March, 82 Calgary.

MECHANICAL ENGINEER'S REFERENCE BOOK

MECHANICAL ENGINEER'S REFERENCE BOOK

Edited by

A. PARRISH, M.B.E., C.Eng., M.I.Mech.E.

With specialist contributors

BUTTERWORTHS

LONDON   –   BOSTON

Sydney – Wellington – Durban – Toronto

The Butterworth Group

United Kingdom **Butterworth & Co (Publishers) Ltd**
London: 88 Kingsway, WC2B 6AB

Australia **Butterworths Pty Ltd**
Sydney: 586 Pacific Highway, Chatswood,
NSW 2067
Also at Melbourne, Brisbane, Adelaide and Perth

Canada **Butterworth & Co. (Canada) Ltd**
Toronto: 2265 Midland Avenue, Scarborough,
Ontario M1P 4S1

New Zealand **Butterworths of New Zealand Ltd**
Wellington: T & W Young Building,
77–85 Customhouse Quay, 1, CPO Box 472

South Africa **Butterworth & Co (South Africa) (Pty) Ltd**
Durban: 152–154 Gale Street

USA **Butterworth (Publishers) Inc**
Boston: 10 Tower Office Park, Woburn, Mass. 01801

First published as *Newnes Engineers Reference
Book* 1946
Eleventh edition (completely revised and reset) 1973
Reprinted 1977, 1978, 1980

© Butterworth & Co (Publishers) Ltd, 1973

ISBN 0 408 00083 X

Printed in the United States of America

PREFACE

The introduction of SI Metric Units has necessitated considerable modifications to British Standards and other National Standards and it is essential that technical literature should reflect or anticipate these changes. This is the principal objective of the eleventh edition of the Mechanical Engineer's Reference Book which has been completely revised; all quantities have been given in SI units only, even though in some instances the original sources of the information are in the inch or MKSA systems.

It is hoped that this book will meet the needs of technologists and other management personnel in the Mechanical, Chemical and Production engineering disciplines. It should also prove valuable to Metallurgists, Welding Engineers and College Lecturers and Students.

The contributors who are drawn from Industry, Trades and Research Associations and Educational bodies have dealt with the technology in sufficient depth to give a thorough coverage of the subject which will be readily understood by the reader. Further information is given by means of numerous tables and illustrations. An extensive bibliography is included in each section which gives reference to national and international recommended literature and standards covering the subject matter.

Much of the technology is common to many countries, furthered by the work of the International Standards Bodies and the actions now being taken for the removal of barriers to trade. Many of the British Standards which are referred to in the text may therefore be compared with the standards of other National bodies.

Absolute units at first present difficulty to those accustomed to gravitational inch and metric systems and it takes use to become familiar with the concept of a newton or a pascal as a force or pressure/stress unit. This is probably because the gravitational systems do not differentiate between the units for loads (mass) and forces. The expressions, however, are more readily understood when the units are in the coherent (SI) system and are not confused by the many special names used in the older systems. No attempt has been made to convert empirical data to another system of units and, indeed, it is not advisable to do this unilaterally. A case in point is gearing, where the British Standards are not yet fully converted to SI. Some of the product designations are based on the former unit systems as, for example, pipe sizes (nominal) which retain the inch nominal sizes in the metric standards and ISO metric bolts and nuts which are based on the material stresses in kgf units. These descriptions do not prevent the use of SI units for expressing the properties of the products.

This book describes a wide variety of engineering subjects and reflects the established practices and technology in metric terms in the U.K. and at International level so far as it is at present practicable to do so.

Extracts from British Standard publications are reproduced by permission of the British Standards Institution, 2 Park Street, London W1A 2BS from whom copies of the complete publication may be obtained.

A. PARRISH

CONTENTS

3 DRAWING OFFICE PRACTICE AND METROLOGY

4 SCREW THREADS, THREADED FASTENERS AND STANDARD SIZES OF MATERIALS

5 METALLURGY AND NON-METALS TECHNOLOGY

6 WELDING AND SURFACE FINISHES

10 PIPEWORK

INDEX

INDEX TO ADVERTISERS

CONTRIBUTORS

H. H. ANDERSON, B.Sc., C.Eng., F.I.Mech.E., Mem.A.S.M.E., Assoc.M.Inst.C.E.,
Design Consultant,
Weir Pumps Ltd. (Section 10)

E. N. ANDREWS, C.Eng., F.I.Gas.E., P.A.I.W.E.,
Manager, Technical Services,
Stanton and Stavely Group, British Steel Corporation. (Section 10)

D. H. BACON, B.Sc., C.Eng., M.I.Mar.E., M.I.Mech.E.,
Lecturer in Thermofluid Mechanics
Plymouth Polytechnic. (Section 2)

A. I. BIGGS, M.Sc., F.R.I.C.,
Chief Technical Adviser, Company Affairs Directorate,
Confederation of British Industry. (Section 15)

G. A. BROWN, C.Eng., M.I.Mech.E.,
Quality Control Manager,
Glenfield and Kennedy Ltd. (Section 10)

R. G. COOKE, B.A.,
Scientific Officer,
The British Ceramic Research Association. (Section 5)

D. M. COWLEY (Mrs.),
Senior Technical Officer,
British Standards Institution. (Section 3)

G. R. DARBY, C.Eng., M.I.Mech.E.,
Secretary, Metric Steering Committee,
C.E.G.B. (Sections 1, 4 and 7)

G. E. DODD, C.Eng., M.I.Mech.E.,
Senior Technical Officer,
British Standards Institution. (Section 4)

T. V. DUGGAN, A.R.T.C.S.(Hons), C.Eng., M.I.Mech.E., M.I.E.D., F.R.S.A.,
Senior Lecturer, Department of Mechanical Engineering and Naval Architecture,
Portsmouth Polytechnic. (Section 7)

W. EDWARDS-SMITH, C.Eng., M.I.Prod.E.,
Principal Lecturer,
Department of Mechanical and Production Engineering,
Leeds Polytechnic. (Sections 10 and 20)

A. A. FIELD, F.I.H.V.E., A.I.Inst.Sc., M.I.L.,
Consulting Environmental Engineer,
Formerly Chief Research Officer,
Brightside WTF Ltd. (Section 15)

W. FIRTH,
Technical Services Executive,
Yorkshire Imperial Metals. (Section 10)

R. E. FISCHBACHER,
Technical Director,
GEC—Elliott Process Instruments Ltd. (Section 18)

G. D. GALVIN, B.SC.,
Senior Research Chemist,
Shell Research Ltd. (Section 14)

D. P. GOLCH, F.I. Plant.E.,
Group Chief Engineer,
Spirax-Sarco Ltd. (Section 10)

DR. M. M. HALL,
Rubber and Plastics Research Association of Great Britain. (Section 16)

D. A. HAMMOND,
Technical Officer,
British Standards Institution. (Section 3)

S. B. HARRISON, C.Eng., M.I.Mech.E.,
Technical Director,
S.A.M. Equipment Ltd and John Tonks & Co. Ltd.
Formerly with I.C.I. (Agricultural Division) Ltd. (Section 2)

D. HERRELL, B.SC.,
Scientific Officer,
British Ceramic Research Association. (Section 5)

S. W. JONES, Assoc.Eng.(Sheff), C.Eng., M.I.Mech.E.,
Principal Lecturer, Mechanical Engineering,
Portsmouth Polytechnic. (Section 13)

H. M. MOSS, B.SC., A.R.T.C., C.Eng., M.I.Mech.E., M.I.E.E., A.M.B.I.M.,
Engineering Services Department,
I.C.I. Ltd. (Sections 15 and 20)

R. A. MOTTRAM, A.I.M.,
Formerly with Agricultural Division,
I.C.I. Ltd. (Section 9)

R. G. NORMAN, M.SC., C.Eng., F.I.Prod.E., M.I.Mech.E., A.M.B.I.M.,
Head of Department of Management Studies,
Sunderland Polytechnic. (Section 20)

F. T. PALIN, Dip. Ceram., L.I. Ceram.,
Experimental Officer,
The British Ceramic Research Association. (Section 5)

A. PARRISH, M.B.E., C.Eng., M.I.Mech.E.,
Consultant,
Formerly with I.C.I. Ltd. (Sections 1, 2 and 10)

L. POWELL, B.SC., F.I.M.,
Formerly with Nobel Division,
I.C.I. Ltd. (Section 12)

J. G. REES, Grad. M.NDT.S.,
Non-destructive Testing Engineer,
Engineering Services (Wilton) Ltd. (Section 8)

DR. G. P. ROTHWELL, M.A., Ph.D., A.I.M., A.R.I.C., A.M.I.Corr.T.,
Department of Metallurgy and Materials Science,
University of Cambridge. (Section 6)

C. G. SCARBOROUGH,
Consulting N.C. Engineer. (Section 19)

J. M. SYKES, M.A., Ph.D., A.I.M., A.R.I.C.,
Department of Chemistry,
The City University. (Section 6)

H. F. TREMLETT, A.R.S.M., B.SC., F.I.M.,
Consulting Metallurgist (Welding). (Sections 5 and 6)

K. B. WARWICK, C.Eng., F.I.Mech.E.,
Consulting Engineer,
Formerly General Manager and Local Director, Geo. W. King Ltd. (Section 12)

P. D. WEBSTER, M.Met., A.I.M., M.I.B.F.,
Senior Lecturer,
West Bromwich College of Commerce and Technology. (Section 5)

F. G. WHITE,
Consultant,
Formerly Chief Designer, Reavell & Co. Ltd. (Section 17)

J. DE WIT,
Shell International Petroleum,
Maatschappij N.V.
The Hague. (Section 11)

D. C. WRIGHT, M.SC.,
Rubber and Plastics Research Association of Great Britain. (Section 16)

1 UNITS, SYMBOLS AND CONSTANTS

UNITS, SYMBOLS AND CONSTANTS

1

1 UNITS, SYMBOLS AND CONSTANTS

METRICATION

G. R. DARBY AND A. PARRISH

THE SYSTÈME INTERNATIONAL D'UNITES

Introduction

SI is the accepted abbreviation for Système International d'Unités (International System of Units) the modern form of the metric system agreed at an international conference in 1960. This system has been adopted by the ISO[19] and the IEC[18] and its use is recommended wherever the metric system is applied. It is already in the process of being adopted in the legislation of twenty-three countries. The indications are that SI Units will supersede the units of existing metric systems and of all systems based on Imperial Units. The SI is now being adopted throughout most of the world and is likely to remain the primary world system of units of measurement for a very long time.

SI units and the rules for their application are contained in ISO Resolution R1000[10] and a BIPM[15] informatory document 'SI–Le Système International d'Unités'. An abridged version of the former is available as BSI publication PD 5686[6]. BS 3763[5] incorporates information from the BIPM document including matters which deal with units outside the International System which are recognised by the CIPM[17] for use in conjunction with it. The BIPM document is based on resolutions of the CGPM[16] or decisions of the CIPM.

Basic SI units

SI comprises seven basic units from which a wide range of quantities can be derived in the form of products and quotients of these units which are:

Quantity	Name of unit	Unit symbol
Length	metre	m
Mass	kilogramme	kg
Time	second	s
Electric current	ampere	A
*Thermodynamic temperature	kelvin	K
Luminous intensity	candela	cd
Amount of substance	mole	mol

* Note: Temperature difference is commonly expressed in degrees Celsius instead of degrees kelvin. The unit of the temperature interval for these scales is the same; $0 K = -273 \cdot 15°C$; $273 \cdot 15°K = 0°C$.

The definition of these units as given in BSI/PD 5686[6] are as follows.

Metre (m). The metre is the length equal to 1 650 763·73 wavelengths in vacuum of the radiation corresponding to the transition between the levels $2 p_{10}$ and $5 d_5$ of the krypton-86 atom. (11th CGPM[16] (1960) Resolution 6).

Kilogramme (kg).   The kilogramme is the unit of mass; it is equal to the mass of the international prototype of the kilogramme. (1st and 3rd CGPM[16] 1889 and 1901).

Second (s).   The second is the duration of 9 192 631 770 periods of the radiation corresponding to the transition between the two hyperfine levels of the ground state of the caesium-133 atom. (13th CGPM[16] (1967) Resolution 1).

Ampere (A).   The ampere is that constant current which, if maintained in two straight parallel conductors of infinite length, of negligible circular cross-section, and placed 1 metre apart in vacuum, would produce between these conductors a force equal to 2×10^{-7} newton per metre of length. (CIPM[17] (1947) Resolution 2 approved by the 9th CGPM[16] (1948).)

Kelvin (K).   The kelvin, unit of thermodynamic temperature, is the fraction $1/273 \cdot 16$ of the thermodynamic temperature of the triple point of water. (13th CGPM[16] (1967) Resolution 4.)

Candela (cd).   The candela is the luminous intensity in the perpendicular direction, of a surface of 1/600 000 square metre of a black body at the temperature of freezing platinum under a pressure of 101 325 newtons per square metre. (13th CGPM[16] (1967) Resolution 5.)

The supplementary base units are defined in 'The International System of Units' as follows.

Plane angle (radian).   The angle subtended at the centre of a circle of radius 1 m by an arc of length 1 m along the circumference.

Solid angle (steradian).   The solid angle subtended at the centre of a sphere of radius 1 m by an area of 1 m² on the surface.

The proposed seventh unit, the *mol* corresponding to the quantity 'amount of substance' is recommended by IUPAP[21], IUPAC[20] and ISO/TC12[22] but needs to be endorsed by the CPGM[16].

The *mole* (symbol 'mol') is defined as an amount of substance of a system which contains as many elementary units as these are carbon atoms in 0·012 kg (exactly) of the pure nuclide 12C.

The elementary unit must be specified and may be an atom, a molecule, an ion, an electron, a proton etc. or a group of such entities according to a stated formula (58th CIPM[17] (1969) Recommendation 1).

Derived units

SI is a rationalised and coherent system because for any one physical quantity it admits of only one measurement unit with its entire structure derived from no more than seven arbitrarily defined basic units. It is coherent because the derived units are always the products or quotients of two or more of these basic units. Thus the SI unit for velocity is m/s (metre per second) and for acceleration is m/s² (metre per second every second).

Special names as shown in Table 1.1(a) have been given to some derived units as an aid to communication.

Table 1.1(b) shows the relationship of some of the quantities.

Although SI is complete in itself, certain non-SI units are recognised for use in conjunction with it where for traditional, commercial or practical purposes it is difficult to discard them.

For example it is impracticable to disregard the minute (in SI — 60 seconds) and the hour (in SI — 3600 seconds) which are non-coherent units.

Gravitational and absolute systems

There may be some difficulty in understanding the difference between SI and the Metric Technical System of units which has been used principally in Europe. The main difference is that whilst mass is expressed in kg in both systems, weight (representing a force) is

Table 1.1(a) SOME DERIVED UNITS HAVING SPECIAL NAMES

Physical quantity	SI unit	Unit symbol
Force	newton	$N = kg \, m/s^2$
Work, energy quantity of heat	joule	$J = N \, m = kg \, m^2/s^2$
Power	watt	$W = J/s = kg \, m^2/s^3$
Electric charge	coulomb	$C = A \, s$
Electric potential	volt	$V = W/A = kg \, m^2/As^3$
Electric capacitance	farad	$F = A \, s/V = A^2s^4/kg \, m^2$
Electric resistance	ohm	$\Omega = V/A = kg \, m^2/A^2s^3$
Frequency	hertz	$Hz = \dfrac{1}{s}$
Magnetic flux	weber	$Wb = kg \, m^2/A \, s^2$
Magnetic flux density	tesla	$T = Wb/m^2 = kg/A \, s^2$
Inductance	henry	$H = kg \, m^2/A^2s^2$
Lummons flux	lumen	$lm = cd \, sr*$
Illumination	lux	$lx = lm/m^2$

*Note: One steradian (sr) is the solid angle which, having its vertex at the centre of a sphere, cuts off an area of the surface of the sphere equal to that of a square with sides of length equal to the radius of the sphere.

The SI unit of electric dipole moment (A s m) is usually expressed as a coulomb metre (C m).

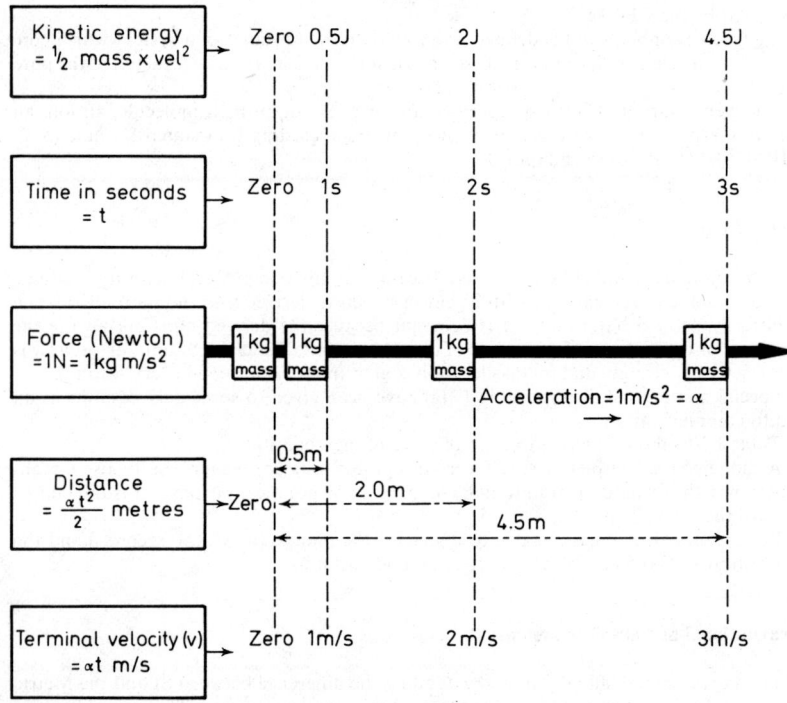

Figure 1.1. Absolute Unit of Force (SI)

Table 1.1(b). SI UNITS—FAMILY TREE

1–5

Electrical current — A

Plane angle — rad

rad/s — rad/s² Angular acceleration

rad/s Angular velocity

As = c Quantity of electricity

A/m Magnetic field strength

Time — s

s⁻¹ = Hz Frequency

m²/s Kinematic viscosity

m/s Speed

m/s² Acceleration

kg m/s² = N Force

Nm = J Work heat

N/m² Pressure

J/s = W Power heat flow rate

Ns/m² Dynamic viscosity

W/A = V Potential difference

As/V = F Electric capacitance

V/A = Ω Resistance

$\frac{1}{\Omega}$ = S Conductance

Vs = Wb Magnetic flux

Vs/A = H Inductance

V/m Electric field strength

Wb/m² = T Magnetic flux density

Length — m

m² Area

m³ Volume

kg/m³ Mass density

Mass — kg

Luminous intensity — cd

Solid angle — sr

cd sr = lm Luminous flux

mol/kg Molarity

J/mol Molar internal energy

lm/m² = lx Illumination

cd/m² Luminance

Amount of substance — mol

J/mol K Molar entropy

Temperature — K

Names of units

A	ampere
cd	candela
F	farad
J	joule
K	kelvin
H	henry
lm	lumen
T	tesla
V	volt
W	watt
Wb	weber
N	newton
Ω	ohm
mol	mole
rad	radian
sr	steradian
S	siemen
C	coulomb
Hz	hertz

expressed as kgf, a gravitational unit, in the MKSA system and as N is SI. An absolute unit of force differs from a gravitational unit of force because it induces unit acceleration in a unit mass whereas a gravitational unit imparts gravitational acceleration to a unit mass. This is illustrated in Figures 1.1 and 1.2.

A comparison of the more commonly known systems and SI is shown in Table 1.1(c).

Table 1.1(c)   COMMONLY USED UNITS OF MEASUREMENT

	SI (absolute)	FPS (gravitational)	FPS (absolute)	cgs (absolute)	Metric technical units (gravitational)
Length	metre (m)	ft	ft	cm	metre
Force	newton (N)	lbf	poundal (pdl)	dyne	kgf
Mass	kg	lb or slug	lb	gram	kg
Time	s	sec	sec	sec	sec
Temperature	°C K	°F	°F °R	°C K	°C K
Energy {mech. / heat}	joule*	ft lbf / Btu	ft pdl / Btu	dyne cm = erg / calorie	kgf m / k cal.
Power {mech. / elec.}	watt	hp / watt	hp / watt	erg's	metric hp / watt
Electric current	amp	amp	amp	amp	amp
Pressure	N/m^2	lbf/ft^2	pdl/ft^2	$dyne/cm^2$	kgf/cm^2

* 1 joule $\doteq$ 1 newton metre or 1 watt second

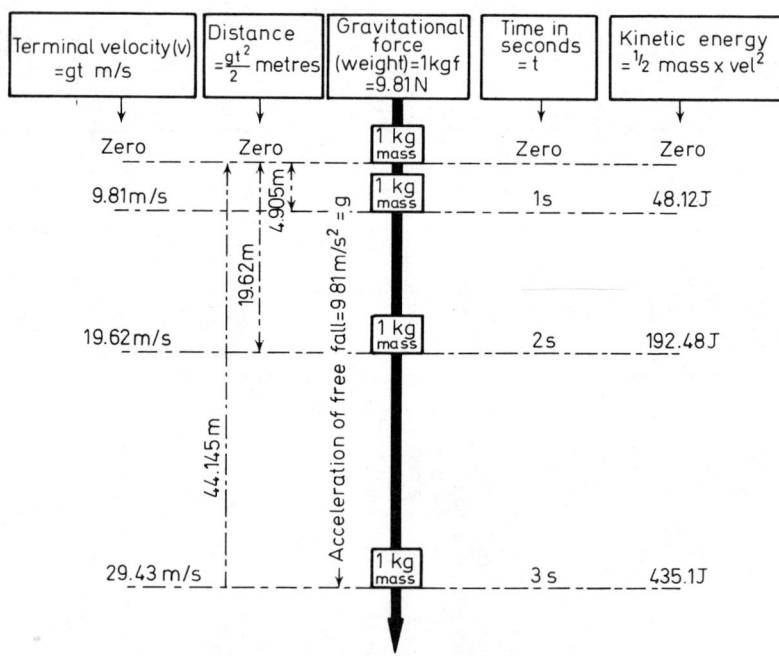

Figure 1.2. Gravitational Unit of Forces (Metric)

It should be noted in particular how all energy and power whether from a mechanical, electrical or heat source share a common derived unit in the SI.

Indices

The metric system is a decimal system in which calculations using the numeral 10, multiplied or divided by itself a number of times, is used. Thus quantities containing many zeros, either before or after the decimal sign appear and these can lead to errors. The calculations can be simplified if the index rules are applied.

The number of times by which 10 is multiplied or divided by itself is referred to as the index of the power of 10. These indices may be whole, fractional, or zero, positive or negative and conform to the following rules:

$$10^m = 10 \times 10 \times 10 \times \ldots \text{ to } m \text{ factors} \tag{1}$$
$$\text{e.g. } 10^3 = 10 \times 10 \times 10 = 1000$$

$$10^m \times 10^n = [10 \times 10 \times \ldots \text{ to } m \text{ factors}] \times 10 \times 10 \times \ldots \text{ to } [n \text{ factors}] \tag{2}$$
$$= 10 \times 10 \times \ldots \text{ to } (m+n) \text{ factors}$$
$$= 10^{m+n}$$
$$\text{e.g. } 10^2 \times 10^3 = 10^{(2+3)} = 10^5$$

$$10^m \times 10^n \times 10^p \times \ldots = 10^{(m+n+p)} \tag{3}$$
$$\text{e.g. } 10^1 \times 10^2 \times 10^3 = 10^{(1+2+3)} = 10^6 = 1\,000\,000$$

$$\frac{10^m}{10^n} = 10^{(m-n)} \text{ when } m>n \text{ and } 10^{(n-m)} \text{ when } m<n \tag{4}$$

$$\text{e.g. } \frac{10^3}{10^2} = 10^{(3-2)} = 10^1$$

$$\text{and } \frac{10^2}{10^3} = \frac{1}{10^{(3-2)}} = \frac{1}{10^1} = 10^{-1} = 0{\cdot}1$$

$$(10^m)^n = 10^{mn} = (10^n)^m \tag{5}$$
$$= 10^m \times 10^m \times 10^m \times \ldots \text{ to } n \text{ factors}$$
$$\text{or } 10^n \times 10^n \times 10^n \times \ldots \text{ to } m \text{ factors}$$
$$\text{e.g. } (10^2)^3 = 10^{(2 \times 3)} \text{ or } 10^{(3 \times 2)} = 10^6 = 1\,000\,000$$

$$10^0 = 1 \tag{6}$$

$$10^{-n} = \frac{1}{10^n} \tag{7}$$

$$\text{e.g. } 10^{-3} = \frac{1}{10^3} = \frac{1}{1000} = 0{\cdot}001$$

$$(10a)^m = 10^m a^m \tag{8}$$
$$\text{e.g. } (10 \times 6)^2 = 10^2 \times 6^2 = 3600$$
$$\text{and } (10 \times 6)^{-2} = \frac{1}{10^2 \times 6^2} = \frac{1}{3600} = 3{\cdot}6 \times 10^{-3}$$
$$= 0{\cdot}000\,277$$

$$\left(\frac{10^x}{10^y}\right)^m = \frac{10^{xm}}{10^{ym}} \tag{9}$$

$$\text{e.g. } \left(\frac{10^3}{10^5}\right)^2 = \frac{10^6}{10^{10}} = 10^{(6-10)} = 10^{-4} = 0{\cdot}0001$$

$$10^{m/n} = \sqrt[n]{(10^m)} \tag{10}$$
$$\text{e.g. } 10^{4/2} = \sqrt[2]{(10^4)} = 10^2 = 100$$

Example.   To convert 1 horsepower to watts

$$1 \text{ hp} = 550 \text{ ft lbf/s} = 5 \cdot 5 \times 10^2 \text{ ft lbf/s}$$

$$1 \text{ ft} = \frac{12 \times 25 \cdot 4 \text{ m}}{1000} = 3 \cdot 048 \times 10^{-1} \text{ m}$$

$$1 \text{ lbf} = 0 \cdot 4536 \text{ kgf} = 4 \cdot 536 \times 10^{-1} \text{ kgf}$$

$$1 \text{ kgf} = 9 \cdot 81 \text{ N}$$

$$1 \text{ W} = 1 \text{ N m/s} = 1 \text{ J/s}$$

$$\therefore 1 \text{ hp} = [5 \cdot 5 \times 10^2] \times [3 \cdot 048 \times 10^{-1}] \times [4 \cdot 536 \times 10^{-1}] \text{ kgf m/s}$$

$$= 5 \cdot 5 \times 3 \cdot 048 \times 4 \cdot 536 \times 9 \cdot 81 \text{ N m/s}$$

$$= 745 \cdot 70 \text{ W}$$

Table 1.1(d) shows the more commonly met terms expressed in coherent units.

Table 1.1(d)   THE USE OF INDICES—MULTIPLES RAISED TO A POWER SHOWN IN TERMS OF COHERENT UNIT

Term	Process	Coherent unit
1 cm^3	$(10^{-2} \text{ m})^3$	10^{-6} m^3
$1 \ \mu \text{ s}^{-1}$	$(10^{-6} \text{ s})^{-1}$	$\dfrac{1}{10^{-6} \text{ s}}$
$1 \text{ mm}^2/\text{s}$	$(10^{-3} \text{ m})^2/\text{s}$	$10^{-6} \text{ m}^2/\text{s}$
1 N/mm^2	$\text{N}/(10^{-3} \text{ m})^2$	$\text{N}/10^{-6} \text{ m}^2 = 10^6 \text{N/m}^2$
1 cm^4	$(10^{-2} \text{ m})^4$	10^{-8} m^4
1 mm^4	$(10^{-3} \text{ m})^4$	10^{-12} m^2
1 mm^2	$(10^{-3} \text{ m})^2$	10^{-6} m^2
1 mm^3	$(10^{-3} \text{ m})^3$	10^{-9} m^3

Expressing magnitudes of SI units

To express magnitudes of a unit, decimal multiples and submultiples are formed using the prefixes shown in Table 1.1(e). This method of expressing magnitudes ensures complete adherence to a decimal system.

Table 1.1(e)   THE INTERNATIONALLY AGREED MULTIPLES AND SUB-MULTIPLES

Factor by which the unit is multiplied		Prefix	Symbol	Common everday samples
One million million (billion)	10^{12}	tera	T	
One thousand million	10^9	giga	G	gigahertz (GHz)
One million	10^6	mega	M	megawatt (MW)
One thousand	10^3	kilo	k	kilometre (km)
One hundred	10^2	hecto*	h	
Ten	10^1	deca*	da	decagramme (dag)
UNITY	1			
One tenth	10^{-1}	deci*	d	decimetre (dm)
One hundredth	10^{-2}	centi*	c	centimetre (cm)
One thousandth	10^{-3}	milli	m	milligramme (mg)
One millionth	10^{-6}	micro	μ	**microsecond (μs)**
One thousand millionth	10^{-9}	nano	n	nanosecond (ns)
One million millionth (one millionth)	10^{-12}	pico	p	picofarad (pF)
One thousand million millionth	10^{-15}	femto	f	
One million million millionth	10^{-18}	atto	a	

* To be avoided wherever possible

Examples of use – Length

$$1 \text{ millimetre } (1 \text{ mm}) = \frac{1}{1000} \text{ metre} \qquad = 10^{-3} \text{ m}$$

10 millimetres	(10 mm)	= 1 centimetre (1 cm)	= 10^{-2} m
10 centimetres	(10 cm)	= 1 decimetre (1 dm)	= 10^{-1} m
10 decimetres	(10 dm)	= 1 metre (1 m)	= 10^{0} m
10 metres	(10 m)	= 1 decametre (1 dam)	= 10^{1} m
10 decametres	(10 dam)	= 1 hectometre (1 hm)	= 10^{2} m
10 hectometres	(10 hm)	= 1 kilometre (1 km)	$=$ 10^{3} m

Rules for use of SI units and the decimal multiples and sub-multiples

The following rules are based on guide rules laid down in ISO R1000[10].

1. The SI units are preferred but it is impracticable to limit usage to these, therefore their decimal multiples and sub-multiples are also required. (For example, it is cumbersome to measure road distances or the breadth of a human hair in metres.)
2. In order to avoid errors in calculations it is preferable to use coherent units. Therefore, it is strongly recommended that in calculations only SI units themselves are used and not their decimal multiples and sub-multiples. (Example: Use $N/m^2 \times 10^6$ not MN/m^2 or N/mm^2 in a calculation.)
3. The use of prefixes representing 10 raised to a power which is a multiple of 3 is especially recommended. (Example: For length, km ... m ... mm ... μm. Thus hm; dam; dm; cm are non-preferred.)
4. When expressing a quantity by a numerical value of a unit it is helpful to use quantities resulting in numerical values between 0 and 1000. Examples:
    $12 \text{ kN} = 12 \times 10^3$ N instead of 12 000 N
    $3 \cdot 94$ mm $= 3 \cdot 94 \times 10^{-3}$ m instead of 0·00394 m
    $14 \cdot 01 \text{ kN}/m^2 = 14 \cdot 01 \times 10^3 \text{ N}/m^2$ instead of 14010 N/m^2
5. Compound prefixes are not used. (Example: Write nm not mμm.) Where, however a name has been given to a product or a quotient of a basic SI unit, for example the bar ($10^5 \text{ N}/m^2$) it is correct practice to apply the prefix to the name e.g. millibar (10^{-3} bar).
6. In forming decimal multiples and sub-multiples of a derived SI unit preferably only one prefix is used. The prefix should be attached to the unit in the numerator. (Example: MW/m^2 not W/mm^2.) The exception is stress where BSI recommend the use of N/mm^2.
7. Multiplying prefixes are printed immediately adjacent to the SI unit symbol with which they are associated. The multiplication of symbols is usually indicated by leaving a small gap between them. (Example: mN = millinewton. If written as m N this would indicate metre newton.)

SI QUANTITIES AND UNITS

The units shown in Tables 1.2(d) to 1.2(k) generally obey the above guide rules. However the BSI expect that initially a practical attitude will prevail in the interpretation of these recommendations in the United Kingdom, particularly as certain countries still use Metric Technical Units which include such units as kgf/cm^2. These units may appear for some time alongside SI Units in BS publications.

Tables 1.2(a) to 1.2(k) contain a list of SI units and a selection of recommended decimal multiples and sub-multiples together with other units or names of units which may be used. These are based on recommendations given in ISO Recommendation R 1000[10] and BSI publication PD 5686[6].

Explanation of Tables

Tables 1.2(a) to 1.2(k) are divided into 8 columns.

Column 1 indicates the physical quantity name. For physical symbols to be used (e.g. for basic formulae...... M for mass; W for weight, g for gravity) reference should be made to ISO R 31[9] and the latter part of this Section dealing with Symbols and Abbreviations.

Columns 2 and 3 indicate the name and symbol of the SI Unit.

Column 4 indicates the recommended multiple to be used by Industry. The consistent use of 10^n where n is ± 3 or multiples thereof should be noted.

Column 5 indicates multiples of SI units which are listed as non-preferred, but nevertheless will occasionally be met for practical reasons. Although a choice of multiples is shown in the Tables for given quantities, it is recommended that a limited selection should be made within the range stated to suit the engineering discipline.

Column 6 indicates these units other than SI which cannot be omitted for social, traditional or commercial reasons.

Column 7 indicates the value of the units shown in Columns 5 and 6 in terms of the SI Unit, Column 4.

Column 8 gives conversion factors from Imperial units to SI and centimetre-gramme-second (cgs) units to SI together with special remarks applicable to the quantity under discussion. The conversion factors marked thus (+) are exact. Detailed conversion tables for certain selected quantities are given in Tables 1.4(a) to 1.4(r). Conversion factors are based on BS 350 Parts 1,2[3] and PD 6203[7].

Explanatory notes to Tables 1.2(a) to 1.2(k)

Notes on particular units and expressions follow each of the Tables. The notes reflect the concensus of opinion of British Industry and should be checked from time to time for further developments. National policy statements on such units are promulgated through the monthly BSI News.

Table 1.2(a) SPACE AND TIME

Quantity (1)	SI units — Name (2)	Symbol (3)	Decimal multiples and sub-multiples — BSI recommended (4)	Others (5)	Units other than SI (6)	Equivalent value in terms of basic unit (7)	Conversion factors and remarks (8)
Plane Angle (1)	radian	rad			degree (°) minute of arc (') second of arc (")	1 rad $\Pi/180$ rad $\Pi/10\,800$ rad $\Pi/648\,000$ rad	$1° = 0\cdot017\,453\,3$ rad+ $1' = 2\cdot908\,88 \times 10^{-4}$ rad+ $1'' = 4\cdot848\,14 \times 10^{-6}$ rad 1 right angle = $\Pi/2$ rad = $90°$
Solid Angle	steradian	sr				1 sr	
Length (2)	metre	m	km mm μm nm	cm		10^3 m 1 m 10^{-2} m 10^{-3} m 10^{-6} m 10^{-9} m	1 inch = 25·4 mm+ 1 foot = 0·3048 m+ 1 yard = 0·9144 m+ 1 mile = 1·609 344 km+ 1 UK nautical mile = 1·853 18 km 1 angstrom (Å) = 10^{-10} metre or 10^{-1} nm
Area (2)	square metre	m²	km² ha a mm²	cm²		10^6 m² 10^4 m² 10^2 m² 1 m² 10^{-4} m² 10^{-6} m²	1 sq in = 645·16 mm²+ = 6·4516 cm²+ 1 sq ft = 0·092 903 m² 1 sq yd = 0·836 127 m² 1 acre = 0·404 686 ha 1 sq mile = 258·999 ha ha = hectare a = are
Volume (3)	cubic metre	m³	mm³	dm³ (litre) cm³ (mlitre)		1 m³ 10^{-3} m³ 10^{-6} m³ 10^{-9} m³	1 in³ = 16·3871 cm³ 1 ft³ = 0·028 3168 m³ 1 yd³ = 0·764 555 m³ 1 UK fl. oz = 28·4131 mlitre 1 gal = 4·5461 litre

Table 1.2(a)—*continued*

| Quantity | SI units | | Decimal multiples and sub-multiples | | Units other than SI | Equivalent value in terms of basic unit | Conversion factors and remarks |
| | Name | Symbol | BSI recommended | Others | | | |
(1)	(2)	(3)	(4)	(5)	(6)	(7)	(8)
Time	second	s	ms μs ns		day, hour, minute	7·344 Ms, 3600 s, 60 s, 1 s, 10^{-3} s, 10^{-6} s, 10^{-9} s	
Angular Velocity	radian per second	rad/s			rev/min rev/s	1 rad/s, 0·104 720 rad/s, 6·283 19 rad/s	1 degree/s = 0·017 4533 rad/s
Velocity	metre per second	m/s			km/h knot (kn)	1 m/s, 1/3·6 m/s, 0·514 444 m/s	1 km/h = 0·277 778 m/s 1 ft/s = 0·3048 m/s+ 1 mile/h = 0·447 04 m/s+ or 1·609 34 km/h 1 UK knot = 0·514 773 m/s
Acceleration	metre per second squared	m/s^2				1 m/s^2	1 ft/s^2 = 0·3048 m/s^2+
Angular Acceleration	radian per second squared	rad/s^2				1 rad/s^2	1 in/s^2 = 0·0254 m/s^2+

Units for common use

NOTES ON IMPLEMENTATION OF PARTICULAR UNITS AND EXPRESSIONS
Table 1.2(a)   SPACE AND TIME

1. *Plane angle.* The Sumerian division of the circle in 360° (hence degrees) is retained for geometry although dynamicists use the radian.

2. *Length—The centimetre.* In the many engineering disciplines the use of the centimetre is non-preferred. It is not recommended for engineering drawings in BS 308[2], but has been adopted as the basic unit of measurement by primary schools and for commercial purposes. Sometimes the centimetre raised to a power (e.g. cm^2; cm^3; cm^4) is used to maintain a sensible range of numerical values in front of the unit. An instance of this concerns steel sections where the moduli of sections and moment of section will be given in steel tables in terms of cm^3 and cm^4 respectively.

Where accuracy to the nearest millimetre is unwarranted, the centimetres can be used to imply a less precise dimension.

3. *Volume and capacity—The litre.* Before 1964 the 1901 litre was equal to $1 \cdot 000\,028$ dm^3 At the XII International CGPM[16] meeting on units (1964) the litre was redefined to equate exactly to 1 cubic decimetre. The same conference agreed that the litre should not be used to express the results of precise measurements, so as to make sure that where high precision was involved (say greater than 1 part in 20 000) the possibility of confusion between the former (1901) litre and the new (1964) litre would be eliminated. The UK legal definition will probably be revised accordingly.

It is recommended that the results of precise measurements of volume should be given only in terms of m^3, dm^3, cm^3, mm^3 etc. even though the millilitre (ex cm^3) and litre (ex dm^3) will still be used for operational and commercial purposes.

Because of the possible confusion of the symbol for the litre 'l' with the figure '1', it is strongly recommended that the unit name be spelt in full.

Centilitre is sometimes used for arbitrary quantities implying a greater degree of tolerance.

Table 1.2(b) PERIODIC AND RELATED PHENOMENA

Quantity	SI units			Units other than SI	Equivalent value in terms of basic unit	Conversion factors and remarks
	Name	Symbol	Decimal multiples and sub-multiples			
			BSI recommended	Others			
(1)	(2)	(3)	(4)	(5)	(6)	(7)	(8)
Frequency	hertz	Hz	GHz MHz kHz			10^9 Hz 10^6 Hz 10^3 Hz 1 Hz	1 c/s (or c.p.s.) = 1 Hz
Rotational Frequency	reciprocal second	s^{-1}		rev/s rev/min		$1\ s^{-1}$	
Wavelength	metre	m		cm		1 m 10^{-2} m	

NOTES ON TABLE 1.2(B)

1. *Rotational frequency (rev/min).* The quantity rev/min is favoured for rotating machinery in place of the rev/s or the SI unit rad/s.

Units for common use

(1) Quantity	SI units — (2) Name	SI units — (3) Symbol	Decimal multiples and sub-multiples — (4) BSI recommended	Decimal multiples and sub-multiples — (5) Others	(6) Units other than SI	(7) Equivalent value in terms of basic unit	(8) Conversion factors and remarks
Mass	kilogramme	kg	Mg		tonne	10^3 kg	1 ton = 1016·05 kg or 1·016 05 tonne
(1)			g			1 kg	1 cwt = 50·8023 kg
(2)			mg			10^{-3} kg	1 lb = 0·453 592 37 kg+
			µg			10^{-6} kg	1 oz = 28·3495 g (avoir)
						10^{-9} kg	
Mass density	kilogramme per cubic metre	kg/m³	Mg/m³	g/cm³ kg/dm³	g/mlitre; kg/litre g/litre	10^3 kg/m³ 1 kg/m³	1 lb/ft³ = 16·0185 kg/m³ ; 1 lb/in³ = 27·6799 g/cm³
Specific volume	cubic metre per kilogramme	m³/kg			litre/kg	1 m³/kg 10^{-3} m³/kg	1 ft³/lb = 0·062 428 m³/kg ; 1 cm³/g = 10^{-3} m³/kg+
Momentum	kilogramme metre per second	kg m/s				1 kg m/s	1 lb ft/s = 0·138 255 kg m/s ; 1 g cm/s = 10^{-5} kg m/s+
Angular momentum	kilogramme square metre per second	kg m²/s				1 kg m²/s	1 lb ft²/s = 0·042 1401 kg m²/s ; 1 g cm²/s = 10^{-7} kg m²/s+
Moment of inertia	kilogramme square metre	kg m²				1 kg m²	1 lb ft² = 0·042 140 1 kg m² ; 1 g cm² = 10^{-7} kg m²+
Force	newton	N	MN kN mN µN			10^6 N 10^3 N 1 N 10^{-3} N 10^{-6} N	1 tonf = 9·964 02 kN ; 1 lbf = 4·448 22 N ; 1 ozf = 0·278 014 N ; 1 pdl = 0·138 255 N ; 1 dyne = 10^{-5} N+ ; 1 kgf or kilopond = 9·806 65 N+
Weight					See Note (1)		

Table 1.2(c)—*continued*

	SI units		Decimal multiples and sub-multiples		Units other than SI	Equivalent value in terms of basic unit	Conversion factors and remarks
Quantity	Name	Symbol	BSI recommended	Others			
(1)	*(2)*	*(3)*	*(4)*	*(5)*	*(6)*	*(7)*	*(8)*
Moment of force (torque)	newton metre	N m	MN m kN m μN m	GN m daN m		10^9 N m 10^6 N m 10^3 N m 10 N m 1 N m 10^{-6} N m	1 tonf ft = 3·037 03 kN m 1 lbf ft = 1·355 82 N m 1 pdl ft = 0·042 140 1 N m 1 lbf in = 0·112 985 N m 1 dyne cm = 10^{-7} N m+ 1 kgf m = 9·806 65 N m+
Mass per unit length	kilogramme per metre	kg/m			tonne/km	1 kg/m 1 kg/m	1 ton/1000 yds = 1·111 16 kg/m 1 ton/mile = 0·631 342 kg/m 1 lb/in = 17·8580 kg/m 1 lb/ft = 1·488 16 kg/m 1 lb/yd = 0·496 055 kg/m
Mass per unit area	kilogramme per square metre	kg/m²			kg/ha	1 kg/m² 10^{-4} kg/m²	1 lb/acre = 1·120 85 kg/ha 1 ton/sq. mile = 3·922 98 kg/ha 1 lb/1000 ft² = 4·882 43 kg/m²
Mass rate of flow	kilogramme per second	kg/s			kg/h	1 kg/s 3600 kg/s	1 lb/s = 0·453 592 kg/s 1 lb/h = 1·259 98×10^{-4} kg/s 1 UK ton/h = 0·282 235 kg/s or 1·016 05 tonne/h
Volume rate of flow	cubic metre per second	m³/s			litre/s litre/min	1 m³/s 10^{-3} m³/s $\dfrac{10^{-3}\ \text{m}^3/\text{s}}{60}$	1 ft³/s (cusec) = 28·3168 litre/s 1 gal/s = 4·546 09 litre/s
Mass flow rate per unit area	kilogramme per square metre	kg/m²s				1 kg/m²s	1 lb/ft²h = 1·356 23×10^{-3} kg/m²s

Quantity	Unit (name)	SI unit	Sub-multiples	Other units	Bar units	Equivalent values	Conversions
Pressure and stress	newton per square metre (pascal)	N/m² (Pa)	MN/m²; kN/m²; mN/m²; μN/m²	daN/mm²; N/mm²; N/cm²	hbar; bar; mbar	10⁷ N/m²; 10⁶ N/m²; 10⁵ N/m²; 10³ N/m²; 10² N/m²; 1 N/m²; 10⁻³ N/m²; 10⁻⁶ N/m²	= 68·9476 mbar; 1 torr = 1·333 22 mbar; 1 in Hg = 33·863 9 mbar; 1 in W.G. = 2·490 89 mbar; 1 kgf/cm² = 0·980 665 bar = 98·0665 kN/m²; 1 tonf/in² = 15·4443 N/mm²; 1 pieze = 10³ N/m²; 1 std atmosphere = 1013·25 + mbar+ = 1·033 23 kgf/cm² = 14·695 9 lbf/in² = 760 torr+ = 29·921 3 in Hg
(3)							
Second moment of area	metre to the power of four	m⁴	mm⁴	cm⁴		1 m⁴; 10⁻⁸ m⁴; 10⁻¹² m⁴	1 in⁴ = 41·6231 cm⁴; 1 ft⁴ = 863 097 cm⁴
Section modulus	cubic metre	m³	mm³	cm³		1 m³; 10⁻⁶ m³	1 in³ = 16·3871 cm³
Dynamic viscosity	newton second per square metre	N s/m²	mN s/m²	P (poise); cP (centipoise)		1 N s/m²; 10⁻¹ N s/m²; 10⁻³ N s/m²	1 lbf s/ft² = 47·8803 N s/m² or 47 880·3 cP; 1 pdl s/ft² = 1·488 16 N s/m² or 1488·16 cP or lb/ft s
(4)							
Kinematic viscosity	square metre per second	m²/s	mm²/s	St (stokes); cSt (centistokes)		1 m²/s; 10⁻⁴ m²/s; 10⁻⁶ m²/s	1 ft²/h = 2·580 64×10⁻⁵ m²/s+ or 25·806 cSt+; 1 ft²/s = 0·092 903 m²/s or 9·2903×10⁴ cSt
(4)							
Surface tension	newton per metre	N/m	mN/m			1 N/m; 10⁻³ N/m	1 lbf/ft = 14·5939 N/m; 1 dyne/cm = 10⁻³ N/m
Energy, work	joule	J (= N m)	GJ; MJ; kJ	MW h; kW h; eV (electron volt)		10⁹ J; 10⁶ J; 10³ J; 1 J; 3·6×10⁹ J; 3·6×10⁶ J; (1·602 10 ± ·00007)×10⁻¹⁹ J	1 kgf m = 9·806 65 J+; 1 ft lbf = 1·355 82 J; 1 erg = 10⁻⁷ J; 1 ft pdl = 0·042 140 1 J; 1 hp h = 2·684 52×10⁶ J; 1 kcal = 4186·8 J+; 1 Btu = 1055·06 J
(5)							

Table 1.2(c)—*continued*

| Quantity | SI units | | | | Units other than SI | Equivalent value in terms of basic unit | Conversion factors and remarks |
| | Name | Symbol | Decimal multiples and sub-multiples BSI recommended | Others | | | |
(1)	(2)	(3)	(4)	(5)	(6)	(7)	(8)
Power	watt	W (= J/s)	GW MW kW mW μW	daJ/cm² J/cm²		10^9 W 10^6 W 10^3 W 1 W 10^{-3} W 10^{-6} W 10^3 J/m² 10^4 J/m²	1 hp = 745·7 W 1 ft lbf/s = 1·355 82 W 1 metric hp = 735·499 W 1 kgf m/s = 9·806 65 W + 1 erg/s = 10^{-7} W
Impact strength	joule per square metre	J/m²	kJ/m²			10^3 J/m² 1 J/m²	
Fuel consumption				litre/km km/litre			1 gal/mile = 2·825 litre/km 1 mile/gal = 0·354 km/litre
Specific fuel consumption	kilogramme per joule	kg/J		kg/MJ	kg/kW h	1 kg/J 10^{-6} kg/J	1 lb/hp h = 0·168 97 kg/MJ
	cubic metre per joule	m³/J		litre/MJ	m³/kW h	1 m³/J 10^{-9} m³/J	1 pint/hp h = 0·211 68 litre/MJ

NOTES ON TABLE 1.2(C)

1. *Mass (Mass by Weight) (kilogramme)*. BSI is considering the spelling of the SI unit for mass so as to replace 'kilogramme' by 'kilogram'.

Confusion sometimes arises over the measuring of the terms 'mass' and 'weight'. Commonly, and in many branches of engineering, it has been the custom to refer to quantities of mass as weights, e.g. weight of coal in kilogrammes.

Weight, however, is dependent upon the gravitational force acting upon the mass. Thus for a mass (M), weight (W) = Mg where g is the local acceleration due to gravity which varies slightly from point to point on the earth's surface. For practical purposes an approximated figure of 9·81 or 9·807 metres per second squared (m/s^2) is used for 'g'.

The force unit in SI is the newton (N) and by using consistent units becomes the force applied to unit mass (kg) to impart unit acceleration (m/s^2) to the mass (as distinct from gravitational acceleration which equals 9·806 65 m/s^2).

Thus it can be more readily understood by comparing the SI system with other systems for mass, weights and measures as shown in Table 1.2(d).

Table 1.2(d) SYSTEMS OF WEIGHTS AND MEASURES

Quantity	Foot pound second	Metric, technical	SI
Mass	1 lb	1 kg	1 kg
Length	1 ft	1 m	1 m
Force	1 lbf	1 kgf	1 N
Definition of Force	$lbf = \dfrac{lb \times ft/s^2}{g}$	$kgf = \dfrac{kg \times m/s^2}{g}$	$N = kg \times m/s^2$.
Definition of Weight (gravitational force)	1 lbf per 1 lb	1 kgf per 1 kg	9·806 65 N per 1 kg

Consider a load of a mass M kg which is to be lifted up by a crane from a shop floor. This mass (M kg) will be referred to as the weight in kilogrammes so the crane hook will be marked with a safe working load in kilogrammes. The designer however is concerned with the forces exerted by that load on the lifting appliances which will be in newtons. Thus although the crane picks up a load of say 1000 kg the designer will calculate in terms of a force of 9806·65 newtons (equals 9·806 65 kN) assuming that the load is lifted with an acceleration of 1 m/s^2 and this is the minimum force needed to overcome the gravitational pull. It should be noted that if the load is snatched, that is, lifted with greater acceleration, the force becomes proportionally greater and allowances should be made for these larger forces in the design. If, however, the load of 1000 kg is pushed horizontally along the floor, and if the friction is ignored, the force required to move it with an acceleration of 1 m/s^2 is reduced to 1000 N (equals 1 kN) (see also Fig. 1.1 and 1.2).

2. *Mass (Tonne/Megagramme)*. The tonne and kilogramme are generally accepted as replacement units for ton and pounds. In particular most lifting equipment already marked in tons can be considered as adequate for lifting the same number of tonnes because of the small excess (1·6 %) of the ton over the tonne.

However in soil mechanics, the megagramme (Mg) rather than the tonne is recommended. This accords with the recommendation of the Institution of Civil Engineers[14] because with large masses involved in work on soil mechanics, confusion between the ton and the tonne could prove very expensive.

3. *Metric Units for Pressure and Stress*. The SI derived unit for force per unit area is the newton per square metre (N/m^2), referred to as the pascal (Pa) and this unit with suitable multiples is favoured as the unit for stress. There are differences of opinion regarding the unit for pressure, but BSI has recommended that, although some flexibility will have to be allowed in the expression of pressure values, the following practice should be adopted[8]:

(a) For the statement of stress property use, without deviation, N/m^2 and appropriate multiples of it e.g. MN/m^2 or this, if preferred, expressed as N/mm^2 or, if essential for non-metallic materials, kN/m^2.

(b) For pressure statements use either N/m^2 (and suitable multiples and sub-multiples of it) or bar or mbar. In such cases the conversion 1 bar $= 10^5 N/m^2$ will always be quoted for reference.

(c) The unit of pressure should be the same in all related publications issued under the authority of any one Industry Standards Committee. In cases where Industry Standards Committees have no strong views, the secretaries concerned will seek to secure conformity with the practice of closely related committees.

(d) Pressures or pressure differences measured by manometer tube may often conveniently be expressed as a height of a column of fluid, the nature of the fluid being stated. Such readings must be converted to terms of N/m^2 if they are to be used in calculations of flow, etc. On the other hand, manometers are sometimes used merely as indicators that a prescribed operating condition has been met. Judgement is therefore required as to when it can be of advantage to use mmH_2O, mmHg etc. or when it is of advantage to calibrate and read manometers in a suitable multiple of N/m^2 or in mbar. It is understood that manometers calibrated in mbar are becoming increasingly available and it is recommended that pressures expressed as a height of a column of fluid should progressively give place to a suitable multiple of the SI unit or to the millibar.

(e) It has been internationally recommended that pressure units themselves should not be modified to indicate whether the pressure value is 'absolute' (i.e. above zero) or 'gauge' (i.e. above atmospheric pressure). If, therefore, the context leaves any doubt as to which is meant, the work 'pressure' must be qualified appropriately.

e.g. '... at a gauge pressure of 12·5 bar'

or '... at a gauge pressure of 1·25 MN/m^2'

or '... at an absolute pressure of 2·34 bar'

or '... at an absolute pressure of 234 kN/m^2'

Table 1.2(e) illustrates some of these practices. Note that this table works on a gauge pressure basis (atmosphere = 0), thus vacuum is shown measured in negative millibars. This continues the custom of associating the higher numerical readings with greater vacuum.

Table 1.2(e) GAUGE VACUUM AND PRESSURE

	Vacuum			Gauge pressure			
Ins. Hg	*30″ Hg*	*20″ Hg*	*10″ Hg*	*10 lbf/in²*	*100 lbf/ in²*	*1000 lbf/in²*	*2000 lbf/in²*
kN/m^2	−101·3	−67·73	−33·86	68·94	689·4	6894	13789·5
	−1·01325	−0·6773	−0·3386				
bar	(−1013·25 mbar)	(−677·3 mbar)	(−338·6 mbar)	0·6894	6·894	68·94	137·895
MN/m^2	−0·1013	−0·0677	−0·0338	0·0689	0·6894	6·894	13·7895

Notwithstanding previous practices of referring to pump performances in terms of pressure, the BSI recommends that the pump total head should be specified in linear measure (metres). The symbol N/m^2 will be recognised internationally as the Pascal (Pa).

4. *Viscosity* (*Centipoise*; *Centistokes*). The recognised derived SI units for Dynamic and Kinematic Viscosity are $N s/m^2$ and m^2/s respectively. However the existing units, centipose (cP) and centistokes (cSt) are so well established internationally, particularly for oils, that the operational use of these units will continue.

5. *Energy* (*MJ/kW h*). The choice of a suitable commercial energy unit common to all energy producing concerns has still to be resolved. The SI unit is the joule and its multiples.

However electrical interests favour the adoption of the kW h ($3 \cdot 6$ MJ $= 1$ kW h).
The following are the probable commercial fuel quantities:

Coal   — tonne
Electricity — kW h
Gas   — 100 MJ (the 'new' therm)
Oil   — $\begin{cases} \text{litre; m}^3 \\ \text{kg; tonne} \end{cases}$

6. *Hardness Values* (*kgf*).   The ISO Technical Committee TC17[23] has agreed to retain the hardness unit kgf to express the load applied by the indenter and this will ensure that most hardness testing machines and empiracally based formulae are not made obsolete. Thus the Rockwell, Vickers and Brinell hardness numbers will be retained. This number is arbitrary and dimensionless and is dependent upon the resistance offered by the material under test to a definite load.

7. *Concentration.*   Concentration should preferably be expressed on a mass/mass basis (i.e. kg/kg; mg/mg; mg/kg) or a volume/volume basis (i.e. m^3/m^3; litre/m^3; millilitre/m^3). It may also be expressed in parts per million (ppm) or as a percentage 'by mass' or 'by volume' respectively.

8. *'ph' Scale.*   This is a number based on the logarithm, to the base 10 of the reciprocal of the concentration of hydrogen ions in aqueous solution. It is used as a method of expressing small differences in the acidity or alkalinity of nearly neutral solutions in biological and electrolytic processes.

This number will remain unchanged.

Table 1.2(f) HEAT

Quantity (1)	SI units				Units other than SI (6)	Equivalent value in terms of basic unit (7)	Conversion factors and remarks (8)
	Name (2)	Symbol (3)	Decimal multiples and sub-multiples BSI recommended (4)	Others (5)			
Absolute temperature (1)	kelvin	K				1 K	K = °C+273·15 K = 1·8°R (Rankine)
Customary temperature (2)					°C (Celsius)		°C = 5/9 (°F−32)
Temperature interval	kelvin	K			°C		1°C = 1 K = 1·8°F (alternative form 1 deg C = 1 deg K = 1·8 deg F)
Temperature coefficient (linear or volumetric)		1/K			1/°C	1/K	
Heat, quantity of heat, internal energy, enthalpy	joule	J	GJ MJ kJ mJ			10^9 J 10^6 J 10^3 J 1 J 10^{-3} J	1 Btu = 1055·06 J 1 cal (IT) = 4·1868 J + 1 CHU = 1899·2 J 1 kW h = 3·6 MJ 1 therm = 105·506 MJ 1 erg = 10^{-7} J
Heat flow rate	watt	W	kW			10^3 W 1 W	1 Btu/h = 0·293 071 W 1 kcal/h = 1·163 W + 1 cal/s = 4·1868 W + 1 frigorie = 4·186 W
Density of heat flow rate	watt per square		MW/m² kW/m²	W/mm²		10^6 W/m² 10^3 W/m²	1 Btu/ft² h = 3·154 59 W/m² 1 cal/cm² s = 41 868 W/m²

Quantity	Unit name	SI unit	Other units	Multiples	Conversions
Thermal conductivity	watt per metre kelvin	W/m K	W/m °C		1 Btu/ft h °F = 1·730 73 W/m °C 1 kcal/m h °C = 1·163 W/m °C+ known as 'k' value
Coefficient of heat transfer	watt per square metre	W/m² K	W/cm² K; kW/m² K; W/m² °C	10^4 W/m² K 10^3 W/m² K 1 W/m² K	1 Btu/ft² h °F = 5·678 26 W/m² °C 1 cal/cm² s °C = 41 868 W/m² + 1 kcal/m² h °C = 1·163 W/m² °C+ known as 'U' value
Heat capacity	joule per kelvin	J/K	kJ/K; J/°C	10^3 J/K 1 J/K	1 Btu/deg R = 1899·11 J/°C 1 cal/°C = 4·1868 J/K+
Specific heat capacity (1)	joule per kilogramme kelvin	J/kg K	kJ/kg K; kW h/kg °C; kJ/kg °C; J/kg °C	10^3 J/kg K 1 J/kg K	1 Btu/lb °F = 4·1868 kJ/kg °C+ 1 cal/g °C = 4·1868 J/kg °C+
Entropy	joule per kelvin	J/K	kJ/K	10^3 J/K 1 J/K	1 Btu/°R = 1899·11 J/K
Specific entropy	joule per kilogramme per kelvin	J/kg K	kJ/kg K; J/g K	10^3 J/kg K 1 J/kg K	1 Btu/lb °F = 4·1868 kJ/kg K + 1 cal/g K = 4·1868 kJ/kg K +
Specific energy	joule per kilogramme	J/kg	MJ/kg; kJ/kg; J/g	10^6 J/kg 10^3 J/kg 1 J/kg	1 Btu/lb = 2·326 kJ/kg+ 1 cal/g = 4·1868 kJ/kg+
Specific enthalpy, specific latent heat	joule per kilogramme	J/kg	MJ/kg; kJ/kg	10^6 J/kg 10^3 J/kg 1 J/kg	1 Btu/lb = 2·326 kJ/kg+
Specific heat content (i) Mass basis	joule per kilogramme	J/kg	MJ/kg; kJ/kg	10^6 J/kg 10^3 J/kg 1 J/kg	1 kcal/kg = 4·1868 kJ/kg+ 1 Btu/lb = 2·326 kJ/kg+ 1 CHU/lb = 4·186 816 kJ/kg 1 therm/ton = 103·84 kJ/kg

Table 1.2(f)—*continued*

		SI units					
			Decimal multiples and sub-multiples			Equivalent value in terms of basic unit	Conversion factors and remarks
Quantity	*Name*	*Symbol*	*BSI recommended*	*Others*	*Units other than SI*		
(1)	*(2)*	*(3)*	*(4)*	*(5)*	*(6)*	*(7)*	*(8)*
(ii) Volume basis	joule per cubic metre	J/m^3	MJ/m^3 kJ/m^3			$10^6\ J/m^3$ $10^3\ J/m^3$ $1\ J/m^3$	1 Btu/ft^3 = 37·2592 kJ/m^3 1 Btu/gal = 0·232 08 kJ/litre 1 therm/UK gal = 23·208 GJ/m^3 1 cal/cm^3 = 4·1868 MJ/m^3 1 kcal/m^3 = 4·1868 kJ/m^3
(2) Heat release rate	watt per cubic metre	W/m^3	kW/m^3			$1\ W/m^3$ $10^3\ W/m^3$	1 Btu/ft^2 s = 37·2589 kW/m^3 1 cal/cm^3 h = 1·163 kW/m^3 +

NOTES ON TABLE (1.2(F)

1. *Temperature* $(K; °C)$.   The basic unit of thermodynamic temperature is the kelvin. The unit for customary or operational temperature is the °C. The °C is now known as Celsius instead of Centigrade because the latter name can be confused with $\frac{1}{100}$ of a right angle which, in some continental countries, is known as a grade. Thus all temperature readings will be in °C. K $(K = °C + 273.15)$ is used for temperatures involved in heat calculations which are equated to absolute zero (i.e. 0 K).

Table 1.2(f) shows heat quantities expressed in both K and °C. Where the choice is given the following quantities are those which, for practical purposes, will be measured in customary temperature i.e. °C.

Linear expansion co-efficient	$°C^{-1} = \dfrac{1}{°C}$
Volumetric expansion co-efficient	$°C^{-1} = \dfrac{1}{°C}$
Thermal conductivity	$W/m °C$
Thermal resistivity	$m °C/W$
Thermal resistance	$m^2 °C/W$
Heat capacity	$kJ/°C$
Specific heat capacity	$kJ/kg °C$

(Originally the basic heat unit was defined in terms of the heat required to raise a unit mass of water through one degree interval of temperature and specific heats were compared with water, taken as unity. The SI joule is not dependent upon the above definition so that the term 'specific heat' is no longer applicable. The word capacity has been added to distinguish the old and new terms so that, for example:

Specific heat capacity of water $=$	4.2×10^{-3} J/kg K
Specific heat – volume basis	$MJ/m^3 °C$
Specific heat – mass basis	$kJ/kg °C$
Heat transfer co-efficient	$W/m^2 °C$

2. *Specific heat content* (*ex calorific value*).   This quantity (previously known as CV) is used for establishing the heat content of fuels. The term 'calorific' is now a misnomer as calories are replaced by joule, thus the quantities should be known as specific heat content or 'joulerific value'.

To give manageable numbers the following magnitudes are preferred:

Mass	Volume
kJ/kg	MJ/m^3

Table 1.2(g) ELECTRICITY AND MAGNETISM

Quantity	SI units — Name	SI units — Symbol	Units for common use — Decimal multiples and sub-multiples BSI recommended	Units for common use — Decimal multiples and sub-multiples Others	Units other than SI	Equivalent value in terms of basic unit	Conversion factors and remarks
(1)	(2)	(3)	(4)	(5)	(6)	(7)	(8)
Electric current	ampere	A	kA			10^3 A	1 emu = 10 A
						1 A	1 esu = $1/3 \times 10^{-9}$ A
			mA			10^{-3} A	
			μA			10^{-6} A	
			nA			10^{-9} A	
			pA			10^{-12} A	
Electric charge	coulomb	C	kC			10^3 C	1 Ah = 3600 C
						1 C	
			μC			10^{-6} C	C = A s
			nC			10^{-9} C	
			pC			10^{-12} C	
Charge density	coulomb per cubic metre.	C/m^3	MC/m^3	C/cm^3		10^6 C/m^3	1 emu = 10^7 C/m^3
			kC/m^3			10^3 c/m^3	1 esu = $1/3 \times 10^{-3}$ C/m^3
						1 C/m^3	
Surface density of charge	coulomb per square metre	C/m^2	MC/m^2	C/mm^2		10^6 C/m^2	
			kC/m^2	C/cm^2		10^4 C/m^2	
						10^3 C/m^2	
						1 C/m^2	
Electric field strength	volt per metre	V/m	MV/m			10^6 V/m	
			kV/m			10^3 V/m	
				V/mm		10^2 V/m	
				V/cm		1 V/m	
			mV/m			10^{-3} V/m	
			$μV/m$			10^{-6} V/m	

Quantity	Unit	Symbol	Multiples / sub-multiples		Powers of ten	Relations
Electric potential	volt	V	kV	mV, µV	10^3 V, 1 V, 10^{-3} V, 10^{-6} V	
Displacement	coulomb per square metre	C/m²	kC/m²	C/cm²	10^4 m², 10^3 m², 1 m²	C/m² = A s/m²
Electric flux	coulomb	C	MC, kC	mC	10^6 C, 10^3 C, 1 C, 10^{-3} C	
Capacitance	farad	F	mF, µF, nF, pF		1 F, 10^{-3} F, 10^{-6} F, 10^{-9} F, 10^{-12} F	F = A s/V = C/V
Permittivity	farad per metre	F/m	µF/m, mF/m, pF/m		1 F/m, 10^{-6} F/m, 10^{-3} F/m, 10^{-12} F/m	$\varepsilon_0 = 8 \cdot 854 \times 10^{-12}$ F/m
Electric polarisation	coulomb per square metre	C/m²	MC/m², kC/m²	C/cm²	10^6 C/m², 10^4 C/m², 10^3 C/m², 1 C/m²	
Electric dipole moment	coulomb metre	C m			1 C m	
Current density	ampere per square metre	A/m²	MA/m², kA/m²	A/mm², A/cm²	10^{-6} A/m², 10^{-4} A/m², 10^{-3} A/m², 1 A/m²	
Linear current density	ampere per metre	A/m	kA/m	A/mm, A/cm	10^3 A/m, 10^2 A/m, 1 A/m	

Table 1.2(g)—*continued*

Quantity	SI units — Name	SI units — Symbol	Decimal multiples and sub-multiples — BSI recommended	Decimal multiples and sub-multiples — Others	Units other than SI	Equivalent value in terms of basic unit	Conversion factors and remarks
(1)	(2)	(3)	(4)	(5)	(6)	(7)	(8)
Magnetic field strength	ampere per metre	A/m	kA/m	A/mm A/cm		10^3 A/m 10^2 A/cm 1 A/m	1 oersted $= 10^3/4\,\Pi$ A/m
Magnetic potential difference		A	kA mA			10^3 A 1 A 10^{-3} A	1 gilbert $= 10/4\,\Pi$ A
Magnetic flux density	tesla	T	mT μT nT			1 T 10^{-3} T 10^{-6} T 10^{-9} T	Wb/m^2 = T 1 gauss $= 10^{-4}$ T
Magnetic flux	weber	Wb	mWb			1 Wb 10^{-3} Wb	V s = Wb 1 maxwell $= 10^{-8}$ Wb
Magnetic vector potential	weber per metre	Wb/m	kWb/m	Wb/mm		10^3 Wb/m 1 Wb/m	1 maxwell/cm $= 10^{-6}$ Wb/m
Mutual inductance, self inductance	henry	H	mH μH nH pH			1 H 10^{-3} H 10^{-6} H 10^{-9} H 10^{-12} H	H = V s/A
Permeability	henry per metre	H/m	μH/m H/m			1 H/m 10^{-6} H/m 10^{-9} H/m	$\mu_0 = 4\,\Pi \times 10^{-7}$ H/m

metre

Quantity	Unit	Symbol				Values	Conversions
Magnetisation	ampere per metre	A/m	kA/m	A/mm		10^3 A/m 1 A/m	1 oersted $= 10^3/4\,\Pi$ A/m
Magnetic polarisation	tesla	T	mT			1 T 10^{-3} T	1 gauss $= 10^{-4}$ T
Magnetic dipole moment	newton square metre per ampere	N m²/A				1 N m²/A	
Resistance	ohm	Ω	GΩ MΩ kΩ mΩ μΩ			10^9 Ω 10^6 Ω 10^3 Ω 1 Ω 10^{-3} Ω 10^{-6} Ω	
Conductance	reciprocal ohm	1/Ω			kS S mS μS	10^3 1/Ω or 10^3 S 1 1/Ω or 1 S 10^{-3} 1/Ω or 10^{-3} S 10^{-6} 1/Ω or 10^{-6} S	S = mho = Siemen
Resistivity	ohm metre	Ωm	GΩ m MΩ m kΩ m mΩ m μΩ m nΩ m	μΩ cm		10^9 Ω m 10^6 Ω m 10^3 Ω m 1 Ω m 10^{-3} Ω m 10^{-6} Ω m 10^{-8} Ω m 10^{-9} Ω m	
(2)							
Conductivity	reciprocal ohm metre	1/Ω m			MS/m kS/m S/m	10^6 1/Ω m or S/m 10^3 1/Ω m or S/m 1 1/Ω m or S/m	
Reluctance	reciprocal henry	1/H				1 1/H	
Permeance	henry	H				1 H	

Table 1.2(g)—continued

Quantity	SI units		Decimal multiples and sub-multiples		Units other than SI	Equivalent value in terms of basic unit	Conversion factors and remarks
	Name	Symbol	BSI recommended	Others			
(1)	(2)	(3)	(4)	(5)	(6)	(7)	(8)
Impedance Reactance	ohm	Ω					
			$M\Omega$			$10^6\Omega$	
			$k\Omega$			$10^3\Omega$	
						1Ω	
			$m\Omega$			$10^{-3}\Omega$	
Conductance	reciprocal ohm	$1/\Omega$					
					kS	$10^3\ 1/\Omega$ or S	
					S	$1\ 1/\Omega$ or S	
					mS	$10^{-3}\ 1/\Omega$ or S	
					μS	$10^{-6}\ 1/\Omega$ or S	
Active power	watt	W					
			TW			10^{12} W	
			GW			10^9 W	
			MW			10^6 W	
			kW			10^3 W	
						1 W	
			mW			10^{-3} W	
			μW			10^{-6} W	
			nW			10^{-9} W	
Apparent power	volt ampere	VA					
			TVA			10^{12} VA	
			GVA			10^9 VA	
			MVA			10^6 VA	
			kVA			10^3 VA	
						1 VA	
			mVA			10^{-3} VA	
			μVA			10^{-6} VA	
			nVA			10^{-9} VA	

Reactive power				T var	10^{12} var
				G var	10^9 var
				k var	10^3 var
				var	1 var
				m var	10^{-3} var
				μ var	10^{-6} var
				n var	10^{-9} var
Electric stress	volt per metre	V/m	kV/mm	1 V/m	1 kV/in = 0·039 370 1 kV/mm
					or 1 V/mil = 39·370 1 kV/m

NOTES ON TABLE 1.2(G)

1. *General note.* SI is an extension of the Giorgi rationalised metre – kilogramme – second – ampere (MKSA) system which has been the preferred system of electrical units in use since 1935, thus engineers should be familiar with most of the units.

Changes which are introduced in SI apart from revised definitions for certain base units are:

(a) The tesla (T) as the unit for magnetic flux density.

(b) The siemen (S) for the reciprocal ohm. This has been adopted by IEC[18] and ISO[19] but has not yet been approved by CGPM[16].

(c) A name is required for the unit to describe periodic phenomena, i.e. the hertz (Hz) for cycles per second.

In electricity and magnetism the SI units align with the MKSA rationalised form of the equations between quantities. Thus the magnetic space constant or permeability of free space constant is $4\Pi \times 10^{-7}$ henry per metre and the electric space constant or permittivity of free space constant is $8·854 \times 10^{-12}$ farad/metre.

2. *Resistivity.* It is the present practice for chemists to work in micro ohms/cm and existing practice is to use dionic instruments etc. calculated in this unit. It will be some time before instruments are calibrated in micro ohm/metre.

Table 1.2(h) LIGHT

Quantity	SI units				Units other than SI	Equivalent value in terms of basic unit	Conversion factors and remarks
	Name	Symbol	Decimal multiples and sub-multiples				
			BSI recommended	Others			
(1)	*(2)*	*(3)*	*(4)*	*(5)*	*(6)*	*(7)*	*(8)*
Luminous intensity	candela	cd				1 cd	
Luminous flux	lumen	lm				1 lm	lm = cd sr (candela steradian)
Illumination	lux	lx		1 m/cm²		1 lx 10⁴ lx	lx = 1m/m²
Luminance	candela per square metre	cd/m²				1 cd/m²	stilb = 1 cd/cm² apostilb = 1/π cd/m² 1 cd/in² = 1550 cd/m² 1 foot lambert = 3426 cd/m² 1 lambert = 3183 cd/m²

Quantities and units of light

The following definitions are based on the International Lighting Vocabulary and on BS 233.

Luminous flux (symbol ϕ)	The light emitted by a source such as a lamp, or received by a surface, irrespective of direction.
Lumen (abbreviation lm)	The SI unit of luminous flux used in describing the total light emitted by a source or received by a surface. (A 100 watt incandescent lamp emits about 1200 lumens.)
Illumination	The process of lighting an object.
Illumination value (symbol E)	The luminous flux incident on a surface, per unit area. The term 'illuminance' has been proposed for international usage. In the Code, the colloquial terms 'illumination' and 'illumination level' have also been used as in current British practice.
Lux (abbreviation lx)	The SI unit of illumination value; it is equal to one lumen per square metre.
Lumen per square foot (abbreviation lm/ft²)	A non-metric unit of illumination value, equal to 10·76 lux. (Previously called the foot-candle, a term still used in the USA.)
Service value of illumination	The mean value of illumination throughout the life of and installation and averaged over the working area.
Initial value of illumination	The mean value of illumination averaged over the working area before depreciation has started, i.e. when the lamps and fittings are new and clean and when the room is freshly decorated.
Mean spherical illumination (scalar illumination)	The average illumination over the surface of a small sphere centred at a given point; more precisely, it is the flux incident on the surface of the sphere divided by the area of the sphere. The term 'scalar' illumination (symbol Es) has been proposed. The unit of scalar illumination is the lux: care is needed to avoid confusing the unit with the illumination on a plane which is measured in the same unit.
Illumination vector	A term used to describe the flow of light. It has both magnitude and direction. The magnitude is defined as the maximum difference in the value of illumination at diametrically opposed surface elements of a small sphere centred at the point under consideration. The direction of the vector is that of the diameter joining the brighter to the darker element. The ratio of the magnitude of the illumination vector to the scalar illumination has been proposed as an index of modelling.
Luminous intensity (symbol I)	The quantity which describes the illuminating power of a source in a particular direction. More precisely it is the luminous flux emitted within a very narrow cone containing that direction divided by the solid angle of the cone.
Candela (abbreviation cd)	The SI unit of luminous intensity. The term 'candle power' designates a luminous intensity expressed in candelas.

Table 1.2(j) SOUND

Quantity	SI units		Decimal multiples and sub-multiples		Units other than SI	Equivalent value in terms of basic unit	Conversion factors and remarks
	Name	Symbol	BSI recommended	Others			
(1)	(2)	(3)	(4)	(5)	(6)	(7)	(8)
Sound intensity	watt per square metre	W/m^2				$1 \ W/m^2$	$1 \ erg \ s^{-1} \ cm^{-2} = 10^{-3} \ W/m^2$
Sound intensity (logarithmic)					decibel bel	$1/10 \ bel$ $1 \ bel$	$20 \log_{10}\left(\dfrac{P}{P_0}\right)$ decibel (dB) where P = measured sound pressure and P_0 = reference sound pressure of $2 \times 10^{-5} \ N/m^2$ (ref. BS. 4196)
Loundness					phon	$1 \ phon$	
Attenuation					neper per metre	np/m	

Table 1.2(k) PHYSICAL CHEMISTRY AND MOLECULAR PHYSICS

Quantity	SI units		Decimal multiples and sub-multiples BSI recommended	Others	Units other than SI	Equivalent value in terms of basic unit	Conversion factors and remarks
	Name	Symbol					
(1)	(2)	(3)	(4)	(5)	(6)	(7)	(8)
Amount of substance					kmol mol	10^3 mol 1 mol	1 lb mol = 0.453 592 37 k mol +
Molar volume					m³/mol	1 m³/mol	
Molar mass					kg/mol g/mol	1 kg/mol 10^{-3} kg/mol	
Molar internal energy					J/mol J/kmol	1 J/mol 10^{-3} J/mol	1 erg/mol = 10^{-7} J/mol
Molar heat capacity					J/mol K; J mol °C J/kmol K; Jkmol °C	1 J/mol K 10^{-3} J/mol K	1 erg mol^{-1} °C^{-1} = 10^{-7} J mol^{-1} K^{-1} Universal gas constant 8·3143 kJ/k mol K
Molar entropy					J/mol K	1 J/mol K	
Molality					kmol/l kmol/m³ mol/l mol/dm³ mol/m³ kmol/kg mol/kg	10^6 mol/m³ 10^3 mol/m³ 10^3 mol/m³ 10^3 mol/m³ 1 mol/m³ 10^3 mol/kg 1 mol/kg	

Table 1.2(k)—*continued*

Quantity	SI units		Units for common use		Units other than SI	Equivalent value in terms of basic unit	Conversion factors and remarks
	Name	Symbol	Decimal multiples and sub-multiples				
			BSI recommended	Others			
(1)	*(2)*	*(3)*	*(4)*	*(5)*	*(6)*	*(7)*	*(8)*
Diffusion coefficient	square metre per second	m²/s				1 m²/s	
Thermal diffusion coefficient	square metre per second	m²/s				1 m²/s	
Molar flow rate					kmol/h mol/h	10^3 mol/h 1 mol/h	1 lb mol/h = 0·453 592 37 kmol/h

Nuclear engineering

It has been the practice to use special units with their individual names for evaluating and comparing results. These units are usually formed by multiplying a unit from the cgs or SI system by a number which matches a value derived from the result of some natural phenomenon. The adoption of SI both nationally and internationally has created the opportunity to examine the practice of using special units in the nuclear industry, with the object of eliminating as many as possible and using the pure system instead.

As an aid to this, ISO draft Recommendations 838[11] and 839[12] have been published, giving a list of quantities with special names, the SI unit and the alternative cgs unit. It is expected that as SI is increasingly adopted and absorbed, those units based on cgs will go out of use. The values of these special units illustrate the fact that a change from them to SI would not be as revolutionary as might be supposed. Examples of these values together with the SI units which replace them are shown in Table 1.2(l).

Table 1.2(l)   NUCLEAR ENGINEERING

	Special unit		SI Replacement
Name		Value	
Angstrom	(Å)	10^{-10}m	m
Barn	(b)	10^{-28}m^2	m^2
Curie	(Ci)	$3 \cdot 7 \times 10^{10}$s^{-1}	s^{-1}
Electronvolt	(eV)	$(1 \cdot 602\,10 \pm 0 \cdot 000\,07) \times 10^{-19}$J	J
Röntgen	(R)	$2 \cdot 58 \times 10^{-4}$C/kg	C/kg

APPLICATION OF UNITS

Calculations

In applying SI units to calculations care should be taken to see that the formulae and references are applicable to an absolute system of units.

Much of the data in existing literature applies to the imperial foot-pound-second system which is a gravitational system.

The following guide rules may prove useful in calculations or design work.

1. Dynamic formulae written W/g will become M (in kilogrammes); conversely gravitational formulae written in W will become Mg (newtons).
2. To avoid errors convert all quantities to consistent units at the outset i.e. keep all quantities in unity multiplied by relevant powers of 10.
3. Use conversion factors based on BS 350[3].

In formulae which need to be transposed, take care that the constants are consistent with the units to be used in SI. Formulae which are expressions of natural laws are always valid and constants are usually independent of units.

Empirical formula should be carefully checked as the constants invariably depend on the units used in the expression.

Consider the deflection (d) of a cantilevered beam with a load (W) supported at a distance (L)

$$d = \frac{WL^3}{3EI}$$

where E = modulus of elasticity for the material
and I = moment of inertia of the section

This expression is valid in any system provided the units are coherent.

These values may be in metric data sheets expressed in the following magnitudes; W in N; kN: L in m; E in N/mm^2; GN/m^2; I in cm^4; d in mm.

In SI the above formula is arranged thus:

$$d(\text{m}) = \frac{W(\text{N})L^3(\text{m}^3)}{3E(\text{N}/\text{m}^2)I(\text{m}^4)}$$

$$\therefore \ (\text{m}) = \frac{(\text{N})(\text{m}^3)}{(\text{N})(\text{m}^4)} \times (\text{m}^2) = \text{m}^{(5-4)} = \text{m}$$

Thus, for example, taking:

$$W = 1 \text{ tonf}$$
$$L = 10 \text{ ft}$$
$$E = 30 \times 10^6 \text{ lbf/in}^2$$
$$I = 69\cdot2 \text{ in}^4$$

(It should be noted that the units for W and L are inconsistent with the units for E and I).

Converting to SI units:

$$W = 1 \text{ tonf} = 1016 \text{ kg} = 1016 \times 9\cdot8605 \text{ N} = 9964\cdot02 \text{ N}$$
$$L = 10 \text{ ft} = 3\cdot048 \text{ m}$$
$$\begin{aligned}E &= 30 \times 10^6/\text{lbf/in}^2 = 30 \times 10^6 \times 6\cdot895 \times 10^3 \text{ N/m}^2\\ &= 206\cdot85 \times 10^9 \text{ N/m}^2\\ &= 206\cdot85 \text{ GN/m}^2\end{aligned}$$
$$\begin{aligned}I = 69\cdot2 \text{ in}^4 &= 69\cdot2 \times 0\cdot254^4 \text{m}^4\\ &= 69\cdot2 \times (2\cdot54 \times 10^{-2})^4\\ &= 69\cdot2 \times 41\cdot61 \times 10^{-8}\\ &= 287\cdot94 \times 10^{-7}\text{m}^4\end{aligned}$$

Putting in these values for the calculations:

$$d = \frac{[9\cdot964 \times 10^3] \times [3\cdot048^3]}{3 \times [206\cdot85 \times 10^9] \times [28\cdot794 \times 10^{-6}]} \text{ m}$$

$$\frac{9\cdot964 \times 3\cdot048^3}{3 \times 206\cdot85 \times 28\cdot794} \text{ m}$$

$$d = 0\cdot0158 \text{ m}$$
$$d = 15\cdot8 \text{ mm}$$

Note how the powers of 10 readily cancel out and how the I value has been brought to the power 10^{-6} to obey the rule of 10^n where $n = \pm3$.

The need for this might be questioned but if 10^{-7} had been used the cancelling out would not be so easily carried out.

Sometimes the conversion of formulae is not so straightforward and in such cases it is suggested that the new metric formulae be proved by comparing the metric calculations with the results obtained from the imperial formula using imperial units which should be converted to metric units when resolved. When using imperial formulae containing constants which apply to several units, the calculation should be made using accurate conversion factors and rounding to the required degree of precision as a final step.

Table 1.3(a)   METRIC TO IMPERIAL CONVERSION FACTORS

SI units	British units
SPACE AND TIME	
Length:	
1 μm (micron)	$= 39 \cdot 37 \times 10^{-6}$ in
1 mm	$= 0 \cdot 039\ 370\ 1$ in
1 cm	$= 0 \cdot 393\ 701$ in
1 m	$= 3 \cdot 280\ 84$ ft
1 m	$= 1 \cdot 093\ 61$ yd
1 km	$= 0 \cdot 621\ 371$ mile
Area:	
1 mm²	$= 1 \cdot 550 \times 10^{-3}$ in²
1 cm²	$= 0 \cdot 1550$ in²
1 m²	$= 10 \cdot 7639$ ft²
1 m²	$= 1 \cdot 195\ 99$ yd
1 ha	$= 2 \cdot 471\ 05$ acre
Volume:	
1 mm³	$= 61 \cdot 0237 \times 10^{-6}$ in³
1 cm³	$= 61 \cdot 0237 \times 10^{-3}$ in³
1 m³	$= 35 \cdot 3147$ ft³
1 m³	$= 1 \cdot 307\ 95$ yd³
Capacity:	
10⁶m³	$= 219 \cdot 969 \times 10^{6}$ gal
1 m³	$= 219 \cdot 969$ gal
1 litre (1)	$\begin{cases} = 0 \cdot 219\ 969 \text{ gal} \\ = 1 \cdot 759\ 80 \text{ pint} \end{cases}$
Capacity flow:	
10³/m³/s	$= 791 \cdot 9 \times 10^{6}$ gal/h
1 m³/s	$= 13 \cdot 20 \times 10^{3}$ gal/min
1 litre/s	$= 13 \cdot 20$ gal/min
1 m³/kW h	$= 219 \cdot 969$ gal/kW h
1 m³/s	$= 35 \cdot 3147$ ft³/s (cusecs)
1 litre/s	$= 0 \cdot 588\ 58 \times 10^{-3}$ ft³/min (cfm)
Velocity:	
1 m/s	$= 3 \cdot 280\ 84$ ft/s $= 2 \cdot 236\ 94$ mile
1 km/h	$= 0 \cdot 621\ 371$ mile/h
Acceleration:	
1 m/s²	$= 3 \cdot 280\ 84$ ft/s²
MECHANICS	
Mass:	
1 g	$= 0 \cdot 035\ 274$ oz
1 kg	$= 2 \cdot 204\ 62$ lb
1 t	$= 0 \cdot 984\ 207$ ton $= 19 \cdot 6841$ cwt
Mass flow:	
1 kg/s	$= 2 \cdot 204\ 62$ lb/s $= 7 \cdot 936\ 64$ klb/h
Mass density:	
1 kg/m³	$= 0 \cdot 062\ 428$ lb/ft³
1 kg/litre	$= 10 \cdot 022\ 119$ lb/gal
Mass per unit length:	
1 kg/m	$= 0 \cdot 671\ 969$ lb/ft $= 2 \cdot 015\ 91$ lb/yd
Mass per unit area:	
1 kg/m²	$= 0 \cdot 204\ 816$ lb/ft²
Specific volume:	
1 m³/kg	$= 16 \cdot 0185$ ft³/lb
1 litre/tonne	$= 0 \cdot 223\ 495$ gal/ton
Momentum:	
1 kg m/s	$= 7 \cdot 233\ 01$ lbft/s
Angular momentum:	
1 kg m²/s	$= 23 \cdot 7304$ lbft²/s

Table 1.3(a)—*continued*

SI units	British units
Moment of inertia:	
1 kg m²	= 23·7304 lbft²
Force:	
1 N	= 0·224 809 lbf
Weight (force) per unit length:	
1 N/m	= 0·068 521 8 lbf/ft = 0·205 566 lbf/yd
Moment of force (or torque):	
1 Nm	= 0·737 562 lbf/ft
Weight (force) per unit area:	
1 N/m²	= 0·020 885 lbf/ft²
Pressure:	
1 N/m²	= 1·450 38 × 10⁻⁴ lbf/in²
1 bar	= 14·5038 lbf/in²
1 bar	= 0·986 923 atmosphere
1 mbar	= 0·401 463 in H₂O
	= 0·029 53 in Hg
Stress:	
1 N/mm²	= 6·474 90 × 10⁻² tonf/in²
1 MN/m²	= 6·474 90 × 10⁻² tonf/in²
1 hbar	= 0·647 490 tonf/in²
Second moment of area:	
1 cm⁴	= 0·024 025 in⁴
Section modulus:	
1 m³	= 61 023·7 in³
1 cm³	= 0·061 023 7 in³
Kinematic viscosity:	
1 m²/s	= 10·762 75 ft²/s = 10⁶ cSt
1 cSt	= 0·038 75 ft²/h
Energy, work:	
1 J	= 0·737 562 ft lbf
1 MJ	= 0·3725 hph
1 MJ	= 0·277 78 kW h
Power:	
1 W	= 0·737 562 ft lbf/s
1 kW	= 1·3410 hp = 737·562 ft lbf/s
Fluid mass:	
(Ordinary) 1 kg/s	= 2·204 62 lb/s = 7936·64 lb/h
(Velocity) 1 kg/m² s	= 0·204 815 lb/ft²s

HEAT

SI units	British units
Temperature:	
(Interval) 1 degK	= 9/5 deg R (Rankine)
1 degC	= 9/5 deg F
(Coefficient) 1 degR⁻¹	= 1 deg F⁻¹ = 5/9 deg C
1 degC⁻¹	= 5/9 deg F⁻¹
Quantity of heat:	
1 J	= 9·478 17 × 10⁻⁴ Btu
1 J	= 0·238 846 cal
1 kJ	= 947·817 Btu
1 GJ	= 947·817 × 10³ Btu
1 kJ	= 526·565 CHU
1 GJ	= 526·565 × 10³ CHU
1 GJ	= 9·478 17 therm
Heat flow rate:	
1 W(J/s)	= 3·412 14 Btu/h
1 W/m²	= 0·316 998 Btu/ft²h
Thermal conductivity:	
1 W/m°C	= 6·933 47 Btu in/ft² h °F

Table 1.3(a)—*continued*

SI units	British units
Coefficient and heat transfer :	
1 W/m^2 C	= 0·176 110 Btu/ft^2 h °F
Heat capacity:	
1 J/°C	= 0·526 57 × 10^{-3} Btu/°R
Specific heat capacity :	
1 J/g°C	= 0·238 846 Btu/lb °F
1 kJ/kg°C	= 0·238 846 Btu/lb °F
Entropy:	
1 J/K	= 0·526 57 × 10^{-3} Btu/°R
Specific Entropy :	
1 J/kg degC	= 0·238 846 × 10^{-3} Btu/lb °F.
1 J/kg degK	= 0·238 846 × 10^{-3} Btu/lb °R
Specific energy/Specific latent heat	
1 J/g	= 0·429 923 Btu/lb
1 J/kg	= 0·429 923 × 10^{-3} Btu/lb
Calorific value :	
1 kJ/kg	= 0·429 923 Btu/lb
1 kJ/kg	= 0·773 861 4 CHU/lb
1 J/m^3	= 0·026 839 2 × 10^{-3} Btu/ft^3
1 kJ/m^3	= 0·026 839 2 Btu/ft^3
1 kg/litre	= 4·308 86 Btu/gal
1 kJ/kg	= 0·009 630 2 therm/ton
ELECTRICITY	
Permeability:	
1 H/m	= $10^7/4 \, \Pi \, \mu o$
Magnetic flux density :	
1 tesla	= 10^4 gauss = 1 Wb/m^2
Conductivity:	
1 mho	= 1 reciprocal ohm
1 Siemen	= 1 reciprocal ohm
Electric stress :	
1 kV/mm	≐ 25·4 kV/in
1 kV/m	= 0·0254 kV/in

Conversion of existing imperial terms

If it is necessary to convert existing imperial terms to a metric equivalent, care should be taken to ensure that the converted value implies the same degree of accuracy. The conversion factor must convey the same order of precision as the original value.

Thus to translate 1 in as 25·4 mm or 1000 ft as 304·8 m conveys a tolerance which in most cases would be too precise.

Particular care is needed when converting machined tolerances. With a simple dimension such as 0·836 in, it is reasonable to assume that this dimension can be met because an imperial micrometer can measure to 0·001 in. The conversion factor from BS 350[3] for 0·836 in is 21·2344 mm. Thus 0·004 mm which is the last figure in our conversion represents an accuracy of 0·000 016 in which is beyond the scope of most toolroom measuring devices. In such cases it should be borne in mind that a metric micrometer can measure to 0·01 mm and with a vernier attachment to 0·002 mm, thus for the greatest possible accuracy our converted reading should be 21·234 mm. BS 2856[4] provides a procedure for converting tolerances which will give the essential accuracy required for precise dimensional interchangeability and gives guidance on the optimum fineness of rounding the converted sizes.

Table 1.3(b)   UNIVERSAL CONSTANTS IN SI UNITS

The digits in parentheses following each quoted value represent the standard deviation error in the final digits of the quoted value as computed on the criterion of internal consistency. The unified scale of atomic weights is used throughout ($^{12}C = 12$). C = coulomb; G = gauss; Hz = hertz; J = joule; N = newton; T = tesla; u = unified nuclidic mass unit; W = watt; Wb = weber. For result multiply the numerical value by the SI unit.

Constant	Symbol	Numerical value	SI unit
Speed of light in vacuum	c	2·997 925(1)	10^8 m s^{-1}
Gravitational constant	G	6·670(5)*	10^{-11} N m^2 kg^{-2}
Elementary charge	e	1·602 10(2)	10^{-19} C
Avogadro constant	N_A	6·022 52(9)	10^{26} kmol^{-1}
Mass unit	u	1·660 43(2)	10^{-27} kg
Electron rest mass	m_e	9·109 08(13)	10^{-31} kg
		5·485 97(3)	10^{-4} u
Proton rest mass	m_p	1·672 52(3)	10^{-27} kg
		1·007 276 63(8)	u
Neutron rest mass	m_n	1·67 482(3)	10^{-27} kg
		1·008 6654(4)	u
Faraday constant	F	9·648 70(5)	10^4 C mol^{-1}
Planck constant	h	6·625 59(16)	10^{-34} J s
	$h/2\pi$	1·054 494(25)	10^{-34} J s
Fine-structure constant	α	7·297 20(3)	10^{-3}
	$1/\alpha$	137·0388(6)	
Charge-to-mass ratio for electron	e/m_e	1·758 796(6)	10^{11} C kg^{-1}
Quantum of magnetic flux	hc/e	4·135 56(4)	10^{-11} Wb
Rydberg constant	R_∞	1·097 3731(1)	10^7 m^{-1}
Bohr radius	a_0	5·291 67(2)	10^{-11} m
Compton wavelength of electron	$h/m_e c$	2·426 21(2)	10^{-12} m
	$\lambda c/2\pi$	3·861 44(3)	10^{-13} m
Electron radius	$e^2/m_e c^2 = r_e$	2·817 77(4)	10^{-15} m
Thomsen cross section	$8\eta r_e^2/3$	6·6516(2)	10^{-29} m^2
Compton wavelength of proton	$\lambda c, p$	1·321 398(13)	10^{-15} m
	$\lambda c, p/2\pi$	2·103 07(2)	10^{-16} m
Gyromagnetic ratio of proton	γ	2·675 192(7)	10^8 rad s^{-1} T^{-1}
	$\gamma/2\pi$	4·257 70(1)	10^7 Hz T^{-1}
(Uncorrected for diamagnetism H$_2$O)	γ'	2·675 123(7)	10^8 rad s^{-1} T^{-1}
	$\gamma'/2\pi$	4·257 59(1)	10^7 Hz T^{-1}
Bohr magneton	μB	9·2732(2)	10^{-24} J T^{-1}
Nuclear magneton	μN	5·050 50(13)	10^{-27} J T^{-1}
Proton moment	μ_p	1·410 49(4)	10^{-26} J T^{-1}
	$\mu_p/\mu N$	2·792 76(2)	
(Uncorrected for diamagnetism in H$_2$O sample)		2·792 68(2)	
Gas constant	R_0	8·314 34(35)	J deg^{-1} mol^{-1}
Boltzmann constant	k	1.380 54(6)	10^{-23} J deg^{-1}
First radiation constant ($2\eta hc^2$)	c_1	3·741 50(9)	10^{-16} W m^2
Second radiation constant (hc/k)	c_2	1·438 79(6)	10^{-2} m deg
Stefan-Boltzmann constant	σ	5·6697(10)	10^{-8} W m^{-2} deg^{-4}

* The universal gravitational constant is not, and cannot in our present state of knowledge, be expressed in terms of other fundamental constants. The value given here is a direct determination by P. R. Heyl and P. Chrzanowski, J. Res. Natl. Bur. Std. (U.S.) 29, 1 (1942).

The above values are extracts from 'Review of Modern Physics' Vol. 37 No. 4 October 1965 published by the American Institute of Physics.

CONVERSION TABLES

Description of Tables 1.4(a) to (r)

Although conversion factors are given in Tables 1.2(a) to 1.2(k), it is sometimes desirable to refer to a conversion table for the more readily used engineering units. Such conversioning tables from Imperial to SI units are shown in Tables 1.4(a) to (r) as follows:

Table	Description
1.4(a)	Length – fractions of an inch to millimetres
1.4(b)	Length – feet and inches to metres
1.4(c)	Length – yards to metres
1.4(d)	Length – miles to kilometres
1.4(e)	Area – square feet to square metres
1.4(f)	Area – square yards to square metres
1.4(g)	Volume – cubic inches to cubic centimetres
1.4(h)	Volume – cubic feet to cubic metres
1.4(j)	Volume – cubic yards to cubic metres
1.4(k)	Capacity – gallons to litres
1.4(l)	Mass (Mass by weight) – pounds to kilogrammes
1.4(m)	Mass (Mass by weight) – tons to tonne
1.4(n)	Mass (Mass by weight) – cwt. qr. to tonne
1.4(p)	Pressure – lbf/in^2 to bar
1.4(q)	Stress – $tonf/in^2$ to N/mm^2
1.4(r)	Stress – kgf/mm^2 to N/mm^2

Reading of tables

It will be noted that each table is shown in two parts i.e. main table and auxiliary table. This is to enable all whole number quantities within the range of the table to be expressed in SI units.

Consider the following example from Table 1.4(b), feet and inches to metres. To convert 83 ft 7 in, to metres.

From Table 1.4(b); main table   3 ft 7 in =   1·092 m
           auxiliary table 80 ft   = 24·384 m

                   83 ft 7 in = 25·476 m

For other detailed conversions see BS 350[3] Part 2 and Supplement No. 1 (BSI/PD 6203)[7].

1-44

Table 1.4(a) LENGTH: FRACTIONS OF AN INCH TO MILLIMETRES (Correct to the nearest millimetre)

in	0	1	2	3	4	5	6	7	8	9	10	11
						mm						
—	—	25	51	76	102	127	152	178	203	229	254	279
$\frac{1}{16}$	2	27	52	78	103	129	154	179	205	230	256	281
$\frac{1}{8}$	3	29	54	79	105	130	156	181	206	232	257	283
$\frac{3}{16}$	5	30	56	81	106	132	157	183	208	233	259	284
$\frac{1}{4}$	6	32	57	83	108	133	159	184	210	235	260	286
$\frac{5}{16}$	8	33	59	84	110	135	160	186	211	237	262	287
$\frac{3}{8}$	10	35	60	86	111	137	162	187	213	238	264	289
$\frac{7}{16}$	11	37	62	87	113	138	164	189	214	240	265	291
$\frac{1}{2}$	13	38	64	89	114	140	165	191	216	241	267	292
$\frac{9}{16}$	14	40	65	90	116	141	167	192	217	243	268	294
$\frac{5}{8}$	16	41	67	92	117	143	168	194	219	244	270	295
$\frac{11}{16}$	17	43	68	94	119	144	170	195	221	246	271	297
$\frac{3}{4}$	19	44	70	95	121	146	171	197	222	248	273	298
$\frac{13}{16}$	21	46	71	97	122	148	173	198	224	249	275	300
$\frac{7}{8}$	22	48	73	98	124	149	175	200	225	251	276	302
$\frac{15}{16}$	24	49	75	100	125	151	176	202	227	252	278	303

AUXILIARY TABLE

ft	1	2	3	4	5	6	7	8	9
m	0·305	0·610	0·914	1·219	1·524	1·829	2·134	2·438	2·743

Table 1.4(b) LENGTH: FEET AND INCHES TO METRES (Correct to the nearest mm)

ft	0	1	2	3	4	5	in 6	7	8	9	10	11
							m					
0	—	0·025	0·051	0·076	0·102	0·127	0·152	0·178	0·203	0·229	0·254	0·279
1	0·305	0·330	0·356	0·381	0·406	0·432	0·457	0·483	0·508	0·533	0·559	0·584
2	0·610	0·635	0·660	0·686	0·711	0·737	0·762	0·787	0·813	0·838	0·864	0·889
3	0·914	0·940	0·965	0·991	1·016	1·041	1·067	1·092	1·118	1·143	1·168	1·194
4	1·219	1·245	1·270	1·295	1·321	1·346	1·372	1·397	1·422	1·448	1·473	1·499
5	1·524	1·549	1·575	1·600	1·626	1·651	1·676	1·702	1·727	1·753	1·778	1·803
6	1·829	1·854	1·880	1·905	1·930	1·956	1·981	2·007	2·032	2·057	2·083	2·108
7	2·134	2·159	2·184	2·210	2·235	2·261	2·286	2·311	2·337	2·362	2·388	2·413
8	2·438	2·464	2·489	2·515	2·540	2·565	2·591	2·616	2·642	2·667	2·692	2·718
9	2·743	2·769	2·794	2·819	2·845	2·870	2·896	2·921	2·946	2·972	2·997	3·023
10	3·048	3·073	3·099	3·124	3·150	3·175	3·200	3·226	3·251	3·277	3·302	3·327
11	3·353	3·378	3·404	3·429	3·454	3·480	3·505	3·531	3·556	3·581	3·607	3·632
12	3·658	3·683	3·708	3·734	3·759	3·785	3·810	3·835	3·861	3·886	3·912	3·937
13	3·962	3·988	4·013	4·039	4·064	4·089	4·115	4·140	4·166	4·191	4·216	4·242
14	4·267	4·293	4·318	4·343	4·369	4·394	4·420	4·445	4·470	4·496	4·521	4·547
15	4·572	4·597	4·623	4·648	4·674	4·699	4·724	4·750	4·775	4·801	4·826	4·851
16	4·877	4·902	4·928	4·953	4·978	5·004	5·029	5·055	5·080	5·105	5·131	5·156
17	5·182	5·207	5·232	5·258	5·283	5·309	5·334	5·359	5·385	5·410	5·436	5·461
18	5·486	5·512	5·537	5·563	5·588	5·613	5·639	5·664	5·690	5·715	5·740	5·766
19	5·791	5·817	5·842	5·867	5·893	5·918	5·944	5·969	5·994	6·020	6·045	6·071
20	6·096	—	—	—	—	—	—	—	—	—	—	—

Table 1.4(b)—*continued*

AUXILIARY TABLE

ft	30	40	50	60	70	80	90	100	150	200
m	9·144	12·192	15·240	18·288	21·336	24·384	27·432	30·480	45·720	60·960

Table 1.4(c) LENGTH: YARDS TO METRES (Correct to the nearest 0·01 m)

yd	0	10	20	30	40	50	60	70	80	90
					m					
0	—	9·14	18·29	27·43	36·58	45·72	54·86	64·01	73·15	82·30
100	91·44	100·58	109·73	118·87	128·02	137·16	146·30	155·45	164·59	173·74
200	182·88	192·02	201·17	210·31	219·46	228·60	237·74	246·89	256·03	265·18
300	274·32	283·46	,292·61	301·75	310·90	320·04	329·18	338·33	347·47	356·62
400	365·76	374·90	384·05	393·19	402·34	411·48	420·62	429·77	438·91	448·06
500	457·20	466·34	475·49	484·63	493·78	502·92	512·06	521·21	530·35	539·50
600	548·64	557·78	566·93	576·07	585·22	594·36	603·50	612·65	621·79	630·94
700	640·08	649·22	658·37	667·51	676·66	685·80	694·94	704·09	713·23	722·38
800	731·52	740·66	749·81	758·95	768·10	777·24	786·38	795·53	804·67	813·82
900	822·96	832·10	891·25	850·39	859·34	868·68	877·82	886·97	896·11	905·26
1000	914·40	—	—	—	—	—	—	—	—	—

AUXILIARY TABLE

yd	1	2	3	4	5	6	7	8	9
m	0·914	1·829	2·743	3·658	4·572	5·486	6·401	7·315	8·230

Table 1.4(d) LENGTH: MILES TO KILOMETRES (Correct to nearest 10 m)

miles	0	1	2	3	4	5	6	7	8	9
					km					
0	—	1·61	3·22	4·83	6·44	8·05	9·66	11·27	12·87	14·48
10	16·09	17·70	19·31	20·92	22·53	24·14	25·75	27·36	28·97	30·58
20	32·19	33·80	35·41	37·01	38·62	40·23	41·84	43·45	45·06	46·67
30	48·28	49·89	51·50	53·11	54·72	56·33	57·94	59·55	61·16	62·76
40	64·37	65·98	67·59	69·20	70·81	72·42	74·03	75·64	77·25	78·86
50	80·47	82·08	83·69	85·30	86·90	88·51	90·12	91·73	93·34	94·95
60	96·56	98·17	99·78	101·39	103·00	104·61	106·22	107·83	109·44	111·05
70	112·65	114·26	115·87	117·48	119·09	120·70	122·31	123·92	125·53	127·14
80	128·75	130·36	131·97	133·58	135·19	136·79	138·40	140·01	141·62	143·23
90	144·84	146·45	148·06	149·67	151·28	152·89	154·50	156·11	157·72	159·33
100	160·94	—	—	—	—	—	—	—	—	—

AUXILIARY TABLES

miles	0·1	0·2	0·3	0·4	0·5	0·6	0·7	0·8	0·9
km	0·161	0·322	0·483	0·644	0·805	0·966	1·127	1·288	1·448

furlongs	1	2	3	4	5	6	7
km	0·201	0·402	0·604	0·804	1·006	1·207	1·408

Table 1.4(e) AREA: SQUARE FEET TO SQUARE METRES (Correct to the nearest 0·01 m²)

ft²	0	10	20	30	40	50	60	70	80	90
						m²				
0	—	0·93	1·86	2·79	3·72	4·65	5·57	6·50	7·43	8·36
100	9·29	10·22	11·15	12·08	13·01	13·94	14·86	15·79	16·72	17·65
200	18·58	19·51	20·44	21·37	22·30	23·23	24·15	25·08	26·01	26·94
300	27·87	28·80	29·73	30·66	31·59	32·52	33·45	34·37	35·30	36·23
400	37·16	38·09	39·02	39·95	40·88	41·81	42·74	43·66	44·59	45·52
500	46·45	47·38	48·31	49·24	50·17	51·10	52·03	52·95	53·88	54·81
600	55·74	56·67	57·60	58·53	59·46	60·39	61·32	62·25	63·17	64·10
700	65·03	65·96	66·89	67·82	68·75	69·68	70·61	71·54	72·46	73·39
800	74·32	72·25	76·18	77·11	78·04	78·97	79·90	80·83	81·76	82·68
900	83·61	84·54	85·47	86·40	87·33	88·26	89·19	90·12	91·05	91·97
1000	92·90	—	—	—	—	—	—	—	—	—

AUXILIARY TABLES

ft²	1	2	3	4	5	6	7	8	9
m²	0·09	0·19	0·28	0·37	0·46	0·56	0·65	0·74	0·84

Table 1.4(f) AREA: SQUARE YARDS TO SQUARE METRES (Correct to the nearest 0·01 m²)

yd²	0	10	20	30	40	50	60	70	80	90
						m²				
0	—	8·36	16·72	25·08	33·45	41·81	50·17	58·53	66·89	75·25
100	83·61	91·97	100·34	108·70	117·06	125·42	133·78	142·14	150·50	158·86
200	167·23	175·59	183·95	192·31	200·67	209·03	217·39	225·75	234·12	242·48
300	250·84	259·20	267·56	275·92	284·28	292·65	301·01	309·37	317·73	326·09
400	334·45	342·81	351·17	359·54	367·90	376·26	384·62	392·98	401·34	409·70
500	418·06	426·43	434·79	443·15	451·51	459·87	468·23	476·59	484·95	493·32
600	501·68	510·04	518·40	526·76	535·12	543·48	551·84	560·21	568·57	576·93
700	585·29	593·65	602·01	610·37	618·73	627·10	635·46	643·82	652·18	660·54
800	668·90	677·26	685·62	693·99	702·35	710·71	719·07	727·43	735·79	744·15
900	753·52	760·88	769·37	777·60	785·96	794·32	802·68	811·04	819·41	827·77
1000	836·13	—	—	—	—	—	—	—	—	—

AUXILIARY TABLE

yd²	1	2	3	4	5	6	7	8	9
m²	0·84	1·67	2·51	3·34	4·18	5·02	5·85	6·69	7·53

Table 1.4(g)   VOLUME: CUBIC INCHES TO CUBIC CENTIMETRES OR MILLILITRES
(Correct to the nearest cm³)

in^3	0	10	20	30	40	50	60	70	80	90
					cm^3					
0	—	164	328	492	655	819	983	1 147	1 311	1 475
100	1 639	1 803	1 966	2 130	2 294	2 458	2 622	2 786	2 950	3 114
200	3 277	3 441	3 605	3 769	3 933	4 097	4 261	4 425	4 588	4 752
300	4 916	5 080	5 244	5 408	5 572	5 735	5 899	6 063	6 227	6 391
400	6 555	6 719	6 883	7 046	7 210	7 374	7 538	7 702	7 866	8 030
500	8 194	8 357	8 521	8 685	8 849	9 013	9 177	9 341	9 505	9 668
600	9 832	9 996	10 160	10 324	10 488	10 652	10 816	10 979	11 143	11 307
700	11 471	11 635	11 799	11 963	12 126	12 290	12 454	12 618	12 782	12 946
800	13 110	13 274	13 437	13 601	13 765	13 929	14 093	14 257	14 421	14 585
900	14 748	14 912	15 076	15 240	15 404	15 568	15 732	15 896	16 059	16 223
1000	16 387	—	—	—	—	—	—	—	—	—

AUXILIARY TABLE

in^3	0	1	2	3	4	5	6	7	8	9
ml.	—	16	33	49	66	82	98	115	131	147

Table 1.4(h)   VOLUME: CUBIC FEET TO CUBIC METRES (Correct to the nearest 0·01 m³)

ft^3	0	10	20	30	40	50	60	70	80	90
					m^3					
0	—	0·28	0·57	0·85	1·13	1·42	1·70	1·98	2·27	2·55
100	2·83	3·11	3·40	3·68	3·96	4·25	4·53	4·81	5·10	5·38
200	5·66	5·95	6·23	6·51	6·80	7·08	7·36	7·65	7·93	8·21
300	8·50	8·78	9·06	9·34	9·63	9·91	10·19	10·48	10·76	11·04
400	11·33	11·61	11·89	12·18	12·46	12·74	13·03	13·31	13·59	13·88
500	14·16	14·44	14·72	15·01	15·29	15·57	15·86	16·14	16·42	16·71
600	16·99	17·27	17·56	17·84	18·12	18·41	18·69	18·97	19·26	19·54
700	19·82	20·11	20·39	20·67	20·95	21·24	21·52	21·80	22·09	22·37
800	22·65	22·94	23·22	23·50	23·79	24·07	24·35	24·64	24·92	25·20
900	25·49	25·77	26·05	26·33	26·62	26·90	27·18	27·47	27·75	28·03
1000	28·32	—	—	—	—	—	—	—	—	—

AUXILIARY TABLE

ft^3	0	1	2	3	4	5	6	7	8	9
m^3	—	0·03	0·06	0·08	0·11	0·14	0·17	0·20	0·23	0·25

Table 1.4(j) VOLUME: CUBIC YARDS TO CUBIC METRES (Correct to the nearest 0·01 m³)

yd³	0	1	2	3	4	5	6	7	8	9
					m³					
0	—	0·76	1·53	2·29	3·06	3·82	4·59	5·35	6·12	6·88
10	7·65	8·41	9·17	9·94	10·70	11·47	12·23	13·00	13·76	14·53
20	15·29	16·06	16·82	17·58	18·35	19·11	19·88	20·64	21·41	22·17
30	22·94	23·70	24·47	25·23	25·99	26·76	27·52	28·29	29·05	29·82
40	30·58	31·35	32·11	32·88	33·64	34·41	35·17	35·93	36·70	37·46
50	38·23	38·99	39·76	40·52	41·29	42·05	42·82	43·58	44·34	45·11
60	45·87	46·64	47·40	48·17	48·93	49·70	50·46	51·23	51·99	52·75
70	53·52	54·28	55·05	55·81	56·58	57·34	58·11	58·87	59·64	60·40
80	61·16	61·93	62·69	63·46	64·22	64·99	65·75	66·52	67·28	68·05
90	68·81	69·57	70·34	71·10	71·87	72·63	73·40	74·16	74·93	75·69
100	75·46	—	—	—	—	—	—	—	—	—

Table 1.4(k) CAPACITY: GALLONS TO LITRES (Correct to the nearest 10 ml)

gal	0	1	2	3	4	5	6	7	8	9
					litres					
0	—	4·55	9·09	13·64	18·18	22·73	27·28	31·82	36·37	40·91
10	45·46	50·01	54·55	59·10	63·65	68·19	72·74	77·28	81·83	86·38
20	90·92	95·47	100·01	104·56	109·11	113·65	118·20	122·74	127·29	131·84
30	136·38	140·93	145·48	150·02	154·57	159·11	163·66	168·21	172·75	177·30
40	181·84	186·39	190·94	195·48	200·03	204·57	209·12	213·67	218·21	222·76
50	227·31	231·85	236·40	240·94	245·49	250·04	254·58	259·13	263·67	268·22
60	273·77	277·31	281·86	286·40	290·95	295·50	300·04	304·59	309·13	313·68
70	318·23	322·77	327·32	331·87	336·41	340·96	345·50	350·05	354·60	359·14
80	363·69	368·23	372·78	377·33	381·87	386·42	390·96	395·51	400·06	404·60
90	409·15	413·69	418·24	422·79	427·33	431·88	436·43	440·97	445·52	450·06
100	454·61	—	—	—	—	—	—	—	—	—

AUXILIARY TABLE

Pint litre	1 0·568	2 1·136	3 1·705	4 2·273	5 2·841	6 3·410	7 3·978

Table 1.4(l) MASS: POUNDS TO KILOGRAMMES (Correct to nearest gramme or 0·001 kg)

Pounds	0	1	2	3	4	5	6	7	8	9
					kilogrammes					
0	—	0·454	0·907	1·361	1·814	2·268	2·722	3·175	3·629	4·082
10	4·536	4·990	5·443	5·897	6·350	6·804	7·257	7·711	8·165	8·618
20	9·072	9·525	9·979	10·433	10·886	11·340	11·793	12·247	12·701	13·154
30	13·608	14·061	14·515	14·969	15·422	15·876	16·329	16·783	17·237	17·690
40	18·144	18·597	19·051	19·505	19·958	20·412	20·865	21·319	21·772	22·226
50	22·680	23·133	23·587	24·040	24·494	24·948	25·401	25·855	26·308	26·762
60	27·216	27·669	28·123	28·576	29·030	29·484	29·937	30·391	30·844	31·298
70	31·752	32·205	32·659	33·112	33·566	24·019	34·473	34·927	35·380	35·834
80	36·287	36·741	37·195	37·648	38·102	38·555	39·009	39·463	39·916	40·370
90	40·823	41·277	41·731	42·184	42·638	43·091	43·549	43·999	44·452	44·906
100	45·359	—	—	—	—	—	—	—	—	—

Table 1.4(m) MASS: TON TO TONNE

Tons	0	1	2	3	4	5	6	7	8	9
0	—	1·0160	2·0321	3·0481	4·0642	5·0802	6·0963	7·1123	8·1284	9·1444
10	10·1605	11·1765	12·1926	13·2086	14·2247	15·2407	16·2568	17·2728	18·2888	19·3049
20	20·3209	21·3370	22·3530	23·3691	24·3851	25·4012	26·4172	27·4333	28·4493	29·4654
30	30·4814	31·4975	32·5135	33·5295	34·5456	35·5616	36·5777	37·5937	38·6098	39·6258
40	40·6419	41·6579	42·6740	43·6900	44·7061	45·7221	46·7382	47·7542	48·7703	49·7865
50	50·8023	51·8184	52·8344	53·8505	54·8665	55·8826	56·8986	57·9147	58·9307	59·9468
60	60·9628	61·9789	62·9949	64·0110	65·0270	66·0430	67·0591	68·0751	69·0912	70·1072
70	71·1233	72·1393	73·1554	74·1714	75·1875	76·2035	77·2196	78·2356	79·2517	80·2677
80	81·2838	82·2998	83·3158	84·3319	85·3479	86·3640	87·3800	88·3961	89·4121	90·4282
90	91·4442	92·4603	93·4763	94·4924	95·5084	96·5245	97·5405	98·5566	99·5726	100·589
100	101·605	—	—	—	—	—	—	—	—	—

Table 1.4(n) MASS: CWT; QTR TO TONNE

qtr	0	1	2	3	qtr	0	1	2	3
cwt		tonne			cwt		tonne		
0	—	0·0127	0·0254	0·0381	10	0·5080	0·5207	0·5334	0·5461
1	0·0508	0·0635	0·0762	0·0889	11	0·3588	0·5715	0·5842	0·5969
2	0·1016	0·1143	0·1270	0·1397	12	0·6096	0·6223	0·6350	0·6477
3	0·1524	0·1651	0·1778	0·1905	13	0·6604	0·6731	0·6858	0·6985
4	0·2032	0·2159	0·2286	0·2413	14	0·7112	0·7239	0·7366	0·7493
5	0·2540	0·2667	0·2794	0·2921	15	0·7620	0·7747	0·7874	0·8001
6	0·3048	0·3175	0·3302	0·3429	16	0·8128	0·8255	0·8382	0·8509
7	0·3556	0·3683	0·3810	0·3937	17	0·8636	0·8763	0·8890	0·9017
8	0·4064	0·4191	0·4318	0·4445	18	0·9144	0·9271	0·9398	0·9525
9	0·4572	0·4699	0·4826	0·4953	19	0·9652	0·9779	0·9906	1·0033

Table 1.4(p) PRESSURE: LB/IN²F TO $10^5 N/m^2$ (bar)

lb/in^2f	0	1	2	3	4	5	6	7	8	9
0	—	0·069	0·138	0·207	0·276	0·345	0·414	0·483	0·552	0·621
10	0·689	0·758	0·827	0·896	0·965	1·034	1·103	1·172	1·241	1·310
20	1·379	1·448	1·517	1·586	1·654	1·724	1·793	1·862	1·931	1·999
30	2·068	2·137	2·206	2·275	2·344	2·413	2·482	2·551	2·620	2·689
40	2·758	2·827	2·895	2·965	3·034	3·103	3·172	3·241	3·309	3·378
50	3·447	3·516	3·585	3·654	3·723	3·792	3·861	3·930	3·999	4·068
60	4·137	4·206	4·275	4·344	4·413	4·442	4·551	4·619	4·688	4·757
70	4·826	4·895	4·964	5·033	5·102	5·171	5·240	5·309	5·378	5·447
80	5·516	5·585	5·654	5·723	5·792	5·861	5·929	5·998	6·067	6·136
90	6·205	6·274	6·343	6·412	6·481	6·550	6·619	6·688	6·757	6·826
100	6·895	—	—	—	—	—	—	—	—	—

AUXILIARY TABLE

lb/in^2f	150	200	250	300	350	400	450	500	550	600
$10^5 N/m^2$ (bar)	10·342	13·79	17·24	20·68	24·13	27·58	31·03	34·47	37·92	41·37

lb/in^2f	650	700	750	800	850	900	950	1 000	—	—
$10^5 N/m^2$ (bar)	44·82	48·26	51·71	55·16	58·61	62·05	65·50	68·95	—	—

To convert from lb/in^2f to kN/m^2 multiply above factors by 100; thus for $15 lb/in^2f = 1·034$ bar $= 103·4$ kN/m²

Table 1.4(q) STRESS: TONS/IN²F TO MN/m² (N/mm²)

ton/in²f	0	1	2	3	4	5	6	7	8	9
0	—	15·44	30·89	46·33	61·78	77·22	92·67	108·11	123·55	139·00
10	154·44	169·89	185·33	200·78	216·22	231·66	247·11	262·55	278·00	293·44
20	308·89	324·33	339·77	355·22	370·66	386·11	401·55	417·00	432·44	447·88
30	463·33	478·77	494·22	509·66	525·11	540·55	555·99	571·44	586·88	602·33
40	617·77	633·22	648·66	664·10	679·55	694·99	710·44	725·88	741·32	756·77
50	772·21	787·66	803·10	818·55	833·99	849·43	864·88	880·32	895·77	911·21
60	926·66	942·10	957·55	972·99	988·43	1003·88	1019·32	1034·77	1050·21	1065·65
70	1081·10	1096·54	1111·99	1127·43	1142·87	1158·32	1173·77	1189·21	1204·65	1220·10
80	1235·54	1250·98	1266·43	1281·87	1297·32	1312·76	1328·21	1343·65	1359·09	1374·54
90	1389·98	1405·43	1420·87	1436·32	1451·76	1467·21	1482·65	1498·09	1513·54	1528·98
100	1544·43	—	—	—	—	—	—	—	—	—

Table 1.4(r) STRESS: kgf/mm² TO MN/m² (N/mm²)

kgf/mm²	0	1	2	3	4	5	6	7	8	9
0	0·00	9·81	19·61	29·42	39·23	40·03	58·84	68·65	78·45	88·26
10	98·07	107·87	117·68	127·49	137·29	147·10	156·91	166·71	176·52	186·33
20	196·13	205·94	215·75	225·55	235·36	245·17	254·97	264·78	274·59	284·39
30	294·20	304·01	313·81	323·62	333·43	343·23	353·04	362·85	372·65	382·46
40	392·27	402·07	411·88	421·69	431·49	441·30	451·11	460·91	470·72	480·53
50	490·33	500·14	509·95	519·75	529·56	539·37	549·17	558·98	568·79	578·59
60	588·40	598·21	608·01	617·82	627·63	637·43	647·24	657·05	666·85	676·66
70	686·47	696·27	706·08	715·89	725·69	735·50	745·31	755·11	764·92	774·73
80	784·53	794·34	804·15	813·95	823·76	833·57	843·37	853·18	862·99	872·79
90	882·60	892·41	902·21	912·02	921·83	931·63	941·44	951·25	961·05	970·86
100	980·67	—	—	—	—	—	—	—	—	—

AUXILIARY TABLE

kgf/mm²	150	200	250	300	350	400	450	500	550	600
MN/m²	1 471·00	1 961·33	2 451·66	2 942·00	3 432·33	3 922·66	4 412·99	4 903·33	5 393·66	5 883·99

IMPLEMENTATION OF SI IN ENGINEERING

The use of SI units in the engineering sectors of industry, particularly in the redesign of existing equipment, is guided by a series of programmes which have been co-ordinated and published by the British Standards Institution.

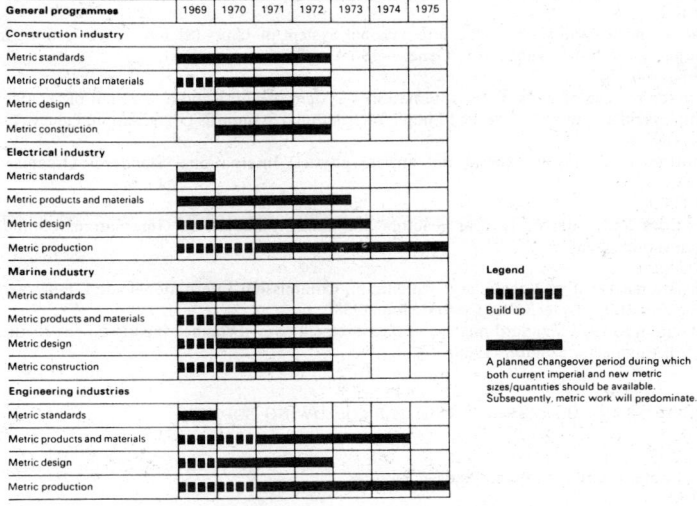

Figure 1.3. General Metrication Programmes

These programmes (see Figure 1.3) are orientated, sector by sector to achieve a substantial amount of metric working by 1975 and are phased to allow for a natural time sequence of metric development within the various engineering disciplines. Thus the building and construction industries metric programme may be completed well before the heavy engineering industry programme.

These programmes are:

PD 6030   Programme for the change to the metric system in the construction industry. BSI (1967).

PD 6424   The adoption of the metric system in engineering: basic programme and guide BSI (1968).

PD 6427   The adoption of the metric system in the electrical industry: basic programme BSI (1969).

PD 6430   The adoption of the metric system in the marine industry: reprt; basic programme and guide BSI (1969).

BIBLIOGRAPHY

1. BS 233
   Glossary of terms used in illumination and photometry; BSI (1953)
2. BS 308
   Engineering drawing practice; BSI (1964)
3. BS 350
   Conversion factors and tables, Parts 1 and 2; BSI (1959 and 1962)
4. BS 2856
   Precise conversion of inch and metric sizes engineering drawings; BSI (1957)

5. BS 3763
   International System (SI) Units; BSI (1964)
6. PD 5686
   The use of SI Units; BSI (1969)
7. PD 6203
   Supplement No. 1 to BS 350, Additional tables for SI conversions; BSI (1967)
8. FIELDEN, G. B. R. Metric units for pressures and stresses; BSI News January 3 (1971)
9. ISO R.31
   Basic quantities and units of the International System of Units (SI units) 2nd edition, International Standards Organisation, Geneva (1965)
10. ISO R.1000
    Rules for the use of units of the International System of Units and a selection of the decimal multiples and sub-multiples of the SI units, International Standards Organisation, Geneva (1969)
11. ISO DR.838
    Quantities and units of atomic and nuclear physics, International Standards Organisation, Geneva (1965)
12. ISO DR.839
    Quantities and units of nuclear reactors and ionizing radiations, International Standards Organisation, Geneva (1965)
13. IEC 50(45)
    International Lighting Vocabulary, 3rd edition, Commission Internationale de L'Eclairage and International Electrotechnical Commission (1958)
14. Addendum to 1967 Standard method of measurement of civil engineering quantities, Institution of Civil Engineers, London (1968)

ABBREVIATIONS IN THE TEXT REFER TO THE FOLLOWING BODIES:

15. BIPM
    The Bureau International des Poids et mesures
16. CGPM
    The Conference Generale des Poids et mesures
17. CIPM
    The Comité International des Poids et mesures
18. IEC
    The International Electrotechnical Commission
19. ISO
    The International Standards Organisation
20. IUPAC
    The International Union of Pure and Applied Chemistry
21. IUPAP
    The International Union of Pure and Applied Physics
22. ISO/TC 12
    ISO Committee, Quantities, Units, Symbols
23. ISO/TC 17
    ISO Committee, Steel

GENERAL REFERENCE PUBLICATIONS

These are additional to those which are referred to in the text.

PD 6031
Use of the metric system in the construction industry; BSI (1968)
PD 6249
Dimensional co-ordination in building. Estimates of timing for BSI work; BSI (1967)
PD 6432
Recommendations for the co-ordination of dimensions in building, arrangements of building components and assemblies within functional groups; BSI (1969)
PD 6444
Recommendations for the co-ordination of dimensions in building. Basic spaces for structures, external envelope and internal sub-division; BSI (1969)

BS 3643
Part 1. ISO metric screw thread. Thread data and standard thread series; BSI (1963)
BS 4318
Recommendations for preferred metric basis sizes for engineering; BSI (1968)
BS 4391
Recommendations for metric basis sizes for metal wire, sheet and strip; BSI (1969)
BS 4500
ISO limits and fits; BSI (1969)

ANDERTON, P. and BIGG, P. H., *National Physical Laboratory Changing to the Metric System. Conversion Factors Symbols and Definitions*, HMSO London (2nd edition 1967)

PARRISH, A., *SI Conversion Charts for Imperial and Metric Quantities*, Iliffe, London (1969)

BSI, *Readimetric*, British Standards Institution, London

MULLIN, J. W., SI Units in Chemical Engineering, The Chemical Engineer, September 1967

Ministry of Public Building and Works, Metrication in the Construction Industry; No. 1 Metrication in Practice; No. 2 Calculations in SI Units Structural, Civil, Heating & Ventilating; No. 3 Craftsman's Pocket Book, HMSO London (1970)

Ministry of Public Building and Works, Going Metric in the Construction Industry; Part 1 Why and When, HMSO London (1967); Part 2 Dimensional Co-ordination, HMSO London (1968)

Public Health Act 1961. The Building Regulations 1965, Metric Equivalents of Dimensions, (Ministry of Housing and Local Government and Welsh Office) HMSO London (1968)

Public Health Act 1961. The Building Regulations 1965, Metric Values. Consultative Proposals, (Ministry of Housing and Local Government and Welsh Office) HMSO London (1969)

FAIRWEATHER, L. and SWILA, JAN. A., *Architects Journal Metric Handbook*, The Architectural Press Limited (1969)

Standard Method of Measurement of Building Works, The Royal Institution of Chartered Surveyors and The National Federation of Building Trades Employers, 5th edition, Metric, July 1968

SI Units for the Compressed Air Industry, Fluid Power International, September 1970

Change to Metric. Reference Manual, The Institution of Heating & Ventilating Engineers 1970

Metrication Guide, Scientific Instrument Manufacturers' Association of Great Britain 1970

Manual on Metrication, British Plastics Federation, June 1969

Metric Units with reference to Water, Sewerage and Related Subjects, Report of Working Party, Ministry of Housing and Local Government, HMSO London (1965)

HADDOCK, A., Going Metric in the UK Petroleum Industry, Journal of Institute of Petroleum No. 548 Vol. 56, March 1970

SYMBOLS AND ABBREVIATIONS

G. R. DARBY

Table 1.5 QUANTITIES AND UNITS OF PERIODIC AND RELATED PHENOMENA
(Based on ISO recommendation R31)

Symbol	Quantity
T	periodic time
$\tau, (T)$	time constant of an
f, ν	exponentially varying quantity frequency
η	rotational frequency
ω	angular frequency
λ	wave length
$\sigma(\tilde{\nu})$	wave number
k	circular wave number
$\log e \, (A_1/A_2)'$	natural logarithm of the ratio of two amplitudes
$10 \log_{10}(P_1/P_2)$	ten times the common logarithm of the ratio of two powers
δ	damping coefficient
Λ	logarithmic decrement
α	attenuation co-efficient
β	phase co-efficient
γ	propagation co-efficient

Table 1.6 QUANTITIES AND UNITS OF MECHANICS
(Based on ISO recommendation R31)

Symbol	Quantity
m	mass
e, ρ	density (mass density)
d	relative density
v	specific volume
p	momentum
b, p_0, p_θ	moment of momentum (angular momentum)
I, J	moment of inertia (dynamic moment of inertia)
F	force
$G(P, W)$	weight
γ	specific weight (weight density)
M	moment of force
M	bending moment
T	torque, moment of a couple
p	pressure
σ	normal stress
τ	shear stress
e, ε	linear strain (relative elongation)

Table 1.6—*continued*

Symbol	Quantity
γ	shear strain (shear angle)
$\Theta\theta$	volume strain (bulk strain)
μ, v	{ Poisson's ratio / Poisson's number
E	Young's modulus (modulus of elasticity)
G	shear modulus (modulus of rigidity)
K	bulk modulus (modulus of compression)
x, κ	compressibility (bulk compressibility)
I, I_a	second moment of area (second axial moment of area)
I_p, J	second polar moment of area
$Z, W\left(\dfrac{I}{v}\right)$	section modulus
$\mu(f)$	co-efficient of friction (factor of friction)
$\eta(\mu)$	viscosity (dynamic viscosity)
γ	kinematic viscosity
$\sigma(\gamma)$	surface tension
A, W	work
E, W	energy
Ep, U, V, Φ	potential energy
E_k, K, T	kinetic energy
p	power

Table 1.7 SYMBOLS FOR QUANTITIES AND UNITS OF HEAT
(Based on ISO recommendation R31)

Symbol	Quantity
T, Θ	{ thermodynamic temperature / absolute temperature
$t, 0\Theta$	customary temperature
α, λ	linear expansion co-efficient
α, β, γ	cubic expansion co-efficient
β	pressure coefficient
Q	heat, quantity of heat
$\Phi(q)$	heat flow rate
$q(\phi)$	density of heat flow rate
$\lambda(k)$	thermal conductivity
h, k, U, α	co-efficient of heat transfer
$a(\alpha, x, k)$	thermal diffusivity
C	heat capacity
c	specific heat capacity
cp	specific heat capacity at constant pressure
cp	specific heat capacity at constant volume
γ, x, k	ratio of the specific heat capacities
S	entropy
s	specific entropy
$U(E)$	internal energy
$H(I)$	enthalpy
F	free energy
G	Gibbs function
$u(e)$	specific internal energy
$h(i)$	specific enthalpy
f	specific free energy
g	specific Gibbs function
L	latent heat
l	specific latent heat

Table 1.8 SYMBOLS FOR QUANTITIES AND UNITS OF ACOUSTICS
(Based on ISO recommendation R31)

Symbol	Quantity
T	period, periodic time
f, v	frequency, frequency interval
ω	angular frequency, circular frequency
λ	wavelength
k	circular wave number
ρ	density (mass density)
P_s	static pressure
p	(instantaneous) sound pressure
$\varepsilon, (x)$	(instantaneous) sound particle displacement
u, v	(instantaneous) sound particle velocity
a	(instantaneous) sound particle acceleration
q, U	(instantaneous) volume velocity
c	velocity of sound
E	sound energy density
$P, (N, W)$	sound energy flux, sound power
I, J	sound intensity
$Z_s, (W)$	specific acoustic impedance
$Z_a, (Z)$	acoustic impedance
$Z_m, (w)$	mechanical impedance
$L_p, (L_N, L_w)$	sound power level
$L_p, (L)$	sound pressure level
δ	damping co-efficient
Λ	logarithmic decrement
α	attenuation co-efficient
β	phase co-efficient
γ	propagation co-efficient
δ	dissipation co-efficient
r, τ	reflection co-efficient
γ	transmission co-efficient
$\alpha, (\alpha_a)$	acoustic absorption co-efficient
R	{ sound reduction index / sound transmission loss
A	equivalent absorption area of a surface or object
T	reverberation time
$L_N, (\Lambda)$	loudness level
N	loudness

Table 1.9 SYMBOLS FOR QUANTITIES AND UNITS OF ELECTRICITY AND MAGNETISM
(Based on ISO recommendation R31)

Symbol	Quantity
I	electric current
Q	electric charge, quantity of electricity
e	volume density of charge, charge density
σ	surface density of charge
$E, (K)$	electric field strength
V, ϕ	electric potential
$U, (V)$	potential difference, tension
E	electromotive force
D	displacement (rationalised displacement)
D'	non-rationalised displacement
Ψ	electric flux, flux of displacement (flux of rationalised displacement)
Ψ'	flux of non-rationalised displacement
C	capacitance
ε	permittivity

Table 1.9 *continued*

Symbol	Quantity
ε_0	permittivity of vacuum
ε'	non-rationalised permittivity
ε'_0	non-rationalised permittivity of vacuum
ε_r	relative permittivity
χ_e	electric susceptibility
χ'_e	non-rationalised electric susceptibility
P	electric polarisation
$p, (p_e)$	electric dipole moment
$J, (S)$	current density
$A, (\alpha)$	linear current density
H	magnetic field strength
H'	non-rationalised magnetic field strength
U_m	magnetic potential difference
F, F_m	magnetomotive force
B	magnetic flux density, magnetic induction
Φ	magnetic flux
A	magnetic vector potential
L	self inductance
M, L_{12}	mutual inductance
$k, (x, k)$	coupling co-efficient
σ	leakage co-efficient
μ	permeability
μ_0	permeability of vacuum
μ'	non-rationalised permeability
μ'_0	non-rationalised permeability of vacuum
μ_r	relative permeability
x, k	magnetic susceptibility
x', k'	non-rationalised magnetic susceptibility
m	electromagnetic moment (magnetic moment)
H_i, M	magnetisation
B_i, J	magnetic polarisation
J'	non-rationalised magnetic polarisation
ω	electromagnetic energy density
S	Poynting vector
c	velocity of propagation of electromagnetic waves in vacuo
R	resistance (to direct current)
G	conductance (to direct current)
e	resistivity
γ, σ	conductivity
R, R_m	reluctance
$A, (P)$	permeance
N	number of turns in winding
m	number of phases
p	number of pairs of poles
ϕ	phase displacement
Z	impedance (complex impedance)
$[Z]$	modulus of impedance (impedance)
X	reactance
R	resistance
Q	quality factor
Y	admittance (complex admittance)
$[Y]$	modulus of admittance (admittance)
B	susceptance
G	conductance
P	active power
$S, (P_s)$	apparent power
$Q, (P_q)$	reactive power

Table 1.10 MATHEMATICAL SIGNS AND SYMBOLS FOR USE IN TECHNOLOGY
(Based on ISO recommendation R31)

Sign, symbol	Quantity
$=$	equal to
$\neq$	not equal to
$\equiv$	identically equal to
$\triangle$	corresponds to
$\approx$	approximately equal to
$\rightarrow$	approaches
$\simeq$	asymptotically equal to
$\sim$	proportional to
∞	infinity
$<$	smaller than
$>$	larger than
$\leq$ $\leqslant$ $\leqq$	smaller than or equal to
$\geq$ $\geqslant$ $\geqq$	larger than or equal to
$\lll$	much smaller than
$\ggg$	much larger than
$+$	plus
$-$	minus
$\cdot$ $\times$	multiplied by
$\dfrac{a}{b}$ a/b	a divided by b
$\lvert a \rvert$	magnitude of a
a^n	a raised to the power n
$a^{\frac{1}{2}}$ $a^{1/2}$ $\sqrt{a}$ $\sqrt{a}$	square root of a
$a^{1/n}$ $a^{\frac{1}{n}}$ $\sqrt[n]{a}$ $\sqrt[n]{a}$	n'th root of a
$\bar{a}$ $\langle a \rangle$	mean value of a
$p!$	factorial p, $1 \times 2 \times 3 \times \dots \times p$
$\binom{n}{p}$	binomial co-efficient, $\dfrac{n(n-1)\dots(n-p+1)}{1 \times 2 \times 3 \times \dots \times p}$
Σ	sum
Π	product
$f(x)$ $f(x)$	function f (of f) of the variable x
$[f(x)]_a^b$ $f(x)/_a^b$	$f(b) - f(a)$
$\lim\limits_{x \to a} f(x)$; $\lim_{x \to a} f(x)$	the limit to which $f(x)$ tends as x approaches a
Δx	delta x = finite increment of x
δx	delta x = variation of x
$\dfrac{df}{dx}$; df/dx; $f'(x)$	differential co-efficient of $f(x)$ with respect to x
$\dfrac{d^n f}{dx^n}$; $f^{(n)}(x)$	differential co-efficient of order n of $f(x)$
$\dfrac{\partial f(x, y, \dots)}{\partial x}$; $\left(\dfrac{\partial f}{\partial x}\right)_{y\dots}$	partial differential co-efficient of $f(x, y, \dots)$ with respect to x, when $y, \dots$ are held constant
df	the total differential of f
$\int f(x) dx$	indefinite integral of $f(x)$ with respect to x
$\int_a^b f(x)dx$; $\int_a^b f(x)dx$	definite integral of $f(x)$ from $x = a$ to $x = b$
e	base of natural logarithms
e^x; $\exp x$	e raised to the power x
$\log_a x$	logarithm to the base a of x
$\ln x$; $\log_e x$	natural logarithm (Napierian logarithm) of x
$\lg x$; $\log x$; $\log_{10} x$	common (Briggsian) logarithm of x
$\mathrm{lb}\, x$; $\log_2 x$	binary logarithm of x
$\sin x$	sine of x
$\cos x$	cosine of x

Table 1.10—*continued*

Symbol	Quantity
tan x; tg x	tangent of x
cot x; ctg x	cotangent of x
sec x	secant of x
cosec x	cosecant of x
arcsin x	arc sine of x
arccos x	arc cosine of x
arctan x, arctg x	arc tangent of x
arccot x, arcctg x	arc cotangent of x
arcsec x	arc secant of x
arccosec x	arc cosecant of x
sinh x	hyperbolic sine of x
cosh x	hyperbolic cosine of x
tanh x	hyperbolic tangent of x
coth x	hyperbolic cotangent of x
sech x	hyperbolic secant of x
cosech x	hyperbolic cosecant of x
arsinh x	inverse hyperbolic sine of x
arcosh x	inverse hyperbolic cosine of x
artanh x	inverse hyperbolic tangent of x
arcoth x	inverse hyperbolic cotangent of x
arsech x	inverse hyperbolic secant of x
arcosech x	inverse hyperbolic cosecant of x
i, j	imaginary unity, $1^2 = -1$
$Re\ z$	real part of z
$Im\ z$	imaginary part of z
$\lvert z \rvert$	modulus of z
arg z	argument of z
z^*	conjugate of x, complex conjugate of z
$\tilde{A}$	transpose of matrix A
A^*	complex conjugate matrix of matrix A
$A\dagger$	Hermitian conjugate matrix of matrix A
Aa	vector
$\lvert A \rvert$, A	magnitude of vector
A · B	scalar product
A × B, A ∧ B	vector product
∇	differential vector operator
∇ϕ, grad ϕ	gradient of ϕ
∇ · A, div A	divergence of A
∇ × A, ∇ ∧ A curl A, rot A	curl of A
$\nabla^2\phi$, $\Delta\phi$	Laplacion of ϕ

Table 1.11 ABBREVIATIONS OF COMMON ENGINEERING TERMS

ab	prefix attached to electrical units to obtain names for electromagnetic units e.g. abampere
abs	absolute
abstr	abstract
AC	alternating current
AD	air dried timber
AF	audio frequency
A/F	across flats
AFC	automatic frequency control
alt	alternating

Table 1.11—*continued*

aml	amplitude modulation with noise limiter
anhyd	anhydrous
amu	atomic mass unit
ani	atmosphère normale internationale (international normal atmosphere)
API	American Petroleum Institute
aq	aqueous
AS	air seasoned timber
asb	asbestos
ASME	American Society of Mechanical Engineers
ASSY	assembly
AT	ampere turn
at	atomic
atm	standard atmosphere
at. wt.	atomic weight
Av	atomic volume
AVC	automatic gain control
avp	avoirdupois (weight)
aw ⎫ AW ⎭	atomic weight
AWG	American Wire Gauge
BA	British Association Screw Thread
BAR	blade area ratio
BDV	breakdown voltage
bev	bevelled
B ext	breadth (extreme)
BG	Birmingham Gauge (Wire)
BHN	Brinell Hardness No.
BHP	brake horse power
BM	bench mark
BMEP	brake mean effective pressure
B mld	breadth (moulded)
bmp	brake mean power
BOD	bio-chemical oxygen demand
BOT	Board of Trade
b.p.	boiling point
bp	brake power
brz	bronze
BFB	broad flange beam
BS	British Standard
BSB	British Standard beam
BSC	British Standard channel
BSC	British Standard cycle (screw thread)
BS COND	British Standard conduit (screw thread)
BS Fine ⎫ BSF ⎭	British Standard Fine (screw thread)
BSG	British Standard gauge
BS Pipe	British Standard pipe (screw thread)
BSPF	British Standard pipe parallel screw thread (fastening)
BSPTr	British Standard pipe taper screw thread
BSW	British Standard Whitworth (screw thread)
CAT	cooled anode transmitting valve
cath	cathode
CB	centre of buoyancy
C'BORE	counterbore
ccb	cubic capacity of bunkers
c to c	centre to centre
cg ⎫ CG ⎭	centre of gravity
cgc	corrugated galvanised iron

Table 1.11—*continued*

cgs	centimetre, gramme second (units system)
CH HD	cheese head
CHAM	chamfered
c.i.⎫ CI⎭	cast iron
CIPM	Conference International des Poids et Mesures
CL	centre line
CLR	centre of lateral resistance
Co	condensation number
conc	concentrated
constr.	constructed
cop	copper
cp	candle power
crit	critical
CRT	cathode ray tube
Cr	chromium
CRS	centres
CS	cast steel
CSK	countersunk
cub⎫ cu⎬ c⎭	cubic
CV	calorific value
CV	cheval vapeur (horsepower)
cy	capacity
cyl.	cylinder
d	density
D	Debye unit
DAR	disc area ratio
DC	direct current
d ext	draught (extreme)
D ext	depth (extreme)
dia⎫ DIA⎭	diameter
DO	Drawing Office
dp	deep
DPN	Diamond Pyramid Hardness Number
DRG	drawing
dw	dead weight
dwg	drawing
E	Eotras unit
E	Youngs' modulus
EHP	effective horse power
emf	electro motive force
emu	electro magnetic unit
equiv	equivalent
ESU	electrostatic unit
F	froude number
fao	finish all over
fb	flat bar
FHP	friction horse power
FFL	finished floor level
FIG	figure
fp	freezing point
FP	forward perpendicular
FS	factor of safety
g	gravity
galv⎫ GALV⎭	galvanised

Table 1.11—*continued*

GL	ground level
Gr	Grashof number
gr. wt.	gross weight
gv	gravimetric volume
Gz	Graetz number
HEX	hexagon
hf	high frequency
HMD	hydraulic mean depth
HP	horse power
HT ⎱ ht ⎰	high tensile
hts	high tensile steel
HYD	hydraulic
hv	high voltage
I/D	inside diameter
IF	intermediate frequency
ig	ignition
IHP	indicated horse power
IMEP	indicated mean effective pressure
insol.	insoluble
INSUL	insulated
I.P.	indicated power
k	1000
Le	Lewis number
LH	left hand
LT	low tension
Lubr	lubricant
lv	low voltage
LU	loudness unit
LVN	limiting viscosity number
M	1000 000
MATL	material
MAX	maximum
M/C	machine
mci	malleable cast iron
mcp	mean effective pressure
mf	machine finish
MIN	minimum
MK	mark
MKSA (or Georgi)	metric, kilogramme, second, ampere (units system)
ml	mean level
mmf	magnetomotive force
Mo	molybdenum
mp	melting point
MW	molecular weight
N	normal concentration
N	modulus of rigidity
NA	neutral axis
Nb	niobium
n. br	naval brass
ND ⎱ NP ⎰	nominal pressure
ndp	normal diametric pitch
NHS	horological screw thread
Ni	nickel
No	number
NPS	American straight pipe screw thread
NPT	American taper pipe screw thread

Table 1.11—*continued*

NPT	American conduit electrical screw thread
NPTG	American taper pipe gas cylinder screw thread
nt	net tonnage
NTP	normal temperature and pressure
nts, NTS	not to scale
Nu	Nurselt number
OASM	ohm, ampere, second, metre (system of units)
OBM	Ordnance Bench Mark
OD	Ordnance Datum (Newlyn)
O/D	outside diameter
o f	oil fired
p	pitch
PCD, pcd	pitch circle diameter
pd	potential difference
Pe	Peclet number
pH	hydrogen ion concentration
pm	phase modulation
PNEU	pneumatic
p–p	peak to peak
ppm	parts per million
Pr	Prandtl number
PRESS	pressure
PS	proof stress
QPC	Quasi propulsive coefficient
rad	radius
R	Rayleigh (luminous intensity)
R, Re	Reynolds number
RF	radio frequency
RH	right hand
RMS	root mean square value
RSJ	rolled steel joist
SAE	American Society of Automotive Engineers
S	Spat
S	Svedberg
SBC	small bayonet cap
SCR	screwed
S'FACE	spot face
Sh	Sherwood number
SI	International System of Units
SIT	spontaneous ignition temperature
S/N curve	stress number curve
SOL	soluble
SP	single pole
SPEC	specification
Sp Gr	specific gravity
Sp ht	specific heat
SQ	square
St	Stanton number
STD	standard
stp	standard temperature and pressure
SWG	Imperial (Standard) Wire Gauge
SWL	safe working load
rh	relative humidity
rms	root mean square
TEMP	temperature
TPI, tpi	threads per inch

Table 1.11—*continued*

TU	transmission unit
T unit	Trichomatic unit
UB	universal beam
UC	universal column
U'CUT	undercut
UN	Unified Thread form
UNC	Unified Coarse Series screw thread
UNEF	Unified extra Fine series screw thread
UNF	Unified Fine series screw thread
UNS	Unified Selective series screw thread
uts	ultimate tensile stress
Va	vanadium
VAC	vacuum
VERT	vertical
VOL	volume
VPM	Vickers Pyramid Number
VU	volume unit
WI	wrought iron
WT	weight
y.p.	yield point

Table 1.12 ABBREVIATIONS OF COMMON UNITS

a	$\begin{cases} \text{are} \\ \text{year} \end{cases}$
Å	ångstrom
A	ampere
asb	apostilb
AU	astronomical unit
AT	assay ton
b	barn
bar	bar
Bi	Biot (unit of current in electromagnetic CGS system)
Btu ⎫ BthU⎭	British thermal unit
c	curie
C	coulomb
°C	degree Celsius
cal	calorie
cc	cubic centimetre
cd	candela
CHU	Centigrade heat unit
Ci	curie
cl	centilitre
cm	centimetre
CM	carat
cP	centipoise
c/s	cycle per second
cSt	centistoke
ct	carat
cu. cm.	cubic centimetre
cu ft	cubic foot
cu in	cubic inch
cusec	cubic foot per second
cwt	hundredweight
d	day

Table 1.12—*continued*

dB	decibel
dm	decimetre
dwt	pennyweight
dyn	dyne
e unit ⎫ E unit ⎭	X-ray doseage
erg	erg
eV	electronvolt
f	force
F	farad
°F	degree Fahrenheit
fc	foot candle
ft	foot
ft L	foot Lambert
ft lb	foot pound
g	gramme
G	gauss
gal	gallon
Gb	gilbert
g cal	gramme calorie
gl	gill
gm	gramme
g.p.m.	gallons per minute
g.p.s.	gallons per second
gr	grain
Gs	gauss
h	hour
H	henry
ha	hectare
hp	horse power
hp hr	horse power hour
Hz	hertz
in	inch
in Hg	inch of mercury
J	joule
K	kelvin
kc	kilocycle
k cal	kilocalorie
kc/s	kilocycle per second
kg	kilogramme
kgf	kilogramme force
km	kilometre
kn ⎫ kt ⎭	knot
kV	kilovolt
kVA	kilovolt ampere
kW	kilowatt
kW h	kilowatt hour
L	lambert
l	litre
lb	pound
lbf	pound force
lea	league
lm	lumen
ly	light year
lx	lux
m	metre
m	⎧ molality ⎩ molal concentration

Table 1.12—*continued*

M	molar concentration
mA	milliampere
mbar	millibar
mcps	mega cycles per second
MEV	mega electron volt
mF	millifarad
micron	$\begin{cases} \text{length} & -10^{-6} \text{ metre} \\ \text{pressure} - 10^{-3} \text{ mm Hg} \end{cases}$
mil	$\begin{cases} \text{angular} & -\frac{1}{1000} \text{ rt. angle} \\ \text{length} \ \frac{1}{1000} \text{ inch} \\ \text{volume} - \text{millilitre} \end{cases}$
min	minute (time)
mks	metre kilogramme second
ml	millilitre
mL	millilambert
mm	millimetre
mm fd	micromicrofarad
mm Hg	millimetre of mercury
mmm	millimicrons
mol	mole (amount of substance)
mpg	miles per gallon
mpm	metres per minute
$\left.\begin{array}{l} \text{m/s} \\ \text{mps} \end{array}\right\}$	metres per second
mt	metric ton
mV	millivolt
mW h	megawatt hour
Mx	maxwell
N	newton
$\left.\begin{array}{l} \text{n. mile} \\ \text{nm} \end{array}\right\}$	nautical mile
Np	neper
nt	nit
ntm	net ton mile
n unit	neutron dose
Oe	oersted
oz	ounze (avoirdupois)
oz. t	ounce (troy)
p	perch
P	poise
P	phon
Pa	pascal
pc	parsec
pdl	poundal
ph	phot
psi	pounds per square inch
pwt	pennyweight
$\left.\begin{array}{l} \text{q} \\ \text{ql} \end{array}\right\}$	quintal
qts	quart
$\left.\begin{array}{l} \text{r} \\ \text{R} \end{array}\right\}$	Röntgen
R	Réaumier
°R	degree Rankine
rad	radian
rpm	revolutions per minute
rps	revolutions per second
s	second (time)

Table 1.12—*continued*

S	Siemen
S St	stokes
sb	stilb
sn	sthéne
sr	steradian
T	tesla
t	tonne
th	thermie
V	volt
VA	volt ampere
W	watt
Wb	weber
yd	yard

Table 1.13   ABBREVIATIONS OF SOME ENGINEERING BODIES

AA	Architectural Association 34 Bedford Square, London WC1
ABBF	Association of Bronze and Brass Founders 69 Harborne Road, Birmingham 15
ABCM	Association of British Chemical Manufacturers 86 Strand, London WC2
ABMAC	Association of British Manufacturers of Agricultural Chemicals 93 Albert Embankment, London SE1
ABOCF	Association of British Organic and Compound Fertilizers 23 St. Mary Axe, London EC3
ABT	Association of Building Technicians 22 London Bridge Street, London SE1
ACE	Association of Consulting Engineers 2 Victoria Street, London SW1
ACMA	Asbestos Cement Manufacturers Association 89 Cornwall Street, Birmingham 3
AEA	Agricultural Engineers Association Ltd. 6 Buckingham Gate, London SW1
AEE	Atomic Energy Establishment Winfrith, Dorchester
AERE	Atomic Energy Research Establishment Harwell, Didcot
AEU	Amalgamated Engineering Union 110 Peckham Road, London SE15
AHEM	Association of Hydraulic Equipment Manufacturers Ltd. 54 Warwick Square, London SW1
AMEME	Association of Mining, Electrical and Mechanical Engineers 62 Talbot Road, Manchester 16
APCM	Association of Plastic Cable Makers 381 Salisbury House, London Wall, London EC2
APLE	Association of Public Lighting Engineers 78 Buckingham Gate, London SW1
ASEE	Association of Supervising Electrical Engineers 26 Bloomsbury Square, London WC1
AWRE	Atomic Weapons Research Establishment Harwell, Didcot

Table 1.13—*continued*

BABS	British Aluminium Building Service
	Norfolk House, St. James' Square, London SW1
BAS	Building Advisory Service
	82 New Cavendish Street, London W1
BCAS	British Compressed Air Society
	11 Ironmonger Lane, London EC2
BCPMA	British Chemical Plant Manufacturers Association
	14 Suffolk Street, London SW1
BCSA	British Constructional Steelwork Association
	Hancock House, 87 Vincent Square, London SW1
BEAB	British Electrical Approvals Board
	Mark House, 153 London Road, Kingston-on-Thames
BEAMA	British Electrical and Allied Manufacturers Association
	Leicester House, Leicester Street, London WC2
BECM	British Electrical Conduit Manufacturers
	96 Hagley Road, Birmingham 16
BEDA	British Electrical Development Association
	2 Savoy Hill, London WC2
BEPC	British Electrical Power Convention
	30 Millbank, London SW1
BHRA	British Hydromechanics Research Association
	Cranfield, Bedford
BICEMA	British International Combustion Engine Manufacturers Association
	6 Grafton Street, London W1
BIM	British Institute of Management
	80 Fetter Lane, London EC4
BIMCAM	British Industrial Measuring and Control Apparatus Manufacturers Association
	23/24 Margaret Street, London W1
BIPM	Bureau International des poids et Mesures (International Bureau of Weights and Measures)
	Paris
BISF	British Iron and Steel Federation
	Steel House, Tothill Street, London SW1
BISRA	British Iron and Steel Research Association
	24 Buckingham Gate, London SW1
BMTFA	British Malleable Tube Fittings Association
	78 Buckingham Gate, London SW1
BNFMP	British Non-Ferrous Metals Federation
	6 Vicarage Road, Birmingham 15
BNFMRA	British Non-Ferrous Metals Research Association
	81 Euston Street, London NW1
BPF	British Plastics Federation
	47 Piccadilly, London W1
BPMA	British Pump Manufacturers Association
	Glen House, Stag Place, London SW1
BSI	British Standards Institution
	2 Park Street, London W1
BVMA	British Valve Manufacturers Association
	25 Victoria Street, London W1
BWRA	British Welding Research Association
	Abington Hall, Abington
CBI	Confederation of British Industry
	21 Tothill Street, London SW1
CDA	Copper Development Association
	55 South Audley Street, London W1
CERA	Civil Engineering Research Association
	1–7 Great George Street, London SW1
DATA	Draughtsmens and Allied Technicians Association
	Drayton House, Gordon Street, London WC1

Table 1.13—*continued*

DOMMDA	Drawing Office Material Manufacturers and Dealers Association
	157 Victoria Street, London SW1
EAA	Electrical Appliance Association Ltd.
	19–21 Conway Street, London W1
EDA	British Electrical Development Association
	2 Savoy Hill, London WC2
EEUA	Engineering Equipment Users Association
	20 Grosvenor Place, London SW1
EFCE	European Federation of Chemical Engineering
	16 Belgrave Square, London SW1
EG	Engineers' Guild Ltd.
	62 Oxford Street, London W1
EIJC	Engineering Institutions Joint Council
	1 Birdcage Walk, London SW1
ELFA	Electric Light Fittings Association
	89 Kingsway, London WC2
ELIC	Electric Lamp Industry Council
	25 Bedford Square, London WC1
ELMA	Electric Lamp Manufacturers Association
	25 Bedford Square, London WC1
ERA	Electrical Research Association
	Cleeve Road, Leatherhead, Surrey
EUROPUMP	European Committee of Pump Manufacturers
	10 Avenue Hoche, Paris 8
FAMEM	Federation of Associations of Mine Equipment Manufacturers
	301 Glossop Road, Sheffield 10
FEPEM	Federation of European Petroleum Equipment Manufacturers
	3 Rue Frey cinet, Paris 16
FMA	Fan Manufacturers Association Ltd.
	414 Chiswick High Road, London W4
FMCEC	Federation of Manufacturers of Construction Equipment and Cranes
	8 St. Brides' Street, London EC4
GTMA	Gauge and Tool Makers Association Ltd.
	2 Old Bond Street, London W1
HA	Hydraulic Association
	Glen House, Stag Place, London SW1
HMSO	Her Majesty's Stationery Office
	Holborn Viaduct, London EC1
HVCA	Heating and Ventilating Contractors Association
	Coastal Chambers, 172 Buckingham Palace Road, London SW1
HVRA	Heating and Ventilating Research Association
	Old Bracknell Lane, Bracknell
IBE	Institution of British Engineers
	46 Victoria Street, London SW1
IBE	Institute of Building Estimators Ltd.
	10 Cromwell Place, London SW7
ICE	Institution of Chemical Engineers
	16 Belgrave Square, London SW1
ICE	Institution of Civil Engineers
	1 Great George Street, London SW1
ICMA	Independent Cable Makers Association
	381–399 Salisbury House, London Wall, London EC2
IE	Institution of Electronics
	Pennine House, Shaw Road, Rochdale
IED	Institution of Engineering Designers
	38 Portland Place, London W1
IEE	Institution of Electrical Engineers
	Savoy Place, London WC2
IEI	Institution of Engineering Inspection
	616 Grand Buildings, Trafalgar Square, London WC2

Table 1.13—*continued*

IERE	Institution of Electronic and Radio Engineers
	8–9 Bedford Square, London WC1
IES	Illuminating Engineering Society
	York House, Westminster Bridge Road, London SE1
IGE	Institution of Gas Engineers
	17 Grosvenor Crescent, London SW1
IHE	Institution of Highway Engineers
	14 Queen Anne's Gate, London SW1
I Loco E	Institution of Locomotive Engineers
	30 Buckingham Gate, London SW1
Inst Met	Institute of Metals
	17 Belgrave Square, London SW1
I Mar E	Institute of Marine Engineers
	76 Mark Lane, London EC3
I Mech E	Institution of Mechanical Engineers
	1 Birdcage Walk, London SW1
I Min E	Institution of Mining Engineers
	3 Grosvenor Crescent, London SW1
I Mun E	Institution of Municipal Engineers
	22 Eccleston Square, London SW1
IOP }	Institute of Petroleum
IP	61 New Cavendish Street, London W1
IOP	Institute of Plumbing
	81 Gower Street, London WC1
I Plant E	Institution of Plant Engineers
	138 Buckingham Palace Road, London SW1
I Prod E	Institution of Production Engineers
	10 Chesterfield Street, London W1
Inst R	Institute of Refrigeration
	30 New Bridge Street, London EC4
IRSE	Institution of Railway Signal Engineers
	21 Avalon Road, Earley, Reading
IRTE	Institute of Road Transport Engineers
	1 Cromwell Place, London SW7
ISI	Iron and Steel Institute
	4 Grosvenor Gardens, London SW1
ISME	Institute of Sheet Metal Engineering
	John Adam House, Adelphi, London WC2
Inst W	Institute of Welding
	54 Princes Gate, London SW7
IWE	Institution of Water Engineers
	11 Pall Mall, London SW1
MFA	Metal Finishing Association
	St. Dunstans' House, Carey Lane, London EC2
MHEA	Mechanical Handling Engineers Association
	Glen House, Stag Place, London SW1
MTIRA	Machine Tool Industry Research Association
	Hulley Road, Hurdsfield, Macclesfield
MTTA	Machine Tool Trades' Association
	25 Buckingham Gate, London SW1
NAMI	National Association of Malleable Iron Founders
	Chamber of Commerce Officers, Tudor House, Bridge Street, Walsall
NBA	National Brassfoundry Association
	5 Greenfield Crescent, Edgbaston, Birmingham 15
NERC	National Electronic Research Council
	8–9 Bedford Square, London WC1
NFETM	National Federation of Engineers' Tool Manufacturers
	Light Trades House, Melbourne Avenue, Sheffield 10
NFI	National Federation of Ironmongers
	20 Harborne Road, Birmingham 15

Table 1.13—*continued*

NLCIF	National Light Castings Ironfounders Federation 30 St. James' Square, London SW1
NPL	National Physical Laboratory Queen's Road, Teddington
OCMA	Oil Companies' Materials Association Cecil Chambers, 86 Strand, London WC2
PERA	Production Engineering Research Association Melton Mowbray
PETMA	Portable Electrical Tool Manufacturers Association Glen House, Stag Place, London SW1
PNEUROP	European Committee of Manufacturers of Compressed Air Equipment 25 Victoria Street, London SW1
SCI	Society of Chemical Industry 14 Belgrave Square, London SW1
SE	Society of Engineers Abbey House, Victoria Street, London SW1
SMMT	Society of Motor Manufacturers and Traders Forbes House, Halkin Street, London SW1
SMRA	Spring Manufacturers' Research Association Doncaster Street, Sheffield 3
SSFA	Stainless Steel Fabricators' Association Chamber of Commerce House, PO Box 360, 75 Harborne Road, Birmingham 15
TEMA	Telecommunication Engineering and Manufacturing Association Stafford House, Norfolk Street, London WC2
TICA	Thermal Insulation Contractors' Association Alderman House, 37 Soho Square, London W1
TIPA	Tank and Industrial Plant Association 197 Knightsbridge, London SW7
TUC	Trades Union Congress 23–28 Great Russell Street, London WC1
VDE	Society of German Electrical Engineers 6 Frankfurt/9, Stressemann, Allee 19
VDI	Society of German Engineers 4 Dusseldorf No. 1, PO Box Poste F CH 1139
WTBA	Water Tube Boilermakers' Association 8 Waterloo Place, London SW1
ZADCA	Zinc Alloy Die Casters' Association 34 Berkeley Square, London W1
ZDA	Zinc Development Association 34 Berkeley Square, London W1

Table 1.14 ABBREVIATIONS OF INTERNATIONAL STANDARDS ORGANISATIONS

ABNT	Brazilian Standards Institute
AFNOR	French Standards Association
API	American Petroleum Institute
ASA	American Standards Association
ASTM	American Society for Testing Materials
CECC	CENEL Electrical Components Committee
CEE	International Commission on Rules for the Approval of Electrical Equipment
CENEL	Committee for European Standardisation in the Electrical Field
CERTICO	ISO Committee on Certification
CETUP	European Oil Hydraulic and Pneumatic Committee
COVENIN	Venezuelan Standards Institute
CSA	Canadian Standards Association
CSK	Committee for Standardisation—Korea
CSN	Czechoslovakia Standards Institute
DGN	Mexican Standards Bureau
DIN DNA	German Standards Association
DNI	Indonesian Standards Association
DS	Danish Standards Bureau
ENO	Greek Standards Office
EOS	Egyptian Organisation for Standisation
GOSI	USSR Standards Association
IBN	Belgium Standards Institute
IEC	International Electrotechnical Commission
IGPAI	Portugese Standards Office
IIRS	Irish Institute for Industrial Research and Standards
INANTIC	Peru Institute of Standards
INDITECHNOR	Chilean Standards Institute
INFCO	ISO Committee for study of scientific and technical information on standardisation
INORCOL	Columbia Standards Institute
IRAM	Argentina Standards Institute
IRATRA	Spanish Standards Institute
ISMIU	Bulgarian Standards Institute
ISI	Indian Standards Institution
ISO	International Standards Organisation
JISC	Japanese Industrial Standards Committee
JZS	Yugoslavian Standards Association
LIBNOR	Lebanese Standards Institution
MSZH	Hungarian Standards Office
NNI	Netherlands Standards Institute
NSF	Norwegian Standards Bureau
NZSI	New Zealand Standards Institute
ONA	Austrian Standards Association
OSS	Rumanian Standards Office
PKN	Polish Standards Institute
PSI	Pakistan Standards Institution
SAA	Standards Association of Australia
SABS	South African Bureau of Standards
SFS	Finland Standards Institute
SII	Standards Institute of Israel
SIS	Swedish Standards Commission
SNIMA	Morocco Standards Association
SNV	Swiss Standards Association
SOI	Standards Organisation of Iran
STACO	Standing Committee for the Study of Principles of Standardisation
STASH	Albanian Standards Bureau
TSE	Turkish Standards Institute
UBARI	Burma Department of Standards
USAISI	USA Standards Institute
UNI	Italian Standards Association

PHYSICAL AND CHEMICAL CONSTANTS

G. R. DARBY

The following Tables provide the physical and chemical constants which are of interest to engineers:

Full coverage of all physical and chemical constants can be obtained by reference to 'Tables of Physical and Chemical Constants' by Kaye and Laby (13th Edition, 1965). The notes below are applicable to the Table heading references:

Atomic number.	Radon (Ra) has an atomic number of 86 (International Atomic Weights Commission). This is the name for one of the isotopes of this element.
Atomic weights.	These are based on the atomic mass of $^{12}C = 12$. Atomic weights are known to be variable because of the natural variation in composition.
Density.	Values are taken at 20°C.
Melting point.	Figures are for a pressure of 1 Atmosphere.
Linear coefficient of expansion.	Known as the 'α' value. Normally worked out as mm expansion over 100 m length using the equation $L = l \ (1 + \alpha t)$ for expansion and $1 = L/1 + \alpha t$ for a contraction (where $1 =$ increase or decrease; $L =$ original length; $t =$ temperature change and $\alpha =$ linear coefficient of expansion).
Heat conductivity.	Sometimes expressed as 'thermal conductivity' and known as 'k' factor. Values are taken at 20°C.
Electric resisitivity.	Values are taken at 20°C.

Table 1.15 TABLE OF ELEMENTS

Element	Symbol	Atomic No.	Atomic weight	Density kg/m³	Melting point °C	Linear coefficient of expansion/°C at normal temp. × 10⁻⁶	Heat conductivity W/m°C	Electric resistivity microhm cm
Actinium	Ac	89	227·00	10 100	1 230	—	—	—
Aluminium	Al	13	26·98	2 700	657	24	217·7	2·655
Antimony	Sb	51	121·75	6 619	630·5	11·29	18·59	39
Argon	A	18	39·944	1·663 (liquid)	−187·9	—	0·017	—
Arsenic	As	33	74·92	5 733	813·8	3·86	—	35
Barium	Ba	56	137·34	3 500	710	—	—	60
Beryllium	Bc	4	9·012	1 822	1 285	12·3	161	2·85
Bismuth	Bi	83	209·00	9 802	271	13·45	8·37	115
Boron	B	5	10·81	2 300	2 030	2	—	$1·8 \times 10^{12}$
Bromine	Br	35	79·9	3 119	−28·3	—	—	—
Cadmium	Cd	48	112·40	8 652	321	29·8	90·86	7·59
Calcium	Ca	20	40·08	1 550	851·3	25	—	4·6
Californium	Cf	98	251					
Carbon (Graphite)	C	6	12·01	2 220	3 500	1·2	23·87	1 000
Cerium	Ce	58	140·12	6 901	775·2	—	—	78
Caesium	Cs	55	132·91	1 899	26	97	—	20
Chlorine	Cl	17	35·457	1 560 (liquid)	−101	11·44	0·0072	10×10^{15}
Chromium	Cr	24	52·01	7 139	1 900	8·1	69·08	13·1
Cobalt	Co	27	58·94	8 904	1 490	12·08	69·08	9·7
Copper	Cu	29	63·54	8 941	1 082	16·42	386·5	1·682
Curium	Gm	96	247					
Dysprosium	Dy	66	162·46	8 500	1 500	—	—	89
Erbium	Er	68	167·20	9 000	1 525	—	—	81
Europium	Eu	63	152·00	5 200	830	—	—	—
Fermium	Fm	100	257					
Fluorine	Fl	9	19·00	1 100 (liquid)	−223	—	—	—
Gadolinium	Gd	64	156·90	7 900	1 320	—	—	126
Gallium	Ga	31	69·72	5 910	29·79	18·3	—	57·1
Germanium	Ge	32	72·60	5 363	958·3	14·4	—	89×10^{3}
Gold	Au	79	197·20	19 310	1 063	14·4	296·1	2·42

Element	Symbol	At. no.	At. wt.	Density	M.P.			
(row cut off)					−253		0·170	77
Hydrogen	H	1	1·008 1	0·083 8	−253	—	0·170	9
Indium	In	49	114·76	7 308	161·2	33	23·87	1·3 × 10^{15}
Iodine	I	53	126·92	4 927	113·5	93	0·043 5	6·08
Iridium	Ir	77	193·10	22 400	2 409	6·41	59	9·8
Iron	Fe	26	55·84	7 861	1 536	11·9	79·56	—
Krypton	Kr	36	83·70	2 160 (liquid)	−170·5	—	0·009	
Lanthanum	La	57	138·92	6 146	826·4	—	—	59
Lead	Pb	82	207·21	11 320	327·3	29·5	34·75	20·65
Lithium	Li	3	6·94	534·3	186	56·0	71·14	8·5
Lutecium	Lu	71	175	9 870	1 700	—	—	54
Magnesium	Mg	12	24·32	1 739	651·3	25·7	154·9	4·46
Manganese	Mn	25	54·93	7 418	1 243	23	—	—
Mercury	Hg	80	200·61	13 540	−38·9	—	8·37	95·8
Molybdenum	Mo	42	95·95	10 190	2 620	5·49	146·6	4·77
Neodymium	Nd	60	144·27	7 058	1 024	—	—	79
Neon	Ne	10	20·183	0·839	−248·5	—	0·046	—
Neptunium	Np	93	239	1 900	—	—	—	—
Nickel	Ni	28	58·70	8 915	1 452	13·7	58·61	6·9
Niobium	Nb	41	92·91	8 571	2 420	7·2	52	15·2
Nitrogen	N	7	14·008	1 165	−209·5	—	0·025	—
Osmium	Os	76	190·2	22 480	3 000	5·7	—	9
Oxygen	O	8	16·000	1·332	−218	—	0·025	—
Palladium	Pd	46	106·7	11 990	1 555	11·60	67·41	10
Phosphorus	P	15	31·02	1 819 (yellow)	44·12	11·25	—	10^{17}
Platinium	Pt	78	195·23	21 420	1 774	8·8	69·5	9·83
Plutonium	Pu	94	239	19 800	640	—	—	150
Polonium	Po	84	209	9 320	254	—	—	—
Potassium	K	19	39·096	858·4	62·29	8·3	99·2	7
Praseodymium	Pr	59	140·92	6 616	940·1	—	—	88
Promethium	Pm	61	147	—	—	—	—	—
Protactinium	Pa	91	231	15 400	1 000	—	—	—
Radium	Ra	88	226·05	5 005	960	—	—	—
Radon	Rn	86	222	4 400	−71	—	—	—
Rhenium	Re	75	186·31	20 000	3 000	8·9	—	21
Rhodium	Rh	45	102·91	12 430	1 966	90·0	89·15	4·93
Rubidium	Rb	37	85·48	1 531	38·3	8·5	—	12·5
Ruthenium	Ru	44	101·7	12 210	2 300	—	—	10

Table 1.15—*continued*

Element	Symbol	Atomic No.	Atomic weight	Density kg/m³	Melting point °C	Linear coefficient of expansion/°C at normal temp. ×10⁻⁶	Heat conductivity W/m°C	Electric resistivity microhm cm
Samarium	Sm	62	150·4	7 752	1 050	—	—	91·4
Scandium	Sc	21	45·10	2 434	1 204	—	—	50·5
Selenium	Se	34	78·96	4 816	220·1	87·0	—	8×10^6
Silicon	Si	14	28·08	2 408	1 427	—	83·74	85×10^3
Silver	Ag	47	107·88	10 520	960·5	18·9	407·9	1·62
Sodium	Na	11	22·997	969	97·52	71·0	135·1	4·6
Strontium	Sr	38	87·63	2 602	77·1	—	—	22·76
Sulphur	S	16	32·06	2 076	113	67·48	0·263 7	$1·9 \times 10^{17}$
Tantalum	Ta	73	180·88	16 620	3 017	6·5	54·43	15·5
Technetium	Tc	43	99	11 400	2 100	—	—	—
Tellurium	Te	52	127·61	6 200	452·3	16·8	6·016	$1·6 \times 10^5$
Terbium	Tb	65	159·2	11 850	310	—	—	—
Thallium	Tl	81	204·39	11 850	303·8	28·0	38·9	18·1
Thorium	Th	90	232·12	11 520	1 700	12·3	—	18
Thulium	Tm	69	169·4	9 330	1 600	—	—	—
Tin	Sn	50	118·70	7 308	232	21	65·73	11·5
Titanium	Ti	22	47·90	4 512	1 680	7·14	—	—
Tungsten	W	74	183·92	19 320	3 370	4·0	199·3	5·48
Uranium	U	92	238·07	18 710	1 133	—	—	60
Vanadium	V	23	50·95	5 675	1 920	8	—	26
Xenon	Xe	54	131·3	152·7	−112	—	—	—
Ytterbium	Yb	70	173·04	6 900	824	—	519·2	27·7
Yttrium	Y	39	88·905	5 509	1 482	—	—	80
Zinc	Zn	30	65·38	7 142	419·5	30	112	5·5
Zirconium	Zr	40	91·22	6 366	1 850	6·3	—	41

Table 1.16 TABLE OF PRINCIPAL ELEMENTS ARRANGED IN ORDER OF VALENCY

	Name	Symbol	Atomic weight
Monovalent	Bromine	Br	79·9
	Chlorine	Cl	35·457
	Fluorine	Fl	19·00
	Hydrogen	H	1·008 1
	Iodine	I	126·92
	Potassium	K	39·096
	Silver	Ag	107·88
	Sodium	Na	22.997
Divalent	Barium	Ba	137·34
	Cadmium	Cd	112·40
	Calcium	Ca	40·08
	Copper	Cu	63·54
	Magnesium	Mg	54·93
	Mercury	Hg	200·61
	Oxygen	O	16·000
	Zinc	Zn	65·38
Trivalent	Aluminium	Al	26·98
	Bismuth	Bi	209·00
	Boron	B	10·81
	Cobalt	Co	58·94
	Gold	Au	197·20
	Iron	Fe	55·84
	Nickel	Ni	58·70
Tetravalent	Lead	Pb	207·21
	Platinum	Pt	195·23
	Silicon	Si	28·08
	Tin	Sn	118·70
Pentavalent	Antimony	Sb	121·75
	Arsenic	As	74·92
	Nitrogen	N	14·008
	Phosphorus	P	31·02
Hexavalent	Chromium	Cr	52·01
	Manganese	Mn	54·93
	Sulphur	S	32·06

Table 1.17 SURFACE TENSION OF SOME COMMON METALS

(This quantity is generally measured in millinewtons per metre (mN/m), equivalent to the free surface energy in millijoules per square metre (mJ/m^2))

Metal	Temperature °C	Surface tension (mN/m)
Antimony	640	350
Bismuth	269	378
Cadmium	320	630
Copper	1 131	1 103
Gold	1 120	1 128
Iron (acc. to C. content)	1 300–1 420	1 150–1 500
Lead	327	452
Mercury	20	465
Silver	998	923
Tin	232	526
Zinc	419	758

Surface Tension of water is 78·5 mN/m at 0"C and decreases by 0·152 mN/m for each degree rise of temperature

Table 1.18 SPECIFIC HEATS AND MELTING POINTS

Material	Specific heat kJ/kg °C	Average watt hours kg/°C	Heat of fusion watt hours/°C	Melting Point °C	Density kg/m³ (approx)
Air (20°C)	0·996 8	0·275	—	—	1·2
Aluminium	0·914 9	0·255	45·9	657	2 700
Brass	0·395 3	0·105	—	850–950	8 400
Carbon	0·854 3	0·235	—	—	2 220
Cobalt	0·448	—	—	1 480	8 904
Copper	0·393 6	0·109	22·8	1 083	8 941
German silver	0·398	—	—	—	8 400
Graphite	0·837 4	0·231	—	—	2 300
Iron, cast	0·46–0·67	0·127–0·255	12·2	1 200	7 000
Lead (solid)	0·131 3	0·359	3·1	327·4	11 320
Lead (liquid)	0·197 2	0·546	—	—	—
Mercury	0·134	—	—	−38·9	13 540
Molybdenum	0·275 9	—	—	2 620	10 190
Nickel	0·454	0·127	2·45	1 452	8 915
Paraffin (solid)	2·6–2·9	0·718–0·800	—	38–56	900
Paraffin (liquid)	29·7	0·821	18·5	—	800
Pitch	—	—	—	—	1 100
Platinum	0·138 1	—	—	1 774	21 420
Silver	0·232 7	—	—	960·8	10 520
Solder	—	—	5·3–9·0	205–185	8 300
Tin (solid)	0·234 5	0·065	7·4	232	7 308
Tin (liquid)	0·268	0·074	—	—	—
Tungsten	0·146 6	—	—	3 370	19 320
Type metal	0·163 3	0·045	—	—	—
Water (20°C)	4·186 8	1·16	42·3	0	—
Zinc (cast)	0·39–0·50	0·108–0·138	14·8	419·5	7 142

Table 1.19 SECTIONAL PROPERTIES OF METALS

Substance	E N/mm²	G N/mm²	σ	K N/mm²	Tensile Strength N/mm²
Aluminium	70 300	26 100	0·345	75 500	90–150
Brass	101 000	37 300	0·350	111 800	280–730
Copper	129 800	48 300	0·343	137 800	120–400
Iron (cast)	152 000	60 000	0·270	109 000	100–230
Iron (Wrought)	211 400	81 000	0·293	170 000	260–450
Lead	16 100	5 600	0·440	45 700	12–17
Magnesium	44 700	17 000	0·291	25 600	60–190
Silver	82 700	30 200	0·366	103 600	300
Platinum	168 000	61 000	0·377	228 000	330–370
Tantalum	185 700	69 200	0·342	196 300	800–1 100
Tin	49 900	18 400	0·357	58 200	20–35
Tungsten	411 000	160 000	0·280	311 500	1 500–3 500
Steel (mild)	211 900	82 200	0·291	169 200	430–690
Steel (hardened)	201 400	77 800	0·295	165 200	1 800–2 300

Notes: 1 N/mm² = 1 MN/m²; E is known as Young's modulus or longitudinal elasticity; G is known as Shear or Rigidity modulus; σ is known as Poisson's ratio; K is known as Bulk modulus.

2 THEORY AND DESIGN DATA

THEORY AND DESIGN DATA 2

2 THEORY AND DESIGN DATA

ENGINEERING THERMODYNAMICS AND HEAT TRANSFER

D. H. BACON

BASIC CONCEPTS

Engineering thermodynamics is concerned with the evaluation of energy transfers in the working substance of a machine. The particular part of the working substance being considered is known as the *system* and is separated from its *surroundings*. Two types of system are required, the *closed* or *non flow* system enclosed by a *boundary* enveloping *constant mass* and the *open* or *flow* system within a *control surface* through which there is a *mass flow rate*. The *state* of a system is defined by *properties* such as pressure, temperature, volume and energy. Properties are usually expressed *specifically*, that is, per unit mass. Two independent properties are required to determine the state of a simple or pure substance.

There are two recognised modes of energy transfer, work (transfer) and heat (transfer). Work and heat are not properties, are not contained in a system and cannot be stored, they are transient phenomena which may appear when a system changes its state. Work and heat may also be expressed specifically.

The first law of thermodynamics

Both open and closed systems obey the law of conservation of energy. Formally titled 'The first law of thermodynamics' it states that the final energy of a system is equal to the initial energy plus that added or subtracted by heat and work transfers. It embraces a sign convention; work from a system is positive and heat to a system is positive and is written

$$Q - W = \Delta E \tag{1}$$

where Q is heat transfer, W is work transfer and ΔE is the change in energy in the sense final energy minus initial energy.

All forms of energy are considered, but when applied to the non flow system only internal energy U is relevant and the *non flow energy equation* is obtained

$$Q - W = \Delta U \tag{2}$$

This may be written specifically

$$q - w = \Delta u \tag{3}$$

where q is the specific heat transfer, w the specific work transfer and Δu the specific internal energy change.

For the flow system, kinetic energy and potential energy are also considered and the *steady flow energy equation* is obtained

$$\dot{Q} - \dot{W}_x = \dot{m}\Delta(h + V^2/2 + gz)$$

or specifically

$$q - w_x = \Delta(h + V^2/2 + gz) \tag{4}$$

In this equation $\dot{Q}$ is the heat transfer rate, $\dot{W}_x$ is the work transfer rate or power, $\dot{m}$ is the mass flow rate, h is the specific enthalpy given by $h = u + pv$, p is the pressure, v is the specific volume, g is gravitational acceleration, z is elevation above some datum and V is velocity. The suffix x is attached to the work term to indicate that this is the *useful* work obtained from the system, the work necessary to maintain flow is accounted for in the specific enthalpy term. All terms in the non flow energy equation will be expressed in joules(J) or if specific, J/kg and in the steady flow energy equation in watts(W) or if specific, J/kg. The mass flow rate $\dot{m}$ may be obtained from the *continuity equation*

$$\dot{m} = \frac{AV}{v} \tag{5}$$

where A is flow area.

In order to use these equations to solve problems it is necessary to be able to evaluate all or some of (i) w (ii) w_x (iii) q and (iv) the property changes Δu and Δh. To obtain w or w_x analytically, the processes involved in a change of state of a system are idealised and the corresponding real processes assigned a *process efficiency* with which the ideal work may be modified to give the actual work. Idealised processes are frictionless and have heat transfers over infinitely small temperature differences, they are termed *reversible* and may be represented by mathematical relations between properties. They may also

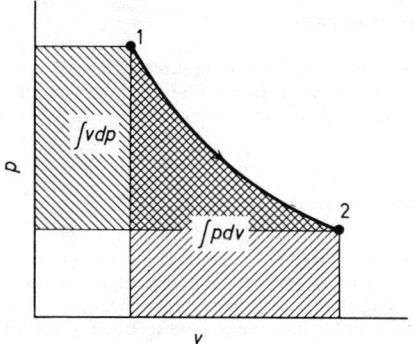

Figure 2.1

be represented on property graphs. The properties usually chosen for the relation are pressure and specific volume and the graph in Figure 2.1 shows a non flow process from state 1 to state 2 for which the specific work is given by

$$w = \int_1^2 p\,dv \tag{6}$$

The same graph could be used for a flow process representation in which case the specific work is given by;

$$w_x = -\int_1^2 v\,dp \tag{7}$$

Equation 7 is only valid if $\Delta V^2/2$ and Δgz are small. The mathematical relationship which suffices for all common processes is

$$pv^n = \text{constant}.$$

It is called the *polytropic* process law. When the index n has particular values, the process has a special name. Table 2.1 shows these processes and the solution to the w and w_x integrals in these cases.

Table 2.1   WORK TRANSFER IN PROCESSES

Process	Law	$w = \int p\,dv$	$w_x = -\int v\,dp$
Constant volume	$v = $ constant	0	$v(p_1 - p_2)$
Constant pressure	$p = $ constant	$p(v_2 - v_1)$	0
Hyperbolic (isothermal for a perfect gas)	$pv = $ constant	$p_1 v_1 \log_e \dfrac{v_2}{v_1}$	$p_1 v_1 \log_e \dfrac{p_1}{p_2}$
Adiabatic*	$pv^k = $ constant	$\dfrac{p_2 v_2 - p_1 v_1}{1 - k}$	$\dfrac{k(p_2 v_2 - p_1 v_1)}{1 - k}$
Polytropic	$pv^n = $ constant	$\dfrac{p_2 v_2 - p_1 v_1}{1 - n}$	$\dfrac{n(p_2 v_2 - p_1 v_1)}{1 - n}$

* For a perfect gas the adiabatic index k may be shown to be equal to the ratio of the specific heat capacity at constant pressure c_p to the specific heat capacity at constant volume c_v. Thus the relation becomes $pv^\gamma = $ constant where $\gamma = c_p/c_v$. Specific heat capacity at constant pressure is defined by $c_p = (\partial h/\partial T)_p$ and specific heat capacity at constant volume by $c_v = (\partial u/\partial T)_v$.

The second law of thermodynamics

When a series of processes brings a working substance back to its initial state a *cycle* has been performed. In a cycle there is no resultant change of state so that there cannot be any change in the value of the property energy and the first law becomes

$$\sum_{\text{cycle}} Q = \sum_{\text{cycle}} W$$

where $\sum\limits_{\text{cycle}}$ means 'the sum around the cycle'.

The primary object of the engineer is to have a continuous work transfer from a continuous heat transfer. The second law of thermodynamics states that this cannot be achieved without wasting a proportion of the initial heat transfer by a second heat transfer to a lower temperature sink. 'It is impossible to build a machine which will give continuous positive work transfer whilst exchanging heat with a single reservoir.'

Figure 2.2 represents a *heat engine* with heat transfer Q_1 from a hot reservoir and heat transfer Q_2 to a cold reservoir giving work transfer W. By the first law, $W = Q_1 - Q_2$ and the measure of success of any heat engine is called *thermal efficiency*, η where

$$\eta = \frac{\text{net work work transfer from the engine}}{\text{heat transfer to the engine}} = \frac{W}{Q_1} \qquad (8)$$

In an ideal heat engine all processes and all energy transfers will be reversible; the engine is termed reversible and it will have the maximum possible efficiency. This is known as the *Carnot efficiency*.

Mathematical consideration of the second law[1] leads to the discovery of two fundamental properties, *thermodynamic temperature T* and *entropy S* related in such a way that for the reversible process from state 1 to state 2 shown in Figure 2.3 on a thermodynamic temperature-specific entropy graph the specific heat transfer

$$q = \int_1^2 T\,ds.$$

Further mathematical consideration shows that the Carnot efficiency for the reversible heat engine may be expressed by

$$\eta = 1 - T_2/T_1 \qquad (9)$$

where T_1 is the maximum temperature in the cycle (the hot reservoir) and T_2 is the minimum temperature (the cold reservoir). Equation 9 is an important relation as it tells the engineer the maximum possible efficiency that could be attained from given maximum and minimum temperatures from any power-producing plant using *thermal energy*. A

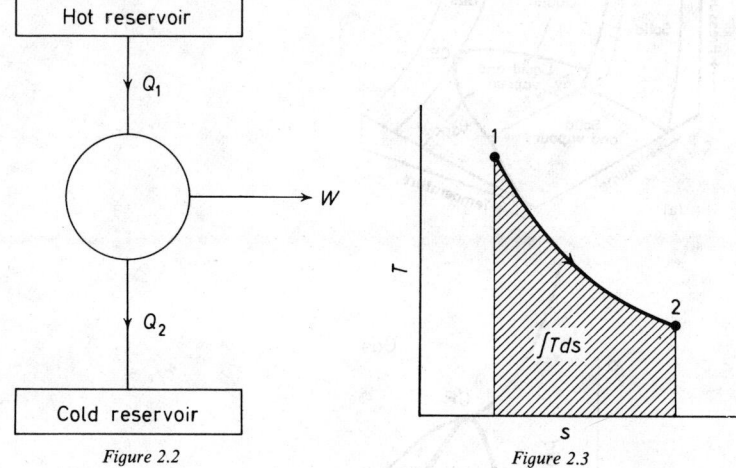

Figure 2.2   Figure 2.3

fuel cell utilising chemical energy converted directly to work is not restricted by the Carnot efficiency and is therefore an attractive proposition.

Further consideration of the process in which q is zero, which is called *adiabatic*, shows that if the process is reversible then $\int T\,ds$ is zero and the process must be one of constant entropy.

This is a very useful concept for the engineer as many real processes are approximately adiabatic and may be idealised for analysis as reversible adiabatic processes. In the real process, friction will cause the temperature of the working substance to reach a higher

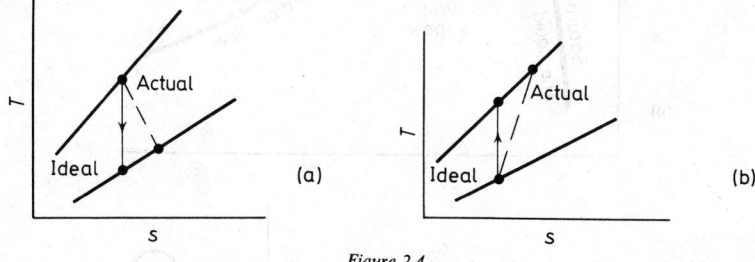

Figure 2.4

value than in the ideal process and Figure 2.4 shows the situation for (a) an expansion and (b) a compression. In both cases it can be seen that the (specific) entropy has increased and this leads to the principle of increase of entropy; 'In real adiabatic processes the entropy always increases'. The process efficiency for an adiabatic process is called *isentropic efficiency* and is used to determine real work from ideal work obtained by analysis. For an expansion isentropic efficiency

$$\eta_{\text{isen}} = \frac{\text{actual work transfer}}{\text{ideal work transfer}} \qquad (10)$$

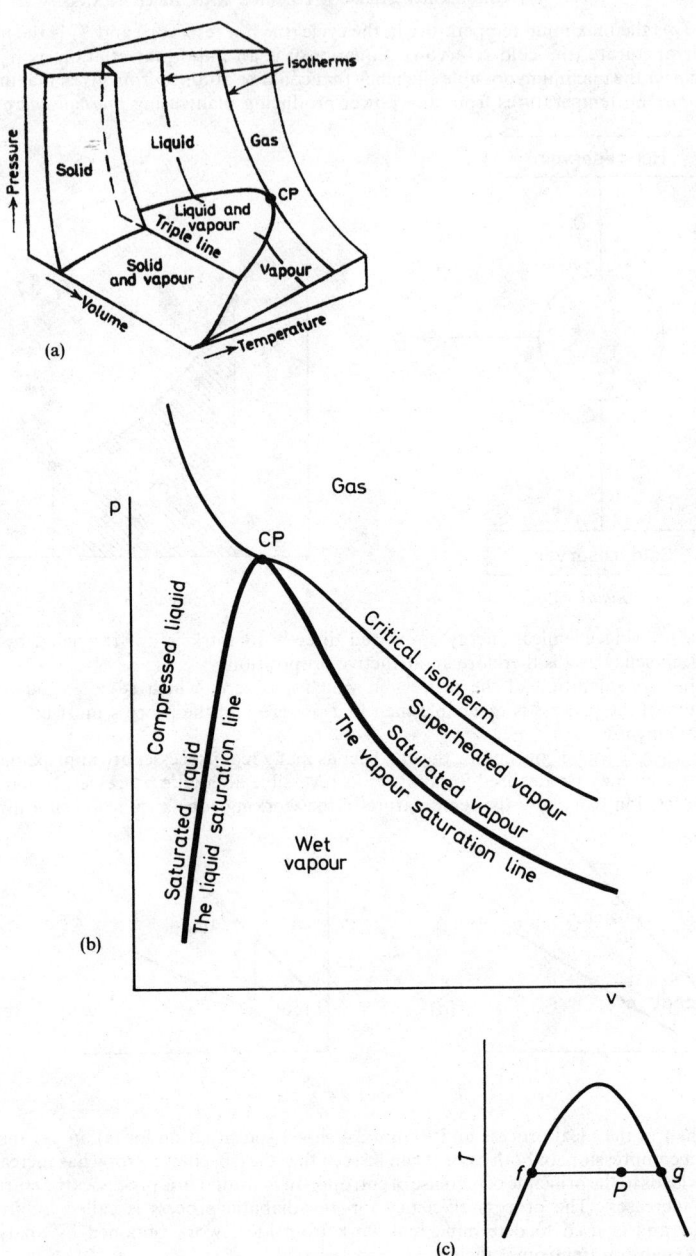

(a)

(b)

(c)

Figure 2.5

and for a compression

$$\eta_{isen} = \frac{\text{ideal work transfer}}{\text{actual work transfer}} \tag{11}$$

Machines which are usually considered adiabatic include steam and gas turbines, gas compressors and nozzles.

PROPERTY DATA

Substances may exist in three phases, solid, liquid and vapour. The solid phase is of little interest in this context. Figure 2.5(a) shows the distribution of possible states for a typical working substance which expands on freezing. Terminology is shown in Figures 2.5(a) and (b). In practice two distinct groups of machine are used (i) those in which the working substance experiences the liquid-vapour phase change and (ii) those in which the working substance remains a gas at all times.

For common working substances, empirical data tables and charts are readily available. The bibliography lists useful sources for these and other less common substances. Values of enthalpy, entropy and energy changes in any process may be determined from these tables. The normal entry to a table is by pressure and temperature but in the case of the *wet vapour* these are not independent properties and the *dryness fraction* x defined as

$$\frac{\text{mass of dry vapour}}{\text{mass of mixture}}$$

is used with pressure or temperature to establish data. In general terms for property z;

$$z = (1 - x)z_f + xz_g \tag{12}$$

e.g. in Figure 2.5(c), which shows the shape of the $T-s$ envelope for steam

$$s_p = (1 - x_p)s_f + x_p s_g.$$

The most useful property chart available is the Mollier $(h-s)$ chart and Figure 2.6 shows its normal form for steam. Examination of equations (17) and (18) (p. **2**–12) shows that the vertical distances represent the work and heat transfers in a steam plant.

Although similar charts and tables are available for gases and common mixtures of gases, it is often more convenient to use a simple equation of state for a gas and to obtain changes in property by analytical methods. When this method is chosen it is simple to programme a computer to store gas properties whereas for steam complex equations are required. These may be obtained from the 1967 IFC formulation for industrial use[2].

The bibliography (p. **2**–37) contains further sources of property data.

Perfect gases

A perfect gas is defined as one having constant specific heat capacity which obeys the simple equation of state

$$pv = RT \tag{13}$$

R is the specific gas constant which may be obtained from the relation $R = R_o/M$, where M is the molecular weight of the gas and R_o is the molar (or universal) gas constant, 8·3143 kJ/kmol K. The specific heat capacities were defined in Table 2.1 and it can be

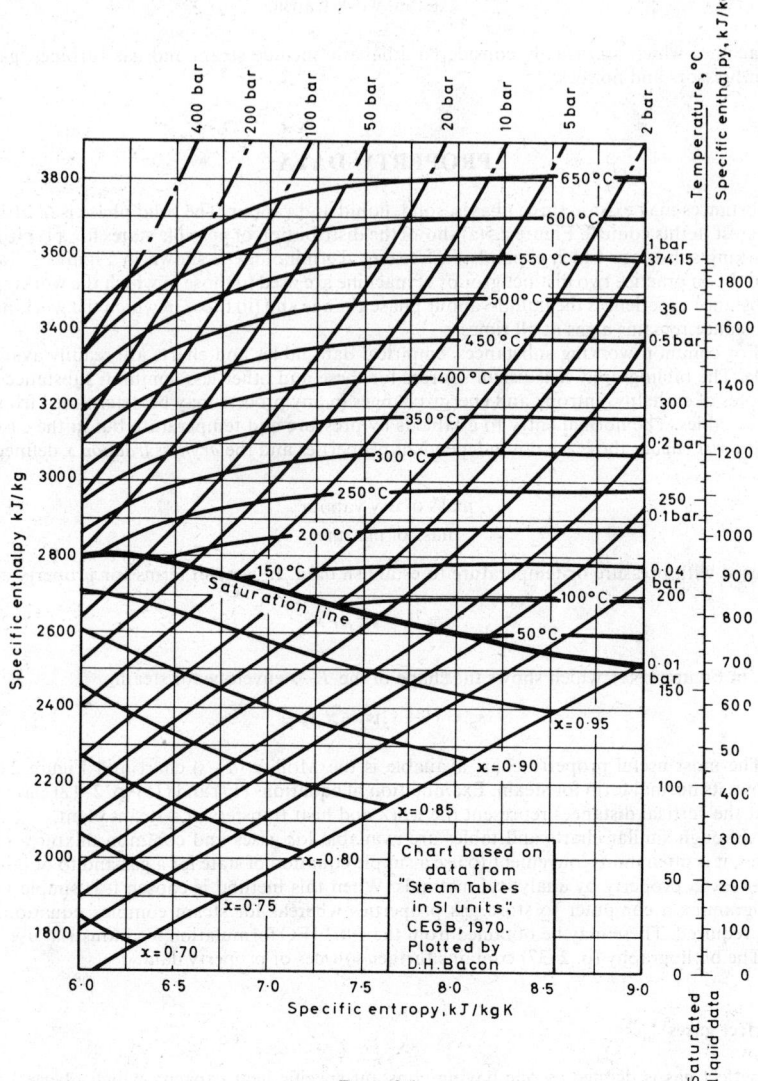

Figure 2.6

seen that if these are considered constant, then enthalpy and energy changes will be obtained by integration and be given by

$$\Delta h = c_p \Delta T \quad \text{and} \quad \Delta u = c_v \Delta T \tag{14}$$

The specific entropy change may be shown to be obtained from

$$\Delta s = c_p \log_e \frac{T_2}{T_1} - R \log_e \frac{p_2}{p_1} \tag{15}$$

Although real gases may be assumed to obey the equation of state $pv = RT$ provided that the temperature is at least twice the critical temperature and the pressure is moderate (for gases used in most engineering problems, air, H_2, O_2, N_2, CO and CO_2 this may be interpreted as 200 to 3000 K and up to 50 bar), their specific heat capacities are not constant and it is necessary to determine these empirically as functions of temperature. These empirical equations, together with the equation of state, enable relations for enthalpy, entropy and energy changes to be calculated or programmed into a computer.[3] A gas which obeys the relation $pv = RT$, but which has specific heat capacities which are functions of temperature is known as a semi-perfect or ideal gas. Some values of specific heat capacity are included in Table 2.13(b).

For the reversible adiabatic perfect gas process equation 15 may be equated to zero to yield

$$\frac{T_2}{T_1} = \left(\frac{p_2}{p_1}\right)^{R/c_p}$$

and if this is compared with the expression of Table 2.1 for the same process it may be seen that

$$\frac{R}{c_p} = \frac{\gamma - 1}{\gamma}$$

or

$$c_p - c_v = R \tag{16}$$

Table 2.2 contains a summary of perfect gas relations.

Table 2.2   PERFECT GAS RELATIONS

Equation of state	$pv = RT$
Specific enthalpy change	$\Delta h = c_p \Delta T$
Specific internal energy change	$\Delta u = c_v \Delta T$
Specific entropy change	$\Delta s = c_p \log_e \frac{T_2}{T_1} - R \log_e \frac{p_2}{p_1}$
Specific heat capacity relations	$c_p - c_v = R$
Polytropic process law	$p_1 v_1^n = p_2 v_2^n, \ T_1 v_1^{n-1} = T_2 v_2^{n-1}, \ \frac{T_2}{T_1} = \left(\frac{p_2}{p_1}\right)^{n-1/n}$
Adiabatic process law	$p_1 r_1^\gamma = p_2 v_2^\gamma, \ T_1 v_1^{\gamma-1} = T_2 v_2^{\gamma-1}, \ \frac{T_2}{T_1} = \left(\frac{p_2}{p_1}\right)^{\gamma-1/\gamma}$
Isothermal process law	$p_1 v_1 = p_2 v_2$

Mixtures of gases (e.g. air) may be treated as a single perfect gas by using the Gibbs–Dalton Law which states

(i) the total pressure exerted by a mixture of gases is equal to the sum of the partial pressures which each component would exert if it alone occupied the volume of the mixture at the temperature of the mixture and

(ii) the enthalpy, energy and entropy of a mixture of gases is the sum of the respective enthalpies, energies and entropies which the components would have if each alone occupied the volume of the mixture at the temperature of the mixture. This gives

$$p = \Sigma p_i, \quad m = \Sigma m_i, \quad u = \frac{1}{m}\Sigma m_i u_i, \quad h = \frac{1}{m}\Sigma m_i h_i \quad \text{and} \quad s = \frac{1}{m}\Sigma m_i s_i$$

where suffix i represents the i'th component.

Further analysis shows that

$$R = \frac{1}{m}\Sigma m_i R_i, \quad M = \frac{m}{\Sigma \dfrac{m_i}{M_i}},$$

$$c_p = \frac{1}{m}\Sigma m_i c_{p_i}, \quad c_v = \frac{1}{m}\Sigma m_i c_{v_i} \quad \text{and} \quad \gamma = \frac{c_{p_{\text{mixture}}}}{c_{v_{\text{mixture}}}}$$

One particular mixture of interest is atmospheric air which contains superheated steam. This may be tackled analytically by treating steam as a perfect gas of molecular weight 18 or by using special psychrometric charts and tables. The approximation made in treating steam as a perfect gas is only valid at partial pressures below 0·07 bar and is not valid when there is water present.

Stagnation properties

When a thermometer is placed in a moving gas stream, the gas will impinge on the thermometer and come to rest. The kinetic energy will be destroyed and the temperature recorded by the thermometer will rise to the stagnation value. The steady flow energy equation shows that

$$T_t = T + \frac{V^2}{2c_p}$$

where T_t is the stagnation temperature.

Similarly, the pressure of the stream of gas will be caused to rise by the destruction of the kinetic energy and if the process is considered reversible the stagnation pressure p_t will be given by

$$p_t = p\left(\frac{T_t}{T}\right)^{\gamma/\gamma-1} \quad \text{or} \quad p_t = p\left(1 + \frac{V^2}{2c_p T}\right)^{\gamma/\gamma-1}$$

Stagnation pressure and temperature are of importance in gas turbines where velocities are considerable and in this application the term in the steady flow energy equation

$$\Delta\left(h + \frac{V^2}{2}\right) = \Delta\left(c_p T + \frac{V^2}{2}\right)$$

may be written $c_p \Delta T_t$ with considerable saving in work.

COMBUSTION

The prime source of thermal energy today is the combustion of hydrocarbon fuels in air. An average solid or liquid fuel contains about 85% carbon and 15% hydrogen by mass and simple chemistry determines the correct (*stoichiometric*) air fuel ratio by mass to be about 15 to 1. Gaseous fuels have varying composition and are often man made.

Exact thermochemical data to energy changes in combustion reactions may be calculated, but for many engineering purposes a simpler criterion is used. This is the amount of energy released when unit mass of fuel is burned under specified conditions and is known as the *calorific value* of the fuel. In experimental determinations of calorific value the H_2O product is constrained to be liquid and the *higher calorific value* is obtained. *Lower calorific value* is the energy release when the H_2O is in the vapour phase which is the normal situation in processes in engines and combustors. The determination of calorific value is made in a *bomb calorimeter* for solid and liquid fuels and in a *gas calori-*

Table 2.3   CALORIFIC VALUES AT $15.5°C$. 1.013 BAR, kJ/kg (gases kJ/m³)

Fuel	Lower C.V.	Higher C.V.	Fuel	Lower C.V.	Higher C.V.
Coal	33 200	34 000	Kerosene	43 300	46 200
Wood	14 300	16 000	Bunker C	40 000	42 000
Coke	30 500	30 800	Town gas	18 000	20 000
Petrol	43 800	46 800	Producer gas	6 000	6 100
Diesel oil	42 700	45 400	North Sea gas	34 500	38 500

meter for gaseous fuels. The procedure for these determinations may be obtained from British Standards[4]. Table 2.3 gives calorific values.

In any particular engine or combustor the products of combustion may be analysed by Orsat apparatus or chromatograph so that the success of the combustion process may be measured. Combustion efficiency may be defined in three ways

(i) $\quad \eta_{comb} = \dfrac{\text{Actual volumetric percentage of } CO_2}{\text{Maximum possible percentage of } CO_2}$ (in non flow combustors)

(ii) $\quad \eta_{comb} = \dfrac{\text{Actual temperature rise in the combustion chamber}}{\text{Maximum possible rise with the same air fuel ratio}}$ (in flow combustors)

(iii) $\quad \eta_{comb} = \dfrac{\text{Actual heat transfer to working substance}}{\text{Maximum possible heat transfer in an isothermal reaction}}$

(in heating devices)

This last expression may be approximated to

$$\eta_{comb} = \frac{q}{\text{calorific value}}$$

Alternatively, the performance of a steam generator or combustor may be continuously metered by a CO_2 diffusion pot indicator. Other available methods of indication include infra red analysis, katharometers, density balances, ionisation detectors and oxygen detectors.

In fields where combustion data is continuously required data is tabulated[5] or plotted. This enables the effects of changes in air fuel ratio or reactant temperatures to be rapidly examined.

In high temperature combustion work the prediction of temperatures and energy release is a complex calculation due to the instability of and *dissociation* of the combustion products. Each reaction has an empirically determined *equilibrium constant* which establishes the analysis of the products at any temperature and the calculation process is iterative. In other applications the product compounds may be expanding during the reaction and the changes in temperature due to the expansion alter the equilibrium constituents continuously. A knowledge of chemical reaction rates is necessary to the combustion engineer.

POWER PLANT CYCLE ANALYSIS

The processes of Table 2.1 may be used to form cycles for power plant. The cycle for the ideal heat engine is known as the *Carnot cycle* but has little use in real plant as it is not composed of the steam or gas processes which are found suitable for practical machinery.

The thermal efficiency of the Carnot cycle is of use to the engineer as it gives him the maximum value that he could attain between given temperature limits.

Steam plant

The ideal cycle for steam plant is the Rankine cycle. It is shown in Figure 2.7(a). The processes forming the cycle are shown in Table 2.4 and the steady flow energy equation is applied to each in turn.

Table 2.4   RANKINE CYCLE PROCESSES

Process	Name and component	Application of energy equation assuming $\Delta V^2/2$ and Δgz negligible
1–2	reversible adiabatic compression in a feed water pump	$-w_x = h_2 - h_1 \doteq v_1(p_2 - p_1)$
2–3	reversible constant pressure heat transfer in a steam generator	$q = h_3 - h_2$
3–4	reversible adiabatic expansion in a turbine	$-w_x = h_4 - h_3$
4–1	reversible constant pressure heat transfer in a condenser	$q = h_1 - h_4$

The thermal efficiency of the cycle

$$\eta = \frac{(h_3 - h_4) - v_1(p_2 - p_1)}{h_3 - h_2} \doteq \frac{h_3 - h_4}{h_3 - h_1} \tag{17}$$

The specific work from the cycle

$$w = (h_3 - h_4) - v_1(p_2 - p_1) \tag{18}$$

Although Figure 2.7(a) shows saturated steam entering the turbine it would be unusual if this were true for real plant. The steam at point 3 would be superheated as shown in Figure 2.7(b). Superheating may be used to the metallurgical limit of the materials used and Figure 2.7(c) shows that efficiency is improved. The specific work from the plant is also improved by the use of superheated steam so that the size of plant for a given duty is reduced. When the limiting temperature is reached, the steam generator pressure may be increased and Figure 2.7(d) shows the efficiency again improved. This has an adverse effect on the dryness of the steam at exit from the turbine and reheating may be used to counteract this undesirable effect. Reheating to the metallurgical limit may be used and, although two turbines are required with this arrangement, the specific work from the plant is increased.

Figure 2.7(e) shows a reheat cycle. Further improvement in efficiency may be achieved by regenerative feed heating in which steam is bled from the turbine to heat the feed water in a feed heater.

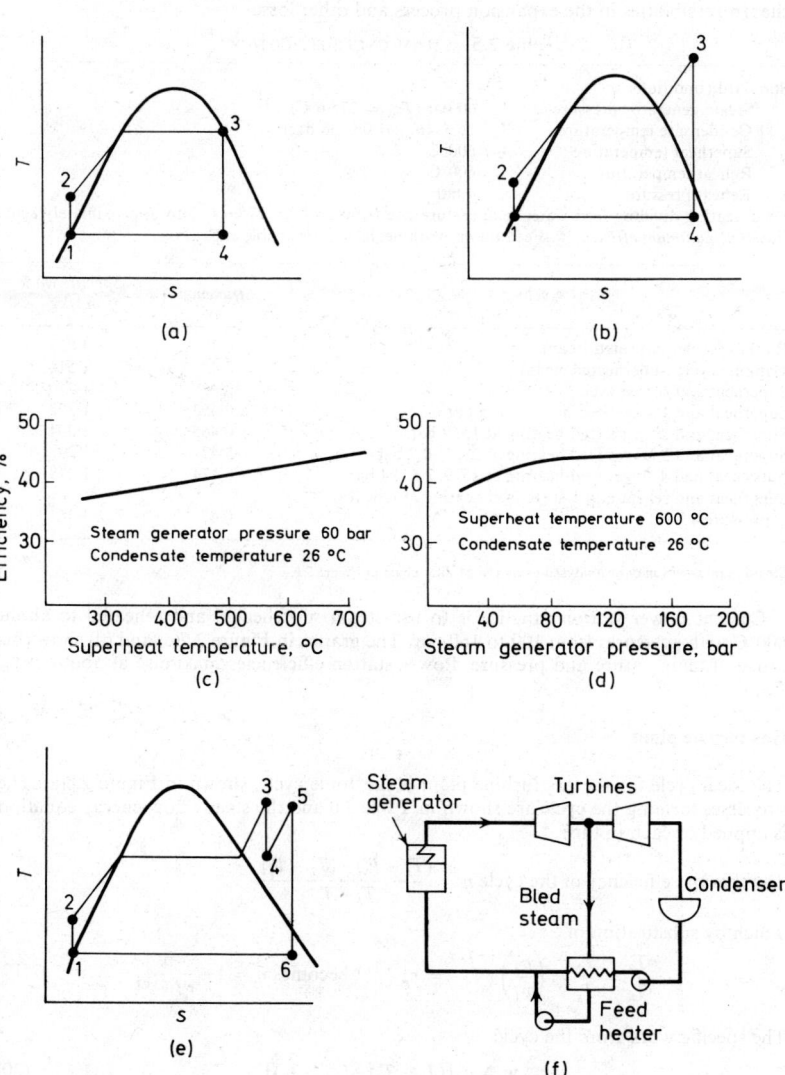

Figure 2.7

Figure 2.7(f) shows a plant with a regenerative feed heater. Several stages of feed heating may be used to advantage and Table 2.5 shows the ideal efficiencies for several schemes. The efficiencies achieved by real plant will be about 15% less than these figures, due to the irreversibilities in the expansion process and other losses.

Table 2.5   STEAM CYCLE EFFICIENCY

Standard conditions:

Steam generator pressure	60 bar ($T_{sat} = 275 \cdot 6°C$)
Condensate temperature	26°C ($p_{sat} = 0 \cdot 0336$ bar)
Superheat temperature	600°C
Reheat temperature	600°C
Reheat pressure	4 bar

Feed heating dividing feed water temperature rise from 26°C to 275·6°C into approximately *equal steps for maximum efficiency*[6]. Feed pump work neglected, reversible expansion.

Cycle	Efficiency	Specific work kJ/kg
Rankine cycle, saturated steam	0·38	1 027
Rankine cycle, superheated steam	0·43	1 516
Superheat and reheat cycle	0·45	1 970
Superheat and 1 stage feed heating at 5 bar	0·453	1 372
Superheat and 2 stages feed heating at 15, 2 bar	0·465	1 304
Superheat and 3 stages feed heating at 20, 5, 0·7 bar	0·47	1 293
Superheat and 4 stages feed heating at 25, 9, 2·4, 0·4 ba	0·474	1 278
Superheat and reheat and 1 stage feed heating at reheat pressure 4 bar	0·47	1 761

Steam plant cycles may be analysed using the Mollier chart of Figure 2.6.

Current power station practice is to use steam superheated and reheated to about 600°C with pressures from 100 to 160 bar. The graphs in Figure 2.7(c) and (d) show this range of temperature and pressure. Power station efficiencies maximise at about 36%.

Gas turbine plant

The ideal cycle for the gas turbine plant is the Joule cycle, shown in Figure 2.8(a). The processes forming the cycle are shown in Table 2.6 and the steady flow energy equation is applied to each in turn.

The thermal efficiency of the cycle $\eta = \dfrac{(T_3 - T_4) - (T_2 - T_1)}{T_3 - T_2}$

which by substitution of

$$\frac{T_3}{T_4} = \frac{T_2}{T_1} = \left(\frac{p_2}{p_1}\right)^{(\gamma - 1)/\gamma} = r_p^{(\gamma - 1)/\gamma} \text{ becomes } \eta = 1 - \frac{1}{r_p^{(\gamma - 1)/\gamma}} \tag{19}$$

The specific work from the cycle

$$w = c_p[(T_3 - T_4) - (T_2 - T_1)] \tag{20}$$

In real gas turbine plant, the cycle above with heater and cooler is rarely used. Instead, an open cycle plant with a combustion chamber replacing the heater and exhausting to atmosphere after the expansion is employed. Figure 2.8(b) shows a simple design of plant where a single turbine drives the compressor and the load. This type of plant has a rising torque characteristic and is, therefore, only suited to constant speed and load applications such as power generation or pumping.

Table 2.6 JOULE CYCLE PROCESSES

Process	Name and component	Application of energy equation assuming $\Delta V^2/2$ and Δgz negligible
1–2	Reversible adiabatic compression in a rotary compressor	$-w_x = h_2 - h_1 = c_p(T_2 - T_1)$
2–3	Reversible constant pressure heat transfer in a heater	$q = h_3 - h_2 = c_p(T_3 - T_2)$
3–4	Reversible adiabatic expansion in a turbine	$-w_x = h_4 - h_3 = c_p(T_4 - T_3)$
4–1	Reversible constant pressure heat transfer in a cooler	$q = h_1 - h_4 = c_p(T_1 - T_4)$

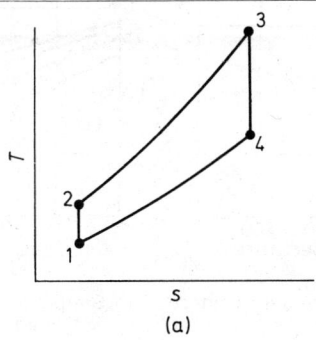

(a)

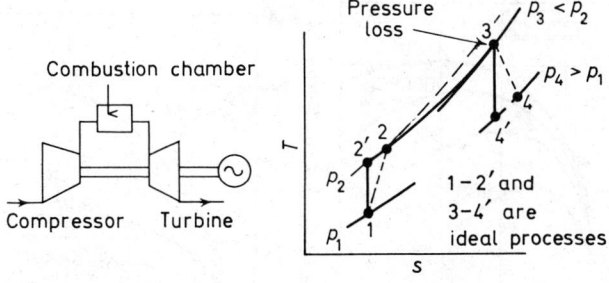

(b)

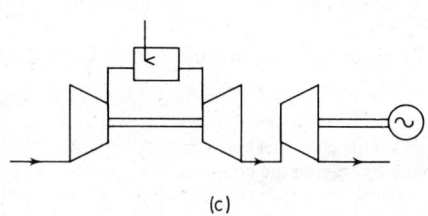

(c)

Figure 2.8

For traction purposes the separate power turbine design of Figure 2.8(c) which has a falling torque characteristic is preferred. The overall efficiency of these simple gas turbines is low but the high power output for a given size or weight is enough to make the plant attractive for some tasks. The efficiency may be improved by fitting a regenerative heat exchanger which heats the air before it enters the combustion chamber utilising the high

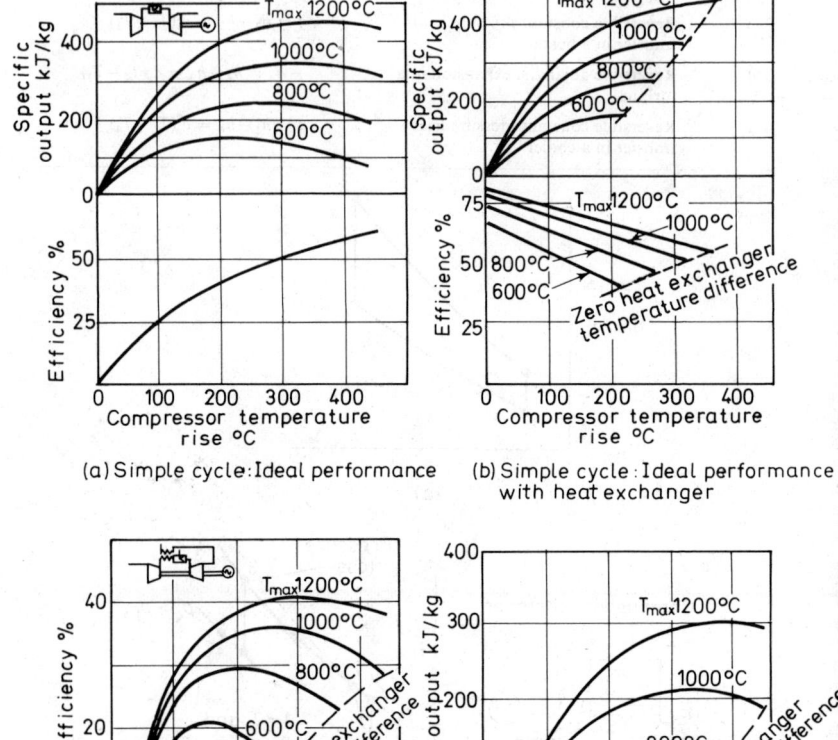

(a) Simple cycle: Ideal performance

(b) Simple cycle: Ideal performance with heat exchanger

(c) Simple cycle: Standard conditions (table 2·7) with heat exchanger

Figure 2.9

temperature exhaust gases. This saves fuel but adds bulk to the plant. Further improvement in efficiency may be made by staging the compression process with intercooling between the stages. To improve specific output the expansion process may be performed in stages with reheat between the stages. For each of these arrangements analysis may be made to find the optimum design conditions[7,8]. Figure 2.9(a) and (b) show performance curves for the ideal cycle[7] and Figure 2.9(c) shows the effect of process efficiencies, pressure losses and heat exchanger effectiveness on the ideal performance of Figure 2.9(b). Table 2.7 makes a comparison at one running condition for several schemes.

Table 2.7   GAS TURBINE CYCLE EFFICIENCY AND OUTPUT

Cycle	Efficiency %	Specific output kJ/kg
Simple cycle	24·1	133
Simple cycle with heat exchanger	26·6	122
Simple cycle with intercooler	27·1	188
Simple cycle with reheat	22·7	181
Simple cycle with intercooler and heat exchanger	33·1	176
Simple cycle with heat exchanger and reheat	29·6	166
Simple cycle with intercooler and reheat	25·6	250
Simple cycle with intercooler, heat exchanger and reheat	35·5	235

Standard conditions for calculations: T_{max} 800°C, T_{min} 15°C, compressor efficiency 87%, turbine efficiency 85%, leakage 2½% of compressor flow, pressure loss in cycle 10% but 15% in heat exchanger cycles, combustion efficiency, 100% intercooler effectiveness 90%, heat exchanger effectiveness 75%, for air $c_p = 1·005$ kJ/kg K, $R = 0·287$ kJ/kg K, for gas $c_p = 1·1$ kJ/kg K, $R = 0·287$ kJ/kg K.

When the design conditions are settled, the off design performance of a plant must be considered by examining the compressor and turbine characteristics and finding a series of matching conditions where the two units will operate together[9].

Reciprocating internal combustion engines

It is not possible to perform cycle analysis calculations for these engines as the processes are not clearly separated, as in steam and gas turbine plant. An alternative approach is to make a computer model to the processes which will enable performance to be predicted[10].

In reciprocating engines, thermodynamic problems are concerned with the combustion process and the rate of energy release in the reaction. In spark ignition engines an approximately stoichiometric mixture is induced. After the spark is made, there is a *delay period* before the developing combustion process causes a detectable change from the motoring curve shown in Figure 2.10(a).

When combustion is established and delay ends, there is a progressive combustion as the flame front moves across the combustion chamber. Correct choice of mixture strength, ignition timing, fuel (octane number) and good combustion chamber design will allow smooth combustion without *knock* which occurs if the end gas reaches the condition where self-ignition causes an explosion of all the mixture remaining in the chamber.

In compression ignition engines, the air fuel ratio varies, since just enough fuel is injected to meet the load demand. The pattern of combustion is shown in Figure 2.10(b). There is a delay period after the injection commences whilst combustion becomes established. This is followed by an explosion of the fuel injected during delay which gives a rapid, often noisy pressure rise. Diffusive combustion of the atomised fuel droplets continues until the reaction ceases. Combustion is smooth, provided the correct fuel (cetane number) is chosen and the combustion chamber design is good.

It may be seen that combustion and combustion chamber design are important topics in this field. Many of the requirements have been well-known for a considerable time[11, 12]

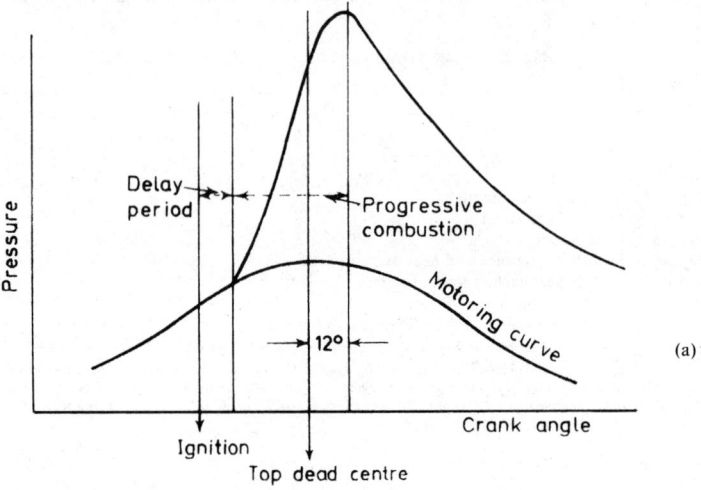

(a)

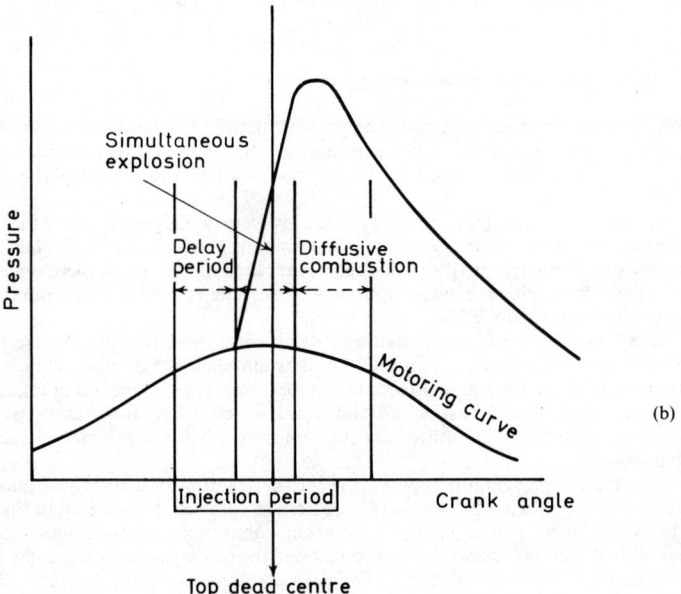

(b)

Figure 2.10

and engines have shown steady improvement in output as experience, better fuels and better materials became available.

COMPRESSORS AND REFRIGERATORS

Reciprocating compressors

Figure 2.11 shows a pressure-volume diagram for a reciprocating compressor. In process 1–2 air is compressed, 2–3 delivered, 3–4 re-expanded and 4–1 induced. The *free air delivery* of a compressor is the volume of air delivered measured at atmospheric pressure and temperature. As the delivery pressure is increased, the free air delivery becomes less and it becomes beneficial to use two or more stages of compression. Intercooling may be used between the stages.

The *volumetric efficiency* of the compressor is a measure of the need to introduce staging, as it is defined as the ratio of the free air delivery to the swept volume. The

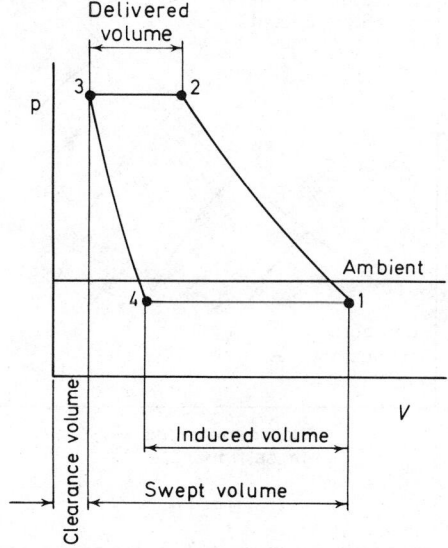

Figure 2.11

specific work required to compress and deliver unit mass of air may be obtained from either energy equation. For a single stage

$$w = RT_1 \cdot \frac{n}{1-n}\left[\left(\frac{p_2}{p_1}\right)^{n-1/n} - 1\right]$$ (21)

For N stages with equal pressure ratios for minimum work and perfect intercooling

$$w = NRT_1 \cdot \frac{n}{1-n}\left[r_p^{n-1/n} - 1\right]$$ (22)

where r_p is the stage pressure ratio, n is the index of polytropic compression (and re-expansion of the clearance volume air) and T_1 is the air inlet temperature.

Reciprocating compressors are capable of giving a large pressure ratio (by using many stages) but are for low mass flow rates. If high mass flow rates are required it is necessary to use a rotary turbocompressor or rotary positive displacement compressor.

Rotary compressors

For pressure ratios less than 4, a centrifugal compressor may be chosen. Higher ratios can be achieved by using two or more in series but the flow path is tortuous and it is preferable to use an axial flow machine in which as many stages as are required can follow each other without difficulty. The design of these machines is an aerodynamic problem and will not be discussed here[13, 14, 15].

Figure 2.12 shows the general form of characteristic that is obtained from rotary compressors. The points on the curves to the left of the *surge line* are regions of unstable flow and the compressor should not be operated in these regions. Surge occurs when

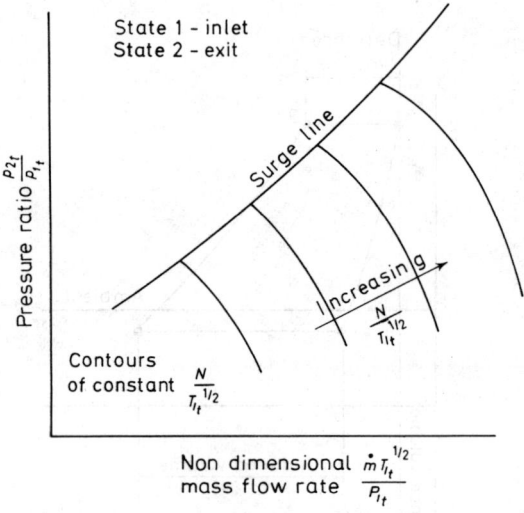

Figure 2.12

the flow reverses temporarily in some part of the compressor and vibrations occur in the flow which are audible and may cause damage to the machine in extreme cases. When designing a compressor installation it is essential to ensure that the machine will not move to surge conditions at off design running states and during acceleration.

The work necessary for compression may be obtained by the steady flow energy equation. Compressors are ideally adiabatic and the ideal specific work is given by

$$-w_x = c_p T_{1_t}[r_{p_t}^{\gamma - 1/\gamma} - 1]$$

where T_{1_t} is the *stagnation* temperature at inlet (to account for the inlet velocity which is not negligible in this type of machine) and r_{p_t} is the stagnation pressure ratio. The actual specific work is

$$-w_x = \frac{c_p T_{1_t}}{\eta_c}\left[r_{p_t}^{\gamma - 1/\gamma} - 1\right] \tag{23}$$

Where η_c is the process efficiency.

The work may also be obtained by considering the momentum changes of the air as it passes through the rotor. For this, the velocity triangles at inlet and outlet are drawn and the force in the direction of motion computed. Figure 2.13 shows triangles for an axial flow stage. The torque at the rotor will be the force multiplied by the radius and the work will be the torque multiplied by the angular velocity. Thus,

$$w_x = \omega(r\,V_{w_2} - r\,V_{w_1}) = u\Delta V_w \tag{24}$$

Equations 23 and 24 may be combined to determine the pressure ratio that will be given

V = Absolute velocity
W = Relative velocity

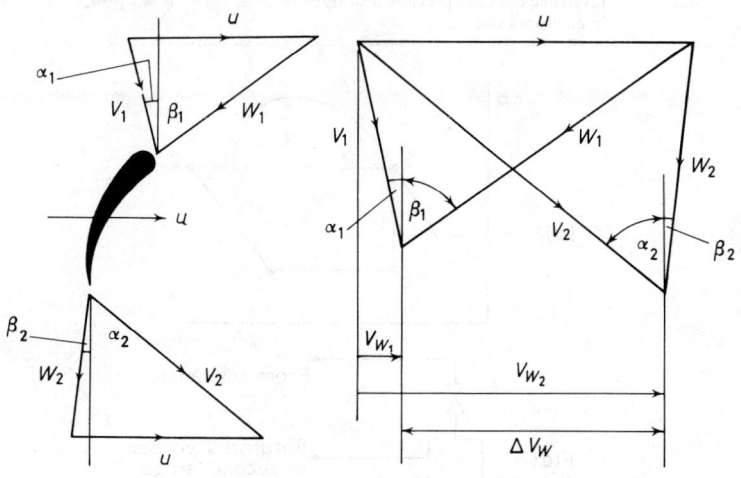

Figure 2.13

by a given blade shape. It must be realised that this simple one dimensional approximate calculation does not solve the aerodynamic problem mentioned earlier.

Rotary positive displacement compressors

This class of machine includes Roots blowers and vane type compressors. These machines achieve their object by scooping air from the atmosphere and transferring it to a receiver. The pressure in the receiver builds up and the compression occurs as each new packet of air is exposed to the receiver pressure. In general terms they are suited to higher flow rates than reciprocating machines and they are more compact. One advantage they offer over the turbocompressors described above is the inability of the flow to reverse for there is a *positive* displacement of the air.

Refrigerators

The ideal vapour compression refrigeration cycle is shown in Figure 2.14(a). The processes forming the cycle are shown in Table 2.8 and the steady flow energy equation is applied to each in turn.

Table 2.8 VAPOUR COMPRESSION REFRIGERATION CYCLE PROCESSES

Process	Name and component	Application of energy equation assuming $\Delta V^2/2$ and Δgz negligible
1–2	reversible adiabatic compression in a reciprocating compressor	$-w_x = h_2 - h_1$
2–3	reversible constant pressure heat transfer in a condenser	$q = h_3 - h_2$
3–4	irreversible adiabatic expansion in a throttle valve	$h_4 = h_3$
4–1	reversible constant pressure heat transfer in an evaporator	$q = h_1 - h_4$

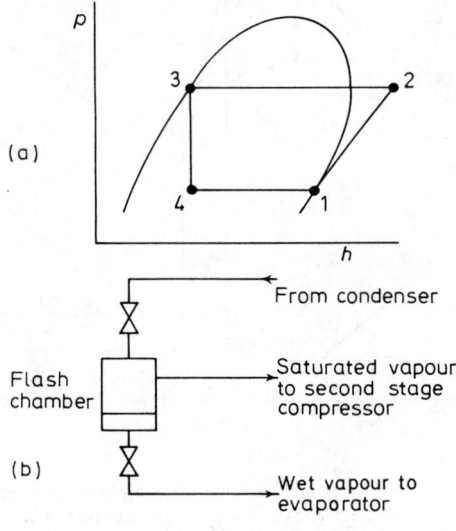

(a)

(b)

Flash chamber

From condenser

Saturated vapour to second stage compressor

Wet vapour to evaporator

Figure 2.14

Table 2.9 PROPERTIES OF REFRIGERANTS

Refrigerant	Evaporator pressure bar	Condenser pressure bar	Refrigeration effect kJ/kg	Volume flow rate per MW of cooling m³/s	cop
Freon 12 (or Refrigerant 12)	1·825	7·45	116·4	0·783	4·68
Methyl chloride	1·46	6·53	349·3	0·8	4·87
Ammonia	2·36	11·67	1 102·9	0·458	4·72
Carbon dioxide	22·9	72·1	132·0	0·127	2·7

The figures in this table are based on the ideal cycle with an evaporator temperature of $-15°C$, a condenser temperature of $30°C$, saturated liquid at condenser exit and saturated vapour at evaporator exit. Of the refrigerants listed, Freon 12 is attractive as a non toxic, non corrosive substance. The other three substances have undesirable physical properties; ammonia forms an explosive mixture in air and is corrosive, methyl chloride has anaesthetic properties, carbon dioxide requires high pressures, and has a critical temperature of $31·05°C$.

The measure of success of the cycle is called the *coefficient of performance*,

$$cop = \frac{\text{cooling heat transfer}}{\text{work supplied}} = \frac{h_1 - h_4}{h_2 - h_1} \tag{25}$$

The cooling heat transfer is called the *refrigeration effect*.

The cycle performance may be improved by fitting a regenerative heat exchanger between the liquid leaving the condenser and the gas leaving the evaporator or by staging the compression and expansion processes. In the staged expansion process, the wet vapour leaving the first throttle valve is separated into dry vapour and saturated liquid by a *flash chamber*, Figure 2.14(b). The dry vapour which has no evaporative cooling potential is returned to the high pressure compressor and the liquid passes to the second throttle valve. The size of compressor required by a plant is determined by the load and by the refrigerant chosen. A refrigerant with small specific volume and large enthalpy vaporisation will lead to a smaller plant. Table 2.9 gives refrigerant properties.

THE MEASUREMENT OF POWER

The application of the laws of thermodynamics to the working substance of an engine or plant will give a measure of the internal (or indicated) power. Due to friction in mechanisms and transmissions the external (or brake) power will be less than this value. The ratio (external power/internal power) is the *mechanical efficiency* η_{mech} of the plant. This ratio is inverted for power absorbing devices such as a compressor.

Internal power may be measured for reciprocating machines with an instrument called an *indicator*. External power is measured in situ by a *transmission dynamometer* or on a test bed by an *absorption dynamometer* or *brake*.

The overall efficiency η_0 of a plant is defined by the ratio

$$\eta_0 = \frac{\text{brake power}}{\text{energy supplied in fuel}}$$

and will allow for mechanical efficiency, combustion efficiency and process efficiencies in the cycle.

Tests may be made to determine these efficiencies and the specification of a selection of relevant trials is contained in the British Standards in the bibliography. The result of these trials is often conveniently presented in a tabular form of the energy equation.

THE MEASUREMENT OF TEMPERATURE

Thermodynamic temperature (a concept linked to the second law of thermodynamics) is a property. It is measured as a number of kelvins (K). It cannot be measured in engineering situations and there is a need for a method of approximation. This is answered by the International Practical (Kelvin) Scale of Temperature, IPKS[16, 17] which has exact numerical values assigned at reproducible conditions and specified methods of interpolation between these assigned values. This International Scale is essential to interchange of data and reproduction of identical standard temperatures in any situation. If 273·15 is subtracted from an IPKS value then an International Practical (Celsius) Scale temperature is obtained.

Temperatures obtained from various thermometers in everyday casual use give an approximation to International Scale values. To obtain an exact value, careful experimental work with a calibrated thermometer is required. Calibration is explained by and will be undertaken by the National Physical Laboratory[18].

Details of methods of temperature measurement are found in BS 1041:1943 'Code for Temperature Measurement' and BS 1900:1952 'Secondary reference Thermometers'.

In many applications where extreme accuracy is not required a thermocouple will be found adequate and convenient[19].

THE MEASUREMENT OF COMPRESSIBLE FLUID FLOW RATE

Common methods available are given in BS 1042:1943 'Code for Flow Measurement' and BS 726:1957 'Measurement of air flow for compressors and exhausters'. Both these standards use methods involving the measurement of a pressure drop through an orifice, venturi or nozzle and details are given of dimensions, installation and calibration. In cases where flow is pulsating a reservoir of adequate capacity to damp the pulsations is specified. For other flow methods or for special applications reference should be made to specialist texts.

COMPRESSIBLE FLUID FLOW IN DUCTS

If flow in a duct is examined by steady flow energy equation and continuity equation,[20] it is found that when the velocity at some point in the flow becomes equal to the local sonic velocity, the duct is passing its maximum mass flow rate and is said to be *choked*. It is also found that if the duct is designed to increase the kinetic energy of a stream of fluid (a nozzle), it must be convergent to accelerate a subsonic flow to sonic conditions and divergent to accelerate a sonic flow to supersonic conditions.

Three cases are of interest;

(i) The adiabatic convergent nozzle which becomes choked with sonic velocity at the exit plane when the ratio of the ambient pressure at the exit plane to the inlet (stagnation) pressure becomes less than or equal to

$$\left(\frac{2}{k+1}\right)^{k/(k-1)},$$

where k is the isentropic index of the fluid.

(ii) The adiabatic convergent-divergent nozzle which becomes choked with sonic velocity at the point of minimum cross sectional area (the *throat*) when the ratio of the throat pressure to the inlet (stagnation) pressure becomes equal to

$$\left(\frac{2}{k+1}\right)^{k/(k-1)}.$$

This condition is not always easy to establish when the nozzle is away from the design condition but will certainly be satisfied if the ambient exit plane pressure is less than or equal to

$$\left(\frac{2}{k+1}\right)^{k/(k-1)}.$$

Similarly, the exit plane velocity in a correctly designed and operated convergent-divergent nozzle will be supersonic but, if incorrectly operated, the exit velocity may be subsonic due to shock wave formation in the divergent portion of the nozzle even though the velocity at the throat is sonic and the nozzle passes the maximum mass flow rate.

The term

$$\left(\frac{2}{k+1}\right)^{k/(k-1)}$$

is called the *critical pressure ratio* and for a perfect gas $k = \gamma$. Table 2.10 gives values of this parameter.

Figure 2.15 shows nozzle flow regimes.

Table 2.10 VALUES OF CRITICAL PRESSURE RATIO

Substance	k	critical pressure ratio
Air	1·4	0·528
Superheated steam	1·3	0·546
Saturated steam	1·135	0·577

Table 2.11 THERMAL CONDUCTIVITY OF MATERIALS

Thermal conductivity values at 20°C, (W/m K)

Aluminium	204	Bakelite	0·232
Duralumin	164	Common brick	0·692
Cast iron	51·9	Facing brick	1·32
Lead	34·6	Concrete	0·8–1·4
Copper	386	Glass	0·762
Bronze (75–25)	26·0	Gypsum plaster	0·485
Brass (70–30)	111	Cotton	0·059
Mercury	8·7	Loose cork	0·036
Constantan	22·7	Loose earth	0·521
Magnesium	171	Glass wool	0·04
Nickel	90	Silk	0·036
Silver	407	Water	0·597
Tungsten	163	Freon 12	0·073
Zinc	112	Methyl chloride	0·163
Tin	64	Engine oil	0·145

Thermal conductivity values at quoted temperature (W/m K)

Diatomaceous earth	204°C	0·242
Diatomaceous earth	872°C	0·312
Fire clay	500°C	1·28
Fire clay	1 100°C	1·4
Magnesite	204°C	3·81
Magnesite	1 204°C	1·9
Asbestos	0°C	0·151
Asbestos	100°C	0·192
Wool felt	30°C	0·0519
Ice	0°C	2·22

Thermal conductivity of steel at 20°C, W/m K
The thermal conductivity of steel depends on the composition of the steel and tests should be made to ascertain the correct value. Some indication of the order of value is given below.

Nickel steel	12–50
Chrome steel	30–60
Carbon steel	35–55
Cr-Ni steel	16
Ni-Cr steel	14
Manganese steel	38
Tungsten steel	62
Silicon steel	31

The data in this table has been adapted from tables in An Introduction to Engineering Heat Transfer, J. R. Simonson, McGraw-Hill London 1967. Further information may be obtained from references 23 and 24.

(iii) In adiabatic pipe flow, the velocity becomes sonic at exit from the pipe when the pipe reaches the *limiting length*. At the limiting length the mass flow rate through the pipe is a maximum. If the length is increased or decreased the mass flow rate falls and at no stage can the flow become supersonic. Values of limiting length are obtained from tables of gas flow[21].

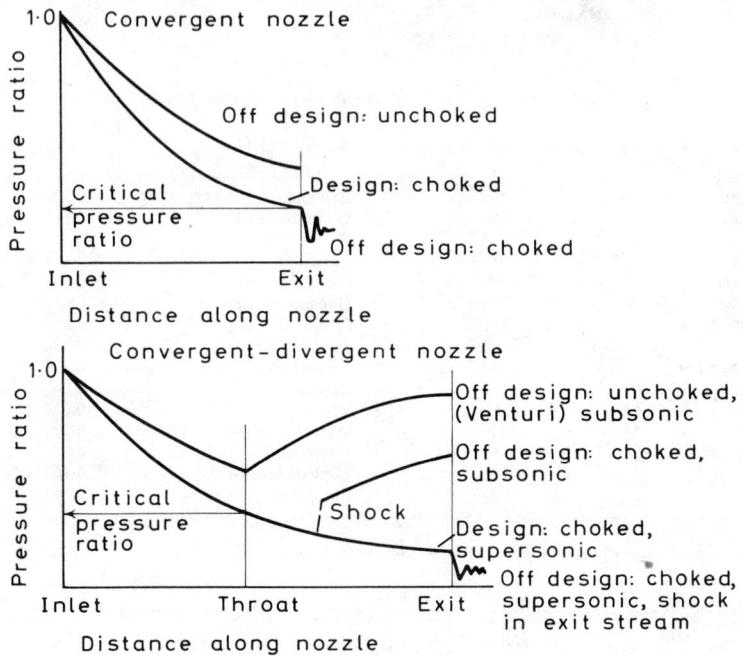

Figure 2.15

HEAT TRANSFER

Conduction

One dimensional conduction is governed by Fourier's law

$$\dot{Q}'' = -k\frac{dT}{dx} \qquad (26)$$

where $\dot{Q}''$ is the heat transfer rate per unit area, k is the thermal conductivity of the material and dT/dx is the temperature gradient in the direction of the heat transfer.

Thermal conductivity is not constant but is usually averaged over the temperature range considered unless the range is very large. Table 2.11 gives some values of k.

Fourier's equation is now applied to four common situations and the solutions obtained by integration are shown in Table 2.12.

The electrical analogy to conduction

The expressions in Table 2.12 may be compared to an Ohm's law type of relation in which $\dot{Q}''$ or $\dot{Q}$ is analogous to I, the current, $(T_1 - T_2)$ to V the potential difference and

$$\sum\left(\frac{\Delta x}{k}\right) \quad \text{or} \quad \sum \frac{\log_e \frac{r_{outer}}{r_{inner}}}{2\pi k_{layer}}$$

Table 2.12 and Figure 2.16 SOLUTION TO CONDUCTION PROBLEMS

Situation	Solution	Notes
Single layer Plane surface 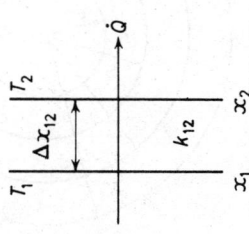	$$\dot{Q}'' = \left[\frac{T_1 - T_2}{x_2 - x_1}\right] k_{12} = \frac{T_1 - T_2}{\left(\dfrac{\Delta x}{k}\right)_{12}}$$	The solution is for heat transfer rate per unit *area*.
Multi layer plane surface 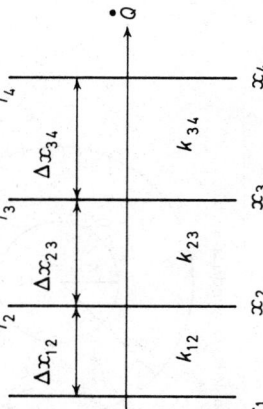	$$\dot{Q}'' = \frac{T_1 - T_4}{\left(\dfrac{\Delta x}{k}\right)_{12} + \left(\dfrac{\Delta x}{k}\right)_{23} + \left(\dfrac{\Delta x}{k}\right)_{34}} = \frac{T_1 - T_4}{\sum\left(\dfrac{\Delta x}{k}\right)}$$	The solution is for heat transfer rate per unit *area*. The heat transfer rate through each layer is the same.

continued

Table 2.12 and Figure 2.16 *continued*

Situation	*Solution*	*Notes*
Single layer cylindrical surface 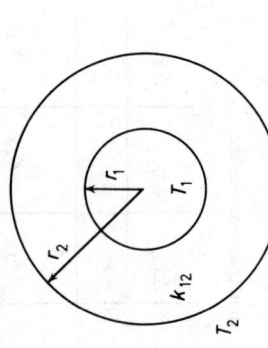	$$\dot{Q}' = \frac{T_1 - T_2}{\dfrac{\log_e \dfrac{r_2}{r_1}}{2\pi k_{12}}}$$	The solution is for heat transfer rate per unit *length* of cylinder as heat transfer area varies with radius.
Multi layer cylindrical surface 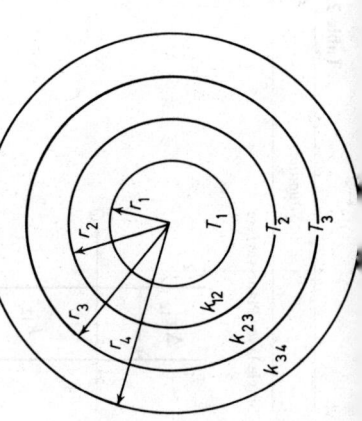	$$\dot{Q}' = \frac{T_1 - T_4}{\dfrac{\log_e \dfrac{r_2}{r_1}}{2\pi k_{12}} + \dfrac{\log_e \dfrac{r_3}{r_2}}{2\pi k_{23}} + \dfrac{\log_e \dfrac{r_4}{r_3}}{2\pi k_{34}}}$$ $$\dot{Q}' = \frac{T_1 - T_4}{\sum \left(\dfrac{\log_e \dfrac{r_{outer}}{r_{inner}}}{2\pi k_{layer}} \right)}$$	The solution is for heat transfer rate per unit *length* of cylinder as heat transfer area varies with radius. The heat transfer rate through each layer is the same.

Table 2.13(a) CONVECTIVE HEAT TRANSFER DATA

Mode		Determining condition	Average value of Nu over a length l		
			Plane	Vertical cylinder	Horizontal cylinder
Free Convection	Laminar	$10^4 < Ra_l = \dfrac{\rho^2 l^3 c_p \beta g \theta}{k\mu} < 10^9$	Vertical $Nu_l = 0.59 Ra_l^{\frac{1}{4}}$	$Nu_l = 0.59 Ra_l^{\frac{1}{4}}$	—
	Laminar	$10^4 < Ra_d = \dfrac{\rho^2 d^3 c_p \beta g \theta}{k\mu} < 10^9$	—	—	$Nu_d = 0.53 Ra_d^{\frac{1}{4}}$
	Turbulent	$10^9 < Ra_l < 10^{12}$	Vertical $Nu_l = 0.13\, Ra_l^{\frac{1}{3}}$	$Nu_l = 0.13 Ra_l^{\frac{1}{3}}$	—
	Turbulent	$10^9 < Ra_d < 10^{12}$	—	—	$Nu_d = 0.13 Ra_d^{\frac{1}{3}}$
Forced Convection	Plane Turbulent	$Re_l = \dfrac{\rho V l}{\mu} > 500\,000$	$Nu_l = 0.036 Re_l^{0.8} Pr^{0.33}$	—	
	Cylinder Turbulent	$Re_d = \dfrac{\rho V d}{\mu} > 2300$	—	$Nu_d = 0.023 Re_d^{0.8} Pr^{0.4}$	

$$Nu_l = \frac{\alpha l}{k}$$
$$Nu_d = \frac{\alpha d}{k}$$
$$Pr = \frac{\mu c_p}{k}$$

Notes

1. The fluid properties in plane surface applications should be extracted from data tables at the mean film temperature of the boundary layer. Film temperature $T_f = (T_w + T_b)2$ where T_w is the surface temperature and T_b is the bulk temperature of the fluid.

2. The fluid properties in cylindrical surface applications should be extracted from the data tables at the mean bulk temperature of the fluid.

3. Nomenclature; μ viscosity
   ρ density
   c_p specific heat capacity
   k thermal conductivity
   l length
   β cubical expansion coefficient
   g gravitational acceleration
   θ temperature difference
   d diameter

The data above is adapted from reference 23.

to R the resistance. Accordingly these latter terms are sometimes called the *thermal resistance* of the situation. It may be seen that thermal resistances in series (as in the multi layer wall) may be added as are electrical resistances in series. This method often gives a quick solution and also enables the dominating resistance to be easily identified.

Problems in unsteady conduction or three dimensional conduction are also solved by Fourier's law but the integral solutions are more complex.[22]

Convection

The heat transfer rate per unit area from a surface due to convection is given by

$$\dot{Q}'' = \alpha\theta \tag{27}$$

where θ is the temperature difference between the surface and the bulk of the fluid above it. Again an electrical analogy may be made and it may be seen that $\dot{Q}'' = \theta/(1/\alpha)$ for a plane wall and

$$\dot{Q}' = \frac{\theta}{\left(\dfrac{1}{2\pi r\alpha}\right)}$$

for a cylindrical surface and the thermal resistances are $(1/\alpha)$ and $(1/2\pi r\alpha)$ respectively.

Table 2.13(b)   TRANSPORT PROPERTIES FOR AIR AND WATER

Temperature °C	Specific volume m³/kg	Specific heat at constant pressure kJ/kg K	Thermal conductivity W/m K	Dynamic viscosity kg/s m	Prandtl number
AIR AT ATMOSPHERIC PRESSURE					
0	0·773	1·01	0·024	$1·7 \times 10^{-5}$	0·72
100	1·057	1·02	0·032	$2·2 \times 10^{-5}$	0·70
200	1·341	1·03	0·039	$2·6 \times 10^{-5}$	0·69
300	1·624	1·05	0·045	$3·0 \times 10^{-5}$	0·69
400	1·908	1·07	0·051	$3·3 \times 10^{-5}$	0·70
500	2·191	1·10	0·056	$3·6 \times 10^{-5}$	0·70
SATURATED WATER					
10	0·00100	4·193	0·587	$130·1 \times 10^{-5}$	9·29
50	0·00101	4·181	0·643	$54·4 \times 10^{-5}$	3·53
100	0·00104	4·216	0·681	$27·8 \times 10^{-5}$	1·723
200	0·00116	4·497	0·665	$13·3 \times 10^{-5}$	0·902
250	0·00125	4·867	0·616	$10·65 \times 10^{-5}$	0·841
300	0·00140	5·762	0·541	$8·97 \times 10^{-5}$	0·955
350	0·00174	10·10	0·437	$6·48 \times 10^{-5}$	1·5

The data above is extracted from Thermodynamic Tables in SI(metric) Units, R. W. Haywood, Cambridge University Press, 1968.

The determination of α, the *surface heat transfer coefficient* is the main problem in convective heat transfer and several methods are available to make reasonable estimates. Several modes of convection are recognised:

(i) *Free or natural* convection in which fluid motion is by buoyancy forces due to density gradients;

(ii) *Forced* convection in which fluid motion is induced by a pump, and

(iii) Both free and forced convection may be associated with *laminar* or *turbulent* flow.

In turbulent flow, heat transfer rates are greater than those in laminar flow and, in forced convection, heat transfer rates are greater than in free convection. Thus *forced turbulent* convection is used when high rates are required. The determination of whether flow is laminar or turbulent in any situation is determined by the value of the Rayleigh Number Ra in free convection and by the value of the Reynolds Number Re in forced convection.

Dimensional analysis and experiment is one of the methods used to form equations in which α is involved in the non-dimensional Nusselt Number Nu. Table 2.13(a) gives results which may be applied to plane and cylindrical surfaces and also indicates the correct method of application. Any relations obtained by empirical methods should be treated with great care since the equations only apply in the exact conditions of the original work. The table shows average values over finite lengths. For other configurations specialist texts should be consulted[23, 24]. Table 2.13(b) shows transport properties for air and water for use in the relations in Table 2.13(a).

Radiation

Heat transfer by radiation becomes increasingly important as temperature rises. This is because the Stefan-Boltzmann Law determining the amount of radiation contains the fourth power of thermodynamic temperature.

Radiation theory is based on an ideal *black body* which absorbs all radiation incident upon it. A *grey body* is also considered which absorbs the same constant proportion of incident radiation at all wavelengths. The amount absorbed is determined by the absorptivity α of the body ($\alpha < 1$). Kirchoff's Law shows that the emissivity ε of a grey body is equal to its absorptivity. Real surfaces have emissivities (and absorptivities) which vary with wavelength and temperature and values are determined empirically and tabulated. Table 2.14 gives radiation laws and Table 2.15 some values of emissivity.

To solve a radiation problem, it is first necessary to decide how much of the radiation that is emitted by the radiator is intercepted by the receiver. This is a problem of geometry and is accounted for by a geometrical or *configuration factor* F for any situation. Configuration factors are not easy to calculate except in simple cases and may be obtained by experiment or from graphs[24, 25] drawn for all common situations. If the configuration

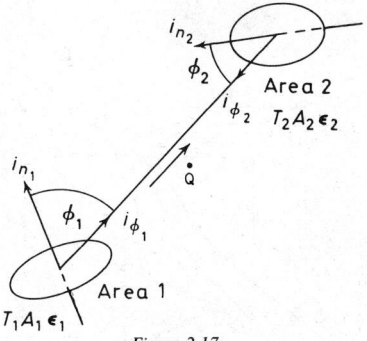

Figure 2.17

factor is defined for the situation in Figure 2.17 by F_{12} = the fraction of energy emitted by 1 that is intercepted by 2, then the energy intercepted is $A_1 \sigma \varepsilon_1 T_1^4 F_{12}$ and of this $\alpha_2 (A_1 \sigma \varepsilon_1 T_1^4 F_{12})$ will be absorbed by 2. The remainder is reflected. A similar expression for the energy emitted by 2 that is intercepted and absorbed by 1 is $\alpha_1 (A_2 \sigma \varepsilon_2 T_2^4 F_{21})$. It may be shown that

$$A_1 F_{12} = A_2 F_{21} \qquad (28)$$

Table 2.14 and Figure 2.18 THE LAWS OF RADIATION

Law	Equation	Notes
Stefan-Boltzmann	$\dot{E}''_b = \sigma T^4$	$\dot{E}''$ is the emissive power in W/m^2 σ is the Stefan-Boltzmann constant; $5.66 \times 10^{-8}\ W/m^2\ K^4$
Planck	$\dot{E}''_{b_\lambda} = \dfrac{c_1 \lambda^{-5}}{e^{c_2/\lambda T} - 1}$ where $c_1 = 3.74 \times 10^{-16}\ Wm^2$ and $c_2 = 0.0144\ mK$	Suffix b refers to black body radiation. This gives the emissive power at wavelength λ and hence $\displaystyle\int_0^\infty \dot{E}''_{b_\lambda}\, d\lambda = \sigma T^4$
Wien	$\lambda_{max} T = \text{constant} = 2.9 \times 10^{-3}\ mK$	
Kirchhoff	$\alpha = \varepsilon$ for a grey body	$\varepsilon = \left[\dfrac{\dot{E}''}{\dot{E}_b}\right]$

(a)

For real bodies the monochromatic emissivity at wavelength λ and temperature T $\varepsilon_\lambda = \left[\dfrac{\dot{E}''_\lambda}{\dot{E}''_{b_\lambda}}\right]_T$. Since this varies with both wavelength and temperature it is often convenient to eliminate the wavelength effect by integrating the radiation over all λ. The total emissivity obtained can then be used to determine the total radiation from a surface at a given temperature T.

$$\text{Thus } \varepsilon = \left[\frac{\displaystyle\int_0^\infty \varepsilon_\lambda \dot{E}''_{b_\lambda}\, d\lambda}{\sigma T^4}\right],$$

that is required coupled with the emissivity at the wavelength of the emitted radiation. If the radiation is at another wavelength either ε or α may be used as they equal total emissivities or total absorptivities.

Lambert

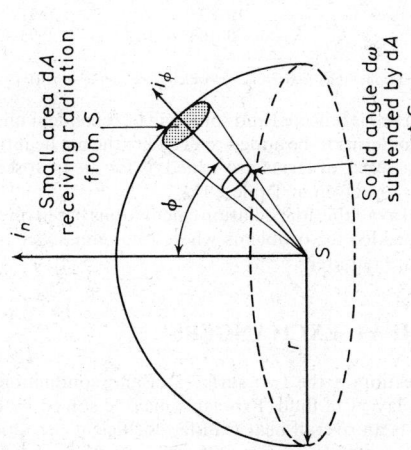

$i_\phi = i_n \cos \phi$

where $i = \left(\dfrac{\mathrm{d}E''}{\mathrm{d}\omega}\right)$

ω being a solid angle

This law determines the spatial distribution of radiation from a point. It is expressed in terms of the intensity of radiation in a direction normal to the surface i_n, where $i_n = \varepsilon \sigma T^4 / \pi$.

(b)

Table 2.15   TOTAL EMISSIVITY OF SURFACES AT THE GIVEN TEMPERATURE

Material	Temperature, °C	Total emissivity
Aluminium		
Commercial sheet	100	0·09
Heavily oxidised	93–505	0·20–0·31
Brass		
Polished	100	0·06
Rolled plate	22	0·06
Chromium		
Polished	100	0·075
Copper		
Polished	100	0·052
Iron		
Cast and polished	200	0·21
Rusted plate	19	0·69
Molten cast iron	1 300–1 400	0·29
Steel		
Polished	100	0·066
Sheet, oxidised	24	0·82
Plate, rough	38–372	0·94–0·97
Molten	1 522–1 650	0·43–0·40
Nichrome wire	49–1 000	0·65–0·79
Platinum filament	27–1 230	0·036–0·192
Silver		
Polished	227–627	0·02–0·032
Tungsten filament	3 320	0·39
Zinc (galvanised iron)	28	0·23
Asbestos board	23	0·96
Brick		
Red, rough	21	0·93
Building	1 000	0·45
Fireclay	1 000	0·75
Magnesite (refractory)	1 000	0·38
Candle soot	97–272	0·952
Lampblack	50–1 000	0·96
Glass	22	0·94
Flat black lacquer	38–94	0·96–0·98
Oil paints, various colours	100	0·92–0·96
Aluminium paints	100	0·27–0·67
Plaster, rough lime	10–87	0·91
Water	0–100	0·95–0·963

The data in this table has been adapted from tables in reference 23 which were compiled by H. C. Hottel.

The radiation that is not absorbed is reflected and some will be reincident on the emitter. This effect will result in a series of terms to be added to evaluate the net heat transfer rate. One case of interest, in which a series of terms are added, is for an enclosed body 1 in an enclosing body 2. Four cases are given in Table 2.16.

There is an alternative method available for radiation calculations based on an electrical analogy[26]. When radiation is tackled for problems where total emissivity is an invalid approximation specialist texts are required[27, 28].

HEAT EXCHANGERS

Heat exchangers involve convection at the two surfaces of and conduction through a solid boundary separating two layers of fluid. Problems may be solved by considering these resistances in series. If U is an overall heat transfer coefficient per unit area for a

Table 2.16   RADIATION BETWEEN AN ENCLOSED BODY AND THE ENCLOSURE

Enclosed body small compared with enclosure

$$\dot{Q} = A_1 \varepsilon_1 \sigma (T_1^4 - T_2^4)$$

Enclosed body

$$\dot{Q} = A_1 \left[\cfrac{1}{\cfrac{1}{\varepsilon_1} + \cfrac{A_1}{A_2}\left(\cfrac{1}{\varepsilon_2} - 1\right)} \right] \sigma(T_1^4 - T_2^4)$$

Concentric spheres or infinitely long concentric cylinders

$$\dot{Q} = A_1 \left[\cfrac{1}{\cfrac{1}{\varepsilon_1} + \cfrac{A_1}{A_2}\left(\cfrac{1}{\varepsilon_2} - 1\right)} \right] \sigma(T_1^4 - T_2^4)$$

Parallel planes of infinite size

$$\dot{Q} = A \left[\cfrac{1}{\cfrac{1}{\varepsilon_1} + \cfrac{1}{\varepsilon_2} - 1} \right] \sigma(T_1^4 - T_2^4)$$

plane heat exchanger and U' an overall heat transfer coefficient per unit length for a cylindrical heat exchanger then

$$\dot{Q} = UA\theta \quad \text{or} \quad \dot{Q} = U'L\theta \tag{29}$$

From earlier considerations

$$\frac{1}{U} = \frac{1}{\alpha_1} + \left(\frac{\Delta x}{k}\right)_{12} + \frac{1}{\alpha_2}$$

and

$$\frac{1}{U'} = \frac{1}{2\pi r_1 \alpha_1} + \frac{\log_e \dfrac{r_2}{r_1}}{2\pi k_{12}} + \frac{1}{2\pi r_2 \alpha_2}$$

The values of α_1 and α_2 are determined for mean conditions and therefore θ in equation

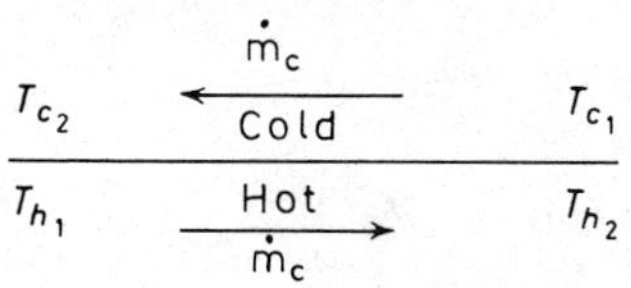

$$\theta_1 = T_{h_1} - T_{c_2}$$

$$\theta_2 = T_{h_2} - T_{c_1}$$

Figure 2.19

(29) must be for mean conditions. It may be shown that θ must be the *log mean temperature difference* given by

$$\theta = \frac{\theta_1 - \theta_2}{\log_e \dfrac{\theta_1}{\theta_2}}$$

where θ_1 and θ_2 are the differences in the two stream temperatures at the ends of the heat exchanger. Figure 2.19 illustrates the calculation of θ. The figure shows a counter flow situation which is normally preferred since it results in a smaller unit for a given heat transfer rate.

It may be seen that if the steady flow energy equation is applied to each stream in Figure 2.19, then the heat transfer rate may also be written

$$\dot{Q} = \dot{m}_h c_{p_h}(T_{h_1} - T_{h_2}) = \dot{m}_c c_{p_c}(T_{c_2} - T_{c_1}) \tag{30}$$

This equation enables heat exchanger design to be completed but the solutions often require trial and error. A more direct approach[29] is given in Table 2.17.

Table 2.17   COUNTER FLOW HEAT EXCHANGER DESIGN

If $\dot{m}_h c_{p_h} > \dot{m}_c c_{p_c}$	If $\dot{m}_c c_{p_c} > \dot{m}_h c_{p_h}$
Let the capacity ratio $C(<1)$ be	
$$C = \frac{\dot{m}_c c_{p_c}}{\dot{m}_h c_{p_h}}$$	$$C = \frac{\dot{m}_h c_{p_h}}{\dot{m}_c c_{p_c}}$$
Let the effectiveness E be the ratio of the actual energy transfer to the maximum possible energy transfer, i.e.	
$$E = \frac{T_{c_2} - T_{c_1}}{T_{h_1} - T_{c_1}}$$	$$E = \frac{T_{h_1} - T_{h_2}}{T_{h_1} - T_{c_1}}$$
but $\dot{Q} = \dot{m}_c c_{p_c}(T_{c_2} - T_{c_1})$ hence $\dot{Q} = EC(T_{h_1} - T_{c_1})\dot{m}_h c_{p_h}$	but $\dot{Q} = \dot{m}_h c_{p_h}(T_{h_1} - T_{h_2})$ hence $\dot{Q} = EC(T_{h_1} - T_{c_1})\dot{m}_c c_{p_c}$
Let the number of transfer units NTU be	
$$NTU = \frac{UA}{\dot{m}_c c_{p_c}}$$	$$NTU = \frac{UA}{\dot{m}_h c_{p_h}}$$
Then it may be shown that for either case	
$$E = \frac{1 - e^{-NTU(1-C)}}{1 - Ce^{-NTU(1-C)}}$$	

Notes: (i) *For parallel flow* similar work yields $E = \dfrac{1 - e^{-NTU(1+C)}}{1 + C}$

(ii) These results are based on U rather than U'. For tubular exchangers U is $U'/2\pi r_2$ where r_2 is the outer radius of the exchange surface. In this case $\dfrac{1}{U} = \dfrac{r_2}{r_1 \alpha_1} + \dfrac{r_2 \log_e \dfrac{r_2}{r_1}}{k_{12}} + \dfrac{1}{\alpha_2}$ and if the tube is thin so that $r_1 \doteq r$

$\dfrac{1}{U} = \dfrac{1}{\alpha_2} + \dfrac{1}{\alpha_1}$.

(iii) Plotted figures of E versus NTU with C as contour are available for various configurations in; Compact Heat Exchangers, W. M. Kays and A. L. London, McGraw-Hill London 1964.

REFERENCES

1. ZEMANSKY, M. W., *Heat and Thermodynamics,* McGraw-Hill, New York (1951)
2. *The 1967 IFC Formulation for Industrial use printed in 1967 Steam Tables,* Electrical Research Association, Arnold, London (1967)
3. MCBRIDE, B. J., *Fortran IV program for calculation of Thermodynamic data,* NASA N67–31592, (1967)
4. BS 526
   Definitions of the calorific values of Fuels (1961)
   BS 1016
   Part 5. Gross calorific values of coal and coke (1967)
   BS 3804
   Part 1. Methods for the determination of the calorific values of fuel gases (1964)
   BS 4379
   Determination of the calorific value of liquid fuels (1969)

5. PEARSON, J. D. and FELLINGER, R. C., *Thermodynamic Properties of Combustion Gases*, Iowa State University Press, Iowa (1965)
6. HAYWOOD, R. W., 'A generalised analysis of the regenerative steam cycle for a finite number of heaters', *Proc. I. Mech. E.*, 161, 157 (1949)
7. HODGE, J., *Cycles and Performance Estimation*, Butterworth, London (1955)
8. PALMER, J. R., *The 'Turbocode' scheme for the programming of thermodynamic cycle calculations on an electronic digital computer*, CoA Report Aero 198, College of Aeronautics, Cranfield (1967)
9. MALLINSON, D. H. and LEWIS, W. G. E., 'The part load performance of various Gas Turbine Engine Schemes', *Proc. I. Mech. E.*, 159, 198 (1948)
10. 'Computers in Internal Combustion Engine Design', *Proc. I. Mech. E.*, 3L, 182 (1968)
11. RICARDO, Sir H. and HEMPSON, J. C. G., *The High Speed Internal Combustion Engine*, Blackie, Glasgow (1968)
12. BROEZE, J. J., *Combustion in Piston Engines*, De Technishe Vitgeverij, Haarlem (1963)
13. HORLOCK, J. H., *Axial Flow Compressors*, Butterworth, London (1958)
14. FERGUSON, T. B., *The Centrifugal Compressor Stage*, Butterworth, London (1963)
15. HAWTHORNE, W. R. (Ed), *Aerodynamics of Turbines and Compressors*, Oxford University Press, London (1964)
16. HAYWOOD, R. W., 'The Rational Treatment of Temperature and Temperature Scales', *Proc. I. Mech. E.*, 182, 501 (1968)
17. *Units and Standards of Measurement employed at the National Physical Laboratory, IV Temperature*, HMSO (1962)
18. HALL, J. A. and BARBER, C. R., *Notes on Applied Science, No. 12 Calibration of Temperature Measuring Instruments*, HMSO (1964)
19. BILLING, B. F., 'Thermocouples, their instrumentation, selection and use', *Institute of Engineering Inspection* (1964)
20. LIEPMANN, H. W. and ROSHKO, A., *Elements of Gas Dynamics*, Wiley, New York (1960)
21. JORDAN, D. P. and MINTZ, M. D., *Air Tables*, McGraw-Hill, McGraw-Hill, New York (1965) or HOUGHTON, E. L. and BROCK, A. G., *Tables for the compressible flow of dry air*, Arnold, London (1970)
22. KUTATELADZE, S. S., et al, *Unsteady State Heat Transfer*, Iliffe, London (1966)
23. MCADAMS, W. H., *Heat Transmission*, McGraw-Hill, New York (1954)
24. CHAPMAN, A. J., *Heat Transfer*, Macmillan Company, New York (1960)
25. HAMILTON, D. C. and MORGAN, W. R., *Radiant Interchange Configuration Factors*, NACA TN-2836 (1952)
26. HOLMAN, J. P., *Heat Transfer*, McGraw-Hill, New York (1968)
27. WIEBELT, J. A., *Engineering Radiation Heat Transfer*, Holt, Rinehart and Winston, New York (1966)
28. HARRISON, T. R., *Radiation Pyrometry and its underlying principles of Radiant Heat Transfer*, Wiley, New York (1960)
29. KAYS, W. M. and LONDON, A. L., *Compact Heat Exchangers*, McGraw-Hill, London (1964)

BIBLIOGRAPHY

KEENAN, J. H., *Thermodynamics*, Wiley, New York (1941)
LAY, J. E., *Thermodynamics*, Pitman, London (1964)
REISS, H., *Methods of Thermodynamics*, Blaisdell, New York (1965)
BAIN, R. W., *Steam Tables 1964, National Engineering Laboratory*, HMSO, Edinburgh (1964)
KEENAN, J. H. and KAYE, J., *Gas Tables*, Wiley, New York (1948)
DIN. F. (Ed), *Thermodynamic Functions of Gases (3 vols.)*, Butterworth, London (1962)
STULL, D. R. (Ed), *Janaf Thermochemical Tables, PB 168–370*, Dow Chemical Company, Midland, Michigan (1965–8)
MCHARNESS, R. C., EISEMAN, B. J. and MARTIN, J. J., 'The New Thermodynamic Properties of Freon 12', *Refrigerating Engineering*, 63, 9 (Sept. 1955)
SCHICK, H. L., *Thermodynamic properties of certain Refractory Compounds*, (2 Vols.), Academic Press, New York (1966)
BENSON, R. S., *Advanced Engineering Thermodynamics*, Pergamon, Oxford (1967)
LEWIS VAN ELBE, *Combustion, Flame and Explosion of Gases*, Academic Press, London (1961)
LICHTY, L. C., *Combustion Engine Processes*, McGraw-Hill, New York (1967)
SMITH, I. E. (Ed), *Combustion in Advanced Gas Turbine Systems*, Pergamon, Oxford (1967)
KALAFATI, D. D., *Thermodynamic Cycles of Nuclear Power Stations*, Israel Programme for Scientific Translations, Jerusalem (1965)

WEIR, C. D., 'An Analytical Study of the Regenerative Multiple Reheat Steam Turbine Cycle', *I. Mech. E. monograph No. 5*, (1967)

BENDER, R. J., 'Steam Generation', *Power magazine*, New York (1960)

'Package Boilers and Liquid Phase Heaters, 1970; Steam and Heating Engineer Survey', *Steam and Heating Engineer*, **39**, No. 358, London (Jan. 1970)

KEARTON, W. J., *Steam Turbine Theory and Practice*, Pitman, London (1961)

Modern Power Station Practice, (5 vols.), CEGB (1963)

HORLOCK, J. H., *Axial Flow Turbines*, Butterworth (1966)

HAWTHORNE, W. R. and OLSEN, N. T. (Eds), *Design and Performance of Gas Turbine Power Plants*, Oxford University Press, London (1960)

SAWYER, J. W., *Gas Turbine Engineering Handbook*, Gas Turbine publications Limited, Stamford, Conn. (1966)

ALCOCK, J. F., 'Some more light on Diesel Combustion', *Proc. Auto. Div. I. Mech. E.*, 179 (1962-3)

ANNAND, W. J. D., 'Heat Transfer in the cylinders of Reciprocating Internal Combustion Engines', *Proc. I. Mech. E.*, 177, 973 (1963)

LYN, W.-T., Calculations of the effect of Rate of Heat 'Release on the shape of cylinder-pressure diagram and cycle efficiency', *Proc. Auto. Div. I. Mech. E.*, 177, 25 (1960-61)

CSANADY, G. T., *Theory of Turbomachinery*, McGraw-Hill, London (1964)

CHLUMSKY, V., *Reciprocating and Rotary Air Compressors*, E. & F. Spon, Ltd., London (1965)

JONES, E. B., *Instrument Technology* (2 vols), Butterworth, London (1953-56)

CONSIDINE, D. M. and ROSS, S. D., Handbook of Applied Instrumentation, McGraw-Hill, New York (1964)

The Instrument Manual, United Trade Press, London, (3rd ed 1960)

SHAPIRO, A. H., *Compressible Fluid Flow* (2 vols), Ronald Press, New York (1953-54)

SIMONSON, J. R., *An Introduction to Engineering Heat Transfer*, McGraw-Hill, London (1967)

POWELL, R. W., HO, C. Y. and LILEY, P. E., *Thermal Conductivity of Selected Materials*, NSRDS-NBS 8

KUTATELADZE, S. S., *Fundamentals of Heat Transfer*, Arnold, London (1963)

NEL, *Heat Bibliography*, HMSO, Edinburgh (annually)

FRAAS, A. P. and OZISIK, M. N., *Heat Exchanger Design*, Wiley, New York (1965)

HERINGTON, E. F. G., 'The measurement of Thermodynamic Properties at the National Chemical Laboratory', *Proc. Chem. Soc.* (June 1964)

BRITISH STANDARDS

BS 132
Steam turbines (1965)
BS 752
Test code for acceptance tests for steam turbines (1958)
BS 845
Code for acceptance tests for industrial type boilers, and steam generators (1961)
BS 1113
Water tube steam generating plant (1969)
BS 1894
Electrode boilers of riveted, seamless, welded and cast iron construction for water heating and steam generating (1954)
BS 2079
Steam receivers and separators (1954)
BS 2885
Acceptance tests for steam generating units (1957)
BS 3285
Methods of sampling superheated steam from steam generating units (1960)
BS 3812
Recommendations for estimating the dryness of saturated steam (1964)
BS 3135
Test code for gas turbines (1959)
BS 3863
Outline specification for gas turbines (1965)
BS 412
Fittings for cylinder pressure indicators for reciprocating engines (1961)
BS 765
Internal combustion engines, spark ignition type (1961)

BS 1701
Filters for air supply to internal combustion engines and compressors (1970)
BS 2637
Method for the determination of knock rating of fuel (1956)
BS 2869
Petroleum fuels for oil engines and burners (1970)
BS 3109
Gas and dual fuel engines (1959)
BS 4040
Petrol for motor vehicles (1971)
BS 526
Definitions of calorific values of fuels (1961)
BS 1016
Part 5. Gross calorific value of coal and coke (1967)
BS 3804
Methods for the determination of the calorific value of fuel gases (1964)
BS 4379
Determination of the calorific value of liquid fuels (1969)
BS 4056
Method of test for ignition temperatures of gases and vapours (1966)
BS 1756
Methods for the sampling and analysis of flue gases
BS 3048
Code for the continuous sampling and automatic recording of flue gases, indicators and recorders (1958)
BS 3156
Methods for the analysis of fuel gases
BS 726
Measurement of air flow for compressors and exhausters (1957)
BS 1042
Code for flow measurement (1943)
BS 1041
Code for temperature measurement (1943)
BS 1900
Secondary reference thermometers (1952)
BS 2082
Code for disappearing filament pyrometers (1954)
BS 3403
Indicating tachometers for general industrial use (1961)
BS 1571
Acceptance tests for positive displacement compressors and exhausters (1949)
BS 2009
Code for acceptance tests for turbocompressors and exhausters (1953)
BS 3122
Rating and testing of refrigeration compressors (1959)
BS 1584
Glossary of terms used in refrigeration (1949)
BS 1586
Methods for testing refrigerant condensing units (1964)
BS 4434
Requirements for refrigeration safety (1969)
BS 1588
The use of thermal insulating materials in the temperature range 95°C to 230°C (1964)
BS 3708
The use of thermal insulating materials between 230°C and 650°C (1969)
BS 2972
Methods of determining thermal properties (1961)
BS 3958
Thermal insulating materials
BS 3274
Tubular heat exchangers for general purposes (1960)

FLUID MECHANICS

D. H. BACON

HYDROSTATICS

Pressure

Hydrostatics is concerned with the determination of forces acting on bodies immersed in or containing fluids at rest. The forces are due to gravitational attraction of the mass of fluid above the level considered. The force per unit area is called *pressure*. Pressure acts in a direction normal to a surface and is obtained from the relation

$$\text{(absolute) pressure, } p = p_o + \rho g h \tag{1}$$

where p_o is the pressure at the free surface of the fluid, h is the depth below the free surface, ρ is the fluid density and g is gravitational acceleration. In many cases p_o is atmospheric pressure and the term $\rho g h$ is termed the *gauge pressure*, that is, pressure above atmospheric value. Atmospheric pressure is measured with a barometer and is approximately 1·013 bar or 101·3 kN/m^2.

Buoyancy

Consider the forces acting on the submerged body of Figure 2.20. All horizontal forces cancel and there will be an upthrust due to the difference between the forces acting on ABCD and EFGH. The force on ABCD = $(p_o + \rho g h_1)A$ and the force on EFGH = $-(p_o + \rho g h_2)A$, thus the net force is $-\rho g(h_2 - h_1)A$ (upwards). This may be seen to be equal to the weight of displaced fluid. Archimedes' principle states that the upthrust on a submerged or partially submerged body is equal to the weight of displaced fluid.

Submerged plane surfaces

Figure 2.21 shows the plane surface. By considering the forces on an elementary area due to the *gauge pressure* and integrating for the whole area, it found that the force on the submerged area is equal to the area multiplied by the pressure at the centre of area. In Figure 2.21,

$$F = \rho g h_G A \text{ or } \rho g x_G \sin \theta \, A. \tag{2}$$

Since pressure is proportional to depth this force will act at a level *below* the centre of area at a point known as the *centre of pressure*. By taking moments of the elementary forces about 00 it may be shown that the position of the centre of pressure is given by

$$x_p = \frac{\text{Second moment of area about 00}}{\text{First moment of area about 00}} = \frac{I_{00}}{A x_G} \tag{3}$$

By use of the theorem of parallel axes this becomes

$$x_p = \frac{I_G + A x_G^2}{A x_G} \quad \text{or} \quad x_p = x_G + \frac{I_G}{A x_G} \tag{4}$$

Where I_G is the second moment of area about an axis through G parallel to 00.

2–40

Equation 4 is the most useful form of the expression since data for second moments of area about the centre of area are often available. Table 2.18 gives some values.

Submerged curved surfaces

Curved surface problems are analysed by considering the equilibrium of a wedge of real or imaginary fluid which has one boundary on the curved surface and others con-

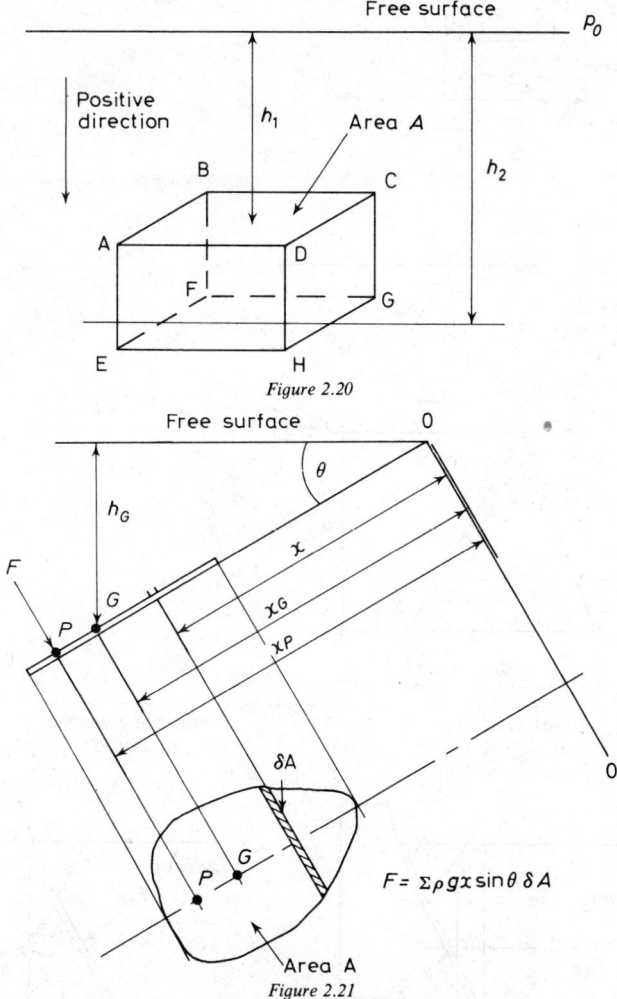

Figure 2.20

$$F = \Sigma \rho g x \sin\theta \, \delta A$$

Figure 2.21

veniently chosen to give a simple solution. This may be achieved by choosing plane vertical and horizontal surfaces and the problem may be solved by the principles of static equilibrium.

If the forces are coplanar and can be reduced to three, one vertical one horizontal and one other the triangle of forces gives a rapid solution. If the curve involved is spherical

Table 2.18 and Figure 2.22 CENTRES OF AREA AND SECOND MOMENTS OF AREA

Figure	Centre of area	Second moment of area

I axis — rectangle of width b and depth d with centre G, $\bar{x}$ measured from base:

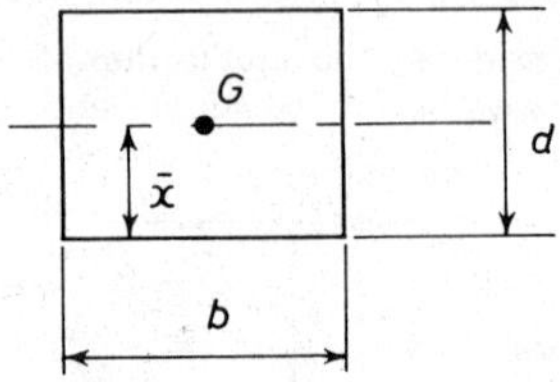

$$\bar{x} = \frac{d}{2}$$

$$I_G = \frac{bd^3}{12}$$

I axis — circle of diameter d with centre G:

—

$$I_G = \frac{\pi d^4}{64}$$

I axis — triangle of base b and height d with centre G, $\bar{x}$ measured from base:

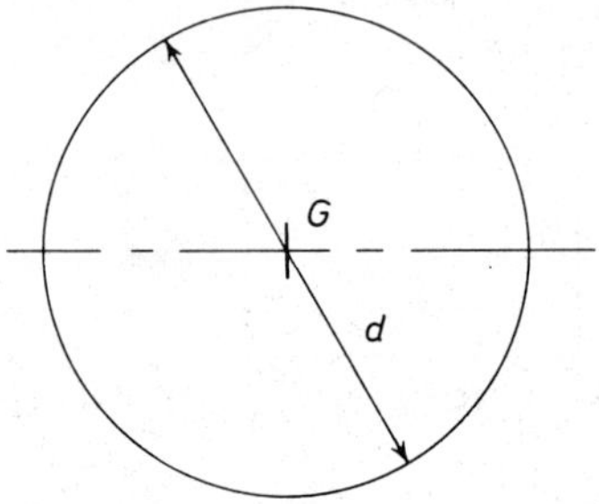

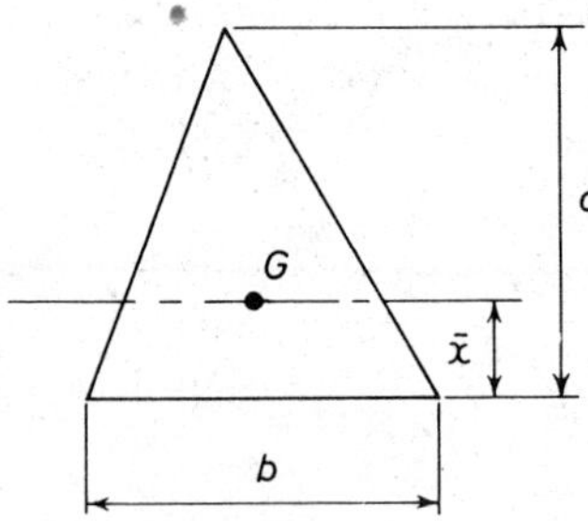

$$\bar{x} = \frac{d}{3}$$

$$I_G = \frac{bd^3}{36}$$

Real wedge of liquid

Imaginary wedge of liquid

or

$$F_{AB} - W' = W'$$

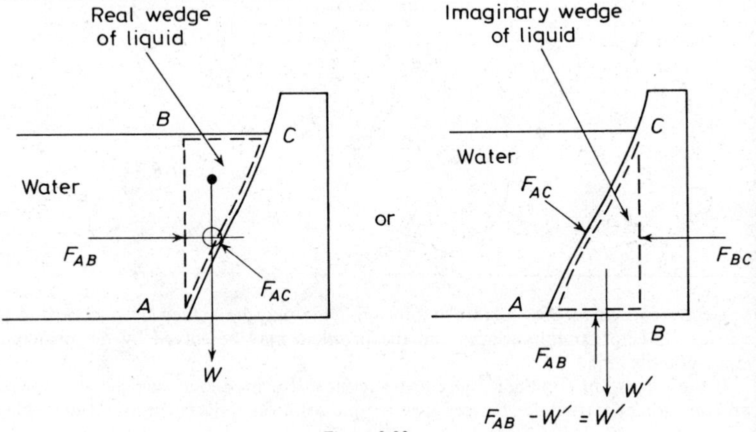

Figure 2.23

or cylindrical the line of action of the force acting on it must pass through the centre of curvature. Figure 2.23 illustrates the principle of choice of wedge.

Gases and liquids. Both gases and liquids are fluids and give rise to hydrostatic forces. In the case of a gas which is within some container, the change of pressure with depth is ignored and pressure is assumed constant. The centre of pressure will then be at the centre of area of the submerged surface. An exception to this approach is in the consideration of the change of pressure with altitude in the atmosphere. For this an International Standard Atmosphere is defined by the equation

$$\frac{p}{p_o} = (1 - 0.000\ 022\ 54\ h)^{5.256}$$

where p_o is the sea level pressure, 1·013 25 bar (101·325 kN/m²) and h is the altitude in metres. This equation is based on a sea level temperature of 288·15 K and values of p are tabulated up to 11 000 m. Between 11 000 m and 20 000 m temperature is constant at 216·7 k [1, 2].

STEADY INCOMPRESSIBLE FLUID FLOW

Fluid flow is analysed as it passes through a fixed region in space defined by a *control surface*. The properties of the fluid are considered as it enters and leaves the volume bounded by the control surface together with mass flow rates and energy transfers by work and heat. Three equations are obtained as follows.

The continuity equation which states that the mass flow rate entering the control surface must equal the mass flow rate leaving the control surface;

$$\dot{m} = (\rho A V)_1 = (\rho A V)_2 = \text{constant} \tag{5}$$

or if the fluid is of constant density,

$$Q = AV = \text{constant.}$$

In these equations A is the flow area measured normal to V the flow velocity. ρ is the fluid density and Q is the volume flow rate.

The energy equation which states that the energy entering the control surface must equal the energy leaving it. Three forms of energy are normally considered, that due to pressure, kinetic energy and potential energy. Situations in fluid mechanics are often adiabatic but there may be work transfer w to or from the control surface in pumps or turbines, whence in frictionless flow,

$$\left(\frac{p}{\rho} + \frac{V^2}{2} + gz\right)_1 = \left(\frac{p}{\rho} + \frac{V^2}{2} + gz\right)_2 + w_{12} \tag{6}$$

In adiabatic, workless, frictionless flow equation 6 becomes

$$\frac{p}{\rho} + \frac{V^2}{2} + gz = \text{constant.} \tag{7}$$

Equations 6 and 7 are expressed as energy per unit mass or specific energy in kJ/kg. Equation 7 is often rearranged to give energy per unit weight or *head* in *m*

$$\frac{p}{\rho g} + \frac{V^2}{2g} + z = \text{constant} \tag{8}$$

Table 2.19 and Figure 2.24 SIMPLE APPLICATIONS OF THE FLOW EQUATIONS[3]

Application	Continuity	Energy	Momentum	Notes
Venturimeter 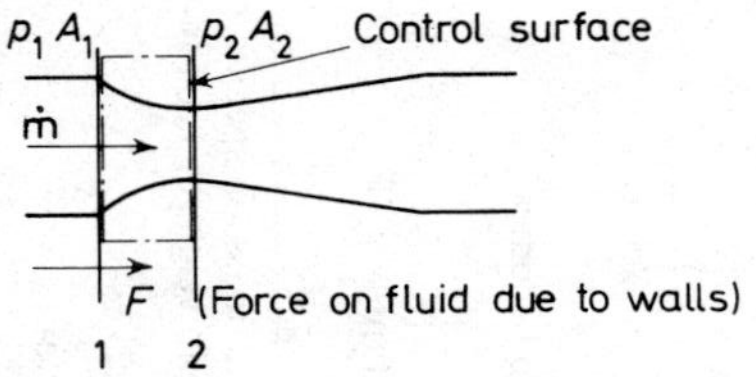 $P_1 A_1$, $P_2 A_2$ Control surface $\dot{m}$ F (Force on fluid due to walls) 1 2 (a)	$Q = A_1 V_1$ $Q = A_2 V_2$	$\dfrac{p_1 - p_2}{\rho} = \dfrac{V_2^2 - V_1^2}{2}$ and $h = \dfrac{p_1 - p_2}{\rho g}$ If H is the reading of a manometer connected between 1 and 2 then $h = \left(\dfrac{pF}{p} - 1\right) H,$ where pF is the density of the manometric fluid	$p_1 A_1 - p_2 A_2 + F = \dot{m}(V_2 - V_1)$ where F is force due to meter walls (not often developed)	Gives the ideal Q as $Q = \dfrac{A_1 A_2 (2gh)^{\frac{1}{2}}}{(A_1^2 - A_2^2)^{\frac{1}{2}}}$ but actual Q is less due to losses. The coefficient of discharge c_d (approximately 0·97) allows for losses, thus $Q = \dfrac{c_d A_1 A_2 (2gh)^{\frac{1}{2}}}{(A_1^2 - A_2^2)^{\frac{1}{2}}}$
Orifice (sharp edged) 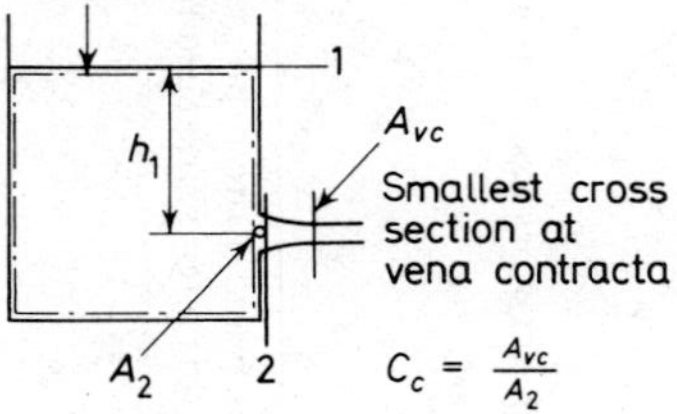 Control surface A_{vc} Smallest cross section at vena contracta A_2 h_1 1 2 $C_c = \dfrac{A_{vc}}{A_2}$ $C_v = \dfrac{V_2}{\text{ideal } V_2}$ $C_d = C_c\,C_v$ (b)	$Q = A_2 V_2$	$\dfrac{V_2^2}{2g} = h_1$		Gives the ideal Q as $Q = A_2 (2gh_1)^{\frac{1}{2}}$ but actual Q is less due to losses and contraction of stream after exit from orifice. The coefficient of discharge c_d (approximately 0·64) allows for both effects. Thus $Q = c_d A_2 (2gh_1)^{\frac{1}{2}}$

Flow between pump or turbine blades (axial flow)

$$Q = V_a A$$

V_a is normal to the flow area A.

$$\frac{p_1 - p_2}{\rho} = \frac{V_2^2 - V_1^2}{2}$$

$$F_w = \dot{m}(V_{2_w} - V_{1_w})$$
$$= \dot{m}\Delta V_w$$

Δ means "out value – in value"

The work transfer rate *to* the fluid due to F_w is

$$\dot{W} = \dot{m}u\Delta V_w$$

In a radial flow machine u is not constant and

$$\dot{W} = \dot{m}\Delta u V_w$$

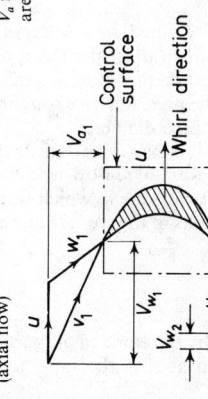
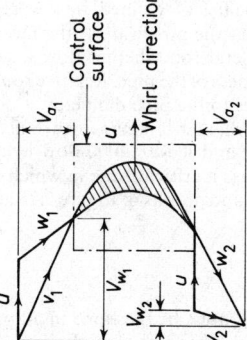
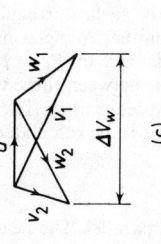

(c)

or energy per unit volume or *pressure* in kJ/m^3 or kN/m^2:

$$p + \tfrac{1}{2}\rho V^2 + \rho gz = \text{constant} \qquad (9)$$

Equations 7, 8 or 9 are known as Bernoulli's equation.

The momentum equation is obtained from Newton's second law of motion and states that the sum of the external forces acting *on* the control surface in a given direction is equal to the rate of change of fluid momentum in that direction

$$F_x = \dot{m}(V_{2_x} - V_{1_x}) \qquad (10)$$

Table 2.19 shows some applications of the flow equations.

Effect of friction on the validity of the flow equations

The continuity and momentum equations are not invalidated by friction losses in the flow but the Bernoulli equations need modification to allow for losses. Equation 11 illustrates the necessary change.

$$\left(\frac{p}{\rho g} + \frac{V^2}{2g} + z\right)_1 = \left(\frac{p}{\rho g} + \frac{V^2}{2g} + z\right)_2 + \text{losses between 1 and 2} \qquad (11)$$

Viscosity

The main cause of loss in incompressible fluid flow is turbulence caused by the *viscosity* of the fluid. Viscosity is defined by the relation $\tau = \mu(\partial V/\partial y)$ where τ is the viscous shear stress at a point in the flow, $\partial V/\partial y$ is the velocity gradient the point and μ is the *coefficient of dynamic viscosity* of the fluid. The coefficient of viscosity is not a constant but may be represented as a function of temperature in many applications. Table 2.13(b) in the thermodynamics section gives viscosity values for air and water. In some applications the coefficient of *kinematic viscosity* μ/ρ may be preferred to μ.

FLUID FLOW IN PIPES

Losses due to friction

The loss of head h_f in circular pipes may be estimated by the D'Arcy equation

$$h_f = \frac{4fLV^2}{2gd} \qquad (12)$$

where f is an experimentally determined friction coefficient defined by $f = \tau_w/\tfrac{1}{2}\rho V^2$. L is the pipe length, τ_w the mean viscous shear stress at the pipe wall, V the mean fluid velocity and d the pipe diameter. The value of f depends on whether flow is laminar (smooth) or turbulent (rough) and on the relative roughness of the pipe. Relative roughness ε/d is the ratio of the height of the surface irregularities to the pipe diameter.

The value of the Reynolds number Re determines whether laminar or turbulent flow prevails. Reynolds number is defined by $Re = \rho Vd/\mu$ and if $Re < 2000$ flow is laminar or if $Re > 2300$ flow is turbulent. Between these values is a critical zone in which flow is transitional and f is not precisely determined. The data and graphs of Figure 2.25 adapted from reference 4 enable head loss to be calculated.

Flow in pipes in parallel

Figure 2.26 shows two pipes in parallel. The head at B must be the same in all particles of fluid irrespective of the route they choose. Since the head at A is the same for all

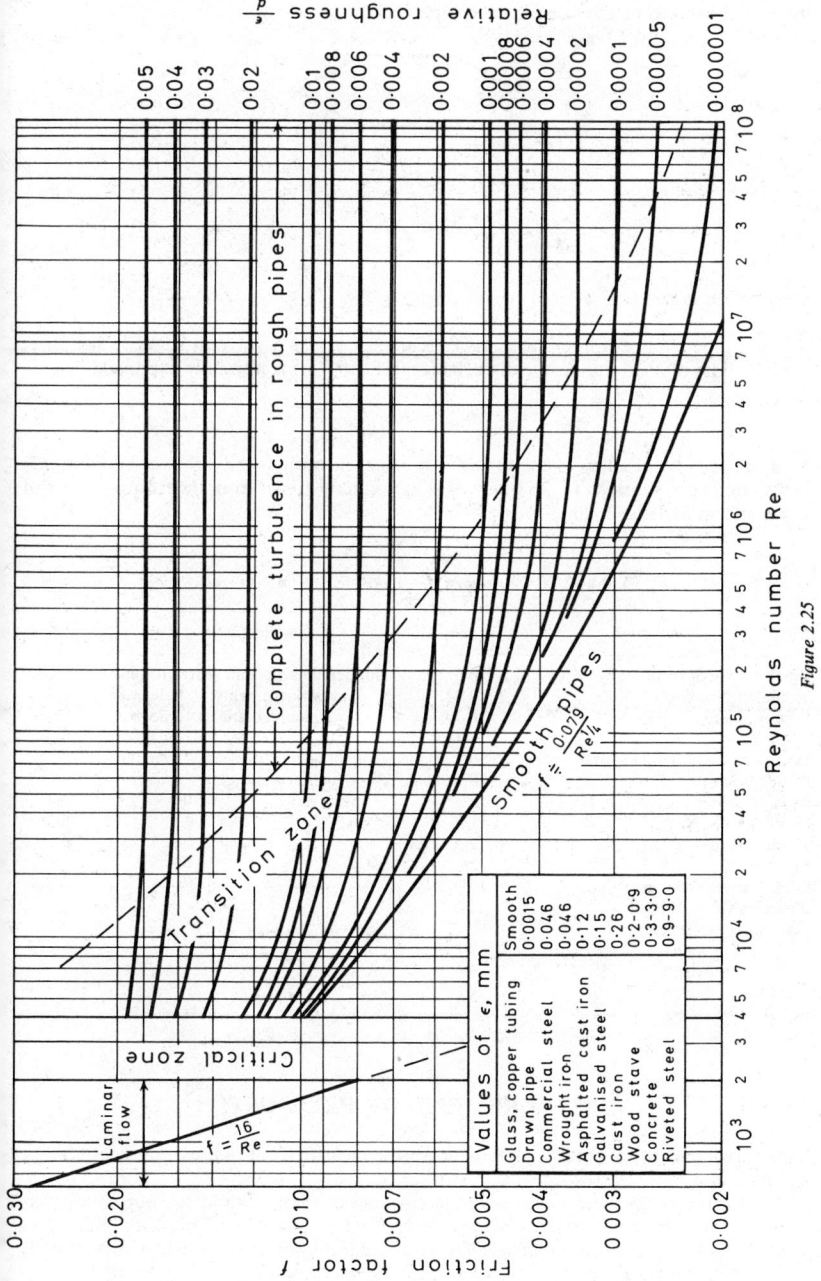

Figure 2.25

particles the distribution of flow between the branches is determined by considering the loss of head to be the same in each pipe.

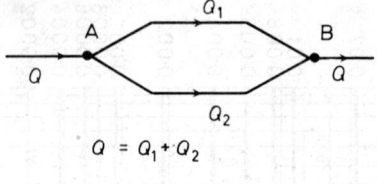

$$Q = Q_1 + Q_2$$

Figure 2.26

Obstruction losses

When a pipe changes direction, changes diameter or has a valve or other fitting there will be a loss of head due to the disturbance in flow. This is normally expressed by

$$h_{\text{loss}} = K \cdot \frac{V_1^2}{2g}$$

V_1 is the velocity at entry to the fitting. An enlargement loss is expressed $K(V_1 - V_2)/2g$ and a contraction loss $K . V_2^2/2g$ where V_2 is the velocity at exit from the fitting. Values of K are shown in Table 2.20.

Table 2.20   OBSTRUCTION LOSSES IN PIPE FLOW

Obstruction	K	Obstruction		K
Pipe entry (exit from tank)	0·5	Sudden enlargement (entry to tank)		0·10
Rounded pipe entry	0·04	Conical enlargements:	6 degree	0·13
Smooth bend	0·30	(total included angle)	10 degree	0·16
Mitre bend	1·1		15 degree	0·30
Mitre bend with guide vanes	0·2		25 degree	0·55
90 degree elbow	0·9	Sudden contractions:		
45 degree elbow	0·42	area ratio	0·2	0·41
Standard T	1·8	(A_2/A_1)	0·4	0·30
Return bend	2·2		0·6	0·18
Strainer	2·0		0·8	0·06
Globe valve, wide open	10·0			
Angle valve, wide open	5·0			
Gate valve, wide open	0·19			
¾ open	1·15			
½ open	5·6			
¼ open	24·0			

The data in this table is adapted from *Fluid Mechanics*, D. A. Gilbrech, Butterworth (1966).

MEASUREMENT OF PRESSURE[5]

There are a number of methods available for the measurement of pressure. These can be grouped into three classes:

(i) Balancing the pressure against the weight of a column of liquid; a group collectively known as *manometers*. Various designs cover a range of pressure differences between 0 and 2 atm. The method is accurate but only suited to substantially constant pressures.

(ii) Balancing the pressure against an elastic stress; this group includes Bourdon gauges, bellows, etc. The method not usually very precise and is only suited to substantially constant pressures.

(iii) Pressure transducing to an electric signal by a piezoelectric, magnetic, capacitance, resistance or inductance transducer. The method is suited to varying or constant pressures and the electrical output may be used by a data logger, computer, pen or other distant recorder.

MEASUREMENT OF INCOMPRESSIBLE FLUID FLOW RATE

Methods involving the use of orifices, nozzles and venturis are given in BS 1042:1943 'Code for flow measurement'. Other methods are given in BS 3680: 'Methods of measurement of liquid flow in open channels'. For continuous metering of low flow rates the 'Rotameter' type of instrument (a variable area orifice meter) may be chosen. It is available in a variety of ranges, the flow rate is indicated by a float whose position is determined by a balance between buoyancy force, weight and pressure difference force across the annular orifice. For accurate metering of steady flow rates a stop watch and calibrated burette will give excellent results particularly if the burette is necked at the timing marks. Reference 5 gives details of a variety of methods.

PRINCIPLES OF SIMILARITY AND MODEL TESTS

Geometric similarity between objects is not sufficient to enable test results from the one to be used to predict performance of the other; *dynamic similarity* is also required between the running conditions of the two objects.

The conditions for dynamic similarity may be obtained from dimensional analysis. Model testing is widely used in pump and turbine, heat exchanger and ship design. Pumps and turbines are discussed in Section 10 and the following paragraphs will be devoted to the similarity for submerged or partially submerged bodies.

Submerged bodies, which include pump and turbine *blades*, pipes and submarines, are found by dimensional analysis to have a resistance to motion R given by

$$R = \rho l^2 V^2 \phi(Re) \tag{13}$$

In equation 13, ρ is density, V is a representative velocity, l is a representative length dimension for the object and Re is Reynolds number.

If tests are made of a model which is geometrically similar to the prototype, dynamic similarity is achieved if the Re of the model and the prototype running conditions are identical. In these circumstances

$$R_p = \left[\frac{\rho_p l_p^2 V_p^2}{\rho_m l_m^2 V_m^2} \right] R_m$$

where R_m is the measured model resistance.

The test must be conducted at the *corresponding speed* such that

$$\left(\frac{\rho V l}{\mu} \right)_m = \left(\frac{\rho V L}{\mu} \right)_p$$

Variation in fluid properties may be achieved by using wind or water tunnels.

Floating bodies present a different problem since their resistance is not only due to viscous effects but also to wavemaking, i.e. lifting water in a gravitational field. Dimensional analysis shows that

$$R = \rho l^2 V^2 \phi(Re, Fr) \tag{14}$$

The new dimensionless group is the Froude number, $Fr = V/(lg)^{\frac{1}{2}}$ and, for dynamic similarity, both the Reynolds number and the Froude number of model and prototype must be equal. This cannot be achieved as the corresponding speeds for these two conditions are different. Model tests are conducted at the corresponding speed for wavemaking given by

$$\left[\frac{V}{(lg)^{\frac{1}{2}}}\right]_p = \left[\frac{V}{(lg)^{\frac{1}{2}}}\right]_m,$$

and the measured resistance at this speed is considered to comprise two parts one due to wavemaking R_w and one due to viscous effects R_v such that $R_m = R_w + R_v$.

The measured resistance of the model is therefore reduced by an amount considered equal to the viscous resistance and the remaining wavemaking resistance is 'scaled up'. The scaled up wavemaking resistance is then increased by an amount considered to be equal to the viscous resistance of the prototype. The viscous resistances to be added or subtracted are obtained from submerged, smooth, flat plate data for which $R_v = \frac{1}{2}\rho V^2 A C_D$. In this relation A is the wetted surface area and C_D is a drag coefficient given (in turbulent flow, $Re > 500\,000$) by the Schoenherr relation

$$\frac{0.242}{C_D^{\frac{1}{2}}} = \log_{10}(Re C_D)$$

This expression may be replaced by the Prandtl–Schlichting relation if

$$Re < 10^9; \quad C_D = \frac{0.455}{(\log_{10} Re)^{2.58}}.$$

Thus

$$R_p = (R_m - R_{v_m})\left[\frac{\rho_p l_p^2 V_p^2}{\rho_m l_m^2 V_m^2}\right] + R_{v_p},$$

where R_m is the measured model resistance at the corresponding speed for wavemaking.

These two examples show that considerable care and experience are necessary before using model test results to predict performance in complex situations.

PUMPS FOR INCOMPRESSIBLE FLUIDS

Reciprocating pumps[6]

The performance characteristics in Figures 2.27(a) and (b) show that the reciprocating pump is suited to applications which require flow rate and efficiency to be functions of speed alone so that the head demanded does not influence performance. Large reciprocating pumps running at speeds in the order 2 rev/s are suited to flow rates up to $0.05 \text{ m}^3/\text{s}$ and heads up to 1000 m whereas small pumps running at speeds up to 25 rev/s can give heads of 10 000 m. The head-swept volume diagram of Figure 2.27(c) shows that the ideal rectangular diagram is modified by the changes in head necessary to accelerate and decelerate the fluid and by friction losses so that the pressure at A is a minimum.

Local boiling or cavitation will occur if p_A is less than the saturation pressure corresponding to the fluid temperature. At higher speeds the acceleration decrease will be larger, and it is at this condition that vapour is most likely to appear. The fitting of an air cushion chamber on the suction side of the pump will smooth the fluctuations in pressure along AB and reduce the chance of cavitation. On the delivery side, an air cushion chamber will smooth the pulsating flow (as will a double-acting or multi-cylinder design). The delivery of the pump is reduced by leakage at valves and the difference between the swept volume and the delivered volume is known as the slip. Alternatively, the volumetric efficiency of the pump

$$\eta_{vol} = \frac{V_{delivered}}{V_{swept}}$$

may be used. η_{vol} is about 95 %.

Rotary positive displacement pumps

There are a considerable number of designs and a specialist handbook[6] must be consulted to decide which type; gear, vane, screw, etc. is best suited to the problem. In general terms, the pump scoops fluid from the suction side and moves it to the delivery side raising the pressure by forcing the extra mass into the delivery space. They are suited to small flow rate or medium head applications. Leakage tends to be greater than in a reciprocating design but as they are often used for viscous fluids such as oil, this effect may be reduced to give a volumetric efficiency of 95 %. Often the rotary pump presents a more compact unit than the reciprocating pump and, if a choice of design is available for a given application, this may be an important factor.

Turbopumps[7, 8]

In a turbopump, the fluid passes through a bladed rotor to a stator. The energy supplied to the rotor by the driving motor is transferred to the fluid which undergoes energy and momentum changes of the type discussed in Table 2.19. The flow through the pump may be radial (in the centrifugal pump), axial or mixed radial and axial. At low heads up to $100 \text{ m}^3/\text{s}$ may be delivered or at low flow rates a head of 500 m may be achieved by a single stage pump. Speeds of up to 500 rev/s may be used. Figure 2.28 shows the shape of turbopump characteristics. It should be noted that head and efficiency are functions of the flow rate. The slope of the H–Q curve in different designs is of particular importance in stability considerations (see 'Matching a pump to a system, page 2–54).

Specific speed

Dimensional analysis may be applied to turbomachines[9] to give, in a range of geometrically similar machines in which viscosity effects are negligible,

$$\frac{H}{N^2D^2} = \phi\left(\frac{Q}{ND^3}\right), \quad \frac{P}{\rho N^3 D^5} = \phi\left(\frac{Q}{ND^3}\right) \quad \text{and} \quad \eta = \phi\left(\frac{Q}{ND^3}\right).$$

Thus dynamic similarity for model tests is achieved if $Q/(ND^3)$ is the same for model and prototype. Another useful parameter for engineers is the specific speed of the pump which relates N, Q and H for all pumps of similar geometric shape. The specific speed $K_n = NQ^{\frac{1}{2}}/(gH)^{\frac{3}{4}}$ is evaluated at the point of maximum efficiency and if N is in rev/s, Q in m^3/s, H in metres and $g = 9{\cdot}81 \text{ m/s}^2$ then K_n is a number of revolutions.

$$\left(K_n \text{ in revolutions is } \frac{1}{15\,700} \times \frac{\text{r.p.m. (g.p.m.)}^{\frac{1}{2}}}{(\text{feet})^{\frac{3}{4}}}\right)$$

Figure 2.29 shows that K_n is related to particular types of pump and may be used as a guide to selection. The larger K_n, the smaller the pump diameter.

Cavitation

Cavitation occurs when the static pressure of the liquid falls below the saturation pressure corresponding to the liquid temperature. The liquid boils and the vapour bubbles cause

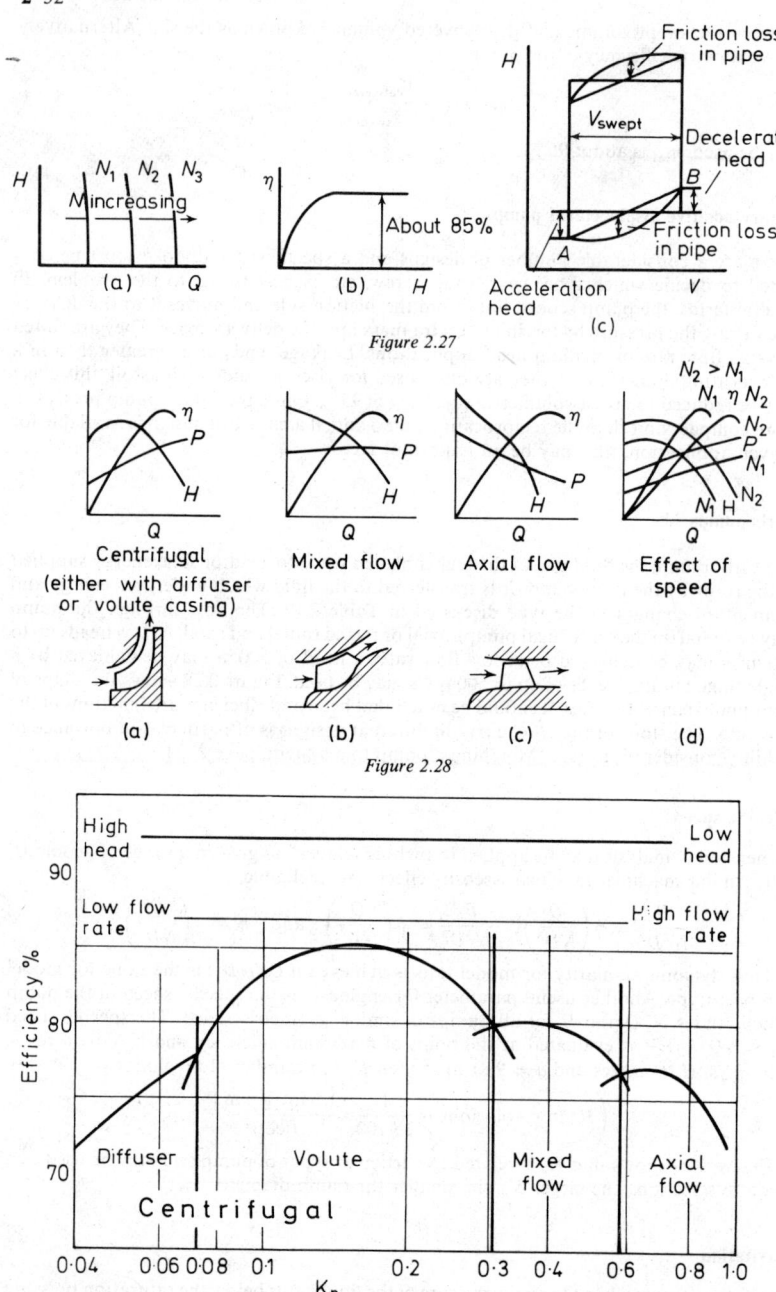

Figure 2.27

Figure 2.28

Figure 2.29

pressure waves as they condense in regions of higher pressure. Cavitation results in pitting of blades, vibration and a sharp fall in efficiency (this latter effect may be used to identify its occurrence). The point of lowest pressure in a pump is at impeller inlet and this is, therefore, the most likely situation for cavitation to occur. Referring to Figure 2.30, if p_{min} is this pressure and V is the velocity at impeller inlet then

$$\frac{p_a}{\rho g} = \frac{p_{min}}{\rho g} + z + h_f + \frac{V^2}{2g}.$$

Now let

$$\frac{V^2}{2g} = \sigma_c H$$

$$\sigma_c = \frac{\dfrac{p_a}{\rho g} - \left[\dfrac{p_{min}}{\rho g} + z + h_f\right]}{H} \qquad (15)$$

The value of σ_c may be determined experimentally by observing the point at which

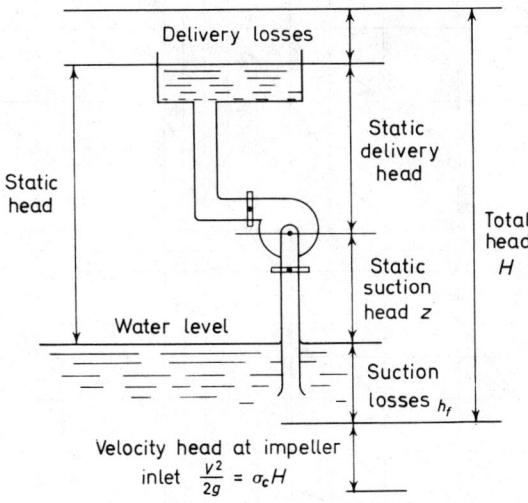

Figure 2.30

cavitation occurs in pumps of various K_n. Then for no cavitation $p_{min} > p_{sat}$ or $\sigma > \sigma_c$ where

$$\sigma = \frac{\dfrac{p_a}{\rho g} - \left[\dfrac{p_{sat}}{\rho g} + z + h_f\right]}{H}.$$

Sigma (σ) is the *Thoma cavitation parameter* and the numerator

$$\frac{p_a}{\rho g} - \left[\frac{p_{sat}}{\rho g} + z + h_f\right]$$

is called the *net positive suction head*. The only variable in this expression is z, the static suction head and this must be chosen to avoid cavitation. As specific speed is increased it may be seen in Figure 2.31, drawn for a single stage centrifugal pump, that z must be decreased to avoid cavitation.

Matching a pump to a system

The losses in the system due to viscous and obstruction effects (proportional to Q^2) are added to the required static head to give a system characteristic. If this is plotted on the

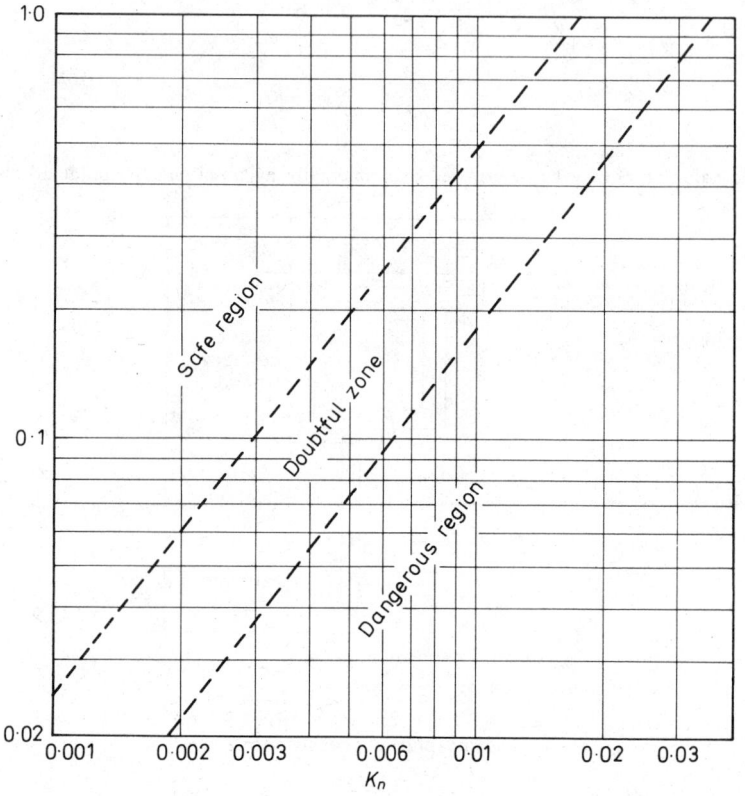

Figure 2.31

pump characteristic the intersection of the two curves will give a matching point. This is shown in Figure 2.32. The match should be at the maximum efficiency point of the pump characteristic for economic operation.

The slope of the H–Q curve is of importance in system matching. If the curve is of negative slope then the pump is stable at all operational conditions but, if partly positive, then unstable conditions may result. The characteristics of mixed and axial flow pumps are usually negative in slope and no problems arise, but the centrifugal pump may have a characteristic which slopes in either direction. With backward facing exit angles (β_2) from the impeller the centrifugal pump characteristic slope is negative and stable but with forward facing exit angles required for high head applications the slope is positive at low flow rates, see Figure 2.33.

Pumps used in series are treated by adding head at constant Q and pumps used in parallel by adding Q at constant head. If pumps are used in parallel there should not be

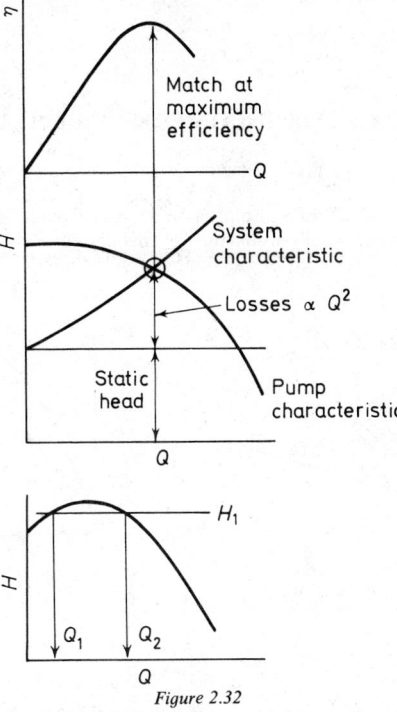

Figure 2.32

a change from positive to negative slope as this will give two head values for a single Q again promoting instability of operation.

Hydraulic efficiency

The hydraulic efficiency of a pump relates the ideal work transfer to the fluid (obtained from the velocity triangles by the method shown in Table 2.19) to the head that is measured

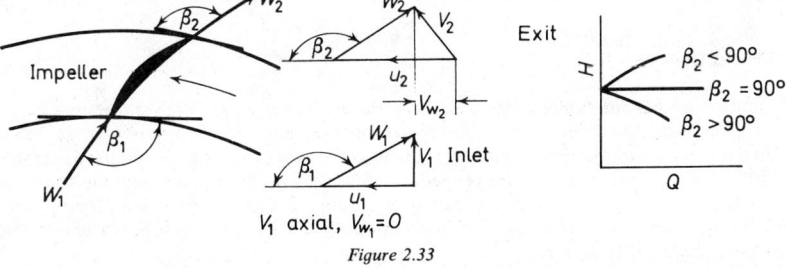

Figure 2.33

between the inlet and outlet of the pump. In Figure 2.33 drawn for an impeller with backward facing blades and axial inlet velocity, the ideal or Euler head is equal to the specific work transfer to the fluid $(u_2 V_{w_2})/g$, as V_{w_1} is zero. The pump head

$$H = \eta_h \times \frac{u_2 V_{w_2}}{g} \tag{16}$$

where η_h is the hydraulic efficiency.

TURBINES FOR INCOMPRESSIBLE FLUIDS[10]

Impulse turbines

In an impulse turbine, a jet of fluid impinges on a series of buckets attached to the periphery of a rotor. The most common machine of this design is known as a *Pelton wheel*. The velocity of the jet is produced by the fluid supply at high head ($V_1 = c_v(2gH)^{\frac{1}{2}}$) and

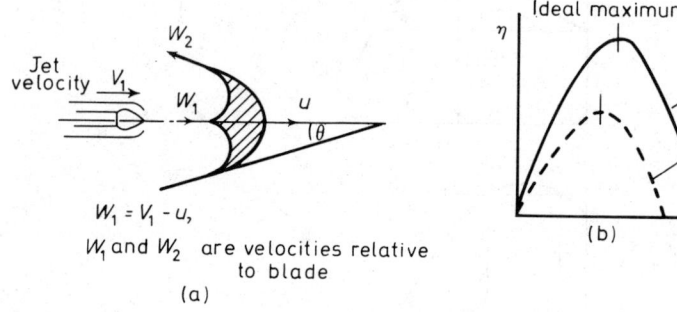

Figure 2.34

the torque on the rotor is produced by the force due to the rate of change of momentum of the fluid passing around the bucket (blade).

Figure 2.34(a) shows a single bucket and the momentum equation gives the specific work transfer from the blade as $u(W_1 + W_2 \cos \theta)$. The efficiency of the rotor will be obtained by the ratio of this specific work to the inlet specific kinetic energy $V_1^2/2$ and is plotted in Figure 2.34(b). The maximum efficiency of a Pelton wheel is about 90%.

Pelton wheels are simple, compact and relatively cheap and are best suited to large heads. It is seen in Figure 2.35 that Pelton wheels have low specific speed and are therefore not suited to all applications.

Reaction turbines

Reaction turbines run full of water and utilise the available head in both stator and rotor. This gives low fluid velocity throughout the machine and efficiencies as high as 95%.

In the *Francis turbine*, the fluid enters radially and leaves in an axial direction so that the whirl velocity at exit is zero. In the *propeller turbine* the flow is axial throughout. In order to obtain optimum running conditions over a range of flow or load conditions both the stator (guide vane or gate) and rotor (runner) blade angles may be varied in the *Kaplan* (propeller) turbine.

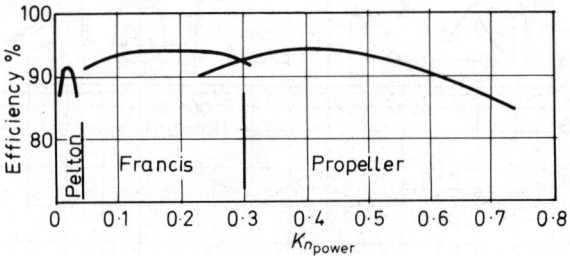

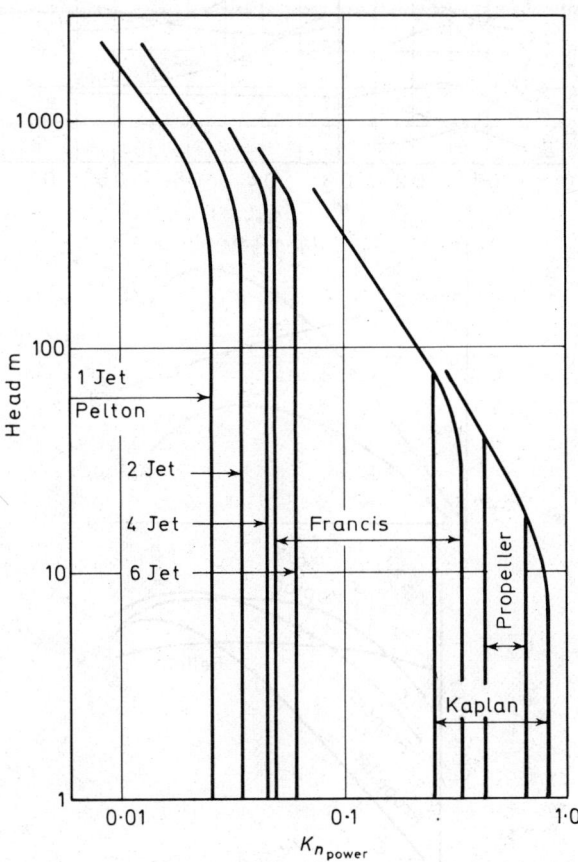

Figure 2.35

2-58

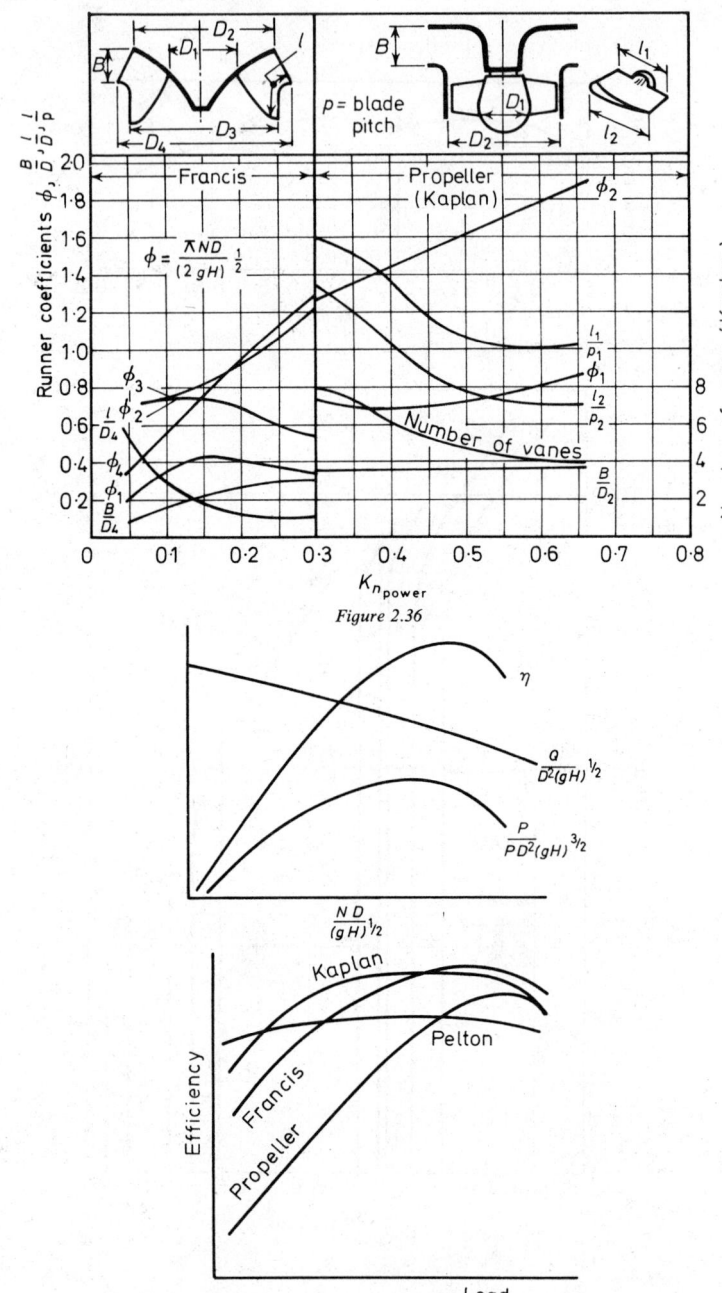

Figure 2.36

Figure 2.37

In all cases the momentum equation may be applied to the flow through the rotor to determine the specific work *from* the fluid by

$$w = u_1 V_{w_1} - u_2 V_{w_2} \text{ (see Table 2.19).}$$

Specific speed

Dimensional analysis applied to a turbine will yield the same groups as those discussed in the previous pump section. The specific speed defined earlier is also applicable to turbines. It is, however, customary to use a separately defined power specific speed $K_{n_{power}}$ (also evaluated at maximum efficiency) relating N, P and H rather than N, Q and H.

$$K_{n_{power}} = \frac{NP^{\frac{1}{2}}}{\rho^{\frac{1}{2}}(gH)^{\frac{5}{4}}}$$

and if N is in rev/s, P in watts, ρ in kg/m³, H in metres and $g = 9{\cdot}81$ m/s² then $K_{n_{power}}$ will be a number of revolutions

$$\left(K_{n_{power}} \text{ in revolutions with } \rho = 1000 \text{ kg/m}^3 \text{ is } \frac{1}{273} \times \frac{\text{rpm(hp)}^{\frac{1}{2}}}{(\text{feet})^{\frac{5}{4}}}\right)$$

Figure 2.35 shows that $K_{n_{power}}$ is related to particular types of turbine and may be used as a guide to selection. Selection may be followed by design and model tests.

Since the design of turbine runners gives shapes related to specific speed a preliminary

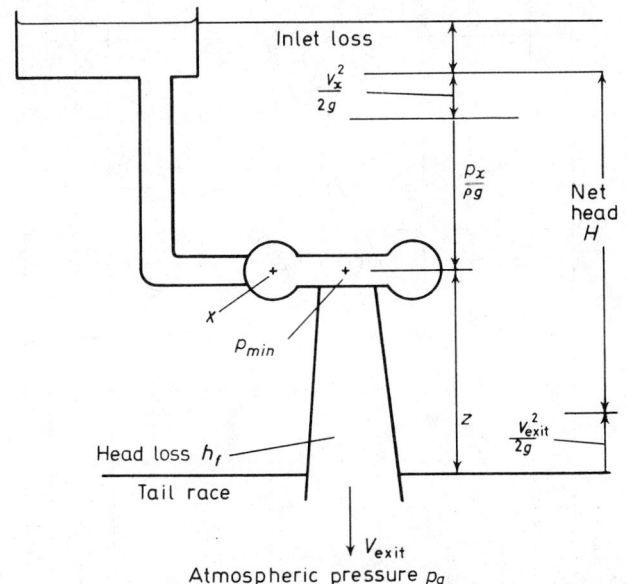

Figure 2.38

idea of the runner dimensions may be obtained by using the curves of Figure 2.36. In these curves ϕ is the turbine runner coefficient given by

$$\phi = \frac{u}{(2gH)^{\frac{1}{2}}} = \frac{\pi DN}{(2gH)^{\frac{1}{2}}} \tag{17}$$

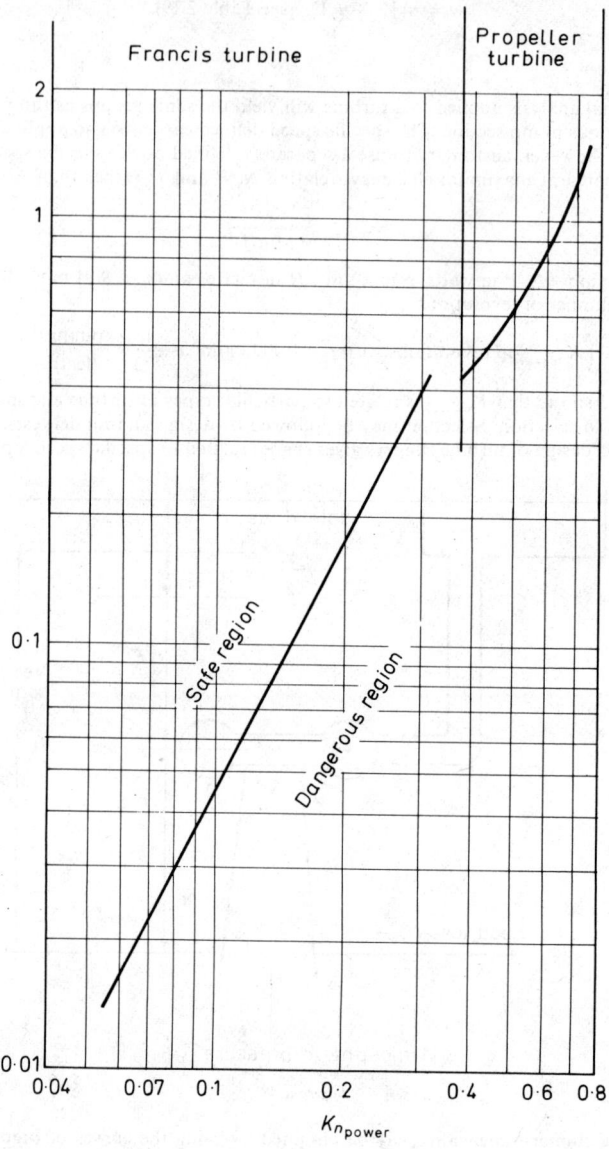

Figure 2.39

(u is the runner peripheral velocity). The curves enable ϕ (and hence D) to be obtained at four points on the Francis turbine runner or at two points on the propeller turbine runner.

The characteristics of various turbine designs are shown in Figure 2.37 in non-dimensional form and at various loads.

Cavitation

Figure 2.38 shows a turbine and the net head

$$H = \frac{p_x}{\rho g} + \frac{V_x^2}{2g} + z - \frac{V_{\text{exit}}^2}{2g}.$$

Cavitation will occur if the static pressure of the liquid falls below the saturation pressure corresponding to the temperature of the liquid. Cavitation results in pitting of runners, vibration and loss in efficiency and should be avoided. The point of minimum pressure will be at runner exit where the velocity is V and

$$\left(\frac{p_{\min}}{\rho g} + z + \frac{V^2}{2g} \right) - h_f = \frac{p_a}{\rho g}$$

where p_a is atmospheric pressure.

Let

$$\frac{V^2}{2g} - h_f = \sigma_c H$$

then

$$\sigma_c = \frac{\dfrac{p_a}{\rho g} - \left[\dfrac{p_{\min}}{\rho g} + z \right]}{H} \tag{18}$$

Now σ_c may be determined experimentally by observing the point at which cavitation occurs in turbines of various $K_{n_{\text{power}}}$, then for no cavitation $p_{\min} > p_{\text{sat}}$ or $\sigma > \sigma_c$ where

$$\sigma = \frac{\dfrac{p_a}{\rho g} - \left[\dfrac{p_{\text{sat}}}{\rho g} + z \right]}{H}.$$

Sigma (σ) as the *Thoma cavitation parameter*. The only variable in this expression is z and this must be chosen to avoid cavitation. The position of the turbine above the tail race surface is critical. As specific speed is increased it may be seen in Figure 2.39 that σ_c increases and the value of z must be decreased. For large head it may be found that the turbine must be submerged below the tail race surface.

REFERENCES

1. Standard Atmosphere; Tables and data for altitudes up to 65 800 ft. NACA 1235 (1955)
2. MAYHEW, Y. R. and ROGERS, G. F. C., *Thermodynamic and Transport Properties of Fluids*, SI units, 2nd ed, Blackwell, Oxford (1969)
3. MASSEY, B. S., *Mechanics of Fluids*, 2nd ed, Van Nostrand-Reinhold, London (1970)
4. MOODY, L. F., *Friction Factors for Pipe Flow, Trans.*, A.S.M.E. 66, 671 (1944)
5. *The Instrument Manual*, United Trade Press, London (1960)
6. *Hydraulics Handbook*, Trade and Technical Press, London
7. STEPANOFF, A. J., *Centrifugal and Axial Flow Pumps*, Wiley, New York (1957)
8. SHERWELL, T. Y. and PENNINGTON, R., 'Centrifugal Pump Characteristics', *Proc. I. Mech. E.*, 123, 621 (1932)
9. NORRIE, D. H., *Incompressible Flow Machines*, Arnold, London (1963)

10. GUTHRIE-BROWN, J. (Ed), *Hydroelectric Engineering Practice*, Vol. 2, Part A, Blackie, Glasgow (1965)

BIBLIOGRAPHY

BARNA, P. S., *Fluid Mechanics for Engineers* (SI edition), Butterworth, London (1971)
LANGHAAR, H. L., *Dimensional Analysis and Theory of Models*, Wiley, New York (1951)
SEDOV, L. I., *Similarity and Dimensional Methods in Mechanics*, Infosearch, London (1959)
KAYE, G. W. C. and LABY, T. H., *Tables of Physical and Chemical Constants*, Longman, London (1966)
PAI, S., *Viscous Flow Theory*, Van Nostrand, New York (1956)
SCHLICHTING, H., *Boundary Layer Theory*, McGraw-Hill, New York (1955)
Hydraulics Handbook, Trade and Technical Press, London
STREETER, V. L. (Ed), *Handbook of Fluid Dynamics*, McGraw-Hill, New York (1961)
DAVIS, C. V. (Ed), *Handbook of Applied Hydraulics*, McGraw-Hill (1952)
CONSIDINE, D. M. and ROSS, S. D., *Handbook of Applied Instrumentation*, McGraw-Hill (1964)
POLLACK, F., *Pump User's Handbook*, Trade and Technical Press, London
Pumping Manual, Trade and Technical Press, London
KARASSIK, I. and CARTER, R., *Centrifugal Pumps*, McGraw-Hill (1962)
KOVATS, A., *Design and Performance of Centrifugal and Axial Flow pumps and Compressors*, Pergamon, Oxford (1964)
KARASSIK, I. J., *Engineers Guide to Centrifugal Pumps*, McGraw-Hill, New York (1964)
CSANADY, G. T., *Theory of Turbomachinery*, McGraw-Hill, London (1964)
NECHLEBA, M., *Hydraulic Turbines*, Constable, London (1957)
WISLISCENUS, G. F., *Fluid Mechanics of Turbomachinery*, McGraw-Hill, New York (1947)

BRITISH STANDARDS

BS 188
Methods for the determination of the viscosity of liquids (1957)
BS 353
Methods of testing water turbine efficiency (1962)
BS 599
Methods of testing pumps (1966)
BS 1042
Code for flow measurement (1943)
BS 1780
Bourdon tube pressure and vacuum gauges (1960)
BS 3680
Methods of measurement of liquid flow in open channels
BS 3829
Principal external dimensions of centrifugal pumps
BS 4082
External dimensions for vertical in line centrifugal pumps
BS 4617
Methods for testing hydraulic pumps and motors for hydrostatic power transmission (1970)
BS 4622
Grey iron pipes and fittings (1970)

MECHANICS

D. H. BACON

In general terms a study of mechanics involves an examination of the state of bodies at rest (statics) or in motion (dynamics) under the action of forces. Before the start of analysis it is useful to sketch a *free body diagram* which isolates the system to be considered from the surroundings. All external forces acting *on* the system are shown in the diagram and the appropriate laws of statics or dynamics applied.

STATICS

The forces that act on a body at rest may be in equilibrium or may give a resultant force or moment. To determine which event obtains the forces shown on the free body diagram are resolved into components in three mutually perpendicular directions x, y and z. The components are summed to give ΣF_x, ΣF_y and ΣF_z and the moments of the components about some convenient point are also summed to give ΣM_x, ΣM_y and ΣM_z.

If $$\left. \begin{array}{ll} \Sigma F_x = 0 & \Sigma M_x = 0 \\ \Sigma F_y = 0 & \Sigma M_y = 0 \\ \Sigma F_z = 0 & \Sigma M_z = 0 \end{array} \right\} \quad \text{then the body is in equilibrium.} \qquad (1)$$

If equation 1 is not satisfied then the magnitude of the resultant force

$$R = [(\Sigma F_x)^2 + (\Sigma F_y)^2 + (\Sigma F_z)^2]^{\frac{1}{2}}$$

and the resultant moment

$$M = [(\Sigma M_x)^2 + (\Sigma M_y)^2 + (\Sigma M_z)^2]^{\frac{1}{2}}.$$

When coplanar forces are involved, the problem is reduced to one of two dimensions and an alternative graphical solution using vector addition of forces and moments may be used. When concurrent forces are involved the sum of the moments is always zero and the forces alone require investigation.

These methods will give solution to situations which are *statically determinate*, i.e. situations which have the minimum number of constraints to support an equilibrium position. Other texts[1] should be consulted for statically indeterminate situations.

DYNAMICS

Basic concepts and definitions

Newton's laws of motion
1. A particle will remain at rest or continue to move in a straight line with uniform velocity until acted on by an external force.
2. The sum of the components of the external forces acting on a particle in a given direction is equal to the rate of change of momentum in that direction.

3. When particles interact the forces of action and reaction are equal in magnitude and line of action but opposite in direction.

Mass (*m*) is a measure of the amount of matter in a body. It may also be considered as a quantitative measure of the *inertia* of a body, i.e. the property causing resistance to changing the motion of the body.

Velocity (*v*) is the rate of change of distance (*x*) with time (*t*),

$$v = \frac{\mathrm{d}x}{\mathrm{d}t} \text{ or } \dot{x}$$

Acceleration (*a*) is the rate of change of velocity with time,

$$a = \frac{\mathrm{d}v}{\mathrm{d}t} \text{ or } \frac{\mathrm{d}^2 x}{\mathrm{d}t^2} \text{ or } \ddot{x}$$

Force (*F*) is the product of mass with acceleration, $F = ma$.

Momentum (*M*) is the product of mass with velocity $M = mv$. The momentum of a system in a given direction is conserved unless the system is acted upon by an external force which has a component in that direction. (Newton's second law)

Impulse (*I*) is the product of force by time, $I = Ft = mat = m(v_2 - v_1)$.

Energy (*E*) is a property of a system which is conserved. In the study of mechanics kinetic energy due to motion, potential energy due to position and strain energy due to elastic deformation are normally examined. Kinetic energy $= \frac{1}{2}mv^2$ in linear motion (angular motion follows), potential energy $= mgz$ (*z* is elevation above a datum, *g* is gravitational acceleration) and strain energy is equal to the work transfer required to cause the strain.

Work (*W*) is a method of energy transfer evaluated in linear motion by the product force times distance moved in the direction of the force.

Heat (*Q*) is a method of energy transfer often ignored in simple mechanics in which systems are considered adiabatic. If energy appears to have been lost careful examination will often reveal that there has been a heat transfer.

Power (*P*) is the rate of work transfer.

Weight (*W*) is the force due to gravitational attraction on the mass of a body: $W = mg$.

Centre of gravity (*G*) is the point at which the distributed weight of a system may be considered to act as a single force. The position of a centre of gravity is found by experiment or analysis. In the latter case moments are taken about three mutually perpendicular axes to give the coordinates of *G*,

$$\bar{x} = \frac{\Sigma \delta mg \cdot x}{\Sigma \delta mg}, \bar{y} = \frac{\Sigma \delta mg \cdot y}{\Sigma \delta mg} \text{ and } \bar{z} = \frac{\Sigma \delta mg \cdot z}{\Sigma \delta mg},$$

where δm is an elementary mass at distance *x*, *y* or *z* from the respective axis. Table 2.21 gives some centre of gravity data.

Moment of Inertia (*I*) about an axis XX is defined by $I_{XX} = \Sigma \delta m x^2 = m k_{XX}^2$ where *x* is the perpendicular distance of the elementary mass *m* from the axis XX and k_{XX} is the *radius of gyration* about axis XX. Table 2.21 gives some moment of inertia data.

Angular velocity (*ω*) is the rate of change of angular distance (*θ*) with time $\omega = \mathrm{d}\theta/\mathrm{d}t$ or $\dot{\theta}$. Angular velocity is also given by linear velocity divided by radius of rotation, thus in Figure 2.41, $v/x = \omega$.

Angular acceleration (*α*) is the rate of change of angular velocity with time $\alpha = \mathrm{d}\omega/\mathrm{d}t$ or $\mathrm{d}^2\theta/\mathrm{d}t^2$ or $\ddot{\theta}$. Angular acceleration is also given by linear acceleration divided by radius of rotation, thus in Figure 2.41, $a/x = \alpha$.

Torque (*T*) is the moment of force about the axis of rotation. In Figure 2.41 the force required to accelerate $\delta m = \alpha x \delta m$ and thus the torque to accelerate the body $T = \Sigma x(\alpha x \delta m) = \alpha \Sigma(\delta m x^2)$ or $T = I_o \alpha$. A torque may also be due to a *couple* (two equal in magnitude but opposite in direction forces whose points of application are separate).

Table 2.21 and Figure 2.40   CENTRES OF GRAVITY AND MOMENTS OF INERTIA
(see also Table 2.18)

—	G	I

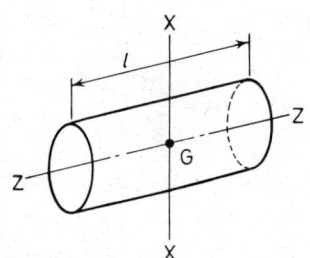

$$\bar{x} = \frac{2r \sin \theta}{3\theta}$$

(*Centroid*)

$$I_{XX} = \frac{r^4}{4}\left(\theta - \frac{\sin 2\theta}{2}\right)$$

$$I_{YY} = \frac{r^4}{4}\left(\theta + \frac{\sin 2\theta}{2}\right)$$

(*Second moments of area*)

plane circular sector

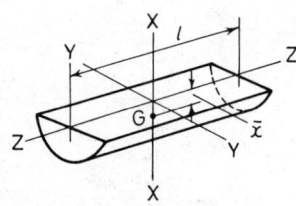

—

$$I_{XX} = \frac{mr^2}{4} + \frac{ml^2}{12}$$

$$I_{ZZ} = \frac{mr^2}{2}$$

cylinder radius r

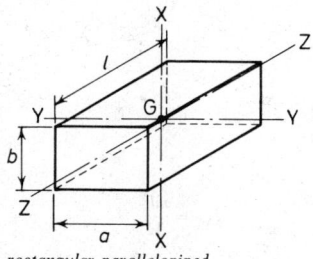

$$\bar{x} = \frac{4r}{3\pi}$$

$$I_{XX} = I_{YY}$$

$$= \frac{mr^2}{4} + \frac{ml^2}{12}$$

$$I_{ZZ} = \frac{mr^2}{2}$$

half cylinder radius r

—

$$I_{XX} = \frac{m}{12}(a^2 + l^2)$$

rectangular parallelopiped

continued

Table 2.21 *continued*

—	G	I

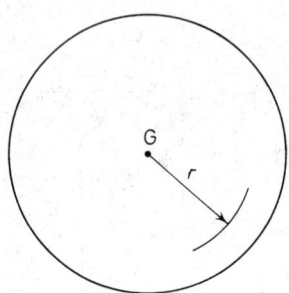

sphere

$$I = \frac{2mr^2}{5}$$

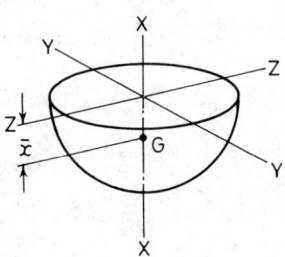

hemisphere radius r

$$\bar{x} = \frac{3r}{8}$$

$$I_{XX} = \frac{2mr^2}{5}$$

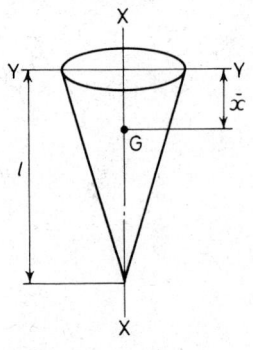

circular cone, basic radius r

$$\bar{x} = \frac{l}{4}$$

$$I_{XX} = \frac{3mr^2}{10}$$

$$I_{YY} = \frac{3mr^2}{20} + \frac{ml^2}{10}$$

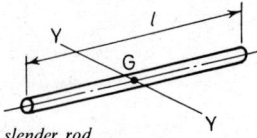

slender rod

$$I_{YY} = \frac{ml^2}{12}$$

Table 2.21 *continued*

—	G	I

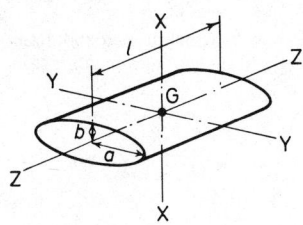

$$I_{xx} = \frac{ma^2}{4} + \frac{ml^2}{12}$$

$$I_{zz} = \frac{m}{4}(a^2 + b^2)$$

elliptical cylinder

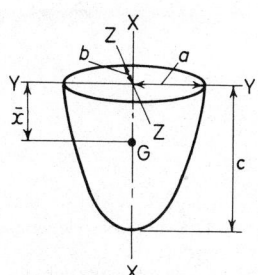

$$\bar{x} = \frac{3c}{8}$$

$$I_{xx} = \frac{m}{5}(a^2 + b^2)$$

$$I_{YY} = \frac{m}{5}(b^2 + c^2)$$

half ellipsoid

The data in this table is adapted from reference 2.

Moment of Momentum or *angular momentum* of a body is the moment of the linear momentum of the body about the axis of rotation, O. In Figure 2.41, angular momentum

$$= \Sigma \delta m \omega x \cdot x = \Sigma \delta m \omega x^2 = \omega I_O$$

and by the theorem of parallel axes

$$I_O = I_G + m \times OG^2$$

so that angular momentum

$$= \omega(I_G + m \times OG^2).$$

The angular momentum of a system remains constant unless the system is acted on by an external torque.

Angular impulse is the product of torque by time, i.e. angular impulse

$$= Tt = I\alpha \cdot t = I(\omega_2 - \omega_1).$$

Angular kinetic energy about an axis O is given by

$$= \Sigma \tfrac{1}{2} \delta m (\omega x)^2 = \tfrac{1}{2} I_O \omega^2$$

but

$$I_O = I_G + m \times OG^2$$

so that angular kinetic energy is

$$\tfrac{1}{2}\omega^2(I_G + m \times OG^2) = \tfrac{1}{2}I_G\omega^2 + \tfrac{1}{2}mv_G^2$$

Work due to torque is the product of torque by angular distance.
Power due to torque is the rate of work:

$$P = \frac{T\theta}{t} = T\omega.$$

Friction. The motion of one body relative to another body with which it is in contact is accompanied by a friction force. The force acts tangentially to the surfaces in contact in

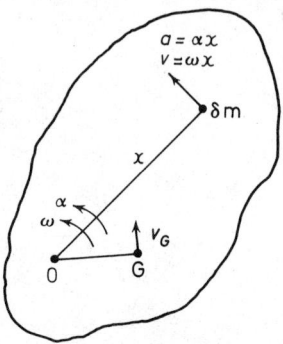

Figure 2.41

a direction opposing motion. The value of the force $F = \mu R$ where R is the normal reaction between the surfaces and μ is the coefficient of friction. The value of μ depends on the nature of the surfaces in contact.

Simple machines are devices for the transmission of energy. They usually allow the deployment of a large force at one point by the application of a small force at some other point. The *mechanical advantage* of a machine is the ratio of the load to the effort and the *velocity ratio* of the machine is the ratio of the distance moved by the effort to the distance moved by the load. The *efficiency* of a machine is the ratio of the energy transfer to the load to the energy transfer by the effort, and is equal to the mechanical advantage divided by the velocity ratio.

LINEAR AND ANGULAR MOTION IN TWO DIMENSIONS[2]

The basic equations of motion were contained in the concepts above

$$v = \dot{x}, a = \ddot{x} = \dot{v}, \omega = \dot{\theta}, \alpha = \ddot{\theta} = \dot{\omega}$$

Integration or differentiation of these relations may be achieved mathematically if formal equations to the displacement, velocity or acceleration are available or graphically if experimental data is being evaluated.
Several cases will be examined:

(i) CONSTANT ACCELERATION

Linear motion

The equations (2) below are obtained suffix 1 refers to values when $t = O$ suffix 2 refers to values when $t = t$

$$\left.\begin{array}{l} x = v_1 t + \tfrac{1}{2}at^2 \\ v_2 = v_1 + at \\ v_2^2 = v_1^2 + 2ax \end{array}\right\} \quad (2)$$

Angular motion

Similar equations (3) obtain:

$$\left.\begin{array}{l} \theta = \omega_1 t + \tfrac{1}{2}\alpha t^2 \\ \omega_2 = \omega_1 + \alpha t \\ \omega_2^2 = \omega_1^2 + 2\alpha\theta \end{array}\right\} \quad (3)$$

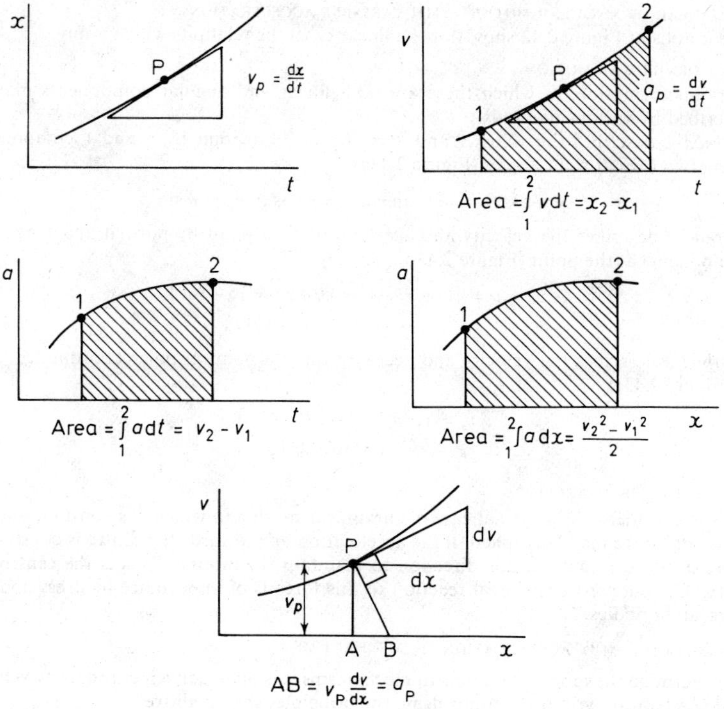

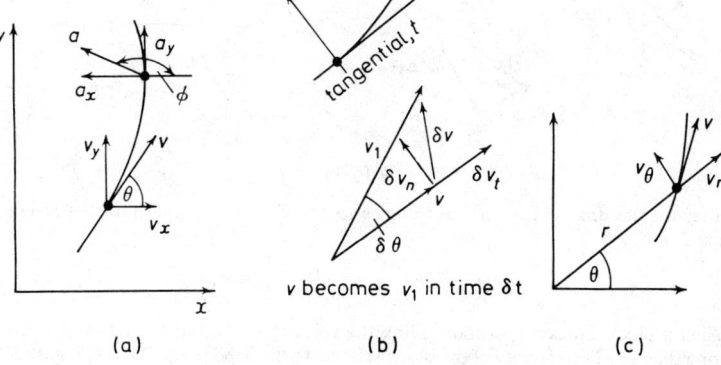

x
P
$v_p = \frac{dx}{dt}$
t
v
2
P
1
$a_p = \frac{dv}{dt}$
t
Area $= \int_1^2 v\,dt = x_2 - x_1$
a
2
1
t
Area $= \int_1^2 a\,dt = v_2 - v_1$
a
2
1
x
Area $= \int_1^2 a\,dx = \frac{v_2^2 - v_1^2}{2}$
v
P
dv
dx
v_p
A
B
x
$AB = v_P \frac{dv}{dx} = a_P$

Figure 2.42

y
a
a_y
a_x
ϕ
v_y
v
θ
v_x
x
(a)
v_1
v
normal, n
tangential, t
v_1
δv
δv_n
δv_t
v
$\delta\theta$
v becomes v_1 in time δt
(b)
v
v_θ
v_r
r
θ
(c)

Figure 2.43

(ii) LINEAR OR ANGULAR MOTION WITH VARIABLE ACCELERATION

The graphs of Figure 2.42 show (for the linear case) the relations which apply.

(iii) CURVILINEAR MOTION

Curvilinear motion, in which there are both linear and angular components may be described by various methods.

Method 1 describes the velocity and acceleration of a point by x and y components along two axes at right angles (Figure 2.43a),

$$v_x = v \cos \theta, v_y = v \sin \theta, a_x = a \cos \phi, a_y = a \sin \phi.$$

Method 2 describes the velocity and acceleration of a point by normal and tangential components at the point (Figure 2.43b);

$$\left. \begin{array}{l} v_t = v = r\theta = r\omega, a_t = r\ddot{\theta}+\dot{r}\dot{\theta} = r\alpha + \dot{r}\omega \\ v_n = 0 \qquad , a_n = v\dot{\theta} = \qquad r\dot{\theta}^2 = r\omega^2 \end{array} \right\} \tag{4}$$

Method 3 describes the velocity and acceleration of a point in polar coordinates r and θ, (Figure 2.43c);

$$\left. \begin{array}{l} v_r = \dot{r}, \quad a_r = \ddot{r}-r\dot{\theta}^2 \\ v_\theta = r\dot{\theta}, a_\theta = r\ddot{\theta}+2\dot{r}\dot{\theta} \end{array} \right\} \tag{5}$$

(iv) MOTION IN A CIRCLE

This is considered as a special case of curvilinear motion in which r is constant and the methods above may be applied. If the acceleration $\omega^2 r$ towards the centre is considered, there must be a force in this direction to maintain the motion. This is the centripetal force. The outward centrifugal reaction to this force is of importance in stress analysis of rotating bodies.

(v) VELOCITY AND ACCELERATION IN MECHANISMS

To determine the motion of points in mechanisms it is often convenient to draw velocity and acceleration vector diagrams using the principles shown above.

Velocities. In a link of fixed length, r is constant and the velocity of one point relative to

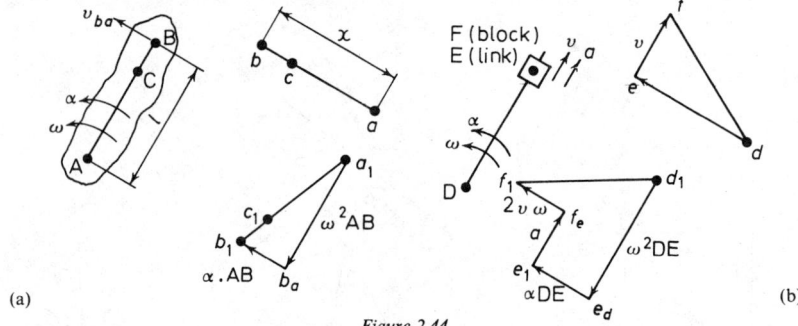

(a)   (b)

Figure 2.44

another on the link will be at right angles to the line joining the points. Referring to Figure 2.44a

$$\omega = \frac{v_{ba}}{AB} = \frac{ab}{AB} = \frac{x}{1} \text{ also } \frac{AC}{AB} = \frac{ac}{ab}$$

When a block slides on a rotating link the velocity of the block v relative to the link is along the link. The velocity of the block relative to the fixed point D is df (Figure 2.44b).

Accelerations. In a link of fixed length (Figure 2.44a), B will have a centripetal acceleration

$\omega^2 AB = a_1 b_a$ towards A and a tangential acceleration αAB. The acceleration of B relative to A is $a_1 b_1$.

When a block slides along a rotating link it may have a linear acceleration a along the link relative to the link. Whether a is positive, negative or zero there will also be a tangential acceleration $2v\omega$. This is known as the Coriolis acceleration and its direction is determined by rotating the sliding velocity vector through 90 degrees in the direction of the link angular velocity ω. Figure 2.44b shows the four acceleration components of the sliding block F.

(vi) GYROSCOPIC EFFECTS

Figure 2.45 shows a rotor with moment of inertia I_{XX} about axis OX. The rotor has angular velocity ω_X about OX which itself rotates (precesses) about axis OY (at 90° to OX) with angular velocity ω_Y. Due to the change (in direction) of the angular momentum

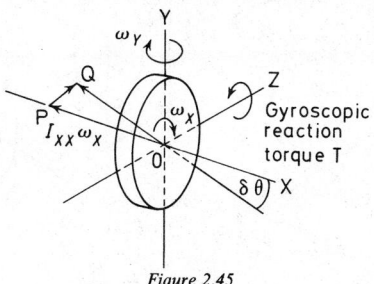

Figure 2.45

vector OP to OQ in time δt there must be a torque which may be represented in direction by the vector PQ. The magnitude of the torque is PQ divided by δt. The gyroscopic reaction torque T is opposite in sign to this torque and must be supported by the bearings of the rotor shaft.

$$T = I_{XX}\omega_X \frac{\delta\theta}{\delta t} = I_{XX}\omega_X\omega_Y \qquad (6)$$

Gyroscopic reaction torque will occur in all situations where a rotor is precessed e.g. automobile wheels on corners, ship turbine rotors in pitch or turns, aircraft turbine rotors in turns or loops.

In situations where the precession is not conveniently in the OZX plane the angular momentum vectors may be resolved. For more complex motion in three dimensions specialist texts should be consulted[3]. Gyroscopic effects are useful in navigation and in the design of missile and spacecraft systems.

BALANCING

At high rotational speeds rotating and reciprocating parts of machines should be balanced to avoid excessive stresses and vibration.

Rotating masses

The forces to be balanced are the centrifugal reactions to the circular motion. When all the rotating masses are in the same plane (Figure 2.46a) balance may be achieved by placing a single extra mass in the plane. This may not be convenient so that more than one mass may be used. Masses are usually placed at the greatest convenient radius to reduce their volume.

The conditions for balance for this two dimensional problem are:

$$\Sigma F_x = \Sigma m\omega^2 r \sin\theta = 0, \Sigma F_y = \Sigma m\omega^2 r \cos\theta = 0$$

where ω is the angular velocity of the shaft, r is the radius of the mass m and θ its angular position relative to the axis y. Axes x and y are at right angles in the plane considered. The equations may be solved graphically with a single force polygon.

When all the masses are not in the same plane (Figure 2.46b) the centrifugal reactions will have moments about any plane which must also be balanced. Two masses in different

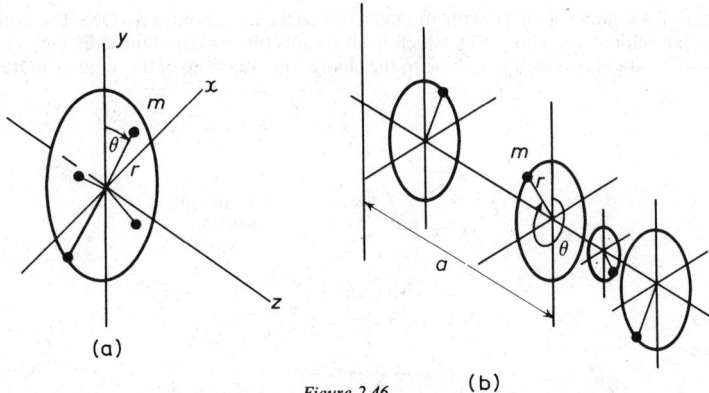

(a)

Figure 2.46   (b)

planes are required to achieve complete balance. Convenient planes are chosen for these masses and moments taken to give the conditions for balance;

$$F_x = \Sigma m\omega^2 r \sin\theta = 0 \qquad F_y = \Sigma m\omega^2 r \cos\theta = 0$$
$$M_x = \Sigma m\omega^2 r \sin\theta \cdot a = 0 \quad M_y = \Sigma m\omega^2 r \sin\theta \cdot a = 0$$

where a is the distance of the mass m from the chosen plane. These equations may be solved graphically by a moment or couple polygon and a force polygon.

Reciprocating masses. Single cylinder machines

The acceleration of the reciprocating piston P of Figure 2.47 may be determined analytically[4] to be a Fourier series of the form:'

$$a = \omega^2 r (\cos\theta + A\cos 2\theta + B\cos 4\theta + c\cos 6\theta \ldots)$$

where A, B and C are functions of $n = l/r$.

Associated with this acceleration will be forces

$$m\omega^2 r (\cos\theta + A\cos 2\theta + B\cos 4\theta + C\cos 6\theta \ldots) \qquad (7)$$

where m is the reciprocating mass. Table 2.22 gives values of A, B, C for various values of n. In simple cases only the first two terms are considered with the approximation

$$F = m\omega^2 r \left(\cos\theta + \frac{\cos 2\theta}{n} \right)$$

This may be seen to be equal to the components in the line of reciprocating motion of the centrifugal reactions to masses rotating at the crankshaft A. Partial balance may therefore be achieved by placing extra rotating masses at the crankshaft. The force is considered in two parts; the primary force $m\omega^2 r \cos\theta$ for which partial primary balance is achieved in a single cylinder machine by a mass Km at radius r_1 rotating at crankshaft speed and the secondary force

$$m(2\omega)^2 \frac{r}{4n} \cos 2\theta$$

which could be partially balanced in a single cylinder machine by a mass rotating at 2ω. Secondary balance is not attempted. Partial primary balance leaves a transverse

Table 2.22   CONSTANTS FOR RECIPROCATING MOTION

n	3·0	3·5	4·0	4·5	5·0
A	0·343 1	0·291 8	0·254 0	0·225 0	0·202 0
B	0·010 1	0·006 2	0·004 1	0·002 8	0·002 1
C	0·000 3	0·000 13	0·000 07	0·000 04	0·000 02

force at the crankshaft which has magnitude $Km\omega^2 r_1 \sin\theta$. The balancing mass and radius are chosen so that a suitable compromise is achieved between the unbalanced reciprocating component and the transverse component.

Reciprocating masses. Multi cylinder in-line machines

In multi-cylinder machines there will additionally be moments due to both primary and secondary forces to be considered. The conditions for complete primary balance are

$$F = \Sigma m\omega^2 r \cos\theta = 0, \quad M = \Sigma m\omega^2 r \cos\theta . a = 0$$

where a is the distance of the reciprocating mass m from the chosen plane. They may be investigated by drawing a force polygon of the centrifugal reactions of the equivalent

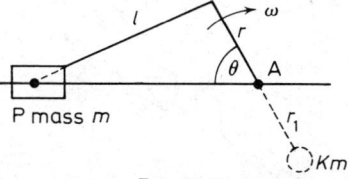

Figure 2.47

rotating masses m at the crankshaft radius r and a moment polygon of the moments due to these forces. Both polygons should close for primary balance. The cranks of a multi-cylinder in line engine are arranged in such a way as to assist primary balance (see below). If the disposition of the cranks does not give complete primary balance extra masses may be introduced at the crankshaft to give the desired effect but unbalanced transverse components will result.

   Secondary balance of multicylinder in line engines will be achieved if

$$F = \Sigma m\omega^2 \frac{r}{n} \cos 2\theta = \Sigma m(2\omega)^2 \frac{r}{4n} \cos 2\theta = 0$$

and

$$M = \Sigma m(2\omega)^2 \frac{r}{4n} \cos 2\theta . a = 0.$$

This will occur if the force polygon and the moment polygon for the equivalent rotating masses m at *radius $r/4n$ drawn at angles 2θ* close. The addition of extra masses to give secondary balance is not attempted.

If it is necessary to consider higher harmonics, then equation 7 must be used and analytical rather than graphical methods may be chosen.

For N identical cyclinders in an in-line engine the resultant m^{th} harmonic force is the series

$$F_m = K_m \left\{ \cos m\theta + \cos m(\theta + \alpha) + \cdots \cos m(\theta + (N-1)\alpha) \right\} \tag{8}$$

where α is the angular (equal) crank spacing for even firing and K_m is given by

$$K_1 = m\omega^2 r, \qquad K_2 = m\omega^2 r \times A, \qquad K_3 = m\omega^2 r \times B \text{ etc.}$$

The sum of the series is

$$S = K_m \frac{\cos m\left(\theta + \left[\dfrac{N-1}{2}\right]\alpha\right) \sin \dfrac{mN\alpha}{2}}{\sin \dfrac{m\alpha}{2}}$$

For a four-stroke engine $N\alpha = 4\pi$ and $\sin mN\alpha/2 = \sin 2\pi m = 0$. Thus $S = 0$ unless $\sin m\alpha/2 = 0$ which occurs when $m\alpha$ is a multiple of 2π when $S = NK_m \cos m\theta$. Since $N\alpha = 4\pi$ unbalanced forces only occur in harmonics which are multiples of half the number of cylinders.

For a two-stroke engine, unbalanced forces occur when m is a multiple of the number of cylinders.

As mentioned above, the disposition of the cranks may be arranged to assist balance. Couples are balanced by arranging that cranks are placed symmetrically about the

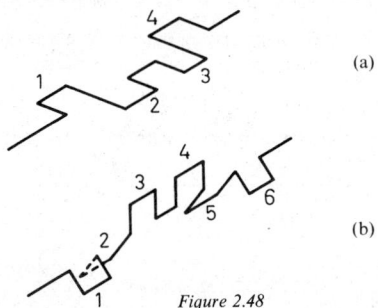

(a)

(b)

Figure 2.48

centre line. Examples; 4 cylinder, 4 stroke engine $\alpha = 4\pi/4 = 180°$. Unbalanced harmonics 2,4,6,8 etc. Cranks 1 and 4, 2 and 3 paired for moment balance (Figure 2.48a).

Six cylinder, 4 stroke engine $\alpha = 4\pi/6 = 120°$. Unbalanced harmonics 6,12,18 etc. Cranks 1 and 6, 2 and 5, 3 and 4 paired for moment balance (Figure 2.48b).

The principles of balancing may be applied to other configurations such as V-engines and radial engines[5]. In the latter case, complete balance of a radial engine with an even number of cylinders N may be achieved by the addition of a single mass $Nm/2$ at crank radius, rotating at crank speed.

VIBRATIONS

Engineering systems subject to vibration may be investigated by forming analytical models which consist of spring and mass arrangements for linear vibrations or shaft and rotor arrangements for torsional vibrations. The value of the analysis depends on the effectiveness of the model chosen. An analogue computer may be used to give visual indication of the effect of varying the model. Tables 2.23 and 2.24 show models, equations and solutions for *single degree of freedom systems*.

Torsional vibrations

The equations of motion in Tables 2.23 and 2.24 may be rewritten in terms of moment of inertia (for mass) and torsional stiffness and torsional damping to yield similar results. However it is unlikely that real torsional systems can be represented by such simple single rotor models and multi rotor models with several degrees of freedom must be analysed; for this specialist texts should be consulted[7, 8].

Transverse vibrations of beams

A light beam with a single concentrated load deflects statically by an amount given by

$$\delta = \frac{Wl^3}{B \times EI}$$

where B is a constant which depends on the support and loading positions. The stiffness of the beam is

$$\frac{B \times EI}{l^3}$$

and if treated as a vibrating body the motion of the load will be described by

$$\left(\frac{W}{g}\right)\ddot{y} + \left(\frac{B \times EI}{l^3}\right)y = 0.$$

This is a case of simple harmonic motion and the fundamental frequency

$$f = \frac{1}{2\pi}\left(\frac{B \times EI \cdot g}{Wl^3}\right)^{\frac{1}{2}} = \frac{1}{2\pi}\left(\frac{g}{\delta}\right)^{\frac{1}{2}},$$

where δ is the static deflection under the load.

Beams with uniformly distributed loads or beams under their own weight exhibit similar vibrations for which the frequency will depend on the method of support[9]. Table 2.25 shows fundamental frequencies for various situations. For beams with varying cross section the determination of the deflection at a load or under the weight of the beam may be obtained by the usual graphical methods[10].

Beams with multiple loads may be solved by Dunkerley's empirical method. In this method, the complex system is considered as a number of simple systems for which the fundamental frequency is easily determined. If f_0 is the frequency of vibration of the beam due to its weight *alone*, f_1 is the frequency with load 1 *alone*, f_2 the frequency with load 2 *alone* etc., then the fundamental frequency of the fully loaded beam f is given by

$$\frac{1}{f^2} = \frac{1}{f_0^2} + \frac{1}{f_1^2} + \frac{1}{f_2^2} + \cdots \frac{1}{f_n^2} \tag{9}$$

Analytical justification for equation 9 is available[9].

Whirling of shafts

Consider a shaft rotating at angular velocity ω which has a rotor of mass m with a centre of gravity of eccentricity e. Figure 2.52 shows the shaft with deflection y due to centrifugal reaction on the mass of the rotor.

Table 2.23 and Figure 2.49 LINEAR VIBRATIONS WITH FREE RESPONSE

Model	Equation of motion and solution	Notes
Undamped* (a) 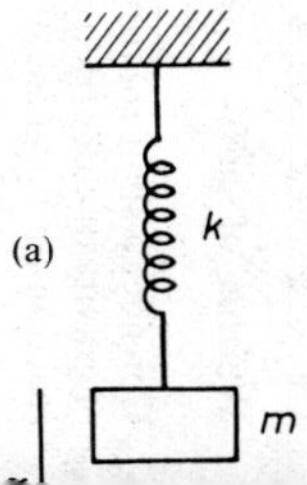	$m\ddot{x} + kx = 0$ or $\ddot{x} + \omega^2 x = 0$ where $\omega^2 = k/m$ Solution: $x = C \sin \omega t + D \cos \omega t$ or $x = A \sin (\omega t + \alpha)$ A, C and D are determined from the initial conditions and α is a phase angle.	The motion is *simple harmonic* with a period $T = 2\pi/\omega$ and a frequency $f = 1/T$.

on the amount of damping. Three cases arise and only one is oscillatory.

$$\ddot{x} + \left(\frac{c}{m}\right)\dot{x} + \omega^2 x = 0$$

where c is the viscous damping coefficient

(i) If $c > 2m\omega$ the system is *overdamped*. The response to a disturbance is a creep back to the undisturbed state.

(ii) If $c = 2m\omega$ the system has *critical damping*, the motion is a periodic and the displacement
$x = e^{-(c/2m)t(A+Bt)}$ where A and B are constants.

(iii) If $c < 2m\omega$ the system has a transient oscillatory motion determined by

$$x = e^{-(c/2m)t}\left[C\sin\left(\omega^2 - \frac{c^2}{4m^2}\right)^{\frac{1}{2}}t + D\cos\left(\omega^2 - \frac{c^2}{4m^2}\right)^{\frac{1}{2}}t\right]$$

The period $T = \dfrac{2\pi}{\left(\omega^2 - \dfrac{c^2}{4m^2}\right)^{\frac{1}{2}}}$ and frequency $f = \dfrac{1}{T}$

The ratio of successive amplitudes (on the same side of the equilibrium position)

$$\frac{a_n}{a_{n+1}} = e^{(c/2m)T} \text{ and } \left(\frac{c}{2m}\right)T \text{ is the logarithmic decrement.}$$

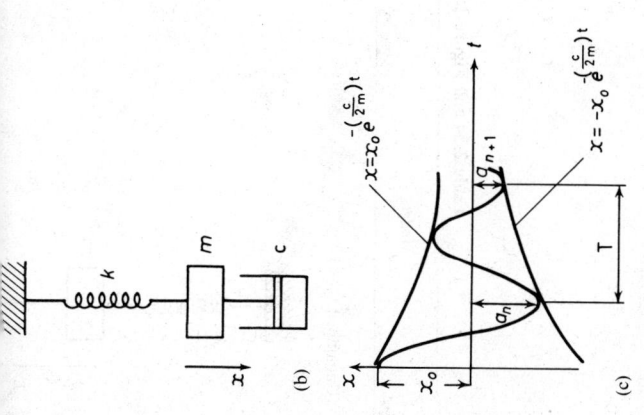

(b)

(c)

$x = x_0 e^{-\left(\frac{c}{2m}\right)t}$

$x = -x_0 e^{-\left(\frac{c}{2m}\right)t}$

Table 2.24 and Figure 2.50   LINEAR VIBRATIONS WITH FORCED RESPONSE DUE TO REGULAR PERIODIC FORCES

Model	Equation of motion and solution	Notes
Undamped	$m\ddot{x} + kx = F_0 \cos pt$ or $\ddot{x} + \omega^2 x = \dfrac{F_0}{m} \cos pt$ Solution: $x = C \sin \omega t + D \cos \omega t + \dfrac{F_0 \cos pt}{m(\omega^2 - p^2)}$	The first two terms of the solution are the free response which dies out leaving an oscillation at the forcing frequency of amplitude $\dfrac{F_0}{m(\omega^2 - p^2)} = \dfrac{F_0}{k}\left(\dfrac{\omega^2}{\omega^2 - p^2}\right)$

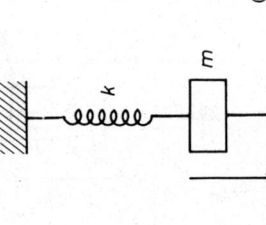

(a)

The fraction $\omega^2/(\omega^2 - p^2)$ is the *dynamic magnifier* and gives the ratio of the amplitude of the vibration to the static deflection under the load F_0. When $\omega = p$ the amplitude becomes infinite as *resonance occurs*

$$m\ddot{x} + c\dot{x} + kx = F_0 \cos pt$$

or

$$\ddot{x} + \left(\frac{c}{m}\right)\dot{x} + \omega^2 x = \frac{F_0}{m}\cos pt$$

Solution: (omitting transients)

$$x = \frac{F_0}{m}\cdot\frac{1}{\left[(\omega^2-p^2)^2 + \left(\frac{cp}{m}\right)^2\right]^{\frac{1}{2}}}\cdot\cos(pt-\alpha)$$

The *dynamic magnifier* is

$$\frac{\omega^2}{\left[(\omega^2-p^2)^2 + \left(\frac{cp}{m}\right)^2\right]^{\frac{1}{2}}}$$

The solution is similar to the undamped case in that the transient free response is damped out leaving a sustained vibration at the forcing frequency.

and resonance occurs when ω is approximately equal to p unless damping is moderate when the maximum amplitude occurs with $p < \omega$. There is a phase shift as p increases tending to a maximum of 180 degrees. It can be seen (Figure 2.50c) that vibration amplitude is minimised by ensuring that the free response (natural frequency) is well below the forcing frequency. If $F_0\alpha p^2$ then it is found that the maximum amplitude occurs with $p > \omega$ and the free response is better well above the forcing frequency.

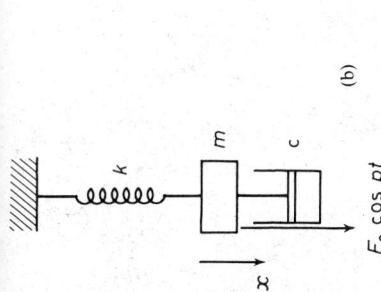

(b)

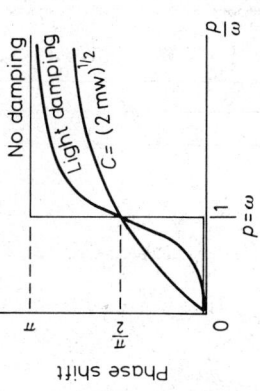

(c)

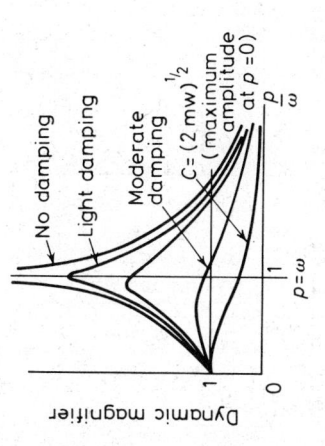

Table 2.25 and Figure 2.51 VIBRATIONS OF BEAMS

Beam loading and support	Frequency Hz

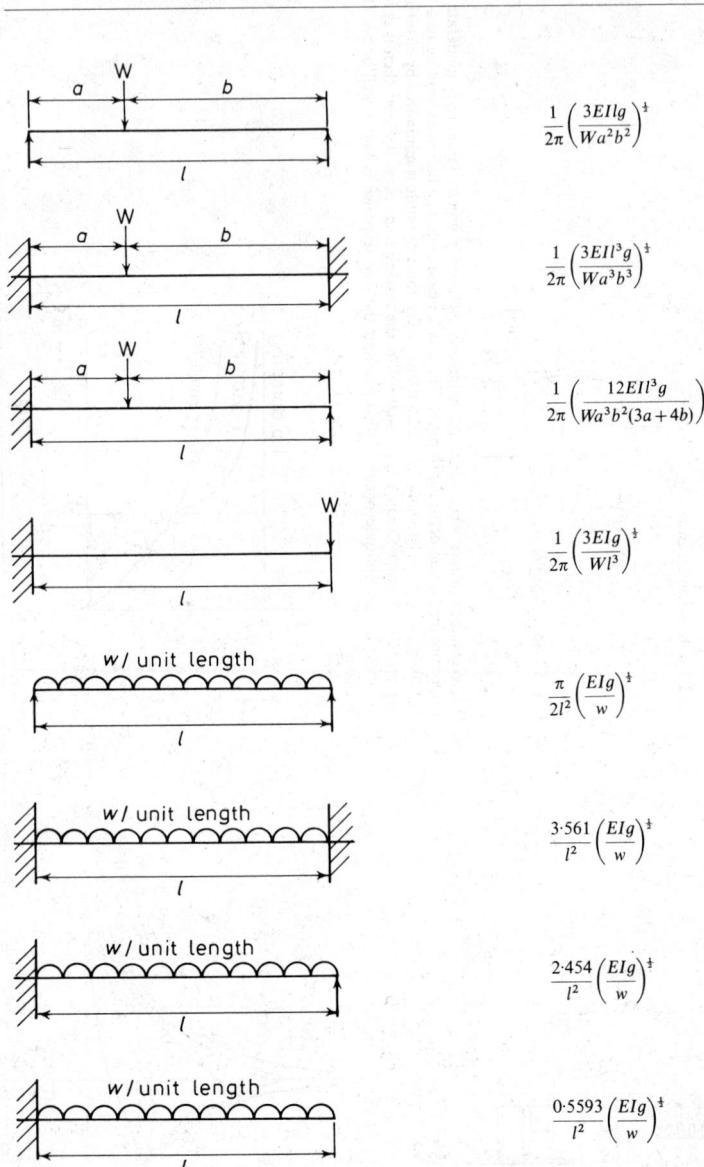

Frequency Hz column:

$$\frac{1}{2\pi}\left(\frac{3EIlg}{Wa^2b^2}\right)^{\frac{1}{2}}$$

$$\frac{1}{2\pi}\left(\frac{3EIl^3g}{Wa^3b^3}\right)^{\frac{1}{2}}$$

$$\frac{1}{2\pi}\left(\frac{12EIl^3g}{Wa^3b^2(3a+4b)}\right)^{\frac{1}{2}}$$

$$\frac{1}{2\pi}\left(\frac{3EIg}{Wl^3}\right)^{\frac{1}{2}}$$

$$\frac{\pi}{2l^2}\left(\frac{EIg}{w}\right)^{\frac{1}{2}}$$

$$\frac{3\cdot561}{l^2}\left(\frac{EIg}{w}\right)^{\frac{1}{2}}$$

$$\frac{2\cdot454}{l^2}\left(\frac{EIg}{w}\right)^{\frac{1}{2}}$$

$$\frac{0\cdot5593}{l^2}\left(\frac{EIg}{w}\right)^{\frac{1}{2}}$$

Thus

$$m\omega^2(y+e) = \left(\frac{B \times EI}{l^3}\right)y \text{ or } y = \frac{e}{\dfrac{B \times EI}{m\omega^2 l^3} - 1} \tag{10}$$

When

$$\frac{B \times EI}{m\omega^2 l^3} = 1,$$

y is infinite and the shaft is said to have a *critical speed of whirling*, ω_c. At this state the shaft is rotating about the line of bearing centres in a bow shape and is also rotating about its own deformed axis. It is seen that

$$\omega_c = \left(\frac{B \times EI}{ml^3}\right)^{\frac{1}{2}}$$

Expressed in rev/s the whirling speed is

$$\frac{\omega_c}{2\pi} = \frac{1}{2\pi}\left(\frac{B \times EI}{ml^3}\right)^{\frac{1}{2}}$$

which is identical to the transverse frequency of vibration of the shaft. By substitution in equation 10;

$$y = \left(\frac{\omega^2}{\omega_c^2 - \omega^2}\right)e$$

thus when $\omega < \omega_c$, y is of the same sign as e and y and e add until when $\omega = \omega_c$ deflection is (theoretically) infinite. When $\omega > \omega_c$, y is of opposite sign to e and if the shaft is run at a speed considerably greater than the whirling speed y tends to $-e$ and the centre of

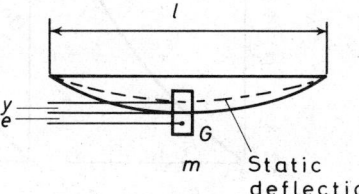

Figure 2.52

gravity of the rotor mass runs on the static deflection curve. This is a desirable running condition but care must be taken to avoid the higher harmonics of transverse vibration.

This simple explanation ignores the phase relation of the centrifugal reaction vector to the bending plane. At low speeds, the vector is in line with the bending plane in an outwards direction, at whirling it is 90° ahead of the bending plane and at speeds well above whirling it approaches 180° ahead of the bending plane i.e. $y \rightarrow -e$. This may be seen to be analogous to the phase shifts shown in Figure 2.50c.

Vibration elimination and reduction

Vibration problems may be limited by various methods. In brief those available are:
  (i) By balancing i.e. removing the disturbing force.
  (ii) By choosing an elastic support such that the dynamic magnifier assists. For this to be a viable method the natural frequency of the support should be small compared with the forcing frequency. This may produce problems.

   (iii) By avoiding resonant conditions.
   (iv) By increasing the damping.
   (v) By introducing a vibration absorber which may mean introducing an equal but
     opposite vibration to the system[7].

CONTROL SYSTEMS

Centrifugal governors

A centrifugal governor is a simple control system for angular speed in which the difference
between the centrifugal reaction of a rotating mass and a gravity or spring force is used to
control the energy supply to an engine. The calculation of the equilibrium conditions for a
given configuration is simple.

A governor is stable if the radius of rotation r of the mass increases as speed increases.
In Figure 2.53a the controlling force F supplied by springs or gravity must be equal to the
centrifugal reaction. Thus

$$F = m\omega^2 r \text{ or } \omega = \left(\frac{1}{m} \times \frac{F}{r}\right)^{\frac{1}{2}}$$

and for stability F/r must increase as r increases. This is achieved if in the F–r curves of
Figure 2.53b $dF/dr > F/r$. For the gravity controlled governor this is always true but is

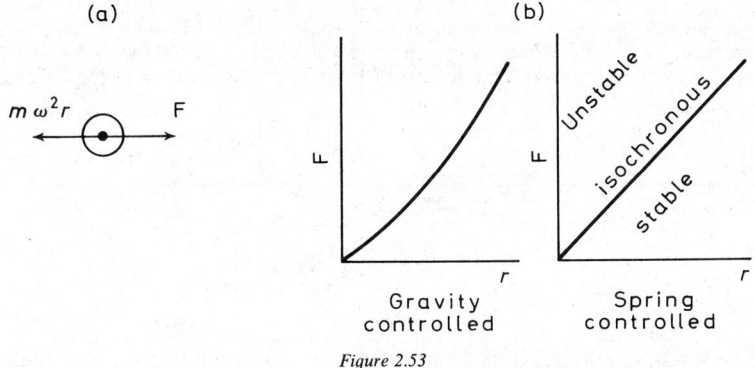

Figure 2.53

only true for the spring controlled governor if the F–r curve lies below the isochronous
line shown. At the isochronous condition there is no control action since $dF/dr = F/r$ at
all radii of rotation.

Automatic control

An automatic control system consists of a number of connected elements. For each
element the ratio of the output signal θ_o to the input signal θ_i is called the transfer operator.
The transfer operator is obtained from the differential equation describing the action of
the element and is expressed in terms of the operator $D \equiv d/dt$; $\theta_o/\theta_i = KG(D)$ where K
is a scalar multiplier and $G(D)$ is some function of the operator.

The control system may be arranged as an open path or as a closed path (feedback[11])
system. Figure 2.54 shows block diagrams of each arrangement. In the feedback system

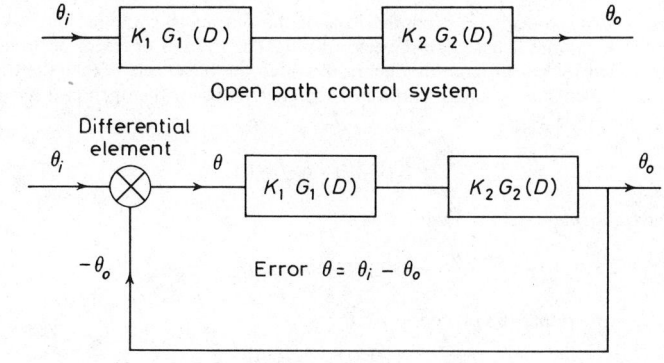

Open path control system

Closed path (feedback) control system

Figure 2.54

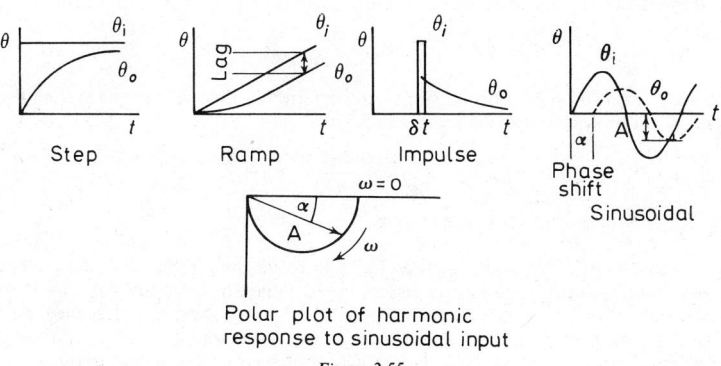

Step   Ramp   Impulse   Phase shift   Sinusoidal

Polar plot of harmonic
response to sinusoidal input

Figure 2.55

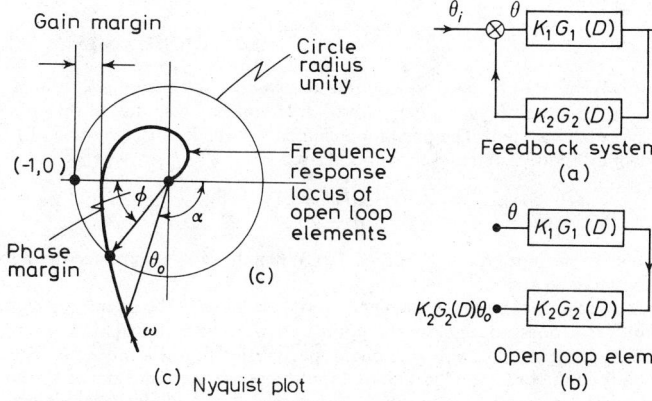

Gain margin

Circle radius unity

Frequency response locus of open loop elements

(-1,0)

Phase margin

(c)

(c) Nyquist plot

Feedback system
(a)

Open loop elements
(b)

Figure 2.56

the control action depends on a comparison of the output θ_o with the input θ_i by a differential element and is based on the error $\theta_i - \theta_o$. The transfer operator of the system may be obtained by block diagram algebra of which the basic rule is that the transfer operators of elements in series are multiplied together. Thus for the open path system of Figure 2.54 the transfer operator

$$\frac{\theta_o}{\theta_i} = K_1 G_1(D) \times K_2 G_2(D)$$

and for the feedback system of Figure 2.54,

$$\frac{\theta_o}{\theta} = K_1 G_1(D) \times K_2 G_2(D)$$

where $\theta = \theta_i - \theta_o$ so that the transfer operator

$$\frac{\theta_o}{\theta_i} = \frac{K_1 G_1(D) \times K_2 G_2(D)}{1 + K_1 G_1(D) \times K_2 G_2(D)} \tag{11}$$

For the control system θ_o represents the instantaneous value of the controlled variable and θ_i the desired value of the controlled variable. Equation 11 may be simplified to

$$\frac{\theta_o}{\theta_i} = \frac{G(D)}{1 + G(D)}.$$

The highest power of D in transfer operator determines the order of the system thus a simple first order system may be represented by

$$\frac{\theta_o}{\theta_i} = \frac{1}{1 + TD}$$

where T is the time constant of the system.

Control system elements may have proportional, integral, derivative or combined action. Analysis of the response of these actions to various inputs will show whether particular actions are suited to given disturbances. Time domain analysis[12] investigates the magnitude of the response to step, ramp and impulsive inputs and frequency domain analysis[12] the magnitude and phase of the response to sinusoidal inputs over a range of frequencies. Figure 2.55 shows the response of a simple first order element to these inputs. For the sinusoidal case the response to various frequencies is best shown on a polar plot where the input $\theta_i = \sin \omega t$ the output

$$\theta_o = \frac{1}{1 + TD} . \sin \omega t = A \sin (\omega t - \alpha).$$

Polar plotting may also be used to examine the stability of a feedback system. The Nyquist[13] locus (Figure 2.56c) is a polar plot of the frequency response of the *open loop* elements of a feedback system. The open loop elements of the feedback system of Figure 2.56a are shown in Figure 2.56b and the ratio plotted is

$$\frac{K_2 G_2(D)\theta_o}{\theta}$$

which for unit error becomes $K_1 G_1(D) \times K_2 G_2(D)$. When the plot is examined the stability criterion below may be applied.

'A feedback system is stable if the frequency response locus of the open loop elements drawn for unit error does not enclose the point $(-1, 0)$.' The stability may be assessed quantitatively by the gain margin or the phase margin shown in Figure 2.56c. The gain margin is usually of the order 0·6 for stability and the phase margin ϕ not less than 30°.

Control systems utilise pneumatic, hydraulic, mechanical and electrical elements to give the desired actions. The bibliography should be consulted for further information.

REFERENCES

1. PARCEL, J. I. and MOORMAN, R. B. D., *Analysis of Statically Indeterminate Structures*, Wiley, New York (1955)
2. MERIAM, J. L., *Dynamics*, Wiley, New York (1966)
3. ARNOLD, R. N. and MAUNDER, L., *Gyrodynamics*, Academic Press, New York (1961)
4. BEVAN, T., *The Theory of Machines*, Longmans, London (1964)
5. TOFT, L. and KERSEY, A. T. J., *Theory of Machines*, Pitman, London (1946)
6. BISHOP, R. E. D. and JOHNSON, D. C., *The Mechanics of Vibration*, Cambridge University Press (1960)
7. KER WILSON, W., *Practical Solution of Torsional Vibration Problems* (5 vols), Chapman and Hall (1965-68)
8. NESTORIDES, E. J. (Ed), *A Handbook of Torsional Vibration*, Cambridge University Press (1958)
9. COLE, E. B., *Theory of Vibrations*, Crosby Lockwood, London (1950)
10. MORLEY, A., *Strength of Materials*, Longmans, Green and Company, London (1956)
11. D'AZZO, J. J. and HOUPIS, C. H., *Feedback Control System Analysis and Synthesis*, McGraw-Hill, New York (1966)
12. WEBB, C. R., *Automatic Control*, McGraw-Hill, London (1964)
13. NYQUIST, H., *Regeneration Theory*, Bell System Tech. Journal, Vol. 11 (1932)

BIBLIOGRAPHY

MERIAM, J. L., *Statics*, Wiley, New York (1966)
JENKINS, W. M., *Matrix and Digital Computer Methods in Structural Analysis*, McGraw-Hill, (1969)
GENNARO, J. J., *Computer Methods in Solid Mechanics*, Macmillan, New York (1965)
EASTHOPE, C. E., *Three Dimensional Dynamics*, Butterworth, London, (1964)
SCARBOROUGH, J. B., *The Gyroscope*, Interscience (1958)
GRABBE, E. M., RAMO, S. and WOOLDRIDGE, D. E., *Handbook of Automation, Computation and Control*, (3 vols), Wiley, New York (1961)
PRIME, H. A., *Modern Concepts in Control Theory*, McGraw-Hill (1969)
SAVANT, C. J., *Control System Design*, McGraw-Hill (1964)
COALES, J. F. (Ed), *Pneumatic Components and Computing Devices for Control Systems*, Butterworth, London (1963)
KAY, F. X., *Pneumatic Circuit Design*, Machinery (1966)
GUILLON, M., *Hydraulic Servo Systems*, Butterworth, London (1969)
TAYLOR, P. L., *Servomechanisms* (Electrical Engineering Series), Longmans (1960)
MINORSKY, N., *Theory of Non-Linear Control Systems*, McGraw-Hill, New York (1969)

BRITISH STANDARDS

BS 1523
Glossary of terms used in automatic controlling and regulating systems (1958)
BS 3015
Glossary of terms used in vibration and shock testing (1958)
BS 3238
Graphical symbols for components of servomechanisms
BS 3318
Locating the centre of gravity of heavy objects (1961)
BS 3851
Glossary of terms used in mechanical balancing of rotary machinery (1969)
BS 3852
Dynamic balancing machines (1965)

STEAM TABLES IN SI UNITS

D. H. BACON AND A. PARRISH

Tables 2.26 and 2.27 have been prepared from The 1967 International Formulation Committee Formulation for Industrial Use. These are reproduced from 'Steam Tables in SI Units' with the permission of the Central Electricity Generating Board who do not accept responsibility for the consequences of any errors.

Table 2.26 'Saturation Line (Pressure) for Range 0·01 to 221·2 bar' is based entirely on CEGB Table 2.

Table 2.27 'Properties of Water and Steam for Range 0·01 to 400·0 bar and 0·01 to 690°C' has been abridged from CEGB Table 3.

A Mollier chart is available on page 2–8.

UNITS AND NOTATION

Variable	Symbol	Unit	
Pressure	p	bar	(10^5 N/m^2)
Temperature	t	°C	(t °C + 273·15
			= T kelvin)
Specific enthalpy	h	kJ/kg	(10^3 J/kg)
Specific entropy	s	kJ/kg K	(10^3 J/kg K)
Specific volume	v	dm^3/kg	(10^{-3} m^3/kg)

SUFFIXES FOR SATURATION DATA:

s	saturation
f	saturated liquid
g	saturated vapour
fg	evaporation increment

Table 2.26 SATURATION LINE (PRESSURE)

Abs. Press. bar p_s	Temp. °C t_s	Specific enthalpy kJ/kg h_f	h_{fg}	h_g	Specific entropy kJ/kg K s_f	s_{fg}	s_g	Specific volume dm³/kg v_f	v_{fg}	v_g	Abs. Press. bar p_s
0·010	6·983	29·34	2 485·0	2 514·4	0·106 0	8·870 6	8·976 7	1·000 1	129 208·0	129 209·0	0·010
0·011	8·380	35·20	2 481·7	2 516·9	0·126 9	8·814 9	8·941 8	1·000 1	118 042·2	118 043·2	0·011
0·012	9·668	40·60	2 478·7	2 519·3	0·146 1	8·764 0	8·910 1	1·000 2	108 696·0	108 697·0	0·012
0·015	13·036	54·71	2 470·7	2 525·5	0·195 7	8·633 2	8·828 8	1·000 6	87 981·1	87 982·1	0·015
0·020	17·513	73·46	2 460·2	2 533·6	0·260 7	8·463 9	8·724 6	1·001 2	67 005·1	67 006·1	0·020
0·025	21·096	88·45	2 451·7	2 540·2	0·311 9	8·332 1	8·644 0	1·001 9	54 255·2	54 256·2	0·025
0·030	24·100	101·0	2 444·6	2 545·6	0·354 4	8·224 1	8·578 5	1·002 7	45 666·3	45 667·3	0·030
0·035	26·694	111·4	2 438·5	2 550·4	0·390 7	8·132 5	8·523 2	1·003 3	39 477·7	39 478·7	0·035
0·040	28·983	121·4	2 433·1	2 554·5	0·422 5	8·053 0	8·475 5	1·004 0	34 801·2	34 802·2	0·040
0·045	31·035	130·0	2 428·2	2 558·2	0·450 7	7·982 7	8·433 5	1·004 6	31 139·8	31 140·8	0·045
0·050	32·898	137·8	2 423·8	2 561·6	0·476 3	7·919 7	8·396 0	1·005 2	28 193·3	28 194·3	0·050
0·055	34·605	144·9	2 419·8	2 564·7	0·499 5	7·862 6	8·362 1	1·005 8	25 769·7	25 770·7	0·055
0·060	36·183	151·5	2 416·0	2 567·5	0·520 9	7·810 4	8·331 2	1·006 4	23 740·0	23 741·0	0·060
0·065	37·651	157·6	2 412·5	2 570·2	0·540 7	7·762 2	8·302 9	1·006 9	22 014·9	22 015·9	0·065
0·070	39·025	163·4	2 409·2	2 572·6	0·559 1	7·717 6	8·276 7	1·007 4	20 530·0	20 531·0	0·070
0·075	40·316	168·8	2 406·2	2 574·9	0·576 3	7·676 0	8·252 3	1·007 9	19 238·1	19 239·1	0·075
0·080	41·534	173·9	2 403·2	2 577·1	0·592 5	7·637 0	8·229 6	1·008 4	18 103·6	18 104·6	0·080
0·085	42·689	178·7	2 400·5	2 579·2	0·607 9	7·600 3	8·208 2	1·008 9	17 099·1	17 100·1	0·085
0·090	43·787	183·3	2 397·9	2 581·1	0·622 4	7·565 7	8·188 1	1·009 3	16 203·3	16 204·3	0·090
0·095	44·833	187·7	2 395·3	2 583·0	0·636 1	7·533 0	8·169 1	1·009 8	15 399·3	15 400·3	0·095
0·10	45·833	191·8	2 392·9	2 584·8	0·649 3	7·501 8	8·151 1	1·010 2	14 673·6	14 674·6	0·10
0·11	47·710	199·7	2 388·4	2 588·1	0·673 8	7·443 9	8·117 7	1·011 1	13 415·0	13 416·1	0·11
0·12	49·446	206·9	2 384·3	2 591·2	0·696 3	7·390 9	8·087 2	1·011 8	12 360·9	12 361·9	0·12
0·13	51·062	213·7	2 380·3	2 594·0	0·717 2	7·342 0	8·059 2	1·012 6	11 464·7	11 465·7	0·13
0·14	52·574	220·0	2 376·7	2 596·7	0·736 7	7·296 7	8·033 4	1·013 3	10 693·2	10 694·2	0·14

continued

Table 2.26—*continued*

Abs. Press. bar p_s	Temp. °C t_s	Specific enthalpy kJ/kg h_f	h_{fg}	h_g	Specific entropy kJ/kg K s_f	s_{fg}	s_g	v_f	Specific volume dm³/kg v_{fg}	v_g	Abs. Press. bar p_s
0·15	53·997	226·0	2 373·2	2 599·2	0·754 9	7·254 4	8·009 3	1·014 0	10 021·8	10 022·8	0·15
0·16	55·341	231·6	2 370·0	2 601·6	0·772 1	7·214 8	7·986 9	1·014 7	9 432·1	9 433·1	0·16
0·17	56·615	236·9	2 366·9	2 603·8	0·788 3	7·177 5	7·965 8	1·015 4	8 909·9	8 911·0	0·17
0·18	57·826	242·0	2 363·9	2 605·9	0·803 6	7·142 4	7·945 9	1·016 0	8 444·2	8 445·2	0·18
0·19	58·982	246·8	2 361·1	2 607·9	0·818 2	7·109 0	7·927 2	1·016 6	8 026·1	8 027·2	0·19
0·20	60·086	251·5	2 358·4	2 609·9	0·832 1	7·077 3	7·909 4	1·017 2	7 648·7	7 649·8	0·20
0·21	61·145	255·9	2 355·8	2 611·7	0·845 3	7·047 2	7·892 5	1·017 8	7 306·3	7 307·3	0·21
0·22	62·161	260·1	2 353·3	2 613·5	0·858 1	7·018 4	7·876 4	1·018 3	6 994·1	6 995·1	0·22
0·23	63·139	264·2	2 350·9	2 615·2	0·870 2	6·990 8	7·861 1	1·018 9	6 708·3	6 709·3	0·23
0·24	64·082	268·2	2 348·6	2 616·8	0·882 0	6·964 4	7·846 4	1·019 4	6 445·7	6 446·7	0·24
0·25	64·992	272·0	2 346·4	2 618·3	0·893 2	6·939 1	7·832 3	1·019 9	6 203·4	6 204·5	0·25
0·26	65·871	275·7	2 344·2	2 619·9	0·904 1	6·914 7	7·818 8	1·020 4	5 979·3	5 980·3	0·26
0·27	66·722	279·2	2 342·1	2 621·3	0·914 6	6·891 2	7·805 8	1·020 9	5 771·3	5 772·4	0·27
0·28	67·547	282·7	2 340·0	2 622·7	0·924 8	6·868 5	7·793 3	1·021 4	5 577·8	5 578·8	0·28
0·29	68·347	286·0	2 338·1	2 624·1	0·934 6	6·846 6	7·781 2	1·021 9	5 397·2	5 398·2	0·29
0·30	69·124	289·3	2 336·1	2 625·4	0·944 1	6·825 4	7·769 5	1·022 3	5 228·3	5 229·3	0·30
0·32	70·615	295·5	2 332·4	2 628·0	0·962 3	6·785 0	7·747 4	1·023 2	4 921·2	4 922·3	0·32
0·34	72·029	301·5	2 328·9	2 630·4	0·979 5	6·747 0	7·726 5	1·024 1	4 649·3	4 650·4	0·34
0·36	73·374	307·1	2 325·5	2 632·6	0·995 8	6·711 1	7·707 0	1·024 9	4 406·8	4 407·8	0·36
0·38	74·658	312·5	2 322·3	2 634·8	1·011 3	6·677 1	7·688 4	1·025 7	4 189·0	4 190·0	0·38
0·40	75·886	317·7	2 319·2	2 636·9	1·026 1	6·644 8	7·670 9	1·026 5	3 992·4	3 993·4	0·40
0·42	77·063	322·6	2 316·3	2 638·9	1·040 2	6·614 0	7·654 2	1·027 3	3 814·0	3 815·0	0·42
0·44	78·194	327·3	2 313·4	2 640·7	1·053 7	6·584 6	7·638 3	1·028 0	3 651·3	3 652·3	0·44
0·46	79·282	331·9	2 310·7	2 642·6	1·066 7	6·556 4	7·623 1	1·028 7	3 502·4	3 503·4	0·46
0·48	80·332	336·3	2 308·0	2 644·3	1·079 2	6·529 4	7·608 6	1·029 4	3 365·5	3 366·5	0·48
0·50	81·345	340·6	2 305·4	2 646·0	1·091 2	6·503 5	7·594 7	1·030 1	3 239·2	3 240·2	0·50
0·55	83·737	350·6	2 299·3	2 649·9	1·119 4	6·442 8	7·562 3	1·031 7	2 962·6	2 963·6	0·55
0·60	85·954	359·9	2 293·6	2 653·6	1·145 4	6·387 3	7·532 7	1·033 3	2 730·7	2 731·8	0·60

0·85	95·152	398·6	2 269·8	2 668·4	1·251 8	6·162 9	7·414 7	1·040 0	1 970·8	1 971·9	0·85
0·90	96·713	405·2	2 265·6	2 670·9	1·269 6	6·125 8	7·395 4	1·041 1	1 868·1	1 869·2	0·90
0·95	98·204	411·5	2 261·7	2 673·2	1·286 5	6·090 6	7·377 1	1·042 3	1 775·9	1 777·0	0·95
1·0	99·632	417·5	2 257·9	2 675·4	1·302 7	6·057 1	7·359 8	1·043 4	1 692·7	1 693·7	1·0
1·1	102·317	428·8	2 250·8	2 679·6	1·333 0	5·994 7	7·327 7	1·045 5	1 548·2	1 549·2	1·1
1·2	104·808	439·4	2 244·1	2 683·4	1·360 9	5·937 5	7·298 4	1·047 5	1 427·1	1 428·1	1·2
1·3	107·133	449·2	2 237·8	2 687·0	1·386 8	5·884 7	7·271 5	1·049 5	1 324·0	1 325·1	1·3
1·4	109·315	458·4	2 231·9	2 690·3	1·410 9	5·835 6	7·246 5	1·051 3	1 235·3	1 236·3	1·4
1·5	111·372	467·1	2 226·2	2 693·4	1·433 6	5·789 8	7·223 4	1·053 0	1 158·0	1 159·0	1·5
1·6	113·320	475·4	2 220·9	2 696·2	1·455 0	5·746 7	7·201 7	1·054 7	1 090·1	1 091·1	1·6
1·7	115·170	483·2	2 215·7	2 699·0	1·475 2	5·706 1	7·181 3	1·056 3	1 029·9	1 030·9	1·7
1·8	116·933	490·7	2 210·8	2 701·5	1·494 4	5·667 8	7·162 2	1·057 9	976·17	977·23	1·8
1·9	118·617	497·8	2 206·1	2 704·0	1·512 7	5·631 4	7·144 0	1·059 4	927·94	929·00	1·9
2·0	120·231	504·7	2 201·6	2 706·3	1·530 1	5·596 7	7·126 8	1·060 8	884·38	885·44	2·0
2·1	121·780	511·3	2 197·2	2 708·5	1·546 8	5·563 7	7·110 5	1·062 2	844·84	845·90	2·1
2·2	123·270	517·6	2 193·0	2 710·6	1·562 7	5·532 1	7·094 9	1·063 6	808·78	809·84	2·2
2·3	124·705	523·7	2 188·9	2 712·6	1·578 1	5·501 9	7·080 0	1·065 0	775·75	776·81	2·3
2·4	126·091	529·6	2 184·9	2 714·5	1·592 9	5·472 8	7·065 7	1·066 3	745·38	746·45	2·4
2·5	127·430	535·3	2 181·0	2 716·4	1·607 1	5·444 9	7·052 0	1·067 5	717·37	718·44	2·5
2·6	128·727	540·9	2 177·3	2 718·2	1·620 9	5·418 0	7·038 9	1·068 8	691·44	692·51	2·6
2·7	129·984	546·2	2 173·6	2 719·9	1·634 2	5·392 0	7·026 2	1·070 0	667·37	668·44	2·7
2·8	131·203	551·4	2 170·1	2 721·5	1·647 1	5·367 0	7·014 0	1·071 2	644·97	646·04	2·8
2·9	132·388	556·5	2 166·6	2 723·1	1·659 5	5·342 7	7·002 3	1·072 3	624·05	625·13	2·9
3·0	133·540	561·4	2 163·2	2 724·7	1·671 6	5·319 3	6·990 9	1·073 5	604·49	605·56	3·0
3·1	134·661	566·2	2 159·9	2 726·1	1·683 4	5·296 5	6·979 9	1·074 6	586·14	587·22	3·1
3·2	135·753	570·9	2 156·7	2 727·6	1·694 8	5·274 4	6·969 2	1·075 7	568·91	569·99	3·2
3·3	136·819	575·5	2 153·5	2 729·0	1·705 9	5·253 0	6·958 9	1·076 8	552·68	553·76	3·3
3·4	137·858	579·9	2 150·4	2 730·3	1·716 8	5·232 2	6·948 9	1·077 8	537·38	538·46	3·4
3·5	138·873	584·3	2 147·4	2 731·6	1·727 3	5·211 9	6·939 2	1·078 9	522·92	524·00	3·5
3·6	139·865	588·5	2 144·4	2 732·9	1·737 6	5·192 1	6·929 7	1·079 9	509·24	510·32	3·6
3·7	140·835	592·7	2 141·4	2 734·1	1·747 6	5·172 9	6·920 5	1·080 9	496·28	497·36	3·7
3·8	141·784	596·8	2 138·6	2 735·3	1·757 4	5·154 1	6·911 6	1·081 9	483·97	485·05	3·8
3·9	142·713	600·8	2 135·7	2 736·5	1·767 0	5·135 8	6·902 8	1·082 9	472·27	473·36	3·9

continued

Table 2.26—*continued*

Abs. Press. bar p_s	Temp. °C t_s	Specific enthalpy kJ/kg			Specific entropy kJ/kg K			Specific volume dm³/kg			Abs. Press. bar p_s
		h_f	h_{fg}	h_g	s_f	s_{fg}	s_g	v_f	v_{fg}	v_g	
4·0	143·623	604·7	2 133·0	2 737·6	1·776 4	5·117 9	6·894 3	1·083 9	461·14	462·22	4·0
4·2	145·390	612·3	2 127·5	2 739·8	1·794 5	5·083 4	6·877 9	1·085 7	440·41	441·50	4·2
4·4	147·090	619·6	2 122·3	2 741·9	1·812 0	5·050 3	6·862 3	1·087 6	421·51	422·60	4·4
4·6	148·729	626·7	2 117·2	2 743·9	1·828 7	5·018 6	6·847 3	1·089 4	404·19	405·28	4·6
4·8	150·313	633·5	2 112·2	2 745·7	1·844 8	4·988 1	6·832 9	1·091 1	388·27	389·36	4·8
5·0	151·844	640·1	2 107·4	2 747·5	1·860 4	4·958 8	6·819 2	1·092 8	373·58	374·68	5·0
5·2	153·327	646·5	2 102·7	2 749·3	1·875 4	4·930 6	6·805 9	1·094 5	359·99	361·08	5·2
5·4	154·765	652·8	2 098·1	2 750·9	1·889 9	4·903 3	6·793 2	1·096 1	347·36	348·46	5·4
5·6	156·161	658·8	2 093·7	2 752·5	1·904 0	4·876 9	6·780 9	1·097 7	335·61	336·71	5·6
5·8	157·518	664·7	2 089·3	2 754·0	1·917 6	4·851 4	6·769 0	1·099 3	324·64	325·74	5·8
6·0	158·838	670·4	2 085·0	2 755·5	1·930 8	4·826 7	6·757 5	1·100 9	314·37	315·47	6·0
6·2	160·123	676·0	2 080·9	2 756·9	1·943 7	4·802 7	6·746 4	1·102 4	304·75	305·85	6·2
6·4	161·376	681·5	2 076·8	2 758·2	1·956 2	4·779 4	6·735 6	1·103 9	295·70	296·81	6·4
6·6	162·598	686·8	2 072·7	2 759·5	1·968 4	4·756 8	6·725 2	1·105 3	287·19	288·30	6·6
6·8	163·791	692·0	2 068·8	2 760·8	1·980 2	4·734 8	6·715 0	1·106 8	279·16	280·27	6·8
7·0	164·956	697·1	2 064·9	2 762·0	1·991 8	4·713 4	6·705 2	1·108 2	271·57	272·68	7·0
7·2	166·095	702·0	2 061·1	2 763·2	2·003 1	4·692 5	6·695 6	1·109 6	264·39	265·50	7·2
7·4	167·209	706·9	2 057·4	2 764·3	2·014 1	4·672 1	6·686 2	1·111 0	257·59	258·70	7·4
7·6	168·300	711·7	2 053·7	2 765·4	2·024 9	4·652 2	6·677 1	1·112 3	251·13	252·24	7·6
7·8	169·368	716·3	2 050·1	2 766·4	2·035 4	4·632 8	6·668 3	1·113 7	244·99	246·10	7·8
8·0	170·415	720·9	2 046·5	2 767·5	2·045 7	4·613 9	6·659 6	1·115 0	239·14	240·26	8·0
8·2	171·441	725·4	2 043·0	2 768·5	2·055 8	4·595 3	6·651 1	1·116 3	233·57	234·69	8·2
8·4	172·448	729·9	2 039·6	2 769·4	2·065 7	4·577 2	6·642 9	1·117 6	228·26	229·38	8·4
8·6	173·436	734·2	2 036·2	2 770·4	2·075 3	4·559 4	6·634 8	1·118 8	223·19	224·30	8·6
8·8	174·405	738·5	2 032·8	2 771·3	2·084 8	4·542 1	6·626 9	1·120 1	218·33	219·45	8·8
9·0	175·358	742·6	2 029·5	2 772·1	2·094 1	4·525 0	6·619 2	1·121 3	213·69	214·81	9·0
9·2	176·294	746·8	2 026·2	2 773·0	2·103 3	4·508 3	6·611 6	1·122 6	209·24	210·36	9·2
9·4	177·214	750·8	2 023·0	2 773·8	2·112 2	4·492 0	6·604 2	1·123 8	204·98	206·10	9·4

11·0	184·06?	781·1	1996·3		2186·0	4371·1	6·542 7	1·135 1		
11·5	186·048	789·9	1991·3	2781·3	2197·7	4336·6	6534·2	1·135 9	168·86	169·99
12·0	187·961	798·4	1984·3	2782·7	2216·1	4303·3	6519·4	1·138 6	162·06	163·20
12·5	189·814	806·7	1977·4	2784·1	2233·8	4271·2	6505·0	1·141 2	155·79	156·93
13·0	191·609	814·7	1970·7	2785·4	2251·0	4240·3	6491·3	1·143 8	149·98	151·13
13·5	193·350	822·5	1964·2	2786·6	2267·6	4210·4	6477·9	1·146 4	144·59	145·74
14·0	195·042	830·1	1957·7	2787·8	2283·7	4181·4	6465·1	1·148 9	139·57	140·72
14·5	196·688	837·5	1951·4	2788·9	2299·3	4153·3	6452·6	1·151 4	134·89	136·04
15·0	198·289	844·7	1945·2	2789·9	2314·5	4126·1	6440·6	1·153 8	130·50	131·66
15·5	199·850	851·7	1939·2	2790·8	2329·2	4099·6	6428·9	1·156 3	126·39	127·55
16·0	201·372	858·6	1933·2	2791·7	2343·6	4073·9	6417·5	1·158 6	122·53	123·69
16·5	202·857	865·3	1927·3	2792·6	2357·6	4048·9	6406·5	1·161 0	118·89	120·05
17·0	204·307	871·8	1921·5	2793·4	2371·3	4024·5	6395·7	1·163 3	115·46	116·62
17·5	205·725	878·3	1915·9	2794·1	2384·6	4000·7	6385·3	1·165 6	112·22	113·38
18·0	207·111	884·6	1910·3	2794·8	2397·6	3977·5	6375·1	1·167 8	109·15	110·32
18·5	208·468	890·7	1904·7	2795·5	2410·3	3954·8	6365·1	1·170 0	106·24	107·41
19·0	209·797	896·8	1899·3	2796·1	2422·8	3932·6	6355·4	1·172 3	103·48	104·65
19·5	211·099	902·8	1893·9	2796·7	2434·9	3911·0	6345·9	1·174 4	100·86	102·03
20·0	212·375	908·6	1888·6	2797·2	2446·9	3889·8	6336·6	1·176 6	98·359	99·536
20·5	213·626	914·3	1883·4	2797·7	2458·5	3869·0	6327·6	1·178 7	95·980	97·158
21·0	214·855	920·0	1878·2	2798·2	2470·0	3848·7	6318·7	1·180 9	93·709	94·890
21·5	216·060	925·5	1873·1	2798·6	2481·2	3828·8	6310·0	1·183 0	91·540	92·723
22·0	217·244	931·0	1868·1	2799·1	2492·2	3809·3	6301·5	1·185 0	89·467	90·652
22·5	218·408	936·3	1863·1	2799·4	2503·0	3790·1	6293·1	1·187 1	87·482	88·669
23·0	219·552	941·6	1858·2	2799·8	2513·6	3771·3	6284·9	1·189 1	85·580	86·769
23·5	220·676	946·8	1853·3	2800·1	2524·1	3752·8	6276·9	1·191 2	83·757	84·948
24·0	221·783	951·9	1848·5	2800·4	2534·3	3734·7	6269·0	1·193 2	82·006	83·199
24·5	222·871	957·0	1843·7	2800·7	2544·4	3716·8	6261·2	1·195 2	80·325	81·520
25·0	223·943	962·0	1839·0	2800·9	2554·3	3699·3	6253·6	1·197 2	78·708	79·905
25·5	224·998	966·9	1834·3	2801·2	2564·0	3682·1	6246·1	1·199 1	77·153	78·352
26·0	226·037	971·7	1829·6	2801·4	2573·6	3665·1	6238·7	1·201 1	75·655	76·856
26·5	227·061	976·5	1825·1	2801·6	2583·1	3648·4	6231·5	1·203 0	74·212	75·415
27·0	228·071	981·2	1820·5	2801·7	2592·4	3632·0	6224·4	1·205 0	72·820	74·025

continued

Table 2.26—continued

Abs. Press. bar p_s	Temp. °C t_s	Specific enthalpy kJ/kg			Specific entropy kJ/kg K			Specific volume dm³/kg			Abs. Press. bar p_s
		h_f	h_{fg}	h_g	s_f	s_{fg}	s_g	v_f	v_{fg}	v_g	
27·5	229·066	985·9	1 816·0	2 801·9	2·601 6	3·615 8	6·217 3	1·206 9	71·477	72·684	27·5
28·0	230·047	990·5	1 811·5	2 802·0	2·610 6	3·599 8	6·210 4	1·208 8	70·180	71·389	28·0
28·5	231·014	995·0	1 807·1	2 802·1	2·619 5	3·584 1	6·203 6	1·210 7	68·927	70·138	28·5
29·0	231·969	999·5	1 802·6	2 802·2	2·628 3	3·568 6	6·196 9	1·212 6	67·716	68·928	29·0
29·5	232·911	1 004·0	1 798·3	2 802·2	2·637 0	3·553 3	6·190 2	1·214 5	66·544	67·758	29·5
30·0	233·841	1 008·4	1 793·9	2 802·3	2·645 5	3·538 2	6·183 7	1·216 3	65·410	66·626	30·0
30·5	234·759	1 012·7	1 789·6	2 802·3	2·653 9	3·523 3	6·177 3	1·218 2	64·312	65·530	30·5
31·0	235·666	1 017·0	1 785·4	2 802·3	2·662 3	3·508 7	6·170 9	1·220 0	63·247	64·467	31·0
31·5	236·561	1 021·2	1 781·1	2 802·3	2·670 5	3·494 2	6·164 6	1·221 9	62·216	63·438	31·5
32·0	237·445	1 025·4	1 776·9	2 802·3	2·678 6	3·479 9	6·158 5	1·223 7	61·215	62·439	32·0
32·5	238·319	1 029·6	1 772·7	2 802·3	2·686 6	3·465 7	6·152 3	1·225 5	60·244	61·470	32·5
33·0	239·183	1 033·7	1 768·6	2 802·3	2·694 5	3·451 8	6·146 3	1·227 3	59·302	60·529	33·0
33·5	240·037	1 037·8	1 764·4	2 802·2	2·702 3	3·438 0	6·140 3	1·229 2	58·386	59·615	33·5
34·0	240·881	1 041·8	1 760·3	2 802·1	2·710 1	3·424 4	6·134 4	1·231 0	57·497	58·728	34·0
34·5	241·715	1 045·8	1 756·3	2 802·1	2·717 7	3·410 9	6·128 6	1·232 7	56·632	57·865	34·5
35·0	242·541	1 049·8	1 752·2	2 802·0	2·725 3	3·397 6	6·122 8	1·234 5	55·791	57·025	35·0
35·5	243·357	1 053·7	1 748·2	2 801·8	2·732 7	3·384 4	6·117 1	1·236 3	54·973	56·209	35·5
36·0	244·164	1 057·6	1 744·2	2 801·7	2·740 1	3·371 4	6·111 5	1·238 1	54·176	55·415	36·0
36·5	244·963	1 061·4	1 740·2	2 801·6	2·747 4	3·358 5	6·105 9	1·239 8	53·401	54·641	36·5
37·0	245·754	1 065·2	1 736·2	2 801·4	2·754 7	3·345 8	6·100 4	1·241 6	52·646	53·888	37·0
37·5	246·536	1 069·0	1 732·3	2 801·3	2·761 8	3·333 2	6·095 0	1·243 4	51·910	53·154	37·5
38·0	247·311	1 072·7	1 728·4	2 801·1	2·768 9	3·320 7	6·089 6	1·245 1	51·193	52·438	38·0
38·5	248·077	1 076·4	1 724·5	2 800·9	2·775 9	3·308 3	6·084 2	1·246 9	50·494	51·741	38·5
39·0	248·836	1 080·1	1 720·6	2 800·8	2·782 9	3·296 1	6·078 9	1·248 6	49·812	51·061	39·0
39·5	249·588	1 083·8	1 716·8	2 800·5	2·789 7	3·284 0	6·073 7	1·250 3	49·147	50·397	39·5
40·0	250·333	1 087·4	1 712·9	2 800·3	2·796 5	3·272 0	6·068 5	1·252 1	48·497	49·749	40·0
41·0	251·800	1 094·6	1 705·3	2 799·9	2·809 9	3·248 3	6·058 3	1·255 5	47·244	48·500	41·0
42·0	253·241	1 101·6	1 697·8	2 799·4	2·823 1	3·225 1	6·048 2	1·258 9	46·048	47·307	42·0

47·0	260·074	1 135·3	1 661·1	2 796·4	2·885 5	3·114 9	6·000 4	1·275 8	40·805	42·081	47·0
48·0	261·373	1 141·8	1 653·9	2 795·7	2·897 4	3·093 9	5·991 3	1·279 2	39·882	41·161	48·0
49·0	262·652	1 148·2	1 646·8	2 794·9	2·909 1	3·073 3	5·982 3	1·282 5	38·995	40·278	49·0
50·0	263·911	1 154·5	1 639·7	2 794·2	2·920 6	3·052 9	5·973 5	1·285 8	38·143	39·429	50·0
51·0	265·151	1 160·7	1 632·7	2 793·4	2·931 9	3·032 8	5·964 8	1·289 1	37·322	38·611	51·0
52·0	266·373	1 166·8	1 625·7	2 792·6	2·943 1	3·013 0	5·956 1	1·292 4	36·532	37·824	52·0
53·0	267·576	1 172·9	1 618·8	2 791·7	2·954 1	2·993 5	5·947 6	1·295 7	35·770	37·066	53·0
54·0	268·763	1 178·9	1 611·9	2 790·8	2·965 0	2·974 2	5·939 2	1·299 0	35·035	36·334	54·0
55·0	269·932	1 184·9	1 605·0	2 789·9	2·975 7	2·955 2	5·930 9	1·302 3	34·326	35·628	55·0
56·0	271·086	1 190·8	1 598·2	2 789·0	2·986 3	2·936 4	5·922 7	1·305 6	33·641	34·947	56·0
57·0	272·224	1 196·6	1 591·4	2 788·0	2·996 7	2·917 9	5·914 6	1·308 9	32·979	34·288	57·0
58·0	273·347	1 202·3	1 584·7	2 787·0	3·007 1	2·899 5	5·906 6	1·312 1	32·339	33·651	58·0
59·0	274·455	1 208·0	1 578·0	2 786·0	3·017 2	2·881 4	5·898 6	1·315 4	31·719	33·035	59·0
60·0	275·550	1 213·7	1 571·3	2 785·0	3·027 3	2·863 5	5·890 8	1·318 7	31·119	32·438	60·0
61·0	276·630	1 219·3	1 564·7	2 784·0	3·037 2	2·845 8	5·883 0	1·321 9	30·538	31·860	61·0
62·0	277·697	1 224·8	1 558·0	2 782·9	3·047 1	2·828 3	5·875 3	1·325 2	29·975	31·300	62·0
63·0	278·750	1 230·3	1 551·5	2 781·8	3·056 8	2·810 9	5·867 7	1·328 5	29·429	30·757	63·0
64·0	279·791	1 235·7	1 544·9	2 780·6	3·066 4	2·793 8	5·860 1	1·331 7	28·899	30·230	64·0
65·0	280·820	1 241·1	1 538·4	2 779·5	3·075 9	2·776 8	5·852 7	1·335 0	28·384	29·719	65·0
66·0	281·837	1 246·5	1 531·9	2 778·3	3·085 3	2·760 0	5·845 2	1·338 2	27·885	29·223	66·0
67·0	282·842	1 251·8	1 525·4	2 777·1	3·094 6	2·743 3	5·837 9	1·341 5	27·399	28·741	67·0
68·0	283·835	1 257·0	1 518·9	2 775·9	3·103 8	2·726 8	5·830 6	1·344 8	26·927	28·272	68·0
69·0	284·818	1 262·2	1 512·5	2 774·7	3·112 9	2·710 5	5·823 3	1·348 0	26·468	27·817	69·0
70·0	285·790	1 267·4	1 506·0	2 773·5	3·121 9	2·694 3	5·816 2	1·351 3	26·022	27·373	70·0
71·0	286·751	1 272·5	1 499·6	2 772·2	3·130 8	2·678 2	5·809 0	1·354 6	25·588	26·942	71·0
72·0	287·702	1 277·6	1 493·3	2 770·9	3·139 7	2·662 6	5·802 0	1·357 9	25·164	26·522	72·0
73·0	288·643	1 282·7	1 486·9	2 769·6	3·148 4	2·646 5	5·794 9	1·361 1	24·752	26·113	73·0
74·0	289·574	1 287·7	1 480·5	2 768·3	3·157 1	2·630 9	5·788 0	1·364 4	24·351	25·715	74·0
75·0	290·496	1 292·7	1 474·2	2 766·9	3·165 7	2·615 3	5·781 0	1·367 7	23·959	25·327	75·0
76·0	291·408	1 297·6	1 467·9	2 765·5	3·174 2	2·599 9	5·774 2	1·371 0	23·578	24·949	76·0
77·0	292·311	1 302·5	1 461·6	2 764·2	3·182 7	2·584 6	5·767 3	1·374 3	23·205	24·580	77·0
78·0	293·205	1 307·4	1 455·3	2 762·8	3·191 1	2·569 5	5·760 5	1·377 6	22·842	24·220	78·0
79·0	294·091	1 312·3	1 449·1	2 761·3	3·199 4	2·554 4	5·753 8	1·380 9	22·487	23·868	79·0

continued

Table 2.26—continued

Abs. Press. bar p_s	Temp. °C t_s	Specific enthalpy kJ/kg h_f	h_{fg}	h_g	Specific entropy kJ/kg K s_f	s_{fg}	s_g	Specific volume dm³/kg v_f	v_{fg}	v_g	Abs. Press. bar p_s
80·0	294·968	1 317·1	1 442·8	2 759·9	3·207 6	2·539 5	5·747 1	1·384 2	22·141	23·525	80·0
81·0	295·836	1 321·9	1 436·6	2 758·4	3·215 8	2·524 6	5·740 4	1·387 6	21·803	23·190	81·0
82·0	296·697	1 326·6	1 430·3	2 757·0	3·223 9	2·509 9	5·733 8	1·390 9	21·472	22·863	82·0
83·0	297·549	1 331·4	1 424·1	2 755·5	3·232 0	2·495 2	5·727 2	1·394 2	21·149	22·544	83·0
84·0	298·394	1 336·1	1 417·9	2 754·0	3·239 9	2·480 7	5·720 6	1·397 6	20·834	22·231	84·0
85·0	299·231	1 340·7	1 411·7	2 752·5	3·247 9	2·466 3	5·714 1	1·400 9	20·525	21·926	85·0
86·0	300·060	1 345·4	1 405·5	2 750·9	3·255 7	2·451 9	5·707 6	1·404 3	20·223	21·627	86·0
87·0	300·882	1 350·0	1 399·3	2 749·4	3·263 6	2·437 6	5·701 2	1·407 7	19·927	21·335	87·0
88·0	301·697	1 354·6	1 393·2	2 747·8	3·271 3	2·423 5	5·694 8	1·411 1	19·638	21·049	88·0
89·0	302·505	1 359·2	1 387·0	2 746·2	3·279 0	2·409 4	5·688 4	1·414 4	19·355	20·769	89·0
90·0	303·306	1 363·7	1 380·9	2 744·6	3·286 7	2·395 3	5·682 0	1·417 9	19·078	20·495	90·0
91·0	304·100	1 368·3	1 374·7	2 743·0	3·294 3	2·381 4	5·675 7	1·421 3	18·806	20·227	91·0
92·0	304·887	1 372·8	1 368·6	2 741·4	3·301 8	2·367 6	5·669 4	1·424 7	18·540	19·964	92·0
93·0	305·668	1 377·2	1 362·5	2 739·7	3·309 3	2·353 8	5·663 1	1·428 1	18·279	19·707	93·0
94·0	306·443	1 381·7	1 356·3	2 738·0	3·316 8	2·340 1	5·656 8	1·431 6	18·023	19·455	94·0
95·0	307·211	1 386·1	1 350·2	2 736·4	3·324 2	2·326 4	5·650 6	1·435 0	17·773	19·208	95·0
96·0	307·973	1 390·6	1 344·1	2 734·7	3·331 5	2·312 9	5·644 4	1·438 5	17·527	18·965	96·0
97·0	308·729	1 395·0	1 338·0	2 733·0	3·338 8	2·299 4	5·638 2	1·442 0	17·286	18·728	97·0
98·0	309·479	1 399·3	1 331·9	2 731·2	3·346 1	2·285 9	5·632 1	1·445 5	17·049	18·494	98·0
99·0	310·222	1 403·7	1 325·8	2 729·5	3·353 4	2·272 6	5·625 9	1·449 0	16·817	18·266	99·0
100·0	310·961	1 408·0	1 319·7	2 727·7	3·360 5	2·259 3	5·619 8	1·452 5	16·589	18·041	100·0
102·0	312·420	1 416·7	1 307·5	2 724·2	3·374 8	2·232 8	5·607 6	1·459 6	16·145	17·605	102·0
104·0	313·858	1 425·2	1 295·3	2 720·6	3·388 9	2·206 6	5·595 5	1·466 8	15·717	17·184	104·0
106·0	315·274	1 433·7	1 283·1	2 716·9	3·402 9	2·180 6	5·583 5	1·474 0	15·304	16·778	106·0
108·0	316·669	1 442·2	1 270·9	2 713·1	3·416 7	2·154 8	5·571 5	1·481 3	14·904	16·385	108·0
110·0	318·045	1 450·6	1 258·7	2 709·3	3·430 4	2·129 1	5·559 5	1·488 7	14·517	16·006	110·0
112·0	319·402	1 458·9	1 246·5	2 705·4	3 444 0	2·103 6	5·547 6	1·496 2	14·143	15·639	112·0
114·0	32?·7?	1 4?7·?	1 234·?	2 70?·?	3 457 4	2 078 3	5 535 7	1·50? 7	13·781	15·284	114·0

124·0	13·664	12·121	1·542 6	5·476 5	1·953 3	3·523 2	2·666 6	1 172·2			
126·0	13·367	11·816	1·550 7	5·464 6	1·928 6	3·536 0	2·676 1	1 160·1	1 516·0	328·401	126·0
128·0	13·078	11·519	1·558 9	5·457 7	1·903 9	3·548 8	2·671 6	1 147·6	1 524·0	329·621	128·0
130·0	12·797	11·230	1·567 2	5·440 8	1·879 2	3·561 6	2·667 0	1 135·0	1 532·0	330·827	130·0
132·0	12·523	10·948	1·575 6	5·428 8	1·854 6	3·574 2	2·662 3	1 122·3	1 540·0	332·018	132·0
134·0	12·256	10·672	1·584 2	5·416 8	1·830 0	3·586 8	2·657 4	1 109·5	1 547·9	333·194	134·0
136·0	11·996	10·404	1·592 8	5·404 7	1·805 3	3·599 3	2·652 5	1 096·7	1 555·8	334·357	136·0
138·0	11·743	10·141	1·601 7	5·392 5	1·780 7	3·611 8	2·647 5	1 083·8	1 563·7	335·506	138·0
140·0	11·495	9·884	1·610 6	5·380 3	1·756 0	3·624 2	2·642 4	1 070·7	1 571·6	336·641	140·0
142·0	11·253	9·634	1·619 7	5·367 9	1·731 3	3·636 6	2·637 1	1 057·6	1 579·5	337·764	142·0
144·0	11·017	9·388	1·629 0	5·355 5	1·706 6	3·649 0	2·631 8	1 044·4	1 587·4	338·874	144·0
146·0	10·786	9·148	1·638 5	5·343 1	1·681 8	3·661 3	2·626 3	1 031·0	1 595·3	339·972	146·0
148·0	10·561	8·913	1·648 1	5·330 5	1·656 9	3·673 6	2·620 7	1 017·6	1 603·1	341·057	148·0
150·0	10·340	8·682	1·657 9	5·317 8	1·632 0	3·685 9	2·615 0	1 004·0	1 611·0	342·131	150·0
152·0	10·125	8·457	1·667 9	5·305 1	1·607 0	3·698 1	2·609 2	990·3	1 618·9	343·193	152·0
154·0	9·913 6	8·235 5	1·678 2	5·292 2	1·581 9	3·710 3	2·603 3	976·5	1 626·8	344·243	154·0
156·0	9·707 2	8·018 6	1·688 6	5·279 3	1·556 7	3·722 6	2·597 3	962·6	1 634·7	345·282	156·0
158·0	9·505 3	7·805 9	1·699 3	5·266 3	1·531 4	3·734 8	2·591 1	948·5	1 642·6	346·311	158·0
160·0	9·307 6	7·597 3	1·710 3	5·253 1	1·506 0	3·747 1	2·584 9	934·3	1 650·5	347·328	160·0
162·0	9·114 1	7·392 6	1·721 5	5·239 9	1·480 6	3·759 4	2·578 5	920·0	1 658·5	348·335	162·0
164·0	8·924 8	7·191 6	1·733 1	5·226 7	1·455 0	3·771 7	2·572 1	905·6	1 666·5	349·332	164·0
166·0	8·738 5	6·993 8	1·744 7	5·213 2	1·429 0	3·784 2	2·565 5	891·0	1 674·5	350·319	166·0
168·0	8·553 5	6·796 6	1·756 9	5·199 4	1·402 1	3·797 4	2·558 6	875·6	1 683·0	351·295	168·0
170·0	8·371 0	6·601 5	1·769 6	5·185 5	1·374 8	3·810 7	2·551 6	859·9	1 691·7	352·262	170·0
172·0	8·191 2	6·408 6	1·782 6	5·171 3	1·347 3	3·824 0	2·544 4	844·0	1 700·4	353·220	172·0
174·0	8·014 0	6·217 9	1·796 1	5·157 0	1·319 8	3·837 2	2·537 1	828·1	1 709·0	354·168	174·0
176·0	7·839 5	6·029 3	1·810 1	5·142 5	1·292 2	3·850 4	2·529 5	811·9	1 717·6	355·106	176·0
178·0	7·667 4	5·842 7	1·824 7	5·127 8	1·264 3	3·863 5	2·521 8	795·6	1 726·2	356·036	178·0
180·0	7·497 7	5·657 9	1·839 9	5·112 8	1·236 2	3·876 5	2·513 9	779·1	1 734·8	356·957	180·0
182·0	7·330 2	5·474 6	1·855 6	5·097 5	1·207 9	3·889 6	2·505 8	762·3	1 743·4	357·868	182·0
184·0	7·164 7	5·292 6	1·872 1	5·082 0	1·179 2	3·902 8	2·497 4	745·3	1 752·1	358·771	184·0
186·0	7·000 9	5·111 7	1·889 3	5·066 1	1·150 1	3·916 0	2·488 8	727·9	1 760·9	359·666	186·0
188·0	6·838 6	4·931 4	1·907 2	5·049 8	1·120 5	3·929 4	2·479 8	710·1	1 769·7	360·552	188·0

continued

Table 2.26—*continued*

Abs. Press. bar P_s	Temp. °C t_s	Specific enthalpy kJ/kg h_f	h_{fg}	h_g	Specific entropy kJ/kg K s_f	s_{fg}	s_g	Specific volume dm³/kg v_f	v_{fg}	v_g	Abs. Press. bar P_s
190·0	361·431	1 778·7	692·0	2 470·6	3·942 9	1·090 3	5·033 2	1·926 0	4·751 5	6·677 5	190·0
192·0	362·301	1 787·8	673·3	2 461·1	3·956 6	1·059 4	5·016 0	1·945 8	4·571 5	6·517 3	192·0
194·0	363·162	1 797·0	654·1	2 451·1	3·970 6	1·027 8	4·998 3	1·966 6	4·390 9	6·357 6	194·0
196·0	364·017	1 806·6	634·2	2 440·7	3·984 9	0·995 1	4·980 0	1·988 6	4·209 2	6·197 9	196·0
198·0	364·863	1 816·3	613·5	2 429·8	3·999 6	0·961 4	4·961 1	2·012 0	4·025 7	6·037 8	198·0
200·0	365·701	1 826·5	591·9	2 418·4	4·014 9	0·926 3	4·941 2	2·037 0	3·839 6	5·876 7	200·0
202·0	366·533	1 837·0	569·2	2 406·2	4·030 8	0·889 7	4·920 4	2·063 9	3·649 9	5·713 8	202·0
204·0	367·356	1 848·1	545·1	2 393·3	4·047 4	0·851 0	4·898 4	2·093 1	3·455 3	5·548 5	204·0
206·0	368·173	1 859·9	519·5	2 379·4	4·065 1	0·809 9	4·875 0	2·125 2	3·254 2	5·379 4	206·0
208·0	368·982	1 872·5	491·7	2 364·2	4·084 1	0·765 7	4·849 8	2·160 9	3·044 1	5·205 1	208·0
210·0	369·784	1 886·3	461·3	2 347·6	4·104 8	0·717 5	4·822 3	2·201 5	2·821 9	5·023 5	210·0
212·0	370·580	1 901·5	427·4	2 328·9	4·127 9	0·663 9	4·791 7	2·248 8	2·582 6	4·831 4	212·0
214·0	371·368	1 919·0	388·4	2 307·4	4·154 3	0·602 6	4·756 9	2·306 1	2·317 8	4·623 9	214·0
216·0	372·149	1 939·9	341·6	2 281·6	4·186 1	0·529 3	4·715 4	2·379 3	2·012 6	4·391 9	216·0
218·0	372·924	1 967·2	280·8	2 248·0	4·227 6	0·434 6	4·662 2	2·483 2	1·632 0	4·115 2	218·0
220·0	373·692	2 011·1	184·5	2 195·6	4·294 7	0·285 2	4·579 9	2·671 4	1·056 5	3·727 9	220·0
221·2	374·150	2 107·4	0·0	2 107·4	4·442 9	0·0	4·442 9	3·170 0	0·0	3·170 0	221·2

Interpolation in Table 2.27 (see pages 2-86 to 2-103)

In the abridged table linear interpolation between pressures for specific enthalpy and specific entropy may be made except at values close to the critical point. Interpolation for specific volume may be obtained by linear interpolation of density values, viz: put

$$\rho_1 = \frac{1}{v_1} \quad \text{and} \quad \rho_2 = \frac{1}{v_2}$$

then at pressure p,

$$\rho = \rho_1 + \frac{p - p_1}{p_2 - p_1}(\rho_2 - \rho_1) \tag{1}$$

and the intermediate value of specific volume

$$v = \frac{1}{\rho} \tag{2}$$

Table 2.27 PROPERTIES OF WATER AND STEAM FOR THE RANGE 0·01—400 BAR AND 0·01—600°C

(Abridged from *Steam Tables in SI Units*, Central Electricity Generating Board, 1970 by D. H. Bacon)

p (abs.) bar	0·01			0·05			0·10			0·20			0·50			1·0		
t_s °C	7·0			32·9			45·8			60·1			81·3			99·6		
t °C	h kJ/kg	s kJ/kg K	v dm³/kg	h kJ/kg	s kJ/kg K	v dm³/kg	h kJ/kg	s kJ/kg K	v dm³/kg	h kJ/kg	s kJ/kg K	v dm³/kg	h kJ/kg	s kJ/kg K	v dm³/kg	h kJ/kg	s kJ/kg K	v dm³/kg
*	29·34	0·1060	1·0001	137·8	0·4763	1·0052	191·8	0·6493	1·0102	251·5	0·8321	1·0172	340·6	0·9912	1·0301	417·5	1·3027	1·0434
†	2514·4	8·9767	129209·0	2561·6	8·3960	28194·3	2584·8	8·1511	14674·6	2609·9	7·9094	7649·8	2646·0	7·5947	3240·2	2675·4	7·3598	1693·7
0·00	0·00	-0·0000	1·0002	0·00	0·0000	1·0002	0·01	0·0000	1·0002	0·02	0·0000	1·0002	0·05	0·0000	1·0002	0·10	0·0000	1·0002
10	2520·0	8·9966	130604·2	42·00	0·1510	1·0002	42·00	0·1510	1·0002	42·01	0·1510	1·0002	42·04	0·1510	1·0002	42·09	0·1510	1·0002
20	2538·6	9·0611	135227·9	83·86	0·2963	1·0017	83·87	0·2963	1·0017	83·88	0·2963	1·0017	83·91	0·2963	1·0017	83·95	0·2963	1·0017
30	2557·2	9·1236	139850·4	125·66	0·4365	1·0043	125·67	0·4365	1·0043	125·68	0·4365	1·0043	125·71	0·4365	1·0043	125·75	0·4365	1·0043
40	2575·9	9·1842	144471·9	2574·9	8·4390	28854·4	167·45	0·5721	1·0078	167·46	0·5721	1·0078	167·49	0·5721	1·0078	167·53	0·5721	1·0078
50	2594·6	9·2430	149092·6	2593·7	8·4981	29783·0	2592·7	8·1756	14869·2	209·26	0·7035	1·0121	209·29	0·7035	1·0121	209·33	0·7035	1·0121
60	2613·3	9·3001	153712·5	2612·6	8·5555	30710·8	2611·6	8·2334	15335·5	251·10	0·8310	1·0171	251·12	0·8310	1·0171	251·16	0·8309	1·0171
70	2632·1	9·3556	158331·7	2631·4	8·6112	31638·0	2630·6	8·2894	15801·2	2628·8	7·9656	7882·7	292·99	0·9548	1·0228	293·03	0·9548	1·0228
80	2650·9	9·4096	162950·5	2650·3	8·6655	32564·7	2649·5	8·3439	16266·4	2648·0	8·0206	8117·2	334·91	1·0752	1·0292	334·92	1·0752	1·0292
90	2669·7	9·4622	167568·7	2669·2	8·7183	33490·9	2668·5	8·3969	16731·1	2667·1	8·0740	8351·1	2663·0	7·6421	3322·9	376·96	1·1925	1·0361
100	2688·6	9·5136	172186·6	2688·1	8·7697	34416·7	2687·5	8·4486	17195·4	2686·3	8·1261	8584·7	2682·6	7·6953	3418·1	2676·2	7·3618	1695·5
120	2726·5	9·6125	181421·3	2726·3	8·8690	36267·3	2725·6	8·5481	18123·0	2724·6	8·2262	9050·8	2721·6	7·7972	3607·4	2716·5	7·4670	1792·7
140	2764·6	9·7070	190655·0	2764·3	8·9636	38116·8	2763·9	8·6430	19049·5	2763·1	8·3215	9515·8	2760·6	7·8940	3795·5	2756·4	7·5662	1888·6
160	2802·9	9·7975	199887·9	2802·6	9·0542	39965·6	2802·3	8·7337	19975·2	2801·6	8·4126	9980·1	2799·6	7·9861	3982·9	2796·2	7·6601	1983·8
180	2841·4	9·8843	209120·2	2841·2	9·1412	41813·7	2840·9	8·8208	20900·4	2840·3	8·5000	10443·7	2838·6	8·0742	4169·7	2835·8	7·7495	2078·3
200	2880·1	9·9679	218352·1	2879·9	9·2248	43661·5	2879·6	8·9045	21825·1	2879·2	8·5839	10906·9	2877·7	8·1587	4356·0	2875·4	7·8349	2172·3
220	2919·0	10·0484	227583·7	2918·8	9·3054	45508·9	2918·6	8·9852	22749·5	2918·2	8·6647	11369·8	2917·0	8·2399	4542·0	2915·0	7·9169	2266·0
240	2958·1	10·1261	236815·0	2957·9	9·3832	47356·0	2957·8	9·0630	23673·6	2957·4	8·7426	11832·4	2956·4	8·3182	4727·7	2954·6	7·9958	2359·5
260	2997·4	10·2014	246046·1	2997·3	9·4584	49203·0	2997·2	9·1383	24597·6	2996·9	8·8180	12294·9	2995·9	8·3939	4913·3	2994·4	8·0718	2452·7
280	3037·0	10·2742	255277·1	3036·9	9·5313	51049·8	3036·8	9·2113	25521·4	3036·5	8·8910	12757·2	3035·7	8·4671	5098·6	3034·4	8·1454	2545·8

Table 2.27 —continued

p (abs.) bar	0·01			0·05			0·10			0·20			0·50			1·0		
t_s °C	7·0			32·9			45·8			60·1			81·3			99·6		
	h kJ/kg	s kJ/kg K	v dm³/kg	v dm³/kg	h kJ/kg	s kJ/kg K	v dm³/kg	h kJ/kg	s kJ/kg K	v dm³/kg	h kJ/kg	s kJ/kg K	v dm³/kg	h kJ/kg	s kJ/kg K	h kJ/kg	s kJ/kg K	v dm³/kg
*	29·34	0·1060	1·0001	1·0052	137·8	0·4763	1·0102	191·8	0·6493	1·0172	251·5	0·8321	1·0301	340·6	1·0912	417·5	1·3027	1·0434
†	2514·4	8·9767	129209·0	28194·3	2561·6	8·3960	14674·6	2584·8	8·1511	7649·8	2609·9	7·9094	3240·2	2646·0	7·5947	2675·4	7·3598	1693·7
t °C																		
300	3076·8	10·3450	264507·9	52896·5	3076·7	9·6021	26445·1	3076·6	9·2820	13219·3	3076·4	8·9618	5283·9	3075·7	8·5380	3074·5	8·2166	2638·7
320	3116·9	10·4137	273738·7	54743·1	3116·8	9·6708	27368·6	3116·7	9·3508	13681·4	3116·5	9·0306	5469·1	3115·9	8·6070	3114·8	8·2857	2731·6
340	3157·2	10·4805	282969·4	56589·6	3157·1	9·7377	28292·1	3157·0	9·4177	14143·4	3156·9	9·0975	5654·2	3156·3	8·6740	3155·3	8·3529	2824·4
360	3197·8	10·5456	292200·1	58436·1	3197·7	9·8028	29215·6	3197·6	9·4828	14605·4	3197·5	9·1627	5839·2	3196·9	8·7392	3196·0	8·4183	2917·2
380	3238·6	10·6091	301430·6	60282·5	3238·6	9·8663	30139·0	3238·5	9·5463	15067·2	3238·3	9·2262	6024·2	3237·8	8·8028	3237·0	8·4820	3009·8
400	3279·7	10·6711	310661·2	62128·9	3279·7	9·9283	31062·4	3279·6	9·6083	15529·1	3279·4	9·2882	6209·1	3279·0	8·8649	3278·2	8·5442	3102·5
420	3321·1	10·7317	319891·7	63975·3	3321·0	9·9888	31985·7	3321·0	9·6689	15990·9	3320·8	9·3488	6394·1	3320·4	8·9255	3319·7	8·6049	3195·1
440	3362·7	10·7909	329122·2	65821·6	3362·7	10·0480	32909·0	3362·6	9·7281	16452·7	3362·5	9·4080	6578·9	3362·1	8·9848	3361·4	8·6642	3287·7
460	3404·6	10·8488	338352·6	67667·9	3404·6	10·1060	33832·3	3404·5	9·7860	16914·5	3404·4	9·4660	6763·8	3404·0	9·0428	3403·4	8·7222	3380·3
480	3446·8	10·9056	347583·1	69514·1	3446·7	10·1627	34755·5	3446·7	9·8428	17376·2	3446·6	9·5228	6948·7	3446·2	9·0996	3444·6	8·7791	3472·8
500	3489·2	10·9612	356813·5	71360·4	3489·2	10·2184	35678·8	3489·1	9·8984	17838·0	3489·0	9·5784	7133·5	3488·7	9·1552	3488·1	8·8348	3565·3
520	3531·9	11·0157	366043·9	73206·6	3531·9	10·2729	36602·0	3531·9	9·9530	18299·7	3531·8	9·6330	7318·3	3531·4	9·2098	3530·9	8·8894	3657·8
540	3574·9	11·0693	375274·2	75052·9	3574·9	10·3265	37525·2	3574·9	10·0065	18761·4	3574·8	9·6865	7503·1	3574·5	9·2634	3574·0	8·9431	3750·3
560	3618·2	11·1218	384504·6	76899·1	3618·1	10·3790	38448·4	3618·0	10·0591	19223·1	3618·0	9·7391	7687·8	3617·8	9·3160	3617·3	8·9957	3842·8
600	3705·6	11·2243	402965·3	80591·4	3705·6	10·4815	40294·7	3705·5	10·1615	20146·4	3705·4	9·8416	8057·4	3705·2	9·4185	3704·8	9·0982	4027·7

*Sat. liquid
†Sat. vapour

Table 2.27 PROPERTIES OF WATER AND STEAM FOR THE RANGE 0.01—400 BAR AND 0.01—600°C

t °C	2.0 — h kJ/kg	s kJ/kg K	v dm³/kg	5.0 — v dm³/kg	h kJ/kg	s kJ/kg K	10.0 — v dm³/kg	h kJ/kg	s kJ/kg K	20.0 — v dm³/kg	h kJ/kg	s kJ/kg K	40.0 — v dm³/kg	h kJ/kg	s kJ/kg K	60.0 — h kJ/kg	s kJ/kg K	v dm³/kg
p (abs.) bar	2.0			5.0			10.0			20.0			40.0			60.0		
t_s °C	120.2			151.8			179.9			212.4			250.3			275.5		
*	504.7	1.5301	1.0608	1.0926	640.1	1.8604	1.1274	762.6	2.1382	1.1766	908.6	2.4469	1.2521	1087.4	2.7965	1213.7	3.0273	1.3187
†	2706.3	7.1268	885.44	374.68	2747.5	6.8192	194.29	2776.2	6.5828	99.536	2797.2	6.3366	49.749	2800.3	6.0685	2785.0	5.8908	32.438
0.01	0.20	0.0000	0.99996	1.0000	0.51	0.0000	0.99971	1.02	0.0001	0.99921	2.04	0.0002	0.99822	4.08	0.0003	6.11	0.0004	0.99722
10	42.19	0.1510	1.0000	1.0001	42.48	0.1509	0.99977	42.97	0.1509	0.99930	43.94	0.1508	0.99836	45.89	0.1506	47.84	0.1505	0.99743
20	84.05	0.2963	1.0015	1.0016	84.33	0.2962	1.0013	84.80	0.2961	1.0008	85.74	0.2959	0.99990	87.62	0.2955	89.49	0.2950	0.99900
30	125.84	0.4364	1.0041	1.0042	126.12	0.4364	1.0039	126.57	0.4362	1.0034	127.48	0.4359	1.0025	129.30	0.4353	131.12	0.4347	1.0016
40	167.62	0.5720	1.0076	1.0077	167.89	0.5719	1.0074	168.33	0.5717	1.0069	169.22	0.5713	1.0060	170.98	0.5706	172.75	0.5698	1.0052
50	209.42	0.7034	1.0119	1.0120	209.68	0.7033	1.0117	210.11	0.7031	1.0112	210.97	0.7026	1.0103	212.69	0.7017	214.41	0.7007	1.0094
60	251.24	0.8309	1.0169	1.0171	251.49	0.8307	1.0167	251.91	0.8305	1.0162	252.75	0.8299	1.0153	254.43	0.8289	256.10	0.8278	1.0144
70	293.11	0.9547	1.0226	1.0228	293.36	0.9545	1.0224	293.76	0.9542	1.0219	294.58	0.9536	1.0210	296.21	0.9524	297.84	0.9512	1.0201
80	335.04	1.0752	1.0290	1.0291	335.28	1.0750	1.0287	335.67	1.0746	1.0282	336.47	1.0740	1.0273	338.06	1.0726	339.64	1.0713	1.0263
90	377.04	1.1924	1.0359	1.0361	377.27	1.1922	1.0357	377.66	1.1919	1.0352	378.43	1.1911	1.0342	379.98	1.1897	381.52	1.1883	1.0332
100	419.14	1.3068	1.0435	1.0436	419.36	1.3066	1.0432	419.74	1.3062	1.0427	420.49	1.3054	1.0417	421.99	1.3038	423.49	1.3023	1.0406
120	503.72	1.5276	1.0604	1.0606	503.93	1.5273	1.0602	504.28	1.5269	1.0596	504.99	1.5260	1.0584	506.39	1.5242	507.81	1.5224	1.0573
140	2747.8	7.2298	934.88	1.0800	589.20	1.7388	1.0796	589.52	1.7383	1.0790	590.17	1.7373	1.0777	591.47	1.7352	592.78	1.7332	1.0764
160	2789.1	7.3275	984.00	383.47	2766.4	6.8631	1.1019	675.70	1.9420	1.1012	676.28	1.9408	1.0997	677.46	1.9385	678.65	1.9361	1.0983
180	2830.0	7.4196	1032.4	404.51	2811.4	6.9647	194.36	2776.5	6.5835	1.1266	763.62	2.1379	1.1249	764.64	2.1352	765.68	2.1325	1.1232
200	2870.5	7.5072	1080.4	424.96	2855.1	7.0592	205.92	2826.8	6.6922	1.1560	852.55	2.3300	1.1540	853.37	2.3268	854.21	2.3237	1.1519
220	2910.8	7.5907	1128.0	444.97	2898.0	7.1478	216.93	2874.6	6.7911	102.09	2819.9	6.3829	1.1878	944.14	2.5147	944.72	2.5110	1.1853
240	2951.4	7.6707	1175.3	464.67	2940.1	7.2317	227.55	2920.6	6.8825	108.43	2875.9	6.4943	1.2280	1037.7	2.7006	1037.9	2.6962	1.2248
260	2991.4	7.7477	1222.4	484.14	2981.9	7.3115	237.88	2965.2	6.9679	114.38	2928.1	6.5941	51.716	2835.6	6.1353	1134.7	2.8813	1.2729
280	3031.7	7.8219	1269.3	503.43	3023.4	7.3879	248.01	3009.0	7.0485	120.04	2977.5	6.6852	55.400	2902.0	6.2576	2804.9	5.9270	33.173

Table 2.27—*continued*

p (abs.) bar	2·0			5·0			10·0			20·0			40·0			60·0		
t_s °C	120·2			151·8			179·9			212·4			250·3			275·5		
	v dm³/kg	h kJ/kg	s kJ/kg K	v dm³/kg	h kJ/kg	s kJ/kg K	v dm³/kg	h kJ/kg	s kJ/kg K	v dm³/kg	h kJ/kg	s kJ/kg K	v dm³/kg	h kJ/kg	s kJ/kg K	v dm³/kg	h kJ/kg	s kJ/kg K
*	1·0608	504·7	1·5301	1·0928	640·1	1·8604	1·1274	762·6	2·1382	1·1766	908·6	2·4469	1·2521	1087·4	2·7965	1·3187	1213·7	3·0273
†	885·44	2706·3	7·1268	374·68	2747·5	6·8192	194·29	2776·2	6·5828	99·536	2797·2	6·3366	49·749	2800·3	6·0685	32·438	2785·0	5·8908
t °C																		
300	1316·2	3072·1	7·8937	522·58	3064·8	7·4614	257·98	3052·1	7·1251	125·50	3025·0	6·7695	58·833	2962·0	6·3642	36·145	2885·0	6·0692
320	1362·9	3112·6	7·9632	541·63	3106·1	7·5322	267·82	3094·9	7·1984	130·81	3071·2	6·8487	61·996	3017·5	6·4593	38·744	2954·2	6·1880
340	1409·5	3153·3	8·0307	560·59	3147·4	7·6008	277·58	3137·4	7·2689	136·00	3116·3	6·9235	64·994	3069·8	6·5461	41·105	3016·5	6·2913
360	1456·1	3194·2	8·0963	579·50	3188·8	7·6673	287·27	3179·7	7·3368	141·10	3160·8	6·9950	67·872	3119·9	6·6265	43·304	3074·0	6·3836
380	1502·7	3235·4	8·1603	598·35	3230·4	7·7319	296·90	3222·0	7·4027	146·14	3204·9	7·0635	70·658	3168·4	6·7018	45·385	3128·3	6·4679
400	1549·2	3276·7	8·2226	617·16	3272·1	7·7948	306·49	3264·4	7·4665	151·13	3248·7	7·1295	73·376	3215·7	6·7733	47·379	3180·1	6·5462
420	1595·6	3318·3	8·2835	635·94	3314·0	7·8561	316·04	3306·9	7·5287	156·07	3292·4	7·1935	76·039	3262·3	6·8414	49·306	3230·3	6·6196
440	1642·1	3360·1	8·3429	654·68	3356·1	7·9160	325·55	3349·5	7·5893	160·98	3336·0	7·2555	78·660	3308·3	6·9069	51·181	3279·3	6·6893
460	1688·5	3402·1	8·4011	673·40	3398·4	7·9745	335·05	3392·2	7·6484	165·86	3379·7	7·3159	81·247	3354·0	6·9702	53·016	3327·4	6·7559
480	1734·9	3444·5	8·4581	692·10	3441·0	8·0318	344·52	3435·1	7·7062	170·72	3423·4	7·3748	83·806	3399·6	7·0314	54·817	3375·0	6·8199
500	1781·2	3487·0	8·5139	710·78	3483·8	8·0879	353·96	3478·3	7·7627	175·55	3467·3	7·4323	86·341	3445·0	7·0909	56·591	3422·2	6·8818
520	1827·6	3529·9	8·5686	729·44	3526·8	8·1428	363·40	3521·6	7·8181	180·37	3511·3	7·4885	88·857	3490·4	7·1489	58·343	3469·1	6·9417
540	1873·9	3573·0	8·6223	748·09	3570·1	8·1967	372·81	3565·2	7·8724	185·18	3555·5	7·5435	91·354	3535·8	7·2055	60·076	3515·9	7·0000
560	1920·2	3616·4	8·6750	766·72	3613·6	8·2496	382·22	3609·0	7·9256	189·96	3599·9	7·5974	93·837	3581·4	7·2608	61·793	3562·7	7·0568
600	2012·9	3704·0	8·7776	803·95	3701·5	8·3526	400·98	3697·4	8·0292	199·50	3689·2	7·7022	98·763	3672·8	7·3680	65·184	3656·2	7·1664

*Sat. liquid
†Sat. vapour

Table 2.27 PROPERTIES OF WATER AND STEAM FOR THE RANGE 0·01—400 BAR AND 0·01—690°C

p (abs.) bar	100·0			150·0			200·0			220·0			300·0			400·0		
t_s °C	311·0			342·1			365·7			373·7			Supercritical			Supercritical		
	h kJ/kg	s kJ/kg K	v dm³/kg	h kJ/kg	s kJ/kg K	v dm³/kg	h kJ/kg	s kJ/kg K	v dm³/kg	h kJ/kg	s kJ/kg K	v dm³/kg	h kJ/kg	s kJ/kg K	v dm³/kg	h kJ/kg	s kJ/kg K	v dm³/kg
*	1408·0	3·3605	1·4525	1611·0	3·6859	1·6579	1826·5	4·0149	2·0370	2011·4	4·2947	2·6714						
†	2727·7	5·6198	18·041	2615·0	5·3178	10·340	2418·4	4·9412	5·8767	2195·6	4·5799	3·7279						
t °C																		
0·01	10·16	0·0007	0·99526	15·19	0·0009	0·99282	20·18	0·0010	0·99042	22·16	0·0010	0·98947	30·05	0·0010	0·98571	39·78	0·0006	0·98112
10	51·71	0·1501	0·99558	56·52	0·1495	0·99329	61·31	0·1489	0·99103	63·22	0·1486	0·99013	70·81	0·1475	0·98660	80·19	0·1459	0·98228
20	93·23	0·2942	0·99722	97·88	0·2931	0·99502	102·52	0·2919	0·99285	104·36	0·2914	0·99199	111·72	0·2895	0·98860	120·85	0·2870	0·98446
30	134·74	0·4334	0·99989	139·26	0·4319	0·99774	143·77	0·4303	0·99562	145·56	0·4296	0·99478	152·73	0·4271	0·99146	161·63	0·4238	0·98742
40	176·28	0·5682	1·0034	180·68	0·5663	1·0013	185·06	0·5643	0·99917	186·82	0·5635	0·99833	193·81	0·5604	0·99504	202·51	0·5565	0·99104
50	217·84	0·6989	1·0077	222·13	0·6966	1·0055	226·41	0·6943	1·0034	228·12	0·6934	1·0026	234·95	0·6897	0·99927	243·46	0·6852	0·99526
60	259·45	0·8257	1·0126	263·63	0·8230	1·0105	267·81	0·8204	1·0083	269·48	0·8194	1·0075	276·15	0·8153	1·0041	284·47	0·8102	1·0000
70	301·10	0·9489	1·0182	305·18	0·9459	1·0160	309·26	0·9430	1·0138	310·89	0·9419	1·0129	317·41	0·9373	1·0095	325·55	0·9317	1·0054
80	342·82	1·0687	1·0244	346·80	1·0655	1·0221	350·78	1·0623	1·0199	352·37	1·0610	1·0190	358·73	1·0560	1·0155	366·70	1·0498	1·0112
90	384·62	1·1854	1·0312	388·49	1·1819	1·0288	392·37	1·1784	1·0265	393·92	1·1770	1·0256	400·13	1·1716	1·0219	407·91	1·1649	1·0175
100	426·50	1·2992	1·0386	430·27	1·2954	1·0361	434·05	1·2916	1·0337	435·56	1·2902	1·0327	441·62	1·2843	1·0289	449·22	1·2771	1·0244
120	510·63	1·5188	1·0551	514·18	1·5144	1·0523	517·75	1·5101	1·0496	519·18	1·5084	1·0486	524·91	1·5017	1·0445	532·12	1·4935	1·0395
140	595·40	1·7291	1·0739	598·71	1·7241	1·0709	602·03	1·7192	1·0679	603·37	1·7173	1·0667	608·74	1·7097	1·0621	615·51	1·7004	1·0566
160	681·03	1·9315	1·0954	684·04	1·9258	1·0919	687·09	1·9203	1·0886	688·31	1·9181	1·0872	693·27	1·9095	1·0821	699·55	1·8991	1·0759
180	767·76	2·1272	1·1199	770·42	2·1208	1·1159	773·13	2·1145	1·1120	774·23	2·1120	1·1105	778·68	2·1022	1·1046	784·38	2·0905	1·0976
200	855·92	2·3176	1·1480	858·14	2·3102	1·1432	860·43	2·3030	1·1387	861·36	2·3001	1·1369	865·20	2·2891	1·1300	870·20	2·2758	1·1220
220	945·93	2·5039	1·1805	947·57	2·4953	1·1748	949·32	2·4869	1·1693	950·04	2·4837	1·1671	953·10	2·4710	1·1590	957·23	2·4560	1·1495
240	1038·4	2·6877	1·2187	1039·2	2·6775	1·2115	1040·3	2·6677	1·2047	1040·7	2·6639	1·2021	1042·8	2·6492	1·1922	1045·8	2·6320	1·1808
260	1134·2	2·8709	1·2647	1134·0	2·8585	1·2553	1134·0	2·8468	1·2466	1134·0	2·8423	1·2432	1134·7	2·8250	1·2307	1136·3	2·8050	1·2166
280	1235·0	3·0563	1·3221	1232·9	3·0407	1·3090	1231·4	3·0262	1·2971	1230·9	3·0207	1·2927	1229·7	2·9998	1·2763	1229·2	2·9761	1·2583

Table 2.27—continued

p (abs.) bar	100.0			150.0			200.0			220.0			300.0			400.0		
t_s °C	311.0			342.1			365.7			373.7								
	h kJ/kg	s kJ/kg K	v dm³/kg	h kJ/kg	s kJ/kg K	v dm³/kg	h kJ/kg	s kJ/kg K	v dm³/kg	h kJ/kg	s kJ/kg K	v dm³/kg	h kJ/kg	s kJ/kg K	v dm³/kg	h kJ/kg	s kJ/kg K	v dm³/kg
													Supercritical			Supercritical		
*	1408.0	3.3605	1.4525	1611.0	3.6859	1.6579	1826.5	4.0149	2.0370	2011.1	4.2947	2.6714						
†	2727.7	5.6198	18.041	2615.0	5.3178	10.340	2418.4	4.9412	5.8767	2195.6	4.5799	3.7279						

t °C

t °C	h kJ/kg	s kJ/kg K	v dm³/kg	h kJ/kg	s kJ/kg K	v dm³/kg	h kJ/kg	s kJ/kg K	v dm³/kg	h kJ/kg	s kJ/kg K	v dm³/kg	h kJ/kg	s kJ/kg K	v dm³/kg	h kJ/kg	s kJ/kg K	v dm³/kg
300	1343.4	3.2488	1.3978	1338.3	3.2278	1.3779	1334.3	3.2089	1.3606	1332.9	3.2018	1.3543	1328.7	3.1757	1.3316	1325.4	3.1469	1.3077
320	2783.5	5.7145	19.256	1454.3	3.4267	1.4736	1445.6	3.3998	1.4451	1442.8	3.3901	1.4351	1433.6	3.3556	1.4012	1429.9	3.3193	1.3677
340	2883.4	5.8803	21.468	1593.3	3.6571	1.6324	1572.4	3.6100	1.5704	1566.2	3.5947	1.5516	1547.7	3.5447	1.4939	1532.9	3.4965	1.4434
360	2964.8	6.0110	23.305	2770.8	5.5677	12.562	1742.9	3.8835	1.8269	1722.0	3.8449	1.7619	1678.0	3.7541	1.6285	1650.5	3.6856	1.5425
380	3035.7	6.1213	24.926	2887.7	5.7497	14.282	2660.2	5.3165	8.2458	2504.5	5.0560	6.1105	1837.7	4.0021	1.8737	1776.4	3.8814	1.6818
400	3099.9	6.2182	26.408	2979.1	5.8876	15.661	2820.5	5.5585	9.9470	2738.8	5.4102	8.2510	2161.8	4.4896	2.8306	1934.1	4.1190	1.9091
420	3159.7	6.3056	27.793	3057.0	6.0016	16.857	2932.9	5.7232	11.197	2874.6	5.6091	9.5883	2557.9	5.0705	4.9215	2145.7	4.4285	2.3709
440	3216.2	6.3861	29.107	3126.9	6.1010	17.940	3023.7	5.8523	12.236	2977.5	5.7556	10.645	2754.0	5.3499	6.2270	2399.4	4.7893	3.1997
460	3270.5	6.4612	30.365	3191.5	6.1904	18.946	3102.7	5.9616	13.154	3064.0	5.8753	11.552	2887.7	5.5349	7.1891	2617.1	5.0906	4.1365
480	3323.2	6.5321	31.580	3252.4	6.2724	19.893	3174.4	6.0581	13.991	3141.0	5.9789	12.368	2993.9	5.6779	7.9849	2779.8	5.3097	4.9414
500	3374.6	6.5994	32.760	3310.6	6.3486	20.795	3241.1	6.1456	14.771	3211.7	6.0716	13.119	3085.0	5.7972	8.6808	2906.8	5.4762	5.6156
520	3425.1	6.6640	33.912	3366.8	6.4204	21.662	3304.2	6.2262	15.507	3278.0	6.1563	13.822	3166.6	5.9014	9.3098	3013.7	5.6128	6.2053
540	3475.1	6.7261	35.040	3421.4	6.4885	22.499	3364.7	6.3015	16.208	3341.0	6.2347	14.488	3241.7	5.9949	9.8901	3108.0	5.7302	6.7346
560	3524.5	6.7863	36.149	3475.0	6.5535	23.313	3423.0	6.3724	16.881	3401.6	6.3083	15.124	3312.1	6.0805	10.433	3193.4	5.8340	7.2185
600	3622.7	6.9013	38.320	3579.8	6.6764	24.884	3535.5	6.5043	18.161	3517.4	6.4441	16.327	3443.0	6.2340	11.436	3346.4	6.0135	8.0884

*Sat. liquid
†Sat. Vapour

Absolute and Gauge pressure

Absolute pressure = Gauge pressure + 1·013 25 bar, see Figure 2.57. It may be noted from Figure 2.57 that negative millibar are recommended in BS 1780: Part 2[1], for vacuum gauge readings.

REFERENCE

BS 1780: Part 2 'Bourdon tube pressure and vacuum gauges' (1971).

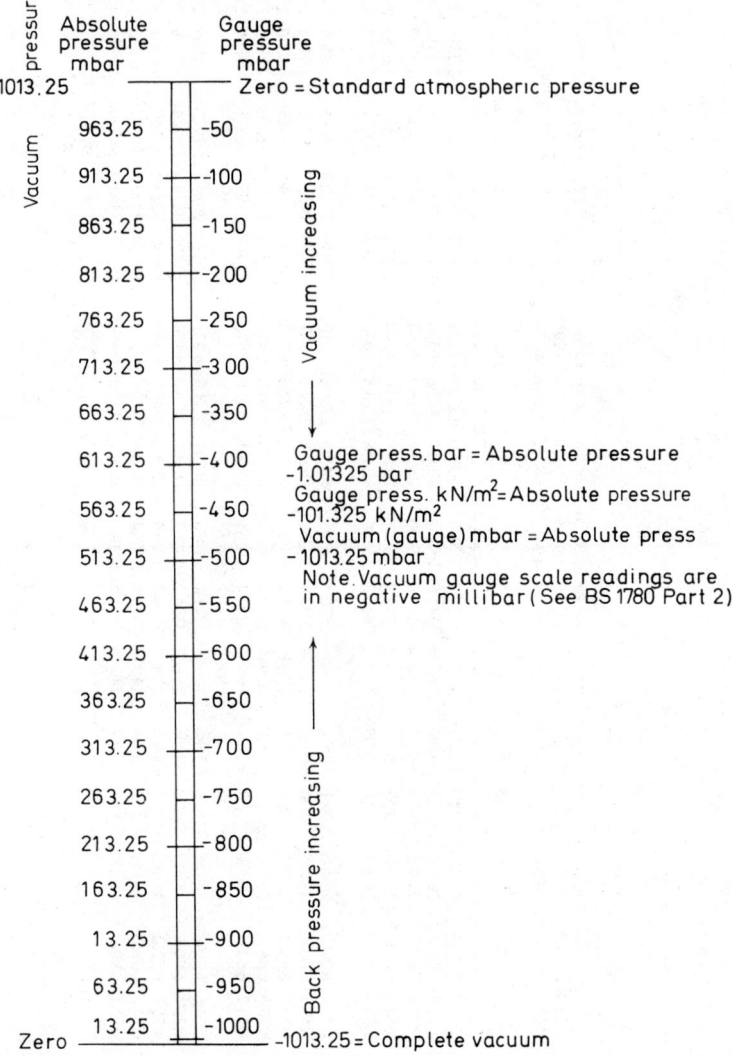

Figure 2.57. Pressure and Vacuum. Absolute and Gauge. (Prepared by A. Parrish)

SPRING DESIGN

S. B. HARRISON

INTRODUCTION

All material bodies are, to some degree, elastic, and will deform under an applied load. Provided that the elastic limit of the material is not exceeded, the body will return to its original shape when the applied load is removed. This inherent elasticity does not imply that all bodies can, or should, be considered as a spring.

In structures, for example, there will be some acceptable deformation under the design loading conditions. Their fundamental requirement, however, is to remain rigid. In contrast, the basic function of a spring is to store energy elastically by virtue of its relatively large displacement.

TYPES OF SPRINGS AND THEIR APPLICATIONS

Many types of springs are available commercially in a variety of shapes and from a wide range of materials. Their extensive applications in the engineering and industrial fields can broadly be classified as follows:
1. To absorb or store energy and to mitigate shock and vibration, e.g. buffers, vehicle suspensions, etc.
2. To apply a definite force or torque, e.g. valves, pipe supports, governors, etc.
3. To indicate or control load or torque, e.g. weighing machines, dynamometers, etc.
4. To provide an elastic pivot or guide, e.g. balancing machines, expansion bends, etc.
Various kinds of springs are used for these purposes but the helical spring is usually chosen in preference to other types because
1. It has practically linear load/deflection characteristics, i.e. a constant rate.
2. It has a relatively wide range of movement.
3. It is compact, which is important in springs which have to absorb energy.
The internal friction of helical type springs is very small so that they return a high proportion of any stored energy. This is a disadvantage in some applications, particularly if there is resonant vibration, when the lack of damping allows large amplitudes to develop. In those applications where damping is an important consideration, other types of springs, e.g. laminated springs, disc springs, etc. may be more suitable than the helical spring.

Helical compression springs

The compression spring is an open coiled helical spring which resists a compressive load. It is made in various forms and from a variety of wire shapes and is probably the most widely used type of spring. In its most common form the coil diameter is uniform throughout its length and it is made from circular section material. Typical compression springs are illustrated in Figure 2.58.

Springs of the open-ended type should only be used where the axial length is restricted. Springs with ends closed and ground flat and square are used much more extensively because they provide a better distribution of end load. Compression springs of square or rectangular section material may be used for those applications where space is limited and relatively large amounts of energy must be absorbed.

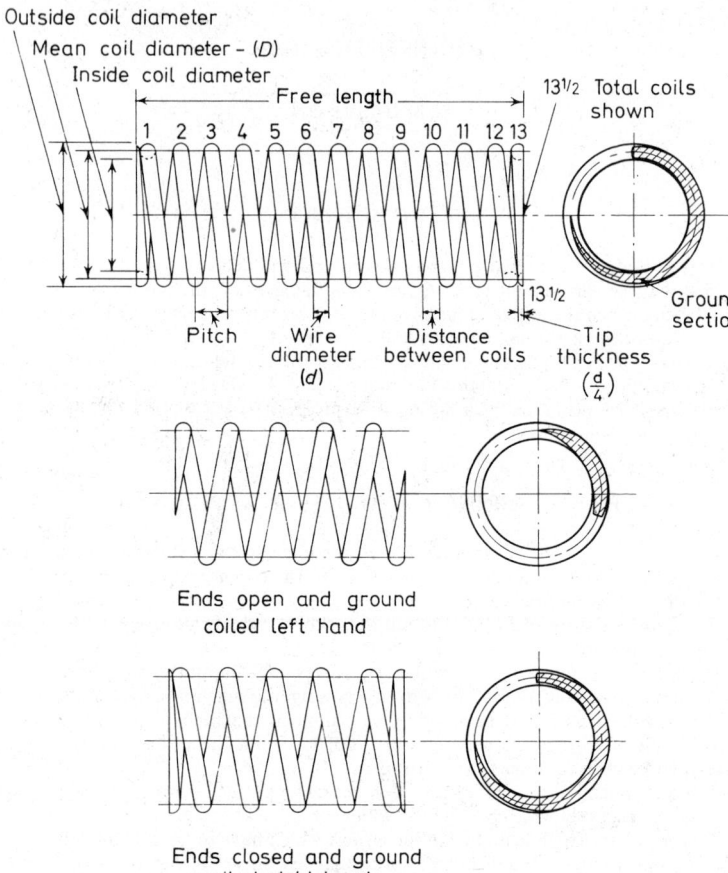

Outside coil diameter
Mean coil diameter – (D)
Inside coil diameter
Free length
13½ Total coils shown

Pitch   Wire diameter (d)   Distance between coils   Tip thickness ($\frac{d}{4}$)   Ground section

Ends open and ground coiled left hand

Ends closed and ground coiled right hand

Figure 2.58. Compression springs

Where possible, however, circular section material should always be used in preference to square or rectangular section, although it may be necessary to use a nest of springs in order to obtain the required performance.

Helical tension springs

The helical tension spring is usually a close-coiled spring of circular section material which resists a tensile load applied by means of a suitable end form, normally a hook or

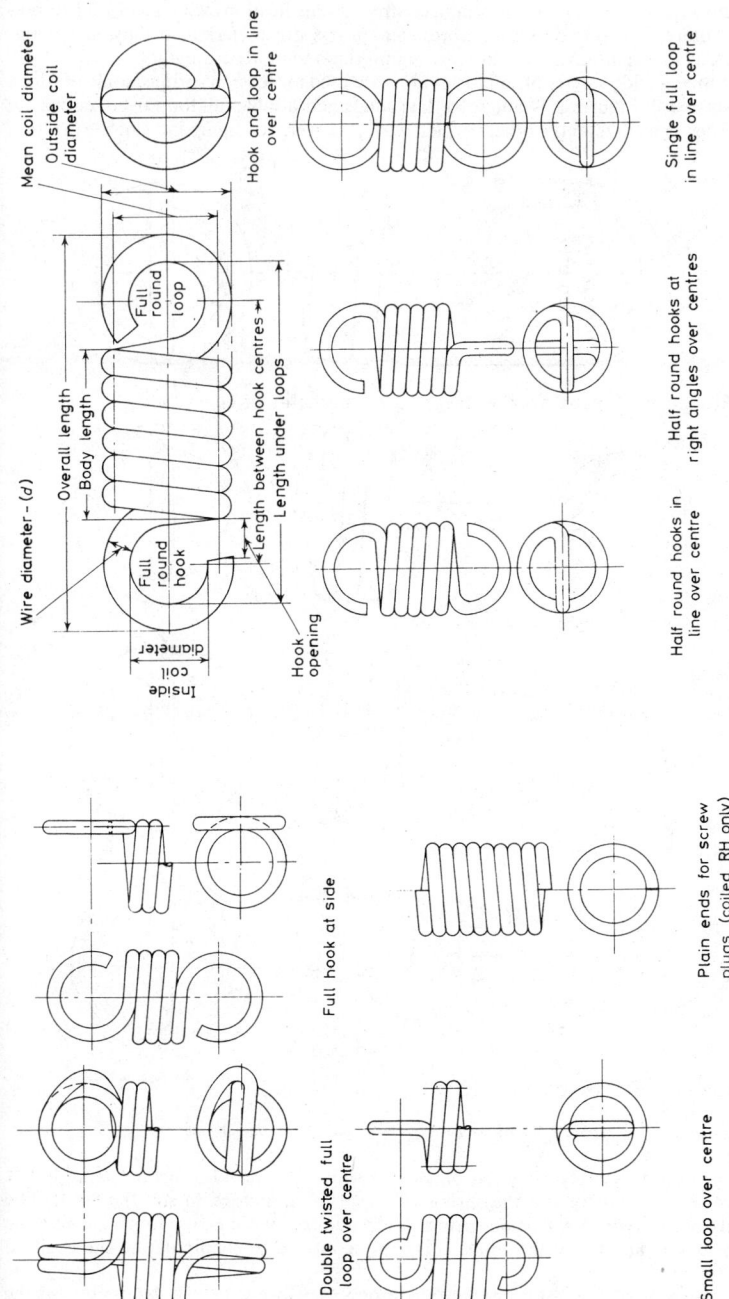

Figure 2.59. Tension springs

loop. This type of end formation produces stress intensifications which can lead to pre-mature failure. For this reason compression springs are preferred to tension springs, and there is often an advantage in rearranging the design accordingly.

In practice, a wide variety of end formations are used for tension springs, some of which are illustrated in Figures 2.59 and 2.60. The single and double full loop over centres are much easier to form than the others, which may, however, be required to suit the position

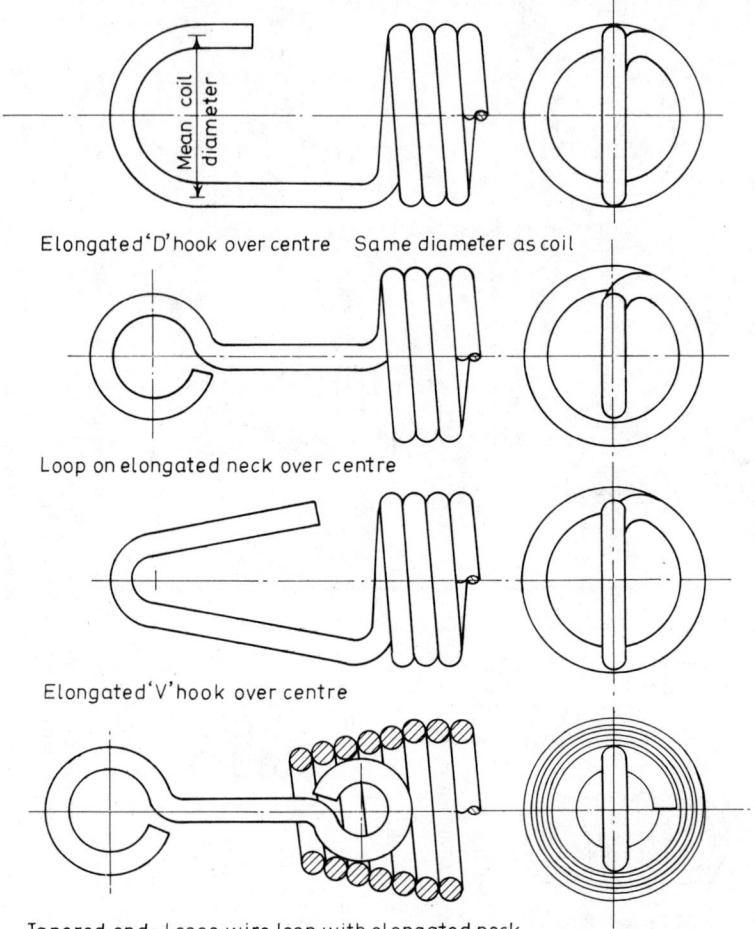

Elongated 'D' hook over centre   Same diameter as coil

Loop on elongated neck over centre

Elongated 'V' hook over centre

Tapered end - Loose wire loop with elongated neck

Figure 2.60. Alternative types of ends for tension springs

of the spring in assembly. Screwed plugs, or arbors, instead of loops are reliable but comparatively expensive and the spring will need to be handed to suit the arbor. The flexibility of this type of extension spring can be adjusted within close limits because the effective length, and hence the active coils can be altered by manipulating the position of the arbors.

Since the coils of a cold-formed tension spring are close-wound it is possible for the

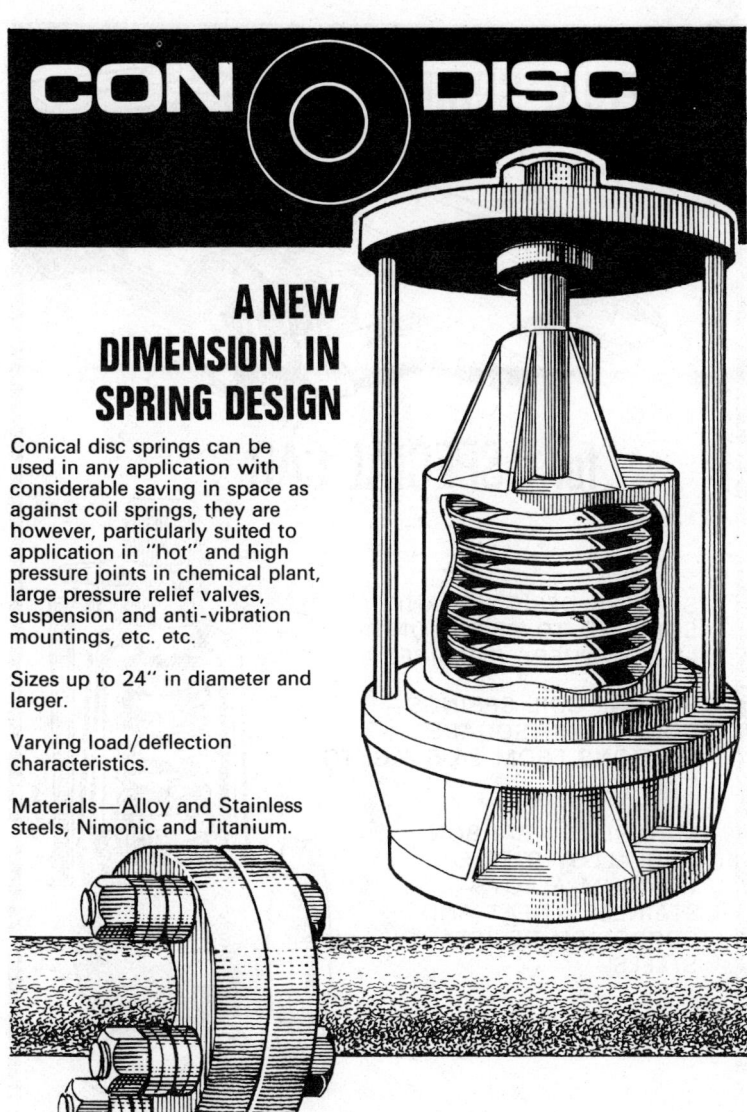

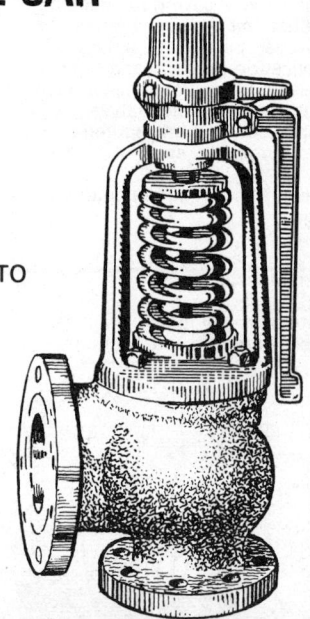

adjacent coils to override during manufacture, thus inducing initial tension in the spring, i.e. an initial load must be applied to the spring before the coils begin to separate. In many applications this effect is desirable but where initial tension is not required it can be removed during manufacture by methods in which the coils become slightly separated.

Helical torsion springs

Torsion springs resist an applied torque when the ends are subjected to an angular displacement and the spring material is stressed in bending by the applied moment. They may be either close-coiled or open-coiled helical springs or torsion bars. Open-coiled helical torsion springs are recommended because the close-coiled type produces friction effects which are difficult to predict.

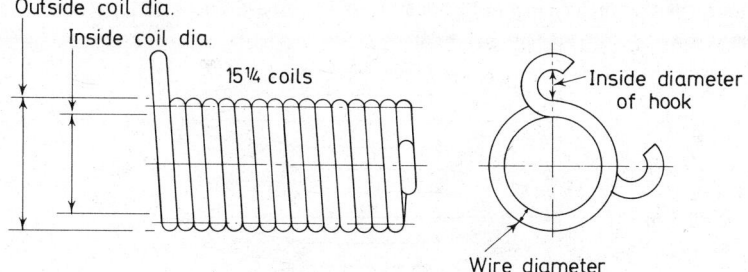

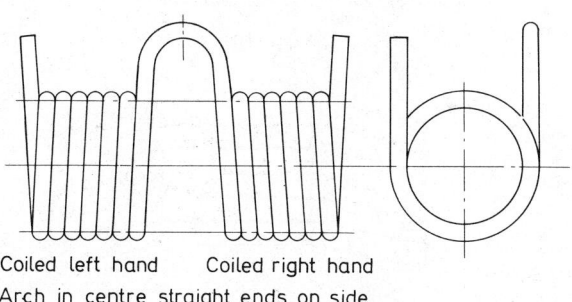

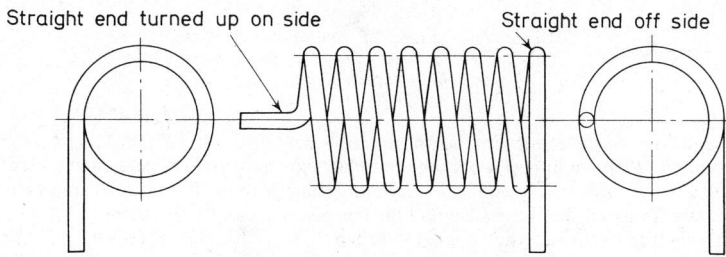

Figure 2.61. Helical torsion springs

Torsion springs are best arranged so that they are 'wound-up' by the application of the load. When this is not done, any residual stresses due to coiling will be in addition to the stress induced by the load. The ends of torsion springs are formed to transmit an external torque or moment, some examples of which are shown in Figure 2.61.

Conical disc springs (or Belleville washers)

Belleville washers are annular coned disc springs of uniform thickness. They are particularly suitable for those applications where high loads are required and limited space is available, and in this respect they possess advantages over the conventional helical spring. By suitable variation in the height to thickness ratio they also offer a wide range of load/deflection characteristics.

Stacks of discs as indicated in Figure 2.62(a) and (b) can be used for those applications requiring either high energy or high load capacity respectively.

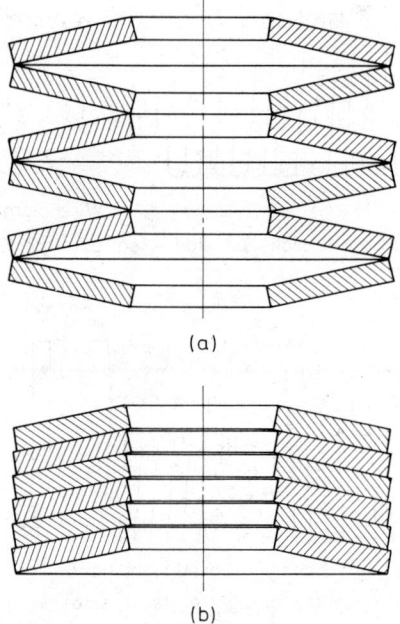

(a)

(b)

Figure 2.62. Stacks of conical disc springs.
(a) Stacked in series. Stack of single pairs.
(b) Stacked in parallel

The maximum deflection obtained from any single disc will depend on how the disc is mounted. When compressed between flat plates the maximum deflection will be when the disc is flat and will be equal to the initial cone height. If the disc is mounted on shoulders, however, deflections beyond the flat position can be obtained.

Usually the deflection required is greater than that which can be obtained from a single disc when it is then necessary to use a number of discs stacked in series. Figure 2.62(a). Similarly, discs stacked in parallel, Figure 2.62(b), may be used to increase the load capacity of a particular disc. To obtain special variations in load/deflection characteristics, discs may be stacked in both series and parallel, either singly or in multiples.

Flat or leaf springs

Flat springs, or leaf springs, generally refer to springs made from flat steel strip or bar. Unlike helical tension or compression springs, however, in which the material is twisted in operation and therefore subjected mainly to torsional stresses, the flat spring absorbs its energy by means of a bending action and the resultant material stresses are mainly tensile or compressive.

There are many types of flat springs with a wide variety of shapes. including cantilever springs. laminated springs. spring clips. circlips. etc.

MATERIALS FOR HELICAL SPRINGS

In general the materials available for the manufacture of round wire helical springs may be divided into two groups:
 1. Patented cold drawn (or hard drawn) materials – in which the required mechanical properties are induced by the drawing process.
 2. Annealed materials – in which the required mechanical properties are obtained by subsequent heat-treatment.
In 1 the springs are cold-formed to the required shape, and normally only require a low temperature stabilising treatment after manufacture. The wire size is therefore a practical limitation to their use. In 2 the springs are usually hot-formed.
Group 1 can be sub-divided into the following classes:
 (a) Carbon steels.
 (b) Alloy steels.
 (c) Austenitic stainless steels.
 (d) Non-ferrous materials.
 (e) Specials. e.g. titanium alloys.
Group 2 is almost exclusively confined to alloy materials. notably
 (f) Silicon-manganese steels (En 45. En 45A).
 (g) Chromium-vanadium steels (En 47, En 50).
 (h) Precipitation hardening steels, e.g. FV 520 (B) steel.
 (i) Specials. e.g. nimonics. titanium alloys.
From the wide range available, the choice of material will depend on its suitability for a particular application. Factors which may have to be considered before a selection is made include: type of loading, i.e. static or dynamic, operating temperature and stresses, nature and effect of a corrosive environment. electrical and/or magnetic requirements. It may be necessary on occasions to use other materials, such as the high nickel alloys (nimonics, etc.) or heat-treatable stainless steels, in order to meet special requirements.

The properties of several spring materials are summarised in Tables 2.28 and 2.29. New designations for these materials are given in the latest edition of BS 970, Parts 4 and 5; see Tables 2.40 and 2·41 at the end of this section.

Cold drawn carbon steel wire to BS 1408[13]

This material can be obtained in three different qualities (BS 1408 B. C and D) and to three strength ranges. Within each strength range the tensile strength varies with wire diameter, increasing as the wire diameter decreases.

The recommended quality is BS 1408 D. which is a high duty ground wire. and the material is therefore free from decarburisation and surface defects. In this condition it is

Table 2.28 CARBON AND LOW ALLOY STEEL SPRING MATERIALS*

Material description	Specification	BS 970 En Ref. No.	Type	Comments	Wire diameters mm	Typical UTS Values GN/m²
Patented cold drawn carbon steel wire for cold-formed springs	BS 1408 B		0·45–0·85 C	Commercial quality Wire	Over 0·230–0·390	2·162–2·472 min
					Over 0·390–0·560	2·008–2·317 min
					Over 0·560–0·840	1·854–2·317
	BS 1408 C		0·55–0·85 C	High duty wire (unground)	Over 0·840–1·300	1·699–2·162
					Over 1·300–2·160	1·545–2·008
					Over 2·160–3·430	1·390–1·854
	BS 1408 D		0·55–0·85 C	High duty wire (ground)	Over 3·430–5·100	1·236–1·699
					Over 5·100–6·680	1·081–1·545
					Over 6·680–10·50	1·081–1·390
Oil hardened and tempered carbon steel wire for cold-formed springs	BS 2803[17] Grade III		0·55–0·75 C	Commercial quality wire	0·914–12·700	1·236–2·008
	BS 2803 Grade II		0·55–0·75 C	High duty wire (unground)	0·914–12·700	1·236–2·008
	BS 2803 Grade I		0·55–0·75 C	High duty wire (ground)	1·626–12·700	1·236–1·776
Annealed carbon steel and low alloy steel wires for oil hardened and tempered springs	BS 1429	En 42B, C, D	0·60–0·90 C	All available as	9·525–19·050	1·081–2·085
		En 44B, C	0·90–1·20 C	Grade 1. Ground	9·525–25·400	1·081–2·781
		En 45, 45A	Si-Mn steel	Grade 2. Unground	9·525–38·100	1·390–1·699
		En 47	Cr-V steel		9·525–38·100	1·390–1·699
		En 50	Cr-V steel		9·525–38·100	1·390–1·699

* A new BS 970, Pts. 1–5 is now available. For details, see Appendix.

Table 2.29 CORROSION RESISTANT SPRING MATERIALS*

Material description	Specification	BS 970 En Ref. No.	Type	Comments	Wire Diameters mm	Typical UTS Values GN/m²
Rust, acid and heat-resisting steel wire for springs	BS 2056	En 56A, B, C, D	13% Cr.	Martensitic qualities available in cold drawn or softened condition		1·545–1·776
		En 57	18/2 Cr. Ni			1·390–1·699
		En 58A	18/8 Cr. Ni	Austenitic qualities available only in cold drawn condition	0·254–0·584	2·008 min
					Over 0·584–0·864	1·854–2·162
					Over 0·864–1·321	1·699–2·008
		En 58J	18/8 Cr. Ni Ti or Nb optional		Over 1·321–2·184	1·545–1·854
					Over 2·184–3·454	1·390–1·699
					Over 3·454–5·131	1·236–1·545
					Over 5·131–6·706	1·081–1·313
					Over 6·706–10·160	1·081 min
Cold drawn titanium alloy (318) wire			90 Ti-6A1-4V	Pickled smooth finish	0·508–7·620	1·158–1·390
Titanium alloy (318) rod			90 Ti-6A1-4V	Annealed and centreless ground	Over 7·620–38·100	0·927 (annealed condition) 1·236 (heat-treated condition)
FV 520 (B) steel rod	BS S143 BS S144 BS S145		14-5-5-1-8-1·7 Cr-Ni-Cu-Mo	Centreless ground	9·525–50·800	0·850–1·081 (overaged condition) 1·158–1·468 (precipitation hardened condition)

* A new BS 970, Pts 1–5 is now available. For details, see Appendix

particularly suitable for those applications where the maximum resistance to fatigue is essential.

Annealed steel wire to BS 1429[14]

This specification covers four grades of low alloy steel and several grades of plain carbon steel for producing springs which may be either hot or cold-formed, but which must be finally quenched and tempered to give the desired mechanical properties. Of the alloy steels the chromium-vanadium types are preferred because of the somewhat greater susceptibility of the silicon-manganese steels to cracking during heat-treatment. Plain carbon steels are even more susceptible to this type of failure.

Recommended permissible design stresses are 0.7 GN/m^2 for the plain carbon steels and 0.85 GN/m^2 for the low alloy steels.

Austenitic steel wire to BS 2056[16]

This specification covers austenitic quality steel wires of 18/8 Cr. Ni types to BS 970[9], En 58, which are available only in the cold-drawn condition. En 58A is a plain 18/8 steel, i.e. unstabilised, whereas EN 58B (18/8/Ti), En 58 F (18/8/Nb) and En 58 J (18/8/Ti or Nb) have the additional stabilising element. In general the stabilised materials can be produced with a better surface finish.

The tensile strength increases as the wire diameter decreases, varying between 1.081 and 2.008 GN/m^2 within the available size range. The maximum design stress for helical compression springs should be limited to 40% of the tensile strength.

Cold drawn titanium 318 alloy wire

Titanium is an expensive material and its use should therefore be limited to those applications where advantage can be taken of its superior corrosion resistance properties. The low density and elastic moduli of titanium, compared with steel, also provide advantages, particularly when used for valve springs in reciprocating machines, e.g. compressors.

Within the available size range the tensile strength is 1.158–1.390 GN/m^2. The maximum design stress for helical compression springs should be limited to 35% of the tensile strength.

Annealed titanium 318 alloy rod for hot-formed springs

This material is used when the wire size required is beyond the available range of the cold drawn titanium alloys (i.e. approx. 9.5 mm upwards). After forming the springs are heat-treated to produce a tensile strength of 1.236–1.313 GN/m^2.

The maximum design stress for helical compression springs should be limited to 45% of the tensile strength.

Hard drawn brass wire to BS 2786[11]

Available in coil in sizes between 0.508 mm and 6.401 mm inclusive. For wire diameters between 0.508 mm and 2.642 mm inclusive the tensile strength is 0.741–0.819 GN/m^2, and for wire diameters from 2.642 mm up to and including 6.401 mm the tensile strength is 0·695–0·772 GN/m². The maximum design stress for helical compression springs should be limited to 35% of the tensile strength.

Characteristics include high conductivity, which is useful for electrical applications, and good resistance to many forms of corrosion. It is unsuitable for use at temperatures above 80°C. Low cost.

Hard drawn phosphor bronze wire to BS 384[10]

This alloy is available, in coil, in sizes between 0·508 mm and 6·401 mm inclusive. For sizes up to and including 2·642 mm diameter the tensile strength is 0·896 GN/m² (min.). For the remaining wire sizes the tensile strength is 0·850 GN/m² (min.). The maximum design stress for helical compression springs should be limited to 35% of the tensile strength.

This material is used where good electrical conductivity is required and where corrosion resistance is important. Unsuitable for use at temperatures above 100°C but withstands higher stresses than brass.

Copper-beryllium

This copper alloy, containing about 2% beryllium, is available in wire sizes up to and including 6·350 mm, and in four different tempers, namely solution-annealed, quarter hard, half hard and hard. The solution-annealed grade is the softest condition available while the harder tempers are produced by cold work. All types can, however, be further hardened by heat-treatment after forming. The properties of Cu.Be. can be varied considerably by heat-treatment. In the cold worked, half hard (reduced, 20% R.A.) condition, the tensile strength is in the range 0·587–0·772 GN/m². After precipitation hardening this tensile range is increased to approximately 1·158–1·313 GN/m².

The maximum design stress for helical compression springs should be limited to 40% of the tensile strength. The mechanical properties are considerably improved as the temperature falls below normal and it is particularly suitable for sub-zero temperature applications. It is, however, unsuitable for use at temperatures above 100°C. Cu.Be. has good corrosion resistance, is non-sparking and non-magnetic. It has good fatigue properties and excellent electrical conductivity.

Monel

A nickel-copper alloy containing approximately two-thirds nickel and one-third copper. It attains its spring properties only after cold work and cannot be hardened by heat-treatment. This alloy is available, in coil, in sizes between 0·711 mm and 6·401 mm inclusive.

The tensile strength is in the range 0·896–1·081 GN/m² and the maximum design stress for helical compression springs should be limited to 40% of the tensile strength. It has good corrosion resistance and good resistance to stress-relaxation up to 200°C. It is slightly magnetic.

K Monel

This material is a precipitation-hardened material and is essentially monel with about 3% of aluminium added. It may be formed in the soft or cold worked condition. Higher tensile strengths (up to 1·158–1·313 GN/m²) can be obtained by an age-hardening treatment after forming.

The maximum design stress for helical compression springs should be limited to 40% of the tensile strength. It is non-magnetic at temperatures down to $-100°C$ and can be used at temperatures up to 260°C. It is available in the same size range as monel.

Inconel

This material is a high-strength, non-magnetic nickel-chromium-iron alloy having good resistance to corrosion. It cannot be hardened by heat-treatment and depends for its spring properties on the degree of cold work put into it. The tensile strength is in the range $1.158-1.313$ GN/m^2 and the maximum design stress for helical compression springs should be limited to 40% of the tensile strength. Available in the same size range as monel.

The maximum operating temperature may be taken as 340°C.

Nimonic 90

This material is a nickel-chromium-cobalt alloy which may be hardened by heat-treatment. The superior creep properties of this alloy make it suitable for use in spring applications at temperatures above those where most available spring materials suffer considerable relaxation. In addition it is non-magnetic. For service below 350°C springs in this material should be coiled from cold drawn wire, and for service above this temperature solution-treated wire should be used.

The tensile strength obtainable ranges between 1.236 and 1.70 GN/m^2 and the maximum design stress for helical compression springs should be limited to 35% of the tensile strength. It is available, in coil, in sizes between 0.711 mm and 6.401 mm inclusive.

FV 520 (B) steel

This material is a high strength, precipitation hardened, chromium-nickel-copper-molybdenum alloy which is more resistant to corrosion than other martensitic steels of the 18 Cr 2 Ni type. It is available in rod form, and in the overaged 550°C condition has a tensile strength of $0.927-1.081$ GN/m^2. The tensile strength can be increased to within the range $1.236-1.468$ GN/m^2 by a precipitation hardening treatment.

The maximum design stress for helical compression springs should be limited to 40% of the tensile strength. It is suitable for use up to 400°C and retains high strength at elevated temperatures. Since the corrosion resistance of FV 520 (B) steel is comparable to that of austenitic quality 18/8 steels it is used as an alternative material when the wire diameter required is beyond the range available for cold drawn austenitic steel.

Physical properties of spring materials

Recommended values of E (modulus of elasticity), G (modulus of rigidity) and density for use in design calculations are given in Table 2.30.

ALLOWABLE STRESSES IN HELICAL SPRINGS

General considerations

The previous data regarding permissible design stresses refer only to statically loaded helical springs, where the stress can safely be related to the nominal elastic limit of the material, either in torsion or in bending. Under cyclic, or fatigue loading conditions, however, the spring will operate between a minimum and maximum load or stress, and the maximum permissible fatigue stress will depend largely on the stress range, i.e. the difference between the initial and final loads or stresses, and the resultant mean stress.

The maximum allowable operating stress will also depend on whether the spring is required to have limited or unlimited life, i.e. say between 10^4 and 10^5 cycles or more than say 10^6 cycles.

Fatigue data are available for most of the usual spring materials, in various conditions, which indicate the limiting stress ranges. These are generally in the form of Goodman type diagrams which have been prepared from S—N curves. This type of diagram may,

Table 2.30 PHYSICAL PROPERTIES OF SPRING MATERIALS

Material	E GN/m^2	G GN/m^2	Density kg/m^3
Carbon steels	206·8	79·3	7 833·4
Silicon-manganese steel	206·8	79·3	7 452·7
Chromium-vanadium steel	206·8	79·3	7 833·4
Martensitic stainless steels	206·8	79·3	7 971·8
Austenitic stainless steels	182·7–193·2	65·5–75·9	7 971·8
FV520(B) steel (overaged 550°C)	200·9	77·2	7 833·4
FV520(B) steel (precipitation hardened)	196·1	75·9	7 833·4
Phosphor bronze	103·5	43·1	8 912·9
Hard drawn brass wire	103·5	37·9	8 525·4
Monel	179·3	65·5	8 802·2
K Monel	179·3	65·5	8 636·1
Inconel	213·8	75·9	8 553·1
Copper-beryllium ⎱ depending on	124·2	48·3	8 248·6
Copper-beryllium ⎰ heat treatment	110·4	41·4	8 248·6
Nimonic 90 (solution treated and aged)	213·8	79·3	8 276·3
Nimonic 90 (cold drawn and aged)	234·5	86·2	8 276·3
Titanium alloy 318	103·5–117·2	34·5–41·4	4 428·8

however, be difficult to use directly for design purposes[1] since the operating stresses will not be known until the design of the spring is complete. A more convenient method, which uses the ratio of the loads corresponding to the minimum and maximum operating stresses is that published by the Ministry of Supply[2], which is summarised below.

Design considerations

The following recommendations are based mainly on research work carried out at the National Physical Laboratory:

1. Higher stresses are permissible in springs for limited endurance life (say 10^4 to 10^5 cycles) than in springs (e.g. valve springs) requiring long endurance life.

2. The permissible maximum stress in a spring depends on the ratio of maximum to minimum stress as well as on the maximum stress at full compression. Thus a spring which will work satisfactorily over a range of say 552 to 1069 MN/m^2 will not give the same service over a range of 0 to 1069 MN/m^2; in the latter case the maximum stress should be limited to 690 MN/m^2, as shown in Figure 2.64. Although the majority of springs are probably designed with an initial compression about half the final compression, the stress range should always be computed and the maximum stress determined accordingly from Figures 2.63, 2.64, 2.65 or 2.66, in which maximum permissible stresses are plotted against the ratio of initial to final load.

3. If the stress range is less than 25 % of the maximum stress, i.e. the initial load is more than 75 % of the final load, high stresses should not normally be used, especially if the

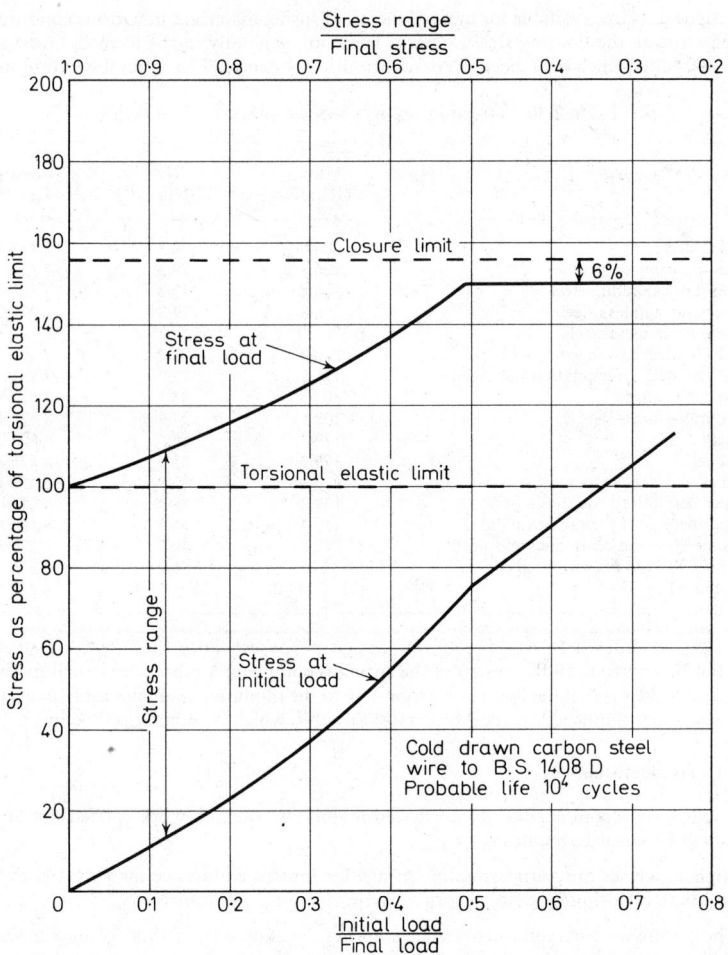

Figure 2.63. (a) Highest possible stresses (category A)

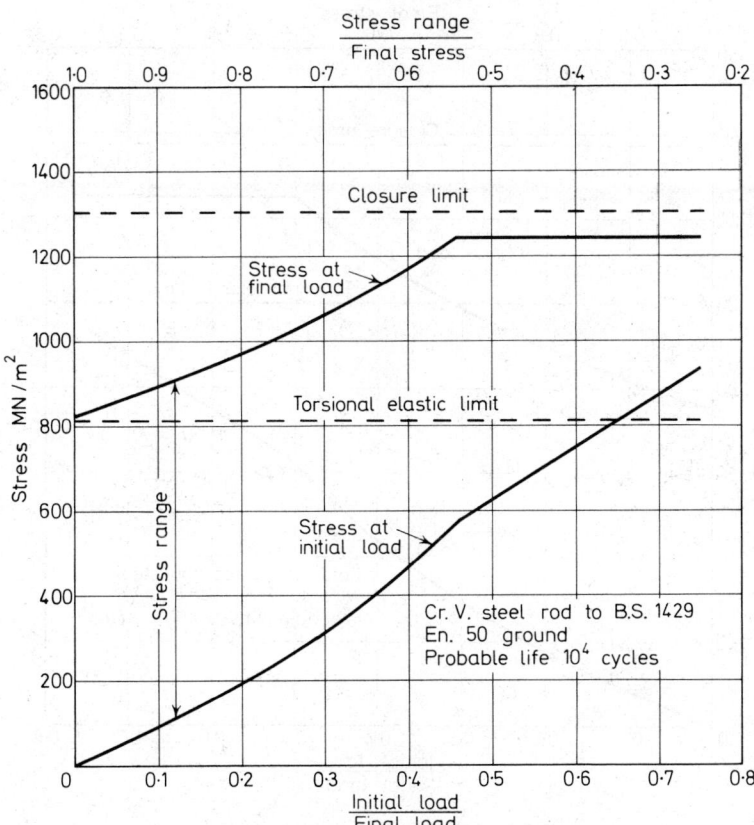

Figure 2.63. (b) Highest possible stresses (category A)

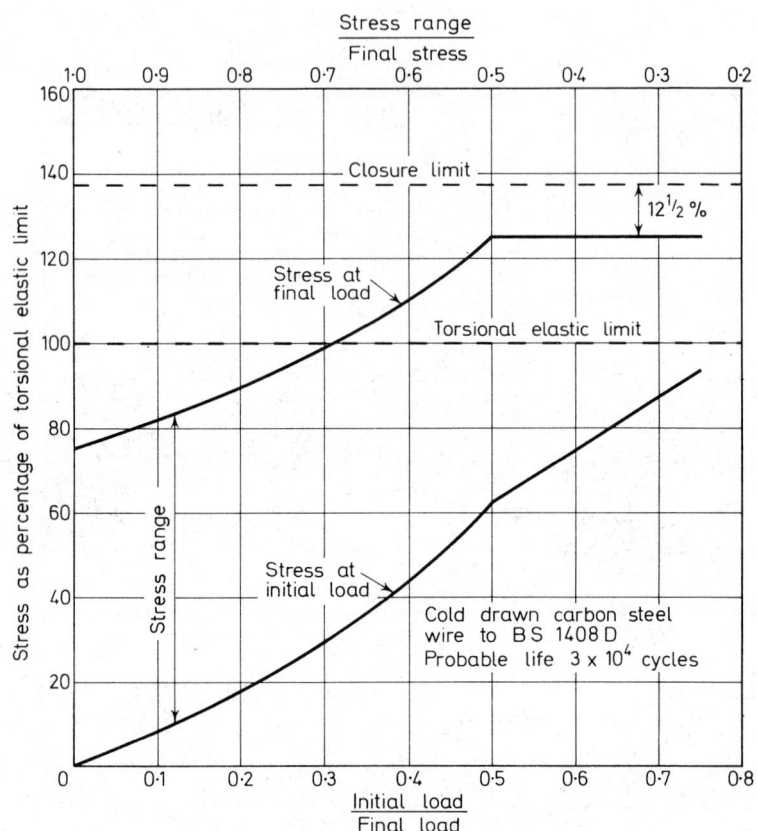

Figure 2.64. (a) High stresses (category B)

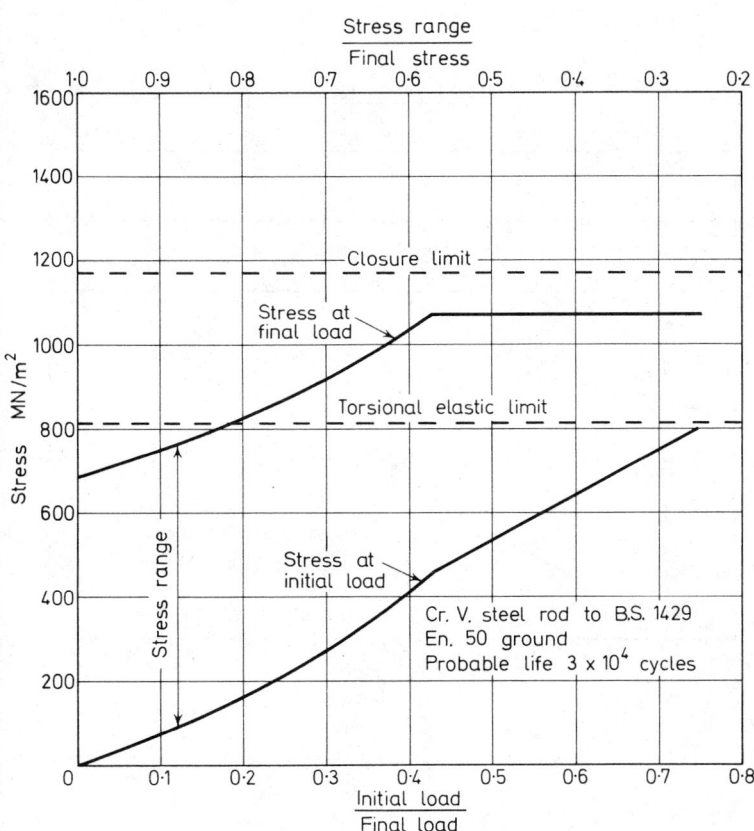

Figure 2.64. (b) High stresses (category B)

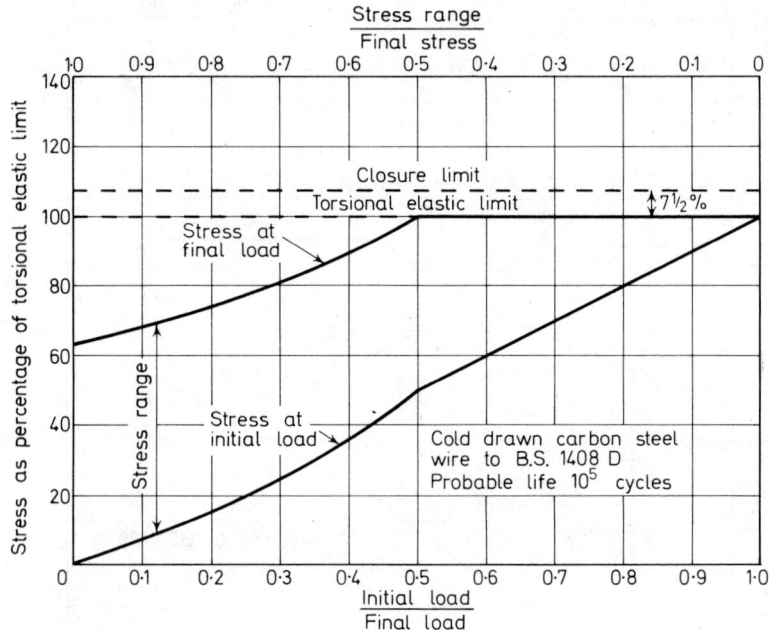

Figure 2.65. (a) Normal stresses (category C)

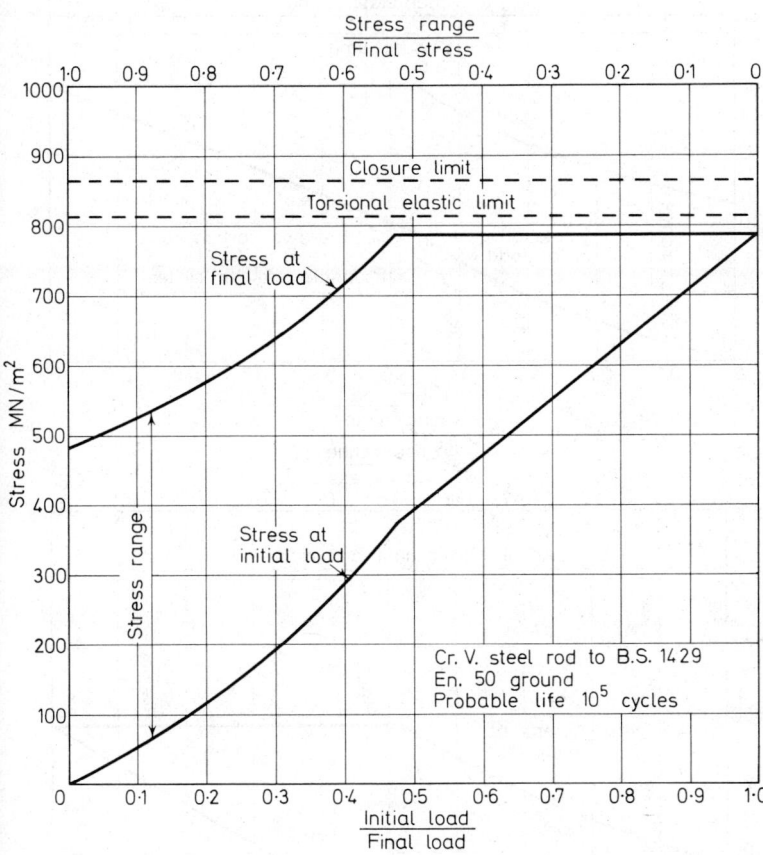

Figure 2.65. (b) Normal stresses (category C)

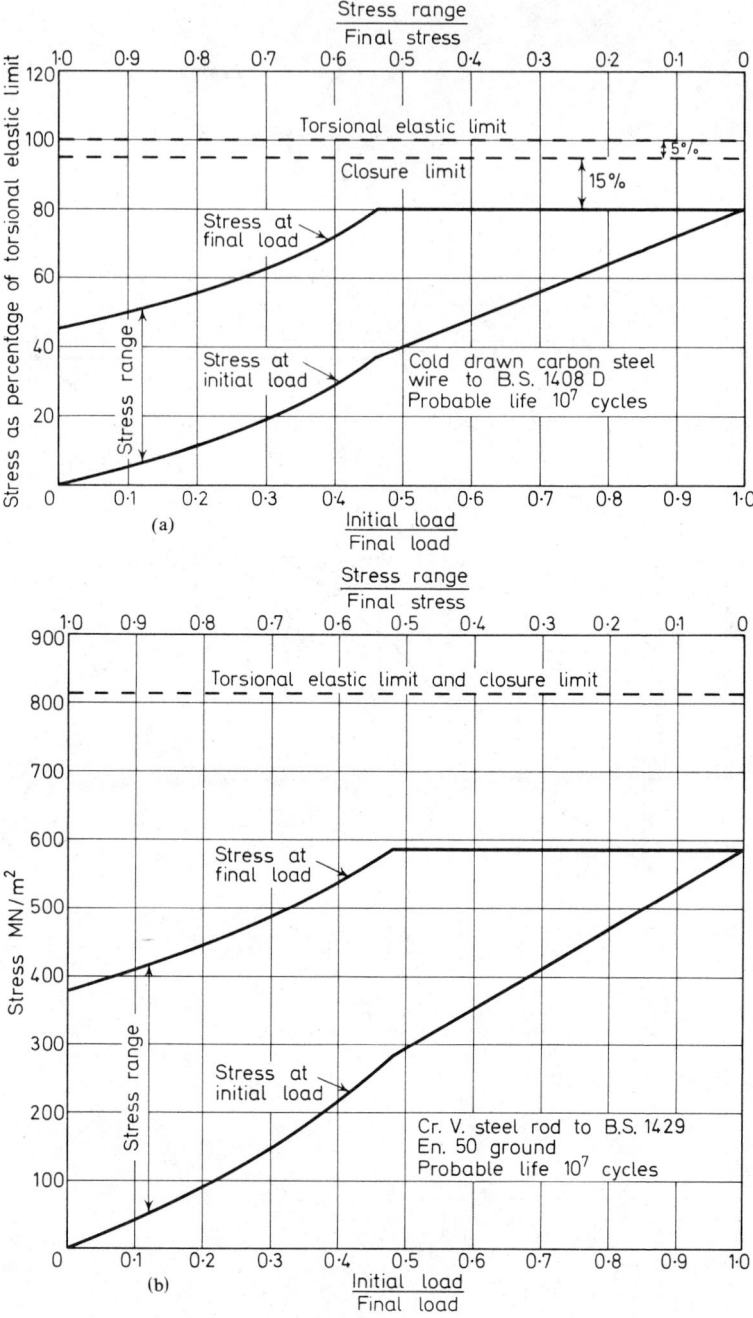

Figure 2.66. Low stresses (category D)

design requires the spring to be maintained over long periods in the state of full compression This restriction is indicated in Figures 2.63 and 2.64.

4. The maximum stresses for cold drawn carbon steel wires are given in Figures 2.63 to 2.66 as percentages of the torsional elastic limits. The torsional elastic limit has been taken as 40°₀ of the UTS. which depends on the wire diameter (see Table 2.28 and Figure 2.67). The maximum stresses for carbon and alloy steel rods are also related to the torsional elastic limit of the material.

5. The pre-stressing (or scragging) process enables the torsional elastic limit to be exceeded for springs of limited endurance life as shown in Figures 2.63 and 2.64.

6. The permissible maximum stresses for carbon and alloy steel rods are somewhat greater for rectangular sections than for circular sections.

7. The closure limits given in Figures 2.63–2.66 represent the limits which must not be exceeded and are not necessarily the limits to be adopted in design. For designs where it is desired to keep the overall length to a minimum, the calculated stress required to close the spring solid will frequently be only about 5% above the maximum working stress. Where very small clearances are specified accurate winding and the use of ground rod are essential.

Table 2.31

Category	A	B	C	D
Stresses	Highest possible	High	Normal	Low
MATERIALS				
Cold drawn carbon steel	BS 1408 D	BS 1408 D	BS 1408 D	BS 1408 D
Low alloy steel rod	En 50	En 50	En 50	En 50
Endurance Life	Low	Medium	High	High
Limits of $\dfrac{\text{Initial Load}}{\text{Final Load}}$	0 to 0·75	0 to 0·75	0 to 1·0	0 to 1·0

The above recommendations show that the permissible stresses depend both on the material used and on the performance required. Accordingly four categories of stresses, A, B, C and D have been allocated which are shown in Table 2.31. Diagrams of maximum stresses and stress ranges for each category and material are given in Figures 2.63–2.66. These are based on empirical formulae and data given in Tables 2.32, 2.33 etc.

Empirical formulae and data

Table 2.32   COLD DRAWN CARBON STEEL WIRE TO BS 1408 D[13]

$$q_{max} = \left[\frac{F_1}{F_2 - r}\right] \% \text{ of torsional elastic limit}$$

$$r = \frac{W_{min}}{W_{max}} \text{ where } \begin{array}{l} W_{min} = \text{minimum working load kg} \\ W_{max} = \text{maximum working load kg} \end{array}$$

Probable life in cycles	10^7 (Cat. D)	10^5 (Cat. C)	3×10^4 (Cat. B)	10^4 (Cat. A)
F_1	48·75	85·13	93·75	148
F_2	1·075	1·352	1·25	1·48
Max. value of q_{max}	80%	100%	125%	150%

Table 2.33   ROUND CR. V STEEL ROD TO BS 1429[14], EN 50 (GROUND)

$$q_{max} = 6\cdot896 \left[\frac{F_1}{F_2 - r} \right] \text{MN/m}^2$$

$$r = \frac{W_{min}}{W_{max}} \text{ as above}$$

Torsional elastic limit assumed to be 813·7 MN/m²

Probable life in cycles	10^7 (Cat. D)	10^5 (Cat. C)	3×10^4 (Cat. B)	10^4 (Cat. A)
F_1	74·61	86·62	120	163·3
F_2	1·357	1·237	1·2	1·36
Max. value of q_{max} (MN/m²)	586·1	786·1	1069	1242

Torsional elastic limit of BS 1408[13] carbon steel wire

Some spring specifications imply the same value of the torsional elastic limit for a wide range of wire sizes. Since the torsional elastic limit varies inversely as the wire diameter

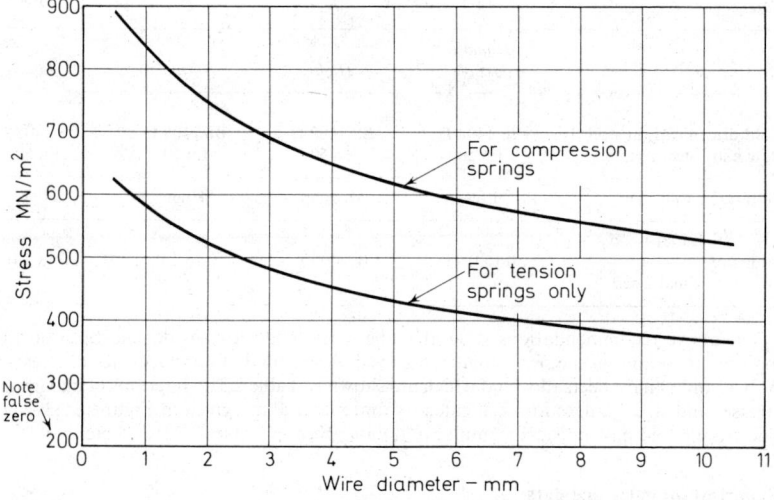

Figure 2.67. Torsional elastic limit of cold drawn carbon steel, BS 1308 D

it is probable that the values for the torsional elastic limit will lie on a smooth curve (see Figure 2.67).

The values in Figure 2.67 are based on the empirical formula:

$$\text{Torsional elastic limit} = \frac{961}{4\sqrt{d}} - \frac{126}{d} \text{ MN/m}^2$$

where d is the wire diameter in mm.

This formula applies to compression springs only. The effective values for tension springs given by the lower curve of Figure 2.67 are 70% of those given by this formula.

Selection of maximum permissible stresses

CATEGORY A—HIGHEST POSSIBLE STRESSES, FIGURES 2.63(a) and (b)

These working stresses should only be adopted in very exceptional circumstances. The production difficulties are considerable but with careful manufacture and close inspection

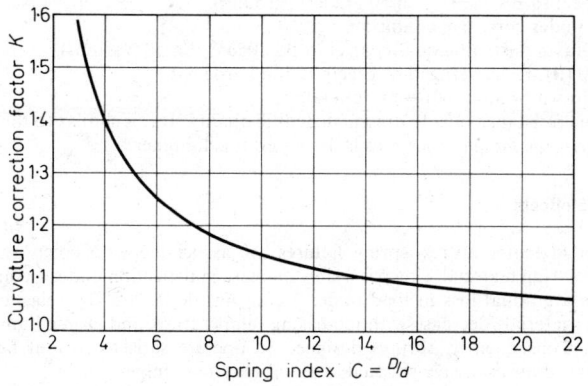

Figure 2.68. The curvature factor K for stress in helical springs

at every stage, a limited life of, say 10^4 to 3×10^4 cycles may be anticipated. As the maximum stress is, in all cases, above the normal torsional elastic limit for the material, the springs must be efficiently pre-stressed.

CATEGORY B—HIGH STRESSES, FIGURES 2.64(a) and (b)

For designs where space or weight limitations justify the use of stresses higher than the normal. In this category also the maximum stress often exceeds the torsional elastic limit, necessitating efficient pre-stressing. With careful manufacture, a life of 3×10^4 to 10^5 cycles may be expected.

CATEGORY C—NORMAL STRESSES, FIGURES 2.65(a) and (b)

For normal designs where the saving in size and weight justifies the use of high quality material and manufacture. Springs designed with these stresses should present no difficulty in manufacture. Pre-stressing may be necessary in some cases.

CATEGORY D—LOW STRESSES, FIGURES 2.66(a) and (b)

For springs designed for normal duty where size and weight are not restricted. In these circumstances the use of high stresses and the highest quality steel are not justified.

Pre-stressing is not necessary and the use of rectangular section material should be very exceptional.

Adverse conditions or long life

In cases where springs are required to operate under adverse conditions e.g. corrosive environment, or where an infinite life is required, stresses as low as possible should be imposed.

For severe duty and/or fatigue conditions, the recommended materials are:
1. Cold drawn carbon steel to BS 1408D[13].
2. Cr.V steel to BS 1429[14], En 50 grade 1 (ground).

For duty under corrosive conditions:
1. Cold drawn austenitic quality steel to BS 2056[16], En 58 (various).
2. FV 520 (B) steel, overaged or precipitation hardened.
3. Cold drawn or annealed Titanium alloy.

The fatigue resistance of cold drawn austenitic quality steel is similar to that of carbon steel: the *corrosion fatigue* resistance is, however, much higher.

Temperature effects

Under elevated temperatures, spring failures are associated with creep or relaxation effects in the spring material. Creep is a slow increase in deflection under a constant load; relaxation is a gradual loss in load under a constant deflection. The elastic moduli of most spring materials decrease with increasing temperature, and elastic limits also tend to decrease. Consequently, springs designed to operate satisfactorily at normal temperatures may show excessive residual strain at elevated temperatures.

Selection of spring materials for high temperature or corrosive service is generally limited to stainless steels or high nickel alloys. Carbon and low alloy steels have a maximum service temperature of about 200°C. In addition they are not normally resistant to corrosion and other forms of high temperature attack. If, in addition, it is desirable that the spring retain maximum load capacity at the elevated temperatures, the choice of materials is further restricted and only the nickel alloys may be satisfactory.

Recent researches[3] on the effect of elevated temperatures on springs have led to the following general conclusions:
1. Plain carbon steels are reliable up to 180°C and 540 MN/m² if 6 % set is acceptable. In addition, if 16 % set can be tolerated 200°C and 772 MN/m² may be used.
2. Alloy steels are reliable up to 200°C and 540 MN/m² with 3 % permanent set. 230°C and 772 MN/m² may be used if 10 % set is permissible.
3. Beyond 230°C freedom from excessive residual strain can be maintained only with such materials as stainless steels and high nickel alloys.

Torsional elastic limit

The torsional elastic limit for cold drawn carbon steel wire varies inversely as the wire diameter (see Table 2.28). Figure 2.67 gives a graph of torsional elastic limit against wire diameter for material to BS 1408 D. The upper curve gives true values which apply only to compression springs. The lower curve gives approximate effective values for tension springs (see Empirical formulae and data).

Design procedure

The final choice of materials for steel springs is linked with the determination of the appropriate working stress. This design procedure combines both.

STEP 1

Select the appropriate stress category from Table 2.31. The choice between cold drawn carbon steel or alloy steel rod will depend on the wire diameter required to satisfy the loading conditions. This can be found to a first approximation by using a suitable trial value of the spring index C and the design charts Figures 2.71 and 2.72.

STEP 2.

If alloy steel rod is used the maximum permissible stress can be obtained directly from Figures 2.63(b) to 2.66(b) for the selected stress category and ratio of initial load to final load.

For cold drawn carbon steel Figures 2.63(a) to 2.66(a) give the maximum permissible stress as a percentage of the torsional elastic limit so a further step is necessary.

STEP 3.

For cold drawn carbon steel only. Find the torsional elastic limit for the chosen wire diameter from Figure 2.67 and use this to find the maximum permissible stress from the percentage found in step 2.

Example

A return spring is required to support a load of 45 kg when compressed 47·5 mm from its initial position at which it must support a load of 11·25 kg. It will be required to have a life of up to 10^7 cycles and its outside diameter must not exceed 48 mm.

First find a suitable working stress.

STEP 1.

Since a long life is required, the stress category should be D (Table 2.31). A suitable trial value of the spring index C is say 8.

Since the outside coil diameter is limited to 48 mm

$$\frac{48-d}{d} = 8 \text{ from which } d = 5\cdot333 \text{ mm}$$

The nearest standard wire diameter is 5·3 mm.

This diameter is within the range of sizes specified for cold drawn carbon steel to BS 1408D[13] (see Table 2.28) and this material will be selected.

STEP 2.

$$\text{Ratio} \frac{\text{initial load}}{\text{final load}} = \frac{11\cdot25}{45} = 0\cdot25$$

From Figure 2.66(a) the maximum permissible stress at final load is 59 % of the torsional elastic limit when the above ratio is 0.25.

STEP 3.

From Figure 2.67 the torsional elastic limit of 5·3 mm diameter cold drawn carbon steel wire to BS 1408D is 610 MN/m². The maximum permissible stress at final load will

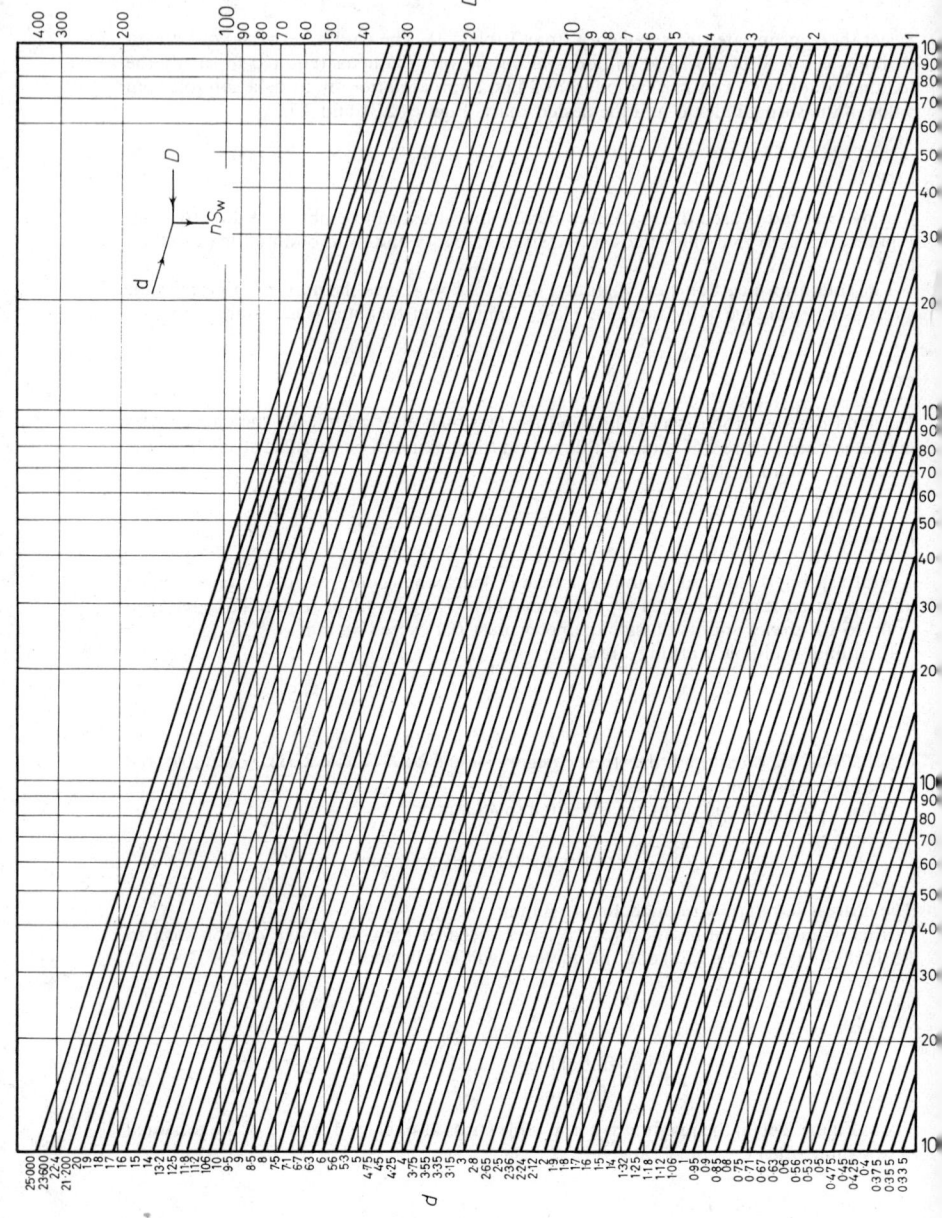

Figure 2.69. Rate and number of active coils •

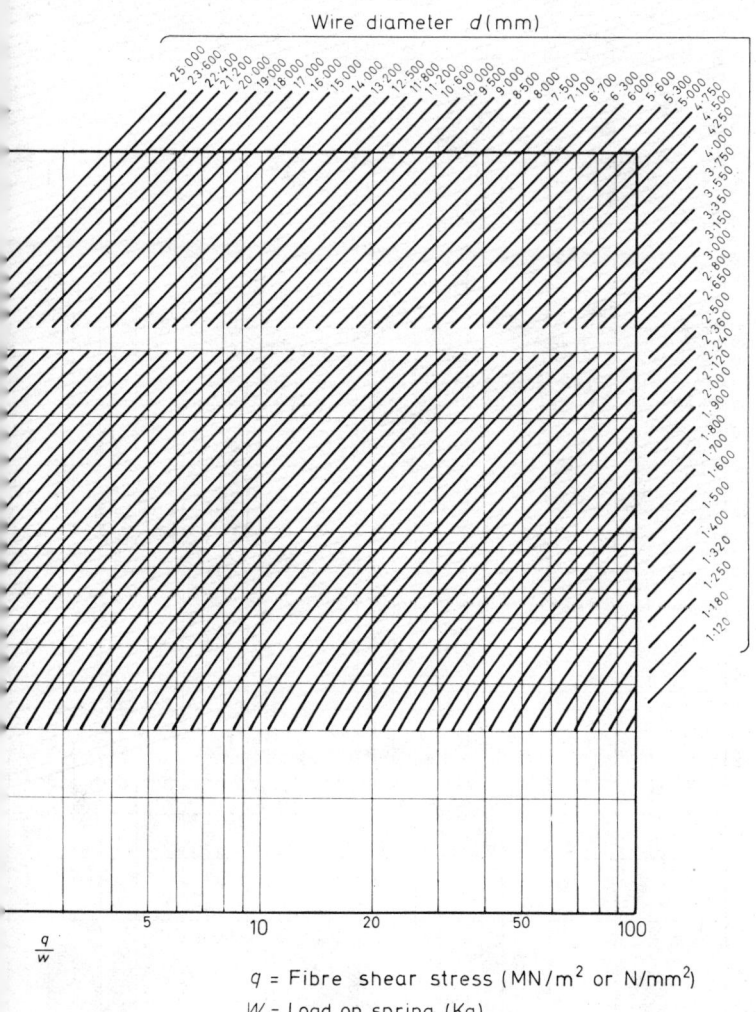

Wire diameter d (mm)

$\frac{q}{w}$

q = Fibre shear stress (MN/m^2 or N/mm^2)

W = Load on spring (Kg)

tress in helical tension and compression springs

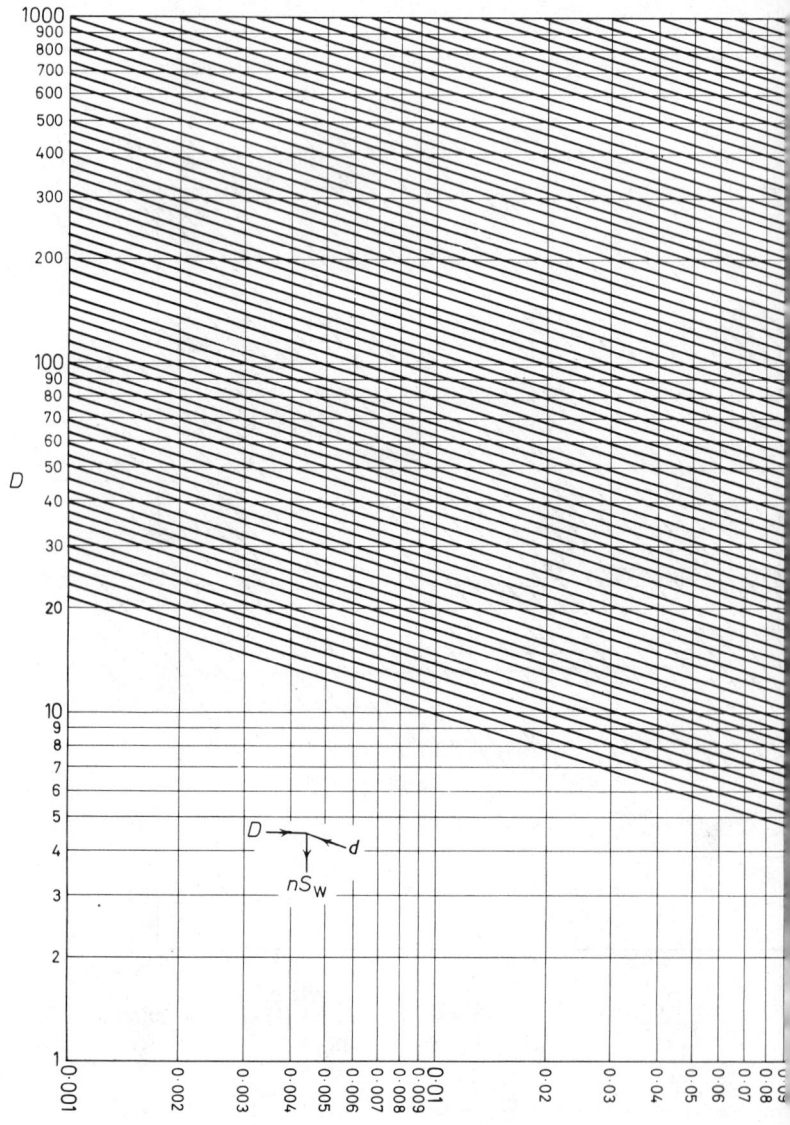

Figure 2.70. Rate an

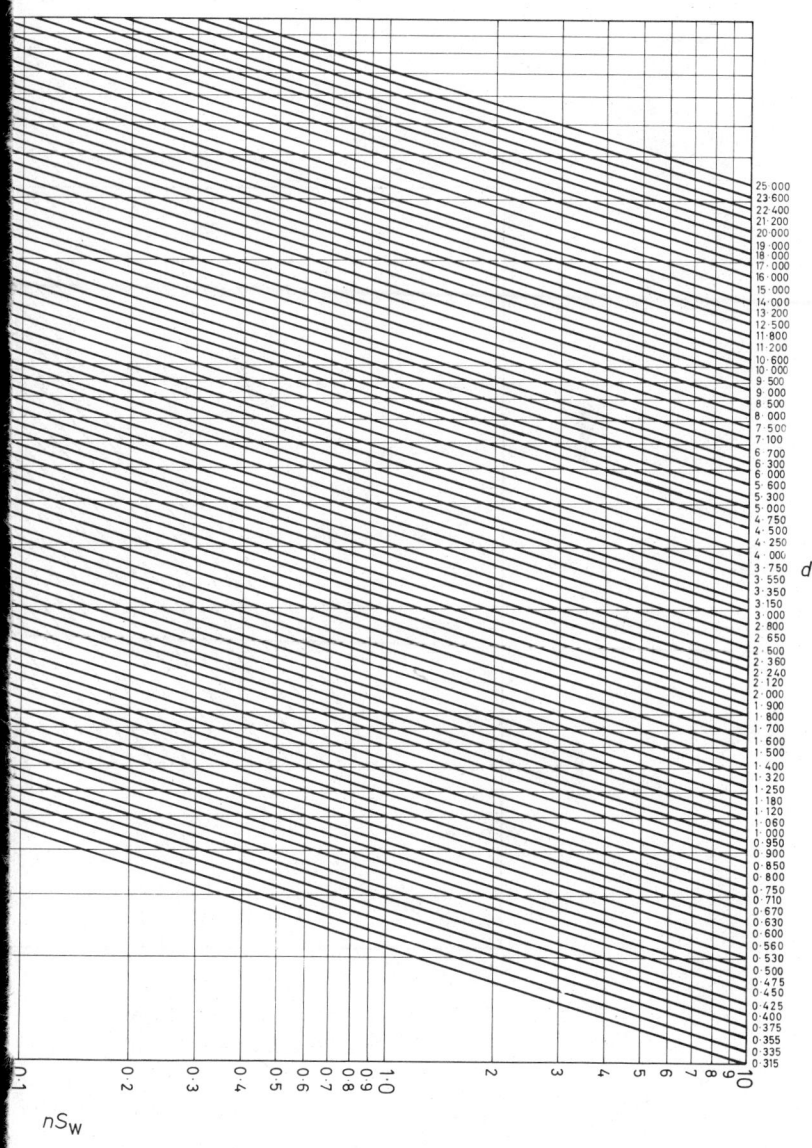

nS_w

number of active coils

d

25·000
23·600
22·400
21·200
20·000
19·000
18·000
17·000
16·000
15·000
14·000
13·200
12·500
11·800
11·200
10·600
10·000
9·500
9·000
8·500
8·000
7·500
7·100
6·700
6·300
6·000
5·600
5·300
5·000
4·750
4·500
4·250
4·000
3·750
3·550
3·350
3·150
3·000
2·800
2·650
2·500
2·360
2·240
2·120
2·000
1·900
1·800
1·700
1·600
1·500
1·400
1·320
1·250
1·180
1·120
1·060
1·000
0·950
0·900
0·850
0·800
0·750
0·710
0·670
0·630
0·600
0·560
0·530
0·500
0·475
0·450
0·425
0·400
0·375
0·355
0·335
0·315

0·1   0·2   0·3   0·4   0·5   0·6   0·7   0·8   0·9   1·0   2   3   4   5   6   7   8   9   10

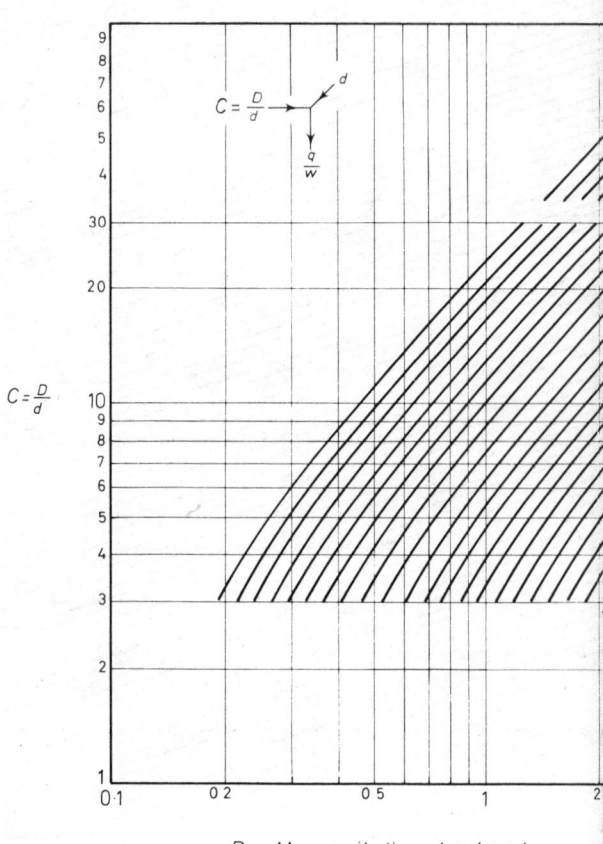

D = Mean coil diameter (mm)

d = Wire diameter (mm)

Figure 2.71. Corrected shear s

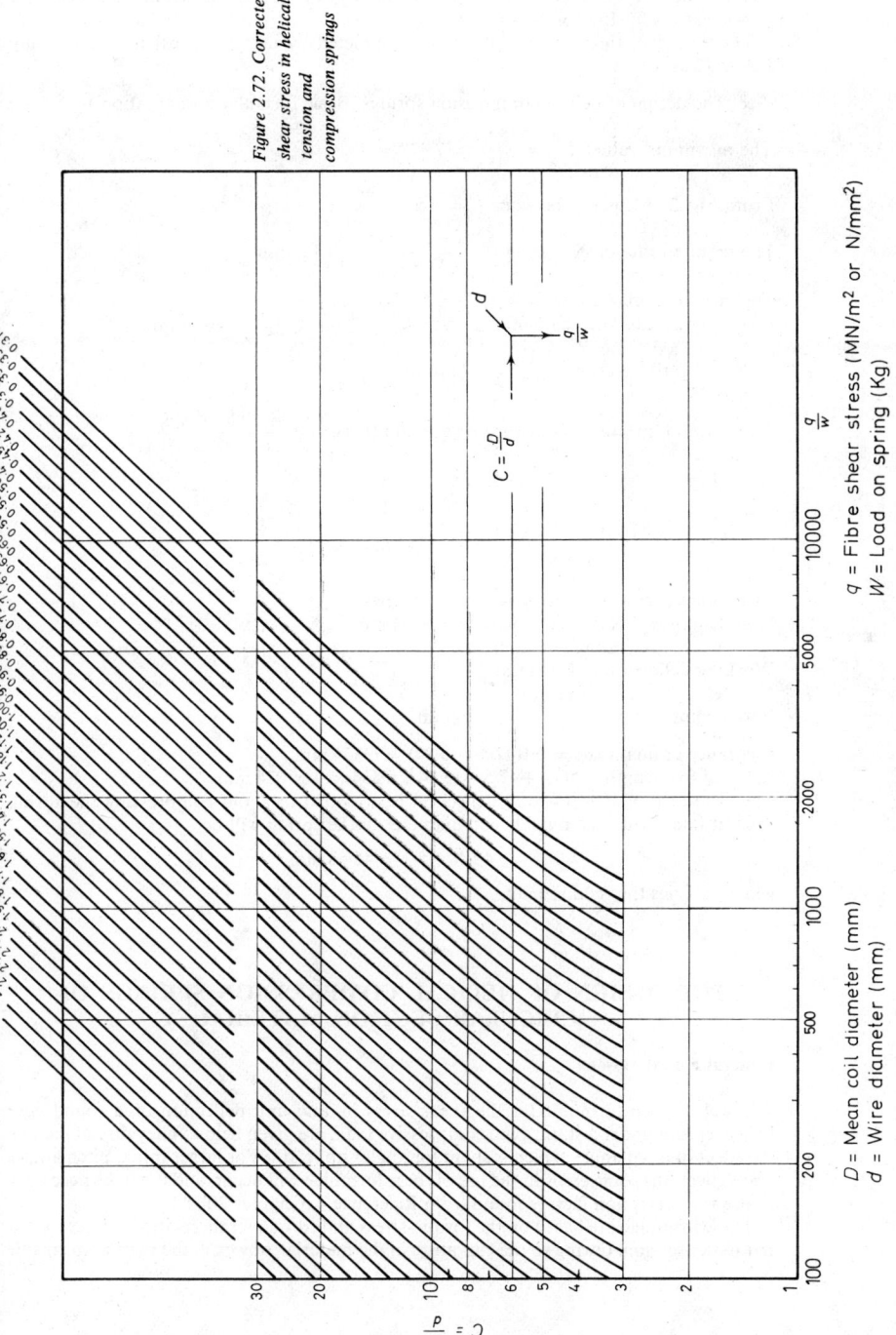

Figure 2.72. Corrected shear stress in helical tension and compression springs

q = Fibre shear stress (MN/m² or N/mm²)
W = Load on spring (Kg)

D = Mean coil diameter (mm)
d = Wire diameter (mm)

therefore be $0.59 \times 610 = 359.9$ MN/m^2 = 360 MN/m^2. This is the maximum permissible *working* stress for the condition.

The design of this spring can now be completed by using the design charts Figure 2.69–2.72 as follows:

(See 'The design of helical compression springs. Basic formulae and notation')

The maximum value of $\dfrac{q}{w} = \dfrac{359.9}{45} = 7.998 = 8$

From Fig. 2.71 this will be when $C = 7.6$

The required rate of the spring $= \dfrac{45 - 11.25}{47.5} = 0.71$ kg/mm

The mean coil diameter D will be $7.6 \times 5.3 = 40.28$ mm
From Figure 2.69 the value of nS_w when $d = 5.3$ mm and $D = 40.28$ mm is 12.

$\therefore$ No. of active coils $n = \dfrac{12}{0.71} = 17$ approx.

A check on q can be obtained from equation (1) i.e. $q = \dfrac{8 \times 9.81 \; WDK}{\pi d^3}$

From Figure 2.68 when $C = 7.6 \quad K = 1.194$

and $q = \dfrac{8 \times 9.81 \times 45 \times 40.28 \times 1.194}{\pi \times 5.3^3} = 363$ MN/m^2

The dimensions of the spring will be as follows:
Solid length $= 5.3(17 + 1) = 5.3 \times 18 = 95.4$ mm
$\qquad\qquad\qquad$ (including end coils)
Working deflection $= 47.5$ mm

Initial compression $= \dfrac{11.25}{0.71} = 15.84$ mm

Clearance at final load $= 0.1(47.5 + 15.84) = 6.334$ mm
$\therefore$ Total free length $= 95.4 + 47.5 + 15.84 + 6.334 = 165.074$ mm
The clearance of 10% of total deflection is provided to avoid compressing the spring solid at final load. The outside coil diameter of the spring will be

$$40.28 + 5.3 = 45.58 \text{ mm}$$

which satisfies the requirement.

THE DESIGN OF HELICAL COMPRESSION SPRINGS OF CIRCULAR SECTION MATERIAL

General considerations

The well known formulae for the shear stress in, and unit deflection of, a round wire helical spring under a static axial compressive load, are given below. They do not include the effect of pitch angle, which is slight for angles up to 10°. For larger initial pitch angles and deflections some error in the use of these formulae will occur, and it will be necessary to use the correction factors found from Reference 4 (Chapter 20).

These formulae are based on the torque-stress and torque-strain relations for a straight bar under torsion. Owing to the curvature of the material, however, the twist at the inside of the coil takes place over a shorter length of material than at the outside of the coil, and as a result, the shear stress at the inside of the coil is greater than that of a corresponding straight bar. The shear stress equation must therefore be corrected by multiplying by a factor K (greater than unity) which depends entirely on the spring index $C = D/d$. The correction factor most widely adopted is that derived by A. M. Wahl[4]. The curvature also affects the deflection but its effect is much smaller and may safely be neglected.

Basic formulae and notation

The axial load W in kilogrammes (kg) should be considered as a force P (kilogrammes–force, kgf) and, to maintain the consistency of SI units, should be expressed in newtons (N).

Thus
$$1 \text{ N} = 1 \text{ kg} \times 1 \text{ m/s}^2$$
$$1 \text{ kgf} = 1 \text{ kg} \times 9 \cdot 81 \text{ m/s}^2 \text{ (approx.)}$$
$$\therefore P = 9 \cdot 81 \, W \text{ newtons}$$

The maximum corrected shear stress, at inside of coil, is given by

$$q = \frac{8 \, PDK}{\pi d^3} = \frac{8 \times 9 \cdot 81 \, WDK}{\pi d^3} = \frac{8 \times 9 \cdot 81 \, WCK}{\pi d^2} \tag{1}$$

The spring rate, or load per unit deflection, is

$$S_W = \frac{W}{\delta} \text{ kg/mm} \tag{2}$$

and the force rate or force per unit deflection is

$$S_p = \frac{P}{\delta} \text{ N/mm} \tag{3}$$

i.e.

$$S_p = \frac{9 \cdot 81 \, W}{\delta} = \frac{Gd^4}{8nD^3} = \frac{Gd}{8nC^3} \tag{4}$$

where

W	kg	= axial load applied to spring.
P	N	= equivalent force applied to spring, i.e. $P = 9 \cdot 81 \, W$.
q	MN/m² or N/mm²	= maximum corrected fibre shear stress.
S_W	kg/mm	= spring rate, or load per unit deflection, i.e. $S_W = W/\delta$.
S_p	N/mm	= force rate, or force per unit deflection, i.e. $S_p = 9 \cdot 81 \, W/\delta$.
δ	mm	= axial deflection of spring.
D	mm	= mean coil diameter of spring.
d	mm	= diameter of wire or rod.
G	MN/m² or N/mm²	= modulus of rigidity (taken as 79323·6 MN/m² for carbon steel).
n		= number of active coils.
N		= total number of coils.
C		= spring index, i.e. $C = D/d$.
K		= Wahl curvature correction factor,

$$K = \frac{4C - 1}{4C - 4} + \frac{0 \cdot 615}{C}$$

Values of K for C values within the probable range can be obtained from Figure 2.68.

For springs with ends closed and ground the number of active coils should be taken as the total number of coils minus $1\frac{1}{2}$, i.e. $n = N - 1\frac{1}{2}$.

The solid length of the spring is then given by $d(n + 1)$.

Use of design charts

Equations (1) and (4) can be rewritten as follows:

$$\frac{q}{w} = \text{shear stress per unit load} = \frac{8 \times 9\cdot81\ DK}{\pi d^3}$$

$$= \frac{8 \times 9.81\ CK}{\pi d^2} \tag{5}$$

or

$$\frac{q}{P} = \text{shear stress per unit force} = \frac{8\ DK}{\pi d^3} = \frac{8\ CK}{\pi d^2} \tag{6}$$

and

$$nS_W = \text{load rate} \times \text{number of active coils}$$

$$= \frac{Gd^4}{9\cdot81 \times 8D^3} = \frac{Gd}{9\cdot81 \times 8C^3} \tag{7}$$

or

$$nS_p = \text{force rate} \times \text{number of active coils}$$

$$= \frac{Gd^4}{8D^3} = \frac{Gd}{8C^3} \tag{8}$$

To facilitate calculations, the design charts reproduced in Figures 2.69 to 2.72 have been prepared from the above equations taking $G = 79\ 323\cdot6$ MN/m². Figures 2.69 and 2.70 are provided for the purpose of obtaining either the rate or the number of active coils, whichever is required. Similarly, Figures 2.71 and 2.72 are used to determine either the corrected stress or the corresponding load.

For materials other than carbon steel the value of S_W obtained from Figures 2.69 and 2.70 must be multiplied by G'/G, where G' is the modulus of rigidity of the chosen material. Thus, for phosphor bronze ($G = 43\ 115$ MN/m²) the actual value of S_W will be

$$\frac{43\ 115}{79\ 323\cdot6}$$

or 0·544 times that obtained from the design charts.

For wire or rod sizes outside the range covered by the design charts, equations (1) and (4) and Figure 2.68 must be used.

EXAMPLE (TO ILLUSTRATE USE OF DESIGN CHARTS)

A carbon steel spring, with ends closed and ground square, has the following dimensions:

Free length	90 mm
Wire diameter	5 mm
Outside coil diameter	38 mm

Determine a suitable rate, and the corrected stress under a working load of 45 kg.
Mean coil diameter $(D) = 38 - 5 = 33$ mm.
From Figure 2.69, with $D = 33$ mm and $d = 5$ mm, $nS_W = 17\cdot6$
i.e. using 10 active coils ($n = 10$), Rate $(S_W) = 17\cdot6/10 = 1\cdot76$ kg/mm.
Solid length $= d(n+1) = 5 \times 11 = 55$ mm
Deflection when solid $= 90 - 55 = 35$ mm

And corresponding load $= 35 \times 1.76 = 61.6$ kg

$$C = \frac{D}{d} = \frac{33}{5} = 6.6$$

From Figure 2.71 with $D/d = 6.6$ and $d = 5$ mm, $q/W = 8.08$.
Hence, corrected solid stress $= 8.08 \times 61.6 = 497.7$ MN/m²
And corrected stress at working load $= 45/61.6 \times 497.7 = 363.6$ MN/m²
    As a check, and to confirm the accuracy of the design charts, equations (1) and (7)
and Figure 2.68 may be used.
    From equation (7).

$$\text{Rate } S_W = \frac{Gd^4}{9.81 \times 8nD^3} = \frac{79\,323.6 \times 5^4}{9.81 \times 8 \times 10 \times 33^3} = 1.76 \text{ kg/mm}$$

And from Figure 2.68, with $D/d = 6.6$, $K = 1.228$.
Hence, corrected stress at working load of 45 kg, from equation (1)

$$q = \frac{8 \times 9.81 \times 45 \times 33 \times 1.228}{\pi \times 5^3}$$
$$= 364 \text{ MN/m}^2 \text{ (or N/mm}^2\text{)}$$

THE DESIGN OF HELICAL COMPRESSION SPRINGS OF SQUARE AND RECTANGULAR SECTION MATERIAL

General considerations

The main advantage offered by helical compression springs of rectangular section
material is their capacity to absorb relatively large amounts of energy in a limited space.
They are, however, more difficult to produce than round wire springs and the quality

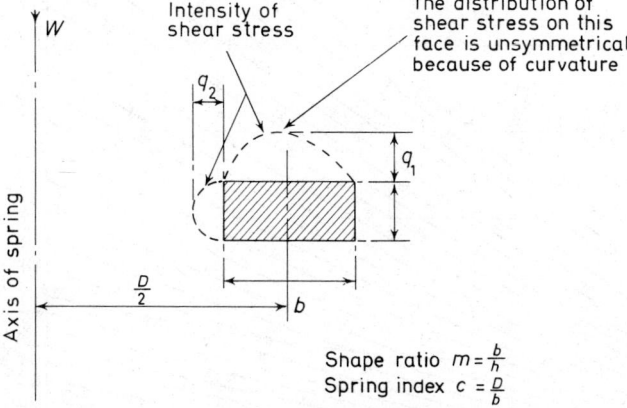

Figure 2.73. Section of one coil of rectangular wire spring

of material available in this type of section is generally below the standard associated with
circular section material.
    The formulae given will enable the maximum stress and rate to be evaluated for a
given spring in order to replace it with an equivalent spring (or nest of springs) of circular
section material. If the conditions allow, round material should always be used in preference to square or rectangular section material.
    Figure 2.73 illustrates the geometry and stress distribution in a single coil of a rect-
angular section spring. The maximum shear stress is either q_1 or q_2 depending on the

spring index and the shape ratio. Which will be the greater in a given spring is indicated in Figure 2.74. Because of the non-uniformity of stress distribution the rectangular section spring is not particularly suitable for fatigue or repeated loading conditions.

Basic formulae and notation

The maximum corrected shear stress is given by

$$q_1 \text{ (or } q_2) = (C+1)(m+1)\lambda_1 \frac{P}{bh} = \lambda \frac{P}{bh} = \lambda \frac{9 \cdot 81 W}{bh} \tag{9}$$

The spring rate, or load per unit deflection, is

$$S_W = \frac{W}{\delta} \text{ kg/mm} \tag{10}$$

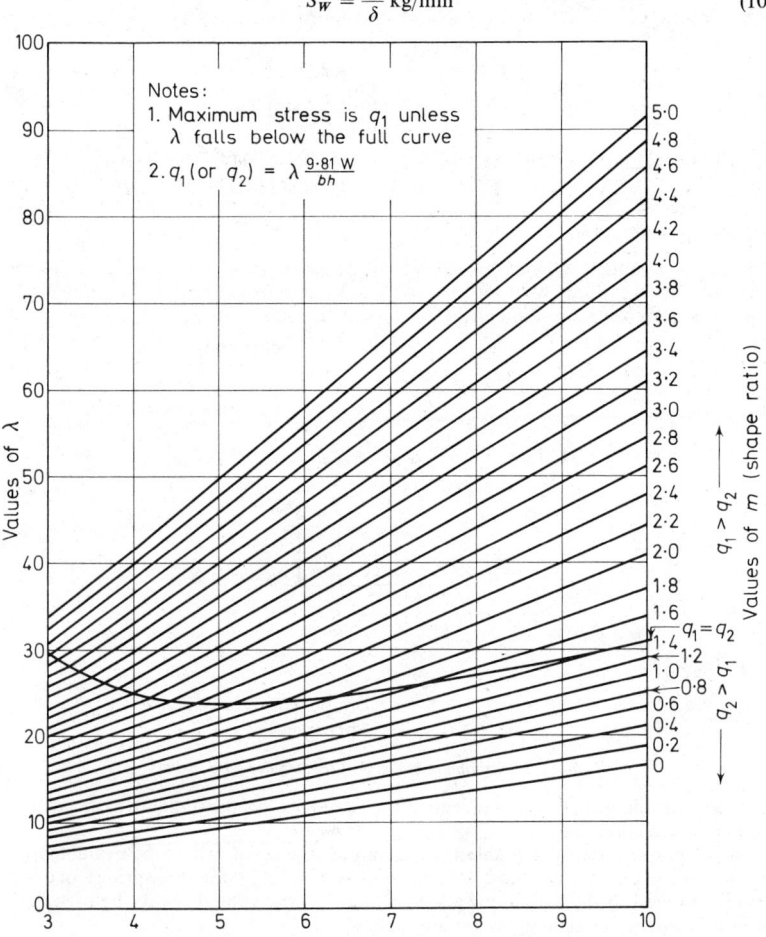

Notes:
1. Maximum stress is q_1 unless λ falls below the full curve

2. q_1 (or q_2) = $\lambda \frac{9 \cdot 81 \, W}{bh}$

Figure 2.74. Design chart for stress — rectangular section springs

And the force rate, or force per unit deflection, is

$$S_P = \frac{P}{\delta} \, \text{N/mm} \tag{11}$$

$$\text{i.e. } S_P = \frac{9 \cdot 81 W}{\delta} = \frac{K_1 G b^2 h^2}{n D^3} = \frac{K_1 G h}{m n C^3} \tag{12}$$

$$\text{or } S_W = \frac{W}{\delta} = \frac{K_1 G b^2 h^2}{9 \cdot 81 n D^3} = \frac{K_1 G h}{9 \cdot 81 m n C^3} \tag{13}$$

where

W	kg	= axial load applied to spring.
P	N	= equivalent force applied to spring i.e. $P = 9 \cdot 81 \, W$.
q_1 (or q_2)	MN/m² or N/mm²	= maximum corrected fibre shear stress.
S_W	kg/mm	= spring rate, or load per unit deflection i.e. $S_W = W/\delta$.
S_P	N/mm	= force rate or force per unit deflection i.e. $S_P = 9 \cdot 81 \, W/\delta$.
δ	mm	= axial deflection of spring.
D	mm	= mean coil diameter of spring.
b	mm	= radial width of rectangular section.
h	mm	= axial thickness of rectangular section.
G	MN/m² or N/mm²	= modulus of rigidity (taken as 79 323·6 MN/m² for carbon steel).
C		= spring index, i.e. $C = D/b$ for rectangular section.
n		= number of active coils.
N		= total number of coils.
λ_1		= function of C and m in the stress formula (Table 2.35).
λ		= stress factor, i.e. $\lambda = \lambda_1(C+1)(m+1)$ (Table 2.36).
m		= shape ratio of cross-section i.e. $m = b/h$.
K_1		= curvature correction factor as a function of C and m in rate formula (Table 2.34).

Values of stress factor λ can be obtained from Figure 2.74 and values of rate factor K_1 can be obtained from Figure 2.75.

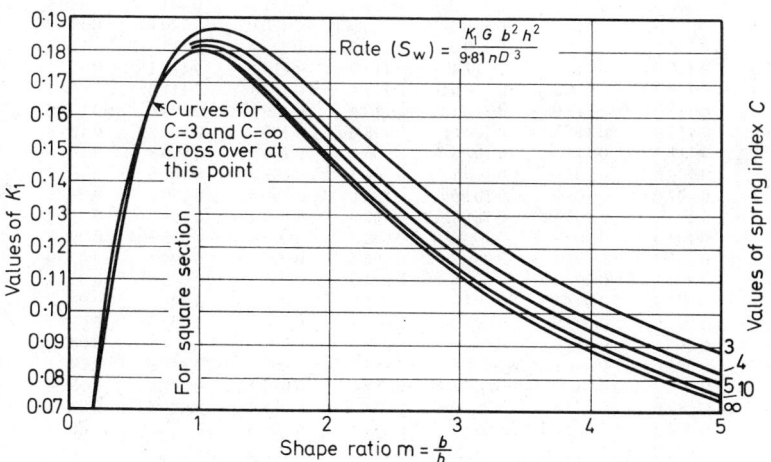

Figure 2.75. Design chart for rate—rectangular section springs

C	3·0	3·2	3·4	3·6	3·8	4·0	4.2
m							
0·0	0·000	0·000	0·000	0·000	0·000	0·000	0·000
0·1	0·038	0·038	0·038	0·038	0·038	0·039	0·039
0·2	0·071	0·071	0·072	0·072	0·072	0·072	0·073
0·3	0·100	0·100	0·100 9	0·100 9	0·100 9	0·101 7	0·101 7
0·4	0·122 6	0·124 3	0·125 2	0·125 2	0·125 2	0·125 2	0·125 2
0·5	0·143 5	0·143 5	0·144 3	0·144 3	0·144 3	0·144 3	0·144 3
0·6	0·159 1	0·159 1	0·159 1	0·159 1	0·159 1	0·159 1	0·159 1
0·7	0·169 6	0·169 6	0·169 6	0·169 6	0·169 6	0·169 6	0·169 6
0·8	0·177 4	0·177 4	0·176 5	0·176 5	0·176 5	0·176 5	0·176 5
0·9	0·182 6	0·181 7	0·181 7	0·180 9	0·180 9	0·180 9	0·180 9
1·0	0·185 2	0·184 3	0·184 3	0·183 5	0·182 6	0·182 6	0·182 6
1·1	0·186 1	0·185 2	0·184 3	0·184 3	0·183 5	0·182 6	0·182 6
1·2	0·186 1	0·185 2	0·184 3	0·183 5	0·182 6	0·181 7	0·181 7
1·3	0·185 2	0·183 5	0·182 6	0·181 7	0·180 9	0·180 0	0·179 1
1·4	0·182 6	0·180 9	0·180 0	0·179 1	0·177 4	0·177 4	0·176 5
1·5	0·180 0	0·178 3	0·176 5	0·175 7	0·174 8	0·173 9	0·173 0
1·6	0·176 5	0·174 8	0·173 0	0·172 2	0·171 3	0·170 4	0·169 6
1·7	0·173 0	0·171 3	0·169 6	0·168 7	0·167 8	0·167 0	0·166 1
1·8	0·169 6	0·167 8	0·166 1	0·165 2	0·163 5	0·163 5	0·161 7
1·9	0·166 1	0·164 3	0·162 6	0·160 9	0·160 0	0·159 1	0·158 3
2·0	0·162 6	0·160 0	0·158 3	0·157 4	0·155 7	0·154 8	0·153 9
2·1	0·158 3	0·156 5	0·154 8	0·153 0	0·152 2	0·151 3	0·150 4
2·2	0·154 8	0·153 0	0·151 3	0·149 6	0·148 7	0·147 8	0·146 1
2·3	0·151 3	0·149 6	0·147 8	0·146 1	0·145 2	0·143 5	0·142 6
2·4	0·147 8	0·146 1	0·144 3	0·143 5	0·141 7	0·140 0	0·139 1
2·5	0·144 3	0·142 6	0·140 9	0·139 1	0·138 3	0·136 5	0·135 7
2·6	0·140 9	0·139 1	0·137 4	0·135 7	0·134 8	0·133 9	0·133 0
2·7	0·138 3	0·135 7	0·133 9	0·133 0	0·131 3	0·130 4	0·129 6
2·8	0·134 8	0·133 0	0·131 3	0·129 6	0·128 7	0·127 0	0·126 1
2·9	0·132 2	0·129 6	0·127 8	0·127 0	0·125 2	0·124 3	0·123 5
3·0	0·128 7	0·127 0	0·125 2	0·123 5	0·122 6	0·121 7	0·120 9
3·1	0·126 1	0·124 3	0·122 6	0·120 9	0·120 0	0·120 0	0·117 4
3·2	0·123 5	0·121 7	0·120 0	0·118 3	0·117 4	0·115 7	0·114 8
3·3	0·120 9	0·119 1	0·117 4	0·115 7	0·114 8	0·113 1	0·112 1
3·4	0·118 3	0·116 5	0·114 8	0·113 1	0·112 1	0·111 3	0·110 4
3·5	0·115 7	0·113 9	0·112 1	0·110 4	0·109 6	0·108 6	0·107 8
3·6	0·113 9	0·111 3	0·109 6	0·108 6	0·107 8	0·107 8	0·105 2
3·7	0·111 3	0·109 6	0·107 8	0·106 1	0·106 1	0·104 3	0·103 5
3·8	0·109 6	0·107 0	0·106 1	0·104 3	0·103 5	0·102 6	0·101 7
3.9	0·107 0	0·105 2	0·103 5	0·102 6	0·100 9	0·100 0	0·099 1
4·0	0·105 2	0·103 5	0·101 7	0·100 0	0·099 1	0·098 3	0·097 4
4·1	0·103 5	0·100 9	0·100 0	0·098 3	0·097 4	0·096 5	0·095 7
4·2	0·101 7	0·099 1	0·098 3	0·096 5	0·095 7	0·094 8	0·093 9
4·3	0·099 1	0·097 4	0·096 5	0·094 8	0·093 9	0·093 0	0·092 2
4·4	0·097 4	0·095 7	0·094 8	0·093 0	0·092 2	0·091 3	0·090 4
4·5	0·096 5	0·093 9	0·093 0	0·091 3	0·090 4	0·089 6	0·088 7
4·6	0·094 8	0·093 0	0·091 3	0·090 4	0·088 7	0·087 8	0·087 0
4·7	0·093 0	0·091 3	0·089 6	0·088 7	0·087 8	0·087 0	0·086 1
4·8	0·091 3	0·089 6	0·087 8	0·087 0	**0·086 1**	0·085 2	0·084 4
4·9	0·089 6	0·087 8	0·087 0	0·085 2	0·084 4	0·083 5	0·082 6
5·0	0·088 7	0·087 0	0·085 2	0·084 4	0·083 5	0·082 6	0·081 7

C	4·4	4·6	4·8	5·0	6·0	10·0	∞
m							
0·0	0·000	0·000	0·000	0·000	0·000	0·000	0·000
0·1	0·039	0·039	0·039	0·039	0·039	0·040	0·040
0·2	0·073	0·073	0·073	0·073	0·073 9	0·073 9	0·073 9
0·3	0·101 7	0·101 7	0·101 7	0·102 6	0·102 6	0·102 6	0·103 5
0·4	0·125 2	0·126 1	0·126 1	0·126 1	0·126 1	0·127	0·127
0·5	0·144 3	0·145 2	0·145 2	0·145 2	0·145 2	0·145 2	0·146 1
0·6	0·159 1	0·159 1	0·159 1	0·159 1	0·159 1	0·159 1	0·160
0·7	0·169 6	0·169 6	0·169 6	0·169 6	0·169 6	0·168 7	0·168 7
0·8	0·175 7	0·175 7	0·176 5	0·176 5	0·175 7	0·174 8	0·174 8
0·9	0·180 0	0·180 9	0·180 0	0·180 0	0·179 1	0·178 3	0·178 3
1·0	0·181 7	0·181 7	0·181 7	0·181 7	0·180 9	0·180 0	0·179 1
1·1	0·181 7	0·181 7	0·181 7	0·180 9	0·180 0	0·179 1	0·178 3
1·2	0·180 9	0·180 9	0·180 0	0·180 0	0·179 1	0·177 4	0·176 5
1.3	0·179 1	0·178 3	0·178 3	0·177 4	0·176 5	0·174 8	0·173 9
1·4	0·175 7	0·175 7	0·174 8	0·174 8	0·173 0	0·171 3	0·170 4
1.5	0·172 2	0·172 2	0·171 3	0·171 3	0·169 6	0·167 8	0·166 1
1·6	0·168 7	0·168 7	0·167 8	0·167 8	0·166 1	0·163 5	0·162 6
1·7	0·165 2	0·164 3	0·164 3	0·163 5	0·161 7	0·159 1	0·158 3
1·8	0·160 9	0·160 9	0·160 0	0·160 0	0·158 3	0·155 7	0·153 9
1·9	0·157 4	0·156 5	0·155 7	0·155 7	0·153 9	0·151 3	0·149 6
2·0	0·153 0	0·153 0	0·152 2	0·151 3	0·149 6	0·147 0	0·146 1
2·1	0·149 6	0·148 7	0·147 8	0·147 8	0·146 1	0·143 5	0·141 7
2·2	0·146 1	0·145 2	0·144 3	0·144 3	0·141 7	0·139 1	0·137 4
2·3	0·141 7	0·141 7	0·140 9	0·140 0	0·138 3	0·135 7	0·133 9
2·4	0·138 3	0·138 3	0·137 4	0·136 5	0·134 8	0·132 2	0·130 4
2.5	0·134 8	0·134 8	0·133 9	0·133 0	0·131 3	0·128 7	0·127 0
2·6	0·132 2	0·131 3	0·130 4	0·129 6	0·127 8	0·125 2	0·123 5
2·7	0·128 7	0·127 8	0·127 0	0·127 0	0·125 2	0·121 7	0·120 9
2·8	0·125 2	0·125 2	0·124 3	0·123 5	0·121 7	0·119 1	0·117 4
2·9	0·122 6	0·121 7	0·120 9	0·120 9	0·119 1	0·115 7	0·114 8
3·0	0·120 0	0·119 1	0·118 3	0·118 3	0·115 7	0·113 1	0·111 3
3·1	0·117 4	0·116 5	0·115 7	0·114 8	0·113 1	0·110 4	0·108 6
3·2	0·113 9	0·113 9	0·113 1	0·112 1	0·110 4	0·107 8	0·106 1
3·3	0·112 1	0·111 3	0·110 4	0·109 6	0·107 8	0·105 2	0·103 5
3·4	0·109 6	0·108 6	0·107 8	0·107 8	0·106 1	0·102 6	0·101 7
3·5	0·107 0	0·106 1	0·106 1	0·105 2	0·103 5	0·100 0	0·099 1
3·6	0·105 2	0·104 3	0·103 5	0·103 5	0·100 9	0·098 3	0·097 4
3·7	0·102 6	0·101 7	0·101 7	0·100 9	0·099 1	0·096 5	0·094 8
3·8	0·100 9	0·100 0	0·099 1	0·099 1	0·097 4	0·094 8	0·093 0
3·9	0·098 3	0·098 3	0·097 4	0·097 4	0·095 7	0·093 0	0·091 3
4·0	0·096 5	0·096 5	0·094 8	0·094 8	0·093 9	0·090 4	0·089 6
4·1	0·094 8	0·093 9	0·093 9	0·093 0	0·091 3	0·088 7	0·087 8
4·2	0·093 0	0·092 2	0·092 2	0·091 3	0·089 6	0·087 8	0·086 1
4·3	0·091 3	0·091 3	0·090 4	0·089 6	0·087 8	0·086 1	0·084 4
4·4	0·089 6	0·089 6	0·088 7	0·088 7	0·087 0	0·084 4	0·082 6
4·5	0·088 7	0·087 8	0·087 0	0·087 0	0·085 2	0·082 6	0·081 7
4·6	0·087 0	0·086 1	0·085 2	0·085 2	0·083 5	0·080 9	0·080 0
4·7	0·085 2	0·084 4	0·084 4	0·083 5	0·081 7	0·080 0	0·078 3
4·8	0·083 5	0·083 5	0·082 6	0·082 6	0·080 9	0·078 3	0·077 4
4·9	0·082 6	0·081 7	0·081 7	0·080 9	0·079 1	0·077 4	0·075 7
5·0	0·080 9	0·080 0	0·080 0	0·079 1	0·077 4	0·075 7	0·073 9

C	3·0	3·2	3·4	3·6	3·8	4·0	4·2
m							
0·0	1·61	1·60	1·59	1·58	1·57	1·56	1·56
0·1	1·53	1·52	1·51	1·51	1·50	1·50	1·50
0·2	1·50	1·49	1·48	1·48	1·47	1·46	1·46
0·3	1·48	1·47	1·46	1·46	1·45	1·44	1·43
0·4	1·47	1·46	1·45	1·44	1·43	1·42	1·42
0·5	1·45	1·44	1·43	1·42	1·42	1·41	1·40
0·6	1·42	1·41	1·40	1·39	1·38	1·38	1·38
0·7	1·40	1·39	1·38	1·37	1·36	1·35	1·35
0·8	1·38	1·37	1·36	1·35	1·34	1·33	1·33
0·9	1·36	1·35	1·34	1·33	1·32	1·31	1·31
1·0	1·34	1·33	1·32	1·31	1·30	1·29	1·29
1·1	1·33	1·32	1·31	1·30	1·29	1·28	1·28
1·2	1·32	1·31	1·30	1·29	1·28	1·27	1·27
1·3	1·31	1·30	1·29	1·28	1·27	1·26	1·26
1·4	1·31	1·30	1·28	1·27	1·26	1·25	1·25
1·5	1·30	1·29	1·27	1·26	1·25	1·25	1·24
1·6	1·30	1·29	1·27	1·25	1·24	1·24	1·24
1·7	1·30	1·29	1·27	1·25	1·24	1·24	1·23
1·8	1·30	1·29	1·27	1·25	1·24	1·24	1·23
1·9	1·29	1·28	1·27	1·25	1·24	1·24	1·23
2·0	1·29	1·28	1·27	1·25	1·24	1·24	1·22
2·1	1·29	1·28	1·27	1·25	1·24	1·24	1·22
2·2	1·29	1·28	1·27	1·25	1·24	1·24	1·22
2·3	1·30	1·28	1·27	1·25	1·24	1·24	1·22
2·4	1·30	1·28	1·27	1·25	1·24	1·24	1·22
2·5	1·30	1·29	1·27	1·25	1·24	1·23	1·23
2·6	1·30	1·29	1·27	1·26	1·25	1·23	1·23
2·7	1·31	1·29	1·27	1·26	1·25	1·24	1·23
2·8	1·31	1·29	1·27	1·26	1·25	1·24	1·23
2·9	1·31	1·29	1·28	1·27	1·25	1·24	1·24
3·0	1·31	1·30	1·28	1·27	1·26	1·25	1·24
3·1	1·31	1·30	1·28	1·27	1·26	1·25	1·25
3·2	1·31	1·30	1·29	1·27	1·27	1·26	1·26
3·3	1·31	1·30	1·29	1·28	1·27	1·27	1·27
3·4	1·31	1·31	1·29	1·28	1·28	1·28	1·28
3·5	1·33	1·31	1·29	1·28	1·28	1·29	1·29
3·6	1·33	1·31	1·30	1·29	1·29	1·29	1·29
3·7	1·34	1·32	1·30	1·29	1·29	1·30	1·30
3·8	1·34	1·32	1·31	1·30	1·30	1·31	1·31
3·9	1·34	1·32	1·31	1·31	1·31	1·32	1·32
4·0	1·34	1·32	1·32	1·32	1·32	1·32	1·32
4·1	1·35	1·33	1·33	1·33	1·33	1·33	1·33
4·2	1·35	1·34	1·34	1·33	1·33	1·33	1·34
4·3	1·35	1·35	1·35	1·34	1·34	1·34	1·34
4·4	1·36	1·35	1·35	1·35	1·35	1·35	1·35
4·5	1·37	1·36	1·36	1·36	1·36	1·36	1·36
4·6	1·37	1·36	1·36	1·37	1·36	1·36	1·36
4·7	1·38	1·37	1·37	1·37	1·37	1·37	1·37
4·8	1·39	1·38	1·38	1·38	1·37	1·37	1·37
4·9	1·39	1·38	1·38	1·38	1·38	1·38	1·38
5·0	1·40	1·39	1·39	1·39	1·38	1·38	1·38

C	4·4	4·6	4·8	5·0	6·0	10·0	∞
m							
0·0	1·55	1·55	1·54	1·54	1·53	1·51	1·50
0·1	1·49	1·49	1·48	1·48	1·47	1·45	**1·45**
0·2	1·45	1·45	1·45	1·45	1·44	1·42	1·43
0·3	1·43	1·43	1·43	1·43	1·41	1·40	1·41
0·4	1·41	1·41	1·41	1·41	1·39	1·38	1·39
0·5	1·39	1·39	1·39	1·39	1·37	1·36	1·35
0·6	1·37	1·37	1·36	1·36	1·34	1·33	1·32
0·7	1·34	1·34	1·33	1·33	1·31	1·30	1·29
0·8	1·32	1·32	1·31	1·31	1·29	1·27	1·26
0·9	1·30	1·30	1·29	1·29	1·27	1·25	1·23
1·0	1·28	1·28	1·27	1·27	1·25	1·23	1·20
1·1	1·27	1·26	1·26	1·25	1·23	1·21	1·22
1·2	1·26	1·25	1·25	1·24	1·22	1·20	1·25
1·3	1·25	1·24	1·24	1·23	1·21	1·19	1·27
1·4	1·24	1·23	1·23	1·22	1·20	1·18	1·28
1·5	1·23	1·22	1·22	1·21	1·19	1·17	1·30
1·6	1·23	1·22	1·22	1·21	1·19	1·18	1·32
1·7	1·22	1·21	1·21	1·20	1·19	1·19	1·33
1·8	1·22	1·21	1·21	1·20	1·18	1·20	1·34
1·9	1·22	1·21	1·21	1·20	1·18	1·22	1·35
2·0	1·22	1·21	1·20	1·19	1·18	1·24	1·36
2·1	1·22	1·21	1·20	1·19	1·19	1·25	1·36
2·2	1·22	1·21	1·20	1·20	1·20	1·26	1·37
2·3	1·22	1·21	1·20	1·20	1·21	1·27	1·38
2·4	1·22	1·22	1·21	1·21	1·22	1·28	1·38
2·5	1·22	1·22	1·21	1·21	1·23	1·28	1·39
2·6	1·23	1·23	1·23	1·22	1·24	1·29	1·39
2·7	1·23	1·23	1·23	1·23	1·25	1·29	1·39
2·8	1·24	1·24	1·24	1·24	1·26	1·30	1·40
2·9	1·24	1·24	1·25	1·25	1·27	1·30	1·40
3·0	1·25	1·25	1·26	1·26	1·27	1·31	1·40
3·1	1·25	1·26	1·27	1·27	1·28	1·31	1·40
3·2	1·26	1·27	1·27	1·28	1·28	1·32	1·41
3·3	1·27	1·28	1·28	1·29	1·29	1·33	1·41
3·4	1·28	1·28	1·28	1·29	1·30	1·33	1·41
3·5	1·29	1·29	1·29	1·30	1·30	1·34	1·41
3·6	1·29	1·29	1·30	1·30	1·31	1·34	1·41
3·7	1·30	1·30	1·31	1·31	1·32	1·35	1·42
3.8	1·31	1·31	1·32	1·32	1·33	1·35	1·42
3·9	1·32	1·32	1·32	1·32	1·33	1·36	1·42
4·0	1·33	1·33	1·33	1·33	1·34	1·36	1·42
4·1	1·33	1·33	1·33	1·33	1·34	1·37	1·42
4·2	1·34	1·34	1·34	1·34	1·35	1·37	1·42
4·3	1·35	1·35	1·35	1·35	1·35	1·38	1·42
4·4	1·35	1·35	1·35	1·35	1·36	1·38	1·43
4·5	1·36	1·36	1·36	1·36	1·36	1·39	1·43
4·6	1·36	1·36	1·36	1·36	1·37	1·39	1·43
4·7	1.37	1·37	1·37	1·37	1·37	1·39	1·43
4·8	1·37	1·37	1·37	1·37	1·37	1·39	1·43
4·9	1·38	1·38	1·38	1·38	1·38	1·39	1·43
5·0	1·38	1·38	1·38	1·38	1·38	1·39	1·43

Table 2.36 VALUE$

C	3·0	3·2	3·4	3·6	3·8	4·0	4·2
m							
0·0	6·440	6·720	6·996	7·268	7·536	7·800	8·112
0·1	6·732	7·022	7·308	7·641	7·920	8·250	8·580
0·2	7·200	7·510	7·814	8·170	8·467	8·760	9·110
0·3	7·696	8·026	8·351	8·731	9·048	9·360	9·667
0·4	8·232	8·789	8·932	9·274	9·610	9·940	10·338
0·5	8·700	9·288	9·438	9·798	10·224	10·575	10·920
0·6	9·088	9·475	9·856	10·230	10·598	11·040	11·482
0·7	9·520	9·925	10·322	10·713	11·098	11·475	11·934
0·8	9·936	10·357	10·771	11·178	11·578	11·970	12·449
0·9	10·336	10·773	11·202	11·624	12·038	12·445	12·943
1·0	10·720	11·172	11·616	12·052	12·480	12·900	13·416
1·1	11·172	11·642	12·104	12·558	13·003	13·440	13·978
1·2	11·616	12·104	12·584	13·055	13·517	13·970	14·529
1·3	12·052	12·558	13·055	13·542	14·021	14·490	15·070
1·4	12·576	13·104	13·517	14·021	14·515	15·000	15·600
1·5	13·000	13·545	13·970	14·490	15·000	15·625	16·120
1·6	13·520	14·087	14·529	14·950	15·475	16·120	16·765
1·7	14·040	14·629	15·088	15·525	16·070	16·740	17·269
1·8	14·560	15·170	15·646	16·100	16·666	17·360	17·909
1·9	14·964	15·590	16·205	16·675	17·261	17·980	18·548
2·0	15·480	16·128	16·764	17·250	17·856	18·600	19·032
2·1	15·996	16·666	17·323	17·825	18·451	19·220	19·666
2·2	16·512	17·203	17·882	18·400	19·046	19·840	20·301
2·3	17·160	17·741	18·440	18·975	19·642	20·460	20·935
2·4	17·680	18·278	18·999	19·550	20·237	21·080	21·570
2·5	18·200	18·963	19·558	20·125	20·832	21·525	22·386
2·6	18·720	19·505	20·117	20·866	21·600	22·140	23·026
2·7	19·388	20·047	20·676	21·445	22·200	22·940	23·665
2·8	19·912	20·588	21·234	22·025	22·800	23·560	24·305
2·9	20·436	21·130	21·965	22·784	23·400	24·180	25·147
3·0	20·960	21·840	22·528	23·368	24·192	25·000	25·792
3·1	21·484	22·386	23·091	23·952	24·797	25·625	26·650
3·2	22·008	22·932	23·839	24·536	25·603	26·460	27·518
3·3	22·532	23·478	24·407	25·318	26·213	27·305	28·397
3·4	23·056	24·209	24·974	25·907	27·034	28·160	29·286
3·5	23·940	24·759	25·542	26·496	27·648	29·025	30·186
3·6	24·472	25·309	26·312	27·296	28·483	29·670	30·857
3·7	25·192	26·057	26·884	27·890	29·102	30·550	31·772
3·8	25·728	26·611	27·667	28·704	29·952	31·440	32·698
3·9	26·264	27·166	28·244	29·527	30·811	32·340	33·634
4·0	26·800	27·720	29·040	30·360	31·680	33·000	34·320
4·1	27·540	28·489	29·845	31·202	32·558	33·915	35·272
4·2	28·080	29·266	30·659	31·814	33·197	34·580	36·234
4·3	28·620	30·051	31·482	32·669	34·090	35·510	36·930
4·4	29·376	30·618	32·076	33·534	34·992	36·450	37·908
4·5	30·140	31·416	32·912	34·408	35·904	37·400	38·896
4·6	30·688	31·987	33·510	35·291	36·557	38·080	39·603
4·7	31·464	32·798	34·360	35·921	37·483	39·045	40·607
4·8	32·248	33·617	35·218	36·818	38·141	39·730	41·319
4·9	32·804	34·196	35·825	37·453	39·082	40·710	42·338
5·0	33·600	35·028	36·696	38·364	39·744	41·400	43·056

$$\text{stress } q_1 \text{ (or } q_2) = \frac{\lambda 9 \cdot 81 W}{bh}$$

C	4·4	4·6	4·8	5·0	6·0	10·0
m						
0·0	8·370	8·680	8·932	9·240	10·710	16·610
0·1	8·851	9·178	9·442	9·768	11·319	17·545
0·2	9·396	9·744	10·092	10·440	12·096	18·744
0·3	10·039	10·410	10·782	11·154	12·831	20·020
0·4	10·660	11·054	11·449	11·844	13·622	21·252
0·5	11·259	11·676	12·093	12·510	14·385	22·440
0·6	11·837	12·275	12·621	13·056	15·008	23·408
0·7	12·301	12·757	13·114	13·566	15·589	24·310
0·8	12·830	13·306	13·676	14·148	16·254	25·146
0·9	13·338	13·832	14·216	14·706	16·891	26·125
1·0	13·824	14·336	14·732	15·240	17·500	27·060
1·1	14·402	14·818	15·347	15·750	18·081	27·951
1·2	14·969	15·400	15·950	16·368	18·788	29·040
1·3	15·525	15·971	16·542	16·974	19·481	30·107
1·4	16·070	16·531	17·122	17·568	20·160	31·152
1·5	16·605	17·080	17·690	18·150	20·825	32·175
1·6	17·269	17·763	18·398	18·876	21·658	33·748
1·7	17·788	18·295	18·949	19·440	22·491	35·343
1·8	18·446	18·973	19·650	20·160	23·128	36·960
1·9	19·105	19·650	20·352	20·880	23·954	38·918
2·0	19·764	20·328	20·880	21·420	24·780	40·920
2·1	20·423	21·006	21·576	22·134	25·823	42·625
2·2	21·082	21·683	22·272	23·040	26·880	44·352
2·3	21·740	22·361	22·968	23·760	27·951	46·101
2·4	22·399	23·229	23·861	24·684	29·036	47·872
2·5	23.058	23·912	24·563	25·410	30·135	49·280
2·6	23·911	24·797	25·682	26·352	31·248	51·084
2·7	24·575	25·486	26·396	27·306	32·375	52·503
2·8	25·445	26·387	27·330	28·272	33·516	54·340
2·9	26·114	27·082	28·275	29·250	34·671	55·770
3·0	27·000	28·000	29·232	30·240	35·560	57·640
3·1	27·675	28·930	30·201	31·242	36·736	59·081
3·2	28·577	29·870	30·937	32·256	37·632	60·984
3·3	29·489	30·822	31·923	33·282	38·829	62·909
3·4	30·413	31·539	32·666	34·056	40·040	64·372
3·5	31·347	32·508	33·669	35·100	40·950	66·330
3·6	32·044	33·230	34·684	35·880	42·182	67·804
3·7	32·994	34·216	35·711	36·942	43·428	69·795
3·8	33·955	35·213	36·749	38·016	44·688	71·280
3·9	34·927	36·221	37·514	38·808	45·619	73·304
4·0	35·910	37·240	38·570	39·900	46·900	74·800
4·1	36·628	37·985	39·341	40·698	47·838	76·857
4·2	37·627	39·021	40·414	41·808	49·140	78·364
4·3	38·637	40·068	41·499	42·930	50·085	80·454
4·4	39·366	40·824	42·282	43·740	51·408	81·972
4·5	40·392	41·888	43·384	44·880	52·360	84·095
4·6	41·126	42·650	44·173	45·696	53·704	85·624
4·7	42·169	43·730	45·292	46·854	54·663	87·153
4·8	42·908	44·498	46·087	47·676	55·622	88·682
4·9	43·967	45·595	47·224	48·852	56·994	90·211
5·0	44·712	46·368	48·024	49·680	57·960	91·740

For springs with ends closed and ground the number of active coils should be taken as the total number of coils minus 2, i.e. $n = N - 2$.
The solid length of the spring is then given by $h(N - \frac{1}{2})$ or $h(n + 1\frac{1}{2})$.
For further information see M.O.S. data sheet DEI[2] and BS 1726: Part 1[15].

Example. A low alloy steel compression spring of rectangular section material has the following dimensions:

Outside coil diameter	= 150 mm
Inside coil diameter	= 90 mm
Axial thickness of section	= 12.5 mm
No. of active coils	= 16

Determine the rate of the spring, and the maximum stress under a load of 1600 kg.

$$b = \frac{150 - 90}{2} = 30 \text{ mm and } D = 120 \text{ mm}$$

$$m = \frac{b}{h} = \frac{30}{12 \cdot 5} = 2 \cdot 40 \text{ and } C = \frac{D}{b} = \frac{120}{30} = 4 \cdot 00$$

From Table 2.34 or Figure 2.75, $K_1 = 0 \cdot 140$ when $C = 4 \cdot 00$ and $m = 2 \cdot 40$

$$\text{Hence rate } (S_W) = \frac{K_1 G b^2 h^2}{9 \cdot 81 n D^3} = \frac{0 \cdot 140 \times 79\,323 \cdot 6 \times 30^2 \times 12 \cdot 5^2}{9 \cdot 81 \times 16 \times 120^3} = 5 \cdot 755 \text{ kg/mm}$$

From Table 2.36 or Figure 2.74, $\lambda = 21 \cdot 08$ when $C = 4 \cdot 00$ and $m = 2 \cdot 40$

$$\text{Hence, maximum stress } (q_2) = \frac{\lambda \times 9 \cdot 81 W}{bh} = \frac{21 \cdot 08 \times 9 \cdot 81 \times 1600}{30 \times 12 \cdot 5}$$

$$\therefore q_2 = 882 \cdot 2 \text{ MN/m}^2$$

THE BUCKLING OF LONG HELICAL COMPRESSION SPRINGS OF CIRCULAR, SQUARE AND RECTANGULAR SECTION

General considerations

A long helical compression spring subjected to an increasing axial compressive load is liable to buckle in the same way as a long strut, at a critical load which depends on its slenderness ratio and end fixing conditions. To avoid buckling, therefore, it is necessary to choose spring proportions such that this load is greater than the maximum working load on the spring.

Notation

L_o	mm =	Initial free height of spring (including end coils).
Δ	mm =	Deflection at which buckling occurs.
D	mm =	Mean coil diameter of spring.
b	mm =	Radial breadth of rectangular section.
h	mm =	Axial depth of rectangular section.
m	=	Shape ratio, i.e. $m = b/h$.

A, B = Numerical coefficients depending on the shape of the section.
F = Factor determined by the end fixing conditions.

Formula for critical deflection

In the design of springs it is convenient to consider the relative critical deflection for buckling rather than the critical load. The relative critical deflection, Δ/L_o, for springs not subjected to lateral loads, derived by Haringx[5], is given by

$$\frac{\Delta}{L_o} = A \left[1 \pm \sqrt{\left(1 - B \left(\frac{D}{L_o F} \right)^2 \right)} \right] \qquad (14)$$

where A and B are coefficients depending on the shape of the section, and F is a factor determined by the end fixing conditions (Figure 2.76 and Table 2.38).

Values of A and B for steel springs (Poisson's ratio = 0·3) of circular, square and rectangular section are given in Table 2.37, and values of F are given in Table 2.38 (F has

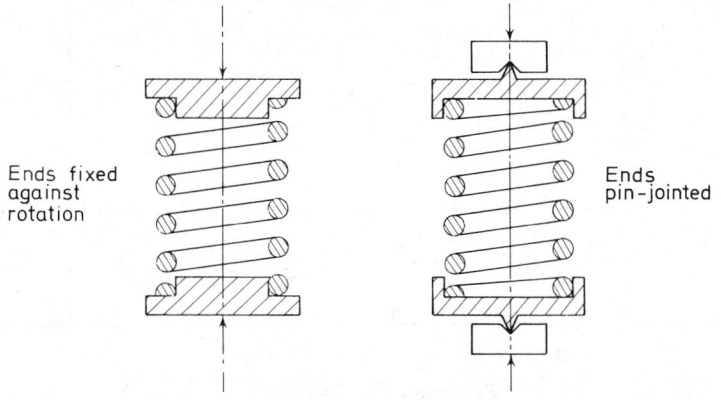

Ends fixed against rotation

Ends pin-jointed

Figure 2.76. Springs with fixed and pin-jointed ends

Table 2.37   VALUES OF A AND B

m	Circular section	Square section	Rectangular section					
		1	*1·5*	*2*	*2·5*	*3*	*4*	*5*
A	0·8125	0·741	0·626	0·576	0·551	0·536	0·521	0·514
B	6·87	8·08	8·27	8·34	8·34	8·31	8·24	8·19

the same value as in the ordinary thin strut formulae.) The variation of $D/(L_o F)$ with Δ/L_o is graphically represented in Figure 2.77.

Design procedure

Springs having a length/diameter ratio less than $2·62/F$ for circular section material or $2·84/F$ for rectangular section material can be fully compressed without buckling and it is unnecessary to proceed further if the length/diameter ratio is less than these values. In

Table 2.38

End conditions			Diagrammatic illustration	F
For rotation		Lateral		
One end	Other end			
Fixed	Fixed	Guided	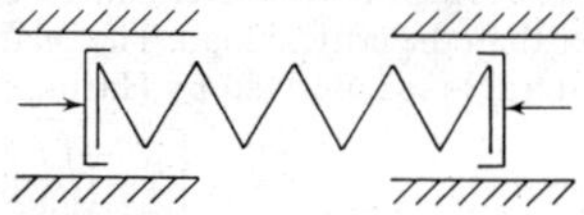	0·5
Fixed	Pin jointed	Guided	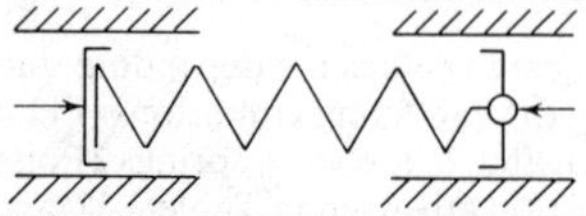	0·67
Pin jointed	Pin jointed	Guided	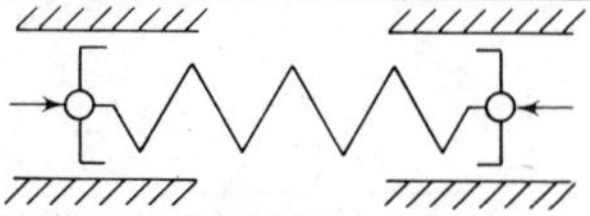	1·0
Fixed	Fixed	Unguided		1·0
Fixed	Pin jointed	Unguided		2·0

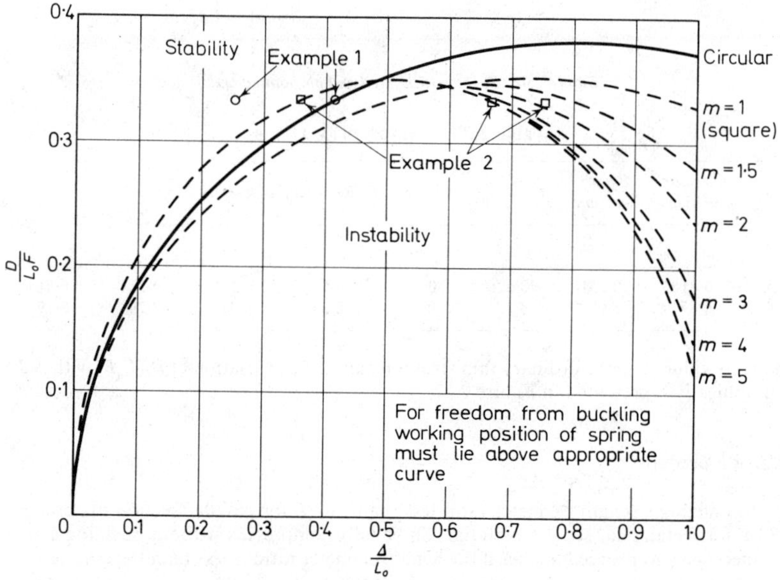

Figure 2.77. Stability chart for helical compression springs

the usual application where both ends are fixed and guided the value $F = 0.5$ applies to a spring with theoretically ideal fixation, a condition which, for various reasons, is unlikely.

The practice of assuming L_o as the overall free length, including end coils, tends to compensate however, and in this particular case the limiting values of L_o/D for circular and rectangular section material may be taken as 5 and 5.5 respectively.

For longer springs, the curves of Figure 2.77 divide the diagram into two fields, one of stability and the other of instability depending on the values of critical deflection Δ/L_o and slenderness ratio $D/(L_oF)$. The full curve applies to circular section springs and the dotted curves apply to rectangular section springs. The working position of the spring must always correspond to a point above the appropriate curve. A horizontal line on the diagram through the value of $D/(L_oF)$ cuts the appropriate curve at the critical deflection of the spring.

For rectangular section springs in which Δ/L_o is less than 0.7, values of $D/(L_oF)$ for intermediate values of m can be estimated accurately by linear interpolation between the curves for $m = 1$ and $m = 5$.

Figure 2.77 also indicates that some rectangular section springs similar to that in example 2, have two critical deflections. Such springs are stable if used at a deflection exceeding the greater of these values but are liable to buckle during fitting. They should not be used if the minimum working deflection is less than the greater critical deflection.

Examples

1. Check the stability of a circular section spring of 25 mm mean coil diameter, free length 150 mm, compressed length 112.5 mm, with both ends fixed and guided.

From Table 2.38, $F = 0.5$ and $\dfrac{D}{L_oF} = \dfrac{25}{150 \times 0.5} = 0.333$

$$\frac{\Delta}{L_o} = \frac{150 - 112.5}{150} = \frac{37.5}{150} = 0.25$$

Plotting these values as indicated in Figure 2.77, shows that the spring will be stable. The critical deflection for buckling is given by $\Delta = 0.415\,L_o = 0.415 \times 150 = 62.25$ mm.

2. A rectangular section spring $m = 5$ is to exert a constant force at a compressed length of 18.75 mm with both ends fixed but unguided. Check its stability if the mean coil diameter is 25 mm and the free length 75 mm.

From Table 2.38 $F = 1.0$ and $\dfrac{D}{L_oF} = \dfrac{25}{75 \times 1} = 0.333$

$$\frac{\Delta}{L_o} = \frac{75 - 18.75}{75} = \frac{56.25}{75} = 0.75$$

Plotting these values in Figure 2.77 shows that the spring will be stable in operation but liable to buckle during assembly at deflections between $75 \times 0.36 = 27$ mm and $75 \times 0.67 = 50.25$ mm.

THE DESIGN OF HELICAL TENSION SPRINGS OF CIRCULAR SECTION MATERIAL

Design considerations

Helical tension springs can be designed using either the basic formulae or the design charts given previously for round wire helical compression springs. They are, however, subject to limitations which do not affect compression springs of similar proportions.

These are:

1. Tension springs are generally coiled with initial tension, which affects the maximum torsional stress in the spring and also its deflection for a given load.
2. Tension springs are easily overstressed by being over-extended, in contrast to compression springs which cannot be compressed beyond the solid position.
3. The various types of end formations used for tension springs often require sharp bends which cause local bending stresses and stress concentrations.
4. The maximum stress and the deflection for a given load in tension springs having hooks or loops at the side (see Figure 2.59) are greater than for those with hooks or loops over centres.

Because of these limitations the proportions of tension springs differ from those of compression springs having the same rate and maximum static load. For these reasons also, the allowable working stresses in statically loaded tension springs are usually considerably lower than those in compression springs.

Figure 2.78 shows a tension spring with full hooks over centres and indicates the usual proportions of the hooks and loops used in forming the ends of tension springs of the

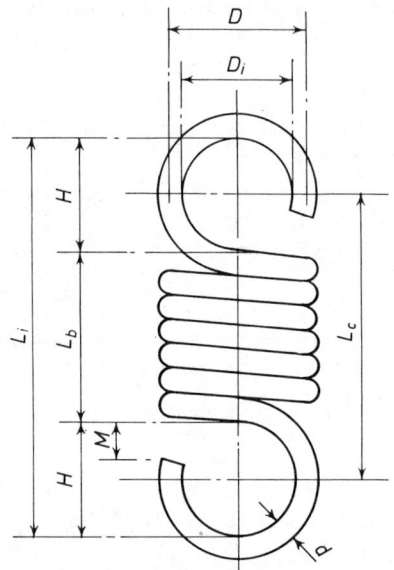

Figure 2.78. Tension spring with full hooks over centres

types illustrated in Figure 2.59. (The difference between a hook and a loop is that there is a definite gap between the ends of a hook and the body coils of the spring, whereas a loop is almost entirely closed upon itself.) The relationship between the dimensions of hooks and loops is approximately as follows:

$$D_i = D - d$$
$$H = D_i \text{ for a full hook or loop, or } D_i/2 \text{ for half hooks or loops}$$
$$M = D_i/3 \text{ for full hooks, or } D_i/6 \text{ for half hooks}$$
$$L_b = d(n' + 1)$$
$$L_i = L_b + 2(D - d) = L_b + 2D_i \text{ for springs with full hooks}$$

$L_i = L_b + (D - d) = L_b + D_i$ for springs with half hooks
$L_c = L_b + D_i$ for full hooks, or $L_b + D_i/2$ for half hooks

Effect of end coils on deflection

To al!' `w for the end effect, the total number of active coils for rate computations can be taken as the number of body coils plus one i.e. $n = n' + 1$, where n' is the exact number of coils between the points where the hooks or loops begin.

Notation

Same as for helical compression springs of circular section material with the following additions:

n'	= number of active body coils.
n	= total number of active coils, i.e. $n = n' + 1$.
W_1 kg	= initial tension load.
σ_A MN/m² or N/mm²	= maximum bending stress in hook end.
q_3 MN/m² or N/mm²	= uncorrected torsional stress induced by W_1.
r_o mm	= mean radius of bend between coil and hook.
r_i mm	= inside radius of bend between coil and hook.
K_2	= stress concentration factor, i.e. $K_2 = r_o/r_i$.
δ mm	= axial deflection (i.e. load W applied axially),

$$\delta = \frac{9 \cdot 81 \; W \times 8nD^3}{Gd^4}.$$

δ' mm	= side deflection (i.e. load W applied at side),

$$\delta' = \frac{9 \cdot 81 \; W \times 12 \, nD^3}{Gd^4}.$$

Initial tension

The initial tension induced by cold-forming a tension spring is a force in the close-wound coils which must be overcome by an applied load before the coils will separate and the spring begin to extend. The rate of such a spring, for loads greater than the initial tension load, is the same as that for a spring with no initial tension, i.e. as obtained from Figures 2.69 or 2.70.

Theoretically, a tension spring has no deflection for loads less than the initial tension load, but in actual springs some coils tend to separate before the others so that the load-deflection curve is non-linear for loads approaching the initial tension load.

The initial tension load in a spring must be allowed for in design and is obtained from

$$W_1 = \frac{\pi d^3 q_3}{9 \cdot 81 \times 8D} = \frac{\pi d^2 q_3}{9 \cdot 81 \times 8C} \tag{15}$$

where q_3 is the uncorrected torsional stress due to a load of W_1, i.e. the torsional stress estimated without regard to the effect of curvature. The initial tension load depends on the spring index $C = D/d$ and on the methods of spring manufacture. For design purposes the maximum *permissible* value of q_3 can be found from Figure 2.79[6].

This curve is derived from practical maximum values obtained under average manufacturing conditions (see example for a comparison between the permissible initial tension stress found from this curve and a measured value). Greater values of initial tension can be produced, if required, by using slower and more costly production methods, and lesser values can be obtained by machine adjustments.

The initial tension in an existing spring can be found as follows:

1. Extend the spring until all the coils are clearly separated. Measure this extension (δ_t) and the load required to produce it (W_2).
2. Extend the spring further to a distance $= 2\delta_t$ and measure the load (W_3) at this extension.
3. The initial tension load W_1 is given by $W_1 = W_2 - (W_3 - W_2) = 2W_2 - W_3$.

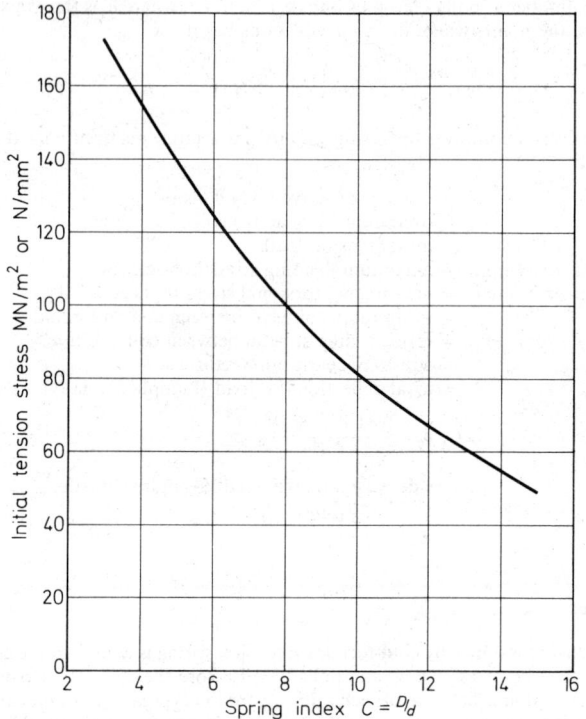

Figure 2.79. Permissible torsional stress values resulting from initial tension in extension springs

The rate of the spring for an extension of δ_t is equal to $W_3 - W_2$ and the initial tension load is equal to the load when all the coils are clearly separated less the load required to separate them.

Stress concentrations in end hooks and loops

Most tension springs have end hooks and loops 'over centres' as indicated in Figure 2.59. A typical spring with half hooks over centres is shown in Figure 2.80. The applied load W produces two types of stress due to:

1. The maximum bending moment $\frac{1}{2} WD$ at A where the sharp bend in the hook section begins.
2. The maximum twisting moment $\frac{1}{2} WD$ at B where the sharp bend from the body section begins.

The *nominal* bending stress at A will be $16 \times 9{\cdot}81\, WD/\pi d^3$ which is numerically twice the maximum torsional stress at B. The maximum stresses are enhanced by the stress concentrations due to the sharp bend and also by the direct tensile stress at A. Hence, the derived maximum stresses are as follows:

$$\sigma_A = \frac{16 \times 9{\cdot}81\, WD\, k_2}{\pi d^3} + \frac{4 \times 9{\cdot}81 W}{\pi d^2} \qquad (16)$$

$$q_B = -\frac{8 \times 9{\cdot}81\, WD\, k_2}{\pi d^3} \qquad (17)$$

where K_2 is a stress concentration factor equal to r_o/r_i (see Figure 2.80).

This value is a practical approximation which slightly overestimates the maximum stresses. Reference 4 gives separate expressions for the stress concentration factors for bending and torsion. These give similar but slightly smaller values.

The ratio r_o/r_i should not normally be allowed to exceed 1·25. This means that the inside radius of the sharp bend should be at least twice the diameter of the spring wire.

Stresses in springs with end hooks and loops at side

The properties of tension springs with end hooks or loops at the side (see Figure 2.59) differ from those of axially loaded springs (with hook or loop over centres) because of eccentric loading. The moment arm of the load on which the maximum torsional stress

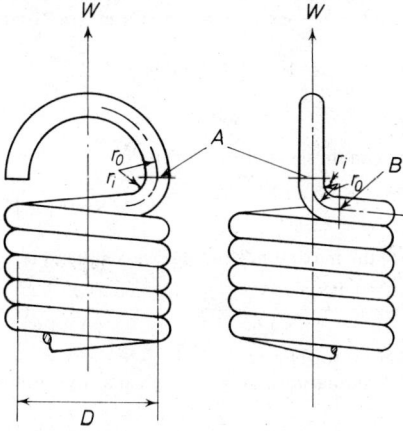

Figure 2.80. Torsion spring with half hooks over centres

depends is practically doubled so that the maximum torsional stress in the coils will be twice that for an axially loaded spring. It can also be shown that the extension of such a spring due to an applied load is given by

$$\delta' = \frac{9{\cdot}81 W \times 12 n D^3}{G d^4},$$

i.e. $1\frac{1}{2}$ times the deflection produced by the same load applied axially.

Springs with ends of this type can be designed from the charts as follows:

1. The rate of the spring is $\frac{2}{3}$ times the value obtained from Figure 2.69 or Figure 2.70.
2. The maximum torsional stress will be twice the value obtained from Figure 2.71 or Figure 2.72.

The initial tension load can be obtained from equation (15). The stresses at the sharp bend of the hook or loop are found from equations (16) and (17) except that the maximum torsional stress at the position corresponding to B in Figure 2.80 is doubled.

Example

A carbon steel tension spring has ten active coils of 1·6 mm diameter wire and a mean coil diameter of 12·5 mm. If the spring is wound with an initial tension load of 1 kg, determine the initial torsional stress, the spring rate, the load required to extend the spring 12·5 mm, and the maximum shear stress at this load. If $r_o = 4$ mm, determine also the maximum shear and bending stresses in the hook ends, which are 'over centres' as in Figure 2.80.

$$C = \frac{D}{d} = \frac{12\cdot5}{1\cdot6} = 7\cdot81$$

From Figure 2.71 with $C = 7\cdot81$ and $d = 1\cdot6$ mm, $q/W = 92$ and when $W_1 = 1$ kg, $q = 92 \times 1 = 92$ MN/m² (or N/mm²). (Note from Figure 2.79 that for $D/d = 7\cdot81$ the permissible uncorrected shear stress resulting from initial tension is 102 MN/m² (or N/mm²). From Figure 2.70 with $D = 12\cdot5$ mm and $d = 1\cdot6$ mm, $nS_W = 3\cdot3$. And when $n = 10$, $S_W = 0\cdot33$ kg/mm.

Hence the load required to extend spring 12·5 mm
$$= 1 + 0\cdot33 \times 12\cdot5 = 1 + 4\cdot125 = 5\cdot125 \text{ kg}$$
and the stress at this load $= \dfrac{5\cdot125}{1} \times 92 = 471\cdot5$ MN/m² (or N/mm²)

$$K_2 = \frac{r_o}{r_i} \text{ and } r_i = r_o - \frac{d}{2} \quad \therefore r_i = 4 - 0\cdot8 = 3\cdot2 \text{ mm}$$

Hence $K_2 = \dfrac{4}{3\cdot2} = 1\cdot25$

From equation (17) maximum shear stress at bend is
$$\frac{8 \times 9\cdot81 \times 5\cdot125 \times 12\cdot5 \times 1\cdot25}{\pi \times 1\cdot6^3} = 488 \text{ MN/m}^2 \text{ (or N/mm}^2)$$

And from equation (16) the maximum bending stress at bend is
$$\frac{16 \times 9\cdot81 \times 5\cdot125 \times 12\cdot5 \times 1\cdot25}{\pi \times 1\cdot6^3} + \frac{4 \times 9\cdot81 \times 5\cdot125}{\pi \times 1\cdot6^2}$$
$$= 976 + 25 = 1001 \text{ MN/m}^2 \text{ (or N/mm}^2)$$
(Note that the *nominal* bending stress is twice as great as the nominal torsion stress.)

HELICAL TORSION SPRINGS

General considerations

Because a torsion spring is subject to a torque about the coil axis (Figure 2.81) the primary stress is a bending stress instead of the shear stress found in helical tension or compression springs. The maximum bending stress occurs at the inside of the coil.

Relative position of ends

The load taken by a torsion spring is affected by the relative positions of the ends which are subject to manufacturing tolerances depending on the spring index (C) and the number of active coils (n).

For values of C between 5 and 10, which are common proportions, a reasonable tolerance is $\pm 10°$ for $n = 4$, increasing to $\pm 20°$ for $n = 12$. These tolerances apply to the angle between the directions of the two end loops (see Figure 2.61).

Clearances

Torsion springs change in diameter as they are wound up or unwound by the applied load. The usual type, which winds up with increasing load, is guided by an arbor or rod as shown in Figure 2.81. High stresses can be induced by binding of the spring on the

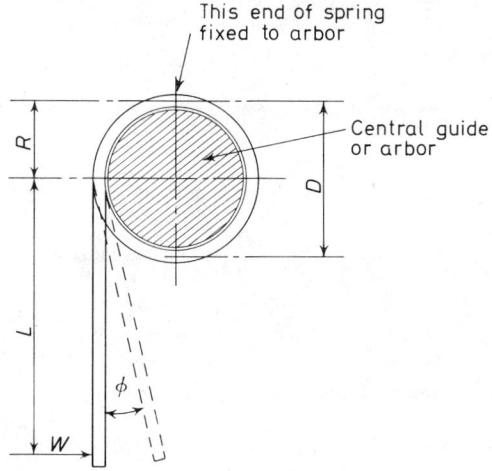

Figure 2.81. Torsion spring subject to a load W at radius L.

arbor. A torsion spring arranged to unwind on loading and guided by a tube can bind in the same way.

The minimum radial clearances to avoid binding for a torsion spring with n active coils which is wound or unwound by N_t turns by the maximum torque are as follows:

Winding up: $y = \dfrac{N_t}{n}(D-d)$ — on the inside

Unwinding: $y = \dfrac{N_t}{n}(D+d)$ — on the outside

N_t is often a fraction of a turn, e.g. if a spring with 8 active coils is wound through 90°, or $\frac{1}{4}$ turn,

$$\frac{N_t}{n} = \frac{0.25}{8} = 0.03 \text{ or approximately } 3\,\% \text{ (Ref. 4)}$$

Basic formulae and notation

The design moments for a torsion spring are as follows:

$M = PL$ N.mm $= 9.81\ WL$ N.mm for the load tending to wind-up the spring (as in Figure 2.81)   (18)

$M = P(L+R)$ N.mm $= 9 \cdot 81 \ W(L+R)$ N.mm for the load tending to unwind the spring (19)

The effect of curvature on the angular rate of a torsion spring is negligible. For round wire springs tending to wind-up under load the angular rate is given by

$$\text{Angular load rate } (S_{Wt}) = \frac{\text{applied moment}}{\text{angular displacement}} = \frac{M}{\phi}$$

$$= \frac{WL}{\phi} \text{ kg.mm/radian} \tag{20}$$

$$\text{Angular force rate } (S_{Pt}) = \frac{\text{applied moment}}{\text{angular displacement}} = \frac{M}{\phi} = \frac{PL}{\phi}$$

$$= \frac{9 \cdot 81 \ WL}{\phi} \text{ N.mm/radian} \tag{21}$$

$$\text{i.e. } S_{Pt} = \frac{9 \cdot 81 \ WL}{\phi} = \frac{Ed^4}{64 \ Dn} \text{ N.mm/radian}$$

$$= \frac{Ed^4 \pi}{32 Dn} \text{ N.mm/turn} \tag{22}$$

$$\text{or } S_{Wt} = \frac{WL}{\phi} = \frac{Ed^4}{9 \cdot 81 \times 64 Dn} \text{ kg.mm/radian}$$

$$= \frac{Ed^4 \pi}{9 \cdot 81 \times 32 Dn} \text{ kg.mm/turn} \tag{23}$$

$$\text{Angular displacement } (\phi) = \frac{9 \cdot 81 \ WL \times 64 Dn}{Ed^4} \text{ radians} \tag{24}$$

The maximum corrected bending stress at inside of coil is given by:

$$\sigma_t = \frac{32 \ PL}{\pi d^3} K_3 = \frac{32 \times 9 \cdot 81 \ WL}{\pi d^3} K_3 \text{ MN/m}^2 \text{ (or N/mm}^2) \tag{25}$$

Rectangular section

$$S_{Pt} = \frac{9 \cdot 81 \ WL}{\phi} = \frac{Ebh^3}{12 \pi Dn} \text{ N.mm/radian} = \frac{Ebh^3}{6 \ Dn} \text{ N.mm/turn} \tag{26}$$

$$\text{or } S_{Wt} = \frac{WL}{\phi} = \frac{Ebh^3}{9 \cdot 81 \times 12 \pi Dn} \text{ kg.mm/radian} = \frac{Ebh^3}{9 \cdot 81 \times 6 Dn} \text{ Kg.mm/turn} \tag{27}$$

$$\text{Angular displacement } (\phi) = \frac{9 \cdot 81 \ WL \times 12 \pi Dn}{Ebh^3} \text{ radians} \tag{28}$$

$$\sigma_t = \frac{6 \ PL}{b \ h^2} K_4 = \frac{6 \times 9 \cdot 81 \ WL}{bh^2} K_4 \text{ MN/m}^2 \text{ (or N/mm}^2) \tag{29}$$

When the load is tending to unwind the spring, the appropriate design moment, as given by equation (19), should be substituted as required in the above expressions.

Where

W	kg	= load, applied at right-angles to arm of spring.
P	N	= equivalent force, applied at right-angles to arm of spring i.e. $P = 9\cdot81\ W$.
σ_t	MN/m^2 (or N/mm^2)	= maximum corrected bending stress.
S_{Wt}	Kg.mm/radian	= angular load rate, i.e. $S_{Wt} = M/\phi$.
S_{Pt}	N.mm/radian	= angular force rate, i.e. $S_{Pt} = 9\cdot81\ M/\phi$.
ϕ	radians	= angular displacement.
D	mm	= mean coil diameter of spring.
d	mm	= diameter of wire.
b	mm	= radial width of rectangular section.
h	mm	= axial thickness of rectangular section.
E	MN/m^2 (or N/mm^2)	= modulus of elasticity.
M	N.mm or Kg.mm	= applied moment or torque.
n		= number of active coils.
L	mm	= length of arm at which W or P is applied.
R	mm	= mean coil radius, i.e. $R = D/2$.
C		= spring index, i.e. $C = D/d$ or D/b.
K_3		= stress correction factor for circular section material, i.e.

$$K_3 = \frac{4C^2 - C - 1}{4C(C-1)}.$$

K_4 = stress correction factor for rectangular section material,

$$\text{i.e. } K_4 = \frac{3C^2 - C - 0\cdot8}{3C(C-1)}.$$

N_t = number of turns required by applied moment or torque,

$$\text{i.e. } N_t = \frac{\phi}{2\pi}.$$

The stress correction factors K_3 and K_4 allow for the effect of curvature and apply to circular and rectangular section material respectively. The theoretical values of K_3 and K_4 (Ref. 4, page 297 and page 296) are

$$\frac{4C^2 - C - 1}{4C(C-1)} \text{ and } \frac{3C^2 - C - 0\cdot8}{3C(C-1)}$$

respectively. Values of K_3 and K_4 as functions of the spring index C are plotted in Figure 2.82.

Allowable stresses

The maximum bending stress (σ_t) should not exceed 50% of the minimum UTS of the wire. In addition, any local bending stresses at end fastenings or hooks not subject to the full design moment should be limited to

$$\frac{\text{UTS}}{2 \times \text{stress concentration factor}}$$

where the stress concentration factor is obtained as for the ends of tension springs (ratio r_o/r_i, see Figure 2.80). It is important to avoid sharp bends in the formation of end hooks or loops and to round off the edges of any clamps or abutments securing the ends of the spring.

Example

A carbon steel torsion spring is loaded as shown in Figure 2.81. It has the following dimensions:

Length of load arm	= 20 mm
Mean coil diameter	= 10 mm
Wire diameter	= 1·06 mm
Applied load at end of arm	= 0·5 kg
Number of active coils	= 6·5

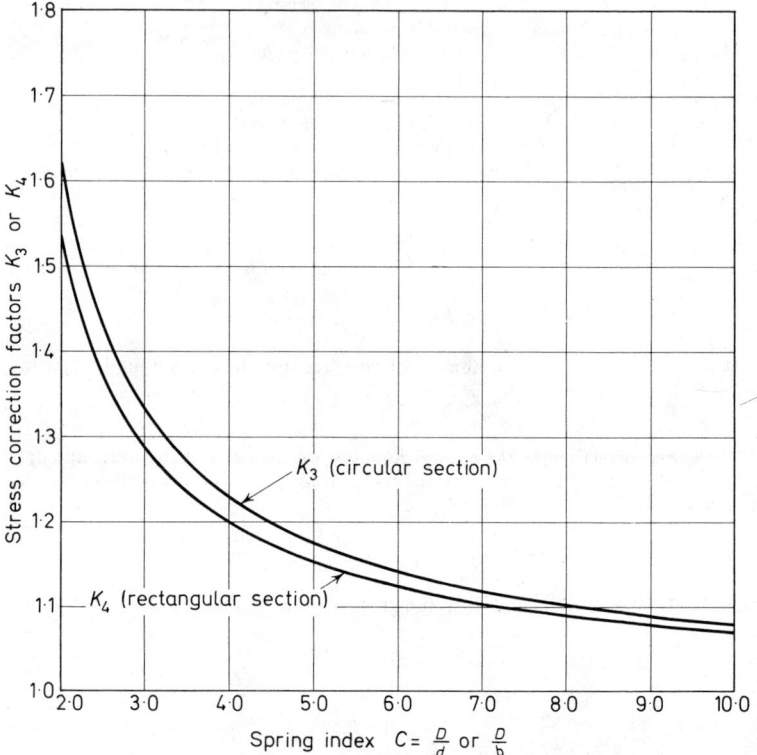

Figure 2.82. Stress correction factors for torsion springs of circular and rectangular section material

Determine the maximum bending stress, the angular displacement and the angular load rate.

$$C = \frac{D}{d} = \frac{10}{1·06} = 9·43 \text{ and from Figure 2.82 } K_3 = 1·086$$

From equation (25) $\sigma_t = \dfrac{32 \times 9·81 \times 0·5 \times 20 \times 1·086}{\pi \times 1·06^3} = 911 \text{ MN/m}^2$

Angular displacement, from equation (24)

$$= \frac{9 \cdot 81 \times 0 \cdot 5 \times 20 \times 64 \times 10 \times 6 \cdot 5}{206800 \times 1 \cdot 06^4} = 1 \cdot 563 \text{ radians}$$

i.e. $N_t = \dfrac{1 \cdot 563}{2\pi} = 0 \cdot 248$ turns (i.e. approx. $90°$)

From equation (20) $S_{W_t} = \dfrac{WL}{\phi} = \dfrac{0 \cdot 5 \times 20}{1 \cdot 563} = 6 \cdot 397$ kg.mm/radian

or $\qquad\qquad S_{P_t} = 9 \cdot 81 \times 6 \cdot 397 = 62 \cdot 75$ N.mm/radian

THE DESIGN OF CONICAL DISC SPRINGS (OR BELLEVILLE WASHERS)

General considerations

The following expressions for load, stress and deflection in conical disc springs have been derived from the theoretical analyses given in References 4.7 and 8. The basis of the recommended design procedure is to use convenient expressions for the load (W_f) on,

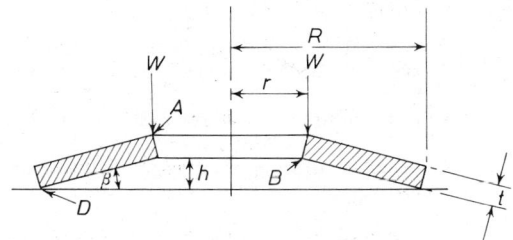

Figure 2.83. Typical conical disc spring

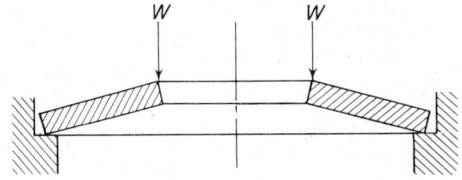

Figure 2.84. Conical disc spring mounted on shoulders

and the maximum stress (S_f) in the disc when in its *flattened* position. The load and maximum stress in any other position can then be expressed in terms of W_f and S_f. Characteristic load-deflection and stress-deflection curves for particular discs can then be readily obtained.

Typical single disc spring arrangements are shown in Figures 2.83 and 2.84. In the arrangement shown in Figure 2.83, the disc can only be deflected to its flattened position.

Single discs mounted on shoulders, however, as in Figure 2.84, can be deflected beyond the flat position.

Notation and basic formulae

R	mm	= outside radius of disc spring.
r	mm	= inside radius of disc spring.
t	mm	= thickness of disc spring.
h	mm	= initial cone height (h = overall free height-thickness, approx.).
δ	mm	= axial deflection of disc spring from free position.
α		= R/r (ratio of outer to inner radius).
β	radians	= initial base angle of cone = $h/(R-r)$.
C_1, C_2, C_3, C_4 and C_5		= stress factors depending on α.
		(Factor C_1 is obtained from Figure 2.85, and Factors C_2, C_3, C_4 and C_5 may be obtained from Figure 2.86.
k		= deflection factor (obtained from Figure 2.85).

$$k = \frac{\alpha - 1}{\log_e \alpha} - 1$$

y		= parameter $\dfrac{t}{h}\left(\dfrac{\alpha - 1}{k}\right)$.
W	kg	= axial load, uniformly applied around edges.
W_f	kg	= axial load when disc spring is in flattened position.
S	MN/m² or N/mm²	= maximum stress in disc spring (at upper inner edge A).
S_f	MN/m² or N/mm²	= maximum stress in disc spring in flattened position (at upper inner edge A).
S_B	MN/m² or N/mm²	= stress at lower inner edge B
S_D	MN/m² or N/mm²	= stress at lower outer edge D
E	MN/m² or N/mm²	= modulus of elasticity (taken as 206 800 MN/m² for carbon and low alloy steels).
v		= Poisson's ratio, taken as 0·3.
x		= Ratio δ/h.

Load-deflection formulae

Load W at deflection δ is given by

$$W = \frac{E\delta C_1}{9\cdot 81\,(1-v^2)R^2}\left[(h-\delta)\left(h-\frac{\delta}{2}\right)t+t^3\right] \qquad (30)$$

and when $\delta = h$ (i.e. disc flat) this reduces to

$$W_f = \frac{1\cdot 1\,ECht^3}{9\cdot 81\,R^2}\,(f\,=\,\text{flat}) \qquad (31)$$

Equation (31) can also be re-written in the form

$$W_f = \frac{1 \cdot 1 \, EC_1 \beta^2}{9 \cdot 81} \left(\frac{\alpha - 1}{\alpha}\right)^2 \frac{t^3}{h} \tag{32}$$

Stress-deflection formulae

Stress (max) at upper inner edge A (compressive), see Figure 2.83, is given by

$$S = \frac{1 \cdot 1 \, E\delta C_1}{R^2} \left[C_2 \left(h - \frac{\delta}{2}\right) + C_3 \, t \right] \tag{33}$$

Stress at lower inner edge B (normally tensile) is given by

$$S_B = \frac{1 \cdot 1 \, E\delta C_1}{R^2} \left[C_2 \left(h - \frac{\delta}{2}\right) - C_3 t \right] \tag{34}$$

Stress at lower outer edge D (normally tensile) is given by

$$S_D = \frac{1 \cdot 1 \, E\delta C_1}{R^2} \left[-C_4 \left(h - \frac{\delta}{2}\right) - C_5 \, t \right] \tag{35}$$

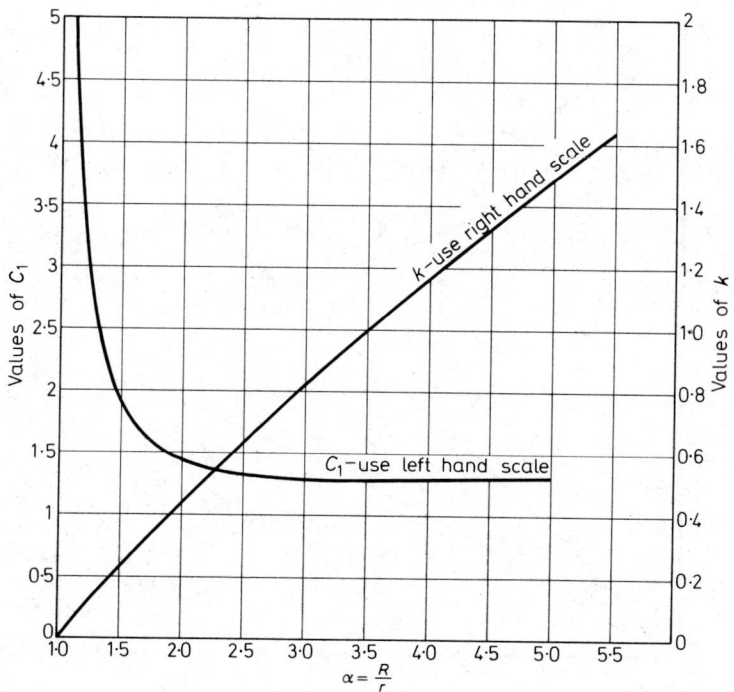

Figure 2.85. Factors C_1 and k for conical disc springs

Expressions for the various factors are as follows:

$$C = \frac{\pi \log_e \alpha}{6}\left(\frac{\alpha}{\alpha - 1}\right)^2$$

$$C_2 = \left(\frac{\alpha - 1}{\log_e \alpha} - 1\right)\frac{6}{\pi \log_e \alpha}$$

$$C_3 = \frac{3(\alpha - 1)}{\pi \log_e \alpha}$$

$$C_4 = \left(\frac{1 - \dfrac{1}{\alpha}}{\log_e \alpha} - 1\right)\frac{6}{\pi \log_e \alpha}$$

$$C_5 = \frac{3\left(1 - \dfrac{1}{\alpha}\right)}{\pi \log_e \alpha}$$

where $\alpha = \dfrac{R}{r}$

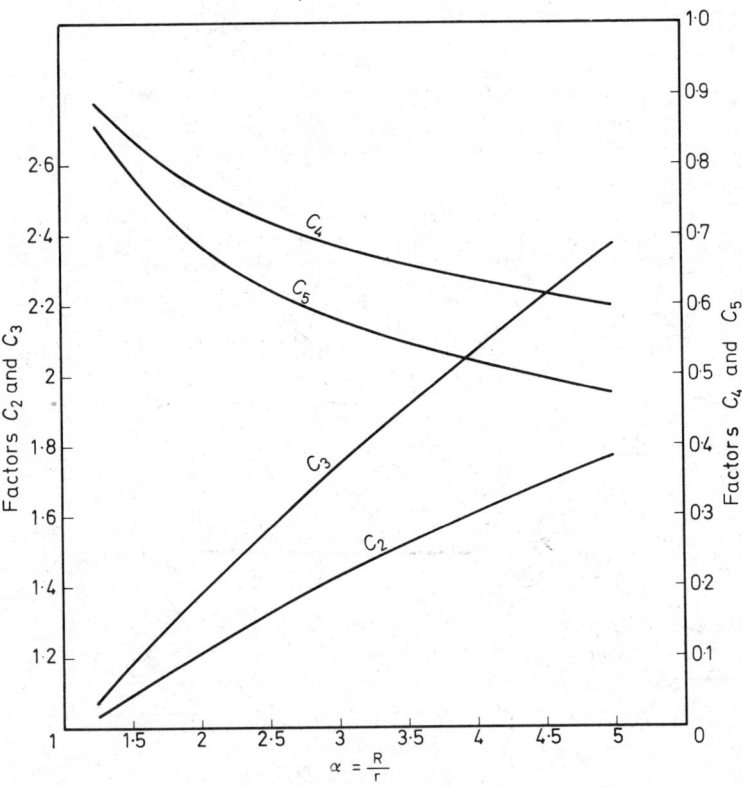

Figure 2.86. Curves for finding factors C_2, C_3, C_4 and C_5

Values of C_1 and k can be obtained from Figure 2.85. Values C_2, C_3, C_4 and C_5 are shown graphically in Figure 2.86.

The stresses obtained from equations (33), (34) and (35) will be compressive when the bracketed quantity is positive. Conversely, the stresses will be tensile when the bracketed quantity is negative.

When $\delta = h$ (i.e. disc flat), equation (33) in terms of β becomes

$$S_f = \frac{1 \cdot 1 \; E\beta^2}{2}\left[k + \frac{t}{h}(\alpha - 1)\right] \tag{36}$$

When the disc deflection is δ, the load and maximum stress are given by

$$\frac{W}{W_f} = x\left[1 + (1-x)\left(1 - \frac{x}{2}\right)\frac{h^2}{t^2}\right] \tag{37}$$

and

$$\frac{S}{S_f} = x\left[\frac{2-x+y}{1+y}\right] \tag{38}$$

where $x = \dfrac{\delta}{h}$

Curves of W/W_f against δ/h for particular values of h/t are shown in Figure 2.87, which indicates the wide variety of flexibility characteristics it is possible to obtain by suitable variation of the height to thickness ratio. From Figure 2.87, it follows that:

if $h/t < \sqrt{2}$, the disc spring rate is always positive, i.e. slope upwards to the right.
if $h/t = \sqrt{2}$, the disc spring rate is zero in the flattened position
if $h/t > \sqrt{2}$, the disc spring rate becomes negative beyond a certain deflection
if $h/t > 2\sqrt{2}$, buckling occurs and the disc spring will snap into a new position, beyond the flattened position, when deflected beyond a certain point.

Curves of S/S_f against δ/h for $\alpha = 2$ and particular values of h/t are given in Figure 2.88, which shows that the stress deflection characteristics for all disc springs of practical proportions are very similar up to the flattened position.

Design limitations

The design of a conical disc spring for a particular requirement is subject to the following considerations:
(a) The value of α should be about 2 for efficient design. Values of α less than 2 result in a disc containing too little spring material. Values of α greater than 2 tend to give disproportionately high stresses at the inner edge.
(b) In order to utilise the disc to its best advantage it should be designed for deflections approaching the flattened position. In most cases the effect of this is to confine the value of h/t to within the range $1 - 1 \cdot 5$. Values of h/t outside this range are possible but such values give disc proportions suitable for the more exceptional requirements. Values of h/t less than 1 represent shallow discs while values of h/t greater than $1 \cdot 5$ represent discs which would either be limited in deflection or would possess negative rate characteristics over a certain range of deflection.
(c) The stress in a disc spring in the flattened position depends mainly on the base angle (β) of the conical face. The proportions of the disc will usually be such that equation (36) gives a value of S_f between $0 \cdot 6 \; E\beta^2$ and $0 \cdot 8 \; E\beta^2$, i.e. for carbon steels and low alloys between $124\,000\beta^2$ and $165\,500\beta^2$ MN/m^2 or N/mm^2. The angle β cannot therefore be made much greater than $0 \cdot 1$ radians without inducing very high stresses in the flattened

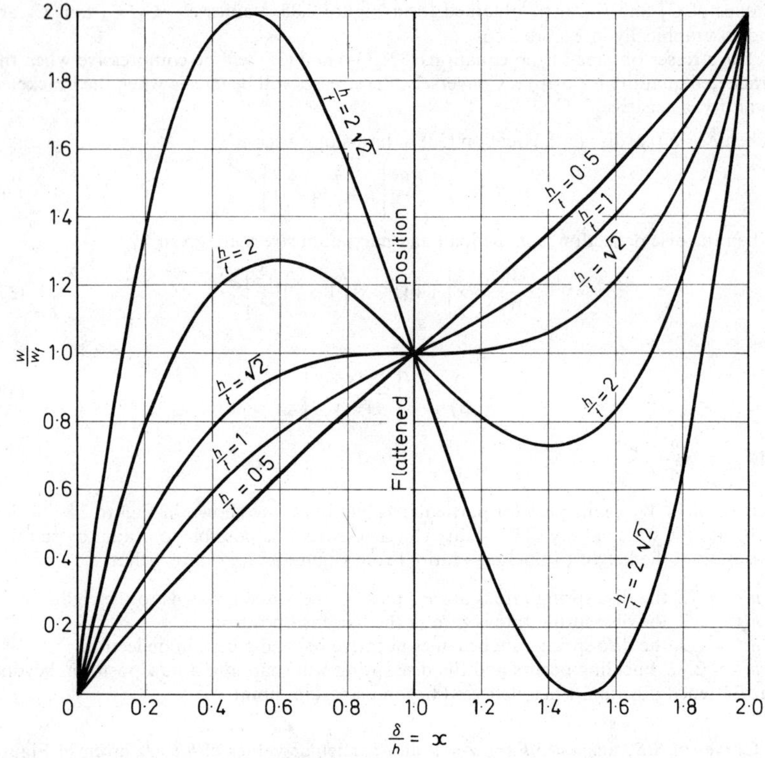

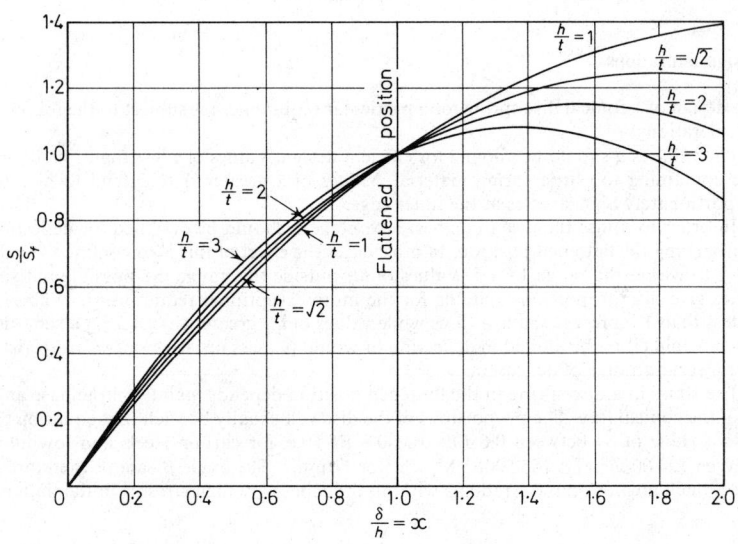

Figure 2.87. Curves of W/W_f against δ/h for particular values of h/t

Figure 2.88. Curves of S/S_f against δ/R for particular values of h/t

position. The actual value of β is determined from the maximum permissible stress S_f in the material and the chosen values of α and h/t using equation (36) which gives

$$\beta = \left[\frac{2S_f}{1 \cdot 1 \, E \left[k + \frac{t}{h}(\alpha - 1) \right]} \right]^{\frac{1}{2}} \tag{39}$$

Having decided on the values of α, h/t and β the proportions of the disc are fixed. The thickness of the disc is then given by equation (32), and the remaining dimensions follow automatically from the relationships

$$\alpha = \frac{R}{r} \quad \text{and} \quad \beta = \frac{h}{R-r}$$

i.e.

$$r = \frac{h}{\beta(\alpha - 1)} \text{ and } R = \alpha r$$

The final values will probably require adjustment to give convenient dimensions for which the load and stress characteristics should be checked.

Materials and allowable stresses

Conical disc springs are usually made from carbon and low alloy steels although stainless alloys and non-ferrous materials can be used if necessary. In general, however, the steels mainly used for this purpose, include the following:

Carbon spring steel, to BS 970, En 44
Silicon-manganese steel to BS 970, En 45, 45A
Chrome Vanadium steel to BS 970, En 47, En 50
3% Cr steel to BS 970, En 29A or En 29B

Suitably hardened and tempered the above materials will provide the following properties:

Hardness   380 Brinell
UTS   1060 MN/m² (min)
Yield   925 MN/m² (min)

Using these materials in the condition described, the following maximum stress values are recommended for design purposes:

Static conditions   1540 MN/m²
Moderate fatigue conditions 1230 MN/m²

These values may be used in spite of the fact that the yield point of the material found by means of a tensile test would be lower than either value. Although the calculated stresses may appear excessive it should be remembered that they are localised at the upper inner edge and subsequent local plastic strain will therefore allow a redistribution of stress.

Example

Determine the height, thickness and maximum stress of a conical disc spring having a zero rate characteristic. The details are as follows:

Outside diameter of disc spring = 150 mm
Inside diameter of disc spring = 75 mm
Load when flat   = 450 kg

Since a 'zero rate' characteristic is required $h/t = \sqrt{2}$

$$\alpha = \frac{R}{r} = \frac{75}{37\cdot5} = 2, \text{ and from Figure 2.85, } C_1 = 1\cdot452 \text{ and } k = 0\cdot443.$$

From equation (31), $ht^3 = \dfrac{9\cdot81\, W_f R^2}{1\cdot1 E C_1}$ and $h = t\sqrt{2}$

$$\therefore\ t^4 \sqrt{2} = \frac{9\cdot81\, W_f R^2}{1\cdot1 E C_1} \qquad \therefore\ t^4 = \frac{9\cdot81\, W_f R^2}{1\cdot1\sqrt{2} E C_1}$$

Hence $\qquad t^4 = \dfrac{9\cdot81 \times 450 \times 75^2}{1\cdot1\sqrt{2} \times 206\,800 \times 1\cdot452} = 53\cdot2 \text{ mm}^4$

from which $t = 2\cdot7$ mm
and $h = 2\cdot7 \sqrt{2} = 3\cdot82$ mm
From Figure 2.86 for $\alpha = 2$, $C_2 = 1\cdot220$ and $C_3 = 1\cdot378$
From equation (33), when $\delta = h$ (i.e. disc flat)

$$\text{Max. stress } (S_f) = \frac{1\cdot1 \times 206\,800 \times 3\cdot82 \times 1\cdot452}{75^2}[1\cdot220 \times 1\cdot91 + 1\cdot378 \times 2\cdot7]$$

$$\therefore\ S_f = 225[2\cdot33 + 3\cdot72] = 225 \times 6\cdot05 = 1360 \text{ MN/m}^2$$

As a check on S_f, from equation (36), with $\beta = \dfrac{3\cdot82}{37\cdot5} = 0\cdot102$ radians

$$S_f = \frac{1\cdot1 \times 206\,800 \times 0\cdot102^2}{2}\left[0\cdot443 + \frac{2\cdot7}{3\cdot82}\right]$$

$$= 1183\,[0\cdot443 + 0\cdot707] = 1183 \times 1\cdot150 = 1360 \text{ MN/m}^2 \text{ (as above)}$$

Finally, as a check on W_f, using equation (32)

$$W_f = \frac{1\cdot1 \times 206\,800 \times 1\cdot452 \times 0\cdot102^2}{9\cdot81}\left[\frac{1}{2}\right]^2 \frac{2\cdot7^3}{3\cdot82} = 450 \text{ kg}$$

SINGLE LEAF SPRINGS

Basic formulae and notation

The simplest form of leaf spring is the straight or flat cantilever with a load applied at the free end as shown in Figure 2.89.

In this case the well known formula for deflection at the free end is

$$\delta = \frac{PL^3}{3EI} = \frac{9\cdot81 WL^3}{3EI} = \frac{4 \times 9\cdot81 WL^3}{bt^3 E} \tag{40}$$

The maximum bending stress at the fixed end, derived from simple beam theory, is

$$\sigma = \frac{M}{Z} = \frac{PL}{Z} = \frac{6PL}{bt^2} = \frac{6 \times 9\cdot81 WL}{bt^2} \tag{41}$$

The stress decreases in proportion to the distance x from the fixed end and becomes zero at the free end where $x = L$. The magnitude of the stress (σ_x) at a distance x from the fixed end is given by

$$\sigma_x = \left(\frac{L-x}{L}\right)\sigma = \sigma\left[1 - \frac{x}{L}\right] \tag{42}$$

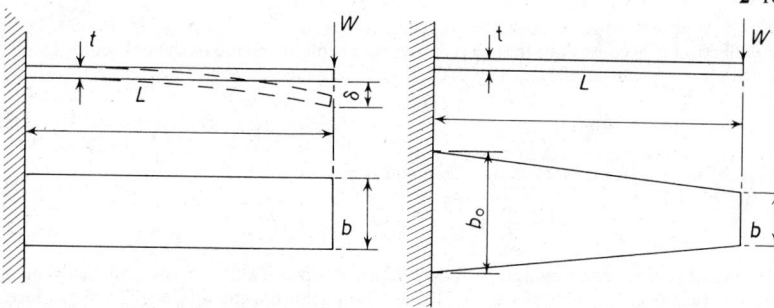

Figure 2.89. Rectangular-uniform thickness—
cantilever type

Figure 2.90. Trapezoidal—uniform thickness—
cantilever type

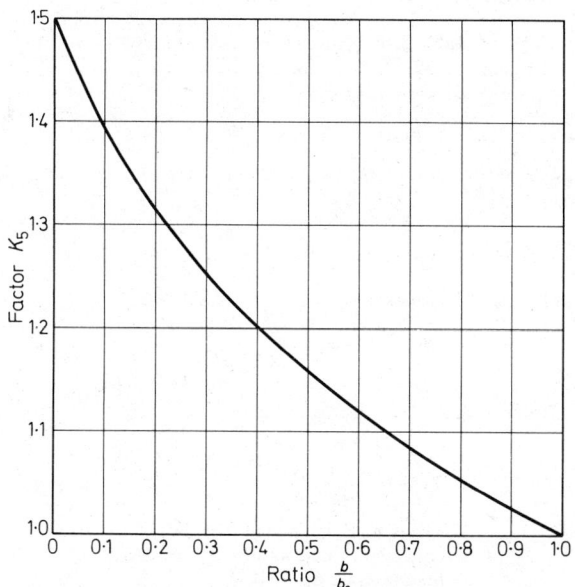

Figure 2.91. Curve for factor K_5 in trapezoidal leaf spring

Table 2.39 DEFLECTION FACTOR K_5 FOR TRAPEZOIDAL CANTILEVER (see Figure 2.91)

$\dfrac{b}{b_0}$	0·000	0·100	0·200	0·300	0·400	0·500	0·600	0·700	0·800	0·900	1·000
K_5	1·500	1·391	1·315	1·254	1·203	1·159	1·120	1·086	1·054	1·026	1·000

$$K_5 = \frac{3}{\left(1 - \dfrac{b}{b_0}\right)^3}\left[0{\cdot}5 - 2\frac{b}{b_0} + \left(\frac{b}{b_0}\right)^2\left(1{\cdot}5 - \log_e\frac{b}{b_0}\right)\right]$$

For a simple cantilever spring of trapezoidal profile, as illustrated in Figure 2.90, the deflection is greater than that of rectangular profile by virtue of the deflection factor K_5, which is a function of b/b_0. The deflection at the free end is given by

$$\delta = \frac{PL^3}{3EI_0} K_5 = \frac{9 \cdot 81 W L^3}{3EI_0} K_5 = \frac{4 \times 9 \cdot 81 W L^3}{b_0 t^3 E} K_5 \tag{43}$$

The maximum bending stress at the fixed end is given by

$$\sigma = \frac{M}{Z_0} = \frac{PL}{Z_0} = \frac{6PL}{b_0 t^2} = \frac{6 \times 9 \cdot 81 W L}{b_0 t^2} \tag{44}$$

Values of K_5 for various values of ratio b/b_0 are given in Table 2.39 and are also shown graphically in Figure 2.91. Referring to Table 2.39, a value of ratio $b/b_0 = 0 \cdot 2$ will produce a deflection $31 \cdot 5\%$ greater than that of a corresponding cantilever spring of rectangular profile.

The elastic properties of the rectangular and trapezoidal cantilever type single leaf springs are summarised in Table 2.40.

Table 2.40   CANTILEVER LEAF SPRINGS

Type of leaf	Maximum bending stress at fixed end	Deflection at free end	Spring load rate
Rectangular with uniform thickness	$\sigma = \dfrac{6 \times 9 \cdot 81 W L}{b t^2}$	$\delta = \dfrac{4 \times 9 \cdot 81 W L^3}{b t^3 E}$ $= \dfrac{2L^2 \sigma}{3tE}$	$S_W = \dfrac{b t^3 E}{4 \times 9 \cdot 81 L^3}$
Trapezoidal with uniform thickness	$\sigma = \dfrac{6 \times 9 \cdot 81 W L}{b_0 t^2}$	$\delta = \dfrac{4 \times 9 \cdot 81 W L^3}{b_0 t^3 E} K_5$ $= \dfrac{2L^2 \sigma}{3tE} K_5$	$S_W = \dfrac{b_0 t^3 E}{4 \times 9 \cdot 81 L^3 K_5}$

Notation

W	kg	= spring load, applied at free end.
P	N	= equivalent force, applied at free end, i.e. $P = 9 \cdot 81 \ W$.
S_W	kg/mm	= load rate, i.e. $S_W = W/\delta$.
S_P	N/mm	= force rate, i.e. $S_P = 9 \cdot 81 W/\delta$.
δ	mm	= spring deflection at free end.
σ	MN/m² (or N/mm²)	= maximum bending stress at fixed end.
σ_x	MN/m² (or N/mm²)	= bending stress at distance x mm from fixed end.
L	mm	= length of spring.
b	mm	= width of leaf of rectangular profile.
b_0	mm	= width of leaf at fixed end of trapezoidal profile.
t	mm	= thickness of leaf.
M	N mm	= bending moment.
E	MN/m² (or N/mm²)	= modulus of elasticity.
I	mm⁴	= moment of inertia ($= bt^3/12$ for rectangular section).
I_0	mm⁴	= moment of inertia of section at fixed end of trapezoidal profile ($= b_0 t^3/12$).
Z	mm³	= section modulus ($= bt^2/6$ for rectangular section)

Z_0 mm^3 = section modulus at fixed end of trapezoidal profile ($= b_0 t^2/6$).
K_5 = deflection factor for cantilever springs of trapezoidal profile.

APPENDIX

Table 2.40   NUMERICAL LIST OF STEELS SHOWING BS REFERENCES AND NEW DESIGNATIONS OF EN STEELS REFERRED TO IN TABLE 2.28

Steel	BS and page reference	En steel replaced	Type
070 A 72	970 Part 5 p. 16	42 B, C, D.	'72' carbon
070 A 78	970 Part 5 p. 16		'78' carbon
060 A 96	970 Part 5 p. 16	44 B, C.	'96' carbon
250 A 53	970 Part 5 p. 16	45	Si Mn '53' carbon
250 A 58	970 Part 5 p. 16	45 A	Si Mn '58' carbon
250 A 61	970 Part 5 p. 16	45 A	Si Mn '61' carbon
735 A 50	970 Part 5 p. 16	47, 50	1% Cr. V.

Table 2.41   NUMERICAL LIST OF STEELS SHOWING BS REFERENCES AND NEW DESIGNATIONS OF EN STEELS REFERRED TO IN TABLE 2.29

Steel	BS and page reference	En steel replaced	Type
410 S 21	970 Part 4 p. 24	56 A	13 Cr. C 0·12
420 S 29	970 Part 4 p. 26	56 B	13 Cr. C 0·17
420 S 37	970 Part 4 p. 28	56 C	13 Cr. C 0·24
420 S 45	970 Part 4 p. 30	56 D	13 Cr. C 0·32
431 S 29	970 Part 4 p. 38	57	17 Cr 2·1/2 Ni C 0·15
302 S 25	970 Part 4 p. 42	58 A	Cr. Ni. 18/9 C 0·12
320 S 17	970 Part 4 p. 48	58 J	Cr. Ni. Mo. 17/12/2·1/2 + Ti C 0·08

REFERENCES

1. BERRY, W. R., *Spring Design*, Emmott (1961)
2. Data Sheets. *Design of Helical Compression Springs*, H.M.S.O. (1951)
3. *Manual on Design and Application of Helical and Spiral Springs*, S.A.E. TR9 (1957)
4. WAHL, A. M., *Mechanical Springs*, 2nd ed, McGraw-Hill (USA)
5. HARINGX, J. A., *On highly compressible helical springs and rubber rods, and their application for vibration-free mountings*, Part 1. H. V. PHILIPS' Research Report, pp. 401–449 (1948)
6. *The Mainspring*, Wallace Barnes Co. (April 1941)
7. ALMEN, J. O. and LASZLO, A., 'Uniform section desk spring', *Trans. A.S.M.E.* **58**, 305–314 (1936)
8. ASHWORTH, GRAHAM, 'Disc spring or Belleville washer'. *Proc. Inst. Mech. Eng.*, **155**, 93 (1946)

BRITISH STANDARDS

9. BS 970
   Wrought steels for automobile and general engineering purposes En series. (1955). See also revised issues. Parts 1 to 5. (1970, 1971, 1972), for new designations

10. BS 384
    Phosphor bronze wire (1963). See also BS 2873
11. BS 2786
    Brass wire for springs (1963). See also BS 2873
12. BS 2873
    Copper and copper alloys (1969). Metric Standard
13. BS 1408
    Patented cold drawn steel spring wire (1964)
14. BS 1429
    Annealed steel wire for oil-hardened and tempered springs (1948)
15. BS 1726
    Guide to the design and specification of coil springs. Helical compression springs. Part 1 (1964)
16. BS 2056
    Rust, acid and heat resisting steel wire for springs (1953)
17. BS 2803
    Oil hardened and tempered steel wire for springs for general engineering purposes (1956)

3 DRAWING OFFICE PRACTICE AND METROLOGY

DRAWING OFFICE PRACTICE
AND METROLOGY 3

3 DRAWING OFFICE PRACTICE AND METROLOGY

ENGINEERING DRAWING PRACTICE

D. A. HAMMOND

INTRODUCTION

The British Standard for Engineering Drawing Practice is BS 308.[1] This section does not in any way attempt to replace the standard, but sets out those aspects of greatest importance for the furtherance of good drawing practice. It must be remembered that the British Standard has adopted those principles agreed internationally within ISO (International Organisation for Standardisation) the first of which were published in 1959. The U.K. has taken a very active interest in all the discussions leading up to the publication of the ISO Recommendations. Indeed, many of the ISO documents have originated from suggestions and papers submitted initially by U.K. representatives.

The 1964 edition of BS 308 took into account the ISO work[2, 3, 4] available at that time, but as far back as 1966 work was commenced on a further edition in order to include the ISO method of specifying geometrical tolerancing by the use of symbols[5]. This revision gave the opportunity of presenting the standard in the metric system in accordance with the changeover gradually taking place throughout the country at that time by government legislation.

As far as engineering drawing practice was concerned, in changing to the metric system, no alterations were made either to the principles of drafting or to the existing interpretations of the text of BS 308 : 1964. Sufficient information had been given in that edition for the production of drawings in the metric system. Every effort was made by the BSI Technical Committee to provide a clear and unambiguous document which would maintain a high standard of work throughout industry in this country. In so doing, it has also maintained its high reputation in the major industrial countries throughout the world.

Means of communication

The engineering drawing is the fundamental means of conveying information from the designer to the manufacturer. It should provide all the details of how a part is to be made or assembled. A draughtsman must be trained to give the information clearly and in such a way that its meaning cannot be misunderstood. It is a Company record.

Clarity of presentation

The drawing should be neatly and carefully executed and thought should be given to the possible ways in which it may be reproduced for future use and reference. For uniformity of density of the print-out the drawing should be produced entirely in ink or entirely in pencil because the reflectance of each is different on the drawing sheet. The microfilming equipment used to reduce the original for filing and storage, and subsequently the enlargement of the film, demands certain qualities and techniques in drawing. These requirements should be incorporated in general principles of engineering drawing in order to have one discipline for all drawings.

Desirability of a common language

Engineering drawing practice seems to have developed into a very sophisticated set of rules in only a relatively short time. The reason for this is twofold. Firstly, the requirements of mass production in industry demands greater care and thought to be given in design drawing to ensure a greater assurance of interchangeability. Secondly, with worldwide associations between the industrial nations, an international convention is necessary to ensure that drawings can be easily read and understood without the barrier of language.

Graphical symbols have been in use for some time to represent complicated pieces of equipment or assemblies and to save valuable drafting time. Such symbols have been rather restricted to the processing industries and some national standardisation has been achieved in these fields, see Table 3.5.

In recent years international work has reached agreement on graphical symbols to take the place of some of the precise instructions formerly applied to drawings as written notes. These symbols indicate the geometrical tolerances of form and position, other than the normal tolerances of size, for the purposes of manufacture. This series is published as an ISO Recommendation and has been widely adopted in industry.

Layout

Care in setting out the drawing is most important. Draughtsmen develop individual styles but strict adherence to the conventions of projection, line thicknesses, methods of sectioning, etc., are important to make the drawing readily understood. Although recommendations are made on the style and size of written characters, the method of applying notes is the responsibility of the individual. Notes properly and expertly set out greatly enhance the drawing and use should be made of lightly ruled lines or backing paper.

Some attempt has been made to standardise drawing sheet format but no success has been achieved because the requirements of individual companies can differ so much. A clear preprinted format is a great asset to a Company in saving much time in both the drawing office and production departments. A typical layout of a drawing sheet format is shown in Figure 3.1.

DRAWING SHEET SIZES

The ISO A series of paper sizes is the only series recognised internationally for use as drawing sheets. The series is not a new one, international documentation goes as far back as 1938, it has been adopted widely in this country. The sizes of drawing sheets are covered in BS 3429[6] and sizes of paper and board in BS 4000[7].

The basic size sheet AO has dimensions of 841 mm × 1189 mm and has an area of 1 square metre. The proportions of the sides are $1 : \sqrt{2}$. In order to obtain the smaller sheets, the longest side is halved – and so on down the range; the proportions of the sides remains $1 : \sqrt{2}$ each time. The five most popular sheets are AO, A1, A2, A3 and A4. Some

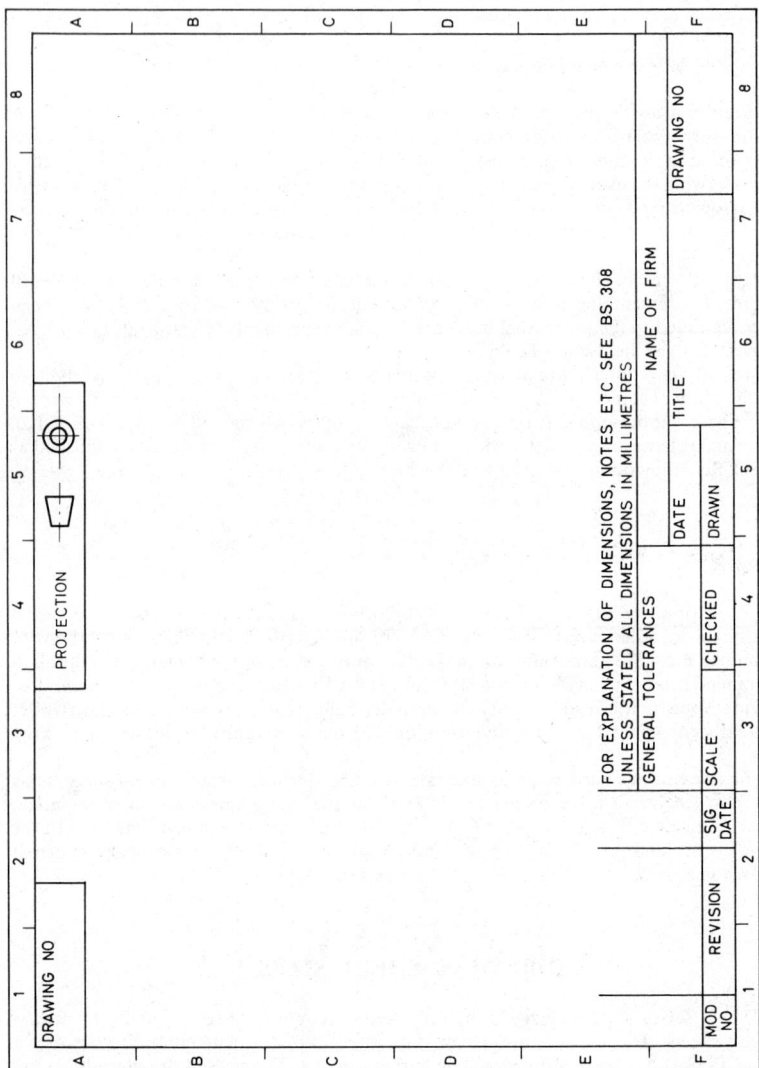

Figure 3.1. A typical layout of a drawing sheet. Showing some of the information which may be of help if printed permanently on the sheet.

need, however, has been found for two extra sizes above and below this group and the recognised list is as follows:

Designation	Dimensions of the trimmed sheet (mm)	
4AO	1682 × 2378	
2AO	1189 × 1682	
AO	841 × 1189	The sizes AO–A4
A1	594 × 841	inclusive are
A2	420 × 594	considered
A3	297 × 420	adequate for most
A4	210 × 297	purposes.
A5	148 × 210	
A6	105 × 148	

DRAWING FORMAT

It is general practice to include the drawing number in the bottom right-hand corner and it is often repeated in the top left-hand corner of the drawing sheet.

Most drawing offices use preprinted drawing sheets. The drawing sheet format is termed as including the border within the trimmed edge of the drawing sheet and information or provision for information necessary to accompany the drawn details. This information contains the name of the Company, the drawing number and various job references, general notes concerning materials, tolerances and quality, amendment procedures and the like. Some offices include a form of grid reference along the edge of the sheet to assist in location of detail and with the advance of microfilm reproduction of engineering drawings, machine centring marks are also being included on the drawing sheet format.

Borders

It has already been stated that due to the many and differing requirements of individual companies, standardisation has not been carried out on drawing format. However, due to the particular requirements of microfilm reproduction, a series of border widths has been established in BS 4210[8]. The recommended widths are as follows:

    A0.   25 mm
    A1.   20 mm
    A2.   20 mm
    A3.   15 mm
    A4.   15 mm

At first sight these sizes may seem rather large but bearing in mind the reduction often made from A0 or A1 originals to A3 and A4 size prints, the resultant borders are of reasonable dimension. Furthermore, the reproduction equipment must have sufficient material on which to clamp and the machine clamps must not of course encroach upon the drawing area.

Drawing revision

It is very important to record carefully on the drawing the revisions which have been found necessary either to correct any error in drafting to improve design or to simplify manufacture. A table is often included in the format to record such modifications.

General tolerances

In many types of engineering much drawing time is saved by the use of general tolerances and a table of values is included in the format. When general tolerances are employed, it must not be forgotten that the responsibility is with the designer to determine that such general size limits are suitable for those dimensions which are not individually controlled.

Scales

An essential piece of information which should be included in the format of the drawing is the original scale. This indicates the scale to which the drawing was made. When the drawing is reproduced, the print size may be reduced considerably from the original dimensions and it is therefore necessary to give an indication of the actual size of the component represented. It must however be borne in mind that a drawing should never be scaled if some necessary piece of information is omitted. Therefore, fundamentally, the record of the scale is for the benefit of the designer.

DRAWING SCALES

In the preparation of an engineering drawing in the metric system, scale multipliers and divisors of 2 and 5 and 10 should be used. The resultant scale representative fractions which are in accordance with both ISO/R 1047[9] and BS 308[1] are as follows:

100 : 1	2 : 1	1 : 1	1 : 100
200 : 1	5 : 1	1 : 2	1 : 200
500 : 1	10 : 1	1 : 5	1 : 500
1000 : 1	20 : 1	1 : 10	1 : 1000
	50 : 1	1 : 20	etc.
		1 : 50	

Scale rules are specified in BS 1347[10], Part 3 'Metric Scales', and in this scales are grouped together in order to establish a fairly short series of scale rules embodying all those scales most likely to be in common use. There may have been instances in the imperial system where a special scale has been found to be of advantage, and in these cases truly metric equivalents or alternatives are impossible to find. Such special dividings may never find their way into the Standard, but it is expected that manufacturers will continue to meet the demand and converted imperial scales should be obtainable for some time to come.

Types of lettering

Lettering should be bold, clear and upright. Characters should be of such a size that they can be easily read if the print of the drawing is reduced in size following microfilm reprography. Capital letters should be used unless lower case characters are more appropriate, i.e. in abbreviations for units or as specified in other standards.

Stencils are used extensively and these produce very clear characters but there is no reason why normal handwriting should not be used. Providing the writing conforms to the general requirement of legibility and drawing printout it does not matter which methods are used for its application. Written information is often applied by means of a typewriter. This has proved quite satisfactory since only the smaller size drawing sheets can be inserted into the machine. The size of the characters on the typewriter is restricted by the machine itself, therefore they may not be suitable for use on the largest drawing

ABCDEFGHIJKLMN
OPQRSTUVWXYZ

1234567890

1234567890

Figure 3.2. Recommended style of lettering

sheet, even if there were machines that could accept them. The recommended style of lettering is shown in Figure 3.2.

Types of lines

Lines should be bold and black and of even density and thickness throughout. All lines on the same drawing sheet may be entirely in ink, or entirely in pencil; a mixture of pencil

Table 3.1. TYPES OF LINES AND THEIR APPLICATIONS

TYPE OF LINE	EXAMPLE OF APPLICATION
Continuous (thick)	Visible outlines and edges
Continuous (thin)	Imaginery outlines and edges Dimension and leader lines Hatching lines for sections Outlines of adjacent parts Outlines of revolved sections
Continuous irregular (thin)	Limits of partial views or sections, if the line is not an axis
Short dashes (thin)	Hidden outlines and edges
Chain (thin)	Centre lines Extreme positions of movable parts.
Chain (thick at ends, thin elsewhere)	Cutting planes
Chain (thick)	Indication of surfaces which have to meet special requirements

and ink may be used, provided uniform density and reflectance is maintained. Mainly for the purposes of microfilming and subsequent print-out of the drawing, adjacent lines should be spaced not less than 1·5 mm apart. It is appreciated that in observing this rule, in some instances the scale of the drawing will be violated.

Two thicknesses of lines are used on engineering drawings and these are referred to in general terms as thick and thin. More precisely, the thicknesses are defined as Thick = 0·7 mm wide and Thin = 0·3 mm. Chain lines comprise long dashes alternated with short dashes and they should start and finish with a long dash. Centre lines should extend beyond the outline of the feature for a short distance, unless they are to be used for dimensioning or other purposes. Centre lines should not terminate on another line, nor should they extend through the spaces between views.

Types of lines and their application are indicated in Table 3.1.

PROJECTIONS

Two methods of projections are accepted internationally and are of equal status. First Angle Projection (Method E) and Third Angle Projection (Method A), (E for European and A for American). There does not seem to be a clear advantage of one method over the other and their use is evenly divided in this country. A statement of which method of projection is used should always appear on the drawing—perhaps preprinted in the format or title block. Figure 3.3 shows how the two methods are developed.

The normal method of drawing showing plan, elevation and side views in single plane is known as orthographic projection. Some help in understanding what is required by the drawing, can be obtained in showing a pictorial view of the part or assembly. There are several types of pictorial projection[11] and the most well known and often used of these is the isometric projection.

The isometric projection is obtained by drawing the normal 'front view' of the orthographic projection at an angle of 30° to the horizontal, vertical lines remaining vertical. It is easier to show by a diagram rather than explain in words how this is done and Figure 3.4 shows an isometric projection of the block used to illustrate 1st and 3rd angle projections in Figure 3.3(a) and (b).

Views

In addition to the normal views: elevation, side views and plan provided by the 1st and 3rd angle methods of orthographic projection, certain other details may be necessary in order to give all the information on the drawing. Auxiliary views giving the detail of a sloping surface or of one which may not be seen clearly in any of the orthographic views are very helpful, see Figure 3.5. Sections through the part assist in appreciating hidden detail of the physical features of the part or how it is assembled, Figure 3.6. In both these techniques it is essential that a clear indication is given on the drawing to show the direction in which the view is to be seen.

Sections

Hatching may be omitted where the meaning of the drawing can be made clear without it. However where it is necessary, sections should be hatched by means of thin parallel lines at approx. 45° to the usual horizontal plane and not less than 4 mm apart. Hatching lines should change direction on adjacent parts in an assembly in order to show a clear separation of the units, see Figure 3.7. In the case of components of very thin section, e.g. sheet metal, the section may be shown as a thick line, not necessarily to scale, see Figure 3.8,

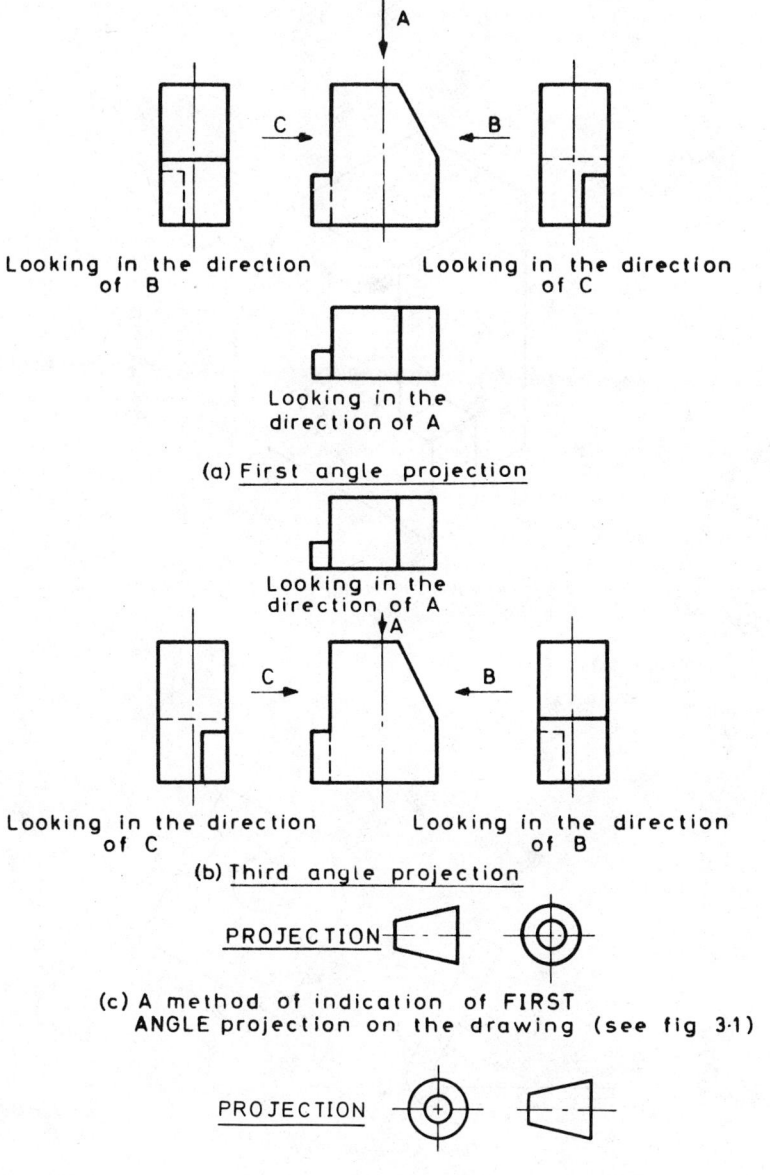

Looking in the direction of B

Looking in the direction of C

Looking in the direction of A

(a) First angle projection

Looking in the direction of A

Looking in the direction of C

Looking in the direction of B

(b) Third angle projection

PROJECTION

(c) A method of indication of FIRST ANGLE projection on the drawing (see fig 3·1)

PROJECTION

(d) A method of indication of THIRD ANGLE projection on the drawing

Figure 3.3. First and third angle projections

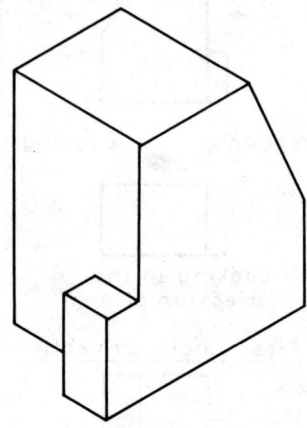

Figure 3.4. Isometric projection of the block shown in Figure 3.3

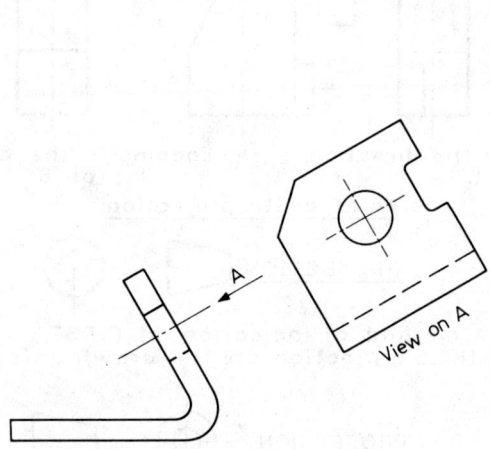

Figure 3.5. Auxiliary view

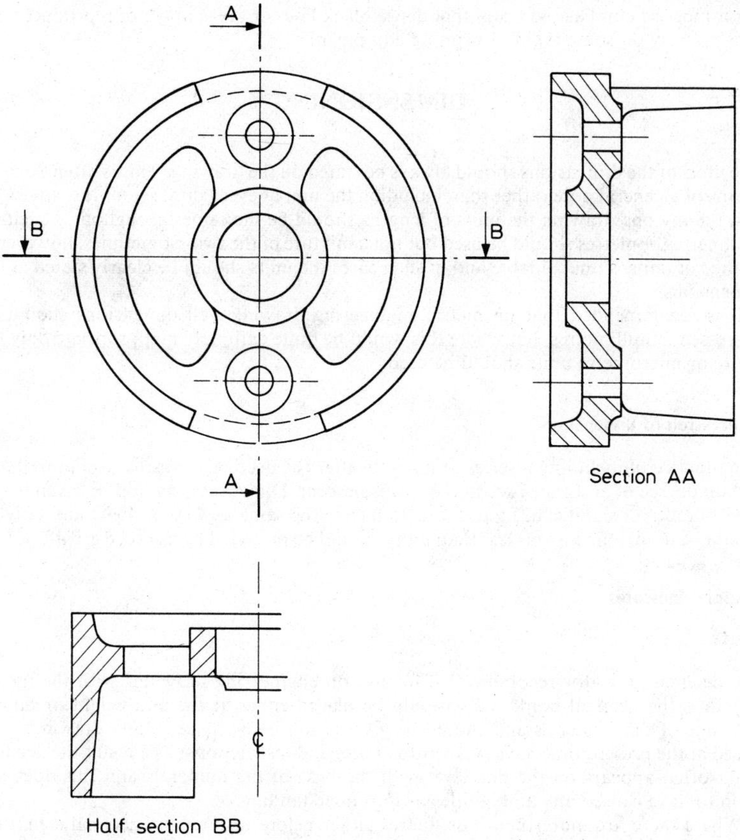

Section AA

Half section BB

Figure 3.6. Sections

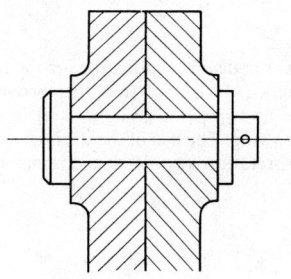

Figure 3.7. Adjacent parts in section

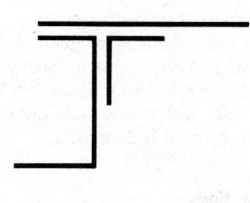

Figure 3.8. Thin sections

but it must be emphasised again that dense black lines or areas may not reproduce very satisfactorily on some types of printing equipment.

DIMENSIONING

Units

The units of the dimensions should always be stated on the drawing. This is often done by means of a general note rather than including the unit at every dimension. It is advisable that on any one drawing the units of lengths should be the same throughout, i.e. either millimetres or metres should be used but not a mixture of the two. Sometimes, however, a change of units is unavoidable and in such cases the units should be clearly stated at the dimensions.

It is recommended that on metric engineering drawings, all dimensions should be expressed in millimetres, but where this would be quite definitely inappropriate only the metre or micrometre units should be used.

Expression of dimensions

The practice of including a series of noughts after the decimal indicator to imply that a certain degree of accuracy is required is obsolescent. Dimensions should be given to the least number of significant figures, e.g. 45 means the same as 45·0 or 45·00 and only 45 should be used. Dimensions less than unity should be preceded by a zero, e.g. 0·45.

Decimal indicator

POINT

The decimal indicator recommended for use on engineering drawings with the metric system is the decimal point. This should be placed either at the mid-height or on the base line of the numerals and should be given a full letter space. Many drawings produced at the present time have typewritten notes and instructions; as a result the decimal point often appears on the line level with the base of the numerals and this does not seem to have caused any ambiguities or misunderstandings.

Where there are more than four figures either before or after the decimal sign, they should be separated into groups of three by a full letter space starting from the decimal indicator. Where there are less than four figures before or after the decimal indicator no separation should be made.
Examples: 35 250
                       4 500
                            0·002 58

COMMA

Some companies have chosen to use the comma as the decimal indicator for the metric system in agreement with general European practice. It is said that this provides a ready means of identification between the metric and imperial systems and in effect provides a common language on the engineering drawing. The use of the decimal comma will generally not create any difficulties in interpretation provided a comma is not retained as a thousands marker.

Dimension lines

Dimension lines should be thin continuous and clear. Arrow heads should be bold and filled in; the point of the arrow should be sharp and should just touch the line to which the dimension refers. Only when it is necessary by reason of shortages of space should a dimension line be broken for the insertion of the figures.

Dimension lines should preferably be placed outside the component being drawn and

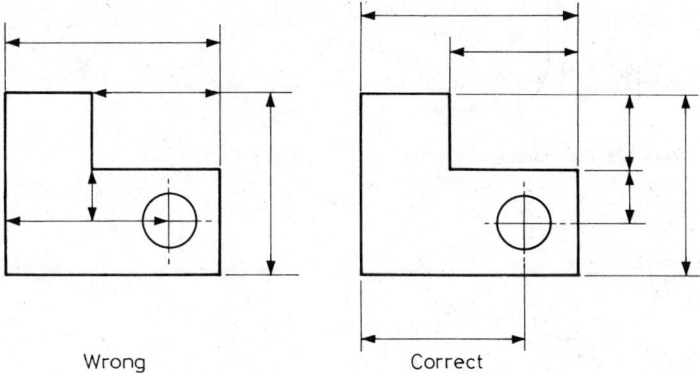

Wrong   Correct

Figure 3.9. Dimension lines

should not be part of a centreline or an extension of the outline or another dimension line, see Figure 3.9.

Arrangement of dimensions

Dimensions should be placed so as to be easily read either from the bottom or from the right hand side of the drawing sheet. Figure 3.10 illustrates how the numerals of the dimension may be applied.

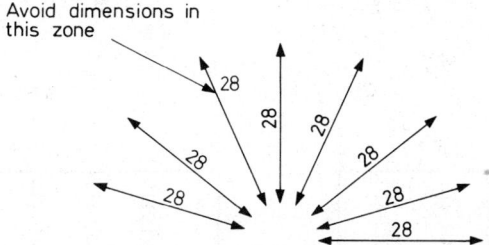

Figure 3.10. Arrangement of dimensions

The methods of dimensioning a radius and a diameter are shown in Figures 3.11 and 3.12 respectively.

Larger dimensions should be placed outside smaller dimensions and where there are several parallel dimension lines in close proximity, the figures should be staggered, see Figure 3.13.

In order to save space on a drawing, dimensioning from a common datum can be achieved effectively by continuing a dimension line through a series of features and

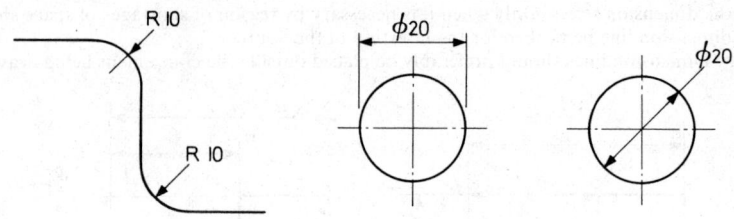

Figure 3.11. Dimensioning of a radius   Figure 3.12. Dimensioning a diameter

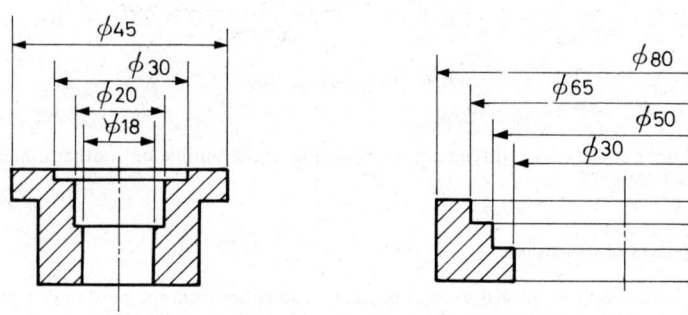

Figure 3.13. Staggered dimensioning

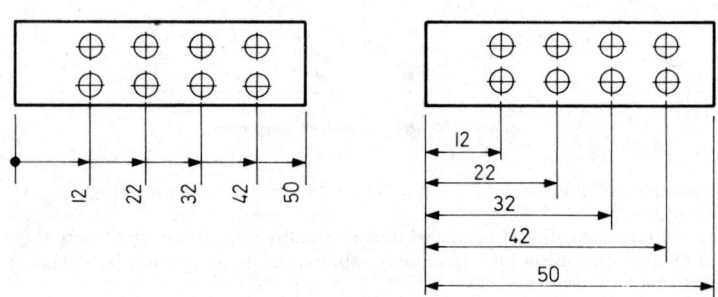

Figure 3.14. Dimensioning from a common datum

placing the values of the dimensions at each appropriate arrow head. Figure 3.14 shows the effect of this method of saving space.

Methods of dimensioning

It is important that the component detailed on an engineering drawing is functionally dimensioned. Dimensions which are not essential to the manufacture of the part are termed redundant. An example of a redundant dimension is given in Figure 3.15 where, with the inclusion of the overall length of the component, one of the dimensions of the chain is not required.

It often occurs however, that certain additional dimensions are given to assist in manufacture or the understanding of the design requirement. Such are called Auxiliary dimensions and should be enclosed in parentheses (.....). Where for functional requirements the overall length of a part is not essential and therefore redundant, it may be given in some

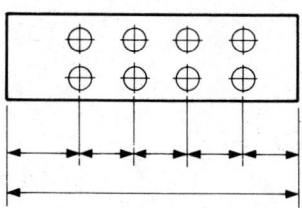

Figure 3.15. Redundant dimensioning

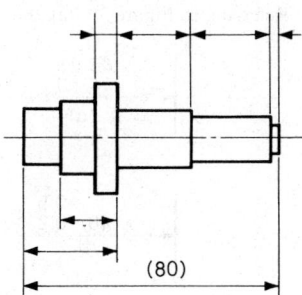

Figure 3.16. Auxiliary dimensioning

cases as an auxiliary dimension to give assistance either for other purposes of design or for early requirements of manufacture, e.g. the cut-off length of raw material. Figure 3.16.

Auxiliary dimensions should not be toleranced and when they relate to position they should refer to the theoretical true position of the features being dimensioned.

The part detailed should be drawn to scale. However, it is often inconvenient and uneconomical to carry out modifications on a drawing to the correct scale and it is sometimes helpful to exaggerate on a very small feature in order to show clearly what is required. It is inevitable therefore, that some dimensions will not be to scale. Such dimensions should be underlined with a bold straight line. The abbreviation NTS, although used widely in the past is not now recommended mainly because it would not be recognised by non English speaking countries. The concept that engineering drawing practice should be universal language is rapidly being accepted and developed by international discussions.

Dual dimensioning

The practice of dual dimensioning is deprecated but in the changeover from the imperial to the metric systems, many concerns have found it a necessary intermediate step in re-education. Dual dimensioning should not be encouraged because of the difficulties in rounding off from one system of units to the other. If extreme care is not taken in the expression of the equivalent units the same parts made by different workshops to the alternative systems may not provide exact interchangeability. The internationally agreed method of conversion is given in BS 2856[15].

When the practice is used, the system of units to which the part was designed and to which it is to be inspected should be given first and the equivalent value of the dimensions in other units should be enclosed in parentheses (.....).

TOLERANCING

Linear tolerancing

INDIVIDUAL DIMENSIONS

There are two main methods of applying tolerances to linear dimensions, either by stating both limits of size directly, Figure 3.17(a) or by specifying the basic size with limits of tolerance either side of the basic value, Figure 3.17(b).

Referring to Figure 3.17(a), the inclusion of an additional nought is perhaps the only

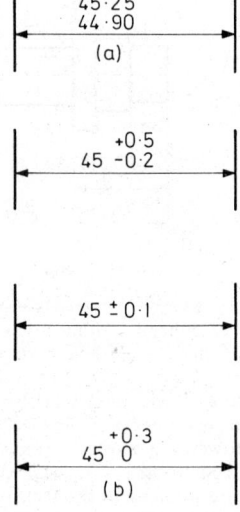

Figure 3.17. Linear tolerancing

case where the general rule is broken. To add the nought makes it absolutely clear that another value has not been forgotten or inadvertently erased.

It should be noticed from Figure 3.17(b) that when one of the limits of tolerance is zero it should be expressed by the single cipher '0' and an equivalent series of decimal places is unnecessary. Furthermore, as explained earlier, the nominal size should not be quoted to the same number of decimal places as the tolerance values.

GENERAL TOLERANCES

It is recognised that the use of a general tolerance note can save a considerable amount of time in preparing a drawing and will simplify the dimensions. The practice, however, tends to take some of the responsibility of design away from the designer and place it with the production and inspection department.

In the imperial system it is possible to express fractions by both the numerator/denominator method and by decimals. These two methods make it possible for a certain degree of accuracy to be implied by whichever type of fraction is used. The numerator/denominator fractions e.g. $\frac{1}{16}$, $\frac{1}{8}$, $\frac{1}{4}$ etc. are assumed to be of a lesser degree of accuracy than decimal fractions e.g. 0·0625, 0·125, 0·25 etc.

The metric system however, being a purely decimal system does not provide a means of implied accuracy. It is a more difficult matter therefore to establish a general tolerance structure. A number of companies have established a series of general tolerances according to the size of the basic dimensions, but the values in such a series would differ according

TOLERANCES ON MILLIMETRE DIMENSIONS EXCEPT WHERE OTHERWISE STATED	
Size	Tolerance
Up to X	± 0·X
Over X up to XX	± 0·X
Over XX up to XXX	± 0·X
Over XXX "	± 0·X
On angle	± X°
On cast thicknesses	± 0·X
On forging "	± 0·X

Figure 3.18. Indication of general tolerances on a drawing

to the type of industry concerned, i.e. the general tolerance values may look entirely different in the constructional engineering industry compared to those essential to precision engineering.

It may take many years of working experience with the metric system before an acceptable general tolerance system is established, one which would be acceptable to all related sectors of industry. Until development is complete, national standardisation cannot be attempted; standardisation should reflect the requirements of the user and it should not invent or try to impose undesirable controls and restrictions on individuals. A typical general tolerance frame structure without numerical values is given in Figure 3.18 to illustrate how such a series may appear with the drawing format.

AN INTRODUCTION TO GEOMETRICAL TOLERANCING

A geometrical tolerance is applied to a feature when there is a requirement to control its variation of form or position. It specifies the size and shape of a tolerance zone which may contain the axis, median place or surface of a feature. It is applied when dimensions and tolerances of size could not in themselves ensure adequate control over the physical shape of the feature, and may even be specified when no special size tolerance is given, e.g. a flatness tolerance on a surface table.

In many cases geometrical tolerances are to be observed regardless of the actual finished sizes of the features concerned. There are, however, other instances in which the allowable deviation depends on whether or not the feature is in its maximum material condition. The permissible error will be least when the feature is in its maximum condition and greatest when it is not.

Certain geometrical tolerancing requirements involve the establishment of other features as datums for those being toleranced. Those datum features should be chosen by consideration of the function of the part and they may themselves be controlled by geometrical tolerances of form and of position.

The principle of geometrical tolerancing is internationally accepted and established in ISO Recommendation 1101[5]. This document specifies the method by the use of symbols and these are well known. Part 3 of BS 308[1] is devoted to the presentation of geometrical tolerancing by the use of symbols and although it gives slightly more explanation in the

SYMBOL	CHARACTERISTIC TOLERANCE
——	Straightness
▱	Flatness
○	Roundness (circularity)
⌀	Cylindricity
⌒	Profile of a line
⌓	Profile of a surface
//	Parallelism
⊥	Squareness (perpendicularity)
∠	Angularity
↗	Run out
⊕	Tolerance of position
◎	Tolerance of concentricity
⚌	Tolerance of symmetry

OTHER SYMBOLS USED IN THIS PRINCIPLE

Ⓜ	Maximum material condition
▭	Boxed dimension — a dimension which defines true position

Figure 3.19. Tolerance symbols

text, it is in agreement with the ISO publication. The complete series of geometrical tolerance symbols is shown in Figure 3.19.

The concept of geometrical tolerancing is complex and many aspects become apparent in its use. As ISO/R1101 has been accepted by international agreement and adopted by all the major engineering manufacturing countries in the world, it would be inadvisable to paraphrase in this handbook what is already said both in the ISO document and in

BS 308[1], Part 3, thus possibly giving rise to further interpretations to the symbols and their applications.

Tolerance frame

A rectangular frame is used to indicate the characteristic and the value of the tolerance with its relationship to any particular datum feature. The tolerance frame is divided into compartments into which the various pieces of information are placed in order to specify the required control.

The symbol for the characteristic is indicated in the extreme left compartment and

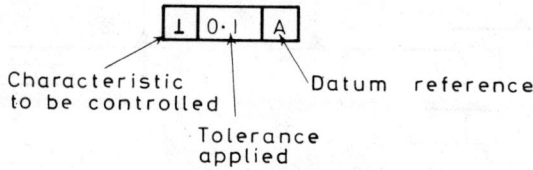

Figure 3.20. Tolerance frame

moving from left to right, the total value of the tolerance is given in the second compartment (if the tolerance value depicts a circular or cylindrical zone it is preceded by the symbol ϕ). Where a datum has to be identified the datum identification letter is placed in the third compartment, see Figure 3.20.

The feature controlled

The feature toleranced is indicated by a leader line which is connected to the tolerance frame. The leader line is terminated in an arrow head either at the surface or at a dimension line dependent on whether the surface itself or the axis of the feature is being controlled, see Figure 3.21.

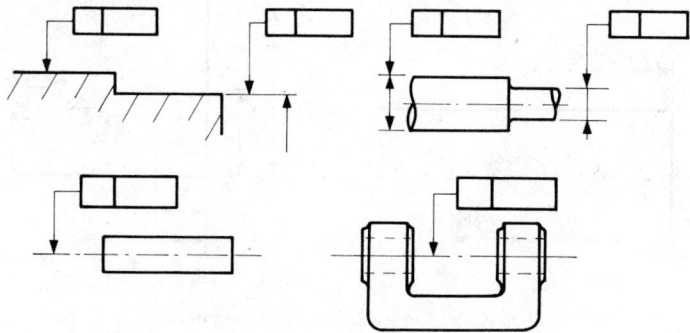

Figure 3.21. The feature controlled

The datum feature

The datum feature is indicated by a leader line which is connected to the tolerance frame but which terminates in a solid triangle either at the surface or at a dimension line dependent on whether the surface or the axis of the feature is designated the datum, see Figure 3.22.

In cases where a tolerance frame cannot be conveniently connected to the datum feature, the datum indentification, a capital letter, in a frame, is connected to the datum feature, the leader line terminating in a solid triangle, see Figure 3.23.

Dimensions which define true position

If tolerances of position, profile or angularity, are to be specified, the dimensions specifying the true position of the features cannot be toleranced. The nominal dimensions of

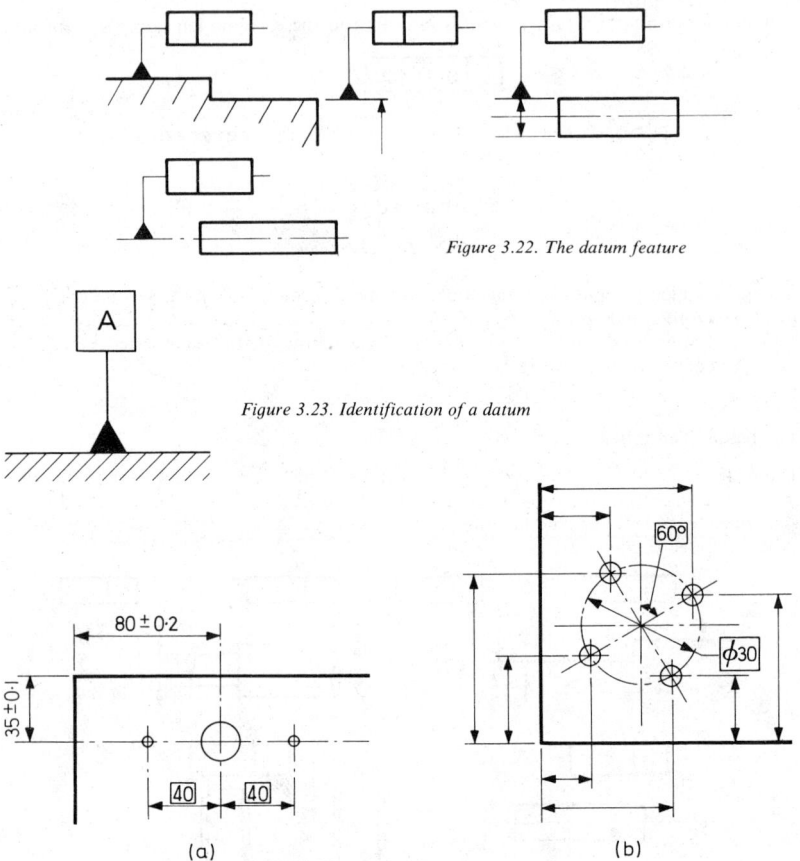

Figure 3.22. The datum feature

Figure 3.23. Identification of a datum

Figure 3.24. Nominal dimensions of true position

true position are placed in a 'box' or rectangular frame to indicate that they are theoretically exact or basic dimensions, see Figures 3.24(a) and 3.24(b).

INDICATION OF SURFACE TEXTURE

Machining and surface texture symbols should be shown on the actual surface under consideration or on an extension to the surface or on a leader line associated with the surface. A symbol should normally occur once on the surface or other line. Where it is necessary to indicate that a particular surface texture or the process to be used is undefined, the symbol of the form of Figure 3.25(a) should be used and it should be drawn normal to the surface as shown in Figure 3.25(b).

Where the component is to be machined all over an additional note can be used, see Figure 3.25(c).

Where a particular surface is to be machined to a specified surface texture, the required surface texture value should be placed above the general symbol, see Figure 3.25(d);

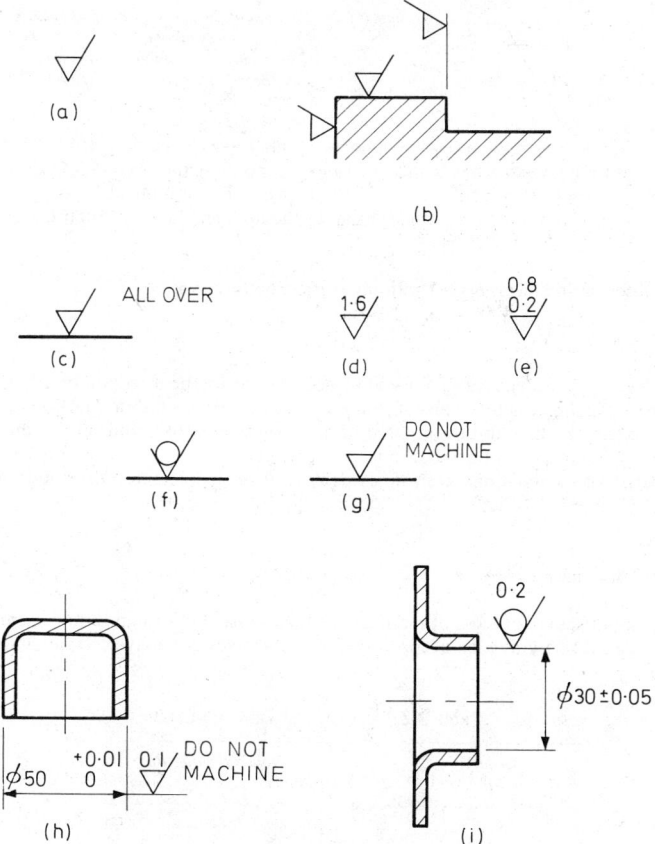

Figure 3.25. Indications of surface texture

the value of the surface texture is expressed numerically in micrometres. If a range of surface texture is permitted the symbol will indicate it accordingly, see Figure 3.25(e).

Where a particular surface *must not* be machined then the symbol shown in Figure 3.25(f) may be used, or a note can be used in association with the general symbol, see Figure 3.25(g).

When a particular surface texture is required but in addition the surface must not be machined, the requirement would be expressed by either of the methods shown in

Figures 3.25(h) and (i). This method of indicating surface texture on drawings which is specified in BS 308 is in agreement with ISO/R 1302[14].

Position of the specifications of surface texture used in conjunction with the symbol

Quite apart from the numerical value of the surface texture, a number of other instructions may be required in order to define completely the quality of the surface. It is

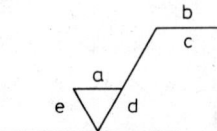

Figure 3.26. Position of the specifications of surface texture:
(a) Roughness number;
(b) Production method treatment or coating;
(c) Sampling length;
(d) Direction of lay;
(e) Machining allowance

important that a convention is adhered to in stating these requirements in order to eliminate numerous written instructions on the drawing. The specifications of surface texture should be placed in relation to the symbol in the manner shown in Figure 3.26.

Other information incorporated with the surface texture symbol

DIRECTION OF LAY

If it is necessary for special functional reasons to specify the direction of lay, it should be shown on the drawing by means of a note or by the use of a symbol. The direction of lay is described as the direction of the predominant surface pattern and this is generally controlled by the method of production.

A series of symbols has been devised by ISO to denote the most common of these surface patterns, see Figure 3.27.

PRODUCTION PROCESS

Where a particular process of manufacture is an essential design requirement, then the surface should be indicated accordingly. Machined surfaces have certain characteristics

Table 3.2   ROUGHNESS GRADE NUMBERS

Grade No.	Nominal value Ra	
	micrometre	microinch
N12	50	2 000
N11	25	1 000
N10	12·5	500
N 9	6·3	250
N 8	3·2	125
N 7	1·6	63
N 6	0·8	32
N 5	0·4	16
N 4	0·2	8
N 3	0·1	4
N 2	0·05	2
N 1	0·025	1

Lay symbols	Indication in conjunction with the surface texture symbol and the interpretation	
=	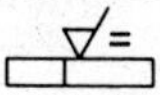 Drawing indication 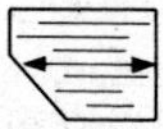 / Interpretation	Parallel to the plane of projection of the view in which the symbol is used is used
⊥	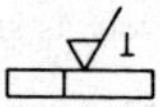 Drawing indication 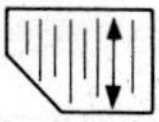 / Interpretation	Perpendicular to the plane of projection of the view in which the symbol is used
X	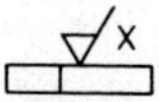 Drawing indication  / Interpretation	Crossed in two slant directions with regard to the plane of projection of the view in which the symbol is used
M	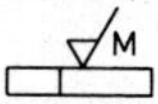 Drawing indication 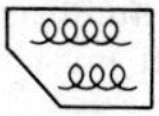 / Interpretation	Multi-directional
C	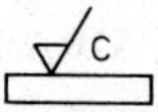 Drawing indication  / Intepretation	Approximately circular relative to the centre of the surface to which the symbol is applied
R	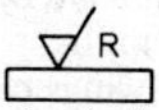 Drawing indication  / Interpretation	Approximately radial relative to the centre of the surface to which the symbol is applied

Figure 3.27. Lay symbols

dependent on the type of machining used and the correct functioning of the part may depend on the correct machining process being applied. The designer should therefore state if it is necessary to use a particular machining method, e.g. grinding, lapping, etc., suitably abbreviated, in conjunction with the surface texture symbol, see Figure 3.28. A machining allowance may also be given if this is necessary.

SAMPLING LENGTH

The length of the surface to be used in the inspection of the surface texture is known as the sampling length and a standard value is established in BS 1134[12]. If a length other than that in the standard is required, it should be clearly stated by a note in conjunction with the surface texture symbol as in Figure 3.29(a), or alternatively as shown in Figure 3.29(b).

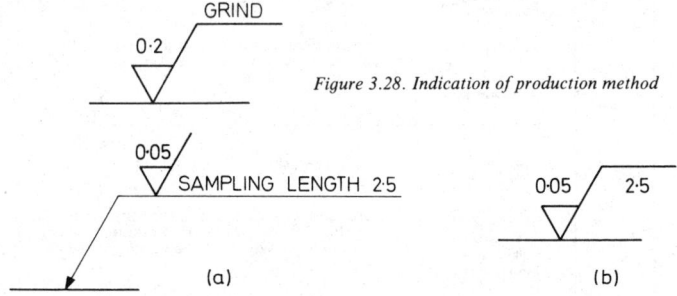

Figure 3.28. Indication of production method

Figure 3.29. Sampling length

Ra values. A comparison of the Roughness numbers, Ra, between the metric and imperial series is given in Table 3.2. The N series of numbers in the internationally agreed notation for the identification of the values. (It must be noted that the term MICROMETRE is to be used. NOT MICRON.)

ABBREVIATIONS FOR USE ON DRAWINGS

Abbreviations are the same in the singular and in the plural. Full stops should not be used unless the abbreviation makes a recognised word in itself e.g. FIG., the abbreviation for Figure. Most engineering draughtsmen write their Notes in capital letters and generally speaking this would make no difference to the meaning of the abbreviations, but there are some cases where a ruling on the use of lower case letters must be observed. Units of length in the metric system make use of the significant difference between upper case and lower case letters, e.g. M- for mega- and m- for milli-.

A set of abbreviations appropriate to engineering drawings can be found in BS 308[1] and further abbreviations covering many fields both academic and industrial are published in BS 1991[13]. It would not be the intention to override these, but the list of abbreviations in Table 3.3 is given to illustrate those most commonly used, and others which may still be seen on older drawings.

SIMPLIFIED DRAFTING

Much drawing office time can be saved by omitting all repetitive work and by the representation of commonly known or proprietary features in a simplified form. For instance,

Table 3.3 GENERAL ENGINEERING TERMS

Term	Abbreviation	Term	Abbreviation
Across flats	A/F	Kilogram	kg
Angles—Degree	°	Left hand	LH
Minute	′	Long	LG
Second	″	Machine	M/C
Assembly	ASSY	Machined	M/CD
British Standard	BS	Material	MATL
Centimetre	cm	Maximum	MAX
Centres	CRS	Metre	m
Centre line	CL or ℄	Millimetre	mm
Chamfered	CHAM	Minimum	MIN
Cheese head	CH HD	Not to scale	NTS
Countersunk	CSK	Number	NO.
Countersunk head	CSK HD	Outside diameter	O/D
Counterbore	C'BORE	Pattern number	PATT NO.
Cylinder or cylindrical	CYL	Pitch circle diameter	PCD
Datum system	DATUM	Pneumatic	PNEU
Diameter—preceding a dimension	φ	Radius—preceding a dimension (capital letter only)	R
—in a note	DIA	Required	REQD
Drawing	DRG	Revolutions per minute	r/min
External	EXT	Right hand	RH
Figure	FIG.	Round head	RD HD
Galvanised	GALV	Screw threads	See
Hardness—Brinell	HB		relevant BS
Rockwell		Screwed	SCR
A scale	HRA	Sheet, when preceding a material	SH
B scale	HRB	Sketch	SK
C scale	HRC	Specification	SPEC
D scale	HRD	Sphere—preceding a dimension (or spherical)	SPHERE
E scale	HRE	Spotface	S'FACE
Vickers	HV	Square—preceding a dimension	□
Hexagon	HEX	Standard	STD
Hexagon head	HEX HD	Undercut	U'CUT
Hydraulic	HYD	Volume	VOL
Insulated or insulation	INSUL	Weight	WT
Internal	INT		
Internal diameter	I/D		

Table 3.4 TERMS RELATING TO DIMENSIONS AND TOLERANCES

Term	Abbreviation	Term	Abbreviation
Basic dimension	BASIC	Flatness tolerance	FLAT TOL
Datum		Parallelism tolerance	PAR TOL
Datum system	DATUM	Positional tolerance	POSN TOL
Datum dimension		Roundness tolerance	RD TOL
True position, or true profile,		Straightness tolerance	STR TOL
dimension in conjunction with		Squareness tolerance	SQ TOL
positional or profile tolerances	TP	Symmetry tolerance	SYM TOL
Angularity tolerance	ANG TOL	Tolerance zone (profiles)	TOL ZONE
Concentric tolerance	CONC TOL	Maximum material condition	MMC
Cylindricity tolerance	CYL TOL	Full indicated movement	FIM

it should not be necessary to draw all the teeth in a set of intermeshing gears or in the detail of a spur wheel, nor should the threads be drawn on a conventional nut and bolt, see Figure 3.30.

A spring may also be drawn schematically rather than in complete detail. A splined shaft may need rather more thought given to it to detail the form of the splines, but only one need be shown. Special features and components may be simplified and repetition should be avoided, see Figure 3.31. Furthermore, if a drawing comprises a series of units of exactly the same specification, only one unit need be drawn in sufficient detail to identify it, and the others should be referred to by notes, or by an outline and location details, see Figure 3.32 and 3.33. In some instances freehand drawing should be permitted, although of course this will be dependent upon the artistic temperament of the draughtsman.

Having simplified the nut and bolt type representation, it should not be difficult to develop the principles of simplified drafting to individual details which are either symmetrical or otherwise have repeated features within them.

Part sections and part views are used to illustrate symmetry or perhaps to show a particular detail in another view for clarification, see Figures 3.34 and 3.35. In drawing only part of a component which is symmetrical about one or two centre-lines, care should be taken to avoid any possibility of misinterpretation resulting in the production of 'half' and 'quarter' parts. Outlines should be extended beyond the axis of symmetry in such a manner as to show clearly what is intended, see Figures 3.36 and 3.37.

If sectioning involves large areas, partial hatching, if it is necessary, need only be drawn at the edges. The indication of common items can effectively be made by note, see Figure 3.38, and unnecessary views and drawings should be avoided by effective brief notes and instructions as illustrated in Figures 3.39 and 3.40.

GRAPHICAL SYMBOLS FOR ENGINEERING DRAWINGS AND DIAGRAMS

A very important aspect of simplified drafting which is applicable to some types of engineering drawings but is appropriate generally to engineering diagrams is the representation of systems and equipment by means of graphical symbols. A number of British Standards have been published to cover the requirements of coal and chemical processing, and heating and ventilating industries, and in recent years a great deal of unification has been found possible in the establishment of a series of graphical symbols for the processing industries. Whereas BS 308 describes how an engineering drawing should be produced, and sets out very precisely the drawing practice to be followed, it does not include within its scope reference to diagrams which to some fields of engineering are just as necessary a function of design and installation. An engineering drawing will show how a part is made or how it is assembled, an engineering diagram on the other hand is used to plan a system or process and perhaps in another form show how the system works or may be operated.

Installation diagrams of heating and air-conditioning systems in buildings and other structures are more correctly termed drawings because they are drawn to scale (location of features if not the features themselves) and the parts are drawn in the correct physical relationship with each other. Diagrams in general however, are drawn in single plane form and the parts may not often be shown in the positions relative to each other in which they would be in reality.

This difference between a drawing and a diagram may help to explain away some of the problems which have been encountered in the representation of wiring/pipes crossing unconnected, and wiring/pipes connected. It is very important that in the preparation of

Title	Convention
External screw threads	
Internal screw threads	
Assembly of screw threads	External threads are shown covering internal threads
Spur gear	
Assembly of spur gears	

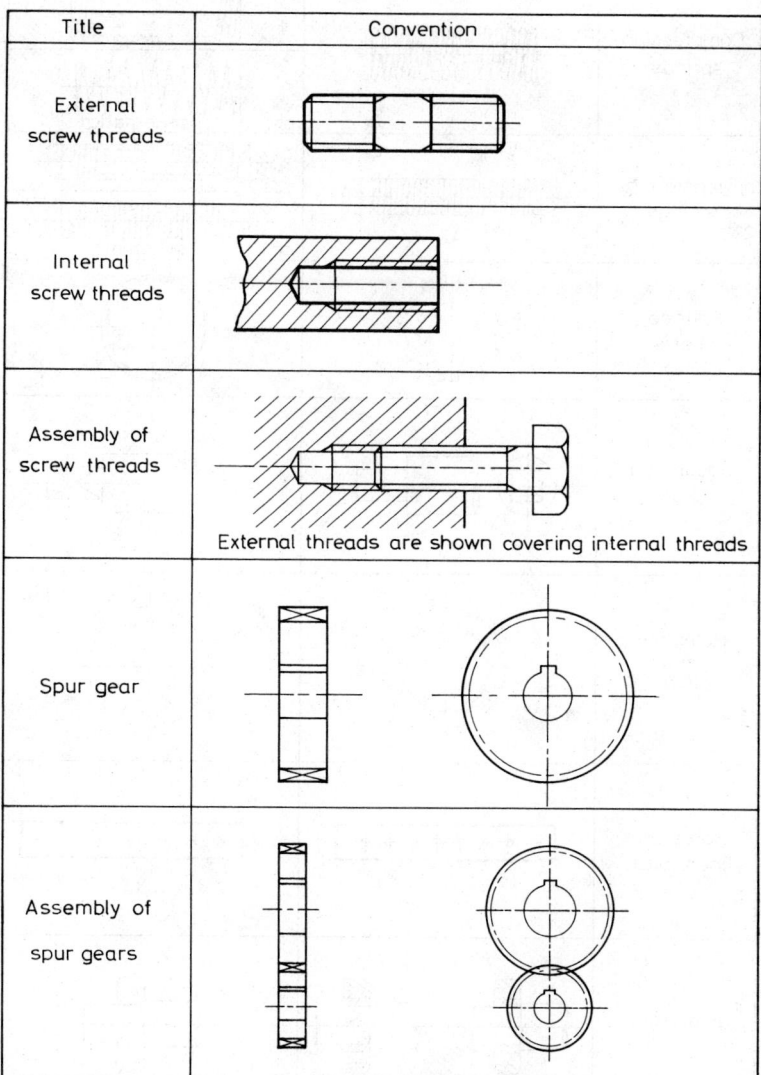

Figure 3.30. Simplified drafting of common features

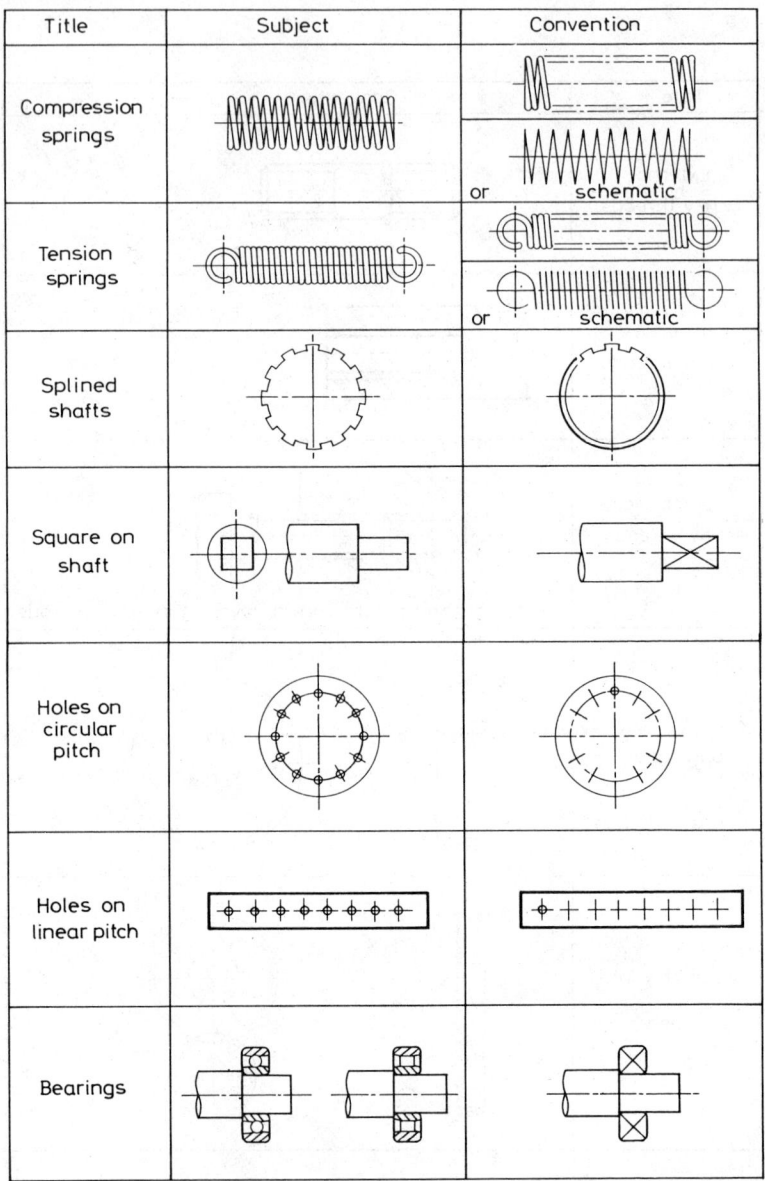

Title	Subject	Convention
Compression springs		or schematic
Tension springs		or schematic
Splined shafts		
Square on shaft		
Holes on circular pitch		
Holes on linear pitch		
Bearings		

Figure 3.31. Simplified drafting of common features

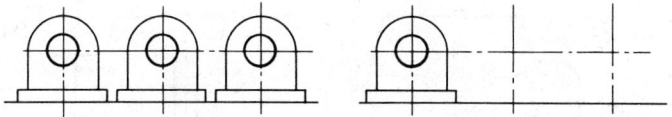

Figure 3.32. Repeated parts shown once only

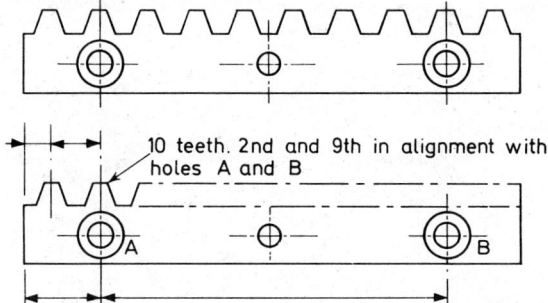

10 teeth. 2nd and 9th in alignment with holes A and B

Figure 3.33. A small number of parts to suggest a pattern

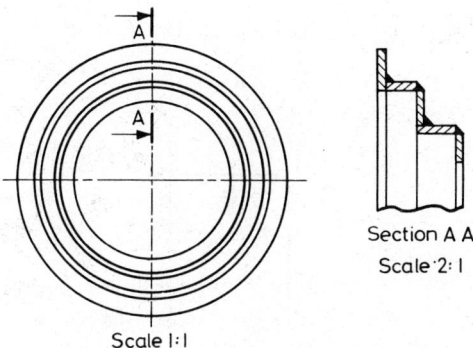

Section A A
Scale 2:1

Scale 1:1

Figure 3.34. Enlarged detail not at too large a scale

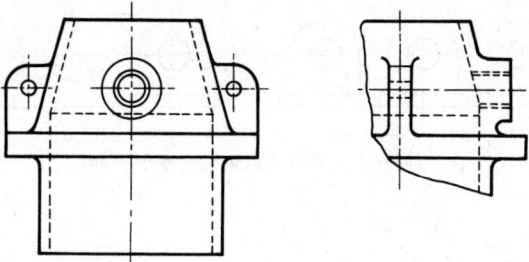

Figure 3.35. Part only of additional view drawn to show certain information

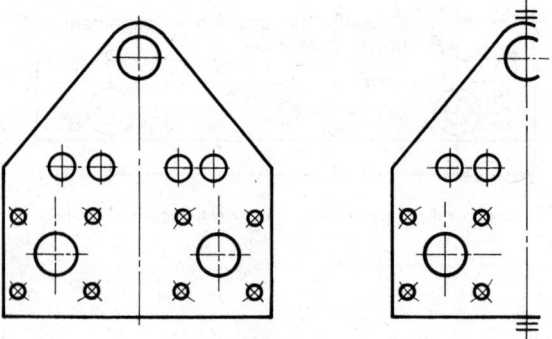

Figure 3.36. Item symmetrical about a centre line

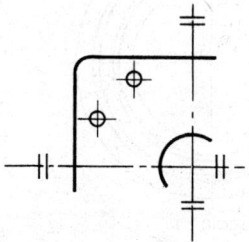

Figure 3.37. Item symmetrical about a centre line

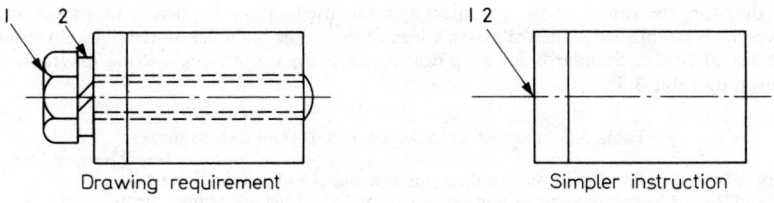

Figure 3.38. Indication of common items by note only

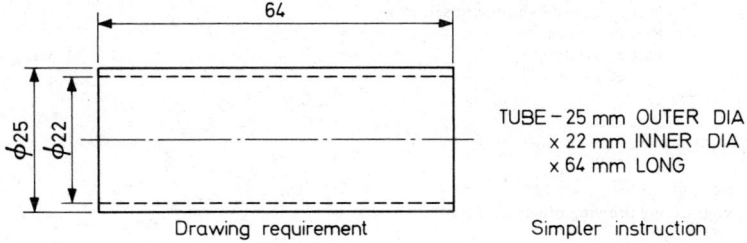

Figure 3.39. Specification by description

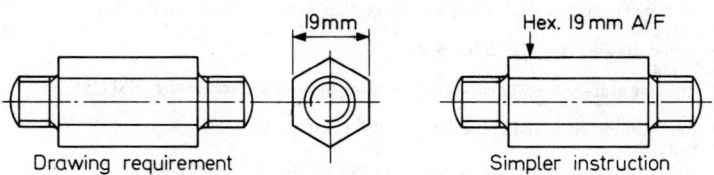

Figure 3.40. Unnecessary views omitted

a diagram, the source of the graphical symbols used should be clearly indicated, and where no established standard exists, a legend should be included on the diagram sheet. A list of British Standards for graphical symbols for use on engineering diagrams is given in Table 3.5.

Table 3.5 BRITISH STANDARDS FOR GRAPHICAL SYMBOLS

BS 974	Symbols for use on flow diagrams of chemical and petroleum plant
BS 1523	Glossary of terms used in automatic controlling and regulating systems.
	Part 1: Process and kinetic control
BS 1553	Graphical symbols for general engineering
	Part 1: Graphical symbols for pipes and valves.
	Part 2: Graphical symbols for power generating plant.
	Part 3: Graphical symbols for compressing plant.
	Part 4: Graphical symbols for heating and ventilating installations.
BS 1635	Graphical symbols for fire protection drawings.
BS 1646	Graphical symbols for process measurement and control functions.
BS 2917	Graphical symbols for use in diagrams for hydraulic and pneumatic systems.
BS 3238	Graphical symbols for components of servo-mechanisms.
BS 3553	Graphical symbols for coal preparation plant.
BS 3939	Graphical symbols for electrical power, telecommunications and electronics diagrams.
BS M24	Graphical symbols for aircraft hydraulic and pneumatic systems.
BS MA1	Graphical symbols representing pipeline systems in ships
	Part 1: General shipboard services.
	Part 2: Ventilation.
	Part 3: Sanitation.

REFERENCES

1. BS 308
   Engineering drawing practice; BSI (1972)
2. ISO R.128
   Engineering drawing, principles of presentation; ISO Geneva (1959)
3. ISO R.129
   Engineering drawing, dimensioning; ISO Geneva (1959)
4. ISO R.406
   Inscription of linear and angular tolerances; ISO Geneva (1964)
5. ISO R.1101
   Tolerances of form and of position, Part 1. Generalities, symbols, indications on drawings; ISO Geneva (1969)
6. BS 3429
   Sizes of drawing sheets; BSI (1961)
7. BS 4000
   Sizes of paper and board; BSI (1968)
8. BS 4210
   35 mm microcopying of engineering drawings and associated data; BSI (1967)
9. ISO R.1047
   Architectural and building drawings: presentation of drawings: scales; ISO Geneva (1969)
10. BS 1347
    Architects', engineers' and surveyors' scales, Part 3. Metric Scales; BSI (1969)
11. BS 1192
    Building drawing practice; BSI (1969)
12. BS 1134
    Centre line average height method for the assessment of surface texture; BSI (1961)
13. BS 1991
    Letter symbols, signs and abbreviations; BSI (1961 to 1967)
14. ISO R.1302
    Method of indicating surface texture on drawings; ISO Geneva (1971)
15. BS 2856
    Precise conversion of inch and metric units. BSI (1972)

LIMITS AND FITS

MRS. D. M. COWLEY

The British Standard BS 4500[1] supersedes BS 1916[2] which was based on the I.S.A. system embodied in ISA Bulletin 25[3]. BS 4500 is in line with ISO Recommendation R/286, the up-to-date version of ISA Bulletin 25.

Very few changes have been made to the previous tolerance values but the system has been extended to cover sizes below 1 mm and above 500 mm and to include two finer grades of tolerance and several intermediate deviations. It now covers the size range from 0 to 3150 mm and has eighteen grades of standard tolerances suitable for all classes of work from the finest – as required for gauge making and horology, for example – to the coarsest. These can be combined with any of twenty-seven fundamental deviations for holes and shafts to give every type of fit from extreme clearance to extreme interference.

It is also important to note that, although the standard refers explicitly only to cylindrical parts described as holes and shafts, and the emphasis is on fits, the tolerances are intended for the most general use and can be applied to other sections and to features which are not members of a fit, e.g. tolerances on bar stock, cutting diameters of drills and reamers, lengths, widths etc. The disposition of the tolerance is indicated by the designating letter and limits of tolerance not tabulated in the general purpose tables are very easily calculated by simple arithmetical means (see below).

Definitions

The basic terms used in the standard are given below and illustrated in Figures 3.41 and 3.42.

Actual size (of a part). The size of a part as obtained by measurement.

Limits of size. The maximum and minimum sizes permitted for a feature.

Maximum limit of size. The greater of the two limits of size.

Minimum limit of size. The smaller of the two limits of size.

Basic size. The size by reference to which the limits of size are fixed. The basic size is the same for both members of a fit.

Deviation. The algebraical difference between a size (actual, maximum, etc.) and the corresponding basic size.

Actual deviation. The algebraical difference between the actual size and the corresponding basic size.

Upper deviation. The algebraical difference between the maximum limit of size and the corresponding basic size. This is designated 'ES' for a hole and 'es' for a shaft, these letters standing for the French term 'écart supérieur'.

Lower deviation. The algebraical difference between the minimum limit of size and the corresponding basic size. This is designated 'EI' for a hole and 'ei' for a shaft, these letters standing for the French term 'écart inférieur'.

Zero line. In a graphical representation of limits and fits, the straight line to which the deviations are referred. The zero line is the line of zero deviation and represents the basic size. By convention, when the zero line is drawn horizontally, positive deviations are shown above and negative deviations below it.

Tolerance. The difference between the maximum limit of size and the minimum limit of size (or in other words, the algebraical difference between the upper deviation and the lower deviation). The tolerance is an absolute value without sign.

Tolerance zone. In a graphical representation of tolerances, the zone comprised between the two lines representing the limits of tolerance and defined by its magnitude (tolerance) and by its position in relation to the zero line.

Fundamental deviation. That one of the two deviations, being the one nearest to the zero

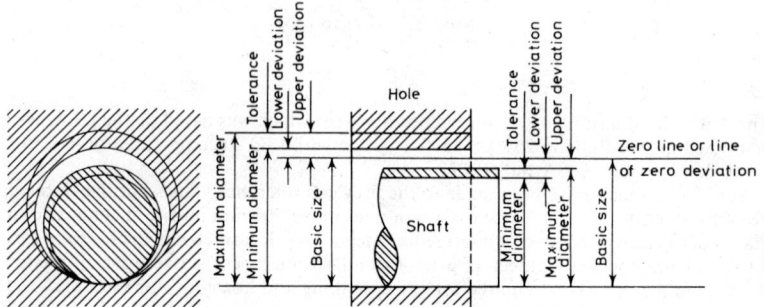

Figure 3.41. Diagrammatic illustration of definitions

line, which is conventionally chosen to define the position of the tolerance zone in relation to the zero line.

Grade of tolerance. In a standardised system of limits and fits, a group of tolerances considered as corresponding to the same level of accuracy for all basic sizes.

Fit. The relationship resulting from the difference, before assembly, between the sizes of the two parts which are to be assembled.

Clearance. The difference between the sizes of the hole and the shaft, before assembly, when this difference is positive.

Interference. The magnitude of the difference between the size of the hole and the shaft, before assembly, when this difference is negative.

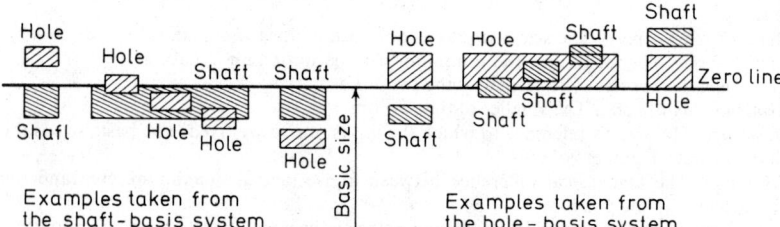

Figure 3.42. Shaft-basis and hole-basis systems of fits

Clearance fit. A fit which always provides a clearance. (The tolerance zone of the hole is entirely above that of the shaft.)

Interference fit. A fit which always provides an interference. (The tolerance zone of the hole is entirely below that of the shaft.)

Transition fit. A fit which may provide either a clearance or an interference. (The tolerance zones of the hole and the shaft overlap.)

Shaft-basis system of fits. A system of fits in which the different clearances and interferences are obtained by associating various holes with a single shaft (or, possibly, with shafts of different grades, but having the same fundamental deviation). In the ISO System, the basic shaft is the shaft the upper deviation of which is zero.

Hole-basis system of fits. A system of fits in which the different clearances and interferences are obtained by associating various shafts with a single hole (or, possibly, with holes of different grades, but always having the same fundamental deviation). In the ISO System, the basic hole is the hole the lower deviation of which is zero.

Hole-basis system of fits. A system of fits in which the different clearances and interferences are obtained by associating various shafts with a single hole (or, possibly, with holes of different grades, but always having the same fundamental deviation). In the ISO System, the basic hole is the hole the lower deviation of which is zero.

Tolerances, limits and fits. The character of a fit depends on the initial clearance (or lack of it) between the mating parts and the magnitude of the tolerance on each of them. The very numerous possible combinations of tolerances and deviations provide a most comprehensive and flexible system of fits and the principle of increasing the magnitude of the tolerances and deviations with diameter ensures that a hole and shaft combination giving a particular quality of fit at 50 mm diameter, for example, will give the same quality of fit at 300 mm.

Standard tolerances. There are eighteen tolerances grades designated IT 01, IT 0, IT 1 . . . IT 16 (IT stands for 'ISO series of tolerances'). In each grade the tolerance increases with diameter in accordance with an empirical formula derived from extensive practical investigations. For a given size in any grade, the magnitude of the tolerance is approximately 1.6 times as great as the tolerance for the next finer grade; above IT 6, the tolerance is multipled by 10 at each fifth step and the same rule will apply if the standard tolerance grades are extended beyond IT 16. For practical purposes, the total size range is subdivided into appropriate steps and tolerances are standardised for the mean size in each step. The standard values are given in Table 3.6 and the formulae from which they are

Table 3.6 STANDARD TOLERANCES   Tolerance unit 0·001 mm

Nominal sizes Over	To	IT 01	IT 0	IT 1	IT 2	IT 3	IT 4	IT 5	IT 6*	IT 7	IT 8	IT 9	IT 10	IT 11	IT 12	IT 13	IT 14†	IT 15†	IT 16†
mm	mm																		
—	3	0·3	0·5	0·8	1·2	2	3	4	6	10	14	25	40	60	100	140	250	400	600
3	6	0·4	0·5	1	1·5	2·5	4	5	8	12	18	30	48	75	120	180	300	480	750
6	10	0·4	0·6	1	1·5	2·5	4	6	9	15	22	36	58	90	150	220	360	580	900
10	18	0·5	0·8	1·2	2	3	5	8	11	18	27	43	70	110	180	270	430	700	1100
18	30	0·6	1	1·5	2·5	4	6	9	13	21	33	52	84	130	210	330	520	840	1300
30	50	0·6	1	1·5	2·5	4	7	11	16	25	39	62	100	160	250	390	620	1000	1600
50	80	0·8	1·2	2	3	5	8	13	19	30	46	74	120	190	300	460	740	1200	1900
80	120	1	1·5	2·5	4	6	10	15	22	35	54	87	140	220	350	540	870	1400	2200
120	180	1·2	2	3·5	5	8	12	18	25	40	63	100	160	250	400	630	1000	1600	2500
180	250	2	3	4·5	7	10	14	20	29	46	72	115	185	290	460	720	1150	1850	2900
250	315	2·5	4	6	8	12	16	23	32	52	81	130	210	320	520	810	1300	2100	3200
315	400	3	5	7	9	13	18	25	36	57	89	140	230	360	570	890	1400	2300	3600
400	500	4	6	8	10	15	20	27	40	63	97	155	250	400	630	970	1550	2500	4000
500	630	—	—	—	—	—	—	—	44	70	110	175	280	440	700	1100	1750	2800	4400
630	800	—	—	—	—	—	—	—	50	80	125	200	320	500	800	1250	2000	3200	5000
800	1000	—	—	—	—	—	—	—	56	90	140	230	360	560	900	1400	2300	3600	5600
1000	1250	—	—	—	—	—	—	—	66	105	165	260	420	660	1050	1650	2600	4200	6600
1250	1600	—	—	—	—	—	—	—	78	125	195	310	500	780	1250	1950	3100	5000	7800
1600	2000	—	—	—	—	—	—	—	92	150	230	370	600	920	1500	2300	3700	6000	9200
2000	2500	—	—	—	—	—	—	—	110	175	280	440	700	1100	1750	2800	4400	7000	11000
2500	3150	—	—	—	—	—	—	—	135	210	330	540	860	1350	2100	3300	5400	8600	13500

* Not recommended for fits in sizes above 500 mm.
† Not applicable to sizes below 1 mm.

derived are given below. Note that grades below IT 6 do not apply to sizes above 500 mm. Grades 5 to 16 are based on the tolerance unit I as follows: For I expressed in micrometres and D in millimetres:

$I = 0.45 \sqrt[3]{D} + 0.001 D$ for sizes up to and including 500 mm

$I = 0.004 D + 2.1$ for sizes above 500 mm

Table 3.7 STANDARD TOLERANCES IN TERMS OF I

	IT5	IT6	IT7	IT8	IT9	IT10	IT11	IT12	IT13
Values	7 I	10 I	16 I	25 I	40 I	64 I	100 I	160 I	250 I

	IT 14	IT 15	IT 16
Values	400 I	640 I	1000 I

Grades IT 01, IT 0 and IT 1 are calculated as follows:

	IT 01	IT 0	IT 1
Values in micrometres for D in millimetres	$0.3 + 0.008D$	$0.5 + 0.012D$	$0.8 + 0.020D$

Grades IT 2 to IT 4: Values are scaled approximately geometrically between the values of IT 1 and IT 5 (see Table 3.6).

Limits of size for a part are obtained by associating the standard tolerances with the fundamental deviations and these can be combined in ways that will give any required fit condition between two mating parts.

Fundamental deviations. The system provides twenty-eight deviations for sizes up to and including 500 mm and 14 for larger sizes. As with the tolerances, the deviations vary with size to maintain the fit characteristics and the values are standardised for the mean size in each range.

The fundamental deviation is that nearest the zero line and establishes the position of the tolerance zone in relation to the basic size; each deviation is designated by a letter or letters as follows (see also Figure 3.43):

For holes: A B C CD D E EF F FG G H J_s J K M N P R S T U V X Y Z

ZA ZB ZC

For shafts: a b c cd d e ef f fg g h j_s j k m n p r s t u v x y z   za zb zc

As can be seen from Figure 3.43, deviations A to H are wholly positive and deviations N to ZC are wholly negative while deviations a to h are wholly negative and deviations n to zc are wholly positive. Deviations J, j, K, k, M, m are partly positive and partly negative (see Tables 3.8 and 3.9). J_s and j_s are symmetrical about the zero line, thus having no fundamental deviation in the strict sense of the term but providing a convenient series of bilateral tolerances for general (non fit) applications.

For holes, when positive, the fundamental deviation is known as the lower deviation (EI) and represents the lower limit of the part, the upper limit being obtained by adding the standard tolerance required. When negative, the fundamental deviation is known as the upper deviation (ES) and represents the upper limit of the part, the lower limit being obtained by subtracting the standard tolerance required.

For shafts, the converse applies and the upper and lower deviations are designated 'es' and 'ei' respectively.

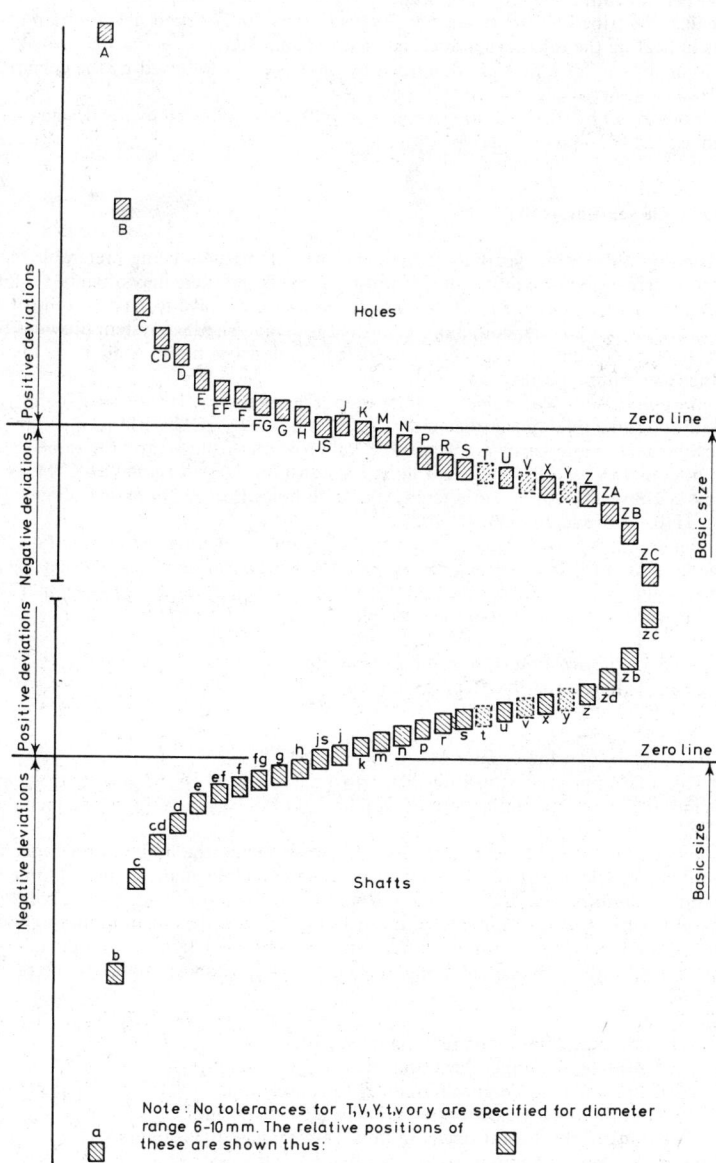

Figure 3.43. Positions of Grade 7 tolerance zones—range 6 to 10 mm diameter

Designation. A hole is described by the capital letter indicating its fundamental deviation and number indicating the tolerance grade, e.g. H7 (see Table 3.10).

A shaft is described by the small letter indicating its fundamental deviation and the number indicating the tolerance grade, e.g. p6 (see Table 3.11).

The limits of size for a part are described by the basic size followed by the deviation and tolerance designations, e.g. 50 H7 or 50 p6.

A fit is described by the basic size common to both parts followed by the designations for each, e.g. 50 H7 – p6 or 50 H7/p6.

Hole basis and shaft basis fits

It has been usual for some years to recommend the hole basis as being preferable for a standard system of limits and fits but it is now recognised that there may often be reasons for preferring a shaft basis, e.g. in the case of shafts which have to have a number of components such as collars, couplings, etc. fitted to them. The ISO system provides for both systems of fits and it is easy to convert from hole basis fits to shaft basis fits by using the rules summarised below.

Clearance fits: With shaft tolerance finer than hole tolerance: H7 – f6 = F7 – h6

With the same tolerance for shaft and hole: H11 – c11 = C11 – h11

Transition and interference fits: Where the total tolerance is large and the same grade of tolerance is used for shaft and hole, i.e. transition fits with holes J, K, M, N in grades above IT8 and for interference fits with holes P to ZC in grades above IT7: e.g. H10 – s10 is identical with S10 – s10.

Where the total tolerance is not very large, and a tolerance one grade finer for the hole than for the shaft is used, when converting from hole basis to shaft basis, the Δ value is employed to calculate the hole deviation. Δ values are given in Table 3.8 and are used as shown in the example below.

In converting the hole basis fit H7 – p6 for a size of 25 mm to the shaft basis fit P7 – h6, the upper limit (ES) for hole P7 = ei(p) + Δ.

From Table 3.8, ei(p) = 22 micrometres

From Table 3.9, Δ for grade 7 = +8

$$ES(P7) = -22 + 8$$
$$= -14 \text{ micrometres}$$

This rule applies to transition fits with holes J, K, M, N in grades up to and including IT8 and to interference fits with holes J to ZC in grades up to and including IT7.

Non tabulated limits of tolerance. Although the tables giving the limits of tolerance for commonly used holes and shafts are very extensive and include many values suitable for application to non-mating parts, they are primarily intended for fits and there will thus be occasions when tolerances indicated on drawings or schedules for non-mating parts cannot be found in them. Such limits of tolerance can easily be derived from the table of fundamental deviations concerned (Table 3.8 and 3.9) and the table of standard tolerances (Table 3.6).

Examples

1. Limits for a part 25 mm in size designated g10:
    From Table 3.8, the upper deviation (es) is – 0·007 mm
    From Table 3.6, the tolerance (Grade IT10) is 0·084 mm
    By addition, the lower deviation is therefore – 0·091 mm
    The limits of size for this part are thus 24·993 mm and 24·909 mm.
2. Limits for a part 125 mm in size designated F11:
    From Table 3.9, the lower deviation (EI) is + 0·043 mm
    From Table 3.6, the tolerance (Grade IT11) is 0·250 mm
    By addition, the upper deviation is therefore 0·293 mm
    The limits of size for this part are thus 125·043 mm and 125·293 mm.

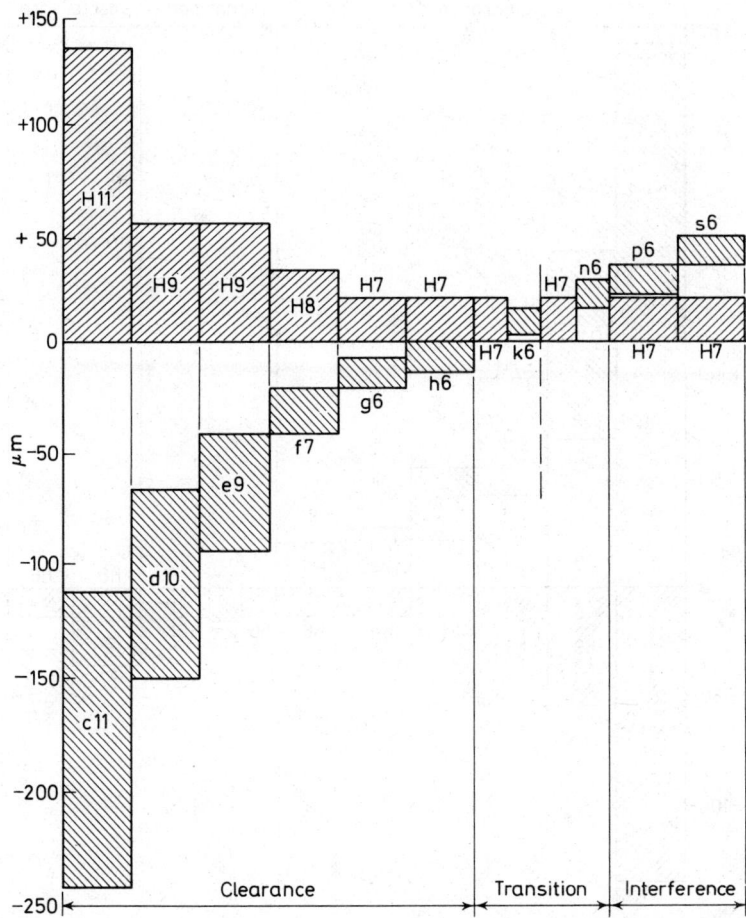

Figure 3.44. Selected fits; hole basis

Recommended fits

The ISO System is so comprehensive and flexible that it can provide an enormous range of fits based on an almost infinitely variable combination of deviations and tolerances. In practice, however, most normal engineering requirements can be met by a much more limited selection of hole and shaft tolerances.

The advantages of standardising on a limited range cannot be over-emphasised; if a user selects three or four standard holes and perhaps ten standard shafts (or three or four standard shafts and perhaps ten standard holes in the case of shaft basis fits) he can combine them in numerous ways to provide clearance, transition and interference fits

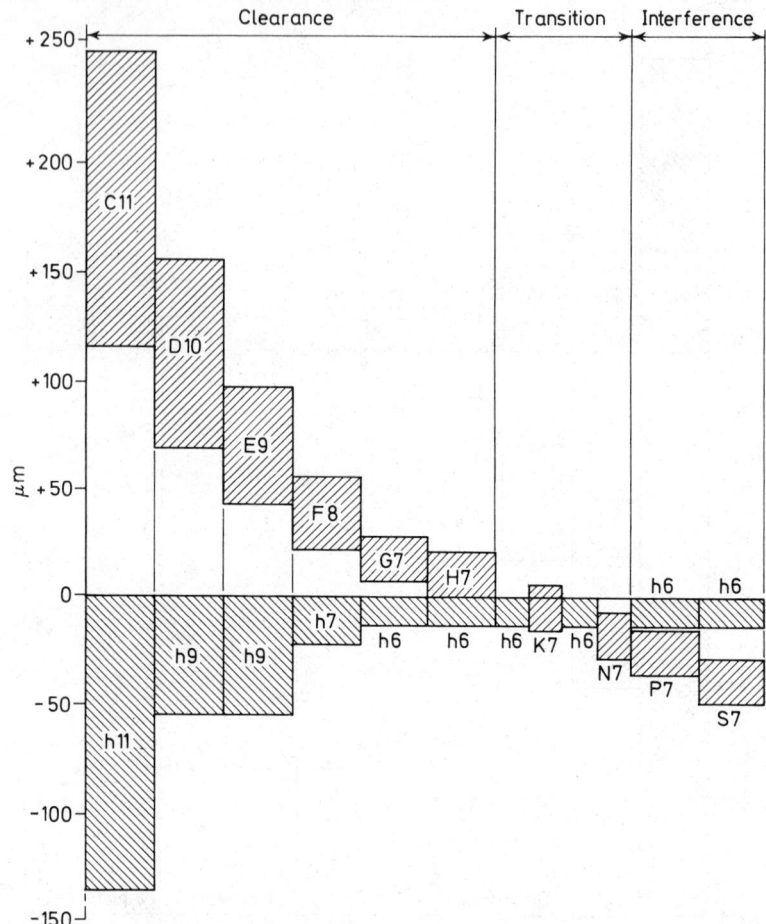

Figure 3.45. Selected fits; shaft basis

but will be able to use the same restricted range of tools and gauges for all of them. For sizes up to about 500 mm the following hole and shaft tolerances provide a range of fits which will be adequate for most normal purposes;

Selected hole tolerances: H7 H8 H9 H11.

Selected shaft tolerances: c11 d10 e9 f7 g6 h6 k6 n6 p6 s6.

Many users may be able to make an even more restricted selection from these to meet their particular requirements.

Figure 3.44 shows the selected fits on a hole basis. Figure 3.45 shows the same selection converted directly to a shaft basis, except for H9–d10 which has been adjusted to D10–h9 instead of D9–h10 to avoid introducing the additional shaft tolerance h10.

Sizes above 500 *mm.* The system now provides tolerances, limits and fits for sizes above 500 mm. The extension is based on principles closely similar to those of the original system and is designed to blend with it. It should be noted, however, that there are certain special problems applying to the manufacture and measurement of large components to small tolerances on a universally interchangeable basis. The effects of temperature are of particular importance and BS 4500[1] explains them in some detail. Where interchangeable manufacture is not practicable, the technique of matched fits may be employed. Briefly summarised, the procedure is as follows:

1. For the nominal dimension concerned, select the fit tolerance thought to be most suitable for the functional requirements of the fit. As an example, assume a nominal size of 2500 mm for which an interference fit of 0·25 to 0·64 mm is required.
2. Select that one of the two mating components which is considered to be the more difficult to manufacture to a small tolerance but which can be measured to the higher standard of accuracy. In most cases this will be the internal diameter.
3. Allocate an easily attainable tolerance to the component selected in (2) above which should be manufactured first. Assume that it is a bore of 2500 mm manufactured to a tolerance of H10.
4. Select from the tables the tolerance most nearly corresponding to the fit tolerance selected in (1) above. The nearest tolerance available for the example given is p8 which gives an interference of 0·25 mm to 0·56 mm.
5. State the complete specification for the matched fit tolerance as follows:
   2500 mm H10 MF p8.
6. Machine the bore to H10 limits, i.e. 2500·7 mm
   2500 mm
7. Measure the bore to the highest possible standard of accuracy. Assume that the measured size is 2500·038 mm.
8. Declare the tolerance for the shaft, which will be calculated as follows:

2500·038 mm	2500·038 mm
0·250 mm	0·560 mm
2500·288 mm	2500·598 mm

9. Manufacture the shaft to the limits of tolerance
   2500·288 mm
   2500·598 mm.

REFERENCES

1. BS 4500
   ISO limits and fits; BSI (1969)
2. BS 1916
   Limits and fits for engineering; BSI (1953)
3. ISA Bulletin 25
   ISA System of limits and fits; ISA (1940)
4. ISO R.286
   ISO System of limits and fits; ISO (1962)

Table 3.8 FUNDAMENTAL DEVIATIONS FOR SHAFTS

Unit = 0·001 mm

Fundamental deviation		Upper deviation es												Lower deviation el					
Letter		a*	b*	c	cd	d	e	ef	f	fg	g	h	js†	j			k		
Grade		01 to 16												5–6	7	8	4–7	≤3	>7
Nominal sizes																			
Over mm	To mm																		
–	3	−270	−140	−60	−34	−20	−14	−10	−6	−4	−2	0	± IT/2	−2	−4	−6	0	0	0
3	6	−270	−140	−70	−46	−30	−20	−14	−10	−6	−4	0		−2	−4	—	+1	0	0
6	10	−280	−150	−80	−56	−40	−25	−18	−13	−8	−5	0		−2	−5	—	+1	0	0
10	14	−290	−150	−95		−50	−32		−16		−6	0		−3	−6	—	+1		0
14	18	−290	−150	−95															
18	24	−300	−160	−110		−65	−40		−20		−7	0		−4	−8	—	+2		0
24	30	−300	−160	−110															
30	40	−310	−170	−120		−80	−50		−25		−9	0		−5	−10	—	+2		0
40	50	−320	−180	−130															
50	65	−340	−190	−140		−100	−60		−30		−10	0		−7	−12	—	+2		0
65	80	−360	−200	−150															
80	100	−380	−220	−170		−120	−72		−36		−12	0		−9	−15	—	+3		0
100	120	−410	−240	−180															
120	140	−460	−260	−200		−145	−85		−43		−14	0		−11	−18	—	+3		0
140	160	−520	−280	−210															
160	180	−580	−310	−230															
180	200	−660	−340	−240		−170	−100		−50		−15	0		−13	−21	—	+4		0
200	225	−740	−380	−260															
225	250	−820	−420	−280															
250	280	−920	−480	−300		−190	−110		−56		−17	0		−16	−26	—	+4		0
280	315	−1 050	−540	−330															
315	355	−1 200	−600	−360		−210	−125		−63		−18	0		−18	−28	—	+4		0
355	400	−1 350	−680	−400															
400	450	−1 500	−760	−440		−230	−135		−68		−20	0		−20	−32	—	+5		0

					±IT/2
630	800	−290	−160	−80	−22	0
800	1 000	−320	−170	−86	−24	0
1 000	1 250	−350	−195	−98	−26	0
1 250	1 600	−390	−220	−110	−28	0
1 600	2 000	−430	−240	−120	−30	0
2 000	2 500	−480	−260	−130	−32	0
2 500	3 150	−520	−290	−145	−34	0
					−38	0

* Not applicable to sizes up to 1 mm.

† In grades 7 to 11, the two symmetrical deviations ±IT/2 should be rounded if the IT value in micrometres is an odd value by replacing it by the even value immediately below.

Fundamental deviation

Lower deviation ei

Letter		m	n	p	r	s	t	u	v	x	y	z	za	zb	zc
Grade							01 to 16								
—	3	+2	+4	+6	+10	+14	—	+18	—	+20	—	+26	+32	+40	+60
3	6	+4	+8	+12	+15	+19	—	+23	—	+28	—	+35	+42	+50	+80
6	10	+6	+10	+15	+19	+23	—	+28	—	+34	—	+42	+52	+67	+97
10	14	+7	+12	+18	+23	+28	—	+33	—	+40	—	+50	+64	+90	+130
14	18	+7	+12	+18	+23	+28	—	+33	+39	+45	—	+60	+77	+108	+150
18	24	+8	+15	+22	+28	+35	—	+41	+47	+54	+63	+73	+98	+136	+188
24	30	+8	+15	+22	+28	+35	+41	+48	+55	+64	+75	+88	+118	+160	+218
30	40	+9	+17	+26	+34	+43	+48	+60	+68	+80	+94	+112	+148	+200	+274
40	50	+9	+17	+26	+34	+43	+54	+70	+81	+97	+114	+136	+180	+242	+325
50	65	+11	+20	+32	+41	+53	+66	+87	+102	+122	+144	+172	+226	+300	+405
65	80	+11	+20	+32	+43	+59	+75	+102	+120	+146	+174	+210	+274	+360	+480
80	100	+13	+23	+37	+51	+71	+91	+124	+146	+178	+214	+258	+335	+445	+585
100	120	+13	+23	+37	+54	+79	+104	+144	+172	+210	+254	+310	+400	+525	+690
120	140	+15	+27	+43	+63	+92	+122	+170	+202	+248	+300	+365	+470	+620	+800
140	160	+15	+27	+43	+65	+100	+134	+190	+228	+280	+340	+415	+535	+700	+900
160	180	+15	+27	+43	+68	+108	+146	+210	+252	+310	+380	+465	+600	+780	+1 000
180	200	+17	+31	+50	+77	+122	+166	+236	+284	+350	+425	+520	+670	+880	+1 150
200	225	+17	+31	+50	+80	+130	+180	+258	+310	+385	+470	+575	+740	+960	+1 250
225	250	+17	+31	+50	+84	+140	+196	+284	+340	+425	+520	+640	+820	+1 050	+1 350

continued overleaf

Table 3.8—*continued*

Unit = 0·001 mm

Fundamental deviation — *Lower deviation ei*

Letter	m	n	p	r	s	t	u	v	x	y	z	za	zb	zc
Grade							01 to 16							

Nominal size Over mm	To mm	m	n	p	r	s	t	u	v	x	y	z	za	zb	zc
250	280	+20	+34	+56	+94	+158	+218	+315	+385	+475	+580	+710	+920	+1 200	+1 550
280	315	+20	+34	+56	+98	+170	+240	+350	+425	+525	+650	+790	+1 000	+1 300	+1 700
315	355	+21	+37	+62	+108	+190	+268	+390	+475	+590	+730	+900	+1 150	+1 500	+1 900
355	400	+21	+37	+62	+114	+208	+294	+435	+530	+660	+820	+1 000	+1 300	+1 650	+2 100
400	450	+23	+40	+68	+126	+232	+330	+490	+595	+740	+920	+1 100	+1 450	+1 850	+2 400
450	500	+23	+40	+68	+132	+252	+360	+540	+660	+820	+1 000	+1 250	+1 600	+2 100	+2 600

Grade					6 to 16									

Over	To	m	n	p	r	s	t	u
500	560	+26	+44	+78	+150	+280	+400	+600
560	630	+26	+44	+78	+155	+310	+450	+660
630	710	+30	+50	+88	+175	+340	+500	+740
710	800	+30	+50	+88	+185	+380	+560	+840
800	900	+34	+56	+100	+210	+430	+620	+940
900	1 000	+34	+56	+100	+220	+470	+680	+1 050
1 000	1 120	+40	+66	+120	+250	+520	+780	+1 150
1 120	1 250	+40	+66	+120	+260	+580	+840	+1 300
1 250	1 400	+48	+78	+140	+300	+640	+960	+1 450
1 400	1 600	+48	+78	+140	+330	+720	+1 050	+1 600
1 600	1 800	+58	+92	+170	+370	+820	+1 200	+1 850
1 800	2 000	+58	+92	+170	+400	+920	+1 350	+2 000
2 000	2 240	+68	+110	+195	+440	+1 000	+1 500	+2 300
2 240	2 500	+68	+110	+195	+460	+1 100	+1 650	+2 500
2 500	2 800	+76	+135	+240	+550	+1 250	+1 900	+2 900
2 800	3 150	+76	+135	+240	+580	+1 400	+2 100	+3 200

Fundamental deviation

Lower deviation EI (Grades 01 to 16) — values positive (+). *Upper deviation ES.* Js† = ±IT7/2.

Nominal sizes Over (mm)	To (mm)	A*	B*	C	CD	D	E	EF	F	FG	G	H	Js†	J 6	J 7	J 8	K ≤8	K >8	M ≤8‡	M >8	N ≤8	N >8§
—	3	270	140	60	34	20	14	10	6	4	2	0	±IT7/2	2	4	6	0	0	−2	−2	−4	−4
3	6	270	140	70	46	30	20	14	10	6	4	0	±IT7/2	5	6	10	−1+Δ	—	−4+Δ	−4	−8+Δ	0
6	10	280	150	80	56	40	25	18	13	8	5	0	±IT7/2	5	8	12	−1+Δ	—	−6+Δ	−6	−10+Δ	0
10	14	290	150	95	—	50	32	—	16	—	6	0	±IT7/2	6	10	15	−1+Δ	—	−7+Δ	−7	−12+Δ	0
14	18	290	150	95	—	50	32	—	16	—	6	0	±IT7/2	6	10	15	−1+Δ	—	−7+Δ	−7	−12+Δ	0
18	24	300	160	110	—	65	40	—	20	—	7	0	±IT7/2	8	12	20	−2+Δ	—	−8+Δ	−8	−15+Δ	0
24	30	300	160	110	—	65	40	—	20	—	7	0	±IT7/2	8	12	20	−2+Δ	—	−8+Δ	−8	−15+Δ	0
30	40	310	170	120	—	80	50	—	25	—	9	0	±IT7/2	10	14	24	−2+Δ	—	−9+Δ	−9	−17+Δ	0
40	50	320	180	130	—	80	50	—	25	—	9	0	±IT7/2	10	14	24	−2+Δ	—	−9+Δ	−9	−17+Δ	0
50	65	340	190	140	—	100	60	—	30	—	10	0	±IT7/2	13	18	28	−2+Δ	—	−11+Δ	−11	−20+Δ	0
65	80	360	200	150	—	100	60	—	30	—	10	0	±IT7/2	13	18	28	−2+Δ	—	−11+Δ	−11	−20+Δ	0
80	100	380	220	170	—	120	72	—	36	—	12	0	±IT7/2	16	22	34	−3+Δ	—	−13+Δ	−13	−23+Δ	0
100	120	410	240	180	—	120	72	—	36	—	12	0	±IT7/2	16	22	34	−3+Δ	—	−13+Δ	−13	−23+Δ	0
120	140	460	260	200	—	145	85	—	43	—	14	0	±IT7/2	18	26	41	−3+Δ	—	−15+Δ	−15	−27+Δ	0
140	160	520	280	210	—	145	85	—	43	—	14	0	±IT7/2	18	26	41	−3+Δ	—	−15+Δ	−15	−27+Δ	0
160	180	580	310	230	—	145	85	—	43	—	14	0	±IT7/2	18	26	41	−3+Δ	—	−15+Δ	−15	−27+Δ	0
180	200	660	340	240	—	170	100	—	50	—	15	0	±IT7/2	22	30	47	−4+Δ	—	−17+Δ	−17	−31+Δ	0
200	225	740	380	260	—	170	100	—	50	—	15	0	±IT7/2	22	30	47	−4+Δ	—	−17+Δ	−17	−31+Δ	0
225	250	820	420	280	—	170	100	—	50	—	15	0	±IT7/2	22	30	47	−4+Δ	—	−17+Δ	−17	−31+Δ	0
250	280	920	480	300	—	190	110	—	56	—	17	0	±IT7/2	25	36	55	−4+Δ	—	−20+Δ	−20	−34+Δ	0
280	315	1050	540	330	—	190	110	—	56	—	17	0	±IT7/2	25	36	55	−4+Δ	—	−20+Δ	−20	−34+Δ	0
315	355	1200	600	360	—	210	125	—	62	—	18	0	±IT7/2	29	39	60	−4+Δ	—	−21+Δ	−21	−37+Δ	0
355	400	1350	680	400	—	210	125	—	62	—	18	0	±IT7/2	29	39	60	−4+Δ	—	−21+Δ	−21	−37+Δ	0
400	450	1500	760	440	—	230	135	—	68	—	20	0	±IT7/2	33	43	66	−5+Δ	—	−23+Δ	−23	−40+Δ	0
450	500	1650	840	480	—	230	135	—	68	—	20	0	±IT7/2	33	43	66	−5+Δ	—	−23+Δ	−23	−40+Δ	0

continued overleaf

Table 3.9—continued

Unit = 0·001 mm

Fundamental deviation			Lower deviation EI												Upper deviation ES		
Letter			A*	B*	C	CD	D	E	EF	F	FG	G	H	Js†	J	K	M	N
Grade									6 to 16									
Nominal sizes																		
Over mm	To mm		+	−	−	−	+	+	+	+	+	+						
500	630		—	—	—	—	260	145	76	—	22	0				0	−26	−44
630	800		—	—	—	—	290	160	80	—	24	0				0	−30	−50
800	1 000		—	—	—	—	320	170	86	—	26	0				0	−34	−56
1 000	1 250		—	—	—	—	350	195	98	—	28	0				0	−40	−66
1 250	1 600		—	—	—	—	390	220	110	—	30	0				0	−48	−78
1 600	2 000		—	—	—	—	430	240	120	—	32	0				0	−58	−92
2 000	2 500		—	—	—	—	480	260	130	—	34	0				0	−68	−110
2 500	3 150		—	—	—	—	520	290	145	—	38	0				0	−76	−135

For columns H, Js†, J: ±IT/2

* Not applicable to sizes up to 1 mm.
† In grades 7 to 11, the two symmetrical deviations ±IT/2 should be rounded if the IT value in micrometres is an odd value by replacing it by the even value immediately below.
‡ Special case: for M6, ES = −9 from 250 to 315 (instead of −11).
§ Not applicable to sizes up to 1 mm.

Fundamental deviation

Upper deviation ES

Values for Δ*

Letter		P	R	S	T	U	V	X	Y	Z	ZA	ZB	ZC						
	P to ZC															Grades:			
Grade	≤7	>7												3	4	5	6	7	8
Nominal sizes																			
Over	To																		
mm	mm																		
—	3		—	—	—	—	—	—	—	—	—	—	—	—	0	0	0	0	0	0
3	6		6	10	14	—	18	—	20	—	26	32	40	60	1	1·5	1	3	4	6
6	10		12	15	19	—	23	—	28	—	35	42	50	80	1	1·5	2	3	6	7
10	14		15	19	23	—	28	—	34	—	42	52	67	97	1	2	3	3	7	9
14	18		18	23	28	—	33	39	40	—	50	64	90	130	1	2	3	3	7	9
18	24	Same deviation as for grades above 7 increased by Δ	18	23	28	—	33	47	45	63	60	77	108	150	1·5	2	3	4	8	12
24	30		22	28	35	41	41	55	54	75	73	98	136	188	1·5	2	3	4	8	12
30	40		22	28	35	48	48	68	64	94	88	118	160	218	1·5	3	4	5	9	14
40	50		26	34	43	54	60	81	80	114	112	148	200	274	1·5	3	4	5	9	14
50	65		26	34	43	66	70	102	97	144	136	180	242	325	2	3	5	6	11	16
65	80		32	41	53	75	87	120	122	174	172	226	300	405	2	3	5	6	11	16
80	100		32	43	59	91	102	146	146	214	210	274	360	480	2	4	5	7	13	19
100	120		37	51	71	104	124	172	178	254	258	335	445	585	2	4	5	7	13	19
120	140		37	54	79	122	144	202	210	300	310	400	525	690	3	4	6	7	15	23
140	160		43	63	92	134	170	228	248	340	365	470	620	800	3	4	6	7	15	23
160	180		43	65	100	146	190	252	280	380	415	535	700	900	3	4	6	7	15	23
180	200		43	68	108	166	210	284	310	425	465	600	780	1 000	3	4	6	9	17	26
200	225		50	77	122	180	236	310	350	470	520	670	880	1 150	3	4	6	9	17	26
225	250		50	80	130	196	258	340	385	520	575	740	960	1 250	3	4	6	9	17	26
250	280		50	84	140	218	284	385	425	580	640	820	1 050	1 350	4	4	7	9	20	26
280	315		56	94	158	240	315	425	475	650	710	920	1 200	1 550	4	4	7	9	20	29
315	355		56	98	170	268	350	475	525	730	790	1 000	1 300	1 700	4	5	7	11	20	29
355	400		62	108	190	294	390	530	590	820	900	1 150	1 500	1 800	4	5	7	11	21	32
400	450		62	114	208	330	435	595	660	920	1 000	1 300	1 650	2 100	5	5	7	13	21	32
450	500		68	126	232	360	490	660	740	1 100	1 250	1 600	2 100	2 600	5	5	7	13	23	34
			68	132	252		540		820	1 000									23	34

* In determining K, M, N up to Grade 8 and P to ZC up to Grade 7, add the Δ value appropriate to the grade as indicated, e.g. for P7 from 18 to 30, Δ = 8 therefore ES = −14.

continued overleaf

Table 3.9—*continued*

Unit = 0·001 mm

Fundamental deviation		Upper deviation ES				
Letter		*P*	*R*	*S*	*T*	*U*
Grade				6 *to* 16		
Nominal sizes						
Over	*To*					
mm	mm					
500	560	78	150	280	400	600
560	630	78	155	310	450	660
630	710	88	175	340	500	740
710	800	88	185	380	560	840
800	900	100	210	430	620	940
900	1 000	100	220	470	680	1 050
1 000	1 120	120	250	520	780	1 150
1 120	1 250	120	260	580	840	1 300
1 250	1 400	140	300	640	960	1 450
1 400	1 600	140	330	720	1 050	1 600
1 600	1 800	170	370	820	1 200	1 850
1 800	2 000	170	400	920	1 350	2 000
2 000	2 240	195	440	1 000	1 500	2 300
2 240	2 500	195	460	1 100	1 650	2 500
2 500	2 800	240	550	1 250	1 900	2 900
2 800	3 150	240	580	1 400	2 100	3 200

Table 3.10 LIMITS OF TOLERANCE (UPPER AND LOWER DEVIATIONS) FOR COMMONLY USED HOLES

ES = Upper deviation
EI = Lower deviation

Holes A to C
Tolerance unit = 0·001 mm

Nominal sizes		A			B				C			
		9	11	9 and 11	8	9	11	8–9 and 11	8	9	11	8–9 and 11
Over	To	ES +	ES +	EI +	ES +	ES +	ES +	EI +	ES +	ES +	ES +	EI +
mm	mm											
—	3	295	330	270	154	165	200	140	74	85	120	60
3	6	300	345	270	158	170	215	140	88	100	145	70
6	10	316	370	280	172	186	240	150	102	116	170	80
10	14	333	400	290	177	193	260	150	122	138	205	95
14	18											
18	24	352	430	300	193	212	290	160	143	162	240	110
24	30											
30	40	372	470	310	209	232	330	170	159	182	280	120
40	50	382	480	320	219	242	340	180	169	192	290	130
50	65	414	530	340	236	264	380	190	186	214	330	140
65	80	434	550	360	246	274	390	200	196	224	340	150
80	100	467	600	380	274	307	440	220	224	257	390	170
100	120	497	630	410	294	327	460	240	234	267	400	180
120	140	560	710	460	323	360	510	260	263	300	450	200
140	160	620	770	520	343	380	530	280	273	310	460	210
160	180	680	830	580	373	410	560	310	293	330	480	230
180	200	775	950	660	412	455	630	340	312	355	530	240
200	225	855	1 030	740	452	495	670	380	332	375	550	260
225	250	935	1 110	820	492	535	710	420	352	395	570	280
250	280	1 050	1 240	920	561	610	800	480	381	430	620	300
280	315	1 180	1 370	1 050	621	670	860	540	411	460	650	330
315	355	1 340	1 560	1 200	689	740	960	600	449	500	720	360
355	400	1 490	1 710	1 350	769	820	1 040	680	489	540	760	400
400	450	1 655	1 900	1 500	857	915	1 160	760	537	595	840	440
450	500	1 805	2 050	1 650	937	995	1 240	840	577	635	880	480

continued overleaf

Table 3.10—*continued*

ES = Upper deviation
EI = Lower deviation

Holes D to E
Tolerance unit = 0·001 mm

Nominal sizes		D								E					
		5	6	7	8	9	10	11	5 to 11	5	6	7	8	9	5 to 10
Over	To	ES +	ES +	ES +	ES +	ES +	ES +	ES +	EI +	ES +	ES +	ES +	ES +	ES +	EI +
mm	mm														
—	3	24	26	30	34	45	60	80	20	18	20	24	28	39	14
3	6	35	38	42	48	60	78	105	30	25	28	32	38	50	20
6	10	46	49	55	62	76	98	130	40	31	34	40	47	61	25
10	14	58	61	68	77	93	120	160	50	40	43	50	59	75	32
14	18														
18	24	74	78	86	98	117	149	195	65	49	53	61	73	92	40
24	30														
30	40	91	96	105	119	142	180	240	80	61	66	75	89	112	50
40	50														
50	65	113	119	130	146	174	220	290	100	73	79	90	106	134	60
65	80														
80	100	135	142	155	174	207	260	340	120	87	94	107	126	159	72
100	120														
120	140	163	170	185	208	245	305	395	145	103	110	125	148	185	85
140	160														
160	180														
180	200	190	199	216	242	285	355	460	170	120	129	146	172	215	100
200	225														
225	250														
250	280	213	222	242	271	320	400	510	190	133	142	162	191	240	110
280	315														
315	355	235	246	267	299	350	440	570	210	150	161	182	214	265	125
355	400														
400	450	257	270	293	327	385	480	630	230	162	175	198	232	290	135
450	500														

Table 3.10—*continued*

ES = Upper deviation
EI = Lower deviation

Holes F to G
Tolerance unit = 0·001 mm

Nominal sizes		F							G					
		4	5	6	7	8	9	4 to 10	4	5	6	7	8	4 to 8
Over	To	ES +	ES +	ES +	ES +	ES +	ES +	EI +	ES +	ES +	ES +	ES +	ES +	EI +
mm	mm													
—	3	9	10	12	16	20	31	6	5	6	8	12		2
3	6	14	15	18	22	28	40	10	8	9	12	16		4
6	10	17	19	22	28	35	49	13	9	11	14	20		5
10	14	21	24	27	34	43	59	16	11	14	17	24		6
14	18													
18	24	26	29	33	41	53	72	20	13	16	20	28		7
24	30													
30	40	32	36	41	50	64	87	25	16	20	25	34		9
40	50													
50	65	38	43	49	60	76	104	30	18	23	29	40		10
65	80													
80	100	46	51	58	71	90	123	36	22	27	34	47		12
100	120													
120	140	55	61	68	83	106	143	43	26	32	39	54		14
140	160													
160	180													
180	200	64	70	79	96	122	165	50	29	35	44	61		15
200	225													
225	250													
250	280	72	79	88	108	137	186	56	33	40	49	69		17
280	315													
315	355	80	87	98	119	151	202	62	36	43	54	75		18
355	400													
400	450	88	95	108	131	165	223	68	40	47	60	83		20
450	500													

continued overleaf

Table 3.10—*continued*

ES = Upper deviation
EI = Lower deviation

Hole H
Tolerance unit = 0·001 mm

Nominal sizes H

Over	To	1	2	3	4	5	6	7	8	9	10	11	12	13	14*	15*	16*	1 to 16
		ES +	ES +	ES +	ES +	ES +	ES +	ES +	ES +	ES +	ES +	ES +	ES +	ES +	ES +	ES +	ES +	EI +
mm	mm																	
—	3	0·8	1·2	2	3	4	6	10	14	25	40	60	100	140	250	400	600	0
3	6	1	1·5	2·5	4	5	8	12	18	30	48	75	120	180	300	480	750	0
6	10	1	1·5	2·5	4	6	9	15	22	36	58	90	150	220	360	580	900	0
10	14	1·2	2	3	5	8	11	18	27	43	70	110	180	270	430	700	1 100	0
14	18																	
18	24	1·5	2·5	4	6	9	13	21	33	52	84	130	210	330	520	840	1 300	0
24	30																	
30	40	1·5	2·5	4	7	11	16	25	39	62	100	160	250	390	620	1 000	1 600	0
40	50																	
50	65	2	3	5	8	13	19	30	46	74	120	190	300	460	740	1 200	1 900	0
65	80																	
80	100	2·5	4	6	10	15	22	35	54	87	140	220	350	540	870	1 400	2 200	0
100	120																	
120	140	3·5	5	8	12	18	25	40	63	100	160	250	400	630	1 000	1 600	2 500	0
140	160																	
160	180																	
180	200	4·5	7	10	14	20	29	46	72	115	185	290	460	720	1 150	1 850	2 900	0
200	225																	
225	250																	
250	280	6	8	12	16	23	32	52	81	130	210	320	520	810	1 300	2 100	3 200	0
280	315																	
315	355	7	9	13	18	25	36	57	89	140	230	360	570	890	1 400	2 300	3 600	0
355	400																	
400	450	8	10	15	20	27	40	63	97	155	250	400	630	970	1 550	2 500	4 000	0
450	500																	

* Grades 14 to 16 do not apply to sizes up to and including 1 mm.

Table 3.10—*continued*

ES = Upper deviation   *Hole J*
EI = Lower deviation   *Tolerance unit* = 0·001 mm

Nominal sizes   J

Over	To	6		7		8	
		ES +	EI −	ES +	EI −	ES +	EI −
mm	mm						
—	3	2	4	4	6	6	8
3	6	5	3	6	6	10	8
6	10	5	4	8	7	12	10
10	14	6	5	10	8	15	12
14	18						
18	24	8	5	12	9	20	13
24	30						
30	40	10	6	14	11	24	15
40	50						
50	65	13	6	18	12	28	18
65	80						
80	100	16	6	22	13	34	20
100	120						
120	140						
140	160	18	7	26	14	41	22
160	180						
180	200						
200	225	22	7	30	16	47	25
225	250						
250	280	25	7	36	16	55	26
280	315						
315	355	29	7	39	18	60	29
355	400						
400	450	33	7	43	20	66	31
450	500						

continued overleaf

Table 3.10—*continued*

ES = Upper deviation
EI = Lower deviation

Hole J_s
Tolerance unit = 0·001 mm

Nominal sizes		1		2		3		4		5		6		7		8	
Over	To	ES +	EI −	ES +	EI −	ES +	EI −	ES +	EI −	ES +	EI −	ES +	EI −	ES +	EI −	ES +	EI −
mm	mm																
—	3	0·4	0·4	0·6	0·6	1	1	1·5	1·5	2	2	3	3	5	5	7	7
3	6	0·5	0·5	0·75	0·75	1·25	1·25	2	2	2·5	2·5	4	4	6	6	9	9
6	10	0·5	0·5	0·75	0·75	1·25	1·25	2	2	3	3	4·5	4·5	7	7	11	11
10	14	0·6	0·6	1	1	1·5	1·5	2·5	2·5	4	4	5·5	5·5	9	9	13	13
14	18																
18	24	0·75	0·75	1·25	1·25	2	2	3	3	4·5	4·5	6·5	6·5	10	10	16	16
24	30																
30	40	0·75	0·75	1·25	1·25	2	2	3·5	3·5	5·5	5·5	8	8	12	12	19	19
40	50																
50	65	1	1	1·5	1·5	2·5	2·5	4	4	6·5	6·5	9·5	9·5	15	15	23	23
65	80																
80	100	1·25	1·25	2	2	3	3	5	5	7·5	7·5	11	11	17	17	27	27
100	120																
120	140	1·75	1·75	2·5	2·5	4	4	6	6	9	9	12·5	12·5	20	20	31·5	31·5
140	160																
160	180																
180	200	2·25	2·25	3·5	3·5	5	5	7	7	10	10	14·5	14·5	23	23	36	36
200	225																
225	250																
250	280	3	3	4	4	6	6	8	8	11·5	11·5	16	16	26	26	40	40
280	315																
315	355	3·5	3·5	4·5	4·5	6·5	6·5	9	9	12·5	12·5	18	18	28	28	44	44
355	400																
400	450	4	4	5	5	7·5	7·5	10	10	13·5	13·5	20	20	31	31	48	48
450	500																

Table 3.10—*continued*

ES = Upper deviation
EI = Lower deviation

Hole J_s
Tolerance unit = 0·001 mm

Nominal sizes		9		10		11		12		13		14*		15*		16*	
		J_s															
Over	To	ES +	EI −	ES +	EI −	ES +	EI −	ES +	EI −	ES +	EI −	ES +	EI −	ES +	EI −	ES +	EI −
mm	mm																
—	3	12	12	20	20	30	30	50	50	70	70	125	125	200	200	300	300
3	6	15	15	24	24	37	37	60	60	90	90	150	150	240	240	375	375
6	10	18	18	29	29	45	45	75	75	110	110	180	180	290	290	450	450
10	14	21	21	35	35	55	55	90	90	135	135	215	215	350	350	550	550
14	18																
18	24	26	26	42	42	65	65	105	105	165	165	260	260	420	420	650	650
24	30																
30	40	31	31	50	50	80	80	125	125	195	195	310	310	500	500	800	800
40	50																
50	65	37	37	60	60	95	95	150	150	230	230	370	370	600	600	950	950
65	80																
80	100	43	43	70	70	110	110	175	175	270	270	435	435	700	700	1 100	1 100
100	120																
120	140	50	50	80	80	125	125	200	200	315	315	500	500	800	800	1 250	1 250
140	160																
160	180																
180	200																
200	225	57	57	92·5	92·5	145	145	230	230	360	360	575	575	925	925	1 450	1 450
225	250																
250	280	65	65	105	105	160	160	260	260	405	405	650	650	1 050	1 050	1 600	1 600
280	315																
315	355	70	70	115	115	180	180	285	285	445	445	700	700	1 150	1 150	1 800	1 800
355	400																
400	450	77	77	125	125	200	200	315	315	485	485	775	775	1 250	1 250	2 000	2 000
450	500																

* Grades 14 to 16 do not apply to sizes up to and including 1 mm.

continued overleaf

Table 3.10—*continued*

ES = Upper deviation								Hole K
EI = Lower deviation							*Tolerance unit* = 0·001 mm	

Nominal sizes		K							
		5		6		7		8	
Over	To	ES +	EI −	ES +	EI −	ES +	EI −	ES +	EI −
mm	mm								
—	3	0	4	0	6	0	10	0	14
3	6	0	5	2	6	3	9	5	13
6	10	1	5	2	7	5	10	6	16
10 14	14 18	2	6	2	9	6	12	8	19
18 24	24 30	1	8	2	11	6	15	10	23
30 40	40 50	2	9	3	13	7	18	12	27
50 65	65 80	3	10	4	15	9	21	14	32
80 100	100 120	2	13	4	18	10	25	16	38
120 140 160	140 160 180	3	15	4	21	12	28	20	43
180 200 225	200 225 250	2	18	5	24	13	33	22	50
250 280	280 315	3	20	5	27	16	36	25	56
315 355	355 400	3	22	7	29	17	40	28	61
400 450	450 500	2	25	8	32	18	45	29	68

Table 3.10—*continued*

ES = Upper deviation								*Hole M*
EI = Lower deviation						*Tolerance unit* = 0·001 mm		

Nominal sizes					M			

Over	To	5		6		7		8*	
		ES	EI	ES	EI	ES	EI	ES	EI
mm	mm								
—	3	2	6	2	8	2	12	—	—
3	6	3	8	1	9	0	12	+2	16
6	10	4	10	3	12	0	15	+1	21
10	14	4	12	4	15	0	18	+2	25
14	18								
18	24	5	14	4	17	0	21	+4	29
24	30								
30	40	5	16	4	20	0	25	+5	34
40	50								
50	65	6	19	5	24	0	30	+5	41
65	80								
80	100	8	23	6	28	0	35	+6	48
100	120								
120	140	9	27	8	33	0	40	+8	55
140	160								
160	180								
180	200	11	31	8	37	0	46	+9	63
200	225								
225	250								
250	280	13	36	9	41	0	52	+9	72
280	315								
315	355	14	39	10	46	0	57	+11	78
355	400								
400	450	16	43	10	50	0	63	+11	86
450	500								

* Where M8 deviations are missing, N8 deviations (with their own sign) should be substituted for them.

continued overleaf

Table 3.10—*continued*

ES = Upper deviation
EI = Lower deviation

Hole N
Tolerance unit = 0·001 mm

Nominal sizes		5		6		7		8		9 to 11	9	10	11
Over	To	ES	EI	ES	EI	ES	EI	ES	EI	ES	EI	EI	EI
mm	mm												
—	3	4	8	4	10	4	14	4	18	4	29	44	64
3	6	7	12	5	13	4	16	2	20	0	30	48	75
6	10	8	14	7	16	4	19	3	25	0	36	58	90
10	14	9	17	9	20	5	23	3	30	0	43	70	110
14	18												
18	24	12	21	11	24	7	28	3	36	0	52	84	130
24	30												
30	40	13	24	12	28	8	33	3	42	0	62	100	160
40	50												
50	65	15	28	14	33	9	39	4	50	0	74	120	190
65	80												
80	100	18	33	16	38	10	45	4	58	0	87	140	220
100	120												
120	140	21	39	20	45	12	52	4	67	0	100	160	250
140	160												
160	180												
180	200	25	45	22	51	14	60	5	77	0	115	185	290
200	225												
225	250												
250	280	27	50	25	57	14	66	5	86	0	130	210	320
280	315												
315	355	30	55	26	62	16	73	5	94	0	140	230	360
355	400												
400	450	33	60	27	67	17	80	6	103	0	155	250	400
450	500												

Table 3.10—*continued*

ES = Upper deviation
EI = Lower deviation

Hole P
Tolerance unit = 0·001 mm

Nominal sizes									P	
		5		6		7		8 and 9	8	9
Over	To	ES −	EI −	ES −	EI −	ES −	EI −	ES −	EI −	EI −
mm	mm									
—	3	6	10	6	12	6	16	6	20	31
3	6	11	16	9	17	8	20	12	30	42
6	10	13	19	12	21	9	24	15	37	51
10	14	15	23	15	26	11	29	18	45	61
14	18									
18	24	19	28	18	31	14	35	22	55	74
24	30									
30	40	22	33	21	37	17	42	26	65	88
40	50									
50	65	27	40	26	45	21	51	32	78	106
65	80									
80	100	32	47	30	52	24	59	37	91	124
100	120									
120	140	37	55	36	61	28	68	43	106	143
140	160									
160	180									
180	200	44	64	41	70	33	79	50	122	165
200	225									
225	250									
250	280	49	72	47	79	36	88	56	137	186
280	315									
315	355	55	80	51	87	41	98	62	151	202
355	400									
400	450	61	88	55	95	45	108	68	165	223
450	500									

continued overleaf

Table 3.10—*continued*

ES = Upper deviation
EI = Lower deviation

Hole R
Tolerance unit = 0·001 mm

Nominal size		R							
		5		6		7		8	
Over	To	ES	EI	ES	EI	ES	EI	ES	EI
mm	mm								
—	3	10	14	10	16	10	20	10	24
3	6	14	19	12	20	11	23	15	33
6	10	17	23	16	25	13	28	19	41
10	14	20	28	20	31	16	34	23	50
14	18								
18	24	25	34	24	37	20	41	28	61
24	30								
30	40	30	41	29	45	25	50	34	73
40	50								
50	65	36	49	35	54	30	60	41	87
65	80	38	51	37	56	32	62	43	89
80	100	46	61	44	66	38	73	51	105
100	120	49	64	47	69	41	76	54	108
120	140	57	75	56	81	48	88	63	126
140	160	59	77	58	83	50	90	65	128
160	180	62	80	61	86	53	93	68	131
180	200	71	91	68	97	60	106	77	149
200	225	74	94	71	100	63	109	80	152
225	250	78	98	75	104	67	113	84	156
250	280	87	110	85	117	74	126	94	175
280	315	91	114	89	121	78	130	98	179
315	355	101	126	97	133	87	144	108	197
355	400	107	132	103	139	93	150	114	203
400	450	119	146	113	153	103	166	126	223
450	500	125	152	119	159	109	172	132	229

Table 3.10—*continued*

ES = Upper deviation
EI = Lower deviation

Holes S and T
Tolerance unit = 0·001 mm

Nominal sizes		S						T*		
		5		6		7		6		7	
Over	To	ES	EI	ES	EI	ES	EI	ES	EI	ES	EI
mm	mm										
—	3	14	18	14	20	14	24	—	—	—	—
3	6	18	23	16	24	15	27	—	—	—	—
6	10	21	27	20	29	17	32	—	—	—	—
10	14	25	33	25	36	21	39	—	—	—	—
14	18										
18	24	32	41	31	44	27	48	37	50	33	54
24	30										
30	40	39	50	38	54	34	59	43	59	39	64
40	50							49	65	45	70
50	65	48	61	47	66	42	72	60	79	55	85
65	80	54	67	53	72	48	78	69	88	64	94
80	100	66	81	64	86	58	93	84	106	78	113
100	120	74	89	72	94	66	101	97	119	91	126
120	140	86	104	85	110	77	117	115	140	107	147
140	160	94	112	93	118	85	125	127	152	119	159
160	180	102	120	101	126	93	133	139	164	131	171
180	200	116	136	113	142	105	151	157	186	149	195
200	225	124	144	121	150	113	159	171	200	163	209
225	250	134	154	131	160	123	169	187	216	179	225
250	280	151	174	149	181	138	190	209	241	198	250
280	315	163	186	161	193	150	202	231	263	220	272
315	355	183	208	179	215	169	226	257	293	247	304
355	400	201	226	197	233	187	244	283	319	273	330
400	450	225	252	219	259	209	272	317	357	307	370
450	500	245	272	239	279	229	292	347	387	337	400

* Where T6 and T7 deviations are missing, U6 and U7 deviations should be substituted for them.

continued overleaf

Table 3.10—*continued*

ES = Upper deviation
EI = Lower deviation

Holes U to Y
Tolerance unit = 0·001 mm

Nominal sizes Over	To	U 6 ES	U 6 EI	U 7 ES	U 7 EI	U 8 and 9 ES	V* 6 ES	V* 6 EI	V* 7 ES	V* 7 EI	X 6 ES	X 6 EI	X 7 ES	X 7 EI	Y† 7 ES	Y† 7 EI
mm	mm															
—	3	18	24	18	28		—	—	—	—	20	26	20	30	—	—
3	6	20	28	19	31		—	—	—	—	25	33	24	36	—	—
6	10	25	34	22	37		—	—	—	—	31	40	28	43	—	—
10	14	30	41	26	44		—	—	—	—	37	48	33	51	—	—
14	18	30	41	26	44		36	47	32	50	42	53	38	56	—	—
18	24	37	50	33	54		43	56	39	60	50	63	46	67	55	76
24	30	44	57	40	61		51	64	47	68	60	73	56	77	67	88
30	40	55	71	51	76		63	79	59	84	75	91	71	96	85	110
40	50	65	81	61	86		76	92	72	97	92	108	88	113	105	130
50	65	81	100	76	106		96	115	91	121	116	135	111	141	133	163
65	80	96	115	91	121		114	133	109	139	140	159	135	165	163	193
80	100	117	139	111	146		139	161	133	168	171	193	165	200	201	236
100	120	137	159	131	166		165	187	159	194	203	225	197	232	241	276
120	140	163	188	155	195		195	220	187	227	241	266	233	273	285	325
140	160	183	208	175	215		221	246	213	253	273	298	265	305	325	365
160	180	203	228	195	235		245	270	237	277	303	328	295	335	365	405
180	200	227	256	219	265		275	304	267	313	341	370	333	379	408	454
200	225	249	278	241	287		301	330	293	339	376	405	368	414	453	499
225	250	275	304	267	313		331	360	323	369	416	445	408	454	503	549
250	280	306	338	295	347		376	408	365	417	466	498	455	507	560	612
280	315	341	373	330	382		416	448	405	457	516	548	505	557	630	682
315	355	379	415	369	426		464	500	454	511	579	615	569	626	709	766
355	400	424	460	414	471		519	555	509	566	649	685	639	696	799	856
400	450	477	517	467	530		582	622	572	635	727	767	717	780	897	960
450	500	527	567	517	580		647	687	637	700	807	847	797	860	977	1 040

* Where V6 and V7 deviations are missing, X6 and X7 deviations should be substituted for them.
† Where Y7 deviations are missing, Z7 deviations should be substituted for them.

Table 3.10—*continued*

ES = Upper deviation
EI = Lower deviation

Holes Z to ZC
Tolerance unit = 0·001 mm

Nominal sizes		Z				ZA				ZB				ZC			
		7		8		7		8		8		9		8		9	
Over	To	ES	EI	ES	EI	ES	EI	ES	EI	ES	EI	ES	EI	ES	EI	ES	EI
mm	mm																
	3	26	36	26	40	32	42	32	46	40	54	40	65	60	74	60	85
3	6	31	43	35	53	38	50	42	60	50	68	50	80	80	98	80	110
6	10	36	51	42	64	46	61	52	74	67	89	67	103	97	119	97	133
10	14	43	61	50	77												
14	18	53	71	60	87												
18	24	65	86	73	106												
24	30	80	101	88	121												
30	40	103	128	112	151												
40	50	127	152	136	175												
50	65	161	191	172	218												
65	80	199	229	210	256												
80	100	245	280	258	312												
100	120	297	332	310	364												
120	140	350	390	365	428												
140	160	400	440	415	478												
160	180	450	490	465	528												
180	200	503	549	520	592												
200	225	558	604	575	647												
225	250	623	669	640	712												
250	280	690	742	710	791												
280	315	770	822	790	871												
315	355	879	936	900	989												
355	400	979	1 036	1 000	1 089												
400	450	1 077	1 140	1 100	1 197												
450	500	1 227	1 290	1 250	1 347												

3–64

Table 3.11 LIMITS OF TOLERANCE (UPPER AND LOWER DEVIATIONS) FOR COMMONLY USED SHAFTS

es = Upper deviation
ei = Lower deviation

Shafts a to c
Tolerance unit = 0·001 mm

Nominal sizes		a*			b*				c			
		9 and 11	9	11	8–9 and 11	8	9	11	8–9 and 11	8	9	11
Over	To	es	ei	ei	es	ei	ei	ei	es	ei	ei	ei
mm	mm											
—	3	270	295	330	140	154	165	200	60	74	85	120
3	6	270	300	345	140	158	170	215	70	88	100	145
6	10	280	316	370	150	172	186	240	80	102	116	170
10	14	290	333	400	150	177	193	260	95	122	138	205
14	18											
18	24	300	352	430	160	193	212	290	110	143	162	240
24	30											
30	40	310	372	470	170	209	232	330	120	159	182	280
40	50	320	382	480	180	219	242	340	130	169	192	290
50	65	340	414	530	190	236	264	380	140	186	214	330
65	80	360	434	550	200	246	274	390	150	196	224	340
80	100	380	467	600	220	274	307	440	170	224	257	390
100	120	410	497	630	240	294	327	460	180	234	267	400
120	140	460	560	710	260	323	360	510	200	263	300	450
140	160	520	620	770	280	343	380	530	210	273	310	460
160	180	580	680	830	310	373	410	560	230	293	330	480
180	200	660	775	950	340	412	455	630	240	312	355	530
200	225	740	855	1 030	380	452	495	670	260	332	375	550
225	250	820	935	1 110	420	492	535	710	280	352	395	570
250	280	920	1 050	1 240	480	561	610	800	300	381	430	620
280	315	1 050	1 180	1 370	540	621	670	860	330	411	460	650
315	355	1 200	1 340	1 560	600	689	740	960	360	449	500	720
355	400	1 350	1 490	1 710	680	769	820	040	400	489	540	760
400	450	1 500	1 655	1 900	760	857	915	1 160	440	537	595	840
450	500	1 650	1 805	2 050	840	937	995	1 240	480	577	635	880

* Deviations a and b do not apply to sizes up to and including 1 mm.

Table 3.11—*continued*

es = Upper deviation

ei = Lower deviation

Shafts d to e

Tolerance unit = 0·001 mm

Nominal sizes		d								e					
Over	To	5 to 11	5	6	7	8	9	10	11	5 to 10	5	6	7	8	9
		es	ei	ei	ei	ei	ei	ei	ei	es	ei	ei	ei	ei	ei
mm	mm														
—	3	20	24	26	30	34	45	60	80	14	18	20	24	28	39
3	6	30	35	38	42	48	60	78	105	20	25	28	32	38	50
6	10	40	46	49	55	62	76	98	130	25	31	34	40	47	61
10 14	14 18	50	58	61	68	77	93	120	160	32	40	43	50	59	75
18 24	24 30	65	74	78	86	98	117	149	195	40	49	53	61	73	92
30 40	40 50	80	91	96	105	119	142	180	240	50	61	66	75	89	112
50 65	65 80	100	113	119	130	146	174	220	290	60	73	79	90	106	134
80 100	100 120	120	135	142	155	174	207	260	340	72	87	94	107	126	159
120 140 160	140 160 180	145	163	170	185	208	245	305	395	85	103	110	125	148	185
180 200 225	200 225 250	170	190	199	216	242	285	355	460	100	120	129	146	172	215
250 280	280 315	190	213	222	242	271	320	400	510	110	133	142	162	191	240
315 355	355 400	210	235	246	267	299	350	440	570	125	150	161	182	214	265
400 450	450 500	230	257	270	293	327	385	480	630	135	162	175	198	232	290

continued overleaf

Table 3.11—continued

es = Upper deviation
ei = Lower deviation

Shafts f to g
Tolerance unit = 0·001 mm

Nominal sizes		f							g				
Over	To	4 to 10	4	5	6	7	8	9	4 to 8	4	5	6	7
		es	ei	ei	ei	ei	ei	ei	es	ei	ei	ei	ei
		−	−	−	−	−	−	−	−	−	−	−	−
mm	mm												
—	3	6	9	10	12	16	20	31	2	5	6	8	12
3	6	10	14	15	18	22	28	40	4	8	9	12	16
6	10	13	17	19	22	28	35	49	5	9	11	14	20
10	14	16	21	24	27	34	43	59	6	11	14	17	24
14	18												
18	24	20	26	29	33	41	53	72	7	13	16	20	28
24	30												
30	40	25	32	36	41	50	64	87	9	16	20	25	34
40	50												
50	65	30	38	43	49	60	76	104	10	18	23	29	40
65	80												
80	100	36	46	51	58	71	90	123	12	22	27	34	47
100	120												
120	140	43	55	61	68	83	106	143	14	26	32	39	54
140	160												
160	180												
180	200	50	64	70	79	96	122	165	15	29	35	44	61
200	225												
225	250												
250	280	56	72	79	88	108	137	186	17	33	40	49	69
280	315												
315	355	62	80	87	98	119	151	202	18	36	43	54	75
355	400												
400	450	68	88	95	108	131	165	223	20	40	47	60	83
450	500												

Table 3.11—*continued*

es = Upper deviation
ei = Lower deviation

Shaft h
Tolerance unit = 0·001 mm

Nominal sizes — h

Over	To	1 to 16	1	2	3	4	5	6	7	8	9	10	11	12	13	14*	15*	16*
		es	ei	ei	ei	ei	ei	ei	ei	ei	ei	ei	ei	ei	ei	ei	ei	ei
mm	mm																	
—	3	0	0·8	1·2	2	3	4	6	10	14	25	40	60	100	140	250	400	660
3	6	0	1	1·5	2·5	4	5	8	12	18	30	48	75	120	180	300	480	750
6	10	0	1	1·5	2·5	4	6	9	15	22	36	58	90	150	220	360	580	900
10 14	14 18	0	1·2	2	3	5	8	11	18	27	43	70	110	180	270	430	700	1 100
18 24	24 30	0	1·5	2·5	4	6	9	13	21	33	52	84	130	210	330	520	840	1 300
30 40	40 50	0	1·5	2·5	4	7	11	16	25	39	62	100	160	250	390	620	1 000	1 600
50 65	65 80	0	2	3	5	8	13	19	30	46	74	120	190	300	460	740	1 200	1 900
80 100	100 120	0	2·5	4	6	10	15	22	35	54	87	140	220	350	540	870	1 400	2 200
120 140 160	140 160 180	0	3·5	5	8	12	18	25	40	63	100	160	250	400	630	1 000	1 600	2 500
180 200 225	200 225 250	0	4·5	7	10	14	20	29	46	72	115	185	290	460	720	1 150	1 850	2 900
250 280	280 315	0	6	8	12	16	23	32	52	81	130	210	320	520	810	1 300	2 100	3 200
315 355	355 400	0	7	9	13	18	25	36	57	89	140	230	360	570	890	1 400	2 300	3 600
400 450	450 500	0	8	10	15	20	27	40	63	97	155	250	400	630	970	1 550	2 500	4 000

* Grades 14 to 16 do not apply to sizes up to and including 1 mm.

continued overleaf

Table 3.11—*continued*

es = Upper deviation							*Shaft j*
ei = Lower deviation					*Tolerance unit* = 0·001 mm		

Nominal sizes		*j*					
		5		6		7	
Over	To	es	ei	es	ei	es	ei
		+	−	+	−	+	−
mm	mm						
—	3	2	2	4	2	6	4
3	6	3	2	6	2	8	4
6	10	4	2	7	2	10	5
10	14	5	3	8	3	12	6
14	18						
18	24	5	4	9	4	13	8
24	30						
30	40	6	5	11	5	15	10
40	50						
50	65	6	7	12	7	18	12
65	80						
80	100	6	9	13	9	20	15
100	120						
120	140	7	11	14	11	22	18
140	160						
160	180						
180	200	7	13	16	13	25	21
200	225						
225	250						
250	280	7	16	16	16	26	26
280	315						
315	355	7	18	18	18	29	28
355	400						
400	450	7	20	20	20	31	32
450	500						

Table 3.11—*continued*

es = Upper deviation
ei = Lower deviation

Shaft j_s
Tolerance unit = 0·001 mm

Nominal sizes j_s

Over	To	1		2		3		4		5		6		7		8	
		es +	ei −	es +	ei −	es +	ei −	es +	ei −	es +	ei −	es +	ei −	es +	ei −	es +	ei −
mm	mm																
—	3	0·4	0·4	0·6	0·6	1	1	1·5	1·5	2	2	3	3	5	5	7	7
3	6	0·5	0·5	0·75	0·75	1·25	1·25	2	2	2·5	2·5	4	4	6	6	9	9
6	10	0·5	0·5	0·75	0·75	1·25	1·25	2	2	3	3	4·5	4·5	7	7	11	11
10 / 14	14 / 18	0·6	0·6	1	1	1·5	1·5	2·5	2·5	4	4	5·5	5·5	9	9	13	13
18 / 24	24 / 30	0·75	0·75	1·25	1·25	2	2	3	3	4·5	4·5	6·5	6·5	10	10	16	16
30 / 40	40 / 50	0·75	0·75	1·25	1·25	2	2	3·5	3·5	5·5	5·5	8	8	12	12	19	19
50 / 65	65 / 80	1	1	1·5	1·5	2·5	2·5	4	4	6·5	6·5	9·5	9·5	15	15	23	23
80 / 100	100 / 120	1·25	1·25	2	2	3	3	5	5	7·5	7·5	11	11	17	17	27	27
120 / 140 / 160	140 / 160 / 180	1·75	1·75	2·5	2·5	4	4	6	6	9	9	12·5	12·5	20	20	31	31
180 / 200 / 225	200 / 225 / 250	2·25	2·25	3·5	3·5	5	5	7	7	10	10	14·5	14·5	23	23	36	36
250 / 280	280 / 315	3	3	4	4	6	6	8	8	11·5	11·5	16	16	26	26	40	40
315 / 355	355 / 400	3·5	3·5	4·5	4·5	6·5	6·5	9	9	12·5	12·5	18	18	28	28	44	44
400 / 450	450 / 500	4	4	5	5	7·5	7·5	10	10	13·5	13·5	20	20	31	31	48	48

continued overleaf

Table 3.11—continued

es = Upper deviation
ei = Lower deviation

Shaft j_s
Tolerance unit = 0·001 mm

Nominal sizes		9		10		11		12		13		14*		15*		16*	
Over	To	es +	ei −	es +	ei −	es +	ei −	es +	ei −	es +	ei −	es +	ei −	es +	ei −	es +	ei −
mm	mm																
—	3	12	12	20	20	30	30	50	50	70	70	125	125	200	200	300	300
3	6	15	15	24	24	37	37	60	60	90	90	150	150	240	240	375	375
6	10	18	18	29	29	45	45	75	75	110	110	180	180	290	290	450	450
10	14	21	21	35	35	55	55	90	90	135	135	215	215	350	350	550	550
14	18																
18	24	26	26	42	42	65	65	105	105	165	165	260	260	420	420	650	650
24	30																
30	40	31	31	50	50	80	80	125	125	195	195	310	310	500	500	800	800
40	50																
50	65	37	37	60	60	95	95	150	150	230	230	370	370	600	600	950	950
65	80																
80	100	43	43	70	70	110	110	175	175	270	270	435	435	700		700	
100	120														700	1 100	1 100
120	140	50	50	80	80	125	125	200	200	315	315	500	500	800			
140	160														800	1 250	1 250
160	180																
180	220	57	57	92	92	145	145	230	230	360	360	575	575	925			
200	225														925	1 450	1 450
225	250																
250	280	65	65	105	105	160	160	260	260	405	405	650	650	1 050	1 050	1 600	1 600
280	315																
315	355	70	70	115	115	180	180	285	285	445	445	700	700	1 150	1 150	1 800	1 800
355	400																
400	450	77	77	125	125	200	200	315	315	485	485	775	775	1 250	1 250	2 000	2 000
450	500																

* Grades 14 to 16 do not apply to sizes up to and including 1 mm.

Table 3.11—*continued*

es = Upper deviation
ei = Lower deviation

Shafts k to m
Tolerance unit = 0·001 mm

Normal sizes		k					m					
		4	5	6	7	4 to 7	4	5	6	7	8	4 to 8
Over	To	es +	es +	es +	es +	ei +	es +	es +	es +	es +	es +	ei +
mm	mm											
—	3	3	4	6	10	0	5	6	8	12		2
3	6	5	6	9	13	1	8	9	12	16		4
6	10	5	7	10	16	1	10	12	15	21		6
10	14	6	9	12	19	1	12	15	18	25		7
14	18											
18	24	8	11	15	23	2	14	17	21	29		8
24	30											
30	40	9	13	18	27	2	16	20	25	34		9
40	50											
50	65	10	15	21	32	2	19	24	30	41		11
65	80											
80	100	13	18	25	38	3	23	28	35	48		13
100	120											
120	140	15	21	28	43	3	27	33	40	55		15
140	160											
160	180											
180	200	18	24	33	50	4	31	37	46	63		17
200	225											
225	250											
250	280	20	27	36	56	4	36	43	52	72		20
280	315											
315	355	22	29	40	61	4	39	46	57	78		21
355	400											
400	450	25	32	45	68	5	43	50	63	86		23
450	500											

continued overleaf

Table 3.11—continued

es = Upper deviation
ei = Lower deviation

Shafts n to p
Tolerance unit = 0·001 mm

Nominal sizes		n					p				
		4	5	6	7	4 to 8	4	5	6	7	4 to 9
Over	To	es +	es +	es +	es +	ei +	es +	es +	es +	es +	ei +
mm	mm										
—	3	7	8	10	14	4	9	10	12	16	6
3	6	12	13	16	20	8	16	17	20	24	12
6	10	14	16	19	25	10	19	21	24	30	15
10	14	17	20	23	30	12	23	26	29	36	18
14	18										
18	24	21	24	28	36	15	28	31	35	43	22
24	30										
30	40	24	28	33	42	17	33	37	42	51	26
40	50										
50	65	28	33	39	50	20	40	45	51	62	32
65	80										
80	100	33	38	45	58	23	47	52	59	72	37
100	120										
120	140	39	45	52	67	27	55	61	68	83	43
140	160										
160	180										
180	200	45	51	60	77	31	64	70	79	96	50
200	225										
225	250										
250	280	50	57	66	86	34	72	79	88	108	56
280	315										
315	355	55	62	73	94	37	80	87	98	119	62
355	400										
400	450	60	67	80	103	40	88	95	108	131	68
450	500										

Table 3.11—*continued*

es = Upper deviation

ei = Lower deviation

Shafts r to s

Tolerance unit = 0·001 mm

Nominal sizes		r					s				
		4	5	6	7	4 to 9	4	5	6	7	4 to 9
Over	To										
		es +	es +	es +	es +	ei +	es +	es +	es +	es +	ei +
mm	mm										
—	3	13	14	16	20	10	17	18	20	24	14
3	6	19	20	23	27	15	23	24	27	31	19
6	10	23	25	28	34	19	27	29	32	38	23
10	14	28	31	34	41	23	33	36	39	46	28
14	18										
18	24	34	37	41	49	28	41	44	48	56	35
24	30										
30	40	41	45	50	59	34	50	54	59	68	43
40	50										
50	65	49	54	60	71	41	61	66	72	83	53
65	80	51	56	62	73	43	67	72	78	89	59
80	100	61	66	73	86	51	81	86	93	106	71
100	120	64	69	76	89	54	89	94	101	114	79
120	140	75	81	88	103	63	104	110	117	132	92
140	160	77	83	90	105	65	112	118	125	140	100
160	180	80	86	93	108	68	120	126	133	148	108
180	200	91	97	106	123	77	136	142	151	168	122
200	225	94	100	109	126	80	144	150	159	176	130
225	250	98	104	113	130	84	154	160	169	186	140
250	280	110	117	126	146	94	174	181	190	210	158
280	315	114	121	130	150	98	186	193	202	222	170
315	355	126	133	144	165	108	208	215	226	247	190
355	400	132	139	150	171	114	226	233	244	265	208
400	450	146	153	166	189	126	252	259	272	295	232
450	500	152	159	172	195	132	272	279	292	315	252

continued overleaf

Table 3.11—continued

es = Upper deviation
ei = Lower deviation

Shafts t to u
Tolerance unit = 0·001 mm

Nominal sizes		t*				u				
		5	6	7	5 to 9	5	6	7	8	5 to 9
Over	To									
		es +	es +	es +	ei +	es +	es +	es +	es +	ei +
mm	mm									
—	3	—			—	22	24	28	32	18
3	6	—			—	28	31	35	41	23
6	10	—			—	34	37	43	50	28
10	14	—			—	41	44	51	60	33
14	18									
18	24	—			—	50	54	62	74	41
24	30	50	54	62	41	57	61	69	81	48
30	40	59	64	73	48	71	76	85	99	60
40	50	65	70	79	54	81	86	95	109	70
50	65	79	85	96	66	100	106	117	133	87
65	80	88	94	105	75	115	121	132	148	102
80	100	106	113	126	91	139	146	159	178	124
100	120	119	126	139	104	159	166	179	198	144
120	140	140	147	162	122	188	195	210	233	170
140	160	152	159	174	134	208	215	230	253	190
160	180	164	171	186	146	228	235	250	273	210
180	200	186	195	212	166	256	265	282	308	236
200	225	200	209	226	180	278	287	304	330	258
225	250	216	225	242	196	304	313	330	356	284
250	280	241	250	270	218	338	347	367	396	315
280	315	263	272	292	240	373	382	402	431	350
315	355	293	304	325	268	415	426	447	479	390
355	400	319	330	351	294	460	471	492	524	435
400	450	357	370	393	330	517	530	553	587	490
450	500	387	400	423	360	567	580	603	637	540

* Where t5 to t7 deviations are missing, u5 to u7 deviations should be substituted for them.

Table 3.11—*continued*

es = Upper deviation
ei = Lower deviation

Shafts v to y
Tolerance unit = 0·001 mm

Nominal sizes		v†				x					y		
		5	6	7	5 to 7	5	6	7	8	5 to 8	6'	7	6 and 7
Over	To	es	es	es	ei	es	es	es	es	ei	es	es	ei
		+	+	+	+	+	+	+	+	+	+	+	+
mm	mm												
—	3	—	—	—	—	24	26	30	34	20	—	—	—
3	6	—	—	—	—	33	36	40	46	28	—	—	—
6	10	—	—	—	—	40	43	49	56	34	—	—	—
10	14	—	—	—	—	48	51	58	67	40	—	—	—
14	18	47	50	57	39	53	56	63	72	45	—	—	—
18	24	56	60	68	47	63	67	75	87	54	76	84	63
24	30	64	68	76	55	73	77	85	97	64	88	96	75
30	40	79	84	93	68	91	96	105	119	80	110	119	94
40	50	92	97	106	81	108	113	122	136	97	130	139	114
50	65	115	121	132	102	135	141	152	168	122	163	174	144
65	80	133	139	150	120	159	165	176	192	146	193	204	174
80	100	161	168	181	146	193	200	213	232	178	236	249	214
100	120	187	194	207	172	225	232	245	264	210	276	289	254
120	140	220	227	242	202	266	273	288	311	248	325	340	300
140	160	246	253	268	228	298	305	320	343	280	365	380	340
160	180	270	277	292	252	328	335	350	373	310	405	420	380
180	200	304	313	330	284	370	379	396	422	350	454	471	425
200	225	330	339	356	310	405	414	431	457	385	499	516	470
225	250	360	369	386	340	445	454	471	497	425	549	566	520
250	280	408	417	437	385	498	507	527	556	475	612	632	580
280	315	448	457	477	425	548	557	577	606	525	682	702	650
315	355	500	511	532	475	615	626	647	679	590	766	787	730
355	400	555	566	587	530	685	696	717	749	660	856	877	820
400	450	622	635	658	595	767	780	803	837	740	960	983	920
450	500	687	700	723	660	847	860	883	917	820	1 040	1 063	1 000

* Where v5 to v7 deviations are missing, x5 to x7 deviations should be substituted for them.

continued overleaf

Table 3.11—*continued*

es = Upper deviation
ei = Lower deviation

Shafts z to zc
Tolerance unit = 0·001 mm

Nominal sizes		z			za			zb			zc		
		6	7	6 and 7	6	7	6 and 7	7	8	7 and 8	7	8	7 and 8
Over	To												
		es +	es +	ei +	es +	es +	ei +	es +	es +	ei +	es +	es +	ei +
mm	mm												
—	3	32	36	26	38	42	32	50	54	40	70	74	60
3	6	43	47	35	50	54	42	62	68	50	92	98	80
6	10	51	57	42	61	67	52	82	89	67	112	119	97
10	14	61	68	50									
14	18	71	78	60									
18	24	86	94	73									
24	30	101	109	88									
30	40	128	137	112									
40	50	152	161	136									
50	65	191	202	172									
65	80	229	240	210									
80	100	280	293	258									
100	120	332	345	310									
120	140	390	405	365									
140	160	440	455	415									
160	180	490	505	465									
180	200	549	566	520									
200	225	604	621	575									
225	250	669	686	640									
250	280	742	762	710									
280	315	822	842	790									
315	355	936	957	900									
355	400	1 036	1 057	1 000									
400	450	1 140	1 163	1 100									
450	500	1 290	1 313	1 250									

4 SCREW THREADS, THREADED FASTENERS AND STANDARD SIZES OF MATERIALS

SCREW THREADS, THREADED FASTENERS AND STANDARD SIZES OF MATERIALS

4

4 SCREW THREADS, THREADED FASTENERS AND STANDARD SIZES OF MATERIALS

SCREW THREADS AND FASTENERS

G. E. DODD

SCREW THREADS

Introduction

The work of the ISO (International Standards Organisation) and the BSI in preparing British Standards for metric screw threads and fasteners, has allowed the British Fastener Industry to provide the necessary screwed components to assist in the conversion to the metric system. This system enables the industries of the world to align usage to internationally agreed standards. Therefore, the declared UK national policy recommends the adoption of the ISO metric screw thread system and fasteners to the eventual exclusion of the BSW, BSF, BA forms, etc.

However, some sections of industry have already changed to the ISO inch (unified) screw thread system. These include a number of large chemical, automotive and aerospace manufacturers, particularly those with major interests in the USA.

It is therefore necessary to provide ISO inch screw threads for some time, but when a change is made from Whitworth and BA threads, ISO metric threads should be used to avoid an intermediate change to ISO inch threads. Both ISO metric and ISO inch screw threads and principal types of fasteners with these threads are included in this section and are recognised by the ISO.

The information in this section covers functional and dimensional data for screw threads used in most applications.

Applications of screw threads

Screw threads are used for many purposes, the principal being as follows:

Means of attachment, between a bolt and nut of a threaded fastener in an assembly.

Means of clamping or tightening between two components such as by means of a bolt and nut if a threaded fastener.

Means of adjustment or travel by translating rotary motion into linear motion e.g. lead screw of machine tool.

Means of applying pressure in a vertical or horizontal component by (linear) movement relating to a piece of machinery or equipment, e.g. lifting jack, screw down gear of metal rolling mills, pug mill or extrusion mill for plastics or compounds.

Means of coupling between one part and another, e.g. fluid coupling compression fittings.

In some of the applications the screw thread is subject to tensile or shear stresses additional to those set up during clamping or torquing by the interaction between flanks of male and female threads. The form or profile of the thread varies according to the appli-

4-2

cation and in general consists of the following: triangular symmetrical; triangular asymmetrical; buttressed; trapezoidal (Acme); and rounded.

Screw thread selection

The selection of a screw thread will depend on a combination of one or more of the following features:

Dimensional.   General; diameter and Pitch relationship.

Functional.

Strength.   Tension; shear; torsion; combination of these.

Variety reduction.

DIMENSIONAL. GENERAL

The Whitworth series of screw threads are obsolescent. In new designs, the Whitworth dimensional thread system is being superseded by the ISO metric system.

The ISO inch system (unified) is recognised when requirements of interchangeability with screw thread in the United States or areas where American National threads predominate. British Standard pipe threads are recognised by ISO and are maintained in the inch system with fractional designations for pipe joints. The combination of a taper internal thread with a parallel internal thread is used together with jointing compound for joints which are to be pressure tight by the mating of the threads. These threads are described in ISO Recommendation R 7[14] and BS 21[1]. Thread combinations where the joint is not made on the thread but on a washer or gasket are described in ISO Recommendation R 228[16] and BS 2779[7].

The dimensions of the threads where strength requirements are secondary will be determined by the dimensions of the components or the space limitations. A thin walled compound will necessitate the use of fine pitch threads and the number of screwed studs or bolts in a flanged pipe joint will depend on space and strength limitations.

Environmental conditions will also need consideration, e.g. combinations of bolt and nut subject to heat or dirty conditions will require a freer fit and wider manufacturing tolerances.

Miniature screw threads used in fine instrument work and horological applications have special profiles with rounded roots in the screws in order to provide strength during assembly rather than strength during attachment.

DIMENSIONAL, DIAMETER AND PITCH RELATIONSHIP

It is necessary to consider the relative merits of screwed components with coarse or fine screw threads. Coarse screw threads have less pitches per unit length than fine screw threads and are usually referred to as screw threads with graded pitches. In other words the number of pitches vary as the diameter increases.

In the selection of a graded pitch series of screw threads it must be borne in mind that a fine pitch thread has a smaller helix angle than a coarse pitch, and therefore, may not tend to loosen during service. Also a fine pitch thread has a greater stress area than a coarse pitch and therefore may, subject to other conditions such as correct length of engagement and selection of proper bolt and nut materials, carry greater loads. The decision to use fine threads or coarse threads needs careful choice and it is necessary to have a complete understanding of the function of the screws with respect to the whole assembly or component before the correct selection can be made.

As most bolted assemblies are made with relatively high tension produced by wrenching the bolt head into a tapped hole, slightly less torquing effort is required to tighten a fine threaded assembly due to the smaller helix angle, but this is negatived by the higher frictional losses of the fine thread.

In assemblies using castle or slotted nuts where the slot in the nut must be stopped to coincide with the hole in the bolt in order to insert a cotter pin, fine threads are an advantage. This is because the linear travel which takes place during rotation of the nut is smaller and allows a finer control of nut tension for a given rotation.

FUNCTIONAL

In applications where the purpose of the screw thread is to perform some operational function rather than carry out an attachment or fastening function a number of features of the thread will be important. Pitch and profile will play an important part in the design of the thread; the former determining the linear travel of the female or nut part of the assembly and the latter needs to be carefully selected in order to withstand the load exerted by the screw. A typical example of the former of these applications is a lead screw in a screw cutting lathe in which one revolution corresponds to the linear travel equal to one pitch.

It is claimed that fine threads are less likely to loosen in a product under vibration than coarse threads. Research has shown that neither coarse nor fine threads will loosen under vibration if the joint is properly designed to provide rigidity by correct initial tightening. Research has also demonstrated that, provided a close fit is adopted between bolt and nut thread, there is little difference between the stripping strength in tension of a fine threaded assembly as compared with a coarse threaded assembly. Wider tolerances and a looser fit between bolt thread and nut thread tend to cause a load concentration on the first threads inside the nut which, with a fine thread and corresponding reduced depth of engagement, will be proportionately greater, thus precipitating a progressive failure of the succeeding threads. Thus coarse threads are used especially when the materials tend to be brittle such as in tapped holes in cast iron or plastics or soft non-ferrous materials.

It is thus necessary to provide correct length of engagement of threads with correctly graded material strengths, and to eliminate such brittleness as may be expected when decarburisation is present, in order to utilise the increased core strength of components with fine threads.

Another aspect which favours the choice of coarse threads as against fine threads is the effect of corrosion which is less damaging to coarse threads in machinery and structures subject to constant outdoor exposure in marine or chemical laden atmospheres. Corrosion preventive coatings such as hot dip galvanising and chromium, cadmium or zinc plating are more effective when applied to coarse threads. Moreover, coarse threads both coated or plain are less prone to damage during assembly than fine threads.

Fine threaded components are more difficult and time consuming to assemble because a fine threaded fastener requires on average two-thirds more turns between nut and bolt during assembly. Moreover a fine thread bolt entering a nut or tapped hole tends to cross thread especially when mating components are not positioned in line one with the other. Coarse threads are less likely to gall or seize in high temperature applications.

Therefore, for the majority of fastener applications coarse threaded fasteners are quite suitable and only special features of the assembly will require the selection of a fine thread series.

STRENGTH OF SCREW THREADS

It is desirable, when possible, to design a threaded connection or fastener so that the failure under tensile load will occur by breakage at the core of the bolt rather than by stripping of the threads in engagement with the nut. Threaded stripping commences by bending of the first threads in engagement and then by progressive bending and stripping of the successive threads. This condition is likely to be accentuated when conditions of repeated assembly and dismantling take place.

BS 3580[8] and NBS Handbook H 28[22] give more detailed methods of assessing the stripping strength, fatigue strength and impact strength of a threaded connection and the beneficial effects likely to be obtained in the strength by specifying certain methods of manufacture. In general, the information contained in these publications refers to triangular profile screw threads only.

The following information has been prepared as a general guide to the design of stressed screwed assemblies where the terms 'bolt and nut' refer to externally and internally

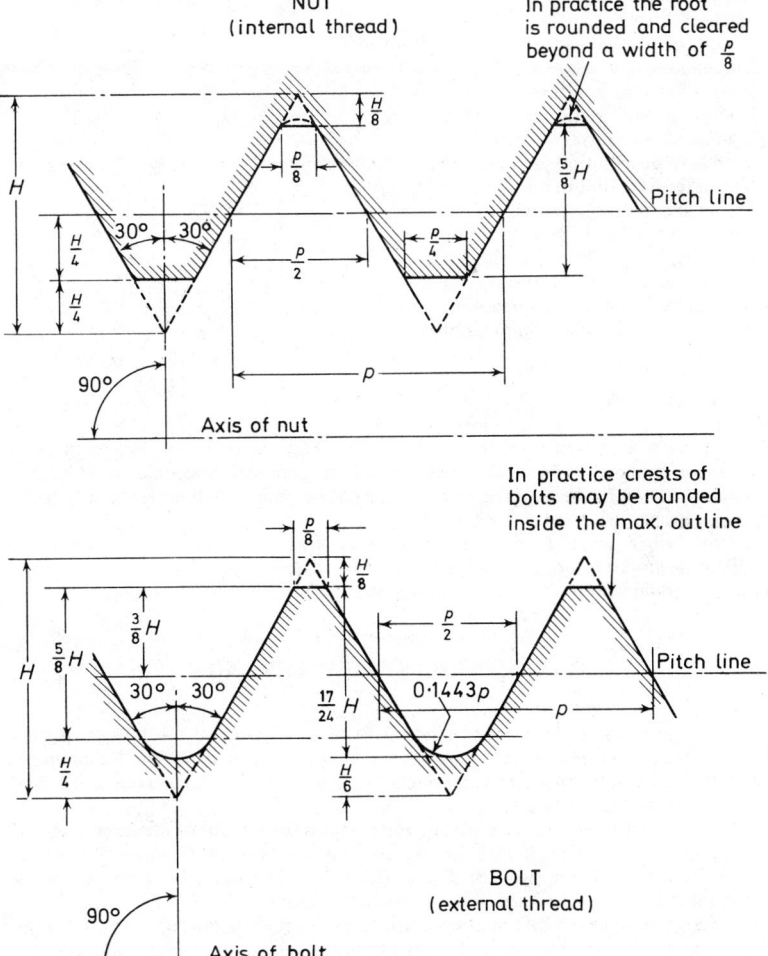

Figure 4.1. Design forms of internal and external threads (maximum metal conditions)

threaded components respectively. For more detailed information on the strength of screw threads reference should be made to NBS Handbook H 28, Part 1.

A threaded assembly should be designed to withstand axial loads of a fluctuating or static nature, together with supplementary shear or bending loads and torsional loads arising from friction due to the helix relationship when clamping, or to the force exerted or due to load imposed or to be carried.

The strength factors will depend on the following:

Materials of the assembly.

Method of production particularly of bolt threads such as cold rolling after heat treatment.

Thread form.

Types of loading.

Diameter of thread.

Pitch of thread.

Length of engagement or number of threads.

Fit.

Form of root in bolt thread.

Friction or anti-friction conditions.

Truncation of crests.

Thread load distribution can occur in connections under the following circumstances:

Nut in tension, bolt in tension.

Nut in shear, bolt in tension.

Nut in compression, bolt in compression.

Nut in tension, bolt in compression.

Nut in compression, bolt in tension.

Nut in shear, bolt in compression.

Nut in tension, bolt in compression.

VARIETY REDUCTION

With regard to standard threaded fasteners, the economical production requires that the number of items produced for stock under mass production techniques be kept to a minimum. Therefore, the designer should select coarse or fine pitch series threads because tools and gauges are more easily available.

Threads which are different from the standard graded series require the design and manufacture of special taps dies and gauges with increased costs and delay in procurement. Selection should also be made from the standard tolerance grades.

ISO METRIC SCREW THREADS

BS 3643[9] relates to single-start, parallel screw threads from 1·0 to 300 mm diameter, having the ISO basic profile for triangular screw threads as given in ISO Recommendation R 68[15], and metric diameters and pitches as given in ISO Recommendation R 261[17]. For overall plan see Table 4.2.

From amongst these diameter/pitch combinations, the standard identifies two series of diameters with graded pitches for use with screws, bolts, nuts and other common threaded fasteners. These two series, one with coarse and the other with fine pitches, are the same as those scheduled in ISO Recommendation R 262[18].

The design profiles for ISO metric internal and external threads are shown in Figure 4.1. These represent the profiles of the threads in their maximum metal condition. It will be noted that the root of each thread is deepened so as to clear the basic flat crest of the other thread. The contact between the threads is thus confined to their sloping flanks.

Table 4.1 THREAD DATA (mm)

Pitch p	p/8	p/4	H (0·866 03p)	H/8 (0·108 25p)	H/6 (0·144 34p)	H/4 (0·216 51p)	3H/8 (0·324 76p)	5H/8 (0·541 27p)	17H/24 (0·613 44p)	3H/4 (0·649 52p)	1¼H (1·082 54p)	1 5/12 H (1·226 88p)
0·25	0·031 2	0·062 5	0·216 5	0·027 1	0·036 1	0·054 1	0·081 2	0·135 3	0·153 4	0·162 4	0·270 6	0·306 7
0·35	0·043 8	0·087 5	0·303 1	0·037 9	0·050 5	0·075 8	0·113 7	0·189 4	0·214 7	0·227 3	0·378 9	0·429 4
0·4	0·050 0	0·100 0	0·346 4	0·043 3	0·057 7	0·086 6	0·129 9	0·216 5	0·245 4	0·259 8	0·433 0	0·490 8
0·45	0·056 2	0·112 5	0·389 7	0·048 7	0·065 0	0·097 4	0·146 1	0·243 6	0·276 0	0·292 3	0·487 1	0·552 1
0·5	0·062 5	0·125 0	0·433 0	0·054 1	0·072 2	0·108 3	0·162 4	0·270 6	0·306 7	0·324 8	0·541 3	0·613 4
0·6	0·075 0	0·150 0	0·519 6	0·065 0	0·086 6	0·129 9	0·194 9	0·324 8	0·368 1	0·389 7	0·649 5	0·736 1
0·7	0·087 5	0·175 0	0·606 2	0·075 8	0·101 0	0·151 6	0·227 3	0·378 9	0·429 4	0·454 7	0·757 8	0·858 8
0·75	0·093 8	0·187 5	0·649 5	0·081 2	0·108 3	0·162 4	0·243 6	0·405 9	0·460 1	0·487 1	0·811 9	0·920 2
0·8	0·100 0	0·200 0	0·692 8	0·086 6	0·115 5	0·173 2	0·259 8	0·433 0	0·490 8	0·519 6	0·866 0	0·981 5
1	0·125 0	0·250 0	0·866 0	0·108 2	0·144 3	0·216 5	0·324 8	0·541 3	0·613 4	0·649 5	1·082 5	1·226 9
1·25	0·156 2	0·312 5	1·082 5	0·135 3	0·180 4	0·270 6	0·406 0	0·676 6	0·766 8	0·811 9	1·353 2	1·533 6
1·5	0·187 5	0·375 0	1·299 0	0·162 4	0·216 5	0·324 8	0·487 1	0·811 9	0·920 2	0·974 3	1·623 8	1·840 3
1·75	0·218 8	0·437 5	1·515 6	0·189 4	0·252 6	0·378 9	0·568 3	0·947 2	1·073 5	1·136 7	1·894 4	2·147 0
2	0·250 0	0·500 0	1·732 1	0·216 5	0·288 7	0·433 0	0·649 5	1·082 5	1·226 9	1·299 0	2·165 1	2·453 8
2·5	0·312 5	0·625 0	2·165 1	0·270 6	0·360 8	0·541 3	0·811 9	1·353 2	1·533 6	1·623 8	2·706 4	3·067 2
3	0·375 0	0·750 0	2·598 1	0·324 8	0·433 0	0·649 5	0·974 3	1·623 8	1·840 3	1·948 6	3·247 6	3·680 6
3·5	0·437 5	0·875 0	3·031 1	0·378 9	0·505 2	0·757 8	1·136 7	1·894 4	2·147 0	2·273 3	3·788 9	4·294 1
4	0·500 0	1·000 0	3·464 1	0·433 0	0·577 4	0·866 0	1·299 0	2·165 1	2·453 8	2·598 1	4·330 2	4·907 5
4·5	0·562 5	1·125 0	3·897 1	0·487 1	0·649 5	0·974 3	1·461 4	2·435 7	2·760 5	2·922 8	4·871 4	5·521 0
5	0·625 0	1·250 0	4·330 2	0·541 2	0·721 7	1·082 6	1·623 8	2·706 4	3·067 2	3·247 6	5·412 7	6·134 4
5·5	0·687 5	1·375 0	4·763 2	0·595 4	0·793 9	1·190 8	1·786 2	2·977 0	3·373 9	3·572 4	5·954 0	6·747 8
6	0·750 0	1·500 0	5·196 2	0·649 5	0·866 0	1·299 1	1·948 6	3·247 6	3·680 6	3·897 1	6·495 2	7·361 3

Table 4.2 ISO METRIC THREADS

| Basic major diameters | | | Coarse series graded pitch | Constant pitches | | | | | | | | | | | |
First choice	Second choice	Third choice		6	4	3	2	1·5	1·25	1	0·75	0·5	0·35	0·25	0·2
1	—	—	0·25	—	—	—	—	—	—	—	—	—	—	—	0·2
—	1·1	—	0·25	—	—	—	—	—	—	—	—	—	—	—	0·2
1·2	—	—	0·25	—	—	—	—	—	—	—	—	—	—	—	0·2
—	1·4	—	0·30	—	—	—	—	—	—	—	—	—	—	—	0·2
1·6	—	—	0·35	—	—	—	—	—	—	—	—	—	—	—	0·2
—	1·8	—	0·35	—	—	—	—	—	—	—	—	—	—	—	0·2
2	—	—	0·4	—	—	—	—	—	—	—	—	—	—	0·25	—
—	2·2	—	0·45	—	—	—	—	—	—	—	—	—	—	0·25	—
2·5	—	—	0·45	—	—	—	—	—	—	—	—	—	0·35	—	—
3	—	—	0·5	—	—	—	—	—	—	—	—	—	0·35	—	—
—	3·5	—	0·6	—	—	—	—	—	—	—	—	—	0·35	—	—
4	—	—	0·7	—	—	—	—	—	—	—	—	0·5	—	—	—
—	4·5	—	0·75	—	—	—	—	—	—	—	—	0·5	—	—	—
5	—	—	0·8	—	—	—	—	—	—	—	—	0·5	—	—	—
—	—	5·5	—	—	—	—	—	—	—	—	—	0·5	—	—	—
6	—	—	1	—	—	—	—	—	—	—	0·75	—	—	—	—
—	—	7	1	—	—	—	—	—	—	—	0·75	—	—	—	—
8	—	—	1·25	—	—	—	—	—	—	1	0·75	—	—	—	—
—	—	9	1·25	—	—	—	—	—	—	1	0·75	—	—	—	—
10	—	—	1·5	—	—	—	—	—	1·25	1	0·75	—	—	—	—
—	—	11	1·5	—	—	—	—	—	—	1	0·75	—	—	—	—
12	—	—	1·75	—	—	—	—	1·5	1·25	1	—	—	—	—	—
—	14	—	2	—	—	—	—	1·5	1·25	1	—	—	—	—	—
—	—	15	—	—	—	—	—	1·5	—	1	—	—	—	—	—
16	—	—	2	—	—	—	—	1·5	—	1	—	—	—	—	—
—	—	17	—	—	—	—	—	1·5	—	1	—	—	—	—	—
—	18	—	2·5	—	—	—	2	1·5	—	1	—	—	—	—	—
20	—	—	2·5	—	—	—	2	1·5	—	1	—	—	—	—	—
—	22	—	2·5	—	—	—	2	1·5	—	1	—	—	—	—	—
24	—	—	3	—	—	—	2	1·5	—	1	—	—	—	—	—
—	—	25	—	—	—	—	2	1·5	—	1	—	—	—	—	—
—	—	26	—	—	—	—	—	1·5	—	—	—	—	—	—	—
—	27	—	3	—	—	—	2	1·5	—	1	—	—	—	—	—
—	—	28	—	—	—	—	2	1·5	—	1	—	—	—	—	—
30	—	—	3·5	—	—	3	2	1·5	—	1	—	—	—	—	—
—	—	32	—	—	—	—	2	1·5	—	—	—	—	—	—	—
—	33	—	3·5	—	—	3	2	1·5	—	—	—	—	—	—	—
—	—	35	—	—	—	—	—	1·5	—	—	—	—	—	—	—
36	—	—	4	—	—	3	2	1·5	—	—	—	—	—	—	—
—	—	38	—	—	—	—	—	1·5	—	—	—	—	—	—	—
—	39	—	4	—	—	3	2	1·5	—	—	—	—	—	—	—
—	—	40	—	—	—	3	2	1·5	—	—	—	—	—	—	—
42	—	—	4·5	—	4	3	2	1·5	—	—	—	—	—	—	—
—	45	—	4·5	—	4	3	2	1·5	—	—	—	—	—	—	—
48	—	—	5	—	4	3	2	1·5	—	—	—	—	—	—	—
—	—	50	—	—	—	3	2	1·5	—	—	—	—	—	—	—
—	52	—	5	—	4	3	2	1·5	—	—	—	—	—	—	—
—	—	55	—	—	4	3	2	1·5	—	—	—	—	—	—	—
56	—	—	5·5	—	4	3	2	1·5	—	—	—	—	—	—	—
—	—	58	—	—	4	3	2	1·5	—	—	—	—	—	—	—
—	60	—	5·5	—	4	3	2	1·5	—	—	—	—	—	—	—
—	—	62	—	—	4	3	2	1·5	—	—	—	—	—	—	—
64	—	—	6	—	4	3	2	1·5	—	—	—	—	—	—	—
—	—	65	—	—	4	3	2	1·5	—	—	—	—	—	—	—

Table 4.2—*continued*

First choice	Second choice	Third choice	Coarse series graded pitch	6	4	3	2	1·5	1·25	1	0·75	0·5	0·35	0·25	0·2
—	68	—	6	—	4	3	2	1·5	—	—	—	—	—	—	—
—	—	70	—	6	4	3	2	1·5	—	—	—	—	—	—	—
72	—	—	—	6	4	3	2	1·5	—	—	—	—	—	—	—
—	—	75	—	—	4	3	2	1·5	—	—	—	—	—	—	—
—	76	—	—	6	4	3	2	1·5	—	—	—	—	—	—	—
—	—	78	—	—	—	—	2	—	—	—	—	—	—	—	—
80	—	—	—	6	4	3	2	1·5	—	—	—	—	—	—	—
—	—	82	—	—	—	—	2	—	—	—	—	—	—	—	—
—	85	—	—	6	4	3	2	—	—	—	—	—	—	—	—
90	—	—	—	6	4	3	2	—	—	—	—	—	—	—	—
—	95	—	—	6	4	3	2	—	—	—	—	—	—	—	—
100	—	—	—	6	4	3	2	—	—	—	—	—	—	—	—
—	105	—	—	6	4	3	2	—	—	—	—	—	—	—	—
110	—	—	—	6	4	3	2	—	—	—	—	—	—	—	—
—	115	—	—	6	4	3	2	—	—	—	—	—	—	—	—
—	120	—	—	6	4	3	2	—	—	—	—	—	—	—	—
125	—	—	—	6	4	3	2	—	—	—	—	—	—	—	—
—	130	—	—	6	4	3	2	—	—	—	—	—	—	—	—
—	—	135	—	6	4	3	2	—	—	—	—	—	—	—	—
140	—	—	—	6	4	3	2	—	—	—	—	—	—	—	—
—	—	145	—	6	4	3	2	—	—	—	—	—	—	—	—
—	150	—	—	6	4	3	2	—	—	—	—	—	—	—	—
—	—	155	—	6	4	3	—	—	—	—	—	—	—	—	—
160	—	—	—	6	4	3	—	—	—	—	—	—	—	—	—
—	—	165	—	6	4	3	—	—	—	—	—	—	—	—	—
—	170	—	—	6	4	3	—	—	—	—	—	—	—	—	—
—	—	175	—	6	4	3	—	—	—	—	—	—	—	—	—
180	—	—	—	6	4	3	—	—	—	—	—	—	—	—	—
—	—	185	—	6	4	3	—	—	—	—	—	—	—	—	—
—	190	—	—	6	4	3	—	—	—	—	—	—	—	—	—
—	—	195	—	6	4	3	—	—	—	—	—	—	—	—	—
200	—	—	—	6	4	3	—	—	—	—	—	—	—	—	—
—	—	205	—	6	4	3	—	—	—	—	—	—	—	—	—
—	210	—	—	6	4	3	—	—	—	—	—	—	—	—	—
—	—	215	—	6	4	3	—	—	—	—	—	—	—	—	—
220	—	—	—	6	4	3	—	—	—	—	—	—	—	—	—
—	—	225	—	6	4	3	—	—	—	—	—	—	—	—	—
—	—	230	—	6	4	3	—	—	—	—	—	—	—	—	—
—	—	235	—	6	4	3	—	—	—	—	—	—	—	—	—
—	240	—	—	6	4	3	—	—	—	—	—	—	—	—	—
—	—	245	—	6	4	3	—	—	—	—	—	—	—	—	—
250	—	—	—	6	4	3	—	—	—	—	—	—	—	—	—
—	—	255	—	6	4	—	—	—	—	—	—	—	—	—	—
—	260	—	—	6	4	—	—	—	—	—	—	—	—	—	—
—	—	265	—	6	4	—	—	—	—	—	—	—	—	—	—
—	—	270	—	6	4	—	—	—	—	—	—	—	—	—	—
—	—	275	—	6	4	—	—	—	—	—	—	—	—	—	—
280	—	—	—	6	4	—	—	—	—	—	—	—	—	—	—
—	—	285	—	6	4	—	—	—	—	—	—	—	—	—	—
—	—	290	—	6	4	—	—	—	—	—	—	—	—	—	—
—	—	295	—	6	4	—	—	—	—	—	—	—	—	—	—
—	300	—	—	6	4	—	—	—	—	—	—	—	—	—	—

Basic numerical data for the various standard pitches of ISO metric threads are given in Table 4.1.

The two standard series of diameters with their related pitches which are recommended for screws, bolts and nuts and other common fasteners with ISO metric screw threads are given in Table 4.3.

Table 4.3 SELECTED 'COARSE' AND 'FINE' SERIES FOR SCREWS, BOLTS AND NUTS AND OTHER COMMON THREADED FASTENERS (mm)

Basic major diameters		Pitches	
First choice	Second choice	Coarse series	Fine series
1·0	—	0·25	—
—	1·1	0·25	—
1·2	—	0·25	—
—	1·4	0·30	—
1·6	—	0·35	—
—	1·8	0·35	—
2	—	0·4	—
—	2·2	0·45	—
2·5	—	0·45	—
3	—	0·5	—
—	3·5	0·6	—
4	—	0·7	—
—	4·5	0·75	—
5	—	0·8	—
6	—	1	0·75
—	7	1	—
8	—	1·25	1
10	—	1·5	1·25
12	—	1·75	1·25
—	14	2	1·5
16	—	2	1·5
—	18	2·5	1·5
20	—	2·5	1·5
—	22	2·5	1·5
24	—	3	2
—	27	3	2
30	—	3·5	2
—	33	3·5	2
36	—	4	3
—	39	4	3

Fine or constant pitch screw threads to BS 3643 are designated by the letter M followed by the nominal diameter and the pitch, e.g. M6 × 0·75. In the absence of an indication of pitch a coarse thread is specified, thus, a coarse thread M6 × 1 is designated 'M6'.

Tolerance zones and classes of fit

A tolerance zone must be specified both in magnitude and position in relation to the basic size of the fit of which it is a part. The nature of a fit is dependent on both the magnitudes of the tolerances and the positions of the tolerance zones for the two members. The position of a tolerance zone is defined by the distance between the basic size and the nearest end of the tolerance zone. This distance is known as the 'fundamental deviation'. In the ISO metric screw thread system fundamental deviations are designated by letters, capitals for internal threads and small letters for external threads. The magnitudes of tolerance zones are designated by tolerance grades (figures). A combination of a tolerance grade (figure) and a fundamental deviation (letter) forms a tolerance class designation, e.g. '6g'.

In BS 3643[9] 'class of fit' indicates the degree of fit between external (bolt) threads and internal (nut) threads. ISO definitions designate this condition as 'tolerance quality', but to maintain conformity with British practice the designation 'class of fit' is preferred. Thus 'fine tolerance quality' conforms to the 'close' class of fit and 'coarse tolerance quality' conforms to the 'free' class of fit. The 'medium' designation is common to both methods.

The complete ISO metric screw thread tolerancing system provides many combinations of tolerance grades and fundamental deviations to cater for most applications. However,

Table 4.4   CLASSES OF FIT FOR ISO METRIC SCREW THREADS

	Tolerance class	
Class of fit	Internal threads (nuts)	External threads (bolts)
Close	5H	4h
Medium	6H	6g
Free	7H	8g

the tolerance classes specified in BS 3643 include tolerances and minimum clearances similar to the Unified thread classes 1A/1B; 2A/2B; 3A/3B.

For external threads (bolts): 8g, 6g, 4h. See 1A, 2A and 3A.
For internal threads (nuts): 7H, 6H, 5H. See 1B, 2B and 3B.
These are shown in Table 4.4 to give close, medium and free thread fits.

Application and classes of fit

Medium fit (6H/6g).   The medium class of fit is appropriate for most general engineering purposes. The minimum clearance associated with this fit assists in free assembly, and this minimises galling and seizing in high speed assembly. (For tolerance details for coarse and fine threads see Tables 4.5 to 4.8 and Figure 4.2).

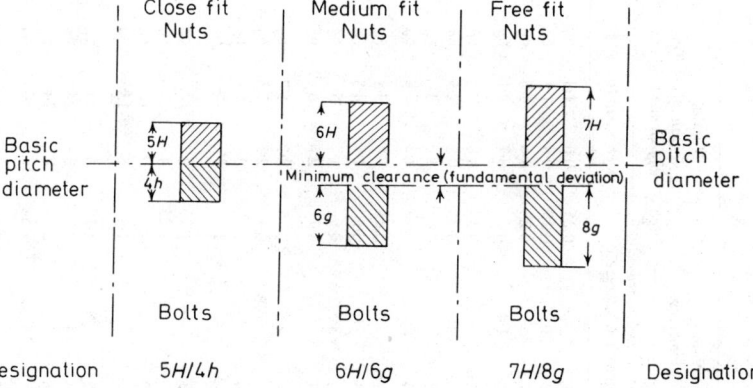

Figure 4.2. Diagram showing relationship between tolerance zones and classes of fit

Table 4.5 LIMITS AND TOLERANCES FOR UNCOATED THREADS, COARSE SERIES THREADS— BOLTS (mm)
Tolerance class 6g

Nominal diameters			Pitch	Normal length of engagement		Tolerance class	Fund. dev.	Major diameter			Pitch diameter			Minor diameter		
1st choice	2nd choice	3rd choice		over	up to			max	tol	min	max	tol	min	max	tol	min
1			0·25	0·6	1·8	6g	0·018	0·982	0·067	0·915	0·820	0·053	0·767	0·675	0·071	0·604
	1·1		0·25	0·6	1·8	6g	0·018	1·082	0·067	1·015	0·920	0·053	0·867	0·775	0·071	0·704
1·2			0·25	0·6	1·8	6g	0·018	1·182	0·067	1·115	1·020	0·053	0·967	0·875	0·071	0·804
	1·4		0·3	0·7	2·1	6g	0·018	1·382	0·075	1·307	1·187	0·056	1·131	1·014	0·078	0·936
1·6			0·35	0·8	2·6	6g	0·019	1·581	0·085	1·496	1·354	0·063	1·291	1·151	0·088	1·063
	1·8		0·35	0·8	2·6	6g	0·019	1·781	0·085	1·696	1·554	0·063	1·491	1·351	0·088	1·263
2			0·4	1	3	6g	0·019	1·981	0·095	1·886	1·721	0·067	1·654	1·490	0·096	1·394
	2·2		0·45	1·3	3·8	6g	0·020	2·180	0·100	2·080	1·888	0·071	1·817	1·628	0·103	1·525
2·5			0·45	1·3	3·8	6g	0·020	2·480	0·100	2·380	2·188	0·071	2·117	1·928	0·103	1·825
3			0·5	1·5	4·5	6g	0·020	2·980	0·106	2·874	2·655	0·075	2·580	2·367	0·111	2·256
	3·5		0·6	1·7	5	6g	0·021	3·479	0·125	3·354	3·089	0·085	3·004	2·743	0·128	2·615
4			0·7	2	6	6g	0·022	3·978	0·140	3·838	3·523	0·090	3·433	3·119	0·140	2·979
	4·5		0·75	2·2	6·7	6g	0·022	4·478	0·140	4·338	3·991	0·090	3·901	3·558	0·144	3·414
5			0·8	2·5	7·5	6g	0·024	4·976	0·150	4·826	4·456	0·095	4·361	3·995	0·153	3·842
6			1	3	9	6g	0·026	5·974	0·180	5·794	5·324	0·112	5·212	4·747	0·184	4·563
		7	1	3	9	6g	0·026	6·974	0·180	6·794	6·324	0·112	6·212	5·747	0·184	5·563
8			1·25	4	12	6g	0·028	7·972	0·212	7·760	7·160	0·118	7·042	6·438	0·208	6·230
		9	1·25	4	12	6g	0·028	8·972	0·212	8·760	8·160	0·118	8·042	7·438	0·208	7·230
10			1·5	5	15	6g	0·032	9·968	0·236	9·732	8·994	0·132	8·862	8·128	0·240	7·888
		11	1·5	5	15	6g	0·032	10·968	0·236	10·732	9·994	0·132	9·862	9·128	0·240	8·888
12			1·75	6	18	6g	0·034	11·966	0·265	11·701	10·829	0·150	10·679	9·819	0·276	9·543
	14		2	8	24	6g	0·038	13·962	0·280	13·682	12·663	0·160	12·503	11·508	0·304	11·204
16			2	8	24	6g	0·038	15·962	0·280	15·682	14·663	0·160	14·503	13·508	0·304	13·204
	18		2·5	10	30	6g	0·042	17·958	0·335	17·623	16·334	0·170	16·164	14·891	0·350	14·541
20			2·5	10	30	6g	0·042	19·958	0·335	19·623	18·334	0·170	18·164	16·891	0·350	16·541
	22		2·5	10	20	6g	0·042	21·958	0·335	21·623	20·334	0·170	20·164	18·891	0·350	18·541

39	4	18	53	6g	0·060	38·940	0·475	38·465	36·342	0·224	36·118	34·033	0·512
42	4·5	21	63	6g	0·063	41·937	0·500	41·437	39·014	0·236	38·778	36·416	0·561
45	4·5	21	63	6g	0·063	44·937	0·500	44·437	42·014	0·236	41·778	39·416	0·561
48	5	24	71	6g	0·071	47·929	0·530	47·399	44·681	0·250	44·431	41·795	0·611
52	5	24	71	6g	0·071	51·929	0·530	51·399	48·681	0·250	48·431	45·795	0·611
56	5·5	28	85	6g	0·075	55·925	0·560	55·365	52·353	0·265	52·088	49·177	0·662
60	5·5	28	85	6g	0·075	59·925	0·560	59·365	56·353	0·265	56·088	53·177	0·662
64	6	32	95	6g	0·080	63·920	0·600	63·320	60·023	0·280	59·743	56·559	0·713
68	6	32	95	6g	0·080	67·920	0·600	67·320	64·023	0·280	63·743	60·559	0·713

Final column (33·521):

39	33·521
42	35·855
45	38·855
48	41·184
52	45·184
56	48·515
60	52·515
64	55·846
68	59·846

For limits and tolerances of other diameters and pitches, see BS 3643, Part 3.

Table 4.6 LIMITS AND TOLERANCES FOR UNCOATED THREADS, COARSE SERIES THREADS— NUTS (mm)

Tolerance class 6H

Nominal diameters			Pitch	Normal length of engagement		Tolerance class	Fund. dev.	Major diameter min	Pitch diameter			Minor diameter		
1st choice	2nd choice	3rd choice		over	up to				max	tol	min	max	tol	min
1			0·25	0·6	1·8	6H								
	1·1		0·25	0·6	1·8	6H								
1·2			0·25	0·6	1·8	6H								
	1·4		0·3	0·7	2·1	6H	0	1·400	1·280	0·075	1·205	1·160	0·085	1·075
1·6			0·35	0·8	2·6	6H	0	1·600	1·458	0·085	1·373	1·321	0·100	1·221
	1·8		0·35	0·8	2·6	6H	0	1·800	1·650	0·085	1·573	1·521	0·100	1·421
2			0·4	1	3	6H	0	2·000	1·830	0·090	1·740	1·679	0·112	1·567
	2·2		0·45	1·3	3·8	6H	0	2·200	2·003	0·095	1·908	1·838	0·125	1·713
2·5			0·45	1·3	3·8	6H	0	2·500	2·303	0·095	2·208	2·138	0·125	2·013
3			0·5	1·5	4·5	6H	0	3·000	2·775	0·100	2·675	2·599	0·140	2·459
	3·5		0·6	1·7	5	6H	0	3·500	3·222	0·112	3·110	3·010	0·160	2·850
4			0·7	2	6	6H	0	4·000	3·663	0·118	3·545	3·422	0·180	3·242
	4·5		0·75	2·2	6·7	6H	0	4·500	4·131	0·118	4·013	3·873	0·190	3·688
5			0·8	2·5	7·5	6H	0	5·000	4·605	0·125	4·480	4·334	0·200	4·134
6			1	3	9	6H	0	6·000	5·500	0·150	5·350	5·153	0·236	4·917
		7	1	3	9	6H	0	7·000	6·500	0·150	6·350	6·153	0·236	5·917
8			1·25	4	12	6H	0	8·000	7·348	0·160	7·188	6·912	0·265	6·647
		9	1·25	4	12	6H	0	9·000	8·348	0·160	8·188	7·912	0·265	7·647
10			1·5	5	15	6H	0	10·000	9·206	0·180	9·026	8·676	0·300	8·376
		11	1·5	5	15	6H	0	11·000	10·206	0·180	10·026	9·676	0·300	9·376
12			1·75	6	18	6H	0	12·000	11·063	0·200	10·863	10·441	0·335	10·106
	14		2	8	24	6H	0	14·000	12·913	0·212	12·701	12·210	0·375	11·835
16			2	8	24	6H	0	16·000	14·913	0·212	14·701	14·210	0·375	13·835
	18		2·5	10	30	6H	0	18·000	16·600	0·224	16·376	15·744	0·450	15·294
20			2·5	10	30	6H	0	20·000	18·600	0·224	18·376	17·744	0·450	17·294
	22		2·5	10	30	6H	0	22·000	20·600	0·224	20·376	19·744	0·450	19·294
24			3	12	36	6H	0	24·000	22·316	0·265	22·051	21·252	0·500	20·752

42	4·5	21	63	6H	0	42·000	39·392	0·315	39·077	37·799	0·670	37·129
45	4·5	21	63	6H	0	45·000	42·392	0·315	42·077	40·799	0·670	40·129
48	5	24	71	6H	0	48·000	45·087	0·335	44·752	43·297	0·710	42·587
52	5	24	71	6H	0	52·000	49·087	0·335	48·752	47·297	0·710	46·587
56	5·5	28	85	6H	0	56·000	52·783	0·355	53·428	50·796	0·750	50·046
60	5·5	28	85	6H	0	60·000	56·783	0·355	56·428	54·796	0·750	54·046
64	6	32	95	6H	0	64·000	60·478	0·375	60·103	58·305	0·800	57·505
68	6	32	95	6H	0	68·000	64·478	0·375	64·103	62·305	0·800	61·505

For limits and tolerances of other diameters and pitches, see BS 3643, Part 3.

Table 4.7 LIMITS AND TOLERANCES FOR UNCOATED THREADS, FINE SERIES THREADS— BOLTS (mm)

Tolerance class 6g

Nominal diameters		Pitch	Normal length of engagement		Tolerance class	Fund. dev.	Major diameter			Pitch diameter			Minor diameter		
1st choice	2nd choice		over	up to			max	tol	min	max	tol	min	max	tol	min
6		0·75	2·4	7·1	6g	0·022	5·978	0·140	5·838	5·491	0·100	5·391	5·058	0·149	4·909
8		1	3	9	6g	0·026	7·974	0·180	7·794	7·324	0·112	7·212	6·747	0·184	6·563
10		1·25	4	12	6g	0·028	9·972	0·212	9·760	9·160	0·118	9·042	8·439	0·208	8·231
12		1·25	4·5	13	6g	0·028	11·972	0·212	11·760	11·160	0·132	11·028	10·439	0·222	10·217
	14	1·5	5·6	16	6g	0·032	13·968	0·236	13·732	12·994	0·140	12·854	12·127	0·248	11·879
16		1·5	5·6	16	6g	0·032	15·968	0·236	15·732	14·994	0·140	14·854	14·127	0·248	13·879
	18	1·5	5·6	16	6g	0·032	17·968	0·236	17·732	16·994	0·140	16·854	16·127	0·248	15·879
20		1·5	5·6	16	6g	0·032	19·968	0·236	19·732	18·994	0·140	18·854	18·127	0·248	17·879
	22	1·5	5·6	16	6g	0·032	21·968	0·236	21·732	20·994	0·140	20·854	20·127	0·248	19·879
24		2	8·5	25	6g	0·038	23·962	0·280	23·682	22·663	0·170	22·493	21·508	0·314	21·194
	27	2	8·5	25	6g	0·038	26·962	0·280	26·682	25·663	0·170	25·493	24·508	0·314	24·194
30		2	8·5	25	6g	0·038	29·962	0·280	29·682	28·663	0·170	28·493	27·508	0·314	27·194
	33	2	8·5	25	6g	0·038	32·962	0·280	32·682	31·663	0·170	31·493	30·508	0·314	30·194
36		3	12	36	6g	0·048	35·952	0·375	35·577	34·003	0·200	33·803	32·271	0·416	31·855
	39	3	12	36	6g	0·048	38·952	0·375	38·577	37·003	0·200	36·803	35·271	0·416	34·855

For limits and tolerances of other diameters and pitches, see BS 3643, Part 3.

Table 4.8 LIMITS AND TOLERANCES FOR UNCOATED THREADS, FINE SERIES THREADS—NUTS (mm)

Tolerance class 6H

Nominal diameters		Pitch	Normal length of engagement		Tolerance class	Fund. dev.	Major diameter	Pitch diameter			Minor diameter		
1st choice	2nd choice		over	up to			min	max	tol	min	max	tol	min
6		0·75	2·4	7·1	6H	0	6·000	5·645	0·132	5·513	5·378	0·190	5·188
8		1	3	9	6H	0	8·000	7·500	0·150	7·350	7·153	0·236	6·917
10		1·25	4	12	6H	0	10·000	9·348	0·160	9·188	8·912	0·265	8·647
12		1·25	4·5	13	6H	0	12·000	11·368	0·180	11·188	10·912	0·265	10·647
	14	1·5	5·6	16	6H	0	14·000	13·216	0·190	13·026	12·676	0·300	12·376
16		1·5	5·6	16	6H	0	16·000	15·216	0·190	15·026	14·676	0·300	14·376
	18	1·5	5·6	16	6H	0	18·000	17·216	0·190	17·026	16·676	0·300	16·376
20		1·5	5·6	16	6H	0	20·000	19·216	0·190	19·026	18·676	0·300	18·376
	22	1·5	5·6	16	6H	0	22·000	21·216	0·190	21·026	20·676	0·300	20·376
24		2	8·5	25	6H	0	24·000	22·925	0·224	22·701	22·210	0·375	21·835
	27	2	8·5	25	6H	0	27·000	25·925	0·224	25·701	25·210	0·375	24·835
30		2	8·5	25	6H	0	30·000	28·925	0·224	28·701	28·210	0·375	27·835
	33	2	8·5	25	6H	0	33·000	31·925	0·224	31·701	31·210	0·375	30·835
36		3	12	36	6H	0	36·000	34·316	0·265	34·051	33·252	0·500	32·752
	39	3	12	36	6H	0	39·000	37·316	0·265	37·051	36·252	0·500	35·752

For limits and tolerances of other diameters and pitches, see BS 3643, Part 3.

Close fit (5H/4h). The close class of fit is applied to threads requiring a closer fit than that normally obtained with the 'medium' class of fit and should only be used when close accuracy of thread form and pitch is particularly required. Consistent production of threads of this fit demands the use of high quality production equipment and particularly thorough inspection.

Free fit (7H/8g). The free class of fit is primarily intended for applications in which quick and easy assembly is needed even when the threads have become dirty and/or slightly damaged.

Method of designation

Coarse threads are designated in accordance with the following examples:

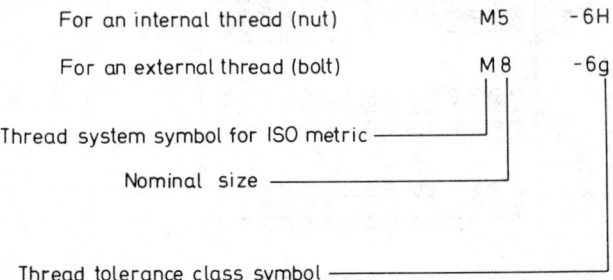

A fit between a pair of threaded parts is indicated by the internal thread (nut) tolerance class designation followed by the external thread (bolt) class designation, the two separated by a stroke, e.g.

$$M8 \qquad -6H/6g$$
$$M5 \qquad -6H/6g$$

Fine threads are designated in accordance with the following examples:

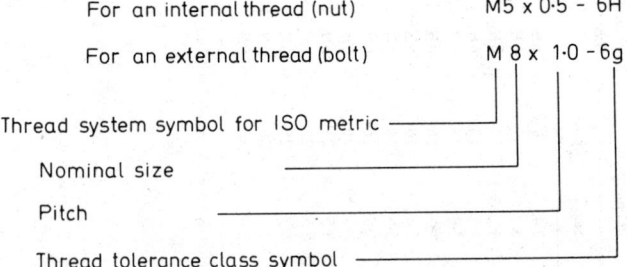

A fit between a pair of threaded parts is indicated by the internal thread (nut) tolerance class designation followed by the external thread (bolt) class designation, the two separated by a stroke, e.g.

$$M8 \times 1{\cdot}0 - 6H/6g$$
$$M16 \times 1{\cdot}5 - 6H/6g$$

Tensile strength related to stress area

A misconception often exists as to the definition of the stress area of a screw thread bolt. The accepted international formula for stress area, (A_s), where it is expected that the failure due to tension will occur in the core of the bolt, uses as the basis the mean of the minor diameter and the pitch diameter.

Thus

$$A_s = \frac{\pi}{4} \, (\text{Mean of pitch and minor diameters})^2.$$

For data on tensile stress areas, see Table 4.9. For areas of other diameters and pitches not shown in the table, see BS 3643, Pt 3.

ISO INCH (UNIFIED) SCREW THREADS

BS 1580 Parts 1 and 2[3] specify particulars of single start clearance fit threads in inch units having the ISO basic profile given in International Standard R 68[15] (see Figure 4.3) and the diameter/pitch combinations for $\frac{1}{4}$ inch diameter and above given in International Standard R 263[19].

There are a number of graded pitch/diameter combinations for use with bolts, nuts, screws and other common threaded fasteners. Two of these series are the UNC and UNF for general purpose use and referred to in International Standard R 263. In addition there are various constant pitch series e.g. 4UN, 8UN etc. intended for special designs requiring larger core area, fine adjustment and small depths of thread. For details of General Plan, see Table 4.10.

The design forms of the threads in their maximum material condition according to whether external (bolt) thread or internal (nut) thread are shown in Figure 4.3. It is

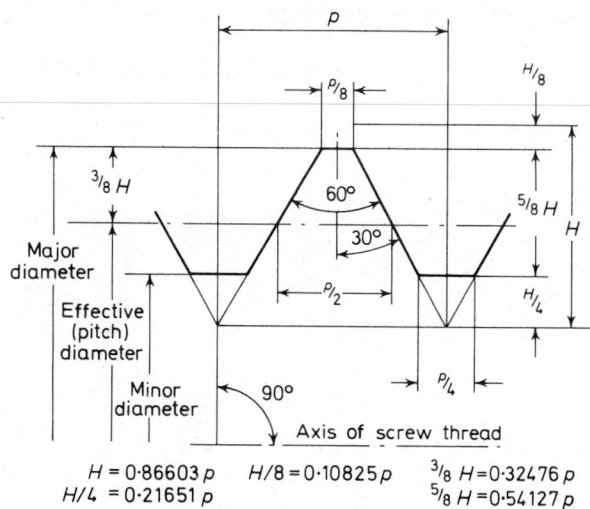

$$H = 0 \cdot 86603\,p \qquad H/8 = 0 \cdot 10825\,p \qquad {}^3\!/_8\,H = 0 \cdot 32476\,p$$
$$H/4 = 0 \cdot 21651\,p \qquad\qquad\qquad\qquad {}^5\!/_8\,H = 0 \cdot 54127\,p$$

Figure 4.3. ISO Basic thread form

important that the roots of external threads should always be rounded within the limiting profiles specified. The basic thread data is given in Figure 4.3.

The Coarse Thread series (UNC) is suitable for mass production of bolts, screws and nuts. A coarse pitch gives good resistance to stripping, consequently the series is suitable for use with lower strength materials such as cast iron and soft materials such as copper and copper alloys, aluminium and other light metals and plastics, see Tables 4.11 and 4.12.

The Fine Thread series (UNF), although less resistant to stripping under conditions of repeated loosening and tightening, is suitable where a greater tensile stress area and smaller depth of thread are necessary, see Tables 4.13 and 4.14.

The tolerancing system adopted by the ISO inch (Unified) screw series is such that combinations of bolt and nut of uncoated threads provide for three grades of fit i.e. free, medium and close. The classes of thread tolerances are 1A, 2A, and 3A for bolts, and 1B, 2B and 3B for nuts. 1A/1B combination gives a free fit, 2A/2B medium fit and 3A/3B close fit conditions. 2A/2B fit corresponds substantially to the 'medium' class external thread and 'normal' class of the Whitworth system.

It should be noted that the bulk of ISO inch (Unified) fasteners are made to 2A/2B tolerances and provide an allowance (clearance) under maximum material conditions which allows for flash plating of the bolt thread without creating undue interference during assembly. For thicker coatings, such as electro galvanising, it is necessary further to reduce the screw thread sizes to allow coating to the finished sizes.

A typical screw thread designation is

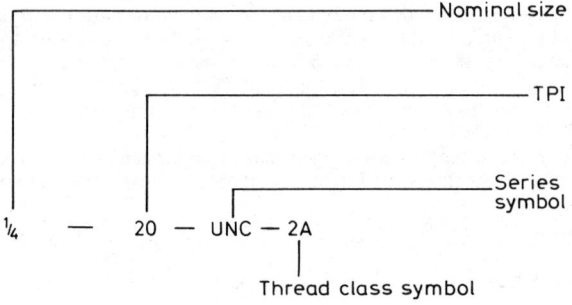

Coarse series

Nominal diameters (mm)			Pitch mm	Section at minor dia (mm²)	Tensile stress area (mm²)
1st choice	2nd choice	3rd choice			
1·6			0·35	1·08	1·27
	1·8		0·35	1·48	1·70
2			0·4	1·79	2·07
	2·2		0·45	2·13	2·48
2·5			0·45	2·98	3·39
3			0·5	4·47	5·03
	3·5		0·6	6·00	6·78
4			0·7	7·75	8·78
	4·5		0·75	10·1	11·3
5			0·8	12·7	14·2
6			1	17·9	20·1
		7	1	26·2	28·9
8			1·25	32·8	36·6
		9	1·25	43·8	48·1
10			1·5	52·3	58·0
		11	1·5	65·9	72·3
12			1·75	76·2	84·3
	14		2	105	115
16			2	144	157
	18		2·5	175	192
20			2·5	225	245
	22		2·5	282	303
24			3	324	353
	27		3	427	459
30			3·5	519	561
	33		3·5	647	694
36			4	759	817
	39		4	913	976
42			4·5	1 050	1 120
	45		4·5	1 220	1 300
48			5	1 380	1 470
	52		5	1 650	1 760
56			5·5	1 910	2 030
	60		5·5	2 230	2 360
64			6	2 520	2 680
	68		6	2 890	3 060

Fine series

Nominal diameters (mm)		Pitch mm	Section at minor dia (mm²)	Tensile stress area (mm²)
1st choice	2nd choice			
6		0·75	20·30	22·00
8		1·0	36·00	39·20
10		1·25	56·30	61·20
12		1·25	86·00	92·10
	14	1·50	116	125
16		1·50	157	167
	18	1·50	205	216
20		1·50	259	272
	22	1·50	319	333
24		2·00	365	384
	27	2·00	473	496
30		2·00	596	621
	33	2·00	733	761
36		3·00	320	865
	39	3·00	980	1 030

Table 4.10 ISO INCH (UNIFIED) SCREW THREADS—GENERAL PLAN

| Sizes | | Basic major diameter | Series with graded pitches | | | Threads per inch — Series with constant pitches | | | | | | | | Sizes |
1st choice in	2nd choice in	in	Coarse UNC	Fine UNF	Extra fine UNEF	4 UN	6 UN	8 UN	12 UN	16 UN	20 UN	28 UN	32 UN	in
1/4		0·250 0	20	28	32	—	—	—	—	—	UNC	UNF	UNEF	1/4
5/16		0·312 5	18	24	32	—	—	—	—	—	20	28	UNEF	5/16
3/8		0·375 0	16	24	32	—	—	—	—	UNC	20	28	UNEF	3/8
7/16		0·437 5	14	20	28	—	—	—	—	16	UNF	UNEF	32	7/16
1/2		0·500 0	13	20	28	—	—	—	—	16	UNF	UNEF	32	1/2
9/16		0·562 5	12	18	24	—	—	—	UNC	16	20	28	32	9/16
5/8		0·625 0	11	18	24	—	—	—	12	16	20	28	32	5/8
	11/16	0·687 5	—	—	24	—	—	—	12	16	20	28	32	11/16
3/4		0·750 0	10	16	20	—	—	—	12	UNF	UNEF	28	32	3/4
	13/16	0·812 5	—	—	20	—	—	—	12	16	UNEF	28	32	13/16
7/8		0·875 0	9	14	20	—	—	—	12	16	UNEF	28	32	7/8
	15/16	0·937 5	—	—	20	—	—	—	12	16	UNEF	28	32	15/16
1		1·000 0	8	12	20	—	—	UNC	UNF	16	UNEF	28	32	1
	1 1/16	1·062 5	—	—	18	—	—	8	12	16	20	28	—	1 1/16
1 1/8		1·125 0	7	12	18	—	—	8	UNF	16	20	28	—	1 1/8
	1 3/16	1·187 5	—	—	18	—	—	8	12	16	20	28	—	1 3/16
1 1/4		1·250 0	7	12	18	—	—	8	UNF	16	20	28	—	1 1/4
	1 5/16	1·312 5	—	—	18	—	—	8	12	16	20	28	—	1 5/16
1 3/8		1·375 0	6	12	18	—	UNC	8	UNF	16	20	28	—	1 3/8
	1 7/16	1·437 5	—	—	18	—	6	8	12	16	20	28	—	1 7/16
1 1/2		1·500 0	6	12	18	—	UNC	8	UNF	16	20	28	—	1 1/2
	1 9/16	1·562 5	—	—	18	—	6	8	12	16	20	—	—	1 9/16
1 5/8		1·625 0	—	—	18	—	6	8	12	16	20	—	—	1 5/8
	1 11/16	1·687 5	—	—	18	—	6	8	12	16	20	—	—	1 11/16
1 3/4		1·750 0	5	—	—	—	6	8	12	16	20	—	—	1 3/4

Size	Decimal (in)	4 UN / UNC	6 UN	8 UN	12 UN	16 UN	20 UN
$2\frac{3}{8}$	2·375 0	—	6	8	12	16	20
$2\frac{1}{2}$	2·500 0	UNC	6	8	12	16	20
$2\frac{5}{8}$	2·625 0	4	6	8	12	16	20
$2\frac{3}{4}$	2·750 0	UNC	6	8	12	16	20
$2\frac{7}{8}$	2·875 0	4	6	8	12	16	20
3	3·000 0	UNC	6	8	12	16	—
$3\frac{1}{8}$	3·125 0	4	6	8	12	16	—
$3\frac{1}{4}$	3·250 0	UNC	6	8	12	16	—
$3\frac{3}{8}$	3·375 0	4	6	8	12	16	—
$3\frac{1}{2}$	3·500 0	UNC	6	8	12	16	—
$3\frac{5}{8}$	3·625 0	4	6	8	12	16	—
$3\frac{3}{4}$	3·750 0	UNC	6	8	12	16	—
$3\frac{7}{8}$	3·875 0	4	6	8	12	16	—
4	4·000 0	UNC	6	8	12	16	—
$4\frac{1}{8}$	4·125 0	4	6	8	12	16	—
$4\frac{1}{4}$	4·250 0	4	6	8	12	16	—
$4\frac{3}{8}$	4·375 0	4	6	8	12	16	—
$4\frac{1}{2}$	4·500 0	4	6	8	12	16	—
$4\frac{5}{8}$	4·625 0	4	6	8	12	16	—
$4\frac{3}{4}$	4·750 0	4	6	8	12	16	—
$4\frac{7}{8}$	4·875 0	4	6	8	12	16	—
5	5·000 0	4	6	8	12	16	—
$5\frac{1}{8}$	5·125 0	4	6	8	12	16	—
$5\frac{1}{4}$	5·250 0	4	6	8	12	16	—
$5\frac{3}{8}$	5·375 0	4	6	8	12	16	—
$5\frac{1}{2}$	5·500 0	4	6	8	12	16	—
$5\frac{5}{8}$	5·625 0	4	6	8	12	16	—
$5\frac{3}{4}$	5·750 0	4	6	8	12	16	—
$5\frac{7}{8}$	5·875 0	4	6	8	12	16	—
6	6·000 0	4	6	8	12	16	—

Table 4.11 UNIFIED COARSE SCREW THREADS, UNC—EXTERNAL THREADS (BOLTS), CLASS 2A
Limits and tolerances for finished uncoated threads (in)

Designation	Major diameter			Effective diameter			Minor diameter		
	max	min	tol	max	min	tol	max	min	tol
1/4 – 20 UNC–2A	0·248 9	0·240 8	0·008 1	0·216 4	0·212 7	0·003 7	0·187 6	0·180 3	0·007 3
5/16 – 18 UNC–2A	0·311 3	0·302 6	0·008 7	0·275 2	0·271 2	0·004 0	0·243 1	0·235 1	0·008 0
3/8 – 16 UNC–2A	0·373 7	0·364 3	0·009 4	0·333 1	0·328 7	0·004 4	0·297 0	0·288 1	0·008 9
7/16 – 14 UNC–2A	0·436 1	0·425 8	0·010 3	0·389 7	0·385 0	0·004 7	0·348 5	0·338 7	0·009 8
1/2 – 13 UNC–2A	0·498 5	0·487 6	0·010 9	0·448 5	0·443 5	0·005 0	0·404 1	0·393 6	0·010 5
9/16 – 12 UNC–2A	0·560 9	0·549 5	0·011 4	0·506 8	0·501 6	0·005 2	0·458 7	0·447 5	0·011 2
5/8 – 11 UNC–2A	0·623 4	0·611 3	0·012 1	0·564 4	0·558 9	0·005 5	0·511 9	0·499 9	0·012 0
3/4 – 10 UNC–2A	0·748 2	0·735 3	0·012 9	0·683 2	0·677 3	0·005 9	0·625 5	0·612 4	0·013 1
7/8 – 9 UNC–2A	0·873 1	0·859 2	0·013 9	0·800 9	0·794 6	0·006 3	0·736 8	0·722 5	0·014 3
1 – 8 UNC–2A	0·998 0	0·983 0	0·015 0	0·916 8	0·910 0	0·006 8	0·844 6	0·828 8	0·015 8
1 1/8 – 7 UNC–2A	1·122 8	1·106 4	0·016 4	1·030 0	1·022 8	0·007 2	0·947 5	0·930 0	0·017 5
1 1/4 – 7 UNC–2A	1·247 8	1·231 4	0·016 4	1·155 0	1·147 6	0·007 4	1·072 5	1·054 8	0·017 7
1 3/8 – 6 UNC–2A	1·372 6	1·354 4	0·018 2	1·264 3	1·256 3	0·008 0	1·168 1	1·148 1	0·020 0
1 1/2 – 6 UNC–2A	1·497 6	1·479 4	0·018 2	1·389 3	1·381 2	0·008 1	1·293 1	1·273 0	0·020 1
1 3/4 – 5 UNC–2A	1·747 3	1·726 8	0·020 5	1·617 4	1·608 5	0·008 9	1·501 9	1·478 6	0·023 3
2 – 4 1/2 UNC–2A	1·997 1	1·975 1	0·022 0	1·852 8	1·843 3	0·009 5	1·724 5	1·699 0	0·025 5
2 1/4 – 4 1/2 UNC–2A	2·247 1	2·225 1	0·022 0	2·102 8	2·093 1	0·009 7	1·974 5	1·948 8	0·025 7
2 1/2 – 4 UNC–2A	2·496 9	2·473 1	0·023 8	2·334 5	2·324 1	0·010 4	2·190 2	2·161 8	0·028 4
2 3/4 – 4 UNC–2A	2·746 8	2·723 0	0·023 8	2·584 4	2·573 9	0·010 5	2·440 1	2·411 6	0·028 5
3 – 4 UNC–2A	2·996 8	2·973 0	0·023 8	2·834 4	2·823 7	0·010 7	2·690 1	2·661 4	0·028 7
3 1/4 – 4 UNC–2A	3·246 7	3·222 9	0·023 8	3·084 3	3·073 4	0·010 9	2·940 0	2·911 1	0·028 9
3 1/2 – 4 UNC–2A	3·496 7	3·472 9	0·023 8	3·334 3	3·323 3	0·011 0	3·190 0	3·161 0	0·029 0
3 3/4 – 4 UNC–2A	3·746 6	3·722 8	0·023 8	3·584 2	3·573 0	0·011 2	3·439 9	3·410 7	0·029 2
4 – 4 UNC–2A	3·996 6	3·972 8	0·023 8	3·834 2	3·822 9	0·011 3	3·689 9	3·660 6	0·029 3

Table 4.12 UNIFIED COARSE SCREW THREADS, UNC—INTERNAL THREADS (NUTS), CLASS 2B

Limits and tolerances for finished uncoated threads (in)

Designation		Minor diameter			Effective diameter			Major diameter
		min	max	tol	min	max	tol	min
1/4 – 20	UNC – 2B	0·195 9	0·207 4	0·011 5	0·217 5	0·222 3	0·004 8	0·250 0
5/16 – 18	UNC – 2B	0·252 4	0·265 1	0·012 7	0·276 4	0·281 7	0·005 3	0·312 5
3/8 – 16	UNC – 2B	0·307 3	0·321 4	0·014 1	0·334 4	0·340 1	0·005 7	0·375 0
7/16 – 14	UNC – 2B	0·360 2	0·376 0	0·015 8	0·391 1	0·397 2	0·006 1	0·437 5
1/2 – 13	UNC – 2B	0·416 7	0·433 6	0·016 9	0·450 0	0·456 5	0·006 5	0·500 0
9/16 – 12	UNC – 2B	0·472 3	0·490 4	0·018 1	0·508 4	0·515 2	0·006 8	0·562 5
5/8 – 11	UNC – 2B	0·526 6	0·546 0	0·019 4	0·566 0	0·573 2	0·007 2	0·625 0
3/4 – 10	UNC – 2B	0·641 7	0·662 7	0·021 0	0·685 0	0·692 7	0·007 7	0·750 0
7/8 – 9	UNC – 2B	0·754 7	0·777 5	0·022 8	0·802 8	0·811 0	0·008 2	0·875 0
1 – 8	UNC – 2B	0·864 7	0·889 7	0·025 0	0·918 8	0·927 6	0·008 8	1·000 0
1 1/8 – 7	UNC – 2B	0·970 4	0·998 0	0·027 6	1·032 2	1·041 6	0·009 4	1·125 0
1 1/4 – 7	UNC – 2B	1·095 4	1·123 0	0·027 6	1·157 2	1·166 8	0·009 6	1·250 0
1 3/8 – 6	UNC – 2B	1·194 6	1·225 2	0·030 6	1·266 7	1·277 1	0·010 4	1·375 0
1 1/2 – 6	UNC – 2B	1·319 6	1·350 2	0·030 6	1·391 7	1·402 2	0·010 5	1·500 0
1 3/4 – 5	UNC – 2B	1·533 5	1·567 5	0·034 0	1·620 1	1·631 7	0·011 6	1·750 0
2 – 4 1/2	UNC – 2B	1·759 4	1·795 2	0·035 8	1·855 7	1·868 1	0·012 4	2·000 0
2 1/4 – 4 1/2	UNC – 2B	2·009 4	2·045 2	0·035 8	2·105 7	2·118 3	0·012 6	2·250 0
2 1/2 – 4	UNC – 2B	2·229 4	2·266 9	0·037 5	2·337 6	2·351 1	0·013 5	2·500 0
2 3/4 – 4	UNC – 2B	2·479 4	2·516 9	0·037 5	2·587 6	2·601 3	0·013 7	2·750 0
3 – 4	UNC – 2B	2·729 4	2·766 9	0·037 5	2·837 6	2·851 5	0·013 9	3·000 0
3 1/4 – 4	UNC – 2B	2·979 4	3·016 9	0·037 5	3·087 6	3·101 7	0·014 1	3·250 0
3 1/2 – 4	UNC – 2B	3·229 4	3·266 9	0·037 5	3·337 6	3·351 9	0·014 3	3·500 0
3 3/4 – 4	UNC – 2B	3·479 4	3·516 9	0·037 5	3·587 6	3·602 1	0·014 5	3·750 0
4 – 4	UNC – 2B	3·729 4	3·766 9	0·037 5	3·837 6	3·852 3	0·014 7	4·000 0

Table 4.13 UNIFIED FINE SCREW THREADS, UNF—EXTERNAL THREADS (BOLTS), CLASS 2A
Limits and tolerances for finished uncoated threads (in)

Designation	Major diameter			Effective diameter			Minor diameter		
	max	min	tol	max	min	tol	max	min	tol
1/4 – 28 UNF – 2A	0·249 0	0·242 5	0·006 5	0·225 8	0·222 5	0·003 3	0·205 2	0·199 3	0·005 9
5/16 – 24 UNF – 2A	0·311 4	0·304 2	0·007 2	0·284 3	0·280 6	0·003 7	0·260 3	0·253 6	0·006 7
3/8 – 24 UNF – 2A	0·373 9	0·366 7	0·007 2	0·346 8	0·343 0	0·003 8	0·322 8	0·316 0	0·006 8
7/16 – 20 UNF – 2A	0·436 2	0·428 1	0·008 1	0·403 7	0·399 5	0·004 2	0·374 9	0·367 1	0·007 8
1/2 – 20 UNF – 2A	0·498 7	0·490 6	0·008 1	0·466 2	0·461 9	0·004 3	0·437 4	0·429 5	0·007 9
9/16 – 18 UNF – 2A	0·561 1	0·552 4	0·008 7	0·525 0	0·520 5	0·004 5	0·492 9	0·484 4	0·008 5
5/8 – 18 UNF – 2A	0·623 6	0·614 9	0·008 7	0·587 5	0·582 8	0·004 7	0·555 4	0·546 7	0·008 7
3/4 – 16 UNF – 2A	0·748 5	0·739 1	0·009 4	0·707 9	0·702 9	0·005 0	0·671 8	0·662 3	0·009 5
7/8 – 14 UNF – 2A	0·873 4	0·863 1	0·010 3	0·827 0	0·821 6	0·005 4	0·785 8	0·775 3	0·010 5
1 – 12 UNF – 2A	0·998 2	0·986 8	0·011 4	0·944 1	0·938 2	0·005 9	0·896 0	0·884 1	0·011 9
1 1/8 – 12 UNF – 2A	1·123 2	1·111 8	0·011 4	1·069 1	1·063 1	0·006 0	1·021 0	1·009 0	0·012 0
1 1/4 – 12 UNF – 2A	1·248 2	1·236 8	0·011 4	1·194 1	1·187 9	0·006 2	1·146 0	1·133 8	0·012 2
1 3/8 – 12 UNF – 2A	1·373 1	1·361 7	0·011 4	1·319 0	1·312 7	0·006 3	1·270 9	1·258 6	0·012 3
1 1/2 – 12 UNF – 2A	1·498 1	1·486 7	0·011 4	1·444 0	1·437 6	0·006 4	1·395 9	1·383 5	0·012 4

Table 4.14   UNIFIED FINE SCREW THREADS, UNF—INTERNAL THREADS (NUTS), CLASS 2B
Limits and tolerances for finished uncoated threads (in)

Designation		Minor diameter			Effective diameter			Major diameter
		min	max	tol	min	max	tol	min
$\frac{1}{4}$ – 28	UNF – 2B	0·211 3	0·219 7	0·008 4	0·226 8	0·231 1	0·004 3	0·250 0
$\frac{5}{16}$ – 24	UNF – 2B	0·267 4	0·277 1	0·009 7	0·285 4	0·290 2	0·004 8	0·312 5
$\frac{3}{8}$ – 24	UNF – 2B	0·329 9	0·339 6	0·009 7	0·347 9	0·352 8	0·004 9	0·375 0
$\frac{7}{16}$ – 20	UNF – 2B	0·383 4	0·394 9	0·011 5	0·405 0	0·410 4	0·005 4	0·437 5
$\frac{1}{2}$ – 20	UNF – 2B	0·445 9	0·457 4	0·011 5	0·467 5	0·473 1	0·005 6	0·500 0
$\frac{9}{16}$ – 18	UNF – 2B	0·502 4	0·515 1	0·012 7	0·526 4	0·532 3	0·005 9	0·562 5
$\frac{5}{8}$ – 18	UNF – 2B	0·564 9	0·577 6	0·012 7	0·588 9	0·594 9	0·006 0	0·625 0
$\frac{3}{4}$ – 16	UNF – 2B	0·682 3	0·696 4	0·014 1	0·709 4	0·715 9	0·006 5	0·750 0
$\frac{7}{8}$ – 14	UNF – 2B	0·797 7	0·813 5	0·015 8	0·828 6	0·835 6	0·007 0	0·875 0
1 – 12	UNF – 2B	0·909 8	0·927 9	0·018 1	0·945 9	0·953 5	0·007 6	1·000 0
$1\frac{1}{8}$ – 12	UNF – 2B	1·034 8	1·052 9	0·018 1	1·070 9	1·078 7	0·007 8	1·125 0
$1\frac{1}{4}$ – 12	UNF – 2B	1·159 8	1·177 9	0·018 1	1·195 9	1·203 9	0·008 0	1·250 0
$1\frac{3}{8}$ – 12	UNF – 2B	1·284 8	1·302 9	0·018 1	1·320 9	1·329 1	0·008 2	1·375 0
$1\frac{1}{2}$ – 12	UNF – 2B	1·409 8	1·427 9	0·018 1	1·445 9	1·454 2	0·008 3	1·500 0

ISO MINIATURE SCREW THREADS. METRIC SERIES

Designation

A complete designation of a thread gives information on the thread system, the size and pitch of the thread and the tolerances applicable to it. ISO/R1501[21] omits the pitch from the designation of miniature threads but a BS[13] soon to be introduced requires its inclusion in all cases. For basic dimensions, see Table 4.15 and Figure 4.4.

Screw threads to this standard are designated by S- followed by the nominal size in mm, the pitch diameter tolerance grade number, the tolerance position letter (capital for internal threads, small for external threads) and the tolerance grade number for the nut minor diameter or bolt major diameter. Example for ISO Miniature External Screw Thread S-0·6 mm diameter: S-0·6 × 0·15 h.

A fit between mating threads is designated by a combination of the nut and bolt designations in that order. Example for ISO Miniature Screw Thread S-0·6 mm diameter:

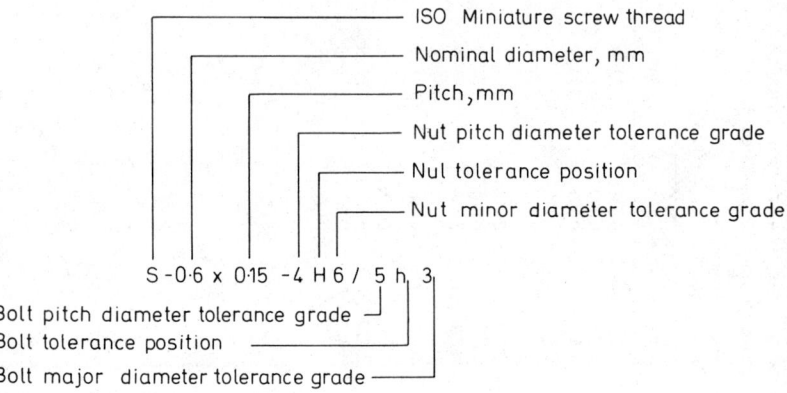

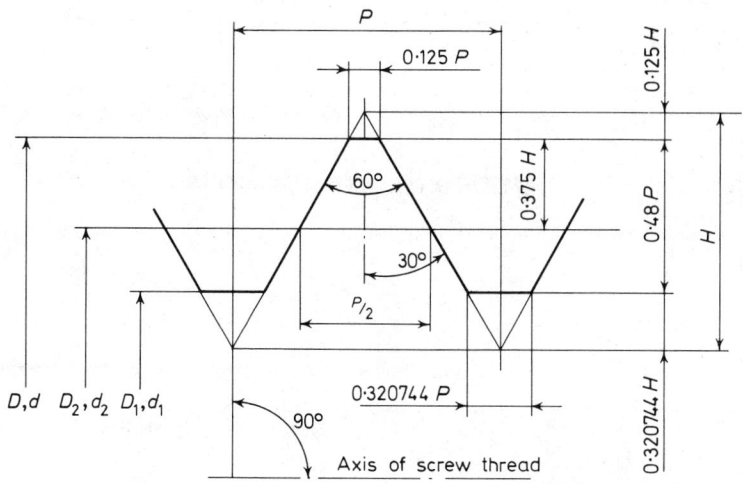

Figure 4.4. Basic profile of ISO metric miniature screw threads

D. Major diameter of internal thread D_1 Minor diameter of internal thread
d. Major diameter of external thread d_1 Minor diameter of external thread
D_2 Pitch diameter of internal thread   p. Pitch
d_2 Pitch diameter of external thread   H. Height of fundamental triangle

Table 4.15   BASIC DIMENSIONS OF STANDARD DIAMETER/PITCH COMBINATIONS
ISO MINIATURE SCREW THREADS (mm) (see Figure 4.4)

		Values in millimetres		
Thread Size	Pitch P	Major diameter D.d	Pitch diameter D_2d_2	Minor diameter D_1d_1
S-0·3	0·08	0·300 000	0·248 038	0·223 200
(S-0·35)	0·09	0·350 000	0·291 543	0·263 600
S-0·4	0·1	0·400 000	0·335 048	0·304 000
(S-0·45	0·1	0·450 000	0·385 048	0·354 000
S-0·5	0·125	0·500 000	0·418 810	0·380 000
(S-0·55)	0·125	0·550 000	0·468 810	0·430 000
S-0·6	0·15	0·600 000	0·502 572	0·456 000
(S-0·7)	0·175	0·700 000	0·586 334	0·532 000
S-0·8	0·2	0·800 000	0·670 096	0·608 000
(S-0·9)	0·225	0·900 000	0·753 858	0·684 000
(S-1)	0·25	1·000 000	0·837 620	0·760 000
(S-1·1)	0·25	1·100 000	0·937 620	0·860 000
(S-1·2)	0·25	1·200 000	1·037 620	0·960 000
(S-1·4)	0·3	1·400 000	1·205 144	1·112 000

Second choice diameters in brackets

DIMENSIONAL COMPARISONS

The thread sizes which are currently in use can be compared by reference to Tables 4.16–4.19.

Table 4.16 DIMENSIONAL COMPARISON OF CURRENT BRITISH SIZES WITH FIRST CHOICE ISO METRIC

Dimensions in brackets are in inches

8BA	(0·086)	M2·5	(0·098 4)
6BA	(0·110)	M3	(0·118 1)
4BA	(0·142)	M4	(0·157 5)
2BA	(0·185)	M5	(0·196 8)
$\frac{3}{16}$ in	(0·187 5)		
0BA	(0·236)	M6	(0·236 2)
$\frac{1}{4}$ in	(0·250)		
$\frac{5}{16}$ in	(0·312)	M8	(0·314 9)
$\frac{3}{8}$ in	(0·375)	M10	(0·393 7)
$\frac{7}{16}$ in	(0·437)	M12	(0·472 4)
$\frac{1}{2}$ in	(0·500)		
$\frac{9}{16}$ in	(0·562 5)	M14	(0·551 2)
$\frac{5}{8}$ in	(0·625 0)	M16	(0·629 9)
$\frac{3}{4}$ in	(0·750 0)	M20	(0·787 4)
$\frac{7}{8}$ in	(0·878 0)	M22	(0·886 1)
1 in	(1·000 0)	M24	(0·944 8)
$1\frac{1}{8}$ in	(1·125 0)	M30	(1·181 1)
$1\frac{1}{4}$ in	(1·250 0)	M36	(1·417 3)
Or $1\frac{3}{8}$ in	(1·375 0)		
$1\frac{1}{2}$ in	(1·500 0)	M42	(1·653 5)
$1\frac{3}{4}$ in	(1·750 0)	M48	(1·889 7)
Or 2 in	(2·000 0)		
$2\frac{1}{4}$ in	(2·250 0)	M56	(2·204 7)
$2\frac{1}{2}$ in	(2·500 0)	M64	(2·519 6)

Table 4.17 DIMENSIONAL COMPARISON OF ISO METRIC, B.A. AND UNIFIED SCREW THREADS

ISO metric threads Dia. mm	Pitch mm	BA threads (BS 93)[2] Desig-nation number	Dia. mm	Pitch mm	Unified screw threads in* Dia. mm	Pitch mm	Designation
ISO metric screw threads to BS 3643 (ISO R261)[9,17]							
6·00 × 1·0		0	6·00 × 1·0		6·350 × 0·907		¼–28 UNF
5·00 × 0·8		1	5·30 × 0·9		5·486 × 0·907		12–28 UNF
4·50 × 0·75*		2	4·70 × 0·81		4·826 × 0·794		10–32 UNF
4·00 × 0·70		3	4·10 × 0·73		4·166 × 0·706		8–36 UNF
3·50 × 0·60*		4	3·60 × 0·66		3·505 × 0·635		6–40 UNF
		5	3·20 × 0·59		3·175 × 0·577		5–44 UNF
3·00 × 0·50		6	2·80 × 0·53		2·845 × 0·529		4–48 UNF
					2·515 × 0·454		3–56 UNF
2·50 × 0·45		7	2·50 × 0·48		2·515 × 0·529		3–48 UNC
2·20 × 0·45*		8	2·20 × 0·43		2·184 × 0·454		2–56 UNC
2·00 × 0·40		9	1·90 × 0·39		1·854 × 0·397		1–64 UNC
1·80 × 0·35*		10	1·70 × 0·35		1·854 × 0·353		1–72 UNF
1·60 × 0·35		11	1·50 × 0·31		1·524 × 0·318		0–80 UNF
Unified miniature screw threads to BS 3369 (excepting the 0·25 mm dia. thread). All these diameter/pitch combinations are included in ISO R68[15].							
1·40 × 0·3*		12	1·30 × 0·28				
1·20 × 0·25		13	1·20 × 0·25				
1·00 × 0·25		14	1·00 × 0·23				
0·90 × 0·225		15	0·90 × 0·21				
0·80 × 0·20		16	0·79 × 0·19		In the Unified miniature screw thread system (BS 3369) diameter/pitch combinations are given in the range from 1·40 mm–0·30 mm. These are identical with those listed in column 1.		
0·70 × 0·175*		17	0·70 × 0·17				
0·60 × 0·15		18	0·62 × 0·15				
0·55 × 0·125*		19	0·54 × 0·14				
0·50 × 0·125		20	0·48 × 0·12				
0·45 × 0·10*		21	0·42 × 0·11				
0·40 × 0·10		22	0·37 × 0·10				
0·35 × 0·09*		23	0·33 × 0·09				
0·30 × 0·08		24	0·29 × 0·08				
0·25 × 0·075		25	0·25 × 0·07				

*Unified screw threads to BS 1580, Part 3, except the ¼–28 UNF which is to BS 1580 Parts 1 and 2. All diameters are in ISO R26[2,19]

Table 4.18 ISO METRIC/B.A. AND INCHES COMPARATIVE DIAMETERS

BA	mm	BS 3643 1st choice	BS 3643 2nd choice	Differences in diameters between ISO metric and BA etc.	
12	1·3				
		1·6			
10	1·7		1·8		
		2·0			
8	2·2		2·2	M2·5 = +0·012 in	
		2·5			
6	2·8			M2·5 = −0·012 in	M3 = +0·008 in
		3			
			3·5		
4	3·6			M3 = −0·024 in	M4 = +0·016 in
		4			
			4·5		
2	4·7			M4 = −0·027 in	M5 = +0·012 in
3/16 in	4·76			M4 = −0·03 in	M5 = +0·009 in
		5			
		6			
1/4 in	6·35			M6 = −0·018 in	
			7		
5/16 in	7·94	8		M8 = +0·0024 in	
3/8 in	9·53			M10 = +0·021 in	
		10			
7/16 in	11·12	12		M12 = +0·035 in	
1/2 in	12·70			M12 = −0·028 in	

Table 4.19 COMPARISON OF INCH AND METRIC THREADS

Diameter in	BSW TPI	BSW Stress area in²	BSF TPI	BSF Stress area in²	ISO metric coarse Metric diameter mm	ISO metric coarse TPI approx.	ISO metric coarse Stress area mm	UN coarse TPI	UN coarse Stress area in²	UN fine TPI	UN fine Stress area in²
1/4	20	0·032 0	26	0·035 6	6	25·4	20·1	20	0·032 4	28	0·036 8
5/16	18	0·052 7	22	0·056 7	8	20·3	36·6	18	0·053 2	24	0·058 7
3/8	16	0·077 9	20	0·083 9	10	17·0	58·0	16	0·078 6	24	0·088 6
7/16	14	0·106 9	18	0·118	12	14·5	84·3	14	0·107 8	20	0·119 8
1/2	12	0·138 5	16	0·152 0				13	0·143 8	20	0·161 2
9/16	12	0·183	16	0·198	14	12·7	115	12	0·184	18	0·205
5/8	11	0·227	14	0·243	16	12·7	157	11	0·229	18	0·258
3/4	10	0·336	12	0·352	20	10·1	245	10	0·338	16	0·375
7/8	9	0·463	11	0·494	22	10·1	303	9	0·467	14	0·513
1	8	0·608	10	0·642	24	8·5	353	8	0·612	12	0·667

PIPE THREADS FOR GAS LIST TUBES AND SCREWED FITTINGS WHERE PRESSURE TIGHT JOINTS ARE MADE ON THE THREADS

The basic forms of the British Standard taper and parallel pipe threads to BS 21[1] are based on that of the British Standard Whitworth thread. British Standard taper pipe threads are designated by the letters R or RE together with the thread size.

It is recommended that these screw threads are referred to on drawings and related documents as follows. Internal taper $RE\frac{1}{2}$; External taper $R\frac{1}{2}$.

British Standard parallel internal threads to BS 21 are designated by the letters RP together with the thread size. It is recommended that these screw threads are referred to on drawings and related documents as follows:

$$RP\frac{1}{2}$$

For terms relating to pipe threads, see Figure 4.5 and for basic dimensions, see Table 4.21.

PIPE THREADS WHERE PRESSURE TIGHT JOINTS ARE NOT MADE ON THE THREADS—METRIC UNITS

The basic form of threads to BS 2779[7] is the Whitworth form and is identical with that of the parallel thread specified in BS 21[1] (ISO/R.7)[14].

Table 4.20 BASIC SIZES OF PIPE THREADS TO BS 2779[7]

			Metric dimensions			
Size Nominal bore	Number of threads per unit	Pitch mm	Depth of thread mm	Major diameter mm	Pitch diameter mm	Minor diameter mm
$\frac{1}{16}$	28	0·907	0·581	7·723	7·142	6·561
$\frac{1}{8}$	28	0·907	0·581	·9·728	9·147	8·566
$\frac{1}{4}$	19	1·337	0·856	13·157	12·301	11·445
$\frac{3}{8}$	19	1·337	0·856	16·662	15·806	14·950
$\frac{1}{2}$	14	1·814	1·162	20·455	19·793	18·631
$\frac{5}{8}$	14	1·814	1·162	22·911	21·749	20·587
$\frac{3}{4}$	14	1·814	1·162	26·441	25·279	24·117
$\frac{7}{8}$	14	1·814	1·162	30·201	29·039	27·877
1	11	2·309	1·479	33·249	31·770	30·291
$1\frac{1}{8}$	11	2·309	1·479	37·897	36·418	34·939
$1\frac{1}{4}$	11	2·309	1·479	41·910	40·431	38·952
$1\frac{1}{2}$	11	2·309	1·479	47·803	46·324	44·845
$1\frac{3}{4}$	11	2·309	1·479	53·746	52·267	50·788
2	11	2·309	1·479	59·614	58·135	56·656
$2\frac{1}{4}$	11	2·309	1·479	65·710	64·231	62·752
$2\frac{1}{2}$	11	2·309	1·479	75·184	73·705	72·226
$2\frac{3}{4}$	11	2·309	1·479	81·534	80·055	78·576
3	11	2·309	1·479	87·884	84·405	84·926
$3\frac{1}{2}$	11	2·309	1·479	100·330	98·851	97·372
4	11	2·309	1·479	113·030	141·551	110·072
$4\frac{1}{2}$	11	2·309	1·479	125·730	124·251	122·772
5	11	2·309	1·479	138·430	136·951	135·472
$5\frac{1}{2}$	11	2·309	1·479	151·130	149·651	148·172
6	11	2·309	1·479	153·830	168·351	150·772

Table 4.21 BASIC DIMENSIONS AND LIMIT(s)

(Basic gauge lengths and limits

Size (nominal bore) in	No. of threads per in	Pitch mm	Depth of thread mm	Basic diameters at gauge plane			Number of turns		
				Prior gauge diameter mm	Pitch mm	Minor mm	Basic	Tolerance plus and minus	Max
$\frac{1}{16}$	28	0·907	0·581	7·723	7·142	6·561	$4\frac{3}{8}$	1	$5\frac{3}{8}$
							4·0	0·9	4·9
$\frac{1}{8}$	28	0·907	0·581	9·728	9·147	8·566	$4\frac{3}{8}$	1	$5\frac{3}{8}$
							4·0	0·9	4·9
$\frac{1}{4}$	19	1·337	0·856	13·157	12·301	11·445	$4\frac{1}{2}$	1	$5\frac{1}{2}$
							5·0	1·3	7·3
$\frac{3}{8}$	19	1·337	0·856	16·662	15·806	14·950	$4\frac{3}{4}$	1	$5\frac{3}{4}$
							6·4	1·3	7·7
$\frac{1}{2}$	14	1·814	1·162	20·955	19·793	18·631	$4\frac{1}{2}$	1	$5\frac{1}{2}$
							8·2	1·8	10·0
$\frac{3}{4}$	14	1·814	1·162	26·441	25·279	24·117	$5\frac{1}{4}$	1	$6\frac{1}{4}$
							9·5	1·8	11·3
1	11	2·309	1·479	33·249	31·770	30·291	$4\frac{1}{2}$	1	$5\frac{1}{2}$
							10·4	2·3	12·7
$1\frac{1}{4}$	11	2·309	1·479	41·910	40·431	38·952	$5\frac{1}{2}$	1	$6\frac{1}{2}$
							12·7	2·3	15·0
$1\frac{1}{2}$	11	2·309	1·479	47·803	46·324	44·845	$5\frac{1}{2}$	1	$6\frac{1}{2}$
							12·7	2·3	15·0
2	11	2·309	1·479	59·614	58·135	56·656	$6\frac{7}{8}$	1	$7\frac{7}{8}$
							15·9	2·3	18·2
$2\frac{1}{2}$	11	2·309	1·479	75·184	73·705	72·226	$7\frac{9}{16}$	$1\frac{1}{2}$	$9\frac{1}{16}$
							17·5	3·5	21·0
3	11	2·309	1·479	87·884	86·405	84·926	$8\frac{15}{16}$	$1\frac{1}{2}$	$10\frac{7}{16}$
							20·6	3·5	24·1
4	11	2·309	1·479	113·030	111·551	110·072	11	$1\frac{1}{2}$	$12\frac{1}{2}$
							25·4	3·5	28·9
5	11	2·309	1·479	138·430	136·951	135·472	$12\frac{3}{8}$	$1\frac{1}{2}$	$13\frac{7}{8}$
							28·6	3·5	32·1
6	11	2·309	1·479	163·830	162·351	160·872	$12\frac{3}{8}$	$1\frac{1}{2}$	$13\frac{7}{8}$
							28·6	3·5	32·1

SIZE FOR BRITISH STANDARD PIPE THREADS

are based on turns of thread)

	Number of turns					Tolerance on position of gauge plane relative to face of internally taper threaded parts (plus and minus)	Diametral tolerance on parallel internal threads (plus and minus) mm
	Useful thread						
min	*basic*	*max*	*min*	*Fitting allowance*	*Wrenching allowance*		
$3\frac{3}{8}$	$7\frac{1}{8}$	$8\frac{1}{8}$	$6\frac{1}{8}$	$2\frac{3}{4}$	$1\frac{1}{2}$	$1\frac{1}{4}$	—
3·1	6·5	7·4	5·6	2·5	1·4	1·1	0·071
$3\frac{3}{8}$	$7\frac{1}{8}$	$8\frac{1}{8}$	$6\frac{1}{8}$	$2\frac{3}{4}$	$1\frac{1}{2}$	$1\frac{1}{4}$	0·071
3·1	*6·5*	*7·4*	*5·6*	*2·5*	*1·4*	*1·1*	
$3\frac{1}{2}$	$7\frac{1}{4}$	$8\frac{1}{4}$	$6\frac{1}{4}$	$2\frac{3}{4}$	$1\frac{1}{2}$	$1\frac{1}{4}$	0·104
4·7	*9·7*	*11·0*	*8·4*	*3·7*	*2·0*	*1·7*	
$3\frac{3}{4}$	$7\frac{1}{2}$	$8\frac{1}{2}$	$6\frac{1}{2}$	$2\frac{3}{4}$	$1\frac{1}{2}$	$1\frac{1}{4}$	0·104
5·1	*10·1*	*11·4*	*8·8*	*3·7*	*2·0*	*1·7*	
$3\frac{1}{2}$	$7\frac{1}{4}$	$8\frac{1}{4}$	$6\frac{1}{4}$	$2\frac{3}{4}$	$1\frac{1}{2}$	$1\frac{1}{4}$	0·142
5·4	*13·2*	*15·0*	*11·4*	*5·0*	*2·7*	*2·3*	
$4\frac{1}{4}$	8	9	7	$2\frac{3}{4}$	$1\frac{1}{2}$	$1\frac{1}{4}$	0·142
7·7	*14·5*	*16·3*	*12·7*	*5·0*	*2·7*	*2·3*	
$3\frac{1}{2}$	$7\frac{1}{4}$	$8\frac{1}{4}$	$6\frac{1}{4}$	$2\frac{3}{4}$	$1\frac{1}{2}$	$1\frac{1}{4}$	0·180
8·1	*16·8*	*19·1*	*14·5*	*6·4*	*3·5*	*2·9*	
$4\frac{1}{2}$	$8\frac{1}{4}$	$9\frac{1}{4}$	$7\frac{1}{4}$	$2\frac{3}{4}$	$1\frac{1}{2}$	$1\frac{1}{4}$	0·180
9·4	*19·1*	*21·4*	*16·8*	*6·4*	*3·5*	*2·9*	
$4\frac{1}{2}$	$8\frac{1}{4}$	$9\frac{1}{4}$	$7\frac{1}{4}$	$2\frac{3}{4}$	$1\frac{1}{2}$	$1\frac{1}{4}$	0·180
9·4	*19·1*	*21·4*	*16·8*	*6·4*	*3·5*	*2·9*	
$5\frac{7}{8}$	$10\frac{1}{8}$	$11\frac{1}{8}$	$9\frac{1}{8}$	$3\frac{1}{4}$	2	$1\frac{1}{4}$	0·180
5·6	*23·4*	*25·7*	*21·1*	*7·5*	*4·6*	*2·9*	
$6\frac{1}{16}$	$11\frac{9}{16}$	$13\frac{1}{16}$	$10\frac{1}{16}$	4	$2\frac{1}{2}$	$1\frac{1}{2}$	0·216
4·0	*26·7*	*30·2*	*23·2*	*9·2*	*5·8*	*3·5*	
$7\frac{7}{16}$	$12\frac{15}{16}$	$14\frac{7}{16}$	$11\frac{7}{16}$	4	$2\frac{1}{2}$	$1\frac{1}{2}$	0·216
7·1	*29·8*	*33·3*	*26·3*	*9·2*	*5·8*	*3·5*	
$9\frac{1}{2}$	$15\frac{1}{2}$	17	14	$4\frac{1}{2}$	3	$1\frac{1}{2}$	0·216
1·9	*35·8*	*19·3*	*32·3*	*10·4*	*6·9*	*3·5*	
$0\frac{7}{8}$	$17\frac{3}{8}$	$18\frac{7}{8}$	$15\frac{7}{8}$	5	$3\frac{1}{2}$	$1\frac{1}{2}$	0·216
5·1	*40·1*	*43·6*	*36·6*	*11·5*	*8·1*	*3·5*	
$0\frac{7}{8}$	$17\frac{3}{8}$	$18\frac{7}{8}$	$15\frac{7}{8}$	5	$3\frac{1}{2}$	$1\frac{1}{2}$	0·216
5·1	*40·1*	*43·6*	*36·6*	*11·5*	*8·1*	*3·5*	

The thread series specified in this Standard is designated by the letter G. It is recommended that these screw threads be referred to on drawings and related documents as follows: Internal, $G\frac{1}{2}$; External Class A, $G\frac{1}{2}A$; External Class B, $G\frac{1}{2}B$.

Where external threads are specified the class reference, i.e. A or B, is added to the designation. The designation of truncated Whitworth pipe threads is by the addition of the abbreviation 'trunc'.

Example: $G \frac{1}{2}$ trunc.

Example: $G \frac{1}{2}$ B trunc.

The basic sizes are shown in Table 4.20.

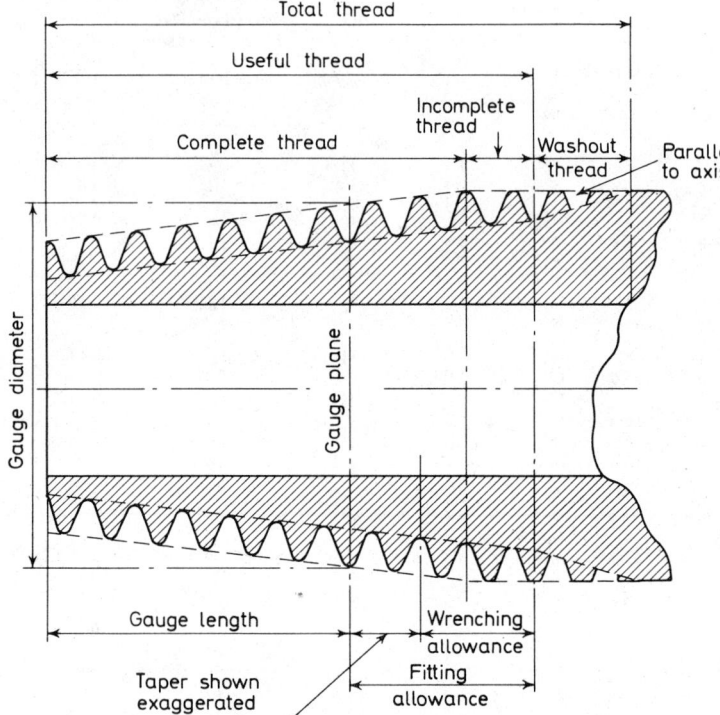

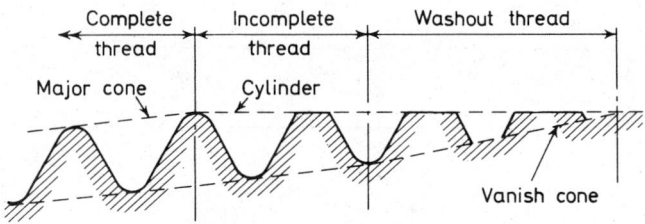

Figure 4.5. Terms relating to pipe threads

THREADED FASTENERS

The following notes are given for guidance in the selection of threaded fasteners for any particular design application. It consists of an aide-memoire and list of the various parameters which need to be considered when designing and drawing up the necessary list of parts. The parameters consist of the following:
Environmental and general consideration.
Head form.
Shank form.
Type of screw thread.
Point form.
Driving media.
Nut selection.
Accessories.

Environmental and general consideration

The general and environmental conditions under which a fastener is applied in service necessitate an examination of:
Choice of basic size to suit assembly.
Choice of material.
Strength versus size.
Corrosion resistance — special material.
— plating or coating.
Resistance to elevated or sub zero temperatures.
Toughness. Resistance to impact or damage.
Availability Stock lines in general purpose materials.
Special materials.

Head form

The head form then needs to be selected according to the following requirements: Appearance; space criterion; weight criterion; availability of torqueing tools; accessibility of torquing tools; bearing area versus pull-through in large clearance holes.

Shank form

The next consideration is the shank form according to the following features and circumstances.
Locking facility whilst tightening:
Square.
Nib.
Clearance.
Avoidance of stress concentration — reduced shank.
Strength in shear; shear/tension; tension; torsion.

Type of screw

Next consider the type of screw thread to be used and the following points will assist in applying the best type.

System. ISO Metric; ISO Inch Unified.

Form Triangular. Symmetric; asymmetric (modified flank angle); rounded root UNJ (inch only).

Pitch/Diameter relationship. Coarse graded series—Plastics; fine graded series; fine constant pitch system. Thin wall; extra fine graded series; special pitches; special diameters; susceptibility to damage and cross threading.

Strength of Screw Thread in: Tension—core strength, stripping strength; tension—shear; shear; torsion; fatigue; vibration.

Fit between male and female thread. Close (better stripping strength); medium (general purpose); free (long lengths of engagement. Dirty conditions).

Method of Forming Thread. Pre-cut rolled or ground (bolt and nut); pre-cut or rolled, bolt only, tapped hole in attachment member; Cut or formed during assembly. Female thread in nut or attachment member.

Point form

It is then necessary to select the type of point which can be according to the following: normal uses; special points, such as: ease of engagement or entry into tapped hole and locking facility of mating component.

Having selected the basic size head form, shank form, and type of point, the final considerations are the method of driving or tightening, the type of nut and type of accessory according to the tightening or locking features required. These should be selected from the undermentioned headings.

Driving media

Facility for tightening.

Facility for untightening. Normal; Dirty conditions; Corroded conditions; Elevated temperature conditions.

Torque effectiveness criterion in following order: Double hexagon or spline; hexagon; square; hexagon socket or spline; recessed head; slotted head; manual.

Availability of tools. These can be classified as: special to suit expediences e.g. flat metal, coin, or no tools (manual).

Nuts

Normal, plain or faced.

Heavy.

Thin. Locking.

Prevailing torque. Thread distortion; top insert; side insert; slotted or castle.

Accessories

Washers. plain; spring; lock; tab.

Pins. Split; taper; split taper; grooved.

Summary of types of fastener

HEAD FORMS

The varieties in the main fastener features are:
Hexagon; hexagon washer faced; hexagon washer head.
Double hexagon; double hexagon washer head.
Square; square washer head.
Countersunk (flat); truncated countersunk (undercut flat).
Raised countersunk (oval); truncated countersunk (undercut oval).
Round (cup).
Pan; pan washer (flanged).
Mushroom (truss).
Cheese (flat fillister); raised cheese (fillister).
Binding; undercut binding.

SHANK FORMS

Full diameter (shank dia. $\simeq$ major dia.).
Scant (shank dia. $\simeq$ pitch dia.).
Reduced (shank dia. < minor dia.).
Fitted (shank dia. > major dia.).
Shoulder.

THREAD TYPES

Screw (cut, rolled, ground).
Tapping screw; thread forming; thread cutting; drive.
Wood screw.

POINT FORMS

Plain rolled.
Round.
Chamfered.
Cone.
Cup.
Dog; full dog.

DRIVING MEDIA

Hexagon, double hexagon, square.
Socket, hexagon socket, double hexagon socket, square socket, spline socket.
Slot.

Recess, Pozidriv; Phillips; Tri-wing; Hi-torque.
Manually operated; wing; thumb; knurl.

NUTS

Hexagon; washer faced hexagon; hexagon flanged; hexagon slotted; hexagon castle.
Double hexagon; double hexagon flanged.
Square.
Wing.
Circular.
Hexagon prevailing torque.
Anchor (plate).

ACCESSORIES (WASHERS)

Plain; Chamfered.
Square; square tapered.
Spring; single coil; double coil; crinkle.
Toothed lock.
Tab.

ACCESSORIES (PINS)

Taper.
Split cotter.

MECHANICAL PROPERTIES OF BOLTS, SCREWS, STUDS AND NUTS—METRIC DESIGNATION

The designation system for property classes of bolts, screws and studs in ISO R 898 1 and 2[20] is shown in Table 4.22. The abscissae show the tensile strength values while the ordinates show those of the elongation after fracture.

The symbol consists of two figures. The first figure indicates 1/10 of the minimum tensile strength value in kgf/mm^2; the second figure indicates 1/10 of the ratio, expressed as a percentage, between minimum yield stress and the minimum tensile strength. The multiplication of these two figures will give the minimum yield stress in kgf/mm^2.

The property classes of nuts are designated by the numbers 4, 5, 6, 8, 10, 12 and 14 as shown in Table 4.23. The designation number is equal to one-tenth of the specified proof load stress in kgf/mm^2. This proof load stress corresponds to the minimum tensile strength of a bolt or screw with which the nut should be assembled, so as to ensure the loading capacity of the bolted connection up to the minimum tensile strength of the bolt.

Table 4.22 BOLTS SCREWS AND STUDS. SYSTEM OF CO-ORDINATES

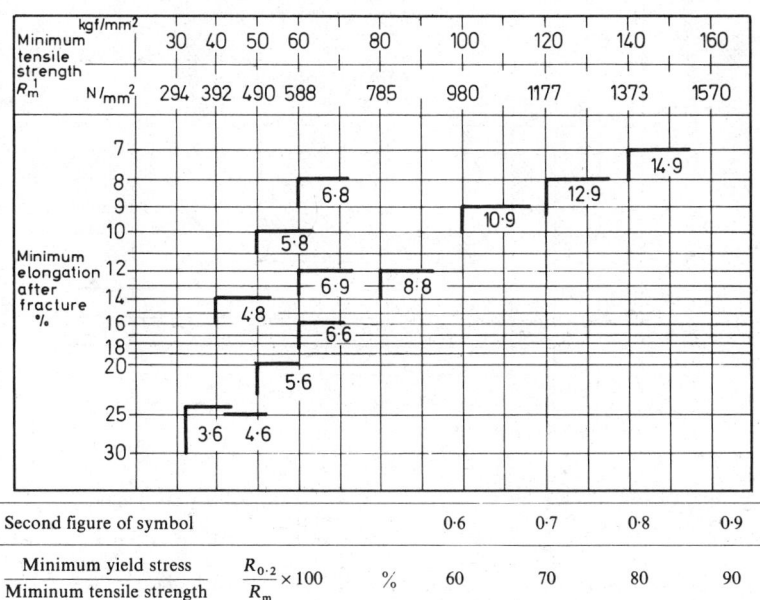

Second figure of symbol		0·6	0·7	0·8	0·9
$\dfrac{\text{Minimum yield stress}}{\text{Miminum tensile strength}}$ $\dfrac{R_{0·2}}{R_{\mathrm{m}}} \times 100$ %		60	70	80	90

Table 4.23 DESIGNATION OF PROPERTY CLASSES OF NUTS IN TO PROOF LOAD STRESS

Property class		4	5	6	8	10	12	14
Proof load stress:	kgf/mm^2	40	50	60	80	100	120	140
	N/mm^2	392	490	588	785	980	1177	1373

Table 4.24   STRENGTH GRADE DESIGNATIONS OF STEEL NUTS

Strength grade designation	4	5	6	8	12	14
Proof load stress kgf/mm²	40	50	60	80	120	140

Table 4.25   RECOMMENDED BOLT AND NUT COMBINATIONS

Grade of bolt	4·6	4·8	5·6	5·8	6·6	6·8	8·8	10·9	12·9	14·9
Recommended grade of nut	4	4	5	5	6	6	8	12	12	14

Note. Nuts of a higher strength grade may be substituted for nuts of a lower strength grade.

Table 4.26

Minimum elongation after fracture	Strength grade designation						
	4·6 to 6·8			8·8	10·9	12·9	14·9
	Tensile strength kgf/mm²						
	40	50	60	80	100	120	140
7							14·9
8			6·8			12·9	
9			6·8		10·9		
10		5·8					
12				8·8			
14	4·8						
16	4·8		6·6				
18							
20		5·6					
25	4·6						
30	4·6						

▨ Free cutting steel allowed   ▥ Hardened and tempered

It is recommended that the grades of nut to be used with each grade of bolt or screw should be as shown in Table 4.25.

The steel used shall be such that the finished product possesses the mechanical properties appropriate to the strength grade quoted. Table 4.26 illustrates the general requirements.

ISO METRIC PRECISION HEXAGON HEAD BOLTS, SCREWS AND NUTS

BS 3692[10] gives the general dimensions and tolerances of precision hexagon bolts, screws and nuts with ISO metric threads in diameters from 1·6 to 68 mm.

Mechanical properties are given only in respect of carbon or alloy steel bolts, screws and nuts, which are not to be used for special applications, such as those requiring

Table 4.27   BASIS FOR STANDARD THREAD LENGTHS

Nominal length of bolt l	Length of thread b
Up to and including 125 mm	$2d + 6$ mm
Over 125 mm up to and including 200 mm	$2d + 12$ mm
Over 200 mm	$2d + 25$ mm

Table 4.28   STRENGTH GRADE DESIGNATIONS OF STEEL BOLTS AND SCREWS

Strength grade designation	4·6	4·8	5·6	5·8	6·6	6·8	8·8	10·9	12·9	14·9
Tensile strength R_m min kgf/mm²	40	40	50	50	60	60	80	100	120	140
Yield stress R_6 min kgf/mm²	24	32	30	40	36	48	—	—	—	—
Stress at permanent set limit R_{m3} min kgf/mm²	—	—	—	—	—	—	64	90	108	126

weldability, corrosion resistance or ability to withstand temperatures above 300°C or below − 50°C.

The dimensional requirements of this standard also apply to non-ferrous and stainless steel bolts, screws and nuts.

Steel precision bolts, screws and nuts are normally supplied with the following finishes.

Heat-treated bolts and screws.   Components heat-treated after manufacture are customarily dull black, although the manufacturer may machine some of the surfaces of the larger size bolts and screws after heat treatment.

Bright finished bolts and screws.   This term is used to describe bolts and screws which are finished bright on all surfaces or which have a finish on the hexagon, produced by bright drawing.

Nuts.   These may be bright on all surfaces or dull black when heat-treated.

The form of thread and diameters and associated pitches of standard ISO metric bolts, screws and nuts are in accordance with BS 3643[9] and are made to the tolerances for the medium class of fit (6H/6g).

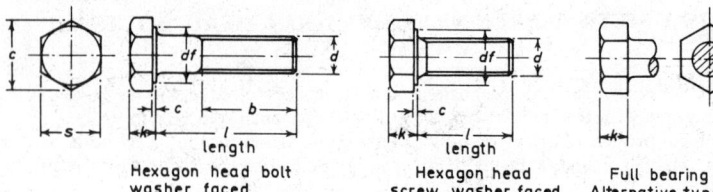

Hexagon head bolt
washer faced

Hexagon head
screw, washer faced

Full bearing head
Alternative type of head
permissible on bolts and
screws

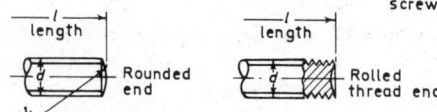

Rounded
end

1¼ d rad. approx.

Rolled
thread end

Alternative types of end permissible on bolts and screws

Figure 4.6. Dimensions of hexagon bolts and screws, precision (BS 3692).

Table 4.29 ISO PRECISION BOLTS AND SCREWS (see Figure 4.6)

Nominal size and thread dia. d	Pitch of thread (coarse pitch series)	Pitch of thread (fine pitch series)	Diameter of unthreaded shank d max	min	Width across flats s max	Width across corners e max	Diameter of washer face d_f max	min	Depth of washer face c	Height of head k max
M1·6	0·35	—	1·6	1·46	3·2	3·7	—	—	—	1·225
M2	0·4	—	2·0	1·86	4·0	4·6	—	—	—	1·525
M2·5	0·45	—	2·5	2·36	5·0	5·8	—	—	—	1·825
M3	0·5	—	3·0	2·86	5·5	6·4	5·08	4·83	0·1	2·125
M4	0·7	—	4·0	3·82	7·0	8·1	6·55	6·30	0·1	2·925
M5	0·8	—	5·0	4·82	8·0	9·2	7·55	7·30	0·2	3·650
M6	1	0·75	6·0	5·82	10·0	11·5	9·48	9·23	0·3	4·15
M8	1·25	1·00	8·0	7·78	13·0	15·0	12·43	12·18	0·4	5·65
M10	1·5	1·25	10·0	9·78	17·0	19·6	16·43	16·18	0·4	7·18
M12	1·75	1·25	12·0	11·73	19·0	21·9	18·37	18·12	0·4	8·18
(M14)	2	1·50	14·0	13·73	22·0	25·4	21·37	21·12	0·4	9·18
M16	2	1·50	16·0	15·73	24·0	27·7	23·27	23·02	0·4	10·18
(M18)	2·5	1·50	18·0	17·73	27·0	31·2	26·27	26·02	0·4	12·215
M20	2·5	1·50	20·0	19·67	30·0	34·6	29·27	28·80	0·4	13·215
(M22)	2·5	1·50	22·0	21·67	32·0	36·9	31·21	30·74	0·4	14·215
M24	3	2·0	24·0	23·67	36·0	41·6	34·98	34·51	0·5	15·215
(M27)	3	2·0	27·0	26·67	41·0	47·3	39·98	39·36	0·5	17·215
M30	3·5	2·0	30·0	29·67	46·0	53·1	44·98	44·36	0·5	19·26
(M33)	3·5	2·0	33·0	32·61	50·0	57·7	48·98	48·36	0·5	21·26
M36	4	3·0	36·0	35·61	55·0	63·5	53·86	53·24	0·5	23·26
(M39)	4	3·0	39·0	38·61	60·0	69·3	58·86	58·24	0·6	25·26
M42	4·5	—	42·0	41·61	65·0	75·1	63·76	63·04	0·6	26·26
(M45)	4·5	—	45·0	44·61	70·0	80·8	68·76	68·04	0·6	28·26
M48	5	—	48·0	47·61	75·0	86·6	73·76	73·04	0·6	30·26
(M52)	5	—	52·0	51·54	80·0	92·4	—	—	—	33·31
M56	5·5	—	56·0	55·54	85·0	98·1	—	—	—	35·31
(M60)	5·5	—	60·0	59·54	90·0	103·9	—	—	—	38·31
M64	6	—	64·0	63·54	95·0	109·7	—	—	—	40·31
(M68)	6	—	68·0	67·54	100·0	115·5	—	—	—	43·31

Sizes shown in brackets are non-preferred.
*This size is included because of its use in the automotive industry.

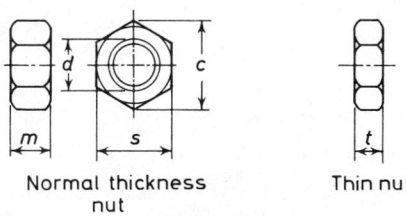

Normal thickness nut   Thin nut

Figure 4.7. Dimensions of hexagon nuts and thin nuts, precision (BS 3692).

Table 4.30 ISO PRECISION HEXAGON NUTS (see Figure 4.7)

Nominal size and thread diameter d	Pitch of thread (coarse pitch series)	Pitch of thread (fine pitch series)	Width across flats s max	Width across corners c max	Thickness of normal nut m max	Thickness of thin nut t max
M1·6	0·35	—	3·20	3·70	1·30	—
M2	0·4	—	4·00	4·60	1·60	—
M2·5	0·45	—	5·00	5·80	2·00	—
M3	0·5	—	5·50	6·40	2·40	—
M4	0·7	—	7·00	8·10	3·20	—
M5	0·8	—	8·00	9·20	4·00	—
M6	1	0·75*	10·00	11·50	5·00	—
M8	1·25	1·00	13·00	15·00	6·50	5·0
M10	1·5	1·25	17·00	19·60	8·00	6·0
M12	1·75	1·25	19·00	21·90	10·00	7·0
(M14)	2	1·50	22·00	25·4	11·00	8·0
M16	2	1·50	24·00	27·7	13·00	8·0
(M18)	2·5	1·50	27·00	31·20	15·00	9·0
M20	2·5	1·50	30·00	34·60	16·00	9·0
(M22)	2·5	1·50	32·00	36·90	18·00	10·0
M24	3	2·0	36·00	41·60	19·00	10·0
(M27)	3	2·0	41·00	47·3	22·00	12·0
M30	3·5	2·0	46·00	53·1	24·00	12·0
(M33)	3·5	2·0	50·00	57·70	26·00	14·0
M36	4	3·0	55·00	63·50	29·00	14·0
(M39)	4	3·0	60·00	69·30	31·00	16·0
M42	4·5	—	65·00	75·10	34·00	16·0
(M45)	4·5	—	70·00	80·80	36·00	18·0
M48	5	—	75·00	86·60	38·00	18·0
(M52)	5	—	80·00	92·40	42·00	20·0
M56	5·5	—	85·00	98·10	45·00	—
(M60)	5·5	—	90·00	103·90	48·00	—
M64	6	—	95·00	109·70	51·00	—
(M68)	6	—	100·00	115·50	54·00	—

Sizes shown in brackets are non-preferred.
*This size is included because of its use in the automotive industry.

Table 4.31 PROOF LOADS FOR STEEL BOLTS AND SCREWS (ISO PRECISION)
Coarse pitch series

Nominal size of bolt or screw mm	Tensile stress area mm²	Strength grade designation									
		4·6	4·8	5·6	5·8	6·6	6·8	8·8	10·9	12·9	14·9
		Stress under proof load kgf/mm²									
		22·6	29·1	28·2	36·4	33·9	43·7	58·2	79·2	95·0	111·0
		Proof load metric tonne force 1 000 kgf									
M1·6	1·27	0·029	0·037	0·036	0·046	0·043	0·055	0·074	0·101	0·121	0·141
M2	2·07	0·047	0·060	0·058	0·075	0·070	0·090	0·120	0·164	0·197	0·230
M2·5	3·39	0·077	0·099	0·096	0·123	0·115	0·148	0·197	0·268	0·322	0·376
M3	5·03	0·114	0·146	0·142	0·183	0·171	0·220	0·293	0·398	0·478	0·558
M4	8·78	0·198	0·255	0·248	0·320	0·298	0·384	0·511	0·695	0·834	0·975
M5	14·2	0·321	0·413	0·400	0·517	0·481	0·621	0·826	1·12	1·35	1·58
M6	20·1	0·454	0·585	0·567	0·732	0·681	0·878	1·17	1·59	1·91	2·23
M8	36·6	0·827	1·07	1·03	1·33	1·24	1·60	2·13	2·90	3·48	4·06
M10	58·0	1·31	1·69	1·64	2·11	1·97	2·53	3·38	4·59	5·51	6·44
M12	84·3	1·91	2·45	2·38	3·07	2·86	3·68	4·91	6·68	8·01	9·36
M14	115	2·60	3·35	3·24	4·17	3·90	5·03	6·69	9·11	10·9	12·8
M16	157	3·55	4·57	4·43	5·71	5·32	6·86	9·14	12·4	14·9	17·4
M18	192	4·34	5·59	5·41	6·99	6·51	8·39	11·2	15·2	18·2	21·3
M20	245	5·54	7·13	6·91	8·92	8·31	10·7	14·3	19·4	23·3	27·2
M22	303	6·85	8·82	8·54	11·0	10·3	13·2	17·6	24·0	28·8	33·6
M24	353	7·98	10·3	9·95	12·8	12·0	15·4	20·5	28·0	33·5	39·2
M27	459	10·4	13·4	12·9	16·7	15·6	20·1	26·7	36·4	43·6	50·9
M30	561	12·7	16·3	15·8	20·4	19·0	24·5	32·7	44·4	53·3	62·3
M33	694	15·7	20·2	19·6	25·3	23·5	30·3	40·4	55·0	65·9	77·0
M36	817	18·5	23·8	23·0	29·7	27·7	35·7	47·5	64·7	77·6	90·7
M39	976	22·1	28·4	27·5	35·5	33·1	42·7	56·8	77·3	92·7	108
M42	1120	25·3	32·6	31·6	40·8	38·0	48·9	65·2	88·7	106	124
M45	1300	29·4	37·8	36·7	47·3	44·1	56·8	75·7	103	124	144
M48	1470	33·2	42·8	41·5	53·5	49·8	64·2	85·6	116	140	163
M52	1760	39·8	51·2	49·6	64·1	59·7	77·0	102	139	167	195
M56	2030	45·8	59·1	57·2	73·9	68·8	88·7	118	161	193	225
M60	2360	53·3	68·7	66·6	85·9	80·0	103	137	187	224	262
M64	2680	60·6	78·0	75·6	97·6	90·9	117	156	212	255	297
M68	3060	69·2	89·0	86·3	111	104	134	178	242	291	340

Proof load = $\dfrac{\text{stress under proof load} \times \text{tensile stress area of bolt}}{1000}$

Table 4.32 PROOF LOADS FOR STEEL NUTS (ISO PRECISION)

Coarse pitch series

Nominal size of nut mm	Tensile stress area of bolt mm^2	Strength grade designation					
		4	5	6	8	12	14
		Stress under proof load kgf/mm^2					
		40	50	60	80	120	140
		Proof load metric tonne force 1 000 kgf					
M1·6	1·27	0·051	0·064	0·076	0·102	0·152	0·178
M2	2·07	0·083	0·103	0·124	0·165	0·248	0·290
M2·5	3·39	0·136	0·170	0·203	0·271	0·407	0·475
M3	5·03	0·201	0·250	0·302	0·402	0·604	0·704
M4	8·78	0·351	0·440	0·527	0·702	1·05	1·23
M5	14·2	0·568	0·710	0·852	1·14	1·70	1·99
M6	20·1	0·804	1·000	1·21	1·61	2·41	2·81
M8	36·6	1·46	1·830	2·20	2·93	4·39	5·12
M10	58·0	2·32	2·90	3·48	4·64	6·96	8·12
M12	84·3	3·37	4·21	5·06	6·74	10·1	11·8
M14	115	4·60	5·75	6·90	9·20	13·8	16·1
M16	157	6·28	7·85	9·42	12·6	18·8	22·0
M18	192	7·68	9·60	11·5	15·4	23·0	26·9
M20	245	9·80	10·2	14·7	19·6	29·4	34·3
M22	303	12·1	15·1	18·2	24·2	36·4	42·4
M24	353	14·1	17·6	21·2	28·2	42·4	49·4
M27	459	18·4	23·0	27·5	36·7	55·1	64·3
M30	561	22·4	28·0	33·7	44·9	67·3	78·5
M33	694	27·8	34·7	41·6	55·5	83·3	97·2
M36	817	32·7	40·8	49·0	65·4	98·0	114
M39	976	39·0	48·8	58·6	78·1	117	137
M42	1120	44·8	56·0	67·2	89·6	134	157
M45	1300	52·0	65·0	78·0	104	156	182
M48	1470	58·8	·73·5	88·2	118	176	206
M52	1760	70·4	88·0	106	141	211	246
M56	2030	81·2	101·5	122	162	244	284
M60	2360	94·4	118	142	189	283	330
M64	2680	107	134	161	214	322	375
M68	3060	122	153	184	245	367	428

Proof load = $\dfrac{\text{stress under proof load} \times \text{tensile stress area of bolt}}{1000}$

Table 4.33 PROOF LOADS FOR STEEL BOLTS AND SCREWS (ISO PRECISION)

Fine pitch series

Nominal size × pitch mm	*Tensile stress area* mm²	*Strength grade designation*									
		4·6	*4·8*	*5·6*	*5·8*	*6·6*	*6·8*	*8·8*	*10·9*	*12·9*	*14·9*
		Stress under proof load kgf/mm²									
		22·6	*29·1*	*28·2*	*36·4*	*33·9*	*43·7*	*58·2*	*79·2*	*95·0*	*111·0*
		Proof load metric tonne force 1 000 kgf									
M6 × 0·75	22·00	0·497	0·640	0·620	0·800	0·746	0·961	1·270	1·742	2·090	2·442
M8 × 1·00	39·20	0·885	1·140	1·110	1·430	1·330	1·710	2·280	3·100	3·270	4·350
M10 × 1·25	61·20	1·380	1·780	1·730	2·230	2·070	2·670	3·560	4·850	5·800	6·790
M12 × 1·25	92·10	2·080	2·680	2·600	3·350	3·120	4·020	5·360	7·290	8·750	10·200
(M14 × 1·50)	129	2·820	3·640	3·520	4·550	4·240	5·460	7·280	9·900	11·900	13·900
M16 × 1·50	167	3·770	4·860	4·710	6·080	5·660	7·300	9·720	13·200	15·900	18·500
(M18 × 1·50)	216	4·880	6·290	6·090	7·860	7·320	9·440	12·600	17·100	20·500	24·000
M20 × 1·50	272	6·150	7·920	7·670	9·900	9·220	11·900	15·800	21·500	25·800	30·200
(M22 × 1·50)	333	7·530	9·690	9·390	12·100	11·300	14·600	19·400	26·400	31·600	37·000
M24 × 2·00	384	8·700	11·200	10·800	14·000	13·000	16·800	22·300	30·400	36·500	42·600
(M27 × 2·00)	496	11·200	14·400	14·000	18·100	16·800	21·700	28·900	29·300	47·100	55·100
M30 × 2·00	621	14·000	18·100	17·500	22·600	21·100	27·100	36·100	49·200	59·000	68·900
(M33 × 2·00)	761	17·200	22·100	21·500	27·700	25·800	33·300	44·300	60·300	72·300	84·500
M36 × 3·00	865	19·500	25·200	24·400	31·500	29·300	37·800	50·300	68·500	82·200	96·000
(M39 × 3·00)	1030	23·300	30·000	29·000	37·500	34·900	45·000	59·900	81·600	97·800	114·000

$$\text{Proof load} = \frac{\text{stress under proof load} \times \text{tensile stress area of bolt}}{1000}$$

Table 4.34 PROOF LOADS FOR STEEL NUTS (PRECISION)

Fine pitch series

Nominal size × pitch mm	*Tensile stress area of bolt* mm²	*Strength grade designation*					
		4	*5*	*6*	*8*	*12*	*14*
		Stress under proof load kg/mm²					
		40	*50*	*60*	*80*	*120*	*140*
		Proof load metric tonne force 1 000 kgf					
M6 × 0·75	22·0	0·88	1·10	1·32	1·76	2·64	3·08
M8 × 1·00	39·20	1·57	1·96	2·35	3·10	4·70	5·50
M10 × 1·25	61·20	2·40	3·06	3·70	4·90	7·35	8·55
M12 × 1·25	92·10	3·50	4·40	5·30	7·00	10·60	12·30
(M14 × 1·50)	129	5·00	6·22	7·50	10·00	15·00	17·50
M16 × 1·50	167	6·70	8·35	10·00	13·40	20·00	23·40
(M18 × 1·50)	216	8·60	10·80	12·90	17·30	25·90	30·20
M20 × 1·50	272	10·90	13·60	16·30	21·80	32·60	38·00
(M22 × 1·50)	333	13·30	16·60	20·00	26·60	40·00	46·60
M24 × 2·00	384	15·40	19·20	23·00	30·70	46·00	53·80
(M27 × 2·00	496	19·90	24·80	29·80	39·70	59·50	69·50
M30 × 2·00	621	24·80	31·00	37·30	49·70	74·50	87·00
(M33 × 2·00)	761	30·40	38·00	45·60	60·80	91·40	106·50
M36 × 3·00	865	34·60	43·20	51·90	69·20	104·00	121·00
(M39 × 3·00)	1030	41·20	50·10	61·80	82·50	124·00	144·00

$$\text{Proof load} = \frac{\text{stress under proof load} \times \text{tensile strength area of bolt}}{1000}$$

Table 4.35 STRENGTH GRADE DESIGNATION MARKING OF BOLTS AND SCREWS

Strength grade designation	4·0	4·6	5·6	5·8	6·6	6·8	8·8*	10·9*	12·9*	14·9*
Marking alternatives		4·6	5·6	5·8	6·6	6·8	8·8	10·9	12·9	14·9
		4·8	56	58	66	68	88	109	129	149
		46								
		48								

*Marking of strength grade is mandatory.

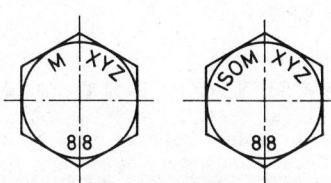

Figure 4.8. Examples of marking of forged products M or ISO M. ISO metric identification XYZ. Manufacturer's identification (trade) marking

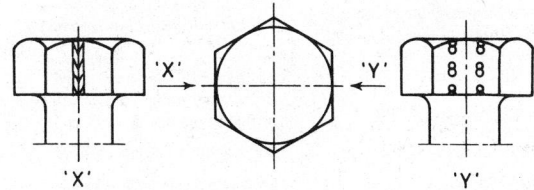

Figure 4.9. Example of marking of bar turned product. For more complete details of marking, see ISO R898[20]

Table 4.36 STRENGTH GRADE DESIGNATION MARKING OF NUTS

Strength grade	4	5	6	8*	12*	14*
Symbol	4	5	6	8	12	14
Alternative 'clock face' marking system	–	–				

* Marking of strength grade is mandatory

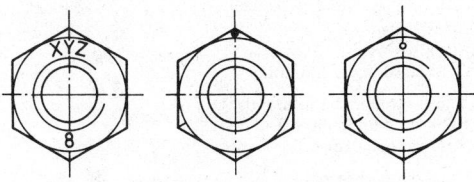

Figure 4.10. Examples of marking of forged nuts

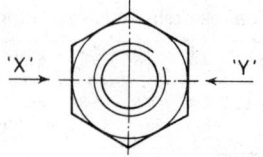

Figure 4.11. Example of marking of bar turned nut

ISO METRIC BLACK HEXAGON BOLTS, SCREWS AND NUTS

BS 4190[12] gives the general dimensions and tolerances of black hexagon bolts, screws and nuts with ISO metric threads, in diameters from 5 mm to 68 mm inclusive.

The term 'black' does not necessarily relate to the appearance of the products since these may be of bright appearance or black in the finished state, but implies the comparatively wider tolerances to which these products are usually made.

Table 4.37 PRODUCT CATEGORIES

Non-machined products (finished black all over)	Partially-machined products
Black bolts	Bolts faced under head only
	Bolts faced under head and turned on shank
Black screws	Screws faced under head only
Black nuts	Nuts faced one side
	Thin nuts (faced both sides)

Screw threads

The form of thread, diameters and associated pitches of standard ISO metric bolts, screws and nuts shall be in accordance with BS 3643, Part 1[9]. Coarse pitch series threads only are specified in this standard.

TOLERANCES

The tolerances on the screw threads are in accordance with BS 3643, Part 2[9] as detailed in Table 4.38.

Table 4.38 THREAD TOLERANCE CLASSES

Product	Tolerance class
Black bolts / Black screws / Bolts faced under head only	8g
Screws faced under head only / Bolts faced under head and turned on shank	6g
Nuts (black or faced)	7H

Table 4.39 MECHANICAL PROPERTIES

Strength grade designation	4·6	4·8	6·9
Tensile strength, kgf/mm²	40	40	60
Yield stress, kgf/mm²	24	32	—
Stress at permanent set limit, kgf/mm²	—	—	54

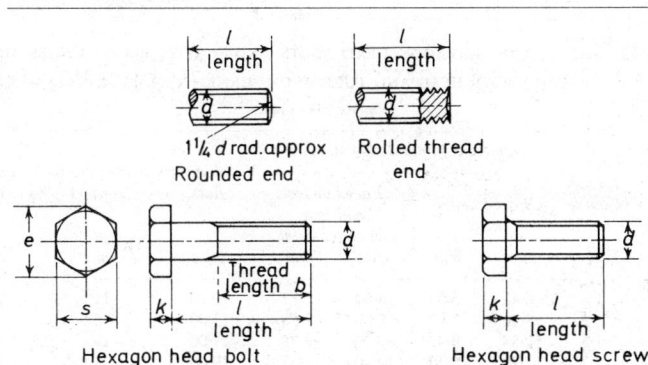

Figure 4.12. Dimensions of black bolts and screws

Table 4.40 ISO METRIC BLACK BOLTS AND SCREWS (see Figure 4.12)

Nominal size and thread diameter	Pitch of thread	Diameter of unthreaded shank		Width across flats	Width across corners	Height of head
d	(Coarse pitch series)	d max	min	s max	e max	k max
M5	0·8	5·48	4·52	8·00	9·2	3·875
M6	1	6·48	5·52	10·00	11·5	4·375
M8	1·25	8·58	7·42	13·00	15·0	5·875
M10	1·5	10·58	9·42	17·00	19·6	7·45
M12	1·75	12·70	11·30	19·00	21·9	8·45
M16	2	16·70	15·30	24·00	27·7	10·45
M20	2·5	20·84	19·16	30·00	34·6	13·90
(M22)	2·5	22·84	21·16	32·00	36·9	14·90
M24	3	24·84	23·16	36·00	41·6	15·90
(M27)	3	27·84	26·16	41·00	47·3	17·90
M30	3·5	30·84	29·16	46·00	53·1	20·05
(M33)	3·5	34·00	32·00	50·00	57·7	22·05
M36	4	37·00	35·00	55·00	63·5	24·05
(M39)	4	40·00	38·00	60·00	69·3	26·05
M42	4·5	43·00	41·00	65·00	75·1	27·05
(M45)	4·5	46·00	44·00	70·00	80·8	29·05
M48	5	49·00	47·00	75·00	86·6	31·05
(M52)	5	53·20	50·80	80·00	92·4	34·25
M56	5·5	57·20	54·80	85·00	98·1	36·25
(M60)	5·5	61·20	58·80	90·00	103·9	39·25
M64	6	65·20	62·80	95·00	109·7	41·25
(M68)	6	69·20	66·80	100·00	115·5	44·25

Table 4.41 THREAD LENGTHS (see Figure 4.12)

Nominal length of bolt *l*	Length of thread *b*
Up to and including 125 mm	2*d* + 6 mm
Over 125 mm up to and including 200 mm	2*d* + 12 mm
Over 200 mm	2*d* + 25 mm

Table 4.42 ISO METRIC HEXAGON HEAD BOLTS AND SCREWS FACED UNDER HEAD OR FACED UNDER HEAD AND TURNED ON SHANK (see Figure 4.12)

Dimensions in mm

Nominal size and thread diameter *d*	Pitch of thread (Coarse pitch series)	Diameter of unthreaded shank *d*				Width across flats *s* max	Width across corners *e* max	Height of head *k* max
		Faced under head		Faced under head and turned on shank				
		max	min	max	min			
M6	1	6·48	5·52	6·00	5·82	10·00	11·5	4·24
M8	1·25	8·58	7·42	8·00	7·78	13·00	15·0	5·74
M10	1·5	10·58	9·42	10·00	9·78	17·00	19·6	7·29
M12	1·75	12·70	11·30	12·00	11·73	19·00	21·9	8·29
M16	2	16·70	15·30	16·00	15·73	24·00	27·7	10·29
M20	2·5	20·84	19·16	20·00	19·67	30·00	34·6	13·35
(M22)	2·5	22·84	21·16	22·00	21·67	32·00	36·9	14·35
M24	3	24·84	23·16	24·00	23·67	36·00	41·6	15·35
(M27)	3	27·84	26·16	27·00	26·67	41·00	47·3	17·35
M30	3·5	30·84	29·16	30·00	29·67	46·00	53·1	19·42
(M33)	3·5	34·00	32·00	33·00	32·61	50·00	57·7	21·42
M36	4	37·00	35·00	36·00	35·61	55·00	63·5	23·42
(M39)	4	40·00	38·00	39·00	38·61	60·00	69·3	25·42
M42	4·5	43·00	41·00	42·00	41·61	65·00	75·1	26·42
(M45)	4·5	46·00	44·00	45·00	44·61	70·00	80·8	28·42
M48	5·0	49·00	47·00	48·00	47·61	75·00	86·6	30·42
(M52)	5·0	53·20	50·80	52·00	51·54	80·00	92·4	33·50
M56	5·5	57·20	54·80	56·00	55·54	85·00	98·1	35·50
(M60)	5·5	61·20	58·80	60·00	59·54	90·00	103·9	38·50
M64	6	62·50	62·80	64·00	63·54	95·00	109·7	40·50
(M68)	6	69·20	66·80	68·00	67·54	100·00	115·5	43·50

Sizes shown in brackets are non-preferred.

Table 4.43 STRENGTH GRADE DESIGNATIONS FOR STEEL NUTS

Strength grade designation	4	6
Proof load stress kgf/mm^2	40	60

It is recommended that the grades of nut to be used with each grade of bolt or screw should be as shown in Table 4.44.

Table 4.44 RECOMMENDED BOLT AND NUT COMBINATIONS

Grade of bolt	4·6	4·8	6·9
Recommended grade of nut	4	4	6

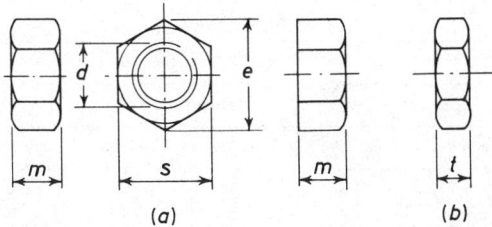

Figure 4.13. (a) Alternative forms of normal thickness nut (b) Thin nut

Table 4.45 ISO METRIC HEXAGON NUTS AND HEXAGON THIN NUTS (see Figure 4.13)

Dimensions in mm

Nominal size of thread diameter	Pitch thread (Coarse pitch series)	Widths across flats s max	Width across corners e max	Thickness of nut m Black max	Thickness of nut m Faced one side max	Thickness of thin nut (faced both sides) t max
M5	0·8	8·00	9·2	4·375	4·0	—
M6	1	10·00	11·5	5·375	5	—
M8	1·25	13·00	15·0	6·875	6·5	5·00
M10	1·5	17·00	19·6	8·45	8	6·00
M12	1·75	19·00	21·9	10·45	10	7·00
M16	2	24·00	27·7	13·55	13	8·00
M20	2·5	30·00	34·6	16·55	16	9·00
(M22)	2·5	32·00	36·9	18·55	18	10·00
M24	3	36·00	41·6	19·65	19	10·00
(M27)	3	41·00	47·3	22·65	22	12·00
M30	3·5	46·00	53·1	24·65	24	12·00
(M33)	3·5	50·00	57·7	26·65	26	14·00
M36	4	55·00	63·5	29·65	29	14·00
(M39)	4	60·00	69·3	31·80	31	16·00
M42	4·5	65·00	75·1	34·80	34	16·00
(M45)	4·5	70·00	80·8	36·80	36	18·00
M48	5	75·00	86·6	38·80	38	18·00
(M52)	5	80·00	92·4	42·80	42	20·00
M56	5·5	85·00	98·1	45·80	45	—
(M60)	5·5	90·00	103·9	48·80	48	—
M64	6	95·00	109·7	51·95	51	—
(M68)	6	100·00	115·5	54·95	54	—

Sizes shown in brackets are non-preferred.

Table 4.46 LIMITING DIMENSIONS FOR CLEARANCE HOLES (see Figure 4.14)

Dimensions in mm

Thread diameter d	Clearance hole D					
	Close fit series		Medium fit series*		Free fit series	
	min dia	max dia	min dia	max dia	min dia	max dia
1·6	1·7	1·8	1·8	1·94	2·0	2·25
2·0	2·2	2·3	2·4	2·54	2·6	2·85
2·5	2·7	2·8	2·9	3·04	3·1	3·3
3·0	3·2	3·32	3·4	3·58	3·6	3·9
4·0	4·3	4·42	4·5	4·68	4·8	5·1
5·0	5·3	5·42	5·5	5·68	5·8	6·1
6·0	6·4	6·55	6·6	6·82	7·0	7·36
7·0	7·4	7·55	7·6	7·82	8·0	8·36
8·0	8·4	8·55	9·0	9·22	10·0	10·36
10·0	10·5	10·68	11·0	11·27	12·0	12·43
12·0	13·0	13·18	14·0	14·27	15·0	15·43
14·0	15·0	15·18	16·0	16·27	17·0	17·43
16·0	17·0	17·18	18·0	18·27	19·0	19·52
18·0	19·0	19·21	20·0	20·33	21·0	21·52
20·0	21·0	21·21	22·0	22·33	24·0	24·52
22·0	23·0	23·21	24·0	24·33	26·0	26·52
24·0	25·0	25·21	26·0	26·33	28·0	28·52
27·0	28·0	28·21	30·0	30·33	32·0	32·62
30·0	31·0	31·25	33·0	33·39	35·0	35·62
33·0	34·0	34·25	36·0	36·39	38·0	38·62
36·0	37·0	37·25	39·0	39·39	42·0	42·62
39·0	40·0	40·25	42·0	42·39	45·0	45·62
42·0	43·0	43·25	45·0	45·39	48·0	48·62
45·0	46·0	46·25	48·0	48·39	52·0	52·74
48·0	50·0	50·25	52·0	52·46	56·0	56·74
52·0	54·0	54·3	56·0	56·46	62·0	62·74
56·0	58·0	58·3	62·0	62·46	66·0	66·74
60·0	62·0	62·3	66·0	66·46	70·0	70·74
64·0	66·0	66·3	70·0	70·46	74·0	74·74
68·0	70·0	70·3	74·0	74·46	78·0	78·74
72·0	74·0	74·3	78·0	78·46	82·0	82·87
76·0	78·0	78·3	82·0	82·54	86·0	86·87
80·0	82·0	80·35	86·0	86·54	91·0	91·87
85·0	87·0	87·35	91·0	91·54	96·0	96·87
90·0	93·0	93·35	96·0	96·54	101·0	101·87
95·0	98·0	98·35	101·0	101·54	107·0	107·87
100·0	104·0	104·35	107·0	107·54	112·0	112·87
105·0	109·0	109·35	112·0	112·54	117·0	117·87
110·0	114·0	114·35	117·0	117·54	122·0	123·0
115·0	119·0	119·35	122·0	122·63	127·0	128·0
120·0	124·0	124·4	127·0	127·63	132·0	133·0
125·0	129·0	129·4	132·0	132·63	137·0	138·0
130·0	134·0	134·4	137·0	137·63	144·0	145·0
140·0	144·0	144·4	147·0	147·63	155·0	156·0
150·0	155·0	155·4	158·0	158·63	165·0	166·0

*Shall be countersunk to accommodate fillet radius under head of bolts and screws.

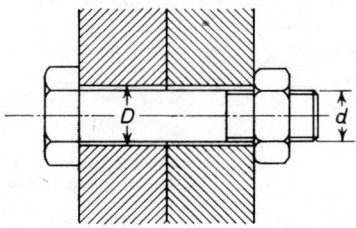

Figure 4.14. Dimensions for clearance holes for metric bolts and screws.

UNIFIED BOLTS, SCREWS AND NUTS

Unified bolts, screws and nuts are based on American standards with which they are generally in alignment. The standards are inch based and although the Normal Precision Series is covered by an ISO Recommendation the Unified standards have no corresponding metric values. These standards are needed for a transition period, principally for use with flanged pipe joints which are based on American standards. The principal British standards are:

BS 1768[4]. Unified Precision Hexagon, Normal Series (See Tables 4.47 to 4.51 and Figures 4.15 and 4.16).
BS 1769[5]. Unified Black Hexagon, Heavy Series.
BS 2708[6]. Unified Black Square and Hexagon, Normal Series.

Mechanical properties

The properties given in Tables 4.49, 4.50 and 4.51 are in inch system units as specified. It should be noted that equivalent properties for ISO metric bolts are expressed in cgs units and are not SI units throughout. The following factors may be used to convert the Imperial system values to align with ISO metric practice:

Tensile strength.	$\text{tonf/in}^2 \times 1\cdot575 = \text{kgf/mm}^2$	(cgs)
	$\text{tonf/in}^2 \times 15\cdot44 = \text{MN/m}^2 \ (\text{N/mm}^2)$	(SI)
Impact.	$\text{ft lbf} \times 1\cdot3558 = \text{J}$	(SI)
Proof Load.	$\text{tonf} \times 1\cdot016 = \text{kgf}$	(cgs)
	$\text{tonf} \times 1\cdot016 = \text{tonnef}$	(cgs)
Proof Stress.	$\text{tonf/in}^2 \times 1\cdot575 = \text{kgf/mm}^2$	(cgs)

BOLTS AND SCREWS (CGS)

$$\text{Proof Load (tonnef)} = \frac{\text{Stress Area (mm}^2) \times \text{Proof Stress (kgf/mm}^2)}{1\ 000}$$

NUTS (CGS)

$$\text{Proof Load (tonnef)} = \frac{\text{Stress Area (mm}^2) \times \text{Min. Tensile Strength of bolt/(kgf/mm}^2)}{1\ 000}$$

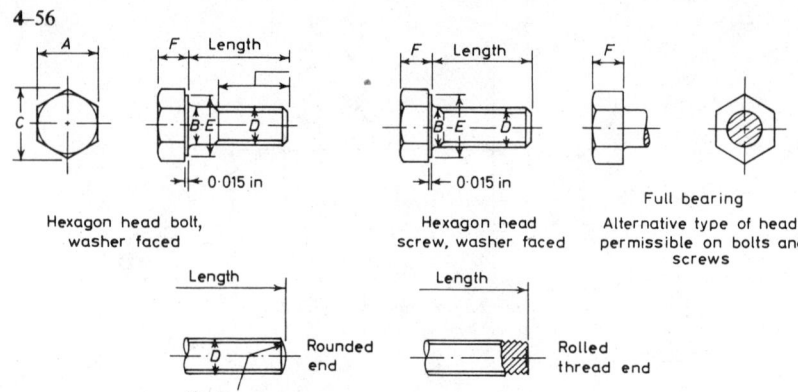

Hexagon head bolt, washer faced

Hexagon head screw, washer faced

Full bearing
Alternative type of head permissible on bolts and screws

Rounded end

1¼ D rad. approx

Rolled thread end

Alternative types of end permissible on bolts and screws

Figure 4.15. Dimensions for unified hexagon head bolts and screws, normal series

Table 4.47 UNIFIED HEXAGON HEAD BOLTS AND SCREWS, NORMAL SERIES (see Figure 4.15)

Dimensions in inches

Nominal size D	Number of threads per inch		Diameter of unthreaded portion of shank B		Width across flats A	Width across corners C	Diameter of washer face E		Thickness of head F
	UNC	UNF	max	min	max	max	max	min	max
¼	20	28	0·250 0	0·246 5	0·437 5	0·505	0·421	0·411	0·163
5⁄16	18	24	0·312 5	0·309 0	0·500 0	0·577	0·483	0·473	0·211
3⁄8	16	24	0·375 0	0·371 5	0·562 5	0·650	0·545	0·535	0·243
7⁄16	14	20	0·437 5	0·433 5	0·625 0	0·722	0·605	0·595	0·291
½	13	20	0·500 0	0·496 0	0·750 0	0·866	0·730	0·720	0·323
9⁄16	12	18	0·562 5	0·558 5	0·812 5	0·938	0·792	0·782	0·371
5⁄8	11	18	0·625 0	0·619 0	0·937 5	1·083	0·918	0·908	0·403
¾	10	16	0·750 0	0·744 0	1·125 0	1·300	1·100	1·090	0·483
7⁄8	9	14	0·875 0	0·867 0	1·312 5	1·515	1·285	1·275	0·563
1	8	12	1·000 0	0·992 0	1·500 0	1·732	1·473	1·463	0·627
1⅛	7	—	1·125 0	1·117 0	1·687 5	1·948	1·641	1·625	0·718
1¼	7	—	1·250 0	1·242 0	1·875 0	2·165	1·813	1·797	0·813
1⅜	6	—	1·375 0	1·365 0	2·062 5	2·382	2·001	1·985	0·878
1½	6	—	1·500 0	1·490 0	2·250 0	2·598	2·188	2·172	0·974
1¾	5	—	1·750 0	1·740 0	2·625 0	3·031	2·543	2·527	1·134
2	4½	—	2·000 0	1·990 0	3·000 0	3·464	2·918	2·902	1·263

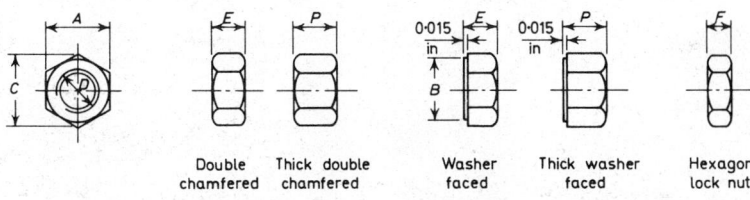

Double chamfered · Thick double chamfered · Washer faced · Thick washer faced · Hexagon lock nut

Alternative types of hexagon ordinary and thick ordinary nuts permissible

Figure 4.16. Dimensions for unified hexagon ordinary nuts, thick ordinary nuts and locknuts, normal series

Table 4.48 UNIFIED HEXAGON ORDINARY NUTS, THICK ORDINARY NUTS AND LOCK NUTS, NORMAL SERIES
Dimensions in inches

Nominal size, D	Width across flats, A	Width across corners, C	Diameter of washer face, B		Thickness		
					Ordinary nuts, E	Thick ordinary nuts, P	Locknuts, F
	max	max	max	min	max	max	max
$\frac{1}{4}$	0·437 5	0·505	0·421	0·411	0·224	0·286	0·161
$\frac{5}{16}$	0·500 0	0·577	0·483	0·473	0·271	0·333	0·192
$\frac{3}{8}$	0·562 5	0·650	0·545	0·535	0·333	0·411	0·224
$\frac{7}{16}$	0·687 5	0·794	0·668	0·658	0·380	0·458	0·255
$\frac{1}{2}$	0·750 0	0·866	0·730	0·720	0·442	0·567	0·317
*$\frac{9}{16}$	0·875 0	1·010	0·855	0·845	0·489	0·614	0·349
$\frac{5}{8}$	0·937 5	1·083	0·918	0·908	0·552	0·724	0·380
$\frac{3}{4}$	1·125 0	1·300	1·100	1·090	0·651	0·822	0·432
$\frac{7}{8}$	1·312 5	1·515	1·285	1·275	0·760	0·916	0·494
1	1·500 0	1·732	1·473	1·463	0·874	1·015	0·562
$1\frac{1}{8}$	1·687 5	1·948	1·641	1·625	0·989	1·176	0·629
$1\frac{1}{4}$	1·875 0	2·165	1·813	1·797	1·087	1·275	0·744
*$1\frac{3}{8}$	2·062 5	2·382	2·001	1·985	1·197	1·400	0·806
$1\frac{1}{2}$	2·250 0	2·598	2·188	2·172	1·311	1·530	0·874
$1\frac{3}{4}$	2·625 0	3·031	2·543	2·527	1·530	—	0·999
2	3·000 0	3·464	2·918	2·902	1·754	—	1·129

In the case of the $\frac{7}{16}$ in and $\frac{9}{16}$ in nuts, the width across flats is $\frac{1}{16}$ in larger than that of the corresponding bolt heads. This is in accordance with the American Standards from which these sizes are derived.
*To be dispensed with wherever possible.

Table 4.49 MECHANICAL PROPERTIES OF FINISHED STEEL HEXAGON HEAD BOLTS AND SCREWS AND HEXAGON NUTS

	Properties of finished bolts						
				Minimum Izod impact value			
			Minimum elongation on a gauge length $4\sqrt{area}$	*Diameter or width across flats*			*Brinell hardness numbers*
Grade	*Minimum tensile strength*	*Stress for proof load*		*Up to $\frac{3}{4}$ in*	*Over $\frac{3}{4}$ in up to and including $1\frac{1}{8}$ in*	*Over $1\frac{1}{8}$ in*	
	tonf/in^2	tonf/in^2	%	ft lbf	ft lbf	ft lbf	
							HB10/3000
A	28	—	10/14	—	—	—	—
B	28	—	17	—	—	—	—
P	35	20	15	20	15	10	152/240
S	50	38	18	35	30	20	223/310
T	55	41	16	35	30	20	248/335
V	65	49·5	14	35	30	20	293/370
X	75	60	12	15	10	15	341/410

Table 4.50 PROOF LOADS FOR BOLTS AND SCREWS

Proof load, ton

Nominal size of bolt or screw (in)	Stress area in² UNC	Stress area in² UNF	Grade P 20 tonf/in² UNC	Grade P 20 tonf/in² UNF	Grade S 38 tonf/in² UNC	Grade S 38 tonf/in² UNF	Grade T 41 tonf/in² UNC	Grade T 41 tonf/in² UNF	Grade V 49·5 tonf/in² UNC	Grade V 49·5 tonf/in² UNF	Grade X 60 tonf/in² UNC	Grade X 60 tonf/in² UNF
1/4	0·032 4	0·0036 8	0·648	0·736	1·231	1·398	1·328	1·509	1·603	1·822	1·944	2·208
5/16	0·053 2	0·0058 7	1·064	1·174	2·021	2·230	2·181	2·406	2·633	2·905	3·192	3·522
3/8	0·078 6	0·0088 6	1·572	1·772	2·986	3·367	3·222	3·633	3·890	4·385	4·716	5·316
7/16	0·107 8	0·119 8	2·156	2·396	4·097	4·552	4·420	4·911	5·335	5·929	6·468	7·188
1/2	0·143 8	0·161 2	2·876	3·224	5·466	6·127	5·896	6·610	7·119	7·980	8·628	9·672
9/16	0·184	0·205	3·68	4·10	6·992	7·791	7·544	8·407	9·107	10·15	11·04	12·30
5/8	0·229	0·258	4·58	5·16	8·702	9·804	9·389	10·58	11·33	12·77	13·74	15·48
3/4	0·338	0·375	6·76	7·50	12·84	14·25	13·86	15·38	16·73	18·57	20·28	22·50
7/8	0·467	0·513	9·34	10·26	17·74	19·49	19·14	21·03	23·12	25·39	28·02	30·78
1	0·612	0·667	12·24	13·34	23·26	25·34	25·09	27·35	30·30	33·01	36·72	40·02
1 1/8	0·771	—	15·42	—	29·30	—	31·61	—	38·17	—	46·26	—
1 1/4	0·978	—	19·56	—	37·16	—	40·10	—	48·41	—	58·68	—
1 3/8	1·166	—	23·32	—	44·31	—	47·80	—	57·72	—	69·96	—
1 1/2	1·418	—	28·36	—	53·89	—	58·15	—	70·20	—	85·08	—
1 3/4	1·92	—	38·4	—	72·97	—	78·72	—	95·04	—	115·2	—
2	2·52	—	50·4	—	95·76	—	103·3	—	124·7	—	151·2	—

Stress area × proof stress = proof load

Table 4.51 PROOF LOADS FOR NUTS

Nominal size of nut (in)	Stress area of bolts in²		Grade 0 nuts — For use with bolts of grades A, B, P — Min. tensile of grade P bolts 35 tonf/in²		Grade 1 nuts — For use with bolts of grade S — Min. tensile of grade S bolts 50 tonf/in²		Grade 3 nuts — For use with bolts of grade T — Min. tensile of grade T bolts 55 tonf/in²		Grade 5 nuts — For use with bolts of grades V, X — Min. tensile of grade X bolts 75 tonf/in²	
			Proof load, ton							
in	UNC	UNF	UNC	UNF	UNC	UNF	UNC	UNF	UNC	UNF
$\frac{1}{4}$	0·032 4	0·036 8	1·134	1·288	1·620	1·840	1·782	2·024	2·430	2·760
$\frac{5}{16}$	0·053 2	0·058 7	1·862	2·054	2·660	2·935	2·926	3·228	3·990	4·402
$\frac{3}{8}$	0·078 6	0·088 6	2·751	3·101	3·930	4·430	4·323	4·873	5·895	6·645
$\frac{7}{16}$	0·107 8	0·119 8	3·773	4·193	5·390	5·990	5·929	6·589	8·085	8·985
$\frac{1}{2}$	0·143 8	0·161 2	5·034	5·643	7·190	8·060	7·911	8·869	10·78	12·09
$\frac{9}{16}$	0·184	0·205	6·440	7·176	9·200	10·25	10·12	11·28	13·80	15·37
$\frac{5}{8}$	0·229	0·258	8·015	9·031	11·45	12·90	12·60	14·19	17·17	19·35
$\frac{3}{4}$	0·338	0·375	11·83	13·12	16·90	18·75	18·59	20·63	25·35	28·12
$\frac{7}{8}$	0·467	0·513	15·97	17·96	23·35	25·65	25·68	28·21	35·02	38·47
1	0·612	0·667	21·42	23·34	30·60	33·35	33·67	36·68	45·90	50·02

Stress area × min. tensile strength of bolt = proof load of nuts

Table 4.52 COMPARISON OF HEAD SIZES ACROSS FLATS

Dimensions in inches

Dia.	ISO metric (BS 3692) max	ISO metric (BS 3692) min	Dia.	ISO unified inch (BS 1768) max	ISO unified inch (BS 1768) min	B.S.W. (BS 1083) max	B.S.W. (BS 1083) min
M6	10·0	9·78	$\frac{1}{4}$	0·437 5	0·430 5	0·445	0·438
(0·236 2)	(0·393 4)	(0·385)					
M8	13·0	12·73	$\frac{5}{16}$	0·500 0	0·493 0	0·525	0·518
(0·314 9)	(0·511 8)	(0·501)					
M10	17·0	16·73	$\frac{3}{8}$	0·562 5	0·554 5	0·600	0·592
(0·393 7)	(0·669 2)	(0·658)					
			$\frac{7}{16}$	0·625 0	0·617 0	0·710	0·702
M12	19·0	18·67	$\frac{1}{2}$	0·750 0	0·742 0	0·820	0·812
(0·472 4)	(0·748 0)	(0·735)					
M14	22·0	21·67	$\frac{9}{16}$	0·812 5	0·804 5	0·920	0·912
(0·551 2)	(0·866 1)	(0·853)					
M16	24·0	23·67	$\frac{5}{8}$	0·937 5	0·929 5	1·010	1·000
(0·629 9)	(0·944 8)	(0·931)					
M20	30·0	29·67	$\frac{3}{4}$	1·125 0	1·115 0	1·200	1·190
(0·787 4)	(1·181 1)	(1·168)					
M22	32·0	31·61	$\frac{7}{8}$	1·312 5	1·300 5	1·300	1·288
(0·866 1)	(1·259 8)	(1·244)					
M24	36·0	35·38	1	1·500 0	1·488 0	1·480	1·468
(0·944 8)	(1·417 32)	(1·393)					
M30	46·00	45·38	$1\frac{1}{8}$	1·687 5	1·657 5	1·670	1·640
(1·181 1)	(1·811 0)	(1·786 6)					
M36	55·00	54·26	$1\frac{1}{4}$	1·875 0	1·830 0	1·860	1·815
(1·417 3)	(2·165 3)	(2·136 2)	$1\frac{3}{8}$	2·062 5	2·017 5		
M42	65·00	64·26	$1\frac{1}{2}$	2·250 0	2·205 0	2·220	2·175
(1·653 5)	(2·559 0)	(2·529 9)					

Inch sizes in brackets.
Different spanner required for each thread type.

SELF-TAPPING SCREWS AND METALLIC DRIVE SCREWS

Types of screws and type designations

The various types of thread forming, thread cutting and drive screw covered by BS 4174 are shown in Table 4.53.

Thread forming screws. These screws are for application in materials which allow sufficient plastic deformation to enable the thread to be formed by displacement without the removal of any material.

Type AB. These screws have widely spaced threads and gimlet points. The thread pitches are the same as those of Type B. They are primarily intended for use in light sheet metal, metal-clad and resin-impregnated plywood, soft plastics, etc. Type AB is suitable for applications in which Type A has previously been used.

Type B. These screws have widely spaced threads and blunt, slightly tapered points. These screws are primarily intended for use in light and heavy sheet metal, non-ferrous castings, soft plastics, metal-clad and resin-impregnated plywood, etc.

Type A. These screws have widely spaced threads and gimlet points. The pitch is coarser than for Types AB and B. They are primarily intended for use in light sheet metal, metal-clad and resin-impregnated plywood, soft plastics, etc. This type of screw is now obsolescent and it is recommended that Type AB be substituted where Type A screws are now specified.

Table 4.53 TYPES OF SCREWS AND TYPE DESIGNATIONS FOR TAPPING SCREWS AND METALLIC DRIVE SCREWS

Thread contour	Type of screw	Type designation for tapping screws and metallic drive screws
	Thread forming	AB
	Thread forming	B (formerly Z)
	Thread cutting	T (formerly 23)
	Thread cutting	Y
	Thread cutting	D (formerly 1)
	Thread cutting	BT (formerly 25)

REFERENCES

1. BS 21
   Pipe threads (1957)*
2. BS 93
   British Association (BA) screw threads (1951)
3. BS 1580
   Unified screw threads (1962)
4. BS 1768
   Unified precision hexagon bolts, screws and nuts, normal series (1963)
5. BS 1769
   Unified black hexagon bolts, screws and nuts, heavy series (1951)

6. BS 2708
   Unified black square and hexagon bolts, screws and nuts, normal series (1956)
7. BS 2779
   Fastening threads of BSP sizes (1956)*
8. BS 3580
   Guide to design considerations on the strength of screw threads (1964)
9. BS 3643
   ISO metric screw threads. Part 1 Thread data and standard thread series (1963); Part 2 Limits and tolerances for coarse pitch series threads (1966)*; Part 3 Limits and tolerances for fine pitch threads (1967)
10. BS 3692
    ISO metric precision hexagon bolts, screws and nuts (1967)
11. BS 4174
    Self-tapping screws and metallic drive screws (1972)
12. BS 4190
    ISO metric black hexagon bolts, screws and nuts (1967)
13. BS 0000
    ISO miniature screw threads, metric series
14. ISO/R7
    Pipe threads for gas list tubes and screwed fittings (1954)
15. ISO/R68
    ISO basic profile for triangular screw threads (1958)
16. ISO/R228
    Pipe threads where pressure-tight joints are not made on the threads (1961)
17. ISO/R261
    ISO metric screw threads. General plan (1962)
18. ISO/R262
    ISO metric screw threads for screws, bolts and nuts, diameter range 6–39 mm (1962)
19. ISO/R263
    ISO inch screw threads. General plan and selection for screws, bolts and nuts (1962)
20. ISO/R898
    Mechanical properties of fasteners. Part 1. Bolts screws and studs (1968); Part 2. Nuts with specified proof load values (1969); Part 3. Marking (1969)
21. ISO/R1501
    ISO miniature screw threads (1971)
22. NBS Handbook H 28. Screw thread standards for general services, Part I (1957)

*These standards are under revision.

STANDARD SIZES OF MATERIALS

G. R. DARBY

WIRE GAUGE SIZES

Introduction

Many wire gauge standards have been used in the past in the UK but the principal ones have been the Standard Wire Gauge (SWG), sometimes referred as the British Imperial Wire Gauge and the Birmingham Gauge (BG).

The SWG and BG series were legalised for use in the UK in 1883 and 1914 respectively, although both orders ceased to have any legal enforcement after 31 January 1964. Table 4.54 summarises the wire gauge systems used throughout the world.

ISO metric wire series

With the advent of metrication, the opportunity has been taken of not only agreeing an International series of wire, sheet and strip sizes but also to adopt a preferred series of increments based on the R10, R20 and R40 preferred numbers in ISO Recommendation R3[1]. The basic diameters for wires are shown in ISO Recommendation R388[2] and BS 4391[3] and cover the size range 0·020–25 mm.

These wire sizes are shown in three columns headed R10, R20 and R40, each giving an increasing number of basic sizes. In selecting sizes, preference should be given to the R10, R20 and R40 series in that order. In particular, the use of the R40 series is to be avoided except where application requires fine differentiation of sizes. The method of designating such thicknesses or diameters is to state the basic size in millimetres followed, if desired, by the letter U to indicate that this size belongs to the ISO metric series, for example 0·020 U. Table 4.55 lists the metric sizes and compares these with the obsolete SWG and BG series.

Steel wire rope sizes

Wire diameters for metric ropes are in accordance with the basic metric wire sizes. These millimetre sizes were introduced in May 1968. In addition to the diameter, the other principal characteristics required for wire ropes as given on the relevant British Standards[5, 6, 7, 8] are:

Wire tensile strength in kg/mm²
Breaking load   in kg/mm²
Weight   in kg/100 m

Breaking strengths of wires can be obtained from Tables of Physical Properties of Steel Wire in Metric Units[9].

METRIC DRILL SIZES

The British Standard metric drill sizes are shown in Table 4.56 together with tapping drill sizes and a correlation to the old number and letter drill gauge sizes.

The tapping drill sizes are based on providing a 70% thread engagement between bolt and nut.

Table 4.54 SUMMARY OF WIRE GAUGE SYSTEMS

Name	Abbreviation	Application
Alhoff & Muller Music Wire Gauge	—	Music wire.
American Screw & Wire Co. Music Wire Gauge	—	Music wire, spring wire.
American Zinc Gauge	—	Zinc sheet.
ASA B32.1[4]	—	Uncoated thin flat metals (USA).
Birmingham Gauge	BG	⎰ Iron hoop and strip. ⎱ Saw blade material. ⎰ Steel sheet.
Birmingham Wire Gauge	BWG	Iron and steel telephone wire.
Birmingham Wire Gauge for Silver and Gold	—	Silver and gold.
Brown & Sharpe Wire Gauge	B & S	Non-ferrous sheet and strip.
Brunton Music Wire Gauge	—	Music wire.
Card Wire Gauge	CWG	Fibre yarns.
Continental Zinc Gauge	—	Sheet and strip.
English Music Wire Gauge	—	Music wire.
English Zinc Wire Gauge	—	Zinc sheet.
Felten & Guilleaume Music Wire Gauge	—	Music wire.
German Sheet Gauge	—	Sheet metal.
Instrument Wire Gauge	—	Instrument wire.
ISO Metric Preferred Series (R388)	—	All applications.
Junge de Paris	—	Copper wires.
Lancashire Pinion Wire Gauge	LPG	Cast steel drill rods.
Mathieson & Hegeler Gauge	M and H	Zinc sheet.
Millimetre Wire Gauge	—	Steel wire.
National Wire Gauge	NWG	Steel wire.
Poehlimann Music Wire Gauge	—	Music wire
Printers Wire Gauge	—	Printers wire.
Roebling Wire Gauge	—	Steel wire.
Sewing and Darning Needle Series	—	Sewing and darning needle.
Standard Wire Gauge	—	Steel and non-ferrous wire and strip. Steel tube thickness.
Stitching Wire Gauge	—	—
Stubs Iron Wire Gauge	—	Iron wire.
Stubs Steel Wire Gauge	—	Steel drill rod.
Tinplate	—	Finished tinplate.
Twist Drill and Steel Wire Gauge	—	Drill rod.
US Galvanised Sheet Gauge	GSG	Galvanised sheet.
US Manufacturers' Standard Gauge	MSG	Iron and steel sheets and plates.
US Standard Gauge	USG	Stainless steel.
US Steel Wire Gauge	USSWG	Steel wire.
Washburn and Moen Gauge	WG	Steel wire.
Warrington Wire Gauge	—	Steel wire.
Yorkshire Wire Gauge	—	Steel wire.

METRIC BAR SIZES

Metric bar sizes, based on BS 4229 are given in Table 4.57 for ferrous materials and Table 4.58 in respect of non-ferrous materials.

Table 4.55 COMPARISON OF SWG, BG AND ISO METRIC WIRE SERIES

	Inch based systems					ISO metric series			
	SWG series			BG series		BS 4391 R.10	ISO R.20	Metric preferred series	
SWG	in	mm	BG	in	mm	mm	mm	R.40 mm	inch equivalent
						0·020			0·000 79
52	0·000 8	0·020 3							
								0·021	0·000 83
							0·022		0·000 87
								0·024	0·000 94
			52	0·000 95	0·024 13				
						0·025			0·000 98
50	0·001 0	0·254							
								0·026	0·001 02
			51	0·001 07	0·027 18				
							0·028		0·001 10
								0·030	0·001 18
49	0·001 2	0·030 5	50	0·001 2	0·030 5				
						0·032			0·001 26
								0·034	0·001 34
			49	0·001 35	0·038 6				
						0·036			0·001 42
								0·038	0·001 50
			48	0·001 52	0·038 6				
						0·040			0·001 58
48	0·001 6	0·040 6							
								0·042	0·001 65
			47	0·001 70	0·043				
							0·045		0·001 77
								0·048	0·001 89
			46	0·001 92	0·048 8				
						0·050			0·001 97
47	0·002 0	0·050 8							
								0·053	0·002 09
			45	0·002 15	0·057 2				
							0·056		0·002 20
								0·060	0·002 36
46	0·002 4	0·061 0							
			44	0·002 42	0·061 5				
						0·063			0·002 48
								0·067	0·002 64
			43	0·002 72	0·069 1				
45	0·002 8	0·071					0·071		0·002 80
								0·075	0·002 95
			42	0·003 06	0·077 7				
						0·080			0·003 15
44	0·003 2	0·081 3							
								0·085	0·003 35
			41	0·003 43	0·087 1				
							0·090		0·003 54
43	0·003 6	0·091 4							
								0·095	0·003 74
			40	0·003 86	0·098				
						0·100			0·003 94
42	0·004 0	0·101 6							
								0·106	0·004 17
			39	0·004 3	0·109 2				

Table 4.55—*continued*

	Inch based systems					ISO metric series			
SWG	SWG series in	mm	BG	BG series in	mm	BS 4391 R.10 mm	ISO R.20 mm	Metric preferred series R.40 mm	inch equivalent
41	0·004 0	0·111 8							
						0·112			0·004 41
								0·118	0·004 65
40	0·004 8	0·121 9	38	0·004 8	0·129 1				
							0·125		0·004 92
39	0·005 2	0·132 1						0·132	0·005 20
			37	0·005 4	0·137 2				
						0·140			0·005 52
								0·150	0·005 91
38	0·006 0	0·152 4							
			36	0·006 1	0·155 0				
						0·160			0·006 30
								0·170	0·006 69
37	0·006 5	0·172 7							
			35	0·006 9	0·175 3				
							0·180		0·007 09
								0·190	0·007 48
36	0·007 6	0·193 0							
			34	0·007 7	0·195 6				
						0·200			0·007 87
								0·212	0·008 35
35	0·008 4	0·212 4							
			33	0·008 7	0·221 0				
							0·224		0·008 82
34	0·009 2	0·233 7							
								0·236	0·009 2
			32	0·009 8	0·248 9				
						0·250			0·009 84
33	0·010 0	0·254 0							
								0·265	0·010 43
32	0·010 8	0·274 3							
			31	0·011 0	0·278 4				
							0·280		0·011 02
31	0·011 6	0·294 6							
								0·300	0·011 81
			30	0·012 3	0·312 4				
30	0·012 4	0·315 0				0·315			0·012 40
29	0·013 6	0·345 4							
								0·335	0·013 19
			29	0·013 9	0·353 1				
							0·355		0·013 98
								0·375	0·014 76
28	0·014 8	0·375 9							
			28	0·015 625	0·396 2				
						0·400			0·015 75
27	0·016 4	0·416 6							
								0·425	0·016 73
			27	0·017 45	0·444 2				
							0·450		0·017 72
26	0·018	0·457 2							
								0·475	0·018 70
			26	0·019 61	0·497 8				
						0·500			0·019 69

continued

Table 4.55—*continued*

	Inch based systems					ISO metric series			
SWG	SWG series in	mm	BG	BG series in	mm	BS 4391 R.10 mm	ISO R.20 mm	Metric preferred series R.40 mm	inch equivalent
25	0·020	0·508 0							
								0·530	0·020 87
24	0·022	0·558 8							
			25	0·022 04	0·569 0				
							0·560		0·022 05
								0·600	0·023 62
23	0·024	0·609 6							
			24	0·024 76	0·628				
						0·630			0·024 80
								0·670	0·026 38
			23	0·027 82	0·706 8				
							0·710		0·027 95
22	0·028	0·711 2							
								0·750	0·029 53
			22	0·031 25	0·793 7				
						0·800			0·031 50
21	0·032	0·812 8							
								0·850	0·033 46
			21	0·034 9	0·886 5				
							0·900		0·035 43
20	0·036	0·914 4							
								0·950	0·037 40
			20	0·039 2	0·995 7				
						1·000			0·039 4
19	0·040	1·016							
								1·06	0·041 7
			19	0·044	1·117 6				
							1·12		0·044 1
								1·18	0·046 5
18	0·048	1·219							
						1·25			0·049 2
			18	0·049 5	1·257 3				
								1·32	0·052 0
							1·40		0·055 1
			17	0·055 6	1·412 2				
17	0·056	1·422							
			16	0·062 5	1·587 5				
						1·60			0·063 0
16	0·064	1·626							
								1·70	0·066 9
			15	0·069 5	1·775 5				
							1·80		0·070 9
15	0·072	1·829							
								1·90	0·074 8
			14	0·078 5	1·993 9				
						2·00			0·078 7
14	0·080	2·032							
								2·12	0·083 5
			13	0·088 2	2·240 3		2·24		0·088 2
13	0·092	2·337							
								2·36	0·092 9
						2·50			0·098 4
			12	0·099 1	2·517 1				

Table 4.55—*continued*

	Inch based systems					ISO metric series			
SWG	SWG series in	mm	BG	BG series in	mm	BS 4391 R.10 mm	ISO R.20 mm	Metric preferred series R.40 mm	inch equivalent
12	0·104	2·642						2·65	0·104 3
							2·80		0·110 2
			11	0·111 3	2·820 0				
11	0·116	2·946							
								3·00	0·118 1
						3·15			0·124 0
			10	0·125 0	3·175 0				
10	0·128	3·251							
								3·35	0·131 9
			9	0·139 8	3·540 0		3·55		0·139 8
9	0·144	3·658							
								3·75	0·147 6
			8	0·157 0	3·987 8				
						4·0			0·157 5
8	0·160	4·064							
								4·25	0·167 3
7	0·176	4·470							
			7	0·176 4	4·480 0				
							4·5		0·177 2
								4·75	0·187 0
6	0·192	4·877							
						5·0			0·196 9
			6	0·198 1	5·029 2				
								5·3	0·208 7
5	0·212	5·349							
							5·6		0·220 5
			5	0·222 5	5·650 0				
4	0·236	5·994							
								6·0	0·236 2
						6·30			0·248 0
			4	0·250	6·350 0				
3	0·252	6·400							
								6·70	0·263 8
2	0·276	7·01							
							7·10		0·279 5
			3	0·280 4	7·213 6				
								7·50	0·295 3
1	0·300	7·620							
			2	0·314 7	7·990 0				
							8·00		0·315 0
0	0·324	8·23							
								8·50	0·334 6
2/0	0·348	8·84							
			1	0·355 2	8·970 0				
							9·00		0·354 3
3/0	0·372	9·45							
								9·50	0·374 0
						10·000			0·393 7
			0	0·396 4	10·065				
4/0	0·400	10·16							
								10·6	0·417 3

continued

Table 4.55—*continued*

	Inch based systems					ISO metric series			
SWG	SWG series in	mm	BG	BG series in	mm	BS 4391 R.10 mm	ISO R.20 mm	Metric preferred series R.40 mm	inch equivalent
5/0	0·432	10·97							
							11·2		0·440 9
			2/0	0·445 2	11·303				
6/0	0·464	11·79							
								11·8	0·464 6
						12·5			0·492 1
7/0	0·500	12·7	3/0	0·500	12·700				
								13·2	0·519 7
			4/0	0·541 6	13·75				
							14·0		0·551 8
			5/0	0·588 3	14·95				
								15·0	0·590 6
			6/0	0·625	15·875				
						16·0			0·629 9
			7/0	0·666 6	16·92				
								17·0	0·669 3
			8/0	0·708 3	17·985				
							18·0		0·708 7
								19·0	0·748 0
			9/0	0·750	19·05				
						20·0			0·787 4
			10/0	0·791 7					
			11/0	0·833 3	21·16				
								21·2	0·834 6
			12/0	0·875 0	22·225				
							22·4		0·881 9
			13/0	0·916 7	23·285				
								23·6	0·929 1
			14/0	0·958 3	24·34				
						25·0			0·984 3
			15/0	1·000	25·4				

Table 4.56   DRILL SIZES

BS 328[10] BS Drill Metric series mm	BS 1157[11] Screw thread series Tapping	BS 328A[12] Old drill gauge and letter size Designation	Decimal in	Equivalent mm
0·20				
0·22				
0·25				
0·28				
0·30				
0·32				
0·35		80	0·013 5	0·342 9
0·38		79	0·014 5	0·368 3
0·40		78	0·016 0	0·406 5
0·42				
0·45		77	0·018 0	0·457 2
0·48				

Table 4.56—*continued*

BS 328[10] BS Drill Metric series mm	BS 1157[11] Screw thread series Tapping	BS 328A[12] Old drill gauge and letter size Designation	Decimal in	Equivalent mm
0·50		76	0·020 0	0·508
0·52				
0·55				
0·58		74	0·022 5	0·571 5
0·60	16 BA	73	0·024 0	0·609 6
0·62				
0·65		72	0·025 0	0·635 0
		71	0·026 0	0·660 4
0·68				
0·70	15 BA	70	0·028 0	0·711 2
0·72				
0·75	M1	69	0·029 2	0·737 0
0·78	14 BA			
0·80				
0·82		67	0·032 0	0·812 8
0·85		66	0·033 0	0·838 2
0·88				
0·90		65	0·035 0	0·889 0
0·92		64	0·036 0	0·914 4
0·95		63	0·037 0	0·939 8
0·98	13 BA	62	0·038 0	0·965 2
1·00		61	0·039 0	0·990 6
		60	0·040 0	1·016 0
1·05	12 BA	59	0·041 0	1·041 4
		58	0·042 0	1·066 8
1·10		57	0·043 0	1·092 2
1·15				
1·20	11 BA			
1·25	M1·6, 0/30			
1·30		55	0·052 0	1·320 8
1·35				
1·40	10 BA	54	0·055 0	1·397 0
1·45				
1·50				
1·55	9 BA, 1/64, 1/72			
1·60	M2	52	0·063 5	1·612 9
1·65				
1·70		51	0·067 0	1·701 8
1·75				
1·80	8 BA	50	0·070 0	1·778 0
1·85	2/56	49	0·073 0	1·854 2
1·90	2/64			
1·95		48	0·076 0	1·930 4
2·00		47	0·078 5	1·993 9
2·05	7 BA	46	0·081 0	2·057 4
2·10	3/48	45	0·082 0	2·082 8
2·15	3/56			
2·20		44	0·086 0	2·184 4
2·25		43	0·089 0	2·260 6
2·30	6 BA			
2·35	4/40			
2·40	4/48			
2·45		41	0·096 0	2·438 4

continued

Table 4.56—*continued*

BS 328[10] BS Drill Metric series mm	BS 1157[11] Screw thread series	BS 328A[12] Old drill gauge and letter size		
	Tapping	Designation	Decimal in	Equivalent mm
2·50	M3	40	0·098 0	2·489 2
2·55	1/8 BSW	39	0·099 5	2·527 3
2·60		38	0·101 5	2·576 0
2·65	5 BA, 5/40	37	0·104 0	2·641 6
2·70	5/44	36	0·106 5	2·700 0
2·75				
2·80		35	0·110 0	2·794 0
		34	0·111 0	2·819 4
2·85	6/32	33	0·113 0	2·870 2
2·90				
2·95	6/40	32	0·116 0	2·946 4
3·00	4 BA	31	0·120 0	0·048 0
3·10				
3·20				
3·30	M4	30	0·128 5	3·260 0
3·40	3 BA			
3·50	8/32; 8/36	29	0·136 0	3·454 4
3·60				
3·70	3/16 BSW	27	0·144 0	3·657 6
		26	0·147 0	3·733 8
3·80		25	0·149 5	3·790 0
3·90	3/16 BSF, 10/24	24	0·152 0	3·860 8
		23	0·154 0	3·911 6
4·00	2 BA	22	0·157 0	3·987 8
		21	0·159 0	4·038 5
4·10	10/32	20	0·161 0	4·089 4
4·20	M5	19	0·166 0	4·216 4
4·30		18	0·169 5	4·300
4·40		17	0·173 0	4·394 2
4·50	1 BA, 12/64	16	0·177 0	4·749 8
4·60	7/32 BSF	15	0·180 0	4·572 0
		14	0·182 0	4·622 8
4·70	12/28	13	0·185 0	4·699 0
4·80		12	0·189 0	4·800 6
4·90		11	0·191 0	4·851 4
		10	0·193 5	4·914 4
5·00	M6	9	0·196 0	4·978 4
5·10	1/4 BSW, OBA	8	0·199 0	5·054 6
		7	0·201 0	5·105 4
5·20	1/4 UNC	6	0·204 0	5·181 6
		5	0·205 5	5·219 0
5·30	1/4 BSF	4	0·209 0	5·308 6
5·40		3	0·213 0	5·410 2
5·50	1/4 UNF			
5·60		2	0·221 0	5·613 4
5·70				
5·80		1	0·228 0	5·791 2
5·90				
6·00	M7	B	0·238 0	6·045 2
6·10		C	0·242 0	6·146 8
6·20		D	0·246 0	6·248 4
6·30				
6·40				

Table 4.56—*continued*

BS 328[10] BS Drill Metric series mm	BS 1157[11] Screw thread series	BS 328A[12] Old drill gauge and letter size		
	Tapping	Designation	Decimal in	Equivalent mm
6·50	5/16 BSW	F	0·257 0	6·527 8
6·60	5/16 UNC	G	0·261 0	6·629 4
6·70				
6·80	M8, 5/16 BSF			
6·90	5/16 UNF	I	0·272 0	6·908 8
7·00		J	0·277 0	7·035 8
7·10				
7·20				
7·30				
7·40		L	0·290 0	7·366 0
7·50		M	0·295 0	7·493 0
7·60				
7·70		N	0·302 0	7·670 8
7·80	M9			
7·90	3/8 BSW			
8·00	3/8 UNC	O	0·316 0	8·026 4
8·10				
8·20		P	0·323 0	8·204 2
8·30	3/8 BSF			
8·40	1/8 BSP	Q	0·332 0	8·432 8
8·50	M10, 3/8 UNF			
8·60		R	0·339 0	8·610 6
8·70				
8·80		S	0·348 0	8·839 2
8·90				
9·00				
9·10		T	0·358 0	9·093 2
9·20				
9·30	7/16 BSW	U	0·368 0	9·347 2
9·40	7/16 UNC			
9·50	M11			
9·60				
9·70	7/16 BSF			
9·80		W	0·386 0	9·804 4
9·90	7/16 UNF			
10·00				
10·10		X	0·387 0	10·083 8
10·20	M12			
10·30		Y	0·404 0	10·261 6
10·40				
10·50	1/2 BSW	Z	0·413 0	10·490 2
10·60				
10·70				
10·80	1/2 UNC			
10·90				
11·00				
11·10	1/2 BSF			
11·20	1/4 BSP			
11·30				
11·40				
11·50	1/2 UNF			
11·60				
11·70				

continued

Table 4.56—*continued*

BS 328[10] BS Drill Metric series mm	BS 1157[11] Screw thread series Tapping	BS 328A[12] Old drill gauge and letter size		
		Designation	Decimal in	Equivalent mm
11·80				
11·90				
12·00	M14			
12·00	9/16 BSW			
12·20	9/16 UNC			
12·30				
12·40				
12·50				
12·60	9/16 BSF			
12·70				
12·80				
12·90	9/16 UNF			
13·00				
13·10				
13·20				
13·30				
13·40				
13·50	5/8 BSW, 5/8 UNC			
13·60				
13·70				
13·80				
13·90				
14·00	M16, 5/8 BSF			
14·25				
14·50	5/8 UNF			
14·75	3/8 BSP			
15·00				
15·25				
15·50	M18			
15·75				
16·00				
16·25	3/4 BSW			
16·50	3/4 UNC			
16·75	3/4 BSF			
17·00				
17·25				
17·50	M20			
17·75				
18·00				
18·25	1/2 BSP			
18·50				
18·75				
19·00	⎰7/8 UNC			
19·25	⎱7/8 BSW			
19·50	M22			
19·75	7/8 BSF			
20·00				
20·25				
20·50				
20·75				
21·00	M24			

Table 4.56—*continued*

BS 328[10] BS Drill Metric series mm	BS 1157[11] Screw thread series Tapping	BS 328A[12] Old drill gauge and letter size Designation	Decimal in	Equivalent mm
21·25				
21·50				
21·75				
22·00	1 BSW			
22·25	1 UNC			
22·50				
22·75	1 BSF			
23·00				
23·25				
23·50	1 UNF			
23·75	3/4 BSP			
24·00				
24·25				
24·50				
24·75	1·1/8 BSW			
25·00	1·1/8 UNC			

Thence in 0·50 mm increments up to and including 50·50 mm dia.
Thence in 1 mm increments from 51–100 mm.
Note. The above table also includes references to the following screw thread systems.
    BA, BSW, BSF in BS 57 and 84 (these are now obsolete) 4/40 etc. BS 1981.
    UNC. BS 1580 (ISO/R68).
    UNF. BS 1580 (ISO/R68).
    M. BS 3643 (ISO/R262).

Table 4.57   METRIC BAR SIZES—FERROUS MATERIALS
(*Based on BS 4229, Part 2*)[10]

Nom. metric size mm	Hot rolled Non-alloy			Bright Non-alloy			Alloy		
	Steel rounds	Steel squares	Steel hexagons	Steel rounds	Steel bars	Steel hexagons	Steel rounds	Steel squares	Steel hexagons
3·2									√
5·0							√	√	
7·0		√					√	√	√
8·0	√	√					√	√	√
10·0	√	√	√				√	√	√
11·0		√	√				√	√	√
12·0	√	√	√		√		√	√	√
13·0		√	√	√	√	√	√	√	√
13·0		√	√	√	√	√	√	√	√
14·0		√	√	√	√	√	√	√	√
15·0		√	√	√	√	√	√	√	√
16·0	√	√	√	√	√	√	√	√	√
17·0		√	√				√	√	√
18·0		√	√	√	√		√	√	√
19·0		√	√	√					
20·0	√	√	√	√	√		√	√	√

continued

4–76

Table 4.57—*continued*

Nom. metric size mm	Hot rolled Non-alloy				Bright		Alloy		
	Steel rounds	Steel squares	Steel hexagons	Steel rounds	Steel bars	Steel hexagons	Steel rounds	Steel squares	Steel hexagons
21·0							√	√	
22·0	√	√	√	√	√	√	√	√	√
23·0							√	√	
24·0		√	√	√		√	√	√	√
25·0	√	√	√	√	√	√	√	√	√
26·0				√			√	√	
27·0			√			√	√	√	√
28·0		√		√	√		√	√	
30·0	√	√	√	√	√	√	√	√	√
32·0	√	√	√	√	√	√	√	√	√
35·0	√	√	√	√	√				
36·0		√				√	√	√	√
38·0		√	√	√					√
40·0	√	√	√	√	√				
41·0			√			√	√	√	√
42·0	√	√		√					
45·0	√	√		√	√				√
46·0						√			
48·0	√		√	√					√
50·0	√	√	√	√	√	√	√	√	√
52·0			√						
55·0	√	√	√	√		√	√	√	√
60·0	√	√	√	√	√	√	√	√	√
65·0	√	√	√	√	√	√	√	√	√
70·0	√	√	√	√	√	√	√	√	√
75·0	√	√	√	√	√	√	√	√	√
80·0	√	√	√	√	√	√	√	√	√
85·0		√	√	√	√	√	√	√	√
90·0	√	√	√	√	√	√	√	√	√
95·0			√	√	√				√
100	√	√		√	√	√	√	√	√
110	√	√		√	√		√	√	
120	√	√		√	√		√	√	
130	√	√		√	√		√	√	
140	√	√		√	√		√	√	
150	√	√		√	√		√	√	
160	√			√	√				
170	√								
180	√			√	√				
190	√								
200	√			√	√				
220	√								
240	√								
260	√								
280	√								
300	√								

√ indicates preferred size selection.

Table 4.58 METRIC BAR SIZES—NON-FERROUS MATERIALS
(based on BS 4229, Part 1)[10]

Metric size mm	Round bar			Square bar			Hexagonal bar		
	Aluminium	Copper	Nickel	Aluminium	Copper	Nickel	Aluminium	Copper	Nickel
3·2							√	√	√
3·0	√	√	√	√	√	√			
4·0	√	√	√	√	√	√	√	√	√
5·0	√	√	√	√	√	√	√	√	√
5·5							√	√	√
6·0	√	√	√	√	√	√			
7·0	√	√	√				√	√	√
8·0	√	√	√	√	√	√	√	√	√
9·0	√	√	√						
10·0	√	√	√	√	√	√	√	√	√
11·0							√	√	√
12·0	√	√	√	√	√	√	√	√	√
13·0							√	√	√
14·0	√	√	√	√	√	√	√	√	√
16·0	√	√	√	√	√	√			
17·0							√	√	√
18·0	√	√	√	√	√	√			
19·0							√	√	√
20·0	√	√	√	√	√	√			
22·0	√	√	√	√	√	√	√	√	√
24·0							√	√	√
25·0	√	√	√	√	√	√			
27·0							√	√	√
28·0	√	√	√						
30·0	√	√	√	√	√	√	√	√	√
32·0	√	√	√	√	√	√	√	√	√
35·0	√	√	√						
36·0				√	√	√	√	√	√
40·0	√	√	√	√	√	√			
41·0							√	√	√
45·0	√	√	√						
46·0				√	√	√	√	√	√
50·0	√	√	√	√	√	√	√	√	√
55·0	√	√	√	√	√	√	√	√	√
60·0	√	√	√	√	√	√	√	√	√
65·0	√	√	√	√		√	√	√	√
70·0	√	√	√	√	√	√	√	√	√
75·0	√		√	√		√	√	√	√
80·0	√	√	√	√	√	√	√	√	√
85·0				√		√	√	√	√
90·0	√	√	√	√	√	√	√	√	√
95·0							√	√	√
100·0	√	√	√	√	√	√	√	√	√
110·0	√		√	√		√			
120·0	√			√		√			
130·0	√			√		√			
140·0	√			√		√			
160·0	√		√	√		√			
180·0	√		√	√		√			
200·0	√			√		√			

√ indicates preferred size selection.

METRIC PLATE SHEET AND STRIP SIZES
(FERROUS AND NON-FERROUS)

Standard thicknesses for sheet and strip

Metric wire sizes are also recommended for sheet and strip thicknesses for both ferrous and non-ferrous products. Proposals are still tentative and are based on a BS Draft for Development DD5[11]. These thicknesses are shown in Tables 4.59 and 4.60.

Table 4.59 STANDARD THICKNESSES FOR STEEL PLATE AND SHEET (mm)

0·50	12	65
0·60	15*	70
0·70	16*	75
0·80	18*	80
0·90	20*	85
1·0	22*	90
1·2	25	95
1·6	28	100
2·0	30	110
2·5	32	115
3·0	35	120
4·0	38	125
5·0	40	130
6·0	45	135
8·0	50	140
10	55	145
	60	150

*These sizes should be regarded as non-preferred.

Table 4.60 STANDARD THICKNESSES FOR NON-FERROUS (mm)

Plate and sheet					
1st choice	2nd choice	1st choice	2nd choice	1st choice	2nd choice
0·25		4·0		40	
0·30		5·0			45
0·40		6·0		50	
0·50		8·0			55
	0·55 (copper only)	10		60	
0·60		12			70
	0·70		14		75
0·80		16		80	
	0·90		18		90
1·0		20		100	
1·2			22		110
1·6		25		120	
2·0			28		130
2·5		30		140	
3·0			35		150

Standard width/length combinations are shown in Tables 4.61 and 4.62.

Table 4.61   WIDTH/LENGTH COMBINATION FOR STEEL PLATES, (mm)

Width	Length
1 000	2 000
1 250	2 500
1 500	3 000
1 500	4 000
1 500	5 000
1 500	6 000
1 500	8 000
1 750	4 000
1 750	5 000
1 750	6 000
1 750	8 000
2 000	4 000
2 000	5 000
2 000	6 000
2 000	8 000
2 000	10 000
2 250	6 000
2 250	8 000
2 250	10 000
2 500	4 000
2 500	5 000
2 500	6 000
2 500	8 000
2 500	10 000

Table 4.62   WIDTH/LENGTH COMBINATIONS FOR NON-FERROUS PLATE AND SHEET (mm)

	Width	Length
Copper and copper alloy	600	1 200
	900	1 800
	1 000	2 000
Light metals and alloys	1 000	2 000
	1 250	2 500
Nickel and nickel alloys	1 000	2 000
	1 000	2 500
	1 000	3 000
	1 200	3 000

REFERENCES

1. ISO/R3
Preferred number, series of preferred numbers (1953)
2. ISO/R388
ISO metric series for basic thicknesses of sheet and diameter of wire (1964)
3. BS 4391
Recommendations for metric basic sizes for metal wire, sheet and strip (1969).
4. ASA B32.1
Preferred thicknesses for uncoated thin flat metal, American National Standards Institute, (1952)
5. BS 302
Wire ropes for cranes, excavators and general engineering purposes (1968, AMD 583–1970)
6. BS 365
Galvanised steel wire ropes for ships (1968)

7. BS 3530
   Small wire ropes (1968)
8. BS 2763
   Round steel wire for ropes (1968)
9. Tables of physical properties of steel wire in metric units, Institute of Iron and Steel Wire Manufacturers, (Nov. 1970)
10. BS 328
    Twist drills and countersinks, Part 1 (1970)
11. BS 1157
    Recommendations for tapping drill sizes (1965)
12. BS 328A
    Twist drills superseding drill gauge and letter sizes (1963)
13. BS 4229
    Recommendations for metric sizes of non-ferrous and ferrous bars. Part 1, Non-ferrous bars, (1967 AMD 831). Part 2, Ferrous bars, (1967 AMD 832)
14. BS DD5
    Draft for development. Recommended metric plate and sheet thicknesses and width-length combinations for metallic materials (1971)

5 METALLURGY AND NON-METALS TECHNOLOGY

METALLURGY AND
NON-METALS TECHNOLOGY

5

5 METALLURGY AND NON-METALS TECHNOLOGY

METALLURGY OF IRON AND STEEL

H. F. TREMLETT

INTRODUCTION

Design engineers are concerned with three main aspects of their materials, apart from cost and delivery. These are:
(a) The mechanical and physical properties as supplied by the manufacturer.
(b) The ability to meet service conditions without failure.
(c) Reliability in respect of properties and performance from beginning to the end of each delivery and from one delivery to the next.

All these three aspects are affected by metallurgical control, in extraction of the metal from the ore, its subsequent forming to shape and the conferring of preferred mechanical properties suited to the particular needs of the application.

Conventional mechanical tests on test samples will reveal the basic properties such as tensile strength, ductility, hardness, toughness etc., of the test sample but the engineer must be aware of the limitations inherent in test sample information, in so far as it may not properly apply to the bulk material which is his concern. These limitations will be discussed in the appropriate sub-section but details of conventional testing methods will be omitted, as these are covered in standard textbooks.

The ability to meet service conditions without failure should follow automatically if the information concerning the conditions and the properties of the material is sufficiently detailed and the design engineer is competent. Unfortunately, from time to time, the occurrence of major failures in bridges, pressure vessels, storage tanks etc., indicates that the correct information may sometimes be lacking or, as happens in welded fabrication, the shop floor practice lacks proper control resulting in dangerous defects. The test sample mechanical properties supplied to the design engineer may well be irrelevant to the safety and performance in service of the finished structure, if it contains defects which are undetected. Where appropriate, these matters, which are of particular interest to design engineers, will be considered in the context of the metallurgy from which they stem and the engineering action necessary to prevent their occurrence.

Reliability in respect of properties and performance in a continuing supply of material is not something which can be covered simply by conventional testing of samples except in a limited range of applications. A degree of variability is inherent in steel products manufactured to the same specification by one steelmaker using the same melting, hot working and heat treatment practices and facilities; the engineer must work within this variability range. Where the steel is to be used in applications not involving welding, the purchase of material from different steelworks to the same specification may well be justified. However, weldability, loosely defined as the relative freedom from welding defects for a standardised welding procedure, is more sensitive to steelmaking factors, e.g. composition range, melting and deoxidation practice, residual element content (copper, arsenic), casting-pit practice and heat treatment than conventional mechanical properties, so that purchasing material from several sources to the same nominal specification will tend to increase the chance of obtaining material outside the engineer's variability range.

Furthermore, it is possible to purchase material where specification control of chemical analysis is restricted to so.few elements that weldability becomes an unknown quantity. An example is mild steel to BS 4360, grade 43A where there is no limit on the manganese content. The use of higher manganese steels (grades 43B, C, D) for notch toughness increases the risk that such steels, which require special welding procedures, will be purchased as mild steel for which the welding shops will apply normal welding procedures. The design engineer should make sure of the reliability of his source of material supply.

Extraction metallurgy of ferrous materials is too wide a subject to be considered in detail here but is covered in brief outline. However, the engineer should note that the conventional mechanical tests used to assess material properties do not necessarily reveal the presence or absence of important properties such as resistance to fatigue failure in service, resistance to brittle fracture in service, reliability in forming and fabrication by welding. The latter may include freedom from cracking, lamination in plates, lamellar tearing in plates, etc. These properties are influenced to a considerable degree by the steelmelting or extraction metallurgy practice. The conventional mechanical tests when applied to steel or iron castings are a more reliable guide. This aspect will receive attention in considering steelmaking practice. The metallurgical principles in forming to shape and in obtaining suitable mechanical properties by heat treatment will then be considered. Foundry practice is dealt with later in this section.

IRON AND STEEL MANUFACTURE (EXTRACTION METALLURGY)

Iron making by blast furnace

Low carbon iron can be obtained by direct reduction of the iron ore but the processes are not as yet generally economic compared with the established method of producing high carbon iron in a blast furnace and then refining this iron to produce steel. Direct reduction processes will not be considered here.

The blast furnace is a vertical cylindrical stack, steel shell, brick lined, tapering to the top (Figure 5.1) with a bosh below the stack, tapering towards the hearth, and a hearth

Table 5.1 MAIN DIMENSIONS OF AMERICAN BLAST FURNACE* (see Figure 5.1)

Foundation	3·429 m thick
Hearth diameter	8.534 m
Hearth depth	3·785 m
Bosh depth	3·480 m
Bosh diameter	9·525 m
Belt depth	2·261 m
Inwall	17·145 m
Throat diameter	6·553 m
Top of foundation to top ring	37·643 m

* No. 3 Furnace; Fairless Works of United States Steel Corporation. *(The Making, Shaping and Treating of Steel, USS Corporation)*

below the bosh. Preheated air is blown into the furnace via tuyeres which encircle the bosh. The hearth collects the molten iron which is run off at intervals. Blast furnaces vary in size; the main dimensions of an American furnace are given in Table 5.1. The furnace is operated and charged continuously once it has been lit. Charging is at the top through a hopper and double bell system, either by skip hoist or bucket hoist. Each charge must be accurately weighed so that the ingredients, coke, iron ore and flux, if used, are maintained in the desired proportions.

The product of the blast furnace, molten iron, is the raw material for the iron foundry and steelmaking plant, see Table 5.2; the former taking it as cold pig iron, the latter either as pig or as partly refined hot metal from a mixer. The composition of the blast furnace product is dependent on the composition of the iron ore, coke and limestone. The iron in the ore body is usually present either as an oxide, hydrated oxide or a carbonate of iron with varying amounts of gangue the ore being either acid or basic in character.

Since the slag must flow from the furnace, the ore may need fluxing unless it is self-

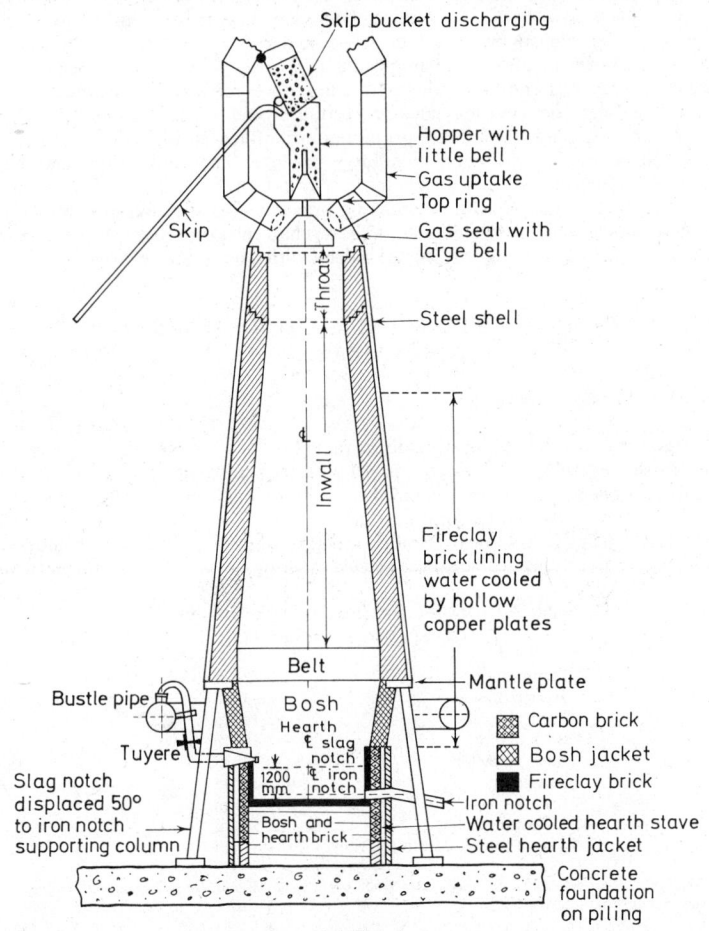

Figure 5.1. Blast furnace

fluxing or it is convenient to mix acid and basic ores in suitable proportions. The reduction of iron from the ore is effected by carbon monoxide from the coke; this monoxide being produced by reaction between coke and carbon dioxide. This latter, in turn, is produced by reaction between iron ore and carbon monoxides. A CO–CO_2 mixture in equilibrium with carbon is consequently present in the furnace and is cooled from a high temperature to below 700°C as it ascends the stack causing the reaction $2CO$

$= C + CO_2$ to occur and this leads to carbon deposition. As the ore is reduced to metallic iron carbon from the gas phase is absorbed into the iron.

Absorption of carbon during the early stages of smelting causes partial melting of the iron and the iron-carbon melt flows over the unreduced ore which is then directly reduced by dissolved carbon. The reduced iron is saturated with carbon by the time it reaches the hearth. It also contains varying amounts of silicon, manganese, phosphorus, sulphur and arsenic depending on burden composition (Table 5.2).

Smelting of extremely pure ore leads to iron with relatively very low sulphur and phosphorus content; phosphatic ores produce iron with a high phosphorus content.

Table 5.2   CHIEF PRODUCTS OF THE BLAST FURNACE
Hot Metal or Cold Pig Iron (British Practice)*

Iron for steelmaking (Furnace type)	Typical composition, %				
	Si	S	P	Mn	Total C
OH. Basic Fixed (cold)	0·7–1·7	0·04–0·07	0·07–1·4	1·0–1·75	3·4–3·8
OH. Basic Fixed (hot)	0·07–1·5	0·04–0·07	0·09–1·7	0·8–1·8	3·1–4·1
OH. Basic Tilting (hot)	0·3–0·7	0·05–0·065	1·4–1·5	0·7–1·0	3·4–3·7
OH. Acid Fixed (cold)	1·6–2·5	0·018–0·06	'0·023–0·06	0·4–2·0	3·5–4·2
Bessemer. Acid	1·0–1·8	0·03–0·08	0·085–0·100	0·6 max	3·8–4·0
Bessemer. Basic	0·3–1·0	0·2	1·9–2·5	0·7–2·5	3·5–4·0

* I.S.I. Special Report No. 22. Symposium on Steelmaking (1938).

High silicon, approximately 2%, iron is necessary for use in converters, or pneumatic steelmaking processes, which derive the heat for refining the iron from the exothermic reactions between oxygen in the air or oxygen blast and the silicon. Basic open hearth steelmaking processes can accept pig or molten iron with phosphorus around 0·25% because it is reduced to acceptable levels, less than 0·06%, in the refining process. The acid open hearth process requires low phosphorus and sulphur iron as neither are reduced in refining. Thus the blast furnace, depending upon the availability of suitable ores, coke and limestone can be worked to produce the raw material for the steelmaking processes.

Steelmaking

Steelmaking is the conversion of non-workable high carbon (3–4%) iron into a lower carbon product which can be worked hot by forging or rolling. Carbon is removed from the molten metal by reaction with oxygen, the oxygen being supplied to the metal by iron oxide (ore or scale) in the Acid Open Hearth process, by ore or oxygen lance in the Basic Open Hearth process, by air or air/oxygen mixtures in the Basic and Acid Bessemer processes, bottom blown converters and by oxygen lance in the LD and Kaldo converters (top blown converters). In the open hearth processes some oxygen is also derived from the furnace gases.

The carbon is removed from the molten metal in gaseous form according to the reversible reaction $FeO + C \rightleftharpoons Fe + CO$. In the open hearth processes the carbon removal is seen as bubbles which rise through the slag during the 'boil'; in the blown processes the escaping CO burns to form the carbon 'flame'. It should be noted that the carbon reduction is effected, in all steelmelting processes where reduction of carbon is required, by iron oxide, which is soluble in the molten metal.[1] In molten steel equilibrium between dissolved FeO and carbon is maintained by increasing the amount of dissolved oxide as the carbon content decreases. Thus at about 0·8% C in the liquid metal the percentage of FeO is about 0·025% whereas at 0·06%C it is about 0·45%, (see Figure 1, ref. 2). Thus at the stage in refining when the desired carbon content has been reached, the molten metal will contain an amount of dissolved oxide which is dependent on the carbon content; such metal, allowed to solidify, would exhibit hot shortness (i.e. cracking in hot forming by forging or rolling) and lack of toughness under impact. The acid open hearth process can provide an exception to this statement if the heat is worked to produce a reduction of silicon from the silicate slag into the metal at the end of the refining period. In such circumstances the ferrous oxide content of the steel will be very low and not directly related to the carbon content.

Control over the amount of dissolved oxygen in the metal is necessary and is effected in the open hearth processes by control of the oxide content in the slag and in the pneumatic processes by cutting off the air or oxygen supply at the appropriate point in the carbon removal process. Control over the amount and distribution of dissolved oxygen in the solidified ingot is handled in one of three ways:

1. By adding elements to the molten metal such as silicon or aluminium which have a strong affinity for oxygen and form oxides, thus reducing the level of dissolved oxygen in the molten metal by forming separate solid or viscous oxides which, under suitable conditions, tend to float out of the steel into the slag. Steel so treated is said to be 'killed'.
2. By allowing, where composition permits (generally carbon content less than 0·15%), the $FeO + C \rightarrow Fe + CO$ reaction to continue in the moulds to produce 'rimmed steel'. This steel is characterised by a pure iron 'rim', with a core which contains iron oxide, because the rimming (evolution of CO gas) action is intentionally suppressed after a suitable rim has been formed. Such steel rolls hot or cold satisfactorily, albeit the notch toughness is poor.
3. By adding a small amount of Silicon the $FeO + C \rightarrow Fe + CO$ reaction is partly suppressed so that a limited amount of gas is formed and the steel is said to be 'semi-killed' or 'balanced'. The latter term refers to the behaviour of the steel in the ingot mould; a 'killed' steel sinks in the mould, a 'rimming' steel rises slightly and a 'semi-killed' steel solidifies without sinking or markedly rising. Whereas a silicon killed steel will contain more than 0·1% approx. silicon a semi-killed steel will contain less than this amount.

The presence of dissolved oxygen in solid steel does not prevent it being used satisfactorily for a very wide range of applications. On the other hand, where steel must be able to stand up to very severe stressing conditions and/or is required to show considerable toughness the level of oxygen should be extremely low. Rimming steel (high oxygen) goes into low carbon products such as wire and sheet, which are the raw materials for products such as nails, screws, fencing wire and most sheet metal products. Semi-killed steel is mainly used in the U.K. for plates and structural sections to produce tanks, pressure vessels, pipes, girders, ships etc. Killed steel is used for forgings, plates for pressure vessels particularly those working in the creep temperature range, pipes, plates for tanks and pressure vessels working at low temperatures, tools, etc.

Rimming and semi-killed steels are produced more cheaply than killed steel because the ingot shrinkage cavity (pipe) in the top of a killed steel ingot requires a feeder head to fill it, otherwise the required 'crop' off the end of the bloom to avoid the pipe would produce excessive waste of good metal in the shoulders of the ingot. Rimming and semi-killed steel ingots are not normally fed. Also killed steels are normally run 'uphill', moulds being set round a central trumpet into which the metal is teemed, because there is less risk of defective ingot surface than with direct teeming into the mould. Forging ingots, which are usually of large size, are teemed direct using an intermediate tun-dish to prevent splashing.

The engineer selecting steel for a particular application must primarily obtain the right quality, price being considered when comparison is made between equal qualities. Specifications covering chemical analysis and mechanical tests on samples should be supported by experience of a steelmaker's product. Significant changes in melting and production practice should not be introduced without the facts being communicated to the purchasing engineer. In applications where experience is lacking some judgement is required to assess the quality that will be needed and, indirectly, the preferred melting practice. Melting practice will now be considered with particular reference to the quality of steel produced.

MELTING PRACTICE

Steel melting practice is conveniently classified according to the type of refining equipment employed and the type of refining slag.

Raw materials

Equipment	Practice	Materials
Open hearth	Cold metal. Acid	Acid pig iron (low S and P), selected scrap, ore, limestone.
Open hearth	Cold metal. Basic	Basic pig iron, scrap, ore, limestone (lime), fluorspar.
Open hearth	Hot metal. Basic	Basic molten iron, scrap, ore, limestone (lime), fluorspar.
Bessemer converter	Hot metal. Acid	Acid molten iron, scrap (for temperature control).
Bessemer converter	Hot metal. Basic	Basic molten iron, scrap (for temperature control), lime.
LD converter	Hot metal. Basic	Basic molten iron, scrap (for temperature control), lime.
Kaldo converter	Hot metal. Basic	Basic molten iron, scrap (for temperature control), lime.
Induction	Cold metal	Selected scrap.

Equipment

Open hearth furnace	Acid slag.
	Basic slag.
Converter	Acid slag.
	Basic slag.
Electric arc furnace	Acid slag.
	Basic slag.
Induction furnace	Acid or basic lining.

Open hearth furnaces operate on the reverberatory principle with the heating gases sweeping the surface of the melted charge. In fixed open hearth practice the average depth of the bath of molten metal is within a range of 250–425 mm, in tilting open hearth practice it is in the range 500–750 mm. The capacity at tapping of fixed open hearth furnaces varies; in acid practice from around 20 tonnes to around 100 tonnes and in basic practice from 50 tonnes to 120 tonnes. Tilting open hearth furnaces have a tapping capacity generally between 120 and 300 tonnes. For details of design and construction see Reference 3, page 373. The chief features of the furnace are shown in Figures 5.2(a), (b) and (c). Converters are brick lined steel vessels which operate either on the principle of air (Bessemer type) or oxygen/steam (VLN) mixtures being blown into the molten metal from the bottom of the vessel or by injecting oxygen through a lance onto the top of the molten metal (LD and Kaldo converters). The time required to refine molten iron in a bottom blown converter is extremely short, of the order of 15–17 min. Consequently, although such converters have a nominal capacity of about 50 tonnes of metal output is potentially greater than that of the open hearth furnace.

The oxygen top blown converters are similar to the bottom blown vessels being supported on trunnions which allow them to be turned from the vertical position, for charging and blowing, into a position just below the horizontal for emptying. The Kaldo vessel is also supported in rings which allow it to be rotated about the longitudinal axis of the vessel, permitting greater contact efficiency between oxygen, slag and metal (see Figure 5.3).

Arc furnaces, which occasionally have a melting capacity up to 100 tonnes or more are more normally of 50–80 tonnes capacity (see Figure 5.4a). Induction furnaces are usually rated at less than 5 tonnes (see Figure 5.4b). The main steelmaking processes are indicated in Figure 5.5.

The functions of slag in steelmaking

In addition to a high carbon content, the product of the blast furnace contains silicon, manganese, sulphur, phosphorus and arsenic in varying amounts depending on the content of these elements in the raw materials of the burden. Sulphur must be reduced in refining to a low level and, if it is desired to produce high sulphur free cutting steel, the metal must be resulphurised. Phosphorus is less harmful but induces notch brittleness and should be below 0·06% for most steel usage. Arsenic is similar to phosphorus in its action but occurs in substantially lower amounts in steel. To reduce these elements, except arsenic which cannot be removed, it is necessary to carry a suitable refining slag on the surface of the metal.

In the processes operated in the open hearth and electric arc furnaces the oxidation of carbon for its reduction in the molten bath is effected by adding iron oxide to the slag (built-up by slag making additions to the charge), from whence it passes into the bath. This ferrous oxide also oxidises silicon and manganese in the bath and these oxides also pass into the slag. This also applies to the LD and Kaldo vessels.

The Bessemer converter which is operated by blowing oxygen through the molten metal either as air blast or air/oxygen mixture or oxygen/steam mixture, does not require additions of iron oxide as oxidation is not dependent on slag metal reactions.

Silicon and manganese are oxidised in addition to carbon and form a slag. However to reduce phosphorus in the molten metal, a slag with a high lime content is required, (basic slag processes), as well as the oxidising conditions needed to convert the phosphorus in the metal to oxide (P_2O_5). The lime ensures the formation of a relatively stable phosphate of lime in the slag, provided the iron oxide and lime content of the slag is high and the silica and phosphoric acid low. To retain these slag conditions with high phosphorus metal, it is necessary to remove the first slag formed and then build up a new one.

Removal of sulphur from the metal bath can only be effected by slag-metal reactions and is favoured by low ferrous oxide, low silica and high lime in the slag, conditions opposite in respect to ferrous oxide content to those needed for reduction of phosphorus and carbon. Thus, in the processes operated with acid slags, high silica content, the slag is the means whereby carbon is reduced by the carbon/ferrous oxide reaction, but there is no worthwhile reduction of sulphur or phosphorus and these processes can only be applied to a metal bath relatively low in sulphur and phosphorus. Where the metal is high in sulphur but low in phosphorus the fixed open hearth and LD or Kaldo converters using a basic slag, high lime, will effectively reduce the sulphur. Where the metal is high in sulphur and phosphorus, the phosphorus must be reduced first, by removing the first slag high in phosphate, again using a high lime basic slag.

The blown metal processes (Bessemer converters) are restricted in the choice of hot metal compositions because they rely on the heat generated by exothermic reactions

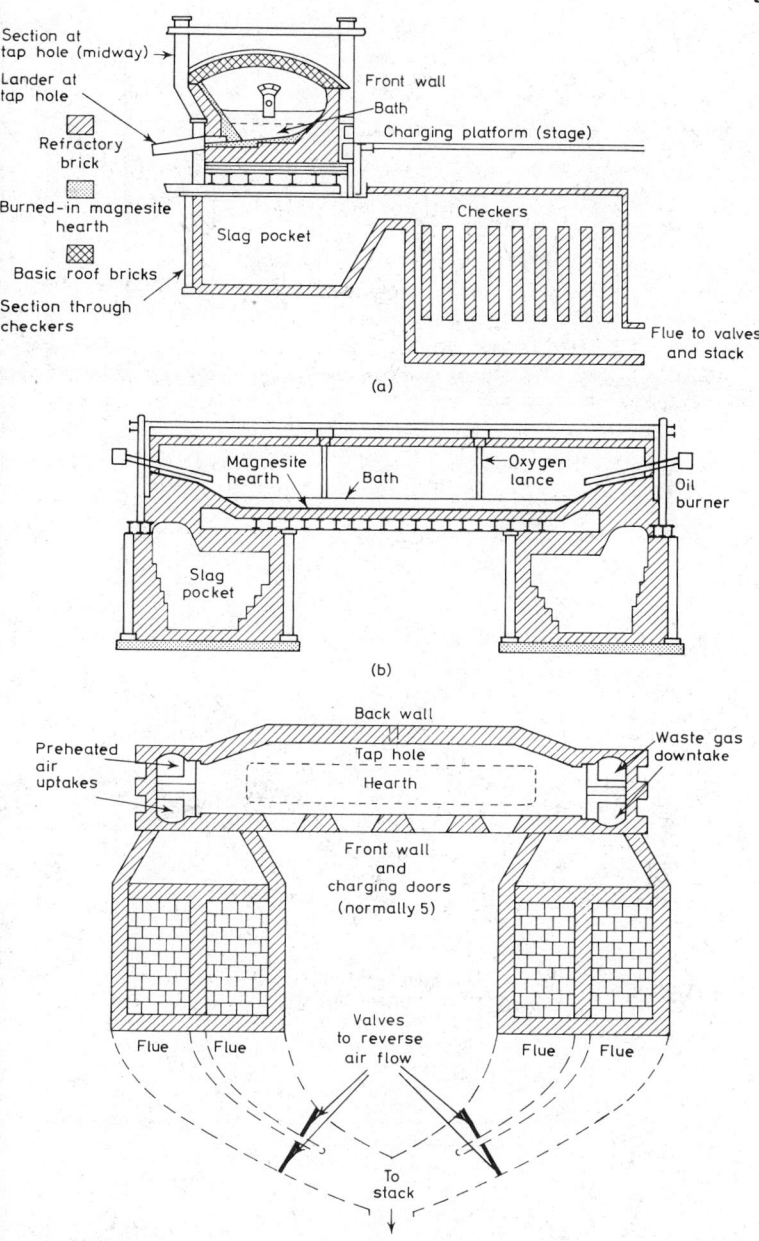

Section at
tap hole (midway)

Lander at
tap hole

Refractory
brick

Burned-in magnesite
hearth

Basic roof bricks

Section through
checkers

Front wall

Bath

Charging platform (stage)

Slag pocket

Checkers

Flue to valves
and stack

(a)

Magnesite
hearth

Bath

Oxygen
lance

Oil
burner

Slag
pocket

(b)

Back wall

Preheated
air
uptakes

Tap hole

Hearth

Waste gas
downtake

Front wall
and
charging doors
(normally 5)

Flue   Flue

Flue   Flue

Valves
to reverse
air flow

To
stack

(c)

Figure 5.2. Fixed basic open hearth furnace—schematic (a) Transverse section, (b) Longitudinal
section, (c) Plan

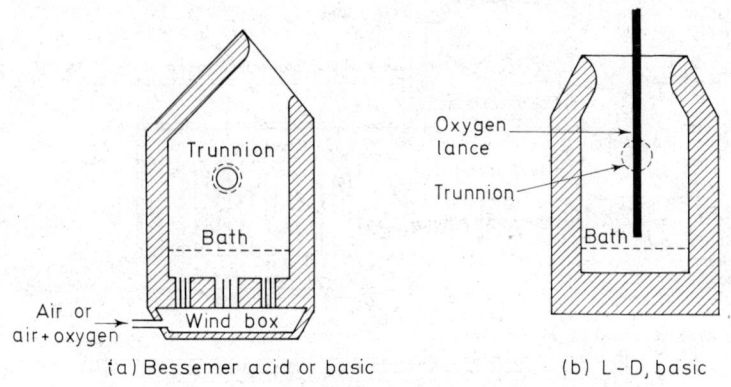

(a) Bessemer acid or basic

(b) L-D, basic

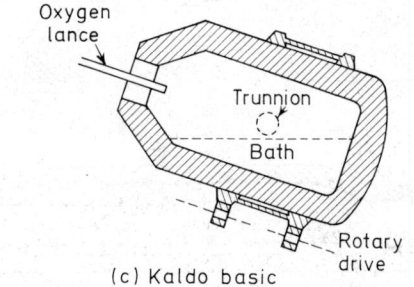

(c) Kaldo basic

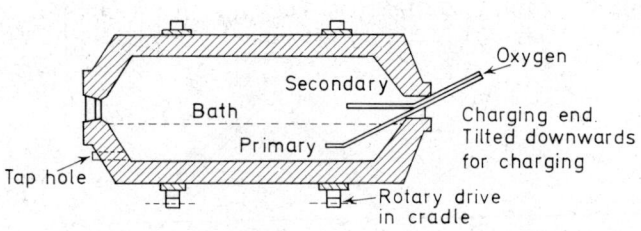

(d) Rotor (Oberhausen) basic

Figure 5.3. Types of converter

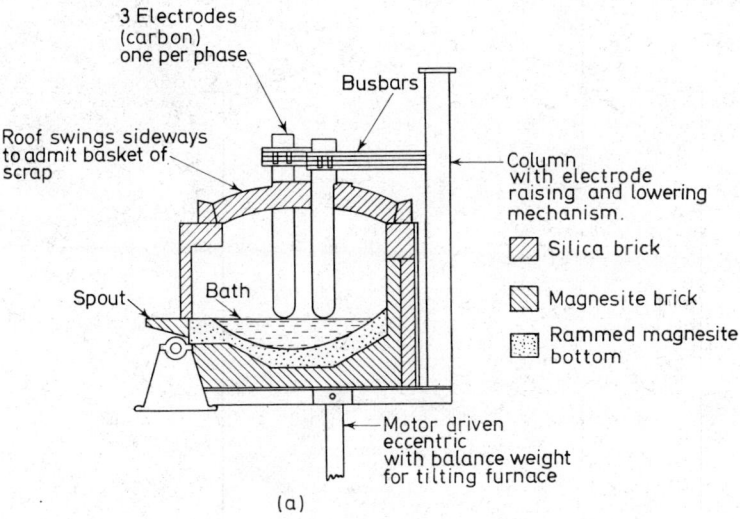

3 Electrodes
(carbon)
one per phase

Busbars

Roof swings sideways
to admit basket of
scrap

Column
with electrode
raising and lowering
mechanism.

▨ Silica brick

◨ Magnesite brick

▤ Rammed magnesite
bottom

Spout

Bath

Motor driven
eccentric
with balance weight
for tilting furnace

(a)

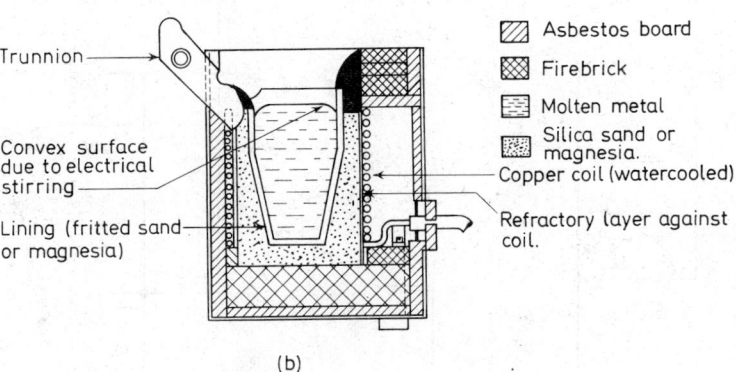

Trunnion

▨ Asbestos board

▩ Firebrick

▦ Molten metal

▨ Silica sand or
magnesia.

Copper coil (watercooled)

Convex surface
due to electrical
stirring

Lining (fritted sand
or magnesia)

Refractory layer against
coil.

(b)

Figure 5.4. Electric furnaces
(a) Basic lined arc furnace
(b) Induction furnace

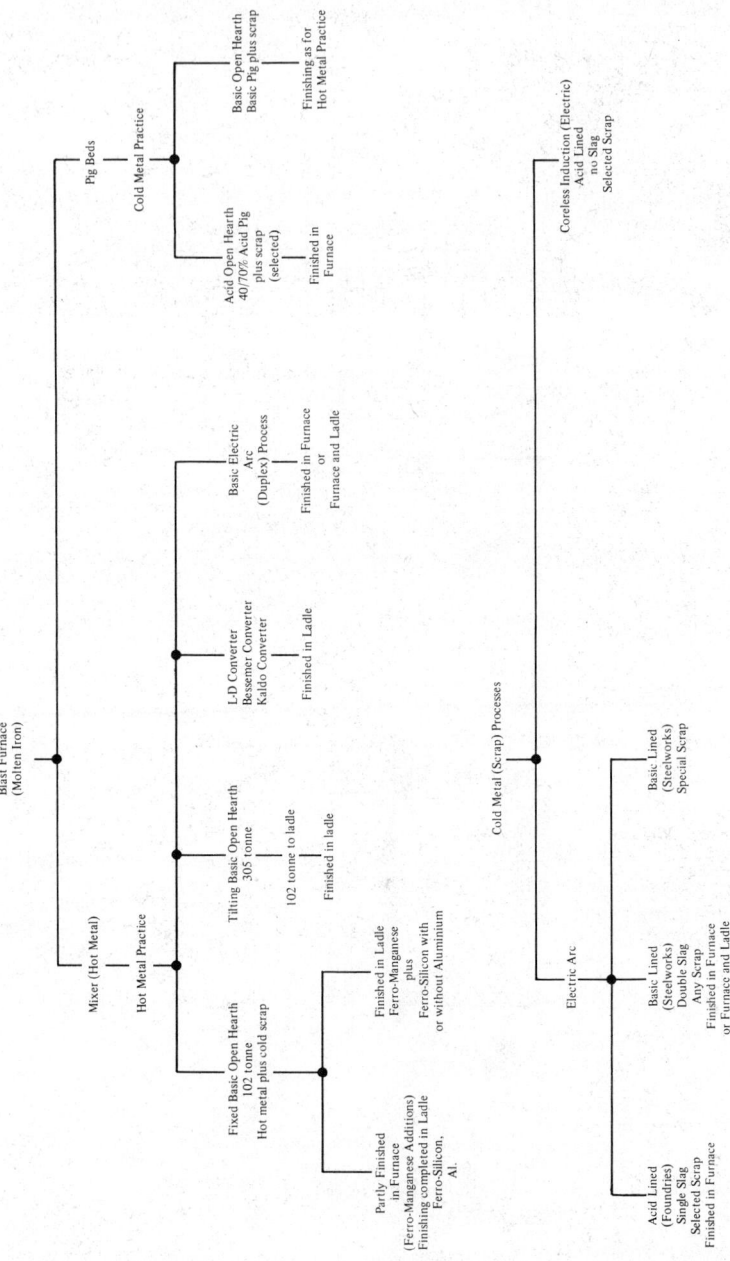

Figure 5.5. Steelmaking processes

between silicon and oxygen to provide the heat required to undertake refining. Carbon reduction means a substantial increase in metal solidification temperature. The silica slag in the acid Bessemer is a product of reaction not a tool in refining. Summarising, the function of slag in refining the blast furnace product into steel, is to supply oxide in the acid and basic processes and to remove sulphur and phosphorus in the basic processes.

Deoxidation

Reduction of carbon and/or phosphorus by reaction with ferrous oxide implies excess dissolved oxide which, at completion of refining, should be reduced to a level which experience shows produces steel with properties adequate for specific purposes.

Low carbon 'soft' steel will necessarily be high in iron oxide at the end of refining. Conversely, medium or high carbon steel intended for forgings or rolled products requiring high mechanical properties will be lower in iron oxide at the end of refining in the acid open hearth process, assuming the carbon is 'caught on the way down'. The basic open hearth, LD and Kaldo processes, because of the need to reduce sulphur, tend to be operated to produce a fairly low carbon content at the end of refining which again means a relatively high oxide content. In the latter case, to raise the carbon to the desired level it is necessary to reduce this oxide in the refined metal by addition of deoxidising elements before adding carbon as pig iron or anthracite.

Removal of ferrous oxide from molten refined steel, is effected in various ways and the method adopted influences the quality of the steel and its reliability for the intended service.

There are three[4] fundamentally different ways of removing ferrous oxide dissolved in molten steel:

1. *Diffusion deoxidation.* In this method the steel is covered by a slag largely free from ferrous oxide which can dissolve the oxide out of the metal as the result of migration (diffusion) of oxide to areas of lower concentration. This method is used in electric arc furnace practice by addition of carbon and silicon to the slag.
2. *Precipitation deoxidation.* The most commonly used method, in which elements with a stronger affinity for oxygen than iron, are added to the metal to decompose the ferrous oxide and form insoluble compounds. This necessarily results in a heterogeneous mixture of molten metal and the products of deoxidation (i.e. oxides of manganese, silicon, aluminium). In the course of time, owing to differences in density, the insoluble oxides rise to the surface of the molten metal. This resembles precipitation of a substance from solution as practised in chemical processes hence the term 'precipitation' deoxidation.
3. *Carbon deoxidation.* In this method carbon is used to react with the ferrous oxide to produce gaseous oxides of carbon which are removed by vacuum treatment.

Diffusion deoxidation, because of the slow rate of diffusion of the ferrous oxide, requires extended periods of time, dependent on the size of furnace which controls the diffusion path, and the fact that ferrous oxide which entered the furnace lining during the first oxidising slag (removal of carbon) period must also be removed to the second white refining slag to produce fully deoxidised steel. High quality steel, made for severe duty, must be fully deoxidised. It is not practicable to carry out diffusion deoxidation in the open hearth or converter vessels. Basic electric arc steel is always deoxidised in this manner.

Precipitation deoxidation is used in open hearth, acid electric and converter processes. In fixed open hearth and electric arc practice the deoxidising elements can be added to the furnace bath or to the ladle, or to both, but in converter practice they are added to the ladle as is the case in open hearth tilting furnace practice.

The addition of silicon to molten steel containing ferrous oxide produces silica, which, being solid at the steel temperature, floats very slowly to the surface. The products of deoxidation should be in the form of a viscous liquid so that they assume a spherical shape, the velocity of movement to the surface being greater the larger the mass. Manganese has a relatively mild deoxidising action compared with silicon or aluminium but, in combination with silicon in suitable proportion, produces viscous liquid deoxidation products. Manganese should be added first or simultaneously with the silicon and in basic open hearth practice is frequently added to the bath and silicon to the ladle.

The acid open hearth process, owing to economic considerations, has almost disappeared. It has an inherent advantage over other steelmaking methods in that a large part of the iron oxide passing into the slag from the bath in the later stages of refining. will not reenter the bath, as it forms stable silicate with the manganese in the slag. Hence the metal is much less oxidised than in the basic open hearth process for a given amount of oxide in the slag.[5] A further and most important point is that the percentage of FeO + MnO in the acid process slag is automatically regulated by the fact that iron-manganese-silicates will dissolve up to about 60% of silica from the furnace banks (silica sand). Hence by regulating the amount of MnO in the slag, the FeO can be closely controlled. Thus the melter is enabled to control more consistently the quality of steel being made.[5] Acid steel practice in fixed open hearth furnaces also permits 'finishing' or deoxidation by 'boiling back' manganese and silicon into the bath at the end of the refining period to effect complete deoxidation. Thus in Swedish open hearth practice (1938) high quality steel (e.g. 1% carbon and 1·2% carbon–1% Cr roller bearing steels) was finished by boiling back 0·2–0·3% silicon from the slag with no additions of ferro-silicon; this practice was considered a very essential factor in making good steel.[6]

Steel quality

High quality steel (carbon steels for wire ropes, alloy engineering steels etc. where a homogeneous steel is essential) can be produced in the fixed basic open hearth process if special melting and deoxidation practice is followed. Rate of carbon elimination is important for quality steel from the basic furnace; after an initial fairly rapid rate, the rate is slowed down and the 'finishings' for deoxidation added to the bath. Such heats, in a 50-tonne furnace, will take twelve to fifteen hours. The same weight of metal can be tapped in 5½ hours using a different melting practice but producing lower quality steel.[7]

Deoxidation in the ladle is not generally considered to produce as good quality steel as furnace deoxidation. However semi-killed steels are normally ladle deoxidised. Low quality steels, where a high inclusion content is not of any consequence, may be deoxidised with aluminium in the moulds. Manganese is essential in steel to prevent hot shortness and must be added whether or not silicon is used as the main deoxidant. If added in the ladle, which is done by shovelling into the stream entering the ladle, as solid ferro-manganese or Spiegel iron (20% Mn pig iron), there is generally insufficient time for the ferro alloy to dissolve and diffuse uniformly throughout the molten metal,[8] with consequent variation in deoxidation and manganese content in different parts of the ladle.

This applies particularly to medium and higher carbon steels which must not be cast at high temperature; the normal heat losses to the ladle and through heating and melting the ferro alloy additions prevent sufficiently rapid solution and diffusion of the deoxidant.[9] The use of molten ferro-manganese as a ladle addition tends to overcome this problem.[10] However ladle deoxidation using the precipitation method has the disadvantage, whether solid or molten ferro alloys are used as deoxidants, that the products of the deoxidation reaction must, to give fully clean steel, rise a distance up to about 4 m in a time of about 75 min maximum. Consequently, the amount of non-metallic deoxidation products in the solid ingots will vary from the start to the end of teeming, increasing as the end of teeming approaches.[11] In general ladle 'finishing' cannot be expected to give as uniform a product as 'finishing' in the furnace. It is particularly difficult to achieve a uniform distribution of manganese in heats where manganese is used as an alloying element (say

1·2–2%) when cold additions of ferro-manganese are made as a ladle addition. However this is frequently practiced.

For the reasons indicated above, in making quality steel, where maximum freedom from non-metallic inclusions is an important objective, as much deoxidation as possible is done in the furnace.[12] In basic open hearth practice there is the possibility that ferrous oxide will revert from the slag to the metal if held in the furnace for the elimination of the rising deoxidation products for too long a period;[13] consequently a common practice for such steels is to add ferro-manganese to the bath and ferro-silicon to the ladle.

In the acid open hearth process, because the oxide will not revert once it is combined in the slag, time can be allowed for elimination of the siliceous non-metallic matter in the furnace.[14] This feature, together with deoxidation by silicon reduction from the slag, enables the acid open hearth to produce steel of very high quality and reliability. Although the basic open hearth process will produce steel with lower sulphur content and consequently a smaller amount of sulphide particles in the steel this does not enable the basic open hearth furnace to produce killed steel for severe duty, such as ball bearings, roller bearings etc. to a better standard than the acid open hearth. This is because the sulphides play an insignificant role in fatigue compared with products of deoxidation of a siliceous or aluminous type.

Unfortunately acid open hearth steel is expensive to produce; special raw materials must be used and it is normal practice to work the heats down from a high carbon content necessitating a long refining process in the furnace. Economic pressure has gradually eliminated many acid open hearth melting shops, to be replaced by the basic electric arc process. Basic electric steel is normally very low in sulphur content and can be as low as 0·006% with very low oxygen content. However, compared with acid open hearth steel, basic electric steel is significantly more 'tender', in that it is more susceptible to overheating and possible burning during subsequent heat treatment. It also produces higher strength alloy steels which are more susceptible to cracking when welded and to cracking during the process of rolling blooms and billets. The latter appear as flakes or hair line cracks.

The acid electric arc process is not used for ingot production but has several advantages for producing castings; where castings must have machined surfaces with superior finish general experience suggests a preference for acid electric steel.[15]

Bulk steel, semi-killed and killed, for plates and sections is deoxidised by the precipitation method. The advent of oxygen blowing for converters improves the economics of steelmaking and substantially lowers the nitrogen content of the steel. However, ladle deoxidation is necessary and the above observations regarding the difficulties in the elimination of non-metallic matter and uniformity of solution of the deoxidising and alloying media (ferro-alloys) apply.

It should be noted that 'dirt' in steel can be the result of failure to remove, by coalescence and flotation, the particles which enter the bath from the scrap, pig iron and furnace banks. These solid silica particles must be liquefied by combination with manganous oxide and ferrous oxide so that they can coalesce. The bath needs active oxidation by ferrous oxide at a suitable stage in refining; this can be achieved in the open hearth and converter processes making low carbon products but medium and high carbon steels made in the open hearth or electric furnace may present difficulties. Experience shows that such heats made from a 'virgin' charge seldom suffer from slag suspension defects.[17] The scrap in such a charge is usually understood to be heavy rolling mill scrap from a hot metal melting shop steelworks. Again one notes the influence of melting practice on end-product quality.

Ladle reactions

Ladle reactions are important to the quality of the metal running into the ingot moulds. The silica lining of the ladle can react with slag from basic processes to promote a return of ferrous oxide into the metal;[18] this oxidation will tend to be most marked in the metal in contact with the slag during teeming and will affect the quality of the ingots in the latter stage of teeming. On the other hand the ladle lining has little or no effect on acid slags,[19] which promotes the production of steel of consistent quality.

Reaction between steel and slag in basic processes which have not yet attained equilibrium in the furnace or converter will continue in the ladle. Also the change of slag composition resulting from solution of the ladle lining, the fall in temperature and the deoxidising and manganese additions combine to create completely new equilibrium conditions which affect practically all reactions;[20] note that this is not applicable to acid steel processes. This is particularly the case in respect of increased total oxygen content of the steel at the end of teeming and the loss of manganese.

In the practice of precipitation deoxidation, comparison between the acid open hearth and the basic processes, open hearth and converter, centres on the attainment of slag-metal equilibria. In the acid open hearth process the slag components are undisturbed in the ladle and consequently produce ingots of relatively uniform quality; in the converter process an oxidised metal passes into the ladle for the critical stage in steelmaking (removal of oxygen) where conditions of non-equilibrium exist so that ingots vary from start to finish of teeming and heats to the same nominal specification vary from one to another.

The steel purchaser must accept that bulk steel produced by basic processes is a variable commodity in respect of certain properties which may or may not matter according to the end usage. For example, the silicate inclusions which are commonly associated with lamellar tearing when killed or semi-killed steel plates are welded under conditions of restraint, are more likely to be found in plates rolled from slabs from the 'back end' of the cast. These inclusions are often a cause of serious financial loss to the steel fabricator but steel affected in this manner passes all normal specification requirements and would be perfectly satisfactory for non-welded applications. To obtain steel which is consistently free from this type of defect resort is made to electric arc melting practice with reducing slags, wherein the reactions in the ladle are completely suppressed[21] combined with vacuum degassing during casting.

Carbon deoxidation (by vacuum degassing)

Removal of carbon from the melt is by the ferrous oxide-carbon reaction forming carbon monoxide gas and metallic iron, This reaction can therefore also be regarded as a deoxidising reaction if ferrous oxide is not added to the system and the gaseous products are withdrawn as in vacuum degassing.

During teeming, the ladle stream is in contact with the atmosphere. Oxygen and nitrogen are absorbed. Excess of deoxidants will prevent the carbon-ferrous oxide reaction and aluminium will combine with the nitrogen but the products, i.e. non-metallic particles will be present in the steel ingot. This is the situation in non-vacuum casting.

For the production of alloy steels of the highest quality for special purposes melting under vacuum, by lowering the partial pressure of the gases formed by the carbon-ferrous oxide reaction, ensures a continuous reduction in the oxide content provided carbon is present and removal of the oxygen in gaseous and therefore harmless form. Nitrogen and hydrogen are also removed.

The deoxidation of steel by vacuum degassing can also be conducted by tapping the furnace into a ladle and then subjecting the ladle and its contents to reduced pressures in a vacuum chamber. For this purpose the chambers may have built-in induction stirring coils to ensure maximum, rapid and uniform exposure of the metal to the reduced pressures.[22] Vacuum deoxidation and vacuum degassing produces steel with improved mechanical properties (ductility, notch toughness) as judged by conventional testing but, more importantly, a very substantial improvement in the performance of end products subject to fatigue failure such as ball and roller bearings.[23] In studying the reasons for this improvement in performance under fatigue stressing, it was concluded that improved toughness of the matrix was a contributory factor.[24]

While there are several methods of vacuum deoxidation and degassing of molten steel,[25] the Design Engineer is primarily interested in the benefits of this deoxidation method in respect of end-product quality. Vacuum degassing reduces hydrogen to low levels which promotes less likelihood of defective rolled or forged products (indicated by hair line cracks) in alloy engineering steels reaching the user. Vacuum deoxidation, which necessarily includes degassing, is primarily applied to processes where sufficient time is available for the reactions to reach stability, i.e. those such as vacuum induction melting and has beneficial effects on cleanness and matrix toughness. The benefits of vacuum deoxidation may prove economically worthwhile when this method is applied to mild and low alloy structural and pressure vessel steels, which, particularly when made in killed steel, are susceptible to lamellar tearing during welding, the most costly defect in such materials.

Scrap melting processes

A considerable quantity of steel is made by remelting scrap, without the addition of pig iron, in electric furnaces either of the arc or induction type. The former carry out refining, usually under double slags of basic composition, but the latter do not ordinarily aim at slag refining. The former can use a wide variety of scrap qualities but the latter require scrap selected for low impurities.

Special alloy steels, stainless steel, high speed steel in the induction furnace, are products of the scrap melting processes. Extra high quality steel is produced in induction furnaces operated under vacuum and it is common practice, in order to minimise hydrogen induced troubles, to vacuum degas alloy steels made in the basic electric arc furnace. In areas where scrap is abundant large basic electric arc furnaces (150 tonnes) are used to produce bulk steels, such as rimming steel, semi-killed and killed mild steel. Such steels can be produced to exceptionally low sulphur levels which can be advantageous for certain applications.

Casting pit practice. Ingots

Casting pit practice is determined by the type of steel being produced and the desired end product. It has important effects on steel quality.

RIMMING STEEL

The steel is teemed with sufficient ferrous oxide dissolved in it, to cause the $C + FeO \rightleftharpoons Fe + CO$ to proceed to the right. In casting a rimming ingot the amount of free oxide must be adjusted so that gas evolution (CO and H) does not begin until a solid rim has formed adjacent to the mould walls and the mould has been filled with metal.

The ingot metal solidifies in the mould with a rim of comparatively pure metal and a core containing blow holes and a concentration of sulphides and oxides. The depth of rim-in is controlled either by chemical means, i.e. addition of aluminium to remove free FeO, sometimes referred to as 'chemical capping', or by placing a steel punching on the top of the ingot when sufficiently rimmed in, known as mechanical capping which seals the surface and prevents further gas evolution and CO reaction. The choice of method can be significant for the suitability of the product for its intended purpose.

Chemical analysis specification, which normally defines rimming steel products does not indicate the method used to control rim-in. A typical analysis would be 0·12% C max, 0·3–0·6% Mn. Rimming steel chemically capped may cause difficulty in subsequent welding operations; furthermore welding wire for mild steel coated electrodes, should be made to a consistent capping practice as a change can cause serious difficulties to the electrode maker.

The rimmed steel ingot contains a large amount of occluded gas (CO) which prevents it from sinking on solidification or indeed may cause it to rise. The behaviour is dependent upon the rate of gas evolution; steel with too high an oxide concentration will evolve gas at such a high rate that the mould cannot be properly filled and the apparent top sinks down to give a 'box-hat' ingot. Slightly lower oxide concentration allows the mould to be filled by an early slow gas release but subsequently more rapid evolution causes the liquid steel to rise in the mould indicating a 'rising' cast. There is virtually no shrinkage pipe in rimmed steel ingots so that practically the whole ingot can be converted to a rolled product the gas holes being welded up during rolling.

Rimmed steel is made up to a carbon content of around 0·15%. Rimmed steel ingots are almost universally cast narrow end up, which is the easiest position for stripping the mould off the ingot, either by direct teeming into each mould or by 'up-hill' teeming through a central trumpet which feeds several ingots set on a plate.

Apart from proper control of the rimming action to ensure the required depth of rim (too shallow a rim can cause trouble because the core can be forced to the surface during rolling), there is a requirement for some rimming steel qualities to contain low sulphur contents, e.g. welding electrode wire, to avoid subsequent hot cracking problems. Other rimmed grades must have completely deoxidised cores, chemically controlled with aluminium or titanium, to avoid trouble due to ageing which causes difficulty in cold working, i.e. deep drawing, pressing etc.

Rimming steel manufacture requires skill in controlling oxidation and rim-in. The user should carry out trials if changing from one source to another and should request notification of any change in practice as can happen when a steelmaker transfers steelmaking from one plant to another.

SEMI-KILLED STEEL

In making semi-killed steel, sufficient silicon is added to the ladle metal to partially suppress the carbon-iron oxide reaction so that oxide concentration is less than is required for the production of a rimmed ingot. However gas evolution is not entirely prevented in the late stages of solidification. The ingots do not show effervescence. The volume of blow holes developed is insufficient to prevent some piping due to contraction but this pipe is situated well to the top of the ingot and is normally of very moderate extent. The silicon content of the semi-killed steel would generally be in the range 0·02–0·07%. Structural and pressure vessel steels in the range 0·12–0·25% C are frequently supplied as semi-killed.

KILLED STEEL

The addition of sufficient deoxidising elements (silicon, aluminium) to suppress completely the carbon-iron oxide reaction produces a 'dead' steel which shrinks on solidification producing a cavity or 'pipe'. If this pipe is 'bridged' by solidifying metal the walls of the pipe may remain unoxidised and will be satisfactorily welded together in subsequent rolling.

Generally, however, killed steel goes into higher quality steel products and the piped area of the ingot must not be included in the rolled product supplied to the customer. Therefore the casting pit practice will be arranged to recover the maximum volume of unpiped metal from each ingot having in mind the following principles:

1. Ingots can be cast in moulds set big end up or big end down, the latter being more economical in pit-side cost but increasing the length of the pipe.

2. The top of the ingot can be 'hot topped' by means of refractory bricks or exothermic powders to prevent early solidification and permit the steel to feed down into the body of the ingot and hence raise the level at which piping starts. This practice increases pit side costs but generally improves quality.

The variation in extent of piping is shown in Figure 5.6. Quality steel is always cast as in A. All forging grade steels are killed and, in general, all steels with carbon content exceeding 0·25% carbon. Ingot structures obtained by controlling the extent of the iron oxide-carbon reaction are shown in Figure 5.7.

Gases in steel. Vacuum degassing

Oxygen, nitrogen and hydrogen are absorbed by liquid steel from the atmosphere and hydrogen and oxygen from hydrated oxides on scrap. Hydrogen present in calcined lime

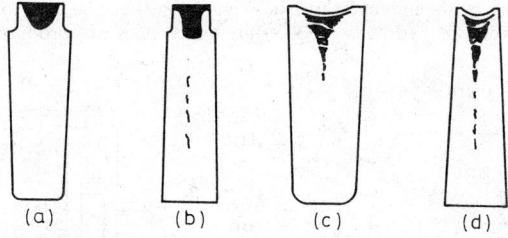

(a)   (b)   (c)   (d)

Figure 5.6. Killed steel.[26] Relation between type of mould and extent of pipe (a) big-end up; hot topped, (b) big-end down; hot topped, (c) big-end up; not hot topped, (d) big-end down; not hot topped

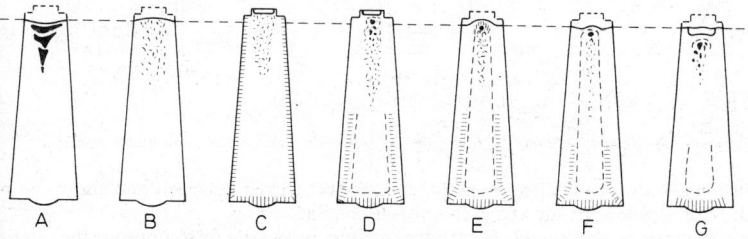

A   B   C   D   E   F   G

Figure 5.7. Ingot structures.[26] Effect of ferrous oxide concentration and gas evolution on solidification structures
A. *Killed steel. No free ferrous oxide. Extensive pipe*
B. *Semi-killed steel. Small FeO. Small gas evolution. Blowholes eliminate pipe*
C. No silicon. Rimming. Capped (mechanical). Larger gas evolution than B. Honeycomb blowholes in skin. No rim
D. *No silicon. Rimming. Capped (mechanical). Larger gas evolution than C. Solid rim internal gas holes. Minor pipe.*
E. *No silicon. Rimming. Not capped. Larger gas evolution than D. Solid rim. Honeycomb holes in lower half*
F. *No silicon. Rimming. Not capped. Larger gas evolution than E. Solid rim. Honeycomb holes almost eliminated.*
G. No silicon. Rimming. Not capped. Larger gas evolution than F. Solid rim. Honeycomb holes eliminated but top discard increased
    **Typical commercial ingots from big-end down moulds*
C. Unsatisfactory. Honeycomb holes in skin. Insufficient gas evolution
E. Rising rimmed ingot. Insufficient gas evolution
G. Box-hat rimmed ingot. Gas evolution too large

and in ferro-alloys also passes into the steel. As indicated in the previous consideration of steelmaking practice, oxygen is used to generate properly positioned internal blow holes in ingots to increase ingot yield but high quality steels for severe service should contain the minimum amount of gases. Gases can cause flaking (hydrogen), embrittlement (hydrogen, oxygen) voids and inclusions (oxygen).

In general there are many qualities of steel suitable for the purpose for which they are intended, which are manufactured without the need to prevent absorption of gases in melting. Where this absorption is undesirable, either melting under vacuum or inert gas is practised or the undesirable gases are extracted from the molten steel before solidification. The former practice is necessarily confined to dealing with relatively small quantities of steel, typically vacuum induction and consumable-electrode vacuum melting, but the latter can be adapted to the degassing of very large quantities since it can be used to remove gases from the running stream of metal. The various possibilities are shown schematically in Figure 5.8 (vacuum melting) and in Figure 5.9 (vacuum degassing).

Vacuum melting by the induction furnace Figure 5.8(a) can be used to produce steel deoxidised by carbon or hydrogen a procedure which does not produce non-metallic

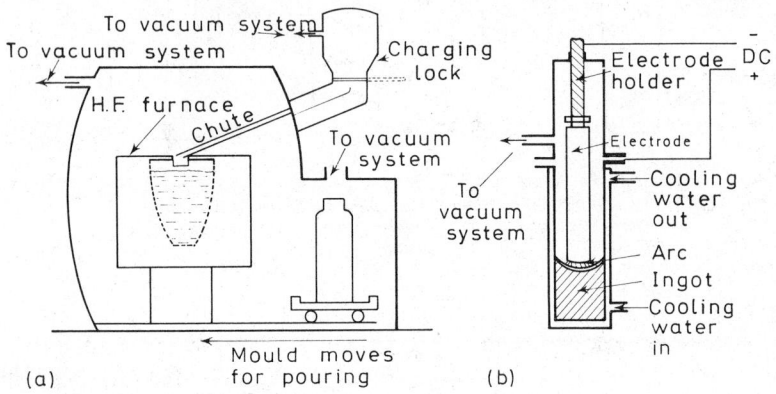

Figure 5.8. Vacuum melting[27]*—schematic (a) Induction furnace, (b) Consumable electrode*

inclusions; additionally nitrogen is excluded so that chromium steels and alloys can be made with very low nitride and carbo-nitride content.

The purpose of consumable electrode remelting in vacuum is to improve the quality of limited weights of steel made in the conventional manner. The steel is cast or forged to make a cylindrical electrode which is then melted by electric arc directly into a water-cooled mould. The melting in vacuo substantially reduces hydrogen, oxygen and nitrogen with improvement in cleanliness, freedom from central porosity and segregation in the ingot. The hot working properties, particularly of high alloy steels, and the mechanical properties, i.e. impact, ductility, fatigue, creep and rupture, at room and elevated temperatures are improved.

Vacuum melted steels, which are necessarily costly to produce, are used for special applications such as ultra-high strength steel for missiles, aircraft steels, special roller bearings and heat resisting alloys where quality is of prime significance. Ingots up to about 20 tonnes have been produced by the vacuum consumable electrode process.

Vacuum degassing is applied to the molten product of conventional steelmaking processes to lower the content of gases in the metal. Hydrogen is markedly reduced by all the following methods; oxygen can be reduced if deoxidising elements such as silicon or aluminium have not been added before degassing and if either the lifting method (Dortmund-Hörder process. Figure 5.10(b)) or the R-H (Ruhrstahl-Heraeus, Figure

5.10(c)) process is used. Both these processes ensure the maximum exposure of the steel, in the form of droplets, to the vacuum.

Ladle degassing (Figure 5.9(a)), in which the contents of the ladle are teemed into moulds after degassing, is possibly the least effective method; an inert gas is sometimes injected into the ladle to promote agitation.

In stream degassing (Figure 5.9(b)) the contents of the tapping ladle from the furnace

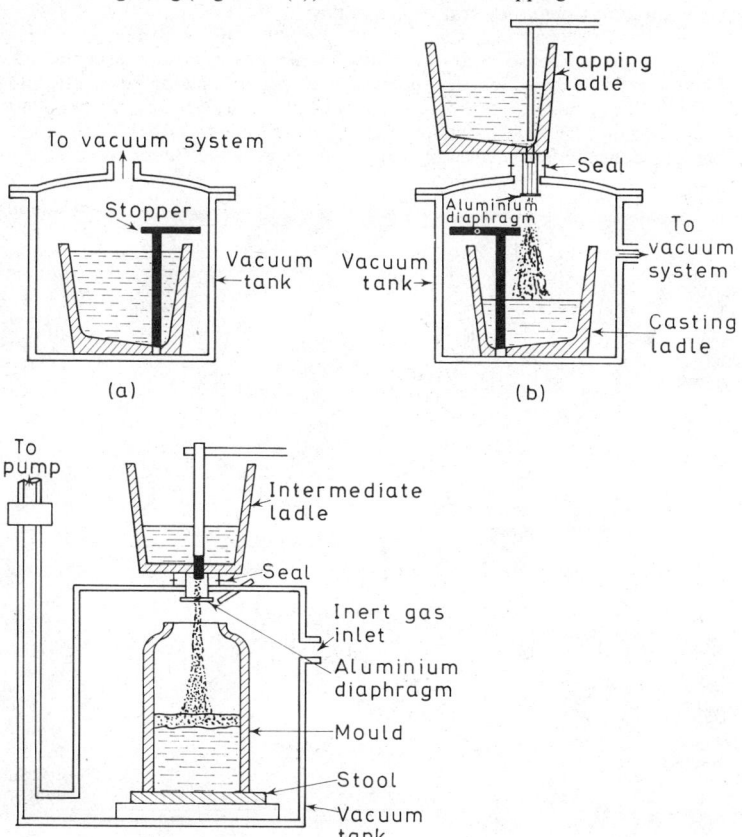

Figure 5.9. *Vacuum degassing—schematic (a) Ladle degassing, (b) Stream degassing, (c) Vacuum-casting using stream degassing*

are passed into a second casting ladle in a vacuum tank, thus exposing the stream to the vacuum. However more heat is lost than in Figure 5.9(a).

Large ingots for forgings can be stream degassed by the method shown in Figure 5.9(c); an intermediate ladle is partly filled by metal from the tapping ladle, the stopper is lifted allowing the stream to perforate the aluminium diaphragm and pass into the ingot mould in the vacuum tank. The intermediate ladle can be continuously fed from tapping ladles so that very large ingots can be made. The method is attractive because it reduces the hydrogen content, to less than 3 mls/100 gms H_2, which means that flakes are not formed during the hot working of alloy steels. Prior to the advent of degassing techniques, bursting of rotor shafts in electrical generating installations was not an unknown occurrence due to the extreme difficulty in preventing flakes or hair-line cracks forming in conventional forging and heat treatment practice. Heavy forgings in alloy steels susceptible to hair line cracking from hydrogen are almost always produced nowadays from degassed ingots. Degassing methods having particular advantages are the following:

DEGASSING DURING TAPPING BY STREAM DEGASSING[28]

By tapping the metal from an electric arc tilting furnace directly into the degassing system, there is little or no increased loss of temperature compared with tapping directly into the casting ladle. The system, Figure 5.10(a), consists of a special removable vacuum type ladle top with tundish which is attached to the ladle while metal is running from the furnace and is then removed when the ladle starts teeming the ingots.

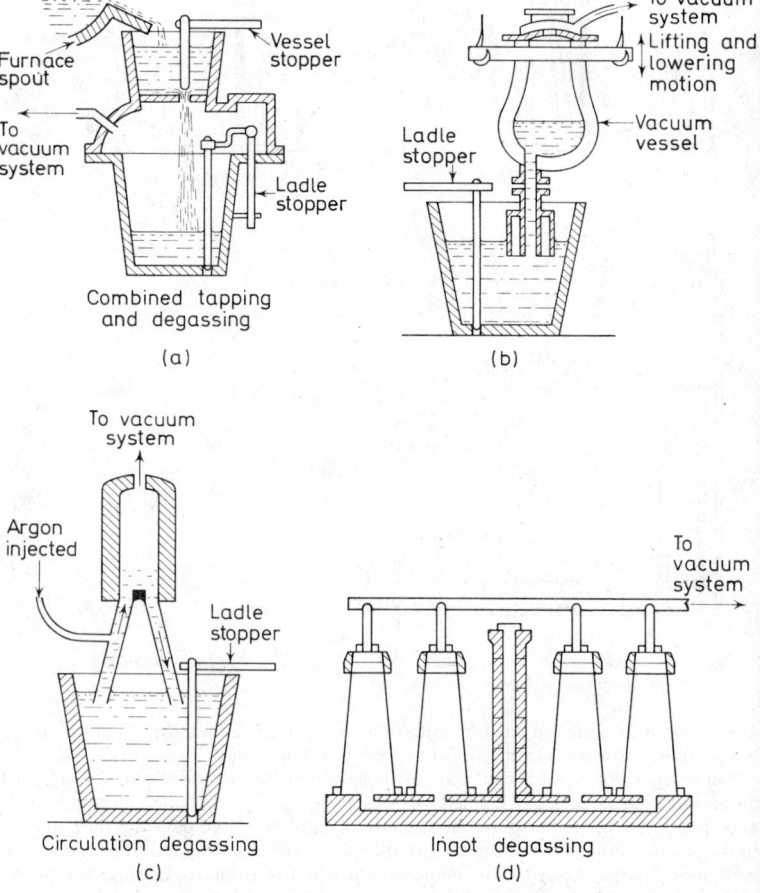

Figure 5.10. Vacuum degassing processes—schematic (a) Combined tapping and degassing, (b) Dortmund-Hörder process, (c) Ruhrstahl-Heraeus process

DEGASSING IN SMALL QUANTITIES BY VACUUM LIFTING DEGASSING[28]

The steel in the ladle is caused to circulate through a separate vacuum vessel held above the casting ladle, Figure 5.10(b). Any quantity of steel can be so treated. Degassing efficiency is high and thorough mixing is achieved. Temperature losses are reduced by heating the vacuum vessel.

DEGASSING BY CIRCULATION[28] R-H PROCESS

By suspending a vacuum vessel over the casting ladle, with two pipes connecting the vessel to the liquid steel and injecting inert gas, the steel can be made to flow continuously through the vacuum vessel, Figure 5.10(c). Efficiency is good and quantities handled can be large (100 tonnes of steel taking about 23 min). Oxide content is lowered in non-killed steels, the metal rising into the vacuum chamber with a boiling action that releases oxygen, hydrogen and nitrogen.

INGOT DEGASSING[28]

Immediately after teeming is finished, ingot moulds set on a plate and fed from a central trumpet, can be capped with vacuum tight covers connected to a flexible pipe leading to the vacuum system, Figure 5.10(d). Under vacuum the carbon-iron oxide reaction in an oxidised steel proceeds almost to completion as the CO product is continuously withdrawn from the system; the ingots effervesce, as in conventional rimmed steel practice, but the process does not lead to the segregated core containing oxides and sulphides typical of conventional rimmed ingots. The sulphide inclusions are more uniformly distributed within the core.[29]

Ingot degassing is specifically aimed at improving 'soft' steel products by reducing the total non-metallic content and producing smaller non-metallic inclusions more uniformly distributed; for the removal of hydrogen from killed steel, ladle or stream de-gassing is more applicable.

Degassing applications

Degassing adds to the cost of steel but the improvement in quality is usually justified by the increased reliability. The applications have defined objectives according to the type of steel and end product:

ENGINEERING ALLOY STEELS

Reduction of hydrogen ensures freedom from hair line cracking in billets, blooms and heavy forgings. From a frequency curve peak value of 4 mls/100 gms of hydrogen before degassing (range 2 mls/100 gms–6·5 mls/100 gms) the hydrogen content was reduced to 2 mls/100 gms by ladle degassing and to 1·5 mls/100 gms by stream degassing[30] both values being less than the critical amount necessary for crack development.

PRODUCTS REQUIRING SPECIALLY CLEAN STEEL. BEARINGS, ROLLS FOR COLD ROLLING ETC.

Under severe duty involving cyclic loading, the stress concentrating effect of aluminous cyclic loading, the stress concentrating effect of aluminous inclusions, the products of conventional deoxidation practice, can lead to unreliable service life due to fatigue failure. But the steel must be fully deoxidised which involves the use of aluminium.

Degassing by an appropriate method, can deoxidise the steel to a degree sufficient to complete the removal of residual oxygen by silicon additions[31], thus avoiding the harmful aluminous inclusions.

PLATES FOR WELDED PRESSURE VESSELS

Rolled plates produced by conventional steelmaking methods are liable to show lamellar tearing at welded joints, associated with non-metallic inclusions and low ductility in the through-thickness direction.

Susceptibility to lamellar tearing is markedly reduced or eliminated by appropriate degassing practice.

NOTCH DUCTILE STEELS FOR WELDED FABRICATION

At low service temperatures, welded fabrication in ordinary quality mild steel, owing to residual stresses and weld defects causing stress concentration, are liable to brittle fracture. To reduce the risk of this occurring, notch ductile steels not susceptible to embrittlement by quench or strain ageing should be used, thus eliminating the significance of the weld stress concentrating defects.

Reliable age resisting steels must be manufactured to be oxygen free and to have the nitrogen combined as nitride, usually aluminium nitride. Such steels, made by conventional bulk steel manufacturing methods in which the deoxidising method is by precipitation in the ladle have an abnormally high non-metallic content, aluminous inclusions being particularly evident. These steels appear to be relatively susceptible to lamellar tearing and the benefits of appropriate deoxidation by degassing followed by aluminium treatment in reducing the non-metallic content are self evident.

NON-AGEING STEELS FOR SHEET PRODUCTS

Using stream degassing, partial deoxidation takes place by the $C + FeO \rightarrow CO + Fe$ reaction, which does not produce non-metallic inclusions. During teeming an accurately calculated addition of aluminium is made, of which 25% is used to combine with residual oxygen and 75% alloys with nitrogen and the steel. About 0·5 kg/ton is added.[32] The product has improved uniformity of properties important in forming operations.

GENERAL UP-GRADING OF STEELS BY VACUUM MELTING OR DEGASSING

As indicated in the sections dealing with steel melting practice, steel quality in any specific category of steel, whether rimmed, semi-killed or killed, depends on the control over deoxidation and extent of deoxidation. Lack of proper control over the oxygen removal conditions in a rimming steel will produce an unsatisfactory product; equally, inefficient oxygen removal from a killed high carbon alloy engineering steel will produce unreliable products.

The use of appropriate vacuum degassing techniques promotes control of oxygen removal and enhances reliability of products of this type, i.e. high carbon engineering and low carbon sheet steels. However, the category of steels which could benefit considerably from this technique is that which covers low carbon low alloy structural and pressure vessel steels where the low carbon content is necessary for resistance to heat affected zone cracking when welding.

In steelmaking, the lower the carbon content of the molten steel the higher the oxygen content, Figure 5.11, so that such steels contain relatively high oxygen contents before deoxidation. They therefore tend to have a high non-metallic content if made by the basic open hearth furnace and deoxidised by precipitation in the ladle, a practice applied to the bulk production of steels for pressure vessel plates and high tensile structural steels, because efficient removal of the non-metallic products of deoxidation cannot be consistently attained. However, if the pressure of the atmosphere is reduced the $C + FeO \rightarrow CO + Fe$ reaction proceeds until equilibrium between carbon and oxygen is re-established at a lower oxygen level. This is shown in Figure 5.12[33] where the line A for 1013 mbar (1 atm) is the line A in Figure 5.11 over the carbon range 0·03–0·13%; the effect of reducing the pressure (e.g. by vacuum degassing) to 133·3 and 13·33 mbar abs appears in curves B and C respectively. Since carbon is lost to the system as well as oxygen the theoretical effect on a steel initially at 0·05%C is shown in line 1 connecting

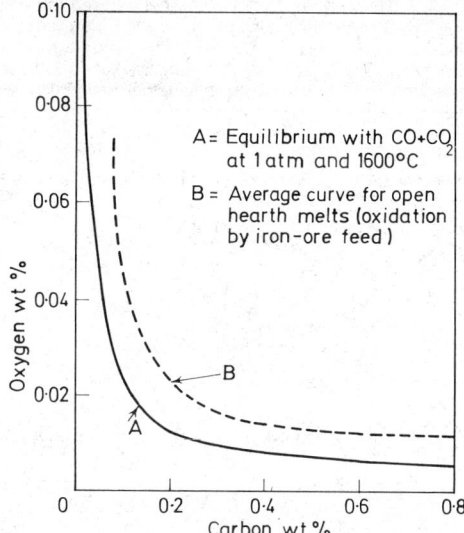

Figure 5.11. Relation between oxygen and carbon contents under equilibrium condition (pressure CO + pressure CO₂ = 1 atm. at 1600°C) and in open hearth melts

the three curves. The actual effect on steels in the 0·05–0·13% C range, as determined on commercial steels with various vacuum degassing techniques is shown in lines 2–5.

Vacuum degassing thus effects reduction of the oxygen content, provided it is applied before the addition of deoxidising elements such as silicon or aluminium, to levels less than half those obtained in the best practice in steels air melted and refined at atmospheric pressure. Consequently, final residual deoxidation can be effected with the minimum amount of aluminium or silicon to produce clean steels. This carbon deoxidation is beneficial in the production of forgings for tube sheets, where non-metallics in the steel can lead to defective welds in the thin-walled butt joints between tubes and tube sheet if conventional de-oxidation practice is followed.

In general, for the reason illustrated in Figure 5.11, vacuum degassing techniques properly applied to permit carbon deoxidation, are the most economical way of upgrading steels and, in particular, low carbon steels in bulk quantity. It should be noted that nitrogen removal by vacuum degassing presents difficulties compared with hydrogen removal since the diffusion velocity of nitrogen is very much lower (100 times). However oxygen-blown steel is low in nitrogen *ab initio* and degassing applied to this steel provides a product low in oxygen, hydrogen and nitrogen.

A welding fabrication problem that can be costly to correct is the appearance, on the radiograph of butt welds in silicon killed pressure vessel steel, of numerous small defects in the heat affected zone. These are small fissures associated with silicate inclusions. Though the cause is not generally agreed, experience suggests atomic hydrogen from the weld metal passes into a pre-existing discontinuity between the inclusion and the matrix steel and the pressure developed by conversion of hydrogen to the molecular state in the cavity promotes tearing which is analogous to blistering. Fissuring is not observed at sulphides. This type of defect is not generally seen at welds in acid open hearth steel, which is a low hydrogen steel due to vigorous and prolonged boiling from a high carbon dead-melt, but occurs in plates in pressure vessel grades made by the basic

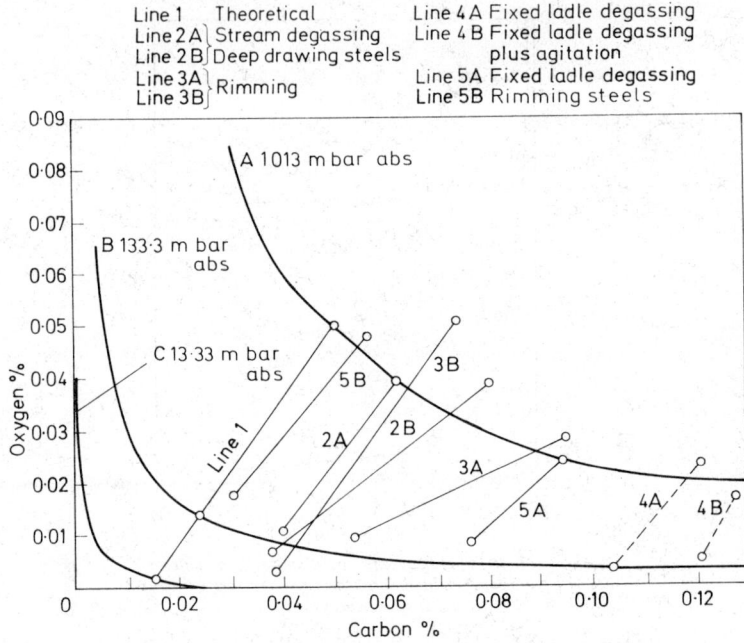

Figure 5.12. Decrease in oxygen and carbon caused by degassing

open hearth where hydrogen can be relatively high. Degassing techniques applied to these pressure vessel heats would be expected to upgrade quality by reducing the total content of silicate and by minimising the size of the silicate inclusions and thus reducing the susceptibility to this type of welding defect.

METALLURGY OF STEEL IN THE SOLID STATE

Aspects of the metallurgy of steel in the solid state of interest to design engineers are those primarily concerned with the interpretation of specifications defining steels and their properties, the factors which affect reliability of specifications and those which determine the dependability of steel in respect of expected service performance.

The metallurgy of the solid state starts at the ingot stage, proceeds through hot and cold working and heat treatment and finishes with the steel possibly undergoing further heating (i.e. welding, hot forming), cold forming or heat treatment at the hands of the purchaser.

Ingots. Solidification behaviour. Macro and micro-segregation

A pure liquid metal freezes to crystals of uniform composition within each crystal and each crystal has the same composition as all others. When the liquid metal, e.g. iron has dissolved other metals or metalloids (C, Mn, Si, S, P), although it is uniform in composition when solution and diffusion of these added elements has been completed, upon solidification the behaviour is not as in the pure metal; instead selective freezing occurs and the resultant solid metal is non-uniform in composition.

Solidification begins with the formation of crystals rich in the element or elements which raise the freezing temperature of the alloy. As these crystals grow, the liquid left between the crystals becomes rich in those elements which lower the melting point of the alloy. This segregation must take place following the physical laws governing crystallisation of an impure liquid. In effect the primary, secondary and tertiary arms of the dendrites are of purer material than the interstitial liquid which is the last to freeze. The latter, in the case of steel, will have a high concentration of C, S and P. This segregation must occur at all positions in the ingot. The iron-carbon constitution diagram (see Figure 5.13) is an indication of the constituents which form from the cooling liquid under near equilibrium conditions of extremely slow heating or cooling. However, there is evidence to show that, in addition to the above segregation, the carbon, sulphur and phosphorus are concentrated in the uppermost regions of the ingot, to an extent which increases as the size of the ingot increases.

Although there is not general agreement as to the cause of this macro-segregation, its occurrence is not in dispute. Some evidence as to the extent of segregation appears in the First Report of the Heterogeneity of Steel Ingots Committee,[34] and suggests that apart from the highly segregated zone at the top of the ingot, C, S and P show horizontal concentration as less in the centre of the ingot than at some intermediate position between centre and outside and vertical concentration as an increase from bottom to top. In general, the cast analysis, determined on the ladle stream, shows higher impurities than the ingot at positions below the middle (Table 5.3). This pattern of segregation is typical of C, S and P in silicon-killed ingots, furnace de-oxidised for high quality forgings.

In contrast to this mode of segregation, the behaviour of silicon is interesting; gravimetric analysis shows that silicon does not segregate markedly as determined by analysis i.e. silicon dissolved in the iron. However, there is marked segregation of silicates[34] in such ingots, these being concentrated at or near the central axis at the bottom of the ingot with a maximum concentration about one-third the distance up the ingot. Later work[35] showed that the silicate concentration area depended on the steelmaking conditions and deoxidising method there being three distribution types which are shown in Figure 5.14 and are listed below.

Distribution type	Silicate type	Finishing slag
1	Infusible 50–75% silica	Highly oxidising. 30% FeO
3	Fusible. High FeO (50%) Low silica (30%), MnO (5%)	Siliceous with low FeO (15%)
2	Intermediate between 1 and 3	

Segregation of C, S and P in semi-killed steels, for plates and sections, is of the same order as in killed steel.[36] Silicon concentrates in the lower middle part of the ingot, as in some killed steels, but to an even more marked degree;[37] this must indicate a concentration of silicates in this region. Free-cutting steel containing high sulphur shows no greater sulphur segregation than a steel of normal sulphur content.[38] The addition of lead to high sulphur free-cutting steel tends towards less sulphur segregation.[39]

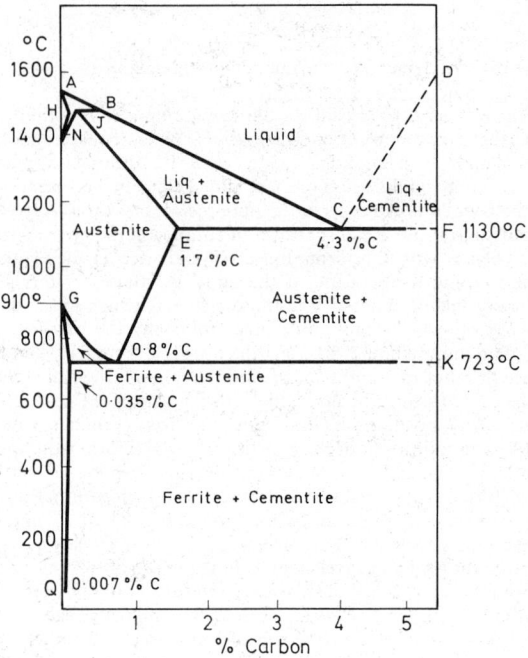

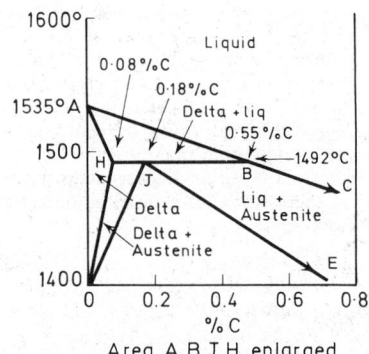

Area A,B,J,H enlarged

Ferrite = alpha = delta = body centred cubic lattice
Austenite = gamma = face centred cubic lattice

Figure 5.13. Iron-carbon constitution diagram

Table 5.3   INGOT SEGREGATION, TYPICAL DATA

HORIZONTAL. ONE-THIRD DOWN FROM TOP

Ingot		Outside		Intermediate	Centre	
No 4	2·8 tonnes	C	0·34	—	0·43	—	0·34
		S	0·039	—	0·045	—	0·039
		P	0·048	—	0·052	—	0·041
No 10	10·7 tonnes	C	0·29	0·29	—	0·32	0·29
		S	0·019	0·019	—	0·022	0·017
		P	0·01	0·01	—	0·011	0·011
No 12	12·5 tonnes	C	0·43	0·43	0·45	0·40	0·39
		S	0·028	0·03	0·033	0·024	0·026
		P	0·036	0·041	0·043	0·032	0·028

VERTICAL. CENTRE LINE

Ingot		Bottom end						Top	
No 10	10·7 tonnes	C	0·24	0·22	0·22	0·29	0·34	0·42	—	—
		S	0·011	0·011	0·008	0·017	0·026	0·042	—	—
		P	0·009	0·009	0·009	0·011	0·011	0·015	—	—
		Cast Analysis C 0·3%, S 0·017%, P 0·010%								
No 11	20·3 tonnes	C	0·17	0·17	0·16	0·17	0·19	0·22	0·23	0·25
		S	0·027	0·025	0·024	0·022	0·026	0·025	0·034	0·035
		P	0·042	0·038	0·037	0·037	0·036	0·049	0·055	0·051
		Cast Analysis C 0·21%, S 0·032%, P 0·04%								
No 15	112 tonnes	C	0·17	0·15	0·22	0·26	0·28	0·35	0·37	0·45
		S	0·025	0·021	0·029	0·033	0·34	0·036	0·045	0·059
		P	0·025	0·024	0·026	0·026	0·031	0·036	0·045	0·050
		Cast Analysis C 0·27%, S 0·033%, P 0·024%								
No. 16	175 tonnes	C	0·19	0·24	0·29	0·37	0·43	0·67	0·49	0·57
		S	0·016	0·021	0·027	0·038	0·046	0·058	0·068	0·085
		P	0·024	0·022	0·03	0·033	0·044	0·055	0·069	0·080
		Cast Analysis C 0·32%, S 0·027%, P 0·033%								

(First Report of Committee on Heterogeneity of Steel Ingots. JISI 1926 No. 1, pp. 39–151. Silicon Killed Steels for Severe Duty.)

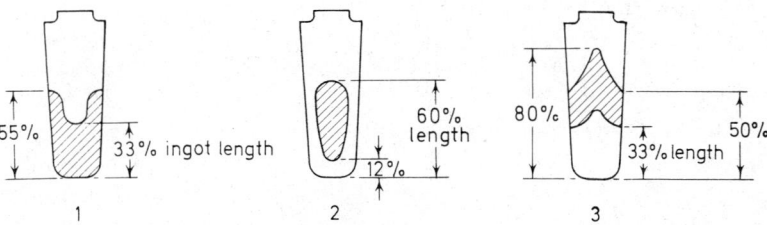

Figure 5.14. Silicate inclusion distribution in ingots (silicon killed). Three distribution types

Since semi-killed steels for plates are of particular interest to engineers concerned with welded fabrication and such steels, with higher manganese and niobium treated are used for weldable high tensile steels, the degree of segregation of C, Mn, Si, S and P observed in six ingots studied by the Committee on the Heterogeneity of Steel Ingots is given in Table 5.4.[40]

The test sample analysis corresponds to the analysis quoted on the mill sheets and the range per cent of C, Mn, Si, S and P is indicative of the variation to be expected in the plates rolled from the ingots. The carbon variation is particularly important in respect of weldability, bearing in mind that manganese shows a tendency to micro-segregation (i.e. segregation in the inter-dendritic material)[41, 42, 43 and 44] which produces alternate bands of high and low manganese content. Consequently the higher tensile steel plates with manganese in the range 1·2–1·8%, with or without niobium, may consist of layers of high and low manganese bands with carbon substantially higher or lower than the cast analysis. Unfortunately welding will affect the bands as if they were separate entities

Table 5.4   RANGE OF COMPOSITION IN REPRESENTATIVE INGOTS FROM SEMI-KILLED PLATE STEELS. ANALYSES FROM DRILLED SAMPLES AT THE SEVEN STANDARD POSITIONS, EXCLUDING E

	Carbon				Manganese		
Test L.S.	Max.	Min.	Range %	Test L.S.	Max.	Min.	Range %
0·10	0·18	0·12	65	0·49	0·496	0·475	4
0·14	0·231	0·14	65	0·52	0·578	0·558	4
0·13	0·19	0·11	61	0·48	0·47	0·45	4
0·15	0·23	0·16	47	0·49	0·53	0·5	6
0·19	0·21	0·15	32	0·51	0·52	0·51	3
0·13	0·16	0·1	46	0·5	0·51	0·49	3

	Silicon				Sulphur		
Test L.S.	Max.	Min.	Range %	Test L.S.	Max.	Min.	Range %
0·012	0·016	0·004	100	0·037	0·065	0·032	89
0·024	0·01	Trace	42	0·039	0·058	0·02	97
0·02	0·018	0·01	40	0·039	0·051	0·024	69
—	—	—	—	0·059	0·078	0·048	51
0·075	0·083	0·061	29	0·041	0·047	0·032	36
0·024	0·025	0·022	12·5	0·033	0·041	0·029	36

	Phosphorus		
Test L.S.	Max.	Min.	Range %
0·036	0·05	0·03	56
0·024	0·035	0·017	75
0·024	0·038	0·021	71
0·061	0·078	0·053	41
0·042	0·055	0·036	45
0·02	0·023	0·016	35

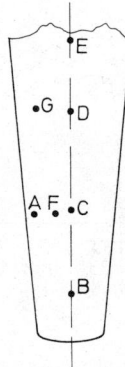

Range % = % of amount of element in ladle stream test sample.

with strong hydrogen cracking susceptibility in the high manganese high carbon bands. In general, macro-segregation results in increasing amounts of C, S and P towards the top of the ingot, so that plates rolled from this part of the ingot may differ materially in weldability, i.e. resistance to hardened zone cracking, from those obtained from the bottom end. On the other hand, silicate inclusions tend to segregate towards the bottom so that defects such as lamellar tearing, fissuring in the heat affected zones in non-hardening steels which tend to be associated with such inclusions are more likely to be experienced in plates from the bottom end.

Micro-segregation of the inter-dendritic type, which is present in varying degree in all ingots, is an additional factor in determining the variation in composition, on a micro scale, of the dendritic and interdendritic parts of the ingot. On forging or rolling, this causes compositional variation to adopt a banded formation, the degree of variation differing from one rolled or forged product to another depending on time and temperature in reheating for hot working and subsequent heat treatment; these factors influencing diffusion of the segregating elements. Micro segregation in an alloy steel ingot affects the hardenability of rolled bar,[41] producing material from the bottom end of the ingot having low hardenability in the centre of the bar, but this effect is almost absent in material from the top end.

It is important to bear in mind, that, as a result of the combined effects of macro segregation and micro segregation, ingots are not uniform in composition on a macro scale (segregation of C, S and P) nor micro scale (Mn, Ni, Mo, P) and, although hot working reduces the degree of compositional variation, it tends to persist, particularly where the amount of hot work imposed is not large. The resulting banded products will have properties associated with those of high and low chemistry bands, which may cause a departure from the properties anticipated from a product of assumed uniform composition as given in the cast analysis.

In the absence of analysis and metallographic examination of the particular finished product of interest to the purchasing engineer, he should note that inferences derived from cast analyses must be conditional upon the following:

(a) The cast analysis is usually taken after one-third of the ingots have been teemed; those teemed before and after this sample will not necessarily have the same composition.

(b) Each ingot will produce material with differing properties according to the position in the ingot from which it came when subjected to the same heat treatment.

(c) Specifications must necessarily give 'making' ranges of permitted chemical composition, so that each specification in reality involves variation in cast analysis, variation in steel analysis from start to finish of each cast and variation in macro and micro-segregation according to casting pit and melting practice. Consequently the specification may not be an adequate and full description of an individual rolled steel product in terms of special properties such as weldability and hardenability. It is however based on experience and the mechanical property ranges are normally obtained from casts meeting the specified composition range.

Gases in solidifying ingots

HYDROGEN

Unless vacuum melted or vacuum degassed liquid steel contains hydrogen. This is derived from the raw materials and the atmosphere in the furnace. Hydrogen is soluble in liquid iron to the extent of approximately 28 millilitres per 100 grammes of metal at 1550°C, the solubility increasing as the temperature rises above 1550°C[45] (Figure 5.15). This is the quantity of hydrogen which can be measured in mild steel weld metals derived from coated electrodes which contain hydrated minerals in the coating. Analysis of liquid steels does not yield such high figures, the range being 2–6·5 ml/100 gm with 4 ml/100 gm occurring most frequently.[46]

However, evolution of hydrogen during solidification, so called 'hydrogen wildness', can be observed in certain circumstances (e.g. 4% Silicon steel) which, according to the solubility of 28 ml/100 gm might be assumed to infer this content in the steel. This is not the case, much lower contents being present. Schenck[47] proffers an explanation, based on the reduction in solubility of hydrogen in the crystals forming in a solidifying ingot. Reference to Figure 5.15 and the iron-carbon diagram Figure 5.13 shows that,

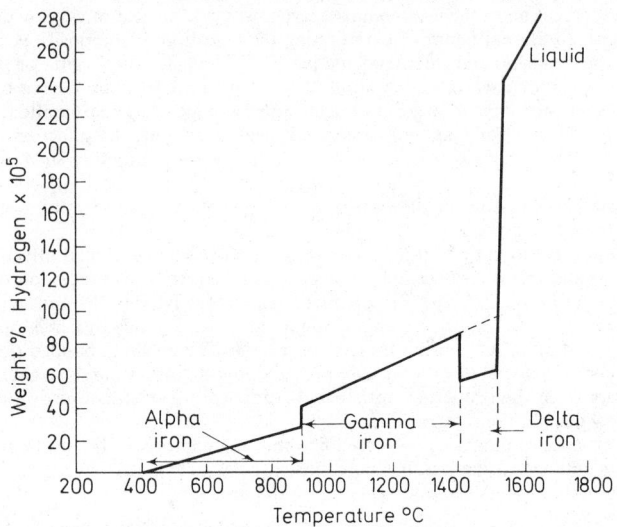

Figure 5.15. Solubility of hydrogen in pure and low-alloyed iron (Physical Chemistry of Steelmaking. H. Schenck, p. 154)

for steels up to 0·5% C, delta iron crystals (body centred cubic lattice) separate first from the melt and the hydrogen solubility in delta iron is about one-quarter the solubility in the liquid. As gamma iron (austenite) is formed, the solubility increases then decreases with falling temperature. On changing from gamma (face centred cubic lattice) to alpha (same as delta, body centred cubic lattice) there is a sudden drop in solubility and this continues to fall as temperature falls.

At the first separation of delta dendrites or gamma dendrites (steels above 0·55% C) from the liquid, the liquid will, if not saturated, dissolve more hydrogen passing out from the dendrites. Consequently, just before complete solidification, the liquid to solidify last (high in elements which lower the melting point, C, S, P etc.) will also be high in hydrogen and possibly saturated. Evolution of hydrogen as gas from pockets of late solidifying liquid is therefore possible, leading to 'looseness' of the structure.

It is also concluded that, at complete solidification, the distribution of hydrogen in the ingot will not be uniform because it will have been constantly passing into the liquid from the solid phases during the solidification process. There will, therefore, be a concentration of hydrogen towards the core of the ingot and the top. Killed steel ingots which are kept open by a feeder head would be expected to retain less hydrogen than semi-killed.

Hydrogen only diffuses through steel when in the atomic form. The diffusion rate above 350°C is controlled by the lattice (i.e. whether body or face centred cubic) and the temperature, but below about 100°C it is primarily determined by release from hydrogen 'traps', such as dislocations or lattice defects which substantially reduces the rate. Between 100°C and 350°C it is a mixture of 'trap release' and lattice diffusion.[48] The diffusion constant for various temperatures is shown in Figure 5.16; in spite of the much higher diffusion rate at high temperatures and the comparatively long time at these temperatures, in heating for hot working and during subsequent heat treatment, leading to diffusion of hydrogen, the room temperature hydrogen content at the centre of large forgings (Figure 5.17[49]) can be approximately 3 ml/100 gm. This is the solubility (equilibrium with 1 atm H_2) in gamma iron at about 900°C and well above the solubility in alpha iron at room temperature.

The liquid steel from which the forgings were made had a hydrogen content of 3·4 ml/100 gm suggesting that there would be no interdendritic hydrogen segregation because this is below the solubility for delta iron. However, if the molten steel contains six or more ml/100 gm hydrogen, interdendritic concentration would occur (reference 50 Figure 4, gives 4–8 ml/100 gm for basic open hearth and 3–7 ml/100 gm for basic electric steel).

It might be supposed that hydrogen diffusion should lead to lower amounts of gas in small solidified ingots compared with large, however this does not appear to be the

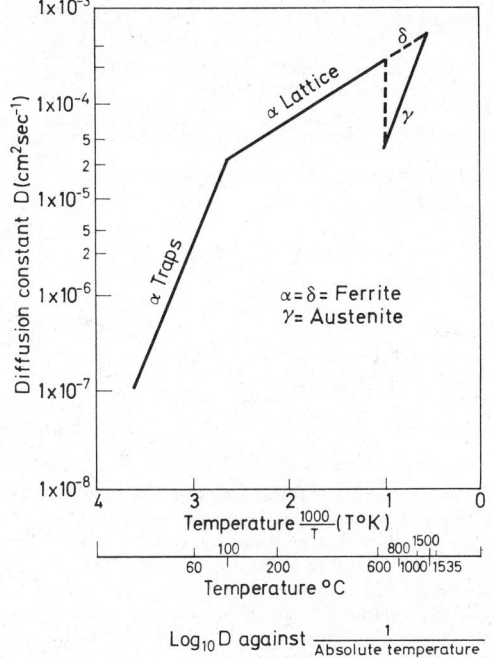

Figure 5.16. Diffusion of hydrogen in iron

case as shown in Table 5.5.[50] Low carbon steels do not show a lower molten metal hydrogen content than high carbon steels, for example[51] fully killed Basic Electric forging steels above 0·25% C had an average hydrogen content of 4 ml/100 gm (range 2–6·5 ml/100 gm) and fully killed basic electric low carbon tube steels with carbon less than 0·2% had 4·5 ml/100 gm.

Although there is little or no data on the hydrogen content of killed and semi-killed steel ingots for plate production the above information suggests a liquid steel hydrogen content range of 4–8 ml/100 gm (as for basic open hearth steel,[50] Figure 4) and a hydrogen content in the centre of the solidified ingot (15–30 tonnes) of 3–7 ml/100 gm. Thus, heats at the top end of the hydrogen range would produce interdendritic and central hydrogen segregation.

The practice of sending ingots, for plates and sections, hot to the soaking pits to conserve heat would tend to maintain a high degree of hydrogen segregation in the rolled product, which will be interdendritic and highly local in character. The effect of this hydrogen on material properties depends on composition and end-usage. In general, problems

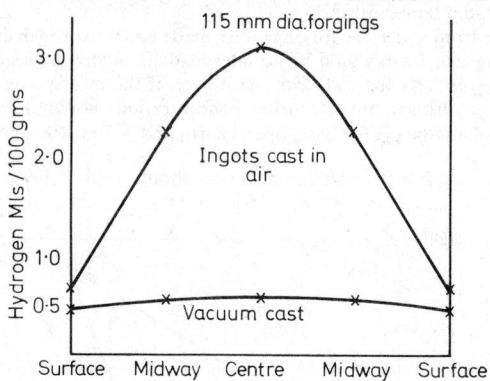

Figure 5.17. Hydrogen in Ni-Mo-V forgings after heat treatment

Table 5.5 HYDROGEN IN SOLIDIFIED INGOTS (ml/100 gm)

	8-tonne ingots			21-tonne ingots		
	Surface	Midway	Centre	Surface	Midway	Centre
Top	1·25	2·25	2·8	1·0	1·6	1·6
Middle	2·25	4	5	1·0	2·5	3·5
Bottom	0·8	1	2	1·0	1·6	1·6

12% Cr-Mo-V Steel.   0·35% C Steel.

arise only where the end-product has thickness exceeding about 50 mm. Sheet and wire are not ordinarily affected by ingot hydrogen, but blooms, billets and forgings may develop hairline cracks if allowed to cool out too quickly in hardenable compositions because transformation of austenite to alpha iron takes place in the presence of hydrogen segregated in the interdendritic austenite region.

Transformation of alloy gamma to alpha produces acicular micro structures (martensite) requiring an expansion or volume increase which, when occurring at low temperatures in the interdendritic regions in otherwise transformed steel, sets up severe internal stress. Stress in the presence of hydrogen above the solubility limit promotes cracking in alpha iron. Hairline cracks develop after an 'incubation' period during which hydrogen is diffusing out and, by lowering the concentration in the segregated region which has transformed to martensite at a lower temperature than the dendritic areas, due to higher C and P, and has therefore generated the higher internal micro-stress system causing the martensite to be subjected to severe stress in the presence of diffusing hydrogen with consequent susceptibility to cracking.

Incubation depends on ambient temperature and other factors but may take hours, days, weeks or months. When the period for a particular steel product (i.e. billet, forging etc.) is reliably known from experience, it is possible to prevent cracking by ensuring that the customer, assuming further hot work to a smaller size is required, carries out this working and heat treatment well before the steel is due to crack. Although the descriptive composition of steel indicates the likelihood or otherwise of hairline cracks developing (alloy engineering steels, particularly those in the air hardening grades being more susceptible than 'softer' steels), the most significant factor is the degree of segregation of carbon, manganese and hydrogen which is not covered by specification.

Steels which, with low alloy and carbon content, would not be expected to crack may do so if the segregation effects produce a banded structure of high carbon high alloy bands alternating with low carbon low alloy bands. Some variability in hairline cracking susceptibility must therefore be taken into account unless the steel has been produced by a vacuum melting or de-gassing route. Hairline cracks, because of preferred orientation in the bloom or billet may not be welded-up in subsequent hot working and, if left in the finished product, may have disastrous consequences under cyclic loading or severe stressing.

Hydrogen dissolves in molten slag; acid furnace slag remelted under an atomic hydrogen welding torch absorbs approximately twice and basic furnace slag about three times the amount soluble in molten steel. Manganese silicate slag particles (from deoxidation) are fluid at temperatures between 1000°C and the solidus. During the early stages of slabbing a plate ingot, the deformation will produce a discontinuity between silicate and surrounding metal which will persist when the silicate becomes viscous. During cooling, hydrogen will diffuse into the discontinuity from the silicate to form molecular hydrogen exerting stress which is concentrated by the geometrical form of the laminar silicate. Hydrogen from the supersaturated matrix metal will also diffuse into these discontinuities.

Internal laminar defects, or fissures, in heavy plate of silicon killed mild steel can be observed by ultrasonic test and are associated with excessive silicate concentrations. Defective areas, associated with inclusions and not detectable by ultrasonic test, which may eventually be revealed as lamellar tears when the plate is welded, are revealed by low ductility (i.e. less than 10% elongation) in a 'through-thickness' tensile test. It is possible that both these defects arise from segregation of hydrogen in the solidifying ingot in conjunction with segregated non-metallic matter which conjointly lower the ductility.

OXYGEN

Oxygen is found alloyed with iron in varying amounts in most steels. It exists in steel either as iron oxide (FeO) in solid solution, in small amounts of manganese oxide (MnO) in solid solution or as non-metallic matter, or inclusions, where it exists as manganese oxide, silica and alumina as such or in various combinations of these oxides with each other. The oxygen in the inclusions is not alloyed with iron.

Iron oxide is dissolved in the molten steel in all steelmaking processes operating a refining slag; in the open hearth and Bessemer 0·5–0·6% FeO may be dissolved.[53] Upon deoxidation the amount is reduced to an extent depending on the type and quantity of deoxidants used and the efficiency of combination, which is dependent on uniformity of diffusion. The solubility of FeO in steel at the melting point is 0·94%[53] whereas at 1600°C 1·37% is soluble. Commercial steel contains less FeO in the molten state, the 'finished' liquid steel containing from traces to 0·30% FeO.[53]

Carbon greatly reduces the solubility of oxygen in liquid iron. The solid solubility of FeO in iron is 0·11% at 715°C.[53] However, the solubility decreases with falling temperature being approximately 0·01% at room temperature in pure alpha iron.[54] Consequently, the oxide is in forced solution in the alpha phase after the steel has cooled down to room temperature. When standing at room temperature, or slight heating, the oxide tends to precipitate out to attain an equilibrium value. This so called ageing process may take weeks or months at room temperature and causes embrittlement.

The correct use of strong deoxidants such as Al or Ti for deoxidation of the liquid steel will ensure a sufficiently low residual oxide content in the solid steel that ageing is prevented but the steel will be 'dirty', to a degree dependent on the original oxide content in the molten steel. Vacuum treatment, as indicated in the earlier section on steelmaking vacuum treatment, when correctly applied can lower the residual oxygen to satisfactory levels without producing a dirty steel.

In rimming steels (high iron oxide in molten metal) a concentration of dissolved oxide in the centre of the solidified ingot is to be expected, but in semi-killed and silicon killed steel ingots oxide dissolved in the alpha phase would not show segregation. Oxygen, in contrast to hydrogen, does not diffuse out from solid steel.

NITROGEN

Nitrogen is soluble in liquid steel; open hearth steel generally contains $0.004–0.006\%$ N_2 while Bessemer (air blown) may contain over 0.02% N_2. The solubility of nitrogen (as iron nitride) in alpha iron at room temperature is not more than 0.001%, having fallen from 0.02% at about $420°C$.[55] Consequently iron nitride may be precipitated under suitable conditions. It can be seen under the microscope as needles in low carbon open hearth and Bessemer steels and in weld metals. It is a cause of 'ageing' embrittlement and increases the hardness of soft steels.

The nitrogen dissolved in the liquid steel would not be expected to segregate in the solidifying ingot. To prevent the solution of iron nitride in the solid steel it must be combined with an element which forms an insoluble compound in the molten metal; Al or Ti are effective in this respect. Alternatively, by melting and refining the steel under vacuum nitrogen is not available to dissolve in the liquid steel and clean steels very low in nitrogen are obtained. For certain products, nitrogen is intentionally retained or added, for others it must be 'fixed', to improve toughness and reduce notch sensivity, or absent.

The solidified ingot will contain the same amount of nitrogen as the liquid steel. The nitrogen content of steel made by various processes is indicated below:

Basic oxygen blown converter	Less than 0.004% N_2
Side blown converter	$0.003–0.008\%$ N_2
Duplex (converter + open hearth)	$0.005–0.008\%$ N_2
Duplex worked down from high carbon bath (vigorous boil)	$0.004–0.006\%$ N_2
Basic open hearth	$0.004–0.006\%$ N_2
Bessemer (air blown)	$0.012–0.02\%$ N_2

Nitrogen, in contrast to hydrogen does not diffuse out from solid steel.

MECHANICAL WORKING OF STEEL

In attempting an understanding of plasticity in metals it is convenient to consider plasticity from two points of view:

1. Physical, i.e. the relation between plasticity and crystal structure.
2. Practical, i.e. the effect of factors such as applied forces, temperature and deformation rate on plastic behaviour.

Plasticity and crystal structure

Solid metals are composed of grains or crystals, in which the atoms have a regular geometrical pattern, separated from neighbouring grains by irregular shaped grain boundaries. The regular pattern of atoms is known as the crystal lattice and is made up of repeated groups of atoms, which are fundamental to the structure, known as unit cells. The lattice, or more correctly space lattice, is of course a three dimensional lattice-work of imaginary lines connecting the atoms in space. The type and dimensions of unit cells of most elements have been determined in the course of the development of X-ray crystallography.

There are seven crystal systems giving fourteen possible space lattices. Most metals crystallise in one or other of three of the fourteen lattices, i.e. the body-centred and face-centred of the cubic system and the close-packed of the hexagonal system. Some metals are polymorphic, meaning they can occur in more than one lattice system, iron being an example of this (Table 5.6).

The plasticity of metals is generally attributed to movements in certain directions along specific crystallographic planes, as a result of applied forces; all the atoms in one series of planes moving as a block relative to a similar series of parallel planes.[56] The planes of

Table 5.6   METALS HAVING SIMPLE SPACE-LATTICES

Lattice	Metals
Face-centred cubic	Aluminium, Titanium*, Lead, Copper, Iron*, Cobalt*, Nickel*, Gold, Silver
Body-centred cubic	Iron*, Vanadium, Chromium*, Niobium, Molybdenum, Tungsten, Zirconium
Hexagonal close-packed	Titanium*, Chromium*, Cobalt*, Nickel*, Magnesium, Zinc

* Metals having polymorphic habit.

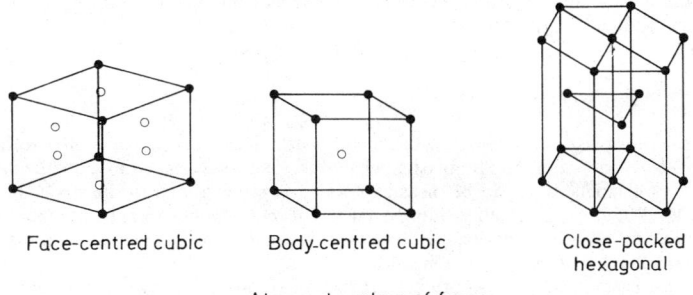

Face-centred cubic   Body-centred cubic   Close-packed hexagonal

o Atoms at centres of faces or body of cube

Figure 5.18. Three simple space-lattice systems unit cells

easiest slip generally have the greatest atomic concentration and distance between the intervening planes, the restraining forces acting between consecutive planes being a minimum. The gliding movement along these planes allows the solid metal to alter its external shape. The three simple lattices have more lines of atoms and atoms per plane than the complex lattice structures and hence show easier slip, i.e. greater plasticity. Within the simple systems the hexagonal close-packed metals are less plastic than the face-centred cubic.[57] The three simple systems are shown in Figure 5.18.

When an external force is applied to a crystal, it is transmitted across the internal crystallographic planes and can be resolved into components so that each slip system in the crystal will have a definite force component working in its direction. It has been proved experimentally that slip occurs when the force per unit area on a system exceeds a critical value, known as the Critical Shear Stress; this varies with the slip system. Most metals are used in a polycrystalline habit, although single crystals of large size have been produced for research purposes. Slip will occur first of all in the crystal and system in which this critical stress is exceeded; subsequently slip occurs in other systems in each of the many crystals making up the polycrystalline metal.

The theoretical force necessary to cause slip in a pure metal is much greater than is observed experimentally, and the explanation appears to derive from the fact that crystals are never perfect but contain dislocations (imperfections of the lattice structure) which allow slip to occur in successive steps by setting up internal forces between the atoms. The theoretical force necessary to cause one dislocation to slip is extremely small but this is opposed by the internal stress set up by other dislocations and imperfections such as foreign atoms and grain boundaries so that in a polycrystalline metal containing multiple imperfections the stress necessary for plastic slip, or yield stress, is very much greater and is dependent on the density of the imperfections. Experimental evidence obtained in the electron microscope shows the marked increase in dislocation density with increasing plastic straining.

Plastic straining beyond yielding, creates new imperfections which, in turn, require the applied force (stress) to be increased to continue the deformation as long as the total of imperfections increases. This is the situation at low temperatures with consequent increase in tensile strength (strain hardening) during cold deformation. However, at high temperatures the thermal energy reduces the imperfection density and also assists the movement of the dislocation, tendencies operating to oppose strain hardening. Thus in hot deformation, the net result of these opposing tendencies depends on rate of deformation, a high rate leading to an increase in imperfections but with a low rate the recovery effect due to the thermal energy may reduce the level of imperfection density to that approaching the annealed condition. The essential difference between hot and cold deformation, in this concept of metal plastic behaviour, is that the stress necessary to produce plastic flow at low temperatures is independent of rate of straining but is very dependent at high temperatures.

Plasticity in practical application

In considering plastic behaviour of the metal 'in the mass' one is concerned with the effect of stress and strain rate on plastic behaviour. If stress is applied to a ductile metal, a degree of elastic strain is to be measured which is proportional to the stress up to a certain level of stress at which yielding occurs. This corresponds to reaching maximum shear stress on the 45° plane; behaviour after this point is reached depends on the type of plastic behaviour typical of the metal.

In certain cases plastic flow continues at constant nominal stress (true stress increasing taking into account reduction of cross-sectional area); in others increased stress is necessary to continue plastic flow owing to 'work hardening'. As stress is increased, fracture eventually occurs when the true breaking stress is reached. It should be noted that a proper understanding of the plastic behaviour of a particular metal sample under tensile loading requires a recorded stress-strain diagram and information on the rate of straining. A high strain rate can increase the true breaking load above the value obtainable at normal rates of straining.

Plasticity is an important aspect of the deformation of metals at elevated temperature. It is intimately bound up with the precipitation phenomena which accompany and influence 'creep' resistance and 'creep' ductility.

Principles of hot working

Mechanical working of steel is based on certain aims and principles. One requires deformation without fracture, fracture occurring when the tensile stress exceeds the cohesive strength of the atoms in the lattice.

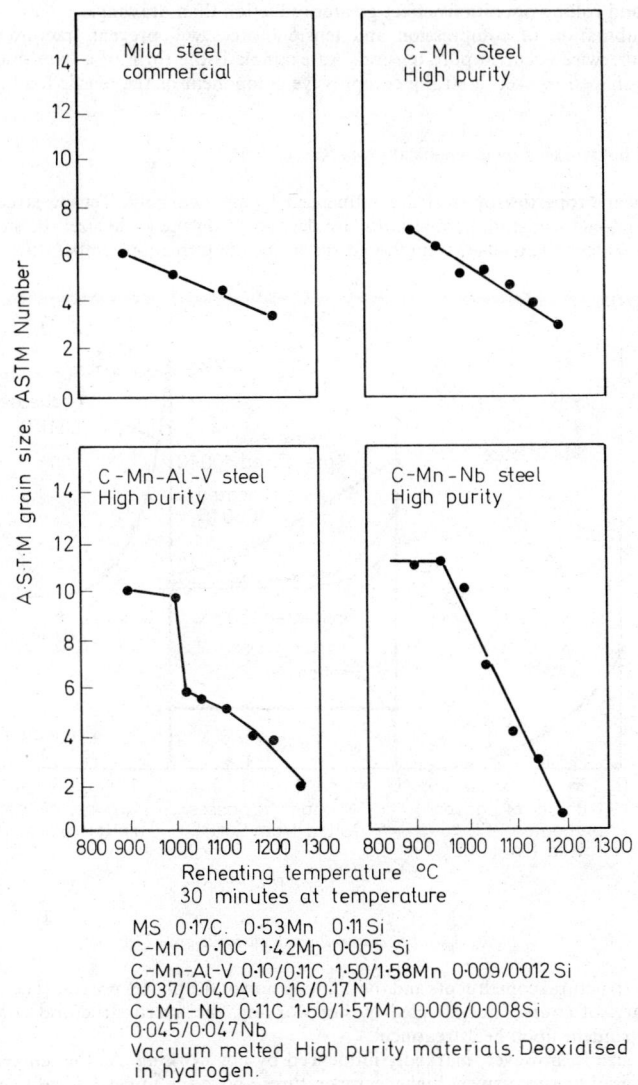

Figure 5.19. Austenite grain size related to reheating temperature

By working the metal under compressive loading, the tensile components of the stress are less likely to reach the critical level necessary for fracture initiation so that compressive loading generally achieves greater plastic work before fracture. For example, forging and rolling operations effect greater reduction than drawing.

A combination of compression and tension forces will prevent fracture where it would otherwise occur in pure tension. An example is the form of a wire drawing die which is shaped to exert a strong compressive component of the tensile load.

Effect of hot working on mechanical properties of steel

Mechanical properties of steel are influenced by hot working. Tensile strength and notched impact transition temperature are dependent on the grain size, the smaller the grain the higher the tensile strength and the lower the transition temperature, the type

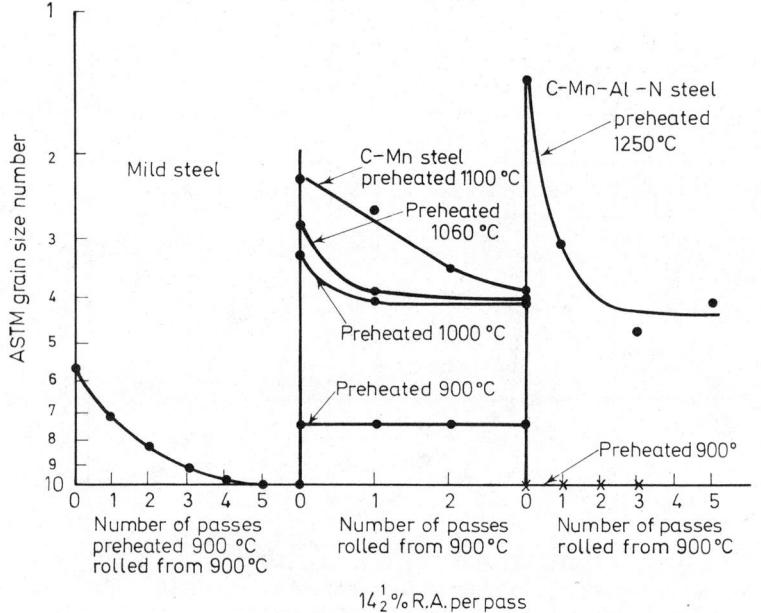

Figure 5.20. Effect of rolling on austenite grain size

of micro-structural constituents and the composition of the phases present. The two latter factors are, of course, not affected by hot working, the micro structural constituents being dependent upon heat treatment.

Grain size is, however, markedly influenced by hot working. As the temperature of steel is raised to successively higher temperatures above the upper critical temperature into the gamma phase field (Figure 5.13) the gamma grain size increases unless inhibited by the addition of aluminium. This inhibiting effect is not observed up to all temperatures and growth, having been restrained up to about 1050°C, starts at a greater rate than in a steel not treated with aluminium; the net result is a similar gamma grain size at 1200°C in the treated steel (Figure 5.19[59]).

Hot working of steel is done at the highest permissible temperature, mild steel ingot temperatures for cogging being in excess of 1200°C. If the grain size shown in Figure 5.20 at 1200°C persisted after forging or rolling, the mechanical properties would be substantially affected adversely. Hot working, however, reduces the austenite grain size, as shown in Figure 5.20[60]; a fine grain austenite produces a fine grain ferrite after transformation. The results quoted in Reference 59 also suggest that an adequately hot worked coarse grained austenite produces a finer grain ferrite than austenite not hot worked. The improvement in mechanical properties arising from this diminution in grain size is indicated by a fall in Charpy impact transition temperature from −40°C, when not worked, to −60°C for hot worked C-Mn steel reheated to 1060°C, rolled 5 passes at $14\frac{1}{2}$% reduction per pass.

The above values apply specifically to the particular steels but the principle that hot work improves the mechanical properties is generally applicable; rolled plates and structural sections which finish rolling at high temperatures and/or with a relatively small amount of work produce a relatively coarse grain structure with inferior impact properties. The importance of degree of working is illustrated in Figure 5.20, the steels tending to reach a limiting grain size which depends upon composition. If, during hot working, it is necessary to reheat for further working, the reheating eliminates the original grain structure and the final structure will depend on the amount of work put in after the last re-heating.

Where the rolling mill stands are sufficiently strong, mild and low alloy steels for structural work (i.e. plates and sections) may be rolled to finish rolling at relatively low temperatures thereby inducing a fine ferritic grain structure with improved tensile strength and notch toughness. This practice is known as 'controlled rolling'.

Methods of hot working

Hot working can be effected in the following ways:

Forging, either by hammers or presses.
Rolling.
Extrusion.
Hot piercing and rolling (tubes).

THE FORGING PROCESS

Hammer forging is used either to break down an ingot into a bloom or bar, or to work down a billet to a rough finished shape, or to shape a piece from a blank in a closed die (drop forging). For breaking down an ingot in certain steels, the hammer is essential as rolling is found to cause fracture; under skilled control the hammer can impose a higher ratio of compression to tension forces than is possible in rolling but it is important that the hammer weight be suitable for the work (too light sets up uneven stresses while too heavy may cause cracking).

Steels that are hammer cogged are typically high speed and tool steels generally, some stainless and heat resisting steels with compositions which produce a two-phase micro structure, the difference in plasticity between the gamma and alpha phases causing ruptures under the strong tension forces induced during rot rolling.

Ease of rolling or forging is determined by composition and ingot casting practice and varies from one plant to another. To assist in hot workability so called 'minor' elements may be added to steel, such elements not being included in normal specifications. For example, boron is commonly added to certain stainless steels, and other stainless steels may have additions of nitrogen which stabilises the gamma phase thus preventing development of a two-phase structure. Forging or rolling is most easily carried out on a single phase structure because plasticity is uniform, so ingots or blooms are heated to temperatures which, as far as possible, ensure the steel is in the gamma (austenitic) phase field (Figure 5.13).

At ultra high temperatures, delta phase is formed from austenite so there is generally a preferred hot working temperature range, the lower end being determined by resistance to deformation and risk of cracking and the upper end by onset of delta formation, particularly in certain stainless steels, and excessive grain growth leading to 'overheating'. The forging or rolling temperature ranges which avoid defects in the product are established from experience.

Where semi-finished steel is purchased from the steelmaker for further hot working by forging, it is most important to keep closely to the recommended hot working temperature range. Heating for forging is an important aspect of successful forging practice. The rate of heat input must not be excessive otherwise there is risk of 'burning' the outside of the ingot or bloom and the heat must soak right through. The furnace atmosphere is important; for example, sulphur content must be a minimum for reheating nickel bearing stainless steels otherwise nickel sulphide is formed at the surface. This is liquid at forging temperatures and, following the grain boundaries, induces cracking or bursting of the forging.

The same considerations apply to heating for rolling. High carbon steels (over 1% C) should be preheated 700–760°C before placing in the forging furnace. In general, the higher the carbon content of the steel, the lower the forging temperature range and the smaller this range. Maximum forging temperatures for American carbon steels are shown in Table 5.7. The safe forging temperature depends on the steel composition, furnace

Table 5.7   MAXIMUM FORGING TEMPERATURES[58]

% carbon	Maximum forging temperature °C
Plain carbon steels	
0·1	1 315
0·3	1 286
0·5	1 259
0·7	1 215
0·9	1 176
1·1	1 133
Alloy engineering steels*	
0·1	1 286
0·2	1 259
0·3	1 231
0·4	1 231
0·5	1 204
0·6	1 204

* SAE 2300, 3100, 3200, 3300, 3400, 4100, 4600, 5100, 6100.

atmosphere, temperature measuring equipment (accuracy and reliability), furnace control equipment (accuracy and reliability) and the type of forging operation (cogging, drop forging). The actual temperatures used may be well below the maxima indicated in Table 5.7.

Forging temperature ranges for plain carbon, alloy, and high speed tool and die steels are given in Table 5.8. The American temperatures are generally higher than British practice; high temperatures are avoided to minimise grain coarsening but the rate and extent of coarsening is partly determined by the steelmaking practice so the difference may be due to inherent steel properties.

Forging by press is usually applied on large masses of steel. Ingots of 305 tonnes have been press forged in Britain to make hollow vessels (pressure drums). The essential difference between press and hammer forging is the slow pressure application of the

Table 5.8 FORGING TEMPERATURE RANGES, TOOL AND HOT WORK DIE STEELS (AMERICAN AND BRITISH PRACTICE)

Type of steel	C	Cr	Mo	W	V	American Start °C	American Finish °C	British Start °C	British Finish °C	Remarks
Carbon Tool (Water hardening)	0·7/1·2 0·85/1·1	— —	— —	— —	— —	981–1 093	— 815	950 —	750 —	General tools and dies. Drill and saw steel.
Tungsten (Oil hardening)	0·9/1·15 1·25	0·5 0·5	— —	0·5/1·6 1·5	— —	981–1 063	871 —	— 950	— 750	
Tungsten Finishing Steel (Water hardening)	0·1/0·3	—	—	3·5/3·75	—	1 008–1 063	871	950–1 010	800	Preheat slowly to 842°C. Slow cool in insulating material. For fine finishing cuts, cold dies, punches, gauges, etc.
Tungsten HSS. (Pil or Salt Bath hardening)	0·8	4	—	18	1	1 120–1 176	926	1 050–1 100	900	Slow cool and anneal after forging.
Hot Work Die Steel Cr-Mo	0·3	5	1	1·25	—	1 093–1 149	898	1 050–1 090	850	Preheat 650–700°C, slow cool, anneal.
Hot Work Steel Tungsten	0·35	3	—	10	0·5	1 120–1 176	898	1 100	900	Preheat 800–850°C, slow cool, anneal.

American data from Sub-Committee on Tool Steels, Metals Handbook 1939 Edition, pp. 991–1031.

former. The softer action of the press is advantageous in upsetting operations which are frequently employed on large ingots to improve central zone structures and distribute the non-metallic matter in preferred orientation. Press forging temperatures are similar to those of hammer forging.

Forgings have, over many years, gained a reputation for dependability under the most severe service conditions. This is largely based on superior quality, toughness and strength characteristics, achieved by the sequence of operations necessary and the care taken to achieve a correct amount of grain refinement and internal soundness. It is normal practice to apply special care to all phases of the forging operation including heating and, given equal attention to rolling variables, it is claimed the rolled product can be of comparable quality.

However, rolling is normally a unidirectional flow process whereas in forging it is common practice to upset the metal once or more times so producing a product with less

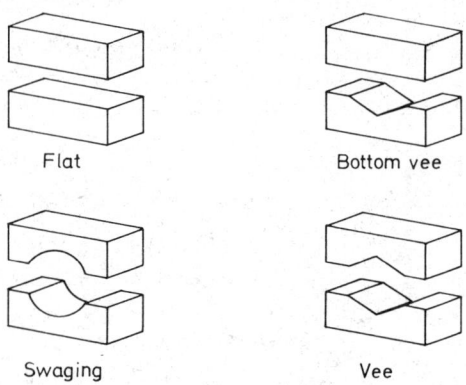

Flat   Bottom vee

Swaging   Vee

Figure 5.21. Die shapes for open-die forging

marked directional properties. The 'through-thickness' properties of rolled products, such as plates, are distinctly inferior to those obtained by testing longitudinally or transversely to the rolling direction.

Press forging is always 'open-die' forging, the metal being shaped between suitable dies and the accuracy of the finished forging being dependent entirely on the skill of the press operator. The usual die shapes are indicated in Figure 5.21. The ingot or forging is held between the press tools in a porter bar, supported by a sling at the centre of gravity of the load, the sling providing the rotation necessary during forging.

Forging begins with 'cogging' the ingot, which is usually cast in a round corrugated or fluted mould, the corrugations preventing ingot surface cracking. Small reductions are used all round the ingot to remove the corrugations and then heavy working is applied, generally to produce an octagon shape, then a square, back to an octagon and so on. This operation is known as drawing-out and produces rounds, rectangular blooms, shafts, etc. For forgings which are made into rotating machinery, such as turbine discs, turbine shafts, etc. where the properties of the steel are of major importance in withstanding severe radial and tangential stresses, upsetting is used to produce circumferential flow. The ingot is forged to a bloom (octagonal) and a slug is parted off; this slug is then upended and forged down to about one-quarter its length. It is then reworked in the original direction.

Hollow forgings (i.e. shells for pressure vessels, rings etc.) are produced by piercing or punching a hole in the centre of the forged workpiece into which is passed a mandrel bar. The wall thickness is reduced and the piece elongated along the bar in stages starting at the centre. As there is a limit to the size (diameter) of a mandrel bar, shells of greater internal diameter are produced by expanding the hollow cylinder; reducing the wall thickness but not increasing the length. This is done by using a stiff bar inside the cylinder as a bottom die; the length of the cylinder that can be expanded is limited by the need to maintain bar stiffness.

APPLICATIONS OF FORGINGS

Forgings are necessarily expensive compared to rolled or cast steel products. For certain applications the shape of the finished article precludes the use of a rolled steel product, an obvious example being a crankshaft; here the choice between a casting and a forging depends upon the level of imposed stresses and the extent to which overloading can be expected. In most crankshaft applications forgings are selected for reliability. Where the shape does not rule out a rolled product, for example, a wide range of shapes can be gas cut from rolled plate and fabricated by welding, to simulate products usually made from forgings, the choice again depends on the nature of the service and the significance of reliability.

For severe duty where failure could be catastrophic, such as chemical pressure vessels (ammonia synthesis columns etc.) forged or forged and welded vessels are frequently chosen. The forged and welded vessel, with forged end-closures circumferentially welded to the body which may be a single forging or a series of forged rings circumferentially welded to produce the body, has the advantage over vessels fabricated from rolled plate that the forgings are not susceptible to lamellar tearing and longitudinal welded seams are avoided.

Lamellar tearing at welded seams is always a risk in rolled plate made by conventional bulk-steel air melting practice with ladle deoxidation and longitudinal welds are subject to the full hoop stress so that detection, by non-destructive tests of all critical size defects is vital in highly stressed vessels.

Forgings in alloy and stainless steels are widely used in nuclear engineering, undercarriage and engine parts of high performance aircraft and for the moulds used in producing centrifugally cast pipes where service conditions are particularly onerous. Forgings are normally used for the turbine and generator shafts of hydro-electric power plants and for the high speed rotor and turbine shafts of steam or gas turbine generators.

ROLLING

In hot working by rolling, the material (i.e. ingot, billet, slab or sheet), is passed between rolls turning in opposite directions, the space between the rolls being less than the thickness of the entering material. The rolls grip the material, reduce its thickness and increase its length; they may be flat as required for rolling slabs or slab ingots into sheets or plates or grooved as needed for rolling shapes.

The products of hot rolling mills are either semi-finished, i.e. hot rolled sheet, billets, rods, or finished (i.e. sections, round and flat bar, plates), a characteristic metallurgical feature being the difference in ductility and notch toughness determined longitudinally or transversely to the rolling direction.

Starting with the ingot, the first stage is to cog down to a suitable thickness for subsequent rolling. Cogging mills produce blooms, also known as blooming mills, for re-rolling into billets. For flat products, such as plate, the ingot is cast rectangular instead of square and is known as a slab ingot being cogged down in a slabbing mill. The rolls for the blooming mill have collars to produce a square or rectangular bloom, those for the slabbing mill are flat. Cogging and slabbing mills are normally two high reversing, Figure 5.22(a). In the next stage the bloom is reheated and rolled to a billet in a billet mill and a slab is rolled to plate in a plate mill or to sheet in a sheet mill.

Reduction in thickness must be effected in stages. The rolls, mounted in a strong rigid housing, are screwed down after each pass of the material, the amount of screw-down depending on the 'bite' of the rolls. Instead of using one pair of rolls in a stand and taking separate passes, it is possible to set several roll stands in tandem, each stand having rolls rotating faster than the previous set. This arrangement is known as continuous rolling; substantially greater output is achieved at much higher capital cost.

Some notes on the rolling mill types indicated in Figure 5.22 are given below:

Type A. Two high reversing. Blooming and slabbing mills. Plate mills. Billet mills.

Type B. Two high non-reversing. Billet mills, bar mills. The mill stands are arranged over a relatively wide mill floor to allow for transfer tables to move the material from one stand to another. Sometimes known as cross country mills. Stands set 'in-line' are nowadays only used for light sheets and pack-rolling as the material must be returned to the front of the stand by hand over the top roll. This is mainly applicable to alloy steels such as high speed steel sheets which require many light passes.

Type C. Two high continuous. Non-reversing. Roll stands are set tandem. Peripheral speed of rolls increases down the line. Billets, bars, sheet, strip.

Type D. Three high. Non-reversing. The material is returned to the front of the stand by mechanically operated lifting tables. Mostly for sheet mills but almost out-of-date.

Type E. Four high. Generally non-reversing and stands set tandem. For plate, sheet and strip. The combination of large diameter back-up and small diameter work rolls minimises deflection and therefore reduces the camber in flat products.

Type F. Planetary mill. Non-reversing. Rolls feed slabs into the work rolls of the mill. These small diameter rolls are retained in a cage around the periphery of the large diameter back-up rolls. They rotate around the back-up rolls at about half the speed of the latter. The work roll stand is followed by a skin pass finishing stand to improve the surface. The mill converts slabs into hot rolled strip in one pass.

HOT EXTRUSION

Hot extrusion is a process of hot working which is widely applied in the non-ferrous field. It has only a limited application in steel manufacture. The presses necessary for steel extrusion must be rigid and strongly constructed, perfect alignment of the die and centre line of the press chamber being essential. Pressures up to 2030 tonnes are in order.

Hot extrusion of steel depends upon glass as a lubricant. Owing to the high cost of the press and ancillary equipment, and the fact that the billets used must have machined surfaces, the process tends to be used for stainless and other expensive alloy steels which may be required in small batches or to unusual shapes.

Hot extrusion produces a good surface finish. Tubular products can be produced by extruding over a central mandrel using billets with axis holes drilled slightly larger than the bore of the finished tube.

SEAMLESS HOT FORMED TUBES

Seamless tubes are hot formed by causing a solid round billet to produce a hole at its axis as a result of rolling between two tapered rolls set side by side and inclined at opposite angles of about 10° to the centre line of the billet axis.

As the hole develops, the billet passes over a mandrel bar which sizes the hole. This is known as a Mannesmann piercing mill. The re-heated tube from the piercing mill is then put through a plug rolling mill in which grooved rolls cause the tube to be forced over a plug (mandrel) which thins the wall and enlarges the bore.

Welded tubes and pipes are widely used. The process entails passing coils of strip through forming rolls which turn the edges towards each other to form a seamed tube, the seam then passing into a welding machine or furnace. Welding may be by electric

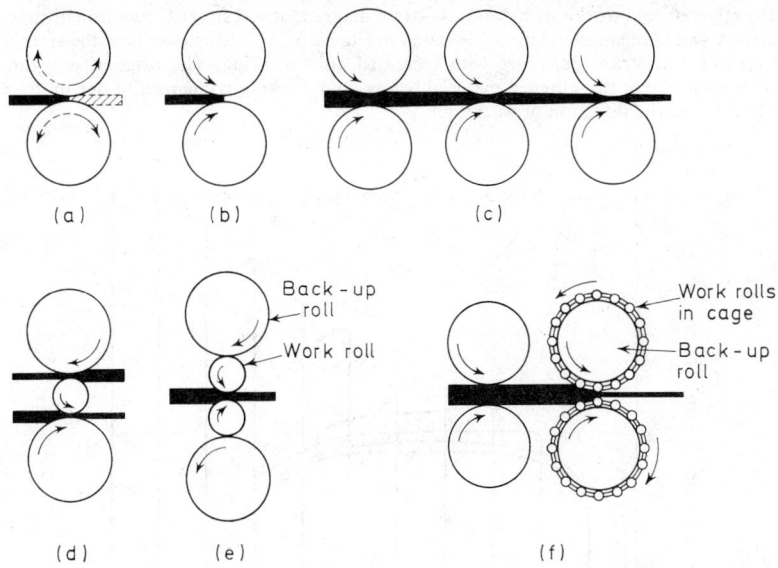

Figure 5.22. Hot rolling mills (schematic)

resistance or induction or fusion welding. In the former, welding rolls force the edges together at the point of heating so making a pressure weld, in the latter the bevelled edges provide a suitable vee which is fused with the addition of filler wire.

Practically all types of steel can produce fusion welded steel tubing. For lower pressure tube applications the strip is passed through forming rolls and a heating furnace and immediately it emerges it enters welding rolls which force the butting edges together to form an upset pressure butt weld. This process is mostly applied to mild steel strip.

These processes are not strictly within the category of hot forming of steel, but they are mentioned because they are widely used in the manufacture of steel tubing.

Cold working. Metallurgical principles

Although cold working is widely applied as a forming process for finished steel products (i.e. pressing, spinning, etc), it is proposed to consider here only cold working as applied to products which are cold worked in the steel mill. These cover cold rolled products (i.e. sheet, strip) and cold drawn products (i.e. round, flat and special shape wire, tubes).

Cold work, involving deformation, is essentially the same mechanism as hot working but being carried out at temperatures below the recrystallisation temperature there is no accompanying recovery. Providing the temperature can be maintained high enough there is no limit to the amount of hot work but there is a strictly limited amount of cold working which can be applied to steel. Cold work causes the yield point of soft steel to disappear, the tensile strength to increase and the ductility to diminish. In contrast, deformation or working of lead, tin or zinc at room temperature is accompanied by recrystallisation so that it resembles the hot working of steel.

The increases in hardness (tensile strength) in cold worked steel is related to the amount of work; several light reductions in wire drawing, for example, having the same effect as one or two heavy reductions given the total reduction is the same in each case. The effect of cold working on the stress-strain diagram of soft steel (Armco iron ultra low carbon and manganese steel) is illustrated in Figure 5.23 which shows how the strength increases and strain decreases with increased cold working. The original polygonal grain structure of the softened material before cold working is changed to one of grains elongated in the direction of working.

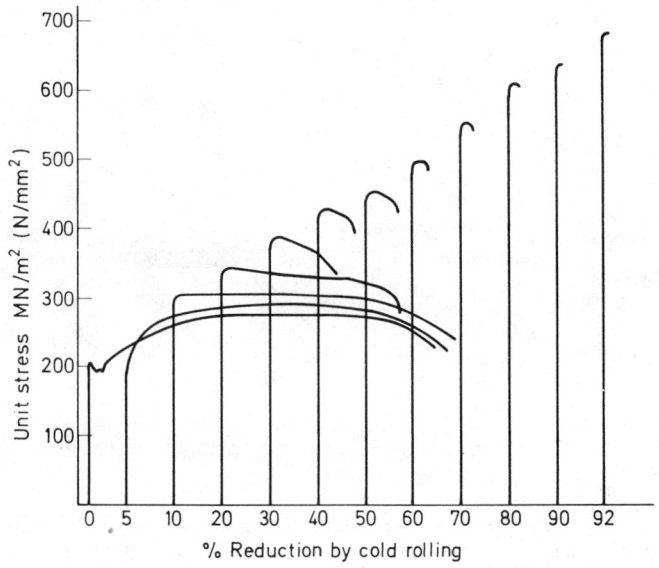

Figure 5.23. Effect of cold rolling on the shape of the stress-strain curve. Armco ingot iron (Kenton and Burns)[61]

The ability to take cold working depends on the plasticity of the phases present in the structure; a small content of non-plastic phase does not prevent cold working, as the plastic grains flow round the non-plastic. Iron carbide (cementite) is a constituent or phase in mild steels, which are readily cold worked, but itself is brittle and non-deformable.

In high carbon steel, which forms the basis of wire ropes, containing about 0·85% C, the cementite forms a major constituent of the structure. Provided the heat treatment has been correct before drawing, the ferrite grains can flow round the fine cementite. The annealed structure with coarse large cementite cannot be drawn, the steel must be cooled at a fairly fast rate to produce very fine cementite particles by patenting. The extent to which steel may be cold worked before fracture and the amount of change in tensile strength varies considerably with the composition and heat treatment, both affecting the phases present, and the initial grain size. In practice, drawing and cold rolling practice is determined by experience.

The above-mentioned change in mechanical properties of soft steel due to cold work is not permanent. These properties change slowly if the material is held at room temperature and more rapidly at slightly higher temperatures, known as ageing. The effect is seen in an increase in yield point, tensile strength and hardness and reduction in ductility and notch impact values.[62] Above 350°C the effect decreases. For lower temperatures the maximum effect is achieved by holding for longer times. The disappearance of the yield point (Figure 5.23) is observed if the test is carried out shortly after cold working but ageing restores it.

Ageing is the result of precipitation of a solute element or compound from a super saturated solid solution. Ageing in cold worked mild steel is considered to be caused by super saturation of very small amounts of carbon or nitrogen precipitated along slip planes induced by the cold working. Ageing after cold working is known as strain-ageing, to distinguish it from quench-ageing, the latter being produced by precipitation of carbon or nitrogen on standing at room temperature or heating to low temperatures after rapid cooling from an elevated temperature. Larger amounts of carbon and nitrogen are required to induce quench-ageing than for strain-ageing. Steels which have been strongly deoxidised with aluminium or titanium tend to be non-ageing; oxygen itself promotes strain-ageing but not quench-ageing. This may be because oxygen reduces the solubility of carbon and nitrogen in ferrite.[63]

Strain-ageing susceptibility is an important factor in various customer applications for steel products which leave the steel mills in a cold worked or slightly cold worked condition. For example, cold rolled mild steel sheet and strip for deep drawing and stamping commonly receive a temper rolling (a light skin pass) as the last operation which may render it susceptible to ageing if the melting practice and production history are such as to pre-dispose it to ageing. Unless the pressing or stamping operations are completed reasonably soon after rolling there is a risk that ageing, with consequent reduction in ductility, will have set in and the material prove unsatisfactory. Deep drawing grade steel, in the annealed condition (fully softened), has a strong inflection in the stress-strain curve at the yield point and this causes stretcher-strain marks in stamping or drawing operations.

As indicated above, cold working eliminates this and also the stretcher-strain marks but, after ageing, the return of this yield point inflection again may promote the appearance of stretcher-strain marks. Thus, storing the lightly rolled coils at the customer's plant may induce stretcher-strain marking during cold deformation.

Structural steel (i.e. sections, angles, etc), although not cold worked in the steelworks, if made from rimming or semi-killed steels, may show decided embrittlement from strain-ageing if they are cold worked sufficiently during fabrication at the customer's plant. Heating to critical ageing temperatures (i.e. galvanising, welding, etc) in the vicinity of cold worked areas (i.e. punched hole, sharp bends, etc) may promote definite embrittlement. Where embrittlement arising from ageing must be avoided the use of non-ageing steels for cold worked applications should be considered.

Objectives in cold working

Hot formed steel products are cold worked for one or more of the following reasons:

To obtain thinner material.
To obtain better surface finish.
To obtain increased mechanical properties.
To obtain products of greater dimensional accuracy.

Methods of cold working

Steel is cold worked in the steelworks either by rolling or drawing through dies.

COLD ROLLING

Rolling mills for cold reduction of hot rolled sheet or strip (after cleaning by pickling and oiling) are either single stands of reversing two high, four high or Sendzimir type (Figure 5.24) or continuous tandem stands of four high type. The single stand type

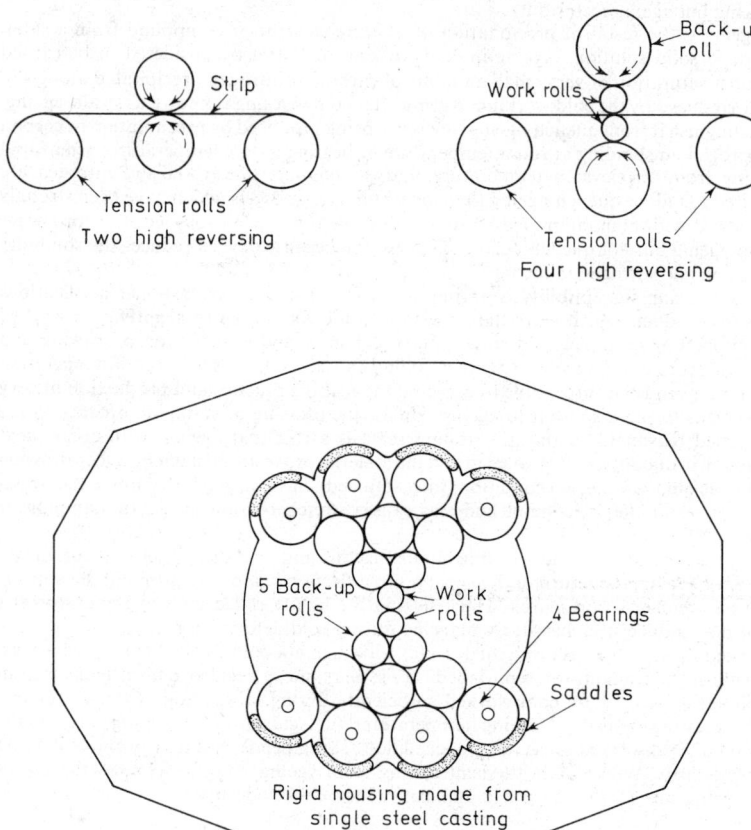

Figure 5.24. Cold rolling mills (schematic)

generally uses tension rolls at front and back. The cluster mill produces strip of exceptional dimensional accuracy.

Steels for cold rolling to thin, flat products e.g. strip, or sheet may be made rimmed (open top, mechanically or chemically capped), semi-killed or killed and melted in the basic open hearth, basic electric furnace or basic converter. Oxygen blown steel is widely used. Because of its good ingot surface, rimmed steel is particularly suited to the manufacture of cold rolled flat products and the bulk are made from this type of steel. A typical composition will be in the range 0·04–0·12% C 0·2–0·6% Mn 0·04% S max. 0·04% P max. Residual elements should be low, Cu and Ni less than 0·15%, Cr less than 0·05%.

Mechanical capping, which suppresses the rimming action at the desired point and leaves the core of the ingot with a uniformly distributed oxide content, gives a more uniform hardness in the cold rolled product across the width and depth of the section than ingots cast open top where variation in rim-in occurs. This may be important to some users.

Chemical capping with aluminium suppresses the rimming action by removing the active oxygen in the core; the steel, tending to be oxygen free also tends to be non-ageing after cold working and, if made correctly, is particularly suitable for extra-deep-drawing applications.

The hot rolled strip, which is the raw material for the cold rolling mill, is rolled from slabs, reheated after surface defects have been removed, or in some mills the hot rolled strip is rolled direct from the ingot. Obviously, in the latter practice, any strip surface defects arising from faulty ingot surface will appear in the hot rolled coil where rectification is practically impossible. The former practice, owing to inherent steelmaking variables, is likely to lead to a more uniform and reliable product.

COLD DRAWING

Cold drawing through dies is used to produce wire (i.e. round, flat, special shapes), tube and bright bar.

Apart from special alloy steels such as heat resisting (for electric resistance elements), stainless etc, in which the alloying elements confer particular properties of oxidation or corrosion resistance, wire production is confined to plain carbon steel, since alloying elements are not required for depth hardening, and low alloy steels for use as consumable welding electrode wires. Combinations of cold working and heat treatment applied to plain carbon steel wire enable mechanical properties to be developed which will meet practically all the demands of the customer.

Heat treatment during wire drawing. Annealing and patenting

The increase in tensile strength as the amount of drawing increases for three carbon ranges is shown in Figure 5.25.[64] Ductility falls as the tensile strength increases (Figure 5.26[64]). When the limit of reduction has been reached the wire must be heat treated to remove the hard drawn microstructure and replace it with a suitable structure for further reduction. For low carbon steels this treatment is a sub-critical anneal, just below the lower critical temperature, which recrystallises the ferite grains to an equiaxed form.

Medium and high carbon wires are generally patented (fairly fast cooling from above the upper critical point by air cooling or quenching in lead) to give a fine pearlitic structure which will draw to very high tensile strengths. Additional to subcritical annealing and patenting, the heat treatments used in wire production include normalising, annealing, hardening and tempering and austempering, all designed to confer structures and properties which have particular relevance to the requirements of specific wire applications.

The tensile strength obtainable depends on carbon content and an approximate indication of the relationship for annealed, patented and hardened and tempered wire is shown in Figure 5.27. Wire has a relatively large surface to volume ratio so that any decarburisation due to heat treatment has a proportionately more significant effect than in heavier steel products. Consequently wire heat treatment is conducted in specialised

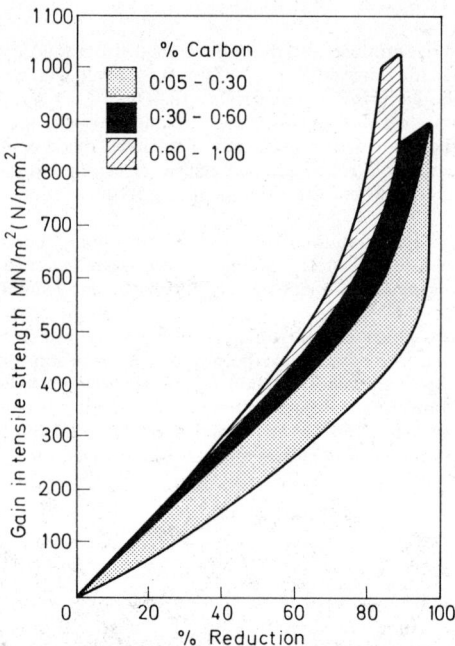

Figure 5.25. Increase in tensile strength related to reduction in wire drawing

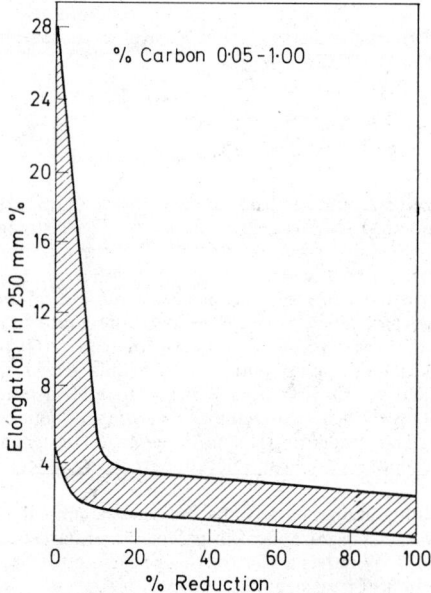

Figure 5.26. Decrease in ductility related to reduction in wire drawing

equipment (i.e. salt baths, atmosphere controlled furnaces, etc) aimed at minimising any such difficulties.

Wire drawing

Cold drawing through dies, while simple in principle, is only successful with considerable experience in:

    (a) Die design with particular reference to the die angles.
    (b) Die lubricants.
    (c) Rod cleaning and baking.

The quality of wire surface is primarily affected by the surface condition of the cleaned (acid pickled) rod, which depends on the type of scale produced in the rod mill and the efficiency of the pickling bath. Careless pickling can leave the centre waps of the hot rolled coil inefficiently clean; the adherence of scale varies even in coils off the same mill. Defects in the surface of billets from which the rod is rolled can persist through the rod

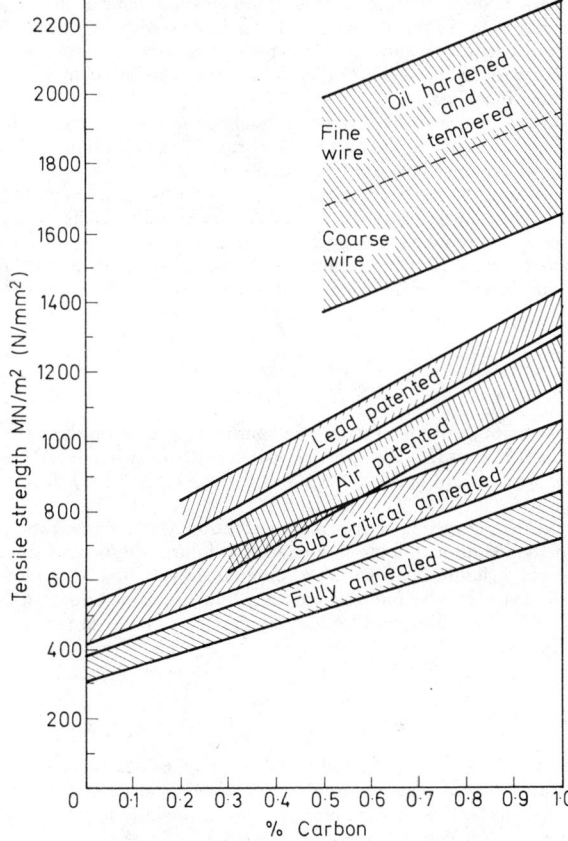

Figure 5.27. Tensile strength—carbon relationship (approx) for wire[64]

stage and will be drawn into the wire surface causing seams or slivers. Any scale left on the rod surface after cleaning will induce scratches in the wire surface.

Billet surface conditioning (grinding, chipping) is practically essential for subsequent rod rolling to produce high tensile wire, where surface defects may endanger service performance. Cleaning of hot rolled rod in pickling baths is usually followed by acid removal by high pressure water wash and neutralising in hot lime solution (milk of lime). The hydrogen absorbed in pickling must then be removed, otherwise the wire would break immediately it entered the first hole, by heating at 200°–250°C in a baking stove. The rod then passes either to dry drawing or wet drawing.

DRY DRAWING

The lime coated wire, the coating thickness is increased for increasing number of holes, is drawn through a die box, ahead of the die, which contains special soap lubricant. The composition of the soap can affect the behaviour of welding wire if it is used in certain gas shielded processes.

Heavy wire (12–25 mm) is drawn on 'bull blocks', one hole usually with a horizontal block. Wire from about 5–12 mm is usually drawn on standard blocks with vertical spindles with one hole. From about 5–3·25 mm blocks with two holes are generally used and for wires requiring three or more holes it is common practice to use multi-holing machines with up to seven blocks in tandem. The speed of successive blocks on these machines increases.

Soft steel rod can be drawn from rod size (oval about 6 mm) direct to about 1·6 mm on such machines. Coils of rod are butt welded, trailing end to leading end, to provide a continuous supply of wire to the drawing machines.

High carbon (usually considered above 0·7% C) rod is given a 'patenting' heat treatment before drawing starts, otherwise the dry drawing method has the same steps as for soft rod except that reduction per hole must be less. Also the wire die boxes are water or air blast cooled because the wire is embrittled if allowed to build up heat from drawing. High carbon wire must be repatented after about 75% reduction, whereas soft steel can go to about 90% before sub-critical annealing.

WET DRAWING

In wet drawing, the objective is to put a thin coating of copper on the wire which, after drawing, produces a better finish to the wire than with dry drawing. It is therefore usually applied to fine wires (less than 0·75 mm) and consequently to dry drawn wire rather than hot rolled rod.

The wire is pickled, dipped in copper plating solution, water washed and either stored under water or baked after dipping in a coating solution. Drawing is done through a lubricated die using liquid soap solution, compared with dry soap powder for dry drawing, which may also be pumped over the blocks, using single or multi-holing machines, the latter commonly having varying diameter blocks in tandem on a single spindle.

WIRE DRAWING DIES

Wire drawing dies are almost universally made of a cylindrical steel case containing a core or 'nib' of tungsten carbide or diamond (for drawing very fine wires in carbon and special steels). On the entry side there is a taper to the die hole size, the taper consisting of three separate angles. The exit side is also tapered.

Where the service conditions of the wire are unusual and not covered by a standard grade of wire, the full facts of the situation should be made clear to the wire drawer who can then apply special care in rod selection and drawing practice to avoid harmful defects. Inferior quality rods cannot produce a consistently high quality wire.

CONSTITUTION AND HEAT TREATMENT OF STEEL

In spite of competition from other materials, steel remains the dominant engineering material today. This is due partly to economic factors, but primarily to its versatility and adaptability. These properties, in turn, depend on the ability to vary and control the constitution, the distribution and nature of the micro-structural constituents, of steel at will by varying chemical composition, heat treatment and hot and cold working, with the resultant wide variation in the mechanical properties of the material over a temperature range extending from very low temperatures to around 1000°C.

To indicate how this variation and control is achieved, it is necessary to consider as briefly as possible matters such as allotropy of iron, metallic solid solutions, constituents of steel, phases, equilibrium, and the iron-carbon constitution diagram.

Allotropy of iron

Solid iron is crystalline in structure (cubic). Pure iron can exist in either of two allotropic modifications, the body-centred cubic or face-centred cubic (Figure 5.18); the important feature being that each is only stable within certain ranges of temperature as shown below.

Modification	Crystal structure	Temperature range of stability °C[66]
Delta	body-centred	1 535–1 400
Gamma	face-centred	1 400– 910
Alpha	body-centred	910 and below

On heating or cooling through the critical temperatures 910°C and 1400°C the change from one modification to the other is accompanied by recrystallisation.

Solid solutions

Solid solutions are formed when a metallic solid dissolves one or more elements in a manner similar to the solution of substances in liquids. The main features are that solubility is not determined by atomic proportionality and that the solute elements can diffuse as in liquid solutions. If the atomic size of the solvent and solute atoms is about the same, the solute atoms tend to replace the solvent atoms in an unchanged space-lattice but, if they are substantially smaller, it is considered likely that they are located in between the solvent atoms.

Theoretically pure iron at room temperature is composed of grains of alpha iron (ferrite). Ferrite is capable of dissolving in limited quantity elements such as silicon, phosphorus, nickel, copper, arsenic, etc, so that commercial steel of ultra-low carbon content will have ferrite grains with these elements in solid solution. It is a convention to refer to alpha iron in commercial steel as ferrite. Gamma iron (austenite) also dissolves elements to form solid solutions and not necessarily to the same extent as alpha or delta iron. If gamma iron dissolves an element to a greater extent than alpha iron, which is usually the case, on cooling through the critical temperature where the change to alpha iron occurs, the alpha must then accommodate more atoms of the element than its solubility allows and the alpha becomes super-saturated. Under suitable conditions precipitation of the excess atoms will occur in order to reach a stable condition.

A great deal of the heat treatment practice which is used to confer particular properties on steel is based on this effect arising from the combination of allotropy with the properties of solid solution. Metallic solids will dissolve compounds as well as elements to form solid solutions. The high solubility of iron carbide (Fe_3C), termed 'cementite' in gamma iron and its relatively low solubility in alpha (ferrite) is the fundamental basis for the heat treatment accorded to non-austenitic steels.

Stability in solid solutions may be difficult to achieve. In the iron-carbon system, the stable form of carbon is graphite. This is only produced in steel when silicon, nickel or aluminium are present and particular heating conditions obtain. For example, the cementite in aluminium killed steels for high temperature service can break down into graphite in the zone near welds. Graphite is a normal constituent of high silicon cast irons. Although steels containing cementite are strictly speaking in a metastable condition, for all practical purposes the condition is one of stability.

Phases, equilibrium and the iron-carbon. Constitution diagram

Constituents are the different homogeneous components of a metal alloy revealed by the microscope after suitable etching procedure. Phases are the parts which are physically and chemically homogeneous and are separated from the rest of the alloy by definite bounding surfaces. In solid steel they are austenite, ferrite, cementite and graphite. All these phases can be termed constituents but some constituents are not necessarily phases, e.g. pearlite is termed a constituent but it consists of the phases cementite and ferrite in a particular arrangement.

The term constitution diagram, commonly used, should properly be called a phase diagram since it represents the phases present, at given temperatures, which are in equilibrium with each other. Equilibrium is attained by slow heating or cooling to allow constitutional changes to be completed.

The iron-carbon constitution diagram, Figure 5.28 (repeat of Figure 5.13), shows the effect of adding carbon to molten iron (Fe) on the temperature range of stability of the delta, gamma, and alpha phases in the solidified alloy and the effect of varying solubility of the iron carbide (cementite) on the phases present at temperatures from solidification downwards. The diagram represents nominally pure alloys; similar diagrams for other iron alloy systems (Fe-Cr, Fe-Mn, etc[67]) show the effect of elements such as chromium, manganese, etc on the range of stability of the delta, gamma and alpha phases. Since commercial plain carbon steels contain manganese and other elements in varying degree Figure 5.28 does not accurately represent the critical temperatures but is a useful starting point for heat treatment.

Equilibrium decomposition of Austenite. Consideration of the iron-carbon constitution diagram

Point A (Figure 5.28) is, at 1535°C, the highest temperature for the existence of solid delta-phase in carbon-free iron; line ABC shows the fall in the temperature, at which the first phase to separate from the solidfying liquid appears, as carbon content is increased. Line AHJEF (the solidus line) shows the temperatures below which all alloys are solid. Pure iron changes from face-centred cubic (austenite) to body-centred cubic (ferrite) at 910°C; adding carbon lowers this temperature as shown by line GS and, as the solubility of cementite in ferrite is less than in austenite the change results in an increase in the carbon content of the remaining austenite, which, on further cooling, then forms more ferrite with again an increase in carbon content of the remaining austenite. This process continues until the last small amount of austenite contains 0·80% carbon which changes at the fixed temperature of 723°C to a mixture of cementite and ferrite called Pearlite or Eutectoid, PSK being the Eutectoid line.

Alloys below 0·8% carbon (hypo-eutectoid) will finish, at room temperature containing the phases ferrite and cementite, as will those above 0·8% carbon (hyper-eutectoid) but the mode of occurrence (the microstructural constituents of hypo-eutectoid steels being ferrite and pearlite, those of hyper-eutectoid being cementite and pearlite) of the phases is different with markedly different effect on mechanical properties. Descriptive sketches or actual photomicrographs may be used to illustrate the mode of occurrence of the

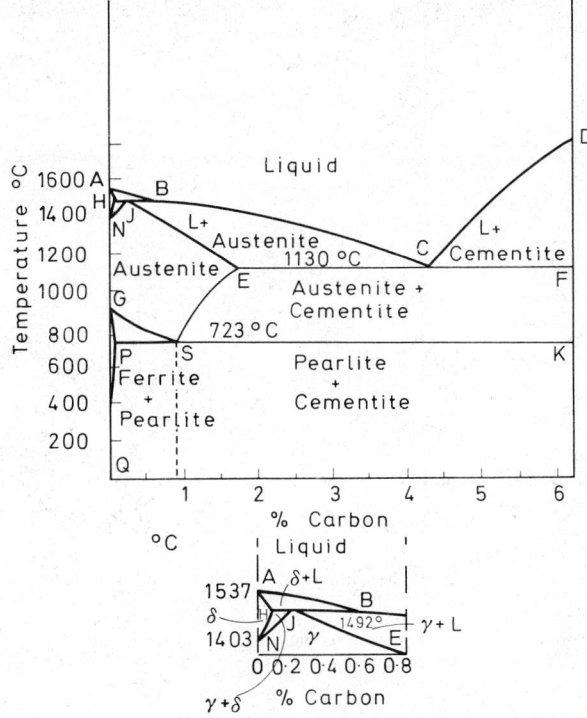

Figure 5.28. Iron-carbon constitution diagram

phases. The constituents in the phase fields and the changes in constituents on cooling for pure iron, 0·10% C and 0·80% C steel are shown schematically in Figure 5.29 when equilibrium conditions exist (diffusion of carbon not inhibited), i.e. extremely slow cooling or heating.

The solid solubility of cementite in gamma iron is 1·7% (point E and eutectic line ECF). Hyper-eutectoid steels between 1·7% and 0·8% C cooling down to line SE precipitate cementite (6·8% C) which reduces the carbon level of the remaining austenite thus requiring a further lowering of temperature before more cementite can precipitate. Continuous cooling down to the eutectoid line PSK results therefore in continuous precipitation of cementite until, at the eutectoid, the austenite contains 0·8% C and precipitates ferrite and cementite simultaneously to produce Pearlite. Steels contain up to 1·7% C, cast irons more than 1·7% C.

It is common practice to use terms such as 'primary austenite', to denote austenite which freezes out of hypo-eutectoid alloys from austenite formed by heating up into the austenite phase field. Also ferrite formed along line GS from steels with less than 0·8% C is called pro-eutectoid ferrite and cementite formed along SE is called pro-eutectoid cementite. The ferrite phase in the eutectoid constituent pearlite is called pearlitic ferrite and the cementite pearlitic cementite.

The microstructural constituents shown schematically in Figure 5.29 are the basis of most commercial steels but the proportions in which they occur and the mode of their occurrence (distribution and form) are profoundly influenced by rate of cooling

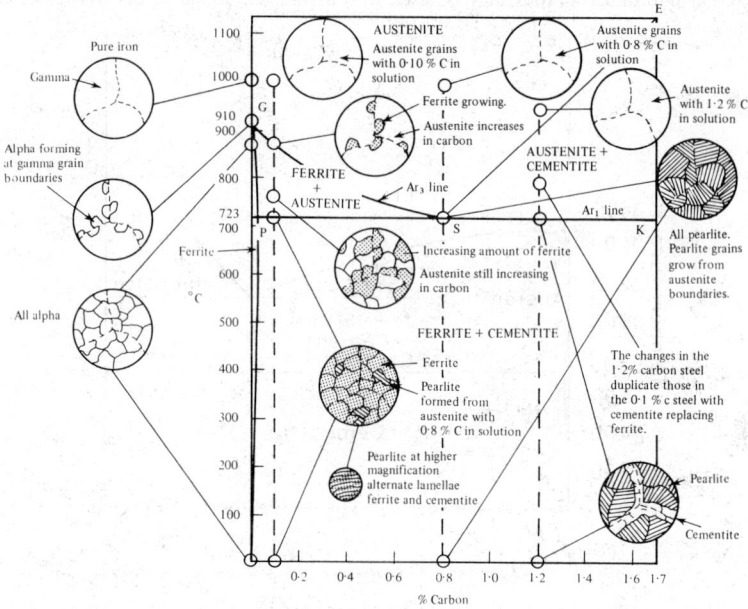

Figure 5.29. Steel portion of iron-carbon constitution diagram. Changes in grain structure and constituents with falling temperature
PURE IRON *Steel at 0·10% carbon (hypo-eutectoid)*
Steel at 0·8% carbon (eutectoid)
Steel at 1·2% carbon (hyper-eutectoid)

from the austenite phase field. The equilibrium conditions necessary to the behaviour shown in Figure 5.29 at the approximate temperatures indicated, (there are no universally agreed figures) do not obtain in steel heat treatment practice except possibly in very large forging ingots when being annealed.

Moreover there is a thermal lag in the allotropic change so that the eutectoid line PSK is at a slightly (3°C approx.) higher temperature during heating up than on cooling down. With normal rates of heating and cooling of steel the eutectoid change is raised or lowered the faster the heating or cooling; consequently in practice a plain carbon steel has a temperature range in which the eutectoid temperature can vary according to heating or cooling rates. This also applies to the position of the austenite to ferrite change line GS.

However, Figure 5.29 illustrates an essential feature of steel heat treatment, namely that austenite in transforming through the lines GS, SE or PSK develops a multiple number of grains of the new constituents in each austenite grain thereby refining the grain structure. The mechanical properties of steels consisting of ferrite and pearlite are strongly influenced by the average grain size of ferrite as well as the amount and type of pearlite (coarse lamellar, fine lamellar, etc). The yield stress varies linearly with the reciprocal of the square root of the grain size.[68]

On heating steel back through the critical temperatures into the austenite field the behaviour observed on cooling is reversed in the following manner.

Steel with 0·1% C. On passing through the lower critical temperature (Ac_1), which is higher than Ar_1, the pearlite areas first transform to austenite of 0·8% C content. This austenite increases by gradually dissolving the surrounding ferrite grains as the temperature is raised and becomes lower in carbon. However, the austenite areas developing from the pearlite consist of numerous crystals[69] so that just above line GS, when the structure is wholly austenitic containing 0·1% C, it consists of numerous small austenite grains. Heating to higher temperatures in the austenite field causes grain growth, some grains growing by absorbing smaller ones around them.

Eutectoid steel. 0·8% C. On heating above the lower critical temperature Ac_1 (which coincides with the upper critical temperature Ac_3 at the eutectoid composition), theoretically the pearlite should transform to austenite of 0·8% C content. In practice it does so over a temperature range, the ferrite lamellae absorbing cementite to form a lower carbon austenite which then dissolves the remaining cementite. Grain growth follows heating to higher temperatures in the austenite field.

Hyper-eutectoid steel. 1·2% C. At the eutectoid line the pearlite starts to transform to austenite of 0·8% C content. As the temperature is raised through the austenite plus cementite phase field pro-eutectoid cementite is gradually dissolved by the austenite adjacent to it and eventually, by carbon diffusion, above the upper critical temperature the austenite is of uniform carbon content at 1·2%. Grain growth follows heating to higher temperatures in the austenite field.

The above simple behaviour of carbon steel can be demonstrated by laboratory heating and cooling procedures tending towards equilibrium conditions;[70] it relies on adequate time being allotted for diffusion of carbon.

When time at temperature is reduced, the diffusion is inhibited in varying degree with pronounced effect on the transformation changes. Thus, in plain carbon steels, increasing the rate of cooling through the critical temperature range Ar_3–Ar_1 lowers this range and alters the proportions of ferrite and pearlite. Steels with less than 0·25% C show refinement of ferrite grains, the growth of individual grains being suppressed, and the pearlitic constituent has finer cementite lamellae. Steels with more than 0·25% C show an increased amount of pearlite and decreased ferrite.

If the steel contains 0·35% C or higher it is possible by sufficient increase in cooling rate to produce a structure entirely consisting of pearlite. This pearlite under high magnification, will differ from the equilibrium pearlite in having very thin cementite lamellae separated by wide ferrite lamellae. Since pearlite (with hard cementite lamellae) is the main contributor to tensile strength in ferrite-pearlite steels, its proportion and morphology in the structure are a prime consideration for heat treatment practice. If cooling rate through the critical range is increased still further, the austenite transformation may be entirely suppressed, and the steel remains as unstable austenite down to a lower temperature where transformation begins for the formation of lower temperature products, e.g. martensite, bainite. When this happens the steel has been cooled at its 'critical rate'.

Martensite has special characteristics which are of great importance in the heat treatment of steel. The essential difference between the mode of formation of martensite and that of pearlite is that the change from the face-centred cubic austenite lattice to the body-centred alpha iron lattice occurs in martensite formation without carbon diffusion, whereas to form pearlite, carbon diffusion must take place producing cementite and ferrite. The effect of this is that the carbon atoms strain the alpha martensite lattice producing micro-stresses and considerable hardness. The higher the carbon content the greater the hardness of martensite (Figure 5.30) and the lower the temperature at which the change to martensite begins.

Furthermore martensite, which is characterised by an acicular appearance, forms progressively over a temperature range as the temperature falls; if the temperature is held constant after the start no further action takes place. Martensite formation produces an expansion, related to the carbon content. The mechanical properties of martensite

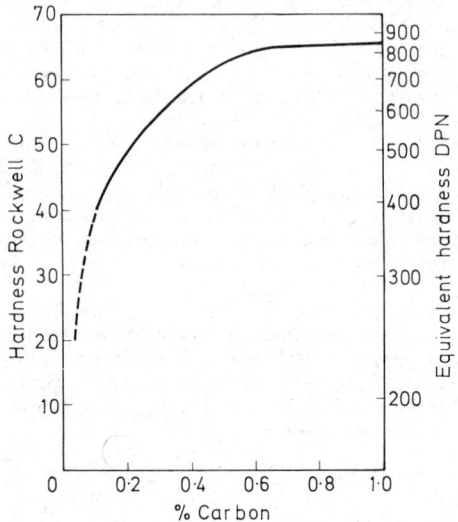

Figure 5.30.   Hardness of martensite related to carbon content

depend on the carbon content; low carbon martensites (less than 0·08% C) have reasonable ductility and toughness, high carbon have no ductility or toughness and extreme hardness and, because of the state of internal stress are very liable to spontaneous cracking. Thus low carbon martensite can be used for industrial purposes, e.g. welded 9% Ni steels for low temperature applications have low carbon marensitic heat affected zones but high carbon. Martensite must be tempered before it is allowed to cool out to room temperature, e.g. carbon tool steels are water quenched to exceed 'critical rate' but the tool is withdrawn from the bath while still hot and immediately tempered.

Isothermal decomposition of austenite

Reference was made in the previous section to the fact that, by suppressing the gamma to alpha transformation by fast cooling, the austenite is an unstable condition. If, before reaching the temperature at which martensite begins to form, the cooling is arrested and the steel held at a constant temperature, the unstable austenite will transform over a period of time to a product which differs markedly from pearlite and has some visual resemblance to martensite in being acicular. This structure is called bainite; it is formed over a range of temperatures (about 550–250°C) and the properties depend to some degree on the transformation temperature, that formed at the lower temperatures being harder. It is tougher than pearlite and not as hard as martensite. It differs fundamentally from the latter by being diffusion dependent as is pearlite.

This type of transformation, at constant temperature, is important in the heat treatment of steel and is called isothermal transformation. It is characterised by an induction period, a start and then a gradual increase in speed of decomposition of the austenite which reaches a maximum at about 50% transformation and then slow completion.

By quenching small specimens of the steel from a fixed temperature, at constant time, in the austenite field above Ar_3 to the temperature at which transformation is desired and holding for various times and determining from microstructure, the proportion of transformed austenite, it is possible to construct an isothermal transformation diagram which gives a summary of the progress of isothermal decomposition of austenite at all temperatures between A_3 and the start of martensite formation (M_s). Such a diagram provides information on the possibilities of applying isothermal heat treatment to bring about complete decomposition of the austenite just below A_1 (isothermal annealing) or just above M_s (austempering) or of holding the steel at subcritical temperatures for a suitable period to reduce temperature gradients set up in quenching without break down of the austenite as in martempering or stepped quenching.[71] Furthermore, if the steel is air hardening or semi air hardening the cooling rate during most welding processes exceeds the 'critical rate' so, by using the isothermal diagram, preheat temperatures and time necessary to hold at temperature to avoid martensite and obtain a bainitic structure can be assessed.

The principle of the isothermal diagram, also known as T-T-T diagrams, is illustrated schematically in Figure 5.31. The dotted lines showing estimated start and finish of transformation indicate the uncertainty of determining with accuracy the start and finish. The main feature of isothermal transformation, the considerable difference in time required to complete transformation at different temperatures within the pearlitic and bainitic temperature ranges is to be noted. These diagrams vary in form for different steels.[72] They also vary according to austenitising temperature (coarseness of gamma grains) and extent to which all carbides are dissolved in the austenite.

In alloy engineering steels containing chromium, molybdenum or tungsten, segregation and carbide banding (size of carbides) varies and can affect extent of carbide solution. In applying these diagrams it is usual to allow considerably longer time for completion of transformation than the time indicated on the diagram to cover the inherent uncertainties in individual consignments of steels.

Effect of carbon and alloying elements on austenite decomposition rate

As carbon content increases the isothermal diagram is moved to the right; in other words austenite transformation is rendered more sluggish.

Alloying elements increase the induction period thus delaying the start and they also increase the time necessary for completion. Furthermore, the effect of adding alloy elements is cumulative but because they have different specific effects on transformation in the pearlitic or bainitic ranges it is not generally possible to predict behaviour of multi-alloy steels.

Decomposition of austenite under continuous cooling conditions

It will be appreciated that the isothermal transformation diagram, while giving the basic information about the characteristics of isothermal transformation for austenite of given composition, grain size and homogeneity (carbide solution efficiency), the common heat treatments used in steel manufacture such as annealing, normalising or quenching are processes which subject the austenite to continuous cooling. This does not necessarily invalidate the use of isothermal diagram data for continuous cooling conditions because, as the steel passes through successively lower temperatures, the microstructures appropriate to transformation at the different temperatures are formed to a limited degree dependent on the time allowed instead of proceeding to completion. The final structure consists of a mixture which is determined by the tendency to form specific structures on the way down, this tendency being indicated by the isothermal diagram.

The time allowed for transformation in the ferrite-pearlite and intermediate (bainite)

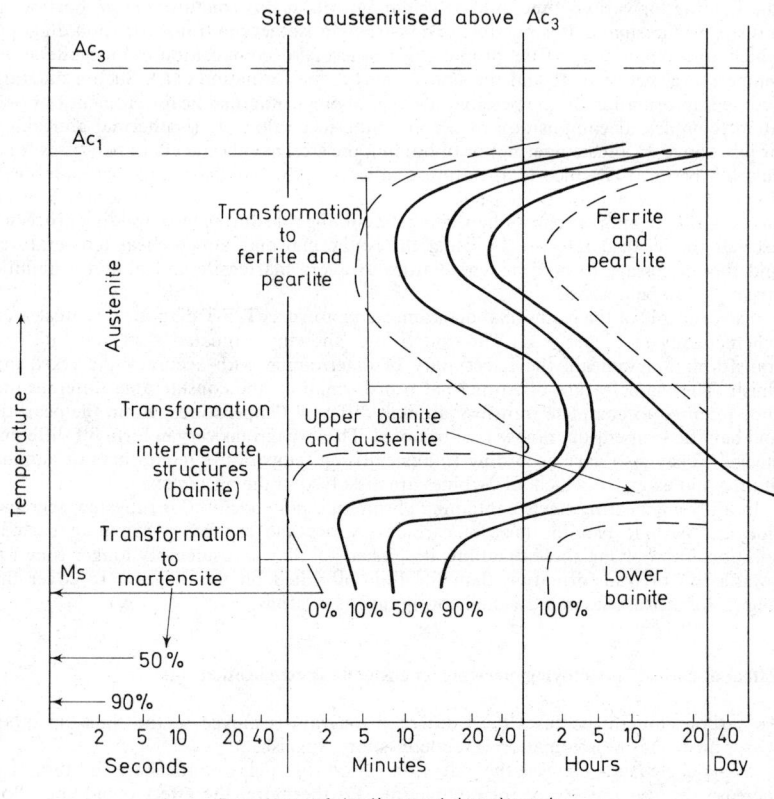

Figure 5.31. Schematic isothermal transformation diagram

regions obviously depends on cooling rate. A continuous cooling transformation diagram will therefore have as its essential features means for indicating the amount of ferrite, pearlite, bainite and martensite which is obtained at various defined cooling rates; these are usually appropriate to heat treatment or selected welding cooling rates. Such a diagram is shown schematically in Figure 5.32.

The effect of continuous cooling is to lower the start temperatures and increase the incubation period so the diagram tends to be below and to the right of the isothermal diagram for the same steel, these effects increasing with increasing cooling rate. As indicated in Figure 5.32, the time axis may be expressed in any suitable form; transformation time in Figure 5.32(a), bar diameter for bars as in Figure 5.32(b).

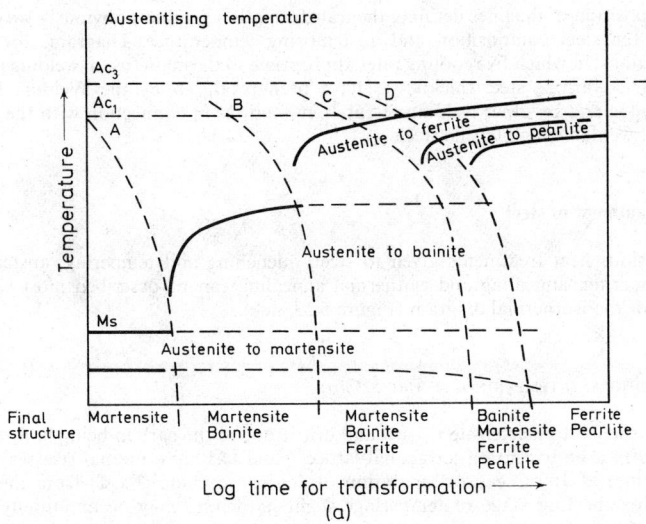

Figure 5.32. (a) Continuous cooling time—temperature—transformation diagram
applicable to forgings, plates and sections
Cooling rate A. Quenching
Cooling rate B. Air cooling light sections. Normalising
Cooling rate C. Air cooling heavy sections. Normalising
Cooling rate D. Slow furnace cooling. Annealing

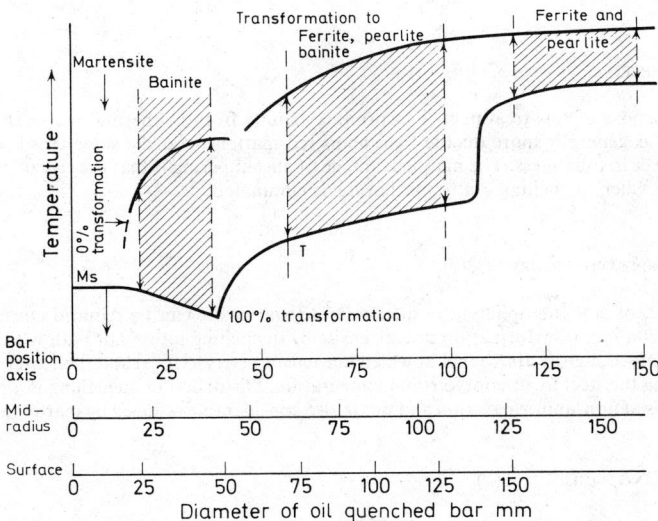

Figure 5.32. (b) Continuous cooling time—temperature—transformation diagram
applicable to heat treatment of bars

The position of the lines defining the transformation products obviously vary according to the steel composition and austenitising temperature. Diagrams for welding applications, in which five cooling rates appropriate to the main fusion welding processes applied to various steel thicknesses have been produced by the Welding Institute, Cambridge and by other welding research institutions in connection with the development of weldable high tensile steels.

Heat treatment of steel

The various heat treatments given to steel, quenching and tempering, austempering, martempering, annealing and isothermal annealing can be described most simply by means of the isothermal diagram (Figure 5.33, a–e).

QUENCHING AND TEMPERING, FIGURE 5.33(a)

Steel quenched to martensite is hard and brittle due to the carbon being in forced solid solution in a body-centred tetragonal lattice[73] and has high internal stresses. Heating (tempering) at 100°C causes separation of eta-iron carbide (Fe_3C) from the matrix, this being the first stage of tempering; slight hardening may occur initially. As the temperature is increased, relief of stress and softening occurs due to cementite formation and release of carbon from the matrix. Marked toughening occurs.

Steels of suitable composition quenched fully to martensite and tempered at appropriate temperatures give the best combination of strength and toughness obtainable from steel. There is a tendency, varying with different steels, for a degree of embrittlement to occur when tempering in the range 250–450°C so steels are either tempered below 250°C, for maximum tensile strength, or above about 550°C for a combination of strength, ductility and toughness due to increasing coalescence of carbides. For more detailed information see Reference 74.

AUSTEMPERING, FIGURE 5.33(b)

The purpose of this treatment is to produce bainite from isothermal treatment; lower bainite is generally more ductile than tempered martensite at the same tensile strength but lower in toughness. The main advantage of austempering is that the risk of cracking, present when quenching out to martensite, is eliminated.[75]

MARTEMPERING, FIGURE 5.33(c)

The risk of cracking inherent in quenching to martensite can be reduced considerably while retaining transformation to martensite by quenching into a salt bath which is at a temperature slightly above that at which martensite starts to form and then after soaking allowing the steel to air cool to room temperature. Distortion in quenching is a problem in pieces of non-uniform section and this is also considerably reduced by martempering.[76]

ANNEALING, FIGURE 5.33(d)

Maximum softness is attained by annealing, involving slow cooling through the ferrite-pearlite field. The pearlitic structure developed provides optimum machinability in medium carbon steels.

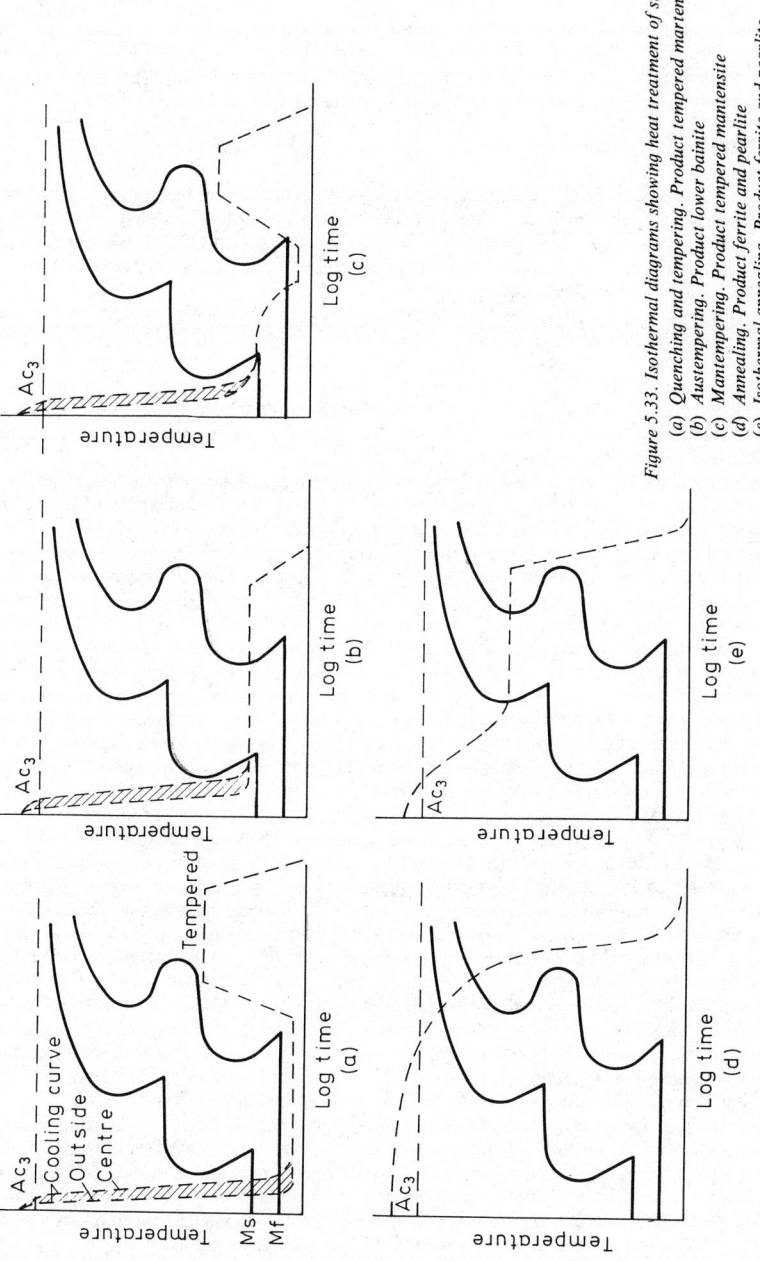

Figure 5.33. Isothermal diagrams showing heat treatment of steel
(a) Quenching and tempering. Product tempered martensite
(b) Austempering. Product lower bainite
(c) Martempering. Product tempered martensite
(d) Annealing. Product ferrite and pearlite
(e) Isothermal annealing. Product ferrite and pearlite

ISOTHERMAL ANNEALING, FIGURE 5.33(*e*)

This treatment is used to produce a soft ferrite-pearlite structure. Its advantage over annealing is that, with appropriate steels and temperatures it takes less total time because cooling down to and following the isothermal treatment may be done at any suitable rate, provided the material is not too bulky or being treated in large batches.[78]

Hardenability of steel

Hardenability in this context refers to the depth of hardening not the intensity. Hardening intensity in a quench is dependent on the carbon content. Plain carbon steels show relatively shallow hardening; they are said to have low hardenability. Alloy engineering steels show deep hardening characteristics to an extent depending primarily on the alloy or alloying elements and austenite grain size.

Hardenability is a significant factor in the application of steels for engineering purposes. Structural steels applied in the as-rolled, normalised or normalised and tempered condition develop sensibly the same microstructure, ferrite and pearlite or bainite, throughout the full thickness for which the properties are claimed by the manufacturer. The tensile or yield strength of such steels is controlled by cooling rate (rolling mill practice in as-rolled condition) with transformation to ferrite and pearlite of fineness adjusted to tensile strength or to bainite.

Similar steels are used as forgings for engineering applications in the normalised condition with yield strength generally below 280 N/mm², the thickness of the section being of minor significance. However, most engineering steels for forgings are used in the oil quenched and tempered condition to achieve optimum properties of strength and toughness based on tempered martensite. It is in this connection that hardenability is important; in general, forgings are required to develop the desired mechanical properties through the full section thickness.

Since the cooling rate in a quench must be slower at the centre of a section than at the surface, the alloy content must be such as to induce sluggishness in the austenite transformation sufficient to inhibit the ferrite-pearlite transformation at the cooling rate obtaining at the centre of the section. It follows that, for a given steel composition and quenching medium, there will be a maximum thickness above which the centre of the section will not cool sufficiently quickly except in those steels which have sufficient alloy content to induce transformation to martensite in air cooling (air hardening steels).

The practical usefulness of engineering steels, ignoring differences in toughness, can therefore be compared on the basis of this maximum thickness or ruling section which must be taken into account when considering selection of steel for any specific application. BS 970, Part 2, Section 4, gives ruling sections for a range of engineering steels. The ductility and toughness depend, not only on achieving adequate hardening in the quench, but also on the particular tempering temperature. Since the steel must be made to a composition range, tempering treatment may need to be adjusted to material composition as delivered.

BS 970 requires the indicated mechanical properties to be obtained, in sections thicker than 62·5 mm, at a position midway between the surface and the centre. The quenched microstructure at the bar centre may be a mixture of martensite and bainite (the cooling curve cutting the bainite nose, Figure 5.33(a)) with not less than 50% martensite. For a full description of ruling section and its application see BS 970, Appendix C.

Instead of expressing hardenability as the limiting bar diameter for given mechanical properties which rely on adequate quenching, it may be regarded as a straight comparison between steels in their ability to harden over progressively slower cooling rates. Thus a series of bars of increasing diameter and standard length are quenched from a constant austenitising temperature and then sectioned for microstructural observation or hardness test at the centre. An alternative method requiring less material, is to cool a bar of standard diameter and length by water jet applied to one end only. The cooling rate at any position along the bar will progressively decrease as one proceeds away from the water sprayed end. The hardness is determined on flats ground on the bar surface located at 180°. The greater the hardenability the further along the bar a fully martensitic structure is developed. This method of assessment is known as the Jominy end-quench test; for full details, see BS 4437.

BS 970 Part 2, Section 3, lists steels which can be purchased to hardenability requirements as described by the Jominy test. They are engineering alloy steels for forgings, etc of medium carbon 1% Ni, $1\frac{1}{2}$% Mn-Mo, $1\frac{1}{4}$% Ni-Cr and 1% Cr-Mo. The range of Jominy hardenability (hardenability band) is given for each steel specification. In this connection the following extract from 'Transformation characteristics of direct − hardening nickel alloy steels'[79] is pertinent:

'The work of the hardenability sub-committee[80] showed that the hardness of quenched bars could not be calculated with accuracy from end-quench hardenability curves and demonstrated that the discrepancies are due mainly to transverse and longitudinal variations of hardenability within steel billets of normal commercial quality.[81]

'The standard 25 mm dia end-quench test piece is normally machined from 29 mm bar and the hardness reflects only the response of the material near the

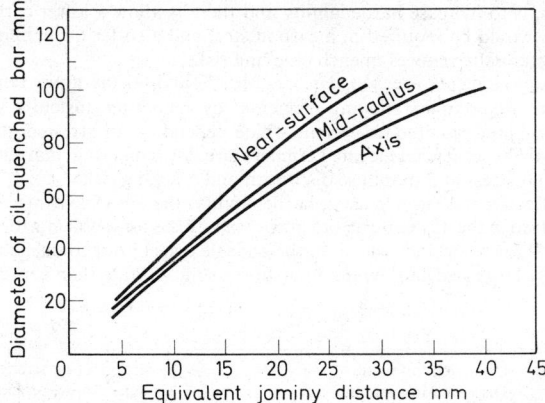

Figure 5.34. Relationship between end-quench hardenability curves and oil-quenched bars

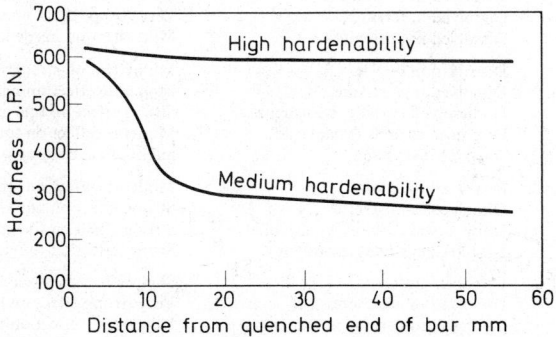

Figure 5.35. End-quench (Jominy) curves for steels of medium and high hardenability

surface of the bar. End-quench hardenability curves should therefore be used only semi-quantitatively to forecast the behaviour of quenched bars. For example, the curve can be used to decide whether a bar of given size is likely to harden fully, partially or to transform completely to soft transformation products when oil quenched. The relationship between end-quench hardenability curves and oil quenched bars (Figure 5.34)[82] is used for this purpose; thus, using the medium hardenability end-quench curve in Figure 5.35,[83] a 12·5 mm dia bar (equivalent Jominy distance 3·81 mm) will harden fully, a 25 mm dia bar (EJD = 7·62 mm) will harden partially being on the slope of the curve and a 50 mm bar (EJD = 15·75 mm) will not harden at the axis.'

The function of alloying elements in engineering alloy steels

Aside from specialised functions, corrosion resistance, abrasion resistance etc, alloying elements are most widely used in engineering alloy steels with carbon in the range 0·25–0·55% or less than 0·15% for case hardening. Their function is to improve the mechanical properties, compared with carbon steel, and, in particular, to make possible the attainment of these properties at section thicknesses which preclude the use of shallow hardening carbon steels, water quenched. In other words their main, but not sole function, is to increase hardenability and thereby allow a lower carbon content to be used than would be required in a carbon steel and a softer quenching medium, e.g. oil. This substantially reduces quench cracking risks.

The alloying elements are Mn, Ni, Cr, Mo, V and Al (as grain refining element). An important function of alloying elements, by rendering austenite transformation sluggish, is to make possible treatments which depend on an arrested quench followed by a timed hold at somewhat elevated temperature (austempering, martempering) which reduce internal stress and minimise distortion and cracking risks.

For full effectiveness in increasing hardenability, the alloy elements should be completely dissolved in the austenite before quenching; this is no problem with Mn and Ni but Cr, Mo and V form carbides which, in the annealed steel prior to quenching, may be of comparatively large size and, owing to a slower solution rate than cementite, are more

Element		Effect
Mn	Dissolved in ferrite.	Strong effect on strength.
	Dissolved in austenite.	Moderate effect on hardenability.
	Undissolved carbide in austenite.	Mild effect on fine grain. Toughness.
	As oxide dispersion.	Slight effect on toughness.
Ni	Dissolved in ferrite.	Mild effect on strength.
	Dissolved in ferrite.	Very strong effect on toughness.
	Dissolved in austenite.	Mild effect on hardenability.
Cr	Dissolved in ferrite.	Mild effect on strength.
	Dissolved in austenite.	Moderate effect on hardenability.
	Undissolved carbide in austenite.	Strong effect on fine grain. Toughness.
	Dispersed carbide (tempering).	Moderate effect on toughness.
	As oxide dispersion.	Slight effect on toughness.
Mo	Dissolved in ferrite.	Moderate effect on strength.
	Dissolved in austenite.	Strong effect on hardenability.
	Undissolved carbide in austenite.	Strong effect on fine grain. Toughness.
	Dispersed carbide (tempering).	Strong effect on toughness.
V	Dissolved in ferrite.	Slight effect on strength.
	Dissolved in austenite.	Very strong effect on hardenability.
	Undissolved carbide in austenite.	Very strong effect on toughness.
	Dispersed carbide (tempering).	Very strong effect on toughness.
	As oxide dispersion.	Strong effect on toughness.

difficult to dissolve. Solution temperatures may therefore be increased and/or times increased.

The effect of alloying elements when tempering is important; in general they retard the rate of softening during tempering compared with carbon steel but the effect, in this respect, of the carbide formers (Cr, Mo, V) is much greater than that of the other elements. They increase the tempering temperatures required for a given degree of softening, which is beneficial for ductility and toughness. Mo and V, at higher levels, show an increase in hardness at higher tempering temperatures due to alloy carbide precipitation; this is 'secondary hardening' and is the basis of hardness in heat treated alloy tool steels. The effect of individual elements is briefly given in the table on page 5–68.

METALLURGY OF STAINLESS AND HEAT RESISTING STEELS

In considering the metallurgical basis for the various commercial chromium and chromium-nickel steels the effect of additions of these alloying elements on the stability of the face-centred cubic (austenite) and body-centred cubic (ferrite) allotropic forms of iron must be noted.

When alloys are made by adding other elements to metals which undergo polymorphic changes, the conditions under which these take place are affected by the added elements and the solubility of the added element in the basis metal is affected by the crystallographic changes.[84] For example nickel, carbon, manganese, copper and cobalt increase

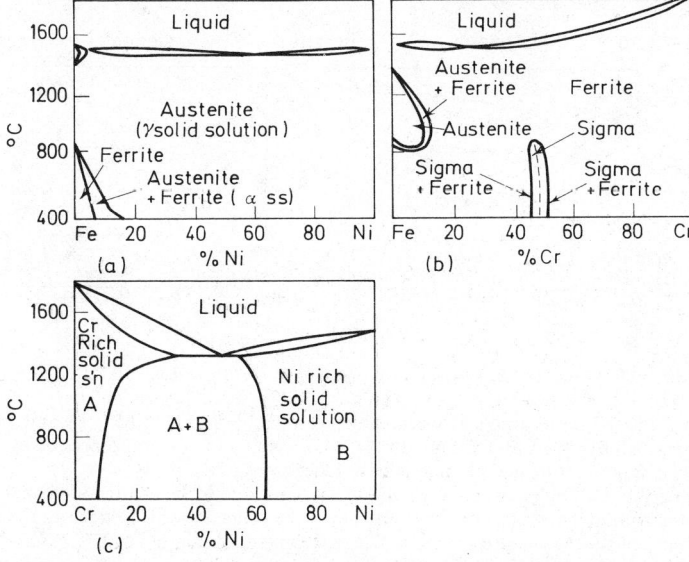

Figure 5.36. Constitution diagrams for the binary Fe-Ni, Fe-Cr and Cr-Ni systems

the range of stability of the face-centred cubic gamma iron at the expense of the body-centred cubic alpha and delta forms; in effect they raise the temperature of the reversible delta to gamma and lower that of the reversible gamma to alpha change. The elements chromium, silicon, phosphorus, tungsten, molybdenum, vanadium and aluminium have the reverse effect in increasing the range of stability of the body-centred cubic alpha and delta forms.

Iron forms a complete series of solid solutions with nickel and with chromium; the alpha or delta form (ferrite) will form solid solutions with chromium up to nearly 100%

of the alloying element without a new phase appearing but will dissolve only a limited amount of nickel and the gamma form (austenite) will dissolve up to almost 100% of nickel without a new phase appearing but can dissolve only limited amounts of chromium. Apart from one significant intermetallic compound sigma phase (FeCr) the iron-chromium-nickel system must consist of face-centred cubic (austenite) and body-centred cubic (ferrite) phases which will vary greatly in composition according to temperature. An intermetallic compound chi phase is of secondary significance.

The above comments are reflected in the phase diagrams for the binary Fe-Ni, Fe-Cr and Cr-Ni systems (Figure 5.36).[85] Of particular significance in the Fe-Cr system (Figure 5.36b) is the small austenite field known as the gamma loop; alloys to the right of this loop are ferritic and undergo no allotropic changes in heating or cooling consequently grain refinement by such changes is not possible. The amount of chromium which closes this loop is 12·8%. Above this figure the Fe-Cr alloys are ferritic and subject to grain growth as temperatures are raised to the liquidus. This is reflected in ultra-large grain formation in the heat affected zone of welds, particularly those made by slower heat-input processes such as oxy-acetylene.

Stainless steels

CHROMIUM STEELS

Carbon is present in all chromium steels. Its effect on the temperature range of stability of austenite and ferrite Fe-Cr alloys is shown in Figure 5.37[86] for two carbon levels.

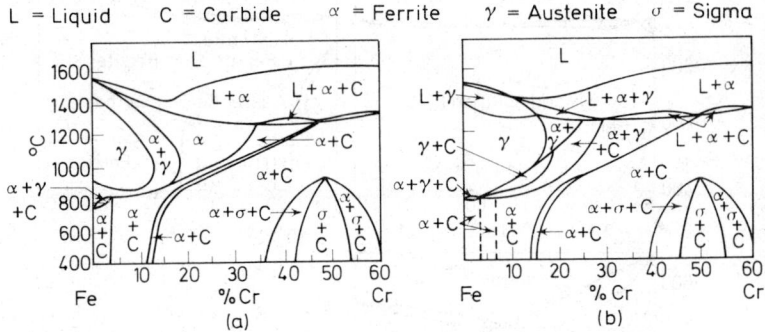

Figure 5.37. Iron-chromium-carbon phase diagrams (a) at 0·10% carbon, (b) at 0·50% carbon

Two carbides of chromium are formed which can dissolve 25% or 55% Fe and an iron carbide which dissolves up to about 15% Cr; to simplify the diagram these distinctions have been ignored. For the full diagram see Reference 86.

The addition of 0·1% carbon to the Fe-Cr system extends the gamma loop slightly to higher chromium contents, carbon being one of the elements which increases the range of stability of the face-centred cubic lattice, austenite disappearing at 17·5% Cr. This is the chromium level at which chromium steels become fully ferritic with consequential grain growth problems when welded. The constitution of Fe-Cr alloys at less than about 2% Cr is similar to Fe-C alloys but as Cr is increased the form of the austenite loop becomes important from the standpoint of heat treatment. Oil quenching from the loop produces a fully martensitic structure owing to the sluggishness of the austenite to ferrite change induced by chromium. The limit of the loop at this carbon content is about 14% Cr but this is extended to approximately 16% Cr when the carbon is increased to 0·5%.

Quenching from the austenite plus ferrite field produces a mixed ferrite-martensite structure. The position of the gamma loop is also affected by the manganese and nitrogen content (both tend to stabilise austenite) and the silicon content (ferrite stabiliser). Chromium steels supplied as bars or billets for hot working will be supplied in the annealed condition for full treatment after working. It is important to adhere strictly to the steelmaker's recommended forging temperature range and quenching temperature to ensure that the steel is austenitised in the desired phase field.

Chromium stainless steels are manufactured in two chromium ranges:

12·5–14·5% Cr with low, medium or high carbon content according to application and desired mechanical properties. These steels are used in the annealed condition if low carbon (ferrite and pearlite structure) and in the oil hardened and tempered condition (troostite, sorbite or martensite according to tempering temperature) if of higher carbon content.

16·5–17·5% Cr with or without nickel or molybdenum additions, with low, medium or high carbon contents. The low carbon grades (C less than 0·1%) are ferritic single phase alloys, supplied and used in the annealed condition, the medium and high carbon grades are quenched and tempered to appropriate products of tempered martensite, sorbite troostite or martensite (which has been tempered at 100–200°C).

Forging and heat treatment practice for chromium stainless steels is indicated by the data quoted for the Remanit 15, 16 and 17 Series of Chromium Steels supplied by Deutsche Edelstahlwerke (Table 5.9). Chromium and chromium-nickel stainless steels are included in BS 1501–1506.

AUSTENITIC CHROMIUM-NICKEL STEELS

The practical importance of these stainless steels rests upon their corrosion resistance, conferred primarily by chromium, and their formability (hot and cold) which is due to the alloys having austenite as the major phase at room temperature. This is attributable to a combination of sufficient nickel and chromium to render the high temperature

is retained down to low temperatures.

The most widely used combination of nickel and chromium to give retained austenite is 8% nickel with 18% chromium. All alloys with higher nickel and 18% chromium are austenitic; those with lower nickel at this chromium level will produce a less stable austenite phase which may more readily transform to alpha (martensite) and carbide. The 18% Cr 8% Ni austenite is not truly stable at room temperature; as a result cold work will induce partial transformation to alpha iron (martensite) and therefore these steels show considerable work hardening capacity,

When chromium is lowered from 18%, austenite can be retained by increasing the nickel content above 8%. For example a 12% Cr steel with 12% Ni, cooled at a suitable rate, is fully austenitic at room temperature. There is no single diagram that will represent accurately the range of compositions within which austenite is retained at room temperature under practical conditions because its retention and relative stability is influenced by factors in addition to the chromium and nickel content, e.g. rate of cooling, initial temperature and carbon content as well as other alloy elements (Mo, Si, etc[87]).

However, a general idea of the composition limits within which austenite may be retained by cooling at ordinary rates from temperatures above 800°C is indicated in Figure 5.38(a).[88] Alloys above line C but below line B are 'persistent' at room temperature after ordinary rates of cooling but liable to become partially ferritic if heated into the range 600–900°C because the stable or equilibrium phase at room temperature for alloys in the range 0–18% nickel and 0–30% chromium is ferrite and continued heating in the range 600–900°C tends to promote an attempt to reach equilibrium.

'Persistent' is a term used to distinguish unstable austenite (retained by ordinary cooling rates), which reaches equilibrium by breakdown to the body-centred cubic structure during cold working, from unstable retained austenite which will not break

Table 5.9 STAINLESS STEELS

Steel	Nominal composition			Forging		Annealing		Hardening or quenching			Tempering	Structure
Remanit	Cr%	C%	Other	Temp. °C	Cooling	Temp. °C	Cooling	Temp. °C	Cooling	For tensile N/mm²	Temp. °C	
1 510	14·3	<0·08	—	1 150–750	Air (still)	750–800	Furnace	—	—	—	—	Ferrite, pearlite
1 510H	13·5	0·10	—	1 150–750	Air (still)	750–800	Furnace	950–1 000	Oil	590–735	700–750	Ferrite, troostite
1 510Al	13·0	<0·08	Al	1 100–750	Air (still)	750–800	Air	—	—	—	—	Ferrite
1 515	13·5	0·15	—	1 150–750	Ashes	750–800	Furnace	960–1 010	Oil	685–835	700–750	Sorbite, troostite
1 520	13·5	0·2	—	1 150–750	Ashes	750–800	Furnace	1 000–1 050	Oil or air	635–785	700–750	Sorbite, troostite
1 540	13·5	0·42	—	1 100–800	Furnace	750–800	Furnace	1 000–1 050	Oil or air	—	100–200	Martensite, ferrite
1 610	16·5	<0·1	—	1 050–750	Air	750–850	Air	—	—	—	—	
1 620	17·0	0·22	1·7Ni	1 150–750	Ashes	—	—	1 000–1 050	Oil	785–930	630–720	Sorbite, troostite
1 740	16·5	0·38	1·2Mo	1 100–750	Ashes	750–800	Furnace	970–1 020	Oil	785–930	670–750	Sorbite, troostite
1 790V	18·0	0·9	1·2Mo + V	1 100–800	Furnace	800–850	Furnace	1 050–1 075	Oil	—	100–200	Martensite

Table 5.9 STAINLESS STEELS (AUSTENITIC)

Steel	Nominal composition				Other	Forging		Quenching*
Remanit	Cr%	C%	Ni	Mo		Temp. °C	Cooling	Temp. °C
1880	18	<0·1	9	—	—	1 150–750	Air	1 050–1 100
1880SW	18	<0·07	10	—	—	1 150–750	Air	1 050–1 100
1880EW	18	0·03 max.	10	—	—	1 150–750	Air	1 050–1 100
1880S	18	<0·1	10·5	—	Nb	1 150–750	Air	1 020–1 070
1880ST	18	<0·1	10·5	—	Ti	1 150–750	Air	1 020–1 070
1880SSW	17·5	0·07 max.	11	2·2	—	1 150–750	Air	1 050–1 100
1880SS	17·5	<0·1	11·5	2·20	Nb	1 150–750	Air	1 020–1 070
1880SST	17·5	<0·1	11·5	2·2	Ti	1 150–750	Air	1 020–1 070
1820S	17·5	<0·07	20	2·2	Cu + Nb	1 150–750	Air	1 050–1 100
1820SST	17·5	<0·07	20	2·2	Cu + Ti	1 150–750	Air	1 050–1 100
1810SSW	17·5	<0·07	12	2·8	—	1 150–750	Air	1 050–1 100
1810SS	17·5	<0·1	12·5	2·8	Nb	1 150–750	Air	1 020–1 070
1810SST	17·5	<0·1	12·5	2·8	Ti	1 150–750	Air	1 020–1 070
1813SSW	17	0·07 max.	13·5	4·5	—	1 150–750	Air	1 050–1 100
2525	25	<0·06	25	2·2	Ti	1 150–850	Air	1 050–1 100

down on cold working. Compositions for 'persistent' alloys are indicated in Figure
5.38(*b*), the austenite line being the same line as A in Figure 5.38(*a*).

Extremely rapid cooling by water quenching tends to move line B, Figure 5.38(*a*) to the
position B1, so that water quenched alloys with lower chromium (11–14%) and higher
nickel content (11–14%) than 18% Cr 8% Ni possess a relatively stable austenite which
remains non-magnetic after cold work and shows minimum work hardenability. These
alloys possess the best deep drawing and stamping properties of all austenitic Cr-Ni
steels. They are, however, rust resisting rather than corrosion resisting being attacked
in acid.

The degree of stability induced in the austenite has important practical implications
depending upon the application. Stainless steels (as distinct from heat resisting steels)

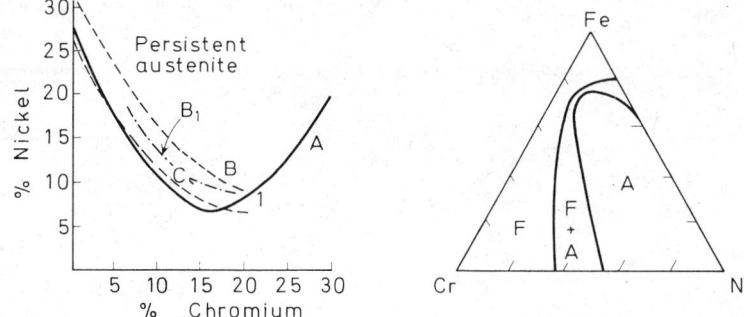

*Figure 5.38. (a) Effect of Nickel and Chromium on the retention of Austerite, (b) Range of
compositions with 'Persistent' Austerite at room temperature. Ordinary cooling rates*

are used at temperatures lower than required to promote partial change to alpha iron
by heating so that they may be considered stable from this point of view. However
fabrication of plate and sheet into vessels involves cold working (i.e. bending, pressing,
etc) plus welding. If partial breakdown is caused the notch toughness is reduced which
may be significant in cryogenic applications. Thin sheets are usually air cooled from the
annealing, carbide solution, temperature but thicker sections are water quenched to
promote stability and retention of carbides in solution.

Carbide solution and precipitation are important aspects of the metallurgy of austenitic chromium-nickel steels. The solubility of chromium carbide in austenite decreases
with decreasing temperature and increasing nickel content (Figure 5.39[90]). At room
temperature the solubility in 18% Cr 8% Ni austenite (solid line) is approximately 0·03%.
If an 18% Cr 8% Ni alloy containing, say, 0·06% C is heated to 1050–1100°C which is a
typical annealing treatment temperature, all chromium carbide is in solution at this
temperature; upon quenching to room temperature all chromium the carbide remains in forced solution in the austenite. When reheating, the alloy attempts to move towards equilibrium
carbide solubility levels and therefore excess carbide at the given temperature must be
precipitated.

The mode of precipitation of carbide is dependent upon whether or not the austenite
has been worked (cold or warm) before precipitation. If the quenched, but not worked,
alloy is heated in the temperature range 450–750°C chromium carbide is precipitated at
the grain boundaries, the lower the temperature the longer the time required for precipitation to be observed. Thus at around 450°C the time taken for precipitation can be about
two years whereas at 700°C it is a matter of minutes. If the alloy has been cold worked,
carbide is precipitated within the grains and at grain boundaries and the speed of the
process is increased.

When the annealed 18% Cr 8% Ni alloy is heated in the above temperature range, precipitation of chromium carbide is effected by uniform diffusion of carbon atoms to the grain boundaries where they combine with approximately sixteen times the number of chromium atoms. Diffusion of carbon is relatively fast at these temperatures but that of chromium extremely slow. Consequently, the chromium atoms are almost entirely supplied by the grain boundaries so that the grain boundary chromium content is substantially lowered. This local depletion of chromium causes loss of passivity in acid corrodants with consequent attack along grain boundaries (weld decay).

Complete disintegration of the metal is possible. The fundamental cause of 'weld decay' is the difference in diffusion rates exhibited by carbon and chromium in this temperature

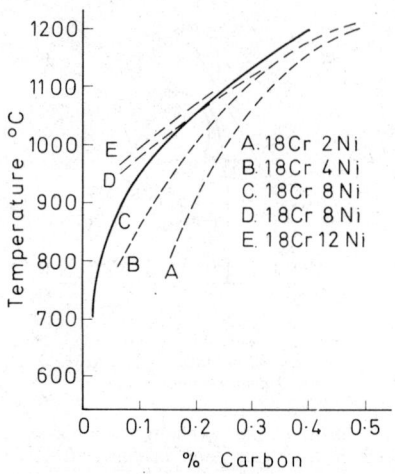

A. 18 Cr 2 Ni
B. 18 Cr 4 Ni
C. 18 Cr 8 Ni
D. 18 Cr 8 Ni
E. 18 Cr 12 Ni

Figure 5.39. Solubility of carbon in 18% Chromium-Nickel alloys

range. When precipitation occurs in cold worked material it takes place along slip planes as well as grain boundaries, consequently the distance that the chromium atoms must diffuse is small. Hence, although the same amount of chromium is removed as carbide, the depletion is more uniformly distributed with a consequential lowering of general corrosion resistance but a lower tendency to intergranular failure.

The heating time in the 450–750°C range is important. Cold worked alloys heated for a short period are very susceptible to corrosion because chromium diffusion is not sufficiently rapid to compensate the locally depleted areas. To assist chromium diffusion heating for some time at 700–750°C or for short times at 750–800°C can be beneficial. Above 800°C carbide redissolves in the austenite and the metal again becomes susceptible to intergranular corrosion on subsequent heating in the 450–750°C range.

Cold rolled 18% Cr 8% Ni is annealed above 800°C to soften it for further working and is therefore susceptible. An increase in nickel content to 15%[89] lowers the recrystallisation temperature (the annealing temperature) to about 750°C so that this alloy in the annealed condition is not susceptible.

Although of no general practical significance it is an interesting confirmation of the importance of slow chromium diffusion as the cause of 'weld decay' that 18% Cr 8% Ni steel with about 0·06% C, heated for five minutes at 650°C will show intergranular corrosion in the standard boiling copper sulphate-sulphuric acid test but if the heating is prolonged to about three months there will be no corrosion although a complete carbide network is present.

The methods relying on chromium diffusion for preventing intergranular corrosion are not widely used. Instead there are two alternative approaches to the problem; either the steel is made to contain 0·03% C where precipitation of carbide in sufficient quantity to cause trouble is impossible or an element such as titanium or niobium, which has a stronger affinity for carbon than chromium, is added to form the appropriate carbide.

Sufficient of the element must be added to combine with all carbon in excess of 0·02% which is in solid solution and steps must be taken to ensure that the carbon is combined with the added element. The theoretical amounts required are titanium 4 x excess carbon, niobium 8 x excess carbon but, in practice, allowance must be made for nitrogen combining with the added element (particularly Ti) and for the effect of mass action on the efficiency of combination, carbon levels below 0·06% requiring a higher titanium or niobium to carbon ratio for complete combination than those above 0·08% C. Furthermore, when a 'stabilised' (Ti or Nb treated) steel is heated to successively higher temperatures above 950°C up to 1250°C the carbide enters solution and is broken down into its constituent elements to an increasing extent so that above 1100°C a relatively small amount of carbon remains combined.

The free carbon is then available to form chromium carbide on subsequent re-heating in the 450–750°C range. Combination of carbon with titanium occurs in the range 850–950°C given adequate time. The whole question of 'stabilisation' is concerned with time, temperature and amount of free carbon.

Steels with apparently the correct titanium addition and passing the standard half hour at 650°C treatment without failure in the copper sulphate-sulphuric acid test may fail the test after extended heating (months or years) at, say 550°C or 450°C, if 'stabilisation' has not been effective. For applications involving service at such temperatures, consideration should be given to either using extra low carbon steels (0·03% C max) or 'stabilised' steels having carbon in excess of 0 08%, preferably the former since no reliance is then placed upon the efficiency of formation of titanium or niobium carbide.

Mass action principles indicate that the amount of alloy carbide formed under particular conditions of time and temperature (in the temperature range of combination of carbon with the alloy element) will be proportional to the product of the total carbon and the amount of available element not combined with nitrogen. Thus in low carbon steels, with less than 0·06% C, manufacturing conditions are more critical in respect of achieving efficient 'stabilisation' but the amount of uncombined carbon must necessarily be low. For most normal applications the small amount of free carbon which *may* be present in such steels does not lead to any risk of intergranular corrosion. It becomes a risk in such partially 'stabilised' steels when service conditions lead to precipitation of this carbon as chromium carbide in circumstances where chromium diffusion cannot occur. This is the case with low temperatures such as 450°C for extended periods of time.

Corrosion resistance of chromium and austenitic chromium-nickel stainless steels

Corrosion resistance of stainless steels is generally considered to depend on a condition of surface passivity arising from the formation of an oxide film which is insoluble, non-porous and, under suitable conditions, self-healing if damaged. Passivity of stainless steel is not a constant condition but it prevails under certain environmental conditions. The environment (atmosphere or liquid) should be oxidising in character.

The chromium content of stainless steels is the primary means whereby passivity is obtained under attack from a given medium. However, other factors are significant, including composition, heat treatment, initial surface condition, variation in corrosion conditions, state of stress in service, welding and service temperature.

COMPOSITION

Stainless steels may be divided into four groups with regard to corrosion:

13% Cr (ferrite-pearlite and martensitic). Rust resisting only. Used for conditions where corrosion is relatively light e.g. atmospheric, steam, oxidation resistance up to 500°C. Applications include cutlery, oil-cracking, turbine blades, surgical instruments, etc.

17% Cr (ferritic and martensitic). Corrosion and light acid resisting. Improved general corrosion resistance compared with 13% Cr steels. Applications for the ferritic grade include domestic and catering equipment, automobile trim, industrial heater parts. The martensitic grade is used in general engineering, for pump and valve parts in contact with non-ferrous metals or graphitic packings. The Mo bearing 17% Cr steel is resistant to sea water, crevice and galvanic corrosion. For wear and acid resistance (valve seats, etc.) the high carbon Mo bearing grade is selected.

18% Cr 8% Ni (austenitic). Corrosion and acid attack. For general corrosion resistance in acids, etc. Applications include domestic, shop and office fittings, food, dairy, brewery, chemical and fertiliser industries.

18% Cr 8% Ni + Mo (austenitic). Resistance to sulphuric acid. Used in chemical, cellulose and textile industries. 18% Cr, 11–13% Ni + Mo + Cu (austenitic). Specially resistant to sulphuric acid, 18% Cr 13% Ni + 4·5% Mo (austenitic) is specially resistant to pitting corrosion.

Note that to ensure an austenitic structure, the nickel content of the Mo bearing steels increases above 8% Ni as the content of Mo + other ferrite stabilising elements (Ti, Nb, etc) increases. The Mo bearing alloys show increasing resistance to specially severe conditions as the Mo increases, the 4·5% Mo alloy being used in sodium chlorite bleaching baths and other very severe environments in the textile industry.

HEAT TREATMENT

Heat treatment, by affecting the microstructure, has a significant influence on corrosion resistance. Maximum resistance is offered when the carbon is completely dissolved in a homogeneous single-phase structure. The 12–14% Cr steels are heated treated to desired combinations of strength, ductility and toughness and are generally satisfactory unless tempered in the range 500–600°C owing to low carbon content.

Austenitic steels (18/8) are most resistant when quenched from 1050–1100°C, the normal condition of supply. Such steels, when used for welded fabrications, should be in a Ti or Nb stabilised grade or better still in the extra low carbon grades, because quenching after welding is usually impracticable. Apart from the amount of carbon, intergranular carbide precipitation (the cause of local intergranular attack) is determined by time and temperature and attack depends on the type of corroding medium (Figure 5.40[91]), which shows the effect of carbide precipitation on an unstabilised 18/8 steel.

SURFACE CONDITION

For maximum resistance to corrosion, the passive film must be properly formed; this is ensured by removing all scale, embedded grit, metal pick-up from tools and other surface contaminants. Polishing improves resistance. Passivating in oxidising acid (10–20% HNO_3 by weight solution at 25°C for 10–30 min) gives maximum resistance. The ferritic-martensitic grades are passivated in nitric acid-potassium dichromate solution (0·5% nitric acid + 0·5% potassium dichromate at 60°C for 30 minutes).

VARIATION IN CORROSION CONDITIONS

If the conditions as to oxidising character, acidity or temperature vary, the passivity of the film may be affected. Concentration of the corroding medium in crevices or narrow gaps in equipment can lead to oxygen starvation with consequent breakdown of the film (design of equipment and absence of concentrating conditions handles this problem).

In the absence of experience samples of the proposed steels should be tested in the condition in which they will be used (i.e. welded, if fabricated) in the intended environment taking full note of the possible variation in service conditions.

STRESS CORROSION

A combination of high residual stress with a certain limited number of corroding solutions, mostly containing halides, will cause local cracking which is usually transgranular (a point for identification). Sufficient stressing may arise from cold working

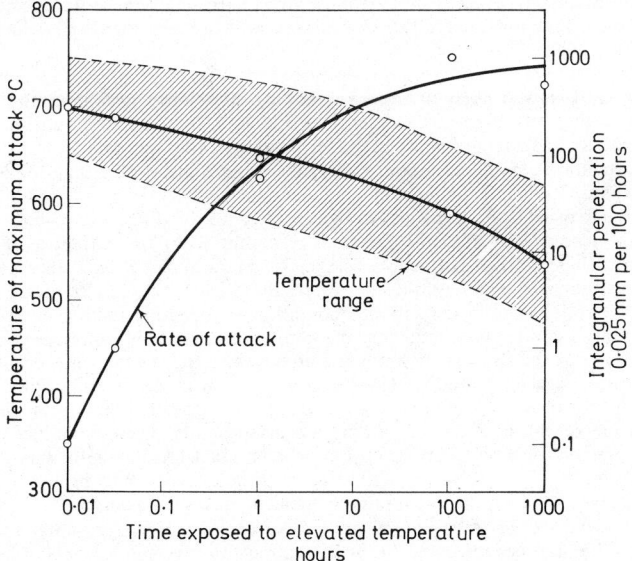

Figure 5.40. Effect of time and temperature on the maximum sensitivity to intergranular attack of low carbon 18% Cr 8% Ni steel

(i.e. bending, forming, pressing, etc) or from differential expansion, set up by local heating, welding, etc.

The cracking is usually observed in solutions operating at elevated temperatures and is typically of a branching type at right angles to the stress. The higher Mo bearing steels offer more resistance to stress corrosion cracking than 18/8; stress relief treatment (two hours at 870°C) after fabrication considerably reduces the risk of cracking.

EFFECT OF WELDING ON CORROSION RESISTANCE

Welding is the normal method of fabricating stainless steel vessels etc. The heat affected zones are raised to incipient fusion temperature but time at temperature varies with different welding processes. Argon arc and spot welding are most satisfactory in heating for minimum time; metal arc welding, inert-gas metal arc and submerged arc are less so in that order from this point of view.

Time at temperature affects the extent of re-solution of the titanium and niobium carbides present in stabilised steels; titanium carbide dissolves more rapidly than niobium carbide. Re-solution takes place at temperatures in excess of approximately 1200°C under welding conditions. The extent depends on carbide particle size as well as time at temperature. If reheated within the sensitisation temperature range, around 650°C, this narrow zone immediately adjacent to the weld metal precipitates intergranular chromium carbide, because combination of Ti or Nb with Cr cannot occur at this temperature.

Thus, although the stabilised steels will not precipitate chromium carbide in the region of the heat affected zone raised to 650°C by welding, there is the possibility, in conditions where the edge of the weld metal is reheated to 650°C, that intergranular attack can occur. This is because the steelmaker ensures the carbon is combined wholly or partially with Ti or Nb. The existence of such conditions depends on the welding practice but in most fabricated articles, as distinct from samples with single run welds, positions must arise at weld junctions where these conditions will obtain; note that welded samples should always have crossed welds.

This particular type of intergranular attack at the weld metal edges is known as 'knife-line' attack; it is most likely to be seen in boiling dilute nitric acid solutions. The composition of the steel affects its incidence; steels with lower nickel to chromium ratio, which produce a greater amount of delta ferrite in the knife-line zone are less susceptible. Fully austenitic Ti stabilised grades appear to be more susceptible than fully austenitic Nb stabilised.

Where corrosion conditions are known to offer a knife-line hazard, heat treatment of the fabrication at 870°C will promote precipitation of the carbon as titanium or niobium carbide with consequent resistance to intergranular attack. It is pertinent to note that steelmaker's and 'weld-decay' test results (Figure 5.40) carried out by subjecting samples of rolled steel in the softened condition to reheating for half-hour at 650°C, Figure 5.40 (a control test frequently made to confirm stabilisation) does not indicate susceptibility of the material to knife-line attack since the steel has not been heated to temperatures exceeding 1200°C. The extra low carbon grades would appear to be the best choice for resisting knife-line attack but it should be noted that although the time-temperature reheating effects of welding are insufficiently severe to induce dangerous carbide precipitation, this may not apply to continued heating, as in service, at temperatures in the 450–750°C range. Therefore these steels should be used for service temperatures above 300°C only on the steelmaker's recommendation.

The unstabilised 17% chromium grade is susceptible to intergranular attack after welding. This can be prevented by heat treating for two hours at 600–800°C which coalesces the carbide films.

Brief summary notes on corrosion and welding aspects of stainless steels are given in Table 5.10.

SERVICE TEMPERATURE

Because stainless steels which are unstabilised or partly stabilised with Ti or Nb may show chromium carbide precipitation when subjected to service temperatures above 350°C, this should be the upper limit for service in corrosive environments. Fully stabilised steels are not restricted in this manner.

Heat resisting steels

The main requirements of heat resisting steels are:

  Resistance to oxidation or corrosion at service temperature in various gaseous atmospheres.
  Structural stability to minimise embrittlement at service temperatures.
  Ductility, formability and weldability for ease of fabrication.
  Resistance to rupture at service temperature.

Table 5.10 CORROSION RESISTANCE AND WELDABILITY OF STAINLESS STEELS

Type	Nominal composition					Corrosion notes
	C	Cr	Ni	Mo	Other	
13% Cr	0·08–0·4	12–14	—	—	—	Rust resisting. Higher carbon grades for engineering applications, turbine blades, cutlery, etc.
13% Cr	0·08 max.	13	—	—	Al	Weldable grade.
17% Cr	0·1 max.	16·5	—	—	—	Resists mild acids. Special feature is resistance to nitric acid. May require heat treatment after welding (600–800°C) to avoid intergranular attack. Forming of sheets up to 3 mm at room temperature; greater thicknesses at 200–350°C.
17% Cr	0·1 max.	17·5	—	—	Ti	Weldable grade not requiring heat treatment. Argon arc (gives minimum grain growth) preferred. Both grades, if welded, should not be applied under conditions of shock loading or vibration.
18/8	0·1 max.	18	9	—	—	Rust and acid resistant. Not generally suitable for welding.
18/8	0·07 max.	18	10	—	—	Lower carbon therefore more resistant to integranular corrosion. Suitable for welding in certain applications.
18/8 ELC	0·03 max.	18	10	—	—	Extra low carbon. Very resistant to integranular corrosion. Weldable for practically all applications.
18/8 Nb } 18/8 Ti }	0·1 max.	18	10·5	—	Nb or Ti	Not susceptible to intergranular attack (but see reference to knife-line attack in text). Applicable above 300°C. Weldable.
18/8/2 Mo	0·07 max.	17·5	11	2·2	—	Resistance to chemical attack better than 18/8 (e.g. severe acid attack). Resists intergranular attack up to 6 mm thickness. Applicable below 300°C. Weldable for most applications.
18/8/2 Mo ELC	0·03 max.	17·5	11	2·2	—	Superior resistance to intergranular corrosion, suitable for thicknesses greater than applications.
18/8/2 MoNb } 18/8/2 MoTi }	0·1 max.	17·5	11·5	2·2	Nb or Ti	Not susceptible to intergranular attack (but see knife-line attack). Applicable above 300°C. Suitable for strong acids at elevated temperatures. Weldable.
18/8/3 Mo	0·07 max.	17·5	12	2·8	—	Resists intergranular attack up to 6 mm thickness. Applicable below 300°C. Corrosion resistance superior to 2% Mo alloys. Weldable for most applications.
18/8/3 MoNb } 18/8/3 MoTi }	0·1 max.	17·5	12·5	2·8	Nb or Ti	For strong acids at high temperatures. Applicable above 300°C. Weldable.
18/8/4½ Mo	0·07 max.	17	13·5	4·5	—	Resistance to strong organic acids at elevated temperatures. Increased resistance to pitting. Applicable below 300°C. Resists intergranular attack up to 6 mm thickness. Weldable for most applications.
25/25 Mo	0·06 max.	25	25	2·2	Ti	More resistant to non-oxidising acids (Viscose manufacture). Weldable.

Creep resisting steels, which are also necessarily heat resisting, are not included. Heat resisting steels can logically be separated into three groups according to composition, microstructure and mechanical properties:

Structure	Alloys	Mechanical property features at temperature
Ferritic	Cr-Si-Al	Lower strength than austenitic alloys.
Ferritic-austenitic	Cr-Si-Ni (low)	Lower strength than austenitic alloys.
Austenitic	Cr-Si-Ni (high)	High strength.

OXIDATION RESISTANCE

Oxidation resistance is determined by composition and the nature of the gaseous environment, i.e. whether oxidising, carburising, reducing or sulphur bearing.

Scaling resistance is related primarily to the chromium content but silicon and aluminium are also important contributors. Tungsten has no beneficial effect. The effect of chromium alone on oxidation resistance is shown in Figure 5.41.[92] Oxidation resistance is not uniform up to all temperatures for the following reason. Resistance is based on the formation of a tightly adherent oxide scale which, having been formed by combination of oxygen from the hot gases with the elements having higher affinity for oxygen than iron present in the steel, prevents further ingress of oxygen after a given thickness has been established.

When a critical temperature is exceeded, the resistance of the scale first formed to

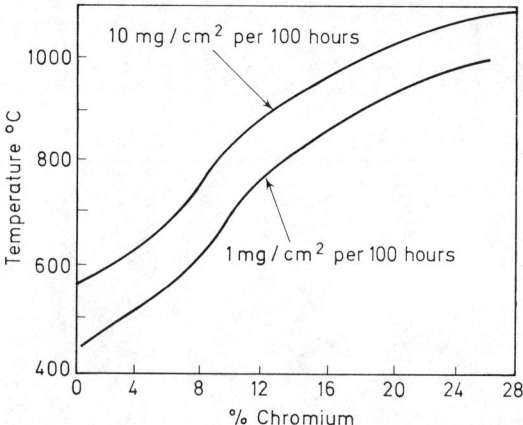

Figure 5.41. Effect of Chromium on oxidation resistance

permeation by oxygen is lowered and multi-layer scale formation (alloy oxide with iron oxide) begins and continues indefinitely. This critical temperature depends on a number of factors such as composition of steel and gases and surface condition (rolling scale, seams, laps, etc). The critical temperature limits the service temperature (see Figure 5.42).

STRUCTURAL STABILITY (ABSENCE OF CHANGE IN MICROSTRUCTURE)

This is relative between the different steels and complete stability is not achieved. Whether or not this is of practical significance depends on the application and particularly upon the design in so far as this determines the extent of additional stressing due to restraint or temperature differences. Although heat resisting steels are based on the iron-chromium-nickel alloy system and therefore have much in common with stainless steels, for example stabilised stainless steels can be used for certain heat resisting applications, the fundamental difference between them is that stainless steels are normally put into service at temperatures below 600°C, and heat resisting steels at service above this temperature.

Stainless steels are, therefore, operated in a theoretically meta-stable condition which for most practical purposes, other than cold work, is a stable state. Austenitic heat resisting steels operate at temperatures sufficiently high for the alloy to approach

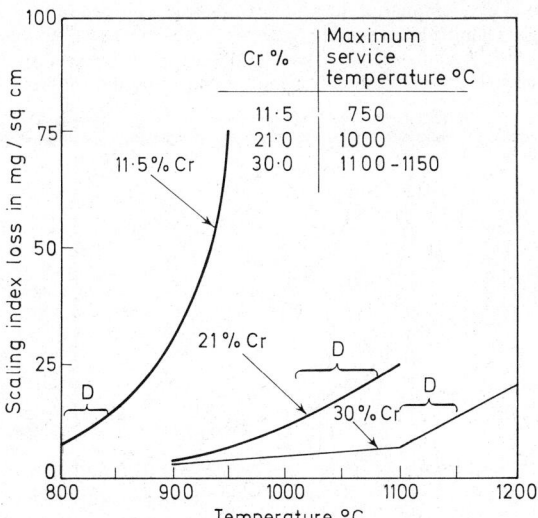

	Cr %	Maximum service temperature °C
	11·5	750
	21·0	1000
	30·0	1100-1150

Figure 5.42. Relationship between scaling loss, temperature and chromium content. Chromium steels

equilibrium so that, having been quenched by the steelmaker to obtain maximum ductility, the carbon is rejected to reach the true solubility level at the particular service temperature (see Figure 5.39). Sigma phase is formed if the service temperature is within the sigma formation temperature range.

The precipitation of carbides and sigma phase produces a marked reduction in ductility and toughness at room temperature but has only a minor effect at the working temperature. However the carbide and sigma, in fully austenitic alloys, will form preferentially at the grain boundaries and there may be corrosive attack on these boundaries in appropriate environments and substantial reduction in notch toughness. Local stressing, due to design faults leading to insufficient accommodation for expansion (a common occurrence in welded fabrication), can produce cracking in steels which have undergone sigma formation.

The straight chromium steels do not undergo sigma phase embrittlement until the chromium content exceeds 27%; it is formed if the steel operates in the range 520–700°C for long periods of time (this situation is unusual however as 27% Cr steels are used at temperatures substantially exceeding 700°C). Reheating above 820°C dissolves the precipitated sigma phase and restores ductility.

A less harmful form of embrittlement (known as '475C') occurs if straight chromium steels at or above about 15% Cr are held in or slowly cooled through the range 425–525°C; it may be removed by reheating above 650°C and cooling rapidly. Sigma phase is non-magnetic. Ferritic 27% Cr steels are magnetic in the absence of sigma precipitation and precipitation of this phase produces a readily detectable change in response to Alnico magnets. Sigma phase may develop in austenitic chromium-nickel steels when subjected to heating in the range 590–925°C; the time required for formation of sigma depends on composition and structure. Silicon, tungsten and particularly molybdenum reduce the time required. A duplex structure of austenite with ferrite also has the same effect. Embrittlement along weld joints frequently precedes general embrittlement owing to delta ferrite formation in the heat affected zones. Sigma formation also occurs in fully austenitic alloys of the 25% Cr 15–20% Ni type particularly when the silicon exceeds 1·5%.

It is possible to over-emphasise the significance of sigma phase embrittlement because, when the design is handled in a manner to prevent local stressing from differential expansion, this embrittlement has slight effect on service performance, although, at room temperature the steel may have ductility and toughness similar to glass.

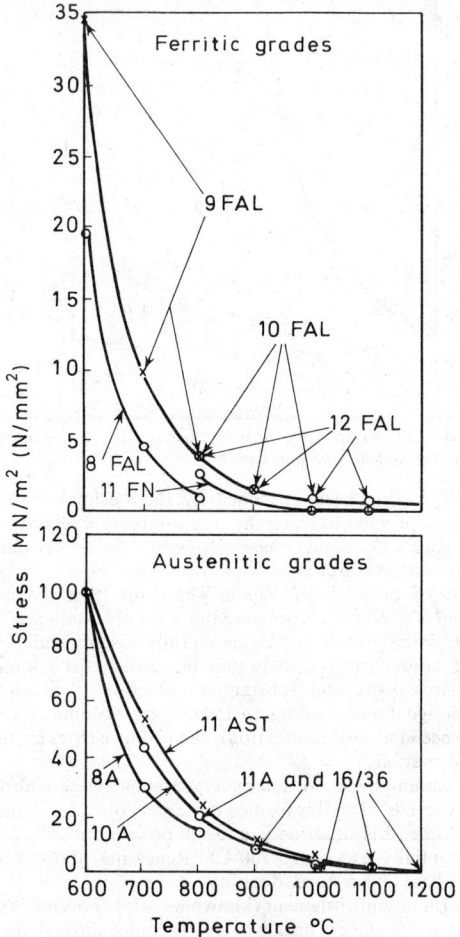

Figure 5.43. Minimum stress for 1% elongation in 1000 hours. Heat resisting steels

Embrittlement due to :-
Carbide precipitation

Sigma formation from austenite

Sigma formation from ferrite

Austenite transformation to α iron

- ○ 0 to 30% embrittlement
- ◑ 30% to 50% embrittlement
- ◕ 50% to 70% embrittlement
- ● 70% to 100% embrittlement

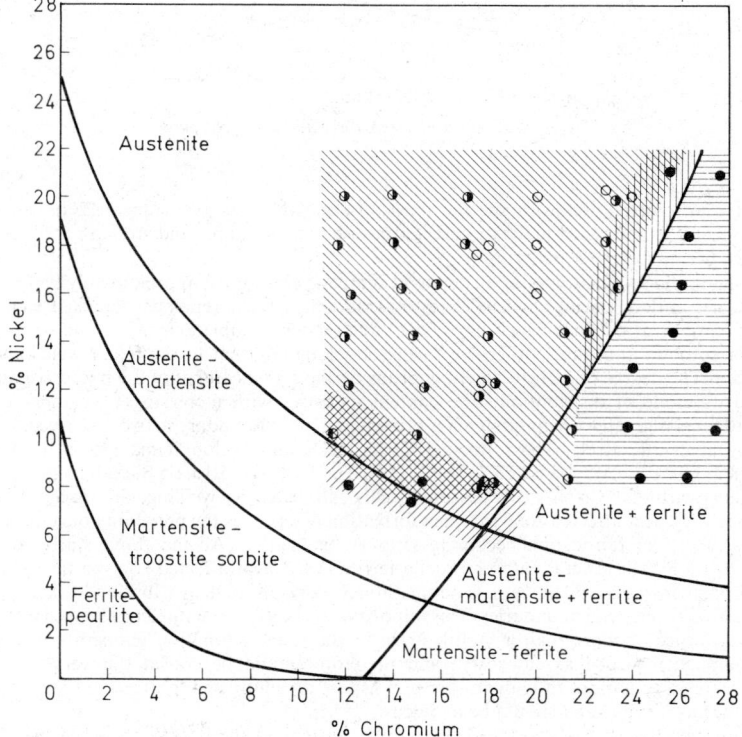

Figure 5.44. Embrittlement of Cr-Ni steels after 1000 hours at 800°C related to microstructure and composition. Dots, etc., show steels examined (After G. Hock)

Designing for the application of heat resisting steels involves selecting a steel which (a) has a maximum permissible working temperature above the service temperature, (b) is suitable for the atmospheric conditions e.g. whether oxidising, reducing or sulphur bearing and (c) has high or low tendency to sigma embrittlement according to the scope offered by the particular design for practical steps to be taken to reduce stress concentrations and local stressing. Where sigma embrittlement has been found to be a significant

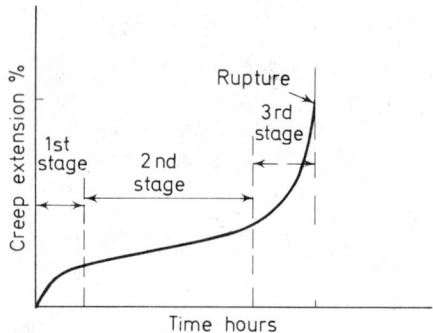

Figure 5.45. Typical creep curve for relatively pure metals

factor in joint failures with specific steels, alternative steels, see Table 5.12(a), *may* be applicable at increased material cost if atmospheric conditions and strength considerations permit (see later paragraph).

Sigma phase redissolves in austenite at temperatures above approximately 950°C so that a solution anneal at 1050°C restores ductility. Weld repairs to cracked welded fabrications or castings which have failed due to sigma embrittlement should be made after solution annealing, otherwise further cracking adjacent to the repair welds is induced. The degree of embrittlement depends on the sigma content of the microstructure and its mode of distribution. The content increases with increasing Cr content and brittleness is greater when the sigma is present as a grain boundary network derived from fully austenitic steels. The following Cr-Ni steels all develop sigma phase; 18/8/Ti, 21/8/1·5 Si/4W, 21/8/1·5 Si, 23/11/1·5 Si/3W, 25/12/1·5 Si, 25/15/1·5 Si and 25/20/1·5 Si.

Since wrought Cr-Ni steels are universally fabricated by welding, the susceptibility of the weld heat affected zone to sigma formation determines the possibility or otherwise of a particular fabrication becoming seriously embrittled. All the above steels except 25/20/1·5 Si will generally develop delta ferrite in the heat affected zone so that sigma although formed, will have a random though concentrated distribution rather than localised at the grain boundaries. This improves resistance to fracture but all fabrications, after service of more than a month or so in the sigma formation temperature range should be regarded as embrittled and therefore should be cooled out very slowly. Repeated fluctuations in temperature pre-dispose an embrittled fabrication to cracking by thermal fatigue and should be avoided.

For applications where the 25% chrome austenitic alloys must be used, serious embrittlement by sigma can be avoided by ensuring that the steel is not held for extended periods of time in the temperature range 600–900°C; if this cannot be avoided the maximum service temperature of around 1050–1150°C for these grades allows, from time to time, a solution anneal at 1000°C (assuming furnace capability) which restores ductility.

Another alternative is to use a steel with 20% chromium 12% Ni which has the same tensile properties up 1000°C as to 25% Cr 20% Ni and differs from it only in being less expensive, in having a lower maximum working temperature (about 1050°C) and being susceptible to only slight embrittlement from sigma phase. For more information on sigma phase formation in Cr and Cr-Ni steels see Reference 93, bearing in mind that, since 1940, experience has shown that sigma forms in fully austenitic 25% chromium steels[94] with 1·5% silicon a point queried in Reference 93.

DUCTILITY, FORMABILITY AND WELDABILITY

The straight chromium steels, which are ferritic, show brittleness at room temperature when the chromium content exceeds about 17%. If the service temperature exceeds 900°C the steels develop grain growth in service which cannot be removed except by hot working; also they suffer grain growth in the weld heat affected zone. Thus, although ductile with adequate cold forming properties up to about 3 mm sheet thickness as supplied by the steelmaker, after welding the fabrication is embrittled at the weld joints and elsewhere if service temperatures exceed 900°C.

This embrittlement disappears at temperatures over 200°C; hence unused plates over 3 mm thickness should be formed hot at temperatures above 200°C and fabrications taken out of service should be similarly preheated before any straightening or forming is attempted.

The austenitic and austenitic-ferritic alloys have good formability, hot and cold. Weldability is also good but, as with the straight chromium grades, arc welding (metal arc and inert gas shielded arc) is preferred since the heat affected zones are narrower.

RUPTURE STRENGTH AT SERVICE TEMPERATURES

The engineering use of steels at elevated temperatures requires the design engineer to decide whether the particular application must maintain a pre-calculated variation from dimensional stability or whether it can be permitted, by proper allowance, to undergo considerable dimensional change. The former requires the use of creep resisting steels, which may also be heat resisting, but the latter allows the use of heat resisting steels which are not necessarily intended as creep resisting. The difference lies in the background metallurgical formulation of the steels.

Most creep resisting materials, whether in steel or the nickel alloys, are based on the solution of a compound in a single phase structure at high temperatures and the precipitation, due to reduced solubility of the solute, within a lower temperature range. The increase in resistance to deformation due to the precipitate, however, is affected by the size and distribution of this precipitate, which is partly determined by time, but mainly by the temperature. Note that an initially fine precipitate coarsens with time but the higher the temperature the coarser the precipitate and the lower the resistance to deformation. Consequently a design based on creep resistance relies on adequate control over temperature and loading. If by accident, the temperature for which relevant creep data have been used, is temporarily exceeded the creep resistance is exceeded, the material being overaged, and the calculated dimensions will not be maintained.

Where temporary accidental overheating occurs in a design not based on creep resistance, with normal safety factors and expansion allowances no permanent damage is done.

Stainless steels of the 18/8 type having carbide stabilising elements, particularly Mo, have greater resistance to creep at temperatures up to 800°C than the heat resisting 25 Cr-12 Ni or 25 Cr-20 Ni alloys. In this context they may be regarded as creep and heat resisting steels (see section on creep resisting steels). Rupture strength may be used as the criterion for designs not based on creep resistance and is determined as the time required for fracture to occur under various constant loads. Design calculations based on stress-rupture tests at various temperatures, which determine either the service stresses or temperatures, are satisfactorily applied provided there is no attempt to extrapolate.

Extrapolation assumes constant metallurgical behaviour which is uncertain in alloys which in service are moving towards structural stability under stress; early fracture may occur. The rupture strengths of four heat resisting steels are shown in Table 5.11.

Table 5.11 RUPTURE STRENGTH OF SOME HEAT RESISTING STEELS
Stress (N/mm²) *for rupture in 1 000 hours*

Temp. °C	27% Cr	18 Cr-8 Ni-Nb	24 Cr-12 Ni	25 Cr-20 Ni
538	—	—	—	220·0
593	41·2	206·0	—	166·0
648	27·4	117·8	104·0	117·8
704	12·3	75·5	80·5	75·5
760	10·8	51·0	—	48·0
815	7·75	30·9	34·3	31·4
871	—	—	18·65	20·6
981	—	—	6·9	13·7

The above notes on oxidation resistance, structural stability, formability, weldability and strength at temperature are illustrated by Thermanit heat resisting steels (Deutsche Edelstahlwerke). The properties and characteristics are given in Table 5.12 and Figure 5.43.

Selection of wrought heat resisting steels

Failure in heat resisting steel applications may be due to several causes; occasionally the wrong steel is chosen for the prevailing conditions on the argument that the more expensive the steel the better it must be. Thus the 25% Cr and 25% Cr 20 Ni steels may be used for service below 900°C which ensures maximum embrittlement under conditions where lower chromium steels have adequate properties.

Assuming correct selection for strength (weak ferritic or strong austenitic) and for scaling resistance (having in mind maximum service temperature, effect of sulphur and carburisation on austenitic steels etc) the remaining factor is embrittlement. The plain chromium steels with more than 25% Cr are embrittled by sigma phase if used below 900°C and by grain growth above 950°C but the latter is primarily only of consequence under shock loading at room temperature which must be avoided. These steels should be used at 1000°C or above.

The embrittlement aspect of chromium-nickel steels is complex but is pictorially represented by Figure 5.44 which shows its dependence on composition. Some degree of embrittlement must be expected in steels with more than 16% Cr; provided the steel is correctly selected and the design takes this into account adequate service is obtainable at less cost than for nickel base alloys. Selection starts with full information on the application being given to the steelmaker. Because of the accentuation of embrittlement problems due to welding, particularly in respect of the straight chromium steels exceeding 25% Cr there is considerable advantage to be gained by using castings if these are available and suited to the application.

CREEP RESISTING STEELS

The standard tensile test as a means of determining the mechanical properties of a steel for design calculation is entirely empirical. The limit of proportionality, the common yardstick of strength, depends on the accuracy of the extensometer and in determining

Type thermix Grade	Nominal composition					Resistance to oxidation under conditions*					Embrittlement and cause where present	Fabrication and welding
	C	Si	Cr	Al	Ni	(a)	(b)	(c)	(d)	(e)		
Ferritic												
8 FAl	0·1	0·75	6·5	0·75	—	800°C	Good	V.G.	Fair	Low	Grain growth above 950°C, i.e. at welds only.	Cold forming up to 3 mm (preferably pre-heat 200–300°C). Hot form 600–800°C above 3 mm. Use austenitic weld metal, capped with 17 Cr in sulphur atmospheres, and inert gas arc process. For welding pre-heat 200–300°C plates thicker than 3 mm. Straightening or bending after welding use 200–300°C pre-heat. Cold forming up to 3 mm after welding anneal fabrication 730–780°C.
9 FAl	0·1	1·0	13·0	1·0	—	950°C	Fair	V.G.	Fair	Low		
10 FAl	0·1	1·0	18·0	1·0	—	1 050°C >900°C	Fair	V.G.	Fair	Low		
12 FAl	0·1	1·5	24·0	1·5	—	1 200°C	Fair	V.G.	Good	Low	General grain growth. Sigma if used in 600°C/900°C range. '475 C' embrittlement.	
Ferritic-Austenitic												
11 FN	0·2	1·0	25·0	—	4·0	1 100°C >900°C	Fair	Fair	Good	Fair	Sigma after long period in 600–900°C range. '475C' also. No grain growth.	Cold forming up to 5 mm. For heavier heat to 900°C for hot forming. Welding, no special precautions.
Austenitic												
8 A	0·08	0·4	18·0	Ti	10·0	800°C	Fair	Fair	Fair	Good	Slight from sigma	Cold forming up to 6 mm. Hot form heavier plates above 900°C. Weldable without precaution.
10 A	0·12	2·0	20·0	—	12·0	1 050°C >900°C	Low	Low	Low	Good	Slight from sigma	
11 A	0·12	2·0	25·0	—	20·0	1 200°C >900°C	Low	Low	V. low	Good	Sigma long periods 600–900°C	
11 AST	0·1	2·0	20·0	Mo, Nb or Ti.	15·0	1 100°C >900°C	Low	Low	V. low	Good	Slight sigma	
16/36	0·12	1·8	16·0	—	36·0	1 100°C >900°C	Fair	Low	V. low	Good	None	

* (a) Air or oxidising gases, maximum service temperature. (b) Reducing or carburising gases. (c) Oxidising sulphur containing gases (SO_2). (d) Reducing sulphur containing gases (H_2S). (e) Low oxygen, nitrogen containing gases.

Table 5.12(b) STRENGTH OF THERMAX HEAT RESISTING STEELS AT ELEVATED TEMPERATURES

Steel	Heat treated condition	Tensile strength (N/mm²)		Min elongation L = 5d	Minimum stress for 1% elongation in 1000 hours at °C (N/mm²)						
		Room temperature			600	700	800	900	1000	1100	1200
		Min yield	Tensile								
8 FAL	Annealed 750–800°C	240	400–590	20	1·9	0·48	0·08	—	—	—	—
9 FAL	Annealed 800–850°C	295	490–635	15	3·4	0·95	0·38	0·14	0·048	—	—
10 FAL	Annealed 800–850°C	295	490–635	12	—	—	0·38	0·14	0·063	—	—
12 FAL	Annealed 800–850°C	295	490–635	10	—	—	0·38	0·14	0·063	0·03	—
11 FN	Quenched 1000–1050°C	390	585–735	25	3·4	0·63	0·29	0·08	0·038	0·019	—
8 A	Quenched 1020–1070°C	263	540–735	40	9·5	2·9	1·45	—	—	—	—
10 A	Quenched 1050–1100°C	293	587–735	40	9·5	4·4	1·9	0·88	0·38	0·14	—
11 A	Quenched 1050–1100°C	293	587–735	40	—	—	1·9	0·88	0·38	0·14	0·048
11 AST	Quenched 1050–1100°C	290	587–735	40	9·5	5·5	2·2	0·9	0·44	0·14	—
16/36	Quenched 1050–1100°C	263	540–735	40	—	—	1·9	0·88	0·38	0·11	0·048

the beginning of permanent strain, the load being increased in steps with corresponding extensions, the test ignores time.

An extensometer which can only detect an extension of 2·5 μm per 25 mm will not record the start of permanent set unless it takes place at the *rate* of 2·5 μm per 25 mm per minute assuming each loading step takes one minute. Thus, when conditions of test (e.g. increase in temperature above ambient) permit considerable deformation with time the figures for limit of proportionality become meaningless in the accepted sense of Elastic Limit. This is illustrated in Table 5.13.[95]

Table 5.13 EXTENSOMETER SENSITIVITY AND LIMIT OF PROPORTIONALITY
(TEST AT 400°C)

Sensivity of Extensometer Limit of strain measurement μm per 25 mm	Deduced value of limit of proportionality from given Stress-Strain curve N/mm²
0·05	154
0·25	173
0·5	192
2·5	260

Also the Limit of Proportionality, determined at a given extensometer sensitivity, tends to decrease with increase in temperature and in the time taken to make the measurement. Short time tensile tests in general are not suitable for assessing the resistance to deformation under constant load of steel at temperatures above 300°C, primarily because the behaviour of the metal under the two different loadings is not the same. The effect of stress and temperature on steels which are not in a microstructurally stable condition is to promote changes towards stability. Such changes may affect the mode of failure under test, promoting an intergranular type with brittle fracture and will certainly affect strength. Above 300°C steel should be subjected to a creep test to determine likely behaviour under service loading and temperature conditions.

When a constant tensile load at constant temperature is applied to a heated steel bar, the bar gradually elongates. The rate of creep will vary considerably according to the load and temperature. The elongation obtained when plotted against time for tests carried out between the two extremes of negligibly small creep at low temperatures with low loads to short time rupture at high loads produces, in relatively pure metals, curves of the form shown in Figure 5.45. The first stage consists of initial rapid creep decreasing with time, the second stage shows a constant creep rate, the curve being linear and the last stage shows a gradual increase in creep rate until rupture.

In practice commercial steels and high temperature alloys do not produce this type of curve with three distinct stages. Instead the curves are irregular with inflections caused by metallurgical changes taking place as the steel approaches a stable condition, though the feature of an approximation to linearity over the middle portion of the curve is generally found. The slope of this line (second stage) is the creep rate, a parameter which is commonly used in the US. In the UK and Continental countries, the generally used parameter is the load (expressed as tensile stress) on the bar required to produce a definite amount of creep (expressed as percent strain) in a given time at a given temperature.

Factors affecting creep

The main factors which, from experience, appear to affect creep are chemical composition, melting process and deoxidation practice, heat treatment, grain size and the length of time of the test. Austenite grain size is a key factor[96] affecting creep rate. If a steel is given a coarse grain treatment and its creep characteristics determined in comparison with those obtained in a fine grain condition the coarse grain bars will show

Table 5.14 TYPICAL CREEP AND RUPTURE DATA FOR FERRITIC CREEP RESISTING STEELS

Type	Nominal composition				Condition* A, N and T, OH and T	Stress ( ) to give 1% strain in T hours at °C (N/mm²)			Stress to give rupture in (a) 1000 hours (b) 10 000 hours (c) 100 000 hours at °C (N/mm²)
	C	Cr	Mo	V		T = 3000	T = 10 000	T = 100 000	
1 Cr ½ Mo	0·12 max	1·0	0·55	—	N	510°C (190) 538°C (150) 565°C (70)	—	—	500°C (a) = 28 (b) = 22 (c) = 17 540°C (a) = 18 (b) = 11 (c) = 7 560°C (a) = 15 (b) = 9 (c) = 4
2¼ Cr Mo	0·15 max	2·25	1·0	—	A or N and T	—	550°C (110) 600°C (50) 650°C (25)	550°C (72) 600°C (25) —	550°C (b) = 126 (c) = 80 600°C (b) = 65 (c) = 35 650°C (b) = 26 (c) = 10
3 Cr Mo	0·25	3·0	0·5	—	OH and T	500°C (180)	—	—	500°C (a) = 290
Mo-V	0·18	—	0·6	0·3	N and T	500°C (260) 550°C (140) 600°C (70)	—	—	500°C (a) = 355 550°C (a) = 220 600°C (a) = 110

* A = Annealed. N = Normalised. N and T = Normalised and tempered. OH and T = Oil hardened and tempered.

lower creep rates. For a given composition, coarse grained steel has the greater resistance to deformation at elevated temperature. The fine structure produced by quenching and tempering has inferior creep resistance to the coarser grain produced by normalising and tempering.[97]

The resistance of the body-centred cubic lattice structure to deformation at elevated temperature is inherently less than that of the face-centred lattice. The resistance of both is increased by precipitation of solute atoms from solid solution. Ideally the steel should enter service at elevated temperature in a stable condition; the body-centred cubic (ferritic) steels are either annealed or normalised and tempered but face-centred cubic (austenitic) steels will be in a meta-stable state. Resistance to deformation is increased to the maximum extent if the ferritic steel precipitate is slow to coalesce at the service temperature. Molybdenum and vanadium, in forming carbides which have this property, confer exceptional creep resistance and are the major alloying elements in creep resisting steels.

The upper limit of service temperature is determined by increasing rate of growth of carbides as the temperature increases with consequential fall in resistance to deformation. Tempering at too high a temperature before service will seriously affect subsequent performance in service as the carbide growth process has been partly completed. Vanadium, in addition to retarding carbide growth, may act as a precipitation hardening element which still further enhances resistance to deformation at elevated temperature.

Creep strength, however, is not usually the sole criterion, ductility also being a necessary feature. Ductility under creep conditions is conferred by chromium. C-Mo and C-Mo-V steels, though possessing excellent creep resistance are lacking creep ductility which somewhat restricts their application. Cr-Mo and Cr-Mo-V steels are widely used for creep resisting applications; optimum chromium addition for ductility plus creep strength is 1·25–1·5%. Increasing Cr up to 6–7% reduces the creep resistance but further addition up to 12–13% improves resistance. The 6% Cr-Mo steel, used mainly for resistance to scaling, has about the same creep strength as a coarse grained carbon steel. Typical creep and rupture strength values for some ferritic creep resisting steels are given in Table 5.14.

As indicated in Table 5.14, the upper temperature for use for loading with ferritic creep resisting steels is around 650°C. Above this temperature austenitic steels are required. Stress rupture data[98] for 18% Cr 11% Ni Nb and 17% Cr 12% Ni 2·5 Mo steels at 650°C and 700°C are given in Table 5.15..Stress to rupture data obtained from

Table 5.15

Steel, AISI alloy content	Stress* (N/mm²) to cause rupture in 100 000 hours	
	at 650°C	at 700°C
347 18 Cr 12 Ni 1 Nb	48	25
316 17 Cr 12 Ni 2·5 Mo	70	25

*Not extrapolated. Values obtained by extrapolation from short-time tests up to 10 000 hours may be substantially higher.

tests of shorter duration (1000 hours) are given in Table 5.16.[99] Comparing Tables 5.15 and 5.16, the considerably reduced rupture strengths of the Nb and Mo bearing steels obtained by extending testing from 1000 hours to 100 000 hours should be noted, indicating that caution is required in applying short time creep and rupture data. A further observation from Table 5.16 is the increase in strength obtained by additions of Ti, Nb or Mo to 18%Cr 8%Ni steel, which is associated with the formation of carbides by these elements with comparatively small tendency to growth.

In addition to strengthening by carbide precipitation, the resistance to deformation at high temperature can be increased by solid solution strengthening (the introduction of foreign atoms into the crystal lattice which do not precipitate out but cause lattice strain by virtue of difference in atomic size). Cobalt is added to Cr-Ni steels and to nickel alloys for this purpose.

A further method of strengthening at high temperature single phase face-centred cubic alloys, i.e. austenitic steels and nickel alloys, is by Precipitation Hardening. The classical example of this is the Nimonic nickel based alloys containing titanium and

Table 5.16

Temperature °C	Stress-to-rupture data for 1000 hours at stated temperatures (N/mm²)			
	18 Cr 8 Ni AISI 304	18 Cr 8 Ni Ti AISI 321	18 Cr 12 Ni 1 Nb AISI 347	17 Cr 12 Ni 2·5 Mo AISI 316
648	100	180	200	162
704	60	120	110	176
760	40	68	70	110

Table 5.17

Alloy	Nominal composition			Temp. °C	Stress to produce rupture (N/mm²) at stated temperature °C in	
	Ni	Cr	Other		1000 hours	10 000 hours
Incoloy 800	32	20		704	90	60
Nimonic 80 A	76·4	20	2·3 Ti 1·3 Al	700	290	170
Incoloy 800	32	20	—	760	60	39
Nimonic 90	59	20	2·5 Ti 1·5 Al 17 Co	750	240	130

Table 5.18 EFFECT OF HEAT TREATMENT ON GRAIN SIZE AND RUPTURE STRENGTH OF INCOLOY 800

Grade	Treatment	Average grain size mm	Stress for rupture in 100 000 hours (N/mm²) at	
			650°C	760°C
1	Annealed 1 hour at 980°C	1 × 0·025	68	11
2	Annealed 1 hour at 1150°C	3·5 × 0·025	81	29

aluminium. These elements form compounds with nickel which, under suitable heat treatment, form very fine stable precipitates. The rupture strength of such alloys is compared with a straight nickel-chromium alloy in Table 5.17.[100]

The attainment of increasing hot strength at higher temperatures is thus achieved by progressing from precipitation hardened ferritic steels (Mo, V) precipitating a fairly stable carbide, through carbide hardened austenitic steels to precipitation hardened nickel alloys precipitating stable compounds.

Grain size has a significant effect on hot strength because creep takes place by a combination of grain distortion and grain boundary sliding, the larger grain size reducing the second mechanism. The solution temperature and time at temperature therefore affect the hot strength as shown in Table 5.18.[101] However, the rupture ductility of coarse grained alloys is normally lower than that of finer grained and in many applications is a significant factor. Cast high carbon austenitic Cr-Ni alloys of the 25/20 and 18/37 type have been used fairly widely, having good hot strength (due to extensive carbide precipitates and coarse grain structure inherent in casting) but they are sensitive to rapid temperature changes because the ductility below 650°C is extremely low (grain boundary carbides).

Most applications of wrought or cast creep resisting alloys involve welded fabrication and the heat affected zone of welded joints is usually the point of lowest ductility after service. Low carbon materials, if fully austenitic, present fewer problems and wrought material offers the possibility of checking quality by ultrasonic inspection; coarse grain in castings prevents this. Assuming environmental conditions are not deterimental, the ultra-low carbon wrought higher nickel alloys (Ni greater than 32%) with 20% Cr seem to have an attractive combination of hot-ductility, resistance to embrittlement and hot strength, which is difficult to match by other materials. Welding is also comparatively straightforward. The low carbon higher nickel alloys also appear comparatively resistant to Thermal Fatigue and Thermal Shock, possibly because of their resistance to embrittlement.

However failure by cracking from thermal fatigue and thermal shock can often arise from faulty design which does not allow for expansion; simple rules[102] which minimise the risks are:

Avoid corner welds (weld toes produce severe stress concentration).

Use rounded corners.

Members of a fabrication which expand in the same direction should have the same cross-section.

If subject to temperature fluctuation, the fabrication should be made from members of the same thickness and section dimensions, whether or not the same sections are under-stressed by virtue of this rule.

Where differential expansion between members is unavoidable consider assembly by bolting with slotted holes to permit movement. Most heat resisting alloys expand approximately 1 mm per 60 mm from room temperature up to 900°C.

STEEL FOR LOW TEMPERATURE USE

The essential property required in steels for low temperature use is resistance to brittle fracture. When ferritic steels are subjected to mechanical tests at temperatures below room temperature they generally show an increase in tensile strength and decrease in ductility and toughness, particularly notched impact toughness. Steels having a fully austenitic structure show increase in strength but not reduction in ductility or notched toughness.

Design engineers use arbitrary factors of safety based on the tensile or fatigue properties of steel to provide a safe margin of strength in a structure over and above the maximum calculated design stresses. This margin of safety is based on experience or tradition; the principle is sound when the service temperature is high enough to prevent the possibility of brittle fracture. The higher the safety factor the thicker the section and the greater the risk that the service temperature will be within the brittle fracture temperature range. The conventional design by safety factor procedure cannot take into account failure by brittle fracture because the calculations do not consider notch effects caused by design, defective material surfaces, defects at welds, etc.

In the presence of notches, cracks or linear defects generally, such as local lack of fusion at weld joints and lamellar tears at weld joints, the breaking strength of the structure, as distinct from that of the steel components from which it is made depends upon the ability of the material at the tip of some critical defect to yield plastically as the load on the structure is increased. Still more important is the question whether the structure will fail in a brittle or ductile manner, the former leading to extensive cracking at loads which may stress the steel well below the calculated 'safe' stress level. Moreover brittle facture is 'fast fracture' so that repairs, as for a slowly developing ductile fracture, are out of the question.

A common feature of failure of *structures* by brittle fracture is that subsequent testing of the steel by conventional tensile tests indicates the material meets the required levels of strength and ductility as determined by elongation and reduction of area. Assessment of the notch ductility of the steel by a notched test is a step in the right direction but it is not possible to evaluate quantitatively and forecast the overall resistance to brittle fracture of a structure by any single test. Machined castings and forgings, properly heat treated, probably represent the situation where the variability of the factors affecting brittle or ductile behaviour are at a minimum. In a structure fabricated by welding and not subsequently stress relieved the variability is probably at a maximum.

Testing for brittle fracture

The design engineer is concerned with avoiding brittle fracture in specific structures. Brittle or ductile behaviour in a particular test on material which is to be used in a structure is of no interest to him unless he can be certain of a significant relationship between the test result and structure behaviour. Unfortunately such a relationship must be established from experience of service behaviour. Suitable tests on the steel to be used will then determine whether control of steelmaking variables, which determine test sample behaviour, has been sufficiently precise to give ductile behaviour in the test and the assumption is then made that the steel quality is the same as that previously found to be satisfactory.

Ductile or brittle behaviour in a test depends on the particular test conditions, such as type of loading (i.e. tensile, bend or impact), type of stress (i.e. round notch or sharp notch having varying degrees of sharpness), thickness of specimen, etc. These different tests may or may not give the same result and indeed may not even place different steels in the same order of resistance to brittle fracture.

In the course of research into the complex subject of brittle fracture many different tests have been devised with each inventor claiming special merit for his particular test. Many tests are suitable as research tools but too difficult to be applied as a quality control test. This function, which demands ease and cheapness of specimen preparation and speed of testing, is handled most successfully by the notched impact test carried out on small specimens. Of the variety of notched impact specimens (i.e. Izod, Charpy Keyhole, Charpy V, Mesnager, D V M, Schnadt) the Charpy is more widely used than any other, partly because it lends itself particularly to testing at sub-zero temperatures.

A branch of engineering science, fracture mechanics, has been developed which aims to avoid relying on establishing a significant relationship between test sample and behaviour of structures in service. It is claimed[103] that fracture mechanics permits the conditions of local stress at fracture in a structure to be related to the size of initiating defect and overall applied stress.

The conditions of local stress at fracture may be described in terms of either a critical elastic stress intensity (K_1c) or, where appreciable local deformation occurs, in terms of a critical displacement of the crack faces to cause fracture (δc). Local stress means the effective stress acting at the ends of the defect which causes failure to occur. For the former condition which is known as plane strain, i.e. elastic deformation at the crack tips, the equation $K = \sigma\sqrt{\pi a}$ relates applied stress (assuming a plate of finite thickness but infinite dimensions in its plane with a sharp through thickness crack of length $2a$) at infinity to half the crack length (a) and an elastic stress intensity K (a measure of stress intensification under these conditions of overall stress and crack length).

The critical point of fracture is given by $K_1c = \sigma c \sqrt{\pi a}$ where σc is the stress to cause failure and K_1c (the critical elastic stress intensity) is called the Plane Strain Fracture

Toughness. Where extensive yielding occurs at the ends of the crack, extension of the elastic analysis to account for plastic deformation leads to the equation

$$\delta = \frac{\delta \varepsilon ya}{\pi} \log_e \sec \frac{(\pi \sigma)}{2\sigma y}$$

where

δ = Crack Opening Displacement,

εy = yield strain,

σy = yield stress,

σ = applied stress at infinity,

a = half the crack length.

It is claimed that if $K_1 c$, measured in the laboratory is chosen to be greater than the K value at a defect in a structure, calculated from the above equation, knowing applied stress and defect size, failure will not occur. Similarly, if a measured value of critical COD δc at fracture is chosen to be greater than the calculated δ value at a crack in a structure based on defect size and applied stress, fracture will not occur. In both instances there is the proviso that internal stresses and design stress (for $K_1 c$) or design strain (for δ) concentrations are taken into account.

This approach to brittle fracture may be compared with conventional methods; the latter uses experience or engineering judgment to select steels for particular service conditions, where structures may be at risk of brittle fracture, using, in the main, the Charpy test as an indicator of steel suitability because it indicates control of steelmaking variables. The assumption is that the structure will be free of linear defects such as cracks.

The Fracture Mechanics approach assumes that the structure will contain linear defects and sets out to determine the acceptable size of defect for given conditions and steel grade using a specially instrumented and controlled notched bend test, to give data relating load to fracture, deformation (either elastic or plastic) and temperature for a standard notch.

Various types of notch are used, one that is favoured being produced as a short fatigue crack from a precut notch. Since notch acuity is known to affect behaviour (brittle or ductile) the natural fatigue induced crack tends to be used for calculations relating to a given structure because no worse instance of stress concentration can be conceived; this will produce, if such cracks are absent, a margin of safety. For comparing one steel with another, standard notches without cracks may be used. Thus the 'margin of error' factor which is accepted in the conventional is also present in the Fracture Mechanics approach.

However, it has shown that the size of cracks which are acceptable under specific conditions of thickness, stressing and temperature for a given material can be calculated from the bend test data and thus, in theory, the cost of removing defects up to a specific size may be avoided. Since considerable expertise is necessary in selecting the test conditions appropriate to a particular structure situation (testing of weld metal, heat affected zone and parent metal may all be required, test thickness determined, notch type decided, etc. i.e. use of Fracture Mechanics judgment), the tendency is for such work to be carried out by a very limited number of organisations with adequate know-how. Small cracks, which may or may not lead to brittle fracture, are commonly associated with welding; thus, in the UK, the Welding Institute has built up the know-how necessary to apply Fracture Mechanics to brittle fracture and other problems arising from linear defects in welded fabrication.

Metallurgical aspects of brittle fracture

Low temperature brittle failure arises from cleavage fracture in steel when in the body-centred cubic lattice form. It occurs by separation of any of the cube faces of the unit cell cube, whereas plastic deformation occurs by slip along planes containing a diagonal of the unit cube. This results in a discontinuous crack connected by plastic deformation at grain boundaries, and means that energy required for propagation of a brittle crack is very low.

A characteristic feature of ductile ferritic steels is a change in fracture mode from ductile to brittle as the temperature is lowered. For this to occur in unnotched specimens extremely low temperatures are required, Figure 5.46.[104] However, in a notched tensile

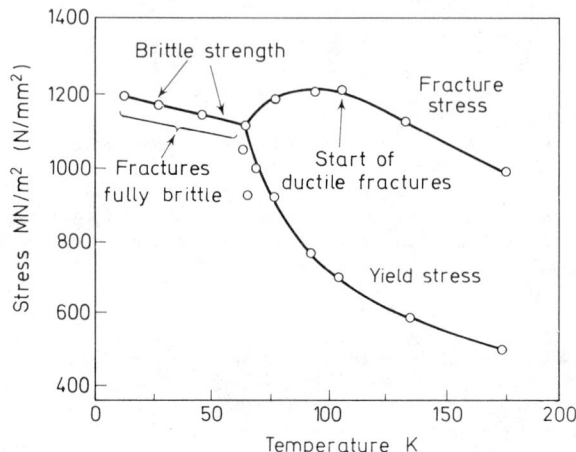

Figure 5.46. Unnotched tensile tests on AISI 1020 *steel showing evidence of a brittle strength curve. Below* 61·5 K *no reduction of area* (After Eldin and Collins)

test bar, e.g. a plate test subjected to axial stress sigma y, there will be a biaxial state of stress at the root of the notch (sigma y, sigma x) with maximum stressing just ahead of the notch root until plastic flow starts. Then the state of stress becomes tri-axial (sigma y, sigma x, sigma z) with lateral contraction at the notch root due to the third stress sigma z acting in the thickness direction. This contraction is opposed by the mass of metal beyond the root of the crack which is under lower stress thus causing constraint against lateral contraction. In an ideally sharp notch the maximum constraint factor is 3 which implies that the stress at the crack tip must rise to about three times the normal uni-axial yield stress of the material before plastic flow occurs.

The yield stress on unnotched tensile test bars increases as the temperature falls. Assuming there is an inherent brittle strength of a material, if the temperature is low enough to cause the stress at the notch tip (3 × the uni-axial yield stress) to reach the brittle strength fracture will be by the brittle mode as will be the case at all lower temperatures. Thus, if the test bars in Figure 5.46 had been notched one would have expected brittle failure at 175° K or lower temperatures. The point is illustrated in Figure 5.47.[105] In this illustration all notched tensile bars broken below $T2$ will fail in the brittle mode. In practice the transition from fully ductile to fully brittle failure takes place over a temperature range not at a single temperature albeit the range may be quite narrow for some steels.

The position of the transition temperature range on the temperature axis depends on

(a) *Metallurgical factors:* steelmaking, strain rate, microstructure, degree of stability, ageing, etc.

(b) *Mechanical factors:* type of loading, type of notch, type and size of test piece.

Reliability in service at low temperatures implies reproducibility of steel quality in respect of low temperature properties from cast to cast which also implies a quality controlled method acceptable to steelmaker and user. The notched impact test has become the recognised tool for this purpose, where testing of the steel mill product is desired but proof of control over melting, deoxidation and heat treatment practice by any other method not involving physical tests could be satisfactory. The effect of metallurgical and mechanical factors on shifting the position of the Charpy transition temperature range will now be considered.

METALLURGICAL FACTORS

These may be conveniently separated into: those which determine the stability of the ferrite structure; those which affect the toughness of stable ferrite; and those which determine ferrite grain size and the microstructural constituents.

Ferrite stability. In this context an unstable ferrite contains carbon, oxygen, nitrogen or other elements which are in supersaturated solid solution at the service temperature. Precipitation in submicroscopic form of compounds of the elements with iron, if suitable circumstances arise, to attain the stable state is therefore potentially possible. Stable

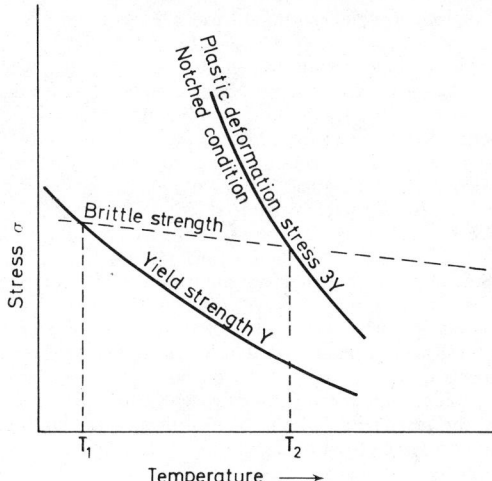

Figure 5.47. Explanation of transition temperature in notch brittle material. Above T_2; fully ductile notched or unnotched. Below T_1; fully brittle notched or unnotched. Between T_1 and T_2; ductile if unnotched; brittle if notched (After E. Orowan)

ferrite is free from these dissolved elements. The unstable ferrite, in moving towards stability by rejection of the compounds, becomes embrittled, the embrittlement process and recovery depending on time and temperature. Ageing of mild steels is an example. Steel melting and deoxidation practice largely determine the stability or otherwise of ferrite in mild steel as it determines the availability or otherwise of the elements for solution. The effect of cold deformation on unstable ferrite (ageing steel) is to markedly increase the rate of progress towards stability, thus ageing Open Hearth and Duplex steels after cold deformation of 5%, 9% and 17% showed a shift to higher temperatures in the impact energy absorption curve of roughly 16°C, 50°C and 70°C respectively.[106] The design engineer will note that welding fabrication will commonly involve some degree of cold deformation (bending, flanging etc) which is followed by ageing; slow when remote from welded joints but faster in the weld heated areas.

Deformation before service is not a prior condition for embrittlement, which takes place during straining the unstable ferrite. This is illustrated by results obtained on large scale tests on box girder beams[107] made from plates. A silicon-aluminium killed steel (stable ferrite) girder tested at −17·8°C showed a similar load-deflection curve to fracture to a semi-killed steel (unstable ferrite) girder tested at 26·7°C.

Elements promoting instability in ferrite are:

Carbon (from fast cooling as in a weld heat affected zone).
Phosphorus. Transition temperature range is raised about 5°C per 0·01% P.
Arsenic. Similar to P.
Nitrogen. High nitrogen steels (e.g. Bessemer air-blown and electric arc) have strong ageing characteristics and relatively high notched impact transition temperature range unless aluminium killed.
Oxygen. Promotes ageing embrittlement by reducing solubility of nitrogen and carbon in ferrite.

Elements promoting stability in ferrite are:
Silicon. By combining with oxygen to form silica whence part of the oxygen is rendered harmless.
Aluminium. By combining with oxygen to form alumina whence the oxygen remaining after silicon killing is rendered harmless. Lowers the transition temperature range by this action but excess aluminium dissolved in ferrite promotes brittleness. A critical amount of Al is necessary for fully effective production of notch tough steels.[108]. By combining with nitrogen (provided the steel is normalised) Al removes this element from solution in ferrite so promoting ferrite stability and improving notch toughness.
Titanium. Combines with oxygen, nitrogen and carbon, having a triple effect on ferrite stability. Prevents ageing in low carbon steels.
Vanadium. Similar to Al in inhibiting grain growth and combining with oxygen and nitrogen but also combines with carbon. Therefore has strong influence towards stabilising ferrite. Has a marked effect in lowering the impact transition temperature range of normalised and heat treated high tensile low alloy steels.[109] Added to rimming steel, it prevents ageing while still allowing a rimming action.
Niobium. Combines with carbon and nitrogen; after normalising these elements are fixed as carbides and nitrides thus ensuring a minimum level of C and N for precipitation on ageing from the ferrite.

Toughness of stable ferrites. Nickel has the most significant effect in lowering the transition temperature of normalised, or normalised and tempered, or heat treated steels, because it markedly improves the toughness of ferrite at low temperatures. The effect is progressive up to 13% nickel in normalised low carbon steels, Figure 5.49(b).[110] It is not predominantly due to its effect on microstructure.

Ferrite grain size and microstructural constituents. The finer the ferrite grain size the lower the transition temperature range. Fine grain ferrite is obtained by normalising, which also eliminates the variations in microstructure inherent in steel supplied in the 'as-rolled' condition, or by 'fine grain' treatment with aluminium or vanadium followed by normalising.

Normalising, while improving grain size, will not remove the instability of the ferrite in a partially killed steel (silicon deoxidation). Although at equal grain size, the plate from a normalised silicon killed steel may show an impact transition temperature range similar to that of a silicon-aluminium fully deoxidised steel; the latter is to be preferred for fabrication involving cold deformation and welding but not stress relieving.

This follows from the situation that weld joints in fabrications which are not stress relieved have high residual stresses combined with the possibility of laminar or linear defects. Therefore, the weld joint should have a greater resistance to brittle fracture than the parent metal. However, welding causes heating above (a) the normalising temperature and (b) the limiting temperature, for inhibiting austenite grain growth by aluminium treatment so the benefit of fine grain structure is not available in a narrow zone adjacent to the weld runs, unless the whole fabrication is renormalised after welding (generally impracticable).

The silicon-killed steel which, by normalising, has attained a fine grain structure and good notch ductility will have a coarse grain in this zone with only a partially stabilised ferrite and lower notched impact properties; the silicon-aluminium killed steel will also have a coarse grain structure but, by fixing all the oxygen as oxide and nitrogen as nitride the aluminium allows more carbon to remain in solution in the ferrite with consequent better approach to stability. A similar effect is obtained with vanadium in low carbon ferritic steels.

Where notch toughness in the heat affected zone is of paramount importance the use of a nickel steel seems indicated (Mn-Ni-V or MN-Ni-Ti) vacuum melted to give minimum oxygen and nitrogen, V or Ti to minimise carbon solution in ferrite and Ni to confer maximum ferrite toughness.

Microstructural constituents have a significant effect on notch toughness in as-rolled or normalised mild and carbon steels. Toughness decreases as the pearlite content increases (with increasing carbon content) but increases as the pearlite becomes finer. The lower the carbon and the higher the Mn content (up to about 1·2%) the greater the notched impact toughness.

Steels which are predominantly bainitic, or a mixture of bainite and tempered martensite tempered to the same hardness level, give lower transition temperature ranges than pearlitic steels.[111] However, full hardening to a martensitic structure followed by tempering can produce the best combination of strength and toughness. The higher the tensile strength in each type of structure (bainitic, tempered martensite) the higher the transition temperature range.

Notch toughness tends to decrease as section thickness increases because thick material is finished in the rolling mill at higher temperatures with consequent coarser grain in the as-rolled condition. Normalising rectifies this and is the first simple step towards obtaining more uniform and reliable toughness in mild steel products.

Within all wrought steel products there is a particular distribution and orientation of non-metallic matter and phases according to production history. This affects the behaviour of the material in the notched impact test according to location of notch relative to the preferred direction of these constituents.

MECHANICAL FACTORS, EFFECT ON POSITION OF TRANSITION TEMPERATURE RANGE ON TEMPERATURE AXIS

The effect of mechanical factors can conveniently be considered under two headings:

Mode of stressing, i.e. whether concentrated and to what degree, whether loading is fast or slow.

Thickness of test piece. Geometrical effect as distinct from the effect of thickness on microstructure.

Mode of stressing. Transition temperature range is purely arbitrary but is a convenient criterion for defining the comparative resistance to brittle fracture of different steels. However, it involves crack initiation followed by crack propagation whichever test is used.

In an unnotched tensile test, by sufficiently lowering the temperature (Figure 5.46) the yield stress is raised to the brittle strength to give a brittle fracture; the yield stress of ferritic steels can also be raised by increasing the rate of straining at any given temperature. This is effectively achieved at the notch root in a notched impact test, but the

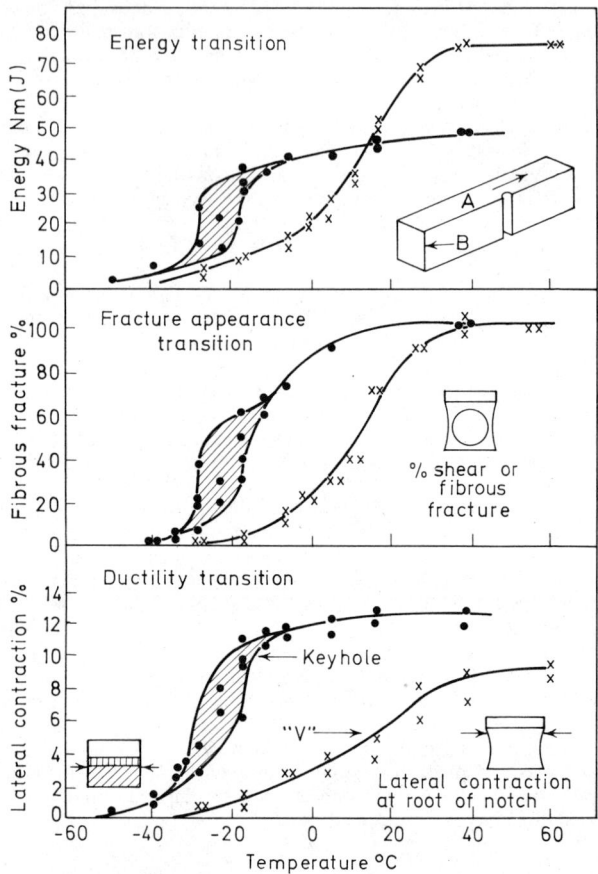

Figure 5.48. Relationship between various transition temperatures. Semi-killed mild steel 0·18% C, 0·54 Mn, 0·07% Si. *A. Rolling direction. B. Plate thickness*

degree of effectiveness depends on notch root radius. By measuring the energy absorbed to fracture and the amount of lateral contraction at the notch the test gives a record of the crack initiation toughness parameters. By recording the fracture appearance of the broken test piece (% fibrous) at each of a series of temperatures the effect of temperature on the crack propagation behaviour of the steel can be set out. Thus three impact fracture behaviour criteria are recognised, i.e. energy, fracture appearance and ductility.

The effect of notch acuity on the position of the various transition curves on the temperature axis is illustrated in Figure 5.48. The diagram also shows the characteristic scatter of results in the transition temperature range. The fracture appearance transition appears at the same or higher temperatures than the ductility transition, generally the latter.

Brittle failure in structures can be related either to the ductility transition or the fracture appearance transition (by empirical correlation). Below the ductility transition temperature, an extremely small amount of local plastic deformation is required to start brittle fracture in the structure but, over a range of temperature just below the fracture appearance transition, a considerable amount of plastic deformation will occur before starting a brittle fracture. Thus the fracture appearance transition temperature range is an indication of the possibility of brittle fracture in the structure at temperatures below it but, if the temperature is above the ductility transition range considerable plastic deformation at the defect or notch is required to *start* a brittle fracture. Below the ductility transition very little plastic deformation at the defect is required to initiate fracture and, once started, it will be by cleavage mode as the temperature is below the fracture appearance transition.

The ductility transition is affected by notch sharpness and rate of loading, being raised to higher temperatures by these factors; this does not ease the problem of relating test to likely behaviour of a structure. However, if one assumes the structure will contain cracks (maximum notch acuity) a notch of maximum acuity would also be used in the test piece and, by breaking the test piece under maximum loading speed (as by an impact load) the test will be more suitable for correlating with crack initiation in structures than if other test conditions are selected.

Thickness of test specimen is also a factor in behaviour; a blunt notch, e.g. keyhole, tends to give a higher transition range in thicker specimens but behaviour with a V notch with impact loading appears to be unaffected by thickness. The fracture appearance transition is also affected by thickness being higher for greater thicknesses. Here there is an argument in favour of determining fracture appearance transition by slow notched (V)—bend specimens of full plate thickness unless a correlation already exists between Charpy fracture appearance and structure behaviour.

Steels for low temperature service

Engineering judgment is required in selecting steels for new applications or ordering steel of a stated quality which has proved satisfactory in the past.

MILD STEELS

Considering the latter aspect first, the degree of reproducibility of behaviour depends upon the extent of control of steelmaking practice required by the grade of steel. A semi-killed steel of the low carbon–high manganese type, which is supplied as-rolled, has little or no control over the factors affecting ferrite stability so that the brittle fracture resistance will alter according to time and temperature experienced in the fabrication by cold working and welding. In this respect there is likely to be variability from cast to cast and particularly from one steelmaker to another. Such steels would be restricted to temperatures at the higher end of the low temperature range. On the other hand, silicon-killed fine grain steels made non-ageing with aluminium treatment and normalised will have a relatively stable ferrite and, providing the steelmaker controls very precisely the fine-grain treatment, reproducibility should be high. However, with both grades the stability of the ferrite in weld heat affected zones is not as satisfactory as in the parent metal.

Carbon and nitrogen dissolved in ferrite are possibly the most potent factors in promoting instability; the addition of niobium is very beneficial in dealing with these elements because the niobium carbides and nitrides are relatively difficult to dissolve in austenite so that the short heating times in the weld heat affected zones do not cause as great a change in stability as in the above-mentioned steels which do not contain carbide formers. Such steels should have good reproducibility when the nitrogen content is under control; when produced as semi-killed and normalised steels their application temperature range would be intermediate between the two aforementioned grades.

The closest approach to stability in the weld zone as well as parent steel and, therefore, to obviating the brittle fracture risks attendant on embrittlement from strain-ageing and quench-ageing is obtained in silicon-killed steels which have fine grain treatment with aluminium and vanadium whereby oxygen, nitrogen and carbon are largely removed from the ferrite. These non-ageing steels will have good reproducibility. An alternative silicon-killed grade of similar quality in this respect is fine grain treated with vanadium only.

From the point of view of reproducibility of resistance to brittle fracture from cast to cast non-ageing steels have the advantage that a strain-ageing test on the product can indicate the success or otherwise of the melting and deoxidation practice applied to the particular heat. The quality control Charpy test applied to ageing steel parent metals is not an indicator of reproducibility of heat affected zone behaviour.

Selecting steels for new applications is a complex operation. Where a choice exists between ageing and non-ageing steels the use of steels subject to ageing embrittlement, yet having relatively high energy absorption due to a fine grain structure achieved by various means, e.g. 'controlled' rolling down to lower temperatures, is justified on economic grounds in lower material cost. However, normal engineering caution, in recognising the added risks at weld joints, would seek to minimise these by careful non-destructive examination to ensure removal of defects above a critical size. Assessment of what is 'critical', by Fracture Mechanics techniques, and the application of NDT (particularly on large structures) tends to be costly so that the material cost advantage may prove illusory. A straight choice between ageing and non-ageing steels evidently exists only towards the higher temperature end of the low temperature range. At the lower end there may be a choice between vanadium bearing non-ageing steels and those based solely on aluminium treatment; the former would appear preferable for applications involving welded fabrication.

Where stress relief heat treatment is properly applied, the risk of brittle fracture in a welded mild steel fabrication is markedly reduced. Where fracture has not occurred during fabrication (such cracking may be possible simply due to residual welding stresses, linear weld defects and strain-ageing embrittlement at the defect tips) the stress relief treatment lowers the residual stress level and reduces, by over-ageing, the condition of embrittlement.

It should be noted that, weldable low-alloy high tensile steels containing vanadium in sufficient amount to confer additional strength (as distinct from its use as a fine-grain non-ageing element) require stress relief treatment over a comparatively narrow and critical temperature range, otherwise, at temperatures lower than this range serious embrittlement will be produced in the weld heat affected zone.

Examples of mild structural steels for low temperature service are given in Table 5.19. The steel FG 26 T is non-ageing.

NICKEL STEELS

The next stage down the low temperature range is covered by the low carbon nickel steels. Nickel is the most potent single element producing good low temperature impact properties. For welded plate fabrication it is used at $3\frac{1}{2}\%$ with carbon under 0.12% giving good welding and cold forming properties, supplied non-ageing and suitable for service temperatures down to minus 100°C. Plates are usually supplied normalised and tempered; stress relief after welding is normal. Charpy data are shown in Figures 5.49a[112] and 5.49b.[110] For lower temperatures down to minus 180°C the choice is between 9% nickel steel and austenitic chormium-nickel steels.

Table 5.19 EXAMPLES OF STRUCTURAL MILD STEELS FOR LOW TEMPERATURE SERVICE

Standard	Composition (cast analyses)				Deoxidation	Condition	Charpy V not aged temp. °C	Nm(J) min.	DVM (DIN 50115)		
	C max	Si	Mn	Other					Not aged		Aged min.
									°C	Nm/cm² min	Nm/cm²
BS 4360 40 C	0·18	—	1·5 max.	—	Semi-killed	As-rolled	0	27			
BS 4360 40 D	0·16	—	1·5 max.	Nb	Semi-killed	Normalised	−20	27			
BS 4360 40 E	0·16	0·1/0·5	1·5 max.	Al	Killed plus Fine grain	Normalised	−50	27			
							−20	61			
DIN StE 26 FG 26*	0·18	0·4 max.	0·4/1·3	Al or Nb or V	Killed plus Fine grain	Normalised	—	—	−20	68·6	
DIN TTStE 26 FG 26T*	0·16	0·4 max	0·5/1·3	Al or Nb or V	Killed plus Fine grain	Normalised	—	—	−20	78·5	58·9
									−50	53·9	

* Mannesmannrohren-Werke, Dusseldorf. Designation.
† Strain Ageing Test. 10% strain. Aged ½ hour at 250°C. Tested at 20°C.

The nickel steel has 0·13% carbon max, and is supplied double normalised and tempered. Mechanical properties at sub-zero temperatures are given in 'Nickel Alloy Steels—Properties at Sub-Zero Temperatures, International Nickel Co. (Mond) Ltd., London' on which Figure 5.50 is based.

Austenitic chromium-nickel steels of the 18% Cr 8% Ni composition (somewhat unstable austenite at low temperatures) partially transform to alpha iron during tensile

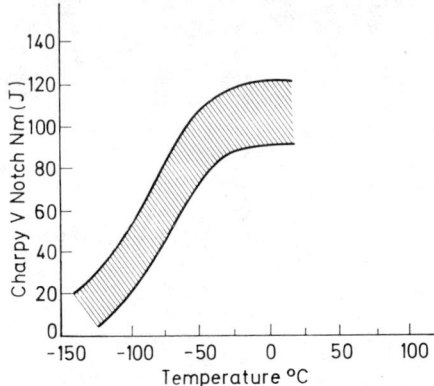

Figure 5.49 (a) Scatter band for 3½% Nickel steels (American; commercial)

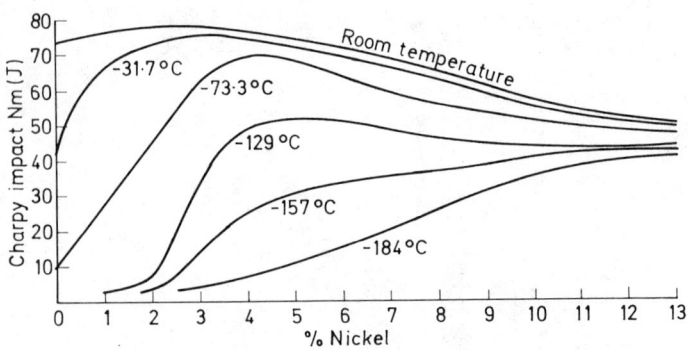

Figure 5.49. (b) Influence of Nickel on the impact strength of normalised low Carbon (0·10–0·20%) steels at low temperature (keyhole notch)

testing at low temperatures, consequently the increase in the tensile strength is not representative of the strength of a fabrication at such temperatures. The heat resisting grade 25% Cr 20% Ni, with a more stable austenite, shows only a small increase in tensile strength at low temperature. An interesting feature is that the fracture of notched impact test pieces, although energy absorption decreases with fall in testing temperature, remains ductile.

Transition curves for several grades of austenitic chromium-nickel steels are shown in Figure 5.51,[113] and for welding electrodes (weld metal) in Figure 5.52.[113] The latter shows that stainless steel weld metals have inferior energy absorption compared with the parent steel at sub-zero temperatures. This is due to a small amount of delta

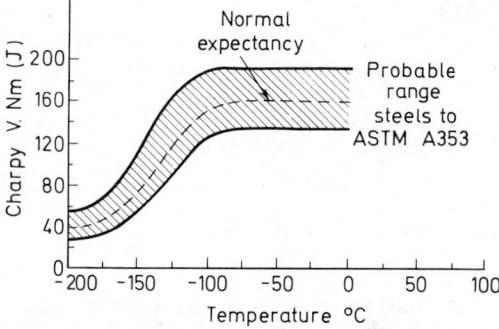

*Figure 5.50. 9% Nickel Steel. Impact-transition curves 25 mm
thick plate*

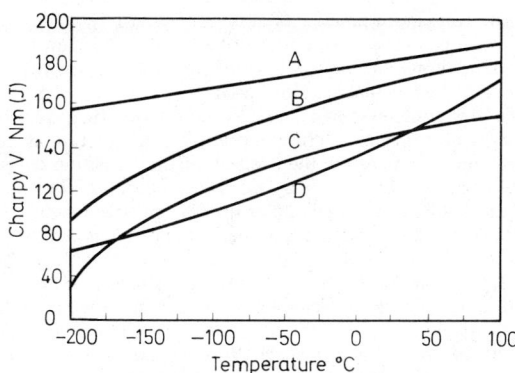

Figure 5.51. Impact transition curves. Stainless steels
A. 12·5 mm plate. 18% Cr-10% Ni + Nb
B. 12·5 mm plate. 18% Cr-8% Ni
C. 12·5 mm plate. 18% Cr-8% Ni-3% Mo + Ti
D. 12·5 mm plate. 23% Cr-21% Ni

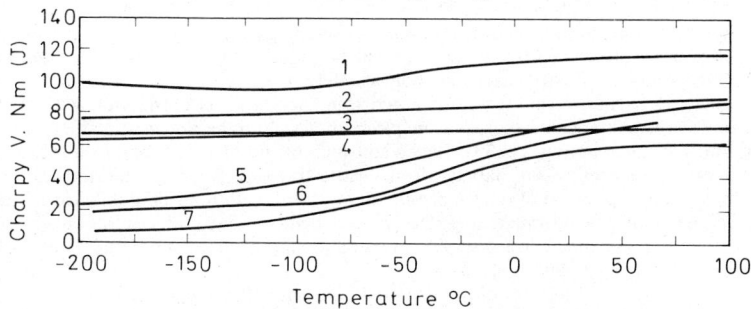

Figure 5.52. Impact transition curves. Weld metals (Stainless Steel)
1. Inco-weld A weld metal. 75 Ni-15 Cr-9 Fe + Mo + Nb
2. Weld joint. Inco weld A on 18 Cr-10 Ni + Nb steel
3. Weld joint. Inco weld A on 18 Cr-8 Ni steel
4. Weld joint. Inco weld A on 18 Cr-8 Ni-3 Mo + Ti steel
5. 18 Cr-8 Ni weld metal
6. 18 Cr-10 Ni + Nb weld metal
7. 18 Cr-8 Ni + 3 Mo + Nb weld metal

ferrite in the weld metal microstructure. On the other hand the energy absorption of Inco weld A-weld metal is practically unaffected down to minus 200°C, either as an all-weld-metal deposit or as weld metal in a joint in the three steels; furthermore the low temperature tensile-ductility properties of the Inco weld A-welded joints were appreciably superior to those of the joints welded with stainless steel electrodes while the room temperature strength is similar.[113]

Selection of steel for low temperature service

The degree of notch toughness required in a steel to ensure ductile behaviour for specific service conditions depends on the temperature, whether steady or fluctuating, the care taken in fabrication and the extent to which designing can be used to eliminate potential stress concentration and other factors conducive to brittle fracture.

The best guide is, of course, past experience but, in this case, it is important that no change in design or fabrication practice even in small details be made that might affect steel behaviour. To some extent selection must be based on the consequences of brittle fracture, in economic terms and risk of loss of life; where a tougher steel is thought necessary, such is generally available at increased steel cost.

The degree of notch toughness in a steel, as indicated earlier, may be expressed in several ways (energy absorption, fracture appearance, contraction at notch root) but, fundamentally one requires to know the position of the transition curve in relation to temperature axis whatever test is being used. Since energy absorption in a notched impact test is most readily determined and as the Charpy specimen is best suited to the techniques required to cool and fracture specimens at known temperatures, the Charpy V or keyhole specimen is widely used.

The determination of the transition curve is time consuming. However, provided the type of steel and hence the general characteristics of the curve is known, the brittle fracture resisting capability may be described either by the temperature at which the change from ductile to brittle fracture starts (lowest uppershelf temperature) or by the temperature at which the change has proceeded half way or by the temperature at which a defined energy absorption just above the lower shelf is obtained.

For quality control and indication of degree of notch toughness, 20 Nm, or 27 Nm or 40 Nm V-notch may be specified to be obtained at a stated temperature. The figure to be attained depends on the type and method of manufacture of the steel. Thus, for semi-killed mild steel 20 Nm is commonly used. Quenched and tempered steels will usually be required to give 40 Nm minimum. The energy absorption required from a less sharp notch at the same temperature will be higher; 20 Nm Charpy V notch in semi-killed steel has 27 Nm Charpy keyhole as alternative.

To establish a datum for degree of notch toughness and its correlation with past experience, the American experience in more or less eliminating brittle fracture in ships' hulls is worth noting. Prior to 1948 a considerable number of Liberty ships broke in half or experienced severe damage from brittle fracture. Although the fabrication techniques, under pressure of output requirements, were, with hindsight, a factor in the failures, the quality of steel plate was clearly not suited to welded fabrication on a mass production scale for the intended service stressing (dynamic) and temperature (fluctuating over a wide range) conditions.

It was observed that failures did not originate in plate 12·5 mm or thinner, so in the 1948 Revised Rules of the American Bureau of Shipping, new specifications for steel manufacture for hull construction were adopted; thus Class A (up to 12·5 mm) no change, Class B (12·5 mm–25 mm) and Class C (over 25 mm) as follows. The carbon in Class B was restricted and the manganese specified to give a higher Mn to C ratio than in the wartime steels and in Class C; additionally the silicon content was specified to give a silicon killed steel with fine grain treatment (impact tests were not called for). The effect of these revised rules on average transition curves is shown in Figure 5.53.[114]

Checking the actual notched impact properties of plates delivered to shipyards over an extended period showed[115] that, for steels supplied to the pre-1948 specifications, the 20 Nm temperature could occur between minus 18°C and plus 66°C with the majority occurring between 4·5°C and 38°C, whereas for the 1948 specification steel the corresponding temperature range was plus 4·5°C to minus 40°C with the majority occurring between minus 18°C and minus 34°C. Experience therefore shows that for mild steel

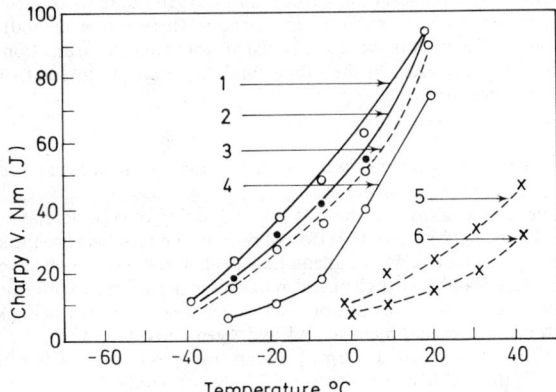

Figure 5.53. Average Charpy V notch impact curves on ship plates
(American Bureau of Shipping)
  1. 25–35 mm. *ABS 1948 Class C.* Mn/C, 4·22
  2. 35–38 mm. *ABS 1948 Class C.* Mn/C, 3·89
  3. Over 38 mm. *ABS 1948 Class C.* Mn/C, 3·74
  4. 12·5–25 mm. *ABS 1948 Class B.* Mn/C, 3·74
  5. 12·5–25 mm. *Pre 1948.* Mn/C, 1·79
  6. 25–35 mm. *Pre 1948.* Mn/C, 1·79

(semi-killed, killed and fine grain treated) the 20 Nm temperature yardstick is an indicator of a satisfactory degree of toughness which correlates well with practical experience of brittle fracture in shipbuilding practice. This has since been widely extended to structural mild steel applications, predominantly semi-killed steel up to 25 mm thickness, so that, where a design in such materials is not covered by existing code requirements the design engineer can accept that the steel must meet 20 Nm minimum Charpy V at the lowest operating temperature.

For semi-killed steels with high Mn/C ratio (low carbon-high Mn) and killed steels made to fine-grain practice a 27 Nm temperature has become accepted. This is largely based on research results, since sufficient experience of brittle failure in such materials has not been built-up to give a correlation. A 40 Nm temperature appears to be regarded as desirable for high tensile alloy structural steels but, it should be emphasised, selection of a criterion to use is a matter of judgment in the absence of information correlating with experience.

Significant experimental contributions to data on brittle fracture

Although experience must be the final yardstick of the brittle fracture resistance of steels, controlled experiments using notched impact or tensile specimens have provided valuable information on the diversity of factors affecting toughness. Contributions of particular interest are briefly noted below.

EFFECT OF ALUMINIUM ADDITIONS ON STEEL PROPERTIES[116]

The transition temperature (27 Nm Charpy V) is shown to be related to the aluminium nitride, aluminium in solid solution, carbon and oxygen contents of steels made from small high frequency melts in carbon-Mn, carbon-Mn (vacuum melted) and $3\frac{1}{2}\%$ Ni steel. Increasing AlN contents were beneficial in lowering the transition temperature by reducing ferrite grain size. On the other hand Al, C and O in solid solution raised the transition temperature.

THE NOTCH DUCTILITY AND TENSILE PROPERTIES OF SOME SYNTHETIC MILD STEELS[117]

The authors, emphasising the fact that the notch ductility of commercial mild steels can vary over wide limits, have studied the effect of minor elements, because the known effects of C, Mn, heat treatment, etc, do not adequately explain some of the observed variations. Synthetic steels were made from high purity materials in the normal carbon and manganese ranges for mild steel but silicon was not used as deoxidant, oxygen being kept at a low content (constant) by vacuum melting and hydrogen deoxidation.

The effect of nitrogen on the Charpy V notch transition curve of C-Mn and that of Al, Si, Zr and Ti on a C-Mn steel with 0.007% N was studied. It was concluded that nitrogen broadens the range of temperatures over which the steel changes, from fully ductile to fully brittle behaviour but it does not effect the temperature at which it is fully brittle. The nitride formers (Al, Zr, Ti) slightly narrow this range. Al up to 0.08% lowered the transition temperature but raised it between 0.8 and 1.0% again lowering it at higher levels. Titanium refines the grain and lowers the transition temperature. Zirconium had little effect on either the range or transition temperature. Up to 0.3% Si lowered the transition temperature but above this level (up to 1%) progressively raised it. In steels containing Al raising the austenitising temperature raised the transition temperature, but had no effect on titanium bearing steels.

MECHANISM OF DEFORMATION AND FRACTURE IN THE CHARPY TEST AS REVEALED BY DYNAMIC RECORDING OF IMPACT LOADS[118]

By determining load-time curves in the Charpy V notch test on 'pedigree' steels (Admiralty Advisory Committee on Structural Steel) down to very low temperatures, the critical stress for cleavage fracture was found to vary between 896 N/mm^2 and 1205 N/mm^2, the semi-killed steel having lower values than the killed. It is concluded that brittle fracture occurs when this critical cleavage stress is reached below the notch root and it can be attained either before or after general yield is reached. At low temperatures fracture occurs at loads below upper yield as a result of twinning but at higher temperatures it initiates from slip. This contribution tends to support the concept of brittle strength, Figure 5.47.

NOTCH BRITTLENESS IN MILD STEEL[119]

Using slow notched-bend tests on Bessemer steel down to very low temperatures, it was shown that cleavage cracking initiated from twinning but from slip at higher temperatures. The connection between the results and the effect of specimen size on fracture behaviour is noted.

INFLUENCE OF NIOBIUM ADDITIONS ON CARBON MANGANESE STEELS[120]

The effect of Nb on tensile and notch toughness of C-Mn steels is recorded. None of the steels examined, which were commercial grades, were non-ageing in the as-rolled or normalised condition. This suggested that Nb did not combine efficiently with nitrogen. In the as-rolled condition, the addition of Nb raises the transition temperature (precipitation hardening by Nb compounds). After normalising, a fine grain structure is obtained which raises the tensile strength and lowers the transition temperature to give notch toughness similar to that of an Al treated steel of similar grain size.

It should be noted that weld heat affected zones will not ordinarily be normalised by heat treatment so that ageing and precipitation hardening in these zones may be anticipated in steels such as BS 4360: 40D, 43D and 50C. The Nb treated (0·10% Nb max) mild and high tensile steels, being semi-killed and hence less expensive to produce, are widely used. Welding processes which involve longer time at high temperature in the heat affected zones, such as Electro-slag, would, unless followed by furnace normalising treatment be expected to produce a substantial reduction in notch toughness in these zones. However, it is in these zones that maximum resistance to brittle fracture initiation is desirable. Further information on Nb bearing mild steels is given in References 121 and 122.

EFFECT OF SMALL ADDITIONS OF VANADIUM ON THE AUSTENITIC GRAIN SIZE, FORGEABILITY, AND IMPACT PROPERTIES OF STEEL[123]

Small additions of vanadium, up to 0·1%, are shown to have effects on impact transition temperature and ferrite grain size similar to treatment with Al. Vanadium nitride formation is shown to take preference over vanadium carbide. This should indicate, as suggested elsewhere,[124] that these vanadium treated steels should be non-ageing and therefore should be advantageous in welded fabrication compared with the Al treated non-ageing steels which, owing to the affinity of Al for oxygen, are generally 'dirty'. Vanadium treated mild steels should be resistant to brittle fracture initiation in the heat affected zone and also resistant to lamellar tearing.

A NOTE ON BRITTLE FRACTURE INITIATION IN MILD STEEL BY PRIOR COMPRESSIVE STRAIN[125]

It is common experience that brittle fractures in service occur without any evidence of plastic deformation such as is seen at the root of notched laboratory test pieces. However if precompression, leading to a small amount of straining, is applied to a notched tensile test piece the subsequent fracture on tensile straining will be brittle with no general plastic deformation and at stresses well below the yield strength.[126, 127, 128]

The effect of prior-compressive straining on pre-notched test pieces cut from plate, (Al killed, oxygen-Bessemer, rimming and silicon-killed steels), was to cause brittle fracture at stresses well below yield in the oxygen-Bessemer steel when tested in the ductile to brittle temperature zone of the Charpy test. No measurable plastic contraction in thickness occurred.

In other respects such as tensile strength the pre-compression straining did not apparently have real effect. Thus a relatively notch ductile steel showed completely different behaviour in the notched tensile test at its transition temperature after pre-straining in compression. The results reported may well have implications for welded fabrication involving cold bending, etc.

LOW ALLOY HIGH TENSILE STRUCTURAL STEELS

Advances in design and the increasing size of steel structures demand the availability of structural steels of higher strength than structural mild steel. Welding, as the major method of joining, imposes a requirement that such steels should be capable of being welded satisfactorily without the need for expensive and possible impracticable, precautions, e.g. site erection conditions, to prevent weld hardened zone cracking.

It is evident, since welding involves rapid cooling of the steel from grain coarsening temperatures which are predisposing to hardening and possibly embrittlement, with concomitant build-up of residual stress that there will be considerable limitation on the increasing strength that can be attained at a commercial price. This is because rectification of unsatisfactory metallurgical situations by post welding heat treatment is not normally possible. The term 'weldable' has no precise meaning but the fabricator considers a 'weldable' high tensile structural steel should be as readily welded as mild steel. Such steels form a restricted category and may not be readily available to UK fabricators who, for the general run of weldable high-tensile steels, are expected to apply special precautions which may involve pre-heating, low hydrogen welding electrodes and special attention to good fit-up. In other words, part of the weldability of such steels must be supplied by the expertise of the fabricator on whom rests the onus of making satisfactory use of the steels. In the USA, where the greatest development of weldable high tensile structural steel has taken place, there is a wide choice which includes many steels with yield strength around 340 N/mm² which are satisfactorily welded by recognised techniques applied to mild steel. Such steels require particular care and skill in manufacture, the expertise being at the steel mill.

Metallurgical

The least expensive method of increasing tensile strength is by increase in carbon, manganese and silicon; but cold forming properties and weldability are adversely affected so that such steels are preferably confined to rivetting practice.

Carbon is the most harmful element in respect of weld cracking susceptibility; therefore, using minimum practicable carbon levels, alloys must be added to attain the highest strength consistent with weldability, formability and notch toughness. The addition of effective amounts of such elements, when the steel is normalised, implies that to avoid martensite or bainite the practical limits for carbon are 0·07–0·1%.[129] Increased strength is obtained by one or more of the following methods:

(a) Strengthening by transformation hardening.
(b) Strengthening by ferrite hardening (solid solution hardening).
(c) Strengthening by fine ferrite grain.
(d) Strengthening by precipitation hardening.
(e) Strengthening by changing the nature and distribution of the carbide constituent.

Elements vary in their effects but the majority have a primary influence on one or two methods with secondary effect on others. This is seen in Table 5.20.

The cooling rate at a metal arc welded joint on cold plate will generally be faster than normalising but not as fast as oil quenching. Because welding involves quenching a weldable steel will make maximum use of (b) to (d) and minimum use of (a). Steels at lowest cost per ton of plate will make maximum use of (a) plus (c) with C and Mn in semi-killed steels. Allowing for the cost of proper welding procedures required to avoid cracking and repair costs, steels made to methods (a) plus (c) may not prove most economical in the long run.

Commercial high tensile structural steels

American high tensile structural steel practice is illustrated in Table 5.21[130] giving ASTM standard steels; in respect of A242 the analysis in Table 5.21 is not fully informative as commercial steels to this specification commonly have carbon below 0·15% (range 0·08–0·2%) with consequent use of alloy elements such as chromium (up to 0·5%), nickel (up to 0·5%) or approximately 0·04% vanadium. Furthermore they are normally fine grained.

Typical compositions and mechanical properties of some American structural steels are shown in Table 5.22.[130] An indication of the differences in weldability of these steels is given in Table 5.23[130] by reference to the temperature levels required for preheat and inter-pass. Steels to A440 require higher temperatures than would be applied, at given thicknesses, to structural mild steel (BS 4360: 43A) particularly if normal hydrogen

Table 5.20 THE STRENGTHENING OF STEEL. INFLUENCE OF ALLOYING ELEMENTS DECREASING ORDER OF INTENSITY

Transformation hardening	Solid solution hardening	Ferrite grain refinement	Precipitation hardening	Carbide nature and form
Carbon P	Phosphorus P	Aluminium P	Copper P	Vanadium S
Manganese P	Silicon P	Vanadium P**	—	Chromium P
Chromium P	Manganese P	Molybdenum S	—	Molybdenum S
Molybedenum P	Nickel P			Manganese M
Vanadium P	Chromium* M			
Nickel S	Molybdenum* M			
Silicon M	Vanadium* M			
	Copper M			

* Over and above the amount required to form carbides.
** In conjunction with nitrogen.
P = Primary effect. S = Secondary effect. M = Minor effect.

level electrodes are used. Steels to A242, particularly when welded with low hydrogen electrodes (which is normal practice) do not require higher temperatures.

British weldable structural steels, Table 5.24, are covered by BS 4360: 1968. The softest grade (40 A-E) is readily weldable, with plate yield strength depending on thickness in the range 210–230 N/mm² or 220–265 N/mm². Grade 43 A-E, at a tensile level of 430/510 N/mm², includes the standard mild structural steel 43A; depending on thickness the yield strength range for the two steels with highest yield strength (D and E—notch ductile lower carbon high Mn) is 240/280 N/mm² and that for 43A the same as the American A373 at 214 N/mm².

The lower of the two high tensile grades, grade 50 A-D, has a yield strength range of 320/350 N/mm² again depending on thickness. As in grade 43, the steels C and D are notch ductile. Grade 50 steels, as regards yield strength at equal thickness, are about in the middle of the American commercial steel range (see Table 5.22). Welding procedures for steels in BS 4360: 1968 are covered by BS 2642: 1968; the essential aim is to provide sufficient heat input (dependent on pre-heat, electrode size and weld bead size) to prevent, for given cooling conditions (dependent on combined plate thickness), the critical cooling rate at 300°C, which produces a hardness exceeding 350 DPH in the heat affected zone, being exceeded.

The critical cooling rate depends on composition, which, in hardening propensity is quantified in the Carbon Equivalent. Using a Carbon Equivalent % formula[131]

$$CE \% = C + \frac{Mn}{6} + \frac{Cr + Mo + V}{5} + \frac{Ni + Cu}{15}$$

the CE ranges, calculated on ladle analyses, for steel to Grade 50B (Hyplus 23)[132] are:

For plates (normal rolling) 0·35–0·46%.
For plates (controlled rolling) 0·35–0·41%.
For sections 0·34–0·47%.
For flats 0·33–0·39%.

Table 5.21 AMERICAN STRUCTURAL AND HIGH TENSILE STRUCTURAL STEELS

ASTM Standard Steels

	A36		A373				A242			A440			A441		
	18–37 mm	37–100 mm	To 12 mm	12–25 mm	25–50 mm	50–100 mm	To 18 mm	18–37 mm	37–100 mm	To 18 mm	18–37 mm	37–100 mm	To 18 mm	18–37 mm	37–100 mm
*C max	0.28	0.28	0.26	0.25	0.26	0.27		0.22			0.28			0.22	
Mn	0.8/1.1	0.85/1.2	—	0.5/0.9	0.5/0.9	0.5/0.9		1.25 max.			1.1/1.6			1.25 max.	
Si	—	0.15/0.3	—	—	0.15/0.3	0.15/0.3		—			0.3 max.			0.3	
Cu	—	—	—	—	—	—		—			0.2			0.2	
V	—	—	—	—	—	—		—			—			0.02	
Yield min.: N/mm²	215	215	215				340	305	290	340	305	305	340	305	290
Tensile N/mm²	400–510	400–510	400–510				480	465	430	480	465	430	480	465	430

Commercial Steels (290–480 N/mm² yield strength)**

	A373	A242	A440	A441
Type (nominal yield strength)	290 or 310	340	400	480

*C   0.15 max. to 0.26 max.
Mn   1.0 max. to 1.35 max.
Si   0.3 max.

Plus one or more of the elements Nb, V, N at about 0.01% each.

	A373	A242	A440	A441
Yield N/mm²	290 or 310	340	400	480
Tensile N/mm²	450	480	510	620

* Ladle analysis.

** Commercial steels to A 242 usually have C in range 0.08–0.22%. Alloys added additional to Mn; 0.5% Cr, 0.5% Ni or 0.04% V. Usually fine grained.

Table 5.22 TYPICAL COMPOSITIONS AND MECHANICAL PROPERTIES OF AMERICAN HIGH-TENSILE STRUCTURAL STEELS

	Nominal composition										Mechanical properties		
	C	Mn	Si	P	Cu	Ni	Cr	Mo	Other		Charpy V 20 Nm(J) Transition temp. °C	N/mm²	
												To 12·5 mm / 12·5–37 mm / 37–100 mm	
1.	0·22	1·4	0·07	0·02	0·27	—	—	—	—	A440 Low material cost, Inferior weldability	−12 to +4°		
2.	0·27	1·5	0·09	0·02	0·26	—	—	—	—				
3.	0·22	1·3	0·03	0·015	0·03	—	—	—	0·02 Nb				
4.	0·24	1·48	0·22	0·013	0·06	—	—	—	—			Yield 330–380 / 300–370 / 270–360	
												Tensile 480–585 / 465–510 / 420–570	
5.	0·09	0·38	0·48	0·09	0·41	0·28	0·84	—	—	A242 Corrosion resisting, Good weldability	−25 to −18°		
6.	0·12	0·75	0·70	0·015	—	—	0·58	—	0·06 Zr				
7.	0·10	0·9	0·09	0·10	1·1	0·6	—	0·15	—				
8.	0·17	1·25	0·35	0·01	0·25	—	—	—	—				
9.	0·15	1·0	0·2	0·022	0·27	—	—	—	0·04 V	A441. Good weldability	−18 to −7°		
10.	0·15	0·85	0·8	0·017	—	—	—	—	0·06 Zr				
												To 10 mm / 10–18 mm	
11.	0·25	1·6	0·25	0·03	—	—	—	—	0·01 V	Extra high yield, Variable weldability		Yield 400–570 / 380–480	
12.	0·2	1·3	0·05	0·02	—	—	—	—	0·03 V, 0·02 Nb			Tensile 570–620 / 480–585	

Adding 0·04% C for difference[133] between ladle and product analyses these ranges become:

For plates (normal rolling) 0·39–0·50%.
For plates (controlled rolling) 0·39–0·45%.
For sections 0·38–0·51%.
For flats 0·37–0·43%.

Histograms[132] show the majority of casts fall into the following carbon equivalent ranges:

Plates (normal rolling) 0·43–0·48%.
Plates (controlled rolling) 0·41–0·45%.
Sections 0·42–0·46%.
Flats 0·40–0·42%.

Depending on the risks the engineer is prepared to accept, selection of welding procedure can be based on the top figure of the full CE range, the top figure of the narrower higher frequency range or any other CE value. Alternatively the procedure can be

Table 5.23   WELDING PRACTICE FOR AMERICAN HIGH TENSILE STRUCTURAL STEELS IN TABLE 5.22

Steel No.	Thickness mm		Minimum preheat and interpass temperature °C	Minimum preheat and interpass temperature °C
			non-low hydrogen electrodes	low hydrogen electrodes
1 to 4	To	10	10	10
A 440		10–18	93	10
		18–31	150	38
		31–50	150	93
	Over	50	Not recommended	150
5 to 10	To	10	10	10
A 242		10–18	10	10
		18–31	38	10
		31–50	93	38
	Over	50	150	93
11 and 12	To	10	38	10
		10–18	93	10
		18–31	150	38

calculated when the cast analysis or analyses for material for a given project is known. The Grade 50 B preheat temperatures for CE of 0·52%, 0·47% and 0·41% are given in Table 5.25[134] showing the drop in preheat temperature with reduction in CE to 0·41%. Still lower carbon equivalents (about 0·38%) would bring the preheats into line with those recommended for A242 steels of the same thickness. Grade 50B is similar to A440 steel No. 3 (Table 5.22) in composition and pre-heat requirements.

Apart from the inferior weldability of the A440 and BS 4360: 1968 Grade 50B steels compared with A242 steels, based on the preheat requirements to avoid cracking in the weld heat affected zone, it is evident the higher carbon steels (carbon above 0·2%) will have greater hardness and hence lower ductility in the heat affected zone, even if not actually cracked, compared with A242 steels. Unless stress relieved after welding, the notch ductility of these zones in the higher carbon steels must be inferior to that of the

Grade	Composition				Condition	Tensile	Minimum yield			
	C (max.)	Si	Mn (max.)	Nb			To 16 mm	16–37 mm	37–63 mm	63–100 mm
40B	Ladle 0·2 Product 0·25	—	1·5 1·6	—	Semi-killed	400/480	230	225	220	210
43A	Ladle 0·25 Product 0·3	—	—	—	Semi-killed	430/510	250	240	230	215
50B	Ladle 0·2 (0·22)* Product 0·24 (0·26)*	—/0·5 —/0·55	1·5 1·6	0·1 —	Semi-killed plus Nb	495/620	350	345	338	320
55C	Ladle 0·22	—/0·6	1·6	0·1	Silicon-killed plus Nb plus fine grain	555/695	450	430	415	—

Mechanical Properties N/mm²

* Over 16 mm.

Table 5.25 WELDING PROCEDURES FOR BS 4360: 1968 GRADE 50 B STEEL. COMPARISON WITH AMERICAN STEELS

Grade	Carbon equivalent %	BS 4360 steels. Welding preheat °C* for thickness (mm)					American A 242 steels. Preheat and interpass °C for thickness (mm)	Normal hydrogen electrodes	Low hydrogen electrodes
		10	18	31	50	Over 50			
50B	0·52	75–100 (20)	100–150 (20–100)	150–200 (125–175)	200 (150–200)	200 (200)**	To 10 mm	10	10
	0·47	20	100–130	150–175	160–190	200	10–18 mm	10	10
	0·41	20	20–100	100–130	110–140	110–140	18–31 mm	38	10
							31–50 mm	93	38
							Over 50 mm	150	93

* Normal Hydrogen level electrodes or Low Hydrogen electrodes baked at 150°C by fabricator. Single run fillets, 5·9 mm or 4·9 mm electrodes.

** Figures in brackets are from BS 2642: 1965. Remaining figures from Reference 134.

parent plate and also inferior to that obtainable from the same zones in A242 steels, because the effect of fine-grain treatment is damaged by the high welding temperatures. Furthermore steels which make use of substantial amounts of manganese are not as suitable for welding as those limited to approximately 1·2% Mn maximum, particularly if welded in the as-rolled condition. This arises from the tendency for carbon and manganese segregation in these steels which leads to bands of alternately high and low chemistry with consequent enhanced weld cracking susceptibility in the former. Reduction of carbon with compensating increase in manganese increases rather than decreases this particular problem.

The most economical way to produce higher yield strength steels is undoubtedly the use of niobium to produce a fine-grain in semi-killed steel containing high manganese (exceeding 1·2%) but such steels inherently introduce uncertainties of welding behaviour which may cause unforeseen expense to the fabricator.

Higher strength steels

The limit of yield strength to be obtained from weldable ferrite-pearlite steels by a combination of fine pearlite, fine ferrite grain and ferrite solid solution strengthening is around 400–450 N/mm^2, bearing in mind the differing concepts of what is the meaning of 'weldable'.

To increase the yield strength from this level, without quenching and tempering, it is necessary to pass from a predominantly ferrite-pearlite to a bainitic structure. Using low carbon content for weldability low carbon bainitic steels can most conveniently be made commercially by the addition of boron and molybdenum to retard the formation of ferrite and by the addition of other alloying elements the bainite formation temperature can be progressively lowered. Developments[135] in this direction have lead to two steels, one having a bainitic structure formed at the top of the bainite formation temperature

Table 5.26 EFFECT OF TEMPERING ON MECHANICAL PROPERTIES OF A LOWER BAINITE HIGH STRENGTH STEEL (NOMINAL COMPOSITION 0·15% C; 1·5% MN, 1·5% CR; 0·5% MO-B)

Tempering		Tensile N/mm^2	Charpy V
Temperature °C	Time hours		Temperature °C for energy absorption of 40 J
—	—	1 160	20
600	1	940	−20
650	1	720	−80
650	4	665	−90

range, known as upper bainite, and the second at the bottom of this range, lower bainite. The bainitic structures which are formed in the middle of the range have unsatisfactory mechanical properties. The upper bainite has a composition (nominal) 0·15% C 0·5% Mn 0·5% Mo-B, the latter[136] 0·15%C 1·5% Mn 1·5% Cr 0·5% Mo-B.

The yield and tensile strength of the upper bainite in the as-rolled or normalised condition are 450–460 N/mm^2 and 620 N/mm^2 respectively. It has marked resistance to hydrogen induced cracking in the weld heat affected zone but may be somewhat susceptible to heat affected zone cracking at aluminous slag stringers; it is not a notch ductile steel. The lower bainite steel[135, 136] is normalised and tempered (to improve notch ductility). In the normalised condition the tensile strength is around 1160 N/mm^2. Tempering to progressively lower strength improves notch ductility (Reference 136,

Figure 15) as shown in Table 5.26. Although this steel has good weldability it has not found wide usage.

HIGH STRENGTH ALLOY STEELS

Where weldability in metal arc welding is not a factor limiting carbon content, quenching and tempering applied to steels with sufficient carbon and alloy content (for hardenability) provides a wide range of combinations of strength with toughness and ductility, the structure being tempered martensite. Strengthening is obtained from the inherently fine structure of the martensite and from the uniformly dispersed carbide particles.

The hardness of untempered martensite increases with carbon content. Alloy elements which promote hardenability increase the martensite hardness but by a constant amount at all carbon levels up to about 0·6% C; above which figure, softening is observed with

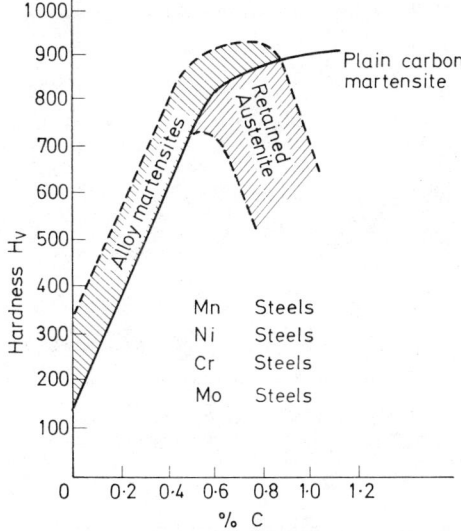

Figure 5.54. Effect of alloying elements on the hardness of Martensite

certain elements due to the partial retention of austenite to room temperature, Figure 5.54.[135] Alloyed deep hardening steels do not contâin more than 0·6% C unless special wear resistance properties, derived from carbides, is required. As the carbon content increases the hardness of martensite, tempered at a given temperature, increases with decrease in ductility and notch toughness; the hardness of tempered martensite at constant carbon content is dependent, over a wide range of alloy steels, on the tempering temperature, Figure 5.55[136].

If the carbon is higher or lower than 0·45%, the band shown in Figure 5.55 is displaced by rotation about the bottom right hand corner. Somewhat longer times at tempering temperature than one hour would not markedly affect hardness. As a rough approximamation, the hardness of steels similar to those indicated in Figure 5.55, in the quenched and tempered condition at carbon contents above or below 0·45% can be gauged from Figure 5.56.[137]

Steels. All at 0·45 %C
Carbon

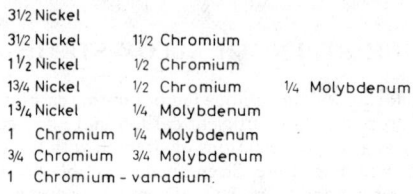

3½ Nickel
3½ Nickel   1½ Chromium
1½ Nickel   ½ Chromium
1¾ Nickel   ½ Chromium   ¼ Molybdenum
1¾ Nickel   ¼ Molybdenum
1 Chromium   ¼ Molybdenum
¾ Chromium   ¾ Molybdenum
1 Chromium - vanadium.

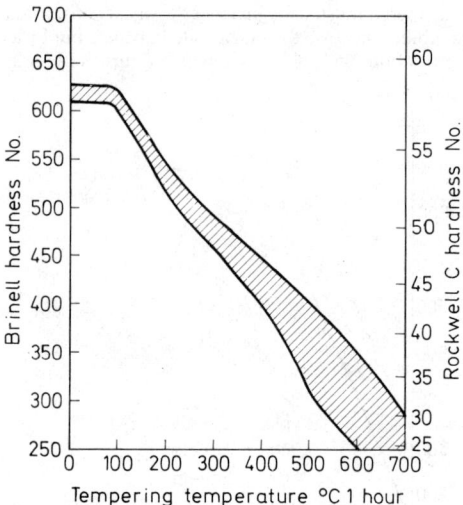

Figure 5.55. Tempered martensite hardness related to tempering temperature at constant carbon content for a range of quenched and tempered steels

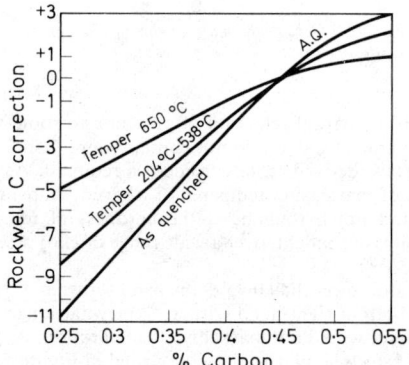

Figure 5.56. Effect of carbon content on hardness of tempered steel

Ductility (elongation %) is related to the hardness obtained by tempering, Figure 5.57.[137] Notch toughness is generally highest with tempering temperatures above about 580°C and lowest for intermediate tempering temperatures (between 220–550°C); the energy absorption level increases with decreasing carbon content and increasing nickel content.

Given sufficient alloys for full hardening (ruling section) to martensite, an alloy steel with 0·4/0·45% C will give up to about 1400 N/mm² tensile strength when tempered

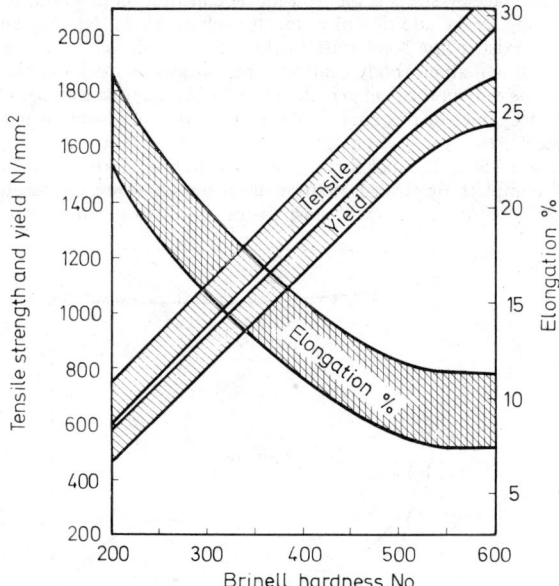

Figure 5.57. Range of tensile properties for several quenched and tempered steels related to hardness

above the temper brittleness range (220/550°C) and up to 1800 N/mm² tensile strength when tempered at 200°C. Tempered alloy steels at this carbon level can give a useful combination of strength, ductility and toughness depending on tempering temperature.

ULTRA-HIGH STRENGTH MAR-AGEING STEELS

Where a fabricator is prepared to adopt welding procedures requiring preheating up to 150/200°C, close control of low hydrogen electrode drying and precise fit-up the limit of yield strength that can be expected from a quenched and tempered low alloy steel suitable for such procedures is in the range 550–750 N/mm², the limit on carbon for weldability setting the limiting strength.

The low alloy medium carbon engineering steels, mentioned earlier, are not regarded as 'weldable'. However they are readily welded by solid phase welding processes, e.g. electric resistance butt and flash butt, and can be satisfactorily welded by fusion welding methods but special techniques are needed which are not readily applicable to large structures. Furthermore these steels are not readily machinable if tempered to a tensile strength above about 1300 N/mm². Therefore, they have to be machined in the annealed condition and subsequently quenched and tempered with consequent risk of distortion.

These difficulties, welding and heat treatment, are largely overcome, with a bonus in strength, ductility and notch ductility, by mar-ageing steels, which have been developed by the International Nickel Co.

Metallurgical

The basis for mar-ageing steels is the iron-nickel constitution diagram, Figure 5.58,[138] and age hardening by the addition of elements such as Al, Ti, Nb, Mo and Co.

Three phases exist in the solid state in the pure iron-nickel system, gamma (face-centred cubic lattice), alpha (body centred cubic lattice) and delta, which is the same as alpha but exists only in the field *acb*. Below 34·4% Ni gamma transforms, on cooling, to alpha; the mechanics of this change is similar to the formation of martensite from austenite in steel. However it is characterised by marked temperature hysteresis. For example in Figure 5.58 a 20% nickel alloy, on slow cooling, starts to transform at 200°C but does not complete the transformation until about −50°C is reached. On slow heating, the 20% Ni alpha does not start to transform to gamma until about 450°C

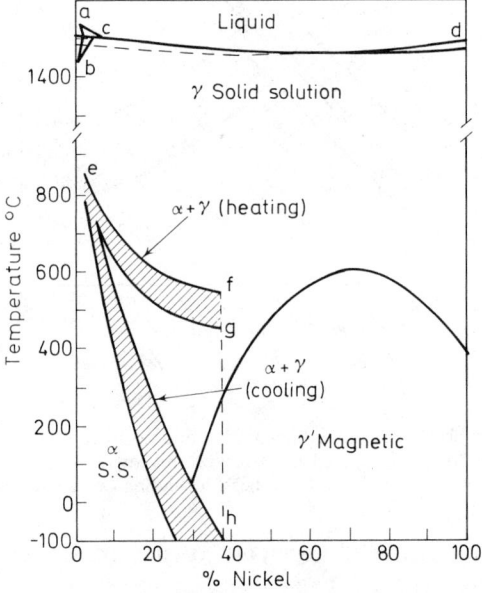

Figure 5.58. Iron-Nickel constitution diagram

is reached and transformation is not completed until the temperature has reached 600°C. Thus. at room temperature the microstructure is predominantly alpha phase.

There are important differences between iron-carbon martensites and iron-nickel martensites. Iron-carbon maretensites are formed only on falling temperature and have properties of hardness, ductility and toughness which are strongly dependent on carbon content. Iron-nickel martensites are formed on falling or constant temperature, a wide range of cooling rates being effective, and the hardness is not appreciably affected by nickel content. The hardness of iron-nickel martensites is around 260 DPN with high ductility (elongation on $4 \times D\%$ of 20–30%, RA 65–80%) and toughness. This contrasts with the brittleness of iron-carbon martensites of similar hardness.

The iron-nickel alloys in the range 18–25% nickel thus provide a material with a remarkable combination of toughness, strength and ductility on which to build a precipitation hardening system. By addition of Mo-Co-Al or Al-Ti-Nb a gamma phase with these elements in solution is formed which on cooling to room temperature is soft (300 DPN) but produces age hardening when it is reheated to the 450–500°C range. Hardness up to 540 DPN can be reached depending on composition, time and temperature.

The alloys with 18–20% nickel are martensitic in the solution annealed condition at room temperature; increasing the nickel to 25% stabilises the gamma phase which then becomes the phase present at room temperature. To cause this to transform (for mar-ageing), the alloy must either be cold-worked and then reheated into the precipitation hardening temperature range (430–450°C) or heated in the range 650–700°C (gamma phase field) which, by formation of Ni_3Ti, lowers the matrix nickel content to produce a less stable gamma phase.

Heat treatment of mar-ageing steels is simple; solution of the age hardening elements at temperatures above 800°C (normally at 820°C) followed by ageing for 1–4 hours at 400–480°C. Welding does not require preheat but specially pure filler materials are required to avoid weld metal cracking if the weld metal is of mar-ageing grade. The heat affected zone of weld joints is in the solution-annealed condition and therefore softer than the parent plate if welding is done on aged material. The most straightforward practice is to form and fabricate in the solution-annealed condition and then age thereby eliminating the softer heat affected zones.

Typical compositions and properties of some mar-ageing steels are given in Tables 5.27 and 5.28.[139]

The production costs of mar-ageing steels based on 18/25% nickel are necessarily high, requiring special melting practice, materials of high purity and very close control

Table 5.27   COMPOSITIONS AND PROPERTIES OF MARAGEING STEELS

	18% Ni-Co-Mo			20% Ni-Ti-Al	25% Ni-Ti-Al
	1400 N/mm²	1700 N/mm²	1930 N/mm²		
Ni %	17–19	17–19	18–19	18–20	25–26
Co %	8–9	7–8·5	8·5–9·5	—	—
Mo %	3–3·5	4·6–5·1	4·7–5·2	—	—
Ti %	0·15–0·25	0·3–0·5	0·5–0·7	1·3–1·6	1·3–1·6
Al %	0·05–0·15	0·05–0·15	0·05–0·15	0·15–0·35	0·15–0·35
Nb %	—	—	—	0·3–0·5	0·3–0·5

of composition. However, bearing in mind the combination of strength and toughness, ease of fabrication by welding and of post-welding heat treatment there are no alternative materials of comparable properties.

The field of application to which they are naturally applicable is in rocket, missile and aircraft production where strength to weight ratio and reliability of fabrication are major considerations. However, cryogenic applications particularly for pressure vessels would also be covered the material being produced as plates, sheets, bars, etc.

STEELS FOR RESISTANCE TO HYDROGEN ATTACK
UNDER PRESSURE

When steel is exposed to hydrogen-hydrocarbon gas mixtures at elevated temperatures and pressures, there is a tendency for decarburisation and loss of strength. This governs the selection of steel compositions for service involving such conditions.

Table 5.28 MECHANICAL PROPERTIES OF MARAGEING STEELS, MAR-AGED CONDITION

	18% Ni-Co-Mo			20% Ni-Ti-Al	25% Ni-Ti-Al
	1400 N/mm² PS	*1700* N/mm² PS	*1930* N/mm² PS		
0·2% PS N/mm²	1 312–1 451	1 605–1 822	2 038–2 099	1 652–1 868	1 652–1 883
Maximum stress N/mm²	1 374–1 513	1 652–1 883	2 069–2 130	1 698–1 930	1 791–2 007
Elongation on 4·5 $\sqrt{A}$%	14–16	10–12	12	10–13	10–15
RA%	65–70	45–60	60	45–60	40–60
Notched tensile strength N/mm²	2 161–2 393	2 393–2 655	2 933–3 088	2 161–2 624	1 930–2 470
Charpy V at: +20°C Joules (Nm)	90–159	20–37	—	16–27	—
−196°C Joules (Nm)	40–81	11–24	—	—	—
Nil ductility transition temperature °C	—	—	—	20 to −60	150–200
Melting practice	Air induction	Air induction	Vacuum induction	Vacuum induction plus vacuum arc remelting	Vacuum induction plus vacuum arc remelting

Carbon steel, exposed to these conditions, suffers decarburisation by hydrogen, but steels in which the carbon is fully combined with an element having a strong affinity for carbon (Ti, V, Nb, Cr, Mo, W) are not readily attacked. Gradations of this situation exist according to the efficiency of combination between carbon and carbide forming element and the operating pressure and temperature.

The attack by hydrogen takes two forms; decarburisation of the outer surfaces caused by migration of carbon atoms to the surface to combine with hydrogen and secondly cracking and rupturing of the steel from internal pressures built-up at grain boundaries

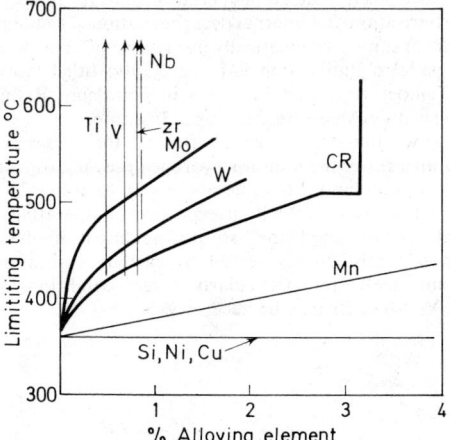

Figure 5.59. Influence of alloying elements on resistance of 0·1% C steels to hydrogen. (Hydrogen pressure 29 N/mm², period 100 hrs)

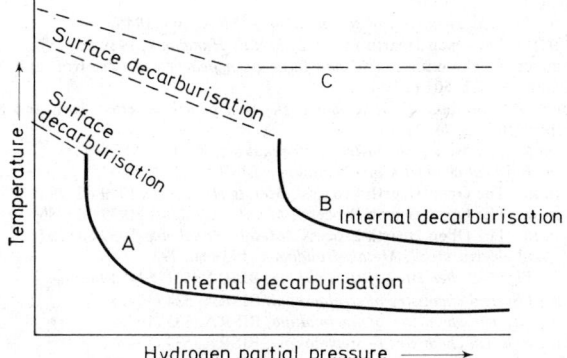

Figure 5.60. Decarburisation by hydrogen. Composition-pressure temperature relationship (schematic)

by methane produced by reaction between the atomic hydrogen diffusing through the steel and carbon derived from the cementite. The internal decarburisation requires more severe conditions of temperature and pressure than surface decarburisation and the effect on mechanical properties is much greater. Steel with slag concentrations is likely to produce blistering as a result of methane gas pressure.

The influence of carbide forming elements in steel on resistance to hydrogen attack is shown in Figure 5.59[140] from which it is seen that Ti, V, Zr and Nb are the most effective elements in promoting resistance. The factors governing the extent of attack by hydrogen, other than composition and heat treatment of the steel as affecting efficiency of combination with carbide forming elements where present, are temperature, pressure and time.

From practical experience over twenty-five years and research data, G. A. Nelson[141] compiled data for carbon, 0·25% Mo, 0·5% Mo, 1·25% Cr–0·5 Mo, 2% Cr–0·5 Mo, 3% Cr–0·5 Mo and 6% Cr–0·5 Mo steels, indicating operating limits for temperature or pressure to avoid surface decarburisation and internal decarburisation. The temperature-pressure relationship is of the form shown schematically in Figure 5.60. For detailed information reference should be made to Publication 941: 1970 July: titled 'Steels for Hydrogen Service at Elevated Temperatures and Pressures in Petroleum Refineries and Petrochemical Plants' published by American Petroleum Institute.

It is to be noted that welding has a harmful effect on the resistance of carbon steel. At normal ambient temperatures the limit for hydrogen pressure for carbon steel should be set at 89·6 N/mm^2 absolute and this value obtains up to about 220°C. At pressures less than 1·5 N/mm^2 there is no attack, irrespective of operating temperature, on unwelded carbon steel, this pressure being halved if welding is involved.

Many applications are satisfactorily served by carbon, carbon-molybdenum and chromium-molybdenum steels; for particularly severe conditions steels containing carbide stabilisers Ti, V, Nb or Zr may be necessary.

REFERENCES

1. HERTY, C. H. JR, 'Alloys of Iron and Oxygen', *Metals Handbook,* 1939 ed, 392
2. SARGEANT, R. J., 'The Determination of FeO in steel from the carbon drop', *Journal of the Iron and Steel Institute,* 2, **CLIV** (1946)
3. 'Symposium on Steelmaking', *Acid and basic open hearth practice,* Iron and Steel Institute Special Report No. 22 (1938)
4. SCHENCK, H., *Physical chemistry of steelmaking,* BISRA, 505 (1945)
5. HERTY, C. H. JR, 'The Open Hearth Process', *Metals Handbook,* 1939 ed, 787
6. 'Symposium on Steelmaking', *Acid and basic open hearth practice,* Iron and Steel Institute Special Report No. 22, 501 (1938)
7. 'Symposium on Steelmaking', *Acid and basic open hearth practice,* Iron and Steel Institute Special Report No. 22, 74 (1938)
8, 9, 10. SCHENCK, H., *Physical chemistry of steelmaking,* BISRA, 517 (1945)
11. SCHENCK, H., *Physical chemistry of steelmaking,* BISRA, 538 (1945)
12. HERTY, C. H. JR, 'The Open Hearth Process', *Metals Handbook,* 1939 ed, 793
13. HERTY, C. H. JR, 'The Open Hearth Process', *Metals Handbook,* 1939 ed, 786
14. HERTY, C. H. JR, 'The Open Hearth Process', *Metals Handbook,* 1939 ed, 787
15. GOTTS, F., 'Acid electric steel', *Metals Handbook,* 1939 ed, 799
16. SCHENCK, H., *Physical chemistry of steelmaking,* BISRA, 531–534 (1945)
17. SCHENCK, H., *Physical chemistry of steelmaking,* BISRA, 534 (1945)
18. SCHENCK, H., *Physical chemistry of steelmaking,* BISRA, 535 (1945)
19. SCHENCK, H., *Physical chemistry of steelmaking,* BISRA, 537 (1945)
20. SCHENCK, H., *Physical chemistry of steelmaking,* BISRA, 536 (1945)
21. SCHENCK, H., *Physical chemistry of steelmaking,* BISRA, 537 (1945)
22. PERRY, T. E., 'High vacuum ladle degassing with induction stirring at the Republic Steel Corporation', *JISI,* 977–979 (Oct 1965)
23. PERRY, T. E., 'High vacuum ladle degassing with induction stirring at the Republic Steel Corporation', *JISI,* 981 (Oct 1965)
24. PERRY, T. E., 'High vacuum ladle degassing with induction stirring at the Republic Steel Corporation', *JISI,* 981 (Oct 1965)
25. MUND, A., 'Methods of degassing and vacuum treating liquid steel', *JISI,* 804–814 (Oct 1962)
26. *The making, shaping and treating of steel,* United States Steel Corporation, 8th ed (page 548)

27. *The making, shaping and treating of steel*, United States Steel Corporation, 8th ed, 544–545, 553–556

28. MUND, A., 'Methods of degassing and vacuum treating liquid steel', *JISI*, 804–814 (Oct 1962)

29. MUND, A., 'Methods of degassing and vacuum treating liquid steel', *JISI*, 811 (Figure 9) (Oct 1962)

30. MUND, A., 'Methods of degassing and vacuum treating liquid steel', *JISI*, 810 (Figure 6) (Oct 1962)

31. MUND, A., 'Methods of degassing and vacuum treating liquid steel', *JISI*, 813 (Oct 1962)

32. MUND, A., 'Methods of degassing and vacuum treating liquid steel', *JISI*, 808 (Oct 1962)

33. MUND, A., 'Methods of degassing and vacuum treating liquid steel', *JISI*, 810 (Figure 8) (Oct 1962)

34. DICKINSON, J. H. S., 'A note on the distribution of silicates in steel ingots', *JISI*, 1, 177–196 (1926)

35. 'Fourth report on the heterogeneity of steel ingots, Iron and Steel Institute Special Report No. 2, 56–57

36. 'Second report on the heterogeneity of steel ingots', *JISI*, 1, 526 (1928)

37. 'Second report on the heterogeneity of steel ingots', *JISI*, 1, 528 (1928)

38. GREGORY, E. and WHITELEY, J. H., 'Examination of a high sulphur-free cutting steel ingot', *JISI*, 2, 13p (1941)

39. GRAHAM, C. S., 'Examination of two ingots of free-cutting steel, one containing lead and the other lead-free', *JISI*, 1, 273p (1945)

40. 'Second report on the heterogeneity of steel ingots', *JISI*, 11, 527 (1928)

41. STEVEN, W. and THORNEYCROFT, D. R., 'Variation of transformation characteristics within samples of an alloy steel', *JISI*, 15–32 (Sept 1957)

42. WARD, R. G., 'Effect of annealing on the dendritic segregation of manganese in steel', *JISI*, 930–932 (Sept 1965)

43. NIELD, B. J., 'Investigation of abnormal structure in a 1·5% Mn steel', *JISI*, 22–26 (Sept 1961)

44. WARD, R. G., 'The dendritic segregation of manganese in steel ingots', *JISI*, 337–342 (April 1958)

45. *Metals Handbook*, 1939 ed, 356

46. MUND, A., 'Methods of degassing and vacuum treating liquid steel', *JISI*, 810 (Figure 6) (Oct 1962)

47. *Physical Chemistry of Steelmaking*, 552–553

48. COE, F. R. and MORETON, J., 'The diffusion of hydrogen in ferritic steels in the temperature range 100–350°C, *Metal Science Journal*, No. 3 (1969)

49. STOLL, J. H., 'Vacuum pouring of ingots for heavy forgings', *JISI*, 191, Pt 1 (Figure 23) (Jan 1959)

50. TIX et al., 'Vacuum treatment of steel', *JISI*, 191, Pt 3, 260–266 (March 1959)

51. MUND, A., 'Methods of degassing and vacuum treating liquid steel', *JISI*, 200, Pt 10, 810 (Oct 1962)

52. ANDREW, J. H. et al., 'The formation of hair line cracks', *JISI*, 2, 262p (1942) and HONEYMAN, contribution to discussion

53. HERTY, C. H. JR, 'Alloys of iron and oxygen', *Metals Handbook*, 1939 ed, 392

54, 55. EPSTEIN, S. and MILLER, H. L., 'Ageing in iron and steel', *Metals Handbook*, 1939 ed, 602

56, 57. VAN HORN, K. R., 'The crystal structure of metals and alloys', *Metals Handbook*, 1939 ed, 74

58. 'Heating bars for forging', *Metals Handbook*, 1939 ed, 833

59. PRIESTNER et al., 'Observations on the behaviour of austenite during the hot working of some low carbon steels', *JISI*, 206, pt 12, 1252–1262 (Dec 1968)

60. PRIESTNER, et al., 'Observations on the behaviour of austenite during the hot working of some low carbon steels', *JISI*, 206, pt 12, 1252–1262 (Figures 4, 5, 7) (Dec 1968)

61. KENYON, R. L. and BURNS, R. S., 'Autographic stress-strain curves of deep drawing sheets', *ASST*, 21, 577 (1933)

62. GREAVES, R. H. and JONES, J. A., 'The effect of temperature on the behaviour of iron and steel in the notched bar impact test', *JISI*, 112, 123 (1925)

63. EPSTEIN, S. and MILLER, H. L., 'Ageing in iron and steel', *Metals Handbook*, 1939 ed, 603

64. LEGGE, E. E., 'Cold drawing steel wire', *Metals Handbook*, 1939 ed, 878

65. *The making, shaping and treating of steel*, United States Steel Corporation, 8th ed, 803 (Figures 29–31)

66. ARCHER, R. S., 'Constitution of iron-carbon alloys', *Metals Handbook*, 1939 ed, 366

67. ARCHER, R. S., 'Constitution of iron-carbon alloys', *Metals Handbook*, 1939 ed, 374–422

68. IRVINE, K. J. and PICKERING, F. B., 'Low-carbon steels with ferrite-pearlite structures', *JISI*, 201, pt 11, 944 (Nov 1963)

69. CARPENTER, SIR H. and ROBERTSON, J. M., *Metals*, 2, 922

70. CARPENTER, SIR H. and ROBERTSON, J. M., *Metals*, 2, 825–879

71. *Atlas of isothermal transformation diagrams*, ISI Special Report No. 40, 9

72. *Atlas of isothermal transformation diagrams*, ISI Special Report No. 40, 15
73. *Transformation characteristics of direct hardening nickel alloy steels*, 3rd ed, The Mond Nickel Co. Ltd., 58
74. *Transformation characteristics of direct hardening nickel alloy steels*, 3rd ed, The Mond Nickel Co. Ltd., 23–29, 56–59
75. *Transformation characteristics of direct hardening nickel alloy steels*, 3rd ed, The Mond Nickel Co. Ltd., 36
76. *Transformation characteristics of direct hardening nickel alloy steels*, 3rd ed, The Mond Nickel Co. Ltd., 31
77. *Transformation characteristics of direct hardening nickel alloy steels*, 3rd ed, The Mond Nickel Co. Ltd., 31
78. *Transformation characteristics of direct hardening nickel alloy steels*, 3rd ed, The Mond Nickel Co. Ltd., 36
79. *Transformation characteristics of direct hardening nickel alloy steels*, 3rd ed, The Mond Nickel Co. Ltd., 54
80. 'Symposium on the hardenability of steel', Iron and Steel Institute Special Report No. 36 (1946)
81. 'Symposium on the hardenability of steel', Iron and Steel Institute Special Report No. 36, 199–252 (1946)
82. *Transformation characteristics of direct hardening nickel alloy steels*, 3rd ed, The Mond Nickel Co. Ltd. (Figure 27)
82, 83. *Transformation characteristics of direct hardening nickel alloy steels*, 3rd ed, The Mond Nickel Co. Ltd. (Figure 26)
84. CARPENTER, SIR H. and ROBERTSON, J. M., *Metals*, **1**, 302
85. BAIN, E. C. and ABORN, R. H., 'The iron-nickel-chromium system, *Metals Handbook*, 1939 ed, 418
86. CRAFTS, W., 'Constitution of iron-chromium-carbon alloys', *Metals Handbook*, 1939 ed, 407A
87. CARPENTER, SIR H. and ROBERTSON, J. M., *Metals*, **2**, 1011
88. CARPENTER, SIR H. and ROBERTSON, J. M., *Metals*, **1**, 1011 (Figure 455)
89. PFEIL, L. B. and JONES, D. J., *JISI*, 127, 337 (1933)
90. CARPENTER, SIR H. and ROBERTSON, J. M., *Metals*, **2**, 1014 (Figure 456)
91. BINDER, W. O., 'The resistance of wrought stainless steels to corrosion', *Metals Handbook*, 1948 ed, 558 (Figure 1)
92. BINDER, W. O., 'The resistance of wrought stainless steels to corrosion', *Metals Handbook*, 1948 ed, 562 (Figure 4)
93. MONYPENNY, J. H. G., 'The brittle phase in high chromium steels', *Metallurgia*, pt 1 (March 1940); pt 2 (July 1940)
94. Wrought heat resisting alloys', *Metals Handbook*, 1948 ed, 563
95. TAPSELL, H. J., *Creep of metals*, O.U.P. (1931)
96. *Report of Joint Research Committee on effect of temperature on the properties of metals*, Preprint No. 29, ASTM (1938)
97. MILLER, CAMPBELL, ABORN and WRIGHT, 'Influence of heat treatment on creep of carbon-molybdenum and chromium-molybdenum-silicon steel, *Trans ASM*, **26**, 81 (March 1938)
98. MURRAY, J. D. and TRUMAN, R. J. *The high temperature properties of Cr-Ni-Nb and Cr-Ni-Mo austenitic steels*, Joint International Conference on Creep, 61 (Aug 1963)
99. 'Wrought heat resisting alloys', *Metals Handbook*, 1948 ed, 565 (Table 6)
100. *Data from Wiggin heat resisting alloys*, Henry Wiggin & Co. Ltd.
101. *Data from Huntington alloys*, Publication T40
102. *Wiggin heat resisting alloys*, Henry Wiggin & Co. Ltd.
103. SAUNDERS, G. G., 'Fracture mechanics and the metallurgist', pt 1, *Welding Institute Research Bulletin*, **10**, (Sept 1969)
104. *Control of steel construction to avoid brittle fracture*, Welding Research Council, New York (Figures 1–7) (1957)
105. *Control of steel construction to avoid brittle fracture*, Welding Research Council, New York (Figures 1–6) (1957)
106. EPSTEIN, S. and MILLER, H. L., 'Ageing in iron and steel', *Metals Handbook*, 1939 ed, 605 (Figure 5)
107. *Control of steel construction to avoid brittle fracture*, Welding Research Council, New York, 40–42 (Figures 1–16 and 1–17) (1957)
108. ERASMUS, L. A. 'Effect of aluminium additions on forgeability, austenite grain coarsening temperature and impact properties of steel', *JISI*, **202**, pt 1, 32–41 (1964)
109. STRAUSS, J. and FRANKLIN, F. F., 'Vanadium in steel', *Metals Handbook*, 1948 ed, 485–488
110. SANDS, J. W., 'Nickel in steel', *Metals Handbook*, 1948 ed, 473–476 (Figure 6) 51 (Figures 1–21(a) and 1–21(b))

111. HOLLOMAN, J. H. *et al.*, 'The effect of microstructure on the mechanical properties of steel., *Trans. ASM*, **38**, 807–844 (1947)
112. *Nickel alloy steels*, 2nd ed, The International Nickel Co. (1949)
113. *Nickel alloy steels, properties at sub-zero temperature*, International Nickel Co.
114. *Control of steel construction to avoid brittle fracture*, Welding Research Council, New York, 51 (1957)
115. *Control of steel construction to avoid brittle fracture*, Welding Research Council, New York, 49 (Figures 1–20(a)) (1957)
116. ERASMUS, L. A., 'Effect of aluminium additions on steel properties', *JISI*, **202**, pt 1, 32–41 (Jan 1964)
117. SAGE, A. M. and COPLEY, F. E. L., 'The notch ductility and tensile properties of some synthetic mild steels', *JISI*, **195**, pt 4, 422–438 (Aug 1960)
118. FERNEHOUGH, G. D. and HOY, C. J., 'Mechanism of deformation and fracture in the charpy test as revealed by dynamic recording of impact loads', *JISI*, **202**, pt 11, 912–920 (Nov 1964)
119. KNOTT, J. F. and COTTRELL, A. H., 'Notch brittleness in mild steel', *JISI*, **201**, pt 3, 249–260 (March 1963)
120. MORRISON, W. B., 'The influence of small niobium additions on the properties of carbon-manganese steels', *JISI*, **201**, pt 4, 317–325 (April 1963)
121. WEBSTER, D. and WOODHEAD, J. H., 'Effect of 0·03% Nb on the ferrite grain size of mild steel, *JISI*, **202**, pt 12 (Dec 1964)
122. PHILIPS *et al.*, 'Effect of Nb and Ta on the tensile and impact properties of mild steel', *JISI*, **202**, pt 7, 593–600 (July 1964)
123. ERASMUS, L. A., 'Effect of small additions of vanadium on the austenitic grain size, forgability and impact properties of steel', *JISI*, **202**, pt 2, 128–134 (Feb 1964)
124. JONES, W. and COOMBES, G., 'Effects of vanadium or chromium on the strain ageing of rimming steels', *JISI*, **174**, 9–15 (1953)
125. TURNER, C. E., 'A note on brittle fracture initiation in mild steel by prior compressive strain', *JISI*, **197**, pt 2, 131–135 (Feb 1961)
126. MYLONAS, C. *et al.*, *Welding Journal Research Supplement* (American Welding Society), 9_s–17_s (Jan 1957)
127. MYLONAS, C. *et al.*, *Welding Journal Research Supplement* (American Welding Society), 473_s–480_s (Oct 1958)
128. MYLONAS, C. *et al.*, *Welding Journal Research Supplement* (American Welding Society), 414_s–424_s (Oct 1959)
129. GILLETT, H. W., 'High strength low-alloy structural steels', *Metals Handbook*, 1948 ed, 534
130. *Welding Handbook*, 5th ed, American Welding Society
131. BS 4360: 1968, Weldable structural steels. Clause 6.5
132. *Hyplus 23 and Hyplus 29*, British Steel Corporation
133. BS 4360: 1968, Weldable structural steels (Table 6)
134. BAILEY, N., *Welding procedures for low alloy steels*, Members' Report, Welding Institute, Cambridge (Figure 32)
135. IRVINE, K. J., 'Development of high strength steels', *JISI*, **200**, pt 10 (Figure 11) (Oct 1962)
136. HODGE, J. M. and BAIN, E. C., 'Functions of the alloying elements in steel', *Metals Handbook*, 1948 ed, 455 (Figures 3–6)
137. HODGE, J. M. and BAIN, E. C., 'Functions of the alloying elements in steel', *Metals Handbook*, 1948 ed, 456 (Figure 7)
138. MERICA, PAUL D., 'Constitution of iron-nickel alloys', *Metals Handbook*, 1939 ed, 386
139. *Nickel alloy steels* (marageing steels for ultra-high strength applications), International Nickel Co. (Mond) Ltd
140. NAUMANN, F. K., 'Influence of alloy addition to steel on resistance to hydrogen under high pressure', *Tech. Mitteilungen Krupp*, 1 (12), 223–234 (1938)
141. NELSON, G. A., 'Action of hydrogen on steel at high temperature and high pressures', Section 2, American Welding Society, Welding Research Council Bulletin 145 (Oct 1969)

CASTING AND FOUNDRY PRACTICE

P. D. WEBSTER

THE ADVANTAGES OF MAKING COMPONENTS AS CASTINGS

This section summarises the ways in which the design engineer can make the best use of cast components and outlines some of the difficulties which may be encountered during the production of castings. If such difficulties are described at length with no other comment, the reader might be given the impression that the production of castings is always a hazardous business giving rise to products of uncertain quality. This is far from the truth; the majority of castings are made without difficulty and the quality of cast components can be extremely high.

Before describing the casting process and the ways in which the design engineer can assist the foundryman, the advantages of this method of production will be outlined so that later discussion of possible defects will not give the wrong emphasis.

Economic advantages

One advantage of castings is very often economy. Making a pattern out of an easily machined material like wood, forming a mould round it with an easily shaped material like moulding sand, pouring liquid metal into this mould and allowing it to solidify is a simple and inexpensive sequence of operations.

If the operation is looked at in slightly more detail, the ways in which the economics of the process are controlled and the advantage of consultation between foundryman and design engineer will be seen. Forgetting such complicating factors as overheads and profits, the cost of a casting may be divided into four sections. These are the cost of producing the pattern equipment, the cost of making the mould, the cost of melting and casting the metal and the cost of finishing and machining the casting.

It is assumed that the design engineer has chosen the most economic material to give him the properties necessary to withstand the service conditions. Also that he has designed the component such that the minimum quantity of metal is required. This leaves pattern cost, moulding cost, and finishing cost. It is interesting to see how these are related to the number of castings to be produced from any one pattern and the dimensional accuracy required of the final finished component. If five castings are to be made, each casting must carry one-fifth of the pattern cost and the total order will only have to include five times the cost of making a mould. If ten thousand castings are required, assuming one casting per mould is still to be produced then the pattern cost per casting will only be one ten-thousandth part of the total pattern cost while the order must carry ten thousand times the cost of making a mould. In the first case it will be seen that any means of saving on pattern making costs, even if this entails more time and skill during moulding, is worthwhile. In the second instance, the pattern equipment will not only have to be of better quality to withstand the wear of the production of ten thousand moulds but also if it can be geared to automated mould production even the saving of a few seconds on the time spent in making each mould will be worthwhile.

The above aspects are related to the dimensional accuracy to which the castings can be made. Low cost pattern equipment and hand moulding do not give the best results from the point of view of close dimensional tolerances. Castings of close dimensional accuracy can be obtained but the techniques needed to do this usually involve expensive pattern equipment or dies. Therefore, although the manufacture of components by the casting process can be the production method with the lowest cost it is essential, if the full economic benefits are to be obtained, that there is full consultation between the purchaser and producer of the castings. The foundryman must be given the most accurate forecast of the number of castings required and only those dimensional tolerances which are really necessary should be imposed.

Casting of complex shapes

The next advantage of producing components as castings is that it is possible to produce extremely complex shapes. Castings can be made not only with complicated external shapes, but hollow castings can be produced with complex internal cavities. If so desired, both the external profile and the inside dimensions can be held to close tolerances.

It should be remembered, however, that if close tolerances are required, special equipment and techniques may have to be employed. Also, that although within itself the casting process can deal with complicated shapes, certain geometric features will involve the foundryman in extra expense. If the full economic advantage of the process is to be gained, slight changes in design can often overcome the necessity for such expense without really curtailing the freedom of design which the casting process gives.

More will be said about this later, but it should be stressed at this point that the foundryman has to look at the problem from many points of view. He will have to think of how the pattern and core boxes are to be made, how the mould is to be made, how the metal is to be poured into the mould, what is likely to happen as the casting solidifies and how he will fettle and finish the casting once it is knocked out of the mould.

Quite often none of these give rise to problems, but deciding which of them may do so takes a great deal of experience, so that only an indication of the kind of difficulty can be given here, it being emphasised that there should always be full consultation between the engineer and foundryman.

MECHANICAL PROPERTIES

Certain mechanical properties of cast products can be of advantage in the correct application. In the production of wrought components by forging, rolling, extruding, drawing etc, grain flow usually takes place giving the finished parts mechanical properties which are superior in certain directions. If these directions coincide with those through which service stresses occur all is well, but when stresses of random direction may take place the non-directional properties of cast components can be an advantage.

Many engineers believe that superior properties will always be obtained from wrought products due to the grain refinement which takes place during the working which occurs in their production. It should be pointed out that correct alloy section, ladle treatment to give fine grained solidification, control of solidification to give sound castings, heat treatment and final inspection to ensure the success of these steps can give rise to products of the highest quality. Certain other cast materials have inherent properties which may be of advantage; the cast irons with their free machining, good damping and good bearing properties being an example of this.

It should also be remembered that certain alloys have been developed so that it is extremely difficult to form them in any other way than by casting. Some of the alloys used for turbine blades have been developed to be very hard at very high temperatures so that they can resist the abrasion of the hot gases. This makes both hot forming processes and machining difficult. This example illustrates a case where not only does the process prove to be of advantage but the high quality of component required indicates that castings can be produced to very rigorous specifications.

Finally it must be stressed that there is a large variety of casting processes available. These are capable of manufacturing parts ranging in size from a few grammes to several hundred tonnes.

HAND MOULDING

Most readers will have a general idea of the way in which a mould is made, but a description of the operation is repeated here in order to introduce many of the terms used in moulding. Also, since the general principles apply to most forms of mould the description will assist with an understanding of the way in which components should be designed when produced as castings.

In order to illustrate the making of a mould a simple component has been chosen. This is the pipe connection elbow illustrated in Figure 5.61(*a*). Note that no dimensions

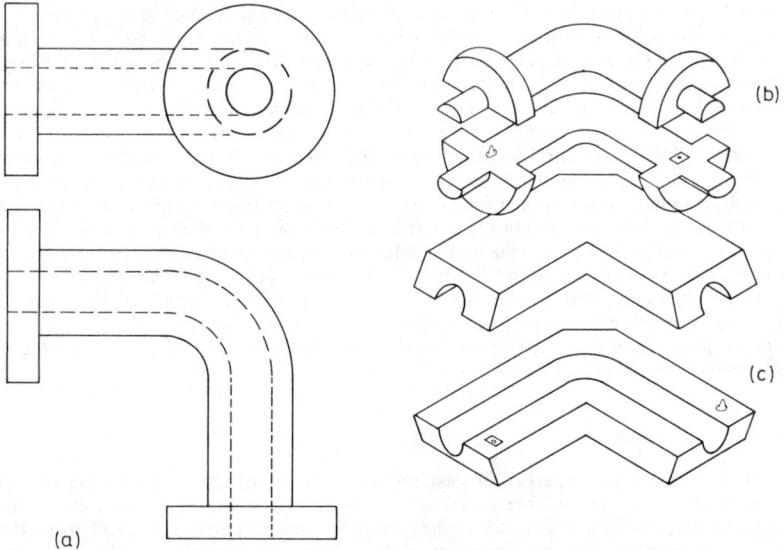

Figure 5.61. Patterns for hand moulding (a) Pipe connection elbow to be moulded by hand. (b) The two halves of the pattern needed to produce A. (c) Core box for the centre hole in A

have been marked in the drawing since the production of this component will be described by a number of different methods and in a number of different alloys and sizes. If this casting were to be produced by hand moulding, a pattern would be required to produce the external shape and this would probably be of the type shown in Figure 5.61(*b*). In order to form the hole which runs through the centre of the casting a core will be required and this could be produced from a core box such as is shown in Figure 5.61(*c*).

Materials

The moulding materials required would probably be a naturally bonded sand for the mould and some special core sand for the core. Moulding sands are silica sands with certain additions to stick them together and form them into a mould, other additions are sometimes made to prevent the high temperature of the molten metal from affecting the mould and causing defects.

Clay bonded moulding sands, as their name suggests, make use of clay to give them enough strength to form the mould. There are two main types of clay bonded sands, the first being synthetic sands which are produced by taking silica sands and adding special clays to them. These give very good properties but tend to need very careful control of the moisture content. The second are natural sands which are quarried with the clay already present; these are more flexible as regards water content and so are usually preferred for hand moulding.

Clay bonded moulding sands are prepared for moulding by the addition of water and milling. Previous use will usually have caused the sand to become dry and the clay then becomes a powder. Simply mixing water with the sand and clay does not produce the desired properties since the water must be kneaded into the clay before it develops its full strength. This is done in a mill in which heavy rollers mix and work the sand to the required consistency.

Cores

Cores are separate portions of the mould which are placed into it and must therefore be hard enough to handle. At one time these were produced by ramming sand mixed with oil into a core box, carefully turning the compacted core out of the core box so as not to damage it and baking it in an oven until it was hard enough to handle. This is now usually done by using other bonding agents than oil.

Sands with additions of sodium silicate can either be instantaneously hardened in the core box by passing carbon dioxide gas through them, or hardened by other chemical agents which are added just before use. Special resins have been developed which harden either when subjected to heat from a heated core box, or by the addition of chemical agents added just before use. The advantages and disadvantages of these various methods of hardening cores will not be discussed here. It is worth mentioning, however, that similar techniques are being employed for the production of moulds.

The first operation would probably be the production of the core. This would be done in the core box illustrated in Figure 5.61(c) probably using one of the modern self-hardening core sands. The two halves of the box locate on each other by the use of dowels, they would be clamped together and the sand compacted into the cavity. In some cases it might be necessary to provide channels through the centre of the core out of which hot gases and moisture could pass during the cooling of the casting; these are called core vents.

In order to strengthen certain cores it is also sometimes necessary to reinforce them with metal wires or grids. Once the core had hardened it would be extracted from the core box ready for use.

Use of patterns

Next the mould would be produced; in this case a loose moulding board has been provided although sometimes patterns are placed directly on the floor or bench and sometimes they are fixed to a moulding board. Half the pattern shown in Figure 5.61(b) is placed on the board and over it is placed a moulding box part. This is a metal frame of steel, cast iron or aluminium usually provided with lugs on its sides with holes through which steel rods known as box pins can pass. The box pins accurately locate one box part to another and also ensures that when one box part is lifted off another the movement is absolutely vertical. This can be important if this lifting also involves extracting the pattern from the mould.

Box parts are usually furnished with handles for lifting and turning over, although in the case of very large moulds these may be trunnions to fit crane slings. There is often some means of clamping moulding box parts together. The inside face of a moulding box is usually not straight but has some form of projection to help lock the moulding sand in. Larger moulding boxes often have strips of metal running across the inside such that they do not take up the space where the pattern will go but do give support to the sand, these are known as box bars.

Figure 5.62(d) shows the half pattern placed on the board with the box part placed over it ready to receive the sand. First, the pattern is dusted over with a parting powder which will prevent the moulding sand from sticking to it. Sometimes fine dry silica sand is used or more often proprietary brands of powder are utilised.

Since the first moulding sand into the box will be that which forms the face of the mould with which the metal will come in contact, it is most important that this should be of high quality. For this reason specially prepared sand called facing sand is usually added first, and is passed through a sieve to give a lump free sand capable of uniform compaction. This is achieved with the use of a peg rammer, which the moulder jabs into the sand in a series of strokes all over its surface. The ramming is carried out so that the sand becomes uniformily consolidated around the pattern without forming strata, a little more sand being added to the mould at a time.

Once the face surface of the mould has been formed and the carefully prepared facing sand is no longer required, so called backing sand may be used. Figure 5.62(a) is a pseudo section of the moulding sand being compacted around the half pattern. When the box part has been completely filled with sand such that it is protruding over the top of the box, this would be scraped off flat with a strip of metal known as a strickle. This box part could then be picked up and turned over. The moulder would then examine the face of the mould to ensure it was not damaged and might mark out the running and feeding system, but more of this later. For the time being the mould making operation only will be considered.

Having turned the first half of the mould over, a second box part is now placed on to the first one, being located with box pins. The second half of the pattern is then placed on top of the first half which is at this time in the compacted first half of the mould Figure 5.62(b). Location of the two halves of the pattern being achieved by dowels.

After clamping the boxes together, parting powder is then dusted on to the pattern and the already formed face of the first half of the mould. The second half of the mould can then be filled and compacted with sand as was the first half. When this has been completed, the clamps are released and the second half of the mould lifted off the first half. At this point the running and feeding system would usually be cut into the mould, but this will be described later.

Removing the pattern from the mould

The next operation is to remove the two halves of the pattern from the two halves of the mould. In some cases, in order to extract the pattern, a spike is driven into the face of it which is showing through the joint face of the mould. A better practice is to provide a special screw location in the pattern to allow a bolt to be screwed into it which can be used for lifting and rapping the pattern.

Having got some means of lifting the pattern from the mould secured, the pattern cannot simply be lifted straight out of the mould as the sand will be too tightly compacted round the pattern. The moulder has now to loosen the pattern by rapping it. This entails striking the bolt first in one direction and then in the other until it is loosened in the mould. The degree to which this is done is one of the skills of hand moulding, and if not carried out with care it can result in loss of dimensional accuracy.

Next the pattern is lifted out of the mould, skill again being required since if the lift is not completely vertical, the sides of the mould cavity will be damaged. Note this is particularly important with vertical faces, an example of which can be seen on the parallel front and back flat faces of the flanges of the pipe connection elbow. This difficulty could be greatly reduced by providing these faces with taper such as that indicated in Figure 5.62(*c*). This is a pseudo section of the mould with the pattern just being lifted from it. Note that the pattern has now left an impression of the outside shape of the casting in the mould. Note also that on the pattern at the points through which the internal hole in the casting will pass there are two projecting pieces, these are known as core prints.

As can be seen from the core box, Figure 5.61(*c*), the core is longer than is necessary just to form the central hole. This extended length fits into the cavities formed by the core prints of the pattern. Half the mould with the core in position is shown in Figure 5.62(*e*). The core prints have not only to support the weight of the core in the mould before the metal enters it, but also have to withstand the pressure caused by the tendency of the core to float once the molten metal is in the mould. This second effect is usually more serious than the weight of the core and, in the case of the pipe connection elbow, the core does not balance very well so it might be necessary to obtain extra support by

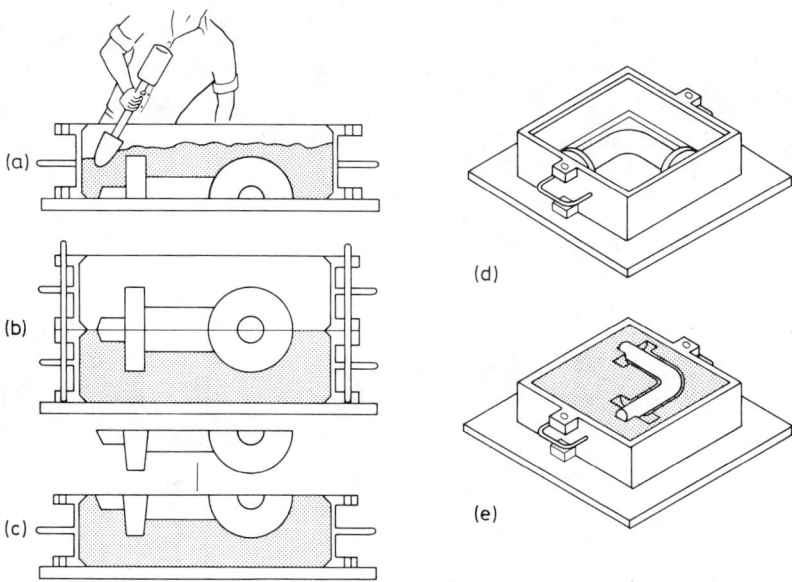

Figure 5.62. Sand casting technique of hand moulding (a) Pseudo-section shows the consolidation of sand round the half pattern to form part of the mould. (b) The first half of the mould has been made and turned over, the second box part and half of the pattern are now in position ready for the other half to be made. (c) Pseudo-section showing half the pattern being withdrawn and leaving the appropriate cavity. (d) Half the pattern placed on a board with a box part in position ready for the introduction of sand. (e) The drag half of the mould with the centre core in position

pinning the core into its prints with special nails used in moulding called sprigs or by extending the core prints. In some instances, cores are supported by putting small studs called chaplets between the mould wall and the core, when the metal enters it partially melts the chaplet so that it becomes fused into the casting. This practice should only be used where the design is such that there is no alternative and where the properties required of the casting permit it.

If the two half moulds, with its core in position, are now put together there is a cavity which will exactly produce the required casting. The top half of the mould is called the cope and the bottom half the drag. A series of channels must be made in the mould in order that the metal can be introduced, this is the running system. Also since the metal undergoes volume changes during cooling, this must be allowed for.

These changes are not only the normal expansion and contracting which takes place on heating and cooling in the solid state and is allowed for by making the pattern slightly larger than the final casting needs to be (called the pattern makers contraction allowance), but also the sudden decrease in volume which often occurs on solidification. The latter has to be compensated for by having reservoirs of metal attached to the casting while it cools. These are called feeder heads and are cut from the casting when cold. Both running and feeding are dealt with later.

Note that the pattern provided to make this mould was produced in two halves so that it could be laid flat on the board which produced the mould joint line. When very few castings are required it may not be economical to do this. There are also instances where splitting a pattern through the joint line makes it so frail that it is not possible to work with. In some cases also, castings have to be made using another casting as a pattern so that splitting is not convenient. In these cases a dummy half mould is first rammed up, the pattern is then roughly buried down to the joint line in this. The first half of the mould can then be made, its joint line cleaned up and then the second half of the mould made on that. The dummy half mould which is discarded is called an odd-side. The necessity for the box pins ensuring that boxes are lifted off each other in a vertical direction can now be seen, since when they are taken apart half the pattern must be extracted from one half of the mould.

DESIGN FEATURES ASSOCIATED WITH PATTERN WITHDRAWAL

In the section on hand moulding it was seen that, after consolidating the moulding material round the pattern, it was necessary to extract the pattern from the mould. This sequence of operations is common to most sand casting techniques and should, where possible, be catered for in the design of the casting. If a casting is designed with

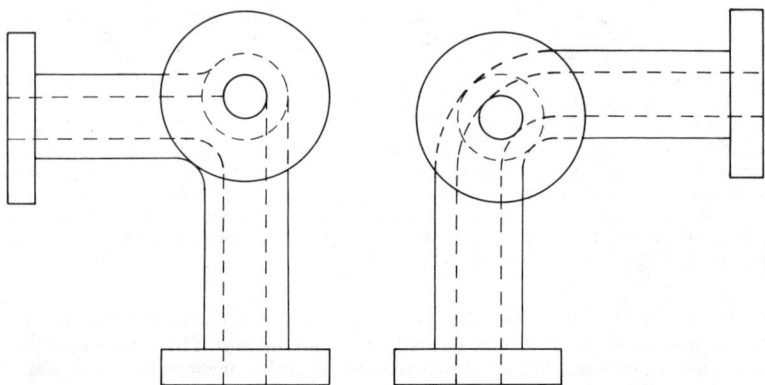

Figure 5.63. Pipe connection elbow with three arms at right angles so that one of the flanges will give difficulty in moulding

projections on it which tend to prevent the pattern from being withdrawn from the mould the difficulty can be overcome in a number of ways.

The pipe connection elbow will again be used as an example and a description of the methods of overcoming this problem will enable the reader to appreciate it. Figure 5.63 shows the elbow, but another flanged arm has been added at right angles to the two previous flanged arms. This means that in whichever direction the casting is moulded, one of the flanges will tend to prevent the pattern from being withdrawn from the mould. The alternative means of overcoming this will be discussed later.

The use of a core

Since it is the sand underneath the flange which is preventing the pattern from being removed from the mould, the problem can be overcome by constructing the pattern as in Figure 5.64(*a*) with the area under the flange blocked off by a core print. A core must then be placed in this print which has a cavity in it which will form the missing part of the mould.

It should be remembered that pattern-makers are very ingenious so that there are a number of ways that the box to produce this core could be made. One simple method is

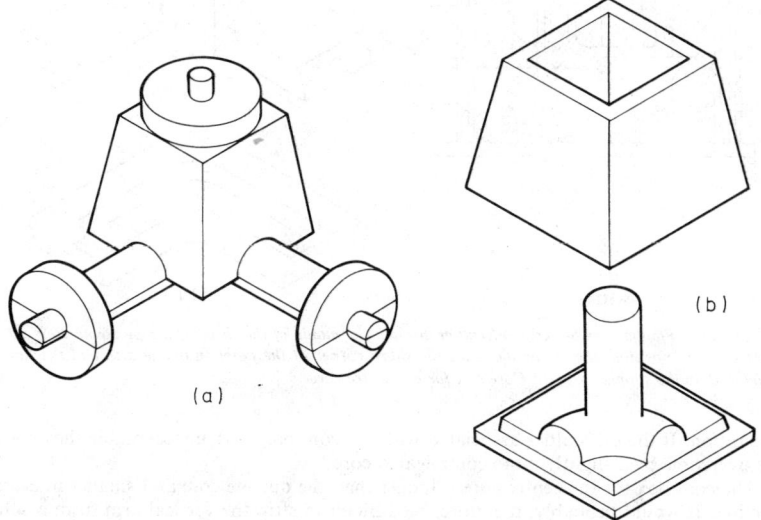

Figure 5.64. Using a core (a) Pattern with a core point to overcome the difficulty in moulding the flange. (b) Two parts of a simple core box to produce the core

illustrated in Figure 5.64(*b*). This is shown with the parts of the box separated so that the inside can be seen. The base part of the box forms the shape of the mould cavity and the top corebox part fits down onto it so that the box can be filled with core sand through the opening in the top. The completed core then fits into the core print as shown in the mould section Figure 5.65(*a*). Note that it is now necessary to use two cores, the one which forms the centre hole and the one which cores out the area under the flange.

The simple construction of the flange core box has certain disadvantages. One of these is that it might be difficult to extract the core from the part of the box forming the mould cavity for the vertical arm of the casting. This would depend on the size of the casting and the amount of taper which was permissible. There is also a possibility that this part might get damaged in use.

The other difficulty is that, since the box is to be filled through the top and excess sand scraped off to give a flat top, while the portion forming the cavity extracts in the opposite direction, the core will form a sharp corner at the junction of the flange and the arm, as indicated by the arrow in Figure 5.65(*a*). Such junctions should have a radiused fillet as shown at the horizontal flange. This is not only because it prevents stress concentration in the final casting, but also because the sharp corner might cause cracking or the formation of shrinkage cavities as the casting cools.

If the elbow were small in size, and to be produced by machine moulding, such a core box might well be used however. The advantage of this is that the machine moulder would have patterns which could be fixed to the machine, and would strip from the mould without difficulty. This could be followed by a relatively simple core setting

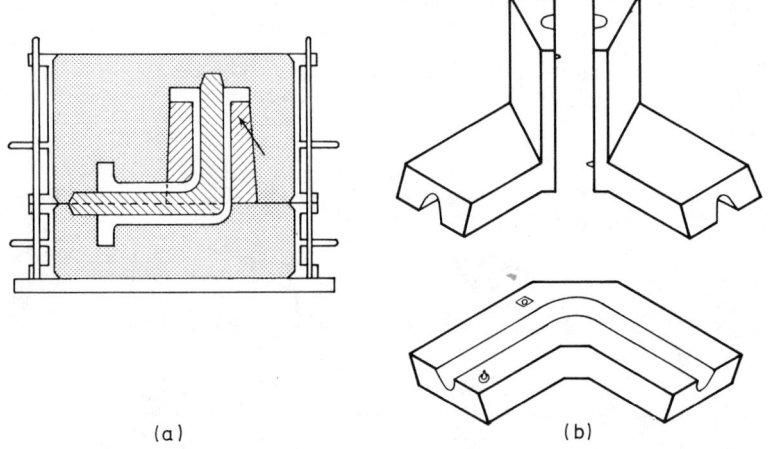

(a)   (b)

Figure 5.65. Flange core boxes. (a) Section through the mould of the three-arm pipe connection elbow with cores in position, the arrow indicates the sharp corner at the junction of the arm and the flange produced by the simple core. (b) Core box for the centre core

operation. If the difficulties associated with the core box were unacceptable they could be overcome by a slightly more complicated core box.

The core box for the centre core is longer than the outside core and smaller in cross-section. It would probably, therefore, be difficult to strip the vertical arm from a solid portion of the box, which would thus be made in three parts as shown in Figure 5.65(*b*). This centre core-box would be suitable for all the methods of production to be described.

The use of a three part mould

The flange on the pattern can also be extracted from the mould by using three moulding box parts instead of two. In this case the pattern is also made in three parts and the mould is produced with two joints. Figure 5.66(*a*) is a section through the mould with the pattern not yet removed. Arrows indicate the direction in which the three parts of the pattern will be withdrawn when the mould is taken apart. Note that the pattern has been split away from the junction between the vertical arm and its flange so that a fillet can be incorporated there.

Although this appears to be an extremely simple answer to the problem, there are limitations to its use. The first is that it is difficult to produce multipart moulds on moulding machines. This is because the centre mould part has to be made with a joint at its top and bottom face which is difficult to do on a machine. The other limitation is that a centre box part of the correct depth is needed. There is a certain degree of latitude

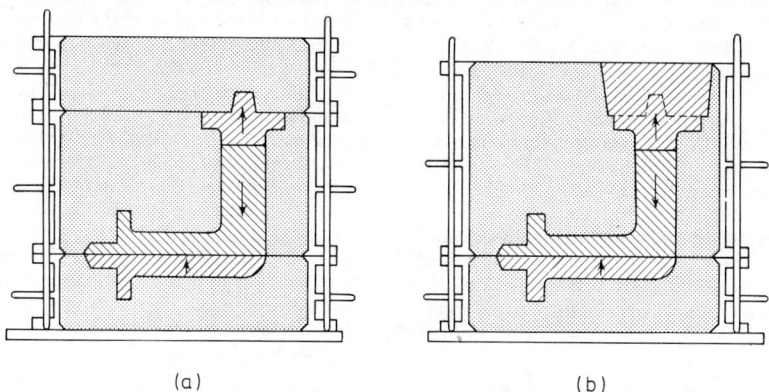

(a)   (b)

Figure 5.66. (a) Three-part pattern and mould. The arrows show the direction of pattern part removal.
(b) Three-part pattern and two-part mould. Arrows again indicate the direction of removal and the broken line shows where the cover core will be placed

here in that a skilled hand moulder can cut a joint to allow for the wrong depth of box to a certain degree. The overall size of the casting is also important, very small castings with very shallow centre box parts are difficult to handle so that this method would most probably be used to produce medium to large castings in fairly small numbers by a skilled moulder.

The use of a cover core

This means of extracting the pattern is very similar to making a three part mould in that the pattern is made in three pieces and extracted in the same directions as before. In this case it is not necessary to have a centre box part of exactly the right depth, since the top two sections of the pattern go into one box part and the depth of these parts of the pattern is adjusted to the depth of the cope box by a core print.

Thus the core print ensures that the pattern comes just to the top of the mould. This print is large enough to ensure that the part of the pattern forming the flange can be withdrawn through the top of the mould, while the part of the pattern forming the vertical arm is withdrawn downwards through the jointline.

The cavity formed by the core-print is afterwards filled in by a so-called 'cover core'. This itself contains a core-print so that the centre core can be located. Figure 5.66(*b*) shows a section through the mould with the pattern still in it. The arrows show the direction in which the parts of the pattern will be extracted, while the broken line indicates the position which will be taken up by the 'cover core'.

While being similar to the three part mould method, this method has the advantage that only two box parts are needed and that a centre box of exact size is not required. It may also be a little more easy when assembling the mould to make sure the centre core is correctly located. The disadvantage is that an extra core box is needed and once again a three part pattern cannot easily be used on a machine. This method would tend to be used in rather similar circumstances to a three part mould but perhaps on slightly smaller castings.

The use of an expendable pattern

There are special moulding techniques where the mould does not need to contain a joint, but in which the pattern is melted, burnt or dissolved out of the mould. In the past, these techniques have been confined to the production of statuary, special precision castings and dental castings, but recently they have found more general uses.

The material which has brought the method into more general use is exfoliated polystyrene. This can very easily be cut to form a pattern and has a very low melting point. One of the advantages is that it can be used to form parts of the pattern which will not withdraw from the mould. The pattern material can then be melted out either by heating the mould prior to casting or by allowing the heat of the metal to drive it in the form of vapour. The technique is also used for the whole of the pattern but this is described in the section on special mould processes.

This method has the advantage of simplicity but since it is still relatively new, it is not as yet being used in many foundries. It has the disadvantage that a new pattern part must be provided each time and even although exfoliated polystyrene can very easily be formed into a pattern shape, this is still an extra operation.

The use of segmented loose pieces

In this method the pattern is made with the part to form the flange in segments. These fit onto the centre part of the pattern which forms the vertical arm and the core-print for the centre core. A detail of this portion of the pattern is shown in Figure 5.67(a) and

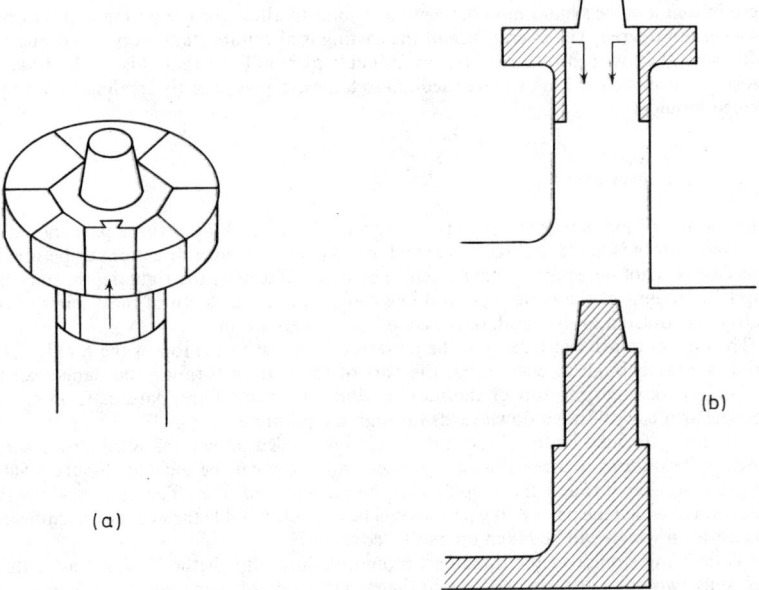

(a)

(b)

Figure 5.67. Use of segmented loose pieces.   (a) Detail of pattern with the flange made as segmented loose pieces. (b) Detail showing a section through the mould with the main part of the pattern removed but the segments still in position. The arrows indicate the direction of segment removal

it can be seen in Figure 5.67(*b*) that, when the main part of the pattern is withdrawn, the segments are left behind in the mould. The moulder then extracts each segment by pulling them into the centre of the mould cavity and then downwards as indicated by the arrows in Figure 5.67(*b*). The segment to which the arrow in Figure 5.67(*a*) is pointing would be withdrawn first after which the others would be free to move.

The limitations of this method are that the mould cavity must be large enough for the moulder to get his hand in and there must be enough clearance to extract the loose

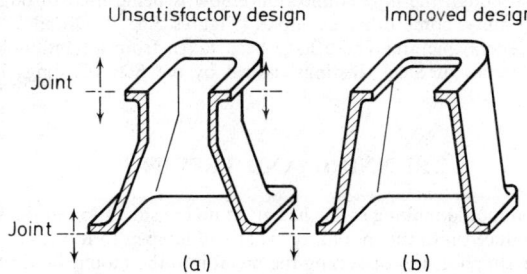

Figure 5.68. *Design modification to eliminate a joint*

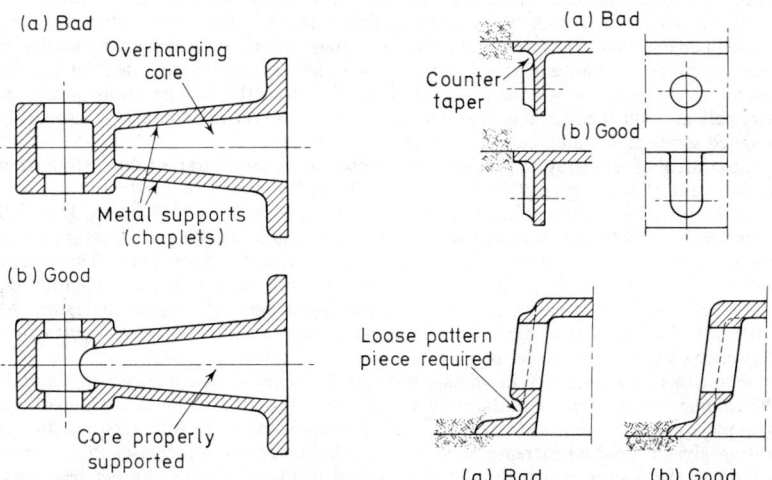

Figure 5.69. *Altering the design not only gives more support to the core so that chaplets are eliminated, but also improves the core venting*

Figure 5.70. *Modifications to simplify moulding of small projecting bosses*

pieces. In this case there is only the minimum clearance, so that if the diameter of the flange were a little larger in relation to the diameter of the vertical arm, the method could not be used. Skill is also required both in pattern making and moulding; nevertheless this method is a very useful one when a number of large castings are being produced.

Overcoming the problem by redesigning the casting

Since a rather simple shape has been chosen to illustrate difficulties associated with pattern withdrawal, the modification of design needed to overcome the problem is also simple.

If the pipe connection had been designed in the form of a T with all the flanges in the same plane it is obvious to see how much trouble would have been avoided. Note, however, that since one of the advantages of producing components as castings is freedom of design, if there was some advantage to the original design, foundries would be perfectly willing to produce it in this way. It can now be appreciated, however, that this will require more time and skill, which of course costs money.

The difficulties which may be encountered with pattern extraction are, of course, many and varied and although the pipe connection elbow is being used throughout for the purposes of continuity, some other examples (Figures 5.68, 5.69 and 5.70) may help to give a more general picture. These have been taken from a handbook called 'The Design and Properties of Steel Castings' issued by the Steel Castings Development Committee.[1]

RUNNING AND FEEDING

The discussion on mould making so far has given no consideration of the way in which the metal is introduced into the mould, or what will happen to it once it starts to cool and solidify. The simplest way of getting the metal into the mould would be to make a small opening in the top and pour it straight in. Unfortunately this is seldom practical because, the bottom of the mould cavity would probably not withstand the eroding effect of a stream of molten metal impinging directly on it, the metal would enter the mould in an extremely turbulent fashion causing the formation of much dross and oxide. There would be no chance of any entrained slag in the metal being separated out and the control of temperature gradients which determine how the casting cools would be extremely difficult. The metal is, therefore, introduced into the mould through a specially designed series of channels called the running system.

The control of temperature gradients is necessary in order to prevent the casting from cracking and that the correct feeding technique can be employed. Feeding of castings is required because of the volume changes which occur during solidification. It is well known that, as metals are heated, they expand. Thus when castings cool from the point at which they are just solid to room temperature a contraction takes place. This means that if the mould were made to the correct size the final casting would be slightly too small. This problem is easily overcome by making the pattern slightly oversize, the dimensions to which it is made being measured with a pattern makers contraction allowance rule, different scales being used for different alloys.

On melting, the majority of metals undergo another sudden increase in volume. The volume change varies with the alloy but 5% is fairly common. This change in volume takes place at the temperature at which the metal melts or solidifies. After melting, further heating gives a gradual increase in volume as the temperature increases in a similar fashion to the way in which the metal expanded in the solid state. Feeder heads are placed on the casting being of such a size that they stay liquid until after the casting has solidified. Thus, as the casting solidifies, the decrease in volume known as solidification shrinkage takes place and liquid metal from the feeder head runs into the casting compensating for it.

The final result should be a solid casting and feeder heads containing all the shrinkage cavities. These heads are then cut off the casting and can be used for remelting. Feeding will be discussed in more detail later while first the method of introducing the metal into the mould which is called running or gating will be considered.

Running systems

In order that the terms used for running and feeding can be understood, Figure 5.71 illustrates the pipe connection elbow with a running and feeding system as it would

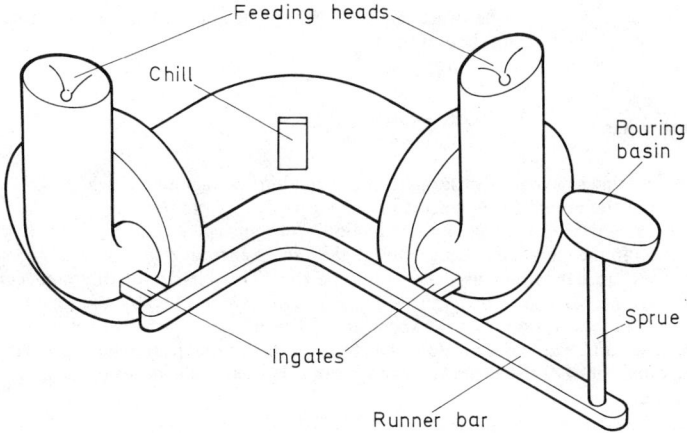

Figure 5.71. Components of a running and feeding system

appear after it had been knocked out of the mould and before these had been cut off. Note that this is not the recommended system for the casting since the needs of the casting would vary with the alloy in which it was to be made and even with its size.

The system merely illustrates what the various components are called. The functions of each of these components are described below.

Pouring basin

This is sometimes called the pouring cup and, indeed, all these parts of running and feeding systems are likely to be given different local names. The functions of the pouring basin is to receive the metal as it is poured from the ladle, and to keep the sprue, which is the vertical channel, completely filled.

If there were no pouring basin, the pourer would have to pour at the exact rate at which the sprue could accept the metal to keep it full, if poured too slowly the sprue would not be filled, if too fast it would immediately overflow. The volume of the pouring basin allows the pourer to adjust the pouring rate since pouring too fast or slowly only produces a relatively slow rise or fall in the level in the pouring basin.

The pouring basin is also a place in which any slag or dross which might be in the metal can float to the surface and not enter the mould. It also serves to break the fall of the metal from the ladle so that the metal enters the mould at a lower velocity and with less turbulence.

The sprue

This is sometimes called the downgate and it is the vertical channel which conveys the metal from the top of the mould down to the level at which it has been decided it should enter the mould cavity which is usually at the joint line.

Because the metal is falling down the sprue and is therefore accelerating under the influence of gravity during its passage through it, there is a tendency for a reduction of pressure to occur which can give rise to mould gases being pulled into the metal stream. This is unimportant except where the metal concerned is one which is easily oxidised and forms dross which would cause defects in the casting. When this is the case, the reduction of pressure can be overcome by tapering the sprue such that it is smaller in cross-section at the bottom than at the top.

The runner bar

This is sometimes called the cross-gate and is the horizontal channel which conveys the metal round the mould to the various points at which it is required.

When the metal stream changes direction from sprue to runner bar, a hydraulic contraction in the stream can cause mould gases to be entrained in it. Specially designed sprue bases are sometimes used to overcome this. This change of direction can also give rise to a decrease in velocity due to friction. This can be aided by making the runner bar larger in cross-section than the sprue and the gentle flow in the runner bar will then enable dross and slag to separate out before the metal enters the casting cavity. The relative dimensions of the sprue and runner can be varied to the needs of the alloy which is being cast.

Slag and dross are trapped in many ways, ranging from runner bars which are made with their top sections in the form of dragons teeth to the use of so-called spinners. These are cylindrical chambers in which separation of the slag is achieved by the metal being made to spin as in a cyclone. Strainers are also used. These are made in refractory or core sand and take the form of plates with small holes in them and may be placed in almost any part of the gating system.

Ingates

These are the openings through which the metal enters the mould cavity and their correct size and positioning are vital to the success of the casting. They must be placed so that the metal will enter the mould cavity rapidly enough to complete filling before the metal goes solid, without having to resort to unreasonably high casting temperatures.

Rapid filling should not, however, result in turbulence or mould erosion. For this reason many alloys which tend to form dross are gated through streamline systems with large ingates giving rise to rapid filling at a low velocity. The ingates are placed low down in the mould so that the metal wells quickly, but quietly, up to the mould cavity. Such systems work well for non-ferrous alloys where the main problem is the formation of oxides which cause dross and scum on the metal. With cast irons, however, very few problems are encountered with oxides; here problems may arise from slag suspended in the form of droplets in the metal. This has come from the reaction of the metal with refractories from the furnaces, ladle or mould.

If the streamline gating systems used for non-ferrous alloys were employed with cast iron, any slag in the metal would be carried straight through the system suspended in the metal stream. In order that this shall not happen, ingates are made with a small cross-section in relation to the rest of the system This so-called 'choking' gives high velocity flow through the ingates, which can be tolerated in the case of cast iron, but means that the metal passes more slowly through the rest of the system enabling the slag to be trapped.

Ingates must also be carefully placed so that they can be conveniently cut or knocked off the casting after it has been removed from the mould. They should be in such a position that grinding the area where they have been, can be carried out with ease, that any marks left by them does not spoil the appearance of the casting, and if machining is to be carried out, porosity which can occur under ingates must not be revealed.

Finally, the positioning of the ingates will in part control the way in which the casting cools, and it is for this reason that running and feeding are usually considered together.

Since the two are not easily separated the way in which the positioning of the ingates affects cooling will be discussed as part of feeding.

Feeding cast irons

It has been mentioned that the design of the running system depends upon the alloy which is to be cast, this is also true of feeding systems. It is, therefore, not possible to discuss the way castings are fed in general but if the alloys are first classified into types, these can be dealt with in a general fashion. The main classifications are the grey cast irons, alloys which solidify over a short range of temperature and those which solidify over a long range of temperature. The grey cast irons are a group of alloys different in their feeding requirements from all others and will be discussed first.

Grey cast iron differs from other metals in that it requires much less feeding indeed many castings require no feeding at all. This is due to the fact that grey irons contain a large amount of carbon (around 3% by weight) and before solidification this is in solution in the molten iron. When solidification takes place most of it comes out of solution to form the graphite flakes which give the iron many of its characteristic properties. Thus, in the molten state, the carbon occupies no volume but on solidification, the graphite takes up space thus compensating for the shrinkage the alloy would otherwise undergo. This being the case, one might expect grey iron castings never to suffer from shrinkage defects; in practice, shrinkage defects are possible but are not nearly the same problem that is experienced with other alloys.

Shrinkage defects can occur in grey iron castings for three reasons. The first is due to pouring at too high a temperature; this can give defects due to the change in volume in the liquid which occurs after melting. If very hot metal fills the mould quickly, because the mould is initially cold, a solid skin of metal can quickly form an envelope inside which is still very hot metal. The fall in temperature to solidification is accompanied by a decrease in volume causing cavities to form. This can, in part, be combated by designing the gating system to give a controlled filling rate. The second cause of shrinkage in grey iron castings is due to the fact that graphite sometimes only starts to form late in the solidification sequence. This problem is most serious in high duty irons containing less carbon. Finally, iron castings sometimes show shrinkage defects due to a tendency for the mould to swell while subjected to the temperature and pressure of the molten iron. Once solidification has proceeded to any extent, swelling of the mould cavity no longer matters since the casting will then be solid enough to support itself. More rigid moulds overcome this problem and this is one of the advantages of high-pressure moulding machines.

Note that all the causes of shrinkage defects are early on in the solidification sequence, so that if feeding is needed at all, it is in the early stages of solidification. Thus grey cast iron is different from other alloys needing either no feeding or feeding in which most importance is during the first stages of solidification. In all other alloys it is most important to be certain that feeders will remain liquid until after the casting is solid, while with grey iron the early stages of solidification are more important.

The effect of the range of temperature on feeding characteristics

Unlike pure metals, alloys do not necessarily solidify at a single temperature. Depending on their composition they may undergo solidification at a single temperature or start to solidify at one temperature and continue solidification with falling temperature which is finally completed at a much lower temperature. The way in which this affects their solidification is illustrated in Figure 5.72.

The top portion of (*a*) is a graph in which the temperature of a solidifying casting has been plotted vertically against a horizontal plot of the distance from the mould walls. This gives a curve which shows the temperature is lower next to the mould which is conducting the heat away, and higher at the centre. There is thus a gradient within the solidifying metal; note that it is steepest where the metal is in contact with the mould

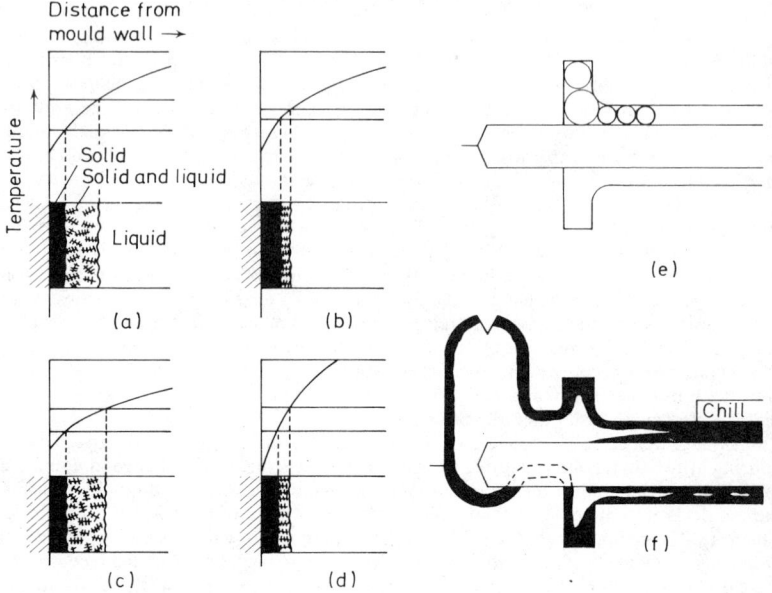

Figure 5.72. Effect of temperature on solidification of alloys. (*a*), (*b*), (*c*) *and* (*d*) *show how the interaction between the temperature gradient and range of solidification temperatures control the way in which solidification takes place.* (*e*) *shows the inscribed circle method of assessing the heat content of sections, while* (*f*) *illustrates a stage in the solidification of a casting being fed by a feeder enclosed in the mould. The black areas are those which have become solid*

wall. Superimposed on to this and on the same temperature scale are two horizontal lines, the top one indicating the temperature at which solidification commences while the bottom indicates where it is completed. If the intersections between the temperature gradient and these two lines are projected down to the bottom half of the diagram the various stages of solidification which have taken place may be illustrated.

The cross hatched area is the mould wall, the back area is completely solid metal, the area containing growing metal crystals in the form of fir-tree like formations called denrites is a mixture of solid and liquid and the clear area is all liquid. The interaction between this particular temperature gradient and solidification temperature range has produced a broad band of semi-solid metal. Thus, in the final stages of solidification, the metal becoming solid is isolated from completely liquid metal by metal which is undergoing solidification itself. There is, therefore, no chance of the liquid metal filling the cavities formed by the decrease in volume which takes place thus producing a casting containing minute shrinkage cavities throughout in the form of microscopic porosity.

In Figure 5.72(*b*) there is a very similar temperature gradient. Here the horizontal lines are closer together denoting that the alloy solidifies over a much narrower range of temperature. This has resulted in a narrow band of semi-solid metal so that, as solidification is completed, liquid metal is adjacent and thus the decrease in volume is continually compensated for until finally a single central shrinkage cavity is located in the last part of the casting to solidify, preferably the feeder head.

Since it is the interaction between the range of temperature over which the alloy solidifies and the temperature gradient, the range of temperature over which solidification takes place cannot be thought of in isolation. In Figure 5.72(c) and 5.72(d) this has been kept constant but Figure 5.72(c) shows a shallow gradient giving a broad band of semi-solid metal and dispersed shrinkage while Figure 5.72(d) has a steep gradient giving centralised shrinkage. It is thus impossible to say in what fashion an alloy will behave by knowing the temperature range over which it solidifies if the temperature gradient is not also known, in fact many alloys cannot be classified in one way or the other only tending in either direction depending on the temperature conditions.

The temperature gradient will be controlled by such things as the thermal properties of the mould, the casting temperature and the section thickness of the casting. Foundrymen know from experience what sort of behaviour to expect from various alloys under various conditions, and design their feeding systems accordingly.

Feeding skin forming alloys

The most straightforward castings to deal with are alloys in which the alloy tends to solidify over a short temperature range and the temperature gradient tends to be steep so that shrinkage is concentrated at the last heat centre to solidify. These castings require very careful feeding since, if it is not carried out properly, very obvious shrinkage defect will occur.

The alloys in such castings are often referred to as skin forming alloys since a skin of solid metal first forms round the outside surface of the casting and then thickens progressively towards the casting heat centre. The object is to promote progressive directional solidification such that thin sections remote from the feeders go solid first. These are fed from intermediate sections which, in turn, solidify and are fed from the heavy sections which are connected to the feeder heads which eventually contain all the shrinkage. Directional solidification can be aided by the runner system. If ingates are provided into, or next to, the feeder head the hottest metal enters here and cools as it flows through the casting. This has been done with the casting in Figure 5.71 and the feeders have been connected to the casting at the joint line in order to facilitate this.

Some foundrymen might prefer to place them on top of the casting to allow gravity to produce a higher pressure to force the feed metal into the casting. Note that, in Figure 5.71, open topped feeder heads are illustrated. These go right to the top of the mould so that the metal at the top of the head is not in contact with the mould and only looses heat by radiation. This top section should therefore remain liquid for a longer period and insulating or exothermic materials are sometimes added to make certain of this. In cases where the mould is too deep or where machine moulding makes it inconvenient for feeder heads to be open topped, heads which are completely enclosed in the mould are used.

With skin forming alloys, feeder heads of the type illustrated in Figure 5.72(f) may be used. In these a skin of metal first forms round the outside, but because the head has been designed with a sharp piece of moulding sand projecting down into its top, this skin is not continuous. This is because this piece of sand is surrounded by hot metal and so becomes heated in such a way that the metal does not solidify round it. As the solidification proceeds, the decrease in volume which is taking place within the envelope of solid metal creates a vacuum. Between the grains of sand, there are pores which makes the mould permeable to the passage of gases. Thus at the tip of the projecting piece of moulding sand atmospheric pressure is able to enter the feeder head and force metal into the casting.

It can now be seen why it is important to keep the metal at the top surface of an open topped feeder liquid, and, also, why the sharp corner at the flange junction should be provided with a fillet. At the bottom of Figure 5.72(*f*) this has not been done and because hot metal is on both sides of the sharp corner solidification has been slower here so that there is a danger of the atmospheric pressure puncturing the surface of the casting and initiating a shrinkage cavity. The way in which this can be overcome is shown in the top filleted part of the flange. For this reason sharp re-entrant angles should be avoided in casting design and it is sometimes preferable to form small slots and holes by machining after the casting is solid.

One or two large feeder heads are much more efficient than a large number of smaller ones. Therefore, it is best to promote the progressive directional solidification towards one or two centres. In order to do this, the natural heat centres of the casting should first be decided on and these may not always be too apparent from the drawing. A technique often employed is to inscribe circles into the drawing such that they just fill the section. For correct feeding these circles should be smallest furthest from the feeder and increase in size towards it. This has been done in Figure 5.72(*e*), note that the technique has emphasised the fact that the true heat centre is at the junction of the flange and the rest of the casting, rather than in the flange generally.

The parallel section which forms the wall of the pipe like part of the casting is long and thin, so that it might be uncertain that directional solidification within it could be achieved. In the bottom of Figure 5.72(*f*) it can be seen that the solid has bridged across this section and that isolated shrinkage cavities will result. This problem can be overcome by the use of a chill. This is a piece of metal which is placed in contact with the pattern during moulding and which remains embedded in the mould after the pattern has been withdrawn. It thus forms a small section of the mould which comes into contact with the cast metal which has higher thermal conductivity and therefore promotes more rapid solidification. Sometimes pieces of metal in the form of spirals of wire or horseshoe nails are allowed to project into the mould cavity itself so that the cast metal fuses on to them and goes solid more rapidly, these are called denseners. The chill in Figure 5.72(*f*) can be seen to have achieved the desired results. In Figure 5.7(*a*) the chill has been shown on the casting to give some idea of the sort of position in which it might be placed, in practice it would fall off the casting when it was knocked out of the mould.

Feeder heads are usually designed to stay liquid until after the casting is solid by making them larger than the sections they have to feed. There are now methods of calculating their size and position. In order to save metal, their efficiency is sometimes increased by surrounding them with insulating or exothermic sleeves. The insulating sleeves simply prevent heat from being conducted away while the exothermic ones ignite and give off heat. On very large casting feeder heads may be heated by topping up with hot metal or striking arcs on them with electrodes.

Alloys which give rise to dispersed porosity

The majority of castings, which are to be fed, are dealt with as described under skin forming alloys. If there is a tendency for dispersed porosity to occur, this is usually overcome by affecting the temperature gradients by control of the casting temperature, gating system and the use of chills.

There are a few alloys, however, in which it is impossible to promote directional solidification when they are cast into certain shapes. The most important of these are the tin bronzes. Even with these since the first metal to touch the cold mould wall goes solid very quickly, the outside shell of the casting solidifies with such a steep temperature gradient that it is free from shrinkage.

The central portion of the casting contains finely dispersed micro porosity, and since this cannot be isolated in a feeder head the foundryman tries to ensure that this is evenly distributed throughout the casting rather than being concentrated in areas of gross porosity. In this case, rather than attempting to promote directional solidification an even rate of solidification is encouraged throughout the casting.

Note that this would not be regarded as a defect and the properties of such alloys are quoted on the basis of their containing a certain amount of dispersed porosity. It should be remembered, however, that if excessive machining allowances are placed on castings in these alloys, the material with the best properties may all be machined off the outside of the casting. It is possible to increase the depth of this sound material by the use of chills, but these will usually tend to concentrate the porosity in some other part of the casting.

Design features associated with feeding

With the exception of the grey cast irons and, in a very few instances the long freezing range alloys, the foundryman will be trying to promote directional solidification so that the design engineer should bear this in mind. Ideally section thickness should progressively increase towards some convenient point which can be fed. Thick sections which are isolated in the centre of thinner sections will cause difficulties. Sharp corners and re-

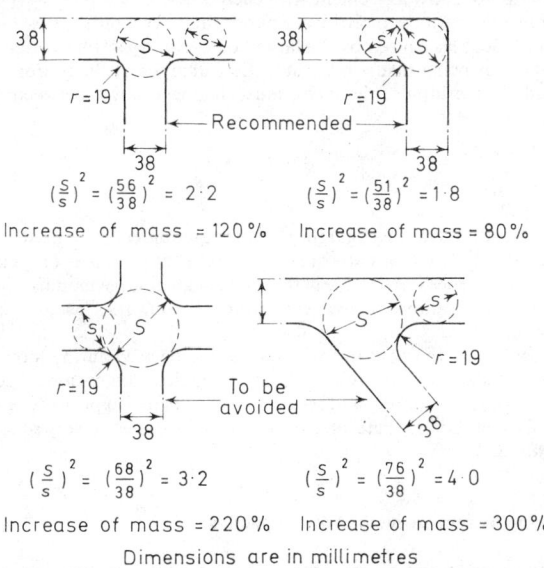

38
S
s
r = 19
Recommended
38
$(\frac{S}{s})^2 = (\frac{56}{38})^2 = 2{\cdot}2$
Increase of mass = 120%
38
s
S
r = 19
38
$(\frac{S}{s})^2 = (\frac{51}{38})^2 = 1{\cdot}8$
Increase of mass = 80%
s
S
r = 19
To be avoided
38
$(\frac{S}{s})^2 = (\frac{68}{38})^2 = 3{\cdot}2$
Increase of mass = 220%
S
s
r = 19
38
$(\frac{S}{s})^2 = (\frac{76}{38})^2 = 4{\cdot}0$
Increase of mass = 300%
Dimensions are in millimetres

Figure 5.73. Increase in mass at various types of junction

a
b
c
c
c
c

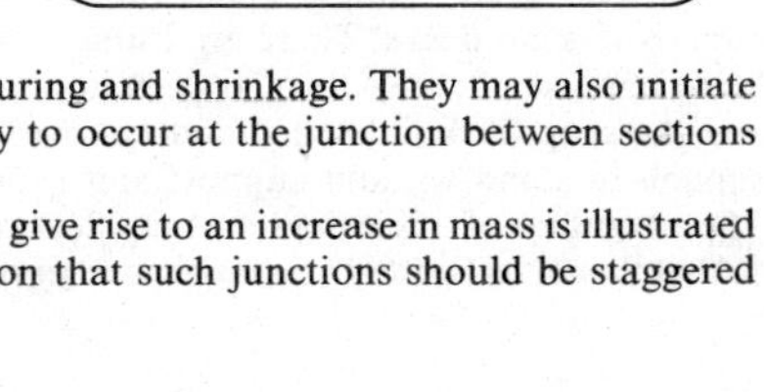

Figure 5.74. Staggering of ribs prevents the increase of section at junctions reducing shrinkage problems and also allows easier deformation to reduce to tendency for cracking

entrant angles may cause atmospheric puncturing and shrinkage. They may also initiate cracks during cooling, and this is most likely to occur at the junction between sections of different thicknesses.

The way in which various types of junction give rise to an increase in mass is illustrated in Figure 5.73. This leads on to the suggestion that such junctions should be staggered as is shown in Figure 5.74. The staggering not only prevents concentration of mass but also enables the casting to withstand stresses created during cooling without cracking.

Since it requires experience to know the way in which any particular alloy will behave in a given situation consultation between the designer and the foundryman at an early stage is most advisable.

CASTING PRODUCTION METHODS

It is extremely difficult to give guidance on the relative costs of production methods since these not only depend on the detailed design and quantity of the castings concerned, but also because competitive markets and technical developments give a constantly changing picture. In a similar fashion although it is desirable that some standard of expected dimensional tolerance be laid down for the various casting processes available, this has as yet not been done in the UK. The standards which have been laid down in other countries do not show agreement with each other.

A report from the Institute of British Foundrymen Technical Sub-Committee 71 [2], emphasises the difficulties caused by the large number of operating variables and shows that some progress is being made in the statistical analysis of them. For the time being both price and dimensional tolerances must be negotiated between supplier and purchaser.

Hand moulding

Hand moulding has the advantage that the skill of the moulder can be used to manufacture moulds from relatively low cost pattern equipment. The overheads on plant and machinery are low, and the process is flexible in that very small or large moulds can be produced and a large number of alloys catered for, although foundries usually specialise in a limited range of castings.

The disadvantages are the need for time and skill to be used during mould production and that, in some instances, it is difficult to maintain close dimensional accuracy. Hand cutting into the mould of running and feeding systems can also produce variable quality of castings. This method is, therefore, now only used when a very small number of castings is required.

Machine moulding by squeeze method of compaction

The simplest form of moulding machine is that in which compaction of the sand is carried out by squeezing. On all moulding machines the patterns must be mounted on plates, a double sided plate to produce the two halves of the mould or two separate plates, one producing copes (top halves) and one drags (bottom halves) being used. Patterns with loose pieces should be avoided, cores being used to overcome difficult projections. Extraction of the pattern is carried out by the machine withdrawing the plate in an absolutely vertical direction, so that pattern rapping is not needed, although in some cases a vibrator is used during the draw. This gives much better dimensional accuracy.

An example of a simple squeeze machine is that which is used for the production of moulds in snap flasks. These are frames which act as a moulding box while the sand is being compacted; once this has been done, the two halves of the mould are put together and the snap flask unlatched and removed from the mould. The mould is then strong enough to stand without support and to withstand the pressure of the molten metal. The advantage of this is that one snap flask can be used to produce any number of moulds so that the capital outlay on a large number of moulding boxes is not required. The method can only be used for the production of light castings and in fact simple squeeze machines in general are only suitable for light castings.

The fact that squeezing is limited to light castings is due to the limitation of compacting sand by squeezing with anything but very high pressures. Squeezing is carried out by placing the moulding box on top of the plated pattern filling it with moulding sand and forcing a ram down on the top surface. The disadvantage of this is the sand becomes very well compacted where it is in contact with this ram which is called the squeeze head but due to friction within the sand grains the squeeze pressure is not transmitted deeper down next to the pattern. This means that the working face tends to be soft.

This disadvantage has been combated by various methods with limited success. These include the use of squeeze heads which are contoured to fit round the pattern, and the so-called 'down sand-frame machine' which uses the pattern and pattern plate as a squeeze head which are pushed up into the box so that the mould is hardest on its working face.

The advantage of compacting by squeezing is that it is very rapid and so it is most useful for light castings, not requiring very hard moulds, which are produced in shallow moulding boxes.

Jolting

In order to compact sand by jolting; the pattern plate, moulding box and sand are lifted and allowed to drop through a short distance when the pattern plate is brought to a halt by striking an anvil. At this instant the moulding sand has gained a certain velocity in its fall and thus compacts itself against the pattern and pattern plate. The mould is thus hardest next to the pattern.

This method has the advantage that it can be used for very large moulds. Unfortunately it is not very good for very deep patterns with tall vertical sides, since the moulds tend to be soft at the top and vertical faces do not compact well.

Jolt squeeze pinlift and rollover machines

Probably the most common form of moulding machine set up being used in this country at the moment utilises machines which use both jolting and squeezing to compact the sand. The machines usually work in pairs, one machine making the top half of the mould, the other the bottom. They are fed with moulding sand from an overhead hopper and usually use mould box sizes which can be handled by hand or small hoists. The bottom half of the moulds go on to a short length of roller track where the cores can be placed in them and then the top halves closed on to them from where they pass on to a conveyer track. On this track the metal is poured into them and, while they cool, they are carried to the knock-out station. From here the moulding boxes are returned to the machines, sand for reconditioning and castings to the fettling shop.

The extraction of the pattern on such machines may be by lifting the compacted mould off the pattern plate with pins which locate on to the edges of the moulding box. These so-called pin lift machines are adequate for shallow patterns but with deep patterns the increased area offers greater resistance to stripping so that there is a danger of cracking the mould. This is overcome, in a roll-over machine, by a mechanism which inverts the pattern and mould so that the mould is stripped by lowering it from under the pattern. The weight of the mould is then self-supporting during the strip which reduces the stress put on it by a considerable amount.

Mechanised foundries using equipment of this type usually concentrate on medium sized castings up to about 150–200 kg. The equipment has a reasonably high rate of production combined with reasonable flexibility in case of pattern change so that both short and long run orders can be tackled. Running and feeding systems will normally be formed by the pattern equipment which both standardises the quality of the castings produced and increases the yield of weight of finished casting to weight of casting plus runners and feeding heads.

Sand slingers

Although some machines using jolting mechanisms are used on quite large moulds and a combination of jolting and squeezing are used in an attempt to get uniform compaction round deep patterns, there is still a problem of obtaining uniform hardness with very large moulds on jolt squeeze machines. One method of overcoming this is the use of the sand slingers in which sand is fed into the centre of rotating impeller blades. These blades throw the sand with great force into the moulding box building up a very hard evenly compacted mould.

The method takes a little longer than other machine moulding methods and tends to give rise to more wear and tear on equipment. However it is extremely useful for the production of larger castings.

Automated moulding lines producing high pressure moulds

Modern mechanised units can achieve very high production rates. Here core production, sand preparation, melting and moulding units must all be geared together and the utmost care with preventative maintenance must be taken since a breakdown of any one of these facilities will bring the whole plant to a halt. Automation may not only include mould making but also pouring. With such fully mechanised systems rates of 300 complete moulds per hour can be obtained from a moulding line incorporating a pair of moulding machines.

In order that more densely compacted moulds would yield sounder and more dimensionally accurate castings the pressures that moulding machines work at have increased in the last twenty-five years. In the late 1940s the pressure exerted on the mould by jolt squeeze machines was of the order of $175-280$ kN/m^2, in the 1950s $350-420$ kN/m^2 became common while the modern high pressure moulding machine exerts squeeze pressures of $700-1\ 400$ kN/m^2

Such machines need very large numbers of the same component to manufacture before they can operate economically, but when they have this they can produce a very high quality component at a low cost. This is not only because high quality pattern equipment and high pressure moulds are being employed, but also because the long production run enables the foundry to spend time in developing exactly the correct running and feeding system.

LOAM AND DRY SAND MOULDING

These two methods of mould production have been grouped together because they both require skilled moulders and tend to be used on large castings made in small quantities and are becoming used less often. They are also similar in that the main agent used to bond them is clay which is the case with all the other processes so far described

With loam moulding, the mould is made with the use of a strickle. The castings to be made must be, in the main, cylindrical in shape, and the mould is made by forming the wet loam with a profile which is rotated on a spindle. Although simple cylindrical shapes are easiest, by ingenious use of this method, quite complicated shapes can be made. Ships' propellers are cast in moulds made this way although not usually in loam sand. Once the loam mould has been struck up, and any cores formed by rotation on a spindle against a profile, the mould and core must be dried to remove excess moisture and to harden the mould material. The advantage is that the only pattern equipment needed is profiled strickle boards.

Dry sand moulding is similar to all other clay bonded sand moulding except that it is usually confined to moulds which are to have large castings made in them and so have to be hard to withstand the extra pressure of metal. Dry sand moulding used to be the only way to achieve this. A mould containing the normal moisture, which is called a green sand mould, is baked in a stove until it becomes hard. Since large moulds are being made, large stoves are required and the drying takes a long time so that expensive space, time and equipment are tied up. These disadvantages have led to the development of other methods of producing hard moulds, one of which is the high pressure moulding machine.

Sand moulds using bonds other than clay

Clay has the limitation that a considerable amount of compaction and in some cases stoving are needed if a hard mould is to be obtained. In order to avoid this, foundrymen have looked towards other bonding agents.

Sand cement

Sand which is hardened by the addition of cement and water is usually used on larger castings. The method is sometimes called the Randupson Process; the moulds take about twenty-four hours to harden. After use, the mould material can be crushed and used again. Cement has also been used as a bond in the fluid sand process which is mentioned later.

Shell moulding

This is often called the Croning Process and utilises a fine sand which has been coated with resin. This is dumped or blown on to a heated pattern plate. The heat from the plate penetrates a certain depth into the sand and causes it to stick together and to the pattern plate. The pattern plate is inverted so that excess sand drops off leaving behind the shell of heat affected sand. The pattern with the shell on it is placed in an oven until the shell hardens. The shell is then stripped from the pattern plate with ejector pins. Two halves of the shell mould are then stuck together to form a mould. Any cores which are needed can be made in a similar fashion in a hot core box.

The advantages of the process are that, since the mould is formed in contact with a high quality metal pattern very good dimensional tolerances are achieved. The fine moulding sand gives good surface finish, and the thin mould allows gases to escape rather than being entrapped in the castings. Disadvantages are that the high cost of the pattern equipment demands long run orders and that the resin sands are more expensive than clay bonded sands.

Mould processes using sodium silicate

If carbon dioxide gas is passed through a mixture of moulding sand and sodium silicate the mixture becomes hardened.

This so-called CO_2 process was first used for the production of cores, it had the advantage that the core could be hardened inside the core box instead of in the core stove. This overcame the danger of the core being distorted as it was being removed from the box and placed in the stove so that the accuracy of cores was much improved. It was soon realised that the process could be applied to the production of large moulds which would then not need so much compaction and no stoving at all.

CO_2 moulds are still used in considerable numbers; the very large ones are, however, difficult to pass the CO_2 gas through evenly and economically. This has led to the development of sodium silicate mixes which are hardened by adding agents to them just before they are used. These include ferrosilicon in the so-called Nishiyama Process and various forms of dicalcium silicate. These bonds are also used in the fluid sand process.

Self hardening resins

In the last few years an extraordinary variety of resins have been developed for the production of moulds and cores. Hot box resins produce cores in a hot core box in a similar fashion to shell moulding sands but are cheaper, more rapid and give off less gas.

Large moulds are made from furane binders which are mixed with catalysts in special mixers which then dispense the mixture into the moulding box where it self hardens to give moulds which are even more rigid than CO_2 moulds.

Fluid sand moulds

In this process, the moulding sand can be poured like a liquid into the moulding boxes. This is done at relatively low moisture contents by adding a foaming agent to the sand. The moisture that is present forms a foam in the mixer. The particles of moulding sand become suspended on this foam and the mixture acts just like a liquid, so that it can be poured round the pattern. It is then hardened by one of the self hardening processes already mentioned.

This process obviously does away with the need for any compaction and again has been applied to the manufacture of larger castings ingot moulds for the steel industry in particular.

Moulding methods using expendable patterns

Having to extract the pattern from the mould before the metal can be poured in, has two disadvantages. The first is that complicated shapes become difficult and the second is that there tends to be a loss in dimensional accuracy. The loss of dimensions is caused by the need for taper, by rapping of the pattern and by the need for the mould to be made in at least two parts with a joint. All these problems are overcome if instead of pulling the pattern out of the mould it is melted, burnt or dissolved out.

The most recent development of this method is in the use of foamed polystyrene and similar plastics. These vapourise at very low temperatures so that the pattern can be left in the mould and, when the metal is poured in, the heat of the metal drives the pattern from the mould in the form of vapour. An interesting form of this is called the 'full mould process' in which the pattern is surrounded by completely unbonded sand. The reason why this dried silica sand does not collapse is not fully understood but it is thought that it is because the mould cavity is always occupied either by the pattern or the metal and that as the metal enters the mould there is an interface between it and the pattern so that collapse cannot take place.

Investment precision casting uses the expendable pattern technique to produce castings of very close dimensional tolerance. To do this the patterns must also be precise, frozen mercury, plastics and wax patterns have been used, wax being by far the most popular.

The patterns are produced by injecting the wax into metal dies. These patterns are then mounted on wax running and feeding systems which are then called assemblies, these are then invested in ethyl silicate bonded moulds, from which the wax can be melted.

The moulds were originally solid and their preparation was rather lengthy, the time taken and the materials used have been much reduced with the introduction of the ceramic shell process. In this the assembly is dipped into a slurry of fine refractory in ethyl silicate so that it becomes coated in this material which then hardens. Further coats are applied until a shell which is strong enough to form a mould has been built up. The wax is then melted out and the mould fired, usually at 1000°C, to give it extra strength.

Pouring the metal while the mould is still hot ensures that the finest detail is reproduced. The process is most useful when machining is difficult due to intricate shape or very hard metals, and is used for alloys which melt at higher temperatures.

For aluminium and magnesium alloys with a relatively low melting temperature precision castings are made in plaster of paris moulds using both expendable and solid patterns. Stripping of the solid patterns is achieved by making them from flexible rubber or plastic.

The 'Shaw Process'

High melting temperature alloys are precision cast into moulds made from solid patterns by the use of the 'Shaw Process'. In this, finely divided refractory such as sillimanite is bonded with ethyl silicate. Hydrolysis produces a rubbery bond which enables the mould to be stripped from the pattern with little taper. The mould is then flash fired to drive off excess alcohol and then taken up to 800°C–1000°C and poured. The advantage being that precision castings can be made without the expense of producing wax injection dies.

The use of permanent metal moulds

PRESSURE DIE CASTING

Because of the high degree of accuracy required by the dies this is mainly limited to lower melting point alloys, alloys of aluminium and zinc being most popular. The metal is injected into the dies under high pressure at temperatures very close to its solidification temperature. The high speed of injection enables very thin sections to be cast while maintenance of the pressure during solidification enables the casting to obtain very fine detail and good surface finish.

The high velocity of metal entry does not enable all the air to escape from the die cavity so that although the skin of the castings are sound their interior usually contains entrapped air. Recently machines which apply vacuum to the die cavity in an attempt to overcome this problem have been developed.

GRAVITY DIE CASTING

When the metal is poured into the dies under gravity they tend to be more robust so that, as well as aluminium and magnesium, higher melting point alloys including copper base alloys and even grey irons can be cast. Completely sound castings can be obtained from gravity dies but because the only pressure being applied during solidification is that of gravity the dimensions are not held quite so close. Sand cores can be used for intricate internal cavities.

LOW PRESSURE DIE CASTING

This process tries to overcome the disadvantages of the other two die casting processes by filling the die at lower velocities than high pressure die casting. Also by maintaining some positive pressure on the metal during solidification so that the tendency for it to draw away from the die does not spoil dimensional accuracy.

The molten metal is held in a sealed sump underneath the die and is introduced into it through a cast iron tube under the pressure of compressed air. The process is at the moment limited to fairly simple shapes which can be filled through one or two openings.

Both low pressure die castings and gravity die castings have very good mechanical properties because the rapid solidification in metal moulds produces a very fine grain size in the castings. All die castings have excellent surface finish and good dimensional accuracy, the high cost of the die equipment needing to be spread over orders of large numbers of castings.

REFERENCES

1. *The design and properties of steel castings*, The Steel Castings Research and Trade Association, Sheffield
2. *First Report of Technical Sub-Committee TS71 — Dimensional tolerances in castings*, The British Foundryman, **LXII**, pt 5, 179 (May 1969)

BIBLIOGRAPHY

'Moulding materials the next 24 years', *The British Foundryman*, **LXIV**, pt II, 423 (Nov 1971)
Publications of the Council of Ironfoundry Association:
    A guide to the Engineering Properties of Iron Castings
    Iron Castings in Engineering Design
    Advantages of Producing Components as Iron Castings
    An Introduction to the Cast Irons

MECHANICAL AND THERMAL PROPERTIES OF HIGH-TEMPERATURE NON-METALLIC MATERIALS

R. G. COOKE

INTRODUCTION

Over the past two decades, a rapid growth in requirements of materials for use in hostile environments has occurred in many branches of technology. One of the most frequent necessities is for articles for use at very high temperatures, where common refractories and ordinary metals cannot be used. Such demands have led to a great improvement in our knowledge of the behaviour of high temperature materials and of the possibility of predicting performance by analysing the design of the article, although it must be emphasised that the design considerations must take into *detailed* account the widely different aspects of such materials from more traditional substances.

There are a number of general features common to these materials and the first part of this section will consider the theoretical approaches to property prediction and also the problems of devising satisfactory parameters from experimental data for use in design. The second part will contain data on selected materials and comment on the implications of the most recent work being carried out on them.

GENERAL THEORIES OF PROPERTIES

The most outstanding difference between high temperature non-metallic materials and metals, is the inherent brittleness of the former and it is this which imposes the greatest difficulty when designing to allow for them. In nearly all cases the stress strain relationship is linear and reversible until fracture, as shown in Figure 5.75, with no observable yielding. It is this form of non-ductile behaviour which requires very careful analysis if mechanical reliability is to be achieved.

To understand the mechanism of failure, the approach first followed by Griffith in his classic work on fracture in glass[1] is taken. His starting point was the well-known observation that the theoretical strength (i.e. the force required to physically separate the constituent molecules of a body) is several orders of magnitude greater than the measured strengths. It occurred to Griffith that the presence of flaws is inevitable in real materials and that these are the sources of weakness, because they act as strong sources of stress concentration, whilst the mean stress remains small. It is this concentrated stress which exceeds the molecular strength at failure.

In the analysis, a crack is regarded as the limiting case of an elliptical hole of major axis 2l with an average tensile stress σ_f perpendicular to the major axis (Figure 5.76). At failure, taking the molecular strength as σ_m, this will be equal to the concentrated stress so that:

$$\sigma_m = 2\sigma_f \left(\frac{1}{\rho}\right)^{\frac{1}{2}} \tag{1}$$

where ρ is the radius of curvature at the ends of the major axis and for a very thin crack this will be of the order of the intermolecular spacing a.

5–155

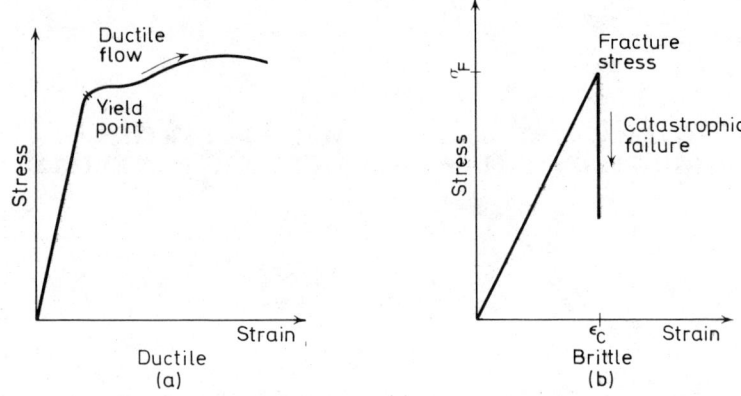

Figure 5.75. Form of stress-strain curves

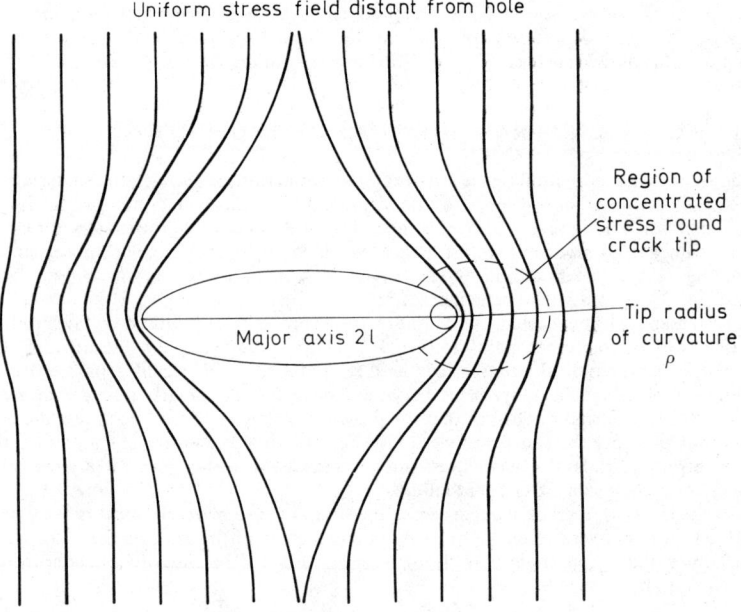

Figure 5.76. Stress concentration round elliptical hole (diagrammatic)

For the crack to propagate, the Griffith criterion must be satisfied, which is that the surface energy of the new surfaces (2γ per unit area) formed must be exceeded by the volume of strain energy released by the fracture, which is of the order of that stored in the volume of the crack, i.e. for unit area:

$$W = a\frac{\sigma_m^2}{2E} \tag{2}$$

so that

$$\sigma_m = 2\left(\frac{E\gamma}{a}\right)^{\frac{1}{2}} \tag{3}$$

hence with $\rho = a$

$$\sigma_f = \left(\frac{E\gamma}{1}\right)^{\frac{1}{2}} \tag{4}$$

From this we see that a critical crack length is required and cracks of this length in a body ('Griffith' microcracks) cause failure at the observed stress.

So far, this analysis applies to all cases of fracture but now the distinctive features of failure in ceramics and other brittle materials must be considered. Primarily, the most usual mode of resistance to crack propagation in metals, i.e. blunting of crack ends by plastic deformation, is not significantly useful in these materials in which propagation is by cleavage and grain boundary fracture. Again, once crack propagation starts, the elastic stored energy is generally so high that failure is catastrophic because insufficient energy is dissipated by surface formation. As the crack proceeds it rapidly gains kinetic energy and, in the case of certain whiskers, this process can even be explosive. It is thus of crucial importance when dealing with these materials to prevent any crack from *starting* to propagate.

Considerations of the type of body and its method of fabrication will often give some guide to the nature of the flaws which may be encountered in a given material and whether there is any possibility of circumventing the difficulties mentioned above, for example by crack blocking. Where these occur in specific materials will be commented upon in the second part of this section. However, from the design point of view an important general aspect of the flaws in an article is their statistical distribution and the effect of this on the prediction of its behaviour.

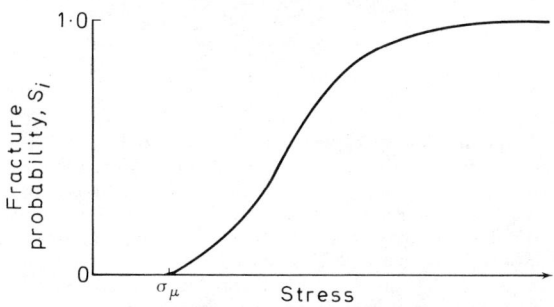

Figure 5.77. Probability of fracture

The most useful form of statistical analysis used for brittle materials is that due to Weibull.[2] His approach was from the point of view of probability theory and he assumed a random distribution of unspecified flaws. The probability of fracture for a sample of unit length and cross sectional area is taken to be shown in Figure 5.77. Let S_i be the probability of failure occurring before a stress level σ_i is reached. Then $1 - S_i$ will be the probability of the sample surviving. If two samples be stressed together, for the chance of both surviving, the product of probabilities is required. Hence the v unit volume elements in a volume V have a survival probability given by:

$$1 - S_{i,v} = (1 - S_i)(1 - S_i) \dots (1 - S_i) = (1 - S_i)^v = (1 - S_i)^V \tag{5}$$

or

$$\log(1 - S_{i,v}) = V \log(1 - S_i) \tag{6}$$

Weibull defined the 'risk of rupture' R as:

$$R = -\log(1 - S_{i,v}) \tag{7}$$

i.e.

$$R = -V \log(1 - S_i) \tag{8}$$

for an arbitrary element of volume

$$dR = -dV \log(1 - S_i) \tag{9}$$

i.e.

$$dR = f(\sigma_i) \, dV \tag{10}$$

and so

$$R = \int_V f(\sigma_i) \, dV \tag{11}$$

Weibull also proposed that

$$f(\sigma_i) = \left(\frac{\sigma - \sigma_u}{\sigma_o} \right)^m \tag{12}$$

where σ_u is the stress below which no failure is likely and σ_o is a material constant relating to the molecular strength. m is the homogeneity factor relating to the flaws and as $m \to \infty$ the material behaves as if it were flawless.

These formulae may be used to calculate mean strengths, deviations and so on but, much more importantly, they may be used to compare strengths obtained from different specimen geometries. As a simple example, consider two specimens of a given material broken in tension, having volumes V_1 and V_2. The stressing is uniform so that:

$$\left(\frac{\sigma_1 - \sigma_u}{\sigma_o} \right)^m \int_{V_1} dV_1 = \left(\frac{\sigma_2 - \sigma_u}{\sigma_o} \right)^m \int_{V_2} dV_2 \tag{13}$$

hence

$$\frac{\sigma_1 - \sigma_u}{\sigma_2 - \sigma_u} = \left(\frac{V_2}{V_1} \right)^{1/m} \tag{14}$$

This method of equating risk of ruptures for different configurations is obviously of considerable importance when considering the possibilities of failure of a material in service when data obtained from particular tests must be used for calculations, although some difficulty may occur in evaluating the volume integrals over complicated shapes. The Weibull method as a whole has the advantage of not requiring a detailed knowledge of the internal flaws themselves. However, its most serious disadvantage is the assumption that the most dangerous flaws are randomly distributed, whereas in many cases the most serious flaws are those at the surface of the body, particularly as this is the region most subject to chemical and mechanical attack.

Determining strength of materials

The most generally used method of determining strength is the three or four point loading bend test, shown in Figure 5.78, giving a value of the 'modulus of rupture'.

Weibull analysis shows that the calculated result will be of order twice the tensile strength, which is not itself determined directly, as it is in metals, because the deviations, due both to inherent properties and also experimental difficulty, are so large that the results are not useful. A variety of more complex disc and ring configuration tests have been used which all raise problems of loading and sample preparation. At least ten specimens must be tested to achieve any reasonable precision and, in general, design must be such that tensile loading in service is *well below* the range of strength values determined, particularly because of the previously noted catastrophic nature of brittle failure.

The strength of these materials defines the limit of loading in given situations, but it is also required to know their deformations under lesser loads and determinations of elasticity are required for this, either by dynamic (forced resonance) or static methods

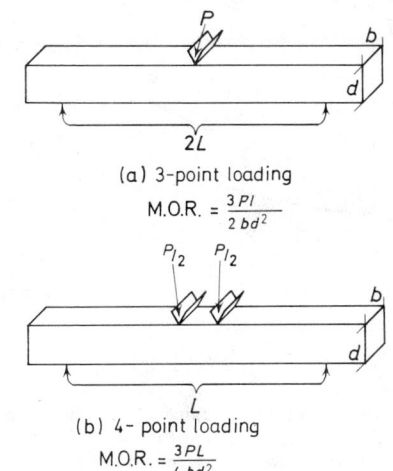

(a) 3-point loading

$$\text{M.O.R.} = \frac{3Pl}{2bd^2}$$

(b) 4- point loading

$$\text{M.O.R.} = \frac{3PL}{4bd^2}$$

Figure 5.78. Bend tests

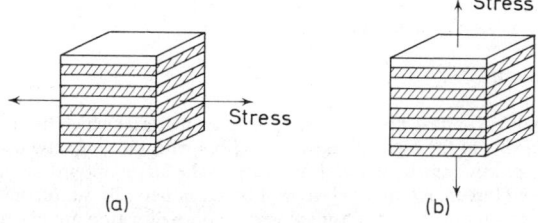

Figure 5.79. Model composites (a) Parallel, constant strain model; (b) Series, constant stress model

which usually agree very well. The strain to failure, i.e. the ratio of modulus of rupture to Young's modulus, which has been termed 'critical strain' by Astbury[3] is fairly constant, ~1 millistrain (0·1 %) for most ceramics and represents a deeper physical insight into material behaviour which may lead to improved performance.

The prediction of bulk properties (e.g. modulus, conductivity, dielectric constant and co) of multiphase materials is generally very complicated, but useful upper and lower bounds may be obtained from simple models. Thus the parallel and series models, shown in Figure 5.79, may be used to give upper and lower bounds of elasticity as

$$E_p = E_1 V_1 + E_2(1 - V_1) \tag{15}$$

$$E_S = \frac{1}{\dfrac{V_1}{E_1} + \dfrac{1 - V_1}{E_2}} \tag{16}$$

with obvious generalisations to multiphase materials. The predicted moduli follow the curves shown in Figure 5.80. The observed moduli fall inside these limits and closer

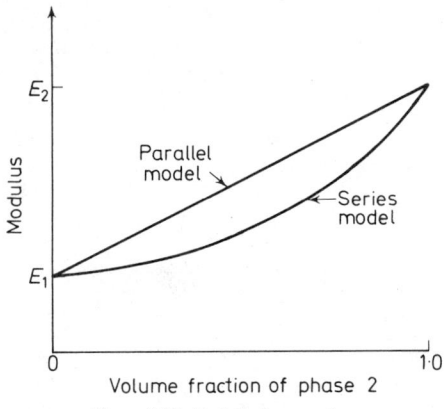

Figure 5.80. Moduli of composites

bounds (obtained for example by the calculus of variations) have been obtained by several authors. Several formulae have been proposed for the effect of porosity, of which the following has been found to give fair agreement with experiment for alumina and magnesia:

$$E = E_o e^{-bp} \tag{17}$$

where p is the volume pore fraction and b is an empirical constant.

Fibre reinforcement

A special form of two-phase body is fibre reinforced material which has the basic aim of utilising the benefits of the high strength of the fibre with the generally useful properties of the metal or plastic matrix material. Because of the large amount of work which has been done on reinforcement, no analysis will be given here. By simple mixture laws the strength and elasticity in the composite along the fibre direction are given by

$$\sigma = V_f \sigma_f + (1 - V_f)\sigma_m \tag{18}$$

where V_f, σ_f are the volume fraction and strength of the fibres and σ_m is the *yield* stress of the matrix, and

$$E = V_f E_f + (1 - V_f)E_m \tag{19}$$

Realisation of the potential of fibre reinforcement is, however, more difficult than simply embedding fibres in a matrix, for conditions of compatibility must obtain such that the fibres are not attacked (and so weakened) by the matrix, whilst the interfaces must be sufficiently well bonded to have useful force transmission, otherwise no benefit is gained. A further useful feature of composites with good bonding is that when the fibre breaks it is still partially adherent to the matrix and work must be done to slide the fibre out, hence usefully increasing the work of fracture. In most recent work it appears that this improvement in work of fracture may be more fruitful than the simpler idea of merely strengthening the body. One other aspect of fibre reinforcement of considerable importance in design is the alignment of the fibres along the direction of the major load, for because of their obvious asymmetry, their efficiency is much less across their orientation and it has also been shown clearly that random distributions of short fibres are very much less effective as reinforcing agents.

The prediction of the build-up of stress in a body and its subsequent failure is the realm of fracture mechanics which has been discussed in detail elsewhere (see bibliography). Here it suffices to point out that the very high stress concentrations caused by sharp edges and re-entrant angles can be particularly dangerous because of the lack of relieving ductility in ceramics.

Recommended reading

There is available a large amount of literature concerning the above topics. A list of books and papers will be found at the end of the section, but it is felt that it will be of value to the reader to give some further details.

Professor Gordon's book *The New Science of Strong Materials* presents a light-hearted, though nevertheless penetrating, view of the whole field and its development. Another book *Materials* contains a collection of articles by many leading scientists, covering all aspects of the discipline, which previously appeared in the *Scientific American Magazine*.

The 6th Volume in the 'Proceedings of the British Ceramics Society' has, in addition to numerous research papers, a useful group of papers on the interpretive problems of ceramics.

MATERIAL PROPERTIES

In this section a number of more commonly used special ceramic materials, and those with potential for development, will be described in terms of available knowledge of their properties, and some indication of problems of fabrication and usage will be given where relevant.

Alumina, Al_2O_3

Alumina is widely used in the special ceramics industry and the results of numerous investigations on it give good agreement where the impurity content is low.

Pure alumina crystallises in the hexagonal system, of theoretical density 3.98 g cm^{-3}. It melts at 2040°C and thermal expansion of 9×10^{-6}°C^{-1}. At room temperature the single crystal thermal conductivity is 43.0 W/m°C, dropping to 6.90 W/m°C by 1000°C (the polycrystalline material effectively follows this behaviour).

Single crystal alumina (sapphire) bend strengths of 0.62 GN/m^2 have been reported with a Young's modulus of 450 GN/m^2.[4, 5] Whiskers have been grown of strength tensile strength of order 0.1 E, which is the best known whisker strength so far, although this may be because of the intensity of the investigation of the material, rather than a distinct property.[6]

Pure, dense polycrystalline alumina can be made with strengths of 0.45 GN/m^2, but increase of grain size and porosity both reduce the strength very markedly. Some impurities have very marked strength reduction dependancies, for example as little as 0·2 wt. % V_2O_5 can halve the bend strength.[4] However, alumina of very high purity is readily available and with normal precautions, fabrication need introduce no significant impurity.

Alumina is extensively used in the electronics industry and can be prepared with high degree of control by a variety of methods. The variation of its properties has been very well documented and the references given here are to data books compiled from a selection of the very large amount of information available on the material. Other uses include refractory applications and optical devices (a translucent polycrystalline form, Lucalox, is available as well as transparent single crystals). Recently, some use has been made of the sintered material in armour, but it is not as efficient as boron carbide because of its lower density.

Beryllia, BeO

Beryllia crystallises in the hexagonal system with a theoretical density of 3·008 g cm^{-3} and a melting point of 2560°C. Its thermal expansion coefficient is $10·3 \times 10^{-6}$°C^{-1} and it has a thermal conductivity at room temperature of 242 W/m°C. This high value is one of the most useful features of beryllia and makes it most attractive for electronic applications where a high rate of heat transfer occurs, although it declines rapidly as the temperature is increased

Polycrystalline bend strengths of 0·276 GN/m^2, but with large variations due to purity, and Young's moduli 400 GN/m^2 have been reported.[4] Single crystal whiskers of strengths up to 0·048 E have been prepared.[6]

From an engineering point of view the most important difficulty in using beryllia is its extreme toxicity in powder form which implies that although it can be prepared by the conventional ceramic forming processes, highly specialised precautions are required to ensure that no health hazard to operators, or by pollution, occurs. This restricts its use to those applications where the benefits of its use outweigh the extra costs of preparing it.

Magnesia, MgO

Cubic magnesia has a theoretical density of 3·57 g cm^{-3} and melts at 3850°C and a thermal expansion of 15×10^{-6}°C^{-1}. Its thermal conductivity is 61 W/m°C at room temperature, showing a rapid decrease with temperature rise. Magnesia has been found to have a high vapour pressure at temperatures greater than 1500°C, such that losses due to evaporation may be of considerable importance. It is also susceptible to attack by water vapour in the atmosphere, although dense materials do not suffer from this to a large extent, and iron oxides have been found to act as inhibitors.

The mechanical properties of single crystal and polycrystalline magnesia have been studied in considerable detail. Bend strengths of polycrystalline magnesia have been found up to 0·30 GN/m^2 for hot pressed material, and Young's modulus of 390 GN/m^2 was found for the same material.[4] However, the most significant feature of magnesia is the presence of ductility in the single crystal form at room temperature. This is because the cubic structure of the material gives an adequate number of independent slip systems, which can lead to microcrack growth from dislocation movement, as opposed to brittle cleavage fracture usually encountered in ceramics.[8]

At low temperatures slip is in $\langle 1\bar{1}0 \rangle$ directions on $\{110\}$ planes and gives two independent slip systems, which is not sufficient for the von Mises criterion to be satisfied for polycrystalline magnesia and so it behaves·as a brittle solid. However, fairly easy slip occurs in the $\langle 110 \rangle$ directions over $\{100\}$ planes above 1500°C and the combination of the three independent slip systems of this set and the two of the above system give the five independent systems required to satisfy the von Mises criterion and so a brittle-ductile transition occurs. Other factors, such as grain boundary migration, occur which further contribute to the plastic flow of polycrystalline magnesias at high temperature. The cheapness of magnesia makes it potentially attractive, and it is used in combination with other materials and in, the iron oxide inhibited form, alone in refractories. The pure oxide has found uses in electronic and optical applications, especially the transparent polycrystalline form.

Zirconia, ZrO_2

Zirconia crystallises in the monoclinic form at room temperatures, but at about 1100°C it undergoes a phase change to the tetragonal form which is stable at higher temperatures. The volume change involved, $\sim 9\%$, is sufficient to shatter any bodies prepared from zirconia when cooled from the sintering temperature, and so the pure oxide cannot be used. However, it readily forms cubic solid solutions with alkaline-earth and rare-earth oxides of which the commonest used are MgO, CaO and Y_2O_3.[9] Such solid solutions do not undergo any phase changes before their melting points (~ 2600°C) and are therefore called 'stabilised zirconia' although it should be borne in mind that this does not necessarily mean chemically stable. The properties given below refer to dense calcia stabilised zirconia (CSZ), but it should here be pointed out that in some applications, much more desirable properties may be obtained by using mixtures of stabilised and unstabilised phases, at the cost of a reduction in strength.[10]

The thermal conductivity, 0·18 W/m°C, is extremely low for ceramic material, and the thermal expansion of CSZ is 12×10^{-6}°C^{-1} (note that the thermal expansion of unstabilised and partially stabilised zirconias are nonlinear because of the phase change).

Strength appears to depend on the quantity of stabiliser added, and with 10% CaO is of the order of 0·240 GN/m^2 and elasticity 156 GN/m^2, but these are very much representative values.[4]

Zirconia has considerable attractiveness as a refractory because of its low reactivity with metals. It is also of interest in applications where its electrical conductivity, which becomes significant above 1000°C, may be used, for example in high temperature fuel cells.

Other Oxides

The oxides of calcium, barium, strontium and the rare-earths can be fabricated by normal ceramic techniques and yield bodies having properties similar to the other oxides discussed above.[4] As the most important uses are often nuclear (e.g. gadolinia as a burnable poison) stringent requirements on purity must often be met but these do not cause any technical problems in principle, although the expense may be high. Many of these oxides are, however, susceptible to attack by atmospheric water vapour.

The radioactive oxides thoria, urania and plutonia are often used as sources of nuclear fuels because of the higher reactor temperature they allow. Again, they behave as typical ceramics[4] and may be made in the usual ways, provided precautions against radiation hazards are taken. The high melting point of thoria (3200°C) and its low radioactivity have made it of interest for refractory applications as well.

Recently, complex oxides have been used as glass-ceramics, in which forming is done in the glassy state and then controlled heat treatment is used to obtain crystallisation, and hence the production of a dense ceramic body.[11]

Non-oxide materials

The methods of preparing oxides, by shaping and sintering, are not available for the non-oxides, none of which sinter readily when heated and which are also susceptible to oxidation during heating in air. Outlined below are the three important fabrication methods which may be used to prepare non-oxides from appropriate raw materials. The properties of the specific materials will be detailed after this.

The first technique which may be used is hot pressing in which the powdered compound, possibly with additives to promote densification, is compressed in a carbon die which is inductively heated. Very high densities may be obtained and the atmosphere can be controlled, but obvious difficulties in die preparation mean that the shapes available are very restricted. Isostatic hot pressing, in which the powder, in a thin walled metal container, is compacted by hot argon gas at high pressure, offers some greater benefit, at much higher cost.

Secondly, reactive methods may be used, in which a solid material (e.g. in the case of silicon carbide, carbon) is shaped in a conventional way and then reacted at high temperature with a penetrating component (in the above case, with molten silicon; similarly nitrogen gas is allowed to penetrate silicon in the preparation of silicon nitride by reaction sintering). This method has the advantage of ease of preparation of complex shapes, but some porosity must be initially present to allow the fluid component to penetrate the whole sample and react, particularly as thermal expansion mismatch between the starting compound and final material may cause very severe internal stresses. The time of penetration of the fluid can also be quite long.

Finally, non-oxides may be obtained by pyrolytic deposition. In this method, appropriate gaseous compounds, often dispersed in a carrier gas, are passed over a heated substrate, where they decompose and deposit a layer of the simpler materials on the substrate. This method can yield very dense, high purity material but very careful control of the system is required, for very minor alterations of flow of gas and temperature can alter the deposition rate and form of crystal grown and also lead to difficulties of non-stoichiometry, all of which may result in high internal stresses.

These methods yield bodies of controlled purity and their relative importance will depend on the particular applications. Methods yielding less pure materials (e.g. clay bonding) will not be discussed in this section.

Silicon carbide, SiC

Cubic silicon carbide has a density of $3 \cdot 21$ g cm^{-3}. It dissociates at atmospheric pressure although rapid heating has been reported to give a melting point of 2600°C.

Self-bonded silicon carbide has a thermal expansion of $4 \cdot 4 \times 10^{-6}$°C^{-1} and a thermal conductivity of 173 W/m°C, although a very pure crystal has been found to be almost three times as conducting.

Very wide ranges of strengths have been found for various forms of single crystal silicon carbides, especially with boron doping. Polycrystals with strengths up to $0 \cdot 410$ GN/m^2 and moduli of 650 GN/m^2 and fibres with strengths of $0 \cdot 024$ E have been reported.[4, 6]

Silicon carbide has generally good resistance to oxidation because of the formation of a protective layer of silica on it, but is susceptible to attack by carbon monoxide, halogens and nitrogen above ~ 1000°C. It finds use as heating elements because of its electrical conductivity, and as an abrasive because of its hardness. More recently, pyrolytic coatings on fuel spheres in dispersed fuel elements have found use in nuclear applications.

Silicon nitride, Si$_3$N$_4$

Interest in silicon nitride has recently grown because of the possibility of using the material in gas turbine applications. It may be made in any of the ways considered above, although for hot pressing some MgO must be added, with the resulting transformation from α to β silicon nitride. The recent work by Jack et al[12] has done much to elucidate the mechanisms of this process, and has demonstrated that the useful α-silicon nitride is in fact an oxynitride. The theoretical density is $3 \cdot 2 \text{ g cm}^{-3}$.

Silicon nitride sublimes above 1900°C. The thermal conductivity and expansion of reaction sintered material of ~85% theoretical density have been given as $9 \cdot 5 \text{ W/m°C}$ and $2 \cdot 5 \times 10^{-6} °\text{C}^{-1}$.[4] Somewhat different values obtain for hot pressed material and the same is true for the mechanical properties given below.

Bend strengths of $0 \cdot 300 \text{ GN/m}^2$ have been obtained, but hot pressed material has been developed with strengths above $1 \cdot 00 \text{ GN/m}^2$. The Young's modulus, of 270–290 GN/m^2, is low for ceramics, implying that silicon nitride has a high critical strain. Its good high temperature strength properties have made it the subject of intensive investigations as a possible successor to the present generation of high temperature alloys.

Boron nitride, BN

Hexagonal boron nitride has a density of $2 \cdot 27 \text{ g cm}^{-3}$ and decomposes above 2700°C at atmospheric nitrogen pressure (but much lower in vacuo). Its structure is closely similar to that of graphite, and it shows the same anisotropic features, but it is a very useful electrical insulator. This anisotropy is of special importance in the pyrolytically deposited form. The cubic form, borazole, analagous to diamond, needs special conditions of high temperature and pressure to be formed, and is the hardest known compound.

Hot pressing gives somewhat aligned material, with a thermal conductivity perpendicular to the pressing axis of $34 \cdot 6 \text{ W/m°C}$ and half the value across this axis. The thermal expansion is $\sim 1 \cdot 0 \times 10^{-6} °\text{C}^{-1}$ perpendicular to the axis and effectively 0 across it, whereas the pyrolytic material has an expansion of $3 \cdot 6 \times 10^{-6} °\text{C}^{-1}$ along the 'c' axis, and again 0 along the 'a' axis.[4] Strength and modulus both vary if taken along the pressing axis, with strength of $0 \cdot 117 \text{ GN/m}^2$ and modulus 97 GN/m^2, and across the pressing axis, strength of $0 \cdot 062$ GN/m^2 and modulus 35 GN/m^2.[4]

Boron nitride has higher oxidation resistance to graphite and is of interest both for its similar properties to graphite and for components in nuclear engineering because of its neutron cross-section of the boron in it. The pyrolytic material has better resistance to oxidation than pyrolytic graphite, but is very slow to deposit.

Other non-oxide materials

Many metal carbides and nitrides occur as precipitated phases in alloys, but their performance is properly considered in metallurgical aspects, rather than ceramic. The hard nature of tungsten carbide is well known in its application for machine tools. Tantalum carbide is of interest as having the highest known melting point, ~4000°C. Boron carbide has found use (together with silicon carbide, silicon nitride and some other non-oxides) as a promising lightweight armour material, an application which exploits the inherent brittleness of ceramics by using the kinetic energy of a projectile to shatter the armour material and thus protect the user.[7]

Sulphides, borides and silicides may be prepared in the same way as ceramics and have similar mechanical properties but do not, as yet, find any commercial application as components. Elemental boron and silicon are also similar, but their mechanical properties do not usually have the importance their electrical properties warrant. Graphite is sometimes considered as a ceramic, but its properties are sufficiently unique and important to have been considered in detail elsewhere.[12]

Conclusions

In the present state of technology, brittle materials are beginning to find more and more applications as their disadvantages become outweighed by the attractiveness of their various other properties. Not surprisingly, the main impetus for improvements in knowledge of them has come from those branches of engineering dealing with the most hostile environments, such as aerospace and nuclear.

In this section, the general aspects of design with ceramics and the properties of a number of particular materials have been outlined and it should be emphasised that the essence of design in ceramics is to use them so that the attractive properties required are brought out and the difficulties with their use minimised. Thus at all times notice must be taken of their brittleness, but good design will also take into account the problems of fabrication and particularly machining such hard materials so that the simplest possible shapes should be used and load bearing minimised. Again, by using novel methods of fabrication and engineering, a ceramic component may be incorporated in an article such that only the property required is used (thus, for example, much weaker foamed materials may be used as thermal insulation, provided all load is carried by some other component).

With care in design and further development, ceramic materials offer a new higher temperature range for engineering, which will probably represent a final limit to which solid components may be used. The improvements of our understanding of these materials, and their consequent greater availability, are making them an attractive and economic alternative to metals in newer technology.

REFERENCES

1. GRIFFITH, A. A., *Phil. Trans. Roy. Soc. A*, **104**, 538 (1920)
2. WEIBULL, W., *Inger. Vetensk. Akad. (Stokholm), Proc.*, No. 151 (1939)
3. ASTBURY, N. F., *Advances in materials research in the Nato countries*, Pergamon Press (1961)
4. LYNCH, J. K., RINDERER, C. G. and DUCKWORTH, W. H., *Engineering properties of selected ceramic materials*, American Ceramic Society (1966)
5. POPPER, P., *Use of ceramics in valves*, B. Ceram. R.A. Spec. Pub. 48 (1965)
6. MACREIGHT, L. R., RANCH, H. W. and SUTTON, W. H., *Ceramic and graphite fibres and whiskers*, Academic Press (1965)
7. WILKINS, M., *An approach to the study of light armour*, URCI 50284 (1967)
8. LANGDON, T. G. and PRICE, J. A. (Ed. Alper, A. M.), *High temperature oxides*, pt III, 53–128, Academic Press (1970)
9. WEBER, B. C., *An annotated bibliography on zirconia*, Aerospace Res. Labs., WPAFB, USAF, ARL64–205 (1964)
10. GARVIE, R. C. (Ed. Alper, A. M.), *High temperature oxides*, pt II, 117–166, Academic Press (1970)
11. MCMILLAN, P. W., *Glass-ceramics*, Academic Press (1964)
12. JACK, K. H., GRIEVESON, P. and WILDE, S. (Ed. Popper, P.), *Special ceramics 5*, B. Ceram. R.A. (June 1972)
13. MANTELL, C. L., *Carbon and graphite handbook*, Interscience (1968)

BIBLIOGRAPHY

GORDON, J. E., *The New Science of Strong Materials*, Penguin Books (Pelican Series) (1968)
Materials, A Scientific American book, W. H. Freeman & Co. (1967)
KINGERY, W. D., *Textbook of Ceramics*, J. Wiley & Sons (1960)
HOLLIDAY, L. (ed), *Composite Materials*, Elsevier (1966)
Advances in Materials Research in Nato Nations, Pergamon (1963)
'Mechanical Properties of Non-Metallic Crystals and Polycrystals', *Proc. Brit. Ceramic Soc.* (6) (June 1966)
'Fracture Toughness Testing', ASTM Special Publication 381 (1964)

REFRACTORIES

D. HERRELL and F. T. PALIN

DEFINITION AND CLASSIFICATION

Broadly speaking, a refractory (also known as a refractory material or a refractory product) is a material that will withstand high temperatures sufficient to permit its use in a furnace lining or other location where it will be exposed to severe heating. The official definition states that a refractory is a non-metallic material or product (but not excluding those containing a proportion of metal) having a pyrometric cone equivalent corresponding to not less than 1500°C. (See later for definition of pyrometric cone equivalent.)

The classifications have been set out by the ISO and refractory materials have been subdivided into three main groups.

Group 1. Dense shaped refractory products

Products classified in this group may be supplied in either the fired or unfired state, and magnesite and dolomite can have additions of tar.

High-alumina, Group I	$Al_2O_3 \geqslant 56\%$
High-alumina, Group II	$45\% \leqslant Al_2O_3 < 56\%$
Fireclay	$30\% \leqslant Al_2O_3 < 45\%$
Low-alumina fireclay	$10\% \leqslant Al_2O_3 < 30\%$
	$SiO_2 < 85\%$
Siliceous	$85\% \leqslant SiO_2 < 93\%$
Silica	$SiO_2 \geqslant 93\%$
Magnesite	$MgO \geqslant 80\%$
Magnesite-chrome	$55\% \leqslant MgO < 80\%$
Chrome-magnesite	$25\% \leqslant MgO < 55\%$
Chrome	$Cr_2O_3 \geqslant 25\%$
	$MgO < 25\%$
Forsterite	
Dolomite	

Special products based on carbon; graphite; zircon; zirconia; silicon carbide; etc.

Group 2. Shaped refractory insulating materials

An insulating material must have a low thermal conductivity and a low thermal capacity, both of which are directly related to the true porosity. By definition, the true porosity of an insulating material must be not less than 45%. Generally the materials in this group may be subdivided in a similar manner to Group 1. The high alumina, fireclay and low alumina fireclay insulating refractories are further subdivided according to the following criteria:

> Temperature at which the permanent linear change does not exceed 2%.
> The limiting values of bulk density. If this is lower than the limits given below, the material belongs to a class designated L.

(See later for definition of permanent linear change and bulk density.)

Note that although the material from which the brick is manufactured has a pyrometric cone equivalent corresponding to 1500°C, the brick itself usually cannot be held at such a temperature without severe slumping.

Designation	Temperature at which the permanent linear change does not exceed 2%	Upper limits of bulk density for class L
110	1 100	65
125	1 250	75
140	1 400	85
150	1 500	95
160	1 600	110
170	1 700	130

Group 3. Prepared but unshaped refractory materials

By definition, unshaped refractory materials are prepared mixtures for use either as delivered or after the addition of an appropriate liquid. This definition covers the refractory cements, mouldable and castable materials, and ramming and gunning mixes. These materials may be further classified as follows:

(a) On the basis of utilisation; e.g. jointing materials, coatings and washes, materials used for monolithic construction or repairs.

(b) On the basis of chemical and mineralogical nature of the major constituents. This is based on the classification given for Group 1 after the separation of the major constituent from the bond.

(c) According to the hardening process; e.g. ceramic binder, hydraulic binder, mineral or organic/mineral binder or organic binder.

Norton gives a classification generally accepted in the United States. This is similar to the ISO classification but is slightly more detailed with respect to high alumina and firebrick refractories.

COMPOSITION AND CONSTITUTION

The composition and constitution of dense refractories found in Group 1 are described below. It is more convenient to discuss the composition of the materials in the other two groups in conjunction with their properties and uses (see later).

High alumina products, Group I ($Al_2O_3 \geqslant 56\%$)

The principal raw materials available for refractories manufacture in this group are bauxite and the sillimanite group. These minerals consist mainly of alumina and silica. Firing bauxite results in the formation of corundum. The sillimanite group consists of sillimanite, kyanite and andalusite. All three minerals have the same empirical formula Al_2SiO_5 and theoretically contain 63% of Al_2O_3. Firing this group results in the formation of mullite and silica. No workable deposits occur in this country and material has to be imported from South Africa and India.

Bauxite is used in this country to manufacture refractories with an alumina content in the range 80-86%. Even though a content of 60-70% would be adequate for a large number of applications, the high cost of the sillimanite group of materials has precluded their use.

Serious firing problems can be encountered when attempting to mix bauxite and fireclay to produce materials with an alumina content in the range 70-85%, some of which may be overcome by the use of phosphoric acid or phosphates to bond the materials together and so dispense with firing the product.

Mullite is a material not commonly found in the natural state. It has the formula $3Al_2O_3 . 2SiO_2$ and contains 72% alumina. Synthetic mullite is manufactured in this country by mixing the requisite amounts of bauxite and fireclay, and either fusing or sintering. Bricks are then manufactured from the fused or sintered grain.

High alumina products, Group II ($45\% \leqslant Al_2O_3 < 56\%$)

The principal raw material available for refractories manufacture in this group is flint clay. This contains 43-57% alumina, consisting of kaolinite with varying amounts of diaspore. South Africa provides most of the material imported into this country at present but it is probable that the USA will provide increasing quantities in the future. Firing this material results in the formation of mullite and silica.

Fireclay products and low-alumina fireclay products ($10\% \leqslant Al_2O_3 < 45\%$)

Fireclays are associated with coal measures. They consist mainly of kaolinite associated with impurities, such as quartz, iron minerals and mica. During firing the kaolinite breaks down to form mullite and silica, and the impurities determine the amounts of glass present in the finished article.

Silica products ($SiO_2 \geqslant 93\%$)

The most commonly used material for silica refractories is quartz rock known as ganister. Deposits are mined in this country where it is found associated with coal seams, some

		% volume expansion
Quartz ↓ 870°C		
Tridymite ⇅ 1470°C	Quartz → tridymite	17
Cristobalite ↓ 1625°C	Quartz → cristobalite	13
Silica glass	Quartz → silica glass	20

material also being imported from Sweden and South Africa. Quartzite consists of approximately 98% silica. Heating quartz results in the conversions to tridymite and cristobalite, but all three forms are relatively stable at room temperature.

The problems with fired silica bricks arise from the large volume expansions associated with the inversions of the three minerals. These are reversible and great care must be exercised when heating these bricks in the low temperature regions.

	Volume increase %
α Quartz $\overset{573^\circ C}{\rightleftharpoons}$ β Quartz	0·86
α Cristobalite $\overset{220^{\circ\circ}C-260^\circ C}{\rightleftharpoons}$ β Cristobalite	2·8
α Tridymite $\overset{117^\circ C}{\rightleftharpoons}$ β₁ Tridymite $\overset{163^\circ C}{\rightleftharpoons}$ β₂ Tridymite	0·14 (117°C)
	0·2 (163°C)

These inversions occur very rapidly in contrast to the slow conversion of one principal form of silica to another.

Magnesite products (MgO ≥ 80%)

The name 'magnesite' refers to magnesium carbonate which occurs naturally in Austria, Greece and Russia, etc. This material is calcined or dead burned to produce magnesium oxide which is used to manufacture bricks, and these are still termed magnesite. No deposits occur in this country although magnesium oxide is extracted from sea-water using, as one constituent, the double carbonate dolomite $MgCO_3 \cdot CaCO_3$ (see later section), deposits of which are mined in this country. This extraction may be represented by the following equations:

$$CaMg(CO_3)_2 \xrightarrow{\text{heated}} CaO + MgO \xrightarrow{H_2O} Ca(OH)_2 + Mg(OH)_2$$
$$\text{Slaked Dolomite}$$

$$\begin{bmatrix} Mg(OH)_2 \\ Ca(OH)_2 \end{bmatrix} + \begin{bmatrix} MgCl_2 \\ MgSO_4 \end{bmatrix} \longrightarrow 4\,Mg(OH)_2 + \begin{bmatrix} CaCl_2 \\ CaSO_4 \end{bmatrix}$$

Slaked   Sea-water   Dissolved in
Dolomite   sea-water

The magnesium hydroxide is calcined or dead burned, graded and pressed into bricks, which after firing consist of magnesite grains and a silicate phase.

Bricks may be used in either the unfired or fired state. In the former phosphoric acid, chromic acid, magnesium sulphate or tar may be used to bond the magnesite grain, and the bricks are fired *in situ* during use. Fired materials may also be impregnated with tar by allowing liquid tar or pitch to penetrate into the pore structure of a warm fired brick held under vacuum. (See also Dolomite.)

Magnesite-chrome and chrome-magnesite (25% ≤ MgO < 80%)

These materials are produced by mixing the requisite amounts of magnesite and chrome ore. This ore consists of a complex solid solution of spinels containing chromium, iron, aluminium, magnesium and oxygen which are associated with impurities or gangue usually consisting of serpentine, feldspars, silica and carbonates. The high firing temperatures used today result in the production of the 'direct-bonded brick' in which the bond is formed directly between the magnesia and chromite grains.

The silicate phase present is usually forsterite. Magnesia and chrome ore may be fused together to form a liquid pool which is cast into moulds. Blocks are then cut from these using a diamond saw. This is known as 'fusion-cast' magnesite-chrome.

Forsterite products (2 MgO . SiO_2)

These refractories are usually made from natural olivines or mixtures of serpentine and magnesia. The refractoriness of the commercial product is much less than that of the pure mineral. These refractories have a lower resistance to slag than the chrome-magnesite products.

Dolomite products

As mentioned above, dolomite is the double carbonate of magnesium and calcium and it is mined in this country. After calcination or dead burning the material consists of a mixture of magnesia and lime with various impurities.

A difficulty with dolomite is that it is subject to hydration and all products need to be protected by some means, usually by coating with tar. A large tonnage of graded grain, with tar, is used as a fettling or ramming material. Alternatively, bricks are produced by bonding the graded material with tar in a similar manner to that described for the magnesite material. It is possible to obtain heat treated tar bonded dolomite which is known as tempered, stoved or toughened, the object being to avoid slumping during the initial heating up. It became apparent that the presence of tar conferred improved properties, and this lead to the use of tar as an impregnating material in fired dolomite and also in magnesite, in which it was not required as a protection against hydration.

Special products

Carbon and graphite. Carbon is usually obtained by coking coal, the high purity material being made from petroleum coke. Graphite occurs naturally in many places in both amorphous and crystalline form. It is an extremely good conductor of both heat and electricity but tends to oxidise at high temperature. Bricks are manufactured by mixing the graded material with tar, pressing and firing under reducing conditions.

Zircon ($ZrO_2 . SiO_2$). This material has a theoretical composition of 67% ZrO_2 and 33% SiO_2 and is found in Brazil, Australia and India. It has a high resistance to slags but when heated for long periods above 1550°C it tends to dissociate into zirconia and silica glass.

Zirconia (ZrO_2). This is commonly found as baddeleyite, deposits of which occur in Brazil. It has a very high melting point (about 2700°C) but like silica, it undergoes an inversion accompanied by a large volume change at 1000°C. The material may be stabilized by the addition of, for example, lime.

Silicon carbide (SiC). This is produced synthetically by heating silica sand and coke. It is a very hard refractory material. When manufactured into shaped articles it is either bonded with clay, bonded with silicon nitride or self-bonded.

Silicon nitride (Si_3N_4). This is prepared synthetically by heating silicon powder in nitrogen or ammonia under pressure. It is a hard material with a good oxidation resistance.

TESTING AND GENERAL PROPERTIES

The testing of a refractory material is often done for one or more of the following reasons:

The assessment of the suitability and uniformity of a raw material.
The control of manufacturing processes and the quality of the products.
To determine within reasonable limits the suitability of a particular product to a specific application.
To supply information for, or to check specifications.
To provide information on the uniformity of wear and to predict campaign lives of long term usage in for example a blast furnace lining.
To examine the effects of service conditions on the refractory lining on completion of particular campaigns.

This data derived from testing must be used with care since it can often be misleading and, in any specific application, certain properties are of prime importance.

The testing methods normally employed in the UK are laid down by the BSI although specifications are often demanded which are based on German (DIN) or American (ASTM) standards. Increased co-operation with European countries could in the future result in a unified testing structure. The normal testing procedures of each of these bodies applied to the different types of refractory materials are shown below.

	Test procedure		
	British	*American*	*German*
Group 1	1902 Pt. 1A, 1B, 2A, 2B, 2C, 2D, 2E		
Group 2	1902 Pt. 1A, 2A, 2B, 2C, 2D, 2E	ASTM Pt. 13	
	2973		DIN Normblatt
Group 3	1902 Pt. 1C, 2A, 2B, 2C, 2D, 2E		Verzeichnis
	915 1881		1971 501 502; 503

These testing procedures give values for the following properties.

Bulk volume. The volume of the solid material plus the volume of the sealed and open pores present.

Apparent porosity. The ratio of the volume of the open pores to the bulk volume expressed as a percentage.

Closed or sealed porosity. The ratio of the volume of the sealed pores to the bulk volume, expressed as a percentage.

True porosity. The ratio of the total volume of the open and sealed pores to the bulk volume, expressed as a percentage.

Apparent solid density. Ratio of the mass of the material to its apparent solid volume.

Bulk density. Ratio of the mass of the material to its bulk volume.

True density or powder density. Ratio of the mass of the material to its true volume.

Apparent specific gravity. For a porous ceramic, the ratio of the mass to the mass of a quantity of water that, at 4°C, has a volume equal to the apparent solid volume of the material at the temperature of measurement.

Permeability. The rate of flow of a fluid (usually air) through a porous ceramic material per unit area and unit pressure gradient.

Pyrometric cone equivalent. A measure of refractoriness: the identification number of the standard pyrometric cone that bends at the temperature nearest to that at which the test cone bends under the standardised conditions of test.

Refractoriness. The ability of a material to withstand high temperatures: it is evaluated in terms of pyrometric cone equivalent.

Refractoriness under load. The ability of a material to withstand specified conditions of load, temperature and time.

Permanent linear change. This property is often referred to as after contraction or expansion and is the permanent contraction or expansion (usually expressed as a linear percentage) which may occur if a refractory product is fired or refired under specified conditions of test.

Reversible thermal expansion. The increase in dimensions of a material when it is heated which is reversible on cooling. This is normally quoted either as a percentage or as a coefficient over a stated temperature range.

Crushing strength. The maximum load per unit area, applied at a specified rate, that a material will withstand before it fails.

Resistance to thermal shock (spalling). The stresses set up by the differential expansion or contraction between the outside and inside of a thermally shocked ceramic, i.e. one which has been subjected to sudden heating and cooling may cause it to crack or spall.

Thermal spalling. Spalling caused by stresses resulting from non-uniform dimensional changes of a brick or block produced by a difference in temperature.

Differential thermal analysis (DTA). A means for the identification and approximate quantitative determination of minerals. The technique is based on the determination of the temperatures at which endothermic and/or exothermic reactions take place when a test sample is heated at a specified rate.

Modulus of rupture (MOR). The maximum transverse breaking stress, applied under specified conditions, that a material will withstand before fracture. When determined at temperatures other than room temperature, it may be called hot modulus of rupture.

Modulus of elasticity. The ratio of stress to strain within the elastic range.

Abrasion resistance. The ability of a material to withstand the action of mechanical rubbing or impact.

Other properties

The standard definitions apply to other properties, e.g. thermal conductivity, specific heat, etc.

Table 5.29 ENGINEERING PROPERTIES OF SELECTED REFRACTORIES

	Pyrometric cone equivalent °C	Refractoriness under load % deformation 2 hr at 1650°C under pressure 0.193 MN/m² unless otherwise stated	Permanent linear change % contraction 2 hr 1700°C unless otherwise stated	Reversible thermal expansion % 20–1400°C	Spalling resistance	Thermal conductivity W/m°K 1300°C unless otherwise stated	Hot modulus of rupture MN/m² 1300°C	Specific heat J/Kg°K 20–1000°C	Cold crushing strength MN/m²
80% Al_2O_3	1 850	3·7–7·8 (1700°C)	0–0·75	0·95	Good to excellent	1·87	11·7	1 130	55–130
Mullite	1 710–1 770	0·5–2·0 (1700°C)	0–0·1	0·6	Good	1·73–2·16	11·7	1 090	62–83
Fireclay 42% Al_2O_3		1 545–1 575°C 5% subsidence under 0·193 MN/m²	0·3–1·3 (1650°C)	0·9	Fair to good	2·02			41–90
Silica	1 710–1 730	1 730°C 5% subsidence under 0·345 MN/m²	0·5 (exp)– 0·5 (1500°C)	1·2	Excellent above 1000°C	1·44 (1000°C)	5·9	1 100	28–34
Magnesite	>1 730	0·5–3·0	0·1–1·0	1·95	Fair to good	2·88	15·2	1 200	34–55
Magnesite chrome		0·5 (1 hr)	0·2	1·60	Excellent	1·87	5·5		38–55
Chrome magnesite	>1 730	0·5 (1 hr)	0·8	1·50	Fair to good	1·73	5·9	1 050	28–38
Chrome		Fails below 1600°C under 0·193 MN/m²	0·5	1·30		1·87		950	55–62
Forsterite	>1 730	10 (1650°C)	1·0–2·5	1·55	Fair	1·44			28–61
Dolomite	>1 730		1–1·5	1·40	Poor	2·31 (1200°C)		1 090	48
Carbon			0–1·0 (1500°C)		Excellent	5·77–7·57 (800°C)			28–62
Silicon carbide	>1 750°C Silicon nitride bonded >1 800°C	0 (50 hr 1500°C) 0·345 Mn/m²)		0·4	Excellent	16·3 (1200°C)	35		
Zircon				0·65	Excellent	3·46		760	83

Other important properties for which there is as yet no accepted standard method are usually determined, e.g. resistance to slag.

PHYSICAL PROPERTIES AND CHEMICAL ANALYSES

Values for some of the above physical properties of a range of refractories may be seen in the Tables 5.29 to 5.32. Typical chemical analyses are given in Table 5.33. The bulk density of a refractory may be affected by the amount of impurity material present,

Table 5.30 PHYSICAL PROPERTIES OF GROUP I REFRACTORIES

	True specific gravity	Apparent porosity %	Bulk density kg/m^3
Corundum	3·75–3·95	20–29	2 880
90% Alumina	3·55–3·65	13–25	2 700–2 900
80% Alumina	3·35–3·45	17–30	2 550–2 750
70% Alumina	3·15–3·25	19–30	2 300–2 400
60% Alumina	2·90–3·05	17–28	2 300–2 550
50% Alumina	2·75–2·85	18–28	2 100–2 350
Mullite	3·00–3·20	15–26	2 400–2 450
Fireclay (42%)	2·65–2·75	20–22	2 050–2 350
Silica	2·30–2·38	20–25	1 680–1 870
Magnesite	3·40–3·60	17–24	2 670–2 950
Magnesite Chrome	3·50–3·80	16–22	2 950–3 050
Chrome Magnesite	3·80–4·00	18–20	2 900–3 100
Chrome	3·80–4·10	18–25	2 950–3 150
Forsterite	3·30–3·40	18–25	2 360
Dolomite	3·25–3·40	16–22	2 500–2 800
Carbon	—	13–21	1 500–1 650
Silicon Carbide	3·17	10–15	2 640–2 710
Zircon	4·68	17–22	3 500

Table 5.31 TYPICAL PROPERTIES OF INSULATING REFRACTORIES

Classification ASTM C155	Bulk density kg/m^3	Permanent linear change ASTM C210 24 hours	strength MN/m^2	Thermal conductivity W/m°K 1100°C
2 000°F (1 100°C)	420–580	0·3–0·4	1·65–3·10	
2 300°F (1 260°C)	770–960	0·6–0·7	4·14–4·48	0·40–0·45
2 500°F (1 370°C)	770–860	0·4	2·76–4·83	0·38–0·40
2 600°F (1 425°C)	740–900	0·3	2·76–4·14	0·35–0·40
2 800°F (1 540°C)	850–900	0·4	4·14–4·83	0·39–0·42
3 000°F (1 650°C)	1 070	0·4	4·83	0·49

the temperature of firing, the method of making and material grading, and the density of the grain.

Although the refractoriness is an indication of the temperature at which the sample will slump, it gives no indication of the ability of the material to bear a load at temperatures below the slumping temperature (see Figure 5.81).

Generally, the ability of a brick to withstand a load is dependent upon the amount and softening point of the glass phase. In this area the direct bonded basic brick has an advantage since the crystals are not bonded together by a glass phase. Fireclay refractories tend to have reduced load bearing capacities at temperature much below the

Table 5.32 TYPICAL PROPERTIES OF PREPARED BUT UNSHAPED REFRACTORIES

Type		Refractoriness °C	Maximum service temperature °C	Density after firing to service limit kg/m³	PLC after firing to service limit (%) contraction	Thermal conductivity 600°C W/m°K
Basic jointing		1 730–1 750	1 700			
High alumina jointing		1 730–1 750	1 700			
Aluminous jointing		1 690–1 770	1 650			
Silica jointing		1 700–1 720	1 650–1 700			
Basic concrete	Chrome	1 600	1 450	2 820	0·1 (exp)	0·94
	Chrome-mag.	1 710+	1 400	2 755	1·94	1·28
	Mag.-chrome	1 710+	1 400	2 565	1·23	1·14
	Mag.	1 710+	1 400	2 490	2·06	1·29
Basic gun mix	Chrome	1 650+	1 500	2 660	0·2	0·97
	Magnesite	1 710+	1 400	2 620	2·52	1·07
Basic ram mix	Magnesite	1 750+	1 700	2 190	1·59	1·66
	Magnesite	1 750+	1 600	2 560	1·98	
Aluminous Hydraulic concrete		1 600+	1 600	2 490	0·5	0·85
		1 750+	1 700	2 570	1·89	1·38
		1 800+	1 800	2 620	1·03	0·92
Aluminous hydraulic Gun mix		1 500+	1 400	1 920	0·5 (exp)	0·90
		1 710+	1 600	2 630	1·99	1·05
Insulating concrete		1 700+	1 700	1 475	2·6 (exp)	—

Table 5.33 TYPICAL CHEMICAL ANALYSES OF VARIOUS REFRACTORIES

	SiO_2	TiO_2	Al_2O_3	Fe_2O_3	CaO	MgO	K_2O	Cr_2O_3	*Loss*	SiC	ZrO_2
High-alumina, *Group I*	12	5	81	1	0·5	0·5			0·5		
High-alumina, *Group II*	42	0·5	55	1			0·3				
Firebrick	51	1	42	3	0·5	0·5	0·5				
Siliceous	88		10	1			0·8				
Silica	96		0·7	0·7	2·3						
Magnesite	2·5		0·8	1·8	2·3	92					
Magnesite-chrome	2·7		11	6	1·6	67		11			
Chrome-magnesite	4·2		19	10	0·8	43		22			
Chrome	5		25	12	0·8	30		26			
Forsterite	24		8	10	0·5	50		9			
Dolomite	3		2	1·5	55	37					
Zircon	34	1·5									65
Zirconia	0·5										97
Silicon carbide	25		4							70	
Diatomite	92		3	1		0·5	0·7				
Vermiculite	43	1	10	6	4	31	5				
China clay	48		38	1·2					13		
Aluminous cement	33	1·5	60	2		0·6	0·9		2·7		
Fireclay	44	1	37	2					14		

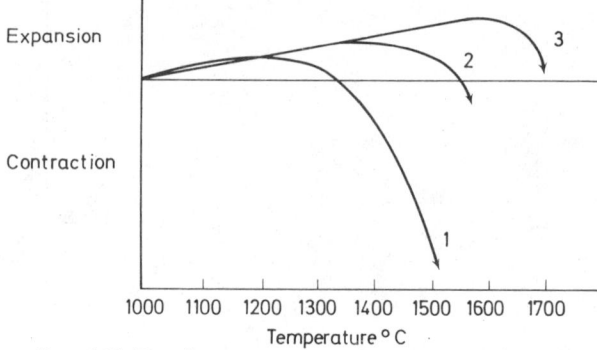

Figure 5.81. Typical rising temperature refractoriness-under-load tests
0. 345 MN/m²–10°C per min. rate of rise
1. Fireclay, 10% deformation at 1500°C
2. Magnesite
3. Silica, failure by shear

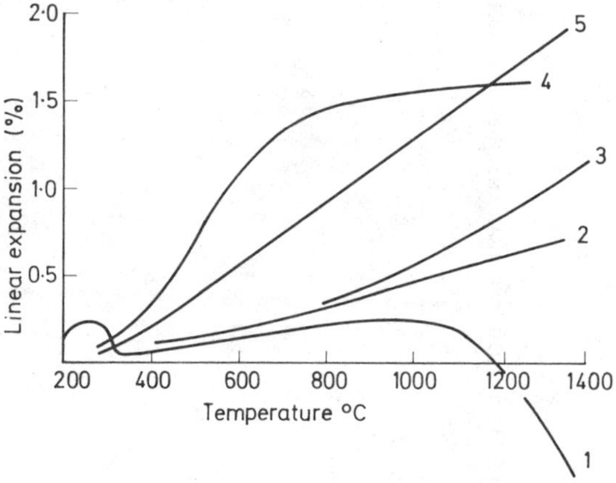

Figure 5.82. Typical linear thermal expansion of various refractories
1. Castable
2. Zircon
3. Fireclay
4. Silica
5. Magnesite

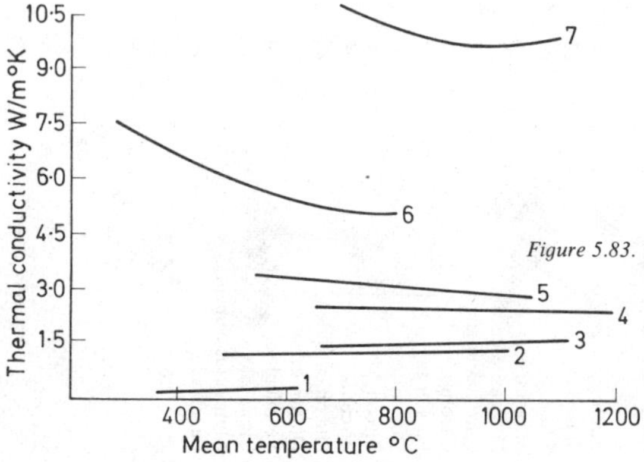

Figure 5.83. Typical thermal conductivities of various refractories
1. Insulating brick
2. High alumina
3. Firebrick
4. Zircon
5. Magnesite-Chrome
6. Magnesite
7. Silicon carbide

refractoriness temperature whilst silica bricks retain their capacities to within a few degrees of the temperature. Consideration must be given to the fact that under service conditions only a narrow zone of the brick is at the top temperature, the main part of the brick is at a lower temperature.

The reversible thermal expansion and permanent change in dimensions of refractories are important properties in connection with calculations of expansion allowances necessary during the building of a furnace structure. Typical curves of reversible thermal expansion for a range of materials may be seen in Figure 5.82. The large expansion of the silica material is due to the inversions mentioned earlier.

The linear coefficient of thermal expansion (Figure 5.82) and the thermal conductivity (Figure 5.83) are factors which are included in equations developed in an attempt to explain the mechanism of spalling. The true mechanism is as yet not well understood but empirical laboratory tests give a useful indication of the ability of a material to undergo rapid changes in temperature without cracking.

The spalling resistance is of importance in a variety of applications. Some steelmaking furnaces are subjected to severe thermal shock during loading and tapping while blast furnace stoves and checkers have hot gas and cold air blown through alternately. A large number of furnaces and kilns are of the intermittent variety and work a room temperature to firing temperature to room temperature cycle, and the speed at which these kilns may be heated and cooled is of primary importance economically.

Slag attack is another major cause of refractory destruction. This slag may be formed from materials added, e.g. during the refining of iron into steel, it may be produced by reaction between the furnace linings and the charge being fired, e.g. volatile constituents of a glaze during glost firing or condensation of alkali vapour in certain areas of a blast furnace lining, called in this case alkali attack. Alternatively, it may be produced by reaction between constituents in the fuel and the refractories, i.e. oil and coal ash. The mechanism of attack in either case is the same. Some constituent or constituents of the slag reacts with a phase of the refractory to produce a low melting point compound which is not refractory at the temperature of operation, and this usually drastically alters the physical properties of the material. A very important factor is the rate of reaction between the refractory and slag, since if the rate is slow other factors may make it economically feasible to use what, at first glance, appears to be an unsuitable refractory.

A high resistance to abrasion is necessary for certain applications, e.g. blast furnace stack and skid rails for reheating furnaces. In these instances the major cause of wear is the constant rubbing away of the material by either the solid material moving down the stack (blast furnace) or the steel billets moving along the skid rails (reheating furnace).

It should be mentioned here that certain types of refractories react if in contact with each other at high temperatures. The temperatures at which these reactions occur are well documented.

SPECIFIC PROPERTIES AND USES

Group I. Dense fired refractories

HIGH ALUMINA

The specialised demands for alumina products are not being dealt with in this treatment since they are often associated, not with its refractory properties, but rather its physical and chemical characteristics.

Generally, high alumina materials are more expensive and have a greater refractoriness than a firebrick. They show good spalling resistance, can be produced with high cold crushing strengths and abrasion resistance, are more resistant to slag than firebrick but are attacked by alkalis.

>80% Al_2O_3. Electric furnace roofs, aluminium production, kilns and furnace structures up to 1850°C.

70% Al_2O_3. Blast furnace stacks, soaking pits, tunnel kilns, crowns and car decks up to 1700°C.

65% Al_2O_3. As above at lower temperature, reheater furnace hearths, skid rails for pusher type furnaces.

Fused mullite. Many areas of glass production, lower stack of blast furnace.

Sillimanite/Kyanite 63% Al_2O_3. Recuperator chambers, electric arc furnace roofs, enamelling muffles and frit melting furnaces, copper production furnaces.

Sillimanite/Kyanite 57% Al_2O_3. High temperature zones of reheating furnaces and soaking pits, regenerator chambers, parts of glass furnaces.

Good electrical resistivity and high thermal conductivity make 'pure' alumina suitable for furnace tubes. Its purity and stability in contact with platinum makes it suitable as thermocouple sheaths and insulators. Other applications include induction furnace crucibles, kiln furniture for firing certain electronic ceramics and it is used in sparking plugs and turbine blades.

FIREBRICK

Generally, firebricks show good resistance to acid slags but resistance to basic slags and alkalis is poor. Resistance depends on its physical properties and improvement may be found by increasing the Al_2O_3 content. Hard firing increases the abrasion resistance, reduces the possibility of carbon monoxide attack and reduces the porosity, possibly resulting in a lower spalling resistance which may be partially overcome by the incorporation of coarse grog grains during making.

Firebricks are used in blast furnace stack, bosh, hearth and ladles, cowper stoves and linings of hot blast mains, areas of the open hearth where temperatures are low. Ladle stoppers and nozzles, in casting bays as runners and guide tubes. Soaking pits and reheating furnaces, small foundry furnaces and steam raising plants. In the continuous casting process the tundishes are often lined with firebrick.

SILICA

A well-fired silica brick will have an excellent spalling resistance between 600 and 1700°C and no contraction up to its melting point, although if the quartz conversion is incomplete, an after-expansion may occur. An obvious disadvantage is the poor spalling resistance at low temperatures. It has good slag resistance, low permeability and a fairly low bulk density.

The use of silica bricks is slowly diminishing and will continue to do so since two of the main uses now appear to have a limited life. These are the basic and acid open hearth steelmaking furnaces and gas raising plants. Silica bricks last indefinitely in coke ovens. Domes and upper layers of cowper stoves often consist of silica bricks.

SILICEOUS OR SEMI-SILICA BRICKS

The properties of this material lie in between that of a firebrick and a silica brick. It has little or no after contraction and a better spalling resistance than silica bricks.

When attacked by slags a glazed reaction zone is formed which tends to be limited to a narrow surface layer.

These materials are used as a backing to silica bricks since they are cheaper and often used in areas normally requiring a silica material where the conditions are less severe. They are used in open-hearth furnace checkers, coke ovens, kiln roofs and flues.

MAGNESITE

Magnesite has a high thermal conductivity (~ 5 W/m°K at 500) which falls with increasing temperature, the spalling resistance is good with the correct grading and analysis. The electrical conductivity is low at room temperature but this increases at higher temperature. Its resistance to ferruginous slag is extremely good and this is improved when tar additions are made. A good quality magnesite will show little or no movement in the refractoriness under load test at 1650°C. The after expansion will depend very much on the firing temperature during manufacture. High cold crushing strengths are obtained on this material.

Application of this material depends very much on the quality, for example in the Kaldo where the demands made on the brick are arduous, a high quality material is required. Magnesite would be used in back and front linings of open-hearth furnaces, linings of LD and rotor furnaces (normally impregnated with tar) and glass tank checker bricks. The lower quality magnesite may be used in the upper walls of electric arc furnaces and safety linings in LD and Kaldo converters.

CHROME

The use of straight chrome bricks is somewhat limited because of the low refractoriness of the gangue. They have poor spalling resistance and show excessive bursting expansion in the presence of iron oxide. They are used as a separating medium between basic and siliceous materials. Excessive expansions of certain chrome materials during kiln firing and subsequent use produces bricks that are mechanically weak. These problems in firing are considerably reduced by the additions of magnesia which also improves its refractoriness. They are used in forge and reheating furnace hearths.

CHROME MAGNESITE

Although the refractoriness under load value of chrome magnesite is improved with firing temperature, the unfired chemically bonded form of this material is now favoured in certain fields. A favourable characteristic of this type of brick is the low after contraction or expansion.

The chemically bonded form is used in open-hearth furnace roofs and back, front and end linings, roofs of copper reverberatory and refining furnaces and, with increased MgO content, it can be used in the upper walls of electric arc furnaces. The fired or direct bonded form has further additional uses as linings of large induction steel melting furnaces and copper converters.

MAGNESITE CHROME

As would be expected, the properties of this material fall in between chrome magnesite and magnesite. Generally it has a good volume stability and refractoriness under load, a higher cold crushing strength and reduced bursting expansion in the presence of iron oxide at temperatures in excess of 1000°C than chrome magnesite. This material is produced in the fired state or chemically bonded with or without metal plates. The fusion cast form has a reduced spalling resistance improved slag resistance and is more expensive.

Chemically bonded bricks are used in roofs, back, front and end linings and suspended uptakes of open-hearth furnaces, and the upper walls of electric arc furnaces. Fired bricks are used in glass tank regenerator superstructure, hot zone linings of rotary cement kilns, linings of copper converters, reverberatory and refining furnaces. Fusion cast magnesite chrome has shown an improved slag resistance when used in a Kaldo converter, compared with magnesite and in areas of maximum wear in electric furnace side walls.

FORSTERITE

Refractoriness under load values are similar to basic materials, thermal expansion is low and a reasonable resistance to acid and basic slags is shown. Its spalling resistance is only moderate.

It is used as lining and regenerator packing in open-hearth furnaces, in glass tank regenerators, in copper melting furnaces and the top courses of checkers.

DOLOMITE

The refractoriness of dolomite is high but its load bearing properties are not good and the spalling resistance is generally poor. The resistance to slag attack is poor, but like many other properties this is improved considerably by the presence of tar or carbon. The properties of dolomite are often improved by additions of MgO.

With additions of tar it is used in LD and LDAC steel converters. Sub hearth and hearth of an open-hearth and electric arc furnaces, basic ladles and casting pits. Dolomite in the crushed form is used for fettling of numerous furnaces and converters.

CARBON

The refractoriness of carbon is extremely high but its use as a hot face lining is limited because it is easily oxidised. A carbon brick gives an excellent refractoriness under load value, good abrasion resistance, good thermal shock resistance and a relatively low thermal expansion. This material can be easily machined which enables complex shapes to be made. Its thermal conductivity depends on its making methods and the degree of graphitisation. It is used as linings for furnaces making aluminium, magnesium, calcium carbide and phosphorus, as electrodes and for conducting hearths of some types of arc furnace, bosh and hearth of blast furnaces. Carbon in the form of rods, tubes or granules is used as the electrical resistor in resistance furnaces.

GRAPHITE

This material tends to have higher electrical and thermal conductivities and a higher density than carbon. The properties depend on the nature and grading of the raw material and the manufacturing process. The uses of graphite are similar to carbon but graphite is used where a high current density is required. It is also used as moulds or plungers for hot pressing, as a resistor in a carbon tube furnace and heating element or inductor in high frequency furnaces. The non-wetting properties of graphite make it useful as a crucible for heating slags in reducing conditions.

ZIRCON

This material has a good spalling resistance, low thermal expansion and high thermal conductivity. It reacts slowly with acid glasses but is very reactive with FeO, fluorspar and some phosphates. Zircon is used in the glass industry, and in pottery kilns because of its resistance to attack from glazes. Its high thermal conductivity enables it to be used as a moulding sand. It is used in the remelting of aluminium and in certain boilers because of its resistance to oil ash.

ZIRCONIA

This is a highly refractory oxide (2677°C) which reacts with basic slags but is stable in both oxidising and reducing gases up to 2200°C. It has a high density, has to be stabilised to avoid inversions and thermal shock.

It is used as inserts in nozzles in the continuous casting of steel. At high temperatures it has a good electrical conductivity and can thus be used as in inductor in high frequency furnaces.

SILICON CARBIDE

Under certain conditions silicon carbide can be used up to 2000°C although in the presence of air, the silicon oxidises to SiO_2 which normally forms a glazed layer to protect the remaining material. The spalling resistance is very good, it has a high thermal conductivity, and a high abradability index. It can be attacked by FeO at moderate temperatures but is resistant to slags of the coal ash type. It is used as tubes in recuperators and electrical resistors (Crusilite rods, Silit rods or Globars), as refractory batts and muffle chambers in the pottery industry and skids in billet heating furnaces. Also used as gas-fired radiant tubes for firing industrial furnaces of controlled atmosphere.

Group II. Shaped refractory insulating materials

These materials are not normally used in areas where slags are likely to occur. They have a poor refractoriness under load. Some products are highly resistant to gas flow and carbon monoxide attack.

Insulating bricks are commonly known as low or high temperature insulators depending upon the composition. Those made from diatomite or vermiculite are in the the low temperature category and are not used above 900–1000°C, and are usually used as a backing material. The high temperature bricks are made from fireclay, china clay or alumina and may be used up to 1650°C as hot face insulation.

A recent development in insulation is the so-called ceramic fibre materials, whose composition approximates to 55% SiO_2, 45% Al_2O_3. These are available in the form of loose fibre, paper or blanket. The high temperature limit of use has been set at 1250°C due to excessive shrinkage. The advantages of fibre are the extremely low bulk density

(50–130 Kg/m³ for formed blanket), low thermal conductivity (0·22–0·12 W/m°K for formed blanket) and ease of installation.

Group III. Prepared but unshaped refractory materials

All unshaped materials are supplied with directions for mixing and method of application. Large variations in properties can be experienced if these directions are not adhered to.

MOULDABLE MATERIAL

This is a mixture of graded refractory aggregates and a plasticiser, which is usually clay, supplied mixed with water in a workable condition. Chemical bonding agents may also be incorporated. The workability of the material is such that it can readily be placed by hand malleting. These materials show good spalling resistance but the compressive strength is not too good although it does improve on the development of a ceramic bond. The thermal conductivity is lower than its equivalent fired material.

Mouldables are used in cement, glass, lime and non-ferrous industries, ships and marine installations, coke ovens, rotary kilns, tunnel kilns and blast furnaces.

RAMMING MATERIAL

This is a mixture of graded refractory aggregates with or without the addition of a plasticiser and with or without water, usually supplied at a consistency which requires a mechanical method of application. Chemical bonding agents may also be incorporated.

Ramming materials are used in copper production furnaces, hot metal mixers, reheat· furnaces, soaking pits and spouts of torpedo ladles.

CASTABLE MATERIAL

This is a mixture of graded refractory aggregates and a hydraulic cement. The material is usually supplied dry and at the appropriate moisture content it may be rammed or cast. The castable usually contains 15–25% of the hydraulic cement.

The aggregate can be alumina, fired kyanite, mullite, zirconia, insulating brick, diatomite or vermiculite. During heating, dehydration occurs which affects the thermal expansion, increases porosity and reduces compressive strength. There is a minimum strength after the hydraulic bond is lost and before the ceramic bond is formed (900–1100°C). Additives are made in an attempt to avoid this. Thermal conductivity is lower than its equivalent fired material. Very little or no shrinkage occurs after heat treatment. The cement may be replaced by vegetable bonds, organic bonds, acids or mineral oils.

Castable materials are used to produce special shapes, as monolithic furnace linings, covers of soaking pits and burner blocks.

GUNNING MATERIALS

Many materials which fall into the above groups may be suitably prepared for application by gunning techniques.

Material applied by gunning techniques is used for general resurfacing of refractory brickwork, re-contouring ladle linings, installation of flue linings, and repiping tapholes in electric furnaces.

BIBLIOGRAPHY

DODD, A. E., *Dictionary of Ceramics*, 2nd ed, Newnes (1967)
CHESTERS, J. H., *Steelplant Refractories*, 2nd ed, United Steel Companies Ltd. (1963)
NORTON, F. H., *Refractories*, 4th ed, McGraw Hill (1968)
GILCHRIST, J. D., *Fuels and Refractories*, Pergamon (1963)
Modern Refractory Practice, Harbison-Walker Refractories Co. (1961)
1970 Annual Book of A.S.T.M. Standards, Part 13, American Society for Testing and Materials (1970)
DIN Normblatt Verzeichnis, 500–502, p. 295–297 (1971)
BS 1902
Methods of testing refractory materials. Pt 1 A (1966); Pt 1B (1967); Pt 1C (1967)
BS 1902
Methods of chemical analysis. Pt 2A (1964); Pt 2B (1967); Pt 2C (1969); Pt 2D (1969); Pt 2E (1970)
BS 915
High alumina cements (incorporating amendments) (1947)
BS 2973
Classification and methods of sampling and testing insulating refractory bricks (1961)
BS 3446
Glossary of terms relating to the refractories industry (1962)
ISO Recommendation R1109
Classification of dense refractory products (1969)
ISO Recommendation R1927
Classification of prepared unshaped materials (dense and insulating) (1971)

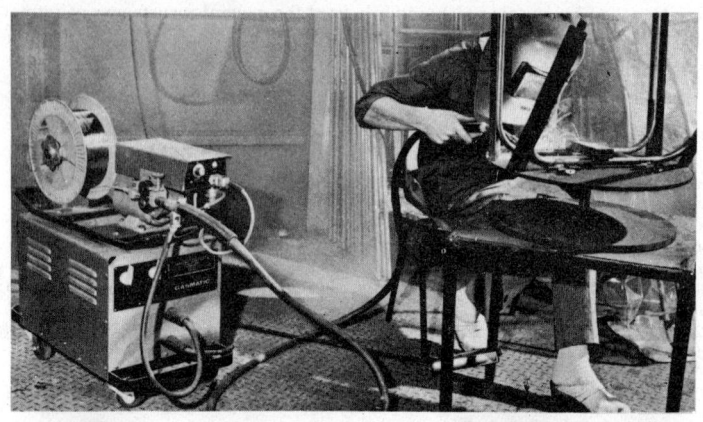
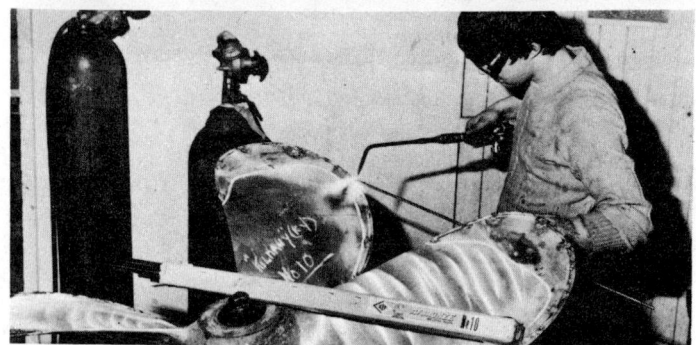

6 WELDING AND SURFACE FINISHES

WELDING AND SURFACE FINISHES 6

6 WELDING AND SURFACE FINISHES

WELDING

H. F. TREMLETT

This section is intended as a source of information for the engineer who is primarily responsible for design. When using mechanical joints, the knowledge of the materials being handled may not need to exceed an understanding of their physical and mechanical properties, resistance to corrosion and wear, hot and cold formability and general response to modern forming methods. If, however, the engineer intends using joining methods such as welding, brazing, soldering, then his knowledge should be broadened to cover a working understanding of these processes. Ideally this is achieved by practical experience in these trades and there is no doubt that the traditional system of industrial engineering organisation which separates the shop-floor tradesman from the design office engineer is responsible for unnecessary financial losses.

Although losses occur as a result of variability in materials handled by the engineer, the most important single cause of early failure in the life of welded components is incorrect design which is the responsibility of the engineer and draughtsman. Errors in design frequently arise from lack of adequate understanding of the welding processes and their effect on the engineering properties of the materials being welded and also from lack of experience of practical welding which leads to defective welded joints.

Welding and Brazing are joining processes entirely dependent on the metallurgical properties and behaviour of metals, whereas mechanical joints depend on the engineering properties (strength, ductility, toughness etc.,). There has been widespread acceptance of the idea that design criteria based on engineering property concepts can be applied without modification or reservations to welded fabrication. This is not to be recommended for universal application although over limited areas it may be acceptable.

Welding involves a highly specialised field of metallurgy which is constantly growing in depth of knowledge. It is impossible for anyone to cover the entire spectrum fully. The design engineer needs an understanding of the broad principles involved in welding practice while leaving the minutiae to the welding engineer.

The information given in this section will, therefore, aim to provide a background understanding of the welding methods and their effect upon the engineering properties of the materials being welded. Where the significance of such effects is of paramount importance to the satisfactory service behaviour of welded components, the matter will be treated in some depth but evidently, in such a wide metallurgical field, many aspects can only be accorded light treatment within the limited space available. References and a bibliography are provided for information in greater detail.

WELDING AND THE DESIGN (DRAWING) OFFICE

Customer loyalty depends, among other things, on anticipated reliability of a firm's products in respect of delivery date and behaviour in service. Both are dependent upon

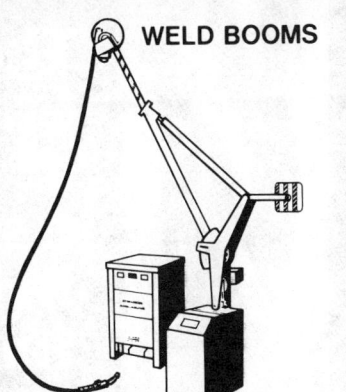
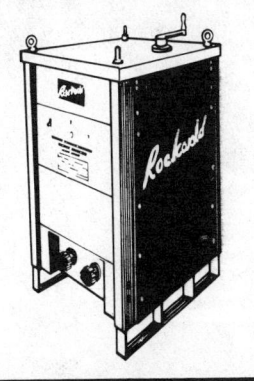
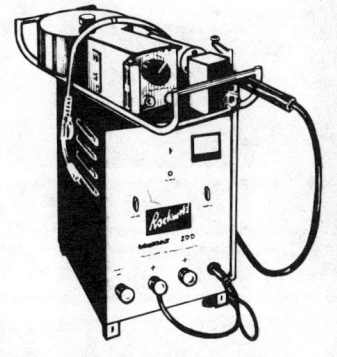

control of weld design and shop-floor practice allied to the manual skills and experience of the welders. This control should be exercised from the Chief Engineer via the Design Office (Engineer responsible for welded design), an Engineer responsible for welded production to ensure that the requirements of the Drawing Office are met and a shop-floor welding supervisor who is responsible for the quality of work produced by the welders.

It is obvious that a standard system of communication between the Drawing Office and the shop-floor is required. Most welding consists of a repetition of butt and fillet joints. All such work can be handled by standard procedures which record the information necessary to ensure that the method of making the weld is clearly defined to the welder. Where unusual joints or joints in unusual materials must be made, co-operation between the Drawing Office and the shop floor is essential to ensure that likely difficulties are foreseen and avoided but the same principle of setting out the procedures on paper should be followed. This has the great merit that if trouble does occur later on there is a record of how the joints were made.

The plater and welder must obtain instructions from the shop floor drawing, the former to deal with cutting, forming and joint edge preparation, preheat etc., the latter to use correct welding materials and procedures.

There are several methods of communicating this information; frequently it is presented on the drawing as a detailed sub-drawing, which tends to be costly in drawing office time. A system of standard symbols, combined with standard procedure or data sheets in booklet form, in which only the symbols appear on the shop floor drawings, is economical once the data sheets have been prepared.

The British Standards Institution have issued BS 499; Part 2: 1965 'Symbols for Welding', and it should be noted that this system is to be used in conjunction with procedure sheets. Details such as the size of gap in an open joint, the amount or position of the root face and the size of the included angle where the plates are edge bevelled are not included in the symbol system but must be given separately on the drawing or in the procedure sheets. The American Welding Society have issued Standard A.2.0–68 'Standard Welding Symbols'. This system includes all the information excluded from the British system, but in other respects the two systems are basically similar.

The correct use of either of these systems promotes precise control of welding operations on the shop floor which gives obvious technical and economic advantages. The use of such systems also tends to eliminate misunderstanding between contractor and sub-contractor. Precision in matters of this sort is the responsibility of the engineering staff but it is also the basic requirement for proper control and forecasting of welding costs. For most organisations involved in large and medium scale fabrication an engineer with practical welding experience is required to ensure that instructions from the Drawing Office are feasible within the limits set by the available equipment and that they will allow the welders to produce work of the required quality.

Where work of a varied and changing nature, in terms of materials and products, is undertaken, it is generally advisable to carry out initial procedure trials before work begins on the shop floor so that the risk of expensive mistakes is reduced. The symbol system combined with procedure or data sheets provides a convenient record of the satisfactory results of initial procedure trials and automatically introduces the new procedure in to the standard data book for future reference.

Standard data sheets should include the following information for each joint:
Material. Defined by specification or trade name.
Material thickness.
Welding position.
Joint preparation.
Pre-heat.
Welding process.
Welding consumables, i.e. electrodes, wire and flux for submerged arc welding, wire for metal inert gas welding, wire for CO_2 welding, filler wire for tungsten inert gas

welding etc. For the shop floor the data sheet should give the consumable manufacturer's trade name but it is also useful to give a national specification number if this is quoted so that potential customers will be informed of the grade and type of consumable. A sketch of the joint giving: Electrode or wire diameter. Current range. Number of weld runs and size of weld if fillet or partial penetration butt. The order in which the runs should be deposited.

Where desirable include special notes such as back chipping or gouging and, for materials which require it, an indication of the interpass temperature. If required, the post-heating (carried out without allowing the joint to fall below the interpass temperature) temperature and time at this temperature must be stated. If required the stress relief temperature and time must also be given.

CLASSIFICATION OF WELDING PROCESSES

There are several ways of classifying the various welding processes. The application for which each is suitable is a practical method. Welding processes can also be classified by the source of heat, i.e. electrical or gas flame or the manner in which joining is achieved, either by fusion or by welding together in the solid state (solid phase welding).

Since the processes are described in detail in numerous reference books and publications by commercial firms, it is sufficient to note here the main features or principles on which they are based and the common applications, advantages and limitations. From such data the engineer may select the process which is best suited to his requirements. The widely established principal welding processes are shown in Table 6.1. Additional fusion welding processes, having special applications, are electron-beam, plasma, micro-plasma, and laser welding and in the solid phase field there are ultrasonic welding and cold welding.

The first question that generally arises is whether or not any given process is suitable for automatic welding, in the sense that semi-skilled labour can be used. This assumes the welding machine controls can be pre-set by a skilled operator and the repetition carried out by semi-skilled or unskilled labour. For the purpose of describing the welding processes in Table 6.1 this situation is termed automatic welding.

The so-called automatic welding processes are 'automated' to the extent that two of the main variables in all fusion welding processes, i.e. the speed of travel of the heat source along the joint and the rate of feed of the consumable wire are controlled by a motorised welding head instead of the movement of the welder's hand. Although the manual element is thereby removed there is still the need to employ skilled welders to ensure that correct control of the welding machine is obtained. This situation is also described as automatic welding in Table 6.1. It is common practice to refer to the situation where wire feed is controlled by a machine but deposition is by manual operation as semi-automatic welding.

METAL ARC WELDING

All fusion welding processes require that atmospheric oxygen and nitrogen be excluded from the weld pool surface. These gases are soluble in molten metals and produce defective welding in varying degree, porosity and embrittlement. The metal arc process achieves this protection by using mineral coatings on the electrode wire which either provide a protective slag blanket or a combination of gaseous shield from combustion of cellulosic materials or calcium carbonate with a slag blanket (see Tables 6.2)[1, 2, 3, 4].

The *manual* metal arc process is more widely used than any other. It is a most versatile process and requires only low capital expenditure. The *automatic* metal arc process, where the forward travel and feed down of the coated wire (from a coil) are made by a

Table 6.1 PRINCIPAL ESTABLISHED WELDING PROCESSES

Category	Fusion							Solid phase		
Process name	Metal arc	Sub-merged arc	Electro-slag	Inert gas metal arc	Inert gas tungsten arc	Oxy-acetyl-ene	Spot, seam, etc.,	Forge and pressure gas	Flash butt and upset butt	Friction
Heat source	Arc	Arc	Resistance heated slag	Arc	Arc	Gas com-bustion	Electric resistance	Fire or gas	Electric resistance	Friction
Welding by:	Fusion	Fusion	Fusion	Fusion	Fusion	Fusion	Fusion	Pressure	Pressure	Pressure
Shielding by:	Slag or slag and gas	Slag	Slag	Gas	Gas	Flame	—	—	Inert gas rarely	Inert gas rarely
Operation*	M	M, A	A	M, A	M, A	M, A	AC	M	AC	AC
Usual operation	M	A	A	M	M	M				

* M. Manual. Requiring skilled welder.
A. Automatic. Machine welding requiring skilled welder.
AC. Automatic. Machine welding by semi-skilled or unskilled labour after skilled setting of machine.

Table 6.2(a) SHIELDED MANUAL METAL ARC WELDING ELECTRODE CLASSIFICATION
(Principal Classes in the AWS A 5–1 and BS 1719. Part 1. Specifications)

Code class (Table 2b) AWS BS 1719	Coating type	Shielding	Current	Positions V = vertical, O = overhead F = flat, H-V = horiz. vert.	Typical applications	Remarks (See table 6.2b)
EXX10 1XX	Cellulose. Sodium	Gas	d.c.+	All. Specially V and O	All-position. X-ray. Ship building, bridges, storage tanks, 'stovepipe' welding, general pipe welding etc.,	
EXX11 1XX	Cellulose. Potassium	Gas	a.c. d.c.+	All. Specially V and O	General. Downhand positioned shop fabrications.	
EXX12 2XX	Titinia. Sodium	Slag	d.c.+ a.c.	All. Mainly F and H-V	General work on mild steel up to medium weight.	
EXX13 3XX	Titinia. Potassium	Slag	a.c. d.c.+	All. Some makes very suitable for V and O	General.	
EXX14 9XXJ	Titinia. Iron pwdr.	Slag	a.c. d.c.+	All. Mainly F and H-V		
EXX24 9XXK	Titinia. Iron pwdr.	Slag	a.c. d.c.+	F. H-V		
EXX15 6XXH	Calcium carb. Sodium. Fluoride.	Slag+ Gas	d.c.+	All	General work in downhand and H-V positions.	
EXX16 6XXH	Calcium carb. Sodium potassium.	Slag+ Gas	a.c. d.c.+	All	General. For low hydrogen welds. High sulphur steels. Root runs (thick steel).	
EXX18 6XXHJ	Calcium carb. Sodium. Iron pwdr.* Fluoride.	Slag+ Gas	a.c. d.c.+	All		
EXX28 6XXHK	Calcium carb. Sodium. Iron pwdr.** Fluoride	Slag+ Gas	a.c. d.c.+	F. H-V	As for EXX15. High deposition rates in flat fillets, and butts.	
EXX27 4XX	Iron oxide-iron pwdr.** Mn oxide-silica	Slag	a.c. d.c.+	F. H-V	Positioned joints in pressure vessels. Deep butt joints (good X-ray).	

* Up to about 130% recovery.
** 130–150% recovery.

Table 6.2(b)

AWS	BS 1719	Remarks
EXX10	1XX	Strong spray type arc. Good root fusion. No under-cutting. Thin slag giving clean high quality welds. Medium weld hydrogen content. Fair to good X-ray. Good notch toughness (E6010. 271 at −28°C). Good for high quality weld outdoors.
EXX11	1XX	Similar to EXX10 but suitable for a.c. Usability not as good as EXX10. Slightly heavier slag.
EXX12	2XX	For single run H-V fillets at high currents. Easier to handle than EXX10. High hydrogen deposit therefore not generally used on low alloy hardenable steels. Weld ductility lower and strength higher than EXX10. X-ray poor to fair. Notch toughness not generally as good as E6010. Subject to weld cracking in restrained joints therefore generally avoided for root runs in heavy mild steel fabrication.
EXX13	3XX	Makers' products in this group vary considerably in usability. Certain British electrodes particularly suitable for V and O welding (these have better X-ray than EXX12). Root crack susceptibility as for EXX12. High hydrogen. Not generally suitable for high current welding.
EXX14	9XXJ	Addition of iron powder increases deposition rates and recovery compared with EXX13, otherwise similar.
EXX24	9XXK	Heavier iron powder addition than EXX14. Recovery exceeds 130%. Very high deposition rates. Crack susceptibility similar to EXX13. Ductility slightly reduced at high currents (normally used). A sub-group has relatively low hydrogen and improved ductility producing mitre H-V fillets which reduces grinding costs when magnetic particle examination required.
EXX15	6XXH	Useful for root runs in heavy restrained joints, not susceptible to cracking. Useful on high sulphur steels. When properly dried are useful on steels subject to hydrogen cracking. Low hydrogen content when fully dried. Tendency to start porosity. Useful for steels to be enamelled. High quality deposits. Good notch ductility and X-ray. Convex deposits may give spurious magnetic particle test result. Short arc essential otherwise porosity present.
EXX16	6XXH	As for EXX15. – Most U.K. Class 6 electrodes are in EXX16 group.
EXX18	6XXHJ	As for EXX15, but handling easier. Electrodes giving more than 1% Mn have very good notch ductility.
EXX28	6XXHK	As for EXX18. but recovery and suitable only for H-V and F positions. Ideal for single run fillets.
EXX27	4XXK	Generally good X-ray standard. Concave profile therefore ideal for deep downhand multi-run butts. Fast deposition. Touch welding electrode. Good for H-V fillets. Not a low hydrogen deposit.

Table 6.2(c) COMPARISON BETWEEN AWS AND BS ELECTRODE CODING SYSTEMS

	AWS	BS
Prefix letter	Designates a manual electrode	Designates method of manufacture E = Solid extrusion R = Extrusion with reinforcement D = Dipping
First digit	Indicates minimum tensile strength of weld deposit undiluted (lbs/sq. inch × 1000)	Indicates type of coating.
Second digit Third digit		Indicates welding position.
Fourth digit	Indicates type of coating.	Indicates type of current (a.c. or d.c.)
Additional letters		Indicates iron powder in coating or low-hydrogen type coating.

welding machine, is useful for long continuous seams so that the stops and starts of the manual method, which can be sources of defects, are eliminated. It is used for the high quality work required on pressure vessel seams and for seams on fabricated girders where its continuous deposition characteristics promote higher footage welded per hour. It is also widely used in shipbuilding where its comparative immunity to deterioration from atmospheric conditions is an advantage.

In manual metal arc welding weld quality depends on the degree of success achieved by the welder in avoiding slag inclusions and porosity and obtaining fusion of the parent metal. Quality therefore is in relation to manual skill, but the degree of skill necessary to obtain a given level of quality depends on the position in which welding must be performed and on the type of coating on the electrode wire.

Coatings are designed to assist the attainment of high quality by varying the constituents in a manner to influence slag behaviour; thus electrodes specifically designed for welding down-hand (to give maximum rate of metal deposition) are not suitable for vertical or overhead welding. Correct electrode selection for any application is important. In pressure vessel joints usually welded down-hand and subject to X-ray examination, it is an advantage in minimising the risk of trapped slag in multi-run welds to use electrodes which produce a flat or concave welded bead surface contour so that slag removal is efficient and easy. In welding mild and medium tensile steel line pipe (for fluid and gas transmission), where the requirements are completion of the maximum number of joints per day with the pipeline axis horizontal and the pipe fixed, about 90 per cent of the world's pipelines are welded with one type of electrode coating because this alone combines the various requirements of slag control, deep penetration, quick freezing metal etc. This makes it possible to weld a large number of joints per day to radiographic standards.

Unfortunately, the electrodes described above, which are primarily gas shielded, promote a fairly high hydrogen absorption into the weld metal and this makes them less suitable for use on steels which harden on welding and may be subject to hydrogen embrittlement and cracking. For pipe welding in such steels, typically chromium–molybdenum qualities for high temperature service, electrodes having coatings which produce a low hydrogen weld metal are preferred. These electrodes are coated to give high quality welds in pipes in the fixed horizontal position but the number of joints which can be welded per day is much less than in the case of line-pipe welding.

It is, therefore, important for the engineer to appreciate that maximum economic benefit from manual metal arc welding is only obtained when the correct electrode is used for the particular application. The process is applicable to all thickness of steel, nickel alloys and most non-ferrous metals although it is not suitable for material thinner than about 2 mm. Control of penetration on thinner material is achieved more easily by other welding methods.

Metal arc welding owes part of its versatility to the successful addition of alloying elements to the coatings which permits readily obtainable core wires of mild steel, low alloy steel or stainless steel to be coated to produce complex and special weld deposit composition which would be commercially unobtainable or extremely expensive as alloyed core wire. Consequently it is the most widely used process for surfacing and hard-facing as well as for fabrication in the general engineering, high temperature, stainless and cryogenic fields.

Manual operation

EQUIPMENT

Power sources: A.C. Specially designed transformer O.C. Voltage 50–100. Normal A.C. input.
D.C. Specially designed generator. Drooping characteristic. O.C. voltage 60 min. Coupled to A.C. motor or I.C. engine.

Electrodes: Core wires, with a normal range of diameters from 2.5 mm to 8 mm, coated with protective gas producing materials (cellulosic) or slag making minerals or combination of gas (carbonates) and minerals.

Electrode Holder and Cables.

MANNER OF OPERATION

An arc is struck between the parent plates and the electrode, parent metal earthed to power source producing molten weld metal pool shielded from contamination from the atmosphere by the slag and products of combustion of the coating. The arc length (18/22 V) is kept constant by the welder manually feeding the electrode in to the pool to match the burn-off rate. Slag is removed this being self-detaching in many instances.

JOINT PREPARATION

Depth of fusion into parent plate normally does not exceed 3.25 mm; therefore butt joints in material thicker than about 6 mm require bevelling to a single or double V. U or J. Multi-run procedures must be used to fill joints thicker than about 4 mm.

DATA SUMMARY

Types of Joint. All.
Welding Position. All. Selection of correct electrode coating type important.
Weld deposition rate. Approximately 1·0–5·0 kg/hr.
Electrodes. Tables 6.2 and References 1, 2, 3 and 4.

PRACTICAL CONSIDERATIONS

The power sources used for manual metal arc welding are robust, reliable workshop tools requiring minimum expenditure on maintenance and, in the case of transformers, of relatively low capital cost. Selection of power source as between a.c. or d.c. depends on the type of application which determines the type of electrode, d.c. being required for certain electrode coatings. In the U.K., for general work, a.c. is widely used. The arc produced by a properly designed d.c. generator or rectifier is more stable than the a.c. arc even on electrodes coated for a.c. and d.c. The weld quality in respect of radiographic soundness is therefore generally superior. For applications where arc blow is not experienced, d.c. is to be preferred for highest quality. Where arc blow is met, as in fabrication work involving welding into corners, a.c. gives less trouble.

The fact that two electrodes may be classified under the same code number does not imply that they will behave alike in respect of 'usability'. For example, in using the stovepipe welding method of welding line-pipe, special usability characteristics are required which are not equally obtainable from all cellulosic electrodes; American cellulosic electrodes have a world wide reputation for consistent behaviour and suitability for this work. However, where the code stipulates a low hydrogen content and the drying conditions specified by the manufacturers are followed, electrodes classified in a code grouping tend to have similar levels of hydrogen.

A particular workshop advantage of the manual metal arc process is that any length of welding cable can be run between power source and the point of application of the electrode so that problems of accessibility are limited to human accessibility.

Automatic operation (shielded metal arc welding)

EQUIPMENT

(a) Welding head. This is essentially a welding head for automatically striking the arc and thereafter holding a constant arc length while burning the electrode. Travel of the head along the seam may be achieved either by motorising the head or the workpiece. Since voltage across the arc varies with the length of the arc it can be used as a control over the speed of the motor used to drive the electrode down to the weld pool. This is a common method of wire feed control. Controlling the speed by a potentiometer type circuit has practical workshop advantage when repairs are necessary by average electrical maintenance staff.

(b) Electrode wire (coil). The important feature of any automatic shielded arc welding method is the means used to pass current into the electrode core. For continuous automatic feeding of the electrode it is obvious that a wire coil should be used and it must be coated with the same type of flux as used in manual metal arc welding.

A reliable practical solution to the current pick-up problem is to introduce a secondary helical wire winding and wire mesh which holds the coating in place during coiling in manufacture and straightening on the welding head and allows simple sliding contacts for current pick-up. The coating is usually of the high titania or calcium–carbonate low hydrogen type and the latter may be alloy bearing. A variation on this theme is to add carbon dioxide gas shielding to the slag shielding of the rutile type coating.

SPECIAL FEATURES OF THE PROCESS

The current pick-up is located near the arc and therefore higher currents can be used than is possible with manual stick electrodes. This gives increased deposition rates and usually more consistent penetration. Welding can be continuous from start to finish of the seam eliminating intermediate craters thus increasing the level of weld soundness.

Although the welder can adjust the head as welding progresses there is a general need to improve accuracy of plate edge preparation and fit up compared with manual welding. The process is primarily applicable to welding in the flat (butts and fillets) position but small leg length fillets (6 mm) can be placed in the H–V position. The process is particularly suitable, compared with automatic submerged arc for those applications where, owing to pick-up of undesirable elements from the parent metal, deep penetration should be avoided. Thus it can tolerate, with a suitable choice of electrode (calcium carbonate type), a comparatively high sulphur parent steel. The process, compared with submerged arc welding, is tolerant to atmospheric conditions and is widely used on outdoor work in shipyards.

Shielded metal arc welding, scope of the process

MANUAL METAL ARC

Manual metal arc welding is applicable to the fabrication of mild, medium and high carbon steels, low and high alloy steels, stainless steels, cast iron (but of limited usefulness on this material), heat resisting steels, nickel and nickel alloys and to the building-up and hard surfacing of these materials with a wide range of hard facing alloys. By suitable coating formulation practically any element except aluminium can be transferred from the coating to the weld metal in controlled amounts. However, since weld metal from coated electrodes has inherently higher strength than wrought or cast steel of the same or similar composition, close matching of parent steel analysis is generally only applicable to corrosion and heat resisting compositions. In most cases the electrode is selected on the basis of its mechanical properties, resistance to weld cracking and general suitability for producing the required degree of weld quality in the particular type of joint, welding position and deposition conditions.

In certain circumstances the over-riding requirement that the weld metal shall be free from cracking, necessitates a miss-match between weld and parent metal compositions, an example being the welding of fully austenitic stainless and heat resisting steels with austenitic-ferritic weld metals. In general, there are few situations in which the weld metal cannot be produced to match or over-match the properties desired in the parent metal. The scope of shielded metal arc manual electrode welding for structural and high tensile steels and hard surfacing is indicated in Table 6.3.

AUTOMATIC SHIELDED METAL ARC

The automatic shielded metal arc welding process makes possible maximum economic benefit from increased deposition rate compared with the manual method. Electrode sizes vary from 3·5 mm to 7 mm with currents in the range 500–900 A. In Europe d.c. is frequently used, and in the U.K. either d.c. or a.c. depending on the type of coating (basic coatings requiring d.c.). Electrodes are either of the class 6, class 2 (rutile) or class 4 (iron oxide–manganese oxide–silica) type. The two latter are suitable for a.c. at 50 V minimum open circuit, or d.c. but the class 6, although a.c. at 80 V minimum o.c. can be used produces highest quality welds on d.c. The process is widely used for fabrication of medium and heavy plate in mild steel, typically ship building, storage tanks, pressure vessels etc. The electrode range covers mild steel, high tensile, stainless and creep resisting steels of the 0·5% Mo and 1% Cr—0·5% Mo types. The low hydrogen basic coated electrode (class 6) produces higher quality welds than the rutile type but the operating characteristics are not as attractive for fast economical welding. The British Oxygen Company have developed an extension of the automatic continuous coated electrode process in which a rutile type coated electrode is deposited under a carbon dioxide gas shield.

The CO_2 gas takes the place of the gas given off by the carbonate in the class 6 manual electrode coating and consequently to some degree the desirable operating characteristics of the rutile type coating are combined with the benefits of the low hydrogen weld (class 6) electrode. The process is known as Fusarc/CO_2 welding. Electrodes are mild steel.

SUBMERGED ARC WELDING

General

With the exception of the manual metal arc process, submerged arc welding is probably the most widely used welding method. It is, however, limited in application being suitable only for butt and fillet welds in the flat or horizontal–vertical position and in positions where there is plenty of room. With properly selected flux and wire it is capable of producing welds to highest radiographic standards.

As the process operates with a continuous coil of electrode wire, downhand butt welds requiring multiple runs to fill the joint can be made with minimum stops and starts. Thus, circumferential joints in cylindrical bodies such as pressure vessels, pipes etc., can be made with one stop and start per revolution of the workpiece, this stop being necessary to re-set the position of the welding head. Consequently, the possibility of stop and start defects is minimised; a most important consideration when reliability in costing is required. Although most widely applied to welding of seams and joints in mild steel, low alloy high tensile steel, creep resisting steels, and to a lesser extent on stainless steels

Table 6.3(a) METAL ARC WELDING (MECHANICAL PROPERTIES OF TYPICAL SHIELDED ARC ELECTRODES (ALL-WELD-METAL DEPOSITS))

Weld metal Type	Electrode class AWS	Electrode class BS 1719	Current	Yield N/mm²	Ultimate N/mm²	Charpy V J	at °C	El%	RA%	Remarks
Mild steel	E6010	E101	d.c. +	350	430	27 min	−29	35	60	Low yield sometimes advantage.
½% Mo. MS.	E7010	E101	d.c. +	380	480	95	−10	30	40	
Mild steel	E6011	E104P	d.c.+a.c.90	400	480	95	−10	25	60	
½% Mo. MS.	E7011	E104P	d.c.+a.c.90	450	570	80	+18	28	60	
Mild steel	E6012	E217	a.c. d.c.	430	510	80	+18	26	40	Fair toughness and ductility. High yield.
Mild steel	E6013	E317	a.c. d.c.	400	500	75	−10	28	60	
Mild steel	E7016	E619H	d.c.+a.c.80.	480	540	80	−30	28	70	Good toughness and ductility.
Mild steel	E7018	E614HJ	d.c.+a.c.80.	410	530	74	−30	30	70	
Low alloy HT.										
1% Mn.	E7018	E611HJ	d.c.+a.c.80.	500	580	60	−30	28	70	
1½ Mn.	E7018	E614HJ	d.c.+a.c.80.	510	600	60	−30	28	70	
1% Mn-1% Ni.	E8018–C3.	E606	d.c.+a.c.70.	530	620	61	−50	35	70	Very good low temperature toughness and good ductility.
1½% Mn-3% Ni.	E8018–C1.	E606	d.c.+a.c.80.	550	670	55 / 20	−80 / −100	30	70	
1½ Ni-Mo.	E9016-G.	E619H.	d.c.+a.c.80.	550	690	40	−30	23	60	
1Mn-2Ni-Cr-Mo.	E12016-G.	E619H.	d.c.+a.c.80.	660	840	20	−20	20	60	
Creep resisting										
1 Cr-½ Mo	E8018–B2.	E614HJ	d.c.+a.c.80.	620	720	—	—	20	70	Good creep strength and ductility.
2¼ Cr-1 Mo	E9018-B3**	E614HJ	d.c.+a.c.80.	680	770	—	—	20	60	
5 Cr-½ Mo	E502-16***	E614HJ	d.c.+a.c.80.	680	770	—	—	20	60	

*Mechanical Properties** (header span over Yield, Ultimate, Charpy V J at °C, El%, RA%)

* The mechanical properties indicated are only typical. They depend on welding procedure; particularly notched impact energy absorption.
** AWS/ASTM.A316—58T.
*** AWS/ASTM.A298—55T.

Table 6.3(b) SOME TYPICAL MANUAL METAL ARC WELDING ELECTRODES FOR HARD FACING

Description	Coating R = rutile B = basic	Alloy type	Current	Hardness Brinell	Machinable	Remarks
Pearlitic air-hardening alloy	R	LC. 1 Cr–Mo	a.c., d.c.	200–250	Yes	Moderate impact and abrasion resistance.
Martensitic-pearlitic A.H. alloy	B	LC. 3 Cr	d.c., a.c.	300	Yes (carbide)	
Martensitic-pearlitic A.H. alloy	R	LC. 3 Cr–Mn	a.c., d.c.	360		Medium impact and abrasion resistance.
Martensitic-pearlitic A.H. alloy	B	LC. 2 Cr–Mn	d.c., a.c.	360		
Martensitic-ferritic A.H. alloy	R	LC. 13 Cr	a.c., d.c.	35–400		Hardness independent of cooling rate. Corrosion resistant.
Martensitic A.H. alloy	B	MC. 13 Cr	d.c., a.c.	500	No	
Martensitic A.H. alloy	B	MC. 7 Cr–Mo–V	d.c., a.c.	560	No	Good impact and abrasion resistance.
Martensitic A.H. alloy	R	MC. 5 Cr–Mo–V	a.c., d.c.	600	No	Moderate impact. Good abrasion resistance
Martensitic and austenitic A.H. alloy	B	HC. 10 Cr	d.c., a.c.	600	No	Good abrasion resistance.
Martensitic A.H. alloy	B	MC. 7 Cr–Mn–Mo–V	d.c., a.c.	650	No	Moderate impact. Good abrasion resistance
Martensitic-austenitic A.H. alloy	R	VHC. 35 Cr–Mn	a.c., d.c.	600 (800WH)	No	Moderate impact. Very good abrasion resistance.
Tool steel alloys	B	MC. 8 W–Co–Cr–Nb	d.c., a.c.		No	Resists softening up to 600°C.
Tool steel alloys	B	HC. 8 Mo–5 Cr–W–V	d.c., a.c.		No	Resists softening up to 550°C.
Work hardening alloy	B	HC. 13 Mn	d.c., a.c.	180 A.D. 650 W.H.	Yes	Must be cold worked in service. Very good impact resistance.
Work hardening alloy (tubular)	—	H.C. 12 Cr–4 Mn	d.c., a.c.	340 A.D. 540 W.H.	Just	Poor abrasion until cold worked. V.G. impact resistance. Moderate abrasion resistance.
Carbide in austenitic matrix (tubular)	—	VHC. 23 Cr. 2 Mn	d.c., a.c.	630 Matrix 1600 carbides	No	Severe abrasion and impact resistance at low to medium temperatures.
Co–Cr–W alloy carbide in non-ferrous matrix	—	VHC. 45Co30Cr12W3Mo	d.c.	630 A.D.	Just (carbide)	Red hard. Resists severe abrasion and moderate impact.
Co–Cr–W alloy carbide in non-ferrous matrix	—	HC. 54Co30Cr5W3Mo3Ni	d.c.	450	Yes (carbide)	Resists severe impact and moderate abrasion. Red hard.
Tungsten carbide	—	Approx. 60% tungsten carbide	d.c.	680	No	Best abrasion and moderate impact resistance at all temperatures.
Carbide in austenitic matrix (tubular)	—	VHC. 27Cr1·5Mn4Mo	d.c., a.c.	700 (matrix) 1600 (carbides)	No	Most severe abrasion and good impact resistance up to medium temperatures.
Bronze. Surfacing	B	Cu–7% Sn	d.c. +	—	Yes	For resistance to corrosion and galling.

LC = Low Carbon. Less than 0·15%
MC = Medium Carbon 0·25–0·5%

AD = As deposited.
WH = Work hardened

it is also widely used for building-up work, either for reclamation or replacement of defective parent metal or for hard surfacing.

Submerged arc welding is a slag shielded process in which, as with manual metal arc welding using slag shielded deposits, slag-metal reactions have a pronounced effect on weld quality and chemistry. There are broadly two main methods of attaining the desired weld chemistry. The first is by employing a neutral flux with de-oxidation and alloying of the weld performed by the wire composition and the second by using an alloyed flux with a mild steel electrode wire. The latter requires more precise control of the welding parameters but provides the possibility of depositing a wider range of compositions. The former is preferred by some because it is less dependent upon control of operator and electrical variables, but the range of compositions is restricted to those which are readily available in wire form.

Although electrode suppliers market a range of such wires which, combined with appropriate fluxes cover almost all normal requirements, special situations may arise requiring non-standard wire compositions which may not be available unless ordered in vast quantity accompanied by extremely long delivery. In these circumstances the alloy flux principle is useful because fluxes are readily adjustable and the necessary wire also readily available. The engineer handling welding matters should, as a first step to using submerged arc welding, assess which method best meets his requirements. The choice of expensive equipment is determined to some extent by this assessment.

Shielding of the weld pool in submerged arc welding is effected by a heavy slag blanket. formed from fusion of the flux powder which is gravity fed into the joint ahead of the moving arc. Submerged arc welding is essentially a machine operation either fully automatic or semi-manual. The latter finds very limited application, (common use is for repairs to steel castings), because the risk of defects on straight seams is high. This is because the welder cannot see the joint line under the flux blanket and may fail to obtain correct fusion in the manual traversing of the welding head.

Submerged arc welding normally uses high current densities (compared with manual metal arc) with consequent substantially higher welding rates and therefore appears attractive from an economic point of view. However, these advantages are accompanied by very much greater mixing of welding wire metal and parent metal so that contamination of the resulting weld metal by undesirable elements in the parent metal is also much greater. This can lead to defects such as cracks and porosity which would not have occurred with the manual metal arc process. An example is the harmful effect of sulphur segregation in steel plate in promoting cracking. On parent steel of satisfactory quality, the process can produce weld metal of the highest radiographic standard at an economic cost (taking into account repair costs as well as initial deposition costs).

The process is widely used for welding the main longitudinal and circumferential seams of vessels where weld quality is an over-riding consideration such as pressure vessels, boiler drums, penstocks, high pressure pipe etc. It is normally carried out 'indoors' since weld quality deteriorates if the flux becomes damp; since welding is virtually confined to the gravity position manipulators and rotators are required to position and move the workpiece.

Adding the capital cost of this equipment to that of the welding head and power source means that the capital outlay is substantially higher than is required for manual metal arc welding. Given an adequate work load plus savings on repairs on good quality steel and cost savings per foot the submerged arc process is attractive.

Data summary

TYPE OF OPERATION, MECHANISED OR AUTOMATIC

The mechanised method, in which the wire feed is motorised but the movement along the seam is manually controlled is not widely used. The automatic method uses either a mechanically propelled welding head or stationary head with mechanically moved workpiece.

EQUIPMENT

The welding head, with or without control panel, carries an electrode wire drive unit and coil spool, flux hopper and dispenser. There is also a traversing mechanism if the head moves over a stationary workpiece.

POWER SOURCE

Either direct or alternating current may be used. Direct current may be supplied either by a variable voltage generator or rectifier or by a flat characteristic generator or rectifier. Generators will deliver up to about 650 A continuous output, rectifiers up to about 1200 A at 100% duty cycle. Alternating current is supplied by welding transformers which may be designed to give a drooping characteristic. Transformer output may range up to 2000 A.

Choice of a.c. or d.c. supply

Technical considerations alone should decide the type of current supply because it is important to the economic working of the process that it should operate smoothly. The following points should be considered:

1. D.C. can be used on positive or negative polarity, substantially affecting bead shape and depth of penetration. D.C. positive provides maximum penetration and smoothest blend-in between weld and parent metal therefore minimising trapped slag in deep multi-run joints. D.C. provides much easier arc striking, so much so that where the type of work requires repeated starts, d.c. is practically mandatory for economic operation.

2. Weld quality depends on arc length control. If the arc snuffs out or shortens unduly without immediate current increase the conducting slag cools and the working of the process is temporarily suspended causing defects. D.C. provides closer arc control; where the weld deposit must be laid on varying contour levels, particularly at high welding speed the very rapid response of the d.c. arc control is virtually essential to success. On the contrary where minimum penetration with maximum stand-up of the weld metal is desired, as may be the case in building-up and hard facing, the d.c. negative arc is preferred. d.c. negative also provides maximum deposition rates.

3. The drawback to d.c. is the tendency to produce arc blow. This increases as the current increases, so that d.c. applications tend to operate at less than 700 A. However, too much emphasis should not be placed on arc blow as there are many common situations where it is non-existent or can be avoided by suitable placement of the earth return lead e.g. in circumferential joints on cylindrical vessels. It is most noticeable at the start and finish of longitudinal seams and its effects can be eliminated by run-on and run-off plates.

4. A.C. minimises arc blow being used for high current application and situations where a d.c. arc would give unacceptable arc blow. Examples are longitudinal welds inside relatively small diameter pipe, and short external longitudinal welds on cylinders.

5. A major requirement in submerged arc welding is to obtain control of arc length adequate to give welds of the specified soundness. With a.c., control is usually obtained by using the voltage across the arc to determine the voltage in the electrode feed motor armature giving a downward feed rate dependent upon arc voltage. Current and voltage fluctuate due to the inherent inertia in this system. With d.c. it is possible to have a downward movement control and also an upward movement (retraction) so

that maximum response to varying arc length is obtained combined with the ability to obtain standing starts at any selected position; a built-in exciter on the generator usually energises the system.

6. An alternative and highly effective way of maintaining a constant arc voltage is to use a constant potential d.c. power source, especially designed to give instant high short circuit current characteristics. Using a constant wire feed speed motor, any variation in burn-off rate or other cause of arc length variation produces an immediate change in the current to bring the arc length (arc voltage) back to its original setting thus balancing the resistance across the arc. The efficiency of this method depends upon the power source electrical characteristics but, given instant response, the high surge currents available make for simple and reliable starting when the electrode is moving relative to the workpiece. This power source is particularly effective on thin material.

7. Fluctuation in output from welding power sources can affect weld quality. Non-rotary sources (transformers and rectifiers) coupled direct to the input line produce a fluctuating output if the input side is subject to this trouble. Generators will smooth out normal line fluctuation.

8. Submerged arc welding can be used to economic advantage over a wide range of applications, providing there is adequate access for the welding head and power source and arc control system together with the wire-flux combination are correctly selected. Some points to be given attention in making the selection are summarised in Table 6.4[5].

Electrodes

The electrode for submerged arc welding is a bare wire in coil form usually copper coated. Two types are available (a) solid wire or (b) tubular wire. The solid wire is widely used for general fabrication of mild and low alloy steels, stainless steels and non-ferrous metals. For welding mild steel and low alloy steels it is either a low carbon ultra-low silicon steel or a silicon-killed steel with manganese addition and sometimes low alloy additions, the selection of either type depending upon the type of flux to be used with it i.e. a flux with manganese or manganese and alloy additions or a neutral flux respectively. The tubular wire (made by forming narrow strip into a tube) carries allow powders which permit the economical production of a wider range of weld compositions than is possible by using the solid wire type. The tubular wire is widely used for hard facing.

Electrode wire is commonly copper coated to obtain a good shelf life and assist current pick-up. However, when welding steels susceptible to hydrogen embrittlement or when depositing high tensile weld deposits, low hydrogen weld metal should be the objective and inferior copper coating can be a cause of undesirably high hydrogen levels. Uncoppered wire can be kept by packing in containers which carry moisture absorbents and storing in heated storerooms. A dull coppered surface is a cause for suspicion.

Classification of electrode wires for submerged arc welding has been set out by the American Welding Society in Specification AWS A5.17–69. With coated manual electrodes, wire and coating are one unit so that such electrodes can be classified according to the type of coating and its effect on weld mechanical properties. In submerged arc welding, any wire may be used with a number of different fluxes with substantially different results in respect of weld quality and mechanical properties. Consequently the AWS system is based on defining the chemical composition range for the electrode wire and the fluxes are classified according to the mechanical properties which they will confer when used with one or other of the wires. Accordingly the completed classification information is provided by a combination of letters and numbers for the flux and letters and numbers for the wires.

Apart from special wires with alloying elements or with special de-oxidation additions, the whole range of wires for submerged arc welding is covered by low carbon (less than

Table 6.4 SUBMERGED ARC WELDING. SELECTION OF POWER SOURCE AND ARC CONTROL SYSTEM

Application requires:	Current	Power sources	Arc Length control
Fast, accurate arc striking. Close arc control.	d.c.	d.c. motor-generator. Variable voltage and current with control excitation.	Voltage control with wire retraction.
Difficult contours.	d.c.		Most sensitive for high speed welding on thin material.
Control of bead shape.	d.c. +		
Maximum deposition rate on sheet metal.	d.c. −	Subject to line voltage fluctuation.	Best for arc striking.
Maximum penetration.	d.c. +	Particularly for: High speed starts.	Best for contour changes.
Minimum penetration.	d.c. −	Line voltage fluctuation.	
Maximum control of burn through	d.c. −		Voltage control without wire retraction.
Smallest pool as on small diameter circumferential welds.	d.c. −	Modified constant Potential M.G. set.	Simplest for a.c.
Short welds. Good arc striking.	d.c.	Rectifier. Constant potential.	Current control.
Starts with moving work-piece.	d.c.	Particularly for: Short arc length (thin metal).	Constant potential power source.
Arc blow. control.	d.c.	Moving starts.	
Butts in round or square bar stock.	a.c.	Intermittent welds. Line voltage fluctuation.	C.P. Power source plus wire retraction and voltage control:
Internal bead on small diameter pipe.	a.c.	a.c. transformer.	For sheet metal and low voltage applications.
Multiple arcs.	One arc a.c.	Drooping characteristic, variable voltage and current.	For moving starts particularly on sheet metal.

0·2%) steel with either low Mn (0·3–0·6%), medium Mn (0·8–1·5%) or high Mn (1·5–3·0%) content. Wires supplied by U.K. and continental manufacturers are broadly in line with the A.W.S. compositions and groups. The A.W.S. classification for wires and fluxes is given in Table 6.5 (Tables 1 and 4 of A.W.S. Specification).

Fluxes

The flux must be an electrical conductor when molten. After many years of development, commercial submerged arc welding fluxes for fabrication of mild and low alloy steels have become reasonably standardised in the U.S., U.K. and on the Continent. There are two main groups; fused and agglomerated. The fused fluxes are powders produced from pre-melted flux mixes; they are sub-divided into (a) the Calcium or Magnesium Silicate type which does not add Mn to the weld metal and (b) the Manganese–Silicate type which adds manganese.

Agglomerated fluxes contain alkaline earth oxides, plus de-oxidising agents, plus a binder and are produced as small pellets or granules of uniform composition. The majority transfer Mn to the weld metal. However, Oerlikon Electrodes Ltd., Hayes, Middlesex, produce a non-manganese bearing agglomerated flux with 3% basicity which is particularly advantageous when good notch toughness is required at abnormally low temperatures (British patent No. 1 080 337).

Manganese is an essential element in the weld metal and it will be appreciated that, in normal circumstances, the manganese free fluxes are used with medium or high manganese wires. The manganese bearing fluxes may be used with medium manganese wires for applications requiring a high manganese weld metal but are most commonly used with wires of low manganese type. The ratio of basic oxides to silica in the flux mix, the basicity of the flux, affects usability and weld metal mechanical properties, particularly the notch toughness. Thus, notch toughness at low temperatures improves as basicity increases.

Highly basic fluxes can only be used satisfactorily with d.c. supply (positive polarity). Alloy bearing fluxes tend to be basic rather than acid in character to obtain control over alloy recovery. For high quality, clean weld metal with optimum notch toughness for the composition basic fluxes with d.c. plus polarity are frequently chosen. Less basic fluxes have increased slag viscosity, an important asset when making circumferential welds on small diameter cylinders or when welding thin material.

The A.W.S. method of classification of fluxes by the mechanical properties achieved in combination with defined wire compositions ignores the above-mentioned practical aspects of the effect of flux formulation. Therefore, it is necessary for the engineer to state his requirements in terms of the application to the electrode wire supplier so that he can obtain recommendations on flux and wire selection.

Because submerged arc welding was based on acid fluxes in its early development stages with consequent rather inferior notch toughness, it became accepted practice to require manual arc welding for notch tough applications. This situation has altered with the development of basic fluxes. Comparison notch toughness figures are given in Table 6.6.

Multiple-arcs

Increased deposition rates and welding speeds can be obtained by using multiple arcs, either as two wires powered from the same source or two or more wires with separately controlled power sources.

Procedures

Submerged arc welding, with its high current density and/or high speed, cannot accommodate variation in joint fit-up as easily as manual metal arc welding. Plate preparation and fit-up should be to a high standard.

Table 6.5 SUBMERGED ARC WELDING
(Classification of Wire and Fluxes for Mild Steel. AWS Specification A5. 17–69).

AWS. class.	Electrode wires — Wire composition % Single figures = maxima			Fluxes					
	C	Mn	Si	AWS class	Yield strength min. N/mm²	Ultimate strength N/mm²	El. % in 50 mm min.	Charpy V J*	Temp. °C.
Low manganese				F60xxxx				None required	
EL8	0·10	0·3–0·55	0·05	F61xxxx				27	−18
EL8K	0·10	0·3–0·55	0·10–0·20	F62xxxx	350	430–550	22	27	−29
EL12	0·07–0·15	0·35–0·6	0·05	F63xxxx				27	−40
Medium manganese				F64xxxx				27	−51
EM5K	0·06	0·9–1·4	0·4–0·7						
	0·03SS, 0·03P, 0·05–0·15Ti, 0·02–0·12Zr, 0·05–0·15Al.								
EM12	0·07–0·15	0·85–1·25	0·05	F70xxxx				None required	
EM12K	0·07–0·15	0·85–1·25	0·15–0·35	F71xxxx				27	−18
EM13K	0·07–0·19	0·9–1·4	0·45–0·7	F72xxxx	410	500–650	22	27	−29
EM15K	0·12–0·2	0·85–1·25	0·15–0·35	F73xxxx				27	−40
High manganese				F74xxxx				27	−51
EH14	0·1–0·18	1·75–2·25	0·05						

E = Electrode.
L = Low manganese.
M = Medium manganese.
H = High manganese.
Numerals indicate nominal carbon level.
K = Silicon killed steel.
Where Si = 0·05% maximum the steel may be made by rimming practice.

F = Flux.
First digit = Tensile strength (minimum) in 10 000 lb/in² e.g. 60 000 lb/in².
Second digit = Minimum impact strength e.g. 0 = none. 4 = 27J at −51°C.
xxxx = The electrode wire code number which, in combination with the flux gives the stated minimum mechanical properties.
* J = Minimum average of three tests with 20J minimum individual test.

Table 6.6 TYPICAL NOTCHED IMPACT VALUES FOR SUBMERGED ARC WELDING MILD STEEL

| AWS classification | | Current | Weld composition (nominal) | | | Flux | | Charpy V |
Wire	Flux		C.	Si.	Mn.	Type	Basicity	J. at °C.
EL12	F62–EL12	a.c., d.c.	0·1	0·5	0·9	Fused Mn–Si.	Acid	40 −20
EM12K	F72–EM12K	a.c., d.c.	0·1	0·5	1·2	Fused Mn–Si.	Acid	40 −20
EL12	F62–EL12	a.c., d.c.	0·08	0·5	1·2	Agg.	Basic	49 −20
EM12K	F72–EM12K	a.c., d.c. 90	0·1	0·5	1·5	Agg.	Basic	49 −20
EL12	F72–EL12	d.c.	0·1	0·2	1·1	Agg.	Highly basic	59 −20
EM12K	F72–EM12K	d.c.	0·1	0·2	1·5	Agg.	Highly basic	59 −20
*		d.c.	0·08	0·2	1·4	Agg.	Highly basic	68 −60

* Silicon killed low carbon—high manganese plus Ti-Al-Zr.

Procedures should be established before undertaking work where experience is not available as a guide. Typical procedures are given in Welding Handbook[7].

ELECTRO-SLAG WELDING

Electro-slag welding is an automatic process for welding material 18 mm or thicker in the vertical position. Square edge plate preparation is used. Joints are limited to butts and T butts. Heat for fusion is obtained by the current in the consumable electrode wire which passes through the molten slag formed by melting of a flux formulated to have high electrical resistance in the molten state.

The joint is set up with a wide gap (approximately 25–36 mm depending on plate thickness), the very large weld pool being contained in the joint by water cooled copper shoes. It is filled in one pass. One, two or three wires are fed into the joint with or without a reciprocating motion to ensure uniform heat generation, the plate thickness determining the number of wires. One head with three wires can weld plate 450 mm thick. The copper shoes rise up the joint to prevent spilling of metal and slag (Figure 6.1). These shoes form part of the welding head.

A variation of the electro-slag process is known as the Consumable Guide process. Here the head remains stationary and feeds one or more wires down a tube which melts into the pool. This method is not applicable to long joints; the equipment (Figure 6.2) is substantially cheaper than the conventional slag welding machine.

Data summary

POWER SOURCE

Either d.c. feeding one electrode wire (Constant potential).
Or d.c. feeding two electrode wires in parallel (Constant potential).
Or a.c. three-phase, one electrode wire per phase. Flat characteristic.
A typical E.S. power source would supply 900 A per wire continuous welding.

WIRE FEEDING UNIT

The wire is fed at constant speed, adjustable to suit requirements.

WELDING HEAD

For the conventional electro-slag process a motorised vertical traversing welding head incorporates an oscillating traverse for the wire guide tubes, the water cooled copper shoes and the wire feeding unit.

For consumable guide welding, where joint length is short, traversing copper shoes are not used; instead copper blocks may be piled on top of each other or a stationary water cooled shoe may be used for repetition work.

ELECTRODES

It is customary to use a standard wire size of 3 mm; the wire is normally copper coated. Solid wire is normally used but cored wire containing flux material has advantages in special situations.

Typical electrode wire compositions are the same as for submerged arc welding

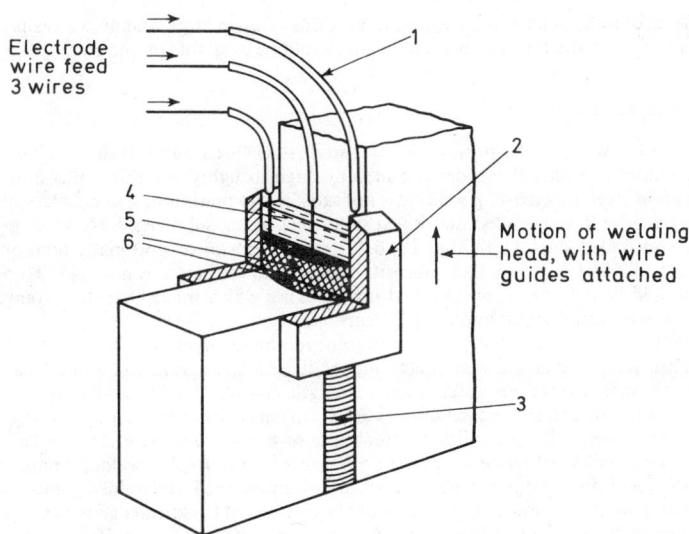

Figure 6.1. Schematic of conventional electro-slag process: 1. Curved wire-guides with oscillating motion; 2. Water-cooled copper shoes; 3. Completed weld; 4. Molten slag pool; 5. Molten weld metal; 6. Solidified weld metal

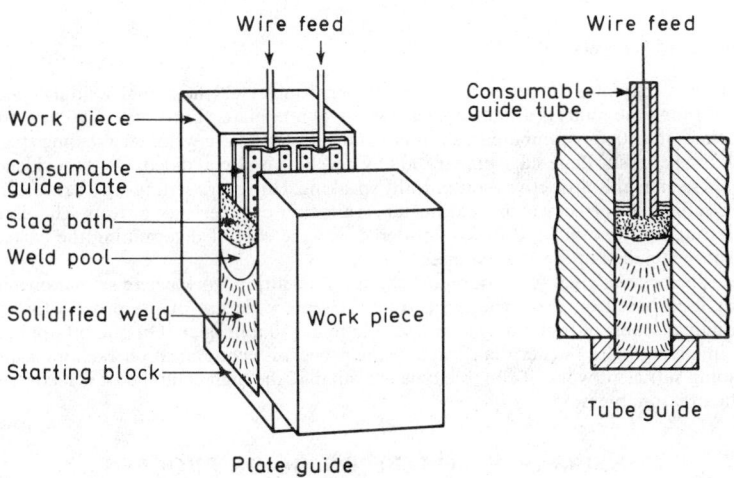

Figure 6.2. Consumable guide electro-slag welding
Note. Copper retaining blocks have been omitted

with the medium or high manganese content. Coils of extra large weight are available as sufficient wire for the full length of the seam should be available in one coil.

Features of the process

In electro-slag welding the proportion of parent steel fused and mixed with electrode wire or guide tube plus electrode wire metal is high, roughly one-third filler and two-thirds parent steel. Electro-slag welding is not satisfactory on rimming steel unless special wires are used. On semi-killed and killed mild steel, using submerged arc wires of low carbon semi-killed steel with either 1% or 2% Mn, welds of exceptionally high quality in respect of freedom from slag, porosity and similar defects are possible. The weld metal should be low in carbon to avoid hot cracking which infers that the parent steel should not generally exceed about 0·25% carbon.

The ultra slow cooling rate of the weld joint minimises hydrogen cracking susceptibility in the parent steel and weld metal but produces a large grain size weld. This tends to give a comparatively poor notch impact strength as determined by conventional tests. Consequently, for pressure vessel applications, current codes require normalising after welding. However, full scale brittle fracture tests of the Wells Wide Plate type have not indicated any increased tendency to brittle fracture in electro-slag welded joints in the as-welded condition compared with metal arc or submerged arc welded joints. Stress relief treatment after welding would generally be considered beneficial in line with normal welding practice applied to heavy plate fabrication. On material up to 50 mm thickness stress relief treatment is normally considered unnecessary.

Fluxes

Electro-slag welding fluxes produce complex silicate slags containing SiO_2, MnO, CaO, MgO, Al_2O_3. Calcium fluoride is added to increase electrical conductivity and lower slag viscosity. Slags based on CaF_2–CaO have a strong de-sulphurising action which assists the welding of steels higher in carbon than 0·25% without hot cracking in the weld metal. Fluxes must be kept dry.

Economics of the process

On heavy steel plate, for pressure vessels, boiler drums etc., the actual welding speed (rate of filling the joint) is about twice as fast in 40 mm plate, four times in 90 mm and eight times in 150 mm compared with multi-run submerged arc welding. Welding speed is 1–1.7 m per hour. Plate-edge preparation by bevelling is also avoided. On this evidence it would seem highly attractive economically speaking. However, 'setting-up' the machine and selecting the correct welding parameters is a matter of experience or tests. Therefore the 'setting-up' time must also be considered and the cost of determining the correct procedure included in the cost estimate.

Thus the process gives full economic benefit on repetition work where set parameters can be used based on experience. An example is the wide use of the process for the longitudinal seams of boiler drums in steel 125 mm–150 mm thick. On one-off applications involving plate thicker than 150 mm the process is economical on each joint and providing sufficient work of suitable type is available, the high capital cost is recovered within a reasonable time.

CONSUMABLE NOZZLE OR GUIDE PROCESS

The consumable nozzle process, while theoretically suitable for very thick sections, is more usually applied to metal up to, say, 150 mm because it is more manageable in these thicknesses. The equipment needed is much simpler than for the conventional process consisting of a constant potential generator or rectifier or transformer with flat characteristic and a wire feed unit. These are also the essential ingredients of submerged arc equipments and, by slight modification, they can be adapted for consumable nozzle welding.

Consumable nozzle welding is widely used in welding massive components of complex profile and contour in the joint where the portable nozzle equipment can be mounted on the work. Various designs of machines of this type are available[8]. An example of such a machine (Russian A–545) operates up to four nozzles. This is clamped to a removable plate (tack welded to the work). The electrode wires are fed to the machine from reels in any convenient position. The wire drive unit consists of motorised driving rolls mounted on a common splined shaft and can be spaced to suit the section thickness. Four wires enable plate 200 mm thick to be welded. All wires (3 mm) are fed simultaneously and at constant speed, adjustable over a range of 60–120 m/hr. None of the wires are insulated from each other; a single phase flat characteristic transformer is the power source, four electrodes taking up to 2000 A. Flexible steel hose takes the wire from the reels into the drive rolls and again, on emerging from the rolls, into the current pick-up and nozzle. This flexibility is important in decreasing setting-up time as the machine need not be accurately located over the joint.

This machine illustrates the essential simplicity of consumable nozzle welding.

Mechanical properties and applications

The weld metal mechanical properties are determined to a considerable degree by the composition and cleanliness of the parent steel. However, proper selection of wire and flux and suitable parent metal confer strength and ductility equal to or better than the parent metal.

In the as-welded condition degradation of notched impact properties in the heat-affected zone results in values lower than the weld metal. Where notched impact requirements must be met, it is necessary to normalise the completed joints. In general, electro-slag welding is an acceptable and economic method of welding thick steel. It finds application for ships' hulls, boiler drums, press frames, nuclear reactors, turbine shafts, rolling mill housings and similar heavy fabrications[10–13].

GAS METAL ARC WELDING (MIG)

Description

This is a manual semi-automatic welding process which can be made fully automatic, although this is not usual. An arc is maintained between the bare electrode wire, continuously fed to the arc via a hand manipulated gun (torch) supplied from a motorised wire drive unit. The wire is of fine gauge. A flow of shielding gas passes into the gun and surrounds the arc and the weld pool. Wire is fed at pre-selected constant speed, the speed determining the current. The arc length is controlled by the power source.

Mig welding is either operated at high current density, when the arc gives a fine 'spray' type of metal transfer or at lower density when the metal passes to the pool as individual globules of comparatively large size. The former is known as spray transfer, the latter as dip transfer. The relationship between current, wire feed speed, wire diameter and type of transfer is shown is Figure 6.3.

Spray transfer must be used in the flat or horizontal position when welding steel. Dip transfer is normally used for vertical or overhead welding; Dip transfer using 'pulsed' arc (in which the globules are pinched off at regular intervals) allows welding of thin sheet metal in the flat position. It also assists dip transfer in general work.

The equipment, power source plus gas supply, is bulky and the length of tubing connecting between the gun and the wire drive unit must necessarily be limited. Consequently accessibility to the working point is not as good as with manual metal arc. For joints with open access, a radius of about 3 m is available from the wire drive unit. This may be installed above the shop floor to conserve space.

Characteristic features of the process

The spray transfer arc permits weld metal to be deposited at up to 9 kg/hr, which is about double the maximum rate obtainable with the manual metal arc process. However, there are several features to bear in mind in comparing the two welding methods. Mig welding uses fine wire, normally in the range 0·6–1·6 mm for manual operation. This wire is

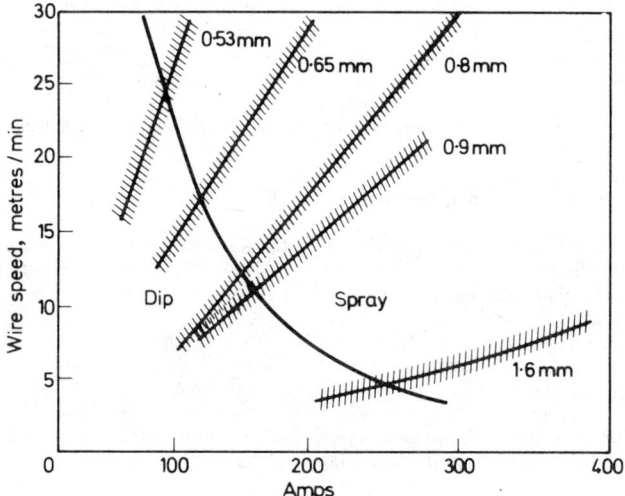

Figure 6.3. Mig welding process. Burn-off curves. Steel. DC positive. Argon or Argon/2 per cent O_2

necessarily expensive to produce so that the higher cost per kilo of the wire must be offset by the greater deposition rate, which implies the need for a high ratio of arcing time to over-all time. This is not necessarily obtainable on all types of application.

It is sometimes suggested that, as there is no flux and welding is from a continuous coil, the arcing time to over-all time ratio must be high compared with manual metal arc welding. However, the slag from modern coated electrodes is virtually self-detaching and the Mig process, particularly with CO_2 gas, requires occasional removal of adherent slag in multi-run joints and stoppages for removal of spatter from the nozzle tip. Nevertheless on down-hand welding of heavy fabrications under suitable conditions cost savings of up to 50% are possible compared with manual metal arc welding.

Wire drive units are generally of the push type but the technical difficulties of pushing the fine wires through the hose and gun have led to the use of pull type units with the unit located on the gun. Such units are only used for the smallest wire gauges. This increases the weight of the gun and makes balancing more difficult in a situation where, using very fine wire, the best balance is desired.

Spray transfer, using the higher currents in the 250–500 A range generally requires water cooling of the current cable and gun. If not supplied from the mains, this requires a recirculating system using a portable tank. Water cooling adds to the weight of the gun and the complexity of the equipment. Compared with manual metal arc equipment the equipment required for the Mig process needs considerably more attention and maintenance for satisfactory operation.

All gas shielded processes should be operated in still air to achieve consistent quality levels since the gas shield issuing from the gun can be disturbed by draughts. The distance between weld pool and gun tip is usually held at not less than 12·5 mm during welding and the issuing gas under still air conditions can cause turbulence with some mixing with entrained air unless conditions are correct. Turbulence depends on nozzle-to-work distance, nozzle diameter, gas velocity, the physical properties of the gas and the design of nozzle and gun body.

Mig wires for welding steels with argon/oxygen gas mixtures have excess silicon, manganese and/or aluminium over and above the amount necessary to kill the steel to combine with oxygen picked-up from the gas and from entrainment of air. When welding in draughty or windy conditions the shielding efficiency may be reduced even with increased gas supply so that more than normal oxygen is absorbed. This need not necessarily induce porosity; it depends on the excess of de-oxidising elements in the wire but it does produce welds with a greater non-metallic content, particularly when the wire is aluminium killed. In outdoor Mig or Tig (Tungsten Inert Gas) welding it is normal practice to erect a windbreak. In metal arc welding with coated electrodes these shielding uncertainties are less pronounced, consequently the chemistry of such welds tends to be more consistent than that of Mig welding under site-welding conditions.

It is normal practice to copper-coat Mig welding wire to prevent corrosion and assist current pick-up. Although Mig welding is inherently a low hydrogen process, inferior copper-coating introduces hydrated oxides, from under the copper skin, into the weld pool with consequent hydrogen pick-up. As in fine wire the ratio of surface area to wire volume is substantially higher than in the heavier gauge wire used in submerged arc welding, the effect of inferior coppering is proportionally greater in the Mig process. In welding low alloy high tensile steels the quality of the copper coating on Mig welding wire is of considerable importance.

When welding vertical and over-head joints in steel, obtaining consistent fusion of the parent metal can be a problem; the welders must have a high degree of manipulative skill. Unfortunately the lack of fusion defect, particularly as it tends to occur on the side walls of a joint, is not necessarily revealed by radiography. Ultrasonic testing is generally more successful but small defects of this type are extremely difficult to detect with certainty.

The problem arises from the fusion method inherent in gas metal arc welding; the parent metal is properly fused when it is directly exposed to the arc. The technique of gas metal arc welding is designed to obtain direct impingement of the arc on the unflooded parent steel. The spray transfer arc is 'stiff', enabling the welder to control the pool in the flat and horizontal positions without difficulty but in vertical and over-head welding in Vee butt joints where dip transfer must be used the type of arc makes control of the pool to avoid adhesion by molten metal running away considerably more difficult. Some of the factors affecting lack of fusion are discussed by A. A. Smith[15].

Power sources

D.C., positive polarity·current gives the best metal transfer and arc handling characteristics. Either generators or rectifiers are used. The drooping characteristic d.c. source, commonly used for manual metal arc welding, may be used for gas metal arc welding but is distinctly less satisfactory than constant potential or rising potential·type. These are produced in various output ratings, some being suitable for submerged arc and having a built-in inductance to give suitable arc characteristics for gas metal arc welding. Lightweight equipment suitable for dip transfer welding at up to 200 A is widely used.

Materials, wires and gases

The Mig process is widely applied to the welding of aluminium and its alloys for which, in conjunction with Tungsten Inert Gas Welding(Tig) it is the most successful technical and economic method. The process is also suitable for copper and nickel and their alloys. For the welding of steel, the carbon dioxide process and its variants (see below) is ideally suited to a wide range of applications, particularly if the welding shop is equipped to position work downhand.

For welding stainless steels the spray transfer arc is commonly used, the gas being argon/1% oxygen (the addition of oxygen from 1–5% being necessary for satisfactory 'wash' action in welding steel and stainless steels). Aluminium-bronze is successfully welded by the Mig process using an argon gas shield.

Gases, electrode, and polarity for several materials are given in Table 6.7.

Selection of shielding gas for Mig welding

Three main gases are used for shielding; argon, carbon dioxide and helium (in U.S.A.). A much wider range of gas mixtures is however used, i.e. mixtures of argon and oxygen, argon and CO_2, argon and helium etc.

The gases and mixtures have effect on arc characteristics, weld quality, welding economics and, most importantly, on weld chemistry and hence mechanical properties. Recommendations of the wire supplier should be followed because the wire composition has been selected to work with particular gases or mixtures.

Gas metal arc welding chemistry

Any fusion welding method is a miniature steel making operation. A significant factor in steel 'quality', particularly in respect of notched impact toughness, is the amount and mode of distribution, whether random or at grain boundaries, of the products of the necessary de-oxidation practice.

In fusion welding processes using fluxes, such as manual metal arc, submerged arc, electro-slag etc., the flux is a major factor in controlling the composition of the products of de-oxidation and hence in deciding whether they may be harmless as a random distribution or harmful as a grain boundary impurity. In gas metal arc welding with solid wire electrodes control has to be exercised by balancing a number of variable factors, not all of which are under the control of the welder but all are significant. The amount and composition of parent metal fused, the type of arc used, the technique used all influence weld chemistry, apart from the electrode wire composition and gas composition.

It would be realistic to expect to have to exert a higher degree of supervision on the shop floor, if consistent weld quality is a prime requirement, with gas metal arc welding than with the other fluxed processes, where the welder is concerned only with preventing trapped slag and is not also responsible for the de-oxidation practice. The complex nature of the inter-acting variables is described by A. A. Smith[23]. The development of the flux-cored wire for use with the CO_2 gas shield eliminates these difficulties, but introduces problems associated with weld hydrogen content. It is also difficult to produce a cored wire which is uniformly fluxed from start to finish of the coil; any breaks in the flux core will produce porosity.

Apart from de-oxidation control the major practical and technical advantage of the flux cored wire is that it produces a smooth blending of weld and parent metal which is

or alloy	Gas	type	sizes used mm.	Volts	Type	Amps.	type	Remarks
Aluminium and Al alloys.	Argon (99.9%)	99.9+ % Al / Al-5% Si. / Al-2½% Mg.	1·6 / 2·4 / 3·2		d.c.+ / d.c.+ / d.c.+	160-350 / 220-400 / 350-450	Spray / Spray / Spray	To prevent porosity use shaved wire properly spooled. Welding techniques Ref. 16.
Copper (Copper alloys) Cu–Ni	Argon (99.9%)	De-oxidised Cu (P, or Si, or B) / 70 Cu-30 Ni (Titanium)	1·6 / 1·6		d.c.+ / d.c.– / d.c.+	300-475 / 250-300	Spray / Spray / Spray	Preheat. TPC not suitable.
Bronze. Silicon / Bronze. Aluminium / Bronze. Manganese		Cu-3% Si / Cu-10 Al-1·5Fe / Cu-11 Al-3 Fe	1·6 / 1·6 / 2·0		d.c.– / d.c.+ / d.c.+	250-300 / 250-300 / 250-300	Spray / Spray / Spray	Keep interpass temp. low. Downhand welding preferred. Stranded wires. Welding technique Ref. 17.
Nickel and Nickel alloys	Argon (99.9%)	Similar to parent metal.	1·2 / 1·6		d.c.+ / d.c.+	150-300 / 200-400	Spray / Spray	See Ref. 18 for wires for nickel alloys and nickel.
Stainless steels (Ni–Cr)	Argon + 1% oxygen	Ultra low carbon (0·025 %*) or Nb stabilised. With or without Mo to suit parent metal.	0·9 / 1·2 / 1·6		d.c.+ / d.c.+ / d.c.+	60-100 / 150-260 / 175-350	Dip / Spray / Spray	Wire composition nominally matches parent steel but is balanced to give 4%–10% ferrite in weld deposit see Ref. 19 for procedures.
Mild steel	CO_2	Low carbon plus excess de-oxidants. / 1·2 Mn-1·0 Si / 1·2 Mn-0·6 Al / 1·2 Mn-Al, Ti, Zr / Flux cored wire rutile or basic	1·2 / 1·6 / 2·4 / 0·6 / 0·8 / 1·0 / 1·2 / 1·6 / 2·0 / 2·4	25-40 / 28-40 / 34-40 / 16-20 / 18-22 / 18-24 / 18-27 / 23-30 / 28-32 / 29-35	d.c.+ / d.c.+ / d.c.+ / d.c.+ / d.c.+ / d.c.+ / d.c.+ / d.c.+ / d.c.+ / d.c.+	150-280 / 225-480 / 330-500 / 50-80 / 60-140 / 80-175 / 120-200 / 150-350 / 250-450 / 350-500	Spray / Spray / Spray / Dip / Dip / Dip / Dip / Spray / Spray / Spray	Electrode wires see Ref. 15 Chap. 4. Mn–Si wire for notch toughness. Mn–Al wire for scaled plate. See Ref. 20. Downhand joints. HV fillets. Good blending. High speed. See Ref. 21.
Low alloy steel	Argon + 1%–5% O_2 / Argon 80% CO_2 20% CO_2	Special low alloy wires with Mn, Cr, Mo, V. / Low alloy wires with Mn, Mo, Ni. Flux cored. Basic	As for mild steel		d.c.+ As for mild steel.			For wire compositions, weld mechanical properties and procedures see Manufacturers' data sheets. Ref. 22.

* Maximum.

important in high tensile steel fabrication and in fatigue situations. The unfluxed weld tends to give a notch effect at the toe of the weld.

Applications of gas metal arc welding

The Mig process is the most widely applied welding method for aluminium and its alloys. With CO_2 gas shielding it is widely applied to steel, particularly in light repetition work.

Flux cored CO_2 welding is used for positioned work of a heavier type.

Further reading

The literature dealing with Mig welding is voluminous. A selection of useful articles is given in the following bibliographies, available from The Welding Institute, Abington, Cambridge:

Bibliography on Mig and Tig Welding 1965–70. Ref. No. Bib. 5/2/71. Compiled by M. C. Szaz

CO_2 Welding Bibliography. Compiled by T. D. Stephens.

GAS TUNGSTEN ARC WELDING (TIG)

Description

An arc is established between a tungsten electrode and the parent metal forming a weld pool into which filler wire is fed, generally by hand. Mechanised systems using wire feed and motorised electrode travel are also widely used on repetition work. Air contamination of the pool is minimised by a shielding gas flow, normally argon.

Materials for which tungsten inert gas (Tig) welding is suitable are as follows. Killed mild steel, killed low alloy steels, stainless and heat resisting steels, aluminium and aluminium alloys, magnesium and its alloys, copper and copper alloys, nickel and nickel alloys, titanium, zirconium, and others.

Features

The process is particularly suited to welding light gauge materials and for the root runs of butt joints in heavy gauges. The clear clean weld pool, high concentration of heat, absence of spatter, precise control of heat input, ability to weld with or without filler metal together with comparative ease of control over penetration in all positions make for reliable welding.

Tungsten inert gas welding is widely used for fusing the root runs in butt joints where exceptionally high quality is essential e.g., stainless steel piping for nuclear energy applications. Fully de-oxidised filler wire is essential and the composition should be balanced to give a self-fluxing weld pool. Wires suitable for oxyacetylene welding are generally satisfactory for Tig welding.

Equipment

An a.c. or d.c. power source with standard generators, rectifiers or transformers is used. The electrode holder (or torch) is either air or water cooled.

The filler wire is selected to match parent metal composition, except with materials which give hot cracking in the weld pool when a single phase alloy is produced for example high nickel stainless steels etc. Special compositions are then required. The wires must be clean.

The gas supply is usually provided from bottled gas but can be piped. Argon of 99·99% purity normally used.

The electrodes are usually of thoriated tungsten. This gives better arc starting, stability, longer life and higher current capacity than ordinary tungsten.

Welding practice

Either d.c. negative polarity or a.c. are used, depending on the metal to be welded, thickness and welding speed. D.C. negative polarity is normally employed, but metals which readily form oxides, such as aluminium, magnesium, aluminium-bronze or beryllium-copper are welded with a.c. On work where exceptional weld quality is required a super-imposed high frequency current may be used to assist arc striking when a.c. welding. When a.c. current is used the positive cycle has a cleaning action in breaking up and dispersing oxide films. A.C. transformer systems for Tig welding must be specially designed to overcome current rectification, this frequently being accomplished with high-frequency stabilisation.

In manual welding practice the arc can be struck by scratching the work lightly but in machine welding high-frequency striking is usual. Many manual equipments use high-frequency arc striking because it eliminates the risk of contaminating the tungsten electrode with parent metal which is particularly important for high quality work.

For most applications of Tig welding, the current ranges from 20–130 A (air cooled holders) but heavier water cooled types can take currents up to 800 A. The electrode size varies with the current and several other factors. Typical electrode sizes for various currents with d.c. and a.c. are given in Table 6.8[24].

Selection of current levels and hence electrode size depends on type of parent metal, joint preparation, thickness and welding position. When fusing the root run of a pipe

Table 6.8 GAS TUNGSTEN ARC WELDING. TYPICAL CURRENT RANGE FOR THORIATED TUNGSTEN ELECTRODES

Electrode dia mm	Gas	d.c. − ve amps	a.c. Unbalanced amps	a.c. Balanced amps
0·5		5–20	5–20	5–20
1·0		15–80	15–80	20–60
1·6		70–150	70–150	60–120
2·5	Argon 99·99%	150–250	140–235	100–180
3·25		250–400	225–325	160–250
4·0		400–500	300–425	200–320
5·0		500–750	400–525	290–390

joint in the fixed-horizontal position (pipe axis horizontal) the current level has to be gradually varied in moving from the downhand to the overhead position and a remote current control connected to the generator is advantageous where high quality welding is required. Procedures must be based on experience and the ability of the individual welder.

Further reading

Bibliography on Mig and Tig welding 1965–1970. Bib. February 5th 1971. The Welding Institute, Abington, Cambridge

OXYACETYLENE GAS WELDING

In this process, acetylene gas is mixed with oxygen in a welding torch and burnt to produce a high temperature flame which melts most commercial metals. Variation in melting capacity is obtained by using torch tips having different sizes of orifice. The type of flame, whether forcing or soft, can be selected by varying the gas flow.

Control of weld oxidation is effected by varying the ratio of oxygen to fuel gas, most welding being done with a neutral flame (1 to 1). It is possible to produce a carburising flame with excess acetylene or an oxidising one with excess oxygen. Gas is usually supplied from bottles but for substantial shop welding installations, piped supplies are used.

Features of the process

The parent metal is first melted and then bare filler metal added. Penetration is determined by tip size in relation to the heat sink from the surrounding parent metal. The type of flame is selected according to the metal to be welded and the joint position. Gas welding is applicable in all positions. Due to its low rate of heat input, it is valued for welding thin material and for welding in situations tending to produce burn-through; for this reason it is particularly suitable for welding the root runs of butt joints in piping especially if fit-up is poor.

The process is applicable to all common ferrous and non-ferrous materials, requires low capital expenditure and employs simple and readily mobile equipment. It is, therefore, the most widely used process for repairing ferrous and non-ferrous materials. A unique feature of the process is the use of a carburising flame to produce a thin layer of molten high carbon steel on the parent steel surface ('sweating') by carbon absorption into the austenite with consequent lowering of melting temperature. This 'sweating' method enables surfacing to be done without melting the parent steel in depth so allowing close control over the deposited metal composition.

The ease of control of the weld pool and slow cooling rate arising from the wide spread of heat make it ideal for surfacing operations with ferrous hard facing alloys, cobalt–chromium alloys, tungsten carbide, nickel and cobalt based hard facing alloys by the 'sweating' method. The process is also the best method for welding cast irons with cast iron filler rods. Bronze welding (in which the parent steel is raised to 'wetting' temperature with bronze filler addition) using the oxyacetylene process is widely employed for cast iron repairs and steel surfacing.

Flux is not required except for cast iron, stainless steel, aluminium, bronze and brass. Aluminium and stainless steel are seldom oxyacetylene welded, argon arc and Mig welding being preferable.

The slow heating and cooling cycles of oxyacetylene welding lead to difficulties with some materials (due to metallurgical damage) and to more distortion than with other fusion welding methods. On the other hand the slow cooling is an advantage on materials tending to harden or embrittle.

Filler rods

For oxyacetylene welding of steel, the filler rod should have sufficient silicon to deal with oxygen pick-up and sufficient manganese to produce a self-fluxing weld pool in which the fluid manganese–silicate slag rises to the surface. A composition around 0·3% silicon, 1% manganese and carbon less than 0·15% would be fairly typical of high quality gas welding rods.

Owing to the slow cooling rate oxyacetylene welding produces welds with yield strengths appreciably lower than in metal arc welding deposits.

Welding practice

Considerable practice is necessary to develop the manual manipulation motions of torch and filler wire combined with travel speed necessary for consistent attainment of full fusion. There is no substitute for a training course under skilled tuition.

The process is seldom used on steel thicker than 2·5 mm. unless special factors warrant it because it is not economically viable. See Reference 25.

SPOT SEAM AND PROJECTION WELDING

Coalescence is obtained in spot, seam and projection welding by electric resistance heating in the two overlapped parent metal parts between copper or copper alloy electrodes which induces the development of an internal melted nugget, half in the top and half in the bottom part. The electrodes apply pressure. In spot welding two conical electrodes (tips) produce a single spot weld.

In seam welding the electrodes are discs or wheels which, rotating continuously along the seam, produce a series of overlapping spot welds by virtue of a pre-set cycled current input. In projection welding, single or multiple projections are mechanically raised in one parent part and the assembly of top and bottom parts placed between flat platens in the welding machine. Passage of current causes resistance heating in the projections which produces a melted nugget at each projection. Thus, one pressure cum current cycle of the process can produce a multiple number of projection welds.

Applications

Spot, seam and projection welding are indispensable for the mass production of all types of sheet metal fabrication which do not involve corrosion resistance.

Joints

Joints are almost 100% of the lapped type. It is technically possible by using a seam welding wheel on a joint with a very small lap to squash the seam flat so producing a form of butt joint.

However, in practice the necessary accuracy of lap is difficult to achieve consistently and the process, known as mash seam welding, is not widely used.

Materials suitable for spot, seam and projection welding

The quenching rate from the molten state is higher in these methods of welding than in any other. If the material will undergo quench-hardening, some embrittlement of the weld nugget must be expected. However spot, seam and projection welds should not be intentionally subjected to tensile loading. The shear strength of the welds depends primarily on the weld size (the spot weld periphery), so that internal defects such as porosity are not significant and cracks are significant only where fatigue strength demands consideration.

Mild steel with normal residual alloy content is not hardened unduly up to approximately 0·15% carbon but an above average Cu and Ni content (say 0·3% Cu plus Cr and Ni = 0·3%) will induce hardening in steel at 0·12% carbon. Most mild steel sheet used for spot welding has less than 0·12% carbon. The austenitic stainless steels are suitable but spot welds in ferritic and hardenable (other than the precipitation hardening grades) stainless steels are brittle and the process is not well suited to them.

Aluminium is readily spot welded but the welding parameters must be closely controlled because the metal has a short plastic temperature range.

Surface preparation

Consistent spot welding results depend on maintaining a standard sheet surface condition, that is to say, one free from any oxide or scale.

Equipment

The range of spot, seam and projection welding machines is almost unlimited. General purpose single spot and seam welding machines find wide application but special machines having multiple spot welding heads for specific applications are also widely used in automobile, railway carriage and similar manufacturing industries.

Practically all units use a single-phase transformer with low secondary voltage and output from several hundred amperes to 50 000 depending on the application. Capacitor discharge stored energy systems from a.c. are sometimes used as also are d.c. stored energy systems from capacitors or batteries. For general information on spot, seam and projection welding reference can be made to the Welding Handbook, Fourth Edition, Section 2, Chapter 30 and to Chapter 32 for information on equipment.

Quality control

Although in highly critical applications, radiographs of spot welds have been undertaken, it is generally impracticable to undertake non-destructive tests on welded fabrications. Therefore quality control is imposed by prior testing of samples accompanied by rigid control over sheet surface condition, and welding machine variables (current, pressure, time).

Since the most important defect is non-formation of a nugget and consequently mere adhesion, a common test is to separate the welded parts by pulling apart (peel test). If the test causes metal to be torn out instead of separating at the original mating surface line a fluid nugget has been obtained and the remaining factor to check is the size of the nugget.

SOLID PHASE WELDING

In solid phase welding, coalescence is effected by pressure applied at a temperature sufficiently high to promote crystal growth, while solid, across the interface between the separate parts.

Processes

1. FORGE WELDING

Blacksmith or Forge Welding.

2. PRESSURE GAS WELDING

Pressure gas welding is an automatic process developed for butt welding solid sections and hollow cylindrical sections. The butting faces, under pressure, are heated by multiple oxyacetylene torches (gas pressure welding) up to around 1200°C. Upsetting and welding occur when the combination of pressure and temperature promote deformation. The slow heating and cooling impart minimum hardening on hardenable steels but steels subject to overheating, such as some electric arc qualities, may suffer damage. The process has been used fairly extensively to make continuous welded tracks on the railways. It has also been used for line-pipe welding. Machined faces are required, free of rust, with good fit-up[26].

3. FLASH BUTT WELDING

This is a hand operated or automatic process in which pressure and upsetting of the butting faces follows pre-heating from electrical resistance heating and flashing. Upsetting should extrude the melted and oxidised materials. A properly made flash butt weld differs from a pressure gas weld or upset butt weld in having a very small swelling with sharp angles at the joint, the extent of heated zone behind the joint being markedly less.

A further difference is that, whereas in the upset butt and pressure gas processes pressure is applied at the start and maintained during heating-up until upset occurs, in the flash butt process, the parts are not in contact during flashing and the upsetting force is applied suddenly when a sufficient plastic zone has been developed. Although a machined face to the parts is not necessary, and indeed is undesirable as it makes flashing less easy to establish, the welding process parameters require more precise control for consistent weld quality than is needed for the other processes.

Flash welding machines are either hand operated or automatic; the movement of the sliding platen being cam operated in the former and by hydraulic cylinder or cam in the latter. Machines range from hand operated, typically for sections up to 2500 mm², to very large automatic type for welding steel sheet and strip in steel mills.

The process is not well suited to jobbing work because the procedure must be established by trial and error. For repetition work of suitable section, geometry and reliable quality material it is extremely economic. Quality depends on ensuring reliable current pick-up by having clean metal surfaces in the zone covered by the contacts; that there is no slipping in the clamps and, after making sure that the welding process parameters are controlled, by ensuring that the input line voltage is not subject to undue variation. A disadvantage is that, owing to the extremely heavy intermittent electrical load imposed on the line by flash welding machines, electrical supply authorities may restrict their use to certain areas in the supply system.

The fact that heating of the faces to be butt welded is effected by the electric flashing and therefore establishes a comparatively narrow zone of plastic metal makes the process suitable for welding wide sheet and strip material which cannot be accomplished with other solid phase welding processes. The heating rate is also considerably faster. Against these advantages is the need to make sure that current is cut off at precisely the correct moment so that upsetting forces all residual molten metal and oxide out of the joint. Retained metal is in a burnt condition and produces defective welds; assuming this is extruded the resulting welds are of high quality. The short heating time tends to reduce the possibility of metallurgical damage by overheating, which is an advantage over upset-butt and pressure gas welding.

Since there is no hydrogen pick-up into the weld practically any steel or non-ferrous alloy can be flash butt welded; hardenable steels may need preheating (by butting the parts lightly before starting flashing) and post-heating. Thus, in flash welding tool steels to carbon steel shanks it is normal practice to stress relief anneal at 830°C-870°C immediately the weld is made otherwise cracks will develop.

Flash welding machines may be general purpose i.e. suitable for rounds, hollows or narrow strip of moderate cross-sectional area or they may be designed for specific jobs. One important consideration is that the mating faces should be identical in cross section geometry and area, heavy sections being machined or swaged down to match a lighter section. The welding machine must be designed with sufficient rigidity to withstand the

Table 6.9 RELATIVE FORGING STRENGTHS OF STEELS

Typical Upset Platen force* N/mm²	Strength category	Typical steel
70	Low	Low carbon (up to 0·2% C) steel bars, forgings, strip, tubes, rods etc., Low carbon (up to 0·12% C) bars, free cutting and Bessemer free cutting etc.,
100	Medium	0·45% Carbon (BS 970 EN43B bars, forgings); rods, tubes, strip. 0·65% Carbon (BS 970 EN42B bars); forgings, rods, tubes, strip. 0·35% Carbon ($1\frac{1}{4}$Ni–$\frac{1}{2}$ Cr BS 970 EN111 bars, forgings) 0·3% C (1 Cr–$\frac{1}{4}$ Mo) forgings. EN19 etc.,
180	High	0·4% C 2 Ni–$\frac{1}{4}$ Mo. Forgings. EN 160 High chromium steels (stainless chromium and austenitic chromium–nickel, high carbon), tungsten and molybdenum high speed steels, die steels, Cr–Ni–Si high carbon valve steels etc.,
250	Ultra High	Heat resisting steels, high temperature Cr–Co–W alloys etc.,

* Weld cross-sectional area.

upset forces. These vary considerably according to the hot strength of the material being welded. The relative compressive strength (forging strength) of steels is given in Table 6.9.

To set up a flash welding machine for a production run it is necessary to produce test samples which should be examined for weld soundness unless previous experience dictates the choice of parameters. The variables to be considered and mode of procedure are detailed in reference No. 27.

4. UPSET BUTT WELDING

In this process, usually applied to comparatively small sectional areas, the parts, with machined faces, are brought into contact under pressure, current is passed to generate heat and the weld produced by upsetting. There is no flashing action.

The process is mainly applied either by quite small machines typically for joining wires and narrow strip (continuous wire drawing) or quite large equipment for continuous welding of longitudinal seams in forming tubes (thin walled) from strip.

5. FRICTION WELDING

In this process one part is held stationary while the mating part, under pressure, is rotated in contact with it so producing heat by friction. Upsetting occurs when a sufficient hot plastic zone has been generated.

Friction welding has the considerable merit of not imposing large transient loads on the supply system. Metallurgically it is attractive, since welding is effected without risk of melting the parent metals; it is possible to weld a wide range of dissimilar metals, for example, aluminium to steel, aluminium to copper etc.

Weld quality depends on the amount of upset achieved, which depends on the axial pressure and temperature reached, the latter being dependent on the speed and time of rotation and the pressure. A constant axial pressure may be applied until rotation stops or it may be at lower level during rotation and higher for upsetting. Rotation must cease immediately before upsetting otherwise oxidised welds are produced. The upset pressures used for friction welding are similar to those required for flash butt welding (Table 6.9). Rotation speed must be such as to develop frictional heat at a substantially higher rate than it is conducted away into the parent metal parts. Welding time therefore is determined by the coefficient of friction, heat conductivity and relative sliding velocity between the parts; the coefficient of friction varies throughout the heating cycle. As with the other solid phase butt welding processes, trial and error methods are used to set up the machines for production runs; the process is not well suited to jobbing work due to preliminary trial cost. A rough indication of representative friction welding parameters is given in Table 6.10[28]. Speed and pressure are not critical in welding carbon steel;

Table 6.10   REPRESENTATIVE FRICTION WELDING PARAMETERS

Material	Diameter mm	Relative rotational Speed rev/min	Pressure kg/mm² Heating	Forging	Total time secs
Carbon steel bar	12·5	3000	3·5	3·5	7
Carbon steel bar	25	1500	5·2	5·2	15
Stainless bar	25	3000	8·2	11·2	7
Copper bar	25	6000	3·5	7·0	18
Stainless to carbon bar	18	3000	5·2	10·4	10
Aluminium bar	18	3800	2·83	4·6	6
Stainless tube	137 OD 125 ID	800	14·0	14·0	35

copper requires high relative speeds; stainless steels, copper and aluminium are usually given a higher forging than heating pressure. These points are illustrated in Table 6.10. Heating and cooling rates in friction welding are fast. Therefore consideration must be given to post-weld heat treatment where appropriate. Welding in an inert gas atmosphere has been used where oxidation can cause trouble.

Further reading

Development of friction welding was pioneered in the U.S.S.R. Thirteen articles on friction welding, theory and application, appear in the Russian Welding Journal Avtomaticheskaya Svarka (Automatic Welding) 1965 No. 3. Translation in English published by The Welding Institute, Abington, Cambridge.

HOLLANDER, M. B., CHENG, C. J. and WYMAN, J. C. 'Friction Welding Parameter Analysis', *Welding Journal*, **42**, (11), 495s–501s (1963)

HAZLETT, T. H. 'Properties of Friction Welds between Dissimilar Metals', *Welding Journal*, **41**, (10), 448s–450s (1962)

HAZLETT, T. H. 'Properties of Friction Welded Plain Carbon and Low Alloy Steels', *Welding Journal*, **41**, (2), 49s–52s (1962)

HOLLANDER, M. B. 'Welding Metals by Friction', *Materials Engineering and Design*, February 1962

HOFFMANN, W. and SCHILDWACHTER, M. 'Friction Welding of Metals', *Schweissen und Schneiden*, **11**, (9), 345–352 (1959)

Friction Welding. Four articles in Svarochnoe Proizvodstvo (Welding Production), October 1959, translated into English by the Welding Institute, Abington, Cambridge. These articles deal with energy distribution, power and heating parameters.

WELDING DESIGN

The following notes emphasise certain aspects of welding design which the design engineer should bear in mind.

Allowable stress

Effective size of Full Penetration Butt Welds = throat thickness less the excess metal.
Effective size of Partial Penetration Butt Welds = $\frac{5}{8} \times$ plate thickness (thinner member)
and penetration of the weld metal must not be less than $\frac{7}{8} \times$ plate thickness (BS 449).
Effective size of Fillet Welds = throat thickness (specified on drawings as leg length).
Leg Length = $1\cdot4 \times$ throat thickness.
Effective size of Deep Penetration Fillets = leg length plus $2\cdot5$ mm.

Plate edge preparation

The need for preparation in fusion welding arises where a square edge joint cannot give
adequate access for depositing weld metal through the required thickness of material.
Unless a backing bar or part of the structure can be used to support the weld metal, the
maximum thickness of mild steel plate that can be welded by single runs from each side is
6 mm, using other than deep penetration electrodes.
   Preparation is single or double bevel or J with root face and gap. Choice of preparation
is determined by accessibility of the process (Mig welding needs V preparation), distortion
(double better than single) and economics (amount of weld metal greater for single than
double, machining or gas cutting a double greater than single and cost of turning over
double sided work greater than single).

Design stresses

STATIC LOADING

Welded joints regarded as having 100% strength of parent metal in steel fabrication. The
allowable tension stress in butt welds is the same as for the parent material; typical
maximum design stresses (buildings BS 449, bridges BS 153) based on 2/3 rds. yield are
given in Table 6.11. The allowable shear stress on the throat area of fillet welds is:
105 N/mm² for electrodes having 308 N/mm² yield strength (BS 639 parts 1 and 2)
131 N/mm² for electrodes having 354 N/mm² yield strength (BS 639 parts 1 and 4)
Values in terms of loading per linear millimetre of fillet weld are given in Table 6.12.
Fillet weld sizes should be not less than:

Parent metal thickness, mm	Minimum fillet, leg length, mm
9·6–19	5
Over 19–31	6
Over 31	8
See References 29–33.	

FATIGUE LOADING

Relevant features of welded fabrications under fatigue loading to minimise risk of
fatigue failure are as follows.

Choice of weld form. Downhand butt welds preferable to standing fillets. Change of
section between connecting members is not desirable, therefore taper thicker material in
butts and replace fillets by butts where possible.

Stress concentration. Eliminate areas of stress concentration such as under cutting at
weld toes by welding downhand or by grinding. Eliminate gussets and stiffeners.

Material. High tensile steel, welded, has the same fatigue life as welded mild steel. Under fatigue loading high tensile steel must therefore be used with circumspection in welded fabrication.

Stressing under fatigue loading. See BS 153. Note that the permissible stresses in BS 153 also apply to high tensile steel.

Weld defects. Avoid all planar defects such as cracks, lack of fusion and incomplete penetration, particularly in transverse joints.

See References 34–36.

DESIGN OUTSIDE THE LIMITS OF BS 153

Material. Where static design stress for mild steel is exceeded, high tensile steel is beneficial under fatigue loading only under short life or high mean stress conditions. Changing

Table 6.11 BUTT WELDS IN STRUCTURAL STEEL
TYPICAL MAXIMUM DESIGN STRESSES. STATIC LOADING (N/mm^2)

Description				Yield stress	
	231	246	323	339	354
Parts in axial tension or compression.	135	146	180	195	208
Parts in bending; tension or compression:					
plates, flats, tubes	150	162	211	220	230
rolled sections	146	150	200	209	218
Parts in bending; tension or compression:					
plate girders	135–	146–	180–	195–	210–
	146	150	200	209	218
Parts in shear:					
maximum	100	105	140	146	150
average	85	92	120	124	131

Note. Where applicable relevant standard will give precise values.

Table 6.12 STRENGTH OF FILLET WELDS. (Loading in kg per mm.
For design shear stresses of 105 N/mm^2 and 131 N/mm^2)

Leg length mm	105 N/mm^2	131 N/mm^2
5	36·4	44·4
6	48·8	59·6
8	61·2	74·4
9·6	73·6	89·6
11	85·6	104·0
12·7	98·0	118·8
16	122·8	149·2
19	138·8	178·0
22·4	171·2	208
25·4	196	238

from mild steel to high tensile steel to overcome fatigue failures occurring in service is not recommended.

Stress relief. If the stress cycle is tensile only, stress relief treatment does not improve fatigue strength but is beneficial if loading is wholly or partially compressive. Design stresses for joints in heat treated components are not available but a 50% increase in fatigue strength could be obtained on non-load bearing fillets under alternating loading.

Improvement procedures. Eliminate the notch effect at weld toes or introduce compressive residual stress e.g. grinding, peening, spot-heating (see Reference 34, chapter 11).

Brittle fracture

Susceptibility to brittle fracture is an attribute of ferritic materials having a body-centred cubic structure. It is temperature dependent, all BCC ferritic materials exhibit brittle fracture below a temperature which is characteristic of the composition, method of steel making, fracture conditions, thickness etc. The change from ductile fracture (slip) to brittle fracture (no slip) is gradual over a transition range of temperature, a possible cause of the change being precipitation of carbon and nitrogen from solution causing blocking of dislocations.

Brittle fracture can occur at very low applied stress, well below yield or design stress or in the absence of a working stress, i.e. from residual stress or change of stress. It is more likely to occur in large structures, or thick plate or sections. It characteristically exhibits a bright cleavage fracture which propagates at extremely high speeds.

Brittle fracture requires a *tensile stress* acting at a *stress concentration,* internal or surface, on material which is 'notch-brittle' at the temperature at which the stress is acting.

The area of stress concentration, typically a linear defect such as cracks or slag inclusions in welds, or geometrical discontinuity causes localised high strain and the ability of the material to withstand this determines its susceptibility to crack *initiation.* Susceptibility to *propagation* of the initiated crack is also temperature dependent, requiring a higher temperature to inhibit it than that required to prevent initiation.

ASSESSMENT OF BRITTLE FRACTURE SUSCEPTIBILITY

This may be a complex and difficult matter. For individual machined forgings in quenched and tempered steels, made from notched-impact quality controlled steel with care taken to remove surface stress concentrators, the assessment of service performance within the temperature range is reliably based on the quality control test and method of steelmaking. For welded fabrication the same principles should apply, but the structure may be composed of plates produced from different casts of steel.

It is important to appreciate that the notched-impact or any other similar test used for quality control has a value only as applied to products made by the same steelmaking practice. If steel made to a rimming practice has been shown by experience to be resistant to brittle fracture for certain applications, it is uneconomic to use a killed steel having a lower transition temperature range. Where experience shows semi-killed steels or silicon-killed steels are susceptible under given conditions a change to a different steel-making practice is indicated where lower transition-temperatures are to be expected. Assessment as to whether or not the actual plates or sections to be used have been properly manufactured is most conveniently and economically done by quality control tests of the notched-impact or tensile or bend type.

Resistance to brittle fracture depends on steel melting practice. Melting process, de-oxidation practice, casting pit practice, steel composition with particular reference to carbon, manganese, aluminium, nitrogen, oxygen, rolling mill finishing temperatures, heat treatment (if any) and final thickness all affect brittle fracture susceptibility. Satisfactory experience with, for example, steel made to fine grain practice with aluminium de-oxidation, where the quality control has been say, 'X' J impact, may not be duplicated by using steel made to a different practice but also giving 'X' J impact at the same temperature. The engineer should satisfy himself as to quality control test result and steel-making practice.

A variety of quality control tests are available but the Charpy test is most widely used and has a strong backing from its correlation with field experience. Where field experience is lacking, assessment of brittle fracture suceptibility in welded fabrications is difficult. Full scale tests on welded joints (wide plate tests etc.) can provide guidelines but are expensive. Increased resistance to brittle fracture means increased cost of material, testing, fabrication, and inspection.

One method of approaching the problem is to obtain a representative Charpy V notched-impact transition curve for the intended material and to apply the criteria of nil ductility transition temperature (NDT), fracture transition for elastic loading (FTE) and fracture transition for plastic loading (FTP) to select a safe minimum working temperature. Judgement is needed in applying these or any other criteria. The selected minimum working temperature increases from NDT to FTE and then to FTP, in mild steels the FTP being about 40°C higher than NDT. These criteria are based on fracture data from full scale welded joints by the drop weight and explosion-bulge crack starter tests with natural cracks initiating the fracture process. (See Figures 6.4 and 6.5 and Table 6.13.) Examples of these criteria are given in Table 6.14.

Design and service determine the conditions of loading while notches and temperature determine the response of the steel to the imposed conditions. The critical fracture transition temperature concept establishes three points on the temperature scale at which, as temperature is raised or lowered, major changes in the fracture behaviour and characteristics of the steel are to be expected.

Welding and brittle fracture

Brittle fracture can occur in unwelded steel components. Welding, however, tends to render a structure more susceptible to brittle fracture than mechanical joining methods (a) because, unless stress relieved, local residual stresses remain after welding (b) some metallurgical damage, such as ageing after rapid cooling or straining, may occur and (c) welded joints may contain linear defects such as slag stringers, fissures or cracks which cause stress concentration. Susceptibility to ageing embrittlement (manifested by drastic reductions in notched-impact toughness) depends on the steelmaking practice and handling of individual heats. Steel can be manufactured to be resistant to ageing. Mylonas[39] showed that 'the properties responsible for brittle fracture are those of damaged not virgin steel', the damage to the mechanical properties being caused by ageing after straining. He inferred that some degree of strain ageing damage must be the primary cause of brittle fracture at welded joints. This tends to be confirmed by Wells[40] who found that, in notched wide plate tests, placing the saw-cut notch in the bevelled joint faces *before* welding, but welding so as not to fill up the saw-cut, could induce low stress brittle fracture, even spontaneous fracture. Cutting a notch *after* welding prevented low stress fractures. Welding over the notch without filling it induces straining and subsequent ageing at the notch tip in materials susceptible to strain ageing.

Steel which is resistant to strain or quench ageing is therefore to be preferred where brittle fracture is an unacceptable risk. Unless it is stated that Charpy tests are carried out in the aged condition, it can be assumed they represent the properties in the unaged condition. Stress relief treatment after welding, properly carried out, effects over-ageing with consequent improvement in toughness, but a quenched and tempered steel, though non-ageing as delivered, can show ageing in the weld heat affected zone unless stress relieved or re-tempered because welding effects solution of the damaging elements. Strain ageing is strongly dependent on de-oxidation practice; resistance to strain ageing increases from rimming steel, through semi-killed and silicon killed to special strain ageing resistant steels. Resistance to strain ageing is judged by Charpy V notched test on samples strained, usually 10%, and aged in the range 200–300°C for half-an-hour. Ageing is a time-temperature dependent process and proceeds at maximum speed in this temperature range. A strain ageing resistant steel will be relatively tolerant of linear welding defects. However, steels not guaranteed resistant to strain ageing may be perfectly satisfactory for the intended service conditions providing they are properly selected

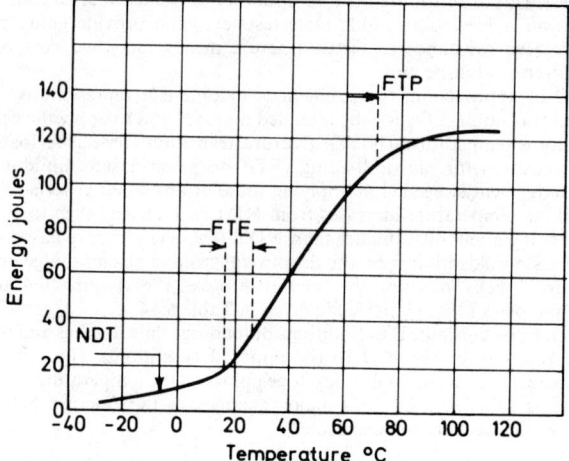

Figure 6.4. *General relationships of crack-starter test results of ABS-A type semi-killed mild steel to Charpy V energy transition curve*

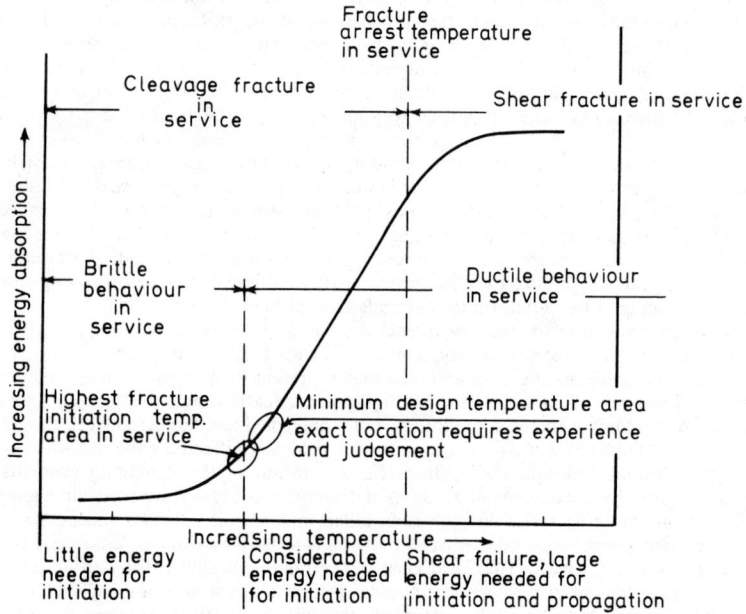

Figure 6.5. *Schematic of minimum design temperature related to Charpy impact energy curve (cf. Ref. 38)*

Table 6.13 (cf. Ref. No. 37) CORRELATION BETWEEN CHARPY V ENERGY ABSORPTION AND NIL DUCTILITY TRANSITION TEMPERATURE FOR VARIOUS STEELS DETERMINED BY DROP WEIGHT TEST

Steel	Observed range of Charpy V correlation with NDT Temp.	Average 'fix' J	Conservative 'fix' J	Remarks
Rimmed mild steel.	4 to 9·5	6·75	9·5	Includes 7 ship fracture steels.
Semi-killed MS. A–7 or equivalent.	6·75 to 13·5	9·5	13·5	Includes 11 ship fracture steels.
ABS–B	9·5 to 43·5	15	20·5	Of 17 steels tested only 3 were higher than 20·5 J
ABS–C (fully killed MS–norm.)	20·5 to 32·5	23	27	Seven tested. One higher than 27 J
High tensile (Norm. HTS)	17·5 to 61	31	41	
A–302	15 to 41	27	41	Fifteen tested. Four higher than 34 J
Quenched and tempered steel. 250 Brinell.	27 to 61	34	47·5	
Quenched and tempered steel 375 Brinell.	15 to 22	17·5	20·5	

Note 1. The Nil Ductility Temperature is the temperature below which there is no ability to deform in the presence of a sharp crack.

Note 2. To 'fix' the NDT temperature on a Charpy curve for silicon killed mild steels, fine grain mild steels and high strength Q and T steels requires higher energy levels than for semi-killed ship steel.

Note 3. The actual NDT temperature was always within ±17°C of that indicated by the tabled 'average fix' value.

Table 6.14 CRACK-STARTER EXPLOSION BULGE TEST ON VARIOUS STEELS. NDT, FTE AND FTP TEMPERATURES. (Ref. No. 37)

Steel	NDT °C	FTE °C***	FTP °C****	Remarks
Ship plate of ship fracture type	−6·7 to −1·1(A)	+21·1(A)	+60(A)	NDT.
A 302. Pressure vessel steel (Casualty)*	+21·1			Below NDT temperature brittle crack initiation is easy at low loads 'no forcing' required. Crack propagation easy at low loads.
A 302. Pressure vessel steel (Casualty)**	+37·8			
A 302.	−12·2			FTE.
ABS–C (Normalised, semi-killed)	−28·9(L)	−6·7(L)	+26·7(L)	At the FTE temperature 'forcing' (plastic deformation at crack tip) required for initiation but propagation of a brittle fracture is a possibility.
HTS (Norm.) Moderate high tensile.	−28·9(A)	−1·1(A)	+26·7(A)	FTP.
HY 80. Quenched and tempered. Low carbon	−90(L)	−67·8(L)	−40(L)	Above the FTP temperature brittle fractures cannot propagate even through severely plastically deformed material.
$2\frac{1}{4}$ Ni–Cr–Mo				A.
				Average value from large number of tests.
				L.
				Average value from limited number of tests.
				Note.
				Quality variation in production steels denoted by an NDT range of up to 33°C.

* Occurred during hydrostatic test.
** Occurred during fabrication from residual welding stress at man-way reinforcement.
*** Over a wide range of steels tested FTE was approximately 22·25°C (16·7 to 27·8°C) above NDT temperature.
**** Over a wide range of steels tested FTP was approximately 55·5°C above NDT temperature.

in relation to the lowest expected service temperature. Assessment by Charpy V notch tests is widely used for both types.

Selection of steel to resist brittle fracture

Selection of a suitable steel for given conditions is simplified if an existing Code is available. Charpy V tests at defined temperatures will then assess the quality of the material. Where Codes are not applicable, selection is a complex matter of inter-acting requirements and factors. Economics must be a prime consideration to be weighed against cost and significance of failure. As the risks are reduced by various means, such as special care in design to avoid stress concentration or special care to eliminate harmful linear defects or the use of special notch ductile or ageing resistant steel or stress relief after welding, separately or jointly in different combinations, the cost will rise and judgement in balancing opposing factors is required.

The main points to bear in mind are:

(1) Brittle fracture is concerned with modes of fracture and temperatures.

(2) The design engineer should have some idea of the Nil Ductility Transition temperature of the material he proposes using if Code practice is not applicable.

(3) It will be necessary to check material quality for (a) steel making practice and (b) notch-impact or other notched type of test performance.

(4) Given a working temperature range, the fabrication practice and type of steel will be selected by judging the relative importance of margins of safety, cost etc., and then using the NDT temperature or some equivalent criterion to establish the suitability of the material for the given application. Factors to consider in design and steel selection are discussed in reference numbers 37, 38 and 40. See also Bibliography.

(5) Correlation between the fracture behaviour of full scale crack containing welded test joints and the Charpy V notch test allows energy absorption in the test to indicate the expected critical temperatures for change in fracture mode, *provided* the *level of energy absorption* used is correct for the *type of steel*. It increases as the strength of the steel increases (see Table 6.15).

Suggested minimum working temperatures for several structural and other steels for low temperature service are given in Table 6.16 together with suggested Charpy energy absorption requirements at the minimum working temperature (abstracted from reference 41). Examples of German high tensile steels made to two grades for resistance to brittle fracture and suitable for welding are given in Table 6.17.

Stress relief heat treatment

This treatment after welding, markedly improves resistance to brittle fracture in welded mild steel (reference 40, section 6) and in low alloy high tensile steels provided that in the latter case, it is correctly carried out within precise temperature limits. This improvement is thought to be due to removal of the strain aged condition at the notch tip.

Weld metal. Resistance to brittle fracture

ASSESSMENT

Unless weld metal is deposited in vacuo (e.g. electron-beam welding) it is subject to absorption of oxygen and nitrogen in varying degree depending on the efficiency with which the welding process excludes these gases. The mode of occurrence of oxygen and nitrogen in the deposited metal depends on the de-oxidation reactions and slag-metal reactions. Except in single run welds the deposition process causes reheating over all

Table 6.15 BRITTLE FRACTURE. (Minimum Working Temperatures for Carbon Manganese Pressure Vessel Steels. Tentative Recommendations by The Welding Institute, Cambridge. Assumes Static Loading Only. Weld Metal having 41 J minimum Charpy V Notch Impact Energy at 0°C.)

Yield strength of steel N/mm²	Condition AW* SR**	Charpy V test Temp. for 27J	Charpy V test Temp. for 41J	Plate thickness mm	Minimum working temp. °C approximate
Up to 27	AW	0		25·4	0
				37·9	+10
		−20		25·4	−10
				37·9	+4
		−40		25·4	−20
				37·9	−5
	SR	+20		75	−45
				37·9	−35
		0		75	−53
				37·9	−45
		−20		75	−65
				37·9	−55
		−40		75	−75
				75	−65
From 29 to 45	AW		0	As above	As above
			−20		
			−40		
	SR		+20		
			0		
			−20		
			−40		

* As welded.
** Stress Relieved (Thermally).

Table 6.16 TENTATIVE MINIMUM WORKING TEMPERATURES FOR VARIOUS STEELS
(Assumes reasonable Design, Fabrication and Inspection standards)

Steel	Type	Thickness mm	Minimum temperature °C. as welded. Stress relieved.		Charpy V requirements for minimum temperature, J
BS 15	MS semi-killed as rolled	25	0	−75	20–34
BS 4360. 43A.		50	+10	−55	
ND1 ⎫ 27J at 0°C	MS semi-killed as rolled	25	−15	−75	20–34
LT0 ⎭		50	−5	−55	
BS 4360.40C, 43C.					
ND4 ⎫ 27J at −50°C	MS silicon killed plus fine grain	25	−25	−80	34–47
LT50 ⎭	plus normalised	50	−15	−60	
BS 4360.40E, 43E		25	−15		
968.27J at −15°C ⎫	Semi-killed plus Nb. high tensile	50	−5		34–61
BS 4360.50C. ⎭					
3½% Nickel		6–25	−100 to −80		27–54
LT 100					
9% Nickel		6–25	−200 to −100		34–68
LT 190					

Table 6.17 GERMAN WELDABLE NOTCH DUCTILE STRUCTURAL STEELS. PLATES.
(FOR PRESSURE VESSELS, HIGH PRESSURE PIPING, BRIDGES ETC.) MANNESMANNROHREN—WERKE AG.

Steel	C% max.	Mn%	V%	Ni%	Cu%	Minimum yield strength† N/mm²	Ultimate strength‡ N/mm²	Impact** (DVM) un-aged at °C. J									Impact (DVM) aged*** at °C. J
								+20	+10	0	−10	−20	−30	−40	−50	−60	+20
FG26	0·18	0·4/1·3	—	—	—	250*	360–480	88·5	88·5	88·5	78·5	68·5	—	—	—	—	—
FG26T	0·16	0·5/1·3	—	—	—	250*	360–480	88·5	88·5	88·5	83·5	78·5	68·5	59	54	49	59
FG32	0·18	0·6/1·5	—	—	—	310*	440–560	88·5	88·5	88·5	78·5	68·5	—	—	—	—	—
FG32T	0·16	0·7/1·5	—	—	—	310*	440–560	88·5	88·5	88·5	83·5	78·5	68·5	59	54	49	59
FG36	0·2	0·9/1·6	—	—	—	350*	490–625	88·5	88·5	88·5	78·5	68·5	—	—	—	—	—
FG36T	0·18	0·9/1·6	—	—	—	350*	490–625	88·5	88·5	88·5	83·5	78·5	68·5	59	54	49	59
FG39	0·18	1·1/1·6	0·1/0·16	—	—	380†	500–645	88·5	88·5	88·5	78·5	68·5	—	—	—	—	—
FG39T	0·16	1·1/1·6	0·1/0·16	—	—	380†	500–645	88·5	88·5	88·5	83·5	78·5	68·5	59	54	49	59
FG47	0·2	1·2/1·7	0·1/0·18	0·4/0·7	—	460†	560–725	88·5	83·5	78·5	73·5	68·5	—	—	—	—	—
FG47T	0·2	1·2/1·7	0·1/0·18	0·4/0·7	—	460†	560–725	88·5	83·5	83·5	78·5	73·5	64	54	49	44	59
FG47C	0·15	1·1/1·5	0·08/0·18	0·5/0·7	0·5/0·7	460†	560–725	88·5	83·5	78·5	73·5	68·5	—	—	—	—	—
FG47CT	0·15	1·1/1·5	0·08/0·18	0·5/0·7	0·5/0·7	460†	560–725	88·5	83·5	83·5	78·5	73·5	64	54	49	44	59
FG51	0·21	1·3/1·7	0·1/0·2	0·4/0·7	—	500†	608–785	78·5	78·5	83·5	78·5	68·5	—	—	—	—	—
FG51T	0·21	1·3/1·7	0·1/0·2	0·4/0·7	—	500†	608–785	88·5	83·5	78·5	73·5	68·5	59	49	44	39	59

* Up to and including 35 mm thickness.
† Up to and including 16 mm thickness.
‡ Up to and including 70 mm thickness.
** Longitudinal. Minimum average of 3 tests. Guaranteed.
*** Strained 10%. Aged half-hour at 250°C.
Note. Additional steels are FG29 and 29T, FG43 and 43T. Notch ductility same as FG26 or 26T.

temperatures on metal which has been deposited at a high rate of cooling. The metal is subject to straining over a wide temperature range. This situation suggests that mild steel weld metal will be as susceptible to ageing as parent metal, but will be dependent on the wide variation in welding procedures and that the welding position (flat, vertical or overhead) must have a marked effect since it influences efficiency of weld pool protection.

The possible variation in weld toughness is shown in reference 42, where energy absorption in different positions of a double V joint varied from 23J to 150J, testing in groups of 18 Charpy V tests per position but welding the joints in one position. This variation in Charpy V test values depending on location of test sample would suggest a fairly wide temperature band for NDT temperatures in welded joints. Temperature ranges for different processes and electrodes are indicated in Figure 6.6. (Reference 43. Figure 2).

Charpy V impact tests, carried out under standard testing conditions (welding procedure, location of notch etc.) on all-weld-metal deposits according to, for example,

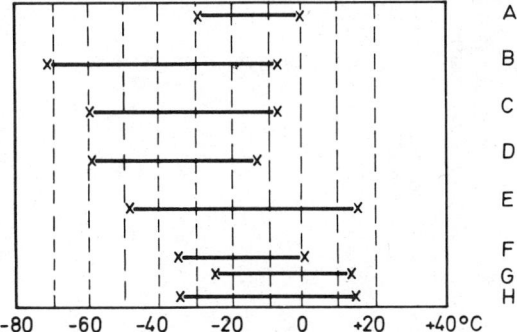

Figure 6.6. Nil ductility temperature ranges for steels and welds. Range of available NDT results: A. Mild steel plates; B. High yield steel plates; C. Low hydrogen (Class 6XX) multi-run welds. The electrode used in Ref. 42 was in this group; E. Submerged-are 2- and 3-run welds; F. CO_2 dip transfer multi-run welds; G. Rutile multi-run welds; H. Electro-slag welds as welded

BS 639 indicate that weld metals vary in notched-impact toughness depending on chemical composition, type of coating and factors associated with manufacturing variables (Table 6.18). Therefore the Charpy test can be used as a manufacturer's quality control test. Its application to welds in fabrications should take account of the large scatter[42] inherent in such welds. The scatter is likely to be less marked in welds from class 1, 2 or 3 electrodes.

SELECTION OF WELD METALS FOR NOTCH DUCTILE APPLICATIONS

Where Code requirements specify a minimum energy absorption level for the weld metal taken from a joint it is important to carry out procedure trials with the selected electrode or consumable because the values attainable in testing to BS 639 may not be achieved 'on the job'. This applies particularly to submerged arc welds and CO_2 welds where substantial mixing with parent plate may alter the weld properties.

The better notch ductility electrodes with class 6 coating (see Table 6.18) have a tendency to break with a mainly cleavage or mainly fibrous fracture at the same temperature when this falls in the transition range between the upper and lower shelf.

Table 6.18 CHARPY V NOTCH ENERGY ABSORPTION FOR VARIOUS MILD AND HIGH TENSILE STEEL WELD METALS. ALL-WELD-METAL DEPOSITS. BS 639: 1969. TYPICAL VALUES FOR COMPOSITION AND PROPERTIES

Manual electrode or wire class	Weld C	Metal Mn	Analysis Si	Submerged arc flux AWS/ASTM	Mechanical yield N/mm²	Properties ultimate N/mm²	Charpy V J at °C		Remarks
BS 1719 Pt. 1									
E104	0·08	0·4	0·2		400	460	95	−10	Upper shelf approximately.
E247	0·08	0·8	0·5		480	550	61	0	Upper shelf approximately.
E317	0·08	0·45	0·2		420	500	68	−10	Upper shelf approximately.
E614	0·08	0·7	0·6		420	520	75	−30	Upper shelf equals 170+J. 75J equals 50% approximate crystallinity.
E611	0·08	1·0	0·6		500	580	81	−30	
E614	0·08	1·5	0·6		510	600	81	−30	
AWS									
EL12	0·1	0·9	0·5	F62. EL12*	340	460	39	−20	
EL12	0·1	1·1	0·2	F72. EL12*	380	490	58	−20	

* Fused Manganese Alloying Silicate.
** Agglomerated Basic Manganese Alloying.

This is known as bi-modal behaviour. Thus low values can be grossly misleading as to the true position of the transition curve on the temperature axis. Mixed low and high values in a set of three tests would suggest the bi-modal zone.

Where there is no Code to follow, a general rule is to match or over-match the energy absorption requirements specified for the plate. This principle can be applied on steels which have a lower yield strength than the weld metal (mild and high tensile steels). If the yield strength of the plate is higher the weld metal energy absorption should be higher than that demanded for the steel.

Further reading

G. M. BOYD (ed), *Brittle Fracture in Steel Structures*, Navy Department Advisory Committee on Structural Steels, Butterworth and Co. (1970)

HALL, KIHARA, SOETE and WELLS, *Brittle Fracture of Welded Plate*, Prentice-Hall

Control of Steel Construction to avoid Brittle Failure, Plasticity Committee of the Welding Research Council, The Welding Research Council (1957)

WELLS, A. A. 'The Brittle Fracture Strengths of Welded Steel Plates', *Trans. Inst. Naval Architects*, **98**, 296–326 (1956)

WELLS, A. A. 'Brittle Fracture Strength of Welded Steel Plates. Tests on Five further Steels', *British Welding Journal*, 259–277 (May 8th 1961)

WELLS, A. A. 'Brittle Fracture Strengths on Welded and Notched Three Inch Thick Steel Plates', *Welding Journal*, **8**, Supplement 32–36 (1961)

WELLS, A. A. 'Pressure Vessel Brittle Fracture and Information required to design against it', *British Welding Journal*, **8**, Supplement 32–36 (1961)

WELLS, A. A. 'Influence of Residual Stresses and Metallurgical Changes on Low Stress Brittle Fracture in Welded Steel Plates', *Welding Research Supplement*, **26**, 182s–192s (May 4th 1961)

'Welded Ferritic Steel Construction for Intermediate Low Temperature Service', *Symposium on Evaluation of Metallic Materials in Design for Low Temperature Service*, A.S.T.M. Special Technical Publication, **302**, 21–40 (1962)

WELLS, A. A. and BURDEKIN, F. M. 'Effects of Thermal Stress Relief and Stress Relieving Conditions on the Brittle Fracture of Notched and Welded Wide Plates', *British Welding Journal*, **10**, 270–276 (May 5th 1963)

BURDEKIN, F. M. 'Steelwork and Brittle Fracture', *Civil Engineering and Public Works Review*, (May 1963)

WOODLEY, C. C., BURDEKIN, F. M. and WELLS, A. A. 'Mild Steel for Pressure Equipment at Sub-Zero Temperatures', *British Welding Journal*, **11**, 123–136 (March 3rd 1964)

BURDEKIN, F. M. 'The Fracture Strength of Welded $3\frac{1}{2}\%$ and 9% Nickel Steels at Low Temperature', *British Welding Journal*, **11**, 586–592 (November 11th 1964)

WEOK, R. 'Failure of Steel Structures: Causes and Remedies', *Proc. Royal Society*, A, **285**, 3–9 (1965)

WELLS, A. A. 'The Application of Fracture Mechanics to Yielding Materials', *Proc. Royal Society*, A, **285**, 34–45

BURDEKIN, F. M. 'The Effect of Thermal Straining during Welding and the Fracture Toughness of Mild Steel', *British Welding Journal*, (February 1967)

BURDEKIN, *et al.*, 'Selection of Weldments to avoid Fracture Initiation', *British Welding Journal*, **15**, 12, 590–600 (1968)

'The Use of Critical Crack Opening Displacement Techniques for the Selection of Fracture Resistant Materials', *First Report of the (CODA) Panel of the Navy Department Advisory Committee on Structural Steels. Symposium on Fracture Toughness*, UKAEA, Culcheth (April 1969)

PELLINI, W. S. and PUSAK, P. P. 'Fracture Analysis Diagram Procedures for the Fracture-Safe Engineering Design of Steel Structures', *Naval Research Laboratory Report 5920*, (March 1963)

MACGREGOR, GROSSMAN and SHEPLER, 'Correlated Brittle Fracture Studies of Notched Bars and Simple Structures', *Welding Research Supplement*, (January 1947)

PELLINI, GOODE, PUSAK, LANGE and HUBER, 'Review of Concepts and Status of Procedures involving Metals of Low to Ultra-High Strength Welds', *Naval Research Laboratory Report 6300*, (June 1965)

PELLINI, W. S. 'Advances in Fracture Toughness Characterisation Procedures and in Quantitative Interpretations to Fracture-Safe Design for Structural Steels', *Welding Research Council Bulletin*, (May 1968)

Lamellar tearing

Lamellar tearing is a characteristically stepped fracture occurring under welded joints
where a combination of straining due to weld contraction and inferior mechanical
properties of the steel in the 'through-thickness' direction leads to a series of tears which
link-up in stepped form. The effect is:

(a) Associated with non-metallic inclusions.
(b) Associated with rolled steel.
(c) Highly local. Susceptible areas are almost impossible to detect before welding.
    However, special ultrasonic techniques may be successful[44].
(d) Not necessarily associated with areas of plate containing detectable defects such
    as laminations.
(e) Most likely to occur where the weld contraction 'pull' is through the plate or
    section thickness (Figure 6.7).
(f) More common in killed steels than semi-killed.
(g) More common in plates over 30 mm thickness (possibly due to higher restraint).
(h) Highly variable in incidence; steel casts, plates from the same cast and areas in a
    plate all vary in susceptibility.

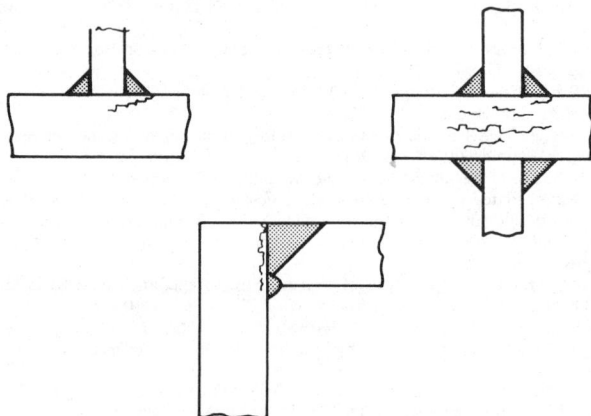

Figure 6.7. Examples of lamellar tearing

REMEDIAL MEASURES

Joints should be used which apply the contraction stress across or tangential to the fibre
direction rather than at right angles to the plate surface.

Pre-buttering with small gauge electrodes (pre-heating if necessary) may prevent
tearing[44].

Selection of plates, by agreement with the steel maker, from the centre of the ingot
should be beneficial.

Forgings, which are virtually free from this defect, can sometimes be used in place of
Tee butts, which are a common cause of trouble[44].

Forged nozzles are preferable to rolled plate nozzles and, for critical applications in
pressure vessel fabrication practice, forged shells are preferable to rolled plate. Apart
from the use of forgings there is no certain method of avoiding lamellar tears other than
welding through the full plate thickness[44] (Figure 6.8) before making the main joint,
so that weld contraction is not applied directly to the plate surface.

Pre-heating to high temperatures (above 200°C), maintaining high interpass temperatures (250°C minimum) and using low yield strength weld metal (BS 1719: Part 1. E101, AWS. E6010 or E6011) are beneficial steps in minimising the risk of Lamellar tearing.

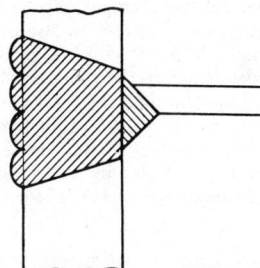

Figure 6.8. Avoiding lamellar tearing at tee joint

Further reading

GOODGER, A. H. *Fissuring along the Flow Structure of a plate under Fillet Welds,* British Standards Institution News (September 1966)

WORMINGTON, H. 'Lamellar Tearing in Silicon killed Boiler Plate', *Welding and Metal Fabrication,* 370 (September 1970)

NICHOLS, D. M. 'Lamellar Tearing in Hot Rolled Steels', *British Welding Journal,* 103–112 (March 1968)

FERRIS, I. J. 'Cracking in Ferrous Welds', *Australian Welding Journal,* **6,** 1, 36–42 (September 1962)

MEYER, H. J. 'Sub-surface Cracks in Plates at the point of application of Stresses perpendicular to the plate surface', *British Welding Research Association Bulletin,* **6,** 263–268 (October 1965)

BURDEKIN, F. A. 'Lamellar Tearing in Bridge Girders—A Case History', *Metal Fabrication and British Welding Journal,* **3,** 5, 205–209 (May 1971)

NOTES ON WELDING VARIOUS MATERIALS

Weld hardening steels

When the rate of cooling at the weld (determined by heat input, speed of travel and heat sink) is sufficient to induce the development of martensitic structures in the parent steel there is a tendency for cracking to occur in the Heat Affected Zone. At lower cooling rates the hardening may not induce cracking but may be unacceptable because ductility is impaired. Cracking is caused by a combination of:

1. A micro-structure sensitive to hydrogen embrittlement.
2. A sufficient amount of hydrogen diffusing in to the steel from the weld metal.
3. A sufficiently low temperature for hydrogen embrittlement to occur (below approximately 185°C).
4. A macro-stress set up by weld contraction against restraint in the structure and micro-stress set up by transformation to martensite to initiate the crack.

Steps to minimise the risk of hardening cracks developing which may be taken, separately or in combination, according to requirements involve:

(a) Preventing the development of the sensitive micro-structure by slower cooling (pre-heating, use of large heat input, slow travel).
(b) Minimising the available Hydrogen (low hydrogen processes, electrodes or consumables).
(c) Pre-heating to above the critical Hydrogen embrittlement temperature. This is beneficial in several ways; Hydrogen diffusion[45], reduction of stress by tempering,

tempering of martensite formed near M_s, etc. Holding times of one hour or more in the range 200–350°C would be selected.

(d) Pre-heat to above hydrogen embrittlement temperature before welding, maintain this temperature throughout welding and, without loss of temperature, raise to stress relieving temperature.

(e) Minimising contraction strains by pre-heating, good fit-up in fillet welds or by using weld metal with low yield or proof stress e.g. austenitic steel.

Weld metals giving the same tensile strength as the high strength steel usually have lower carbon (less than 0·1% C) contents than the steels and produce less hydrogen sensitive micro-structures but the hydrogen level of fusion welds is always higher than in the Heat Affected Zone, (HAZ). In general, the requirements for preventing HAZ cracking will also be advantageous in respect of the weld metal in preventing fissuring or major transverse contraction cracking.

Steel hardenability

Choice of measures (a) to (d) depends on the hardenability of the steel; a shallow hardening steel having a high critical cooling rate, a deep hardening one having a sufficiently slow rate that welding processes, even with pre-heating at reasonably practical temperature levels, produce martensitic hardened zones. At the cooling rates experienced in manual metal arc and other arc processes steels can be roughly classified[46] into:

Class 1. Those which do not harden to a critical degree in arc welding.

Class 2. Those having low hardenability (requiring high cooling rate to harden) and producing micro-structures with low crack susceptibility.

Class 3. Those having low hardenability and producing micro-structures having high crack susceptibility.

Class 4. Those having high hardenability with low crack susceptibility.

Class 5. Those having high hardenability with high crack susceptibility.

Precautions (a) to (d) need not be applied to Class 1. Class 2 may respond to (a) in respect of a modest increase in heat input and will certainly respond to pre-heating. Some, at lower cooling rates, will respond to low hydrogen consumables without recourse to pre-heating. Class 3, typically plain carbon steels above 0·3% carbon, will respond to substantial heat input increases to reduce cooling rate up to 0·4% carbon. At higher carbon contents pre-heating is effective. Class 4 are typically low carbon (less than 0·17% carbon) structural high tensile steels with small alloy additions (Mn, Cr, Mo, V) to give yield strengths up to about 450 N/mm², at the heavier end of the alloy element range, or higher strength precipitation hardened steels (USS Carilloy T 1 for example) or 9% Nickel steel. Low carbon contents with medium alloy or ultra-low carbon with a large alloy content characterise this class. They are steels which produce low carbon martensitic micro-structures at low cooling rates but having auto-tempering characteristics, the low carbon content tending to promote low crack susceptibility. Precautions (b) and (c) are frequently necessary. Class 5 steels produce martensitic micro-structures at low cooling rates which do not have adequate auto-tempering characteristics; examples are the low carbon Cr–Mo high temperature steels and medium to high carbon low alloy engineering steels. Precautions include (b) and (c) or (b) and (d). Where practicable precaution (e) is accepted as for all weld hardenable steels.

Unfortunately the division between classes is not clear-cut. Indeed, a given steel specification can overlap from one class to another in its composition range. The concept of Carbon Equivalent is an attempt to calculate weld hardenability from chemical composition. It may be used as a guide only, provided it is applied to steels within a given type e.g. high tensile structural, because variation in hardenability within alloyed steels due to carbon and alloy segregation precludes more positive prediction. The use of a Carbon Equivalent formula is illustrated in the preparation of a Nomogram[47] for welding Carbon–Manganese steels of the C—Mn—Nb High Tensile type (BS 4360:

1968 grade 50) and Carbon Manganese H.T. structural steels generally. When dealing with steels of a restricted composition range and type such as BS 968 the C.E. concept can be used to predict conditions under which pre-heating can be omitted[48]. For the general run of engineering alloy, carbon, high temperature and cryogenic steels pre-heating practice is based on experience, accepted good practice being assumed. This includes proper control of hydrogen to low levels where required, good fit-up, pre-heating through the full thickness avoiding stress concentration as at under-cutting and avoiding markedly convex weld profiles by downhand welding.

A guide to pre-heating and post-heating requirements for the manual metal arc process is given in Table 6.19. Weld metals of matching strength are used for multi-run welds but for single runs, mild steel electrodes are frequently effective with reduced weld metal cracking risk. Low hydrogen consumables are normally used (assumed in Table 6.19 except where indicated). Low hydrogen-iron powder electrodes are widely used in the downhand position to avoid stress concentration at weld toes.

The procedures in Table 6.19 apply to the general welding, repair or build-up welding of alloy engineering steels, examples being forgings for heavy engineering applications. Repair welding not properly carried out can lead to brittle fracture or fatigue fracture.

Welding hardenable steel castings

The pre-heat temperatures and post-heat treatment for welding castings, either for repairs or fabrication, are given in Table 6.20[50].

The pre-heat temperatures indicated are for guidance only as variations in composition, mass, design (stress concentration areas, rigidity etc.,) and extent of welding determine the required amount of pre-heat.

Welding hardenable steels. Processes other than manual metal arc

CO_2 welding (clean wire) is applicable as a low hydrogen fusion welding process but the deep penetration of the spray arc and quick cooling rate leads to increased risk of weld cracking and porosity on steel over 0·3% carbon. Pre-heating should be as given in Table 6.19 working well above the minimum figures.

Submerged arc welding, at normal current levels and speeds with dry fluxes, requires pre-heating in the 100–250°C range for low alloy and carbon steels in the 0·25–0·4% carbon range. At higher carbon levels the pre-heat should be 200–350°C (the deep penetration on positive polarity or on a.c. leads to substantial carbon and alloy pick-up). For build-up welding on carbon and alloy steels negative polarity is frequently used to avoid this pick-up.

Flash-butt welding does not require pre-heating, but steels having high hardenability and crack susceptibility to quench cracking (e.g. tungsten high speed steel) should be stress-relieved or tempered or given full heat treatment without allowing the joint to cool out.

Stainless steels

Most stainless steels are readily welded by metal arc, argon arc (Tig), resistance and friction welding. Free cutting stainless steels cause problems of weld cracking. Oxy-acetylene welding tends to produce unacceptably high carbon contents; inert gas metal-arc welding (argon) is subject to 'cold-start' difficulties. Cracking may occur from (a) weld metal hot cracking (b) weld hardening (c) weld embrittlement (d) stress corrosion. Hot cracking is mainly confined to fully austenitic micro-structures; hardening and embrittlement cracking (usually in high ferrite content welds) are controlled by pre- and

Table 6.19 PRE-HEAT AND POST HEAT PROCEDURES FOR WELD HARDENING STEELS. MANUAL METAL ARC PROCESS

Steel	Nominal composition cast analysis						Thickness mm	Electrode tensile	Type class (BS 1719)	Welding procedure Pre-heat and interpass Temp. °C	Welding procedure Post-heat treatment
	C	Mn	Ni	Cr	Mo	V					
Mild steel	0·3	0·7	—	—	—	—	Over 75	MS MS	6 Root runs 2 fill	100 min 100 min	SR 650
C–Mn Structural. HT. (BS 4360: 1968)	0·23 max	1·6 max					25 25 50	MS MS MS	6 6 6	150 min Nil 150 min	
Carbon BS 970											
EN8	0·4	0·8				—	25 75	MS MS	2 2	100–150 200–250	
EN43A	0·5	0·8				—	25 25	MS MS	2 2	200 min 225 min	
EN9	0·55	0·8				—	Over 25 Over 25	MS MS	2 2	250 min 250 min	
Cr–Mo (high temperature)											
1 Cr–½ Mo	0·15 max	0·4		0·85/1·2	0·4/0·6		Over 12	M***	6	100 min	630–730 SR
½ Cr–½ Mo–¼ V	0·15 max	0·4		0·5	0·5	0·25	Over 12	1 Cr–½ Mo	6	100 min	630–730 SR
2¼ Cr–1 Mo	0·15 max	0·4		2·0/2·5	0·9/1·1	—	All	1 Cr–½ Mo	6	200–250	700–760 SR*
3 Cr–1 Mo	0·15 max	0·4		2·75/3·25	0·8/1·0	—	All	1 Cr–½ Mo	6	320–370	730–760 SR*
5 Cr–½ Mo	0·15 max	0·4		4·0/6·0	0·45/0·65	—	All	1 Cr–½ Mo	6	320–370	730–760 SR*
Cryogenic											
3½% Nickel	0·12	0·85	3·5	—	—	—	All	M	6	150 min	SR if code requirement
9% Nickel	0·10 max	0·4	9·0	—	—	—	All	Ni base	6	150 min	565–580 SR
Engineering											
EN 13	0·18	1·6					25 50	MS MS	6 6	None 100–125	550–660 SR 550–660 SR
EN 14B	0·25	1·5					25	HT	6	100 min	550–660 SR

1 Cr–Mo	0·4	0·5	—	1·0	0·25	25–50	HT	6	250–300 590–670 SR
						12	HT	6	200–250 590–670 SR
						12–50	HT	6	300–350 Hold 1 hr 300–350** 590–670 SR

* Transfer directly to Stress Relief without allowing to cool below pre-heat temperature.
** Low alloy Engineering Steels containing over 0·39% Carbon should be pre-heated 300–350°C and held 1 hr minimum at this temperature before Stress Relief if over 12 mm thick.
*** M = Matching tensile strength.
Electrodes Class 2 = Rutile with high hydrogen.
            Class 6 = Basic with low hydrogen.

Steel		Remarks
Mild steel		Pre-heat to prevent root run cracks and fissures in fill-up runs when using rutile electrodes
C–Mn	25 mm	Small gauge electrodes 1·25 kJ/mm
	25 mm	Large gauge electrodes 3·25 kJ/mm ⎫ Groups 2 and 4
	50 mm	Large gauge electrodes 3·25 kJ/mm ⎭
Carbon EN8	25 mm	Medium gauge electrodes 2 kJ/mm
	75 mm	Medium gauge electrodes 2 kJ/mm ⎫ Class 2 electrodes give less mixing than Class 6 therefore with high carbon
EN43A	25 mm	Large gauge electrodes 3 kJ/mm ⎬ parent steel there is less risk of weld cracking if pre-heating is as indicated.
	25 mm	Medium gauge electrodes 2 kJ/mm Group 3
EN9	Over 25 mm	Large gauge electrodes 3 kJ/mm ⎫
	Over 25 mm	Large gauge electrodes 3 kJ/mm ⎭
Cr–Mo		
1 Cr-½ Mo		Group 4
½ Cr-½ Mo-¼ V		Group 4
2¼ Cr-1 Mo		Group 5
3 Cr-1 Mo		Group 5
5 Cr-½ Mo		Group 5
Cryogenic		
3½% Ni		Pre-heat if rigidity is high. Group 2
9% Ni		Group 4. Reference 49

(continued overleaf)

Table 6.19—*continued*

	Steel	Remarks
Engineering		
EN 13	25 mm	Electrodes at 3 kJ/mm. Group 2 if carbon less than 0·2% otherwise Group 5.
	50 mm	Electrodes at 3 kJ/mm
EN14B	25 mm	Electrodes at 3 kJ/mm
	25 mm	Electrodes at 2 kJ/mm ⎫
EN17	50 mm	Electrodes at 3 kJ/mm ⎬ Group 5
		Electrodes at 3 kJ/mm ⎭
1 Cr–Mo 0·3C	12–25 mm ⎫	
	20–50 mm ⎬ First run with mild steel electrodes. Group 5.
0·4C	12 mm ⎫	
	12–50 mm ⎭	

Table 6.20 WELDING CARBON AND LOW ALLOY STEEL CASTINGS. MANUAL METAL ARC PROCESS

Steel	Nominal composition						Pre-heat minimum °C*	Post-heat treatment	Electrodes	Remarks
	C	Mn	Ni	Cr	Mo	V				
BS 592B	0·35	—	—	—	—	—	100	600–650 SR	BS 639 Medium H.T.	Immediate post-heat treatment often desirable.
BS 592C	0·45	—	—	—	—	—	150	600–650 SR	BS 639 Medium H.T.	
Carbon–Mo	0·15/0·25	—	—	—	0·4/0·7	—	150	630–680 SR	BS 2493 A	
Carbon–Mn	0·18/0·25	1·2/1·7	—	—	—	—	150	600–650 SR	BS 639 Medium H.T.	
	0·25/0·33	1·2/1·7	—	—	—	—	200	600–650 SR	BS 639 Medium H.T.	
	0·4/0·5	1·0	—	—	—	—	250	600–650 SR	BS 639 Medium H.T.	
	0·55/0·65	1·0	—	—	—	—	300	600–650 SR**	BS 639 Medium H.T.	Weld preparation = chipping, grinding, machining.
C–Mn–Cr	0·45/0·55	1·0	—	1·0	—	—	300	600–650 SR***	BS 639 Medium H.T.	
	0·55/0·65	1·0	—	1·0	0·3	—	350	600–650 SR***	BS 639 Medium H.T.	
3½% Ni	0·15	—	3·5	—	—	—	300	580–630	Matching	
Mo–V	0·15	—	—	—	0·6	0·25	300	675–700 SR**	BS 2493 C	
Cr–Mo–V	0·15	—	—	1·0	1·0	0·25	300	675–700 SR***	BS 2493 C	
1¼ Cr–Mo	0·2 max	—	—	1·25	0·5	—	250	600–650 SR	BS 2493B	
2¼ Cr–Mo	0·18 max	—	—	2·25	1·0	—	275	640–690 SR**	BS 2493C	
3 Cr–Mo	0·2 max	—	—	3·0	0·5	—	275	640–690 SR**	BS 2493 D	
5 Cr–Mo	0·2 max	—	—	5·0	0·5	—	300	650–700 SR***	BS 2493 D	

* Preheat entire casting in furnace. Maintain temperature until welding completed. Generally desirable to post-heat treat without casting cooling out.

** Essential for post-heat treatment to be applied immediately welding is completed without allowing to cool below preheat temperature.

post-heating, stress corrosion cracking by using preferred compositions and more rarely by stress relief treatment.

To avoid weld metal hot cracking the predominantly austenitic weld metal should contain not less than 3% ferrite (controlled by the ratio of ferrite forming elements Cr, Si, Mo to austenite forming elements C, Mn, Ni, N). Excepting the 25 Cr–20 Ni alloys, manual metal-arc electrodes provide weld analyses which, allowing for normal dilution, take account of this when welding their 'type' composition. Given a knowledge of the degree of mixing, the diagrams, Figure 6.9 and 6.9(a) to 6.9(f), offer a guide to the likely weld metal micro-structures from the calculated resultant weld composition in terms of Cr and Ni equivalents. Stainless weld metals fall into four broad groups (1) Martensitic (2) Ferritic and Ferritic–Martensitic (3) Ferritic–Austenitic and (4) Austenitic with or without Ferrite up to about 30%. Procedures to minimise the above cracking possibilities are indicated in Tables 6.21(a), 6.21(b) and 6.21(c). The steels in groups 1, 2 and 3 normally require pre-heating and post-heating. General welding information for Chromium Irons and Steels and for Austenitic Chromium–Nickel Stainless and Heat Resisting Steels is given in Reference 51, chapters 64 and 65 respectively.

Fully ferritic chromium heat-resisting weld metal is exceedingly brittle below 200°C in the as-welded condition and must be heat treated before cooling out. The 25 Cr 12 Ni and 25 Cr 20 Ni weld metals and alloys develop in varying degree, carbide and sigma phase embrittlement in service at temperatures in the 500°C to 950°C range; therefore freedom for expansion is an important aspect of design. Weld repairs to material embrittled and cracked in service is generally unsuccessful unless the material is given a prior heat treatment at 1050°C to 1150°C.

INERT GAS TUNGSTEN–ARC WELDING (TIG)

This is the ideal welding process for austenitic stainless steel sheet and for root runs in butt joints in pipe. Accessibility of the torch to deep joints can be a problem. Argon gas is used. Austenitic filler wires are made to compositions which give a small ferrite content in an undiluted all-weld-metal deposit; Table 6.21(b). (BS 2901: part 2: 1970. part extraction) lists the various grades of austenitic stainless and heat resisting compositions in common usage. Ferritic filler wires (group 2) are not generally available.

The composition of the diluted weld metal depends on amount of penetration, composition of parent steel and wire composition; in root runs the parent steel has a predominant effect on composition and as it is not made to impart a given minimum ferrite content to the root run, it is common practice to use an insert (EB insert) of suitable composition.

INERT GAS METAL–ARC WELDING (MIG)

This process is suitable for a wide range of general fabrication work in stainless steels using argon–oxygen gas mixtures. However, there is a tendency for 'adhesion' rather than fusion to occur at the start of the weld run; full fusion of the parent metal requires direct exposure to the arc and, at the start, establishing the arc results in flooding the pool and blanketing the underlying metal. Once forward movement has started the blanketing effect disappears provided correct travel speed is used. The 'adhesion' defective areas are not readily detected by radiography and the inherently coarse grain of austenitic weld deposits makes inspection by ultrasonic methods difficult.

For critical applications (pressure vessels etc.) and particularly for multi-run welds, it is questionable whether the Mig process is really suitable. However, provided the defects are not exposed to a surface corrosive medium by subsequent machining operations, their significance on static mechanical properties is low because they are sited in metal of extremely good notch toughness. This comment however does not necessarily

Susceptibility to:—

☐ Martensitic cracking below 400°C (see Table 6.21c)

○ Hot cracking above 1250°C (see Table 6.21c)

⊕ Sigma phase embrittlement after heat treatment or service at 500°C–900°C (see Table 6.21c)

■ Cold brittleness after grain growth due to high temperatures (above 1150°C). Ductile above 400°C (see Table 6.21c)

Overlapping areas, e.g. ■⊕, show characteristics of both.

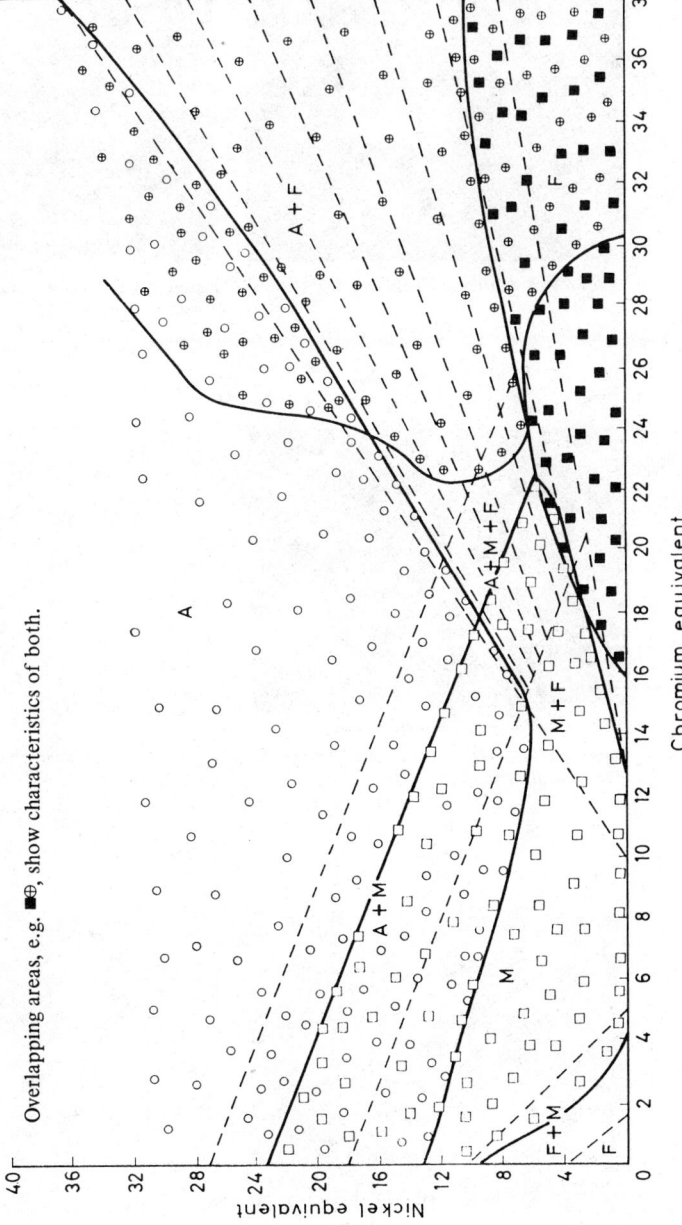

Figure 6.9. Stainless weld metals. Susceptibility to embrittlement etc. related to composition

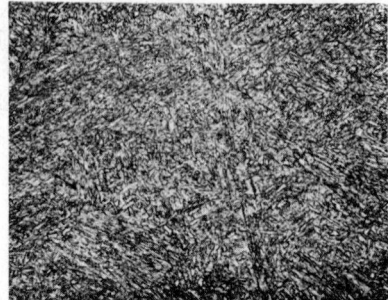

Figure 6.9(a). Martensitic weld metal

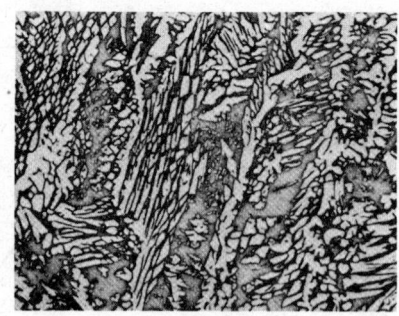

Figure 6.9(d). Ferritic-Austenitic weld metal (Ferrite = dark areas)

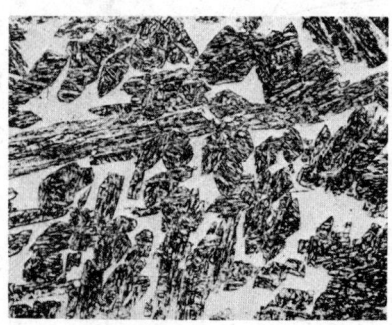

Figure 6.9(b). Ferritic-Martensitic weld metal. (Ferrite = light areas)

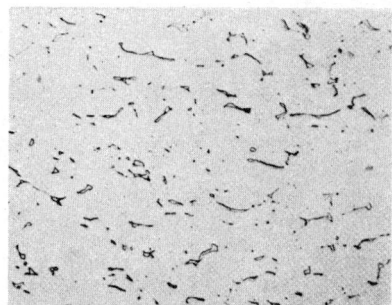

Figure 6.9 (e). Austenitic weld metal (with small ferrite content)

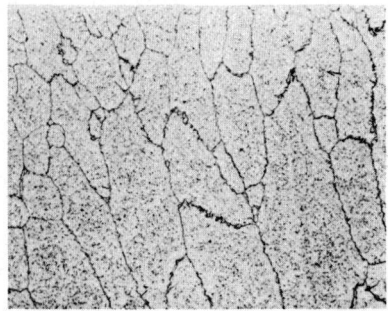

Figure 6.9(c). Ferritic weld metal (with fine carbides)

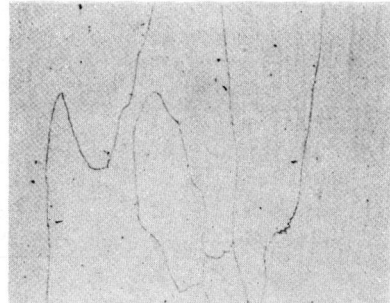

Figure 6.9 (f). Fully austenitic weld metal

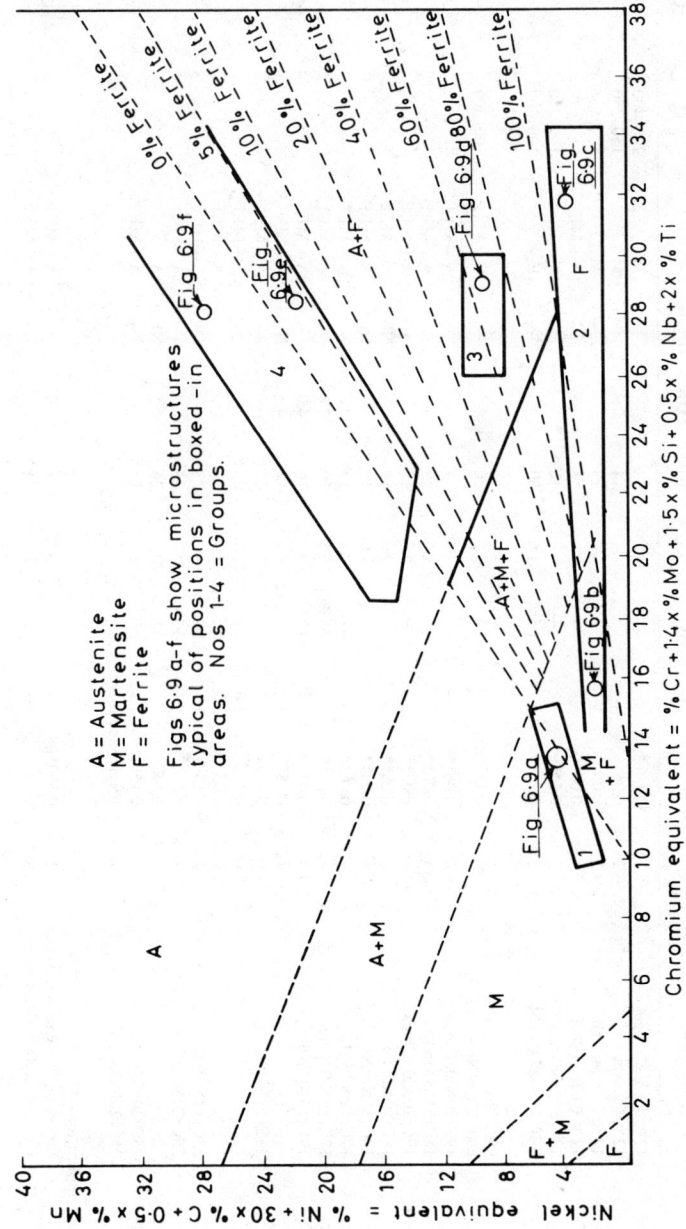

Figure 6.9(g). Stainless weld metal constitution diagram. Modified Schaeffler diagram 1 Deutsche Edelstahlwerke. Note 1. No. 4 extends to 30 per cent Ferrite. Note 2. Composition of fabrication welds depends on extent of mixing with parent metal. Calculate accordingly

Table 6.21(a) WELDING PROCEDURES FOR STAINLESS AND HEAT RESISTING STEELS. MANUAL METAL ARC PROCESS. TYPICAL ALL-WELD METAL COMPOSITIONS

A.I.S.I.	Parent steel Type	Group (see Fig. 6.9)	Weld metal nominal composition							Micro-structure***	Welding procedure	
			C	Mn	Si	Ni	Cr	Mo	Others		Pre-heat °C	Post-heat °C
—	12%–18% Cr steel	1	0·1/0·2	—	—	—	13	—	—	M	200–400	720–760
410, 403	12% Cr iron	2	0·06	—	1·5	—	14	—	—	M+F	150–250	730–780A*
405	12% Cr Iron (Al)	2	0·06	—	—	—	13	—	—	M+F	—	—
405	12% Cr Iron (Al)	4	0·06**	—	—	9	18	2·5	—	A+MF	—	—
430	14–18% Cr Iron	2	0·1**	—	1·5	—	18	—	—	F	100–250	700–800A*
446	0·35C 23–27 Cr (N)	2	0·1	—	1·5	1·6	30	—	—	F	100–300 sheets	750–800A*
—	0·4C 29Cr / 0·4C27Cr4Ni	3	0·15**	—	1·5	4·5	26	—	—	F+A	100–300	—
304	19Cr10Ni08Cmax	4	0·06**	—	—	9·5	19	—	—	A+MLF	—	—
202	18Cr5Ni8Mn (N)	4	0·06**	—	—	9·5	19	—	—	A+MLF	—	—
304L	19Cr10Ni03Cmax	4	0·03**	—	—	9·5	19	—	—	A+MLF	—	—
321	18Cr10Ni (Ti)	4	0·08**	—	—	9·0	19	—	—	A+MLF	—	—
347	18Cr11Ni (Nb)	4	0·08**	—	—	9·0	19	—	Nb	A+MLF	—	—
316	17Cr12Ni2½Mo	4	0·06**	—	—	10·5	19	2·6	—	A+MF	—	—
316L	17Cr12Ni03Cmax	4	0·03**	—	—	10·5	19	2·6	—	A+MF	—	—
317	19Cr13Ni3½Mo	4	0·03	0·7	0·8	13·5	19	4·0	Nb	A+MF	—	—
		4	0·08**	—	—	13·5	18	4·5	—	A+MF	—	—
309	23Cr13Ni20C	4	0·1	1·0	1·2	11	22	—	Nb	A+MF	—	—
and 309S	23Cr13Ni08C	4	0·1	3·0	0·8	12	23·5	—	—	A+MLF	—	—
310	25Cr20Ni25C	4	0·05	1·5	0·5	12	23	0·5	3W	A+MLF	—	—
and 310S	25Cr20Ni08C	4	0·15	1·7	1·0	11	23·5	—	—	A+HF	—	—
		4	0·1	7	0·3	20	26	—	Nb	A	—	—
314	25Cr20Ni2Si25C	4	0·15	1·5	0·8	20	26	0·5	3W	A	—	—
			Electrodes as for 310.							A+MF		
Special applications												
347	Steam piping. Creep resisting	4	0·05/0·08	—	—	8	17	2·0	—	A+MLF	—	SR
—	Cr–Iron or 18/8 cladding	4	0·10	—	0·8	12	23·5	—	—	A+MLF	250 (13% Cr)	—
—	Dissimilar metals (stainless/MS etc.), armour plate	4	0·1	—	—	8·5	20·5	3·5	—	A+HF	—	—

A.I.S.I.	Remarks: Applicable to weld metal
410, 403	Pre-heat not generally necessary on light structures.
405 Austenitic electrodes	Pre-heat not required.
405	Resists sulphur attack at higher temperatures.
430	Preferable to use 18 Cr 9 Ni 2½ Mo electrode for root and fill runs up to last layer.
446	—
— 29 Cr	—
— 27 Cr 4 Ni	—
304	Extra low carbon for intergranular corrosion resistance without stabilisation.
202	Nb stabilised weld metal for Ti stabilised steel.
304L	Nb stabilised weld metal for Nb stabilised steel.
321	—
347	Extra low carbon Mo bearing.
316	—
316L	Also for group 2 and 3 steels except surface runs.
317	Also for group 2 and 3 steels except surface runs.
309	Susceptible to hot cracking (26 Cr 20 Ni weld metal).
309S	Less susceptible to hot cracking (26 Cr 20 Ni 7 Mn, Nb weld metal).
310 and 310S	For root runs in 25 Cr 20 Ni joints (23·5 Cr 11 Ni 3 W weld metal).
310 and 310S	As for 310.
310 and 310S	
314	
Special applications	
347 Creep resisting	Controlled ferrite (4 %–8 %) Mo bearing weld metal reduces risk of stress relief cracking in heat affected zone.
— Cr–Iron or 18/8 cladding	For first run on to mild steel backing material.
— Dissimilar metals	—

Table 6.21(b) FILLER RODS AND WIRES FOR INERT-GAS TUNGSTEN-ARC AND INERT-GAS METAL-ARC WELDING OF AUSTENITIC STAINLESS AND HEAT RESISTING STEELS. (BS. 2901: Part 2. 1970)

| Type | Chemical composition | | | | | | | | | | | | | | Mig process undiluted weld metal structure* |
| | C | | Si | | Mn | | Ni | | Cr | | Mo | | Nb | | |
	min	max	min	max	min	max	min	max	min	max	min	max	min	max	
308S92	—	0·03	0·25	0·60	1·0	2·5	9·0	11·0	19·5	22·0	—	—	—	—	A+F (8%)**
308S96	—	0·08	0·25	0·60	1·0	2·5	9·0	11·0	19·5	22·0	—	—	—	—	A+F(5%)**
308S93	—	0·03	0·7	1·0	1·5	2·0	9·5	10·5	20·0	21·0	—	0·3	—	—	A+F(15%)**
347S96	—	0·08	0·25	0·6	1·0	2·5	9·0	11·0	19·0	21·5	—	—	10×C	1·0	A+F (7%)**
309S94	—	0·12	0·25	0·6	1·0	2·5	12·0	14·0	23·0	25·0	—	—	—	—	A+F (7%)**
311S94	—	0·12	0·25	0·6	1·0	2·5	12·0	14·0	23·0	25·0	—	—	10×C	1·3	A+F(10%)**
310S94	0·08	0·15	0·25	0·6	1·0	2·5	20·0	22·5	25·0	28·0	—	—	—	—	A
310S98	0·35	0·45	0·8	1·3	1·0	2·5	20·0	22·5	25·0	28·0	—	—	—	—	A
313S94	0·06	0·13	0·25	0·60	1·0	2·5	20·0	22·5	25·0	28·0	—	—	10×C	1·3	A
316S92	—	0·03	0·25	0·6	1·0	2·5	11·0	14·0	18·0	20·0	2·0	3·0	—	—	A+F(10%)**
316S96	—	0·08	0·25	0·6	1·0	2·5	11·0	14·0	18·0	20·0	2·0	3·0	—	—	A+F (2%)**
316S93	—	0·03	0·7	1·0	1·2	1·8	10·0	13·5	18·0	20·0	2·5	3·0	—	—	A+F(10%)**
317S96	—	0·08	0·25	0·6	1·0	2·5	13·0	15·0	18·5	20·5	3·0	4·0	—	—	A+F(10%)**
318S96	—	0·08	0·25	0·6	1·0	2·5	11·0	14·0	18·0	20·0	2·0	3·0	10×C	1·0	A+F (6%)**

* Middle of range. Assumes 30% loss of C (above 0·03 and Nb and 25% loss of Mn and Si.
** % Ferrite ± 2½%

Table 6.21(c) STAINLESS WELD METAL CHARACTERISTICS (Ref. Fig. 6.9(g))

Group	Microstructure	Characteristic	Remarks
1.	Martensitic (strongly hardening)	Susceptibility to cracking under normal weld cooling rates.	Pre-heating and interpass (150°–400°C) and holding pre-heat four hours minimum prevents cracking (annealing necessary for full toughness) in joints in hardenable chromium steels. Hard facing deposits not usually annealed. Electrodes must be fully dried. Anneal heavy fabrications without cooling out from welding.
2. (a)	Ferritic–Martensitic (hardening)	Lower susceptibility to cracking than Group 1	Pre-heat 100°–250°C for joints in chromium irons (anneal for full toughness unless for heat resisting service). Facing deposits not usually annealed. Dry electrodes fully.
2. (b)	Ferritic (non-hardening)	Very notch brittle below 200°C. Welds having more than 28% equivalent Cr (80% ferrite+) develop sigma phase embrittlement in service (500°C–900°C).	Pre-heating required to prevent brittle fractures. Annealing improves notch ductility and removes residual stress. Sigma embrittled steels relatively tough at furnace operating temperatures. Fully dry electrodes. Risk of brittle fracture if fabrication cooled out quickly after embrittlement in service. For non-heat resisting applications ferritic chromium steels frequently welded with Group 4 weld metals and capped on working face with Group 2b (gives tougher joints without annealing).
3.	Ferritic–Austenitic (non-hardening) 40–80% Ferrite	Relatively notch tough as-welded compared with Group 2b but ferrite areas subject to sigma-phase embrittlement in service (500°C–900°C).	Pre-heating avoids brittle fracture. Cool out slowly after welding or after service in embrittlement temperature range. Fully dry electrodes.
4. Austenitic (a)	Fully Austenitic	Subject to hot cracking and fissuring.	Weld metals with high carbon (e.g. 25 Cr 20 Ni 0·4 C) resistant to hot cracking and fissuring but only suitable for high temperature use. For low carbon welds use root runs with 5%–25% ferrite.
(b)	4–15% Ferrite	Not subject to cracking, fissuring or service embrittlement.	
(c)	15–30% Ferrite	Subject to sigma embrittlement (high temperature service) or in Mo bearing corrosion resisting welds.	Embrittlement not generally of practical significance in heat resisting welds since tough at working temperatures. Mo bearing welds may suffer loss of corrosion resistance (sigma-phase formed by multi-run heating); select lower ferrite weld metals.

apply to heat resisting steels in an embrittled condition. Their main significance lies in the tendency, with metal of low resistance to hot cracking, to generate hot tears of an intergranular nature which can propagate under dynamic loading conditions. The 'adhesion' defects are most serious in root runs where the root of the weld is exposed to the corrosive medium because rapid corrosion can occur in the crevices; Mig welding is unsuited to welding root runs of pipe joints.

Nickel and nickel-base alloys

The majority of these alloys are readily welded by the manual metal arc, tungsten arc (argon), inert-gas metal arc (argon), plasma arc, electron-beam and resistance welding processes. Oxyacetylene and submerged arc welding have only limited application. A feature of these alloys is susceptibility to serious embrittlement from sulphur and lead contamination. For all welding processes the surfaces to be fused or heated must be completely free of any contaminants containing these elements, such as cutting oils, oxide, grease, films from atmospheric pollution etc. Nickel 200, nickel 201 and monel 400 require most attention in this respect, abrasive cleaning being necessary to remove air pollution films[52].

High nickel weld metal is more sluggish than mild or stainless steel weld metals and penetration is less in manual metal arc and inert-gas metal arc processes. Therefore room for manipulation must be increased by wider V angles and root radii than are used for mild steel. Recommended joint preparation, welding conditions, filler and electrodes etc., for MMA, Tig, Mig, submerged arc, oxy-acetylene and resistance welding of nickel base alloys are given in Reference 52. Since the weld metal composition must, with certain alloys, contain additional elements to control porosity or hot cracking the 'matching analysis' concept is frequently not applicable. Proprietary electrodes and filler wires for various nickel base alloys are given in Table 6.22[52]. Other proprietary electrodes of similar composition are available for Monel and Inconel 600. Filler wires for Tig and Mig welding are covered in BS 2901: Part 5: 1970. For detailed information on the welding of nickel and nickel base alloys see also Reference 53.

Aluminium and aluminium alloys

These materials are usually welded by the inert-gas tungsten arc and inert-gas metal arc processes or by resistance spot and seam welding.

DESIGN OF JOINTS

Butt, lap and tee joints are all used; edge and corner joints are less satisfactory. Aluminium has a considerable advantage from the welding design point of view in that extruded sections are generally available which reduce joint preparation costs to a minimum by eliminating bevelling for butts, tees and thick-to-thin section joints. Where even root penetration is required 'all position' (e.g. a pipe joint welded in the fixed horizontal position), a typical joint design is shown in Figure 6.10.

The main problem in welding aluminium alloys is weld metal hot cracking, caused by crack sensitive compositions and restraint. Methods of avoiding this involve controlling composition by minimising mixing with parent metal (bevelled butts are preferable to close square butts) and using a filler wire with higher alloy content, silicon or magnesium, than the parent metal e.g. for N4 alloy (aluminium, $2\frac{1}{4}\%$ Mg) NG6 wire (aluminium, 5% Mg) is used, and for Al—Mg—Si alloys, either NG6 or NG 21 (aluminium, 5% Si) is used. A minor problem is porosity caused by hydrogen. Properly degassed wires and parent materials and clean dry (proper storage) wire surfaces minimise porosity but absolute soundness is difficult to achieve and seldom necessary. Work hardened or heat treated alloys may pose problems in attaining matching joint strength but given correct weld metal compositions sound as-welded joints match or over-match the parent metal strength in the non-heat treatable alloys. Filler wires for Tig and Mig welding are given in Table 6.23. Tig welding of aluminium is usually done on a.c. current to

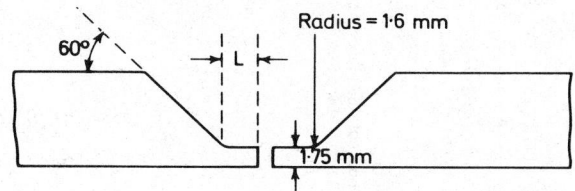

Figure 6.10. Butt weld joint design in aluminium for complete penetration
for all-position welding (cf. Ref. 54)
L = 3·25 mm. Minimum inert gas tungsten-arc welding
= 1·6 mm. Minimum inert gas metal-arc welding

break up and disperse oxide film. Weld quality in inert-gas metal arc welding is dependent on the quality of the electrode wire in terms of reliability of freedom from gas and non-metallics, surface cleanness and reliability of spooling. For detailed information on fusion welding aluminium and aluminium alloys, see Reference 54.

Further reading

BAYSINGER, F. R. and ROBINSON, I. B. 'Welding aluminium Alloy 7039', *Welding Journal, Research Supplement*, **45**, 10, 433s–444s (October 1966)
HOLMES, A. W. and ROGERSON, J. H. 'Pulse Arc Welding of Aluminium', *Welding and Metal Fabrication*, **34**, 9, 349–354 (September 1966)
'The Welding of Aluminium and its Alloys', *Welding News*, 129, 15–17 (January 1967). 130. 17–18 (May 1967). 131, 9–11 (September 1967). 132, 17–18 (January 1968). 133, 10–12 (April 1968). A general study of the properties, alloying elements, surface and joint preparation, welding techniques for Tig and Mig processes, including equipment and filler metals.
'Refuelling Tanks for Concorde', *Welding and Metal Fabrication*, **36**, 10, 362–366 (October 1968). Mig Fabrication in Aluminium Alloys. Equipment, procedure, operations before and after welding. Control of dimensions. Pressure testing.
CLEMENTS, H. W. 'Power Sources for Gas Shielded welding of Aluminium', *Welding and Metal Fabrication*, **36**, 12, 440–446 (December 1968). Tests to select suitable power sources for Tig welding aluminium.
YOUNG, J. G. 'B.W.R.A. experience in the welding of Al-Zn-Mg alloys', *Welding Journal, Research Supplement*, **47**, 10, 451s–461s (October 1968). Influence of base metal composition and heat treatment on metallurgical problems in gas shielded arc welding of Al-Zn-Mg alloys.
HIRSCHFIELD, J. A. and WATERBURY, W. C. 'Mig welding of Aluminium', 'Tig welding of Aluminium', Papers given at Aluminium Welding Seminar on 13th-14th November 1969. San Francisco. Sponsored by Aluminium Association and American Welding Society.
CLEMENTS, H. W. 'Power Sources for Gas Shielded Welding of Aluminium. (Mig Process)', *Welding and Metal Fabrication*, **37**, 3, 93–99 (March 1969)
KENT, K. G. 'Weldable Al-Zn-Mg Alloys', Proceedings of Second Commonwealth Welding Conference, 239–245 (May 1965). Institute of Welding, Abington, Cambridge

Copper and copper alloys

For fusion welding copper and copper alloys the inert-gas tungsten arc or inert-gas metal arc are preferred. The high heat conductivity of copper makes fusion of the parent

Table 6.22 ELECTRODES AND FILLER RODS FOR SOME NICKEL AND NICKEL BASE ALLOYS

Material	Manual metal arc Electrode	Drying time at 260°C hrs	Tungsten arc argon filler	Mig argon wire	Oxy-acetylene Wire	Oxy-acetylene Flux	Submerged arc Wire	Submerged arc Flux	BS 2901, Part 5
Nickel 200	Nickel 141*	2	Nickel 61*	Nickel 61*	Nickel 41	—	—	—	Nickel 61 = NA 32
Nickel 201	Nickel 141*	—	Nickel 61*	Nickel 61*	—	—	—	—	—
Monel 400	Monel 190*	1	Monel 60*	Monel 60*	Monel 40*	Monel	Monel 60*	Inco-flux 5	Monel 60 = NA 33
Inconel 600	Inconel 132*	2	Inconel 82*	In'l 82*	NC 80/20	In'l**	In'l 82*	Inco-flux 4	Inconel 82 = NA 35
	Inconel 182	2	NC 80/20						
Inconel 718			Inconel 718	In'l 718					—
Inconel X-750		—	Inconel 69	—					—
Incoloy DS	Inco weld A*	2	NC 80/20	NC 80/20	NC 80/20	In'l**	—	—	NC 80/20 = NA 34
Incoloy 800	Inco weld A*	2	Inconel 82*	In'l 82*	NC 80/20	In'l**	—	—	—
Incoloy 825	Incoloy 135	2	Incoloy 65	In'ly 65	—	—	—	—	Incoloy 65 = NA 41
Nimonic 75	Inconel 132	2	NC 80/20	NC 80/20	NC 80/20	In'l**	—	—	—
	Inconel 182	2	Inconel 82*	In'l 82*					
Niloalloys	Inco weld A*		Inconel 92*	In'l 92*	—	—	—	—	Inconel 92 = NA 39
			Nickel 61*	Nickel 61*					—

* Weld deposit or wire differs from parent metal in respect of content of Nb, Ti, Al etc.
** Flux is boron-free.

BS*	Parent material Nominal composition	Filler rod or wire BS 2901, Pt 4 1970	Filler rod or wire Nominal composition	Remarks
1, 1A, 1B, 1C, LMO	Pure aluminium, various grades	G1B	Pure aluminium	
N3	$1\frac{1}{4}$% Manganese	NG3	$1\frac{1}{4}$% Mn	Preferred.
		G1B	Pure aluminium	
N4	$2\frac{1}{4}$% Mg	NG6	5% Mg	Preferred.
		NG61	$5\frac{1}{4}$% Mg, $\frac{3}{4}$% Mn	
		NG4	$2\frac{3}{4}$% Mg	May give cracking.
		N51	$2\frac{3}{4}$% Mg, $\frac{3}{4}$% Mn	
		NG52	$2\frac{3}{4}$% Mg, $\frac{3}{4}$% Mn	
		NG6	5% Mg	
		NG61	$5\frac{1}{4}$% Mg, $\frac{3}{4}$% Mn	
N6	5% Mg	NG6	5% Mg	
		NG61	$5\frac{1}{4}$% Mg, $\frac{3}{4}$% Mn	
N8	$4\frac{1}{2}$% Mg, $\frac{1}{2}$% Mn	NG6	5% Mg	
		NG61	$5\frac{1}{4}$% Mg, $\frac{3}{4}$% Mn	
H20	1% Mg, $\frac{1}{2}$% Si	NG6	5% Mg	Strongest welds but possible cracking difficulties.
H30	1% Mg, 1% Si	NG21	5% Si	Use to avoid cracking.
		NG2	10% Si	Use when cracking is exceptionally severe.
H14	4% Cu	NG2	10% Si	Weak joints.
H15	$4\frac{1}{4}$% Cu			
H17	$4\frac{1}{4}$% Zn, $1\frac{1}{4}$% Mg	NG6	5% Mg	
		NG61	$5\frac{1}{4}$% Mg, $1\frac{1}{4}$% Mn	Al–Zn–Mg filler could be considered in exceptional cases.
—	Stronger Al–Zn–Mg alloys e.g. 4% Zn 2% Mg	NG6	5% Mg	Fillet welds and low stress butt welds.
		NG61	$5\frac{1}{4}$% Mg, $1\frac{1}{4}$ Mn	(Fillet welds and low stress butt welds.)
		—	$4\frac{1}{4}$% Zn, $2\frac{1}{4}$% Mg	Hiduminium 48
			$3\frac{3}{4}$% Mg, $2\frac{3}{4}$% Zn	ASA 5039
LM6	13% Si	NG2	10% Si	
LM9, LM20	13% Si	NG2	10% Si	
LM8, LM18	5% Si	NG21	5% Si	
LM25	7% Si	NG21	5% Si	
LM5	$4\frac{1}{2}$% Mg	NG6	5% Mg	
LM10	10% Mg	NG61	$5\frac{1}{4}$% Mg, $\frac{3}{4}$% Mn	

* BS 1470–1477. Sheet, plate, bar etc.
BS 1490. Castings (LM numbers).
N51. Supplementary series.

metal the major problem necessitating preheating other than light gauge sheet metal. Too high a preheat temperature causes oxidation of the joint faces with consequent cracking in the weld metal. Filler wires for Tig and wires for Mig welding are given in Table 6.24.

TIG WELDING

Copper parent metal should be de-oxidised or oxygen-free. Argon gas is normally used with d.c. negative polarity. Heavy sections are unsuited to Tig welding. Preheating is required for copper over 3 mm thickness. Copper alloys should only be preheated in exceptional circumstances (i.e. castings).

MIG WELDING

For welding copper, wide angle preparation is necessary. Over 9 mm thickness copper should be preheated. Copper–Zn need less wide angles when welding with silicon–bronze. The high zinc brasses and manganese bronze are welded with aluminium–bronze wires. The copper–silicon alloys tend to show weld hot cracking; use small weld pool and low inter-pass temperature not exceeding 100°C. Wide angle preparation is required for the aluminium–bronzes. Downhand welding is preferable. Argon gas is generally used for Mig welding copper alloys, which are not normally pre-heated. For detailed welding data on copper and copper alloys see References 56, 57.

MANUAL METAL ARC PROCESS

Suitable coated electrodes give satisfactory results on aluminium–bronze, tin–bronze, and cupro–nickel alloys.

DISSIMILAR METAL WELDS

There are many situations in engineering practice where it is advisable to use filler metal (coated electrode or filler wire) which deposits metal of dissimilar composition to that of one or both of the parent metals.

Examples include strongly hardening alloy steels welded with austenitic chromium nickel molybdenum electrodes (cf. Table 6.21), ferritic chromium steels welded with austenitic Cr–Ni weld metals etc. Copper is frequently joined to mild steel, stainless steel and to nickel alloys; to prevent excessive heat loss to the copper side of the joint during the closing weld the copper should firstly be overlayed with nickel pre-heating where necessary. High nickel alloy or stainless steel, depending on the other parent metal, are frequently used to complete the joint. Where heat loss is more balanced than in copper to other metals, direct welding is possible. Aluminium–bronze is suitable for joining several copper alloys to steel as it is not affected by normal amounts of iron dilution from the steel.

Further reading

CLEWS, K. J. and YOUNG, Y. G. *Processes for Copper Welding—recent developments and their Applications*, Copper Development Association Conference on 'Modern Aspects of Copper in Electrical Engineering', London, 7, (April 30th 1968)
DAWSON, R. J. C. 'Inert Gas Shielded Arc Welding of Copper and Copper Alloys; the dissimilar joint', *Welding and Metal Fabrication*, **36**, 2, 60–66 (February 1968)

Table 6.24 FILLER RODS, WIRES AND GASES FOR INERT-GAS SHIELDED ARC WELDING OF COPPER AND COPPER ALLOYS

(Filler designation ex BS 2901 : Part 3 : 1970)

Material	Filler or wire	Gas	Remarks
TUNGSTEN-ARC			
Copper, PDO	C7	He or A	{ Helium preferable, allows lower pre-heat, produces superior quality welds. C7 wire should
	C7	He or A	be tin free for resistance to stress corrosion cracking in certain circumstances.
	C21	He or A	For applications requiring high conductivity weld metal.
Cu-Zn alloys	C12	50A/50He or A	Pre-heat castings up to 100°C.
Cu-Si alloys	C9	A	
Cu-Sn alloys	C10	A	Selected to match or over-match parent metal Sn content.
	C11	A	Selected to match or over-match parent metal Sn content.
Cu-Al alloys	C12	A	Better corrosion resistance than C13. For 7% Al alloys.
	C12Fe	A	To match BS copper alloy CA106.
	C13	A	General purpose filler for Cu-Al alloys.
	C20	A	Weld metal has high hot plasticity. Good corrosion resistance.
Cu-Ni alloys	C16	A	Selected to match or over-match Ni in parent metal.
	C18	A	Selected to match or over-match Ni in parent metal.
INERT-GAS METAL ARC			
Copper, PDO } Copper, oxygen free }	C7	50/50He or A or N	C7 general purpose wire. C21 high conductivity. A/He gas preferred. Welding grade N = excessive spatter and rough but sound deposits.
Cu-Zn alloys	C12	50A/50He or A	Base metal must be lead free. Vee grooves 70–90 degrees.
Cu-Si alloys	C9	50A/50He or A	
Cu-Sn alloys	C10	50A/50He or A	
	C11	50A/50He or A	
Cu-Al alloys	C12	50A/50He or A	See tungsten-arc.
	C12Fe	50A/50He or A	
	C13	50A/50He or A	
	C20	50A/50He or A	
Cu-Ni alloys	C16	50A/50He or A	
	C18	50A/50He or A	

TAYLOR, E. A. and BURN, A. H. *Development of improved Fillers for Inert-Gas Arc Welding Cupro-Nickel Alloys,* Institute of Welding, Autumn meeting London, **3**, 41–51 (1967)

GARRIOTT, F. E. 'Welding Design and Process Manual—Copper and its Alloys', *Welding Engineer,* **52**, 6, 59–66 (June 1967)

DAWSON, R. J. C. 'Inert-Gas Shielded Arc Welding of Copper and Copper Alloys', *Welding and Metal Fabrication,* **34**, 12, 545–458 (December 1966) and **35**, 2, 59–63 (February 1967). Subjects include weldability of copper and various copper alloys by Tig and Mig processes. Filler metals. Plasma-arc welding.

Welding Cupro-Nickel Alloys, International Nickel Co. London (1965)

CLEWS, K. J. *Metallurgical considerations in welding Aluminium Bronze,* Institute of Welding. Second Commonwealth Welding Conference, London, 213–217 (1965). Includes weldable and non-weldable alloys, cracking and brittleness of welds and base metal etc.

CLEWS, K. J. 'Cracking of Aluminium-Bronze Plate during Welding', *British Welding Journal,* **12**, 6, 301–309 (June 1965)

ANON, 'Welding Naval Brass', Welding and Metal Fabrication, **32**, 10, 398–399 (October 1944)

NOLAN, M. J. and WINTERTON, K. 'Metal Inert-Gas Welding of Tin Bronze Castings', *Copper Abstract,* 54, (August 1963), *Canada Dept. of Mines Technical Surveys* (June 1962)

COTTON, H. 'Seal Welding of Heat Exchanger Tubulars by the Metal Arc Process', *British Welding Journal,* **10**, 5, 250–257 (May 1963)

THE SIGNIFICANCE OF WELD DEFECTS

The design engineer is responsible for the effectiveness of his design. The significance of defects in welds, which can be assumed to occur in normal fabrication, is therefore of importance to design engineers.

The tolerance of welded joints to defects ranges from acceptance of gross defects (lack of penetration, gross slag inclusions, lack of fusion) under static loading at low stress levels to sensitivity to extremely small defects, such as cracks, which are almost impossible to detect by non-destructive testing. The significance of individual defects depends on the micro-structure in which the defect occurs, the mechanical properties of the material with particular reference to notch toughness, the type of general loading (static, cyclical or shock), the environment (corrosive or non-corrosive), section thickness, type and size of defect and stress pattern local to the defect etc. The subject is complex. Unnecessarily high standards are wasteful but adequate standards essential. It is also essential that customer, designer and fabricator are fully agreed on the standard to be applied *before* the design is started. Various aspects of the subject are considered in papers given at two Conferences[55]. The papers given to the Second Conference attempt to assess the significance of defects in respect of fatigue strength, behaviour under shock loading, behaviour under static loading and on susceptibility to brittle fracture and consider acceptance standards for weld defects based on suitability for service, judged by mechanical and physical property assessment rather than subjective opinions and historical Codes. Although weld defects due to faulty consumables or techniques may or may not be acceptable on this basis the relatively large number of failures in welded fabrication due to low stress high cycle fatigue show that many designers unwittingly introduce 'design defects' in to their products. The care needed in design and execution of welded fabrication in this loading situation is well illustrated in Reference 35 and the significance of 'design' and 'workmanship' defects is clear. However, design data relevant to the significance and definition of design and workmanship defects under high stress low cycle loading is meagre; acceptance criteria for defects influencing creep failure and lamellar tearing have yet to be established.

Further reading

LINDH, D. V. and PESLAK, G. M. 'The influence of weld defects on performance', *Welding Journal,* **48**, 2, 45s–56s (February 1969). Experimental work on effect of defects on static and fatigue properties of welds in titanium alloys.

HOLT, D. 'The influence of weld defects on service performance', *British Engine Technical Report*, **9**, 9–26 (1969). Acceptability of defects in particular fabrications. Importance of different types of defect. Failures due to fatigue and brittle fracture.

PENSE, A. W. and STOUT, R. D. 'Influence of weld defects on the mechanical properties of Aluminium Alloy weldments', *Welding Research Council Bulletin*, 152, 1–16 (June 1970). Literature survey. Information on effect of weld defects on tensile and fatigue strength and notch toughness in aluminium alloys.

Defects in welds, Machinery Lloyd (European Edition), 21–22 (December 1968)

HERSH, M. S. 'Effect of discontinuities on fatigue properties of aluminium welds', *Welding Journal*, **48**, 9, 389s–394s (September 1969). Effect of porosity and inclusions on fatigue properties.

WELDING REFERENCES

1. BS 1719
   Part 1. Classification and Coding of Covered Electrodes for Metal-Arc Welding (1969)
2. American Welding Society Specifications, *Mild Steel Covered Arc-Welding Electrodes*, AWS. A5.1–69
3. American Welding Society Specification, *Low Alloy Steel Covered Arc-Welding Electrodes*, AWS.A5. 5–69
4. *Welding Handbook*, Section 5, 5th ed, Chapter 95, 4–18, American Welding Society
5. *Welding Handbook*, Section 2, 4th ed, Chapter 28, 2–39, American Welding Society
6. *Welding Handbook*, Section 2, 4th ed, Chapter 28, 6–7, American Welding Society
7. *Welding Handbook*, Section 2, 4th ed
8. 'Fabrication and Equipment Guide', *Welding and Metal Fabrication*
9. PATON, B. E. 'Electro-Slag Welding', Translated from the Russian. Published by American Welding Society Inc. New York, 2nd ed, 231
10. *Welding Handbook*, Section 3, 5th ed, Chapter 55, American Welding Society
11. DANKIEN, F. G. 'Western European Technique in Electro-slag Welding', *Welding Journal*, **41**, (1), 17–23 (1962)
12. PATON, B. E. Electro-slag Welding: of very thick material, *Welding Journal*, **41**, (12), 1115–1123 (1962)
13. THOMAS, R. D. Jnr. 'Electro-slag Welding—A new process for heavy fabrication', *Welding Journal*, **39**, (2), 111–117 (1960)
14. HOULDCROFT, P. T. *Welding Processes*, 91, Cambridge University Press (1967)
15. SMITH, A. A. *CO₂ Welding*, 143, The Welding Institute
16. *Welding Handbook*, Section 4, 4th ed, Chapter 69a, 35–42, American Welding Society
17. *Welding Handbook*, Section 4, 4th ed, Chapter 68, American Welding Society
18. *Welding, Bracing and Soldering Wiggin Nickel Alloys*, Henry Wiggin and Co. Ltd., Hereford
19. *Welding Handbook*, Section 4, 4th ed, Chapter 65, 21–23, American Welding Society (Cleaver-Hume Press, London)
20. SMITH, A. A. *CO₂ Welding*, Chapter 4, The Welding Institute
21. SMITH, A. A. *CO₂ Welding*, Chapter 11, The Welding Institute
22. 'Fabrication and Equipment Guide', *Welding and Metal Fabrication*
23. SMITH, A. A. *CO₂ Welding*, Chapters 2,5,6,7, The Welding Institute
24. *Welding Handbook*, Section 2, 4th ed, Chapter 29, 29, American Welding Society
25. *Welding Handbook*, Section 2, 4th ed, Chapters 21 and 23, American Welding Society
26. *Welding Handbook*, Section 2, 4th ed, Chapter 22, American Welding Society
27. *Welding Handbook*, Section 2, 4th ed, Chapter 31, American Welding Society
28. *Welding Handbook*, Section 3, 5th ed, Chapter 52, 6, American Welding Society
29. BS 1856
   General requirements for the metal-arc welding of mild steel
30. BS 449
   The use of structural steel in buildings
31. BS 153
   Steel girder bridges
32. *Handbook for Welding Design*, The Welding Institute
33. *The use of Welding in Steel building structures*, British Constructional Steel Work Association
34. GURNEY, T. R. *Fatigue of Welded Structures*, Cambridge University Press
35. RICHARDS, K. G. *Fatigue Strength of Welded Structures*, The Welding Institute

36. HARRISON, J. D., BURDEKIN, F. M. and YOUNG, J. G. 'A proposed acceptance standard for weld defects based upon suitability for service'. *Paper to Second Conference on Significance of defects in Welds*. The Welding Institute

37. PUZAK, P. P. and PELLINI, W. S. 'Evaluation of the Significance of Charpy Tests for quenched and tempered steels, *Welding Journal*, 275s–290s (June 1956)

38. *Control of Steel Construction to avoid Brittle Failure*, Manual prepared by Plasticity Committee of The Welding Research Council. Part 3: Design 115–144. Welding Research Council, 29 West Thirty-Ninth Street, New York, N.Y.

39. MYLONAS, C. 'Exhaustion of Ductility and Brittle Fracture of Project E Steel caused by Prestrain and Ageing.' Report number 1 to S.S. Committee. Dept. Navy: Bureau of Ships: Brown University, Providence R. I. (December 1963). See also Ref. 38: 219

40. HALL, W. J., KIHARA, H., SOETE, W. and WELLS, A. A. *Brittle Fracture of Welded Plate*, Prentice-Hall Inc. Englewood Cliffs, N.J. 216–218, 90–91

41. *Brittle Fracture and General Engineering*, British Welding Research Association Bulletin, **7**, No. 2, (February 1966)

42. HIPKINS, M. G. and DICKINSON, S. F. 'Influence of Notch Orientation and position on Notch Ductility of Automatic and Manual Welds in Mild Steel.' *Welding and Metal Fabrication*, **33**, No. 5, 216–227 (May 1965)

43. HERBERT, D. C. 'The Significance of Weld Defects for Brittle Fracture of Steel.' *Proceedings of the Second Conference on The Significance of Defects in Welds*. The Welding Institute, 81–83 (1969)

44. MEYER, H. J. 'The Influence of Defects in Plates on Weld Failure under Stresses Perpendicular to the Plate Surface. *Proceedings of the Second Conference on The Significance of Defects in Welds*. The Welding Institute, 15–24 (1969)

45. COE, F. R. and MORETON, J. 'Hydrogen Removal during the Heat Treatment of Welded Structures.' The Welding Institute

46. BAKER, R. G., WATKINSON, F. and NEWMAN, R. P. The Metallurgical Implications of Welding Practice as Related to Low Alloy Steels, *B.W.R.A. Bulletin*, **6**, 8 (August 1965) or *Proceedings of Second Commonwealth Welding Conference*, 125–131, The Welding Institute (1966)

47. BAILEY, N. Welding Carbon-Manganese Steels, *Metal Construction and British Welding Journal*, (October 1970)

48. WATKINSON, F. and BAKER, R. G. Welding of Steel to BS 968: 1962, *British Welding Journal*, 603–613 (November 1967)

49. BAILEY, N. Welding Procedures for Nine Per Cent Nickel Steel, *Metal Construction and British Welding Journal*, 419 (October 1970)

50. *A Recommended Procedure for the Welding of Steel Castings by the Metal-Arc Process*, The British Steel Castings Research Association (September 1964)

51. *Welding Handbook*, 4th ed, Section 4, Chapter 64, American Welding Society

52. *Welding, Brazing and Soldering Wiggin Nickel Alloys*, Henry Wiggin and Co.

53. *Welding Handbook*, 4th ed, Section 4, Chapter 67

54. *Welding Handbook*, Section 4, Chapter 69, Fig. 69a, 11

55. 'Significance of Defects in Welds', *Proceedings of the First Conference*, (February 1967). *Proceedings of the Second Conference*, (May 1968). The Welding Institute

56. *Welding Handbook*, 4th ed, Section 4, Chapter 68, American Welding Society

57. *Welding Handbook*, 5th ed, Section 5, 95, 49, American Welding Society

BRAZING

H. F. TREMLETT

Brazing is a process of joining metals in which, after heating to a suitable temperature, molten filler metal is drawn by capillary attraction into the space between closely adjacent surfaces of the parts to be joined. In general, the melting point of the filler metal is above 500°C but is always below that of the parent metals.

Braze welding[1] is the joining of metals by a technique similar to fusion welding but using a filler metal with a lower melting point than the parent metal. This process has no practical resemblance to brazing and will not be dealt with here.

Principles of brazing

Brazing methods rely upon bonding taking place between the unmelted parent metals and the filler metal after the latter has solidified. This only occurs if the molten filler metal 'wets' the surface of the parent metal. Wetting will not take place satisfactorily if the surfaces are oxidised or contaminated with oil or grease or if the temperature of the surfaces is too low or too high.

In brazing, oxidised surfaces are avoided by pre-cleaning chemically or mechanically, and by using fluxes which dissolve oxides or prevent oxidation during the brazing process. Temperature control must be precise and is determined by experience.

Scope and applications

As a non-mechanical method of joining metals, brazing probably has wider scope than any other. It is applicable to any material having a melting point not less than 50°C higher than that of the brazing alloy.

Since brazing alloys have been developed to cover a brazing temperature range from approximately 620°C to 930°C the process is applicable to cast and wrought irons, steels, galvanised iron or steel, copper and copper alloys, aluminium and aluminium alloys, magnesium and magnesium alloys, cupronickels, nickel and nickel alloys, stainless and heat resisting steels, Ti, Zr, Be, ceramics and refractory metals and to dissimilar metal joints.

In practice, brazing of metals is easy where the metal oxide is readily fluxed (e.g. iron oxide, copper oxide) but may be difficult where the metal produces a refractory oxide during heating, for example, materials containing chromium and/or aluminium or titanium as alloying elements or has a refractory oxide skin as in stainless steels. Brazing obviously cannot be applied to metals which depend upon a specific heat treatment for mechanical properties if the brazing heating adversely affects the prior treatment.

Selection of filler metals may be restricted for some metals by metallurgical considerations, for example, phosphorous bearing brazing alloys must not be used on high nickel alloys or steels. Furthermore, again due to metallurgical problems, nickel and its alloys must be absolutely free of any surface contaminants containing lead or sulphur such as drawing and cutting lubricants, paint, oil, grease etc. This illustrates a general principle that advice should be sought from the manufacturers of materials where full brazing information is not already to hand. Such information includes suitability of the brazed joint for the service conditions in respect of strength at temperature and corrosion resistance and suitability of the type of joint necessary for high quality brazed joints to the particular application.

Successful and economical brazing depends on careful control of fit-up, attention to precise control of fluxing and temperature and selection of the correct brazing alloy and method for the particular materials being brazed. Brazing materials and methods have been developed for a very wide range of materials, including almost all ferrous and non-ferrous metals, refractory metals, ceramics and non-metals. Fluxing is as important to the success of the operation as selection of correct filler alloy and heating method; for example incorrect flux may cause poor wetting action and flux retention in the joint with consequent severe corrosion.

The above points indicate the need to consider carefully all aspects and seek advice where doubt exists when assessing any potential brazing application.

Joint design and preparation

In brazing, a controlled close fit between the surfaces to be joined is essential. Preferred types of joint are used which facilitate this control and also the addition of the filler alloy which may be effected by using manual application of wires or pre-set application of the alloy in the form of shims, wire rings, washers etc.

Correct joint design is extremely important. The principles operating in designing brazed joints are illustrated in References 2, 3 and 4. Design should also cater for inspection which is usually visual; the filler should be introduced into the joint on the side opposite to that which is most readily accessible for visual inspection.

Joint preparation for most metals involves (a) cleaning to a bright metallic surface by chemical or mechanical methods, removing all oxide, oil and grease and (b) where necessary, assembling the parts in jigs or simple fixtures which hold them in correct position and (c) where necessary, preplacing the filler and flux. Stop-off paints are sometimes used to prevent the brazing alloy from running off the desired brazing areas[5].

Joint preparation also includes setting up the gap clearance, which is an important factor in joint quality. It depends on a number of variable factors such as filler alloy fluidity and wetability, susceptibility to slag entrapment (refractory oxides) or voids which are metallurgical in origin but optimum mechanical strength is attained by using the minimum clearance which avoids the above-mentioned defects. The clearance desired must be obtained at the brazing temperature and, in the case of dissimilar metal joints, adjustment to the room temperature gap for difference in expansion may be required. Recommended joint clearances at brazing temperature, (extracted from the Welding Handbook) are given in Table 6.25.

Filler metals

Brazing is one of the oldest metal joining methods known to man. It has been based for many years on copper or silver, or alloys of these. Technological advances have lead to the use of metals such as Titanium, Zirconium, Molybdenum, and Tungsten as engineering materials. The welding of many of these materials presents considerable difficulties, and this has caused the development of suitable brazing alloys. Materials for high temperature service must obviously be brazed with filler metal having suitable high temperature properties yet this requirement must be combined with a brazing temperature lower than the melting temperature of the parent metal and coupled with adequate wetting characteristics.

The properties of brazed joints depend on metallurgical reactions and, given no restriction on cost, a metallurgical solution to most apparently intractable problems has been found. Consequently an enormous range of filler metals is currently available for existing materials and there is no reason to suppose that, as new metals and alloys are developed, brazing filler alloys cannot be produced to meet the new requirements.

Table 6.25   RECOMMENDED JOINT CLEARANCES AT BRAZING TEMPERATURE

Filler metal AWS–ASTM Group classification	*Joint clearance* mm	
B.Al Si	Lap less than 6 mm	0·153–0·25
	Lap greater than 6 mm	0·25 –0·625
B.Cu P		0·025–0·125
B.Ag		0·050–0·125
B.Cu Au		0·05 –0·125
B.Cu		0·00 –0·05
B.Cu Zn		0·05 –0·125
B.Mg		0·10 –0·25
B.Ni		0·05 –0·125

It is impracticable to list all available filler alloys, but the main groups of alloys are given in Tables 6.26. Reference to BS 1723: 1963 and BS 1845. and to the Welding Handbook will indicate the standard filler alloys for various materials. Information on these and commercial filler alloys for normal and special situations is obtainable from alloy suppliers[6].

Certain combinations of filler and parent metals produce unsatisfactory joints due to metallurgical factors leading to excessive erosion (solution of parent metal by filler alloy), brittleness, lack of proper wetting, intergranular cracking of parent metal, poor corrosion resistance or susceptibility to stress corrosion cracking. In the absence of satisfactory experience of a particular combination of filler and parent metal under given working conditions it is advisable to consult the alloy suppliers.

Parent metals, cautionary notes

ALUMINIUM AND ALUMINIUM ALLOYS.

Certain aluminium alloys are not suitable for brazing owing to their low melting range. Aluminium alloys containing a high magnesium content are particularly difficult to braze due to poor fluxing and wetting. Only fluxes specifically intended for brazing aluminium should be used for this purpose.

COPPER

Generally available as oxygen bearing e.g. electrolytic and tough-pitch or as oxygen free which includes phosphorous de-oxidised and oxygen free high conductivity copper. The oxygen bearing coppers are not generally suitable for brazing owing to loss of strength and ductility due to oxide migration; furthermore any hydrogen bearing

Table 6.26(a) BRAZING FILLER METALS. BASED ON A.S.T.M. SPECIFICATION B260 AND NEAREST BS 1845 EQUIVALENT
(THE WELDING HANDBOOK. AMERICAN WELDING SOCIETY.)

Filler metal A.S.T.M. BS 1845	Cu	Ag	P	Zn	Cd	Cr	Au	Ni	Al	Si	B	Other	Liquidus °C
B.Al Si 2 Al 3									92·5	7·5			612
3									86·0	10·0			585
4 Al 2									88·0	12·0			582
5									90·0	10·0			590
B.Cu P 1	95		5										898
2 CP 3	92·75		7·25										793
3	89	5	6										807
4	86·75	6	7·25										723
5 CP 1	80	15	5										801
B.Ag 1	15	45·0		16·0	21·0								617
1a AG1	15·5	50·0		16·5	18·0								635
2 AG3	26·0	35·0		21·0	18·0								701
3	15·5	50·0		15·5	16·0			3					687
4	30·0	40·0		28·0				2					779
5	30·0	45·0		25·0									743
6	34·0	50·0		16·0									773
7	22·0	56·0		17·0								Sn 5·0	650
8	28·0	72·0											779
8a	27·8	72·0										Li 0·2	765
13	40·0	54·0		5·0				1·0					857
18	30·0	60·0										Sn.10·0	718
19	7·3	92·5										Li 0·2	890
B.Au 1 AU 3	62·5						37·5						1 015
2 AU 1	20·0						80·0						890

Designation					4·0	3·5		M.P.
RB.Cu.Zn.								
A CZ 5	59·25	40·0					Sn. 0·75	898
D CZ 8	48·0	42·0		10·0			C. 0·75	935
B.Ni								
1 NI 7			14	73·25	4·5	3·1	Fe 4·5	1 037
2 NI 3			7	82·4	4·5	3·1	Fe 3·0	998
3 NI 4				90·9	4·5	3·1	Fe 1·5 Max	1 037
4 NI 5				93·4	3·5	1·6	Fe 1·5 Max	1 065
5 NI 8			19·0	70·9	10·1			1 135
6 NI 1		11·0		89·0				876
7 NI 2		10·0	13·0	77·0				887
PD 1	26·5	68·5					Pd 5·0	810
PD 2	31·5	58·5					Pd 10·0	850
PD 3	22·5	67·5					Pd 10·0	860
PD 4	20·0	65·0					Pd 15·0	900
PD 5	28·0	52·0					Pd 20·0	900
PD 6	2·0	54·0					Pd 25·0	950
PD 7		95·0					Pd 5·0	1 010
PD 8	82·0						Pd 18·0	1 090
PD 9		75·0					Pd 20·0 / Mn 5·0	1 120
PD 10		64·0					Pd 33·0 / Mn 3·0	1 200
PD 11				48·0			Pd 21·0 / Mn 31·0	1 120
PD 12	55·0			15·0			Pd 20·0 / Mn 10·0	1 110
PD 13	35·0			20·0			Pd 30·0 / Mn 15·0	1 090
PD 14				40·0			Pd 60·0	1 235

Table 6.26(b)

A.S.T.M.	BS 1845	Remarks
B.Al Si		
2	Al 3	Only as cladding on aluminium core sheet. For dip and furnace.
3		General purpose for aluminium alloys suitable for brazing. Dip and furnace.
4	Al 2	General purpose. For torch brazing also dip and furnace.
5		Filler for furnace and dip brazing. Also as cladding.
B.Cu P		
1		For preplacing in joints. Resistance and furnace.
2	Cp 3	}
3		} Highly fluid. Clearance 0·025–0·075 mm.
4		}
5	Cp 1	For wider clearance of 0·075–0·125 mm. The Cu P group used for copper and copper alloys. Self fluxing on Cu. Flux needed on copper alloys and other non-ferrous alloys.
B.Ag		
1		}
1a	AG 1	}
2	AG 3	General purpose.
3		General purpose. For wider clearance.
4		For corrosion resisting joints in stainless steels. For carbide tips.
5		} For carbide tips at higher brazing temperature than 3.
6		} General purpose. Used where cadmium-free filler mandatory.
7		For furnace brazing.
8		For applications requiring no volatiles. Dry hydrogen or vacuum preferred for ferrous metals.
8a		Self-fluxing on ferrous metals and alloys in protective atmosphere. Not for torch with flux.
13		High melting point for aircraft and aero engine construction.
18		For fluxless furnace brazing where volatiles not accepted.

Designation	Description
B.Cu 1 1a 2 4 AU 5	General purpose for ferrous metals, nickel and copper nickel alloys. Widely used for furnace brazing of steel without flux. If steel contains Cr, Mn, Si, V or Al in excess of 2% flux probably necessary. Furnace brazing (flux free) requires reducing atmosphere.
RB. Cu Zn A CZ 5 D Cz 8	General purpose for copper alloys, nickel alloys, cast iron and steel. Alloy A especially for torch brazing. Alloy D for dip, torch or furnace.
B.Ni 1 NI 7	Erosive. Therefore unsuitable for very thin sheet.
2 NI 3	Less erosive. Useful where higher melting point alloy unsuitable. Free flowing.
3 NI 4	For wider clearances.
4 NI 5	For high strength and oxidation resistance at high temperature up to 1093 °C. Low erosion.
5 NI 8	Free flowing. Minimum erosion.
6 NI 1	Strong joints produced at low temperatures. Useful for honeycomb structures.
7 NI 2	The nickel alloy fillers are used for heat and corrosion resistance. Used on stainless and heat resisting steels for high temperature and cryogenic applications.
PD 1/14	These palladium bearing silver alloys may be regarded as high temperature silver brazing alloys. The absence of the volatile zinc and cadmium constituents (present in low temperature silver brazing alloys) makes them suitable for vacuum brazing and for joining parts for vacuum applications. Most of the alloys are available with lithium addition to promote self-fluxing where this property is required.

heating atmosphere embrittles this material. Oxygen free copper is entirely satisfactory and widely used for brazed components.

COPPER–ZINC ALLOYS (BRASS)

Flux should generally be used with protective atmospheres to reduce loss of zinc (vaporisation) and consequent reduction in strength. Slow heating required on stressed sheets to prevent stress cracking.

COPPER–SILICON ALLOYS (SILICON BRONZES)

Susceptible to intergranular penetration of filler metal when in a stressed condition Should be stress relieved before brazing or heating must be very slow.

COPPER–ALUMINIUM ALLOYS (ALUMINIUM BRONZES)

Special fluxes are essential for these alloys owing to refractory oxides.

COPPER–NICKEL ALLOYS

Readily brazed but susceptible to inter-granular penetration and cracking if stressed. Slow heating or stress relief before brazing required.

LOW CARBON AND LOW ALLOY STEELS

Low carbon steels present no problems. With low alloy steels which tend to harden, the brazing should be done below the critical temperature or slow cooling to below the critical rate should be applied.

STAINLESS AND HEAT RESISTING STEELS

It is advisable to confirm suitability of proposed filler-parent metal-brazing process combination by test joints submitted to intended service conditions. Special fluxes are essential owing to refractory oxides and difficult wetting. Stabilised (Ti or Nb) or extra low carbon stainless steels are suitable for brazing by the nickel group fillers but unstabilised steels with carbon exceeding about 0·05% may suffer intergranular carbide precipitation with loss of corrosion resistance. Stressing during brazing should be avoided by supporting the parts. Precipitation hardened stainless steels must be brazed in a manner to avoid damaging the effects of the hardening heat treatments. It is essential to check with the steel supplier.

NICKEL AND HIGH NICKEL ALLOYS

These materials are seriously embrittled by traces of sulphur, lead, bismuth and antimony. To avoid pick-up in brazing any trace of cutting oils, grease etc., must be removed and furnace atmospheres must be sulphur free. Nickel and nickel alloys are susceptible to stress corrosion cracking in the presence of brazing metals and consequently should be stress relieved before brazing.

TITANIUM, ZIRCONIUM AND BERYLLIUM BASE ALLOYS, CARBIDES, CERMETS, CERAMICS ETC.

Notes on brazing the above materials are given in the Welding Handbook, Section 3, 5th edition, Chapter 43 (pp. 61–72).

FLUXES

Unless an inert or reducing atmosphere is used during heating, the brazing process will cause oxidation of the metal surfaces. The oxides so formed must be removed to permit wetting. The function of fluxes is two-fold (a) to dissolve the oxide in to the molten flux and remove it to the surface of the molten filler metal and (b) to spread ahead of the filler metal and so protect the metal surface from oxidation. No single flux composition is universally applicable. Modern commercially available fluxes, tailored to perform correctly for the given application, should be used in preference to 'home-made' mixtures based on borax.

General information on commercial fluxes (The Welding Handbook, Section 3, 5th edition, chapter 43, table 3) is reproduced in Table 6,27. The table provides a guide for classification. Reference should also be made to the American Welding Society Brazing Manual (1963 ed).

FLUX REMOVAL

Flux removal is essential immediately after brazing. This is usually effected with hot water or, if obstinate, by chemical dips or mechanically.

Heating (furnaces)

Heating for brazing should be done in such a manner as to slowly and uniformly raise the temperature of the component to the brazing heat. This should be done in a heating atmosphere which inhibits oxidation and/or reduces oxides.

Manual (torch brazing) methods and furnace methods will achieve the required conditions but, naturally, the former tend to be less reliable being dependent on the operator's judgement. Modern brazing furnace equipment provides better control over the composition of the atmosphere and uniformity of temperature. Consequently, this method is to be preferred where output justifies the capital cost. Where this situation obtains, the economics of brazing are attractive.

Hydrogen bearing furnace atmospheres are widely used for batch or continuous type furnaces for the copper brazing of steel articles, these atmospheres being obtained from cracked ammonia, town's gas etc. Refractory metals, such as tungsten, molybdenum etc., are commonly brazed in an inert gas atmosphere, generally argon. Brazing is widely used in the manufacture of components for the aerospace industry and, by suitable metallurgical ingenuity combined with availability of satisfactory protective gases, inert or reducing, it seems that there are few combinations of metals and metals to non-metals which cannot be joined by brazing, this particular industry being notable for the range of metals and materials it handles. Useful information on heating methods for brazing is obtainable from references in the appended bibliography.

Inspection

The design department, by co-operation with the inspection department, can reduce the cost of inspection and ensure its reliability. It should be regarded as a cardinal rule

Table 6.27 GENERAL INFORMATION ON COMMERCIAL BRAZING FLUXES.
(The Welding Handbook. American Welding Society.)

A.W.S. flux type	Flux applications Parent metals	Filler metals (see Table 6.26)	Effective temperature range flux °C	Major constituents	Form	Application methods (see below)
1	Aluminium, aluminium alloys	B.Al Si	370–643	Fluorides, chlorides	Powder	1, 2, 3, 4
2	Magnesium alloys	B.Mg	482–648	Fluorides, chlorides	Powder	3, 4
3a	Copper, copper base alloys (except with aluminium), iron based alloys, cast iron, carbon and alloy steel, nickel and nickel based alloys, stainless steels, precious metals	B.Cu P B.Ag	555–871	Boric acid, borates, fluorides, fluoborate, wetting agent	Powder, paste, liquid	1, 2, 3, 5
3b	As for 3a	B.Cu, Cu P, Ag, Au, Ni, RB., Cu Zn	732–1148	As for 3a	Powder, paste, liquid	1, 2, 3, 5
4	Aluminium-bronze, aluminium-brass	B.Ag B.Cu Zn B.Cu P	565–871	Borates, fluorides, chlorides	Powder, paste	1, 2, 3
5	As for 3a	B.Cu, B.Cu P, B.Ag (8–19), B.Au, B.Cu Zn, B.Ni.	750–1204	Borax, boric acid, borates	Powder, paste, liquid	1, 2, 3, 5

1. Dry powder sprinkled on joint.
2. Heated filler rod dipped in powder or paste.
3. Mix to paste.
4. Molten flux bath.
5. Liquid sprayed on joint or via fuel gas (torch brazing).

that filler is introduced to the joint on the side opposite to that which will be visually inspected if inspection all round the joint is not possible. This ruling may require careful consideration of joint design. Where flux is not used, visual inspection is frequently all that is required for example copper brazing in furnaces. Pressure tests on completed components is an economical method of ensuring sound joints.

REFERENCES

1. BS 499
   Welding terms and symbols. Part 1 (1965)
2. BS 1723
   Specification for Brazing (1963)
3. *The Welding Handbook*, Section 3, 5th ed, Chapter 43, 7-9, American Welding Society
4. *Welding, Brazing and Soldering Wiggin Nickel Alloys*, Publication No. 3367 A. Henry Wiggin and Co. Ltd., Hereford
5. *The Welding Handbook*, Section 3, 5th ed, Chapter 43, 44-47
6. 'Fabrication Equipment Guide', *Welding and Metal Fabrication*

BIBLIOGRAPHY

(a) Brazing and Design
DENT, H. C. 'Vacuum brazing as a production technique.' 'Metal Construction and British Welding Journal', 1, 12, 560-562 (December 1969). Fundamentals of brazing. Joint configurations. Table of brazing fillers and their uses. Advantages of vacuum furnaces.
DILLEY, D. C. 'Brazing Trouble—Its prevention and cure', *Welding Engineering*, 54, (9), 68-70,72,74, 76 (September 1968). Importance of cleanliness, clearance, personnel training.
POTAK, S. 'Brazing in shielding atmosphere for the precision mass production Industry.' (German) V.D.I.-Z 111, 9, 630-32 (May 1969). Furnace brazing using cracked ammonia for mass production of copper alloy, steel and stainless steel components. Design of Components and inspection.
MEHRKAM, Q. D. 'Salt—Bath Brazing Techniques', *Welding Journal*, 49, (4), 273-274, 277-280 (April 1970). Salt-bath brazing of Aluminium alloys. Pre-treatment, flux and fillers, fixtures and Joint design for Al, brass and ferrous brazing operations.

(b) Furnace Brazing
WALKER, A. 'Vacuum Brazing', *Metals and Materials*, 4, 5, 190, 192-193, 196 (May 1970)
'Copper Brazing Carbon Steel Parts', *Metal Treating*, 70, (5), 42-43, (October-November 1969)
HERR, HARRY K. 'Fluxless Vacuum Brazing of Aluminium', *Metal Progress*, 96, (6), 68-70 (December 1969)
TERRILL, J. R. Brazing Aluminium in a dry air atmosphere, *Metal Progress*, 96, (6), 70-73 (December 1969)
Vacuum Brazing Easily joins Hard-to-braze Materials Iron Age Metalwork Int. 8, (7), 20-21 (July 1969). Furnaces, Filler Alloys for Al, Be, superalloys, Ti, refractory metals, stainless steel and dissimilar metal combinations.
WATSON, C. A. Jr. 'Cathodic cleaning removes flux from Furnace Brazed Assemblies', *Welding Journal*, 48, (9), 721-725 (September 1969)
'Precious Metals Boost Brazing uses', *Welding Engineering*, 55, (1), 38-41 (January 1970). Use of precious metals and alloys in controlled atmosphere furnace brazing.
PETRAK, F. Newly developed and improved brazing and soldering machines (German). *Schw— Techn*, 19, No. 11, 506-8 (November 1969). Production of cans by induction, furnace, wave and resistance brazing and soldering.
HEAP, H. R. and RILEY, C. C. Furnace Brazing of Commercially Pure Titanium, *Metal Forming*, 36, (7), 200-209 (July 1969)
Weld-brazing and Brazing of Aluminium and its Alloys (French), *Revue Aluminium*, (372), 294-306 (March 1969). Furnace, salt-bath and low temperature brazing. Suitable metals, equipment, joints, methods and applications. Induction brazing. Diffusion brazing in vacuum.
DUMEZ, B. and L. Furnace Brazing (French), *Soud. Techn. Connexes*, 22, (1-2), 21-33 (January-February 1968). Furnace design. Controlled atmospheres. Prevention of oxidation. Vacuum furnace problems. Brazing fillers. General brazing design factors.

SOLDERING

H. F. TREMLETT

Soldering is a method of joining metals in which the parent metals are heated to a suitable temperature less than about 500° and are joined by non-ferrous filler metals (solders) which have melting temperatures below those of the parent metal. The molten filler metal dissolves a very small amount of the parent metal to produce, when solid, a layer of an intermetallic compound at the filler metal/parent metal interfaces thus effecting union between the parent metals.

As in brazing, the soldering process depends upon the molten filler metal wetting the parent metal.

Principles

For wetting to take place, the parent metals must be chemically cleaned and at the proper temperature. To meet these requirements a flux must be used for most applications. The flux will remove existing oxide film and prevent new film being formed during heating. Where a metal, such as aluminium, has a tenacious and refractory oxide film it is possible to break this up mechanically in the presence of molten solder to effect wetting by the solder.

The ease with which the oxide film on a metal can be removed and prevented from reforming during heating for soldering varies with the composition of the metal so that fluxes of varying composition and activity are required for different parent metals. Correct selection of flux is a mandatory requirement for satisfactory soldering.

Soldering alloys are of low strength in relation to the parent metals so joint design is commonly based on mechanical joints with the solder acting as a seal and stiffening agent. Solders will not withstand movement during cooling down from the molten condition so that firm clamping or jigging is essential. Heating to the required temperature is achieved by any means which simultaneously ensures that the parent metal reaches the temperature necessary for wetting by the solder and the solder is melted.

Practical aspects

JOINT DESIGN

Joints are either lap or butt. The former is the most commonly used, since it is preferred for strength and for sealing purposes. For sheet metal work, the joints used are typically lock seaming in its various forms; riveting; spot welding and bolting; and various socketed joints for pipe or tubular sections.

Clearance is important since the solder must be drawn in to the joint by capillary action. Overlarge clearances must be avoided e.g. they should not generally exceed 0·25 mm. The optimum range of clearances is 0·075 mm to 0·125 mm. For typical joint designs, see References 1 or 2.

PREPARATORY CLEANING

All oil, grease, paint, residual cutting lubricants, atmospheric dirt, oxide and rust films must be removed for effective wetting. This is achieved by either mechanical or chemical means. Mechanical cleaning involves shot blasting (which is better than sand), grinding, sanding, filing etc. Oily or greasy surfaces should be degreased before mechanical cleaning.

Chemical cleaning includes degreasing in solvents of the trichlorethylene type (vapour or liquid) or hot alkali type such as trisodium phosphate used hot. The cleaning solution must be thoroughly washed off by soft water or steam.

Before soldering it is advisable to check the efficiency of the proposed method. Removal of scale, oxides etc. by chemical methods involves acid cleaning, usually by hydrochloric and sulphuric acid, followed by hot water washing and *immediate* drying.

FLUXES

From a practical point of view, fluxes can be classified according to the degree of corrosiveness of the residue left after soldering since, if it is corrosive, it is most important

Table 6.28 RELATIVE EASE OF SOFT SOLDERING BASED ON FLUX ACTIVITY REQUIREMENTS

Parent metal or finish	Non-corrosive	Flux corrosive	Special flux and/or solder	Soldering not recommended
Aluminium			X	
Aluminium-bronze			X	
Beryllium				X
Beryllium copper		X		
Brass	X	X		
Cadmium	X	X		
Cast iron			X	
Chromium				X
Copper	X	X		
Copper-chromium		X		
Copper-nickel		X		
Copper-silicon		X		
Copper-silicon-manganese				X
Gold	X			
Inconel			X	
Lead	X	X		
Magnesium			X	
Manganese-bronze				X
Monel		X		
Nickel		X		
Nichrome			X	
Palladium	X			
Platinum	X			
Rhodium		X		
Silver	X	X		
Stainless steel			X	
Steel		X		
Tin	X	X		
Tin-bronze	X	X		
Tin-lead	X	X		
Tin-nickel	X	X		
Titanium				X
Zinc		X		
Zinc die castings			X	

Table 6.29 COMPOSITION OF COMMON SOLDERS AND THEIR APPLICATIONS

Solder alloy type	Classification ASTM B 32	BS 219 1959	Sn	Pb	Fluxes	Heating methods	Comments, application etc.
Tin-lead	5A		5	95	Active preferred	Not bit	Relatively poor wetting and flow. Sealing pre-coated containers. High soldering temperature.
	10A		10	90			
	15A		15	85	All	All	Good wetting and flow. General application. Dipping.
	20A	V	20	80			
	25A	J	25	75	All	All	General. Plumbing. Wiping cable and lead pipe joints.
	30A	H	30	70		T, M	Dipping baths.
	35A	G	35	65			General purpose solders. Best combination of wetting, strength and economy. Widely used on sheet metal work.
	40A	R	40	60	All	All	
	45A	F	45	55		B, T	Machine soldering. Resin cored wire for electrical work.
	50A		50	50			
	60A	K	60	40	All	All	Free running. For components liable to heat damage (electrical, radio etc.). For zinc and coating.
	70A		70	30	All	All	
Tin-lead-antimony	20C	N	20	79·0	All	All	The antimony bearing solders are not recommended for zinc, galvanised metals or aluminium. The ASTM, B series contain up to 0·5% Sb as impurity and may normally be used as for series A above. Strength of C series higher than A but somewhat inferior flow.
	25C		25	73·7	All	All	
	30C	D	30	68·4	All	All	
	35C	L	35	63·2	All	All	
	40C	C	40	58·0	All	All	
		M	45	52·5	All	All	
		B	50	47·3	All	All	
Tin-antimony	95A		95	5 Sb	All	All	For somewhat higher temperature service conditions than A series above. Excellent soldering and strength. Used where lead must be avoided (food containers). Copper tubing joint.
Tin-silver			96·5	3·5 Ag	All, usually non-corrosive	Bit, resistance	For fine instrument work.
			95·0	5·0 Ag			

	60 30	40 Zn 70 Zn			70/30 and 80/20 useful for pre-coating.
Cadmium–silver	95 Cd	5 Ag	Depends on parent metal		For higher temperature service applications. High strength joints at room temperature. Can be used on aluminium. Fumes are toxic.
Cadmium–zinc	82·5 Cd 40 10	17·5 Zn 60 90	Depends on parent metal		The 82·5/17·5 solder (eutectic) has no pasty range. These solders are useful for aluminium. They have intermediate strength and corrosion resistance is good with suitable flux. Fumes are toxic.
Zinc–aluminium	95 Zn	5 Al	Active only		Specifically for aluminium. High strength. Good corrosion resistance. No pasty range. Parent metal must withstand soldering temperature above 370°C.
Indium alloys	Various alloys of tin and indium with or without bismuth, lead and cadmium		All	All	For special applications. The 50% indium–50% tin alloy may be used for glass to metal and glass to glass joints. The bismuth alloys need active fluxes or pre-coating.

M = machine.   T = torch.   B = bit.   (these methods particularly suitable).

to remove all traces. Selection of the flux depends on the parent metals and filler alloy Metals requiring a strongly active flux for wetting must be fluxed with corrosive fluxes Flux requirements for various metals are given in Table 6.28 (Reference 1, chapter 44 table 1).

Corrosive fluxes are either highly corrosive, based on inorganic acids and salts, on mildly corrosive, based on organic acids and derivatives of these. The highly corrosive fluxes (used as solutions, pastes or dry salts) can be used with all types of heating but the mildly corrosive fluxes tend to volatilize with torch or flame heating leading to difficulty in wetting under continued heating. The latter tend to be used for quick spot soldering applications when flame heating is used. Non-corrosive fluxes, based on resin, are mainly protective and therefore highly dependent on efficient preparatory cleaning but so-called activated resin fluxes are available with a more active fluxing action.

Corrosive and non-corrosive fluxes are generally available as pastes. Application as paste is usually more convenient than in other forms. It is also possible to obtain fluxes either of the inorganic and organic acid types, blended with finely divided solder metal and a neutral carrier, commonly referred to as solder pastes.

Active fluxes are:

Zinc chloride/ammonium chloride eutetic (70% $ZnCl_2$ 30% NH_4Cl).

Zinc chloride (25%–50%), ammonium chloride (5%–20%) balance water.

Paste flux. 20% anhydrous zinc chloride 74% petroleum jelly 5% ammonium chloride and 1% water.

Non-active (non-corrosive) fluxes are:

Resin, tallow, stearine.

Activated resins (addition of aniline hydrochloride for example).

The non-corrosive fluxes are widely used in electrical applications and must be applied to clean parent metal surfaces. They are available as paste, soldering fluid, solder cream and resin cored solder wire.

SOLDERS[3,4]

Solders are tin–lead, tin–antimony–lead, tin–antimony, tin–silver, tin–zinc, lead–silver, cadmium–silver, cadmium–zinc, zinc–aluminium alloys etc. The tin–lead alloys have the widest general application of all solders, the others being for special application, for example the zinc–aluminium solder is specifically intended for soldering aluminium.

An indication of the compositions of common solders and their applications is given in Table 6.29 but many others are available from commercial sources for special applications and duties[5].

Heating for soldering

Good soldering practice requires proper application of heat. The surface of the parent metal must reach the melting temperature of the solder and should not greatly exceed this temperature. The heated surfaces should only be exposed to the atmosphere for the minimum time possible before the molten solder makes contact, ideally the melting and parent metal heating should take place simultaneously.

Various methods of heating are used according to the practical aspects of the application, the parent metals and the economics of the application. Methods used include:

Soldering irons with copper bits.

Torch or blow pipe.

Dipping in a bath of molten solder.

Induction.

Electric resistance.

Oven (with or without protective gas atmosphere).

The soldering iron is probably the most widely used heat source followed by torch soldering. Whichever method is selected the heat source must be correctly designed and, in particular, of adequate heat capacity for the work involved. Expert advice should be sought[5].

Flux residue removal

ACTIVE TYPE FLUXES. ZINC CHLORIDE BASE

The residue left after soldering should be completely removed by washing in warm acidified (2% concentrated hydrochloridic acid) water, followed quickly by a clean hot water rinse. An extra precaution to ensure removal of any acid after the acidified water wash is to wash in water made slightly alkaline with a few crystals of sodium carbonate followed by a rinse in clean water.

INTERMEDIATE TYPE FLUXES

Mild organic acids (stearic, oleic or tallow) can be regarded as non-corrosive. Urea with organic hydrochlorides should be regarded as corrosive. Double rinsing in hot water is effective.

NON-CORROSIVE FLUXES

The resin based non-corrosive fluxes need not be removed. If removal is desirable, it can be effected with alcohol, petrol, turpentine or most organic solvents. The activated resin fluxes can be treated as non-activated resins.

PRE-COATING

If active corrosive fluxes must be used yet removal by the acidified water treatment is impracticable the parts should be pre-coated with solder using the active flux and then all residue completely removed before the parts are assembled for soldering with a non-active flux.

BIBLIOGRAPHY

MANKO, H. H. *Solders and Soldering*, New York, McGraw-Hill (1964)
'Joining and Fastening of Materials. Brazing and Soldering Alloys', *Materials in Design Engineering*, **58**, 486–489 (October 1963)
'Soldering Aluminium', *New Zealand Eng.*, **24**, (2), 72–74 (February 15th 1969)

REFERENCES

1. *The Welding Handbook*, Section 3, 5th ed, Chapter 44, 5, American Welding Society
2. *Welding, Brazing and Soldering Wiggin Nickel Alloys*, Henry Wiggin and Co., Hereford
3. *The Welding Handbook*, Section 3, 5th ed, Chapter 44, 17–21, American Welding Society
4. BS 219 (1959)
5. 'Fabrication Equipment Guide', *Welding and Metal Fabrication*

METAL FINISHING

G. P. ROTHWELL and J. M. SYKES

SURFACE PREPARATION

If an article is to have an attractive and sound surface finish after electroplating, anodising or any of the processes discussed below, it will in general be necessary to pay close attention to the state of the surface at the outset of the finishing process. The amount of effort involved at this stage will be directly dependent on the methods of manufacture adopted. A die-casting with a sound surface may need only a mild flash-removal before the finishing cycle is commenced, whereas a sand-cast or machined article will often require a more complicated series of surface preparation operations which will depend on the type of finishing process selected and the quality of finish required. Grinding, polishing and brushing processes are typical of such operations: similar results may often be obtained by barrel finishing, and the process may also be arranged to effect flash-removal, deburring and radiusing. Sand- and grit-blasting, vapour blasting and similar processes may also be considered as surface preparation operations, but probably belong more properly to the engineering works than the finishing shop, and are not discussed here.

The distinction between grinding and polishing operations, which is relatively clear in metallographic applications, for example, is by no means so distinct in the context of industrial metal finishing. In the former case, both grinding and polishing are considered to be cutting operations, the object being to disturb the basis metal as little as possible, and the distinction is made that grinding takes place on fixed (coarse) abrasives which produce long scratches and remove metal rapidly, whereas polishing occurs on free (fine) abrasive particles, which thus produce much shorter, finer scratches, leading finally to a reflective surface. In the metal finishing context however, grinding is frequently performed on relatively loosely bound abrasive particles from grease-bonded compositions, and much of the polishing process involves a burnishing action, causing the surface to flow, in addition to the cutting action of the polishing compound.

We may take it however, that grinding is, broadly, that process which removes macroscopic irregularities and grooves from the surface while the polishing process removes the grinding scratches and produces a mirror finish.

Grinding

The most commonly used abrasives are emery, natural and synthetic corundum, and silicon carbide. Naxos or Turkish emery was once the standard grinding abrasive. It is a compound of iron and aluminium oxides, containing about 60–75% Al_2O_3. It is somewhat less hard than the aluminas or silicon carbide and has more rounded grains, which reduces the degree of cutting. Emery is on the decline as a grinding medium, its place being taken by the synthetic products.

Natural alumina, or corundum, contains 90–95% Al_2O_3. It has a hardness of 9 on Moh's scale and cleaves readily in use, maintaining new, sharp cutting edges. Its principal use is for bonding into grinding wheels. Synthetic alumina is harder and more uniform than the natural material. It contains up to 95% Al_2O_3; material of higher purity is available but too costly. Synthetic alumina is sharp, tough and fast cutting, and is readily bonded onto polishing wheels and belts. It is thus today probably the most widely used abrasive.

Silicon carbide is harder than Al_2O_3 (approaching 10 on Moh's scale) and more brittle in use, retaining its cutting edge well. It is thus an excellent abrasive, and economically competitive with the aluminas, and finds very extensive use. Earlier problems with the bonding of silicon carbide to grinding wheels and belts have been overcome by improved adhesive formulations. These materials are available in screened sizes from 8 mesh to 600 mesh, but grades between 60 and 240 are most commonly used in the grinding process.

In addition Tripoli, and Kieselguhr, both diatomaceous silicates with a hardness of 5-6 on Moh's scale, may occasionally be used for fine grinding particularly of brasses and aluminium alloys.

For most purposes, the grinding abrasive is used on abrasive bands, dressed wheels or in grease-bonded compositions on mops or brushes. The abrasive band is driven by a contact wheel on the polishing motor, and tension is maintained by an idler wheel. The work is applied to the portion of the belt which passes over the contact wheel, which thus replaces a dressed wheel. The abrasive band has the advantage of a larger contact area, cooler running and longer life. Bands can be obtained from about 24 mesh to 280 mesh, bonded with glue or synthetic resin, the latter giving a waterproof band with better heat resistance than glue-bonded bands. Contact wheels may be of hard, compressed canvas or woven sisal for flat surfaces, or softer calico may be used for grinding contoured work. Alternatively, light alloy wheels with rubberised canvas vanes may be used, in a variety of grades.

Abrasive wheels may be of solid felt, of felt- or leather-covered wooden hubs, of clamped leather or canvas sections, or stitched mops, the latter being the most flexible and often used as scurfing mops for initial descaling of tubular and contoured work. They are dressed with abrasive grains, using glue or resin adhesive, and must be balanced before use. They may be used on a variety of machines, ranging from single and two-spindle manual machines to more complicated automatic and centreless grinders. Sisal mops and fibre rotary brushes are used dressed with grease-bonded grinding compositions for finish-grinding of steel components. They may be built up in sections to give very wide contact surfaces for surface dressing of steel strip and similar applications.

Polishing

Many materials may be used for the final polishing stages, depending upon the specific application. Among the most common are Vienna lime, rouge, chromic oxide and alumina. Vienna lime is calcined dolomite, with a grain size below about 1 μm. It is an excellent finishing abrasive for aluminium and nickel, but has a strong affinity both for carbon dioxide and for the fatty acid compounds used in polishing compositions, forming calcium and magnesium carbonates and soaps respectively and becoming useless for polishing. Scrupulous care is therefore needed in the storage and use of Vienna lime compositions.

Rouge is a very fine ferric oxide. Ferric oxide may be produced by many routes, and a compound with different characteristics is formed in each case. A few of these rouges are useful polishing agents, and considerable skill is exercised in blending these to produce suitable polishing materials.

Chromic oxide, *green composition*, is very hard, and suitably prepared makes an excellent and extremely fine polishing abrasive for hard materials such as chromium and stainless steel. It is somewhat expensive in comparison with other materials.

Alumina exists in two modifications, both of which find use in the polishing of metals. If bauxite is dehydrated at about 750-850°C, very fine gamma alumina is produced. This cubic modification of the oxide has a hardness of about 8 on Moh's scale, and is used extensively for the final stages of metallographic polishing. If γ-Al_2O_3 is heated above about 1200°C or if the original dehydration takes place at 1000–1200°C, the hexagonal alpha modification is produced. This has a hardness of 9, and is usually somewhat coarser than the γ-Al_2O_3. It is extensively used for industrial and metal finishing.

For use, the polishing abrasive is compounded into a composition with fats and waxes blended to give the required melting point. The media are mainly neutral, but some fatty acids may be used to control the pH of the composition and minimise staining of the workpiece. Synthetic silicone materials are also used, and are said to give a better gloss with a shorter polishing time. In the selection of compounding media it is most important to remember that the residual grease or wax on the workpiece must be removed during the subsequent cleaning stages, before electrodeposition or other chemical finishing processes can take place. For this reason, greaseless compositions often find favour, since their medium is water soluble and cleaning is facilitated. Another approach to ease of operation is the use of liquid compositions. In these, the abrasive is suspended in an oil-in-water emulsion, facilitating spray application in large automatic machines. The composition is formulated so that most of the water evaporates on atomisation, and leaves the oil to bind the abrasive to the polishing wheel and lubricate the cutting process.

The polishing composition is used on mops or brushes, which may be larger and are usually softer than those used for grinding. There is a wide variety of types of brushes and mops, and a detailed discussion would be out of place. Endless woven belts of wool gauze are also available, as are endless felt belts, so that polishing may be carried out on a standard belt grinder. The types of polishing machine available range from single-spindle hand machines to large multi-station machines which can polish all surfaces of complex articles automatically with little attention from the operator.

With material which is to be electroplated, final polishing may take place after deposition, depending on the finish required and the availability of plating processes. Satin finishes may also be obtained after plating by mopping or brushing with very fine grinding compounds.

Barrel finishing

An alternative method of surface preparation is barrel finishing, which provides an attractive low-labour, high productivity method of treating small parts, such as gear wheels and small castings, particularly in cases when no sharp corners are to be kept on the articles. It can be arranged to provide treatments equivalent to any process from grinding to honing, and the coarser treatments can be arranged to provide radiusing and deburring at the same time, although really deep damage cannot be removed, since there is no possibility of providing concentrated local treatment for bad areas. Processing of a single charge can take many hours, but since a large number of components are treated at one time the productivity is high, and since all parts receive identical treatment the reject rate is virtually nil.

Either closed polygonal horizontal barrels or open inclined barrels are used, usually constructed of steel plate with a hardwood, rubber or plastic lining, although inclined barrels may be of all wooden construction. Rotational speeds may vary between 15 and 50 rev/min depending on the type of charge. The work is charged with between two and ten times its own volume of barrelling media and usually with barrelling compound. The media may consist of pebbles, shaped ceramic pieces, steel shapes, steel, brass or bronze balls or wooden pegs; the choice of material and size depends on the particular situation.

Sizes and types of media may be mixed, for example to enable internal and external surfaces to be prepared in one operation. The purpose of the media is to separate the workpieces, and prevent damage by direct impingement as the load tumbles, to increase the weight of the load and thus increase the grinding or burnishing pressure, and either to perform the grinding or burnishing, or to carry added abrasive evenly to the surface of the workpiece.

Work may be barrelled dry, but the use of barrelling compound leads to a more uniform finish, and the solution keeps the work clean and lubricates the process. The compounds are usually aqueous solutions containing wetting agents or soaps, anti-rust compounds, and buffering agents, and where necessary additional abrasive such as silicon carbide, alumina, rouges or emery. Control of water softness and pH is most important. Most compounds are neutral, but acid compounds are used to give more rapid descaling of ferrous components. For burnishing, foaming compounds may be chosen, but for most procedures foam is to be avoided.

The majority of barrelling processes are carried out in two stages, scouring and deburring, followed by burnishing or polishing, with appropriate changes of media and compounds between. However, sometimes a single process can suffice, and, again, a barrelling procedure may be combined with one or more conventional grinding and polishing stages to optimise the surface preparation process. By the use of fixtures to keep the work stationary while the media move, the process can be extended to quite large parts, which could never be allowed to strike each other directly while rotating freely in a barrel.

Vibratory finishing. The vibrator provides an alternative to conventional rotary barrelling processes. It consists of an open rubber-lined work container which is vibrated on rubber supports by a suitable arrangement of motor and eccentric. The vibration, the amplitude of which may be adjusted, provides the necessary motion between the work and the media. Process times are considerably shorter than conventional barrelling procedures and the degree of control somewhat superior.

Electropolishing

An alternative method of polishing which finds some industrial applications is electropolishing, which may be used as a finishing process itself, or as a preliminary to other metal finishing processes with most pure metals and high-quality single-phase alloys, in particular aluminium, single-phase copper alloys, plain and stainless steels and jewellery and semiconductor materials. It has been used for articles as diverse as domestic sinks, automobile trim, lamp reflectors, surgeons needles and large gear wheels. As the metal is removed electrochemically, electropolishing can in a sense be thought of as electroplating in reverse, and as such it must have a particular attraction to the electroplater since it requires the same tank lines, jigs and power supplies as are already available in the plating shop. Again, it is a batch process, and can often be included in an electrofinishing line to provide polishing without the necessity for separate handling and rejigging of the workpieces. It can also be used to provide very rapid polishing after electrodeposition without rejigging.

Since electropolishing is a batch process, it is frequently much cheaper than manual polishing of individual items. It may be used to produce a bright and level surface from material prepared by 180-grit abrasion, and typically some 25 μm of metal are removed. The throwing power of electropolishing baths is generally good, so that a satisfactory finish may be produced over quite complex surface contours with relatively simple counter electrodes. The finish produced by electropolishing is not quite as level as that produced by the best mechanical polishing on flat products, but the colour of the metal is often superior, particularly with white metals, such as aluminium. A real advantage is that electropolishing produces a different type of surface from that produced by mechanical finishing processes, in that surface deformation and polishing debris are absent, and the polished surface retains the true crystal structure of the material, which makes it ideal for subsequent electrochemical finishing operations. However, since electropolishing is an oxidation process, many materials exhibit passivity after polishing and an intermediate activation may be necessary before subsequent operations.

The electropolishing process can be considered as the combination of two separate phenomena; *levelling*, consisting in the preferential removal of metal from prominences on the surface, thus reducing the macroscopic (>1 μm) roughness of the surface, and *brightening*, which involves the suppression of crystallographic etching, thus preventing the surface becoming rough on a microscopic scale. Levelling is associated with diffusion controlled transfer of ions through the viscous Jacquet layer at the metal/solution interface. (The ions involved may be those of the dissolving metal moving outwards to the bulk of the solution, or alternatively may be ions moving inwards to complex the metal as it dissolves). For this reason, most polishing solutions are relatively viscous, and many formulae call for the addition of dissolved metal ion (which increases viscosity) before the solution is ready for use, or else polish much better when some metal has been allowed to accumulate in the solution. Again, failure to level may often be traced to factors which affect the stability and thickness of the diffusion layer, such as viscosity, temperature, stirring rate or concentration of dissolved metal.

Brightening on the other hand is generally thought to be due to dissolution of the metal through a thin (a few nm) *solid* film of oxide or basic salt of the metal, so that the selection of atoms dissolving depends on the random location of (rare) vacant lattice sites in the film, through which the ions can diffuse to the solution, rather than on the energy differences between the metal atoms as a result of their crystallographic location. Thus, random rather than preferential solution is assured and the surface remains bright. Faults in brightening may often be traced to factors affecting the stability of the solid film (pH, concentration of dissolved metal ion, electrode potential, impurity ions present).

Most of the electropolishing baths in common use are viscous, being based on phosphoric or sulphuric acid, or containing glycerol or other additives to maintain viscosity. In addition they usually contain a strong acid (or alkali for amphoteric metals), to ensure ready dissolution of metal, and may also contain additional oxidising agents, to facilitate setting up the correct polishing regime. Solutions based on perchloric acid,

Table 6.30 CONSTITUENTS OF COMMONLY USED CHEMICAL POLISHING BATHS

Aluminium	or	Sodium carbonate, sodium phosphate
		fluoboric acid
α-Brasses	or	Phosphoric acid, chromic acid
		phosphoric acid, sulphuric acid
Copper		Phosphoric acid, Cr^{3+}, Al^{3+}
Carbon steels	or	Phosphoric acid, sulphuric acid,
		phosphoric acid, sulphuric acid, chromic acid
Gold and silver		Potassium cyanide, potassium carbonate
Nickel		Sulphuric acid
Stainless steels		Phosphoric acid, glycerine

which find extensive use in research applications, are generally thought to be too hazardous for industrial use. The baths fall into two types, those which are used to exhaustion and then discarded, and those which are virtually self-maintaining by plating out of dissolved metal onto the cathode, or by sludging-out when the solubility of the metal is exceeded. Replenishment may, of course, be needed, and control is usually by measurement of specific gravity or dissolved metal concentration.

All the solutions commonly used have high conductivity, and a series rheostat circuit is used to control the polishing current. Current densities of 1000–4000 A m^{-2} are used, with cell voltages between 5 and 15 V. The current must be kept to about 2–3 A per litre of the bath to avoid over-heating. Polishing is usually carried out with copious gas evolution, which serves to stir the solution and prevent pitting and streaking.

Baths for industrial electropolishing are well documented in proprietary and patent literature; constituents of the most commonly used are listed in Table 6.30.

Chemical polishing

Chemical polishing provides an alternative to electrolytic polishing for aluminium and its alloys, copper and the α-brasses and for a few other non-ferrous materials, but is not usually very successful for ferrous materials. Many of the baths loosely referred to as chemical polishes are in fact bright dips, in that the levelling function of the polishing process is absent. They are referred to further below.

The mechanism of chemical polishing is similar to that of electropolishing, save that the driving force for dissolution of the metal is provided by the reduction of an oxidising agent present in the bath, rather than electrically. Such baths generally therefore contain nitric acid or chromic acid or sometimes hydrogen peroxide. Solutions for aluminium and its alloys are based on orthophosphoric-nitric-acetic acid mixtures or ortho-phosphoric-nitric-sulphuric acid mixtures. The first type of solution is also useful for copper and its alloys, and for nickel. Sulphuric acid-chromic acid mixtures or sulphuric acid-hydrogen peroxide mixtures are useful for cadmium and zinc. The hazardous nature of these solutions is apparent, and stringent precautions should be observed in their use.

Measurement of surface roughness

Stylus recording instruments are most commonly used for measurement of surface finish. In operation, a stylus with a tip radius which may be between 2 and 20 μm depending on application, is traversed across the specimen. Different instruments employ mechanical, pneumatic or electro-magnetic amplification of the vertical movement of the stylus by a factor up to 10 000–40 000 times, and the resulting profile is recorded on a chart or film. Alternatively, meters or dial gauges may be used to record the movement, particularly in workshop instruments.

Stylus instruments are suitable for measurement of surface roughness between about 100 and 0·1 μm. Such values are reported as CLA (Centre Line Average) values in this country, the figure being the arithmetic mean of excursions from the centre-line of the surface profile. In the USA, r.m.s. roughness values are preferred. BS 1134[1] lays down conditions to which measuring instruments must conform, and specifies the methods of measurement to be adopted.

Interference microscopy may be used to study surfaces finer than those accessible to stylus instruments, and might typically be used to study profiles from 1 to 0·005 μm. Such instruments are however, essentially for laboratory rather than workshop use. Electron microscopy can also be used for the study of surfaces on a still finer scale, and in addition will give information about surface deformation and inclusion of polishing

Table 6.31   TYPICAL ROUGHNESS VALUES
FOR VARIOUS SURFACE FINISHES, μm CLA

Coarse grinding	10–20
Sandblasting	4
Fine grinding	2·5–0·25
Mechanical polishing	0·1–0·01
Honing, lapping, superfinishing	0·05–0·01
Electropolishing	0·6–0·1
25 μm levelling nickel on previously ground surface	0·1

materials, but it is even further away from the realm of practical metal finishing.

Frequently, the properties of a polished surface which are of most importance to the metal finisher are *gloss* or *lustre*, and *truth* (undistorted image reflection). Lustre is assessed by a series of photometric instruments which measure directly or indirectly the amount of specular reflection from the surface. Such instruments are extremely sensitive, and capable of detecting differences between surfaces which cannot be distinguished by eye. Surface truth is assessed from the quality of the reflected image of a strongly illuminated pattern of black and white stripes or squares. It is primarily a subjective assessment, depending on the distance the polished surface can be moved away from the test pattern before the image becomes unacceptably blurred.

Typical CLA roughness values for a variety of surface finishes are given in Table 6.31. Actual values depend greatly on the material involved and the precise manner in which the treatment is carried out. Again, values for levelling electrodeposits and electropolished surfaces refer to long-range waviness, or imperfect smoothing of the surface, whereas those for mechanical finishing are short-range roughness values.

Cleaning

The contaminants which are present on a soiled metal surface as it is processed may be organic, such as natural or compounded oils, inorganic, such as smuts, oxide or machining debris, or mixed, such as polishing residues or pigmented drawing compounds in which an inorganic phase is dispersed in an organic binder. Before electrodeposition can proceed, all of this material must be removed to give a clean metal surface.

The cleaning process may be divided into three stages, preliminary cleaning or degreasing, pickling or bright dipping, and final cleaning. The necessity for all of these stages, the methods at each stage and the order in which the stages are carried out will depend on the precise plating system involved, convenience and cost-effectiveness. In some processes several preliminary cleaning steps may be used, at various stages in the preparation of the workpiece. The treatments discussed below are intended to be illustrative rather than exhaustive. Final surface cleanliness may be assessed in a variety of ways; most of the practical methods involve observation of wettability as shown by water-break or spray-pattern tests. The final assessment of cleanliness must, of course, be the success of subsequent plating processes. Preliminary cleaning may be carried out by solvent or emulsion cleaning processes or by hot alkaline cleaning.

SOLVENT CLEANING

Solvents may be employed to remove greasy soil from the surface of the workpiece. Inhibited trichloroethylene or perchloroethylene are the solvents usually used; they may be boiled in a special vat, with heating coils at the bottom and cooling coils at the top to prevent vapour loss. The work is suspended in the vapour region and clean, distilled solvent condenses onto it, washing away oil and grease. Condensation continues until the work reaches the temperature of the solvent vapour, when cleaning ceases.

Vapour degreasing may be inadequate for components with particularly hard and compact greasy soil, or for thin sections which reach the vapour temperature very rapidly. Immersion in boiling liquid solvent may be used in such instances, to give a greater cleaning action, assisted by the stirring motion of the boil. Plants are available with double vats to allow components to be rinsed in clean vapour after degreasing in boiling liquor. Alternatively, articles may be soaked in a self-emulsifying solvent system, consisting of suitable mixtures of solvents and emulsifying agents, which allows the work to be water-rinsed after soaking, to flush away soil and residues. Self-emulsifying compositions are usually used cold, but may be warmed.

EMULSION CLEANING

Heavy soils and polishing residues may be treated by emulsion cleaners. These are mixtures of solvents, soaps and wetting agents, dispersed in an aqueous phase. They may be used hot, applied by soak or spray.

Spray application is particularly effective, the pressure of the jet facilitating reaction with the soil, and markedly reducing cleaning time. Spray booths may conveniently be integrated into automated process lines. Ultrasonic cleaning in emulsion cleaners is also used successfully in a number of applications.

The action of these cleaners is twofold, the solvent phase dissolves the oils and greases present, while the aqueous phase modifies the inorganic soils so that they are more readily removed by rinsing or by subsequent alkaline cleaning.

HOT ALKALINE CLEANING

An alternative approach to preliminary degreasing is hot alkaline cleaning. Alkaline cleaners function by saponification of animal and vegetable oils and emulsification of most other organic soils. They consist of mixtures of hydroxides and carbonates, condensed phosphates and silicates, often with added detergents, although the presence of some detergents may inhibit subsequent processes.

Cleaners with a variety of pH values are available, ranging from strongly alkaline cleaners for steel to nearly neutral for aluminium and zinc alloys. They may be used hot as soak or spray cleaners, although low foaming formulations must be chosen for spray applications, and are also frequently used as electrolytic cleaners, or in a two stage, soak-plus-electrolytic scheme.

Electrocleaning is very effective, since the evolution of gas at the surface produces a scouring action which aids in removal of soils. Cathodic cleaning is preferred for non-ferrous metals, since the volume of hydrogen evolved at a cathode is twice the volume of oxygen evolved at an anode operating at the same current density, and the scouring action is correspondingly greater. Anodic cleaning is employed for most ferrous metals however, to avoid the risk of embrittlement due to hydrogen pick-up. An attendant disadvantage is that anodic cleaning of steel frequently leaves the surface in a passive condition, necessitating an intermediate acid dip before further processing.

FINAL CLEANING

After preliminary cleaning, many articles undergo further machining or polishing stages, pickles or dips to remove oxide films, or other processes, and before the surface is ready for plating a final cleaning stage is generally necessary.

This may be a cyanide electrocleaning bath, or if cyanide is unwelcome in the plating line, further electrolytic cleaning in an appropriate alkaline cleaner. An exception to this occurs with steel articles which are required to take a heavy nickel deposit, when anodic acid etching may be used to clean and roughen the surface.

Pickling and bright dipping

At various stages in the preparation sequence, articles may be pickled to remove scale and oxide films. Typically, this is carried out after preliminary cleaning, but obviously before polishing and final cleaning. The problem which arises in the removal of heavy scale by acid pickling is attack on the bare metal causing roughening of the surface, and change in dimensions. The difficulty may be overcome by the use of a pickling restrainer to mitigate attack on the metal but, nevertheless, it is usual to carry out such operations when necessary as early as possible in the process.

For iron and steels, hot dilute sulphuric acid is used to remove heavy oxide films although cold dilute hydrochloric acid may be used to remove thin oxide films, and leaves a cleaner surface than sulphuric acid at short treatment times. Stainless steels require a more vigorous treatment for the removal of oxide films, and a nitric acid–hydrofluoric acid mixture is often used. Plain dilute hydrofluoric acid is also used for pickling aluminium and its alloys, and for the treatment of sand castings in a variety of materials, where its ability to remove residual sand particles is particularly advantageous with castings for hydraulic equipment.

Copper and its alloys may be picked in hot sulphuric acid solutions similar to those used for steels, but it is often an advantage to use bright dips for small parts in copper, brasses and nickel silver. Bright dips perform a brightening process similar to that found in electrolytic and chemical polishing, but with markedly inferior levelling. Nevertheless, with many articles, bright dipping effectively replaces lengthy individual mechanical polishing processes by a rapid chemical batch process. *Aqua fortis* bright dip is often used, a mixture of sulphuric acid and nitric acid with a little sodium chloride added; alternatively *red dip*, a chromic-sulphuric acid mixture may be used, particularly for plumbers' brassware and similar articles. Aqua fortis bright dip evolves noxious brown fumes, so that efficient extraction equipment is necessary; both dips are exceedingly corrosive and appropriate precautions must be taken.

In addition to the pickling processes, most materials receive a rapid activating or destaining dip in dilute acid or dilute cyanide solution immediately before plating. Copious and efficient water rinsing is necessary at all stages of cleaning and surface preparation.

Surfaces requiring special treatment

Despite suitable pickling and cleaning procedures, some surfaces reform their oxide films so rapidly that special treatments are required to ensure efficient subsequent deposition. Typical of this group is aluminium, which usually receives an alkaline cleaning treatment followed by immersion plating of a thin layer of zinc from an alkaline zincate solution; alternatively an acid etch may be followed by plating in a strongly acid copper or nickel strike bath.

Magnesium-based alloys and titanium also require vigorous etching followed by immersion plating with zinc. Stainless steels carry a very tenacious surface film, but they may usually be treated successfully by a simultaneous activation and strike process in strongly acid nickel sulphate or chloride, or copper sulphate baths.

ELECTROPLATING

Fundamentals

ELECTROLYTE SOLUTIONS

Metal salts are ionic compounds, consisting of positively-charged metallic cations and negatively-charged anions arranged in a geometric lattice. When a salt is dissolved in in water, the structure is broken down and the salt dissociates into separate ions. The dissociation involves coordination of the cations with water molecules to produce hydrated ions. Alternatively, other ions or molecules may be added which will replace this bound water and form more stable complexes. In many respects, such solutions of complex ions behave like less concentrated solutions of simple (i.e. hydrated) ions. To take account of such effects, the behaviour of ions in solution is discussed in terms of their *activity*, that is, their effective concentration under the relevant conditions.

Acids and alkalis also dissociate in solution, yielding hydrated hydrogen ion (H_3O^+) plus an anion, or hydroxyl ion (OH^-) plus a cation respectively. In addition, the water itself is slightly dissociated, giving small quantities of hydrogen ion and hydroxyl ion in equilibrium with the water molecules. This equilibrium is preserved in acid or alkaline solutions, so that strongly acid solutions contain much hydrogen ion and little hydroxyl ion, whereas in alkaline solutions the reverse is true. Numerically $[H^+][OH^-] \simeq 10^{-14}$ at 25°C where $[H^+]$, $[OH^-]$ are the activities of hydrogen ion and hydroxyl ion, respectively, in gram ions per litre. Thus hydrogen ion activity is a measure of the strength of both acid and alkaline solutions. We define

$$pH = -\log_{10}[H^+],$$

where $[H^+]$ is the hydrogen ion activity in gram ions per litre. Therefore a 0·1 molar solution of HCl, a strong acid which will be completely dissociated, has a pH of 1, whereas a 0·1 molar solution of NaOH, a strong alkali, contains 0·1 grams per ion per litre of OH^-, and has a pH of 13. In neutral solutions, there are equal quantities of hydrogen and hydroxyl ions, so that the pH is 7. The situation is less simple in the case of weak acids and bases, where incomplete dissociation makes the calculations more complicated.

Electrolytic conduction. If a potential difference is applied between electrodes in an electrolyte solution, movement of ions will occur in the field, producing an electric current. The positively-charged cations will be drawn towards the negative electrode, the cathode, and the anions to the anode. The current depends on the mobility of the ions, and is proportional to the field strength. For any solution the conductivity depends on the concentrations of the different ions in solution, their sizes and the charges they carry.

ELECTRODE PROCESSES

So that a continuous current can flow, charge must be transferred to the solution at the electrodes by an electrochemical reaction involving either species in the solution, the electrodes themselves, or both. Typical reactions in the electroplating context might be:

At the anode:
$$Cu \rightarrow Cu^{2+} + 2e^-,$$
$$2H_2O \rightarrow 4H^+ + O_2 + 4e^-,$$
$$Fe^{2+} \rightarrow Fe^{3+} + e^-,$$

At the cathode:
$$Ni^{2+} + 2e^- \rightarrow Ni,$$
$$2H^+ + 2e^- \rightarrow H_2.$$

The weight of product in such reactions is proportional to the quantity of charge passed and inversely proportional to the number of electrons involved in the unit reaction. According to Faraday's Law, 96,540 coulombs, or 26·8 Ah (a quantity of electricity known as the Faraday) will deposit one gram equivalent of metal. The gram equivalent weight of a metal is the atomic weight in grams divided by the number of electrons involved in the unit reaction. If side reactions can occur simultaneously with the desired reaction, the ratio of current usefully consumed to the total current passing is termed the *current efficiency* of the process. This is usually obtained by comparing the actual yield with that calculated from Faraday's Law.

THE ELECTROCHEMICAL SERIES

In the absence of an external current, the potential difference between electrode and solution comes to a value at which the anodic and cathodic processes occur at equal rates. This *electrode potential* cannot be measured directly, but must be obtained by comparison with a reference electrode. The zero point of the electrode potential scale is arbitrarily taken as the potential of the standard hydrogen electrode. For a single electrode reaction the electrode potential, e, can be calculated from the Nernst equation, which for a pure metal in a solution of its ions gives

$$e = e^\circ + \frac{RT}{zF} \log [M^{z+}],$$

where $[M^{z+}]$ is the activity of M^{z+} ions in solution and R, T, F and z are the gas constant, temperature (°K), Faraday's constant and the number of electrons involved in the reaction, respectively. The *standard electrode potential, e°*, is the potential of the electrode in a solution of its ions at unit activity. Metal electrodes thus undergo anodic dissolution if held at potentials higher than their equilibrium electrode potential in a particular solution, and cathodic deposition will occur at potentials lower than this value.

The electrochemical behaviour of different metals may be discussed in terms of their standard electrode potentials. Such a list of metals in terms of their standard electrode potentials is known as the Electrochemical Series; typical values are given in Table 6.32. Metals at the noble end of this series are more readily deposited than the more base metals. During deposition of metals below hydrogen in the series, there is always the possibility of hydrogen evolution at the cathode, and the more base the metal deposited the greater the tendency towards hydrogen reduction.

Fortunately hydrogen evolution is very sluggish on many metal cathodes, and so the current efficiency of most deposition processes is economically high.

POLARISATION

To drive a deposition reaction at any desired rate, the potential of the electrode must be brought to some value below the equilibrium potential by an amount η_c, which is called the overpotential or polarisation of the electrode. At its equilibrium potential, a metal

Table 6.32   THE ELECTROCHEMICAL SERIES.
STANDARD ELECTRODE POTENTIALS OF SOME METALS
AT 25°C, e^0, VOLTS VS. STANDARD HYDROGEN
ELECTRODE

NOBLE END		
Gold/Au^{3+}		$+1\cdot50$
Silver/Ag$^+$		$+0\cdot80$
Copper/Cu^{2+}		$+0\cdot34$
Hydrogen/H$^+$		$0\cdot00$ arbitrary zero
Lead/Pb^{2+}		$-0\cdot13$
Tin/Sn^{2+}		$-0\cdot14$
Nickel/Ni^{2+}		$-0\cdot25$
Cadmium/Cd^{2+}		$-0\cdot40$
Iron/Fe^{2+}		$-0\cdot44$
Chromium/Cr^{3+}	approx.	$-0\cdot70$
Zinc/Zn^{2+}		$-0\cdot76$ V
BASE END		

is in a state of dynamic equilibrium, in which the small rate of dissolution of metal atoms is balanced by an equal rate of spontaneous deposition of metal ions from the solution.

To obtain a net rate of deposition, the energy barrier at the metal/solution interface has to be tilted in the sense that favours deposition, by the application of the overpotential. Provided that the current is low enough for the metal/solution energy barrier to be the rate determining step, the current density is related to the overpotential by the Tafel equation

$$\eta_c = a + b \log i_c,$$

where η_c is the cathodic overpotential, i_c the cathodic current density and a and b are constants. The values of a vary widely, but values of b are frequently about 0·06 V per decade for monovalent cations, 0·12 V per decade for divalent cations. Polarisation in the regime in which Tafel's equation is obeyed is called *activation polarisation*.

As the current density is increased, the energy barrier at the metal/solution interface ceases to be rate-determining, and instead the rate at which ions can pass through the diffusion layer to the electrode controls the rate of deposition. This is the *concentration polarisation* regime, and ultimately a limiting current density is reached, which cannot be increased by increasing the polarisation. The limiting current density can however be increased by increasing the stirring rate in the bath, the temperature, or the concentration of dissolved ions. The depletion of ions at a cathode working in the concentration polarisation region gives rise to the so-called *back emf* when the current is switched off.

Figure 6.11(a) summarises the polarisation behaviour of a single electrode, and Figure 6.11(b) emphasises the point that if an electrode is held at a potential at which several reactions can occur, the total current density is to be found by adding all the current densities from the various individual reactions.

Further, if the bath has a high resistance, or if gas films or solid layers impede current flow to the electrode, more driving force is required to overcome these effects. The extra polarisation required is called *resistance polarisation*.

The plating process

PLATING SOLUTIONS

Plating solutions are nearly always solutions of metal salts in water, but it is possible to produce useful metal deposits from molten salts or from solutions in organic solvents. The primary function of the solution is to carry metal ions to the workpiece, but it may also act as a source of metal if inert anodes are used. In either case it is important that the electrolyte contains a high concentration of the metal which is to be deposited, since this

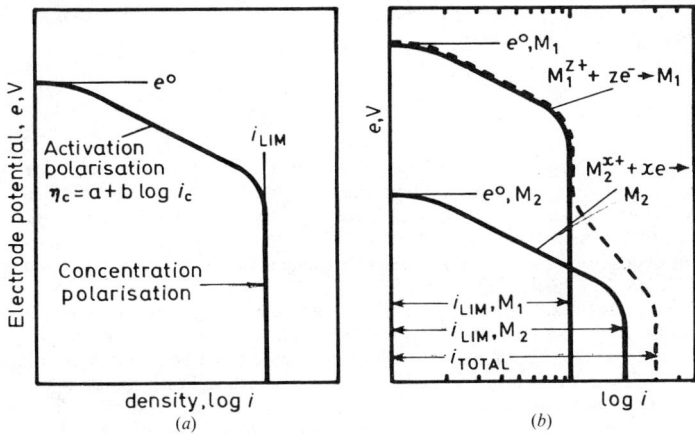

Figure 6.11. Schematic cathodic polarisation curves showing (a) activation and concentration polarisation regimes and (b) addition of curves for two individual reactions to give the net cathodic polarisation curve

allows higher currents and therefore faster plating rates to be used. If the concentration is too low for the plating rate used, the cathode will become heavily concentration polarised and rough, loose, or discoloured deposits may result. The main constituent of a plating bath will therefore be a readily soluble salt of the metal to be plated.

A further desirable property of the electrolyte is high conductivity, so that ohmic power losses will be minimised. Conductivity can often be improved by adding ions of high mobility such as halide or hydrogen ion. Halides may also be beneficial in ensuring that the anode dissolves at a low overpotential.

In many electrolytes it will be important to control pH; if it is too low excessive amounts of hydrogen will be liberated at the cathode, but if it is too high hydroxides may be precipitated, especially at the cathode, where there will be a local pH increase. For strongly acid or alkaline solutions, pH will be fairly stable, but near-neutral solutions require the addition of a buffer agent to take account of excess alkali or acid generated in the cell.

In addition to these main constituents there will often be addition agents to control pitting or internal stress, or to improve the smoothness or brightness of the deposit.

COMPLEX PLATING BATHS

In aqueous solutions the metal ions are associated with surrounding water molecules, or other ions or molecules which can displace the water and form a complex in which the metal ion will be more tightly bound. Many types of complex are so stable that they cannot be broken down by electrolysis, but others provide the basis for successful plating baths.

Because the ion is more strongly bonded to the surrounding species the activity of the metal in solution is lowered causing the deposition potential of the metal to become less noble, and the polarisation to be increased. Nevertheless complex plating baths have many advantages; they usually yield a finer-grain deposit, have less tendency to form loose immersion deposits by displacement reactions with the basis metal, and tend to give better metal distribution.

Two widely used complexing agents in plating solutions are halide and cyanide ions, which tend to form neutral or anionic species but are nevertheless discharged at the cathode to give good metal deposits.

ALLOY PLATING

In some cases the deposition potentials of two metals may be so close together that a mixture of the two ions will give a deposit containing both metals, but the deposition potentials are often far apart. If the difference is not too large and the more noble metal shows a limiting current density as the potential of the work is made more negative, then when the deposition potential of the less noble metal is reached, increasing amounts of it will be obtained in the deposit.

In many cases the situation is more difficult and a complexing agent must be added. This can lower the activity of the more noble metal so that its deposition potential is sufficiently negative for both metals to plate out together, or if both ions form complexes the deposition potentials may still be brought close together. For instance, in the silver-zinc system addition of cyanide brings the difference in deposition potential from 1·56 V to 0·3 V. In some baths the two ions can be present as different complexes. The ratio of the metals in the electroplate will depend on their activities in the plating bath and will, therefore, depend not only on their relative concentrations, but on the amount of complexing agent present.

Some alloy systems show anomalous effects, in that a noble metal plates more slowly than a base metal; deposits can also contain metals which are not normally obtainable from aqueous solutions.

HYDROGEN EVOLUTION

Most of the common metals are deposited at negative potentials on the hydrogen scale and, depending on the hydrogen overpotential of the metal, more or less hydrogen is evolved at practical plating current densities.

This presents three main difficulties. The current efficiency of the process is decreased, in that only part of the current is used for the reduction of metal ions, the rest being for hydrogen production. The gas evolved causes a spray of solution above the bath which may be a serious health or corrosion hazard. Finally, the gas can lead to pitting of the electrodeposit, due to bubbles obscuring the work. To prevent this, wetting-agents may be added to the bath to decrease the adhesion between the bubbles and the deposit, or oxidising agents may be added to oxidise nascent hydrogen before bubbles form.

LEVELLING

One important function that electroplating can perform is levelling of the workpiece surface; thus a scratched, poorly polished surface can be transformed into a smooth one. One might expect an increase in surface relief during the deposition, with protuding ridges attracting most of the current, but in fact even without the addition of levelling agents some smoothing can occur. This is *geometric levelling* and occurs providing that the current distribution is not controlled by concentration polarisation.

In this case the thickness deposited in a fixed time will be the same in a groove or on a flat surface and successive layers will eliminate grooves (Figure 6.12(b)). Levelling can be

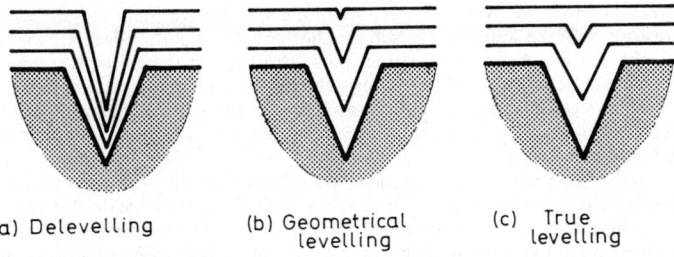

(a) Delevelling   (b) Geometrical levelling   (c) True levelling

Figure 6.12. Delevelling, geometric levelling and true levelling

greatly improved by the addition of small quantities of organic molecules which can absorb on the work and polarise the deposition reaction. The addition agent diffuses to the metal surface but is consumed by incorporation in the deposit; the greatest concentration of leveller is found on ridges (short diffusion path), the least in grooves (long diffusion path).

Because the leveller inhibits metal deposition, the plating rate is thus highest in grooves and lowest on ridges, bringing about *true levelling* (Figure 6.12(c)). In this case, the sides of the groove do not remain parallel to the original groove but become progressively less steep.

Levellers are particularly important in producing suitable nickel undercoats for bright chromium, and in this context various sulphur-containing compounds have been successful.

BRIGHTENING

Even with levellers in the plating bath, the deposit will usually lack lustre and the final surface must be polished. A deposit which is bright as-plated is obviously most desirable. Brightness is often associated with fine grain size, but this is fortuitous, since bright deposits can be obtained on large single-crystal substrates. What is required is a surface without crystallographic facets which scatter reflected light, that is, a surface which is flat on a sub-microscopic scale. To effect this the normal preference for growth of certain crystallographic planes must be suppressed. Brighteners, organic addition agents, are thought to act by adsorbing strongly on to the growing crystal surface as a labile barrier layer and thus ensuring a random distribution of metal.

Brighteners, particularly those for nickel plating baths, are again often sulphur-containing compounds, such as thiourea or saccharin, which adsorb strongly, and traces of sulphur become incorporated in the deposit. Unfortunately such traces decrease the corrosion resistance of the metal.

Bright deposits can sometimes be obtained without the addition of brighteners, notably in the case of chromium. In this case, deposition appears to take place through a film of hydroxide or chromate on the chromium surface.

Degradation of addition agents. For many addition agents the active species in the levelling and brightening processes are not the addition agents themselves, but breakdown products formed by electrolysis or reaction with the solution.

THROWING POWER

For long life in service it is important that an object to be plated receives an even thickness of deposit over the whole surface, but the amount of metal on different areas depends on the current distribution. Areas nearer the anode will receive more metal because the ohmic resistance of the solution is less in the shorter current path. Edges will also attract high current densities because of current paths available in the electrolyte beyond the extent of the anode and cathode. It is very difficult to obtain adequate deposits on the inside of articles or on surfaces facing away from the anode. These considerations may also effect the quality of the deposit if there is only a limited range of acceptable cathode current densities.

Fortunately there may be considerable polarisation at the workpiece, particularly with complex electrolytes, which can be comparable with the voltage drop in the electrolyte. This will tend to even out the distribution of metal. The property of producing an even distribution of metal on an object of irregular shape is known as *throwing power*. It depends not only on the solution but on the operating conditions. The best throwing power is obtained with a solution of high conductivity operated so that the slope of the polarisation curve (dE/di) is large.

Various methods can be used to measure throwing power, the simplest is to plate on to a bent strip cathode of known dimensions, which will give a qualitative comparison between different baths. Alternatively a Hull cell can be used in which a trapezoidal cell

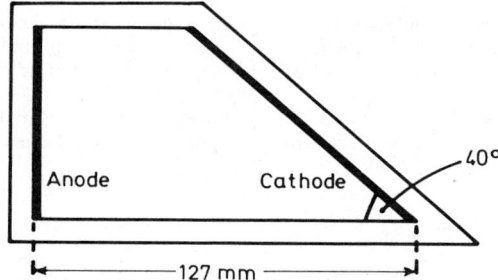

Figure 6.13. The 267 ml Hull cell (plan)

contains a cathode inclined at a known angle to the anode (Figure 6.13), again a qualitative test. The first definition of throwing power was made with the aid of the Haring cell, a standard rectangular cell with a cathode at either end and a gauze anode placed much nearer to one cathode than the other (Figure 6.14). The cathodes are connected together and plated, more metal being deposited on the nearer cathode. If the ratio of the lengths of the two halves of the cell is L, throwing power was defined as $100(L-M)L\%$, where M is the ratio of the masses of metal deposited at the two cathodes. L is usually set at five. There are good theoretical reasons for this derivation, but it has a disadvantage that for perfect throwing power, that is, $M=1$, TP$=80\%$. The British Standards Institution adopted a more practical definition, $100(L-M)/(L+M-2)$. Here, when $M = 1$, TP $= 100\%$; if M $= 5$, TP $= 0$; or for M greater than five TP is negative.

This quantity can be determined for any plating solution and set of plating conditions by weighing the cathodes before and after plating for a fixed length of time. The ratio of the two currents can of course be determined rapidly by direct measurement (with suitable equipment), or by potential measurements in the solution, but this has the

$$L = l_2/l_1$$

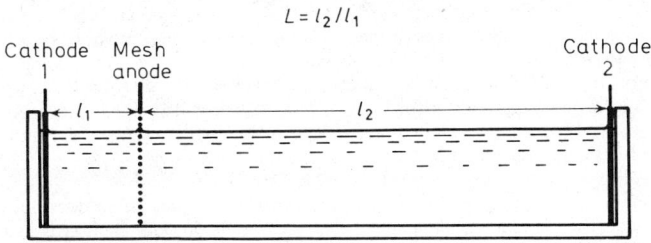

Cathode 1   Mesh anode   Cathode 2

$\leftarrow l_1 \rightarrow$ $\longleftarrow \quad l_2 \quad \longrightarrow$

Figure 6.14. The Haring cell (longitudinal section). The wetted area of the electrodes is usually 50 mm square

disadvantage that variations of current efficiency with current density are not taken into account. For instance, at low current densities in an acid copper plating bath no hydrogen is evolved, but much may be evolved at high current density; conversely with Watt's nickel bath, at low current density mainly hydrogen evolution will occur.

The effects of poor throwing power can often be offset by careful distribution of anodes so that all points on the workpiece are roughly equidistant from one or other anode. If there are projections or corners which would attract too high a current, leading to poor deposits, a shield of non-conducting material may be used to block the current flow, or an electrode connected to the cathode can be used to draw current away. This is known as a *robber* or *thief*.

COVERING POWER

This is an effect of current distribution, as is throwing power, but refers to the extreme case in which no metal is deposited in low current density areas. It is a property not only of the electrolyte and plating conditions, but also of the substrate material. Poor covering power is obtained on materials bearing oxide films, where the voltage drop across the film is large enough to make initiation of a metal deposit difficult.

It is particularly difficult with aluminium and nickel alloys. One way of ensuring good covering is initially to pass a very high current to ensure that the potential at all points on the workpiece is sufficiently cathodic to start deposition; further growth can then proceed on this metal film under normal conditions. An alternative method is to pre-treat the workpiece to remove the oxide film, then either to plate at once, or to give a quick flash plating with a metal giving good covering power.

The simplest way of assessing covering power is to use a Hull cell with a cathode of the appropriate material.

Vat plating

The basic equipment required for vat plating is a tank of suitable size, anodes, which may be consumable or inert, and a supply of direct current. The tank must resist the corrosive action of the solution, and it is usually constructed of a cheap material such as steel or timber, with an inert lining of lead, rubber or plastic.

ANODES

Consumable anodes. These are usually cast or rolled from the metal to be plated. The purity necessary varies from metal to metal, and additions may be made to decrease anode polarisation. It is particularly important that the anodes should be free from insoluble inclusions that may become suspended in the electrolyte. In some cases the anode may consist of small pieces of metal contained in an inert metal basket.

Because there are usually some insoluble particles in the anodes, and metal grains may drop out of an anode if the grain boundaries are preferentially dissolved, it is preferable that the anodes be contained in fine-woven bags which will prevent these getting into the solution and becoming occluded in the deposit. The bags may be of cotton twill, but synthetic fibres are preferred for their better chemical resistance.

Inert anodes. Inert anodes are preferred if the metal to be deposited does not dissolve readily, or in the correct valence state. In such cases the anode reaction is usually the evolution of oxygen, although it is also possible for species in the solution to be oxidised. Inert anodes may be used in conjunction with soluble anodes to prevent a build-up of metal in the solution, if the current efficiency at the anode is greater than that at the cathode.

JIGGING

The simplest way of supporting the objects to be plated is to suspend them by copper wire. This is adequate for one-off jobs, but for routine production, jigs are used. These are frames onto which the objects are wired or hung, and are coated with insulating plastic, wax or lacquer except for the points of contact. This facilitates handling of large numbers of small objects as they are transferred from tank to tank. The work must be fixed on the racks so that all pieces receive an equal share of the current.

With irregularly-shaped objects it is often possible to devise an arrangement in which neighbouring pieces draw current away from projections on each other, and so produce a more nearly uniform current distribution (Figure 6.15).

TEMPERATURE CONTROL

Many plating baths are operated at elevated temperatures to give higher rates of deposition, but for reproducible results, the temperature must be carefully controlled. Thus provision must be made for controlled heating of the bath. This may be achieved by the provision of steam coils, or electrical immersion heaters, or by circulating the solution through an external heat exchanger. Alternatively, the bath may be enclosed in a water jacket heated in the above manner or by steam injection or gas burners.

Again, during high current operation, cooling may sometimes be needed. Coils carrying recirculating cooling water will usually suffice, but plate-type radiators or block-types heat exchangers may be necessary for large tanks. Most conventional types of thermostat can be used to control the heat balance. For work at elevated temperatures, lagging the plating tanks to reduce heat loss will often prove to be worthwhile.

FILTRATION

Despite precautions to keep particles from getting into the baths, after lengthy operation sludge is bound to accumulate. To maintain optimum performance, particularly with bright deposits, it is necessary to remove all such solids by filtration.

The simplest method is gravity filtration, in which the solution is raised by an air-lift into a filter bag above the tank, and allowed to drain back slowly into the tank. This method allows only limited throughput and for rigorous filtration a pressure filter must be used. Particles of sizes down to 1 μm can be removed by this means. Activated carbon may be added to the filter to remove organic impurities, particularly when these are

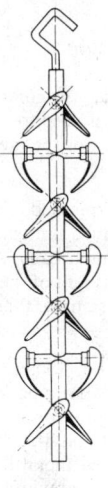

Figure 6.15. Jigging of irregularly-shaped objects to give more uniform current distribution and minimise deposit build-up

generated by breakdown of addition agents. It is often convenient to include other components in such a circuit, for example heat exchangers, or facilities for conditioning or purifying the solution.

AGITATION

Various problems can arise if electroplating solutions are not agitated. Hydrogen bubbles may cling to the cathode, becoming occluded in the deposit and leaving a pitted surface, the metal-rich solution at the anode may sink to the bottom of the tank and form a discrete layer, local pH changes may occur at the cathode, and the maximum operating c.d. will be decreased. For these reasons, it is important that the solution is thoroughly agitated, and the simplest, most widely used method is by air bubbles from a compressed air line in the bottom of the tank. These bubbles should be diverted towards the cathode or disturbance in the solution will help sediment to penetrate the anode bags. This method should not be used when a wetting agent is included in the bath formulation, or excessive frothing may occur.

An alternative method is to move the work back and forth by reciprocating the cathode bar with a motor-driven crank. In automatic plant in which the work is moved along the tank by conveyor belt the motion will serve to stir the solution around the cathodes.

FUME EXTRACTION

With tanks operating at high temperature, or where there is a lot of gas evolved, fumes can constitute a health risk and lead to serious corrosion of overhead steelwork, particularly with strongly acid solutions, such as the chromic acid bath for chromium plating.

To minimise fuming, hollow balls of plastic or glass may be floated on the top of the solution; these part easily to admit work but condense vapour as it tries to leave the surface. This can also substantially reduce heat loss. Fume extraction must also be used, and in most cases ducts are fixed around the lip of the tank to draw fume off directly.

SOLUTION CONTROL

Specific Gravity.   Regular measurement of the specific gravity of the plating solution, by hydrometer, is used to monitor the solute content of the bath, which may be used as a convenient index of the metal concentration.

Acidity.   As explained above it is usually important to maintain the solution pH within well-defined limits and it must be checked regularly, by measurement with a pH meter or colorimetrically. Adjustments are usually made with free acid or a basic salt of the metal constituent of the bath in acid solutions, or with alkali. Corrections are usually made on a test sample, then the addition needed to the tank is calculated.

Soluble Impurities.   There are many ways in which the plating bath can become contaminated over a long period of operation. There may be dissolution of the workpiece, particularly on interior surfaces, or work may be dislodged from the jig and fall to the bottom of the tank. If possible this should be recovered at once. There may also be attack on exposed metal in the equipment, and there will be *drag-out,* that is solution carried over from previous stages of the plating operation. Drag out can be minimised by careful jigging to ensure that solution will not be retained in hollows, and by thorough rinsing. The dissolution of work will be minimised if the current is switched on before loading the tank, and active metals such as zinc die-castings should always be preplated from a *strike bath* before loading them into an acid solution.

It is far easier to prevent contamination than to cure it. If, however, the bath begins to yield poor deposits (they may become dull or burnt) the solution must either be purified or replaced. In any event, the solution should be analysed from time to time, or the deposits produced may have poor corrosion resistance or poor mechanical properties as a result of codeposition of impurities. Some metals can be removed adequately by *dummying* the bath, that is plating out at low current density onto dummy cathodes. Others, depending on the type of plating solution, can often be precipitated by adjusting pH. The filter circuits can conveniently be used to run off the solution into a treatment tank, and will then remove any precipitate or suspension as the electrolyte is returned.

Organic contaminants.   Organic contamination can only be detected by test plating, or by recognising the onset of deterioration in deposit quality. Filtration over active carbon will help to minimise contamination, but when it becomes excessive it is normal practice to oxidise organic material in the bath, for instance with H_2O_2, then absorb the products on charcoal. Any addition agents must then be replenished.

AUTOMATIC PLATING LINES

An automatic line will consist of a series of tanks for the various stages of the pretreatment and plating processes with a conveyor belt to transfer racks of work from tank to tank. The work may be moved along the tanks by conveyor during the plating operation, or the conveyor may be arranged to set down work into a tank for a specified period and then to collect it again for transfer to the next stage of the process.

During routine operation, the timing of the process and the current settings can be controlled automatically.

ELECTRICAL POWER SUPPLIES

Early plating shops used d.c. generators as power sources, but modern rectifiers have superseded these. Metal-plate rectifiers can be used but germanium, or particularly silicon, semi-conductor rectifiers have greater power capacities. The voltage of the supply is usually regulated by an autotransformer before rectification, but if one unit is supplying electricity to several tanks, then individual resistance boards must be used to regulate the current to each tank.

Constant current density operation.   Units are now available which can be set to give a constant plating current density independent of the size of the load in the tank; in fact these rely on operating at a fixed voltage and unless the anode polarisation is small and constant, problems may be encountered. It may be that by using a reference electrode and fixing the *potential* of the workpiece that constant current density operation may find wider application.

Barrel plating

With small articles it is both laborious and expensive to mount them individually on jigs, and to avoid this the barrel plating process has been devised. The barrel, which may be horizontal or tilted, is filled with the work which is then tumbled over and over. Contacts connected to the negative supply touch some of the pieces in the barrel and many other pieces will make contact indirectly. Not all the work will be plating at any one time but as the load is moved about metal will be deposited over all the work. The rubbing together of the objects also helps to burnish the deposit giving a good quality finish.

There are two main types of barrels, closed ones which are tilted and actually hold the solution, and mesh or perforated barrels which are suspended in plating tanks with the anodes outside. In the closed type the anode is fixed inside the barrel and contact with the load made with studs on the inside of the barrel. Generally, best results can be obtained with the perforated, horizontal, fully immersed barrel and this has the advantage that the work can be transferred automatically inside the barrel through all the stages of the plating process. Contact is made through *danglers*, wires or chains hanging down from the spindles.

There are difficulties with barrel plating, first if the objects are of irregular shape they may become entangled (or flat objects may stick together), so that an even deposit cannot be produced, also there is difficulty in determining the effective area of load. Care must be taken that the barrel is not over- or under-loaded, and experience is needed to judge the correct amount of any particular article.

With barrel plating higher voltage power supplies are needed to maintain the same metal deposition rates.

Specific metal deposits

COPPER

Acid copper solutions.   Copper can readily be deposited as a satisfactory coating from acidified solutions of many of its salts; cupric sulphate is most commonly used but the fluoborate is more soluble and allows faster plating rates. A typical formulation for an acid copper sulphate solution is:

$$CuSO_4 . 5H_2O \qquad 200$$
$$H_2SO_4 \qquad 75 \text{ gram litre}^{-1}$$

In the temperature range 18–60°C this bath can be operated at up to 2000 A m^{-2}.
The fluoborate bath contains:

$$Cu(BF_4)_2 \qquad 225\text{--}450$$
$$HBF_4 \qquad 15\text{--}30$$
$$H_3BO_3 \qquad 15\text{--}30 \text{ gram litre}^{-1}$$

At 40–50°C this can be operated at up to 4000 A m^{-2}.

Thick deposits from copper acid solutions tend to be rough and coarse-grained, but addition agents such as glue, gelatin, thiourea and many others can be used to improve this. Acid solutions are widely used for electro-forming and electro-typing.

Cyanide copper baths.   With acid copper solutions many metals, particularly steel, form loose immersion deposits by a displacement reaction, but by complexing the cuprous ions with cyanide this can be prevented. Cyanide copper gives good fine-grained deposits and operates with high throwing power; in particular it can be used as a strike bath to prevent immersion deposits on steel or zinc die-castings in other plating solutions, and to limit corrosion of the workpiece.

A typical cyanide copper solution contains:

$$CuCN \qquad 25$$
$$NaCN \qquad 34$$
$$Na_2CO_3 \qquad 15 \text{ gram litre}^{-1}$$

This is operated at high temperature (60–80°C) at up to 1000 A m^{-2}. For strike plating, special baths containing higher cyanide concentrations are needed to prevent immersion deposits; alternatively Rochelle salt baths containing tartrates may be used.

Pyrophosphate copper.   Cyanide baths are, of course, more hazardous to operate than simple acid solutions, and the pyrophosphate solution offers a less toxic alternative. It gives fine-grained deposits but is not suitable as a strike bath. One formula is:

$$Cu_2P_2O_7 . 3H_2O \qquad 110$$
$$K_4P_2O_7 . 1OH_2O \qquad 400$$
$$\text{Ammonia } (0.880) \qquad 3 \text{ gram litre}^{-1}$$

at 50–60°C this gives plating rates up to 8000 A m^{-2}. An addition of 5–10 gram litre^{-1} of nitrate ion is usually made to suppress hydrogen evolution.

Electroless copper.   In electroless plating, metal ions are deposited by the action of a reducing agent in solution instead of an electric current. Strictly speaking those solutions which can plate out onto any surface, e.g. silver-mirror solutions, or those forming immersion deposits by reaction with the substrate are not included, but only those where the metal surface serves as a catalyst for the electrochemical reaction. It is difficult to obtain satisfactory copper deposits of any great thickness by this method, but it is widely used to prepare non-conductors for electroplating, particularly in the electronics industry.

Solutions are usually based on a Rochelle salt formulation with formaldehyde as the reducing agent. Stabilisers such as EDTA are added but even so solutions should be mixed just before use.

NICKEL

The basis of the nickel plating industry is the traditional Watt's nickel bath from which most of the bright and semi-bright solutions are derived. The basic formula is:

$$NiSO_4 \cdot 7H_2O \qquad 240-340$$
$$NiCl_2 \cdot 6H_2O \qquad 30-60$$
$$H_3BO_3 \qquad 30-40 \text{ gram litre}^{-1}.$$

The chlorides may be added as an alkali metal salt but the nickel salt is preferred. The bath can be operated over a range of conditions but pH must be rigorously maintained between five and six. The maximum permissible current will depend on operating temperature, but typical operating conditions are 600 A m^{-2} and 50 °C. The physical properties of the deposit can be varied by adjusting the sulphate:chloride ratio and variants of the bath are used for special purposes. A slightly different bath for electrotyping uses ammonium chloride both to provide chloride ions and to buffer pH.

Addition agents.   Wetting agents can be used to ensure that hydrogen bubbles do not stick to the cathode and form pits. These should be used sparingly as they lead to loss in deposit ductility.

Hydrogen peroxide may be added to the solution to prevent gas pitting by oxidising hydrogen back to water. It must be replenished daily and cannot be used with brighteners which are degraded by it.

When nickel is used as a substrate for decorative chromium it is important that it be able to mask scratches in the underlying surface. Coumarin and acetylenic alcohols are widely used to improve the levelling power of Watt's nickel bath and many other types of compound can act as levellers. Proprietary levellers containing different mixtures of individual agents are claimed to give superior results.

If cobalt ions are added to the Watt's solution the deposit, which contains some cobalt, is found to be semi-bright, but to obtain a fully-bright nickel plate organic addition agents must be used. Brighteners have been classified into two types. Brighteners of the first type can be characterised by the group $=C-SO_2^-$, for instance sulphonic acids, sulphonates or sulphonamides, and these can produce brightness without causing any brittleness in the deposit. Brighteners of the second type cannot be used alone or they would produce brittle deposits, however, they can be used together with brighteners of the first type. They are characterised by C—C, C—N, C—O, C—S or N—N double or triple bonds. Common types are aldehydes, ketones and acetylene compounds, and safranine and fuchsin are also well known examples.

Solution maintenance.   Because nickel plates out at a high overpotential, impurities will tend to be codeposited with it. For this reason it is important to prevent contamination, and to purify the solution if poor deposits are produced. Copper can be removed by displacement with nickel powder, but iron must be precipitated as hydroxide by raising the pH to about 5·5 in a treatment tank and allowing the sludge to settle. Organic impurities are normally oxidised with H_2O_2 and adsorbed onto charcoal. This also removes addition agents, which must be replenished.

Nickel anodes.   Nickel anodes should be at least 99% pure to prevent grain boundary attack and to ensure even dissolution. They may be cast or rolled. Pure nickel tends to passivate under anodic polarisation so that various additions are needed to prevent this. *The depolarised* nickel anodes containing NiO, and carbon-nickel anodes resist passivation in low pH solutions, but suffer grain boundary attack at high chloride concentrations. Sulphur is very beneficial, and can be introduced into electrolytic nickel anodes.

Small pieces cut from electrolytic nickel sheet may be used as anodes, by containing them in titanium anode baskets.

Sulphamate nickel. Nickel sulphamate has a very high solubility especially at elevated temperatures so that it is an ideal basis for a high speed plating bath. It is fortunate that deposits from nickel sulphamate solutions, even when plated at high current densities have exceptional ductility, giving a bath which is excellent for applications such as electroforming, where high plating speed is essential. A typical nickel sulphamate solution is:

$$Ni(NH_2SO_3)_2 \qquad 400$$
$$NiCl_2 \cdot 6H_2O \qquad 10$$
$$H_3BO_3 \qquad 30\text{--}40 \text{ gram litre}^{-1}.$$

this can be used at pH 3–4 and gives current densities of up to 5000 A m^{-2} at between 30 and 50 °C.

Electroless nickel. This process has had great success and in many applications can compete with electroplated nickel. In particular it has very high throwing power, depositing well wherever solution reaches the metal.

The best reducing agent for nickel ions is the hypophosphite ion, the electrochemical reactions being:

$$Ni^{++} + 2e^- \rightarrow Ni,$$
$$H_2PO_2^- + H_2O \rightarrow H_2PO_3^- + 2H^+ + 2e^-.$$

Solutions contain 20–30 gram litre^{-1} of $NiCl_2 \cdot 6H_2O$ with 10–30 gram litre^{-1} of $NaH_2PO_2 \cdot H_2O$ and an addition of either an organic acid, e.g. hydroxyacetic or lactic, to buffer the pH and complex the nickel, or ammonia and sodium citrate sometimes with pyrophosphate ions. Both types of bath give deposits containing 4–8% of phosphorous at rates of 10–30 μm h^{-1}.

With the acidic solutions deposition rates rise rapidly with temperature, usually kept between 90–100 °C, and also with agitation. The level of hypophosphite is not critical. To stabilise the solutions or reduce the plating rate additions of cadmium, lead or thiocyanate ions may be made. Fluoride ion or acetic or propionic acids, on the other hand, act as accelerators.

In ammoniacal solutions the effect of agitation is negligible but the concentration of hypophosphate ion needs to be carefully controlled.

Electroless deposits form on many metals spontaneously, but on others, for instance, copper, need to be initiated by making workpieces cathodic for an instant. The alternative is to activate the work by dipping it in palladium chloride solution immediately before plating. For non-metals, proper activation must be carried out by immersing in a reducing agent such as acid stannous chloride before dipping in the palladium solution.

CHROMIUM

Chromium is used in great quantities as a decorative coating and the only solution in general use is based on chromic acid. Pure chromic acid alone will not yield deposits and additions of small quantities of catalyst are needed. The best catalysts are sulphate and fluosilicate ions which are added in the ratio of about 1 part to 100 of the chromate ions. Very strong solutions containing 200–400 gram litre^{-1} Cr_2O_3 give good deposition rates, but normally operation is governed by the conditions which lead to a bright deposit. At 25 °C there is only a very small range of current density which will give brightness, but at 50°C bright deposits can be obtained from 500–2500 A m^{-2}. Because current density varies over the work-piece the most practical temperature for plating is 40–45 °C which gives a useful range of 300–1500 A m^{-2}. Bright deposits cannot be obtained at higher temperatures. Normally baths will operate at a fairly low current efficiency (10–25%), and current efficiency will fall as chromic acid concentration is increased. Maximum efficiency is found at about 75 gram litre^{-1} Cr_2O_3. Deposits of chromium from chromic acid baths although hard and very corrosion resistant are often brittle and highly stressed, leading to patterns of fine cracks in the coating.

Anodes.   Chromium anodes are not suitable, and inert lead-tin or lead-antimony alloy anodes must be used. During correct operation a chocolate-coloured film of lead peroxide forms on the anodes, and this ensures that all chromic ions are oxidised to chromate. It is advisable to run new anodes at maximum current to form the film and to ensure that anodes are not left idle for long periods or the film will be converted to yellow lead chromate, preventing proper operation.

Fume.   Because of the heavy gassing at both anode and cathode, a spray of chromic acid is produced which must be eliminated. A blanket of plastic balls on the electrolyte will help to reduce spray (and heat loss), but good fume extraction is essential.

Self-regulating high speed baths.   The control of catalyst levels in chromium plating baths is of major importance; however, if the catalyst is added as an excess of sparingly soluble salts, the concentration of dissolved catalyst will be maintained constant. This procedure allows mixtures of catalysts giving optimum performance to be used, without the difficulties of constant solution maintenance. Such formulae can give crack-free deposits at high temperature.

Hard chromium plating.   Chromium gives a hard corrosion-resistant coating which is excellent for engineering purposes. It can be used to render surfaces of tools, dies and gauges wear-resistant or for building up worn parts. It is best used on hard steels since soft substrates allow the electroplate to become damaged. Accurate hard chromium plating calls for special skills from the electroplater.

Trivalent chromium plating solutions. The conventional chromic acid plating solution has several disadvantages, notably toxicity and low current efficiency. These can be overcome with a Cr^{3+} solution. This can be less acid, giving less hydrogen evolution, and in any event only half the current is required for an equivalent amount of chromium deposition. A successful solution has recently been devised using a mixture of water and dimethyl-formamide as solvent, containing 220 gram litre^{-1} of $CrCl_2 \cdot 6H_2O$ and small quantities of NH_4Cl, NaCl and boric acid. This can operate at 2500 A m^{-2} at room temperature and a pH of 1·3, giving a current efficiency of 35%, an improvement of four to five times over deposition from chromic acid. The deposits are micro-discontinuous but darker in colour than conventional chromium. It appears that this solution will find use as a decorative coating, but cracks too readily for engineering applications.

NICKEL-CHROMIUM DECORATIVE PLATING

Because chromium deposits usually contain cracks, or being brittle can become damaged easily, chromium alone does not constitute a satisfactory protective coating. On a base metal such as steel or zinc alloy the corrosive attack at defects in the electroplate would be greatly accelerated by galvanic coupling leading to serious pitting. To prevent this chromium must have an undercoat, generally of nickel or of copper plus nickel. Even when coupled with chromium the rate of attack on pure nickel is quite small and the corrosion product is not as unsightly as that of iron. With bright or semi-bright nickel the corrosion resistance is less good because of the sulphur in the deposit, but this property can be used with benefit if a *duplex* nickel coating is formed, with bright nickel on top of semi-bright nickel. Penetration of the semi-bright nickel is suppressed because preferential attack on the bright nickel affords a measure of cathodic protection. This difference in performance is reflected in the requirements of BS 1224[2].

Penetration of the nickel undercoat can also be delayed by the use of special chromium plating. If the crack pattern is modified to give more smaller cracks, the so-called micro-cracked chromium will give longer life. This change in crack pattern is normally induced by plating two layers of chromium, the stresses in the first layer inducing the micro-crack pattern to form in the second. It may also be induced by the post-nickel strike technique where an extra strike of nickel induces the fine crack pattern.

An alternative to micro-cracked chromium is micro-porous chromium obtained by plating on top of the normal nickel undercoat a layer of nickel containing 1–5 μm particles of a non-conducting solid. The chromium does not deposit on the solid particles, and so small holes are left in the chromium deposit which prevents the onset of cracking and delocalises the corrosive attack.

In all cases the chromium should be applied to the nickel undercoat as soon as possible after its deposition, and in normal practice the two operations occur in the same plating line. If for any reason there is a delay, the nickel will passivate and must be reactivated by a suitable pickle such as is used for preparing nickel alloys for plating, to secure good adhesion.

Regardless of service conditions, BS 1224[2] specifies that thickness of chromium shall be >0·3 μm for regular or microporous chromium or >0·8 μm for crack free or micro-cracked chromium. Microporous chromium should have >10 000 pores per 100 mm^2 with no pores visible to the naked eye, and microcracked chromium should have not less than 250 cracks per 10 mm in a closed network. The type of service environment determines the minimum thickness of nickel undercoat, 40 μm of duplex or semi-bright nickel being required on steel for severe environments with regular chromium, but only 30 μm beneath microdiscontinuous chromium. For mild conditions, 10 μm nickel suffices. Substrates such as copper alloys and zinc-base alloys require rather thinner nickel coatings, but aluminium alloys require the same thickness as steel substrates, and in addition crack-free chromium is not permitted.

ZINC

Zinc is one of the best protective coatings for steel because of its ability to provide cathodic protection to the steel at discontinuities in the coating. Zinc can be applied to steel by hot-dip galvanising, but electrogalvanising can give thinner, more consistent coatings and is superior and cheaper for many applications. The mean thickness of zinc specified by BS 1706[3], varies from 8 μm to 37 μm, depending on conditions.

Acid zinc plating.   Adequate deposits can be obtained from a simple solution of zinc sulphate, and although quality and throwing power are not good the process is cheap and useful for thicker coatings. The solutions contain 200–400 gram litre^{-1} $ZnSO_4 . 7H_2O$ usually with chlorides added ($ZnCl_2$, NaCl and NH_4Cl). Sodium acetate or $AlCl_3$ are added as buffers to give a pH of about 4. This solution, operating at room temperature, can be used at current densities up to 1000 A m^{-2}. The quality of the electroplated metal may be improved by additions of dextrin, gelatin, β-naphthol or a wide variety of other addition agents.

With plating on moving strips very high current densities are obtained and a special all-chloride solution is often used at up to 5000 A m^{-2}. This solution contains 250 gram litre^{-1} $ZnCl_2$ plus NH_4Cl and is used at elevated temperatures.

Zinc anodes can be used although impurities in solution will often plate out while the tank is idle and must be removed. Silver-lead anodes are also suitable.

Cyanide zinc.   This plating bath contains zinc both as sodium zinc cyanide complex and as zincate ion. One formula is:

$Zn(CN)_2$	60
NaCN	60–140
NaOH	70–120 gram litre^{-1},

This is operated at 250–600 A m^{-2} at 20–35°C, and has much better throwing power than acid zinc solutions. It also produces a more attractive deposit which can be brightened by the use of addition agents.

CADMIUM

Cadmium, like zinc, can give good protection of steel by cathodic protection and itself has excellent tarnish resistance. However, because of its greater expense it is only used instead of zinc for special applications such as aircraft parts, or anywhere where corrosion resistance is of paramount importance. The plating process is similar to that for zinc, using a simple cyanide solution of the formula:

CdO	23–35
NaCN	90–120 gram litre^{-1},

which will operate up to 500 A m^{-2} at about 30°C. Organic brighteners are usually added. Anodes of cobalt (99·9% pure) or steel can be used.

With the large amounts of hydrogen liberated from these cyanide solutions, parts made from high-strength steel may suffer from hydrogen embrittlement after plating and must be annealed to allow hydrogen to diffuse out. Even so, it is difficult to recover the full strength of the metal and for this application it is advisable to use an acid cadmium plating bath, which liberates less hydrogen. The following fluoborate solution is preferred:

$Cd(BF_4)_2$	150–300
NH_4BF_4	60–120 gram litre^{-1},

which can plate at 600 A m^{-2} at 25°C. BS 1706[3] specifies a mean thickness of 6 μm for interior use, increasing to 13 μm for marine environments.

TIN

Tin is widely used as a protective coating for steel, despite the fact that no protection is conferred on the steel at discontinuities except in the case of tin cans containing fruit, in which tin can act as a sacrificial anode due to complexing by the organic acids. Electro-tinning has now largely superseded the hot-dipping process.

For utensils for food or water, BS 1872[4] specifies a minimum local thickness of 30 μm of >99·5% tin. For other applications, recommended thicknesses vary between 5 and 15 μm on copper alloys and between 10 and 20 μm on steel substrates, depending on severity of conditions. However, for electro-tin plate the specifications are much less severe, and BS 2920[5] indicates thicknesses between 4·9 and 25·8 g m^{-2} (0·4–2μm) depending on end use.

Alkaline tin bath. Tin is deposited from an alkaline stannate solution, the potassium salt being preferred for its greater solubility, conductivity and current efficiency. High or low strength solutions may be used.

	A *(low strength)*	B *(high strength)*
$K_2SnO_3 . 3H_2O$	100	420
KOH	15	22 gram litre^{-1}
Max c.d.	1000	4000 A m^{-2}

Both are used hot at 70–90°C and pH $\sim$ 13. It is important that the concentration of stannite ion is kept low or spongy deposits will result, and H_2O_2 may be added to correct this, but if tin anodes are used it is more important to ensure proper anode operation. The number of anodes must be small enough to ensure a high enough current density to produce all stannite ion, but if the c.d. becomes too high passivity of the anodes will ensue, leading to oxygen evolution.

Normally, the current density is raised till the onset of passivity, then lowered to the correct working level (200–300 A m^{-2}), when the anode forms a visible green/yellow film. If an anode makes poor contact with the anode rail it can idle and introduce stannite into the bath. Anodes should never be left standing in the solution. Anodes containing 1% aluminium (high speed anodes) may be used; these can sustain nearly double the anode current density of ordinary tin anodes. The use of steel anodes eliminates the problem of stannite formation.

Acid tin plating. This solution has poor throwing and covering power and requires addition agents to give acceptable deposits; nevertheless it is adequate for routine electro-tinning and gives good current efficiency. Tin anodes can be used without difficulty. A typical formula is: $SnSO_4$ 50–100, H_2SO_4 50 to 100 gram litre^{-1}. Addition agents may be gelatin, naphthols, cresols or sulphonic acids, and help to refine grain size and prevent dendritic growth. The bath is used at room temperature without air agitation to minimise the oxidation of stannous ion to stannic ion, and at a current density of 100–1000 A m^{-2}.

The plating rate can be improved by using tin fluoborate solutions which can contain up to 80 gram litre^{-1} of tin.

Flow-Brightening of tin coatings. To improve the appearance and integrity of thin electro-deposited films, tin coatings may be rapidly melted to give a bright level surface on solidification. This is normally achieved by immersion in hot palm oil or tallow at 240°C with free organic acids as flux, but radiant or induction-heating may also be used. For good results the tin must be clean and freshly plated, and its thickness must lie between 2 and 7 μm.

IRON

Iron is not used as a surface coating but it can be used as a cheap material for building up worn parts or for electro-forming. It can be deposited easily from acid solutions of ferrous salts, but ferric salts lead to poor deposits.

GOLD

Besides its decorative application in jewellery, gold plating is widely used for high quality electronic parts, particularly for contact areas of switches or connectors. It exhibits good resistance to tarnishing and has excellent solderability. Plating baths are mostly based on cyanide solutions, such as the two examples below, used for thin flash deposits or for heavier plate.

	Flash deposits	*Heavy plate*
Au (as $KAu(CN)_2$)	2	8
free KCN	15	20
K_2HPO_4	15	20
K_2CO_3	—	20 gram litre^{-1}.

These solutions are used hot at 55–70°C with stainless steel anodes and operate at current densities between 10 and 100 A m^{-2}. Barrel plating solutions use lower gold content. The pH is buffered at 11 to 11·5 by the phosphate. For high speed plating addition of 200 gram litre^{-1} $K_4Fe(CN)_6$ allows gold concentrations up to 22 gram litre^{-1}. For electrical purposes, BS 4292[6] specifies deposits of 0·5 to 20 μm of gold (>90% Au) on an 8 μm undercoat of copper or nickel.

For use in jewellery the majority of gold is not plated pure but with additions of baser metals, and in decorative plating good colour matching is essential. Nickel, cobalt, cadmium, silver, copper or zinc can all be added to the cyanide baths as cyanides, or tin may be added as stannate. BS 4292 allows up to 37·5% of other metals (15 carat gold), and coatings in the range 4–16 μm are recommended.

Bright acid gold. To avoid the use of cyanide, various processes have been devised. Gold may be deposited from a hot solution containing 75 to 125 millilitre litre^{-1} HCl and 15–40 gram litre^{-1} of gold (as chloride), or various proprietary solutions have been formulated with gold complexed by organic acids. Sulphite solutions are now also used and offer excellent performance with suitable brighteners. These solutions all require gold anodes.

SILVER

Silver, like gold, is used mainly for decoration or in electronics but can be also used as a bearing material. It is normally plated from a cyanide bath, despite formulations for non-cyanide solutions. Typically this bath contains about 30 gram litre^{-1} of silver, 30–45 gram litre^{-1} KCN and 30–90 gram litre^{-1} K_2CO_3. High speed baths contain more KCN and up to 100 gram litre^{-1} of KNO_3. These solutions can be used at 25°C at up to 500 A m^{-2}. Pure silver anodes (>99·9%) should be used. BS 4290[7] covers deposits between 2 and 50 μm thick.

Silver can form immersion deposits on many metals and this must be prevented either by flash plating from a high cyanide solution, or by amalgamating (*quicking*) the substrate by dipping in a solution containing a mercuric salt.

Sulphur-containing brighteners are very effective for silver and thiourea is widely used. The traditional brightener is prepared by mixing a portion of plating solution with CS_2 for several days then adding this mixture to the bath.

THE PLATINUM METALS

With the exception of rhodium, these metals are only soluble as complexes and therefore tend to be difficult to plate despite their noble character. Many solutions have to be strongly acid to prevent hydrolysis and the main product at the cathode is hydrogen, giving poor current efficiency. Rhodium alone is easily deposited from simple salt solutions. A solution containing 2 gram litre^{-1} rhodium (as sulphate or phosphate) plus 20 millilitre litre^{-1} conc. H_2SO_4 or 40 millilitre litre^{-1} H_3PO_4 can be operated at 40–45°C and 100–1000 A m^{-2}. These processes are widely used in jewellery and for searchlight reflectors to give enhanced tarnish resistance. Platinum and palladium are also plated commercially, commonly from ammine- or chloro-complexes using platinum anodes. A typical bath is:

$Pt(NH_3)_2(NO_2)_2$	16·5
NH_4NO_3	100
$NaNO_2$	10 gram litre^{-1}
NH_4OH (28%)	50 millilitre litre^{-1};

the diammine complex is dissolved in the ammonia to form a tetrammine in solution, $Pt(NH_3)_4(NO_2)_2$. The bath plates at 40–100 A m^{-2} near boiling point. Alternatively:

Pd (as H_2PdCl_4)	50
NH_4Cl	20–50 gram litre^{-1}
HCl	50 millilitre litre^{-1}

This operates at 50°C, again at only 100 A m^{-2}.

The platinum group metals find their main application in plating jewellery or electronic components, which may be plated not by electrolysis but by charging with hydrogen before dipping into the solution. Because of the high cost of materials it is important that pore free deposits be produced even at very low thicknesses and careful specimen preparation and solution maintenance are essential.

There is much interest in the development of solutions for the other, cheaper platinum metals.

Deposition of alloys

Alloys are widely used in metallurgy for many applications. Many of these alloys have been prepared successfully as electroplated coatings, but the majority are little used in practice. One important reason for this is that control of deposit quality and composition is too difficult on all but the simplest shapes, and the benefits gained are not always economically worthwhile. Again because of the capital and space required to set up any particular plating line, proliferation of plating systems is not encouraged.

A number of alloy plating baths do however find limited use; brass is used for promoting rubber-to-metal adhesion or for decoration and corrosion resistance, as is bronze. Tin-zinc alloy gives a protective coating for steel deriving good properties from both the parent metals, and tin-nickel alloy shows extraordinary resistance to severe corrosion environments. Thin alloy films containing iron, nickel and cobalt are used in computer technology for their magnetic properties.

Solution control.   To control the composition of the deposit, not only must the ratio of metals in the bath be maintained, but the concentration of complexing species must also be kept constant.

Anodes.   Anodes may be inert, of the constituent metals or of the alloy. If alloy anodes are used they should have the same ratio of constituents as the deposit, not the bath. This may of course pose difficulties; with two-phase anodes, one phase may dissolve preferentially, or an intermetallic compound may remain unattacked. Pure metal anodes may be used if the areas of the two types of anode are suitably adjusted, or if the current ratio is regulated by using separate supplies. If inert anodes are used, chemical control of the solution becomes more difficult.

Plating parameters. If current density is increased, the concentration of less noble metal will be increased, whereas agitation tends to increase that of the more noble metal. The effect of temperature varies from bath to bath.

Throwing power. On all but the simplest shapes cathode current density must vary from point to point and with it the composition of the alloy. In some cases this can affect colour or decrease corrosion resistance. It is apparent therefore that the throwing power of the alloy plating bath is of the greatest importance. It is fortunate that most of these baths contain complex ions which in general give better throwing power than simple salts.

BRASS

Brass plating is widely used as a decorative finish particularly for use indoors. It can be used as-plated, or with a bronze-type finish but should be lacquered to prevent tarnishing. White brass, containing up to 70% zinc has been suggested as a substitute for a nickel undercoat with chromium. Brass can also be used to improve the bonding of rubber to ferrous or non-ferrous metals.

Brass plating baths contain both zinc and copper as cyano-complexes with an excess of free cyanide as follows:

$CuCN$	35
$Zn(CN)_2$	13
$NaCN$	50
Na_2CO_3	15
$NaHCO_3$	15
NH_4OH	12·5–25 gram litre^{-1}.

The carbonate-bicarbonate pair act as a buffer, with ammonia added for colour control. This alloy composition depends on operating conditions but these are typically 80°C and up to 160 A m^{-2} with 80:20 brass anodes. Baths for white brass contain more zinc, 60 gram litre^{-1} of NaOH instead of the carbonates, with 0·5 gram litre^{-1} Na_2S as an addition agent. An ether-aldehyde brightener may be used.

BRONZE

Bronze is plated from a solution containing copper as a cyano-complex and tin as stannate, for instance:

$CuCN$	36
$NaCN$	26
Na_2SnO_3	50 gram litre^{-1}.

The pH is adjusted to 12·5 and the bath operates at 65°C and up to 500 A m^{-2}. Separate copper and tin anodes make it easier to ensure that tin dissolves as stannate and not stannite. Decorative deposits are plated with about 10% tin.

IRON-ZINC

Iron-zinc alloys can be plated at high rates from simple salt solutions and offer a cheap alternative to zinc coatings for steel. With only 16% zinc the alloy itself is resistant to rusting but 65% is needed to give a coating which will protect steel in outdoor exposure. Bright deposits may be obtained from a bath such as:

$FeSO_4$	250
NH_4SO_4	118
$ZnSO_4$	9
KCl	10
Citric acid	0·5 gram litre^{-1}.

The bath operates at pH 1·7 and 50°C at up to 250 A m^{-2}.

TIN-ZINC

Deposits containing 20–25% zinc combine some of the advantages of both metals. Deposits are easily soldered but give cathodic protection to steel; in some environments the protection is better than that given by pure zinc. The appearance is better than zinc alone, being whiter and finer grained. Solutions contain stannate ion and zinc as cyano-complex:

K_2SnO_3	120
$Zn(CN)_2$	90
KCN	21
KOH	6·5 gram litre^{-1}.

This bath operates at 65°C and 100–250 A m^{-2}.

TIN-NICKEL

This alloy has extraordinary properties, resisting attack by the most aggressive solutions far better than either of its constituent metals. Its behaviour is more typical of an inter-metallic compound than an alloy and the equiatomic composition NiSn is closely adhered to even when plating variables are changed. The bath which produces this deposit contains nickel (II) and tin (II) ions together with at least enough fluoride ion to complex all tin as SnF_4^{2-} or SnF_6^{2-}:

$SnCl_2·2H_2O$	50
$NiCl_2·6H_2O$	250
$NH_4F.HF$	40 gram litre
$NH_4OH(0·880)$	35 millilitre litre^{-1}.

The operating conditions are 65–70 C and about 250 A m^{-2}. The alloy has not found wide use despite its unique properties, but has been applied to balance weights, electrical components and watch and clock parts. It requires only an 8 μm copper undercoat on steel, and BS 3597[8] covers three grades, with between 8 and 25 μm of electrodeposit.

IRON-NICKEL-COBALT

The binary and ternary alloys of iron, nickel and cobalt exhibit a wide range of magnetic properties. The Ni-21% Fe permalloy has a very high permeability and alloys with more iron or cobalt have excellent coercivity, which makes them suitable for computer memory units. Furthermore, for this type of application very thin films of the alloys are required and electro-plating offers a good method of manufacture. In fact the deposition potentials of iron, nickel and cobalt in simple salt solutions are very close together and additions of iron or cobalt to Watt's nickel plating will result in codeposition. The variety of processes used in this type of application arise from the need to carefully control many properties of the deposits.

LEAD ALLOYS

Lead and lead alloys can be electroplated as bearing materials, lead itself from fluoborate, fluosilicate or sulphamate solutions, and Pb-10% Sn alloys with improved performance from the fluoborate bath. Lead-indium alloys are also used for this purpose.

60:40 lead-tin alloys have also been electrodeposited on to electrical components from the fluoborate bath as protection and to improve solderability.

Plating on non-conductors

If a non-conductor is coated with metal or graphite powder, if this will adhere, or a paint containing powder mixed with just enough lacquer or varnish to act as a binder, it can then be electroplated. Metal coatings may also be applied as a preparation for electroplating by vacuum evaporation or chemical deposition.

Graphite may be brushed on dry, or in suspension, and paints can contain graphite, copper or silver powder. Proprietary silver paints which can be fused on to ceramic materials after application give extremely good adhesion. Evaporation in vacuum gives good adhesion, good surface reproduction and a high purity deposit, but it is too expensive for most applications. It has been used for electronic components.

Chemical plating is now widely carried out particularly in preparing plastics for plating. Silver or nickel are suitable, but electroless copper is satisfactory and is the cheapest. Normal silver mirror or electroless copper or nickel solution may be used, but the surfaces will usually require activation. Electroplating of the coated articles can be carried out by vat or barrel plating as normal but lower current densities should be used, particularly at first, to prevent excessive build-up of deposit near the electrical connection to the coating. With plastic or other light materials some difficulty may be encountered in barrel plating as they float in the solution.

Because chemically or vacuum deposited coatings are usually very thin, care must be taken that the plating solution is not allowed to dissolve the coating away completely at any point before electrodeposition commences.

PLATING ON PLASTICS

Chromium plated injection mouldings of ABS, polypropylene, or other plastics are now displacing light-alloy die-castings particularly for trim on cars and domestic equipment. The success of these components relies on a proper pre-treatment of the surface to obtain good adhesion, and effective activation to give an even metal deposit. A typical preparation sequence for ABS would be:
1. Etch in chromic acid.
2. Sensitise with acid stannous chloride solution.
3. Activate in palladous chloride solution to form palladium nuclei.
4. Coat with electroless copper or nickel.

More nickel (or copper) and chromium would then be plated from a normal plating bath.

This sequence may be modified for other plastics. The etching step is needed to obtain adequate adhesion, but not all plastics form the deep pores formed by the action of chromic acid on ABS. To promote acid attack on other polymers, such as high-impact polystyrene and polypropylene, they can be treated with an emulsion of solvent, for example turpentine or toluene, and water.

Non-aqueous plating solutions

Many metals cannot be deposited from aqueous solutions because electrolytic decomposition of water takes place more easily than metal deposition. In fact many other metals can only be plated because of the high overpotential for hydrogen evolution. With other active metals deposits contain oxide or hydroxide which are produced by reaction with the water.

It is therefore only possible to electrodeposit reactive metals such as aluminium, magnesium or titanium from a non-aqueous solution or from molten salts. This can have other advantages; current efficiency can be improved and if the substrate passivates easily in aqueous solution, this is prevented in molten salts or non-aqueous solutions.

MOLTEN SALTS

Molten salts can be inconvenient to work with, but eutectic mixtures with melting points below 200°C are not uncommon. The technology of metal extraction from molten salts to give a liquid product is well established, and processes for producing refractory metals by electro-reduction in melts are proving competitive. These can give coherent deposits, but the melts may cause deterioration of the workpiece.

So far a process for plating aluminium from a 20% solution of $AlCl_3$ in sodium chloride at 175°C has been developed and the deposition of iridium and ruthenium from cyanide melts has also been reported.

ORGANIC SOLUTIONS

There remains the possibility of using organic solvents as a substitute for water, and a great variety have been investigated. Fundamental research with metals such as copper, silver, zinc and cadmium has shown that deposits can be obtained from many solvents, for instance, tetramethylurea, formamide or mixtures of pyridine and acetic acid. In most cases complete exclusion of water is needed. Much more work is needed in this field before plating with reactive metals becomes commonplace, but a successful solution based on $AlCl_3$ and $LiAlH_4$ in dimethyl ether offers good prospects.

Properties of the deposits

A large number of testing methods are available for the assessment of visual, mechanical and chemical properties of electrodeposited coatings. The properties which will be important in any particular case will obviously depend on the specification and end-use which the plated article has to meet. The precise parameters to be measured, and recommended methods for measurement are usually included in the relevant specifications.

APPEARANCE

We have discussed above the examination of surface appearance in terms of *gloss* and *truth*. Visual assessment of these characteristics, together with colour and evenness of deposit may suffice for articles which are to serve a purely decorative purpose. More complete analysis in terms of true surface profile and measured optical properties of the finish is also possible, but it is more appropriate to laboratory investigations than production line control.

METAL THICKNESS AND DISTRIBUTION

The problem of obtaining relatively even metal distribution and uniform thickness was discussed under throwing power. It is important to be able to determine how closely the ideal of completely uniform deposition has been achieved; the chemical properties of the deposit in particular are affected by non-uniform deposits. The thicknesses of deposits involved may range from as little as 0·3 μm for bright chromium to as much as 30–50 μm for nickel, zinc or hard chromium and it is sometimes required that the specification thickness of deposit be present at all surfaces which can be touched with a 20 mm sphere (BS 1224[2]).

Methods for measurement of thickness may be destructive or non-destructive; the latter are normally preferred. In principle, the simplest non-destructive methods are those involving direct measurement of weight gain or increase in thickness. In practice, such methods are only suitable for thick coatings, and the weight-gain test assumes a uniform coating and gives a figure for mean thickness.

A large number of instruments utilise the magnetic properties of the substrate and measure the pull-off force required to remove the testing device from either a ferromagnetic substrate with a non-ferromagnetic coating, or (with more sensitive devices) vice versa. These may usually be used in the range of deposit thickness between 3 and 30 μm. Other instruments measure the magnetic flux in a ferromagnetic basis metal. Further instruments are based on the principles of induction, of eddy currents, of conductance measurements and of thermoelectric effects between deposit and substrate when a heated probe is applied to the surface. Each type of instrument has been designed for some particular application, and this list is by no means exhaustive.

Destructive methods are equally varied. Sectioning methods are used, although many individual readings are necessary to obtain meaningful mean thickness values. Taper sectioning may be used to enhance accuracy, as may the method of Mesle of either grinding a flat on a curved specimen or grinding through a flat coating with a grinding wheel, and calculating the deposit thickness from the geometry of the exposed section. None of these methods is very good for routine measurement, since all require good metallographic and measuring facilities.

Other destructive methods involve chemical or electrochemical coating removal followed by weighing, coulometric dissolution of the deposit, or chemical dissolution with visual observation of time to first appearance of basis metal or hydrogen gas as an index of coating thickness. The simplest methods are best suited to production control, but require calibration. A test much used to assess thickness of non-ferrous coatings is the jet test, in which specified solutions flow from a standard jet onto a sample of the deposit, and the time for first appearance of substrate is calibrated against coating thickness.

MECHANICAL PROPERTIES

The mechanical properties which may be important in electrodeposited coatings are adhesion, ductility, hardness and internal stress. Quantitative tests for the measurement of adhesion are complicated by the difficulty of gripping the deposited coating, and so usually apply only to relatively thick coatings on specially designed specimens. Qualitative tests useful in the plating shop involve such procedures as burnishing with a hard object such as the edge of a copper coin, filing across the edge of the substrate and deposit, bend tests or hammering tests. Ductility is also only measured with ease on specially prepared specimens, which may be subjected to straightforward tensile tests while observing the deposit surface for cracking, or wound around a special spiral jig to determine the minimum radius to which the specimen can be bent before cracking. The measurement of hardness of electrodeposited coatings is a problem with all but the thickest coatings, due to interference by the substrate material. Micro-hardness measurements at very low loads are usually required, with either Vickers or Knoop diamond indenters. However, scratch hardness measurements with calibrated needles have been used successfully. If wear-resistance is a more relevant parameter than hardness *per se,* accelerated wear tests may be used.

Most coatings are electrodeposited with tensile or compressive internal stresses, which if not monitored and controlled to within certain limits detract from the performance of the coatings. Internal stresses are usually measured with prepared shim specimens in specially designed instruments. If such a thin strip restrained at one end is plated on one side only, the curvature of the strip is a measure of the stress in the deposit. Instruments are designed to measure the displacement of strip or spiral specimens, or the restraining force required to prevent the specimen from bending. An alternative design allows a flat disc to bow under the influence of the internal stress of the deposit, and measures the volume of fluid forced out of a sealed chamber behind the disc.

CORROSION BEHAVIOUR

The measurement of corrosion properties of electrodeposits may be done in various ways. A series of tests involving filter paper soaked in suitable solutions may be used to assess porosity in the deposits. For example, porosity in nickel on steel may be measured by covering with a filter paper soaked in ferroxyl indicator (phcnolphthalein and potassium ferricyanide in salt solution) for 10 min, and then observing the number of blue spots on the paper, which indicate pores through which ferrous ions have been able to dissolve.

Immersion in boiling water may also be used, when rust spots form over pores in the coating. Again, some coatings, e.g. tin on steel may be tested by immersion in 1% hydrochloric acid solution, when evolution of hydrogen occurs at individual pores. The evaluation of all such tests involves visual examination and usually some sort of weighting procedure to give more importance to discontinuities of sizes which are considered more harmful.

Another type of test is the accelerated corrosion test. Salt-spray tests were first used for this application, but plain salt-spray tests tend to be irreproducible and are now seldom used. Acetic acid may be added to reduce the pH of the 5% salt solution to about 3·3, and if the test is run at about 35°C, a much more corrosive environment is produced which gives a pattern of attack similar to that observed in outdoor service on plated zinc die-castings and plated steel. A more rapid and reproducible test still is the copper-accelerated acetic acid-salt spray (CASS) test. The spray medium is 5% sodium chloride, with 0·25 grams/litre^{-1} cupric chloride and with pH lowered to 3·1 with glacial acetic acid. The test is conducted at 49°C, and articles are usually tested for two periods of 18 h. Reproducible results are obtained on (typically) plated motor car parts which correlate well with service performance.

Another method employs a moist sulphur dioxide environment to evaluate coatings. Under these conditions nickel-chromium deposits can be tested in as little as 24 h. One final chemical test worthy of mention is the Corrodkote test, in which a kaolin paste containing ammonium and ferric chlorides and cupric nitrate is applied as a slurry before parts are tested for 20 h at a relative humidity of 100%. BS 3749[9] discusses the application and analysis of accelerated corrosion tests for these and other applications in considerable detail and considers the expression of criteria of corrosion resistance in numerical terms.

Despite accelerated test procedures, the final analysis of corrosion resistance of electro-deposited coatings still rests with long-term exposure tests in typical service environments (e.g. rural, industrial, marine) with periodic examination. The procedure is time consuming, but the most accurate method of predicting suitability for particular purposes.

DIP-, DIFFUSION- AND SPRAYED COATINGS

Hot-dipped coatings

For the formation of satisfactory coatings of metals by hot-dipping, it is essential that the coating metal is able to wet the basis metal readily. This involves some reaction between the coating metal and the substrate, and the formation of alloy layers or intermetallic compounds at the interface promotes good wetting and adhesion. Thus zinc, tin and aluminium all readily form compounds with iron, and produce sound hot-dipped coatings on iron and steel. Lead, on the other hand, does not dissolve in iron and forms no compounds with it, and it is virtually impossible to produce satisfactory coatings of pure lead by hot-dipping. Small additions of tin or antimony to the lead promote wetting and adhesion, for both are soluble in lead and form compounds with iron.

Unfortunately the intermetallic compounds at the coating/substrate interface tend to be hard and brittle, and their thickness must be minimised by the use of short coating times, so that the mechanical properties of the coated article are not unduly impaired.

ZINC

Hot dip galvanising consists simply in immersion of a clean substrate, suitably fluxed to remove residual oxide, into a bath of molten zinc at 430–460°C. This may be performed manually or mechanically depending on the type and quantity of articles to be dipped. The majority of material so treated is low carbon steel.

There are two processes which may be considered. In the so-called *wet* process, the article to be galvanised is passed into the zinc pot through a layer of molten ammonium chloride flux. In the *dry* process, the pickled work is dipped in molten zinc ammonium chloride flux and the flux coating then dried. As the work is immersed in the zinc bath, the flux melts and dissociates, liberated hydrochloric acid removing the last traces of oxide from the metal, and ensuring complete wetting by the zinc. After coating, wire or sheet may be roller-wiped, other articles are drained or wiped manually to produce a coating of the desired thickness, which might typically be 75 μm. Up to half of this thickness is usually intermetallic compound, the remainder pure zinc.

The purity of the zinc used for galvanising is of prime importance. Most metallic impurities are detrimental to the quality of the coating. However, up to 0·25% aluminium may be added to the zinc to increase the fluidity and give thinner coatings (and therefore less intermetallic compound). Aluminium also improves the brightness and surface appearance of the coating. Up to 1% tin may also be added to improve the uniformity and adherence of the coating.

Steel sheet and strip may be continuously galvanised. The steel is first oxidised in air at between 300 and 500°C, then passed through an atmosphere of cracked ammonia to reduce the thin oxide coating, and immersed in the zinc bath while still shielded by the inert atmosphere.

Since the zinc wets the substrate readily, the coating has very few pinholes, but in use the zinc pot tends to build up *dross*, FeZn intermetallic compound, and if this becomes included in the coating, poor appearance and inferior performance result. Coatings which are to be stored or used in a humid environment are frequently given a chromate passivating dip, to minimise the incidence of *white rust*.

BS 729[10], Part 1 and BS 2989[11] specify required weights of coating, adherence and continuity for galvanised products; the former refers to hot-dip galvanised articles, the latter to continuously galvanised sheet and strip.

TIN

Low carbon steels and copper-based alloys may be hot-dip tinned successfully by immersion in molten tin at 235–300°C through a suitable flux. In a single-bath process, the tinning bath is covered with a zinc chloride-ammonium chloride-hydrochloric acid flux, to which tin chloride may be added. In a two bath process the first bath is usually flux-covered and held at a relatively high temperature in the above range; the second is frequently covered with palm oil and held at a slightly lower temperature.

After tinning, the work is allowed to drain for a few seconds in a bath of hot palm oil at between 235 and 250°C, to reduce the thickness of coating to the required value. The thickness depends markedly on the temperature of the palm oil; a fully-drained coating may be only 5–7 μm, but significantly thicker coatings are required in many instances, particularly for cooking utensils etc., where 30–50 μm may be necessary. Intermetallic compounds ($FeSn_2$ or Cu_6Sn_5 and Cu_3Sn, depending on the substrate) comprise perhaps 10% of the total thickness. After cooling, the oil is removed from the tinplate surface with sawdust, or by more conventional degreasing methods.

BS 3788[12] specifies thicknesses of tin required for cooking utensils, and discusses the permissible purity of the coating.

Tinplate. Low carbon steel sheet and strip is continuously tinned in large quantities. It is a particularly valuable material because the coating is sufficiently ductile to allow forming operations to take place after coating, and it is very easy to solder. Much of the world's tinplate is used for food cans, the tin having no toxic effect on foodstuffs.

Steel for tinplate, typically 0·25–0·5 mm thick, is taken either as continuous strip, or more often as individual sheets, pickled and fed through a layer of flux into the tin pot at 300–340°C. It leaves the tin pot through the grease pot, at 250–280°C, where squeeze rolls reduce the molten coating to the required thickness. The tinplate passes through a wet alkaline cleaning bath and then through a dry cleaning machine in which it is mopped with sawdust or wood meal.

The thickness of coating used to be expressed in pounds of tin per basis box, where a basis box was 112 plates 14 in × 20 in, or the equivalent area of tinplate (31 360 in²). Nowadays coating weights are expressed in g m^{-2} of tinplate. (1lb per basis box is 22·4 g m^{-2} of tinplate, or 11·2 g m^{-2} of *coated surface*). Typical coating weights range between 5·6 and 56 g m^{-2} (a thickness range of approximately 0·5 to 5 μm). BS 2920[5] discussed coating weights and mechanical properties of tinplate, and indicates appropriate testing methods.

LEAD ALLOYS

Pure lead is bonded to clean steel by hot-dipping only with the greatest difficulty, since it neither dissolves nor forms compounds and therefore does not wet the substrate. There is a very strong tendency to form pinholes, and the coating detaches easily if the steel is deformed. However, if 2% tin or up to 2% antimony are added to the lead, the wetting problem is alleviated and steel sheet can be coated by hot dipping in a manner very similar to that used for zinc and tin. A zinc ammonium chloride flux may be used, followed by either one or two lead baths at 340–380°C. Sheets are withdrawn through squeeze rolls in an oil bath. Typical coating weights are in the region of 3 g m^{-2}.

Terne plate. Terne plate is of more commercial interest. It is tinplate-grade steel coated either with 93% lead 7% tin, or 80% lead 20% tin (tin-terne), and may be manufactured in a similar manner to tinplate. Light coatings up to 8 μm may be prepared by a single dip, and coatings up to 25 μm are prepared by multiple dipping. Terne plate is used for roofing, for petrol tanks and hardware, and in the fabrication of deep-drawn components, where the coating acts as an excellent lubricant for the drawing operation.

ALUMINIUM

Aluminium coatings on iron and steel may be prepared by hot dipping, and commercial processes are available for manual dipping and for continuous treatment of wire and strip. There are, however, considerable difficulties. The substrate needs to be exceptionally clean, to prevent the formation of aluminium oxide films at the interface. Further, the intermetallic compound $FeAl_3$ forms particularly readily and it is extremely brittle, so that it is very easy to impair the properties of the steel. Nevertheless coatings of 50 μm and more are produced in pure aluminium and in aluminium-silicon baths. The silicon reduces the rate of compound formation and also makes the coating more suitable to resist high temperature oxidation.

Diffusion processes

Diffusion processes can be used to produce surface layers of many metals on many substrates. The most common processes are those producing aluminium- and zinc-rich coatings, but processes are available for materials such as tungsten, vanadium, zirconium and boron. Any substrates are suitable provided that alloys or compounds can be formed with the coating metal.

The hard and brittle nature of such compounds produces considerable hardening of the surface of components treated in this way. The process can be either one of gas-phase transport of a compound of the coating metal or of physical contact with particles of the metal. Diffusion- or cementation processes are particularly suitable for castings, where porosity would lead to bleeding of solution after electrodeposition, but is generally confined to small parts, such as can be packed into retorts or boxes of a reasonable size.

ZINC

Sherardising, the production of zinc coatings by cementation, is particularly suited to the coating of relatively small parts. They are packed into sealed drums with a charge of zinc powder, zinc oxide and a filler, and heated to 320–370°C for between 2 and 6h, while the drum is rotated, causing the charge and parts to tumble. At the end of the treatment, the drum must be cooled before it can be opened, or the heated zinc powder will ignite.

The charge may conveniently be blue powder from a zinc smelter, which contains about 80–90% of zinc and up to 10% of zinc oxide. Sand or alumina are often added to prevent caking. The thickness of the diffusion layer is about 25 μm, and it provides protection against corrosion which is about equivalent to electroplated zinc but rather inferior to hot-dipped coatings. The coating is ideal for threaded parts, since the thickness is uniform, independent of the geometry of the part. It consists of a mixture of the compound Fe Zn and zinc and usually contains networks of fine cracks due to the difference of expansion coefficients between coating and substrate.

BS 729[10]. Part 2, specifies continuity criteria for sherardised coatings, and recommends tests for the evaluation of the coating.

ALUMINIUM

Calorising is the process of producing aluminium-rich cementation coatings on the surface of metal parts, which are generally of iron or steel, but may be of copper-based alloys, to protect against high-temperature oxidation. It is used to protect exhaust-systems, furnace tubes, aero-engine parts, heat treatment pots and similar articles up to temperatures of about 900°C. In one version of the process the parts are packed in a rotating drum with a charge of up to 50% aluminium powder, up to 5% ammonium chloride and alumina powder to prevent coalescence, and heated to 850–950°C in an inert atmosphere for about 5 h. (Copper alloys are treated at 750°C). They are then removed and given a further treatment of about 24 h at 900°C, to reduce the aluminium concentration at the surface and improve adhesion and ductility of the coating. Smaller more fragile parts may be pack-calorised in a sealed box for up to 24 h.

The layer is between 100 μm and 1 mm thick, with about 25% aluminium in the surface. A vapour-transport method also exists, using aluminium chloride vapour, which produces thinner coatings.

CHROMIUM

Chromising may be used to provide a corrosion-resistant surface for steel parts. It is only applicable to low carbon steels, or to higher carbon steels after surface decarburisation. In pack chromising the steel is heated at between 1000 and 1450°C with a charge of 55% chromium or ferrochrome and 45% alumina, in hydrogen, for up to 4 h. This treatment gives a layer up to 200 μm thick with 10–20% chromium at the surface.

Alternatively, vapour chromising may be used. The work is sealed with a mixture of chromium, alumina and ammonium chloride in an atmosphere of hydrogen. At 1000°C, chromous chloride is generated and serves to carry the chromium to the substrate. Layers up to 400 μm may be formed, but thicknesses between 75 and 125 μm are usual, for which treatment for about 4 h is required. The chromium concentration in the surface may be as high as 50%. The performance of these coatings is roughly equivalent to that of a 25% Cr steel.

SILICON

If low carbon steel is packed with ferrosilicon and treated with a mixture of hydrogen and chlorine at 1000°C, silicon is transported by silicon tetrachloride vapour and a silicon diffusion coating is formed on the work. Coatings from 125 μm to 1 mm can be produced with up to 14% silicon in the surface. The precise depth of coating depends upon the time and temperature of treatment. The process is known as *ihrigising*, and is useful in producing acid-resistant surfaces.

Sprayed metal coatings

Metal spraying is quite a different method of applying metallic coatings, in that it relies on purely mechanical bonding, rather than attempting to attain chemical bonding to clean surfaces. The process thus uses no fluxes, and can be used for virtually any metals and alloys, and indeed can be modified to apply some ceramic coatings. Metal spraying can be used to build up very thick coatings if required, and is the only method available for applying aluminium coatings thicker than 0·1 mm.

Surface preparation is most important, the more so since a particular type of rough surface is required rather than the smooth finish required for other types of coating. Steel surfaces must be grit blasted to white metal with angular chilled-iron grit to obtain a sharp contour to which the sprayed metal will adhere. Sharp sand may also be used as an abrasive, but can only be used in particular circumstances due to the silicosis hazard. The white metal surface is dusted and inspected and sprayed as soon as possible after cleaning, to avoid adhesion problems arising due to reformation of oxide films and absorption of moisture. It is recommended that if grit blasted work must be handled before spraying, gloves are worn. An alternative type of preparation in reclamation and repair work is rough machining to give a coarsely grooved surface.

The original metal spraying process used molten metal, but modern guns use wire or powder. The powder method is the less popular, although it is used to some extent for zinc, which may be obtained particularly cheaply as powder. The wire type of pistol is most often used for metal spraying. It is fed with wire which may range from very small sizes up to about 5 mm in diameter. The wire is fed automatically to the pistol, and is fused by an oxygen-fuel gas (usually oxy-propane) flame. The molten metal is atomised and accelerated towards the work surface by a compressed air blast. (An inert gas blast must be used for spraying reactive metals like titanium and zirconium). Electric-arc pistols have also been developed, but have generally been thought to be too cumbersome. They do however, make possible the spraying of high melting point metals like tungsten and molybdenum.

When the molten droplets are flattened as they strike the surface, they key into the angular surface profile to give good mechanical adhesion. Subsequent droplets adhere to the initial layers in a similar manner and may become welded to them. The precise strength of the bond between coating and substrate depends on the temperature and velocity of the spray, but may be in the order of $7-10$ MN m^{-2}. Up to 24 Kg h^{-1} of zinc or 6 Kg h^{-1} of aluminium can be sprayed from a wire pistol. The metal contains up to 3% oxide, and has a density of 85–90% of the theoretical density. The lower density is due to porosity within the coating.

Since adhesion is purely mechanical, there are few restrictions on which coatings can be sprayed onto which substrates. Thus for reclamation work high carbon steels or stellites can be sprayed onto low carbon steels to provide extra wear resistance. Nickel, monel and stainless steels can be sprayed to give improved corrosion resistance, although the porosity may cause disappointing results. Porosity can however be reduced by mechanical treatments. Bearings can be reclaimed by spraying copper alloys or babbitts. when the porosity may be an advantage, giving improved oil retention. The pistols are small enough to be used on lathes or in production lines, and so spray coating has become an integral part of the production engineering cycle.

The metals most commonly sprayed are zinc and aluminium for corrosion resistance and aluminium for high temperature oxidation resistance. Coatings of 100–300 μm may be used for corrosion resistance, and coatings between 150–200 μm for oxidation resistance. Aluminium coatings sprayed for oxidation resistance are subsequently annealed to improve adhesion. Zinc or aluminium coatings may be sprayed to protect ships hulls, storage tanks, bridges, structural steelwork and similar structures, and, for example, sprayed zinc may be used to repair galvanised structures after welding. In these instances the pores become plugged by corrosion product, after which the corrosion rate falls to a very low value. In a few instances, an alloy of 65% zinc 35% aluminium has been used, and it is claimed to give a harder and more protective film of corrosion product.

BS 2569[13] is in two parts; Part I is concerned with aluminium and zinc as protection against atmospheric corrosion, Part II with aluminium for protection against oxidation of elevated temperatures. The standard specifies compositions, surface preparation, methods of application and thickness of sprayed coatings, and detailed testing methods.

With structures where first coating cost is less important than maintenance costs, for example bridges and similar civil engineering structures, protection of an extremely high order can be obtained even in adverse conditions by painting sprayed coatings with specially formulated advanced paint systems.

ANODISING AND CHEMICAL CONVERSION COATINGS

Anodising

Anodising is extensively used for the production of protective and decorative coatings on aluminium and many of its alloys. During the process, the natural air-formed film on the metal is grown to a substantial thickness by anodic treatment in acid solutions. The resulting film is coherent, adherent to its substrate, hard and electrically insulating. The type of film obtained depends on the solution used and the conditions of treatment. Anodic treatment in, for example, boric acid or ammonium tartrate solutions at pH 5–6, results in parallel-sided *barrier layer* growth, in which the film thickens to about 1·5 nm per volt applied. Voltages up to 250 V may be used, to produce films approaching 350 nm thick. These barrier layer processes are useful for the treatment of evaporated aluminium mirrors, or in the manufacture of electrolytic condensers, although high purity substrates are necessary if the necessary electrical integrity is to be maintained.

Most industrial anodising, on the other hand, is carried out in much lower pH solutions of sulphuric acid, chromic acid or oxalic acid. In these solutions, a barrier layer is first formed, but then dissolution reactions become important and a relatively thick porous film appears above the barrier layer. A substantial proportion of the aluminium converted to oxide is dissolved in the solution. The structure of the anodised film approximates to a hexagonal array of cells each with a more or less cylindrical pore down the centre (Figure 6.16). The thickness of the pore-base layer increases with forming voltage, as does the

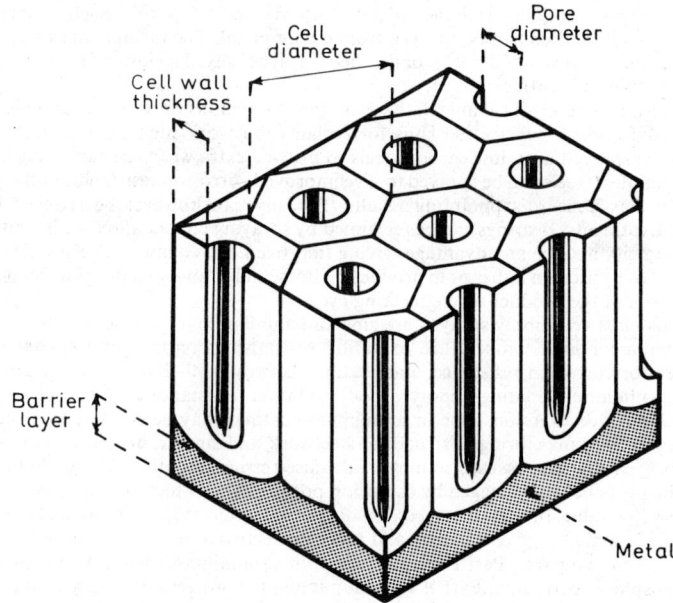

Figure 6.16. Schematic sketch of porous oxide film on anodised aluminium

cell size and the wall thickness. The pore diameter is a function of anodising solution, temperature and current density. These porous films are usually grown to thicknesses up to 30–50 μm, but over 100 μm of oxide may be used in hard-anodising processes.

Since the alumina film is electrically insulating, the anodising process has excellent throwing power and current is automatically diverted to thinner or less perfect areas of the film until these have a coating of uniform thickness.

After anodic treatment the porous film may be *sealed* in boiling water, steam or other reagents. The amorphous oxide is converted to böhmite, AlO(OH), and the resultant swelling of the pore walls closes the pores and increases the resistance of the film. Prior to sealing, the anodised article may be coloured with organic dyes, which are absorbed into the pore structure, or by inorganic pigments precipitated in situ in the pores, to give a wide range of decorative effects.

BS 1615[14] specifies thicknesses of anodic films on aluminium, and discusses sealing requirements, light-fastness and leaching resistance of dyes, corrosion and abrasion resistance and electrical breakdown. Between 15 and 35 μm of anodic coating are recommended for architectural anodising, 10 or 15 μm for general corrosion protection and other exterior applications, whereas 5 μm is usually satisfactory for interior use and 1 or 3 μm as a basis for paints and lacquers. BS AU89[15] covers the use of anodised aluminium and its alloys for motor-vehicle trim. 5 to 8 μm of coating are recommended for exterior, 2·5 to 5 μm for interior use, and the specification indicates alloys for use where specific appearance is required in the finished article.

After aluminium, titanium and its alloys, and tantalum are the most commercially important applications of anodising, but a variety of other metals can be treated in a similar manner, notably niobium, zirconium, hafnium and tungsten.

SULPHURIC ACID ANODISING

The sulphuric acid process is most widely used for decorative and protective anodising, although it is not suitable for fabricated or riveted parts, where traces of acid retained in interstices after treatment may lead to corrosion. It produces a transparent film which is colourless on pure aluminium, but darker on some alloys, particularly those with a high silicon content. The lustre of the underlying metal is retained. Decorative applications are thus usually based on pure aluminium, chemically polished or bright-dipped before treatment, to give a brilliant metallic finish after anodising, and bright clear colours when dyed. The films are hard and abrasion resistant, but not ductile. They contain appreciable quantities of sulphate.

A general purpose solution contains 10% by volume of concentrated sulphuric and is operated at 110–160 A m^{-2} and 21–24°C. To achieve this current density, a voltage of the order of 15 V is required, but for most consistent results the current density must be maintained at a constant value by varying the voltage as anodising proceeds. Aluminium or lead cathodes are used, or if lead-lined vats are used the lining may serve as cathode. Typical anodising times vary from 10 to 60 min, the shorter times being used for decorative applications, the longer for architectural and outdoor purposes. Thorough boiling-water or steam sealing must be provided for aluminium for architectural purposes.

CHROMIC ACID ANODISING

The chromic acid process is the other major industrial process for protective anodising, and it also finds a few decorative uses. It provides very high resistance to corrosion, and is preferred for anodising assemblies or stressed parts, where any traces of anodising solution retained are inhibitive rather than corrosive. The film produced is virtually free from chromate. It is opaque, and varies in colour from light grey on pure aluminium to dark blue-grey on alloys containing high percentages of silicon. It is not quite so hard as the film produced in sulphuric acid, but is considerably more ductile. Dyeing may be used only to produce pastel or very dark shades, so that the process is only used for decoration where opaque enamel-like finishes are desired, without metallic lustre. The chromic acid process is not suitable for alloys containing more than 5% copper.

The original process uses a solution of 50 gram litre^{-1} chromic acid with a voltage cycle in which the anodising voltage is raised from 0 to 40 V in 10 min, held for 20 min, then raised to 50 V in 5 min and held for a further 5 min. A modified cycle for castings replaced the final voltage increase by a further 10 min at 40 V. The current density is about 30 A m^{-2} for pure aluminium, rising to about 200 A m^{-2} for high silicon alloys.

The process is normally run at about 40°C, with stainless steel cathodes, which should have a total area about one eighth of the anode area.

An alternative process which is somewhat easier to control uses 100 gram litre^{-1} chromic acid at 55°C. A constant 30 V is used, giving a current density of the order of 150 A m^{-2}. This rate corresponds to film thickening at about 0·3 μm min^{-1}.

Boiling water or steam sealing may be used, often preceded by a neutralising dip in an ammoniacal or bicarbonate solution. After sealing, the films may be waxed or greased to give additional protection, particularly for military specifications.

HARD ANODISING

Where a more abrasion resistant finish is required, as may be the case in engineering and aircraft applications, conditions can be modified to produce harder, thicker coatings than the normal anodised films. The hardness of the film may approach 500 VPN, and the thickness of the film may be increased to between 50 and 100 μm, or more, depending on the alloy. This type of finish has made possible the use of aluminium in hydraulic equipment, gears, automobile pistons and cylinders etc., where previously a much harder (and heavier) material would have been specified. It can compete in many applications with hard chromium plating, to which it is superior in adhesion and oil retention and often also on grounds of cost of application.

Hard anodising produces a surface growth equal to about half the thickness of the anode film, so that it may sometimes be used to reclaim over-machined parts. The surface of the work is roughened by hard anodising, so that if a smooth finish is required particular care must be taken to produce very smooth substrates. The hard anodised surface can be honed and lapped to improve the finish. All hard anodising coatings have fine cracks in the anodic film, which are accentuated at corners and edges. It is important therefore that such regions are correctly radiused and deburred before anodising. For this reason, hard anodising is restricted to the specific areas which need to be wear resistant, the remainder of the work being given a normal sulphuric acid or chromic acid treatment.

There are two main methods of hard anodising. The 10% sulphuric acid solution mentioned above may be used between −10 and 5°C, at 220 to 1100 A m^{-2}. The voltage required may rise to as high as 75 V by the end of the process. Because of the greater power dissipation in this type of process, adequate agitation and refrigeration must be provided. The rate of film formation depends on current density; it is of the order of 1·5 μm min^{-1} at 500 A m^{-2}.

An alternative hard anodising process uses a sulphuric acid-oxalic mixture containing:

H_2SO_4(conc.)   100 millilitre litre^{-1}
oxalic acid   50 gram litre^{-1}.

This solution may be used at 150 to 250 A m^{-2} at between 5 and 10°C. In both these processes, the voltage is raised slowly at the start of anodising, so that the required current density is not exceeded.

The films may be sealed in water or steam, but this may produce some softening of the surface. For engineering applications it is more usual to oil or wax the film without sealing. The thickest films are usually dark coloured, and are not dyed, but thinner films are sometimes dyed for architectural applications.

Chemical conversion coatings

PHOSPHATING PROCESSES

Phosphate conversion coatings are most commonly applied to iron and steel, but aluminium, zinc, cadmium and tin may also be treated. *Coslettising* was the first version of the process to be developed. *Parkerising* and *Bonderising* are well-known trade names for versions of the modern process, and there are many other proprietary versions. The process can be used by barrelling, dipping or spraying, although only *rapid* solutions are suitable for spraying, and intricate parts present no problem.

A typical solution would consist of acid and normal phosphates of zinc in phosphoric acid, balanced so that the normal zinc phosphate is near to saturation. Steel articles immersed in the solution are attacked by the phosphoric acid, and the pH near to the surface rises as hydrogen is evolved, the solution becomes supersaturated with zinc phosphate, which is precipitated onto the surface as a crystalline layer. Some iron from the solution is also included in the precipitate. The rate of formation is controlled by the rate of hydrogen evolution during metal dissolution, and is very slow in plain solutions. Many baths contain accelerators, which are usually oxidising agents to depolarise the hydrogen evolution reaction, or copper salts. Phosphates of manganese and of iron itself are also used as the basis of phosphating baths. Baths are classified as *standard*, in which the process time is greater than 30 min, *accelerated*, in which less than 30 min is necessary, and *rapid*, which will produce coatings in less than 5 min.

The coatings produced have excellent adhesion, although they are not very protective on their own and would usually be covered with oil or paint. The most corrosion-resistant coatings have small grain sizes, which usually decrease as the treatment temperature increases. However satisfactory cold solutions do exist, as do alkaline solutions.

For corrosion resistance, BS 3189[16] classifies five qualities of coating, ranging from thick manganese or manganese-iron phosphate coatings of over 7 g m^{-2} for maximum protection, through zinc phosphate coatings of the same weight, to medium coatings of over 4 g m^{-2} of zinc phosphate for general use, with thin and extra thin coatings between 30 and 1000 mg m^{-2} for use under highly protective paint or lacquer coatings on light gauge sheet. The standard also discusses procedure for coating, rinsing and testing.

Thin coatings produced by spray application of rapid solutions are used as a basis for paint on car bodies, bicycle frames, domestic appliances and plain steel sheet and strip. The coating improves adhesion of the paint and reduces under-rusting. Thick manganese or iron phosphate coatings are used to give improved lubrication and wear resistance on piston rings and cylinder liners, diesel engine parts etc. The thick ($10-40 \text{ g m}^{-2}$) relatively coarse grained films serve to retain grease and lubricant, and the grain size is varied to suit the application. Thinner zinc phosphate coatings are also used to retain lubricant and prevent galling in cold forming operations.

Aluminium, zinc and cadmium receive light zinc phosphate coatings to improve paint adhesion, and light amorphous phosphate coatings can be produced on aluminium for protection from corrosion.

CHROMATING PROCESSES

Chromate conversion coatings are usually applied to zinc, cadmium, aluminium and magnesium. They are normally chemically produced, although electrochemical processes are available. The coatings are very thin, of the order of 1 μm, and unlike phosphate coatings they are amorphous rather than crystalline, and very fragile when freshly formed. The coatings act as both a barrier against the environment and as a source of inhibiting ions, and may be used alone or as undercoats for paint systems. Many proprietary processes are available.

Zinc and cadmium are normally treated in acid baths containing hexavalent chromium. The mechanism is similar to that of the phosphating process. On immersion, the metal is attacked and the interface pH rises due to hydrogen evolution. Some hexavalent chromium is reduced to the trivalent state, and when the pH rises to a critical value a basic chromium chromate is precipitated. Treatment times range from a few seconds to several minutes at about 40°C. The thickest coatings are olive in colour; the thinner yellow-bronze coatings can be dyed while wet, but the colours are not light-fast. A post-treatment bleach may render the thinnest coatings clear and colourless, and these may be lacquered for decorative purposes.

Aluminium may be treated in similar solutions, with added fluoride, to produce similar films. The electrical resistance of the thinner chromate coatings is not high, unlike that of anodised coatings, and chromated aluminium may be used successfully in electrical work, although the coatings are more frequently used as paint substrates. An alternative process uses a chromic acid-phosphoric acid bath to produce a mixed film of chromium and aluminium phosphates. Aluminium may also be treated in an alkaline chromating bath:

$$Na_2CO_3 \qquad\qquad 50$$
$$Na_2CrO_4 \qquad\qquad 15 \text{ gram litre}^{-1}.$$

This is the MBV (Modified Bauer-Vogel) solution. It is used at 90–95°C, and times of 3–5 min may be used to produce coatings of the order of 2 μm, or longer times up to 30 min to produce more substantial coatings. The dark grey coatings are a mixture of aluminium oxide and chromium oxide, and may be further improved by sealing in sodium silicate solution. There are many modifications of the original MBV formula.

Virtually all magnesium sand- and die-castings are given a chromate passivation treatment for protection during storage, or as a basis for subsequent painting The work is first pickled in hydrofluoric acid or anodically treated in ammonium fluoride, then treated in either boiling sodium dichromate solution or a boiling ammoniacal dichromate-sulphate bath. The coatings vary from matte grey to dark red in colour, and give good protection. For maximum protection, an alternative process involves treatment in a sodium dichromate-nitric acid-sodium acid fluoride bath. Treatment is for 1 min at 35°C, and very coherent deep brown coatings are produced.

Tin plate may also be treated in an alkaline chromate dip, to increase corrosion resistance during storage and to prevent black sulphur-staining when used with foodstuffs.

SELECTION OF COATINGS

Metal coatings

Coatings can perform a variety of functions, all of which are intended to improve some property of the surface. Whatever the particular function of a coating it must always resist deterioration in its service environment, however mild, so that corrosion-resistance is always important. The importance of other properties will depend on the particular application, and discussion of coating materials and techniques is simplified by the consideration of several separate types of application:

1. Decoration and protection from corrosion.
2. Reclamation and repair.
3. Lubrication and wear-resistance.
4. Electrical applications.
5. Special applications.

PROTECTION FROM CORROSION

The first requirement of a protective coating is that it should resist service conditions better than the basis metal. Decorative coatings must remain lustrous and free from unsightly corrosion products in use.

Protective coatings must be able to isolate the substrate from its surroundings, or if they cannot do this completely, must be able to protect exposed metal. Limited initial porosity is acceptable if exposure leads to rapid blocking of pores by corrosion products. In general, metal coatings can be divided into two categories, those which confer cathodic protection, dissolving in preference to the substrate, and those which do not. The behaviour of a particular metal cannot always be predicted from the electro-chemical series but will depend upon exposure conditions.

Zinc and cadmium normally act as sacrificial coatings, as also may tin plate in many of the organic acids found in foodstuffs, where tin corrodes without attack on the steel. Aluminium can also behave anodically on steel in an aggressive environment such as an SO_2-containing industrial atmosphere, and makes an excellent corrosion-resistant coating under these conditions. Alloys of tin and zinc or iron and zinc applied by electroplating also give cathodic protection. Tin-zinc has better appearance than zinc but unlike tin confers cathodic protection on steel. Iron-zinc is a cheap alternative to pure zinc.

Cadmium is preferred to zinc in tropical or humid conditions where *white rust* can occur, but zinc is better in marine environments. Cadmium is often used in engineering applications where small thicknesses are required to maintain tolerances but here sherardised zinc coatings are also important, particularly for small parts like nuts and bolts. Care must be taken with electrodeposits of both metals on high strength steels to ensure that hydrogen enbrittlement is prevented by a suitable heat-treatment after plating. Cadmium is toxic and should not be used with articles which may contain foodstuffs. For maximum life it is common practice to apply a simple paint system over these coatings.

With coatings that are cathodic to underlying metal there is a danger of accelerated corrosion and rapid penetration at any discontinuities, particularly with noble metals. To give adequate protection relatively thick, pore-free coatings will be needed. For moderately corrosive environments thick nickel electrodeposits or hot dipped tin or lead give good service. Bright nickel alone can be used on much domestic equipment, and resists most natural waters. Thin electrotin gives limited protection which is improved by lacquers, but thicker hot dipped coatings are widely used for food containers such as milk churns. Lead is used as terne plate, which performs well as a roofing material in industrial conditions, or in contact with flue gases. SO_2 in the air causes pores to become blocked with $PbSO_4$. Copper-tin and copper-zinc alloys can also give limited protection but are generally for interior use.

The metals which resist corrosion well and do not tarnish, such as chromium and the noble metals, tend to be expensive so that it is often not economical to apply them sufficiently thickly to give complete exclusion of aggresive species. In this case it is normal practice to apply a comparatively thick undercoat of another metal which is pore-free, and a much thinner top-coat which may well contain discontinuities. The nickel-chromium system is most widely used in aggresive environments, the chromium increasing resistance to attack and preventing fogging of the nickel. Nickel-tin alloy can also be used under severe conditions. Silver, gold, platinum and rhodium are used as thin coatings for jewellery and luxury goods, but also find other use in specialised applications. In addition to electronic and aerospace uses thicker coatings of gold and silver, applied by cladding, may be used on equipment for the pharmaceutical industry.

For particular applications, the type of plating and thickness of deposits must largely be determined by previous experience; expert advice can however be obtained from a number of advisory institutions. Particularly in the case of corrosion-resistant coatings, the nature of the substrate and of any undercoat will be of vital importance. The type of environment must also be considered, but as a general guideline the British Standards Institution suggest specifications for many decorative coatings.

LUBRICATION AND WEAR

In tribological applications, hard coatings may be used to resist abrasion, or softer materials to resist seizing, a major cause of wear.

Hard chromium applied by electroplating is greatly favoured as a coating for tools, dies and gauges, but for heavy duty applications such as agricultural and civil engineering machinery metal spraying with alloy steels or stellite gives a more durable surface. This type of hard-facing is not suitable for soft substrates which can yield and cause cracking of the coating. A new process particularly useful for light alloy parts is the electro-deposition of nickel or cobalt containing ceramic particles from a plating bath containing these particles in suspension.

Electrodeposited coatings of soft metals such as lead or indium alloys can improve the life of journal bearings and silver plating is also used for heavy duty applications. Tin may be used on engine pistons to prevent seizing during running-in.

RECLAMATION AND REPAIR

Electrodeposits or sprayed coatings can be used to replace metal worn from machinery, tools and dies. In this case mechanical strength will be the most important property, but there must also be good adhesion to the substrate. Flame-spraying can be carried out with the same metal as the substrate, but a hard-facing alloy will give a greater extension of life. Likewise steel parts can be rebuilt with electrodeposited iron but nickel, or even chromium, are preferred despite the extra expense.

ELECTRICAL APPLICATIONS

Many electroplating processes are used in the electrical and electronic industries. Copper conductors can be electrodeposited onto printed circuit boards, and waveguides may be electroformed. Silver and gold coatings are used to protect both circuit boards and wave-guides. One of the main uses is to ensure that switches and connectors have low contact resistance. Silver alone may be used but a gold or rhodium top-coat will prevent tarnishing. Gold and silver deposits also give good solderability. The use of platinum metals is particularly important for contacts where sparking may occur, which would erode coatings with lower melting point. Tin and tin-lead alloy are widely used to make connectors and wires easily solderable, and cadmium is used to plate steel chassis because it protects well and has good solderability.

Vacuum evaporated aluminium is used widely to make polymer surfaces electrically conducting for the manufacture of condensers.

SPECIAL APPLICATIONS

There are many uses of coatings which depend on special properties of the coating material. One of the best known is the use of brass plating to make rubber bond to steel. Sulphur in the rubber reacts chemically with the brass to give a strong bond, but with copper alone yields a loose film of the sulphide at the interface.

Metals have always been used to produce mirrors, traditionally by the chemical reduction of silver solutions, but also nowadays by vacuum deposition of aluminium. On front-silvered precision equipment, rhodium plating of the silver prevents tarnishing. The high reflectivity of metals is invaluable for protecting space equipment from solar radiation, and gold-plating is widely used.

Electrodeposits of copper can also be used for masking parts before carburising, limiting hardening to the area where it is required. Tin deposits can be used in a similar way to mask in nitriding processes.

METHODS OF COATING

The number of application techniques which can be used for any coating is usually quite limited. Many metals cannot be electroplated because deposition potentials are too low, or because they react too readily with water. Hot-dipping can only be used for low melting point metals on high melting point substrates; there are usually similar restrictions on other methods. From the methods available for any metal there are particular cases, described above, where certain techniques are preferred but there are also general guide lines.

Electroplating accounts for the majority of coatings on small parts. Hot dipping tends to be used for heavy-duty applications where thick coatings are needed, but electro-deposits are much more economical where thin deposits are adequate, when the improved control over thickness can lead to big savings in coating metal. Hot dipping can be used for large sections; there are tanks in operation which can coat sections up to 10 m long, but to electroplate this size of part would need a much bigger plating tank and is not really practical. It is on large sections in particular that metal spraying becomes important, even large bridges are coated in this way, although usually before final assembly. It can also compete with hot dipping for corrosion protection of many objects in outdoor exposure.

For precision work electrodeposition can deal with most problems, but it is in this area that vapour deposition and vacuum evaporation become important, particularly for very thin coatings.

Electroless plating, chiefly of nickel and copper is being used increasingly for a variety of objects. It is more expensive than electroplating but has extremely good throwing power, giving even deposits even in small holes.

In general, where size or accuracy are not controlling factors, decisions must be based on economic considerations.

Non-metallic coatings

Selection of inorganic coatings is much more restricted than is the case with metallic coatings. In the first place, the nature of the substrate limits the types of coating which may be made, and further, the applications for which various coatings are suitable may be quite specific. Thus, anodising is largely limited to aluminium, but the films may be used as paint substrates when thin, as an excellent protection against corrosion, or if prepared on polished substrates and perhaps dyed and sealed, for decorative purposes. Hard anodised films are used for wear resistance and other engineering applications. Phosphate and chromate coatings on aluminium may also be considered for paint substrates, and chromate coatings for corrosion resistance.

Chromate coatings also give corrosion resistance to zinc, cadmium and magnesium, and sometimes tin, and may be used as bases for paints and lacquers.

Many different phosphate coatings can be applied to iron and steel, to achieve corrosion resistance, wear resistance, improved lubrication or paint bonding. For these and most other inorganic treatments, the processes are usually proprietary, and the manufacturers have wide experience in their application. Ample advice is therefore available, and should be sought. The precise choice of a process will rest on the scale of application, availability of plant, and ultimately, on economic considerations.

Organic coatings

Although the above treatment refers solely to metallic and other inorganic coatings in metal finishing, decorative and corrosion resistant coatings can also be provided by organic systems. Bitumen and plastics coatings are widely used, and protect the substrate by isolating it from the environment. Properly formulated paint systems while not providing complete isolation can still give excellent protection, either by providing inhibiting ions, metal pigments which give cathodic protection to the substrate, or by introducing a high resistance between anodic and cathodic areas on the metal. Used alone on properly prepared substrates, or used over sprayed or deposited metal coatings where economic considerations warrant, organic systems provide highly competitive coatings for both decoration and corrosion protection.

REFERENCES

1. BS 1134
   Centre-line-average height method for the assessment of surface texture (1961)
2. BS 1224
   Electroplated coatings of nickel and chromium (1970)
3. BS 1706
   Electroplated coatings of cadmium and zinc on iron and steel (1960). (Will be revised in line with ISO/R2081 and ISO/R2082 (1971))
4. BS 1872
   Electroplated coatings of tin (1964)
5. BS 2920
   Cold-reduced tinplate and cold-reduced blackplate (1957)
6. BS 4292
   Electroplated coatings of gold and gold alloy (1968)
7. BS 4290
   Electroplated coatings of silver for cutlery, flat-ware and holloware (1968)
8. BS 3597
   Electroplated coatings of 65/35 tin-nickel alloy (1963)
9. BS 3745
   Evaluation of results of accelerated corrosion tests on metallic coatings (1970)
10. BS 729
    Zinc coatings on iron and steel articles. Parts 1 and 2 (1961)
11. BS 2989
    Hot-dip galvanised plain steel sheet and coil (1967)
12. BS 3788
    Tin coated finish for culinary utensils (1964)
13. BS 2569
    Sprayed metal coatings. Parts 1 and 2 (1964/1965)
14. BS 1615
    Anodic oxidation coatings on aluminium (1961)
15. AU 89
    Anodised aluminium for automobile use (1965)
16. BS 3189
    Phosphate treatment of iron and steel for protection against corrosion (1959)
17. BS 4758
    Electroplated coatings of nickel for engineering purposes (1971)

7 MECHANICS AND PROPERTIES OF SOLIDS AND MATERIALS

MECHANICS AND PROPERTIES
OF SOLIDS AND MATERIALS

7

7 MECHANICS AND PROPERTIES OF SOLIDS AND MATERIALS

MECHANICS OF SOLIDS

TERANCE V. DUGGAN

INTRODUCTION

In considering the design of a structure or machine component, it is essential to ensure that the response of the elements is such that they can perform their required function with a high degree of confidence and reliability. This requires particular attention to establishing that any external forces, whether static or dynamic, can be safely accommodated for the desired life without failure. Failure in this sense may be defined as a condition which causes the component or structure to become unsafe, or one in which the component ceases to perform its required function.

The major factors which need to be considered in order to assess the integrity of a component from a solid mechanics point of view (neglecting tribological situations and the like) are those of

- (i) strength
- (ii) elastic deformation, and
- (iii) elastic stability

Environmental factors may also considerably influence the integrity of a component or structure, particularly where creep and fatigue considerations are involved.

It is consequently necessary for the designer to be able to calculate stresses, deflections and buckling loads, and to be able to relate them to the appropriate limiting safe values in practice. The decision as to which of the above factors is the critical one will depend upon the nature of the load, the shape of the structure or component and the mechanical properties of the material selected.

STRESSES IN MACHINE PARTS OR STRUCTURES

The type of stress induced in a machine or structure may be either static or dynamic, and one, or a combination, of three basic types, namely

- (i) mechanical,
- (ii) thermal, or
- (iii) chemical.

A mechanical stress is caused by the application of an external force; a thermal stress by differential expansion due to temperature gradient; and a chemical stress by corrosion or some deliberately introduced chemical process such as nitriding or case hardening.

In real situations the designer will be concerned with members which have changes in section, and the effect of such a discontinuity or stress raiser is to cause a local concentration of stress[1]. This local concentration effectively means that the nominal or average stress is amplified by means of a stress concentration factor, usually designated K_t. Thus by definition:

$$K_t = \frac{\text{maximum localised elastic stress}}{\text{nominal or average stress}} \qquad (1)$$

The nominal or average stress can be calculated in the usual way using the equations derived from applied mechanics, and the maximum localised stress can then be obtained by multiplying this nominal stress by the appropriate value for K_t. Values for K_t for a wide variety of common geometries and loading situations have been presented in the

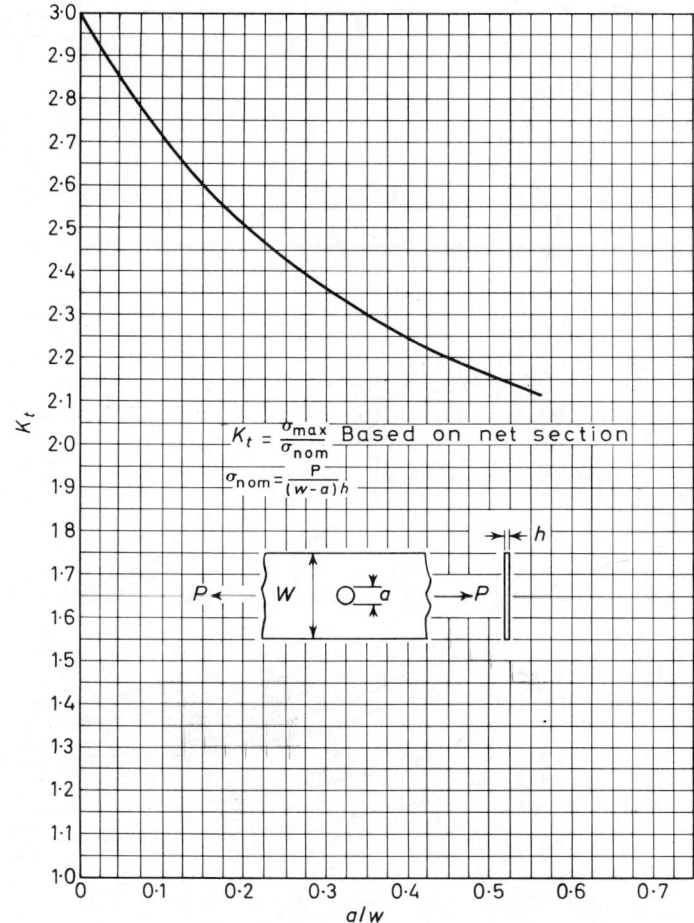

Figure 7.1. Stress concentration factor for axial loading of a plate of finite width with a tranverse hole: nominal stress $= P/[(w-a)h]^2$

form of design curves by Peterson[2], and those of most use are reproduced in Figures 7.1 to 7.12.

Under static loading conditions, unless the material is a brittle homogeneous one, stress concentration is usually of little or no significance. However, for dynamic loading situations, stress concentration is of paramount importance, and should always be considered.

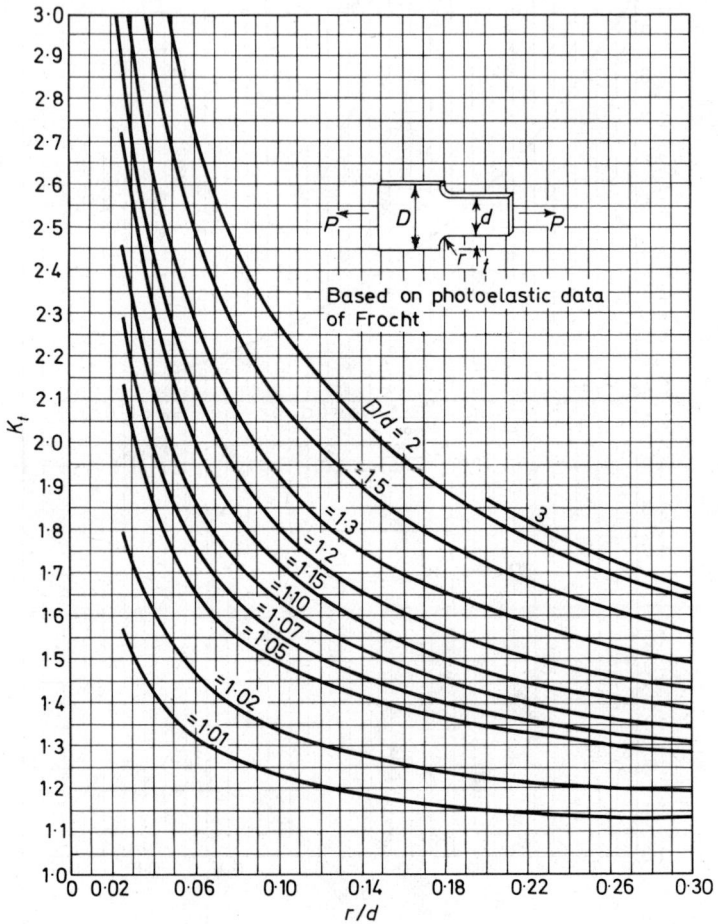

Figure 7.2. Stress concentration factor for the tension of a flat bar with a shoulder fillet[2]

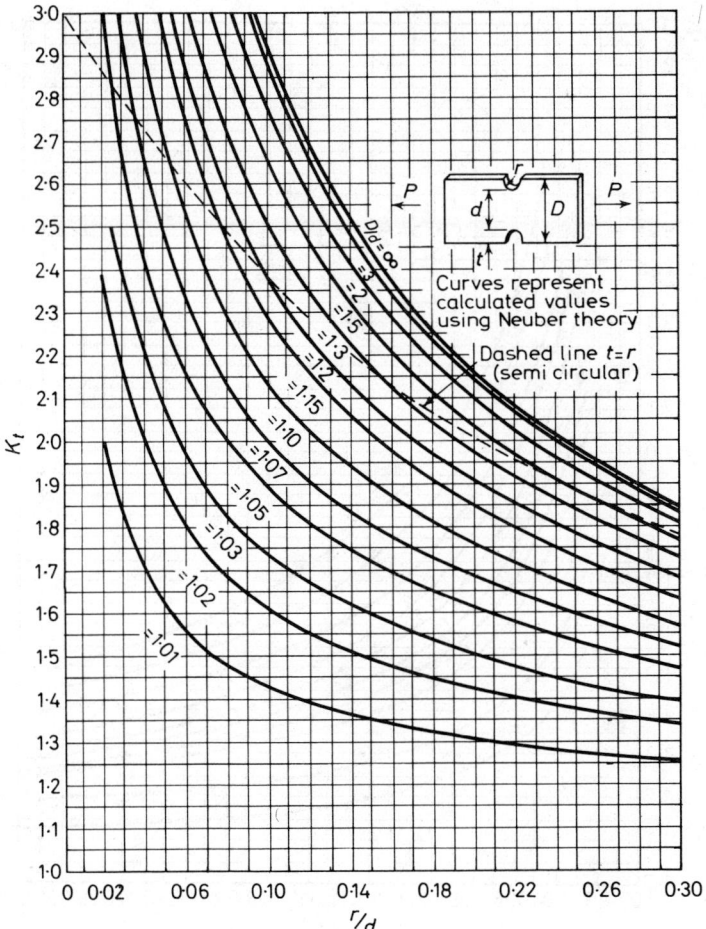

Figure 7.3. Stress concentration factor for a notched flat bar in tension[2]

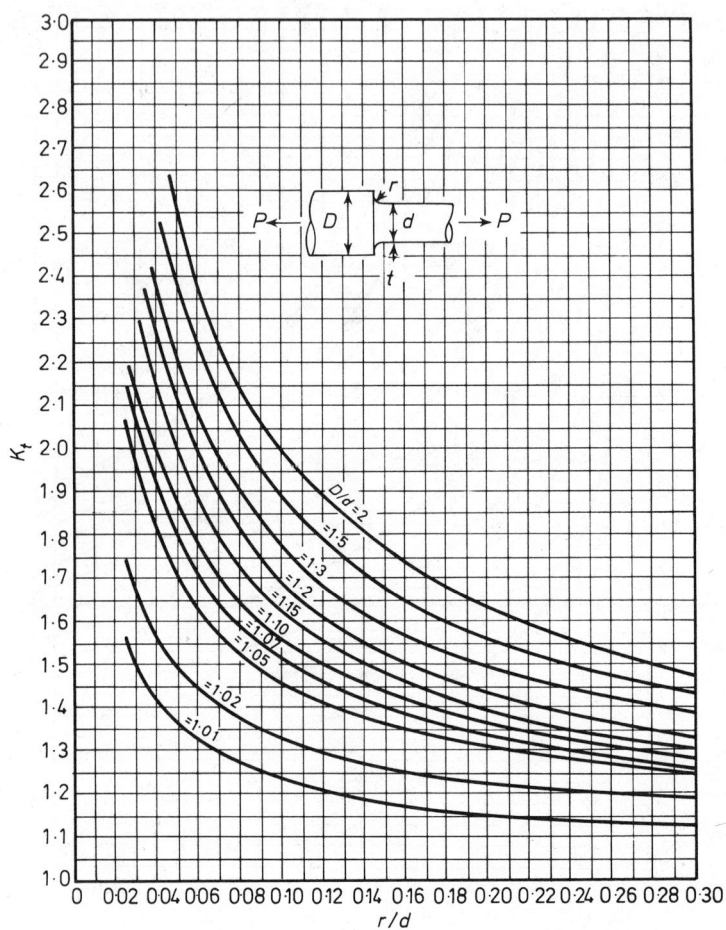

Figure 7.4. Stress concentration for tension of a shaft with shoulder fillet[2]

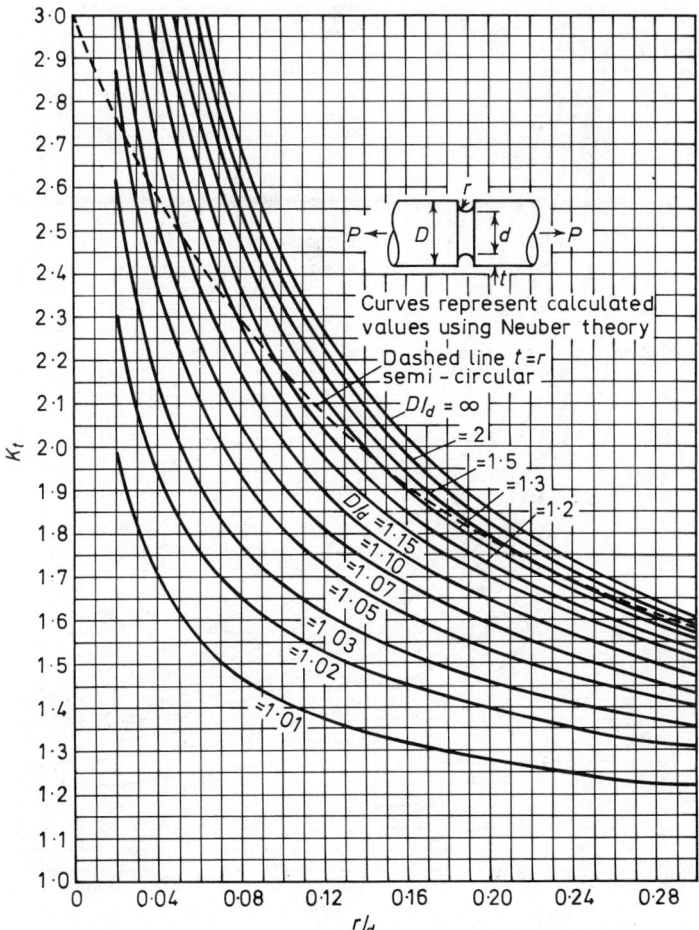

Figure 7.5. Stress concentration factor for a grooved shaft in tension[2]

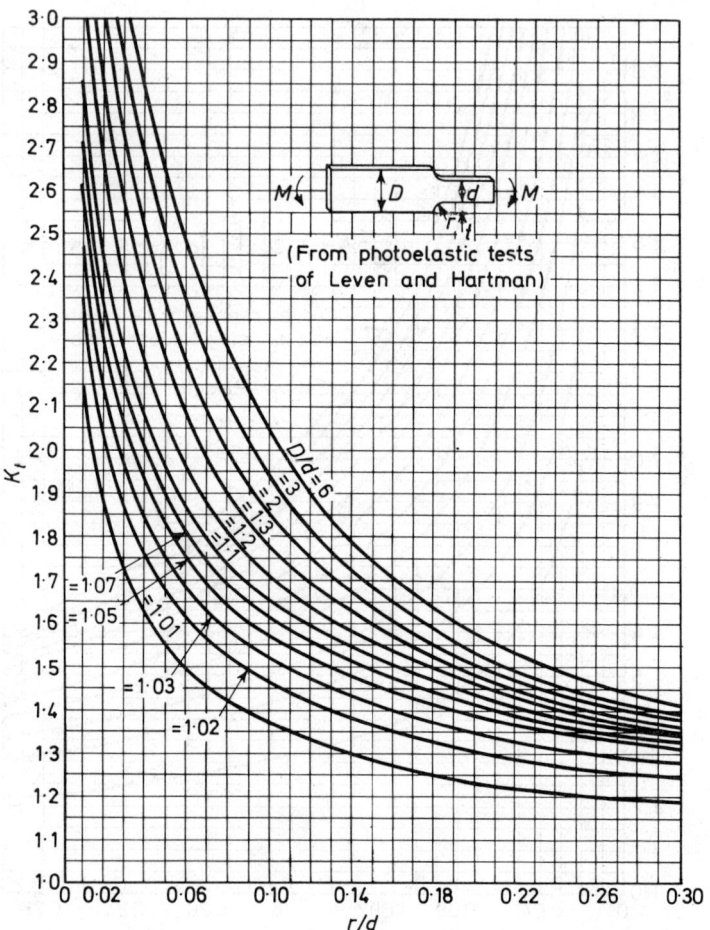

Figure 7.6. Stress concentration for the bending of a bar with a shoulder fillet[2]

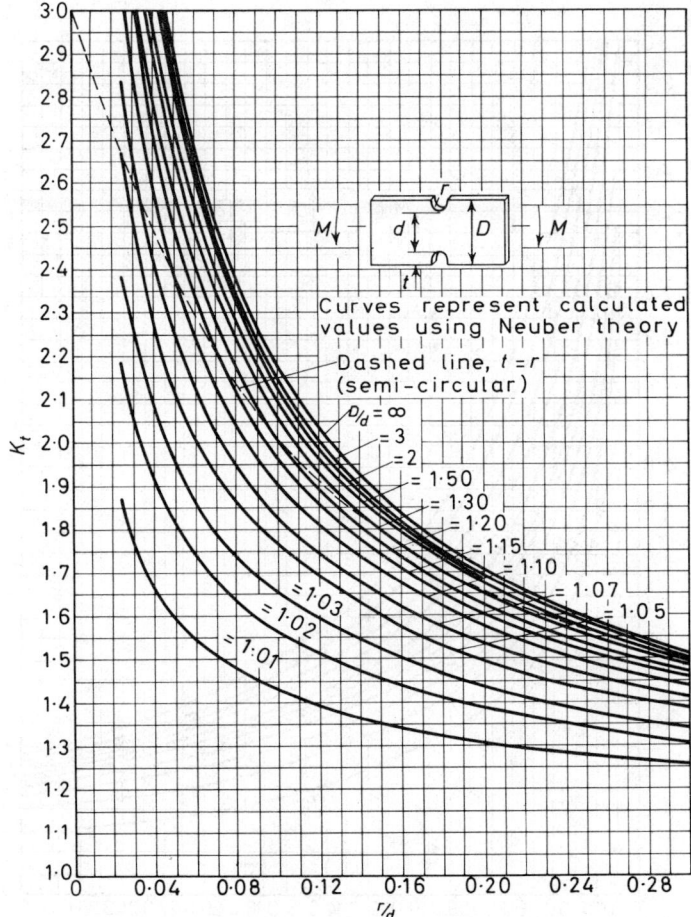

Figure 7.7. Stress concentration factor for a notched flat bar in bending[2]

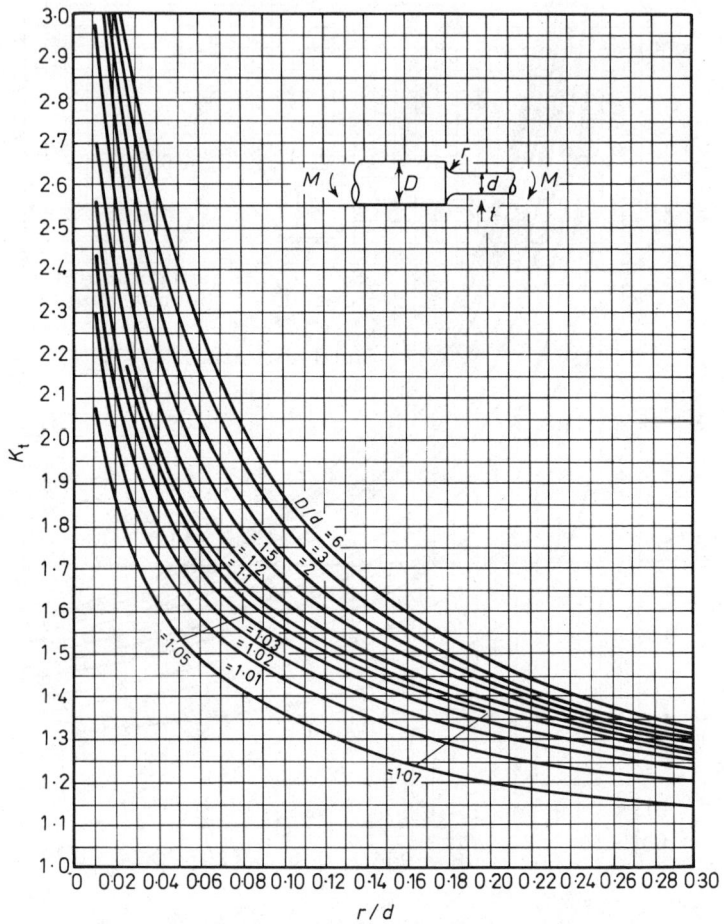

Figure 7.8. Stress concentration factor for bending of a shaft with shoulder fillet[2]

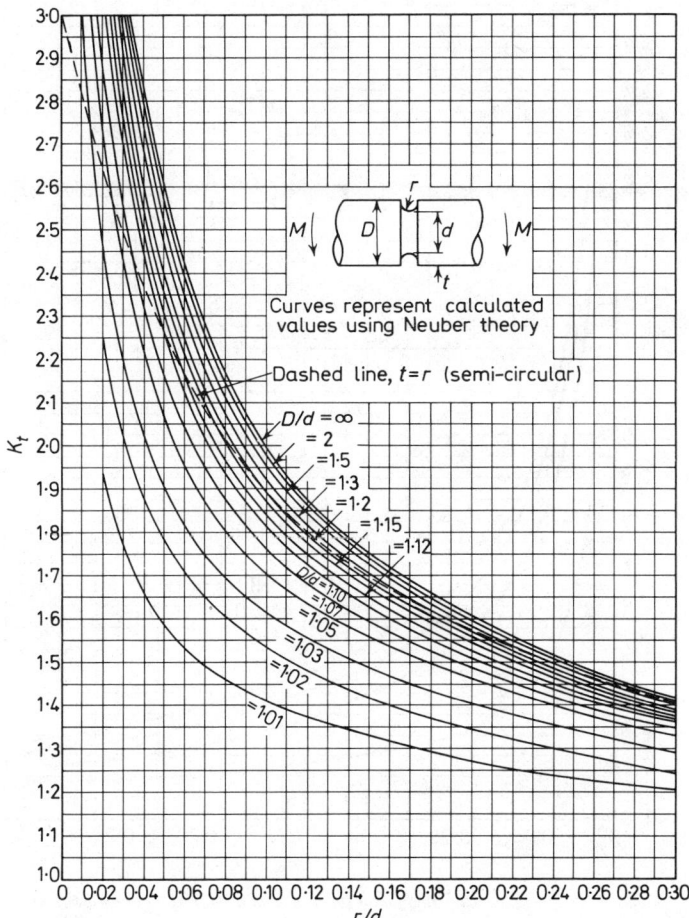

Figure 7.9. Stress concentration factor for a grooved shaft in bending[2]

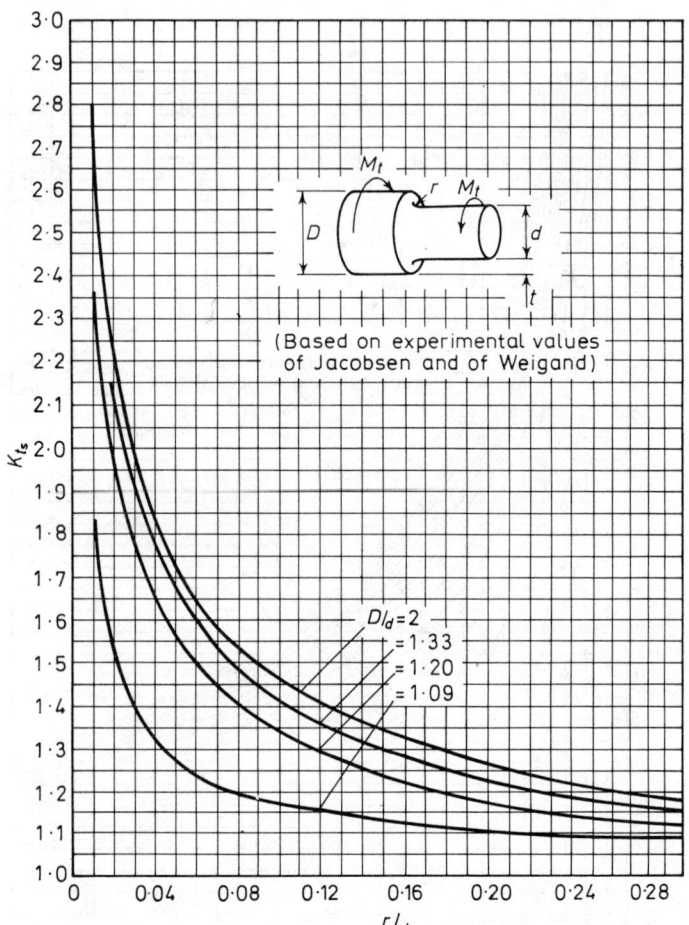

Figure 7.10. Stress concentration factor for torsion of a shaft with a shoulder fillet[2]

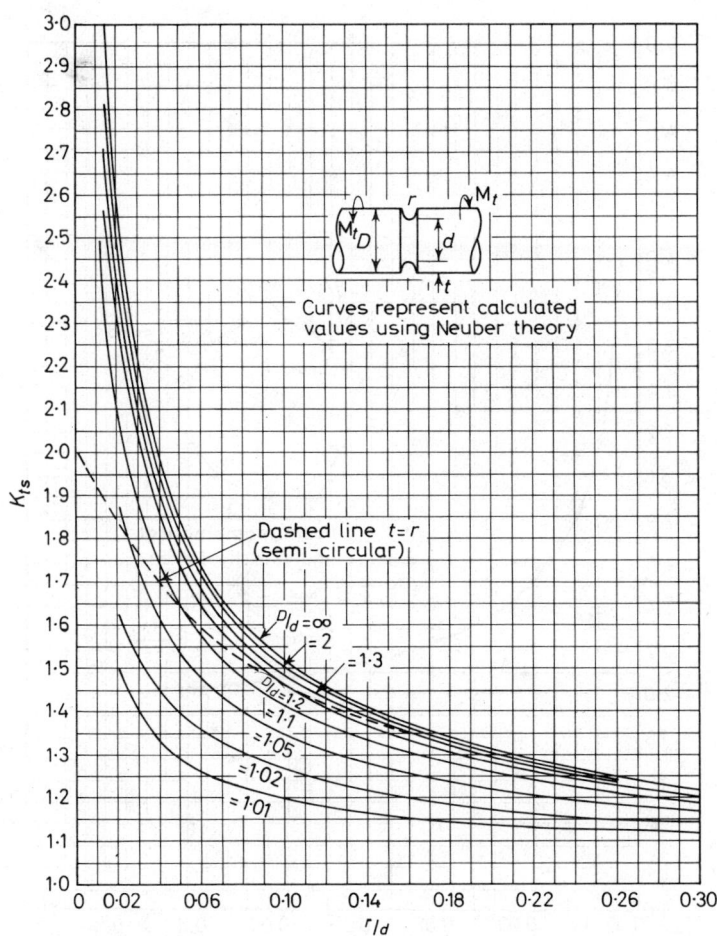

Figure 7.11. Stress concentration factor for a grooved shaft in torsion[2]

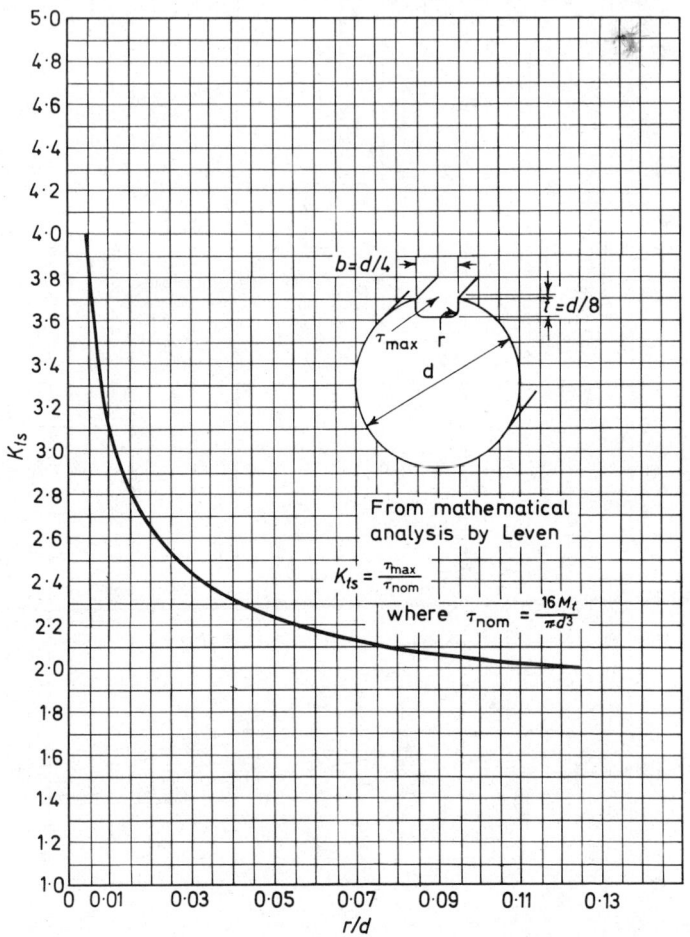

Figure 7.12. Stress concentration factor for the straight portion of a keyway in a shaft in torsion,[2] nominal torsional stress $= 16M_t/\pi d^3$

(Figures 7.1 to 7.12 are reproduced from "Stress Concentration Design" (R. Petersen). Copyright © 1953 John Wiley & Sons Inc. by permission of John Wiley & Sons Inc.

The following sections indicate some of the more useful equations for both mechanical and thermal stress calculations, and are given without proof[1–6].

Mechanical stresses

Mechanical stresses due to external loading can be divided into the following types:
  (a) Axial loading stresses;
  (b) Bending stresses;
  (c) Direct shear stresses;
  (d) Torsional shear stresses;
  (e) Transverse or longitudinal shear stresses;
  (f) Contact, bearing or Hertzian stresses.

AXIAL LOADING STRESSES

The simplest type of stress encountered in a machine element is due to axial loading, and may be calculated from the equation

$$\sigma_a = \frac{P}{A} \tag{2}$$

where σ_a = axial loading stress, MN/m²,
  P = applied external load, N,
  A = cross-sectional area, mm².

Apart from all the usual assumptions about the material, this equation assumes that the load P is axially applied without any eccentricity. It further assumes that the section under consideration is remote from any change in section. If the stress is required at a discontinuity, the nominal stress must be multiplied by K_t to give the peak stress.

BENDING STRESSES

When a beam (which may be part of a structure or a machine part) is loaded by transverse loads, the bending stress may be calculated from the equation

$$\sigma_b = \frac{M}{I} y = \frac{M}{Z} \tag{3}$$

where σ_b = bending stress, MN/m²
  M = bending moment at section considered, N mm,
  Z = section modulus = I/y, mm³
  I = second moment of area of section, mm⁴
  y = distance from neutral axis to outermost fibre, mm.

Values for I and Z for various sections may be calculated using the equations listed in Table 7.1. Note that if all forces are expressed in units of N, and all lengths in mm, the stresses calculated will automatically be in MN/m². If the stress is required at a point of discontinuity, the peak stress is obtained by multiplying the nominal stress by the value of K_t.

Thick curved beams. In the case of thick curved beams, such as encountered in crane hooks and the like, the neutral axis and the centroidal axis are not coincident, and the stresses for such beams may be calculated[1,3] using the equation

$$\sigma_b = \frac{My}{Ae(r+y)} \tag{4}$$

Table 7.1

PROPERTIES OF SECTIONS

y = Distance from neutral axis to outermost fibre

A = Cross-sectional area

I = Second moment of area

Z = Section modulus = I/y

k = Radius of gyration = $\sqrt{I/A}$

Section	y	A	I	Z	k
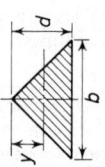	$d/2$	bd	$\dfrac{bd^3}{12}$	$\dfrac{bd^2}{6}$	$d/\sqrt{12}$
	$b/\sqrt{2}$	b^2	$b^4/12$	$b^3/6\sqrt{2}$	$b/\sqrt{12}$
	$\dfrac{2d}{3}$	$\dfrac{bd}{2}$	$\dfrac{bd^3}{36}$	$\dfrac{bd^2}{24}$	$d/\sqrt{48}$

Table 7.1 (*cont.*)

Section	y	A	I	Z	k
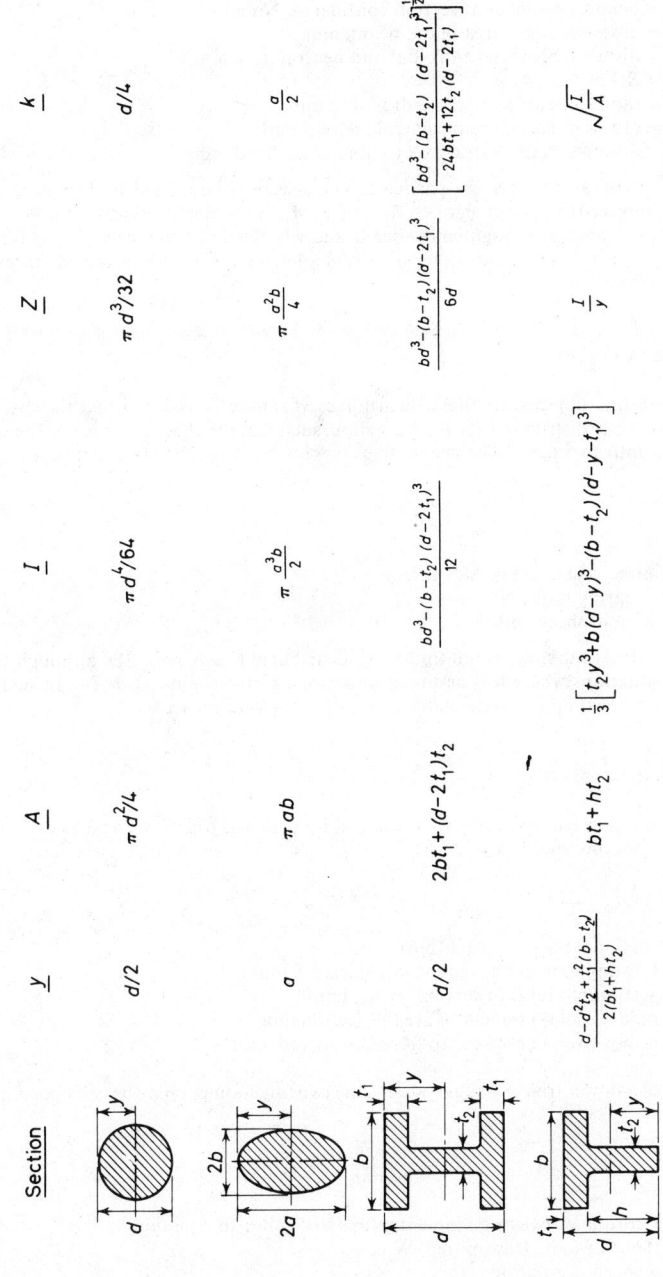	$d/2$	$\pi d^2/4$	$\pi d^4/64$	$\pi d^3/32$	$d/4$
	a	πab	$\pi\dfrac{a^3 b}{2}$	$\pi\dfrac{a^2 b}{4}$	$\dfrac{a}{2}$
	$d/2$	$2bt_1+(d-2t_1)t_2$	$\dfrac{bd^3-(b-t_2)(d-2t_1)^3}{12}$	$\dfrac{bd^3-(b-t_2)(d-2t_1)^3}{6d}$	$\left[\dfrac{bd^3-(b-t_2)(d-2t_1)^3}{24bt_1+12t_2(d-2t_1)}\right]^{\frac{1}{2}}$
	$\dfrac{d-d^2t_2+t_1^2(b-t_2)}{2(bt_1+ht_2)}$	bt_1+ht_2	$\dfrac{1}{3}\left[t_2 y^3+b(d-y)^3-(b-t_2)(d-y-t_1)^3\right]$	$\dfrac{I}{y}$	$\sqrt{\dfrac{I}{A}}$

where σ_b = bending stress, MN/m^2
M = bending moment at section considered, N mm,
A = cross-sectional area of the beam, mm^2,
e = distance between centroidal and neutral axis, mm,
  = $R - r$
r = radius of curvature of neutral axis, mm,
R = radius of curvature of centroidal axis, mm,
y = distance from neutral axis to fibre considered, mm.

The radii of curvature for some particular cross-sections are listed in Table 7.2. It is important to give the correct signs to M and y. M is considered negative if it acts in a direction which tends to straighten the beam, and y is positive if measured away from the centre of curvature. This results in tensile stresses having positive signs and compressive stresses negative signs.

DIRECT SHEAR STRESSES

For components subjected to direct shear forces, it is usually assumed that the shearing stress is uniformly distributed across the section, such that the effect is to slide or shear the component into two parts. The shear stress on this basis is calculated from the simple equation

$$\tau = \frac{Q}{A} \qquad (5)$$

where τ = direct shear stress, MN/m^2,
Q = shearing force, N,
A = area in shear, mm^2

In practice, the shear stress is not uniformly distributed (see page 7–21), although there are many situations where it is assumed to be so, e.g. rivets, pins, keys, etc. In such instances, ensure that appropriate material values for safe stress are used.

TORSIONAL SHEAR STRESSES

For a solid circular bar, subjected to a twisting moment, the torsional shear stress can be calculated from the equation

$$\tau = \frac{M_t}{J} r = \frac{M_t}{Z_t} \qquad (6)$$

where τ = torsional shear stress, MN/m^2,
M_t = twisting moment at section considered, N mm,
Z_t = section modulus in torsion = J/r, mm^3,
J = polar second moment of area of section, mm^4,
r = radius from axis of bar to fibre considered, mm.

In the case of a power transmission system[7], the twisting moment may be calculated from the equation

$$M_t = \frac{9 \cdot 55\,W}{rev/min} \qquad J(Nm) \qquad (7)$$

where M_t = torque or twisting moment, Nm, (Note units in equation 6),
W = power to be transmitted, W or J/s,
rev/min = speed of rotation.

Table 7.2 SOME RADII OF CURVATURE FOR THICK CURVED BEAMS

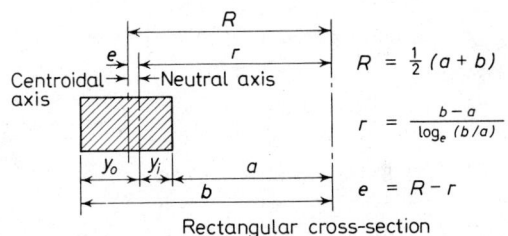

$$R = \tfrac{1}{2}(a+b)$$

$$r = \frac{b-a}{\log_e(b/a)}$$

$$e = R-r$$

Rectangular cross-section

$$R = \tfrac{1}{2}(a+b)$$

$$r = \frac{R}{1+(d/4R)^2}$$

$$e = R-r$$

Circular cross-section

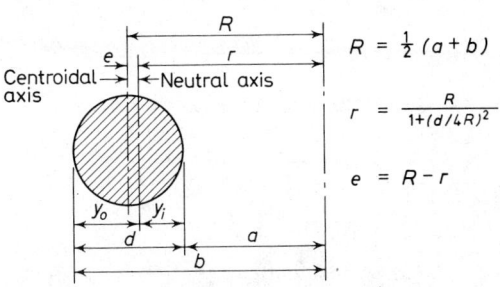

$$R = a + \frac{h(w_2+2w_1)}{3(w_1+w_2)}$$

$$r = \frac{h^2(w_1+w_2)}{2\left[(bw_2-aw_1)\log_e(b/a)-h(w_2-w_1)\right]}$$

$$e = R-r$$

Trapezoidal cross-section

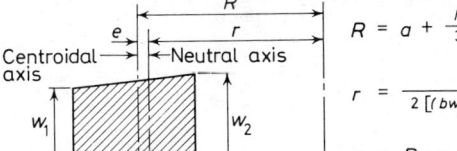

$$R = a + \frac{h^2 t_o + t_i^2(w-t_o)}{2\left[ht_o+(w-t_o)t_i\right]}$$

$$r = \frac{t_i(w-t_o)+ht_o}{(w-t_o)\log_e\frac{a+t_i}{a}+t_o\log_e(b/a)}$$

$$e = R-r$$

Tee cross-section

Table 7.3 MAXIMUM SHEAR STRESS AND ANGLE OF TWIST
FOR SOME SECTIONS UNDER TORSION

$$\tau_{max} = \frac{16 M_t}{\pi d^3}$$

$$\theta = \frac{32 M_t l}{G \pi D^4}$$

(1) Solid circular section

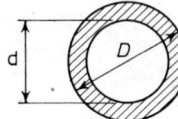

$$\tau_{max} = \frac{16 M_t D}{\pi (D^4 - d^4)}$$

$$\theta = \frac{32 M_t l}{G \pi (D^4 - d^4)}$$

(2) Hollow circular section

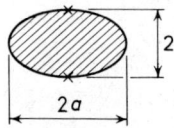

$$\tau_{max} = \frac{2 M_t}{\pi a b^2}$$

$$\theta = \frac{M_t l (a^2 + b^2)}{G \pi a^3 b^3}$$

(3) Solid elliptical section

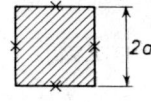

$$\tau_{max} = \frac{3 M_t}{5 a^3}$$

$$\theta = \frac{M_t l}{2 \cdot 25 G a^4}$$

(4) Solid square section

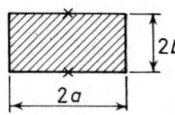

$$\tau_{max} = \frac{M_t (15 a + 9b)}{40 a^2 b^2}$$

$$\theta = \frac{M_t l}{G a b^3 [(\frac{16}{3} - 3 \cdot 36 \frac{b}{a} (1 - \frac{b^4}{12 a^4})]}$$

(5) Solid rectangular section

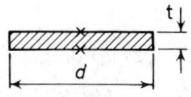

$$\tau_{max} = \frac{3 M_t}{d t^2}$$

$$\theta = \frac{3 M_t l}{G d t^3}$$

(6) Thin rectangular section (plate)

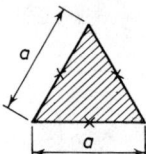

$$\tau_{max} = \frac{20 M_t}{a^3}$$

$$\theta = \frac{80 M_t l}{\sqrt{3} a^4 G}$$

(7) Solid triangular section

If the torsional shear stress is required at a discontinuity, the peak stress is calculated by multiplying the nominal stress by the value for K_t.

It is important to realise that equation (6) is only applicable to solid bars of circular cross-section. With non-circular sections, plane sections do not remain plane but warp. Table 7.3 lists equations for the maximum shear stress and angle of twist for a variety of sections [4, 5, 6].

LONGITUDINAL SHEAR STRESSES

When a beam is subjected to transverse loading, in general, at any point along the beam there exists both bending moments and shearing forces. The stresses due to bending can be evaluated using the equations already discussed, but if the shearing force is assumed to act uniformly over the cross-sectional area, i.e. equation (5) is applied, some error will result. In fact, the shear stress at any section can be calculated using the equation[3]

$$\tau = \frac{Q}{b\,I} \Sigma A_1 \bar{y} \tag{8}$$

where τ = longitudinal shear stress, MN/m^2,

Q = total shear force at cross-section considered, N,

I = second moment of area of cross-section about neutral axis, mm^4,

b = width of cross-section at the point where the shear stress is required, mm,

A_1 = area of that portion of the cross-section between the point where the stress is required and the outer fibres, mm^2,

$\bar{y}$ = distance from the neutral axis to the centroid of area A_1, mm.

The application of equation (8) is best illustrated by reference to Figure 7.13, which indicates the cross-sectional area of a beam subjected to a shear force Q. Study of equation

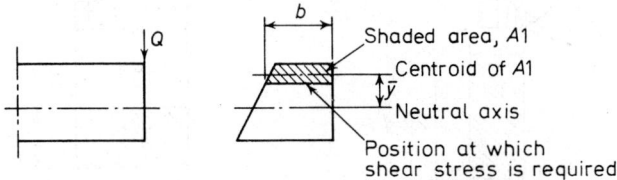

Figure 7.13. Longitudinal shear stress in a beam

(8) indicates that the maximum value of the longitudinal shear stress will occur when $\Sigma A_1 \bar{y}$ is a maximum, and this obviously occurs at the neutral axis. At the outer fibres, where $\Sigma A_1 \bar{y}$ is zero, the shear stress is obviously zero. Consequently, the maximum shear stress occurs where the bending stress is zero, and the maximum bending stress occurs where the shear stress is zero. It is usual, therefore, to consider the bending stresses and shear stresses separately, and if these values are less than the permissible working stresses, then the beam is likely to be satisfactory from the point of view of stress.

CONTACT, BEARING OR HERTZIAN STRESSES

When two elastic bodies are in contact under the action of an externally applied load, local compressive stresses are set up, and deformation of the surfaces occurs. Even though the external load may be quite small, the consequential contact stresses may be quite high owing to the very small area of contact. The classical solution for the local stresses and deformations which occur when two elastic bodies having point contact under load are pressed together is due to Hertz[8], and for this reason the contact stresses are often referred to as Hertzian stresses.

For the sake of convenience, two types of contact can be considered, namely point contact and line contact. Naturally, these hypothetical conditions are obtained only under the action of zero force since, when pressure is developed between the contacting surfaces, some deformation occurs and a finite area of contact is established. Figure 7.14

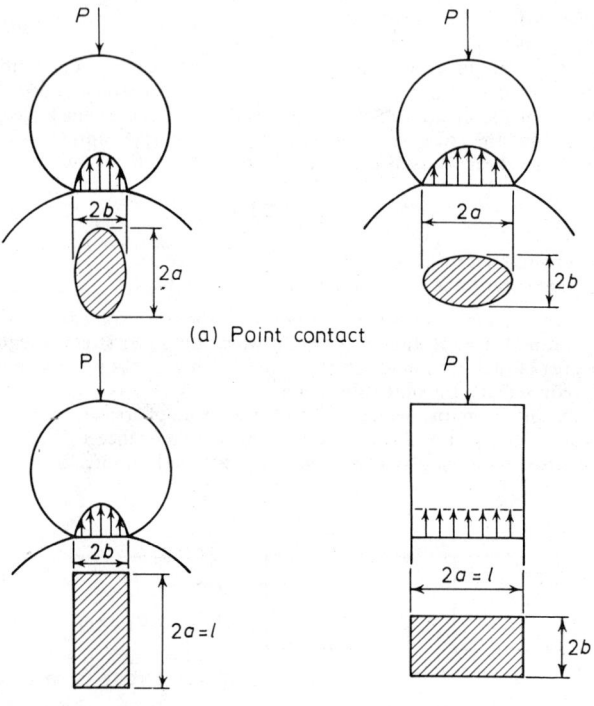

(a) Point contact

(b) Line contact

Figure 7.14. Surface compressive stress distribution, (a) point contact, (b) line contact

indicates the surface compressive stress distribution which occurs for the two conditions of contact, and Table 7.4 lists some equations for calculating Hertzian stresses.

Where
$$k_1 = \frac{1 - v_1^2}{\pi E_1} \text{ and } k_2 = \frac{1 - v_2^2}{\pi E_2}$$

v and E being Poisson's ratio and the elastic modulus respectively.

Combined stresses and theories of failure

Under the action of combined stresses, the elementary equations for direct, torsional, bending and longitudinal stress, although of fundamental importance, are not in themselves sufficient. Many problems in design are concerned with components which are

Table 7.4 HERTZIAN STRESS EQUATIONS FOR SOME SPECIAL CASES

Sphere and flat plate

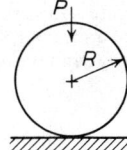

$$\sigma_{cmax} = 0{\cdot}269 \left[\frac{P}{R^2 (k_1 + k_2)^2} \right]^{1/3}$$

Two spheres

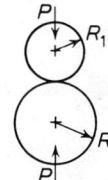

$$\sigma_{cmax} = 0{\cdot}269 \left[\frac{P (R_1 + R_2)^2}{R_1^2 R_2^2 (k_1 + k_2)^2} \right]^{1/3}$$

Sphere and spherical socket

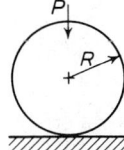

$$\sigma_{cmax} = 0{\cdot}269 \left[\frac{P (R_2 - R_1)^2}{R_1^2 R_2^2 (k_1 + k_2)^2} \right]^{1/3}$$

Cylinder and flat plate

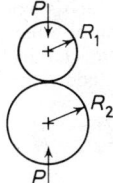

$$\sigma_{cmax} = 0{\cdot}318 \left[\frac{P}{Rl (k_1 + k_2)} \right]^{1/2}$$

Two cylinders

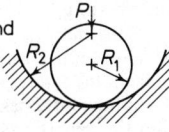

$$\sigma_{cmax} = 0{\cdot}318 \left[\frac{P (R_1 + R_2)}{R_1 R_2 l (k_1 + k_2)} \right]^{1/2}$$

Cylinder and cylindrical groove

$$\sigma_{cmax} = 0{\cdot}318 \left[\frac{P (R_2 - R_1)}{R_1 R_2 l (k_1 + k_2)} \right]^{1/2}$$

subjected to combined stresses, and it is necessary to know how to deal with these situations. Further, one must consider the manner in which materials fail, and appropriate failure theories must be adopted.

If an element from a component, as shown in Figure 7.15, is subjected to direct stresses σ_x and σ_y in two mutually perpendicular planes, and a shear stress τ_{xy} in the x-plane, then it can be shown[1] that the greatest resultant direct stress and the minimum resultant direct

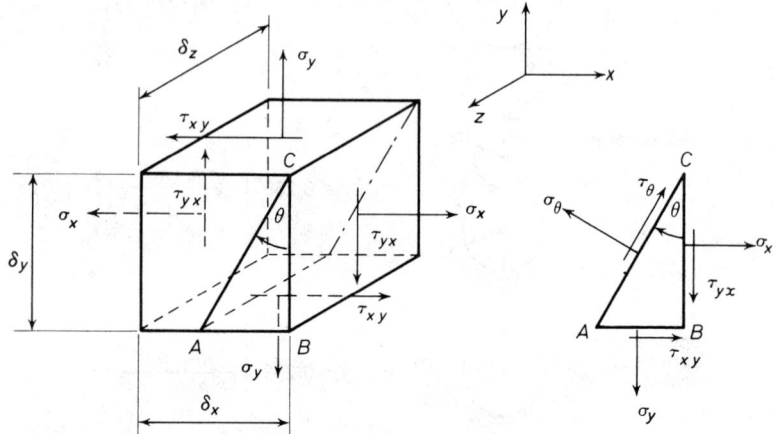

Figure 7.15. A two-dimensional stress system

stress occur on planes 90° apart. These stresses are termed the *maximum* and *minimum* *principal stresses*, and the planes on which they act are termed the *principal planes*. The magnitude of these principal stresses, designated σ_1 and σ_2 can be calculated from the equations

$$\sigma_1 = \tfrac{1}{2}(\sigma_x + \sigma_y) + \tfrac{1}{2}[(\sigma_x - \sigma_y)^2 + 4\tau_{xy}^2]^{\tfrac{1}{2}} \tag{9a}$$

$$\sigma_2 = \tfrac{1}{2}(\sigma_x + \sigma_y) - \tfrac{1}{2}[(\sigma_x - \sigma_y)^2 + 4\tau_{xy}^2]^{\tfrac{1}{2}} \tag{9b}$$

These results are extremely important in stress analysis work, since the application of any failure theory usually involves the calculation of principal stresses. It can also be shown[1] that there is a plane on which the shear stress has its maximum value, known as the plane of maximum shear stress, and the magnitude of the maximum shear stress can be calculated from the equation

$$\tau_{max} = \pm\tfrac{1}{2}[(\sigma_x - \sigma_y)^2 + 4\tau_{xy}^2]^{\tfrac{1}{2}} \tag{10}$$

There are numerous situations where combined stresses are encountered, such as in a shaft which is subjected to combined bending and twisting, and in such instances it is essential that the principal stresses be calculated. This may be done using the above equations, or by means of a convenient graphical solution first suggested by Mohr[9], as shown in Figure 7.16. The construction is evident from the figure.

When an elastic body is subjected to a system of combined stresses, it may fail according to one of a number of limiting conditions. Failure is generally implied, under static loading conditions with the load gradually applied, when the material reaches its yield point or proof stress. Since in many cases yielding does not necessarily mean that a component can no longer perform its required function, it is preferable to consider failure to have occurred when some critical property of the material is reached. Several failure theories have been proposed, each assuming a different limiting quantity as the criterion, but in practice, only three theories are in general use, namely,

   (i) maximum principal stress theory,
   (ii) maximum shear stress theory, and
   (iii) shear strain energy theory.

The *maximum principal stress theory* assumes that failure occurs when the maximum principal stress reaches the limiting acceptable stress for the material, i.e. the yield strength, S_p, the ultimate strength S_u, or some other critical stress value, S_c. Stated in equation form, failure is deemed to occur when

$$\sigma_1 \geq S_c \tag{11}$$

The *maximum shear stress theory* assumes that failure occurs when the maximum shear stress reaches the limiting acceptable shear stress for the material, i.e. the yield point in

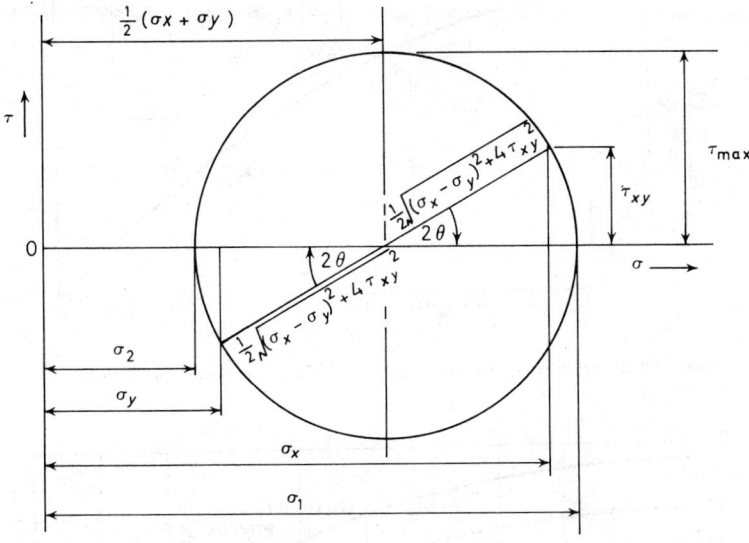

Figure 7.16. Mohr's stress circle

shear, S_{sp}, the ultimate shear strength S_{su}, or some other limiting shear strength S_{sc}. Stated in equation form, failure is deemed to occur when

$$\tau_{max} - \geq S_{sc} \tag{12}$$

or, in terms of the principal stresses,

$$\tfrac{1}{2}\left| \sigma_1 - \sigma_3 \right| \geq S_{sc} \tag{13}$$

where the convention $\sigma_1 > \sigma_2 > \sigma_3$ is used, due account being taken of sign. The *shear strain energy theory*, sometimes referred to as the *distortion energy* or *von-Mises-Hencky* criterion, assumes that failure occurs when the shear strain energy for the complex stress system is equal to the shear strain energy at some critical value in tension. In equation form, failure is deemed to occur when

$$(\sigma_1 - \sigma_2)^2 + (\sigma_2 - \sigma_3)^2 + (\sigma_3 - \sigma_1)^2 \geq 2S_c^2 \tag{14}$$

The assumption is usually made, based upon experimental observation, that $S_{sc} = S_c/2$, so that all the failure theories may be expressed in terms of a single material property, more usually S_c being assumed equal to S_p.

In applying these theories of failure in real design situations, it is necessary to decide which failure theory applies to which material. There is no universal method of knowing this, but generally it has been found that the maximum principal stress is applicable to brittle materials, whilst both the maximum shear stress and shear strain energy theories are applicable to ductile materials. In this latter case, the maximum shear stress theory is more conservative, but the maximum shear strain energy gives the better correlation with experimental data. In distinguishing between brittle and ductile materials, it is suggested that an arbitrary division be made at about 5 per cent elongation.

In applying these theories, use can be made of the design charts shown in Figures 7.17 and 7.18, for the special case where the stress in the y direction is zero. If the intercept of

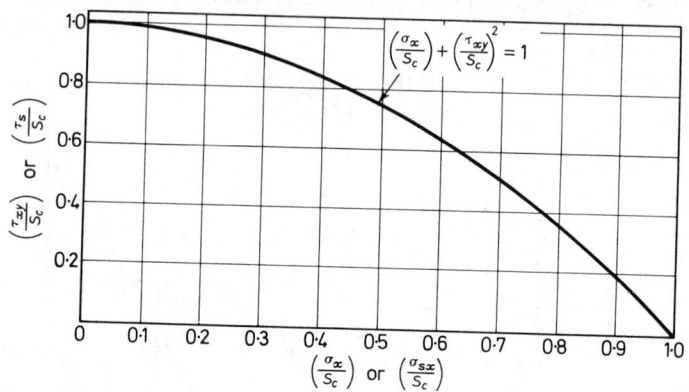

Figure 7.17. Graphical representation of failure as predicted by the maximum principal stress theory

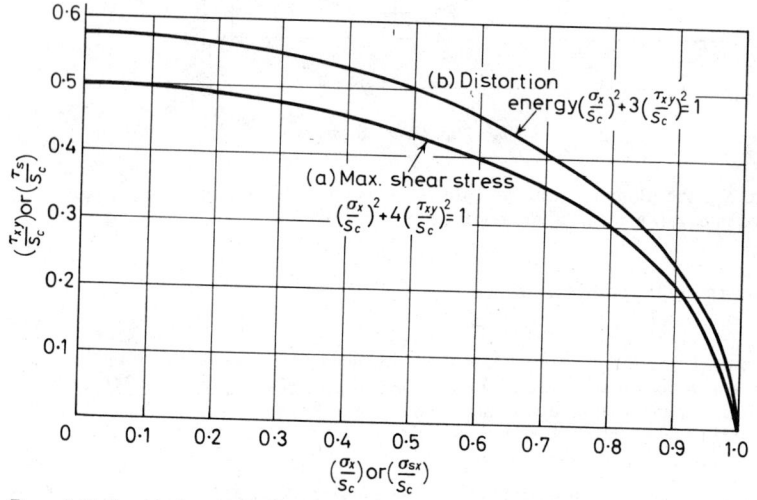

Figure 7.18. Graphical representation of failure as predicted by (a) maximum shear stress theory and (b) maximum shear strain energy theory

the two ratios lies within the curve the design is safe, but if outside the curve it is unsafe. No factors of safety have been incorporated, and these should be considered separately. Of course, in using these design curves, it may be convenient to incorporate an appropriate factor of safety or reserve factor in the value for S_c. Thus, if yielding was to be considered the criterion of failure, and a reserve factor was also to be used, then it would be possible to calculate S_c from

$$S_c = \frac{S_p}{RF} \qquad (15)$$

CIRCULAR SHAFTS SUBJECTED TO COMBINED LOADING

For the situation of a circular shaft subjected to combined bending and torsion, provided that the loads are statically applied or a generous factor of safety is incorporated to allow

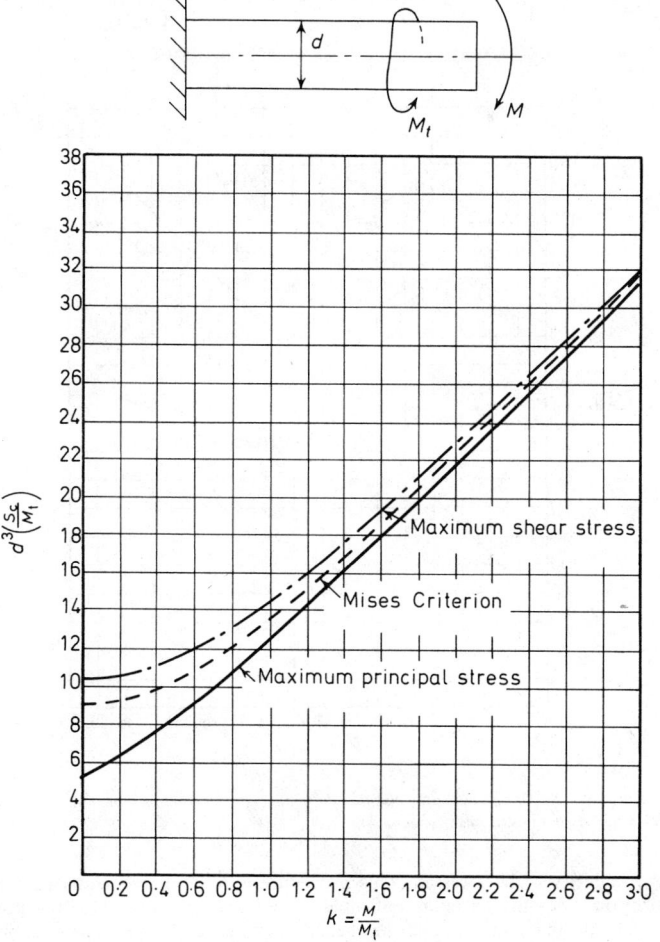

Figure 7.19. Design curves for circular shaft subjected to combined bending and twisting (static case)

for shock and fatigue, the principal stresses may be calculated from equations (9), where

$$\sigma_x = \frac{M}{Z} = \frac{32\,M}{\pi d^3} \tag{16}$$

$$\tau_{xy} = \frac{M_t}{Z_t} = \frac{16\,M_t}{\pi d^3} \tag{17}$$

and

$$\sigma_y = 0$$

If an axial load acts at the same time, the direct stress may be calculated using equation (2), and this value added to the bending stress calculated from equation (16), before determining the principal stresses. The application of an appropriate failure theory

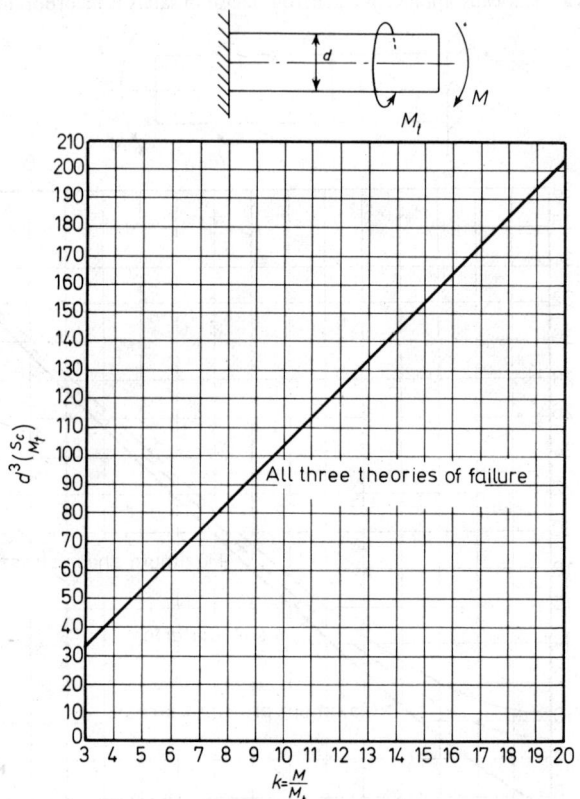

Figure 7.20. Design curve for circular shaft subjected to combined bending and twisting (static case)

enables the integrity to be assessed. This procedure may be conveniently carried out using appropriate design charts, as indicated in Figures 7.19 to 7.21. If the point of intercept of the ordinate and abscissa lies beneath the curve, the design is safe, but if it lies outside the curve it is unsafe. These curves may be used to determine an appropriate shaft diameter, a reserve factor or factor of safety, a limiting value for either M or M_t, or

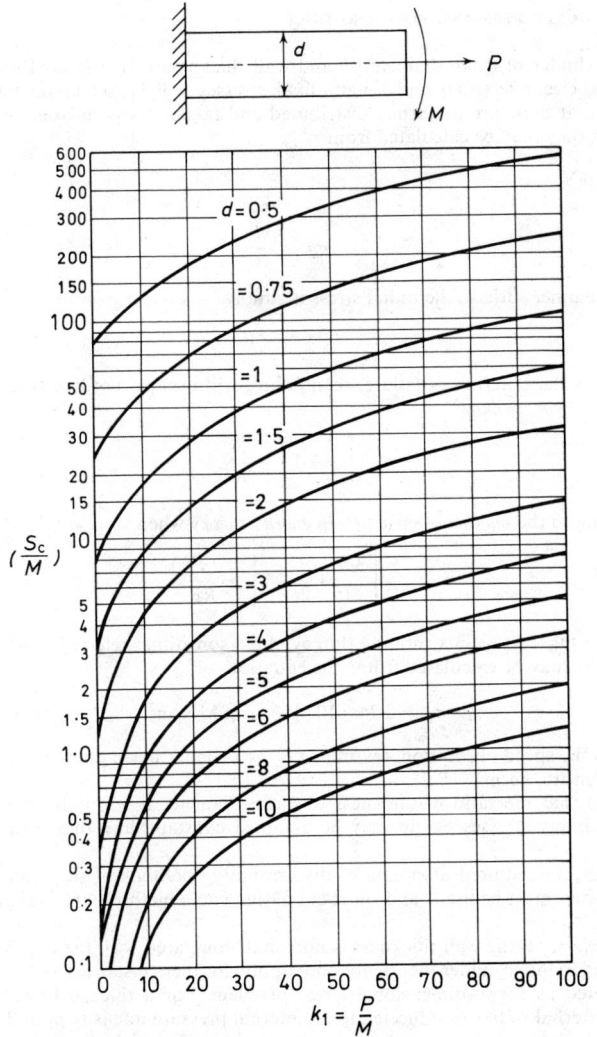

Figure 7.21. Design curve for circular shaft subjected to combined bending
and axial load (static case)

to aid in deciding on an appropriate material. The case of fatigue loading is considered in the section dealing with cyclic loading and fatigue (page 7–33).

STRESSES IN CYLINDERS AND ROTATING DISCS

If a thin cylinder of mean diameter d, and wall thickness t, is subjected to an internal pressure p, circumferential and longitudinal stresses will be set up in the wall, and assuming that these are uniformly distributed and taken at a point remote from a discontinuity, they may be calculated from[3, 4]

$$\sigma_1 = \frac{pd}{2t} \tag{18}$$

$$\sigma_2 = \frac{pd}{4t}$$

Also, at the inner surface, the radial stress is simply

$$\sigma_3 = -p$$

If yielding is the criterion of failure, then failure will occur according to the *maximum shear stress theory* when

$$p\left[\frac{d}{2t} + 1\right] \geq \frac{S_p}{RF} \tag{19}$$

or according to the *maximum shear strain energy theory* when

$$p^2\left[\frac{3d^2}{8t^2} + \frac{d}{2t} + 1\right] \geq \frac{2S_p^2}{RF} \tag{20}$$

The centrifugal stress in a rotating thin cylinder, sometimes referred to as the *bursting* or *skin stress* may be calculated from[3] the equation

$$\sigma_1 = 2\cdot74 \times 10^{-15} N^2 d^2 \rho \ \text{MN/m}^2 \tag{21}$$

where N is the speed of rotation, rev/min, d is the cylinder mean diameter, mm, and ρ is the mass density, kg/m^3.

Provided that the ratio of thickness to mean diameter is less than about $\frac{1}{10}$ (more correctly $\frac{1}{20}$), any stresses set up may be assumed constant, and thin cylinder theory applies.

If stresses are required at points of discontinuity, methods of slope and deflection compatibilities must be used[5] and short cut design curve methods are available[9].

Thick cylinders   If the wall thickness is not small compared with the internal diameter of the cylinder, thin cylinder theory does not apply; for these instances, the problem must be considered as a two-dimensional stress problem. For a thick cylinder of internal radius a, external radius b, subjected to an internal pressure intensity p_1 and an external pressure intensity p_2, it may be shown[3] that the circumferential stress σ_1 and the radial stress σ_2, at any radius r within the wall thickness are given by the equations

$$\sigma_1 = A + \frac{B}{r^2} \tag{22}$$

$$\sigma_2 = A - \frac{B}{r^2} \tag{23}$$

where

$$A = \frac{p_1 a^2 - p_2 b^2}{b^2 - a^2} \tag{24}$$

and

$$B = \frac{a^2 b^2 (p_1 - p_2)}{b^2 - a^2} \tag{25}$$

From these expressions the stress distribution across the cylinder wall can be plotted. In the special case where the external pressure is zero, the maximum stresses occur at the bore and are given by

$$\sigma_{1\,max} = \frac{p_1(a^2 + b^2)}{b^2 - a^2} \tag{26}$$

and

$$\sigma_{2\,max} = -p_1$$

Applying an appropriate theory of failure, the integrity of a cylindrical vessel may be assessed. The procedure may be conveniently carried out using the design chart shown in Figure 7.22. Thus, for any value of the parameter (p/S_c), the required ratio $R(= a/b)$ can be read from the appropriate curve, and hence for a given bore the necessary thickness is obtained. Alternatively, for any given cylinder the maximum safe pressure for a given value of S_c (usually $S_c = S_p/RF$) can be obtained. This curve applies only if low cycle fatigue is of no consequence[10,11]. Note that Figure 7.22 uses non-dimensional parameters, and therefore any consistent units may be used.

Where compound cylinders are to be used to increase efficiency where very high pressures are involved, the magnitude of the contact pressure corresponding to a particular interference fit can be determined[3] using the equation

$$p = \frac{\Delta}{bK} \qquad \text{MN/m}^2 \tag{27}$$

where

$$K = \frac{1}{E_1}\left(\frac{b^2 + c^2}{c^2 - b^2} + v_2\right) + \frac{1}{E_1}\left(\frac{a^2 + b^2}{b^2 - a^2} - v_1\right) \tag{28}$$

and a = inner radius of inner cylinder, mm,
     b = outer radius of inner cylinder, mm,
       = inner radius of outer cylinder, mm,
       = radius of mating surface, mm,
E_1, E_2 = elastic modulus for the material of the inner and outer cylinder respectively, MN/m^2,
v_1, v_2 = Poisson's ratio for the material of the inner and outer cylinder respectively,
  Δ = radial interference, mm.

The magnitude of the interference fit will depend upon the conditions required, i.e. axial, torsional or radial holding ability, but should not create stresses greater than those permissible. In practice, an interference fit of about 0·001 mm per mm of diameter at the mating surface is used, i.e. as an empirical guide only

$$\Delta = 0·000\ 5\ b \text{ mm} \tag{29}$$

Having determined the interference pressure from equation (27), the stresses due to this pressure can be calculated using equations (22) and (23), to which must be added any other stresses due to internal or external pressures.

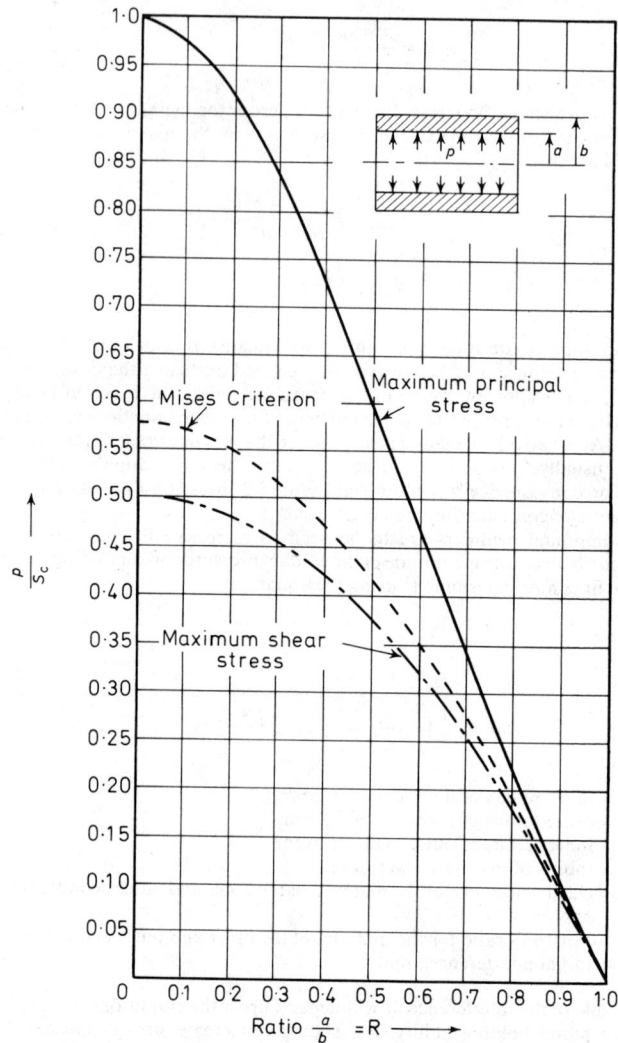

Figure 7.22. Design curve for thick cylinder subjected to internal pressure (static case)

Rotating discs and cylinders   For a disc or cylinder with a central hole rotating freely, it may be shown[3] that the circumferential stress σ_1 and radial stress σ_2 at any radius r are given by the equations

$$\sigma_1 = A + \frac{B}{r^2} - k_1 r^2 \ \text{MN/m}^2 \tag{30}$$

$$\sigma_2 = A - \frac{B}{r^2} - k_2 r^2 \ \text{MN/m}^2 \tag{31}$$

where k_1 and k_2 are rotational constants having the following values

$$k_1 = 1 \cdot 37 \times 10^{-15} N^2 (1 + 3v) \rho \ \text{for a disc} \tag{32a}$$

$$k_1 = 1 \cdot 37 \times 10^{-15} N^2 \left(\frac{1 + 2v}{1 - v} \right) \rho \ \text{for thick cylinder} \tag{32b}$$

$$k_2 = 1 \cdot 37 \times 10^{-15} N^2 (3 + v) \rho \ \text{for a disc} \tag{33a}$$

$$k_3 = 1 \cdot 37 \times 10^{-15} N^2 \left(\frac{3 - 2v}{1 - v} \right) \rho \ \text{for a thick cylinder} \tag{33b}$$

The stresses calculated from equations (30) and (31) will be in MN/m^2, if all dimensions are in mm, and the mass density, ρ is in kg/m^3. The constants A and B are determined from the boundary conditions, and for a freely rotating disc having an inner radius a mm, and an external radius b mm, these constants are calculated from the equations

$$A = k_2 (a^2 + b^2) \tag{34}$$

$$\text{and } B = k_2 a^2 b^2 \tag{35}$$

Cyclic loading and fatigue

When a component is subjected to a fluctuating load, it is possible for failure to occur at a stress level considerably less than that estimated on a static basis, and it is estimated that about 80 % of all machine failures are due to *fatigue*. In calculating the stresses under dynamic loading, the basic equations previously presented in this section can be used, but due account must be taken of stress concentration. However, the calculated stresses will not be constant in value, and the material properties considered must be those obtained under dynamic loading, namely the fatigue strength. Further, many factors affect the fatigue strength of a component, and these must be carefully considered in any analysis.

The fatigue strength of a material must be determined experimentally, and this is normally done by subjecting test specimens to repeated loads of specified magnitude and determining the number of cycles required to produce failure. Such results are plotted graphically with the alternating stress as the ordinate and the number of cycles to produce failure as the abscissa, as shown in Figure 7.23. There is a great deal of scatter associated with the fatigue curve obtained in the laboratory, even for apparently identical test specimens, and the so-called *intrinsic* or *basic material fatigue curve* shown in Figure 7.23 is obtained by fitting the best curve to the test data. It is not possible in a discussion such as this to consider all the important aspects associated with fatigue design, and only design for (assumed) infinite life will be considered. For a more comprehensive discussion the reader should consult the literature available[1, 10-14].

Even accepting that an intrinsic fatigue curve for the material of interest can be obtained by statistical methods to give probability fatigue curves with a high confidence level, before this can be applied to a real component, it must be modified to account for those factors not included but nevertheless affecting fatigue life. Simplifying the situation, the fatigue strength of a component, corresponding to any particular life N_f, might be represented by the equation

$$S'_a = \frac{Sa}{K_f} K_s C_s C_L \tag{36}$$

The lower curve in Figure 7.23 represents the modified fatigue curve for the component, obtained by the application of the factors indicated[1]. In equation (36), the factors are as follows,

(a) K_f is the fatigue strength reduction factor, depending upon the geometry and type of loading, as well as the material,
(b) K_s is the surface finish factor, depending upon type of finish and tensile strength or grain size,
(c) C_s is the size factor, associated with stress gradient, and
(d) C_L is the loading factor, depending upon whether the component is subjected to bending, axial or torsional loading.

S'_a represents the fatigue strength of the component corresponding to any particular life N_f, and S_a is the intrinsic fatigue strength of the material for the same life. The factors mentioned above all assume their greatest effect in the high life region, their influence diminishing somewhat as the life is reduced. Many materials exhibit a so-called *endurance*

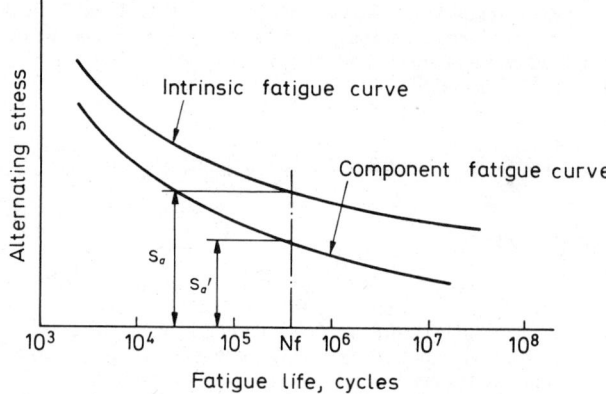

Figure 7.23. Intrinsic and modified fatigue curve

limit, i.e. a stress below which failure will not occur, no matter how many times the load is applied. For steels and ferrous alloys generally, this endurance limit occurs at about one million cycles, whilst for aluminium and non-ferrous alloys, it is generally assumed that an endurance limit exists at about 10^8 or 5×10^8 c.

Fatigue analysis can be conveniently divided into two parts, namely that of determining the component or modified fatigue strength, and that of calculating the stresses, allowing for the combination of mean and alternating components. These two must be equated to assess the fatigue integrity.

For steels having tensile strengths not greatly exceeding about 1200 MN/m², the endurance limit is approximately given by

$$S_e = 0.5 \, S_u \tag{37}$$

and for cast irons and many aluminium alloys the approximation that

$$S_e = 0.3 \, S_u \tag{38}$$

is often used. These relationships are at the best only approximate, and it is recommended that they are only used in the absence of more precise fatigue data.

DETERMINATION OF MAJOR FACTORS AFFECTING FATIGUE STRENGTH

The factors incorporated in equation (36) are the major ones affecting fatigue strength (excluding environmental considerations), and corresponding to the endurance limit, i.e. when $S_a = S_e$, they may be estimated as follows.

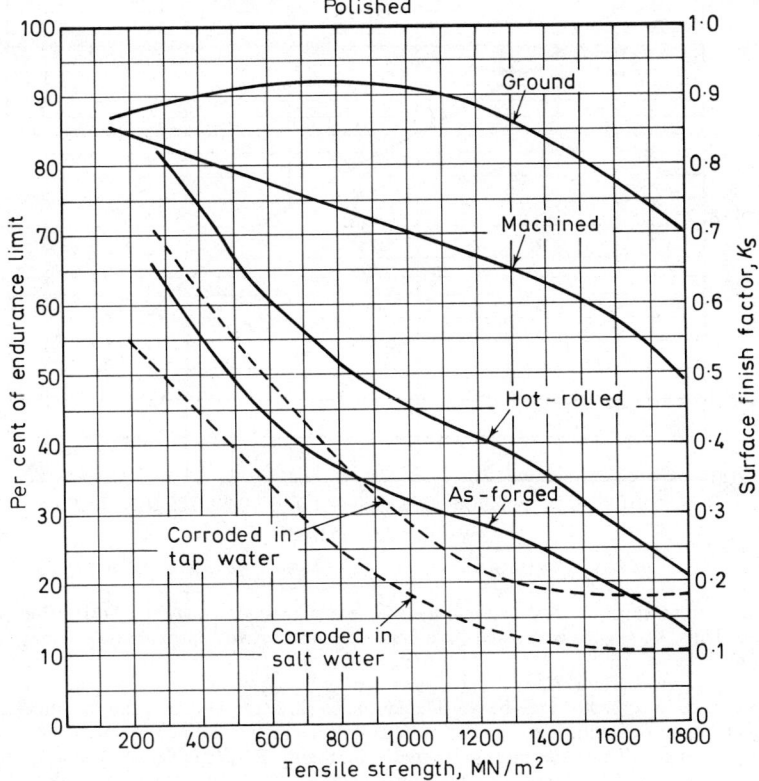

Figure 7.24. Reduction of endurance strength due to surface finish[13]

(a) *Fatigue strength reduction factor, K_f*
The value for K_f may be obtained using the equation[13]

$$K_f = 1 + q(K_t - 1)K_s \tag{39}$$

where K_t = stress concentration factor (Figures 7.1 to 7.12),
K_s = surface finish factor (obtained from Figure 7.24),
q = notch sensitivity index.

The notch sensitivity index is dependent upon the size of the notch (i.e. the radius of a fillet, hole, groove, keyway, or other stress concentration feature), and the material. It may be calculated[15] from the equation

$$q = \frac{1}{1+(a/r)} \tag{40}$$

where r = notch radius, mm, and a = material constant having dimensions mm. The material constant a may be obtained from the curve[16] given in Figure 7.25 for axial or

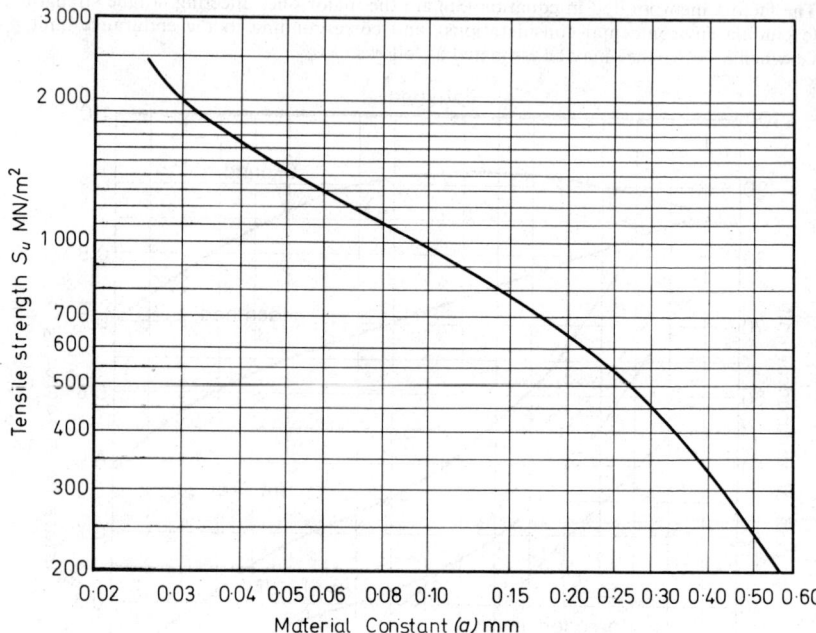

Figure 7.25. Variation of material constant with tensile strength for steels[16]

bending conditions, but for torsional loading it is suggested[15] that 0·6 of this value be taken. Thus, K_f may be calculated. As a first and conservative approximation, it may be assumed that $K_f = K_t$.

(b) Surface finish factor, K_s

The surface finish factor may be considered due to surface roughness, and its influence depends upon the grain size, which in turn (for steels) may be approximately related to tensile strength. Thus, the data presented[13] in Figure 7.24, enables the value of K_s corresponding to a particular combination of surface finish and tensile strength to be obtained.

(c) Size factor, C_s

It has been established that for components subjected to a stress gradient, an increase in component size above the size of the test specimen (usually about 7 to 8 mm) has the effect of reducing the fatigue strength[11, 17]. Up to about 13 mm diameter the fatigue strength decreases by about 15%, and thereafter remains sensibly constant up to about 50 mm or more. Thus, between this range of sizes it is suggested that

$$C_s = 0·85 \text{ for bending and torsion,}$$
$$C_s = 1·0 \text{ for axial loading.}$$

There is evidence[17,20] that reductions of up to 25 or 30% can be obtained for larger components, but sufficient experimental evidence is not available to generally confirm this.

(d) *Loading factor, C_L*

The intrinsic fatigue curve is usually obtained under conditions of rotating bending. If the component being designed is subjected to some other type of loading, its fatigue strength will be different, and must be suitably modified. The following values may be assumed[11,13] for the loading factor

$$C_L = 1 \text{ for rotating or reversed bending,}$$
$$C_L = 0.85 \text{ for axial loading, and}$$
$$C_L = 0.6 \text{ for torsional loading.}$$

Thus, the factors K_f, K_s, C_s and C_L may be obtained as indicated, and the fatigue strength of the component S'_e corresponding to the endurance limit can be obtained using equation (36), where $S'_a = S'_e$

FATIGUE ANALYSIS OF ONE-DIMENSIONAL STRESS SYSTEMS

For a constant amplitude periodic stress, as depicted in Figure 7.26, the system may be completely defined in terms of a mean stress component σ_m, and an alternating component σ_{alt}. These components may be calculated using the equations

$$\sigma_m = \tfrac{1}{2}(\sigma_{max} + \sigma_{min}) \tag{41}$$

and

$$\sigma_{alt} = \tfrac{1}{2}(\sigma_{max} - \sigma_{min}) \tag{42}$$

where σ_{max} and σ_{min} are the maximum and minimum stresses calculated for a one dimen-

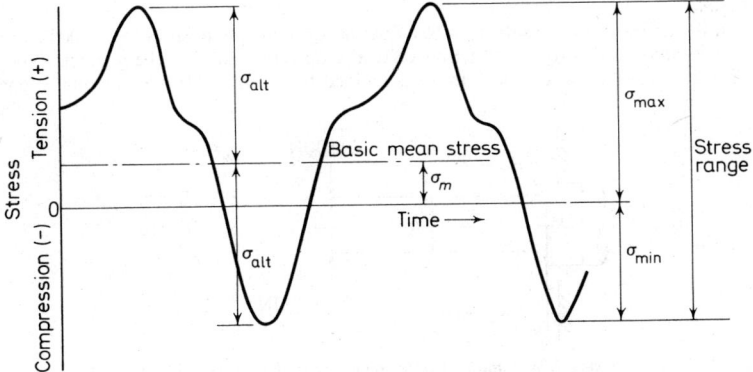

Figure 7.26. Constant amplitude periodic stress cycling

sional stress system using the elementary equations presented in the section on mechanical stresses (page 7-2).

The intrinsic fatigue curve is based on the assumption that the stresses are completely reversed, i.e. that the mean stress is zero. However, the effect of a mean stress is to reduce the fatigue strength, and this must obviously be taken into account. There are a number of suggestions for doing this[1,11,13,18], perhaps the most popular being the use of the modified Goodman relationship. This method is represented in Figure 7.27, and may be expressed mathematically as

$$\sigma_{alt} = S'_e\left[1 - \frac{\sigma_m}{S_u}\right] \tag{43}$$

or, introducing a general reserve factor for both the mean and alternating components, and transposing,

$$RF = \frac{S_u}{\sigma_m + (S_u/S'_e)\sigma_{alt}} \qquad (44)$$

Thus, if σ_m and σ_{alt} are calculated using equations (41) and (42), and S'_e is obtained from equation (36), the value for the reserve factor or factor of safety can be determined from equation (44). Alternatively, the stresses σ_m and σ_{alt} may be in terms geometry, such as

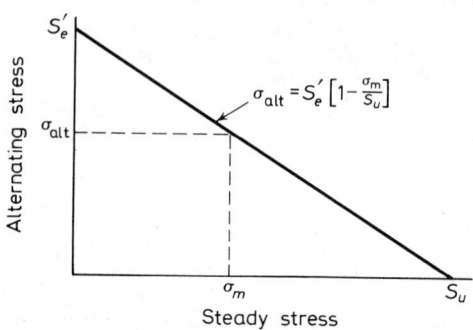

Figure 7.27. Modified Goodman relationship

shaft diameter, section modulus, etc., and if the material has been decided and an appropriate value for RF has been estimated, the unknown geometry can be determined. Another variation on the same theme is the selection of a suitable material.

Example

A cantilever beam is to be subjected to a fluctuating load which varies from 500 N acting vertically downwards to 200 N acting vertically upwards, and has the geometry shown in Figure 7.28. The cantilever is to be machined from En 24(U) steel, having a tensile

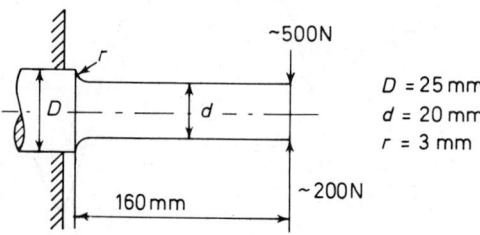

$D = 25\,\text{mm}$
$d = 20\,\text{mm}$
$r = 3\,\text{mm}$

Figure 7.28. Cantilever beam subjected to fluctuating bending

strength of 930 MN/m². Determine the reserve factor (factor of safety), for a life in excess of 10^6c.

Solution

In the absence of more precise information assume (equation 37) that

$$S_e = 0.5\,S_u = 465\,\text{MN/m}^2$$

Reference to the section on fatigue strength (page 7–35) will enable the reader to verify that the major factors to be applied in correcting the endurance strength for the component are

Loading factor (reversed bending) $C_L = 1$
Size factor $C_s = 0.85$

Surface finish factor (Figure 7.24) $K_s = 0.71$
Stress concentration factor (Figure 7.8) $K_t = 1.48$

Using equation (40), and noting that $a = 0.1$ mm (from Figure 7.25),

$$q = \frac{1}{1+(1/3)} = 0.75$$

Using equation (39)

$$K_f = 1 + 0.75(1.48 - 1)0.71 = 1.256$$

By equation (36)

$$S'_e = \frac{465}{1.256} \times 0.71 \times 0.85 \times 1 = 224\ \text{MN/m}^2$$

The maximum and minimum bending stresses can be calculated using equation (3), and noting that $Z = \pi d^3/32$ from Table 7.1. The maximum and minimum bending moments at the critical section are

$$M_{max}\ \ 500 \times 160 = 80\,000\ \text{N mm}$$
$$M_{min} = (-200) \times 160 = -32\,000\ \text{N mm}$$

and the bending stresses are

$$\sigma_{max} = \frac{32 \times 80\,000}{\pi \times (20)^3} = 102\ \text{MN/m}^2$$

$$\sigma_{min} = \frac{32 \times (-32\,000)}{\pi \times (20)^3} = -40.6\ \text{MN/m}^2$$

Using equations (41) and (42) it may be verified that

$$\sigma_m = 30.7\ \text{MN/m}^2, \text{ and}$$
$$\sigma_{alt} = 71.3\ \text{MN/m}^2$$

The value for the reserve factor may now be calculated from equation (44) as

$$RF = \frac{930}{30.7 + (930/224) \times 71.3} = 2.84$$

which is satisfactory.

FATIGUE ANALYSIS FOR COMBINED STRESSES

The above simple procedure may be extended to cope with those situations involving combined fluctuating stresses. This necessitates replacing the dynamic stresses by an equivalent static stress which would have the same damaging effect[21], and assuming that the static failure theories are then applicable. For the general case of two-dimensional fluctuating stresses, the equivalent static stresses can be calculated from the following

$$\sigma_{sx} = \sigma_{mx} + c_1\sigma_{altx} \qquad (45a)$$

$$\sigma_{sy} = \sigma_{my} + c_2\sigma_{alty} \qquad (45b)$$

$$\tau_s = \tau_m + c_3\tau_{alt} \qquad (45c)$$

where σ_{mx}, σ_{my}, σ_{altx}, σ_{alty} represent the mean and alternating components in the x and y directions, and may be calculated using equations (41) and (42). The quantities c_1, c_2 and c_3 represent combined material and geometrical constants, defined as

$$c_1 = \frac{S_{ux}}{S'_{ex}}; c_2 = \frac{S_{uy}}{S'_{ey}}; c_3 = \frac{S_{su}}{S'_{se}} \qquad (46)$$

where S_{ux} and S_{uy} represent the ultimate tensile strengths of the material in the directions x and y, S_{su} represents the ultimate strength of the material in shear (note that S_{su} may be assumed approximately equal to $0.75 \, S_u$ for a wide range of steels); S'_{ex} and S'_{ey} represent the corrected fatigue strengths in the x and y directions, and S'_{se} the corrected torsional fatigue strength. For a component which is isotropic, $S_{ux} = S_{uy}$. Thus, if the equivalent static stresses are calculated using equations (45), the principal stresses may be calculated in the usual way, and the static failure theories applied.

According to the *maximum principal stress theory*

$$\left(\frac{\sigma_{sy} + \sigma_{sy}}{S_c}\right) + \left(\frac{\tau_s}{S_c}\right)^2 - \left(\frac{\sigma_{sx}\sigma_{sy}}{S_c}\right)^2 = 1 \tag{47}$$

If $\sigma_{sy} = 0$, equation (47) reduces to

$$\left(\frac{\sigma_{sx}}{S_c}\right) + \left(\frac{\tau_s}{S_c}\right)^2 = 1 \tag{48}$$

According to the *maximum shear stress theory*

$$\left(\frac{\sigma_{sx} - \sigma_{sy}}{S_c}\right)^2 + 4\left(\frac{\tau_s}{S_c}\right)^2 = 1 \tag{49}$$

If $\sigma_{sy} = 0$, equation (49) reduces to

$$\left(\frac{\sigma_{sx}}{S_c}\right)^2 + 4\left(\frac{\tau_s}{S_c}\right)^2 = 1 \tag{50}$$

According to the *maximum shear strain energy theory*

$$\left(\frac{\sigma_{sx}^2 + \sigma_{sy}^2}{S_c^2}\right) - \left(\frac{\sigma_{sx}\sigma_{sy}}{S_c}\right)^2 + 3\left(\frac{\tau_s}{S_c}\right)^2 = 1 \tag{51}$$

If $\sigma_{sy} = 0$, equation (51) reduces to

$$\left(\frac{\sigma_{sx}}{S_c}\right)^2 + 3\left(\frac{\tau_s}{S_c}\right)^2 = 1 \tag{52}$$

Equations (48), (50) and (52) can be represented graphically as design curves. Such curves will be identical with those shown in Figure 7.17 and 7.18 for static stresses, but the static stresses are replaced by the equivalent static stresses used in a fatigue analysis.

The procedure for assessing the integrity of a component subjected to combined fluctuating stresses can be summarised as follows

1. Determine the values for K_f, K_s, C_s and C_L (see major factors affecting fatigue strength (page 7–35))
2. Calculate the corrected endurance strengths S'_e and S'_{se} using equation (36)
3. Determine the values for the combined material and geometric constants using equations (46)
4. Calculate the maximum and minimum stresses in the x and y directions, and the maximum and minimum shear stresses; using equations (41) and (42) determine the mean and alternating components in each direction.
5. Determine the equivalent static stresses using equation (45)
6. Using an appropriate failure theory, assess the fatigue integrity.

The above procedure may be used to determine geometry, reserve factor or to enable a decision regarding a suitable material to be made, depending upon the particular circumstances. Where the geometry is initially unknown, it will be necessary to make an initial assumption concerning K_f, and it is suggested that $K_f = K_t$ be used. A useful value in the absence of other information is $K_t = 2$. It is suggested that for the general two-dimensional case either equation (49) or (51) be used, equation (49) being somewhat more conservative. In the special case where the direct stress in one direction is zero, the curves indicated in Figure 7.18 may be used, provided that equivalent static stress values are used.

Thermal stresses and creep

In the previous discussions the effect of environment, more particularly that of temperature, has not been considered, and this imposes certain limitations on the results presented. In the first instance, the effect of temperature will, in general, be to alter the mechanical properties of materials from those normally obtained at room temperature; secondly, it may cause thermal stresses to be set up due to external restraint or to differential expansion caused by internal restraint; and thirdly, if the temperature is sufficiently high (in excess of about 0·35 to 0·7 times the absolute melting point of the base metal) progressive deformation of a material under constant load may occur, known as *creep*.

THERMAL STRESSES

Thermal stresses are set up in a structure or component whenever the expansion or contraction which would normally occur due to heating or cooling is prevented. This may be due to either external restraint, e.g. a bar which is uniformly heated clamped rigidly at each end to prevent expansion; or to internal restraint caused by non-uniform temperature distribution giving rise to differential expansion, e.g. a cylindrical pressure vessel with a non-uniform temperature distribution through its wall thickness. In a handbook discussion such as this, it is not possible to consider thermal stresses in detail, and the reader who requires a more comprehensive and detailed discussion should consult the references[5, 22, 23, 24]. It is probably important to point out that many of the difficulties associated with thermal stress calculations are actually due to establishing temperature distributions, which will require heat transfer considerations. For convenience, thermal stress equations for some of the more commonly encountered situations are given below. In all cases, the material properties involved should be taken as those corresponding to the temperature of interest. In all the equations given, α is the coefficient of thermal expansion, mm per mm per deg. C, E is the elastic modulus, MN/m^2, and v is Poisson's ratio. The stresses will then be in units of MN/m^2, with temperature differences in deg. C.

Thermal stresses in beams   For a straight beam, rigidly clamped at each end, and subjected to a temperature change $\Delta\theta$ throughout, the thermal stress is

$$\sigma = E\alpha\Delta\theta \tag{53}$$

For the same beam, but this time subjected to a linear variation of temperature across any two opposite faces (say temperature θ_1 on one face and θ_2 on the other) the maximum thermal stress due to bending is

$$\sigma = \tfrac{1}{2}E\alpha(\theta_1 - \theta_2) \tag{54}$$

Thermal stresses in plates   The theory of plates and shells generally has been treated very comprehensively[25], and equations useful in calculating thermal stresses are given below.

For a uniform flat plate of thickness t, rigidly clamped at its edges, and subjected to a temperature change $\Delta\theta$ throughout, the thermal stress is

$$\sigma = \frac{E\alpha\Delta\theta}{1-v} \tag{55}$$

For the same plate, but this time subjected to a linear variation of temperature across its thickness t (say temperature θ_1 on one face and θ_2 on the other), the thermal stress is

$$\sigma = \frac{E\alpha(\theta_1 - \theta_2)}{2(1-v)} \tag{56}$$

For a flat equilateral triangular plate, rigidly clamped at its edges, and subjected to a linear temperature gradient through its thickness t (say θ_1 on one face and θ_2 on the other), the thermal stress is greatest at the corners and is

$$\sigma = \frac{3E\alpha(\theta_1 - \theta_2)}{4} \tag{57}$$

Thermal stresses in thin cylindrical shells   In the case of a long thin cylindrical shell, subjected to a linear temperature variation across its wall thickness, the circumferential and longitudinal thermal stresses at points remote from the ends, are given by the equation for a restrained plate, i.e.

$$\sigma = \frac{E\alpha(\theta_1 - \theta_2)}{2(1 - v)} \tag{56}$$

Thermal stresses in thick cylinders   In the case of long thick cylinders, a linear distribution of temperature through the wall thickness is no longer justified, and it is more usual to assume a logarithmic distribution. For a cylinder of internal radius a, external radius b, temperature θ_1 at the inner surface and zero at the outer surface, the temperature θ at any radius r from the centre is

$$\theta = \frac{\theta_1}{ln(b/a)} \, ln\frac{b}{r} \tag{58}$$

The thermal stresses in the radial, circumferential (or hoop) and axial directions respectively can be shown[5, 23] to be given by the equations

$$\sigma_r = \frac{E\alpha\theta_1}{2(1-v)ln(b/a)}\left[-ln\frac{b}{r} - \frac{a^2}{b^2-a^2}\left(1 - \frac{b^2}{r^2}\right)ln\frac{b}{a}\right] \tag{59}$$

$$\sigma_h = \frac{E\alpha\theta_1}{2(1-v)ln(b/a)}\left[1 - ln\frac{b}{r} - \frac{a^2}{b^2-a^2}\left(1 + \frac{b^2}{r^2}\right)ln\frac{b}{a}\right] \tag{60}$$

$$\sigma_a = \frac{E\alpha\theta_1}{2(1-v)ln(b/a)}\left[1 - 2\,ln\frac{b}{r} - \frac{2a^2}{b^2-a^2}\,ln\frac{b}{a}\right] \tag{61}$$

The equations presented here should prove adequate for many instances, but for cases not dealt with, the reader should consult the references.

CREEP BEHAVIOUR

Creep is a distortional phenomenon which may be described as progressive deformation under the action of a constant external load. It is only of significance if the operating temperature is in excess of about 0·35 to 0·7 times the melting point of the base metal (depending upon the material).

Basic creep data is usually obtained by testing specimens under conditions of constant stress and temperature, either to produce definite amounts of deformation or until actual rupture. These types of test are known as *creep* and *creep-rupture* tests respectively. Having obtained such basic data, it then has to be applied to design situations, usually very different in conditions from those of the actual test. This handbook discussion permits only a superficial treatment, and the reader is well advised to refer to the literature for a better understanding[23, 26, 27].

Creep in tension   Basic creep data is usually obtained under simple tension, and the creep strain with time is found to consist of the stages indicated in Figure 7.29; the stage most important in design is the second stage.

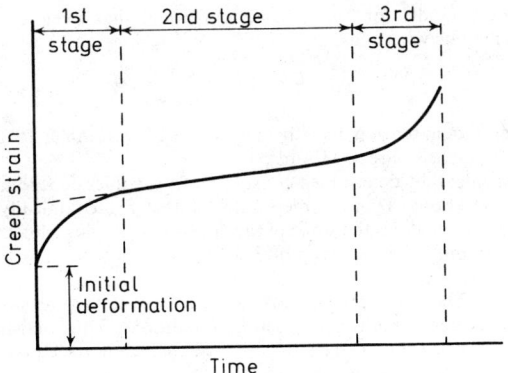

Figure 7.29. Typical creep curve

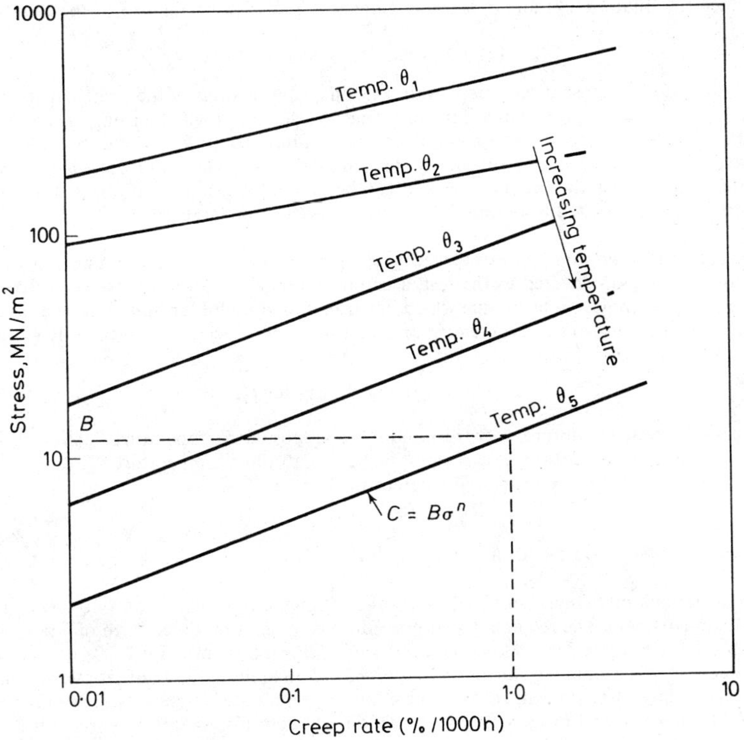

Stress, MN/m² (vertical axis)

Temp. θ_1

Temp. θ_2

Temp. θ_3

Temp. θ_4

Temp. θ_5

Increasing temperature

B

$C = B\sigma^n$

Creep rate (%/1000h)

Figure 7.30. Logarithmic plot of creep rate (%/1 000 h) against stress (MN/m²)

A mathematical model for creep behaviour which has been extensively verified experimentally may be written as

$$C = \frac{\varepsilon}{t} = B\sigma^n \tag{62}$$

where C = creep rate in tension, usually expressed as % strain per 1000 h,
    σ = applied tensile stress, MN/m^2,
B and n are experimentally determined constants whose values depend upon the material and operating conditions. Typical creep data, plotted logarithmically, is indicated in Figure 7.30. The exponent n is the slope of this line on the log-log plot, and B is the stress intercept corresponding to a log-creep rate of 1.0.

Multi-axial creep   The data obtained from simple creep tests in tension can be used to predict creep rates under multi-axial loading conditions. Thus, it can be shown[23, 27] that creep rates for any state of stress can be predicted using the equations

$$C_1 = B\sigma_e^{n-1}[\sigma_1 - \tfrac{1}{2}(\sigma_2 + \sigma_3)] \tag{63a}$$

$$C_2 = B\sigma_e^{n-1}[\sigma_2 - \tfrac{1}{2}(\sigma_1 + \sigma_3)] \tag{63b}$$

$$C_3 = B\sigma_e^{n-1}[\sigma_3 - \tfrac{1}{2}(\sigma_1 + \sigma_2)] \tag{63c}$$

where B and n are experimentally determined constants obtained in simple tension, as defined by equation (62), and σ_e is the so-called *equivalent stress*, defined by the equation

$$\sigma_e = (\sigma_1^2 + \sigma_2^2 + \sigma_3^2 - \sigma_1\sigma_2 - \sigma_2\sigma_3 - \sigma_1\sigma_3)^{\frac{1}{2}} \tag{64}$$

C_1, C_2 and C_3 represent the creep strain rates in the directions of the principal stresses σ_1, σ_2 and σ_3 respectively. Thus, if the principal stresses are calculated using the methods outlined in 'Combined Stresses and Theory of Failure' (page 7–22), and the material constants B and n are experimentally determined values which are obtained for the material of interest, the creep strain rates can be calculated using equations (63), and the values obtained related to permissible values to assess creep integrity.

Extrapolated creep data   In extrapolating creep data curves to values of time and stress beyond the region covered in the test, a certain amount of caution must be exercised. One such extrapolation technique which has been found useful for quite a wide application, is the Larson-Miller parameter[28]. In equation form this can be conveniently written as

$$\theta_1(20 + \log t_1) = \theta_2(20 + \log t_2) \tag{65}$$

where t_1 = time to rupture, h, corresponding to an absolute temperature θ_1, K,
    t_2 = time to rupture, h, corresponding to an absolute temperature θ_2, K.
The constant 20 is an experimentally determined value.

COMBINED CREEP AND FATIGUE

For situations involving combined creep and fatigue, any analysis can only be, at the best, approximate. The recommended procedure is to use a modified form of Goodman diagram (see Fatigue Analysis of One-dimensional Stress Systems page 7–37), combining fatigue and creep data, as shown in Figure 7.31. The limiting value of fatigue strength corresponding to zero mean stress is the endurance limit at the temperature of interest. The limiting value of mean stress corresponding to zero alternating stress is either the creep rupture stress, or the stress corresponding to an acceptable creep elongation after a certain period of time.

The linear relationship is generally over-conservative, and an elliptical approximation has been suggested which, in equation form, may be written as

$$\left(\frac{\sigma_{alt}}{S'_e}\right)^2 + \left(\frac{\sigma_m}{S_c}\right)^2 = 1 \tag{66}$$

where S'_e is the modified fatigue strength for the component (see Cyclic Loading and Fatigue, page 7-33), at the temperature of interest, and S_c is either the creep rupture stress or the stress corresponding to an acceptable creep elongation, both in MN/m^2.

The procedure for one-dimensional stress systems is straight forward, and for multi-axial stress systems involving fluctuating stresses and high temperatures, the methods

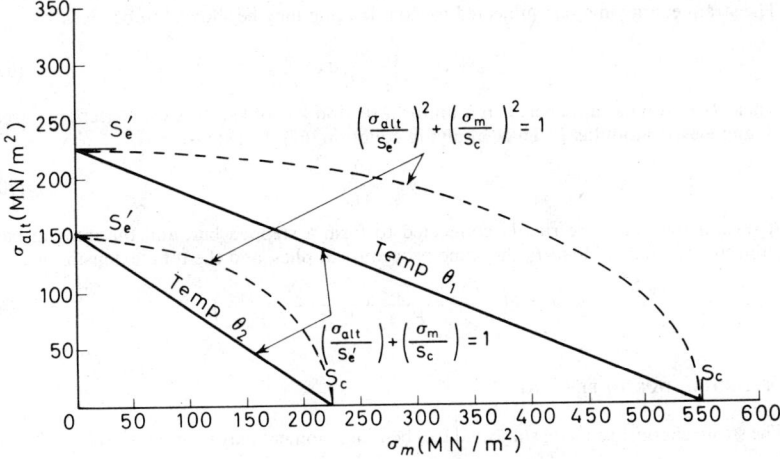

Figure 7.31. Combined creep-fatigue curves ($\theta_2 < \theta_1$ for same material)

outlined in 'Fatigue analysis for combined stress systems and creep behaviour' (pages 7-39 and 7-42) can be combined to assess integrity.

Elastic deformation

When a component is subjected to applied forces some deformation takes place, the actual amount being dependent upon shape and elasticity of the component, as well as the nature of the load. It is important to ensure that such deformations as do occur do not interfere with the satisfactory operation of the system, and it is consequently necessary to calculate such deflections and relate these to safe values. In the present discussion, the emphasis is on general methods of solution, rather than simply quoting numerous standard equations which would, in the main, be of only academic interest. For complex systems, a very powerful method of solution is that using strain energy considerations. Stated briefly, the deflection under any load (real or dummy), can be obtained by writing the mathematical equation for strain energy of the system, and then applying Castigliano's first theorem. This states that if the total strain energy in a body is partially differentiated with respect to a particular load, then the partial derivative so obtained will give the deflection corresponding to the load in the direction in which the load acts. Stated mathematically,

$$\frac{\partial U}{\partial P} = y \tag{67}$$

Similarly, the partial derivative of the total strain energy of an elastic structure, with respect to any particular twisting moment or torque, will give the angular displacement of the twisting moment in the direction in which it acts, thus,

$$\frac{\partial U}{\partial M_t} = \theta \tag{68}$$

Further discussions on these methods can be found in the references [1, 3, 4]

DEFLECTION DUE TO AXIAL LOADING

The strain energy in a bar subjected to axial loading may be shown[3] to be given by

$$U = \int_0^l \frac{P^2}{2AE}\,dx \tag{69}$$

where U = total strain energy in a uniform section bar of length l, cross-sectional area A, and elastic modulus E. Thus, applying equation (67) the extension is

$$y = \frac{Pl}{AE} \tag{70}$$

If several such bars are rigidly connected to form a stepped bar, and the integral bar subjected to an axial load P, the same procedure applies, and the total extension is

$$y = P\left(\frac{l_1}{E_1 A_1} + \frac{l_2}{E_2 A_2} + \frac{l_3}{E_3 A_3} + \cdots \frac{l_n}{E_n A_n}\right) \tag{71}$$

DEFLECTION DUE TO BENDING

The strain energy in a beam subjected to a bending moment may be shown[3] to be given by

$$U = \int_0^l \frac{M^2}{2EI}\,dx \tag{72}$$

where U = total strain energy of a uniform section beam of length l, second moment of area I, and elastic modulus E. To determine deflection for any type of transverse loading, simply write down the equation for the bending moment in terms of the loads, substitute into equation (72), and partially differentiate with respect to the load (or dummy load) at which the deflection is required. The reader should be able to obtain the standard equations for beams of constant flexure in this way. If the deflection is required at a point where no load acts, simply introduce a dummy load at this point, and after partially differentiating, put this load equal to zero.

In the majority of real design situations concerned with beams, shafts and the like, the flexure (EI) is not likely to be constant throughout, nor is the force system likely to be simple. In such instances, a number of powerful methods are available[1, 7], and the discussion here will be limited to a modification of the area moment method[3]. This method is illustrated in Figure 7.32, and is based on the principle that the deflection of the elastic line relative to the datum line (which may be the original unbent centre line of the beam in elementary cases where the point of zero slope is known), at any point, is obtained by taking the first moment of area of the (M/EI) diagram about that point and the point of zero slope. Thus, referring to Figure 7.32, the distance Δ_1 can be obtained by working between the points B and A, i.e.

$$\Delta_1 = \int_B^A \frac{M}{EI} \times dx = A_1 \bar{x}_1$$

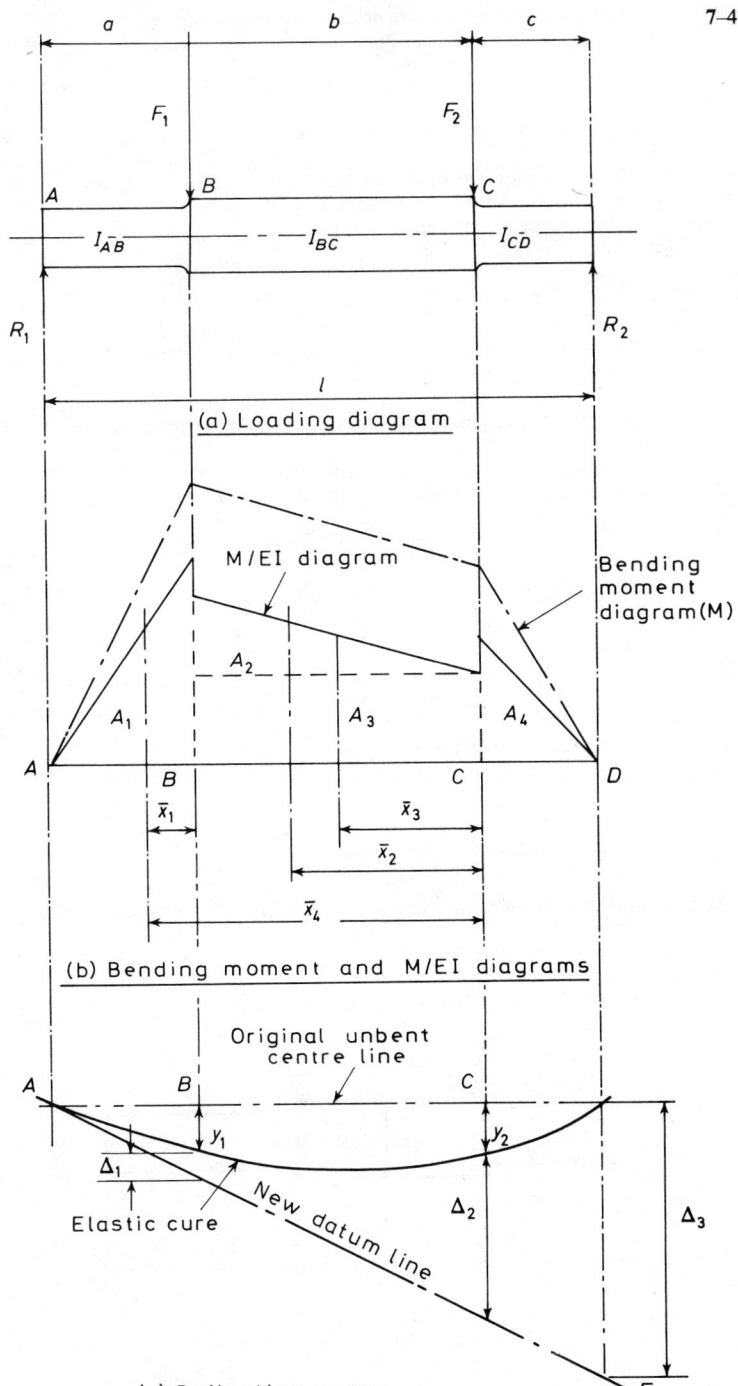

Figure 7.32. *Modified area moment method for beam of non-constant flexure*

Similarly,
$$\Delta_2 = \int_c^A \frac{M}{EI} \times dx = A_1\bar{x}_4 + A_2\bar{x}_2 + A_3\bar{x}_3$$

where A_1, A_2 and A_3 represent the elemental areas of the (M/EI) diagram (as indicated in Figure 7.32), and $\bar{x}_1$, $\bar{x}_2$, $\bar{x}_3$, etc., represent the appropriate distances from the point where the deflection is required to the centroid of the elemental areas of the (M/EI) diagram. The distance Δ_3 can be obtained in a similar manner. The actual deflections y_1, y_2, etc., can be obtained from similar triangles, i.e.

$$\frac{\Delta_1 + y_1}{a} = \frac{\Delta_2 + y_2}{a+b} = \frac{\Delta_3}{l}$$

The method is an extremely powerful one, permitting deflection calculations to be made for beams of non-constant flexure having complex loadings. No scale drawings are required, it being necessary only to calculate the bending moments and sketch the (M/EI) diagram. Note that where a change in section occurs on the beam, we have a discontinuity in the M/EI diagram, and the beam should be divided up into sections of constant flexure, and the bending moment calculated at each discontinuity.

The natural frequency of vibration for a beam or shaft may be estimated based on the static deflection curve, and the following equations[7] enable the natural frequency to be calculated.

or

$$f = 15 \cdot 76 \left[\frac{\Sigma_1^n m_i y_i}{\Sigma_1^n m_i y_i^2} \right]^{\frac{1}{2}} \text{ Hz} \tag{73}$$

$$N = 945 \cdot 6 \left[\frac{\Sigma_1^n m_i y_i}{\Sigma_1^n m_i y_i^2} \right]^{\frac{1}{2}} \text{ vib/min} \tag{74}$$

where m_i — mass of appropriate lumped parameter mass, kg,

y_i = deflection corresponding to the i'th mass, mm.

A thorough discussion relating to vibration considerations of beams and shafts has already been presented by the author[7], to which the reader is referred for further information.

DEFLECTION DUE TO TORSION

The strain energy in a circular shaft subjected to twisting may be shown[3] to be given by

$$U = \int_0^L \frac{M_t^2}{2JG} dx \tag{75}$$

where U = total strain energy of a uniform section shaft of length L, polar second

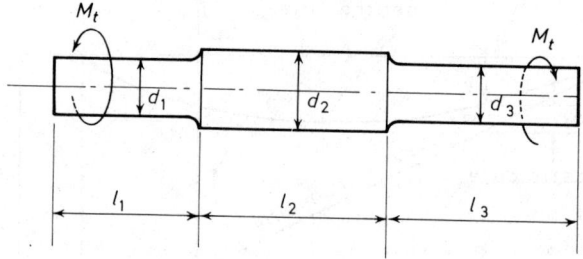

Figure 7.33. Stepped shaft in torsion

moment of area J, and elastic shear modulus G. Thus, applying equation (68), the angle of twist is

$$\theta = \frac{L}{GJ} M_t \qquad (76)$$

In the case of a stepped shaft, as indicated in Figure 7.33, it is usual to replace the stepped shaft by an equivalent uniform diameter shaft of length L, having a diameter

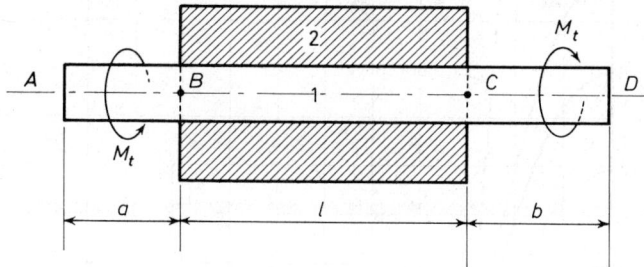

Figure 7.34. Sleeved shaft in torsion

corresponding to the smallest diameter of the stepped shaft. It can be shown[3] that the length L of the equivalent shaft, having diameter d_1 which twists through the same angle as the original stepped shaft is given by the equation

$$L = l_1 + l_2 \left(\frac{d_1}{d_2}\right)^4 + l_3 \left(\frac{d_1}{d_3}\right)^4 \qquad (77)$$

which may be extended for any number of sections.

Where it is advantageous to make a shaft torsionally stiffer by fitting a sleeve, as illustrated in Figure 7.34, if the interference fit between the shaft and the sleeve is sufficient to prevent slipping, then over the section where the sleeve fits, it may be shown[3] that the angle of twist is given by the equation

$$\theta_{BC} = \frac{l}{G_1 J_1 + G_2 J_2} M_t \qquad (78)$$

where the suffix 1 refers to the shaft, and the suffix 2 to the sleeve.

Elastic instability

Structures may fail in some instances not by exceeding limiting stress or deflection values, but by general collapse or instability. Such instances may arise if the cross-sectional dimensions are small compared with the length, and the instability or buckling will always take place about the axis of least resistance. This is usually the axis containing the least second moment of area of the section, but the end conditions play a very important part. Pressure vessels can also fail by collapse due to the application of either internal or external pressure, although the latter is more likely.

BUCKLING OF STRUTS

The classical theory for buckling of long columns or struts is due to Euler, and in equation form[3] may be expressed by the equation

$$P_{cr} = n \frac{\pi^2 EI}{l^2} \qquad (79)$$

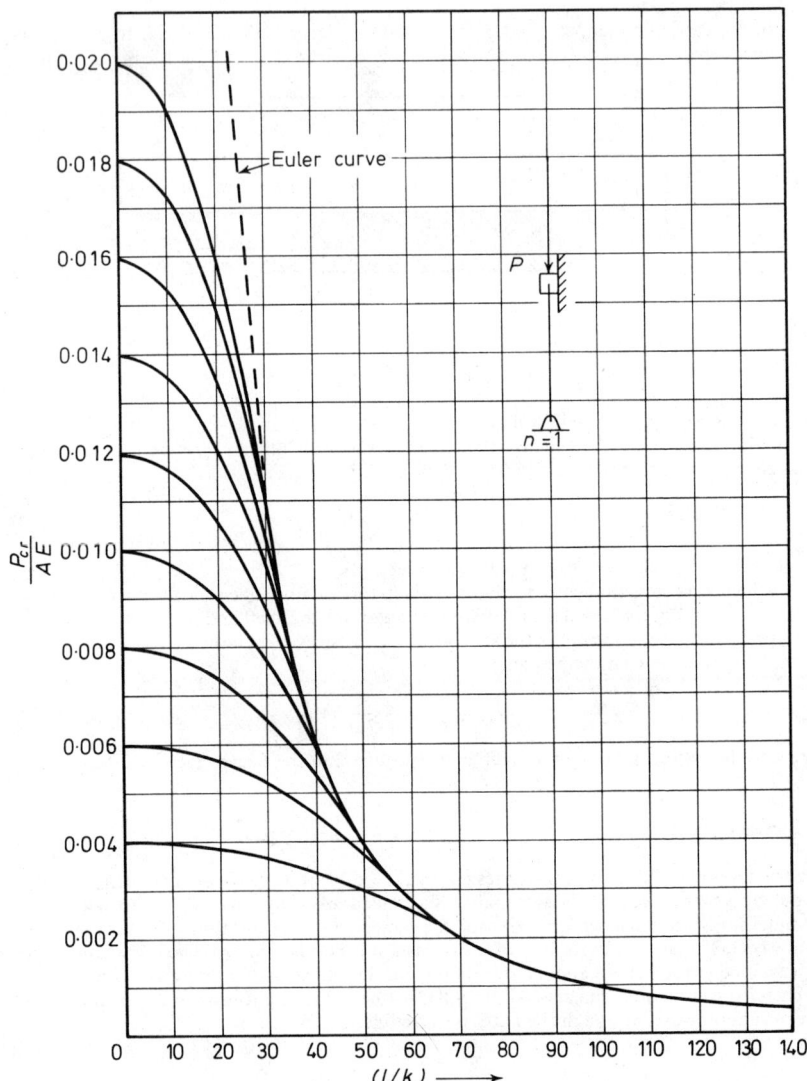

Figure 7.35. Design curves for a strut, position fixed at each end[1]

or as

$$\sigma_E = \frac{P_{cr}}{A} = n\frac{\pi^2 E}{(l/k)^2} \tag{80}$$

where P_{cr} = critical buckling load,
    σ_E = Euler buckling stress,
    n = end-fixity coefficient,
    E = elastic modulus,
    I = second moment of area of section = Ak^2,
    A = cross-sectional area of column,
    k = radius of gyration,
    l = length of strut,
    l/k = slenderness ratio.

For columns whose lengths are less than that necessary for the Euler theory to apply (see equation 82), the Johnson parabolic[29] equation may be used, i.e.

$$\frac{P_{cr}}{A} = S_p - \left(\frac{S_p}{2\pi}\right)^2 \left(\frac{l}{k}\right)^2 \frac{1}{nE} \tag{81}$$

The transition from the Euler to the Johnson equation occurs when

$$\frac{l}{k} = \pi\left(\frac{2nE}{S_p}\right)^{\frac{1}{2}} \tag{82}$$

For convenience, the values obtained by substituting numerical data into equations (80) and (81) have been plotted for a range of values of slenderness ratio for different end conditions[1], and these plots are reproduced in Figures 7.35 to 7.38. The non-dimensional parameter P_{cr}/AE is plotted as the ordinate, and the slenderness ratio as the abscissa. The values of P_{cr}/AE corresponding to l/k equal to zero, represents the yield strength divided by the elastic modulus. Thus, for a given column of known material and dimensions, the appropriate figure may be used to determine the buckling load. The correct curve to use is located by calculating the ratio S_p/E, and for a given slenderness ratio, the corresponding value for P_{cr}/AE is obtained. The critical buckling load can then easily be determined.

To extend the curves shown in Figures 7.35 to 7.38 so that they may be used for brittle materials, such as cast irons, a straight line is drawn between the tangent to the Euler part of the curve and the value for S_u/E corresponding to l/k equal to zero. Although these curves are intended for guidance as working design curves, it may be that in some instances they do not extend into a particular region. In such cases the buckling load can be determined using the appropriate equation (i.e. either equation (80) or (81)), the value for the end-fixity coefficient being as indicated in Figures 7.35 to 7.38.

BUCKLING OF THIN CYLINDERS OR TUBES

For a thin cylinder or tube subjected to external pressure loading p, buckling may occur if the external pressure is equal to the critical pressure, as calculated from the equation

$$p_{cr} = \frac{2t^3 E}{d^3(1-v^2)} \tag{83}$$

or, in terms of stress as

$$\sigma_{cr} = \frac{E}{1-v^2}\left(\frac{t}{d}\right)^2 \tag{84}$$

If the stress exceeds the yield strength, the elastic modulus should be replaced by the

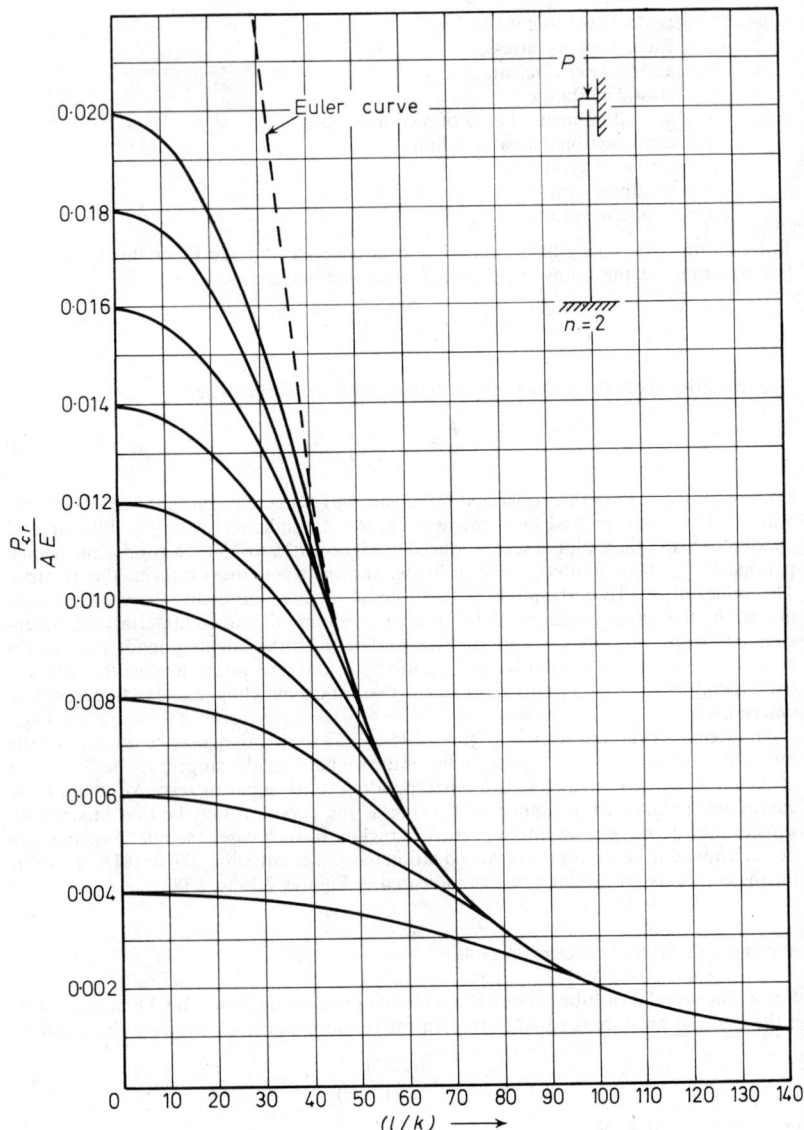

Figure 7.36. Design curves for a strut, built-in at one end, position fixed at the other[1]

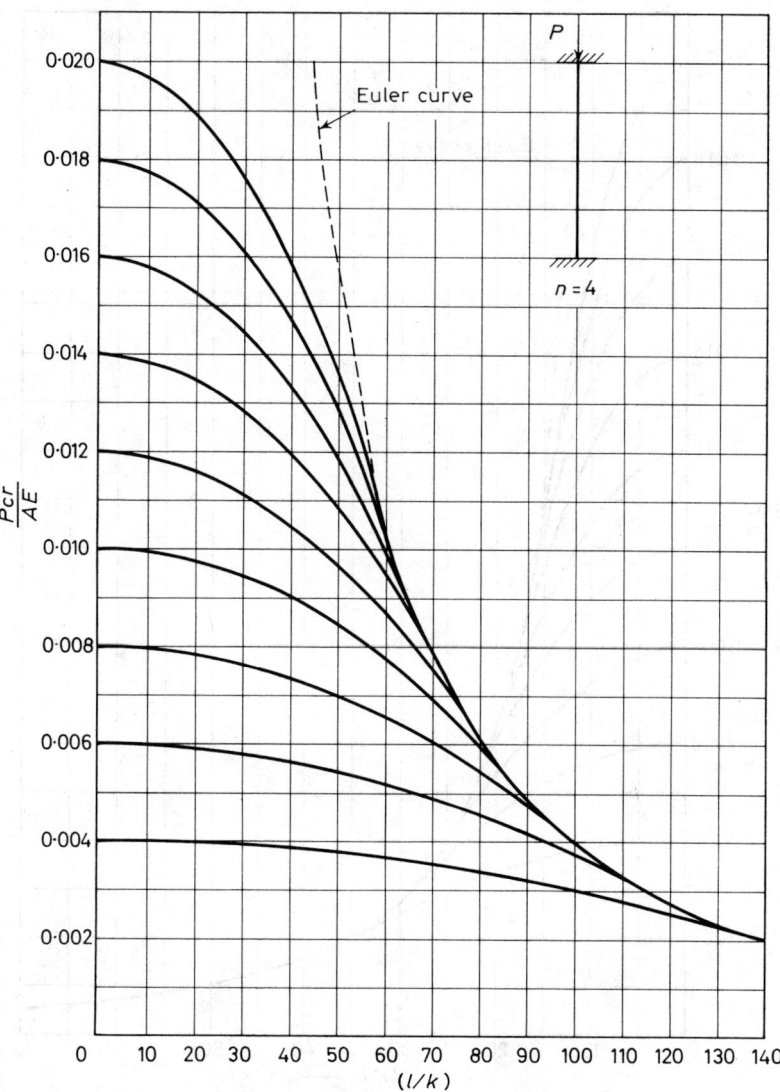

Figure 7.37. Design curves for a strut, built-in at both ends[1]

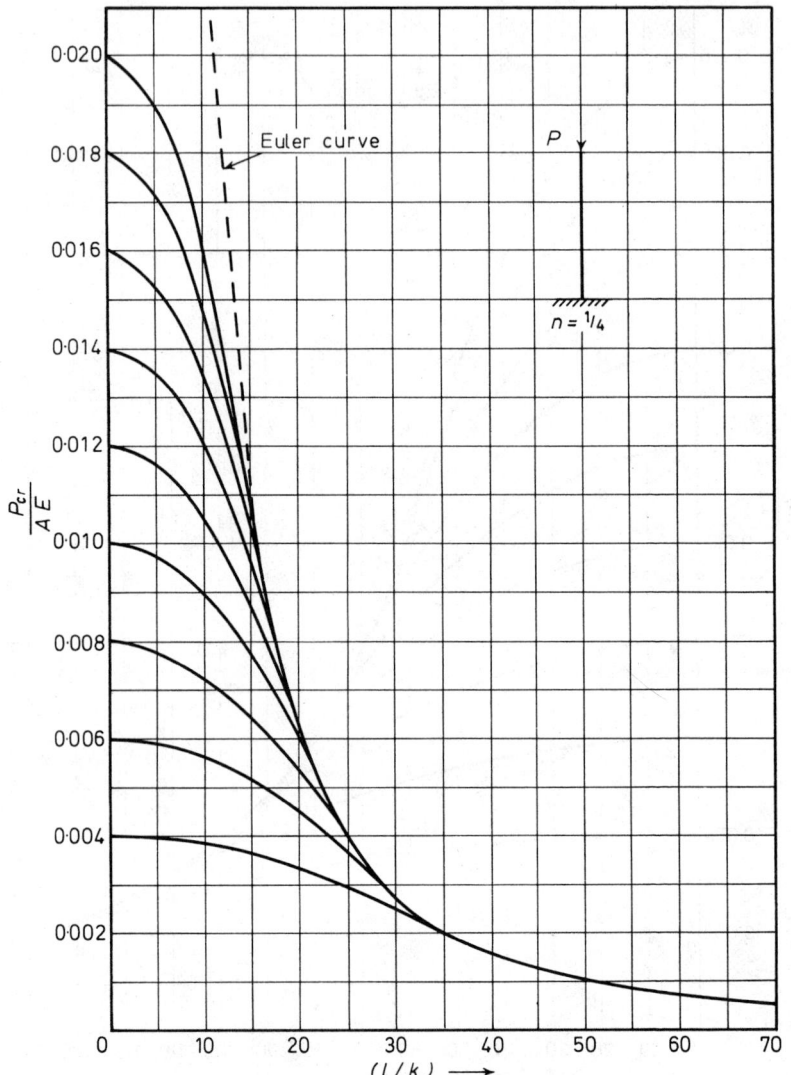

Figure 7.38. Design curves for a strut, built-in at one end, free at the other[2]

tangent modulus to estimate the critical buckling pressure. For a more extensive treatment the reader is referred to the references[23, 30].

RESERVE FACTORS OR FACTORS OF SAFETY

A component is considered to have failed when it can no longer perform its required function, or when it becomes unsafe. Just what is considered to be safe depends upon a number of factors, and it is usual to group these factors together under the general term *factor of safety*, although *factor of ignorance* would be more appropriate in many instances.

A rational approach to determining *reserve factors* or factors of safety has been proposed[31], in which it is recognised that the actual load which a component will experience in service may be somewhat different from that calculated. Also, the load capability of the component may be different from that anticipated, due to variations in material properties and manufacture. The variations of these quantities from the mean or design values are assumed to be reasonably predictable, and using statistical considerations two equations are derived. The first equation recognises that failure occurrence can be tolerated to a limited extent, and the reserve factor is obtained from the equation

$$\left[5 \cdot 4 A_f^{0 \cdot 256} - \left(\frac{\Delta L}{L}\right)^2\right] RF^2 - 10 \cdot 8 A_f^{0 \cdot 256} RF + 5 \cdot 4 A_f^{0 \cdot 256} - \left(\frac{\Delta \lambda}{\lambda}\right)^2 = 0 \qquad (85)$$

For instances where no failure occurrence can be tolerated, the reserve factor is obtained from the equation

$$RF \geq \frac{1 + \Delta \lambda / \bar{\lambda}}{1 - \Delta L / \bar{L}} \qquad (86)$$

where RF = reserve factor,
$\quad \bar{\lambda}$ = mean or design load,
$\quad L$ = mean load capability,
$\quad \Delta \lambda / \bar{\lambda}$ = decimal percentage tolerance band on anticipated actual load,
$\quad \Delta L / \bar{L}$ = decimal percentage tolerance band on load capability,
$\quad A_f$ = decimal percentage failure which can be theoretically tolerated (usually between 0·001 and 0·015)

The decimal percentage tolerance band on anticipated actual load and on load capability, should be based on data obtained from service conditions.

REFERENCES

1. DUGGAN, T. V., *Applied Engineering Design and Analysis*, Butterworths, London (1970)
2. PETERSON, R. E., *Stress Concentration Design Factors*, Wiley, New York (1953)
3. DUGGAN, T. V., *Stress Analysis and Vibration of Elastic Bodies*, Iliffe, London (1964)
4. TIMOSHENKO, S., *Strength of Materials*, Parts 1 and 2, Van Nostrand, Princeton, 3rd ed (1955)
5. TIMOSHENKO, S. and GOODIER, J. N., *Theory of Elasticity*, McGraw-Hill, New York, 2nd ed (1951)
6. JOHNSON, W. and MELLOR, P. B., *Plasticity for Mechanical Engineers*, Van Nostrand, London (1962)
7. DUGGAN, T. V., *Power Transmission and Vibration Considerations in Design*, Butterworths (1971)
8. HERTZ, H. 'On the Contact of Rigid Elastic Solids and on Hardness', *Miscellaneous Papers*, Macmillan, London, 163–183 (1896)
9. DUGGAN, T. V., *Discontinuity Stresses in Pressure Vessels*, DD7, Heywood-Temple Industrial Publications (1968)
10. DUGGAN, T. V., 'Current Trends in Fatigue Research', *The Chartered Mechanical Engineer*, **17**, 10 (1970)
11. DUGGAN, T. V. (Ed), 'Fatigue as a Design Criterion', Butterworths, London (1973)
12. MANSON, S. S., 'Fatigue: A Complex Subject—Some Simple Approximations', *Expt. Mechanics*, S.E.S.A., **5**, 193 (1965)

13. LIPSON, C. and JUVINALL, R. C., *Handbook of Stress and Strength*, The Macmillan Company, New York (1963)
14. FORREST, P. G., *Fatigue of Metals*, Pergamon, Oxford (1962)
15. PETERSON, R. E., 'Notch Sensitivity', *Metal Fatigue* (Eds. SINES, G. and WAISMANN, J. L.), McGraw-Hill, New York, Chap. 13 (1959)
16. LANGER, B. F., 'Application of Stress Concentration Factors', *Bettis Tech. Rev.*, WAPD-BT-18, 1 (1960)
17. CAZAUD, R., *Fatigue of Metals*, Chapman and Hall, London (1953)
18. HEYWOOD, R. B., *Designing Against Fatigue*, Chapman and Hall, London (1962)
19. MANSON, S. S., *Thermal Stress and Low Cycle Fatigue*, McGraw-Hill, New York (1966)
20. COYLE, M. B. and WATSON, S. J., 'Fatigue Strength of Turbine Shafts with Shrunk-on Discs', *Proc. Inst. Mech. Engrs.*, **178**, 147–183 (1964)
21. DUGGAN, T. V., 'Fatigue as a Design Criterion', *Engrg. Matls. and Design*, Part 2, **10** 6 (1967)
22. BOLEY, B. A. and WEINER, J. H., *Theory of Thermal Stresses*, Wiley, New York (1960)
23. FAUPEL, J. H., *Engineering Design—A Synthesis of Stress Analysis and Materials Engineering*, Wiley, New York (1964)
24. ROARK, R. J., *Formulas for Stress and Strain*, McGraw-Hill, New York, 3rd ed (1954)
25. TIMOSHENKO, S. P. and WOINOWSKY-KRIEGER, S., *Theory of Plates and Shells*, McGraw-Hill, New York, 2nd ed (1959)
26. JUVINALL, R. C., *Engineering Considerations of Stress, Strain and Strength*, McGraw-Hill, New York (1967)
27. MARIN, J., *Engineering Materials*, Prentice-Hall, Englewood Cliffs, N. J. (1952)
28. LARSON, F. R. and MILLER, J., 'Time Temperature Relationships for Rupture and Creep Stresses', *Trans. A.S.M.E.*, **74**, 765 (1952)
29. JOHNSON, J. B., *The Theory and Practice of Modern Framed Structures*, New York (1893)
30. TIMOSHENKO, S. and GERE, J. M., *Theory of Elastic Stability*, McGraw-Hill, New York, 2nd ed (1961)
31. JOHNSON, R. C., *Optimum Design of Mechanical Elements*, Wiley, New York (1961)

MECHANICAL PROPERTIES OF FERROUS MATERIALS

G. R. DARBY

Explanation of tables

Tables 7.5 to 7.23 relate to the more popular grades of steel used in the United Kingdom and have been converted to SI Units from data shown in current British Standards[1-7].

Reference should also be made to these British Standards, particularly the BS 970[4] En series, because they are still under revision. The stress value conversions have been taken to the nearest MN/m^2 (N/mm^2).

Hardness values—Table 7.5

Hardness testing is under discussion by ISO to cover the introduction of SI Units in relation to the forces applied to the indenter in a hardness test. Committee ISO/TC17 (ISO Technical Committee on Steel) unanimously endorses the view that regardless of the introduction of SI Units, the hardness values given in the following scales should remain unchanged:

Symbol	Description	Controlling British standard
HRA	Rockwell hardness 'A' scale	BS 891, Part 1[13]
HRB	Rockwell hardness 'B' scale	BS 891, Part 1
HRC	Rockwell hardness 'C' scale	BS 891, Part 1
HRD	Rockwell hardness 'D' scale	BS 891, Part 1
HRE	Rockwell hardness 'E' scale	BS 891, Part 1
HRF	Rockwell hardness 'F' scale	BS 891, Part 1
HRG	Rockwell hardness 'G' scale	BS 891, Part 1
HRH	Rockwell hardness 'H' scale	BS 891, Part 1
HRK	Rockwell hardness 'K' scale	BS 891, Part 1
HR15N HR30N HR45N	Rockwell superficial hardness 'N' scale	BS 4175, Part 1[14]
HR15T HR30T HR45T	Rockwell superficial hardness 'T' scale	BS 4175, Part 1
HB	Brinell hardness No	BS 240, Part 1[15]
HV	Vickers hardness No	BS 427, Part 1[16]
—	Scleroscope No	—

Tables 7.8 to 7.12 and 7.14 give values in the Brinell hardness system (HB) which are dimensionless numbers. Table 7.5 gives a comparison of the more generally used hardness scales. For an exact comparison of these hardness scales reference should be made to BS 860[1] for the formula to use applicable to the metal under examination.

7–57

Impact values—Table 7.6

ISO Technical Committee TC17 has recommended that SI Units should be introduced for expressing the result of impact tests carried out in accordance with ISO R148[8] (V-notch) and ISO/R83[9] (U-notch) for steel. The energy absorbed should be expressed in joules (N m). Tables 7.7 to 7.17 quote Izod values (unless otherwise stated) in N m (joules).

Table 7.6 gives energy conversion factors from ft lbf to N m and N m to ft lbf.

Elongation values

There is as yet no international acceptance of a standard gauge length for elongation values. The present practice in the United Kingdom is to change from $4\sqrt{area}$ to $5\cdot65\sqrt{area}$. Elongation values obtained on gauge lengths of $4\sqrt{area}$ correspond approximately to the following values obtained on a gauge length of $5\cdot65\sqrt{area}$.

$4\sqrt{area}$	12	13	14	15	16	17	18
$5\cdot65\sqrt{area}$	8	9	10	11	12	12	13

Bend tests

The bend test values shown in the Tables are indicated thus 'lt'. This is the product of the radius of bend multiplied by the thickness of section under test. This provides the value for the internal radius over which the test piece shall be bent through 180° whilst the outer convex surface remains free of cracks. Thus for a 16 mm thick plate with a bend test specification of 'lt', the test piece should be bent through 180° around a former of radius 16 mm.

Tensile strength

The values in the Tables are given in newton per square millimetre (N/mm^2) equivalent to meganewton per square metre (MN/m^2). There has been some preference in the steel industry for expressing the tensile strength in hectobar (hbar) equal to 10^7 N/m^2 because the numerical value closely aligns with continental values expressed in kilogramme force per square millimetre (1 hbar = $1\cdot020$ kgf/mm²). The values given in the Tables can be converted to hbar by simply moving the decimal point one place to the left.

Methods of mechanical test

For a summary of the methods and test pieces of mechanical tests, reference should be made to BS 18[7].

Steel plates

Some overlap at present occurs in BS specifications for material grades of Plates. Table 7.7 compares the more popular plate steel properties, using the BS 1501[2] series as the basis. It should be noted that Yield Stress and Elongation values are dependent upon the thickness of plate. The appropriate British Standard should be referred to for more detailed information.

Wrought steel

Existing specifications for wrought steels as billets, blooms, bars and forgings present the user with a formidable choice. The most popular specifications quoted are the BS 970[4] En series, BS 1502[2] and BS 1506[2]. Attempts have been made to correlate such specifications, notably through the issue of BS 3141[6]. At present radical changes are taking place in the revision of the BS 970 series and as a result the 'En numbers are gradually being replaced by a designation system using six digits explained in BS Publication PD 6290[10] PD 6423[11] and PD 6431[12]. A letter has also been introduced at the fourth digit and its use is presently restricted to denote the condition of supply in the case of carbon and alloy steels and material type in the case of stainless steels. The system takes the following form:

	1st three digits			*4th digit*	*5th and 6th digit*	
	1	*2*	*3*	*(letter A, M, H or S)*	*5*	*6*
Carbon { Plain carbon		} 000 to 199		A = Analyses		
Carbon manganese				M = Mechanical properties	100 × mean carbon content	
{ Free cutting			200–240			
Alloy			500–999	H = Hardenability		
Stainless			300–499	S = Stainless	Basic alloy 01 specific alloy 11–99	

Tables 7.8 to 7.14 give the properties of these wrought steels based on the BS 970 series and correlate to the new digit classification (where allocated by British Standards) and also to the BS 1501–6 series of wrought steels.

Forgings

The more readily obtainable forging materials are specified in BS 1503[2]. Table 7.16 lists the mechanical properties of the BS 1503 grades and correlates these to other current British Standards grades.

Castings

The more readily obtainable casting materials are specified in BS 1504[2]. Table 7.17 lists the mechanical properties of the BS 1504 grades and correlates these to other current British Standard grades.

Table 7.5 HARDNESS COMPARISON TABLE

Brinell 3000 kg load 10 mm ball		Diamond pyramid number (Vickers) (HV)	Rockwell number		Scleroscope number	Equivalent tensile strength for steel N/mm²
Diameter of impression	Hardness No (HB)		C scale 150 kg diamond Cone (HRC)	B scale 100 kg 1·6 mm ball (HRB)		
2·05	899	—	—	—	—	3 027
2·10	856	—	—	—	—	2 890
2·15	817	—	—	—	—	2 749
2·20	780	1 150	70	—	106	2 640
2·25	745	1 050	68	—	100	2 517
2·30	712	960	66	—	95	2 409
2·35	682	885	64	—	91	2 286
2·40	653	820	62	—	87	2 193
2·45	627	765	60	—	84	2 116
2·50	601	717	58	—	81	2 023
2·55	578	675	57	—	78	1 946
2·60	555	633	55	120	75	1 869
2·65	534	598	53	119	72	1 792
2·70	514	567	52	119	70	1 730
2·75	495	540	50	117	67	1 670
2·80	477	515	49	117	65	1 606
2·85	461	494	47	116	63	1 544
2·90	444	472	46	115	61	1 498
2·95	429	454	45	115	59	1 452
3·00	415	437	44	114	57	1 405
3·05	401	420	42	113	55	1 359
3·10	388	404	41	112	54	1 297
3·15	375	389	40	112	52	1 266
3·20	363	375	38	110	51	1 220
3·25	352	363	37	110	49	1 174
3·30	341	350	36	109	48	1 143
3·35	331	339	35	109	46	1 112
3·40	321	327	34	108	45	1 081
3·45	311	316	33	108	44	1 050
3·50	302	305	32	107	43	1 019
3·55	293	296	31	106	42	988
3·60	285	287	30	105	40	958
3·65	277	279	29	104	39	927
3·70	269	270	28	104	38	911
3·75	262	263	26	103	37	880
3·80	255	256	25	102	37	849
3·85	248	248	24	102	36	834
3·90	241	241	23	100	35	803
3·95	235	235	22	99	34	788
4·00	229	229	21	98	33	772
4·05	223	223	20	97	32	757
4·10	217	217	18	96	31	726
4·15	212	212	17	96	31	710
4·20	207	207	16	95	30	695
4·25	201	202	15	94	30	680
4·30	197	197	13	93	29	664
4·35	192	192	12	92	28	649
4·40	187	187	10	91	28	633
4·45	183	183	9	90	27	618
4·50	179	179	8	89	27	610
4·55	174	174	7	88	26	602
4·60	170	170	6	87	26	595
4·65	167	166	4	86	25	587

Table 7.5—*continued*

Brinell 3000 kg load 10 mm ball		Diamond pyramid number (Vickers) (HV)	Rockwell number		Scleroscope number	Equivalent tensile strength for steel N/mm²
Diameter of impression	Hardness No (HB)		C scale 150 kg diamond Cone (HRC)	B scale 100 kg 1·6 mm ball (HRB)		
4·70	163	163	3	85	25	579
4·75	159	159	2	84	24	564
4·80	156	156	1	83	24	556
4·85	152	153	0	82	23	541
4·90	149	149	−1	81	23	525
4·95	146	146	−2	80	22	517
5·00	143	143	−3	79	22	510
5·05	140	140	−4	78	21	494
5·10	137	137	−6	77	21	486
5·15	134	134	−7	76	21	479
5·20	131	131	−9	74	20	463
5·25	128	128	−11	73	20	456
5·30	126	126	−12	72	19	448
5·35	123	124	−14	71	19	440
5·40	121	121	−16	70	19	432
5·45	118	118	−17	69	19	417
5·50	116	116	−19	68	19	409
5·55	114	114	−20	67	19	402
5·60	111	112	—	66	19	394
5·65	109	109	—	65	18	386
5·70	107	107	—	64	18	378
5·75	105	105	—	62	18	371
5·80	103	103	—	61	18	363
5·85	101	101	—	60	18	355
5·90	99·2	99	—	59	18	351
5·95	97·3	97	—	57	17	347
6·00	95·5	95	—	56	17	355

Table 7.6 IZOD IMPACT VALUES = ENERGY CONVERSION FACTORS

ft lbf *to* N m

ft lbf	×1	×10	×100	×1000
1	1·3558	13·5582	135·582	1355·82
2	2·7116	27·1164	271·164	2711·64
3	4·0675	40·6745	406·745	4067·45
4	5·4233	54·2327	542·327	5423·27
5	6·7791	67·7909	677·909	6779·09
6	8·1349	81·3491	813·491	8134·91
7	9·4907	94·9073	949·073	9490·73
8	10·8465	108·465	1084·65	10846·5
9	12·2024	122·024	1220·24	12202·4

N m *to* ft lbf

N m	×1	×10	×100	×1000
1	0·7376	7·3765	73·7562	737·562
2	1·4751	14·7512	147·512	1475·12
3	2·2127	22·1269	221·269	2212·69
4	2·9502	29·5025	295·025	2950·25
5	3·6878	36·8781	368·781	3687·81
6	4·4254	44·2537	442·537	4425·37
7	5·1629	51·6294	516·294	5162·94
8	5·9005	59·0050	590·050	5900·50
9	6·6381	66·3806	663·806	6638·06

Table 7.7(a) MECHANICAL PROPERTIES OF ALLOY STEEL PLATES

Material description	BS No	Tensile strength range min. N/mm²	Yield stress (according to section) N/mm²	Elongation % $5.65\sqrt{S_0}$	V longitudinal min. 20°C N m	0°C N m	−40°C N m	−196°C N m	U transverse notch 5 mm N m/cm²	3 mm N m/cm²	Bend test	Limiting section mm	Heat treatment condition	Nearest equivalent BS
Molybdenum boron	1501–261	556/664	417/448	16	20/41	—	—	—	49	59	$1\frac{1}{2}$ t	90	N	—
Manganese chromium Molybdenum Vanadium	1501–271	587/695	386/463	16	41	27	—	—	49	59	$1\frac{1}{2}$ t	150	NT	—
Nickel chromium Molybdenum Vanadium	1501–281	587/695	386/463	16	—	—	27	—	88	98	$1\frac{1}{2}$ t	150	NT	—
Nickel chromium Molybdenum Vanadium	1501–282	571/710	417/494	18	—	—	41	—	69	82	$1\frac{1}{2}$ t	150	NT or NNT	—
3½% Nickel steel	1501–503	448	263(0·2 % P.S)	20	—	—	34 (at −80°C)	—	—	—	1 t	38	NT or QT	—
9% Nickel steel	1501–509	695	525(0·2 % P.S)	18	—	—	—	27	—	—	1/1½ t	50	NNT or QT	—
9% Nickel steel	1501–510	695	587(0·2 % P.S)	18	—	—	—	27	—	—	1/1½ t	50	QT	—
1%Cr ½%Mo	1501–620/27	417/541	286 min	21/19	—	—	—	—	—	—	—	150	NT	—
	1501–620/31	479/602	340 min	18/16	—	—	—	—	—	—	—	150	NT	—
					20°C	0°C	−15°C	−30°C						
1¼% Cr ½% Mo	1501–621	448/602	317/340	18/16	—	—	—	—	—	—	—	150	NT	—
2¼% Cr 1% Mo	1501–622/31	479/602	278 min	18/16	—	—	—	—	—	—	—	—	—	—
	1501–622/45	695/819	556 min	16/15	—	—	—	—	—	—	—	150	NT	—

Heat treatment symbols: N, Normalised; NT, Normalised and tempered; NNT, Double normalised and tempered; QT, Quenched and tempered; AR, As rolled

Table 7.7(b) MECHANICAL PROPERTIES OF CARBON STEEL PLATES

Material description	*BS No*	*Tensile strength range min.* N/mm²	*Yield stress (according to section)* N/mm²	*Elongation* % $5{\cdot}65\sqrt{S_0}$	*Impact test (charpy)* *V longitudinal min.* 20°C N m	0°C N m	–15°C N m	*U transverse notch* 5 mm N m/cm²	3 mm N m/cm²	*Bend test*	*Limiting section* mm	*Heat treatment condition*	*Nearest equivalent BS*
Weldable structural steel	4360 Gd. 40	400/480	210/230	25	27	27	27	—	—	$1\frac{1}{4}$ t			Replacing
	4360 Gd. 43	430/510	220/280	22	27	27	27	—	—	$1\frac{1}{2}$ t	50	AR or N	3706, 2762
	4360 Gd. 50	500/620	325/355	20	—	27	27	—	—	$1\frac{1}{2}$ t			3706, 2762, 15
	4360 Gd. 55	550/700	415/450	19	—	27	—	—	—	$1\frac{1}{2}$ t			968 —
Carbon steel (rimming)	1501–141	355/432	—	25·4/28·9	—	—	—	—	—	1 t	19	AR	—
Carbon steel (semi-killed)	1501–151/23	355/432	186/202	28·9/25·4	—	—	—	—	—	1 t		AR (up to	1633
	1501–151/26	400/485	214/228	26·8/23	—	—	—	—	—	1 t	100	45 mm)	14
	1501–151/28	432/515	228/247	25·4/21·5	—	—	—	—	—	$1\frac{1}{2}$ t		N	—
Carbon steel (semi-killed with aluminium)	1501–154/23	355/432	202	28·9/25·4	—	—	—	—	—	1 t			—
	1501–154/26	400/485	228	26·8/23	—	—	—	—	—	1 t	9·5	AR or N	—
	1501–154/28	432/515	247	25·4/21·5	—	—	—	—	—	$1\frac{1}{2}$ t			—
Carbon steel (silicone killed)	1501–161/23	355/432	186/202	28·9/25·4	—	—	—	—	—	1 t			1633
	1501–161/26	400/485	214/228	26·8/23	—	—	—	—	—	1 t	150	AR or N	—
	1501–161/28	432/515	228/247	25·4/21·5	—	—	—	—	—	$1\frac{1}{2}$ t			—
Carbon manganese (semi-killed)	1501–211/26	400/485	225/236	26·8/23	41	27	—	49	59	1 t			1633
	1501–211/28	432/515	239/255	25·4/21·5	41	27	—	49	59	$1\frac{1}{2}$ t	100	AR	—
	1501–211/30	463/555	256/273	24/19·8	41	27	—	49	59	$1\frac{1}{2}$ t			—
	1501–211/32	494/594	273/290	22·6/18	34	23	—	49	59	$1\frac{1}{2}$ t			—
	1501–211/26	400/485	225/236	26·8/23	47	34	27	59	69	1 t			1633
	1501–211/28	432/515	239/255	25·4/21·5	47	34	27	59	69	$1\frac{1}{2}$ t	45	N	—
	1501–211/30	463/555	256/273	24/19·8	47	34	—	59	69	$1\frac{1}{2}$ t			—
	1501–211/32	494/594	273/290	22·6/18	41	27	—	59	69	$1\frac{1}{2}$ t			—

Table 7.7(b)—*continued*

Material description	BS No	Tensile strength range min. N/mm²	Yield stress (according to section) N/mm²	Elongation % 5·65√S₀	Impact test (charpy)						Bend test	Limiting section mm	Heat treatment condition	Nearest equivalent BS
					V longitudinal min.				U transverse notch					
					20°C N m	0°C N m	−15°C N m	−30°C N m	5 mm N m/cm²	3 mm N m/cm²				
Carbon manganese (semi-killed niobium treated)	1501–213/28	432/515	286/301	25·4/21·5	68	54	41	27	69	78	1½ t		N	—
	1501–213/30	463/555	301/317	24/19·8	68	54	41	27	59	69	1½ t	100		—
	1501–213/32	494/594	324/340	22·6/18	68	54	41	27	49	59	1½ t			—
Carbon manganese (silicone killed)	1501–221/26	400/485	222/236	26·8/23	41	27	—	—	49	59	1 t			1633
	1501–221/28	432/515	239/255	25·4/21·5	41	27	—	—	49	59	1½ t	100	AR	—
	1501–221/30	463/555	256/273	24/19·8	41	27	—	—	49	59	1½ t			—
	1501–221/32	494/594	273/290	22·6/18	34	23	—	—	49	59	1½ t			—
	1501–221/26	400/485	222/236	26·8/23	47	34	27	—	59	69	1 t			1633
	1501–221/28	432/515	239/255	25·4/21·5	47	34	27	—	59	69	1½ t	45	N	—
	1501–221/30	463/555	256/273	24/19·8	47	34	—	—	59	69	1½ t			—
	1501–221/32	494/594	273/290	22·6/18	41	27	—	—	59	69	1½ t			—
Carbon manganese (silicone killed niobium treated)	1501–223/28	432/515	286/301	25·4/21·5	68	54	41	27	69	78	1½ t			—
	1501–223/30	463/555	301/317	24/19·8	68	54	41	27	59	69	1½ t	150	N	—
	1501–223/32	494/594	324/340	22·6/18	68	54	41	27	49	59	1½ t			—
Carbon manganese (silicone killed aluminium treated)	1501–224/26	400/485	232/263	26·8/23	68	68	54	47	69	78	1 t			—
	1501–224/28	432/515	255/286	25·4/21·5	68	68	54	47	69	78	1½ t	150	N	—
	1501–224/30	463/555	278/309	24/19·8	68	68	54	47	59	69	1½ t			—

Heat treatment symbols: N, Normalised; NT, Normalised and tempered; NNT, Double normalised and tempered; QT, Quenched and tempered; AR, as rolled.

Table 7.8 MECHANICAL PROPERTIES OF COLD DRAWN BARS—OTHER THAN FREE CUTTING STEEL BARS—En Series

BS spec	Limiting Section mm	Tensile Strength N/mm²	Elongation $4\sqrt{S_0}$ %	Impact Test N m	Hardness HB	Comparable BS
En 3B	—	432	17			1506/121
En 3D	31·8 and less	541	15			29
	over 31·8 to 63·5	463	15			400
	over 63·5	432	15			1506/121
En 4A	50·8 and less	494/649	12			
En 5D	12·7 and less	695	20	54	255	1 449
	over 12·7 to 19	618	18	27	229	
	over 19 to 63·5	541	15	20/13	229	
En 6 & 6A	19 and less	587/741	12	27		
	over 19 to 63·5	541/695	12	20/13		
	over 63·5	494/695	15	20/13		
En 6K	19 and less	587/741	12	27		
	over 19 to 63·5	541/695	12	20/13		
	over 63·5	494/695	15	20/13		
En 8	31·8 and less	649	10		241	
	over 31·8 to 63·5	602	10		229	
	over 63·5	571	10		229	
En 43A	50·8 and less	695/927	12		201/277	
En 14A	50·8 and less	695	15	20		
En 14B	50·8 and less	695	15	20		
En 9	50·8 and less	772/1004	12		223/302	1506/155
En 9K	50·8 and less	772/1004	12		223/302	1506/155

Table 7.9 MECHANICAL PROPERTIES OF COLD DRAWN FREE CUTTING STEEL BARS

BS spec	Limiting section mm	Tensile strength N/mm²	Elongation $4\sqrt{S_0}$ %	Impact test N m	Hardness HB
En 1A	13·5 and less	494	10		
	over 13·5 to 38	432	14		
	over 38 to 63·5	386	14		
	over 63·5 to 100	355	26		
En 1B	13·5 and less	417	10		
	over 13·5 to 38	386	12		
	over 38 to 63·5	355	12		
En 7	12·7 and less	618/772	15		
	over 12·7 to 44·5	541/695	15	27/13	
	over 44·5 to 63·5	463/618	12	13	
En 7A	28·6 and less	541/695	15	27/20	
	over 28·6 to 63·5	463/618	15	13	
En 8M	38 and less	587	12		229

Table 7.10 MECHANICAL PROPERTIES OF HOT ROLLED OR NORMALISED BARS etc.—
OTHER THAN FREE CUTTING

S spec	Limiting section mm	Tensile strength N/mm²	Yield stress N/mm²	0·2% Proof stress N/mm²	Elongation $4\sqrt{S_0}$ %	Impact test N m	Hardness HB	Comparable BS
2	150	309			28			1 449, 29
2E	150	309			28		120	
3	150	386/541			25			29, 400, 1506/102
3A	150	432			25			29,400 1 506–111
3C	150	432			25			29, 400 1 506–111
4	150	432/587			25		126/179	1 503–151
5	63·5	494	247		25	27	143/193	1 449, 29
5K	63·5	494	247		25	27	143/193	1 449
6, 6A	19	541/695			12	27		
	63·5	541/695			12	20/13		
	150	494/695			15	13		
8	150	541	278		20		152/207	
8K	150	541	278	263	20	13	152/207	
12	150	541	309		20	20	152/207	
14A	150	541	324		20	20	152/207	
14A/1	28·5	541/633	324		18			1 501/221 1 506/221
	63·5	510/602	293		18			1 501/221 1 506/221
14B	150	587	355		20	20	170/223	
43A	28·5	618	324		18		179/229	
9, 9K	100	695	355		18		201/255	1 506/155

Table 7.11 MECHANICAL PROPERTIES OF HARDENED AND TEMPERED BARS, BILLETS, FORGINGS AND DROP FORGINGS

Digit class	BS spec	Condition	Limiting section mm	Tensile strength N/mm² (min)	Yield stress N/mm²	0·2% Proof stress N/mm²	Elongation $4\sqrt{S_0}$ %	Impact test N m	Hardness HB	Comparable BS
	En 5	P	63·5	541	340	—	22	27	152/207	
		Q	19	618	432	—	20	27	179/229	
		R	12·7	695	494	—	20	54	201/255	
	En 5K	P	63·5	541	340	309	22	54	152/207	
		Q	19	618	432	402	20	27	179/229	
		R	12·7	695	494	463	20	54	201/255	
	En 8	Q	63·5	618	432	—	22–17 dependent upon final condition required	13–27 dependent upon final condition required	179/229	1506–162
		R	22	695	494	—	20–17 dependent upon final condition required	13–54 dependent upon final condition required	201/255	
	En 8K	Q	63·5	618	432	386	22	34	179/229	
	En 8M	Q	50	618	432	386	22	34	179/229	
		R	12·7	695	494	463	20	54	201/255	
	En 43A	R	50	695	463	432	18	—	201/255	
		S	28·5	772	510	479	18	—	223/277	
	En 9	R	50	695	463	—	18	—	201/255	
		S	28·5	772	510	—	18	—	223/277	
		T	28·5	849	556	—	15	—	248/302	
	En 9K	R	50	695	463	432	18	—	201/255	1506–221
		S	28·5	772	510	479	18	—	223/277	
		T	28·5	849	556	525	15	—	248/302	
	En 14A	Q	100	618	432	402	20	40	179/229	
		R	28·5	695	494	463	20	34	201/255	
	En 14B	Q	100	618	432	402	20	47	179/229	1045
		R	63·5	695	494	463	20	34	201/255	
	En 15A	Q	100	618	432	—	22	34	179/229	
		R	63·5	695	525	—	20	34	201/255	
		S	22	772	587	—	20	34	223/277	
	En 15 AM	R	63·5	695	525	—	20	34	201/255	

[illegible]	En 12	Q	150	618	463	448	17	34	179/229	
785M19	En 13	Q	150	618	463	448	18	54	179/229	
503M40	En 18	R	100	695	525	510	17	54	201/255	
		S	63·5	772	587	571	15	54	223/277	
		T	28·5	849	680	664	13	54	248/302	
	En 160	R	150	695	525	494	22	54	201/255	
		S	100	772	587	556	20	54	223/277	
		T	63·5	849	680	633	18	54	248/302	
		U	28·5	927	757	710	17	47	269/321	
	En 22	S	100	772	587	556	20	54	223/277	
		T	63·5	849	680	633	18	54	248/302	
526M60	En 11	T	100	849	618	—	11	—	248/302	
		V	63·5	1004	741	—	8	—	293/352	
608M38	En 17	R	254	695	525	479	15	41	201/255	
			150	695	525	510	17	54	201/255	
		S	254	772	587	541	13	34	223/277	
			150	772	587	571	15	54	223/277	
		T	100	849	680	664	13	54	248/302	
		U	63·5	927	757	741	12	47	269/331	
		V	28·5	1.004	84·9	834	12	47	293/352	
709M40	En 19	R	254	695	525	479	15	34	201/255	1506/621A
		S	254	772	587	541	13	27	223/277	
			150	772	587	571	15	54	223/277	
		T	100	849	680	664	13	54	248/302	
		U	63·5	927	757	741	12	47	269/331	
		V	28·5	1 004	849	834	12	47	293/352	
	En 20A/B	T	63·5	849	680	633	18	54	248/302	1506–621B
		U	63·5	927	741	710	17	47	269/321	
		V	28·5	1 004	803	772	16	47	293/341	
816M40	En 110	S	254	772	556	541	15	34	223/277	
			150	772	587	571	15	54	223/277	
		T	100	849	680	664	13	54	248/302	
		U	63·5	927	757	741	12	47	269/331	
		V	28·5	1 004	849	834	12	47	293/352	
817M40	En 24	T	254	849	587	633	13	41	248/302	
			150	849	680	664	13	54	248/302	
		U	100	927	757	741	12	47	269/331	
		V	63·5	1 004	849	834	12	47	293/352	
		W	28·5	1 081	942	927	11	41	311/375	
		X	28·5	1 158	1 019	1 004	10	34	341/401	
		Z	28·5	1 544	1 234	1 127	5	11	444 min	

Table 7.11—*continued*

Digit class	BS spec	Condition	Limiting section mm	Tensile strength N/mm² min	Yield stress N/mm²	0·2% Proof stress N/mm²	Elongation $4\sqrt{S_0}$ %	Impact test N m	Hardness HB	Comparable BS
	En 21	R	100	695	494	494	22	40	201/255	
		S	63·5	772	587	556	20	40	223/277	
605M30	En 16 & 16D	R	254	695	494	479	15	25	201/255	
605M36		S	150	695	525	510	17	40	201/255	
		T	100	772	587	571	15	40	223/277	
		U	63·5	849	680	664	13	35	248/302	
		V	28·5	927	757	741	12	35	269/331	
			19	1 004	849	834	12	40	293/352	1506-621A
708M40	En 19A	R	150	695	525	510	17	40	201/255	
		S	100	772	587	571	15	40	223/277	
		T	63·5	849	680	664	13	40	248/302	
		U	28·5	927	757	741	12	35	269/331	
945M38	En 100	R	254	695	494	479	15	25	201/255	
		S	150	695	525	510	17	40	201/255	
		T	100	772	587	571	15	40	223/277	
		U	63·5	849	680	664	13	40	248/302	
		V	28·5	927	757	741	12	35	269/331	
			285	1 004	849	834	17	35	293/352	
640M40	En 111	R	150	695	525	510	15	40	201/255	
		S	100	772	587	571	13	40	223/277	
		T	63·5	849	680	664	12	40	248/302	
		U	28·5	927	757	741	15	35	269/331	
653M31	En 23	S	150	772	680	664	13	54	223/277	
		T	100	849	757	741	12	54	248/302	
		U	63·5	927	587	571	15	47	269/331	
En 29A	302	R	150	695	525	494	22	54	201/255	
		S	150	772	587	556	20	54	223/277	
		T	150	849	680	633	18	47	248/302	
		U	150	927	757	710	17	47	269/321	
		V	100	1 004	849	772	16		293/341	

Material	En	Pos	t						
826M31	En 25	U	254	927	741	726	12	34	269/331
			150	927	757	741	12	48	269/331
		V	150	1 004	849	834	12	48	293/352
		W	100	1 081	942	927	11	41	311/375
		X	63·5	1 158	1 019	1 004	10	34	341/401
		Z	63·5	1 544	1 236	1 127	5	11	444 min
830M31	En 27	T	254	849	649	633	13	41	248/302
			150	849	680	664	13	54	248/302
		U	150	927	757	741	12	48	269/331
		V	100	1 004	849	834	12	48	293/352
		W	63·5	1 081	942	927	11	41	311/375
	En 28	U	150	927	741	710	17	48	269/321
		V	150	1 004	803	772	14	48	293/341
		W	100	1 004	803	772	16	41	293/341
			100	1 081	896	849	15	34	311/375
		Y	63·5	1 236	1 050	988	14	34	363/415
		U	254	927	741	726	12	48	269/331
		V	150	927	757	741	12	34	269/331
		W	254	1 004	834	819	12	48	293/352
826M40	En 26	V	150	1 004	849	834	11	27	293/352
		W	254	1 081	927	911	11	41	311/375
		X	150	1 081	942	927	10	34	311/375
		Y	150	1 158	1 019	1 003	10	34	341/401
		Z	150	1 236	1 097	1 081	7	13	363/429
835M30	En 30A	Z	150	1 544	1 236	1 127	7	13	444 min
	En 30B	Z	150	1 544	1 313	1 236	10	13	444 min
			150	1 544	1 313	1 127	7	20	444 min

Table 7.12 MECHANICAL PROPERTIES OF HARDENED AND TEMPERED BARS, BILLETS, FORGINGS AND DROP FORGINGS—NITRIDING STEELS

Digit class	BS spec.	Condition	Limiting section mm	Tensile strength N/mm²	Yield stress N/mm²	0·2% proof stress N/mm²	Elongation $4\sqrt{S_0}$ %	Impact test N m	Hardness HB
	En 40A	R	150	695 min	525 min	494 min	22	54	201/255
		S	150	772	587	556	20	54	223/277
		T	150	849	580	633	18	54	248/302
		U	150	927	741	710	17	47	269/321
722 M24	En 40B	T	254 / 150	849	649	633	13	40	248/302
			150	849	680	664	13	54	248/302
		U	150	927	757	741	12	47	269/331
905 M31	En 41A	R	100	695	525	510	17	54	201/255
		S	63·5	772	587	571	15	54	223/277
905 M39	En 41B	R	150	695	525	510	17	54	201/255
		S	100	772	587	571	15	54	223/277
		T	63·5	849	680	664	13	47	248/302
897 M39	En 40C	85 tons/in² (1313 N/mm²) Z	63·5	1 313	1 158	1 112	8	20	375
		Z	28·5	1 544	1 236	1 205	7	13	444

Table 7.13 MECHANICAL PROPERTIES—CASE HARDENING STEEL BAR, BILLETS, FORGINGS

BS spec	Limiting section mm	Tensile strength N/mm^2	Elongation $4\sqrt{S_0}$ %	Impact test N m
En 32A	12·7	494 min	20	54
En 32B, 32C	over 12·7	494	20	54
En 32M	—	494	20	54
En 202	—	587	20	41
En 37	—	618	20	68
En 201	—	618	20	54
En 33	—	695	18	54
En 34	—	695	18	54
En 351	—	695	18	41
En 361	—	849	15	27
En 35	—	849	15	30
En 36A	—	849	15	47
En 325	—	849	15	41
En 352	—	849	15	27
En 362	—	849	15	20
En 36B, 36C	—	1 004	13	41
En 38	—	1 004	13	41
En 353	—	1 004	12	27
En 363	—	1 004	—	—
En 354	—	1 158	12	27
En 39A/B	—	1 313	12	34
En 355	—	1 313	12	34

Table 7.14 MECHANICAL PROPERTIES OF STAINLESS AND HEAT RESISTING STEELS

Digit class	BS	Condition	Limiting section mm	Tensile strength N/mm²	Yield stress N/mm²	0·2% proof stress N/mm²	Elongation A_4/S_0 %	Impact test	Hardness HB	Comparable BS
302S25	En 58A		—	510 min	—	208 min	40	—	183	1506–801
303S21	En 58M		19	864	—	695	12	—	—	1506–821
304S15	En 58E	Softened	25	787	—	556	15	—	—	for En 58
321S12	En 58B/C	Cold drawn	30	726	—	448	20	—	—	B, G, M
347S17	En 58F/G (Austenitic chromium)		38	695	—	340	28	—	—	
			44	649	—	309	28	—	—	
315S16	En 58H (Austenitic chromium)	Softened	—	463	—	170	40	—	183	1 449
										1 554
										2 056
320S17	En 58J (Austenitic chromium)		—	494	—	178	40	—	183	BS 1506–845
		Softened	19	865	—	695	12	—	—	
		Cold drawn	25	788	—	556	15	—	—	
			30	726	—	448	20	—	—	
			38	695	—	340	28	—	—	
			44	649	—	309	28	—	—	
331S40	En 54	As rolled or	—	—	—	—	—	—	—	—
331S42	En 54A (Chromium tungsten)	softened								
401S45	En 52 (Silicon chromium)	Rolled or rolled and stress relieved	—	—	—	—	—	—	225 min	—
410S21	En 56A	Hardened and tempered condition P	150	541/695	371 min	340 min	20/15	34/27	152–207	1506–713
416S21	En 56AM (Martensinic 13% chromium)									

420S15	En 60 (17% chromium)	Softened condition	63	432 min	278 min	247 min	20	—	170 min	1 449
431S29	En 57 (chromium nickel)	Hardened and tempered 'T' condition	150	849/1 004	680 min	633 min	11	34/20 ATS	248–302	1 449
443S65	En 59 (chromium nickel silicon)	Specified by purchaser	—	—	—	—	—	—	269 min	—

(ATS = According to Section)

Table 7.15 MECHANICAL PROPERTIES OF SECTIONS AND BARS (OTHER THAN BOLTING MATERIALS)—BS 1502 SERIES

Material description	BS No	Tensile strength range/ min. value N/mm²	Yield stress (according to section) N/mm²	Elongation $5.65\sqrt{S_0}$ %	Impact test	Bend test	Limiting section		Heat treatment condition
							Section	Bar	
Carbon steel	1502–151	432 min	min 247 232 229	25·4/21·5	No	$1\frac{1}{2}$ t	mm 16 16–32 32–64	25 25–51 51–102	AR
Carbon manganese steel	1502–161	432 min	247 232 229	25·4/21·5	No	$1\frac{1}{2}$ t	16 16–32 32–64	25 25–51 51–102	AR
Carbon manganese steel (semi-killed)	1502/211	432 min	255 247 239	25.4/21·5	No	$1\frac{1}{2}$ t	16 16–32 32–64	25 25–51 51–102	AR
Carbon manganese steel (silicon-killed)	1502/221	432 min	255 247 239	25·4/21·5	No	$1\frac{1}{2}$ t	16 16–32 32–64	25 25–51 51–102	AR
Carbon manganese steel (silicon-killed aluminium treated)	1502–224	432 min	286 270 255	25·4/21·5	No	$1\frac{1}{2}$ t	— — —	25 25–51 51–102	AR

Material description	BS No.	strength range/ min. value N/mm^2	(according to section) N/mm^2	Elongation % $5{\cdot}65\sqrt{S_0}$	Impact test	Bend test	Limiting section	Heat treatment condition	Nearest equivalent BS
Carbon steel (silicon-killed)	1503–161/26	402/479	201 min	23/20		$\frac{1}{2}$ t/$\frac{2}{3}$ t		N	BS 24 Pt.4 Cl.B; BS 29
	1503–161/28	432/510	216	22/19	No	$\frac{1}{3}$ t/$\frac{2}{3}$ t		N & T	BS 29
	1503–161/32	494/556	247	20/17		—/—			
Carbon manganese steel (silicon-killed)	1503–221/32	494/571	247	19/16	No	$\frac{1}{2}$ t/1 t		N	
	1503–221/34	525/602	263	18/15		$\frac{2}{3}$ t/1$\frac{1}{3}$ t		N & T	
Carbon manganese steel (silicon-killed, Niobium-treated)	1503–223/28	432/510	247	22/19	No	$\frac{1}{3}$ t/$\frac{2}{3}$ t		N	
	1503–223/32	494/571	278	19/16		$\frac{2}{3}$ t/1 t		N & T	
								Q & T	
Carbon manganese steel (silicon-killed, aluminium-treated)	1503–224/28	432/510	232	22/19		$\frac{1}{3}$ t/$\frac{2}{3}$ t		N	
					No			N & T	
	1503–224/32	494/571	263	19/16		$\frac{2}{3}$ t/1 t		Q & T	
Low-alloy manganese chromium molybdenum vanadium steel	1503–271	556 min	371 (0·2 % P.S.)	19/15	No	1 t/2 t		N & T Q & T	
3$\frac{1}{2}$ % nickel steel	1503–503	494 min	293 (0·2 % P.S.)	21/16	No	1 t/2 t		N & T Q & T	
9 % nickel steel	1503–509	695 min	541 (0·2 % P.S.)	18/13	No	2 t/3 t		DQ & T Q & T	
1 % Cr $\frac{1}{2}$ % Mo steel	1503–620	417/571	232 (0·2 % P.S.)	23/18	No	1 t/2 t		N & T Q & T	
1$\frac{1}{4}$ % Cr $\frac{1}{2}$ % Mo steel	1503–621	463/618	263 (0·2 % P.S.)	21/16	No	1 t/2 t		N & T Q & T	
2$\frac{1}{4}$ % Cr 1 % Mo steel	1503–622	541/695	371 min (0·2 % P.S.)	19/15	No	1 t/2 t		N & T Q & T	
3 % Cr $\frac{1}{2}$ % Mo steel	1503–623/38	587/741	417 (0·2 % P.S.)	19/15	No	1 t/2 t		N & T	En 29A
	1503–623/47	726/880	541 (0·2 % P.S.)	17/13		1 t/2 t		Q & T	
5 % Cr $\frac{1}{2}$ % Mo steel	1503–625	618	448 (0·2 % P.S.)	18/14	No	2 t/3 t		N & T Q & T	
$\frac{1}{2}$ % Cr $\frac{1}{2}$ % Mo $\frac{1}{4}$ % Va steel	1503–660	463/618	293 (0·2 % P.S.)	21/16	No	1 t/2 t		N & T Q & T	

Heat treatment condition: PS, Proof stress; N, Normalised; N and T, Normalised and tempered; Q and T, Quenched and tempered; DQ and T, Double quenched and tempered.

Table 7.17 MECHANICAL PROPERTIES OF CARBON AND ALLOY CASTING STEELS—BS 1504[2] SERIES

Material description	BS No	Tensile strength range/ min. value N/mm²	Yield stress (according to section) N/mm²	Elongation % 5·65√S₀	Impact test N m	Bend test min. t angle	Heat treatment condition	Comparable BS
			min	min				
Carbon steel for structural purposes	1504–101/A	402/494	201	20		1¼ 120°	fully annealed	592A
	1504–101/B	432/541	216	20		1½ 120°	and normalised	
	1504–101/C	541/618	270	15		1½ 90°		592C
Carbon steel for parts under pressure	1504–161/A	432	216	22	20	1½ 120°	annealed	
	1504–161/B	479	247	22		1½ 120°	annealed or normalised	
Carbon molybdenum steel	1504–240	463	247	20	20	1½ 120°	annealed / annealed or normalised	1398
3½% nickel steel	1504–503	448	270	25	41	1½ 90°	annealed, normalised and tempered	
1¼% Cr Mo steel	1504–621	479	278	20	34	1½ 120°	annealed / annealed or normalised	
2¼% Cr 1% Mo steel	1504–622	479	278	20	27	3 90°	annealed / annealed and normalised	
3% Cr Mo steel	1504–623	618/722	371	18	27	3 120°	air hardened and tempered / oil hardened and tempered	1461
5% Cr Mo steel	1504–625	618	417	18	27	3 90°	air hardened and tempered	1462
9% Cr Mo steel	1504–629	618	417	18	—	3 90°	air hardened and	1463

						on hardened and tempered	
Austenitic chromium nickel steel	1504–801	463	208	20	1½ 120°	heat treated and descaled	1631
Stabilised austenitic Cr Ni steel	1504–821	463	208	20	1½ 120°	heat treated and descaled	1631
Austenitic Cr Ni 2½% Mo steel	1504–845	463	208	15	2 120°	heat treated and descaled	1632
Austenitic Cr Ni 3½% Mo steel	1504–846	463	208	15	2 120°	soften and cool in air or water	1632

Table 7.18 DESIGN STRESSES (TENSILE) FOR FERROUS PLATES (BASED ON IMPERIAL VALUES GIVEN IN BS 1500[3])

Material	Reference BS number	Tensile strength N/mm²	Permissible design stress at the design temperature of the metal (N/mm²)													
			149°C	260°C	288°C	315°C	342°C	371°C	399°C	427°C	454°C	482°C	495°C	510°C	523°C	538°C
Carbon steel (semi-killed)	1501–151/23	355–432	88.9	88.9	88.9	88.9	88.9	87.6	80	—	—	—	—	—	—	—
	1501–151/26	402–486	100	100	100	100	100	98.6	87.6	—	—	—	—	—	—	—
	1501–151/28	432–517	108.2	108.2	108.2	108.2	108.2	104.8	92.4	—	—	—	—	—	—	—
Carbon steel (silicone killed)	1501–161/23	355–432	88.9	88.9	88.9	88.9	88.9	87.6	80	71	57.9	43.4	33.1	23.1	—	—
	1501–161/26	402–486	100	100	100	100	100	98.6	87.6	75.2	60	43.4	33.1	23.1	—	—
	1501–161/28	432–517	108.2	108.2	108.2	108.2	108.2	104.8	92.4	77.9	61.4	43.4	33.1	23.1	—	—
Carbon manganese steel (semi-killed)	1501–211/26	402–486	100	100	100	100	100	98.6	87.6	—	—	—	—	—	—	—
	1501–211/28	432–517	108.2	108.2	108.2	108.2	108.2	104.8	92.4	—	—	—	—	—	—	—
	1501–211/30	463–556	115.8	115.8	115.8	115.8	115.8	114.5	103.4	—	—	—	—	—	—	—
	1501–211/32	494–595	123.4	123.4	123.4	123.4	123.4	121.3	108.2	—	—	—	—	—	—	—
Carbon manganese steel (semi-killed niobium treated)	1501–213/28	432–517	108.2	108.2	108.2	108.2	108.2	104.8	—	—	—	—	—	—	—	—
	1501–213/30	463–556	115.8	115.8	115.8	115.8	115.8	114.5	—	—	—	—	—	—	—	—
	1501–213/32	494–595	123.4	123.4	123.4	123.4	123.4	121.3	—	—	—	—	—	—	—	—
Carbon manganese steel (silicone killed)	1501–221/26	402–486	100	100	100	100	100	98.6	87.6	75.2	60	43.4	33.1	23.1	—	—
	1501–221/28	432–517	108.2	108.2	108.2	108.2	108.2	104.8	92.4	77.9	61.4	43.4	33.1	23.1	—	—
	1501–221/30	463–556	115.8	115.8	115.8	115.8	115.8	114.5	103.4	83.4	63.4	44.1	33.1	23.1	—	—
	1501–221/32	494–595	123.4	123.4	123.4	123.4	123.4	121.3	108.2	86.2	64.8	44.1	33.1	23.1	—	—
Carbon manganese steel (silicone killed, aluminium treated)	1501–224/26	402–484	100	100	100	100	100	98.6	—	—	—	—	—	—	—	—
	1501–224/28	432–517	108.2	108.2	108.2	108.2	108.2	104.8	—	—	—	—	—	—	—	—
	1501–224/30	463–556	115.8	115.8	115.8	115.8	115.8	114.5	—	—	—	—	—	—	—	—
	1501–224/32	494–595	123.4	123.4	123.4	123.4	123.4	121.3	—	—	—	—	—	—	—	—
Carbon molybdenum steel	1501–240	417–510	104.1	104.1	104.1	104.1	104.1	104.1	104.1	100.6	95.1	89.6	84.1	62	35.9	26.2

Intermediate values may be obtained by linear interpolation

Table 7.19 DESIGN STRESSES (TENSILE) FOR FERROUS BARS AND SECTIONS (BASED ON IMPERIAL VALUES GIVEN IN BS 1500[3])

Material	Reference BS number	Tensile strength N/mm²	Permissible design stresses at the design temperature (N/mm²) MN/m²													
			149°C	260°C	288°C	315°C	342°C	371°C	399°C	427°C	454°C	482°C	495°C	510°C	523°C	538°C
Carbon steel	1501–101	402–494	100	—												
Carbon steel	1501–151/A	371–432	92·4	92·4	92·4	92·4	92·4	91	82·7	—	—	—	—	—	—	—
	1501–151/B	402–463	100	100	100	100	100	98·6	87·6	—	—	—	—	—	—	—
	1501–151/C	432–494	108·2	108·2	108·2	108·2	108·2	104·8	92·4	—	—	—	—	—	—	—
Carbon steel (silicone killed)	1501–161/A	371–432	92·4	92·4	92·4	92·4	92·4	91	82·7	72·4	58·6	43·4	33·1	23·1	—	—
	1501–161/B	402–463	100	100	100	100	100	98·6	87·6	75·2	60	43·4	33·1	23·1	—	—
	1501–161/C	432–494	108·2	108·2	108·2	108·2	108·2	104·8	92·4	77·9	61·4	43·4	33·1	23·1	—	—
Carbon manganese steel	1501–221	463–556	115·8	115·8	115·8	115·8	115·8	114·5	103	83·4	63·4	44·1	33·8	23·1	—	—
		510–602	127·6	127·6	127·6	127·6	127·6	124·8	110·3	87·6	65·5	44·1	33·8	23·1	—	—
Carbon molybdenum steel	1501–240	541–633	135·1	135·1	135·1	135·1	135·1	131·7	115·1	90·3	66·9	44·1	33·8	23·1	—	—
		417–510	104·1	104·1	104·1	104·1	104·1	104·1	104·1	100·7	95·1	89·6	84·1	62·1	35·9	26·2

(1) Intermediate values may be obtained by linear interpolation.
(2) Although above table makes reference to BS. 1501[2] classification of steels, note should be taken of the fact that for bars and sections the classification will be BS. 1502[2] and already BS. 1502 Grades 151; 161; 211; 221; 224 are available with the same physical, chemical and mechanical properties of the equivalent grades in the BS. 1501 series.

Table 7.20 DESIGN STRESSES (TENSILE) FOR FERROUS FORGINGS (BASED ON IMPERIAL VALUES GIVEN IN BS 1500³)

Material	Reference BS number	Tensile strength N/mm²	Permissible design stresses at the design temperature (N/mm²) MN/m²													
			149°C	260°C	288°C	315°C	342°C	371°C	399°C	427°C	454°C	482°C	495°C	510°C	523°C	538°C
Carbon steel (silicone killed)	1503–161/A	371–432	92·3	92·3	92·3	92·3	92·3	91	82·7	72·4	58·6	43·4	33·1	23·1	—	—
	1503–161/B	432–494	108·2	108·2	108·2	108·2	108·2	104·8	92·4	77·9	61·4	43·4	33·1	23·1	—	—
	1503–161/C	494–556	124·1	124·1	124·1	124·1	124·1	118·6	102	83·4	64·1	43·4	33·1	23·1	—	—
Carbon manganese steel	1503–221	494–587	124·1	124·1	124·1	124·1	124·1	122	108·2	86·2	64·8	44·1	33·8	23·1	—	—
Carbon molybdenum steel	1503–240/A	402–510	100·7	100·7	100·7	100·7	100·7	100·7	100·7	97·9	92·4	86·9	81·4	62	35·9	26·2
	1503–240/B	494 min	124·1	124·1	124·1	124·1	124·1	124·1	117·2	113·8	108·9	100	86·9	60	32·4	20·7

(1) Intermediate values may be obtained by linear interpolation.
(2) Grade BS. 1503/240 no longer appears in the current BS. 1503; 1969 Specification.

Table 7.21 DESIGN STRESSES (TENSILE) FOR FERROUS CASTINGS (BASED ON IMPERIAL VALUES GIVEN IN BS 1500³)

Material	Reference BS number	Tensile strength N/mm²	Permissible design stresses at the design temperature (N/mm²) MN/m²													
			149°C	260°C	288°C	315°C	342°C	371°C	399°C	427°C	454°C	482°C	495°C	510°C	523°C	538°C
Carbon steel	1504-101/A	402–494	179	179	179	179	179	176	157	134	108	77	60·2	41·7	—	—
	1504-101/B	432–541	193	193	193	193	193	188	166	139	110	77	60·2	41·7	—	—
	1504-101/C	541–618	242	242	242	242	242	232	196	158	119	77	60·2	41·7	—	—
Carbon steel (for parts under pressure)	1504-161/A	432 min	193	193	193	193	193	188	163	139	110	77	60·2	41·7	—	—
	1504-161/B	479 min	215	215	215	215	215	207	179	147	114	77	60·2	41·7	—	—
½% Molybdenum steel	1504-240	463 min	207	207	207	207	207	207	202	195	185	173	156	108	57·9	37

Intermediate values may be obtained by linear interpolation

Table 7.22 DESIGN STRESSES (TENSILE) FOR FERROUS PIPES (FOR USE AS SHELLS, BRANCH PIPES AND SIMILAR PARTS OF PRESSURE VESSELS—BASED ON IMPERIAL VALUES GIVEN IN BS 1500³)

Material	Reference BS number	Tensile strength N/mm²	Permissible design stresses at the design temperature (N/mm²) MN/m²														
			up to 204°C	260°C	288°C	315°C	342°C	371°C	399°C	427°C	454°C	482°C	495°C	510°C	523°C	538°C	566°C
Carbon steel seamless	3601–HFS22	402–494	84·8	84·8	—	—	—	—	—	—	—	—	—	—	—	—	—
	3601–HFS27	355–432	104·1	104·1	—	—	—	—	—	—	—	—	—	—	—	—	—
	3601–CDS22	402–486	84·8	84·8	—	—	—	—	—	—	—	—	—	—	—	—	—
	3601–CDS27	432–517	104·1	104·1	—	—	—	—	—	—	—	—	—	—	—	—	—
	3602–HFS23		88·9	88·9	88·9	88·9	88·9	88·3	81	71	57·9	45·4	33	23·1	—	—	—
	3602–HFS27		104·1	104·1	104·1	104·1	104·1	101·3	89·6	76·5	60·7	43·4	33	23·1	—	—	—
	3602–CDS23		88·9	88·9	88·9	88·9	88·9	88·3	81	71	57·9	43·4	33	23·1	—	—	—
	3602–CDS27		104·1	104·1	104·1	104·1	104·1	101·3	89·6	76·5	60·7	43·4	33	23·1	—	—	—
Carbon steel electric resistance welded	3601–ERW22		84·8	84·8	—	—	—	—	—	—	—	—	—	—	—	—	—
	3601–ERW27		104·1	104·1	—	—	—	—	—	—	—	—	—	—	—	—	—
	3602–ERW23		88·9	88·9	88·9	88·9	88·9	88·3	81	71	57·9	43·4	33	23·1	—	—	—
	3602–ERW27		104·1	104·1	104·1	104·1	104·1	101·3	89·6	76·5	60·7	43·4	33	23·1	—	—	—
Carbon steel low temperature duties	3603–HFS27 LT30/50		104·1	104·1	104·1	104·1	104·1	101·3	—	—	—	—	—	—	—	—	—
	3603–CDS27 LT30/50		104·1	104·1	104·1	104·1	104·1	101·3	—	—	—	—	—	—	—	—	—
1% Chromium ½% Molybdenum steel	3604–CD620		104·1	104·1	104·1	104·1	104·1	104·1	104·1	102·3	100·6	91	84·1	77·2	63·8	51·7	31
	3604–HF620		104·1	104·1	104·1	104·1	104·1	104·1	104·1	102·3	100·6	91	84·1	77·2	63·8	51·7	31

Intermediate values may be obtained by linear interpolation

Legend

HFS—Hot Finished Seamless
EDS—Cold Drawn Seamless

ERW—Electric Resistance Welded
CD —Cold Draw

HF—Hot Finished
LT —Low Temperature

Table 7.23 VALUE OF 'E' FOR FERROUS MATERIALS (BASED ON IMPERIAL VALUES GIVEN IN BS 1500[3])

Material	Permissible design values at design temperature of the metal—GN/m²									
	66°C	145°C	232°C	316°C	371°C	399°C	427°C	454°C	487°C	510°C
Carbon steel	203.3	198.6	191.7	182.7	175.8	172.4	167.5	160.9	151.7	141.4
Carbon manganese steel	203.3	198.6	191.7	182.7	175.8	172.4	167.5	160.9	151.7	141.4
Carbon molybdenum steel	203.3	198.6	191.7	182.7	175.8	172.4	167.5	160.9	151.7	141.4

Intermediate values may be obtained by linear interpolation

REFERENCES

The following references are used in the Tables:

1. BS 860
   Tables for comparison of hardness scales; BSI (1967)
2. BS 1501-6
   Steels for use in the chemical, petroleum and allied industries; BSI (1958. BS 1501 Part 1. Carbon and carbon manganese steel plates; BSI (1964). BS 1501 Part 2. Alloy steel plate; BSI (1970). BS 1502. Sections and bars; BSI (1968). BS 1503. Forgings; BSI (1969). BS 1504. Castings*. BS 1506. Bars for Bolting Materials.*
3. BS 1500
   Fusion welded pressure vessels for general purposes. Part 1. Carbon and low alloy steels; BSI (1958)
4. BS 970
   Wrought steels in the form of bars, billets and forgings up to 6 inch ruling section for automobile and general engineering purposes. En Series; BSI (1955-1972)
5. BS 4360
   Weldable structural steels. Part 1. Imperial units; BSI (1968). Part 2. Metric units; BSI (1969)
6. BS 3141
   Master schedule of steels. Part 1. Carbon steels; BSI (1960)
7. BS 18
   Methods for tensile testing of metals; BSI (1962)
8. ISO R.148
   Beam impact test (V notch) for steel; ISO Geneva (1960)
9. ISO R.83
   Charpy impact test (U notch) for steel; ISO Geneva (1959)
10. PD 6290
    New designation system for stainless steels; BSI (1967)
11. PD 6423
    New designation system for certain steels; BSI (1968)
12. PD 6431
    New designation system for alloy steels; BSI (1969)
13. BS 891
    Method for Rockwell hardness test. Part 1. Testing of metals; BSI (1962)
14. BS 4175
    Method for Rockwell superficial hardness test (N and T scales). Part 1. Testing of metals; BSI (1967)
15. BS 240
    Method for Brinell hardness test. Part 1. Testing of metals; BSI (1962)
16. BS 427
    Method for Vickers hardness test. Part 1. Testing of metals; BSI (1961)

* Still retained in BS 1501–6 BSI 1958

MECHANICAL PROPERTIES
OF NON-FERROUS MATERIALS

G. R. DARBY

Explanation of tables

Tables 7.24 to 7.64 relate to commercially available grades of Aluminium, Copper and Nickel and their respective alloys used in the U.K. These tables give the mechanical properties, converted to the appropriate SI unit, from data shown in current British Standards. All British Standards in this series[1-21] have now been metricated with the stress values shown in hectobars (hbar). These values have been converted in the above tables to MN/m^2 (N/mm^2). Values in hbar can be obtained by dividing the values in N/mm^2 by 10.

Hardness values

Aluminium and Aluminium Alloys: Hardness values are to the Brinell (HB) series.
Copper and copper alloys: Hardness values are to the Vickers (HV) series.
Nickel and nickel alloys: Hardness values are to the Vickers (HV) series

Elongation

Aluminium and aluminium alloys: Elongation is given on two bases, (i) 50 mm length; (ii) $5.65\sqrt{area}$
Copper and copper alloys: Elongation is given on four bases, (i) 100 mm length; (ii) 200 mm length; (iii) $5.65\sqrt{area}$; (iv) $4\sqrt{area}$.
Nickel and nickel alloys: Elongation is given on two bases, (i) $5.65\sqrt{area}$; (ii) 50 mm length.

Bend test

Aluminium and aluminium alloys: Not specified.
Copper and copper alloys: The test piece shall be bent with the skin in tension and the test piece shall not crack when bent (cold) once through the appropriate angle shown in the column 'Angle (°)' where t = diameter or width across flats of test piece. A transverse bend is made on a test piece cut with its major axes at right angles to the direction of rolling; a longitudinal bend test is made as a test piece cut with its major axes parallel to the direction of rolling. Bend tests are in accordance with BS 1639[22].
Nickel and nickel alloys: Not specified.

Aluminium and aluminium alloys

MATERIAL DESIGNATION

The BS designation is given as the primary designation. The designation in brackets, formulated around the chemical composition of the material, is that which conforms to ISO R208[23] and ISO R209[24].

The main grades of material used are 1; 1A; 1B; 1C; NZ; N3; N4; N5; N6; N8; H9; H12; H15; H16; H20; and H30, to which is attached a prefix to indicate form of material. The prefixes adopted are S for plate sheet and strip; E for bars, extruded round tube and sections; F for forging stock and forgings; LM for ingots and castings (above material designations do not necessarily apply to this series of materials); R for rivets; B for bolt and screw stock; G for wire and T for drawn tube.

CONDITION

The designations for heat treatment conditions are also in line with ISO proposals. Such work hardened designations as ¼H, half hard, ¾H etc. have now been replaced by an appropriate symbol selected from the range H1 to H8 (the higher number indicates increase in strength). The conditions of heat treated alloys is also indicated by the letter T followed by a qualifying letter B, D, E, F, or H, indicating the precise condition following the heat treatment.

The legend used in the following tables is as follows:

M	As manufactured. Materials which acquire some temper from shaping processes in which there is no special control over thermal treatment of amount of strain hardening.
O	Annealed Material which is fully annealed to obtain the lowest strength condition.
H1 H2	Strain hardened. Material subject to the application of cold work after
H3 H4	annealing (or hot forming) or to a combination of cold work and partial
H5 H6	annealing/stabilising in order to secure the specified mechanical properties.
H7 H8	The designations are in ascending order of tensile strength.
TB	Solution heat-treated and naturally aged. Material which receives no cold work after solution heat treatment except as may be required to flatten or straighten it. Properties of some alloys in this temper are unstable.
TB 7	Solution heat treated and stabilised.
TD	Solution heat treated, cold worked and naturally aged.
TE	Cooled from an elevated temperature shaping process and precipitation treated.
TF	Solution heat treated and precipitation treated.
TF 7	Full heat treatment plus stabilisation.
TH	Solution heat treated, cold worked and then precipitation treated.

SECTIONS

Aluminium and aluminium alloy section dimensions (metric series) have yet to be agreed internationally.

Copper and copper alloys

MATERIAL DESIGNATION

The BS designation for copper and copper alloys are based on two letters being the main constituent elements followed by a number e.g. CZ 106 . . . 70/73% copper; remainder zinc (with other trace elements). However, ISO Recommendation grades based on ISO R426 etc.[25-29] have also been given where available. These are formulated around the chemical composition of the material.

The copper and copper alloy series have a British Standard for sheet, strip and foil and a further British Standard for plate. Plate in this context is defined as flat material over 10·0 mm thick and over 300 mm wide. The casting alloys are shown in three groups. Group A alloys are alloys in common use (preferred for all general purposes); Group B alloys are special purpose alloys (for applications requiring their particular properties e.g. HCC 1 . . . high conductivity copper for electrical work) and Group C alloys being alloys in limited production.

CONDITION

The material shall be supplied in one of the following conditions as specified by the purchaser:

O	Material in the annealed condition.
$\frac{1}{4}$H $\frac{1}{2}$H H EH	The various harder tempers produced by cold rolling.
H EH	For certain of the materials in this standard, these tempers may be produced by partial annealing.
SH ESH	Spring hard tempers produced by cold rolling of thinner material.
M	Material in the 'as manufactured' condition.
P	Precipitation treatment.
W	Material which has been solution heat treated and will respond effectively to precipitation treatment.
W($\frac{1}{4}$H) W($\frac{1}{2}$H) W(H)	Material which has been solution heat treated and subsequently cold worked to various harder tempers.
WP	Material which has been solution heat treated and precipitation treated.
W($\frac{1}{4}$H)P W($\frac{1}{2}$H)P W(H)P	Material which has been solution heat treated, cold worked and then precipitation treated.

Nickel and nickel alloys

MATERIAL DESIGNATION

The material designation for nickel and nickel alloys are based on a number system used in conjunction with the prefix letters NA. The designations are not as yet the subject of an ISO Recommendation.

CONDITION

The various conditions obtainable are as specified in the Tables 7.59 to 7.64.

Material designation	Condition	Thickness Over mm	Thickness Up to and including mm	0.2% proof stress minimum N/mm²	Tensile strength Minimum N/mm²	Tensile strength Maximum N/mm²	Elongation % on 50 mm; Material thicker than 0.5 mm minimum	0.8 mm minimum	1.3 mm minimum	2.6 mm minimum	3.0 mm minimum	Elongation on 5.65 $\sqrt{S_0}$ over 12.5 mm minimum %
BS 1470/S1 (ISO Al 99.99)	O	0.2	6.0	—	—	65	30	35	40	45	45	—
	H4	0.2	6.0	—	80	95	7	8	10	12	12	—
	H8	0.2	6.0	—	100	—	3	4	5	6	6	—
BS 1470/S1A (ISO Al 99.8)	M	3.0	25.0	—	—	—	—	—	—	—	—	—
	O	0.2	6.0	—	—	90	29	29	29	35	35	—
	H4	0.2	12.5	—	95	120	5	6	7	8	8	—
	H8	0.2	3.0	—	125	—	3	4	4	5	—	—
BS 470/S1B (ISO Al 99.5)	O	0.2	6.0	—	55	95	22	25	30	32	32	—
	H4	0.2	12.5	—	100	135	4	5	6	6	8	—
	H8	0.2	3.0	—	135	—	3	3	4	4	—	—
BS 1470/S1C (ISO Al 99.0)	M	3.0	25.0	—	70	105	20	25	30	30	30	—
	O	0.2	6.0	—	95	120	4	6	8	9	9	—
	H2	0.2	6.0	—	110	140	3	4	5	5	7	—
	H4	0.2	12.5	—	125	150	2	3	4	4	6	—
	H6	0.2	6.0	—	140	—	2	3	4	4	4	—
	H8	0.2	3.0	—	—	—	—	—	—	—	—	—
BS 470/NS3 (ISO Al Mn 1)	O	0.2	6.0	—	90	130	20	23	24	24	25	—
	H2	0.2	6.0	—	120	145	5	6	7	9	9	—
	H4	0.2	12.5	—	140	175	3	4	5	6	7	—
	H6	0.2	6.0	—	160	195	2	3	4	4	4	—
	H8	0.2	3.0	—	175	—	2	3	4	4	—	—
BS 4300/6/NS31 (Al Mn Mg)	O	0.2	3.0	—	110	155	16	18	20	20	—	—
	H2	0.2	3.0	115	130	175	2	3	4	5	—	—
	H4	0.2	3.0	145	160	205	2	2	3	4	—	—
	H6	0.2	3.0	170	185	230	1	1	2	3	—	—
	H8	0.2	3.0	190	215	—	1	1	1	2	—	—
BS 1470/NS4 (ISO Al Mg 2)	M	3.0	25.0	—	—	—	—	—	—	—	—	—
	O	0.2	6.0	60	160	200	18	18	18	20	20	—
	H3	0.2	6.0	130	200	240	4	5	6	8	8	—
	H6	0.2	12.5	175	225	275	3	4	5	5	5	—

continued overleaf

Table 7.24—continued

Material designation	Condition	Thickness Over mm	Thickness Up to and including mm	0.2% proof stress minimum N/mm²	Tensile strength Minimum N/mm²	Tensile strength Maximum N/mm²	Elongation % on 50 mm; Material thicker than 0.5 mm minimum	0.8 mm minimum	1.3 mm minimum	2.6 mm minimum	3.0 mm minimum	Elongation on 5.65 √S_0 over 12.5 mm minimum %
BS 4300/7/NS41 (Al Mg 1)	O	0.2	3.0	—	95	145	18	20	21	22	—	—
	H2	0.2	3.0	80	125	170	4	5	6	8	—	—
	H4	0.2	3.0	100	145	185	3	3	5	6	—	—
	H8	0.2	3.0	165	185	—	1	2	3	3	—	—
BS 1470/NS5 (ISO Al Mg 3–5)	O	0.2	6.0	85	215	275	12	14	16	18	18	—
	H2	0.2	6.0	165	245	295	5	6	7	8	8	—
	H4	0.2	6.0	225	275	325	4	4	6	6	6	—
BS 4300/8/NS51 (Al Mg 3 Mn)	M	3.0	25.0	—	—	—						14
	O	0.2	3.0	80	215	285	12	14	16	18	—	—
	H2	0.2	3.0	180	250	305	4	5	7	8	—	—
	H4	0.2	3.0	200	270	325	3	4	5	6	—	—
BS 1470/NS8 (ISO Al Mg 4.5 Mn)	M	3.0	25.0	—	—	—						14
	O	0.2	25.0	125	275	350	12	14	16	16	16	10
	H2	0.2	6.0	235	310	375	5	6	8	10	8	6
	H4	0.2	6.0	270	345	405	4	5	6	8	6	6
BS 1470/HS 15 (ISO Al Cu 4 Si Mg)	TB	0.2	25.0	245	385	—	13	14	14	14	14	10
	TF	0.2	3.0	375	430	—	6	6	7	7	9	—
	TF	3.0	25.0	380	440	—	—	—	—	—	—	6
	TF	25.0	40.0	360	430	—	—	—	—	—	—	6
	TF	40.0	63.0	345	420	—	—	—	—	—	—	6
BS 1470/HC 15 (ISO Al Cu 4 Si Mg)	TB	0.2	12.5	230	375	—	13	14	14	14	14	10
	TB	12.5	25.0	245	385	—	—	—	—	—	—	—
	TF	0.2	3.0	325	400	—	7	7	8	8	8	—
	TF	3.0	12.5	365	425	—	—	—	—	8	—	8
	TF	12.5	25.0	380	440	—	—	—	—	—	—	6
BS 1470/HS 30 (ISO Al Si Mg Mn)	O	0.2	3.0	—	200	155	16	16	16	18	—	—
	TB	0.2	3.0	120	200	—	15	15	15	15	15	15
	TB	3.0	25.0	115	295	—	8	8	8	8	8	15
	TF		3.0	255								8

Table 7.25 PROPERTIES OF ALUMINIUM AND ALUMINIUM ALLOY DRAWN TUBE

Material designation	Condition	Wall thickness		0·2% Proof stress minimum	Tensile strength		Elongation on 50 mm or $5·65\sqrt{S_0}$ minimum
		Over	Up to and including		Minimum	Maximum	
		mm	mm	N/mm²	N/mm²	N/mm²	
BS 1471/T1B	0	—	12·0	—	—	95	—
(ISO Al 99·5)	H4	—	12·0	—	100	135	—
	H8	—	12·0	—	135	—	—
BS 1471/T1C	0	—	12·0	—	—	105	—
(ISO Al 99·0)	H4	—	12·0	—	110	140	—
	H8	—	12·0	—	140	—	—
BS 1471/NT4	0	—	10·0	60	160	200	18
(ISO Al Mg 2)	H4	—	10·0	175	225	—	5
BS 1471/NT5	0	—	10·0	125	275	350	12
(ISO Al Mg 3·5)	H4	—	10·0	235	310	—	5
BS 4300/10/NT51	0	—	10·0	85	215	260	16
(Al Mg 3 Mn)	H2	—	10·0	165	245	295	8
	H4	—	10·0	225	275	325	4
BS 1471/NT8	0	—	10·0	125	275	350	12
(ISO Al Mg 4·5 Mn)	H2	—	10·0	235	310	—	5
BS 1471/HT9	0	—	10·0	—	—	155	—
(ISO Al Mg Si)	TB	—	10·0	100	155	—	15
	TF	—	10·0	180	200	—	8
BS 1471/HT15	TB	—	10·0	290	400	—	8
(ISO Al Cu 4 Si Mg)	TF	—	10·0	370	450	—	6
BS 1471/HT20	H4	—	6·0	160	185	—	5
(ISO Al Mg 1 Si Cu)	TB	—	6·0	115	215	—	12
	TB	6·0	10·0	115	215	—	14
	TF	—	6·0	240	295	—	7
	TF	6·0	10·0	225	295	—	9
BS 1471/HT30	TB	—	6·0	115	215	—	12
(ISO Al Si 1 Mg Mn)	TB	6·0	10·0	115	215	—	14
	TF	—	6·0	225	310	—	7
	TF	6·0	10·0	240	310	—	9

7-92

Table 7.26 PROPERTIES OF ALUMINIUM AND ALUMINIUM ALLOYS FORGING STOCK AND FORGINGS

Material Designation	Condition	Condition of test sample bar	Size of bar Over mm	Up to and including mm	0.2% Proof stress minimum N/mm²	Tensile strength minimum N/mm²	Elongation % on $5.65\sqrt{S_0}$ minimum
BS 1472/F1B (ISO Al 99·5)	M	Forged or extruded	—	150	—	60	22
BS 1472/NF4 (ISO Al Mg 2)	M	Forged or extruded	—	150	60	170	16
BS 1472/NF5 (ISO Al Mg 3·5)	M	Forged or extruded	—	150	100	215	16
BS 4300/11/NF51 (Al Mg 3 Mn)	O	Forged or extruded	—	150	85	275	18
	M	Forged or extruded	—	150	100	—	16
BS 1472/NF8 (ISO Al Mg 4·5 Mn)	M	Forged or extruded	—	150	130	280	12
BS 1472/HF9 (ISO Al Mg Si)	TB	Forged or extruded	—	150	85	140	16
			150	200	85	125	13
	TF	Forged or extruded	—	150	160	185	10
			150	200	130	150	6
BS 1472/HF12 (ISO Al Cu 2 Ni 1 Mg Fe Si)	TB	Forged	—	150	160	310	13
		Extruded	—	200	145	310	13
	TF	Forged	—	150	300	385	6
		Extruded	—	200	285	385	6
BS 1472/HF15 (ISO Al Cu 4 Si Mg)	TB	Forged	—	150	215	370	13
			—	20	230	370	11
		Extruded	20	75	250	390	11
			75	150	250	390	8
			150	200	230	370	8
	TF	Forged	—	150	395	450	6
			—	20	370	435	7
		Extruded	20	75	435	480	7
			75	150	420	465	7
			150	200	390	435	7
BS 1472/HF16 (ISO Al Cu 2 Mg 1·5 Fe 1 Ni 1)	TF	Forged or extruded	—	200	340	430	5
BS 1472/HF30 (ISO Al Si 1 Mg Mn)	TB	Forged	—	150	120	185	16
	TB	Extruded	—	150	120	190	16
			150	200	100	170	13
	TF	Forged	—	150	225	295	8
			—	20	255	295	8
	TF	Extruded	20	150	270	310	8
			150	200	240	280	5

Table 7.27 PROPERTIES OF ALUMINIUM AND ALUMINIUM ALLOY RIVET, BOLT AND SCREW STOCK

Material designation	Condition of supply	Condition of test	Diameter Over mm	Diameter Up to and including mm	0·2% proof stress minimum N/mm²	Tensile strength Minimum N/mm²	Tensile strength Maximum N/mm²
RIVET STOCK							
BS 1473/R1B (ISO Al 99·5)	H5	H5	—	12	—	110	—
BS 1473/NR5 (ISO AL Mg 3·5)	O or M	O or M	—	25	—	215	—
	H2 annealed and drawn 10–20% reduction in area	H2 annealed and drawn 10–20% reduction in area	—	25	—	245	—
BS 1473/NR6 (ISO Al Mg 5)	O or M	O or M	—	25	—	250	—
	H2 annealed and drawn 10–20% reduction in area	H2 annealed and drawn 10–20% reduction in area	—	25	—	280	—
BS 1473/HR15 (ISO Al Cu 4 Si Mg)	H2 annealed and drawn 20–40% reduction in area	TB	—	12	—	385	—
BS 1473/HR30 (ISO Al Si Mg Mn)	H2 annealed and drawn 20–40% reduction in area	TB	—	25	—	200	—
BOLT AND SCREW STOCK							
BS 1473/NB6 (ISO Al Mg 5)	H4	H4	—	12	240	310	360
BS 1473/HB15 (ISO Al Cu 4 Si Mg)	H2 annealed and drawn 20–40% reduction in area	TF	—	12	390	430	—
BS 1473/HB30 (ISO Al Si Mg Mn)	H2 annealed and drawn 20–40% reduction in area	TF	—	6	255	295	—
			6	12	270	310	—

Table 7.28 PROPERTIES OF ALUMINIUM AND ALUMINIUM ALLOYS BARS AND SECTIONS

Material Designation	Condition	Thickness		0·2% Proof stress minimum N/mm²	Tensile strength		Elongation %	
		Over mm	Up to and including mm		minimum N/mm²	maximum N/mm²	On 5·65 $\sqrt{S_0}$ minimum	On 50 mm minimum
BS 1474/E1B (ISO Al 99·5)	M	—	—	—	60	—	25	23
BS 1474/E1C (ISO Al 99·0)	M	—	—	—	60	—	20	18
BS 1474/NE4	M	—	150	60	170	—	16	14
BS 1474/NE5 (ISO Al Mg 3·5)	O	—	150	35	215	275	18	16
	M	—	150	100	215	—	16	14
BS 1474/NE8 (ISO Al Mg 4·5 Mn)	O	—	150	125	275	—	14	13
	M	—	150	130	280	—	12	11
BS 1474/HE9 (ISO Al Mg Si)	O	—	200	—	—	140	15	13
	M	—	200	—	100	—	13	12
	TB	—	150	70	130	—	16	14
	TB	150	200	70	120	—	13	—
	TE	—	25	110	150	—	8	7
	TF	—	150	160	185	—	8	7
	TF	150	200	130	150	—	6	—
BS 1474/HE15 (ISO Al Cu 4 Si Mg)	TB	—	20	230	370	—	11	10
	TB	20	75	250	390	—	11	—
	TB	75	150	250	390	—	8	—
	TB	150	200	230	370	—	8	6
	TF	—	20	370	435	—	7	—
	TF	20	75	435	480	—	7	—
	TF	75	150	420	465	—	7	—
	TF	150	200	390	435	—	7	—

TB	—	150	120	190	—	16	14
TB	150	200	100	170	—	13	—
TF	—	20	255	295	—	8	7
TF	20	150	270	310	—	8	—
TF	150	200	240	280	—	5	—
BS 4300/12/NE51 (Al Mg 3 Mn) O	—	150	85	215	275	18	16
M	—	150	100	215	—	16	14

Table 7.29 PROPERTIES OF ALUMINIUM AND ALUMINIUM ALLOY WIRE

Material Designation	Condition	Diameter Over mm	Up to and incl. mm	Tensile strength minimum N/mm²	maximum N/mm²
BS 1475/GIA	O	—	10	—	90
(ISO Al 99.8)	M	—	10	—	—
	H8	—	10	125	—
BS 1475/GIB	O	—	10	—	95
(ISO Al 99.5)	M	—	10	—	—
	H8	—	10	135	—
BS 1475/NG2	M	—	10	—	—
(ISO Al Si12)					
BS 1475/NG21	M	—	10	—	—
(ISO Al Si5)					
BS 1475/NG3	O	—	10	—	130
(ISO Al Mn1)	M	—	10	—	—
	H8	—	10	175	—
BS 1475/NG4	O	—	10	170	215
(ISO Al Mg2)	M	—	10	—	—
	H8	—	10	260	—
BS 4300/9/NG41	O	—	10	—	140
	H8	—	10	180	—
BS 1475/NG5	M	—	10	—	—
(ISO Al Mg3.5)					
BS 1475/NG6	O	—	10	250	310
(ISO Al Mg5)	M	—	10	—	—
	H4	—	10	310	360
	H8	—	10	385	—
BS 1475/NG61	M	—	10	—	—
(ISO Al Mg5.2MnCr)					
BS 1475/HG9	M	—	10	—	—
(ISO Al MgSi)	TB	—	10	140	—
	TF	—	10	185	—
	TD	—	6	280	—
	TD	6	10	230	—
BS 1475/HG15	TB	—	10	385	—
(ISO Al Cu4SiMg)	TF	—	10	430	—
BS 1475/HG20	TH	—	6	370	—
(ISO Al Mg1SiCu)	TH	6	10	355	—

Table 7.30 PROPERTIES OF ALUMINIUM AND ALUMINIUM ALLOY INGOTS AND CASTINGS

Designation	Condition	Tensile strength minimum		Elongation minimum		Hardness	Remarks
		Sand cast N/mm²	Chill cast N/mm²	Sand cast %	Chill cast %	HB	
BS 1490/LM 2	M	—	150	—	—	—	
BS 1490/LM 4	M	140	160	2	2	—	
	TF	230	280	—	—	—	
BS 1490/LM 6	M	160	190	5	7	—	
BS 1490/LM 20	M	—	190	—	5	—	General
BS 1490/LM 24	M	—	180	—	1·5	—	purpose
BS 1490/LM 25	M	130	160	2	3	—	alloys
	TE	150	190	1	2	—	
	TB7	160	230	2·5	5	—	
	TF	230	280	—	2	—	
BS 1490/LM 27	M	140	160	1	2	—	
BS 1490/LM 0	M	—	—	—	—	—	
BS 1490/LM 5	M	140	170	3	5	—	
BS 1490/LM 9	M	—	190	—	3	—	
	TE	170	230	1·5	2	—	
	TF	240	295	—	—	—	
BS 1490/LM 10	TB	280	310	8	12	—	Special
BS 1490/LM 12	M	—	170	—	—	—	purpose
BS 1490/LM 13	TE	—	210	—	—	90–130	alloys
	TF	170	280	—	—	100–150	
	TF7	140	200	—	—	65–90	
BS 1490/LM 16	TB	170	230	2	3	—	
	TF	230	280	—	—	—	
BS 1490/LM 18	M	120	140	3	4	—	
BS 1490/LM 21	M	150	170	1	1	—	
BS 1490/LM 22	TB	—	245	—	8	—	
BS 1490/LM 26	TE	—	210	—	—	90–120	Special
BS 1490/LM 28	TE	—	170	—	—	90–130	purpose
	TF	120	190	—	—	100–140	alloys
BS 1490/LM 29	TE	120	190	—	—	100–140	
	TF	120	190	—	—	—	
BS 1490/LM 30	M	—	150	—	—	—	
	TS	—	160	—	—	—	

Table 7.31 DESIGN STRESSES (TENSILE) FOR WROUGHT ALUMINIUM MATERIALS (BASED ON BS 1500: PART 3)

Material	Type	Condition	Minimum tensile strength N/mm²	0.2% proof stress N/mm²	Minimum Elongation on 50 mm %	38°C	66°C	93°C	121°C	149°C	176°C	204°C
1B	99.5% Al	Annealed	62	18	30	13.8	12.8	11.7	10.3	93	82.7	72.4
		½ H	100	69	8	20.7	20.7	20.0	18.6	16.2	13.8	11.0
N3	Al-1¼% Mn	Annealed	98	35	25	23.1	21.7	20.0	18.6	16.5	14.5	12.4
		½ H	139	117	7	34.5	33.4	32.4	30.3	27.6	24.1	21.4
N4	Al-2¼% Mg	Annealed	170	62	18	42.7	42.7	42.7	41.4	37.2	32.1	21.4
		¼ H	200	131	12	53.4	53.4	52.7	49.0	44.1	38.6	24.1
		½ H	232	199	5	58.6	58.6	57.9	58.1	47.6	42.1	24.1
Al-Mg-Mn alloy	Al-2¾% Mg	Annealed	213	83	18	53.4	53.4	53.0	49.6	38.0	29.6	24.1
		¼ H	249	179	12	62.1	62.1	60.7	51.7	38.0	29.6	24.1
(ASA 5454)[41]	¾% Mn	½ H	269	200	10	67.2	67.2	63.8	51.7	38.0	29.6	24.1
N5	Al-3½% Mg	Annealed	216	100	18	53.8	53.8	—	—	—	—	—
N6	Al-5% Mg	Annealed	263	100	20	65.5	65.5	—	—	—	—	—
N8	Al-4½% Mg ¾% Mn	Annealed or as manufactured	278	131	12	69.0	69.0	—	—	—	—	—
		Bars and sections	263			65.5	65.5	—	—	—	—	—
H9	Al-Ng-Si	P	154	116	10	37.9	35.2	33.8	31.7	29.0	21.4	13.8
		WP	200	162	12	51.7	49.0	46.9	42.1	31.0	21.4	13.8
		Annealed	108	—	15	29.3	29.0	27.6	26.2	24.8	19.0	13.1
H30	Al-Mg-Si-Mn	W	200	116	18	51.7	49.6	48.3	46.2	44.1	38.6	27.6
		WP	293	254	10	72.4	70.3	68.3	64.8	54.5	42.7	30.3
LM5	Al-5% Mg	M (Sand test)	139	—	3	23.1	21.4	20.3	19.7	18.6	17.6	16.9
		M (Chill test)	170	—	5	28.3	25.5	24.1	23.4	22.4	21.0	20.3
LM6	Al-12% Si	M (Sand test)	162	—	5	26.9	23.4	22.4	21.0	18.6	16.9	15.1
		M (Chill test)	185	—	7	30.7	26.9	25.5	24.1	21.4	19.3	17.2
LM8	Al-5% Si	M (Sand test)	184		?	20.7						

Table 7.32 DESIGN STRESSES (TENSILE) FOR ALUMINIUM BOLTING ALLOYS
(BASED ON BS 1500: PART 3)

Material	Type	Condition	Minimum Tensile strength on test bars N/mm^2	Permissible design stresses (N/mm^2) for temperature not exceeding						
				38°C	66°C	93°C	121°C	149°C	176°C	204°C
H30	Al-Mg-Si	WP	290	57·9	56·5	54·5	51·7	43·4	33·8	22·8
H15	Al-Cu-Mg	WP	448	89·6	84·1	80·0	71·7	49·6	30·3	20·7
N4	Al-2¼% Mg	½ H	232	58·6	58·6	57·9	53·1	47·6	42·1	24·1
N5	Al-5% Mg	½ H	309	75·8	74·5	—	—	—	—	—

Table 7.33 VALUE OF 'E' (GN/m^2) FOR ALUMINIUM MATERIALS
(BASED ON BS 1500: PART 3) 1 GN/m^2 = 1000 N/mm^2

Material	Design temperature of the metal								
	−216°C	−16°C	38°C	66°C	93°C	121°C	149°C	177°C	204°C
1B	77·2	70·3	68·9	68·3	67·6	66·9	65·5	64·1	62·1
N3	77·2	70·3	68·9	68·3	67·6	66·9	65·5	64·1	62·1
N4	77·2	70·3	68·9	68·3	67·6	66·9	65·5	64·1	62·1
Al-Mg-Mn alloy	77·2	70·3	68·9	68·3	67·6	66·9	65·5	64·1	62·1
N5	77·2	70·3	68·9	68·3	67·6	66·9	65·5	64·1	62·1
N6	77·2	70·3	68·9	68·3	67·6	66·9	65·5	64·1	62·1
N8	77·2	70·3	68·9	68·3	67·6	66·9	65·5	64·1	62·1
H9	73·0	66·9	65·5	64·8	64·1	63·4	62·1	60·7	58·6
H15	82·0	74·5	73·1	72·4	71·7	71·1	69·6	68·3	65·5
H30	77·2	70·3	68·9	68·3	67·6	66·9	65·5	64·1	62·0
LM5	77·2	70·3	68·9	68·3	67·6	66·9	65·5	64·1	62·0
LM6	88·3	80·7	79·3	78·6	77·2	76·5	75·2	73·8	71·7

Table 7.34 PROPERTIES OF COPPER SHEET, STRIP AND FOIL

Material Designation	Description	Condition	Thickness Over mm	Up to and including mm	Tensile strength Up to and including 450 mm (minimum) N/mm²	Over 450 mm (minimum) N/mm²	Elongation on 50 mm %	Transverse bend Angle °	Radius	Longitudinal bend Angle °	Radius
BS 2870/C101	Electrolytic tough pitch high conductivity copper	O	0·6	10·0	210	210	35	180	close	180	close
		½H	0·6	1·3	245	245	10	180	t	180	t
		½H	1·3	10·0	245	245	15	180	t	180	t
		H	0·6	2·7	310	285	—	90	t	90	t
		H	2·7	10·0	290	280	—	90	t	90	t
BS 2870/C102	Fire refined tough pitch high conductivity copper	O	0·6	10·0	210	210	35	180	close	180	close
		½H	0·6	1·3	245	245	10	180	t	180	t
		½H	1·3	10·0	245	245	15	180	t	180	t
		H	0·6	2·7	310	285	—	90	t	90	t
		H	2·7	10·0	290	280	—	90	t	90	t
BS 2870/C 103	Oxygen free high conductivity copper	O	0·6	10·0	210	210	35	180	close	180	close
		¼H	0·6	1·3	245	245	10	180	t	180	t
		½H	1·3	10·0	245	245	15	180	t	180	t
		H	0·6	2·7	310	285	—	90	t	90	t
		H	2·7	10·0	290	280	—	90	t	90	t
BS 2870/C 104	Tough pitch non-arsenical copper	M or O	0·6	10·0	210	210	35	180	close	180	close
		¼H	0·6	1·3	245	245	10	180	t	180	t
		½H	1·3	10·0	245	245	15	180	t	180	t
		H	0·6	2·7	310	285	—	90	t	90	t
		H	2·7	10·0	295	280	—	90	t	90	t
BS 2870/C 105	Tough pitch arsenical copper	M or O	0·6	10·0	210	210	35	180	close	180	close
		¼H	0·6	1·3	245	245	10	180	t	180	t
		½H	1·3	10·0	245	245	15	180	t	180	t
		H	0·6	2·7	310	285		90	t	90	t

arsenical copper	½H	1·3	10·0	245	245	15	180		180	t
	H	0·6	2·7	310	285	—	90	t	90	t
	H	2·7	10·0	295	280	—	90	t	90	close
BS 2870/C 107										
Phosphorus deoxidised non-arsenical copper	M or O	0·6	10·0	210	210	35	180	close	180	t
	¼H	0·6	1·3	245	245	10	180	t	180	t
	½H	1·3	10·0	245	245	15	180	t	180	t
	H	0·6	2·7	310	285	—	90	t	90	t
	H	2·7	10·0	295	280	—	90	t	90	t

Table 7.35 PROPERTIES OF BRASS SHEET, STRIP AND FOIL

Material Designation	Description	Condition	Thickness Over	Thickness Up to and including	Tensile strength Up to and Including 450 mm wide minimum N/mm²	Tensile strength Over 450 mm wide minimum (N/mm²)	Elongation on 50 mm minimum %	Vickers hardness HV Up to and including 450 mm wide min.	max.	Over 450 mm wide min.	max.	Transverse bend Angle °	Radius	Longitudinal bend Angle °	Radius
BS 2870/CZ 101 (R426 Cu Zn 10)	90/10 Brass	O	—	10·0	245	245	35	—	75	—	75	180	close	180	close
		H ½	—	3·5	310	280	7	95	—	85	—	180	close	180	close
		H ½	3·5	10·0	310	280	7	95	—	85	—	90	2 t	90	t
		H	—	10·0	350	325	3	110	75	100	75	90	t	90	t
BS 2870/CZ 102 (R426 Cu Zn 15)	85/15 Brass	O	—	10·0	245	245	35	—	75	—	75	180	close	180	close
		H ½	—	3·5	325	295	7	95	—	85	—	180	close	180	close
		H ½	3·5	10·0	325	295	7	95	—	85	—	180	t	180	t
		H	—	10·0	370	340	3	110	—	100	—	90	2 t	90	t
BS 2870/CZ 103 (R426 Cu Zn 20)	80/20 Brass	O	—	10·0	265	265	40	—	80	—	80	180	close	180	close
		H ½	—	3·5	340	310	10	95	—	85	—	180	close	180	close
		H ½	3·5	10·0	340	310	10	95	—	85	—	180	t	180	t
		H	—	10·0	400	370	5	120	—	110	—	90	2 t	90	t
BS 2870/CZ 106 (R426 Cu Zn 30)	70/30 Brass	O	—	10·0	280	280	50	—	80	—	80	180	close	180	close
		H ¼	—	10·0	325	325	35	75	—	75	—	180	close	180	close
		H ½	—	3·5	350	340	20	100	—	95	—	180	close	180	close
		H ½	3·5	10·0	350	340	20	100	—	95	—	180	t	180	t
		H	—	10·0	415	385	5	125	—	120	—	90	2 t	90	t
BS 2870/CZ 107 (R426 Cu Zn 33)	2/1 Brass	O	—	10·0	280	280	45	—	80	—	80	180	close	180	close
		H ¼	—	10·0	340	325	35	75	—	75	—	180	close	180	close
		H ½	—	3·5	385	350	20	110	—	100	—	180	close	180	close
		H ½	3·5	10·0	385	350	20	110	—	100	—	180	t	180	t
		H	—	10·0	460	415	5	135	—	125	—	90	2 t	90	t
		EH	—	10·0	525	—	—	165	—	—	—	—	—	90	2 t
BS 2870/CZ 108 (R426 Cu Zn 37)	Common brass	O	—	10·0	280	280	40	—	80	—	80	180	close	180	close
		H ¼	—	10·0	340	325	30	75	—	75	—	180	close	180	close
		H ½	—	3·5	385	350	15	110	—	100	—	180	close	180	close
		H ½	3·5	10·0	385	350	15	110	—	100	—	180	t	180	t

21 Al 2)

Standard	Material	Temper													
BS 2870/CZ 112	Naval brass	O	—	10·0	340	340	25	—	—	—	180	t	180	t
		H	—	10·0	400	400	20	—	—	—	—	—	90	t
BS 2870/CZ 118	Leaded brass 64% copper 1% lead	½H	—	6·0	370	370	10	110	—	140	—	—	—	—
		H	—	6·0	430	430	5	140	—	165	—	—	—	—
		EH	—	6·0	510	510	3	165	—	190	—	—	—	—
BS 2870/CZ 119	Leaded brass 62% copper 1% lead	½H	—	6·0	370	370	10	110	—	140	—	—	—	—
		H	—	6·0	430	430	5	140	—	165	—	—	—	—
		EH	—	6·0	510	510	3	165	—	190	—	—	—	—
BS 2870/CZ 120 (R426 Cu Zn 39 Pb 2)	Leaded brass 59% copper 2% lead	½H	—	6·0	—	—	10	110	—	140	—	—	—	—
		H	0·3	0·3	510	510	5	—	—	—	—	—	—	—
		H	—	6·0	510	510	5	140	—	165	—	—	—	—
		EH	—	6·0	575	575	3	165	—	190	—	—	—	—
BS 2870/CZ 123 (R426 Cu Zn 40 Pb)	60/40 Brass	M	—	10·0	370	370	20	—	—	—	—	—	—	—
BS 2870/CZ 125	Cap copper	O	—	10·0	—	—	—	—	75	75	180	close	180	close

Table 7.36 PROPERTIES OF COPPER-NICKEL SHEET, STRIP AND FOIL

Material		Condition	Thickness		Tensile strength	Elongation on 50 mm minimum	Vickers hardness HV	Bend test			
Designation	Description		Over	Up to and including				Transverse bend		Longitudinal bend	
								Angle °	Radius	Angle °	Radius
			mm	mm	N/mm²	%	maximum				
BS 2870/CN 101 (R429 Cu Ni 5 Fe 1 Mn)	95/5 Copper-nickel-iron	M	—	10·0	230	35	—	—	—	—	—
BS 2870/CN 102 (R429 Cu Ni 10 Fe 1 Mn)	90/10 Copper-nickel-iron	M	—	10·0	310	30	90	—	—	—	—
		O	—	10·0	280	40	90	—	—	—	—
BS 2870/CN 103	85/15 Copper-nickel	O	0·6	2·0	280	35	—	180	close	180	close
		O	2·0	10·0	280	38	—	180	close	180	close
BS 2870/CN 104 (R429 Cu Ni 20)	80/20 Copper-nickel	O	0·6	2·0	310	35	—	180	close	180	close
		O	2·0	10·0	310	38	—	180	close	180	close
BS 2870/CN 105 (R429 Cu Ni 25)	75/25 Copper-nickel	O	0·6	2·0	340	30	—	180	close	180	close
		O	2·0	10·0	340	35	—	180	close	180	close
BS 2870/CN 106 (R429 Cu Ni 30)	70/30 Copper-nickel	O	0·6	2·0	370	30	—	180	close	180	close
		O	2·0	10·0	370	35	—	180	close	180	close
BS 2870/CN 107 (R429 Cu Ni 30 Fe Mn)	70/30 Copper-nickel	O	0·6	2·0	370	30	—	180	close	180	close
		O	2·0	10·0	370	35	—	180	close	180	close

Table 7.37 PROPERTIES OF PHOSPHOR BRONZE SHEET, STRIP AND FOIL

Material Designation	Description	Condition	Thickness up to and including (mm)	Tensile strength Up to and including 450 mm wide min N/mm²	Tensile strength Over 450 mm wide min N/mm²	0·2% proof stress Up to and including 450 mm wide min N/mm²	0·2% proof stress Over 450 mm wide min N/mm²	Elongation on 50 mm min %	Vickers hardness (HV) Up to and including 450 mm wide min	Vickers hardness (HV) Up to and including 450 mm wide max	Vickers hardness (HV) Over 450 mm wide min	Vickers hardness (HV) Over 450 mm wide max	Transverse bend Angle °	Transverse bend Radius	Longitudinal bend Angle °	Longitudinal bend Radius
BS 2870/ PB 101 (R427 Cu Sn 5)	3% Phosphor bronze	O	10·0	295	295	—	—	40	—	80	—	80	180	close	180	close
		¼H	10·0	340	340	125	125	30	100	—	100	—	180	close	180	close
		½H	10·0	460	430	390	340	8	150	—	130	—	90	t	180	t
		H	6·0	540	495	480	435	4	180	—	150	—	—	—	90	t
		EH	6·0	620	—	580	—	—	190	—	—	—	—	—	90	t
BS 2870/ PB 102	5% Phosphor bronze	O	10·0	310	310	—	—	45	—	85	—	85	180	close	180	close
		¼H	10·0	350	350	140	140	35	110	—	110	—	180	close	180	close
		½H	10·0	495	460	420	385	10	160	—	140	—	90	t	180	t
		H	6·0	590	525	540	480	4	180	—	160	—	—	—	90	t
		EH	6·0	645	—	615	—	—	200	—	—	—	—	—	90	t
BS 2870/ PR 103 (R427 Cu Sn 7)	7% Phosphor bronze	O	10·0	340	340	—	—	50	—	90	—	90	180	close	180	close
		¼H	10·0	385	385	200	200	40	115	—	115	—	180	close	180	close
		½H	10·0	525	460	440	380	12	170	—	150	—	90	t	180	t
		H	6·0	620	540	560	480	6	200	—	165	—	—	—	90	t
		EH	6·0	695	—	650	—	—	215	—	—	—	—	—	90	t
		SH	0·6	—	—	—	—	—	220	240	Up to 150 max wide		—	—	—	—
		ESH	0·6	—	—	—	—	—	240	—	Up to 150 max wide		—	—	—	—

Table 7.38 PROPERTIES OF ALUMINIUM AND SILICON BRONZE SHEET, STRIP AND FOIL

| Material | | | Thickness | | Tensile strength min. | Elongation on 50 mm | Bend test | |
| Designation | Description | Condition | Over | Up to and including | | | Transverse bend | |
			mm	mm	N/mm²	%	Angle °	Radius
BS 2870/ CA 101 (R428 Cu Al 5)	5 % aluminium bronze	M	—	2·7	—	—	180	close
		M	2·7	10·0	340	40	180	2 t
BS 2870/ BS 101	Silicon bronze	M	—	2 7	—	—	180	close
		M	2·7	10·0	370	50	180	2 t

Table 7.39 PROPERTIES OF NICKEL SILVER SHEET, STRIP AND FOIL

Material		Condition	Thickness up to and including	Vickers hardness (HV)		Bend test			
						Transverse bend		Longitudinal bend	
Designation	Description			min.	max.	Angle °	Radius	Angle °	Radius
BS 2870/NS 103	10% Nickel silver	O	10·0	—	100	180	t	180	t
		½H	10·0	125	—	180	t	180	t
		H	10·0	160	—	90	t	90	t
		EH	10·0	185	—	—	—	90	t
BS 2870/NS 104	12% Nickel silver	O	10·0	—	100	180	t	180	t
		½H	10·0	130	—	180	t	180	t
		H	10·0	160	—	90	t	90	t
		EH	10·0	190	—	—	—	90	t
BS 2870/NS 105	15% Nickel silver	O	10·0	—	105	180	t	180	t
		½H	10·0	135	—	180	t	180	t
		H	10·0	165	—	90	t	90	t
		EH	10·0	195	—	—	—	90	t
BS 2870/NS 106 (R430 Cu Ni 18 Zn 20)	18% Nickel silver	O	10·0	—	110	180	t	180	t
		½H	10·0	135	—	180	t	180	t
		H	10·0	170	—	90	t	90	t
		EH	10·0	200	—	—	—	90	t
BS 2870/NS 107 (R430 Cu Ni 18 Zn 27)	18% Nickel silver	—	—	—	—	—	—	—	—
BS 2870/NS 108	20% Nickel silver	O	10·0	—	110	180	t	180	t
		½H	10·0	140	—	180	t	180	t
		H	10·0	175	—	90	t	90	t
		EH	10·0	205	—	—	—	90	t
BS 2870/NS 109	25% Nickel silver	O	10·0	—	115	180	t	180	t
		½H	10·0	150	—	180	t	180	t
		H	10·0	180	—	90	t	90	t
		EH	10·0	210	—	—	—	90	t

Table 7.40 PROPERTIES OF COPPER-BERYLLIUM STRIP AND FOIL

Material Designation	Description	Condition	Thickness up to and including	Tensile strength N/mm²	Elongation on 50 mm minimum %	Vickers hardness (HV) minimum	maximum	Bend test Transverse bend Angle °	Radius	Longitudinal bend Angle °	Radius
BS 2870/CB 101	Copper beryllium	W	10·0	415	40	85	120	180	close	180	close
		W ($\frac{1}{4}$ H)	10·0	510	15	150	190	90	t	180	t
		W ($\frac{1}{2}$ H)	10·0	590	10	190	225	90	2 t	90	t
		W (H)	10·0	695	4	225	—	—	—	90	2 t
		WP	10·0	—	—	350	—	—	—	—	—
		W ($\frac{1}{4}$ H) P	10·0	—	—	350	—	—	—	—	—
		W ($\frac{1}{2}$ H) P	10·0	—	—	350	—	—	—	—	—
		W (H) P	10·0	—	—	350	—	—	—	—	—

Table 7.41 PROPERTIES OF COPPER AND COPPER ALLOY PLATE

| Material | | | Thickness | | Tensile strength min | Elongation on 200 mm min. |
Designation	Description	Condition	Over mm	Up to and including mm	N/mm²	%
BS 2875/ C 101	Electrolytic tough pitch high conductivity copper	M or O	10	—	210	35
		H	{ 10	16	280	15
			{ 16	25	250	20
BS 2875/ C 102	Fire refined tough pitch high conductivity copper	M or O	10	—	210	35
		H	{ 10	16	280	15
			{ 16	25	250	20
BS 2875/ C 103	Oxygen free high conductivity copper	M or O	10	—	210	35
		H	{ 10	16	280	15
			{ 16	25	250	20
BS 2875/ C 104	Tough pitch non-arsenical copper	M or O	10	—	210	35
		H	{ 10	16	280	15
			{ 16	25	250	20
BS 2875/ C 105	Tough pitch arsenical copper	M or O	10	—	220	35
		H	{ 10	16	280	15
			{ 16	25	250	20
BS 2875/ C 106	Phosphorus deoxidised non-arsenical copper	M or O	10	—	210	35
		H	{ 10	16	280	15
			{ 16	25	250	20
			{ 25	—	240	25
BS 2875 C 107	Phosphorus deoxidised arsenical copper	M or O	10	—	210	35
		H	{ 10	16	280	15
			{ 16	25	250	20
			{ 25	—	240	25
BS 2875/ C 108	Copper cadmium	H	{ 10	16	310	13 } on
			{ 16	25	280	18 } $5{\cdot}65\sqrt{S_0}$

Table 7.42 PROPERTIES OF BRASS PLATE

Material Designation	Description	Condition	Thickness Over mm	Thickness Up to and including mm	Tensile strength min N/mm^2	Elongation on $5 \cdot 65\sqrt{S_0}$ min %
BS 2875/CZ 105	70/30 Arsenical brass	M or O	10	—	280	40
		H	10	16	360	18
			16	25	340	22
BS 2875/CZ 106 (R426 Cu Zn 30)	70/30 brass	M or O	10	—	280	40
		H	10	16	360	18
			16	25	340	22
BS 2875/CZ 110 (R426 Cu Zn 21 Al 2)	Aluminium brass	M	10	—	280	36
		O	10	—	280	40
BS 2875/CZ 112	Naval brass		10	25	360	18
		M or O	25	125	340	18
			125	—	310	18
		H	10	125	400	18
BS 2875/CZ 123 (R426 Cu Zn 40 Pb)	60/40 brass		10	25	340	18
		M	25	125	310	18
			125	—	290	18

Table 7.43 PROPERTIES OF PHOSPHOR BRONZE PLATE

Material Designation	Description	Condition	Thickness Over mm	Thickness Up to and including mm	Tensile strength min N/mm^2	Elongation on $5 \cdot 65\sqrt{S_0}$ min %
BS 2875/PB 101	3% phosphor bronze	M or O	10	—	280	30
		H	10	16	410	12
			16	25	390	16
BS 2875/PB 102	5% phosphor bronze	M or O	10	—	310	35
		H	10	16	430	12
			16	25	400	16

Table 7.44 PROPERTIES OF ALUMINIUM AND SILICON BRONZE PLATE

Material		Condition	Thickness		Tensile strength min	Elongation on $5 \cdot 65\sqrt{S_0}$ min
Designation	Description		Over mm	Up to and including mm	N/mm^2	%
BS 2875/CA 102	7% aluminium bronze	M	10 50	50 —	480 450	31 31
BS 2875/CA 105	10% aluminium bronze	M	10 85	85 —	620 560	10 10
BS 2875/CA 106 (R428 Cu Al 8 Fe 3)	7% aluminium bronze	M	10 50	50 —	480 450	31 31
BS 2875/CS 101	Silicon bronze	M O	10 10	— —	340 320	31 40

Table 7.45 PROPERTIES OF COPPER-NICKEL—IRON PLATE

Material		Condition	Thickness		Tensile strength min	Elongation on $5 \cdot 65\sqrt{S_0}$ min
Designation	Description		Over mm	Up to and including mm	N/mm^2	%
BS 2875/CN 101 (R429 Cu Ni 5 Fe 1 Mn)	95/5 copper nickel-iron	M or O	10	—	250	27
BS 2875/CN 102 (R429 Cu Ni 10 Fe 1 Mn)	90/10 copper nickel-iron	M O	10 10	— —	280 280	27 36
BS 2875/CN 107 (R429 Cu Ni 30 Fe Mn)	70/30 copper nickel	M or O	10	—	310	27

Table 7.46 PROPERTIES OF COPPER AND COPPER ALLOY TUBES

| Material | | Tensile strength | | | Elongation on $4\sqrt{A}$ (annealed condition) % | Tube application British standards |
| | | Annealed (O) | | As drawn (M) | | |
Designation	Description	min N/mm²	max N/mm²	min N/mm²		
BS 2871 C 101	Electrolytic tough pitch high conductivity copper	200	247	263	40	BS 1977
BS 2871/ C 102	Fire refined tough pitch high conductivity copper	200	247	263	40	BS 1977
BS 2871/ C 103	Oxygen-free high conductivity copper	200	247	263	40	BS 1977
BS 2871/ C 106	Phosphorus deoxidised non-arsenical copper	216	263	278	40	BS 61 Part 1 BS 378; BS 659 BS 1306 Pt. 2 BS 1386; BS 2017
BS 2871/ C 107	Phosphorus deoxidised arsenical copper	216	263	278	40	BS 61 Pt. 1 BS 378; BS 659 BS 1306 Pt. 2 BS 1386; BS 2017
BS 2871/ CZ 105	70/30 arsenical brass	278	370	386	—	BS 378; BS 885 BS 1464; BS 2579 Pt. 1
BS 2871/ CZ 110	Aluminium brass	309	402	417	—	BS 378; BS 885 BS 1464; BS 2579 Pt. 1
BS 2871/ CZ 111	Admiralty brass	278	370	386	—	BS 378; BS 1464 BS 2579 Pt. 1
BS 2871/ CN 107	70/30 copper-nickel	417	463	479	—	BS 378; BS 1464 BS 2579 Pt. 1
BS 2871/ CA 102	7% aluminium bronze bronze	386	463	510	50	BS 378; BS 1464 BS 1867 BS 2579 Pt. 1

Note: BS 2871 is under revision. The new version of BS 2871 will incorporate the tube application British Standards as follows.
BS 2871 Part 1 replaces BS 659, 1386, 3931
    Part 2 replaces BS 61 Part 1 Tables 1 and 2
                BS 1306 Part 2 Tables 1, 3, 4 and 6
                BS 1977
                BS 2017
    Part 3 replaces BS 378, BS 1464

Table 7.47 PROPERTIES OF COPPER AND COPPER ALLOY FORGING STOCK AND FORGINGS

Material		Size		Tensile strength minimum N/mm²	0·2% proof stress minimum N/mm²	Elongation on 5·65√S₀ minimum %
Designation	Description	Over mm	Up to and including mm			
BS 2872/CZ 109 (R426 Cu Zn 40)	Lead free 60/40 brass	6 and over	—	310	—	25
BS 2872/CZ 112	Naval brass	6 and over	—	340	—	15
BS 2872/CZ 114	High tensile brass	6 and over	—	460	195	15
BS 2872/CZ 115 (R426 Cu Zn 39 Al Fe Mn)	High tensile brass (soldering quality)	6 and over	—	460	195	15
BS 2872/CZ 116	High tensile brass	6 and over	—	540	295	12
BS 2872/CZ 122	Leaded brass	6 and over	—	310	—	20
BS 2872/CZ 123 (R426 Cu Zn 40 Pb)	60/40 brass	6 and over	—	310	—	25
BS 2872/CA 103	9% Aluminium bronze	6 and over	—	520	215	20
BS 2872/CA 104 (R428 Cu Al 10 Fe 5 Ni 5)	10% Aluminium bronze	6	10	700	400	10
		over 10	70	700	400	12
		over 70	120	700	350	14
		over 120	—	650	320	12
BS 2872/CA 106 (R428 Cu Al 8 Fe 3)	7% Aluminium bronze	6	10	540	240	30
		over 10	70	520	220	30
		over 70	—	460	190	30
BS 2872/NS 101 (R430 Cu Ni 10 Zn 45)	Leaded 10% nickel brass	6 and over	—	460	—	8
BS 2872/CS 101	Copper silicon	6 and over	—	340	—	20

Table 7.48 PROPERTIES OF COPPER AND COPPER ALLOY RODS AND SECTIONS

Material Designation	Material Description	Condition	Size Over (mm)	Size Up to and including (mm)	Tensile strength Round min N/mm²	Round max N/mm²	Square and hexagon min N/mm²	Square and hexagon max N/mm²	Rectangular min N/mm²	Rectangular max N/mm²	Elongation Round min %	Section and hexagon min %	Rectangular min %
BS 2874/ C 101	Electrolytic tough pitch high conductivity copper	O	3	6	—	260	—	—	—	—	32	—	—
			6	10	—	250	—	—	—	—	32	—	—
			10	12	—	240	—	240	—	—	40	40	40
			12	50	—	230	—	230	—	240	45	45	45
			50	80	—	230	—	230	—	230	45	45	45
BS 2874/ C 102	Fire refined tough pitch high conductivity copper	½H	3	6	290	—	—	—	—	—	4	—	—
			6	10	280	—	—	—	—	—	8	—	—
			10	12	260	—	260	—	250	—	12	12	12
			12	25	250	—	250	—	230	—	18	18	18
			25	50	230	—	230	—	230	—	22	22	—
			50	80	230	—	230	—	—	—	22	22	—
BS 2874/ C 103	Oxygen free high conductivity copper	H	3	6	350	—	—	—	—	—	—	—	—
			6	10	330	—	—	—	—	—	—	—	—
			10	12	320	—	310	—	280	—	6	6	6
			12	50	290	—	280	—	260	—	8	8	8
			50	80	260	—	250	—	—	—	12	12	—
BS 2874/ C 106	Phosphorus deoxidised non-arsenical copper	O	6 and over		210	—	210	—	210	—	33	33	33
		M	6 and over		230	—	230	—	230	—	13	13	13
BS 2874/ C 109	Copper tellurium	O	Up to 6		210	—	210	—	210	—	28	28	28
		M	6	50	260	—	260	—	260	—	8	8	8
		M	50		240	—	240	—	240	—	8	8	8
BS 2874/ C 111	Copper sulphur	O	Up to 6		210	—	210	—	210	—	28	28	28
		M	6	50	260	—	260	—	260	—	8	8	8
		M	50		240	—	240	—	240	—	8	8	8

Table 7.49 PROPERTIES OF BRASS RODS AND SECTIONS

Material		Condition	Size		Tensile strength min N/mm²	0.2% Proof stress min N/mm²	Elongation on $5.65\sqrt{S_0}$ min %	Bend test	
Designation	Description		Over mm	Up to and including mm				Angle °	Radius
BS 2874/CZ103 R426 Cu Zn 20	80/20 brass	M	6	—	310	—	24	120	t
BS 2874/CZ104	Leaded 80/20 brass	M	6	—	310	—	22	120	t
BS 2874/CZ106 (R426 Cu Zn 30)	70/30 brass	O M	6 6	— —	280 340	— —	45 28	— —	— —
BS 2874/CZ109 (R426 Cu Zn 40)	Lead free 60/40 brass	M M	6 50	50 —	340 310	160 —	26 26	75 75	t t
BS 2874/CZ112 (R426 Cu Zn Sn1)	Naval brass	M M	6 20	20 —	400 340	— —	18 20	75 75	t t
BS 2874/CZ113	Naval brass (special mixture)	M M	6 20	20 —	400 340	— —	16 18	75 75	t t
BS 2874/CZ114 (R426 Cu Zn 39 Al Fe Mn)	High tensile brass	M M H	6 80 6	80 160 40	460 460 540	240 200 290	18 18 12	— — —	— — —
BS 2874/CZ115 (R426 Cu Zn 39 Al Fe Mn)	High tensile brass (soldering) quantity	M Hot worked M Cold worked and stress relieved	— 6 40 60	40 60 100 100	460 540 500 460	190 280 250 220	20 12 14 16	— — — —	— — — —
BS 2874/CZ116	High tensile brass	M	6	100	540	290	12	—	—
BS 2874/CZ119	Leaded brass 62% Cu 2% Pb	M M	6 50	50 —	350 320	— —	22 22	180 180	t t
BS 2874/CZ121 (R426 Cu Zn 40 Pb 3)	Leaded brass 58% Cu 3% Pb	M M	6 80	80 —	380 340	— —	12 16	— —	— —

Table 7.49—*continued*

| Material | | Condition | Size | | Tensile strength min | 0·2% Proof stress min | Elongation on 5·65√S₀ min | Bend test | |
Designation	Description		Over mm	Up to and including mm	N/mm²	N/mm²	%	Angle °	Radius
BS 2874/CZ122	Leaded brass 58% Cu 2% Pb	M	6	80	380	—	18	—	—
		M	80	—	340	—	18	—	—
		H	6	40	460	240	18	—	—
		H	40	50	380	160	20	—	—
BS. 2874/CZ123 (R426 Cu Zn 40 Pb)		M	6	50	340	—	24	75	t
		M	50	—	310	—	24	75	t

| Material | | | Size | | | | Tensile strength minimum | 0·2% proof stress minimum | Elongation 5·65√S₀ |
| | | | Thickness | | Width or dia. or width A/F | | | | |
Designation	Description	Form and Condition	over mm	up to and including mm	over mm	up to and including mm	N/mm²	N/mm²	%
BS 2874/CZ 124 (R426 Cu Zn 36 Pb3)	Leaded brass 62% Cu 3% Pb	Round and hex. rod — M	—	—	6·0	25	350	130	12
		M	—	—	25	50	300	115	18
		M	—	—	50	—	280	95	22
		½H	—	—	6·0	12	400	160	6
		½H	—	—	12	25	380	160	9
		½H	—	—	25	50	340	130	12
		½H	—	—	50	—	310	95	18
		H	—	—	3·0	5·0	550	290	—
		H	—	—	5·0	8·0	480	220	3
BS 2874/CZ 124 (R426 Cu Zn 36 Pb3)	Leaded brass 62% Cu 3% Pb	Rect. and squares — M	6·0	25	—	150	300	115	18
		M	25	—	—	150	280	95	22
		½H	6·0	12	—	25	340	160	9
		½H	6·0	12	25	150	310	105	12
		½H	12	50	—	50	310	105	18

Table 7.50 PROPERTIES OF PHOSPHOR BRONZE RODS AND SECTIONS

| Material | | Condition | Size | | Tensile strength minimum | 0.2% proof stress minimum | Elongation $5.65\sqrt{S_0}$ |
| Designation | Description | | Over | Up to and including | | | |
			mm	mm	N/mm²	N/mm²	%
BS 2874/ PB 102	5% Phosphor bronze	M	6	20	500	410	12
		M	20	40	460	380	12
		M	40	70	380	320	16
		M	70	120	310	240	20
		M	120	—	270	80	24

Table 7.51 PROPERTIES OF ALUMINIUM BRONZE RODS AND SECTIONS

| Material | | Condition | Size | | Tensile strength minimum | 0·2% proof stress minimum | Elongation $5 \cdot 65 \sqrt{S_0}$ | Bend test | |
Designation	Description		Over mm	Up to and including mm	N/mm²	N/mm²	%	Angle °	Radius
BS 2874/CA 103	9% Aluminium bronze	M	6·0	—	520	215	22	75	t
BS 2874/CA 104 (R428 Cu Al 10 Fe 5 Ni 5)	10% Aluminium bronze	M	6·0	10	700	400	10	—	—
		M	10	70	700	400	12	—	—
		M	70	120	700	350	14	—	—
		M	120	—	650	320	12	—	—
BS 2874/CA 106 (R428 Cu Al 8 Fe 3)	7% Aluminium bronze	O	6	10	460	190	30	—	—
		M	6	10	540	240	30	180	t
		M	10	70	520	220	30	180	t
		M	70	—	460	190	30	180	t

Table 7.52 PROPERTIES OF LEADED NICKEL BRASS AND LEADED NICKEL SILVER RODS AND SECTIONS

Material		Condition	Size	Tensile strength min	Elongation on $5 \cdot 65 \sqrt{S_0}$ min	Bend test	
Designation	Description		mm	N/mm^2	%	Angle °	Radius
BS 2874/ NS 101 (R430 Cu Ni 10 Zn 45)	Leaded 10% nickel brass	M	6 and over	460	8	—	—
BS 2874/ NS 102	Leaded 14% nickel brass	M	6 and over	510	8	—	—
BS 2874/ NS 111	Leaded 10% nickel silver	M	—	—	—	180	t
BS 2874/ NS 112	Leaded 15% nickel silver	M	—	—	—	180	t
BS 2874/ NS 113 (R430 Cu Ni 18 Zn 19 Pb)	Leaded 18% nickel silver	M	—	—	—	180	t

Table 7.53 PROPERTIES OF COPPER SILICON RODS

Material		Condition	Size		Tensile strength minimum	Elongation on $5 \cdot 65 \sqrt{S_0}$ minimum	Bend test	
Designation	Description		Over mm	Up to and including mm	N/mm^2	%	Angle °	Radius
BS 2874/ CS 101	Copper silicon	O	6	70	370	40	75	$\frac{1}{2}$t
		O	70	—	340	40	75	$\frac{1}{2}$t
		M	6	20	540	15	75	$\frac{1}{2}$t
		M	20	70	460	20	75	$\frac{1}{2}$t
		M	70	—	430	25	75	$\frac{1}{2}$t

Table 7.54 PROPERTIES OF COPPER AND COPPER ALLOY WIRE

| Material | | Condition | Diameter | | Tensile strength minimum N/mm² | Elongation on 200 mm minimum % |
Designation	Description		Over mm	Up to and including mm		
BS 2873/C 101	Electrolytic tough pitch high conductivity copper	O	—	0·125	—	10
			0·125	0·50	—	15
			0·50	0·90	—	20
			0·90	1·25	—	25
			1·25	10·0	—	30
BS 2873/C 102	Fire refined tough pitch high conductivity copper	H	—	1·60	455	—
			1·60	2·50	445	—
			2·50	—	—	—
BS 2873/C 103	Oxygen free high conductivity copper	—	—	—	—	—
BS 2873/C 106	Phosphorus deoxidised non-arsenical copper	—	—	—	—	—
BS 2873/C 108	Copper-cadmium	H	—	—	—	—

Table 7.55 PROPERTIES OF BRASS WIRE

| Material | | Condition | Diameter | | Tensile strength | | Elongation on 100 mm min % |
Designation	Description		Over mm	Up to and including mm	Min N/mm²	Max N/mm²	
BS 2873/ CZ 101 (R426 Cu Zn 10)	90/10 brass	O ½ H H	By agreement between purchaser and supplier				
BS 2873/ CZ 102 (R426 Cu Zn 15)	85/15 brass	O	0·5	10·0	290	—	25
		½ H	0·5	10·0	430	590	—
		H	0·5	10·0	590	—	—
BS 2873/ CZ 103 (R426 Cu Zn 20)	80/20 brass	O	0·5	10·0	310	—	30
		½ H	0·5	10·0	460	620	—
		H	0·5	10·0	620	—	—
BS 2873/ CZ 106 (R426 Cu Zn 30)	70/30 brass	O	0·5	10·0	310	—	45
		½ H	0·5	10·0	460	620	—
		H	0·5	10·0	620	—	—
BS 2873/ CZ 107 (R426 Cu Zn 33)	2/1 brass	O	0·5	10·0	320	—	35
		½ H	0·5	10·0	460	620	—
		H	0·5	10·0	620	740	—
		EH	—	2·5	740	820	—
		EH	2·5	6·0	700	770	—
BS 2873/ CZ 108 (R426 Cu Zn 37)	Common brass	O	0·5	10·0	320	—	35
		½ H	0·5	10·0	460	620	—
		H	0·5	10·0	620	770	—
		EH	—	2·5	740	—	—
		EH	2·5	6·0	700	—	—
BS 2873/ CZ 119	Leaded brass	O ½ H H	By agreement between purchaser and supplier				

7.56 PROPERTIES OF PHOSPHOR BRONZE WIRE

Material			Diameter		Tensile strength		Elongation on 100 mm min %
Designation	Description	Condition	Over mm	Up to and including mm	Min N/mm²	Max N/mm²	
BS 2873/	5%	O	0·5	10·0	340	—	40
PB 102	phosphor	½ H	0·5	10·0	540	700	—
	bronze	H	0·5	10·0	700	850	—
		EH	—	2·5	850	—	—
		EH	2·5	6·0	800	—	—
BS 2873/	7%	O	0·5	10·0	370	—	50
PB 103	phosphor	½ H	0·5	10·0	590	740	—
	bronze	H	0·5	10·0	740	900	—
		EH	—	2·5	900	—	—
		EH	2·5	6·0	850	—	—

Table 7.57 PROPERTIES OF COPPER—BERYLLIUM WIRE

Material			Diameter		Tensile strength min. N/mm²	Elongation on 200 mm min. %
Designation	Description	Condition	Over mm	Up to and including mm		
BS 2873/CB 101	Copper-beryllium	W	0·5	10·0	390	30
		W (H)	—	3·0	770	—
		WP	0·5	10·0	1 050	—
		W (H)P	—	3·0	1 240	—

Table 7.58 PROPERTIES OF CC

| Material | | Tensile strength | | | | 0·2% p | |
Designation	Description	Sand N/mm²	Chill N/mm²	Con-tinuous N/mm²	Centri-fugal N/mm²	Sand N/mm²	C/ N/r
GROUP A ALLOYS							
BS 1400/PB 4	Phosphor bronze	190–270	270–370	330–450	280–400	100–160	140
BS 1400/LPB 1	Leaded phosphor bronze	190–250	220–270	270–360	230–310	80–130	130
BS 1400/LB2	80/10/0/10 Leaded bronze	190–270	220–280	280–390	230–310	80–130	140
BS 1400/LB 4	85/5/0/10 Leaded bronze	160–190	200–270	230–310	220–300	60–100	80
BS 1400/LG 2	85/5/5/5 Leaded gunmetal	200–270	200–280	270–340	220–310	100–130	110
BS 1400/LG 4	87/7/3/3 Leaded gunmetal	250–320	250–340	300–370	230–370	130–140	130
BS 1400/SCB 1	Brass for sand casting	170–200	—	—	—	80–110	-
BS 1400/SCB 3	Brass for sand casting	190–220	—	—	—	70–110	-
BS 1400/SCB 6	Brass for brazable casting	170–190	—	—	—	80–110	-
BS 1400/DCB 1	Brass for die casting	—	280–370	—	—	—	90
BS 1400/DCB 3	Brass for die casting	—	300–340	—	—	—	90
BS 1400/PCB 1	Brass for pressure die casting	—	280–370	—	—	—	90
GROUP B ALLOYS							
BS 1400/HCC 1	High conductivity copper	160–190	—	—	—	—	
BS 1400/CC 1	Copper chromium	270–340	—	—	—	170–250	-
BS 1400/PB 1	Phosphor bronze	220–280	310–390	360–500	330–420	130–160	170
BS 1400/PB 2	Phosphor bronze	220–310	270–340	310–430	280–370	130–170	170
BS 1400/CT 1	Copper tin	230–310	270–340	310–390	280–370	130–160	140
BS 1400/LB 5	75/5/0/20 Leaded bronze	160–190	170–230	190–270	190–270	60–100	80
BS 1400/LG 1	83/3/9/5 Leaded gunmetal	180–220	180–270	—	—	80–130	80
BS 1400/AB 1	Aluminium bronze	500–590	540–620	—	560–650	170–200	200
BS 1400/AB 2	Aluminium bronze	640–700	650–740	—	670–730	250–300	250
BS 1400/CMA 1	Copper manganese aluminium	650–730	670–740	—	—	280–340	310
BS 1400/CMA 2	Copper manganese aluminium	740–820	—	—	—	380–470	
BS 1400/HTB 1	High tensile brass	470–570	500–570	—	500–600	170–280	210
BS 1400/HTB 3	High tensile beta brass	740–810	—	—	740–930	400–470	
GROUP C ALLOYS							
BS 1400/LB 1	76/9/0/15 Leaded bronze	170–230	200–270	230–310	220–300	80–110	130
BS 1400/G 1	88/10/2 Gunmetal	270–340	230–310	300–370	250–340	130–160	130
BS 1400/G 3	Nickel gunmetal	280–340	—	340–370	—	140–160	-
BS 1400/G3 WP	Nickel gunmetal (fully heat treated)	430–480	—	430–500	—	280–310	-
BS 1400/SCB 4	Naval brass	250–310	—	—	—	70–110	-

		Elongation on $5{\cdot}65\sqrt{S_0}$				Hardness (HB)				Impact value
s / 1^2	Centri-fugal N/mm^2	Sand %	Chill %	Con-tinuous %	Centri-fugal %	Sand	Chill	Con-tinuous	Centri-fugal	N m
70	140–230	3–12	2–10	7–30	4–20	70–95	95–140	95–140	95–140	—
)0	130–160	3–12	2–12	5–18	4–22	60–90	85–110	85–110	85–110	—
20	140–190	5–15	3–7	6–15	5–10	65–85	80–90	80–90	80–90	11
70	80–110	7–12	5–10	9–20	6–13	55–75	60–80	60–80	60–80	11
40	110–140	13–25	6–15	13–35	8–30	65–75	80–95	75–90	80–95	26
50	130–160	16–25	5–15	13–30	6–30	70–85	80–95	80–95	80–95	26
	—	18–40	—	—	—	45–60	—	—	—	—
	—	11–30	—	—	—	45–65	—	—	—	—
	—	18–40	—	—	—	45–60	—	—	—	—
	—	—	23–50	—	—	—	60–70	—	—	—
	—	—	13–40	—	—	—	60–70	—	—	—
	—	—	25–40	—	—	—	60–70	—	—	—
	—	23–40	—	—	—	40–45	—	—	—	61
	—	18–30	—	—	—	100–120	—	—	—	—
50	170–230	3–8	2–8	6–25	4–22	70–100	95–150	100–150	25–150	—
50	170–200	5–15	3–7	5–15	3–14	75–110	100–150	100–150	100–150	—
70	140–190	6–20	5–15	9–25	6–25	70–90	90–130	90–130	90–130	—
0	80–110	5–10	5–12	8–16	7–15	45–65	50–70	50–70	50–70	—
	—	11–15	2–8	—	—	55–65	65–80	—	—	—
	200–270	18–40	18–40	—	20–30	90–140	130–160	—	120–160	41
	250–310	13–20	13–20	—	13–20	140–180	160–190	—	140–180	24
	—	18–35	27–40	—	—	160–210	—	—	—	41
	—	9–20	—	—	—	200–260	—	—	—	—
	210–280	18–35	18–35	—	20–38	100–150	—	—	100–150	26
	400–500	11–18	—	—	13–21	150–230	—	—	150–230	20
0	130–160	4–10	3–7	9–10	4–10	50–70	70–90	70–90	70–90	—
0	130–170	13–25	3–8	9–25	5–16	70–95	85–130	90–130	70–95	—
0	—	16–25	—	18–25	—	70–95	—	90–130	—	—
0	—	3–5	—	3–7	—	160–180	—	160–180	—	—
	—	18–40	—	—	—	50–75	—	—	—	—

Table 7.59 PROPERTIES OF NICKEL AND NICKEL ALLOY SHEET AND PLATE

Material		Condition	Thickness	Tensile strength minimum	0·2% proof stress	Elongation on 50 mm minimum	Hardness (HV)	
Designation	Description		mm	N/mm²	N/mm²	%	Minimum	Maximum
BS 3072/NA11	Nickel	Cold rolled and annealed	0·25–4·25	380	105	40	—	125
		Hot rolled and annealed	2·0–9·5	380	105	40	—	125
			over 9·5					
BS 3072/NA 12	Low carbon nickel	Cold rolled and annealed	0·25–4·25	850	345	40	—	110
		Hot rolled and annealed	2·0–9·5	850	345	40	—	110
			over 9·5					
BS 3072/NA 13	Nickel copper alloy	Cold rolled and annealed	0·25–4·25	485	195	35	—	130
		Hot rolled and annealed	2·0–9·5	485	195	35	—	130
			over 9·5					
BS 3072/NA14	Nickel-chromium iron alloy	Cold rolled and annealed	0·25–4·25	550	240	30	—	200
		Hot rolled and annealed	2·0–9·5	550	240	30	—	200
			over 9·5					
BS 3072/NA15	Nickel-iron-chromium alloy	Cold rolled and annealed	0·25–4·25	515	205	30	—	200
		Hot rolled and annealed	2·0–9·5	515	205	30	—	200
			over 9·5					
BS 3072/NA 16	Nickel-iron-	Cold rolled and annealed	0·25–4·25	585	240	30	—	200

	alloy		over 9·5					
BS 3072/NA 18	Nickel-copper aluminium alloy	Cold rolled solution treated and precipitation treated	0·25-4·25	895	620	15	255	—
BS 3072/NA 18	Nickel-copper aluminium alloy	Hot rolled solution treated and precipitation treated	over 9·5	895	620	15	255	—
		Hot rolled and precipitation treated	over 9·5	965	690	18	275	—

Also available as sheet or plate in the hot rolled condition but mechanical properties not specified in BS 3072

Table 7.60 PROPERTIES OF NICKEL AND NICKEL ALLOY STRIP

Material		Condition	Tensile strength min N/mm²	0·2% proof stress N/mm²	Elongation on 50 mm min %	Hardness (HV)	
Designation	Description					min	max
BS 3073/NA 11	Nickel	Cold rolled and annealed	380	105	40	—	110
		Cold rolled, half hard				145	170
		Cold rolled hard				185	—
BS 3073/NA 13	Nickel-copper alloy	Cold rolled and annealed	485	195	35	—	130
		Cold rolled, half hard				155	185
		Cold rolled, hard				200	—
BS 3073/NA 14	Nickel-chromium-iron alloy	Cold rolled and annealed	550	240	30	—	200
BS 3073/NA 15	Nickel-iron-chromium alloy	Cold rolled and annealed	515	205	30	—	200
BS 3073/NA 16	Nickel-iron-chromium-copper alloy	Cold rolled and annealed	585	240	30	—	200
BS 3073/NA 17	Nickel-iron-chromium-silicon alloy	Cold rolled and annealed		Not specified		—	200
BS 3073/NA 18	Nickel-copper-aluminium alloy	Cold rolled half hard and precipitation treated	1 000	760	8	290	—
		Cold rolled, hard and precipitation treated	1 170	895	5	330	—
		Cold rolled solution treated and precipitation treated	895	620	15	255	—

Table 7.61 PROPERTIES OF NICKEL AND NICKEL ALLOY TUBES

Material Designation	Description	Condition	Size Over mm	Size Up to and including mm	0.2% proof stress minimum N/mm²	Tensile strength minimum N/mm²	Elongation on 50 mm or 5·65√S₀ %	Outside diameter availability
BS 3074/NA 11	Nickel	Cold worked and annealed	—	—	105	380	40	6·0 mm–115 mm cold worked
		Cold worked and stress relieved	—	125	275	450	15	
			125	—	105	380	40	
		Hot worked and annealed	—	—	85	345	40	60 mm–250 mm hot worked
BS 3074/NA 12	Low carbon nickel	Cold worked and annealed	—	—	205	415	15	
		Cold worked and stress relieved	—	125	85	345	40	
			125	—	70	345	40	
		Hot worked annealed	—	—	195	485	35	60 mm–250 mm hot worked
BS 3074/NA 13	Nickel-copper alloy	Cold worked and annealed	—	—	380	585	15	6·0 mm–115 mm cold worked
		Cold worked and stress relieved	—	125	195	485	35	
		Hot worked and annealed	125	—	170	485	35	60 mm–250 mm hot worked
BS 3074/NA 14	Nickel-chromium-iron-alloy	Cold worked	—	—	Properties not specified			
		Cold worked and annealed	—	—	240	550	30	
		Hot worked	—	125	205	550	35	
			125	—	170	515	35	
		Hot worked and annealed	—	125	205	550	35	60 mm–250 mm hot worked
			125	—	170	515	35	
BS 3074/NA 15	Nickel-iron-chromium alloy	Cold worked and annealed	—	—	205	515	30	6·0 mm–115 mm cold worked
		Cold worked and solution treated	—	—	170	485	30	
		Hot worked	—	—	170	485	30	
		Hot worked and annealed	—	—	170	485	30	
		Hot worked and solution treated	—	—	170	485	30	

continued overleaf

Table 7.61—*continued*

Material			Size		0·2% proof stress minimum N/mm²	Tensile strength minimum N/mm²	Elongation on 50 mm or 5·65√S₀ %	Outside diameter availability
Designation	*Description*	*Condition*	*Over mm*	*Up to and including mm*				
BS 3074/NA 16	Nickel-iron-chromium-molybdenum-copper alloy	Cold worked and annealed	—	—	240	585	30	60 mm–250 mm hot worked
		Hot worked and annealed	—	—	170	515	30	
BS 3074/NA 17	Nickel-iron-chromium-silicon alloy	Cold worked and annealed			Properties not specified			6·0 mm–115 mm cold worked
		Hot worked						60 mm–250 mm hot worked
		Hot worked and annealed						
BS 3074/NA 18	Nickel-copper-aluminium alloy	Cold worked			Properties not specified			6·00 mm–115 mm
		Cold worked and solution treated						
		Cold worked solution treated	—	—	620	895	15	
		Precipitation treated						

Table 7.62 PROPERTIES OF NICKEL AND NICKEL ALLOY RODS

| Material | | Condition | Size | | Tensile strength minimum N/mm² | 0·2% proof stress minimum N/mm² | Elongation on $5.65\sqrt{S_0}$ minimum % | Form of rod R = round, S = square, round or hexagonal |
Designation	Description		Over mm	Up to and including mm				
BS 3076/NA 11	Nickel	Cold worked	—	25·0	550	415	9	R, S
			25·0	55·0	515	345	13	R, S
		Cold worked and annealed	All sizes		380	105	37	R
BS 3076/NA 12	Low carbon nickel	Hot worked and annealed	All sizes		345	70	37	R, S
		Cold worked and annealed	All sizes. Round rods		485	170	32	R, S
BS 3076/NA 13	Nickel-copper alloy	Cold worked and stress relieved	—	40·0	585	415	20	R
			40·0	55·0	585	380	20	
		Cold worked and stress relieved	Square, rectangular and hexagonal rods. All sizes. Round rods		580	345	20	S
BS 3076/NA 13	Nickel-copper alloy	Cold worked and stress equalised	—	40·0	620	485	15	R
			Square, rectangular and hexagonal rods. All sizes		585	380	18	S
		Hot worked and annealed	All sizes		485	170	32	R
BS 3076/NA 14	Nickel-chromium iron alloy	Cold worked	—	12·5	825	620	7	R, S
			12·5	25·0	760	585	9	
			25·0	55·0	725	550	11	R, S
		Cold worked and annealed; Hot worked and annealed	All sizes		550	240	28	R, S; R

continued overleaf

Table 7.62—*continued*

Material Designation	Material Description	Condition	Size Over mm	Size Up to and including mm	Tensile strength minimum N/mm²	0.2% proof stress minimum N/mm²	Elongation on $5.65\sqrt{S_0}$ minimum %	Form of rod R = round, S = square, round or hexagonal
BS 3076/NA 15	Nickel-iron-chromium alloy	Cold worked and annealed		All sizes	515	205	28	R, S
		Cold worked and solution treated		All sizes	485	170	28	R, S
		Hot worked and annealed		All sizes	515	205	28	R
		Hot worked and solution treated		All sizes	485	170	28	R
BS 3076/NA 16	Nickel-iron-molybdenum-copper alloy	Cold worked and annealed		All sizes	585	240	28	Not specified
		Hot worked and annealed		All sizes	585	240	28	
BS 3076/NA 17	Nickel-iron-chromium-silicon alloy	Cold worked and annealed			Properties not specified			Not specified
		Hot worked						
		Hot worked and annealed						
		Cold worked and precipitation treated	Round rods —	25·0	1 000	760	14	
			25·0	55·0	965	690	16	
			Hexagons —	25·0	965	690	14	
BS 3076/NA 18	Nickel-chromium-aluminium alloy	Cold worked solution treated and precipitation treated		All sizes	895	620	18	Not specified
		Hot worked and precipitation treated		All sizes	965	690	18	
		Hot worked solution treated and precipitation treated		All sizes	895	620	18	

Rods also available in hot worked condition only; properties not specified

Table 7.63 PROPERTIES OF NICKEL AND NICKEL ALLOY WIRE

| Material | | Condition | Size | | Tensile strength | | Elongation on 50 mm min |
Designation	Description		Over mm	Up to and incl. mm	Min N/mm²	Max N/mm²	%
BS 3075/ NA 11	Nickel	Cold drawn	—	3·25	540	—	—
			3·25	8·0	465	—	—
		Cold drawn and annealed	—	0·45	385	—	20
			0·45	8·0	385	—	25
BS 3075/ NA 13	Nickel–copper alloy	Cold drawn	—	3·25	770	—	—
			3·25	8·0	695	—	—
		Cold drawn and annealed	—	0·45	485	—	20
			0·45	8·0	485	—	25
BS 3075/ NA 14	Nickel–chromium–iron alloy	Cold drawn	—	3·25	850	—	—
			3·25	8·0	770	—	—
		Cold drawn and annealed	—	0·45	550	—	20
			0·45	8·0	550	—	25
BS 3075/ NA 15	Nickel–iron–chromium alloy	Cold drawn and annealed	—	0·45	515	—	20
			0·45	8·0	515	—	25
BS 3075/ NA 17	Nickel–iron chromium–silicon alloy	Cold drawn and annealed	Properties not specified				
BS 3075/ NA 18	Nickel–copper –aluminium alloy	Cold drawn	—	8·0	760	—	—
		Cold drawn and solution treated	—	8·0	—	760	—
		Cold drawn spring temper	—	1·5	1140	—	—
			1·5	3·0	1070	—	—
			3·0	6·0	1035	—	—
			6·0	8·0	1000	—	—
		Cold drawn precipitation treated	—	8·0	1070	—	—
BS 3075/ NA 18	Nickel–copper –aluminium alloy	Cold drawn solution treated and precipitation treated	—	8·0	897	—	—
		Cold drawn spring temper and precipitation treated	—	3·0	1240	—	—
			3·0	8·0	1170	—	—
BS 3075/ NA 19	Nickel– chromium– cobalt titanium– aluminium alloy	Cold drawn	—	—	—	1545	Reverse bend and wrapping test specified
		Cold drawn and solution treated	—	—	—	1080	30
		Cold drawn and precipitation treated	0·45	1·0	1545	—	Torsion test
			1·0	5·0	1390	—	Specified
			5·0	8·0	1315	—	10
		Cold drawn solution treated and precipitation treated	0·45	1·0	1080	—	15
			1·0	8·0	1080	—	15

7–132

Table 7.64 PROPERTIES OF NICKEL AND NICKEL ALLOY CASTINGS

Material		Tensile strength min. N/mm²	0·1% Proof stress N/mm²	Elongation on 50 mm min %	Hardness (HB) min
Designation	Description				
BS 3071/NA1	Nickel–copper (1% silicon)	432	—	16	—
BS 3071/NA2	Nickel–copper (2·75% silicon)	541	232	10	—
BS 3071/NA3	Nickel–copper (4% silicon)	618	—	—	250

REFERENCES

1. BS 1470
   Wrought aluminium and aluminium alloys. Plate, sheet and strip; BSI (1969)
2. BS 1471
   Wrought aluminium and aluminium alloys. Drawn tube; BSI (1969)
3. Wrought aluminium and aluminium alloys. Forging stock and forgings; BSI (1969)
4. BS 1473
   Wrought aluminium and aluminium alloys. Rivet, bolt and screw stock; BSI (1969)
5. BS 1474
   Wrought aluminium and aluminium alloys. Bars, extruded round tube and sections; BSI (1969)
6. BS 1475
   Wrought aluminium and aluminium alloys. Wire; BSI (1969)
7. BS 1490
   Aluminium and aluminium ingots and castings; BSI (1970)
8. BS 1161
   Aluminium and aluminium alloy sections. (Imperial series); BSI (1951)
9. BS 2870
   Rolled copper and copper alloys—sheet, strip and foil; BSI (1968)
10. BS 2871
    Copper and copper alloys—tubes; BSI (1957).
11. BS 2872
    Copper and copper alloys—forging stock and forgings; BSI (1969)
12. BS 2873
    Copper and copper alloys—wire; BSI (1969)
13. BS 2874
    Copper and copper alloys—rods and sections (other than forging stock); BSI (1969)
14. BS 2875
    Copper and copper alloys—plate; BSI (1969)
15. BS 1400
    Copper alloy ingots and copper alloy castings; BSI (1969)
16. BS 3071
    Nickel—copper alloy castings; BSI (1968)
17. BS 3072
    Nickel and nickel alloys—sheet and plate; BSI (1968)
18. BS 3073
    Nickel and nickel alloys—strip; BSI (1968)
19. BS 3074
    Nickel and nickel alloys—tube; BSI (1968)
20. BS 3075
    Nickel and nickel alloys—wire; BSI (1968)
21. BS 3076
    Nickel and nickel alloys—rods; BSI (1968)

22. BS 1639
Methods for bend testing of metals; BSI (1964)
23. ISO R.208
Composition of aluminium alloy castings; ISO Geneva (1961)
24. ISO R.209
Composition of aluminium alloy castings; ISO Geneva (1961)
25. ISO R.426
Classification of brasses, leaded brasses, special brasses and high tensile brasses; ISO Geneva (1965)
26. ISO R.427
Classification of tin bronzes and special tin bronzes; ISO Geneva (1965)
27. ISO R.428
Classification of aluminium bronzes and special aluminium bronzes; ISO Geneva (1965)
28. ISO R.429
Classification of copper-nickel alloys; ISO Geneva (1965)
29. ISO R.430
Classification of copper-nickel-zinc alloys; ISO Geneva (1965)
30. BS 1977
Copper for electrical purposes—tubes (high conductivity); BSI (1963)
31. BS 61 Part 1
Copper tubes (heavy gauge) for general purposes; BSI (1947)
32. BS 378
Solid drawn copper and copper alloy tubes for condensers, evaporators, heaters and coolers; BSI (1963)
33. BS 659
Light gauge copper tubes for water, gas and sanitation; BSI (1967)
34. BS 1306 Part 2
Seamless copper tubes for steam services; BSI (1948)
35. BS 1386
Copper tubes to be buried underground; BSI (1957)
36. BS 2017
Copper tubes for general purposes; BSI (1963)
37. BS 885
Brass tubes for general purposes 70/30 brass, aluminium brass; BSI (1963)
38. BS 1464
Solid drawn copper alloy tubes for heat exchange equipment in the Petroleum Industry; BSI (1957)
39. BS 2579 Part 1
Tubes for the manufacture of screwed ferrules for condenser, evaporator, heater and cooler tubes; BSI (1955)
40. BS 1867
Solid drawn aluminium bronze tubes; BSI (1952)
41. ASA 5454
Material grade shown in ASTM B241 aluminium alloy seamless pipes and seamless extruded tubes; (American Society for Testing of Materials 1969)
42. BS 1500. Part 3
Fusion welded pressure vessels for general purposes—Aluminium.

8 NON-DESTRUCTIVE TESTING

NON-DESTRUCTIVE TESTING 8

8 NON-DESTRUCTIVE TESTING

J. G. REES

INTRODUCTION

Non-destructive testing may be defined as any test or examination applied to an item or material without impairing its usefulness.

Many physical and other principles are employed in this fast developing science. For example the X-ray, ultra-violet and infra-red bands of the electromagnetic spectrum are utilised. In addition capillary attraction, magnetic flux, eddy currents, and the piezo electric effect are widely employed phenomena. Acoustic techniques and holography are at present in their infancy, but more will be heard of them in the not too distant future.

Acceptance standards

Much non-destructive testing is carried out to customers specifications, but techniques are becoming so sophisticated that, particularly in maintenance work, rejecting every detected flaw is an extremely drastic step.

Subjective judgement alone is not always sufficient so that techniques like *LEO* have been devised, which break the problem down into a small number of manageable factors. In the case of a crack, the critical crack length is denoted L, the anticipated crack growth is E, and the efficiency of the non-destructive testing technique and operator is O. The derivation of the factors and some case histories is described fully by Birchon.[1]

Non-destructive testing on raw material

The start point in the working of any metallic component is some form of casting or ingot. From this stage onwards, non-destructive testing is applied at various stages in the working procedure to eliminate defects which would be unacceptable in the final product and to save the cost of processing this substandard material. The tests carried out in large scale production are usually automated, but the principles and techniques are similar to those applied manually.

The cost of the non-destructive testing is recovered by improving production techniques and the reduction of defective material. As components are being subject to continually higher stresses, reliability is becoming increasingly important. Non-destructive testing is applied either as an instrument of quality control i.e. to maintain a reasonable standard or on a 100% basis to a specified acceptance standard.

Non-destructive testing as a maintenance tool

Once a component has been shown to be manufactured to a satisfactory standard it is placed in service. Non-destructive testing is widely used to ensure that any deterioration caused by service conditions is detected. These inspections may take place during periods of overhaul, but there are considerable advantages if the condition of items can be monitored without costly dismantling or even while actually in service.

Some aspects of visual methods[2]

Visual inspection is the oldest and most widely used of all non-destructive testing techniques. All visual inspections are subjective and so may be influenced by outside factors. It should therefore be carried out under the best possible conditions. There must be adequate lighting with no dazzle or glare, the inspector should be comfortable and protected from disturbing factors such as noise, draughts, extremes of temperature and inclement weather.

A systematic approach is essential; if it is not possible to cover a large area in a natural sequence then a grid system should be used and one square examined at a time. Beware of eye fatigue and plan the work so that there are opportunities for the eyes to be rested.

VISUAL AIDS

Aids to assist the eye

A single or compound lens with a magnification of × 3 to × 5 is a simple but essential aid to visual inspection. The watchmaker type which can be fixed to a pair of spectacles leaving both hands free and an illuminated hand lens are useful variations. There is now available a pair of binoculars which has a range from infinity down to 0·7 m which finds many applications in examining structural items without requiring physical access.

Aids to extend the eye

The simplest aid for the eye is a common mirror. Mirrors mounted on a stick sometimes with a light source and a lens or two are now available in kit form. Introscopes, Endoscopes, Borescopes, etc. (The Trade name depending on the manufacturer) are forms of rigid, narrow, long industrial telescopes which introduce light and permit visual examination through small apertures e.g. down a small bore tube. They range from 2–50 mm in diameter and, in special cases, may be made from small sections up to 50 m long.[3] Illumination is provided by a lamp built in the viewhead of the instrument or by piping 'cold' light through a hollow tube, solid quartz rod or glass fibre strands from a projector type lamp. Instruments may have a fixed, adjustable or infinite depth of field and produce an image of actual size or slightly enlarged. It is important to know the optical characteristics of an instrument before use. The direction of view may be adjustable, fixed forward, or lateral, and some instruments have interchangeable viewheads with forward fore-oblique, right angle or retrograde viewing directions. Special purpose instruments are available for high temperature applications.

It has already been stated that light may be piped down strands of glass fibres. If bundles of these fibres are arranged, precisely positioned with respect to their neighbours, they form a 'coherent bundle'. Such a bundle can be used as a flexible viewing instrument having the advantage of being able to examine otherwise unaccessible places, but the quality of the image does not compare well with a conventional optical system.

A miniaturised TV camera which passes through an aperture of 40 mm enables inspection through pipelines, vessels, sewers etc., at distances up to 60 m from the controls. A useful feature is that the examination may be recorded on a video tape recorder, and studied later at leisure or preserved as records.

Another remarkable visual aid is a self-illuminating shutterless remote-controlled still camera which takes pictures on single frames of 16 mm cine film. One version of this camera passes through a 25 mm hole, and can be used up to 35 m from the controls.

RADIOGRAPHY

X-rays are produced when electrons from a heated filament (the cathode) are accelerated by a high voltage and strike a tungsten target set in copper (the anode) in an evacuated tube. Only about 1 % of the energy is converted to X-rays, the rest being heat. X-rays produced in this manner are limited to around 400 kilovolts (peak), kVp but super voltage or high energy generators are available to produce X-rays with energies between 1 and 31 megaelectron volts, MeV. The best known of these are the betatron and the linear electron accelerator.

X-ray generators have two basic types of circuitry, the half wave or self-rectified, and the constant potential circuit. The self-rectified circuit is ideal for portable applications and usually cost about half of a c.p. unit of similar potential but exposure times are much longer.

Gamma rays are continuously emitted from some radiation materials; for the purpose of industrial radiography, the gamma ray sources are artificial isotopes of iridium, cobalt, caesium or thulium. Gamma and X-rays are the same phenomenon; the voltage on an X-ray set is variable, but the wavelengths of the radiation from a radio-isotope are constant, and in the case of the ones mentioned above generally shorter and therefore more penetrating than conventional X-rays.

Table 8.1   PROPERTIES OF THE ISOTOPES USED IN INDUSTRIAL RADIOGRAPHY

Isotope	Output *R/hr/Ci at 1 m	Half life	Optimum thick range of steel (mm)
Thulium 170	0·002 5	125 days	2·5–12
Iridium 192	0·48	74·4 days	12–60
Caesium 134	0·89	30 years	50–100
Caesium 137	0·33	30 years	50–100
Cobalt 60	1·35	5·26 years	50–200

* R/hr/Ci at 1m = 1 curie equivalent activity at 1 m in rontgens per hour.

The isotopes are right cylinders in sizes from $\frac{1}{2} \times \frac{1}{2}$ mm to 6×6 mm encapsulated in stainless steel.

Table 8.2   PRACTICAL LIMIT OF STEEL THICKNESS OF VARIOUS KILOVOLTAGES FOR SENSITIVE TECHNIQUES

Voltage or energy	Steel thickness (mm)
200 kV	40
300 kV	70
400 kV	100
1 MeV	125
2 MeV	200
31 MeV	350

$1 \text{ eV} = 1 \cdot 602 \times 10^{-19} \text{J}$

The properties of X-rays and gamma rays which are of importance in radiography are:
1. They penetrate all matter depending on thickness and density.
2. They travel in straight lines at the speed of light $(3 \times 10^9$ m/s)
3. They effect photographic emulsions.
4. They cause certain chemicals to fluoresce.
5. They can be scattered.
6. They produce adverse biological effects to living tissues and blood.
7. They ionise gases.

X-ray films

An X-ray film consists of a cellulose acetate base coated with a silver halide emulsion (similar but not identical to a photographic emulsion). There is usually an emulsion on each side of the base, this reduces exposure time and assists in processing. The range of X-ray films available is from fast coarse grain films to slow ultra fine grain films, again as in photography.

Intensifying screens

The X-ray film is placed in a cassette or film holder between a pair of intensifying screens. Fluorescent or salt screens consist of a piece of cardboard coated with a very thin layer of one of the salts which fluoresce when excited by X-rays, like calcium tungstate. There are different grades of salt screens according to grain size, but the nett effect is to reduce exposures by several hundred times but with some loss of resolution. Lead screens around 0·1 mm thick are used for voltage in excess of 120 kV; they reduce exposures between two and four times and enhance the radiographic quality because the lead absorbs much of the scattered soft radiation. For high energy applications copper screens are also used.

A radiograph is produced by exposing radiation through a specimen on to a film placed at the opposite side. The exposed film is then processed according to the manufacturers instructions. Details of techniques for producing radiographs of fusion welds in steel are given in BS 2600[10] and in steel castings BS 4080[11].

Image quality indicators (Penetrameters)

I.Q.I.'s consist of either several wires of increasing diameter or a stepwedge of increasing thickness; they are used to compare techniques and to measure relative sensitivity e.g.

$$\text{sensitivity} = \frac{\text{diameter of smallest wire visible on radiograph} \times 100}{\text{thickness of specimen}}$$

Legislation

The use of ionising radiations for the purpose of industrial radiography is subject to the following legislation.

The Radioactive Substances Act 1960 requires all users of radioactive material to register (or apply for exemption) with the Ministry of Housing and Local Government, now the Department of the Environment. It is also concerned with registration of mobile radioactive apparatus and the right of entry and inspection by the Minister's Inspector.

The Ionising Radiations (Sealed Sources) Regulations 1969 deal principally with industrial radiographic processes and the protection of both users of ionising radiations and the general public. The Factory Inspector for the district must be notified in writing before any work with ionising radiations takes place for the first time in any factory. Workers are classified, which means they are under medical supervision and detailed records are kept of the individual dosages. The regulations also deal with administration, notification and records, basic principal of protection, radiological supervision, organisation of work, monitoring and the schedule of maximum permissible radiation doses.

The Radioactive Substances (Carriage by Road) G.B. Regulations, 1970. These impose requirements on packaging, labelling and procedure involved in the transportation of radioactive substances on roads to which the public have access.

The Radioactive Substances (Road Transport Workers) G.B. Regulations 1970 are concerned with the protection of workers engaged in the transportation of radioactive substances. The Code of Practice for the Carriage of Radioactive Materials by Road is intended to assist all concerned to discharge their obligations under the Law. Notification containing prescribed information on vehicles employed in regulated transport operations must be given to the Local Licensing Authority.

Radiation Protection

The basic principles of protection are distance, shielding and time and in practice combinations of these principles are used. The rem (rontgen equivalent man) is a measure of biological effect and is the unit of dose of ionising radiation which gives the same effect as that due to 1 rontgen of X-rays. The maximum dose rate that may be received

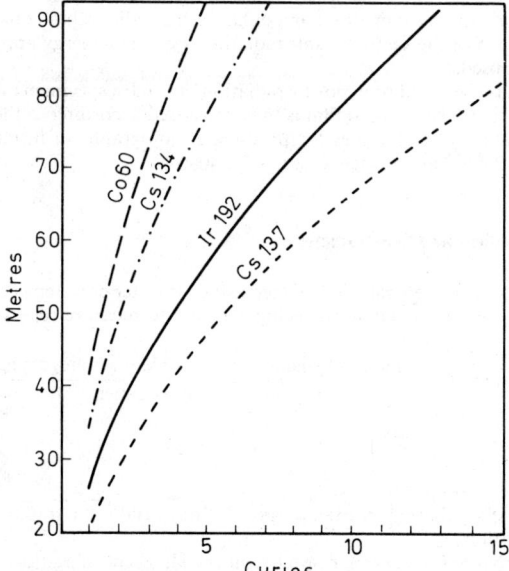

Figure 8.1. Distance at which output of unshielded isotope is reduced to 0·75 m rem/h

by a member of the general public is 0·75 m rem/h averaged over 1 minute and by a classified worker 2·5 m rem/h. The distance from an unshielded gamma ray source at which the output falls to 0·75 m rem/h may be calculated from the inverse square law which is expressed mathematically by the equation:

$$\frac{I_1}{I_2} = \frac{d_2^2}{d_1^2} \tag{1}$$

where I_1 is the activity in curies (or millicuries) at distance d_1 in metres and I_2 is the activity (in the same units as I_1) at distance d_2 in metres. The output of the isotope is given in Table 8.1. For example if it is desired to calculate the distance from an unshielded isotope of Iridium 192, activity 5 curies, at which the output is reduced to 0·75 m rem/h:

$$\frac{0.48 \times 5 \times 1\,000}{0.75} = \frac{d_2^2}{1^2}$$

$$d_2 = \sqrt{\frac{0.48 \times 5 \times 1\,000}{0.75}} = 56.46 \text{ m}$$

The safe distances for cobalt, iridium and caesium are given in Figure 8.1.

A most efficient form of protection is, where it is possible, to use the radiation in a directional beam pointing vertically into the ground. Shielding is either the use of dense materials like steel, lead or spent uranium; or greater thicknesses of cheaper, less dense materials such as concrete and brick. Shielding is usually used in conjunction with distance, this is an effective and economical form of protection. Some useful shielding data has been compiled by Langley[5].

ULTRASONIC TESTING

The use of sound waves at frequencies beyond the audible range (above 20 kHz) for testing purposes is known as ultrasonic testing. The frequencies used range between 100 kHz and 25 mHz, although the most commonly used frequencies are between 0·5 mHz and 10 mHz. Ultrasonic testing is probably the most flexible of all non-destructive testing techniques because it can be adapted for so many different applications.

Piezo-electric effect

When a low alternating voltage is applied to a piezo-electric crystal it vibrates at a similar frequency and conversely the crystal will convert mechanical vibrations into electrical impulses. If the crystal is coupled to a solid media, ultrasonic sound waves are transmitted through the media. The piezo-electric effect was found originally in naturally occurring crystals such as quartz.

Although some quartz is still used most crystals in use today are synthetically produced like barium titanate and lead zirconate. Each crystal thickness has an optimum frequency which is indicated on the outside of the transducer probe which houses the crystal.

Transducers

The piezo-electric crystal is silvered on one side, damped with an epoxy resin loaded with powdered tungsten carbide and fitted into a suitable metallic or plastic container; the crystal usually has a plastic contact piece. Transducers transmit sound waves either normally or at an angle to the surface of the work piece. The plastic contact piece has many functions; it prevents wear on the crystal, can be shaped to suit special applications and is used in wedge form in the production of angle probes; it is one way, using special high temperature plastics of protecting the crystal when in use on hot surfaces. Probes may have a single crystal which transmits and receives on alternate pulses or two crystals one of which acts as a transmitter, the other as a receiver.

Wave forms

The three modes of sound propagations used in ultrasonic testing are longitudinal or compression, transverse or shear and surface waves. The type of surface wave generated by an ultrasonic transducer is controversial but for testing purposes it is sufficient to designate them surface waves.

For longitudinal waves the velocity of propogation V_L is related to the elastic constants of the material.

$$\text{Thus } V_L = \frac{E(1-\delta)}{\rho(1+\sigma)(1-2\sigma)} \text{ m/s} \tag{2}$$

$$\text{For transverse waves } V_T = \frac{\sqrt{G}}{\rho} \text{ m/s} \tag{3}$$

Where   E = Young's Modulus in MN/m^2
     G = Shear (rigidity) modulus in MN/m^2
     σ = Poisson's ratio in MN/m^2
     ρ = Density of material in kg/m^3

Hence

$$V_T = \sqrt{\frac{E\,1}{\rho\,2(2-\sigma)}} \tag{4}$$

Surface waves travel at approximately 0·9 of the speed of a transverse wave. For steel and most metals a transverse wave is roughly half the speed of a longitudinal wave.

Methods of ultrasonic testing

The resonance method is applied to parallel-sided specimens by applying an ultrasonic wave generated by a variable frequency oscillator and adjusting until resonance takes place. This resonance can be detected by an audio technique or on a cathode ray tube. In the former case the thickness of the specimen can be deduced from the reading on the variable frequency oscillator which has previously been calibrated against known standards.

In the pulse echo method, which is the most widely used, short pulses are transmitted until they are reflected by some inhomogeneity or the far surface of the specimen. The pulse rate frequency may be anything from 50 to 2000 pulses per second. The great advantage of this method is that flaws may be accurately located with access required from one side only.

In the transmission method a separate transmitter and receiver probe are placed on opposite sides of the specimen, ultrasonic energy may be pulsed or continuous. The specimen is evaluated by the difference in amplitude between the transmission and receiving signals. This method is used for testing very coarse materials e.g. concrete.

Types of cathode ray display

'A'-scan presentation (see Figure 8.2a) has a horizontal base line (X-axis) on the cathode ray tube which indicates transit time from left to right. The signal amplitude is indicated

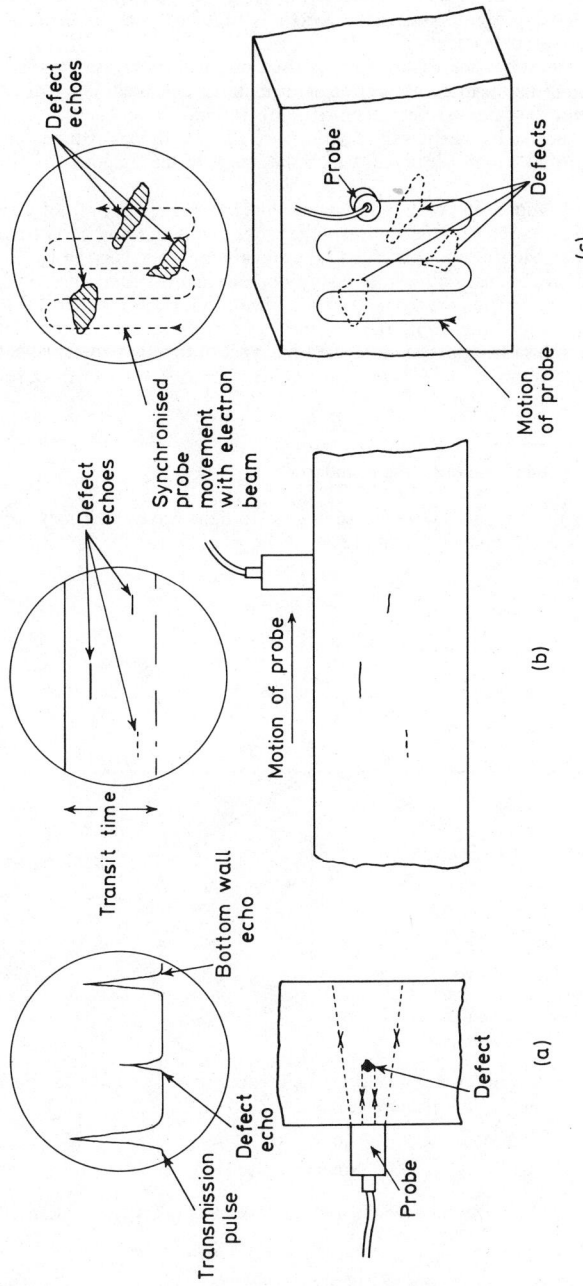

Figure 8.2. Scans. (a) A—Scan, (b) B—Scan, (c) C—Scan

by the vertical deflection (Y axis) and corresponds to the intensity of the reflected ultrasonic energy. On most equipments the X axis is fully rectified. 'A'-scan is the usual display on manual instruments.

'B'-scan presentation (see Figure 8.2b) is used on automatic installation and where the distribution of defects in a cross section of the material is desired. The Y axis represents transit time and the X axis is synchronised with the motion of the probe. The signal amplitude is indicated by the intensities of the trace on the cathode ray tube. A persistence type tube is normally used i.e. one in wnich the image of the signals persist for several seconds.

'C'-scan (see Figure 8.2c) presents an image which is a plan view of the area under test. The tube is synchronised with the motion of the probe scanning along the two coordinates. Again a long persistence tube is used and the signal amplitude is given by the intensity of the trace. 'C'-scan is used almost exclusively on automatic systems.

Special purpose instruments are made where the output is fed into a meter or a digital counter so that, after calibration, the thickness of a material may be read off directly in any desired unit of length. The more effective types of these instruments feature automatic gain control circuitry. This enables use over wide variations of thickness, surface conditions and textures of materials.

Transmission of sound waves across boundaries

The behaviour of ultrasonic waves is analogous to light waves and obeys some of the light laws. When a beam of sound travels at an angle a (to the normal) from solid

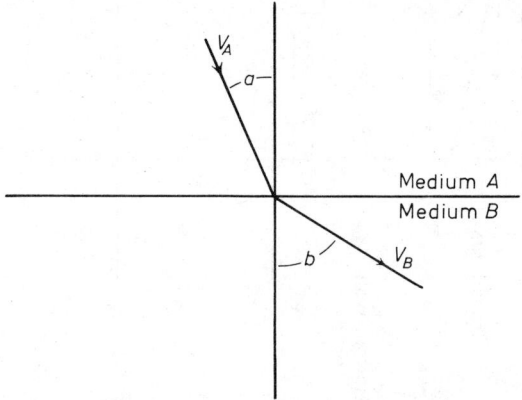

Figure 8.3. Refraction of beam of sound through the boundary between two solid media

medium A to solid medium B at a velocity V_A it is refracted at an angle b and a velocity V_B according to Snell's Law (Figure 8.3).

$$\frac{\sin a}{\sin b} = \frac{V_A}{V_B} \tag{5}$$

This relationship also applies when mode conversion takes place at the boundary.

$$\frac{\sin a(L)}{\sin b(T)} = \frac{V_L}{V_T} \tag{6}$$

An important boundary relationship is the one between perspex and steel because the angle of most angle probes is determined by this relationship. When angle a is between 0 and 29° then both longitudinal and transverse waves are refracted. When angle a is between 30° and 60° shear waves only are refracted, beyond 61° only surface

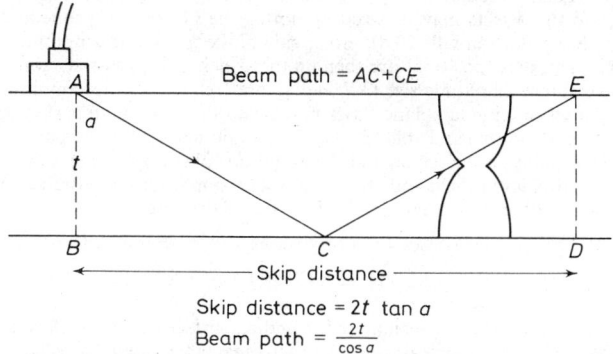

Skip distance = $2t$ tan a

Beam path = $\frac{2t}{\cos a}$

Figure 8.4. Diagram showing skip distance and beam path of shear wave used to scan weld in plate

waves are generated. Commercial angle waves are made at angles (to the normal) of 35°, 45°, 60°, 70° and 80° in steel. These probes may also be used for other materials but then the refracted angle is changed according to Snell's Law.

Table 8.3   REFRACTIVE ANGLES, SKIP DISTANCES & BEAM PATH OF COMMERCIAL
ANGLE PROBES IN STEEL, COPPER & ALUMINIUM

Angle of incidence in Perspex	Refractive angle in steel	Refractive angle in		Skip distance			Beam path		
		Cu	Al	Fe	Cu	Al	Fe	Cu	Al
29·0	35	23·6	33·1	1·4	0·9	1·3	2·4	2·2	2·4
36·7	45	29·7	42·4	2·0	1·1	1·8	2·8	2·3	2·7
46·9	60	37·2	55·5	3·5	1·5	2·9	4·0	2·5	3·5
52·6	70	41·1	63·7	5·5	1·7	4·0	5·8	2·7	4·5
56·4	80	43·6	69·9	11·3	1·9	5·5	11·5	2·8	5·8
57·7	90 (surface wave)	44·4	—	—	2·0	—	—	2·8	—

See Figure 8.4 showing skip distance and beam path of shear wave used to scan weld in plate.

Determination of position and size of defects

Before use, a flaw detector is calibrated over a suitable range using a calibration block.[6] The position of a defect is then ascertained by a direct reading in the case of a compression wave probe and by simple trigonometry or a specially designed slide rule when using a shear wave probe. The size of defects may be estimated by using one or more of the following methods.

(i) The maximum amplitude technique in which the defect is scanned from as many angles and orientations as possible. At the point of maximum amplitude the beam path and surface distance readings are noted and plotted. The plots give a facsimile outline of the defect.

(ii) For the 20 dB technique, the beam profile of each probe is plotted out 20 dB each side of the centre of the beam. This beam is then logged on a slider of transparent plastic. A cross-section of the weld or detail of item under examination is drawn full size on to another piece of plastic. The slider is superimposed on to the latter, duplicating the position of the probe on the item being examined. The origin of any indications are simply identified and the defects may be sized by noting the position of probe and reflection when the defect reflection falls 20 dB either side of the maximum amplitude in any one straight scan and these positions are then on the sketch of the cross-section.

(iii) The distance amplitude size (AVG) diagram[7] relates distance amplitude and size or defects based on circular plane flaws parallel to the examining surface. AVG's are applicable to both normal and angle probes. The amplitude on a sound part of the sample is set at a zero point on the screen and the amplitude of the defect measured in dB after making correction for far field and attenuation. The amplitude and distance of the defect is used to read off the equivalent circular flaw size of the defect.

MAGNETIC PARTICLE TESTING

Magnetic particle testing is a technique for finding surface or near surface cracks, laps, non-metallic inclusions and segregations in ferromagnetic materials. When a magnetic field is interrupted by a discontinuity, stray lines of force can be detected by applying

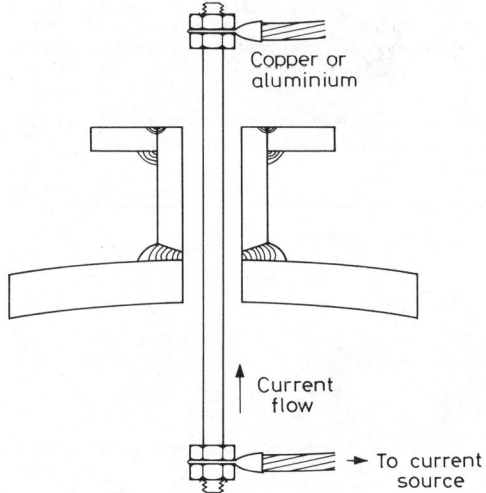

Figure 8.5. Showing threaded bar technique in nozzle weld detail

magnetic iron oxide (Fe_3O_4) either in dry form or in a suspension of kerosene. The stray lines of force attract the particles revealing the outline of the defect.

There are five principle ways of applying a magnetic field, they are:

(i) Permanent magnet.
(ii) Electromagnet, solenoid coil or magnetic yoke.
(iii) Current flow, using prods or contacts.
(iv) Threaded bar (Figure 8.5).
(v) Induction or induced current flow.

The selection of the most suitable magnetising method depends on the size and shape of the object and also the orientation of the defect. It may be necessary to apply a technique for more than one direction or indeed to apply more than one technique in order to achieve a complete examination.

Sensitivity levels vary, but a field strength of between 3 and 6 kA/m (38 and 75 oersteds) is generally acceptable. The field strength (and direction) may be measured with a tangential field strength meter.

Some rules of thumb

The following points are a quick guide to the methods of applying the magnetic field listed above.

  (i) This not adjustable except by varying the distance from the poles.

  (ii) The current is variable but saturation, which is indicated by furring of the particles, should be avoided.

 (iii) 500 A a.c. or 700 A d.c. per 25 mm diameter of specimen.

 (iv) 500 A a.c. per 25 mm diameter of cylinder or hole.

  (v) $AT = \dfrac{40000\,D}{L}$ (7)

Where A = current in amperes T = number of turns
D = Diameter of specimen L = length of specimen.
(only applicable where the ratio of L/D is greater than 5.)

Demagnetisation

Many engineering components require to be demagnetised after testing; this is achieved by passing the components two or three times through a coil carrying alternating current.

Magnetic particles

Particles are coloured black, red or yellow and may also be fluorescent in order to achieve maximum contrast against the background colour of the item being inspected. Specially formulated white paints are also applied in order to achieve good contrast.

Interpretation of indications

Magnetic particle testing requires skill and training in interpreting indications. In addition to discontinuities, particles may be attracted to sharp changes of section, surface scratches and boundaries of dissimilar metals. Subsurface defects are recognised by their slightly blurred outline. A suspected crack may be confirmed or otherwise by removing a film of metal with emery paper or a smooth file and retesting.

LIQUID PENETRANT TESTING

Penetrant testing is the modern version of the old oil and whiting technique and is applicable to all metals and many non-metallic materials. Liquid penetrants are used to find discontinuities such as cracks, laps and pores which are exposed to the surface.

Principles of the technique

Penetrant is applied by spray, brush or by dip to the surface and seeps into any surface defect by capilliary attraction. The surplus penetrant is removed and a developer, a white ink or powder, soaks out the penetrant from the defect so that it is clearly visible against a white background.

Types of penetrant

The basic types of liquid penetrant are:
  (i) Red dye penetrant.
 (ii) Visual fluorescent penetrant which shows red in white light and orange under ultra-violet light.
(iii) A pure fluorescent which shows greeny yellow under ultra-violet light.

Removal of penetrants

Penetrants are formulated in three groups for the purpose of removal.
  (i) Solvent removal penetrants.
 (ii) Post emulsion penetrants where an emulsifier is applied on to the penetrant, left for a predetermined time and the whole removed by water.
(iii) Water washable penetrants.

Types of developer

Developers are also in three basic types:
  (i) A hard drying developer paint which cannot be accidentally removed.
 (ii) A soft drying developer paint which is easily removable.
(iii) A dry powder developer which is applied by blowing or dusting.

Stages in liquid penetrant testing

   (i) Preclean and degrease, a vapour phase degreasing bath is ideal.
  (ii) Preliminary inspection.
 (iii) Application of penetrants.
  (iv) Soaking of penetrant.
   (v) Spraying of post emulsifier (where applicable).
  (vi) Removal of penetrant, (a) excess by tissues or rags; (b) remainder by solvent or water.
 (vii) Dry; hot air is very suitable (except where dry powder developer is used).
(viii) Application of developer.
  (ix) Soaking of developer; this may take anything from 10 minutes to 24 hours depending on size of crack and sensitivity required.
   (x) Final inspection.

Safety in using penetrant materials

Solvents used in penetrant materials may be both inflammable and toxic. When carrying out liquid penetrant testing great care should be taken to ensure good ventilation and that precautions are taken when using inflammable liquids.

It is not generally realised that if penetrant materials, particularly aerosols, are introduced into confined spaces, the Maximum Allowed Concentration (MAC) or Threshold Limit Value (TLV) laid down in 'Dust and fumes in Factory Atmosphere',[12] may be exceeded. To overcome this problem, aerosols should not be used in confined spaces.

Only minimum quantities of materials should be introduced into confined spaces and arrangements made for the testing of the atmosphere for the level of the solvents used.

EDDY CURRENTS

Eddy currents are produced in any electrically conducting material when influenced by a coil in which an alternating current is flowing.[8, 9] For the purpose of non-destructive testing two coils are used, a small alternating current is generated in the primary coil which induces an alternating magnetic field in the specimen; this in turn generates an a.c. current of similar frequency in the receiver coil which induces an opposing alternating magnetic field in the component.

The current in the receiver coil is affected by any change in physical properties or of homogeneity and so may be used for comparing the quality of a manufactured component against a standard, for testing tubes and bars, or for material sorting. In a slightly different form, the two coils mounted in a small transducer may be used for crack detection, the measurement of the thickness of thin coatings on metallic bases or for the comparison of hardness. Heat exchanger tubes in non-ferrous materials can be inspected by using internal coils.

Eddy currents are not as penetrating as either ultrasonics or radiography, but have three outstanding advantages which are:

(i) Physical contact is not required.

(ii) It is a very fast form of inspection.

(iii) It is easily automated.

NEW TESTING TECHNIQUES

As technology improves more techniques are being used for non-destructive testing. Infra-red cameras which are able to discriminate small changes of temperature and are used for detecting faults in electrical contacts and for locating hot spots in pressure vessels and furnaces.

The availability of *Neutron Sources* has made possible a new field of neutron radiography. Neutron absorption is almost the converse of X-ray absorption. This makes possible many applications which are not possible with conventional radiography. For example, neutrons will easily penetrate dense materials to form a sensitive radiograph of less dense material.

The use of unsealed short half-life radioactive tracers is increasing in chemical process investigation, the detection of leaks and the checking of effluent disposal.

Acoustic emission is a technique, still in its infancy which by using acoustic transducers placed in carefully selected positions monitor noise made by defects propagating or about to propagate. The interpretation of the noises is not yet fully understood, but several organisations in both U.K. and U.S.A. are working in this field.

UNITS

Quantities and units of nuclear reactions and ionising radiations are being considered by the International Standards Organisation in terms of the International System of Units (SI) and will appear as part of ISO Recommendation R/31.[13] At present the units of the c.g.s. system are widely used for some of the 'mechanical' quantities and these are the basis for the units used in the foregoing chapter.

REFERENCES

1. BIRCHON, D., 'Engineering Design and Non-Destructive Testing—The LEO technique', *The Engineer*, (27th Sept. 1968)
2. REES, J. G., 'Visual Methods of N.D.T.' Lecture at Sunderland Polytechnic (5th March 1970)
3. MCMASTER, R. C. (Ed), *Non-Destructive Testing Handbook*, Ronald, New York (1959)
4. *Radiation Sources for Industry and Research 1971*. The Radiochemical Centre, Amersham, Bucks.
5. LANGLEY, R., *Gamma Radiography—Review 10*. The Radiochemical Centre, Amersham, Bucks.
6. BS 2704: 1966 *Calibration Blocks and Recommendations for their use in ultrasonic Flaw Detector*.
7. KRAUTKRAMER, J., *The British Journal of Applied Physics*, **10** (June 1959)
8. BLITZ, J., KING, W. G. and ROGERS, D. G., *Electrical, Magnetic and Visual Methods of Testing Materials*, Butterworths (1969)
9. SCHALL, W. E., *Non-Destructive Testing*. The Machinery Publishing Co., Ltd. (1968)
10. BS 2600: 1962. *General recommendations for the radiographic examination of fusion welded butt joints in steel*.
11. BS 4080: 1966. *Methods for non-destructive testing of steel castings*.
12. New Series No. 9, *Dust and fumes in factory atmospheres*, H.M.S.O.
13. ISO R.31, *Quantities and units of nuclear reactions and ionizing radiations*, ISO Geneva

BIBLIOGRAPHY

Ionising Radiations: Precautions for industrial users, HMSO (1969)
BETZ, C. E., *Principles of penetrants*, Magnaflux Corporation Illinois (1963)
HALMSHAW, R., *Physics of Industrial Radiology*, Heywoods (1966)
BANKS, B., OLDFIELD, G. E. and RAWDING, H., *Ultrasonic flaw detection in metals*, Iliffe (1962)
Industrial Radiography, Kodak Ltd., London
BETZ, C. E. and DOANE, *The principles of Magnaflux*, Magnaflux Corporation
ROCKLEY, J. C., *An Introduction to Industrial Radiology*, Butterworths (1964)
KRAUTKRAMER, J. and H., *Ultrasonic Testing of Materials*, Allen & Unwin Ltd., London, Springer-Verlag, Berlin, Heidelberg, New York
FILIPCZNSKI, L., PAWLOWSKI, Z. and WEHR, J. (Ed, BLITZ, J.), *Ultrasonic Methods of Testing Materials*, Butterworths (1966)
Industrial Oils Review. Quarterly published by Burmah-Castrol Industrial Ltd., Special Products Division, Burmah-Castrol House, Marylebone Road, London NW1
SHARPE, R. S. (Ed), *Research techniques in Non-Destructive Testing*, Academic Press, London and New York (1970)
'Sonatest News'. Published by Sonatest Ltd., Old Wolverton Road, Milton Keynes
'The Echo'. Wells Krautkramer Ltd., Blackhorse Road, Letchworth, Herts.
'Non-Destructive Testing'. Research and practice published bi-monthly by Iliffe Science and Technology Publications Ltd. This publication includes N.D.T. Info, which is a list of references drawn from two thousand British and International publications compiled by the Non-Destructive Testing Centre, AERE, Harwell, Didcot, Berks.
'The British Journal of Non-Destructive Testing'. Published bi-monthly by the Non-Destructive Testing Society of Great Britain, Maitland House, Warrior Square, Southend-on-Sea, Essex
'Material Evaluation'. Monthly official journal of the American Society for Non-Destructive Testing, 914 Chicago Avenue, Evanston, Illinois 60202.
'Ultrasonics'. Published quarterly by Iliffe Science and Technology Publications Ltd.

BRITISH STANDARDS CONCERNED WITH OR REFERRING TO NON-DESTRUCTIVE TESTING

BS 487
Part 2. Specification for fusion-welded steel air receiver not exceeding 500 lb/in^2 (1963)
BS 499
Part 3. Terminology of, and abbreviations for, fusion weld imperfections as revealed by radiography (1965)
BS 806
Ferrous pipes and piping installations for and in connection with land boilers (1967)
BS 1113
Water-tube boilers and their integral superheaters (1969)
BS 1500
Fusion welded pressure vessel for general purposes. Part 1. Carbon and low alloy steels (1958).
Part 3. Aluminium (1965)
BS 1500A
Carbon and low alloy steels (1960)
BS 1515
Fusion welded pressure vessels for use in the chemical, petroleum and allied industries. Part 1.
Carbon and ferritic alloy steels (1965). Part 2. Austentic stainless steel (1968)
BS 2079
Steam receivers and separators (1954)
BS 2600
General recommendations for the radiographic examination of fusion welded built joints in steel
(1962)
BS 2633
Class 1 arc welding of ferritic steel pipework for carrying fluids (1966)
BS 2654
Vertical mild steel welded storage tank with butt-welded shells for the petroleum industry (1961)
BS 2704
Calibrations blocks and recommendations for their use in ultrasonic flaw detectors (1966)
BS 2790
Part 1. Shell boilers of welded construction (other than water-tube boilers) (1969)
BS 2791
Class II metal arc welding of steel pipelines and pipe assembling for carrying fluids (1961)
BS 2910
General recommendations for the Radiographic examination of fusion-welded circumferential
butt joints in steel pipes (1965)
BS 3351
Piping systems for the petroleum industry (1961)
BS 3451
Methods of testing fusion welds in aluminium and aluminium alloy (1962)
BS 3889
Part 1a. Ultrasonic testing of ferrous pipes (excluding cast). Part 2a. Eddy current testing, ferrous
pipes and tubes. Part 2b. Eddy current testing of non-ferrous tubes. Part 3a. Penetrant testing
ferrous pipes and tubes. Part 4a. Magnetic particle flaw detection ferrous pipes and tubes
BS 3915
Carbon and low alloy steel pressure vessels for primary circuits of nuclear reactors (1965)
BS 3923
Methods for ultrasonic examination of welds. Part 1. Manual examination of fusion welded butt
joints in ferritic steels (1968). Part 2. Automatic examination of welded seams (1965). Part 3.
Manual examination of nozzle welds (1965)
BS 3971
Image quality indicators for radiography (1966)
BS 4069
Magnetic flaw detection inks and powders (1966)
BS 4080
Methods for non-destructive testing of steel castings (1966)
BS 4094
Recommendation for data on shielding from ionising radiation. Part 1. Shielding from gamma
radiation (1966). Part 2. Shielding from X-radiation (1971)
BS 4097
Gamma radiography exposure container for industrial purposes and their source holders (1966)

BS 4124
Methods for non-destructive testing of steel forgings. Part 1. Ultrasonic flaw detection (1967).
Part 2. Magnetic particle flaw detection (1968). Part 3. Penetrant flaw detection (1968)
BS 4206
Methods for testing fusion welds in copper and copper alloys (1967)
BS 4208
Carbon and low alloy steel containment structures for stationary nuclear power reactors (1967)
BS 4331
Methods for assessing the performance characteristics of ultrasonic flaw detection equipment. Part 1.
Overall performance (1968)
BS 4336
Methods for non-destructive testing of plate material. Part 1a. Ultrasonic detection of laminar
imperfections in ferrous wrought plate (1968)
BS 4397
Methods for magnetic particle testing of welds (1969)
BS 4416
Method for penetrant testing of welded or brazed joins in metals (1969)
BS 4515
Field welding of carbon steel pipelines (1969)

9 FABRICATION OF MATERIALS

FABRICATION OF MATERIALS *9*

9 FABRICATION OF MATERIALS

HIGH ENERGY RATE FORMING TECHNIQUES

R. A. MOTTRAM

INTRODUCTION

In recent years new fabrication techniques have been developed to satisfy the demands which modern technological industries have made for components of complex shape and size which hitherto could not be economically produced by conventional metal forming. In the development of these techniques, high speed forming has played an important part and there is now an ever widening field of application for this method of forming.

High speed forming offers a number of important advantages over conventional techniques and while many claims are made which cannot be substantiated in practice, it is certain that, by high speed forming, large and complex parts can be made with a reduction in plant costs by the elimination of power presses and a reduction in tooling costs, since in general only a female die is required. For short runs the dies which are required can often be made from cheap materials such as concrete or nodular iron. In hot working processes the reduction in time of contact between die and work made possible by high speed forming permits higher working temperatures and large reductions in thickness in one operation.

High speed forming techniques can be classified into three groups:

1. Explosive forming.
2. Electrical discharge forming.
3. High speed mechanical forming.

In explosive forming, components are formed by the sudden release of energy from explosive materials. The explosive may be in close proximity to the metal, as in contact operations, with the explosive acting directly on the metal, or the energy released by the explosive may be transmitted to the component by an intervening medium, as in stand off operations.

Electrical discharge methods use the energy released by the rapid discharge of an electric current either through a coil in which case the magnetic field produced repels the component, as in electro-magnetic forming, or by discharging across a small gap in an ionisable liquid. The latter process produces a shock wave in the fluid which causes deformation of the component, as in electro-hydraulic forming.

High-speed machines developed for extrusion, stamping and pressing of components use either a high pressure gas or the force from an exploding gas to propel the tools of the machine at high speeds.

EXPLOSIVE FORMING

The use of explosives for metal forming has been commercially applied for many years mainly in the U.S.A. and to a lesser extent in Europe. Components such as rocket nose cones, vessel dished ends, dental palates, have been formed by this technique. Explosive

2

welding is exploited by several companies providing metal clads, e.g. titanium on steel, which cannot be made by more conventional cladding techniques. Explosive is also applied to the fabrication of tubular components and is of great value in making heat exchanger tube to tube plate joints.

Nature of explosive materials

Explosives are thermodynamically unstable compounds which react exothermically on initiation to form more stable and usually gaseous products. The reaction is completed in a very short time and is propagated through the explosive by means of a detonation wave in which very high pressures (15 to 30×10^4 bar) exist. The detonation velocity varies with composition of the explosive material but falls within the range of 3000–9000 m/sec as shown in Table 9.1.

Table 9.1   TYPICAL DETONATING VELOCITIES OF EXPLOSIVES

	Density gm/cc	Detonating velocity m/sec
T.N.T.	1·00	5 000
T.N.T.	1·64	6 940
P.E.T.N.	1·00	5 550
P.E.T.N.	1·70	8 300
Tetryl	1·00	5 600
Tetryl	1·71	7 850
RDX	1·00	6 080
RDX	1·65	8 180
Metabel	1·47	6 500–7 500
Trimonite No. 3	1·13	3 900

When a charge has completely detonated, the energy released is transmitted to the surrounding medium by means of a shock wave characterised by a sharp rise in pressure followed by an approximately exponential decay over a comparatively long time. Metal in the path of this shock wave is deformed by it at a high speed within 100 micro seconds. Under these high strain rates the elastic limit of some materials is about $2\frac{1}{2}$ times their normal values.

Most explosive metal working operations can be divided into two groups, contact operations where the charge is in intimate contact with the workpiece and stand off operations where the charge is located some distance from the workpiece and the energy is transmitted through an intervening medium such as air, water, oil, plasticine, clay, etc.

Contact explosive techniques are used to perform a variety of operations such as hardening, welding, powder compaction, metal embossing and cutting. Although the great majority of parts being manufactured by explosive forming operations involve stand off operations, for example, expanding, forming and dishing.

EXPLOSIVE HARDENING

Explosive hardening is achieved when explosive material is fired in contact with the metal surface. The metal workpiece may only experience about 5% plastic strain but at the same time hardening can be achieved to a degree that would require 80% cold reduction by conventional rolling.

The process is used commercially[1] for hardening Hadfields manganese steel where the Brinell hardness at the metal surface can be increased to more than 400 HB as compared with the 200–250 HB of the untreated steel. The increase in hardness to greater than

300 HB may be obtained to depths of more than 150 mm below the metal surface. The effect obtained is determined by the thickness or weight of explosive (of the order of 2500 g/m²), but attempts to increase the depth of hardening by using very large charge weights can result in fracture.

Many metals can be hardened by explosives as shown by Dieter and others[2, 3, 4] in Table 9.2.

Table 9.2   EXPLOSIVE HARDENING OF METALS

Material	Condition	Yield strength N/mm²	Tensile strength N/mm²	Elonga- tion %	Hard- ness HV
Iron	Annealed	130	215	68	—
	Explosive hardened	750	850	16	—
Mild steel	Annealed	240	400	25	120
	Explosive hardened	800	850	7	230
Stainless steel Type 304	Annealed	—	—	—	150
	Explosive hardened	—	—	—	350
Tool steel (H11)	Annealed	1 300	2 000	9	—
	Explosive hardened	2 000	2 350	5	—
Nickel	Annealed	235	420	37	100
	Explosive hardened	540	620	10	230
Copper	Annealed	—	—	—	50
	Explosive hardened	—	—	—	110

EXPLOSIVE WELDING

The explosive welding of two or more metals to produce strong metallurgical bonds has been used for a considerable period of time and is reported in many places. Most metal combinations can be welded together including, for instance, titanium to mild steel, copper to stainless steel, aluminium to stainless steel. The bonds are made with little generation of heat at the weld interface and consequently there is no significant melting or heat affected zone.

Explosion welding is successfully used for the manufacture of clad plates for pressure vessels and chemical reaction vessels, for the welding of tubes into tube plates of heat exchangers, in the joining of pipe lines, the lining of tubular components and the manufacture of clad metal coinage.[5, 6, 7]

Welding mechanism

Bonding is achieved by the oblique high velocity impact between the two members being jointed. The technique commonly used for cladding one small flat plate to another is shown in Figure 9.1 in which the plate is inclined at an angle (α) to the base plate and backed up by explosive.

When the charge is detonated, the detonation front passes through the explosive and the angled plate is forced onto the base plate at high velocity. Figure 9.2 shows the condition during the progressive collapse of the plate and it will be seen that the effective angle between the plates has become (β)

where
$$\beta = \alpha + \frac{V}{U}$$

α = initial angle.   V = velocity of plate.
U = velocity of detonation front propagating through explosive.

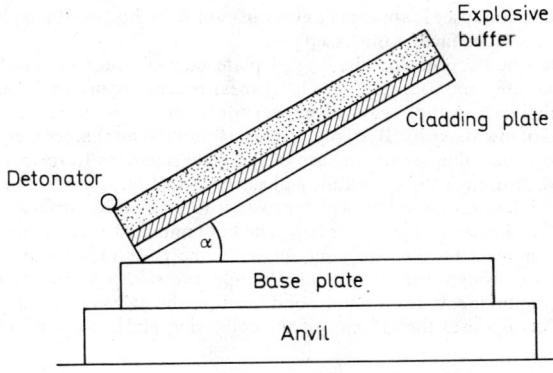

Figure 9.1. Explosive welding of angled plates

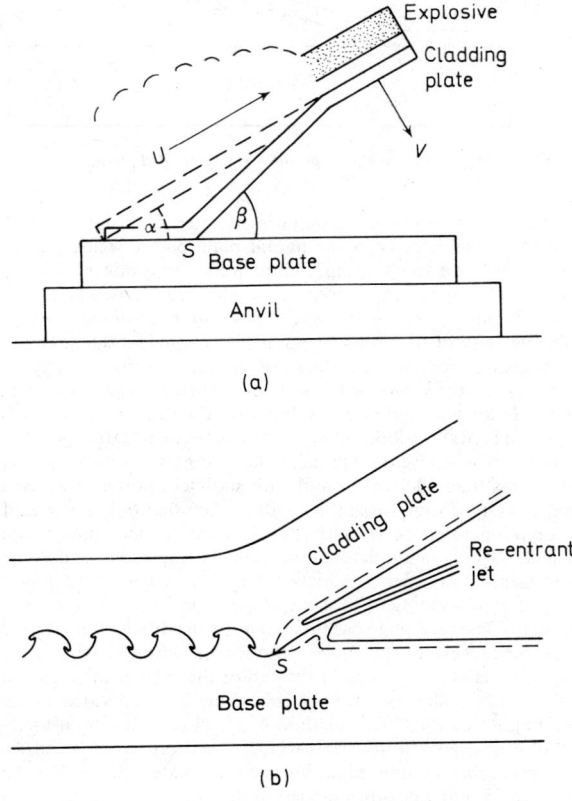

(a)

(b)

Figure 9.2. Collapse of angled plate during explosive welding

As the angled plate collapses it appears at every instant to be hinged at a collision point S which moves across the face of the baseplate.

At the instant when a point on the angled plate impacts the baseplate, very high pressures are generated and as a result the local shear resistance, in compression, of the materials is negligible and the region of impact for a short time behaves in a similar manner to a fluid of low viscosity. By applying the laws of fluid mechanics it can be shown that as the angle plate collapses the surface of the plate separates to form a re-entrant jet, Figure 9.2(b). Although the re-entrant jet has very small mass it has a high velocity which gives it a high scouring ability which causes it to pick up by surface traction the thin contaminated surface of the base plate. The contaminated surface layers of both plates are thus removed by the sweeping action of the re-entrant jet and the newly created surfaces are brought together under very high pressure to form a weld.

The essential requirements for a good bond are that the collision should produce a re-entrant jet. This requires the velocity of the collapsing plate to be sufficiently high

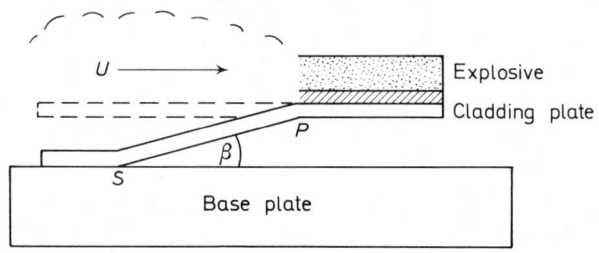

Figure 9.3. Explosive welding parallel plates

to generate high pressure in the region of impact. The collision point (S) velocity must be less than the sonic velocity in the parent plate before welding is achieved and the angle of impact must be above the minimum needed to produce a re-entrant jet.[8]

By using the right type of explosive it is possible to weld two plates which are parallel to each other. Figure 9.3 shows two such plates after the detonation front has reached point P. The velocity of the collision point (S) is equal to the detonation velocity (U) of the explosive and providing this does not exceed the sonic velocity in the material of the parent plates and the values of U and V are sufficient to give an angle (β) greater than the critical angle for welding, then welding of the initially parallel surfaces is possible. In general, the velocity of sound in metals varies between 3100 and 5400 m/sec and it is therefore necessary when welding parallel plates to have explosive material with detonation velocities less than 3000 m/sec and with sufficient power to accelerate the plate to the required velocity. From Table 9.1 it will be seen that only a few explosive materials have the required detonation velocity (e.g. Trimonite) which makes welding of parallel plates possible. Where high detonation velocity explosive is used an angled plate technique is essential in order to achieve the necessary low collision point velocity.

Tube to tube plate welding of heat exchangers can be achieved by a technique shown in Figure 9.4.[9, 10] The tube plate has tapered holes the depth of the taper depending upon the materials being welded. An angle is formed between the tube and the tapered hole and the explosive charge is placed in the end of the tube in the region of the taper. On detonation the tube collapses onto the tube plate and, provided the angle of taper is correct, welding is achieved. This method of attachment has been used successfully for fabrication of exchangers in materials which would otherwise be difficult to weld, e.g. aluminium brass tubes in aluminium bronze tube plates. Figure 9.5 shows one of four exchangers with 35 mm o.d. tubes welded in this way.

A similar technique where the taper is machined onto the outside of the tubes can be used where the ligament spacing between tube holes is such that the tapering of the hole would weaken the tube plate.

By using low detonating velocity explosives, tubes can be welded into tube plates without the need to taper the tube or hole, provided that there is a radial clearance between tube and hole sufficient to allow the formation of the required angle of collapse during detonation of the explosive.[10, 11]

Properties of explosive welds

All explosively welded joints are true metallurgical bonds. They are basically simple pressure welds which rely upon interatomic attraction to develop their strength.

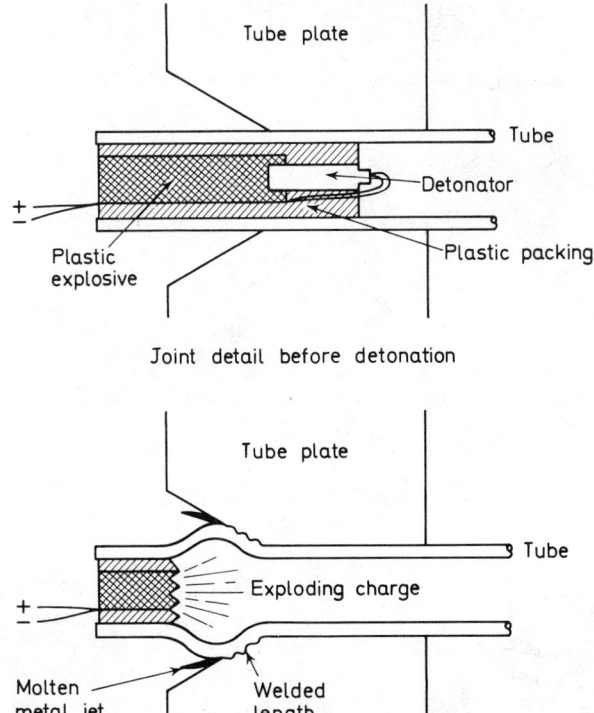

Joint detail before detonation

Formation of explosion weld

Figure 9.4. Tube to tube plate welding technique

Explosive welds generally have a characteristic wavy appearance along the interface although this is not essential to produce a sound weld. Figure 9.6 shows the structure of a typical explosive weld between copper and steel. With some metal combinations, intermetallic phases are formed at the weld interface although these are usually widely dispersed along the interface so that little deterioration of joint strength results.

The strength of explosive welds is, in many cases, equal to the strength of the parent materials; tensile and bend tests invariably lead to failure in the parent metal. The

Figure 9.5. Heat exchanger made by explosive tube to tubeplate welding

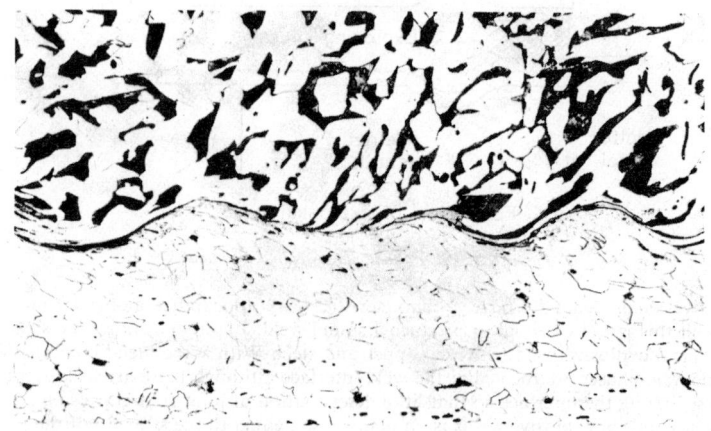

Figure 9.6. Explosive weld between steel and copper

completeness of the bond can be shown by ultrasonic examination, no reflection from the interface indicating a complete weld. This method of testing is used commercially for acceptance testing of explosion clad plate.

The main drawback to explosive welding operations is the heavy charge weight required to produce large clads; the resultant noise problems usually mean that cladding operations have to be carried out in relatively isolated locations. Despite this, clad plate is commercially available in a variety of combinations with clad layers such as Nickel, Copper etc. up to 20 mm thick on base metals such as mild steel 20–250 mm thick in plates 2500 mm by 5000 mm. Tube to tube plate welding only requires a small charge and can therefore be carried out within a normal metal fabrication workshop without too much difficulty.

EMBOSSING AND CUTTING

By placing a stencil between the charge and the base metal an embossed detail can be formed in the base metal. The stencil can be made in a variety of materials including paper or cardboard. The design of the charge must be such that a uniformly flat pressure pulse is obtained over the whole of the base plate; this is best achieved by some form of explosive plane wave generator.[12] Alternatively, the charge may be detonated with the stencil between a metal sandwich, or by using stand off operations under water where the stand off distance gives the pressure pulse the required flat contour but at the expense of a reduction in pressure at the workpiece.

Metal cutting by explosives is a commonly used demolition technique. It is achieved either by detonation of a linear charge in contact with the workpiece, where the explosive width is twice the metal thickness, or by a linear shaped charge consisting of a thin metal angle backed by explosive. As detonation proceeds along the length of the charge, the inner metal liner collapses inwards projecting forwards a high velocity jet of metallic fragments which penetrate the metal (Munroe effect). The cuts obtained by both methods have relatively ragged edges and the material around the cut is highly deformed.

METAL FORMING BY EXPLOSIVES

The great majority of metal forming operations use some form of stand off technique where the charge is located at a distance from the workpiece and the energy released by the charge is transmitted to the workpiece by the intervening medium. The selection of the transmission medium has a significant effect on the energy which is received by the workpiece from the charge, for example for a given charge and stand off distance water will transmit a pressure pulse 2 to 5 times as great as that of air. Plastic, rubber, clay, sand, etc., are also suitable transmission media.

Explosive metal forming has been used for a wide variety of operations, pressing, bulging tube, dishing ends, often in large sizes up to 6000 mm diameter.

For convenience of use many forming operations are carried out in specially constructed water tanks with air bubble screens to protect the walls of the tank against the repeated shock loading. The dies used may either allow the metal to form freely under the action of the charge or be designed to confine the workpiece to the desired shape. Where forming into a closed die is undertaken, steps must be taken to remove the air from between the workpiece and the die as trapped air can distort the workpiece.

The choice of die material is mainly determined by the number of components required, the size and final tolerance to be achieved. For expanding and sizing operations of high strength materials cast or forged steel, dies could be required but zinc base casting alloys can be used for softer materials. For small numbers of components, epoxy resin or concrete dies may be acceptable.

Hot forging by explosives is carried out using transmission materials such as sand, glass beads, powdered alumina. However, as most explosive materials detonate at relatively low temperatures, care must be taken in the heating of blanks and the siting of charges.

No punch or internal tooling is required in explosion forming so it is ideally suited to the forming of tubular components into an outer die. The tube is filled with a transmission media, such as water, and a central charge detonated to form the tube into the die. Complex shapes are possible which could not be produced by conventional

Figure 9.7. Explosive tube to tubeplate expansion

forming techniques. Perforation of holes, formation of stub branches and flanging can be achieved by suitable modification of the outer die.

One aspect of tube forming which has found application in the chemical industry is the explosive expansion of tubes into heat exchanger tube plates.[13, 14] This method of explosive forming has been of considerable success in repairing joints which could not be sealed by conventional roller expansion or welding. The small charge required is located co-axially in the tube end within the tube plate using a plastic transmission medium. As the charge is small it can be detonated in the exchanger on site or in normal fabrication workshops without the need for a specially constructed explosive forming site. Figure 9.7 shows a typical alloy steel exchanger with 20 mm dia. tubes being expanded into position in the tube plate.

All explosive forming techniques can be dangerous if adequate precautions are not taken in the handling of the explosive materials.[15] Many countries have strict legislation on the use of explosives and explosive forming operations may be restricted by the size of charge which can be detonated at one time. Frequently steps are taken to minimise the noise problems arising, but invariably explosive forming operations have to be carried out on isolated sites away from other conventional processes.

ELECTRICAL DISCHARGE-FORMING PROCESSES

For small or medium sized components, some of the difficulties arising from explosive forming techniques can be overcome by using the shock wave produced when a high voltage electrical discharge is made across a small gap usually in water or through a coil.[16, 17, 18]

The energy released by explosives is of the order of 4000–5000 J/gram. For a typical modest explosive charge of 5 g, the equivalent electrical energy would require a bank

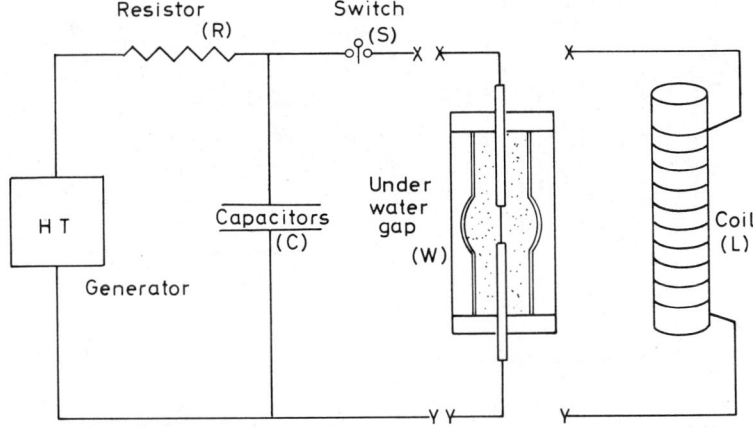

Figure 9.8. Basic electrical circuit—electrical discharge forming

of capacitors capable of storing and delivering 20 000 J. Such energy can be obtained from capacitors of 400 μf at 10 kV, 64 μf at 25 kV or 16 μf at 50 kV and equipment can be constructed of quite small size capable of generating and storing this energy. If a much larger energy requirement of the order 100 000 J is needed, to compare with more sizable explosive charges, much more elaborate electrical equipment with a consequent loss of simplicity in operation is required. Electrical impulse methods are therefore likely to prove most useful in fabrication of small components particularly where the advantages offered by repetitive impulses can be exploited.

The two methods by which electrical discharge can be applied to metal forming both have the same basic electrical circuit as illustrated in Figure 9.8. It consists of an h.t. generator which charges the capacitors C through a current limiting resistor R. A high voltage switch S isolates the capacitors from the underwater gap W or the coil L until the moment of discharge. The h.t. generator should be capable of charging the capacitors in a short time to avoid long delay between discharges, and the choice of the resistor must be sufficiently large that the peak charging current does not exceed the current rating of the generator and yet low enough to keep the charging time short. High working voltages are therefore beneficial since for a given energy requirement the capacitor value and hence the circuit time between charging and discharging can be kept low $(E = \frac{1}{2}CV^2)$.

ELECTRO-HYDRAULIC FORMING

Figure 9.9 shows a typical arrangement for forming tubular or flat plate components by electro hydraulic forming. On discharging the capacitor bank across the gap between the electrodes, which are immersed in water, the fine wire bridging the electrodes is

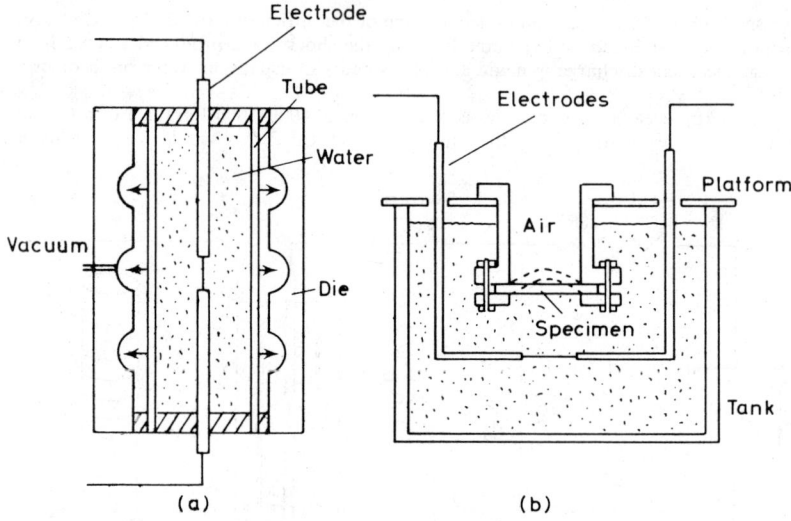

Figure 9.9. Electro-hydraulic forming (a) bulging tube, (b) forming flat plate

heated, melted and vaporised with explosive force. The water vapour produced between the electrodes is highly ionised and forms a plasma, the expansion of which is restricted by the inertia of the surrounding liquid. The surge of current passing across the gap is tens of thousands of amperes over a duration of microseconds and the heating and expansion of the plasma occurs with such speed that a high pressure pulse is transmitted to the water as a shock wave and the tube containing the water and electrodes or a metal plate held in the path of the shock wave will be deformed.

Electro-hydraulic forming operations have been used mainly for the formation of tubular components including bulging into dies, blowing side holes in tubular components, expanding tubes onto flanges, formation of screw threads in tube, etc. The 'fluid punch' imposes little restriction on the shape to be formed and causes little relative movement between die and workpiece which leads to improved surface finish on the formed component.

A medium-size forming machine of 40 000 joules is capable of discharging forty-five times for 1 kWh providing a rapid and economical forming process.[19]

ELECTRO-MAGNETIC FORMING

In electro-magnetic forming, the energy stored in the capacitor bank is discharged rapidly through a coil producing a magnetic field around the coil. If the workpiece is an electrical conductor and is placed in close proximity to the coil a current will be induced in it which in turn produces its own magnetic field around the workpiece. The interaction of the magnetic field around the coil and the induced magnetic field in the workpiece causes the workpiece to be repelled way from the coil in a direction perpendicular to the magnetic field.

The rapid electrical discharge through the coil causes extremely high currents in the workpiece and coil and a high strength magnetic field in the small space between the two. Typical field strengths may be of the order of 30 tesla and surface pressures on the workpiece as high as 3500 bar have been estimated. Machines have been built[20] (which are capable of providing 12 kJ of electrical energy at eight pulses per minute. The coils

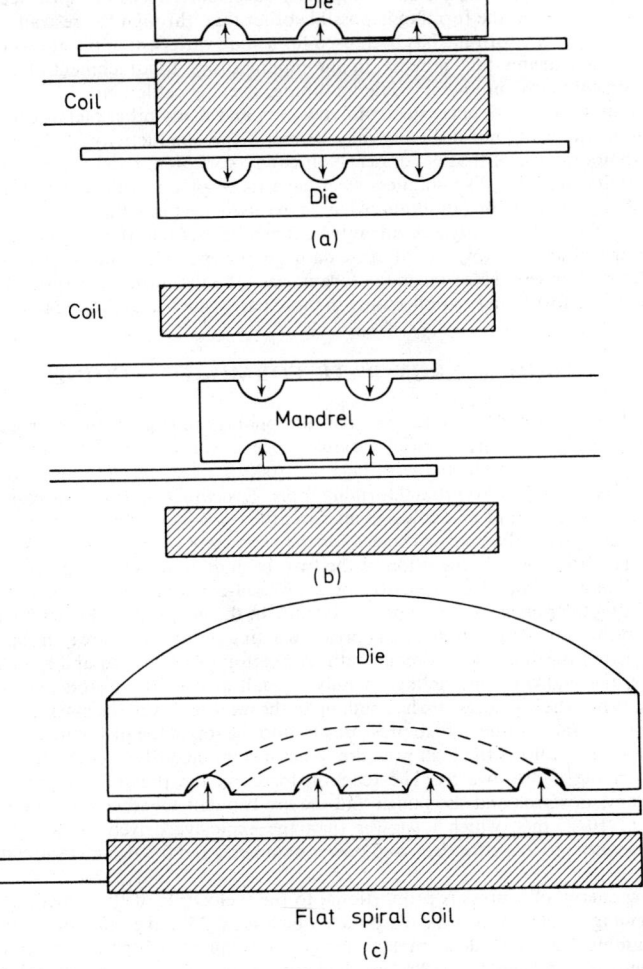

Figure 9.10. Electro-magnetic forming (a) expanding, (b) compression, (c) forming flat sheets

must have adequate strength to withstand the reaction forces but, when intense fields are required, expendable coils may be used. The magnetic field can be used in a number of ways as shown in Figure 9.10. With the induction coil surrounding the workpiece such that the force is exerted radially inwards to collapse the workpiece, possibly onto a mandrel, or conversely, by placing the coil inside, a cylindrical workpiece can be expanded against an external die. In the forming of sheet, flat spiral coils producing a uniform magnetic pressure against the surface of the sheet, can be used to form it into a die.

The most attractive feature of the electro-magnetic forming technique is that no transmitting medium other than air is required. The area, or volume, over which the magnetic force acts is determined solely by the distribution of the induced currents in the workpiece. By suitable design of coil, the induced currents can be localised and some degree of control on the forming is possible other than through the restraint of the die.

Electro-magnetic forming has been used for a wide variety of applications from sizing large diameter drums 1300 mm dia, in aluminium alloys, to attachment of compression fittings on tubular components, bulging tubing, sleeving cables, etc.

Aluminium, brass, copper, steel, molybdenum and many other metals can be formed electro-magnetically, but difficulty may be met if the conductivity of the workpiece is less than about 10% of that of copper. Non-conductors are not deformed at all and low conductivity materials like stainless steel have to be plated with copper, or a 'driver' sheet of light gauge soft aluminium has to be used between the coil and workpiece. While this fact is frequently a disadvantage, it can be useful in those cases where a force on all parts of an assembly would cause damage, for example to non-conducting plastic parts. In such cases electro-magnetic forming may be the most suitable method in that the deforming force can be applied only to those parts which are required to be deformed.

HIGH ENERGY RATE FORMING MACHINES

The application of high speed to more conventional forming processes such as blanking, stamping and forging, has been of variable success. Many types of machine have been developed[21, 22, 23, 24] with energy ratings of 10 000 to over 100 000 J, using a variety of ways of accelerating conventional forming tools. Two which are most successful involve either the detonation of an explosive mixture, the force of which accelerates the punch until it impacts the die and workpiece, or the expansion of a pressurised gas against a piston. The principles of operation of the two techniques are shown in Figure 9.11.

The operation of the explosive press is self-explanatory, the explosive force accelerating the punch to speeds of 25–100 m/s. In the air-pressured machine the piston is held in the starting position by gas pressure acting on the large area on the underside of the piston. A rubber seal is made between the top of the piston and a high pressure cylinder, the high pressure acting on only a small area of the piston until the seal is broken. When the high pressure has built up to the required level, the machine is triggered by injecting a small surge of high pressure air onto the top of the piston in order to unseat the piston. This allows the high pressure to act instantaneously over the full area of the piston driving it downwards. The reaction forces in the pressure chamber drive the press frame upwards and the punch and die are brought together with equal thrust at speeds of 10–25 m/s, which is slower than the explosive driven press. These speeds compare with 3 to 9 m/s for drop and power hammers, and 0·3–1·5 m/s for conventional presses.

As the energy of a press is proportional to the (velocity)^2 of the punch, high energy rate forming machines may have a punch which is 10–25% the mass of a conventional drop hammer for equivalent impact energy. Consequently, high energy rate forming machines can be of much smaller size than comparable conventional machines, for the same energy rating. Since the machines are normally self-reacting or counter-blow, it is not necessary to provide the substantial foundations normally required for forging hammers, thus also saving capital cost. The energy delivered by the high speed machines can be accurately controlled by the setting of the air pressure, thereby removing some of the dependence upon the skills of the operator required by conventional presses.

High energy rate forming machines have been used widely in the hot forging field. The rapid forming allows the workpiece to maintain temperature and thin sections such as fins or webs can be formed without problems caused by rapid chilling met in conventional processes. This is particularly advantageous in materials like titanium and the nickel base alloys. Because of the short forming time, the temperature difference between die and workpiece is large and rapid chilling of the workpiece occurs after the die is filled causing shrinkage away from the die; this allows the forging of products with low draft angles and therefore more accurate dimensions and less subsequent machining. The short forming time, large strains and rapid cooling possible by high

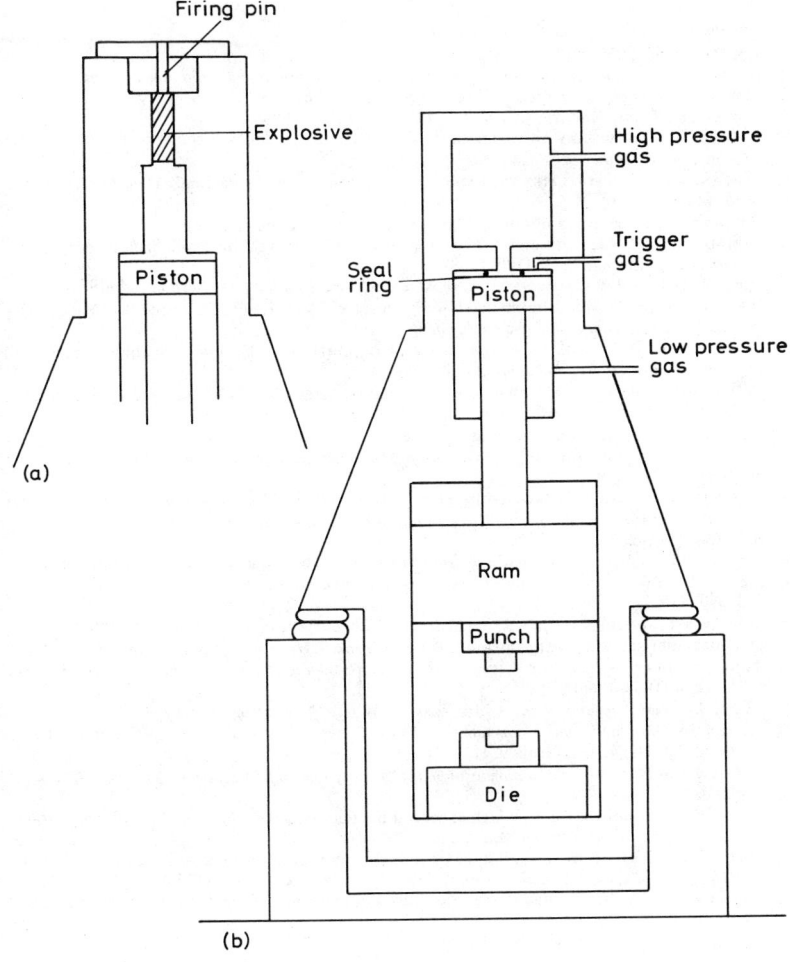

Figure 9.11. High energy rate forming machines (a) explosive press, (b) high pressure gas press

energy rate forming machines, ensures that a fine grain size is obtained in the product with resultant improved mechanical properties, although this effect is probably only marginal as subsequent heat treatment tends to even out any differences between these techniques and the conventional processes. The bulk of the deformation in high energy forming machines is carried out in one die and while this can cause more die wear the greater precision and fewer forging stages can lead to a cheaper product.

Cold forming in high energy rate forming machines is not as extensively used as hot forming and it is unlikely that these machines will be able to compete with modern automated mechanical presses, except perhaps for high speed blanking where the reduction in dishing and edge distortion leads to considerable improvement in quality.

REFERENCES

1. *The use of Metabel for hardening manganese steel*
   UK Patent 765,305
2. DIETER, G. E., 'High rate and high pressure forming', *International conference on international research in production engineering, Pittsburgh*, published by A.S.M.E. (Sept. 1963)
3. SMITH, C. S., *Trans. Met. Soc. A.I.M.E.*, **212**, 574 (1958)
4. NOLAND, M. C., GADBERRY, H. M., LOSER, J. B., SNEEGAS, E. C., 'Shock forming bangs out complex parts accurately, *Materials Engineering* (May 1967)
5. BAHRANI, A. S. and CROSSLAND, B., 'Explosives and their use in engineering', *Metals and materials* (Feb./March 1968)
6. 'Titanium-clad steel by explosion welding', *Engineering* (6 January 1967)
7. CARLSON, R. J., LINSE, V. D. and WITTMAN, R. H., 'Explosive welding bonds metals, large areas, *Materials Engineering*, 70–75 (July 1968)
8. CROSSLAND, B. and WILLIAMS, J. D., 'Explosive welding', *Metallurgical Reviews*, **15** (1970)
9. Yorkshire Imperial Metal Industries UK Patents 1,123,836, 1,149,387, *Impact welding of tubes into tube plates in condensers and exchangers*
10. CHADWICK, M. D., HOWD, D., WILDSMITH, G. and CAIRNS, J. H., Explosive welding of tubes and tube plates', *British Welding Journal* (Oct. 1968)
11. CROSSLAND, B., BAHRANI, A. S., WILLIAMS, J. D. and SHRIBMAN, V., Explosive welding of tubes into tube plates', *Welding and Metal Fabrication* (March 1967)
12. COOK, J. H., 'Engraving on metal plates by means of explosives', *Research*, **1**, 474 (1948)
13. MOTTRAM, R. A. and ANDREW, R. V., 'Tube expansion and welding by explosives', *Engineer* (29 Sept. 1967)
14. MOTTRAM, R. A., 'Explosive tube to tube plate expansion Paper 7', *Proc. of high pressure technology association meeting on use of explosives in forming, welding and compaction*. Queens University of Belfast (March 1968)
15. *Explosives—the sale, storage and conveyance by road*, Nobel Explosives Ltd., 5th Edition (1966)
16. KIRK, J. W., 'Impulse forming by electrical discharge methods', *Sheet Metal Industries*, 39 (424), 533–540 (1962)
17. DUNCAN, J. L. and JOHNSON, W., 'Comparison of the behaviour of different sheet materials formed by underwater spark discharge method. Paper 3, *Proc. I.Mech.E.*, **179**, pt. 1, No. 7 (1964–1965)
18. BAINES, K., DUNCAN, J. L. and JOHNSON, W., 'Electro-magnetic forming', *Proc. I.Mech.E.*, **180**, pt. 1, No. 4 (1965–1966)
19. *Electro hydraulic forming*, Vickers Ltd., Elswick Works, Newcastle upon Tyne
20. *Magneform, Magnetic pulse forming machine*, General Dynamics Corp., General Atomic Division, San Diego, California 92112
21. MANG, W. O., '*Dynapak', a new dimension in high energy forming*, Sheet Metal Industries, 541–554 (1962)
22. DAVIES, R., 'Some effects of very high speeds in impact extrusion', *Proc. of 5th Int. Conf. Mach. Tool Research and Development*, Birmingham (1964)
23. BAKHTAR, F., *et al.*, 'Recent developments in high energy rate forming with petro-forge', *Proc. 2nd Int. Conf. of centre for high energy forming*. Estes Park, Colorado (1968)
24. DOWER, R. J. 'The use of high energy rate forming machines in metal working', *Iron and Steel* (Dec. 1971)

MODERN FABRICATION TECHNIQUES

W. EDWARDS-SMITH

INTRODUCTION

The relatively recent increase in the use of hard, high-strength and temperature-resistant materials in engineering has necessitated the development of new machining techniques. With the exception of grinding, conventional methods of removing material from a workpiece are not readily applicable to these new materials. Even when such machining is possible, it is usually slow and highly inefficient.

Although most of the new machining processes have been developed specifically for materials which are difficult to machine, some of them have found use in the production of complex shapes and cavities in softer, more readily machined materials.

ULTRASONIC MACHINING

Machining by the electro-chemical or electrical discharge processes is confined strictly to metals, whereas ultrasonic machining may be applied to a much wider range of materials. Typical examples of the range of work covered by this process are listed below:

1. Machining of ceramics and porcelain.
2. Cutting glass including thread cutting.
3. Cutting precious and semi-precious stones.
4. Machining silicon, germanium and other semi-conducting materials.
5. Machining ferrite.
6. Machining small punches and dies for presswork.
7. Machining accurately cavities of limited depth in cemented carbide.
8. Drilling holes of varied internal configuration.

The development of this process has opened up several possibilities for the extended use of materials such as ceramics and semi-conducting materials which would hitherto have been impractical. It must be recognised however that at this moment in time there are limitations to the process which have not been overcome namely:

1. Relatively small machining area.
2. Low cutting rates for hard metals relative to other work materials.
3. Relatively high power consumption.
4. High rate of tool wear when cutting hard metals relative to other work materials.
5. Relatively shallow depth of cut.

The cutting of brittle materials by the application of repeated blows of a concentrated nature has long been applied. This principle is applied in ultrasonic machining where the machining of a wide range of brittle materials can be carried out[1-6].

In ultrasonic machining the material is removed by the action of an abrasive which is driven into the work surface by a tool oscillating normal to the surface at high frequency. The tool which is machined to the shape of the desired hole or external profile is made from a tough metal, such as tool bits of alloy steel. A slurry of abrasive powder in liquid is fed to the tool which is oscillated at frequencies between 2000 and 3000 Hz with amplitudes in the region of 0·025–0·075 mm. The tool is pressed onto the work with a force of a few newtons, feeding down as the profile is cut on the work.

Excitation of the tool at these frequencies may be produced by an electro-mechanical transducer of the piezo-electric or magnetostrictive design.

Piezoelectric transducers

It has been known for a considerable time that an electrical charge can be created by certain types of crystal when tensile or compressive stress is applied to the crystal face. It was further shown that the charge was directly proportional to the applied force and that a change from tension to compression reversed the sign of the charge.

Piezo-electric vibrators consist of plates cut from single crystals of piezo-electric material and by applying an alternating voltage of the required frequency to the crystal it will vibrate. The amplitude of vibration being dependent upon the resonant frequency of the crystal, therefore to produce high amplitude the crystal length is matched to the generator frequency. Quartz crystals are used extensively to generate ultrasonic vibrations in solid and liquid media and are capable of producing frequencies up to 25 MHz when vibrating in a fundamental mode.

To produce the resonant condition the crystal length has to be half the wavelength of sound in the material.

The velocity of sound $V_G = (E/\rho)^{\frac{1}{2}}$

$$V_G = \left(\frac{E}{\rho}\right)^{\frac{1}{2}}$$

$$= \left(\frac{N}{m^2\rho}\right)^{\frac{1}{2}}$$

$$= \left(\frac{Nm^4}{m^2Ns^2}\right)^{\frac{1}{2}}$$

$$= ms^{-1}$$

where E = Youngs modulus (k N/m^2), and

ρ = density of crystal (kg/m^3) = (Ns^2/m^4)

In the case of quartz, $E = 5·2 \times 10^{10}$ N/m^2 and $\rho = 2·6 \times 10^3$ kg/m^3

$$V_G = 4472 \text{ m/s}$$

Now as the velocity of propagation of sound waves is proportional to the wavelength and frequency, i.e.
where

λ = wavelength (mm)

f = frequency of vibration (Hz)

Then for a frequency of 20 kHz the wavelength of sound with a quartz crystal is:

$$V_{G/f} = \frac{4472}{20 \times 10^3} = 0·2246 \text{ m}$$

Therefore, if a single crystal were used as transducer at this frequency it would have to be about 0·1123 m long, as it is necessary for the crystal length to be half the wavelength of sound in order to produce resonance. As it is impractical to use large crystals, sandwich type transducers of the type shown in Figure 9.12 have been devised, the alternate tension and compression being applied as a result of the oscillation and the clamping effect of the bolt.

In order to conserve momentum the velocity and amplitude of vibration in the high and low density materials is inversely proportional to the density thus giving a high amplitude at the radiating face.

Magneto-strictive transducers

For any specified magnetic conditions a given material has a unique group of physical and chemical characteristics. The effect of a change in the magnetic field is to produce linear, circular and volumetric changes. Fundamentally a magneto-strictive transducer is a magnetised rod with a coil wrapped around it, a voltage may be induced in it by stretching or compressing the bar. As the bar returns to its original length it induces

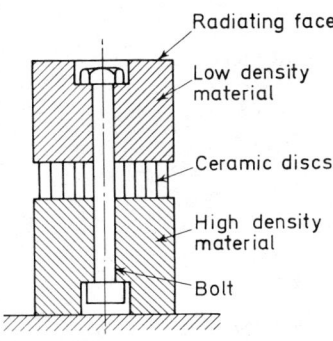

Radiating face

Low density material

Ceramic discs

High density material

Bolt

Figure 9.12. The sandwich type transducer

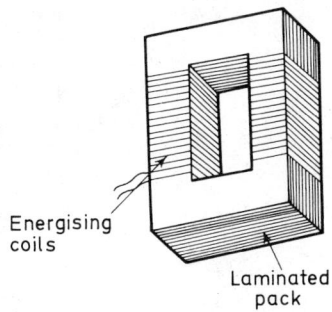

Energising coils

Laminated pack

Figure 9.13. Showing the stacking of nickel laminations of the magnetostrictive transducer

another voltage of opposite polarity, thus by vibrating the rod at some a.c. frequency an a.c. voltage of the same frequency is induced in the coil. The magnitude of the voltage is proportional to the amplitude and therefore at resonant frequency maximum voltage occurs.

If a voltage is applied to the coil the rod will vibrate at the frequency of the applied voltage with an amplitude approximately proportional to the voltage, and ultrasonic waves will be radiated from the rod extremities. A ferro-magnetic material such as nickel is used in the manufacture of magneto-strictive transducers and to reduce eddy current losses the nickel is usually stacked in the manner shown at Figure 9.13, where the laminations are insulated. Current of the required frequency is fed to the energising coils and the heat which is thus generated is dissipated by a surrounding water jacket.

The magneto-strictive vibrator has several advantages over the piezo-electric crystal when applied to a practical ultrasonic machine tool. These advantages being:

1. Generally more robust.
2. Less affected by temperature changes.
3. For a given power output the voltage requirements are modest.[8-11]

The principle of ultrasonic machining

The basic principle of ultrasonic machining is shown in Figure 9.14 and, in order to increase the amplitude of vibration of the transducer, a velocity transformer is fixed to the radiating face of the transducer. The resonant movement of the radiating face of the transducer being in the region of 0·012 mm by attaching a velocity transformer the longitudinal amplitude can be increased to 0·025–0·075 mm. The transformer is made from materials having high fatigue strength and low energy loss and a material such as brass meets this specification. The profile of the transformer is tapered exponentially, Figure 9.15, the increase in the ratio of the amplitude being inversely proportional to the ratio of the areas of the two ends.

The toolholder (or tool cone) is fixed to the end of the transformer and to give minimum damping the vibrating components are clamped at the nodes and are acoustically matched. A transducer-transformer-tool system is thus produced which is a resonant mechanical transmission line two or more half-wavelengths in lengtn. As tool wear

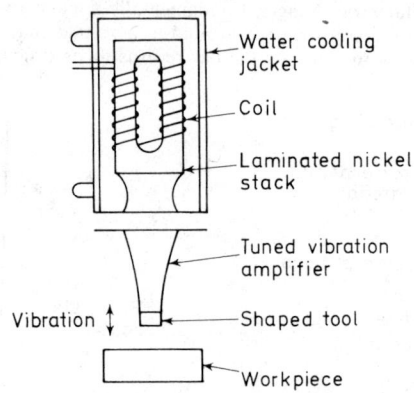

Figure 9.14. The principle of ultrasonic machining

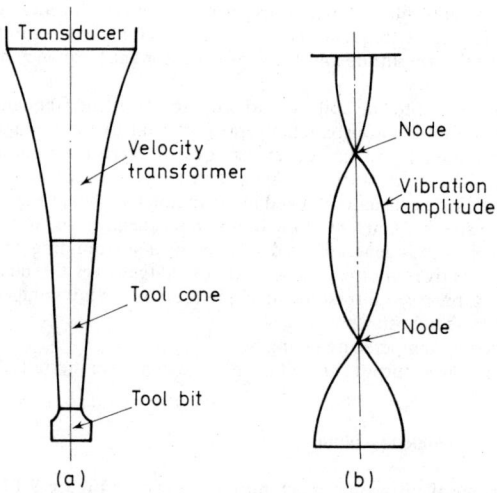

Figure 9.15. Profile of the transformer

produces shortening of the combined length of tool cone and tool, it is usual practice to make them slightly longer than a half-wavelength and thereby increase their effective life.

When machining components by this method the tool is held against the work by a light static force of about 18 N and an abrasive slurry is applied to the work area either by hand or by a circulating pump. As the tool tip is oscillated by the transducer, abrasive particles are trapped between the tool and the work and a chipping action ensues which removes small pieces of material from the workpiece. The vibration of the tool tip accelerates the abrasive at very high rates thus imparting the necessary force to produce the cutting action. Cavitation which occurs in the gap assists removal of the chips and circulation of the abrasive.

The abrasive is suspended in a liquid which serves several functions; it provides an acoustic bond between vibrating tool and work and therefore produces an efficient means of transferring energy between tool and work. It also acts as a coolant at the tool face and assists in carrying the abrasive to the cutting area and to wash the spent abrasive and swarf away.

A number of different types of abrasive have been used for ultrasonic machining but those most commonly used are aluminium oxide, silicon carbide, boron carbide and to a lesser extent, diamond dust. The grit size varies between 200 and 2000 with the coarse grade being used for roughing and the finer grades (say 600–700) used for finishing. The very fine grades (1000–2000) only being used for the final pass on work of the highest accuracy.

The liquid used to carry the abrasive must have the following properties:

1. Its density must be approximately equal to that of the abrasive.
2. It must have good wetting properties.
3. In order to be able to carry the abrasive down between tool and work, it must have low viscosity.
4. For the efficient removal of heat from the cutting zone it must have high thermal conductivity and specific heat.

Water will satisfy most of these requirements with such additives as rust inhibitors and wetting agents.

Factors influencing cutting performance

The manner by which impacted abrasive grains remove metal from a surface is as yet not fully understood and of the several attempts made to construct a model of the process none of them satisfies all the variables of the process.

The experimental work of Neppiras and Goetze provide extensive details of the process which indicate the considerable number of variables involved and therefore the virtual impossibility of formulating an explicit empirical relationship. Thus the work carried out in this field has been limited to a selected number of variables.

The effect of amplitude frequency and grain diameter upon cutting rate appears to be a source of disagreement between the work of Goetze[12, 13] and Neppiras and Foskett.[1, 3] The former claims that cutting rate increases linearly with increase in the three variables and that above a certain critical abrasive/water ratio the relationship R/afd is constant.

Where R = penetration rate of tool,
a = amplitude of vibration,
f = frequency of vibration,
d = mean diameter of abrasive.

The work of Neppiras and Foskett in this area showed, as can be seen from Figure 9.16 (a) and (b), that a nonlinear relationship between amplitude, frequency and cutting rate R exists. They were also able to show that by increasing the static force a rapid increase in cutting rate occurred, this is followed by a broad peak after which the rate decreases.

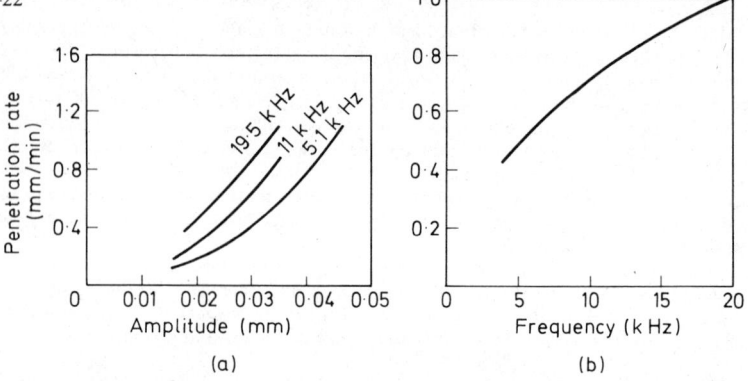

Figure 9.16. Cutting rates in glass (a) as a function of amplitude (peak to peak) of the toolface,
(b) as a function of frequency for a fixed amplitude of 0·03175 mm (after Neppiras and Foskett[2])

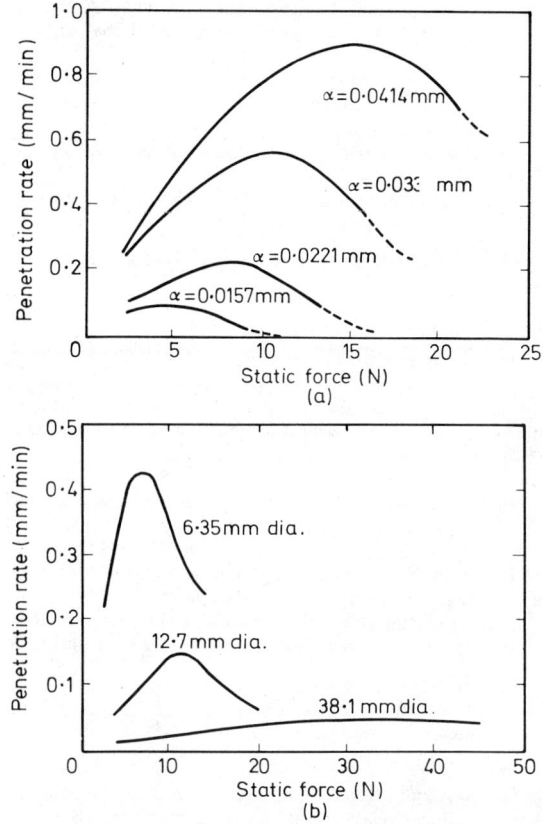

Figure 9.17. Cutting rate as a function of the applied static force
(a) amplitude of vibration as parameter, (b) diameter of toolface as
parameter (after Neppiras and Foskett[1])

An excessive force on the tool may prevent abrasives reaching the tool so that cutting would eventually stop. The position of the maximum cutting rate was influenced by the amplitude of the tool and the cross-sectional area of the tool, as shown in Figure 9.17[1] and the optimum force, which will vary from several to tens of newtons will differ from job to job. Goetze[13] showed that the shape of the tool face would influence the maximum cutting rate, a narrow rectangular cross-section giving a higher maximum cutting rate than a tool having a square cross-section of the same area.

Neppiras and Foskett examined the effect of abrasive concentration in the slurry.

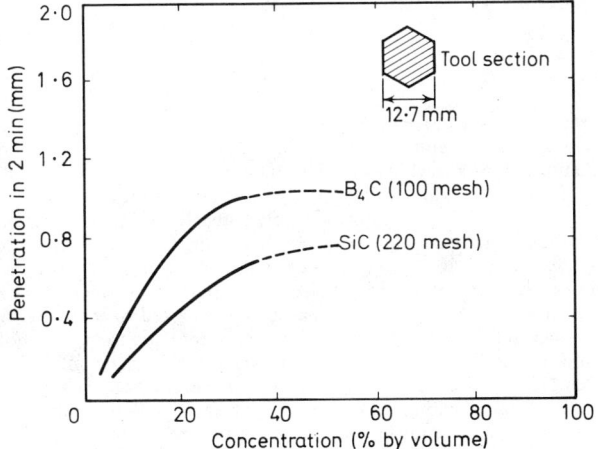

Figure 9.18. Cutting speed in glass as a function of abrasive concentration
(after Neppiras and Foskett[1])

It was shown that violent cavitation occurs over the tool face due to the liberation of absorbed gases and that strong currents occurred in local areas across the cutting face and between the lateral faces of the tool and cavity. The intensity of these currents will vary with end and side clearance, frequency and amplitude of vibration. Cavitation erosion and the abrasive action of the strong current flow of the slurry have little influence upon the cutting action at the tool face. The flow of slurry does however tend to take up specific positions in the clearance space between tool and work and thereby produce local scouring of the work surface, thus reducing surface quality. By providing flutes down the side of the tool a higher flow of cutting media is possible and therefore the cutting speed may be increased.

The viscosity of the slurry has a damping influence on the cutting tool oscillations which is of particular importance when manufacturing work with a large side area, such as splines. If the viscosity is too low, then it will retard the passage of slurry and waste in the side clearance and hence reduce the cutting rate. To eliminate this, the shank of the tool is reduced in size behind the face.

The result of varying the abrasive concentration is shown in Figure 9.18, which indicates a rise in the cutting rate as the abrasive content is increased until a practical limit is reached. It is quite likely that a further increase in the abrasive concentration will decrease the cutting rate as a stage may be reached where the grits will jam together. Neppiras and Foskett have carried out tests with a range of work materials and abrasives, the general trends being shown in chart form as in Figure 9.19.[14]

The success of the abrasive as a cutting media depends not only upon its hardness but also on the number and durability of the cutting edges which are a function of the shape of the particles. The brittle behaviour of the work material will greatly influence the cutting rate, brittle non-metallic materials being cut at very high rates.

One important advantage of this process is that it leaves no residual mechanical stresses in the cut material when operating under normal conditions. It can be shown that even with the most easily fractured materials the chipped workpieces are small even when compared with the abrasive, indicating that the stresses transferred to the work must be very localised. It is important to remember, however, that an adequate supply of liquid coolant must be maintained. If the abrasive is allowed to dry out, high temperatures may develop at the tool tip, and thermal stresses will occur in the workpiece followed by possible work fracture.

Surface finish and accuracy

The abrasive grit size is a major factor which influences surface finish. Neppiras and Foskett[1] showed that the applied static load had no effect upon surface finish. When using, for example, boron carbide on glass the roughness may vary from 5 μm for 100 mesh grit to 1–25 μm when a 600 mesh is used. If tungsten carbide is machined, however, the roughness values are approximately a quarter of those listed above. It was also shown that the roughness obtained on the face of the tool is generally less than that on the sides. As was previously indicated, cavitation streams may produce furrows having a depth roughly equal to the grit size which mark the surface of the sides.

The process, being dependent upon the abrasive being fed to the working face will cause a slight taper to form due to lateral wear of both tool and work. The hole which is produced by this method will tend to be slightly oversize in the region of 0·025–0·40 mm. In order to achieve high accuracy, it will generally be necessary to carry out machining in several stages, initial roughing being done at high speed with an undersized tool and coarse grit. The final sizing requiring one or two operations with tools of increased size and finer grits. It is possible, however, to predict the amount of oversize to within

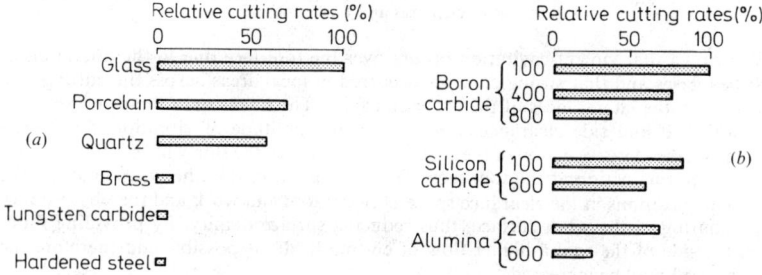

Figure 9.19. The influence of (a) work material and (b) abrasive on the relative cutting rates. Rates expressed as a percentage of the cutting rate for soda glass with 100 grit boron carbide abrasive (after Neppiras[14])

$\pm 0·05$ mm which will permit accurate prediction of the allowance to be made on the tool to achieve a sized hole.

As the ultrasonic machining process is limited by its relatively low metal removal rate, several methods have been suggested for improving this. Pentland and Ektermanis[15] have shown that by refrigerating the abrasive slurry this causes embrittlement of the work material and when machining a AISI 4140 steel the machining rate was increased by 20–30 per cent when the slurry temperature was 0°C. An alternative method of improving the machining rate has been suggested by Rozenberg[16] where he shows that by increasing the pressure at which the slurry is fed into the cutting zone substantial improvements in cutting rate can be obtained. When using a high supply pressure, improvements of the order of 10 times may be obtained.

Advantages

The advantages of ultrasonic machining are as follows:

1. The ease with which accuracy and good surface finish are obtained.
2. No heat generated and therefore no changes in the physical structure of the material occur.
3. Equipment is safe to handle and little skill required in its operation.
4. Cheap abrasives are used.
5. The process is applicable to all brittle materials and is not limited by any other physical property.

A more recent development of ultrasonic machining makes use of a diamond impregnated cutting tool with a rotary transducer which it is claimed gives substantially faster cutting rates and greater accuracy than conventional methods. As the cutter rotates, the process may be likened to that of end milling and hence by traversing the tool, profiles are generated. It is possible to achieve work tolerances in the region of 0·0125 mm in glass and ceramic components and when machining glass cutting rates of the order of 10 mm^3 per minute have been obtained.

The tool is rotated at 100–1800 rev/min with an axial vibration in the region of 20 Hz and an amplitude of 0·005–0·0125 mm.

ELECTRICAL DISCHARGE MACHINING (EDM)

This process is sometimes referred to as electro-erosion or electro-spark machining. It may be described briefly as the removal of material by means of repetitive short-lived electric sparks which occur between the tool (i.e. electrode) and workpiece.

The principle of the process is shown in Figure 9.20 in which the tool is mounted on a chuck or other clamping device attached to the spindle of the spark machine. The down feed of this spindle being controlled by a servo-drive through a reduction gearbox. The workpiece (i.e. cathode) is immersed in a tank filled with a dielectric fluid usually paraffin, in order to reduce the risk of fire a depth of several inches is maintained over the top of the workpiece. Dielectric fluid is circulated under pressure by a pump and filtered to remove vaporised workpiece particles.[17]

The circuit shown in Figure 9.20 is a d.c. relaxation circuit fed usually from a mercury arc or selenium type rectifier. In order to achieve erosion, a gap (approx. 0·025–0·05 mm) between tool and workpiece has to be maintained by the servo-drive. This type of circuit is not particularly efficient, as will be shown later, but does represent the earlier type of circuit used in e.d.m.

When the power supply is switched on a voltage begins to build up in the bank of condensers C, such that the condenser voltage V_c builds up exponentially towards the supply voltage V_s.

Where

$$V_c = V_s \left(1 - e^{-t/R_b C}\right) \tag{1}$$

where t = time,

R_b = ballast resistance,

C = capacitance.

In this initial stage the spark gap behaves as an open circuit and no current flows. As voltage V_c builds up, it eventually reaches the breakdown voltage V_b of the gap (this being determined by the gap width and the dielectric fluid); a spark is produced across the gap, the dielectric ionises, the condenser then being discharged. After the condenser has discharged the surrounding dielectric deionises thus again becoming an effective insulator and the cycle is repeated. Thus a rapid succession of sparks is obtained, the frequence of spark production being of the order of 10 000 per second.

As the current is discharged across the gap temperatures in the region of 10 000–50 000°C are developed, this evaporates part of the surrounding dielectric fluid and

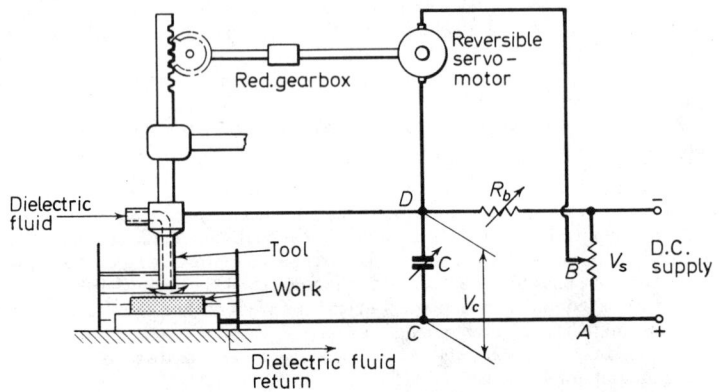

Figure 9.20. Electrical discharge machining

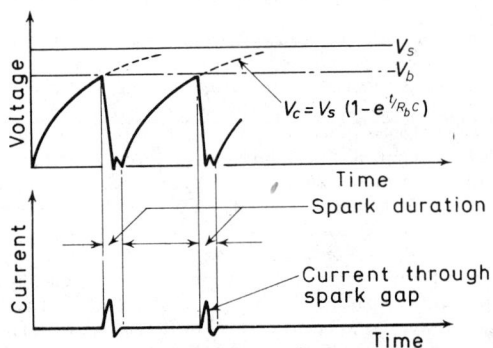

Figure 9.21. Voltage and current in EDM

vaporises the metal thus forming a small crater on the work surface. As the spark always occurs where the tool and work are closest together, the high spots of the work are gradually melted away thus producing the form of the tool on the work. The vaporised metal globules produced during the process are removed by the flowing dielectric and if this were not the case these particles, largely of non-conducting material, would act as insulators thus preventing spark formation.

It will be seen from Figure 9.21 how the voltage builds up in the condenser until it reaches the breakdown voltage of the gap when it is discharged at peak current in a short interval of time. This results in a low efficiency of relaxation circuits as characterised by the very rapid discharge time but long pause and charge time (i.e. approximately

70 per cent pause time and 1 per cent discharge time). A further difficulty is that polarity is reversed at the end of the discharge which results in erosion of the cathode (i.e. the tool now becomes the cathode).

If high production is important, then the high intensity low frequency spark production of the relaxation circuit is unsatisfactory and sparks of high frequency produced by rotary impulse generators are preferable. This will ensure better control of frequency and higher metal removal rates. In addition less tool wear occurs.

Tool wear and tool materials

The fact that wear occurs upon tool (i.e. electrode) is one of the major difficulties in EDM, especially where a relaxation circuit is used. The wear ratio, i.e. metal lost from tool divided by volume of metal removed from work will depend upon the tool and work materials used. Typical examples are that of a brass tool used on a brass workpiece, where the ratio is of the order of 0·5, whereas a brass tool used on a tungsten carbide workpiece the ratio may be as high as 3.[19]

Electrode wear will produce inaccurate machining, the wear taking place predominantly at the leading edges of the electrode producing tapered cavities. Advantage may be taken of this feature when producing press tool blanking dies, the electrode being put through the die from the reverse side thus providing natural die clearance. In most cases, however, electrode wear is not an advantage as it may be necessary to use several electrodes in order to achieve accuracy of an acceptable value.

In order to minimise the cost of electrode manufacture much work has been done to obtain suitable materials for the electrodes and with suitably chosen materials the wear ratio may be as low as 0·1.[20] For many applications brass or copper is used, although copper-graphite and tungsten carbide have been used to advantage. The higher the melting point of the electrode material the less wear that occurs. Thus, by using materials such as copper-graphite and more recently with the pulse generator circuit, graphite, lower wear rates are obtained.

Servo-controller tool feed

The basic outline of the control system is shown in Figure 9.20 where a reversible servo-motor is connected between one terminal of the condenser D and the slider of variable resistance B. A and C are at the same potential and if the mean value of V_c (Figure 9.20) equals the pre-set voltage between A and B, then the potentials at B and D are equal and the servo-motor does not rotate. On the other hand, if the mean value of V_c changes due to alteration in gap width then the potential at D changes and the servo-motor rotates until the desired gap width is restored. Thus a mean spark gap voltage is maintained and hence the gap width remains constant. In the event of overshoot or alternatively build-up of disintegrated work material in the gap, which produces short-circuiting, the motor reverses until correct gap width is restored or the work material cleared.

Characteristics of machined surface

Investigations by Barash[18, 21] indicated that high tensile stresses exist in the top 0·025 mm of the surface, this being due to the high localised surface temperatures which lead to expansion, plastic flow and subsequent contraction. He also states the results of fatigue tests on such surfaces. These tests showed that the endurance limit was substantially less than that of the same material with a turned surface, and whilst stress relieving raised the endurance limit it was still below that of the turned surface.

Opitz[22] has shown that the structure of surface layers formed by discharge machining will, in the case of steels, be a high-carbon layer which may contain as much as 4 per cent carbon as a result of the breakdown of the dielectric fluid at high temperatures. Below this there may be successive layers of austenite and martensite blending through troostite and sorbite to the annealed material. Under finish machining conditions the depth of the heat affected zone is of the order of 0·02–0·1 mm.

Metal removal

The rate at which metal can be removed by the EDM process is largely dependent upon the electrical parameters, which will be discussed later. If the highest quality surface finish is required then lower metal removal rates have to be used so that it will be necessary to achieve a compromise dependent upon whether rough or finish machining is required.

Using modern pulse generator machines metal removal rates of up to 8 mm³/min can be achieved when rough machining. The metal removal rate is affected by the spark duration for a given energy input, for a given power input an optimum discharge time exists thus giving maximum metal removal per discharge. If the discharge time is too small then tool wear becomes excessive with resulting loss of accuracy.[23]

Accuracy in EDM

The principal application of the process is in die and mould manufacture and as no force is applied to the tool then thin or delicate tools can be used without risk of bending or breaking.

In addition to errors occurring due to incorrect discharge times, the accuracy with which the electrode is produced is clearly an important factor. Allowance must be made for the spark gap, and if parallel sided cavities are to be produced, then by using several electrodes or by relieving the tool behind the tip, this can be achieved. As the depth of the cavity increases it becomes more difficult for the dielectric fluid to wash the disintegrated work particles clear of the tool/work gap. This has the effect of reducing the breakdown voltage of the dielectric fluid with the result that sparking occurs between the sides of the electrode and work cavity.

Variation of electrical parameters

The surface geometry produced by EDM consists of small spherical craters resulting from the metal being removed by the individual sparks. The size of these craters depend upon the energy dissipated per discharge which is a function of the gap break-down voltage V_b and the capacitance C.

Now, as the metal removal rate depends on the energy dissipated per discharge and the frequency of discharge then the conditions required to control metal removal must be examined. It has been shown by Barash[18] that the maximum useful power is transmitted through the charging circuit when the conditions are adjusted to give:

$$V_c = 0·72 \, V_s \qquad (2)$$

and the frequency of discharge was approximately shown to be:

$$f = \frac{1}{2\pi\sqrt{(LC)}} \qquad (3)$$

where L = inductance of the leads.

As it is a basic requirement of EDM that no current shall pass through the gap between sparks then upon completion of the spark the gap must be nonconducting. If this condition is not attained current will flow continuously thus preventing build-up of energy in the condenser resulting in an arc which damages the structure and finish of the surface.

In order to increase metal removal rate the supply voltage V_s should be increased whilst the ballast resistance R_b is decreased. On the other hand if this resistance is too low then continuous arcing will take place as the sparking frequency becomes too high. Barash[18] found experimentally that the minimum ballast resistance necessary to avoid arcing was:

$$R_b \min = \frac{60}{\sqrt{C}} \tag{4}$$

This limits the sparking frequency such that only approximately $\frac{1}{8}$ of the working time is used for erosion, the rest being used for charging. A reduction in V_s and therefore of V_c permits a smaller spark gap to be maintained thus giving greater accuracy and improved surface finish. As V_s is decreased, however, metal removal decreases rapidly, but high cutting rates can only be obtained by accepting some arcing and therefore loss of surface finish. Thus it is necessary in most cases to reach a compromise in the adjustment of the electrical parameters.

Surface finish in EDM

The energy of a discharge in a relaxation circuit is given by:

$$Q = \tfrac{1}{2}V_c^2 C \tag{5}$$

but as only part of this energy is used in eroding the surface, the volume of material removed per discharge may be written:

$$V = k_1 V_c^2 C \tag{6}$$

where k_1 is less than half.

If it is assumed that a spherical crater is produced then the volume removed is proportional to the cube of the crater depth. If, in addition, we assume the craters are regularly arranged and having a constant depth-to-diameter ratio, then the CLA value may be taken as proportional to the depth, i.e.

$$(h_{CLA})^3 = k_2 V_c^2 C$$

or

$$h_{CLA} = k_3 V_c^{\frac{2}{3}} C^{\frac{1}{3}} \tag{7}$$

This relationship, which was derived by Barash[18] was found experimentally to be:

$$h_{CLA} = k V_c^{0.5} C^{(0.31 \text{ to } 0.36)} \tag{8}$$

which shows reasonably good agreement with equation (7), it being suggested that the lower power of V_c indicates that the proportion of energy loss in the gap increases with gap length.

When high production rates are required the surface roughness may be as high as 1·75–3·75 μm CLA but by reducing the sparking rate CLA values in the region of 0·25 μm may be achieved.

Applications

Spark machining has its greatest application in the manufacture of press tools, mould forgings and extrusion dies. The material can be spark machined after it has been hardened thus eliminating the effects of distortion due to heat treatment. Tools manufactured from cemented carbide can be machined when sintering has been completed, thereby eliminating the need for an intermediate partial sintering stage and thus eliminating the errors resulting from final sintering after the cavities, etc. have been machined.

The successful application of this method to the drilling of holes with a large length/ diameter ratio, as in the case of the nozzles of diesel engine fuel injectors, has been carried out.

In addition to the conventional applications of EDM, the process has been used for turning and precision band sawing. In the former case the workpiece is rotated as in turning, but the cutting tool is replaced by the electrode. The dielectric fluid is flooded into the spark gap.

When precision band sawing of hard materials a fine wire wound onto two reels is used, with the dielectric fluid flooding into the work. To prevent the wire from breaking and wearing it is continually wound from one reel to the other. In one advanced type of band sawing machine the path moved by the wire is under numerical control.[24]

ELECTROCHEMICAL MACHINING (ECM)

Electrochemical machining is a process of metal removal in which electrolytic action is utilised to dissolve the workpiece material. It is in effect the reverse of electroplating. The term ECM is sometimes used for the electrolytic grinding process, which is a modification of electrochemical machining.

The basic elements of an electrochemical machine are shown in Figure 9.22. The

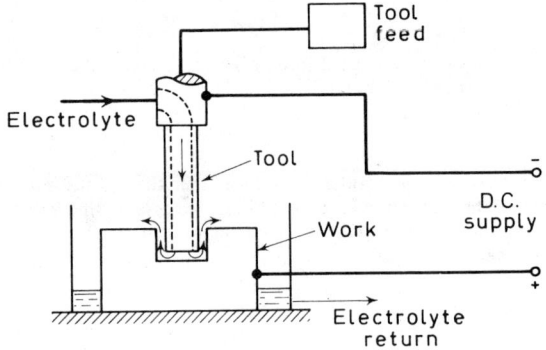

Figure 9.22. Electrochemical machining

workpiece (which must be a conductor of electricity) is placed in a tank on the machine table and connected to the *positive* terminal of a d.c. supply. The tool electrode, which is shaped to form the required cavity in the workpiece is mounted in the tool holder and connected to the negative terminal of the supply. An electrolyte flows through the gap between the tool and workpiece and is then pumped back to the working zone either through the tool or externally, depending on application.

The action of current flowing through the electrolyte is to dissolve the metal at the anode, i.e. the workpiece. The electrical resistance is lowest (and hence the current highest) in the region where the tool and work are closest together. Since the metal is dissolved from the work most rapidly in this region, the form of the tool will be reproduced on the work.

There is no mechanical contact between work and tool and any tendency of the work metal to be plated on the tool (the cathode) is counteracted by the flow of electrolyte which removes the dissolved metal from the working zone. Hence there is neither tool wear nor plating of the work material on the tool, so that one tool can produce a very large number of components in its life.

Electrolyte

The electrolyte serves two important functions in ECM. Firstly, it provides the medium by which electrolysis takes place and, secondly, it removes the heat generated in the working zone due to the flow of a high current through the electrodes and through the electrolyte itself. The flow of electrolyte must be sufficient to prevent the boiling point of the liquid being reached.[20]

Besides the ability of the electrolyte to carry out the above functions, there are two aspects of its chemical nature which must be considered. It must be sufficiently active chemically to cause efficient metal removal, but should not be too corrosive or rapid deterioration of the parts of the machine it contacts will occur. The parts of the machine exposed to the electrolyte are usually manufactured from stainless steel or coated with corrosion-resistant paint.[25]

The electrolyte most commonly used is brine; other salt solutions are used for specific applications.

Nature of machined surface

The fact that metal removal in ECM is not achieved by mechanical shearing or by melting and vaporisation of the metal means that no thermal damage occurs and no residual stresses are produced on the worked surfaces.[20] The only heat generated is that due to electrical resistance, and the temperature cannot be allowed to rise above the boiling point of the electrolyte.

Surface finishes of 0.75–1 μm[26] are readily obtained, while surface finishes better than 0.25 μm[20] have also been achieved. A great advantage of ECM, which is contrary to conventional machining, is that, as the metal removal rate is increased, the surface finish and accuracy are improved.

Merits of ECM

These can be listed as follows:

1. Stress free surface on workpiece.
2. No work-hardening of material.
3. No thermal cracking of work surface.
4. High surface finish.
5. No softening of work surface.
6. Freedom from burrs.
7. High metal removal rates on hard materials.
8. No electrode wear.

ELECTROLYTIC GRINDING

This is a modification of the ECM process. The tool electrode consists of a rotating abrasive wheel which can conduct electricity; this is usually a metal bonded diamond wheel. The electrolyte is fed between the wheel and work surface in the direction of movement of the wheel periphery, so that it is carried past the work surface by the wheel rotation. The abrasive particles help to maintain a constant gap between wheel and work. The current flows between work and wheel as shown in Figure 9.23.

In this process the predominance of the electrolytic action reduces wheel wear to a negligible amount and makes it possible to grind hard materials rapidly. Further, the wheel can be used for long periods without dressing.

In applying electrolytic grinding it is desirable to design the operation in such a way that the area of contact between the wheel and work is as large as possible. This gives

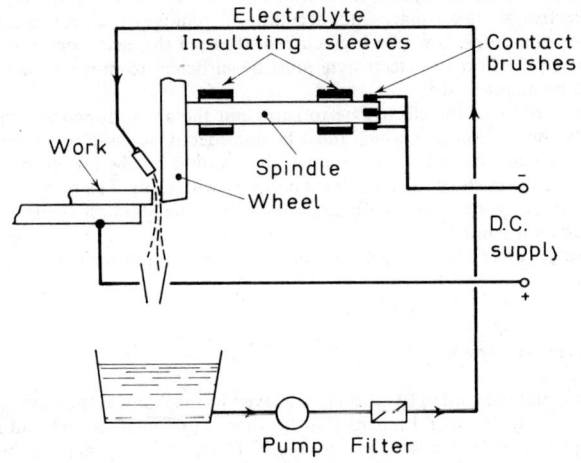

Figure 9.23. Electrolytic grinding

the highest rate of removal for a given current density i.e. i/A and allows full use to be made of the available current capacity. The most notable application of this method is the grinding of carbide tool tips where surface finishes of the order of $0 \cdot 1 \ \mu m$ have been obtained.

Applications

Its main applications are in machining hard metals such as those used in high temperature service. The removal rate is much higher than in conventional machining if the latter can be used at all.

ECM has been applied successfully to the following machine operations.[27]

1. Machining of through holes of any cross-section.
2. Machining blind holes with parallel sides.
3. Machining shaped cavities, such as forging dies.
4. Wire cutting of large slugs of metal.
5. Machining complex external shapes, e.g. turbine blades.

Basic theory

The electrochemical machining process employs the principle of electrolytic anode dissolution.[23, 28, 29, 30] Figure 9.24 shows a typical electrolysis process, two electrodes, the positive (anode) A and the negative (cathode) C are immersed in a solution and connected to a source of potential difference through meter G which measures the current. If the anode is silver, the cathode copper, and the solution silver nitrate in water, then when the current is in the indicated direction, the copper cathode will gradually become plated and the silver anode will decrease in mass. Thus silver atoms are being transferred from the anode to the cathode.

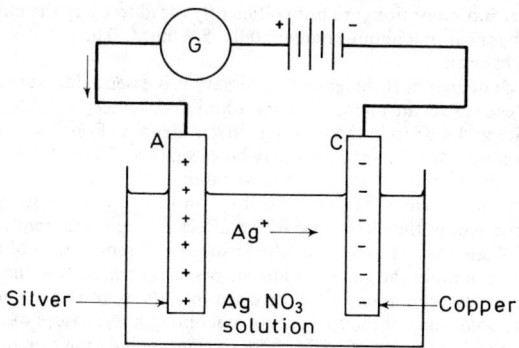

Figure 9.24. Typical electrolysis process

Faraday found that the mass W of material deposited is directly proportional to the charge Q_I passing through the galvanometer.

$$W \propto Q_I \tag{9}$$

(If the current I is constant, then Q_I is I times the time allowed for deposition.)
Thus $Q_I = I_t$ coulombs (i.e. 1 amp sec = 1 coulomb)
Also for a given transfer of charge, the mass deposited is proportional to the atomic mass (atomic weight) of the substance divided by its valence:
Thus

$$W \propto w/n$$

where

w = atomic weight, and

n = valence

The two above results may be summarised by:

$$W \propto Q_I w/n \tag{10}$$

This can be written as an equation if we introduce a constant of proportionality, then

$$W = \frac{1}{F} Q_I \frac{w}{n} \tag{11}$$

The constant F is called the faraday. The faraday can be accurately determined by measuring the mass W deposited by a measured charge Q_I of a substance with known valence and atomic mass.
Thus

$$F = \frac{Q_I/n}{W/w} = \frac{Q_I w}{nW} \text{ coulombs/mole} \tag{12}$$

w and n are dimensionless.

$$F = 9 \cdot 6487 \times 10^4 \text{ coulomb/mole} \simeq 96\,500 \text{ coulombs/mole}$$

Electrolyte flow

In electrochemical machining a high flow of electrolyte is used in order to flush away the metal removed from the workpiece (anode), but also to prevent the formation of an ionised layer on the work surface.

In traditional electrolytic processes the normal mechanism of ion transfer, i.e. convection, diffusion and migration, are not sufficiently rapid to carry the current densities used in electrochemical machining, namely $0.05-5$ A/mm². Thus forced circulation of the electrolyte is essential.

Although a high degree of turbulence is necessary it is essential to avoid cavitation or gas formation, leading to an increase in the ohmic resistance, a reduction of anode-electrolyte contact and a drop in the limiting current density. For this reason electrodes are separated by a very small gap, which may be as small as 0.1 mm, through which the electrolyte passes at a rate of several thousand millimetres/second. Heat and gases are evolved by the process which causes pressure, temperature, viscosity gradients, and decomposition changes in the electrolyte which affect the electrical conductivity.

Two major problems are encountered when using small separations of the electrodes. The first is how to maintain the gap constant for precision machining, the second is how to introduce the necessary amount of electrolyte into the gap. It becomes a matter of compromise. The constancy of the gap during machining is a measure of the correctness of 'machine setting'. In the case of ECM these settings include the rate of circulation of the electrolyte, composition of the electrolyte, its clarification, current and voltage, rate of tool feed, and control of temperature. Provided then, that an optimum balance of setting variables is achieved, the process is self adjusting.

Then from

$$F = \frac{Q_I w}{nW}$$

$$W = \frac{Q_I w}{96\,500\,n} \tag{13}$$

and

$$W = a\rho f$$

Where

a = active area of tool (mm³)

ρ = density of work material (g/mm³)

f = feed rate (mm/s)

Then

$$a\rho f = \frac{Itw}{96\,500\,n}$$

and

$$f = \frac{IW_c}{96\,500\,a\rho} \tag{14}$$

Where W_c = combining weight, i.e. atomic weight/valence.

From Figure 9.25 the gap between tool and work has thickness h and cross-sectional area a, the electrolyte flowing through the gap at Q mm³/s.

Then the heat H to raise m kilograms of electrolyte to boiling point is

$$H = mc(T_b - T_o) \tag{15}$$

where

m = total mass of electrolyte passing through gap (kg),

c = specific heat of electrolyte (J/kg K),

T_b = boiling point of electrolyte (K),

T_o = temperature of electrolyte as it enters gap (K)

Then

$$H = V\rho_e c(T_b - T_o)$$

and

$$H/t = Q\rho_e c(T_b - T_o) = P \text{ (J/s)} \qquad (16)$$

where ρ_e = density of electrolyte (kg/mm³).

Then, if it is permissible for the temperature of the electrolyte to reach its boiling point as it passes through the gap, and this temperature rise is due to power loss (i.e. I^2R loss) in the electrolyte in the gap.

$$I = (P/R)^{\frac{1}{2}} \qquad (17)$$

Now as $R = P_s h/a$
(where P_s = specific resistance of electrolyte (ohm/mm))
Then

$$I = \{[Q\rho_e ca(T_b - T_o)]/P_s h\}^{\frac{1}{2}} \qquad (18)$$

substituting equation (8) in equation (14) then

$$f_\rho = (W_c/96\ 500\ \rho)[(\rho_e Qc/P_s ha)(T_b - T_o)]^{\frac{1}{2}} \qquad (19)$$

Where f_ρ = maximum permissible feed (mm/s).

Thus it is possible, by substituting a given value of F for to determine the temperature rise in the electrolyte.

The metal removal rate varies with the type of operation performed and the composition of the work material. When producing cavities for example, if the tool were sunk into

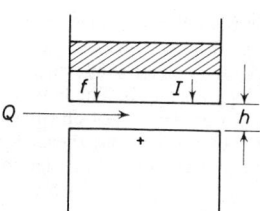

Figure 9.25. Gap setting of electrodes

the work at a rate of 5 mm/min then this would be considered high. Tables giving machining rates for various work materials are available where standard current densities are used.

The type of material used for the cathode depends upon its electrical conductivity, resistance to chemical action, cost and ease of manufacture. Materials such as brass, aluminium, bronze, carbon and stainless steels are commonplace.

ELECTROLYTIC POLISHING

As with electro-chemical machining, the process is essentially the reverse of electro-plating, the material being polished is the anode and is immersed in electrolyte where a cathode is provided to complete the circuit. The process has found reasonable popularity for the industrial polishing of components in recent years.[31, 32, 33, 34]

Electropolishing removes the fine surface scratches on metal as it takes advantage of the fact that the peaks are more easily dissolved than the valleys when subjected to electrolytic action. The surface produced is of high brilliance, it being possible to attain a finish of 0·25 μm to 1 μm if the original surface is not rougher than 2 μm. If on the other hand the surfaces are initially ground then surface finishes of better than 0·05 μm can be achieved.

In order to obtain a high quality finish the correct choice of current density, operating temperature and electrolyte is critical and Walton[35] investigated these characteristics in electrolytic polishing. He considered that the process was dependent upon the formation of a thin viscous layer at the work surface. This layer tends to mask the grooves in the surface thus allowing the electrolyte to selectively attack the high spots, thus progressively improving the surface. As the thin viscous layer spreads over the whole surface electrolytic polishing will not improve the flatness of coarse machined or otherwise imperfect surfaces.

The type of electrolyte used depends on the material being polished. Electrolytes which are mixtures of perchloric acid, acetic anhydride, ethanol, glacial acetic acid, orthophosphoric acid, pyrophosphoric acid or sulphuric acid are commonly used. The electrolyte temperature is in the region of 1 to 90°C with current densities between 0·01 and 7 A/cm^2 with polishing times which rarely exceed 15 min.[32, 34]

THERMO-ELECTRIC PROCESSES

The principal feature of these rather exotic machining processes are the high temperatures and high thermal energy densities which are used. The energy is of such a magnitude as to evaporate the material being machined, two of these processes, electron beam machining (EBM) and laser machining (LBM) will now be considered.

ELECTRON BEAM MACHINING

This process is based on the principle that it is possible to accelerate electrons to very high velocities, i.e. upwards of 160 000 km/sec. The electrons are then focused into a narrow beam by an electric field which can be bent by electrostatic and electromagnetic fields in a similar way to light rays being bent by glass lenses. The high velocity stream of electrons are so focused as to impinge upon a very small spot on the work material, the kinetic energy of the electrons then being transformed into thermal energy which vaporises the material in the local area of the focused spot. In order to prevent the electrons from colliding with gas molecules in the atmosphere, which would scatter the electron beam, the operation is carried out in a vacuum.

A typical system used for generating these high kinetic energy electron streams is shown in Figure 9.26.

Producing the electron beam

An electron gun is used to form the electron beam, the gun consisting of

1. A hot tungsten filament (cathode) which emits high negative potential electrons.
2. A grid cup which is negatively biased in relation to the cathode.
3. An anode at ground potential through which the accelerated electrons pass.

The electron stream emits from the tip of the hot cathode and is accelerated towards the anode by a high accelerating potential between anode and cathode. The extent of the negative bias applied to the grid cup controls the beam current or electron flow, and may be used to turn the beam on or off. The electrostatic field formed by the grid cup prefocuses the electrons which pass as a converging beam through the anode. Immediately the electrons have passed through the anode they have attained maximum velocity for a given accelerating potential, and they maintain this velocity (as they are passing through a collision-free area) until they make impact with the workpiece.

In order to produce a collision-free environment for the transmission of the electron

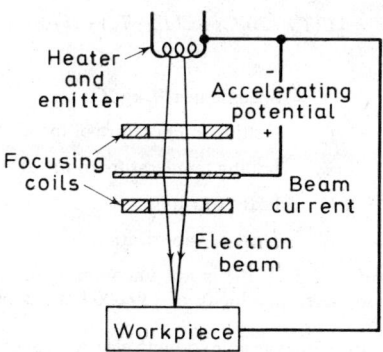

Figure 9.26. Electron beam machining apparatus

beam a vacuum of 1.333×10^{-2} N/m^2 abs. or better is required. Due to the high velocity impingement of electrons, X-rays are emitted thus making it necessary to shield the vacuum chamber in order to absorb this radiation.

The basic equations controlling the performance of the electron are given below:

The kinetic energy per electron is given by

$$KE = \tfrac{1}{2}\,mV^2 = Ee \tag{20}$$

where

$$mg = \text{Weight of electron} = 9.106\,6 \times 10^{-26}$$

$$e = \text{Charge on each electron} = 1.60 \times 10^{-19}\ \text{(Joules)}$$

$$E = \text{Voltage applied}$$

$$V = \text{Electron velocity (cm/s)}$$

$$1\ \text{g}-\text{cm} = 9.807 \times 10^{-5}\ \text{(joules)}$$

The number of electrons/s

$$(N) = In \tag{21}$$

where

$$I = \text{electron beam current (amps)}$$

$$n = 6.3 \times 10^{18}\ \text{electron/s/amp}$$

and the total power $p = EI = EenI$ (watts)

$$e \times n = 1.0 \qquad\qquad 1\ \text{watt} = 1\ \text{joule/sec.}$$

and the metal removal rate may be obtained from:

$$M = \eta P/W \text{ (cm}^3/\text{s)} \tag{22}$$

where

P = power (joules/s),

W = specific energy required to vaporise metal (joule/cm^3),

η = efficiency (cutting).

$$W = [C(T_M - 20°C) + C(T_B - T_M) + H_f + H_v] \tag{23}$$

where:

C = Specific heat J/kg °C

T_M = melting temperature of metal °C

T_B = boiling temperature of metal °C

H_f = heat of fusion

H_v = Heat of vaporisation

A potential of the order of $1·5 \times 10^5$ is not uncommon and by substitution of this value in equation (20) an electron velocity of 230 000 km/s is obtained with a current of $2·5 \times 10^{-5}$ A.

As indicated from Figure 9.26 a variable strength electromagnetic lens is used to refocus the beam to any diameter down to about 0·002 5 mm with precise location on the workpiece. By focusing down the beam to such small dimensions exceptionally high current densities are obtained which immediately vaporise any material in the path of the beam. Mounted below the focusing coils Figure 9.26 is a magnetic deflection coil which is used to bend the beam and thus direct it over the work surface. This facility allows any geometrical profile required to be produced when using the appropriate deflection coil current input.

Advantages and limitations of EBM

This process provides what is perhaps the most precise cutting tool available which can also cut any known material, metallic or non-metallic which can exist in a high vacuum. As all cutting must be done in a vacuum, the degree of vacuum restricts the component size and time must be allowed to evacuate the chamber. This method is very suitable for micromachining as holes down to 0·05 mm diameter can be machined with no cutting tool pressure or wear. Hole diameter to depth ratios of the order of 1 : 200 can be achieved and due to the small beam diameter close tolerances can be held ($\pm 0·012 - 0·05$ mm).

Distortion free machining of thin or hollow parts is possible with no physical or metallurgical damage to the work material. Only relatively small cuts are economically possible as the metal removal rate is approximately only 0·1 mg/s. The holes produced by this method have a slight crater where the beam enters the work and additionally the hole is tapered by about 2 degrees per side. The profile of the hole varies with its depth due to the fact that the beam fans out above and below its focal point thereby tending to produce an hourglass shape.

High capital cost of the equipment plus the need for considerable operator skill and the fact that the process is applicable to thin material rather limits its range of application.

Applications

The process has been applied to such operations as drilling either round or profile shaped holes for metering flow on sleeve valves, rocket fuel injectors of injection nozzles for diesel engines. Further applications have been concerned with producing wire-drawing dies, spinnerets for manufacturing synthetic fibres and drilling gas orifices for pressure differential devices.

LASER BEAM MACHINING

The laser (light amplification by stimulated electron radiation) is a highly mono-chromatic and coherent beam of light having a very high energy density which will cut metal in a similar way to the electron beam.

The light energy emitted by a laser has several important features which distinguish it from other light sources:

(a) Due to the highly monochromatic nature of the laser beam simple lenses may be used for focusing the beam (i.e. colour corrected lenses not required).

(b) The light beam is highly collimated diverging by angles of only 10^{-2} to 10^{-4} radians.

(c) Due to its small beam divergence, all the beam energy can be collected by simple optical system and focused onto a small area.

Basic principles

The orbital electrons of an atom can jump to orbits further away from the nucleus (i.e. higher energy levels) by absorbing quanta of stimulating energy. At this stage the atom is in the 'excited' state and may spontaneously radiate the absorbed energy, at the same time the electron drops back to its original orbit or to some intermediate level. With the atom in the excited condition, if another quanta of energy is absorbed by the electron the two quanta of energy are radiated. The stimulated (radiated energy) and stimulating energies both have the same wavelength, the result being that the stimulating energy is amplified as shown in Figure 9.27 which indicates the principle of laser operation.

In order to contain the stimulating energy in the form of coherent light and intensify it to high power density additional items are required. An early type of laser is shown in

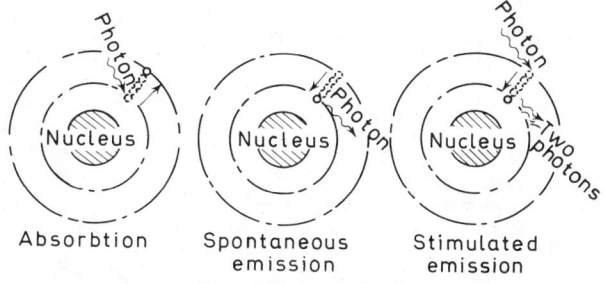

| Absorbtion | Spontaneous emission | Stimulated emission |

Figure 9.27. Laser principle

Figure 9.28 which consists of a xenon-filled tube which provides the excitation source, and a ruby element at the left hand end of which is a reflecting mirror and at the right hand or output end is a partially reflecting mirror. The light radiated by the excited electrons is in phase with the initiating beam and this intensified light is subject to reversals through the laser material (i.e. ruby, gas) stimulating more and more electrons to give up their energy. By suitably adjusting the mirrors there will be a build-up of a single exceedingly powerful clearly defined pattern of standing waves which bursts forth from the partially reflecting end.

Types of laser

The optically pumped solid state and the continuous $N_2 - CO_2$ molecular gas lasers appear to offer the greatest potential for practical industrial application.

Solid state laser

Due to its low beam divergence and the fact that this type of laser produces the highest energy and peak power output of any pulsed laser produced makes it an attractive proposition for machining applications.

Two optically pumped solid state laser materials which find the widest industrial application are ruby and neodymium-in-glass. The ruby laser material being crystalline aluminium oxide (sapphire) whilst the neodymium-in-glass is based on barium crown glass containing between 0·5 and 6 per cent by weight of neodymium (Nd). Both these

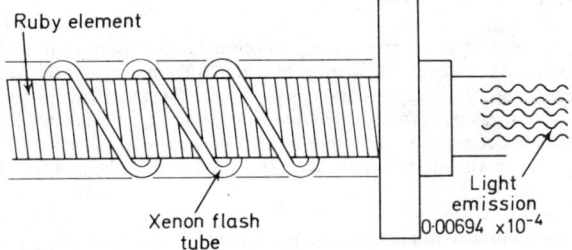

Figure 9.28. An early method of disposing the flash tube (prior to the use of elliptical reflectors)

materials are produced in the form of rods, the ends being finished to a high order of optical tolerance.

Both ruby and Nd-in-glass are electrically insulating materials and consequently must be powered by a method other than simple electrical excitation. The xenon discharge tube is widely used as an optical pumping source which injects energy into the materials by generating a very intense light flux which can be absorbed by the laser material and converted into a collimated laser beam.

The laser rod and lamp are placed in a cylinder of elliptical or circular cross-section in order that the light produced by the lamp is efficiently used. This cylinder focuses the light from the discharge lamp onto the laser rod as shown in Figure 9.29.

The efficiency with which the electrical energy is converted into laser light energy is in the region of 0·5-5 per cent and therefore cooling, usually obtained from air streams, is required to maintain a reasonable temperature in the laser.

The energy output from a typical laser device will be in the region of 20 J, with a pulse of 10^{-3} s duration and a peak power of 20 000 W. Beam divergence for a laser of this type would be in the order of 2×10^{-3} radians indicating the high degree of collimation which can be achieved. The power density attainable is sufficient to vaporise any known metal, but not all of it is removed by vaporisation. The high intensity of the laser beam evaporates a small proportion of the liquid metal so rapidly that a sizeable impulse is transmitted to the liquid causing it to leave the working area at relatively high velocity.

In view of the ability of a laser to cut through optically transparent materials and melt or vaporise any known metal independent of atmospheric environment, it is preferable to EBM in many cases. As in the case of EBM small heat affected zones occur and little thermal damage exists. It is, however, difficult to control hole size, especially when drilling deep holes, due to rewelding of incompletely expelled material, and the holes produced are not always round or straight.

The low efficiency of the laser plus the short life expectancy of the discharge tube and

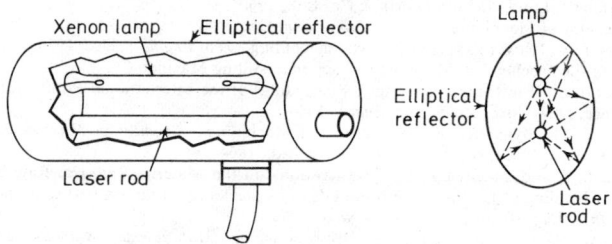

Figure 9.29. Elliptical focusing structure using linear lamp

maintenance costs make the process very costly to operate and is generally most econom-cal when the volume of material to be removed is small. Although lasers have improved considerably in design and performance, further improvements in efficiency, laser material, pumping, cooling and optical systems are required if laser machining is to find a broader field of application to industrial machining operations.

REFERENCES

1. NEPPIRAS, E. A. and FOSKETT, R. D., 'Ultrasonic Machining', *Philips Technical Review* (1956/57)
2. NEPPIRAS, E. A., 'Research' (January 1955)
3. NEPPIRAS, E. A., 'Report on Ultrasonic Machining', *Metalworking Production* (1956)
4. ROZENBERG, L. D., *et al.*, 'Ultrasonic Cutting', Moscow (1962)
5. MARKOV, A. I., 'Machining Intractable Materials with Ultrasonic and Sonic Vibrations', Moscow (1962)
6. KOPS, L., 'Ultrasonic Machining and Machine Tools', Warsaw (1961)
7. BROWN, B., 'Recent developments in the application of ultrasonics to machining processes', *Machinery*, Lond. (1965)
8. BALAMUTH, L., 'Mechanical Impedance Transformers in Relation to Ultrasonic Machining', *Trans. Inst. Radio Engs.*, PGUE-Z (Nov. 1954)
9. HOOVER, R. M., 'High Power Operation of a Magnetostrictive Transducer', *J. of Acoust. Soc. Amer.* (1956)
10. CAMP, L., 'Lamination Designs for Magnetostrictive Underwater Electroacoustic Transducers', *J. Acoust. Soc. Amer.* (1948)
11. NEPPIRAS, E. A., 'Design of ultrasonic machine tools', Conference on Technology of Manufacture, I.Mech.E. (1958)
12. GOETZE, D., 'Effect of vibration amplitude frequency and composition of abrasive slurry on the rate of ultrasonic machining Ketos* Tool Steel', *J. Acoust. Soc. Amer.* (1956)
13. GOETZE, D., 'Effect of Pressure between Tool Tip and Workpiece on the Rate of Ultrasonic Machining in Ketos Tool Steel', *J. Acoust. Soc. Amer.* (1957)
14. NEPPIRAS, E. A., 'Ultrasonic Machining and Forming', *Ultrasonics* (1964)
15. PENTLAND, W., and EKTERMANIS, J. A., 'Improving Ultrasonic Machining Rates', *Trans. Amer. Soc. Mech Engrs. Series B. J. of Eng. for Ind.* (1965)
16. ROZENBERG, L. D., KAZANTSEV, V. F. and MECHETSER, B., 'Improving the Efficiency of Ultrasonic Cutting', *Soviet Physics: Doklady* (1966)
17. Spark Machining, *Machinery's Yellow Back Series, No. 46*
18. BARASH, M., 'Electric Spark Machining', *Inst. J. Mach. Tool Des. and Res.* (1962)

19. LAWRENCE, A. J., 'The application of spark erosion machining', *J.I. Prod. E.*, Nov. 1958.
20. KRABACHER, E. J., HAGGERTY, W. A., ALLISON, C. R. and DAVIS, M. F., 'Electrical methods of machining', International Research in Production Engineering (International Production Engineering Research Conference, Pittsburgh 1963)
21. BARASH, M., 'Some properties of spark-eroded surfaces', *Microtecnic* (1959)
22. OPITZ, H., 'New developments in Spark-Erosion Machining', *Microtecnic* (1959)
23. OPITZ, H., 'Electrical Machining Processes', International Research in Production Engineering (International Production Engineering Research Conference, Pittsburgh 1963)
24. LIVSKITS, A. L., *The Electro-Erosion of Machining of Metals* (Translated by E. Bishop: Editor R. S. Bennett), London: Butterworth & Co. Ltd., 1960
25. 'Electrochemical Machining', *Aircraft Production*, Dec. 1962
26. WILLIAMS, L. A., 'How to apply electrolytic machining', *Tool Engineer*, Dec. 1959
27. Report on Electrochemical Machining, Cincinnati Milling Machine Co.
28. PAHLITZSCH, G., 'Electrolytic Grinding', Proceedings of the International Production Research Conference, New York: *Amer. Soc. Engrs.*, 1964
29. TIPTON, H., 'The dynamics of Electrochemical Machining', *Advances in Machine Tool Design and Research*, p. 509, Oxford: Pergamon Press Ltd., 1964
30. KASZMAREK, J. and ZACHWIEJA, T., 'Investigations on the Material Removal Rate by Electrochemical Grinding of Cutting Tool Materials in Dependence on the Properties of the Grinding Wheel', *Inst. J. of Mach. Tool Des. and Res.* (1966)
31. RILEY, M. W., 'Where and When to Use Electropolishing', *Materials and Methods* (June 1954)
32. 'Conditions for Electrolytic Polishing of Metals and Alloys' *Materials and Methods* (1954)
33. STEER, A. T., 'Electropolishing—Its Influence on the Fatigue Endurance Limit of Ferrous and Non-ferrous Parts', *Aircraft Production* (July 1953)
34. BOYLE, S. C., 'The Principles of Electropolishing, Electrodeburring and Electromachining', *Metal Finishing Journal* (Nov. 1962)
35. WALTON, H. F., 'The Anode Layer in Electrolytic Polishing of Copper', *Journal of the Electrochemical Soc.* (July 1950)
36. Electron beam process. *Aircraft Prod.*, Lond. (1960)
37. *Electron beam techniques for microelectronics*, R.R.E. symposium, Malvern (1964)
38. DE BARR, A. E., 'Modern metal removal techniques', *The Chartered Engineer* (1966)
39. SCHWARZ, H., 'Power density of Optimally focused space-charge limited electron beams', *J. Applied Physics* (1962)
40. BURTON, G. and MATCHETT, R. L., 'Welding, Drilling and Machining by Electron Beam bombardment', *American Machinist* (Feb. 23, Mar. 9, 1959)
41. LEINWALL, S., *Understanding Lasers and Masers*, Iliffe Books (1965)
42. MICROSA, R., 'Laser in metalworking', *Microtecnic* (1965)
43. NEUROTH, N., 'The operation and application of lasers', *Chemie-Ingr-Tech.* (1964)
44. SCHMIDT, A. O., HAM, I. and MOSHI, T., 'Laser Features and applications to micro-machining and micro-welding', *Advances in Machine Tool Design and Research*, Oxford: Pergamon Press, 1964
45. ADAMS, C. M. and HARDWAY, G. A., 'Fundamentals of Laser Beam Machining and Drilling', I.E.E.E. Trans. on Ind. and General Applications, I.G.A.-I. (March–April 1965)

10 PIPEWORK

PIPEWORK 10

10 PIPEWORK

STEEL PIPES AND FLANGES

A. PARRISH

In 1940, ISA Bulletin 5a[1] was prepared and this became the basis of European metric pipework standards. The International Standards Association (ISA) was replaced by the International Standards Organisation (ISO) and recommenced work in 1951 on the standardisation of steel pipes and flanges. The U.K. became a Member Body of ISO and it was thus necessary to consider both inch and metric standards.

International recommendations were agreed for the sizes of pipes and tubes and it was found possible to set 'corresponding values' in inches and millimetres for the outside diameters and wall thicknesses of most pipes and tubes in general use. These sizes ensured the interchangeability of both inch and metric pipes within the limits of the existing manufacturing tolerances.

No agreement was possible for flanges because the European, American and British standards were not interchangeable.

Basic ISO sizes for pipes and tubes

The descriptions 'pipe' and 'tube' are considered to be synonymous and both are used.

Outside diameters. The range of corresponding values in inches and millimetres is listed in ISO/R64[3] (non-screwed tubes) and ISO/R65[4] (screwed tubes).

Wall thicknesses. The range of corresponding values in inches and millimetres is listed in ISO/R221[5].

General dimensions. The outside diameter and wall thickness sizes in ISO/R64 and ISO/R221 are combined in ISO/R336[7] (carbon and alloy steel plain end tubes) and ISO/R1127[9] (stainless steel plain end tubes). Tolerances and masses per unit length are specified.

Pipe threads. Suitable for pipes up to 150 mm (6 in) nom size to ISO/R65 are given in ISO/R7[2]. These are based on British Standard pipe threads (BS 21)[11]. The above apply where a pressure tight joint is made on the thread.

A series of pipe threads where pressure tight joints are not made on the threads is given in ISO/R228[6] (BS 2779)[26].

Where screwed pipes are required in sizes above 150 mm (6 in) it is recommended that the threads should conform to ASA B2.1[41] or API Std. 5B[42] for line pipe threads.

Precision tubes which are cold drawn to close tolerances, suitable for compression fittings, are listed in metric sizes in ISO/R560[8] (BS 4368 Pt. 1)[38].

British Standards

There are two series of British Standards for metric pipes, flanges and fittings which are not interchangeable.

(a) Based on ISO recommendations and German (DIN) standards using ISO metric bolts and nuts.

(b) Based on the standards of the American National Standards Institute (ANSI)[41]. These are inch standards converted to exact metric equivalents using ISO inch (Unified) bolts and nuts. The pipe outside diameter sizes, however, are included in ISO/R64 and ISO/R336 and in the range of nominal sizes above 150 mm are (apart from tolerances) the same as those in (a). The pipe wall thicknesses are not included in ISO recommendations. ANSI standards are established in common use throughout the world (except the USSR and eastern Europe) in producing and refining activities and in many instances ISO is considering their adoption as the bases of ISO standards.

Pipe diameters. See Table 10.1 for the outside diameters of steel pipes and tubes for nominal sizes 6–300 mm ($\frac{1}{8}$–12 in)

Table 10.1 OUTSIDE DIAMETERS OF STEEL PIPES AND TUBES (6–300 mm nom. size)

		Screwed to BS 21 (ISO/R 7) Heavy and medium series BS 1387 (ISO/R 65)		Plain end pipes		
Nominal size				BS 3600 ISO/R 64 ISO/R 336	BS 1600 : Part 2 ISO/R 64 ISO/R 336	Non ISO
mm	in	*Max*	*Min*			
6	$\frac{1}{8}$	10·6	9·8	10·2		
	$\frac{1}{8}$					10·3
8	$\frac{1}{4}$	14·0	13·2	13·5		
	$\frac{1}{4}$					13·7
	$\frac{3}{8}$					17·1
10	$\frac{3}{8}$	17·5	16·7	17·2		
15	$\frac{1}{2}$	21·8	21·0	21·3	21·3	
	$\frac{3}{4}$					26·7
20	$\frac{3}{4}$	27·3	26·5	26·9		
	1					33·4
25	1	34·2	33·3	33·7		
32	1$\frac{1}{4}$	42·9	42·0	42·4	42·4*	
40	1$\frac{1}{2}$	48·8	47·9	48·3	48·3	
50	2	60·8	59·7	60·3	60·3	
	2$\frac{1}{2}$					73·0*
65	2$\frac{1}{2}$	76·6	75·3	76·1		
80	3	89·5	88·0	88·9	88·9	
90	3$\frac{1}{2}$	102·1	100·4	101·6	101·6*	
100	4	115·0	113·1	114·3	114·3	
125	5	140·8	138·5	130·7		
	5					141·3*
150	6	166·5	163·9	165·1		
150	6			168·3	168·3	
175	7	Pipes 168·3 mm outside		193·7*		
200	8	diameter and larger may		219·1	219·1	
225	9	be screwed to ASA B2.1		244·5*		
250	10	or API Std 5B for line		273·0	273·0	
300	12	pipe threads.		323·9	323·9	

*Avoid using these sizes.

Note. The above nominal sizes approximate to the nominal bore sizes. Nominal pipe sizes are used in BS 1600, Pt. 2 which approximate to the nominal bore sizes in inches but are the same as the outside diameters in millimetres.

(a) to BS 1387[14], screwed to BS 21
    BS 3600[30], plain end pipes
(b) to BS 1600, Part 2[19], plain end pipes.

The outside diameters of steel pipes and tubes for nominal sizes from 350–1000 mm (14–40 in) to BS 1600, Part 2 and BS 3600, are given in Table 10.2.

Table 10.2 OUTSIDE DIAMETERS OF STEEL PIPES (350–1000 mm nom. size)

BS 1600 *and* BS 3600 (*Metric*)
ISO/R 64 *and* ISO/R 336

Nominal size		Outside dia
mm	in	mm
350	14	355·6
400	16	406·4
450	18	457·2
500	20	508
550	22	558·8
600	24	609·6
650	26	660·4
700	28	711·2
750	30	762
800	32	812·8
850	34	863·6
900	36	914·4
1 000	40	1 016

Note 1. Welded pipes can be obtained in sizes up to 2120 mm outside diameter and in thicknesses up to 32 mm.

Note 2. The inch nominal sizes are equal to the outside diameters; the metric nominal sizes are rounded from the outside diameters.

Note 3. Nominal pipe sizes are used in BS 1600, Pt. 2 which are the same as the outside diameters in millimetres.

Pipe thicknesses. Table 10.3(a) gives the range of pipe wall thicknesses given in BS 3600 for nominal sizes 6–1000 mm. ($\frac{1}{8}$–40 in.). The detailed sizes available commercially in the various materials are shown in Table 10.3(b).

The pipe thicknesses in BS 1600, Part 2 are based on schedules formulated by the

Table 10.3(a) WALL THICKNESSES OF WELDED AND SEAMLESS STEEL PIPES (BS 3600)

Nominal size		Wall thicknesses	
mm	in	Seamless and alloy steels mm	Austenitic stainless steels mm
6	$\frac{1}{8}$	1·2–2·6	1·2–2·3
8	$\frac{1}{4}$	1·2–3·6	1·2–2·6
10	$\frac{3}{8}$	1·2–4·5	1·2–2·9
15	$\frac{1}{2}$	1.2–5.4	1·2–3·2
20	$\frac{3}{4}$	1·2–7·1	1·2–5
25	1	1·2–8·8	1·2–6·3
32	$1\frac{1}{4}$	1·2–·8·8	1·2–6·3
40	$1\frac{1}{2}$	1·2–11	1·2–6·3
50	2	1·4–14·2	1·6–6·3
65	$2\frac{1}{2}$	1·8–20	1·6–6·3

Table 10.3(a)—*continued*

Nominal size		Wall thickness	
mm	in	Seamless and alloy steels mm	Austenitic stainless steels mm
80	3	1·8–22·2	1·6–6·3
90	3½	2·3–25	1·6–6·3
100	4	2·3–25	1·6–6·3
125	5	3·2–25	1·6–6·3
150	6	3·2–25	1·6–6·3
*175	7	3·2–25	
200	8	3·2–25	2·0–6·3
*225	9	3·2–25	
250	10	3·2–25	2·0–6·3
300	12	3·2–25	2·0–6·3
350	14	3·2–25	2·0–6·3
400	16	3·2–25	2·0–6·3
450	18	3·2–25	2·0–6·3
500	20	3·2–25	2·0–6·3
550	22	3·2–25	
600	24	4·0–25	
650	26	4·5–25	
700	28	4·5–25	
750	30	5·0–25	
800	32	5·4–25	
850	34	5·4–25	
900	36	6·3–25	
1 000	40	6.3–25	

*Avoid using these sizes.

Note 1. Seamless carbon and alloy steel pipes 100 mm up to and including 450 mm nominal size can be obtained in thicknesses up to 25 per cent of the outside diameter.

Note 2. Larger welded carbon and alloy steel pipes can be obtained in sizes up to 2121 mm outside diameter and in thicknesses up to 32 mm.

Note 3. The thicknesses for seamless austenitic stainless steel pipes over 6·3 mm should be selected from the sizes listed for carbon and alloy steel pipes.

Note 4. See Table 10.3(b) for the detailed range of thicknesses.

Note 5. The above are extracted from BS 3600 and conform to ISO/R336 and ISO/R1127.

American Petroleum Institute (API) where it is the practice to refer to the wall thickness by a schedule number. These schedules bear a relation to the pressure rating of the piping and are eleven in number ranging from the lowest at 5 through 10, 20, 30, 40, 60, 80, 100, 120, 140 to schedule no. 160. For tubing 150 mm (6 in) nominal size and smaller, schedule 40 is the lightest which is specified. Only schedules 40 and 80 cover

Table 10.3(b)   ISO THICKNESS SIZES (mm)

Materials	Thickness sizes									
Carbon and alloy steels					1·2	1·4	1·6	1·8	2·0	2·3 2·6
Austenitic stainless steels					1·2		1·6		2·0	2·3 2·6
Pipes for compression couplings	0·5	0·6	0·8	1·0	1·2		1·6		2·0	2·3 2·6

continued

Table 10.3(b)—*continued*

Materials	Thickness sizes									
Carbon and alloy steels	2·9	3·2	3·6	4·0	4·5	5·0	5·4	5·6	5·9	6·3 7·1
Austenitic stainless steels	2·9	3·2	3·6	4·0	4·5	5·0		5·6		6·3
Pipes for compression couplings	2·9	3·2	3·6	4·0	4·5	5·0				

Materials	Thickness sizes									
Carbon and alloy steels	8·0	8·8	10	11	12·5	14·2	16·0	17·5	20	22·5 25

Note. The above are taken from BS 3600 and conform to ISO/R 64, ISO/R 336 and ISO/R 1127. The sizes from 1·2 mm to 5·9 mm are derived from the Imperial Standard Wire Gauges (SWG). All larger sizes (and 5·6 mm) are based on inch fractions.

Table 10.4 STEEL PIPES—API SCHEDULE NUMBERS (BS 1600)

Nominal** pipe size		Wall thicknesses in millimetres							
in	mm	Sch. 10	Sch. 20	Sch. 30	Std. wall	Sch. 40	Sch. 60	Sch. 80	Sch. 100
⅛	10·3				1·73	1·73		2·41	
¼	13·7				2·24	2·24		3·02	
⅜	17·1				2·31	2·31		3·20	
½	21·3				2·77	2·77		3·73	
¾	26·7				2·87	2·87		3·91	
1	33·4				3·38	3·38		4·55	
*1¼	42·4				3·56	3·56		4·85	
1½	48·3				3·68	3·68		5·08	
2	60·3				3·91	3·91		5·54	
*2½	73·0				5·16	5·16		7·01	
3	88·9				5·49	5·49		7·62	
*3½	101·6				5·74	5·74		8·08	
4	114·3				6·02	6·02		8·56	
*5	141·3				6·55	6·55		9·52	
6	168·3				7·11	7·11		10·97	
8	219·1		6·35	7·04	8·18	8·18	10·31	12·70	15·09
10	273·0		6·35	7·80	9·27	9·27	12·70	15·09	18·26
12	323·9		6·35	8·38	9·52	10·31	14·27	17·47	21·44
14	355·6	6·35	7·92	9·52	9·52	11·13	15·09	19·05	23·82
16	406·4	6·35	7·92	9·52	9·52	12·70	16·64	21·44	26·19
18	457·2	6·35	7·92	11·13	9·52	14·27	19·05	23·82	29·36
20	508·0	6·35	9·52	12·70	9·52	15·09	20·62	26·19	32·54
22	558·8	6·35	9·52	12·70	9·52	15·88	22·22	28·58	34·92
24	609·6	6·35	9·52	14·27	9·52	17·48	24·61	30·96	38·89
26	660·4	7·92	12·70		9·52				
28	711·2	7·92	12·70	15·88	9·52				
30	762·0	7·92	12·70	15·88	9·52				
32	812·8	7·92	12·70	15·88	9·52	17·48			
34	863·6	7·92	12·70	15·88	9·52	17·48			
36	914·4	7·92	12·70	15·88	9·52	19·05			

*Avoid using these sizes.
**For corresponding ISO Nominal Sizes in millimetres see Table 10.3(a).

Note. None of these wall thicknesses conforms to ISO/R 221 and ISO/R 336 cf. (Table 10.4).

The above are extracted from BS 1560 and BS 1600 (Pts. 2) and are based on ANSI Std B 36·10 and B 36·19.

the full range from 15 mm ($\frac{1}{2}$ in) up to 600 mm nom. sizes. A selection is shown in Table 10.4

Calculated pipe thicknesses

Formulae are given in BS 806[12] and BS 3351[29] for calculated pipe thicknesses. These can be used with SI units as follows:

BS 806, CLAUSE 2.2

$$tf = \frac{pD}{2fe+p} \tag{1}$$

where tf = minimum thickness in mm.

p = design pressure in N/m^2 (1 bar = 10^5N/m^2).

D = outside diameter of pipe in mm.

f = maximum permissible design stress in N/m^2

e = 1·0 for seamless and ERW pipes.

= 0·95 for EFW pipes to BS 3602 with weld fully radiographed.

= 0·90 for welded pipes to BS 3601 other than ERW.

tf is the minimum thickness for straights and provision should be made for minus tolerances.

BS 3351, CLAUSE 5.2.1

Applicable to ferritic steel pipe at metal temperatures up to 600°C and austenitic steel pipe up to 825°C.

$$t = \frac{pD}{20S+p} \tag{2}$$

where t = internal pressure design thickness in millimetres.

p = internal design pressure in bar, (10^5N/m^2).

D = outside diameter of pipe in millimetres.

S = design stress in N/mm^2, (MN/m^2).

Pipe with t equal to or greater than $D/4$ requires special consideration.

Thicknesses may need to be increased to cover wastage from chemical action, oxidation, scaling, abrasion, erosion and for threaded or grooved connections. Table 3 in BS 3351, (see Table 10.5) gives values of S appropriate to design temperatures applicable to pipes to BS 3601-5 and API 5L. The design stress values for pipe with a longitudinal or spiral welded seam are adjusted for weld joint factors. Conventional mechanical and internal pressure loads are provided for in this design procedure. Reference should be made to BS 1500[15] and BS 1515 for piping design subject to external pressure.

Pipe specifications and materials

BS 3601 to 3605[31-35] specify the materials, properties and tests for carbon, alloy and austenitic stainless steel pipes which are manufactured by various processes, as follows.

CDS *Cold drawn seamless.* Pipe is manufactured by hot working without a welded seam and finished to close tolerances and size by cold working.

HFS *Hot finished seamless.* Pipe is manufactured by hot working without a welded seam.

Table 10.5 DESIGN STRESSES FOR FERROUS PIPES

Values of S in N/mm^2 for metal temperatures in °C not exceeding*

Material	50	100	150	200	250	300	350	400	425	450	475	500	525	550	575	600
BS 3601 HFS, CDS, Steel 22	125·5	114·5	103·9	99·3	94·5	89·7	85·0	75·0	65·7	56·9						
BS 3601 HFS, CDS, Steel 27	154·5	141·3	127·5	121·7	114·2	110·0	104·0	87·7	75·8	62·0						
API 5L Grade A, seamless; steels open hearth electric furnace and basic oxygen	122·5	111·9	101·3	95·8	91·3	86·5	82·0	73·1	64·8	55·8						
API 5L Grade B, seamless, submerged arc, spiral weld; steels open hearth, electric furnace and basic oxygen	153·0	139·8	126·5	119·0	113·8	108·5	102·5	89·3	75·5	62·0						
BS 3602 CDS, HFS and ERW, steel 23	130·0	125·5	120·6	117·3	111·5	96·5	88·9	78·2	67·9	57·7	47·5	35·8				
BS 3602 CDS, HFS and ERW, steel 27	154·5	143·0	132·0	127·0	119·0	110·0	104·2	89·7	75·9	62·3	48·6	35·8				
BS 3602 EFW Grade 28 B	123·5	116·5	109·6	104·0	95·2	90·4	87·0	73·8	62·0	50·3	39·3	24·8				
BS 3603 HFS, CDS, steel 27 LT 30	154·5	143·3	132·0	126·5	119·0											
BS 3603 HFS, CDS, steel 27 LT 50	154·5	143·3	132·0	126·5	119·0											
BS 3603 HFS, CDS, steel 503 LT 100	159·0	147·8	135·3	129·3	121·9											
BS 3604 620 HFS, CDS, steel 27	166·5	144·2	144·0	139·0	124·5	114·0	109·5	104·0	101·5	98·5	92·3	81·0	62·8	43·8	28·9	17·2
BS 3604 621 HFS, CDS, steel 27	144·5	144·2	144·0	139·0	124·5	115·5	110·5	106·0	103·8	100·0	92·8	81·2	64·3	49·3	34·2	24·4
BS 3604 622 HFS, CDS, steel 27	144·5	134·2	129·6	125·0	120·5	115·5	111·0	106·0	103·8	101·0	92·5	82·8	63·8	47·5	36·2	26·8
BS 3604 622 HFS, CDS, steel 31	164·0	164·0	164·0	164·0	156·5	150·5	144·0	137·2	134·0	130·2	107·5	82·8	63·8	47·5	36·2	26·8
BS 3604 660 CDS, HFS	172·0	172·0	167·0	164·0	156·5	150·5	144·0	137·2	134·0	130·2	127·5	115·6	79·9	53·7	36·9	
BS 3604 625 HFS	137·0	131·0	125·5	119·5	114·0	108·5	103·0	97·0	92·8	86·5	81·0	72·4	59·0	43·8	31·3	21·1

Values of S in N/mm² for metal temperatures in °C not exceeding*

Material	−200 to +50	100	150	200	250	300	350	400	425	450	475	500	525	550	575	600	625	650	675	700	725	750	775	800	825
BS 3605 Grade 801 (304)	126·0	113·5	103·5	95·0	87·5	81·9	76·3	71·7	68·9	66·8	64·8	63·3	61·3	59·7											
BS 3605 Grade 811 (304H)															55·8	48·0	37·6	30·7	22·4	17·6	13·4	10·3	7·9	6·2	4·5
BS 3605 Grade 801L (304L)	106·8	102·8	90·3	76·5	68·2	63·3	60·0	57·2	55·8																
BS 3605 Grade 822Ti (321)	129·3	127·5	117·0	108·5	105·5	102·8	102·0	101·3	100·5	99·0	97·5	96·0	94·0												
BS 3605 Grade 832Ti (321H)														91·7	88·5	75·8	51·3	34·5	25·5	19·6	15·2	11·8	9·3	7·6	6·5
BS 3605 Grade 845 (316) and Grade 845Ti and Grade 846 (317)	129·3	128·2	123·5	120·6	118·7	118·0	117·5	116·5	115·5	114·0	111·5	106·0	99·3	90·3	79·3	67·6	56·2	46·6	36·8	28·3	22·0	17·6	14·4	11·8	9·6
BS 3605 Grade 845L (316L)	107·5	106·2	100·0	83·4	79·3	74·3	66·9	62·3	60·7	58·6															

*Intermediate values may be obtained by linear interpolation.

Note. The figures in parenthesis alongside the grades are equivalent AISI types. There is no AISI equivalent of 845Ti. (Extracted from BS 3351. Piping Systems for petroleum industry, 1971, by permission of the British Standards Institution, 2 Park Street, London W1A2BS.)

EFW   *Electric fusion welded.* The abutting edges of steel plate formed into a tube are joined by manual or automatic electric arc welding with or without a filler metal.

SFW   *Spiral seam welded.* Steel strip, sheet or plate is formed into a tube with a helical seam around its circumference. The seam is electric fusion welded by an automatic metal arc welding process, or by electric resistance welding.

ERW   *Electric resistance welded.* Steel strip formed into a tube and resistance welded continuously along the longitudinal seam by passing an electric current across the abutting edges without the addition of filler metal.

BW   *Butt-welded.* A continuous steel strip is heated in a tunnel furnace to forge welding temperature and passed through a series of rolls in which it is formed into a tube and the abutting edges forge welded together.

Electric induction welded. Steel trip or plate formed into a tube and longitudinally induction welded continuously by passing an electric current across the abutting edges without adding filler metal.

Electric flash welded. Steel strip or plate formed into a tube and longitudinally flash welded simultaneously along the full length by passing an electric current across the abutting edges without adding filler metal.

BS 3601[31] applies to ordinary duties for the conveyance of gaseous and liquid fluids at pressures up to 20·7 bar at 260°C max.

BS 3602[32] applies to carbon steel pipes for high duty applications and guidance is given in appendices of special supplementary requirements and the creep and rupture

Table 10.6   SCREWED STEEL PIPE JOINTS WITH BSP (ISO R.7) THREADS
PERMISSIBLE WORKING PRESSURES

Pipes.   BS 1387, heavy weight and medium weight.
API 5L Grade B, seamless (equivalent to BS 3601 Steel 27) with dimensions to BS 1600 Pt 2. Sch. 40 and 80.
Screwing.   BSP threads to BS 21 (ISO/R7).

| Nominal size | | Maximum non-shock service pressures bar | | | | | | | |
| | | R/Rp threads* | | | | R/Rc threads* | | | |
mm	in	Heavy weight	Medium weight	Sch. 80	Sch. 40	Heavy weight	Medium weight	Sch. 80	Sch. 40
up to 25	up to 1	12·06	10·34	12·06	—	20·68	17·23	20·68	—
40	1½	10·34	8·61	10·34	—	17·23	13·78	17·23	—
50	2	8·61	6·89	—	8·61	15·51	12·06	—	15·51
80	3	8·61	6·89	—	8·61	13·78	10·34	—	13·78
100	4	6·89	5·17	—	6·89	10·34	8·61	—	10·34

*R = External taper thread on pipe.
Rp = Internal parallel thread in flange.
Rc = Internal taper thread in flange.

Note. The above are maximum pressures for temperatures up to 260°C assuming a satisfactory joint can be made. If necessary they must be decreased to suit; the rating of flanges or other equipment the fluid being conveyed; conditions, i.e. shock (R/Rc threads are better than R/Rp threads under shock conditions).

properties at elevated temperatures (over 426·6°C). The pipes for use at these elevated temperatures are silicon killed and are further designated by the letter H.

BS 3603[33] is specific to low temperature service and includes appendices giving details of Charpy V notch impact tests at specified sub-zero temperatures.

BS 3604[34] covers low and medium alloy steel pipes and includes appendices giving the long term elevated temperature properties of the materials.

BS 3605[35] specifies austenitic stainless steel pipes for heat and corrosion duties.

BS 3600[30] gives dimensions based on ISO recommendations applicable to pipes in accordance with BS 3601–5.

BS 1387[14] covers the dimensions of service pipes with screwed and socketed joints or with plain ends suitable for screwing with threads to BS 21[11], and is based on ISO recommendations. These pipes are suitable for use with wrought steel pipe fittings to BS 1740: Part 1[40]. Table 10.6 shows recommended non-shock service pressures for the heavy weight and medium weight qualities when used with screwed pipe joints. It should be noted that the 165.1 o.d. size should be used for screwing to BS 21 and not the 168.3 o.d. size (see Table 10.1).

BS 1600 PART 2[19] gives dimensions of steel pipe converted from ANSI standards and applies mainly to pipes manufactured to API and ASTM[43] specifications for materials.

Flanged pipe

FLANGES

Three series of flanges for steel pipes are covered in British Standards:
BS 10[10] in Imperial units.
BS 1560 Pt. 2[18], a metric version of ASA[41] standards.
BS 4504[39], a metric standard based on European practice and ISA Bulletin 5a.
BS 10 is obsolescent and will not be metricated, but will remain for many years.
BS 1560 Pt. 2 is technically the same as BS 1560 Pt. 1 (Imperial units) which has world wide usage. This has been submitted to ISO by BSI as the basis of an ISO standard. Sizes above 609.6 o.d. (24 in nominal) are covered in BS 3293 which is being considered for metrication.

BS 4504 flanges are in many cases identical with the German (DIN) flanges but in some cases there is only partial alignment. Generally, there are no areas of exact agreement between BS 4504 and BS 10, although in some cases it is possible to drill blank flanges in either system to suit the other.

ISO/R2084 (1971) gives the mating dimensions for pipeline flanges for general use.

PRESSURE/TEMPERATURE RATINGS

The flange standards give pressure/temperature ratings for moderate shock duties covering a range of recommended materials. The average trends are shown in Figure 10.1, but reference should be made to the appropriate specification for actual values.

NOMINAL SIZE

This description is based on the pipe and is used to identify the pipes, valves and fittings in a system. It does not define the actual dimensions, which for a pipe are generally governed by the outside diameter, wall thickness and manufacturing tolerances. It should be noted that in BS 4504 and BS 3600 the nominal size in millimetres is approximately equal to the bore size. In BS 1560, Part 2 (appendix) and BS 1600, Part 2, the description 'nominal pipe size' (in millimetres) is a metric conversion of the nominal pipe size in inches and these metric sizes do not line up with those in BS 4504 and BS 3600 for similar pipes.

It is expected, however, that the nominal sizes will continue to be expressed in inches.

BOLTING

BS 1750[20] provides information for a wide range of bolting both ISO metric and ISO inch and gives recommended temperature ranges for the materials. Appendix E refers

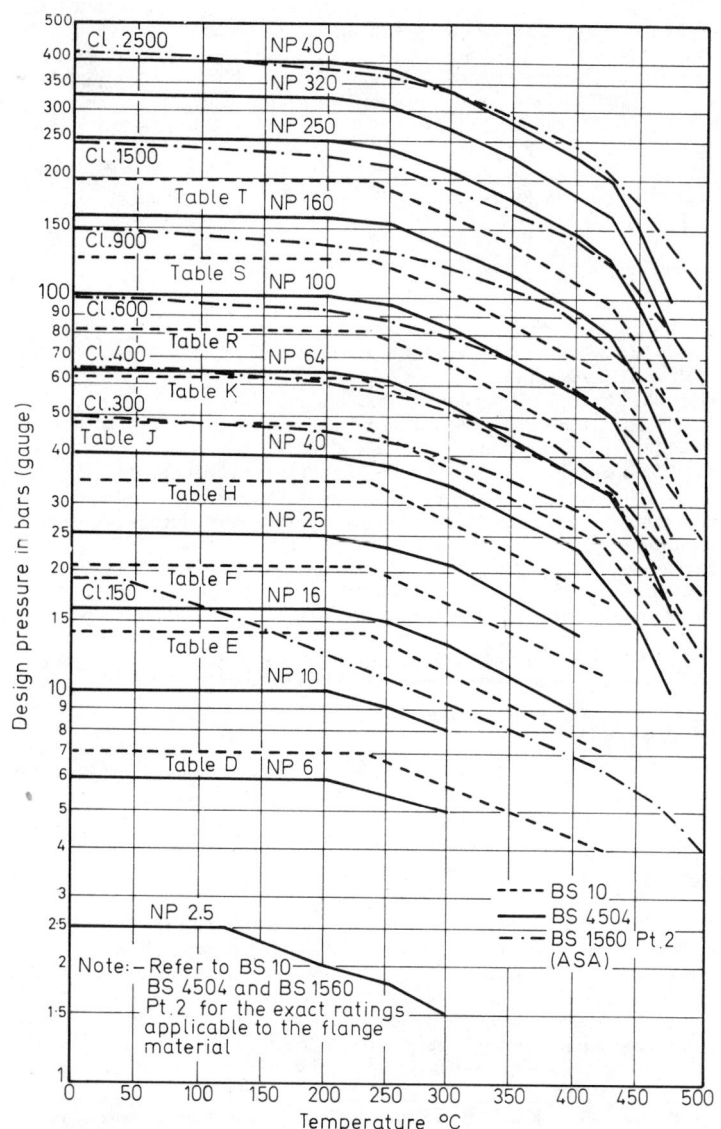

Figure 10.1. Pressure/temperature ratings for steel flanges

to appropriate standards for nuts and bolts in the ISO metric series and Appendix F to the ISO inch series.

ATTACHMENT OF FLANGES

Flanges should be attached to pipes suitable for the pressure/temperature ratings of the flanges. Guidance may be obtained from appropriate application standards such as BS 806[12], 3351[29] and welding standards such as BS 1821[23], 2633[24], 2640[25] and 2971[27].

FLANGED PIPE JOINTS

The flange standards show a variety of flange types and it is necessary to make a selection from these appropriate to the end service for which the pipes are required. The pipe thicknesses may be calculated for the specific duties and an appropriate class of pipe

Table 10.7   LIST OF STEEL FLANGE PIPE JOINTS

Joint table No.	Type	Flange specification			Piping BS No.†	Joint figure No.
		BS No.	Class No.	Table No.		
10·8	Screwed boss	4504	—	6/4	1387	10·2
10·9a	Screwed boss	4504	—	16/4	1387	10·2
10·9b	Screwed boss	4504	—	16/4	1387	10·2
10·10	Screwed boss	4504	—	40/4	1387	10·2
10·11	Screwed boss	1560 Pt. 2	150	—	1600 Pt. 2	10·3
10·12	Screwed boss	1560 Pt. 2	300	—	1600 Pt. 2	10·3
10·13	Plate	4504	—	6/3	3600	10·4
10·13	Slip-on boss	4504	—	6/5	3600	10·5
10·14	Plate	4504	—	16/3	3600	10·4
10·14	Slip-on-boss	4504	—	16/5	3600	10·5
10·15	Plate	4504	—	40/3	3600	10·4
10·15	Slip-on boss	4504	—	40/5	3600	10·5
10·16	Welding neck	4504	—	40/2	3600	10·6
10·17	*Welding neck	4504	—	40/2	3600	10·6
10·18	Welding neck	4504	—	100/2	3600	10·6
10·19	Slip-on-welding	1560 Pt. 2	150	—	1600 Pt. 2	10·7
10·20	Welding neck	1560 Pt. 2	150	—	1600 Pt. 2	10·8
10·21	*Slip-on-welding	1560 Pt. 2	150	—	1600 Pt. 2	10·7
10·21	*Welding neck	1560 Pt. 2	150	—	1600 Pt. 2	10·8
10·22	Slip-on-welding	1560 Pt. 2	300	—	1600 Pt. 2	10·7
10·23	Welding neck	1560 Pt. 2	300	—	1600 Pt. 2	10·8
10·24	*Slip-on-welding	1560 Pt. 2	300	—	1600 Pt. 2	10·7
10·24	*Welding neck	1560 Pt. 2	300	—	1600 Pt. 2	10·8
10·25	Welding neck	1560 Pt. 2	600	—	1600 Pt. 2	10·9

*For use at temperatures above 400°C.
†Dimensional standards. See Joint Tables for materials standards.

selected for the purpose from the pipe standards. In practice it is convenient to prepare standards for flanged pipe joints using pipe thicknesses and qualities which are in normal production.

A draft proposal is being considered by ISO giving pipe outside diameters and associated thicknesses for pipes suitable for use at pressures up to 100 bar at ambient temperature.

Table 10.7 gives a list of suggested standards using materials which should be readily available; these pipe joint standards are detailed in Tables 10.8 to 10.25 inclusive and are illustrated in Figures 10.2 to 10.9 inclusive.

Table 10.8 STEEL FLANGED PIPE JOINTS SCREWED BOSS FLANGES (BS 4504, Table 6.4)
(See Figure 10.2)

(Dimensions are in millimetres unless otherwise stated.)

Nominal size mm	in	Pipe medium weight o.d. ≈	t	D	b	h	Flange R	Bolting	No.	Drilling d	k
15	½	21·3	2·65	80	12	20	40	M10	4	11	55
20	¾	26·9	2·65	90	14	24	50	M10	4	11	65
25	1	33·7	3·25	100	14	24	60	M10	4	11	75
40	1½	48·3	3·25	130	14	26	80	M12	4	14	100
50	2	60·3	3·65	140	14	28	90	M12	4	14	110
80	3	88·9	4·5	190	16	34	128	M16	4	18	150
100	4	114·3	4·5	210	16	40	148	M16	4	18	170

Maximum non-shock service. Pressure/temperature rating*

Nominal size mm	in	Flanges steel 070 M20 up to 120°C bar	200°C bar	250°C bar	Flanges steel 1503–161A and 1504–161A up to 120°C bar	200°C bar	250°C bar
15 to 80	½ to 3	6·0	5·0	4·5	6·0	6·0	5·5
100	4	5·17	5·0	4·5	5·17	5·17	5·17

Flanges. To BS 4504 Table 6/4. Materials carbon steel:
Forgings to BS 1503–161A or 070 M20 (BS 970 En 3).
Castings to BS 1504 161A.
Pipe. To BS 1387, medium weight.
Screwing. To BS 21 (ISO R/7). R/Rp threads. R = external taper thread on pipe.
Rp = internal parallel thread in flange.
Bolting. To BS 1750 Pts 1 and 2 (ISO metric series). Carbon steel.
Dimensions of hexagon headed bolts and nuts to BS 3692 or BS 4190.
(BS 3692 designation, bolts 6·6.)

* The above ratings apply to installations at temperatures up to 250° C subject to moderate shock. If necessary they must be decreased to suit conditions of severe shock and/or the fluid being conveyed.

Table 10.9(a) STEEL FLANGED PIPE JOINTS SCREWED BOSS FLANGES
(BS 4504, Table 16.4, see Figure 10.2)

(Dimensions are in millimetres unless otherwise stated)

Nominal size mm	in	Pipe o.d. ≈	t Heavy Weight	t Medium Weight	Flange D	b	h	R	Bolting	Drilling No.	d	k
15	½	21·3	3·25	2·65	95	14	20	45	M12	4	14	65
20	¾	26·9	3·25	2·65	105	16	24	58	M12	4	14	75
25	1	33·7	4·05	3·25	115	16	24	68	M12	4	14	85
40	1½	48·3	4·05	3·25	150	16	26	88	M16	4	18	110
50	2	60·3	4·5	3·65	165	18	28	102	M16	4	18	125
80	3	88·9	4·85	4·05	200	20	34	138	M16	8	18	160
100	4	114·3	5·4	4·5	220	20	40	158	M16	8	18	180

Ratings. See Table 10.9b for pressure/temperature ratings.
Flanges. To BS 4504 Table 16/4. Materials carbon steel:
Forgings to BS 1503–161B or 070 M20 (BS 970 En 3). Castings to BS 1504–161B
Pipe. To BS 1387, heavy weight or medium weight. See Table 10.9b.
Screwing. To BS 21 (ISO R/7). See Table 10.9b.
Bolting. To BS 1750 Pts. 1 and 2 (ISO metric series), carbon steel.
Dimensions of hexagon headed bolts and nuts to BS 3692 or BS 4190.
(BS 3692 designation, bolts 6·6.)

Table 10.9(b) STEEL PIPES WITH SCREWED BOSS FLANGES (BS 4504, Table 16.4)
Pressure/Temperature Ratings in bars up to 250°C

Nominal size		Steel 070 M20 (flanges)			Steels 1503–161B and 1504–161B (flanges)		
		Up to 120°C	200°C	250°C	Up to 120°C	200°C	250°C
mm	in	bars	bars	bars	bars	bars	bars
		Medium weight pipes – R/Rp threads*					
15 to 25	½ to 1	10·34	10·34	10·34	10·34	10·34	10·34
40	1½	8·61	8·61	8·61	8·61	8·61	8·61
50	2	6·89	6·89	6·89	6·89	6·89	6·89
80	3	6·89	6·89	6·89	6·89	6·89	6·89
100	4	5·17	5·17	5·17	5·17	5·17	5·17
		Medium weight pipes – R/Rc threads*					
15 to 25	½ to 1	16·0	14·0	13·0	16·0	16·0	15·0
40	1½	13·78	13·78	13·0	13·78	13·78	13·78
50	2	12·06	12·06	12·06	12·06	12·06	12·06
80	3	10·34	10·34	10·34	10·34	10·34	10·34
100	4	8·61	8·61	8·61	8·61	8·61	8·61
		Heavy weight pipes – R/Rp threads*					
15 to 25	½ to 1	12·06	12·06	12·06	12·06	12·06	12·06
40	1½	10·34	10·34	10·34	10·34	10·34	10·34
50	2	8·61	8·61	8·61	8·61	8·61	8·61
80	3	8·61	8·61	8·61	8·61	8·61	8·61
100	4	6·89	6·89	6·89	6·89	6·89	6·89
		Heavy weight pipes – R/Rc threads*					
15 to 25	½ to 1	16·0	14·0	13·0	16·0	16·0	15·0
40	1½	16·0	14·0	13·0	16·0	16·0	15·0
50	2	15·51	14·0	13·0	15·51	15·51	15·0
80	3	13·78	13·78	13·0	13·78	13·78	13·78
100	4	10·34	10·34	10·34	10·34	10·34	10·34

* R = External taper thread on pipe. Rp = Internal parallel thread in flange. Rc = Internal taper thread in flange.

Note 1. For dimensions see Table 10.9a and Figure 10.2.

Note 2. The above ratings apply to installations subject to moderate shock. If necessary they must be decreased to suit conditions of severe shock and or the fluid being conveyed.

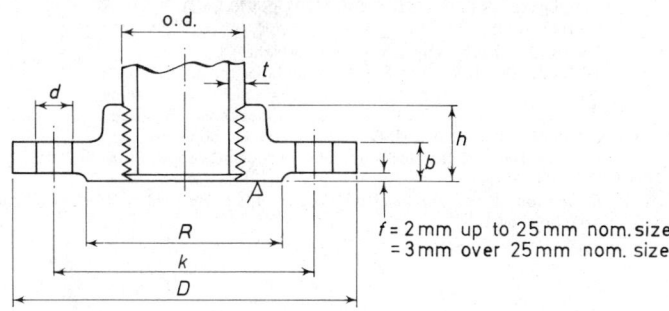

△ Continuous spiral groove to
BS 4504 clause 4.4

Figure 10.2. Steel flange pipe joints with screwed boss flanges (see Tables 10.8, 10.9a, 10.9b and 10.10)

Table 10.10 STEEL FLANGED PIPE JOINTS SCREWED BOSS FLANGES
(BS 4504, Table 40.4, see Figure 10.2)

(Dimensions are in millimetres unless otherwise stated)

Nom. size		Pipe o.d. ≈	t Heavy weight	t Medium weight	D	Flange b	h	R	Drilling	Drilling No.	d	k
mm	in											
15	½	21·3	3·25	2·65	95	16	22	45	M12	4	14	65
20	¾	26·9	3·25	2·65	105	18	26	58	M12	4	14	75
25	1	33·7	4·05	3·25	115	18	28	68	M12	4	14	85
40	1½	48·3	4·05	3·25	150	18	32	88	M16	4	18	110
50	2	60·3	4·5	3·65	165	20	34	102	M16	4	18	125
80	3	88·9	4·85	4·05	200	24	40	138	M16	8	18	160
100	4	114·3	5·4	4·5	235	24	44	162	M20	8	22	190

Maximum non-shock service. Pressure/temperature rating

Nom. Size mm	in	Heavy weight R/Rp threads* bar	Heavy weight R/Rc threads* bar	Medium weight R/Rc threads* bar
15	½	12·06	20·68	17·23
20	¾	12·06	20·68	17·23
25	1	12·06	20·68	17·23
40	1½	10·34	17·23	13·78
50	2	8·61	15·51	12·06
80	3	8·61	13·78	10·34
100	4	6·89	10·34	8·61

The above ratings apply to installations at temperatures up to 250°C subject to moderate shock. If necessary they must be decreased to suit conditions of severe shock and/or the fluid being conveyed.

Flanges. To BS 4504 Table 40/4. Materials carbon steel:
Forgings to BS 1503–161B or 070 M20 (BS 970 En 3).
Castings to BS 1504–161B.

Pipe. To BS 1387, heavy weight or medium weight.

Screwing. To BS 21 (ISO R/7)   * R = external taper thread on pipe
Rp = internal parallel thread in flange
Rc = internal taper thread in flange

Bolting. To BS 1750 Pts. 1 and 2 (ISO metric series). Carbon steel:
Use 1 per cent Cr. Mo. steel stud bolts for pressures above 19 bar at 250° C. Bolt Gr. B7M (BS 1506–621A).
Dimensions of hexagon headed bolts and nuts to BS 3692 or BS 4190. (BS 3692 designation, bolts 6·6.)

Table 10.11 STEEL FLANGE PIPE JOINTS. SCREWED BOSS FLANGES, BS 1560 CLASS 150
(See Figure 10.3)

(Dimensions in millimetres unless otherwise stated.)

| Nom. size | | Pipe | | | | | Flange | | | | Bolt dia. | Drilling | | |
mm	in	BS 1600 Pt. 2 o.d.	t	Sch. No.	BS 1387 Heavy weight o.d. ≈	t	D	b	h	R	in	No.	d	k
15	½	21·3	3·73	80	21·3	3·25	89	11·1	16	35	½	4	15·9	60·3
20	¾	26·7	3·91	80	26·9	3·25	98	12·7	16	43	½	4	15·9	69·8
25	1	33·4	4·55	80	33·7	4·05	108	14·3	17	51	½	4	15·9	79·4
40	1½	48·3	5·08	80	48·3	4·05	127	17·5	22	73	½	4	15·9	98·4
50	2	60·3	3·91	40	60·3	4·5	152	19·0	25	92	⅝	4	19·0	120·6

| Nom. size | | Maximum non-shock working pressures bar Temperatures 0°C up to °C | | | | | | | | | | |
mm	in	38	50	75	100	125	150	175	200	225	250	260
15	½											
20	¾	19·0	18·3	17·3	16·3							
25	1					15·4	14·5	13·5	12·6	11·7	10·7	10·3
40	1½	17·2	17·2	17·0	16·3							
50	2	15·5	15·5	15·5	15·5							

Flanges.	To BS 1560 Pt. 2 except screwing. Material carbon steel:
	Forgings to BS 1503–161 Gr. 28B or Gr. 32B or ASTM A105 Gr. I or Gr. II.
	Castings to BS 1504–161 Gr. B or ASTM A216 Gr. WCB.
Pipe.	To BS 1600 Pt. 2. Material carbon steel:
	API 5L Gr. B or BS 3601 Steel 27 or BS 1387, heavy weight.
Screwing.	To BS 21 R/Rc threads. R = external taper thread on pipe.
	Rc = internal taper thread in flange.
Bolting.	To BS 1750, carbon steel, ISO inch series (Unified), Pts. 1 and 3.
	For lengths see BS 1560 Pt 2 Table 12.
	Dimensions of hexagon headed bolts and nuts to BS 1768 or BS 1769.
Nom. sizes.	ISO nominal sizes are given in millimetres.
	BS 1600 Pt. 2 nom. pipe sizes in millimetres are equal to pipe o.d.

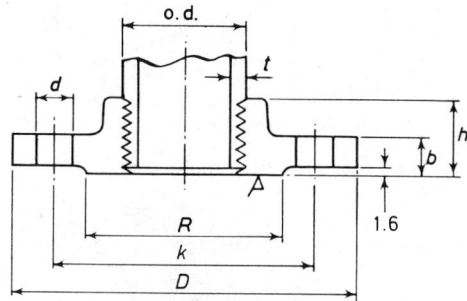

Δ Continuous spiral groove to
BS 1560 pt.2 clause 3.3.1

*Figure 10.3. Steel flange pipe joints with screwed boss flanges
(see Tables 10.11 and 10.12)*

Table 10.12 STEEL FLANGE PIPE JOINTS. SCREWED BOSS FLANGES (BS 1560 Class 300)
(See Figure 10.3)

(Dimensions in millimetres unless otherwise stated.)

| Nom. size | | Pipe | | | | | Flange | | | | Bolt | | Drilling | |
| mm | in | BS 1600 Pt. 2 | | | BS 1387 | | | | | | dia. | | | |
		o.d.	t	Sch. No.	Heavy weight o.d. ≈	t	D	b	h	R	in	No.	d	k
15	$\frac{1}{2}$	21·3	3·73	80	21·3	3·25	95	14·3	22	35	$\frac{1}{2}$	4	15·9	66·7
20	$\frac{3}{4}$	26·7	3·91	80	26·9	3·25	117	15·9	25	43	$\frac{5}{8}$	4	19·0	82·6
25	1	33·4	4·55	80	33·7	4·05	124	17·5	27	51	$\frac{5}{8}$	4	19·0	88·9
40	$1\frac{1}{2}$	48·3	5·08	80	48·3	4·05	156	20·6	30	73	$\frac{3}{4}$	4	22·2	114·3
50	2	60·3	3·91	40	60·3	4·5	165	22·2	33	92	$\frac{5}{8}$	8	19·0	127·0

Maximum non-shock working pressures

Nom. Size	mm in	15 $\frac{1}{2}$	20 $\frac{3}{4}$	25 1	40 $1\frac{1}{2}$	50 2
Pressure bar		20·68	20·68	20·68	17·23	15·51

The above are maximum pressures up to 260°C and are for non-shock conditions. The pressures should be decreased according to the fluid being conveyed and for shock conditions.

Flanges. To BS 1560 Pt. 2 except screwing. Material carbon steel:
Forgings to BS 1503–161 Gr. 28B or 32B, or ASTM A105 Gr. I or II.
Castings to BS 1504–161 Gr. B, or ASTM A216 Gr. WCB.

Pipe. To BS 1600 Pt. 2. Material carbon steel:
API 5L Gr. B, or BS 3601 Steel 27, or BS 1387 Heavy Weight.

Screwing. To BS 21 R/Rc threads. R = external taper thread on pipe.
Rc = internal taper thread in flange.

Bolting. To BS 1750 Pts. 1 and 3, ISO inch (unified) series.
Lengths. See BS 1560 Pt. 2, Table 14.
Use carbon steel bolts and nuts for pressures up to 19 bar at 260°C.
Dimensions of hexagon headed bolts and nuts to BS 1768 or BS 1769.
Use 1 per cent Cr. Mo. steel for pressures above 19 bar at 260°C. Bolt Gr. B7, Bolt Gr. B7 (BS 1506–621A), Stud bolts.

Nom. sizes. ISO nominal sizes are given in millimetres. BS 1600 nom. pipe sizes are equal to pipe o.d. in millimetres.

Table 10.13 STEEL FLANGE PIPE JOINTS. NOMINAL PRESSURE 6 BAR
(For Plate Flanges, see Figure 10.4)
(For Slip-on-welding Flanges, see Figure 10.5)

(Dimensions are in millimetres unless otherwise stated.)

Nom. size		Pipe		Flange						Bolt size	Drilling		
mm	in	o.d.	t	D	R	b_p	b_s	h_s	w		No.	d	k
15	½	21·3	2·6	80	40	12	12	20	6·0	M10	4	11	55
20	¾	26·9	2·9	90	50	14	14	24	6·0	M10	4	11	65
25	1	33·7	3·2	100	60	14	14	24	6·0	M10	4	11	75
40	1½	48·3	3·6	130	80	16	14	26	6·0	M12	4	14	100
50	2	60·3	4·0	140	90	16	14	28	6·0	M12	4	14	110
80	3	88·9	4·0	190	128	18	16	34	6·0	M16	4	18	150
100	4	114·3	4·5	210	148	18	16	40	7·0	M16	4	18	170
150	6	168·3	4·5	265	202	20	18	44	7·0	M16	8	18	225
200	8	219·1	5·0	320	258	22	20	44	7·0	M16	8	18	280
250	10	273	5·9	375	312	24	22	44	9·0	M16	12	18	335
300	12	323·9	6·3	440	365	24	22	44	9·0	M20	12	22	395
350	14	355·6	*	490	415	26			*	M20	12	22	445
400	16	406·4		540	465	28				M20	16	22	495
500	20	508		645	570	30				M20	20	22	600
600	24	609·6		755	670	32				M24	20	26	705

* Calculate for specific duties to save costs. Dimension $w = 1·4\ t$.

Design pressure (bar gauge) at temperature °C (Moderate shock)

Flange material	− 10 to 120	200	250	300
070 M20	6	5	4·5	
BS 1501, 1503, 1504	6	6	5·5	5

Pipes to BS 3601 and 1387 should be limited to 260° max.

Fabrication. To BS 806 or BS 3351.
Flanges. To BS 4504. Steel plate, Table 6/3. Steel slip-on boss, Table 6/5.
Materials: Steel plates to BS 1501–151 Gr. 23A.
Steel forgings to BS 1503–161A or 070 M20 (BS 970 En 3).
Cast steel to BS 1504–161A.
Bolting. To BS 1750 Pts. 1 and 2 (ISO metric series).
Material: Carbon steel (BS 3692 Designation 4.6).
Dimensions of hex. bolts and nuts to BS 3692 or BS 4190.

Pipe.

Nom. pipe size mm	Specification BS 3600 and BS	Process
15–25	3601 Gr. 22 or Gr. 27	Cold drawn seamless
	3602 Gr. 23 or Gr. 27	Cold drawn seamless
15–100	3601 Gr. 22 or Gr. 27	Electric resistance welded
	3601 Gr. 22	Butt welded
	1387	
	3602 Gr. 23 or Gr. 27	Electric resistance welded
25–450	3601 Gr. 22 or Gr. 27	Hot finished seamless
	3602 Gr. 23 or Gr. 27	Hot finished seamless
500 and 600	3601 Gr. 26	Electric fusion welded
	3602 Gr. 28	Electric fusion welded

Table 10.14 STEEL FLANGE PIPE JOINTS. NOMINAL PRESSURE 16 BAR
(For Plate Flanges, see Figure 10.4)
(For Slip-on-welding Flanges, see Figure 10.5)
(*Dimensions are in millimetres unless otherwise stated.*)

Nom. size		Pipe		Flange						Bolt size	Drilling		
mm	in	o.d.	t	D	R	b_p	b_s	h_s	w		No.	d	k
15	$\frac{1}{2}$	21·3	2·9	95	45	14	14	20	6·0	M12	4	14	65
20	$\frac{3}{4}$	26·9	2·9	105	58	16	16	24	6·0	M12	4	14	75
25	1	33·7	3·6	115	68	16	16	24	6·0	M12	4	14	85
40	$1\frac{1}{2}$	48·3	4·0	150	88	16	16	26	6·0	M16	4	18	110
50	2	60·3	4·0	165	102	18	18	28	6·0	M16	4	18	125
80	3	88·9	4·5	200	138	20	20	34	7·0	M16	8	18	160
100	4	114·3	4·5	220	158	20	20	40	7·0	M16	8	18	180
150	6	168·3	5·4	285	212	22	22	44	8·0	M20	8	22	240
200	8	219·1	7·1	340	268	24	24	44	10·0	M20	12	22	295
250	10	273	8·0	405	320	26	26	46	11·0	M24	12	26	355
300	12	323·9	8·8	460	378	28	28	46	13·0	M24	12	26	410
350	14	355·6	*	520	438	32			*	M24	16	26	470
400	16	406·4		580	490	36				M27	16	30	525
500	20	508		715	610	44				M30	20	33	650
600	24	609·6		840	725	52				M33	20	36	770

* Calculate for specific duties to save costs. Dimension $w = 1·4\,t$.

Design pressure (bar gauge) at temperature °C (Moderate shock)						
Flange material	*— 10 to 120*	*200*	*250*	*300*	*350*	*400*
O70 M20	16	14	13			
BS 1501, 1503, 1504	16	16	15	13	11	9

Pipes to BS 3601 and 1387 should be limited to 260°C max.

Fabrication.	To BS 806 or BS 3351.
Flanges.	To BS 4504. Steel plate, Table 16/3. Steel slip-on boss, Table 16/5.
	Materials: Steel plates to BS 1501–151 Gr. 26B.
	Steel forgings to BS 1503–161B or 070 M20 (BS 970 En 3).
	Cast steel to BS 1504–161B.
Bolting.	To BS 1750 Pts 1 and 2 (ISO metric series).
	Materials: Carbon steel up to 300°C max. (BS 3692, Designation 6.6).
	Dimensions to BS 3692 or BS 4190.
	1 per cent Cr. Mo. steel up to 400°C, Bolt Gr. B7 M (BS 1506–621A).
	Dimensions of stud bolts to BS 1750 Pt. 2.

Pipe.	Nom. pipe size mm	Specification BS 3600 and BS	Process
	15–25	3601 Gr. 22 or Gr. 27	Cold drawn seamless
		3602 Gr. 23 or Gr. 27	Cold drawn seamless
	15–100	3601 Gr. 22 or Gr. 27	Electric resistance welded
		3601 Gr. 22	Butt welded
		1387	
		3602 Gr. 23 or Gr. 27	Electric resistance welded
	25–450	3601 Gr. 22 or Gr. 27	Hot finished seamless
		3602 Gr. 23 or Gr. 27	Hot finished seamless
	500 and 600	3601 Gr. 26	Electric fusion welded
		3602 Gr. 28	**Electric fusion welded**

Note. Bessemer steel should be oxygen blown.

Table 10.15 STEEL FLANGE PIPE JOINTS. NOMINAL PRESSURE 40 BAR
(For Plate Flanges, see Figure 10.4)
(For Slip-on-welding Flanges, see Figure 10.5)
(*Dimensions are in millimetres unless otherwise stated.*)

Nom. size		Pipe		Flange						Bolt size	Drilling		
mm	in	o.d.	t	D	R	b_p	b_s	h_s	w		No.	d	k
15	½	21·3	2·9	95	45	16	16	22	6·0	M12	4	14	65
20	¾	26·9	2·9	105	58	18	18	26	6·0	M12	4	14	75
25	1	33·7	3·6	115	68	18	18	28	6·0	M12	4	14	85
40	1½	48·3	4·0	150	88	20	18	32	6·0	M16	4	18	110
50	2	60·3	4·0	165	102	20	20	34	6·0	M16	4	18	125
80	3	88·9	5·4	200	138	24	24	40	8·0	M16	8	18	160
110	4	114·3	5·9	235	162	26	24	44	9·0	M20	8	22	190
150	6	168·3	7·1	300	218	30	28	52	10·0	M24	8	26	250
200	8	219·1	8·0	375	285	34	32	52	11·0	M27	12	30	320
250	10	273	10·0	450	345	42	40	60	14·0	M30	12	33	385
300	12	323·9	11·0	515	410	50	48	67	16·0	M30	16	33	450
350	14	355·6	*	580	465	56			*	M33	16	36	510
400	16	406·4		660	535	64				M36	16	39	585
500	20	508		755	615	72				M39	20	42	670

* Calculate for duties to save costs. Dimension $w = 1·4\,t$.

Design pressure (*bar gauge*) at temperature °C (subject to moderate shock)						
Flange material	*− 10 to 120*	*200*	*250*	*300*	*350*	*400*
O70 M20	40	32	28			
BS 1501, 1503, 1504	40	40	38	33	28	23

Fabrication. To BS 806 or 3351.
Flanges. To BS 4504. Steel plate, Table 40/3. Steel slip-on boss, Table 40/5.
        Materials: Steel plates to BS 1501–151 Gr. 26B.
                Steel forgings to BS 1503–161B or 070 M20 (BS 970 En 3).
                Cast steel to BS 1504–161B.
Bolting. To BS 1750 Pts. 1 and 2 (ISO metric series).
        Materials: Carbon steel up to 300°C max. (BS 3692 designation 6.6).
        Dimensions to BS 3692 or BS 4190.
        1 per cent. Cr. Mo. steel up to 400°C, Bolt Gr. B7 M (BS 1506–621A).
        Dimensions of stud bolts to BS 1750 Pt. 2.

Pipe.

Nom. pipe size mm	Specification BS 3600 and BS	Process
15–25	3602 Gr. 23 or Gr. 27	Cold drawn seamless
15–100	3602 Gr. 23 or Gr. 27	Electric resistance welded
25–400	3602 Gr. 27	Hot finished seamless
500	3602 Gr. 28†	Electric fusion welded

†Plate to BS 1501–151 Gr. 28A. (Bessemer steel should be oxygen blown.)

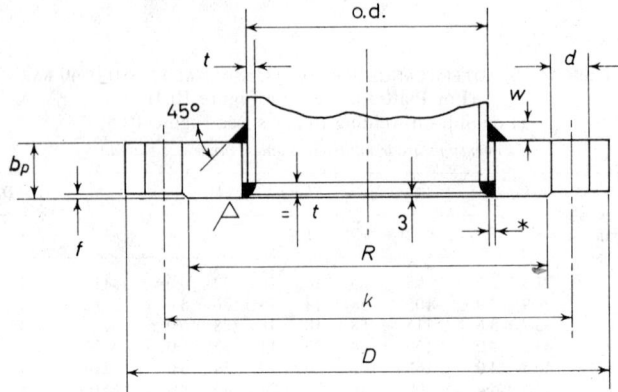

Figure 10.4. Steel flange pipe joint with plate flanges for welding (see Tables 10.13, 10.14 and 10.15)

△ Continuous spiral groove to BS 4504 Cl. 4.4

* Initial clearance required. Maximum allowable is 3 mm at any position but sum of clearance on any diameter should not exceed 5 mm. For nom. sizes up to and including 150 mm the pipe may be expanded, within the specified limits, into the flange to assist assembly before welding

$f = $ 2 mm for nom. sizes up to and including 25 mm
  3 mm for nom. sizes from 40 up to and including 250 mm
  4 mm for nom. sizes from 300 up to and including 500 mm
  5 mm for nom. sizes over 500 mm

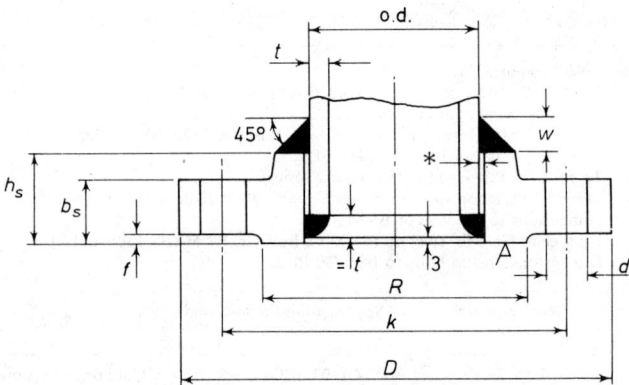

Figure 10.5. Steel flange pipe joint with slip-on boss flanges for welding (see Tables 10.13, 10.14 and 10.15)

△ Continuous spiral groove to BS 4504 Cl. 4.4

* Initial clearance required. Maximum allowable is 3 mm at any position but sum of clearance on any diameter should not exceed 5 mm

$f = $ 2 mm for nom. sizes up to and including 25 mm
  3 mm for nom. sizes from 40 up to and including 250 mm
  4 mm for 300 mm nom. size

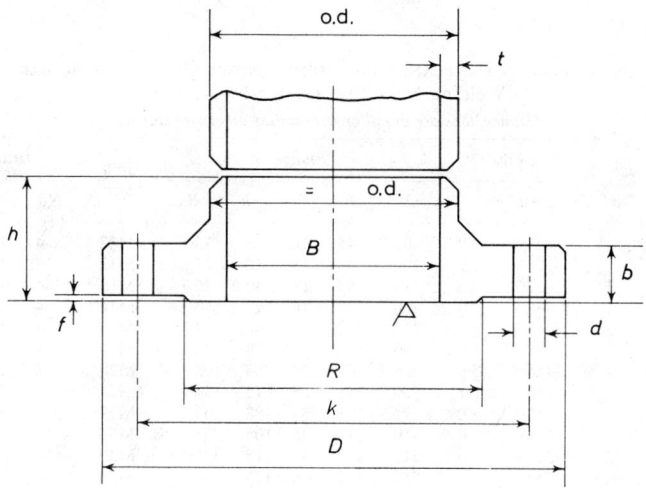

Figure 10.6. Steel flange pipe joint with welding neck flange (see Tables 10.16, 10.17 and 10.18)
△ *Continuous spiral groove to BS 4504 Cl. 4.4*
f = 2 mm for nom. sizes up to and including 25 mm
    3 mm for nom. sizes from 40 up to and including 250 mm
    4 mm for nom. sizes from 300 up to and including 500 mm

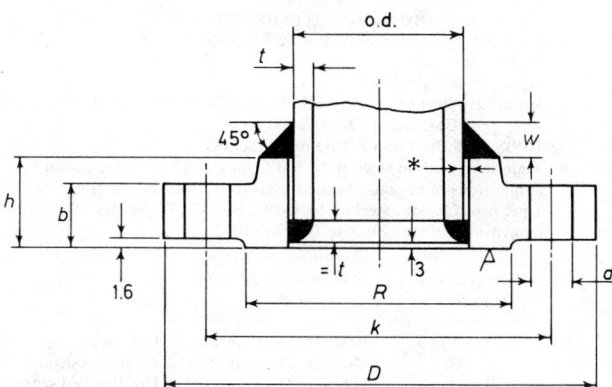

Figure 10.7. Steel flange pipe joint with slip-on welding flange (see Tables 10.19, 10.21, 10.22, and 10.24)
* Initial clearance required. Maximum allowable is 3 mm at any position but sum of clearance on any diameter should not exceed 5 mm
△ *Continuous spiral groove to BS 1560 Pt. 2, clause 3.3.1*

Table 10.16 STEEL FLANGE PIPE JOINTS. NOMINAL PRESSURE 40 BAR
(Welding Neck Flanges, see Figure 10.6)
(*Dimensions are in millimetres unless otherwise stated.*)

Nom. size		Pipe		Flange					Bolt size	Drilling		
mm	in	o.d.	t	D	R	b	h	B		No.	d	k
15	½	21·3	2·9	95	45	16	38	15·5	M12	4	14	65
20	¾	26·9	2·9	105	58	18	40	21·1	M12	4	14	75
25	1	33·7	3·6	115	68	18	40	26·5	M12	4	14	85
40	1½	48·3	4·0	150	88	18	45	40·3	M16	4	18	110
50	2	60·3	4·0	165	102	20	48	52·3	M16	4	18	125
80	3	88·9	5·4	200	138	24	58	78·1	M16	8	18	160
100	4	114·3	5·9	235	162	24	65	102·5	M20	8	22	190
150	6	168·3	7·1	300	218	28	75	154·1	M24	8	26	250
200	8	219·1	8·0	375	285	34	88	203·1	M27	12	30	320
250	10	273	10·0	450	345	38	105	253	M30	12	33	385
300	12	323·9	11·0	515	410	42	115	301·9	M30	16	33	450
350	14	355·6	*	580	465	46	125	*	M33	16	36	510
400	16	406·4		660	535	50	135		M36	16	39	585
500	20	508		755	615	52	140		M39	20	42	670

* Calculate for specific duties to save costs. Dimension $B = (\text{o.d.} - 2\,t)$.

Design pressure (*bar gauge*) at temperature °C (subject to moderate shock)

Flange material	−10 to 120	200	250	300	350	400
070 M20	40	32	28			
BS 1503, 1504	40	40	38	33	28	23

Weld Preparation.   Class I Arc welding to BS 2633.
  Class II Metal arc welding to BS 2971.
  Class I Oxy-acetylene welding to BS 1821.
  Class II Oxy-acetylene welding to BS 2640.

Fabrication.   To BS 806 or 3351.

Flanges.   To BS 4504, Table 40/2.
  Materials: Steel forgings to BS 1503–161B or 070 M20 (BS 970 En 3).
       Cast steel to BS 1504–161B.

Bolting.   To BS 1750 Pts 1 and 2 (ISO metric series).
  Materials: Carbon steel up to 300°C max. (BS 3692 designation 6.6)
  Dimensions of hexagon headed bolts to BS 3692 or BS 4190.
  1 per cent Cr. Mo. steel up to 400°C, Bolt Gr. B7 M (BS 1506–621A)
  Dimensions of stud bolts to BS 1750 Pt. 2.

Pipe.

Nom. pipe size mm	Specification BS 3600 and BS	Process
15–25	3602 Gr. 23 or Gr. 27	Cold drawn seamless
15–100	3602 Gr. 23 or Gr. 27	Electric resistance welded
25–400	3602 Gr. 27	**Hot finished seamless**
500	3602 Gr. 28†	Electric fusion welded

Note. Bessemer steel should be oxygen blown.
†Plate to BS 1501–151 Gr. 28A

Table 10.17 STEEL FLANGE PIPE JOINTS. NOMINAL PRESSURE 40 BAR
(Welding Neck Flanges for temperatures above 400°C to 475°C, see Figure 10.6)

Dimensions.	See Table 10.16.
Fabrication.	To BS 806 or 3351.
Weld Preparation.	Class I Arc welding to BS 2633.
	Class I Oxy-acetylene welding to BS 1821.
	Class II Metal arc welding to BS 2971.
	Class II Oxy-acetylene welding to BS 2640.
Flanges.	To BS 4504, Table 40/2.
	Materials: Steel forgings to BS 1503–161B or BS 1503–240B.
	Cast steel to BS 1504–161B or BS 1504–240.
Bolting.	To BS 1750 Pts 1 and 2 (ISO metric series). Stud bolts.
	Materials: Up to 450°C, 1 per cent Cr. Mo. steel, bolt Gr. B7A M (BS 1506–621B).
	Up to 475 °C, 12 per cent Cr. steel, bolt Gr. B6 M (BS 1506–713).

Pipe.	Nom. pipe size mm	Specification BS 3600 and BS	Process
	15–25	3602 Gr. 27H (silicon killed)	Cold drawn seamless
	25–400	3602 Gr. 27H (silicon killed)	Hot finished seamless
	500	3602 Gr. 28*	Electric fusion welded

* Plate to BS 1501–161 Gr. 28B.

Design pressure (bar gauge) at temperature °C (subject to moderate shock)

Flange material	200	250	300	350	400	425	450	475
BS 1503–161B BS 1504–161B	40	38	33	28	23	20	15	10
BS 1503–240B BS 1504–240		40	35	31	30	29	28	22

Table 10.18 STEEL FLANGE PIPE JOINTS. NOMINAL PRESSURE 100 BAR
(Welding Neck Flanges, see Figure 10.6)
(Dimensions are in millimetres unless otherwise stated)

| Nom. size | | Pipe | | Flange | | | | | Bolt Size | Drilling | | |
mm	in	o.d.	t	D	R	b	h	B		No.	d	k
15	½	21·3	4·0	105	45	20	45	13·3	M12	4	14	75
20	¾	26·9	4·0	130	58	22	58	18·9	M16	4	18	90
25	1	33·7	4·5	140	68	24	58	24·7	M16	4	18	100
40	1½	48·3	5·0	170	88	26	62	38·3	M20	4	22	125
50	2	60·3	5·6	195	102	28	68	49·1	M24	4	26	145
80	3	88·9	8·0	230	138	32	78	72·9	M24	8	26	180
100	4	114·3	8·8	265	162	36	90	96·7	M27	8	30	210
150	6	168·3	11·0	355	218	44	115	146·3	M30	12	33	290
200	8	219·1	12·5	430	285	52	130	194·1	M33	12	36	360
250	10	273	*	505	345	60	157	*	M36	12	39	430
300	12	323·9		585	410	68	170		M39	16	42	500
350	14	355·6		655	465	74	189		M45	16	48	560

*Calculate for specific duties to save costs. Dimension $B = (\text{o.d.} - 2t)$.

Design pressure (bar gauge) at temperature °C (Subject to moderate shock)

Temperature °C	−10 to 120	200	250	300	350	400
Pressure bar	100	100	95	82	70	57

Fabrication.	To BS 806 or 3351
Weld preparation.	Class I Arc welding to BS 2633
	Class I Oxy-acetylene welding to BS 1821
	Class II Metal arc welding to BS 2971
	Class II Oxy-acetylene welding to BS 2640
Flanges.	To BS 4504, Table 100/2.
	Materials: Steel forgings to BS 1503–161B.
	Cast steel to BS 1504–161B.
Bolting.	To BS 1750 Pts. 1 and 2 (ISO metric series). Stud bolts.
	Materials: 1 per cent Cr. Mo. steel, Bolt Gr. B7 M (BS 1506–621A).

Pipe.

Nom. pipe size mm	Specification BS 3600 and BS	Process
15 to 25	3602 Gr. 23 or Gr. 27	Cold drawn seamless
15 to 100	3602 Gr. 23 or Gr. 27	Electric resistance welded
25 to 350	3602 Gr. 27	Hot finished seamless

Table 10.19 STEEL FLANGE PIPE JOINTS (Slip-on-welding Flanges, BS 1560 Class 150)
(Maximum Temperature 400°C, see Figure 10.7)
(Dimensions are in millimetres unless otherwise stated.)

Nom. size		Pipe BS 1600 Pt. 2			Flange					Bolt	Drilling		
mm*	in	o.d.	Sch. No.	t	D	b	h	R	w	Dia. in	No.	d	k
15	½	21·3	40	2·77	89	11·1	16	35	6·0	½	4	15·9	60·3
20	¾	26·7	40	2·87	98	12·7	16	43	6·0	½	4	15·9	69·8
25	1	33·4	40	3·38	108	14·3	17	51	6·0	½	4	15·9	79·4
40	1½	48·3	40	3·68	127	17·5	22	73	6·0	½	4	15·9	98·4
50	2	60·3	40	3·91	152	19·0	25	92	6·0	⅝	4	19·0	120·6
80	3	88·9	40	5·49	190	23·8	30	127	8·0	⅝	4	19·0	152·4
100	4	114·3	40	6·02	229	23·8	33	157	9·0	⅝	8	19·0	190·5
150	6	168·3	40	7·11	279	25·4	40	216	10·0	¾	8	22·2	241·3
200	8	219·1	30	7·04	343	28·6	44	270	10·0	¾	8	22·2	298·4
250	10	273·0	30	7·80	406	30·2	49	324	11·0	⅞	12	25·4	362·0
300	12	323·9	30	8·38	483	31·8	56	381	13·0	⅞	12	25·4	431·8
350	14	355·6	30	9·52	533	34·9	57	413	14·0	1	12	28·6	476·2
400	16	406·4	30	9·52	597	36·5	64	470	14·0	1	16	28·6	539·8
450	18	457·2	Std. Wall	9·52	635	39·7	68	533	14·0	1⅛	16	31·8	577·8
500	20	508·0	20	9·52	698	42·9	73	584	14·0	1⅛	20	31·8	635·0
600	24	609·6	20	9·52	813	47·6	83	692	14·0	1¼	20	34·9	749·3

* ISO nominal sizes are given in mm BS 1600 Pt. 2 nom. pipe sizes in mm = o.d.

Maximum non-shock service. Pressure/Temperature Rating

O C to	38	50	75	100	125	150	175	200
bar	19	18·3	17·3	16·3	15·4	14·5	13·5	12·6

O C to	225	250	260	275	300	325	350	375	400
bar	11·7	10·7	10·3	10·0	9·4	8·7	8·1	7·5	6·9

Fabrication. To BS 3351.

Flanges. To BS 1560 Pt. 2. Material carbon steel:
Forgings to BS 1503–161 Gr. 28B or 32B or ASTM A105 Gr. I or II.
Castings to BS 1504–161 Gr. B or ASTM A216 Gr. WCB.

Bolting. To BS 1750 Pts. 1 and 3, ISO inch (unified) series.
Materials: Carbon steel up to 300°C max.
Stud bolts, 1 per cent Cr. Mo. steel up to 400°C, Bolt Gr. B7 (BS 1506–621A).
Dimensions of hexagon headed to BS 1768 or BS 1769.
Lengths. See BS 1560 Pt. 2, Table 12.

Pipe.	Nom. Pipe size mm*	Specification	Process
	15 to 450	API 5L Gr. B	Seamless
		ASTM A 106 Gr. B	Seamless
	15 and 20	BS 3602 Gr. 23	Electric resistance welded
		BS 3602 Gr. 27	Cold drawn seamless
	40 to 450	BS 3602 Gr. 27	Hot finished seamless
	500 and 600	API 5L Gr. B	Submerged arc welded
		BS 3602 Gr. 28†	Electric fusion welded

Note. Bessemer steel should be oxygen blown.

* ISO nominal sizes are given in mm BS 1600 Pt. 2 nom. pipe sizes in mm = o.d.

† Plate to BS 1501–151 Gr. 28A

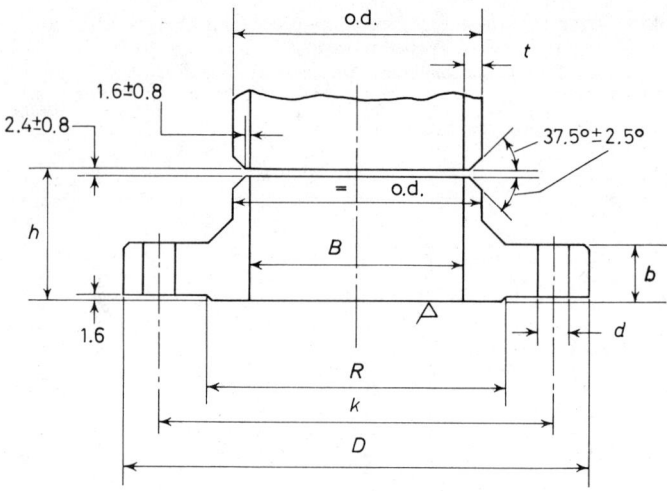

Figure 10.8. Steel flange pipe joint with welding neck flange (see Tables 10.20 to 10.24)
△ Continuous spiral groove to BS 1560 Pt. 2, clause 3.3.1

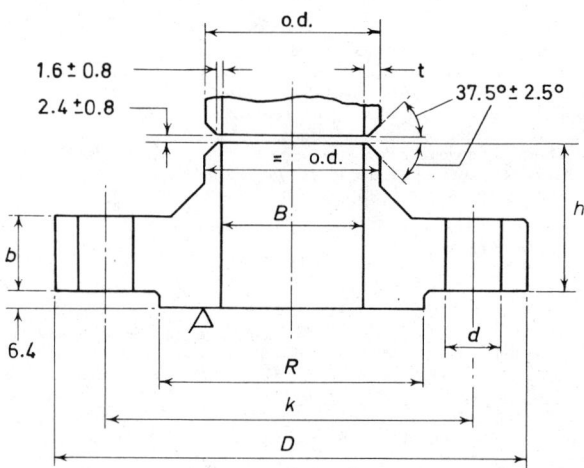

Figure 10.9. Steel flange pipe joint with welding neck flange (see Table 10.25)
△ Continuous spiral groove to BS 1560 Pt. 2, clause 3.3.1

Table 10.20 STEEL FLANGE PIPE JOINTS (Welding Neck Flanges, BS 1560 Class 150)
(Maximum Temperature 400°C, see Figure 10.8)
(Dimensions are in millimetres unless otherwise stated.)

Nom. size		Pipe BS 1600 Pt. 2 Sch.			Flange					Bolt Dia.		Drilling	
mm*	in	o.d.	No.	t	B	D	b	h	R	in	No.	d	k
15	½	21·3	40	2·77	15·8	89	11·1	48	35	½	4	15·9	60·3
20	¾	26·7	40	2·87	21·0	98	12·7	52	43	½	4	15·9	69·8
25	1	33·4	40	3·38	26·6	108	14·3	56	51	½	4	15·9	79·4
40	1½	48·3	40	3·68	40·9	127	17·5	62	73	½	4	15·9	98·4
50	2	60·3	40	3·91	52·5	152	19·0	64	92	⅝	4	19·0	120·6
80	3	88·9	40	5·49	77·9	190	23·8	70	127	⅝	4	19·0	152·4
100	4	114·3	40	6·02	102·3	229	23·8	76	157	⅝	8	19·0	190·5
150	6	168·3	40	7·11	154·1	279	25·4	89	216	¾	8	22·2	241·3
200	8	219·1	30	7·04	205·0	343	28·6	102	270	¾	8	22·2	298·4
250	10	273·0	30	7·80	257·4	406	30·2	102	324	⅞	12	25·4	362·0
300	12	323·9	30	8·38	307·1	483	31·8	114	381	⅞	12	25·4	431·8
350	14	355·6	30	9·52	336·6	**533**	34·9	127	413	1	12	28·6	476·2
400	16	406·4	30	9·52	387·4	597	36·5	127	470	1	16	28·6	539·8
450	18	457·2	Std Wall	9·52	438·2	635	39·7	140	533	1⅛	16	31·8	577·8
500	20	508·0	20	9·52	489·0	698	42·9	144	584	1⅛	20	31·8	635·0
600	24	609·6	20	9·52	590·6	813	47·6	152	692	1¼	20	34·9	749·3

*ISO nominal sizes are given in millimetres. BS 1600 Pt. 2 nom. pipe sizes in millimetres are equal to pipe o.d.

Maximum non-shock service. Pressure/Temperature Rating

O°C to	38	50	75	100	125	150	175	200
bar	19	18·3	17·3	16·3	15·4	14·5	13·5	12·6

O°C to	225	250	260	275	300	325	350	375	400
bar	11·7	10·7	10·3	10·0	9·4	8·7	8·1	7·5	6·9

Fabrication. To BS 3351.
Flanges. To BS 1560 Pt. 2. Material carbon steel:
Forgings to BS 1503–161 Gr. 28B or 32B or ASTM A105 Gr. I or II.
Castings to BS 1504–161 Gr. B or ASTM A216 Gr. WCB.
Bolting. To BS 1750 Pts. 1 and 3, ISO inch (unified) series.
Materials: Carbon steel up to 300°C max. Dimensions to BS 1768 or 1769.
Stud bolts, 1 per cent Cr. Mo steel up to 400°C, Bolt Gr. B7 (BS 1506–621A).
Lengths. See BS 1560 Pt. 2, Table 12.

Pipe.

Nom. pipe size mm*	Specification	Process
15 to 450	API 5L Gr. B	Seamless
	ASTM A 106 Gr. B	Seamless
15 and 20	BS 3602 Gr. 23	Electric resistance welded
	BS 3602 Gr. 27	Cold drawn seamless
40 to 450	BS 3602 Gr. 27	Hot finished seamless
500 and 600	API 5L Gr. B	Submerged arc welded
	BS 3602 Gr. 28†	Electric fusion welded

Note. Bessemer steel should be oxygen blown.

* ISO nominal sizes are given in millimetres. BS 1600 Pt. 2 nom. pipe sizes in mm are equal to o.d.

† Plate to BS 1501–151 Gr. 28A.

Table 10.21 STEEL FLANGE PIPE JOINTS (BS 1560 Class 150)
(Temperatures above 400°C to 500°C)

Dimensions. Slip-on-welding flanges. See Table 10.19 and Figure 10.7.
Welding neck flanges. See Table 10.20 and Figure 10.8.
See Tables 10.19 and 10.20.
The values of *t* for nom. pipe sizes 400 mm and larger should be calculated for specific duties to save costs.
In such cases W (Figure 10.7 $= 1·4t$ and B (Figure 10.8) $= o.d. - 2t$.

Specifications and materials. See Tables 10.19 and 10.20 except for pipe and bolting. (See below.)
Pipes for service above 400°C to 500°C should be selected from the following:

Nom. size mm	Specification BS 1600 Pt. 2 and	Process
15 to 450	ASTM A 106 Gr. B (silicon killed)	Seamless
40 to 450	BS 3602 Gr. 27H (silicon killed)	Hot finished seamless
15 and 20	BS 3602 Gr. 27H (silicon killed)	Cold drawn seamless
500 and 600	BS 3602 Gr. 28* ASTM A 155 Gr. KC60†	Electric fusion welded

* Plate to BS 1501–161 Gr 28B
† Plate to ASTM A 515 Gr. 60.

Bolting. To BS 1750 Pts. 1 and 3. ISO inch (unified) series.
Materials: Up to 450 C 1 per cent Cr. Mo. steel (BS 1506–621B), bolt Gr. B7A
     Up to 500 C 12 per cent Cr. steel (BS 1506–713), bolt Gr. B6.
Lengths: See BS 1560 Pt. 2, Table 12 (stud bolts).

Maximum non-shock service. Pressure/Temperature rating				
400°C to	425	450	475	500
bar	6·4	5·8	5·1	4·0

Table 10.22 STEEL FLANGE PIPE JOINTS (Slip-on-welding Flanges, BS 1560 Class 300) (Maximum Temperature 400°C, see Figure 10.7)
(*Dimensions are in millimetres unless otherwise stated.*)

| Nom. size | | Pipe BS 1600 Pt. 2 Sch. | | | Flange | | | | | Bolt Dia. | | Drilling | |
mm*	in	o.d.	No.	t	D	b	h	R	w	in	No.	d	k
15	½	21·3	80	3·73	95	14·3	22	35	7·0	½	4	15·9	66·7
20	¾	26·7	80	3·91	117	15·9	25	43	7·0	⅝	4	19·0	82·6
25	1	33·4	80	4·55	124	17·5	27	51	7·0	⅝	4	19·0	88·9
40	1½	48·3	80	5·08	156	20·6	30	73	7·0	¾	4	22·2	114·3
50	2	60·3	40	3·91	165	22·2	33	92	7·0	⅝	8	19·0	127·0
80	3	88·9	40	5·49	210	28·6	43	127	8·0	¾	8	22·2	168·3
100	4	114·3	40	6·02	254	31·8	48	157	10·0	¾	8	22·2	200·0
150	6	168·3	40	7·11	318	36·5	52	216	10·0	¾	12	22·2	269·9
200	8	219·1	30	7·04	381	41·3	62	270	10·0	⅞	12	25·4	330·2
250	10	273·0	30	7·80	444	47·6	67	324	11·0	1	16	28·6	387·4
300	12	323·9	30	8·38	521	50·8	73	381	12·0	1⅛	16	31·8	450·8
350	14	355·6	30	9·52	584	54·0	76	413	14·0	1⅛	20	31·8	514·4
400	16	406·4	40	12·70	648	57·2	83	470	19·0	1¼	20	34·9	571·5
450	18	457·2	40	14·27	711	60·3	89	533	21·0	1¼	24	34·9	628·6
500	20	508·0		14·27	775	63·5	95	584	21·0	1¼	24	34·9	685·8
600	24	609·6		15·88	914	69·8	106	692	22·0	1½	24	41·3	812·8

*ISO nominal sizes are given in millimetres. BS 1600 Pt. 2 nom. pipe sizes in mm are equal to o.d.

Maximum non-shock service. Pressure/Temperature Rating.

0°C to	38	50	75	100	125	150	175	200
bar	49·6	49·4	48·7	48·0	47·5	46·8	46·5	46·0

0°C	225	250	275	300	325	350	375	400
bar	45·1	43·7	41·8	39·6	37·4	34·8	32·1	29·1

Fabrication. To BS 3351.

Flanges. To BS 1560 Pt. 2. Material carbon steel:
Forgings to BS 1503–161 Gr. 28B or 32B or ASTM A105 Gr. I or II.
Castings to BS 1504–161 Gr. B or ASTM A216 Gr. WCB.

Bolting. To BS 1750 Pts. 1 and 3, ISO inch (unified) series.
Materials: Carbon steel up to 300°C max. Dimensions to BS 1768 or 1769.
Stud bolts, 1 per cent Cr. Mo. steel up to 400°C, Bolt Gr. B7 (BS 1506–621A).
Lengths. See BS 1560 Pt. 2, Table 14.

Pipe.

Nom. pipe size mm	Specification	Process
15 to 450	API 5L Gr. B.	Seamless
	ASTM A 106 Gr. B.	Seamless
15 and 20	BS 3602 Gr. 23	Electric resistance welded
	BS 3602 Gr. 27	Cold drawn seamless
40 to 450	BS 3602 Gr. 27	Hot finished seamless
500 and 600	API 5L Gr. B.	Submerged arc welded.
	BS 3602 Gr. 28*	Electric fusion welded.

Note. Bessemer steel should be oxygen blown.
*Plate to BS 1501–151 Gr. 28A.

Table 10.23 STEEL FLANGE PIPE JOINTS (Welding Neck Flanges, BS 1560 Class 300)
(Maximum Temperature 400°C, see Figure 10.8)
(*Dimensions are in millimetres unless otherwise stated.*)

Nom. size mm*	in	Pipe BS 1600 Pt. 2 Sch. o.d.	No.	t	B	D	b	h	R	Bolt Dia. in	No.	d	k
15	$\frac{1}{2}$	21·3	80	3·73	13·8	95	14·3	52	35	$\frac{1}{2}$	4	15·9	66·7
20	$\frac{3}{4}$	26·7	80	3·91	18·9	117	15·9	57	43	$\frac{5}{8}$	4	19·0	82·6
25	1	33·4	80	4·55	24·3	124	17·5	62	51	$\frac{5}{8}$	4	19·0	88·9
40	$1\frac{1}{2}$	48·3	80	5·08	38·1	156	20·6	68	73	$\frac{3}{4}$	4	22·2	114·3
50	2	60·3	40	3·91	52·5	165	22·2	70	92	$\frac{5}{8}$	8	19·0	127·0
80	3	88·9	40	5·49	77·9	210	28·6	79	127	$\frac{3}{4}$	8	22·2	168·3
100	4	114·3	40	6·02	102·3	254	31·8	86	157	$\frac{3}{4}$	8	22·2	200·0
150	6	168·3	40	7·11	154·1	318	36·5	98	216	$\frac{3}{4}$	12	22·2	269·9
200	8	219·1	30	7·04	205·0	381	41·3	111	270	$\frac{7}{8}$	12	25·4	330·2
250	10	273·0	30	7·80	257·4	444	47·6	117	324	1	16	28·6	387·4
300	12	323·9	30	8·38	307·1	521	50·8	130	381	$1\frac{1}{8}$	16	31·8	450·8
350	14	355·6	30	9·52	336·6	584	54·0	143	413	$1\frac{1}{8}$	20	31·8	514·4
400	16	406·4	40	12·70	381·0	648	57·2	146	470	$1\frac{1}{4}$	20	34·9	571·5
450	18	457·2	40	14·27	428·7	711	60·3	159	533	$1\frac{1}{4}$	24	34·9	628·6
500	20	508·0		14·27	479·4	775	63·5	162	584	$1\frac{1}{4}$	24	34·9	685·8
600	24	609·6		15·88	577·8	914	69·8	168	692	$1\frac{1}{2}$	24	41·3	812·8

*ISO nominal sizes are given in millimetres. BS 1600 Pt. 2 nom. pipe sizes in millimetres are equal to pipe o.d.

Maximum non-shock service. Pressure/Temperature Rating

0°C to	38	50	75	100	125	150	175	200
bar	49·6	49·4	48·7	48·0	47·5	46·8	46·5	46·0

0°C	225	250	275	300	325	350	375	400
bar	45·1	43·7	41·8	39·6	37·4	34·8	32·1	29·1

Fabrication. To BS 3351.
Flanges. To BS 1560 Pt. 2. Material carbon steel.
  Forgings to BS 1503–161 Gr. 28B or 32B or ASTM A105 Gr. I or II.
  Castings to BS 1504–161 Gr. B or ASTM A216 Gr. WCB.
Bolting. To BS 1750 Pts. 1 and 3, ISO inch (unified) series.
  Materials: Carbon steel up to 300°C max. Dimensions to BS 1768 or 1769
  Stud bolts, 1 per cent Cr. Mo. steel up to 400°C, Bolt Gr. B7 (BS 1506–621A)
  Lengths. See BS 1560 Pt. 2, Table 14.

Pipe.	Nom. size mm	Specification	Process
	15 to 450	API 5L Gr. B	Seamless
		ASTM A 106 Gr. B	Seamless
	15 and 20	BS 3602 Gr. 23	Electric resistance welded
		BS 3602 Gr. 27	Cold drawn seamless
	40 to 450	BS 3602 Gr. 27	Hot finished seamless
	500 and 600	API 5L Gr. B	Submerged arc welded
		BS 3602 Gr 28*	Electric fusion welded

Note. Bessemer steel should be oxygen blown.
* Plate to BS 1501–161 Gr. 28A.

Table 10.24 STEEL FLANGE PIPE JOINTS (BS 1560 Class 300)
(Temperatures above 400°C to 500°C)

Slip-on-welding flanges. See Table 10.22 and Figure 10.7.
Welding neck flanges. See Table 10.23 and Figure 10.8.
Dimensions. See Tables 10.22 and 10.23

The values of t for nom. pipe sizes 400 mm and larger should be calculated for specific duties to save costs.

In such cases w (Figure 10.7) $= 1·4t$ and B (Figure 10.8) $= o.d. - 2t.$

Specifications and materials. See Tables 10.22 and 10.23 except for pipe and bolting. (See below.)
Pipes for service above 400°C to 500°C should be selected from the following:

Nom. size mm	Specification BS 1600 Pt. 2 and	Process
15 to 450	ASTM A 106 Gr. B (silicon killed)	Seamless
40 to 450	BS 3602 Gr. 27H (silicon killed)	Hot finished seamless
15 and 20	BS 3602 Gr. 27H (silicon killed)	Cold drawn seamless
500 and 600	BS 3602 Gr. 28* ASTM A 155 GR.KC 60†	Electric fusion welded

* Plate to BS 1501–161 Gr. 28B
† Plate to ASTM A 515 Gr. 60.

Bolting. To BS 1750 Pts 1 and 3, ISO inch (unified) series.
    Materials. Up to 450°C, 1 per cent Cr. Mo. steel (BS 1506–621B). Bolt Gr. B7A.
               Up to 500°C, 12 per cent Cr. steel (BS 1506–713). Bolt Gr. B6.
    Lengths. See BS 1560 Pt. 2, Table 14. (Stud bolts)

Maximum non-shock service. Pressure/Temperature rating.				
400°C to	425	450	475	500
bar	25·5	21·4	16·8	12·4

Table 10.25 STEEL FLANGE PIPE JOINTS (Welding Neck Flanges, BS 1560 Class 600)
(Maximum Temperature 400°C, see Figure 10.9)
(Dimensions are in millimetres unless stated.)

Nom. size		Pipe BS 1600 Pt. 2			Flange					Bolt	Drilling		
mm*	in	o.d.	No.	t	B	D	b	h	R	Dia. in	No.	d	k
15	½	21·3	80	3·73	13·8	95	14·3	52	35	½	4	15·9	66·7
20	¾	26·7	80	3·91	18·9	117	15·9	57	43	⅝	4	19·0	82·6
25	1	33·4	80	4·55	24·3	124	17·5	62	51	⅝	4	19·0	88·9
40	1½	48·3	80	5·08	38·1	156	22·2	70	73	¾	4	22·2	114·3
50	2	60·3	80	5·54	49·2	165	25·4	73	92	⅝	8	19·0	127·0
80	3	88·9	80	7·62	73·7	210	31·8	83	127	¾	8	22·2	168·3
100	4	114·3	80	8·56	97·2	273	38·1	102	157	⅞	8	25·4	215·9
150	6	168·3	80	10·97	146·3	356	47·6	117	216	1	12	28·6	292·1
200	8	219·1	80	12·70	193·7	419	55·6	133	270	1⅛	12	31·8	349·2

*ISO nominal sizes are given in millimetres. BS 1600 Pt. 2 nom. pipe sizes in millimetres are equal to pipe o.d.

Maximum non-shock service. Pressure/Temperature Rating.

0°C to	38	50	75	100	125	150	175	200
bar	99·2	98·5	97·4	96·0	94·9	94·0	93·1	91·9

0°C to	225	250	275	300	325	350	375	400
bar	90·4	87·8	83·5	79·0	74·5	69·3	63·8	58·2

Fabrication. To BS 3351.
Flanges. To BS 1560 Pt. 2. Material carbon steel:
Forgings to BS 1503–161 Gr. 28B or Gr. 32B or ASTM A105 Gr. I or II.
Castings to BS 1504–161 Gr. B or ASTM A216 Gr.WCB.
Bolting. To BS 1750 Pts. 1 and 3, ISO inch (unified) series.
Material. 1 per cent Cr. Mo. steel. Bolt Gr. B7 (BS 1506–621A).
Lengths. See BS 1560 Pt. 2, Table 18 (stud bolts).

Pipe.	Nom. pipe size mm*	Specification	Process
	15 to 450	API 5L Gr. B	Seamless
		ASTM A 106 Gr. B	Seamless
	15 and 20	BS 3602 Gr. 23	Electric resistance welded
		BS 3602 Gr. 27	Cold drawn seamless
	40 to 450	BS 3602 Gr. 27	Hot finished seamless
	500 and 600	API 5L Gr. B	Submerged arc welded
		BS 3602 Gr. 28†	Electric fusion welded

Note. Bessemer steel should be oxygen blown.
*ISO nominal sizes are given in millimetres. BS 1600 Pt. 2 nom. pipe sizes in mm are equal to o.d.
† Plate to BS 1501–151 Gr. 28A.

Table 10.26 WELDED AND SEAMLESS STEEL PIPES
For use with Compression Fittings

Nom. o.d.	BS 4368 Pt. 1 Table 1 Limits of size		BS 3600 Table 3 Thicknesses Range
	max.	min.	
6	6·1	5·9	0·5 to 1·2
8	8·1	7·9	0·5 to 1·6
10	10·1	9·9	0·6 to 2·3
12	12·1	11·9	0·6 to 3·6
16	16·1	15·9	0·8 to 4·5
20	20·1	19·9	0·8 to 5·0
25	25·1	24·9	0·8 to 5·0
30	30·1	29·9	0·8 to 5·0
38	38·15	37·85	1·0 to 5·0
50	51·2	50·8	1·0 to 5·0

Diameters. To BS 4368 Pt. 1 Table 1.
Thicknesses. To BS 3600 Table 3. See Table 10.3(b) for the standard sizes. Sizes over should be selected from Table 10.3(a).
Materials. To BS 3601 or BS 3602 for carbon steel pipes.
To BS 3605 for austenitic stainless steel pipes.
Application. For use with carbon and stainless steel compression fittings to BS 4368 Pt. 1.
Dimensions are in millimetres.

Table 10.27 STEEL BOILER AND SUPERHEATER TUBES

Outside diameter		Thickness*
mm	in	mm
25·4	1·0	2·0 to 6·3
31·8	1·25	2·0 to 8·0
38	1·5	2·0 to 8·8
44·5	1·75	2·3 to 10·0
51	2·0	2·3 to 10·0
54	2·125	2·6 to 10·0
57	2·25	2·6 to 11·0
60·3	2·375	2·6 to 11·0
63·5	2·5	2·6 to 11·0
70	2·75	2·6 to 11·0
76·1	3·00	2·9 to 11·0
82·5	3·25	3·2 to 11·0
88·9	3·50	3·2 to 11·0
101·6	4·00	3·6 to 11·0
108	4·25	4·0 to 11·0
114·3	4·50	4·0 to 11·0
127	5·0	4·0 to 11·0

These tubes conform to BS 3059 and to ISO/R 64, ISO/R 221 and ISO/R 336
* See Table 10.3(b) for the detailed range of thicknesses to ISO/R 221 and ISO/R 336

Pipes for compression fittings

Pipes for use with carbon and stainless steel compression fittings to BS 4368, Part I[38] are manufactured to close tolerances. These are shown in Table 10.26.

Boiler and superheater tubes

These are specified in several materials and qualities in BS 3059[28] and details are given in this standard of elevated temperature proof stress properties. Table 10.27 gives details of the sizes covered in this specification, ISO/R1129 (1971) specifies the dimensions, tolerances and masses per unit length of boiler tubes and is based on ISO/R336.

REFERENCES

1. Bulletin 5a, *Pipes and connections,* International Federation of the National Standardisation Associations (1940)
2. ISO R.7
   Pipe threads for gas test tubes and fittings (1954)
3. ISO R.64
   Steel tubes. Outside diameters (1971)
4. ISO R.65
   Steel tubes suitable for screwing in accordance with ISO R.7 (1971)
5. ISO R.221
   Steel tubes. Thicknesses (1971)
6. ISO R.228
   Pipe threads where pressure-tight joints are not made on the threads (1961)
7. ISO R.336
   Plain end tubes, welded or seamless. General table of dimensions and masses per unit length (1971)
8. ISO R.560
   Cold drawn precision steel tubes. Metric series. Dimensions and masses per metre (1967)
9. ISO R.1127
   Stainless steel tubes, dimensions, tolerances and conventional masses per unit length (1967)
10. BS 10
    Flanges and bolting for pipes, valves and fittings (1962)
11. BS 21
    Pipe threads (1951)
12. BS 806
    Ferrous pipes and piping installations for and in connection with land boilers (1967)
13. BS 970
    Wrought steels in the form of bars, billets and forgings (1972)
14. BS 1387
    Steel tubes and tubulars suitable for screwing to BS 21 pipe threads (1967)
15. BS 1500
    Part 1. Fusion welded pressure vessels for general purposes (1958)
    BS 1515
    Part 1. Fusion welded pressure vessels for use in the chemical, petroleum and allied industries (1965)
    (These standards are being metricated and will be issued as a single BS probably in 1972. Most of the basic formulae are valid for any consistent system of units).
16. BS 1501–6
    Steels for use in the chemical, petroleum and allied industries (1958)
17. BS 1501
    Steels for fired and unfired pressure vessels. Plates (1964)
18. BS 1560
    Part 2. Steel pipe flanges and flanged fittings for the petroleum industry (1970)
19. BS 1600
    Part 2. Dimensions of steel pipe for the petroleum industry (1970)

20. BS 1750
    Bolting for flanges and valves (1972)
21. BS 1768
    Unified precision hexagon bolts and nuts (1963)
22. BS 1769
    Unified black hexagon bolts and nuts (1951)
23. BS 1821
    Class I oxy-acetylene welding of steel pipelines and pipe assemblies for carrying fluids (1957)
24. BS 2633
    Class I arc welding of ferritic steel pipework for carrying fluids (1966)
25. BS 2640
    Class II oxy-acetylene welding of steel pipelines and pipe assemblies for carrying fluids (1955)
26. BS 2779
    Fastening threads of BSP sizes (1956)
27. BS 2971
    Class II metal-arc welding of steel pipelines and pipe assemblies for carrying fluids (1961)
28. BS 3059
    Steel boiler and superheater tubes (1968)
29. BS 3351
    Piping systems for petroleum refineries and petrochemical plants (1971)
30. BS 3600
    Dimensions of welded and seamless steel pipes for pressure purposes, metric units (1972)
31. BS 3601
    Steel pipes and tubes for pressure purposes. Carbon steel: ordinary duties (1972)
32. BS 3602
    Steel pipes and tubes for pressure purposes. Carbon steel: high duties (1972)
33. BS 3603
    Steel pipes and tubes for pressure purposes. Carbon and alloy steel: low temperature duties (1972)
34. BS 3604
    Steel pipes and tubes for pressure purposes. Low and medium alloy steel (1972)
35. BS 3605
    Steel pipes and tubes for pressure purposes. Austenitic stainless steel (1972)
36. BS 3692
    ISO metric precision hexagon bolts and nuts (1967)
37. BS 4190
    ISO metric black hexagon bolts and nuts (1967)
38. BS 4368
    Carbon and stainless steel compression couplings for tubes. Part 1. Metric sizes (1968)
39. BS 4504
    Flanges and bolting for pipes, valves and fittings, metric series (1969)
40. BS 1740
    Wrought steel pipe fittings (screwed BSP thread) metric units (1971)
41. American National Standards Institute (ANSI) formerly American Standards Association (ASA):
    ANSI B31.3
    Standard code for pressure piping. Section 3. Petroleum refinery piping
    ASA B2.1
    Pipe threads (except Dryseal)
    ANSI B36.10, and B36.19
    Pipe thicknesses
42. API 5B
    'Threading, gageing and thread inspection of casing, tubing and line pipe threads.' *American Petroleum Institute*
43. Book of ASTM standards. Part 1. Steel piping, tubing and fittings. *American Society for the testing of materials*

GREY AND DUCTILE IRON PRESSURE PIPES AND FITTINGS

E. N. ANDREWS

The BSI have recently published British Standards covering metric size iron pipes and fittings and also a code of practice dealing with the design and construction of iron pipelines in land. To realise the maximum benefits from metrication, particular attention was given, during the preparation of these publications, to dimensional co-ordination with products manufactured on the Continent.

ISO/R13 'CAST IRON PIPES, SPECIAL CASTINGS AND CAST IRON PARTS FOR PRESSURE MAIN LINES'

Grey iron pipes are established products in all the major Continental countries and in many metric orientated countries outside Europe.

There has existed since 1955 an adequate ISO Recommendation covering grey iron pipes and fittings, designated ISO/R13. In consequence, the BSI, when preparing the new British Standard for metric size grey iron pipes and fittings, looked to ISO/R13 for guidance.

Although ISO/R13 is a widely used international standard it is not completely satisfactory for use as a national standard. The DNA, the German equivalent of the BSI, has issued a comprehensive series of standards based on ISO/R13 which take into account the specific requirements of users and manufacturers in that country. Similarly, when ISO/R13 was under consideration by BSI, the views of manufacturers and users in the United Kingdom resulted in modifications to the original basic standard. Further, and this was most important, dimensional co-ordination with the parallel British Standard for Metric Size Ductile Iron Pipes and Fittings[2] was a prime objective.

BS 4622 'GREY IRON PIPES AND FITTINGS (METRIC UNITS)'

SIZES

The new British Standard for Grey Iron Pipes and Fittings, BS 4622[3], has adopted the range of nominal internal diameters established by ISO/R13 except for the inclusion of the 450 mm size, also widely used in France, the omission of the 125 mm size and the upper limitation of 700 mm nominal internal diameter against that of the 1000 mm size in ISO/R13. The diameter range is, therefore, 80, 100, 150, 200, 250, 300, 350, 400, 450, 500, 600 and 700 mm (see Table 10.28).

In the inch series BS 1211[4], for all sizes up to and including 10 in, a single mean external diameter is specified for each size, irrespective of class. This feature was extended in ISO/R13 to cover every size, and was incorporated in BS 4622 (see Table 10.29). The three thickness classes of pipe in BS 4622, designated classes 1, 2 and 3

Table 10.28 SPUN IRON PIPES: MEAN INTERNAL DIAMETERS

| Nom int. dia. in | Inch pipe BS 1211 | | | Ductile iron | | Metric pipe BS 4622 | | | BS 4772 | Nom int. dia. mm |
	Class B mm	Class C mm	Class D mm	Class B external mm	Class C external mm	Class 1 mm	Class 2 mm	Class 3 mm	Class K9 mm	
3	81	81	80	—	—	84	82	81	86	80
4	107	106	104	110	—	103	101	100	106	100
5	134	133	130	—	—	—	—	—	—	—
6	161	158	155	166	—	153	152	150	157	150
7	187	184	181	—	—	—	—	—	—	—
8	214	210	207	219	—	204	202	200	209	200
9	240	236	233	246	—	—	—	—	—	—
10	266	262	258	272	—	254	252	250	260	250
12	312	319	315	319	330	304	302	300	312	300
14	363	371	366	—	—	355	352	350	363	350
15	389	397	392	396	409	—	—	—	—	—
16	415	423	418	—	—	404	401	399	413	400
18	466	474	469	474	489	453	451	448	463	450
20	517	526	521	—	—	504	501	498	514	500
21	543	552	547	552	567	—	—	—	—	—
24	620	629	624	629	646	603	600	597	615	600
27	694	705	698	—	—	703	699	696	716	700
30	785	782	779	—	—	—	—	—	—	—
—	—	—	—	—	—	—	—	—	819	800
36	941	937	933	—	—	—	—	—	920	900
—	—	—	—	—	—	—	—	—	1021	1000
42	1096	1091	1087	—	—	—	—	—	—	—
—	—	—	—	—	—	—	—	—	1123	1200
48	1250	1245	1245	—	—	—	—	—	1224	1200

Note. Diameters of inch size pipes are converted to mm for comparison purposes.

Table 10.29 SPUN IRON PIPES: MEAN EXTERNAL DIAMETER OF BARREL

	Inch pipe		Metric pipe	
Nom. int. dia. in	BS 1211 Class B Grey and ductile iron mm	BS 1211 Class C/D Grey and ductile iron mm	BS 4622 BS 4772 Grey and ductile iron mm	Nom. int. dia. mm
3	96	96	98	80
4	122	122	118	100
5‡	150	150	—	—
6	177	177	170	150
7‡	205	205	—	—
8	232	232	222	200
9	259	259	—	—
10	286	286	274	250
12	334	345	326	300
14‡	387	399	378	350
15	413	426	—	—
16‡	439	453	429	400
18	492	507	480	450
20‡	545	560	532	500
21	572	587	—	—
24	650	667	635	600
27‡	729	747	738	700
30*	826	826	—	—
—	—	—	842	800†
36*	985	985	945	900†
—	—	—	1 048	1 000†
42	1 143	1 143	—	—
—	—	—	1 152	1 100†
48*	1 300	1 300	1 255	1 200†

Note. Diameters of inch size pipes are converted to mm for comparison purposes.
*These sizes conform to those specified in BS 78: Part 1 (grey iron only)
†These sizes in ductile iron only.
‡These sizes in grey iron only.

Table 10.30   SPUN IRON PIPES: MEAN THICKNESSES OF PIPE BARREL

Nom. int. dia.	Inch pipe				Metric pipe				Nom. int. dia
	BS 1211			Ductile iron	BS 4622			BS 4772	
	Class B	Class C	Class D		Class 1	Class 2	Class 3	Class K9	
in	mm	mm	mm	mm	mm	mm	mm	mm	mm
3	7·4	7·4	7·6	—	7·2	7·9	8·6	6·0	80
4	7·6	7·9	8·9	5·8	7·5	8·3	9·0	6·1	100
5	7·9	8·6	9·9	—	—	—	—	—	—
6	8·4	9·4	10·9	5·8	8·3	9·2	10·0	6·3	150
7	8·6	10·2	11·7	—	—	—	—	—	—
8	9·1	10·9	12·4	6·4	9·2	10·1	11·0	6·4	200
9	9·4	11·4	13·2	6·6	—	—	—	—	—
10	9·9	11·9	14·0	6·9	10·0	11·0	12·0	6·8	250
12	10·9	13·2	15·2	7·6	10·8	11·9	13·0	7·2	300
14	11·7	14·2	16·5	—	11·7	12·8	14·0	7·7	350
15	11·9	14·7	17·0	8·4	—	—	—	—	—
16	12·4	15·2	17·5	—	12·5	13·8	15·0	8·1	400
18	13·2	16·3	18·8	9·1	13·3	14·7	16·0	8·6	450
20	14·0	17·0	19·6	—	14·2	15·6	17·0	9·0	500
21	14·2	17·5	20·3	9·9	—	—	—	—	—
24	15·2	18·8	21·6	10·7	15·8	17·4	19·0	9·9	600
27	17·3	21·1	24·4	—	17·5	19·3	21·0	10·8	700
30	20·3	21·8	23·4	—	—	—	—	—	—
—	—	—	—	—	—	—	—	11·7	800
36	21·8	23·9	25·9	—	—	—	—	12·6	900
—	—	—	—	—	—	—	—	13·5	1 000
42	23·6	25·9	28·2	—	—	—	—	—	—
—	—	—	—	—	—	—	—	14·4	1 100
48	25·4	27·9	30·5	—	—	—	—	15·3	1 200

Note. Thicknesses of inch size pipes are converted to mm for comparison purposes.

respectively, are identical to those in ISO/R13 where they are designated LA, A and B respectively. The designation change has been made to avoid confusion between Class B in BS 1211, which is the thinnest class in that standard, and Class B in ISO/R13 where it is the thickest class. A comparison of the mean thickness of BS 1211 pipe, Class B, C and D and those of BS 4622 Classes 1, 2 and 3 as shown in Table 10.30. It will be seen that the mean thicknesses of BS 1211 Class B pipe and those of BS 4622 Class 1 pipe, the lightest class in each standard, are almost identical although derived quite independently.

PRESSURE RATINGS

The pipes in BS 1211 are classified according to the specified field test pressure and, as no working pressure is suggested, the pipeline designer using BS 1211 pipe has considerable flexibility of choice. He can, by adjusting the safety factor, use a lighter class of pipe for a given working pressure provided the field test pressure is not higher than that

Table 10.31   HYDRAULIC AND HYDROSTATIC PRESSURES FOR
GREY IRON PIPES AND FITTINGS

Nominal internal diameter DN	Description		Suggested maximum hydraulic working pressure (Inclusive of surge)	Suggested maximum hydro-static site test pressure	Hydrostatic works test
mm			bar	bar	bar
80–700	Spigot and socket centrifugally cast iron pipe	Class 1	10	16	35
80–700	Spigot and socket centrifugally cast iron pipe	Class 2	12·5	20	35
80–700	Spigot and socket centrifugally cast iron pipe	Class 3	16	25	35
80–300	Flanged centrifugally cast iron pipe	Class 3	12·5	20	20
350–600	Flanged centrifugally cast iron pipe	Class 3	10	16	16
80–700	Flanged sand cast pipe	Class 3	10	16	16
80–300	Flanged sand cast pipe	Class 4	12·5	20	20
80–300	Standard fittings (except condensate receivers)		12·5	20	20
350–600	Standard fittings (except condensate receivers)		10	16	16
700	Standard fittings (except condensate receivers)		6	10	10
80–700	Condensate receivers		—	—	7

specified in BS 1211. In practice, however, a safety factor of 2:1 is commonly accepted and the maximum working pressure is usually, but not inevitably, half the specified test pressure. This point was considered by BSI when preparing the metric standard BS 4622 and it was felt that the new standard should give more positive guidance to designers and users by following established American and German practice in actually suggesting maximum hydraulic working pressures, inclusive of surge. As these were already available in a German national standard, they were conveniently abstracted and appear in Table 8 of BS 4622 (see Table 10.31).

BS 4622 also includes flanged pipes and a range of standard fittings, both flanged and all-socket. Suggested maximum hydraulic working pressures, inclusive of surge, are similarly given in Table 10.31.

The fittings are of a single pressure class and, where required to work at pressures higher than those in Table 10.31 ductile iron fittings or strengthened grey iron fittings should be substituted.

However, the standard grey iron fittings are adequate for the large majority of current working conditions for grey iron pipe.

It will be noted that the suggested hydrostatic site test pressures shown in Table 10.31 denote a further change in that these are 1.6 times, and not twice, the suggested maximum hydraulic working pressure, inclusive of surge. This ratio is again based on established German practice. The actual figure is 1.5 times but 1.6 ensures closer accordance with the ISO series of preferred numbers, thus:

$$10 \times 1 \cdot 6 = 16$$
$$16 \times 1 \cdot 6 = 25 \text{ (approx)}$$
$$25 \times 1 \cdot 6 = 40$$
$$40 \times 1 \cdot 6 = 64$$
$$64 \times 1 \cdot 6 = 100 \text{ (approx)}$$

The pressures listed in Table 10.31 are now incorporated in the Code of Practice for the Design and Construction of Iron Pipelines in land, CP 2010, Pt 3,[6] and it should be noted that, as always, these pressures are subject to any limitations imposed by the type of flange specified.

SOCKET FITTINGS

The all socket fitting (see Figure 10.10) has, as a result of certain advantages, superseded the spigot and socket fitting shown in the inch based BS 78, Part 2. In this respect, BS 4622 is now in line with national standards in the USA, Canada and in Europe. The

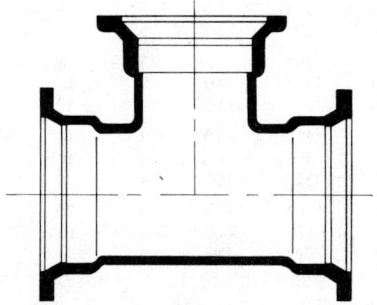

Figure 10.10. All socket fitting

general shape of the castings in ISO/R13 has been followed in BS 4622 but the opportunity has been taken to rationalise the laying dimensions and internal dimensions of the castings with those of the generally more compact metric size ductile iron fittings.

Although certain castings familiar to United Kingdom users, e.g. hydrant duckfoot bends and condensate receivers, which are not included in ISO/R13 have been added to

BS 4622, the multiplicity of types and size combinations of fittings in BS 78, Part 2 has been much reduced.

Metric flanges to BS 4622[4] and BS 4504[8]

Only two types of flange are included in BS 4622, these are:
  NP10 for 80–700 mm nominal internal diameter, (see Table 10.32).
  NP16 for 80–300 mm only nominal internal diameter, (see Table 10.33).
These flanges are not interchangeable with the inch based flanges to BS 10[8] and BS

Table 10.32   STANDARD FLANGES—TYPE NP10

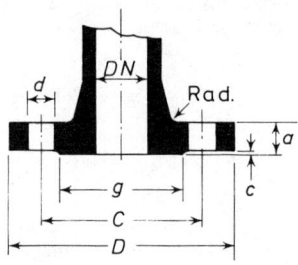

Nominal internal diameter DN	D	C Pitch circle diameter	g	a	c	Bolt holes		Radius	Mass*
						No.	d		
mm	mm	mm	mm	mm	mm		mm	mm	kg
80	200	160	133	24·0	3	8	19	6	3·7
100	220	180	153	25·0	3	8	19	6	4·2
150	285	240	209	26·0	3	8	23	6	6·7
200	340	295	264	27·5	3	8	23	8	9·3
250	400	350	319	29·0	3	12	23	8	12·0
300	455	400	367	31·5	4	12	23	8	14·8
350	505	460	427	33·0	4	16	23	8	19·0
400	565	515	477	34·0	4	16	28	10	23·4
450	615	565	527	35·5	4	20	28	10	27·7
500	670	620	582	37·0	4	20	28	10	32·1
600	780	725	682	41·0	5	20	31	10	44·0
700	895	840	797	43·5	5	24	31	10	59·9

*The mass of the flange is exclusive of the mass of the barrel of the pipe or fitting.

2035[9]. Metric flanges generally incorporate a facing strip which is contrary to current British practice.

Because the maximum pressure rating for flanged pipes and fittings above 300 mm is 10 bar, the NP10 flange only is included for the larger sizes. Details of NP16 flanges and NP25 flanges in grey iron are included in BS 4504 which covers metric size flanges in all materials, these details could be used if required.

There are minor differences between the flanges specified in BS 4622 and those in BS 4504 but it must be emphasised that these do not affect jointing.

Table 10.33 STANDARD FLANGES—TYPE NP16

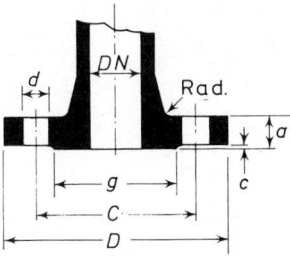

Nominal internal diameter DN	D	C Pitch circle diameter	g	a	c	Bolt holes No.	d	Radius	Mass*
mm	mm	mm	mm	mm	mm		mm	mm	kg
80	200	160	133	24·0	3	8	19	6	3·7
100	220	180	153	25·0	3	8	19	6	4·2
150	285	240	209	26·0	3	8	23	6	6·7
200	340	295	264	27·5	3	12	23	8	9·3
250	400	355	319	29·0	3	12	28	8	12·0
300	455	410	367	31·5	4	12	28	8	14·8

*The mass of the flange is exclusive of the mass of the barrel of the pipe or fitting.

The technical specification BS 4622

The technical specification in BS 4622 is based on ISO/R13 but has been redrafted to meet the requirements of United Kingdom practice and the accepted form of presentation in British Standards.

A direct comparison between the mechanical requirements of BS 4622 and those of BS 1211, BS 78, Part 2, and BS 2035 is complicated but they are generally similar.

BS 4772 'DUCTILE IRON PIPES AND FITTINGS'

SIZES

In common with BS 4622, BS 4772 is based on the proposals of the International Standards Organisation for ductile iron pipe and fittings.

BS 4772 has points in common with BS 4622. The diameter range is similar but has been extended to include nominal internal diameters 800, 900, 1000, 1100 and 1200 mm (see Table 10.28). The mean external diameters of BS 4622 centrifugally cast pipe, taken from ISO/R13, are used for ductile spun iron pipe with obvious advantages particularly in respect of interchangeability (see Table 10.29).

For the majority of working conditions, a single thickness class of ductile spun iron pipe, designated Class K9 in line with the system of designations used in the ISO proposals, is used (see Table 10.30). Class K9 is, in fact, the only standard thickness spigot and socket pipe in the British Standard but provision is made for pipes having either thinner or greater thicknesses for special cases.

The shapes and laying dimensions of the fittings accord with those in BS 4622 and there is a single class of fitting. As for pipes, the class designations of fittings are based on formulae detailed in the ISO proposals, and are directly related to the mean wall thicknesses of the castings. Consequently, tees, which are thicker than other castings,

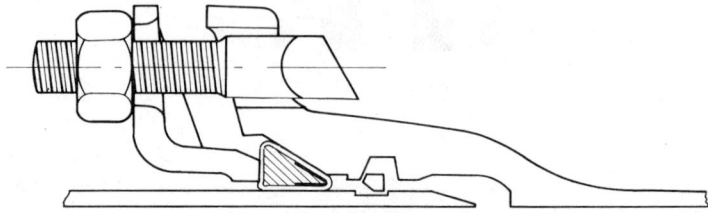

Figure 10.11. The ductile iron Stanlock self-anchoring joint

have the class designation K14 and all other types the designation K12. Provision is made for thicker fittings and there is little doubt that these will be required in the larger diameters, say above 600 mm, to suit the much higher working pressures now being envisaged for ductile iron pipelines.

Ductile iron pipes and fittings are available with flexible joints of the 'push-in' type, generally used for water or sewage mains, and the mechanical type which, for gas mains only, may be fitted with a self-anchoring mechanism (see Figure 10.11).

Pressure ratings for Metric Ductile iron pipes and fittings

The British Standard for Ductile Iron Pipes and Fittings does not include pressure ratings as these are now shown only in Code of Practice, CP 2010 Pt 3, for Design and Construction of Iron Pipelines in Land[5]. This is in line with the principle that a British Standard shall cover the manufacture of the product but that a Code of Practice shall cover the use of the product. The maximum hydraulic working pressures for standard ductile iron pipes and fittings with flexible joints are:

    80–300 mm nominal internal diameter 40 bar
    350–600 mm nominal internal diameter 25 bar
    700–1200 mm nominal internal diameter 16 bar

The Code also gives guidance about pressure ratings for flanged pipes and fittings and makes recommendations in regard to maximum field or site hydrostatic test pressures for standard ductile iron pipes and fittings as well as for grey iron pipes and fittings. This recognises the desirability of differentiating between the test pressures applicable to pipes and fittings of these two materials in order to take advantage of the superior physical strength, toughness and resistance to transit damage of ductile iron.

The field hydrostatic test pressure should always be sufficient to test the structural soundness and leak-tightness of the pipeline but there are considerable advantages in avoiding applying unnecessarily high test pressures which are unlikely, even under surge conditions, to be repeated in service. Consequently, the Code recommends that the test pressure be:

    (a) the maximum sustained working pressure, or the maximum static pressure, plus
        5 bar

or (b) The sum of the maximum sustained working pressure, or the maximum static pressure, and the maximum calculated surge pressure.

The Code suggests that, where the surge pressure may cause the total pressure to exceed the normal working or static pressure by more than 10%, protective devices should be fitted to the pipeline. It follows that (a) will normally be the greater. Consequently, the Code bases its recommended field hydrostatic test pressures on the maximum hydraulic pressure rating, plus 5 bar, giving (for pipelines with flexible joints):

    80–300 mm nominal internal diameter, 45 bar
    350–600 mm nominal internal diameter, 30 bar
    700–1200 mm nominal internal diameter, 21 bar

Works proof test pressures for metric ductile iron pipes and fittings

The works hydrostatic proof test pressures for pipes and fittings follow the provisions in the ISO proposals, those for pipes being:

    80–300 mm nominal internal diameter, 50 bar
    350–600 mm nominal internal diameter, 40 bar
    700–1000 mm nominal internal diameter, 32 bar
    1100–1200 mm nominal internal diameter, 25 bar

The application of high hydrostatic test pressures to ductile iron fittings presents special problems, particularly those of anchorage and sealing. Further, as the end loadings required to effect the seal are applied before the test pressure, there is a possibility of the castings becoming distorted due to the high stresses applied.

In view of these factors, the application of high works hydraulic test pressures to fittings is precluded and both the British Standard and Code, and also the ISO proposals, specify only a leak tightness test with water at moderately high pressure, these pressures being:

    80–300 mm nominal internal diameter, 25 bar
    350–600 mm nominal internal diameter, 16 bar
    700–1200 mm nominal internal diameter, 10 bar

Provision is also made for the air pressure proof testing at works of ductile iron gas pipes and fittings at a pressure of 3·5 bar.

Metric ductile iron flanges

The British Standard[2] includes four types of metric flange.

    NP10 and NP16. 80–1200 mm nominal internal diameter (see Tables 10.34 and 10.35)
    NP25.   80–600 mm nominal internal diameter (see Table 10.36)
    NP40.   80–300 mm nominal internal diameter (see Table 10.37)

As in the case of the metric size grey iron flanges, these flange tables already incorporate a degree of rationalisation. Ductile iron flanges are not included in BS 4504 and this standard will refer readers to BS 4772.

The technical specification

The technical specification included in the new British Standard is based on that in the ISO proposals with the inclusion of certain more stringent requirements; for example the section covering surface defects has been greatly improved and the differences between routine quality control tests and the purchaser's acceptance tests are more clearly defined.

Interchangeability

Grey and ductile iron metric pipes and fittings, whether they be spigot and socket or flanged, are dimensionally different from the corresponding inch sizes and are not interchangeable (see Table 10.29).

In order to facilitate the incorporation of metric components into existing inch pipelines the following procedures have been proposed:

1. Change collars may be used where the differential thrusts, resulting from the action of pressure on the unequal pressure areas inside the collar, are insufficient to cause movement of the collar along the smaller pipe or where the purchaser is

Table 10.34   STANDARD FLANGES—TYPE NP10

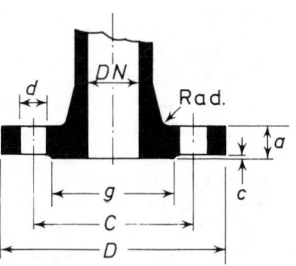

Nominal internal diameter DN	D	C Pitch circle diameter	g	a	c	Bolt holes No.	Bolt holes d	Radius	Mass*
mm	mm	mm	mm	mm	mm		mm	mm	kg
80	200	160	133	19	3	8	19	6	2·9
100	220	180	153	19	3	8	19	6	3·3
150	285	240	209	19	3	8	19	6	4·9
200	340	295	264	20	3	8	23	8	6·8
250	400	350	319	22	3	12	23	8	9·6
300	455	400	367	24·5	4	12	23	8	12·8
350	505	460	427	24·5	4	16	23	8	14·1
400	565	515	477	24·5	4	16	28	10	16·3
450	615	565	527	25·5	4	20	28	10	18·1
500	670	620	582	26·5	4	20	28	10	21·8
600	780	725	682	30	5	20	31	10	30·8
700	895	840	797	32·5	5	24	31	10	40·5
800	1 015	950	904	35	5	24	34	10	54·8
900	1 115	1 050	1 004	37·5	5	28	34	10	64·3
1 000	1 230	1 160	1 111	40	5	28	37	10	81·4
1 100	1 340	1 270	1 221	42·5	5	32	37	10	98·7
1 200	1 455	1 380	1 328	45	5	32	40	10	120·7

prepared to anchor the collar to resist this thrust. Change collars are included in BS 4622 and in the British Standard for Ductile Iron Pipes and Fittings but it is unlikely that grey iron castings will be produced by the major manufacturers.

2. Where pressures and the combination of pressure areas is such that substantial differential thrusts would be developed in a change collar, spigot and socket change pieces of compact design should be used. These castings will not be manufactured in grey iron because of the limitation of the 10 bar rating for fittings above 300 mm.

Table 10.35 STANDARD FLANGES—TYPE NP16

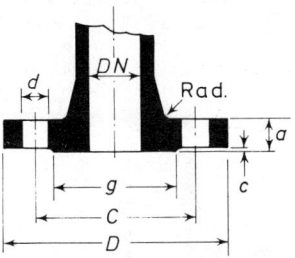

Nominal internal diameter DN	D	C Pitch circle diameter	g	a	c	Bolt holes No.	d	Radius	Mass*
mm	mm	mm	mm	mm	mm		mm	mm	kg
80	200	160	133	19	3	8	19	6	2·9
100	220	180	153	19	3	8	19	6	3·3
150	285	240	209	19	3	8	23	6	4·9
200	340	295	264	20	3	12	23	8	6·6
250	400	355	319	22	3	12	28	8	9·2
300	455	410	367	24·5	4	12	28	8	12·4
350	520	470	432	26·5	4	16	28	8	17·2
400	580	525	484	28	4	16	31	10	21·9
450	640	585	544	30	4	20	31	10	26·7
500	715	650	606	31·5	4	20	34	10	37
600	840	770	721	36	5	20	37	10	57·3
700	910	840	791	39·5	5	24	37	10	55·6
800	1 025	950	898	43	5	24	40	10	74
900	1 125	1 050	998	46·5	5	28	40	10	88·2
1 000	1 255	1 170	1 115	50	5	28	43	10	122·9
1 100	1 355	1 270	1 215	53·5	5	32	43	10	141·4
1 200	1 485	1 390	1 329	57	5	32	49	10	185·1

Generally change collars and change pieces will be provided with flexible joints only at the metric size end. The inch size with a flexible joint or with a lead-caulked joint as the latter may be necessary where a connection is to be made on an old main.

It is anticipated that there will be some demand for metric size flanged fittings with BS 10 flanges and, possible, ASA 125[10] flanges and this requirement can be accommodated but it is expected that the need to supply fittings with BS 10 flanges, already classed as obsolescent by the BSI, will gradually fall off.

Identification of metric pipes and fittings

The introduction of metric pipe into a previously inch-size market could obviously lead to difficulties of identification. Identification markings are based on two colours:

Red indicates 'ductile iron'.

Blue indicates 'metric size'.

The nominal internal diameter will be stencilled near the socket or near one flange of pipes and will be cast on fittings. The class designation will be stencilled on the barrel

Table 10.36 STANDARD FLANGES—TYPE NP25

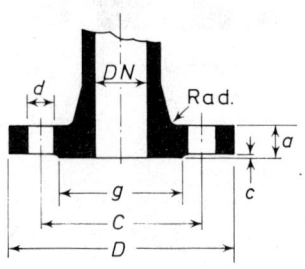

Nominal internal diameter DN	D	C Pitch circle diameter	g	a	c	Bolt holes		Radius	Mass*
						No.	d		
mm	mm	mm	mm	mm	mm		mm	mm	kg
80	200	160	133	19	3	8	19	6	2·9
100	235	190	159	19	3	8	23	6	3·8
150	300	250	214	20	3	8	28	6	5·9
200	360	310	274	22	3	12	28	8	8·7
250	425	370	331	24·5	3	12	31	8	13·1
300	485	430	389	27·5	4	16	31	8	18
350	555	490	446	30	4	16	34	8	25·5
400	620	550	503	32	4	16	37	10	33·2
450	670	600	553	34·5	4	20	37	10	42·2
500	730	660	613	36·5	4	20	37	10	48·7
600	845	770	718	42	5	20	40	10	71·5

of the pipe. In the case of metric grey iron pipe the ISO/R13, class designation may appear in addition to that of BS 4622. Metric sockets will carry a band of blue colour but metric flanges will be identified only by means of the raised face, a noticeable feature of the metric design.

It is particularly necessary to distinguish between inch flanged ductile iron fittings, formerly manufactured to BS 2035 laying dimensions, and the new metric bodied flanged ductile iron fittings. The latter will have the letters 'ND' cast on.

In order to distinguish between the inch and the new metric flexible joint gaskets, the latter are marked with the diameter in millimetres.

Mean bores of metric pipe

The internal dimensions of both grey and ductile iron fittings in sizes up to and including 600 mm are identical. In the case of all-socket fittings, the actual internal diameters of these fittings have been made equal to those of the equivalent Class K9 ductile iron pipe. Consequently, there is no restriction in bore at the fitting, indeed in a grey iron pipeline

Table 10.37 STANDARD FLANGES—TYPE NP40

(*Refer to diagram for Figure 10.36*)

Nominal internal diameter DN	D	C Pitch circle diameter	g	a	c	Bolt holes No.	d	Radius	Mass*
mm	mm	mm	mm	mm	mm		mm	mm	kg
80	200	160	133	19	3	8	19	6	2·9
100	235	190	159	19	3	8	23	6	3·8
150	300	250	214	26	3	8	28	6	7·9
200	375	320	281	30	3	12	31	8	13·4
250	450	385	343	34	3	12	34	8	21·7
300	515	450	406	39·5	4	16	34	8	31·3

the internal diameters of the fittings are slightly larger than those of the pipes. The metric pipe bore is generally slightly smaller than the equivalent inch size (see Table 10.28).

REFERENCES

1. ISO/R13
   Cast iron pipes, special castings and cast iron parts for pressure main lines, ISO, Geneva (1955)
2. BS 4772
   Ductile iron pipes and fittings (1971)
3. BS 4622
   Grey iron pipes and fittings (1970)
4. BS 1211
   Centrifugally cast (spun) iron pressure pipes for water, gas and sewage (1958)
5. CP 2010; Part 3
   Code of practice for the design and construction of iron pipelines in land (1972).
6. BS 78
   Part 2. Cast iron spigot and socket fittings (vertically cast) and spigot and socket fittings. Part 2. Fittings (1965)
7. BS 4504
   Flanges and bolting for pipes, valves and fittings (metric series) (1969)
8. BS 10
   Flanges and bolting for pipes, valves and fittings (inch series) (1962)
9. BS 2035
   Cast iron flanged pipes and flanged fittings (1966)
10. USAS B16.1
    Cast iron pipe flanges and flanged fittings. ASA 125 USAS New York (1967)

METRICATION IN THE VALVE INDUSTRY

G. A. BROWN

The projected change from Imperial units to a metric system of weights and measures was considered by the valve industry in 1966. The change was recognised as a unique opportunity to improve the effectiveness and the efficiency of the industry by ensuring every effort was made to reduce variety, eliminate traditional but uneconomic practices, maximise quantities, and improve export potential and capability.

To achieve such aims, however, it was necessary, not simply to change inches into millimetres, but to accept the challenge of a new philosophy, and to study the rules, procedures and requirements of the European environment. With this in mind a survey of the valve industry in Western Europe was arranged by the valve manufacturers trade association (BVMA) and the Ministry of Technology.

European valve survey

In order to study as wide a field of activity in the valve industry as possible in Western Europe and Scandinavia, all aspects of marketing, manufacturing, stocking and usage of valves for general purposes were considered.

The survey team contacted twenty-four organisations, and detailed questionnaires were sent to nine countries, Germany, France, Sweden, Denmark, Italy, Netherlands, Austria, Switzerland and Belgium. Visits were made to Germany, France and Sweden and discussions held with plant constructors, valve manufacturers, consulting engineers, agents, stockists, trade associations, and standard organisations.

From the information gathered it was apparent that the DIN system of valve standards was predominant throughout Europe and the Nordic countries, and that 80% of all valves used, were made, or based on, German DIN Standards. Of the remainder, some 85% were manufactured to American valve standards for use primarily in the petroleum and petro-chemical industries.

The survey also revealed a need, and desire, for a new system of European valve standards to supersede existing, and in many cases outdated, national standards. Evidence was found of preliminary investigations by German, French and Swedish national standards organisations towards a rationalised system.

British Standards Institution

As the organisation responsible for the compilation and publication of national standards in the U.K., the British Standards Institution was vitally concerned in the results obtained from the valve survey, and took the initiative in contacting the standards organisations of Germany and France with a view to establishing discussions on the subject of European Valve Standards.

A ready-made forum was already in existence for discussions in the form of the European Standards Co-ordinating Committee (CEN) and a sub-committee was formed, consisting of user, manufacturer and standards organisation representatives from Germany, France and the U.K. Observer members from the Nordic countries were also included in the sub-committee which was given the reference CENTRI 13.

European valve standards

Meetings of sub-committee CENTRI 13 were held during 1970/1971 and progress has been made towards the establishment of European valve standards, by the agreement on a number of basic parameters of nominal sizes, nominal pressures, face to face dimensions and dimensional tolerances.

NOMINAL SIZES

The nominal size range for Standard Valves for general purposes is from 10–1000 mm, and in the following millimetre sizes:

10	15	20	25	32	40	50	65	80	100
125	150	200	250	300	350	400	450	500	600
700	800	900	1000						

In the case of certain valve types, e.g. flanged butterfly valves, the range may be extended by the following additional sizes.

1200   1400   1600   1800   2000

NOMINAL PRESSURES

A range of nominal pressures for general purpose valves has been agreed. These nominal pressures are the maximum design working pressures in bars (1 bar $= 10^5$ N/m^2) at 20°C and are:

10   16   25   40   64   100 bar

For low pressure valves 200 mm nominal size and larger an isomorphic series, where the nominal pressure reduces as the size increases is included having ratings of:

1,   1·6,   2·5,   4,   6 bar (see Table 10·38)

FACE TO FACE DIMENSIONS

Agreement has been reached on face to face dimensions of a wide range of valve types. These dimensions are given in tables 10.38 to 10.46.

ACTUAL BORES

In Tables 10.38 to 10.46 the size of the valve has been expressed as 'nominal size', as dimensional differences between the bore of the valve, inferred by the size, and the actual bore, occur in the larger sized valves for the higher nominal pressures.

Table 10.47 gives details of actual bores related to nominal sizes and nominal pressures.

Other requirements

In addition to the basic requirements for valve standards of size, pressure ratings, and face-to-face dimensions, other criteria relating to materials, testing, performance and

Table 10.38 FLANGED GATE VALVES (Face to face dimensions, mm)

Nominal pressure (bar)	1 to 6 (see below)				10 and 16			25 and 40		64 and 100
Basic material	Iron		Steel		Iron and steel			Steel		Steel
Nominal size mm	NP	Face to face	NP	Face to face	Short	Medium	Long		Alternative	
10	—	—	—	—	102	—	—	—	—	—
15	—	—	—	—	108	150	—	140	—	165
20	—	—	—	—	117	160	—	152	—	190
25	—	—	—	—	127	160	—	165	—	216
32	—	—	—	—	140	180	—	178	—	229
40	—	—	—	—	165	190	240	190	203	241
50	—	—	—	—	178	200	250	216	254	292
65	—	—	—	—	190	215	270	241	273	330
80	—	—	—	—	203	230	280	283	292	356
100	—	—	—	—	229	250	300	305	330	432
125	—	—	—	—	254	275	325	381	381	508
150	—	—	—	—	267	300	350	403	394	559
200	6	⎧ 228	—	—	292	350	400	419	445	660
250		⎨ 255	—	—	330	400	450	457	457	787
300		⎩ 285			356	425	500	502	508	838
350	4	⎧ 315	6	⎧ 315	381	475	550	762	508	889
400		⎪ 340		⎪ 340	406	525	600	838	559	991
450		⎨ 360		⎨ 360	432	575	650	914	610	1 092
500		⎩ 380		⎩ 380	457	625	700	991	—	1 194
600	2·5	⎰ 425	4	⎰ 425	508	725	800	1 143	—	1 397
700		⎱ 470		⎱ 470	620	825	900	—	—	—
800	1·6	510	2·5	510	680	925	1 000	—	—	—
900	1	⎰ 555	1·6	⎰ 555	740	1 025	1 100	—	—	—
1 000		⎱ 600		⎱ 600	760	1 125	1 200	—	—	—

marking are necessary. Such details have yet to be discussed and agreed by CENTRI 13 before complete European Valve Standards can be finalised.

World standards for valves

The adoption of European Standards by the member countries of EEC and EFTA will assist greatly towards the preparation and acceptance of International Valve Standards on a world wide basis, by the processing of such European Standards through the International Organisation for Standards (ISO). Such a process is the logical sequel to the current endeavours.

Table 10.39 FLANGED SCREWDOWN STOP VALVES

Type	Straight (globe) Face to face dimensions mm				Angle Centre to face dimensions mm			
Nominal pressure bar	10 and 16		16/25 and 40	64 and 100	10 and 16		16/25 and 40	64 and 100
Basic material	Iron		Steel	Steel	Iron		Steel	Steel
Nominal size mm	Short	Long	—	—	Short	Long	—	—
10	—	120	120	210	—	85	85	105
15	—	130	130	210	—	90	90	105
20	—	150	150	230	—	95	95	115
25	—	160	160	230	—	100	100	115
32	—	180	180	260	—	105	105	130
40	165	200	200	260	82	115	115	130
50	203	230	230	300	102	125	125	150
65	216	290	290	340	108	145	145	170
80	241	310	310	380	121	155	155	190
100	292	350	350	430	146	175	175	215
125	330	400	400	500	165	200	200	250
150	356	480	480	550	178	225	225	275
200	495	600	600	650	248	275	275	325
250	622	730	730	—	311	325	325	—
300	698	850	850	—	350	375	375	—
350	787	980	980	—	394	425	425	—
400	914	1 100	1 100	—	457	475	475	—

Table 10.40 FLANGED LIFT CHECK VALVES

Type	Straight (globe) Face to face dimensions mm				Angle Centre to face dimensions mm			
Nominal pressure bar	10 and 16		16/25 and 40	64 and 100	10 and 16		16/25 and 40	64 and 100
Basic material	Iron		Steel	Steel	Iron		Steel	Steel
Nominal size mm	Short	Long	—	—	Short	Long	—	—
10	—	120	120	210	—	85	85	105
15	—	130	130	210	—	90	90	105
20	—	150	150	230	—	95	95	115
25	—	160	160	230	—	100	100	115
32	—	180	180	260	—	105	105	130
40	165	200	200	260	82	115	115	130
50	203	230	230	300	102	125	125	150
65	216	290	290	340	108	145	145	170
80	241	310	310	380	121	155	155	190
100	292	350	350	430	146	175	175	215
125	330	400	400	500	165	200	200	250
150	356	480	480	550	178	225	225	275
200	495	600	600	650	248	275	275	325
250	622	730	730	—	311	325	325	—
300	698	850	850	—	350	375	375	—
350	787	980	980	—	394	425	425	—
400	914	1 100	1 100	—	457	475	475	—

Table 10.41 FLANGED SWING CHECK VALVES

Type	Straight Face to face dimensions mm				Angle Centre to face mm	
Nominal pressure bar	10 and 16		16/25 and 40	64 and 100	25 and 40	64
Basic material	Iron		Steel	Steel	Steel	Steel
Nominal size mm	Short	Long	—	—	—	—
10	—	120	120	210	—	—
15	—	130	130	210	—	—
20	—	150	150	230	—	—
25	—	160	160	230	—	—
32	—	180	180	260	—	—
40	165	200	200	260	—	—
50	203	230	230	300	—	—
65	216	290	290	340	—	—
80	241	310	310	380	—	—
100	292	350	350	430	—	—
125	330	400	400	500	—	—
150	356	480	480	550	—	—
200	495	600	600	650	275	325
250	622	730	730	775	325	390
300	698	850	850	900	375	450
350	787	980	980	1 025	425	515
400	914	1 100	1 100	1 150	475	575
450	965	1 200	1 200	1 275	500	—
500	1 067	1 250	1 250	1 400	575	700
600	1 219	1 450	1 450	—	675	—
700	—	1 650	1 650	—	775	—
800	—	1 850	1 850	—	875	—
900	—	2 050	2 050	—	975	—
1 000	—	2 250	2 250	—	1 075	—

Table 10.42 BUTTERFLY AND DIAPHRAGM VALVES (Face to face dimensions, mm)

←———————— Butterfly valves ————————				→←— Diaphragm valves —→		
Type	Flanged			Wafer	Flanged	
Nominal pressure bar	10 and 16	25	—	6/10/16	10/16	25/40
Nominal size mm	—	—		Short	Long	
10	—	—	—	95	120	120
15	—	—	—	102	130	130
20	—	—	—	117	150	150
25	—	—	—	127	160	160
32	—	—	—	146	180	180
40	106	140	33	159	200	200
50	108	150	43	190	230	230
65	112	170	46	216	290	290
80	114	180	64	254	310	310
100	127	190	64	305	350	350
125	140	200	70	356	400	400
150	140	210	76	406	480	480
200	152	230	89	521	600	600
250	165	250	114	635	730	730
300	178	270	114	749	850	850
350	190	290	127	—	—	—
400	216	310	140	—	—	—
450	222	330	152	—	—	—
500	229	350	152	—	—	—
600	267	390	178	—	—	—
700	292	430	229	—	—	—
800	318	470	241	—	—	—
900	330	510	241	—	—	—
1 000	410	550	300	—	—	—
1 200	470	630	350	—	—	—
1 400	530	710	—	—	—	—
1 600	600	790	—	—	—	—
1 800	670	870	—	—	—	—
2 000	760	950	—	—	—	—

Table 10.43   FLANGED PLUG AND REDUCED BORE BALL VALVES
(Face to face dimensions, mm)

Nominal pressure bar	10 and 16		25 and 40	64 and 100
Basic material	Iron and Steel		Steel	Steel
Nominal size mm	Short	Long	—	—
10	102	—	—	—
15	108	—	140	165
20	117	—	152	190
25	127	—	165	216
32	140	—	178	229
40	165	—	190	241
50	178	203	216	292
65	190	222	241	330
80	203	241	283	356
100	229	305	305	432
125	254	356	381	508
150	267	394	403	559
200	292	457	419	660
250	330	533	457	787
300	356	610	502	838
350	381	686	762	889
400	406	762	838	991
450	432	864	914	1 092
500	457	914	991	1 194
600	508	1 067	1 143	1 397

Table 10.44 FLANGED FULL BORE BALL VALVES (Face to face dimensions, mm)

Nominal pressure bar	10 and 16		25 and 40		64 and 100
Basic material	Iron and steel		Steel		Steel
Nominal size mm	Short	Long	Short	Long	
10	102	120	—	—	—
15	108	130	140	—	165
20	117	150	152	—	190
25	127	160	165	—	216
32	140	180	178	—	229
40	165	200	190	200	241
50	203	230	216	230	292
65	222	290	241	290	330
80	241	310	283	310	356
100	305	350	305	350	432
125	356	400	381	400	508
150	394	480	403	480	559
200	457	600	502	600	660
250	533	730	568	730	787
300	610	850	648	850	838
350	686	980	762	980	889
400	762	1 100	838	1 100	991
450	864	—	914	—	1 092
500	914	1 250	991	1 250	1 194
600	1 067	1 450	1 143	1 450	1 397

Table 10.45 STAINLESS STEEL GATE VALVES (Face to face dimensions, mm)

Nominal pressure bar	10	16	25	40	64	100
Nominal size mm	—	—	—	—	—	—
10	108	108	108	108	108	108
15	108	108	108	108	108	108
20	117	117	117	117	117	117
25	127	127	127	127	127	127
32	127	127	127	127	127	228
40	136	136	136	136	190	241
50	142	142	142	216	216	292
65	154	154	154	241	241	330
80	160	160	203	283	283	355
100	172	172	229	305	305	432
125	186	186	254	381	381	—
150	200	200	267	403	403	559
200	228	292	292	419	419	660
250	255	330	330	457	457	787
300	285	355	355	501	501	838
350	315	381	381	—	—	—
400	340	406	406	—	—	—
450	360	432	—	—	—	—
500	380	457	—	—	—	—

Table 10.46 FLANGED OBLIQUE VALVES (Face to face dimensions, mm)

Nominal pressure bar	10 and 16	25 and 40	64 and 100
Basic material	Stainless Steel	Stainless Steel	Stainless Steel
Nominal size mm			
10	120	120	210
15	130	130	210
20	150	150	230
25	160	160	230
32	180	180	260
40	200	200	260
50	230	230	300
65	290	290	340
80	310	310	380
100	350	350	430
125	400	400	500
150	480	480	550
200	600	600	650
250	730	730	—
300	850	850	—
350	980	980	—
400	1 100	1 100	—

Table 10.47 VALVE ACTUAL BORE DIMENSIONS

Nominal pressure bar	NP 10	NP 16	NP 25	NP 40	NP 64	NP 100
Nominal size mm			Actual bore mm			
10	10	10	10	10	10	10
15	15	15	15	15	15	15
20	20	20	20	20	20	20
25	25	25	25	25	25	25
32	32	32	32	32	32	32
40	40	40	40	40	40	40
50	50	50	50	50	50	50
65	65	65	65	65	65	65
80	80	80	80	80	80	80
100	100	100	100	100	100	100
125	125	125	125	125	125	125
150	150	150	150	150	150	150
200	200	200	200	200	200	200
250	250	250	250	250	250	250
300	300	300	300	300	300	300
350	350	335	335	335	335	325
400	400	385	385	385	385	375
450	450	440	430	430	430	420
500	500	490	485	485	480	465
600	600	590	585	585	575	560
700	700	690	686	686	675	660
800	800	792	781	781	770	755
900	900	891	876	876	860	840
1 000	1 000	989	960	960	940	915

METRIC PUMPS

H. H. ANDERSON

Metric dimensional and output standards for single stage centrifugal pumps have been formulated in Germany, Britain and Russia with the aim of providing uniformity to assist users and to facilitate manufacture.

The German DIN standards are in general use in Europe with minor local variations in France, Scandinavia etc.

DIN STANDARD

The German standards authority published DIN 24255 for water duties and DIN 24256[2] for chemical duties. These specifications described single stage, single entry centrifugal pumps with horizontal spindle, end suction, and vertical centre line discharge. The pump casing carries support feet and in the chemical specification an additional support bracket is provided at the drive end of the bearing housing.

A dimension is recorded for the distance to be provided by spacer shaft for withdrawal of rotor, bearing housing and end cover so as to dismantle the pump for maintenance without disturbing the pipe joints or driving unit.

Duties

The DIN specifications cover duties from 12·5 m³/hr 20 m head to 250 m³/h, 80 m head at 2900 rev/min., and from 6·3 m³/h, 5 m head to 315 m³/h, 50 m head at 1450 rev/min.

Dimensions

The size increase between successive pumps in the range is approximately 26% or a multiplying factor of 1·26. This has the advantage of providing for most of the range a geometric scaling which helps designing and manufacture. A size increase of 1·26 gives a flow increase of 1.26 cubed equals 2 and a head increase of 1.26 squared equals 1·59.

For the greater part of the range the flow steps increase by 2 and the head steps increase by approximately 1·59.

Pressures

The chemical range is designed for a maximum working pressure of 16 bar whilst the water range is designed for a maximum working pressure of 10 bar.

Designation

DIN pumps are designated by the outlet branch diameter and the nominal impeller diameter; for example, 50-125 indicates 50 mm outlet branch dia and 125 mm impeller diameter.

OTHER EUROPEAN STANDARDS

In general, these follow DIN standards with minor local variations. For example the French standard NF E44-112 for water shows the support feet on the bearing housing, the pump casing being overhung. Minor variations in flow quantities also occur in certain specifications. The Russian standard GOST 5.515-70[4] for a centrifugal canti-levered pump for chemical industries shows one frame with an output of 45 m³/h 54m head. Overall clearance dimensions and limiting weight are specified, rather than detail sizes and the general emphasis is on quality and reliability.

British Standards

The British Standards define single stage, single entry centrifugal pumps with outlet branch axis passing through the pump spindle centre line.

In line pumps

BS 4082[5] covers vertical in line pumps where in the pump impeller is fitted on to the shaft extension of an electric motor.

In the case of the 'I' type, Part 1 of BS 4082 the inlet and outlet branches are on opposite sides of the pump and lie in a common centre line intersecting the vertical spindle axis. The pump set may be supported by the pipes or may be supported on a base by means of a small foot.

BS 4082, Part 2, defines a 'U' type pump wherein both branches are at one side of the pump.

End suction pumps

BS 4519[6] defines medium duty horizontal chemical pumps with end suction and vertical centre line outlet branch. The pump casing carries the support feet with the optional addition of a small support bracket at the drive end of the bearing housing.

Dimensions are given for the distance to be provided by a spacer shaft to withdraw the rotor, end cover and bearing housing without disturbing pipe joints or driving unit.

Duties for BS 4082 and BS 4519

The pump duties for each type range from 12 m³/h, 10 m head to 384 m³/h, 160 m head at 2900 rev/min except for duties of 384 m³/h. 25 and 40 m head which are at 1450 rev/min. Separate charts show 60 Hz duties.

Duty steps

The step of flow between consecutive sizes in the range is two to one whilst the step of head is 1·6 to one thus affording the advantages of geometric ranges as described above.

Designation

On the main output chart at 2900 rev/min. (with two frames at 1450 rev/min as mentioned above) the flow rates have a quantity code of 10 to 60 to define flows of 12 to 384 m³/h in steps of two, whilst the head code of B to H defines respectively heads of 10, 16, 25, 40, 64, 100, 160 m.

COMPARISON OF DIN STANDARDS AND BRITISH STANDARDS

It should be noted that although the head step of approximately 1.6 is the same, the incidence of heads differs in that the DIN code has a series of heads intermediate to the heads in the BS code.

PRESSURE TESTS

The BS 4082 defines a pressure test for Code R of 60 bar and a pressure test for Code L of 25 bar. BS 4519 defines a pressure test of 25 bar for head codes of B to G covering heads of 10 to 100 m and a pressure test of 40 bar for head code H covering 160 m head.

INTERNATIONAL STANDARDS ORGANISATION

ISO committees are formulating international standards for centrifugal pumps using the various national standards as reference documents.

REFERENCES

1. DIN 24255
   Water single stage end suction pumps, Deutsche Normen (1966)
2. DIN 24256
   Chemical single stage end suction pumps, Deutsche Normen (1968)
3. NFE 44–112
   Single stage end suction centrifugal water pumps. Frame Mounted. Norme Francaise en registree (1967)
4. GOST 5.515–70
   Electrically driven centrifugal cantilevered pumps for chemical duties. State Standard of the USSR
5. BS 4082
   External dimensions for vertical in-line centrifugal pumps (Part 1. 1966 and Part 2. 1969)
6. BS 4519
   Part 1. Horizontal end-suction pumps. Medium duty-chemical (1969)

BIBLIOGRAPHY

ANDERSON, H. H., *Centrifugal Pumps,* 2nd ed, Trade and Technical Press (1972)
Monthly Journal. 'Pumps Pompes Pumpen', Trade and Technical Press

CONDENSATE REMOVAL

D. P. GOLCH

Steam is the most common conveyor of heat used in industry. It would be difficult to name any finished product which did not require heat at some stage of its process, either to evaporate moisture or to bring about a change in the state of the product. This is common the world over, and in some parts of the world there is an additional, but perhaps seasonal, demand for space heating.

The heat is contained in the fuel, be it coal, coke, gas, oil, wood or nuclear energy, but in general it is not convenient to extract the heat from the fuel at the processing point where the heat is required. It is, therefore, common practice to provide for combustion of the fuel to take place in a fired heat exchanger, a boiler, and to transfer the heat to some other heat carrying medium.

The steam temperature can be easily controlled which is an essential requirement of so many processes, and it can also easily be generated and distributed to the points of usage. Some useful hints on pipe sizing and general layout are given in 'Steam Distribution'.

The purpose of this section is the removal of condensate. Immediately steam leaves the boiler and passes into the distribution pipework some heat will be given up due to radiation losses, thus causing some of the steam to condense and revert to water. The amount of steam which condenses in this way, may, as a proportion of the total load, be comparatively small, say no more than 1% of the total load in a well-insulated pipe, but it must be remembered that this a continuous operation for just as long as steam is in the pipe. For example, a 100 mm nominal bore line, well lagged and 30 m long, carrying steam at 7 bar gauge with surrounding air at 10°C will condense approximately 15 kg of steam per hour. So at the end of one hour the line will contain about 15 litres of water; at the end of two hours 30 litres, and so on. This must not be allowed to accumulate, otherwise it will cause waterhammer and be the cause of wet steam reaching the unit heat transfer surfaces.

Provision is usually made for getting this water out of the line by installing drain points. A drain point is correctly formed by providing a large pocket in the steam line into which the condensate will drain from whence it can be discharged through a steam trap.

Such drain pockets should be provided at intervals of about 30–45 m and should be formed by using an equal tee on mains up to say 100 mm nominal bore. On larger mains, for economy, it is practical to use a 100 mm pocket welded to a 150 mm line, a 150 mm pocket on a 200 mm line and so on and generally as shown in Figure 10.12.

A steam trap connected directly into the main as Figure 10.13 should never be used, as it is impossible to ensure that the water travelling at speed along the bottom of the main will pass down through the small bore pipe to the trap.

The steam traps most suited to this application are the Thermodynamic (TD), (Figure 10.14), Inverted Bucket (Figure 10.15) and Open Top Bucket (Figure 10.16) types, the TD trap being particularly suitable for external mains where freezing is possible.

All low points where water can collect in the steam mains should be similarly drained

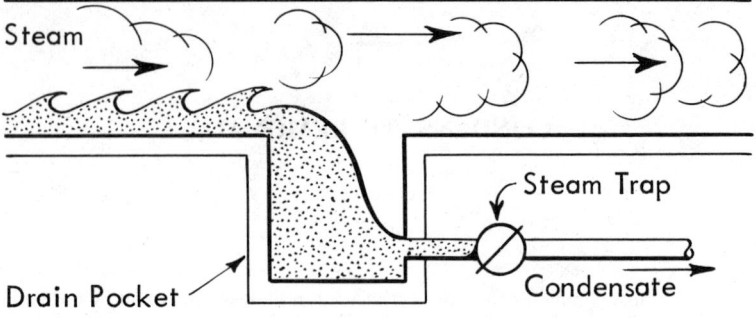

Figure 10.12. Drain pocket in steam line

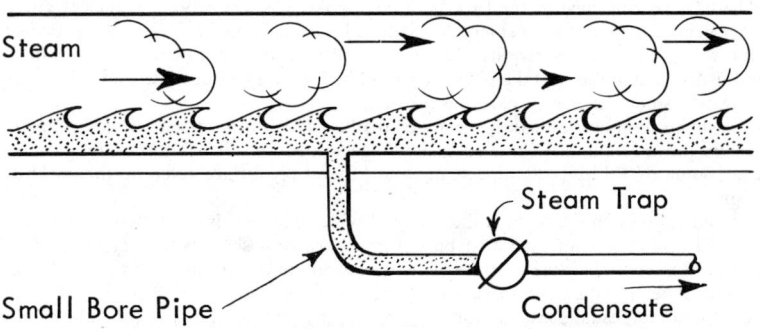

Figure 10.13. Steam trap connected into main (not recommended)

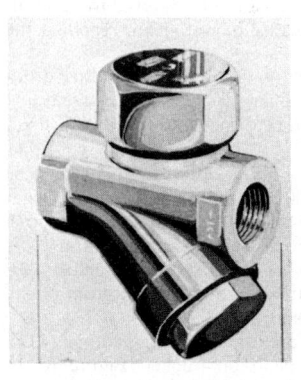

Figure 10.14. TD traps

and trapped. This applies particularly to branch lines which drop to low level before serving plant and are often controlled by some form of automatic valve. The valve may be shut for quite long periods, during which condensate will build up on the upstream side of the valve, so that when next the valve opens as a result of a call for heat in the

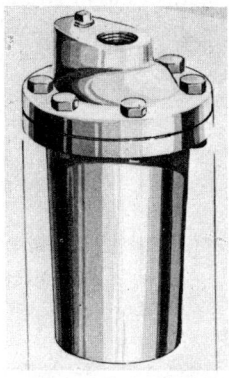

Figure 10.15. Inverted bucket trap

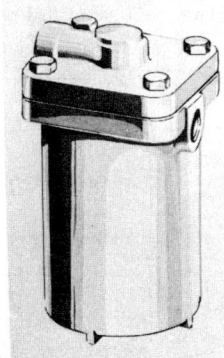

Figure 10.16. Open top bucket trap

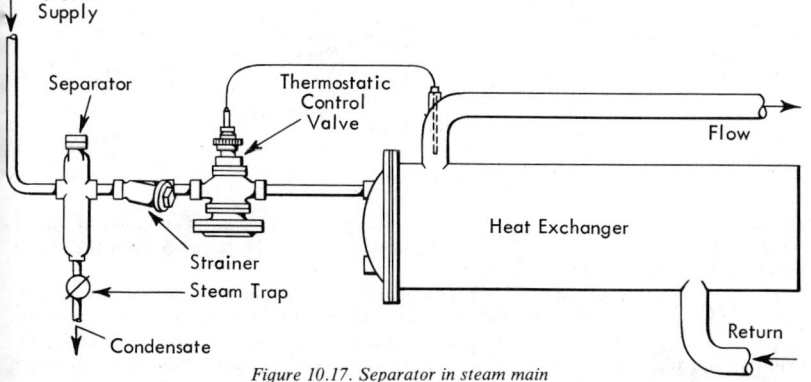

Figure 10.17. Separator in steam main

process this slug of water has to first pass through the valve. This can result in sluggish response of the control, waterhammer and excessive wear on the working parts of the valve.

It is an advantage sometimes to use a separator for this purpose. Apart from the condensate being carried in the steam main, the steam itself may well be wet, carrying particles of water which serve no useful purpose and are detrimental to obtaining maximum heat transfer from the steam. A separator as shown in Figure 10.17 will act as a drain pocket and will also separate the water particles from the steam.

Air venting

There are many steam users operating day shifts only, where the boiler is shutdown each night. The residual steam condenses in the pipelines and air takes its place. So next morning when steam is turned on, it can only pass through the distribution network of pipes at a speed dictated by the release of air, and those systems operating on a daily or even weekly start up can greatly benefit by additional air venting of the steam mains. This can be done automatically using a thermostatic air vent as shown in Figure 10.18.

The purpose of using steam is to provide heat for process or space heating and we now come to the place in the system where that heat is used. In giving up its heat inside the plant, be it a jacketed pan, pipe coil, tracer line, air heater battery or some

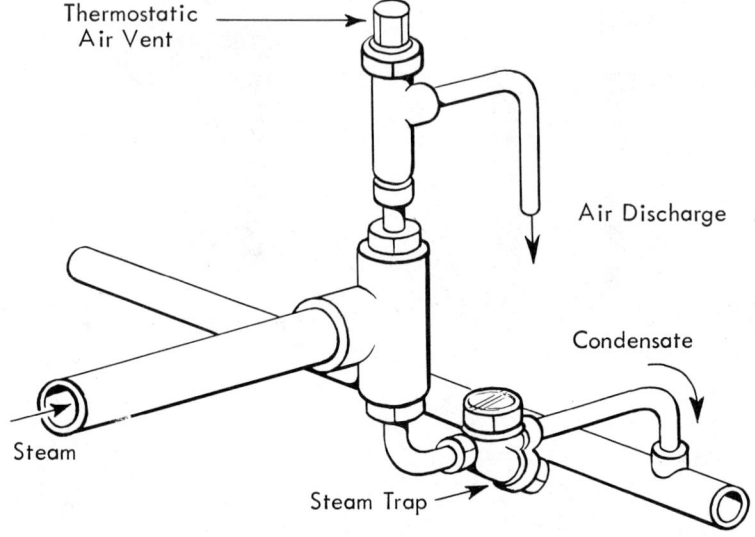

Figure 10.18. Air venting a steam main

specially designed heater unit, the steam condenses to water, each kilogram of steam condensing to a kilogram of water, which obviously must be released from the steam space to make way for more steam.

This, again, is the duty of a steam trap which has an extremely important function. It may have to operate in adverse conditions, such as air binding, steam locking, variations in pressure and load, back pressure, waterhammer, etc. together with possibilities of external and internal corrosion. These are conditions which vary from plant to plant and no single type of steam trap can possibly serve all purposes; for this reason there are many different types of steam trap. It is not the purpose of this chapter to define these traps in detail, and the reader is referred to the specialist literature on the subject, (see 'Steam Trapping and Air Venting'[2] and 'Practical Steam Trapping').

Flash steam

Flash steam is one problem which often confronts a steam user. Most steam traps used to drain steam mains or process or heating plant are of a type which discharge condensate at, or near to, steam temperature. Indeed traps are specially chosen having these characteristics to ensure that the condensate is cleared from the steam spaces as quickly as it forms, so that maximum plant efficiency can be maintained. However, because the traps are handling condensate at or near to steam temperature the resultant discharge will be a mixture of water and steam—flash steam.

A study of the Steam Tables (Tables 2.26 and 2.27) will show that as pressure is increased so the temperature at which water boils is elevated. Conversely a drop in pressure will lower the boiling point. We also know that when steam gives up its latent heat and condenses to water its temperature remains constant.

If we now put these facts into practice we can study why flash steam is formed. For example, a plant, (Figure 10.19) is using steam at 7 bar gauge and is drained by a float type trap which discharges condensate at steam temperature. The steam which is at 7 bar gauge has a temperature of approximately 170°C and condenses to water at the same temperature which drains down into the body of the trap. But it can only remain as water at this temperature so long as it is subjected to a pressure of 7 bar gauge or more. The trap is discharging into a condensate return line at atmospheric pressure at which the boiling point of water is 100°C.

This is now a situation where water at 170°C is discharged through the trap orifice into a condition where it cannot exist as water at a temperature above 100°C. In this case the surplus heat in the high temperature water will revert to latent heat, re-evaporating some of the water into steam, known as flash steam. Just how much water will

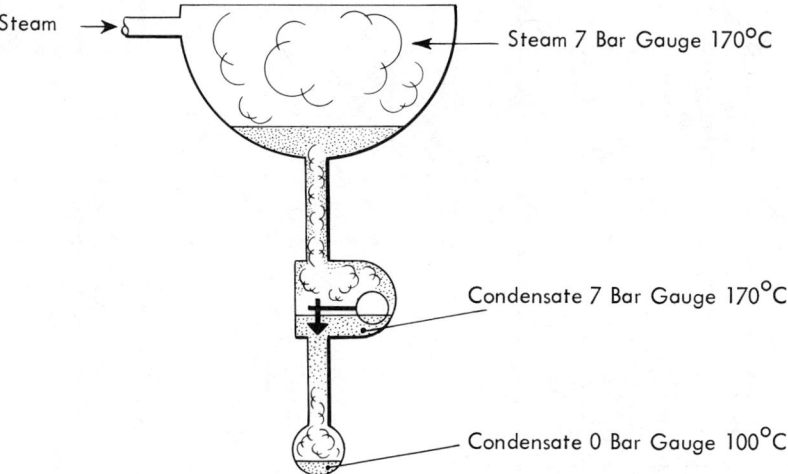

Figure 10.19. Formation of flash steam in condensate line

re-evaporate into steam in this way will depend upon the difference in pressure and the temperature of the condensate at the moment of discharge. Figure 10.20 will give some guidance as to the amount of flash formed from traps discharging condensate at steam temperature.

The flash steam contains valuable heat; it can be returned along with the condensate to the boiler feed tank but over a long distance much of its useful heat may be lost by radiation losses, particularly if the condensate mains are unlagged. Alternatively the flash steam can be separated from the condensate, recovered and used as a low pressure steam supply. In this way greater use is made of the heat content of the flash steam, particularly if the recovery and use are local.[4]

Condensate recovery

When condensate is first discharged from the steam trap it will still contain much of the heat (probably about 25% under average conditions) originally put into it by burning fuel in the boiler.

Furthermore, the water in most cases has been chemically treated to prevent or reduce scale formation inside the boiler and having once removed the offending salts further treatment is seldom necessary. So, because the condensate can be regarded as good hot, treated, water it is ideal for boiler feeding for which reason first choice is to return all condensate to the boilerhouse.

In this way, the temperature of the boiler feed water will be increased by the recovery of heat which otherwise would have been wasted. This can be a decided advantage with modern boiler plant having no economiser, because the nearer the feed water temperature approaches saturation temperature the less work the boiler has to do in converting

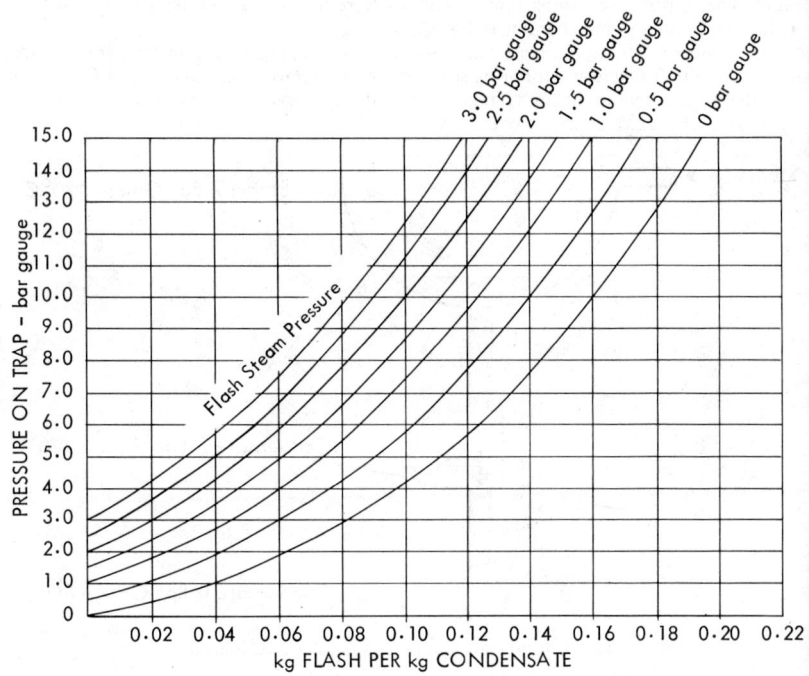

Figure 10.20. Metric flash graph

the water into steam. In fact, for approximately every 5°C rise in feed water temperature 1% less fuel will be used at the boiler.

In some cases the return of all condensate can be an embarrassment by producing water which is too hot to handle. If the problem is simply that due to the high temperature, cavitation takes place at the feed pump, this can be overcome by arranging the feed tank so that it provides a positive head over the pump. The necessary head will depend upon the temperature of the water and the type of pump, but the pump makers will usually provide this information. If it is not practical to put the feed tank at high level an 'automatic pump' or 'pumping trap' can be used to lift the water up to a head tank as Figure 10.21. The automatic pump or pumping trap has the advantage that it fills by gravity from the main feed tank and therefore does not suffer from cavitation no matter how hot the water is.

Another way of overcoming the cavitation problem is to reduce the temperature of the feed water by extracting some of the heat from the condensate before returning it to the boiler house. Flash recovery is an excellent way of doing this.

Condensate pipe sizing

If the condensate is to be collected and returned to the boiler house or to any other point of recovery then there arises the problem of pipe carrying capacity.

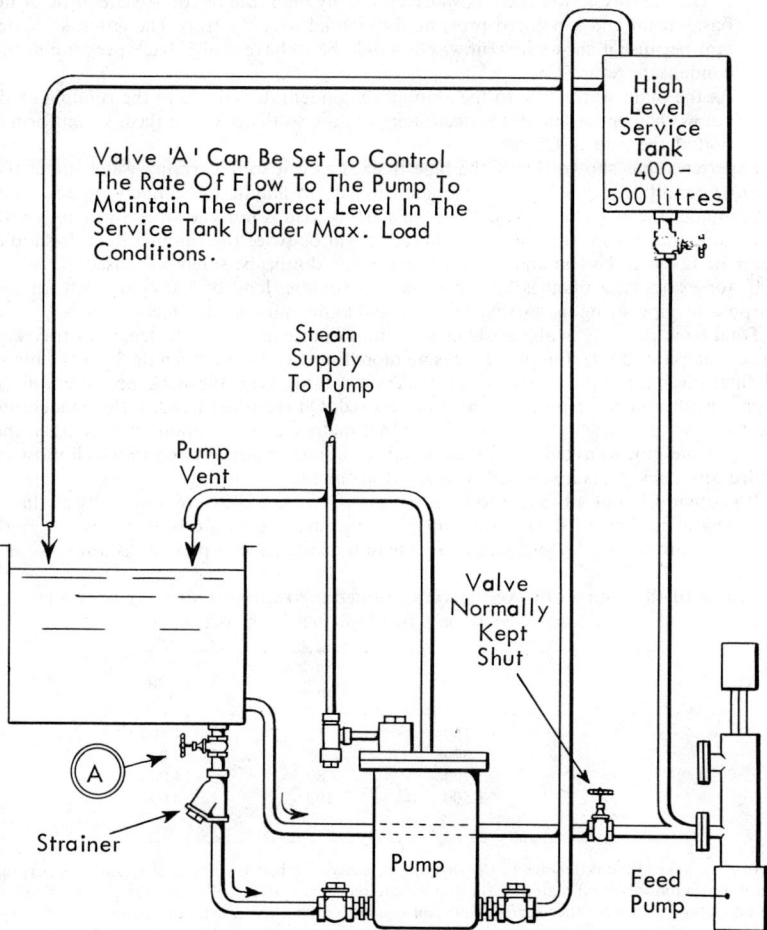

Figure 10.21. *Automatic pumps feeding high level service tank above feed pump*

From time to time many different recommendations have been given about the carrying capacity of condensate pipes, and with some wide differences of opinion about sizes. This is understandable because the conditions vary so much; indeed, to take into consideration the flash steam alone would call for a different chart for every change of pressure/temperature condition. Therefore, a compromise is necessary.

There are basically three stages which have to be considered when deciding the size of pipe:

1. On starting up, some air may be discharged through the trap into the return line.
2. This will be followed by a large amount of cool condensate. When starting up, the plant is cold and this will condense steam very rapidly, so that during this period steam may be condensed at some two or even three times the normal running rate. What is more, because of this heavy condensing rate, the pressure drop in the plant may be considerable.

   During this period there is an exceptionally high rate of condensate, little or no flash steam and a reduced pressure differential over the trap. The latter is important because it shows how unwise it would be to have a high back pressure in the condensate return line.
3. As the plant warms up, so the amount of condensate reduces to the running load, but as the condensate nears steam temperature so there will be flash steam formation at the trap discharge.

Experience has shown that if the pipe is sized as if it were carrying water under the *starting* conditions, it will have adequate capacity to handle the condensate and flash steam under the *running* conditions. If the starting load is not known it is generally safe to assume that for most average conditions it will be twice the running load. It should never be taken to be less and may, if there is any doubt, be safely increased.

If, for example, a plant is known to have a running load of 700 kg/hr then for the purpose of pipe sizing, a starting load of 1400 kg/hr must be assumed.

Total back pressure is obviously of some importance and also, the length of travel of the condensate. If for example, there is no more than say 3 m between the trap and point of final discharge into a receiver or boiler feed tank, then the back pressure will be insignificant unless the pipe is grossly undersized. On the other hand, if the condensate has to travel over a length of several hundred metres, as is common practice, then the sizing of the pipe to avoid excessive resistance is most important and to which must be added any back pressure caused by any lift in the line.

It is obviously not advisable to impose too high a back pressure, especially as this is normally at its highest during start up when the steam pressure is at its lowest. High back pressure cannot be tolerated at all where the initial steam pressure is low or where

Table 10.48   AMOUNT OF CONDENSATE CARRIED IN STANDARD NOMINAL BORE PIPES WITH A RESISTANCE OF 8 TO 10 mb PER 10 m OF TRAVEL

Size mm	Kg/hr	Size mm	Kg/hr
15	160	50	4 500
20	370	65	9 000
25	700	80	14 000
32	1 500	100	30 000
40			

Note. It makes little difference to the carrying capacity whether the pipe is of steel or copper, the two most commonly used materials for condensate recovery systems. Whereas the copper pipe has the advantage of a smoother surface and consequently offers less resistance to flow, the nominal bore of a steel pipe is usually greater so that what is lost in frictional resistance is gained in cross-sectional area.

the plant temperature is regulated by a modulating type of control. Yet, on the other hand, condensate lines must not be oversized otherwise cost of installation can be prohibitive. A compromise to provide a quick simple answer to this problem, suitable for most everyday installations, is to size on a frictional resistance of 8/10 mb per length of 10 m.

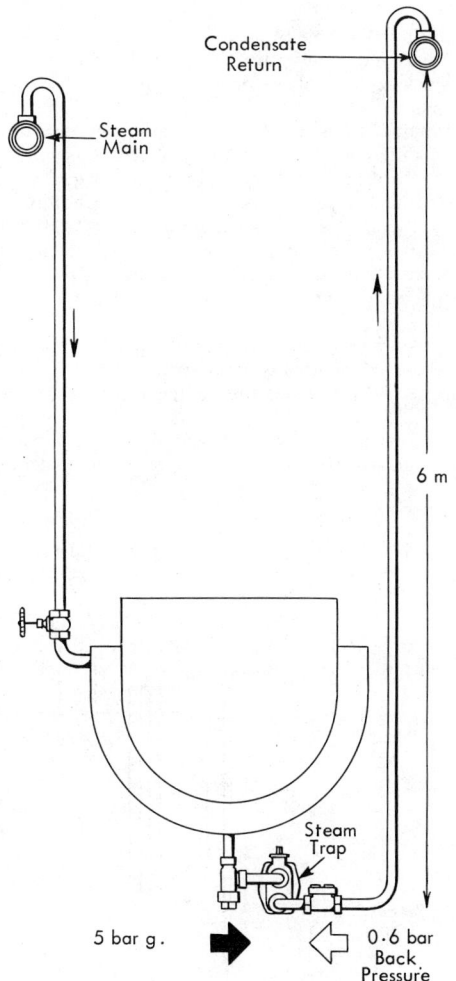

Condensate
Return

Steam
Main

6 m

Steam
Trap

5 bar g .

0·6 bar
Back
Pressure

4·4 BAR DIFFERENTIAL PRESSURE

Figure 10.22. Effect of back pressure

Table 10.48 can be used as a simple sizing chart for most conditions, but it must be remembered that the figures shown represent starting loads.

Lifting the condensate

In many installations it is desirable to lift the condensate from the trap situated at low level up to a high level condensate return line. It is the steam pressure at the trap and not the trap itself which does the lifting. Basically for each 0·1 bar pressure the condensate can be lifted 1 m.

It must be noted that, not only does this present a back pressure, but in so doing it reduces the pressure differential over the trap. When considering back pressure the question must always be asked—will there always be adequate steam pressure *at the trap* to overcome the back pressure and under these conditions will the trap have sufficient capacity. Let us look at a few examples.

A process jacketed pan is operating with steam at 5 bar gauge in the jacket, from which it is reasonable to assume that the pressure at the trap is also in the region of 5 bar gauge when the pan is up to temperature. There is a lift of 6 m after the trap as shown in Figure 10.22.

The 6 m lift imposes a back pressure of approximately 0·6 bar gauge so with 5 bar gauge at the trap there will normally be sufficient pressure to overcome the back pressure of 4·4 bar, i.e. '5-0·6 and this differential will be even lower during start up conditions.

Supposing the same plant was supplied with steam at 1 bar gauge then, whilst theoretically there would still be positive pressure differential available to clear the

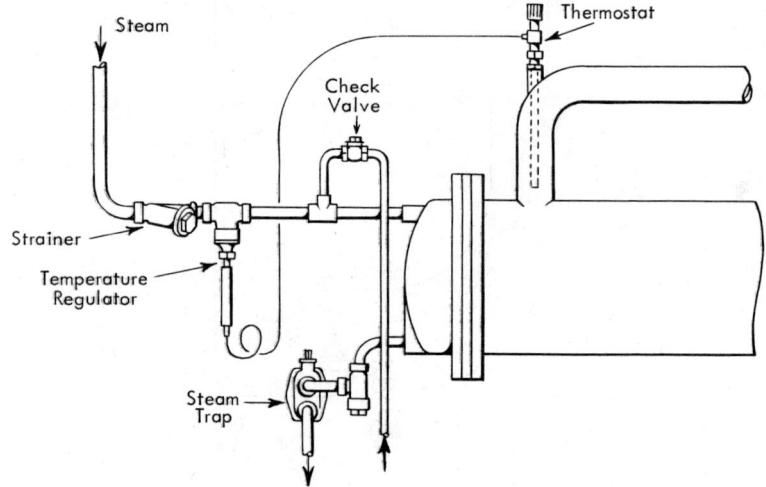

Figure 10.23. Using a check valve as a vacuum breaker

condensate against the back pressure, this could be an extremely risky operation. This is because pressure drop in the plant, especially under starting up conditions, could bring the pressure at the trap down to well below the back pressure, with resultant water-logging of the plant. This is particularly noticeable where plant is thermostatically controlled, especially by valves of the modulating type.

For example, in a steam/water heating calorifier, it is required to meet the maximum heating demand only when the outside air temperature is at its lowest. For all higher outside temperatures the calorifier is usually controlled in such a way that the steam supply is throttled to meet the reduced demand. The effect of this throttling causes a reduction in steam pressure inside the steam space which often falls to zero or below when the load is very light, and because of this it would be quite impossible to lift the condensate. Any attempt to do so would result in waterlogging of the tube bundle with consequent water-hammer and possible tube fracture.

Where because of reduced demand, sub-atmospheric pressures are likely to be encountered, it is obviously impossible under these conditions to get the condensate out

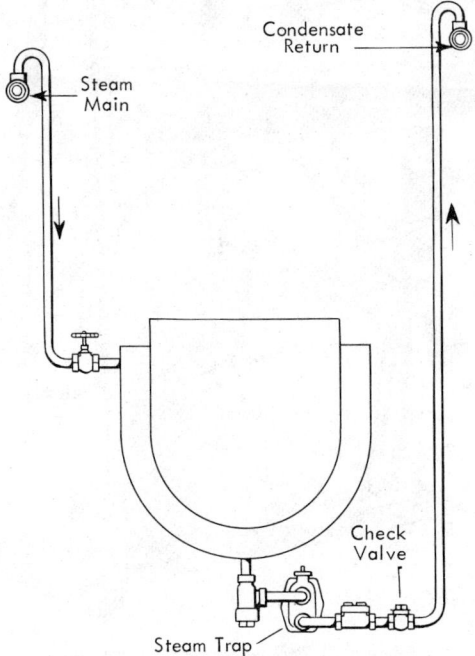

Figure 10.24. Check valve filled between trap discharge and lift

of the plant. A check valve fitted as a vacuum breaker as shown in Figure 10.23 can be an advantage, since it will open when a vacuum tries to form and bleed air into the steam space. Condense can then be discharged through the trap due to the static head. Another use for a check valve is on the discharge side of the trap as shown in Figure 10.24. In this position it will stop water in the rising line from flooding back into the steam space when the plant is shut down.

Pumping the condensate

It will be seen that there are occasions when condensate cannot be lifted and there are certain conditions when it can. But whatever the circumstances it is better not to lift the condensate if it can be avoided. Even under the most favourable conditions, lifting can be an embarrassment on starting up because it causes a back pressure which slows down the clearance of condensate just at a time when this is least wanted. It prevents the clearance of air through the trap and, of course, it means that whenever maintenance has to be carried out it may be necessary to drain down the column of water.

This can be avoided by taking the condensate by natural fall to a receiver and then sending it on its journey back to the boiler house by an independent pump. Electrically

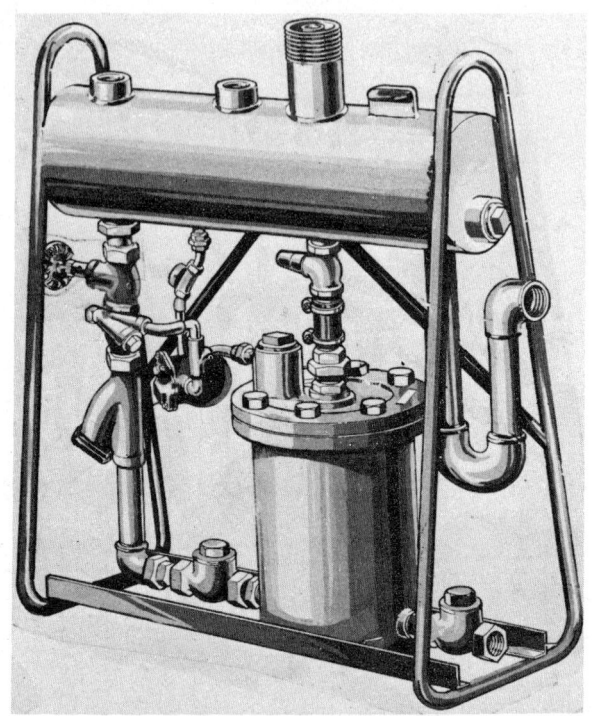

Figure 10.25. Packaged automatic pump unit

driven pumps are frequently used for this purpose and are particularly suitable where the lines are long and tortuous and where the capacities are high. For short and medium runs, one very practical way of delivering the condensate back to the boiler feed tank is by using an 'automatic pump' or 'pumping trap'. These are available as single units or as packaged units complete with receiver, as Figure 10.25.

The pump is operated preferably by steam although compressed air can also be used. This could, however, aerate the condensate and may, in some cases, aggravate corrosion. An advantage of using an automatic pump for this duty is that it can be fitted with a stroke counter so that the amount of condensate handled can be measured.

Dirt

On any new installation, mill scale, building rubble, weld splatter, pipe jointing etc., will normally be present in the pipework system. Even if this is washed out of the

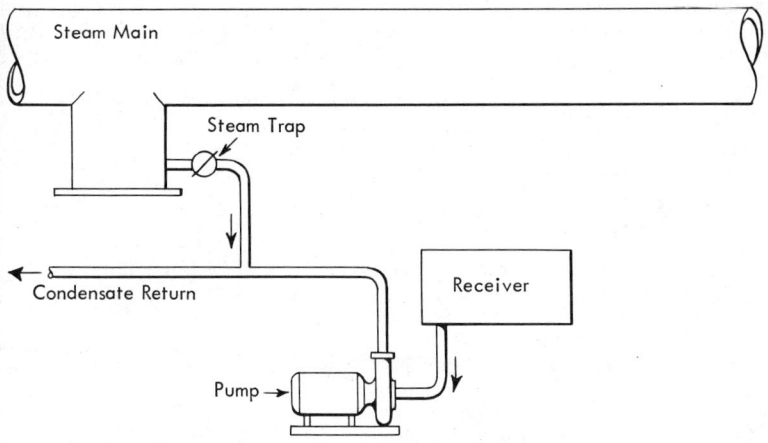

Figure 10.26. A typical layout which may promote waterhammer

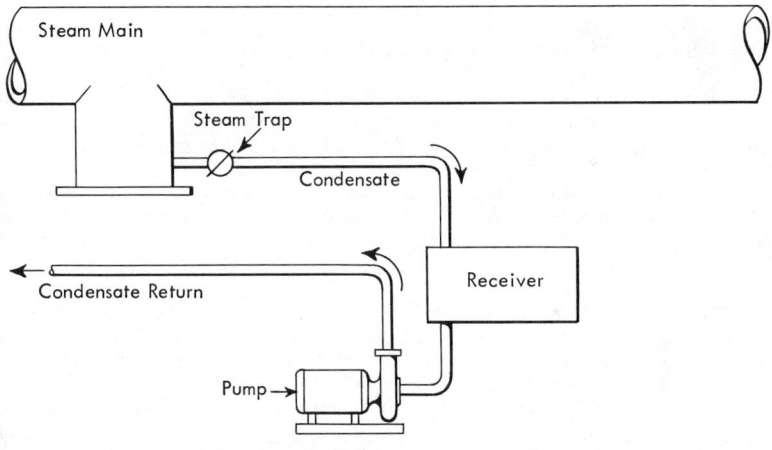

Figure 10.27. A typical layout to prevent waterhammer

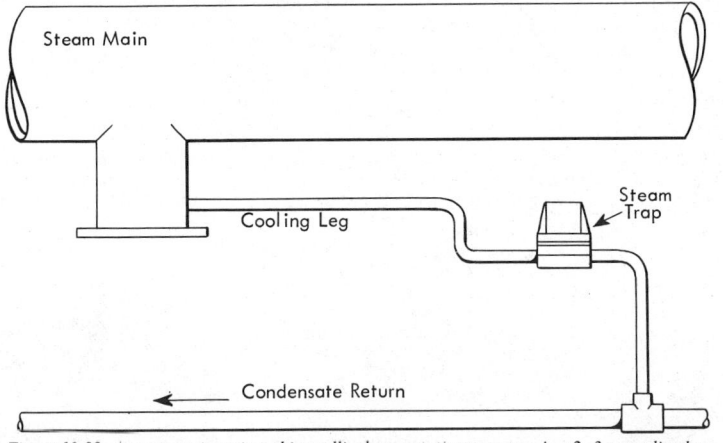

Figure 10.28. A compromise using a bimetallic thermostatic steam trap in a 2–3 m cooling leg

system before commissioning, which is an obvious advantage, unless water treatment and plant management are strictly observed, carry over of solids from the boiler combined with products of corrosion from the pipework will collect to form a mobile sludge.

Steam traps, control valves, reducing valves etc., can only work properly if they are kept clean, so a strainer should always be fitted immediately upstream of all such equipment.

Waterhammer and flooded condensate lines

Taking the discharge from a stream trap into a flooded condensate line can often cause trouble with waterhammer. For example, in Figure 10.26 a pumped return line runs alongside the steam main so it is convenient to take the discharge from the trap draining the line into this return line; but this is almost certain to give trouble.

The function of the trap draining the main is to keep it clear of condensate at all times, so ideally a trap is chosen of a type which will handle condensate at steam temperature. Now this means that, on discharge, some of the condensate will re-evaporate into flash steam and the expanding flash steam must find a place in the return line. But as this is already full of water and under pressure too it can only enter by forcing the water out of the way—a common cause of waterhammer. Another cause of noise under these circumstances is the collapsing of the flash steam bubbles in the water. The obvious answer in a case like this is to extend the trap discharge into the receiver as Figure 10.27.

Extending the trap discharge in this way is not always practical; the nearest pump unit may be much too far away. This is difficult because, as already explained, first choice must be a trap which handles the condensate at steam temperature as a result of which flash steam is formed. It is formed. It is this flash steam which causes the trouble, but to eliminate it entirely would mean using a trap capable of holding back the condensate until it had cooled down to or below the saturation temperature relative to the pressure on the downstream side of the trap. This is not practical because it would inevitably cause waterlogging of the steam main. It could, in fact, simply transfer the waterhammer from one side of the trap to the other.

A compromise is sometimes found by using a trap of the balanced pressure bimetallic type; because this is operating on a balanced pressure principle it is controlling the condensate at a preset temperature below that of the steam. Whilst this cannot, under all circumstances, eliminate the flash steam it does considerably reduce it. As the trap is operating at a temperature below that of the steam some provision must be made for condensate cooling on the upstream side of the trap—this can usually be done by having an adequate drain pocket and a 2 m to 3 m unlagged cooling leg, as Figure 10.28. This is a poor compromise and the only sure way of avoiding trouble is the arrangement as shown in Figure 10.27.

REFERENCES

1. *Steam Distribution*, Spirax-Sarco Ltd., Cheltenham
2. NORTHCROFT. L. G. and BARBER. W. M., *Steam Trapping and Air Venting*, Hutchinson (1968)
3. *Practical Steam Trapping*, Spirax-Sarco Ltd., Cheltenham
4. *Flash Steam Recovery*, Spirax-Sarco Ltd., Cheltenham

COPPER AND COPPER ALLOY TUBES AND FITTINGS —METRIC SIZES

W. FIRTH

The Copper and Copper Alloy Tubes and Fittings Manufacturing Associations have, through the normal BSI channels, been active participants in the appropriate ISO Technical Committee over a long period. Therefore, when the decision to change to metric was taken, the basic agreement on appropriate sizes had largely been achieved.

In accordance with the decision taken previously by BSI to have all wrought copper and copper alloy products incorporated in a series of omnibus standards, previously known as schedules, with each product type in one standard, e.g. BS 2873 'Copper and Copper Alloy Wire', BS 2875 'Copper and Copper Alloy Plate', instead of the numerous individual standards previously in existence, copper and copper alloy tubes are also covered in one standard, namely, BS 2871[1].

Due to the large variety in size, alloy, mechanical properties and preferred sizes of tubes for different applications, it was felt that it would be unwieldy to include all these in one document and probably confusing. BS 2871 is therefore issued in three parts, each part being generally related but not exclusively to the type of end usage.

TUBES FOR WATER, GAS, SANITATION AND HEATING SERVICES

Part 1 covers copper tubes for these services, which is the end usage formerly covered in inch sizes by the individual standards BS 659[2], 1386[3] and 3931[4]. Two grades of copper are specified. These are the phosphorus deoxidised type, i.e. BS 1172[5] non-arsenical and BS 1174[5] arsenical; the arsenical grade would not normally be supplied except by special requirement. Tubes can be supplied in the hard, half hard or annealed temper, each temper of tube being supplied to a specific thickness in relation to outside diameter, and these are detailed in three separate BS tables designated X, Y and Z in order to avoid confusion with letters and numbers which have been quoted previously in standards now superseded.

The outside diameters in all tables are consistent, to ensure compatibility with the appropriate fittings and flanges. For the assistance of designers and users, maximum working pressures at ambient temperature have been calculated using appropriate design stress values, and these are also quoted in each table.

The annealed and half hard temper tubes can be fabricated and bent using the normal machine or spring bending techniques, although it will be appreciated that for either of these methods of bending, the appropriate size springs and/or formers and guides will be necessary. Consultation between the spring and bending machine manufacturers and the tube manufacturers have ensured that these are also available. BS Table X, which is the type of tube in greatest demand, is reproduced as Table 10.49.

TUBES FOR GENERAL PURPOSES

As the title implies, BS 2871 Part 2 'Copper and Copper Alloy Tubes for General Purposes', caters for the many unspecified uses of tubes for a multitude of different purposes.

It specifies five different grades of copper and, in addition, five copper alloys all in regular production. For information purposes, a range of mechanical properties is quoted against each material, and whilst the large majority of tubes supplied would meet these requirements, there may be some at the extremes of the size ranges, i.e., combination of diameters of thickness which could be supplied, which would not necessarily comply but which would, of course, be very near.

Copper alloy tubes in aluminium brass and/or 90/10 copper-nickel-iron alloy are the usual choice of material for salt water circulating systems on board ship. A preferred

Table 10.49

1	2	3	4	5
Size of tube	Outside diameter		Nominal thickness	Maximum working pressures*
	max	min		
mm	mm	mm	mm	bar†
6	6·045	5.965	0·6	133
8	8·045	7·965	0·6	97
10	10·045	9·965	0·6	77
12	12·045	11·965	0·6	63
15	15·045	14·965	0·7	58
18	18·045	17·965	0·8	56
22	22·055	21·975	0·9	51
28	28·055	27·975	0·9	40
35	35·07	34·99	1·2	42
42	42·07	41·99	1·2	35
54	54·07	53·99	1·2	27
76·1	76·30	76·15	1·5	24
108	108·25	108·00	1·5	17
133	133·50	133·25	1·5	14
159	159·50	159·25	2·0	15

* Based on material in ½H condition at 65°C.
† 1 bar = 0·1 N/mm² = 10⁵ N/m².

size range of tubes used for this purpose for many years is listed in BS 2871, Part 2 (Table 3), and reproduced in Table 10.50.

This preferred range of sizes is also the accepted range in the majority of European shipyards and others throughout the world. Consequently, a large rationalisation of pipe sizes is expected from their adoption. Such pipes are, of course, entirely suitable for many other applications, and benefits both for users and manufacturers may be expected in limited ranges for manufacturing and availability of stocks.

TUBES FOR SCREWING AND STEAM SERVICES

Also covered in BS 2871 Part 2, are the requirements for copper tubes suitable for screwing, previously covered by BS 61 Part 2[6], and copper tubes for steam services previously covered by BS 1306 Part 2[7]. Opportunity has been taken to effect further rationalisation in these ranges of sizes quoted in BS tables 5–10 inclusive, by ensuring, whenever possible, similar sizes both of diameter and thickness are utilised or, at least, similar diameters where a different thickness may be necessary.

The design formula for calculation of wall thickness for these applications was amended some time ago to incorporate hoop stress values widely accepted in industry both in the

UK and abroad. The use of this new up-to-date formula has assisted in this rationalisation by enabling the number of tables to be reduced for steam services from six previously specified to four at present.

The outside diameters of tubes suitable for screwing, whilst quoted in millimetres, are the same (direct conversion) as previously quoted in inch units because the pipe thread

Table 10.50

Tube size	O.D. max.	O.D. min.	Preferred standard thickness	Other standard thickness	Other* standard thickness
mm	mm	mm	mm	mm	mm
3	3·045	2·965	0·8	—	—
4	4·045	3·965	0·8	—	—
6	6·045	5·965	0·8	1·0	0·6
8	8·045	7·965	0·8	1·0	0·6
10	10·045	9·965	0·8	1·0	0·6
12	12·045	11·965	0·8	1·0	0·6
16	16·045	15·965	1·0	1·5	—
20	20·055	19·975	1·0	2·0	—
25	25·055	24·975	1·5	2·0	—
30	30·055	29·975	1·5	2·0	—
38	38·07	37·99	1·5	2·0	—
44·5	44·57	44·49	1·5	2·0	—
57	57·20	57·12	1·5	2·0	—
76·1	76·30	76·15	2·0	2·5	—
88·9	89·15	89·00	2·5	3·0	—
108	108·25	108·00	2·5	3·0	—
133	133·50	133·25	2·5	3·0	—
159	159·50	159·25	2·5	3·0	—
193·7	194·50	194·25	3·0	3·5	—
219·1	219·90	218·30	3·0	4·0	—
267	268·00	266·40	3·0	4·0	—
323·9	324·90	323·30	4·0	4·5	—
368	369·00	367·40	4·0	4·5	—
419	420·00	418·40	4·0	4·5	—
457·2	458·20	455·20	4·0	4·5	—
508	509·00	506·00	4·5	5·0	—

Note. Tubes for pipelines are usually required suitable for cold bending or other fabrication processes and unless otherwise specified, are supplied in either the TA, ½H or O condition appropriate to material and size at the manufacturer's discretion.

* Aligns with BS 2871: Part 1.

to BS 21[8] has been adopted internationally by ISO Recommendation No. 7[9] and, in fact, this is the most common type of thread used for pipe purposes in Europe.

Where any requirement for a particular size or application cannot be met by one of the preferred tables of sizes, it can still be supplied to the standard by meeting the tolerances given in BS table 4 or 11, which cover a wider range of unspecified sizes.

HEAT EXCHANGER TUBES

Within the past few years there have been significant changes in methods of manufacture and in some methods of installation of heat exchanger tubes. These are acknowledged in BS 2871 Part 3 'Copper and Copper Alloy Tubes for Heat Exchangers'.

For many years in steam condensers, the predominant method of installation was by packing and/or ferruling into the tubeplate at both ends. For this purpose, a hard tube was necessary to avoid distortion or damage under the packing. Alternatively, packing and/or ferruling at the outlet end and expanded at the inlet end, and in this case one end was locally annealed to enable the tube to be satisfactorily expanded. The hard condition was, in fact, the only one covered in BS 378[10].

In other types of heat exchangers, particularly the smaller equipments, e.g., oil coolers, air coolers, etc., expanding the tube into the tubeplate at both ends was the preferred method. This has now generally been adopted for large steam condensers also, to such an extent that it is now the preferred method. However, the different fitting techniques are still used and likely to be so for many years and it is therefore necessary to cater for them by the provision of tubes in suitable tempers, appropriate for the particular method of fixing adopted. Brass tubes, which were supplied in a temper suitable for expansion into tubeplates at both ends, were usually referred to as the heat-treated condition.

This terminology has, however, created confusion in some quarters and in BS 2871 Part 3 the terminology 'temper annealed' is used, which it is felt is more descriptive of the specific temper referred to. The different tempers in which various alloys can be supplied are clearly designated, together with specific range of hardness applicable to each temper, together with a maximum grain size for the temper annealed, or annealed alloy, where these are appropriate.

The preferred method of specifying the thickness of heat exchanger tubes has for many years been a nominal wall thickness with a manufacturing tolerance of plus and minus, although in some instances a minimum wall only was specified with the tolerance purely plus. Whilst this latter method has decreased in recent years, it is occasionally still required. Both methods are therefore covered in BS 2871 Part 3.

QUALITY CONTROL

A significant advance in quality control methods in recent years for heat exchanger tubes and also some types of copper tubes, has been the development and fairly universal acceptance of eddy current methods of non-destructive testing. There are many proprietary types of equipment available which for operation are required to comply with the requirements of BS 3889 Part 2[11]. The actual levels of calibration for inspection purposes are usually decided by the tube manufacturer, although in some instances this is by agreement between the tube manufacturer and the user, e.g., some of the multi-national electricity equipment manufacturers who have wide experience in this subject. This is now included as a standard method of test for the smaller sizes of copper tubing in BS 2871 Part 1, and also for all tubes in BS 2871 Part 3. In some cases of heat exchanger tubes it may be supplemented by agreement between the manufacturer and purchaser by the addition of either an internal pneumatic or hydraulic pressure test.

The main advantage of the eddy current test is its ability to detect not only defects or discontinuities which may be present on the surfaces of the tubes and which could be detected visually, but those also present within the wall which (except for gross defects) would hitherto have gone undetected. Allied to this advantage is, of course, the elimination of the human interpretation of defects and the ability to mechanise the inspection procedures when eddy current methods are adopted.

FITTINGS

For connecting metric pipes it is essential that the appropriate fittings, flanges, etc., are used and this is being provided for by the issue of British Standards for metric sized fittings and flanges. As in the case of tubes, wherever possible these are based on agreements reached internationally, even though at the time of writing these may not all have been issued as international recommendations or standards.

The largest volume of fittings in copper and copper alloys are those traditionally used by plumbers and heating engineers for connecting copper tubes for domestic water, heating and gas services, and BS 864 Part 2[12] covers these products. It will be noted that the same British Standard number has been maintained as that covering the inch size fittings formerly used. One advantage of this is that the number is widely quoted in Codes of Practice Water By-Laws. Consequently, the retention of the same number eliminates confusion in this context.

The sizes of fittings in this standard are the same basic sizes as those given in ISO Recommendation 2016 and are suitable for use with tubes specified in BS 2871 Part 1. Whilst the basic sizes are the same, the socket diameter tolerances for capillary fittings are a little tighter than in the ISO Recommendation because similar tighter tolerances have been used in the UK for very many years and it is known that tighter tolerances are an aid to sounder joints. Therefore, it was felt desirable to retain these tighter tolerances in the metric standard.

As a result of the availability of metric fittings (stocks are held by most builders' merchants in addition to manufacturer's own stocks) allied to their economic cost, they are being increasingly used for many engineering applications where they are suitable for the service conditions involved. Indeed, many such applications have relatively low pressures and/or temperature requirements, where such fittings are more than adequate. This increased and widening usage is acknowledged in the ISO. Recommendation where a statement is now included indicating to users that they can be used for engineering applications, thereby specifically drawing designers' and users' attention to this point.

Fittings for engineering purposes and particularly those for higher pressures, are covered in BS 2051 Part 1.[9] Initially, it was intended that the range of sizes would be selected from the tube diameters given in Table 3 of BS 2871 Part 2. In view of the acceptance for some applications of the sizes quoted in BS 864 Part 2,[12] the complete range of sizes given in this standard includes some sizes suitable for use with fittings to BS 2871 Part 1, and also some for sizes in BS 2871, Part 2. There is, however, a restricted range of sizes up to and including 42 mm with clearly specified maximum working pressures at ambient temperatures. As with other types of fittings where the connector ends are required to be threaded, these are usually the British Standard parallel thread in accordance with ISO Recommendation R228[14], although the taper thread in accordance with ISO Recommendation R7 or other types of thread by special agreement can be supplied.

In addition to the smaller sizes of fittings covered by BS 864 Part 2 and BS 2051 Part 1[13] metric terms, a number of fittings manufacturers produce proprietary types of fittings in larger sizes. As the vast majority of these larger sizes are required for shipbuilding and engineering applications, such fittings are made in sizes suitable for use with tubes complying with Table 3 of BS 2871 Part 2. Whilst in the very large sizes the variety and types of fittings are not required, frequently tight radius bends are, and these can usually only be produced by special forging components instead of the usual machine bending. These are also available in metric sizes.

FLANGES

Flanges in all materials in metric sizes are covered by BS 4504.[15] At the present time, this only covers copper and copper alloy integral flanges up to and including 108 mm. However, flanges are already available in various types, e.g., integral, composite, weld-neck, etc., suitable for use with the various sized copper and copper alloy tubes in metric sizes, and these will eventually be included in the standard.

In the shipbuilding industry, the metric sizes have been established for some years and this is now being consolidated within ISO and these recommendations are being used as the basis for the sizes in BS 4504. There is already a fair measure of agreement between flanges in inch sizes and those currently supplied in metric sizes, but the ISO work at present in progress will produce greater standardisation particularly as far as number of bolts and bolt diameters are concerned, ensuring greater utilisation and standardisation throughout Europe and other parts of the world.

REFERENCES

1. BS 2871
   Copper and copper alloys tubes (1971 and 1972)
2. BS 659
   Light gauge copper tubes (light drawn) (1967)
3. BS 1386
   Copper tubes to be buried underground (1957)
4. BS 3931
   Hard-drawn thin wall copper tubes (1965)
5. BS 1172–1174
   Included in BS 1035
   BS 1035
   Raw copper (1964)
6. BS 61
   Screw threads for copper tubes. Part 2 (1946)
7. BS 1306
   Seamless copper tubes for steam services (1948)
8. BS 21
   Pipe threads (1957)
9. ISO/R7
   Pipe threads for gas list tubes and fittings; ISO (1954)
10. BS 378
    Solid drawn copper and copper alloy tubes for condensers, evaporators, heaters and coolers (1963)
11. BS 3889
    Eddy current testing of non-ferrous tubes. Part 2B (1966)
12. BS 864
    Capillary and compression tube fittings of copper and copper alloy. Part 2 (1971)
13. BS 2051
    Copper and copper alloy capillary and compression tube fittings (for use with fractional o.d. sizes of tubes). Part 1 (1953)
14. ISO/R228
    Pipe threads where pressure-tight joints are not made on the threads; ISO (1961)
15. BS 4504
    Flanges and bolting for pipes, valves and fittings, metric series (1969)

11 STEEL VERTICAL CYLINDRICAL STORAGE TANKS

STEEL VERTICAL
CYLINDRICAL
STORAGE TANKS 11

11 STEEL VERTICAL CYLINDRICAL STORAGE TANKS

J. de WIT

INTRODUCTION

Crude oil and oil products for the petroleum industry are generally stored in steel vertical cylindrical tanks with flat bottoms. Requirements for the design, fabrication, erection and testing of such tanks are given in BS 2654 'Vertical steel welded tanks'[1]. The tanks can be divided into two main groups: fixed-roof tanks and floating-roof tanks. Two different types of roof are necessary to enable adequate reduction of vapour losses for the various types of oil.

A fixed roof consists of steel plates resting on a supporting framework and attached

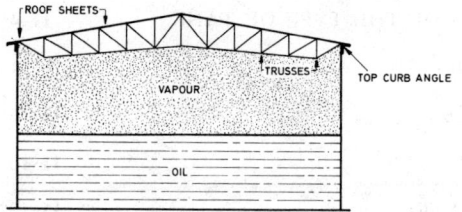

Figure 11.1. Fixed-roof tank

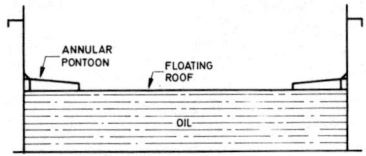

Figure 11.2. Floating-roof tank

to the tank only at the top of the shell, Figure 11.1. A floating roof is a steel disc which floats on the surface of the oil and rises or falls with the oil levels as the tank is filled or emptied, Figure 11.2.

VAPOUR LOSSES

In a fixed-roof tank the space between the oil and the roof is filled with vapour. Owing to the presence of this vapour space, evaporation losses from breathing and filling occur during use.

Breathing losses

These are caused by the different night and day temperatures. During the day the sun heats the tank, raises the temperature of the vapour in the tank's vapour space, causing evaporation. Some vapour is expelled from the tank through the roof vents. At night, as the temperature falls, vapour in the tank cools and condenses to some extent. Air is drawn into the tank and mixes with the internal vapour. The next day, the tank is again

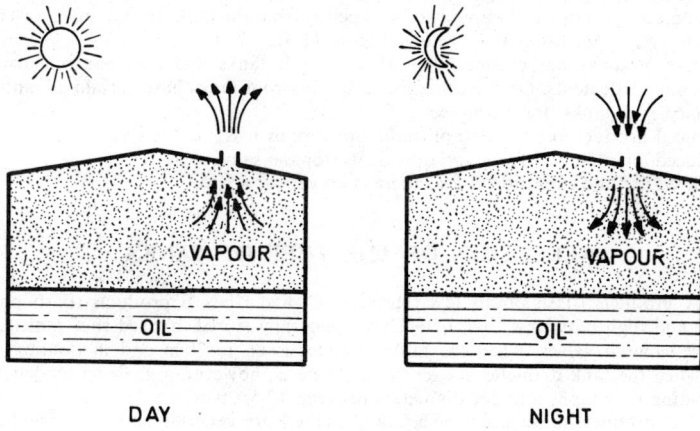

DAY   NIGHT

Figure 11.3. Breathing losses of a fixed-roof tank

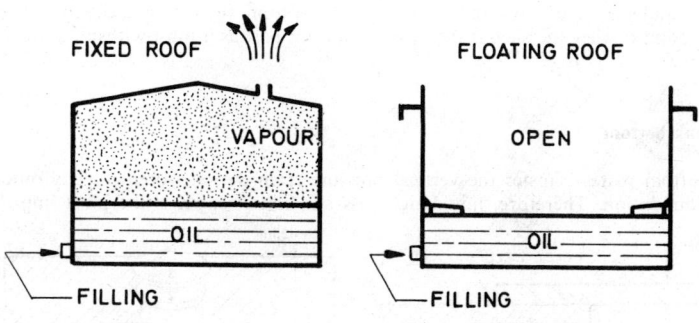

Figure 11.4. Filling losses from fixed and floating-roof tanks

heated by the sun, a quantity of this vapour/air mixture is again expelled from the tank and so on throughout the year (Figure 11.3).

In the case of fixed-roof tanks, these breathing losses can be reduced by installing pressure/vacuum relief valves on the roof. These valves only open at a certain over-pressure (usually 20 mbar or 56 mbar) and at a certain underpressure (usually 6 mbar). These pressure ranges are equivalent to a change in temperature of approximately 8°C to 18°C respectively. This means that breathing does not occur in such tanks when the pressure drop over twenty-four hours is less than the equivalent temperature ranges. In floating-roof tanks breathing losses are virtually eliminated. Evaporation can take place only in the narrow space between the roof and the tank shell (width 100–150 mm).

Filling losses

These occur during the filling of a fixed-roof tank, when the inflowing quantity of oil causes an equal quantity of vapour to be expelled from the tank. In floating-roof tanks there are no vapour losses from filling (Figure 11.4).

Fixed-roof tanks are cheaper than floating-roof tanks and also less sensitive to uneven soil settlements. On the other hand floating-roof tanks have certain advantages over fixed-roof tanks, for example:

Reduced product loss because of minimum vapour loss;
Reduced air pollution for several products, for the same reason;
Reduced fire and explosion risk as there is no vapour space.

SELECTION OF THE TYPE OF TANKS

Class A products (flash points less than 22·8°C) and Class B products (flash points 22·8°C and higher but less than 65·6°C) are generally stored in fixed-roof tanks with pressure/vacuum valves, when the tank diameter is up to 20 m and in floating-roof tanks when the tank diameter is over 20 m. There is, however, a growing tendency to use floating-roof tanks also for diameters between 12·5–20 m.

Class C products (flash point 65·6°C and higher) are generally stored in fixed-roof tanks with open vents.

DESIGN OF THE TANK

A tank can be divided into three main parts: the bottom, the shell and the roof. From a design point of view the shell is the most important part as it must withstand the liquid pressure.

The tank bottom

The bottom plates transfer the vertical pressure from the oil stored directly onto the tank foundation. Therefore, liquid tightness and not strength is the most important

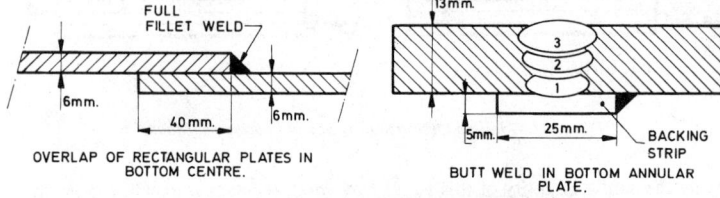

Figure 11.5. Welded joints in bottom plates

point for the tank bottom, except for the bottom annular plates. The bottom annular plates, placed under and fixed to the tank shell are highly stressed by the horizontal liquid pressure acting on the lowest part of the tank shell.

The centre part of the bottom consists of 6 mm thick rectangular plates. These plates are connected with overlaps, welded on the top side only by a full fillet weld, see Figure 11.5. For small tanks ($\leq$ 12·5 m dia) the rectangular plates are also placed under the tank shell; however, the part under the shell is made flush. For the larger tanks (> 12·5 m dia) an annular ring of butt-welded plates is placed under the tank shell.

The bottom annular plates have a thickness of 10–13 mm and are butt-welded on backing strips to obtain full penetration (Figure 11.5).

When welding the tank bottom, the welding sequence should be selected in such a

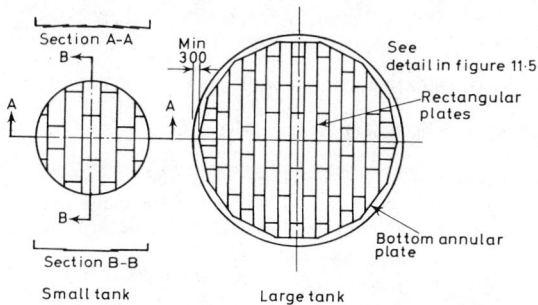

Figure 11.6. Tank bottom layout

manner that the shrinkage is kept to a minimum. Figure 11.6 shows the layout of bottom plates for a large and a small tank.

The tank shell

The height of a tank depends on the allowable soil pressure at the location where the tank is to be built. The tank height generally varies from 14–22 m. In most cases it is preferable to build the tank as high as possible, especially for floating-roof tanks. The soil condition should be checked by soil investigations, especially if large tanks are involved, to ensure that excessive soil settlement is prevented.

The tank shell is made up of a number of courses usually of the same height, whose plate thicknesses gradually increase downwards. Each course has a width of 1·5–2 m for the small tanks and 2–2·5 m for the large ones. Each course is made up of a number of equal plates with a length of approximately 7·5 m for small tanks and approximately 10 m for big tanks.

The thickness of the tank shell increases stepwise downwards, and the plate thickness of courses is calculated for the liquid pressure at the lowest point of the course involved. The shell plate thickness is calculated in accordance with the formula:

$$t = \frac{4 \cdot 9 \times D \times (H - 0 \cdot 3)}{S \times E} \tag{1}$$

where: t = minimum course thickness, in mm;
   S = maximum allowable stress, in N/mm²;
   E = joint efficiency factor for welding;
   D = nominal diameter of tank, in m;
   H = height from top of shell to the lower edge of the course under consideration, in m.

The plate thicknesses are calculated on the assumption that the tank will be filled full of water, since all tanks are hydrostatically tested after construction. This allows for the tank being filled with any type of oil, independent of the specific gravity.

For erection purposes minimum thicknesses are specified for shell plates. For tanks with a diameter up to 33 m the minimum thickness is 6 mm, for diameters between 30–60 m it is 8 mm and over 60 m diameter it is 10 mm. The allowable stress (S) is taken as 2/3 of the minimum guaranteed yield stress of the steel to be used.

Until about 1965 it was general practice to use a joint efficiency factor (E) of 0·85 in view of possible weld defects since tanks were only visually inspected. Since 1965 a joint efficiency factor of 1 is generally applied. However, spot X-ray examination is to be used to maintain a good overall weld quality. All shell weld seams shall have full penetration and full fusion.

EXAMPLE OF A PLATE THICKNESS CALCULATION FOR THE SHELL

Tank diameter 66 m; tank height 20 m consisting of eight courses 2·5 m wide each. Grade 50 material is used for the six lower courses and grade 43 material for the two top courses. Guaranteed minimum yield stress of grade 50 material is 355 N/mm^2 for plates up to and including 16 mm thick, 345 N/mm^2 for plates over 16 mm thick and 245 N/mm^2 for grade 43 material (See BS 4360, 'Weldable structural steels')[2].

$$\text{Course 1.} \quad t = \frac{4\cdot9 \times 66 \times 19\cdot7}{0\cdot666 \times 345 \times 1} = 27\cdot7 \text{ mm, grade 50 D.}$$

$$\text{Course 2.} \quad t = \frac{4\cdot9 \times 66 \times 17\cdot2}{0\cdot666 \times 345 \times 1} = 24\cdot2 \text{ mm, grade 50 D.}$$

$$\text{Course 3.} \quad t = \frac{4\cdot9 \times 66 \times 14\cdot7}{0\cdot666 \times 345 \times 1} = 20\cdot7 \text{ mm, grade 50 D.}$$

$$\text{Course 4.} \quad t = \frac{4\cdot9 \times 66 \times 12\cdot2}{0\cdot666 \times 345 \times 1} = 17\cdot2 \text{ mm, grade 50 C.}$$

$$\text{Course 5.} \quad t = \frac{4\cdot9 \times 66 \times 9\cdot7}{0\cdot666 \times 355 \times 1} = 13\cdot3 \text{ mm, grade 50 C.}$$

$$\text{Course 6.} \quad t = \frac{4\cdot9 \times 66 \times 7\cdot2}{0\cdot666 \times 355 \times 1} = 9\cdot9 \text{ mm (10 mm), grade 50 B.}$$

$$\text{Course 7.} \quad t = \frac{4\cdot9 \times 66 \times 5}{0\cdot666 \times 245 \times 1} = 9\cdot9 \text{ mm (10 mm), grade 43 A.}$$

$$\text{Course 8.} \quad t = \frac{4\cdot9 \times 66 \times 2\cdot2}{0\cdot666 \times 245 \times 1} = 4\cdot4 \text{ mm (10 mm), grade 43 A.}$$

Courses 7 and 8 shall have the minimum thickness of 10 mm for tanks with a diameter over 60 m. The subgrades B, C and D of grade 50 material are required for notch ductility and carbon equivalent requirements of BS 2654[1].

The vertical joints between shell plates should not be in alignment within any three consecutive courses and the distance between vertical joints in adjacent courses should be one-third of the shell plate length, see Figure 11.7.

The vertical welds of the shell are generally hand-welded. The horizontal welds of the shell are generally done automatically.

The shell is attached to the tank bottom by two fillet welds, see Figure 11.8. This bottom to shell connection is almost acting as a hinge, as only a small bending moment can be absorbed.

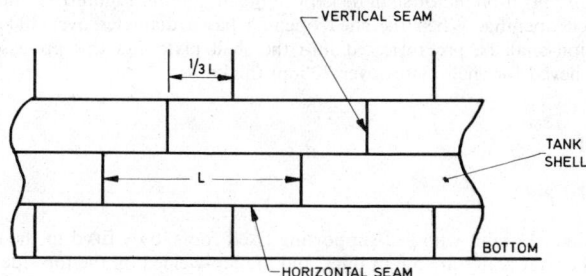

Figure 11.7. Location of vertical shell seams

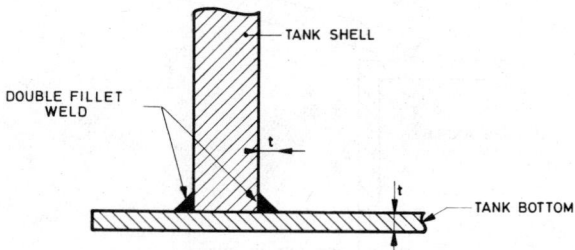

Figure 11.8. Bottom to shell connection

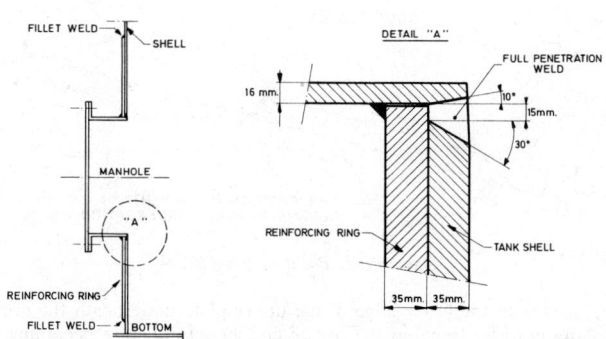

Figure 11.9. Shell nozzle connection

Manholes and nozzle connections for inlets, outlets, product drains, etc., shall be installed in the lowest shell course. When a joint efficiency factor of 1 is applied the stress concentration factor shall be kept to the minimum. Figure 11.9 shows an example of a shell opening. When the shell opening has a diameter over 300 mm the nozzle connection shall be prefabricated into the shell plate and this pre-assembly shall be stress-relieved for shell plates over 20 mm thickness.

The roof

THE FIXED ROOF

Most tanks are built with self-supporting fixed roofs, only fixed to the top of the tank shell. The roof plates are 5 mm thick and are lap-welded on the top side only. The roof

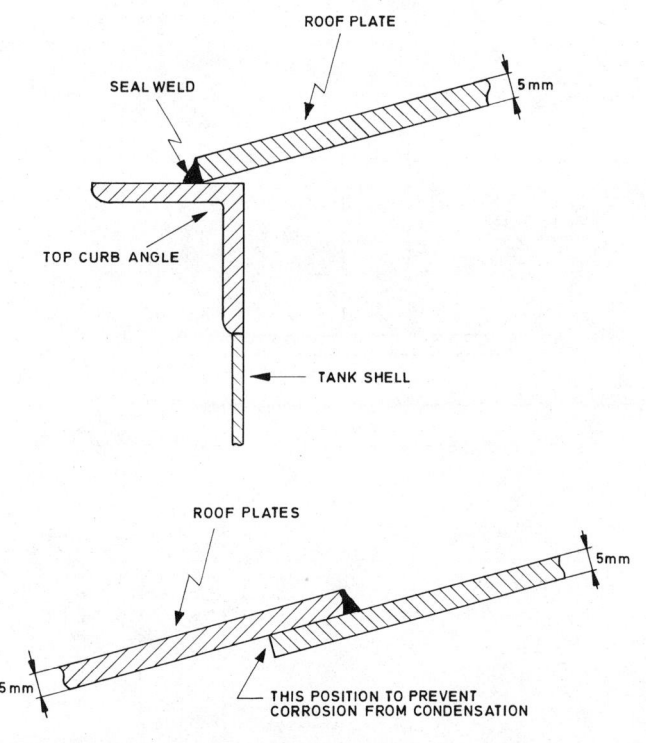

Figure 11.10. Position of roof plates

plates are placed with the lower edge of the upper plate underneath the upper edge of the lower plate, in order to avoid the risk of condensed moisture becoming trapped in the lap joint on the underside of the roof (Figure 11.10).

The roof plates are not fixed to the roof-supporting structure. The roof plates are only welded to the top curbangle of the shell with a seal weld (Figure 11.10). This is done to enable the roof plates to blow away in the case of an explosion in the tank, so that no damage is done to the tank shell and the oil stays in the tank. Manholes, nozzle connections for gauging and dipping and vents or pressure/vacuum valves are placed on the roof. For safety reasons a railing is generally installed around the periphery of the roof.

The floating roof

This floats as a disc on the oil and rises or falls with the oil level. In most cases the pontoon-type roof is used. The buoyancy is supplied by an annular pontoon at the

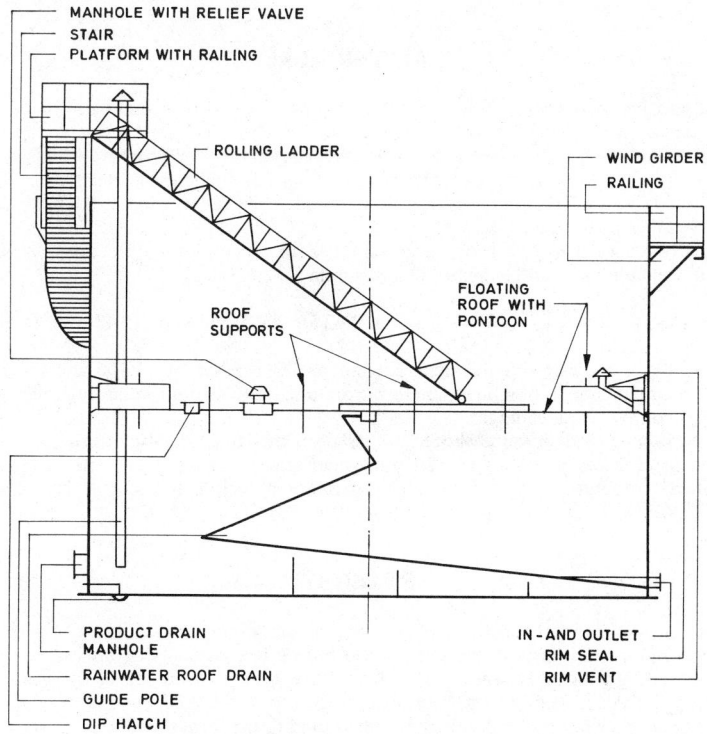

MANHOLE WITH RELIEF VALVE
STAIR
PLATFORM WITH RAILING

ROLLING LADDER

WIND GIRDER
RAILING

ROOF SUPPORTS

FLOATING ROOF WITH PONTOON

PRODUCT DRAIN MANHOLE
RAINWATER ROOF DRAIN
GUIDE POLE
DIP HATCH

IN - AND OUTLET
RIM SEAL
RIM VENT

Figure 11.11. Pontoon-type floating roof

periphery which covers 20–25% of the total roof area. The centre deck, made up of 5 mm thick lap-welded plates, is welded to the inner side of the annular pontoon.

The rainwater is discharged from the roof via an articulated pipe drain in the tank, which follows the level of the roof. An access ladder is provided from the top of the shell to the roof, running over a rail track on the roof. Often these ladders are provided with self-levelling stair treads, which are always in a horizontal position. The roof is provided with roof supports, which can be placed in two positions. The first position is approximately 0·9 m above the bottom to keep the roof free from all accessories on the tank bottom. The second position is approximately 1·8 m above the tank bottom for access under the roof during maintenance.

For large diameter tanks sometimes double-deck roofs (double-deck over the whole liquid surface) or special-type floating roofs are applied. Figure 11.11 shows a pontoon-type roof.

To enable the free movement of the floating-roof along the tank shell, a flexible roof seal is installed between the roof and the shell. There are two types of roof seal: metallic seals and fabric seals. An example of a metallic seal is shown in Figure 11.12; it can make an inward and outward movement of 150 mm. The spring hangers push the continuous sealing ring plate against the shell. The continuous weathershield at the top prevents rainwater, snow or dirt entering the rim space.

Fabric seals are rubber hoses filled with a liquid or foam which can be pressed inward and outward during the movements of the roof, see Figure 11.12.

MATERIALS

Small and medium size tanks are made from mild steels. Large tanks are made from high-tensile steels. These steels are indicated in BS 4360[2]. The mild steels are specified as Grade 43.

The high-tensile steels are specified as grade 50. For high-tensile steels over 20 mm thick a guaranteed carbon-equivalent is required to ensure a good weldability under poor weather conditions.

A carbon equivalent is a formula which takes into account the influence of elements on the weldability of all elements. This formula is:

$$C_{eq} = C + \frac{Mn}{6} + \frac{Cr + Mo + V}{5} + \frac{Ni + Cu}{15} \leqq 0.43 \text{ for 20-25 mm and } \leqq 0.42 \text{ over 25 mm} \quad (2)$$

where: C_{eq} = the carbon equivalent, percent, and C, Mn, Cr, Mo, V, Ni and Cu are the contents of carbon, manganese, chromium, molybdenum, vanadium, nickel and copper, as percentages.

This carbon equivalent value should be based on the ladle analysis of the plates.

For the thicker plates (over 12·5 mm) guaranteed impact values are required at specified temperatures to prevent the possibility of brittle fracture at low ambient temperatures.

STABILITY

The stability of the tank shell, when the tank is empty, must be sufficient to withstand the maximum wind gusts. Wind gusts could buckle the thin upper courses.

For floating-roof tanks where no roof structure is fixed to the top of the shell as for fixed-roof tanks, a wind girder is installed at the top of the shell, see Figure 11.11. The strength of this wind girder is calculated according to the formula:

$$W = 0.058 D^2 H, \quad (3)$$

where: W = section modulus in cm^3;
D = diameter of the tank in m;
H = height of the tank in m.

For large storage tanks one or more intermediate stiffening rings may be required on the top half of the tank shell to prevent buckling. The calculation method is specified in BS 2654[1].

WELD CONTROL

All vertical and horizontal seams are to be checked by spot X-ray examination. The thicker lower shell courses and the T-junctions between horizontal and vertical seams

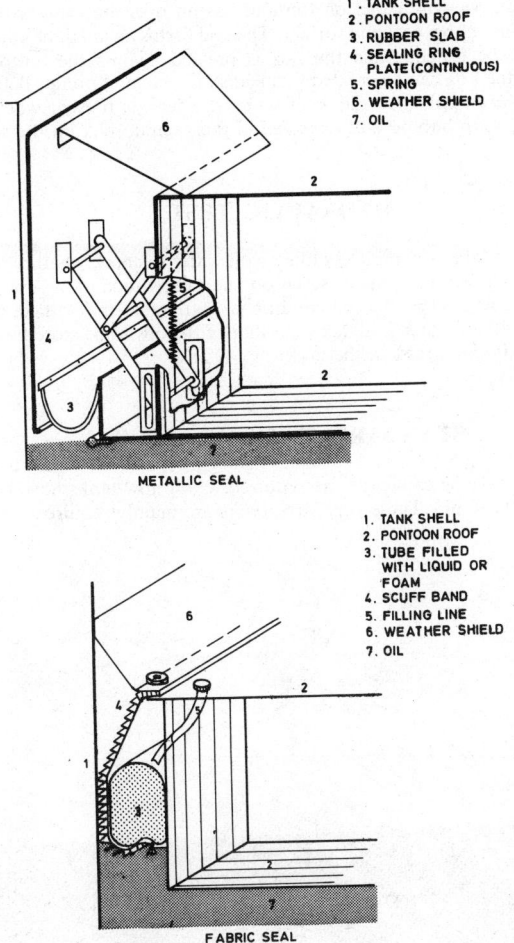

1. TANK SHELL
2. PONTOON ROOF
3. RUBBER SLAB
4. SEALING RING
   PLATE (CONTINUOUS)
5. SPRING
6. WEATHER SHIELD
7. OIL

METALLIC SEAL

1. TANK SHELL
2. PONTOON ROOF
3. TUBE FILLED
   WITH LIQUID OR
   FOAM
4. SCUFF BAND
5. FILLING LINE
6. WEATHER SHIELD
7. OIL

FABRIC SEAL

Figure 11.12. Floating roof seals

are especially checked. Apart from this X-ray examination a thorough visual inspection is always required. The X-ray examination is required to maintain a good weld quality and should be done immediately after the welding of each course.

The bottom and roof fillet welds are leak-tested with a vacuum box, which enables any leaks in the seams to be positively located by visual examination. The vacuum box is fitted with a glass viewing panel on top and has an open bottom, around which is secured a continuous rubber seal and former. The seal forms an air-tight joint around the section of weld to be tested, when the box is pressed against the bottom plates. A partial vacuum in the box can be created by means of a vacuum pump. Before testing is started, soap suds are applied to the weld seam. If a leak is present when the partial vacuum is created, soap bubbles are observed in the vacuum box on the weld seam.

HYDRAULIC TEST

After construction, a tank is completely filled with water to check whether any leaks are present. This also allows the tank to settle on the tank foundation. The filling rates during water testing depend on the soil conditions. Settlement should take place slowly. As the specific gravity of water is higher than that of the oils to be stored, the hydraulic test places the maximum stress on the tank.

STANDARD SIZES OF TANKS

Tanks with greatly varying capacities are required. Until now tanks have been built up to a capacity of 150 000 m³. These very large tanks are mainly required for the storage

Figure 11.13. 100 000 m³ floating roof tank

of crude oil. It is obvious that standardisation of tank dimensions is necessary to obtain a logical tank farm layout and an efficient design and construction (Figure 11.13).

BS 2654 indicates a list of standard diameters, which are followed by all tank manufacturers in the UK and also often on the continent. The standard dimensions are shown in Tables 11.1 and 11.2. The standard diameters are based on the plate lengths

Table 11.1 NOMINAL CAPACITIES OF STANDARD VERTICAL CYLINDRICAL TANKS

| Height (m) | \multicolumn{13}{c}{Tank diameter (m)} | | | | | | | | | | | | |
	3	4	6	8	10	12·5	15	17·5	20	22·5	25	27·5	30
	\multicolumn{13}{c}{Nominal capacities (m³)}												
1	7	12	28	50	78	122	176	240	314	397	490	593	706
2	14	25	56	100	157	245	353	481	628	795	981	1 187	1 413
3	21	37	84	150	235	368	530	721	942	1 192	1 472	1 781	2 120
4	28	50	113	201	314	490	706	962	1 256	1 590	1 963	2 375	2 827
5	35	62	141	251	392	613	883	1 202	1 570	1 988	2 454	2 969	3 534
6	42	75	169	301	471	736	1 060	1 443	1 884	2 385	2 945	3 563	4 241
7		87	197	351	549	859	1 237	1 683	2 199	2 783	3 436	4 157	4 948
8		100	226	402	628	981	1 413	1 924	2 513	3 180	3 926	4 751	5 654
9			254	452	706	1 104	1 590	2 164	2 827	3 578	4 417	5 345	6 361
10			282	502	785	1 227	1 767	2 405	3 141	3 976	4 908	5 939	7 068
11				552	863	1 349	1 943	2 645	3 455	4 373	5 399	6 533	7 775
12				603	942	1 472	2 120	2 886	3 769	4 771	5 890	7 127	8 482
13					1 021	1 595	2 297	3 126	40 84	5 168	6 381	7 721	9 189
14					1 099	1 718	2 474	3 367	4 398	5 566	6 872	8 315	9 896
15					1 178	1 840	2 650	3 607	4 712	5 964	7 363	8 909	10 602
16					1 256	1 963	2 827	3 848	5 026	6 361	7 853	9 503	11 309
17						2 086	3 004	4 088	5 340	6 759	8 344	10 097	12 016
18						2 208	3 180	4 329	5 654	7 156	8 835	10 691	12 723
19						2 331	3 357	4 570	5 969	7 554	9 326	11 285	13 430
20						2 454	3 534	4 810	6 283	7 952	9 817	11 879	14 137
21							3 711	5 051	6 597	8 349	10 308	12 473	14 844
22							3 887	5 291	6 911	8 747	10 799	13 067	15 550
23							4 064	5 532	7 225	9 144	11 290	13 661	16 257
24							4 241	5 772	7 539	9 542	11 780	14 254	16 964
25							4 417	6 013	7 853	9 940	12 271	14 848	17 671

generally used in Europe. The heights are not standardised so that manufacturers can select their most economical plate widths. Therefore, the tank capacity is indicated for every metre of height for all standard diameters. The maximum possible height of a tank is almost invariably determined by the permissible load in view of the nature of the soil. Where this permissible height has been determined the standard diameter and the corresponding height can be determined for every tank capacity desired by means of the capacity table.

REFERENCES

1. BS 2654
     Part 1. Vertical steel welded storage tanks with butt-welded shells for the Petroleum Industry (1970)
2. BS 4360
     Part 2. Weldable structural steels (1969)

Table 11.2 NOMINAL CAPACITIES OF STANDARD VERTICAL CYLINDRICAL TANKS

Nominal capacities (m³)

Height (m)	\ Tank diameter (m) 33	36	39	42	45	48	51	54	57	60	66	72	78	84	90	96	102
1	855	1 017	1 194	1 385	1 590	1 809	2 042	2 290	2 551	2 827	3 421	4 071	4 778	5 541	6 361	7 238	8 171
2	1 710	2 035	2 389	2 770	3 180	3 619	4 085	4 580	5 103	5 654	6 842	8 142	9 556	11 083	12 723	14 476	16 342
3	2 565	3 053	3 583	4 156	4 771	5 428	6 128	6 870	7 655	8 482	10 263	12 214	14 335	16 625	19 085	21 714	24 513
4	3 421	4 071	4 778	5 541	6 361	7 238	8 171	9 160	10 207	11 309	13 684	16 285	19 113	22 167	25 446	28 952	32 685
5	4 276	5 089	5 972	6 927	7 952	9 047	10 214	11 451	12 758	14 137	17 105	20 357	23 891	27 708	31 803	36 191	40 856
6	5 131	6 107	7 167	8 312	9 542	10 857	12 256	13 741	15 310	16 964	20 527	24 428	28 670	33 250	38 170	43 429	49 027
7	5 987	7 125	8 362	9 698	11 133	12 666	14 299	16 031	17 862	19 792	23 948	28 500	33 448	38 792	44 532	50 667	57 198
8	6 842	8 142	9 556	11 083	12 723	14 476	16 342	18 321	20 414	22 619	27 369	32 571	38 226	44 334	50 893	57 905	65 370
9	7 697	9 160	10 751	12 468	14 313	16 285	18 385	20 611	22 965	25 446	30 790	36 643	43 005	49 875	57 255	65 143	73 541
10	8 552	10 178	11 945	13 854	15 904	18 095	20 428	22 902	25 517	28 274	34 211	40 714	47 783	55 417	63 617	72 382	81 712
11	9 408	11 196	13 140	15 239	17 494	19 905	22 470	25 192	28 069	31 101	37 633	44 786	52 561	60 959	69 978	79 620	89 883
12	10 263	12 214	14 335	16 625	19 085	21 714	24 513	27 482	30 621	33 929	41 054	48 857	57 340	66 501	76 340	86 858	98 055
13	11 118	13 232	15 529	18 010	20 675	23 524	26 556	29 772	33 172	36 756	44 475	52 929	62 118	72 042	82 702	94 096	106 226
14	11 974	14 250	16 724	19 396	22 266	25 333	28 599	32 063	35 724	39 584	47 890	57 000	66 896	77 584	89 064	101 335	114 397
15	12 829	15 268	17 918	20 781	23 856	27 143	30 642	34 353	38 276	42 411	51 317	61 072	71 675	83 126	95 425	108 573	122 569
16	13 684	16 285	19 113	22 167	25 446	28 952	32 685	36 643	40 828	45 238	54 739	65 143	76 453	88 668	101 787	115 811	130 740
17	14 540	17 303	20 308	23 552	27 037	30 762	34 727	38 933	43 379	48 066	58 160	69 215	81 232	94 209	108 149	123 049	138 911
18	15 395	18 321	21 502	24 937	28 627	32 571	36 770	41 223	45 931	50 893	61 581	73 286	86 010	99 751	114 510	130 287	147 082
19	16 250	19 339	22 697	26 323	30 218	34 381	38 813	43 514	48 483	53 721	65 002	77 358	90 788	105 293	120 872	137 526	155 254
20	17 105	20 357	23 891	27 708	31 808	36 191	40 856	45 804	51 035	56 548	68 423	81 429	95 567	110 835	127 234	144 764	163 425
21	17 961	21 375	25 086	29 094	33 399	38 000	42 899	48 094	53 586	59 376	71 844	85 501	100 345	116 377	133 596	152 002	171 596
22	18 816	22 393	26 280	30 479	34 989	39 810	44 941	50 384	56 138	62 203	75 266	89 572	105 123	121 918	139 957	159 240	179 767
23	19 671	23 411	27 475	31 865	36 579	41 619	46 984	52 675	58 690	65 030	78 687	93 644	109 902	127 460	146 319	166 479	187 939
24	20 527	24 428	28 670	33 250	38 170	43 429	49 027	54 965	61 242	67 858	82 108	97 715	114 680	133 002	152 681	173 717	196 110
25	21 382	25 446	29 864	34 636	39 760	45 238	51 070	57 255	63 793	70 685	85 529	101 787	119 458	138 544	159 042	180 955	204 281

12 MATERIALS HANDLING EQUIPMENT AND LIFTING APPLIANCES

MATERIALS HANDLING
EQUIPMENT AND
LIFTING APPLIANCES

12

12 MATERIALS HANDLING EQUIPMENT AND LIFTING APPLIANCES

LIFTING APPLIANCES

L. POWELL

INTRODUCTION

A number of significant developments have taken place in the field of lifting appliances since the Second World War. Prior to 1946 most lifting appliances were made either of wrought iron or mild steel (usually rimming steel having a carbon content of about 0·08 %). The strength of these materials is such that a mild steel chain of 1 in nominal diameter will have a minimum breaking load of 30 tons. The breaking load can thus be expressed in the form $30d^2$ tons where 'd' is the nominal diameter of the chain and this method of expressing breaking strength has been adopted as the basis of a system of grading for steel chain. Thus mild steel chain (BS 590) became known as Grade 30 chain.

Subsequent to 1946 there was a growing demand for stronger, more reliable, lifting appliances. At the same time welding techniques, particularly those concerned with the type of repetitive welding required in chain manufacture, had improved significantly so that it became feasible to manufacture lifting appliances from semi-killed or fully killed steels of higher carbon content. Thus in 1950, BS 1663 'Higher Tensile Steel Chain (Grade 40)' was published. This provides for the manufacture of chain from a steel having a maximum carbon content of 0·3 % and there followed a whole series of co-ordinated standards for other lifting gear components made from this steel or from a carbon-manganese steel (1·5 % manganese). The term 'higher tensile steel' has become generally applied to lifting gear of this quality having a minimum breaking load of $40 d^2$ tons (Grade 40).

More recently the need for even greater strength and compactness of lifting appliances has led to the use of low alloy steels. These usually have carbon contents in the range 0·17–0·23 % with small additions of nickel (0·40–0·70 %), chromium (0·30–0·70 %), molybdenum (0·15–0·25 %). British Standards for alloy steel chains Grade 60 and Grade 80 were both published in 1959 (BS 3113 and BS 3114) and are being revised as metric standards.

These developments followed by a number of years satisfactory operating experience with both higher tensile and alloy steel lifting gear have made wrought iron and mild steel gear obsolete. It is logical therefore to deal only with higher tensile and alloy steel lifting appliances and to give guidance for the safe use of lifting appliances. The manufacture, maintenance and use of lifting appliances should comply with the relevant British Standards, codes of practice and, of course, current statutory requirements. A key to these statutory requirements appears below and the prescribed statutory forms are given in detail under 'G. Statutory Forms' (page 12–77) and at the end of Section 12. References appear in curved brackets throughout the text to help the reader appreciate where a legal requirement is involved and in these cases the action called for is of course mandatory. The bibliography at the end of the section indicates some of the authorities and publications consulted and acknowledgement is hereby made of the considerable assistance obtained from these.

KEY TO STATUTORY REQUIREMENTS

FA 1961	Factories Act, 1961.
C(LO)R	Construction (Lifting Operations) Regulations (SI 1961, No. 1581).
DR	Docks Regulations (SR & O 1934, No. 279).
SI	Statutory Instruments.
SR & O	Statutory Rules and Orders.

Other Statutory Rules and Orders and Statutory Instruments to which reference is made in the text are as follows:

Year	Number	Title
1934	279	Docks Regulations.
1938	599	The Chains, Ropes and Lifting Tackle (Register) Order.
1961	1580	The Construction (General Provisions) Regulations.
1961	1581	The Construction (Lifting Operations) Regulations.
1962	225	The Construction (Lifting Operations) Reports Order.
1962	227	The Construction (Lifting Operations) Certificates Order.
1962	272	The Construction (Notice of Accidents, etc.) Order.
1963	1382	The Lifting Machines (Particulars of Examinations) Order.
1962	715	The Hoists Exemption Order.
1968	868	Offices, Shops, and Railway Premises (Hoists and Lifts) Order.

A. DEFINITIONS

A.1 Lifting appliances

The term 'Lifting Appliance' includes both 'lifting tackle' and 'lifting machine'. It should be noted that this differs from the legal definition given in Construction (Lifting Operations) Regulations 1961.

A.2 Lifting tackle

Table 12.1 contains a list of items which will normally be considered as 'lifting tackle'. Obvious exceptions occur, e.g. chains used for fencing, ropes used for towing or as permanent guys to stacks, and turnbuckles used for the suspension of permanently installed pipelines.

A.3 Lifting machine

Table 12.2 shows typical items which will normally be considered as 'lifting machines'.

A.4 Hoist and Lift

A 'hoist' or 'lift' means a lifting machine, whether worked by mechanical power or not, with a carriage, platform or cage, the direction of movement of which is restricted by a guide or guides.
  (FA 1961, S 25(9); SI 1961 No. 1581 4(2))

Table 12.1. LIFTING TACKLE

Item	See Notes below	Details reference in Section F
Boatswain's chair		F.1
Bordeaux connection		F.2
Bulldog grip		F.3
Chain	b, c	F.4
Eyebolt	b	F.5
Girder clip	b	F.6
Hook	a, b, c	F.7
Lifting beam		F.8
Lifting box or skip		F.9
Link	b, c	F.15.1
Plate lifting clamp	b	F.10
Rigging or stretching screw or turnbuckle		F.11
Ring	a, b, c	F.15.1
Rope, fibre		F.12
Rope, wire		F.13
Shackle	a, b, c	F.14
Sling, chain, wire or fibre	a, b	F.15
Socket, wire rope		F.16
Swivel	a, b, c	F.18
Thimble for wire rope		F.20
Weighing machine, suspended type and other items specially designed for particular application		F.22

Notes. (a) These items are defined as 'lifting tackle' in the Factories Act, 1961.
   (b) These items are defined as 'lifting gear' in the Construction (Lifting Operations) Regulations 1961.
   (c) These items are defined as 'lifting gear' in the Docks Regulations, 1934.

Table 12.2. LIFTING MACHINES

Item	See Notes below	Details reference in Section F
Block, chain, or rope	d, e	F.23
Cranes of all types	d, e, f	F.24
Derrick	f	F.24.4
Excavators	e	F.26
Hoists or lifts of all types	e, f	F.27
Jacks of all types		F.29
Pile driver	e	F.30
Pile extractor	e	F.30
Overhead runway beam	d	F.31
Stacking machines and fork lift trucks		F.32
Winch or crab	d, e, f	F.33
and other items specially designed for particular application		

Notes. (d) These items are defined as 'lifting machines' in the Factories Act 1961.
   (e) These items are defined as 'lifting appliances' in the Construction (Lifting Operations) Regulations 1961.
   (f) These items are defined as 'lifting machinery' in the Docks Regulations 1934.

A.5   Competent persons

A 'competent person' shall be a person of such practical and theoretical knowledge and actual experience of the type of lifting appliance he has to examine or test, as will enable him to detect defects or weaknesses which it is the purpose of the examination or test to discover, and to assess their importance in relation to the strength and functions of the particular appliance.

It is recommended that the appointment of such a competent person be made in writing. The written authority should specify the extent of the duties for which the competent person is appointed.

SUPERVISION, OPERATION OR OTHER ACTION

A 'competent person' referred to in any specific regulation dealing with supervision, operation, or other action is a person who is qualified by training and experience to carry out the requirements of the regulation.

A.6   Inspection

An 'inspection' means a visual inspection of a lifting appliance by a competent person, carried out with sufficient care to decide whether or not the lifting appliance is safe for use. The driver or operator of the lifting appliance may, in certain cases, carry out an inspection.

A.7   Examination

'Examination' means thorough examination.

A.8   Thorough examination

A 'thorough examination' means a visual examination by a competent person supplemented, if necessary, by other means such (DR.19(b)(iii)) as a hammer test, carried out as carefully as conditions permit, in order to arrive at a reliable conclusion as to the safety of the parts examined; and if necessary for the purpose, parts of the machines and gear shall be dismantled. In this connection various forms of non-destructive testing may be appropriate.

A.9   Test

A 'test' means the application in a manner specified of a 'proof load' on the actual lifting appliance and if required, a 'test load' on a sample. In each case there shall be a thorough examination after the test.

A.10   Proof load

A 'proof load' is a specified load which a lifting appliance shall withstand without showing permanent set exceeding a specified amount or showing other defect.

A.11   Test load

A test load is the load applied:
1. to test a sample of rope to destruction to obtain the ultimate breaking load: or
2. to test a sample of chain for the purpose of determining both the ultimate breaking load and the elongation at fracture.

A.12   Energy absorption factor

The energy absorption factor, usually of a chain, is the product of the ultimate breaking load and the elongation at fracture.

A.13   Safe working load

The safe working load of an appliance is the maximum load which that appliance is allowed, by statutory regulations, to support under particular conditions of service and is required to be marked on the lifting appliance.
(SI 1961 No. 1581 4(2)).

A.14   Store

The term 'store' used throughout this volume means a Lifting Appliances Store and not a General Store.

B.   DESIGN AND ORDERING

B.1   General

The regulations in this Section deal only with matters of general application. It is intended that reference should be made to Section F for detailed information regarding any particular appliance.

It is obvious that all lifting appliances should comply with statutory requirements, be of good construction, sound material, adequate strength, suitable quality and free from patent defect. They should also be in accordance with the appropriate British Standard.
(FA 1961 S 22(1), 26(1), 27(1); SI 1961 No. 1581 10(1)(a))

It is recommended that all items of lifting gear should be manufactured either of higher tensile steel (Grade 40) or of low alloy steel (Grade 60 or Grade 80). Wrought iron and mild steel (Grade 30) lifting gear should now be regarded as obsolete (see Introduction page 12.1). Much development work is in hand at the present time and over the next few years further grades of alloy steel lifting gear will undoubtedly become available. Generally speaking it will be in order to use these additional grades provided they are covered by a British Standard. In cases of doubt it would be wise to consult a lifting gear specialist.

B.2   Drawings

All drawings of lifting gear components should specify clearly:
   (i)   The material of construction.
   (ii) The appropriate British Standard.
   (iii) The initial heat treatment (see Section E 'Heat Treatment').

(iv) The safe working load of the components.

(v) Details of tests to be carried out.

(vi) Full details for marking the components with:

    (a) means to enable any person using the appliance to ascertain the safe working load. (See sub-section F for detailed requirements.)

    (b) the following symbols to denote the material of construction and the heat treatment:

        '04'   for higher tensile steel, Grade 40, in the hardened and tempered condition.

        '06'   for alloy steel, Grade 60, in the hardened and tempered condition.

        '08'   for alloy steel, Grade 80, in the hardened and tempered condition.

    (c) the identification marking.

The above marking symbols for different grades of components follow the BSI recommendations.

B.3 Approval of design

In cases where lifting appliances are specially designed for a particular purpose the design should be approved by an appropriate appointed person.

B.4 Allowable stresses

It is not possible to particularise on the matter of allowable stresses, and in determining these for any particular application the special conditions under which the appliance will be used should be taken into consideration. In many cases an appropriate British Standard will be available and the stresses recommended in these should not be exceeded. Reference should also be made to items listed in the bibliography at the end of this section.

B.5 Tests

A test certificate relating to each component of lifting gear should always be obtained in accordance with the detailed requirement set out in sub-section F. All test certificates should be filed for reference purposes.

    (FA 1961 3.26(1)(e), 27(b); SI 1961 No. 1581 28(1), 34(1)(a), 35, 46(1)(d); DR 18(a)(b), 19(a)(d), 20(a))

B.6 Guarding

The guarding of all lifting machines should conform to the requirements of the Factories Act 1961.

    (FA 1961 S 12, 13, 14; SI 1961 No. 1581 42, 43 DR 26)

B.7 Certificate from makers

Every order shall specify the certificates which are to be supplied by the manufacturer.

C. ISSUE, MAINTENANCE, AND RECORDS

C.1 Personnel

Every well-run factory making use of lifting appliances will normally have an engineer or technical officer specifically appointed to take full responsibility for the proper organisation and use of the appliances and seeing that all the legal requirements are fully met. In the ordinary way no delegation of this responsibility and authority is permitted but assistants may be appointed to carry out such detail duties as can be safely entrusted to them.

C.2   Responsibilities

The Lifting Appliance Organisation should be responsible for the following:
Storage of lifting appliances.
Checking new appliances against specification.
Testing.
Registration.
Marking.
Records.
Issue.
Periodic Inspection and Examination.
Permissible wear.
Scrapping of worn appliances.
Repairs.
Action in cases of loss of appliances.
Action in cases of failure of appliances.

It may be convenient to sub-contract some of the above work to specialist commercial organisations but it remains the statutory obligation of the persons defined in sub-section C.1 to ensure that such work is properly completed.

The following sub-sections detail the action which should be taken in each case.

C.3   Storage

Wherever practicable all lifting appliances, until they are handed over for use, should be kept in the Lifting Appliances Stores. It is important that the store should be kept dry and warm and provided with facilities to enable each lifting appliance to be stored separately. Arrangements should also be made for cleaning and oiling.

C.4   New appliances

The appropriate certificate of test and examination should be obtained for any newly acquired lifting appliance before it is authorised for use, regardless of the source from which it is obtained.

All lifting appliances should be checked against the ordering specification.

When appliances are designed specially for particular applications, the design should be approved by the appropriate appointed person.

C.5   Testing

Before being used for the first time, and after alteration or repair affecting the safety or stability of the equipment, every lifting appliance should be tested by the competent person and a certificate of test specifying the safe working load, or loads, and signed by the person making the test, should be obtained.

(FA 1961 S 26(1) (e), 27(6); SI 1961 No. 1581 28(1), 34(1) (a), 35, 46(1) (a))

The following items of lifting tackle are exempt from testing: Bulldog grips, thimbles for ropes, fibre ropes, wire ropes, and wire rope slings. Samples of fibre and wire ropes are tested to destruction.

In most cases it is a statutory requirement that all cranes be tested by a competent person before being taken into use, either for the first time or after any substantial alteration or repair, and in the case of cranes used under the Construction (Lifting Operations) Regulations 1961 periodic tests are required at least every four years.

The Construction (Lifting Operations) Regulations 1961 also require, after each erection of a crane or alteration affecting the anchorage or ballasting, tests to be carried out to establish the security of the anchorage and adequacy of the ballasting. These tests require either the imposition of a 25% overload above the appropriate maximum safe working load to be lifted by the crane applied at a position where there is the maximum pull on each anchorage, or the imposition of a reduced load at an increased radius to give an equivalent test of the anchorage or ballasting arrangements. In the latter case caution is necessary to avoid a considerable over-stressing of components. (See also F.24.4).

(SI 1961 No. 1581 19(4))

The schedule of the Docks Regulations 1934, gives details of the manner of testing lifting appliances. Information is also available from H.M. Stationery Office publications[1, 2], and from the BS Code of Practice 'Safe Use of Cranes' [3]

(SR & O 1934 No. 279).

When lifting appliances require periodic test, a recall notice should be sent to the user stating the date on which the appliance is to be returned to store or made available for test.

Particulars and records of all tests are required to be entered on prescribed forms or in the approved register. Details are given in sub-sections F and G.

C.6   Registration

Regulations require that each lifting appliance be given a Factory identification number which must be quoted on all certificates and other documents referring to that particular appliance.

(SR & O 1938 No. 599; SI 1963 No. 1382)

When a lifting appliance consists of an assembly of individual items of tackle not permanently attached to each other, each item must be given an identification number which will provide means for identifying:

1. The individual item, and
2. The assembly to which that item belongs.

It is not necessary to give separate identification numbers to items of tackle permanently attached to each other.

C.7   Marking

Every lifting appliance is required to have clearly marked upon it in plain legible figures and letters not readily obliterated or removed and as detailed in Section F,

1. the identification number
2. the safe working load (see note below)
3. the Symbol 04, 06 or 08 to denote the material of which the appliance is made and the condition of heat treatment (see particularly F.4.1 and F.4.2).

The proof load shall not be marked on any lifting appliance. Note that there is no exception from marking single and multiple leg slings used at various angles and it is recommended that the marking should be the safe working load at an included angle of 90° between the legs.

(FA 1961 S 27(a); SI 1961 No. 1581 29, 34(1)(c); DR 23, 24, 29)

A table showing the safe working loads of every size and every kind of chain, rope, or lifting tackle in use, and in the case of a multiple sling, the safe working load at different angles of the legs shall be posted in the store in which the equipment is kept. This information must also be displayed in prominent positions throughout the occupied premises and on sites where lifting operations are carried out.

In the case of a crane which is on occasion dismantled or partially dismantled, any structural member which is separated from the crane shall be clearly marked to indicate the crane of which it forms a part.

(FA 1961 S 26(b))

C.8   Records

A register is required to be kept containing the prescribed particulars of all lifting appliances. This is called the Lifting Appliances Register, and all certificates and reports of examinations, inspections, heat treatment, repairs or other particulars as may be prescribed should be entered in or attached to it. The official forms are:

Factories Act	Form No. 88
Construction (Lifting Operations) Regulations	Form No. 91
Docks Regulations	Form No. 99

Other statutory forms are dealt with in Section G.

The Factories Act 1961, 22(2), also requires reports of six-monthly thorough examinations of hoists and lifts on Form 54 to be attached to the General Register. The official form for the General Register is No. 31.

The following particulars will need to be recorded:
Name of factory
Address of factory
Identification number or mark and description sufficient to identify the appliance.
Date of each examination and by whom carried out.
Particulars of any defect found which would affect the safe working load and details of the steps taken to remedy the defect.
Date, (if after 30th June 1938) when the appliance was first taken into use in the factory.
Dates and numbers of certificates of tests and examinations.

Provided the permission of the District Inspector of Factories is obtained, the Lifting Appliances Register may be in the form of a Card Register and may be extended to include some or all of the records of lifting appliances. In this case it will not be necessary to use the official forms. This latter method of registration is recommended.

(FA 1961; S 26(1) (g), 27(2) SI 1961; No. 1581 10(1) & (2), 19(4), 30(1) (2) (3), 46(1); SI 1962 No. 225; SR & O 1938 No. 599; SI 1963 No. 1382)

C.9   Issue

Lifting appliances should only be issued for use on a written order signed by an authorised person and only after all the statutory regulations have been complied with in regard to:
Checking against specification
Approval of special designs
Testing
Registration
Marking
Periodic examination

C.10   Periodic inspection or examination

All lifting appliances in use should be inspected or examined by a competent person at prescribed intervals as detailed in sub-section F. Appliances should be recalled for inspection by means of a written notice to the user as described in sub-section C.5.

The whole of the appliances for the anchorage of a Scotch Derrick, Guy Derrick, or Tower Derrick crane should be examined by a competent person on each occasion the crane is erected.

The prescribed particulars and results of every inspection or examination should be entered on the appropriate certificate or register within fourteen days after the inspection has been made. Full details of the frequency and the forms or registers to be used are shown in Table 12.3.

Lifting appliances which are not in use need not be periodically inspected but they should not be issued or used unless they comply fully with sub-section C.9.

If an examination shows that certain lifting appliances cannot continue to be used safely until repairs have been carried out, a copy of the report of the examination should be sent to the Factory Inspector for the district within twenty-eight days of completion of the examination.

(FA 1961 S 27(2))

C.11 Method of inspection or examination and testing of load chains

The following procedure is recommended when a load chain is examined:
1. Examine visually for obvious defects.
2. Check to see whether the chain has stretched during use. The chain should be scrapped if the stretch exceeds 3% of the original length. The original length of the chain is the measured length after application of the proof load and before the chain was put into service.
3. If the chain is stretched up to the allowable 3% then each individual link should be checked separately. Any obviously stretched or distorted links should be replaced by new links (see C.15.1) though in cases where a large number of damaged links are found it may be more economical to fit a complete new chain.
4. The wear on chain links should at no point exceed a 10% reduction in cross-sectional area. Table 12.4 shows minimum reduced diameters of links based upon a 10% reduction of area of cross-section.
   NO GO limit gauges representing the permissible minimum reduced link diameters as given in Table 12.4 may be used.

No chain shall be authorised for use if the links have 'shouldered' or 'socketed' or if there is the slightest tendency to locking of the links because of wear, stretch or other deformation or damage, even though the wear is within the permissible limit.

C.12 Worn or corroded lifting gear

Lifting appliances which have become worn or corroded to such an extent as to be unsafe for use at the normal safe working load should not be authorised for use.

Items of lifting gear which have become worn or corroded should not be re-registered for use at a reduced safe working load.

C.13 Permissible wear

A wire rope which, in any length equal to five rope diameters, contains more than five visible broken wires should not be used for raising, lowering, or suspension purposes. Similarly, a wire rope which shows signs of excessive wear, corrosion or other defect should be regarded as unfit for use.

Table 12.3. PERIODIC EXAMINATION AND INSPECTION FREQUENCY AND FORMS TO BE USED

Examination is required before use by FA, C(LO)R and DR
The official form numbers are:
FA—Form No. 88, Part I; C(LO)R—Wire Ropes—Form No. 87 and
DR—Form No. 86—Wire Ropes—Form No. 87

	Factories Act 1961				Dock Regulations 1934				Construction (lifting operation) Regulations			
	Examination		Inspection		Examination		Inspection		Examination		Inspection	
	Frequency	Form	Frequency	Form	Frequency	Form	Frequency	Form	Frequency	Form	Frequency	Form
Lifting tackle (see A.3)	6 months	88 Part III	Z	Domestic form	12 months	99 Part III or 1952	X 3 months	Domestic form	6 months	91 Part II	Z	Domestic form
Lifting machine (see A.4)	12 months	88 Part IV	Z	Domestic form Y	12 months	99 Part II	Derricks only 12 months	99 Part I	14 months	88 Part IV	weekly	91 Part I
Hoists (see A.5)	6 months	54	Z	Domestic form	12 months	99 Part II	Z	—	6 months	91 Part II	weekly	91 Part I

Z. Inspection only to be carried out when arduous duties make it desirable (see sub-section F for recommended frequencies).
R. See Section E Heat Treatment.
X. After a wire has broken in the rope the inspection frequency shall be at least once in every month.
Y. Except Derricks which are inspected every four years.
All reports shall be issued and attached or entered in the prescribed register within fourteen days of the completion of the examination or inspection.

Chains which on any link exhibit wear greater than that shown in Table 12.4 are unfit for use.

(SI 1961 No. 1581 34(3); DR 20(c))

Table 12.4. PERMISSIBLE WEAR OF STEEL CHAINS

Original nominal dia.		Minimum reduced dia.		Original nominal dia.		Minimum reduced dia.	
in	mm	in	mm	in	mm	in	mm
$\frac{1}{4}$	6·35	$\frac{15}{64}$	5·95	$\frac{15}{16}$	23·81	$\frac{55}{64}$	21·83
$\frac{5}{16}$	7·94	$\frac{9}{32}$	7·14	1·0	25·4	$\frac{56}{64}$	23·41
$\frac{3}{8}$	9·52	$\frac{11}{32}$	8·73	$1\frac{1}{8}$	28·57	$1\frac{3}{64}$	26·59
$\frac{7}{16}$	11·1	$\frac{13}{32}$	10·32	$1\frac{1}{4}$	31·75	$1\frac{11}{64}$	29·77
$\frac{1}{2}$	12·7	$\frac{15}{32}$	11·91				
$\frac{9}{16}$	14·29	$\frac{17}{32}$	13·49				
$\frac{5}{8}$	15·87	$\frac{37}{64}$	14·68	For chains above $1\frac{1}{4}$ in (31·75 mm)			
$\frac{11}{16}$	17·46	$\frac{41}{64}$	16·27	dia. allow 10% reduction in area as a			
$\frac{3}{4}$	19·05	$\frac{45}{64}$	17·86	maximum.			
$\frac{13}{16}$	20·64	$\frac{49}{64}$	19·45				
$\frac{7}{8}$	22·22	$\frac{53}{64}$	21·03				

C.14   Scrapping

Lifting tackle which is beyond repair should be scrapped either by or with the knowledge of a competent person who should enter in the Lifting Appliances Register particulars of the action taken.

Scrapping should be carried out by cutting the tackle into a sufficient number of pieces to make it useless.

C.15   Repairs

GENERAL

The repair of Lifting Tackle should be done only under the supervision of a competent person who should ensure that such work as forming and welding is done by qualified persons. Items of lifting tackle other than chain should not be repaired by fusion welding, though this does not preclude the repair of specially designed items such as lifting beams and lifting boxes.

C.15.1   REPAIR OF CHAIN

Higher tensile steel and alloy steel chains should be repaired only by the removal of the whole of the defective length of chain followed by the insertion of a length of new chain of the appropriate quality. The length of new chain should be in the hardened and tempered condition and should be joined to the existing chain by a link of higher tensile steel or alloy steel as appropriate.

The welding of the joining link should be done on a fully automatic link welding machine. Subsequently all the joining links should be hardened and tempered in a link heater or alternatively the whole chain should be hardened and tempered in a suitable furnace. If the joining link is heat treated in a link heater it is permissible to quench from the hardening temperature into a molten salt bath held at an appropriate temperature.

The equipment needed for carrying out chain repairs by this approved method may not be available to all users and it will then be necessary to send the chains for repair either by the original manufacturer or by an approved chain repairer.

C.15.2   TREATMENT AFTER REPAIR OF ITEMS OTHER THAN CHAINS

After repairs involving the heating of higher tensile steel or alloy steel rings, hooks, shackles, links, eyebolts, swivels, etc., these items shall be heat treated by hardening and tempering.

C.15.3   WIRE ROPE SPLICING

A thimble or loop splice made in a wire rope should always have at least three tucks with the whole strand of the rope and two tucks with one-half of the wires cut out of each strand. The strands in all cases shall be tucked against the lay of the rope. An equally efficient form of splice may be used if preferred.

For joining a wire rope to a bordeaux connection a nine-tuck splice should be used as recommended in BS 461.

C.15.4   TEST AND EXAMINATION

After alteration or repair, all lifting appliances should be thoroughly examined and tested, and an appropriate certificate should be obtained and filed. A list of official forms is given in sub-section G.
(SI 1961 No. 1581 28(2), 6; SI 1962 No. 227)

C.16   Fibre ropes and slings

It is recommended that natural fibre ropes and slings should not be used out-of-doors unless they have been treated with a water repellent, nor should they be used under conditions where there is danger of contamination with corrosive substances. If it is not possible to avoid such usage of natural fibre ropes or slings then they should be carefully examined and smelt at frequent regular intervals and if damage is suspected they should be destroyed.

Under certain corrosive or wet conditions, synthetic fibre ropes made from 'Nylon', 'Terylene', or polyethylene have advantages over natural fibre ropes. In deciding which of the synthetic fibres is most suitable for a particular set of conditions reference should be made to clause F.12, to the appropriate British Standards and to information from manufacturers and suppliers.

C.17   Loss of lifting appliances

A 'black list' of lifting appliances lost or not returned to the Lifting Appliances Store on the date specified, should be issued to all users and should also be posted in prominent positions on the factory premises. Instruction should be given prohibiting the use of any 'black listed' equipment and if found requiring its immediate return to Store. 'Black lists' should be brought up-to-date and re-issued each month.

C.18   Failure of lifting appliances

In all cases of collapse or failure of a crane, derrick, winch, hoist, or other appliance used in raising or lowering persons or goods, or any part of such equipment (except the breakage of chain slings or rope slings), or the overturning of a crane, written notice of the incident, whether personal injury is caused or not, shall be sent to HM District Inspector of Factories.

After any such incident nothing in the vicinity of the incident shall be moved, except for the purpose of rescue, until permission is given by the person in charge of the investigation.

(FA 1961 S 80, 81; SI 1962 No. 272)

D. APPLICATION AND USE

This sub-section deals with the general requirements relating to the use of lifting gear.

D.1   Withdrawal from stores

Lifting appliances should be withdrawn from store only with a written order signed by an authorised person.

D.2   General requirements

Lifting appliances should not be used unless the requirements relating to inspection, examination and test detailed in sub-sections C and F have been fully complied with.

Lifting appliances should be used only for the purpose for which they have been designed.

All lifting appliances must be properly maintained, and defective appliances must not be used.

Wire lashings and hand lines are not lifting appliances and must not be used for lifting purposes or as a direct means of support for lifting or lowering a load.

(FA 1961 S 22(1), 26(1) (a), 27(1); SI 1961 No. 1581 10; SI 1962 No. 225).

D.3   Marking

Every lifting appliance should be clearly marked in plain legible figures and letters, in a manner which is not readily obliterated or removed, with the following information:
1. The identification number.
2. The safe working load (see note below).
3. A symbol in accordance with the appropriate British Standard, to denote the material of which the item has been made and the heat treatment it has been given.

(FA 1961 S 22(8), 26(1) (b), 27(4); SI 1961 No. 1581 29(1), 34(1) (c), 45; DR 23, 24, 29; SI 1961 No. 1581 11(6)).

The proof load must not appear on any lifting appliance. It is recommended that the marking of multiple leg slings should be the safe working load at an included angle of 90° between the legs. A table showing the safe working loads of every size and kind of chain, rope or lifting tackle in use, and in the case of a multiple sling, the safe working load at different angles of the legs shall be posted in the store in which the chains, ropes or lifting tackle are kept, in prominent positions on the premises, and on the site of operations.

(FA 1961; S 26(b)).

In the case of a crane which is occasionally dismantled or partially dismantled, any structural member which is separated from the crane must be clearly marked so as to identify the crane of which it is a part.

(SI 1961 No. 1581 11(6)).

D.4 Lighting

Every place where lifting appliances are used for raising or lowering operations must be adequately and suitably lighted.

Effective provision must be made for securing and maintaining sufficient and suitable lighting, whether natural or artificial, in every part of the premises where persons are either working or passing.

(SI 1961 No. 1580 47; FA 1961 S 5(1)).

D.5 Operators

Lifting appliances must be used only by, or under the direct supervision of, persons who are qualified by training or experience to use appliances in a proper and safe manner. No person under eighteen years of age must be allowed to operate any lifting machine driven by mechanical power or give signals to the operator of any such machine.

(SI 1961 No. 1581 26(1) (2)).

D.6 Cabins

Where reasonably practicable, the cabin of every power driven lifting machine must be adequately heated in cold weather.

For recommendations regarding overhead cranes and a second means of escape from the cab for the driver, see sub-section F.24.

(SI 1961 No. 1581 14).

D.7 Vision and signals

The operator of a lifting appliance must have a clear view of the load being lifted or must be assisted by a specially instructed signaller. It is recommended that a standard code of visual and sound signals be used throughout the works. (Figure 12.1.)

Every signal given for the movement or stopping of a lifting appliance must be distinctive in character and such that the person to whom it is given is able to hear or see it easily and clearly.

Devices or apparatus used for giving sound, colour or light signals for the purposes stated above must be properly maintained, and the means of communication must be adequately protected from accidental interference.

(SI 1961 No. 1581 26(3) (5), 27).

D.8 Access to machines

Safe access must be provided to the cabs and platforms of lifting machines and as far as practicable to all points which require examination, repair or lubrication.

(SI 1961 No. 1581 17).

D.9 Access to area of operation

Every stage, gantry or other place where a lifting machine with a travelling or slewing motion is employed, must be provided with an unobstructed passageway not less than two feet wide. The passageway must be maintained between any part of the appliance liable to move and any guard rails, fencing or other nearby fixture.

If at any time it should prove impracticable to maintain such a passageway at any place or point, all reasonable steps must be taken to prevent the access of any person to

Figure 12.1. Crane signals. The signaller should stand in a secure position where he can see the load and can be seen clearly by the driver. Face the driver if possible. Each signal should be distinct and clear

such place or point at such time. Operators must not operate a lifting machine with persons standing under the jib or within the radius over which the jib may swing. Steps must be taken, by fencing and notices or other positive means to prevent any person from standing or passing under a load while it is being raised, lowered or suspended.

Where any aerial cableway passes above a place where persons work or habitually pass and are thus liable to injury by objects falling from the cableway, then appropriate, effective measures such as the provision of screens or other equipment, so far as is reasonably practicable, must be taken to protect persons from injury.

D.10   Containers

No receptacle must be used for raising or lowering unless it is so enclosed, constructed or designed as to prevent the accidental fall of its contents. This provision will not apply in the case of a grab, shovel or other similar excavating receptacle provided effective steps are taken to prevent any person being endangered by the fall of the contents.

For the requirements of a receptacle to be used for carrying persons. (See sub-section D.11).

(SI 1961 No. 1581 49(3)).

D.11   Lifting of persons

The raising or lowering of persons must whenever practicable, be undertaken by means of an approved suspended scaffold.

Where such an approved scaffold is not practicable, no person must be raised or lowered or carried by a lifting appliance, except in a skip or cage, at least 1 m deep, which is either constructed wholly of, or carried by two strong bands of carbon or low alloy steel properly fastened and carried round the sides and bottom. Alternatively a properly constructed Boatswain's Chair may be used (See sub-section F.1).

Aluminium alloys are not suitable materials of construction for this purpose.

The suspension of approved scaffolds, skips, cages or boatswain's chairs shall be by means of wire ropes or higher tensile or low alloy steel chain. Fibre ropes shall not be used.

When such an approved receptacle is used, suitable measures shall be taken to prevent spinning or tipping and to prevent the occupant falling out.

Materials or tools liable to interfere with a person's hand or foot hold or otherwise cause danger shall not be carried in the receptacle.

Regulations covering the use of appliances for raising or lowering persons are detailed under Hoists, Mobile Cranes, Overhead Cranes and Winches.

The machine used for the raising or lowering of persons in an approved receptacle must have facilities for braking and holding the receptacle, both by means of the prime mover and the brake. The gear clutch engaging the hoist drum to the engine shall always be engaged while raising or lowering persons.

D.12   Securing of loads

Every part of a load must be securely suspended or supported and secured to prevent danger from slipping or displacement.

(SI 1961 No. 1581 49(1)).

The lifting of a load should be halted after the load has been raised a few inches and the security of the slinging attachments should then be checked before proceeding with the operations. (See also section D.14).

The wheel of a barrow must not be used as a means of suspending the barrow unless positive steps are taken to prevent the axle from slipping out of the bearings.

(SI 1961 No. 1581 49(6) (7)).

Adequate steps must be taken by the use of suitable packing or other means to prevent the edges of a load from coming into contact with any sling, rope or chain in such a manner as to cause damage or danger.
(SI 1961 No. 1581, 38; DR 25).

D.13 Suspended loads

No load must be left suspended from a lifting appliance unless a competent person remains in charge of the machine for the whole of the time the load is suspended.
(SI 1961 No. 1581 49(7); DR 33(b)).

D.14 Load shared by two or more lifting machines

When more than one lifting machine is required to lift or lower a single load, the method adopted must be such that no single machine is, at any time loaded beyond its safe working load, or be rendered unstable in the raising or lowering of its load. A competent person, who should be of at least charge-hand status, must be specially appointed to supervise the operation.
(SI 1961 No. 1581 32(2)).

D.15 Loads approaching the safe working load of a lifting machine

When a lifting machine is required to lift a load equal to, or slightly less than, the safe working load of the machine, the lifting must be halted after the load has been raised a few inches and before the operation is proceeded with. When a machine is being used on repetition work it is sufficient that the above provision should apply to the first lift only.
(SI 1961 No. 1581 32(1)).

D.16 Overloading

No lifting appliance must be loaded beyond its safe working load, except for the purpose of making an approved test.
(FA 1961 S 22(8), 26, 27).

D.17 Obstruction to moving load

Steps must be taken to prevent a load coming into contact with and displacing any object, or being itself displaced, while being raised or lowered.
(SI 1961 No. 1581 12).

D.18 Wet paint

No steelwork or structure on which there is wet paint, other than paint used for jointing, must be moved or manipulated with a lifting machine, except for the purpose of painting.
(SI 1961 No. 1581 51).

D.19   Rope anchorages on winding drums

The rope anchorages on winding or derricking drums must be securely fixed and there must be at least two dead turns of rope on the drum in every operating position.
(SI 1961 No. 1581 15).

D.20   Stability of lifting appliances

No lifting appliance must be used on soft or uneven surfaces in circumstances in which the stability of the appliance is likely to be affected, or on a slope unless appropriate precautions are taken to ensure safety.
   No crane must be used for raising or lowering unless its stability has been ensured by
   1. Secure anchorage.
   2. Adequate weighting by suitable ballast which must be properly placed on the crane structure and effectively secured to prevent it being accidentally displaced.
   No part of any rails on which the crane is mounted, or the sleepers supporting the rails must be used as anchorage for this purpose. Where the stability of the crane is secured by means of removable weights, a diagram or notice indicating the position and amount of such weights must be fixed on the crane where it can readily be seen.
   No crane must be used or erected under weather conditions likely to endanger its stability. After exposure to such conditions, the anchorage arrangements and ballast must be examined by a competent person as soon as practicable and before the crane is next used, and all necessary steps must be taken to ensure the stability of the crane. Where outriggers are used arrangements must be made to ensure the continued stability of the appliance.
   (SI 1961 No. 1581 19).

D.21   Repairs and modifications

Repairs or modifications of registered lifting appliances shall be the responsibility of the Lifting Appliances Organisation. In cases where welding or heat treatment is involved the Lifting Appliance Organisation must seek specialist advice.
   The repair of higher tensile and alloy steel chains is a specialised operation and the correct procedure is given in sub-section C.15.1.

D.22   Guarding

All revolving shafts, flywheels, couplings, toothed gearing, belt and pulley drives, chain and sprocket drives, all projecting screws, bolts or keys on any revolving shaft or pinion, every live electric conductor, every steam pipe and every dangerous part of any lifting machine must be securely fenced or so enclosed or so situated as to be safe to every person employed on or working near the machine.
   (FA 1961 S 14(1); SI 1961 No. 1580 42; DR 26).
   Special requirements apply to prime movers.
   (FA 1961 S 12).

D.23   Return to store

Appliances registered by the Lifting Appliances Organisation, which have become worn or defective must, where practicable, be returned to the Lifting Appliances Store or must be reported to the Organisation which shall be responsible for the repair or destruction of the items concerned. (See also sub-section D.20).

Appliances recalled by the Lifting Appliances Organisation for periodic test and examination must not continue to be used beyond the date for return specified on the recall note. If replacement items are required for lifting appliances on recall, arrangements must be made to obtain such replacements in good time.

D.24   Loss of appliances

The loss of any lifting appliance must be reported to the Lifting Appliances Organisation without delay. A lost appliance 'black list' (see sub-section C.17) shall be posted in prominent positions throughout the factory.

D.25   Failure of appliances

In all cases of collapse or failure of a crane, derrick, winch, hoist or other appliance used in raising or lowering persons or goods or any part of such equipment (except the breakage of chain slings or rope slings), or the overturning of a crane, written notice of the incident, whether or not personal injury or disablement is involved, must be sent to the Factory Inspector of the district.

After such a failure nothing in the vicinity of the incident must be moved, except for the purpose of rescue, until permission is given by the person in charge of the investigation. (FA 1961 S 80, 81; SI 1962 No. 272).

D.26   Equipment not owned by the factory occupier or site operator

Although the actual owner of lifting equipment is required to comply with the statutory regulations relating to guarding, testing and examination, and the general safe operation of the equipment, the factory occupier or site operator has certain responsibilities from which he is not released by the failure of the owner to fulfil his duties.

In cases where equipment is hired, it is the hirer's responsibility to see that the law is complied with. Thus, where equipment is either owned or hired by a contractor working in a factory or on a construction site it is the responsibility of the factory occupier or site operator to see that his employees and those of any third party are not endangered.

Appropriate action must be taken at all times to ensure compliance with the law and the scale of action required will depend upon circumstance. For example, a mobile crane on short term hire together with a driver, may require only the inspection of a copy of the current test and examination certificate, but equipment on long term hire and operated by the occupier, employees must be inspected before use and be registered with the Lifting Appliance Organisation, to ensure that it is receiving the statutory periodic examinations.

D.27   Diesel engines in flammable atmospheres

Some lifting machines are powered by diesel engines and, in these cases, there is in addition to any danger from electrical equipment, hot surfaces or exhaust systems a special hazard when the machine is operated in an area where flammable vapours may exist. In the event of an escape of such vapour it is possible that this may be drawn into the air intake of the diesel engine and act as a fuel.

In such cases cutting off the normal fuel supply may not stop the engine. Certain flammable materials may cause the engine to race out of control of its governor and fuel cut off device, leading to valve bounce and back-firing. A diesel engine will not normally run on all flammable vapours without access to its normal fuel, but the availability of such vapours is likely to cause back-firing. Both conditions may well cause ignition of any surrounding flammable atmosphere.

Before diesel driven equipment is taken into areas where there may be a flammable atmosphere, written authorisation should be obtained from the factory or site manager.

E. HEAT TREATMENT

Heat treatment is a specialised technique and should not be attempted without first seeking advice of a recognised authority.

E.1   General

The heat treatment of lifting gear components has previously been given much prominence mainly because most early equipment was made of wrought iron, a material which is susceptible to surface embrittlement under normal operating conditions. This gave rise to the need for periodic annealing to restore normal surface ductility, a process which became a statutory requirement and which is still mandatory under the Factories Act, 1961, (Section 26, para. (1) (f)) for wrought iron gear. Any wrought iron lifting gear still in use must comply with these regulations.

This Section, however, does not cover the use of wrought iron lifting tackle, now considered to be obsolete, and for this reason it does not deal, in detail, with periodic heat treatment.

The type of lifting gear dealt with here, made either of carbon (higher tensile) steel or low alloy steel, must under no circumstances be subjected to periodic annealing.

E.2   Definitions

NORMALISING

Normalising is a treatment applied to items of lifting gear made from carbon steel (mild steel or higher tensile steel). The treatment consists of heating the part in a furnace until the whole of it attains a uniform temperature within 50°C above the upper critical point of the steel used, withdrawing the component and allowing it to cool in still air. The normalising temperature for mild steel lifting gear is in the range 880–910°C while that for higher tensile steel gear is in the range 860–900°C.

HARDENING AND TEMPERING

Hardening and tempering is a treatment applied to both carbon (higher tensile steel) and low alloy steels. The hardening process consists of heating in a furnace until the whole of the component attains a uniform temperature within 50°C above the upper critical point of the steel used, holding at this temperature for an appropriate time interval and then rapidly quenching in either water or oil. Tempering is a toughening treatment and is carried out immediately after hardening. It consists of re-heating the component uniformly in a furnace at a suitable intermediate temperature, holding at this temperature for an appropriate time interval and finally cooling in still air. The hardening temperature is usually within the range 840–890°C but the precise temperature employed will depend upon the actual composition of the steel used. Tempering is carried out in the range 550–650°C depending upon the actual properties required.

E.3 Furnaces

The furnaces used for all heat treatment of lifting appliances shall be properly constructed and have suitable means of indicating and controlling the temperature.

E.4 Link heater

This is a device which enables a single link in a chain to be heat treated without affecting neighbouring links. Its principal use is in the heat treatment of a single link used to join together two lengths of chain. It is possible to carry out normalising or hardening and tempering operations.

Whilst the device is useful it must be used under careful supervision if consistent results are to be obtained from one link to the next.

E.5 Initial heat treatment

HIGHER TENSILE STEEL AND ALLOY STEEL

Lifting tackle components in both these materials shall be given an initial heat treatment consisting of hardening and tempering.

E.6 Heat treatment after repair

If a repair involving heating is applied to an item of lifting tackle which has been given a final heat treatment after manufacture, then the same heat treatment shall be applied after repair.

E.7 Relevant statutory requirements

There are no Statutory Requirements relating to the heat treatment of lifting gear manufactured from higher tensile steel or alloy steel. Lifting gear of this type is exempt from periodic heat treatment by certificate of the H.M. Factory Inspector.

F. DETAILED REGULATIONS

The intention of this section is to give the designer and Lifting Appliance Organisation details of the most commonly used lifting appliances. It is not possible to include the many items of special gear used in industry but sufficient examples are given to allow a proper decision being made in most cases.

When it is necessary for a particular purpose to make up a specialised lifting appliance, it must be properly designed and constructed of suitably specified materials (see subsection C.4). Before it is used it must be inspected and tested, including proof testing, and it must be clearly marked with its safe working load.

The periods within which inspection, examination or test must be made are based on the requirements of the Factories Act, 1961, Construction (Lifting Operations) Regulations, Dock Regulations 1934 and other Statutory Rules and Orders.

Full information regarding the correct use of prescribed forms and registers has been given, though if a card register system is adopted (see section C.8) the use of some of these will not be necessary. Where no form or register for the entry of a statutory record is prescribed, recommendations are in certain cases made regarding a form or register which may appropriately be used.

When an appliance consists of an assembly of separate items it is necessary either:
1. To hold test and examination certificates for the complete assembly, and
2. to ensure that items do not become detached and used apart from the assembly, and
3. to re-test the complete assembly if any part of it is replaced,
or
to hold test and examination certificates for each item of the assembly.

For example, in the case of a boatswains chair, test and examination certificates must be held either for the complete assembly or for the chair, blocks, rope, etc., individually.

Most of the items detailed in this sub-section F are described in BS 3810 'Glossary of Terms used in Materials Handling'. This contains, where appropriate, line drawings illustrating typical examples of the items defined.

F.1   Boatswain's chair

(See also Clause D.11, Lifting of Persons)

Specification: BS 2830 'Suspended Safety Chairs and Cradles for use in the Construction Industry'.

Material: All load bearing parts and supports of the boatswain's chair must be of steel. Wrought iron and aluminium, although included in BS 2830, are not now considered to be suitable materials for this purpose. However, timber seats are permissible provided they are fully supported on a steel member which in turn is connected directly to the hanger.

Test: (i) when new, (ii) after alteration or repair.

Safe Working Load: 113 Kg.

Proof Load: 152 Kg.

Marking: (i) identification number. (ii) safe working load.

Test Certificate: Use Form 86.
(*Note.* It is necessary to hold the maker's certificate on Form 87 for a wire rope).

Inspection: Immediately before use.

Examination: Every six months.

F.2   Bordeaux connection

Specification: BS 461 'Metric Units'.

The form and dimensions of higher tensile steel grab shackles to be employed when connecting a chain to a bordeaux connection are shown in Figure 12–2 and

Table 2 of BS 461. In all other respects grab shackles must comply with BS 3032 'Higher Tensile Steel Shackles'.

A normal five-tuck splice in the attached wire rope is unsuitable for bordeaux connections, and a method of making the recommended nine-tuck splice is given in BS 461.

Material:

Rope fitting. Carbon steel casting to BS 3100, Specification BS 392 Grade B or C in the hardened and tempered condition.

Connecting Link. Steel to BS 970 Part I, 060 A27, 080 A30 or 150 M19 in the hardened and tempered condition.

Heat Treatment: Initially and after alteration or repair and to consist of hardening and tempering.

Test: (i) when new, (ii) after alteration or repair.

Proof Load: 2 × safe working load.

Test Certificates:

	Certificate	Register
FA	Use Form 86	Use Form 88, Pt. I
C(LO)R	Use Form 97	Use Form 91, Pt. II H
DR	Use Form 86	Use Form 99

Marking: (i) identification number, (ii) safe working load, both to be stamped on the rope connection.

Periodic Inspection and Examination: See Table 12.5.

Table 12.5

	Inspection	Form No.	Examination	Form No.
FA	None	—	6 months	88 Pt. II
C(LO)R	None	—	6 months	97
DR	None	—	12 months	1952 or 99 Pt. III

F.3 Bulldog grips

Specification: BS 462 'Metric Units'.

Bulldog grips are used as an alternative to splicing or socketing in cases where skilled labour or facilities for such work are not available. BS 462 details the method to be adopted in applying bulldog grips to wire ropes and recommends the number of grips to be used for various rope diameters.

Material:

U bolts and nuts. Steel to BS 970 Part I, 070 M20 in the normalised condition.

Bridges. Steel drop forgings to BS 970 Part I, 070 M20 in the normalised condition, or steel castings to BS 3100 Specification BS 592 Grade A.

Note that malleable iron castings and spheroidal graphite iron castings included in BS 462 are not now considered suitable materials for this purpose.

Heat Treatment: Initial heat treatment only is required and should consist of normalising.

Test: Not required on finished bulldog grips.

Test Certificate: No statutory form is required but a manufacturer's certificate of compliance with BS 462 must be filed in the lifting appliance register.

Marking: Nominal size to be marked on bridge.

Periodic Inspection and Examination: See Table 12.6(a)

Table 12.6(a)

	Inspection	*Form No.*	*Examination*	*Form No.*
FA	None	—	6 months	88 Pt. II
C(LO)R	None	—	6 months	97
DR	None	—	None	—

Care and discretion are needed in using bulldog grips. If they are wrongly applied the reduction in the tensile strength of the rope may be substantial. The bridge of the grip must always be fitted to the working part of the rope and the U-bolt on the tail or dead end of the rope, see Figures 12.2 and 12.3.

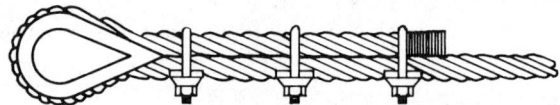

Figure 12.2. Correct method of making a loop in a rope

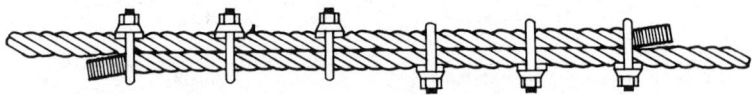

Figure 12.3. Correct method of joining two ropes. When joining two ropes, twice the number of grips is required than is necessary when forming a loop in a single rope

The minimum number of grips to be used for each rope depends on the size of the rope and must not be less than that given in Table 12.6(b).

Table 12.6(b)

Nominal diameter of wire rope mm	Minimum number of bulldog grips
Up to and including 19	3
Over 19 up to and including 32	4
Over 32 up to and including 38	5
Over 38 up to and including 44	6
Over 44	7

F.4   Chain

A load must not be raised, lowered or suspended on a chain which has a knot tied in any part under direct tension.

A chain which has been shortened or joined to another by means of bolts and nuts must not be used for raising, lowering or suspending a load. The use of chain adjusters for shortening a chain is not recommended.

Hand chains must not be used as slings or in any way directly for raising, lowering or suspending a load.

No chain must be issued for service if, because of wear or other agency, the diameter of a right section through any link is less than that shown in Table 12.4 (See sub-section C13).

Chains must not be used if the links have become 'shouldered' or 'socketed' at the bearing points or if there is the least tendency to locking.

Special care should be exercised in the use of lifting appliances which incorporate plate link load chains. This type of chain is flexible only in one plane and it is important to avoid conditions which impose loading in any other plane as this may cause distortion of the chain.

It is possible for high strength chains, including some of those discussed below, to become embrittled by hydrogen. Such embrittlement can arise, for example, by the exposure of chains to an acidic corrosive environment. Although, in direct exposure to hydrogen, susceptibility to embrittlement increases with the tensile strength of the steel, exposure tests in an industrial atmosphere have shown that the risk of embrittlement of Grade 80 chains is no higher than with Grade 40 chains, and these have given satisfactory service for many years.

To guard against embrittlement, high strength chains, when used under conditions where corrosion is likely to occur, should be protected with a coating of a suitable oil. The chains should be examined regularly to ensure that the oil film is adequate and corrosion is not taking place.

(SI 1961 No. 1581 39)

F.4.1   PITCHED OR CALIBRATED LOAD CHAIN OF HIGHER TENSILE STEEL OR ALLOY STEEL

Specifications: BS 1663 'Higher Tensile Steel Chain Grade 40 (short link and pitched or calibrated) for lifting purposes' and BS 3114 'Alloy Steel Chain Grade 80. Polished short link calibrated load chain for pulley blocks'.

Higher Tensile Steel Chain to BS 1663 has a minimum breaking load of 40 d^2 ton (Grade 40) where 'd' is the nominal size of the chain in inches. The recommended safe working load for this type of chain provides a factor of safety of 5 to 1.

Alloy Steel Chain to BS 3114, has a minimum breaking load of 80 d^2 ton and the recommended safe working load provides a factor of safety of 5·75 to 1. This type of chain is unsuitable for general lifting purposes and must be used only on pulley blocks.

Material: Grade 40 chains to BS 1663 are made of low carbon steel having a carbon content less than 0.30%.

Grade 80 chains to BS 3114 are made of low alloy steel containing nickel, chromium and molybdenum.

Heat Treatment: Only required initially and/or after modification or repair. In the case of both Higher Tensile and Alloy Steel Chains the heat treatment consists of hardening and tempering.

Periodic heat treatment must on no account be carried out on these chains.

Proof Load: $2 \times$ safe working load, or $1\frac{1}{2} \times$ safe working load when used with hand operated pulley blocks (Form 1951).

Test: (i) when new according to BS 1663 or BS 3114 as appropriate. (ii) after alteration or repair — with proof load only.

Test Certificate:

	Certificate	Register
FA	use Form 86.	Form 88 Pt. I.
C(LO)R	use Form 97.	use Form 91 Pt. II H.
DR	use Form 86.	Form 99.

Form 86 shall be endorsed for heat treatment for higher tensile and low alloy steel chain.

Marking:
    (a) identification number;
    (b) safe working load.
    (c) BS 1663 requires the manufacturer of higher tensile steel chain to mark each twentieth link, or links at intervals of 3 ft (0·914 m) apart with the symbol ⑭ to indicate that the chain is of higher tensile steel, grade 40 quality, in the hardened and tempered condition.
        BS 3114 precludes marking of chain links themselves but stipulates that substantial metal tabs, tallies or labels should be fitted to the chains. These should be marked with the symbol ⑱ indicating that the chain is of alloy steel grade 80, quality in the hardened and tempered condition.

Periodic Inspection and Examination: See Table 12.7.

Table 12.7

	Inspection	Form No.	Examination	Form No.
FA	None	—	6 months	88, Pt. II
C(LO)R	None	—	6 months	97
DR	Each occasion before use or 3 months	None	12 months	99, Pt. II or 1952 and 99

F.4.2   SHORT LINK AND NON-PITCH LOAD CHAIN OF HIGHER TENSILE STEEL OR ALLOY STEEL

Specifications: BS 1663 'Higher Tensile Steel Chain Grade 40 (short link and pitched or calibrated) for lifting purposes'.
    BS 3113 Alloy Steel Chain Grade 60 short link for lifting purposes.
    BS 1663 provides for a chain having a minimum breaking load of 60 d^2 ton and a factor of safety of 5 to 1.
    Alloy steel chain to BS 3113 has a minimum breaking load of 60 d^2 ton also with a factor of safety of 5 to 1.

    BS 3113 is likely soon to be superseded by a Standard covering a higher grade of chain possibly with a minimum breaking load of around 75 d^2 ton. A considerable quantity of chain of this type is already being manufactured and such chains may be offered in preference to BS 3113. These higher grade chains are of perfectly acceptable quality but in the meantime they are not covered by a British Standard.

It should be noted, particularly, that certain high strength chains are produced with a specifically hardened surface layer. Such chains are considered unsuitable for lifting purposes and should not be employed for these duties.

Material: Grade 40 chains to BS 1663 are manufactured from low carbon steel having a carbon content less than 0·30%.

Grade 60 chains to BS 3113 are made of a low alloy steel containing nickel, chromium and molybdenum.

Heat Treatment: Only required initially and/or after modification or repair. In case of both Higher Tensile and Alloy Steel Chains the heat treatment shall consist of hardening and tempering.

Periodic heat treatment must on no account be carried out on these chains.

Proof Load: 2 × safe working load for both higher tensile and alloy steel chains.

Test: (i) When new according to BS 1663 or BS 3113 as appropriate. (ii) after alteration or repair, with proof load only.

Test Certificate:

	Certificate	*Register*
FA	Use Form 86	Form 88, Pt. I
C(LO)R	Use Form 97	Use Form 91, Pt. II, J
DR	Use Form 86	Use Form 99

Form 86 shall be endorsed for heat treatment for both higher tensile and low alloy steel chain.

Marking:
(a) Identification number.
(b) Safe working load.
(c) BS 1663 requires the manufacturer of higher tensile steel chain to mark each twentieth link, or links at intervals of 3 ft (0·914 m) apart with the symbol ⑭ to indicate that the chain is of higher tensile steel grade 40 quality in the hardened and tempered condition.

BS 3113 requires each twentieth link, or links at intervals of 3 ft (0·914 m) apart, whichever is the lesser distance shall be legibly and permanently marked with the symbol ⑯ to denote Grade 60 chain in the hardened and tempered condition.

Periodic Inspection and Examination: See Table 12.8.

Table 12.8

	Inspection	*Form No.*	*Examination*	*Form No.*
FA	None	—	6 months	88, Pt. II
C(LO)R	None	—	6 months	91, Pt. II, J
DR	Each occasion before use or every 3 months	—	12 months	99, Pt. III

F.5   Eyebolt

Specifications: BS 4278 'Eyebolts for Lifting Purposes, Metric Units'.

BS 4278 covers three types of higher tensile steel eyebolts, namely, collar eyebolts, collar eyebolts with link and dynamo eyebolts. The first two types are suitable for general lifting operations but dynamo eyebolts are intended for vertical loading only.

Collar eyebolts should always be tightened down to ensure full contact between the underside of the collar and the seating, but undue tightening of eyebolts must be avoided. Collar eyebolts are suitable for the application of oblique loads, such as result from the use of multiple leg slings, provided the loads are applied in the plane of the eye.

It is strongly recommended that eyebolts without collars should not be used. Eyebolts of this type were formerly provided for in BS 529 Part II which has now been superseded by the above specification.

When two or more eyebolts are used simultaneously for raising a load, each eyebolt must be loaded by an individual sling leg. When rigid items are lifted on four eyebolts then each eyebolt should be capable of taking half the total load, i.e. the safe working load of each eyebolt should be equal to one half of the total load.

A continuous sling must not be threaded through two eyebolts because of the excessive stress set up in them due to the tension in the horizontal part of the sling.

Material: The whole of the eyebolt must be made in steel to BS 970 Part I Grade 060 A27 (En5A). Eyebolt bodies, including the shank, shall be made as one piece drop forgings. Eyebolt links must be welded and inspected as required by BS 4278.

Heat Treatment: Eyebolts and links must be in the hardened and tempered condition.

Proof Load: 2 × axial safe working load.

Test: When new; according to BS 4278.

Test Certificate:

	Certificate	Register
FA	Use Form 86	Form 88, Pt. I
C(LO)R	Use Form 97	Form 91, Pt. II, J
DR	Use Form 86	Form 99

Marking: The following markings are to be stamped on the eyebolt collar.
  (i) Identification number.
  (ii) Safe working load.
  (iii) The symbol (04) to indicate that the complete eyebolt is made of higher tensile steel in the hardened and tempered condition.
  (iv) Identification of thread on screwed shank.

Periodic Examination and Inspection: See Table 12.9.

Table 12.9

	Inspection	Form No.	Examination	Form No.
FA	None	—	6 months	88, Pt. II
C(LO)R	None	—	6 months	91, Pt. II, J
DR	None	—	12 months	99, Pt. II

Table 12.10 MAXIMUM RECOMMENDED WORKING LOADS FOR COLLAR EYEBOLTS WHEN USED IN PAIRS FOR INCLINED LOADING CONDITIONS

Nominal size of eyebolt		Safe working load of a single eyebolt		*Included angle between legs*					
				30°		60°		90°	
BS 4278	BS 529								
mm	in	tonne	ton cwt	tonne	ton cwt	tonne	ton cwt	tonne	ton cwt
18	$\frac{3}{8}$	1·0	5	1·3	5	800 kg	3	500 kg	2
20	$\frac{1}{2}$	1·2	10	1·6	11	1·0	7	630 kg	5
22	$\frac{5}{8}$	1·6	18	2·0	1 0	1·2	13	800 kg	9
24	$\frac{3}{4}$	2·0	1 8	2·5	1 12	1·6	1 1	1·0	14
27	$\frac{7}{8}$	2·5	2 0	3·2	2 6	2·0	1 10	1·2	1 0
30	1	3·2	2 15	4·0	3 2	2·5	2 0	1·6	1 7
33	$1\frac{1}{8}$	4·0	3 10	5·0	4 0	3·2	2 10	2·0	1 14
36	$1\frac{1}{4}$	5·0	4 10	6·3	5 2	4·0	3 6	2·5	2 4
39	$1\frac{1}{2}$	6·3	6 10	8·0	7 8	5·0	4 16	3·2	3 4
45	$1\frac{3}{4}$	8·0	9 0	10·0	10 4	6·3	6 12	4·0	4 8
52	2	10·0	12 0	12·5	13 12	8·0	8 16	5·0	6 0
56	$2\frac{1}{4}$	12·6	15 0	16	17 0	10·0	11 0	6·3	7 8
64	$2\frac{1}{2}$	16	20 0	20	22 14	12·5	14 14	8·0	9 18
70	—	20	—	25	—	16	—	10·0	—
76	3	25	30 0	32	34 0	20	22 0	12·5	14 16

F.6 Girder clips

Specification: None available.

Material: Clips must be made as single piece forgings in steel to BS 970 Pt. I, steel 060 A27 (En5A).

Rings to be made of the same steel, welded and inspected as required by BS 4278 'Eyebolts'.

Heat Treatment: The clip and ring assembly shall be hardened and tempered.

Proof Load: 2 × safe working load.

Test: When new.

Test Certificate: Use Form 86, FA; 97, DR 86 (C(LO)R)).

Marking: (i) Identification number. (ii) Safe working load.

Periodic Inspection and Examination: Examine every six months.

F.7 Hooks

Specifications: BS 2903 'Higher Tensile Steel Hooks, Metric Units' and BS 3017 'Mild Steel Forged Ramshorn Hooks'.

BS 2903 provides for six types of hand or drop forged higher tensile steel hooks of trapezoidal section. Hooks of both point and c type are included and each form is specified with a screwed shank or alternatively with an eye for use with chain or wire rope thimbles, the c hook in the latter case being supplied with a link. Provision is made in BS 2903 for hooks with forged lugs for the attachment of safety catches. Safe working loads conform with the R.10 series of preferred numbers and shank hooks are available up to a safe working load of 63 tonnes.

BS 3017 provides a range of forged ramshorn hooks of trapezoidal section having safe working loads in the range 20 to 200 ton (20 to 200 tonne, approx.). Crane hooks exceeding 30 tonne safe working load should preferably be of the ramshorn type. (See BS 466 'Electric overhead travelling cranes').

All hooks to these two specifications are required to be one piece forgings and hooks with welded shanks are specifically excluded. Similarly, hooks which have been flame cut from plate material are not permissible for general lifting applications.

It is strongly recommended that all hooks, whether intended for building operations, works of engineering construction, or general use should be provided with a safety catch. For the first two of the above applications the use of hooks fitted with safety catches is mandatory.

(SI 1961 No. 1581 36(a) and (b))

Material: BS 2903 'Steel to BS 970 Pt. I Grades 080 A27, 080 A30, 150 M19 or 150 M28. BS 3017 'Steel to BS 970 Pt. I Grade 070 A20'.

Heat Treatment: Initial heat treatment only is required and must consist of: BS 2903 'Hardening and Tempering' and BS 3017 'Normalising'.

Proof Load:

    (i) For hooks having safe working loads of 25 tonnes or less: proof load = 2 × safe working load.

    (ii) For hooks having safe working loads in excess of 25 tonnes, see BS 2903 Appendix B, and BS 3017.

Test: When new according to appropriate specification.

Test Certificate:

	Certificate	*Register*
FA	Use Form 86	Form 88, Pt. I
C(LO)R	Use Form 97	Form 91, Pt. II, J
DR	Use Form 86	Form 99

Form 86 must be endorsed for heat treatment.

Marking: The marking of hooks must be done at a position near the tip as shown in BS 2903 and must include:

    (a) Identification number.

    (b) Safe working load but see BS 3017 for details of marking the included angle on ramshorn hooks.

    (c) (i) BS 2903 hooks must be marked with the symbol ⑭ to denote that they are of higher tensile steel quality in the hardened and tempered condition. (ii) BS 3017 hooks must be marked with the word 'Steel'.

Periodic Inspection and Examination: See Table 12.11.

Table 12.11

	Inspection	Form No.	Examination	Form No.
FA	None	—	6 months	88, Pt. II
C(LO)R	None	—	6 months	91, Pt. II, J
DR	Each occasion before use or every 3 months	None	— 12 months	1952

F.8 Portable lifting beams (spreader, strongback)

Specification: None available.

Lifting beams may take many forms. Each must be properly designed for its particular application and it must be properly constructed of suitable material.

Material: Structural steel to BS 4360.

Heat Treatment: If the manufacture of the lifting beam involves forging or welding operations then the completed beam must be normalised before use. Otherwise no heat treatment is required.

Proof Load: $2 \times$ safe working load up to 25 tonnes, or in excess of this, according to BS 2903 Appendix B.

Test: Before use or after alteration or repair.

Test Certificate:

	Certificate	Register
FA	Use Form 86	Form 88, Pt. I
C(LO)R	Use Form 97	Form 91, Pt. II, H
DR	Use Form 86	Form 99

Marking: (i) Identification number. The identification number must be marked on each detachable component. (ii) Safe working load.

Periodic Inspection and Examination: See Table 12.12.

Table 12.12

	Inspection	Form No.	Examination	Form No.
FA	None	—	6 months	88, Pt. II
C(LO)R	None	—	6 months	91, Pt. II, H
DR	Each occasion before use or every 3 months	None	12 months	99

F.9   Lifting box or skip for materials only (for lifting of persons, see sub-section D.11)

Specification: None available.

Every lifting box or skip must be properly designed and constructed and all load bearing members must be of a suitable quality of steel. Boxes or skips used for raising or lowering of loose objects such as for example bricks or tiles, must be designed and constructed in such a way as to prevent the accidental fall of such objects.

Heat Treatment: If the manufacture of the box or skip involves forging or welding operations then either the whole completed box or skip or the completed load bearing steel framework must be normalised before use.

Proof Load: 2 × safe working load.

Test: (i) when new. (ii) after alteration or repair.

Test Certificate: FA and DR; Form 86; C(LO)R; Form 87.

Marking: (i) Identification number. (ii) Safe working load.

Periodic Inspection and Examination: Every six months.

F.10   Plate lifting clamp

Specification: None available.

Material: Steel to BS 970 Pt. I Grade 170 A20.

Heat Treatment: Normalise after forging.

Proof Load: 2 × safe working load.

Test: When new and after repair.

Test Certificate:

	Certificate
FA	Form 86
C(LO)R	Form 97
DR	Form 86

Marking: (i) identification number. (ii) safe working load. (iii) plate thickness for which clamp is suitable.

Periodic Inspection and Examination: Every six months.

F.11   Rigging screws, stretching screws and turnbuckles

Specification: BS 4429 'Metric Units'.
This standard provides for the following range of nominal sizes: Rigging Screws, 10 mm to 110 mm; Turnbuckles, 10 mm to 56 mm.

The size of a rigging screw or turnbuckle is the diameter of the eye or other end attachment. The safe working loads in the standard increase in accordance with the R.10 series of preferred numbers.

Material: The tubular bodies must be of longitudinally welded or seamless steel tubing to BS 1775 Grade 13, or Grade 16. Tubular bodies may also be machined from solid material to BS 970 Pt. I Grade 080 A27.

Eyes, fork ends and open bodies must be one piece forgings in steel to BS 970 Pt. I Grade 080 A27. Other forms of end attachment must comply with the appropriate British Standard.

Heat Treatment: Initial heat treatment only is required and consists of normalising.

Proof Load: (i) Up to and including 25 tonnes: Proof Load = 2 × safe working load. (ii) Above 25 tonnes: see BS 4429.

Test: (i) before use. (ii) after alteration or repair.

Test Certificate:

	Certificate	*Register*
FA	Use Form 86	Form 88, Pt. I
C(LO)R	Use Form 97	Form 91, Pt. II, H
DR	Use Form 86	Form 99

Marking: (i) Identification number. (ii) Safe working load. (iii) The symbol ⑭ to denote that the components are of higher tensile steel in the normalised condition.

Periodic Inspection and Examination: See Table 12.13.

Table 12.13

	Inspection	*Form No.*	*Examination*	*Form No.*
FA	None	—	6 months	88, Pt. II
C(LO)R	None	—	6 months	91, Pt. II, H
DR	Each occasion before use or every 3 months	None	12 months	99, Pt. III or 1952

F.12   Fibre rope (natural and synthetic)

Specifications:
BS 2052 'Ropes made from Coir, Hemp, Manila and Sisal'.
BS 4002 'Cotton Ropes'.
BS 3758 'Polyester Filament Ropes (Terylene)'.
BS 3912 'Polythene Filament Ropes'.
BS 3977 'Polyamide (Nylon) Filament Ropes'.
BS 3724 'Glossary of Terms Relating to Fibre Ropes and Cordage'.

Materials: Coir ropes are unsuitable for lifting purposes because their load carrying capacity is low, they chafe readily and have poor resistance to water.

Cotton makes a soft, flexible rope with good shock resistance but its load carrying capacity is low.

Hemp makes a strong, soft, flexible rope but it is not much used for lifting purposes. The most useful of the natural fibres for making lifting ropes are manila and sisal. Manila fibres are long and tough and produce a strong flexible rope which stands up well to wear in both wet and dry conditions. Sisal produces a strong rope but its resistance to both wear and water is not as good as that of Manila. However, at the present time it is used more extensively than any other natural fibre for lifting ropes.

Synthetic fibre ropes are made either of Nylon, Terylene, or polypropylene (Ulstron). As compared with natural fibre ropes they have much higher tensile strengths and shock resistance, better resistance to wear, mildew and attack by some chemicals, and do not absorb water. They are however, easily damaged by contact with hot surfaces or by friction.

Nylon is resistant to alkalis but readily attacked by mineral acids.

Terylene has good resistance to acids but is attacked by hot, strong alkalis and concentrated phenols.

Polypropylene has excellent resistance both to acids and alkalis but is attacked by some hydrocarbons and by halogenated hydrocarbons.

For practical purposes it is essential that all natural fibre ropes must be treated with a suitable water repellant if they are to be used out-of-doors. For such use it is generally preferable to specify ropes made of synthetic fibre.

Ropes should not be allowed to lie on wet ground. Where the ground is wet or there is danger of the rope coming into contact with corrosive material, the loose fall should be collected in a suitable receptacle such as an empty drum or box.

When natural fibre ropes became wet they must be dried naturally and not by the application of excessive heat as this may render them unfit for further use. They should be stored in a warm (10–16°C), dry place, free from fumes, and should be so hung or stocked that air can freely circulate around them.

Under many conditions synthetic fibre ropes made from nylon, Terylene or Ulstron have advantages over natural fibre ropes. Before deciding to use ropes made from synthetic fibres it is recommended that information should be obtained regarding the suitablity of the particular fibre for the actual conditions of use. Such information is normally available from rope manufacturers.

Test: A sample of the rope shall be tested to destruction by the manufacturer strictly in accordance with the appropriate specification.

Proof Test: Fibre ropes are not proof tested because they would be permanently damaged by an overload.

Safe Working Load: Various factors of safety are to be found in the literature depending upon factors such as size of rope and purpose for which it is used. However, the factor of safety to be applied to all fibre ropes must be in accordance with British Standard recommendations for fibre ropes. See F.15.2 and Table 12.24 for safe working loads. 'Recommendations for the use of Fibre Rope Slings'. See Clause F.15 Table 12.24 for safe working loads.

Test Certificate:

	Certificate	Register
FA	Maker's	Use Form 88
C(LO)R	Maker's	Use Form 91, Pt. II
DR	Maker's	Use Form 99

When a length of rope is cut from a stock coil. the Test Certificate Number of the coil must be entered in the Lifting Appliances Register against the cut off length.

Marking: (a) Identification number. (b) Safe working load. The marking must be stamped on metal ferrules firmly attached at each end of the rope.

Periodic Examination: See Table 12.14.

Table 12.14

	Inspection	Form No.	Examination	Form No.
FA	None	—	6 months	88, Pt. II
C(LO)R	None	—	6 months	91, Pt. II, H
DR	None	—	None	—

Although statutory regulations require fibre ropes to be examined at six-monthly intervals it is recommended that because of the special hazards encountered in their use, they should be examined more frequently. The preferred procedure is for the slinger to be trained and instructed always to look over every fibre rope immediately before each occasion on which it is used, looking particularly for cuts and abrasions and signs of chemical attack. In addition, each rope must be examined by the competent person every six months.

F.13 Wire rope

Specifications:
BS 302. Wire Ropes for Cranes, Excavators and General Engineering Purposes (Metric Units).
BS 329. Steel Wire Ropes for Electric Lifts (Metric Units).
BS 3530. Small Wire Ropes.

Wire ropes must not be used as lifting tackle if in any length of 10 diameters the total number of visible broken wires exceeds five.
Ropes must not be used on sheaves or pulleys having groove widths less than the diameter of the rope. The ratio of the diameter of the sheave or pulley to the diameter of the rope must be adequate to avoid excessive bending of the rope (see BS 302 and BS 329).
(SI 1961 No. 1581 34(3); DR 20(c).
A load must not be raised, lowered or suspended by a wire rope which has a knot tied in any part under direct tension.
Ropes must be protected from damage by sharp edges on the load or at fixed anchorages.
Ropes must not be allowed to lie on wet ground. Where the ground is wet or there is danger of the rope coming into contact with corrosive material, the loose fall should be collected in a suitable receptacle such as an empty drum or box.
(SI 1961 No. 1581 39; SI 1961 No. 1581 38)
A thimble or loop splice made in a wire rope must have at least three tucks with a whole strand of the rope and two tucks with one half of the wires cut out of each strand. The strands in all cases shall be tucked against the lay of the rope. An equally efficient form of splice may be used if preferred.
(DR 20(d)).
For making a wire rope bordeaux connection splice a nine-tuck splice should be used and a recommended method for making this is given in BS 461.

Material: Cold drawn carbon steel wire conforming to BS 2763 'Round steel wire for ropes (Metric Units)'. The wire may have tensile strengths in various ranges between 1175 and 1765 MN/m^2. Stainless steel wire ropes are also available for special duties.

Tests:
    (a) A tensile test to destruction on a sample of a production length of new rope carried out as detailed in the appropriate British Standard.
    (b) Tensile, torsion and galvanising tests on samples of individual wires removed from the rope as detailed in the appropriate British Standard.

Safe Working Load: The recommended minimum factors of safety for new ropes shall be as listed in Table 12.15.

Table 12.15   FACTORS OF SAFETY FOR NEW ROPES

Type of crane or lifting appliance	Class*	Factor of safety in the rope	
		Running rope	Standing rope or guys
Power-drive Scotch	1 and 2	5:1	2·5:1
derrick	3	6:1	—
	4	7:1	—
Power-driven travelling	2	4·5:1	3·5:1
jib crane	3	5·5:1	4:1
Overhead travelling	1	5:1	—
gantry crane	2	6:1	—
	3	7:1	—
	4	8:1	—
Power-driven mobile	2	4·5:1	3·5:1
crane	3	5·5:1	4:1
High pedestal or portal	2 and 3	6:1	8:1
crane	4	7·5:1	8:1
Power-driven tower crane	1 and 2	5·5:1	4:1
Goods hoist	—	12:1	—
Passenger hoist	—	12:1	—

* Cranes are classified according to the work they are intended to do—See Table 12.16.

Table 12.16   CLASSIFICATION OF CRANES

Class	Maximum number of hours in service per year assumed for design purposes	Type of duty
1	1000	Light duty.
2	2000	Medium duty—general use in factories.
3	3000	Heavy duty—foundry and intermittent grabbing.
4	In excess of 3000	Extra-heavy duty—continuous grabbing and steelworks use.

Test Certificate:

	Certificate	Register
FA	Use Form 87	Form 88, Pt. I
C(LO)R	Use Form 87	Form 91, Pt. II, H
DR	Use Form 87	Form 99

When a length of rope is cut from a stock coil, the Test Certificate Number of the coil must be entered in the Lifting Appliances Register against the cut off length.

Marking: (a) Identification number. (b) Safe working load.
The marking shall be stamped on metal ferrules one to be firmly attached at each end of the rope.

Periodic Inspection and Examination: See Table 12.17.

Table 12.17

	Inspection	*Form No.*	*Examination*	*Form No.*
FA	None	—	6 months	88, Pt. II
C(LO)R	None	—	6 months	91, Pt. II, H
DR	(1) When no wires are broken – 3 months	None	None	—
	(2) When one or more wires are broken – 1 month			

Although statutory regulations require wire ropes to be examined at six-monthly intervals it is often advisable that they should be examined more frequently, the period depending upon the conditions under which they are used This applies particularly to ropes attached to lifting machines, e.g. ropes on cranes working to C(LO)R would normally be inspected with the crane at weekly intervals, and excavator ropes, being subject to heavy wear should be similarly treated.

F.14 Shackles

Specifications:
BS 3032. 'Higher Tensile Steel Shackles'.
BS 3551. 'Alloy Steel Shackles'.

BS 3032 provides for five types of shackles namely: small and large dee, small and large bow and grab shackles. BS 3551 provides for one range of dee and one range of bow shackles both with four alternative types of shackle pin.

Materials: Shackles to BS 3032 are made from three grades of steel namely BS 970 Pt. I Grades 080 A27, 080 A30 and 150 M19 or 150 M28.
Alloy steel shackles are made of low alloy (nickel, chromium, molybdenum) steels as defined in BS 3551.

Initial Heat Treatment: The bodies and pins of both higher tensile steel and alloy steel shackles must be hardened and tempered as required by the appropriate specification.

Test: When new according to the appropriate specification.

Proof Load: 2 × safe working load.

Test Certificate:

	Certificate	Register
FA	Use Form 86	Form 88, Pt. II
C(LO)R	Use Form 97	Form 91, Pt. II
DR	Use Form 86	Form 99

Form 86 must be endorsed for heat treatment.

Marking: (a) Identification number. (b) Safe working load. (c) The symbol Ⓞ4 or Ⓞ6 to denote higher tensile steel or alloy steel respectively in the hardened and tempered condition.

The identification number and the symbol indicating the material must be stamped both on the pin and on the 'dee' or 'bow' of the shackle; the safe working load must be stamped on the 'dee' or 'bow' only.

Periodic Inspection and Examination: See Table 12.18.

Table 12.18

	Inspection	Form No.	Examination	Form No.
FA	None	—	6 months	88, Pt. II
C(LO)R	None	—	6 months	91, Pt. II, H
DR	Each occasion before use or 3 months	None	After test and every 12 months	99, Pt. III or 1952 and 99

F.15   Slings

Tables showing the safe working loads of all slings at different angles of the legs must be posted in prominent positions on premises where the slings are used.

Single leg slings shall be marked with the safe working load in choke hitch at 120° spread, i.e. 80% of single leg value. Multi-leg slings should be marked with the safe working load at an included angle of 90° between legs.

Note: BS publication PD 6464 (1972) 'Slinging Practice' now makes recommendations for the safe working loads of multi-leg slings but leaves the SWL for single leg slings and for single leg slings used in basket hitch and in-choke hitch to be decided by the appropriate British Standard committees responsible for slings made in various materials. Whilst recommendations for the latter are still awaited, PD 6464 should be consulted in conjunction with the guidance, based on past practice, given in sub-section F.15.

(FA 1961 26(1)B; SI 1961 No. 1581 34(2); DR 24)

Care shall be taken when attaching a sling to a lifting appliance that the method used is not one likely to result in damage to the sling or to the lifting appliance.

When loads with sharp edges are to be lifted, adequate protection from damage to the ropes or chains should be given by the use of suitable packing.

No double or multiple sling must be used for raising or lowering if:

1. The upper ends of the sling legs are not connected by means of a shackle, ring or link of adequate strength.

   Note. A link is preferred to a ring for this purpose because its shape ensures that the stress set up in it is always in one direction, whereas a ring being free to take up any position may be subject to reversal of stress each time a load is applied.

OR

2. The safe working load of any sling leg is exceeded as a result of the angle between the sling legs.

   Slings should be kept on suitable racks or pegs when not in use and not left on the floor where they are liable to be damaged.

When a sling is attached to a hook, care must be taken to ensure that the component placed on the hook is of suitable size and shape to ride freely on the hook.

Four leg slings should not be used for lifting rigid articles (e.g. castings), as in such cases unequal distribution of the load is inevitable and two legs may carry the total load.

'Back hooking' on to the main lifting hook involves risk and this method should not be used. Sling hooks should always be hooked into a ring or shackle.

Care must be taken when 'lowering off' to ensure that the load does not rest on a sling, thereby crushing it, and suitable packing must be placed under the load so that the sling may be withdrawn without injury.

Slings must not be dragged along the floor or ground and must never be pulled from under a load which has been lowered and is resting on the sling.

When a sling is reeved round a load, care must be taken to avoid setting up excessive tension in the inclined legs by forcing the bight down too low. Even when the distance of the bight above the load is as much as one quarter of the length of

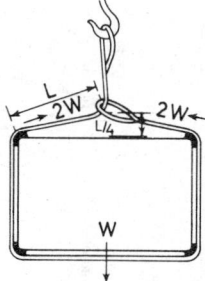

Figure 12.4. Showing tensional forces on a sling

the inclined leg i.e. 150° approximately, as illustrated in Figure 12.4 the tension in each of the inclined legs is equal to twice the load being lifted.

(FA 1961 S26(1)(b); SI 1961 No. 1581 37 and 38; DR 25)

F.15.1   CHAIN SLINGS INCLUDING RINGS AND LINKS (REFER ALSO TO CLAUSE F.4)

Specifications: BS 2902 'Higher Tensile Steel Chain Slings' and BS 3458 'Alloy Steel Chain Slings'.

Both these specifications cover chain slings, rings, links alternative to rings, e.g. links and intermediate links, whilst BS 3458 also covers trapezoidal section eye hooks.

Table 12.19 SAFE WORKING LOADS OF HIGHER TENSILE STEEL GRADE 40 CHAIN SLINGS TO BS 2902

Size of chain		Permissible working load of one leg			Included angle between legs 30°			60°			90°			120°			SWL of single sling reeved in choke hitch angle not to exceed 120°		
mm	in	t(kg)	ton	cwt	t(kg)	ton	cwt	t(kg)	ton	cwt	t(kg)	ton	cwt	t(kg)	ton	cwt	t(kg)	ton	cwt
6	¼	(500)		10	(960)		19	(860)		17	(700)		14	(500)		10	(400)		8
8	5⁄16	(760)		15	1·4	1	8	1·3	1	5	1·0	1	1	(760)		15	(600)		12
10	3⁄8	1·1	1	2	2·1	2	2	1·9	1	18	1·5	1	11	1·1	1	2	(890)		17
11	7⁄16	1·5	1	10	2·9	2	17	2·6	2	11	2·1	2	2	1·5	1	10	1·2	1	4
13	½	2·0	2	0	3·9	3	17	3·5	3	9	2·8	2	16	2·0	2	0	1·6	1	12
16	5⁄8	3·1	3	2	6·0	5	19	5·4	5	7	4·4	4	7	3·1	3	2	2·5	2	9
19	¾	4·5	4	10	8·8	8	13	7·9	7	15	6·4	6	7	4·5	4	10	3·6	3	12
22	7⁄8	6·1	6	2	11·9	11	15	10·7	10	11	8·7	8	12	6·1	6	2	4·9	4	17
25	1	8·1	8	0	15·6	15	9	14·1	13	17	11·4	11	6	8·1	8	0	6·4	6	8
29	1⅛	10·2	10	2	19·8	19	10	17·7	17	9	14·5	14	5	10·2	10	2	8·2	8	1
32	1¼	12·7	12	12	24·5	24	2	21·9	21	13	17·9	17	13	12·7	12	12	10·1	10	0
35	1⅜	15·3	15	2	29·6	29	3	26·5	26	3	21·6	21	7	15·3	15	2	12·2	12	1
38	1½	18·2	18	0	35·2	34	15	31·6	31	3	25·7	25	9	18·2	18	0	14·6	14	8

Where practicable the included angle should not exceed 90°.
Marking of SWL: Single slings as choke hitch.
        Multi-leg slings for 90° included angle.

Table 12.20 SAFE WORKING LOADS OF ALLOY STEEL GRADE 60 CHAIN SLINGS TO BS 3458

| Size of chain | | Permissible working load of one leg | | | Included angle between legs | | | | | | | | | | | | SWL of single sling reeved in choke hitch angle not to exceed 120° | | |
| | | | | | 30° | | | 60° | | | 90° | | | 120° | | | | | |
mm	in	t(kg)	ton	cwt	tonne	ton	cwt	tonne	ton	cwt	tonne	ton	cwt	t(kg)	ton	cwt	t(kg)	ton	cwt
6	1/4	(760)		15	1·4	1	9	1·3	1	6	1·0	1	1	(760)		15	(600)		12
8	5/16	1·1	1	3	2·2	2	4	2·0	2	0	1·6	1	12	1·1	1	3	(930)		18
10	3/8	1·6	1	13	3·2	3	3	2·8	2	17	2·3	2	6	1·6	1	13	1·3	1	6
11	7/16	2·2	2	5	4·4	4	7	3·9	3	18	3·2	3	3	2·2	2	5	1·8	1	16
13	1/2	3·0	3	0	5·8	5	16	5·2	5	4	4·2	4	4	3·0	3	0	2·4	2	8
16	5/8	4·7	4	13	8·9	8	19	8·0	8	1	6·5	6	11	4·7	4	13	3·7	3	14
19	3/4	6·8	6	15	13·2	13	0	11·8	11	13	9·6	9	11	6·8	6	15	5·4	5	8
22	7/8	9·2	9	3	17·9	17	13	16·0	15	17	13·1	12	18	9·2	9	3	7·4	7	6
25	1	12·1	12	0	23·5	23	3	21·1	20	15	17·2	16	19	12·1	12	0	9·7	9	12
29	1 1/8	15·3	15	3	29·7	29	5	26·6	26	4	21·7	21	8	15·3	15	3	12·3	12	2
32	1 1/4	19·0	18	15	36·8	36	5	32·9	32	9	26·9	26	10	19·0	18	15	15·2	15	0
35	1 3/8	23·0	22	13	44·4	43	15	39·8	39	4	32·5	32	0	23·0	22	13	18·4	18	2
38	1 1/2	27·4	27	0	52·9	52	3	47·5	46	15	38·7	38	3	27·4	27	0	21·9	21	12

Where practicable the included angle should not exceed 90°.
Marking of SWL: Single slings as choke hitch.
Multi-leg slings for 90° included angle.

Materials: BS 2902 specifies short link higher tensile steel chain to BS 1663 and other components in steel to BS 970 Pt. I Grades 080 A27, 080 A30, 150 M19 or 150 M28.
BS 3458 specifies short link Grade 60 alloy steel chain to BS 3113 and components of alloy steel as detailed in the specification.

Initial Heat Treatment: Sling assemblies in both higher tensile and alloy steel must be hardened and tempered as detailed in the appropriate specification.

Test: Separate components and complete sling assemblies must be tested.
1. Before use according to the British Standard.
2. After alteration or repair by proof loading only.

Proof Load: Proof loading must be carried out in accordance with the requirements of these two British Standards.

Test Certificate: A certificate is required for the separate components and for the complete sling.

	Certificate	Register
FA	Use Form 86	Form 88, Pt. I
C(LO)R	Use Form 97	Form 91, Pt. II, H
DR	Use Form 86	Form 99

Marking: (a) Identification number. (b) Safe working load. (c) the symbol ⑭ or ⑯ to denote higher tensile steel or alloy steel respectively in the hardened and tempered condition.

Periodic Inspection and Examination: See Table 12.21.

Table 12.21

	Inspection	Form No.	Examination	Form No.
FA	—	—	6 months	88, Pt. II
C(LO)R	—	—	6 months	91, Pt. II, H
DR	Each occasion before use or 3 months	None	After test, heat treatment or repair	99, Pt. IV or 1946

F.15.2   FIBRE ROPE SLINGS (REFER ALSO TO CLAUSE F.12)

Specification: BS. The safe working load of Fibre Rope Slings and Fibre Rope Lifting Blocks (in preparation).

Material:
(a) Manila or Sisal rope in the water repellant condition and to BS 2052.
(b) Synthetic fibre rope to BS 3758, BS 3912 or BS 3977.
Other components of the slings must be as detailed under the appropriate heading in Section F.

Safe Factors Related to Rope Size: See Table 12.22

Table 12.22

Rope size, dia. mm	12	14	16	18	20	24	28	32	40	48
Grade safety factor	12	10	10	8	8	7	7	7	6	6

Test including Proof Test: A test of the complete sling is not required, but a sample of the rope and other components must be tested in accordance with the appropriate British Standard.
1. Before use.
2. At periods specified for the particular item.

Test Certificates: As required for the separate components.

Marking: (a) Identification number. (b) Safe working load.
 In cases where the sling consists entirely of fibre rope the marking must be stamped on a metal ferrule fixed firmly to the rope.

Periodic Inspection and Examination: See Table 12.23.

Table 12.23

	Inspection	Form No.	Examination	Form No.
FA	None	—	6 months	88, Pt. II
C(LO)R	None	—	6 months	91, Pt. II, H
DR	Each occasion before use or 3 months	None	None	—

Table 12.24 SAFE WORKING LOADS OF FIBRE ROPES AND ENDLESS SLINGS

Rope diameter mm	Polyamide (nylon) BS 3977 tonnes (kg)	Polyester (terylene) BS 3758 tonnes (kg)	Poly-propylene* tonnes (kg)	Manila Grade 1 BS 2052 tonnes (kg)	Sisal or Manila Grade 2 BS 2052 tonnes (kg)
			Single vertical fall		
12	(250)	(190)	(165)	(90)	(75)
14	(500)	(315)	(275)	(140)	(125)
16	(800)	(400)	(340)	(200)	(175)
18	1·0	(600)	(550)	(300)	(265)
20	1·7	(800)	(650)	(400)	(350)
24	2·2	1·2	1·0	(650)	(550)
28	2·8	1·7	1·4	(850)	(750)
32	4·9	2·2	1·8	1·0	(950)
40	6·8	3·9	3·1	1·9	1·7
48	9·4	5·5	4·5	2·7	2·4
			Endless sling		
12	(400)	(300)	(265)	(140)	(125)
14	(650)	(500)	(450)	(225)	(200)
16	(850)	(650)	(550)	(325)	(275)
18	1·3	1·0	(900)	(500)	(425)
20	1·6	1·2	1·0	(650)	(550)
24	2·7	2·0	1·7	1·0	(900)
26	—	—	—	1·2	1·0
28	3·5	2·7	2·3	1·3	1·2
32	4·5	3·5	3·0	1·7	1·5
40	8·0	6·2	5·0	3·0	2·7
48	11·0	9·0	7·2	4·5	3·8

* BS in course of preparation.

F.15.3   WIRE ROPE SLINGS (REFER ALSO TO CLAUSE F.13)

Specifications: BS 1290 'Wire Rope Slings and Sling Legs' and BS 3865 'Wire Rope Sling Legs, with Ferrule-Secured Eye Terminals'.

BS 1290 provides for a series of wire rope slings with various terminal fittings secured by looping the rope through them and then splicing it to itself. The wire rope must comply with BS 302. Rings of alternative materials are offered but higher tensile steel rings only are considered acceptable for general lifting applications.

BS 3865 is a companion specification providing for slings with eye terminals formed by looping the rope end back parallel to the main length and joining the two parts with a ferrule secured under pressure.

Test: A test of a complete sling with manual splices is not required, but a sample of the rope (see sub-section F.13) and the other components must be tested before both use and after alteration or repair as specified in this sub-section F, for each particular item.

A sling with machine splices to BS 3865 shall be tested to twice the safe working load before use. The test requirements for a sample of the rope and component items are as for manually spliced slings.

Test Certificate: A test certificate is required for the separate components and for the complete sling.

	Certificate	*Register*
FA	Use Form 86	Form 88, Pt. I
C(LO)R	Use Form 97	Form 91, Pt. II, H
DR	Use Form 86	Form 99

BS 3865 also provides for the testing of complete sample slings to 90% of the breaking strain of the rope, unless otherwise agreed between purchaser and manufacturer. This is intended as a test of the ferrule-secured eye terminal and not of the rope.

Marking: (a) Identification number. (b) Safe working load.

Periodic Inspection and Examination: Table 12.25.

Table 12.25

	Inspection	*Form No.*	*Examination*	*Form No.*
FA	None	—	6 months	88, Pt. II
C(LO)R	None	—	6 months	91, Pt. II, H
DR	Each occasion before use or 3 months	None	M.S. components 12 months	99, Pt. III or 1952

F.15.4   FLAT SLINGS — WIRE

Specification: BS 3481. 'Part I: Wire Coil Flat Slings'.

BS 302. 180 kgf/mm² (115 tonf/in²) fibre core
Round strand 6 × 19 (12/6/1) Table 3, 8 to 16 mm
Round strand 6 × 37 (18/12/6/1) Table 5, 18–38 mm

Included angle between legs

Dia of rope		Permissible working load of one leg		30°		60°		90°		120°		S.W.L of endless sling reeved angle not to exceed 120°		S.W.L of sling reeved choke hitch angle not to exceed 120°	
mm	in	t(kg)	ton cwt	tonne	ton cwt	tonne	ton cwt	t(kg)	ton cwt	t(kg)	ton cwt	t(kg)	ton cwt	t(kg)	ton cwt
8	5/16	(700)	13	1·3	1 5	1·2	1 2	(900)	18	(700)	13	1·1	1 0	(560)	10
9	3/8	(890)	17	1·7	1 12	1·5	1 9	1·2	1 4	(890)	17	1·4	1 7	(710)	13
10		1·1	1 1	2·1	2 0	1·9	1 16	1·5	1 9	1·1	1 1	1·7	1 13	(880)	16
11	7/16	1·3	1 6	2·5	2 10	2·3	2 5	1·8	1 16	1·3	1 6	2·1	2 1	1·0	1 0
12		1·5	1 11	3·0	2 19	2·7	2 13	2·2	2 3	1·5	1 11	2·5	2 9	1·2	1 4
13	1/2	1·8	1 16	3·6	3 9	3·2	3 3	2·6	2 10	1·8	1 16	2·9	2 17	1·4	1 8
14	9/16	2·1	2 2	4·1	4 1	3·7	3 12	2·9	2 19	2·1	2 2	3·4	3 7	1·7	1 13
16	5/8	2·8	2 15	5·4	5 6	4·9	4 15	4·0	3 17	2·8	2 15	4·5	4 8	2·2	2 4
18	11/16	3·4	3 7	6·2	6 9	5·9	5 16	4·8	4 14	3·4	3 7	5·4	5 7	2·7	2 13
19	3/4	3·8	3 15	7·4	7 6	6·6	6 9	5·4	5 6	3·8	3 15	6·1	6 0	3·0	3 0
20	13/16	4·2	4 3	8·1	8 0	7·3	7 3	5·9	5 17	4·2	4 3	6·7	6 13	3·3	3 6
22	7/8	5·1	5 1	9·9	9 15	8·8	8 14	7·2	7 2	5·1	5 1	8·2	8 1	4·1	4 0
24	15/16	6·1	6 0	11·8	11 11	11·2	10 7	8·6	8 9	6·1	6 0	9·8	9 12	4·8	4 16
26	1	7·1	7 1	13·8	13 12	12·4	12 4	10·1	9 19	7·1	7 1	11·4	11 6	5·7	5 12
28	1⅛	8·3	8 3	16·0	15 14	14·4	14 2	11·7	11 10	8·3	8 3	13·2	13 0	6·6	6 10
32	1¼	10·8	10 13	20·9	20 11	18·8	18 8	15·3	15 1	10·8	10 13	17·3	17 1	8·6	8 10
35	1⅜	12·9	12 15	25·0	24 12	22·4	22 1	18·3	18 0	12·9	12 15	20·3	20 8	10·3	10 4
36		13·7	13 10	26·5	26 1	23·8	23 7	19·4	19 1	13·7	13 10	22·0	21 12	11·0	10 16
38	1½	15·3	15 1	29·6	29 1	26·5	26 1	21·6	21 6	15·3	15 1	24·4	24 1	12·2	12 1

Where practicable the included angle should not exceed 90°.
Marking of swl.: Single slings as choke hitch.
Multi-leg slings for 90° included angle.

Material: Flat coil slings shall be made of low carbon steel wire of the composition stated in the British Standard. Terminal attachments shall be made of steel to BS 970 Part I Grade 070 A20.

Initial Heat Treatment: The terminal attachments shall be normalised.

Test: (i) When new, according to British Standard. (ii) After repair.

Proof Load: 2 × safe working load in straight pull.

Safe Working Load in Choke Hitch: If there is any doubt about limiting the load when using the sling in choke hitch it is customary that the load should be within the safe working load in straight pull as given in BS 3481:Part I.

Test Certificate:

	Certificate	*Register*
FA	Use Form 86	Form 88, Pt. I
C(LO)R	Use Form 97	Form 91, Pt. II, H
DR	Use Form 86	Form 99

The test certificate shall give the following information for each sling:
Distinguishing mark, to identify the particular sling.
Designation; whether heavy, medium or light mesh.
Effective length.
Mesh breadth.
Proof load applied.
Safe working load in straight pull.

Marking: (a) Identification number. (b) Safe working load.

Periodic Inspection and Examination: See Table 12.27.

Table 12.27

	Inspection	*Form No.*	*Examination*	*Form No.*
FA	None	—	6 months	88, Pt. II
C(LO)R	None	—	6 months	91, Pt. II, H
DR	Each occasion before use or 3 months	None	None	—

F.15.5   FLAT SLINGS—SYNTHETIC FIBRE

Specification: BS 3481 Part 2.

Size Range: 50 mm to 320 mm wide.

Materials: Synthetic fibre flat slings are available in Nylon, polyester (Terylene) and polypropylene. For normal industrial application Nylon slings are recommended. Extreme care is necessary when selecting synthetic fibre slings for duties which may involve corrosive conditions.

All of these synthetic materials are highly resistant to mildew and other microbiological attack but their resistance to attack under acid or alkali conditions varies considerably. In general, nylon is unsuitable for use under acidic conditions and Terylene is attacked by strong alkalis. Polypropylene on the other hand has appreciable resistance to both acids and alkalis.

All synthetic fibres are easily damaged by contact with hot surfaces or by friction.

Before deciding to use synthetic fibre flat slings it is strongly recommended that information should be obtained from a recognised authority regarding the suitability of the particular fibre for the actual conditions of use.

Proof Load: (a) Sling. Not required for each individual sling (See Clause 1.7 of BS 3481, Pt. 2). (b) End Fittings. All end fittings shall be proof loaded to twice the safe working load.

Test Certificate: The manufacturer shall supply a certificate in accordance with BS 3481, Pt. 2. This shall include a statement that the sling has been examined for quality and compares in every way with a similar sling that has been tested to destruction and complies fully with BS 3481: Pt. 2.

Safe Working Load: See BS 3481: Pt. 2.

Periodic Inspection and Examination: Although, by statute, fibre slings are required to be examined at six-monthly intervals, because of the special hazards encountered in their use, they should be examined more frequently. The preferred procedure is for the slinger to be trained and instructed always to look over every sling immediately before each occasion on which it is used, looking particularly for cuts, abrasions and signs of chemical attack. In addition, each sling shall be examined by the competent person every six months. For further guidance see BS 3481, Pt. 2 Appendix G.

F.16   Socket for wire rope

Specifications:
   BS 463. Pt. 1. Inch units.
   BS 463. Pt. 2. Metric units.

Material: BS 970. Pt. 1. Grades 080 A27 and 070 A20.

Initial Heat Treatment: As required by appropriate British Standards.

Test: Before use; as required by British Standards.

Proof Load: 2 × breaking strength of the rope to be used.

Test Certificate:

	Certificate	Register
FA	Use Form 86	Form 88, Pt. I
C(LO)R	Use Form 97	Form 91, Pt. II, H
DR	Use Form 86	Form 99

Marking: (a) Identification number. (b) Nominal size of rope for which socket has been machined.

Periodic Inspection and Examination. As in Table 12.28.

Table 12.28

	Inspection	Form No.	Examination	Form No.
FA	None	—	6 months	88, Pt. II
C(LO)R	None	—	6 months	91, Pt. II, H
DR	Each occasion	None	12 months	99, Pt. III or
	before use or			1952
	3 months			

F.17 Stretching screw

Refer to Section F.11. 'Rigging Screw'.

F.18 Swivel

Specification: BS 4238. Swivels for Lifting Purposes (Metric Units).
This Standard provides for two types of swivel, namely ball-bearing and plain-bearing. Each comprises a bowpiece with one of three alternative types of end attachment, namely a round eye, an elongated eye or a 'c' type trapezoidal shank hook.
Swivel with safe working loads from 1·0 tonne to 12·5 tonnes are included.

Material: Carbon steel to BS 970 type 080 A27 or type 080 A30 or carbon-manganese steel to BS 970 type 150 M19.

Initial Heat Treatment: All swivel bowpieces, eyes and hooks should be hardened and tempered as required by the specification.

Test: (i) Before use. (ii) After repair.

Proof Load: 2 × safe working load.

Test Certificate:

	Certificate	Register
FA	Use Form 86	Form 88, Pt. I
C(LO)R	Use Form 97	Form 91, Pt. II, H
DR	Use Form 86	Form 99

Marking: (a) Identification number. (b) Safe working load. (c) The symbol (04) to denote higher tensile steel in the hardened and tempered condition.
The markings must be stamped on the swivel in accordance with the requirements of the British Standard.

Periodic Inspection and Examination: See Table 12.29.

Table 12.29

	Inspection	Form No.	Examination	Form No.
FA	None	—	6 months	88, Pt. II
C(LO)R	None	—	6 months	91, Pt. II, H
DR	Each occasion before use or 3 months	None	12 months	99, Pt. III or 1952

F.19   Teagle

Every teagle opening or similar doorway shall be securely fenced and shall be provided with a secure hand-hold on each side of the opening. The fencing shall be properly maintained and shall be kept in position except when the hoisting or lowering of goods or materials is being carried on at the opening.
(FA 1961 S 24).

F.20   Thimbles for wire ropes

Specifications: BS 464. 'Thimbles for Wire Ropes', and BS 3226. 'Thimbles for Natural Fibre Ropes'.

Because of the lower breaking strengths of fibre ropes as compared with wire ropes, a lighter type of thimble is appropriate and these are provided for in BS 3226. This standard specifically relates to natural fibre ropes because it was issued before the specifications for synthetic fibre ropes became available (see F.12). Thimbles to BS 3226 are however suitable for synthetic fibre ropes.

These specifications provide for a variety of forms of thimble but it is recommended that only thimbles made as solid forgings, or machined for rolled bar should be used in lifting operations.

Material: Rolled carbon steel bar to BS 4360 Grades 40 or 43 or BS 970 Grade 070 A20.

Initial Heat Treatment: None.

Test: None required, visual inspection only.

Marking: None.

Periodic Inspection and Examination: To be carried out at the same time as that specified for the item of lifting tackle of which the thimble forms part.

F.21   Turnbuckle

See sub-section F.11. 'Rigging Screws'.

F.22   Weighing machines, suspended type (crane weigher)

This type of weighing machine must be given a thorough examination at least once every six months and the results of the examination must be entered in the Lifting Appliances Register.

F.23   Pulley blocks

Specifications:

BS 3243. 'Hand-operated chain pulley blocks'.

BS 4018. 'Pulley Blocks for use with Wire Rope'. (Max. lift 25 tonnes in combination).

BS 4536. 'Heavy Duty Pulley Blocks for use with Wire Rope. (Metric Units)' (Max. lift approx. 180 tonnes).

BS 408. 'Ships Cargo Blocks for use with Wire Rope'.

BS 1839. 'London Pattern Pulley Blocks for natural fibre ropes'.

BS 4344. 'Pulley Blocks for use with natural and synthetic fibre ropes (Metric Units)'.

BS 1692. 'Gin Blocks'.

No British Standard is available for power operated chain blocks.

Poles or beams supporting pulley blocks or gin wheels: No pulley block or gin wheel attached to a pole or beam shall be used for raising or lowering materials unless the pole or beam:

1. Is adequately secured at least at two points to support the load with safety and to prevent the pole or beam moving into contact with any part of the scaffold.
2. If secured only to a scaffold, is secured to at least two standards or extension poles of that scaffold.

The pole or beam must not be supported on a part of the scaffold which serves as a ledger or putlog.

(SI 1961 No. 1581 18).

Loss of efficiency in rope pulley blocks due to friction: Particularly when handling large loads the tension in the hauling part of a fall is greater than the value of

$$\frac{\text{Load lifted } (W)}{\text{Number of ropes in fall } (n)}$$

due to friction. To allow for the effect of friction it is customary to increase the value of W/n by a given percentage for each sheave in the fall.

For this purpose Tables 12.30 and 12.31 may be used and it is recommended that an allowance of 5% per sheave be made in the case of wire rope and 10% per sheave in the case of fibre rope. Provided the block is of good design and in proper working condition these allowances should be adequate.

Table 12.30   TENSION IN THE HAULING PART WHEN LOAD LIFTED $= 1$.
(HAULING PART, LEADING OFF FIXED BLOCK IN FALL)

Allowance for friction, %	Number of ropes in fall					
	1	*2*	*3*	*4*	*5*	*6*
0	1·00	0·50	0·33	0·25	0·20	0·17
5	1·05	0·55	0·38	0·30	0·25	0·22
10	1·10	0·60	0·43	0·35	0·30	0·27

Similarly, the load that may be lifted by the fall will be less than the value of the safe working load of the rope multiplied by the number of ropes in the fall.

Table 12.31   LOAD LIFTED BY THE FALL WHEN TENSION IN HAULING PART = 1.
(HAULING PART, LEADING OFF FIXED BLOCK IN FALL)

Allowance for friction, %	Number of ropes in fall					
	1	*2*	*3*	*4*	*5*	*6*
0	1·00	2·00	3·00	4·00	5·00	6·00
5	0·95	1·82	2·61	3·33	4·00	4·62
10	0·91	1·67	2·31	2·86	3·33	3·75

Hand-operated Pulley Blocks: It is recommended that manually operated blocks on runways should be geared for travelling with hand chain operation, particularly when accurate positioning is required or when the height from the operating platform is more than 4 m or where the capacity is one tonne or greater. The type which is moved by pushing the load is not recommended. When practicable it is better to use travelling blocks which are constructed integral with the trolley.

Power operated Pulley Blocks: These can be either electrically or pneumatically operated.

Power operated chain blocks must always be provided with means by which the load chain can be guided on to and stripped from the sprocket wheel. In the case of wire rope blocks provision must be made to prevent the wire rope jumping the end flanges of the drum (see sub-section D.19).

(a) *Electrically Operated Blocks.* The electrical equipment including limit switches and pendant controls must be of a type which is known to be safe for the particular operating conditions. Electromechanical brakes must be provided on the lifting and traversing motions. The brakes must be mechanically applied and electrically released and must be applied automatically when the associated controller is in the STOP position so that the brake is on when the associated motor is de-energised.

The brake on the hoisting motion must be capable of preventing the lowering of a load equal to the proof load. A manual brake release must be fitted on the transverse drive.

(b) *Pneumatically Operated Blocks.* Pneumatically operated blocks must be fitted with an automatic brake/safety device which, in the event of a failure of the air pressure, will retain the load at its height at the moment of failure.

Materials: The load chains to be used on pulley blocks are either higher tensile steel chains, Grade 40 to BS 1663 or alloy steel chains Grade 80, both being pitched or calibrated (see section F.4.1)

Wire and fibre ropes are specified to the appropriate British Standard (see sub-sections F.12 and F.13).

Hooks must comply generally with BS 2903 (see sub-section F.7), and other terminal fittings must be made in either higher tensile steel or alloy steel in the hardened and tempered condition, and where possible must comply with the appropriate British Standard.

Various alternative materials are specified for the component parts of pulley blocks and the block manufacturer is required to produce evidence that the materials used have adequate strength.

Aluminium alloys are not suitable and must not be used in the manufacture of load bearing components of pulley blocks.

Test: (i) Before use. (ii) After alteration or repair.

Proof Load:

(a) *Hand operated chain block.* The factor of safety of any part of the chain block (BS 3243) shall be not less than 5. Every chain pulley block must be subjected to a proof load of $1\frac{1}{2}$ times the safe working load through a length of lift which will ensure that every part of the block mechanism and each tooth of the gears comes under load.

(b) *Hand operated wire rope pulley block.* The factor of safety shall be not less than 5 for safe working loads up to and including 100 tonnes and not less than 4 for safe working loads over 100 tonnes. For multiple sheave block capacities up to and including 25 tonnes the proof load must be 2 times the safe working load. For capacities in excess of 25 tonnes the proof load must be on a descending scale from 100% excess load at 30 tonnes capacity down to 33% excess load at 180 tonnes capacity and as laid down in BS 4536.

(c) *Hand operated fibre rope pulley block, single and multiple sheave.* The factor of safety of any part of the rope block must be not less than 5. The proof load must be:

Single sheave blocks.	$4 \times$ S.W.L.
Multiple sheave blocks	
Up to 20 tonnes	$2 \times$ S.W.L.
20 to 40 tonnes	SWL $+$ 20 tonne
Over 40 tonnes	$1\frac{1}{2} \times$ SWL

(d) *Power operated chain or wire rope block complete assembly.*

Up to 20 tonnes	SWL $+25\%$
20 to 50 tonnes	SWL $+5$ tonnes
Over 50 tonnes	SWL $+10\%$

Note that before a block is used for the first time, it is necessary to hold a test certificate for the complete assembly. When an individual part, e.g. chain, hook etc. is renewed it is not necessary to retest the complete assembly provided an individual test certificate for the new part is obtained and retained together with the original certificate.

Test Certificate:

	Certificate	Register
FA	Use Form 86	Form 88
C(LO)R	Use Form 80	Form 91, Pt. II, F
DR	Use Form 86	Form 99

Marking: Every block must have a plate permanently attached to it and clearly marked with an identification number and the safe working load. Detachable fittings such as hooks, chains, shackles etc. must be marked with the identification number of the assembly.

Periodic Inspection and Examination: See Table 12.32.

Table 12.32

	Inspection	Form No.	Examination	Form No.
FA	None	—	14 months	88, Pt. IV
C(LO)R	Weekly	91, Pt. I, B	14 months	91, Pt. II, F
DR	Each occasion before use or 3 months	None	12 months	99, Pt. II

F.24 Cranes

F.24.1 GENERAL REQUIREMENTS

Specifications:
 BS 2573. 'Permissible Stresses in Cranes'.
 'British Standard Code of Practice for the Safe Use of Cranes'. Part 1. Mobile Cranes, Tower Cranes, and Derrick Cranes.
  These specifications which apply to cranes generally contain much useful data relating to the various types of cranes. A number of appendices deal with matters such as information to be supplied by the crane manufacturer, lists of British Standards covering materials and equipment, legislation affecting the use of cranes and other matters.

Materials: Most of the British Standards, listed below under the particular types of cranes, contain appendices detailing the materials and equipment to be used in the manufacture of the cranes. All items of lifting gear associated with cranes must be made in accordance with the requirements detailed in this Section F for each individual item.
  No crane which has any timber structural member shall be used for work which is subject to the construction (Lifting Operations) Regulations 1961.
  (SI 1961 No. 1581 24)

Platform: Crane platforms shall be close planked or plated, provided with guard rails at heights above 1·2 m and toe-boards if more than 2 m from the ground, and shall be of sufficient area for the driver.
  (SI 1961 No. 1581 13(i))

Cabin: Every power-driven crane shall be provided with a cabin for the driver which will afford him adequate protection from the weather, give as clear vision as possible and where reasonably practicable, is heated adequately in cold weather. This regulation shall not apply when the driver is indoors or otherwise adequately protected from the weather. For means of escape from cabins, see clauses covering travelling jib and overhead travelling cranes.
  (SI 1961 No. 1581 14)

Controls: Every control lever, handle or wheel shall have marked upon it or adjacent to it, its purpose and mode of operation.
  Every control lever, handle or wheel shall, where practicable be provided with a

suitable locking device to prevent accidental movement in cases where this is liable to cause danger.

(SI 1961 No. 1581 16(3) 16(2))

Interlocks: On every crane having a derricking jib operated through a clutch there shall be provided and properly maintained an effective interlocking arrangement between the derricking clutch and the pawl sustaining the derricking drum. This must ensure that the clutch cannot be disengaged unless the pawl is in effective engagement with the derricking drum and the pawl cannot be disengaged unless the clutch is in effective engagement with the derricking drum, provided that this regulation shall not apply to any crane in which:

(i) The hoisting drum and the derricking drum are independently driven; or

(ii) The mechanism driving the derricking drum is self locking.

(SI 1961 No. 1581 22)

Brakes: An effective brake or other safety device shall be provided which will prevent the fall of a suspended load and which will effectively control the lowering of the load. Brakes shall be sufficiently powerful to hold the test load.

(SI 1961 No. 1581 16(i))

Guards: All moving parts with which a person is liable to come into contact shall be effectively guarded. It is recommended that the guards should not prevent visual inspection of gear wheels.

(SI 1961 No. 1580 42)

Rope Anchorages: The rope anchorages on winding and derricking drums shall be securely fixed and there shall be at least two dead turns of rope on the drum in every operating position.

(SI 1961 No. 1581 15)

Mounting of Cranes: Every bogie, trolley or wheeled carriage on which a crane is mounted shall be of good construction, sound material and adequate strength and suitable to support the crane having regard to the purposes for which the crane is being used.

(SI 1961 No. 1581 12)

Restriction on the Use of Cranes: The hoisting mechanism of a crane shall not be used for any purpose other than raising or lowering a load vertically unless no undue stress is imposed on any part of the crane structure or mechanism, and the stability of the crane is not thereby endangered, and unless such use is supervised by a competent person.

(SI 1961 No. 1581 23)

Stability of Lifting Appliances: No lifting appliance shall be used on a soft or uneven surface in circumstances in which the stability of the appliance is likely to be affected, or on a slope, unless appropriate precautions are taken to ensure safety.

Outriggers provided to improve stability shall be used in appropriate circumstances. They shall be properly secured when extended and their feet firmly based on a satisfactory bearing surface.

(SI 1961 No. 1581 19)

Notification of H.M. Factory Inspectorate: Where an examination shows that a lifting machine cannot continue to be used with safety unless certain repairs are carried out immediately, or within a specified time; the person making the report of the exami-

nation shall within twenty-eight days of the completion of the examination send a copy of the report to the inspector for the district.
(FA 1961 S 27(2e); SI 1961 No. 1581 46(3))

F.24.2 MOBILE CRANES

Definition: The term 'mobile crane' is intended to mean a wheeled crane on pneumatic or solid tyres, power or hand operated as well as crawler mounted vehicles. Locomotive cranes are not included.

Specification: BS 1757. 'Power-Driven Mobile Cranes'. See also British Standard Code of Practice for the Safe Use of Cranes: Part I, Mobile cranes, tower cranes and derrick cranes (in course of preparation).

Additional Regulations Applying Particularly to Mobile Cranes: (See also sub-section D and F.24.1).

Cab: Adequate ventilation must be provided in the operating cabs of mobile cranes. Where a crane may overturn and block the normal means of exit from the cab, a second means of exit must be provided, e.g. a knock-out panel.

Limit Switches: Upper and lower limit switches shall be provided to prevent over-winding or slack rope.

Back Stops: Whenever it is foreseen that a crane jib may kick backwards, suitable back stops must be fitted.

Access: Safe access must be provided to the cabins and platforms of all cranes, and as far as practicable to all points which require examination, lubrication etc.
　　The tail of all mobile cranes must be marked distinctively and precautions must be taken to prevent the access of persons to any place where it is impossible to maintain an unobstructed passageway not less than 0·6 m wide between any part of a travelling or slewing crane and any guard rail or other fixture.
　　When the crane is travelling the steering gear must be properly secured.
　　(SI 1961 No. 1581 17)

Excavations: Mobile cranes must not be used near the edges of excavations or embankments unless the walls of the excavation have been adequately shored or are of adequate strength.

Level Indicators: Mobile cranes are designed, and their safe working loads assessed, for operation on level ground. If, as often happens, the crane is required to work on ground which is uneven or sloping, the driver should at least be able to check whether the crane is, so far as practicable, level before taking the lift. A level indicator should be fitted which provides information sufficient for the driver to level the crane.

Marking and Indication of Safe Working Load: Derricks and jib cranes shall:
   (a) be marked with the safe working load at various radii of the jib and the maximum radius at which the jib may be worked;
   (b) be provided with an accurate indicator showing the radius of the jib at any time and the safe working load corresponding to that radius; and
   (c) be provided with an approved type of automatic indicator which indicates when the load being carried approaches the safe working load of the crane at

that radius, and gives an effective sound signal when the load being carried is in excess of the safe working load of the crane at that radius.

The requirements of (c) above do not apply:
  (i) in the case of the Factories Act,
        to any crane provided the person in charge is able to ascertain the weight of the load to be handled and the inclination of the jib;
  (ii) in the case of the Construction (Lifting Operations) Regulations,
        to any guy derrick crane; or to any hand crane which is being used solely for erecting or dismantling another crane; or to any crane having a maximum safe working load of 1 tonne or less.
(FA 1961 S 27(4); SI 1961 No. 1581 24, 30(1))

Every such automatic load indicator shall be properly maintained and tested by a competent person other than the crane driver after erection or installation of the crane on a building site and before the crane is taken into use. The indicator shall be inspected thereafter at intervals not exceeding one week, when the crane is in use, by the person carrying out the inspection required under sub-paragraph (c) or paragraph (1) of Regulation 10 of SI 1961 No. 1581, and the results of every such inspection shall be reported in the manner specified in that sub-paragraph.

Cranes fitted with alternative hoisting arrangement require safe load indicators to suit.

Automatic indicators operated by electric batteries shall be checked by the driver each day before the crane is used to ensure that the batteries are charged.

Within the factory area there must always be a second person in attendance to give the appropriate signals when travelling or lifting. Care is necessary to ensure that the jib or hoist rope shall not foul pipelines, overhead wires, plant or structures, especially in confined areas.
(SI 1961 No. 1581 10 and 30)

Test: (i) Before use for the first time. (ii) After substantial alterations or substantial repairs. (iii) Every four years in the case of a crane used subject to C(LO)R.

Proof Load: The proof load shall exceed the safe working load as follows:

Safe Working Load	*Proof Load*
Up to 20 tonnes	25% excess
Over 20 tonnes up to 50 tonnes	5 tonnes excess
Over 50 tonnes	10% excess.

The proof load shall be hoisted and swung as far as possible in both directions. If the jib has a variable radius it shall be tested with a proof load as defined above at both the maximum and minimum radii of the jib. In the case of hydraulic cranes, where, owing to limitations of pressure it may not be possible to hoist a load 25% in excess of the safe working load it shall be sufficient to hoist the greatest possible load.

Test Certificate:

	Certificate	*Register*
FA	Use Form 1945	Form 88
C(LO)R	Use Form 1945	Form 91, Pt. I
DR	Use Form 1945	Form 99

Marking: (a) Identification number, including additional jibs. (b) All items of lifting gear in accordance with the regulations detailed for each particular item in this sub-section F.

Periodic Inspection and Examination: See Table 12.33.

Table 12.33

	Inspection	Form No.	Examination	Form No.
FA	None	—	14 months	88, Pt. IV
C(LO)R	Weekly	91, Pt. I, B	14 months	91, Pt. II, F
DR	Derricks – 12 months	99, Pt. I	Derricks – 4 years	99, Pt. I
			Other cranes – 12 months	99, Pt. II

Tyre pressures must be checked daily. All ropes and other lifting tackle used on any crane must be tested, examined and inspected in accordance with the requirements for each particular item. See also sub-section C.5.

F.24.3 RAIL-MOUNTED CRANES

Specification: BS 357. 'Power driven travelling jib cranes (rail-mounted low carriage type)'.

Additional Regulations Applying Particularly to Rail Mounted Cranes: See also sub-sections D and F.24.1.

Rails: All rails on which a crane moves shall:
   (a) have an even running surface, be sufficiently and adequately supported, and be of adequate section;
   (b) be jointed by fish plates or double chairs;
   (c) be securely fastened to sleepers or bearers;
   (d) be supported on a surface sufficiently firm to prevent undue movement of the rails;
   (e) be laid in straight lines or in curves of such radii that the crane can be moved freely and without danger of derailment;
   (f) be provided with adequate stops or buffers on each rail at each end of the track. All rails and other equipment shall be properly maintained.
   (SI 1961 No. 1581 20; FA 1961 S 27(3))

Excavations: Rail mounted cranes shall not be used near the edges of excavations or embankments unless the walls have been adequately shored and are of adequate strength.
   Rail tracks, when approaching an excavation, shall be given a considerable up-gradient towards the buffer or stop.

Weight and Clearance: The weight and minimum height of each crane must be clearly displayed in the cab in order to assist in safe travelling about the site.
   Care must be exercised by the drivers of locomotive cranes to prevent the jib or main hoist rope coming into contact with overhead electric cables, plant or structures.
   For details of cabs, limit-switches, safe load indicators, back-stops, test and examination, see sub-section F.24.2.

F.24.4 SCOTCH DERRICK CRANES

Specification: BS 327. 'Power Driven Derrick Cranes'. See also Safety Pamphlet No. 15 'The Use of Derrick Cranes' (HMSO).

Additional Regulations Applying Particularly to Scotch Derrick Cranes: (See also sub-section D and F.24.1).

Anchorage, Ballast or Kentledge: No fixed crane shall be used unless it is securely anchored or adequately weighted by suitable ballast properly placed on the crane structure to secure the stability of the crane. The anchorage shall be of sufficient weight or strength to give a margin of 50% above the weight required to balance the over-turning moment on the crane when carrying the safe working load in any position. (SI 1961 No. 1581 19)

When the stability of a crane is secured by means of removable weights, a diagram or notice indicating the position and amount of such weights shall be fixed to the crane in a position where it can be seen readily. The ballast on each leg of the crane structure shall never be less than three times the maximum safe working load; care shall be taken to ensure that the ballast is properly secured and carried on the legs of the crane structure.

It is recommended that the ballast for derrick cranes should consist of heavy items not readily removed by hand. The use of materials such as earth or sand, which are liable to be removed for use in construction works, should be avoided.

No part of any rails on which a crane is mounted, or the sleepers supporting the rails shall be used as anchorage for this purpose.

Arrangements shall be made to prevent the foot of the king-post being lifted out of its socket.

A scotch derrick crane shall not be erected with its jib between the back stays, nor shall it be used for moving any loads lying between the back stays.

The racking gear shall not be relied on for anchoring the jib against movement by the wind. Jibs should be drawn well up and anchored or lowered off when the crane is out of use.

For details of limit-switches and safe load indicators see sub-section F.24.2 'Mobile Cranes'. (SI 1961 No. 1581 33(3); DR 31)

Test and Examination: Erection:

(i) The whole of the appliances for the anchorage or ballasting of a crane shall be examined by a competent person on each occasion before the crane is erected.

(ii) After each erection of a crane on a building site, or any adjustment to any member of a crane, being a removal or adjustment which involves changes in the arrangements for anchoring or ballasting the crane, the security of the anchorage or the adequacy of the ballasting as the case may be, shall, before the crane is taken into use, be tested by a competent person by the imposition either:

(a) of a load of 25% above the maximum load to be lifted by the crane as erected at the positions where there is the maximum pull on each anchorage, or

(b) of a lesser load arranged to provide an equivalent test of the anchorage or ballasting arrangements. (SI 1961 No. 1581 19)

The prescribed particulars of such tests and their results shall forthwith be entered in or attached to the prescribed register. (SI 1962 No. 225)

(iii) If the person making tests under (ii) above considers that the maximum load which may safely be lifted by that crane as erected is less than the safe working load of the crane as defined in (a) above, he shall specify that maximum among the said particulars and a loading diagram appropriate to the stability of the crane as at the time of the test, taking into account, in the case of a crane mounted on wheels, the condition of the track and indicating a modified safe working load or loads shall be fixed in a position where it can be seen readily by the crane driver. Such modified safe working load or loads shall be deemed (for the purpose of the Regulations) to be the safe working load or loads of the crane as erected.

(iv) No crane shall be erected except under the supervision of a competent person. (SI 1956 No. 1581 25)

Test: (i) before use for the first time; (ii) after substantial alteration of substantial repair; (iii) every four years in the case of a crane used subject to C(LO)R.

Test Certificate:

	Certificate	Register
FA	Use Form 1945	Form 88
C(LO)R	Use Form 96	Form 91, Pt. I, D
DR	Use Form 1945	Form 99

Marking:
(a) Identification number; all structural parts of derrick cranes shall be so marked.
(b) All items of lifting gear associated with the crane shall be marked in accordance with the appropriate detailed regulations set out in this Section.

Periodic Inspection and Examination: See Table 12.34.

Table 12.34

	Inspection	Form No.	Examination	Form No.
FA	None	—	14 months	88, Pt. IV
C(LO)R	Weekly	91, Pt. I, B	14 months	91, Pt. II, F
DR	12 months	99, Pt. I	4 years	99, Pt. I

All ropes and other lifting tackle used on any crane shall be tested, examined and inspected in accordance with the requirements for the particular item.

In appropriate cases HM Factory Inspectorate must be notified of inspections and examinations. See sub-section F.24.1.

F.24.5   OVERHEAD TRAVELLING CRANES

F.24.5.1. Additional Regulations Applying Particularly to Overhead Cranes (See also sub-sections D and F.24.1).

Access: Safe access shall be provided to the cabins of all cranes, and as far as is practicable to all points which require examination, lubrication, etc.

Means of Escape: Where there is a risk to the driver of the crane from fire, explosion or escape of gas, provision shall be made for his safe escape.

Persons Working on or near Wheel Tracks: If any person is required to work on or near the wheel track of an overhead travelling crane in any place where he would be liable to be struck by the crane, effective measures shall be taken by warning the driver of the crane or otherwise to ensure that the crane does not approach within 6·6 m of that place.

The driver of an overhead travelling crane is responsible for ensuring that no persons are on the crane when he is about to move it, or that if there are any persons on the crane, they are aware that the driver is going to move the crane before he actually does so. This also applies to people above ground level who are liable to be struck by the load.

Consideration must be given to the provision of an audible warning device to be used by the crane driver.

(FA 1961 S 27(7), (8))

Hoisting Gear: All individual items such as for example, wire rope, shackles, hooks, anchorage to drums, etc. must conform to the specified requirements detailed in Section 12.

Outdoor Cranes: Where cranes operate in the open, proper weather protection must be provided to all parts likely to be affected by rain, damp, snow or ice.

Reliance must not be placed on the crane brakes to hold the crane in position when parked under windy conditions and a positive locking device must be incorporated to hold the crane stationary in this situation.

Rail Stops: Rail stops shall be provided at the long and cross travel to prevent the crane over-running the end of the track. Consideration must also be given to the provision of limit switches before rail stops, in order to control the travel.

For information on limit switches see Clause F.24.2.

F.24.5.2. Electrically Operated Cranes

Specification: BS 466. 'Electric overhead travelling cranes for general use in factories, workshops and warehouses'.

Controls: The two main systems of control are cab and floor operation.

Cab operation is used only where
  (i) the crane is in constant use, or
 (ii) loads are to be placed into positions which preclude the use of floor operations, or
(iii) there is regular use under bad climatic conditions.

All cab controls should be fitted with a 'dead man's handle' type of equipment and be clearly marked to show the type of motion and the direction of operation.

Floor operation may be by a pendant push-button or by radio control. The use of pendant ropes or chains which are pulled to operate the crane, is not recommended.

In all electrically operated push-button control systems, the maximum voltage to earth of the control circuits must not exceed 25 V.

Pendant controls shall include a 'stop' push-button which will isolate all electrical supply to the crane or hoist.

The push-buttons or switches must be clearly labelled to show their relevant motion and direction of travel.

Operating Speeds: For floor controlled cranes the long travel speed should not exceed 36·5 m per min. and the cross travel speed should not exceed 9 m per min.

F.24.5.3 Air Operated Cranes

Specification: There is no British Standard for air-operated cranes but these should comply generally with BS 466 and the various clauses in this Section 12 which are relevant.

Air Supply: The air hose should be so arranged as to allow the crane to operate through all positions and should be fully supported.
   Means must be provided for cutting off the compressed air feed to the equipment and this must automatically vent to the atmosphere all pipes on the air motor of the isolator. It shall be arranged to lock the isolator in the OFF position only.

Brakes: The control of the brakes must be so arranged as to prevent creep and to ensure that the brakes are applied should the air pressure available to the hoisting motor fail or fall below that necessary to sustain the maximum load plus the 25% overload.

F.24.5.4 Hand Operated Cranes

Specification: Hand operated overhead cranes should comply structurally, and where applicable in other respects, with BS 466, the crab being fitted with anti-derailment gear.

Hand chain: Hand chains shall comply with BS 590 'Mild Steel Chain Grade 30', be pitched and polished and shall have one or more links left unwelded as a safety measure to ensure that the chain will part if it is accidentally caught up in another object.
   The following provisions apply to all types of overhead travelling cranes.

Test: (i) Before use for the first time. (ii) After substantial alteration of substantial repair. (iii) Every four years in the case of a crane used subject to C(LO)R.

Proof Load: The proof load shall exceed the safe working load as follows:

Safe Working Load	Proof Load
Up to 20 tonnes	25% excess
Over 20 tonnes up to 50 tonne	5 tonnes excess
Over 50 tonnes	10% excess.

The method of test shall be as detailed in BS 466.

Test Certificate: Use Form 91D.

Marking: (a) Identification number; all structural members of the crane must be marked. (b) All items of lifting gear must be marked in accordance with the appropriate regulations detailed in this sub-section F.

Periodic Inspection and Examination: See Table 12.35.

Table 12.35

	Inspection	Form No.	Examination	Form No.
FA	None	—	14 months	88, Pt. IV
C(LO)R	Weekly	91, Pt. I, B	14 months	91, Pt. II, F
DR	12 months	99, Pt. I	12 months	99, Pt. II

All ropes and other lifting tackle which is an integral part of an overhead crane must be tested in accordance with the detailed regulations given in this sub-section F, for each particular item. For periodic inspection and examination they must be regarded as part of the crane.

F.25 Derrick pole and guy derrick

The following regulations are additional to the general requirements given in sub-section D.

Wherever practicable guys must be equally spaced (in plan view) and the base line or spread of each guy must be as long as possible. When bulldog grips are used for making a loop in a guy rope or for joining two ropes to make a longer guy, the minimum number of grips which must be used depends upon the circumference of the rope and is given in sub-section F.3, which also shows the correct application of the grips.

The selection of suitable anchorages needs particular care. It is advisable if guys are to be fastened to existing structures, to examine these structures to ensure that they are not weakened by corrosion or other forms of decay. Where horizontal girders are used the point of attachment of the guy shall be near one end to prevent bowing of the girder. Hand rails and similar light structural members must never be used even for guys for light derrick poles. If no suitable structure is available for guy anchorages other approved methods should be adopted as shown in Safety Pamphlet No. 15. 'Use of Derrick Cranes'[2].

The foot of the mast must be adequately supported on a solid level base and care taken to ensure that the ground is not overloaded. The mast must be secured so that there is no risk of the foot lifting out of the socket or support when in use.

It is recommended that the angle of a derrick pole should never exceed five degrees from the vertical when loaded.

Timber derrick poles should always be suspect on account of possible hidden deterioration.

(SI 1961 No. 1581)

Test: This is given in full in F.24.4.

Marking: (a) Identification number. (b) Safe working load at maximum inclination from the vertical at which the derrick pole may be used.

F.26 Excavator (fitted with face or drag shovel, skimmer, dragline or grab)

(When fitted with crane equipment see sub-section F.25.)

Specification: BS 1761:1951 'Single Bucket Excavators'.

Materials: Appendix E of BS 1761 lists those British Standards covering materials to be used in the manufacture of Excavators.

Cabin: Every excavator shall be provided with a cabin for the driver which affords him adequate protection from the weather, gives as clear vision as possible and, where reasonably practicable, is adequately heated in cold weather.

(SI 1961 No. 1581 14)

Drums and Pulleys: Every drum or pulley shall be of suitable diameter and construction for the chain or rope which it carries.
   (SI 1961 No. 1581 15)

Rope Anchorages: The rope anchorage on every drum should be fixed by more than one bolt or screw, and there shall be at least two dead turns of rope on the drum in every operating position.

Controls: Every control lever, handle or wheel shall have marked upon or adjacent to it its purpose and mode of operation; it shall, where practicable, be provided with a suitable locking device to prevent accidental movement in cases where this is liable to cause danger.
   (SI 1961 No. 1581 16)

Test of Excavator: (i) if fitted with crane equipment; for details of test see 6.25; (ii) if not fitted with crane equipment; no test required.

Marking: (a) Identification number. (b) Hooks, ropes, etc., in accordance with the appropriate detailed regulations.

Periodic Inspection and Examination: (including Table XXXIV).
   (a) if fitted with crane equipment, see 6.25.
   (b) if not fitted with crane equipment, see Table 12.36.

Table 12.36

	Inspection	Form No.	Examination	Form No.
FA	None	—	None	—
C(LO)R	Weekly	91, Pt. I, B	14 months	91, Pt. II, F
DR	None	—	None	—

It is recommended that all excavators should be inspected and examined in accordance with the requirements of C(LO)R.

All ropes and other lifting tackle used on any excavator shall be tested, examined and inspected in accordance with the Detailed Regulation for the particular item.

F.27 Hoist or lift

General Definitions: No lifting machine or appliance shall be deemed to be a hoist or lift unless it has a platform or cage the movement of which is restricted by a guide or guides.
   (FA 1961 S 25(1))

The term 'hoist' means a lifting machine whether worked by mechanical power or not, with a carriage, platform or cage, the movement of which is restricted by a guide or guides.
   (SI 1961 No. 1581 4(2))

F.27.1   HOISTS FOR PASSENGERS AND MATERIALS USED FOR CONSTRUCTION PURPOSES—
TEMPORARY INSTALLATIONS

Specification: BS 4465 'Electric Hoists for Passengers and Materials'.

The Statutory Regulations covering hoists used on work subject to the Construction (Lifting Operations) Regulations are as follows:

Hoistway: The hoistway of every hoist shall at all points at which persons are liable to be struck by any moving part of the hoist, be efficiently protected by a substantial enclosure fitted with gates. The enclosure and gates shall be at least 2 m high, except where a lesser height is sufficient to prevent any person falling down the hoistway or coming into contact with any moving part of the hoist but in no case shall they be less than 0·9 m high.

The following additional requirement shall apply if the hoist is used for carrying persons. Every gate in the hoistway enclosure shall, so far as is reasonably practicable, be fitted with efficient interlocking or other devices to ensure that the gate cannot be opened except when the cage is at a landing place and that the cage cannot be moved away from the landing place until the gate is closed.

(SI 1961 1581 42(1), 48(2))

Platform or Cage: Efficient devices which will support the platform or cage, together with its safe working load, in the event of failure of the hoist rope or ropes or any part of the hoisting gear shall be provided.

(SI 1961 No. 1581 42(2))

Automatic devices which will ensure that the platform or cage does not over-run the highest point to which it is for the time being constructed to travel, shall be provided.

(SI No. 1581 42(3))

The following additional requirements shall apply if the hoist is used for carrying persons:

Suitable efficient automatic devices which will ensure that the cage comes to rest at a point above the lowest point to which the cage can travel, shall be provided.

(SI 1961 No. 1581 48(3))

The cage shall be so constructed as to prevent, when the cage gate is shut, any person carried from falling out or from being trapped between any part of the cage and any fixed structure or other moving part of the hoist or from being struck by articles or materials falling down the hoistway.

(SI 1961 No. 1581 48(i) (a))

It is recommended that a trap door be provided in the roof of the cage to permit the inspection of the suspension ropes or chains and the balance weight attachments, this inspection being carried out from within the cage. The trap door shall be fitted with a master interlock switch which will stop the lift when the trap door is open.

The cage shall be fitted with a gate or gates with efficient interlocking or other devices to secure that the gate cannot be opened except when the cage is at a landing place and the cage cannot be moved away from the landing place until the gate is closed.

(SI 1961 No. 1581 (i) (b))

Method of Operation: The construction and installation arrangements of a hoist shall be such that the hoist can be operated at any one time only from the cage or only from one other position. Every hoist in which a person is being carried shall be operated from the cage of the hoist only.

(SI 1961 No. 1581 43(1))

Types of Winch: When the platform or cage is raised or lowered by means of a winch, the winch shall be so constructed that the brake is applied when the control lever, handle or switch is not held in the operating position. The winch shall not be a

winch fitted with a pawl and ratchet gear on which the pawl has to be disengaged before the platform or cage can be lowered.
(SI 1961 No. 1581 44)

Rope and Chain Anchorages: The rope or chain anchorages on winding drums should be fixed by more than one bolt or screw, and there shall be at least two dead turns of the rope or chain on the drum in every operating position.

In the case of hoists in certain chimneys exemption from some of these requirements is allowable under Certificate of Exemption No. 1 (Hoists in Certain Chimneys) Form No. 2006.
(SI 1961 No. 1581 15)

Materials: Appendix D of BS 4465 gives a list of British Standards covering the materials and equipment to be used in the manufacture of hoists.

Proof Load: 1·25 × safe working load.

Test: (a) Before the hoist is put into service for the first time. (b) After modification or repair.

Test Certificate:

	Certificate	Register
C(LO)R	Form 75	Form 91, Pt. I, F & H

Marking:
(a) All hoists must be marked with an identification number and the safe working load.
(b) Hoists used for carrying persons must be marked with the maximum number of persons to be carried at any one time.
(c) Hoists not operated from the platform or cage must display a notice, fixed to the platform or cage stating that the carriage of persons is prohibited.

Periodic Inspection and Examination: See Table 12.37.

Table 12.37

	Inspection	Form No.	Examination	Form No.
C(LO)R	1 week	91, Pt. I, B	6 months	91, Pt. II, G

F.27.2   HOISTS FOR MATERIALS ONLY USED FOR CONSTRUCTION PURPOSES—TEMPORARY INSTALLATIONS

Specification: BS 3125 'Power Driven Hoists for Materials'.

Operation: The construction and installation arrangements of every hoist shall, where practicable, be such that it can be operated at any one time only from one position, and a hoist shall not be operated from the cage unless the requirements of SI 1961 No. 1581, 48 are complied with.
(SI 1961 No. 1581 43)

If a person operating a hoist has not a clear and unrestricted view of the platform or cage throughout its travel, except at points where such a view is not necessary for safe working, then effective arrangements shall be made for signals for operating the hoist to be given to the operator from each landing place at which the hoist is used and to enable him to stop the platform or cage at the appropriate level.

No loaded truck or wheelbarrow shall be carried on the open platform of a hoist unless it is so loaded that no part of the load is liable to fall off.

Materials: BS 3125 contains an appendix which gives a list of British Standards covering the materials and equipment to be used in the manufacture of hoists.

Proof Load: See BS 3125.

Test: (a) Before the hoist is put into service for the first time. (b) After modification or repair.

Test Certificate:

	Certificate	Register
C(LO)R	Form 75	Form 91, Pt. H

Marking:
  (i) The winch of the hoist must be marked with an identification number or symbol and with the safe working load.
  (ii) The travelling carriage or platform must be marked with the safe working load and with the maker's name and number.
  (iii) The mast section must have the safe working load stamped at each end.

Periodic Inspection and Examination: See Table 12.38.

Table 12.38

	Inspection	Form No.	Examination	Form No.
C(LO)R	—	—	6 months	91, Pt. II, H

F.27.3   HOISTS FOR PASSENGERS AND GOODS—PERMANENT INSTALLATIONS

Specifications: BS 2655. 'Lifts, Escalators, Passenger Conveyors and Paternosters'. Parts 1 to 10.
(Part 10 is in course of preparation).
CP 407. 'Code of Practice — Electric, Hydraulic and Hand Powered Lifts'.

The statutory requirements relating to hoists covered by BS 2655 will be generally governed either by the Factories Act 1961 or by the Construction (Lifting Operations) Regulations.

Hoistway: Every hoistway or liftway shall be efficiently protected by a substantial enclosure fitted with gates, being such an enclosure as to prevent, when the gates are shut, any person falling down the way or coming into contact with any moving part of the hoist or lift.
  (FA.1961 S 22(4))

Hoistway gates shall be fitted with efficient interlocking or other devices to secure that the gate cannot be opened except when the cage or platform is at the landing and that the cage or platform cannot be moved away from the landing until the gate is closed.
  (FA 1961 S 22(5))

Every hoist or lift and every hoistway enclosure shall be so constructed as to prevent any part of any person or any goods carried in the hoist being trapped between any part of the hoist and any fixed structure or between the counterbalance weight and any other moving part of the hoist.
(FA 1961 S 22(7))

Every hoistway in a building constructed after 1st July, 1938 shall be completely enclosed with fire resisting materials, and all means of access to the hoist shall be fitted with doors of fire-resisting materials. The hoistway, however, shall be enclosed at the top only by some materials easily broken by fire, or provided with a vent at the top.
(FA 1961 S 48(4))

The following additional regulations apply to hoists or lifts used for carrying persons, whether together with goods or otherwise.

Platform or Cage: Efficient automatic devices shall be provided to prevent the cage or platform over-running.
(FA 1961 S 23(1) (a))

Every cage shall, on each side from which access is afforded to a landing, be fitted with a gate, and in connection with every such gate efficient devices shall be provided to secure that, when persons or goods are in the cage, the cage cannot be raised or lowered unless the gate is closed and will come to rest when the gate is opened.
(FA 1961 S 23(1) (b))

In the case of a hoist constructed or reconstructed after 30 July, 1937 where the platform or cage is suspended by rope or chain, there shall be at least two ropes or chains, separately connected with the platform or cage, each rope or chain and its attachments being capable of carrying the whole weight of the platform or cage and its maximum working load, and efficient devices shall be provided and maintained which will support the platform or cage with its maximum working load in the event of a breakage of the ropes or chains or any of their attachments.
(FA 1961 S 23(3))

Exemptions: When subject to Factories Act regulations various classes or descriptions of hoists are exempted, subject to certain conditions, from certain requirements, regarding the hoistway enclosure and gates. These exemptions are given in the Hoists Exemption Order 1962 which refer to Sections 22, 23, and 25 of the Factories Act 1961. These should be consulted for the actual details.
(SI 1962 No. 715)

The class or description of hoists which are subject to such exemptions are as follows:

1. Hoistways of pavement hoists, that is to say, hoists in the case of which the provision of a permanent enclosure at the top landing would obstruct a street or public place, or yard or other open space within a factory where persons are required to pass.
2. Mobile hoists used in various positions for the stacking of goods or materials or for loading or unloading directly to or from vehicles, which have no fixed landings above the lowest landing.
3. Hoists which are fixed in position and which are used for stacking of goods or materials or for loading or unloading directly to or from vehicles which have no fixed landing above the lowest landing and in the case of which the maximum height of the cage or platform above the ground or floor level exceeds 2 m.
4. Platform hoists which are fixed in position and in the case of which the maximum height of platform above ground or floor level does not exceed 2 m.
5. Hoists used solely for lifting material directly into a machine.

6. Hoistways of hoists which are not used for carrying persons and into or from which, goods or materials are not loaded or unloaded except at a height of not less than 0·84 m above the level of the floor or ground where loading or unloading is performed.
7. Hoists which are not connected with mechanical power and which are not used for carrying persons, and the enclosures of the hoistways of such hoists.
8. Hoists mainly used for raising materials for charging lime kilns or for charging blast furnaces in which the process of smelting iron ore is carried on.
9. Hoists used for raising or lowering or tipping standard gauge or broader gauge railway rolling stock.
10. Drop-pit hoists used for raising or lowering wheels or bogies detached from standard gauge or broader gauge railway rolling stock.
11. Hoists in the case of which the doors of the hoistway are of solid construction and the interior surfaces of the said doors and of the hoistway opposite to any side of the cage in which there is an opening are, throughout the height of travel of the cage, smooth and flush with each other save for any recess designed for working purposes and not more than 12 mm in depth, and hand grips not exceeding 25 mm in depth provided for closing doors and so constructed as to prevent trapping.
12. Hoistways of hoists into or from which goods or materials are loaded or unloaded automatically and to the platform or cage of which there is no access for persons.
13. Hoistways of hoists which are not used for carrying persons and on which the goods or materials stacked on the platform or in the cage are loaded or un-loaded with the top layer of the stack at landing level.
14. Hoists and hoistways the landing and cage entrances of which are protected by lattice gates.

Materials: BS 2655 does not detail the materials to be used for various component parts of Lifts, Paternosters, etc., but Appendix D of BS 4465 gives a list of British Standards covering materials and equipment to be used in the manufacture of this equipment.

In general all items of lifting gear associated with lifts, escalators, passenger conveyors, paternosters and hoists shall be made in accordance with the requirements set out in this sub-section F.

Proof Load:
Static Proof Load = 1·25 × safe working load.
Mobile Proof Load = 1·10 × safe working load.

Test: (a) Before use for the first time. (b) After substantial alteration or substantial repair.

Test Certificate: See Table 12.39.

Table 12.39

	Type of hoist	Certificate	Register
FA	Electric lifts, P. and G.*	BS 2655, Part 7 Appendix A	Form F54
FA	Hydraulic lifts, P. and G.*	BS 2655, Part 7, Appendix B	Form F54
FA	Service lift	BS 2655, Part 7, Appendix C	Form F54
FA	Excalators and passenger conveyors	BS 2655, Part 7, Appendix D	Form F54
FA	Paternosters	BS 2655, Part 7, Appendix E	From F54 plus item 5. Appendix F of BS 2655, Part 7

* Passengers and Goods.

Hoists in premises covered by the Offices, Shops and Railway Premises (Hoists and Lifts) Regulations 1968 shall be treated as shown for Factories Act premises.

Marking:
(i) All hoists must be marked with an identification number and the safe working load.
(ii) Hoists used for carrying persons must be marked with the maximum number of persons which can be carried at any one time.
(iii) Hoists not operated from the cage must have a notice fixed on the platform or cage stating that the carriage of persons is prohibited.

Periodic Inspection and Examination: See Table 12.40.

Table 12.40

	Inspection	*Form No.*	*Examination*	*Form No.*
FA	None	—	6 months	F54
DR	Each occasion before use or 3 months	None	12 months	99, Pt. III, or 1952

F.28 Hoists, chain lever

Specification: BS. Chain Lever Hoists (in preparation).

This British Standard provides for lever hoists fitted with alloy steel short link chain to BS 3114 and hooks conforming with BS 2903. (It is recommended that the hooks be fitted with safety catches and that the use of lever hoists fitted with plate link chains be avoided).

Where lever hoists are used for hauling over rough surfaces or uneven ground, care must be taken to ensure that the safe working load is not exceeded on account of friction between the load and the point over which it is being hauled.

Where loads are being handled through distances exceeding 6 m or where low operating efforts are necessary, advice should be sought from the manufacturer of the equipment on the design of the brake mechanisms incorporated in the hoist intended for such use.

If the maximum range of travel is exceeded an excessive and dangerous load is applied to the load chain terminal stop and failure may result.

As chain lever hoists are portable tools and liable to abuse, adequate lubrication of the load chain and the moving parts will assist in ensuring a reasonable working life.

Materials: The materials used in the manufacture of lever hoists shall comply with the appropriate British Standard.

Aluminium alloys are not suitable materials for load bearings components of lever hoists and must not therefore be incorporated in this equipment (see also sub-section F.23 'Pulley Blocks').

Test: (i) Before use. (ii) After alteration or repair.

Proof Load: 1·5 × safe working load.

Note. Before a lever hoist is used for the first time it is necessary to hold a certificate of test and examination issued by the manufacturer. When an individual part of the hoist, e.g. chain, hook, etc. is renewed it is not necessary to retest the complete assembly provided an individual test certificate for the new part is obtained and kept together with the original certificate.

Test Certificate:

	Certificate	*Register*
FA	Use Form 86	Form 88
C(LO)R	Use Form 80	Form 91, Pt. II, F
DR	Use Form 86	Form 99

Marking: (i) Identification number. (ii) Safe working load. (iii) Grade of chain together with the name of maker or supplier.

Periodic Inspection and Examination: See Table 12.41.

Table 12.41

	Inspection	*Form No.*	*Examination*	*Form No.*
FA	None	—	14 months	88, Pt. IV
C(LO)R	Weekly	91, Pt. I, B	14 months	91, Pt. II, F
DR	Each occasion before use or 3 months	None	12 months	99, Pt. III

F.29   Jack

Specification: No British Standard is available.

Operation: Both ram-type and screw-type jacks must have positive means for limiting the outward movement and so prevent the complete ejection of the ram or screw from the body of the jack.

When a load is being raised by a jack it must always be followed up closely by packing so that in the event of the jack slipping or falling the load will have only a short distance to fall on to the packing.

A load must not be left supported entirely by jacks.

Precautions must be taken to prevent the heads of jacks slipping, if necessary by the use of suitable packing between the head of the jack and the load. Care must also be taken to prevent the jack kicking out as a result of an oblique thrust.

When using traversing jacks care must be taken to have the bed of the jack on a solid foundation. When not in use hydraulic jacks should be kept under pressure so that the leather cup washers remain expanded.

Test: (i) When new. (ii) After alteration or repair.

Proof Load: The proof load shall exceed the safe working load as follows.

Safe Working Load	*Proof Load*
Up to 20 tonnes	25% in excess
20 to 50 tonnes	5 tonnes in excess
Over 50 tonnes	10% in excess.

Test Certificate:

	Certificate	Register
FA	Use Form 86	Form 88
C(LO)R	Use Form 97	Form 91, Pt. I
DR	Use Form 86	Form 99

Marking: (a) Identification number. (b) Safe working load.

In the case of a toe-jack the safe working load which may be lifted on the toe must be marked on the jack as well as the safe working load permitted on the head.

Periodic Inspection and Examination: All jacks shall be examined once in every period of 12 months and the information recorded in the register provided for this purpose.

F.30 Pile driver and pile extractor

Specification: BS 1761:1951 'Single Bucket Excavators'.

Parts of this specification deal with the use of equipment for pile driving. There is no available British Standard for Pile Extractor.

Material: BS 1761 has an appendix listing British Standards covering materials and equipment to be used in the manufacture of pile drivers. All items of lifting gear associated with pile drivers shall be made in accordance with the requirements set out in sub-section F for each individual item.

Cabin: Every winch-driven pile driver or extractor shall be provided with a cabin for the driver which affords him adequate protection from the weather, gives as clear vision as possible and where reasonably practicable, is adequately heated in cold weather.

(SI 1961 No. 1581 14)

Drums and Pulleys: Every drum or pulley shall be of suitable diameter and construction for the chain or rope which it carries.

(SI 1961 No. 1581 15)

Rope Anchorages: The rope anchorages on every drum should be fixed by more than one bolt or screw, and there shall be at least two dead turns of rope on the drum in every operating position.

(SI 1961 No. 1581 15)

Controls: Every control lever, handle or wheel shall have marked upon or adjacent to it its purpose and mode of operation; it shall, where practicable, be provided with a suitable locking device to prevent accidental movement in cases where this is liable to cause danger.

(SI 1961 No. 1581 16(2)(3))

Test: No special test of the complete pile driver or extractor is required, but a preliminary test should be made after each erection to ensure that everything is in order for putting the machine to work.

Marking: (a) Identification number. (b) hooks, ropes, etc., in accordance with the appropriate detailed regulation.

Periodic Inspection and Examination: See Table 12.42.

Table 12.42

	Inspection	Form No.	Examination	Form No.
FA	None	—	None	—
C(LO)R	Weekly	91, Pt. I, B	14 months	91, Pt. II, F
DR	None	—	None	—

It is recommended that helmets and crowns used with pile drivers be carefully examined each day before work is commenced. All ropes, other lifting tackle and winches used on any pile driver or extractor shall be tested, examined and inspected in accordance with the detailed regulations for the particular item.

F.31   Overhead runway beam

*Specification:*BS 2853. 'The Design and Testing of Steel Overhead Runway Beams'.

Every overhead runway beam must be properly designed, of adequate size and strength and must have an even running surface. It must be adequately supported or suspended and properly maintained.
(FA 1961 S 27(3))

Test: There is no statutory obligation to test an overhead runway beam before use although BS 2853 calls for a test and, in most circumstances, a test may be advisable. In all cases before use, the installation must be checked against the approved drawing together with the security and fit of all the bolts.

Marking: All runway beams must be marked with a permanent inscription, readily legible from ground level bearing the following information:
  (i)   Identification number.
  (ii) Safe working load.
  (iii) Any limiting conditions.
    It should be noted that the safe working load applies to the runway beam only and does not apply to a travelling trolley or lifting appliance operating on the runway beam.

Periodic Inspection and Examination: All runway beams and tracks must be thoroughly examined at least once in every 14 months.

F.32   Fork lift, platform, stacking and reach trucks

Specifications:
  BS 3726. 'Stability Testing of Fork Lift Trucks'.
  BS 4430. 'Recommendations for the Safety of Powered Industrial Trucks'. Part 1, Manufacture (Metric Units); Part 2, Operation and Maintenance (Metric Units).
  BS 4436. 'Methods for Stability Testing of Reach Trucks and Straddle Trucks (Metric Units)'.

The British Standards contain useful information on stability tests and recommendations for the manufacture, operation including hazardous situations (see sub-section D.27 particularly for diesel operated trucks) and maintenance. None of these standards, however, contain any detailed provisions concerning the specification of industrial trucks.

Certain parts of fork lift trucks and similar machines are lifting appliances. These include lifting chains and ropes, forks and any associated lifting attachments and the mast assembly. Fork lift trucks and similar machines must therefore be registered with the Lifting Appliances Organisation so that the necessary periodic examination of the various lifting components can be provided for (see also sub-section D.26).

Other parts of these machines require routine attention, e.g. lights, tyres, steering and brakes and provision must be made for these items to be inspected regularly. It is strongly recommended that a routine inspection schedule be drawn up for this purpose. Certain items such as tyres will require daily inspection whilst others will need weekly, monthly or 6 monthly attention.

Materials: All attachments and items of lifting gear associated with lift trucks must be manufactured in accordance with the requirements set out in this sub-section F for each particular item.

Special attention is drawn to the lifting forks themselves since experience has shown them to be prone to failure and because they are not dealt with elsewhere. It is recommended that lifting forks should be made as single piece forgings in steel to BS 970, Grade 150 M19 or Grade 150 M28. Each fork shall be clearly forged and finished with a smooth surface and adequately radiused corners.

If lifting or locating lugs are to be fitted to the forks, they shall be made of the same steel as the forks themselves and shall be attached by the manual metallic arc welding process using an adequate degree of preheat and employing electrodes complying with BS 1719 Class E 61.

Heat Treatment: The various standard items of lifting gear associated with fork lift trucks shall be heat treated as laid down in this sub-section F for each particular item.

Lifting forks shall, on completion of forging and welding operations, be hardened and tempered so that their mechanical properties conform to condition Q of BS 970, Grade 150 M19 or Grade 150 M28. Brinell Hardness measurements made on the surface of the forks shall lie within the range 180–230.

Test: Various aspects of the testing of fork lift and other types of industrial trucks are dealt with in the above British Standards.

A test of the complete truck is required before it is first taken into use and also after any substantial alteration or repair.

Proof Load: The complete truck shall be tested under a load of $1.25 \times$ safe working load.

Each lifting fork shall be individually tested to the total safe working load of the truck.

All other components of lifting gear shall be proof tested as laid down for the individual items in this Section F.

Test Certificate:

	Certificate	Register
FA	Form 1945	Form 88
C(LO)R	Form 96	Form 91, Pt. I, D
DR	Form 1945	Form 99

Marking:
  (a) Identification number.
  (b) Safe working load.
     Both of the above shall be painted clearly on the mast assembly.
  (c) Where appropriate, the tyre pressures shall be clearly marked on the truck body above each wheel.

Periodic Inspection and Examination: Fork lift, platform, stacking and reach trucks should be examined at least once in every period of six months. Particular attention should be given to the mast assembly and the lifting forks, and it is recommended that the latter should be carefully examined for the presence of cracks and fissures by means of magnaflux or dye penetrant methods.

Items of lifting gear used on industrial trucks shall be inspected as laid down in this sub-section F for each particular item.

The results of all inspections and examinations shall be entered in the appropriate register.

F.33   Winch or grab

Specification: BS 3701 Hand Operated Plate Sided Winches. No British Standard is available for power operated winches.

Material: Every part of the framework, including bearers shall be of metal, preferably mild steel. All items of lifting gear associated with winches or crabs shall be made in accordance with the requirements set out in this sub-section F for each individual item.

Cabin: Every power-driven winch shall be provided with a cabin for the driver which affords him adequate protection from the weather, gives as clear vision as possible, and where reasonably practicable, is adequately heated in cold weather. This regulation shall not apply when the driver is indoors or otherwise adequately protected from the weather, or when the appliance is used only occasionally or for short periods.
(SI 1961 No. 1581 14)

Access: Safe access shall be provided to the cabins and platforms of all winches, and as far as practicable, to all points which require examination, lubrication, etc.
(SI 1961 No. 1581 17)

Controls: Every control lever, handle or wheel shall have marked upon or adjacent to it, its purpose and mode of operation. This shall not apply to rotating handles provided for raising or lowering the load in the case of a winch not operated by mechanical power.

Every control lever, handle or wheel shall, where practicable, be provided with a suitable locking device, to prevent accidental movement in cases where this is liable to cause danger.
(SI 1961 No. 1581 16(2)(3))

Brakes: An effective brake, or other safety device shall be provided which will prevent the fall of a suspended load and which will effectively control the lowering of the load.
(SI 1961 No. 1581 16(1))

Guards: All moving parts with which a person is liable to come into contact shall be effectively guarded. It is recommended that guards should not prevent visual inspection of gear wheels.
(SI 1961 No. 1580 42)

Rope Anchorages: The rope anchorages on winding drums should be fixed by more than one bolt or screw, and there shall be at least two dead turns of rope on the drum in every operating position.
(SI 1961 No. 1581 15)

Test:
   (i) Before use for the first time.
   (ii) After each erection.
   (iii) After substantial alteration or substantial repair.
   (iv) Every four years in the case of a winch used subject to C(LO)R.

Proof Load: The proof load shall exceed the safe working load as follows:

Safe Working Load	*Proof Load*
Up to 20 tonnes	safe working load + 25%
Over 20 tonnes up to 50 tonnes	safe working load + 5 tonnes
Over 50 tonnes	safe working load + 10%

The proof load shall be applied either by hoisting movable weights, or by means of a spring or hydraulic balance or similar appliance.

Test Certificate:

	Certificate	*Register*
FA	Use Form 1944	Form 88
C(LO)R	Use Form 80	Form 91, Pt. 2, F
DR	Use Form 1944	Form 99

Marking: (a) Identification number. (b) Safe working load.

Periodic Inspection and Examination: See Table 12.43.

Table 12.43

	Inspection	*Form No.*	*Examination*	*Form No.*
FA	None	—	14 months	88, Pt. 4
C(LO)R	Weekly	91, Pt. I, B	14 months	91, Pt. 2, F
DR	None	—	12 months	99, Pt. 2

All ropes and other lifting tackle used on any winch or crab shall be tested, examined and inspected in accordance with the Detailed Regulation for the particular item.

G STATUTORY FORMS

Official forms have been prescribed for most of the records which have to be compiled for the testing, periodic inspections, and examinations of lifting appliances. Where there are no offical forms, records shall be attached to the appropriate Lifting Appliances Register, viz:

FA	Form No. 88
C(LO)R	Form No. 91 (Part II)
DR	Form No. 99

Provided permission is obtained from the District Inspector of Factories, the Lifting Appliance Register may take the form of a card index. In this case the official forms need not be used.

The following forms have been prescribed and are referred to in sub-section F. Detailed Regulations.

Factories Act 1961

FORM NO.

31   General Register—to which shall be attached within 28 days reports of the results of 6-monthly thorough examinations of hoists and lifts.

54   'Report of Examination of Hoist or Lift.'

88   'Register of Chains, Ropes and Lifting Tackle and Lifting Machines.' In four parts, as follows:

Part I   'Test and Thorough Examination of Chains, Ropes or Lifting Tackle (except Fibre Ropes or Fibre Rope Slings) before being taken into use in any Factory for the first time in that Factory.'

Part II   'Six-monthly Thorough Examination of all Chains, Ropes or Lifting Tackle.'

Part III   'Annealing or Other Approved Form of Heat Treatment of Chains or Lifting Tackle Except Rope Slings.'

Part IV   'Thorough Examination of Cranes and Other Lifting Machines.'
The particulars which have to be recorded in Form 88 are given in S.R. & O. 1938, No. 599 'The Chains, Ropes and Lifting Tackle (Register) Order' and S.I. 1963 No. 1382, 'The Machines (Particulars of Examination) Order, 1963'.

661   'Certificate of Exemption No. 1.'

Construction (Lifting Operations) Regulations

FORM NO.

75   'Certificate of Test and Thorough Examination of Hoist.'

80   'Certificate of Test and Thorough Examination of Crabs, Winches, Pulley Blocks and Gin Wheels or Sheer Legs.'

87   'Certificate of Test and Examination of Wire Rope, before being Taken into Use.'

91   Register in two parts as follows:

Part I   'For reports on Weekly Inspection of Lifting Appliances, Anchorage Tests for Derrick Cranes, Test and Examination of Passenger Hoists after erection or alteration of height of travel.'

Part II   'Register for reports on Thorough Examination of Lifting Appliances (Except Hoists) every 14 months, or after substantial alteration or repair:
Six-monthly Examination of Hoists;
Six-monthly Thorough Examinations of Chains,
Ropes and Lifting Gear;
Heat Treatment of Chains and Lifting Gear.'

96   'Certificate of Test and Thorough Examination of Crane.'

97   'Certificate of Test and Examination of Chains and Lifting Gear.'

Docks Regulations

FORM NO.

86   'Certificate of Test and Examination of Chains, Rings, Hooks, Shackles, Swivels and Pulley Blocks, before being Taken into Use.'

87 'Certificate of Test and Examination of Wire Rope, before being Taken into Use.'

99 Register of Machinery, Chains, etc., and Wire Ropes.
In four parts as follows:

Part I 'Annual Inspection and Quadrennial Thorough Examination of Derricks and Permanent Attachments (including Bridle Chains) to the Derricks, Masts and Decks.'

Part II 'Annual Thorough Examination of Cranes, Winches, Hoists and Accessory Gear other than Derricks and Permanent Attachments Thereto.'

Part III 'Annual Thorough Examination of Gear Exempted from Annealing.'

Part IV 'Annealing of Chains, Rings, Hooks, Shackles and Swivels (other than those Exempted).'

1944 'Certificate of Test and Examination of Winches, Derricks and Accessory Gear, before being Taken into Use.'

1945 'Certificate of Test and Examination of Cranes or Hoists and their Accessory Gear, before being Taken into Use.'

1946 'Certificate of Annealing of Chains, Rings, Hooks, Shackles and Swivels.'

1950 'Certificate of Exemption No. 1 — Annealing of Certain Classes of Lifting Gear.'

1951 'Certificate of Exemption No. 2 — Manner of Test of Lifting Gear before being Taken into Use.'

1952 'Certificate of Annual Thorough Examination of Gear Exempted from Annealing!

BIBLIOGRAPHY

1. *The Use of Chains and Other Lifting Gear. Safety Pamphlet No. 3,* 7th ed. H.M.S.O.
2. *The Use of Derrick Cranes. Safety Pamphlet No. 15,* 3rd ed. H.M.S.O.
3. LINDER, L. *Safe Working Loads of Lifting Tackle,* 2nd ed. Issued by the Testing Department of Coubro & Scrutton Ltd., London (1952).
4. MCCAULLY, C. H. A. *The Chain Testers Handbook.* Published by the Chain Testers Association of Great Britain.

MECHANICAL HANDLING

K. B. WARWICK

The importance of Mechanical Handling cannot be overstressed; it remains almost the only tool in the hands of the individual factory manager with which to overcome increasing overheads. Wages, salaries, rates and taxes, cost of raw materials, power and fuel which are all beyond his direct control as are the prices of new machinery and equipment. On the other hand the methods of handling within his plant are very much his concern. Here he can show quite remarkable savings by the application of soundly based schemes of mechanical handling.

It should be realised that all forms of handling are inefficient, they add to the cost of manufacturing but add nothing to the value of the product, in fact poor handling can actually add to the cost if, as often happens, the product is damaged in transit.

Advantages of mechanical handling

Handling should be as simple as possible, particularly in the small concern. If gravity will do the job, give it the opportunity by using inclined sliding surfaces, gravity roller track or similar equipment, particularly for inter-machine transit. Each problem, however, requires study by the company's experts who have been trained for the purpose, by a mechanical handling manufacturer, or by a specialist consulting engineer. The latter is often better because the engineer concerned has, almost certainly, much experience of similar problems and is not inhibited to change established practices.

The advantages of a good mechanical handling scheme are:

Economy of labour by either reducing the labour force or increasing the work of individual operators;

Avoidance of damage to product by manhandling;

Less handling avoiding double handling and stacking in space-wasting heaps.

There are, however, other less well appreciated advantages. The chief of these is generally the reduction of work in progress and stocks. This is such that one large British Automotive Factory, some years back spent well over £1 000 000·00 on mechanical handling plant to reduce its work in progress from a 6 weeks stock level to a 2 days level. This ensured that, in some cases, they were receiving payment for their final product before they had been required to meet the monthly account of their raw material supplier.

A further hidden advantage which is of tremendous aid where mass- or batch-production is carried out is the control of tempo. This needs careful study when the mechanical handling scheme is introduced if trouble is to be avoided. The workers' representatives should be consulted, a 'start' rate agreed and each increase negotiated. Once accepted, a mechanical handling scheme is often welcomed by the operators, as it enables them to develop a sense of rhythm and ensures that the output of the first hour is the same as the others throughout the working shift. This also tends to ensure that supplies are regulated and enables management to forecast output with reasonable precision.

In this section various types of plant are dealt with in some detail. It would be useless to mention prices which fluctuate but these are available from manufacturers or from price lists. Care is needed to ensure that these lists are up to date.

When a large scheme is visualised this will need to be most carefully phased, to fit in with existing production and should be installed, if possible, when the works are closed for annual shut-down. Thus the need arises for a most careful watch on the progress of the work, possibly calling for a critical path analysis chart or at least a carefully drawn series of bar charts, provided and reviewed, week by week, by the parties concerned. Otherwise chaos can set in and delivery be delayed until it is too late for appropriate remedial action.

Having installed mechanical handling plant there are two further important matters for consideration, the amortisation rate and maintenance. The former should be assessed by financial experts for this must, to some extent depend on the market, the effects of discounted cash flow, and upon proposed and anticipated leglislation. However something between 10% to 20% p.a. seems usual, i.e. between 10 and 5 years to write off completely.

Maintenance

With regard to maintenance, the importance of this is paramount. This particularly applies to the more complicated plant and especially where a single item such as an assembly conveyor, or mechanised paint plant can hold up the entire production and hence the output of a factory. In such cases advice should be sought as to the standby spares to hold and the possibility of replacing such items quickly. It is of interest that one large motor car factory debits its plant maintenance department with a sum of £1000·00 per minute for stoppage of the final assembly or feeding conveyors. It must be remembered that a large percentage of the maintenance work on mechanical handling plant serving production equipment must of necessity be done between shifts, at weekends, and at closure periods.

The more complicated the plant, the more maintenance work it will demand and the class of worker employed upon it must be of a highly skilled class. Schemes of preventive maintenance should be drawn up with the utmost care and the schedules for these should give the greatest detail of the work to be done, the time allowed for it, the frequency of the task, the materials required to be to hand, and information as to whether it can be done during normal working hours. A French car factory uses record cards for each task which are mounted on the part concerned and each card is signed by the mechanic who performs the task, duly dated with a brief reference to his formal report. The system has proved most helpful in ensuring regular and thorough service.

In small organisations, particularly where only one mechanic is employed to attend to many duties, it is advisable to arrange a maintenance contract with the makers or a competent firm to inspect and report on all plant at regular visits.

A useful equation was suggested some time ago by Professor Bright in a paper given to the Institution of Mechanical Engineers which reads:

$$What + Where + When = How$$

Interpreted as follows:

> *What* needs to be handled
> *Where* is it to be handled (source, destination and route)
> *When* does it need to be handled (frequency, speed, etc.)

These data will almost certainly reveal the type or types of handling required, i.e. *How*.

ROLLER TRACK

One can only assume that the method of moving articles by mounting them on rollers is as old or even older than the invention of the wheel. It is, however, still a most efficient way of transporting goods.

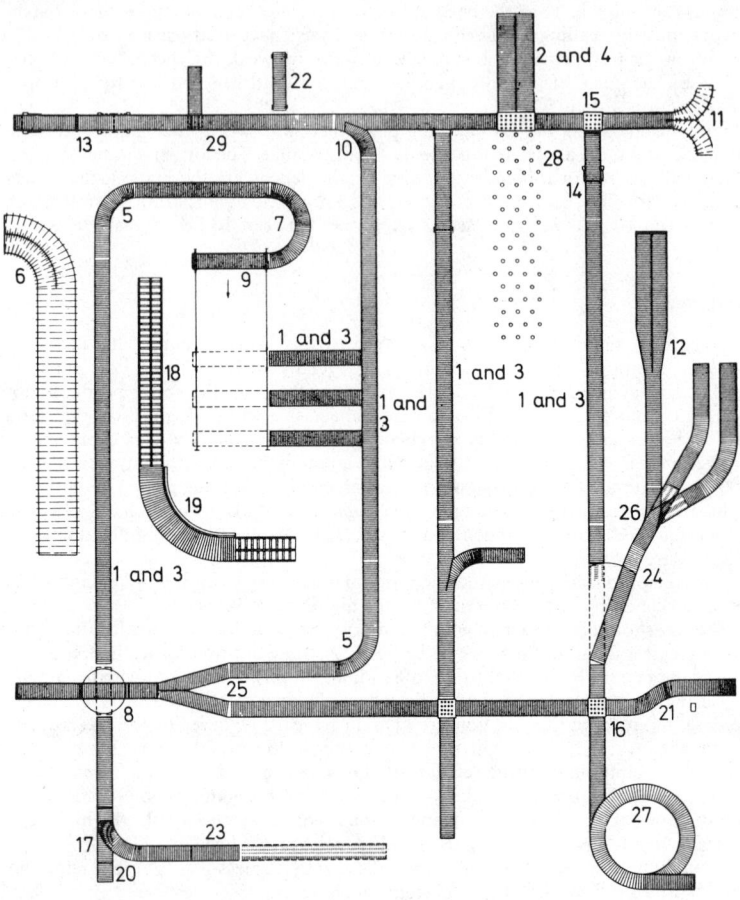

Figure 12.5. Typical roller track layout

1. Single straight track
2. Double straight track
3. Fixed height stands
4. Adjustable height stands
5. Parallel roller bends (single)
6. Parallel roller bends (double)
7. Taper roller bends
8. Turntables
9. Traversing section
10. 90° junction
11. Curved Y junction
12. Breeches section
13. Double counterbalanced gate unit
14. Single counterbalanced gate unit

15. Ball tables
16. Castor tables
17. Turn-over section
18, 19. Vee roller track and curve
20. Hydraulic lift table
21. Off-set section
22. Mobile section
23. Retractable sections
24. Switch section
25. Y section
26. Flexible roller curve
27. Spiral storage
28. Ball pillars
29. Skate wheel transfer

The rollers of modern track are normally of a bright finish steel tubing with simple ball bearings of a non-precision type. These are mounted on steel rod spindles secured in angle section steel 'ladders'. The size of the tube and pitch of rollers will, of course, depend on the duty required. Some manufacturers offer nylon bearings for rollers in their range as an alternative to ball bearings. Normally there exists a great variety of other alternatives within a maker's range, e.g. pressed steel or angle side frames, different widths of rollers, precision bearings, sealed bearings for foundry or similar work, flexible joints, liftable gate sections, ordinary or tapered roller bends, transfer sections and simple turntables. This is illustrated by the typical layout shown

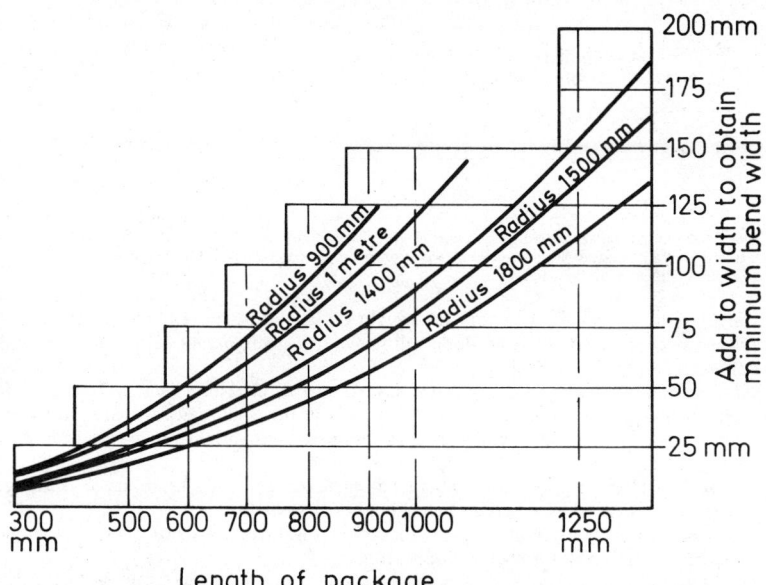

Figure 12.6. Roller track conveyor. Gravity track bend width selection chart. To ensure safe negotiation of bends with guard rails to prevent loads from projecting over sides of track (i.e. gravity falls) width of track must be increased over and above maximum width of loads. This additional width can be read from the graph

Note that on gravity installations spiralling of bend affects even distribution of load and may call for heavier rollers on bend and succeeding straight.

in Figure 12.5. Track is available from Ultra Light with 25 mm rollers through Light, Standard (57 mm rollers) and several intermediate sizes to a Heavy type ranging from 76–89 mm (approx.).

Some useful rules for design check points for the use of non-mechanised roller track are as follows:

Pitch. Under rigid flat based loads there must be full 3-roller contact at all times. Flexible loads such as cardboard cartons require reduced pitch, to reduce sagging to an acceptable minimum. Reduced pitch is also advisable for easily damaged loads, e.g. unpoured foundry moulds.

Width. Rigid flat based loads can be carried on rollers narrower than the loads but flexible based loads must be borne on rollers wider than the loads. Where bends are involved refer to Figure 12.6 for extra width. Care is necessary to ensure that the correct track is used. Some designs of bends are suitable for both horizontal and gravity applications, others must be applied as directed. Taper roller bends allow for faster surface speeds and minimise skidding. These are suitable for both gravity and horizontal applications and most standard bends of this kind allow for fall of up to 76 mm (approx.).

Support. Support is recommended at all joints and at pitches not greater than 1·5 m. Pitch should be reduced where heavy accumulating loads are involved.

Gradient. Table 12.44 gives minimum falls required in mm per 300 mm run, for gravity applications.

Table 12.44   MINIMUM FALLS FOR GRAVITY APPLICATIONS

Loads (kg)	*Fall* (mm/300 mm)	
	Straight runs with ball bearing rollers and taper roller bends	*Straight runs with sealed ball bearings or bushed rollers. Curves with parallel roller bends*
2·25–7·0	15–18	23–27
7·0–23·0	13–15	19–23
23–45	10–13	15–19
45–90	8–10	11–15

Note. Minimum fall figures should be increased by 10% if guard rails are fitted. Minimum fall figure should be increased by 100% if the conveyor is inaccessible, so as to ensure positive travel. The type of load and its shape can also call for modification of these figures.

Care should be taken not to vary the weight of articles by more than 10% to 15% on gravity conveyors, otherwise the slope will prove too steep, or not sufficiently steep for the efficient transport of the goods.

Load projections. Nails, binding wire, seams, straps, etc. impede travel and might impose full load on each individual roller. Loads of this kind call for extra care when specifying rollers. Loads with unsuitable running surfaces should preferably be carried on pallets.

Guard rails. These are desirable where falling loads could cause injury or damage. Their use is considered essential on bends in gravity systems.

Special rollers. Rollers of light alloy or with rubber, plastic or corrosion resistant coatings are normally obtainable for special applications.

Sacks. Only *firm* sacks (as for cement, sugar, etc.) should be carried. Manual assistance is usually required for the first metre or so until a running surface is formed. Consultation with makers is recommended before planning for sack loading.

Power-operated roller installations

The operation of roller conveyor by power is achieved by driving the rollers by cycle type chain or bevel gearing, alternatively by belts driving by friction the underside of the rollers. Such conveyors are most useful for boosting gravity runs or as a main or circulator conveyor feeding of (possibly selectively, using 'plough-off' sections) onto a multiplicity of horizontal or gravity lines.

A high degree of automatic control for sorting articles has been achieved for installations such as airport baggage handling. Anti interference devices can be incorporated to ensure that, at transfer points of multi-line systems, goods are not transferred on a collision course.

OVERHEAD CONVEYORS

The early development of this type of conveyor is reputed to have originated in the pork packing houses in the Mid-West of the USA in the latter part of the 19th century. Starting from a series of trolley/hooks mounted on an overhead track to carry the animals from the killing to dismembering stages, the trollies were linked to a hauling chain which later became the endless chain of a conveyor loop and was finally mechanically driven.

This was a ready tool for the automotive industry as it assumed a mass-production role. The basic constituents remain, namely

  (a) *The Track*. This usually is a rolled steel joist of size varying from approximately 75 × 40 mm to 150 × 75 mm with the popular 'Standard' 100 × 75 mm for loads of around 350 Kg per single trolley;

  (b) *The Chain*, for the standard, most frequently of forged links and pins or of plate link type of the internationally acceptable 458 pattern with a pitch of approximately 100 mm and breaking strain in the region of 20 tonnes;

  (c) *The Drive Unit* of either the sprocket type, or the more modern caterpillar type which has the advantage that it can be inserted at almost any point in the track as required by the pull in the chain and does not involve diverting the line of the conveyor to accommodate a sprocket.

Bends

The horizontal bends in the track are normally provided with a sprocket or plain wheel to divert the chain. Alternatively a bank of rollers is used particularly when the diversion is less than 45°. Vertical bends are rarely truly vertical but the name is intended to imply 'bends in the vertical plane'. These, unless special universally jointed chain is employed, are normally restricted to an angle of 45°, and are often on grounds of loading or load clearance, held to 30° or less.

Chain pull

A factor of 10 is frequently applied, not so much as a factor of safety, (5 or 6 could be justified for this on the basis of crane practice) but to cater for wear of the parts. This dictates the capacity of the drive units at 2000 Kg.

The chain pulls must, therefore, be carefully assessed and balanced as far as practicable, taking into account the following rules

  1. The drive unit(s) should be sited at high, rather than low levels but should not be located too close to the top of vertical or to horizontal bends;

  2. Drive units should not be located immediately after tension or take-up units.

  3. Mounting drive units on vertical bends is not recommended.

When calculating chain pull, a percentage should be applied for tractive effort on straight and level sections, this can vary from 2% to 5% depending on the type of chain rollers, the chain itself, and the track. For vertical bends the number of loads on a bend, must be considered and the horizontal component, based on the angle of the bend, be calculated. Horizontal bends should be treated, for this calculation, on the belt drive formula; the 'slack' side and 'tight' side being the input and output ends of the chain respectively. The calculation is obviously progressive with a selected starting point at the output side of the optimum position for the first drive unit, where the pull in the chain is assumed to be zero.

Multi-drive applications

This limitation of chain pull is further complicated by the pulsation of the conveyor chain which becomes serious when very light or fragile loads are concerned.

There is no easy way of solving this problem, empirical rules have been evolved e.g. 'Never exceed 300 m run on a single drive unit for a standard 458 overhead conveyor'. This rule is really only a guide, factors to be considered are:

Speed, a slow conveyor (say 50 mm/sec) is much more likely to pulsate than one running at three times that speed.

Location of Tension unit. The use of more than one floating tension unit renders a conveyor most vulnerable, but tension units at high levels may also be a cause of jerky operation.

Method and location of Loading. Obviously the loads on down slopes tend to balance loads ascending. Rarely is a true balance achieved due to position of loading and unloading stations, pitch of loads, variation of loading programme during a working shift, variation of weight of individual loads etc. Thus it is unsafe to take this factor into full account beyond assuming that all normally loaded rises are fully laden and all normally loaded dropping runs will be one third loaded.

It will be seen therefore that, despite the amount of study and research devoted to this subject, that the multiplicity of variables still renders it inexact.

Balancing of drive units

In order that the chain pull may be shared amongst the drives and to avoid slack or unduly taut chain developing in the system, drives must be fitted with some device for automatic adjustment.

At first sight it would seem sufficient to synchronise the drives but this will not achieve the balance required. Chain tolerance is difficult to hold to finer limits than (for 458 forged chain) ± 3 mm in 3000 mm. Thus it is possible to have one drive coping with say 30 m of chain *nominal* length, which is actually 30 mm short whilst its partner is driving 30 m, nominal length which is actually 30 mm longer. This is exaggerated if a conveyor is extended after a year or so, the new chain is likely to be on 'top limit' whilst the older, original chain may quite easily be 100 mm long in each 10·000 m length, and this new chain would, unless otherwise specified, be in one isolated length.

The usual means of balancing is to couple a fluid coupling, or electric dynamatic coupling to the high-speed shaft of the drive. The latter can be trimmed by varying resistance and has the additional advantage that it can be used for speed control; the former can be adjusted by adding or extracting up to 30 cc of the fluid at the plug level. High slip motors have also been used successfully for balancing the drives of light (unit loads up to 100 kg) conveyors.

TWIN TRACK CONVEYORS

From the above type of 'Power-Pulled' conveyors, the twin track or 'Dual-Duty' conveyor was developed during the 1930's by Donald M. King. In this application the loads were no longer attached directly to the endless conveyor chain but were suspended from four-wheel trolleys running on a parallel track below the chain track. The chain was fitted with hinged dogs capable of driving or retaining the load trolleys. This was a great advance, for the loads could now be diverted from the main track onto sidings or 'Dead-line' where painting or inspection operations could be undertaken, or could be held for storage.

The full advantages of this new system were not utilised for some 20 years, due mainly to war-time activities; but in 1950, the further development permitted the deadlines to be driven as separate conveyors, indexing to be used on storage lines and the whole system of delivery, transfer, and storage made automatic and to be controlled by punched cards. Further development has enabled the indexing operation to be provided on the individual trolleys. Lowering sections of load trolley track are available, known as drop sections; these can also be made automatic and are used not only for introduction and extraction of loads but also for automatic mounting power units etc. A system of this nature effectively controls the 'Tempo' of the whole plant.

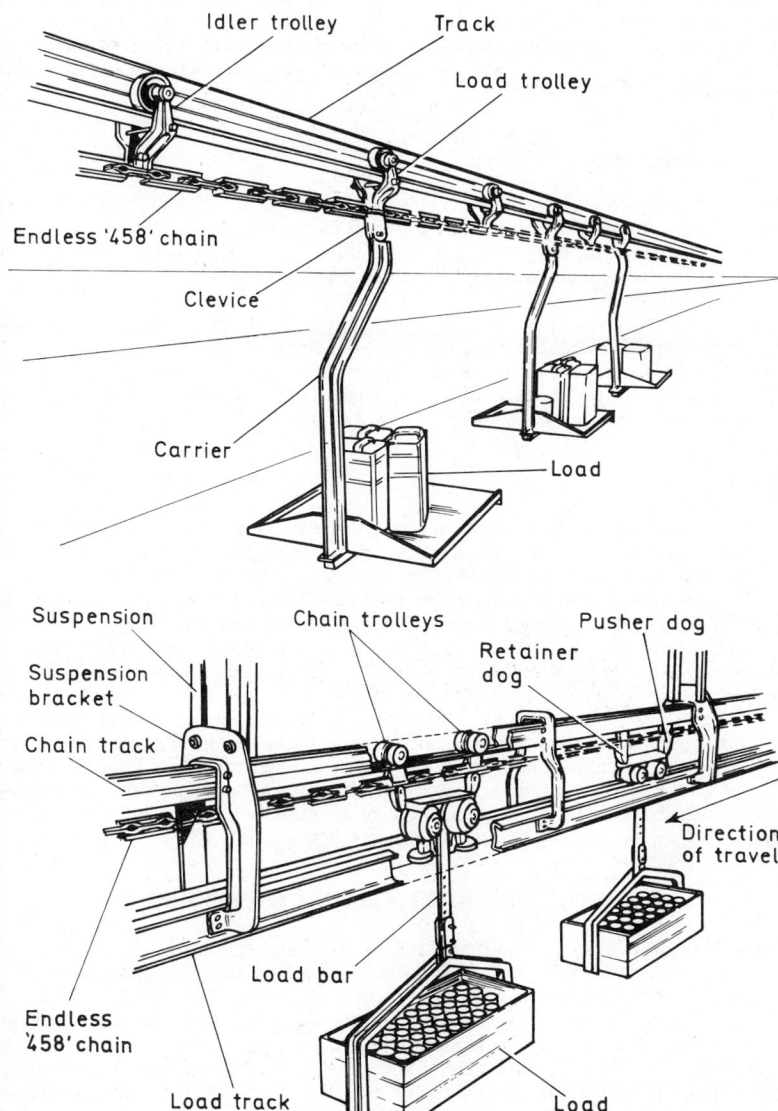

Figure 12.7. Overhead chain conveyors. On the conventional chain conveyor (above) the load trolley is tied to the chain. The dual duty type below allows the trolley to be temporarily disconnected from the chain and arrested on secondary track so that it can be bypassed by other loads

In the main the dual-duty conveyor, or 'Rest and Run', as it is known in USA, is basically the same as the normal power pulled conveyor. The chain, drive technique, tension units, roller bends or wheel corners, are of a similar nature, but naturally the track differs, also the 'furniture' of the main chain. The basic differences can be seen in Figure 12.7.

Loading and unloading

The overhead conveyor is particularly suited to the provision of loading and unloading devices which utilise the motion of the conveyor for their operation. These often incorporate roller conveyor, ramps or air-operated rams, for diversion of the loads and can be made selective by limit switches actuated by pegs on the suspension members on the carriers for the loads.

It is useless to invest in a conveyor installation to reduce labour if almost as many people are employed for loading and unloading it. Suitable devices for this purpose have, almost invariably, to be specially designed for particular applications and this, together with the manufacture of the carriers, should be left to the conveyor makers who have the necessary experience and knowledge.

Guarding

Overhead conveyors are not specifically mentioned in the Factories Act. Guarding is thus dealt with in the general sections dealing with the protection of prime movers, transmission machinery, safe positions of working and means of access.

Generally speaking, drive units which have protruding shaft keyways etc., must be guarded together with exposed gearing; the nipping points between chain and wheels or roller bends must be guarded if less than around 2 m from an access point and any 'dangerous point', e.g. an unloading device at or near ground level, should be duly 'fenced'.

For dual-duty or similar twin track conveyors it is advisable that *all* load trolleys be tested and certified for their Safe Working Loads as the conveyor may be provided with a drop section which, with all its working gear should be treated as a Lifting Machine subject to Section 27 of the Factories Act 1961. The electrical equipment of such installations is subject to the Electrical Regulations of the Act. See also concluding section on Safety.

FLOOR MOUNTED CHAIN CONVEYORS

This type of conveyor with its twin chain and wood or metal slats falls into two categories:
1. *The Service Conveyor*. This is used for moving goods from one position to another in a straight line in a level or inclined plane.
2. *The Process Conveyor*. This application is arranged so that work may be carried out on the goods whilst they are in motion.

The service conveyor is much simpler and straightforward, the chains are driven by two, usually octagonal, sprockets mounted on a common shaft driven by a worm or train of spur gears, the final drive being roller chain in most cases to ensure compactness. The material of the slats is normally dictated by the duty requirements. A tractive effort for this type of conveyor is usually around 5%. The structural members consist of angles for the track supported on intermediate channels forming legs to suit the application.

There exists a great variety of designs of process conveyors to cope with the immense range of applications. The advantage of the slat over the belt conveyor for this type of use is primarily the rigidity of the continuous working surface which can support jigs and fixtures and is readily adaptable to the incorporation of automatic processes.

Return strand of overhead feeder conveyor

Special 'drop section' bringing bodies to floor conveyor.

Operators fitting wheels, after which car transferred to cakestand conveyor.

Cakestand floor conveyor. Note men working below and on level with vehicle.

'Chocks' bolted to slats propel car by front wheels.

Figure 12.8. Cakestand floor conveyor

With frames of the appropriate height, the slats can become a travelling work-bench. The height can be varied at appropriate locations along the run of the conveyor as in the case of the 'Cake-stand Conveyors' of the motor car assembly plants. Here the car frame and transmission units are set up on a single slat conveyor and when the wheels and tyres are added are easily transferred to a twin slat line where chocks act on the car wheels up an incline to a higher level. The vehicle is now on the 'Cakestand' track where the operators can work on platforms carried by the conveyor structure or inside the car, or beneath it from the ground level (Figure 12.8).

Care is required on such conveyors to ensure that the chocks referred to above are protected where they descend below ground level in order to avoid workers' feet being trapped. The descent should be gradual and a device, similar to the 'Combs' provided on escalators, should be fitted.

For flat ground level conveyor runs carrying wheeled vehicles, the level of slats at prearranged pitches can be lowered so that the vehicles are duly spaced.

As an alternative to the bench type conveyor a more recent development is for the slats to run at floor level and support work-carrying pedestals (Figure 12.9). This enables the

Figure 12.9. Diesel engine assembly on single slat floor conveyor with pedestal mountings allowing all-round accessibility

operator to have all-round access to his job and, where necessary, the head of the work carrier can provide a swivel unit.

This type of conveyor may have lengths of 100 m or more, which call for a heavier than normal chain to deal with the increased chain-pull; even so the problem of pulsation may arise (see 'Overhead Conveyors'). As such process conveyors almost invariably require some speed variation provision, it is advised that a second drive unit be provided at the start or tension end of the conveyor. This must be a combined drive and tension and should be fitted with dynamatic couplings to ensure load sharing.

Floor conveyors without slats

When it is required to convey pallets, work-carrying skids or fixtures, slats are not needed and the carrying unit is mounted directly on the chain. Alternatively, if wheeled trucks are involved, dogs are provided on the chain to engage the abutments on the trucks.

This provides two further classifications, namely double strand floor conveyors and drag link floor conveyors.

DOUBLE STRAND CONVEYORS

These could be also described as slat conveyors without slats. The chains are often of the raised link type where the rollers provided at each end of the links are offset to ensure that the links neither drag on the track nor rub on the loads. This greatly reduces the chain-pull, allows for longer length and smoother running.

DRAG LINK CONVEYORS

Drag link conveyors are normally of single strand chain, often of the 458 or 678 pattern as used for overhead conveyors, mounted on the floor in a box type track with guide channel rails for trucks. These conveyors are ideal for paint plants because, there is no chance of dust or grease being dropped from an overhead track. Trucks, however must, be accurately made and constantly maintained and checked particularly where automatic painting is involved. Here, as with slings for overhead conveyors, it is recommended that the conveyor maker be required to design and manufacture the trucks, and that their rigidity be well tested prior to bulk manufacture.

Floor conveyors (Carrousel)

This type of conveyor is frequently used for stores service and small part assembly. The layout is a continuous loop with shaped slats driven by a chain mounted centrally. The slats may require to overlap to prevent screws and small workpieces falling between them which would result in jamming on bends.

The carrousel conveyor has proved to be most useful for baggage handling at airports.

Future developments

The assembly bench type conveyor lends itself to development by the use of linear motors particularly for the return of pallets which, with their complicated work holders and fixtures, represent a high capital outlay on long slat conveyors.

A reciprocating conveyor with a stroke equal to the length of a work station has been used successfully for such applications and the employment of the hovercraft principle is not impossible.

ELEVATORS

The elevator can provide a useful link in a mechanical handling scheme and can be applied to bulk or unit handling. Many manufacturers have standard ranges with capacity quoted from light loads of a few kg to 150 tonnes/hour.

Horsepower calculations are relatively simple as the load of the loaded side can be assessed assuming in the case of bucket elevators, 75 to 85% full buckets. The return belt and buckets or trays will balance the rising ones and the height of lift is settled; therefore a speed must be agreed and an efficiency to cover the gearing etc., applied. The motor rating will be based upon whether the machine will be required for continuous or intermittent operation. In the former case fan cooling must be considered. The method of loading and unloading must be taken into account. In the case of bulk handling this should preferably be automatic from feeder conveyors to storage hoppers or directly into process machinery.

Many makers provide self-supporting elevators. Standard bucket type with discharge height of 30–35 m and facilities are available to fit standard extension lengths up to this limit should the original, shorter height, require to be increased.

'Swing Tray' Elevator for unit load handling calls for more individual design to allow for load size, method of loading and unloading. These trays may actually take the form of a skeleton frame with spaces in the bases to accommodate combs for unloading at different floors. The combs can embody rollers so that unloading can be automatic. The safety record of such equipment is reasonable but care must be taken to ensure that workers cannot put their heads into the track where they may be struck by loads or carriers.

Bucket type elevators vary in design with the loads to be handled. Abrasive materials require special buckets and sealed ball bearings. Where wet, lumpy materials are handled, or powdered materials are liable to build up, special provisions such as self-cleaning or slatted boot pulleys are called for to secure reliability in operation.

Lifting Tables also provide a link in mechanical handling systems. They normally have a vertical lift of less than 300 mm, which is achieved by hydraulic or pneumatic rams or by screw or cam, or a scissors motion driven by a relatively small motor. They are used to gain additional height in gravity roller systems or to effect a transfer in slat or twin chain floor mounted conveyors and for raising or lowering loads for loading or unloading overhead conveyor systems.

For safety, see that apron guards are provided to ensure that there is no trapping zone within reach between the lifting table itself, and any fixed structure, or the floor of the building. Otherwise workers' feet are particularly vulnerable on floor mounted tables.

BELT CONVEYORS

There are two classes of belt conveyor, flat belt and troughed belt. Broadly speaking the flat belt type is used for unit handling and the troughed belt type for bulk handling. Both are driven from the far end by a head drum usually crowned to minimise the danger of the belt tracking or drifting to one side. Self aligning roller sets are also available to assist tracking for long conveyors.

BS 490:Part 1[1] has been metricated and specifies preferred sizes for rubber belting following joint work with the Federation Europeenne de la Manutention and the International Standards Organisation. BS 2890[2] for troughed belt conveyors is being revised as a metric standard and BS 4531[3] for portable belt conveyors is a metric standard.

The calculation for the drive unit is based upon the normal belt drive formula and demands the determination of the effective tension required to move the empty belt, to move the load, and to raise or lower the load. This, however, is not the total belt tension as additional tension is required to prevent belt-slip which must be added and this constitutes the slack side tension of the belt. It is derived by applying the coefficient of friction and allowing for the type of drive, i.e. plain, snubbed or tandem. Allowance is also required where the tension unit is screw, or gravity operated and for the angle of belt wrap and whether the pulley is bare or lagged. The value of the coefficient of friction is normally taken as 0·25 for bare pulleys and 0·35 for lagged. A rough estimation for the take up weights of a gravity tension unit is to double the slack side tension and add a little to counteract the friction in the guides.

Belts are normally rubber- or plastic covered textile material, ranging from cotton, doubled cotton/nylon; rayon; rayon/nylon to all terylene and terylene/nylon of varying plies. The covering is obtainable in differing thickness as demanded by the duty. Rubber tyred idler rollers are also obtainable to suit impact duty. Belting of stainless steel is frequently used in the food industry and woven wire belts are normal when high temperature service is required.

Belt conveyors have a good record of trouble-free running even though high speeds (sometimes in excess of 3 m/s) are common with high, long gradients, and heavy and sharp materials are dealt with. The electrical equipment is usually quite straightforward but interlocks should be provided in any continuous run of conveyors to ensure that if any one of these stops or develops serious belt slip, the whole series will stop together with any automatic loading mechanism.

Much detailed information will be found in British Standards and in the publications of the British Tyre and Rubber Association. The unloading of flat belts can be made to operate automatically by the provision of ploughs which can be brought into position diagonally across the belt at appropriate positions and retracted by air cylinders. Troughed belts, however, present a more difficult problem which can be solved by a tripper. This device incorporates a gradually inclined track with troughed idler pulleys, the conveyor belt is carried up a ramp and over a drum at the highest point so that the material being conveyed tips onto a built-in chute. The belt is then fed over a further drum at conveyor level and thence to its normal operating position.

Trippers are mounted on four rail wheels, fitted with locking devices to retain them in the selected positions. Their travel motion is operated by hand through gearing; by the belt with hand or automatic control for direction; by independent motor; or by a fixed winch. The discharge chutes can be arranged on either or both sides of the conveyor track or forward of it and two three-way valves provided to discharge selectively or simultaneously as required.

Troughed belt conveyors will move a very large amount of material. For example, a conveyor with a 0·9 m belt running at a speed of 2 m/s and with a lift of 24 m has a capacity of 650 tonnes of 200 mm sandstone per hour.

Flat belt conveyors have a multiplicity of applications and can be used for light assembly work. With a spiral track they can form most compact cooling conveyors but a special belt is required. Baggage handling and assisted walkways at airports are typical applications. Composite schemes consisting of belt, roller and slat or roller-flight conveyors are being used in automatic warehouse schemes. The latter comprise a series of rollers slung between twin horizontal chains; the special advantage being that the loads can accumulate on the conveyor and be held stationary although the chains are running normally. Flat belt conveyors have also been used with a travelling swing tray elevator acting as a kind of tripper for 'order-picking' work. In such cases the operator is located on the elevator and his cabin can be made to rise and fall to suit the location of the racking.

The angle at which the belt conveyor can operate is determined by the angle of repose of the material and the friction between belt and material. Where the limitation of space is a determining factor, however, the 'Blanket conveyor' may be employed with advantage. This introduces a second belt covering the loaded lower belt and permits far more inclination of the conveyor. The patented 'Zipper belt' achieves a similar advantage, in that the side edges of the conveyor belt are specially shaped to interlock, when pressed together by training rollers, so that the conveyor belt forms a complete tube on the inclined portion and is unzipped again at the discharge point.

The installation of belt conveyors is extremely important, particularly of long runs. The drive and tension units must be most carefully aligned and all idlers positioned and set up accurately if good belt tracking is to be assured.

PNEUMATIC CONVEYORS

The advantages claimed for this type of conveyor are:

Space saving and accessibility. Pneumatic ducting can be routed along walls and ceilings and enable saving of floor space (a similar claim made for the overhead conveyor.) Operators can transmit materials between inaccessible points without radical structural alterations to existing premises.

Bulk purchasing is made practicable for smaller firms, with resultant saving of expense.

Improved housekeeping. Pneumatic handling virtually eliminates wastage and spillage and is clean and economical.

Lower running costs. Few working parts are involved and reduced maintenance costs are thus achieved.

Positive overall plant efficiency over more usual methods of bulk handling can be achieved.

Many differing systems are available. The simple pressure method with a blower pack induces the pressure into the ducting, the material to be conveyed being fed from hopper via a rotary valve. The final storage bin is fitted with filters to clean the exhaust air. The simple vacuum system uses a similar layout except that no rotary valve is required and the blower pack is fitted at the end of the line connected to the filter on the storage bin; the air involved can then be recycled. Various combinations of the two systems are built up to suit the particular requirements.

It is important that the purchaser should ensure that the supplier has the facility to run tests of samples of his material, to ensure that the system is the best for the particular material and that no damage will be done to the material in rotary valves etc.

Pneumatic conveyor systems lend themselves quite easily to automation and process control. As a bonus it is frequently possible to arrange for actual processing during conveying, for example, Product Cooling and heating and drying which can also be efficiently carried out due to the intimate contact of each particle of the material with the conveying gas. Similarly blending, homogenising and classification can be achieved.

The demands of industry, together with the improvement in compressors, filters and valves over the last decade have made this form of handling one of the most rapidly developing forms of bulk type mechanical handling. Conveyors to handle 500 tonne/hour over approximately 1000 m are now possible and designers are planning installations to unload ships, off-shore, of cargos of powders and grains. This will enable such materials to be landed in areas which, a few years ago would have been impossible.

A disproportionate amount of power is required to convey material up any inclined ducting. Therefore layouts should be arranged for horizontal and vertical components of any slope, thus avoiding the friction forces of material acting on any inclined ducts.

Wear on equipment and pipelines caused by abrasive material is being counteracted by the use of vacuum or pressure vessels which avoid the use of rotary valves or venturis and by reinforced bends. Reverse jets or back-washing filters enable very light or fine powders to be efficiently handled and separated from the conveyor air stream.

Level control in hoppers has been a problem; this has been overcome by the use of rotating paddle controllers which are insensitive to dust and pressure but give positive control.

It can be expected that this type of bulk handling will be increasingly used for heavier work such as the conveying of pulverised fuel, for quarry work and in the iron and steel industry.

SCREW AND 'EN MASSE' CONVEYORS

These conveyors can be used for either horizontal or vertical transport and run under full bore conditions. The screw conveyor which has been called 'an animated mincing machine' certainly employs the same principle. It is dealt with in BS 4409[4] (Parts 1 and 2) which has been revised for metric terminology. The 'En Masse' type conveyor consists of a special endless chain with H-shaped attachments at frequent pitches, housed in rectangular ducting. The forces exerted by the attachments and the chain exceed the friction force of the material on the sides of the housing and thus it is carried along by them.

Both these types of conveyor, being totally enclosed, tend to be dust free, which strongly recommends them to the Food and Chemical Industries. The material conveyed is normally of a powdered or granular nature. In the latter case there exists a risk of crushing of the granules between the conveying medium and the trunking, this must be assessed, in relation to the need to preserve the material, intact or otherwise, in the project stage.

INDUSTRIAL TRUCKS

Industrial trucks fall into two categories, based primarily upon functional requirements. These are transporting trucks and stacking trucks. There are, however, numerous variations and combinations of the two. Trucks used solely for transporting, range from the small and cheap pallet truck for hand or propulsion by tractor-trailer system to platform trucks, skid trucks and huge straddle carriers capable of transporting long loads of timber or steel sections weighing some 30 tonnes at high speeds.

Stacking or high trucks which are used for transportation also are fork lift trucks, reach fork lifts, narrow aisle straddle trucks, side loaders, high lift platform trucks etc.

The optimum advantage of trucking is obtainable when unit loads can be provided. As an example, consider loading a lorry with a capacity of 3 tonnes with 90 loads each of 33 kg. A man could carry only one carton of this weight at a time. Thus, it would require many men to load the lorry quickly or, alternatively, the loading would take several hours. On the other hand, a fork truck with the modest capacity of 1400 kg could load the vehicle in three operations, provided that the cartons were not bulky. Thus if unit loads, duly palletised, can be arranged for in-going and out-going goods the handling problem is simplified.

Within the factory the advantages of trucking are headed by that of flexibility. It is comparatively simple to alter the layout of a machine shop or stores and still use the same method of mechanical handling. Advantage can also be taken of narrower aisles by the introduction of a new truck without any confusion which might arise with other methods of handling. The unit price of trucks tends to be lower than many other materials handling systems but provision should be made for radio control to keep in contact with the drivers when breaks are necessary for meals and refuelling the trucks. One chemical company is reported to have sanctioned an outlay of £10 000 on mechanical handling equipment to replace one operator.

The fork truck carries its load forward of the front wheels, it relies on the weight acting on the centre of gravity of the truck to balance the load tipping effect. There is thus a stability problem of some magnitude but stringent tests are required by BS 3726[5] and BS 4338[6]. Pallets are dealt with in BS 2629[7]. Figure 12.10 shows a fork truck undergoing a stability test.

Normally fork trucks are fitted with two forks but, for specific purposes, special attachments have been developed e.g. a round ram for cylindrical articles such as coils of wire, large pipes etc., a crane jib and rotators or clamps which are power driven. Molten metal, paint, plastics and glass in liquid state can be handled and poured by fork trucks with appropriate attachments.

Straddle trucks are so called because the load is carried within the wheelbase of the truck. The front wheels extend beyond the main structure widely spaced to enable the load to be placed between them. As the centre of gravity is inside the wheelbase the heavy weight needed to counteract the tipping effect of the load on the normal counterbalance type fork truck is not required and a much more compact chassis is permissible. The overall length of a straddle truck tends to be some 1200 mm less than an ordinary fork lift of similar capacity. It is thus possible to reduce the access gangway by this amount. The disadvantage is, however, that the width of the truck is increased by the width of the straddle arms and front wheels.

The reach truck is a combination of the counterbalance truck and the straddle truck. It is more expensive but its space saving qualities can normally more than offset this. The load is carried within the wheelbase of this truck but the reach truck is also able to move its load forward thus making it, in effect, a counterbalance truck and allowing the stacks to be much closer together. When the truck is travelling with its load it can turn through 90° in a stacking aisle in a much shorter length than a comparable counter-balance truck. When it reaches the ready position for stacking the load is moved forward whilst the truck itself remains stationary. Many straddle and reach trucks have the operator standing in the driving compartment but later developments have provision

Figure 12.10. Fork truck on stability test

for the driver to be seated sidewise across the truck. This not only takes up the minimum space but also enables him to have the best visibility in all directions.

Reach trucks can handle loads of 1 tonne up to heights of 8 to 9 m which is a great advantage in modern warehouses. On the other hand the modern demand is for high bay warehouses where stacking heights of some 30 m are required. This calls for the use of the stacker crane, which has often been described as an upside-down fork truck.

CO-ORDINATED SCHEMES

In order to take full advantage of the benefits of mechanical handling the co-ordinated scheme is desirable if not essential. This is particularly so when mass and batch production are involved. The extra unloading and reloading of products at the various stages of the work, the creation of unnecessary storage areas, the damage caused to goods during the inevitable double-handling, are all kept to a minimum, if not completely eliminated. Above all, the conveyors involved can be made to control the tempo of the whole factory.

Naturally the scheme selected will be tailor-made to suit the product and processes involved. It is likely that a twin-track circulating conveyor of light or heavy construction with linked indexing conveyors will prove the solution as has been the case in so many instances, ranging from the production of shoes to heavy commercial vehicles. Automatic transfer mechanisms, drop sections, and load and unload mechanisms are a feature of such conveyor schemes. On the other hand the work may be better handled by floor mounted conveyors as in the modern brewery, where cleaning, and bottling plants are fed by floor mounted conveyors with universal chain feeding circulator conveyors, with automatic restricting arms holding up the transfer of goods on the feeder conveyors until a suitable space exists on the circulator. Such schemes may well incorporate a palletiser unit to stack the cases of bottles onto pallets, and be subsequently conveyed on roller flight or slat conveyors to the transporting vehicles; different conveyors being linked by transfer mechanisms incorporating lifting tables. Alternatively, the final delivery from the palletiser may well be by fork truck.

Preventive maintenance of a high order is necessary on all co-ordinated schemes and it is also desirable to have conveyorised storage banks at strategic points which can be used to feed subsequent conveyors for, say, 15 min in the event of any hold-up. The provision of mimic diagrams with lights on the displayed conveyor routes and at the positions of other allied plant should indicate the stoppage of any vital link in the production chain. In this way the man at the control panel can call up the nearest mechanic and instruct him as to the machine concerned and the nature of the fault.

When installing such a scheme it is considered important to place the responsibility for its design, manufacture, installation and commissioning upon one competent body. This can be a main manufacturer or a consulting engineer. The manufacturer selected may not make all types of plant and the purchaser may be able to purchase at a cheaper price than quoted by the main manufacturer, but it is considered undesirable to expect the plant engineers to cope with such additional work satisfactorily.

SAFETY

The conveyor as such is not specifically mentioned in the Factories Act 1961, and although the track and trollies of an overhead conveyor might be considered to constitute a runway which is included in the scope of 'Lifting Machine' in Section 27 Subsection (9). However, the Act never seems to have been interpreted in this way, despite the fact that a similarly worded clause was in the 1938 Act.

It thus appears that only General Sections are operative for conveyors, elevators, roller track and industrial trucks. Some swing or fixed tray elevators might be thought by some to constitute 'Lifts or Hoists' coming under Section 23. Here again, this interpretation does not seem to have been made officially, presumably because the trays are not considered to be 'Platforms' (See Section 23 Subsection (1)).

Section 12 Subsection (1) and (3) 'Fencing of Flywheels' will apply to motor couplings. Section 13 Subsections (1), (2) and (3) refer to fencing of transmission and also belt requirements. Section 14 deals with the guarding of 'Dangerous parts of Machinery'. This is very wide in scope. Section 17 deals with protection of keys, bolts, setscrews on revolving shafts, also toothed or friction gearing. It should be noted in connection with this section that the responsibility is upon the maker or hirer if the plant was manufactured after 31st July 1937. In the normal way the factory occupier bears responsibility.

The question of underguarding for overhead conveyors is frequently raised and here the matter seems to be left to the inclination of the occupier. It is extremely rare for loads to fall but this can happen particularly when the operation of loading has been carelessly done. Further, the guarding can be useful if boarding, e.g. a plank walkway, is provided over the usual wire meshing; this can provide a useful platform for inspection and maintenance purposes. The fitting of automatic stops to vertical bends is recommended if personnel are working under or near the track of an overhead conveyor or if the conveyor route passes over a gangway. (See also a reference under 'Overhead Conveyors—Guarding'.)

A serious trapping point can exist at the transfer point of two belt conveyors in line. This can be worsened if metallic 'Alligator' type belt jointing is used. Such points can be guarded by a tunnel arch guard over the junction point so that it is no longer possible to reach the danger zone. It is recommended that as many guards as possible be constructed of expanded metal so that any trouble at the transfer point can be clearly seen.

With regard to industrial trucks the main danger is that of stability. This is covered in the relevant British Standards so far as the machines are concerned but, of course, the flooring must be well maintained and a serious danger is present if loaded trucks are required to come out of a building at right angles to a gradient. This should be avoided if possible, otherwise large warning notices should be displayed.

REFERENCES

1. BS 490
   Conveyor and elevator belting (1968)
2. BS 2890
   Troughed belt conveyors (under revision)
3. BS 4531
   Portable troughed belt conveyors (1969)
4. BS 4409
   Trough type screw conveyors (Part 1, 1969)
   Tubular type screw conveyors (Part 2, 1970)
5. BS 3726
   Methods for stability testing of counter-balanced fork lift trucks (1964)
6. BS 4338
   Rated capacities of fork lift trucks (1968)
7. BS 2629
   Pallets for materials handling for through transit (1967)

BIBLIOGRAPHY

BS 2075
Chains and chainwheels for slat conveyors (1971)
BS 2567
Steel roller conveyors (under revision)
BS 4116
Steel roller chains, chainwheels and attachments for conveyors (1971)
BS 4155
Dimensions of pallet trucks (1967)
BS 4337
Principal dimensions of hand-operated stillage trucks (1968)
BS 4339
Rating of industrial tractors (1968)
COPE, J. L., 'Consider the Conveyor', *Concrete Construction* (Dec. 1967)
CAPLAN, B., 'Conveyor beats toughest conditions', *Mechanical Handling* (Feb. 1968)
'Radical Conveyor System', *Modern Manufacturing* (*Factory*) (June 1968)
GRIERSON, A., 'Belt Conveyor Design', *Proc. Inst. Mining and Metallurgy*, **73** (1963)
THRAILKILL, P., 'Overhead Conveyor in Tight Space', *Plant Engineering* (Jan. 1965)
MUMBY, K., 'Minerals in Bulk', *Mechanical Handling* (Aug. 1965)
LE GRAND, R., 'New Angleson Conveyors', *American Machinist* (Dec. 1965)
WINFIELD,   'Applications and Selection of Conveyors', *Plant Engineering* (Jan. 1967)
HUDSON, W. G., *Conveyors and Related Equipment*, Chapman & Hall Ltd.
Overhead Conveyors, McGraw Hill Publishing Co.
SPIVAKOVSKY, A. and DYACHKOV, V., *Conveyors and Related Equipment*, Peace Publications, Moscow
WARWICK, K. B., 'Modern Conveyors', *The Chartered Mechanical Engineer* (Oct. 1969)

ACKNOWLEDGEMENTS

The author wishes to acknowledge, with thanks, permission from the following Organisations for the use of material and assistance given in the preparation of this section. The British Standards Institution; The Mechanical Handling Engineers Association; Douglas Rownson Ltd.; Geo. W. King Ltd. (T.I.); W. & C. Pantin Ltd.; and Lancing Bagnall Ltd.

13 POWER TRANSMISSION

POWER TRANSMISSION 13

13 POWER TRANSMISSION

STANLEY W. JONES

A full consideration of power transmission elements involves reference to numerous British Standards which have not yet been issued in SI units. Many such standards contain empirical formulae which are not valid in a consistent system of units and in these special cases the original form has been quoted. It is undesirable to anticipate action by the British Standards Institution and where Imperial units are currently the major source of reference, a metric conversion has been added.

ENERGY TRANSFER

This section, which deals with methods of energy transfer, outlines the principles which define the range of operation of various shaft mounted elements concerned with the transmission of power. The basis of shaft design for various operating criteria will also be examined.

The diversity of requirements for power transmission connections between prime mover and driven machine has led to the development of a wide range of clutches and couplings.

Couplings

Some manufacturers market more than six different types of coupling, several in combination with clutch and braking capability. Although certain applications are not critical as to the type of connection used, it is wise to compare alternative designs on a feasibility basis.

In the simplest type of *rigid* coupling the shafts rotate as one and any initial errors are transmitted to the driven member. Conventionally one half has a concentric spigot on its flange which fits into a recess in the other half flange; the fit being arranged to prevent lateral loads being taken by the connecting bolts. End thrust cannot be accommodated, neither is rapid disconnection possible.

Oldham couplings are rigid types intended to connect parallel shafts. The flanged ends on each shaft, which have a central grooved face, are connected by a central member having a tenon on each face, offset by 90°.

Hooke's joint or coupling has a star-like configuration made up of four equidistant arms, the pairs of which are pin-jointed into forks carried by the two shafts. Relative angular displacement of the shafts about the perpendicular centres of the pins is thus provided. If there is an appreciable difference between centre lines of parallel shafts, a cardan shaft with a joint at each end can be provided and this reduces the variation in velocity ratio.

A *flexible* coupling specification should differentiate between the ability to tolerate misalignment and the inherent torsional flexibility. The specification should also clearly define the degree of flexibility required, whether this be radial (torsional) or in respect of an axial or end float motion. Data to be assembled should list the nature of torque

13-2

Full coverage

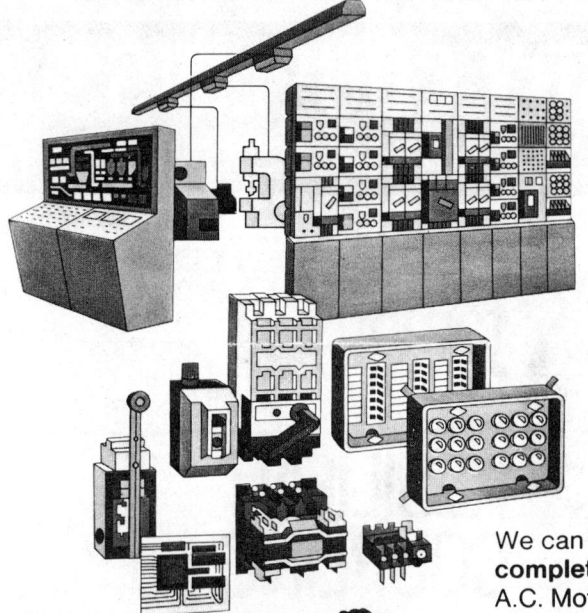

We can supply a **complete** range of A.C. Motor Control and Switchgear from Components to Prewired Load and Control Centres.

fluctuation at particular speeds, the lack of concentricity between the shafts to be connected and any limitations in terms of space availability or environment. Manufacturer's recommendations should be sought in respect of peak loadings and the degree of misalignment that can be permitted.

Coupling characteristics exert considerable influence on the vibratory response of a shaft system and the type chosen should help in controlling response when the system is subjected to various stimuli. Lateral vibrations may be transmitted from adjacent machinery or arise within the system itself due to unbalanced masses. Torsional vibration will be minimised if operating ranges avoid critical speed areas. The degree of response in a 'soft' coupling will permit smooth running at speeds much above critical, alternatively a 'stiff' coupling will place certain orders of frequency outside the operating speed range.

The simplest *spider* type of flexible coupling takes its name from the rubber star or cruciform which engages with the metal jaws of each half coupling. Maximum capacity is 4 hp (3 kW) at 100 rev/min or its equivalent. In *Disc* couplings the half body flanges are fitted with steel drive pins which engage with a rubberised fabric centre disc. Torsional flexibility is small but compensated for by good capacity under impulsive loading and powers up to 700 hp (520 kW) at 1800 rev/min can be accommodated.

In *pin* type couplings, one half body has pins mounted on flexible buffers which engage in holes bored in the other half body. Some variations of this type provide a brake drum facility and incorporate shear pin overload protection. Powers up to 3500 hp (2610 kW) at 100 rev/min are typical.

Couplings which rely upon the power transmitting capability of rubber-like materials usually involve the element in shear or compression loading. Rubber in shear will permit a low stiffness design but is not capable of sustaining a large stress because of its tendency to creep. However, the high bulk modulus of rubber can be utilised by arranging mountings to create compression loading and a progressive increase in stiffness may thus be provided.

Hydrokinetic or fluid couplings transmit power due to the kinetic energy transfer in a fluid filling operation. Essentially, the torque developed depends upon the speed difference between the two halves and the parameter characteristics of the fluid. A variable speed facility is available if a convenient reservoir can be tapped.

Spring type couplings are grooved in the periphery of each half body to permit connection by flat springs which are intended to flex in the process of providing radial flexibility and absorbing shock loads. A totally enclosed lubricated container is typical for powers up to 650 hp (485 kW) at 100 rev/min.

A number of manufacturers have developed *toothed gear* couplings which utilise externally toothed hubs keyed to driving and driven shafts. Connection between them is provided by an internally toothed sleeve and the assembly is enclosed to ensure adequate lubrication between mating teeth. These types have a high power/size ratio and are readily available for capacities up to 500 hp (375 kW) at 100 rev/min.

A simple way of connecting two shafts is by the use of *chain* couplings. This involves mounting single row chain wheels on the end of each shaft and fitting a duplex chain for inter-connection. The assembly is enclosed and grease lubricated. Although misalignment flexibility is small, the simplicity of disconnection is attractive for many applications and powers up to 2500 hp (1865 kW) at 750 rev/min are typical. Effective coupling design depends on the type of fit contemplated for the bolts or other connecting media, since this determines how the load is shared. Methods which depend on an equation of the torsional capacity of shaft with bolt resistance are unlikely to yield accurate results for dynamic situations, although equality of shearing might be used for a preliminary assessment of non-critical dimensions.

Clutches

Clutches are used when a connection to an intermittently rotating member is required. A positive connection involves engagement between projecting jaws with provision for one member to slide out of mesh. This means that shock loads must be accommodated and, if gradual starting of the driven member is required, some form of slippage must be provided.

The *centrifugal* clutch is largely automatic; it incorporates sliding spiders or slippers which do not make contact with the driven member until the speed has reached a sufficient value for centrifugal force to become significant. Spring control ensures that slippers are held clear at low speeds so as to eliminate drag. With this type of connection, an electric motor can be protected from the effects of a high initial starting torque, when trying to actuate a dead load, until its speed is sufficient for the load to be taken up in a smoothly accelerated fashion. Many types of slipping clutch incorporate flat plates or friction discs, held against mating faces by spring pressure; the characteristic of operation is with the clutch in the engaged position. Modern practice for operated clutches favours a linkage to increase mechanical advantage and improve power/space ratio.

Electromagnetic and *pneumatic* clutches are widely used in some quarters, although slip ring control in the former restricts maximum speed, and the latter types need special arrangements for air supply. The above types of clutch have largely superseded the use of conical friction clutches, since the considerable axial force which is then needed for operation poses problems of accommodating end thrust.

BRAKES

The conventional mechanical *brake* interposes a friction material between stationary and rotating members in order to dissipate kinetic energy as heat.

Drum brakes are preferred for high inertia loads because of their increased thermal capacity, although disc types offer faster operation. Both dry and lubricated friction materials are utilised. Wear rate at friction surface can be kept under control provided that the critical temperature is not exceeded.

Fluid brakes dissipate kinetic energy by friction resulting from the agitation of a fluid, while dynamic brakes such as generators or dynamometers are used for absorbing large powers, such as in the testing of prime movers.

Many brakes associated with hoisting machinery are electrically released and spring engaged so as to provide automatic engagement, a fail safe procedure, in the event of a power failure. Perhaps the most important design or selection parameter is to define a limit for the amount of work to be done in one operation so that thermal capacity is not exceeded and the braking surfaces are kept below the temperature beyond which rapid wear rate will lead to early failure.

Chain drives

Chain drives readily satisfy the field of application which would involve unnecessarily large gear wheels, and yet does not permit centre distances at which belt drives would be efficient. The pitch diameter of a chain sprocket is the diameter of a circle on which hinge centres lie and, since the links are rigid, the effect of chordal pitch means that the relationship between chain speeds and angular velocity tends to vary with angular position and makes precise timing difficult.

Current manufacture of chain components is concentrated on bush roller types and BS 228[1] provides details. A reduction ratio of 6/1 is maxima for stock drives, although up to 9/1 is possible with specially designed drives for light powers. The numbers of pinion teeth are conventionally standardised as 17 19 21 23 or 25 and the preferred number range for wheels extends to 114 teeth. The standard suggests details for a range of 15 chains, single, double or triplex form according to size, with pitches ranging from 8 mm to 114·3 mm, although various manufacturers offer a wider range, say 9 to 150 teeth and pitches exceeding 150 mm. The pitch should be chosen to suit both operating power and speed requirements, a range 5000 → 300 rev/min might typify the pitch range quoted earlier.

Attempts have been made to standardise on preferred centre distances but stock drives are usually available for centres which lie between 30 and 100 pitches of chain. Manufacturers' ratings for powers up to 700 hp (520 kW) at 550 rev/min are usually based on 19 tooth pinions and centre distances up to 80 pitches.

An essential requirement for efficient working and to realise a typical design life of 15 000 hours is the provision of a properly lubricated chain case, which should allow for adjustment of chain length, either by take-up or by inclusion of a jockey wheel. Although some tolerance can be permitted on centre distances, shaft alignment is critical. The most effective form of lubrication is by the pumping of a mineral oil on to the inside of chain face, or by running through an oil bath if the wheel is below.

V-belts

BS 1440[2] provides metric specifications for moulded endless industrial V-belt drives. The belts consist of fabric and/or cord suitably reinforced and treated with rubber or its compounds, the whole being moulded together and surface treated to impede the ingress of moisture and render the material suitable for an ambient temperature range 18 C to 60 C. Previous editions of the standard established the practice of referring to various cross-sectional pairings by a letter symbol. Figure 13.1 shows the relationship

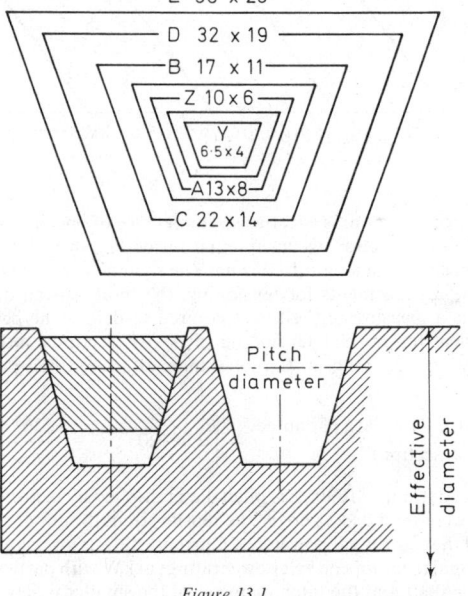

Figure 13.1

between them together with a typical cross section of pulley for A size grooves. It should be noted that the largest section E is now little used and that sections Y and Z, formerly the subject of BS 3548,[3] are intended for light drives where a single belt will suffice.

The included angle of each profile has been shown at a nominal 40° although this angle varies slightly depending on the size of section and the range of pulley pitch diameters for which it is intended. Notice that the included angle for A section is 34° for pitch diameters 75 to 125 mm but 38° for larger sizes and that typically this section is used for the pitch diameter range 75 to 200 mm. Earlier versions of the standard made similar angle stipulations, but based on the effective diameter.

In general, the optimum belt life occurs when the largest possible pulley for a particular section is used, provided that the belt speed is restricted to 30 m/s. Although most manufacturers offer the standard sizes as listed the general commercial range is considerably extended by the provision of narrow, deeper belts in the small size range which can transmit much larger powers. The choice of an appropriate size of cross-section is made from a range chart, Figure 13.2 is typical, thereafter individual drives can be selected from the available pitch lengths quoted in manufacturer's catalogues.

For example, A range belts are available with pitch lengths between 630 and 3 540 mm. Some data is still published in terms of nominal inside length or effective length, the latter

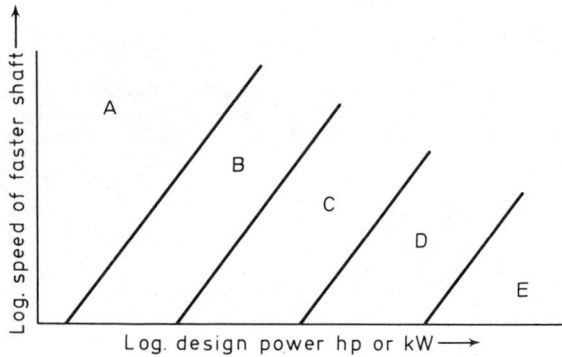

Figure 13.2

being based on twice centres plus effective circumference of one of the two equal measuring pulleys. Thus for A section when nominal inside length is 1 067 mm, the effective length is 1 118 mm and pitch length 1 102 mm. The significance of the reference to pulleys here relates to the arrangements for tensioning, the total tension divided between the working pulleys is a measure of the force required to deflect the belt a given distance, 1 – 10 kgf is typical for normal tensioning of A → E sections, although retensioning may be necessary after running in. Earlier standards quoted belt powers in terms of a formula such as

$$\text{HP per belt for 180° arc of contact (inch units A section)} = \left[2{\cdot}684(S)^{-0{\cdot}09} - \frac{5{\cdot}236}{d_e} - 0{\cdot}013\,6\,S^2 \right] S \quad (1)$$

where S is belt speed ft/min ÷ 1000, d_e is pitch dia multiplied by factor depending on speed ratio. The terms were intended to cover the tensile strength function, the bending effect due to flexing and the centrifugal power loss. Current editions do not quote these formulae and it is conventional to quote safe belt power ratings in kW with particular reference to the speed of the faster shaft and the pitch diameter of the smaller pulley

$$\left[1 \text{ hp} = 0{\cdot}746 \text{ kW or } 1 \text{ kW} = \frac{\text{kgf} \times \text{m/sec}}{102} = 1{\cdot}34 \text{ hp} \right]$$

A small additional power per belt is noted in terms of the operating speed ratio. The A size belt, taking drive from a 75 mm pulley mounted on the shaft of a 1 440 rev/min motor, is rated at 0·91 kW for an arc of contact of 180°. This value is upgraded by over

20 % if a speed ratio of 2/1 is involved, but would then be re-appraised in terms of a correction for the reduced arc of contact on the smaller pulley.

$$\text{This arc is twice the angle whose cosine is } \frac{\text{Difference of pulley diameters}}{\text{twice centre distance}}$$

and the correction multipler is 0·82 when arc of contact falls to 120°. Power correction factors for the pitch lengths already quoted for A section belts increase from 0·80 to 1·16 with the belt lengths.

Some manufacturers offer additionally a range of narrow V belts according to the specifications of BS 3790.[4] These exhibit a larger ratio of top width to depth than the 'classical' belts of BS 1440 and are intended for drives where space is at a premium. Nomenclature such as 3 V, 5 V, 8 V are typical standards and other sizes are available to make up a range capable of transmitting a maximum of 375 kW at 5 000 rev/min.

It is important to make a realistic appraisal of the effect of environment on belt performance and particularly of the load characteristics expected on driven machines. Most manufacturers quote a service factor which allows both for the type of driving unit and the duty to be experienced at driven machine, factors up to 1·8 depending on the operating period are typical for the A B C D range. The factor is used as a multiplier of the nominal motor rating to determine the design power which the drive should be able to transmit.

Pulleys are conventionally cast iron or steel of hardness sufficient to assure adequate wearing properties and to enable a satisfactory surface finish to be achieved. Consideration of the need for dynamic balancing will arise for certain face width/diameter ratios and generally when surface belt speeds exceed 10 m/s. The intention should be to eliminate the resultant and any couple which arise due to the effect of centrifugal force. The later section of this chapter on keyed connections will cover the attachment of V belt pulleys, although the convenience of tapered bush screwed fitments, with or without keyways has been widely recognised as an alternative means of fast and easy fixing. Some data on V flat drives is still available, these methods are a convenience when the larger pulley is needed to have a solid rim because of its flywheel type duties.

One feature of pulley type transmissions is the facility to alter effective diameters by adjusting the working position of the V section and speed variation ratios up to 4/1 can be provided by single pulley pairings of this type. Large scale ratio variations, up to 16/1 with powers up to 130 hp (100 kW), are available in closed cases with fixed centre positions for output and input shaft. These devices utilise the converging/diverging action of taper cones. Precise regulation down to very low speeds is difficult because of belt slip.

One of the problems associated with V belts, and with the chain drives considered earlier, is the variation in velocity ratio which occurs from time to time. Timing belt drives, or *synchronous* belt drives as described in BS 4548[5] are one method of ensuring greater precision. The grip required for driving is provided by the meshing action between the moulding on the underside of the belt and the gaps in the pulleys. The positive nature of this meshing greatly reduces the need for initial tension, lubrication is not required and there is negligible stretch or slip. Typical practice is to use a flanged smaller pulley with both flanged on vertical drives, the five pitch codes approximately 5 mm to 32 mm and the eleven belt widths approximately 6 mm to 127 mm cover power ratings up to 75 kW and conventional speeds up to 30 m/s.

Static balancing is usual for all pulleys and dynamic balancing is desirable for higher speeds. Selection procedures involve the use of service factors rated in terms of driving source, operating periods and the dynamic characteristics of the driven machine, in conjunction with a range power/speed chart of the type used for V belt drives.

Flat belts

In spite of the increased range of application for chain drives and wedge section belt drives there remains a sphere of operations in which leather or balata or composite flat belts, spanning a considerable centre distance, will provide a single stage drive that might demand a countershaft or separate stages if any other means were used.

In long centre drives the maintenance of tension is largely due to sag in the belt (with slack side uppermost), but when centres are short the elasticity of belt and the provision of aids such as pivoted seatings or idler pulleys are necessary factors in ensuring that the arc of contact on the smaller pulley does not fall to a dangerously small value. A belt stretched over pulleys at rest is subjected to initial tension on both sides, the normal force on pulleys which results, sets up a friction force and this, as the driver starts to rotate, acts as a tangential force at the rim. The initial tension must be sufficient to ensure clinging during the process of increasing tension on one side and decreasing it on the other, so ensuring rotation of the driven pulley.

Initial tension is often assumed as half the sum of tensions on tight and slack sides but it is more accurately determined in terms of the active arcs of contact. Effective tension is the tension difference and this multiplied by belt speed enables determination of power transmitted. The simple tension relationship T_1/T_2 is a function of the arc of contact at the smaller pulley, which is calculated from $180°$

$$180° - 2\sin^{-1}\left(\frac{D-d}{2C}\right) \tag{2}$$

Assuming average values for the friction coefficient, T_1/T_2 for a flat belt (from $e^{\mu\theta}$), is unlikely to exceed 2 before slip starts to occur. Recommendations for centre distance are based on the reduction ratio. The maximum ratio is not likely to exceed 6/1 unless special tensioning devices are provided.

The ideal surface belt speed is approximately 18 m/s but this may result in inconveniently large pulleys and 8–10 m/s is more common. Belt drives are designed on the basis of a safe pull per unit width, depending on the thickness of leather or balata involved. Modern belts are capable of considerably higher performance, they typically compose a three-section sandwich in which the central tension member is nylon or polyester, outers are tanned leather or impregnated fabric. Crowned pulleys are typical and much higher speeds up to 50 m/s are possible with powers up to 3750 kW.

As higher speeds are contemplated with conventional materials, the effect of centrifugal force needs to be taken into account. The loss in power due to centrifugal force will be noted at speeds exceeding 15 m/s unless allowance is made; it is common for belts apparently tight when stationary to fail to exert grip when centrifugal effects predominate. It is convenient to derive another tension relationship and with symbols $T_T T_S$ representing the belt tensions, T_C the centrifugal force, we have

$$\frac{T_T - T_C}{T_S - T_C} = e^{\mu\theta} \tag{3}$$

in appropriate units. This is a significant result in terms of the maximum power which a belt can be expected to transmit, typically the surface belt speed should be such that the stress due to centrifugal effects is one third of the maximum working stress permitted in the belt material. It is necessary to distinguish clearly between the effects of Creep and Slip which can occur even when the tension relationship is less than the limiting value.

Both phenomena are sometimes linked in experimental investigations since they combine to produce a small difference in speed between the driving and driven pulleys, the amount of which is a function of the tension difference. It seems likely that creep takes place over an active arc of contact and that over the remaining angle of lap there is no change of tension. The properties of the material are of significance here, although the stress/strain curve for belt materials is not straight the increase of stress at a greater rate than strain can generally be predicted. It should be noted that there is a field of application for long centre, large power drives in which flat belts remain the viable choice and should merit consideration at feasibility level before turning to more expensive alternatives.

Shaft design

Shaft design should proceed by means of an iterative process of analysis and decision to establish geometry and material and to so specify the component that value engineering considerations are incorporated. A typical procedure might involve initial design to satisfy strength requirements with subsequent modifications as required to satisfy limiting values, such as for deflection or a safe operating speed.

One consideration of early concern is the accuracy of the determination made as to an allowable working stress for the available material. In some cases the analysis of imposed stress, due to the loading pattern, may prove difficult to resolve although, in general, a state of multiaxial stress can be represented by a maximum of three forces acting on the faces of a cube and in many design problems the third principal stress may be assumed zero. The methods suggested below will prove effective for the commonly used design criteria but it is important to appreciate the assumptions made in their derivation.

A careful analysis will be wasted unless the material capability is well documented so that the relationship between significant stresses and characteristic material strengths can be defined in terms of a probable failure phenomenon. It will be assumed for our purpose that the postulation of a shaft in pure torsion is likely to be untenable, at the very least self weight and the method of support will produce a stress changing from tensile to compressive in each revolution. Neither, in present day machinery, are we likely to find many examples of the lineshafting type, when considerable lengths of shaft can be safely proportioned by making stiffness the basis of design. The situation most commonly found in practice is a sensibly constant torque with a cyclic reversal of bending moment due to self weight and transverse load. The type of support provided by bearings is significant, an assumption of simple support provided by rolling contact bearings is a convenience but most bearings involving rolling elements or separate bushes will induce constraint or fixing of varying degree.

It seems, then, that the problem of material choice and an estimate of intrinsic load carrying capacity should receive early consideration in any design procedure. Standards such as BS 970/71[6] and the recent PD 6423[7] for carbon steels publish test values for a range of materials. It is necessary to be satisfied that the test procedure on which such values are based is suitable for the application under review and representative of the loading pattern likely to be experienced. A further problem is that the care expended in making the test piece; in terms of accuracy, surface finish and the likelihood that manufacture has removed the predisposition to surface originated failures inherent in ferrous materials, may not be repeated when a real component of substantially larger dimensions is to be made.

It is with this in mind that the standards propose the observation of a ruling section code, an indication of the maximum cross-section for which published test values are likely to be acceptable. An additional factor concerns assumptions as to the likely mode of failure, the various failure theories argue that the stress pattern under combined loading is not likely to prove more damaging than a particular type of stress at yield in a simple tensile test. They are usually stated in terms of the principal stress relationship at a point when failure could occur on the basis of the material strength in simple tension or compression.

Whether an analysis based on maximum principal stress, maximum shear stress or distortion energy is the most significant in shaft design need not be discussed here. It will suffice to say that the formulae should be manipulated to provide an expression for a safe working stress or shaft diameter, having taken the predominant type of loading into account. If, for example, the maximum shear stress theory of failure is accepted as satisfactory for a ductile material, subjected to combined bending (M) and twisting (T) of a non-fluctuating type, a suitable shaft diameter (D) may be extracted from

$$D^3 = \frac{32}{\pi\sigma_w}\sqrt{(M^2 + T^2)} \qquad (4)$$

where σ_w is the allowable working stress. The square root term is sometimes called the equivalent twisting moment.

Assumptions made here do not involve acceptance of Hooke's law if a discernible yield point can be detected, but expect similar capabilities in tension and compression. The choice of an alternative theory depends on what seems likely to constitute a failure occurrence. Static failure theories based on yielding will be acceptable for a few cycles of load application but, if the stress fluctuation is frequent, a criterion which involves fatigue failure should be introduced. For certain dynamic situations, a cumulative damage theory may be most appropriate.

In a typical case of fluctuating stress, the proposals of Goodman or Soderberg may be adopted. The characteristic modified diagram due to the former is based on an assumption as to a safe endurance limit in terms of ultimate stress. It enables any relationship between variable and mean stresses to be determined ready for subsequent insertion in the Goodman hypothesis or in the Soderberg equations when the mean stress axis is referred to a yield criterion. Other important terms to be included are for a factor of safety or reserve factor and for a multiplier of the variable stress term so as to allow for the effect of discontinuities or stress raisers, in the form of a stress concentration factor.

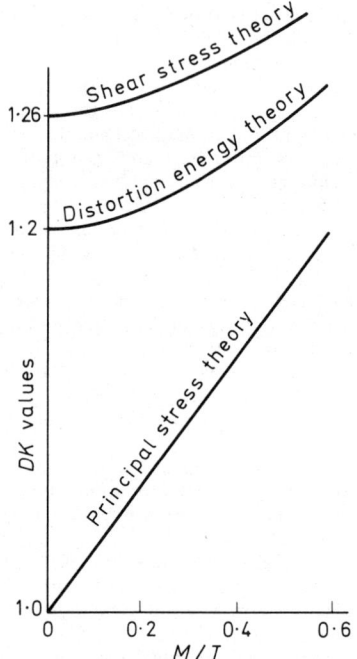

Figure 13.3

Theoretical values K_T, as a ratio of maximum to average stress, are found from mathematical or experimental techniques, although more accurate results for cyclic loading conditions are determined by values of K_F, sometimes called a fatigue strength reduction factor. This takes notch sensitivity into account and attempts to assess the varying response of different materials to the same stress raiser. It is desirable also to account for possible differences between the performance of a test piece and the real component, as was discussed earlier.

Turning now to explicit design procedures, the equivalent twisting moment has been the basis of several design codes intended to cover combinations of bending and twisting in static loading, sometimes with additional multipliers which purport to allow for shock and the fatigue effect due to stress fluctuations. The A.S.M.E. Code is one method of this type in which additional terms allow for the presence of axial loading by introducing a column effect. The allowable stress is based on shear capability, assuming this to be some percentage of the tensile capacity at either yield or fracture conditions. Graphical solutions facilitate the determination of critical areas and may be accurate enough for preliminary calculations. A typical example would be the proportioning of a shaft on the proposition that its lowest critical speed be kept within a particular range or it might be decided that none of the steps in the speed range should approach closer than 20% of a critical harmonic, this will involve determination of shaft deflection at various positions. Similarly the permitted deflection of an attached component such as a gear wheel, in order not to prejudice meshing conditions, might be the basis of a rigidity calculation to determine dimensions. The same analysis in terms of a slope curve could determine the type of bearing support to be provided, since one design criterion limits the permitted slope of shaft through bearings. The use of moment-area methods for deflection calculations should be studied.

The purpose of the foregoing has been to indicate the unlikelihood of an empirical formula or a list of theoretical shaft power capabilities providing little other than an estimate of power availability at critical areas. The distribution and take off of load, the effect of various bearing arrangements and methods of location all conspire to emphasise the importance of a realistic methodology such as analysing the loading pattern and synthesising to an equivalent stress system before determining dimensions, knowing the capability of various materials. For these reasons there is no attempt in this text to include a table of shaft powers or a list of recommended bearing centres.

Reference to Figure 13.3, however, will indicate how a shaft diameter can be determined, knowing the ratio between bending moment M and twisting moment T for each of the failure theories mentioned earlier. The DK ordinates indicate the product of shaft diameter and the constant

$$K = \frac{D}{C(T)^{\frac{1}{3}}}$$

The basis of their plotting when $\quad C = \sqrt[3]{\dfrac{16}{\pi \sigma_w}} \quad$ and $A = M/T$ is as follows

For Principal stress $\qquad DK = [A + \sqrt{(A^2 + 1)}]^{\frac{1}{3}}$ (5)

For Distortion Energy $\qquad DK = [\sqrt{(4A^2 + 3)}]^{\frac{1}{3}}$ (6)

For Shear Stress $\qquad DK = [2\sqrt{(A^2 + 1)}]^{\frac{1}{3}}$ (7)

Thus when A is unity the ordinates are 1·342, 1·383, 1·415 respectively, from which a shaft diameter can be extracted when torque and allowable stress are known.

Keyed connections

The transmission of power from a shaft to its attachment is frequently by a keyed connection. The metric equivalent of BS 46: Part 1[8] is BS 4235: Part 1.[9] The tolerance values proposed were intended to arrange for a key to be retained more securely in the shaft than in the mating-hub, although provision for either top or side fitting has been made. The simplest design basis is to equate the product of resisting area and allowable material stress, for both shearing and crushing criteria, to the tangential force determined from torque to be transmitted. It might appear, when values are substituted, that the safety factor in compression is less than that for shear, but when the effect of fit is allowed for the real compression stress is less than the theoretical value.

Parallel square or rectangular keys are usually side fitting with top clearance, following the arrangement shown in Figure 13.4 which is typical for shafts over one inch diameter. The tolerance arrangements are such as to permit withdrawal of the component over the key in cases where a gib head cannot be accommodated or when there is insufficient clearance to drift out the key. For assemblies likely to require periodic withdrawal a taper key can be specified, the fitting arrangements depending on the loading pattern.

The dimensioning method for shafts and hubs should follow the recommendations of BS 308[10]. Conventional rectangular key proportions are based on a nominal width of shaft diameter/4 and a thickness of shaft diameter/6. In order to rationalise sizes of keybar a series of preferred size ranges each has a particular pair of key dimensions, being based on the largest shaft size in the range. Thus for a shaft over one inch (25 mm) but less than $1\frac{1}{4}$ in (31 mm) diameter the nominal key dimensions are $\frac{5}{16} \times \frac{1}{4}$ in (8 × 6 mm),

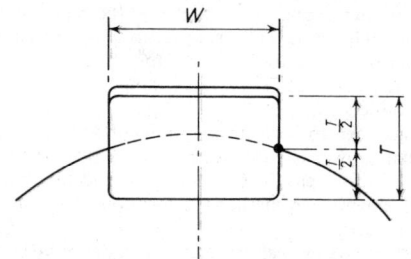

Figure 13.4

whereas for the range 5 in to 6 in (127 to 152 mm) the nominal key is $1\frac{1}{2} \times 1$ in (38 × 25 mm). Similar sizes are suggested for the cross-section of taper keys, with or without gib heads, the fitting portion being provided with a taper of 1 in 100.

When frequent sliding has to be accommodated. single or double keying may be replaced by a splined connection. Alternative schemes are suggested in BS 2059,[11] dimensions and fits being on a hole basis with various grades arranged by variation of shaft root diameter and spline width. Although a variety of spline proportions have at times been suggested it is now conventional to base on six splines, bottom fitting, to either 'deep' or 'shallow' proportions.

PLAIN BEARINGS

A bearing is defined as that part of an assembly which supports another component moving in sliding contact with it, the intention being that the relative movement occurs with the minimum resistance to motion. The degree of failure in performance is reflected in the heat generated and the loss in size of mating components. There is no essential difference between a bearing and a bush, although some authorities reserve the latter name for complete hollow cylinders and speak of bearings as the split outer containers within which rotates the shaft or journal.

Types of plain bearing

The simplest type of rubbing bearing involves the use of an unlubricated bush, usually of non-metallic origin for which a P.V. criterion [(lbf/in² of projected area) (ft/min)] is commonly quoted. This empirical product has little to commend it academically, although it was at one time regarded as an elementary design basis for film lubricated bearings.

Values quoted by manufacturers of non-metallic bearings can be used to plot fields of application, logarithmic ordinates of load and speed are typical, and from these a particular working range can be extracted. It should be realised that a continuous wearing process is taking place so that the mating materials ought to include properties suitable for the maintenance of a low wear rate, test data is frequently based on a wear rate of 0·00025 mm/hour.

Typical dry bearing materials are thermo-plastics (often filled to improve properties and metal-backed for additional strength) or woven fibre reinforced thermosets. More sophisticated types involve a fluorocarbon polymer overlay on an intermediate impregnated bronze with a steel backing. It is intended that a film of overlay be transferred between the asperities of the mating surfaces. Plastic faced materials need larger running clearances than metals, which emphasises the importance of alignment since there is no film of lubricant to spread the load. Many types are supplied finish machined so that shaft and housing dimensions can be prepared to suit the correct running clearance and a suitable mounting fit.

There is a wide field of application for oil impregnated porous metals bushings (essentially bronzes) intended to provide a self-lubricating effect. Powder metallurgy processes are employed in the manufacture of these bearings when the melting points of constituent alloys are so high as to preclude fusion techniques, or when the required composition can not be provided by casting. The intention is to provide compacts of porosity 25–35 per cent and typically 90 per cent of this porosity can be oil impregnated. Retention is by capillary attraction so that there must be a suitable relationship between pore size and surface tension. The P.V. product is frequently used as the application criterion and the main queries which arise are:

   (i) Has the prediction of operating conditions been accurate enough to remove the need for supplementary lubrication?

  (ii) Have arrangements for mounting and supporting allowed for the various expansion coefficients and the stresses imposed during fitting?

As with the types reviewed previously, there are restrictions on load carrying capacity due to problems of heat dissipation and the usual design intention is to operate at a much lower temperature than for full fluid lubrication. The bushes can be supplied finish machined and adherence to the manufacturers recommendations for shaft and housing sizes will ensure correct fit and clearance. Supplementary machining and oil grooving is to be avoided. BS 1131 part 5[12] lists sizes of liners and bushes commonly available in inch units and BS 4480 part 1[13] in the metric series.

Bearing materials

Before considering factors related to the design of hydrodynamic or film lubricated bearings it is appropriate to compare a real and an ideal bearing. The hypothetical situation envisages smooth and parallel surfaces, not subjected to thermal or elastic distortion and separated by a film of lubricant whose properties are at all times able to provide sufficient load carrying capacity to prevent metal contact. This brings in its train assumptions that the lubricant behaves as a simple Newtonian fluid and that material strength is the only criterion of bearing performance. Such a reference or theoretical bearing is impossible of realisation; deficiencies of surface finish, distortion of the shell and housing and journal deflection all occur in a real case with the added risk of penetrating the oil film by surface irregularities. It is in these circumstances that complementary properties of the material pairs become important, decisions taken to reduce hardness and elastic modulus to the minimum acceptable values involve consideration of properties such as;

   Compatibility, a measure of anti-scoring characteristics

   Embeddability, the possibility of inclusions or debris being accepted by the softer
        material.

   Deformability, the hardness/modulus rationalisation intended to compensate for
        misalignment or geometric error.

The selection of one hard and one soft material is the simplest possible design precept for a journal bearing, with the intention of reducing the likelihood of seizure by local melting and in the hope that any wear debris will be implanted so as to eliminate further damage by circulation. The desirable hardness ratio of a metallic bearing pair might range from 5–14 over typical groupings of relatively soft bushes such as

Babbitts or White Metals $\longrightarrow$ Copper/Lead $\longrightarrow$ Lead Bronze
(tin, antimony, copper)   (typically 70/30)   (75/20/5)
(copper-lead-tin)

These values are a useful guide when specifying a suitable journal hardness and a surface finish 0·254–0·381 μm will be appropriate, assuming steel as the universal journal material.

At the weaker end of this range, the bearing material needs a stronger back and present day practice tends towards the use of softer faced bearings in preference to gun metals or phosphor bronzes, particularly when there is a risk that the boundary state may be prolonged. BS 1131, Parts 1–4, specifies ranges of plain and flanged bearings as wrapped metal bushings (from strip) and paired half liners. These are broadly based on copper alloys (BS 1400) available either pre-finished or for machining in situ.

Lubrication

The theory of full fluid or hydrodynamic lubrication assumes that the shaft and journal are rigid and always separated by a film of fluid within which all significant events occur, The key to understanding is an appreciation of variations in the bearing modulus ZN/p where Z is viscosity in centipoises at the operating temperature, N is rotational speed rev/min and p is pressure on the projected bearing area lbf/in^2. Figure 13.5 is drawn to

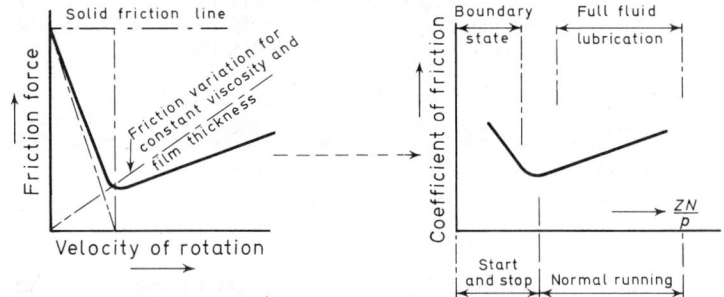

Figure 13.5

indicate the relationship of these properties with friction force or coefficient and indicates how various phases of bearing operation can be isolated.

Since the intention is that the rotating journal shall act as a pump and generate sufficient oil pressure in the clearance space to support the load and prevent metallic contact, it is of some significance to appreciate the effect of changes in separate parameters and of their overall effects in combination. If, for example, the speed is reduced with Z and p originally constant the film becomes thinner, frictional losses are reduced and operating temperature falls, which in due course causes Z to increase. At starting and stopping Z may be small and film is weak in compression, N is small and there is little pumping action, p may be large due to high static loading. These events conspire to low values of ZN/p, a complete film is not formed, only wetting of the surfaces and lubrication relies on the adherence of the molecular constituents, or their 'oiliness' to minimise metallic contact. All journal bearings which are not force fed pass through this boundary lubricated regime during certain phases of their operation. These are periods when viscosity plays little part in frictional behaviour, and it is during these periods that significant surface damage can occur.

At the other end of the scale, a poorly designed assembly may generate an oil pad of such dimensions that churning and sliding of molecular layers in the film generates considerable heat quite apart from the difficulty of ensuring positive location between journal and bearing when film thickness is large. Thus the situation is one of compromise, and the component design process should be sufficiently flexible to allow for the inaccuracies of manufacture and to accept that parameters associated with the third member of the assembly, the lubricant, are of equal if not greater importance if friction and wear is to be minimised.

In terms of lubrication techniques for metallic bearings, oil is the conventional choice although, provided that the greasing regime is reliable, their reduced end leakage makes some self lubricating systems an attractive proposition. Elaborate grooving arrangements are rarely necessary, they impede the formation of an oil film and should certainly be avoided in the lower half of the bearing adjacent to the maximum pressure region. In a one-piece radial bearing, a single longitudinal groove is often sufficient while, in a split bearing, a longitudinal chamber, formed by bevelling the edges of cap and base, will serve. Assuming that dimensions have been properly chosen, the most common type of bearing failure is due to the selection of oil having inappropriate body to satisfy the range of operating conditions which might vary between a temperature extreme due to poor ventilation or a cold environment where fluid approaches its pour point.

Oxidation is always a problem, insoluble products deposited as sediment may clog the oilways and retard cooling due to their insulating effect. Iterative calculations for bearing parameters are simplified by lubricant data which classifies in terms of a tolerance on viscosity at a fixed temperature rather than in a range varying with temperature. The eighteen grade designations of BS 4231[14] bear close relation to kinematic viscosity in centistokes at 38°C, one of the points on the Viscosity – Index system, and the standard lists comparisons with SAE data which is widely quoted for commercial lubricants in this country.

It is instructive for the young designer to surmise as to the lubrication situation in frequently encountered engineering components. A petrol engine connecting rod is one example where totally different regimes occur at opposite ends. At the big end, the speed is sufficient for an oil pad to quickly build up over a relatively wide area, unit pressure is low with load well distributed so that the bearing material can be quite soft. At the small end, however, where reciprocation is through a small angle the speed is too low and pressure too high for bearing modulus to reach a value synonymous with full fluid lubrication. The oil in such locations may not be of much more than molecular proportions and loading is only kept within reasonable bounds by careful attention to the fit of components. In such circumstances it is important that the properties of the bearing material shall allow for as much elastic deformation as can be tolerated. The lubricant has an additional commitment in many applications of this type, that is to provide a cooling facility as part of a distribution system, so that oil quantity is an important parameter if operating temperatures are to be maintained within reasonable limits.

Design procedure

The first priority in this section has been to acquaint the reader with the most commonly used types of journal bearings and with some of the problems involved in their efficient lubrication. It seems appropriate to conclude by reviewing various steps which might form the basis of a design procedure. The simplest method, when dimensions and lubricant properties are known, is to use a bearing modulus curve (or assume a value of ZN/p) which can be substituted in the McKee equation. This yields an expression for Coefficient of Friction (Imperial Units) as

$$\left[\frac{473}{10^{10}} \left(\frac{ZN}{p} \right) \frac{D}{C} \right] + k \qquad (8)$$

where D and C are journal diameter and diametral clearance respectively, k varies with journal size. A balance can then be struck between heat generated and heat dissipated to yield an operating temperature.

It will be noticed that this empirical basis makes no allowance for variation in friction coefficient or for changes in film thickness due to changes in position of the journal centre. Figure 13.6 shows typical relationships and makes the assumption that journal centre

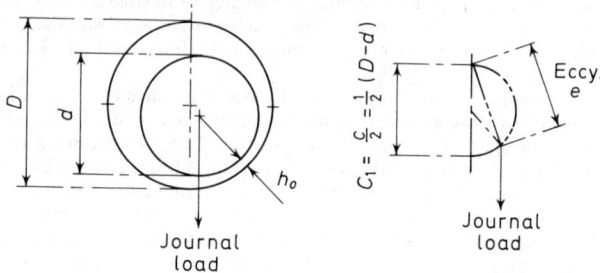

Figure 13.6

lies on an arc locus of diameter $C/2$. These parameters are included in methods of determining horsepower lost in friction by using design charts plotted in terms of the eccentricity ratio or attitude

$$\varepsilon = 1 - \frac{2h_0}{C} = \frac{2e}{C}$$

where h_0 is minimum oil film thickness.

A typical chart for a given length/diameter ratio, but plotted in terms of radial clearance C_1, is shown in Figure 13.7. Enter chart on the dotted line having assumed a value for h_0 and calculating ε so that the viscosity μ_1 can be determined. Subsequent interpolation of the full line values will allow calculation of friction force. More recent work on short bearing approximations is based on the Sommerfeld solution to Reynolds equation and is referenced at the end of this section. Particular attention should be paid to the assumptions made in these proposals.

Finally, mention should be made of the design chart methods typified by Engineering Sciences Data 66023. This recognises the more frequent use of pressure or force feeding and proposes solution methods for full bearings where a hydrodynamic situation may be expected to develop. A particular difference lies in the considerably larger film thicknesses which can be tolerated and the increase in operating pressure requires a more precise evaluation of oil temperature. Reserve should be exercised in the application of test data since some of this is derived from experimental work on test rigs whose rigidity would be quite impracticable in commercial bearings. Such results may predict low friction coefficients or minimum film thickness, based on geometrical accuracy and surface finish, outside conventional manufacturing specifications.

ROLLING CONTACT BEARINGS

The ball or roller bearing journal bearing consists basically of rolling elements which run on curved tracks in the inner and outer races, the elements being guided and separated by a spacer or cage. The most widely used rollers have flat and parallel ends, accuracy to within 0·002 5 mm on diameter and 0·005 mm on length is typical. The accuracy of balls is within 0·002 5 mm on both diameter and sphericity. These values represent conventional manufacturing practice, but the set which makes up a given bearing is graded to considerably finer tolerances in order to assist in uniformity of load sharing. The track profiles vary slightly between one manufacturer and another according to the type of bearing, the load which is to be carried or the preferred mounting arrangement. External dimensions are standardised in BS 292,[15] the module being the outside diameter of the outer race and it is conventional to specify millimetre dimensions for all new bearing layouts.

Race materials are direct hardening chrome-carbon steel bars, or for larger sizes a

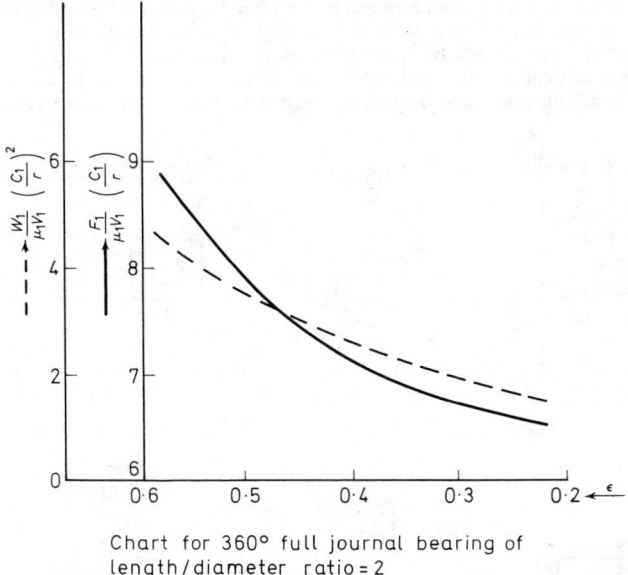

Chart for 360° full journal bearing of length/diameter ratio = 2

Figure 13.7

press forged mild steel which is suitable for subsequent carburising. The alloy chrome steel used for balls and rollers is for the former either cold pressed or hot forged and rough ground before hardening, whereas roller material is centreless ground and parted off before hardening. The object is to achieve hardness values at the working surfaces of Brinell 630 or HD 800, depending upon the manufacturer. Various non-ferrous metals, synthetics and non-metals are used for annular cages, solid construction or pressed in sections and riveted according to the method of guidance or location preferred.

Types of bearing

The general proportions of the types most commonly used are compared in Figure 13.8, the diagrams being to scale for a given series and a common bore size.

The single row rigid ball journal is the most widely used rolling contact bearing, its grooved tracks conform closely to the ball contour so as to provide maximum support and consequently it will tolerate only small errors of misalignment. There is thus a compromise between easy running and adequate capacity. The basic form of this type has remained unchanged for many years but improvements in the control of material properties and in manufacturing methods have allowed progressively larger balls to be used within a given cross-section. For heavier loads, the double row types are specified with each row running on separate tracks. When alignment is a problem, the single row type can be provided with periphery ground in spherical fashion to suit a housing, when two rows of balls are provided there are separate grooved tracks on the inner race and a spherical bore to outer race, so permitting the rolling elements to swivel.

Angular contact ball bearings are typified by arrangements for the balls to make contact with the sides, rather than the bottom of the tracks, so permitting these types to carry a combination of journal (or radial) and axial (or thrust) loads. It is necessary to be precise as to the operating conditions, unless the thrust load is uni-directional and always exceeds the journal load arrangements should be made to use single row bearings in pairs, adjusted against each other to keep the balls up to their tracks.

Last in the series of ball bearings, come those types designed to carry *thrust* only, usually in a vertical direction. For all but the lightest loads, the tracks are grooved. In addition to double row types, which take thrust in both directions, a variety of spherical

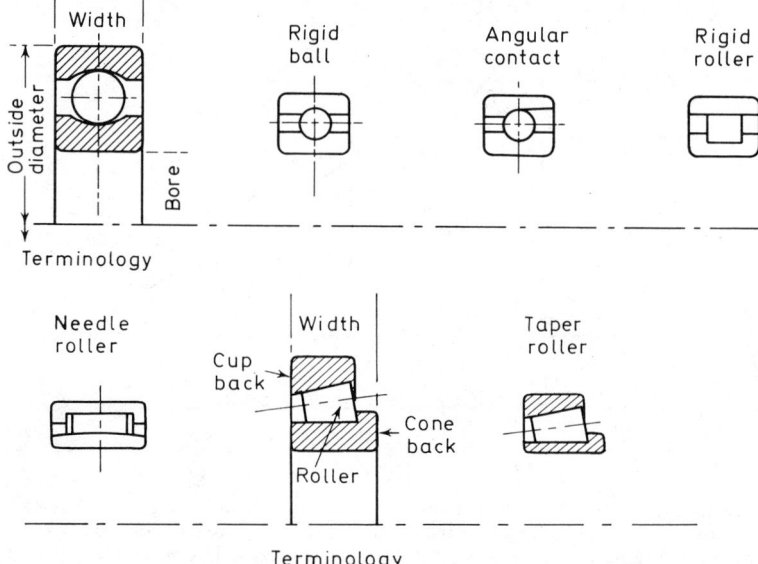

Figure 13.8

seating arrangements are available including parallel mounting sleeves and nuts. By this means the adjustment problem common with double thrust bearings can be avoided.

The single row parallel roller journal is characterised by line contact and so offers a higher load capacity than a ball journal of the same size. The conventional bearing employs rolling elements whose length is equal to their diameter, these are guided on concentric tracks in the inner race, while the outer race is a plain annulus. There is thus no facility for locating the shaft endways, but a variety of alternative lip or circlip arrangements can be provided in the outer race. These are intended to facilitate assembly or dismantling and can be used to provide location in situations where end thrust is intermittent.

The needle roller journal bearing differs from other roller types, in that true running depends on controlled circumferential clearance between the rollers, whose length/diameter ratio might range between 5 and 10. There is no facility for dealing with other than journal duties or for provision of location duties. The most economical arrangement is a complement of loose rollers, axially located between end washers, using existing surfaces as raceways, but it sometimes is not possible to arrange for adequate hardness and fine surface finish on the contacting surfaces. In these circumstances it may be better to specify full complement drawn cup bearings, although this requires the provision of suitably designed housings, retention of rollers being ensured by turned in lips or by a closed end. Load sharing problems and the control of tolerances, which at one time caused the allowable speed range to be restricted, have now been largely overcome and since capacity differs little from other bearings of the same outside diameter the specification of needle rollers usually allows a larger shaft to be used.

The simple construction of the *taper roller bearing* is based on the formation of a cone which envelops the rolling elements and raceways, so ensuring true rolling and an obvious capacity for handling a combination of radial and thrust loads. The inherent property is that of alignment between the outer race or cup and the inner race or cone which can be ensured without the need for guidance from the cage. It is important to determine accurately the relationship between radial and thrust loads since manufacturers specify different cone operating angles for variance in this ratio. Because of the tapered construction the imposition of a purely radial load will induce thrust and it is

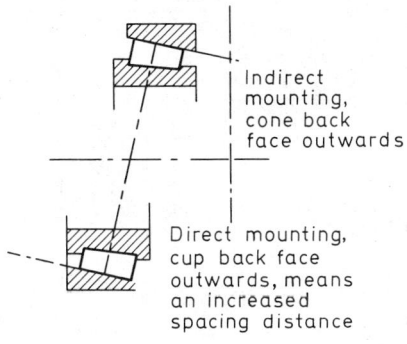

Indirect mounting, cone back face outwards

Direct mounting, cup back face outwards, means an increased spacing distance

Figure 13.9

conventional to mount single row bearings in pairs with their tapers opposed. It must be resolved which bearing has to carry any excess thrust and this suggests the importance of design to achieve maximum rigidity.

The definition of direct and indirect mounting, an important design concept, is explained in Figure 13.9 and the increasing use of these bearings has led to their dimensional standardisation as detailed in BS 3134.[16]

Metric annular bearings

In accordance with international standards, metric annular bearings are made in different *diameter* series, so that for a given bore a range of outside diameters is available. Further, since different widths are available for these diameter series, each set of sizes can be referred to in terms of a *dimension* series, conventionally designated by a two digit number indicating the width and diameter code respectively. Titles such as Light, Medium and Heavy have often been used to classify a dimension series. The complete ISO identification code has three main sections for bore, type and dimension series, with supplementary data indicating internal fit, lubrication, cage type etc. Some manufacturers add a further number, which when multiplied by 5 indicates the bore size mm. Figure 13.10 indicates to scale typical boundary dimensions for various dimension series and while

BS 292[15] lists standardised external dimensions for journal bearings it must be remembered that manufacturers have freedom, within the recommended values, to allow different proportions for internal components.

There should be little difficulty in formulating choice criteria provided that the issues raised in the specification are dealt with methodically. A typical check list might include the following questions;

   (i) What is load, speed and desired life?

  (ii) Is there an incidence of shock, deflection at supports or a requirement for end float?

 (iii) What dimensional limitations exist?

 (iv) Is the environment polluted to the extent of requiring special sealing?

  (v) Will the temperature differential problem necessitate special lubrication arrangements?

If alignment is guaranteed, rigid bearings can safely be specified and these types will cope with the highest speeds likely to be encountered. The centrifugal force due to rota-

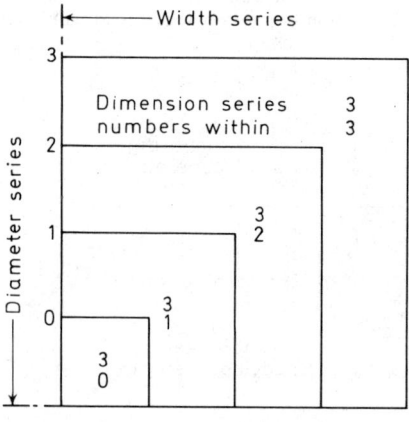

Figure 13.10

tion is sometimes calculated in terms of an equivalent load to be added to the imposed external load and such dynamic effects are sometimes used to set an upper limit to speed. Ball types will be preferred to rollers in all cases where loading is within their capacity unless there is a mounting or dismantling problem.

Life considerations

The investigations of Heinrich Hertz into the deformation of curved elastic bodies in contact were accepted by early researchers as a basis for predicting the behaviour of rolling elements. In their different fields the work of Stribeck, Palmgren and Goodman led to the definition of rating formulae involving the number of balls, the ball size and contact angle or alternatively to the idea of a conformity or relative load capacity factor based on the average stress in the contact area.

For most practical purposes it is sufficient for the designer to calculate a static equivalent load, sometimes called the equivalent bearing load, the simple formula found in manufacturer's catalogues involves multipliers for radial and thrust components intended to allow for dynamic effects. It is at this stage that life formulae need to be taken into account, since the capacity of a rolling contact bearing is determined by the hardness of the contacting surfaces and their ability to retain shape and finish in the working environment. The calculated life of a rolling contact bearing is defined as the period (number of revolutions or working hours at some constant speed) which 90% of a group will reach or exceed before showing signs of fatigue.

$$\begin{array}{l}\text{Typically Life } L \text{ in millions}\\\text{of revolutions}\end{array} = \left\{\frac{\text{load capacity for bearing with life of } 10^6 \text{ revs}}{\text{equivalent load } P}\right\}^3 \quad (10)$$

so that life is inversely proportional to the cube of load.

The basic dynamic capacity shown in manufacturer's tables will therefore typically correspond to the load to which a bearing may be subjected during a life of one million revolutions, such values may be regarded as a means of comparison for the relative strengths of similar types. In practice life is most difficult to forecast; quite apart from the difficulty in rationalising load/life ratings when the loading cycle varies, the commonly accepted theoretical basis for stress repetitions to initiate failure varying inversely up to the ninth power of stress is unrealistic for rolling elements.

Mounting rolling contact bearings

The mounting problem for rolling contact bearings is a relatively simple one, provided the manufacturer's instructions are followed. In the process of manufacture a predetermined amount of diametric or running clearance will be arranged to allow for size variations due to fit or to the effect of temperature. Race seatings must be parallel, truly circular and of sufficiently robust construction to avoid distorting the rolling elements. Similarly, any abutment surfaces must be truly square with the axis of rotation. The suggestions shown for fit should be strictly adhered to, in particular the slight interference recommended for the rotating race in order to prevent creep with respect to the mounting surface.

Difficulties sometimes arise if insufficient attention is paid to alignment and location, a single shaft length should be located endways by not more than one bearing in each direction, although on occasions a single bearing can deal with the location in both

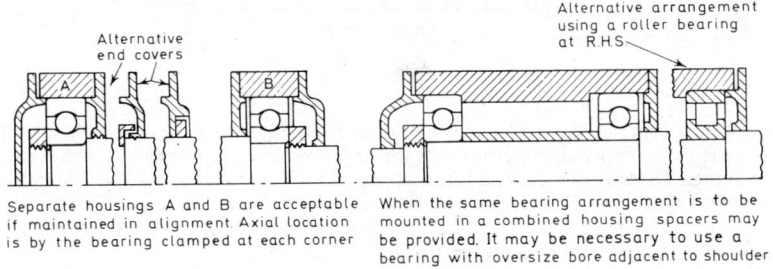

Separate housings A and B are acceptable if maintained in alignment. Axial location is by the bearing clamped at each corner

When the same bearing arrangement is to be mounted in a combined housing spacers may be provided. It may be necessary to use a bearing with oversize bore adjacent to shoulder

Figure 13.11

directions. The most effective method of explaining various solutions of the mounting problem is by reference to Figure 13.11 which shows the alternative arrangements recommended for separate and combined housings. The illustration shows also alternative methods of sealing and protection. A series of grooves turned in the bore of an end cover will form an effective seal when filled with grease, quite adequate for clean and dry conditions. For a wet or dirty environment a labyrinth seal may be effective, although it has become customary to standardise on a spring mounted wiping surface mounted within a lipped metallic frame which can be pushed into a counterbore in the end cover.

Pre-loading of rolling contact bearings is defined as the practice of setting up an initial internal loading between rolling elements, so as to take up low load deflection and remove unwanted clearance before the external load is applied. The intention is to increase the rigidity of the bearing assembly, since peak working accuracy and repeatable performance from a machine tool demands considerable stiffness in each element of the structure. Once the operating and mounting conditions are known, a suitable pre-load can be determined and many standard bearing types, albeit manufactured to extra precision specifications, can be rated for pre-load in terms of length of axial interference. An assembly involving single row angular contact bearings need only be mounted in similar fashion to that for which pre-load of the pair has been determined during manufacture. Alternative arrangements for radial pre-loading involve forcing the bearing along a tapered ring to expand the inner race. The behaviour of a bearing in adjusting to the combined effect of mounting and applied load is sometimes called compliance. It is important to distinguish between the experimental deflection results from a test rig (where the housing may be over-rigid and supporting shaft solid) and those likely to occur in practice, where the effect of pushing outer race into housing or mounting inner race on a hollow shaft will produce a different pattern of load sharing.

A comparison between the load/deflection relationship for conventionally mounted and pre-loaded bearings is shown in Figure 13.12. Notice the non-linear variation in

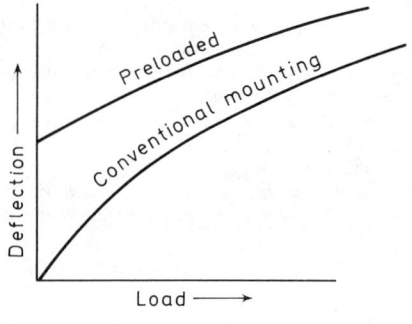

Figure 13.12

the former type; as the load increases the rate of deflection increase declines, which suggests that an operating region exists where pre-loading can be applied to maximum advantage. The key to assessment is the influence of the load sharing factor on deflection, which is a function of load and acceptable bearing play. Excessive pre-load will increase running torque, may cause lubrication problems because of the higher operating temperature and significantly reduce bearing life.

A simple method of explaining the mechanism of pre-load is to consider the assembly as a problem in applied mechanics. Figure 13.13(a) shows in free body outline the rolling elements replaced by springs whose rate or stiffness is K and having a free length $x + 2x_1$. When mounted on a sliding pin, which represents the shaft, the effect of constraint to a length x causes each spring to exert a force W. If subsequently an external load of sufficient magnitude is applied as shown in (b), it is easy to see that the right-hand spring exerts no force on its constraints, equilibrium being maintained by application of force $2W$ to the slider. Notice how the rate of the assembly is increased by the use of helical springs, which have nominally a linear load/deflection relationship. The benefits are even more marked in a rolling bearing assembly where such relationships are not linear. It is not

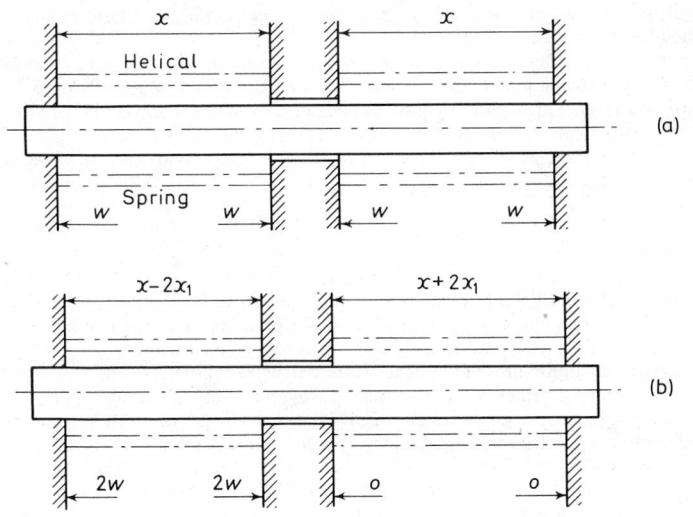

Figure 13.13

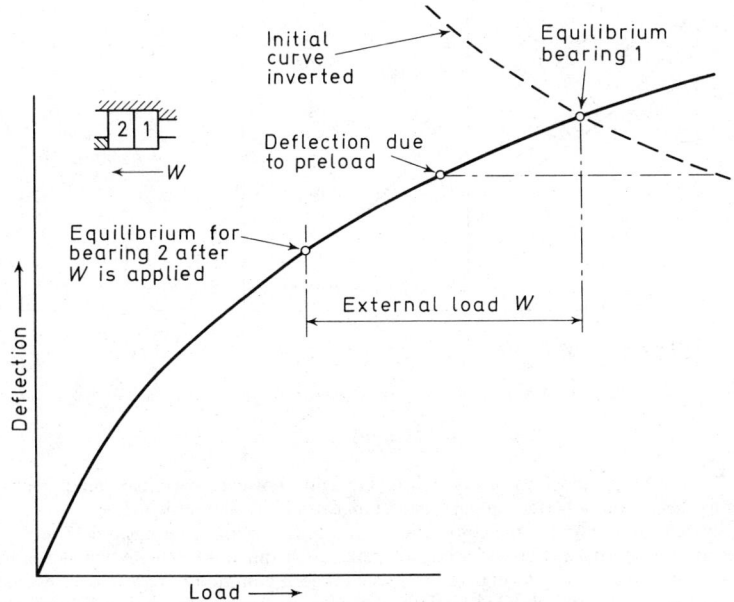

Figure 13.14

difficult to calculate the point of disappearance of pre-load when the working load has reached a certain value.

Figure 13.14 is drawn to scale for an angular contact pair and indicates the variation of individual bearing loads when an external load is applied to a pre-loaded assembly. The use of shims or springs for pre-loading is widely advocated in some quarters but equally effective results can be obtained by using matched pairs in the manner prescribed by the manufacturers; this will also reduce the danger of over-loading during the assembly process.

Lubrication

The prime functions of a lubricant are to reduce friction and prevent the ingress of foreign matter. Since true rolling is unobtainable in the pressure area, the sliding friction due to conformity and deformation must be countered by lubrication, quite apart from the need to protect highly finished surfaces from corrosive influences. The safe speed limits up to which rolling elements can operate, is set by the ability of the lubricant to maintain a film, although since a great variety of oils and greases will perform adequately in the majority of situations the choice is frequently made on the basis of external considerations. Oils are more effective than greases, they carry away heat more efficiently, feed readily

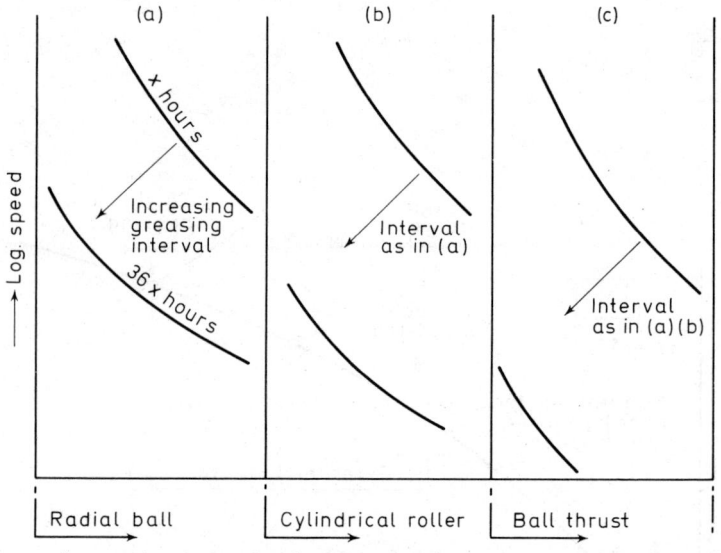

Figure 13.15

into contact areas and flush away debris. Greasing however permits a much simpler housing design and a better seal against the ingress of dirt and moisture.

A straight mineral oil is the conventional choice for rolling bearings, selection being in terms of viscosity at the operating temperature, a range 90–450 Redwood Seconds (21.5–117 cS) might cover the high to low speed range. A common error is to over-estimate the amount of oil required. One fortieth a drop, properly dispersed, will lubricate a small bearing rotating at say 1000 rev/min for one hour. The particular problem in grease manufacture is to control crystallisation of the soap so as to provide a fibrous structure which will immobilise quantities of lubricant and allow it to weep at a desired rate.

There is a close relationship between the oils used for greases and those for conventional lubricants, although the additional refining required tends to reduce Viscosity Index, lower values of which signify a larger change of viscosity for a given temperature variation. A tenacious or clinging nature will assist in adherence but should not be so pronounced as to impede rolling action, or from the opposite view, be unable to cope with the considerable agitation tending to cause separation. Some greases will not regain their initial state after working. The choice is usually resolved in terms of operating temperature and speed.

Cup greases (*lime soap*) and sponge greases (*soda soap*) will cover most applications in which operating temperature maximum is 49°C and 93°C respectively. A particular advantage in using greases is the ease by which inhibitors or additives can be included. Anti-oxidants, corrosion inhibitors and pour depressants are typical, and the well-known extreme pressure additives provide a unique service when contact loads are large.

Figure 13.15 is a typical chart showing how the greasing interval varies according to the type of bearing. This is an important economic consideration, the more often greasing is necessary the more it is desirable to reduce the labour required to perform the task by improved housing design.

TOOTHED GEARING

Involute properties

It is not the intention of this section to provide a design text for toothed gearing, but certain properties of the involute profile will first be explained because their correct understanding will materially assist in defining the theory of gear measurements and the basis of corrected teeth.

The simple involute definition as the locus of a point on a straight line rolling round a circle, or of the curve described by the end of a cord being unwound from a circle is shown in Figure 13.16 to indicate the geometrical construction and certain properties

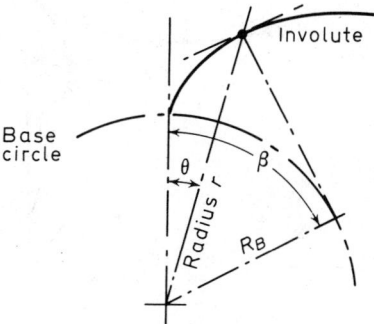

Figure 13.16

related to the base circle whose radius is R_B. When $r\theta\beta$ refer to a point on the involute,

$$\theta = \beta - \tan^{-1}\frac{(r^2 - R_B^2)^{\frac{1}{2}}}{R_B} \tag{11}$$

and the length of the generating line is the same as the part of the base circle circumference subtended by β. Hence the polar equation of the involute curve is found by substituting $(r^2 - R_B^2)^{\frac{1}{2}}/R_B$ for β and, since this value is also the tangent of the angle $\beta - \theta$, we see that the generating line is the normal to the involute curve and the radius of curvature of the involute at a given point is the length of generating line. If, then, the shape of the involute curve is dependent only on the size of base circle the angular motion of one involute will be transmitted to another regardless of their centre distance; the rate of motion being in inverse proportion to size which is the same as saying in inverse proportion to tooth numbers. All points of contact will lie on the path which is the common tangent to the base circles.

An involute does not have a fixed pitch circle; any diameter on it is a potential pitch diameter, with the size of pitch circles and the pressure angle depending on base circle dimensions and centre distance. When the involute form is used for gear profiles, various curves are developed from the same base circle to form the profiles of successive teeth. The distance between these involutes, measured along a tangent to the base circle, is always the same and equal to

   (i) Length of arc of base circle between origins

   (ii) Circumference of base circle divided by number of teeth, called base pitch.

Study of the profile construction will indicate the variation in length of generators and thus of the radii of curvature, so that when two involutes are in contact, a combined rolling and sliding action must occur because of the varying lengths of equal angular increments.

Figure 13.17 shows equal involutes and *ab/gh* are typical contact pairs. Note the distance by which the profiles must slide against each other to make up the difference in lengths

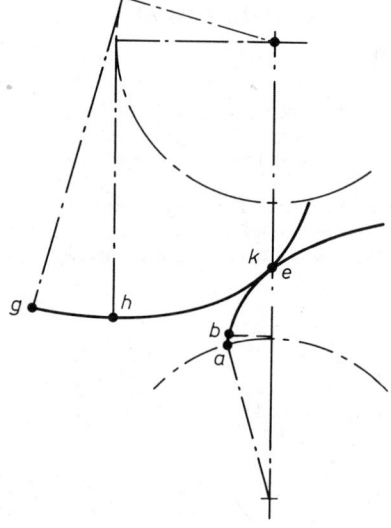

Figure 13.17

between *ab* and *gh*. It should be evident that the rate of sliding is constantly varying, while there is a similar numerical value of sliding velocity for the pair it is distributed over different lengths of profile.

Tooth proportions

The layout shown in Figure 13.18 is typical of a mating pair working at a standard centre distance and indicates basic tooth proportions and nomenclature. It will be seen that the maximum allowable radius for the wheel addendum circle is O_2M if interference is to be avoided. The maximum permitted wheel addendum is

$$O_2M - O_2P = O_2P[\sqrt{(1+(O_1P/O_2P)[(O_1P/O_2P)+2])}\sin^2 \phi)-1] \qquad (12)$$

The actual size of tooth used is determined by the power to be transmitted and may be much smaller than the maximum available. If we equate the expression for wheel adden-

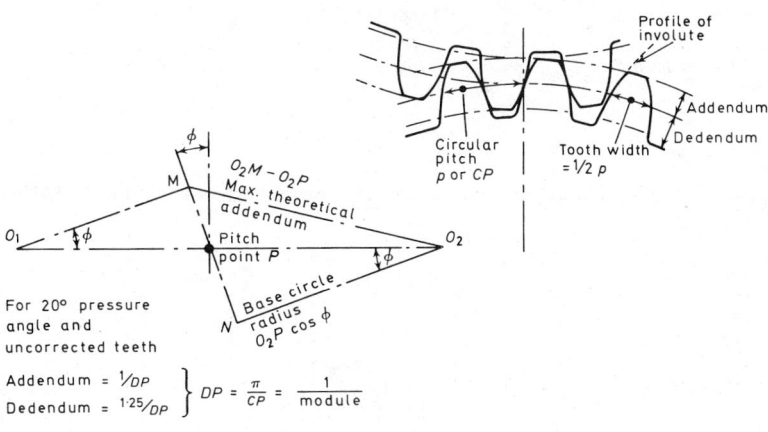

Figure 13.18

dum to $x . m$, where x is a fraction or percentage and m is module, equations for minimum numbers of teeth t or T can be derived, such as

$$T = \frac{2x}{\sqrt{(1+A \sin^2 \phi)-1}} \quad \text{where } A = \left(\frac{t}{T}\right)\left[\left(\frac{t}{T}\right)+2\right] \qquad (13)$$

Therefore, for a pinion of 20° pressure angle having a tooth addendum of one module ($x = 1$), the minimum number of pinion teeth, when meshing with a wheel which provides a velocity ratio of 3:1, is

$$t = \frac{2 \times 1 \times \frac{1}{3}}{\sqrt{(1+(\frac{1}{3})[(\frac{1}{3})+2] \sin^2 (20))-1}} = 15 \cdot 1 \text{ say 16 teeth}$$

Examination of the involute profile will indicate its unsuitability, close to the origin on base circle, as a tooth shape because of the rapidly changing radius of curvature which tends to cause high contact stresses. Some authorities suggest a minimum total of 60 teeth in a standard pair if space permits, and suggest a scheme of centre distance extension for low tooth numbers. There are limitations in the use of tooth generating processes, the effect of which can be assessed by considering the engagement between a pinion and a rack.

The three standard methods of measuring tooth pitch are defined below:
  (i) Circular pitch CP or p is the distance in inches measured round the pitch circle between similar points on successive teeth.
 (ii) Diametral pitch DP or P is the number of teeth per inch of pitch diameter.
(iii) Module m is the reciprocal of DP but based on a millimetre pitch diameter. Since $1/DP$ is also the standard proportion for addendum we speak of a standard addendum of one module, or some proportion of this in the case of corrected teeth.

$$\text{Summarising, } CP = \frac{\pi(PCD)}{\text{No teeth}} \quad DP = \frac{\text{No teeth}}{PCD} \quad m = \frac{PCD}{\text{No teeth}}$$

The relationship between standard pitches is shown in Table 13.1.

In American practice there are four standard spur gear forms, quite apart from a number of other types representing the proposals of gear cutter manufacturers. The basic

Table 13.1

Circular pitch in	Diametral pitch	Module mm	Circular pitch in	Diametral pitch	Module mm	Circular pitch in	Diametral pitch	Module mm
0·250	12·566	2·021	0·618	5·079	5	1·236	2·540	10
0·261	12	2·116	0·625	5·026	5·053	1·250	2·513	10·106
0·309	10·159	2·5	0·628	5	5·079	1·256	2·5	10·159
0·314	10	2·540	0·742	4·233	6	1·360	2·309	11
0·371	8·466	3	0·750	4·188	6·063	1·484	2·116	12
0·375	8·377	3·031	0·785	4	6·349	1·500	2·094	12·127
0·392	8	3·174	0·865	3·628	7	1·570	2	12·699
0·432	7·257	3·5	0·875	3·590	7·074	1·731	1·814	14
0·448	7	3·628	0·897	3·5	7·257	1·750	1·795	14·148
0·494	6·349	4	0·989	3·174	8	1·979	1·587	16
0·500	6·283	4·042	1·000	3·1416	8·084	2·000	1·570	16·169
0·523	6	4·233	1·047	3	8·466	2·226	1·411	18

standards need no other reference than details of the rack which defines their interchangeability and can be summarised as

14½° full depth,   total depth $2·157/DP$
20° full depth,   differs only in respect of pressure angle and fillet shape
14½° composite,   involute above pitch circle, cycloidal below it
20° stub,   total depth $1·8/DP$

Provided that the condition for uniform velocity transfer is maintained i.e. that the common normal of the profiles at the point of contact shall pass through the pitch point, it is possible within limits to choose an arbitrary shape for one profile and then determine a mating or conjugate shape to suit. The practical considerations limiting this choice will be considered later under Gear Cutting. A particular virtue of the involute profile is that a pairing may be run at non-standard centres without destroying conjugate action. Quite apart from unintentional inaccuracy in manufacture or location it is sometimes necessary to mount deliberately at extended centres to simplify gearbox design.

In the case of standard gears, interference will not normally be a problem, although steps taken to prevent it may limit the size of the mating gear and thus the possible speed ratio. Remedial methods may include increase of pitch radius or pressure angle, but the most suitable procedure is to utilise a scheme for corrected profiles. A typical scheme which is based on BS 436[17] ensures full involute action by variation of addendum with pitch and tooth number according to formulae for external gears, such as

$$\text{'a' for pinion} = (p_N/\pi)(1+k_p)$$
$$\text{'A' for wheel} = (p_N/\pi)(1+k_w)$$

where p_N is normal pitch, k_p and k_w are usually equal but of opposite sign and determined by tooth number. Typically when $(t+T) \sec^3\sigma$ is not less than 60, k_p is the greater of $0·4(1-t/T)$ and $0·02(30-t \sec^3\sigma)$ where σ is helix angle.

Simplicity and compactness should be the keynote of gear train design. When the train is to be gearbox enclosed it is often desirable to replace the conventional train of Figure 13.19(a) by the reverted arrangement shown at (b). These layouts are drawn to scale and

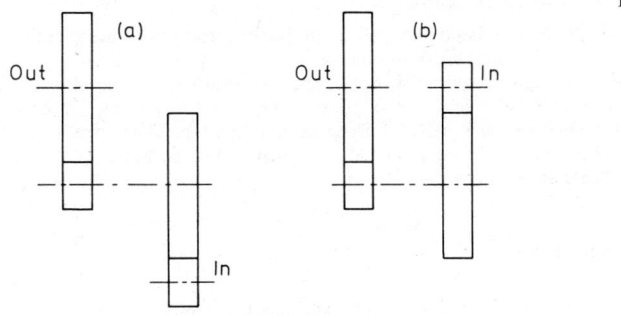

Figure 13.19

V-BELT REDUCTION FROM 2-SPEED MOTOR THROUGH HEADSTOCK PROVIDES 16 SPEEDS AT MAIN SPINDLE. A TYPICAL MAXIMUM REDUCTION IS 4·4/1 FROM MOTOR SPEED TO CHUCK SPEED.

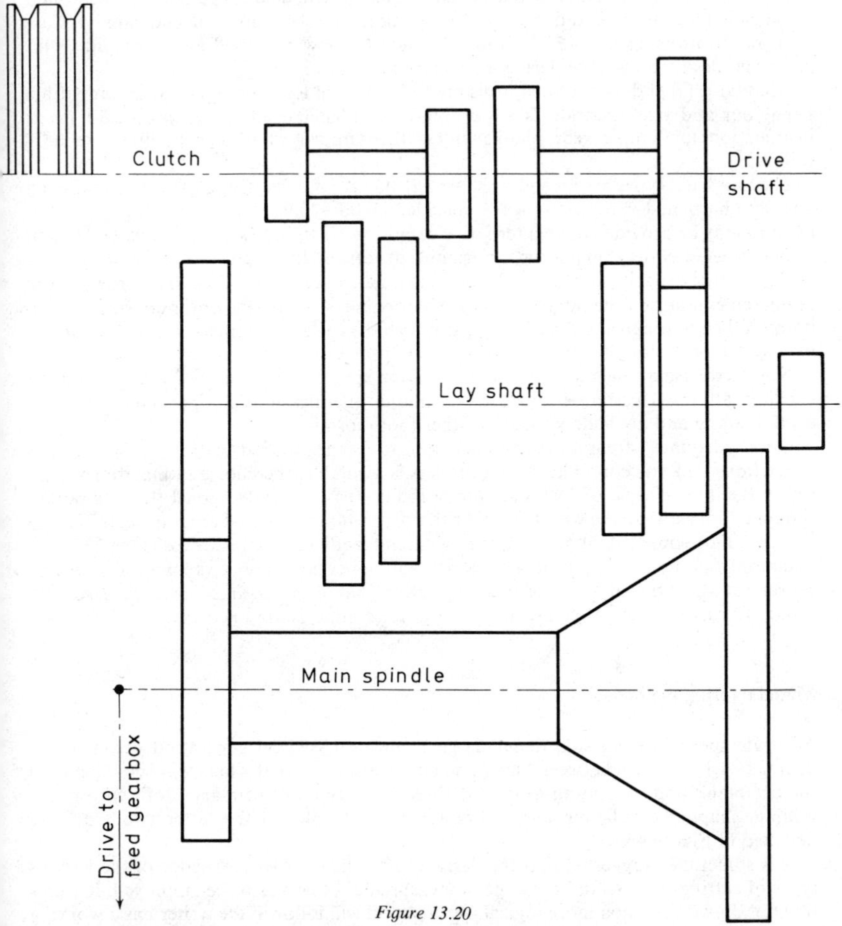

Figure 13.20

the saving in space is obvious, although further restrictions are thereby imposed in the selection of pitch and tooth numbers

Some machines require a range of speeds to satisfy a range of operating conditions, usually involving a selection of output speeds from a given input. The relationship between such speeds is called a progression and a typical arrangement for a lathe head-stock is shown in Figure 13.20. Arrangements must be made to allow speed changing while gears are in motion.

Material considerations

The choice of material for metallic gears is not always simple since it depends upon availability and convenience for raw material shape, the property amendments which subsequent treatment may produce and the manufacturing processes available. Hardness is a significant parameter; easy cutting is possible at 200 Brinell whereas grinding would be employed to finish materials hardened to 500 Brinell. The range 300–350 will often define adequate strength and is a commonly quoted limit for good machineability. Cast iron specifications to BS 1452[18] (grades 12 and 14) are widely used for gear blanks with higher grades to BS 821[19] offering a stronger alternative.

Nodular (SG) and innoculated irons are best reserved for situations where compatible behaviour and size retention is important, since their considerably higher cost needs justification for simple gear blanks unless their improved tensile qualities are fully utilised.

The four basic types of gear steels listed in BS 436 are the cast, forged, through hardened and case hardened varieties typically specified in BS 970/71. Notice that a given specification may be provided in an alternative condition, thus En 8 is a widely available plain carbon steel often used in a forged (normalised) version, in which state it provides maximum bending capacity and gives a better performance than when used in a hardened and tempered condition. The intention of case hardening is to provide optimum surface conditions allied to tough and ductile core, although the surface treatment may subsequently involve the provision of sophisticated finishing processes for the tooth profiles.

Non-ferrous materials are not often used for spur gears, mainly on the grounds of expense, although certain bronzes have unique qualifications for particular types of tooth loading and are widely used for other tooth forms.

Once adequate strength requirements have been satisfied within ruling section restrictions and an economic cutting process is available, the choice of material should reflect the dual criteria of cost and compatibility. It is generally agreed that the pinion should be harder than the wheel, to cover the larger number of load repetitions, otherwise the search for compatibility often means a careful study of case histories. Table 13.2 lists commonly used pairings for metals and gives an indication of relative material costs and power ratings. Optimisation of 'best condition' has been assumed in each case with Grade 12 cast iron values given unity ratings for comparison purposes.

Manufacturing methods

Accurate speed transmission demands gear cutting based on generation or on repro-ducing relative motion between blank and cutter similar to that which will occur between the cut blank and its mating gear. The three main types of primary tooth cutting are milling, shaping or planing and hobbing. The principles of the latter two types are depicted in Figure 13.21.

It is sometimes suggested that the designer need have little knowledge of the various types of cutting process and the scale of measurement that has to be employed. It seems obvious, however, that a more logical specification will follow if the writer has a working knowledge of the problems involved and appreciates the degree of accuracy which it is possible to maintain.

Milling is sometimes used as a preliminary to other processes for large gears, it is typical of intermittent cutting, using a rotating cutter with each tooth repeating the performance of its predecessor. The *Gleason* method is used extensively for generating curved tooth bevel gears involving a tool similar to a milling cutter and the cutting of Hypoid gears follows the same general principles, although a separate cutter may be needed for pinion flanks. There is extensive use of milling procedures in the manufacture of worms.

The planing process (Figure 13.21) is a clear indication as to how a rack shape can be

Table 13.2

Pinion	En number	8		9		27		34		40B
	Power rating	1·4		2·1		3·1		10		7·5
	Cost rating		1·25		1·25		4·5		5	9
Wheel	En or type	CI(12)		8		9		19		19
	Power rating	1·0		1·4		2·1		6·5		6·5
	Cost rating		1·0		1·25		1·25		2	2

Note that an initial pairing for moderate loading En 8/9, can be replaced by a higher strength alternative En 9/27, which provides improved quality at about the same percentage increase in cost.

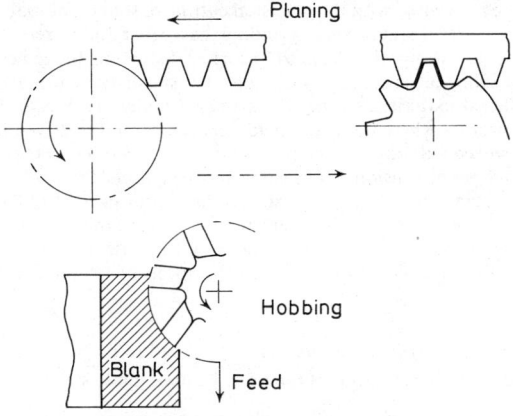

Figure 13.21

used for tooth generation. When the *Sunderland* process is used for spur and helical gears, arrangements are made for blank to rotate at the same linear speed as the longitudinal movement of cutter. During the working stroke horizontal blank rotation and vertical cutter translation proceed, until after a tangential advance of one pitch or thereabouts, the cutter moves out of contact when rotation stops and these phases are repeated until one revolution is completed. The process continues with gradually increasing depths until full depth teeth are provided. In the alternative *Maag* process, where work is mounted with axis vertical and work transverses from side to side, cutting takes place on the downward stroke. The essence of the shaping process is based on the premise of conjugacy between gears generated by the same basic rack. The circular cutter is similar to one member of a meshing pair and toothing is effected by reciprocating cutter across face of work while feeding radially to required depth.

The *Fellows* machine uses a rotating table in conjunction with a vertical reciprocating cutter, a cam mechanism moves the work in and out of mesh. In the *Sykes* machine, with work mounted horizontally, the cutter motion combines horizontal reciprocation with rotation at the speed of a comparable meshing pair. The shaping method is particularly suited to the manufacture of internal gears and BS 2887[20] provides details of pinion type cutters.

The most accurate gears are produced by the hobbing process which is a continuous generation with the hob or cutter carried at an angle to work axis so that its threads mesh with the teeth being cut. The hob constitutes the pinion of a pair, similar to crossed helicals, whose speeds of rotation are in inverse ratio to number of teeth in work and number of threads in hob. The thread profile of the latter makes contact with teeth on work in a series of points as the cutting proceeds; this is due to radial feeding to required depth and traversing across workpiece face in a direction parallel to the axis of work rotation. Worm wheels can be hobbed in conventional fashion, but more commonly by special generating machines. Cylindrical hobs have a helical configuration similar to a screw thread, their cross section being similar to rack teeth of the same pitch. Standard dimensions are shown in BS 2062[21]. The standards of accuracy demanded by modern practice mean that hob teeth are frequently finished by a subsidiary process such as grinding after the initial gashing operation. Accuracy in change gears and in their mounting considerably affects accuracy of the work, for example, profile and pitch errors can be attributed to either the change or the table driving gears.

Grinding and Shaving are the finishing processes most commonly used. Perhaps the best summary of the former lies in its intention to provide, on hardened surfaces, grades of accuracy which are associated with the precision cutting of softer materials. Apart from its use in precision cutter manufacture, grinding has often to be resorted to when hardening follows initial cutting with the need to correct distortion due to heat treatment. The properties of shaving cutter materials do not often satisfy the cutting requirements typical of fully hardened materials, so that the process usually precedes heat treatment. Shaving is another process where the blank held between centres is driven by the cutter as it traverses, the crossed axis angle of the pair is critical depending on the type of finish required. BS 2007 suggests dimensions for circular shaving cutters. It must be remembered that while finishing operations will improve surface finish the amount of metal removal will not significantly correct dimensional errors arising from the initial cutting operation.

Many gear cutting machines are similar to screw cutting lathes in that change gears are used to arrange the necessary ratio between the speed of two shafts. For a screw cutting lathe

$$\frac{\text{Product of driver tooth numbers}}{\text{Product of driven tooth numbers}} = \frac{\text{Lead to be cut}}{\text{Lead of guide screw}}$$

Thus for a milling machine set up to cut a helical groove, the right hand side of the equation reads

$$\frac{\text{Lead of helix to be cut}}{\text{Lead of table feed screw}} \times \frac{1}{N}$$

where N is number of revolutions of first driving gear per revolution of work, usually 40.

The method of continued fractions is a convenient device for determining the numbers of teeth in change gears. Taking the 'fraction' of 0·3286, written 3286/10 000, the procedure is to divide denominator by numerator and then to divide remainder into numerator, continuing until there is no remainder.

10000/3286	yields 3	remainder 142
3286/142	yields 23	remainder 20
142/20	yields 7	remainder 2
20/2	yields 10	

Take these quotients to yield fractions of $\frac{1}{3}$

$$\frac{23(1)+0}{23(3)+1} = \frac{23}{70}$$

$$\frac{7(23)+1}{7(70)+3} = \frac{162}{493}$$

$$\frac{10(162)+23}{10(493)+70} = \frac{1643}{5000}$$

which is the required ratio.

It is necessary then to determine a suitable combination of gear tooth numbers to yield the required ratio, say

$$\frac{31 \times 53}{50 \times 100}$$

When the fraction is larger than unity, say 2·317, the first step is reversed by:

dividing 2317 by 1000	yields 2	remainder 317
1000/317	yields 3	remainder 49
317/49	yields 6	remainder 23

complete solution yields quotients 2 3 6 2 7 1 2 and the fractions are

$$\frac{2}{1}, \frac{3(2)+1}{3(1)+0} = \frac{7}{3} \text{ note the reversal, } \frac{6(7)+2}{6(3)+1} = \frac{44}{19}$$

and similarly 95/41, 709/306, 804/347, 2317/1000.

In some situations where the final fraction has no convenient factors an earlier fraction might be chosen. It is a matter then of deciding whether sufficient accuracy has been achieved.

If $95/41 = 2\cdot31707$ is chosen, a train of $\dfrac{80 \times 53}{30 \times 61} = 2\cdot31693$ might satisfy.

Gear metrology

The increasing attention being paid to component accuracy is reflected in the 1967 and 1970 issues of the revised BS 436[17]. The latest amendments go much further than the original specification of 5 grades of accuracy, as modifications to the basic rack profile, by:

    (i) Suggesting a range of preferred pitches (DP or metric module).

    (ii) Specifying grades of accuracy for elements likely to affect efficient running.

    (iii) Indicating tolerances on pitch, profile and alignment which will define such grades.

Additional proposals for associated features involve tolerances on radial run-out, tooth thickness and tooth to tooth error. Certain measurements, although of little practical importance in running, must be controlled so as not to affect the accuracy of other recordings. A typical example is the outside diameter of a blank, often used as a location for fixture purposes, while its circularity controls the accuracy of depth setting for measurement of chordal addendum or thickness.

The stages or phases of accuracy must be clearly defined, often on the basis of the intended operating speed. Taking machine accuracy as the first phase and a precision hobber as the most demanding, accuracy wise, the minimum tests for machine performance might comprise the following elements:

Departure of table from true horizontal plane.

Truth of the column.

Accuracy of feed screw, master worm, hob saddle.

Efficiency of grating measurement for transmission error.

There is little virtue in manufacturing to an appreciably higher standard than the application demands, but care must be taken to specify the lower limit below which a product will not be acceptable. BS 4500[22] defines appropriate size tolerances and BS 3800[23] the basis of geometrical and practical tests for machine tool accuracy. It should not be forgotten that the optimum requirement of a gear pair is to run in a specified fashion and, since measurements in isolation may not prove sufficient, most practical tests are based on the running of an engaged pair under pre-determined meshing conditions. The parallelism of axes within 0·01 mm might be a typical requirement, or meshing conditions specified as in BS 1807, perhaps as an acceptable percentage of the contact area.

In many gearing calculations the involute function is involved. The involute function of angle ϕ is defined as the angle β radians formed by the radial lines through the point of origin of the involute and the point on the involute from which ϕ is derived. From Figure 13.22

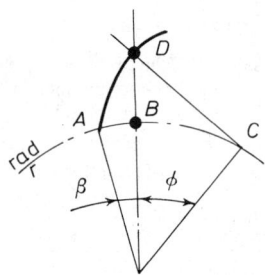

Figure 13.22

$$\beta \text{ radians} = \text{arc/radius} = AB/r = \text{INV. } \phi$$
$$\text{arc } AB = \text{arc } AC - \text{arc } CB$$
$$= CD - \text{arc } CB = r \tan \phi - r\phi = r(\tan \phi - \phi)$$
$$\text{and } \beta \text{ radians} = \tan \phi - \phi = \text{INV. } \phi \tag{15}$$

Tooth thickness and space width are equal when the axis of symmetry of the generating rack coincides with pitch circle of the gear. When displaced to avoid undercutting, the amount of radial rack displacement (value k for unit pitch) has positive sign when rack is moved away from gear and thickness of corrected tooth exceeds a standard tooth by the amount $2k \tan \phi$.

The problem of measuring tooth thickness when it is defined as the length of an arc, can be solved with moderate accuracy by measurement of chord using a gear tooth caliper or by calculating base tangent and using a tooth micrometer.

$G = 2r \sin \theta$ where $H = A + r(1 - \cos \theta)$ as in Figure 13.23.

For corrected teeth
$$\theta = \frac{\frac{1}{2}p + 2k \tan \phi}{2r} \tag{16}$$

Alternatively the existence of a constant chord, which joins the points making contact with mating teeth, can be used to measure chordal thickness. For a given pitch this chord is independent of tooth number and the illustration also shows the arrangement for corrected teeth. The dimensions defining the constant chord are:

$$X = (\tfrac{1}{2}p + 2k \tan \phi) \cos^2 \phi \qquad Y = A - (\tfrac{1}{8}p + \tfrac{1}{2}k \tan \phi) \sin 2\phi \tag{17}$$

Tooth thickness of a spur gear may be obtained from measurement over rollers of a specified diameter. When d is diameter of a cylinder at which the normal pressure angle is ϕ_N and w is the normal thickness of cutter tooth at a plane touching the pitch circle of generation, then for rollers of diameter $w \cos \phi_N$ and an even number of teeth, the distance over rollers when placed in opposite spaces is

$$d + w \cos \phi_N \qquad\qquad (18)$$

When number of teeth t is odd the first term becomes

$$d\left(1 - \frac{1\cdot23}{t^2}\right)$$

Pitch or tooth spacing errors may be assessed by step readings or by direct angular measurement. The former method relies for its accuracy on minimising the error of a chosen datum and is comparative thereafter with respect to successive teeth, whereas

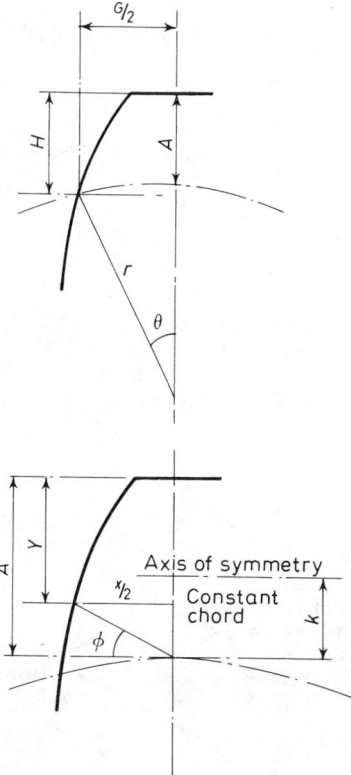

Figure 13.23

an accurate dividing head and a vertical measuring machine is required for spot readings. Profile measurement may involve optical projection or direct measurement of the difference between a true involute and the shape under test. Self-contained involute testing machines utilise a stylus to explore the test profile and record on a chart any differences from the true involute generated by rolling a straightedge round a circle. Concentricity can be assessed by mounting the gear on an arbor and testing a roller placed successively between teeth while comparison of height from a parallel mounting bar will indicate the accuracy of alignment.

As an alternative to the measurement of individual elements, the functional or rolling gear test involves a master gear mounted on a fixed arbor and arranged to mesh with the gear to be measured, which is mounted on a floating carriage. Single and double flank tests are used and, in this way, the important dimension of correct centre distance is best assessed, or centre distance variation used to indicate errors in test gear when the pair are meshed under controlled spring pressure. The care necessary in the manufacture of master gears is recognised in BS 3696[24] which lists proposals for materials and dimensions. Perhaps the most important consideration is that elements involved in the measurement of gear accuracy should not be so sensitive as to reject gears finished to ordinary commercial limits because their industrial use does not demand exceptional accuracy.

Spiral gears

The toothed connection between non-parallel, non-intersecting shafts is best made by spiral gears and their relationship with other gearing types is explained in Figure 13.24

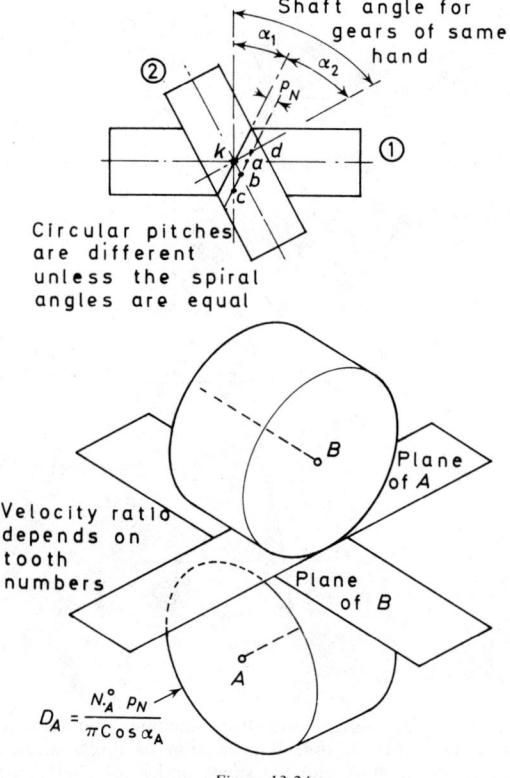

Figure 13.24

where ka and kc are multiples of circular and axial pitches respectively. Coincidence of a and b means the parallel shaft axis case of the helical gear while coincidence of bc and ad means right angle axes for which a worm and wheel connection is appropriate.

The situation unique to a spiral pair is that each gear has two circular pitches and two pitch surfaces, one a cylinder and the other a plane. The former rotates about its own axis while the pitch plane travels in the direction of rotation of the mating gear. Normal pitch is the same for both in a pair but the circular pitches will be different unless the spiral angles are equal. It is only for this latter equality that velocity ratio may be determined from pitch diameters, otherwise it is found from tooth numbers and it is common practice to refer to driver and follower instead of pinion and wheel.

Except when the spiral angles are equal, there can only be a graphical solution for angles and diameters unless the centre distance has been chosen to suit a particular combination. The theoretical condition of mesh is point contact which means that only small

P is diametral pitch p is circular pitch

P_N or NDP is normal diametral pitch $= \dfrac{\text{N}^\circ \text{ of teeth}}{d \cos \alpha} = P \sec \alpha$

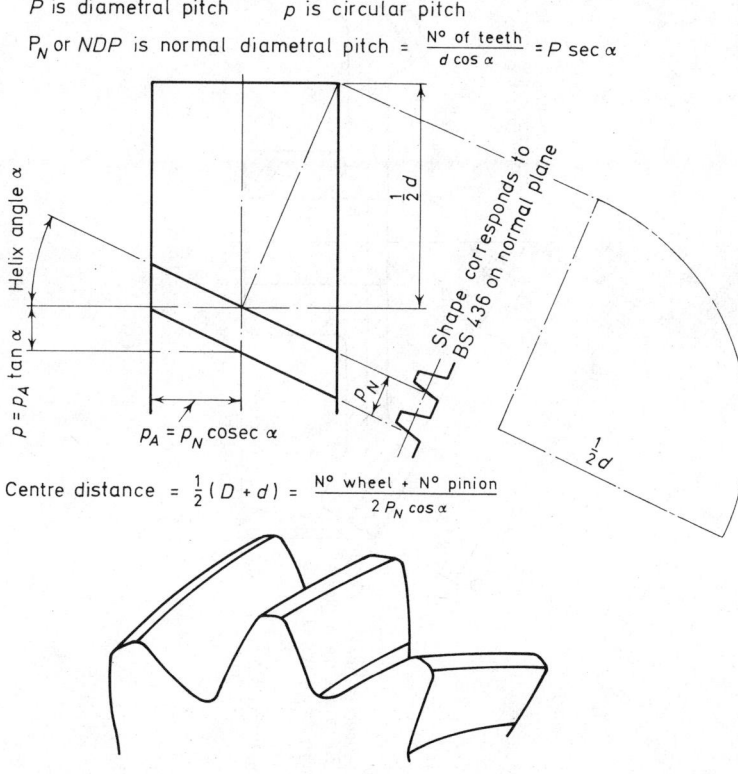

Centre distance $= \frac{1}{2}(D + d) = \dfrac{\text{N}^\circ \text{ wheel} + \text{N}^\circ \text{ pinion}}{2 P_N \cos \alpha}$

Figure 13.25

powers can be transmitted and that the quality of tooth surface finish is an important consideration. The useful width of face depends on the travel of point contact and any additional width does not increase load carrying capacity as is the case with parallel axis gears.

Helical gears

The advantages claimed for helical gears are in terms of increased power capacity for a given size with quietness and smoothness in operation, even at enhanced speeds. Although both circular and axial pitches are quoted, it is usual to refer parameters to the normal plane which is that perpendicular to a tooth element on the pitch cylinder surface at the

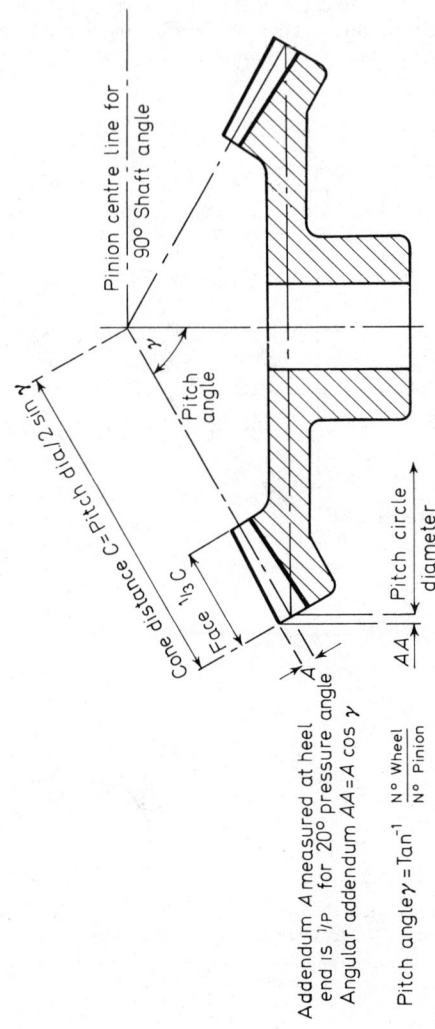

Pinion centre line for 90° Shaft angle

Pitch angle

γ

Cone distance C=Pitch dia/2 sin γ

Face ⅓ C

A

AA

Pitch circle diameter

Addendum A measured at heel
end is 1/P   for 20° pressure angle
Angular addendum AA=A cos γ

Pitch angle γ =Tan⁻¹   N° Wheel
                        N° Pinion

Figure 13.26

pitch point. It is on this plane that the shape of tooth corresponds to the basic rack rooth form BS 436 when p_N is normal circular pitch.

Figure 13.25 shows typical proportions and indicates that the normal plane is a function of the helix angle α, the angle between a tangent to tooth helix and the plane containing the gear axis. This angle cannot be chosen independently of the pitch diameter, since P_N or NDP, the normal diametral pitch, must be standard size for which a cutter is available. α varies in practice from $15°$ to $45°$; minimum values being just sufficient to ensure continuous action while larger values are associated with inconveniently large end thrusts which may necessitate the provision of special bearing arrangements or the use of double helical gears. The interchangeability, for manufacturing purposes, with straight spur gear cutters is found by calculating the formative gear set. This is based on the virtual number of teeth N_F on the surface formed by the intersection of the normal plane and pitch cylinders, N_F = actual No teeth/$\cos^3\alpha$. There are practical limits to face width if proper tooth contact is to be assured. These are often defined in terms of an overlap ratio so that the trailing tip of one tooth does not leave engagement before the leading tip of succeeding tooth comes into mesh. A minimum face width of $4p$, as used for straight spurs, is sometimes specified, alternatively the range from minima of $\frac{1}{12}$ (wheel diameter) $(1 + \tan \alpha)$ to $\leq$ twice pinion diameter is a typical empirical design basis.

Bevel gears

Bevel gear characteristics typify the class of toothed wheels necessary for the connection of shaft axes which intersect. The point of intersection is the apex of a conjugate action in which pitch cones roll over each other. The common pitch surface may be thought of as the pitch plane of a notational crown wheel conjugate with the cutter of two mating wheels. General proportions for the simplest type of straight bevel tooth layout are shown in Figure 13.26. Notice how the teeth of the notational crown wheel conform to radial lines and in normal section the tooth profile approximates to that of a straight spur gear, see BS 545[25] for details. Tooth shape is based on a gear whose virtual number of teeth is the actual number divided by the cosine of the pitch angle.

The smooth running of bevel gears depends greatly upon the accuracy of mounting, and the presence of both radial and thrust loads may demand the provision of rolling bearings in preference to plain bushings, with facilities to shim or adjust for the most effective tooth contact. In the general class of spiral bevel gears, the tooth spirals in the pitch plane are no longer radial but take on the same relationship as do helicals to straight spurs in cylindrical gears. So called curved tooth spiral bevel proportions are based on the spiral angle specified at pitch point at mean radius, although the angle between the tangent to tooth spiral and pitch cone generator at that point varies with radius. The Zerol gear has a zero mean radius spiral angle and the ability to generate this form by rotary cutter (the *Gleason system*) facilitates accuracy since the teeth can subsequently be ground. BS 545 provides proposals for straight and curved teeth having a normal pressure angle of $20°$ at the pitch cone, various classes of manufacturing accuracy being recommended for particular speed ranges.

Arrangements for corrected teeth follow the pattern already explained for spur gears and so does the procedure for calculation of power transmission capacity, which as before is carried out on both a strength and wear basis. The formulae take into account possible ranges of spiral angle, overlap ratio and the relationship between cone distance and face width, which allows for the tapering size of tooth. It is conventional to specify tooth dimensions at the large or heel end of the bevel tooth. The relationship between different types of curved bevel teeth is shown in Figure 13.27.

The offset arrangement of *hypoid* gears means that shaft axes do not intersect. It permits a higher ratio of reduction but care must be taken to combat the severe lubrication conditions due to the combination of loading, the degree of longitudinal sliding increases with the amount of offset. The mating members of a hypoid pair are usually steel and no lubricant which does not contain an extreme pressure additive is likely to be effective. It is a complex matter to define geometrical conjugacy of a hypoid or skew pair, the pitch diameters are not in fixed proportion to the numbers of teeth. In some cases one of the pair is cut as a conventional spiral bevel, the profile of its mate being designed as a conjugate with it. The obvious attribute of spiral type profiles is that due to the large

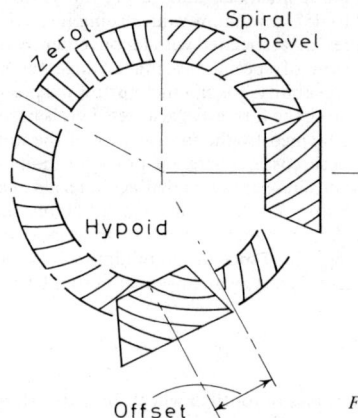

Figure 13.27

number of pairs in mesh the transfer of load is much smoother, noise and vibration are much reduced. The convention for handing is determined by the direction in which teeth curve from view point, a right hand spiral curves off to the right. The handing of a pair should be chosen so as to make the member carrying the greater axial thrust tend to move out of mesh when the load is applied.

Worm gears

In worm gears, the region of mesh is dispersed about a line normal to the shaft axes and in the typical 90° arrangement shown in Figure 13.28, engagement in an axial plane through the shaft axis has an analogy with a helical pinion and rack. The axial pitch of the worm corresponds to the circular pitch of the wheel. In a single enveloping set

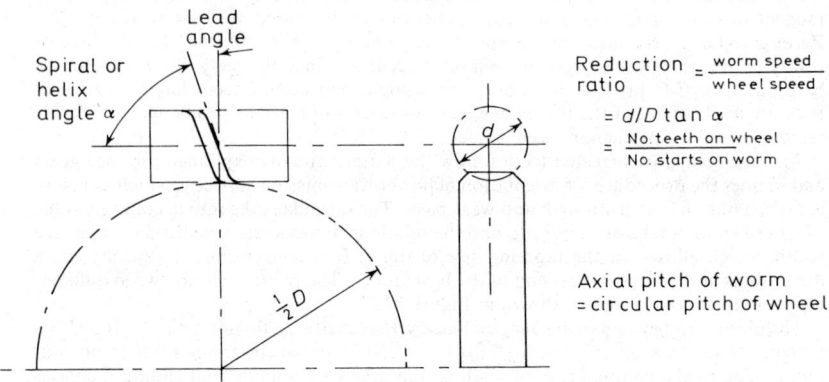

Figure 13.28

(straight worm and throated wheel) the motion involves sliding and rolling in the plane transverse to wheel in addition to sliding in the direction of worm axis.

Worm sets are sensitive to the accuracy of alignment and their supporting bearings are heavily loaded. There is considerable thrust at the worm and high radial loading at wheel bearings. Sliding speeds are a critical feature which demands particular attention to lubrication in order to avoid an excessive temperature rise due to friction. This is one reason for the recommendation in BS 721[26] of a case-hardened steel worm and a phosphor bronze wheel as the best choice of compatible pairings for heat dissipation availability.

The Standard proposes an involute helicoid thread based on the rack profiles previously discussed, having a normal pressure angle of 20° and generated with the plane of rack cutter normal to thread helix on the worm pitch cylinder. Conventional nomenclature of a worm pair is defined by the symbols $t/T/q/m$ meaning number of worm starts/number of wheel teeth/diameter factor/axial module and the specification is completed by the addition of centre distance. There is much less scope for manoeuvre with worm pairs than with other gear sets. The BS lists proposed values for centres, ratios and axial modules based on a reference diameter for the worm which is the product of module and diameter factor. Thus, the nominal worm pitch diameter is that of a spur gear having the same module pitch and q teeth. A basic design concept is to reduce the worm size to the smallest possible consistent with strength and rigidity, than to settle suitable dimensions for the pair and decide on materials to sustain the power transmission requirements.

The basis of calculation for strength and wear ratings is clearly laid out in BS 721. Lead angles should be chosen with care, it is possible to arrange a theoretically self-locking pair when lead angle is small, but the arrangement would be inefficient from the power transmitting point of view. Sometimes the relationship between the helix angle and its complement is chosen in terms of thread shape and the number of starts.

Lubrication

For the gear types discussed in previous sections the anti-frictional properties of their lubricants were not a prime consideration. The continuous rating of worm sets, however, tends to be limited by consideration of temperature rise rather than by bending or wear

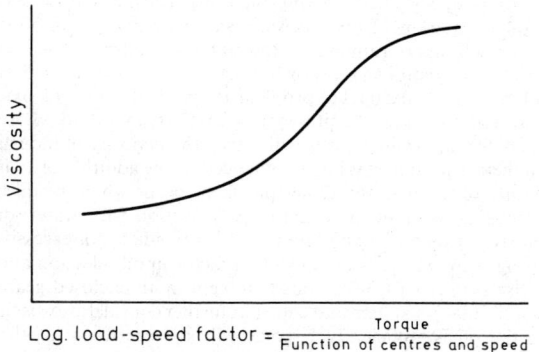

$$\text{Log. load-speed factor} = \frac{\text{Torque}}{\text{Function of centres and speed}}$$

Figure 13.29

capacity and there is some merit in classifying lubricants in terms of their thermal range, particularly when high reduction ratios are involved. There is some discrepancy in the advice for lubricant grades quoted in national standards and by leading manufacturers and care should be taken in the use of empirical formulae to determine viscosity ranges. These are best determined from performance tests although curves, such as Figure 13.29, which are based on typical centres and wheel size relationships, find favour in some quarters. The load/speed factor is a measure of surface stress in the contact region and indicates the likelihood of film generation.

The increasing use of worm reduction gears in enclosed cases for industrial drives has been recognised by the issue of BS 3027[27] which lists principal external dimensions for eight sizes of unit, the centre distances between high and low speed shafts ranging from 4–20 in (101–508 mm). Details of performance, efficiency, materials or allowable temperature rise do not form part of the specification. Careful assessment and some experience is necessary to evaluate the performance of alternative manufacturer's products over the single stage ratio range between 5 : 1 and 70 : 1. This is particularly the case in terms of recommendations for load characteristic multipliers for various types of driven machine, for overhung loads and thermal ratings. A variety of handing and mounting arrangements is available for standard gearboxes.

Lubrication is an important performance factor in power transmission because of its influence on operating efficiency. In general, the efficiency of a set is reduced as the order of sliding velocity increases. The rotation of a journal within a bearing revolving at a different speed is sometimes used as an analogy for the sliding action between engaging teeth, although rolling should also be considered. First contact of a spur gear pair involves maximum sliding and minimum roll, this changes to zero sliding and maximum rolling at pitch line. Such combinations aid film formation in a favourable environment, being most effective early and late in a typical contact and less efficient as the pitch line is approached. There will be a variation in the coefficient of friction μ as the point of contact moves over the profiles. If boundary conditions obtain, the tooth surface conditions will determine μ but if a satisfactory film is present, μ will vary with film thickness, sliding velocity and the lubricant viscosity.

Attempts are sometimes made to distinguish between friction coefficients due to rolling and sliding, although the measurement of an instantaneous μ as a function of the meshing position is prone to misinterpretation. Most data is provided by disc tests where two rollers are run in tangential contact under a loading regime which reproduces the stress conditions under review. A significant situation is that, for irregular contact, the nominal film thickness must exceed a minimum value in respect of surface roughness. Machining specifications for toothed gears rarely include surface finish stipulations although a conventional accuracy specification will imply a particular surface state. Assuming an average accuracy requirement for precision gears, a reduction of 50 per cent in CLA values within a few hours of running ductile materials under load is frequently found.

A necessary characteristic of a gearbox lubricant is its resistance to thermal degradation and chemical change. A particular problem is the probability of oxidation, which is accentuated by agitation, and the presence of catalysts in a fluid whose operating temperature is high. Straight mineral oils will satisfy the majority of industrial applications particularly if their oxidation stability is enhanced by the addition of an inhibitor selected to suit the nature of the base oil. Other problems occur when the oil has poor demulsibility and tends to form sludge due to emulsification with water, or when foaming becomes excessive instead of being minimised by avoidance of excessive churning.

The operating temperature is an important factor in oil selection and there is often a temperature rise between 10–38°C above ambient in an enclosed gear set which is not connected to a circulatory system and cooler. A further consideration is the tooth pressure, which unless the oil is sufficiently viscous, will squeeze it out at the wedge. This is symbolic of the selection of too light an oil. If however the oil is over-viscous, unnecessary friction will result and the operating temperature will be high. The characteristics of open gearing arrangements are the sparsity of lubrication and the infrequency of application. A boundary situation frequently prevails in which oiliness, rather than viscosity, may form the basis of choice. The lubricant properties should enhance the attraction between fluid and metallic surfaces and should not be those likely to permit squeezing out under load or throwing off.

Enclosed gearboxes are more common and sometimes they are a part of a circulatory system with oil sprayed on to the teeth at the point of mesh. When the splash system is used it is more difficult to form an oil film and, in general, a heavier-bodied oil should be used. Usually the larger gear dips in and carries oil to the meshing area. Oil level is all important since too high a level results in churning with unnecessary generation of heat and subsequent thinning of the oil. It is in splash systems where severe service factors such as tendency to atomisation, coupled with the oxidising effect of the atmosphere, combine to cause particular problems for the lubricant engineer.

Bearing loads

Figure 13.30 shows the conventional representation of forces on a spur gear tooth when any errors of profile, spacing or dynamic effects due to maldistribution of load are ignored. When a single spur gear is mounted on a shaft and the bearings are assumed to provide simple support, the bearing loads can be assessed by taking simple moments in terms of the normal tooth load P. This is the resultant of the tangential tooth load T and the separating force S. When more than one gear is involved, the load on each bearing due to T and S should be calculated separately and the total load obtained by combination of these having due regard to direction.

The spacing and positioning of bearings is often determined by the spatial requirements of the application and may not depend on the size or type of gears involved. As a first approximation the empirical proportion for straddle mounting suggests a bearing spacing dimension as 70 per cent of the pitch diameter. If a gear is overhung, the bearing distant from that supporting the overhang should be spaced at least twice the overhang dimension from its partner. Variations in bearing spacing have sometimes to be arranged to satisfy the compromise between calculated load values and the capacity of the bearing it is desired to use.

Considering the straight spur set shown in Figure 13.31.

Tangential force T_1 on A and $B = 126\,000\,(\text{hp})/ND_A$ in English units where N is speed and D is pitch diameter.

Separating component $S_1 = T_1 \tan \phi$

Normal tooth pressure $P_1 = T_1/\cos \phi$

For the C and D pair;

    Tangential $T_2 = T_1 (D_B/D_C)$

    Separating $S_2 = T_2 \tan \phi$

    Normal $P_2 = T_2/\cos \phi$

Radial bearing loads

            (1) $= P_2(k/l)$

            (2) $= P_2(j/l)$

            (3) $= P_1(a/c)$

            (4) $= P_1(b/c)$

Combined loads

$$(5) = \sqrt{([T_1(e/g)+T_2(f/g)]^2+[S_1(e/g)-S_2(f/g)]^2)} \tag{19}$$

$$(6) = \sqrt{([T_1(d/g)-T_2(h/g)]^2+[S_1(d/g)+S_2(h/g)]^2)} \tag{20}$$

Using the same diagram to represent a single helical gear set, the expression for separating force must take into account the helix angle of thread α so $S = T \tan \phi/\cos \alpha$.

Additionally it is necessary to allow for end thrust $E = T \tan \alpha$ which will add or subtract from the load due to separating force according to direction. Using symbols as before

$$\text{Radial load on (5)} = \sqrt{([T_2(k/l)]^2+[S_2(k/l)\pm E_2(D_D/2l)]^2)} \tag{21}$$

As a further example, if the helical gears B and C are mounted within bearings 5 and 6, as shown in Figure 13.32, the radial load on (5) =

$$\sqrt{([T_1(x/z)-T_2(y/z)]^2+[S_1(x/z)+S_2(y/z)\pm E_1(D_B/2z)\pm E_2(D_C/2z)]^2)} \tag{22}$$

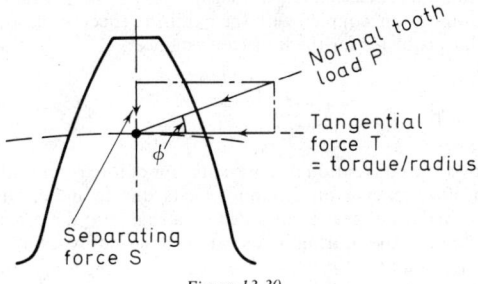

Figure 13.30

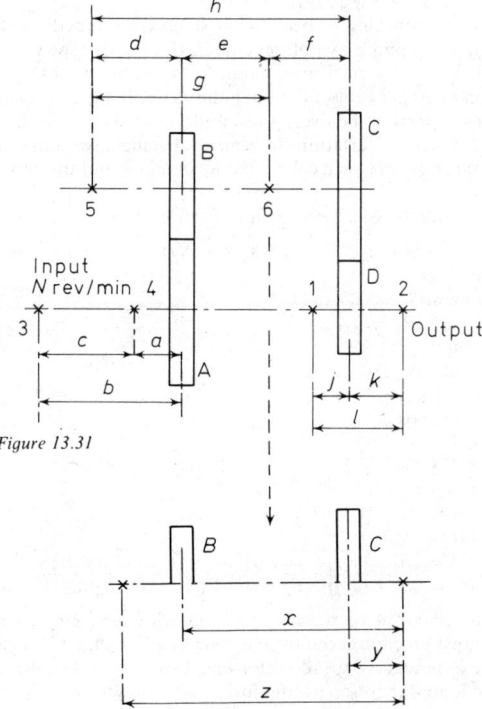

Figure 13.31

Figure 13.32

The + or − sign indicates that thrust load will vary according to the hand of spiral and direction of rotation. One convention for direction is to consider the driver as having a right or left hand thread, the thrust along this shaft is in the same direction as the driver would move if rotated in the same direction as the shaft.

Figure 13.33 shows a straight tooth bevel gear support layout for shafts at 90°. It is

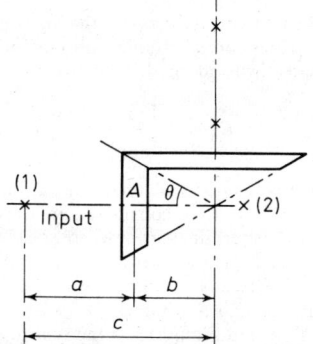

Figure 13.33

necessary to allow for the thrusts acting away from the gearing centre and end thrust must be dealt with at one position on each shaft. The typical expression for radial load on bearing (2) =

$$\sqrt{[T(a/c)]^2 + (E_B(a/c) - (E_A \cdot D_A/2c)]^2} \qquad (23)$$

where E_A is thrust on driving shaft = $T \tan \phi \sin \theta$
E_B is thrust on driven shaft = $T \tan \phi \cos \theta$
T is tangential force
D is mean pitch circle diameter
θ is half pitch cone angle of driver
ϕ is pressure angle.

In hypoid layouts it is necessary to consider the effect of thrust on the crown wheel. This imposes a couple which results in radial loads at right angles and these are resolved

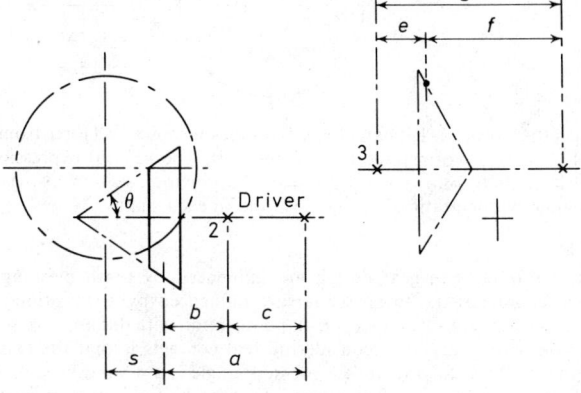

Figure 13.34

with tangential force and pinion thrust. For the layout shown in Figure 13.34 and for anti-clockwise rotation of driver

Radial load on bearing (2) $= \sqrt{([T(a/c)]^2 + [E_B(a/c) - E_A(D_p/2c)]^2)}$ (24)

T is tangential force, E_A and E_B are driving and driven shaft thrust respectively. D_p is mean diameter pinion and ϕ is pressure angle.

Radial load (3) $= \sqrt{[T(f/g) - E_B(x/g)]^2 + [E_A(f/g) + E_B(s/g)]^2}$ (25)

Care must be taken to correctly assess the sign of thrust along the driving or driven shafts. A typical expression, assuming anti-clockwise rotation of driver and left hand helix angle α, means thrust along driving shaft is

$$E_A = T \left[\frac{\tan \phi \sin \theta}{\cos \alpha} + (\tan \alpha \; \cos \theta) \right]$$ (26)

and provision must be made to deal with this at one position.

In worm pairs the forces to be considered are the end thrust on worm or the tangential force on wheel, say A, determined from worm speed and lead. Alternatively the end thrust on wheel or tangential force on worm, say B, can be determined (English units) from 126 000 (HP)/(N.D.) where N is worm speed and D is worm pitch diameter. The force tending to separate the pair $S = \tan \phi \sqrt{A^2 + B^2}$.

For the layout shown in Figure 13.35 the typical expression for radial load on bearing (1) is

$$\sqrt{[B(b/c)]^2 + [S(b/c) \pm A(D/2c)]^2}$$ (27)

Careful attention must be paid to determining the direction of gear forces. In the usual case with the worm as driver, rotating clockwise, the directions are as shown in the

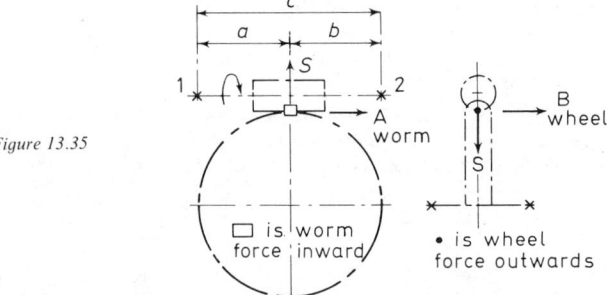

Figure 13.35

diagram when the worm has left hand helix. Symbol ● means wheel force from A outwards and symbol ☐ means worm force from B inwards. The general expression for thrust along the driven shaft follows the pattern of the previous section, provision must be made for dealing with end thrust at one position on each shaft.

Power ratings

Most of the parameters in gear design are influenced by tooth meshing conditions; load carrying characteristics in particular are influenced by assumptions as to where the load is acting and how allowance for shape variation with different sizes can be made. The first lesson to be learned in considering real contacts is that the concept of load coming on to a tooth at the theoretical pressure angle is not valid because of the deformation of the parts and the maldistribution of the load. One method proposed for commercially cut gears is to define a rectangular load region which makes allowance for addendum correction to be included as required, typically to cover the case of non-standard centres or to prevent undercutting.

Figure 13.36 shows one layout when the pitch line of generation is not the line of symmetry. Notice how the load region is used to define the extreme points of involute contact, not the theoretical positions defined by addendum circles but conventionally taken as half the base pitch from coincidence point, an assumption that in the worst case the load is carried by one pair of teeth.

The simplest type of rating formula for spur gears suggests that the allowable tangential load/inch of face is the product of a material strength term S, a rotational speed

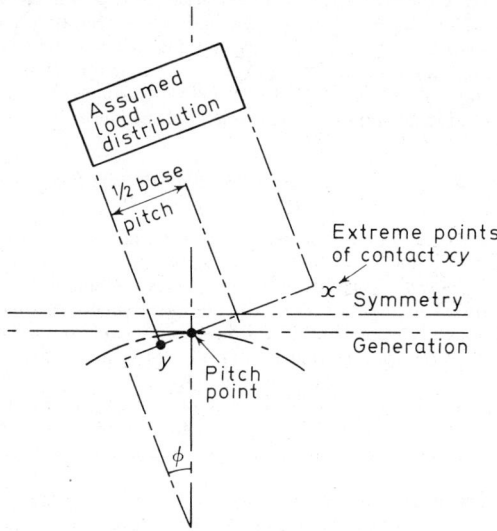

Figure 13.36

factor X and a strength factor Y divided by diametral pitch P. The S term is a comparative value, a function of the material bending stress while the speed factor represents the relationship between permissible loading margins, assuming that the accuracy of particular gears is appropriate for their working speed range.

$$\text{Since torque } T = \frac{33\,000\,(\text{hp})}{2\pi N}$$

where N is speed rev/min, and allowable tangential load/inch of face

$$= \frac{\text{torque}}{\text{pitch radius} \times \text{face width}} = \frac{SXY}{P}$$

the h.p. capacity on a strength basis

$$= \frac{S.X.Y.N \times \text{face} \times \text{number of teeth}}{126\,000\,(P)^2} \tag{28}$$

The problem of surface failure has proved more intractable than that of the bending situation, largely due to the difficulty of correlating conventional stress calculations with the endurance properties of various materials, and data on the fatigue strength of gear teeth is generally inadequate. The expression for a safe tangential load follows the same pattern as before, symbols SXZ/K, where S is a material surface stress factor involving a compressive stress function, often a comparative value which relates tooth load to the radius of curvature. Z compares the load carrying capacity of different tooth number combinations of common pitch while K is a function of diametral pitch. In this way an expression for a wear h.p. capacity can be built up as in the earlier pattern and the safe capacity for a pinion/wheel combination will be the lowest value from the four calculations. Full details are shown in BS 436 which also indicates how to determine the load carrying capacity of single helical gears by variations in the values used for the Y and Z parameters. A particular requirement of rating proposals is that they shall allow for the characteristic of spur gear contacts i.e. for the variation between one and two pairs in mesh, which is one of the factors concerned in causing a fluctuating load cycle.

Nominal ratings are on the basis of a 12 hour/day programme or for an equivalent running time of 26 000 hours, although few industrial operating schedules involve prolonged running without some load variation. The real criterion is the number of cycles at which the gear materials reach their endurance limit. One method of determining the value of a uniform load equivalent to a variable loading cycle is to calculate the equivalent running time (at uniform load) which will produce the same effect on gear life.

Epicyclic gears

Most of the difficulties found in assessing the speed ratio of epicyclic gears arise from the situation that the centre lines of certain elements are not fixed. In an analysis not considering efficiency of power transmitted, the only factors of importance are the relative velocity relationships. The intention should be to derive equations which satisfy all the kinematic possibilities of a given train. A simple method of solution is to determine algebraic relationships on the assumption that all centres are fixed and that all gears are free to rotate. It is however important to differentiate between those wheels which revolve on fixed centres, usually called *sun* gears and those which revolve on moving centres, called *planet* pinions. Other definitions are for *star* gears when the planet carrier is fixed with rotation of sun and annulus, or in *solar* gear systems when only the sun is fixed.

Most epicyclic trains are constructed using a number of planet carriers at equal intervals, and one of the design criteria is to ensure that assembly of the train is feasible. In the simple arrangement shown in Figure 13.37, the sum of tooth numbers in the sun

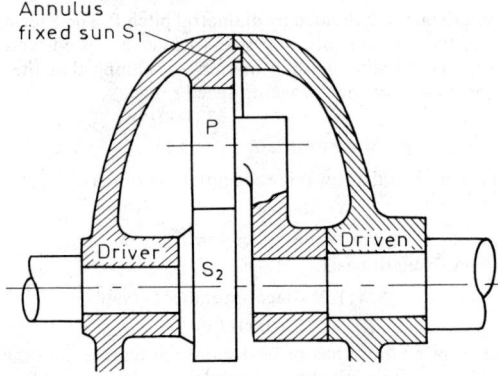

Figure 13.37

gears must be divisible by the number of planet pinions. There is a close similarity between the design problems of internal gears and those of the outer sun wheel or annulus. If numbers of teeth are small it may be necessary to use a corrected tooth system or some form of stub tooth to minimise tooth overhang.

Various methods have been proposed for the determination of velocity ratios. The algebraic and tabular methods will be assessed in terms of the compound train shown in Figure 13.38. It is clear that ABE can rotate about the common axis at different speeds but that only two of these can be arbitrarily assigned. A typical algebraic relationship is

$$\frac{N_E - N_A}{N_B - N_A} = -\frac{T_D \cdot T_B}{T_E \cdot T_C}$$ where N and T refer to speeds and tooth numbers respectively.

When one of the members, say B is fixed the speed ratio of the others can be determined since

$$\frac{N_E}{N_A} = 1 + \frac{T_D \cdot T_B}{T_E \cdot T_C} \tag{29}$$

The alternative tabular method is typified by Table 13.3 and consists of summing a

A	B	C / D	E
+1	+1	+1	+1
0	-1	$-\dfrac{T_B}{T_C}$	$+\dfrac{T_B}{T_C} \cdot \dfrac{T_D}{T_E}$
+1	0	$1 - \dfrac{T_B}{T_C}$	$1 + \dfrac{T_B \cdot T_D}{T_C \cdot T_E}$

series of tabular exercises. One step may be assumed as the revolution of the assembly with all gears locked. The procedure of the second step is then determined according to the information required. In the example shown, it is the velocity ratio between E

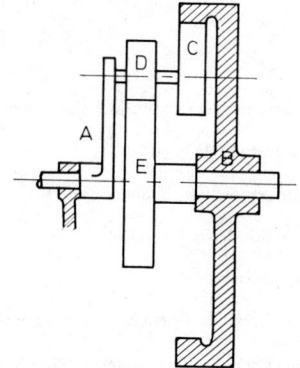

Figure 13.38

and A which is to be determined, the intention of the second step should be to make the sum of A events as unity, so that A is fixed and B is given -1 revolutions. Sometimes it is not convenient to fix any wheel. If the tabular method is to be used for the arrangement shown in Figure 13.39 a solution can be obtained provided that the operating step relationship is pre-determined. With the locked set given x revolutions clockwise, where x is revs of arm, the second stage step for the planet (with arm fixed) would be y revs =

$N_P - N_A$. It is instructive to analyse concentric arrangements, from their end elevations, in terms of their potentiality as speed increasing or reducing units.

The arrangement shown in Figure 13.40 using symbols as diameters or as numbers of teeth, can be used as a speed *increaser* when arm is driver and the largest wheel is fixed. It yields rotation value for the follower or driven member D_1 according to the ratio

$$1 + \frac{x(D_2)}{y(D_1)} \tag{30}$$

A similar arrangement can be used for speed *reduction* when the arm is follower, D_1 is the driver and D_2 is fixed. It yields the same expression for reduction ratio. The solution of problems involving torque and reaction are best assessed on the assumption

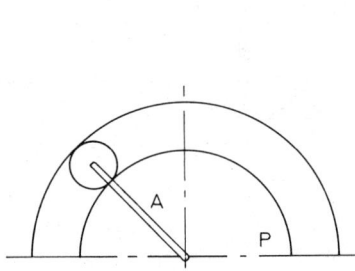

Figure 13.39

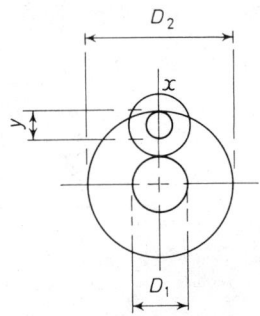

Figure 13.40

of uniform speed of elements, so that no angular accelerations are involved. On the basis that the algebraic sum of all externally applied torques is zero, the minimum involvement is in terms of the three values, input torque at driver; load torque on driven; braking torque at fixing.

REFERENCES

1. BS 228
   Transmission precision roller chains and chainwheels (1970)
2. BS 1440
   Endless V-belt drives (1971)
3. BS 3548
   Endless V-belt drives, sections Y and Z (1970)
4. BS 3790
   Endless narrow V-belt drives for industrial purposes: sections 3V, 5V and 8V (1964)
5. BS 4548
   Synchronous belt drives (1970)
6. BS 970/971
   Wrought steels, bars, billets, forgings, En series (1972)
7. PD 6423
   New designation system for carbon steels (1968)
8. BS 46
   Keys and keyways (1958)
9. BS 4235
   Metric keys and keyways (1967)
10. BS 308
    Engineering drawing practice (1964)
11. BS 2059
    Straight-sided splines and serrations (1953)

12. BS 1131
    Plain bearings (metal) (1955)
13. BS 4480
    Plain bearings metric series (1969)
14. BS 4231
    Viscosity classification for industrial liquid lubricants (1967)
15. BS 292
    Dimensions of ball bearings and cylindrical roller bearings (1969)
16. BS 3134
    Dimensions of tapered roller bearings (1968)
17. BS 436
    Spur and helical gears (1970)
18. BS 1452
    Grey iron castings (1961)
19. BS 821
    Iron castings for gears and gear blanks (1938)
20. BS 2887
    Pinion type cutters for spur gears (1957)
21. BS 2062
    Gear hobs (1959)
22. BS 4500
    ISO limits and fits (1969)
23. BS 3800
    Methods of testing the accuracy of machine tools (1964)
24. BS 3696
    Master gears (1963)
25. BS 545
    Bevel gears (machine cut) (1949)
26. BS 721
    Worm gearing (1963)
27. BS 3027
    Dimensions of worm gear units (1968)

BIBLIOGRAPHY

ISO R.701
International gear notations, symbols for geometric data (1958)
ESD 65007
General guide to the choice of journal bearing type (1965)
ESD 66023
Calculation methods for pressure fed journal bearings (1966)
ESD 68018
Dry rubbing bearings, guide to design and material selection (1968)
ESD 66001
Geometric design of spur and helical gears (1966)
ESD 68028
Design of parallel axis gears, backlash and inspection (1968)
ESD 68040
Choice of materials and estimate of dimensions (1968)

   The above publications are produced by the International Standards Organisation (ISO) and the Engineering Sciences Data Unit (ESD). The ESD publications can be obtained direct from 4 Hamilton Place, London W.1.

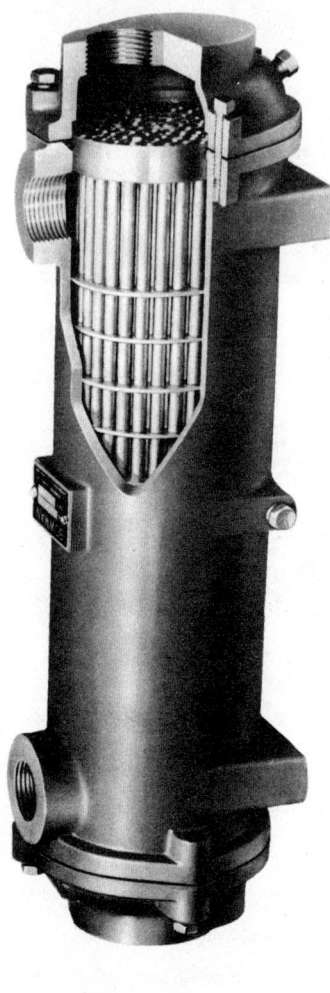

14 LUBRICANTS AND LUBRICATION

LUBRICANTS AND LUBRICATION 14

14 LUBRICANTS AND LUBRICATION

G. D. GALVIN

INTRODUCTION

In recent years the tendency to treat wear, friction, lubrication and bearing materials as distinct and separate topics has been much reduced and the technology of Tribology as a unifying approach to the above to seek the best solution to a problem has led to great advances. The importance of Tribology has been recognised by the publication of many new books and scientific journals. In the U.K. the Institution of Mechanical Engineers has formed a Tribology group open to engineers, chemists, physicists, metallurgists, etc. to discuss problems at all levels from the completely practical to the research level. Of particular value to the practising engineer are the bound volumes of the Tribology Group's Annual Conventions, the reports of specialised meetings and the 1967 Conference 'Lubrication and Wear: Fundamentals and Applications to Design'.[1] The Engineering Sciences Data Unit has produced many relevant publications of direct interest and usefulness.[2, 3]

The aim of Tribology is to prevent wear and control friction in machinery. This is done by separating the moving surfaces by a lubricant film or making one or both of the load bearing surfaces of a self-lubricating material. If a lubricant is used it may be a gas, a liquid or a solid and it may be required to cool the surfaces, carry away swarf and wear debris and prevent the ingress of dirt and contaminants. The load bearing surfaces may take many forms and be made of many materials depending, amongst other things, on load, speed, temperature, direction of load, contamination, availability of lubricant under normal running and at start-up, and corrosivity of the lubricant. If it is suspected that at any stage the lubricant film thickness will be less than several times the sum of the roughnesses of the surfaces, then special care will be needed in the selection of bearing material and lubricant otherwise some form of failure, such as wear, seizure or fatigue-pitting may well occur.

The regimes of lubrication under which bearings and lubricants operate are as follows.

Hydrodynamic lubrication

This is the normally preferred type of lubrication in which the bearing surfaces are completely separated by the lubricant, which may be a liquid or a gas. The pressure in the lubricant film is most commonly generated by the relative movement of the surfaces and their geometry (converging). It is also possible to generate the pressure externally, hydrostatic lubrication, and this has certain advantages such as ensuring the presence of lubricant at start-up and when the relative movement is insufficient to generate an oil film; it does require, however, the supply of a pump and control devices.

In hydrodynamic lubrication the pressures will usually not exceed 7 MN/m² and can be much less. Film thicknesses will be typically 10^{-1}–10^{-3} mm and the coefficient of friction of the order of 10^{-3}.

The subject of hydrodynamic lubrication is dealt with adequately in available published literature[4-14] and these contain formulae for calculating the load carrying capacity, film thickness, friction and heat generation. Whilst the attainment of a full oil film is often the ideal situation, it is sometimes not possible to achieve it. For example, special precautions will be needed if the bearing has to start up under load, if shock loads are encountered or if changes in the direction of the loading or the direction of sliding occur the surfaces may come into contact. These may take the form of dissolving boundary or anti-wear additives in the lubricant (see below) or ensuring that the bearing surfaces themselves can tolerate such conditions.

Elastohydrodynamic lubrication

Where the bearing surfaces do not conform geometrically the load is carried by very small areas and the pressures are very high.[15-18] Classical hydrodynamic lubrication theory would indicate that the surfaces would be effectively in contact because such theory ignores the increase in contact area caused by elastic distortion and the increase of the viscosity of the lubricants at elevated pressures.

Heavily loaded gears, ball and roller bearings, and other devices operate in the elasto-hydrodynamic lubrication regime typically with film thicknesses of the order 1–0·1 μm (comparable to surface roughnesses) and film pressures of the order of 1 GN/m^2. Where the bearing surfaces are very easily deformed, as in the lubrication of rubber coated surfaces, elastohydrodynamic lubrication may occur under much less severe conditions.

Boundary lubrication

Where it is not possible to generate a complete oil film because loads are too high, entraining velocities too low or a great deal of sliding occurs in the contact area, special ways have to be found to control wear and friction.[19-22] The viscosity of the lubricant has little influence and special additives are dissolved in the oil. Under relatively mild conditions boundary lubricating additives are used. These usually take the form of polar, long-chain organic compounds such as fatty acids, fats or soaps which adsorb on the bearing surfaces to form an effectively solid film. This film differs from films of solid lubricants in that although it melts at lower temperatures it is self-repairing. Coefficients of friction about 0·05–0·1 are typical and the melting points of the films roughly 70° to 150°C depending on whether or not the boundary lubricant forms a soap at the bearing surfaces. Mineral oils contain natural mild boundary lubricant compounds.

Under the most severe operating conditions as in metal cutting or the shock loading of hypoid gears extreme-pressure additives are used. These are oil-soluble compounds containing sulphur, chlorine or phosphorus which react with the bearing surfaces under the high temperatures generated where the oil film is ruptured to form high melting point films. In the process of extreme-pressure lubrication some wear is inevitable but it may be very small and may almost cease if the surfaces can run-in.

Solid lubrication

Certain inorganic compounds such as graphite and molybdenum disulphide have the property of forming tenaciously held films on metal surfaces which shear easily and resist penetration by the rubbing surfaces.[23-26] Such compounds can be used as lubricants alone or suspended in liquid carriers or greases with metal bearing surfaces. Plastics such as polytretrafluorethylene or nylon can be used as bearing surfaces which do not require further lubrication.[27-29]

LUBRICANTS

The broad classes of lubricants available which include, gases, liquids, semi-solids and solids are shown in Figure 14.1.[30-38]

Gases

Air or helium are typical gases which can be used to provide hydrodynamic lubrication and because of their very low viscosities give low friction and low heat generation. Rotational speeds as high as 10^5–10^6 rev/min have been achieved. Because of their low viscosities the load carrying capacity of gases is very much smaller than liquids and very thin gas films necessitate closer control of bearing dimensions and finishes.

Gases can be used over a very wide temperature range, their viscosities increase with increasing temperature and often it is the heat resistance of the bearing itself which is the

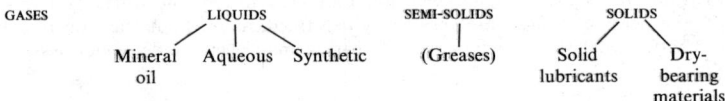

Figure 14.1. Types of lubricant

limiting factor at high temperatures. Air being cheap and non-toxic is frequently used and, as with other gases, it should be free from solid contaminants and preferably dry. Gas lubricated bearings may be of the self-generating or externally pressurised type and it should be noted that gases have no boundary lubricating properties.

Mineral oils

Mineral oils are the almost automatic choice if a liquid lubricant is to be used, being excellent lubricants, cheap and available in a wide range of viscosities. A vast amount of experience exists on their use and properties and compatible metals, rubbers, paints, etc. are readily available.

There are three basic types of mineral oil used to meet practical requirements and each type is available in a range of viscosities. The three types differ in the sort of crude oil and the severity of refining used in their manufacture. They are called High Viscosity Index (HVI), Medium Viscosity Index (MVI) and Low Viscosity Index (LVI). The Viscosity Index is a measure of the change of viscosity with temperature, a high Viscosity Index indicating a smaller change than a low Viscosity Index. The differences in the oil types are illustrated in Table 14.1.

Because of their properties, HVI oils are used where the least change of viscosity with temperature and/or high resistance to oxidation is required because of high operating temperatures or long life requirements. A large proportion of mineral oils contain one or more additives to improve their properties; boundary and extreme pressure additives have already been mentioned. In general it is not wise to mix additive containing oils from different sources unless it is known that they are compatible. Other commonly used additives are:

1. Oxidation inhibitors to prolong the working life of the oil or allow it to be used at high temperatures.
2. Viscosity index improvers to reduce the change of viscosity with changes of temperature.
3. Pour point depressants to lower the setting points of oils made from paraffinic crudes.
4. Dispersant additives to prevent flocculation of small contaminant particles.

Anti-rust and anti-foam additives are self-explanatory.

Aqueous lubricants

Aqueous lubricants are used because of the cooling properties of water and its resistance to burning. Oil-in-water emulsions containing small percentages of oil in the form of finely dispersed droplets are used in metal working lubricants and in large scale hydraulic systems. These are cheap and are excellent coolants but have little lubricating ability.

Table 14.1 CLASSES OF MINERAL OILS

Type	Viscosity index	Source	Refining	Oxidation resistance
HVI	80–100	Paraffinic crude oil	Severe	Greatest
MVI				
(i) Naphthenic (MVI-N)	30–80	Naphthenic crude oil	Severe	Equivalent to HVI
(ii) Paraffinic (MVI-P)	30–80	Paraffinic crude oil	Less severe	Least
LVI	<30	Naphthenic crude oil	Less severe	Least

Water-in-oil emulsions containing about 40%v of water are used as fire-resistant hydraulic fluids.

Aqueous solutions containing no oil are used as metal working lubricants and fire-resistant hydraulic fluids. If it is necessary to restrict water loss, operating temperatures must be kept below about 60°C, and emulsions may separate if they become contaminated. Aqueous lubricants tend to give much shorter fatigue lives of heavily loaded ball and roller bearings than mineral oils.

Synthetic lubricants

Synthetic lubricants are used in specific situations where mineral oils cannot be used. Examples of such situations are

1. Where a lubricant has to be used over a very wide range of temperature and mineral oils are not suitable because, their setting points are too high, or their change of viscosity is too large, or they cannot stand the high temperatures involved.
2. Where the lubricant must be fire-resistant.

The main types of synthetic lubricants are silicone fluids, organic esters, polyalkylene glycol derivatives, phosphate esters and polyphenyl ethers. Some of their more important properties are given in Table 14.2.

Greases

Greases consist essentially of a thickener dispersed in a liquid lubricant in such a way that the required rheological properties are obtained, additives may also be present. Usually the liquid lubricant is mineral oil but synthetic lubricants such as organic esters

Table 14.2 PROPERTIES OF SYNTHETIC LUBRICANTS

Lubricant	Main properties	Uses
SILICONES Methyl Low phenyl High phenyl	Least change of viscosity with temperature, low volatility, oxidation resistant. Very wide range of viscosities available, low freezing points. Poor lubricants, very expensive. Methyl silicones usable in air to 200°C; if air excluded, over 300°C. Low phenyl silicones, greater change of viscosity with temperature; usable in air to 250°C; if air excluded, 320°C. High phenyl usable in air above 250°C; 350°C if air excluded.	Hydraulic dampers, instrument fluids, high temperature applications.
ESTERS Diesters Polyesters	Excellent lubricants. Low setting points, low volatility, high viscosity indices. Respond well to certain oxidation inhibitors. Diesters useful from −50°C to 150°C bulk oil temperature, 250°C in hot spots. Polyesters useful to some 50°C higher temperatures.	Aviation gas-turbine lubricants, low-temperature lubricants, low-volatility lubricants.
POLYGLYCOLS Polyethylene oxide Polypropylene oxide	Ethylene oxide polymers: water-soluble, very high viscosity indices, high setting points. Propylene oxide polymers: water-insoluble, not quite so high viscosity indices but low setting points. Polyglycols respond well to specific oxidation inhibitors and oxidation products are volatile or soluble; usable to about 200°–250°C in absence of air; 130°C in air.	Automotive brake fluids, worm gear lubricants, high temperature lubricants; thickeners for diester lubricants and water-glycol type hydraulic fluids.
POLYPHENYL ETHERS	Outstandingly resistant to thermal breakdown and oxidation. Usable at around 300°C for extensive periods. Physical properties roughly equivalent to those of HVI mineral oils.	High temperature lubricants, potential lubricants for Mach 3 aircraft jet engines.
PHOSPHATE ESTERS	Excellent lubricants. Aromatic phosphates very resistant to burning but poor viscosity-temperature properties. Aliphatic phosphates less resistance to burning but better viscosity-temperature properties.	Fire-resistant hydraulic fluids. Fire-resistant steam turbine lubricants.

may be used for special applications. The thickener may be a metallic soap (e.g. sodium, calcium or aluminium stearate or lithium hydroxystearate), specially treated clay or organic pigments.

Greases are non-Newtonian and have a yield stress. Because of their properties they have less tendency to escape from bearings than mineral oils; consequently bearings can be prepacked for life or fitted with simple provisions for relubrication and sealing. Grease itself forms a seal and protects the bearing from contamination or corrosion. On the other hand since there is no flow of lubricant through the bearing heat is not conducted away and speed and temperature limitations may have to be imposed on bearing operation.

Solid lubricants

The best known solid lubricants are graphite and molybdenum disulphide which are available neat and as suspensions in liquids or greases. They find uses in very high temperature applications, space and ultra-high vacuum lubrication conditions, nuclear energy applications and under conditions where relubrication is difficult and it is possible that the bearing will be starved of lubricant.

Films of solid lubricant must adhere strongly to bearing surfaces and are often baked on in the presence of binders. The thickness of the lubricant film is rather critical—if it is too thin it will be strong but not durable and if it is too thick it will be soft. The limited life of such films can be overcome by making provisions for relubrication if the mechanism is self-cleaning of wear debris.

General guide to selection of lubricants

Lubricants derived from petroleum, i.e. mineral oils and greases, comprise about 90% of all lubricants used and should be considered first. If rolling bearings alone are involved then it is likely that the mechanical simplicity of grease lubrication will offer advantages unless speed and temperature considerations make the use of oil essential. In selecting an oil the temperature of operation, the temperature range to which the oil is subjected and the life expected of the lubricant should first be considered to see if it is possible to use a mineral oil. Consideration should then be given to the selection of the type of oil, the viscosity grade and the incorporation of additives. In general the recommendations of lubricant suppliers and machinery manufacturers should be sought and followed since these are based on special knowledge and wide experience.

If mineral oils are not suitable then the relative merits of gases, synthetic lubricants, solid lubricants and dry bearing materials should be considered. Specialised advice is then even more necessary. The British Standard Institution publication[39] and the paper by Beuerlein and Kara[40] are very helpful in discussing the many factors involved.

PROPERTIES OF LUBRICANTS

Viscosity

Viscosity, the resistance to flow of a fluid, is the most important physical property of a lubricant since it controls load carrying capacity, film-thickness, friction and heat generation in bearings. It is defined as the ratio of shear stress to shear rate, i.e.

$$\eta = \tau \Big/ \frac{\partial v}{\partial \gamma}$$

where η = viscosity; τ = shear stress; $\dfrac{\partial v}{\partial \gamma}$ = shear rate.

In lubrication technology it is common practice to retain the c.g.s. system of units where τ is expressed in dyn/cm^2 (10^{-1} N/m^2), v in cm/sec and γ in cm. The unit of viscosity is then the Poise (P); the sub-unit centipoise (cP) equal to $10^{-2}P$ is frequently used. In the SI system of units

$$IP = 10^{-1}\ Ns/m^2$$
$$IcP = 10^{-3}\ Ns/m^2$$

In many flow problems and in the determination of viscosity by capillary viscometers the ratio of viscosity to density occurs. This is called kinematic viscosity and in the c.g.s. system the unit is the Stokes (St):

$$1\ St. = 10^{-4}\ m^2/s$$
$$1\ cSt = 10^{-2}\ St = 10^{-6}\ m^2/s$$

In lubrication problems it is viscosity rather than kinematic viscosity which is important, but because of ease of precise measurement, data on lubricants are normally quoted in cSt. The densities of most liquid lubricants lie between 0·8 and 1·1 g/cm^3 so that the viscosity and kinematic viscosity expressed in cP or cSt are numerically similar. This is not the case for gases: viscosity of air at 20°C is 0·018 cP, 15 cSt, whereas the viscosities of mineral oils range from approximately 10–1400 cP, 12–1500 cSt.

The viscosities of clear liquid lubricants are normally determined by measuring the time taken for a volume of the liquid to flow under gravity through a vertical glass capillary. The conditions are carefully controlled and given in detail in the Institute of Petroleum 'Standards for petroleum and its products', Method IP 71/66[41] and BS 188. Viscometers having capillaries of different diameters are available so that the flow time is not inconveniently long nor less than 200 s when timing accuracy would be lost and corrections would have to be applied.

Conventional or jet type viscometers of lower accuracy are still used, the Redwood I and II instruments in Great Britain, the Saybolt Universal and Furol in the USA and the Engler in Europe. In the Redwood instruments the time taken for 50 ml of the oil to flow through the jet under gravity is measured. Only two jet sizes are available and the flow time must not be less than 30 s. If flow times exceed 2000 s in the No. I viscometer the No. II is used and gives one-tenth the time of the No. I. The Saybolt instruments are similarly used but in the Engler Viscometer the flow time of the oil at the required temperature is compared with the flow time of water at 20°C, the ratio being reported as degrees Engler, °E.

It is permissible to convert kinematic viscosities to conventional viscosities but it is not recommended to convert the other way as large errors may be introduced. The viscosities of a wide variety of liquids and petroleum products are given in the literature.[42, 43] A viscosity conversion table is given in the Appendix.

Effect of temperature on the viscosities of lubricants

The viscosities of gases increase with increasing temperature whereas those of liquid lubricants fall. The effect of temperature on the viscosity of liquids is strongly dependent on their molecular size and shape. In general, for non-polymeric liquids the larger the molecule the more its viscosity is affected by changes of temperature and liquids with rigid, ring-type structures show larger changes than flexible, chain-like molecules of which the dimethyl silicone fluids are the outstanding example. The change of viscosity with temperature can be of over-riding importance where machinery has to be used over a wide temperature range and extreme conditions, which mineral oils cannot satisfy, have necessitated the development of synthetic lubricants of the silicone fluid and organic ester type.

In the measurement of viscosity of lubricants the standard temperatures used include

100°F and 210°F. Where 37·8°C and 98·9°C are quoted, the measurements were taken at 100°F and 210°F.

VISCOSITY-TEMPERATURE RELATIONSHIPS

The relation between viscosity and temperature is well described for mineral oils by the equation due to Walther:

$$\log \ \log (v + 0\cdot6) = m \log T + b$$

where

v = kinematic viscosity in centistokes; T = temperature, °Rankine ($= °F + 460$); m and b are constants for a given oil.

The American Society for Testing and Materials has produced a series of viscosity-temperature charts with axes scaled to the Walther equation so that the viscosity data of mineral oils plot as straight lines to allow easy interpolation and extrapolation. The most useful of these charts is ASTM Standard Viscosity-Temperature Chart for Liquid Petroleum Products (D 341–39), Chart C. Kinematic Viscosity—High Range, the charts are available from the Institute of Petroleum. The REFUTAS Chart (Baird and Tatlock (London) Ltd.) serves the same function and in addition has scales for converting kinematic viscosity to conventional viscosity and also a blending scale.

A relationship said to be applicable to a wider range of liquids has been proposed by Roelands.[44] It takes the form:

$$\log (1\cdot200 + \log \eta) = \log b - S \log \left(1 + \frac{t}{135}\right)$$

where

η = viscosity, centipoises; t = temperature, °C; b = constant.

S, the Slope Index, is claimed to be constant for families of oils and for homologous series of liquids. It has also been suggested that the original Viscosity Index system described later should be replaced by:

Dynamic Viscosity Index (DVI) $= 220 - 7 \times 10^S$, where S = Slope Index

Viscosity-temperature charts scaled to the Roelands Equation are to be made available commercially by the copyright owners, the Technological University of Delft.

Viscosity index

The change of viscosity with temperature of a mineral oil is dependent on the crude oil from which it was made and the refining processes used. Because of the importance of viscosity-temperature properties the Viscosity Index system was introduced as a means of expressing the effect of temperature on the viscosity of a mineral oil as a single number. The system is based on the comparison of the viscosities of the oil at 37·8°C (100°F) and 98·9°C (210°F) with those of a series of high reference oils to which a viscosity index of 100 is allocated and a series of low reference oils to which a viscosity index of 0 is allocated. The series of oils of VI = 100 and VI = 0 were the oils showing the smallest and largest changes of viscosity respectively which were available at the time the system was introduced.

The viscosity of the test oil of unknown VI is measured at 37·8°C and 98·9°C and from tables are obtained the viscosities at 37·8°C of the high and low reference oils having the same viscosity at 98·9°C as the test oil. If the viscosities at 37·8°C of the test oil, the

Table 14.3 KINEMATIC VISCOSITIES, cSt, OF MINERAL OILS AT 98·9°C AND 37·8°C (210°F AND 100°F)

Kinematic viscosity, cSt, at 98·9°C	2	4	6	8	10	12	14	16	18	20	25	30	35	40	45	50
Kinematic viscosity, cSt, at 37·8°C for viscosity index of																
0	8·34	26·9	62·6	109·2	162·9	223·7	293·1	370·0	456·9	552·4	829	1 159	1 539	1 973	2 459	3 000
20	8·03	25·6	58·2	100·1	148·2	202·3	263·8	331·7	408·1	491·9	733	1 020	1 348	1 721	2 139	2 602
40	7·68	24·4	53·8	91·0	133·4	180·9	234·5	293·4	359·3	431·3	638	880	1 157	1 470	1 819	2 204
60	7·33	23·1	49·5	82·0	118·7	159·5	205·2	255·1	310·6	370·8	542	741	966	1 219	1 499	1 807
80	6·98	21·8	45·1	72·9	104·0	138·1	175·9	216·8	261·8	310·2	446	601	775	967	1 179	1 410
100	6·63	20·6	40·7	63·8	89·3	116·7	146·6	178·5	213·0	249·7	350	462	583	716	859	1 012
For VI$_E$ of																
120	6·36	18·9	36·5	56·3	77·7	100·3	125·0	150·9	178·7	208·2	288·0	375·6	470·3	572·3	682	798
140	6·13	17·6	33·2	50·4	68·7	88·0	108·6	130·2	153·3	177·5	242·7	313·4	389·2	470·0	556	648
160	5·93	16·5	30·5	45·5	61·6	78·2	95·8	114·1	133·6	154·0	208·2	266·7	328·8	394·7	464	538

high reference oil and the low reference oil are U, H and L centistokes respectively, the VI of the test oil is defined as

$$\frac{L-U}{L-H} \times 100$$

In commercial practice it is usual to define High Viscosity Index (HVI) oils as having VI's > 80, Medium Viscosity Index (MVI) oils as having VI's between 80 and 30 and Low Viscosity Index oils as having VI's < 30. The VI system is useful only for comparing oils of about the same viscosity. The absolute change of viscosity for oils of all types increases greatly as their viscosities increase. The viscosity index system has been extended to cover lubricants whose VI's are considerably greater than 100 and for which the original system is unsuitable.[41] The extended system is based on the equation

$$VI_E = \frac{(\text{antilog } N) - 1}{0.0075} + 100$$

where

$$N = \frac{\log H - \log U}{\log V}$$

H and U have their original meanings and V is the kinematic viscosity of the test oil at 98·9°C.

Table 14.3 shows the effect of temperature on viscosities of a range of lubricating oils.

Viscosities of synthetic lubricants

Commercial synthetic lubricants are often complex mixtures of base stocks and additives. Additives may be used to change the viscosity level of the base stocks and to reduce the variation of viscosity with temperature. Polymeric lubricants such as the poly dimethyl silicones and the polyethylene oxide and polypropylene oxide derivitives are available

Table 14.4   SOME PROPERTIES OF TYPICAL FIRE-RESISTANT HYDRAULIC FLUIDS

Property	Tri-tolyl phosphate	Tri-2ethylhexyl phosphate	Industrial F-R fluid	Aviation F-R fluid	Commercial water-glycol based fluid	Commercial water-in-oil emulsion
Kinematic viscosity cSt, at						
98·9°C(210°F)	4·37	2·23	6	3·9	—	—
37·8°C(100°F)	35·1	7·98	49	11·5	42·8	104
−40°C(−40°F)	—	840	—	500	—	—
Pour point °C	−25	< −55	−20	< −65	−50	−20
Specific gravity 15·6°C	1·16	0·926	1·29	1·06	1·055	0·929

Fluid type — *Phosphate esters* spans the first four fluid columns.

in a very wide range of viscosities. For all these reasons it is not possible to give more than a few examples, see Tables 14.4 and 14.5. For further information the manufacturers' literature should be consulted.

The physical properties, including viscosity, of most commercially available fire-resistant hydraulic fluids are listed in References 31–34. The book 'Synthetic Lub-

Table 14.5 SOME PROPERTIES OF SYNTHETIC LUBRICANTS

Property	Fluid type									
	Silicone fluids					Organic esters				Mixed polypropylene oxide-polyethylene oxide derivative
	Dimethyl			Medium phenyl	High phenyl	Di-2ethylhexyl sebacate	Commercial hindered ester	Specification D Eng RD 2847 fluid	Specification MIL-L-23699A fluid	
Kinematic viscosity, cST, at										
98.9°C (210°F)	17	30	145	18	30	3·3	5·1	7·7	5·1	23·1
37.8°C (100°F)	40	80	400	90	300	12·5	25·9	35·6	29·1	144
25°C (77°F)	50	100	500	150	500	—	—	12 500	—	—
−40°C (−40°F)	—	—	—	—	—	1 450	7 000	—	10 650	—
Pour point, °C	−55	−55	−50	−50	−20	< −59	< −57	—	< −62	−32
Specific gravity at										
25°C	0·96	0·97	0·97	1·07	1·11	0·912	0·995	0·947	1·005	1·025
15.6°C	—	—	—	—	—	—	—	—	—	—

ricants'[31] is useful as a source of data on physical and chemical properties of base stock of synthetic lubricants except those containing water.

Viscosities of gases

Table 14.6 gives the viscosities of gases which can be used to provide hydrodynamic lubrication over a temperature range from −50°C to 300°C.

Viscosity of lubricating oils

Table 14.7 gives viscosity classifications for lubricating oils based on BS 4231 and ASTM D2422–68. Other specifications covering oils are as follows.

BS 4475: 1969 Specification for straight mineral oils uses the same Viscosity Grade Nos. and covers oils for total loss systems (TLS grades) and for circulating oil systems (CSA and CSB grades). The TLS oils range from Viscosity Grade No. 10 to 1500 and their

Table 14.6   VISCOSITY OF GASES

Gas	Viscosity, micropoise, at							
	−50°C	0°C	50°C	100°C	150°C	200°C	250°C	300°C
Air	140	171	198	221	241	260	278	294
Ammonia	72	92	109	128	146	165	181	204
Carbon dioxide	110	139	161	186	208	228	247	268
Helium	160	186	210	228	250	267	285	308
Hydrogen	72	83·5	94	104	112	120	129	138
Methane	87	103	120	133	148	160	175	190

Table 14.7   LUBRICATING OIL SPECIFICATIONS BASED ON VISCOSITY
BS 4231: 1967. Viscosity classification for industrial liquid lubricants
ASTM D2422–68. Viscosity system for industrial fluid lubricants

BS viscosity grade No. (cSt)	ASTM viscosity No. (Saybolt Universal Seconds)	Kinematic viscosity (cSt) at 37·8°C (100°F)
2	32	1·98– 2·42
3	36	2·88– 3·52
5	40	4·14– 5·06
7	50	6·12– 7·48
10	60	9·0 –11·0
15	75	13·5–16·5
22	105	19·8–24·2
32	150	28·8–35·2
46	215	41·4–50·6
68	315	61·2–74·8
100	465	90–110
150	700	135–165
220	1 000	198–242
320	1 500	288–352
460	2 150	414–506
680	3 150	612–748
1 000	4 650	900–1 100
1 500	7 000	1 350–1 650

viscosities are specified at one temperature only. The CSA oils range from 2 to 1500 and a minimum VI of 40 is stipulated. The CSB oils range from 68 to 680 and a minimum VI of 70 is stipulated.

BS 489: 1968 Specification for steam turbine oils defines its Light, Medium, Heavy and Extra Heavy oils as having Viscosity Grade Nos. 32, 46, 68 and 100 respectively. A minimum VI of 60 is required.

BS 2626: 1955 Specification for refrigerator oils

Grade	Kinematic viscosity, cSt, at 37·8°C
A	10–30
B	30–55
C	55–80
D	80–115

Crankcase oil viscosity

The viscosity classifications of crankcase oil is given in Table 14.8. The viscosities of the W grades are measured at −17·8°C, by ASTM D 2602 'Method of test for apparent

Table 14.8   CRANKCASE OIL VISCOSITY CLASSIFICATION SAE J300A
(Society of Automotive Engineers)

SAE viscosity number	Viscosity at −17·8°C, cP	Viscosity at 98·9°C, cSt
5W	Less than 1 200	
10W	1 200–2 400	
20W	2 400–9 000	
20		5·7– 9·6
30		9·6–12·9
40		12·9–16·8
50		16·8–22·7

Table 14.9   TRANSMISSION AND AXLE LUBRICANT CLASSIFICATION SAE J306
(Society of Automotive Engineers)

SAE Viscosity Number	Viscosity at −17·8°C		Viscosity at 98·9°C	
	SSU	cSt	SSU	cSt
75	<15 000	< 3 250		
80	15 000–100 000	<21 700		
90			75–120	14·2–25·0
140			120–200	25·0–43·0
250			>200	>43·0

viscosity of motor oils at low temperatures using Cold Cranking Simulator', the results are reported in centipoises. If an oil density at −17·8°C of 0·9 g/cc is assumed the corresponding limits in centistokes are:

less than 1300; 1300–2600; 2400–9600.

The viscosity of all oils in this classification must exceed 3·9 cSt at 98·9°C and the minimum viscosities of the 10W and 20W grades may be waived if the viscosities of the oils are not less than 4·2 and 5·7 cSt respectively at 98·9°C.

Transmission and axle lubricant viscosity

Table 14.9 gives the viscosity classification of transmission and axle lubricant. The classification is based on viscosities expressed in Seconds Saybolt Universal and all oils must have a viscosity not less than 40SSU at 98·9°C. The minimum viscosity of SAE 80 may be waived if the viscosity is not less than 48SSU at 98·9°C. The maximum viscosity of SAE 90 may be waived if the extrapolated viscosity at −17·8°C is not greater than 750 000SSU.

American Gear Manufacturers Association viscosity classifications

Table 14.10   SPECIFICATION FOR LUBRICATION OF INDUSTRIAL ENCLOSED GEARING (AGMA 250.02)

AGMA lubricant number	Viscosity, seconds, Saybolt Universal	
	at 37·8°C	at 98·9°C
1	180–240	—
2	280–360	—
3	490–700	—
4	700–1 000	—
5	—	80–105
6	—	105–125
7 compounded	—	125–150
8 compounded	—	150–190
8A compounded	—	190–250

The oils referred to as compounded contain 3–10% acidless tallow or other suitable animal fat and must have a Viscosity Index > 90. The Viscosity Index of the other oils should exceed 30 (or 60 if the operating temperatures vary more than 44·5°C).

Table 14.11   SPECIFICATION FOR MILD EXTREME-PRESSURE LUBRICANTS FOR INDUSTRIAL ENCLOSED GEARING (AGMA 252.01)

AGMA mild EP lubricant number	Viscosity, seconds, Saybolt Universal	
	at 37·8°C	at 98·9°C
2 EP	280–400	—
3 EP	400–700	—
4 EP	700–1 000	—
5 EP	—	80–105
6 EP	—	105–125
7 EP	—	125–150
8 EP	—	150–190

The oils must have VI's > 60.

Table 14.12 THE EFFECT OF PRESSURE ON THE VISCOSITY OF LUBRICANTS

Lubricant	Gauge pressure, MN/m²	Viscosity, cP, at			Pressure coefficient γ, (cm²/dyne) × 10⁻⁹[(m²/N) × 10⁻⁸] between 0 and 34·5 MN/m² at		
		30°C	60°C	100°C	30°C	60°C	100°C
A	0	250	50·5	12·6	2·50	2·13	1·76
	34·5	590	105	23·2			
	68·9	1 330	206	40·5			
	103·4	2 880	386	70			
B	0	122	26·3	7·3	2·70	2·16	1·75
	34·5	310	55·5	13·3			
	68·9	755	113	23			
	103·4	1 750	221	39			
C	0	107	23·3	6·4	2·96	2·28	1·78
	34·5	298	51	11·8			
	68·9	780	111	21·2			
	103·4	2 000	227	36·5			
D	0	310	44·2	9·4	3·46	2·63	1·95
	34·5	1 025	110	18·5			
	68·9	3 400	265	36			
	103·4	—	645	67			
E	0	30·7	8·6	3·1	2·57	2·03	1·54
	34·5	74·5	17·4	5·3			
	68·9	178	33·7	8·7			
	103·4	450	65	14·0			
F	0	173*	60	15·3	3·10	2·44	1·85
	34·5	460*	139	29			
	68·9	1 240*	319	54			
	103·4	3 180*	705	97			
G	0	204	62·5	22·5	1·76	1·43	1·22
	34·5	380	104	35			
	68·9	685	166	52			
	103·4	1 200	260	73			
H	0	143	84	46·5	1·81	1·81	1·94
	34·5	268	156	90·5			
	68·9	445	255	146			
	103·4	700	385	219			
I	0	246*	80	18·0	1·59	1·44	1·23
	34·5	418*	131	27·4			
	68·9	700*	206	40·5			
	103·4	1 090*	310	56·5			
J	0	43·5	11·6	4·05	2·08	1·66	1·35
	34·5	89	20·5	6·45			
	68·9	175·5	35·5	9·9			
	103·4	329	60·5	14·8			
K	0	535	73	13·9	0·59	0·55	0·36
	34·5	655	88	15·8			
	68·9	815	106	18·2			
	103·4	1 020	125	21·4			

*Measured at 40°C

Effect of pressure on the viscosities of lubricants

With the exception of fluids containing major amounts of water over certain pressure and temperature ranges, all fluids increase in viscosity with increasing pressure. In the case of gases the effect is small and can normally be neglected in the pressure range used in gas bearings. For information on the viscosities of gases at higher pressures, above 10–15 bar, the best sources of information are text books on physics or physical chemistry.[45]
   The viscosities of mineral oils vary with pressure according to the type and viscosity level of the oil. The effect of an increase in pressure follows that of a decrease in temperature, LVI oils showing larger changes of viscosity with both temperature and pressure than HVI oils. Very roughly the viscosity of a mineral oil doubles for each 30 MN/m^2 increase in pressure. The viscosities of synthetic lubricants, except aqueous ones, depend also on structure and viscosity level. Those lubricants which are composed of rigid ring containing structures, e.g. tri-tolyl phosphate, polyphenyl ethers, will resemble mineral oils but flexible polymeric fluids, e.g. silicones, polyglycols will be affected less by pressure than are mineral oils. In classical hydrodynamic lubrication it is assumed that viscosity is not affected by pressure. In systems at intermediate pressures, e.g. hydraulic systems the viscosity of the fluid will be significantly greater than its atmospheric pressure value. In elastohydrodynamic lubrication with pressures of the order of 1 GN/m^2 the viscosity of a mineral oil may be 10^5–10^6 times its atmospheric value although such

Table 14.13   SOME PHYSICAL PROPERTIES OF THE LUBRICANTS TESTED

Lubricant	Kinematic viscosity, cSt, at		Kinematic viscosity index	Specific gravity, at 15·6°C	Specific gravity, relative to water at 15·6°C, at		
	37·8°C	98·9°C			30°C	60°C	100 C
A	175·3	15·36	96	0·891	0·884	0·866	0·843
B	83	8·8	84	0·899	0·890	0·871	0·847
C	77	7·8	63	0·889	0·880	0·862	0·838
D	180·1	10·84	8	0·946	0·938	0·920	0·895
E	21·6	3·5	−6	0·927	0·919	0·898	0·874
F	219·8	17·73	99	0·944	0·931	0·915	0·890
G	143·4	24·10	142	1·018	1·008	0·983	0·955
H	130	53	144	0·974	0·962	0·936	0·903
I	295	20·2	87	0·963	0·955	0·935	0·909
J	30·1	4·32	12	0·988	0·978	0·957	0·928
K	232	11·9	−11	1·265	1·256	1·238	1·212

little published information as exists has been determined at low shear rates. Under the combined action of very high pressures and very high shear rates lubricants will show non-Newtonian behaviour. Over small pressure ranges viscosity varies with pressure according to the equation:

$$\eta_p = \eta_o e^{\alpha P}$$

where η_p and η_o are the viscosities at pressure P and atmospheric pressure respectively and α is the pressure coefficient of viscosity.
   In Table 14.12 the effect of pressure on the viscosities of a variety of lubricants is given. The physical properties of the lubricants are given in Table 14.13. The identification codes in these tables are as follows:

   A. HVI mineral oil.
   B. MVI mineral oil to OM 100 specification.
   C. Highly refined naphthenic white oil.

    D. High viscosity LVI oil.
    E. Low viscosity LVI oil.
    F. LVI oil + methacrylate Viscosity Index Improver.
    G. Ethylene oxide—propylene oxide copolymer.
    H. Polydimethyl silicone fluid.
    I. Castor oil.
    J. Di 2ethylhexyl phthalate.
    K. Glycerol.

This information was taken from Galvin, Naylor and Wilson.[46] Further information on the effect of pressure on the viscosity of a wide range of HVI oils is given in Reference 47.

For calculations of film thickness in elastohydrodynamic lubrication it is usual to take the value of the pressure viscosity coefficient between atmospheric pressure and

Table 14.14   PRESSURE-VISCOSITY COEFFICIENT OF MINERAL OILS

Viscosity, cp	α, cm^2/dyne (10 m^2/N)	
	HVI oils	LVI oils
10	$1 \cdot 70 \times 10^{-9}$	$2 \cdot 00 \times 10^{-9}$
100	$2 \cdot 25 \times 10^{-9}$	$2 \cdot 65 \times 10^{-9}$
1 000	$2 \cdot 75 \times 10^{-9}$	$4 \cdot 15 \times 10^{-9}$

about 30 MN/m^2 since it is the pressures in the inlet region to the elastohydrodynamic contact which control the thickness of the lubricant film. Values of α for MVI oils will be close to those for LVI oils.

The best source of information on the viscosity of mineral oils at very high pressures ≈ 1 GN/m^2, is the Pressure-Viscosity Report 1951[47], which gives information on a wide range of mineral oils and a few synthetic lubricants and pure hydrocarbons. More specific information on the effect of pressure on the properties of proprietary lubricants is usually obtainable from lubricant manufacturers.[48]

Paraffinic mineral oils solidify at high pressures, the pressure at which this occurs depends on the nature of the oil and the refining process. At 25°C signs of solidification were noted[47] at pressures ranging from 76–345 MN/m^2, increasing the oil temperature raises the solidification pressure. Data on the solidification of some synthetic lubricant is given by Kobzova.[49] In the lubrication of gears etc., where very much higher pressures than those noted above occur, solidification does not seem to occur probably because of the very much higher shear stresses and the much shorter exposure times than occur in viscometric determinations.

Density of lubricants

Whilst the preferred units of density are kg/m^3 the alternative g/ml is accepted and is in fact most commonly used in reference to lubricant properties. To convert g/ml to kg/m multiply by 10^3. For mineral oils, specific gravities are usually quoted at the referenc temperature 15·6°C (60°F), the density of water at this temperature is 0·999 g/ml. The densities of mineral oils lie in the range 0·85–0·95 g/ml at room temperature increasin with increasing viscosity and aromatic content. The nature of the crude oil, the way th lubricant has been refined and the presence of additives especially lead, soaps an chlorinated compounds all have an effect on density.

Some typical values for the densities of mineral oils at 15·6°C are quoted in Table 14.1 Values for oils from different sources will lie within ±2% of those quoted.

In America the API Gravity System is still used. The relation between this and specific gravity at 15·6°C with respect to water at 15·6°C is given by

$$\text{Degrees API} = \frac{141\cdot5}{\text{S.G.}_{15\cdot6°/15\cdot6°C}} - 131\cdot5$$

The effect of temperature on the density of mineral oils is illustrated by the data given

Table 14.15   THE DENSITIES OF SOME TYPICAL MINERAL OILS

Kinematic viscosity, cSt, at 37·8°C(100°F)	Density, g/ml, at 15·6°C of		
	HVI oil	MVI(N) oil	LVI oil
10	0·86	0·87	0·91$_5$
50	0·87$_5$	0·89	0·93
100	0·88	0·90	0·94
500	0·90	—	0·95

in Table 14.13. For most purposes the variation of density with temperature is given with sufficient accuracy by the equation

$$d_t = d_{20} - 64 \times 10^{-5}(t-20) + 1\cdot5 \times 10^{-7}(t-20)^2$$

where d_t and d_{20} refer to density, g/ml, at $t°C$ and 20°C respectively.

Compressibility and bulk modulus

At the working pressures of modern hydraulic systems liquid lubricants are compressed by a significant amount, 2%v, and at the very high pressures encountered in elastohydrodynamic lubrication the reduction in their volume may be 15 to 20%. The response of servomechanisms and the speed of propagation of shock waves are related to the isentropic bulk modulus of the liquid, the reciprocal of the isentropic compressibility. Isothermal values of compressibility and bulk modulus are normally obtained from measurements of the effect of pressure on the volume of a fixed amount of liquid under controlled temperature conditions whereas isentropic values are obtained by measurements of the velocity of sound waves. In all cases the temperature and pressure (or pressure range) at which the measurements are made must be specified.

Compressibility and bulk modulus values are expressed in several ways, the most common being given below. Isothermal values are given the subscript T, isentropic the subscript s.

$$\text{Compression} = \frac{V_o - V}{V_o}$$

$$\% \text{ Compression} = 100\,\frac{(V_o - V)}{V_o}$$

$$\text{Compressibility} = \beta = -\frac{1}{V}\frac{\partial V}{\partial P}$$

$$\text{Bulk modulus} = K = -V\frac{\partial P}{\partial V}$$

The last two definitions are often called the tangent compressibility and bulk modulus to differentiate them from secant values defined as

$$\text{Secant bulk modulus} = \bar{K} = \frac{V_o P}{V_o - V}$$

The velocity of sound in a liquid is given by

$$\text{Isentropic bulk modulus } K_s = \rho c^2$$

where ρ is the density of the fluid and c is the velocity of sound.
The isentropic and isothermal bulk moduli are related as follows:

$$\frac{K_s}{K_T} = \gamma$$

where γ is the ratio of specific heats at constant pressure and constant volume.
According to Cameron this can be taken as being 1·1–1·2 for mineral oils.[50]

In general, the compressibilities of liquids decrease with increasing molecular weight (or viscosity) and increase with increasing temperature. Rigid, ring-type structures are less compressible than flexible, chain type structures. The compressibilities of liquid lubricants do not vary widely. At 345 MN/m² the percentage reduction in volume at 20°C for glycerol, typical mineral oils and dimethyl silicone fluids are 6·25, 10–11 and 15 respectively.
Data for a typical mineral oil are given in Table 14.16.
Compressibility data on specific lubricants used as hydraulic fluids are available from lubricant manufacturers. Data on mineral oils at 20° and 100°C and pressures up to 345 MN/m² are given by Galvin, Naylor and Wilson,[46] up to 1 GN/m² between 0° and 200°C in the ASME Pressure-Viscosity Report[47] (in the form of density data)

Table 14.16   COMPRESSIBILITY OF A TYPICAL MINERAL OIL

Pressure MN/m²	% reduction in volume at 20°C	100°C
69	3·3	4·2
138	5·6	7·2
207	7·5	9·5
276	9·1	11·4
345	10·5	12·9

and up to 1·586 GN/m² by Ahmed et al.[51] Hayward has reported isentropic compressibility data on 47 mineral oils and 34 fire-resistant fluids at 20°C.[52] He finds that at 20°C and 69 MN/m² the isentropic secant compressibility, $\bar{\beta}_s$, of mineral oils is related to their kinematic viscosity, η, at 22°C by the expression

$$\bar{\beta}_s = 3·5 - 0·2 \log \eta$$

where $\bar{\beta}_s$ is expressed in 10^{-6} lbf/in² and η in cSt.
When $\bar{\beta}$ is expressed in N/m² and η in cSt this becomes

$$\bar{\beta}_s = \frac{3·5 - 0·2 \log \eta}{6·895 \times 10^{-3}} \, \text{N/m}^2$$

Under the same conditions typical isentropic compressibility values for water-glycol, water-in-oil emulsion and phosphate ester fluids are 2·01, 3·07 and 2·31 × 10^{-6} lbf/in² respectively[53, 54] (13·859, 21·168 and 15·927 × 10^{-3} N/m² respectively).
Peeler and Green[55] give a large number of references to isothermal and isentropic measurements on a wide variety of fluids.

Solubility of gases in lubricating oils

Mineral oils contain approximately 10% v/v of dissolved air under normal conditions. This dissolved air has no measurable effect on the physical properties of the oil but is very advantageous under boundary lubrication conditions. To a first approximation Henry's Law is obeyed, i.e. the weight of gas dissolved in the oil is directly proportional to the partial pressure of the gas. Thus if pressure is reduced below atmospheric, as may happen for example on the inlet side of hydraulic pumps, air will be released from the oil. This air may give rise to cavitation troubles or cause inadequate lubrication. The compressibility of an air-oil mixture is much greater than of oil containing dissolved air. According to Hayward[56] the viscosity of the bubbly oil is related to that of air-free oil as follows:

$$\mu_B = \mu_o(1 + 0.15B)$$

where μ_B and μ_o are the viscosities of the bubbly oil and the bubble-free oil and B is the percentage bubble content.

If pressure is raised above atmospheric then more gas will dissolve in the oil if gas is available. Large amounts of dissolved gas can cause major reductions in oil viscosity. On returning to atmospheric pressure the gas will be released and must be given time to clear if the oil is to be recirculated. With large amounts of gas dissolved in the oil and especially at elevated temperatures chemical reaction between the gas and the oil may occur and in the case of oxygen the reaction may be violent.

The solubilities of gases in liquids may be expressed in terms of the Bunsen or the Ostwald Coefficients. The Bunsen Coefficient, α, is defined as the volume of gas at NTP dissolved in unit volume of the liquid when the partial pressure of the gas is one atmosphere. The Ostwald Coefficient, $\aleph$, is the ratio of the concentration of gas in solution to its concentration in the gas phase. The two coefficients are related as follows:

$$\aleph = \frac{\alpha T}{273}$$

where T is the temperature in K.

The Ostwald Coefficient, which gives directly the volume of gas dissolved in unit volume of the oil at the same temperature and pressure, is the more useful in practical situations.

Daniel and Baldwin[57] show that gas solubility decreases as mineral oil viscosity increases. They give data, Table 14.17, for gases dissolved in two HVI mineral oils having kinematic viscosities at 37.8°C of 35 and 268 cSt respectively. The smaller of each pair of numbers is the solubility in the more viscous oil.

Table 14.17   SOLUBILITY OF GASES IN TWO HVI MINERAL OILS

	Ostwald coefficient at		
	20°C	60°C	100°C
Hydrogen	0·052–0·050	0·067–0·064	0·090–0·081
Carbon dioxide	1·04 –0·92	0·78 –0·70	0·64 –0·57
Oxygen	0·155–0·135	0·16 –0·145	0·18 –0·16
Nitrogen	0·082–0·071	0·096–0·090	0·11 –0·10
Air	0·097–0·085	0·11 –0·10	0·13 –0·11

Campbell[58] found that air dissolved in oil at pressures up to 27·6 MN/m² did not affect the compressibility of the oil and that the ratio of volume of dissolved air to volume of oil was 0·10 at all the pressures and temperatures used. Summers–Smith[59] describing the problems arising from gas solubility and reactivity in compressor lubrication gives further information on gas solubility and the effect of dissolved gas on oil viscosity. Safronova and Zhuze[60] give solubility data up to 31 MN/m² and 100°C for methane, propane, ethylene, carbon dioxide and nitrogen. Swearingen and Redding[61] have measured the reduction in viscosity of mineral oils saturated with natural gas at high pressures. Klaus, Johnson and Fesco[62] give data on the effect of nitrogen on the viscosities of a mineral oil and some synthetic lubricants at high pressures.

Thermal properties of liquids

Most of the published information on the specific heat and thermal conductivity appears to be based on the correlations suggested by Cragoe[63] some forty years ago. These are:

$$\text{Specific heat (Btu/lb °F or cal/gm °C)} = \frac{(0.388 + 0.000\ 45\ t)}{\sqrt{d}}$$

$$\text{Thermal conductivity (Btu in/sq ft h °F)} = \frac{0.821 - 0.000\ 244\ t}{d}$$

where

$$t = \text{temperature °F}$$
$$d = \text{S.G. } 15.6°/15.6°C$$

1 Btu/lb°F = 4.187 kJ/kg °C
1 Btu in/sq ft h °F = 0.144 2 W/m² °C

Gunderson and Hart[31] say that the specific heats of diesters are approximately 5–10% greater than those of mineral oils at ambient temperatures but change rather less with temperature. Those of hindered esters are similar to diesters. The typical data in Table 14.18 for other types of synthetic lubricants have been taken from a variety of source

Table 14.18   SPECIFIC HEATS AND THERMAL CONDUCTIVITIES OF SOME SYNTHETIC LUBRICANTS

Fluid	Specific heat, kJ/kg °C	Thermal conductivity, W/m² °C
Water-glycol F. R. fluids	2.9–3.4	—
Water-in-oil emulsions	2.3–3.4	—
Phosphate esters	1.55	0.0086
Polyglycols	1.8–2.0	0.012–0.013
Chlorinated aromatics	1.2	0.0086
Silicone fluids	1.40–1.55	0.0086–0.012

including the manufacturers' literature. In such literature it is noticeable that data quoted for apparently very similar fluids may differ by suspiciously large amounts.

Operating temperature range of lubricants

The lower level of operating temperature of liquid lubricants is set either by the development of excessively high viscosity or by solidification. Complex mixtures such a lubricating oils do not have sharp solidification temperatures but some indication of the lowest temperature at which an oil will flow under its own weight is given by the pour point determined by the standard methods I.P. 15/67[41] or ASTM D97-66. In this tes LVI and MVI(N) lubricating oils show no phase change but attain a viscosity—about 10^4 Poise—when no movement of the oil can be detected. HVI and MVI(P) oils precipi tate small amounts of wax crystals which effectively form a structure in the oil and prevent it flowing under gravity. This structure is easily broken down by shearing stresses.

The pour points of HVI and MVI(P) oils are controlled by the severity of the dewaxin process used in refining and by the use of additives, pour point depressants. Sinc dewaxing to give very low pour points is very expensive the oils will usually be treated to give a pour point of about $-5°C$ and for use at lower temperatures, e.g. $-30°C$ pour point depressants will be used. The pour points of LVI and MVI(N) oils are not affected by such additives but are governed by their composition and refining. Typically the least viscous LVI and MVI(N) oils will have values of about $-40°C$ rising to $-10°C$ for the most viscous oils.

The upper operating temperature limit for liquid lubricants is set by thermal degradation, oxidation and volatility (especially in the case of aqueous fluids). Exact figures cannot be given as the extent of thermal degradation and oxidation is dependent on many factors including time spent at the high temperature, the availability of air and the presence of pro-oxidant catalytic metals, especially copper. Fowle[35] has collected together experience on solvent refined oils and gives the following figures as a guide for conditions where (a) the life is limited by thermal degradation in the absence of air, (b) the life of a straight mineral oil is limited by oxidation, (c) the life of an oil containing an oxidation inhibitor is limited by oxidation.

Temperature limit,	Life in hours			
°C	10 hrs	100 hrs	1 000 hrs	10 000 hrs
Condition a	380°	360°	330°	300°
Condition b	130°	100°	75°	50°
Condition c	160°	135°	110°	90°

Lansdowne,[64] besides giving the upper temperature range of some unconventional high temperature lubricants such as glass and liquid metals, has estimated life-v-operating temperature for polyphenyl ethers, silicones, esters and phosphate esters. Hatton[65] has given decomposition temperatures and operational temperature ranges for a wide variety of synthetic lubricants.

GREASES

Greases are solid or semi-solid mixtures of a liquid lubricant and a thickener, the thickener being dispersed in such a way as to impart the desired rheological properties to the grease. Any maltreatment such as subjecting the grease to excessively high temperatures or to excessive mechanical working which changes the form of the thickener will change the rheological properties of the grease and may cause it to liquefy.

The most commonly used liquid lubricants in grease manufacture are mineral oils; di-esters and silicone fluids are used where very high or very low temperatures are encountered. Fatty acid soaps of sodium, calcium, lithium and aluminium are commonly used as thickeners; specially treated clays, thermally stable organic pigments and other solids are also used. Additives both similar to those used in liquid lubricants and insoluble additives such as molybdenum disulphide may be added. The nature of the liquid lubricant and of the thickener both have important effects on the properties of the grease, e.g. sodium soap based greases have poor water resistance because of the water solubility of the sodium soap.

The consistency of the grease is adjusted to suit the application. Hard, block greases can be shaped and pressed against slow moving shafts, more fluid greases are required for centralised lubrication systems. Consistency is measured by the cone penetration method of ASTM D.217-68, IP 50/69 (or IP 167/50T for semi-fluid greases having a penetration of more than 475). The grading system proposed by the National Lubricating Grease Institute is widely used and is shown in Table 14.19.

Where a grease is working at temperatures approaching its upper limit frequent relubrication is necessary.

The rheological properties of greases permit simple sealing arrangements to be made and the intervals between relubrication can be very long. However, the fact that there is no flow of lubricant through the bearing means that the cooling function is lost and

Table 14.19

NLGI grade	Worked penetration
000	445–475
00	400–430
0	355–385
1	310–340
2	265–295
3	220–250
4	175–205
5	130–160
6	80–115

APPROXIMATE TEMPERATURE RANGE OF GREASES

1. Limits set by liquid lubricant

	Lower limit, °C	Upper limit, °C
Mineral oils	−25° (−40° with special oils)	150°
Esters	−75°	175°
Silicones	−75°	225°

2. Limits set by thickener

Calcium soap (water stabilised)	60°
Calcium complex	150°
Sodium complex	150°
Lithium hydroxystearate	150°
Aluminium soap	60°
Aluminium complex	150°
Clay	Greater than 150° Limited by waterproofing agent used

the maximum speeds of operation of grease lubricated ball and roller bearings are therefore limited. Details are given in manufacturers' catalogues. Information on the manufacture of greases, their properties and methods of testing are given in References 66–70.

Further topics

References to several important areas of tribology such as selection of type and viscosity grade of lubricant, bearing design, etc. are included in the bibliography.

REFERENCES

General reference and source books

1. 'Lubrication and Wear: Fundamentals and Application to Design', Proceedings of conference, *Proc. Inst. Mech.E.*, London, **182**, Pt 3A (1967–68)
2. O'CONNOR, J. J. (editor), *Standard Handbook of Lubrication Engineering* (sponsored by the American Society of Lubrication Engineers), McGraw-Hill (1968)
3. *Tribology Handbook*, Butterworth (1973)

Hydrodynamic and Hydrostatic Lubrication

4. CAMERON, A., *Principles of Lubrication*, Longmans (1966)
5. BARWELL, F. T., *Lubrication of Bearings*, Butterworth (1962)
6. SHAW, M. C. and MACKS, E. F., *Analysis and Lubrication of Bearings*, McGraw-Hill (1949)

7. GROSS, W. A., *Gas Film Lubrication,* Wiley (1962)
8. GRASSAM, N. S. and POWELL, J. W., *Gas Lubricated Bearings,* Butterworth (1964)
9. STANSFIELD, F., *Hydrostatic Bearings,* Machinery Publishing Co. (1971)
10. 'Externally Pressurised Bearings', Proceedings of Symposium, *Inst. Mech.E.,* London (1972)
11. LLOYD, T. and MCCALLION, H., 'Recent developments in fluid film lubrication theory', Proceedings of Conference, *Proc.I.Mech.E.,* **182,** Pt 3A, Paper 3 (1967–68)
12. CAMPBELL, J., LOVE, P. P., MARTIN, F. A. and RAFIQUE, S. O., 'Bearings for Reciprocating Machinery. A review of the present state of theoretical, experimental and service knowledge', Proceedings of Conference, *Proc.I.Mech.E.,* **182,** Pt 3A, Paper 4 (1967–68)
13. RAMSDEN, P., 'Review of published data and their application to the design of large bearings for steam turbines', Proceedings of Conference, *Proc.I.Mech.E.,* **182,** Pt 3A, Paper 6 (1967–68)
14. GROSS, W. A., 'Gas Bearings: Journal and Thrust', Proceedings of Conference, *Proc.I.Mech.E.,* **182,** Pt 3A, Paper 9 (1967–68)

Elastohydrodynamic lubrication

15. DOWSON, D. and HIGGINSON, G. R., *Elastohydrodynamic Lubrication: The Fundamentals of Gear and Roller Lubrication,* Pergamon Press (1965)
16. Proceedings of Symposium on Elastohydrodynamic Lubrication, *Proc.I.Mech.E.,* **180,** Pt. 3B (1965–66)
17. 'Elastohydrodynamic Lubrication', Proceedings of Symposium, *Proc.I.Mech.E.* (1972)
18. DOWSON, D., 'Elastohydrodynamics', Proceedings of Conference, *Proc.I.Mech.E.,* **182,** Pt. 3A, Paper 10 (1967–68)

Boundary lubrication

19. BOWDEN, F. P. and TABOR, D., *The Friction and Lubrication of Solids, Pt. I,* Oxford University Press (1950)
20. BOWDEN, F. P. and TABOR, D., *The Friction and Lubrication of Solids, Pt. II,* Oxford University Press (1964)
21. LING, F. F., KLAUS, E. E. and FEIN, R. S. (editors), *Boundary Lubrication: An Appraisal of the World Literature.* American Society of Mechanical Engineers (1969)
22. TABOR, D., 'Solid Friction and Boundary Lubrication', Proceedings of Conference, *Proc.I.Mech.E.,* **182,** Pt. 3A, Paper 16 (1967–68)

Solid lubricants

23. BRAITHWAITE, E. R., *Solid Lubricants and Surfaces,* Pergamon (1964)
24. BOYD, J. (editor), *International Conference on Solid Lubrication,* American Society of Lubrication Engineers (1971)
25. ROBBINS, E. J., 'Dry Lubrication', *Tribology,* **3,** No. 2, p. 84 (1970)
26. DEVINE, M. J., LAMSON, E. R., CERINI, B. A. and CARROLL, R. J., 'Solid Lubricants', Proceedings of Conference, *Proc.I.Mech.E.,* **182,** Pt. 3A, Paper 20 (1967–68)
27. SALOMON, G., 'Plastics and Rubbers in Machine Design', Proceedings of Conference, *Proc.I. Mech.E.,* **182,** Pt. 3A, Paper 24 (1967–68)
28. LANCASTER, J. K., 'Self-lubricating Composite for Bearings. Engineering Materials and Methods', *I.Mech.E.,* London (1971)
29. Engineering Sciences Data Item No. 68018, 'Dry Rubbing Bearings—A Guide to Design and Material Selection', Engineering Sciences Data Unit (1968)

Lubricants

30. BRAITHWAITE, E. R., *Lubrication and Lubricants,* Elsevier (1967)
31. GUNDERSON, R. C. and HART, A. W., *Synthetic Lubricants,* Reinhold (1962)
32. HATTON, R. E., *Introduction to Hydraulic Fluids,* Reinhold (1962)
33. WARRING, R. H., *Hydraulic Fluids,* Trade and Technical Press Ltd. (1961)
34. *Fire Resistant Fluids for Fluid Power Systems,* Association of Hydraulic Equipment Manufacturers
35. FOWLE, T. I., 'Lubricants for Fluid Film and Hertzian Contact Conditions', Proceedings of Conference, *Proc.I.Mech.E.,* **182,** Pt. 3A, Paper 19 (1967–68)
36. SCARLETT, N. A., 'Use of Grease in Rolling Bearings', Proceedings of Conference, *Proc.I.Mech.E.,* **182,** Pt. 3A, Paper 21 (1967–68)
37. TANTAM, D. H., 'Cold Environments', Proceedings of Conference, *Proc.I.Mech.E.,* **182,** Pt. 3A, Paper 29 (1967–68)
38. BLANCHARD, P. M., 'Lubricants for Hot Environments with Special Reference to Aero Gas Turbines Operating Under Severe Conditions', Proceedings of Conference, *Proc.I.Mech.E.,* **182,** Pt. 3A, Paper 30 (1967–68)

Selection of lubricants

39. *Information on Factors Affecting the Selection of Lubricants*, British Standards Institution (1969)
40. BEUERLEIN, P. and KARA, W. H., 'Selection of Lubricants', Proceedings of Conference, *Proc.I. Mech.E.*, **182**, Pt. 3A, Paper 36 (1967–68)

Viscosity

41. *I.P. Standards for Petroleum and its Products, Pt. I*, Institute of Petroleum (revised annually)
42. Engineering Sciences Data Item No. 66024, 'Approximate Data on the Viscosity of Some Common Liquids', Engineering Science Data Unit (1966)
43. Engineering Sciences Data Item No. 67015, 'A Guide to the Viscosity of Liquid Petroleum Products', Engineering Science Data Unit (1967)
44. ROELANDS, C. J. A., BLOK, H. and VLUGTER, J. C., 'A New Viscosity-Temperature Criterion for Lubricating Oil', *ASME Paper No. 64—LUB-3*
45. PARTINGTON, J. R., *An Advanced Treatise on Physical Chemistry, Vol. I*, Longmans (1949)
46. GALVIN, G. D., NAYLOR, H. and WILSON, A. R., 'The Effect of Pressure and Temperature on Some Properties of Fluids of Importance in Elastohydrodynamic Lubrication', 2nd Lubrication and Wear Group Convention, *Proc.I.Mech.E.*, **178**, Pt. 3N (1964)
47. *Viscosity and Density of over 40 Lubricating Fluids of Known Composition at Pressures up to 150 000 psi and Temperatures to 425°F*, Vols. I and II, American Society of Mechanical Engineers (1953)
48. *Technical Data on Shell Tellus Oils*, Shell International Petroleum Co. Ltd., 2nd ed (1967)
49. KOBZOVA, R. I., ERSHOVA, T. P., OPARINA, E. M. and TOBYANSKAYA, G. S., 'Solidification of Lubricating Oils at High Pressures', *Chemistry and Technology of Fuels and Oils*, pp 586–8 (1967)

Compressibility

50. CAMERON, A., 'The Isothermal and Adiabatic Compressibilities of Oil', *Journal Institute of Petroleum*, **31**, 263, pp 421–7 (1945)
51. AHMED, N., GOLDMAN, B., VENKATESEN, P. S., and CARTWRIGHT, J. S., 'The Compressibility of Selected Fluids at Pressures up to 230 000 psi', *American Society of Lubrication Engineers*, Pre-Print No. 71 AM, 2E-1 (1971)
52. HAYWARD, A. T. J., MARTINS, R. R. and ROBERTSON, J., 'Compressibility Measurements on Hydraulic Fluids, Pt. I, Isentropic Measurements on 47 Mineral Oils at 20°C and Pressures up to 10 000 lb/in², *National Engineering Laboratory Report No. 173* (1964)
53. HAYWARD, A. T. J., MARTINS, R. R. and ROBERTSON, J., 'Compressibility Measurements on Hydraulic Fluids, Pt. II, Isentropic Measurements on 34 Fire-Resistant Fluids at 20°C and Pressures up to 10 000 lb/in²', *National Engineering Laboratory Report No. 176* (1965)
54. HAYWARD, A. T. J., 'Generalisations for Isentropic and Isothermal Compressibility of Hydraulic Mineral Oils', *National Engineering Laboratory Report No. 443* (1970)
55. PEELER, R. L. and GREEN, J., 'Measurement of Bulk Modulus of Hydraulic Fluids', *ASTM Bulletin*, pp 51–7 (Jan. 1959)

Solubility of gases

56. HAYWARD, A. T. J., 'Air Bubbles in Oil—Their Effect on Viscosity and Compressibility', *National Engineering Laboratory Report No. 5* (1961)
57. BALDWIN, R. R. and DANIEL, S. G., 'The Solubility of Gases in Lubricating Oils and Fuels, *Jour. Inst. of Petroleum*, **39**, No. 350, p 105 (1953)
58. CAMPBELL, J. E., 'Investigation of the Fundamental Characteristics of High Performance Aircraft Hydraulic Systems', *U.S.A.F. Wright-Patterson Base Technical Report* 5997 (June 1950)
59. SUMMERS-SMITH, D., 'Selection of Lubricant Viscosity Grade for Reciprocating Gas Compressors', 1968 Tribology Group Convention, *Proc.I.Mech.E.*, **182**, Pt. 3N (1967–68)
60. SAFRONOVA, T. P. and ZHUZE, T. P., 'The Solubility of Gases in Petroleum Oils at Elevated Pressures and Temperatures', *Khim i Tekhnol. Topliva i Mosel, Pt. 2*, 1958 p 41 (English translation available—NLL ref. SLE-TT 61-10489)
61. SWEARINGEN, J. S. and REDDING, E. D., 'Viscosity Characteristics of Lubricating Oil Saturated with Natural Gas at High Pressures', *Jour. Ind. Eng. Chem.*, **34**, p 1496 (1942)
62. KLAUS, E. E., JOHNSON, R. H., and FRESCO, G. P., 'Development of a Precision Capillary-type Pressure Viscometer', *Trans. ASLE*, **9**, p 113 (1966)

Thermal properties

63. CRAGOE, C. S., Bureau of Standards, *Misc. Publication*, **97**, 1929
64. LANSDOWN, A. R., 'Liquid Lubricants—Functions and Requirements', NASA Symposium on Interdisciplinary Approach to Liquid Lubricant Technology (1972)

65. HATTON, R. E., 'Synthetic Oils', NASA Symposium on Interdisciplinary Approach to Liquid Lubricant Technology (1972)

Greases

66. BONER, C. J., *Manufacture and Applications of Lubricating Greases*, Reinhold (1954)
67. HARRIS, J. H., *The Lubrication of Rolling Bearings*, Shell Mex and B.P. Ltd., London (1967)
68. JONES, E. F., 'General Characteristics of Grease and Grease Lubrication', *Tribology*, **1**, No. 3, p 161 (1968)
69. JONES, E. F., 'Grease Manufacture', *Tribology*, **1**, No. 4, p 209 (1968)
70. MOORE, H. D., 'Laboratory Tests and their Significance', *Tribology*, **2**, No. 1, p 18 (1969)

BIBLIOGRAPHY
Journals publishing papers on tribological matters
ASLE Transactions
Bulletin of the Japanese Society of Mechanical Engineers
Chemistry and Technology of Fuels and Oils (Russian lubrication work)
Industrial Lubricants and Tribology
Journal of the Institute of Petroleum
Journal of Lubrication Technology
Journal of Mechanical Engineering Science
Lubrication Engineering
Proceedings of the Institution of Mechanical Engineers
Schmiertechnic Tribologie
Tribology
Tribos (literature abstracts)
VDI Zeitschrift
Wear

Symposia, conferences and conventions
Institution of Mechanical Engineers, London. Annual Meetings of Lubrication and Wear Group.
1st Annual Meeting: Non-Conventional Lubricants and Bearing Materials such as used in Nuclear Engineering. 3rd Annual Meeting: Iron and Steel Works Lubrication.
Institution of Mechanical Engineers, London. Conventions. Lubrication and Wear Group, 1963–67; Tribology Group, 1968 onwards
Institution of Mechanical Engineers, London. Proceedings of Symposia
    Fatigue in Rolling Contact, 1964
    Journal Bearings for Reciprocating and Turbo Machinery
    Lubrication and Wear in Living and Artificial Human Joints, Proceedings, **181**, Pt. 3J
    Lubrication of Textile Machinery
    Experimental Methods in Tribology, Proceedings, **182**, Pt. 3G (1967–68)
    The Use of Grease as an Engineering Component, Proceedings, **184**, Pt. 3F (1969–70)
    Gearing in 1970, Proceedings, **184**, Pt. 30 (1969–70)
    Lubrication in Hostile Environments, **183**, Pt. 31 (1968–69)
National Aeronautics and Space Administration
    Advanced Bearing Technology, NASA, SP-38 (1964)
    Interdisciplinary Approach to Friction and Wear, NASA, SP-181 (1968)
    Interdisciplinary Approach to the Lubrication of Concentrated Contacts, NASA, SP-237 (1970)

Useful introductory texts
BOWDEN, F. P. and TABOR, D., *Friction and Lubrication*, Methuens Monographs on Physical Subjects (1967)
FREEMAN, P., *Lubrication and Friction*, Pitman (1962)
Industrial Lubrication, British Petroleum Co. Ltd. (1966)
Lubrication Theory and its Application, B.P. Trading Ltd. (1969)
SCHILLING, A., *Motor Oils and Engine Lubrication*, Scientific Publications (GB) Ltd. (1972)
SUMMERS-SMITH, D., *An Introduction to Tribology in Industry*, The Machinery Publishing Company (1969)
The Application of Lubricants, Shell International Petroleum Co. Ltd. (1965)

Bearings, selection and design; gears
Engineering Sciences Data Item No. 65007, 'General Guide to the Choice of Journal Bearing Type', Engineering Science Data Unit (1965)

Engineering Sciences Data Item No. 67003, 'General Guide to the Choice of Thrust Bearing Type', Engineering Science Data Unit (1967)

Engineering Sciences Data Item No. 66023/1, 'Calculation Methods for Steadily Loaded Pressure Fed Hydrodynamic Journal Bearings', Engineering Science Data Unit (1966)

Engineering Sciences Data Item No. 69002A, 'Computer Service for Prediction of Performance of Steadily Loaded Pressure Fed Hydrodynamic Journal Bearings', Engineering Sciences Data Unit (1969)

NEALE, M. J., 'The Selection of Bearings', Proceedings of Conference, *Proc.I.Mech.E.*, **182,** Pt. 3A, Paper 35 (1967–68)

NEALE, M. J., 'The Design of Plain Bearings. Symposium. Iron and Steel Works Lubrication', *Proc.I. Mech.E.*, **179,** Pt. 3D (1964–65)

RIPPEL, H. C., *Cast Bronze Bearing Design Manual*, Cast Bronze Bearing Inst. Inc. (1965)

RIPPEL, H. C., *Cast Bronze Hydrostatic Bearing Design Manual*, Cast Bronze Bearing Inst. Inc. (1965)

RIPPEL, H. C., *Cast Bronze Thrust Bearing Design Manual*, Cast Bronze Bearing Inst. Inc. (1967)

ALLAN, R. K., *Rolling Bearings*, Pitman (1964)

BONER, C. J., *Gear and Transmission Lubricants*, Reinhold (1964)

BUCKINGHAM, E., *Analytical Mechanics of Gears*, McGraw-Hill

DUDLEY, D. W., *Gear Handbook*, McGraw-Hill (1962)

Lubrication of Industrial Gears, Shell International Petroleum Co. Ltd. (1964)

MERRIT, H. E., *Gear Engineering*, Pitman (1971)

TUPLIN, W. A., *Gear Design*, The Machinery Publishing Co. Ltd. (1962)

APPENDIX
VISCOSITY CONVERSION TABLE (see page 14-8)

Kinematic viscosity centistokes	Saybolt universal seconds	Redwood I seconds	Engler degrees	Kinematic viscosity centistokes	Saybolt universal seconds	Redwood I seconds	Engler degrees
5	42	39	1·4	55	255	226	7·3
6	46	41	1·5	56	260	230	7·46
7	49	44	1·55	57	264	234	7·55
8	52	46	1·65	58	269	238	7·7
9	55	49	1·75	59	274	242	7·8
10	59	52	1·85	60	278	246	7·95
11	62	55	1·95	61	283	250	8·05
12	66	58	2·05	62	287	254	8·2
13	70	62	2·15	63	292	258	8·3
14	74	65	2·25	64	297	262	8·45
15	77	68	2·35	65	301	266	8·6
16	81	72	2·45	66	306	271	8·7
17	85	75	2·55	67	311	275	8·85
18	89	79	2·65	68	315	279	9·0
19	94	82	2·75	69	320	283	9·1
20	98	86	2·9	70	324	287	9·25
21	102	90	3·0	72	334	295	9·5
22	106	94	3·1	74	343	303	9·75
23	111	97	3·25	76	352	311	10·05
24	115	101	3·35	78	361	319	10·3
25	119	105	3·45	80	371	328	10·55
26	128	109	3·6	82	380	336	10·8
27	128	113	3·7	84	389	344	11·1
28	132	117	3·85	86	399	352	11·35
29	140	121	3·95	88	408	360	11·6
30	141	125	4·1	90	417	369	11·9
31	146	129	4·2	92	426	377	12·15
32	150	133	4·35	94	436	385	12·4
33	155	137	4·45	96	445	393	12·6
34	159	141	4·6	98	454	401	12·95
35	164	145	4·7	100	463	410	13·2
36	168	149	4·85	102	473	418	13·45
37	173	153	4·95	104	482	426	13·75
38	177	157	5·1	106	491	435	14·0
39	182	161	5·2	108	500	443	14·25
40	186	165	5·35	110	510	451	14·5
41	191	169	5·5	112	519	459	14·8
42	195	173	5·6	114	528	467	15·05
43	200	177	5·75	116	537	476	15·3
44	204	181	5·85	118	547	484	15·6
45	209	185	6·0	120	556	492	15·85
46	214	189	6·5	122	565	500	16·1
47	218	193	6·25	124	575	508	16·35
48	223	197	6·4	126	584	517	16·65
49	228	201	6·5	128	593	525	16·9
50	232	205	6·65	130	602	533	17·15
51	237	209	6·75	132	612	541	17·4
52	241	213	6·9	134	621	549	17·7
53	246	218	7·05	136	630	558	17·95
54	251	222	7·15	138	640	566	18·2

continued overleaf

Viscosity Conversion Table—*continued*

Kinematic viscosity centistokes	Saybolt universal seconds	Redwood I seconds	Engler degrees	Kinematic viscosity centistokes	Saybolt universal seconds	Redwood I seconds	Engler degrees
140	649	574	18·5	180	834	738	23·75
142	658	582	18·75	184	853	754	24·3
144	667	590	19·0	188	871	771	24·8
146	677	599	19·25	192	890	787	25·35
148	686	607	19·55	196	908	804	25·85
150	695	615	19·8	200	927	820	26·4
152	704	623	20·05	204	946	836	26·95
154	714	631	20·35	208	964	853	27·45
156	723	640	20·6	212	983	869	28·0
158	732	648	20·85	216	1 001	886	28·5
160	742	656	21·1	220	1 020	902	29·05
164	760	672	21·65	224	1 038	918	29·55
168	779	689	22·2	228	1 057	935	30·1
172	797	705	22·7	232	1 075	951	30·6
176	816	722	23·25	236	1 094	968	31·15
				Above 236 multiply by 4·634 7		4·10	0·132

15 EFFLUENTS AND ENVIRONMENT

EFFLUENTS AND ENVIRONMENT 15

15   EFFLUENTS AND ENVIRONMENT

WATER AND EFFLUENTS IN INDUSTRY

A. I. BIGGS

INTRODUCTION

For the purpose of this chapter Industrial Pollution will be confined to pollution of inland waters, coastal waters and the atmosphere by discharges of liquid and gaseous effluents from industrial premises and the effect on noise levels due to industrial operations. There are very few industrial manufacturing units which are not involved in at least one of these aspects of pollution and many in all three.

The obligations of industry to abate pollution from industrial sources are laid down in statutory legislation and to some extent by Common Law. The requirements can vary considerably from one area to another since the basic principle of UK legislation is that discharges of effluent, whether gaseous or liquid, must be with the consent of the appropriate authority. When granting this consent, the authority may impose conditions which it considers are relevant to the circumstances of the discharge. It is also advised that these conditions should be reasonable and practicable and should be capable of technical achievement. The system is necessarily elastic and enables conditions to be varied at specified periods of time to cover changes in circumstances and also changes in industrial processes, the conditions and future use of the receiving source and improvements in the technology associated with the abatement of pollution.

LEGISLATION

The CBI handbook on 'Disposal of Trade Effluent in England and Wales[1] describes in some detail the legislative controls which are related to the discharge of liquid effluents but for convenience the principal legislation connected with industrial pollution is listed below.

In the case of discharges of liquid effluents to rivers, streams and tidal waters the principal legislation is
The Rivers (Prevention of Pollution) Acts 1951 and 1961.
The Clean Rivers (Estuaries and Tidal Waters) Act 1960.
The Water Resources Act 1963.
The equivalent legislation in Scotland is the Rivers (Prevention of Pollution) (Scotland) Acts 1951 and 1965. The Water Resources Act 1963 does not apply to Scotland.

In the case of discharges of industrial effluents to the public sewerage system, the main legislation is
The Public Health Act 1936.
The Public Health (Drainage of Trade Premises) Act 1937.
The Public Health Act 1961.
The equivalent legislation in Scotland is the Sewerage (Scotland) Act 1968.

As regards emissions of smoke, grit, dust and gases to the atmosphere the main legislation is
The Alkali etc. Act 1906.
The Clean Air Acts 1956 and 1968.
These Acts apply to England, Scotland and Wales.

In the case of noise the main Act is:
The Noise Abatement Act 1965.
The nuisance aspect of noise is covered by the Public Health Acts. The above summary does not refer to all the legislation connected with pollution due to liquid and gaseous effluents or noise but it does indicate that industrial pollution is subject to extensive legislation. In addition Common Law provides additional protection, particularly to riparian owners. The judgement of Lord Macnaghten[2] that every riparian owner is entitled to river water in its natural state has established the background for Common Law rights in connection with discharges of effluent to water courses. The same rights could be established in respect of pollution of the atmosphere and nuisance due to noise.

Although cases connected with pollution and rights under Common Law are not common, they can be very significant in respect of industrial production.

Discharges of liquid effluent

The main provisions of the legislation dealing with discharges of industrial effluent to watercourses and public sewers are described in the CBI Handbook[1] but for convenience these are summarised below.

DISCHARGES TO WATER COURSES, INCLUDING TIDAL WATERS

(a) Discharges of effluent require the consent of the appropriate river authority.
(b) The Authority may attach conditions to the consent to discharge.
(c) The discharger has a right of appeal to the Secretary of State for the Environment (Secretaries of State for Scotland and Wales) if the conditions are considered unreasonable or if consent is unreasonably withheld. The only exceptions are some pre-1960 discharges of effluent to tidal waters.

DISCHARGES TO PUBLIC SEWERS

(a) Post-1937 discharges of effluent to public sewers are subject to the consent of the appropriate local authority,
(b) The authority may attach conditions to the consent, including a charge for reception and treatment.
(c) The discharger has a right to appeal to the Secretary of State for the Environment (Secretary of State for Wales or Secretary of State for Scotland when provisions of 1968 Act become effective) if the discharger considers that the conditions, including the charge, are unreasonable or if consent is unreasonably withheld. In the case of pre-1937 discharges which are unchanged in terms of volume, rate of flow and nature and composition the authority is entitled to serve a direction on the discharger. This direction may include a charge for reception and treatment but conditions are restricted to pH, temperature, inspection and sampling manhole and the giving of relevant information.

It will be apparent from the above that practically all discharges of liquid industrial effluent to public sewers, watercourses and tidal waters are subject to the control of the appropriate river authority or local authority. The only significant exemptions are pre-

1960 discharges of effluent to tidal waters which are not subject to a Tidal Waters Order. (These are listed in Appendix VI of the CBI Handbook but the Secretary of State for the Environment and the Secretaries of State for Wales and Scotland have powers to make additional orders for estuaries or coastal waters). The exemption for pre-1960 discharges is conditioned by the requirement that the volume and nature and composition of the effluent discharged to tidal waters should not be substantially different from the effluent discharged in the previous year.

Management is responsible for ensuring that the consent of the authority is obtained and that the conditions written into the consent are complied with. It must be stressed that the conditions should be such that they can be met by the discharger. This involves consultations with the authority *before* the consent (and the conditions contained therein) are served on the discharger since this is the most appropriate time when the difficulties involved in meeting the conditions and the need to relate these to the problems envisaged by the authority may be resolved.

In the event of an agreed solution not being possible, or the conditions being impracticable or not capable of being met, the discharger should exercise his right of appeal to the Minister once the formal consent has been served. Very little purpose is served and indeed a great deal of harm can be done by accepting conditions which cannot be met on the assumption that any breach of conditions will not be noticed. Apart from the liability to prosecution, the relations between the discharger and the authority will be soured and a successful prosecution may provide the springboard for a Common Law action and the liability of an injunction.

TECHNICAL CONDITIONS

Although the legislation spells out in some detail the requirements which may be included in a consent, the most important from the industrial point of view are the technical conditions which may be imposed on the discharge and the charge for treatment if discharged to a public sewer. The following paragraphs deal with some of the conditions, but not all, which may be imposed, the probable limits and the reasons for the conditions.

As regards discharges to watercourses, the conditions are intended to maintain the 'wholesomeness of rivers'; this covers two broad types of constituents. The first such as cyanides which may exert a toxic effect in their own right; the second such as oxygen demand which may lead to deoxygenation of the river water and a consequential deterioration of the water, particularly its ability to support aquatic life.

In the case of discharges to public sewers, the conditions imposed are two-fold: firstly, to ensure that effluents do not contain compounds which will inhibit the treatment processes or block the sewers and secondly to enable the treatment authority to meet the conditions imposed by the appropriate river authority on the effluent discharged from the sewage outfall.

The conditions imposed by most river authorities and local authorities may be summarised as follows:

pH

The pH scale applies to aqueous solutions at temperatures between $0°$ and $95\ °C$ and is defined in BS 1647.[11]

The need for limits for pH are readily appreciated but should take into account the buffering action of sewage in the case of discharges to sewers and the dilution effect in rivers. A range of pH 6·0 to pH 10·0 is reasonable for discharges of effluent of significant volumes to sewers but pH 5·0 to pH 12·0 would not be unreasonable for a discharge of effluent to a main trunk sewer or where considerable dilution was effected.

In the case of discharges to rivers, an acceptable pH range would be pH 6·0 to pH 9·0 but wider limits would be acceptable under certain circumstances e.g. rapid dilution, estuarine waters etc.

Temperature

There would seem to be general agreement that effluents discharged to public sewers should not exceed 43·4°C. This could be eased in the case of rapid dilution of the effluent or some similar circumstances. There is not the same degree of agreement in the case of discharges to rivers but the essential requirement should be an insignificant increase in the temperature of the receiving waters.

Many authorities recommend a temperature of 30°C or a temperature which can be calculated by formulae summarised by Dr. L. Klein[3].

Suspended solids

In the case of discharges to watercourses, the main consideration is the effect of suspended solids on the river water and the bed of the watercourse. This depends on the volume of effluent and the dilution in the river but the Royal Commission of 1900–1912 advised that for a dilution of 8 to 1 in an inland river the Suspended Solids content should not exceed 30 ppm. Increased concentrations would be acceptable for greater dilutions and for tidal waters.

For discharges to sewers, the limit must be related to the volume of the effluent, the dilution in the sewerage system and the treatment capacity of the sewage works. There is no justifiable reason for arbitrarily limiting suspended solids in industrial effluent to the content of domestic sewage and there is every good reason for the removal of suspended solids at a central sewage treatment works rather than a series of similar plants on industrial sites.

It is not possible to suggest average limits for suspended solids because of the varying circumstances but the limit should not be less than 400 ppm, the normal design figure for many sewage works. However, the figure may be as high as 10 000 ppm provided the capacity is available and the particles of a suitable size. In addition the term 'Suspended Solids' can mean many things, ranging from large coarse particles to fine materials such as china clay.

Oxygen demand

Oxygen demand is a measure the oxygen uptake of an effluent and, consequently, the deoxygenating effect which an effluent will have on a watercourse or the treatment demand on the biological processes at a sewage works. A number of parameters are in use, the main ones being BOD_5 (five-day biological oxygen demand), PV (permanganate value), the McGowan Factor and COD (Chemical Oxygen Demand). There is an increasing tendency to use COD, particularly for discharges to public sewers.

These parameters are described in the publication of the Ministry of Housing and Local Government entitled 'methods of chemical analysis as applied to sewage and sewage effluents'. This publication is currently being revised.

The oxygen demand of an effluent discharged to a watercourse should be related to the dilution in the watercourse and the Royal Commission on sewage disposal 1901–1915 recommended a limit of 20 ppm for a dilution of 8 to 1 for inland waters. Higher limits were proposed where the dilution factor was greater than this and for tidal waters. The main criterion must be the wholesomeness of the river which is largely dependent on its dissolved oxygen content, its nature and its self purification ability. The circumstances vary considerably and the limits for oxygen demand imposed on discharges of industrial effluent must depend on the circumstances of each discharge.

In the case of discharges to public sewers, the criterion for limits for oxygen demand is similar to that for suspended solids. There is no magic in the oxygen demand of domestic sewage which is about 400 ppm BOD. The important factors are the volume of the effluent, the treatment capacity of the sewage works and the limits imposed by the river authority on the sewage outfall. The limits could range from a BOD_5 of about 300 ppm to a figure of several thousand for an effluent of relatively small volume or where adequate treatment capacity is available. Effluents with high BODs, 10 000 ppm or so, are success-fully treated at sewage works where the capacity is available, the effluent is adequately balanced and the rate of discharge suitably controlled.

Non-ferrous metallic compounds

The metal salts normally included under this heading are those of chromium, copper, nickel, cadmium and zinc. The main objection to the presence of these compounds in effluents is the effect which they may have on the fauna and flora of river waters and a possible inhibitory effect on sewage treatment processes. It is usual for consents for dis-charges of effluent from industrial premises involving metal finishing and allied processes to include limits for these compounds.

The limits for discharges to watercourses should take into account the dilution in the river, other discharges of a similar nature to the river and the further use of the river water. For discharges of a significant volume the limits are likely to be about 2 ppm, very often divided into two categories; 1 ppm for zinc and 1 ppm for other metallic salts in toto. For smaller discharges and for discharges to tidal waters these conditions may be con-siderably more relaxed.

For discharges to public sewers, the concentration of soluble metallic salts in the effluent passing from the primary treatment system to the biological treatment processes of a sewage works is probably the most important factor. There is evidence that a figure of 2 ppm is acceptable but this could depend on the proportion of each metallic com-pound in the composite mixture. From the volume of the effluent, the dilution of the effluent in the sewerage system and discharges from other sources, it should be possible to determine realistic conditions for each discharge.

It is worth noting that, in the Birmingham area, where large volumes of effluent from metal finishing processes are received into the sewerage system the present conditions imposed by the authority are 10 ppm for soluble metallic salts, and 20 ppm for insoluble toxic metallic salts. For most other areas it should be possible to have somewhat more relaxed conditions.

Cyanides

Cyanides and compounds which may produce hydrocyanic acid have a toxic effect on both river flora and fauna and also on the biological processes of sewage treatment. In addition these compounds may cause the release of hydrocyanic gas in sewers and similar confined spaces and thus be hazardous to men working on sewerage systems.

The limit for cyanides in discharges of effluent to watercourses is likely to be 1 ppm, although somewhat less stringent conditions may be imposed for discharges which are subject to considerable dilution or are into tidal waters. In the case of discharges to public sewers, the normal limit is 10 ppm although this may be relaxed in the case of discharges of small volume. Conversely the limit may be reduced for discharges of considerable volume or where dilution is limited.

The conditions imposed on sulphides are, to some extent, similar to those which may be imposed in the case of cyanides.

Oil

Immiscible oil has a deleterious effect on both river water and on sewage treatment processes. Apart from amenity and other aspects the formation of a layer of oil inhibits the oxygenation of river water. In theory, the true measure of the polluting effect of oil should be in the film thickness, but, in practice, River Authorities impose limits which are normally expressed in terms of concentration. These vary a great deal and range from 'no visible signs of oil' (about 5 ppm depending on the type of oil) to about 25 ppm.

In the case of discharges of effluent to public sewers, the limits vary widely and should be related to the volume of effluent and the capacity of the sewage works. Perhaps the most acceptable condition is that the oil content of the effluent 'should not be physically separable'.

Synthetic detergents

The difficulties caused by synthetic detergents are well-known but the introduction of modern 'soft' detergents which are readily biodegradable has eased the problem considerably.

In the case of discharges to public sewers, a limit of 150 ppm expressed in terms of Monoxal O.T. is not uncommon whereas in the case of direct discharges to watercourses, a limit of 10 ppm or less, depending on the volume of effluent and the dilution in the river, should be acceptable.

Sulphates

Many effluents contain sulphates, arising partly from production processes and partly from treatment and neutralisation of effluents. As regards sewerage systems, sulphates are known to have a deleterious effect on concrete structure and this has been reported on by the Water Pollution Research Laboratory.[4] It would seem that a concentration of sulphates equivalent to 1000 ppm as SO_3 *in the sewer* is acceptable: a reasonable limit for sulphates in the discharge of effluent would need to take into account the dilution when the effluent is discharged to the sewer.

Sulphates and other soluble inorganic salts do not have any marked effect on river water except where the total dissolved solids render the water unfit for use for public water supply or for farming purposes or lead to eutrophication of the water.

Other toxic compounds

This includes a very wide range of compounds, many of them organic, which may have a deleterious effect either on river water or on sewage treatment processes. It is not possible in this chapter to deal with these individually, but a continuing study of these compounds is being made by the Water Pollution Research Laboratory. They are listed in the Laboratory's periodic publications of 'Instab'[5].

In the previous paragraphs, some of the compounds which are likely to be controlled by River Authorities and Local Authories and the conditions which may be imposed have been described. The Authorities will be within their rights to impose conditions, provided these are reasonable and justifiable, and it will be the responsibility of management to comply with them.

It is essential, however, that management should ensure:

(i) That the conditions are reasonable.

(ii) That the effluent (or effluents) discharged from the factory site will meet the conditions which have been imposed or that steps can be taken to ensure that the effluent will meet the conditions and

(iii) That an appeal be made to the Secretary of State for the Environment or an extension of the appeal period sought if the effluent does not meet the conditions or can not be treated at reasonable cost, to meet the conditions.

The need for management to initiate consultations with the appropriate authority before the conditions of consent are formally decided is stressed.

PROBLEMS ARISING FROM THE TREATMENT OF LIQUID EFFLUENT

The imposition of conditions by a local authority or a river authority on the discharge of industrial effluent almost invariably involves the company concerned in the segregation and treatment of effluent. This, in turn, may involve the company in redrainage of the factory site, suitable space for a treatment plant, the disposal of sludges and concentrates connected with treatment and possible nuisance arising from the treatment process.

These problems have been discussed in a previous paper[6] but are summarised in the following paragraphs.

Drainage

Any scheme for effluent treatment requires a satisfactory drainage system in terms of capacity, rate in flow, gradients and intake and outfall points. Many factories in the United Kingdom were established before effective legislation on abatement of river pollution and control of discharges of effluent to public sewers was enacted. In addition many large factory sites have old, and in some cases, unknown drainage systems and a number of points of discharge to sewers or to watercourses.

The cost of redraining old factory sites can be as expensive and be more difficult than the installation of treatment plant. These difficulties must be borne in mind when discussing effluent disposal with the appropriate river authority or local authority.

Space for treatment plant

Effluent treatment plants on factory sites require space, and the area involved can be particularly significant if the plants involve the treatment of suspended material or a reduction in the oxygen demand of the effluent. The size of the plant would also be related to the stringency of the conditions imposed by the authority.

Many factory sites in the industrial conurbations were not laid out with this in view and space for this purpose, or any other purpose, is very often at a premium. The treatment plant must also be an integral part of the drainage system and be satisfactorily situated from the point of view of factory drainage and discharge to the public sewer or watercourse. Recent developments in the design of treatment plant have eased this problem to some extent, particularly the use of continuous and automated plant.

The problem of available space is very often critical and must be taken into account when considering conditions to be imposed on discharges of effluent and the facilities required to meet these conditions.

Disposal of sludges and concentrates

The treatment of almost all industrial effluents involves the removal of sludges arising either from the settlement of solids or from biological oxidation. The latter is a variable, but in the case of domestic effluent about 0·5 kg of suspended solids are produced for every 1 kg of BOD removed.

The problem of sludge disposal is well-known in most quarters, but the difficulties being experienced by industry in this respect cannot be over-stressed. Indeed this subject is included in the Report of Ministry of Housing and Local Government Technical Committee on the Disposal of Toxic Solid Wastes which was published in August 1970.[7] It is imperative therefore that management, when considering the treatment of effluent for discharge to a watercourse or treatment before discharge to a public sewer, should take into account the sludge which will be produced and facilities for the disposal of the sludge.

Nuisance

The law on nuisance is complicated but it is probably sufficient to note that the construction of an effluent treatment plant on a factory site may result in a nuisance abatement order, possibly in connection with smell or atmospheric pollution. This problem is less likely to arise if the effluent is discharged to the public sewer.

In determining whether or not treatment should be carried out on site, management should take into account the possible effect on their own operations, (e.g. manufacture of food and drink) and the possibility of creating a nuisance.

TREATMENT

The variables associated with discharges of effluent, both within an industry and from one industry to another, are well known and have been described for a wide sector of Industry by Southgate[8] and Isaac.[9] In addition many books, monographs and papers have been published on the more detailed aspects of treatment of particular effluents or effluents from selected industries. In the wider sphere, the Water Pollution Research Laboratory[5] has a very deep interest in the causes of pollution of watercourses and coastal waters and the treatment processes required for abatement of pollution purposes. The Annual Reports of the Laboratory, and the numerous papers, abstracts, reports and studies produced by the Laboratory contain much information of both a fundamental and detailed nature on effluent treatment. In addition a number of Research Associations, Universities and research and development organisations attached to a large number of manufacturing companies are actively interested in this field.

It is not possible in a chapter of this length to deal with any but the major features of effluent treatment and readers who wish to have detailed knowledge are advised to refer to more authoritative and exhaustive sources of information. It may be helpful, however, to refer to the broader aspects which are common to most effluent treatment systems and to highlight their significance.

Measurement of volume and preliminary treatment

Practically all consents for discharge of an effluent will stipulate a maximum daily volume and a maximum hourly rate of flow. For effluents of considerable volume, say greater than 225 m^3 per day, there are good reasons for the installation of a flow measuring recorder. It must be noted, however, that in many of the older factories, effluents are discharged through a number of points of discharge and the difficulties in installing flow recorders and indeed the need for them are very diverse. In addition, effluents of low flow or variable flow or of unusual composition are likely to give inaccurate readings on flow recorders.

Inter alia, the subject of measurement of volume has been considered by the Local Authority Associations/CBI Trade Effluent Joint Advisory Panel[10] and the Model consents and Directions agreed by the Panel refer specifically to measurement of flow by recorders or by alternative acceptable methods. The latter normally refers to measurement of volume of effluent from records of water intake with agreed deductions for water lost in processing, steam raising etc.

In the case of industrial effluents, preliminary treatment is largely confined to balancing effluents from different sources, adjustment of pH and settling out of suspended solids before the effluent is passed on to the biological treatment stage or for further chemical treatment.

Biological oxidation

A wide range of industrial effluents has an oxygen demand (measured as five-day Biological Oxygen demand, Permanganate Value, McGowan Factor or Chemical Oxygen demand). It is frequently necessary for this to be reduced before discharge to a watercourse, tidal waters or the public sewerage systems. In most cases this is achieved by biological oxidation processes and a number of alternatives is available to industry, either singly or in combination. Considerable use is made of percolating filters or activated sludge processes but increasing use is being made of high rate plastic medium filters, either alone or as a pretreatment technique for final treatment by traditional methods.

Biological oxidation requires an effluent of reasonably consistent quality, corrected for pH and nutrient balance, and free of inhibitory substances. In general practice these conditions are met by admixture of the industrial effluent with domestic sewage and, other things being equal, most industrial effluents requiring biological treatment are more satisfactorily treated at a local authority sewage works, at least for final treatment, before discharge to the receiving watercourse.

Sludge removal and disposal

Sludge arise from several sources in effluent treatment processes, mainly settlement of suspended solids, use of flocculants, precipitates from neutralisation processes and semi-solid materials arising from biological processes. The latter is an important factor since, for domestic effluent, approximately 0·5 kg of dried solids are produced for every 1 kg reduction in BOD.

The settlement of sludges arising from treatment processes is normally carried out in tanks, the design of the tank varying with the type of matter to be settled and the sludges removed by mechanical scrapers and by the hydrostatic head from conical tanks. The importance of designing the tank to meet the physical requirements of the suspended material and fluctuations in rate of flow is stressed. In recent years, extensive use has been made of polyelectrolytes and other similar substances to accelerate the settlement processes and obtain reductions in tank size. Flotation techniques for the removal of fine suspended solids, particularly of the type which do not readily settle have found an increasing use in the treatment of industrial effluents.

The dewatering of sludges has been the subject of considerable study, particularly by Water Pollution Research Laboratory. Sludges arising from treatment of industrial effluents vary considerably and many are amenable to dewatering by mechanical filtration e.g. vacuum filtration, pressure filtration etc. which is more adaptable and normally more economic than the traditional open drainage and drying techniques used for sewage work sludges. Techniques have been developed by Water Pollution Research Laboratory for assessing the dewatering properties of sludges.

Tertiary treatment

The normal biological and settlement methods for reducing the oxygen demand and suspended solids content of both domestic and industrial effluents usually need to be supplemented by tertiary treatment if an effluent of higher quality than Royal Commission Standard i.e. five-day BOD, 20 mg/l and Suspended Solids 30 mg/l is required. Experience of tertiary treatment techniques is still being acquired and a wide range of methods is being examined. In the case of industrial effluents tertiary treatment, where necessary, is normally restricted to sand filtration, land irrigation, or mechanical filtration.

In addition to the main features of effluent treatment described in the previous paragraphs, industrial effluent can require a wide range of supplementary treatment and this can differ from one effluent to another. In particular many industrial effluents are more satisfactorily treated in concentrated form and before admixture with the main effluent stream, e.g. removal of ammonia and toxic metallic compounds; oxidation of cyanides; treatment of chemical concentrates etc. In addition industrial effluents from the same process vary hour by hour, day by day and usually require more sophisticated techniques for balancing, pH adjustment, chemical dosing etc. than effluents from domestic sources.

The treatment of industrial effluent, because of its variable nature, is not amenable to the 'rule of thumb' approach and, in most cases, the design of the main treatment plant and any ancillary units must be determined by the nature and composition of the effluent and the circumstances of the discharge and the receiving watercourse or public sewerage system.

FINANCIAL ASPECTS

River authorities and local authorities have extensive powers to control discharges of industrial effluent to watercourses, tidal waters and public sewers and to require the treatment of effluent. To meet these conditions the treatment of industrial effluent has become a significant item of cost, and as these powers are progressively introduced, is likely to become even more so. The expenditure incurred is:

(a) the cost of treatment plants to bring the effluent to a standard acceptable to a river authority or a local authority;

(b) the charge which may be made by the local authority for this purpose and

(c) the cost of disposal of sludges and similar substances which arise from treatment processes.

It is difficult to predict how much this costs industry, due to the variation in the volume and average strength of industrial effluent discharged to watercourses and public sewers. In addition, the cost must include the sewerage element of the rates paid by industry to local authorities.

During 1970 the Department of the Environment has initiated the River Pollution Survey and the Report should provide, *inter alia*, figures for the total cost of disposing of all liquid effluents to rivers, estuaries and tidal waters. Industry has cooperated with the Department in this Survey and a questionnaire has been sent to a wide sector of industry with a view to obtaining information on industrial costs for effluent treatment and disposal. The Survey is likely to be completed in 1971 and should provide factual information on both present and future costs.

The cost of effluent treatment plant may also be considered as a percentage of the total capital investment by industry in new plant. In evidence to the Lena Jeger Committee[12], the CBI estimated that this ranged from 2% to 15%, the variation being due partly to the nature of the industry and partly to whether the industry was capital intensive or not. It is interesting to note, however, that figures available for some new investments early in 1971 indicate that the capital expenditure on plant for abatement of pollution purposes is approximately 10% of the total capital outlay. This figure is higher than that applicable to most other industrialised countries where 5 to 6% would seem to be the order.

From a management point of view, the main objective in the treatment and disposal of industrial effluent is to comply with the requirements of the legislation and to do this in the most economic manner. It cannot be stressed too often that perhaps the most profitable step which can be taken by industry (and others) is control of individual discharges of effluent at source. This involves, among other things a high degree of 'good housekeeping'; the segregation of highly polluting and noxious effluents from the main volume of the effluent discharge; the reuse and recycling of water where this can be done economically and satisfactorily and the appraisal of possible future changes in processes or treatment of individual effluents which will result in improving the consistency and the quality of effluent discharge.

Perhaps the two most crucial factors are the determination of realistic and acceptable conditions for the discharge of effluent and the development of processes and treatment systems which will result in the most economic treatment and disposal of the effluent. In the case of discharges of effluent to public sewerage systems it is essential that the charge imposed by the local authority should be properly related to the volume and strength of the effluent discharged from the factory and the cost to the local authority for the reception and treatment of the effluent. This requires proper measurement of the volume and composition of the effluent discharged by the factory and a full knowledge of the basis on which the local authority assesses its charge.

In some circumstances, alternatives are available for the disposal of effluent. For example, either full treatment on site before discharge of the effluent to a watercourse or tidal waters; discharge of the untreated or partially treated effluent to the public sewerage system; or disposal of a proportion of the effluent by a contractor. To ensure the most economic method for disposal it is necessary for management to investigate the advantages and disadvantages, both technical and financial, of all possible methods of disposal, both in the short term and the long term. In any of these calculations provision should be made for the disposal of sludges and similar materials which may arise from treatment.

GOVERNMENT ADVICE

The provisions of the legislation and some of the technical aspects of abatement of pollution have been briefly described. It has also been pointed out that, in the event of an authority imposing conditions which are unreasonable or incapable of being fulfilled, the discharger has the right of appeal to the Secretary of State for the Environment or the Secretaries of State for Scotland or Wales. The right of appeal to the Minister is important to industry and care should be taken to make sure that the time allowed in which to appeal is not allowed to lapse.

The appellate functions of the Secretaries of State make it difficult for the Departments to give advice on the implementation of the provisions of the various Acts. Because of the complexity of the subject and the various interpretations which can be placed on parts of the legislation the Secretaries of State have from time to time issued circular letters giving advice on some aspects of this subject. It is essential that those involved in this field should be aware of this advice and the main circulars are briefly described below. These refer specifically to England but similar circulars apply to Scotland and Wales.

The circular letters and memoranda referred to below reflect the views of the Departments on the conditions and charges which could properly be applied to discharges of industrial effluents and effluents from local authority sewage disposal works. These documents should be closely studied by those who are involved, to any great extent, in discharging industrial effluent to watercourses or public sewerage systems or the treatment of domestic effluent.

Rivers (Prevention of Pollution) Act 1961, Circular 39/61

After the Rivers (Prevention of Pollution) Act 1961 became law the Ministry of Housing and Local Government issued a circular letter, No. 39/61 dated 30 August 1961 which briefly described the main purpose of the legislation. In paragraph 8 the following statement is made:

'In this connection (referring to improvements) I am to point out that the Report of the Trade Effluents Sub-Committee contemplated a gradual and progressive campaign against river pollution, during which improvements would be closely related to what it was reasonable and practicable to require from a particular discharger in the circumstances obtaining at any one time. That will be a consideration which the Minister will have in mind when considering appeals'.

This statement which was repeated in a subsequent circular letter clearly indicates that the Minister expects river authorities to seek improvements which are realistic.

Public Health Act 1961, Circular 46/61

Shortly after the Public Health Act 1961 became law the Ministry issued circular No. 46/61 dated 29 September 1961 which described the provisions in the legislation but in paragraph 14 the following statement is made:

'Concern has been expressed also that the trader should be shown how the charge relates to the cost of treating the effluent. The Minister realises that in a number of cases local authorities are content with a charge which approximates to or is even less than the cost of treating ordinary domestic effluent and that in these instances there may be little need for detailed explanation. Where, however, they are making a special charge for an effluent which is difficult to treat, it seems right that they should make their calculations known as plainly as they can to the trader. Similarly where factors of the kind mentioned in the previous paragraph (capital contribution) by a trader or payment being made for another trade effluent) are taken into account the local authority should endeavour to show as plainly as possible what weight has been allotted to them. By so doing they may well be avoiding unnecessary appeals to the Minister'.

This advice clearly indicates that the basis of the determination of trade effluent charges should be explained to industrial dischargers.

Circular 37/66

On the 13 October 1966 the Ministry issued circular No. 37/66 and a technical memorandum entitled 'Technical problems of River Authorities and Sewage Disposal Authorities in Laying down and Complying with Limits of Quality for Effluents more restrictive than those of the Royal Commission'. Although this document refers essentially to discharges of effluent from local authority sewage treatment works to watercourses much of the advice could be regarded as applying to large discharges of industrial effluent to a watercourse.

In particular the need to explain in detail the reason for a choice of standards and carrying out improvements in stages could be important.

Circular 64/68

On 4 December 1968 the Ministry issued circular No. 64/68 and a technical memorandum on 'Standards of Effluents to Rivers with Particular Reference to Industrial Effluents'. This memorandum should be read in conjunction with the earlier technical memorandum but refers particularly to the criteria which should be applied in determining conditions which should be imposed on discharges of industrial effluent to watercourses. The memorandum recognises the fact that industrial effluents vary a great deal more in volume and composition than domestic effluents and the need to deal with each case on its merits. The memorandum should be studied by the management of any company discharging liquid effluents to a watercourse.

The accompanying circular letter issued with the memorandum raises two points which are of particular interest to industry. Firstly, it advises that river authorities should be prepared to justify the conditions imposed on *all* discharges of industrial effluent to watercourses and secondly refers to the need for greater consultation between industry, river authorities and local authorities particularly when industrial effluent is a significant proportion of the total flow to a local authority sewerage works.

Circular 44/70

On 4 June 1970 the Ministry issued circular No. 44/70 which, *inter alia*, reiterates the need for River Authorities to justify the conditions imposed on all discharges of industrial effluents.

FUTURE DEVELOPMENTS

1970 was European Conservation Year and 1972 was chosen as World Conservation Year. There is no doubt that environmental pollution has become a matter of considerable public interest and resulted in a great deal of activity at both national and international levels. These and other pressures will undoubtedly lead to demands for more stringent measures on abatement of pollution which are likely to have an appreciable effect on industry and others. It is not possible in this chapter to deal with these in detail but management should be aware of some of the current studies which may have an effect on abatement of pollution by all dischargers of effluent.

Reference has already been made to the River Pollution Survey being conducted by the Department of the Environment. This Report, together with the Trent Economic Study provides considerable information on the state of river pollution and also guide-lines on the rate of improvement which may be achieved.

The Report of the Ministry of Housing and Local Government Committee on 'Sewage Disposal'[12], covers a much wider field than its name implies since it deals with every form of liquid effluent, the effect of effluent discharges on watercourses and recommendations for future Government action. In particular the Report recommends a much greater public investment in sewage and sewerage systems; more research into the problems involved and the technology associated with effluent treatment; a general tightening up of legislation on this subject; and changes in the form of grant aid in respect of capital investment by industry on effluent treatment plant provided by local authorities.

The Report of the Key Committee on the 'Disposal of Toxic Solid Wastes'[13] covers the problems connected with the disposal of toxic and semi-toxic solids; sludges; and liquid concentrates, many of which arise from industrial processes the use of chemical products or are by-products from effluent treatment. The Report identifies the extent of the problem and recommends fairly extensive action by the Government to control the disposal of these materials.

The management of water resources, water supplies and disposal of effluent is vested in River Authorities, Statutory Water Undertakings and Sewerage Authorities respectively. There are twenty-nine River Authorities, just under two hundred Water Boards and about one thousand four hundred Sewerage Authorities in England and Wales. The Central Advisory Water Committee under the Chairmanship of Sir Alan Wilson is studying the reorganisation of these authorities and has assumed that the demand for water will double by the end of the century. In addition, and in the long term more importantly, a Royal Commission is investigating the problems of Environmental Pollution.

There is no doubt that as a result of these Reports there will be considerable changes in the administrative organisation and control related to the abatement of river pollution and the disposal of industrial effluent. The existing legislation is comprehensive and, in the view of many people, adequate but the facilities for implementing the legislation would seem to require reconsideration. It is likely, therefore, that future legislation (and this may come sooner than many people think), will be aimed at reorganisation and possible amendments to the existing legislation where this is thought necessary. Management should be aware of these trends and, in determining future company policy, take into account an almost certain increase in the stringency of conditions which will be attached to discharges of industrial effluent and the disposal of sludges and similar materials and more rigid control over the abstraction of water for industrial purposes. The Reports should be studied by management and the legislation which will arise from them closely watched.

In addition to national matters there is considerable activity in the international field. International agencies like the Economic Commission for Europe, OECD, Council of Europe, International Maritime Consultative Organisation, to mention a few are actively engaged in considering the need for international agreement and harmonisation of legislation connected with abatement of pollution. Although much of this is related to international matters there is no doubt it will have a considerable bearing on future legislation in the U.K. and possibly on implementation of the legislation.

SUMMARY

The discharge of industrial liquid effluents to watercourses, estuaries and public sewers is already the subject of very extensive legislation and practically all these discharges are required to meet conditions which may be imposed by the appropriate authority. The same applies to discharges of gases, smoke and grit and dust to the atmosphere and, to a lesser extent, the abatement of noise. It is also apparent that there will be increasing pressures to abate pollution whether by liquids, gases or noise in the foreseeable future.

It is the responsibility of industrial management to be aware of the existing requirements of the legislation and also to take into account, in any long-term planning, in possible future legislation. It is also the responsibility of management to comply with this legislation and to adopt the best practices and techniques which will enable this to be done in the most economic and satisfactory manner. It is again stressed that good housekeeping within a factory, sound planning, and an adequate knowledge of the effluents which are being disposed of are essential for good management and that consideration of the legal requirements, technical capabilities and cost involved in abatement of pollution will pay handsome dividends in the years to come.

Table 15.1 summarises the general statute law relating to trade effluent disposal in England and Wales and has been taken from the CBI handbook.[1]

Table 15.1 GENERAL STATUTE LAW RELATING TO TRADE EFFLUENT DISPOSAL IN ENGLAND AND WALES

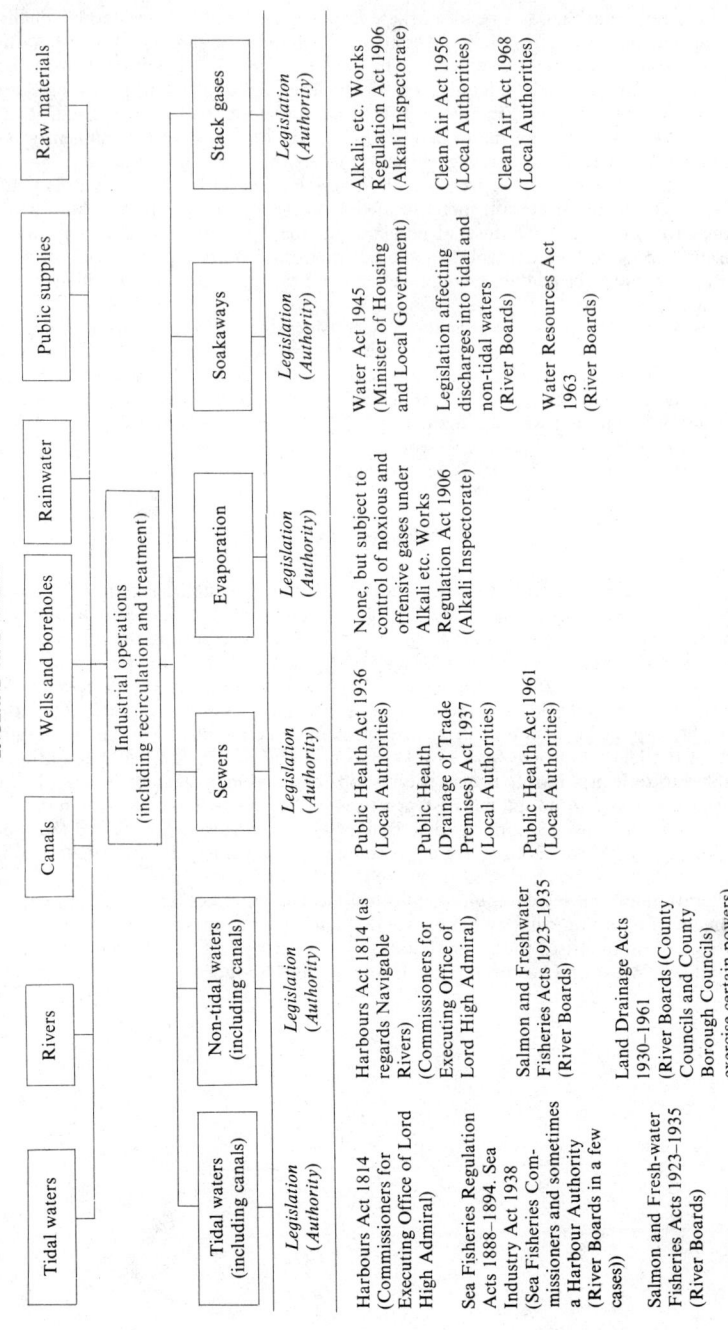

Top-level headings:

Tidal waters | Rivers | Canals | Wells and boreholes | Rainwater | Public supplies | Raw materials

Industrial operations (including recirculation and treatment)

Tidal waters (including canals)	Non-tidal waters (including canals)	Sewers	Evaporation	Soakaways	Stack gases
Legislation (Authority)	*Legislation (Authority)*	*Legislation (Authority)*	*Legislation (Authority)*	*Legislation (Authority)*	*Legislation (Authority)*
Harbours Act 1814 (Commissioners for Executing Office of Lord High Admiral)	Harbours Act 1814 (as regards Navigable Rivers) (Commissioners for Executing Office of Lord High Admiral)	Public Health Act 1936 (Local Authorities)	None, but subject to control of noxious and offensive gases under Alkali etc. Works Regulation Act 1906 (Alkali Inspectorate)	Water Act 1945 (Minister of Housing and Local Government)	Alkali, etc. Works Regulation Act 1906 (Alkali Inspectorate)
Sea Fisheries Regulation Acts 1888–1894. Sea Industry Act 1938 (Sea Fisheries Commissioners and sometimes a Harbour Authority (River Boards in a few cases))	Salmon and Freshwater Fisheries Acts 1923–1935 (River Boards)	Public Health (Drainage of Trade Premises) Act 1937 (Local Authorities)		Legislation affecting discharges into tidal and non-tidal waters (River Boards)	Clean Air Act 1956 (Local Authorities)
Salmon and Fresh-water Fisheries Acts 1923–1935 (River Boards)	Land Drainage Acts 1930–1961 (River Boards (County Councils and County Borough Councils) exercise certain powers)	Public Health Act 1961 (Local Authorities)		Water Resources Act 1963 (River Boards)	Clean Air Act 1968 (Local Authorities)
Land Drainage Acts 1930–1961 (River Boards (County	River Boards Act 1948				

(River Boards)

Coast Protection Act 1949
(Minister of Transport and Harbour Authorities)

Rivers (Prevention of Pollution Acts 1951–1961
(River Boards)

Oil in Navigable Waters Act 1955
(Minister of Transport (Harbour Authorities and Sea Fisheries Committees may prosecute for certain offences))

Clean Rivers (Estuaries and Tidal Waters) Act 1960
(River Boards)

Pollution) Acts 1951–1961
(River Boards)

Oil in Navigable Waters Act 1955 (Inland Navigable Waters for Seagoing Vessels)
(Minister of Transport, Harbour Authorities and Sea Fisheries Committees may prosecute for certain offences)

Clean Rivers (Estuaries and Tidal Waters) Act 1960
(River Boards)

Note: (1) Since the Water Resources Act 1963 the River Board have become River Authorities.
(2) The Water Resources Act 1963 and the Clean Air Act 1968 (Soakaways and Stack Gases respectively) have become law.

REFERENCES

1. CBI Handbook, 'Disposal of Trade Effluent in England and Wales', C.B.I., London, (1962)
2. Judgement of Lord Macnaghten, Young & Bankur Distillery Co. (1893), (A.C. 691:69 L.T. 838;58 JP, 100, H. L. Sc)
3. KLEIN, L., *River Pollution*, Butterworths
4. Notes on Water Pollution No. 6, 'Discharge of Sulphates into concrete sewers', *Water Pollution Research Laboratory*, (Sept. 1959)
5. Information Service on Toxicity and Biodegradibility, *Water Pollution Research Laboratory*
6. *Report of the Committee on Disposal of Solid Toxic Wastes*, H.M.S.O.
7. SOUTHGATE, B. A., *Treatment of Industrial Effluents*, H.M.S.O.
8. ISAAC, P. C. G., *Waste Treatment*, Pergamon Press (1960)
9. Annual Reports, 'Pollution Abstracts', *Water Pollution Research Laboratory*
10. CBI/Local Authority Associations Trade Effluent Joint Advisory Panel, 'Circular Memoranda', C.B.I., London
11. BS 1647, *pH Scale*, British Standards Institution (1961)
12. Report of the Working Party on Sewage Disposal, H.M.S.O. (1970)
13. *Report of the Technical Committee on the Disposal of Solid Toxic Waste*, H.M.S.O. (1970)

BIBLIOGRAPHY

NONHEBEL, G. (Ed), *Gas Purification Processes*, 2nd ed Butterworths (1972)

NOISE

H. M. MOSS

Noise is defined as undesired sound. It is one form of pollution of the environment and interferes with the efficacy of workers and diminishes the pleasure of leisure hours. The range of intensity of noise is such that emotions, from mere annoyance to severe pain can be experienced and in consequence it is regarded as a nuisance in the broadest meaning of the word. At present, therefore, authorities are more aware of the need to regulate noise-producing activities with a view to obviating the nuisance to the public in general and to private individuals in the particular cases.

At Common Law an individual can approach the Courts for relief from noise as a nuisance but such legislation as the Noise Abatement Act 1960 and the Public Health (Recurring Nuisances) Act 1969, has given wider powers to Local Authorities to initiate action for reducing noise on behalf of the community as a whole.

As people age, their hearing is impaired due to natural processes, but living and working in noisy environments accentuates this impairment. The ear is provided with a natural defence mechanism which restores the hearing facility substantially to its original condition when it is no longer subject to extreme noise. If, therefore, working in a noisy environment causes deafness, this deafness may be temporary and hearing will gradually recover once the noise source has been silenced (Threshold Shift). Industry is concerned with the onset of this sort of deafness which through time and age can become permanent and a measurement of the hearing faculty in recruits for various industrial situations is becoming common. Similarly, the possibility of the inclusion of the loss of hearing due to adverse working conditions within the scope of legislation regulating compensation actions has awakened industrial concerns to their responsibilities for protecting workpeople from the effects of noise in work-places.

Two duties fall upon industry. It is now incumbent on industry to prevent, by all practicable means, the emission of noise becoming a nuisance to householders and others who live and have their business in the vicinity of potentially noisy factories; and to prevent the emission of noise from noisy machinery and equipment causing preventable impairment of the hearing of workpeople.

NOISE RATINGS

It has been found that certain noise intensities can be accepted by the vast majority of people. The particular values in decibels have been plotted as a series of curves which are set out in IEC/R226.[1] The basis of these 'noise rating' curves is the sound pressure level at 1000 Hz frequency in decibels relative to $2 \times 10^{-5} N/m^2$. The noise rating numbers of the curves intersecting the 1000 Hz ordinate represent the maximum permissible value of the noise level for the emission of sound at that frequency. If it is agreed that a machine noise is acceptable at 1000 Hz with a rating of 80 dB, the rating at 4000 Hz will be 75 dB and at 250 Hz, 85 dB (Figure 15.1). To assimilate noise, standard instruments have been designed with filter circuits having differing characteristic curves. These filters are referred to as 'weightings' and there were three A B and C (there is now a fourth weighting D which is supposed to come close to a measure of the 'annoyance' component of noise but there is some doubt as to its value).

At one time noise levels were taken on the C weighting and the units were designated 'decibels C' or dbC. It is now thought that the A filter which has a less flat characteristic

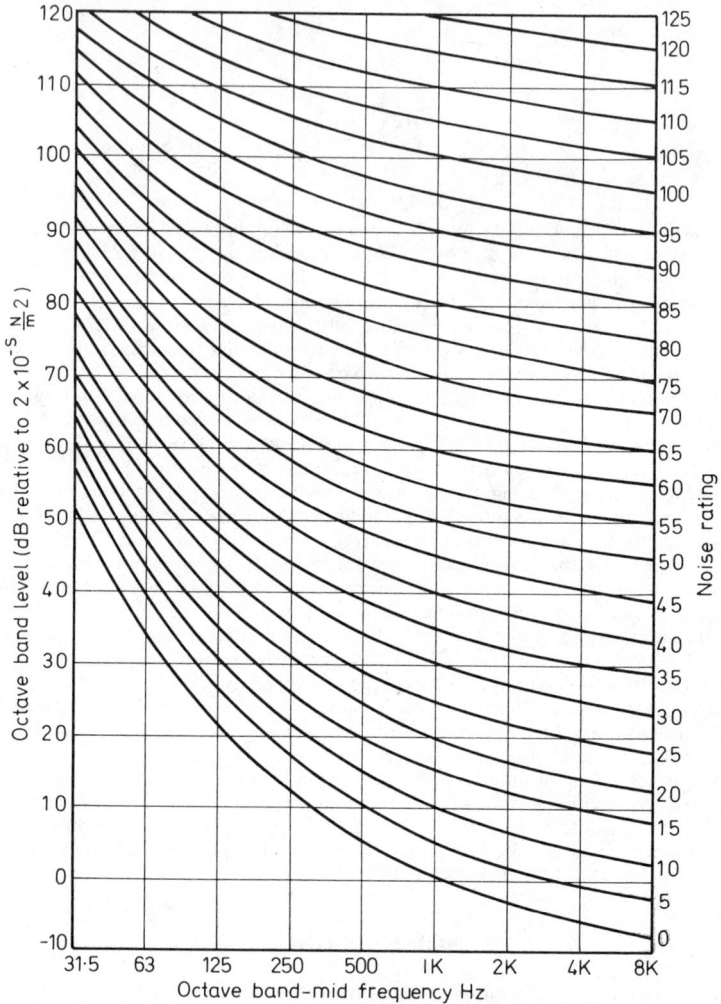

Figure 15.1. Noise Rating Curves

than the C filter gives a more representative measure of the effect of noise on the human ear. Noise levels in consequence are now measured in 'decibels A' or dBA.

Sound, and therefore noise, is the result of the undulated flow of fluid. This flow, which is a disturbance of the fluid, creates pressure differences which in turn affects the fine membranes of the ear, the tympanic and the basilar of the cochlea. These pressures are very small and it has been established that a pressure of $0\cdot00002$ N/m^2 is the smallest pressure which can affect these membranes, i.e. the threshold of hearing, and is by convention 0 dB. In a sound wave the intensity is proportional to the square of the pressure and by convention the decibel scale of pressure is likewise based on the pressure squared thus:

$$\text{Sound Pressure Level in dB} = 10 \log_{10} \left\{ \frac{\text{Sound Pressure}}{0 \cdot 00002 \text{ N/m}^2} \right\}^2 \tag{1}$$

To ensure that machinery will not be objectionable from the noise aspect it is becoming customary for purchasers to stipulate certain noise level values in the purchasing specifications. In addition, as a result of legislation, the noise level at the perimeter of works areas especially in residential districts is more than ever restricted. Sound level meters are used to obtain qualitative assessments of the noise emitted from such areas and monitoring systems have been employed in some of the larger industrial estates. Where new plants are involved or where changes are made to existing plants or workshops, some industrial concerns make recommendations which they expect the contractor firms engaged in the work to follow as closely as practicable. Table 15.2 exemplifies modern outlook in this respect.

Table 15.2   NOISE RATINGS FOR VARIOUS INDUSTRIES

Noise rating No.		Limiting sound pressure level in dB re 2×10^{-5} N/m^2							
	Octave band mid frequency	63 Hz	125 Hz	250 Hz	500 Hz	1000 Hz	2000 Hz	4000 Hz	8000 Hz
85	Plant sites	102	96	91	87	85	83	81	79
65	Workshops	87	78	72	68	65	63	62	60
55	Control rooms and plant offices	79	69	62	58	55	53	51	50
50	General offices	75	65	59	53	50	48	46	45
45	Canteens	72	61	53	48	45	43	40	39
40	Private offices	68	57	49	44	40	37	35	33

The procedure to be adopted in measuring noise levels is set out in BS 4196[2] and the kind of instrument to be employed is specified in BS 4197.[3] The results of the measurement can be assessed on the lines recommended in BS 4198.[4] In residential areas the annoyance caused to the residents by noise from adjacent factories is assessed by the method specified in BS 4142.[5]

EXPOSURE TO NOISE

The effect of long exposure to noise of workpeople is irreversible impairment of hearing. In some countries the legislature has set clearly defined limits to the length of time to which a work person is exposed to noise of a maximum noise level rating without some form of protection. The simplest kind of protection involves ear plugs and wadding and, as an alternative, the wearing of ear defenders.

If the noise level exceeds the limits laid down steps must be taken to reduce the level by sound absorption media, silencing devices or ultimately a redesign of the working environment. A similar approach to the matter is adopted in this country. According to British Occupational Hygiene Society a continuous noise level of 90 dB should not be exceeded habitually, i.e. for 8 hours per day, 5 days per week and 48 weeks per year, without hearing protection.

The American Noise Control Regulations, Section 10, state that 'for an eight hour day a worker shall not be exposed to more than 90 dBA unless protection is provided and used. For less than 8 hours, a worker may be exposed to higher noise levels for specified periods of time.' The following periods are specified.

Sound level dBA	Durations per day in hours
90	8
92	6
95	4
97	3
100	2
102	$1\frac{1}{2}$
105	1
110	$\frac{1}{2}$
115	$\frac{1}{4}$

Impulsive or impact noise shall not exceed 140 dBA measured on the C scale fast response setting of the sound level meter. The regulations set out a method of computation where two or more periods of noise exposure of different levels are involved.

Noise levels of the order set out above may well be adopted internationally despite the lack of such precision in the U.K. and other industrial countries. Where a worker is subject to various noise levels over a normal working day the following method of determining the equivalent noise level has much to recommend it. The formula used is the well known one

$$L_e = 10 \log \frac{1}{T_2 - T_1} \int_{T_1}^{T_2} 10^{(L_p/10)} dt \tag{2}$$

where L_p = Momentary sound level in dB(A) at time t, T_1 and T_2 = Times at beginning and end of 8-hour working day. The uniform equivalent exposure level L_e of a worker who spends his eight-hour working day in the following noise environments is calculated from the data:

2 hours at 100 dB(A)
3 hours at 97 dB(A)
2 hours at 90 dB(A)
1 hour at 80 dB(A)

L_p	$10^{(L_p/10)}$	$T_2 - T_1$	$10^{(L_p/10)}(T_2 - T_1)$		
100	10^{10}	2	$2 \times 10^{10} = 20 \times 10^9 = 20$		20×10^9
97	$10^{9.7}$	3	$3 \times 10^{.7} \times 10^9 = 3 \times 5 \times 10^9 =$		15×10^9
90	10^9	2	$2 \times 10^9 = 2 \times 10^9 =$		2×10^9
80	10^8	1	$1 \times 10^8 = 0.1 \times 10^9 =$		0.1×10^9

$$\Sigma 37.1 \times 10^9$$

$$L_e = 10 \times \log \frac{37.1}{8} \times 10^9$$

$$= 90 + 10 \log 4.636$$
$$= 90 + 6.6$$
$$= 96.6 \text{ dB(A)}$$

Other methods involve the use of tabular data for adjustments according to the length of exposure. Such data is published in books and specifications relating to noise.

TEMPORARY HEARING PROTECTION

For short duration exposure to high noise levels, ear wadding, ear plugs and ear muffs may be useful. Ear wadding is least effective but has the advantage that it can be thrown away when the need for it no longer exists; it is not hygienic, however. Ear plugs are better but require to be retained as the personal property of the user.

Ear muffs represent the best form of hearing protection on a temporary basis. The acoustic properties of ear muffs tend to vary with the designs supported by various manufacturers and the comfort to the wearer depends upon the working environment; the muffs should be regarded as the wearer's personal property.

The effectiveness of hearing protection on a temporary basis is, of course, dependent upon the enforcement of the use of the protector means.

NOISE REDUCTION

To achieve a quiet working environment and to avoid litigation, methods of noise suppression should be considered at the initial stages in the planning of new or altered works sites. Machinery can be silenced by modifications to design and by greater attention to noisy components such as gearboxes, drives and impulsive elements. In some cases successful use of heavier foundations, or enclosure within sound absorbing housings has improved the environment for both work people and the public.

Office machinery can generate much unnecessary noise especially as it wears; where such noise is likely to cause annoyance, consideration should be given either to the use of quieter equipment, even if more expensive, or to the various electronic alternatives existing at present. Certain equipment during its running life will emit high piercing noises which can be reduced by silencers, and similar devices.

It should be remembered that, occasionally, in producing a less noisy environment for the benefit of some people the conditions for others are worsened. The employment of consultants may be recommended when there is no noise engineer on a factory staff but the field is so wide and varied that the experience and field of operation of the consultant firms is a matter of careful consideration before a choice is made.

REFERENCES

1. IEC/R.226
   Normal equal-loudness contours for pure tones and normal threshold of hearing under free field listening conditions; International Electrotechnical Commission (1961)
2. BS 4196
   Guide to the selection of methods of measuring noise emitted by machinery (1967)
3. BS 4197
   A Precision sound level meter (1967)
4. BS 4198
   Methods for calculating loudness (1967)
5. BS 4142
   Method of rating industrial noise affecting mixed residential and industrial areas (1967)

BIBLIOGRAPHY

Useful information regarding noise and methods of suppression are given in numerous relevant publications of which the following is a small selection.

HARRIS, C. M., *Handbook of Noise Control*, McGraw Hill Book Co. (1957)
BURNS, W., *Noise and Man*, John Murray (1968)
BURNS, W. and ROBINSON, D. W., *Hearing and Noise in Industry*, H.M.S.O. (1970)
BSI Code of Practice CP3: Chap. III *Sound Insulation and Noise Reduction*, British Standards Institution (1960)
Noise: Final Report, H.M.S.O. (1963)
EEUA, *Measurement and Control of Noise*, Constable (1968)
DAY, B. P., FORD, R. D. and LORD, P., *Building Acoustics*, Elsevier (1969)
EVANS, E. J. and BAZLEY, E. N., *Sound Absorbing Materials*, H.M.S.O. (1960) with reprints (1961 and 1968)
Various publications on Noise in Buildings, published by the Building Research Station.
Codes of Practice for reducing the exposure of employed persons to noise, H.M.S.O. (1972).

ENVIRONMENTAL ENGINEERING

A. A. FIELD

Environmental engineering normally deals with the mechanical services needed to provide an acceptable environment — thermal, visual and acoustic — in a building. To an increasing extent, however, environmental engineering is receiving a much wider interpretation to include all the ergonomic factors such as sanitation, transportation, electrical services, the building envelope, space design, colour and so on. These are all fields which, until recently, represented completely independent disciplines.

This section deals with environmental engineering in its generally understood sense as first defined.

ENVIRONMENT

The basic question to be answered is what kind of environment do people need for comfort. Comfort in this sense being related to heat, light and sound, and not to space design or aesthetics.

Human beings are capable of living in extreme conditions of heat and cold by variation in clothing and activity. But in the normal built environment with relatively fixed standards of clothing and moderate activity, only a narrow range of temperature is acceptable for comfort.

Heat production

The rate at which heat is evolved in the body depends almost entirely on activity and only to a very small extent on the temperature of the surroundings.

Knowledge of this rate of heat production is necessary in working out the cooling load in densely occupied buildings such as theatres. Table 15.3 gives typical values for design purposes,[1] and Figure 15.2 shows how the rate varies with different kinds of activity.[2] It has been found that 80 % of the heat evolved under comfort conditions for sedentary or moderate activity is sensible, and 20 % is latent (evaporative). As the temperature increases, more heat is lost by evaporation (Figure 15.3).[3]

Parameters of comfort

The body always establishes a balance between its internal heat production and loss, otherwise body temperature would rise or fall. Transient conditions which produce changing skin temperature affect warmth sensations but generally this is not considered in environmental engineering which deals only with the steady state. The parameters of thermal comfort are summarised in Table 15.4.

In the built environment, the most important factors in determining warmth are the air temperature and the mean radiant temperature. Numerous warmth indices relating

15–24

Table 15.3 TOTAL BODILY HEAT PRODUCTION
(SENSIBLE AND LATENT)

Activity	Total heat production (W)
Basal metabolism	70–80
Sitting at rest	100–120
Sedentary activity	120–140
Walking on a level at 4 km/h	250–320
Walking on a level at 7 km/h	350–470
Light industrial work	180–350
Moderate industrial work	350–470

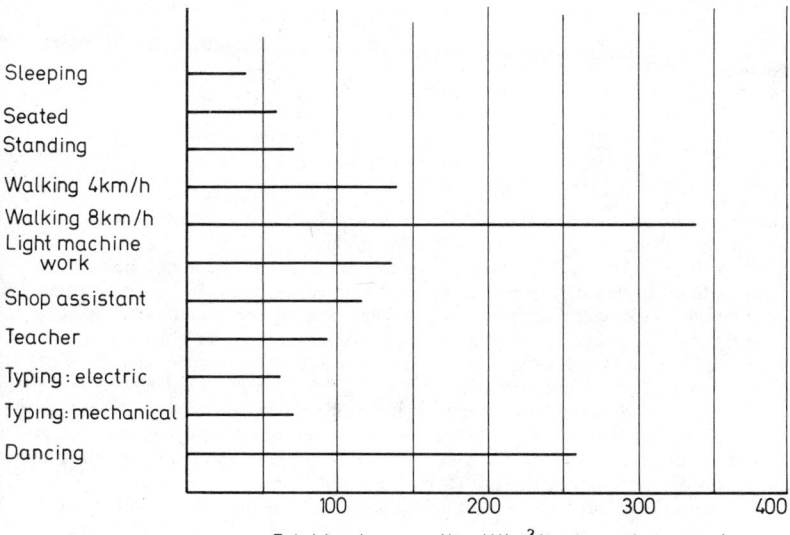

Total heat generation W/m² (body surface area)

Figure 15.2. Variation of body heat production with activity

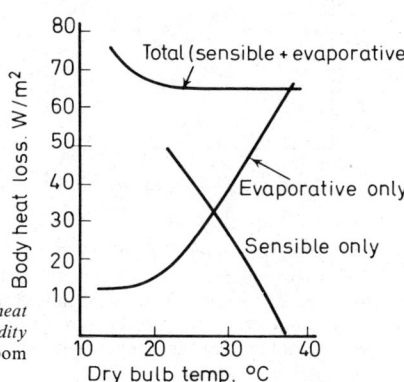

Figure 15.3. Variation of sensible and latent heat production with temperature (Relative humidity 40%). (Based on ASHRAE calorimeter room experiments.)

these two factors, sometimes combined with humidity and air velocity, have been developed over the years e.g. effective temperature, equivalent temperature, equivalent warmth, and operative temperature, but the U.K. is now tending to favour the use of *resultant temperature*. This is a simple concept originated by Missenard.[4] and is sufficiently accurate for normal conditions. The resultant temperature, t_r is defined as

$$(t_a + t_m)/2 \tag{1}$$

where t_a is the air temperature and t_m the mean radiant temperature. Any combination of air temperature and mean radiant temperature, giving the same resultant temperature

Table 15.4   PARAMETERS OF THERMAL COMFORT

Environmental factors		Human factors
Primary	*Secondary*	
Air temperature	Humidity	Age
Mean radiant temperature	Assymetric radiation	Sex
Air velocity	Conduction	Clothing
		Body weight

will therefore produce the same feeling of warmth. Thus the temperature of the surroundings is as important as the temperature of the air.

Mean radiant temperature (mrt) is roughly the same as the area-weighted average surrounding temperature. In rooms with radiant heat emitters, and in large spaces, the calculation is more complicated and a heat balance must be worked out:[2, 4] a computer program exists for the case of high temperature radiants in factories.[6] In practice, resultant temperature is measured with a 150 mm diameter globe thermometer.

Humidity has little effect on warmth in the range 30% to 70%, which is the maximum likely to occur in any case. Air movement in rooms in winter is not likely to exceed 0·15 m/s, which is not significant as far as warmth is concerned. Draughts from ventilating grilles can cause discomfort but the effect depends on the air temperature. In summer, higher air speeds can be tolerated (Figure 15.4).[5]

Assymetric radiation, that is, heat falling more strongly on one part of the body than

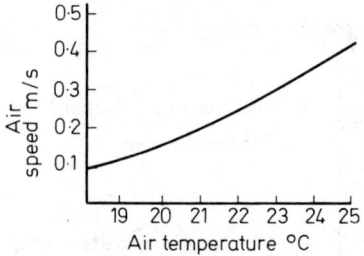

Figure 15.4. Relationship between air temperature and air speed for equal comfort (Based on DIN standards)

another or the presence of nearby heat sinks such as large windows, can be uncomfortable. A particular case is radiation at head level from heated panels. The limit here is the rise of mrt at head level, which should not exceed 2 °C at an air temperature of 18 °C.[7] Graphical solutions to this for different geometries are given by Kollmar.[8]

Vertical gradients of air temperature should not exceed 3°C in the occupied zone, i.e. over a height of about 2 m.

Conduction can be important where prolonged contact with a hot or cold surface occurs, for example standing on a heated floor panel. The maximum floor temperature is generally taken as 27°C for air at 20°C. The 'cold floor' sensation is mainly due to the stratification of cold air, not conduction into the floor. Contact chilling can occur however, with very light footwear. Nevins[9] gives a lower limit of 17°C.

The human factors of age, sex, weight and country of origin do not have much influence. Comfort may be judged perhaps, 1°C higher or lower at the most.[2]

Clothing is not an important variable factor in heated buildings. When necessary the insulating value of clothing can be worked out.[2]

Comfort zones

Individual differences make it impossible to fix an optimum resultant temperature. The choice therefore is a statistical one; the temperature at which the least number of people are feeling too cool or too warm. The range is accepted generally at 19 °C to 23 °C.

Table 15.5   NORMALLY ACCEPTED ROOM TEMPERATURES

Type of building	Environmental temperature °C
FACTORIES:	
Sedentary work	19
Light work	16
FLATS, RESIDENCES AND HOSTELS:	
Living rooms	21
Bedrooms	18
Bathrooms	22
Staircases and corridors	16
HOSPITALS:	
Operating theatre suite	18–21
Wards and patient areas	18
HOTELS:	
Bedrooms (standard)	22
Bedrooms (luxury)	24
Public rooms	21
Staircases, corridors, halls and foyers	18
OFFICES:	
General	20
Private	20
RESTAURANTS AND TEASHOPS	18
SCHOOLS AND COLLEGES:	
Classrooms	18
Lecture rooms	18
SHOPS AND SHOWROOMS:	
Small	18
Large	18
Department store	18

Table 15.5 gives a selection of some of the recommended temperatures for different types of building.[10] In summer the comfort zone shifts upwards by about 2 °C.

Hypothermia can result in underheated rooms. Taylor[11] found body temperature could be appreciably lowered and death by no means uncommon.

The hot industries (steel works, glass works) are not normally the concern of the environmental engineer. Here the exposure is best measured by the heat stress index.[12] Solutions include cold air douche, radiation screening and protective clothing.

Air

The minimum quantity of fresh air necessary to supply sufficient oxygen is extremely small, less than 0·5 litre/s per person, but much greater air quantities are needed in buildings where the main criterion is odour and not oxygen or carbon dioxide. The normally accepted range is 3·5 to 12·0 litre/s per person, rising to 24 litre/s in special cases where there is heavy smoking. Local regulations may specify a minimum fresh air requirement.

Table 15.6 summarises the recommendations of the Institution of Heating and Ventilating Engineers (IHVE). These are of course only for fresh i.e. outside air, requirements;

Table 15.6 IHVE RECOMMENDED MINIMUM FRESH AIR REQUIREMENTS (litre/s, per person)

Air space per person (m³)	Smoking not permitted	Smoking permitted
3	17·0	22·6
6	10·7	14·2
9	7·8	10·4
12	6·0	8·0

ventilation systems have to circulate more than this basic quantity to promote good distribution and to provide heating or cooling.

Light

Window design for natural lighting is traditionally the task of the architect and not the environmental engineer. However, window size related solely to the lighting level may lead to excess solar gain in summer and co-operation between the two disciplines is necessary.

There is now a trend towards the design of deep buildings having only vision-strip windows and permanent artificial light (PSALI). Lighting levels in this case are higher than normal to compensate for window glare and give a better standard of illumination.

Subjective response and lighting standards

The eye will accept a light intensity span of about 10^{12} to 1, but can only resolve detail in an ambient range of 10^3 to 1. Colours are not received with equal sensitivity, the eye being least sensitive to red and violet, and most sensitive to the yellow-green part of the spectrum (Figure 15.5).

Visual acuity, the ability to discern detail, depends on colour and the relative brightness of the object: over the normal room lighting range, acuity varies by about 20%. The The candela (cd) is the unit of light intensity, and the flux per unit steradian from a uniform source is the lumen (lm). The intensity of illumination of a surface lm/m^2 is the lux (lx). A minimum illuminance value of 200 lx is specified by the Code of the Illuminating Engineering Society for all working areas. Table 15.7 gives some typical values for different applications.

An important factor in the design of lighting installations is glare; this is basically a wide disparity in brightness between two areas, for example, the light fitting itself and the illumination level produced. Installations must be designed to meet the IES limiting glare indices.

Noise suppression

The services in a building should be designed so that noise generated by the prime movers, or flow of air and water does not impose on the general background noise level. Equipment fixed outside the building such as cooling towers or dust arresters should not create annoyance.

The environmental engineer is not generally concerned with architectural questions

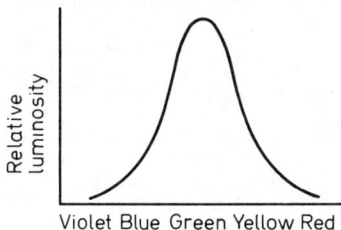

Figure 15.5. Light sensitivity of the eye (Based on work by Gibson and Tyndall)

Table 15.7 TYPICAL ILLUMINATION VALUES FOR VARIOUS TASKS

Situation	Lighting level, lux	
	General	Detail work
Offices	400	600
Shops	200	600
Dwellings	100	600
Industrial	200	900

Illuminance for very fine detail work, e.g. electronic component assembly, can go up to 2000 lx.

such as insulation against the external environment or acoustic treatment of building surfaces.

Sound energy and pressure units

Sound energy (W) needs to be known in the rating of sound sources such as fans and motors, and sound intensity (W/m^2) determines how much energy flows through a given area. Both quantities are measured in decibels (dB) above a standard datum.

$$\text{Energy level (dB)} = 10 \log_{10} W/W_0 \qquad (2)$$
$$\text{Intensity level (dB)} = 10 \log_{10} I/I_0 \qquad (3)$$

The datum value in both cases is 10^{-12} W.

Sound pressure is the most appropriate unit for the subjective effects of sound, and in any case is the easiest to measure. Since sound energy is a function of the square of the pressure

$$\text{Pressure level (dB)} = 10 \log_{10} (P/P_0)^2 = 20 \log_{10} P/P_0 \qquad (4)$$

The datum value in this case, $P_0 = 2 \times 10^{-5}$ N/m².

This in effect brings back the pressure level to the same basis as the energy level.

Decibels cannot be added or subtracted and in order to find the total of two sound sources, the combined decibel level must be recalculated from the actual values of energy or pressure.

Subjective response to sound

Subjective response to sound follows the Weber law of sensation which basically says that a perceptible change in the sound level is always produced by the same proportional increase in sound, whatever the absolute value happens to be. The range of sound pressures which the ear can accommodate is about 10^6 to 1 and therefore the convention of using decibels is seen to be not only convenient but related to the nature of the hearing response.

The ear is not equally sensitive to all frequencies and the datum of 20 μN/m^2 is just below the threshold at 1000 Hz. The limits of normal hearing are about 20 to 20 000 Hz. Figure 15.6 gives curves of equal loudness based on experiments which related the actual

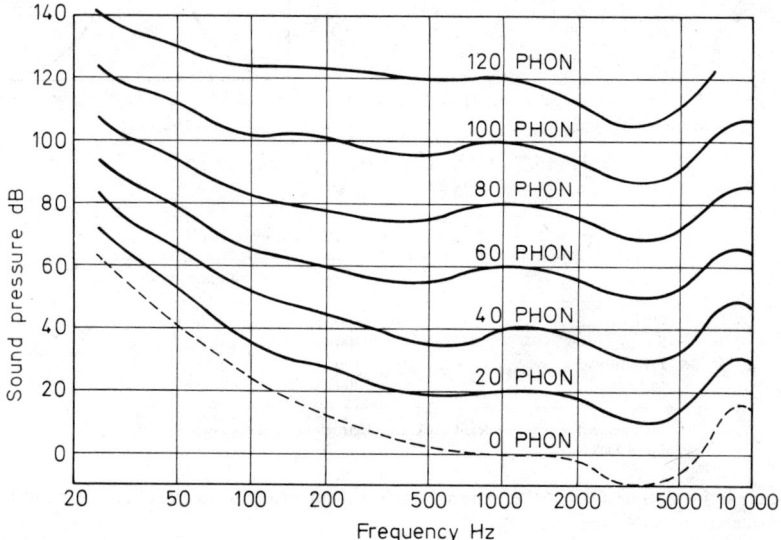

Figure 15.6. Fletcher and Munson's equivalent loudness contours for varying sound pressure and frequency

noise being heard to a standard noise at 1000 Hz. Thus, at 100 Hz, a 50 dB sound would appear to be as loud as a 38 dB sound at 1000 Hz. A sound thus rated on the basis of the 1000 Hz value is known as the *phon*.

The *sone* is another unit introduced to give a value directly related to the loudness i.e. a sound of 60 sones would be twice as loud as one of 30 sones.

Noise criteria

Noise from equipment is generally distributed throughout the spectrum (broad-band or white noise). Because of the variable sensitivity of the ear, noise criteria must relate dB to frequency. From fieldwork, Beranek formulated NC curves (Figure 15.7) which limit the interference to speech (600 to 4800 Hz) and the loudness (high and low frequency).

These curves are now widely accepted as a means of rating an environment. Table 15.8 gives recommended NC values for different situations.

CLIMATE

Architecture and construction do not enter specifically into environmental engineering but knowledge of how the building reacts to the external conditions of temperature, sun and wind is essential for the designer in matching up the internal services.

The most important variable in winter is the air temperature. In spite of its relatively high latitude, the U.K. has a fairly mild climate, and temperatures lower than $-5°C$

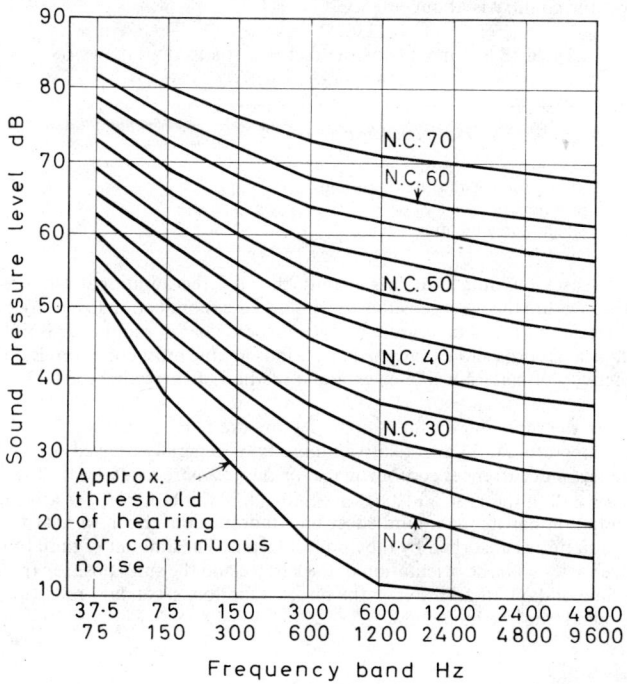

Figure 15.7. Beranek's NC curves

Table 15.8   RECOMMENDED NOISE RATINGS FOR
DIFFERENT SITUATIONS

NC rating	Situation
25	Concert halls, sound studios, hotel bedrooms
30	Theatres and cinemas, TV studios, houses, hotel bedrooms
35	Schools, hospital wards, flats, hotels
40	Circulation areas, restaurants, offices
45	Shops, supermarkets, canteens
50	Offices with machinery
60	Industrial

occur only for one or two days during the year. In general, the temperature is 0°C or higher most of the time. It would be uneconomical, of course, to design the system on the lowest recorded external temperature, and the procedure originally proposed by the Ministry of Works Study Committee[13] was to choose the lowest temperature which on average lasts for a period equal to the time lag of the building. For example, if the building had a time lag of two days, and the lowest temperature which lasted for two consecutive days was −2°C, this would be the basis for designing the heating system; periods of less than two days where the temperature might be −3°C would be neglected.

Non-average temperatures are covered by the normal overload margin on the system (15 to 20%). Table 15.9 gives the IHVE recommendations. The variation from these figures over the country is about ±0·5°C.

Table 15.9   IHVE RECOMMENDATIONS FOR EXTERNAL DESIGN TEMPERATURES IN WINTER °C

Type of building	System margin	
	20 per cent	None
Multi-storey with solid floors and partitions	−1	−4
Single storey buildings	−3	−5

Wind increases building heat loss mainly through the additional air flow through cracks; windows are the most sensitive to wind as far as conduction losses are concerned. The combination of low temperature and wind speeds of 5m/s and over is not frequent. Meteorological records show that northerly winds are the most common. Rain increases the conductivity of materials like brick and a method of correcting for this is described later.

In summer the most important climatic factors are temperature and solar radiation. Dry bulb temperature (and also relative humidity) is generally derived on the basis of a 1% frequency of occurrence; contour maps for the U.K. are available[10]. Typical values for the lower half of the U.K. are 28°C dry bulb, 20°C wet bulb (48% relative humidity).

Solar radiation has its maximum effect on windows, conversion to heat taking place when the radiation is absorbed by the floor and walls. The amount of radiation received by a surface varies with the orientation of the surface and the sun's position (period of the year and time of day). In addition to the direct radiation, secondary radiation from the sky and ground must be included.

Heat losses

Heat is lost from a building by direct conduction through the structure to the outside air (fabric loss), and in the warmer air displaced by air entering from the outside (air change loss). In well insulated buildings, often the air change loss is the greater part of the total.

Fabric loss

The heat conduction rate through the building fabric is characterised by the U-value, defined as the heat flow through unit area per 1°C difference in air temperature between the inside and outside (W/m² °C).

The U-value is compounded of the direct conduction through the material and the so-called surface conductance. The latter substitutes an equivalent conductance for the radiation and convection heat exchange at the surface with its surroundings.

The general solution for U-value is:

$$U = 1/(1/C_i + 1/C_0 + t_1/k_1 + ... t_n/k_n) \tag{5}$$

where C_i and C_o are the surface conductances (inside and outside), and t and k are the thicknesses and conductivities of the various structural elements.

It is more convenient in practice to use resistance and resistivity ($1/C$ or $1/k$), giving $m^2 °C/W$ and $m °C/W$.

Thus the equation is:

$$U = 1/(R_i + R_o + t_1 r_1 + \dots t_n r_n) \tag{6}$$

The total resistance of a material is found by simply multiplying the resistivity by the thickness. The higher the resistance of the material, the better its insulating value.

For non-homogeneous components, e.g. cored blocks or double glazing, the resistance of the whole must be known.

Building materials

In general the resistivity of material is inversely proportional to the density. Figure 15.8 gives the variation for lightweight concretes. In the higher resistivity range, 20 to 40 m °C/W, the relationship is not quite as close but nevertheless gives a good idea of the

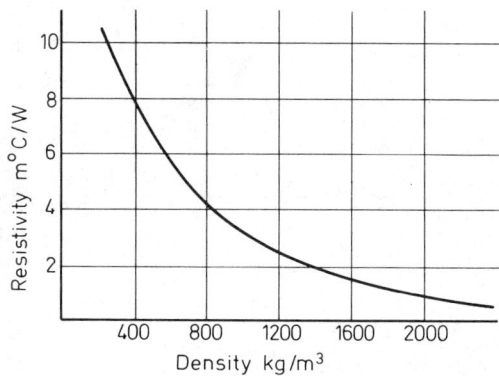

Figure 15.8. Variation of resistivity with density for concretes (Based on data by the Building Research Station)

performance of new materials. Table 15.10 lists resistivity values for a range of common building materials.

When a material becomes damp, the resistivity decreases. Table 15.11 gives correction factors proposed by Jakob[14] which can be applied to the resistivity values in Table 15.10. For design purposes, the moisture content of exposed brick and concrete is taken to be 5%.

Air spaces

Air spaces in a structure, such as occur for example between a fibre building board and a corrugated roof, are good insulators. In general, an air space reaches its maximum resistance at a width of about 20 mm; at less than 10 mm, the resistance falls off fairly quickly.

Where air spaces are created they should be at least 12 mm wide giving 90% of maximum resistance. Table 15.12 gives values proposed by the Building Research Station. The higher resistance of the air space when one of the surfaces has low emissivity (aluminium foil, for example) is due to the fact that radiation transfer across the air space is almost entirely suppressed.

Table 15.10 TYPICAL RESISTIVITIES AND DENSITIES OF SOME COMMON BUILDING MATERIALS

Material	Density kg/m^3	Resistivity m °C/W
Asbestos cement	1 500	3·5
Bitumen	1 100	6·2
Brick	—	0·8
Concretes:		
dense	2 300	0·7
foamed slag	1 100	5·5
foamed slag	1 500	2·6
vermiculite	400	9·2
vermiculite	560	6·3
vermiculite	880	3·9
aerated	320	11·8
aerated	640	6·9
aerated	960	3·9
Felt	200	26
Glass	2 500	1·0
Hardboard	800	9·7
Metals	3 000 to 8 000	0·01 to 0·03
Mortar	1 900	1·1
Plaster	—	2·5
Plasterboard	950	6·3
Plastics	1 000 to 2 200	0·2 to 0·5
Stone	2 200	0·6
Timber	600 to 700	6 to 8

INSULATING MATERIALS	
Cellular glass	18
Compressed straw slab	10
Corkboard	20 to 25
Expanded polystyrene	25 to 28
Expanded PVC	35
Fibreboard	15 to 20
Foamed infills (polyurethane)	40 to 45
Loose fill granules (vermiculite, perlite)	20 to 30
Mineral fibre blanket	20 to 30

Table 15.11 MOISTURE CORRECTION FACTORS FOR THE RESISTIVITY OF MASONRY

Moisture content % vol.	Factor
1	0·77
3	0·62
5	0·57
10	0·48
15	0·43
20	0·39
25	0·36

Multiple glazing is simply a series of air spaces separated by a highly conducting material.

Cold bridges

In certain composite structures like curtain walling or hollow blocks, part of the material forms a directly conducting path to the outside air. The rest of the structure may be well insulated but the 'cold bridge' increases the total heat loss. The combined U-value can be found by dividing each area U-value product by the total area.

Cold bridges can cause condensation and this aspect is dealt with later in this section.

Solid floors

Solid floors represent a special case in calculating the conduction path. Generally, it can be said that most of the heat from a solid floor is lost around the building perimeter. Table 15.13 is from work by Billington[15].

Table 15.12   AIR SPACE RESISTANCES m^2 °C/W, UNVENTILATED (BASED ON BRS DIGEST NO. 108, 1969)

Space between:	Width mm	Surface	Heat flow direction	
			Horizontal or upwards	Downwards
Two flat surfaces	5	plain*	0·11	0·11
	5	alum.	0·18	0·18
	20	plain*	0·18	0·21
	20	alum.	0·35	1·06
Flat and corrugated sheets in contact	—	plain	0·09	0·11

* Plain refers to any surface of high emissivity.

Table 15.13   HEAT LOSS FROM SOLID FLOORS

Dimensions, L	m W	No. of exposed edges	U* W/m^2 °C
Any length	30		0·1
	15	1	0·2
	8		0·3
	3		0·5
30	30		0·3
30	8		0·5
15	15	4	0·4
15	8		0·6
3	3		1·5

* Applies to the total floor area.

Insulation to a solid floor need not cover the whole area to be effective. A vertical perimeter strip of 25 mm glass wool, 0·5 m deep, will reduce the average loss by about 12 %.

Surface conductances

Work by the Building Research Station[16] has established standard values and these are reproduced in Table 15.14. Aluminium, providing it is not painted, will retain its low emissivity even when oxidised.

Table 15.14 SURFACE RESISTANCES (BRS DATA) m^2 °c/w

Building element	Surface	Inside	Outside		
			Sheltered	Normal (standard)	Severe
Wall	plain*	0·12	0·080	0·053	0·027
	alum.	—	0·106	0·062	0·027
Floor (downward flow)	plain*	0·15	—	—	—
Roof	plain*	0·11	0·070	0·044	0·018
	alum.	—	0·088	0·053	0·018

* Any surface of high emissivity.

Definition of the exposure terms in Table 15.14 are as follows:
Sheltered — Single or two-storey buildings in towns.
Normal (standard value) — Third to fifth storeys of buildings in towns and most country buildings.
Severe — Sixth and higher storeys of buildings in towns: buildings in very exposed positions outside towns.

Standard U-values

All U-values calculated from the 'normal' exposure values are said to be standard. Table 15.15 gives standard U-values for a representative range of constructions. Values for sheltered or severe exposure can be derived as follows:

$$U' = U/[1 - U(R_0 - R'_0)] \qquad (7)$$

where the superscripts denote corresponding values of U and R for the non-normal exposure.
More comprehensive tables will be found in the IHVE Guide Book A.

Condensation

In houses and flats, condensation on the inside of the structure is fairly common, more particularly in the older housing stock. There are several reasons: poor standard of insulation, lack of ventilation, dispersal of moisture from cooking and laundering, and only partial heating. Moisture evolved in a warm area is carried by convection currents or diffusion to a lower temperature area where it condenses on the cool surfaces of walls and windows.
Condensation within the structure itself can also occur when moisture diffuses through a permeable inner skin into a low temperature zone. A good example of this is the roof of a sheeted industrial building lined with unpainted fibreboard, where moisture pene-

trates through the board and condenses on the cold inner face of the sheet. The rate of condensation depends on the vapour diffusivity of the material.

The drop of temperature up to any point in a material is proportional to the fraction of the total resistance. Thus, with a U-value of $1 \cdot 0$ ($R = 1 \cdot 0$) and a temperature difference of $20°C$, the temperature drop to the inner surface (facing the room) — resistance $0 \cdot 123$ — is: $(0 \cdot 123/1 \cdot 0) \times 20 = 2 \cdot 4°C$. The inner surface would thus be at the temperature of $20 - 2 \cdot 4 = 17 \cdot 6°C$. This would have to be above the dew point of the air to prevent condensation.

The weakest points in a structure for condensation are cold bridges and single glazing. Where high humidity is normal in a building, e.g. swimming pools, double glazing is

Table 15.15   STANDARD U-VALUES, W/m^2 °C

Construction	Thickness mm	U	Construction	Thickness mm	U
WALLS			ROOFS		
Brick, plastered	110	3·6	Concrete (asphalt and		
	220	2·7	plaster)	150	3·1
Cavity brick, plastered	270	1·7	Hollow concrete beams		
Concrete	100	4·0	(asphalt and plaster)	150	2·5
	150	3·6	Hollow steel decking		
Brick (cavity) lightweight			insulated 25 mm		
concrete inner skin	260	1·6	polystyrene board		1·0
Wood	25	2·8	Cement-bonded wood-		
			wool decking		1·6
Corrugated asbestos-					
cement sheets:			Corrugated asbestos-		
Uninsulated		6·5	cement sheets:		
12 mm fibre board		2·0	Uninsulated		8·0
Corrugated aluminium			Sandwich with 25 mm		
sheets:			glass fibre		1·1
Uninsulated		4·0	Pitched, tiled roof with		
Sandwich with 25 mm			plaster ceiling:		
glass fibre		0·8	Uninsulated		2·2
Single glazing		5·7	25 mm glass fibre		0·8
Partition walls:					
110 mm brick		2·9			
75 mm breeze blocks		2·4			
Double skin steel		2·4			

essential to avoid continuous condensation although even double-glazing will not prevent condensation all the time.

Graphical solutions to the calculation procedures including vapour transmission, are given in the DoE publication 'Design Guide to Condensation in Buildings'.

Air change load

The amount of air entering the building is determined by the area of crackage, the wind pressure and the stack effect caused by the temperature difference. For multi-storey buildings with corridors and stairwells, the calculation is complicated and best solved by computer: a program for this has been developed by the HVRA[17]. For single-space buildings the IHVE calculation procedure can be used[10].

Empirical data in terms of air-change rate (m^3/h per m^3 of building space) can be used. Some typical values are given in Table 15.16. The infiltration heat load can be calculated with sufficient accuracy from the volumetric specific heat of air (1200 J/m³ °C):

$$1200\ N\ (t_i - t_0)/3.6 \times 10^3 = 0.33\ N\ (t_i - t_0)\ \text{W/m}^3 \tag{8}$$

where N is the rate of infiltration, changes /h.

t_i is the internal air temperature °C.

t_0 is the external air temperature.

Environmental temperature

This is a concept developed by Loudon of the Building Research Station to compensate for the error involved in taking the total heat transfer through the fabric as a function of the inside/outside air temperature difference. Radiation exchange is also involved, the mrt not being the same for all surfaces.

Environmental temperature is defined as $0.67\ t_m + 0.33\ t_a$, but for practical purposes

Table 15.16   AIR INFILTRATION RATES; m³/h PER m³ OF SPACE

Space	Rate of infiltration
Houses and flats:	
Living rooms	1·0
Bedrooms	0·5
Hospitals*	1·0 to 2·0
Hotels	1·0 to 1·5
Offices	1·0
Schools	1·0 to 2·0
Shops	0·5 to 1·0
Factories (sheeted, lined construction)†	1·0 to 2·0

* More detailed information is given in the Hospital Building Notes series (HMSO).
† The lower rate applies to large spaces. In existing, unlined factories air change rate can reach three or four changes per hour.

environmental temperature may be equated to resultant temperature. Thus the procedure is to work out the air change load as a function of air temperature, and the fabric heat loss as a function of the resultant temperature.

Calculating heat losses

Heat losses are generally calculated on a standard form which lists room number, dimensions, areas (for walls, glazing, ceiling and floors), U-values, and temperature difference. A complete room heat loss would be:

$$0.33\ NQ\ (t_a - t_0) + [A_w U_w + A_g U_g + A_c U_c + A_f U_f](t_r - t_0) \tag{9}$$

the answer being in watts.

This is the steady state loss assuming that temperature does not change with time. In certain cases, notably with electric storage heating where the heat input falls off during the day, and with solar gain, an idea of the thermal response of the building is necessary. The ability of the walls and floors in a building to smooth out temperature fluctuations is characterised by the admittance (W/m² °C). A calculation procedure is described by Loudon[18].

COMPUTERS

Several heat loss programs exist (HVRA) but the most promising technique so far which eliminates data transfer, is the Plancal desk calculator. A stylus, which is linked to the machine, is used to trace the building outline. The machine integrates the input with height, temperature difference and *U*–value.

Heat gain

The environmental engineer must show whether a case for air conditioning exists. The curves of Figure 15.9 were derived by BRS work with an analogue computer and show clearly how overheating is related to glazing area.

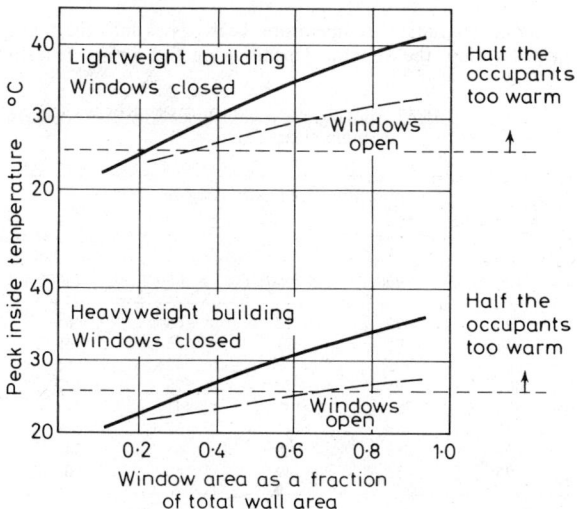

Figure 15.9. Room temperatures without air conditioning (Based on the work of A. G. Loudon, BRS Current Paper CP4/68)

The cooling load of an air-conditioning plant is made up of the following items, roughly in order of importance:
Direct solar transmission through windows.
Internal gains.
Conduction through the building fabric.
Fresh air brought in by the air-conditioning plant.
Infiltration of outside air.
Latent heat.
Detailed techniques of calculation will be found in the IHVE Guide Book A, and the Carrier Handbook of Air-Conditioning Design. Several programs are available for computer calculation of loads, e.g. those by Svenska Flaktfabrikan, and Faber Computer Operations.

Direct solar transmission

The amount of energy entering a window depends on the position of the sun, whether shades or special glazing is used, and the orientation of the window. Table 15.17 gives the transmission through plain, single-glazing for the two peaks in June and September for 52°N latitude. The figures include diffuse radiation from the sky and ground as well as the direct component. Correction factors for different kinds of glazing and blinds are included in a footnote to the table.

External shading can take the form of fabric blinds, louvres, balconies or structural fins. Graphical methods of arriving at shade factors are available[19].

Conduction

In addition to the higher outside temperature, heat passes into the building through solar radiation absorption, the amount depending on the surface reflectivity and the

Table 15.17. SOLAR HEAT TRANSMISSION THROUGH WINDOWS (51°N) w/m^2
(Single clear glazing)

Orientation	Month and time of day (hrs. GMT)					
	June			September and March		
	1000	1400	1600	1000	1400	1600
N	95	95	72	65	65	42
NE	133	95	72	65	65	42
E	383	95	72	342	65	42
SE	467	95	72	528	137	42
S	331	331	137	441	441	205
SW	95	467	441	137	528	414
W	95	383	528	65	342	406
NW	95	133	349	65	65	182

Derived from data in the IHVE Guide Book A

Correction factors:	
Double glazing	0·8
Heat absorbing glass	0·5–0·7
Heat reflecting glass	0·3
White Venetian blinds	0·6
Double glazing with Venetian blinds between panes	0·4

U-value. To avoid having to make a separate heat balance, the concept of sol-air temperature is used. This allows the total heat transfer to be written in terms of a fictitious temperature difference.

Tables of sol-air temperatures are given in the IHVE Guide Book A, also the technique of calculating the time lag through the structure.

HEATING SYSTEMS

The various means of building heating can be divided broadly into direct and indirect systems depending on whether the energy conversion to heat takes place on the spot or is transmitted by a suitable medium (Table 15.18)[20]. To see how various systems fit

Direct			Indirect			
Source	Appliance	Application	Source	Transport medium and temperature range °C	Appliance	Application
Solid fuel	Open fire, closable stove	Domestic	Boiler (solid, liquid, or gas fuel)	Water (100–200) Steam (100–200) Thermal Fluid (100–300)	Unit heater Air heater battery (plenum system or packaged air handling unit)	Industrial Industrial
Liquid fuel	Paraffin fires and convectors.	Domestic				
	Forced-draught air heater	Industrial	Heat exchanger (steam/water or water/water) Waste gas heat exchanger (Incinerator or gas turbine)		Radiant panel	Industrial
Gas	Radiant fire					
	Unflued convector	Domestic				
	Radiant panel	Industrial				
	Direct-mixing air heater	Industrial	Boiler (solid, liquid or gas fuel)	Water (35–120)	Radiator Convector (natural or forced) Skirting Air heater battery Embedded panel Suspended ceiling panel	Domestic/ Commercial Commercial
	Forced draught air heater	Domestic/ Industrial				
Electric	Floor heating	Mostly commercial	Electric thermal storage (solid or liquid medium)			
	Radiators and convectors	Domestic				
	Block storage units	Domestic/ Commercial	Heat exchanger (steam/water, or water/water)			
	Block storage warm air unit	Domestic	Direct-fired oil or gas/air heat exchanger	Air (100–250)	Radiant tubes	Industrial
			Rotary regenerative heat exchanger air/air	Metal	Warm air system	Commercial/ Industrial

the building requirements, they are reviewed below in terms of building sector, i.e. domestic, commercial, public and industrial.

Domestic heating

FIRES AND CONVECTORS

Solid fuel direct systems are exclusively domestic. The simple open fire has the merit of high radiation intensity, giving comfort at much lower air temperatures than normal, but high radiation assymetry, low efficiency (less than 20%), and high secondary air induction. Improved open fires with added convection are much better; efficiency over 50%, lower radiation output but higher air temperature, and less air induction. The closeable stove has the highest efficiency, about 60%, but gives appreciable air temperature gradient.

Paraffin convectors use all the heat in the fuel but also produce moisture and high temperature gradient.

Gas radiant fires have a high radiation output plus convection and low induction rates (efficiency over 60%).

WARM AIR

Gas and oil-fired warm-air units for housing consist of a combustion chamber and secondary hot gas transfer section. Air is blown over the heated surfaces and discharged into the room through short ducts.

Electric warm air systems can use either fan convectors (on-peak), or central block storage units for off-peak power. Air is circulated in the same way as with oil and gas systems, the advantage being that no flue is required.

Central warm-air systems are generally only suitable for new buildings. The advantages are rapid warm up time and no space taken up by appliances: the disadvantages are poor temperature distribution, noise, and increased fire transmission risk.

HYDRONIC SYSTEMS

Domestic systems using water as the heat transport medium are generally called hydronic. Heat can be generated in a boiler using any of the three fossil fuels, or from an electric block storage core with a heat exchanger. Water temperature generally does not exceed 85°C with radiators or similar appliances, but sometimes higher temperatures, up to 110°C are used with finned tube skirting or convectors where the element is hidden and cannot be accidentally touched. Water is pumped through the piping using a variety of distribution techniques Figure 15.10.

The small-bore system uses a constant-size pipe ring, generally 15 mm dia copper, with radiators taken off by tee connections, the flow in the secondary radiator circuit being promoted by the pressure drop across the connections to the ring. A variation of the ring (or single-pipe) system links the radiators in series and uses shunt valves to divert part of the ring flow through the radiator. With the one-pipe system, radiators have to be made larger (for the same emission) as the water cools progressively along the circuit.

In the microbore system much smaller pipes are used (6 and 10 mm dia.) every radiator having a separate set of flow and return connections from a manifold, which may contain up to sixteen tappings. Using coils of soft copper tube, the run-outs to each radiator can be bent by hand and are only jointed at the manifold and the radiator.

The most widely used room heating appliance is the radiator (reviewed later); other types are the convector (natural draught, or fan-assisted), skirting heater using finned pipe, and sometimes a water/air heat exchanger forming part of a warm air system. Most domestic systems have a load range of about 10–20 kW.

ELECTRIC

Electric floor heating was the first system to use off-peak supply, but today it has been largely replaced by the block-storage heater and central warm-air unit. Block storage heaters use a core of concrete or metal (heated by electric resistance elements) and insulated so that storage can be improved by using very high temperatures, up to 500°C. Improved designs of unit use a circulating fan to give a more selective use of the stored heat. The central warm-air unit is a larger version of this.

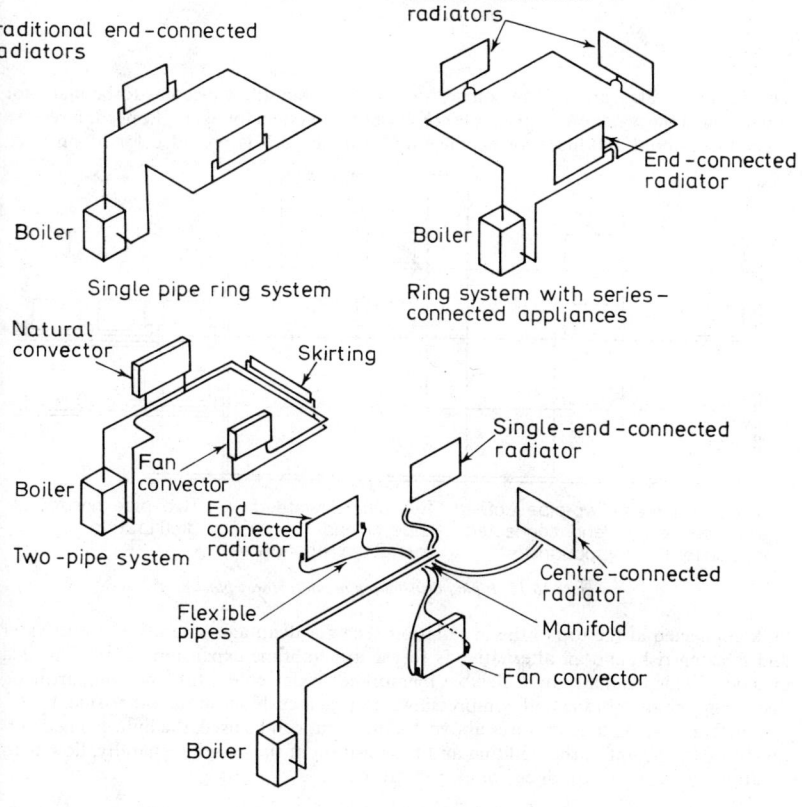

Figure 15.10. Piping distribution systems for domestic heating

On-peak, direct appliances like oil-filled radiators and radiant fires are mainly only used as a back up to other means of heating, or intermittantly for full heating.

HEATING COMMERCIAL AND PUBLIC BUILDINGS

Hot water systems

Buildings such as office blocks, hospitals, hotels, shops, libraries and so on are almost invariably heated by a low temperature water system. This is described below.

PIPING CIRCUITS

The two-pipe principle is used more widely for the distribution system, the one-pipe principle being only applicable to radiators or other low pressure drop appliances Figure 15.11 shows the variations possible.

EXPANSION

The pumped water circuit is of course, a closed one, but allowance has to be made for expansion of the water on heating: a 300 kW radiator system for example, needs a reserve capacity of about 150 litres for expansion. Normally, this is provided for by an open

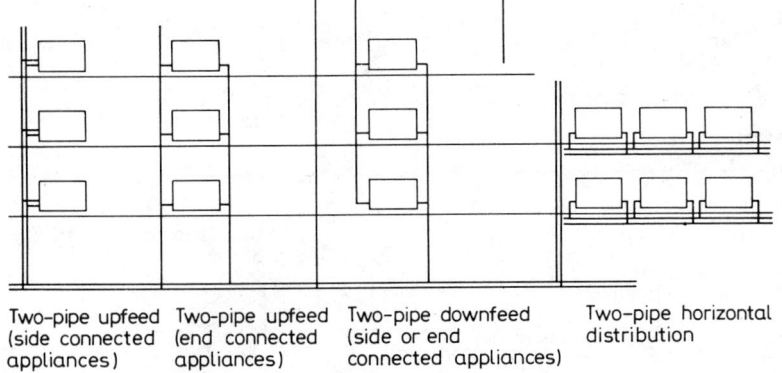

| Two–pipe upfeed (side connected appliances) | Two–pipe upfeed (end connected appliances) | Two–pipe downfeed (side or end connected appliances) | Two–pipe horizontal distribution |

Figure 15.11. Piping distribution in multi storey blocks

tank connected at the top of the installation. The disadvantage is aeration of the water and freezing risk, and an alternative is to use a membrane expansion vessel. This is a cylinder divided radially by a flexible membrane; water enters the one compartment, displacing the membrane and compressing a nitrogen cushion on the other side.

With this system, temperatures above boiling point can be used, the limit being determined by the height of the building and the position of the vessel. Generally, flow temperature with this system does not exceed 110°C.

APPLIANCES

The *radiator* is probably the most versatile of space heating appliances. Cast-iron radiators have almost entirely disappeared from the market, the disadvantage being principally the weight and the limited design variations. The steel radiator is basically a flat panel containing waterways. The outer surface is generally contoured, e.g. ribbing and corrugations to increase the convection output and for the sake of appearance. Higher outputs can be obtained by sandwiching two or three panels together, or by increasing the convective surface on the back of the radiator with welded fins. Generally radiator connections are at the ends; some makes however, have bottom centre connections, and in two of these the control valve is also built in. The advantage of the radiator as a heat emitter is the even air temperature distribution produced, the higher resultant temperature compared with a fully convective appliance, and the adaptability.

Radiator emission is measured in a standard calorimeter room and the rating given in terms of total output. (Formerly this was given in terms of Watts per m² of surface area).

Convectors use a finned pipe or pipe battery as the primary heating surface. By placing the element at the base of a cabinet with openings top and bottom, air flow is induced over the element by natural stack effect. This is the principle used in the continuous convector system; adjoining cabinets forming an unbroken line around the building periphery. Each module in this system can have different heat outputs according to the number of fins, and pipework is hidden. Independent cabinet convectors can work either by natural convection or by forced convection using a fan. A two-position switch can give a higher fan speed to boost output.

Skirting is essentially low level continuous convector surface, more generally applicable to domestic heating.

Something between a radiator and a convector is the high-pressure coil radiator. This is a pipe-grid clipped to a face plate of expanded metal. The advantage is that it can be used with high temperatures and higher pressures than normal.

Panel heating uses part of the ceiling or floor as low-temperature heating surface. Coils of pipe are either embedded directly in the structural concrete, or as a suspended ceiling panel held in contact with the ceiling itself. Figure 15.12 shows the various types. The directly-embedded system suffers from considerable time lag and the tendency today is to use suspended metal ceilings. One maker uses plaster tiles with aluminium spreader plates connecting to the pipes. *Floor heating* is not often sufficient in itself, since the temperature difference floor surface/air is limited to about 10°C for comfort, and needs to be supplemented by convectors or radiators.

Ceiling panels have about twice the emission of floor panels since they are able to run at higher temperatures, the limitation here is the downward radiation which must not exceed the comfort level (see page **15**-26).

Warm air systems

Where mechanical ventilation is required for a space, it is often more economic to combine ventilation and heating in one system.

Air is delivered to the building through a system of ducts and air diffusers. These are discussed more thoroughly under air-conditioning systems. Most of the air is recycled, since it is being used as a heat transfer medium and not simply for ventilation. The air flow rate is calculated from the temperature drop between the entering warm air and the room, normally in the 10 to 20°C region. Thus, air flow rate, m³/s, is given by

$$H_l/1200 \, (t_g - t_a) \tag{10}$$

where H_l is the heat loss in W and $t_g - t_a$ the excess of entering air temperature over room temperature °C.

The amount of fresh air will depend on the leakage rate of the building and the air required for the particular application (see page **15**-28).

Depending on the kind of building, the system can have a central plant room or rooms, with distribution ducts in ceiling voids and core spaces; or separate, packaged air handling units can be used. The heating medium is generally water, 100–120°C with sectional boilers, or up to about 150°C with shell boilers. Warm air systems, which are part of a large complex, can use whatever medium is available.

Electric systems

The running cost of direct electric heating normally excludes it from consideration. The same off-peak techniques can be used as described earlier, but the only satisfactory one is the forced convection block storage unit.

Large buildings can also use water as a storage medium. Power can be supplied either to an electrode boiler on a pumped water circuit to the storage cylinders, or each cylinder

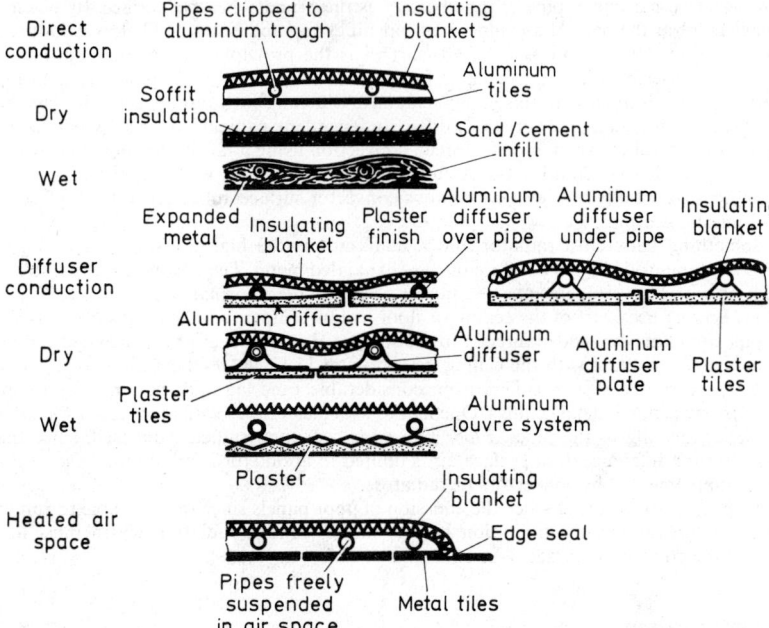

Figure 15.12. Types of low temperature panel heating (Courtesy of Heating, Piping and Air Conditioning, USA)

can have immersion heaters. The efficiency of storage depends on the difference between the storage temperature and the minimum system return temperature. Pressurising the cylinders can raise the storage potential, but the amount is limited by the cylinder cost: generally 120°C is the maximum in practice. An advantage of this technique is that a standard hydronic system can be used for heating.

INDUSTRIAL HEATING

Direct fired air heaters

One of the most widespread techniques in heating factories is by means of the free standing, direct fired air heater. This consists of a combustion chamber with secondary heat transfer tubes for the hot gases, fired by a pressure-jet oil burner or gas burner.

Air is circulated over the outside of the combustion chamber and tubes by a fan and discharged through adjustable louvres at the top of the unit. Several high velocity warm air streams are produced which are directed by means of the louvres to cover as wide an area as possible.

Discharge temperature for such units is about 70°C, which compared to the room temperature of 20°C gives 50°C buoyancy, resulting in overheating at roof level. Initial cost is, of course, low compared with a central hot water or steam system.

High temperature water and steam

Steam was at one time the only medium used for piped heating systems in factories. Now it has been largely superseded by medium or high temperature water (up to 200°C), which compared to steam has numerous advantages. For example, pipes do not need to be graded to run off condensate, and pipework is simplified by the absence of traps and relay points. Ancillaries like condensate tanks, pumps and flash recovery vessels are eliminated.

High temperature water (HTW) is inherently easier to control since any temperature can be produced by mixing at the boiler. Steam, on the other hand, has only a limited temperature variation through changing the pressure; for example, reducing pressure from 5 bar to 1 bar produces a fall from 159°C to 120°C, only about 30% reduction in heat output.

To generate water at temperatures above the atmospheric boiling point, the system is pressurised either from a vessel containing a nitrogen gas cushion, or by using the relatively new technique of continuously-running pressure pumps. Figure 15.13 shows the basic principles of both types of system. Circulating pumps for HTW need to have water-cooled bearings, otherwise they are not essentially different from the types used on low temperature water. A two-pipe distribution supplies the heat emitters.

When steam is used for heating it is always saturated, not superheated. Heat transfer with superheated steam is poor because the steam behaves as a gas, not as a condensing vapour which has a film conductance several hundred times as great. This is not important when the heating surface is a plain pipe, but it can severely affect the performance of heat exchangers which have a much higher transfer rate; desuperheating (spraying water into the steam) is necessary in this case.

Where steam requirements are intermittent and sharply peaked, advantage can be taken of the accumulator technique which stores steam at low load by condensing it in a water storage cylinder and allowing it to flash off as the pressure falls on increase of load.

Radiant heating

Factory heating can be either radiant or convective. In radiant heating over half the total heat output from the emitters is by radiation, which gives considerable advantages over warm air. The resultant temperature is always higher because the radiation has a direct warming effect on the occupants, the temperature gradient from floor to roof, even in buildings 20 m high, is very small. The radiant emitters (Figure 15.14) are generally pipe grids with steel or aluminium radiating plates fixed to them (radiant strip) and insulated on the opposite face so that heat is given off from one side only. The strip is fixed near the roof to radiate downwards. Individual panels are also sometimes used, these normally being fixed to the walls.

Another form of radiant system uses air as the heat transport medium. A direct-fired air heater is used to supply air at high temperature (about 250°C) which is circulated around a hairpin loop made up of circular sheet metal ducts side by side to form radiating

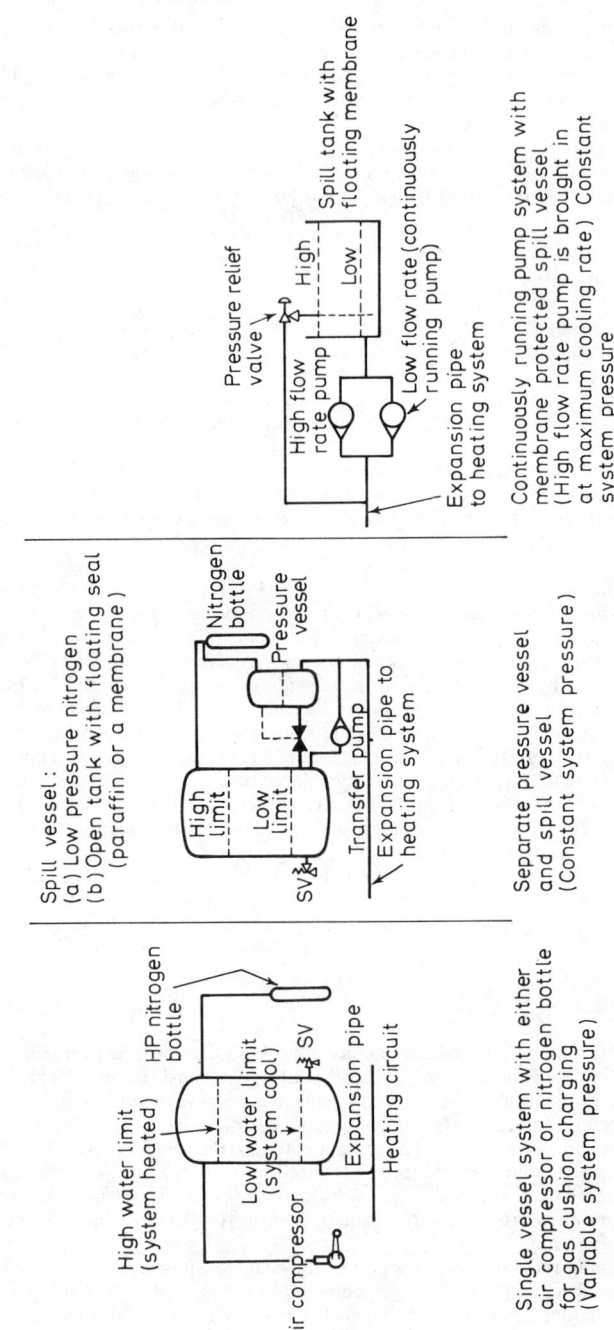

Figure 15.13. Techniques of pressurisation for high temperature water systems

surface. The advantage compared with piped radiant heating is to eliminate the boiler house and at the same time giving a better response rate.

Convection heating

Convection systems normally use unit heaters, which are small recirculating air handling units containing a finned tube battery. They can also be combined with a fresh air inlet. The disadvantages are maintenance of the motors, dust clogging of the battery, and uneven temperature distribution.

Central warm air (plenum) systems are normally only used when a ventilation requirement exists e.g. where the factory has a high air extract rate from process equipment. Heat

Section	Basic construction	Range of downward radiation. percent
	Grooved steel plate clipped or welded to 25–32 mm pipe grid; insulated	60–70
	Grooved steel or aluminum plate clipped to 25–32 mm exposed pipe grid; insulated	60–70
	40 mm pipe welded to heavy gauge steel plate, insulated	60–70
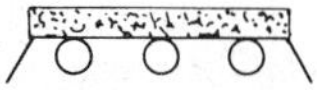	40–50mm pipe grid with steel plate in contact; insulated	60
	Grooved aluminum plate with 20 mm pipe grid; insulated	60–70
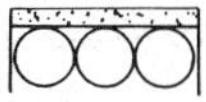	Sheet metal ducts heated by high temp. forced air circuit.	70–75

Figure 15.14. Types of radiant heat emitter

can be supplied by finned batteries from the works' steam or HTW system, or direct fired oil or gas heat exchangers can be used. With natural gas it is possible to directly mix the combustion products with the air providing all the air is taken from outside: the usual application is in high volumetric flow rate for air replacement.

Another type of direct-fired plenum system circulates air at less than half the normal rate, and at much higher temperatures. The technique is to discharge the air at high velocity from nozzles designed to induce air from around the nozzle and mix it with the primary jet. The advantages are smaller ducts, and lower cost compared with systems needing boiler plant.

Figure 15.15 gives the profile diagram[21] summarising the various characteristics and cost advantages of industrial heating systems.

District heating

Several buildings may be heated from the same source using a fluid medium transmitted through a network of pipes buried in the ground. The idea is not basically any different from that of gas and cold water service distribution.

Steam was at one time widely used for district heating, but now because of the disadvantages outlined earlier, it has been largely replaced by hot water. Most small district

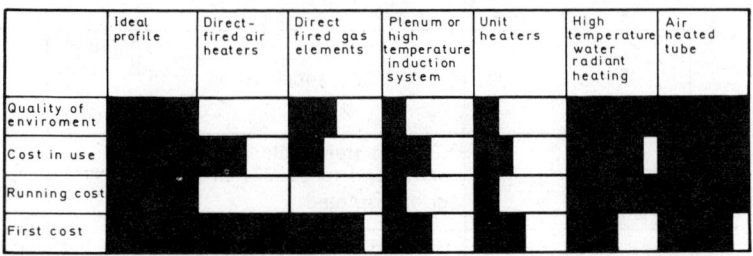

Figure 15.15. Relative merits of various factory heating systems (Courtesy 'The Engineer')

heating schemes, less than about 15 MW, use water at temperatures up to 120°C. Larger schemes, particularly those involving whole towns, use mainly HTW.

A two-pipe distribution is almost always used, although variations do exist, for example, the three-pipe, which uses separate flow pipes for heating and domestic hot water, and a common return. A system developed in the USSR uses only one pipe to supply both services, the water ultimately going to waste. This is similar to the geothermal systems used in Iceland and only applicable when the cost of recirculation is uneconomic. Pumps for district heating are basically the same as with any large heating system except that generally speed reduction is required to save power in intermediate weather. Pumping pressures of 4 bar are not uncommon.

The buried network is the part of a district heating system most susceptible to damage. If the low density heat insulation material becomes saturated with ground water, pipes corrode and heat is wasted. Current insulation techniques (Figure 15.16) range from hydrophobic trench infills to double-wall pipe using steel or plastics as the outer cover.

Any kind of hydronic system can be supplied by district heating but the majority use radiators. Water temperatures do not normally exceed 90°C, although some warm air hydronic systems use higher temperatures. Where the network uses HTW, temperature is broken down through substation heat exchangers, and sometimes with mixing ejectors (Figure 15.17).

Pipe networks

MATERIALS

Piping in all closed circuits, with the exception of domestic heating systems, is of mild steel with screwed or welded joints. There are three grades, light, medium and heavy, but only the medium and heavy grades are in normal use, and medium grade with welded joints covers all the temperature and pressure ranges met with in practice. The range of

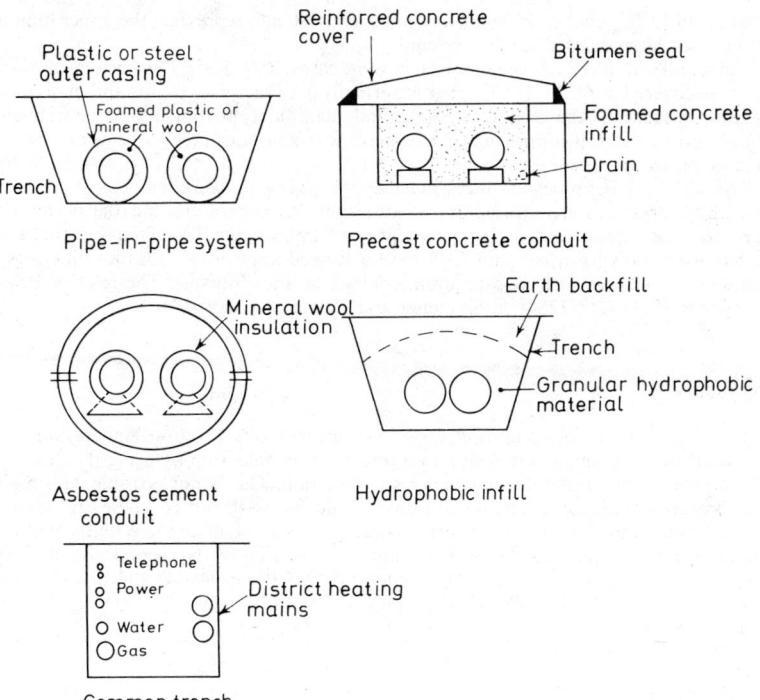

Figure 15.16. Insulation of buried district heating mains

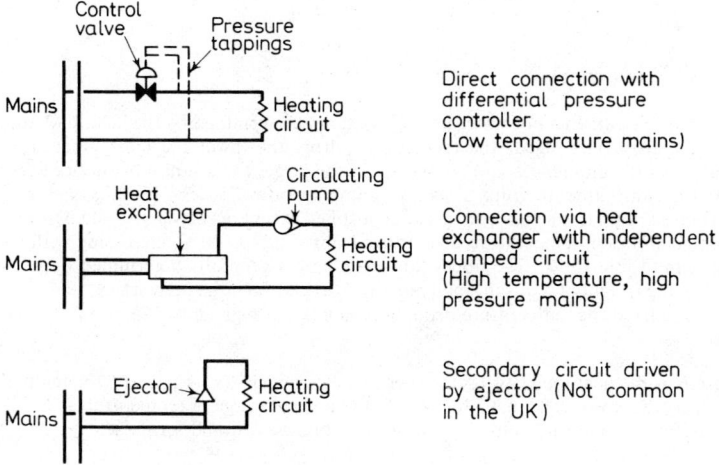

Figure 15.17. District heating network connections

sizes in BS 1387[37] ends at 150 mm, and in general 300 mm represents the upper limit for all but very large district heating systems.

Thin-wall stainless steel is economical in some cases (BS 4127)[38]. Dimensions of copper tube are covered in BS 2871[39]. Copper is normally used on open circuits and in domestic heating because of the ease of jointing and manipulation (brazing or compression couplings). Coiled soft copper tube can be made with a plastic outer sleeve where the tube is to be buried in concrete.

The use of plastics pipework is restricted by the inverse pressure/temperature relationship; high pressures are satisfactory at atmospheric temperature, but the permissible pressure falls rapidly with rising temperature. Unplasticised PVC and polythene are widely used; polypropylene and ABS have a limited application. Plastics pipework is suitable for services such as sanitation, cold water and drainage. The relative British Standards are 1972[40], 3284[41] (polythene), and 3505[42] and 3506 (PVC)[43].

INSULATION

EXPANSION

Axial expansion of pipes is allowed for by various methods. For low temperature water in the smaller size range, expansion movement can be taken up by using the flexibility of the pipe system in the normal changes in direction. This is not possible with higher temperatures and larger pipes, and various techniques are then used; these can be either bellows joints and sliding joints, or articulated bellows and ball joints. With the first two, the pipes move axially, constrained by guides. In the case of the second, the idea is to displace the pipe axially, rotating at the joint itself. With articulated and ball joints, the forces on the anchors (fixed parts of the piping) are much reduced.

INSULATION

Insulation is required on distribution pipework, particularly on high temperature systems, to prevent waste of heat. The economical thickness of the insulation depends on the first cost, the fuel saving, and the amortisation period. Solutions are given in BS 1588[44] and BS 1334[45].

DESIGN CALCULATIONS

The pumping rate in a hot water heating system is determined by the heat load and temperature drop. The higher the temperature drop, the lower the mean operating temperature of the appliances and consequently the larger the amount of surface needed.

At the same time, the pipe sizes and pumping power is less. Finding the economic break-even point is not easy and not often justified except in very large systems. Generally accepted values of temperature drop range from 10°C to 100°C, increasing with the size of system (Table 15.19). The upper flow rate limit is generally determined by noise, the range being roughly 1.5 m/s in small pipes to 4 or 5 m/s in large networks.

Having fixed the temperature drop, flow rate (kg/s) is given by:

$$H/s\theta \tag{11}$$

where H is the heat load (kW), s the specific heat capacity (kJ/kg), and θ the temperature difference °C. s varies from 4.2 to 4.4 over the normal range of temperatures.

Pressure drop in the piping is determined from the rational formula:

$$p = 4flv^2/2gd \tag{12}$$

where p is the manometric head loss (m).

In SI terms of pressure

$$P \,(\mathrm{N/m^2}) = p\rho g \tag{13}$$

Combining (12) and (13) in terms of pressure loss per metre length of pipe,

$$P/1 = 4fv^2\rho/2d \tag{14}$$

where $P/1$ = pressure loss, $\mathrm{N/m^3}$
f = friction factor
v = fluid velocity, m/s
ρ = density, $\mathrm{kg/m^3}$
d = diameter, m

The friction factor is a function of the Reynolds' number and the relative roughness of the pipe, and in most cases has a range of about 0·004 to 0·008. Solutions for f are given

Table 15.19 CIRCUIT TEMPERATURE DROP RANGES
IN PRACTICE

Type of system	Temperature drop range °C
Domestic:	
Single pipe	10–17
Two-pipe	17–22
Commercial (two-pipe)	17–22
Industrial:	
MTW	28–33
HTW	33–56
District heating:	
L and MTW	22–56
HTW	45–83

graphically in the Moody curves, or can be solved from the Colebrook–White expression[22]. Pressure loss in fittings is normally expressed in terms of equivalent length of pipe: typically the pressure loss of fittings varies from 10% to 30% of the straight pressure loss depending on the configuration.

The most detailed flow tables are those published by the IHVE[22]. A short extract valid for steel pipes is reproduced in Table 15.20.

Table 15.20 FLOW OF WATER IN MEDIUM GRADE STEEL PIPES
(Temperature 75°C)
(Extract from the IHVE Guide Book C)

Pressure drop $\mathrm{N/m^3}$	Diameter (mm), and flow rate (kg/s)			
	15	25	50	100
10	0·017	0·073	0·439	2·74
40	0·038	0·157	0·931	5·74
60	0·047	0·196	1·16	7·11
80	0·055	0·228	1·35	8·26
100	0·062	0·258	1·52	9·28
200	0·091	0·373	2·18	13·3
400	0·132	0·537	3·12	19·0
600	0·163	0·664	3·85	23·3

With the two-pipe system where appliances are connected across the flow and return mains, the pressure differential depends on the position of the appliance along the circuit. Nearer the pump, the differential is higher. The pump must, of course, be able to provide the maximum differential pressure required at the furthest point of the circuit To prevent short circuiting, the sub-circuit connections have to be sized so that the differential is absorbed at the required flow rate: this means either smaller pipes or a throttling valve. Appliances with inherently high pressure drop, heat exchangers for example, give better circuit stability. In district heating systems where pressure differences of several bars can exist along the circuit, differential pressure regulators are used to limit flow into sub-circuits. Where differentials are less than one bar, precalibrated valves can be used, set to absorb a given pressure.

For steam pipe sizing, an approximate type formula is generally used. Figure 15.18 gives a graphical solution[23]. The steam flow rate is determined principally from the latent heat, since the sensible heat content represents a very small fraction of the condensing heat potential. Flow rate (kg/s) is given by

$$H/h \qquad (15)$$

where h is the specific enthalpy of the saturated vapour, kJ/kg.

Sub-circuit balancing is not necessary since, providing the pipe is large enough, the flow will be self regulating depending on the condensation rate. The change in temperature with pressure over the typical pressure drop range, is small enough to make little difference to heat output.

COMPUTER AIDED METHODS

Several computer programs exist for sizing hot water pipe networks, possibly the best being the French suite of programs developed by the CoSTIC Laboratory and available in the U.K. through G and M Research Applications Ltd.

HEAT GENERATION

Boilers

There are basically three types of boiler; cast iron or steel sectional, steel shell, and water tube.

The normal cast-iron sectional boiler is limited to a pressure of 4 bar, but those of spheroidal graphite iron can be worked at 12 bar. Shell boilers cover the total pressure range generally met with in environmental engineering, and water tube boilers have an inherently high working pressure because of the tubular construction.

The cast-iron boiler is still the most widely used form of heat generator in the industry. The advantages are long life (because of the substantial wall thickness of the sections), and the facility for erecting a large boiler in a space where the only access is a normal sized door. A sectional boiler has a further advantage that heat output can be increased by adding more sections, thus meeting load increases without having to instal another boiler. Ratings cover the range from domestic (about 10 kW) up to 1200 kW. Working limits defined by BS 779[46] are; systems with open expansion tanks 4 bar, 100°C and pressurised systems 3·5 bar, 120°C.

Originally, sectional boilers were intended for solid fuel where high radiation transfer exists, but the newer designs for oil and gas firing rely on forced convection with gas velocities of about 15 m/s. Some pressurised combustion chamber boilers work at velocities of 45 m/s, giving overall heat transfer rates of about three times that of the traditional sectional boiler[24]. Combustion efficiency with modern sectional boilers is in the 75 to 80% region, with exit gas temperatures of about 250°C.

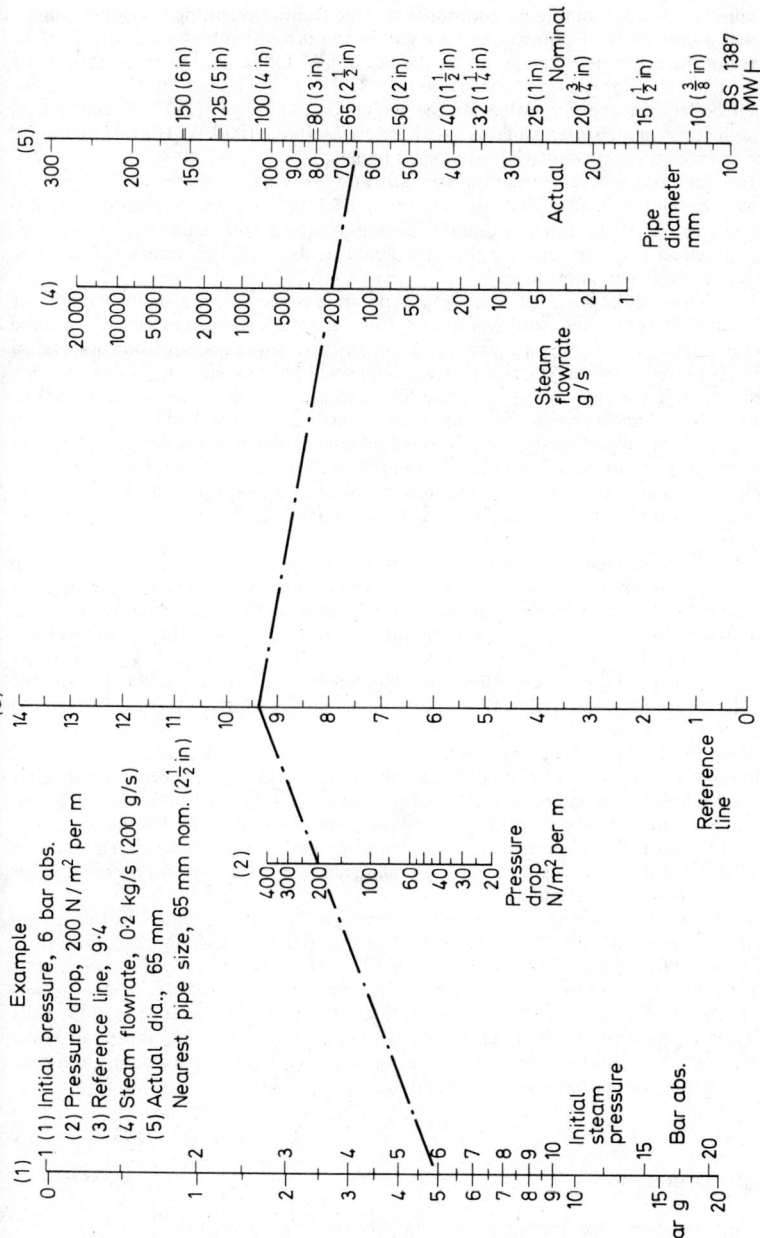

Figure 15.18. Steam flow in pipes (Nomogram by courtesy of Brightside Heating and Engineering Co. Ltd)

Example

(1) Initial pressure, 6 bar abs.
(2) Pressure drop, 200 N/m² per m
(3) Reference line, 9·4
(4) Steam flowrate, 0·2 kg/s (200 g/s)
(5) Actual dia., 65 mm
Nearest pipe size, 65 mm nom. (2½ in)

Sectional boilers in steel are used relatively little. However, one-piece steel boilers, particularly those combining an additional storage compartment for hot water supply, are used widely on the Continent and are gaining an increasing acceptance in the U.K. Ratings of the combined type go up to about 200 kW. Often they have the facility for dual-fuel firing; oil or gas on one side and solid fuel or refuse on the other.

Shell boilers are more suitable for the higher output ranges (250 kW and above) although some models are available rated as low as about 100 kW. High temperature water and steam systems invariably use shell boilers.

Safety requirements, construction strength and working pressures are subjects of BS 759[47] (mountings), BS 2790[48] (general) and BS 1501[49] (steels). Standards are also laid down by the AOTC representing the leading insurance companies.

The practical pressure limit for the shell boiler is about 20 bar: above this a water tube boiler is a better solution.

Shell boilers of the so-called economic type are designed with a main fire tube or combustion chamber, and banks of smoke tubes for the combustion gases. The tube banks are arranged in series to give a number of traverses or passes from one end of the boiler to the other. A typical three-pass boiler will have an efficiency of about 85% (based on the gross calorific value of the fuel) and a gas temperature of about 200°C. The upper size limit in practice for a fire tube boiler is about 6 MW, although sizes can go up to 12 MW. When used on HTW, the difference in temperature between flow and return connections to the system must be limited to about 60°C to avoid thermal stress. Normally all makes of shell boiler are now supplied as a packaged unit i.e. the boiler is complete with induced draught fan, burner, control gear and instruments, and mountings.

Another form of shell boiler, which does not follow the traditional 'economic' pattern of fire tube and smoke tubes, uses instead double steel shells to produce an annular water space. The burner (gas or oil) fires into the main centre shell, which is in effect a water-cooled cylinder, and then passes around the outside of the annulus. In some makes there are two such annular cylinders, arranged concentrically. In another type, the flame is arranged to reverse in the combustion chamber itself giving the effect of a double pass in the primary combustion zone. The rating range is roughly 0·1 to 2·5 MW, and pressures go up to 10 bar.

Water-tube boilers at one time were limited to very large installations like district heating and thermal-electric stations. Packaged models however, have brought the water tube boiler into the range suitable for factories and medium-sized building complexes (2–30 MW). The advantages of the water tube over the shell boiler are basically less weight, higher operating pressure (40 bar for some packaged boilers), differentials up to 120°C tolerated and quicker response to change in firing rate. Cost is generally higher until the rating exceeds about 12 MW.

Also in the water tube boiler field is the down-jet flash steam boiler. In this design there is a helical tube bundle with the core space of the helix forming the combustion chamber. Water is pumped into the lower end of the helix and, as it moves upwards, it is progressively heated and then vaporised into steam. The very low water content means that steam can be generated within a few minutes of start-up. The expansion of the helix is used as a signal to detect water supply failure, and the boiler has the usual safety devices. The application is generally in transport heating, process steam supply and peak load infill; it is not normally used for building heating systems.

Heat exchangers

The shell and tube heat exchanger is probably the most frequently used type in the industry. Some typical applications are; sub-stations on HTW district heating networks to reduce both temperature and pressure for the building heating circuit, MTW generation from a primary steam supply, and oil line heaters using steam or HTW. With condensing steam, the film conductance is extremely high and the arrangement is to pass the water or oil through the tube bundle, with the steam in the shell. Where both media are liquid, baffles are used in the shell to increase the velocity over the tube bundle. With liquid/ liquid heat exchangers, the contra flow technique is preferable to parallel flow. With contra flow the exit temperature of the secondary medium can be higher than the exit temperature of the primary, whereas with parallel flow, the two exit temperatures approach the same value. Other types of heat exchanger such as the parallel plate or spiral plate, spiral coil or graphite block, offer high output in a small space but cost limits their use in the environmental engineering field.

The system designer normally specifies the heat exchanger performance in terms of heat output, temperature difference, working pressure and maximum pressure drop to the manufacturer and is not generally concerned with working out the heating surface or tube geometry.

Heat pumps

A refrigerating machine is called a heat pump when the interest is in using the heat at the condenser instead of the cooling produced at the evaporator. Generally, the heat pump is arranged to take heat from some low-grade source (effluent gas, including air, or water) and to raise it to a usable temperature for heating. The effectiveness of the heat pump cycle is characterised by the coefficient of performance:

$$T_1/(T_1 - T_2) \tag{16}$$

where T_2 is the source temperature and T_1 the required temperature, both in K.

The nearer the two temperatures approach, the higher the coefficient of performance (COP). In practice performance is described by the performance energy ratio (PER), which is the ratio of the actual heat output and input. This works out at about 0·5 of the ideal COP. Generally heat pumps work with a PER of about 4.

Heat pumps are rarely justifiable unless combined with air conditioning. (see air conditioning systems). Few successful installations exist in the U.K.[25].

Total energy

Where electric power is generated on site, and the waste heat reclaimed for heating the building and driving absorption refrigeration machines, the system is called *total energy*.

Development has been greatest in the United States but several installations now exist in the U.K. The gas turbine is the most favoured power source, with diesel engine systems second. The high capital cost of the installation has to be set against power and fuel savings. For large total energy systems natural gas is often offered at extremely low unit rates.

FUELS AND COMBUSTION

Oil and gas take the major share of the fuel market today; the use of coal is slowly declining. The reason for the decline of coal is basically cost and convenience. The heavier grades of oil cost about the same in terms of heat as coal, and coal needs automatic handling equipment for conveying, firing and ash removal. Storage and delivery is not as easy as with oil.

Natural gas at the normal tariff, although more costly than oil or coal, has the added advantage of no storage and a completely clean effluent.

Oil

The various grades of oil range from distillates (domestic vaporising burners and stoves use kerosene) to residuals of high viscosity. Table 15.21 gives the principal physical properties and the grades according to BS 2869.[50]

Table 15.21 GRADES AND PROPERTIES OF FUEL OILS

Class (BS 2869)	Common name	Viscosity Redwood No. 1 seconds (38°C)	Specific gravity	Gross Calorific value kJ/kg	Sulphur % by weight
C	Kerosene	—	0·79	46 400	0·2
D	Gas oil	35	0·84	45 500	0·8
E	Light	220	0·93	43 400	3·2
F	Heavy	950	0·95	42 900	3·5
G	Extra heavy	3 500	0·97	42 500	3·5

Means of burning the fuel vary with the viscosity. Apart from kerosene, which can be directly vaporised, all other grades need to be sprayed into the combustion chamber. In the pressure jet burner, which is the most widely used in the industry, a high pressure pump supplies oil to a nozzle which effectively breaks up (atomises) the oil stream into a fine spray. Combustion air is delivered around the nozzle from a blower and the disposition of the nozzle apertures and air inlets designed to give effective mixing.

Centrifugal force is another atomisation technique used in the spinning cup burner. An advantage is the ability to reduce the oil flow rate without altering the degree of atomisation.

Residual fuels need heating for effecting atomisation: Table 15.22 gives the typical ranges.

Table 15.22 STORAGE AND HANDLING TEMPERATURES FOR FUEL OILS, °C

Class	Tank and distribution		Burner	
	Storage	Pumping	Pressure jet	Spinning cup
E	7	7	60– 80	30
F	20	27	80–100	60
G	32	38	100–130	130

Sectional boilers nearly always use pressure jet burners, and shell boilers can use either pressure jet or spinning cup.

With the higher sulphur content fuels, corrosion of the gas-side surfaces is a danger: where the surface temperature drops below the dew point of 60°C, water will condense out, which in the presence of acid (150°C dewpoint), will rapidly corrode steel and cast iron. To prevent this, the return water temperature must be kept high by using a re-circulating pump or ejector on the boiler. Many packaged boilers now incorporate this feature.

Oil is normally stored in cylindrical steel tanks, the capacity being calculated to provide two or three weeks fuel at full load. The minimum size should be 3500 litres, and the typical range goes up to 50 000 litres. Constructional details are given in BS 799.[51] Although the tank is open to atmosphere through a vent pipe, it must be designed to withstand the pressure which may be produced if the road tanker accidentally pumps over the vent; the developed pressure being a function of the pumping rate (which can be as high as 15 litres/s), the size of the vent and the height of the outlet above the tank.

Kerosene and gas oil can be stored at atmospheric temperature but the residual grades need to be kept heated (Table 15.22). Insulation is advisable for grades F and G, and an efficiency of about 75 % should be aimed at.

Oil can flow direct from the tank to the burner for small installations with one boiler, but for residuals a pumped ring is used. Additionally the piping will have electric tracer heating.

Gas

North sea gas is almost all methane. Some of the properties compared with town gas will be seen from Table 15.23. For a given heat output, natural gas will have about half the flow rate of town gas.

Table 15.23 SOME PROPERTIES OF NATURAL AND TOWN GAS

Property	Natural gas (methane)	Town gas
Gross calorific value (MJ/m³)	37·0	18·6
Specific gravity (air = 1·0)	0·55	0·47
Air requirement (m³/MJ)	0·26	0·23
Maximum burning velocity (m/s)	0·34	1·0
Air/gas ratio	9·5	4·3

The air requirements, also in terms of heat, are about the same. The lower flame speed gives natural gas a tendency to break away from the burner, and with over twice the air requirements per m³ of gas, aeration is not as easy as with town gas. These difficulties have, in general, been overcome in present-day burners.

Coal

Coal is basically carbon, hydrogen and oxygen, with various other components such as ash and sulphur. It is classified by the NCB rank number which is a function of the volatile content and the caking property. Table 15.24 is a selection from NCB data.

Table 15.24 CALORIFIC VALUES AND SULPHUR CONTENTS OF COALS

Rank No.	Type	Gross CV MJ/kg	Sulphur %
101	Anthracite	29·7	1·0
201	Dry steam	30·6	1·0
202	Coking steam	30·8	1·0
206	Low volatile	30·0	1·2
305	Medium volatile	30·4	1·2
501	Strongly caking	29·2	1·7
701	Weakly caking	26·8	1·7
902	Non-caking	23·8	1·7

Only small boilers are now hand-fired with coal. There are several techniques of automatic firing i.e. the underfeed stoker, chain grate stoker, sprinkler stoker and coking stoker.

The underfeed stoker generally has an upper limit of about 1 MW: the other types are suitable for up to 10 MW. The gravity-feed boiler is another example of automatic firing; coal is held in a hopper which forms part of the boiler and is allowed to pass through to the combustion chamber. The upper size limit is about 3 MW. Efficiency on most solid-fuel boilers is at least 75 %, and the automatic types 80 %.

A promising development in solid fuel combustion is the fluidised bed technique which gives much higher combustion intensity than mechanical stoking.

Chimneys

Chimneys are required with boilers having non-pressurised combustion chambers to promote the flow of gases from the boiler, suction pressures up to 1·0 mb being necessary. With pressurised boilers, the chimney is needed to disperse the gases and there is no specific draught requirement.

The Clean Air Act Memorandum on chimney heights requires tall chimneys where high sulphur content fuels are being used; the height of the chimney depends on the rate

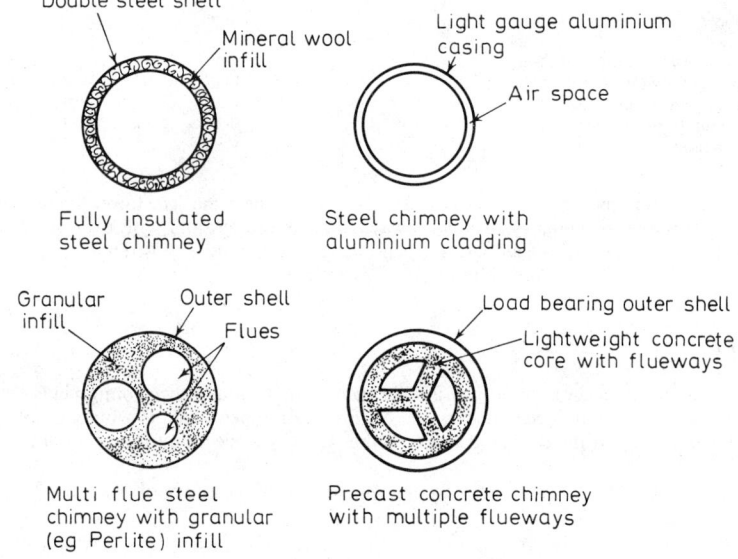

Figure 15.19. Types of insulated chimneys

of emission of sulphur dioxide, the height/length ratio of the building to which the chimney is attached, and the existing background level of pollution.

There are two types of chimney in general use; the insulated steel chimney and the pre-cast concrete chimney. Uninsulated steel chimneys and brick chimneys (except when part of a building) are almost never used. The Memorandum requires each boiler to have a separate flue and this had led to the development of multi-flue chimneys within a single shell (Figure 15.19). Chimney insulation is necessary for sulphur bearing fuels since with a cold chimney acid deposition can take place. The critical temperature is about 130°C, but flue design should aim at keeping the wall temperature above 150°C. Generally speaking this means that with all residual fuel oils, insulation equivalent to about 25 or 50 mm of glass wool is needed to prevent condensation: in concrete chimneys insulation is provided by the low density concrete used to form the inner flues.

The height of a chimney calculated from the Memorandum nomograms to meet pollution requirements is normally more than enough to provide sufficient draught: the chimney designer must, therefore, arrive at the most economical chimney size in terms of diameter. The Memorandum specifies a lower limit of 6 m/s on the gas velocity at the top of the chimney, rising to 15 m/s for very large boilers (up to 140 MW). The combustion gas flow rate can be calculated from the following:

$$Q_{20} = 2 \cdot 7 \times 10^{-4} H(y + 107)/e \qquad (17)$$

where Q_{20} is gas flow rate m^3/s at a standard temperature of 20°C; H, boiler rating kW; y, excess air for combustion, %; e, boiler efficiency, %.

This is valid for coal, oil and natural gas. The gas volume (Q_{20}) is at a standard temperature of 20°C. The actual volume flow can be found from the inverse ratio of the absolute temperatures. For example, with a gas temperature of 200°C and a standard temperature flow rate of 2·0 m^3/s, the actual flow rate would be 2·0 × [(200 + 273)/(20 + 273)] = 3·2 m^3/s.

The gas velocity is then easily found from Q/A, where A is the area of the flue.

The pressure drop in the flue should be checked against the available fan pressure (with induced draught or pressurised boiler), or with a natural draught boiler against the stack effect. This can be done most simply using the set of monograms contained in the Brightside Chimney Design Manual[26]. The Manual also gives graphical solutions for temperature drop in the chimney so that the amount of insulation needed can be worked out.

With natural gas fuel, the Clean Air Act Regulations do not apply and the products of combustion need not be discharged from tall stacks. The technique is to dilute the flue gases with a substantial amount of air, using an induced draught fan, to reduce the concentration of the various components to below the known threshold limits. Gases can then be discharged horizontally from louvres in the side of the boiler house.

WATER SERVICES

Hot water supply

Hot water for sinks, wash-basins, showers and so on can either be drawn from a storage vessel, or generated at the time it is required. With storage high flow rates can be sustained for a short time without loss of temperature: instantaneous methods are limited to a maximum flow depending on the amount of heat available.

Hot water requirements

To assess storage vessel capacity or outflow heating load, the total water demand must be established.

Table 15.25 gives generally accepted figures for temperatures and outflow rates from fittings. Data for specialised services such as laundries and hospitals, and more extensive versions of Table 15.25 and 15.26 can be found in the IHVE Guide Book B and Code of Practice 342[52].

Table 15.26 gives the amount of storage and the associated heat input required from boilers and heat exchangers for different buildings[27].

Hot water storage

In single family dwellings, and in blocks of flats using the individual storage principle, notably with district heating, hot water is stored in a vertical cylinder. Cold water is supplied from an individual tank (in some makes forming the upper part of the cylinder), or from a central tank system. The cold water connection is at the base of the cylinder

Table 15.25 HOT WATER OUTFLOW RATES AND TEMPERATURES

Fitting	Flow rate ml/s	Temp. range °C
Wash basin:		
Ordinary tap	160	40–60
Spray tap	40	40–60
Sink	200–300	60
Bath	300	40–60
Shower:		
Nozzle	100–150	45–60
Rose	350	45–60
Dish washer	250–500	50–80

Table 15.26 HOT WATER STORAGE AND HEAT INPUT (per person)

Building	Hot water storage litres	Heat input kW	Peak daily demand litres
Houses and flats	35–50	0·7–1·2	120–150
Hotels	40–50	0·9–1·2	120–150
Offices and factories	5	0·1	15
Schools and colleges:			
Boarding	25	0·7	125
Day	5	0·1	15

so that entering water displaces the hot water above the connection level instead of mixing with it and thus reducing the effective heat store. The water in the cylinder is normally heated from an immersed pipe coil, or occasionally from an external heat exchanger on a close-coupled circuit.

Sometimes, in single family dwellings, water is heated directly by a boiler without the intermediary of a heat exchanger, but the danger is the formation of deposits or corrosion in the boiler and circuit.

When part of a large heating system, the cylinder is supplied from the pipe network in the same way as a radiator.

A method which dispenses with the primary circulating pipework is the combined boiler-cylinder unit. In all heating circuits, two cold water tanks are normally needed, one for expansion and one to supply water to the cylinder; to prevent intermixing of water in the two circuits. Small systems can take advantage of the single-feed cylinder where only one tank is necessary, the water in the two circuits being separated by an air seal in the cylinder. Figure 15.20 shows diagramatically the various boiler-piping arrangements with storage systems.

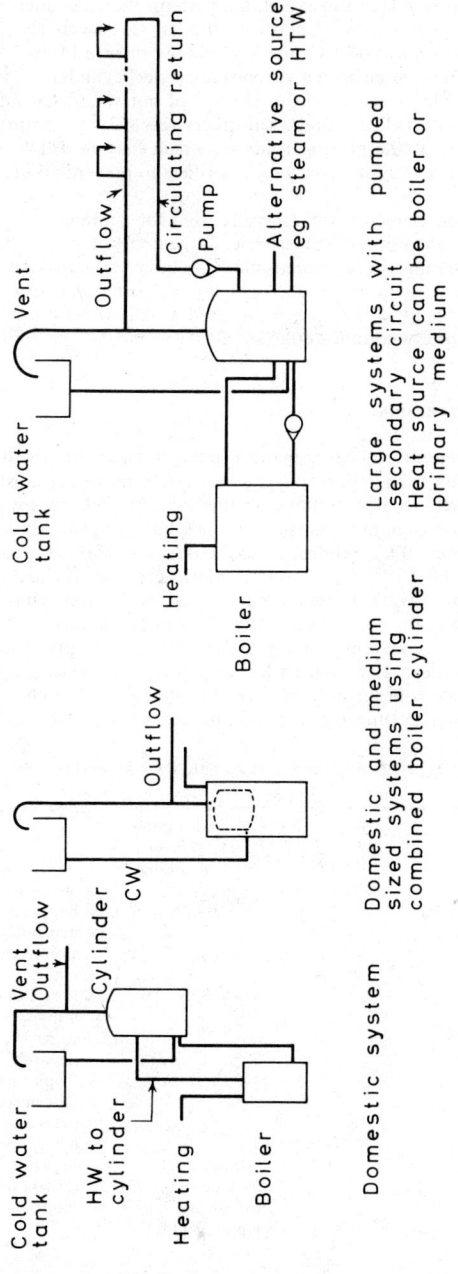

Figure 15.20. Storage hot water systems

Most storage cylinders are in copper, with a few being of galvanised steel. Stainless steel is being used to an increasing extent, but perhaps the most interesting development has been the fibre-glass-reinforced plastics cylinder: this meets the same specification as copper cylinders to BS 1566[53], Grade 3 (working pressure 1·0 bar). Combined boiler-cylinder units are also offered with a vitreous lined steel cylinder.

Large buildings must rely on central storage of hot water. Cylinders with heat exchangers are generally called storage calorifiers (BS 853).[54] The primary medium in this case is the same as the heating medium, which may be HTW or steam. Control devices and safety valves are necessary on calorifiers to prevent accidental over-pressure or over-heating.

Electric immersion elements are generally used for summer hot water supply with boiler systems, but an independent electrically heated hot water service is becoming more common. Running costs comparable with boiler systems can be achieved with off-peak generation using more storage capacity and better insulation. The applications are mainly domestic but the principle can be used equally well for local hot water supply for example, in factories and office blocks.

On-line hot water generation

Generation of hot water by a heat exchanger in the boiler at the rate of draw-off involves much higher heat flow than with storage methods since the boiler must work at the peak rate and not the average. For example, in the domestic field where a draw-off rate of 200 millilitres/s from one outlet is normal, the heat input to give a discharge temperature of 60°C would be over 30 kW, almost twice the heat loss of the average house.

The larger direct-fired heaters are able to meet nearly the normal flow rate required for a single outlet but electric heaters are restricted by the maximum permissible phase loading, and generally do not exceed 6 kW. This means accepting a much lower flow rate than with a storage system. Table 15.27 gives an idea of present-day equipment.

On-line generation can also be used for large hot water heating installations but, in this case, all the boiler output may be diverted into the heat exchanger at the time of maximum draw-off. The time lag of the heating system and building is sufficient to

Table 15.27   LOCAL HOT WATER SUPPLY: DIRECT GAS-FIRED AND ELECTRIC UNITS

Type	Rating kW	Hot water delivered or storage	Characteristics
'Instantaneous' gas-fired unit	10–25	50–150 ml/s	Supply to one or several taps, depending on output. Connected direct to water main.
Electric in-line heater	3–6	15–30 ml/s	Direct on tap or pipeline mounted. Connected direct to water main.
Gas-fired storage units	10–15	Typical storage 200–300 l	Separate feed tank needed (some are built into the unit). One make has a high pressure cylinder (up to 8 bar).
Electric storage units	3–7	Storage 3 l (tap mounted), to 150 l for normal supply. Up to 300 l for off-peak.	Separate feed tanks needed (except on some single tap models).

absorb the changes in boiler output. (The ratio between peak hot water demand and heat loss in large buildings is less than 1·0). The hot water generator is basically a high-output shell and tube heat exchanger using either straight or helical tube bundles, sometimes with additional finning; the hot water can be in the shell or the tubes depending on the design. Combined boiler-cylinder units can also be designed for on-line generation, the reserve capacity in the cylinder being limited to about 15 minutes at the maximum draw-off rate.

Hot water draw-off circuits

Where the draw-off points are not more than about 10 m away from the storage cylinder or hot water generator, a single pipe or 'dead leg' connection can be used. With such connections there is an appreciable time taken for the hot water to reach the draw-off point, depending on the length, diameter and flow rate. On low-flow outlets such as spray taps, the dead leg must be much shorter. Apart from the time factor dead legs may be a source of heat loss if they pass through unheated spaces or ducts.

In large buildings, dead leg feeds are not possible and the hot water is pumped around a closed circuit. The flow pipe is of course, made large enough to take the highest discharge rate and the return pipe need only be sufficient to carry the circulation required to offset the ring main losses.

Sizing the outflow pipe depends on making an assessment of the probable maximum draw-off rate: it would clearly be uneconomic to make the flow pipe large enough to supply all the outlets simultaneously, except on small installations. Techniques of working out the probable number of taps open at the same time are described in the IHVE Guide Book B, and by Burberry and Griffiths[28]. Alternatively, water usage can be analysed on an hour-by-hour basis from known usage pattern and the peak flow identified. A computer program solution is available from Faber Computer Operations Ltd.

The operating pressure is the static head between the level in the cold water tap and the outlet. Piping in hot water supply systems is invariably copper with soldered or compression fittings.

Cold water supply

The amount of cold water storage needed is defined in the Model Water Byelaws and Code of Practice CP 310[55]. Table 15.28 summarises the requirements. Flow outlet requirements are the same as for Table 15.25.

Table 15.28   COLD WATER STORAGE REQUIREMENTS
(BS. CP 310) PER PERSON

Building	litres
Flats	90
Hotels	90
Offices	45
Houses	230 min. total storage (Model Byelaws)

Tanks for cold water storage are generally in galvanised steel but increasing use is being made of plastics. Cold water pipework can also be in plastics, taking advantage of the much greater strength of polythene and UPVC at low temperatures.

In tall buildings, mains pressure may not be sufficient to elevate the water and pressure boosters must be used. These can either by continuously running pump units, or gas pressurised storage cylinders (Figure 15.21). The lower storeys of a building can be supplied direct from the mains and the upper storeys taking water from the basement pressure source or from a header tank at the top of the building. A distinction is sometimes made between drinking water (which is held in closed vessels) and water for other purposes which may be stored in an open tank.

Water treatment

Data on the degree of hardness (permanent and temporary) of water supplies can be obtained from the Water Engineer's Handbook, which lists all the water undertakings in the country.

AIR CONDITIONING

Air conditioning is basically understood to mean a technique of supplying air to a building, heating or cooling it as required, adding moisture or taking it out, and purifying it to a greater or lesser degree.

Air conditioning is becoming increasingly necessary in the U.K., not because of high summer temperatures, which are rare, but because the built environment is changing. Glazing often occupies three-quarters of the total building facade, about three times as much as in the early 1950's. Structures are much lighter in weight and therefore respond more rapidly to the effects of sunshine. More artificial lighting is being used, particularly in offices and shops where loads can approach 30 W/m^2, of floor area. These three factors — glazing, mass and lighting — are the most important in determining air conditioning load. Figure 15.9 is a guide to the designer in deciding whether air conditioning is needed on the basis of glazing alone.

A development made possible through air conditioning is the deep plan building where permanent artificial lighting is combined with vision strip glazing. Air conditioning is essential in places like theatres and cinemas where the population density is high and natural ventilation is not possible. It is also essential in specialised areas such as operating theatres, clean rooms, computer centres, laboratories, lecture theatres and windowless structures.

Range of equipment

Typical air conditioning system components are listed in Table 15.29 and the main characteristics summarised. Figure 15.22 shows the essential stages of the process. Few air conditioning systems use all the possible stages, and many small packaged units only act as air coolers.

Built-up central station plant with distribution ductwork was, at one time, the only technique used in air conditioning, but over the last decade the use of packaged equipment has increased. The present day range covers small self-contained room air conditioners, larger floor-standing or ceiling-mounted units with separate condensers for use with a limited amount of ductwork, and air handling units in sizes going up to about 20 m^3/s; enough for a central station for a building of around 2000 m^2 floor area. Air handling units include fans, heating and cooling coils, coarse filters and silencers. Packaged equipment eliminates expensive site labour and possible design errors which may occur in built-up plant.

Filters

Atmospheric dust has a wide size distribution from less than 0·5 μm to about 50 μm, but the greater mass of particles is concentrated in the size range above 5 μm. In terms of the

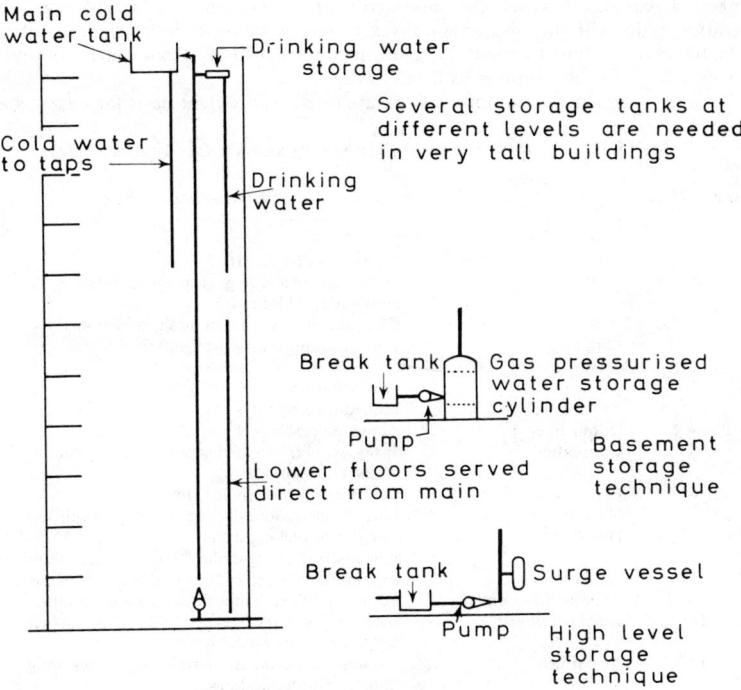

Figure 15.21. Boosting cold water supply

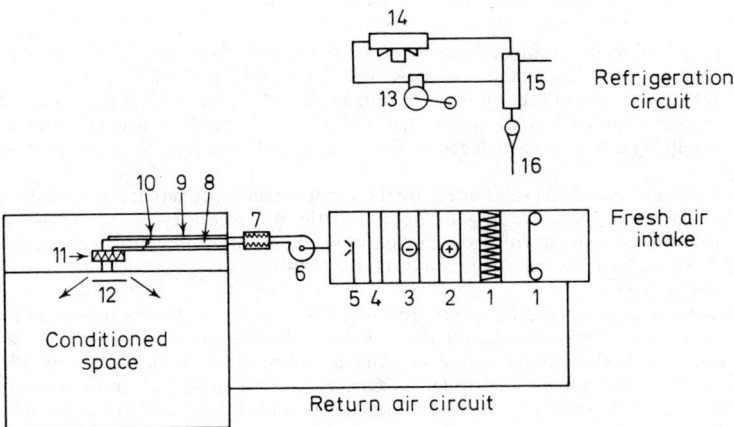

Figure 15.22. Air conditioning components (Not all these would necessarily be used at the same time). See Table 15.29 for the key to the numbers

number of particles however, the sub-micrometre sizes predominate. The mass concentration varies with the area, a typical yearly average value for industrial towns being about 100 μg/m^3, falling to about 20 μg/m^3 in country districts. Peak winter concentrations may reach five to ten times as much as this.

Filters for normal air conditioning work are rated on a weight basis for coarse stage

Table 15.29 AIR-CONDITIONING SYSTEM COMPONENTS (see Figure 15.22)

Key to 15.22	Item	Characteristics
1	Main filters:	
	1st stage	Coarse dust (> 5 μm):
		Disposable media (e.g. glass fibre, cotton wool, automatic roll type).
		Cleanable media: (nylon mesh, coated metallic).
	2nd stage	Fine dust and smoke (< 5 μm):
		electrostatic, paper asbestos.
2	Air heater	Finned tube battery.
3	Air cooler	Finned tube battery.
4	Odour filter	Activated carbon cells.
5	Humidifier	Spray washer, irrigated capillary cell, spinning disc atomiser, steam jet.
6	Fan	Centrifugal, axial, mixed-flow
7	Silencer	Lined duct section, packaged unit with splitters.
8	Ducts	Circular or rectangular: galvanised steel, rigid glass fibre, plastics, aluminium.
9	Insulation	Rigid mineral fibre with external vapour barrier.
10	Volume flow control	Motor operated damper, self-actuated damper.
11	Terminal filter	High efficiency (penetration < 0·1 % for particles > 0·1 μm): paper-asbestos.
12	Air diffuser	Circular or rectangular terminals, nozzles, long slots, perforated ceilings.
13	Refrigerating machine	Compression cycle: reciprocating, centrifugal. Absorption cycle: steam or gas.
14	Condenser	Air cooled: forced draught finned-tube battery. Evaporative: spray cooling tower, pond.
15	Evaporator	Shell and tube heat exchanger
16	Chilled water circuit	Pumped, 2-pipe.

filtration, and on dust stain or methylene blue and sodium flame tests for second stage or terminal filtration. Test methods are described in BS 2831[56] and 3928[57]

The dirtying power of the air is caused almost entirely by the smaller particles, and for effective air cleaning, blackness test figures of 80 % and above are generally necessary: this inevitably means using high density fine or medium fibre filter cells, or an electrostatic filter.

Two main types of first stage filters are the viscous coated media type (i.e. metal louvres, expanded metal, metal wool, glass fibre) and the dry media types (i.e. cotton wool, pleated fibre, continuous roller type, nylon mesh). Some of the viscous coated media are of the disposable type, but others can be cleaned. Most of the dry media filters are of the disposable type.

Second stage filters are either electrostatic plate type, or high density packed media (paper-asbestos), sometimes called absolute filters. Electrostatically charged media can also be used. Both first and second stage filters can be self-cleaning e.g. a fibre blanket can be arranged on two rollers to form a curtain which is slowly driven from one roller to the other as the medium becomes saturated with dust. Electrostatic filters can have an automatic washing sequence.

Wet techniques of filtration such a spray washing were used at one time but have an efficiency lower than most coarse filters.

First stage filtration is used to relieve the dust loading on second stage filters, which have a relatively low holding capacity.

Air speed through filters is limited by the pressure drop and by dynamic effects within the filter. By folding the media, the actual penetration velocity can be made less than the face velocity. The normal range for face velocity is 1·0 to 2·5 m/s, and for pressure drop 0·2 to 2·0 mb.

Air heaters and coolers

A finned tube battery is necessary because of the low outside film conductance of air compared with the liquid or vapour inside the tubes. Batteries are rated according to BS 3208[58] (hot water) and BS 2619[59] (steam). Liquid refrigerant can also be delivered to the coil from the condenser. The number of banks of tubes determines the heat transfer with the air, and in the case of cooling coils the amount of moisture condensed out. Air speeds through the battery are in the range 2·5 to 4·0 m/s, and pressure drop 0·2 to 1·5 mb.

Odour filters and air purification

Adsorption by activated carbon is the technique generally used for eliminating odours in the conditioned air stream. The medium is held in canisters or panels and built up to form the necessary quantity. High temperature reactivation is essential at intervals. Ozone was at one time used for odour control but concentrations at which it is effective can be above the desirable maximum and produce irritation.

Ultra-violet irradiation of the air can destroy micro organisms but equally effective results can be obtained with high efficiency terminal filters. In special cases ultra violet can be used as a second stage bacterial filter.

Humidifiers

Spray washers, which work on a closed circuit with a high pressure water supply to nozzles, are used to a limited extent to humidify the air, sometimes acting as a chilled water-to-air heat exchanger. The efficiency, expressed as a fraction of the total amount of moisture needed to saturate the air, is about 70 to 80%.

Capillary cell humidifiers use cells of glass filaments irrigated from low pressure nozzles. Efficiency is high, approaching 95%. Pumping power is much less than with the spray washer. Face velocity with both types of washer is about 2·5 m/s.

Air heater and cooler batteries can also be used as humidifiers by spraying water onto the tube bank: the efficiency is not more than 50%.

With the spinning disc humidifier, water is atomised by being thrown off the disc edge against a separating rim. The water aerosols produced by this means can give a super-saturated air stream. In the textile industry for example, such humidifiers are arranged to discharge directly into the conditioned space to take advantage of high moisture emission. Duct mounted spinning disc machines are common in air conditioning.

Steam jet humidifiers produce no moisture droplet carry-over as with spray techniques. Being in vapour form already, no sensible heat is required from the air stream. There is no problem with scale deposits or bacterial growths as sometimes occur in closed circuit spray systems.

Fans

Most central station plants and packaged units use centrifugal fans since, in general, they are the only type to give the required combination of high volume, high pressure and

RECTANGULAR GRILLE

Fixed or adjustable vanes,
or perforated metal grille.

LINEAR GRILLE

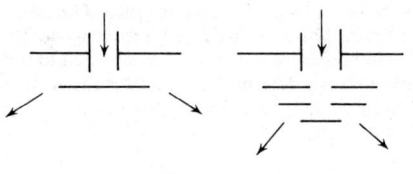

Grilles with high aspect ratio
(greater than 10:1).
Fixed or adjustable vanes.

SLOT DIFFUSER

Metal framed single or
multiple parallel slots.

CEILING DIFFUSERS

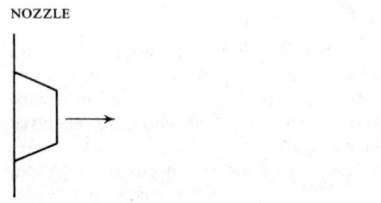

Ceiling fixed terminals with single
circular or rectangular deflector, or
multiple deflectors. Adjustable for
radial jet direction. Can be
combined with exhaust terminal.

NOZZLE

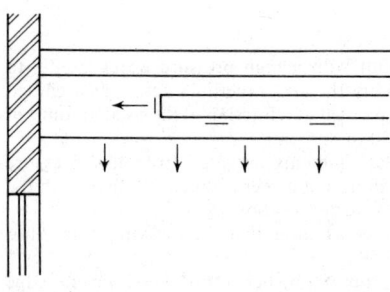

Cylindrical or tapered nozzle
giving single air jet.
Direct adjustable.

PERFORATED CEILING

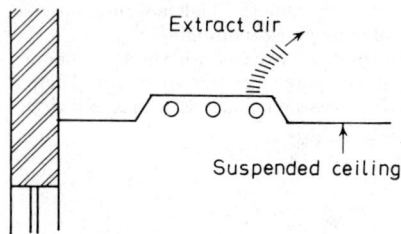

Suspended tile ceiling, part
or all of the area perforated or
slotted to diffuse air from the
plenum space.

LIGHTING TROFFER EXHAUST UNIT

Extract air

Suspended ceiling

Combination lighting troffer and
exhaust terminal designed to pick up
convective heat from the troffer
and lights to prevent it entering
the room. Also available as a dual
inlet-exhaust unit. Part of an
'integrated' heat-from-light system.

Figure 15.23. Characteristics of air diffusers

reasonably high efficiency. There are four types of impeller: forward curved (multivane), backward curved (single blade), backward curve (aerofoil blade), and paddle blade.

The forward curved impeller is a general purpose fan suitable for medium pressure systems and fairly quiet because of the low tip speed: the backward curved fan can produce higher pressures and will not overload the motor if worked against a lower pressure; the aerofoil bladed version of the backward curved fan has the highest efficiency; the paddle blade fan is reserved for industrial dust work, having an efficiency of 50 % or less. The efficiencies of the four types vary from 50 % to 60 % with forward curved, and up to 85 % with the aerofoil bladed.

Axial flow fans are used to a lesser extent than centrifugal fans because of the higher noise level and lower pressures produced. Combined with suitable silencers, however, advantage can be taken of their straight-through design for direct duct work mounting.

Propeller fans are limited to very low pressures, mainly as roof and wall extractors.

Air diffusers

The various characteristics of air diffusers are summarised in Figure 15.23. The basic idea of the air diffuser is to direct the air jet so that it mixes with room air and at the same time gives sufficient turbulence for freshness without causing a draught. Where very high air turnover rates are used (over 30 changes/h), terminals give way to perforated ceilings.

The shape of the terminal, and any induction devices included in it, makes it impossible to predict performance with any accuracy and manufacturers' test data must be used. This is usually in terms of the throw, the point at which the velocity falls to about 0·5 m/s (the radius of diffusion in the case of circular terminals); and the rise or fall of the jet depending on whether it is heated or cooled.

Refrigerating machines

The principal techniques of refrigeration are compression and absorption. Compression refrigeration can use either reciprocating, centrifugal or screw machines depending on the load. The limit for reciprocating machines is about 1000 kW, and screw machines 1500 kW, with centrifugal machines going up to 15 000 kW. Absorption refrigeration machines can go up to 5000 kW, using gas or oil as the heat source.

The standards for refrigerating machines are BS 3122[60] and CP 406.[61]

Condensers and cooling towers

An air-cooled finned battery can be used for the condenser stage but size and noise level limit it in practice to about 400 kW: units in parallel are sometimes used up to 1500 kW.

For large installations, a shell and tube condenser in the refrigeration circuit is supplied with water on a partially closed circuit from a cooling tower. The most frequently used type today is the induced draught, packaged unit designed for contra or cross flow of atmospheric air and cooling water. Water is sprayed into the tower where it cascades over an extended surface medium.

The performance of a tower is determined largely by the atmospheric wet bulb temperature, since this is the ultimate limit to which the condenser water can be cooled.

	Main features	*Variations*

SINGLE DUCT LOW VELOCITY SYSTEM

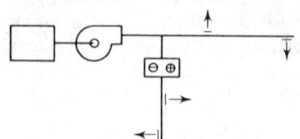

Conditioned air from central plant distributed through duct network. Velocity less than 6 m/s.

Zone reheaters and coolers.

SINGLE DUCT HIGH VELOCITY SYSTEM

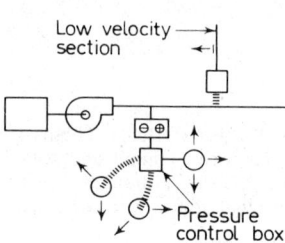

Low velocity section

Pressure control box

Conditioned air from central plant distributed through duct network. Velocity normally less than 20 m/s. Sound-absorbing/pressure-reducing distribution boxes feeding low velocity duct sections or terminals.

Zone reheaters and coolers. Limited control of room temperature by variable volume regulation.

DUAL DUCT HIGH VELOCITY SYSTEM

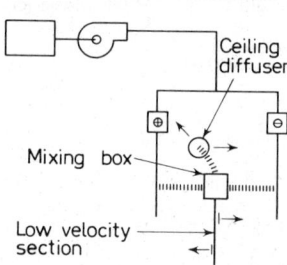

Ceiling diffuser

Mixing box

Low velocity section

Conditioned air from central plant split into two parallel ducts—velocity normally less than 20 m/s—one carrying heated air, the other cooled air. Air is mixed in sound-absorbing/pressure-reducing boxes to meet individual zone temperature requirements. Mixed airflow rate from the box remains constant. To a lesser degree the system also provides humidity control.

HIGH VELOCITY INDUCTION SYSTEM

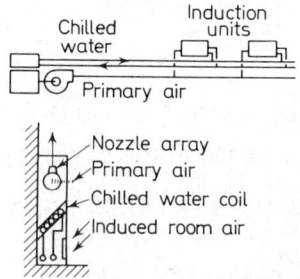

Chilled water

Induction units

Primary air

Nozzle array
Primary air
Chilled water coil
Induced room air

Conditioned, minimum fresh air at fixed dew point from central plant distributed through duct network. Velocity normally less than 20 m/s.
Room terminal is a unit containing chilled water coil, filter and an array of nozzles. Primary air discharge through nozzles induces secondary circulation (from three to six times the primary flow rate) from the room over the filter and coil.

Winter changeover to heated water using the same coil. Non-changeover. Terminal reheat only. 2, 3 and 4 pipe distribution. Overnight background heating with hot water circuit only.

Figure 15.24. Air conditioning systems (From an original diagram by Brightside Heating and Engineering Co. Ltd)

	Main features	*Variations*

CENTRAL REFRIGERATOR/HEAT PUMP SYSTEM

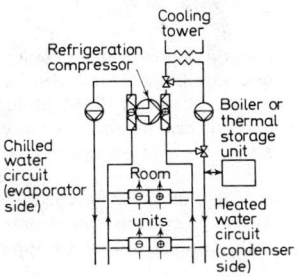

Used with high velocity induction system, fan coil units, or multi-zone air handling units. Heat normally wasted in the condenser of the refrigeration machine is added to the heating circuit. Additional heat from boiler plant or electric thermal storage may be necessary with low winter temperatures. Applicable to buildings with high internal heat gain.

Three pipe distribution for chilled and heated water using the same coil in the room unit.

UNIT REFRIGERATOR/HEAT PUMP SYSTEM

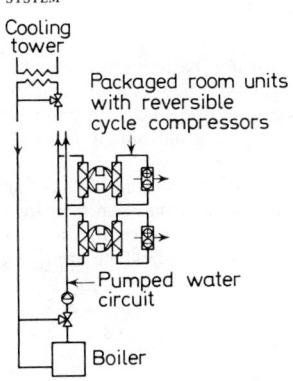

Self-contained fan–coil–refrigerator units with pumped, constant temperature water circuit for the condenser. The refrigeration cycle can be reversed to act as a heat pump, taking the water circuit as a source. Thus the room can be heated or cooled from the same unit. A low-cost fundamental system in new buildings. Also suitable for existing buildings, sometimes using existing radiator pipework.

Can be extended as necessary. Fresh air delivered from central plant, or each room unit can have a fresh air inlet, concealed on the facade.

FAN COIL SYSTEM

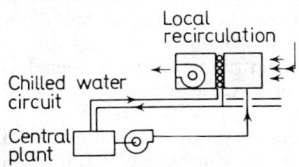

Individual room handling units with chilled water coil. Conditioned, minimum fresh air at fixed dew point from central plant.

Fresh air delivered direct to the room through separate diffusers. Fresh air taken direct from outside. Second coil supplied from independent heating circuit. Winter changeover to heated water using the same coil.

SPLIT PACKAGE SELF-CONTAINED UNIT

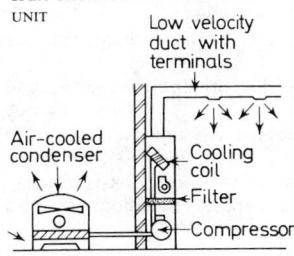

Filter, cooling coil and refrigeration compressor in one cabinet with refrigerant piped outside the building to an air-cooled condenser. Short low velocity distribution duct to terminal units in the room.

Water-cooled condenser. Reverse cycle heat pump operation for winter.

Noise emission from the tower (data is generally available from the manufacturer) must be checked against the background level for possible nuisance.

Air conditioning systems

The simplest form of air conditioning is a direct expansion fan-coil unit with an air cooled condenser. This is the principle used in all small packaged equipment, for example, window mounted air conditioners, floor standing units and so on. The larger the building, the more refined the control system and the more diverse the heat loads, the more complicated the system must become. Figure 15.24 shows the various approaches and outlines the main characteristics.

A broad distinction can be made between all-air-systems, where all the air needed for ventilation and cooling is delivered by the system, and air/water systems where only the fresh air is supplied from the central plant and most of the cooling is done through local fan-coil units or air induction-coil units.

HEAT RECLAIM TECHNIQUES

By pumping the refrigerator condenser water through heater batteries on zone air handling units, it is possible to reclaim heat that would otherwise be wasted in the condenser. The technique is used in buildings where there is a disparity in load; heat removed at one point can be used where heat is required. The refrigerating machine in this case is working as a heat pump so far as the heating circuits are concerned.

If a higher than normal standard of building insulation is used, heat gain from lighting in a building in winter can be made to supply a significant part of the total heat loss. The lighting units are combined with air recirculation points in the ceiling so that convection from lights is removed at its source and the fitting is cooled, which improves the light output.

DESIGN

Having determined the cooling loads, by calculation or by use of the computer, the system air flow rate and temperatures must be determined. This involves a knowledge of the properties of moist air: the various characteristics can be found from tables[10] or charts. Figure 15.25 is an outline of the IHVE psychometric chart[10] generally used.

The temperature of the conditioned air discharged from the terminals is limited in summer to about 10°C less than the room air, but the actual value depends on the kind of outlet and location.

The amount of air needed (with an all-air system) depends on the total cooling load (sensible plus latent). The entry conditions are defined by the sensible heat ratio, which is the ratio of the sensible heat gains to the total i.e. including moisture gains. The IHVE chart contains a sensible heat ratio sector diagram which gives the slope of the line on the chart: any entry state point must fall on this line. The mass flow rate of the air is found from dividing the total heat load by the difference in specific enthalpy between the entering and the room air conditions.

Loading of cooler and heater batteries is found from the enthalpy change required and the total air flow rate.

Figure 15.26 gives an idea of the range of refrigeration loads and air volumes in practice[23].

Air distribution in ducts can be designed on the high or low velocity basis (roughly 6 m/s and 20 m/s). High velocity systems economise in duct space, only about one third

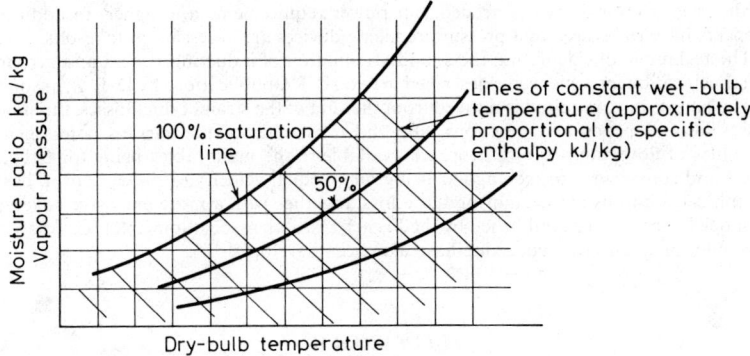

Figure 15.25. Outline psychrometric chart

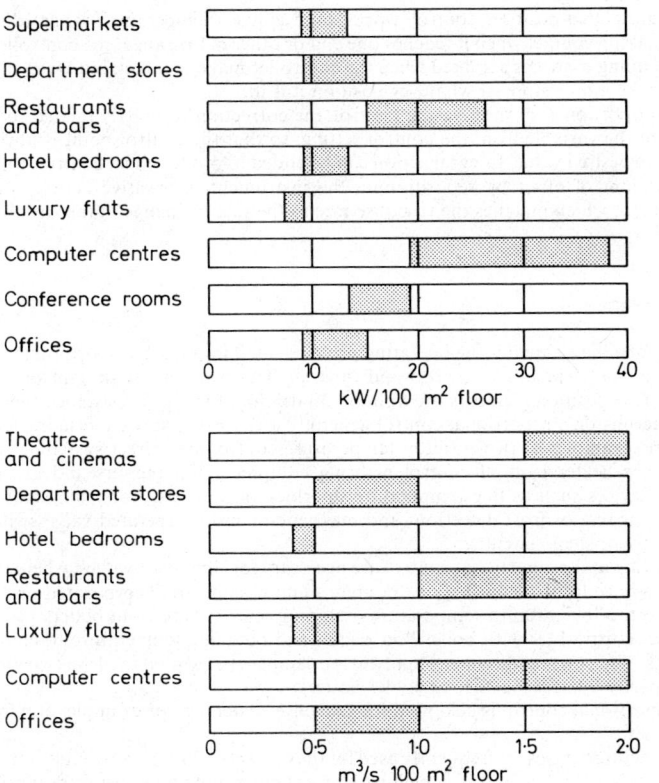

Figure 15.26. Typical refrigeration loads and air flow rates (Courtesy Brightside Heating and Engineering Co. Ltd)

of the cross-sectional area is needed, but power requirements are higher. In addition, silencers have to be used and pressure reducing devices are necessary at take-offs.

The technique of calculating the sound attenuation of a ducting system and arriving at a specification for a silencer is described in two HVRA publications by D. R. Johnson.[29]

The fan pressure required is derived from the sum of the straight duct losses, the shock losses around bends and junctions, and the resistances of the various components. SI tables of flow rate/pressure drop are available,[23] the main tables being for circular ducts and conversions to rectangular being on an equal velocity or pressure drop basis. Graphical solutions for straight pressure loss, together with a more extensive coverage of shock losses has been published by the French research association CoSTIC.[30] Several computer programs on duct sizing have also been developed.[31]

CONTROLS

The majority of control gear used is electric or electronic; very little use is made of pneumatic or fluidic devices. The main control variables are temperature, humidity, pressure and flow. The simplest form of control is the two-position or on-off, used in thermostats.

Floating two-position control works over a fixed temperature range (differential) only making contact when it reaches one end or other of the range; the controlled device, e.g. a pulling motor, is actuated when the controller makes contact, but when the contact breaks the motor stops at whatever position it is in.

In proportional, or modulating control, the correction is directly related to the deviation of the variable from the control setting, so that the control point is always offset from the desired value. Integral action can be added to proportional control; this reduces the amount of offset by repositioning the instrument. Derivative action is a further refinement which matches the response rate to the rate of change in the variable.

Temperature

The bimetallic element is the operating principle used for most two-position temperature control. The element can be rod and tube, or flat strip — either straight or as a helix. Other two-position elements are volatile fluids, liquid in tube, resistance thermometer and thermistor. Proportional control generally uses the resistance thermometer.

Some examples of two-position temperature control are; direct switching of electric resistance heaters; on/off control of firing equipment (oil burners and gas burners); safety devices such as flue temperature interlock on burners, water temperature overshoot warning or frost detection; and magnetic or motor operated valve switching on hot water or steam circuits.

Floating and proportional control are necessary for three-way valves, where two water streams have to be blended. For example, shunt circuits are incorporated on boilers to prevent too low a return temperature, outgoing temperature from boilers produced by mixing return water with boiler flow water, and outgoing temperature to chilled water circuits. The same requirements apply to air dampers being used to blend two air streams, or to partly by-pass a heater or cooler battery.

Proportional control is also used for sequence switching, for example, of refrigerating machines.

Self-acting devices are frequently used for the control of fluids; an example is the thermostatic radiator valve. In this, a bellows element containing a temperature sensitive fluid or vapour acts directly on the valve spindle, positioning it according to the room temperature. The same principle is employed for in-line controllers for HTW and steam.

Humidity

Simple two-position control uses the change in length of a hygroscopic material as a sensor: proportional control uses electric resistance elements having a known resistance/moisture-content relationship.

The control action is directed to switching spinning disc humidifiers, or sequence switching of several units, resetting by-pass dampers or controlling the input to a steam jet humidifier. Dehumidification can also be controlled through positioning the valve or dampers on the chilled water coil.

Pressure

Most applications require only two-position control. The sensors can be bourdon tube, diaphragm or piezoelectric. Typical control linkages are steam pressure switching firing equipment, pressure differential over filters operating a warning signal, limit switches to operate pumps on HTW pressurised systems, and hydrostatic booster pumps on cold water storage vessels.

Pressure is also used as the signal to measure liquid level in fuel oil tanks and water storage tanks.

Flow

Differential devices such as orifice plates, nozzles and Dall tubes can be used to produce a pressure signal, or a direct output obtained from an electromagnetic meter. Air flow is more easily measured in terms of pressure by a pitot-venturi, or in terms of kinetic pressure by a deflecting vane.

The applications are proportional control of metering valves, or low flow indication.

Level

Liquid level in storage tanks can be detected by proximity (inductance) probes, electric resistance probes and mechanical floats, in addition to hydrostatic pressure.

Table 15.30 reviews the main techniques used in control for heating and air-conditioning.

ELECTRIC LIGHTING

The amount and quality of light needed in a building depends on the activities taking place. Excluding decorative and display lighting, the installation must provide a given level of illumination, at the same time limiting the glare and keeping the contrast between the illuminated area and the surroundings within certain limits.

Lighting may be required only for night time, or maybe in continuous use in a window-less building, or a building with vision strip windows designed to rely on supplementary lighting.

Fittings

Where lighting is intended for nearly continuous use, fluorescent tubes give the best operating economy in terms of running costs and initial cost. Light output from fluorescent tubes decreases with rise in ambient temperature and, where fittings contain several tubes in parallel, air cooling of the fitting is desirable, otherwise temperatures may rise to 60°C or more. The hot air extracted can be used as part of a heat pump reclaim system, or can be recirculated, and mixed with fresh air for the ventilating or air conditioning plant.

Noise can be appreciable from badly mounted fittings, or from a large number of fittings. The estimated noise level should be compared with the NC curve.

In order to limit glare from the surface of the tube, egg-crate or louvre shields are used; the desirable cut-off angle being about 45° each side the vertical plane through the fitting.

Lighting system design

The distribution of light from a fitting i.e. the amount of light directed upwards and downwards through the horizontal plane, and the sideways component above the 45° cut off zone, is rated according to the IES British Zonal Method (or BZ method). The class of fitting, (BZ 1 to 10) is given by the manufacturer, and the designer can work out the light intensities on the horizontal and vertical planes from the fitting data, and from the known reflectances of the room surfaces.

A simplified method of calculation has been developed by the IES, who also give a technique for determining the amount of glare related to the BZ class of the fitting and the total light output.[32] Limiting glare indices are defined for various classes of work.

ECONOMICS

Cost-in-use is the only criterion by which various systems can be compared. Low running costs may be obtained for high capital investment, or conversely, low initial cost may involve high running cost. Cost-in-use combines both factors:

$$C = \frac{C_i}{n} + C_i \left[\frac{n+1}{2n} \cdot \frac{P}{100} \right] + E + M + R + I + S \tag{18}$$

where C = annual cost-in-use.
C_i = initial cost.
n = amortisation/depreciation period in years.
P = interest rate on capital %.
E = energy cost.
M = maintenance cost.
R = cost of repairs and replacements.
I = insurance.
S = salaries of operators.

This equation can be used to give a rough idea of the comparative economics of different system designs, the least value of C being the best solution in terms of cost. For more accurate assessment, taking into account the depreciation of the value of money, it is desirable to use the discounted cash flow technique.[33]

First cost

Although trade organisations of the services industry have made attempts from time to time to correlate cost data, they have always failed through inadequate support. There is therefore, no well documented information on cost.

Table 15.31 gives some idea of the cost ranges to be expected for various heating and air conditioning installations. An analysis in terms of building sector will be found in Spon's Mechanical and Electrical Services Price Book. Reference can also be made to the Architects' Journal Building Study Series, or to the HVCA Cost Information Service.

Table 15.30 AN OUTLINE OF CONTROL APPLICATION

Equipment	Control techniques
Boilers and heat exchangers	1. Fixed or variable flow temperature by switching or proportioning the firing rate (or primary medium).
Hot water heating systems	1. Fixed flow temperature direct from boiler or through mixing valve.
	2. Variable flow temperatures, depending on outside temperature, through mixing valve (fixed boiler temperature).
	3. Independent sub-circuit control with two-position valves, or proportional mixing valves (with sub-circuit pumps).
Heating appliances	1. Self-acting controls (radiators, convectors).
	2. Magnetic or two-position valves.
	3. Fan motor switching (convectors, unit heaters).
	4. Direct switching of electric resistance elements.
Warm air systems	1. Two-position or proportional control of heating medium to main air heater battery, and zone batteries.
	2. Direct switching of firing equipment.
Air conditioning	1. Two-position or proportional control of valves to chilled water battery.
	2. Proportional control of mixing dampers on heater and cooler batteries.
	3. Step control of stages in direct expansion battery.
	4. Mixing dampers on dual duct terminal units.
Refrigerating machines	1. Direct switching of compressor (small units).
	2. Capacity control of machine (cylinder unloading, inlet vanes).
	3. Energy source control (absorption cycle).

Table 15.31 AVERAGE INSTALLATION COSTS OF HEATING AND AIR CONDITIONING

Type of system	Cost £ per:	
	m^2 floor	m^3 space
Hydronic:		
LTW, radiators or convectors	5	—
LTW, panel heating	6	—
HTW radiant heating	—	0·7
HTW unit heaters	—	0·6
Warm-air (direct fired):		
Domestic and commercial	3	—
Industrial	—	0·4
Air conditioning:		
Small packaged units	15	—
Central station systems	35	—

Running costs

The cost of energy is the principal part of the running cost component. For heating systems, the amount of fuel used is directly proportional to the inside/outside temperature difference. To assess the total amount of fuel in a year therefore, the average outside temperature and operating cycles of the plant must be known.

An approximate method uses the idea of time under equivalent full load. This works out at between 1500 hours and 2500 hours depending on the particular case. The mass of fuel used is thus:

$$3 \cdot 6 \times 10^5 \, H_l \, n / eC \qquad\qquad (19)$$

where H_l is the design heat load, kW.

     n is the number of equivalent hours on full load.

     e is boiler or heat source efficiency %.

     C is calorific value of the fuel, kJ/kg.

The number of equivalent full load hours can be arrived at by taking the actual operating time and multiplying by a weather factor (approx range 0·6 to 0·7). Weather factors can be derived from a knowledge of the number of degree-days for the area. Monthly figures are published by the Gas Council and a contour map for the U.K. will be found in the Olsen Report.[34] An explanation of the method of calculation will be found in the ASHRAE Guide.

Energy usage for air conditioning in summer cannot be established in the same way since the load is not directly related to outside temperature. The energy used by fans and pumps can be derived for the total estimated hours of operation. The refrigeration power can be derived from an analysis of the probable outside temperature and sunshine pattern. Cost estimation must take into account the additional electricity charges made for maximum demand and connected load.

An analysis of the measured energy distribution in several air conditioned buildings is being made by the Building Research Station.[35]

POLLUTION

Atmospheric pollution is the major by-product of environmental engineering installations. Smoke from chimneys can be avoided by using suitable fuels and firing techniques. Grit can be removed by inertial separators, but sulphur dioxide can only be diluted; no reasonable-cost solution exists for removal from the gases nor from the fuel itself.

Natural gas is the only fuel free from atmospheric contaminants. Over the last thirty years, smoke pollution in the U.K. has fallen by about two-thirds while energy usage at the same time has almost doubled. Sulphur dioxide emission increased up to 1960 and has since remained nearly steady. The pattern can be seen from Figure 15.27.[36]

In the last fifteen years, the increasing number of smokeless zones has halved the average ground level smoke concentration; for London the reduction is three-quarters.

The Warren Spring Laboratory is responsible for the continuous monitoring of atmospheric pollution, and published a survey giving data on dust and sulphur dioxide for over a thousand sites.

The most comprehensive review of all aspects of pollution in the U.K. is the First Report of the Royal Commision on Environmental Pollution, 1971.

Smoke

The Clean Air Act 1956 sets out to reduce smoke emission from industrial chimneys, and where possible to eliminate it altogether. It is an offence under the Act for a chimney to emit *dark* smoke continuously i.e. defined by comparison with No. 2 of the Ringlemann optical scale. A maximum period of four minutes is permitted where dark smoke can be discharged, or an aggregate time of ten minutes in eight hours. All new furnaces must be smokeless.

Grit and dust

The Clean Air Act also requires boiler plant burning between 45 and 23 000 kg of coal equivalent per hour to be equipped with grit arresters. The lower limit corresponds to a heat load of about 150 kW.

All new boiler plant must be equipped with grit sampling points in the flue. Measurement techniques are described in BS 3405.[62]

Sulphur dioxide

Atmospheric dilution with tall chimneys is the only means at present of limiting ground level concentration of sulphur dioxide. The Clean Air Act Memorandum gives the chimney

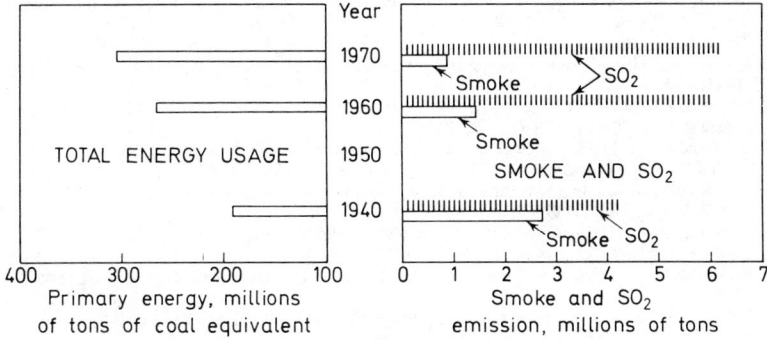

Figure 15.27. Patterns of energy usage: based on an analysis by Dr A. Parker (Courtesy of Heating, Piping and Air Conditioning, USA)

height required for a known rate of SO2 discharge. Nomogram solutions are given in the Memorandum and the Brightside Chimney Design Manual.[26]

The emission of SO2 (kg/h) can be found from:

$$100\,WS \tag{20}$$

where W is the mass of fuel burned, kg/h, and S is the sulphur content %.

The exit velocity of the gases at the top of the chimney should always be over 6 m/s under full load. If the boiler has an induced draught fan the velocity should be 8 m/s, rising to 15 m/s at 130 MW.

The relationship between chimney height and SO2 emission is roughly as follows. Ground level concentration varies directly with the emission, inversely as the square of the height, and inversely as the wind speed. The point of maximum concentration is between 10 to 15 chimney heights away from the base of the chimney. At the moment there is no legal limit on SO2 concentration. In Europe statutory limits vary from 0·2 to 0·4 parts/million.

Noise

Equipment noise is not yet subject to legislation (except generally through the Noise Abatement Act) but obviously the designer will make sure that externally generated noise does not become a nuisance.

Local regulations on noise exist however. For example, the Ministry of Health in the Hospital Design Note on noise control specifies maximum sound pressures from equipment in the 8 frequency bands and a few local authorities specify NC ratings for living zones.

REFERENCES

1. GIVONI, B., *Man Climate & Architecture*, Elsevier (1969)
2. FANGER, P. O., *Thermal Comfort*, Danish Technical Press (1970)
3. ASHRAE *Handbook of Fundamentals*
4. MISSENARD, F. A., *Le Chauffage et le Rafraichissement par Rayonnement*, Eyrolles (1959)
5. RECKNAGEL-SPRENGER, *Taschenbuch fur Heizung Luftung & Klimatechnik*, Oldenbourg (1970)
6. BASNETT, P., 'Space heating by medium temperature panels', *Jour. IHVE*, **36**, 120 (1968)
7. CHRENKO, F. A., 'Heated Ceilings and Comfort', *Jour. IHVE*, **20**, 375 (1953)
8. KOLLMAR and LIESE, *Die Strahlungsheizung*, Oldenbourg, 4th ed (1957)
9. NEVINS, R. G. and FEYERHERM, A. M., 'The effect of floor surface temperatures on comfort', *Trans. ASHRAE*, **73**, Part 2 (1967)
10. *IHVE Guide, Book A* (1970)
11. TAYLOR, G., 'Hypothermia—low body temperature', *IHVE District Heating Symposium Report* (1967)
12. BELDING, H. S. and HATCH, T. F., 'Index for evaluating heat stress in terms of resulting physiological strains', *Trans. ASHRAE*, **62**, 213 (1956)
13. JAMIESON, H. C., 'Meteorological data and design temperatures', *Jour, IHVE*, **22**, 465 (1955)
14. JAKOB, M. *Heat Transfer*, Vol. 1, Chapman & Hall,
15. BILLINGTON, N. S., 'Heat loss through solid ground floors', *Jour. IHVE*, **19**, 351 (1951)
16. *B.R.S. Digest*, No. 108 (1969)
17. JACKMAN, P. J., *HVRA Laboratory Report No. 53* (1969)
18. LOUDON, A. G., *B.R.S. Current Paper No. 47/68*
19. PETHERBRIDGE, P., *Sunpath Diagrams & Overlays for Solar Heat Gain Calculations*, H.M.S.O. (1969)
20. FIELD, A. A., *Paper to Condensation in Buildings Conference*, York University (1972)
21. FIELD, A. A., *The Engineer, Special Report Series* (July 1971)
22. *IHVE Guide Book C.* (1970)
23. *Private Communication*, Brightside Heating & Engineering Co. Ltd.
24. RILEY, J. M., 'The development of the cast iron sectional boiler', *Jour. IHVE*, **38**, 160 (1970)
25. KELL, J. R. and MARTIN, P. L., 'The Nuffield College Heat pump', *Jour. IHVE*, **30**, 333 (1963)
26. *Brightside Chimney Design Manual*, Technitrade Journals Ltd., 2nd ed (1970)
27. *IHVE Guide*, 3rd ed (1965)
28. BURBERRY and GRIFFITHS, 'Service engineering hydraulics', *Arch. Journ.* 1185 (November 1962)
29. JOHNSON, D. R., 'Acoustical design of engineering systems', *HVRA Technical Notes*, Nos. 5 and 7 (1959)
30. *Pertes de Charge Aerauliques*, CoSTIC (1965)
31. REHVA, 'Index of Computer Programs for Heating Ventilating & Air Conditioning', *HVRA* (January 1971)
32. *Illuminating Engineering Society, Technical Report Nos. 2 and 10*,
33. STONE, P., *Building Design Evaluation*, Spon
34. OLSEN, F. E., *Thermal Insulation for Buildings*, OECD, Paris
35. MILLBANK, N. O., 'Operating costs of mechanical services in office buildings', *Jour. IHVE*, **39**, (1971)
36. FIELD, A. A., 'London's clean air', *Heating, Piping and Air Conditioning*, 165 (January 1971)

BRITISH STANDARDS

37. BS 1387
    Steel tubes and tubulars suitable for screwing to BS 21 pipe threads (1967)
38. BS 4127
    Light gauge stainless steel tubes
39. BS 2871
    Copper and copper alloy tubes (1971)

40. BS 1972
Polythene pipe (type 32) for cold water services (1967)
41. BS 3284
Polythene pipe (type 50) for cold water services (1967)
42. BS 3505
Unplasticised PVC pipe for cold water services (1968)
43. BS 3506
Unplasticised PVC pipe for industrial purposes (1969)
44. BS 1588
The use of thermal insulating materials. Temperature range 95°C to 230°C (1963)
45. BS 1334
The use of thermal insulating materials for central heating and hot and cold water supply installations (1966)
46. BS 779
Cast iron boilers for central heating and hot water supply
47. BS 759
Valves, gauges and other fittings for boilers and piping installations
48. BS 2790
Cylindrical land steam boilers (other than water-tube boilers) (1969)
49. BS 1501
Steels for fired and unfired pressure vessels, Plates (1964)
50. BS 2869
Petroleum fuels for oil engines and burners (1967)
51. BS. 799
Oil burning equipment (1970)
52. CP 342
Centralised domestic hot water supply (1970)
53. BS 1566
Copper indirect cylinders for domestic purposes
54. BS 853
Calorifiers for central heating and hot water supply
55. CP 310
Water supply (1955)
56. BS 2831
Methods of test for air filters used in air-conditioning and general ventilation (1957)
57. BS 3928
Method of test for low-penetration air filters (1965)
58. BS 3208
Methods of test and rating for hot-water air-heater batteries (1960)
59. BS 2619
Method of test and rating for steam-heated air-heater batteries (1960)
60. BS 3122
Rating and testing of refrigeration compressors (1959)
61. CP 406
Mechanical refrigeration (1952)
62. BS 3405
Simplified methods for measurement of grit and dust emission from chimneys (1961)

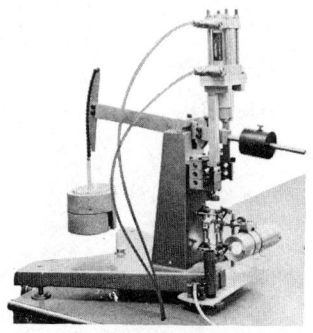 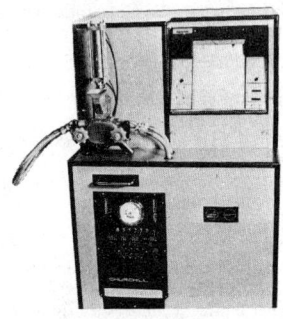

16 PLASTICS

PLASTICS 16

16 PLASTICS

M. M. HALL and D. C. WRIGHT

INTRODUCTION

Plastics complement and overlap the longer established materials which are available to design and applications engineers. They are selected for applications in which they will function efficiently with a reduction in the total costs involved. In general, they are less stiff, and weaker than metals, with a limited service temperature range. However, their ease of processing, low density, attractive appearance, and resistance to corrosion, offer a unique combination of properties.

The ability to control the properties of plastics by using controlled amounts and types of fillers, and by altering the structure of the polymer, gives the engineer a group of materials with an exceptionally wide range of properties. Plastics, combined with other materials into composites, can be used satisfactorily for applications which are subjected to moderate loads.

This section describes the engineering properties of plastics together with their advantages and disadvantages. The methods of conversion of the raw material into finished products are outlined. The physical and mechanical properties of the more important plastics are given, together with guidance on material selection and techniques for designing for specified stiffness and strength. Sources of further information about plastics are listed.

Plastics can be divided into two broad categories which are distinguished by their processing characteristics. These categories, thermoplastics and thermosets, are still valuable classifications although they are tending to break down due to developments in processing machinery and modifications to material performance.

Plastics are often distinguished from polymers, upon which they are based, in the following manner. Additives or bulk fillers are incorporated in the base polymer resin for reasons of ease of processing, enhancement and stability of properties, or economics. These modified polymers are called plastics.

Thermoplastics

Thermoplastics are polymer-based materials which soften on heating and harden on cooling in a reversible manner. They can therefore be readily moulded while hot, but must cool before they become stable in shape.

There is essentially no chemical change associated with the physical change during a heating cycle, and hence scrap or waste thermoplastic can be reused.

Thermosets

Thermosets are polymer-based materials which, upon application of heat, show an initial softening and then go hard or 'cure' in an irreversible manner. Cured mouldings are dimensionally stable while still hot, and remain hard on cooling.

In cooling, the thermoset is undergoing the chemical change of polymerisation and cross-linking. Hence, scrap cured thermoset is different from uncured thermoset, and cannot be reused.

Fillers

Reinforced thermosets dominate the production of polymer-based composites. The primary role of fillers may be as a reinforcement, but alternatively, they may act largely as pigments or extenders. They may be organic or inorganic materials which are either fibrous or particulate. They vary widely in cost, properties, and availability. A major reason for their inclusion is often to reduce the cost of the bulk plastic relative to the cost

Table 16.1 COMMON TYPES OF FILLERS

Filler type	Typical applications	Advantages of composites
Kaolin clay	Polyester sheet and bulk moulding compound	Reduced cost, controlled mouldability.
Asbestos	Thermosetting moulding compounds, vinyl floor tiles	Reduced cost, improved mechanical properties and resistance to heat.
Calcium silicate	Used in many thermoplastics and thermosets	Reduced cost, good resin compatibility, good electrical and thermal resistance, low water absorption.
Mica	Potting compounds.	Good temperature resistance.
Metal powder	Epoxy-based cast forming tools	Controlled and improved physical properties.
Carbon	Pigment and filler for some thermoplastics and thermosets, e.g. black polycthylene sheet	Improved resistance to ultra-violet degradation.
Wood flour	In thermosets	Reduced cost and mould shrinkage.
Papers	Laminated thermosets (sheet, rod, tube) e.g. bakelite (phenolic resin-based)	Cheap (phenolic-based) good combination of mechanical, electrical, and chemical resistance properties.
Glass fibre	Unidirectional, woven, and chopped—strand mat reinforced polyesters and epoxies, short-fibre reinforced thermoplastics	Increased strength and stiffness reduced thermal expansion and contraction.

of the polymer resin upon which it is based. The fibrous reinforcements may be as unwoven or woven mat, or as short fibres (particularly for thermoplastics).

Some common types of fillers are given in Table 16.1, together with their main advantages and typical applications. The detailed effects of the incorporation of the fillers depend upon the amount and degree of dispersion of the fillers as well as its type. The fibre-reinforced thermosets are particularly important engineering plastics.

PROCESSING

Before processing to produce end products, the plastics are in the form of powders, granules, syrups or doughs; or in semi-fabricated forms such as sheet, rod, block, or tube. The principles of the different techniques for converting to end products are as follows.

Extrusion

Extrusion is normally used for the production of continuous lengths of thermoplastics with a wide variety of profiles, including sheet, rod, and tube. Granules or powder are fed through a hopper into the heated barrel of the extruder where they are converted to a melt (plasticised): see Figure 16.1.

The barrel contains a screw, or screws, which transport the melt to the die as well as mechanically working (and therefore heating) and homogenising the polymer. The die is often oversize to allow for shrinkage on cooling, and to allow for the reduction in

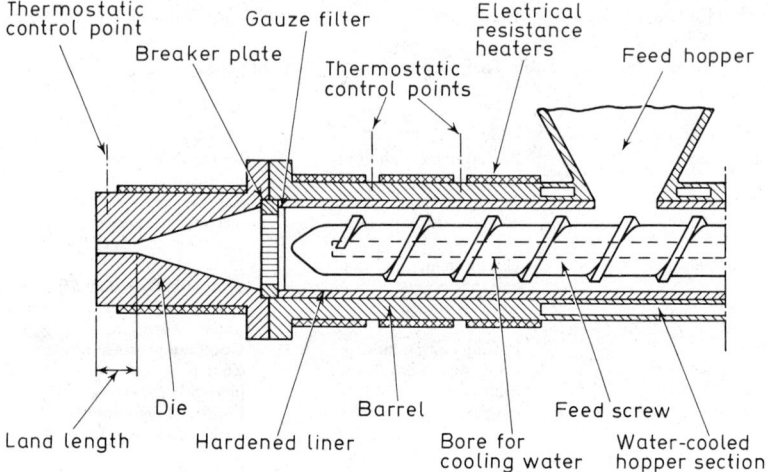

Figure 16.1. The single-screw extruder. (Reproduced from PI monograph Extrusion of Plastics *(Fisher) by courtesy of The Plastics Institute)*

cross section of the extrudate which occurs during take-off. The take-off mechanisms involve conveyor systems which support and transport the extrudate while cooling by air or water. Deliberate draw-down of the extrudate is achieved by having differential take-off and extrusion rates.

The extruder can be used for the continuous production of a coated substrate by combining it with a continuous feed of substrate. Typical substrates are wire, paper, fabrics and roll-formed metal shapes.

Dual extrusion makes it possible to produce profiles formed from the combination of two different types, grades or colours of thermoplastics. Adhesion between different plastics in a dual extrusion depends upon the plastics concerned and can be improved by mechanical keying due to undercuts.

'Lay-flat' films of the polyolefines (polyethylene and polypropylene) and vinyls are formed by extruding tube which is inflated by internal air pressure as it cools before flattening by passing through the nips of a pair of rolls.

Thermosets can be extruded by a discontinuous process in which a high-pressure reciprocating ram forces the thermoset through the die.

Symmetrical components such as bottles or drums can be blow-moulded for thermoplastics. A vertical extruder produces a tube (parison) which passes downwards through a split-mould. Mould closure seals the ends.

Injection moulding

Injection moulding is a particularly important process for producing complex mouldings, primarily from thermoplastic powders or granules but also from thermosetting powders.

The powder or granules are delivered from the hopper into a heated chamber where they are plasticised before injecting into a cool metal mould. The molten plastic is held under pressure in the mould until it solidifies by cooling (thermoplastic) or curing (thermosets).

Typical heated chamber and mould temperature ranges for some of the more common thermoplastics are as in Table 16.2.

Table 16.2 TYPICAL CHAMBER AND MOULD TEMPERATURES FOR INJECTION MOULDING OF THERMOPLASTICS

Plastics	Chamber temperature °C	Mould temperature °C
Acetal copolymers	190–240	80–130
Acrylics	210–240	50–70
ABS	210–275	40–90
Nylon 6	225–280	20–120
Polycarbonate	275–320	85–120
Polyethylene (low density)	190–250	30–50
PPO (modified)	270–300	90–110
Polypropylene	200–280	40–70
High impact polystyrene	170–260	10–70
Rigid PVC	140–180	20–60

The basic parts of injection moulding machines are shown in Figure 16.2. The torpedo spreads the plastics against the inner wall of the heated barrel and, therefore, aids plasticisation. Modifications to the conventional injection moulding machine include ram or extruder systems to plastify the material before it is fed into a holding chamber and injected into the mould by a piston. Plasticisation and injection may alternatively be carried out by a reciprocating screw inside a barrel. These types of modifications can reduce cycle time, broaden the range of materials that can be handled, and improve the homogeneity of the melt plastic.

Machine capacities are rated according to the quantity of general purpose polystyrene that can be injected in one moulding cycle. Capacities range from less than 0·03 kg to more than 84 kg. Thermosets are moulded by a reciprocating screw injection technique, with typical barrel temperature of 55–120°C. and mould temperature of 120–180°C.

Moulds, which may contain more than one cavity, are usually split into two sections which can be clamped together by locking forces of up to about 4×10^5 kg. They contain channels for cooling or heating, and for ventilating entrapped air (Figure 16.3).

The melt polymer enters the mould via the sprue and is distributed to cavities via runners, entering each cavity through the gate. The runners may be heated. It is normally desirable to have all cavities equidistant from the sprue in a balanced manner. The gate is generally smaller than the runner and is a compromise in size between being large enough to allow a suitable flow rate but small enough to prevent back flow. Hollow components can be injection moulded by incorporating cores in the mould cavity. Mouldings can be removed by ejection pins.

Compression moulding

Compression moulding is normally limited to thermosets. A measured quantity of material is placed in the cavity of the mould, which is normally in two parts. The mould

16–6

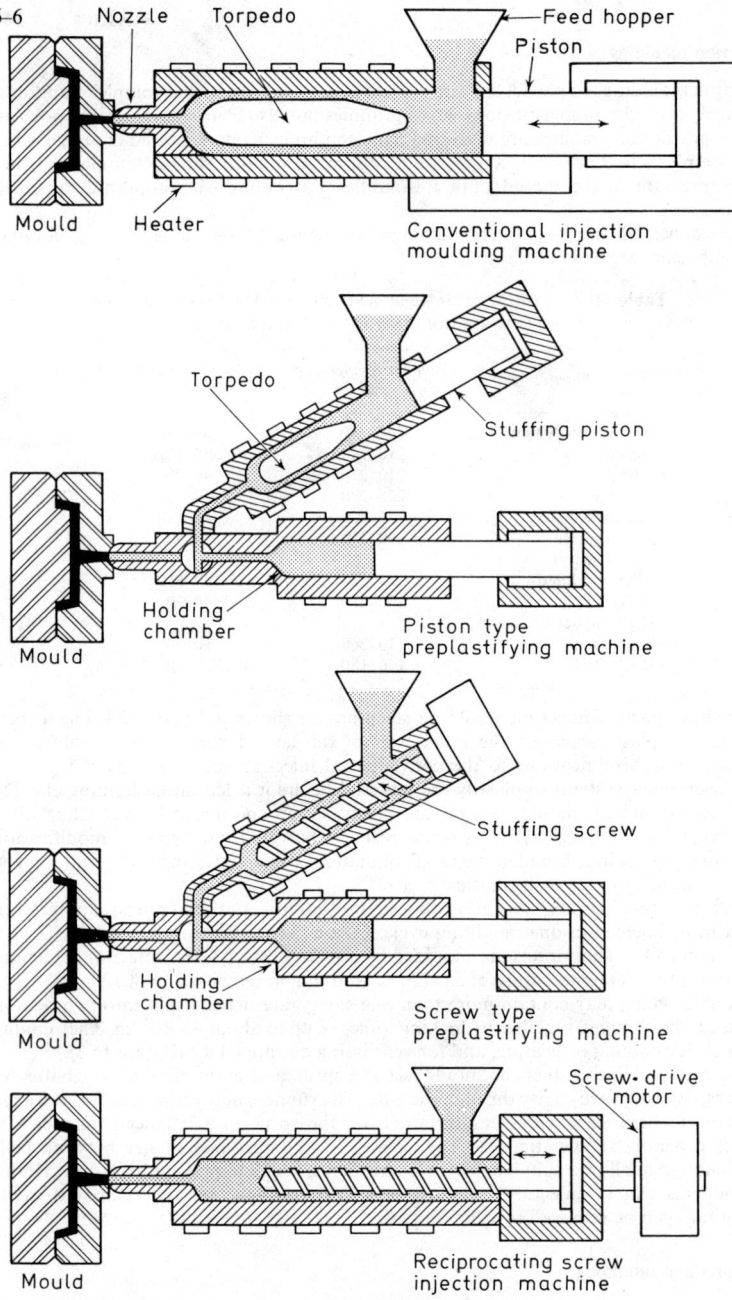

Nozzle Torpedo — Feed hopper / Piston
Mould Heater Conventional injection moulding machine

Torpedo — Stuffing piston
Holding chamber Piston type preplastifying machine
Mould

Stuffing screw
Holding chamber Screw type preplastifying machine
Mould

Screw·drive motor
Mould Reciprocating screw injection machine

Figure 16.2. Basic types of injection moulding machines. (*Reproduced from* Plastic Product Design (*Beck*) *by courtesy of the Allied Chemical Corporation*)

parts are fastened to the platens of a press. Either the mould or the press is heated to temperatures within the range of 140–190°C and the mould closure creates pressure from 14 MN/m² to 70 MN/m². The material softens and flows to fill the mould before curing; the moulding can then be removed while hot.

Compression moulding can be used for deep large-area components of relatively simple shape. Cycle times vary from less than one minute to about 15 minutes. The moulding accurately reproduces the surface finish of the mould cavity but finishing operations are necessary to remove the flash created by the flow of the material along the split lines of the mould. Inserts can be moulded-in, but if the inserts are relatively

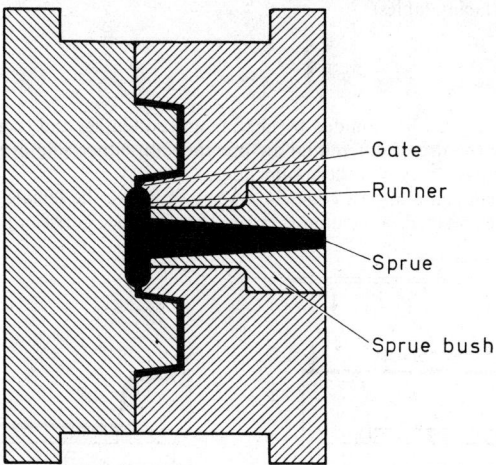

Figure 16.3. A schematic diagram of an injection mould. (Reproduced from Injection Mould Design *(Pye) by courtesy of the Plastics Institute)*

delicate, transfer moulding is necessary, which offers the further advantage that flash is minimised. Transfer moulding involves plasticisation in a separate chamber before transfer into the closed mould. Cycle times can be reduced by the use of pre-heating units.

The thermoset to be moulded may be powder or alternatively preformed by cold compression into a convenient size for handling and fitting into the mould cavity. Fibre-refined thermosets, notably glass-fibre reinforced (grp), can be compression moulded between matched metal moulds.

The moulding material may be in three alternative forms. It may be in the form of a dough moulding compound (DMC), a sheet moulding compound (SMC), or a preform moulding. The moulding compounds have putty-like consistency as a result of the mechanical blending of glass fibres, resin, and additives (DMC); or as a result of the pre-impregnation of glass-fibre sheet (SMC). Preform mouldings are made by air-spraying chopped glass fibres on to a perforated former of the same size as the final moulding. The fibres are held in place on the former by suction and are sprayed with resin which binds them in place. The binding resin is cured to produce a preform which can be handled and transferred to the compression mould. Resin is added and the final moulding produced.

Short production runs of grp components can be produced by simple lay-up techniques. A wood or plaster mould is covered by chopped-strand or woven glass-fibre mats which are impregnated with activated resin. Compaction and the removal of air bubbles can be achieved by hand rolling or a variety of techniques involving the use of pressured flexible bags.

Pressure vessels and tubes of grp can be produced by winding continuous filaments of glass-fibres onto resin-coated mandrels. The mandrel is removed after the moulding is cured. The winding angles are chosen to take advantage of the anisotropy of the mechanical properties of mouldings produced in this way. Fabrics and papers can be impregnated with thermosetting resin and then bonded together by heat and pressure to form sheets, blocks, rods or tubes.

Bakelite, based on phenolic resin, is an example of a well-established, widely-used laminate with excellent electrical properties. The laminate can include a facing ply of metal (copper-clad laminates for the electrical industry) or a patterned paper (decorative laminates for surfacing tables).

Thermoforming

Sheet thermoplastic can be moulded by heating and softening, and then forcing against a cold mould by the use of differential air pressures on either side of the sheet. Figure 16.4 shows three basic stages of thermoforming.

The sheet is clamped along its edges in a position above the female mould and then heated and hence softened. A vacuum is created in the space between the softened sheet

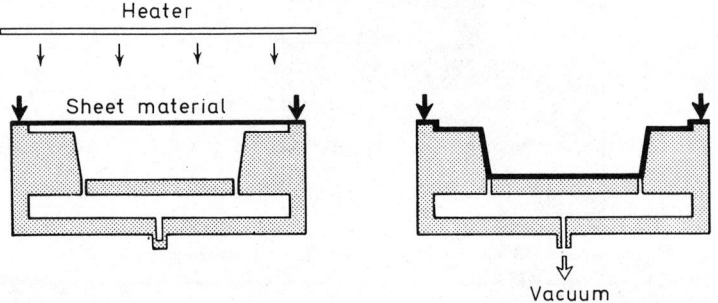

Figure 16.4. Vacuum forming into a female mould. (*Reproduced from* Thermoplastics, effects of processing (*Ogorkiewicz*) *by courtesy of the Plastics Institute*)

and the mould and the sheet is drawn against the cold mould surface. The cold moulding will need trimming. The wall thickness decreases with depth of draw.

There are a number of varieties aimed at producing mouldings with deep draws or more even wall thickness. A male mould may be used; the mould moving up into the heated sheet to be draped by it before the vacuum draws the sheet against the mould. The sheet may be pre-blown into a bubble to achieve a deep draw.

Thermoplastics which are commonly thermoformed are the polystyrenes, ABS and the acrylics. Typical products are refrigerator liners and cups for drink vending machines.

A limited number of plastics are available as liquids suitable for casting. Acrylic sheet is commonly produced by casting into a mould formed from two high-quality glass sheets separated by a gasket spacer. Nylon 6 can also be cast as a monomer to polymerise in the mould. Epoxy and polyester resin can be cast into moulds before curing. Casting is a slow process but can use cheap moulds and is valuable for the encapsulation and protection of delicate components.

Hollow thermoplastic mouldings with a degree of symmetry can be produced by rotational moulding. Powder or paste is placed in a hot hollow mould which is then rotated about two perpendicular axes. The plastic flows and evenly coats the mould walls and is then cooled. Typical mouldings are polyethylene containers and bottles.

Plastics can be used as coatings for a number of materials, either for decoration or as a protection. The component to be coated may be pre-heated and then dipped in a plastic

solution or a fluidised bed of plastic powder. Alternatively the component can be sprayed with a plastic solution before heating.
Detailed descriptions of these processes are given in References 1 to 11.

DESIGN CONSIDERATIONS

In designing for a novel application, or in considering plastics as a replacement for other materials in established applications, three questions must be answered i.e.
   What material?
   What processing techniques?
   What dimensions?
The answers are inter-related and dominated by cost factors. The applications engineer should work closely with the moulders and material suppliers, and if necessary consult impartial authorities such as RAPRA[12] and the BPF[13].

Table 16.3 A CHECK LIST TO AID MATERIALS SELECTION

Economic	Material costs
	Fabricating and finishing costs
Environmental	Influence of temperature on all properties
	Ageing due to heat, light and other radiations
	Weathering
	Flammability
	Chemical attack
	Influence of active environments on rupture, fatigue, ductile-brittle transitions and the effects of notches
	Influence of humidity on creep and electrical properties
Thermal	Shrinkage from the mould
	Thermal expansion
	Specific heat
	Dimensional stability
	Upper and lower temperature limits
Aesthetic	Colours possible
	Optical clarity
	Surface finishes possible
	Stiffness
	Creep: at various temperatures and stress levels
	Stress relaxation: at various temperatures and strain levels
	Dynamic stiffness: at various temperatures, frequencies, and stress levels.
Mechanical	Strength
	Creep rupture: at various temperatures
	Impact strength: notch sensitivity, effect of surface finish. Temperature dependence
	Dynamic fatigue: at various temperatures, frequencies, and stress levels
	The effect of processing conditions on stiffness and strength
	Surface properties: hardness, friction, wear
Electrical	Electrical strength, ac and dc
	Tracking resistance, ac and dc
	Volume and surface resistivity
	Dielectric constant at range of frequencies
	Loss tangent at range of frequencies
Safety	Toxicity
	Flammability

Materials selection

In order to select an appropriate material the designer must establish as closely as possible the anticipated working conditions of the product. Until this is done the available property data sheets on plastics are of little value; the use of data relevant to conditions other than the anticipated service conditions can be positively misleading.

A first survey of the materials is most usefully made from comprehensive lists of materials which, because of the number of materials included, contain only a limited amount of information on each material: (see 'Properties Data' section and References 2, 14, 15). Many materials can be rapidly eliminated if the first search is for materials to satisfy the service temperatures and chemical resistance requirements. A check list such as that given in Table 16.3 is helpful to ensure that all of the appropriate factors are considered.

For a given application the different factors will have a different degree of importance and a limited number of factors will dominate the selection of the material[16].

Dimensions

The dimensions of a component are a compromise between ease of processing and general finish, ergonomics and value engineering, and stressing factors. Special size and tolerance controls may be necessary if the plastic is part of an assembly.

Plastics shrink on cooling: the degree of shrinkage depending upon the nature of the

Table 16.4 SHRINKAGE FROM THE MOULD FOR VARIOUS PLASTICS

	Mould shrinkage %
THERMOSETS	
Alkyd moulding powder	0·5–0·7
Epoxy moulding powder	0·4–1·0
Phenolics	
wood flour-filled	0·4–0·8
mineral-filled	0·1–0·4
Polyesters DMC	
glass fibre-filled	0·2–0·5
'low profile'	±0·05
THERMOPLASTICS	
Acetal copolymer	1·8–2·5
Acrylic	0·1–0·8
ABS	0·4–0·7
Nylon 6	0·5–1·5
Nylon 6·6	1·2–1·8
Polycarbonate	0·7–0·8
Polyethylene	
low density	1–2
high density	1–3·5
PPO (modified)	0·5–0·7
Polypropylene	1·8
Polystyrene (general purpose)	0·2–0·6
Polystyrene (high impact)	0·3–0·7
Rigid PVC	0·7–1·5
Polyethylene Terephthalate ('Arnite')	
amorphous	0·1
crystalline	1·5–2·0

plastic and the processing history. Useful generalisations are that partially crystalline polymers shrink more than non-crystalline (amorphous) polymers, and shrinkage can often be reduced or controlled by the use of fillers. Some typical shrinkage data is given in Table 16.4.

The actual shrinkage of a particular material depends upon moulding conditions, and a requirement for high tolerances can be costly. A system of tolerances for injection-moulded thermoplastics, based upon practical experience, has been drawn up and presented as a British Standard (BS 4042).[43] Tolerances for moulding thermosets are given in BS 2026.[41] Detailed data for a number of plastics are also given by the Society of the Plastics Industry.[17]

Suggested minimum wall thicknesses for moulded articles are given in Table 16.5. The suggestions are based upon the need for ease of processing as well as general structural considerations.

Table 16.5 SUGGESTED WALL THICKNESSES OF MOULDED ARTICLES (mm). (*Reproduced from* Plastics Engineering Handbook *by courtesy of the Society of the Plastics Industry Inc.*)

	Minimum for any article	For small articles	Average for most articles	Large to maximum articles
THERMOSETTING				
Phenolics				
general-purpose and flock-filled	1·27	1·58	3·17	4·75 to 25·4
fabric-filled	1·58	3·18	4·75	4·75 to 9·53
mineral-filled	3·18	3·18	4·75	5·08 to 25·4
Alkyd				
glass-filled	1·02	2·36	3·18	4·75 to 12·7
mineral-filled	1·02	3·18	4·75	4·75 to 9·53
Ureas and Melamines				
cellulose-filled	0·89	1·58	2·54	3·18 to 4·75
fabric-filled	1·27	3·18	3·17	3·18 to 4·75
mineral-filled	1·02	2·36	4·75	4·75 to 9·53
THERMOPLASTIC				
Acrylics	0·64	0·89	2·36	3·18 to 6·35
Cellulose Acetate	0·64	1·27	1·91	3·18 to 4·75
Cellulose Acetate Butyrate	0·64	1·27	1·91	3·18 to 4·75
Ethyl Cellulose	0·89	1·27	1·58	2·36 to 3·18
Polyamide	0·38	0·64	1·53	2·36 to 3·18
Polyethylene	0·89	1·27	1·58	2·36 to 3·18
Polystyrene	0·76	1·27	1·58	3·18 to 6·35
Polyvinyls	2·36	1·58	2·36	3·18 to 6·35

Wall thicknesses should be as even as possible. Sudden changes in section thickness in thermoplastics can lead to unbalanced residual stresses and hence distortion, as well as moulding faults such as sinks or voids. Corners should have radii to reduce the stress concentration factors. The effect of the radius of a fillet upon the stress concentration factor is shown in Figure 16.5; there is limited benefit in increasing the fillet radius to thickness ratio above about 0·6.

The rigidity of plates or walls can be increased by the use of ribs, or by increasing the wall thickness. Ribs are clearly more efficient in terms of material usage, and can also be used as part of the runner system in injection moulding for the efficient distribution of material. Rib dimension can be calculated using the standard 'strength of materials' formulae of classical elasticity provided that the plastic stiffness data used in the formulae are appropriate to the temperature and loading conditions of service. Sinks or voids in the region of the rib and supported wall can be minimised by moulding ribs with section thicknesses less than the corresponding wall thickness. Detailed descriptions of the design of ribs, bosses, inserts, undercuts and mould design in general, can be found in textbooks.[11, 18, 19]

Designing for stiffness

The stiffness (modulus) of plastics depends upon the temperature, the time under load, the rate of application of load. There are non-linear relationships between stress and strain, and hence the stiffness (the ratio of stress to strain) is also dependent upon the strain level at which it is measured. In principle, therefore, the classical laws of elasticity are inapplicable to plastics. In practice, however, the designer can use the standard

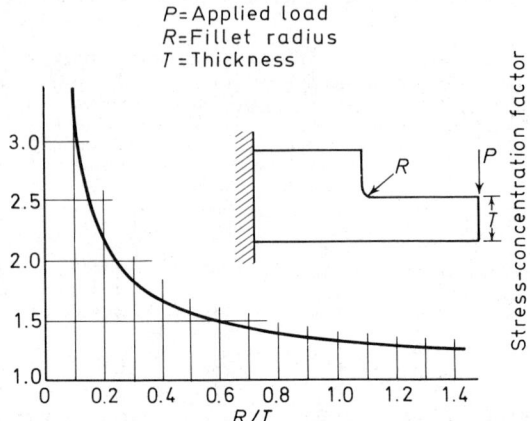

Figure 16.5. Effect of fillet radius on stress concentration factor. (Reproduced from SPI Plastics Engineering Handbook *by courtesy of The Society of the Plastics Industry Inc.)*

'strength of materials' formulae which are based upon classical elasticity, together with specially selected modulus values abstracted from plastics data[20–22].

There are recommended formats for the presentation of plastics design data (see BS 4618).[44] The basic stiffness data are creep data; the total strain response to a constant stress. Uniaxial tensile creep data are most commonly available and are obtained as the strain response to a series of stress levels. They can alternatively be represented as isometric (constant strain) stress relaxation data, or as isochronous (constant time) stress-strain data (Figure 16.6).

Stress relaxation data is useful in applications such as a push-fit insert into a plastic boss, when the insert generates hoop stresses as a result of the constant hoop strains in the bosses. The stresses will relax with time and could become inadequate to support the insert. Isochronous stress-strain data gives information upon the stress level which will give a particular strain response after a life-time corresponding to the time of the isochronous curve. It also demonstrates the strain dependence of the modulus.

The creep modulus of a plastic can be determined by dividing the applied stress by the corresponding creep strain. Substitution of the creep modulus $E(t)$ for Young's modulus in standard elasticity formulae gives an estimate of the long-term deformation of the plastic part. For example $d(t)$, the creep deflection of an end-loaded beam, would be given by:

$$d(t) = \frac{4\,Wl^3}{E(t)bd^3}$$

where W, l, b and d are the load, length, width, and thickness respectively.

Standard formulae can also be used in a more conventional manner. For example, the formula for the bending of a beam can be used to calculate the thicknesses of different materials to give equal resistance to bending. If materials A and B have equal resistance to bending then their Young's moduli and thicknesses are related by:

$$E_A d_A^3 = E_B d_B^3$$

Table 16.6 gives thicknesses of a number of plastics which have been calculated, according to the above formula, to have an equivalent bending resistance to steel and aluminium sheet. The moduli quoted are typical values for standard tensile tests at room temperature.

Thermoplastics can recover from finite strains. The magnitude of these strains depends upon service conditions but is commonly equivalent to tensile strains of several percent[22].

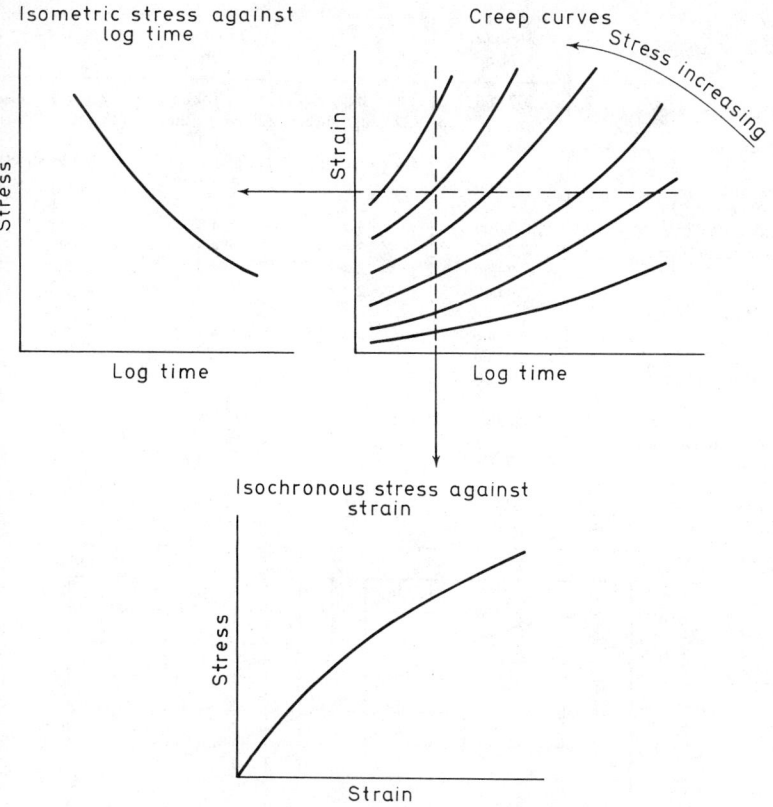

Figure 16.6. Alternative presentation of creep data

Flexibility of this type is an advantage in releasing complex shapes from moulds, as well as for intermittent loading situations.

The use of the appropriate creep modulus in standard formulae accounts for the time dependent behaviour of plastics. Errors resulting from the use of standard formulae because of the non-linear stress-strain behaviour of plastics can be controlled by limiting the allowable strain. A useful method of selecting the maximum allowable strain is the 'modulus accuracy limit' approach[20]. Non-linearity at a given strain level can be described by the ratio of the secant modulus to the initial modulus. The secant modulus is the gradient of the line drawn from a point on the curve to the origin. Hence the ratio has a maximum value of unity when the point is at the origin, and decreases with increasing strain (Figure 16.7).

The use of initial modulus in standard formulae becomes less accurate as the strain increases. The choice of the maximum allowable error is arbitrary; 15% is a commonly used accuracy. The strain corresponding to a secant/initial modulus ratio of 0·85 can be determined from data of the type shown in Figure 16.7. The corresponding stresses to give this strain at the design lifetime can be determined from the corresponding isochronous stress-strain relationship. If the 'modulus accuracy limit' stress or strain is exceeded in a

Table 16.6   THICKNESSES FOR EQUIVALENT RESISTANCE TO BENDING

Material	Modulus MN/m²	Thickness mm		
Steel	207 000	0·8 (22 gauge sheet)	1·0 (20 gauge sheet)	1·3 (18 gauge sheet)
Aluminium	69 000	1·1 (19 gauge sheet)	1·4 (17 gauge sheet)	1·8 (15 gauge sheet)
Acrylic	2 900	3·3	4·1	5·4
PPO (modified)	6 500	2·5	3·1	4·1
Rigid PVC	2 700	3·4	4·3	5·5
Acetal copolymer	2 800	3·4	4·3	5·5
ABS	2 000	3·8	4·8	6·2
Nylon 6·6 (30% glass-fibre)	9 000	2·3	2·9	3·7
Polyester DMC	3 000	3·3	4·1	5·4
Polyester (30% chopped strand mat)	10 000	2·2	2·8	3·6
Polyester (67% woven glass fibre)	20 000	1·7	2·1	2·8
Phenolic (paper laminate)	7 000	2·5	3·1	4·1
Epoxy (67% woven glass fibre)	24 000	1·6	2·0	2·6

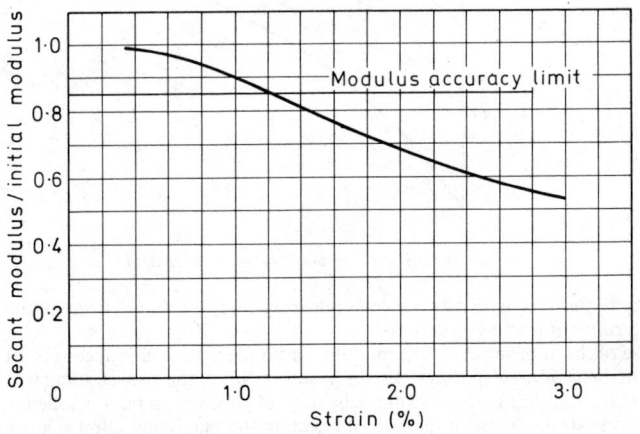

Figure 16.7. Typical strain dependence of the ratio of secant modulus to initial modulus

particular design of a structural application, then the geometry of the component, or the material must be changed.

The complete characterisation of isotropic materials requires a knowledge of Poisson's ratio (or the bulk modulus) as well as the tensile modulus. For design purposes with isotropic plastics the Poisson's ratio is assumed to be independent of time and strain, and to adopt a range of values from 0·35 (rigid plastics) to 0·5 (rubbers). It is not normally a dominant parameter in standard formulae.

Current practices in designing for stiffness with thermoplastics do not consider the influence of anisotropy, although this factor is now receiving increasing attention in research organisations. There have been a number of applications in which controlled fibre directions in fibre-reinforced thermosets have been used to enhance the stiffness of the composite in a specific direction. Anistropy has a more important role in design in its influence upon the strength of thermoplastics and thermosets.

Designing for strength

The important strength properties are yield and rupture although a component may be considered to have failed due to excessive creep, stress relaxation, crazing, discolouration, or other arbitrary criteria which can only be specific for a particular application.

Strength properties of plastics are sensitive to the processing history of the component, in particular the moulded-in molecular orientation and the presence of internal in-homogeneities or external notches. The internal inhomogeneities may be unintentional, e.g. voids or foreign particles; or intentional, e.g. fibres or rubber particles. The direction of the fibres in fibre-reinforced plastics controls the anisotropy of the strength of the

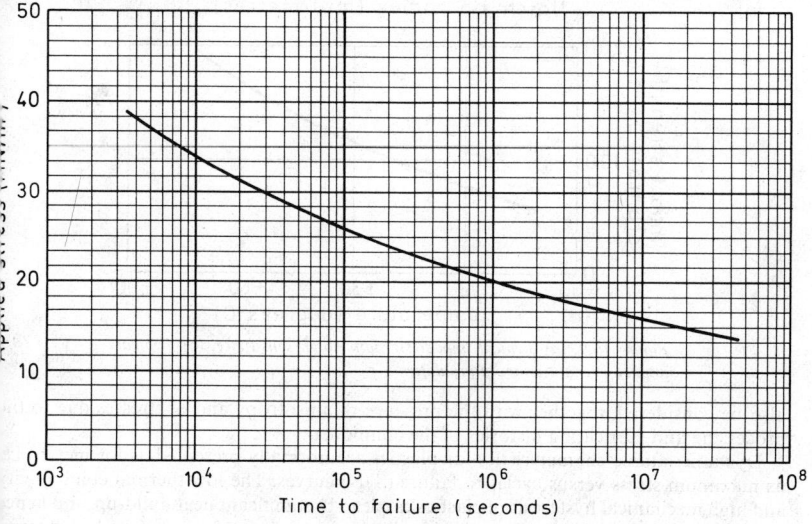

Figure 16.8. Typical static fatigue characteristics

moulding. The nature of the environment is important and, for example, can lead to premature localised cracks and crazes in certain materials under certain conditions[8]. Service temperatures and loading histories are also important.

Rupture under constant load can be considered the end point of a creep curve. Creep rupture (or static fatigue) data represent the relationship between the applied stress and time to failure. Static fatigue data may alternatively represent some other failure criteria such as yield (Figure 16.8).

Standard 'plasticity theory' formulae are used in calculating design stresses to avoid yield failure in plastic structures[20]. At the present time the use of these formulae is largely limited to the design of pipes and pressure vessels. The tensile yield stress corresponding to the design lifetime can be selected from static fatigue data and substituted into the von-Mises yield criteria. The effects of temperature and other environmental conditions should

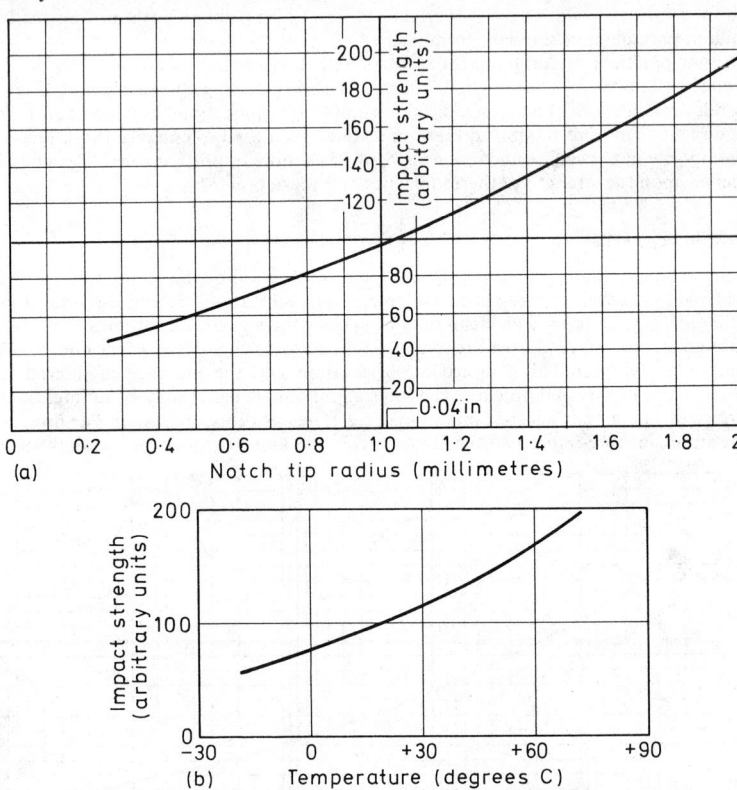

Figure 16.9. (a) Typical impact strength notch sensitivity. (b) Typical impact strength-temperature data

also be considered, together with the presence of anisotropy and of stresses due to the processing and fabricating histories of the component.

Dynamic fatigue characteristics for plastics are normally presented as for metals, i.e. as maximum stress versus cycles to failure (S–N) curves. The low thermal conductivity and high mechanical hysteresis of plastics can lead to significant heat build-up, and hence S–N curves for plastics become sensitive to the test frequency, at frequencies which depend upon the plastic but are typically of the order of 100 Hz.

The impact properties of plastics depend markedly upon sample preparation history and testing procedure as well as upon environmental conditions. The assessment of impact strength as single-point data, i.e. data obtained for a single combination of test conditions, can be totally misleading when comparing the relative merits of different plastics. Impact strength data cannot be used in design calculations; the designer should survey the impact strength as a function of notch sensitivity, temperature and any other factors relevant to the service conditions of the particular application (Figure 16.9).

Prototypes

Careful consideration of all of the factors involved in material selection, and in the choice of dimensions to satisfy processing and engineering design requirements, will reduce design costs by reducing the number of prototypes when compared with an 'ad hoc' approach. Prototypes should be made for a practical evaluation of the design and as an aid to further industrial engineering (aesthetics, ergonomics etc.). They are commonly machined from rod or bar in one or more parts, even when the final product will be injection moulded. It may be possible to take advantage of vacuum forming, casting or compression moulding techniques, using cheap moulds with a short working life-time. For grp systems the prototype can be laid up by hand.

Moulds costs for production can be very high. Injection moulding tools can cost more than the capital costs of the injection moulding machine itself, and extrusion dies and associated equipment may cost 20% of the extruder cost. Vacuum forming tools rarely cost more than 5% of the capital machine costs, and can be considerably less. Prototype testing under service conditions is therefore an important design stage before drawing-up specifications for production. If a prototype is unsatisfactory then an alteration in the dimensions will be necessary. If the prototype proves unsatisfactory during test the designer can modify both dimensions and material. In principle material changes will also involve dimensional changes because of the different processing characteristics of the new material.

Machining

A major advantage of plastics is the ability to mould complex shapes with excellent surface finishes in a single operation. However, machining operations may be necessary and can be carried out successfully provided that consideration is given to the special properties of plastics.

The relatively low stiffness of plastics, associated with an elastic flexibility, will allow the material to flex away from the cutting tool unless it is properly supported. If the feed-rate of the cutting tool is too high then, for example, the part in a lathe may flex away with a consequent reduction in tolerances, or a drilled hole may be smaller than the drill diameter.

Heat generation due to mechanical working must be kept to a minimum because of the high coefficient of expansion and low thermal conductivity of most plastics which could lead to a local degradation or softening. In general, water-soluble cutting fluids can be used but attention must be given to the chemical resistance of the material.

Sharp wood-working equipment can be used for plastics, using cutting fluids, low feed-rates, and high surface speeds for the plastic in the appropriate operations. However, for optimum finishes and high tolerances the geometry of the cutting tool should be specific to each plastic. Detailed description of tool geometries are given elsewhere[2].

Joining

Plastic parts can be joined by welding, riveting, cementing, press and snap fitting, or by the use of self-tapping screws, threaded bolts or threaded inserts. The efficiency of the joint depends upon the nature of the joint and the materials involved. For details see References 2, 17, 23.

Decorating

Thermosets can be decorated by engraving, paints and lacquers, labelling, transfers, vacuum metalising and (with some restrictions) printing. Thermoplastics can be decorated by engraving, hot foiling, paints and lacquers, labelling, electroplating, printing, transfers, and vacuum metalising. The more hygroscopic thermoplastics can be surface dyed.

A number of these forms of decoration will offer protection against chemical attack or ultra-violet damage as well as having aesthetic appeal. For details see References 2, 17, 24.

Costs of plastic material

The total cost of a product can be subdivided into categories of raw material, conversion, installation, and maintenance. This section discussed raw material costs only.

Costing a plastic material for use in a particular application is both an economic and technical exercise, with an emphasis on the technical aspects. Performance/costs ratios are valuable in comparing the relative costs of different materials. They enable the engineer to consider the relevant performance characteristics in terms of costs as an aid to materials selection, and should be used in conjunction with the more conventional cost/unit weight and cost/unit volume information (Table 16.7).

The costs given in Table 16.7 are only for general guidance. They are based on quantities of 1000 kg, but are subject to considerable variation by negotiation with the supplier, superimposed upon general trends of price changes with time. The 'single-point' modulus and tensile strength values used in the cost/performance ratios are typical tensile test values under ambient conditions.

It can be seen from Table 16.7 that the relative costs of different materials depend upon

Table 16.7 RAW MATERIAL COSTS (FOR GENERAL GUIDANCE ONLY)

Material	Specific gravity	Cost per unit weight p/kg	Cost per unit volume £/m³	Tensile strength (TS) MN/m²	Tensile modulus (E) MN/m²	$\dfrac{TS}{cost/unit\ volume}$ MN m £	$\dfrac{E}{cost/unit\ volume}$ MN m £
THERMOPLASTICS							
Polystyrene (high-impact)	1·05	18	210	25	2 500	0·12	12
Polypropylene (copolymer)	0·90	24	230	25	1 000	0·11	4·3
ABS	1·05	33	350	35	2 000	0·10	5·5
Polypropylene (40% asbestos)	1·27	30	380	30	3 000	0·08	7·9
Rigid PVC	1·34	26	380	50	2 400	0·13	6·5
Nylon 6·6	1·13	58	630	69	3 000	0·11	4·8
PPO (modified)	1·06	68	730	80	6 500	0·11	8·9
Polycarbonate	1·2	66	800	60	2 200	0·07	2·7
Nylon 6·6 (30% glass-fibre)	1·36	63	860	130	9 000	0·15	10·1
Acetal copolymer	1·41	70	990	60	2 800	0·06	2·8
Polysulphone	1·24	115	1 470	70	2 500	0·05	1·7
THERMOSETS							
Phenolic (moulding powder)	1·36	18	250	40	6 000	0·16	24
Urea (moulding powder)	1·55	23	330	65	8 000	0·20	24
Melamine (moulding powder)	1·53	33	500	65	8 000	0·13	16
Polyester DMC	1·68	33	580	45	3 000	0·08	5·2
METALS							
Steel (sheet)	7·8	6	500	310	207 000	0·62	410
Aluminium (ingots)	2·6	25	660	150	69 000	0·23	105

Note: 254 £/m³ = 1 pence (d)/in³
0·918 pence (p)/kg = 1 pence (d)/lb.

the method of presenting the cost data. The technical aspects of costing are increased when the effects of temperature and time under load are considered. If, for example, the tensile strength and modulus data in Table 16.7 are replaced by static fatigue and creep modulus data respectively then cost ratios can be calculated as a function of time. The relative costs of materials may be significantly different at times (or temperatures)[25].

PROPERTIES DATA

The tabulated data given in this section are necessarily restricted. For further details on specific materials refer to manufacturers' literature or other publications[2, 15, 22, 26, 27].

Mechanical properties data

It has already been emphasised in the sections on designing for stiffness and strength that plastics as a group are non-linear and viscoelastic. The response of the material to in-service conditions; which obviously includes wide ranges of temperature, humidity, service life, stress levels, chemical and weather environment, and the component history of all these factors; is complex and certainly beyond the scope of this chapter.

Rational design procedures involving plastics must take due account all those factors within a specified service environment which significantly affect behaviour. The simple single-point data provided by Table 16.8 are therefore not recommended for design but as an aid to material selection. Once such a selection has been made it is suggested that the designer contacts the appropriate material supplier (manufacturers' addresses are given at the end of this chapter).

DEFORMATION

The short-time tensile creep modulus of a typical unreinforced engineering thermoplastic is one or two orders down on the comparable elastic modulus of metals. The long-time creep modulus would reveal an even more unfavourable comparison. However, the specific gravity of plastics is low (most fall into the category 1·0–1·5) so the stiffness to weight ratio becomes more attractive. In addition, most plastics exhibit recovery from strains as high as several percent.

The deformation behaviour at elevated temperatures is generally poor, showing in some cases an extremely rapid reduction in stiffness at temperatures not very far removed from ambient. An indication of this effect is given in the thermal properties data. Humidity generally has comparatively little effect upon plastics, with the notable exception of the polyamides (nylons).

STRENGTH

Apart from a few exceptions, plastics will fail in a brittle fashion when strained or stressed at a high rate. The presence of stress raisers which may be external (scratches or sharp corners), or internal (voids, crazes or microcracks), reduce impact resistance. The impact strength data provided have been obtained by using notched specimens, but this does not convey the effect of notch sensitivity nor indeed of temperature (impact strength falls with temperature), see Figure 16.9.

In general the impact strength of stiff plastics is not as high as that exhibited by the flexible grades. Those plastics which have become accepted as engineering materials i.e. acetals, polycarbonates, polyamides, ABS, and rigid PVC, have a good combination of stiffness and impact strength.

Table 16.8 MECHANICAL PROPERTIES DATA

	Tensile strength MN/m²	Flexural strength MN/m²	Impact strength J/25 mm	Elongation at break %	Tensile modulus MN/m²	Hardness 'Scales'
Test method	ASTM D638³⁶	ASTM D790³⁹	BS 2782⁴²	ASTM D638³⁶	ASTM D638³⁶	ASTM D2240–68⁴⁰ ASTM D785–65³⁸
Acetal Homopolymer	70	100	2·0	25 injection 75 extrusion	3 500	R120
Acetal Copolymer	60	90	1·75	60–75	2 800	M80
Acrylic (cast sheet)	55–80	90–120	0·4	2–7	2 600	M80–M100
Acrylonitrile—butadiene-styrene	28–48	35–95	1·5–15	30–60	2 000	R110
Cellulose Acetate	20–60	13–80	4·0	5–50	2 000	R34–R124
Cellulose Butyrate	35	28–65	4·0	50–100	1 600	R31–R116
Cellulose Propionate	12–50	20–75	1–13	30–100	1 600	R10–R122
Cellulose Nitrate	50	65–75	6·5	20–40	1 500	R95–R115
Chlorinated Polyether	40	35	35	130	1 000	R100
Epoxide+GF	70–200	70–400	12–35	4·0	20 000	M100–M110
Ethylene-vinyl Acetate	10	No break	No break	750	28	D35
Fluorinated Ethylene Propylene	20	No break	No break	250–300	600	D57–D60
Ionomer Resin	30	No break	15	400	150–800	D50–D65
Melamine Formaldehyde (paper laminate)	55–80	60–80	0·3	1·0	9 000	M115–M125
Melamine Formaldehyde (wood filled)	55	50–80	0·2	0·6	6 500	E64–E95
Melamine Formaldehyde (asbestos filled)	30	75	0·1	0·3	7 000	E75–E97
Polyamide (nylon 6·6)	70	35–110	1·8	90	3 000	R108–R120
Polyamide (nylon 6)	60	35–110	1·5–5·0	300	2 000	R103–R119
Polyamide (nylon 6·10)	50	No break	3·0	250	2 000	R111
Polyamide (nylon 11)	55	No break	2·5	300	1 300	R108
Polyamide (nylon 6·6+30% GF)	130	130	3·0	1·5	9 000	R116–R120
Polycarbonate	60	90	20	75	2 500	M70
Polycarbonate+30% GF	140	110	4·0	2	10 000	M88–M95
Polyester (rigid casting grade)	55	75	0·3	3·0	3 500	M70–M95
Polyester (flexible casting grade)	10	No break	8	40–300		D84–D94
Polyester+70% satin weave GF	300	400	30	0·5–5·0	10 000–30 000	
Polyester+30% GF mat	150	150	12–25	0·5–5·0	6 000–14 000	M70–M120
Polyester dough moulding compound	50	70	3·5–8·0	3–5	3 000	M110
Polyethylene (low density)	10	No break	20	100–800	100–250	D41–D46
Polyethylene (medium density)	15	35	15	50–600	250	D50–D60
Polyethylene (high density)	30	20	20	250–600	500–1 500	D60–D70

Table 16.8—*continued*

Test method	Tensile strength MN/m² ASTM D638[36]	Flexural strength MN/m² ASTM D790[39]	Impact strength J/25 mm BS 2782[42]	Elongation at break % ASTM D638[36]	Tensile modulus MN/m² ASTM D638[36]	Hardness 'Scales' ASTM D2240–68[40] ASTM D785–65[38]
Polyethylene Terephthalate	60	100	1·0	60–100	—	M106
Polyimide	70	95	1·0	5·0–7·0	3 000	
Poly(4 methyl-pentene-1) TPX	25		1·0	15	1 400	L67–L74
Polphenylene Oxide (modified)	85	140	2·0	5	6 500	M78–M93
Polypropylene Homopolymer	35	50	1·5	200–700	1 400	R85–R110
Polypropylene Copolymer	25	40	10	200–700	1 000	R50–R96
Polystyrene (general purpose)	50	75	0·6	1·5	3 000	M65–M80
Polystyrene (high impact)	30	40	1·5	20	2 500	M20–M80
Polysulphone	70	100	2·0	75	2 500	M69
Polytetrafluoroethylene PTFE	21		3·0	100–400	350–600	D50–D65
Polyurethane Elastomer	40	5–45	No break	10–650	70–2 000	M28
PVC (plasticised)	10–27	No break	No break	300	3÷18	A50–A100
PVC (rigid)	50	70	3·0	2–40	2 500	D65–D85
Silicone (flexible cast)	5	No break	No break	100–300		A15–A65
Styrene Acrylonitrile	75	100	0·7	2·0	3 200	M80–M90
Urea Formaldehyde (paper filled)	70	95	0·25	0·75	8 500	M110–M120
Urea Formaldehyde (wood filled)	70	95	0·25	0·50	8 000	M120

The tensile strength of plastics is generally more sensitive to processing conditions and component environment such as ultra-violet radiation and chemical attack than is stiffness. The effect of loading time is comparable. It must be emphasised that the designing of plastic components to sustain static loads on strength criteria is not as common as with many other materials. The strain at break is generally so high that design on a stiffness criteria to prevent excessive deformation or deflection is most common. There are some exceptions to this; pipe design being a good example.

HARDNESS

Reference to the appropriate testing procedures is necessary if direct comparison between the hardness of plastics is to be made. The Rockwell hardness test (ASTM D785–5)[38] has five scales R, L, M, E and K denoting different combinations of indenter diameter and load. The Shore hardness test (ASTM D2240–68)[40] has two scales, A for the softer materials and D for the harder.

An indication of the hardness of plastics compared with other materials can be seen by comparing scratch resistance. On the Mohs scale plastics falls within the range 2–3 which is approximately the same as aluminium. Steel is quoted in the range 5–8.

Temperature resistance

Plastics operate efficiently over a relatively limited range of temperatures. Thermoplastics soften and flow at elevated temperatures, whereas thermosets degrade with a limited amount of softening. At low temperatures all plastics become increasingly brittle.

The softening of plastics takes place over a range of temperatures; it is therefore not possible to define the resistance of a given plastic by a single characteristic softening temperature. Standards organisations use tests which give the temperature at which a loaded test piece deforms by a given amount. Heat distortion data of this type is arbitrary but nevertheless a useful aid to material selection. The heat distortion data quoted in Table 16.9 is according to ASTM D648[32] in which the test piece is a beam subjected to a 3-point loading.

The crystalline melt temperature T_m is an alternative means of indicating resistance of thermoplastics to elevated temperatures. T_m is useful to processors but of limited value to applications engineers. It represents the temperature at which the partially crystalline plastic finally changes state into a viscoelastic fluid. The glass transition temperature T_g[28, 29, 30] is more useful to the engineer as it represents a temperature at which the degree of softening increases rapidly. It defines the transition of a polymer from a glass-like to a rubber-like material.

Thermoplastics are normally only used at temperatures below T_g. At temperatures increasing below T_g plastics become more brittle. A 'brittle-point' temperature is sometimes observed from impact data (see Figure 16.9b).

Thermal properties data

The thermal properties of plastics are such that they may often prevent the use of the material in critical or sophisticated engineering applications. The important properties are listed in Table 16.9 and the following notes are given in explanation.

THERMAL EXPANSION COEFFICIENT

This is typically an order greater than found in metals. Precise engineering application demanding interlocking parts or instrument-like geometrical stability are not generally recommended. The superior flexibility of plastics however can often compensate for this by allowing the designer to specify less critical tolerances. Plastic bushes or bearings are often capable of deforming to accommodate mis-alignment. Plastic gear teeth profiles need not be constructed precisely to prevent excessive noise, wear or scuffing.

THERMAL CONDUCTIVITY

Plastics are good thermal insulators. The comfortable 'feel' of the material is attributed to a low thermal conductivity. This is one of the properties which helps to explain or justify the use of plastics as handles, knobs, steering wheels, packaging, furnishings and personal accessories. Many other applications demanding thermal protection or insulation can be satisfactorily met by the use of plastic components.

Low thermal conductivity in combination with poor temperature resistance limits the fatigue endurance of plastics at high loading frequencies. At frequencies below about 100 Hz (depending upon component geometry) localised temperature build-up is not as important and fatigue endurance is typically as good if not better than many conventional engineering materials.

Table 16.9 THERMAL PROPERTIES DATA 16–23

Test method	Distortion temperature (fibre stress = 1·8 MN/m²) °C	Thermal conductivity W/m °C	Thermal expansion coefficient 10⁻⁵/°C
	ASTM D648[32]	ASTM C177[31]	ASTM D696[37]
Acetal Homopolymer	125	0·25	8
Acetal Copolymer	110	0·25	8
Acrylic cast sheet PMMA	70–100	0·20	5–9
Acrylonitrile butadiene-styrene ABS	105–120	0·20–0·35	6–9
Cellulose Acetate	45–90	0·15–0·35	10–15
Cellulose Butyrate	45–95	0·15–0·35	11–17
Cellulose Propionate	45–110	0·15–0·35	11–17
Cellulose Nitrate	60–70	0·25	8–12
Chlorinated Polyether	210	0·13	3·5
Epoxide resin + glass fibre	100–250	0·15–0·4	1–3·5
Ethylene Vinyl Acetate EVA	55–65*	0·3	—
Ionomer Resin	40–50	0·25	12
Melamine Formaldehyde (paper filled)	130	0·12	3
Nylon 6·6	70–110	0·25	10
Nylon 6	60–70	0·25	9
Nylon 6·10	—	0·20	15
Nylon 6·10	50	0·25	15
Nylon 12	50	0·25	10
Nylon 6·6 + 30% glass fibre	250	0·20	3
Phenolic (wood filled)	130–170	0·25	3
Polycarbonate	135	0·20	6·5
Polycarbonate + 30% glass fibre	150	0·4–0·6	1·5–4·0
Polyester + 30% glass fibre	220	—	1·5
Polyester DMC	190–250	0·15	5–10
Polyethylene (low density)	30	0·35	22
Polyethylene (high density)	50	0·50	13
Polyethylene Terephthalate	70	0·15	2·5
Polyimide	480	0·40	4–5
Poly(4-methylpentene-1) TPX	125*	0·15	12
Modified Polyphenylene oxide PPO	130–150	0·20	3·5
Polypropylene	50–60	0·12	6–10
Polystyrene (general purpose)	85	0·12	6–7
Polystyrene (high impact)	80	0·10	6–7
Polysulphone	175	0·25	5·5
Polytetrafluoroethylene PTFE	260*	0·25	10
Polyurethane	Variable	0·1–0·3	10–20
PVC (flexible)	60–105*	0·15	5
PVC (rigid)	75	0·13	5
Silicon moulding compound	450	0·15–0·30	30
Styrene Acrionitrile	80–105	0·12	3
Urea Formaldehyde (paper filled)	130–150	0·35	4

* Max. service temperature.

Table 16.10 ELECTRICAL PROPERTIES DATA

Ohms	Volume resistivity cm^3	Dielectric strength kV/cm	Dielectric constant at 1 MHz	Dissipation factor at 1 MHz
Test method	ASTM D257[35]	ASTM D149[33]	ASTM D150[34]	ASTM D150[34]
Acetal Homopolymer	10^{15}	150	3·7	0·005
Acetal Copolymer	10^{14}	200	3·7	0·006
Acrylic (cast)	10^{15}	180–220	2·2–3·2	0·025
Acrylonitrile Butadiene Styrene ABS	$(1–5)\times10^{16}$	150–200	2·4–3·8	0·007–0·015
Cellulose Acetate	$10^{11}–10^{15}$	100–240	4·0–5·0	0·03–0·04
Cellulose Butyrate	$10^{11}–10^{15}$	100–160	4–5	0·02–0·04
Cellulose Propionate	$10^{12}–10^{16}$	120–180	3·3–3·8	0·01–0·05
Cellulose Nitrate	10^{11}	120–240	6·4	0·06–0·09
Chlorinated Polyether	10^{15}	160	2·9	0·01
Epoxide Resin + 30% glass fibre	10^{14}	120–160	3·5–5·0	0·01
Ethylene Vinyl Acetate EVA	$2·5\times10^{16}$	80	—	0·035
Fluorinated Ethylene Propionate FEP	2×10^{18}	200–240	2·1	0·0003
Ionomer Resin	10^{16}	360–440	2·5	0·002
Melamine Formaldehyde (paper)	$10^{11}–10^{13}$	140–160	5–7	0·03–0·06
Phenol Formaldehyde (wood filled)	10^{10}	120	5·0	0·1
Polyamide 6·6	$10^{14}–10^{15}$	155–190	3·5	0·04
Polyamide 6	$10^{13}–10^{15}$	180–200	3·5–4·5	0·03
Polyamide 6·10	$10^{12}–10^{15}$	140–160	3·5	0·04
Polyamide 11	5×10^{13}	170	3·2	0·03
Polyamide 12	$2·5\times10^{15}$	180	3·0	0·03
Polyamide 6·6 + 30% glass fibre	10^{14}	150–190	3·0	0·02
Polycarbonate	2×10^{16}	160	3·0	0·009
Polycarbonate + 30% glass fibre	5×10^{16}	180	3·2	0·007
Polyester Resin	10^{15}	140–200	2·8–4·1	0·006–0·026
Polyester + 30% glass fibre	10^{14}	140–200	4·0–5·5	0·01–0·03
Polyester dough moulding compound	$0·010^{14}–10^{15}$	160	4·2–5·8	0·01–0·02
Polyethylene (low density)	10^{16}	180–400	2·3	0·0005
Polyethylene (high density)	10^{16}	180–200	2·3	0·0005
Polyethylene Terephthalate	—	—	3·0	0·02
Polyimide	10^{16}	220	3·4	0·005
Poly(4-methylpentene-1) TPX	10^{16}	280	2·2	0·00002
Polypropylene Homopolymer	10^{16}	200–250	2·5	0·0005–0·00
Polypropylene Copolymer	10^{17}	200–250	2·3	0·0005–0·00
Polystyrene (general purpose)	10^{16}	200–280	2·5	0·0001–0·00
Polystyrene (high impact)	10^{16}	120–240	2·4–3·8	0·0004–0·00
Polysulphone	5×10^{16}	280	3·1	0·0035
Polytetrafluoroethylene PTFE	10^{18}	200	2·1	0·0002
Polyurethane Elastomer	$10^{11}–10^{13}$	120–360	4–5	0·05–0·075
PVC (flexible)	$10^{11}–10^{15}$	120–400	3·3–4·5	0·4–0·15
PVC (rigid)	10^{16}	170–520	2·8–3·1	0·006–0·019
Styrene Acrylonitrile	10^{16}	160–200	2·6–3·1	0·007–0·01
Urea Formaldehyde (paper filled)	$10^{12}–10^{13}$	120–160	6·8	0·25–0·35

Electrical properties data

Plastics are extensively used in the electrical industry as capacitive, resistive or insulation components. Switchgear, circuit boards and cable insulation being common examples.

D. C. VOLUME RESISTIVITY ASTM D 257

The volume resistivity (between embedded electrodes) is a more repeatable characteristic than any technique which involves surface resistivity as the latter is so sensitive to surface condition. At voltages of either 100 or 500 V the current after 60 seconds of electrification is used to compute resistivity. This is performed under standard conditions of humidity and temperature, the data reported in Table 16.10 being collected at 23°C and 50% RH. The resistivity is sensitive to both temperature and humidity and may reduce by a factor of 10^5 or 10^6 for increases from 25°C to 100°C, or from 25% RH to 90% RH respectively.

DIELECTRIC STRENGTHS AT COMMERCIAL POWER FREQUENCIES ASTM D149[33]

This represents the voltage gradient at which the insulation properties of the plastic begin to fail. The data are applicable to voltage sources with frequencies in the range 50–60 Hz. Breakdown may occur by localised electrical discharge, thermal build-up or intrinsic breakdown.

DIELECTRIC CONSTANT AND DISSIPATION FACTOR AT 1 MH$_z$ ASTM D150[34]

The phase angle θ between an applied voltage and current within a plastic material can be used to provide the loss angle, $(90° - \theta)$. The tangent of the loss angle is known as the dissipation factor. This is a measure of the a.c. loss within an insulator and an indication of the thermal build-up. The dielectric constant (permittivity) is always a maximum at low frequency whilst the dissipation factor is a characteristic relaxation function of the material with maxima and minima at various frequencies.

Chemical resistance data

One of the major materials application criteria in plastics is the ability of a material to resist attack by chemicals associated with that environment. It is not sufficient only to define the normal service conditions but also to account for rare contingencies e.g. the use of detergents, organic cleaning fluids, etc. This is necessary because most common plastics are susceptible to almost immediate and irreversible damage by at least one group of domestic or industrial chemical.

The result of chemical attack may be one or a combination of the following; embrittlement, discolouration, crazing, cracking, disintegration, softening, swelling, loss of strength or complete or partial absorption of the plastic by the chemical. The rate at which these deteriorations proceed is not unnaturally proportional to chemical concentration, temperature and conditions of exposure (e.g. whether or not turbulent). The combination of externally applied stress or processing stresses also significantly modify the response to a chemical environment as in solvent stress cracking or crazing.

Table 16.11 assesses in a qualitative manner the reaction of most of the common plastics to crudely grouped classes of chemicals. These do not discriminate between grades, combination of additives, sensitivity to temperature or the user requirements. For

Table 16.11 WEATHER AND CHEMICAL RESISTANCE DATA

P = Poor   F = Satisfactory   G = Good

Plastic	Weather resistance	Acids dil.	Acids conc.	Alkali dil.	Alkali conc.	Oils and greases	Organic chemicals
Acetal	Chalks slightly	F	P	G	F	G	Generally good
Acrylic PMMA	Excellent	G	P	G	F	F	Poor
Acrylonitrile-Butadene-Styrene	Slight discoloura-tion + Embrittlement	G	F	G	F	G	Attacked by K, AH, CH
Cellulose Acetate	Slight to moderate deterioration	F	P	F	P	G	Resists non-polar chemicals
Cellulose Butyrate and Propionate	Resistant grades available	F	P	F	P	G	Poor
Cellulose Nitrate CN	Poor	F	P	F	P	F	
Chlorinated Polyether	Slight surface embrittlement	G except chromic	F	G	G	G	Attacked by some CH, but otherwise good
Epoxide Resin	Slight	G	F	G	F	G	Attacked by K, CH
Ethylene-vinyl Acetate EVA	Excellent when stabilised	G	F	G	F	G	Attacked by K, AH, CH
Fluorinated Ethylene Propylene FEP	Excellent	G	G	G	G	G	Excellent
Ionomer Resin	Poor to fair depending on formulation	G	P	G	G	G	Generally good
Melamine Formaldehyde MF	Fair	F	P	G	P	F	Generally good
Polyamides (nylon)	Slight discolouration	F	P	G	F	G	Generally good
Polycarbonate PC	Fair to good when stabilised	G	F	P	P	G	Attacked by K, AH, CH
Polyester Resin	Slight	G	F	P	P	G	Generally poor
Phenolic	Darkened by sunlight	F	F	P	P	G	Generally good
Polyester DMC	Slight yellowing	F	P	F	P	G	Attacked by K, CH
Polyethylene	Very poor unless stabilised	G	F	G	F	G	Attacked by AH, CH
Polyethylene Terephthalate	Embrittles	G	F	G	G	G	Generally good
Polyimide	Fair	F	P	P	P	G	Generally good
Poly(4-methyl-pentene-1)	Crazes if not protected	G	G	G	G	G	Attacked by K, AH, CH
Polyphenylene Oxide	Good	G	F	G	G	F	Attacked by K, AH, CH
Polypropylene	Crazes if not protected	G	F	G	F	F	Good at low temperatures
Polystyrene	Slight yellowing	G	F	G	G	F	Attacked by K, AH, CH
Polysulphone	Slight strength loss and swelling	G	G	G	G	G	Attacked by K, AH, CH
Polytetrafluoro-ethylene PTFE	Good	G	G	G	G	G	Excellent
Polyurethane	Slight yellowing	P	P	P	P	G	Generally poor
PVC	Fair to good	G	G	G	F	G	Poor
Silicones	Good	G	G	F	P	G	Moderately good
Urea Formaldehyde	Poor	F	P	F	P	G	Generally good

instance, discolouration may not be important in that the component will still function providing mechanical properties are not adversely affected. A poor (P) Classification is, therefore, not necessarily universally applicable. More comprehensive data[26, 27] can generally be obtained from material manufacturers.

The organic chemicals considered fall into the following groups: alcohols (A), ketones (K), aromatic hydrocarbons (AH) and chlorinated hydrocarbons (CH).

Weather resistance

Table 16.11 includes similarly qualitative information on the weather resistance of common plastics. The interaction between a naturally variable climate consisting of several important factors is complex. These factors include:

RADIATION

Ultra-violet radiation (3×10^{-7} m to 4×10^{-7} m) in sunlight is the most damaging factor to be considered. The photon energy at this wavelength is sufficient to disassociate many chemical bonds in polymerised chains. The rate at which deterioration occurs can be reduced considerably by the addition of u.v. stabilisers or opaque inducing fillers such as carbon black.

TEMPERATURE

Alternatively high and low temperature can lead to surface crazing or cracking initiated by expansion and contraction fatigue. This particularly applies to the more brittle plastics. High temperatures accelerate the rate of degradation in the presence of oxygen.

WATER

Polymers are almost completely insoluble in water. However, stabilisers, plasticisers and other additives which are invariably included in plastic compounds are often soluble. Thin films are especially susceptible to this form of attack.

COMMON PLASTICS

* indicates thermoset.)

ABS
See Acrylonitrile-Butadiene-Styrene

ACETAL RESIN

Both homopolymer and copolymer grades offer an excellent combination of strength, stiffness, and impact resistance. Dynamic fatigue endurance is also notably good. Coupled with dimensional stability and temperature resistance (up to 120°C) makes possible the use of the material in engineering components such as gears, bearings and bushes.

Chemical resistance is not outstanding, particularly in an acid environment. Also some deterioration is likely if exposed to u.v. radiation.

ACRYLICS

Moulding, casting, and extrusion grades are available, all of which offer excellent clarity and weather resistance. Extensively used in external glazing and light fitting applications e.g. sky-light mouldings, subsonic aircraft glazing and automobile reflectors.

Low impact strength and notch sensitivity generally restrict this material to lightly loaded application. Many common solvents and organic chemicals can cause severe deterioration.

ACRYLONITRILE—BUTADIENE—STYRENE (ABS)

Available in extrusion, high impact moulding grade, heat resistant moulding grade, and self extinguishing moulding grade. Similar to the acetals in that the material possesses a good combination of engineering properties. Used extensively in extruded pipe for complex shapes with good dimensional accuracy and stability.

ABS is attacked by strong acids and alkali and some hydrocarbons.

AMINO PLASTICS*

See Melamine and Urea formaldehyde.

CELLULOSE ACETATE

A low cost rigid material with good impact strength, clarity, or colourability. Main disadvantage is a poor dimensional stability caused by both plasticiser loss and water absorption. Chemical and weathering resistance is also not good. Typical application includes extruded tape, switches, knobs and tool handles.

CELLULOSE ACETATE BUTYRATE

Improved dimensional stability, heat and weather resistance compared with cellulose acetate. It is, however, more expensive. Applications include pen and pencil barrels, signal lenses and decorative trim.

CELLULOSE ACETATE PROPIONATE

Similar to the butyrate but with improved hardness and rigidity. Weathering resistance and impact strength is however reduced. Steering wheels and toothbrush handles are notable applications.

CELLULOSE NITRATE

Exhibits excellent dimensional stability and toughness but discolouration and embrittlement occurs when exposed to sunlight. Flammability is extremely high. Mainly used as indoor decorative accessories.

CHLORINATED POLYETHER

Noted particularly for its excellent chemical resistance and low temperature impact strength. The cost of the material, however, is high which restricts its use to specialised

nti-corrosive applications, namely pipe and liquid container linings. Care must be taken
ot to use in the presence of chromic acids or concentrated oxidising acids.

POXIDE RESINS AND COMPOSITES*

 very versatile engineering material. Essentially liquids, they are cured and crosslinked
y the reaction of catalytic hardeners. In this form they are used for encapsulation of
lectrical parts, adhesives, and surface coatings. It is extensively used in combination
ith glass fibre mat, cloth, and roving, to improve stiffness and strength properties.

THYLENE VINYL ACETATE EVA

 very flexible rubber-like plastic with excellent flex-crack resistance even at low tem-
eratures. Typical strain at break 750%. Many potential applications particularly in
ne food packaging industry, flexible tubing, bellows, etc.

LUORINATED ETHYLENE PROPYLENE FEP

ne of the most expensive of all plastics. Similar in properties to polytetrafluoroethylene
TFE.

NOMER RESINS

 tough material particularly relative to other thermoplastics at low temperatures.
esistance to foodstuffs, oils and greases is excellent combined with good sheet drawing
haracteristics explains its use in the packaging industry in the form of coated substrates.
equires internal protection on account of u.v. instability. Dust collection due to the
eneration of high electrostatic charges is a problem.

ELAMINE FORMALDEHYDE MF*

ommonly cellulose filled. Used in the production of moulded tableware and electrical
ccessories where heat resistance is required. The material is noted particularly for its
rface properties which exhibit high gloss and mar-resistance. Also used in the form of a
per or cotton laminate it is often used as an insulator in industrial environments.
 Disadvantages include low impact resistance, high notch sensitivity, and relatively
or dimensional stability.

RYL

modified Polyphenylene Oxide.

LONS

e Polyamides.

RSPEX

e Polymethyl Methacrylate.

PHENOL FORMALDEHYDE*

Unfilled phenolics are restricted in usage to adhesives, surface coatings and various
other sealing applications. Moulding formulations are available principally including
wood, flour, cotton, asbestos, or cellulose as a filler. It is also laminated with a variety
of fibrous materials to form sheet, rod, and bars which can be readily machined. Use
are numerous particularly in the electrical and light engineering industry.

PMMA

See Polymethyl Methacrylate.

POLYACETAL

See Acetals.

POLYAMIDES

A class of materials commonly known as nylons available as nylon 6, 6·6, 6·10, 11 and 1.
All nylons have excellent bearing qualities including low unlubricated friction coefficien
abrasion resistance, chemical resistance, resilience, and heat resistance. A major outle
for these materials is therefore to be found in gears, cams, brushes and slides. The
can be reinforced with glass fibre to improve stiffness and strength. The general purpos
grades are sensitive to water showing a change of mechanical properties and dimension:
stability. Nylon 12 is the least affected grade.

POLYCARBONATE

The toughest of all rigid thermoplastics combined with good dimensional stability
Optical clarity, although not up to the class of acrylics, is good. Apart from impac
strength, and clarity, all properties can be improved by short glass fibre reinforcemen
Applications include, terminal insulators, fuse boxes, safety helmets, air filter bowl
feeding bottles, and high output light fittings.
    Care must be taken to choose a u.v. stabilised grade for outdoor applications. Chemic:
resistance is generally only fair.

POLYESTER RESINS*

Unfilled polyester is generally not used in load bearing applications due to a low
resistance. Uses include, surface coatings, potting and encapsulation, and adhesives.
very versatile dough moulding compound (short fibre) is available which is noted f
low cost, rigidity and thermal stability. Often reinforced by fibres or lamination 1
produce boat hulls, chimney stacks, storage tanks, etc.

POLYETHYLENE

A low cost material with low tensile strength and stiffness. Higher density grades hav
improved load bearing (apart from impact) and heat resisting characteristics. General

ısed in applications where rough handling rather than high static loads are to be expected
ŧ.g. dustbins, blow moulded bottles, milk crates, and pipe liners.
Chemical resistance is generally good but often subject to environment stress cracking.
ᒪong term outdoor use is not recommended without u.v. stabilisation.

POLYETHYLENE TEREPHTHALATE

ᒍossess the qualities required of a bearing material namely good abrasion resistance,
ʟow friction coefficient and good dimensional stability. Chemical resistance is not ex-
ːeptional particularly in the presence of acids and steam.
Possible applications include, gears, rollers, and bearings for electrical machinery.

POLYIMIDES

Noted particularly for its ability to sustain loads at continuously high temperatures
250°C). Good combination of mechanical properties including stiffness, strength and
brasion resistance. Impact resistance however is not as good as many engineering
hermoplastics. Combined with a low friction coefficient these properties make the
 material attractive for special purpose bearing applications especially if graphite filled.
ᒍeneral applications may be ruled out on account of cost.

POLYMETHYL METHACRYLATE

ᒪee Acrylics.

POLY (4-METHYL—PENTENE—1) TPX

A low temperature material of good clarity and temperature resistance. Chemical
ᒪsistance is generally good although oxidising reagents, and surface active agents can
ᒪduce mechanical properties, or cause stress cracking. Impact strength is higher than
ᒪcrylics.
Uses include, high power light fittings, aircraft instrument fascias, measuring cylinders,
ᒪeakers and sterilisable medical ware.

POLYOLEFINS

ᒪee Polyethylene and Polypropylene.

POLYPHENYLENE OXIDE

ᒪvailable in a modified formulation. Notable for the retention of properties and dimen-
ᒪonal stability at working temperatures in the region of 100°C. Resistance to hot or
ᒪoiling water is excellent. Creep resistance is outstanding compared with other thermo-
ᒪastics. Chemical resistance is also good especially in contact with detergents.
Applications include hot water fittings and particularly washing machine parts.

POLYPROPYLENE

ᒪelatively inexpensive general purpose thermoplastic. Stiffness and strength properties
ᒪe not of a high order but, in general, are slightly superior in these respects to Polyethylene.

Flex and impact resistance is good and hence its uses in integral hinges, beer crates safety helmets, etc.

Ultra-violet resistance is poor and even with protection is not recommended fo permanent outdoor use. Adhesion is difficult due to a characteristic waxy surface Dimensional stability is comparatively poor so applications demanding close tolerance are not recommended.

POLYSTYRENE

The least expensive of all thermoplastics. Tensile strength and modulus are comparativel high but impact strength is low. Tougher grades are however available. Chemica resistance particularly to solvent attack is poor, as is resistance to u.v. radiation.

Applications exploit the materials' excellent processibility and clarity. Example include packaging toys, hypodermic syringes, ball point pens, etc.

POLYSULPHONE

Tough and rigid with good creep resistance. Toughness however is considerably reduce by notch sensitivity. Susceptible to stress cracking unless annealed. Retention of mech anical properties over continuous periods of high temperature is a notable qualit; Poor resistance to solvent attack.

Applications include, dishwasher parts, cooker control knobs, appliance handle and many electrical uses.

POLYTETRAFLUOROETHYLENE

Very similar in its chemical, mechanical and electrical properties to FEP and PCTFI but cheaper. Very low coefficient of friction. Often used with fibrous or metallic fille to increase rigidity and abrasion resistance. Unlike FEP it cannot be processed b conventional melt techniques and requires pre-forming as a powder and common sintering.

Applications include, pipe and flexible hose, mechanical seals, wire and cable insulatio bellows, utensil coatings, bushes and bearings.

POLYURETHANE

Available as a flexible or rigid foam, or an elastomer. Flexible foam is used extensive in bedding and furniture cushioning, often replacing latex rubber foam which is mo expensive. Rigid foam is used in applications demanding buoyancy, or thermal insulatic including boat building, and refrigerator units.

The elastomers are tough and resilient with good abrasion resistance. Uses inclu conveyor belting, shoe heels and gaskets.

POLYVINYL CHLORIDE (PVC)

The most widely used of all the thermoplastics. Versatility is accounted for by a wi choice of material stiffness ranging from rigid to plasticised grades. The latter is used upholstery, protective coverings, and coatings, rainware and numerous others. O notable disadvantage with this grade is the migration of plasticiser which eventua (accelerated at high temperatures) leads to brittleness and discolouration.

Rigid PVC has a good combination of strength, stiffness, and impact resistance but is not recommended for applications involving elevated temperatures. Chemical resistance is good. Uses include pipes, decorative cladding, gramophone records, bottles, etc. Calendering is by far the most common processing technique for both the flexible and rigid grades.

PPO

See Polyphenylene Oxide.

PTFE

See Polytetrafluoroethylene.

SILICONES

Noted for exceptional heat resistance (maximum service temperature 300°C) low mould shrinkage, chemical resistance and anti-adhesion properties. Used as a release agent in the encapsulation of active electronic devices. Silicone rubbers are often used to replace natural rubber formulations if extremes of temperature are envisaged. Weathering resistance is also superior.

STYRENE ACRYLONITRILE

Similar in most respects to polystyrene but with improved strength and rigidity. Solvent resistance though not exceptional is much improved compared with polystyrene.

TPX

See Poly (4-methyl-Pentene-1).

UREA FORMALDEHYDE*

Similar to melamine formaldehyde but with inferior temperature resistance. Chemical resistance is also inferior. Uses include domestic electrical fittings knobs and handles.

REFERENCES

1. BERNHARDT, E. C. (Ed), *Processing of Thermoplastic Materials*, Rheinhold (1959)
2. *Modern Plastics Encyclopedia*, Modern Plastics, (1969–70 or latest)
3. WALKER, J. S. and MATTIN, E. R., *Injection Moulding of Plastics*, Butterworth (1966)
4. FISHER, E. G., *Extrusion of Plastics*, 2nd ed., Butterworth (1964)
5. BUTLER, J., *Compression and Transfer Moulding of Plastics*, Butterworth (1959)
6. ELDEN, R. A. and SWAN, A. D., *Calendering of Plastics*, Butterworth (1971)
7. FISHER, E. G., *Blow Moulding of Plastics*, Butterworth (1971)
8. OGORKIEWICZ, R. M. (Ed), *Thermoplastics: Effects of Processing*, Iliffe (1969)
9. THIEL, A., *Principles of Vacuum Forming*, Iliffe (1965)

10. BUTLER, J., *A Moulder's Notebook*, British Industrial Plastics (1961)
11. BECK, R. D., *Plastic Product Design*, Van Nostrand Reinhold (1970)
12. Rubber and Plastics Research Association of Great Britain, Shawbury, Shrewsbury SY4 4NR
13. British Plastics Federation, 47 Piccadilly, London W1V ODN
14. FARAGO, P. J. (Ed), *Plastics Handbook*, Product Journals (1968)
15. *Plastics Materials Guide*, British Plastics, (January 1970)
16. HALL, M. M., 'An Approach to Material Selection', *British Plastics*, **44**, No. 10, 85 (1971)
17. Society of the Plastics Industry Inc., *Plastics Engineering Handbook*, 3rd ed, Chapman and Hall (1960)
18. PYE, R. G. W., *Injection Mould Design*, Butterworth (1968)
19. BEBB, R. H., *Plastics Mould Design: Vol. 1 Compression and Transfer Moulds*, Butterworth
20. *Application Design*, (E. I. du Pont de Nemours) (1961)
21. BAER (Ed), *Engineering Design for Plastics*, Chapman and Hall (1964)
22. OGORKIEWICZ, R. M. (Ed), *Engineering Properties of Thermoplastics*, Wiley-Interscience (1970)
23. NEUMANN, J. A. and BOCKHOFF, F. J., *Welding of Plastics*, Reinhold (1959)
24. GOLDIE, W., *Metallic Coating of Plastics*, Electro-chemical Publications (1969)
25. ABRAHAMS, M. and DIMMOCK, J., *Mechanical and Economic Comparisons of Reinforced Thermoplastics*, Plasticised Polymers, **39**, No. 141, 187 (1971)
26. *RAPRA Data Sheets* (*Chemical*), See Reference 12 for address
27. RAPRA *Ageing and Weathering of Plastics*, see Reference 12 for address
28. EISENBERG, A. and SHEN, M., *Rubber, Chem. Technol.*, **43**, No. 1, 156 (1970)
29. RUBBER REVIEWS, **36**, No. 5, 1303 (1963)
30. ANDREWS, E. H., *Fracture in Polymers*, Oliver and Boyd Ltd., 9 (1968)
31. ASTM C177–63
    Test for thermal conductivity of materials by means of the guarded hot plate (1968)
32. ASTM D648–56
    Test for deflection temperature of plastics under load (1961)
33. ASTM D149–64
    Tests for dielectric breakdown voltage and dielectric strength of electrical insulating materials at commercial power frequencies (1970)
34. ASTM D150
    Tests for A.C. loss characteristics and dielectric constant (permittivity) of solid electrical insulating materials (1970)
35. ASTM D257
    Tests for D.C. resistance or conductance of insulating materials (1966)
36. ASTM D638
    Tests for tensile properties of plastics (1968)
37. ASTM D696
    Test for coefficient of linear thermal expansion of plastics (1970)
38. ASTM D785–65
    Test for Rockwell hardness of plastics and electrical insulating materials (1970)
39. ASTM D790
    Test for flexural properties of plastics (1970)
40. ASTM D2240
    Test for indentation hardness of rubber and plastics by means of a durometer (1968)
41. BS 2026
    Tolerances for mouldings in thermosetting materials (1953)
42. BS 2782
    Methods of testing plastics (1970)
43. BS 4042
    Specification for a system of dimensional tolerances for small injection mouldings in rigid and semi-rigid thermoplastic materials (1966)
44. BS 4618
    Recommendations for the presentation of plastics design data (1970)

U.K.-BASED MANUFACTURERS AND SUPPLIERS OF PLASTICS

ABS

Anchor Chemical
Aquitane-Organice
BASF (UK)
Bayer Chemicals
BXL Plastics

Caplin
Cole Plastics
Consort Plastics
Curtis

Delpha Assoc.
Dow Chemical

Huls (UK)

Jupiter Plastics

Marbon
Monsanto
Muehlstein-Northwestern

Polymer (UK)

SIC Plastics
A. Schulman
Sterling Moulding Materials

Tar Residuals

Uniroyal

Joseph Weil.

Du Pont

East Anglia Plastics

ICI

Lennig

Joseph Weil

ALKYD MOULDING POWDERS

Anglo-Austrian

Bayer
BP Plastics
BXL Plastics
British Celanese

Cole Plastics

East Anglia

Hercules Powder

M & B Plastics

SIC Plastics

Troviplast

Victor Plastics

ACETAL COPOLYMER

BIP

Hoechst

ICI

ACETAL HOMOPOLYMER

Du Pont

ACRYLICS

Bayer Chemicals

Cornelius Chemical

CELLULOSE ACETATE BUTYRATE

Bayer

Jupiter Plastics

M & B Plastics

Telcon

CELLULOSE PROPIONATE

Bayer

CHLORINATED POLYETHER

Hercules Powder

DAP

Dow

Huls (UK)

Kingsley & Keith

Resinous Chemicals

EPOXIDE RESINS

Anglo Austrian

Bayer Chemicals
BXL Plastics
Borden Chemical

Ciba (ARL)

Dow Chemical

Emmersons

Grilon & Plastic Machinery

Hysol Chemicals UK

Shell Chemicals UK

EPOXIDE MOULDING COMPOUNDS

Anglo-Austrian

Ciba (ARL)

Dohm Plastics
Dow Chemical

Grilon & Plastic Machinery

Hysol Sterling

Shell Chemicals

Kingsley & Keith

Sterling Moulding Materials

Telcon

EVA

BASF
BXL
BIP

Caplin
Cole Plastics

Du Pont

Greeff, R. W.

ICI

Monsanto
Muehlstein-Northwestern

US Industrial Chemicals

FLUOROCARBONS—FEP

Du Pont

FLUOROCARBONS—PTFE

Anglo-Austrian

Du Pont

Hoechst

ICI

Kingsley & Keith

Maflon
M. & B. Plastics

Plastic Coatings
Polypenco

Turner Bros Asbestos

Joseph Weil

IONOMERS

Du Pont

MBS

SIC Plastics

MELAMINE FORMALDEHYDE

BIP

Ciba (ARL)
Curtis

ICI

Sterling

Tar Residuals

Joseph Weil

PHENOLICS

Associated Electrical Industries

BP Chemicals
Beck Koller
Birkbys
Borden Chemical
British Resin Products
BXL Plastics Materials

CIBA

Featley Products
Ferguson

ICI

Moulding Powders

Sterling

Troviplast
Turner Bros.

POLYALLOMERS

Anchor
Aquitane-Organico (UK)

BASF (UK)
Bayer Chemical
BIP

Grilon & Plastic Machinery
Huls (UK)

ICI

M & B Plastics
Muehlstein-Northwestern

Polypenco

Joseph Weil

POLYAMIDE 6.6

Anchor

BASF (UK)
BIP
British Chemical Products

Du Pont

ICI

M & B Plastics
Monsanto
Muehlstein-Northwestern

Polypenco

POLYAMIDE 6.10

BASF (UK)

Du Pont

ICI

M & B Plastics
Mobay Plastics

Polypenco

POLYAMIDE 11

Aquitane-Organico (UK)

POLYAMIDE 12

Aquitane-Organico UK

Grilon & Plastic Machinery

Huls (UK)

POLYCARBONATE

Bayer Chemical

POLYESTER DOUGH MOULDING COMPOUNDS

BP Chemicals
BIP Chemicals

BXL Plastics
BRP

Freeman Chemicals

R. W. Greeff
Resinous Chemicals
Spencer Knight

POLYESTER RESINS

Anglo-Austrian
Artrite
Astor Boisselier

BP Chemicals
BIP Chemicals
BXL Plastics
Beck, Koller
BRP

Cellon

Freeman Chemicals

Greeff

Honeywill Atlas
Huls (UK)

ICI

Mitchell & Smith

Pinchin Johnson
Plastanol

Resinous Chemicals

Scott Bader

Tar Residuals

United Coke & Chemicals

Joseph Weil
Witco Chemical

POLYETHYLENE (H.D.)

Anglo-Austrian

British Hydrocarbon
BRP
BASF (UK)
BIP

Caplin
Cole
Curtis

Dow

East Anglia

Greenham

Hellyar
Hoechst
Huls (UK)

Kaylis
Kingsley & Keith

Muehlstein-Northwestern

Phoenix Rubber
Phillips
Polypenco

Shell Chemicals
A. Schulman

Tar Residuals

US Industrial Chemicals

Victor

Joseph Weil

POLYETHYLENE (L.D.)

Anglo-Austrian
Aquitane Organico (UK)

BASF (UK)
BIP
British Chemical Products
BP Chemicals
BXL Plastics

Caplin & Co.
Cole
Curtis

Dow
Du Pont

East Anglia Plastics

Hellyar

ICI

Kaylis
Kingsley & Keith

Monsanto
Muehlstein-Northwestern

Plastic Coatings
Polypenco

Shell
A. Schulman

Tar Residuals
Telcon

US Industrial Chemicals

Joseph Weil

**POLYETHYLENE TEREPHTHALATE
MOULDING POWDER**

BIP

POLYIMIDES

Du Pont

**POLY (4-METHYL
PENTENEY)**

ICI

PPO-MODIFIED

Kingsley & Keith

POLYPROPYLENE

BASF (UK)'

Caplin

Dow

Eastman Chemical
East Anglia Plastics

Hoechst
Huls (UK)

ICI

Jupiter Plastics

May & Baker
Muehlstein-Northwestern

Shell
A. Schulman

Joseph Weil

POLYSTYRENES

Anglo-Austrian
Aquitane Organico UK

BASF (UK)
Bayer Chemical
BP Plastics
BXL Plastics

Caplin
Cole
Curtis

Delpha
Distrene
Dow

East Anglia

Hellyar
Hoechst
Huls (UK)

Industrial Polymers

Kaylis

Monsanto
Muehlstein-Northwestern

Polypenco

Shell
A. Schulman
SIC
Sterling Moulding

Victor

Joseph Weil

POLYSULFONES

BXL Plastics

POLYURETHANES

Anchor Chemical

Bayer Chemical

Dow Chemical

Elastomer Products

Hysol Sterling

ICI

Jacobson van den Berg

Witco Chemical

POLYVINYL CHLORIDE
POLYMER

Anglo-Austrian
Aquitaine Organico UK
Ault & Wiborg

BASF (UK)
BP Chemicals
BXL Plastics Materials
Britona
British Geon

Caplin
Cole

Dow

East Anglia
Elson & Robbins

Hellyar
Hoechst
Huls (UK)

ICI

M & B

Norsk Hydro (UK)

Phoenix Rubber
Plastic Coatings
Polypenco
Scott Bader
Shell
SIC Plastics
Sterling Moulding

Tar Residuals
Troviplast

Victor
Vinatex
Vinyl Compositions

Joseph Weil

UREA FORMALDEHYDE

Anglo-Austrian

BIP Chemicals

CIBA

Ferguson

ICI

Joseph Weil

U.K.-BASED SUPPLIERS' NAMES AND ADDRESSES

Aeropreen Ltd.
Lindsay Avenue, High Wycombe, Bucks.
Anchor Chemical Co. Ltd.
Clayton, Manchester 11
Anglo Austrian Trading Co. Ltd.
1–11 Hay Hill, London W1
Aquitaine-Organico
Colthrop Lane, Thatcham, Newbury, Berks.
Artrite Resins Ltd.
Stanhope Road, Camberley, Surrey
Associated Electrical Ind. Ltd.
Mill Road, Rugby
Astor Boisselier & Lawrence Ltd.
9 Savoy Street, London WC2
Ault & Wiborg Ltd.
Cow Lane, Watford
Bakelite Xylonite Ltd.
Manningtree, Essex
The Baxenden Chemical Co. Ltd.
Paragon Works, Baxenden, Nr. Accrington, Lancs.
Beck, Koller & Co. Ltd.
Research and Technical Service Labs., North Site, Speke, Liverpool 24
B.I.P. Chemicals Ltd.
P.O. Box 6, Popes Lane, Oldbury, Warley, Worcs.
Birkbys Ltd.
P.O. Box 2, Liversedge, Yorkshire
Borden Chemical Co. (UK) Ltd., (The)
North Baddesley, Southampton, Hampshire
B.P. Plastics Ltd.
West Halkin House, West Halkin Street, London SW1
Britiona Ltd.
Mulkern Road Works, London N19
British Celanese Ltd.
Chemicals & Plastics Group, 345 Foleshill Road, Coventry, Warwicks
British Chemical Products & Colours Ltd.
Buckingham House, 6 Buckingham Street, London WC2
British Geon Ltd.
Devonshire House, Piccadilly, London W1
British Hydrocarbon Chemicals Ltd.
Devonshire House, Piccadilly, London W1
British Resin Products Ltd.
Devonshire House, Mayfair Place, Piccadilly, London W1
BTR Industries Ltd.
Silvertown House, Vincent Square, London SW1
BXL Plastics Materials Group Ltd.
12–18 Grosvenor Gardens, London SW1
Caplin Co. (Plastics) Ltd.
Third Avenue, Bletchley, Bucks.
Cellon Div. of Pinchin Johnson & Assoc. Ltd.
Richmond Road, Kingston-on-Thames, Surrey
Ciba (A.R.L.) Ltd.
Duxford, Cambridge
Cole R. H. Plastics Ltd.
Batsford Mill, Lower Luton Road, Harpenden, Herts.
Consort Plastics Ltd.
9 Wimpole Street, London W1

Coolag Ltd.
P.O.Box 3, Charlestown, Glossop, Derbyshire
Coopers Plastic Foams Ltd.
Stamford Works, Crompton Street, Ashton-under-Lyne, Lancs.
Cow, P. B. (Plastics) Ltd.
5 Falmouth Road, Trading Estate, Slough, Bucks.
Croid Ltd.
Berkshire House, 168–173 High Holborn, London WC1
Curtis Plastics Ltd.
Cradock Road, Luton, Beds.
Declon Foam Plastics Ltd.
Cranborne Road, Potters Bar, Herts.
Delfa Associates Ltd.
Belco House, Park Royal Road, London NW10
Distrene Ltd.
Devonshire House, Piccadilly, London W1
Dohm Ltd.
167 Victoria Street, London SW1
Dow Chemical Co. (UK) Ltd.
105 Wigmore Street, London W1
Du Pont Co. (UK) Ltd.
18 Bream's Buildings, Fetter Lane, London EC4
East Anglia Plastics Ltd.
Knight Road, Strood, Kent
Eastman Chemical International AG
246 High Holborn, London WC1
Elastomer Products Ltd.
275 King Street, London W6
Elson & Robbins Ltd
Bennett Street, Long Eaton, Nottingham
Expanded Rubber & Plastics Ltd
Mitcham Road, Croydon, Surrey
Featley Products Ltd.
Farrell Street, Manchester 7
Ferguson, James, & Sons Ltd.
Prince George's Road, London SW19
Fibre Form Ltd.
Garratt Mills, Trewint Street, Earlsfield, London SW18
Foley Packaging & Insulation Ltd.
Hardy Road, Farlington, Portsmouth
Freeman Chemicals Ltd.
Ellesmere Port, Wirral, Cheshire
Grilon & Plastic Machinery Ltd.
Leader House, 119–120 Snargate Street, Dover, Kent
Harrison & Jones (Flexible Foam) Ltd.
Bee Mill, Shaw Road, Royton, Oldham, Lancs.
Hellyar, John, & Co. Ltd.
Colewood Road, Industrial Estate, Swalecliffe, Whitstable, Kent
Hercules Powder Co. Ltd.
1 Gt. Cumberland Place, London W1
Hoechst, UK Ltd.
Hoechst House, Kew Bridge, Brentford, Middx.
Honeywill Atlas Ltd.
Mill Lane, Carshalton, Surrey
Huls (UK) Ltd.
41 Dover Street, London W1
Hysol Sterling Ltd.
Sterling House, Heddon Street, London W1

ICI Dyestuffs Div.
Hexagon House, Blackley, Manchester 9
ICI Plastics Div.
Bessemer Road, Welwyn Garden City, Hertfordshire
Jablo Group Sales Ltd.
Monsanto House, 10–18 Victoria Street, London SW1
Jacobson Van Den Berg & Co. (UK) Ltd.
Marketing House, 2–4 Richbell Place, London WC1
Kay Bros, Plastics Ltd.
Waterhouse Mill, Bollington, Macclesfield, Cheshire
Kaylis Chemical Ltd.
Weston Street, Bolton, Lancs.
Kingsley & Keith (Chemicals) Ltd.
73–76 Jermyn Street, London SW1
Lankro Chemicals Ltd.
P.O. Box 1, Eccles, Manchester
Lenning Chemicals Ltd.
26–28 Bedford Row, London WC1
Maflon Ltd.
Swallowfields, Ditching Road, Keymer, Sussex
Mitchell W. A. & Smith Ltd.
Church Path, Church Road, Mitcham, Surrey CR4 3YE
Monsanto Chemicals Ltd
10–18 Victoria Street, London SW1
Moulding Powders Ltd
Lamberhead Industrial Estate, Wigan, Lancs.
Muehlstein-Northwestern Ltd.
7 Cork Street, London W1
Norsk-Hydro (UK) Ltd.
2 Holly Road, Twickenham, Middx.
Pfizer Ltd.
Ramsgate Road, Sandwich, Kent
Phillips Petroleum UK Ltd.
Portland House, Stag Place, London W1
Phoenix Rubber Co. Ltd.
Buckingham Avenue, Trading Estate, Slough, Bucks.
Pinchin Johnson Paints
380 Richmond Road, Kingston upon Thames, Surrey
Plastanol Ltd.
Crabtree Manorway, Belvedere, Kent
Plastic Coatings Ltd.
Industrial Estate, By-pass, Guildford, Surrey
Polymer (UK) Ltd.
Yorkshire House, Chapel Street, Liverpool 2
Polypenco Ltd.
Gate House, Fretherne Road, Welwyn Garden City, Herts.
Pritex (Plastics) Ltd.
Wellington, Somerset
Resinous Chemicals
Portland Road, Newcastle-on-Tyne 2
Scott Bader & Co. Ltd.
Wollaston, Wellingborough, Northants.
Shell International Chemicals Ltd.
Shell Centre, Downstream Building, London SE1
S.I.C. Plastics Ltd.
58–64 High Street, Epsom, Surrey
Spencer Knight & Co. Ltd.
Britannia Mill, P.O. Box 4, Mirfield, Yorks.

Sterling Moulding Materials Ltd.
Sterling House, Heddon Street, London W1
Telcon Plastics Ltd.
Farnborough Works, Green Street Green, Orpington, Kent
Thermalon
Berkeley Square, London W1
Turner Bros. Asbestos Co. Ltd.
P.O. Box No. 40, Rochdale
Troviplast Ltd.
Greener House, 66–68 Haymarket, London SW1
Union Carbide Ltd.
8 Grafton Street, London W1
Uniroyal Ltd.
Chemicals Division, Stoke Works, Bromsgrove, Worcester
United Coke & Chemicals Co. Ltd.
P.O. Box 136, Handsworth, Sheffield 13
U.S. Industrial Chemicals Co. International
12–16 High Street, Walton-on-Thames, Surrey
Victor Plastics (Mfg) Ltd.
67 Westbourne Grove, London W2
Vinatex Ltd.
Mill Lane, Carshalton, Surrey
Vinyl Compositions Ltd.
Grimshaw Lane, Bolington, Nr. Macclesfield, Cheshire
Joseph Weil & Son Ltd.
39–41 New Broad Street, London EC2

SOCIETIES AND INSTITUTIONS

American Society of Tool and Manufacturing Engineers
20501, Ford Road, Dearborn, Michigan, 48128, U.S.A.
Board of Trade
1 Victoria Street, London SW1
British Hydromechanics Research Association
Cranfield, Beds.
British Numerical Control Society
10 Chesterfield Street, London WIX 8DE
British Standards Institution (Sub Committee MEE/6/25 and technical committee DPE/19)
2 Park Street, London W1
Centre D'Etudes et de Recherches de la Machine Outil
150 Boulevard Bineau, Neuilly-sur-Seine, France
Electronic Industries Association
2001, Eye Street, N.W. Washington DC 20006, U.S.A.
Institution of Electrical Engineers
Savoy Place, London WC2
Institution of Mechanical Engineers
1 Birdcage Walk, London SW1
International Organisation for Standardisation,
1 Rue de Varembe, Geneva, Switzerland
Machine Tool Industry Research Association
Hulley Road, Hurdsfield Macclesfield, Cheshire
Machine Tool Trades Association
25 Buckingham Gate, London SW1
National Engineering Laboratories
East Kilbride, Glasgow
Production Engineering Research Association
Melton Mowbray, Leicestershire
Society of Automotive Engineers
485 Lexington Avenue, New York 10017, U.S.A.

17 INDUSTRIAL GASES

INDUSTRIAL GASES 17

17 INDUSTRIAL GASES

F. G. WHITE

Air, in its natural form, and the separate gases of which it is composed are of great importance in industry. In factories and shipyards, compressed air is necessary in many operations such as drilling, grinding, chipping, riveting, painting, lifting and other services.

The concentated power which can be developed and the ease of control make compressed air particularly convenient for the operation of portable tools and equipment. The working pressure at the tools is usually between 7 and 10 bar (gauge) and there is a wide choice of compressors in standard ranges, available for the duty. Small to medium capacities are covered by both rotary and reciprocating machines; very large capacities are suitable for the dynamic types. The three basic types differ very much in general design and in the actual mechanics of compression. Some, particularly those with fixed cut-off, may vary from the ideal if used under non-standard conditions but the basic laws of thermodynamics are generally applicable to all.

COMPOSITION AND PRESSURE OF THE ATMOSPHERE

Atmospheric air is a mixture of gases varying only slightly according to the locality. Its main constituents and approximate proportions are Nitrogen 78%, Oxygen 21%, Argon nearly 1%, with traces of carbon dioxide, helium and other gases. A further constituent of importance to the compressor designer and user is water vapour. The pressure exerted by the atmosphere and consequently, its density, is affected by temperature, humidity and altitude. The last is a characteristic of the geographical situation while the two former factors can vary widely from day to day or even hourly. For a meaningful assessment of the relative efficiencies of different makes and types of compressor, it is therefore necessary to specify standardised conditions at the compressor inlet and to make corrections to test figures if necessary.

The ISO 'Standard Reference Atmosphere'[1] specifies:

Pressure 1013 mbar; Temperature 20°C; R. Humidity 65%.

whereas another standard gives

Pressure 1013·25 mbar; Temperature 15°C, Dry.

Since there is, at present, no universally accepted standard, it is advisable, where guarantees are concerned, to specify in detail the intake conditions of pressure, temperature and humidity to which the guarantee applies.

Where conditions on test are not too widely different from those specified, correction factors are applicable. It will be clear that pressure and temperature have a direct effect upon the volume of free air drawn into the machine. Humidity increases the volume of the free air, and, as most of it is condensed and extracted as water, this represents a shrinkage of useful volume and in some cases, a measurable waste of power.

SYMBOLS AND UNITS

Use of symbols

In this section the following symbols will be used.

P_1, P_2, initial and final absolute pressure.
V_1, V_2, initial and final volumes.
T_1, T_2, initial and final absolute temperatures, Kelvin.
n, compression index, usually the ratio of specific heats, 1·41 for air under normal conditions of pressure and temperature.
mep, Mean effective pressure, referred to intake conditions.
$\log_e$, Hyperbolic logarithm.

Choice of units

For the sake of clarity in exposition and to take advantage of the coherence of the SI system, the following units are used.

$$\text{Rate of Flow,} \quad m^3/s$$
$$\text{Pressure,} \quad N/m^2$$
$$\text{Power,} \quad \text{watts.} \left(\frac{m^3}{s} \times \frac{N}{m^2} = \frac{Nm}{s} = W \right)$$

Notes. Substituting the pascal for N/m^2 leaves the power equations unchanged, since

$$1 \text{ Pa} = 1 \text{ N/m}^2.$$

Substituting kN/m^2 or kPa gives power directly in kW.
Using the bar for pressure introduces a factor of 10^5, since 1 bar $= 10^5$ N/m^2.

THEORY OF COMPRESSION

Although air is not a perfect gas, it follows closely enough to the standard physical laws to enable these to be safely used for calculations under most normal conditions. When temperature remains constant, pressure and volume are related by the law

$$P \times V = \text{Constant} \tag{1}$$

P and V may be in any self-consistent units but it must be remembered that in this expression and in all calculations relating to the compression of air, P means the total pressure i.e. the gauge reading plus the atmospheric pressure. This total pressure is usually referred to as the absolute pressure meaning the pressure above absolute zero.

Since $P.V$ is constant, a reduction of volume brings a proportionate increase of pressure

$$\frac{P_2}{P_1} = \frac{V_1}{V_2} \tag{2}$$

If the final temperature is higher than the initial, the pressure is further increased in accordance with the law

$$\frac{P_2}{P_1} = \frac{V_1 T_2}{V_2 T_1} \tag{3}$$

Since compression increases the activity of the molecules, it causes the temperature to rise and, in order to achieve constant temperature (isothermal) compression, it reasonably possible to provide sufficient cooling to maintain the air inside the compressor at a constant temperature and a good deal of the heat of compression is carried away with the air as it leaves the machine. Although the ideal of isothermal compression can rarely be achieved, the theoretical isothermal power consumption forms a useful standard when comparing efficiencies. It may not be obvious at first but the normal compressor carries out two separate operations:

1. The compression of the air to the desired working pressure.
2. The expulsion of the volume of compressed air against the back pressure in the delivery main.

Since these functions cannot very easily be separated in practice, they are combined in the equations relating to power. During the process of isothermal compression and delivery, the mean effective pressure (mep) against which the compressor has to operate is given by

$$\text{mep} = P_1 \log_e \frac{P_2}{P_1} \qquad (4)$$

The power absorbed is the product of mep and volume, in suitable units. Using newtons per square metre and cubic metres per second,

$$\text{Power absorbed (watts)} = \text{mep} \times V_1 \qquad (5)$$

The term 'adiabatic' has been in use for many years to describe the process of compression in which all heat is retained. For technical reasons, the expression isentropic' is preferred in connection with dynamic compressors.

With all heat retained, compression will raise the temperature in accordance with

$$\frac{T_2}{T_1} = \left(\frac{P_2}{P_1}\right)^{(n-1)/n} \qquad (6)$$

For power calculations,

$$\text{mep} = \frac{n}{n-1} \times P_1 \left\{ \left(\frac{P_2}{P_1}\right)^{(n-1)/n} - 1 \right\} \qquad (7)$$

and, using the same units as in equation (5),

$$\text{Power absorbed (watts)} = \text{mep} \times V_1/s, \text{ as before.}$$

Using equation (6) it can be calculated that for a working pressure of 8·5 bar (gauge), the theoretical temperature of compression would be, with all heat retained, over 275°C or even higher if frictional heat is included. This might cause difficulties with the oil and with some of the small highly stressed parts. Small machines, and slow running types have a high ratio of cylinder surface to aspired volume which gives them an advantage in getting rid of some of the heat. Forced draught air cooling and water jackets also help, and many machines with cylinders of small diameter operate regularly against a pressure of 8·5 bar (gauge) or even higher, in a single stage.

With modern high speed machines, as the size increases, the cooling surface of the cylinder has less effect upon the temperature of the air while, at the same time, it becomes increasingly important to achieve good efficiency. In oil flooded rotary machines and some other types and also in certain reciprocating machines, a quantity of cooling fluid is injected into the compression space along with the air or gas. The fluid takes up heat directly from the air, reducing the working temperatures and also the net power required for compression.

An alternative method, widely used, is to carry out the compression in stages, the air being cooled back to, or near to, its initial temperature between the stages. In the design of two-stage machines, the object is to divide the work equally between the two stages and thus have equal temperatures. Since T is a function of P_2/P_1, the ideal compression ratio in each stage will be $\sqrt{(P_2/P_1)}$ and the theoretical temperature rise less than half the single stage figure.

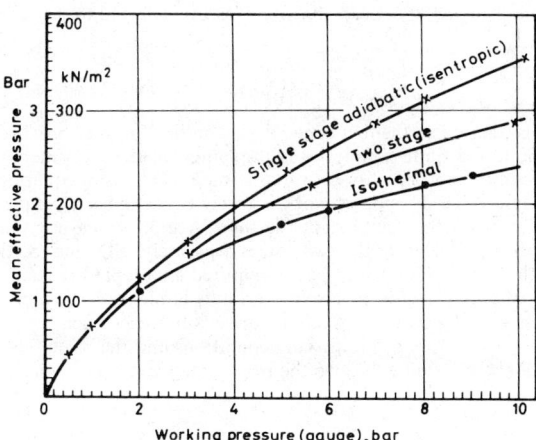

Figure 17.1. Mean effective pressure for air compression, 0–10 bar

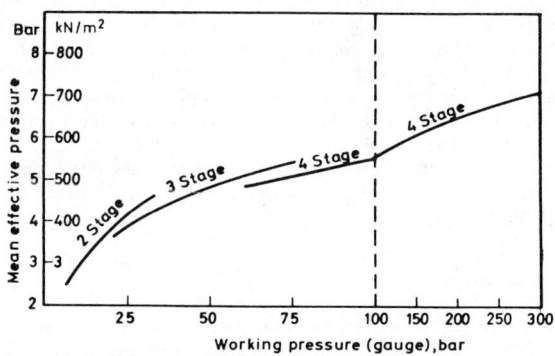

Figure 17.2. Mean effective pressure for multi-stage adiabatic compression of air to higher pressures

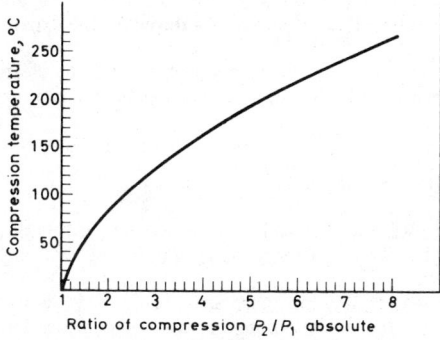

Figure 17.3. Temperature of air compression

For the whole machine, the mean effective pressure, referred to the first stage is given by:

$$2\frac{n}{n-1}\times P_1\left\{\left(\frac{P_2}{P_1}\right)^{(n-1)/2n}-1\right\} \tag{8}$$

Since the mep is referred to the first stage, the theoretical power absorbed by the whole machine can be found quite simply by multiplying together this mep and the intake volume of the compressor, i.e. Power (watts) $= $ mep $\times V_1/s$, as in equation (5).

In actual practice, the division of work is rarely exact and the temperature between stages is usually different from that of the inlet. A more exact computation may be made, if desired, by calculating the two stages separately, allowing for these factors.

To convert theoretical power into power required at compressor shaft, it is necessary to include an efficiency factor to cover thermodynamic and air friction losses, also mechanical losses and power required to drive auxiliaries such as oil pumps. The efficiency varies, according to the working conditions and the basic type and design of machine. The makers' catalogues give the best guidance for particular cases.

VOLUMETRIC EFFICIENCY

In a normal type of positive displacement compressor, there is a difference between the volume swept by the first stage piston (or other displacement unit) and the actual volume of atmospheric air retained, compressed and delivered by the machine. The ratio between these two is known as the volumetric efficiency. Although, in extreme cases, the volumetric efficiency may have an effect upon the overall performance, there is no direct connection and a high VE is not necessarily associated with a high overall efficiency. Other factors are involved, as detailed by White[2].

The reasons why the volumetric efficiency is normally below 100% may be considered under two headings and will perhaps be easier to follow if related first of all to a reciprocating compressor.

Effect of clearance spaces

When the piston reaches the end of the compression and delivery stroke, it does not actually touch the cylinder cover; a small clearance remains. There are also pockets under the valves which are not swept by the piston. All these together form the clearance space which remains charged with compressed air when the delivery valve closes. As the piston returns down the cylinder, the air in these pockets expands and the inlet valve does not open until the pressure has dropped to that of the atmosphere or a little below. In other words, the effective stroke is reduced by the expansion of the trapped air.

The work done in compressing this fraction of air into the clearance space at each stroke is not entirely lost since during its expansion the air returns some work to the piston.

Leakage and thermodynamic losses

The quantity of air drawn into the machine is further reduced by throttling and friction in passageways and by contact with hot surfaces also by internal leakage. All these represent wasted power.

The total of these losses may, in a particular case, be as low as 15% or as high as 40%, depending to a large extent upon the ratio of compression in the first stage cylinder, giving a volumetric efficiency of 85% or 60% respectively.

Although the previous descriptions are particularly related to a reciprocating compressor, the actual processes of re-expansion, throttling, air friction and internal leakage occur together or separately in many types of positive displacement rotary.

Effect of altitude

The pressure of the atmosphere varies according to the altitude above (or below) sea level.

This has two effects upon the compressor:

(i) Where it increases the compression ratio in the first stage, it will reduce the

Table 17.1   SOME USEFUL FORMULAE (adapted from *Industrial Air Compressors*)

Isothermal (constant temperature)

$$P_1 V_1 = P_2 V_2$$

$$\frac{P_2}{P_1} = \frac{V_1}{V_2}$$

$$\text{mep} = P_1 \log_e \frac{P_2}{P_1}$$

Theoretical power = mep × V_1 (rate of flow)

Adiabatic or Isentropic (all heat retained)

$$\frac{P_1 V_1}{T_1} = \frac{P_2 V_2}{T_2}$$

$$\frac{P_2}{P_1} = \left(\frac{V_1}{V_2}\right)^n$$

$$\frac{V_1}{V_2} = \left(\frac{P_2}{P_1}\right)^{1/n}$$

$$\frac{T_2}{T_1} = \left(\frac{P_2}{P_1}\right)^{(n-1)/n}$$

$$\text{mep} = \frac{n}{n-1} \times P_1 \left[\left(\frac{P_2}{P_1}\right)^{(n-1)/n} - 1\right]$$

Theoretical power = mep × V_1 (rate of flow), see 'Symbols and Units'.

Table 17.2   EFFECT OF ALTITUDE ON COMPRESSOR DELIVERY

Altitude, m	250	500	750	1000	1500	2000	2500
Useful output %	97	94·5	91·5	89	83	78	73

Note. At high altitudes, there may be a further reduction due to a change in volumetric efficiency.

Table 17.3   EFFECT OF INLET TEMPERATURE ON COMPRESSOR DELIVERY. BASED ON STANDARD INLET OF 20°C

Inlet temperature °C	0	10	20	30	40	50	60
Useful output %	107	103·5	100	96·5	93·5	91	88

volumetric efficiency. Single-stage machines are usually more sensitive to altitude than two-stage, in this respect.

(ii) As the altitude above sea level increases, the atmospheric pressure decreases and so also does the density. This means that a greater volume of free air is needed to perform a given duty, if the operating pressure is constant, as shown on the pressure gauge.

PRACTICAL COMPRESSORS

The Pneurop international classification of compressors[3] lists a total of sixteen basic designs offered for low and medium air or gas delivery pressures. They may be grouped, according to their operating principles, into three fundamental types:

Reciprocating, in which air is compressed by a piston travelling to and fro in a close-fitting cylinder with inlet and outlet ports or valves to control the flow.

Positive rotary, in which the air is driven forward mechanically by a rotary piston or some other close-fitting rotary unit. In some types, there is more than one rotor.

Dynamic rotary, in which air passes through radial impellers and is compressed partly by centrifugal force, partly by velocity conversion or, in an axial flow type, is accelerated by turbine type blades and has its pressure raised by subsequent conversion of kinetic energy. In the medium pressure models it is usual to have a number of stages of compression.

Reciprocating compressors

One popular form of compressor for use in factories and out-of-doors, is the single-acting reciprocating type.

In a simple form of this design, a lightweight piston of the type used in an internal combustion engine is reciprocated by a crank and connecting rod or other appropriate means. The surface area of the piston is sufficient to deal with the side thrust of the

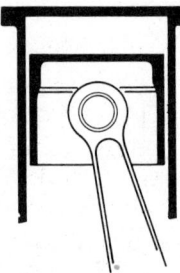

Figure 17.4. Single-acting cylinder, with connecting rod attached directly to piston

drive so that an additional crosshead and guide is not required. The piston is a close fit in an accurately finished cylinder bore and, in order to minimise leakage, is commonly fitted with metallic rings.

During the 'inward' stroke, when the piston is travelling towards the crank, air is admitted to the cylinder by an automatic non-return valve in the cylinder head. During the return stroke, the air already admitted is compressed in the cylinder and finally discharged into the delivery main, through a second non-return valve, operating in the reverse direction. Since there is a considerable variation in resistance during each revolution, a flywheel is usually fitted to the crankshaft to even out the turning moment. The flywheel is often used as a pulley for belt drive.

As many of these machines run at moderately high speeds, practical considerations limit the quantity of air which can be dealt with in a single cylinder. Larger capacities are obtained by using two or more cylinders, arranged in various ways according to the makers' preference, side by side, in line, in V or W formation or in other configurations.

TWO-STAGE SINGLE-ACTING COMPRESSORS

Some makers have a complete range of two-stage compressors, others use a single stage for the lower end of their range and for low pressures, turning over to two stages for larger capacities and higher pressures.

In multi-cylinder machines, the high pressure piston is often on a separate line of parts, driven from the same crankshaft as the low pressure piston. In other designs, the HP piston may be in tandem with the LP and driven from the same crankpin.

Cooling. To carry away some of the heat, air or water cooling may be used. On two-stage machines there is generally an intercooler between the stages and this also may be air or water cooled. Sometimes an air-cooled intercooler is used with water jacketed cylinders or vice-versa.

DOUBLE-ACTING COMPRESSORS

In the single-acting series, only one side of the piston is used so that, from a capacity point of view, alternate strokes are idle.

An obvious way of increasing the capacity is to use both sides. This involves some

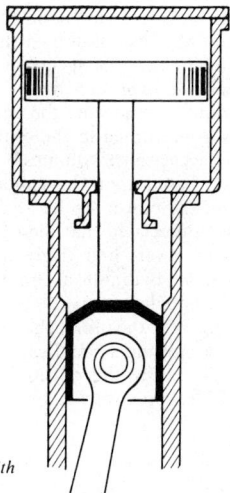

Figure 17.5. Double-acting cylinder and piston with piston rod and crosshead

form of crosshead to take the thrust of the connecting rod (Figure 17.5). An alternative is to use a stepped piston with crosshead combined.

Double-acting compressors may have vertical, horizontal or angled cylinders; sometimes two or more cylinders are mounted in tandem, according to the duty. Another popular arrangement has two or more cylinders grouped around a common crankcase, the angle between the cylinders being carefully chosen to combine a high degree of mechanical balance with a smooth turning moment.

MEDIUM AND HIGH PRESSURE COMPRESSORS

Both single-acting and double-acting types are available for working pressures above 10 bar, with two, three, four or more stages of compression, as may be appropriate. Working pressures in kilobars are not uncommon in the chemical and plastics industries.

A particular field for medium pressure machines is provided by the need for starting air for diesel engines both on land and on shipboard. The usual storage pressure is between 25 and 40 bar. Since compactness is important, several makers produce neat two- and three-stage single-acting compressors for the purpose.

FREE PISTON DIRECT ACTING MACHINE

This is another type of reciprocating compressor. In one form, this consists of a single power cylinder of two-stroke type, arranged horizontally and fitted with opposed pistons, to the outsides of which are attached compressing pistons and also the piston or pistons for providing scavenging air for the power cylinder.

The compressing pistons may be arranged to give one or more stages of compression according to the working pressure required, a single stage model for the lower pressures, two or three stages for intermediate pressures and high pressure versions with four or more stages for 200 bar and upwards. Although there is no crankshaft and connecting rod in the ordinary sense, a link motion is usually provided, to maintain synchronism between the moving pistons and to operate auxiliary equipment.

The operation is as follows. At starting, the power cylinder is charged with air and the pistons brought together to compress the charge. Fuel is injected at the appropriate moment and the explosion drives the pistons apart, carrying the compressor pistons with them and commencing the compression process. At the end of the power stroke, the pistons are returned to the central position by compressed air, using either the air trapped in the clearance volumes or special cushion cylinders or a combination of both. Governing is based upon control of the quantity of fuel injected which in turn controls the length of the power stroke and so regulates the quantity of air delivered.

It will be appreciated that the opposed piston principle makes it possible to produce machines with a very high degree of balance.

The air-loaded hydraulic accumulator calls for pressures of two or three hectobars and upwards. Although a fair volume of air may be stored, it is often possible to allow 24 hours or longer for the initial charge, after which the duty is usually only to make up the leakage so permitting a small capacity compressor to be used. The single-acting design is often employed for this purpose, with three or more stages as required. Where larger volumes of high pressure air are required, double acting machines are convenient.

BOOSTERS

A good deal of misunderstanding attends the expression 'boosting up the pressure'. The meaning used here relates to a special machine which is designed to draw air or gas from a pressurised system at P_1, to compress it further and to deliver into a separate system at a higher pressure P_2.

Standard machines, intended to draw their supplies from the atmosphere cannot often be used for this purpose without being overloaded and the makers should always be consulted. Reference to the theory of compression will show that, for a given pressure ratio, the mean effective pressure is increased in direct proportion to the absolute inlet pressure. The power requirement is similarly increased.

BOOSTER CALCULATIONS

Compressor equations (2) to (8) apply equally to boosters, where P_1 is the absolute inlet pressure and V_1 the intake volume; P_A is the pressure of the atmosphere. At equal temperatures,

$$V_1 = \text{Free Air Volume} \times \frac{P_A}{P_1}.$$

The performance of boosters is affected by the extra densities involved, particularly when P_1/P_A is high and P_2/P_1 is low.

OIL FREE COMPRESSORS

There are many processes for which oil-free air is required. Among reciprocating machines, the double-acting types are particularly suitable, since the cylinders can be sealed off and separated from the crankcase in which the motion work is lubricated in the ordinary way. Modern materials such as certain plastics, carbon graphite and some fluorocarbons can be used under oil-free conditions. The synthetic material known as PTFE has a very low coefficient of dry friction when operated against suitable working surfaces. Such materials, reinforced where necessary, with glass fibre or other substances, make possible the manufacture of piston sealing rings, gland packing rings and guide or bearing rings which prevent metallic contact between piston and cylinder wall.

For smaller capacities, several manufacturers offer oil-free single acting machines with the actual compressing cylinder or cylinders protected from the crankcase oil by a distance piece and oil seal.

In another type of oil-free reciprocating compressor, the problem of cylinder lubrication is solved in a different way. In this example the piston, instead of being fitted with sealing rings, is provided with labyrinthine expansion grooves and is made a close fit in the cylinder but without touching. The bore may have a plain surface or may also be grooved. Metallic contact between piston and cylinder is prevented by the closely controlled clearances of specially designed and accurately machined crosshead and/or piston rod guides which are external to the cylinder and suitably lubricated.

It will be appreciated that, in the absence of oil, polished working surfaces may be affected by rust when the compressor is idle. It is an advantage, therefore, to have rust-resisting materials for these parts. Cylinder inlet and outlet valves must also be of a type which will operate for long periods without lubrication.

It is possible in some cases to obtain conversion kits to enable standard compressors to operate without cylinder lubrication.

OIL FREE ROTARIES

To enable the machines in the crescent group to operate without internal lubrication, it is usual to fit blades or vanes of self lubricating material, this may be combined with special material for the contacting parts, where feasible.

In the twin- or multi-rotor group, those types in which there is no metallic contact of moving parts in the working chamber are particularly suitable for oil-free compression and many are standardised in this form. For larger capacities, the dynamic rotaries are by nature oil free except for the lubrication of the bearings and gears.

Positive rotary compressors

There are many designs of positive rotary compressors in regular use, and the following descriptions are typical. One group contains a single rotor and a reciprocating element, the blades, while another has two or more rotors per machine.

SLIDING BLADE TYPE

This descriptive term is often applied to a group of designs in which the rotor is smaller than the casing bore and offset, thus indicating the alternative description 'Crescent type'. The rotor may have an eccentric motion or, as in the design illustrated in Figure 17.6 it may be mounted concentrically on a shaft which is itself eccentric to the casing.

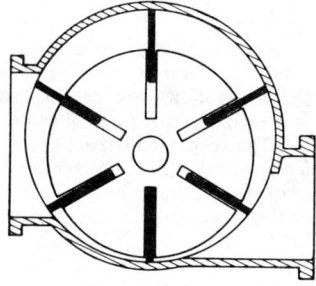

Figure 17.6. Sliding blade type of rotary. (The rotor axis may be above, below or at one side of the axis of the casing, with ports arranged to suit)

In this design, blades of metal or plastic material, fitted to slots in the rotor are free to slide and keep contact with the casing bore, as the rotor turns. The crescent-shaped space between rotor and casing is therefore divided into a series of compartments which increase and decrease in volume during each revolution.

Considering a single compartment and one revolution, there are three phases of operation:

    (i) As the volume increases, air is drawn into the compartment through the suction port.

  (ii) The compartment is cut off when the trailing blade passes the edge of the port. The reduction of volume which follows during rotation causes the trapped air to be compressed to a pressure determined by the position of the delivery port.

 (iii) The delivery port is uncovered and the compressed air is discharged into the pressure main.

LIQUID RING TYPE

In the liquid ring type of compressor, the operating principle is somewhat similar, but the blades or vanes are rigidly attached to the rotor. The cyclical variation of volume in the compartments is brought about by a ring of liquid, carried round with the rotor.

The liquid advances and retreats in the compartments, following the contour of the inside of the casing which may, for example, be eccentric or elliptical. The air enters and leaves through appropriately placed ports.

TWIN ROTOR GROUP

The classification 'twin rotor' is intended to cover a group of designs in which two rotors are arranged side by side in a common casing and so constructed that they mesh with one another. The rotors are not always 'twin' in the sense that they are alike and they may not be even of the same size.

The twin rotor group may be roughly divided into two sub-groups, according to whether the air flow is in a direction substantially at right-angles to the axes of the rotors or parallel to them.

The first sub-group includes the *Roots blowers,* in which the casing is swept by fixed blades, or lobes, carried by the rotors on their separate axes. The spacing of the axes is such that the lobes mesh together closely in the middle of the casing, where the rotors approach one another. Air from the inlet port, trapped between the lobes, is carried round the inner circumference of the casing and expelled into the discharge piping. It is prevented from returning to the inlet side by the meshing of the lobes. In the example described there may be two, three or more lobes on each rotor, according to the duty, and the lobes, although they approach one another closely, do not actually rub against each other. They are phased by accurately machined gearing in a separate compartment so that the air or gas being handled has no contact with lubricating oil. In some, the gearing is omitted. There are several other interesting designs which may be included in this sub-group. Some have their rotors so constructed that the air passing through is compressed before it reaches the discharge port. Sometimes, more than two rotors may be used.

The second group includes the *Screw Type* rotor shown in Figure 17.7. One rotor is formed with a number of deep spiral channels, running from end to end, after the style

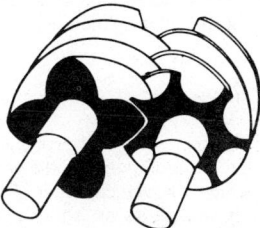

Figure 17.7. Diagram of rotors for a screw type compressor

of a multi-start screw thread. The other rotor carries projecting spiral lobes of opposite hand, so designed and profiled that they fit accurately into the channels of the first at all points of the revolution when the rotors are assembled at their correct centre distance. The casing, in which the rotors are a close fit, is provided with an inlet port at one end through which air enters the spirals and is trapped by the intermeshing lobes. Further rotation drives the air along the channels, compressing it on the way, to an extent depending upon the design of the rotors and the position of the cut-off edges of the discharge port, through which the compressed air passes on its way to the main.

The rotors may, or may not, have an equal number of spirals. In this type, they do not touch one another, but have a very small clearance, being kept in phase by external gears. In some lubricated and oil-flooded types, the rotors make actual contact, except, of course, for the thickness of the oil film, and the drive is transmitted directly from one to the other.

TWO STAGE COMPRESSION

The above rotary types and others are made to compress in more than one stage when required. One of the methods is to couple together a large low pressure casing and a smaller high pressure unit either directly or through gearing but other methods may be used, according to the general design.

An important variant of the machines described and of other rotary designs is the oil flooded version. In this group, a substantial quantity of suitable oil is injected into the compressor with the air and carried right through.

The quantity injected is such that it not only lubricates the parts and seals the clearances (to minimise internal leakage) but also carries away a considerable part of the heat of compression. This, naturally, reduces operating temperatures throughout the machine. On reaching the delivery side, the oil is filtered out, cooled and recirculated.

Dynamic compressors

In these machines, which normally run at high speeds, there is no actual contact between rotor and stator. The air is compressed by dynamic action

CENTRIFUGAL TYPE

The air flows in a substantially radial direction through disc type or similar impellers and is compressed partly by centrifugal force and partly by velocity conversion in diffuser type passage ways which conduct the air from the circumference of one impeller to the central inlet of the next.

AXIAL FLOW TYPE

The air is accelerated and propelled in an axial direction by turbine type blades mounted on the circumference of the rotor. The moving blades are followed by a stationary set in which the velocity energy is converted into pressure and a given machine will usually have a series of sets of moving and fixed blades in order to build up the final pressure required.

Dynamic compressors for medium and high pressures have a number of stages of compression with one or more intercoolers, as may be necessary. As the pressure rises and the volumetric flow shrinks, stage by stage, a point may be reached where aerodynamic considerations make changes necessary in the proportions of the impellers. With this in mind, the impellers for the high pressure stages are often mounted on a separate shaft or shafts. It is then possible to drive them at a different speed where this is appropriate to the design.

CHOOSING A COMPRESSOR

Amongst the basic types described and the variations due to individual manufacturers there are probably several which appear suitable for a particular requirement. The makers' advice should be sought in all cases but it is necessary first to make a few preliminary decisions which may affect the choice except, perhaps, in the case of standard portable sets, where a decision can be based largely on makers' performance figures in relation to cost.

Where a more permanent installation is concerned, the first decision is whether to have individual small compressors in strategic positions or to group the whole compressor capacity in a single power house. There are advantages in either scheme and there is no hard-and-fast rule. The choice will probably be affected by a consideration of the distances involved, it being usually cheaper to run and maintain a long electric cable than a long air line.

Capacity

Individual compressors may be tailored to suit the particular piece of equipment which they serve, with a suitable allowance for loss of efficiency and accidental leakage. Power station compressors, on the other hand, have to deal with widely fluctuating demands, according to how many tools etc. happen to be in use at a particular time.

It is not considered good practice to instal a single machine which will cover the maximum demand if it is so large for the normal requirement that it has to run unloaded for long periods. Some works engineers prefer to have a battery of several compressors in the power house so that they can be shut down one by one as the demand falls off. This arrangement has the further advantage that regular maintenance can be carried out without interfering with production.

Installation

Nearly all the standard machines are available, even in quite large sizes, as 'packaged' units. This usually means compressor with its coolers, motor and starter mounted upon a fabricated framework, ready to drop into position upon minimum foundations and requiring only to be connected to the services. Alternatively, it may be more convenient in some cases, particularly with belt drives, to instal compressor and motor individually on a prepared concrete base.

Many modern machines are so well balanced that deep foundations are not required unless the subsoil is poor. The makers are usually willing to give detailed advice on this aspect.

ACCESSORIES

Among the accessories likely to be required are the following.

Air intake filter

On a packaged unit, an air filter is usually included on the inlet side but there are many installations where the air is drawn through a duct or pipe and a separate filter is required. If this is out of doors, it should be protected from the weather and consideration should also be given to the installation of a silencer since the noise made by the air entering the compressor can, under some circumstances, be very disturbing, particularly if reflected off surrounding buildings or in a narrow space. The filter will require regular servicing, in accordance with makers' instructions.

Aftercooler

Since the air leaving the discharge port of a compressor is usually rather warm, it will carry a heavy charge of water vapour and, perhaps, oil. If the distance is short, as in the case of a portable compressor serving road drills, the vapour and oil content may not matter but where the compressed air has some distance to travel, both will be deposited in the pipe line as the air gradually cools.

The accumulated water may cause damage to expensive tools while oil deposits may constitute a fire hazard. For these reasons, it is considered better in most cases to cool the air as it leaves the compressor.

Oil and moisture separator

After cooling, the air passes through a separator which is designed to extract condensed moisture and excess oil, allowing both to be drained off either by hand or by use of an automatic drain trap.

Dryer

If the purity of the air is of great importance, (e.g. for certain process work) a dryer may be installed, either to follow the separator or at a point close to where the air will be used.

Many modern dryers operate on an adsorption cycle using a medium such as silica gel, activated alumina, molecular sieve, etc. Since these materials are of a porous

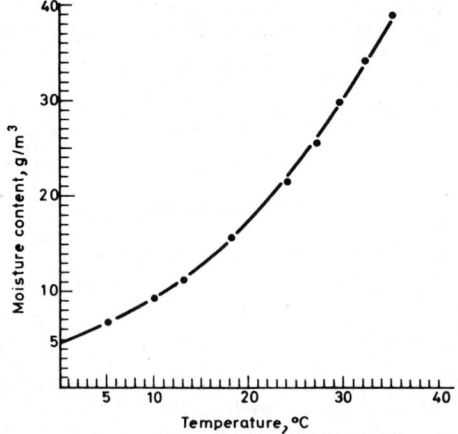

Figure 17.8. Moisture content of air

nature and adsorb moisture by a physical process, they can be regenerated by heating so that a single charge of desiccant will last for a long time.

In one well-known version, the desiccant is contained in two cylinders which are inter-connected in such a way that the air flows through them alternately. Electric heaters, embedded in the material, drive off the water from the idle cylinder, assisted by a small bleed of air from the dry side or from a separate unit. The change-over valves can be operated automatically on a time cycle or through a humidity sensor. As an alternative to heating, the medium may be reactivated by vacuum and/or use of a purge gas.

Air receivers

The air receiver is a valuable accessory for the ordinary general purpose compressor serving a battery of tools and appliances, although some compressors, particularly those used in individual installations and the large rotaries, may be operated without one. Such compressors have to deal with violent fluctuations, owing to intermittent operation of individual tools and the air receiver plays an important part in damping out the peaks and troughs of the demand. If fitted with an internal pipe, it also assists in the removal of oil and moisture.

The ideal size for the receiver depends on many factors local to the installation, some of which may be indeterminate. A long-standing recommendation is that the holding capacity should be equal to one minute's output from the compressor but, in the interests of first costs and economy of space, modern installations may include receivers of three-quarters or even half this figure. On the other hand, where peak demand is high in relation to average rate of consumption, an air receiver of increased capacity is desirable. The size of receiver is related to the functioning of the unloader and reference should also be made to the paragraphs 'Pressure and Volume Control'.

The receiver should be installed in a cool place and should be provided with safety valve, pressure gauge, drainage arrangements, inspection and cleaning doors, in accordance with current standard specifications.

PRESSURE AND VOLUME CONTROL

Since it is not normally feasible to match the output of a general service compressor exactly to the demand, a form of automatic regulator is usually fitted. In many cases this consists of an on/off control operated by air receiver pressure through a relay or equivalent unit. The relay is a pressure sensitive device and is often set with a range of 1 bar. When the demand for air diminishes, the pressure in the air receiver will rise until, at a predetermined figure, the relay operates and cuts out the compressor, which continues to run unloaded. When the pressure has fallen again and the lower limit is reached, the relay operates in the reverse sense and the compressor resumes its load. Alternatively, the relay may shut down the compressor entirely, but this method is not often used on large machines, owing to the effect of frequent switching on the supply mains and on the motor starter.

The method of unloading varies according to the type of machine and the preference of the maker. For example, single-acting reciprocating compressors are, in many cases, unloaded by holding open the cylinder inlet valve so that air drawn in during the suction stroke is discharged freely to the atmosphere when the piston returns. The second stage cylinder if any, is unloaded at the same time.

In double-acting machines, a popular method is to close the inlet so that the compressor runs idly with a degree of vacuum in the cylinders, absorbing very little power. Another method is to provide a clearance chamber in connection with the cylinder. During the compression stroke, the air, instead of being delivered into the main, is compressed into this chamber from which it expands during the suction stroke and prevents a further charge of air from being drawn in. A chamber of smaller capacity will reduce the amount of free air drawn in instead of cutting it off entirely. A further method is to bypass the delivery, allowing the machine to discharge partly or wholly to the atmosphere, a non-return valve being usually provided to retain the pressure in the main.

Closing the suction and/or bypassing the delivery are methods used also on some positive rotaries. On oil flooded types, special arrangements are included for dealing with the oil.

Starting unloader

Many reciprocating and rotary compressors are equipped with means of relieving the load during the starting and running-up period. Often this is combined with the volume control but sometimes a separate device is used.

Proportional unloading

Modifications and combinations of basic principles and also speed changes are used to give step-by-step unloading or continuous variation in certain cases. It is not feasible to include details of all varieties in this brief summary, but mention should be made of a sophisticated type of control which may be all pneumatic or pneumatic and electrical. This system lends itself particularly to the control of a group of compressors feeding a common main.

To simplify the description, it is assumed that there are four in the group, of which only two are running at a particular moment. No. 1 is selected for control. When the air pressure reaches the top limit, the sensitive relay unloads No. 1 in the manner described above. It also sets in motion a timing device, set for an interval of, say 15 minutes. At the end of that time, if the demand is still small, No. 2 compressor shuts itself down. No. 1 continues to supply the reduced requirements, loading and unloading as necessary.

When the demand increases again, and the pressure falls, No. 2 starts up again, followed by No. 3 and No. 4 if needed. The sequence can, of course, be changed weekly or monthly by switching over.

In the case of large dynamic rotaries, problems arise in view of the large power being handled and the inertia of the high speed rotating parts. In some special cases, such as,

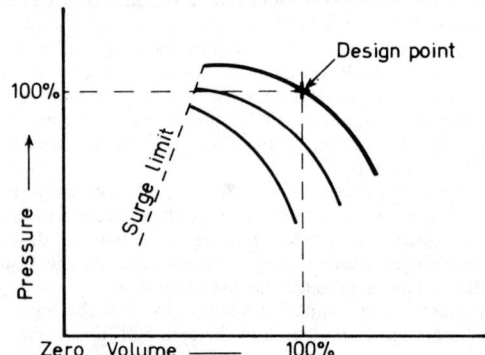

Figure 17.9. Sample characteristic curves for a dynamic compressor

for instance, the supply of air to mines, where the storage capacity is very large, it may be possible to shut off the compressor intermittently in order to control the output, in spite of the big power fluctuations involved.

Where large power fluctuations cannot be faced, flow regulation can be arranged by speed variation, by throttling and by the use of variable diffusers and guide vanes.

In a dynamic compressor, pressure and volume are inter-related in accordance with the characteristic curves appropriate to that particular machine. The curves also indicate the extent to which the flow rate can be varied on that machine, while avoiding the surge point at which the aerodynamic flow pattern becomes disordered. When necessary, the onset of surging can be prevented by an automatic device, such as a controlled bypass to ensure that when the demand for air is low, the compressor will still operate within the stable part of its range.

PIPELINES

For small diameter pipes, a popular material is copper or copper alloy. It is easy to manipulate and several types of patented couplers enable joints to be made quickly, without brazing. It is also resistant to corrosion, for which reason it is strongly recommended for the connection of pressure gauges, relays, unloaders and other delicate

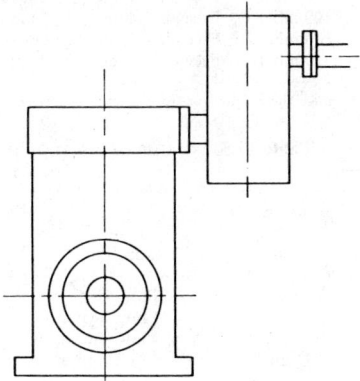

Figure 17.10. Pulsation damper fitted to compressor delivery (Adapted from 'Industrial Air Compressors'[2])

items which might be affected by rust or scale. Larger sizes, below 100 mm, are often made of steam piping, screwed and socketed. Above this diameter, flanged piping is generally used, of steel or cast iron, according to the relative cost.

Piping should be strongly supported but with freedom to expand. It should slope gently downwards in the direction of air flow so that condensed oil and moisture will be carried forward, away from the compressor. Drain legs should be fitted at convenient low points to enable the condensate to be drained off periodically. Sharp bends should be avoided as they have a tendency to encourage pipeline vibrations set up by pulsations and sudden changes in the rate of flow, as indicated by White[4].

As a further precaution against pipeline vibrations, many makers recommend a pulsation damper on the delivery outlet of the compressor. This usually consists of a small pressure vessel of size appropriate to the compressor capacity, see Figure 17.10. The proportions and the relative positions of inlet and outlet connections are important and are usually determined by the compressor manufacturer. A correctly proportioned pulsation damper will often lead to a noticeable improvement in efficiency. Provision for inspection and drainage should not be forgotten.

The size of the air main should be fixed from consideration of the normal maximum flow rates, the length involved and the permissible pressure drop. Allowance should be made for possible extensions and increased use of pneumatic appliances. Where a long run round a building is involved, consideration should be given to the use of a ring main which allows sudden local concentrations of demand to be fed from both directions.

Table 17.4 shows the expected pressure drop in a given length of main for various flow rates. This may be used as a general guide but actual figures will vary according to the internal smoothness of the pipe and the number of obstructions, bends etc.

In laying out the main, one should bear in mind, always, the question of drainage. The condensation problem may be more or less severe according to conditions in the factory and what precautions have been taken at the compressor but, in most installations, there will be occasions when a film of moisture may collect on the lower internal

Table 17.4   FLOW OF AIR IN PIPING

Allowable pressure drop in 100 m	Initial pressure 7 bar gauge (steady flow)	Nominal bore, mm					
		25	50	75	100	150	200
25 mbar	Free air volume	9 dm³/s	56 dm³/s	160 dm³/s	340 dm³/s	1 m³/s	2·1 m³/s
50 mbar	Free air volume	13 dm³/s	80 dm³/s	220 dm³/s	465 dm³/s	1·4 m³/s	2·9 m³/s
100 mbar	Free air volume	17·5 dm/s	110 dm³/s	320 dm³/s	650 dm³/s	2·0 m³/s	4·1 m³/s
200 mbar	Free air volume	26 dm/s	160 dm³/s	460 dm³/s	930 dm³/s	2·8 m³/s	5·8 m³/s
300 mbar	Free air volume	32 dm/s	200 dm³/s	560 dm³/s	1170 dm³/s	3·5 m³/s	7·0 m³/s

Note. These figures are for general guidance. Exact pressure drop will vary from one installation to another.

Table 17.5.   VOLUME OF AIR FLOWING TO THE ATMOSPHERE FROM A ROUND HOLE IN A PRESSURE SYSTEM

Gauge pressure at inlet bar	Flow measure at atmospheric pressure					
	dia. mm	2	4	6	8	10
3	Flow, dm³/s	2·3	9·2	21	37	58
5	Flow, dm³/s	3·5	14	31	55	87
7	Flow, dm³/s	4·7	19	42	75	116
10	Flow, dm³/s	6·5	26	58	100	158

Note. The coefficient of discharge may vary from 0·65 for a sharp opening in thin plate to 0·98 for a scientifically rounded nozzle. The above values are theoretical.

surface of the pipe. If the connection to the tool is taken in what appears to be the easiest way, from the bottom of the main, such moisture will drain into the branch pipe and eventually into the tool. To avoid this, it is recommended that branch connections should be taken from the upper side of the main, in spite of not having quite such a neat appearance.

Leakage should, of course, be kept to a minimum. Table 17.5 gives an idea of the quantity of air which can escape through a small hole.

VACUUM PUMPS

Since a vacuum pump is really a compressor working at a pressure below that of the atmosphere, it may be expected that many of the types already described would be suitable for the duty. This is, in fact, the case and there are many industrial vacuum applications for which standard, or nearly standard, compressors are used, particularly within the inlet pressure range of 50–650 mbar.

It will be appreciated that, when drawing in gas at 50 mbar and discharging against atmospheric pressure, the pump will be operating with a compression ratio of twenty-to-one and the volumetric efficiency may not be very high unless the machine has several stages or is designed specially for the duty. As the absolute inlet pressure is further reduced, additional difficulties arise owing to the feeble force exerted by the air or gas for operating automatic valves etc. and the growing seriousness of even tiny leaks. Machines for achieving very low absolute pressures are therefore usually of specialised design although certain of the more standard types are often used as backing pumps.

Owing to the combined effects of volumetric clearance, slip, leakage etc. the volumetric efficiency, the ratio of quantity of air aspired to volume swept, diminishes as the compression ratio increases, see Figure 17.11. For this reason, the method of specifying

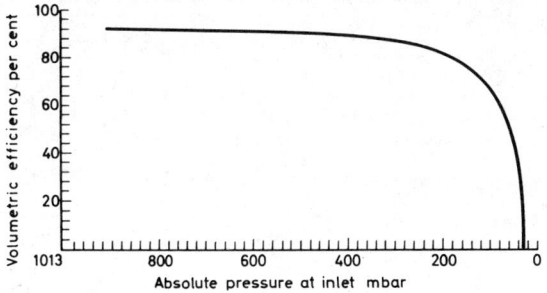

Figure 17.11. Typical volumetric efficiency curve, reciprocating vacuum pump

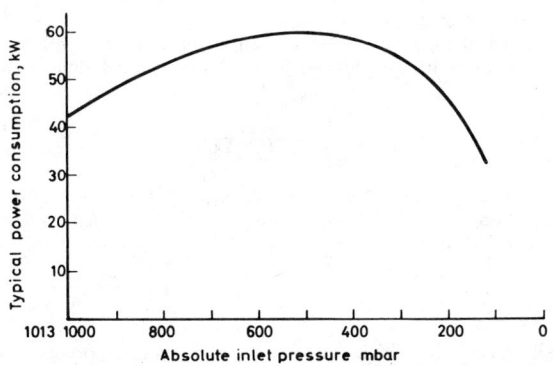

Figure 17.12. Sample power curve for reciprocating vacuum pump
(Piston displacement 1·0 m³/s)

the capacity of a rough vacuum pump differs from that used for a compressor aspiring from the atmosphere at a steady intake pressure. Three methods are in common use, according to duty:

1. To specify the piston displacement (or swept volume) and the working vacuum or minimum absolute pressure required.
2. To state the time allowed to pump a chamber of specified volume down to a specified absolute pressure or vacuum.
3. To maintain a specified vacuum against a steady leakage from outside. Sometimes the leakage rate is specified in grams or standard m³/s. Alternatively, the equivalent diameter of leak hole may be stated. In either case, the actual volume to be aspired by the vacuum pump per second increases as the absolute pressure falls, until equilibrium is reached when the volume of the in-leakage is equal to the volume aspired by the pump at the same absolute pressure.

Many practical applications involve a combination of (2) and (3). In such a case, it is hardly possible to decide upon the piston displacement required without detailed knowledge of the characteristics of the type of pump proposed.

In dealing with any requirement for a vacuum pump, it is important to bear in mind the distinction between piston displacement, which is a matter of mechanical dimen-

Table 17.6 QUANTITY OF AIR FLOWING FROM THE ATMOSPHERE THROUGH A ROUND HOLE INTO A VACUUM SYSTEM

	System pressure mbar	Diameter, mm.				
		2	4	6	8	10
Free air volume	500 or lower	0·57	2·3	5·1	9·1	14·2
Volume at system pressure	500	1·2	4·6	10·3	18·3	28·4
	200	2·9	11·5	26	46	72
	100	5·8	23	52	92	144
	50	11·6	46	104	184	288
	10	58	230	520	920	1440

Note. Theoretical figures are given. Coefficient of discharge is affected by shape of hole, as in the case of pressure flow (see Table 17.5).

sions, and the actual volume aspired at the working vacuum. The ratio between the two is the Volumetric Efficiency.

The graph shown in Figure 17.11 is an example of a curve of volumetric efficiency plotted against absolute intake pressure for a single stage vacuum pump of reciprocating type.

PNEUMATIC TOOLS

Air operated tools are available in such variety and for so many purposes that it is impossible to catalogue them all here. They may be classified roughly into types, according to their principle of operation.

Percussive type

This group includes concrete breakers, tampers, rock drills, chipping chisels and many sizes of hammer, from hand held to the large forging hammers used in steel works. Many of these use the direct acting principle, the energy stored in the compressed air being transmitted to the bit by a comparatively heavy piston, free to reciprocate in its cylinder.

The movement of the piston is controlled by an automatic valve and a series of ports, so arranged that when the piston reaches the end of its stroke, the valve is thrown over into the reverse position and the piston at once returns. This process repeats itself automatically and the piston strikes a series of rapid blows, as long as the trigger is held open, the rate depending upon the design of valves and ports and the inertia of the piston. In single stroke hammers, such as those used for making forgings, the movement of the piston and the force of the blow are frequently controlled by hand.

Rotary type

In this group is found a large family of machines, drills, reamers, trimmers, tapping machines, stud drivers, nut runners, impact wrenches, grinders, polishers etc. The driving member is usually some form of air motor which may be of reciprocating or rotary type. One form has sliding blades and is similar in principle to the sliding blade machine described under rotary compressors.

In rotary tools intended for screwing nuts etc., there is often a built-in device which, when resistance is felt, imparts a series of impacts to the spanner in a rotary sense for final tightening.

Squeeze type

This group is typified by the squeeze rivetter but might also be considered to include varieties of chucks, clamps, air vices and gripping devices of many kinds.

Operating cylinders and air motors

Both these devices are in frequent use for hoisting, lifting and transferring as well as being incorporated into the design of many vehicles, where power assistance is required.

Air motors are also often used for driving pumps etc.

Blow guns and spray guns

In industries where dust, fine shavings and so on are a problem, the dusting gun is a valuable tool, also for general cleaning of small parts. The grit or shot blast is another and nearly all factories have a use for the paint spray, which surely needs no description.

Effect of altitude

When pneumatic tools are operated at an altitude above sea level, the gauge pressure being unchanged, the consumption of free air is increased, owing to its lower density.

Air lift

Amongst other industrial uses, are the transport of solid or granular materials by compressed air and the pumping of liquid by methods which include the *air lift*.

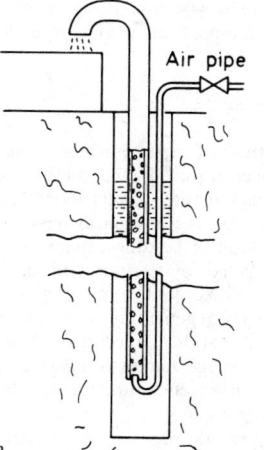

Figure 17.13. Principle of air lift pump

In this process of well pumping, the column in the rising main has its density lowered by aeration and is then forced upwards by the pressure of the surrounding water in the well. Another type of application is found in various pneumatic control and logic systems. Descriptions of some of these will be found in appropriate publications of the British Compressed Air Society[5].

OTHER INDUSTRIAL GASES

Oxygen

Oxygen, chemical symbol O_2, is one of the major constituents of the atmosphere. Its Specific Gravity, relative to air is 1·105, boiling point at atmospheric pressure 90 K, compression index 1·40. It is colourless and odourless, and under suitable conditions, it combines readily with many substances to form oxides and more complicated compounds. A major source of supply is the atmosphere which is a mechanical mixture of gases containing approximately 21% of oxygen by volume. Another source is ordinary water, H_2O, which, as its symbol indicates, consists of two atoms of hydrogen and one of oxygen, chemically combined.

A practical method of separating oxygen from the other components of the atmosphere is by liquefaction and fractional re-evaporation. The air is first compressed and then cooled either by means of an expansion engine or by the Joule-Thomson effect of direct expansion through a nozzle, usually employing pre-cooling to induce a further reduction of temperature. Pressures vary from about 200 bar or so in small direct expansion plants to very much lower pressures in large 'tonnage' plants for steelworks. Since the various gases have different boiling points, control of the temperature of the liquefied air makes it possible to separate the oxygen.

For the smaller sizes of liquid oxygen plant, requiring pressures of the order of 200 bar, it is convenient to use an air compressor of reciprocating type, having three, four or more stages of compression, with inter- and aftercooling carried on a single frame. Many of the reciprocating types described in previous paragraphs are available in compact multi-stage construction, the actual arrangement of the cylinders depending upon the volume of air to be handled. Oil, introduced into the air during compression, is removed by purifiers before the liquefaction column. Large, low pressure, tonnage plants, producing 400 tons or more of oxygen per day, employ multi-stage dynamic machines to compress air in this process.

ADSORPTIVE PROCESS

Another method of separating the atmospheric gases is by selective adsorption in an Oxygen Concentrator.[6] In this method, the air, after compression is passed through a bed of artificial zeolite, a microporous crystalline solid which adsorbs and holds the nitrogen, allowing the oxygen to pass.

When the limit of holding capacity has been reached, the air flow is cut off and the bed is regenerated, by vacuum action, drawing off the nitrogen and discharging it to waste. To provide a continuous flow, two cylinders are employed, cross connected, so that one is being regenerated while the other is on stream, on a timed cycle.

It is claimed that the Concentrator can be adjusted for an oxygen concentration of up to 80%. This range makes it possible to select a degree of enrichment suitable for various applications including medical use, breathing masks, welding, brazing, heating, metal spraying etc.

In the electrolytic process for preparing oxygen, water is first rendered conductive. The passage of an electric current loosens the bonds of some of the molecules and the two components, hydrogen and oxygen, appear separately at cathode and anode respectively.

In addition to its bulk consumption in the manufacture of steel, oxygen plays an important part in many chemical processes, in the treatment of sewage, etc. It is widely used in the constructional industries for intensifying the flame of acetylene, propane and other fuels for cutting, welding etc.

Liquid oxygen is transported in bulk in heat insulated road and rail tankers from which it can be transferred to customers' storage vessels by means of centrifugal pumps. From the insulated storage, it is evaporated off as required. Smaller quantities are transported in gaseous form in high pressure cylinders at 125 bar or, in the new higher pressure cylinders, at 172 bar.

Since oxygen reacts energetically with many substances, it is important to avoid exposing flammable materials such as ordinary lubricating oils to intimate contact, particularly under conditions of high pressure and temperature. Compression is therefore carried out in oil-free machines. For bottling purposes, from atmospheric pressure, special, water lubricated multi-stage reciprocating compressors have been developed. The water is usually removed before storage.

Nitrogen

Nitrogen[7], chemical formula N_2, forms approximately 78 % by volume or 75·5 % by weight, of the atmosphere. It is colourless and odourless and does not readily react with other elements. Its specific gravity is 0·967, relative to air. Its boiling point at an absolute pressure of 1 atmosphere is 77 K, enabling it to be separated from the other gases in air by the process of liquefaction and fractional distillation.

Owing to its comparatively inert nature, nitrogen is in demand as a blanket gas for many processes in which the presence of oxygen might be dangerous or objectionable. It is used in bulk in the manufacture of fertilisers and explosives and for many purposes in the chemical industry. In the liquid form, it is valuable as a refrigerant where low temperatures are required.

The compression of nitrogen offers few difficulties. The compression index, the ratio of specific heats, is 1·41. Owing to its non-reactive nature, it has been compressed in oil lubricated compressors to very high pressures. When necessary, it can also be handled in various types of oil-free machine where it is important to avoid contamination or under conditions where the operating temperatures are so low that lubricating oils would lose their fluidity. Bulk supplies for chemical plant are generally produced near to the point of usage, in conjunction with the production of oxygen and other air gases.

Where pipeline compressors are needed, they are generally of oil-free type. Smaller quantities are normally transported in high pressure cylinders. Bulk transport of liquid nitrogen and its storage in customers' works are as for liquid oxygen Nitrogen may also be obtained from combustion gases, and delivered in a dry condition, using an adsorption technique as described for oxygen.

Argon

Argon[7], chemical formula A, is the third major ingredient of the atmosphere, occupying just under 1 % by volume. Its boiling point of 87 K at atmospheric pressure, lies between those of nitrogen and oxygen, enabling it to be separated by fractional distillation, from the other components of liquid air.

Argon is used as a shield gas during arc welding, both hand and automatic; the choice of the pure form or a mixture containing a percentage of other gases depends upon the base material and the actual welding method. It is also used for other applications where an inert atmosphere is required and is one of the gases used for filling electric light bulbs. Since argon is monatomic, it has a high ratio of specific heats. The compression index is 1·667 and care needs to be taken in the selection of a suitable compressor to avoid difficulties with excessive temperatures. Lubricated or oil-free compressors may be employed, according to the need for purity and the temperature scale. Low temperatures may cause difficulties with lubricating oils.

In process handling for circulation over catalysts for purification, oil-free, non-contact, positive rotary machines are often used. Transport of small and medium quantities is usually in high pressure cylinders as for oxygen and nitrogen. During the production of argon, it is a normal practice to charge the cylinders by means of liquid pumps and evaporators. Liquid argon is also transported in bulk, in a similar manner to oxygen and nitrogen.

MANUFACTURED AND NATURAL GAS[7, 8]

Many industrial installations are conveniently situated for a piped supply of Towns gas which is a clean general purpose fuel. It is used for space heating, the operation of furnaces of many types, the forging, hardening and tempering of steel, the melting and fusing of alloys and non-metallic substances and many other technical processes for which controllable heat is required. In a blow torch with compressed air or oxygen, it can be used for concentrated local heating, brazing, flame cutting, etc. It may be used to generate power and when compressed to 200 or 250 bar in lightweight bottles, has been carried in place of liquid fuel on some types of vehicle.

For a long period Towns gas was produced from coal of selected grades in a specially developed retort. However, rapidly increasing demand for the gas led to changes in manufacturing processes and the employment of liquefied methane and other hydrocarbons from the oil industry as alternative or additional feedstocks.

Properties of manufactured gas

The exact composition of coal gas or its equivalents naturally depends to some extent upon the raw material and the process of manufacture. A typical sample might have a hydrogen content of 50–55%, with other major components methane and carbon monoxide. The high hydrogen content gives the gas a comparatively low specific gravity, usually of the order of 0·45 relative to air. The calorific value is usually controlled in the neighbourhood of 18·6 kJ/dm³ but higher and lower values are sometimes found.

In many parts of the world, natural gas from wells has long been used for domestic and commercial purposes but it was only in the late 1960s and early 1970s that intensive exploration of the Continental Shelf confirmed the presence of large quantities of gas of good quality under the bed of the North Sea and made this gas commercially available for general use in Britain.

Successful off-shore wells were rapidly developed and pipelines laid under the sea to on-shore receiving stations where undesirable moisture and condensibles are stripped off. High pressure feeders carry the supplies across country to link up with existing Towns gas mains for distribution to individual consumers.

Natural gas takes the place of manufactured gas for most industrial processes but burners etc. intended for coal gas or its equivalents are not generally suitable for natural gas owing to its higher calorific value and the higher ratio of air required for combustion. In most cases, the necessary changes can be readily made and large areas of Britain have already been converted to natural gas.

Properties of natural gas

As in the case of manufactured gas, the composition of natural gas varies according to the source from which it is obtained. A typical sample might contain 90% methane with other major components ethane, propane and nitrogen. The specific gravity would normally be about 0·6 relative to air and the calorific value of the order of 38 kJ/dm³. Since the main constituent is methane with a boiling point of 111 K at atmospheric pressure, natural gas can be readily liquefied for transport in ships or vehicles. Surplus gas can also be stored in liquid form and re-evaporated to meet peak demands.

Gas compression

The compression of either manufactured or natural gas can be carried out in normal compressors unless the gas is wet or corrosive. The compression index (ratio of specific heats) for either type is usually in the range 1·3 to 1·35 according to the proportions of the separate components of the mixture. If an exact figure is required, it may be calculated from a knowledge of the mass composition of the sample.

In spite of the low specific gravity, dynamic compressors are widely used. Most of the other compressor types described earlier in this section are also suitable, provided that they incorporate the necessary arrangements to prevent leakage of gas to the atmosphere or inward leakage of atmospheric air. The choice of lubricated or non-lubricated type depends on circumstances.

Where an industrial compressor or engine draws its gas from a piped supply, it is important to take precautions. This is usually achieved by interposing a pulsation damper, to prevent the feed back of pressure pulsations or shock waves which may interfere with the performance of other appliances, and particularly of automatic gas burners, fed from the same main.

Propane and Butane

Propane[9] and Butane are hydrocarbon compounds with chemical formulae C_3H_8 and C_4H_{10} respectively. Large sources of supply are gas and oil wells and refineries. Their uses in industry include the oxy-propane or oxy-butane torch for flame cutting and for other processes requiring the local application of heat. Their high calorific value makes them attractive as fuels for many purposes either independently or mixed in selected proportions. At the same time, their comparatively high boiling points enable them to be transported in liquid form without needing excessively high pressures or very low temperatures for storage. These advantages make them popular for domestic as well as industrial purposes in appropriate cases.

In addition to their usefulness as fuels, or as additives to other fuels, they also form a convenient base for use in the production of more complicated chemicals, plastics, etc. Both propane and butane are practical refrigerants. The compression index for propane is 1·15 and for butane, 1·108. Temperature difficulties are, therefore, not likely to arise during the process of compression unless the operating conditions are high on the temperature scale. Under very cold conditions, problems of lubrication may arise. For these conditions and also where extreme purity is desired, oil-free compressors may be employed with advantage.

In cases where the operating temperature is at or below the dew point of the gas. condensation may take place inside the machine. The deposited liquid may interfere with lubrication and cause hydraulic damage, particularly in reciprocating compressors, Here, automatic cylinder drainage is an advantage as, for example, in the horizontal type.

LEGISLATION RELATING TO PRESSURE VESSELS AND THE CONVEYANCE OF GASES

Information is given in Appendix R of the Report of the Committee of Enquiry on Pressure Vessels[10], from which the following brief extracts are made with the permission of the Controller of Her Majesty's Stationery Office. The principal statutory requirements are contained in or regulated by: the Factories Act 1961, the Mines and Quarries Act 1954, the Petroleum Consolidation Act 1928, the Gas Cylinders (Conveyance) Regulations 1931 (amended 1947 and 1959) and the Boiler Explosion Acts of 1882 and 1890.

Amongst the items controlled by the Factories Act 1961 administered by the DEP, are Air Receivers. The Act defines requirements governing the method of examination and the periods between examinations. It also requires examination 'by a competent person'.

The Petroleum Consolidation Act 1928 controls the conveyance of gas cylinders by road in the UK, in respect of cylinders containing the permanent gases of air, argon, carbon monoxide coal gas, hydrogen, methane, neon, nitrogen and oxygen. The main instrument of control is the Gas Cylinders (Conveyance) Regulations 1931, amended in 1947 and 1959, which lays down in considerable detail specifications for materials, design, construction and testing of cylinders. The Home Office is responsible for the legislation relating to the carriage of gas cylinders on roads in the UK. This does not cover factory sites which come under the Factory Acts (DEP), but every gas cylinder to which the Gas Cylinders (Conveyance) Regulations apply must comply with these regulations because all gas cylinders are, at one time or another, transported by road.

Home Office regulations

The Home Office Regulations cover both approval and certification of each gas cylinder as it is made, as well as the recertification of every gas cylinder (by an approved inspection organisation or by an approved manufacturer with an approved inspection department) at least every 5 years.

It should be noted that, in accordance with para. 22, alterations are proposed to both the Factories Act 1961 and the Petroleum Consolidation Act 1928.

In addition to UK requirements, the appendix also includes information regarding legislation in other countries.

British Standards

Some British Standards of interest to the engineer dealing with compressed air and gases are the following.

BS 1571 deals with the testing of positive displacement compressors, both reciprocating and rotary. A suitable test layout is recommended and standard methods of taking and recording pressures, temperatures, water quantities etc. Suggestions are made regarding the form of the test report and the necessary adjustments where the test conditions do not agree exactly with those of the specification.

BS 2009 provides similar information for testing dynamic compressors, for which the procedure is slightly different owing to the different nature of the machines and of the physical principles involved.

BS 1042 covers orifices, nozzles and venturi tubes used for the measurement of air and gas flow rates and gives details of the calculations.

BS 1339:1965 gives formulae for calculating the humidity of air. When carrying out exact tests, it is important to make allowances for the dryness of the air aspired.

BS 1701 concerns filters for the air intake of compressors and internal combustion engines. It covers many types, including the dry element, the viscous, the oil bath, the centrifugal and velocity change types and various combinations of such units. Performance requirements and methods of proving the efficiency of filtration are included also recommendations regarding the test dust.

BS 4142 and 4196: 1967 relate to the estimation and measurement of noise level and sound power. This is a subject of increasing importance under present conditions, particularly where residential districts may be involved. Certain types of noise, particularly air noise, will sometimes carry for long distances and give rise to complaints from unexpected quarters.

BS 1045 concerns manganese steel cylinders for the storage of air and gases which do not readily condense. It deals with the quality and composition of the materials, manufacture, heat treatment, tensile and bend tests, hydraulic tests etc.

BS 1288 gives similar information for the condensible gases.

BS 428, 429, 430 and 1099 deal with forged and welded air receivers. In the case of welded units, formulae are given for calculating the required thickness of the shell and of the dished ends. Details of welds are also included.

BS 1123 concerns itself with the fittings required on air receivers. Particulars are included of types of valves, pressure gauges and recommendations for the size and pressure range of safety valves.

BS 2915 refers to a special type of safety device known as a bursting disc. These are often used where slight leakage might be a nuisance and where instantaneous release of pressure is desired. As the name implies, the device consists essentially of a disc of foil, which may be flat or domed, secured between hollow flanges. An example of its use is on the outer casing of the water jacket of a particular type of heat exchanger where a tube failure might release a burst of high pressure air or gas into the water chamber. The standard gives recommendations regarding manufacture, the type of foil, methods of testing etc.

BS 3274 is concerned with tubular type heat exchangers. It gives advice on materials, design, construction and testing.

BS 4364, 4365 and 4366: 1968 specify requirements for atmospheric gases for industrial use. The standards cover gases produced from air by fractional distillation; oxygen from the electrolysis of water; or argon as a by-product of ammonia synthesis. Permissible limits for impurities are laid down and recommendations made for the treatment of samples. The standards relate to industrial oxygen, argon and nitrogen, respectively.

It should be noted that British Standards may be revised from time to time. Where dates of issue are included the standards have been classified as metric standards by B.S.I.

REFERENCES

1. *Standard Reference Atmosphere*, International Standards Organisation R.554
2. WHITE, F. G., *Industrial Air Compressors*, chap. 3, G. T. Foulis
3. PNEUROP EUROPEAN COMMITTEE, *Compressors, Classification and Glossary of Technical Terms*, (in six languages) The British Compressed Air Society

4. WHITE, F. G., *Industrial Air Compressors*, chap. 12, G. T. Foulis
5. *Pneumatic Circuits and Low Cost Automation—Fluid Logic in Simple Terms*, British Compressed Air Society
6. Separating Gases by selective adsorption, *Rimer-Birlec Ltd.*
7. WALTERS, W. J., 'North Sea Gas. Its impact on the British Gas Industry', *Inst. of Gas Engineers*, Communication 760
8. ROOKE, D. E., 'The development of new sources of gas for the British Gas Industry', *International Gas Union, 10th International Conference*, Paper IGU/B6-67
9. DIN, F. (Ed), *Thermodynamic Functions of Gases*
   Vol 1: Ammonia, Carbon Dioxide and Carbon Monoxide, (1956)
   Vol. 2: Air, Acetylene, Ethylene, Propane and Argon, (1956)
   Vol. 3: Methane, Nitrogen, Ethane, (1961)
   (Butterworth)
10. *Report of the Committee of enquiry on Pressure Vessels, 1969, Appendix R*, H.M.S.O.

18 INSTRUMENTATION

INSTRUMENTATION 18

18 INSTRUMENTATION

R. E. FISCHBACHER

The range of instrumentation relevant to the mechanical engineer is too great to be covered comprehensively in this section which covers temperature, pressure, level flow, and means of indicating or recording. Information on measurements and techniques beyond the scope of this section can be found in standard text books and reference books devoted to the subject of Instrumentation.

TEMPERATURE

Temperature is probably the most often measured variable encountered in industrial practice. The methods used to measure it depend on the temperature range and on accuracy or repeatability requirements. The span of temperature practically encountered may range from liquefied gases (e.g. hydrogen $-253°C$) to molten metals (blast furnace $1900°C$). For some purposes it is adequate to know temperature to the nearest 10–$20°C$; for others, the $0.01°C$ accuracy which can be achieved by resistance thermometry is necessary. Figure 18.1 shows the temperature ranges of various sensors.

Industrially, thermocouples, filled-system thermometers, and radiation pyrometers are most widely used. Resistance thermometry at its best offers a high degree of stability and repeatability. For approximate temperature indications, especially at high temperatures, pyrometric cones and temperature sensitive crayons and paints may be used. Liquid-in-glass thermometers are sufficiently familiar not to be discussed further.

Thermocouples

Thermocouples use an effect discovered by Seebeck in 1821. When two conductors of dissimilar metals M_1 and M_2 form an electrical circuit as in Figure 18.2 and the temperatures T_1 and T_2 of the junctions differ, an electrical current will flow in the circuit and its magnitude will be related to the temperature difference (T_1-T_2).

In the practical application of this effect, the circuit is not closed and the thermo-electrical potential is measured at points A and B. If the temperature of one junction is stabilised the emf of the terminals is a direct function of the temperature of the other junction, which then forms the temperature sensitive element. The characteristics of some thermocouples are shown in Figure 18.3.

Materials commonly used for thermocouples are copper-Constantan, iron-Constantan, Chromel-Alumel, and platinum/rhodium-platinum. Approximate ranges are given in Table 18.1.

Base metal thermocouples have a relatively high output voltage for a given temperature change and exhibit a virtually linear characteristic. They are more generally used than the rare metal couples. The thermocouple junction is normally housed in a protective sheath or pocket which resists mechanical and thermal shock and corrosion and provides insulation, but has adequate heat transfer properties.

18–2

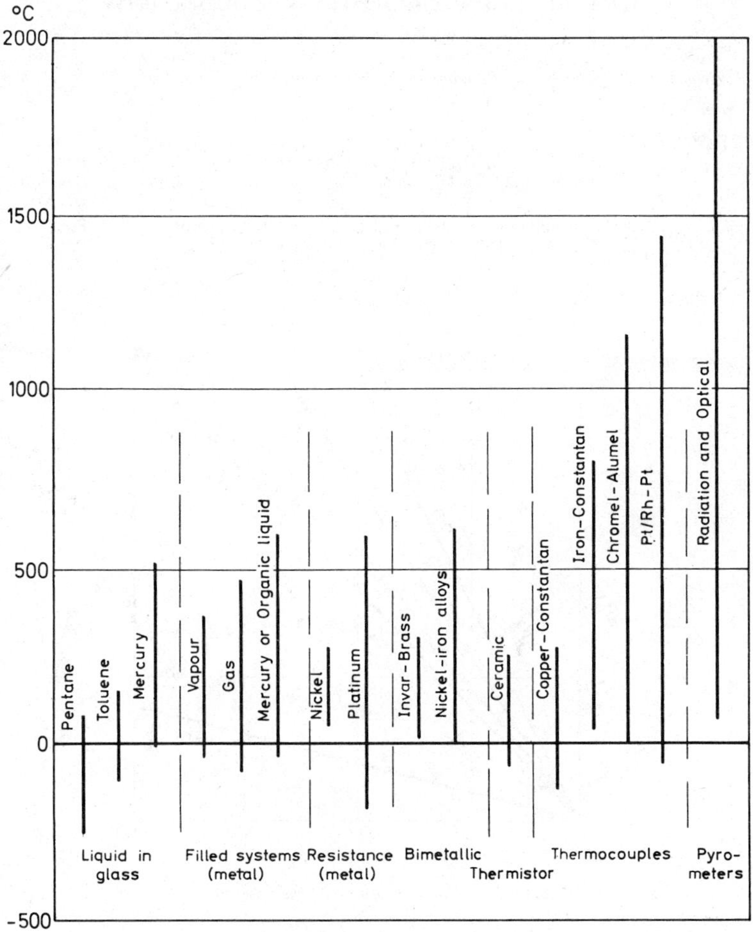

°C

Figure 18.1. Approximate ranges of temperature sensors

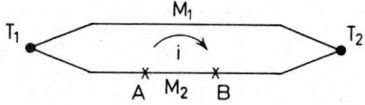

Figure 18.2. Principle of thermocouple

Table 18.1 GENERAL CHARACTERISTICS OF THERMOCOUPLES

Positive element	Negative element	Approx. temp. range °C	Max. temp. °C	Oxidation resistance	Reduction resistance
Pt/Rh (90%/10%)	Pt	0 to 1450	1 700	Very good	Poor
Chromel	Alumel	−200 to 1100	1 200	Good	Poor
Iron	Constantan	−200 to 750	1 000	Good to 400 °C	Good to 400 °C
Copper	Constantan	−200 to 350	600	Good	Poor
Chromel	Constantan	−100 to 1000	1 000	Good	Poor

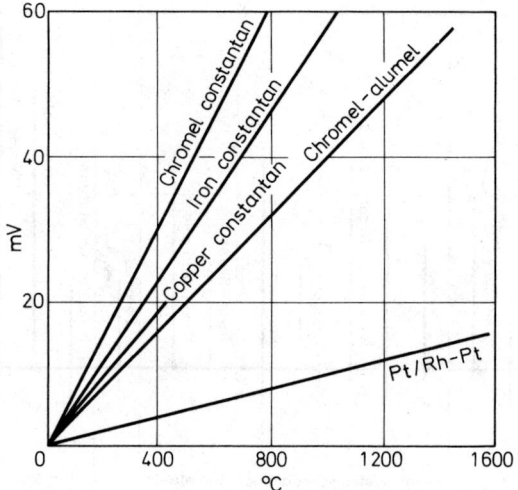

Figure 18.3. Output characteristics of thermocouples

Table 18.2 COMPENSATING CABLE MATERIALS

Thermocouple		Compensating angle		Range °C
Positive element	Negative element	Positive	Negative	
Platinum/Rhodium	Platinum	Copper	Copper-nickel	25 to 200
Chromel	Alumel	Chromel	Alumel	0 to 200
		Copper	Constantan	25 to 100
		Iron	Copper-nickel alloy	25 to 200
Iron	Constantan	Iron	Constantan	0 to 200
Copper	Constantan	Copper	Constantan	−50 to 100

Where measuring distances are significant, cost usually dictates that the hot junction be connected to the cold junction by 'compensating cables' of metals with thermo-electric properties similar to the thermocouple metals. Table 18.2 indicates suitable materials.

Cold junctions may be stabilised by placing them in a controlled oven maintained above ambient temperature, or in an ice bath; alternatively cold junction compensation may be provided by temperature sensitive electrical or mechanical devices. The indicating instrument may be a galvanometer connected as a millivoltmeter or more probably a self-balancing potentiometric recorder.

Resistance thermometers

Resistance thermometers consist of lengths of very fine wire wound on a suitable insulating former, usually of ceramic or similar material, and protected by an insulated metal sheath. Platinum wire is widely used to provide a high degree of accuracy and repeatability of temperature measurement.

The resistance-temperature relationship is given by the Callendar equation

$$T = \left(\frac{R_t - R_0}{R_{100} - R_0}\right) \times 100 + \delta\left(\frac{\tau}{100} - 1\right)\frac{\tau}{100} \tag{1}$$

where T = temperature in °C, R_t = resistance at t°C, R_0 = resistance at 0°C, R_{100} = resistance at 100°C, and δ = constant characteristic of the material (approx 1·5 per °C for platinum). Typically resistance values may range from a few ohms to several hundred ohms.

Resistance is usually measured by bridge methods, including self-balancing potentiometric bridges.

Thermistors

Certain semiconducting materials have properties which lend themselves to temperature sensing. Oxides of heavy metals (cobalt, iron, manganese, nickel, tin, zinc, for example) are pressed into minute pellets and heat treated before encapsulation or enclosure in a suitable protective mounting. These elements have exponential resistance-temperature characteristics, $R = a \exp(b/T)$ where a and b are constants. A typical thermistor characteristic is shown in Figure 18.4.

The thermistor offers short time constants in appropriate conditions (down to a few tenths of a second) and high sensitivity, but is inherently less stable over long periods than metal resistance thermistors. The non-linearity may not be of major importance when it is used to control a preset temperature or indicate deviations from a set point.

Filled-system thermometers

The fragility of liquid-in-glass thermometers is overcome by using metal containers filled with expanding liquids or gases. The essential elements are a bulb filled with the expanding medium, a Bourdon element which distorts with pressure variation, a capillary to connect the two, and an indicator actuated by the Bourdon element (see Figure 18.5).

The expansion medium may be liquid or gas. Liquids include mercury which can be used from −80°C to 700°C, and organic liquids such as pentane, iso-propyl alcohol, and xylene for sectors of the range −85°C to 260°C. For most practical purposes liquid in metal thermometers are linear. Vapour pressure thermometers are filled with a volatile liquid (methyl chloride, ethyl alcohol, ether, toluene). They cover the approximate range −20°C to 340°C, but any individual type will span less than 150°C. Vapour pressure systems are not inherently linear.

Gas-filled systems contain inert gases such as nitrogen or argon under pressure. They have lower thermal capacity than similar volume liquid types and cover the range −200°C to 800°C, being substantially linear over their operating span.

Depending on type and range, filled-system bulbs may be very small, e.g. 6 mm o.d. × 75 mm long. Bulb size will increase in high pressure systems and more so if a short

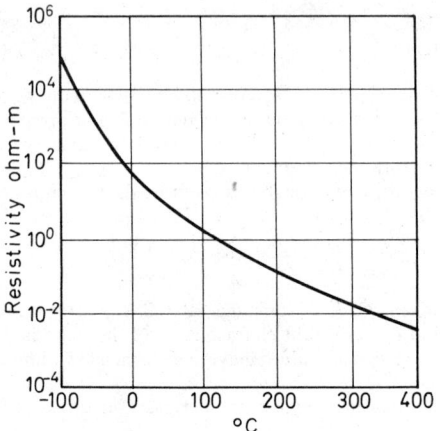

Figure 18.4. Thermistor characteristics

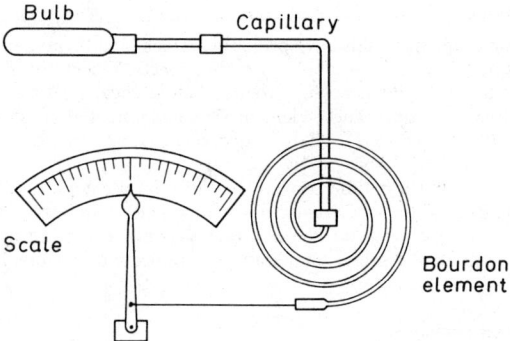

Figure 18.5. Filled-system thermometer

temperature span requires a larger filled volume. Bulb materials commonly used include stainless steel, monel, and copper.

Capillary tubes of up to 50 mm length may be used between the measuring point and the indicator. Variations in temperature along the capillary may require the use of compensation devices.

The Bourdon element may take many forms in practice. The most readily described form is the flattened tube, closed at one end and bent into a helix or a C-shape. As the liquid or gas in the bulb expands with rising temperature the resultant pressure causes a deflection of the end of the Bourdon tube, which is used to actuate the indicator.

Filled-system thermometers are essentially simple and reliable, requiring no ancillary electrical supplies or circuits. These advantages may not apply where temperatures have to be recorded or transmitted to a data logger or computer.

Bimetallic thermometers

Relatively cheap, robust thermometers are constructed using the differential expansion of two metals. Two metals with different thermal coefficients of expansion are laminated into a single strip. Since the expansion coefficients of the two laminae differ, temperature change will cause the strip to distort. To achieve a useful sensitivity a very long strip must be used, and it is common to form the strip into a helix or a flat spiral. The deflection of the end is used to actuate the indicator, or an electrical contact. Typical

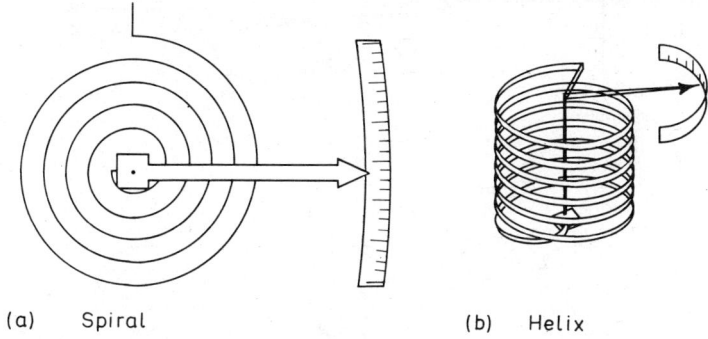

(a)   Spiral   (b)   Helix

Figure 18.6. Bimetal thermometers—basic elements

industrial thermometer configurations of this type are shown in Figure 18.6. For domestic and other water heater control, even simpler construction uses the differential expansion between a rod and an enclosing metal tube which is immersed in the water.

Pyrometers

Pyrometers differ from devices already described in that no contact is made with the body whose temperature is being measured. Pyrometers are most commonly used for measuring very high temperatures such as are encountered in furnaces and kilns.

Classically, pyrometers are of two types; the 'radiation' pyrometer and 'optical' pyrometer. The radiation pyrometer collects a broad spectrum of radiation and focuses it on to a thermopile as in Figure 18.7 which is sensitive to radiation over a broad band. The output of the thermopile may be used to give a local temperature indication, but is frequently transmitted to a remote recorder or indicator. Frequently pyrometers have to be air or water-cooled to protect the thermopile from high ambient temperatures and radiant heating. Radiation pyrometers can be made to operate down to a few tens of °C, but are normally used at higher temperatures up to 2000°C.

Optical pyrometers use only a narrow band of radiation, essentially measuring the 'colour' of the radiating source. The simplest optical pyrometers present to the eye a compound image (an example is shown in Figure 18.8) composed of an image of the measured body superimposed on the image of a filament whose temperature can be varied by controlling the current through it. The current is adjusted until the image of the filament merges into the image of the hot target. Thus the instrument is sometimes known as a 'disappearing filament' pyrometer. Optical pyrometers are widely used in the range 800–2000°C or higher.

More complex designs, such as two-colour pyrometers and other variants provide compensation for smoke and fumes, and for emissivity variations in the observed body. In general pyrometers have to be carefully calibrated against specially designed sources.

Temperature indicators

Where approximate temperature measurements only are required, or the site is difficult of access, relatively crude methods are sometimes of value. Temperature-sensitive paints and crayons change colour or melt respectively at known temperatures. Colour

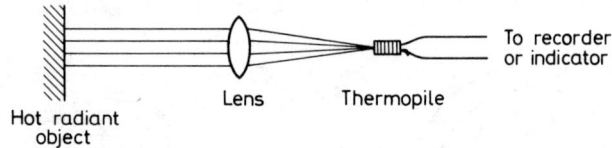

Hot radiant object   Lens   Thermopile

Figure 18.7. Principle of radiation pyrometer

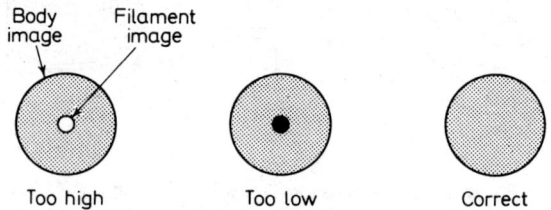

Too high   Too low   Correct

Figure 18.8. Optical pyrometer settings

changes are irreversible and melting can easily be detected. Such materials can be obtained covering 25–1100°C in steps of 8 − 30°C.

Pyrometric cones are used to indicate temperature, mainly in the ceramics industry. These are slim trihedral cones of ceramics material which soften and sag at different temperatures. Some 60 cones cover the range 580–2000°C. Some operator experience is usually required to interpret results since exposure time as well as temperature affects behaviour.

PRESSURE

Pressure measurements are normally made in terms of absolute, gauge, vacuum, or differential pressure. These terms refer to the reference point against which the measurement is made. Absolute pressure is referred to absolute zero (zero molecular activity); gauge and vacuum pressure take atmospheric pressure as the reference point (the former normally above, the latter below, atmospheric pressure); differential pressure is the difference between two pressures which are themselves unknown or irrelevant.

Apart from high-vacuum gauges, pressure instruments fall broadly into two categories. One uses a liquid column of appropriate height and density to counterbalance the measured pressure; the other uses the elastic deformation of metal elements or

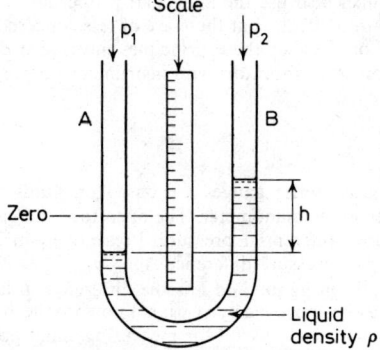

Figure 18.9. Simple manometer

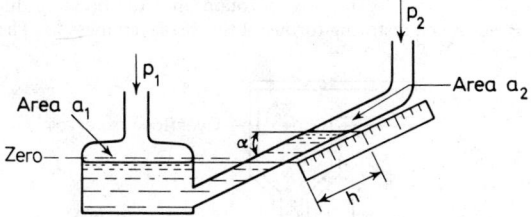

Figure 18.10. Inclined tube manometer

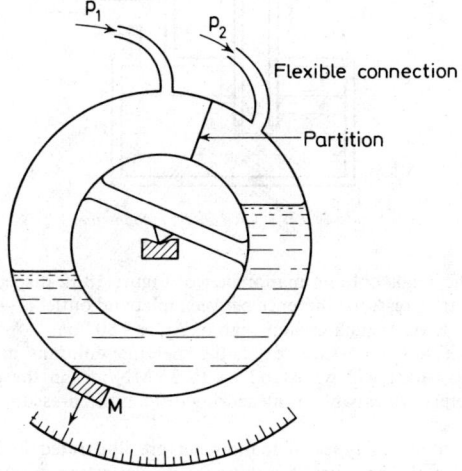

Figure 18.11. Ring balance

structures. Vacuum gauges also use the molecular properties of gases such as thermal conductivity and ionisation effects, but these are of less concern in the present context. Pressure gauges based on each of these principles have been devised in a very large number of configurations for laboratory and instrument purposes.

Gravimetric gauges

The most elementary gravimetric gauges are based on fluids under the influence of gravity and include the simple manometer. The open tube mercury manometer (Figure 18.9) essentially measures differential pressure. Pressure p_1 in limb A is higher than pressure p_2 in limb B. The pressure difference $\Delta p = p_1 - p_2 = 9\cdot807 \, \rho h$ N/m^2 where ρ is the density of the liquid in kg/m^3 and h is the difference in liquid height in metres. The reading accuracy may be improved by a variant known as the inclined-tube manometer (Figure 18.10); this is particularly useful for measuring small pressure differences. The differential pressure $\Delta p = 9\cdot807 \, \rho h \, (1 + a_1/a_2) \sin \alpha$. Where a_1 is very much greater than $a_2 \, \Delta p = 9\cdot807 \, \rho h \sin \alpha$.

The ring balance of Figure 18.11 balances the differential pressure against a mass M attached to the ring which is freely suspended at its centre. Flexible connections to the measured pressure points allow the ring to rotate until the pressure difference on the tube end is balanced by the restoring torque of the displaced mass M. The liquid acts as

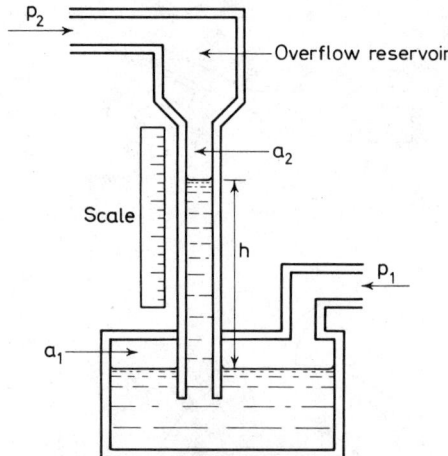

Figure 18.12. Single column manometer

a pressure seal. The single column manometer of Figure 18.12 dispenses with one limb of the U-tube, and the pressure difference between inlet and outlet is given by $\Delta p = 9\cdot807 \, \rho h \, (1 + a_1/a_2)$. If a_2 is very much greater than $a_1 \, \Delta p = 9\cdot807 \, \rho h$.

Where very high static pressures exist, the mercury columns may be enclosed in stainless steel tube which will withstand up to 35 MN/m^2, in the presence of which precision manometers are capable of measuring differential pressure to around 7 N/m^2 (0·07 mbar).

The bell and double-bell types of manometer are illustrated in Figures 18.13 and 18.14. In the former the pressure is opposed by a deflecting spring, and a pressure-tight spindle is required to transmit the motion outside the housing; in the latter

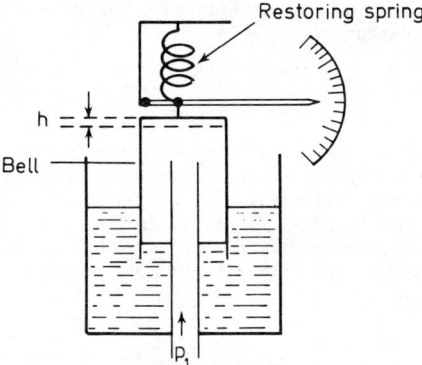

Figure 18.13. Bell manometer

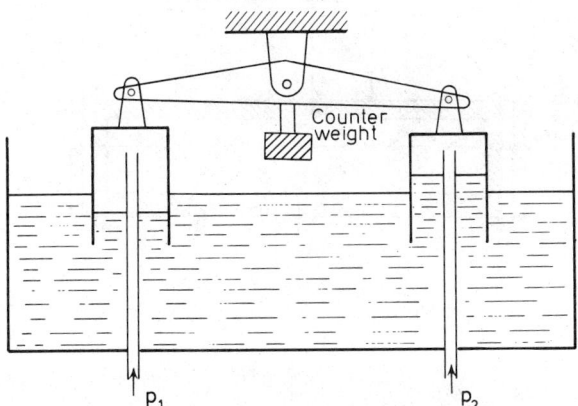

Figure 18.14. Double-bell manometer

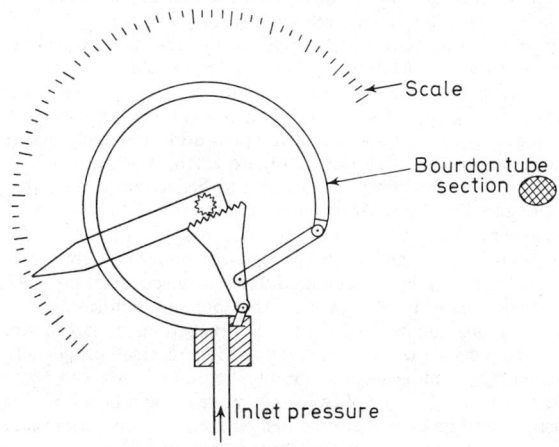

Figure 18.15. Bourdon pressure gauge

displacement is opposed by the mass M. Many other variants of the U-tube manometer may be found in the standard texts.

Elastic elements

The majority of industrial pressure measurements dispense with liquids and employ the pressure to deflect an elastic element, bourdon tube, diaphragm, or bellows.

The bourdon tube in its most elementary form consists of an elliptical section tube closed at one end and bent into the shape of the letter C (Figure 18.15). Pressure applied at the open end tends to deform the tube wall toward circularity. This causes the C shape to uncoil and deflect the pointer over the scale. Sensitivity may be increased by

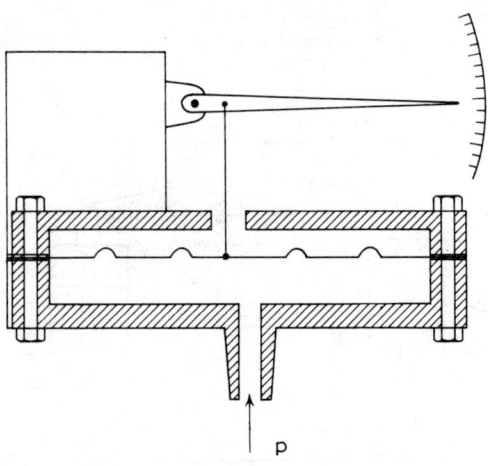

Figure 18.16. Diaphragm pressure gauge

coiling the tube further into a flat spiral, or into a helix. Bourdon gauges most commonly measure 'gauge' pressure (i.e. relative to atmosphere) and may be calibrated to negative (i.e. below atmospheric) as well as positive pressures.

Since this type of mechanism can be made with very thick-walled tube, very high pressures may be measured. Materials employed for the elastic element depend on the application for which it is intended; they include bronze, phosphor-bronze, stainless steel, Monel, beryllium-copper, Ni-span, Inconel, and low-alloy steels.

The metal diaphragm is normally used for measuring relatively low pressures. The simplest form of diaphragm is shown in Figure 18.16. The measured pressure is admitted to one side of the diaphragm causing it to deflect and rotate the pointer via a suitable lever or geared motion. More commonly the single diaphragm is replaced by a capsule of the general form shown in Figure 18.17. Each half of the capsule is pressed from metal strip carefully selected for its homogeneity dimensionally and metallurgically. The two corrugated halves are soldered, brazed, or electron-beam welded round the periphery to form a capsule of reproduceable pressure-deformation characteristics. The thickness of metal, number and depth of corrugations, material, and diameter of the capsule determine its pressure sensitivity. Metallurgical properties (composition and heat treatment) determine secular stability, reproducibility and hysteresis.

The capsule may be used in one of two main modes. It can be allowed to deflect from its neutral position and its deformation detected mechanically, electrically or magnetically. This deflection is a measure of the applied pressure difference between inside and outside of the capsule. Alternatively the diaphragm may be used in the force-balance mode (Figure 18.18) where extremely small deflection from the neutral position causes a restoring force to be applied electromagnetically (or pneumatically). This keeps the diaphragm close to the neutral position and can considerably reduce hysteresis effects sometimes associated with large deflections.

Frequently, multiple capsule elements are constructed by bonding a number of capsules to form a more sensitive stack (Figure 18.19). Bellows elements (Figure 18.20) are formed from tubular material by mechanical rolling or by hydraulic pressure. They

Figure 18.17. Pressure capsule

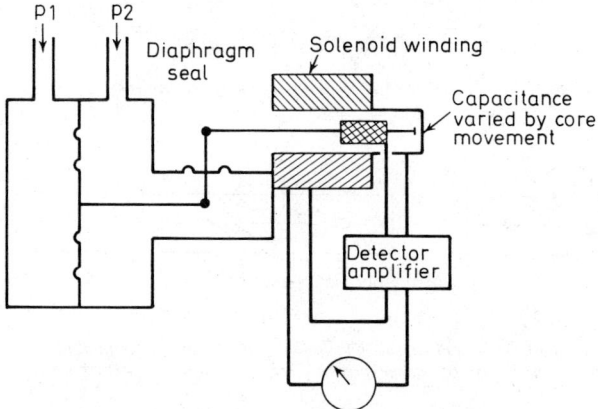

Figure 18.18. Schematic force-balance pressure transducer

are flexible along the tube axis, being expansible and collapsible within defined limits. Sensitivity is determined by material, wall thickness, length, and number and depth of convolutions.

As with all pressure measuring devices, bellows and capsules essentially respond to the difference in pressure between inside and outside. The method of assembly and housing determines whether the resultant instrument measures absolute, gauge, or differential pressure. Figure 18.21a illustrates how a capsule may be mounted to detect absolute pressure by evacuating the inner enclosure; in Figure 18.21b the outer chamber is exposed to atmospheric pressure and the device will indicate gauge pressure; in

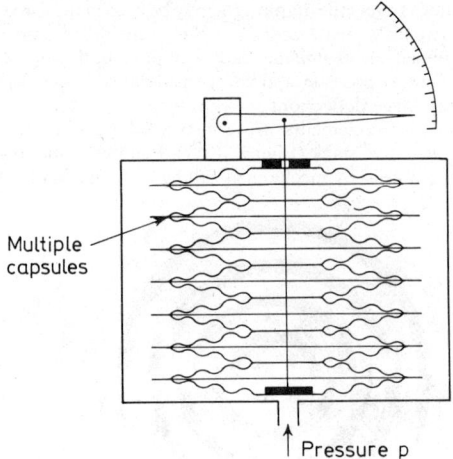

Figure 18.19. Multiple capsule element

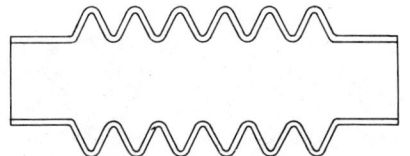

Figure 18.20. Bellows-type pressure element

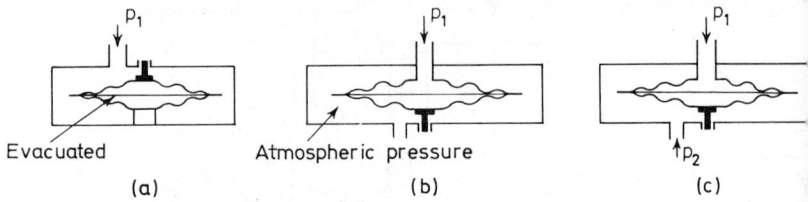

Figure 18.21. Configurations of capsule for absolute, 'gauge', and differential pressure measurement.
(a) Absolute pressure gauge, (b) 'gauge' pressure gauge, (c) differential pressure gauge

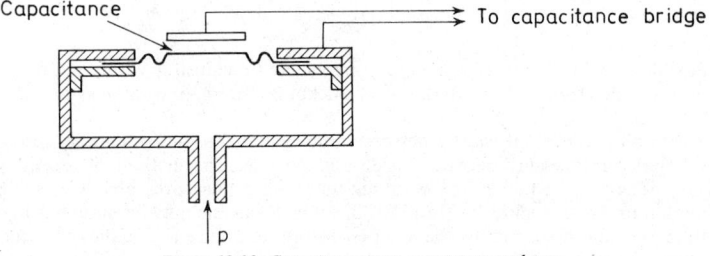

Figure 18.22. Capacitance-type pressure transducer

Figure 18.21c the outer chamber is connected to one pressure source and the capsule to another, so that differential pressure is measured. To protect the bellows or capsule from overload, it is normal to build a restraining 'stop' into the design.

Detectors

Elastic-element pressure devices depend on deformation to indicate pressure. The deformation may be detected in a number of ways. The simplest is by mechanical linkage operating a pointer. In the most sensitive cases, and where the pressure reading has to be transmitted over any distance, other methods are used. Foil or wire strain gauges bonded to the elastic element may be connected to a strain gauge bridge. In Figure 18.22 capacitance between the element and a fixed plate will vary with deformation and may be measured. One of the commonest transducers used with pressure elements to detect diaphragm displacement is the differential transformer (Figure 18.23).

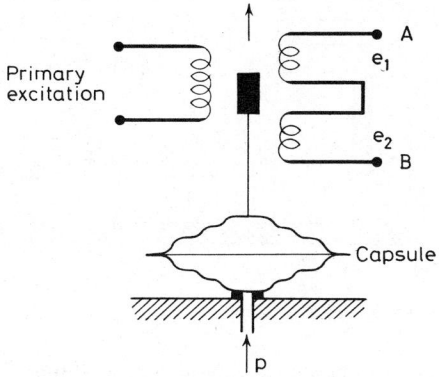

Figure 18.23. Differential transformer

This has the merit of giving a relatively large output signal for a very small displacement of the movable magnetic cone. When this is in the balance position the outputs e_1 and e_2 from the two secondary windings are equal and can be arranged to be opposing in phase. The output between terminals A and B is therefore zero.

As the cone moves from the balance position in the direction shown, the magnetic coupling to the upper secondary winding increases and to the lower decreases, so that e_1 becomes greater and e_2 smaller. A voltage then appears across A and B which may be used to control a solenoid to restore the diaphragm virtually to its initial position, or the voltage may be used to operate an indicator or recorder directly. In the force balance situation the solenoid current will be used as a measure of pressure.

Vacuum gauges

Within limits, pressures below atmospheric may be measured by variants of the devices already mentioned. Among the additional methods used at high vacuum are the Pirani and thermocouple gauges. These depend on the thermal conductivity of the gas, which below about 100 N/m^2 decreases with pressure, linearly below about 50 N/m^2. Up to about 100 N/m^2 the scale is roughly logarithmic. There are two heated elements connected in a Wheatstone bridge, one is evacuated and sealed, the other being exposed to the gas pressure. The conductivity of the gas cools one element unbalancing the bridge, and pressure may be read from the unbalance indicator. The gauge has to be calibrated for gas composition.

Various versions of ionisation gauge are also used at high vacuum. The gas may be ionised by hot filament, electrical discharge, or radiation, but in each case the ionisation current is related to gas pressure.

Dead-weight testers

Apart from calibration against accurately measured heads of liquids of known density, the dead-weight tester (Figure 18.24) offers a good calibration facility over a very wide pressure range. The tester consists of a piston operating in an oil-filled cylinder of known bore and carrying accurately known weights. The gauge under test is attached

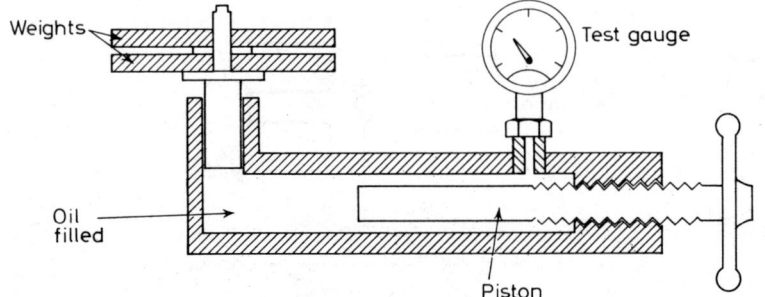

Figure 18.24. Dead-weight tester

to the tester and the required pressure is applied by a screwed piston or ram until the weights are lifted. While the gauge is being read, the weights are spun to reduce friction effects.

Table 18.3   TYPES OF PRESSURE GAUGES WITH PRESSURE RANGE AND APPLICATION

Gauge type	Approximate usable range	Approximate accuracy
PRESSURE GAUGES		
Bourdon tube	Up to 10^4 bar	$\frac{1}{2}\%$
Bellows—general	Vacuum to 10 bar	$\frac{1}{2}$–1%
Nested capsule	0–10 mbar to 0–2 bar	$\frac{1}{4}\%$
Slack diaphragm	0–50 m bar	1%
Liquid-in-glass	20 mbar–2 bar	mbar
Inclined tube	0·2–2 mbar	0·01 mbar
Ball manometer	0–5 mbar	1%
Ring balance	0–5 mbar	1%
Force balance	Very wide	$\frac{1}{2}\%$
Dead weight tester	Very wide	0·05%
VACUUM GAUGES		
McLeod	0·05 N/m^2–50 kN/m^2	1%
Thermocouple	1–10^3 N/m^2	
Pirani	1 N/m^2–2 kN/m^2	
Ionisation	10^{-3}–1 N/m^2	

Note that in all gravimetric types of gauge, where the highest accuracy is required corrections for the local gravitational constant (which varies with geographical location and altitude) may have to be made.

Applications and accuracy

Table 18.3 gives approximate pressure ranges associated with various types of pressure gauge, with some indication of accuracy and application.

LEVEL

The need to determine the level of liquids or granular materials in hoppers or closed vessels frequently arises. The object may be to maintain a constant level in the vessel, or it may be to determine the start and stop sequences for discharging or refilling. Many of the techniques used for liquids may be applicable to granular solids which are fine enough to have some of the properties of liquids.

The most familiar example of liquid level control is the domestic water tank in which a float both senses the liquid level and actuates a water valve to open when the level

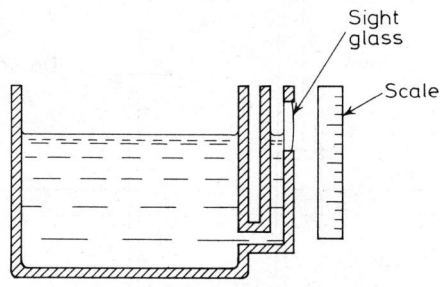

Figure 18.25. Simple gauge glass

falls and close when the set level is regained. Such a device is both simple and effective, but it lacks sufficient accuracy for some purposes and operating range for others.

The most rudimentary level measurement devices are visual. If the vessel is transparent or can have a transparent section inserted a calibrated scale is all that is needed to determine level. A more common variant is the gauge glass which is, as Figure 18.25 shows, an open ended manometer. Many vessels are, however, operating under internal pressure and the gauge glass may take on a more sophisticated form (Figure 18.26), including facilities for isolating the gauge in the event of rupture of the sight glass. Gauge glasses can be used up to 300 bars pressure and up to temperatures of 550°C.

Other basic visual devices used with graduated scales or tapes are shown in Figure 18.27.

Float types

Direct-reading float gauges employ a float on a radius arm which drives a pointer through a suitable mechanical linkage (Figure 18.28). The buoyancy of the float determines the power available to overcome static and dynamic friction. The nature of the mechanical construction of this type of gauge limits the range of level over which it

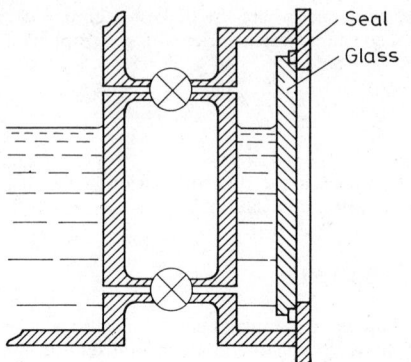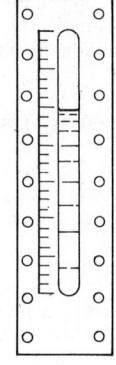

Figure 18.26. Pressure gauge glass

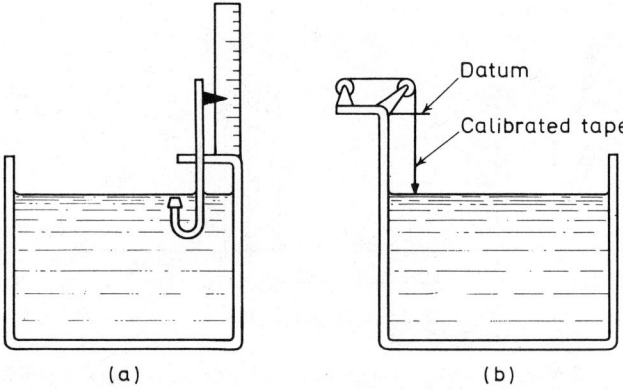

Figure 18.27. Point-contact dip stick and calibrated tape

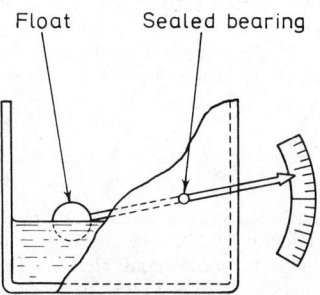

Figure 18.28. Simple float gauge

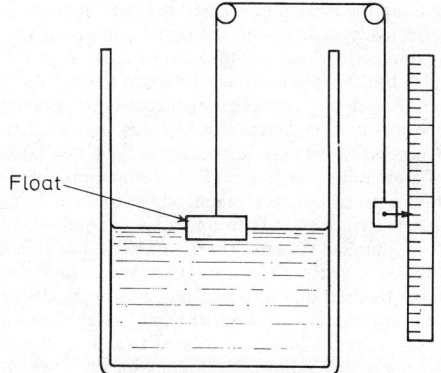

Figure 18.29. Float and tape gauge

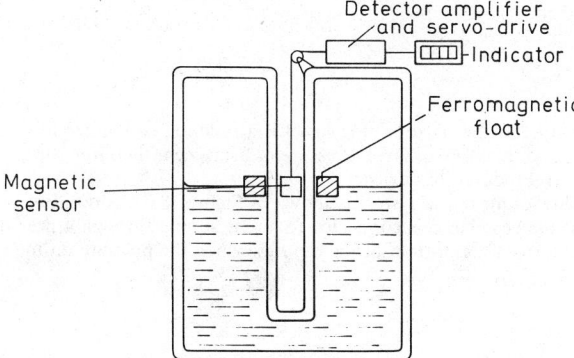

Figure 18.30. Magnetic float level detector

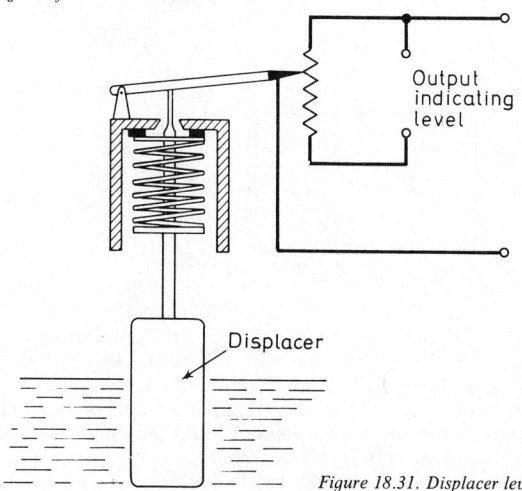

Figure 18.31. Displacer level

can be used. Magnetic coupling is sometimes used between indicator and driver, so that the indicator can be effectively sealed from the liquid content of the vessel.

A more versatile version is the float-and-tape gauge illustrated in Figure 18.29. Here the use of an inextensible but flexible tape permits much greater travel of the float, and such a gauge may operate over several metres displacement. Once again the number of configurations in which this type of device can be recast is considerable.

Where isolation is necessary between the detector and the liquid (which may be highly corrosive or inflammable) a magnetic float is sometimes used (Figure 18.30). A 'follower' magnet within a protective non-magnetic sealed tube follows the position of the annular float. Magnets strong enough to operate a counterbalanced pointer can be used to produce an all-mechanical system. Alternatively the position of a ferromagnetic float can be detected by a servo-driven electromagnetic position sensor within the tube.

The float may however be used in a different mode as in the displacer type of gauge (Figure 18.31). The float in this case is cylindrical and longer than the limits of level to be measured. It is immersed in the liquid suspended from a spring. As the liquid level changes, the changing apparent weight of the float as shown by extension of the spring can be used to operate a pointer or suitable position transducer to produce an electrical output. Again isolation from the vessel contents may be achieved by magnetic coupling to the displacer rod.

Pressure types

The pressure existing at the bottom of a vessel is a function of the height of the fluid column above it, and therefore of level. Pressure measurement therefore forms the basis of a number of level measuring devices.

A simple air flow or air purge system is shown in Figure 18.32. A dip tube is lowered to the bottom of the vessel and is connected to an air supply through a pressure regulator. Air will leak from the bottom of the dip tube when the pressure on the tube side

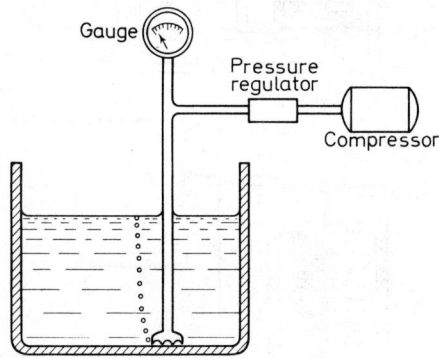

Figure 18.32. Air-purge gauge

of the regulator reaches the pressure at the bottom of the vessel, maintaining the pressure in the dip tube very slightly above that at the bottom of the vessel. A pressure gauge connected to the tube may be calibrated in terms of level. The same principle can be applied to closed pressurised vessels by using a differential pressure gauge connected back to the top of the pressure vessel (Figure 18.33). The supply compressor will have to be powerful enough to supply air pressure well above that in the vessel. Where air is inadmissible, inert gas, or even a liquid may be used.

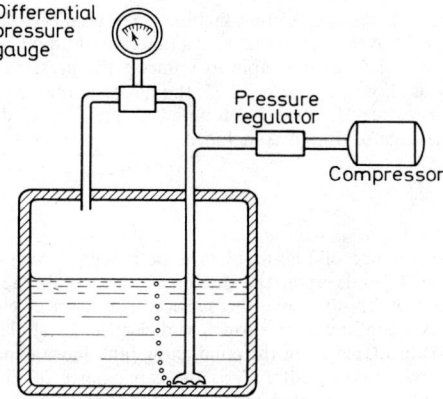

Figure 18.33. Air-purge in pressurised vessel

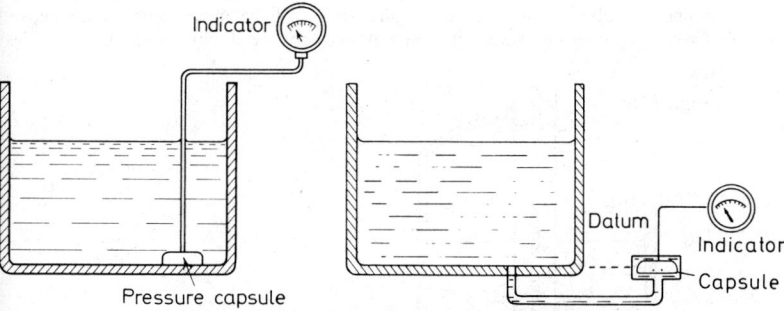

Figure 18.34. Pressure capsule gauge   *Figure 18.35. Pressure capsule—externally mounted*

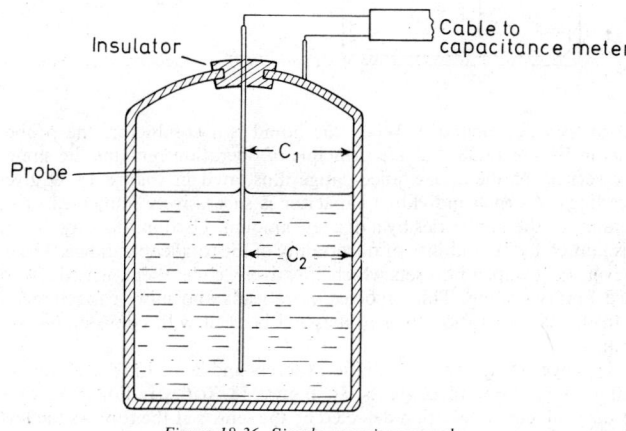

Figure 18.36. Simple capacitance probe

Pressure capsules with remote reading facilities may be placed at the bottom of the vessel as in Figure 18.34. Again the pressure is a function of liquid head and therefore of level. If it is not convenient or desirable to immerse the pressure detector it may be mounted externally as shown in Figure 18.35, though care must be taken to ensure that at installation the pressure detector is mounted at the correct datum level, or that compensation for change of datum is included.

Other types

The capacitance level gauge of Figure 18.36 is potentially a very rugged, yet simple, device. Its operation depends upon the dielectric constant of the liquid being much higher than that of the vapour above the surface. The form of electrode depends on whether the liquid is a conductor or a low-loss dielectric. In the latter case the simple probe (Figure 18.37) insulated from the conducting tank has a capacitance to the tank walls composed of two components. C_1 is the capacitance of the section above the liquid level. C_1 is low because the dielectric content of the vapour is low (close to unity). C_2 is the capacitance of the submerged section which, length for length, will exceed the value of C_1 by a factor close to the dielectric constant of the liquid. As liquid level changes, the total capacitance alters and is measured on a sensitive capacitance bridge.

In practice a probe of the form of Figure 18.37 will be much more sensitive and potentially more accurate, since the capacitances involved will be higher and less

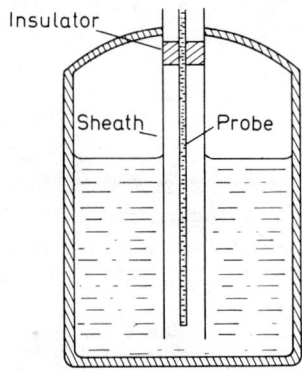

Insulator

Sheath   Probe

Figure 18.37. Capacitance probe in sheath

influenced by stray capacitance. When the liquid is a conductor, the probe must be insulated as in Figure 18.38, but the principle of operation remains the same.

Another version of the capacitance gauge illustrated in Figure 18.39 gives discontinuous readings. As each individual capacitor is successively immersed or uncovered the capacitance of the unit varies by a discrete amount. Used in this way, the number of elements is limited by the stability or uncertainty of the total capacitance. The capacitors may, however, be grouped into sets which effectively 'code' them, usually in some form of modified binary coding. This involves more leads and more capacitance detecting circuitry. In this configuration long probes can be read with considerable accuracy in digital form.

Nucleonic gauges using gamma radiation sources and detectors depend on absorption of the radiation in the tank contents. In Figure 18.40 the radioactive source at the bottom of the tank emits radiation detected by the sensor at the top. As the level rises in the tank more of the radiation is progressively absorbed, and an approximately linear

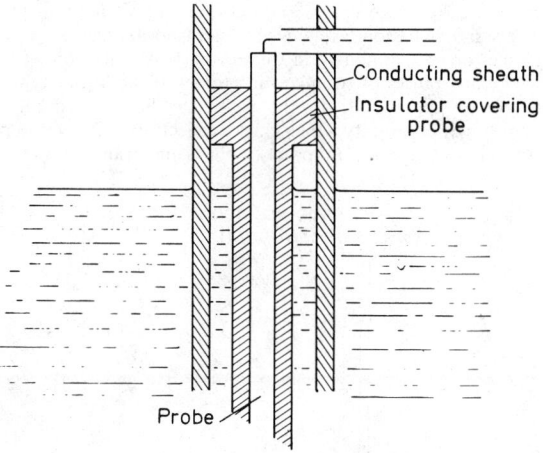

Figure 18.38. Insulated capacitance probe for conducting liquids

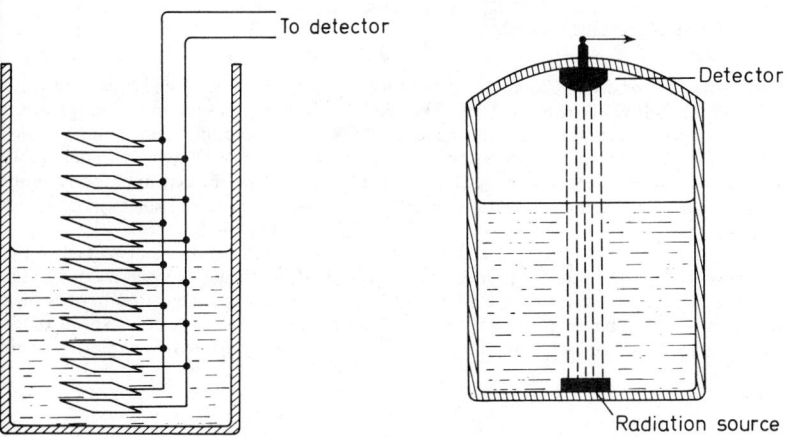

Figure 18.39. Discrete capacitance gauge   *Figure 18.40. Nucleonic level gauge*

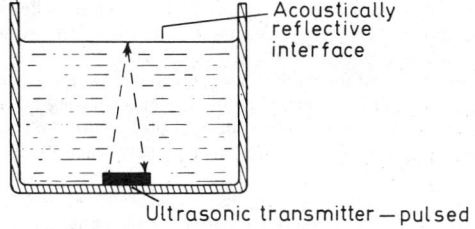

Figure 18.41. Acoustic level gauge—submerged

calibration can be achieved. The source and detector can often be mounted external to steel-walled vessels and radiation levels need present no hazard.

Ultrasonic gauges operate as shown in Figure 18.41. An ultrasonic transducer at the base of the vessel emits pulses of sound which travel to the liquid-vapour interface at the velocity of sound appropriate to the particular liquid medium. The interface represents an acoustic discontinuity which causes reflection of the sound pulse back to a receiving transducer at the base. In practice the same transducer may act as both

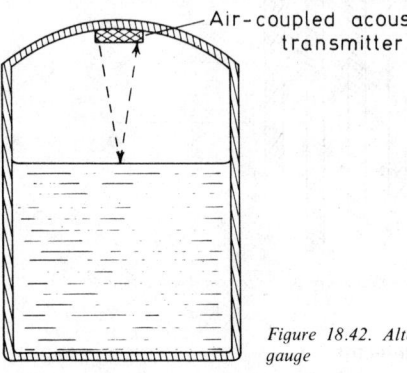

Air–coupled acoustic transmitter

Figure 18.42. Alternative form of acoustic level gauge

transmitter and receiver. The total transit time is a function of the length of path traversed, and therefore of the liquid level. The velocity of sound is a function of temperature and of composition, and such effects have to be taken into account. Aeration or bubbles may also lead to errors or inconsistency. An alternative form is shown in Figure 18.42 where the vapour space above the interface is measured. Although apparently more attractive this arrangement may present difficulties because of the problem of coupling adequate acoustic energy of the required frequency into the vapour space.

Thermal gauges use a heated filament or tape immersed in the liquid. The part of the tape immersed in the liquid will lose heat more rapidly than the exposed section. This causes a change in the resistance of the tape; the resistance is measured by a bridge which gives a reading proportional to level. This type of gauge may be useful in fluids which congeal on cooling, and thus leave a coating on gauge parts, impairing their operation.

Discontinuous gauges

The gauges described so far have been of the continuous type which can be used to give an indication of level over the whole or part of the depth of the vessel. In many cases, especially for level control purposes, it is adequate to know whether the level has reached a predetermined datum or not. Variants of many of the above methods lend themselves to this type of application. Float gauges with a relatively small travel can operate magnetic, electrical and mechanical switches at pre-set level. Simple capacitance gauges with the capacitor plates or probe at the preset level may be used in an on-off mode. Thermal tapes may be replaced by a thermal probe with the sensitive element at the tip. Free-flowing air gauges will develop back pressure when the liquid level reaches the orifice.

Ultrasonic and nucleonic gauges can also be used to detect the pressure or absence of contents at the datum level. In the nucleonics case the liquid content cuts off or attenuates the radiation. On the other hand the ultrasonic beam will be transmitted across the vessel

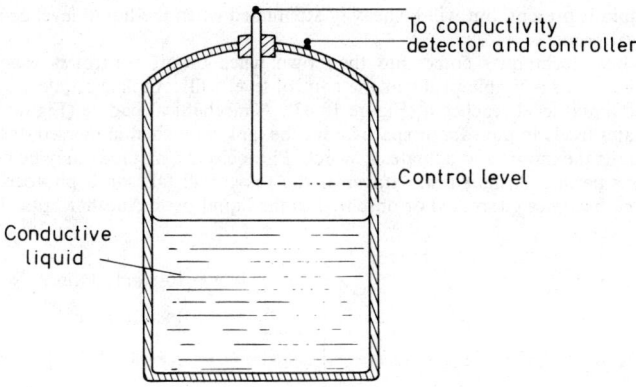

Figure 18.43. Conductive level probe

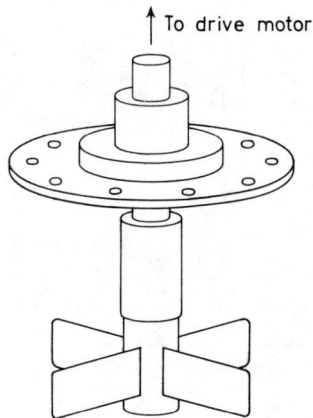

Figure 18.44. Paddle level detector

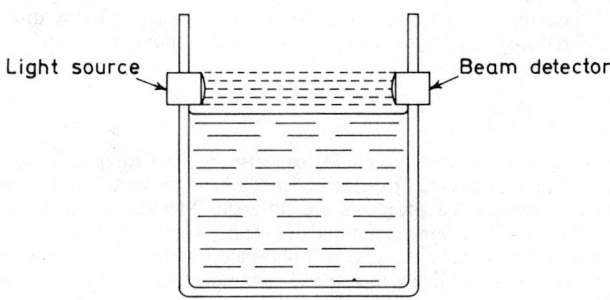

Figure 18.45. Light level detector

when liquid is present, but will be heavily attenuated when the liquid level drops below the critical point.

Additional techniques come into their own when on-off control is sought. With conductive fluids a simple probe at the control level will complete an electrical circuit when the liquid level reaches it (Figure 18.43). A mechanical paddle (Figure 18.44) or vane rotates freely in the vapour space above the tank, but when immersed, the reaction torque stalls the motor and actuates a switch. Photoelectric methods may be used when conditions permit. A light beam traverses the vessel and falls on a photocell (Figure 18.45), but becomes alternated or obscured as the liquid rises. Another optical device is

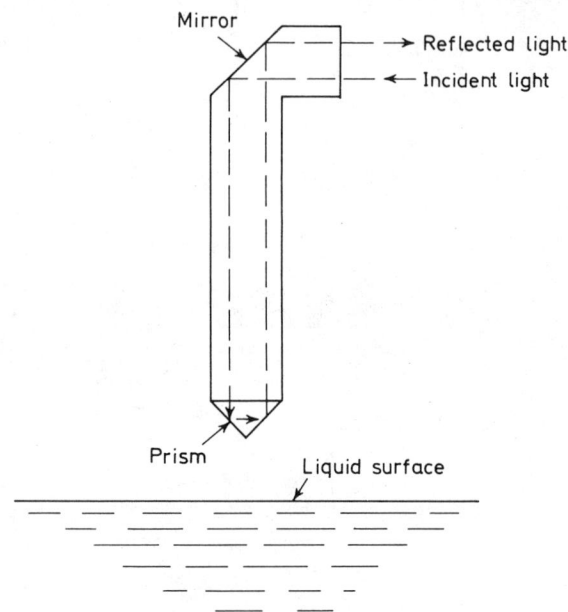

Figure 18.46. Optical dipstick

the optical 'dip-stick'. This is a probe with a prism at the lower end (Figure 18.46). Above liquid level, light passed down the tube is reflected back up onto a photo-detector. When the prismatic end is immersed the refractive index discontinuity is substantially reduced and little light is returned to the photodetector.

Solids

Some of the techniques used for liquids can also be used for powdered or granular solids. Some finely powdered free-flowing solids have properties not dissimilar from liquids, but in general the problems are different. Mechanical methods which are variants of the paddle or vane technique are clearly applicable to solids. Capacitance gauges will perform in many cases, as will photoelectric detectors. Solids with appreciable moisture content are likely to respond to conductivity probes. Nucleonic gauges detect solids as effectively as liquids. Thermal devices and air-flow techniques may also be used where conditions allow.

Floats as such are not likely to be applicable, but a plumb-line technique in which a float-like device is lowered until it touches the surface may be used in conjunction with a servo-system operated by strain in the suspension to maintain it in contact with the solids surface (Figure 18.47). A pressure diaphragm in the vessel wall can be used to detect the presence or absence of material at diaphragm level; a variant of this is to detect the damping of a vibrated diaphragm operating rather like a loudspeaker cone.

FLUID FLOW

Most industrial chemical processes and many others require the measurement or control of fluid flow rates. In the majority of cases, measurement of volume flow will

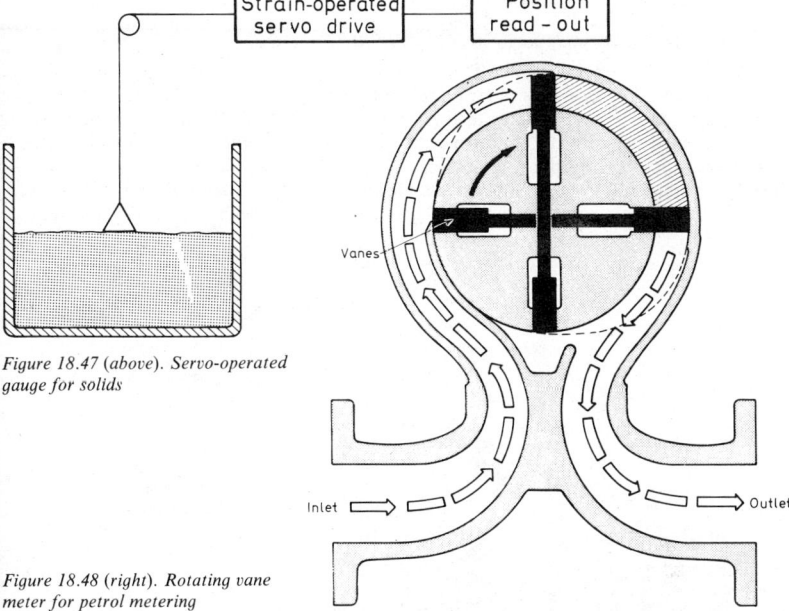

Figure 18.47 (above). Servo-operated gauge for solids

Figure 18.48 (right). Rotating vane meter for petrol metering

suffice, but in some situations it is important to know mass flow. For many purposes repeatability of flow measurement is of more importance than absolute accuracy.

The diversity of requirements for flow measurement in terms of range, type of fluid, accuracy, pressure-loss characteristics, and mass/volume sensing has led to the development of a very large number of flow measuring devices, so that it is only possible here to outline the principles and basic characteristics of the more significant versions.

Positive displacement

Perhaps the most basic and accurate method of flow measurement is by discharging the fluid into a container of known volume, and emptying the container to the outlet when full while the discharge is diverted to a second container. Alternatively, the container can be emptied when a predetermined weight is reached if mass flow is required. Indeed, a refined version of this principle is used for flowmeter calibration.

Other methods of passing known volumes of liquid more or less continuously through closed conduits are in widespread use. The simplest of such devices resembles a two-cylinder internal combustion engine, suitably ported and driven by the measured fluid. Such a piston meter is lubricated by the fluid, and provided the cylinders are efficiently filled and exhausted on each stroke, the flow is a function only of the swept volume; the number of cycles of the meter can be indicated by a suitably geared counter. The rotating vane meter of Figure 18.48 accomplishes the same purpose without the necessity of reciprocating parts. The defined volume is that enclosed between successive vane seals and the annular wall of the meter. Rate of rotation or number of rotations define flow rate and total flow respectively. Piston meters are capable of accuracy around $\frac{1}{2}\%$ and rotating vane meters may achieve $0\cdot1\%$ under favourable circumstances.

The nutating disc flowmeter (Figure 18.49) is another example in common use which depends on an accurately known swept volume. The disc is not free to rotate, being constrained by a radial partition in the working chamber. It is however free to nutate about its vertical axis. Flow of liquid through the meter causes the disc to be driven with a 'swash' action, the pin above the top bearing driving a gear train which actuates a counter through a sealing gland. The viscosity of the liquid takes care both of sealing and lubrication. For gases, a similar principle is used in the nutating bell meter in which a half submerged chamber nutates in the same way as the disc. In this case the working space is formed between the chamber wall and the liquid, which acts as a seal.

The gear-type or rotating-lobe meter of Figure 18.50 is also used for gas flow. The impellers rotate in opposing directions under pressure from the flowing gas. The swept volume is that between each impeller and the chamber wall; this volume is passed

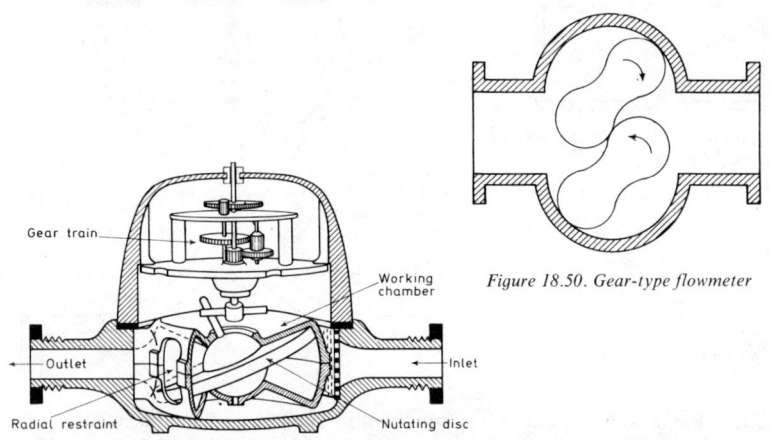

Figure 18.50. Gear-type flowmeter

Figure 18.49. Nutating disc meter

through the meter four times per revolution. The relative positions of the two impellers is maintained by gearing and sealing is effected by close tolerances and scrapers on the impeller tips, which also serve to keep the chamber walls clean.

The oscillating or semi-rotary piston meter of Figure 18.51 looks somewhat different. Liquid enters through the inlet port on one side of the partition, which also acts as a guide for the split cylindrical piston. The liquid flow drives the piston in an anticlockwise direction. The piston is constrained by the radial partition so that the piston slot slides up and down over the partition. The piston is also constrained by bearings which cause its centre to take a rotary path which maintains the outer wall of the cylinder in contact with the piston. The piston thus has an oscillatory motion which uncovers inlet and outlet ports successively and sweeps out a constant volume during each

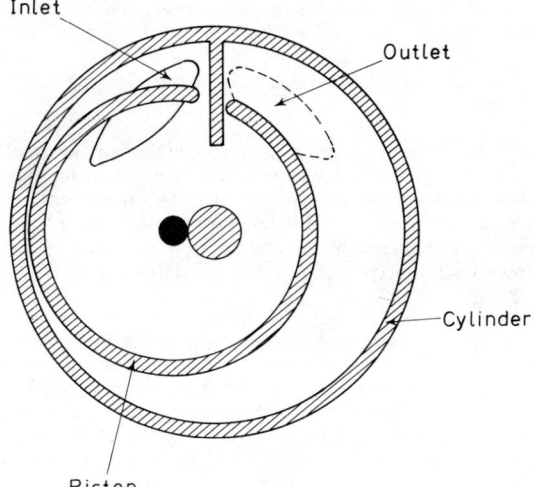

Figure 18.51. Oscillating or semi-rotary piston meter

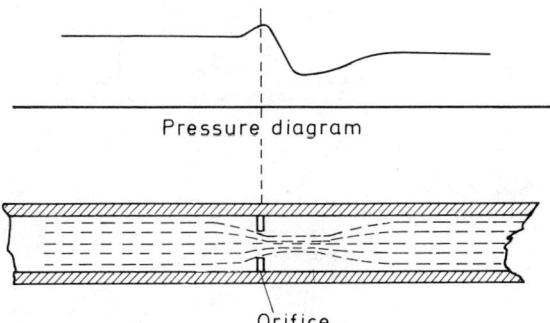

Figure 18.52. Orifice plate—flow and pressure patterns

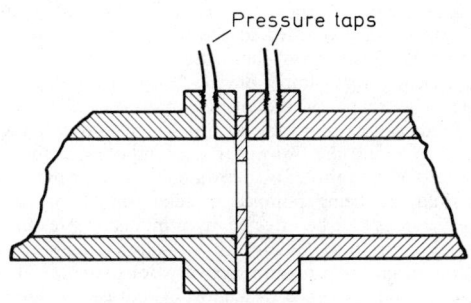

Figure 18.53. Orifice plate—flange taps

revolution. Piston cycles can be counted by a simple gear and counter mechanism. This type of meter is capable of around 2% accuracy in industrial conditions.

Differential pressure

Another genus of flowmeter depends on the principle that any obstruction placed in a pipe or closed conduit alters the pressure distribution within the pipe. Figure 18.52 shows a pipe with the most common type of restriction, the orifice plate. The pressure graph shows the nature of pressure distribution on either side of the restriction. By measuring pressure at points just before and after the restriction it can be shown that the relationship between flow rate Q and pressure difference Δp is of the type

$$Q = k \, (\Delta p / \rho)^{\frac{1}{2}}$$

where ρ is fluid-density. The square-root relationship is a disadvantage which is reflected in the reduced effective range over which any one meter may be used effectively, although the principle is used in pipes from a few millimetres to metres in diameter.

Positioning of the pressure tapping points is of considerable importance. Ideally they should be placed where disturbing factors are least. For example, considerable pressure

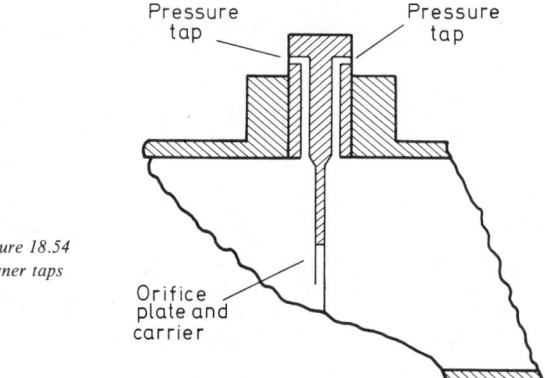

Pressure tap

Pressure tap

Figure 18.54
Corner taps

Orifice plate and carrier

instability is likely to be encountered between $\frac{3}{4}$ and 3 pipe diameters downstream. Flange taps are sometimes used (Figure 18.53), having the advantage that the taps are readily accessible for cleaning. Corner taps (Figure 18.54) have constructional advantages in that they can be incorporated in the orifice plate carrier, but they may be susceptible to disturbance upstream of the orifice.

Orifice plates are clearly very simple devices, easy to install and to maintain, and easily reproducible. The shape of the orifice, however, must be maintained, and corrosive or dirty fluids can cause problems unless constant care is exercised.

Where pressure drop across the flowmeter is of importance the orifice plate is at a disadvantage, especially if the ratio of orifice diameter to pipe diameter is small. Where large volumes of fluid are being pumped at high velocity in industrial conditions considerable power savings can be effected by minimising pressure drop at the restriction. The classic Venturi tube (Figure 18.55) follows the flow-lines of fluid at a constriction, resulting in greater pressure recovery downstream. The practical Venturi consists of two conical sections joined by a short cylindrical section. The shorter of the two cones forms the lead-in section to the restriction, and the longer cone forms the exit or recovery section. Long downstream cone sections (5–7° angle) have better pressure recovery characteristics than shorter sections (14–15° angle) but are more expensive to construct and require more space to install. Because of the smooth flow-lines of the Venturi it is much more suitable for fluids with suspended solids than the orifice plate, which may clog up readily.

The Dall tube (Figure 18.56) has even lower pressure loss than the Venturi, and has the further advantage of being short in length. It consists of two truncated cones as in

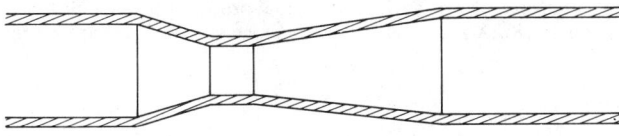

Figure 18.55. Venturi tube

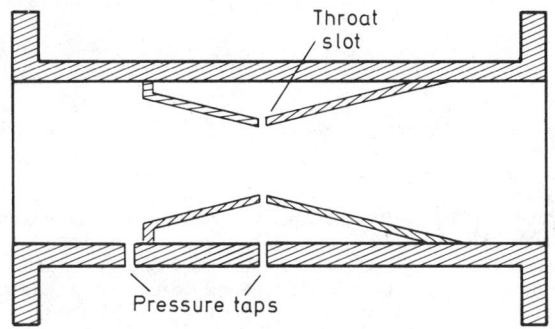

Figure 18.56. Dall tube

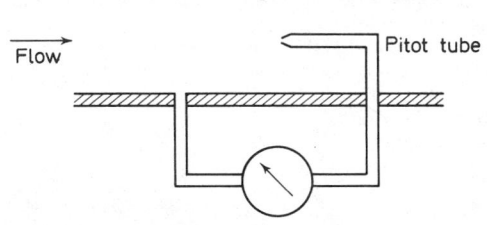

Figure 18.57. Pitot tube

the Venturi, but with steeper cone angles and no cylindrical section between. At the conjunction of the cones there is a slot which acts as a 'wall-attachment' device for the fluid stream, reducing turbulence and pressure loss. The Dall tube is not suitable for fluids containing solids which might settle in the slot.

The Pitot tube is more often used for exploring flow patterns than as a flowmeter in its own right. Essentially it measures flow velocity at a point in the flow stream. The Pitot tube is shown diagrammatically in Figure 18.57. It measures the difference between the 'impact' pressure p_1 and the static pressure p_2 in the conduit. As with orifice and Venturi meters, there is a square-root relationship between the differential pressure and flow.

Area meters

Variable area flowmeters are simple in concept. The flow acts on an obstruction and moves it in such a way as to increase progressively the effective cross-section of the flow path. In the Rotameter type of device the internal profile of the pipe or tube is slightly tapered (Figure 18.58). A plummet within the cone is forced upward by the flowing

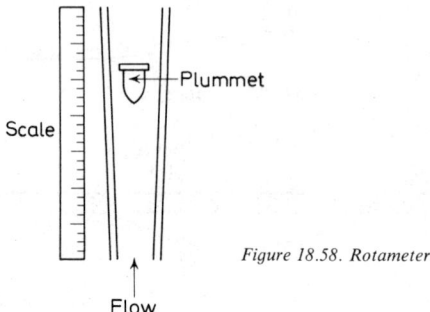

Figure 18.58. Rotameter

fluid and suspended at a point up the taper where the drag of the plummet counterbalances its weight. The flow rate is measured against calibrations on the side of the tapered tube where this is of glass. Earlier plummets or floats were designed to rotate with flow and thus gain an element of stability. Some floats, however, are guided within the tube on stabilising constraints. The flowmeter has a virtually linear characteristic.

For use with opaque liquids or at high pressures the tube may be of metal such as stainless steel or non-ferrous alloy. The position of a magnet in the float can be detected from outside the tube by a simple magnetic coupling to an indicating mechanism external to the tube, or by a servo-controlled external follower mechanism. Many other variants of the principle may be found, mostly concerned with convenience of indicating float position in difficult conditions.

Instead of a free float other variable area meters use spring or gravity constrained pistons, flaps, or discs to expose an increasing cross section of the meter to flow. The displacement of the piston or flap may be detected mechanically or electromagnetically to provide indication of flow rate.

Open channels

Rather different forms of meter have been evolved for fluids flowing in open channels. The simplest is the weir (Figure 18.59). Flow over the weir depends on the head of fluid upstream over the weir plate height. The plate itself is sharp-edged and may be rectangular or V-notch. Other forms are used to give slightly different characteristics. Flow rate over a rectangular weir is related to upstream head h by a $\frac{3}{2}$ law (i.e. $Q \propto h$ exp $\frac{3}{2}$) approximately, while for a V-notch weir a $\frac{5}{2}$ law applies. Ideally, immediately above the weir flow rate should be reduced to a minimum. In practice corrections to the idealised flow relationships have to be applied for discharge coefficient, effect of contraction of flow at notch edges, and other factors.

The weir is analogous to the orifice plate in closed conduits; the flume is the analogue of the Venturi, and is used for the same reason, reduction of head loss. The shape of the

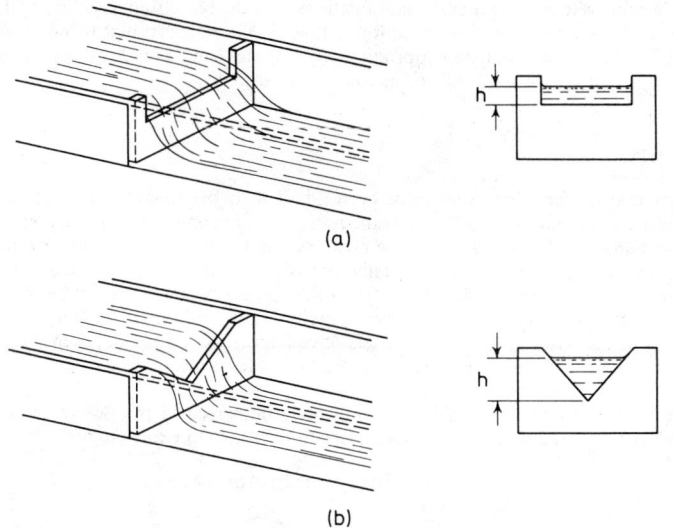

Figure 18.59. Rectangular (a) and V-notch (b) weirs

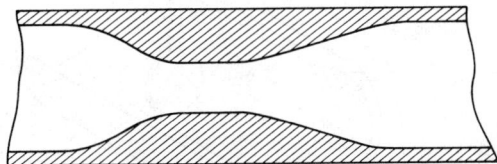

Figure 18.60. Flume: Plan view

Turbine rotor

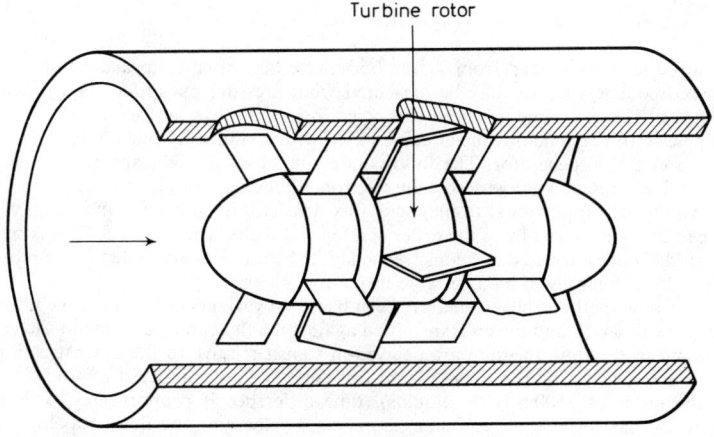

Figure 18.61. Turbine meter

flume is subject to many possible modifications, but the basic flume section is shown in Figure 18.60. Under 'free flow' conditions flow rate is proportional to upstream head with a $\frac{3}{2}$ power law relationship approximately. Flumes can be made as large as required without difficulty and are used to measure very large flows.

Other types

A great many other principles have been adapted to the measurement of flow. The turbine meter (Figure 18.61) is one of the more widely encountered in the measurement of clean liquids and gases. The turbine meter consists of a multi-bladed rotor immersed totally in the fluid stream and usually preceded by flow straighteners. By careful bearing and hydrodynamic design the friction drag may be made very small indeed, and very high accuracy is claimed for turbine meters. The rotor need have no mechanical coupling to the associated tachometer. Electromagnetic coupling between a magnet in the rotor and a pick-up coil outside the pipe provides a drag-free measure of rotation speed.

The pressure of flowing fluid on a drag body immersed in the flow creates a force which has been used with a force transducer to measure flow rate (Figure 18.62). More

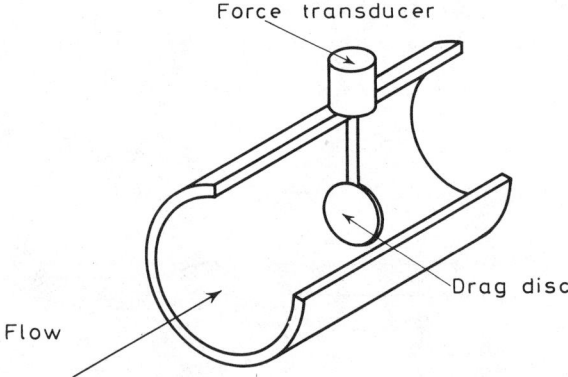

Force transducer

Drag disc

Flow

Figure 18.62. Drag flowmeter

recently the vortices shed from a drag body have been used to measure flow rate; the vortex frequency, which may be measured from pressure pulsations, is a measure of flow rate. This principle has been embodied in several physical forms.

The electromagnetic flowmeter can be used with conducting fluids only. It is essentially an electrical generator. The fluid conductor moves at right angles to a magnetic field and generates a voltage across the electrodes housed in the pipe walls (Figure 18.63). The voltage is proportional to magnetic flux density, pipe diameter and mean velocity of the fluid, and is usually of the order of a few millivolts. The magnetic field is supplied by 'saddle' coils energised at mains frequency; the electrodes are sealed in the pipe walls and isolated from them by a suitable insulating material.

Ultrasonic methods have generally been based on variants of the principle illustrated in Figure 18.64. Sound pulses transmitted against the flow and travelling at the velocity of sound in the fluid medium are delayed in transit relative to the downstream pulses which are sped up by the flow. Where flow rate is slow compared to sound velocity (of the order of 1500 m/s for liquids), time difference is proportional to flow, but correction has to be made for changes in sound velocity in the fluid, which is usually highly temperature sensitive. By making each arriving sound impulse trigger the next

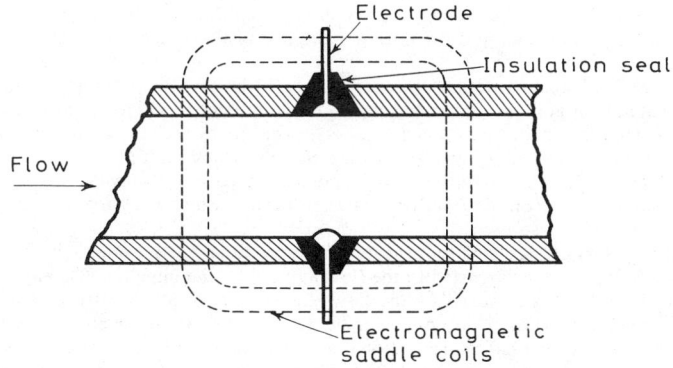

Figure 18.63. Electromagnetic flowmeter

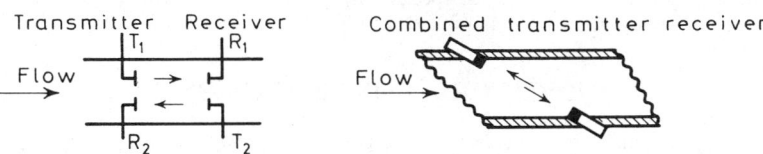

Figure 18.64. Ultrasonic flowmeter

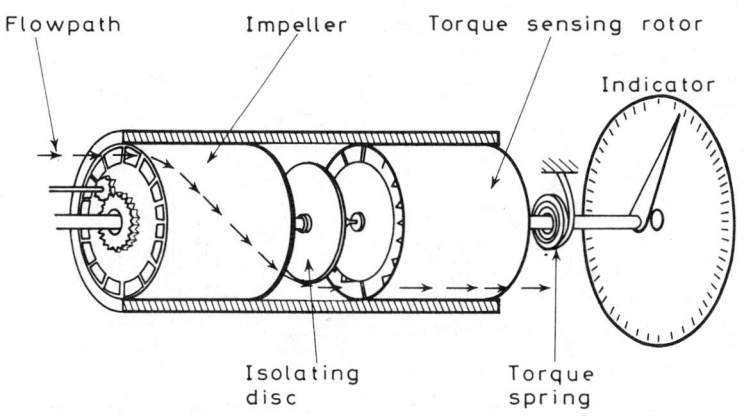

Figure 18.65. Swirl mass flowmeter

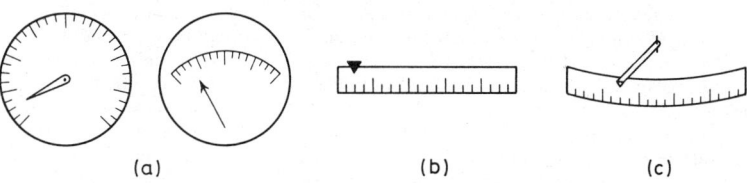

Figure 18.66. Circular, linear and circumferential pointer displays

transmitted pulse the two circuits may be set into oscillation, and the difference or beat frequency is proportional to flow rate. A more practical arrangement is shown in Figure 18.64b.

The cooling effect of a gas stream on a hot wire is the basis of the hot-wire anemometer. It is capable of good accuracy, but has to be calibrated to achieve good results.

Velocity flowmeters can be made to act as mass flowmeters by the use of a continuous density measuring device and a signal processing element which multiplies the velocity signal by the density signal. The simplest form of densitometer is a float whose displacement is measured; more complex and sophisticated devices which are capable of high accuracy are, however, also used.

Certain flowmeters are designed to measure mass flow directly. An example is shown in Figure 18.65. Rotation is imparted to the flowing fluid by the impeller. The angular momentum of the fluid is removed by the flow straightening rotor downstream. The rotor is constrained to act as a torque sensing device, and the torque which is detected by an angular position pick-off is proportional to mass rate of fluid flow. An isolating disc between the impeller and torque sensor 'decouples' the sensor from viscous drive at zero flow.

Further devices and variants for special purposes are too numerous to be covered here. They are extensively covered in available literature.

INDICATORS AND RECORDERS

The most elementary form of indicator is the pointer moving over a scale, which may be of circular, linear, or circumferential configuration (Figure 18.66). Such indicators may be driven directly by mechanical means or actuated remotely by pneumatic or electrical signals.

Mechanical motions also lend themselves to driving a drum counter (Figure 18.67) to give digital readout. Electrical signals may also be displayed digitally on self-luminescent panels (Figure 18.68) after the signal has been digitised by an analogue-to-digital converter.

Analogue displays (e.g. pointer on dial) have two main merits: fairly accurate information can be absorbed very rapidly, at a distance if required, as with a distantly

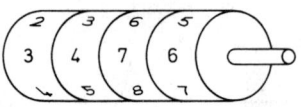

Figure 18.67. Drum counter or cyclometer

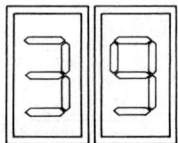

Figure 18.68. Electro-luminescent display

glimpsed clock face; and trends or fluctuations can be detected very rapidly. Digital displays are unambiguous, can be read by the unskilled, and may be more accurate. They can, however, be no more accurate than the input information or the associated circuitry; a factor sometimes masked by the seeming validity of an array of numeric characters. When signals are not constant digital displays may be very confusing.

When temporary or permanent records are required, chart recorders are the most common means of information storage. A circular chart recorder comprises a replaceable circular chart on a backing disc which rotates at a rate chosen to suit the requirements of the situation, and usually electrically driven. A pen fed with ink from an accessible reservoir is mounted on the end of an arm driven by the actuating signal which may be electrical, pneumatic or mechanical. Time indices are marked on the chart.

Several signals may be marked on a single chart using separate pointers designed to allow cross-over, and using coloured inks to differentiate records.

Strip chart recorders operate on the same principle except that the chart record is in roll form fed steadily down the face of the instrument under the recorder pen.

Self-balancing potentiometric recorders operate from electrical signals only. The potentiometric bridge is balanced by driving a carriage along a resistive slidewire until the null point is reached. The same carriage transports the pen or pens across the chart paper. Sensitivity of these recorders is high and may be as little as 100 μV full scale, with resolution to 1 μV. On higher ranges, such recorders give accuracy of $\frac{1}{4}\%$ full scale. Full

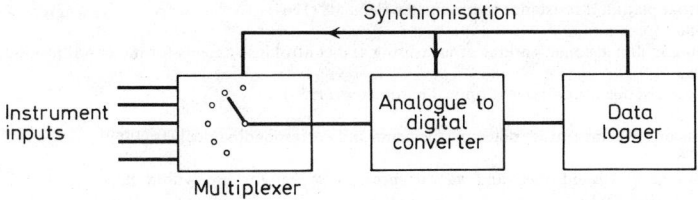

Figure 18.69. Multiple data logging system

scale deflection may be achieved in 0·25 second. Chart speeds are usually adjustable in fixed steps over a wide range.

Much instrumentation is now designed to be connected to computer systems, or to be compatible with computers. Where information has to be stored as well as processed, records from such systems are fed to a data logger which usually prints a numerical record at predetermined intervals, or on request. Since data loggers can handle the output of many instruments at high speed, a multiplexer connects instruments in turn to the logger which assimilates their information sequentially. Most instruments supply their information in analogue form, and an analogue to digital converter is also needed in the system. By placing this between the multiplexer and the logger, one converter can serve many instruments. Figure 18.69 illustrates the system.

REFERENCES

1. CONSIDINE, D. M. (Ed), *Process Instruments and Controls Handbook*, McGraw-Hill (1957)
2. MILLER, J. T., *The Instrument Manual*, 4th ed United Trade Press (1971)
3. DOYLE, F. E., *Instrumentation: pressure and liquid level*, Blackie (1969)
4. DOYLE, F. E. and BYROM, G. T., *Instrumentation: temperature*, Blackie (1970)
5. JONES, E. B., *Instrument technology*, Vol. 1, 2nd ed Butterworths (1965)
6. HALL, J. A., *The measurement of temperature*, Chapman & Hall (1966)

BRITISH STANDARDS

BS 1041
Code for temperature measurement. Section 2.1. Liquid-in-glass expansion thermometers; Part 2. Expansion thermometers; Part 3. Industrial resistance thermometry; Part 4. Thermocouples; Part 5. Optical pyrometers; Part 7. Temperature/time indicators (1972)
BS 1042
Code for flow measurement; Part 1. Orifice plates, nozzles and venturi tubes; Part 2. Pilot tubes; Part 3. Guide to the effect of departure from methods in Part 1 (1965)
BS 1259
Intrinsically safe electrical apparatus and circuits for use in explosive atmospheres (1958)
BS 1704
General purpose thermometers (1951)
BS 1780
Bourdon tube pressure and vacuum gauges (1960)

BS 1794
Chart ranges for temperature recording instruments (1952)
BS 1826
Reference tables for plantinum/rhodium v platinum thermocouples (1962)
BS 1827
Reference tables for nickel/chromium v nickel/aluminium thermocouples (1952)
BS 1828
Reference tables for copper v constantan thermocouples (1961)
BS 1829
Reference tables for iron v constantan thermocouples (1962)
BS 1904
Industrial platinum resistance thermometer elements (1964)
BS 1986
Design and dimensional features of measuring and control instruments for industrial processes
BS 2082
Code for disappearing-filament optical pyrometers (1954)
BS 2765
Dimensions of temperature detecting elements and corresponding pockets (1969)
BS 3166
Thermographs (liquid-filled and vapour pressure types) for use within the temperature range −30°C to 105°C (1959)
BS 3231
Thermographs (bimetallic type) for air temperatures within the range −20°C to 60°C (1960)
BS 3586
Analogue direct current signals for telemetry and control (1970)
BS 3680
Methods of measurement of liquid flow in open channels. Part 1. Glossary of terms; Part 3.Velocity area methods; Part 4. Weirs and flumes (1969)
BS 3693
Recommendations for design of scales and indexes (1964)
BS 3792
Recommendations for the installation of automatic liquid level and temperature measuring instruments on storage tanks (1964)
BS 4161
Gas meters (4 parts) (1968)

19 COMPUTERS IN DESIGN AND OPERATIONAL CONTROL

19 COMPUTERS IN DESIGN AND OPERATIONAL CONTROL

C. G. SCARBOROUGH

NUMERICAL CONTROL

Work in the aerospace industries in the USA[1] in the fifties led to the development of machine tools to which an electronic control system had been retrofitted. Information was fed to the control system by means of a punched paper tape. The information on the tape was in numerical format and thus, the name Numerical Control[2] (N.C.) was evolved. This name has now become accepted as meaning that its only application is in the removal of metal from basic stock with the machine tool under tape control.

Management system[3]. With computers involved in the preparation of control tapes, a much wider concept is now possible, and, indeed, is essential if the full benefits of numerical control are to be achieved.

The current climate

Generally speaking, the product of any company begins its life as an idea which is developed in the design office. The design scheme, the end product of the design office passes to the detail office and from there the necessary drawings find their way to the manufacturing and quality control departments. Only when the planning departments become involved is any consideration given to numerical control.

However, a considerable effort is devoted at this stage to ordering materials, tools, fixtures, etc., and to planning which machines will be employed in manufacturing the piece, together with the time of employment. Much of this type of planning is now using computers to reduce the manual effort required and to give more accurate information more rapidly.

The various aspects of planning systems have generally been developed as separate systems and without consideration of control tape procurement. The inevitable outcome of these isolated systems is that when the scheduled machining time arrives one or more essential items may fail to arrive according to plan. Thus, the planning is completely ineffective. Further, most of the systems employed are based on the assumption that once a plan is made, it is inflexible. No provision is made for feeding back information reflecting current changes in the shop floor situations to enable the computer to re-assess the requirements and so produce dynamic plans.

The requirements of a complete management system

The development of a complete management system must start no later than the design stage, when all dimensional information together with material types, etc., require to be stored on computer file. In this way, the detail office can obtain the information it needs quickly and without error, and, by the use of an N.C. drafting machine can also obtain drawings and accurate layouts as they are required. The ultimate development of such a system means that prints need not be produced and circulated by time-consuming methods. When the detail office has completed its task, the computer records can be up-dated to permit planning and inspection departments in turn to obtain their basic working information from the revised computer record, which at this point contains full information concerning:

(a) The detail drawing requirements.
(b) Material requirements.
(c) Number of pieces per assembly.

The presence of interacting computer systems is essential. Obviously, a system must exist by which the computer can be instructed of the number of assemblies to meet the complete order, and the dates on which deliveries are required. There must also be a system which can collate information of cutting tool requirements from direct instructions and from technological N.C. computer system files. If a mathematical model exists to calculate tool life, the computer is able to calculate the number of tools of each type which will be needed to meet the order.

Computer systems to cope with material requirements are also essential, as are systems for producing N.C. tapes. For a complete system it is necessary to develop a computer system which will simulate machine loading, making due allowance for setting, maintenance etc. The operation of such inter-reactive systems will produce many advantages, such as

(a) Eliminate duplication of effort, e.g. there will be no requirement for each area to produce its own scale drawings, since these will be available via an N.C. drafting machine and will be to a consistent standard. Similarly, dimensions can be related to any datum by a simple computer manipulation thus saving hours of drafting time.
(b) Suitable design of printed computer output can be used in conjunction with transparent overlays which are preprinted with standard block formats, etc. and xeroxed to produce a standard company document, again saving manpower on clerical tasks.
(c) Prepare instructions to (i) buying departments; (ii) tool stores; (iii) foremen.

Dynamic action

A system as outlined above would only produce a picture of events which should happen. Hence, an essential feature of such a system will be the ability to feed back to the computer information about events which have happened, or should have happened but have failed to occur.

In this way, the computer will be in a position to produce a new schedule and so give management a complete and up-to-date statement of the true position of the production cycle. Also, sufficient information exists on computer files to enable a more accurate costing system to be introduced.

Problems

Such a system drastically changes the ways in which things require to be done, and change is the most strongly resisted force in industry. It is important when embarking on any system which involves change to study carefully the problems that such change will produce, and try to reduce their effects on individuals to a minimum. Above all, it is important, once the changes have been decided on, to educate the people involved, directly or indirectly, in the objectives of the systems as well as their use.

Drawing standards

In most companies, the information contained on drawings is designed to suit conventional machines and gives no consideration to the changed requirements of N.C. or to the use of a computer as part of an information system.

Effects on drawing offices

Apart from the need to look carefully at drawing standards, consideration must be given to methods of dimensioning. For instance, many ways can be found by which a pattern of 100 holes equally spaced on a pitch circle may be defined on a detail drawing. In many offices, it is the practice to number the holes consecutively and to tabulate the co-ordinate dimensions X and Y. To achieve the values of these co-ordinate dimensions it is necessary to calculate the angle subtended by each hole at the circle centre relative to one datum

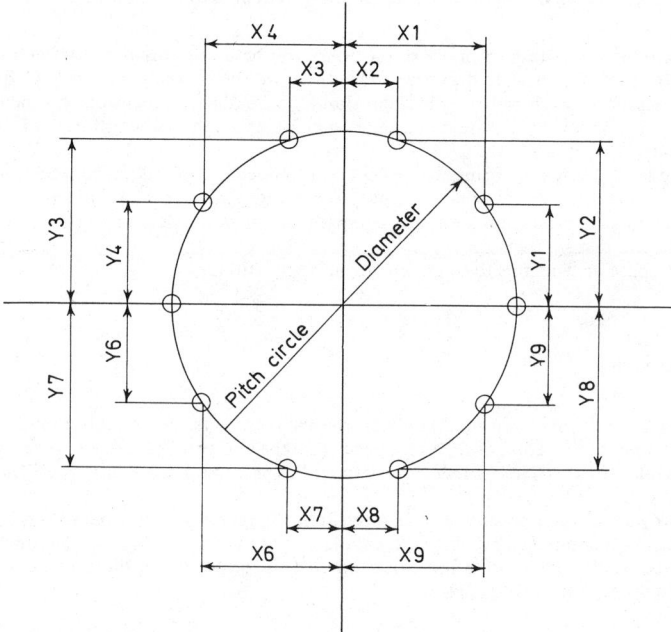

Figure 19.1. Holes on a pitch circle

and to determine the sine and cosine of these angles. Finally, to multiply these trigonometrical values by the pitch circle radius.

Thus, some four hundred calculations have been performed, and there still remains the task of printing the values of the co-ordinates on the drawing sheet. Apart from the time consuming and boring aspects of this method, it is extremely error prone. Furthermore, there is a tacit assumption that the datum selected is suitable to all areas of the company. This assumption may be unfounded, and it may be necessary, for instance, that the co-ordinates be transposed to another datum for inspection purposes. Thus, hundreds of further calculations are required, with more copying on to another drawing.

Using an APT like language (Automatically Programmed Tools) the pattern of holes described above can be described by means of the two simple statements:

CIR 1 = 0, 0, 20

PAT 1 = PATERN/ARC, CIR 1,0, CLW, 100

Such a statement could be used on the drawing to define the hole pattern dimensions precisely and without any possibility of ambiguity. Furthermore, such a definition could be interpreted almost anywhere in the world to mean exactly the same thing, because of the universal use of APT.

The pattern of holes shown in Figure 19.1 can be described by the statement

CIR 2 = 0, 0, 5

PATERN 2 = PATERN/ARC, CIR 2, CLW, 10

This clearly illustrates that the amount of effort required to describe a hole pattern

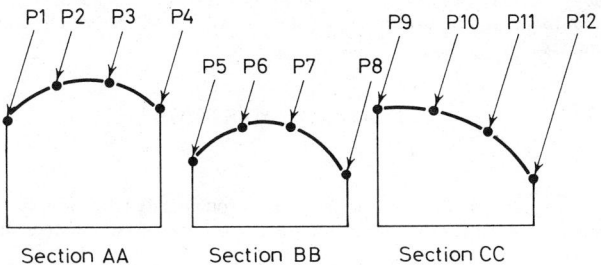

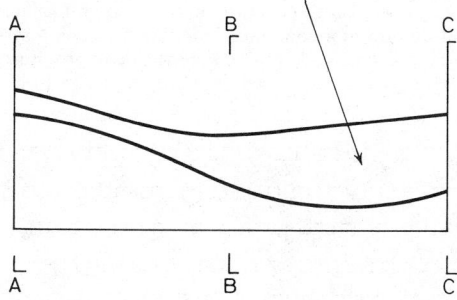

Figure 19.2. Faired sections

remains the same for the same type of pattern, irrespective of the number of holes in the pattern.

To write such a statement takes a few minutes, once the language is understood. If a list of the co-ordinates is required, the computer can supply this, and, if policy still dictates that these are issued with the drawing, consideration should be given to using a photo-reproduction of the computer output as a supporting document for the drawing. This results in maximum utilisation of manpower and a transmission of information at a minimum cost and without error.

Surfaces

The arguments used in connection with hole patterns apply even more in respect of surface definition. In many instances, surfaces are defined by a series of sections in which each section profile is defined by co-ordinates. The drawing then defines the surface by a statement such as FAIR between the given sections (Figure 19.2). This gives licence to anyone in the manufacturing organisation to interpret the surface in their own way. Thus, it is possible for the piece to be produced to one definition and inspected to another, a state of affairs which can scarcely be considered desirable.

The use of an APT statement such as:

SRLD = RLDSRF/SURF1,P1,P2,P3,SURF2,P4,P5,P6

(where SURF1 and SURF2 are previously defined surfaces and P1, P2, P3, P4, P5 and P6 are previously defined points) to define the surface (Figure 19.2) produces a complete and wholly unambiguous situation which means the same to all people, no matter what their native language.

CONTROL SYSTEMS

Tapes

Two types of control tapes are in use in industry today, magnetic tapes[4] and paper tapes. The former has two categories, digital magnetic tapes and phase analogue magnetic tapes.

Phase analogue magnetic tapes

These have been used mostly on three-axis continuous path machines and because of their nature, demand the use of a computer. The tape for such a system is normally a 6 mm wide plastic tape which is oxide coated on one side to give it magnetic properties. It is similar in all respects to the tape used on a tape recorder. In N.C. operation, four

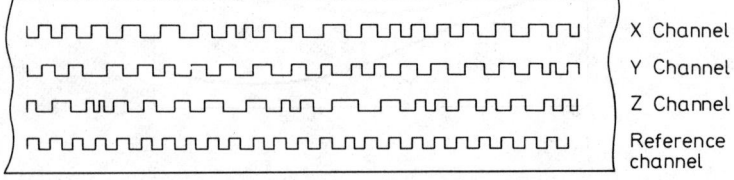

X Channel
Y Channel
Z Channel
Reference channel

Figure 19.3. Diagrammatic representation of phase analogue magnetic tape

tracks are recorded on the tape (diagrammatic representation Figure 19.3) one track controls the X axis, one the Y axis and one the Z. The fourth track is a reference track used for comparison with the other three.

The signals on each track are made up of a square wave form, with the reference track having a constant frequency of 100 Hz. The signals on the X, Y, and Z tracks are frequency modulated according to the relative distances to be moved by the cutting tool in the X, Y and Z axes. The modulated frequency varies between 50 and 100 Hz.

The X, Y, Z channels are continuously compared with the reference channel for differences of phase angle relative to the latter. The control system is so designed that, for any channel, 7·2 degrees difference in phase from the reference channel represents one basic unit (bit) of movement along the particular axis. The bit size is normally of the order of 5 micrometres.

A characteristic of this type of system is that feedrate is actually controlled by virtue of the rate of change of phase angle.

Advantages and disadvantages of phase analogue magnetic tape

Because a phase analogue tape contains all the information necessary to machine the piece, no further calculation, i.e. interpolation, is required in the control system. Hence, the latter should be simpler and cheaper than that required for a digital system.

The requirement to produce the magnetic form on tape involves the use of an expensive piece of equipment, e.g. a Copath, and this tends to offset the lower cost of machine control systems. Other disadvantages are:

1. Since information must be contained in the tape to provide the continuous signals for each basic unit (bit) of movement of the machine tool (generally a movement of the order of micrometres), the amount of calculation required to be performed by the computer is extremely large – hence computing costs.
2. The information contained on the tape cannot be read directly.
3. The continuous information and the need for special recording equipment means that manual editing cannot be undertaken, and to implement even the smallest modification requires a complete reprocessing of the information.
4. Information on the tape can easily be erased if brought into contact with strong random electric or magnetic fields.
5. If stored for long periods without use, the magnetic images tend to induce reflective images in adjacent layers thus producing a faulty tape.

Digital magnetic tapes

It is anticipated that the International Standards Organisation will recommend that coding and format of digital magnetic tape will conform to the standards laid down for paper tape.[8]

Therefore, apart from the fact that on these tapes the signals consist of magnetic spots instead of the holes used in the case of paper tapes, the very nature of magnetic tape permits high density storage of information. This results in shorter tapes than would be required when using paper tapes.

Care of magnetic tapes

It cannot be too strongly emphasised that it is extremely important to take proper care of magnetic tapes, both when in use and in storage. The following should be noted.

1. All control tapes are valuable and tangible assets, and should be treated with at least the same care given to fixtures and tools.
2. When in use on the control system, the lid of the control should always be closed. Tapes should be protected at all times against oils, coolants and swarf which are present on the shop floor.
3. Care should be taken that no undue longitudinal tension is applied to the tape because a stretched tape will undoubtedly be faulty.
4. Non-magnetic leaders and trailers should be attached to all tapes as standard practice.
5. When not in use, the spooled tape should be placed in a plastic bag and stored in a cardboard box.

6. The reel and box should identify the tape and the piece which it will produce.
7. Boxes containing spools of tape should never be stored in piles of more than ten.
8. Boxes of tapes should be stored neatly in metal cupboards devoted exclusively to this purpose.
9. Tapes should be protected from electric and magnetic fields. Care must be taken to avoid contact with magnetic chucks, etc. The chokes of fluorescent lights are a potential source of danger.

Punched tapes

The standard punched tape[5] is 25·4 mm in width although it can be of a number of different materials, e.g. paper, a material known as Durable or Mylar which may be pure Mylar or coated with various materials to produce additional properties. Whatever the material, the tape should conform to BS 3880.[5] On low speed punches or readers, deviation from this standard may not be too important, but at high speed any out of tolerance condition may result in a faulty tape and this can be a costly business if scrap is produced.

The material used should depend upon the tape readers on the various control systems. Paper tape, being cheap, is recommended as the standard for punching by computer and for use as a master tape, but its properties do not make it an ideal material for use on the shop floor. When the master tape has been prepared, it should be transferred to a reproducer which will produce one or more tapes, as required, in a material suitable for shop floor use. A comparator should then be used to compare the tapes so produced with the master tape. This step is considered necessary, because punches, being mechanical, do not have the reliability of other computer hardware. Moreover, it should be standard practice to make every effort to ensure that an error proof tape reaches the machine tool.

Experience has shown that the Durable type of tape operates well on photo-electric readers, having the requisite opacity. Mylar, is strong and is extremely suitable for mechanical readers which can have a tendency to damage the tape. The various coated or bonded Mylars are also extremely reliable in photo-electric readers.

Any tests to compare relative strengths of various tape materials should not be carried out on virgin tape, but on punched tape. The results from virgin tape are completely misleading.

Information on the tape

BINARY

The basis of all coding on a digital system is the mathematical concept of binary numbers. In this concept only the digits 0 and 1 are recognised. The fact that only 0 and 1 are recognised means that these can be related to correspond electrically to 'switch on' and 'switch off', magnetically to 'N' and 'S', on a punched tape to 'no hole' and 'hole', and so on.

The method by which decimal numbers greater than 1 can be represented in binary form is shown in Table 19.1.

A little thought will reveal that the right-hand digit in the binary number represents the decimal number 1, the next column to the left represents the decimal number 2, the next left 4, and the extreme left 8. Consider the binary number 111. This therefore, represents $4+2+1 = 7$ in decimal notation. Greater numbers can be represented in a similar way, but N.C. restricts itself to the numbers in the Figure 19.4, and this restricted use of binary notation is referred to as binary coded decimal. To represent the number 734, it is necessary to use 111 followed by 11 followed by 100, and this lends itself ideally to punched tape.

The standard paper tape has eight channels, and, with binary coded decimal, it is only necessary to use four of these channels. Using a hole to represent 1 the decimal numbers 1 to 9 can be represented on tape (Figure 19.4).

Channel 5 is called the parity channel and is punched to make the number of holes in any row odd or even according to whether the control system is odd or even parity,

Table 19.1. BINARY CODE

Column value →	8	4	2	1
Decimal number ↓	Binary number ↓			
0	0	0	0	0
1	0	0	0	1
2	0	0	1	0
3	0	0	1	1
4	0	1	0	0
5	0	1	0	1
6	0	1	1	0
7	0	1	1	1
8	1	0	0	0
9	1	0	0	1

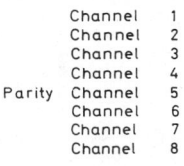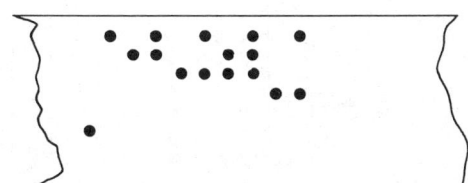

Figure 19.4. Binary digits in tape

i.e. requires an odd or even number of holes to operate. This parity condition is also used as a check on punches.

Alphabetic and special characters can be represented by using channels 6 and 7, (Figure 19.5).

The code discussed here, has been the International Standards Organisation Code (ISO), which is, incidentally, identical with that of the British Standards Institution (BSI).[8] Other codes are in existence, one of the most commonly used being the Electrical Industries Association (EIA) code.

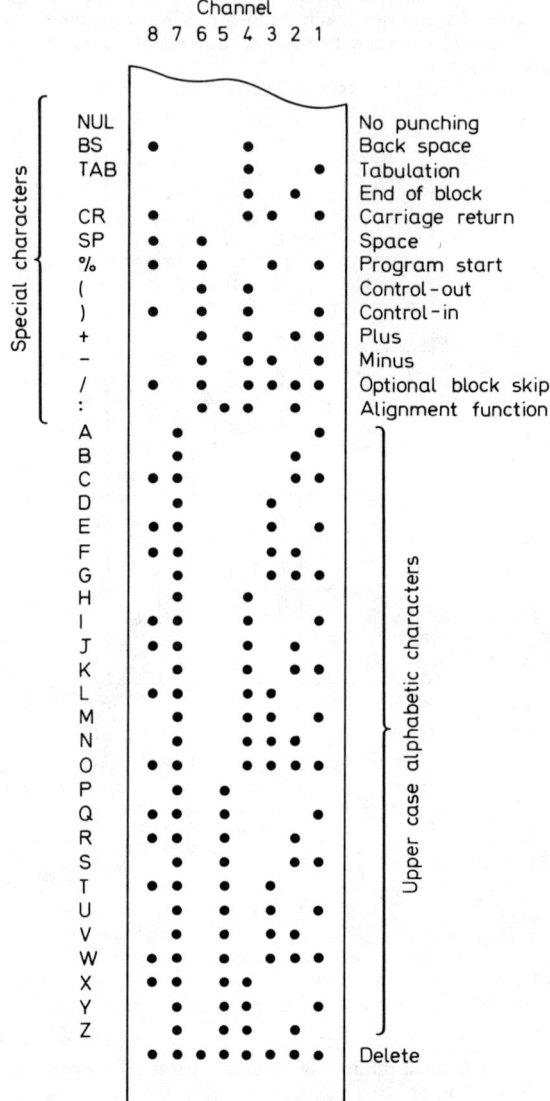

Figure 19.5. Alphabetic and special characters

It is, therefore, important that a company should endeavour to operate one code, although, a computer is easily programmed to produce tapes in any desired code. Obviously, if there is a proliferation of codes, flexibility is reduced since tapes are not interchangeable from one machine to another of the same type. Problems are also likely to arise in connection with mechanical tape punches.

Word address format

The ability to encode alphabetic information as well as numeric information enables us to precede encoded numeric information with alphabetic information e.g. X2759. This is referred to as WORD ADDRESS FORMAT, whilst the X is referred to as the ADDRESS, i.e. it identifies the function to be performed. The alphanumeric information X2759 constitutes a WORD.

Block

Figure 19.6 shows a set of encoded instructions required to perform a drilling operation at a given point. This set of instructions is referred to as a block. It is possible to have fixed block lengths or variable block lengths. The choice will be with the control system manufacturer.

The block may be of fixed format, in which case, the position of information within the block is firmly prescribed. On the other hand, the block may be variable in content. The control manufacturers' instruction manuals should be consulted to ascertain the block format.

Advantages of punched tapes

1. They are comparatively cheap.
2. Punches, reproducers and verifiers are not high cost items.
3. The control systems contain interpolation facilities etc. and are therefore small computers. Consequently, a minimum of information is produced by the main computer, thus reducing costs.
4. The computer can label tapes in man-readable characters.
5. Any information on the tape can be decoded visually.
6. Simple modifications can be carried out on a suitable typewriter without the need for a computer run.
7. Tapes are not affected by magnetic or electric fields.
8. Tapes can be stored for long periods without deterioration.

Disadvantages of punched tape

1. If a large amount of information is required on the tape, it may be necessary to use a number of individual tapes and for the job to be broken down accordingly.
2. The tapes, particularly paper, are easily damaged.
3. Tape punches, being mechanical, are error prone.
4. Mylar tape is extremely tough, and can induce wear on the anvils of the tape punch. It is desirable to use carbide tipped anvils, if Mylar is to be used extensively.

Care of paper tape

Again it cannot be too highly emphasised that whilst a virgin roll of paper tape may be comparatively a petty cash item, a roll of tape containing control information is an expensive and valuable asset and should be treated as such. The following recommendations are made.

1. A master tape should be kept in a secure place.
2. Two working tapes should be made. The first for use on the shop floor and the second tape held in reserve in the stores. Should the first tape become unfit for service,

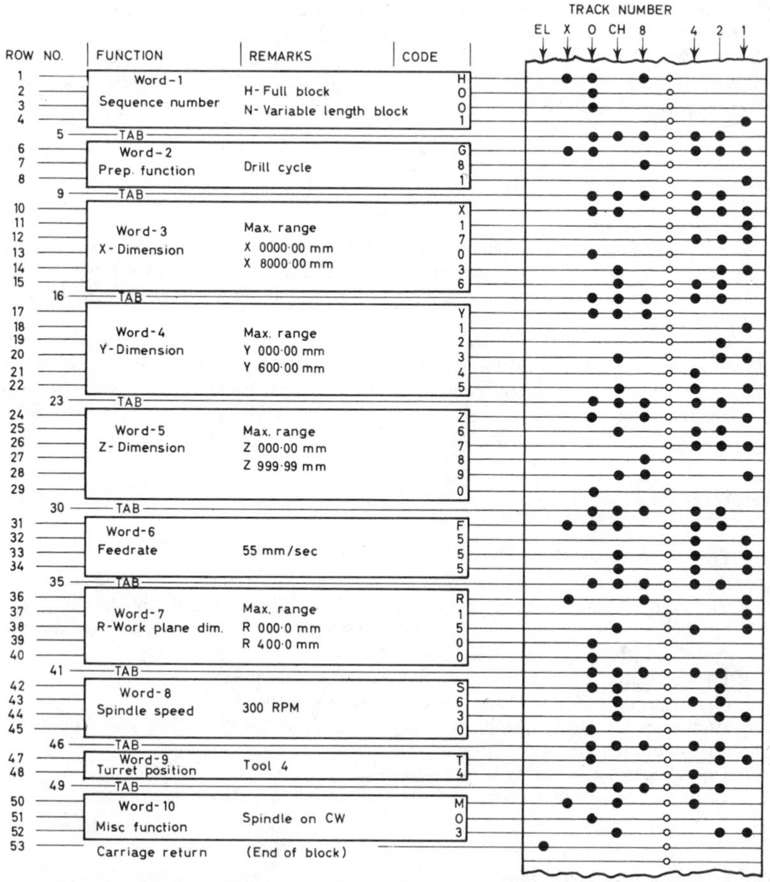

ROW NO.	FUNCTION	REMARKS	CODE
1 2 3 4	Word–1 Sequence number	H–Full block N–Variable length block	H O O 1
5	TAB		
6 7 8	Word–2 Prep. function	Drill cycle	G 8 1
9	TAB		
10 11 12 13 14 15	Word–3 X–Dimension	Max. range X 0000·00 mm X 8000·00 mm	X 1 7 0 3 6
16	TAB		
17 18 19 20 21 22	Word–4 Y–Dimension	Max. range Y 000·00 mm Y 600·00 mm	Y 1 2 3 4 5
23	TAB		
24 25 26 27 28 29	Word–5 Z–Dimension	Max. range Z 000·00 mm Z 999·99 mm	Z 6 7 8 9 0
30	TAB		
31 32 33 34	Word–6 Feedrate	55 mm/sec	F 5 5 5
35	TAB		
36 37 38 39 40	Word–7 R–Work plane dim.	Max. range R 000·0 mm R 400·0 mm	R 1 5 0 0
41	TAB		
42 43 44 45	Word–8 Spindle speed	300 RPM	S 6 3 0
46	TAB		
47 48	Word–9 Turret position	Tool 4	T 4
49	TAB		
50 51 52 53	Word–10 Misc function	Spindle on CW	M O 3
	Carriage return	(End of block)	

TRACK NUMBER — EL X O CH 8 4 2 1

Figure 19.6. One block of information

Figure 19.7. MAN readable leader

the second tape should be brought into use and a further reserve prepared from the master.
3. All tapes should carry a man-readable identification at the beginning (Figure 19.7).
4. Tapes should be protected from coming into contact with coolant, oil or swarf.
5. Proper spools should be used to carry tapes.
6. When not in use, tapes should be kept in boxes which will offer protection.
7. Cupboards should be used for storage.
8. Tapes must not be creased or torn. They should not be allowed to trail the floor.

TAPE PROVING

The fundamental aim in tape preparation is to ensure an error free working tape at the machine tool at the right time. Continuous care is essential to this end, and certain obvious steps should be taken.
1. Every effort should be made to ensure that the part program is correct before it reaches the computer.
2. All equipment such as tape punches, reproducers and verifiers receive adequate preventative maintenance.
3. Output from the computer is thoroughly checked.
4. Computer prepared drawings showing piece and cutter path are prepared wherever possible (Figure 19.8).
5. Tapes are purchased on the understanding that these conform to the specified standards and that specimen tapes are checked from time to time for conformance.

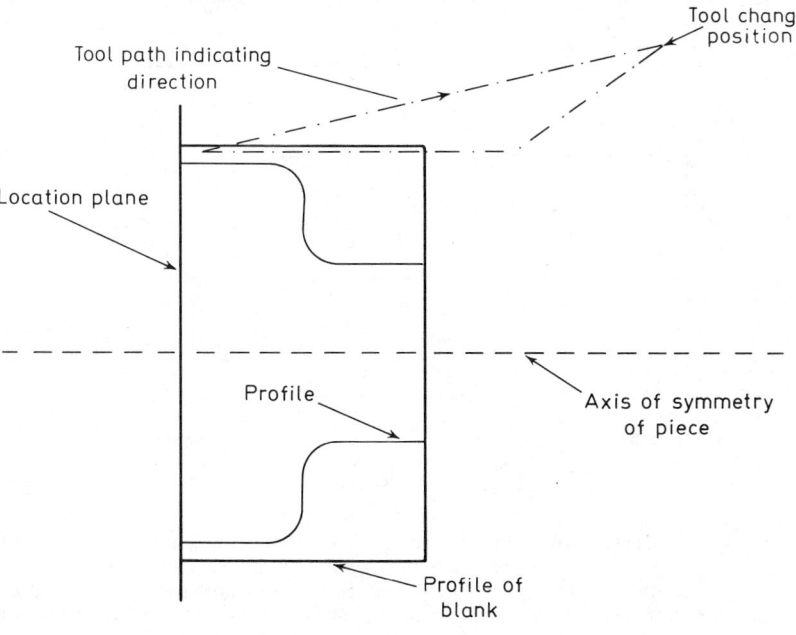

Figure 19.8. Automatically produced cutter path drawing

6. Punches etc. are allowed to warm up before being used for tape preparation.
7. Hole spacing, a common source of error, should be checked regularly while tape production is carried out. Suitable gauges are easily purchased. Provision of adequate systems incorporating the above points will go a long way to producing error free tapes, but despite all precautions, the occasional rogue will slip through and only be detected on the machine tool. It is here that the fast turn round of new tapes on the computer should reduce the resulting problems.

ADAPTIVE CONTROL

Adaptive control is the technique by which such parameters as feeds, speeds, depth of cut, etc. can be controlled during an N.C. machining cycle by a feedback process which is dependent on the variable parameters developed on the machine during that cycle.

Shortcomings of present day numerical control

The preparation of part programs in which such parameters as feeds, speeds and depth of cut have to be estimated or calculated in advance makes it impossible to take into consideration any variations from these forecasts which will inevitably occur during the actual machining process. Among the factors which contribute to the difficulties experienced in assessing the correct machining conditions are:
1. *Variations in width and depth of cut.* The geometry of the piece part or variations in the rough stock configuration can result in a continually changing shape of cross section

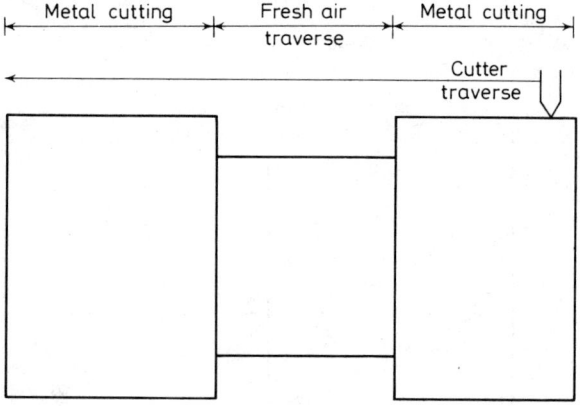

Figure 19.9. Air gap in cutter path

across the actual cut. This results in continual change in the load imposed on the cutter and part programmers must of necessity select conservative feeds and speeds.
2. *Workpiece hardness.* In most materials used on N.C. machines hardness varies considerably in any specimen. Whilst these variations are not significant in light alloys, in other materials they impose conservatism on the part programmer.
3. *Feed and speed ranges.* Most N.C. machines have specified and therefore limited ranges of feeds and speeds. This limitation can make it impossible to select the optimum values for a given operation and it becomes apparent that, for high strength thermal resistant

alloys, incorrect selection of optimum speeds and feeds may have considerable effect on tool life.

4. *Tool wear.* Normal N.C. machine feeds and speeds are not responsive to tool wear, and once feeds and speeds have been built into the part program on the basis of an average tool they may be incapable of being changed without reprogramming. On some machines it is possible to apply manual over-ride, but the effectiveness of this is open to debate, depending, as it does, on human judgement.

5. *Air gaps.* It is quite common when machining a piece to have a condition (Figure 19.9) where the cutter cuts metal for a period, then traverses in fresh air until metal is reached once again. When this condition occurs, it is often impracticable for the part programmer to program maximum speeds to cover the fresh air travel.

6. *Stalling.* If feeds and speeds are programmed such that the spindle motor is producing its maximum horse power, any hard spots in the work piece can result in stall. Again, the part programmer must be conservative.

7. *Inspection.* Variations in workpiece size and hardness, cutter wear, etc., often affect the repeatability of the process and necessitate considerable inspection to control the process. This can significantly increase the cost of the product.

Advantages of adaptive control

Many of the shortcomings listed above can be eliminated by the use of Adaptive Control which ensures:

1. The machine is protected against overload and still performs at the optimum feeds and speeds within the power capacity limits of the machine.
2. The cutting tools are protected, since excessive cutting forces are not generated.
3. The piece itself is protected against tools breaking and remaining in situ in the piece e.g. drills.
4. There is less dependence on human decision or intervention.
5. Part programming is simplified by eliminating the need to guess at requirements for speeds and feeds etc.
6. Optimisation of cutting conditions results in increased productivity.
7. Complete optimisation of cutting conditions reduces machining costs.
8. Inspection costs are reduced because of greater repeatability especially if in-process measuring devices are incorporated.

Survey of systems

A number of Adaptive Control systems are commercially available, although some of them are still in the development stage. Further work is being conducted in various research establishments and whilst most of these have established the lines on which technical research will be directed, not enough progress has been made to enable the viability of the projects to be assessed or to determine their economic value. These factors often make it difficult to obtain technical information on individual products and projects.

A brief review of some of the better known systems is given below.

1. AEG. This unit was exhibited at the 1970 Machine Tool Exhibition at Hanover. Little information is available at the moment, and it is expected that development work will not be completed before 1973. It would appear from information available that feed rate and depth of cut are controlled by feed back from the speed and torque of the spindle.

2. BENDIX. Despite the fact that Bendix units have been seen at the last two Chicago Machine Tool Shows, very little detailed information about the controlling parameters is available. An early system is known to have used torque, temperature, tool vibration and cutter loading as control parameters.

3. CINCINATTI 'ACRAMISER'. The principal control parameters of this system are cutter torque, spindle deflection and horse-power being applied. These cause feedrate and speed to be adjusted automatically, thus optimising within the maximum power of the machine, whilst permitting maximum traverse rates when not cutting metal. The sensors on the spindle are so designed that when a tool wears to the point where economic cutting ceases, a signal light warns the operator that the tool needs changing.

4. GENERAL ELECTRIC. The system offered by General Electric will be available as a retrofit to the Mark Century and also for the 7542 series. The spindle load is measured by a watts transducer and this is used to control feedrates within predetermined limits by adjusting rates which have been supplied to the system via the normal part program.

5. MAKINO. The Makino machining centre, exhibited at Osaka in 1970, offered a feature which used a computer to calculate cutting feeds and speeds. The computer files carried tool numbers, tool life data, and maximum permissible torque. The tool change signals initiated the computer calculation of feeds and speeds, spindle torque sensors fed back signals to optimise these rates, yet spindle torque was prevented from exceeding the machine's capacity.

6. SIEMENS SINUMERIK. This system has been seen at a number of exhibitions. The control parameters are spindle torque and spindle speed, and are used to control the feedrate and depth of cut. A feature of this system is that it is only used for the roughing operation, the finishing operation being performed without adaptive control. For the roughing operation, it is only necessary to define the geometry of the rough finished contour, from which, together with the feed back signals, the system will determine the path of the cutter and depth of feed.

7. PROFESSOR PORTER[7]. A system of adaptive control is being developed by Professor Porter of Salford University and has been fitted to a drilling machine. The system uses a performance factor H which may be varied to produce minimum operation cost or minimum machining time. The controlling parameters measured on the machine tool are torque, tool temperature and vibration (Figure 19.10). Salford University have been given a contract to develop the system and to assess its economic viability.

8. RADAN ASSOCIATES LTD. (BILLET). Research into methods of inducing faster reaction times is being carried out by Radan Associates Ltd. This faster reaction is due to the use of a thyristor stack and trigger unit (Figure 19.11). Because the system gives control of tool wear but does not actually measure it; vibration analysis is used to indicate the need for tool replacement. The system would appear to be most suitable for facing operations, but much work is still required to assess and evaluate the system.

9. RAE, FARNBOROUGH. The RAE has worked on very similar lines to Billet, the chief difference being in the methods of thermal measurement used.

Monitoring techniques

The techniques of monitoring are outside the scope of this document, but a bibliography[6] is included for the benefit of those seeking more detailed information.

Limitations

The foregoing discussion of systems is limited because of the difficulties experienced in obtaining information from various sources which are scattered across the world.

Economics

When equipment becomes commercially available in realistic quantities, engineers will become interested in purchasing. As with all technological developments, it will be

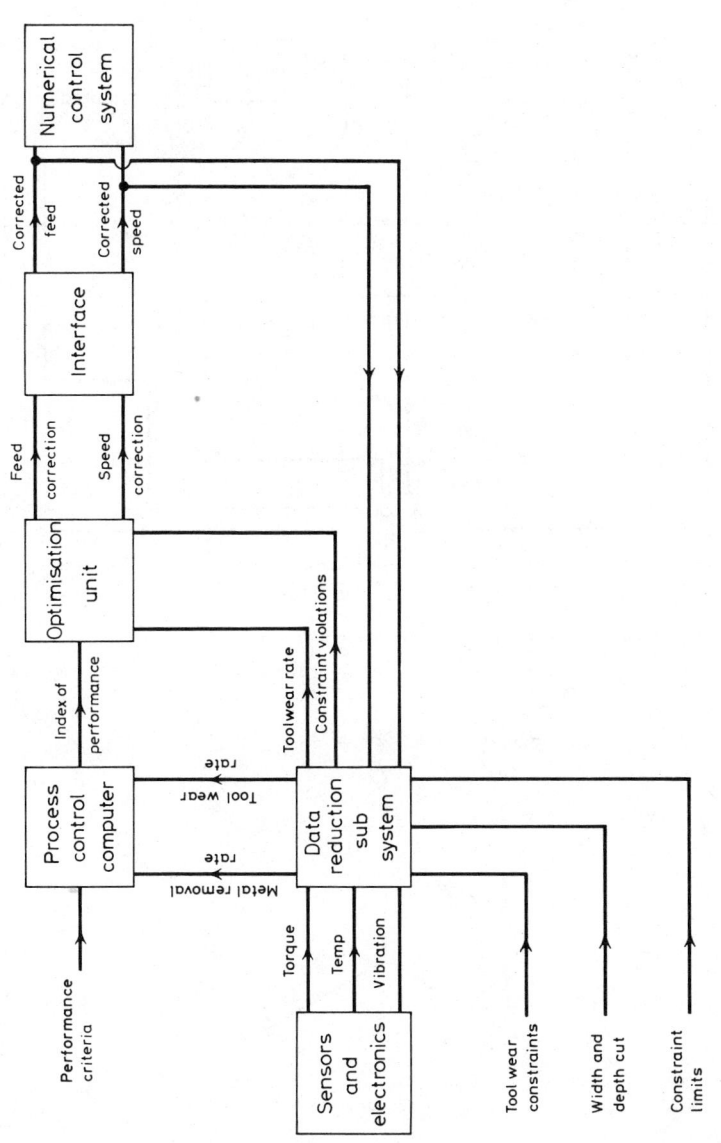

Figure 19.10. Prof. Porter's concept

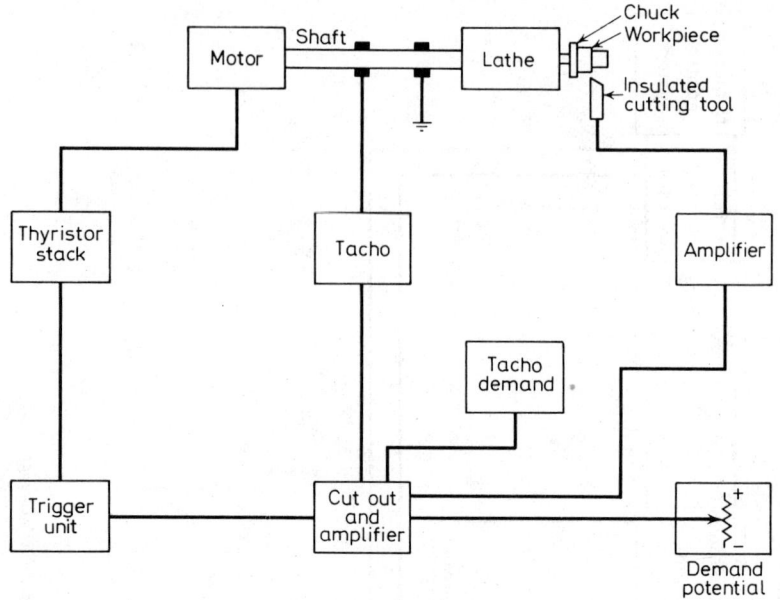

Figure 19.11. Control system developed by billet (Radan Associates Ltd)

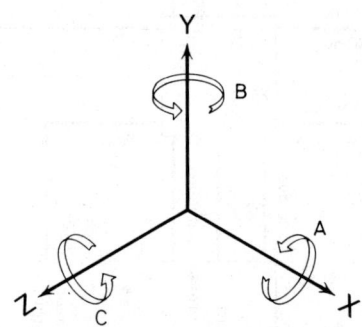

Figure 19.12. Cartesian axes

essential to conduct strict economic appraisals, and not to make purchases based on technical aspects alone.

Future development

Economic and technically viable systems will have a place on individual machines on the shop floor, but, they will certainly become an essential requirement for a really effective system of On Line Control.

AXES[8]

Co-ordinate systems

In order to carry out part programming, it is necessary to have a set of axes, related to the piece, to which geometric features can be related. The standard method of achieving this is to use three mutually perpendicular axes referred to as the X axis, the Y axis and the Z axis (Figure 19.12), thus permitting the use of cartesian geometry to define points, circles, etc., in three dimensional space.

Machine movements

It is also necessary to be able to define the movements of the work piece mounted on the machine tool relative to the cutting tool. Again, a right-handed cartesian system is used. This cartesian system is related to the principal linear slideways of the machine tool, and each of the coordinate axes is labelled X, Y or Z. Movements are thus said to occur in the X, the Y or the Z axis.

Axis relationship

The co-ordinate system of the machine tool remains constant, but, the co-ordinate system used in the part program is dependent upon the requirements of the work piece as decided by the part programmer. It is essential therefore, that the part program contains a facility by which the two sets of co-ordinate axes may be related.

Positive movement

The positive direction of movement is that which will produce an increasing positive direction on the work piece e.g. on a drilling machine a movement in the negative Z direction will cause the drill to penetrate the work piece.

The Z axis of motion

The spindle which supplies cutting power is used to identify the Z axis. On a drilling, boring or tapping machine, this spindle is the one which rotates the tool (Figure 19.13), but on machines generating a turned shape, this spindle rotates the workpiece.

In cases where there is more than one spindle, the one performing dominantly during the machining cycle should be selected as the Z axis and it should preferably be one normal to the work-holding surface. On some machines, the principal axis has a swivelling capability. In such instances or where there is no rotating spindle, the Z axis should be selected to be normal to the work holding surface.

The X axis of motion

It is preferable that the X axis of motion should be horizontal and parallel to the work-holding surface and should be the principal axis of motion in the positioning plane of the tool or workpiece. If the machine possesses neither rotating tools or workpieces, the X

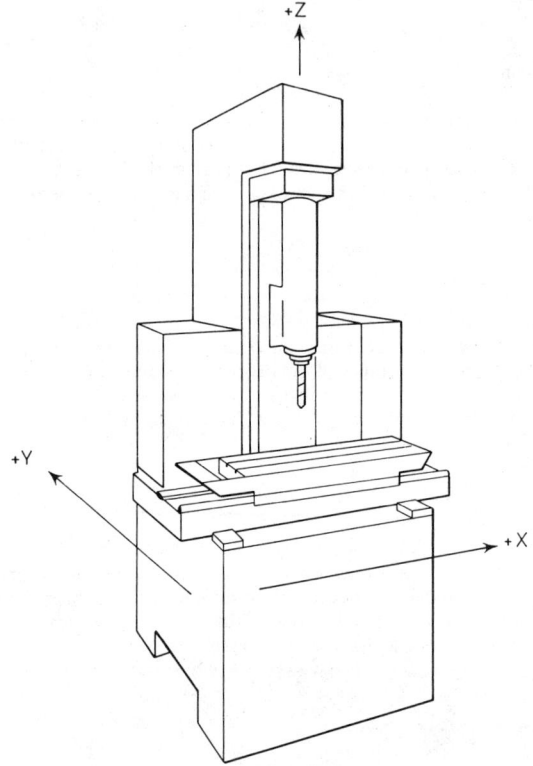

Figure 19.13. Three-axis machine

axis is parallel to the principal direction of cutting. The positive direction of the X axis is in this direction of cutting.

Machines with rotating workpieces (lathes, etc.) have the X axis in a radial direction parallel to the cross slide, and the positive direction of motion is such that a tool mounted on the principal tool post recedes from the axis of rotation.

Machines with rotating tools such as milling machines, required to be considered in two categories:

1. When the Z axis is horizontal, the positive X motion is from left to right when looking from the principal tool spindle in the direction of the work piece.
2. When the Z axis is vertical, the positive X motion is to the right when the view is

from the principal tool spindle to the column, whilst in gantry machines the view from the principal spindle to the left hand column.

The Y axis of motion

The Y axis is that required to complete the right hand cartesian system.

Right-hand rule

A simple method of determining the positive directions of X, Y and Z for a right handed axis system follows. If the thumb, the first and second fingers of the right hand are extended so as to be mutually at right angles to each other, then the thumb points in the positive

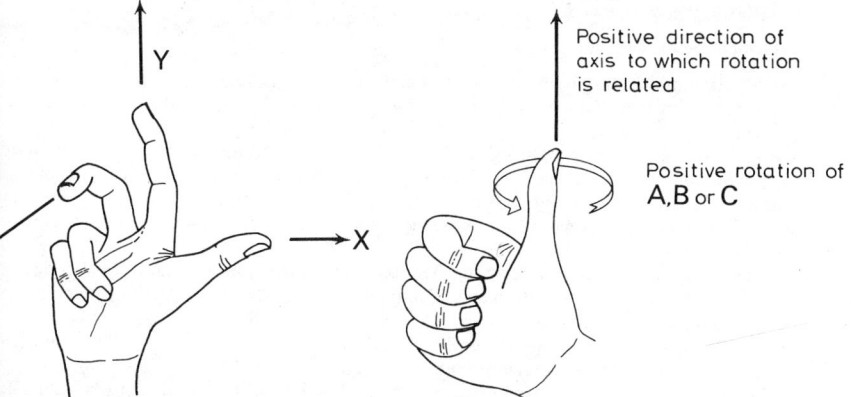

Figure 19.14. Right-hand rule for linear axes   *Figure 19.15. Right-hand rule for rotation*

direction of the Y axis and the second finger in the positive direction of the Z axis (Figure 19.14).

Rotary motions

To define rotary motions which occur about the X, Y and Z axes are called A B and C respectively, their positive directions of motion being such as would advance a right-hand screw in the positive direction of the associated axes.

A right-hand rule is a useful aid in determining the sign attributable to rotation. If the thumb of the right hand is extended, and pointed in the positive direction of the associated axis, then the tips of the slightly curved fingers point in the direction of positive rotation (Figure 19.15).

Additional axes

Many instances of machines with parallel slideways each performing different functions. Motions parallel to X, Y and Z axes are designated U, V and W respectively, whilst additional motions parallel to U, V or W designated P, Q or R respectively.

Should there be more than three rotary motions A, B and C, these will be designated D or E, irrespective of their inter-relationship with other axes.

Moving work

The rules outlined previously define the axes for machines in which power is transmitted to the tool to make it change its position. However, if power is used to move the work holding surface, then the tape signals must be such as to move the surface in the opposite direction to which the tool itself would require to move in order to produce the same result on the workpiece. Such movements are indicated by X^1, Y^1 and Z^1.

The effect of a tape command instructing a positive movement in X is such that on a machine in which the tool moves and the work holding surface remains stationary, the tool moves in the positive direction of X, whilst on a machine with moving holding surface and stationary tool, the table will move in the negative X direction.

Classification of machines

In order to indicate the capability of N.C. machine tools, it has become common practice to classify them by reference to the number of continuously controlled axes which they possess.

Two-axis machines. A draughting machine capable of continuous control in the X and Y axes, with the pen operating on a simple 'pen up', 'pen down' instruction is a typical example of a two-axis machine tool. Similarly, a drilling machine with continuous control on the X and Y axes, but with the Z movement being carried out independently falls into the same category.

Three-axis machines. In this category there is a wide range of machine tools which have the capability of controlled motion in X, Y and Z. Typical examples are, drilling machines, three-axis mills, lathes, grinding machines and filament winding machines.

Four-axis machines. If a three-axis milling machine has a controlled rotary table added, it thus becomes a four-axis machine tool.

Multi-axis machines. In this manner, additional freedom of movement can be added, building up five-, six-axis machines, etc. Obviously, the statement that a machine is a nine-axis machine is of no consequence on its own, but merely a guide to the movements which are provided. The value of these movements depends upon the ability of the machine to perform the operations on the type of work which it is planned to carry out on the machine tool. This is important since each additional axis adds, not only to the cost of the machine tool, but to the cost of the control system.

Machining centres. Multi-axis machines with automatic tool changing facilities are generally referred to as *machining centres*. These are extremely versatile machines, capable of carrying out the most complicated machining, thus enabling the piece part to be completely machined in one setting, whilst, if fitted with automatic pallet changing facilities, work can be set up whilst the first is being machined.

In such installations capital expenditure is high, and the economics of machining centres must receive careful consideration prior to installation. Not only must they be justified on the basis that they can make a profit, but their performance must be monitored to ensure that they do make a profit. The aim must also be to achieve maximum utilisation. The latter statement is true for any N.C. machine, but, in the case of machining centres becomes even more important.

PART PROGRAMMING

Part Programming[9] is normally started from a fully dimensioned drawing of the piece to be manufactured. The subsequent events, however, depend upon the actual method which will be used to prepare the controlling information. Three main methods are available, manual part programming, computer assisted part programming, or with the assistance of a programming centre.

Manual part programming

It is essential to be familiar with the characteristics and capabilities of the machine tool and its associated control system.

The following information should be given via the control tape, to the control system:
(a) The amount to be moved in the X, Y, and Z planes.
(b) The direction relative to the datums in which these movements will occur.
(c) The feed rate of the cutting tool relative to the piece.
(d) The speed of the cutting tool.
(e) The diameter of the cutting tool.
(f) Other operations which the machine tool may be called upon to perform under control of the tape, e.g. automatic switching of the coolant. This type of instruction is referred to as a Miscellaneous Function.

This information is detailed step by step, in the exact order in which it is desired to perform the operations, on a programming sheet (Figure 19.16).

When the programming sheet has been prepared and checked, the information is typed on a special typewriter which automatically produces a punched tape. To provide a check, it is useful to have the programming sheet ruled to correspond with the line spacing on the typewriter. The part program can then be written using every third line. The typist is able to type immediately below the written entry, thus giving a simple method of checking her work. The completed tape can then be run through the typewriter to produce a third entry on the sheet, thus proving tape content. This is a very necessary step because the punches producing the tape can easily malfunction and so produce an invalid tape.

Advantages of manual programming

For simple work manual part programming offers many advantages, chief amongst which may be mentioned:
1. The user does not have to buy expensive equipment.
2. The Production Engineer controls the operation.
3. Tapes can be modified quickly.

Disadvantages of manual programming

Figure 19.6 shows a block of information such as would be required to drill one hole at a specified position on a point to point machine. To prepare this block, however, involves a considerable amount of preparatory work:
1. Calculation of the hole position relative to the datum.
2. Determination of the tool required.
3. Calculation of the feed rates and spindle speeds.
4. Preparation of the actual tape which contains fifty-two rows of information.
5. Checking all the calculations and information.
6. Checking the tapes.

Quite obviously when a large number of operations are required to be performed, tape preparation can become a mammoth task. Indeed, when the requirement is for continuous path operation, manual tape preparation often becomes an impossibility. Manual tape

Sequence number	Preliminary function	X axis	Y axis	Z axis	Feed	Speed	Tool	Miscellaneous function	Remarks

N.C. programming sheet

Figure 19.16. Programming sheet

preparation by its very nature is extremely error prone and is liable to be a boring procedure.

Programming centres

The problems involved in manual part programming have led to the development of machines, often called Programming Centres, designed to produce tapes from a drawing.

Basically, the machine consists of a table with a measuring cursor capable of moving round the table. The position of the cursor at any time can be automatically transferred to the control system. In some cases, the cursor is designed to represent a turret containing the required tools. Because of the ability to rotate this diagrammatic turret it is possible to bring the diagram of the required tool into position relative to a drawing of the piece part positioned on the table. The tip of the diagrammatic tool can thus be moved across the drawing to reproduce the tool movements which are required to be made during actual machining operations. Thus, the position of the cursor at any point of its traverse is recorded.

In addition, the system has an electric typewriter connected to the control system permitting additional information such as feeds, speeds, etc., to be supplied. Accurate positional information can be supplied via this typewriter. The information which is fed to the control unit, in reality a small computer, can be processed to produce a tape.

Advantages of a programming centre

From the method of operation it can be appreciated that the programmer can visualise clearly the tools movements which are being programmed, and at the same time, he can also ensure that no collisions between tools and other parts of the work piece occur in the plane of the drawing.

Disadvantages of a programming centre

Programming Centres are expensive items of capital equipment with certain limitations:
1. Only one part programmer can use the machine at a given time.
2. Additional electronics are likely to be required to cater for various machine tool/ control system combinations.
3. A certain amount of manual calculation may be required.

Computer-assisted part programming[9]

This often starts as in other systems of part programming with the drawing, but more use is being made of computers to produce drawings as well as tapes. The part programmer uses a special N.C. language, of which there are many, to communicate with the computer. In general, these N.C. languages are English like in appearance but, have the capability of conveying complex instructions to the computer in a brief easily understood instruction[5].

For example, the statements

$$CIR = CIRCLE/0, 0, 8.5$$
$$PAT = PATERN/ARC, CIR, 10, CLW, 151$$

processed by the computer will produce the X and Y co-ordinates of 151 holes equally spaced on the pitch circle in the figure and numbered in a clockwise direction.

Advantages of computer-assisted programming

1. More rapid production of tapes.
2. Reduction in the number of errors on first tape.
3. Complex calculations are performed by the computer.
4. More complex surfaces can be produced.
5. Costs of part programming can be reduced by 50% on appropriate work.
6. Proving drawings can often be produced.

Disadvantages of computer-assisted programming

1. High capital outlay or service charges.
2. The need for a more highly qualified part programmer.
3. The introduction of another service which is not normally under the production engineer's direct control and brings with it its own problems and terminology.

In the majority of cases, the advantages of computer assisted part programming far outweigh the disadvantages.

ORGANISATION OF PART PROGRAMMERS

Objectives

The introduction of new systems in industry has been shown to produce social as well as technical problems. Hence, it is extremely important to consider both these factors when establishing part programmers in the organisation, and every effort needs to be made to establish a standard of professionalism into part programming.

This cannot be too highly stressed when so much of the success of N.C. will depend upon the efforts of these members of the organisation.

Skills

A part programmer should ideally possess certain skills and attributes at the time he comes to the job. These should include:

1. The possession of a National Certificate or equivalent.
2. A good working knowledge of Algebra, Geometry and Trigonometry.
3. Understanding of the capabilities of the N.C. machine tools installed in the company.
4. Appreciation of the properties of materials in general use in the company.
5. Knowledge of the properties and capabilities of cutting tools.
6. Experience of the effects of processes on material properties, e.g. heat treatment, work hardening, etc.
7. Capability to design holding fixtures.
8. Ability to specify special tools.

The part programmer task

An employee whose task is solely concerned with preparing the computer input or coding sheet is faced with the prospect of monotony after a very short period and as a result will become bored and error prone. Certainly he will not bring an air of professionalism to the task. It may be argued that part programs should be written by anyone in production engineering, but this approach falls short of the professional, because the individuals concerned do not become sufficiently familiar with the N.C. languages in use and are therefore liable to be error prone.

The first step to professionalism is to clearly identify the task and responsibilities of the individual and to ensure that they are completely involved in N.C. Ideally, the part programmer will commence his work with the detail drawing and thence undertake the following tasks:

1. Decide those operations that are most effectively performed on N.C. machines and those on conventional machines.
2. Inform the personnel responsible for conventional planning of the operations they are required to perform.
3. Prepare the specifications for the holding fixtures which will be required for the N.C. operations and advise the appropriate staff in order that these fixtures are available when required.
4. Decide on the cutting techniques to be employed.
5. Prepare a list of tools, standard and special, which are needed for the operations planned and initiate the actions to ensure delivery of these tools to the machine at the correct time.
6. Prepare the part programs necessary to produce the required tapes and drawings, and submit for computer processing after adequate checking.
7. Check the computer output of printed sheets, drawings and tapes to ensure that there is the maximum chance of first time success on the machine tool.
8. Supervise the production of the first piece on the machine tool and, if necessary, carry out any work required to bring the piece to the standard required by quality control and to achieve optimum results on the machine tool, e.g. modify feeds and speeds, reduce fresh air travel to a minimum, etc.

Selection of part programmers[10]

Once the task of the part programmer has been defined, the personnel having the necessary prerequisites can be considered. There is much to recommend a two part selection process. The first part of this selection process would be ideally an aptitude test. This should be designed to test the qualities required of a part programmer, and would not be the normal intelligence test. Such a system is quite commonplace in the selection of computer programmers.

The second part of the selection process should be by interview. Perhaps the chief part programmer and a production engineer could each interview the candidate separately and informally and compare their findings afterwards. The final selection would be dependent upon the test and the findings of the interviews. It is worthwhile to make some special effort on selection since it is only by selecting the right people that the professional approach to the job can be achieved.

Training part programmers[8]

Having carefully selected the part programmer, some formal training is called for, in order that he can learn the N.C. languages which he will be required to use. Courses in these languages are available from organisations external to the user company and are advertised in the various journals associated with production engineering.

Alternatively, if adequately trained staff are available in-house, a company can arrange its own training, but great care must be taken to ensure this training is adequate. It must not be assumed that having attended a week or a fortnight's course on an N.C. language, that a person has become a part programmer. At this stage, only the basic rules have been imparted and it requires application and experience to develop expertise and efficiency.

It is important that the trainee returning from a training course must immediately be given the opportunity to use his new experience. He should not be left for weeks without practice as he will not only forget his training, but he will lose interest. Continual involvement is important.

Grouping

Part programmers in general are least efficient when each individual works in isolation; the best performances being achieved when a number of part programmers work together in a group. Many of the advantages are fairly obvious. For instance there is a back-up available when, as is inevitable, a part programmer is absent because of illness or holiday. A less obvious benefit is that expertise spreads more rapidly in a group than among a number of isolated staff, and there is less likelihood of know-how being retained by one person in order to maintain a superiority over his fellows.

Training can also be planned more efficiently in the group organisation, and the team spirit can be cultivated much more effectively in such an environment, thus making a good foundation on which to build the professional approach.

Career development

Not only must the part programmer be given a definition of his job and provided with adequate training, but a fully defined career path should be developed, otherwise part programming may appear as a dead end job. In this case, new entrants may be difficult to find and the staff actually employed on part programming may become disinterested and lose their enthusiasm.

It is also important that the status of the part programmer is defined and known, not only to himself, but to the whole organisation.

THE COMPUTER ROLE

Machine classification

Before considering the details of computer systems it is necessary to consider the two major classifications of machine tools. Depending upon the control system, these are:
1. Point to Point[11]
2. Continuous Path[12]

POINT TO POINT

These systems are generally used in connection with drilling and boring operations where it is important that the cutting tool is exactly positioned at a number of specified points, whilst the path taken by the tool from point to point is of no great importance. Figure 19.17 shows how a tool can move between two discrete points A and B.

The cutter leaves point A moving parallel to the X axis, then moves at an angle until it achieves the same Y value as point B, when movement continues parallel to the Y axis until point B is accurately achieved. The amount of movement parallel to the various axes is not precisely predictable, and care may have to be exercised to ensure that there are no projecting features which may be involved in collision with the cutter. Some control systems also offer a facility for milling in straight lines parallel with the axes.

CONTINUOUS PATH

When sculpturing or profiling of the material is required, it becomes important that the cutter follows a precisely defined path which is accurately maintained at any point in time. Therefore, this cutting path must be the result of simultaneous motion in two or more axes. This simultaneous motion means that it is essential to define a large number of

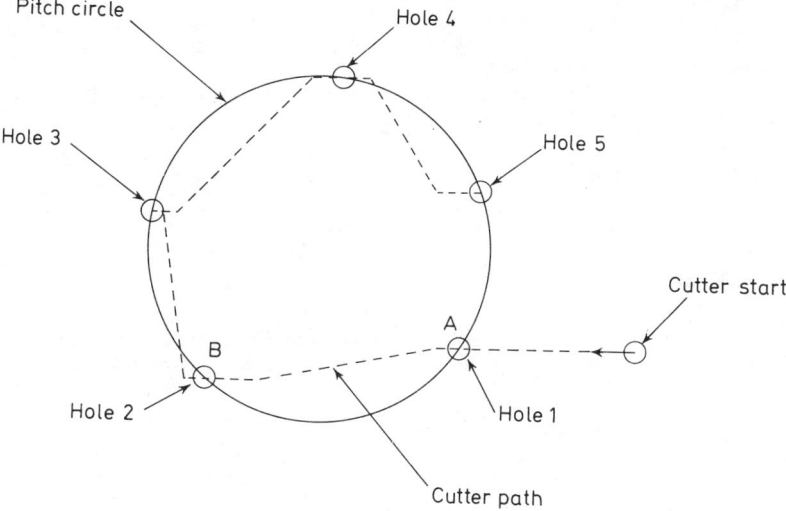

Figure 19.17. Point to point machining

points to the control system. Consequently manual part programming tends to become an impossible task, and the computer becomes an essential aid.

Figure 19.18 illustrates the cutter path which must be achieved to profile a cam.

Processing

The part program which has been written on part programming sheets is sent to the computer department or bureau for processing.

The first step is to punch the information on the part programming sheets to punched cards or tape. This is done by means of a special punching machine in the case of cards, and by a special electric typewriter in the case of tapes. The resulting cards or tapes are fed into a computer which produces the machine tool control tapes, printed output, etc., automatically. The process is represented diagrammatically in Figure 19.19.

Processors[13]

In order that the computer can convert (or process) the part program into a machine tool control tape, it is necessary to have a computer program which reads the information in the part program, carries out the necessary interpretation and calculation to permit the control tape to be produced by the computer.

In N.C., such computer programs are known as processors, and they are a series of instructions for handling part programs written in a form which the computer can understand. Processors may exist in many forms:

1. A Manuscript.
2. On Punch cards or punched tape.
3. On Magnetic Tapes.
4. On Magnetic Drums.
5. On Data Cells.

It is important to appreciate this simple definition.

Types of N.C. computer systems

Basically there are three types of N.C. Computer systems:

1. Special Purpose.
2. Machine Oriented.
3. General Purpose.

Special purpose systems

These are developed to meet special requirements existing within an individual company, and are generally related to a specific product within that company. Systems of this type

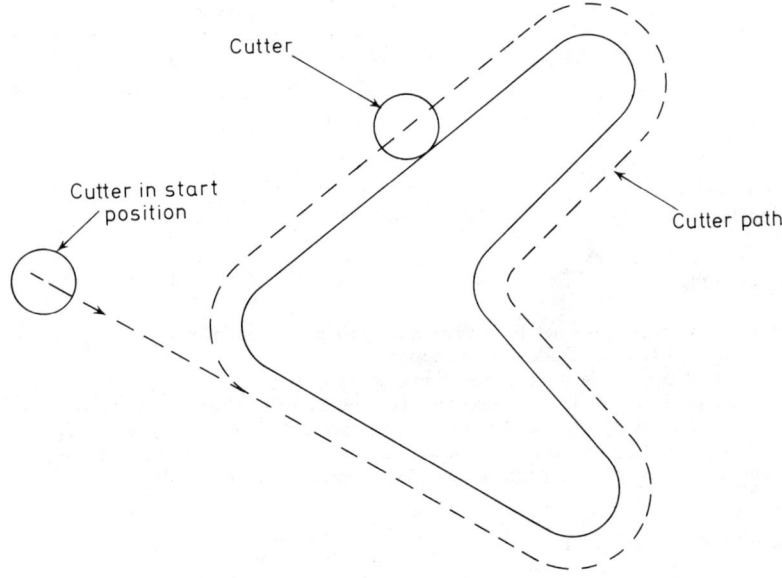

Figure 19.18. Continuous path machining

are expensive to develop, and because of their company orientation may not be worth anything as a marketable commodity.

For this reason, it is important to carry out a detailed survey of estimated costs and benefits before embarking on such a project. However, if assistance is required on a problem or series of problems which occur regularly, and if the extent of the problem can be defined by simple parameters, a special purpose system can become simple to program,

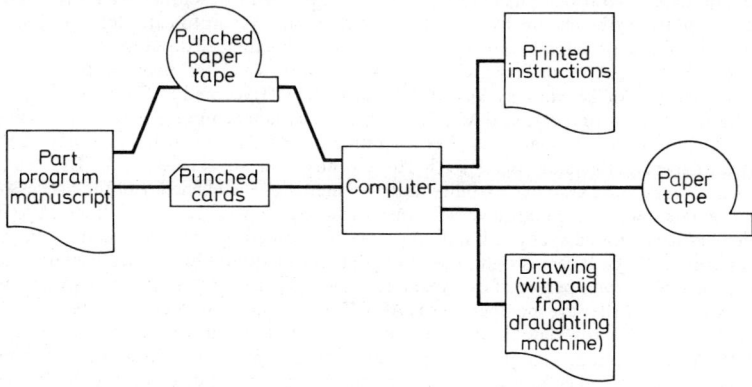

Figure 19.19. The computer process

giving economical use of part programmers' time and computer time and of greatest importance can be highly profitable, producing, as they often do considerable savings in production engineering.

Machine oriented systems

Some machine tool builders provide computer systems which are designed specifically for a given machine tool. The maintenance of such systems is the responsibility of the manufacturer, but there may sometimes be difficulty in obtaining modifications which may be required to meet the special techniques employed by one specific company.

Naturally, such systems cannot be used on other machines, and with an organisation purchasing machines from a number of sources, this may impose the need for part programmers to learn a number of N.C. languages which can create obvious problems.

General purpose systems

The most popular current practice is to use general purpose systems, because their use is not limited to a particular make of machine tool or product. In general, there is a tendency for the language of a general purpose system to be based on APT (Automatically Programmed Tools).

Whilst general purpose systems are independent of the source of manufacture of the machine tool, some are designed to cater for specific types of machine tools, e.g. drilling machines, lathes, mills, etc. e.g. EXAPT 1, 2C etc.

AVAILABLE PROCESSORS

Apt[14]

The most comprehensive processor available is APT which derives its name from the initials of Automatically Programmed Tools. APT was developed by the Massachusetts Institute of Technology as a result of a contract awarded by the U.S. Air Force in 1957 to speed up the preparation of machine tool control tapes. MIT developed APT 1 to handle point-to-point problems but by 1961, had extended the system to cater for two dimensional machining on three-axis machines (APT II). At this stage of development work was transferred to the Armour Research Foundation which was later incorporated into the Illinois Institute of Technology Research Institute (IITRI) located in Chicago.

The new contractors produced APT III to cater for a limited amount of 3D work on five-axis machines. Development has been continuous and additional features have been added to give eight versions of APT III. This has now been superseded by APT IV which further increases the system capability and also increases the efficiency of the processor.

No single company could have developed a processor with the capabilities of APT, which has necessitated over one hundred years of combined effort from machine tool and computing experts. It seems right, therefore, that IITRI should have a proprietary right on the processor for a period of two years. During this period no other company can use the system unless they are members of the APT Club. It is not proposed to quote membership fees, but full up-to-date information can easily be obtained from IITRI in Chicago.

IBM have maintained a policy of supplying computer programs for use on their computers, and numerical control was no exception, and in continuance of this policy implemented a two-year-old version 7 of APT III and have continued to develop the system which is available to users of the larger computers in the S-360 range.

Capabilities

APT is a geometrical processor, so all information regarding feed rates, speeds, use of coolants, machining cycles, etc., has to be supplied by the part programmer. It has the ability to cater for the most complicated machines, from draughting machines, simple metal cutting machines, to the most complicated multi-axis machines. In practice, this versatility tends to make processing more expensive than many simpler systems which are designed to cope with a limited range of problems or with specific machine requirements. Hence, it tends to produce the best results when dealing with complicated multi-axis machines. It must be pointed out that because of the size of the processor, operation on a large computer is essential.

Format of language

The part programming language of APT, which is tending to become the basis of world standards, uses English-like words in its application, e.g.

LINE = a straight line
CIRCLE = a circle
TANTO = tangent to

The total collection of words used is known as the VOCABULARY of the language, whilst a number of characters are used as punctuators so that STATEMENTS can be formulated. The simple statement to describe a circle of radius 2 with its centre at a previously defined point P1 whose co-ordinates are X and Y.

CIRCLE/CENTRE, P1, RADIUS, 2

It is also possible to give this statement a unique identifying symbol C1 and to state

C1 = CIRCLE/CENTRE, P1, RADIUS, 2

and to recall the same circle at any time in the part program by referring only to C1.
The very nature of the language offers many advantages among which are:
1. Ease of standardisation.
2. Ease of understanding.
3. Non ambiguity.
4. Reduces training requirements provided that all systems used by a company are based on this language.

DEVELOPMENTS IN EUROPE

In the mid-sixties, European N.C. machine tool users became aware of a number of problems which were affecting their efficiency. These included
1. A growing shortage of experienced computer staff.
2. Costs could be reduced by using high power computers.
3. Difficulties in transmitting information.
4. Difficulties of adjusting software to meet machine tool/control system combinations.

Many companies in Britain and Europe were participating members of the APT Club but two events in the last half of the last decade brought about many changes of allegiance. In 1965, a numerical control programming group was formed by the National Engineering Laboratory[15] in East Kilbride with its objective the development of systems to cater for European requirements. The second event was the formation in 1967 of the EXAPT Association[16] in West Germany. This Association was based on the universities of Aachen, Stuttgart and Berlin and had the backing of West German industry. Within a very short time, a number of British companies became members and a U.K. EXAPT Users Association was formed to look after the interests of British industry. The association is currently represented on the Council of the EXAPT Association. Membership is by subscription and full up to date information can be obtained from the association headquarters in Aachen.

Both organisations recognised the virtues of APT and aimed to standardise their own work on:
1. The part programming language used in APT.
2. The CLDATA which acts as the interface between the processors and post processors.

Complete standardisation in these two areas would have resulted in real economies being made because the same part programming language could be used for all machine tools and irrespective of the language, it would only be necessary to produce one post processor for a specific machine tool-control system combination.

Unfortunately, 100% standardisation has not yet been achieved, but the national standards organisations of various European countries (including the British Standards Institution) are all working to this end, with the International Standards Organisation playing a co-ordinating role. Awareness of the problem exists with the system developers, and APT, NEL, EXAPT and the French IFAPT group formed the Uniformed Numerical Control Language Group (UNCL) to work on this standardisation.

Unless standardisation is achieved post processor writing will be proliferated. In particular, failure to standardise the CLDATA interface will mean that every combination of machine tool-control system-computer combination will require its own special post processor, with the result that machine tool manufacturers will be reluctant to supply post processors. This is especially so, since good post processor writers are in short supply, and cost effectiveness and machining efficiency of a machine tool can be affected by the efficiency of the post processor.

Work of the EXAPT Association

The EXAPT Association have virtually taken subsets of APT and with extensions to incorporate technology, have developed processors specially suited to specific types of machining operation. Thus three processors have been or are in the process of development:

    (a) EXAPT I
    (b) EXAPT II
    (c) EXAPT III

EXAPT I

This is a geometrical and technological system specially suited to drilling machines. Complete work cycles can be instructed by a simple instruction. For instance, the instruction TAP will result in all predrilling operations being planned by the computer, with the necessary tools, feed rates, spindle speeds and coolant conditions also being determined automatically by the processor. This processor is currently available.

EXAPT II

This is also a technological processor suited to lathes, and probably one which has proved most effective. Whilst tools are not selected automatically, feed rates, spindle speeds are. In addition, the processor determines the depth of cut and tool path.

EXAPT III

For problems concerned with three-axis machines, EXAPT III is being developed, but is not currently available.

National Engineering Laboratory

The laboratory has undertaken considerable work in the field of N.C. which has been aimed at providing assistance to British industry, taking into account the special requirements of Britain. The processor 2CL and its subsets have been made available.

2CL

This processor, which at the present time is purely geometrical, takes its name from the fact that it caters for two continuous path axes and one linear axis. Designed to cater for work on three-axis machines, it is based on the APT part programming language and CLDATA.

2C

As stated previously, the processor 2C is a subset of 2CL and hence is geometric. The subset has been extracted to provide a system which is geared to lathes.

2P

A further subset of 2CL has been evolved for point to point work. Again, this is purely geometric.

IBM

In addition to a version of APT for use on its own computers, IBM have developed a number of geometric processors of lesser complexity for use by their customers.

ADAPT

A subset of APT has been developed by IBM to cater for 2D machining, and to provide some of the arithmetic facilities of APT. It was designed for use on small computers.

AUTOMAP I

Although not so powerful as ADAPT, AUTOMAP I is an IBM subset of APT which can be used for piece parts where profiling can be achieved by a sequence of machining made up of straight lines, circles or for simple point to point motions.

AUTOSPOT II

For the operation of point to point machines with straight line milling capabilities, IBM developed the program AUTOSPOT II. In some instances, this processor can be used for pocketing and machining arcs of circles. Like AUTOMAT and AUTOPROMT, a system in which part-programming is claimed to be reduced to a minimum. The system has been designed for machining with a bale-ended cutter, such that the surface to be machined is always tangential to the cutter, and caters for:
  (a) Planes.
  (b) Spheres.
  (c) Cylinders.
  (d) Cones.
  (e) Quadrics (ellipsoid, paraboloid, hyperboloid).
  (f) Torus.

AUTOPOL

The German office of IBM have produced AUTOPOL. This system is oriented to turning machines, an area of the machine tool industry in which West Germany have achieved remarkable sales.

Other systems

Other systems have been developed by individual production engineering companies, some from machine tool manufacturers, and some from computer manufacturers who use them to boost computer sales. The cost of developing and maintaining general purpose processors is high and co-operatives have resulted with membership recruited from industry, manufacturers, and government and academic institutes. An example of such a co-operative is the German based EXAPT Association.

Production Engineering Research Association

To meet the needs of companies with small computers, PERA have recently produced PICNIC, a non technological system using a language very similar to 2CL and EXAPT I.

This system is designed for point to point machines and does not use the processor/post processor concept. A different version is required for each machine tool/control system combination. Many of the bureaux operating commercially offer PICNIC to their customers who have the option of a direct terminal link. It is of interest to note that PICNIC has been sold behind the Iron Curtain.

GENERAL ELECTRIC INFORMATION SERVICES (GEIS)

A time sharing system in which the user has a terminal in his own office linked directely to a GEIS computer is provided by GEIS. This is a geometrical point to point system using EXAPT I type input.

ACTION is a system which covers point to point, contouring lathes and mills with up to three-axis capability. It is purely a geometrical processor designed for small computers and it does not use an APT like language.

Two main advantages are claimed for the system
1. That it simplifies part programming.
2. That it uses a minimum computer time. In fact, it is claimed to be four times faster than ADAPT.

Terminals are not available at the present time.

Systems not normally available commercially

Many companies have developed their own systems for in-house use and whilst these may not be available commercially, they could possibly become so in the future.

PRESSED STEEL CO. LTD., COWLEY

This company have developed AUTOPRESS for its own use. The system, which has in-built post processing, covers side milling of contours in any specified plane by means of a cylindrical cutter, cutter offset position being automatically calculated.

HAWKER SIDDELEY AVIATION LTD.

The De-Havilland Division of Hawker Siddeley has produced CLAM whilst the AVRO Division has produced AUTOCODE. Each of these systems has been developed for the company's own use.

CLAM

The language of CLAM is, in general, in English-like statements even though not compatible with APT. Designed for two-and-a-half axis work, the system produces a tape which is subsequently used to create a phase analogue magnetic tape for the machine tool control system.

AUTOCODE

Mnemonic abbreviations form the basis of the language of AUTOCODE which is used for two-and-a-half axis machines, and whilst it was designed to cater for points, straight lines and circles, provision has been made for continued expansion.

ICT LTD., ADVANCED SYSTEMS GROUP

This group developed the ICT Milwaukee-Matic Program which was aimed to make planning and part-programming simpler. Capabilities are provided for drilling, reaming and tapping, as well as face and pocket milling and profiling of rectangular shapes parallel with the machine axes. The language uses English-like words. The system is capable of working out tool paths together with feedrates and speeds, and can determine work cycles and tools required from a single instruction such as TAP. In addition, guard areas can be specified such that the cutter is automatically excluded from such an area during the cutting cycle.

INTERNATIONAL COMPUTERS LTD.

In conjunction with Ferranti Ltd., ICL developed the SURFACE FITTING PROGRAM PMT2 for milling the complex shapes encountered in turbine blades and wings for aircraft models. The shape of the required surface is fed into the computer as a matrix of co-ordinate values of the points forming the surface. The control tape resulting is a phase analogue magnetic tape.

ROYAL AIRCRAFT ESTABLISHMENT (RAE)

A system PIDGEON was developed by the RAE to cater for two axis template grinding machines. The language of the system consists of a limited vocabulary of non-abbreviated English words, and hence the geometric capabilities are limited.

FERRANTI LTD.

The control systems developed by Ferranti Ltd use phase analogue tapes and because of the equipment necessary to produce such tapes, it became necessary to develop PRO-FILEDATA, continuous path system for 3-axis milling machines. Since the recent amalgamations in the N.C. control system industry, PROFILEDATA has been acquired by Plessey Ltd. PROFILEDATA permits components to be defined in terms of points, straight lines and circles, and uses symbolic references to define these features, and mnemonics for the remainder of the language. PROFILEDATA is available for use through some bureaux.

ROLLS-ROYCE LTD.

The systems developed by Rolls-Royce Ltd were intended for their own use and are not commercially available at the present time. Among the systems developed are PRO-CONSEL, COCOMAT, SCHAUDT GRINDING, etc.

PROCONSEL

A powerful but highly specialised system used in detailing, planning, manufacturing and inspecting turbine and compressor blades. This system is an excellent example of developing N.C. to cover as wide a range as possible of a company's requirements on a specific product; a big step towards a complete management system.

COCOMAT

Designed to cater for two and a half dimensional machining, COCOMAT caters for straight lines, circles and interpolated curves. Mnemonic abbreviations make up the part programming language.

Characteristics of general purpose processors

There are, in general, four fundamental characteristics:
1. The geometry of the piece, and in general, the machining techniques are specified by the user.
2. This geometrical and machining information is communicated to the computer by means of a special computer part programming language.
3. Two separate processing operations are required, i.e. the *Processor* and the *Post Processor*.
4. In current practice, the communication from the *Processor* to the *Post Processor* is in the form of *Cutter Location Data*. (CLDATA). The CLDATA is usually stored on computer file and can be printed out as required (Figure 19.20).

Processor

This phase is concerned with the geometry of the piece and the cutting tools involved which have been defined in the appropriate part programming language. At the end of this process, the points which define the cutter path, have been calculated and stored in the CLDATA file.

Information which relates to feed, speeds and other machine oriented functions is recorded on the CLDATA file for subsequent processing by the *post processor*. The *processor* is thus completely independent of the machine tool, and its control system and is only required to suit the requirements of the company product and the computer which is used for processing.

Post processors[18]

This phase of operations takes its information from the CLDATA file and carries out all the calculations to ensure that the feeds and speeds etc. which have been specified are corrected to be within the capacities of the machine tool.

This information is stored on a further CLDATA file and contains all the information for the preparation of the punched tape.

It is important to stress that a post processor must be written to suit the format of the CLDATA, the computer in use and the machine tool control system combination. It follows therefore, that if a company has a number of drilling machines of different manufacture and linked to different control systems, it will be necessary to have a separate post processor for each machine tool/control system combination. Even minor non-standard features of a particular machine tool/control system may call for a post processor modification.

Flexibility

The concept of using a part programming language, in conjunction with a processor and post processor results in a very flexible system, thus permitting a common processor to

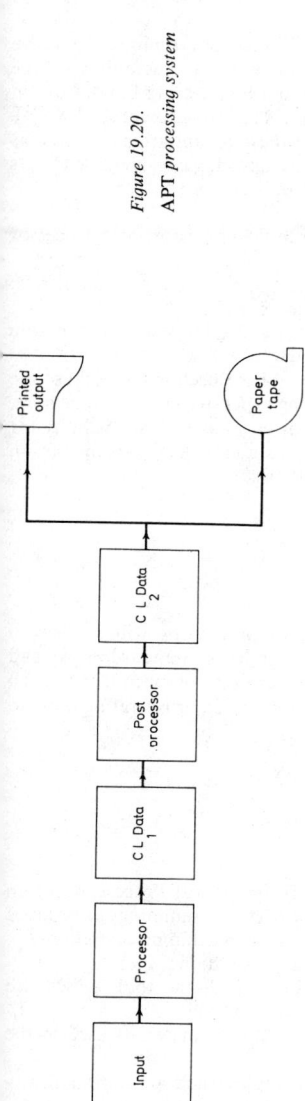

Figure 19.20.
APT processing system

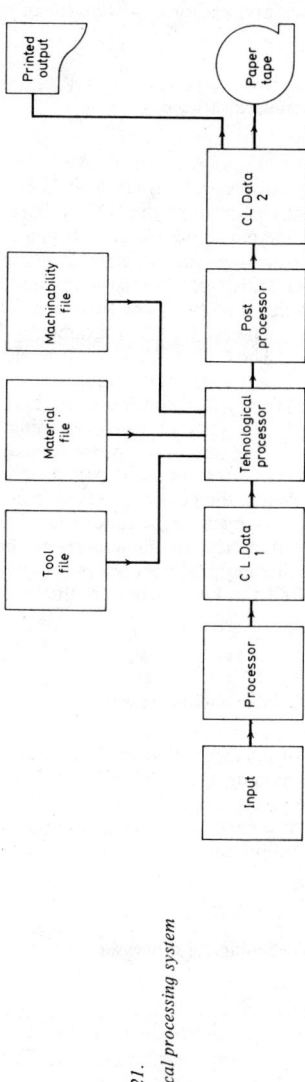

Figure 19.21.
Technological processing system

be used for a wide range of processes, e.g. welding, flame cutting, laser cutting, drafting etc., and embraces a wide range of machine/control combinations.

Standardisation

The development of the APT system in the United States saw the beginning of a move towards standardisation of N.C. languages in that country, but it was only with the establishment of the N.C. Software Group at the National Engineering Laboratory in 1965, that standardisation began to take effect in England. The formation of the EXAPT Association in 1967 spread this trend into Europe. All these organisations, as well as the British and International Standards Organisations are devoting considerable efforts in this direction. The first areas to receive attention were based on:
1. The APT part programming language.
2. The CLDATA which provides the interfaces between the processor, post processor and tape production.

Work is still continuing in this direction with the object of making it possible to use:
1. The same part programming language for all machine tools.
2. The same post processor for a given machine/tool control system combination irrespective of the part programming language used.

When the goal is achieved, considerable economies will have been made because the development and maintenance of post processors is a long expensive task. Furthermore, machine tool manufacturers are beginning to accept that they have a responsibility for providing post processors for their equipment, but unless a complete standardisation of CLDATA is achieved, they are faced with a gigantic task.

Modular post processors

Considerable work is being conducted on modular post processors, with a view to providing a method of building post processors from a number of standard bricks and so giving the ability to a post processor to cater for a wider range of machines. In such a system, is is important to improve efficiency and so keep computer operating costs to a minimum.

Technological processors

The EXAPT Association introduced the concept of Technological Processors, in an attempt (which has been successful to date) to simplify part programming, to produce complicated machining cycles automatically and to optimise machining operations.

To provide these benefits, the computer system requires additional files:
1. A tool library which lists the pertinent dimensions of all the tools which are required. This is a permanent file and can be added to at will.
2. A materials file which lists the cutting properties of all the materials used in the company.
3. A machinability file which determines the size of tools which are automatically selected e.g. selects the drill diameter and depth for a reamed hole.

The system works (Figure 19.21) in very much the same way as any general purpose system, except that there is an additional processor.

THE BATCH PROCESS

Batch processing

The system of supplying part programs on preprinted sheets, punching the information contained thereon on to tape or cards and then processing in a computer is generally referred to as Batch Processing. Such a system obviously pre-supposes access to a suitable computer.

The non-availability of a computer within a company (In House) does not preclude the use of computers for numerical control since, in recent years, a number of organisations (Bureau) have been set up to carry out such work in the bureau on behalf of customers whose works may be spread over a very wide area.

The decision as to which approach is most suited to a particular company is dependent upon many practical factors and also to the economics involved.

In house processing

Most companies would appear to introduce computers to deal with administrative problems such as the preparation of the company payroll etc.

Rarely does one find a computer installed solely for numerical control work, and, indeed, this may not be a practical method for most companies since the total N.C. load on the computer could be a fraction of the total capacity available, and the resulting service would be inordinately expensive. Therefore, if a computer is available within the company, it should be considered whether this will be suitable for numerical control.

Processors

The first decision is to determine the processors which will be used. This decision must be made in the light of the company's technical requirements. The processor must be capable of handling the variety of piece parts which the company makes or is likely to make in the foreseeable future, and this will automatically determine the size of the computer which is essential. The proprietors of the processor and the computer manufacturers are able to advise on this aspect.

The fact that the company computer is of adequate capacity does not automatically imply that this computer provides the right answer. Economics are all important, and many factors must be considered.

Size of computer

Small computers are much cheaper to purchase than large computers. This difference in cost can be very considerable but it is not merely a question of size of the available core storage which must be taken into account. The large computers can process information many times faster than a small computer.

Thus, the cost of processing a job on a small computer can be many times greater than the cost of the same job on a large computer, and it may well be, that although there is a small computer in house, a much more economic service could be obtained from a bureau.

Advantages

Certain advantages of In House service are immediately obvious. Lead time for processing can be kept to a minimum because of the proximity of the computer to the part programmer, whilst computer schedules can easily be adjusted to meet priority requirements.

Moreover, use of an In House computer for all aspects of computing provides greater utilisation of the equipment, with consequent cost reduction per job. All things being equal, In House running costs should easily be competitive, since a commercial profit does not have to be included in the cost. Considerable advantage accrues when a job has to be run three or four times before a correct tape is produced, since little time is wasted in transmitting information from one place to another.

SELECTING A SYSTEM

Difficulties

Once the machine tool-control system configuration has been decided, a further decision must be made as to whether the processing is to be performed 'in house' or at a bureau. Should the decision be the latter, all that remains is to arrange for transfer of input and output to and from the bureau whose systems he will have to accept. However, if an 'in house' service is to be developed many questions must be answered.

New phraseology[20]

To assess the capabilities of an N.C. processor, it is essential to study the part programming language concerned and to decide whether it meets the requirements of the individual company. In general, the engineer experiences difficulties in assessing these languages and finds it difficult to obtain assistance, because very few organisations have detailed knowledge of all systems. A further difficulty is that both computing and numerical control technologies have generated their own terminology and as specialists tend to sprinkle their discourses very liberally with these new terms, the engineer finds himself in a new and confusing world.

In order to prepare himself to meet these new outlooks, the engineer should devote considerable time to reading the technical journals and some of the technical books on the subject which are now on the market. Attendance at seminars is another useful approach, but only if the individual participates. It is not sufficient merely to sit and listen. The engineer must ask questions to settle the doubts and grey areas which exist in his ideas, and if possible, he should use the many intervals during the sessions to buttonhole the speakers to discuss problems to a greater depth.

A very useful organisation which has branches throughout the country, each of which presents local lectures on interesting aspects of numerical control, is the British Numerical Control Society. Membership of this society immediately gives personal contact with hundreds of people involved in numerical control. It is also advisable for the newcomer to attend a course on computer appreciation. In this way, the mysteries of such things as 'Source Decks', 'IO Operations', 'Core Size', 'Bits', 'Bytes', etc. can become clear.

Preliminary survey

Before making a final decision about the approach to be made to N.C. a preliminary survey should be conducted. As part of the survey, discussions should take place with the company's electronics engineers, computing specialists and production staff, to consider:

1. The processors, geometrical and technological, which are currently available.
2. The size and capabilities of the computers installed within the company.

3. The pros and cons of using 'in house' systems as against a bureau.
4. The action which will be needed to implement the system.
5. The lead times which can be expected from part programming to the availability of the finished tape.
6. The organisational changes which will be required to produce the most efficient service.
7. Any new job specifications which may be required in the new organisation.
8. The availability of suitable staff.
9. The training which will be required to prepare staff for the operation of the new system.

In conducting this survey, care should be taken to keep it on as wide a basis as possible. The effects on personnel and systems not directly involved should also be considered, because weaknesses in associated areas can often prevent the best utilisation of numerical control methods.

No one must lose sight of the fact that the overheads and support costs of N.C. machines are higher than conventional machines, and that the benefits of the former may be considerably reduced if sufficient attention is not paid to detail.

Factors to consider

The ultimate factor which decides whether a system is right for a given company is economics, just as is the case in most business decisions. All costs and all savings must be accounted for. There must be no hidden costs or exaggerated savings and the engineer must be sure that he has a true bill before he makes his decision. To this end, further factors require careful consideration, and where possible, a cash value attributing to them:

1. Is it desirable that the part program should conform to some recognised standard such as APT. Future requirements as well as current needs should be taken into account?
2. Should the system be special or general purpose, or is a machine-oriented system adequate?
3. Is standard CLDATA essential to future policy?
4. Are adequate post processors available, and what is to be the source of future post-processors?
5. Does the system contain a good method for checking input information and can any such errors be corrected easily and quickly? Errors can be expensively wasteful of computer time.
6. Does the system supply warning and error messages in a form which is easily understood without ambiguity, or are they supplied in a mysterious code?
7. Can 'remarks' or 'special instructions' be fed into the system in such a way that they can be usefully shown on the printed output?
8. Is the printed output in an intelligible form, and can it be used as the basis of company documentation?
9. Can coded instructions which may be required for the operation of certain machine tool functions be entered easily into the system?
10. Is the system easy to master?
11. What service is available for correction of inadequacies which may be found to exist in the system?
12. What back-up services are available?
13. Is any additional system available which will permit drawings of components and tool paths to be produced?
14. How much formal training is required, not only of personnel directly involved but of management, supervision and others indirectly involved?

15. Is such training readily available, and is it necessary to obtain such training from a bureau, or can an 'in house' service be adequate?
16. Does the system use technological processors? If so, how much effort will be required to supply and maintain the various technological files concerning tooling, materials and machinability?
17. Can the system easily deal with metric and imperial units?
18. Has the system reached the limit of its development or can it be extended to cope with future developments? Are any such extensions currently planned?
19. How much does it cost (a) to secure the rights to use the system? (b) to implement the system 'in house'? (c) to maintain the system?
20. What is the likely cost of producing a tape for a specific operation where the cost of tape preparation by other systems is known?
21. Is the lead time required for tape preparation compatible with the requirements of the company?

COMPUTER GRAPHICS

Reactive graphical display screens have been developed which, in the future, may play a large part in the part programming process. These systems are used by many companies, but as the capital cost tends to be high, extended use tends to be slow. It is, of course, anticipated that the cost of these systems will reduce as sales increase, so permitting even wider usage.

Basically the system consists of a television screen which is linked via a suitable interface to a computer. In addition to the television screen, a fibre optic probe, a typewriter keyboard and a press button panel are included. The fibre optic probe is capable of detecting the electron beam producing the image on the screen, and hence is capable of indicating to the computer the exact point of search.

The press-button panel can be used in such a manner that each button represents a code which is recognised by the computer program. Whilst the typewriter keyboard is available as a means of calling up the required program or entering precise dimensional information.

Having called up the program via the keyboard, the computer displays on the screen instructions for the operator controlling the device. Thus, the operator does not need extensive training, since he is instructed, step-by-step, on his next course of action or option.

In addition, the computer also displays on the screen, when appropriate, a menu. This menu can be a list of parameter names, and in some cases, possible parameter values. The parameter name and value required from the menu can be indicated to the computer by the operator pointing the end of the fibre optic lead to the screen and pressing it gently on the screen to trigger a switch mechanism.

It is possible by these means to produce straight lines, circles, conic sections etc., on the screen, and in this way, build up the image of a piece part on the screen. When building up pictures in this manner, all lines are treated as being of infinite length, but, as there is a facility to delete any portion of a line, this causes no problems.

Many facilities are available to give the system wide flexibility. For instance, the whole drawing may be moved to any part of the screen, and any small detail can be enlarged to the full size of the screen to permit detailed investigation or modification.

Programs have been developed to permit dimension lines to be positioned on the display, together with the appropriate dimensions which are calculated by the computer. Choice of tolerancing presentation is available, e.g. plus and minus tolerances may be shown, or minimum and maximum dimensions.

To obtain the greatest value from such graphical systems, it is necessary to be able to transfer all the information developed on the screen to a drafting machine which will thus be capable of producing the same picture on paper or some similar medium.

Part programming is possible, because, computer programs have been developed which enable the part programmer to generate the cutter path on the screen. This information transmitted to the appropriate post processor results in the production of a control tape for the machine tool.

It is possible with computer graphics to store the information on computer file so that it can be used again, either for the purpose of modification or to develop further information, such as detail drawings, production drawings etc.

Obviously, such devices emphasise the need to regard N.C. as a concept of involvement of the computer from the design board to the finished product.

ON LINE CONTROL[21]

Current problems

The most irksome aspect of N.C. at the present time is tape proofing, for, despite all the precautions which have been taken, there cannot be 100% certainty that the tape will carry out the necessary machining operations correctly.

Should the tape fail to perform correctly, the paramount urgency becomes the production of another tape in which the error has been eliminated. If this involves additional computer processing, this will obviously take time. Thus the machine tool must either remain idle during this period, or be re-set to produce a different piece; either way, this costs money. A possible solution is to use *On Line Control*.

The tape is the normal medium for transferring control information from the computer to the machine tool. The presence or absence of a hole in the tape is a representation of an electrical condition, hence, the simple concept of *On Line Control* in which the control information produced by the computer is transferred directly to the machine tool by conducting wires or radio link (Figure 19.22).

State of the art

During the latter part of the last decade, considerable work was carried out by universities, machine tool manufacturers etc. into the feasibility of *On Line Control*. The

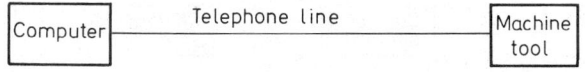

Figure 19.22. Computer to machine tool

major areas of activity were in the United States and Japan, and 1970 saw the advent of the first few systems on the world market. Consequently, there is, as yet, very little practical experience of this approach on the shop floor in industry. England, however, has not been completely idle, for instance Molins Ltd. have designed 'SYSTEM 24'. This takes a revolutionary approach and the system offers
1. A complete new outlook in the machine tool itself.
2. On line control of the complete shop floor process with control signals supplied from a carousel of magnetic tapes.
3. A complete management system.

Process control computers

Computers have been developed for process control work which have a high inherent reliability, because, failure in process control could be disastrous. Not only are such computers extremely reliable, they are designed on a 'Fail safe' basis and to permit planned maintenance to be carried out whilst the computer continues its normal function.

Process Control Computers have been used in the petro-chemical industries for some considerable time with outstanding success. Obviously a failure in these industries cannot be tolerated. Similarly, a pilot scheme in the N.C. field carried out by McDonnell Douglas did not experience a single failure during six month's continuous operation.

Tapeless control

In the present method of operating N.C., one controller serves each machine tool, and each controller is driven by a tape in which the presence or absence of holes signals the instructions to be carried out.

A logical development is to eliminate the tape, which is the most easily damaged piece of tooling. Moreover, its method of preperation, depending as it does on mechanical systems, is the most suspect phase of N.C. operations. This can be achieved by introducing a process control computer on to the shop floor and with control signals from this computer fed to each control system via an electrical feeder (Figure 19.23).

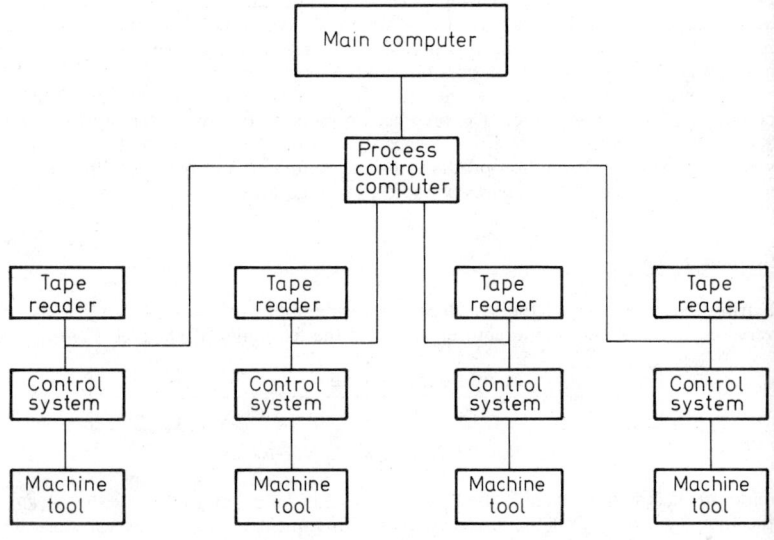

Figure 19.23. Tapeless control

Connection of the feeders from the computer between the tape reader and electronic of the control system produces a system which can be used either in the on line or tape controlled mode, thus providing a back up service in the unusual event of failure of the process control computer or during periods of preventive maintenance in the latter. The process control computer itself is connected by electrical feeders to the main computer which has the necessary capacity to process N.C. part programs.

The permissible distance between the process control computer and the machine tools is generally subject to limitations because of signal strength attenuation in the feeder lines. However, the main computer can transmit over very long distances.

This configuration permits interpolation to be carried out in the individual controllers thus the process control computer can handle a large number of machine tools.

This process, in which interpolation is carried out in the individual control systems is generally referred to as tapeless control.

Shared interpolation

Interpolation can be carried out in the process control system, with the resulting signals passing via the connecting line to the servo controls (Figure 19.24). Known as a

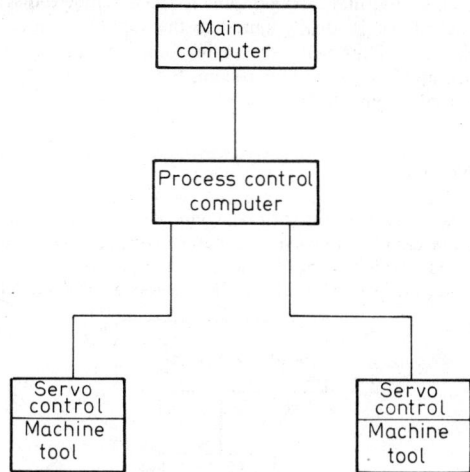

Figure 19.24. Shared interpolation

shared interpolation system, this system has greater limitations, since the large amount of interpolation required will rapidly saturate the capacity of the process control computer.

Fully automated factory

To supply information to the machine tool directly from the main computer via a process control computer brings the fully automated factory for batch processing one step nearer. However, two important steps will still be required to achieve this final goal:

1. Control of movement of stock on the shop floor.
2. Control of the whole process.

Transfer lines

Movement of stock on the shop floor is one area of batch processing which often creates problems such as material which is available failing to reach the right place at the right time. The solution can well lie in the use of transfer lines, the operation of which is commonplace in the motor car industry. In general, the latter systems are fixed mission systems, transferring pieces to predetermined destinations at fixed intervals of time.

For a batch manufacturing factory, a variable mission would appear to be an essential requisite, in order that piece parts may be transferred from one machine to another as required. Such a system offers a facility for rationalising the machine tools sited on the shop floor, to permit operations to be carried out on the cheapest machine in the N.C. complex, using expensive, sophisticated equipment only for absolutely essential operations. In this way, the most economic utilisation of equipment may be achieved.

So far, no mention has been made of the manner in which variable mission operation of the transfer line can be achieved, and it is here that the process control computer can be further utilised. This computer, being on line to the machine tools, and having access to the processing schedules, is ideally suited to the supply, on line of information to operate the transfer line. Thus, with the main computer, process control computer, N.C. machine tools, and a transfer line system, the fully automated batch processing factory can be achieved (Figure 19.25).

Control of the whole process

It has been emphasised that systems for management control are essential for any N.C. systems, but with the capital intensive automated factory such systems become even more essential. Systems which schedule the materials, tools, control information etc., will be vital, but even more important will be the retrieval of facts concerning actual

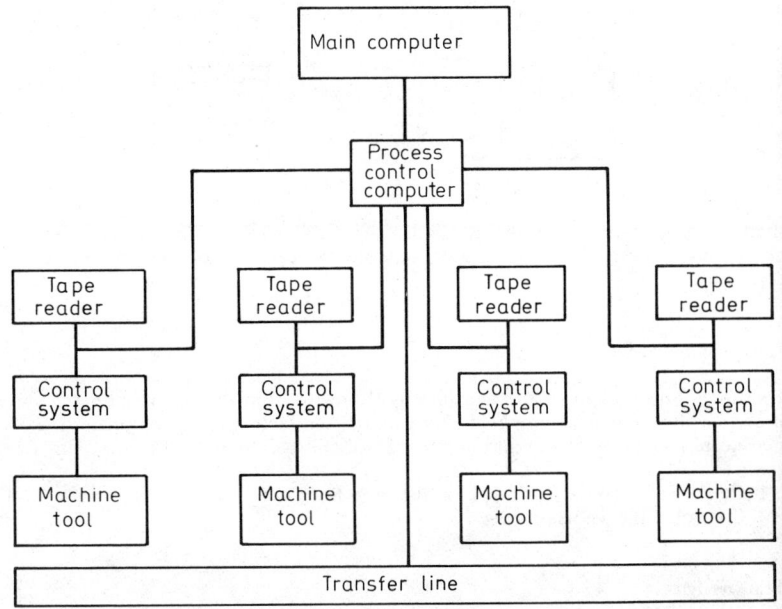

Figure 19.25. Transfer line

achievements related to these schedules. In such an environment, the main computer will be capable of assessing the current state of the art on the shop floor, and, where achievement does not meet the current schedule, will be capable of preparing rapidly new schedules to permit optimum manufacturing.

It must not be thought that such a system takes over the function of management; far from it. As with all good systems, it will give management the information needed for decision making. Management will still, rightly, make the plans and decision, and the system will advise whether or not the plans are going to be achieved. No longer will management be left in ignorance of the precise state of performance, any failure in the manufacturing plan will be signalled *immediately*.

REFERENCES

1. PUCKLE, O. S. and ARROWSMITH, J. R., *An Introduction to Numerical Control of Machine Tools*, Chapman & Hall (1968)

   ROSS, D. T. and others, *Automatic Programming of Numerically Controlled Machine Tools*, Massachusetts Institute of Technology. Interim reports 6873-IR-1 to 11 and final report 6873-FR-3 (1956–60)

   ROSS, D. T. and WARD, J. E., *Investigations in Computer-Aided Design for Numerically Controlled Production*, U.S. Air Force Materials Laboratory, Technical Report. AFML-TR-68-206 (1967)

2. LOCKWOOD, F. B., *Fundamentals of Numerical Control*, Machinery Publishing Company, Brighton (1968)

3. WILSON, R. K., 'The Future Impact of Numerical Control on Management', *Proc. ASTME-WESTEC Engineering Conference*, Technical paper MM67-723 (March 1967)

   BENTLEY, W. H., 'Numerical Control is a Management Tool', *Automation* (Sept. 1965)

4. WALKER, D. F., 'Why Use Magnetic Tape as Information Carrier for Modern Numerically Controlled Machines', *Technica (German)*, (15 December 1967)

   EBERLE, M. and WAIBEL, G., 'Analogue System of Absolute Position Measurement for Numerical Control of Machine Tools', *Siemens Review*, (February 1965)

5. BS 3880 (1971) *Specification for the Dimensions of Punched Paper Tape for Data Processing*

6. KOHL, R., 'Adaptive Control—Towards the Thinking Machine', *Design*, (1 May 1969)

   DYKE, R. M., 'Adaptive Control Optimises N.C. Technology', *Canadian Controls and Instrumentation*, (March 1969)

   HOLMES, L. L., 'Adaptive Control for Metalworking Applications', *Automation*, (August 1967)

7. PORTER, PROF., 'Adaptive Machine Control', *Machinery and Production Engineering*, (5 February 1969)

8. BS 3635. The Numerical Control Machine Tool Control Input Data (1972)

   ISO Standard R840. *Code for Numerical Control of Machines (compatible with the ISO 7-bit character set)*

   ISO Standard R841. *Access on motion nomenclature for N.C. machines*

   U.S.A. Standard X3.8.65. *Interchangeable Perforated Tape Variable Block Format for Contouring and Contouring/Positioning of Numerically Controlled Tools*, U.S.A. Standards Institute

   ISO Standard 1056. *Punched Tape Block Format for the control of N.C. Machines*

   ISO Standards 1057 and 1058. *Interchangeable Punched Tape Variable Block Format for Positioning and Straight Cut N.C. Machines*

   ISO Standard 1059. *Punched Tape for Fixed Block Format for N.C. machines*

   EIA Standard RS 227. *One Inch Perforated Tape*, Electrical Industries of America

   EIA Standard RS 326, RS 273A. *Interchangeable Perforated Tape Fixed Block Format for Positioning and Straight Cut N.C. Machines*, Electrical Industries of America

9. National Engineering Laboratory, *Programming of Numerically Controlled Machine Tools*, NEL Report 187 (May 1965)

   EXAPT, *EXAPT 2. Part Programmers Reference Manual*, EXAPT, Verein, Aachen (1967)

   EXAPT, *EXAPT 1—Part Programmers Reference Manual*, EXAPT, Verein, Aachen (1967)

   Illinois Institute, *APT IV System Manual*, Illinois Institute of Technology Research Institute, Chicago, Illinois (1968)

10. MOORHEAD, J., 'Training for Numerical Control', Paper No. 18, Proc. 6th Annual Meeting and Technical Conference of the U.S. Numerical Control Society (1969)

    MOORHEAD, J., 'Training for Numerical Control', *American Machinist*, (6 June 1966)

11. DYKE, R. M., 'Effective Utilisation of Point to Point Numerical Control', *Machinist*, (April 1967)

    CHILDS, J. J., 'Contour Machining by Point to Point N.C.' *Machinery and Production Engineering*, (Dec. 1966)

    National Engineering Laboratory, *'A Survey of Part Programming for Point to Point Numerically Controlled Machine Tools*, NEL Report 291

12. GOEZ, W. E., *Three Axis Profiling of Wind Tunnel Model Parts*, 6th Technical Conference U.S. Numerical Control Society, Paper No. 23 (1969)

    MCAVOY, W., *Part Program for Three Axis Multi-Tool Machine*, ASTE Southeastern Engineering Conference Collected Papers, Paper No. 781 (1965)

13. WILBURN, J. E., 'Future Marriage of N.C. and Computer Control', *Automation* (1966)

14. DOBE, J. W., *The APT Long Range Programme*, Paper No. 2. The Annual Technical Conference of the U.S. Numerical Control Society (1969)

15. National Engineering Laboratory, *Machine Tools Software*, NEL Report No. 399L5834, (1969)

    National Engineering Laboratory, *Programming of Numerically Controlled Machine Tools*, NEL Report No. 187 (1965)

    LESLIE, W. H. P., 'Features of Computer Programs for Numerical Control', Paper No. 29, *Proc. Conference on Advances in Computer Control. I.E.E.* (1967)

16. Association for Computing Machinery, 'Development of Numerical Control Languages in Europe', *Proc. 22nd National Conf. Assoc. for Computing Machinery*, Thompson Book Company, Washington (1967)

    OPITZ, H. and others, *The EXAPT Programming System for Numerically Controlled Machine Tools*, (German) Industrie-Anzeiger (1967)

    Compagnie Internationale-pour L'information, *IFAPT Language Description*, (German) 68, Route de Versailles a Louveciennes (1966)

17. Illinois Institute, *CLTAPE Format & Vocabulary Codes*, Illinois Institute of Technology Research Institute, Illinois, Chicago (1957)

18. International Business Machines, *Post Processor Program Writing*, IBM Form EZO-8154, International Business Machines Corporation, New York (1964)

    THOMAS, R. A., 'Role of the Post Processors in Programming N.C. Machines', *Automation*, (Oct. 1966)

    CATON, R. W., 'In-house Post Processors', *American Machinery Metalworking Manufacturing*, (26 Aug. 1968)

19. OLDFIELD, L. B., 'Look Before You Leap Into N.C.' *Metalworking Production*, (9 Feb. 1966)

20. THORNHILL, R. B., 'Numerical Control Terminology', *Automation* (Sept. 1968)

21. THORNHILL, R. B., 'Computers and Numerical Control—An Uneasy Marriage', *Data Systems* (Feb. 1968)

    THORNHILL, R. B., 'Computer Aids Flexibility to N.C.' *Tool and Manufacturing Engineer* (March 1968)

    THORNHILL, R. B., 'Goodbye to Tape', *Tool and Manufacturing Engineer* (May 1968)

BIBLIOGRAPHY

KELLEY, R. A., *A Selective Bibliography on Numerical Control*, North Holland Publishing Co. (1970)

HATRICS, *Numerical Control of Machine Tools—A Bibliography*, Hampshire Technical Research Industrial Commercial Service, Central Library, Southampton

REMPP, K. A., *Numerical Control—A Bibliography*, Institute of Science & Technology, Industrial Development Division, University of Michigan

DIRECTORIES

Kensington, London, SW7 4EF
Register of Research in Machine Tools, Machine Tools Branch, Ministry of Technology, London
Membership Directory, Numerical Control Society, Princeton, U.S.A.

PERIODICALS

N.C. News (Irregular) National Engineering Laboratory, Glasgow
B.N.C.S. News (January, March, May, August, November), British Numerical Control Society
N.C. Scene, Numerical Control Society, Princeton, New Jersey, U.S.A.

20 MANAGEMENT

MANAGEMENT 20

20 MANAGEMENT

ENGINEERING PRODUCTION METHODS AND LAYOUT

R. G. NORMAN

Manufacturing resources such as facilities, materials, labour, space and technica know-how are possibly the most vital factors that contribute to the profitability an economic development of an enterprise. Yet, if engineering production practices ar inferior the productive utilisations of these resources will also be inferior. Many tech niques have been developed to assist production decision makers to be more proficien and in this text an attempt has been made to bring as many of the techniques to the notic of both the novice and practising engineer. This section endeavours to integrate tradi tional and orthodox practices with those of modern analytical engineering productior concepts and mathematical applications.

However, it would be impracticable to suggest that proficiency in the use of all sucl techniques could be achieved by reading one author's text, the depth of which is essentiall limited to an introductory level. To compensate for certain inhibitions an extensiv bibliography has been included.

It has long been accepted in engineering production that the most economical method of manufacture occur when high volume and flow-line production techniques can b established. Unfortunately the heterogeneous nature of products and activities i individual companies inhibits this ideal. Many organisations never attain an ideal an are more often faced by the nature of their products and the fluctuations in consume demand, as well as history and tradition, to consider various and changing methods o production.

The circumstances of demand and economic necessities of each firm must be carefull considered before a decision is taken as to which type of manufacture would be justifie for both short and long-term success.[1] For example:

Continuous. Generally related to steelmaking, chemical and allied industries, an which theoretically produce 24 hours per day throughout the year.

Repetitive. Often found in the clothing, food and automotive industries wher products and their component parts are processed in batches, each successiv batch of items passing through the same sequence of operations.

Intermittent. In which single products or small batches of components are made i response to separate customer orders.

Most companies are content to adopt the intermittent method of manufacture simpl because it provides considerable leeway and freedom for management to meet crise as they arise. This method, nevertheless, tends to restrict capital and labour utilisatio and necessitates a relatively high level of working capital in the form of facilities, materia handling, stocks and work-in-progress.

TYPES OF PRODUCTION

Basically there are four types of production method, which are related as shown i Figure 20.1.

Jobbing production. Consists of 'one-off' products characterised by the fact that the whole product is considered as one operation, where work is completed on each product before passing onto the next. Manufacture is usually carried out to customer order only.

Batch production. Involves the processing of discrete groups of a particular component, or product, through a series of operations which are completed through the whole batch before the next operation is undertaken. Usually produced at a rate exceeding the average demand in anticipation of repeat orders, which will justify some provision of special production facilities and possible holding of stocks.

Flow production. Imposes its own rhythm of work on a guaranteed market distribution and causes a continuous and progressive process of material, i.e. as one

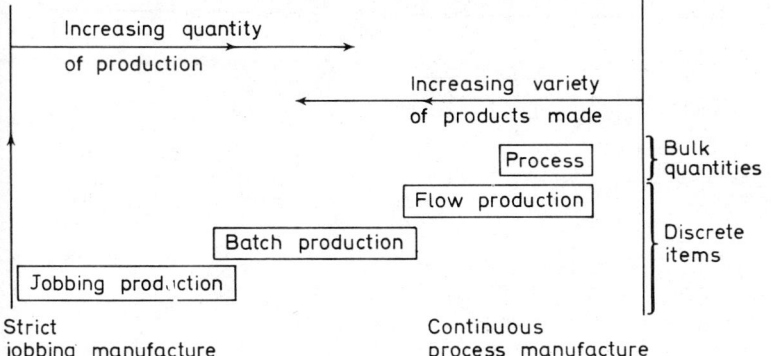

Figure 20.1. Types of production (from 'Production Management' (Wild) published by Holt, Rinehart and Winston)

operation is completed the component, or product, is passed immediately to its next operation without waiting for the operations to be completed on a total batch or order.

Process manufacture. Often carried out by chemical rather than mechanical means and in bulk quantities.

There are four layout arrangements (Figure 20.2) which are associated with the various production methods:

Layout by process (or function). Where facilities are grouped together in the same department or part of a factory according to activities of a similar nature. For example, separate areas may exist for drilling operations, milling, grinding, fitting, assembling, inspection and so on. It is a layout most frequently adopted for its simplicity of supervision and convenience of services. It is highly flexible and permits complex products to be manufactured alongside simple repetitive items. Nevertheless, it has the disadvantages of operating with a comparatively high level of work-in-progress, extended throughput time, low equipment and labour utilisation and high cost materials handling.

Layout by product. Where the facilities are laid out to the operations sequence of the products and where services are usually provided by ring-main type power. This layout is typical of 'flow' production methods, the equipment being arranged in a uni-directional basis. Because it is a specialised production layout designed for individual products in large quantities, or at least a very small range of standardised products, it is relatively inflexible. The advantages of such a layout are: low-level work-in-progress, quick throughput time, high equipment and labour utilisation and relatively low materials handling cost.

Layout by process or function

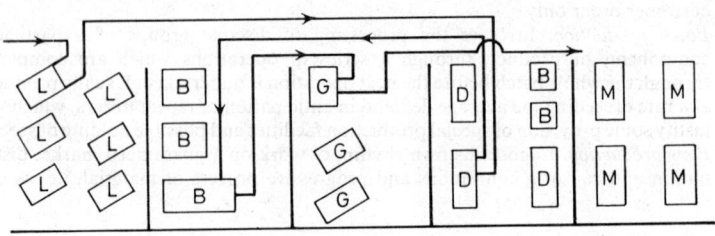

Layout by product

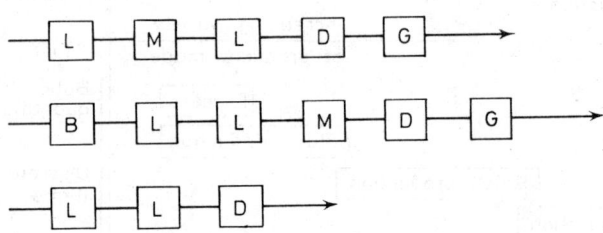

Layout by cell or grouping

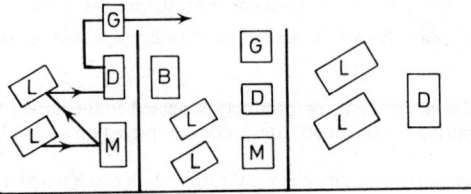

Layout by fixed position of product

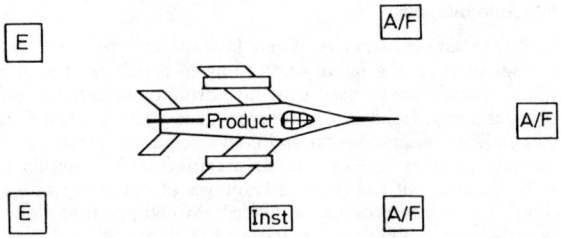

Figure 20.2. Layout arrangements

L.	Lathe	G.	Grind
B.	Borer	A/F.	Airframe
D.	Drill	INST.	Instruments
M.	Mill	E.	Engines

Layout by 'cell' or 'group'. An arrangement that combines a functional layout with that of a product layout in order that 'families' of components can be manufactured on a more profitable and economic basis. The arrangement allows similar and certain dissimilar products and parts to be processed on groups of machines laid out exclusively for performing most, if not all, of the required operations on a product.

The workshop may be composed of a number of such groups laid out in a line, or in a cell.

It is a system that has developed with the application of Group Technology[2, 3] a relatively new philosophy that tends to highlight the need to rationalise variety, reduce tooling set-ups and changeover time. It is a layout that encourages quicker throughput times, more productive utilisation of labour and equipment and involves a relatively low cost absorbed in materials handling.

Layout by fixed position of product. In the previous layout systems the product moves past stationary production equipment. In this case the reverse applies. For example, the product, a ship or aircraft, remains stationary until it is completed. The method usually involves a low content of work-in-progress, low utilisation of labour and equipment, high materials handling cost and a relatively slow throughput time.

FACTORS THAT INFLUENCE METHODS OF PRODUCTION

The exclusive existence in a company of any single layout strategy is rare, most methods of production require a combination of strategies.[4] Whilst many firms prefer jobbing-shop arrangements, policies influenced by demand, standardisation, rationalisation, product diversity as well as other influences as listed in Table 20.1, may encourage 'batch' production methods on a product-line basis and similarly, for some products, flow production methods may be economically justified.

Table 20.1   FACTORS INFLUENCING PRODUCTION METHODS OF MANUFACTURE

Factor	Jobbing production	Batch production	Flow production
Kind of plant	A wide range of general purpose machines and equipment. Usually over-size for flexibility and versatility. High capital investment to rate of output, usually breakdown maintenance.	Mixed general purpose with some special and single purpose machines and equipment. Tooling designed to de-skill the job for ease of interchangeability. Some planned maintenance.	Large range of automatic and transfer machinery, moving tracks, and conveyors. Tools designed to completely de-skill the job. Low capital investment in relation to output per head. Planned maintenance essential.
Personnel: Technical	Substantial planning staff knowledgeable of estimating and quantities. Wide knowledge of profession and competent contract and site engineers.	Competent design and tooling staff knowledge of scheduling and shop loading problems. Some Work Study experience.	Large highly specialised staff trained in flow production. Sophisticated production planning and control department, including Work Study, plant layout and Materials Handling engineers, etc. Highly efficient maintenance pool.

continued

Table 20.1 (*continued*)

Factor	Jobbing production	Batch production	Flow production
Supervisors	Competent general engineers, mostly of skilled craftsmen ability. Knowledge of many crafts not merely organisers.	Apprenticed in a particular craft, some knowledge of other crafts. Good organiser with a flair for planning and control.	Essential to be a good organiser capable of handling mixed labour able to instruct operatives. Capable of planning and control and a competent human relations man.
Operatives	Highly skilled capable of working with the minimum of supervision.	Constant labour force of semi-skill and some skilled machine setters.	Semi-skilled operatives dexterous in repetitive work conditions and competent in paced working condition. Surplus labour essential, able to act as 'floaters' or stand-ins.
Plant layout	Process layout convenient for supervisor's control. Large storage area in relation to machining area required. General purpose handling facilities twists, overhead cranes, etc.	Combined process and product layout. Palletisation necessary. Mechanical handling facilities in the form of Fork Lift trucks and roller conveyors, overhead cranes, etc.	Product layout not necessarily straight line, must have direct access for materials handling. Conveyorised handling facilities and transfer type machinery. Overhead cranes, etc.
Cost of product	Very expensive due to individuality.	Larger the order the cheaper the product, relates to economic batch size.	Relatively cheap due to long-run and large quantities.
Utilisation of capital and labour	Low — large proportion of 'preparation' and 'put-away' time with little productive work.	Improved utilisation relating to batch sizes, 'preparation' time and put-away time reduced.	Constant high utilisation with minimum set-up times in proportion to throughput time.
Design of product	To suit customer requirements.	Designed mostly by manufacturer with variations to suit customer's requirements, limited to resources available.	Standardised design for maximum production.
Store facilities	Large and cumbersome contains excessive variety of materials and tools. Location and coding usually at the discretion of storeman.	Large accommodation palletisation laid out for ease of handling. In-process stores with tool stores separate. Tends to be specialised store-holder with coding and location experience.	Work-in-progress stores with packing facilities. Time table transport facilities. Specialised store-holder fully conversant with coding and location, instructions usually issued by Production Planning and Control.
Quality control	100% inspection — highly skilled inspectors.	Patrol inspection and sampling techniques provided. Skilled inspectors.	Physically located inspection in flow-line, sampling process. Skilled and semi-skilled inspectors.

Company policies, cost information control and organisation behaviour essentially influence the method of production. Policies, for instance, often stem from the origins of 'make' or 'buy' decisions. Theoretically, every item purchased is always under scrutiny for internal manufacture. Conversely every item manufactured internally is a potential for purchase externally. Obviously both decisions will be strongly influenced by the history of the company and the available resources. By making an item internally, dependence on other companies can be reduced and the consequence of their labour disputes avoided. Trade secrets, methods of manufacture and quality standards can be preserved. Otherwise, to purchase items rather than manufacture internally may result in quicker delivery and additionally, may reduce costs such as those associated with special skills and technology, storage, handling, paperwork, etc. as well as releasing facilities for jobs on which they might be more suitably and profitably employed.

Inevitably, 'make' or 'buy' must be an economic decision, but there is no simple and straightforward answer. Costing systems used by most companies are inadequate for such decision making, the essential information for control is missing. For instance, normally, value is added to material (capital + labour + a realistic proportion of over-heads) at each 'work-station' in a sequence of operations, the true value of which is generally ignored; additionally the inventory cost absorbed whilst the material is in progress or delay at and between each 'work-station', is ignored. Generally, no consideration is given to the items that are in production less time than others, that by-pass storage, are handled the least, and have the least number of documents associated with their movement and control: nor is consideration given to the facilities utilisation whether the correct machine for the job by capacity and feeds and speeds is being employed. All facilities are charged as if they were operating under maximum conditions.

Few companies can derive the true production cost and consequently the profit per item manufactured; variety control of non-profitable items appears to be beyond most companies' capabilities. The major reasons for this uncontrollable weakness arises from a popular labour based overhead absorption costing system, one in which the operating costs of an enterprise (excluding direct materials which are costed separately) are allocated to the amount of direct labour hours or cost and distributed over the range of products manufactured. Now that companies are becoming more capitalised with facilities, new and improved costing systems need to be considered. Costing systems, variety control and inventory controls are dealt with later.

Other policies are concerned with whether to manufacture basically for stock, or to customer order. Companies in the same industry, even divisions and departments within companies can have very different policies. For instance, one may decide to manufacture to customer order only, with the risk that as the business slackens the company is forced either to carry the excess cost, work short-time, or dismiss part of the labour force, and still be left to carry the unutilised plant cost. Another may forecast demand and manufacture for stock thereby maintaining a relatively stable labour force and capital utilisation. Whilst a third company may rely on the excellence of its products and marketing to risk a large order back-log at peak seasons, overtime becoming a necessary excess cost in such a company.

In the behaviour of organisations, sales departments generally have a dominating position (not an uncommon feature), they are often the cause of confusion and delays by making frequent demands for changes of production plan, generally influenced by the constraints of unrealistic delivery dates. Even a company with an established planning and control unit finds it frustrating to operate effectively under such conditions. Sales forecasting is an essential attribute to a company's well being. The organisation of responsibility for decision between sales, planning and control and manufacture should create harmonious rather than competitive situations based on a recognition of the need for each other's work and company objectives, for example, optimising the utilisation of a company's resources.

Whatever the policy, it can only be effectively carried out by having a 'well-defined' planning and control department where the allocation of resources can be economically distributed. Without such a department, a firm has to rely on local and intuitive solutions of individual supervisors many of whom are inadequately prepared for planning, and where more than one department is involved in the production of the product the probability of bottlenecks, queues and delays will be extremely high.

WORK STUDY

Work Study is predominantly a British title originating before World War II from Imperial Chemical Industries. It is one aspect of engineering production which has been accepted, and practised in industry and commerce, not altogether favourably, as part of the management function. As a 'management tool' it has been partly responsible for an astonishing productivity improvement record and has proved itself a vital factor in industrial progress and development.

The International Labour Office,[5] defines Work Study as 'a term used to embrace the techniques of Method Study and Work Measurement which are employed to ensure the best possible use of human and material resources in carrying out a specified activity'.

The scope of Work Study extends well beyond the boundaries of engineering industries. For example, its application in commerce, distribution, armed services, agriculture, public services, government and many other organisations has been just as successful as in manufacturing.

With this widening application BS 3138[6] established a common language in the form of a 'Glossary of terms in Work Study' in 1959. However, it is not intended that the use of this terminology should be binding on anyone, nor is it expected that rapid changes

Table 20.2

SUMMARY OF THE AIMS AND TECHNIQUES OF WORK STUDY

QUALITATIVE	QUANTITATIVE
Method Study to improve methods of production	Work Measurement to assess human effectiveness
Resulting in more effective utilisation of labour, materials plant and equipment	Making it possible to improve planning and control and to provide a basis for incentive payments

Improve rate of generating
more output per head i.e.
Net output (or Added Value)
per Employee

in existing language will be affected by the publication of a 'Glossary' alone. It is considered, nevertheless, that an earnest endeavour will be made to introduce the recommended terminology wherever possible.

BS 3138 also gives a definition of Work Study as follows: 'A generic term for those techniques, particularly Method Study and Work Measurement, which are used in the examination of human work in all its contexts, and which lead systematically to the investigation of all the factors which affect the efficiency and economy of the situation being reviewed, in order to effect improvement'.

METHOD STUDY: PRIMARY APPROACH

This sub-section is concerned primarily with the quality aspects, commonly referred to as 'Method Study', whilst the quantitative aspects are dealt with later.

There are usually numerous ways to perform a task, but with the knowledge obtainable at any one time, one method is usually more superior to the others. Mundel[7] states that 'the scientific method of solving problems is more productive of better work methods than is undisciplined ingenuity'. Nevertheless speculation, innovation and creativity provide us with bright ideas which lead towards better methods and ways of doing work. An ordered and systematic approach encourages ideas to emerge more often.

The first man on earth who succeeded in simplifying his job by reason can be considered as the originator of method study, for method study is an inquisitive attitude of mind. Frank Bunker Gilbreth and his wife Dr. Lillian Gilbreth[8] are recognised as the pioneers of the analytical approach, who devised various symbols and techniques for the portrayal and analysis of elements of work which led to the effective layout of work areas.

Definitions

The Gilbreths called their collection of techniques 'Motion Study' and defined it as: 'The division of work into fundamental elements, analysing these elements separately and in relation to one another, and from these studied elements, when timed, building methods of least waste'.

The Industrial Engineering Terminology published by the American Society of Mechanical Engineers (ASME) provides separate definitions for 'motion study' and 'method study' although many other American publications accept them as one and the same. The BSI agreed to use the term 'method study' rather than 'motion study' since it was felt that the former was generally considered to have the wider implication. They define it as: 'The systematic recording and critical examination of existing and proposed ways of doing work as a means of developing and applying easier and more effective methods and reducing costs'.

Where an examination of the process as a whole is justified method study should be applied as a primary approach and motion study as a secondary approach. An improvement resulting from method study in the design of a product or process may eliminate the need for motion study at individual stages under the original conditions. Nevertheless, the best possible improvement cannot emerge without the assistance of work measurement and while it is accepted generally that method study should precede work measurement in many aspects the two are quite often closely linked. However, for the purposes of this text it has been necessary to separate these highly related techniques.

Figure 20.3 gives an outline of method study approach.

Systematic investigation

The systematic investigation in a working situation is concerned with the following points:[9]

> *Activity of the operator.* This covers the examination, improvement, specification and measurement of operatives' working methods. Also, the training and retraining of operators which can only be carried out satisfactorily with the willing and intelligent cooperation of operators and management alike.

20–10

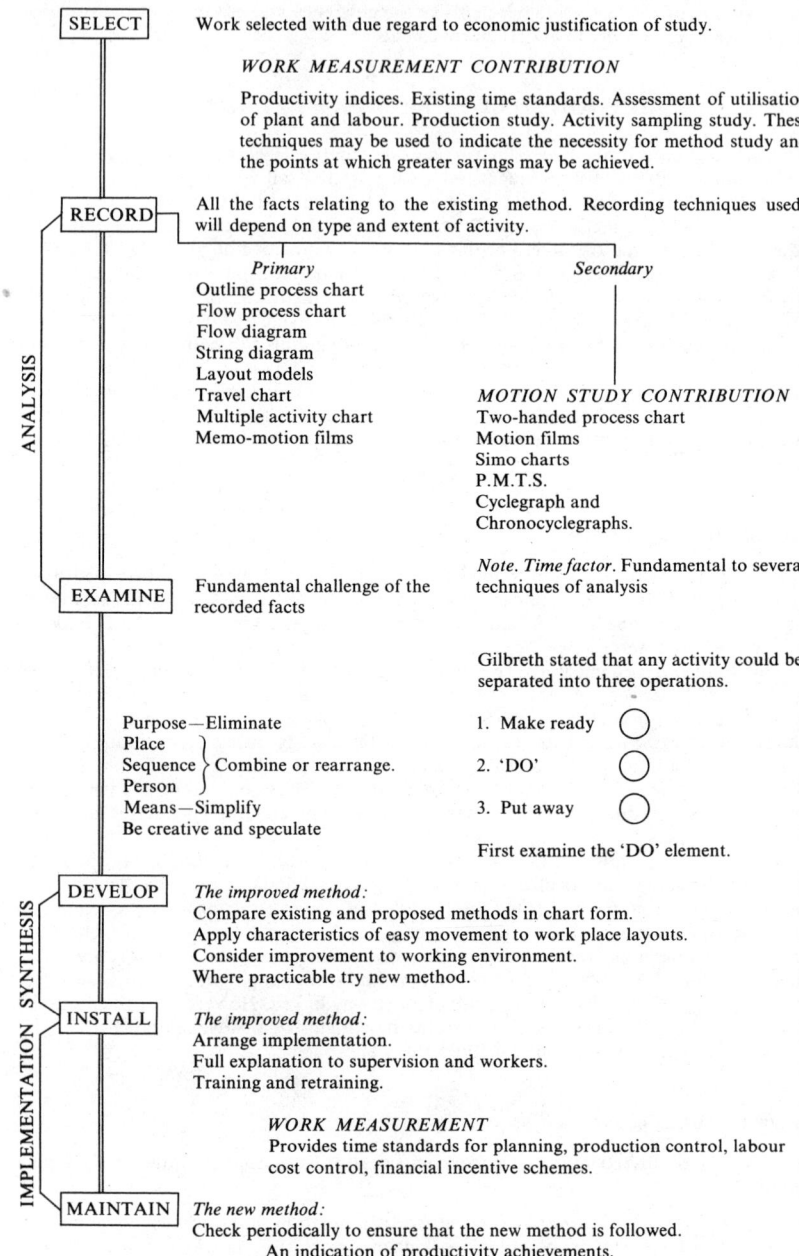

SELECT — Work selected with due regard to economic justification of study.

WORK MEASUREMENT CONTRIBUTION

Productivity indices. Existing time standards. Assessment of utilisation of plant and labour. Production study. Activity sampling study. These techniques may be used to indicate the necessity for method study and the points at which greater savings may be achieved.

RECORD — All the facts relating to the existing method. Recording techniques used will depend on type and extent of activity.

Primary	*Secondary*
Outline process chart	
Flow process chart	
Flow diagram	
String diagram	
Layout models	
Travel chart	*MOTION STUDY CONTRIBUTION*
Multiple activity chart	Two-handed process chart
Memo-motion films	Motion films
	Simo charts
	P.M.T.S.
	Cyclegraph and
	Chronocyclegraphs.

Note. Time factor. Fundamental to several techniques of analysis

EXAMINE — Fundamental challenge of the recorded facts

Gilbreth stated that any activity could be separated into three operations.

Purpose—Eliminate
Place ⎫
Sequence ⎬ Combine or rearrange.
Person ⎭
Means—Simplify
Be creative and speculate

1. Make ready ◯
2. 'DO' ◯
3. Put away ◯

First examine the 'DO' element.

DEVELOP — *The improved method:*
Compare existing and proposed methods in chart form.
Apply characteristics of easy movement to work place layouts.
Consider improvement to working environment.
Where practicable try new method.

INSTALL — *The improved method:*
Arrange implementation.
Full explanation to supervision and workers.
Training and retraining.

WORK MEASUREMENT
Provides time standards for planning, production control, labour cost control, financial incentive schemes.

MAINTAIN — *The new method:*
Check periodically to ensure that the new method is followed.
An indication of productivity achievements.

ANALYSIS

IMPLEMENTATION SYNTHESIS

Figure 20.3. Outline of method study approach

Tools and equipment. Normally the design and behaviour of such items is a technical problem, but work study is concerned with the utilisation and suitability for the purposes and methods used. The methods adopted for maintenance and standardisation, for instance, are an important part of work study.

Materials. All materials at all stages of process must be continually examined for their suitability, utilisation, rate of consumption, waste quality and reliability.

Working conditions. There are at least two kinds of conditions that affect work: physical and psychological. The effectiveness of work depends very much upon environment, services and morale, together with the safety, lighting, general housekeeping, services to and from the operator, in the form of clear and concise orders, material movements, clearing production delays and promptly removing obstructions. Psychological conditions reflect the whole climate of personnel relations and work study can contribute greatly to good psychological conditions needed to obtain cooperation.

Selection

Cost reduction and employee attitudes to a large extent determine whether it is economically justified to select a particular job for method study. Evidence of potential cost reductions and a clear definition of the problem can be gained by carrying out preliminary studies to determine the efficiency levels of working groups and associated plant, equipment and materials. Activity sampling as discussed by Barnes[10] or memo-motion filming can provide an effective way to measure productive and non-productive work. A simpler and more practical method is that of Norman's 'Disc-O-Tec'[11]; this is a novel approach to activity sampling and one that foremen can apply without detriment to their normal duties and responsibilities.

Employee attitudes are the most difficult to foretell, since mental and emotional reactions to investigations and change have to be anticipated. Change is implied by definition and this is very important at supervisory level. The authority of the supervisor inevitably is challenged because it brings in outside control to his department and he is very much concerned in the success or otherwise of the introduction of both method study and work measurement. Operatives will have a natural resistance to change with the attendant fears of redundancy and transfers, so that all reasonable assurances should be given.

For instance, method study may be more readily accepted, if good industrial relations exist, and an explanation is given to all concerned. The latter should state why an investigation is taking place; the scope of the investigation; as much as is known about the probable effects. The first jobs selected for investigation are those recognised as unpleasant and where the undesirable features can be removed.

Record

The success of an investigation depends upon the accuracy with which the facts are portrayed. All the facts must be related to the situation under investigation and should be obtained and arranged for appraisal and comparison. Various recording techniques have been evolved in order to clarify working situations. These comprise Process charts for the analysis of producing goods; Multiple activity charts for the analysis of teamwork and man/machine systems, and Plant layout and movement charts, diagrams and models.

The use of any such techniques must be acceptable to both management and employees.

Process charts

It is essential when constructing charts that consistency and 'scale' be maintained throughout the work cycle being investigated. Otherwise a distorted view results and difficulty arises in comparing activities 'like with like', particularly in succeeding stages of the investigation.

For analytical purposes and to aid in detecting and removing wasted effort, it is preferable to classify the actions which occur during a given process into five activities, standardised by the American Society of Mechanical Engineers (ASME).

When unusual situations outside the range of the definitions are encountered, the intent of the definitions summarised in the following tabulation will enable the analyst to make the proper classification.

Classification		Predominant result
Operation	○	Produces or accomplishes
Inspection	☐	Verifies
Transportation	⇨	Moves
Permanent storage	▽	Keeps
Delay	D	Interferes

○ OPERATION. An operation occurs when an object is intentionally changed in any of its physical or chemical characteristics, is assembled or disassembled from another object, or is arranged or prepared for another operation, transportation, inspection, or storage. An operation also occurs where information is given or received or when planning or calculating takes place.

☐ INSPECTION. An inspection occurs when an object is examined for identification or is verified for quality or quantity in any of its characteristics.

⇨ TRANSPORTATION. A transportation occurs when an object is moved from one place to another, except when such movements are a part of the operation or are caused by the operator at the work station during an operation or an inspection.

▽ PERMANENT STORAGE. Indicates a controlled storage when an object is kept and protected against unauthorised removal

D DELAY or TEMPORARY STORAGE. This occurs to an object when conditions except those which intentionally change the physical or chemical characteristics of the object, do not permit or require immediate performance of the next planned action.

◪ COMBINED ACTIVITY. When it is desired to show activities performed either concurrently or by the same operator at the same work station, the symbols for those activities are combined.

To simplify chart construction and its subsequent analysis, a colour-code is recommended, as listed below, to amplify the differences in portrayal of 'present' and 'proposed' and 'productive' and 'non-productive' methods of working. For example, any job of work can be separated into three broad categories.

○ 1 'make ready'
○ 2 'Do'
○ 3 'put away'.

The 'Do' part being productive while the remaining parts are non-productive. It is suggested that operations ○ concerned with 'Do-ing' should be coloured 'black' and the other ○ operations 'blue'. The remaining symbols may be coloured as recommended in the following table supplied by Cumberland Pencil Co. Ltd.

Pencil No.	Colour	Symbol	Comment
67	Black	○	'DO'
29	Blue	○	make ready and put away
45	Green	⇨	on the move
14	Red	▽	costs money
14	Red diagonals	D	temporarily costs money
6	Yellow	☐	take care

SYMBOL	NAME	USED TO REPRESENT
◎	Origin of form	Form first being made out.
 ◎◎	Origin of form	Form first being made out in duplicate.
 ◎ (stacked)	Origin of form	Form first being made out in triplicate, etc.
◯	Operation	Work being done on form; computations or additional information added, etc.
○	Movement	A change in location of form, not changing it.
▽ (crossed)	Temporary file	Forms waiting to be worked on, such as in desk basket.
▽	File	Forms in a file
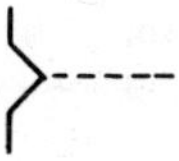 }-----	Information take-off	Information being taken off form for entry onto another or for use by someone. Point of line indicates symbol on other parallel chart where information is going. (Use--------broken line to indicate destination if destination appears on chart and line is aid to clarify).
 ⅄	Disposal	Form or copy destroyed.
◇-----	Inspection Quality	Correctness of information on form checked by comparison with other source of information. (Use----------- broken line drawn to other source if other source appears on chart and line is aid to clarity.)
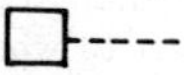 □----	Quantity	
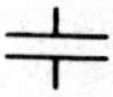 ╪	Item Change	Change in item charted.
 ⤬	Gap	Activities not pertinent to study and hence not charted in detail.

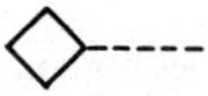

Figure 20.4. Mundel symbols. Combined analysis symbols for use with forms

Although the ASME symbols have achieved wide usage since they were recommended in 1947, other sets of symbols exist. Of these the best known are the Gilbreth originals.

GILBRETH SYMBOLS

○ Operation
□ Inspection for quantity
◇ Inspection for quality
○ Transport
▽ Permanent storage
⩔ Temporary storage or delay.

Mundel[7] symbols are particularly useful in clerical procedures, Figure 20.4.

Outline process charts

Drawings and specifications give us 'what to manufacture', sales and forecasting tell us 'how many to manufacture' and the Outline Process Chart (OPC) summarises the 'sequence of manufacture'.

The chart gives an overall view of a process from which it can be decided whether a further and more detailed record is required.[12] It is a portrayal of operations ○ and inspections □ only, carried out on materials during processing. Provision can be made on the chart for recording units of time, quantities, and a brief description of the activity, but it does not show where, or at what time work takes place, or who performs it; neither does it concern itself with transportations delays or storages.

An example of the machining and assembly of a refrigerator stop valve is shown in Figure 20.5.

The simultaneous occurrence of two different activities can be shown by using combined symbols ⊡. One symbol representing the major activity superimposed on the other.

Numbering each symbol facilitates quick reference and comparison. Like symbols are numbered consecutively from the beginning of the chart, the combined symbol having two numbers.

Flow process chart

This chart provides a more detailed picture of activities in a process and employs five symbols to illustrate the type of activity being carried out. The rules for construction are similar to those for Outline Process Charting.

Simultaneous activities of two or more subjects can be presented alongside each other, but it is essential that only the activities of the particular subject to which the chart refers, either: Man; Material; or Equipment, are recorded on any single chart. Consistency must be strictly adhered to and 'levelling' maintained throughout. Two types of chart are shown in Figures 20.6 and 20.7. Where the job being investigated is simple and requires only one subject to be charted, an OTIS chart may be preferred. Figure 20.8 (O operation; T transport; I inspection; S store).

Multiple activity chart

Often referred to as a man/machine chart, work planning or activity analysis chart. The effect being to illustrate 'balance or utilisation of teamwork', by relating the sequence of activities of men, equipment and/or materials belonging to one team, to a time scale.

Chart begins : Material feeding into process   Present method   Signed

Chart ends   : Inspection after final assembly   Date

Job manufacture and assembly of Stop Valve

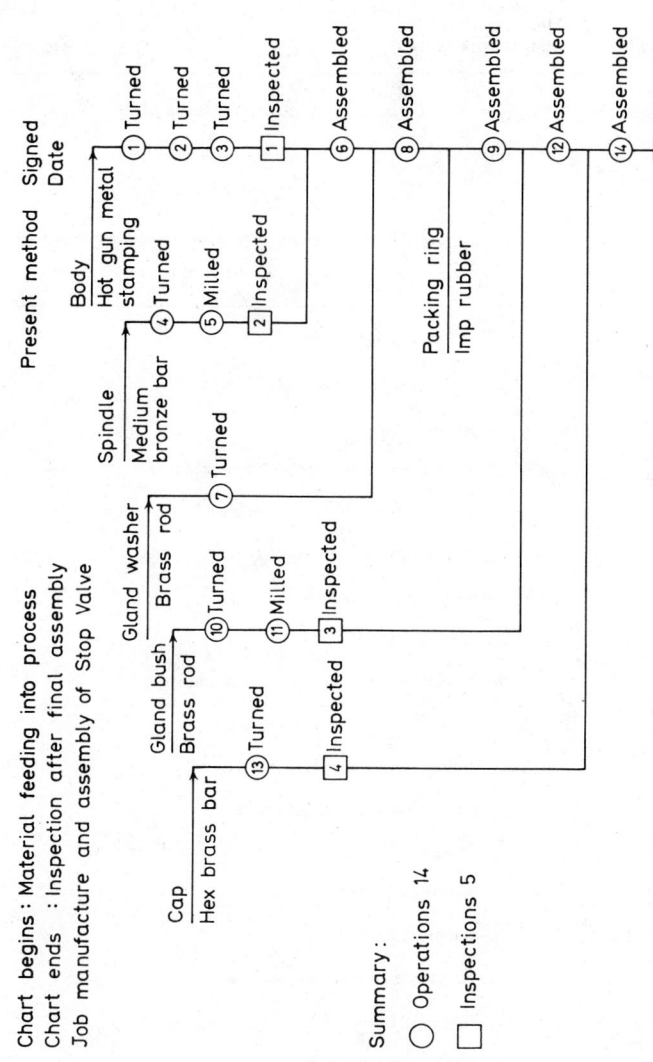

Summary:

◯ Operations 14

▢ Inspections 5

Figure 20.5. Outline process chart

Location:	Push, Pull & Carrie Radio Co.
Process:	Radio screen production.
Material:	Copper and brass sheets.
Chart starts:	Man at work bench.
Chart ends:	Man at work bench.

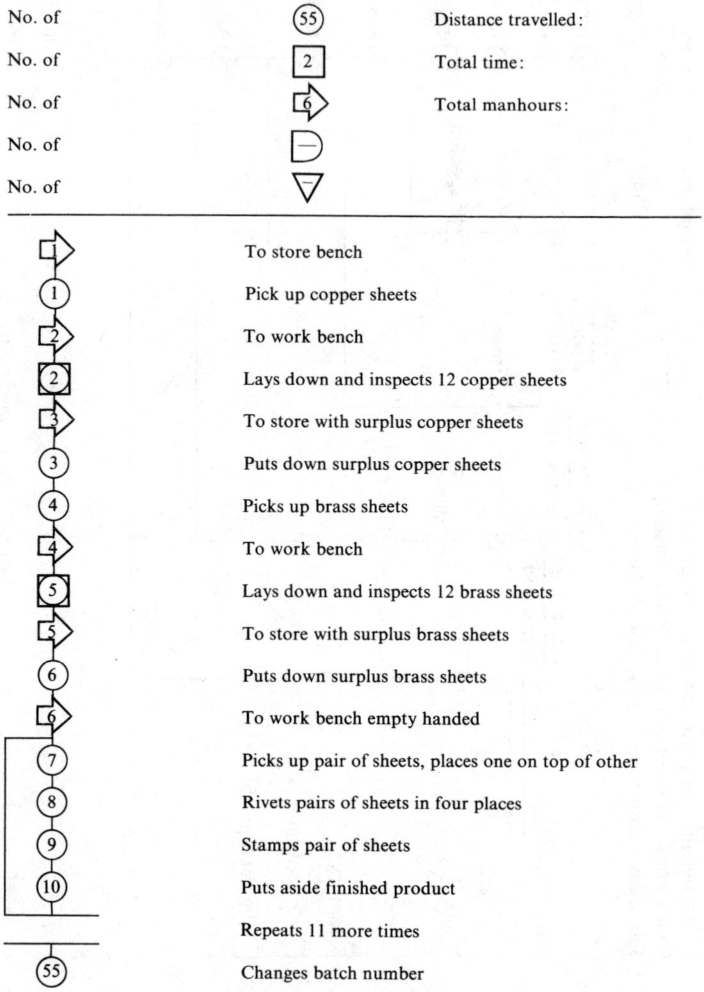

No. of   (55)   Distance travelled:

No. of   [2]   Total time:

No. of   ⑥⟩   Total manhours:

No. of   ⟨—⟩

No. of   ▽

⟩	To store bench
①	Pick up copper sheets
⟩	To work bench
▣ 2	Lays down and inspects 12 copper sheets
⟩	To store with surplus copper sheets
③	Puts down surplus copper sheets
④	Picks up brass sheets
⟩	To work bench
▣ 5	Lays down and inspects 12 brass sheets
⟩	To store with surplus brass sheets
⑥	Puts down surplus brass sheets
⟩	To work bench empty handed
⑦	Picks up pair of sheets, places one on top of other
⑧	Rivets pairs of sheets in four places
⑨	Stamps pair of sheets
⑩	Puts aside finished product
	Repeats 11 more times
(55)	Changes batch number

Figure 20.6 Flow process chart

Location: Push, Pull & Carrie Radio Co.
Process: Radio screen production
Material: Copper and brass sheets
Chart starts: Materials on store bench
Chart ends: Finished product on work bench

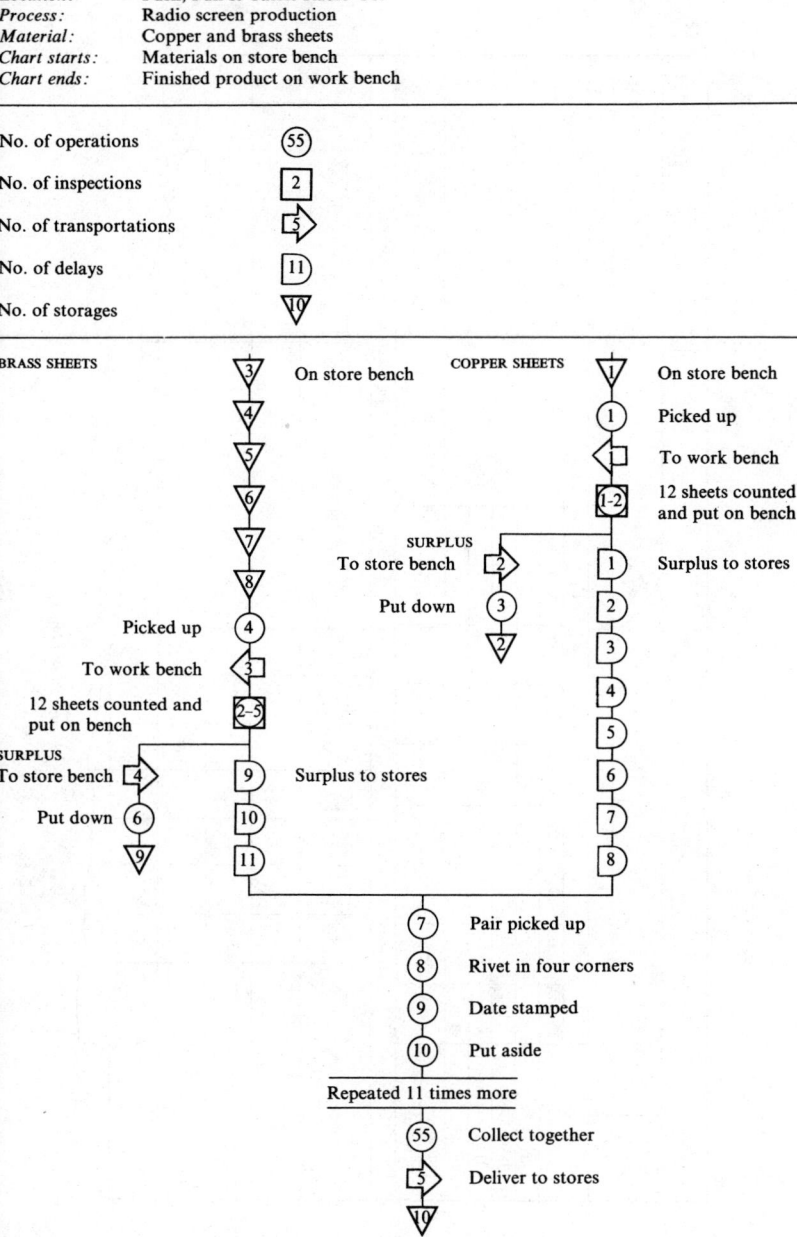

No. of operations   55

No. of inspections   2

No. of transportations   5

No. of delays   11

No. of storages   10

Figure 20.7. Flow process chart—Material type

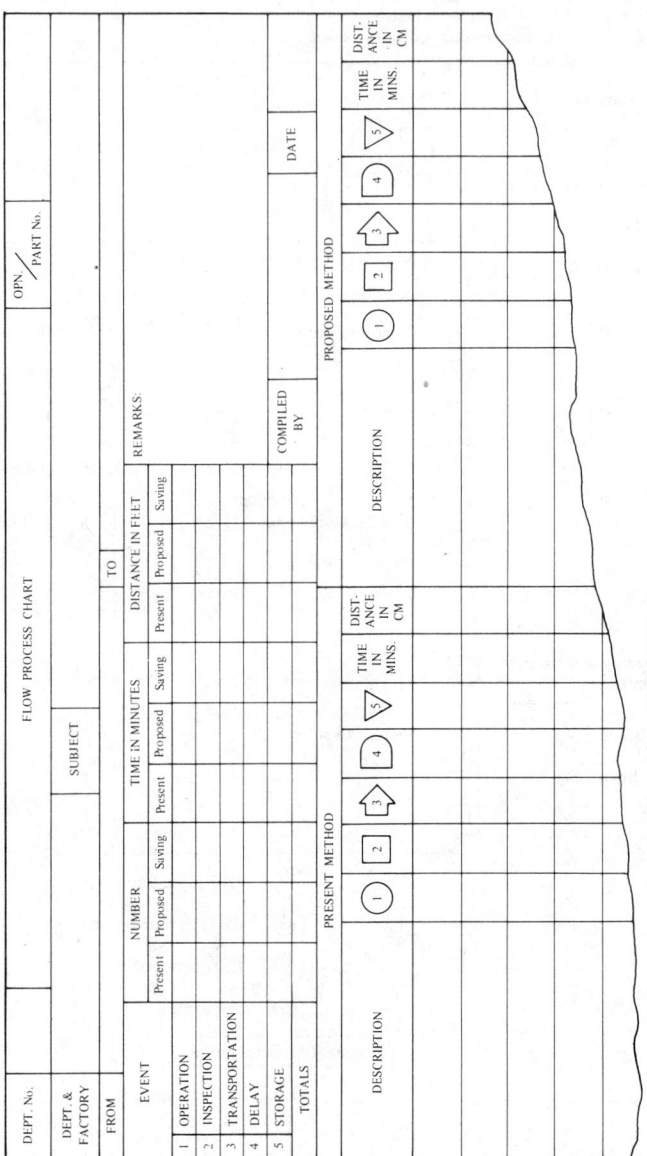

Figure 20.8. Flow process chart

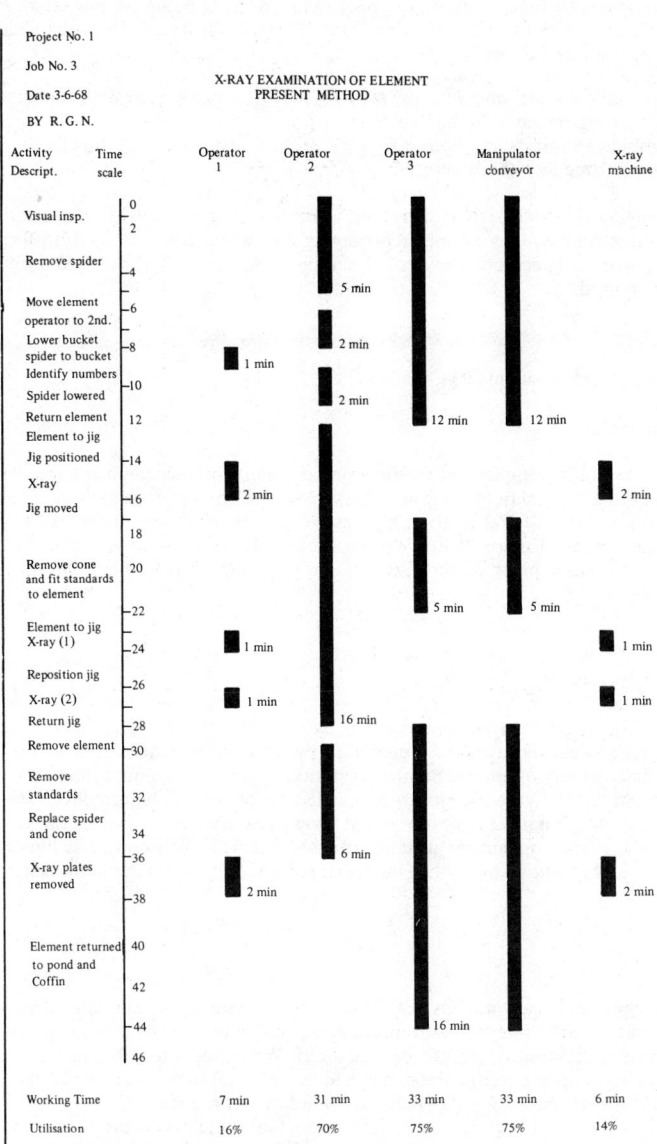

Project No. 1
Job No. 3
Date 3-6-68
BY R. G. N.

X-RAY EXAMINATION OF ELEMENT
PRESENT METHOD

Activity Descript.	Time scale	Operator 1	Operator 2	Operator 3	Manipulator conveyor	X-ray machine

Visual insp.
Remove spider
Move element operator to 2nd.
Lower bucket spider to bucket
Identify numbers
Spider lowered
Return element
Element to jig
Jig positioned
X-ray
Jig moved
Remove cone and fit standards to element
Element to jig
X-ray (1)
Reposition jig
X-ray (2)
Return jig
Remove element
Remove standards
Replace spider and cone
X-ray plates removed
Element returned to pond and Coffin

0 2 4 6 8 10 12 14 16 18 20 22 24 26 28 30 32 34 36 38 40 42 44 46

5 min · 2 min · 2 min · 1 min · 2 min · 12 min · 12 min · 2 min · 5 min · 5 min · 1 min · 1 min · 1 min · 1 min · 16 min · 6 min · 2 min · 2 min · 16 min

		Operator 1	Operator 2	Operator 3	Manipulator conveyor	X-ray machine
Working Time		7 min	31 min	33 min	33 min	6 min
Utilisation		16%	70%	75%	75%	14%

Figure 20.9. Multiple activity chart

It is constructed initially by placing Flow Process Charts (FPC) of individual subjects in the team, side by side, so that they appear as separate columns on one chart. Periods of productive and non-productive time can be clearly illustrated by block shading in order to distinguish between

(a) independent work where the operative(s) performs work other than that connected with the operation of a machine(s),
(b) combined work, where both operative(s) and machine(s) work together,
(c) waiting time by either operative(s) or machine(s).

By challenging the 'DO' activity, it is then possible to eliminate waiting time for operative(s) and machine(s) by simply rearranging the work cycle or by transferring or balancing work between operatives. An example is shown in Figure 20.9 and others are seen in Maynard.[13]

Movement-diagrams and models

FLOW DIAGRAM

It is often helpful to supplement the flow process chart by superimposing the portrayed symbols onto a floor plan of the work area, at specific points where activity takes place, or changes, so that spatial relationships as well as the type of activity can be appropriately recognised (Figure 20.10). When a subject moves about several floors elevated, as well as different parts of the factory, a three dimensional flow diagram may be particularly useful.

STRING DIAGRAMS

These are used to record the routes taken by men, materials and equipment within a defined area. A scaled drawing is prepared of the area under review, pins are inserted at points where activity or change in movement takes place, and coloured thread is wound around the pins following the movement carried out. A string diagram illustrates where delays occur, where unnecessary movement takes place, and on measuring the amount of thread used, a total movement distance can be obtained. An example is illustrated in Figure 20.11 which shows the path of movement of fuel elements undergoing examination.

MODELS

Three dimensional scale models add an element of realism which can sometimes be justified particularly where headroom, access, material flow by overhead crane and environmental conditions need to be visualised. While such models can be made in a woodworking shop to specifications, it is seldom practical to do so since die-cast models are available commercially for virtually all standard equipment.

Models are of vital importance in any intellectual attack on a problem. A model is always an abstraction to some degree of the real life process for which we want to predict performance.[14] Such models need not be elaborate or expensive, as long as the overall dimensions of plant and machinery, etc., are accurate and, for instance, include extended travel of machine tables and spindles.

Two-dimensional templates are by far the most commonly used, they provide simplicity in manufacture and reproduction. One such aid is supplied by Ozalid and known as

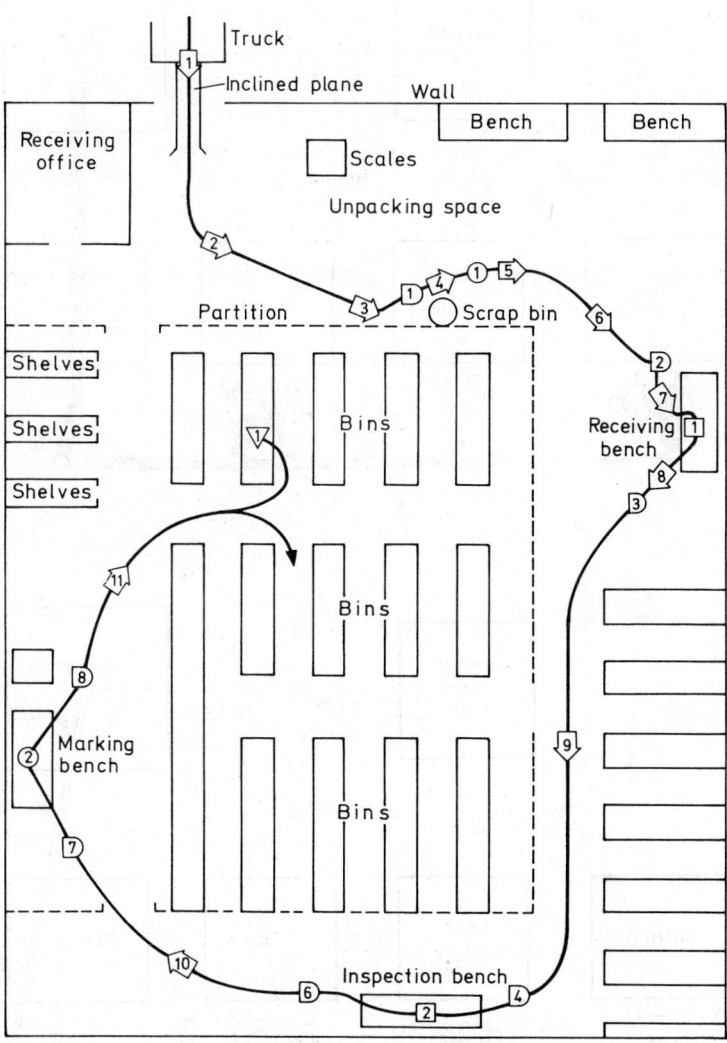

Figure 20.10. Example of a flow diagram

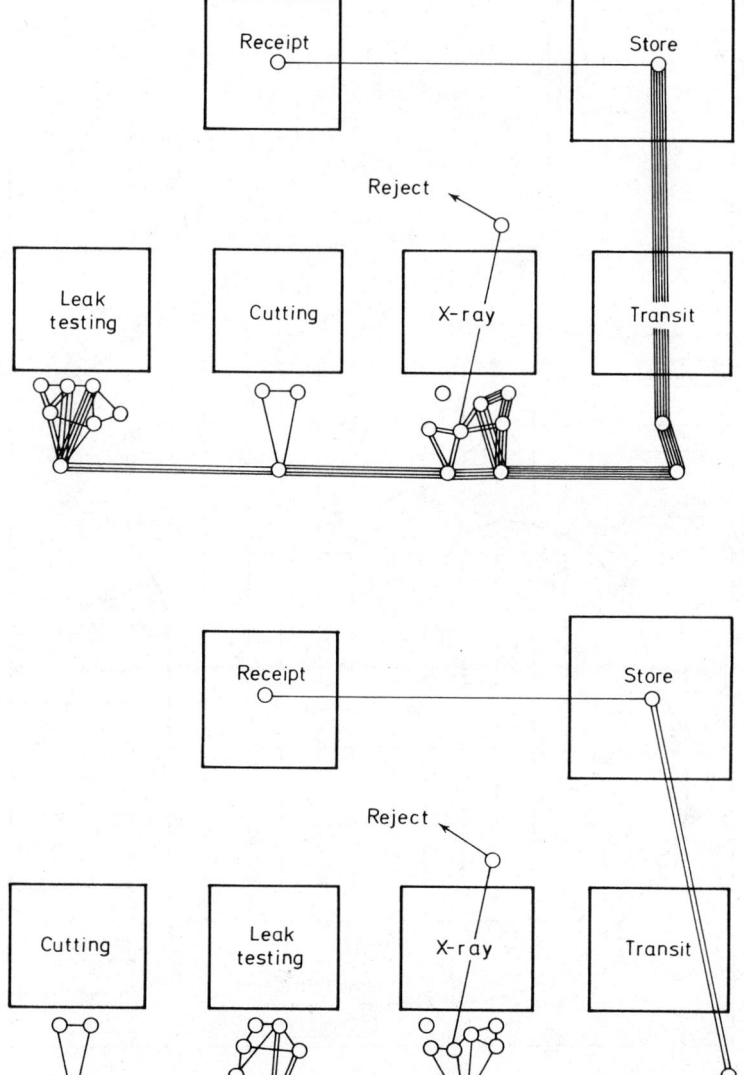

Figure 20.11. String diagrams

'2D Planning System' which includes a transparent grid and aquaprint self adhesive templates and tapes. The original is in colour with reproductions in black and white.

METHOD STUDY: SECONDARY APPROACH

Motion study

Economic justification as the criterion of the depth of investigation, forms the basis of selection of the techniques most suitable for analysis of the situation. Motion study is not only useful in the analysis of industrial and business activities but has been applied with great success to the design of machine tools and their controls, control panel layouts, and numerous examples of man/machine relationships. However where human capacities are involved in 'fitting the job and environment to the worker' it is considered to be the field of the Ergonomist.

Motion study is the detailed analysis of the motions of individual workers during particular operations, with the object of eliminating unnecessary motions and fatigue, and of defining the best working rhythm for the operation at any stage of its development.

OBJECTIVES

The phrase 'Principles of Motion Economy' is considered to be inappropriate and it is recommended by the BSI that 'Characteristics of Easy Movement' (BS 3138: 22301) be adopted.

There are twenty-two principles[9] which form the basis for improving the efficiency and reducing the fatigue in manual work. They fall into three main groups:

1. Use of the human body,
2. Arrangement of the workplace, and
3. Design of the tools and equipment.

By following these rules the operative will achieve a higher rate of generating output, without working harder.

In an attempt to achieve the 'characteristics' of easy movement at the workplace, particular attention should be given to:

1. Areas of reach and visual control.
2. Motion paths in the horizontal and vertical planes.
3. Position and the design of containers, tools, jigs and fixtures.
4. Relative heights of chair, working surfaces, feeding and locating surfaces.
5. Receipt and removal of work.

'Mock-up' models are particularly useful in the development and trial of workplace design.

The following paragraphs give the 'characteristics of easy movement' or 'principles of motion economy' as expounded by R. M. Barnes.[15]

Analysis of methods and movement at the workplace

The study of an operative's activities at his workplace starts, as does Method Study— Primary Approach, with a process chart.

TWO-HANDED PROCESS CHART

This chart provides the basis for the simplest form of motion analysis and assesses where further micromotion analysis is economically justified. The symbols used are a modified version of the ASME process symbols, with ∇ indicating hold, and as the chart is primarily concerned with hand activities ☐ is rarely used, although sometimes it may be useful in drawing attention to the examination of a component by touch perception. The record often includes a sketch of the workplace layout.

Micromotion analysis

Where the justification for extremely detailed study exists, a camera provides the means for obtaining a complete and accurate record of the motions involved in a work cycle. A frame by frame examination of the film enables a complete sequence of events to be recorded onto a 'Simultaneous Motion Cycle Chart' or SIMO Chart as it is often called. Gilbreth states that this chart is a 'device for visualising a process as a means of improving it'.

The symbols used to identify the motions are known as Therbligs (Figure 20.12) and the unit of time is recorded in 'Winks', a term used to denote 1/2000th part of a minute, equivalent to $\frac{1}{2}$ a frame:

i.e.   film speed = 16 frames per sec. = 960 frames per min.
    or   1 frame = 1/960th min $\simeq$ 1/1000th min
        $\frac{1}{2}$ frame $\simeq$ 1/2000th min = 1 wink

Alternatively, a microchronometer, or wink counter having 100 divisions each representing 1/2000th part of a minute is placed near to the subject being filmed so that time measurement relative to motion might be more precise. If the film is required for sound production then 24 frames per second speed is recommended.

To analyse the methods and movements, it is preferred to select one body member and follow through the complete sequence frame by frame before attempting to analyse a second member, otherwise confusion will arise.

Stereoscopic photography

This is the most preferable means of filming the path of movement, with the aid of 'Cyclegraphs' and 'Chronocyclegraphs' (Figure 20.13). The latter is particularly useful for demonstrating accelerations, retardations and hesitations (Ann Shaw 1952). The best effect is obtained by filming in a relatively darkened room, so that the path traced by the light source contrasts with that of the silhouette of the subject.

Nadler (1955) developed a method by which a motion pattern is traced by projecting a cine film, frame by frame, of the work cycle onto a sheet of paper. The points reached in successive frames by the body members are marked on the sheet. Attention is drawn to the acceleration and retardation of motions by the distance between markings. The extent of dwell at any point can be noted by recording the number of frames elapsed at that point. The speed of motion is measured by reference to the film speed.

Memomotion photography

A cine film is exposed at an unusually slow speed at time intervals between $\frac{1}{2}$ s and 4 s depending upon the range of the interval time recorder and the activity being recorded.

Whilst this technique is extensively used for the recording of multiple activities, particularly maintenance and construction work, it can also be used for the recording of individual activities, if economically justified. It involves setting up an 8 mm or 16 mm cine camera with an unobstructive view of the subject.

A wide angle or zoom lens may be preferred, depending on the detail required. An interrupter gear operated by the time lapse unit is attached to the camera to enable single

Symbol		Description	Colour	No.	Explanation
Sh		Search	Black	67	Eye turned as if searching
F		Find	Grey	69	Eye fixed on an object
St		Select	Light grey	70	Reaching for an object
G		Grasp	Lake red	14	Hand open for grasping object
H		Hold	Gold ochre	3	Magnet holding iron bar
TE		Transport empty	Olive Green	51	Empty hand
TL		Transport loaded	Green	45	Hand with something in it
P		Position	Blue	29	Object being placed by hand.
RL		Release load	Carmine red	19	Dropping content from hand
PP		Preposition	Pale blue	39	Set-up nine pin in bowling alley
U		Use	Purple	22	Letter U
A		Assemble	Violet	25	Noughts and crosses
DA		Disassemble	Light violet	26	One line of assembly removed
I		Inspect	Burnt ochre	60	Magnifying lens
UD		Unavoidable delay	Yellow ochre	6	Man bumping nose unintentionally
AD		Avoidable delay	Lemon ochre	1	Man lying down on job voluntarily
Pn		Plan	Brown	55	Man thinking fingers on brow
R		Rest to overcome fatigue	Orange	10	Man seated as if resting

Figure 20.12. Therbligs

Example; Picking up a screwdriver.

From a flat surface
Chronocyclegraph

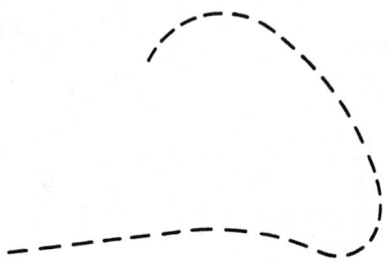

From a raised tool rest
Cyclegraph

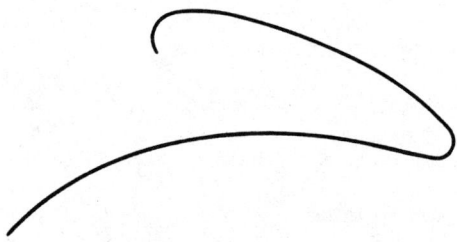

Figure 20.13. Cyclegraph and chronocyclegraph

frame exposures to be taken at pre-determined intervals. Without this technique several observers may have to be used and their results would be very difficult to co-ordinate. Film consumption is economical.

Analysis and implementation

EXAMINE

It is the aim of recording techniques to define all processes into relatively simple activities so that they may be subjected to a question of need or change. Gilbreth stated that any job or process can be separated into three very broad 'operations':

1. Make ready   ○ (blue)
2. 'DO'   ○ (black)
3. Put away   ○ (blue)

An immediate challenge by sequential questioning towards eliminating, combining, rearranging or simplifying the 'DO' operation of the process is then carried out.

The sequence of examination can be summarised as follows:

Purpose: What is done?
          Why is it done?
          What else could be done?
          What should be done?

Place: Where is it done?
          Why is it done there?
          Where else could it be done?
          Where should it be done?

Sequence: When is it done?
          Why is it done then?
          When else could it be done?
          When should it be done?

Person: Who does it?
          Why does that person do it?
          Who else could do it?
          Who should do it?

Means: How is it done?
          Why is it done that way?
          How else could it be done?
          How should it be done?

The means of improvement materialised from 'brain-storming' sessions, objectively challenging—purpose, place, sequence, person and means, providing alternatives (no matter how ludicrous they might lead to bright ideas) and from these selecting the most economically justified.

For instance:

Purpose—where possible eliminate.
Place   ⎫
Sequence   ⎬ combine or change.
Person   ⎭
Means—simplify.

The questioning sequence follows a well established pattern as shown in the example of 'packing a parcel' (Table 20.3) where the primary questions must be answered before continuing to answer the secondary questions (BS 3138: 22201a and b). By eliminating the need or changing the 'DO' operation, the remaining questions will be similarly affected and considerable time and effort will be saved in further examination.

DEVELOP

Comparison of existing and proposed methods is essential in order to show that improvements are factual and it is important, therefore, that all activities are portrayed in the same 'scale' and at the same level. The synthesis of possible improvement will be viewed from the standpoint of practicability and economic justification of change; the full effect of the selected and alternative improvements must be considered carefully.

Table 20.3   METHOD STUDY: CRITICAL EXAMINATION SHEET

Description of Element   Ref. ....
'DO' Operation Packing Contents   Page ....
                                  Date ....

PRIMARY QUESTIONS The Present Facts		SECONDARY QUESTIONS Alternatives	Selected Alternative for Development
Purpose – what is achieved?	Is it necessary? Yes If yes – why?	What else could be done?	What?
Contents of parcel are contained.	To facilitate despatch, min. postal charges.	Post without packing, customer collects. Revise packing method. Deliver ourselves.	New packing material.
Place – where is it done?	Why there?	Where else could it be done?	Where?
Postroom packing bench.	Facilities are there.	Warehouse. Any available space. Re-arranged post room.	Re-arranged post room
Sequence – when is it done?	Why then?	When else could it be done?	When?
After receipt of assembly note and goods. Before 4.30 p.m. when parcels are collected.	Cannot be done before to meet delivery time.	Before leaving warehouse. After 4.30 p.m. Shift work.	As at present.
Person – who does it?	Why that person?	Who else could do it?	Who?
Two full time operatives and one part time. Male packers. Day workers.	Specification males available. Day work only.	Any other employee (other than C/hand). Females – Shift employees.	Two packers. Males.
Means – how is it done?	Why that way?	How else could it be done?	How?
Wrap in corrugated paper. Pack in carton. Wrap in brown paper. Stick on label. Tie with string. Wax for reg. Manual and Op. standing.	To prevent damage cheap postal rate. Easier to handle. Clear address. Postal requirements.	Leave in unwrapped condition. Pack in straw or rubber. Wrap in P.V.C. Contain in wood box. Strap with steel strip. Seal with Sellotape.	Wrap in shredded paper. Pack in carton. Seal with Sellotape. Stick on label. Reg. and unreg. packed the same way.

Impartial and objective reports are usually demanded by senior management and these must be written in a manner not to recriminate, otherwise a tendency to do this will antagonise management and employees and defeat the purpose of the investigation. To ensure cooperation throughout, it might be useful and indeed most beneficial to persuade the head of the department under investigation to join as co-author of the report which will give him the opportunity to comment on the evidence presented.

INSTALL

The first stage of any installation is to plan and programme the content of work and the resources available and to:

1. The running down of the old method.
2. Introduce the change-over to the new as quickly as possible.
3. Set up necessary administrative controls and ensure those concerned know their operation. Remove old systems otherwise paper builds up into a 'slow paper moderated reactor'.
4. Ensure the service departments are informed of any changes that may affect them, e.g. transport and canteens.
5. Anticipate wage and working condition problems, try to assess these before they arise.

If time and facilities are available, rehearse the new system using supervisors, work study officers, or ideally operatives who are undergoing training in work study laboratories, training schools, or at work place.

Where production line work is involved, new changes should generally be made outside normal working time. Progress thoroughly through during the first few weeks of operating the new system. More than likely amendments will be necessary; attend to these immediately and notify departmental managers who are involved, correct the job specifications where necessary. Sort out weaknesses in team work, either replace or retrain. Finally give credit due to operatives and supervision for full cooperation under difficult conditions.

The installation is the prime responsibility of the Work Study Officer actively responsible for the project, or as advisor and coordinator. He should make sure that implementation is carried out truly to the accepted recommendations.

MAINTAIN

A method must be clearly defined and specified. This is especially important where it is used for setting time standards for financial incentives. Operatives should not be allowed to slip back into old methods and it is therefore necessary to deal with conditions of work and wage rate problems immediately they arise and to arrange periodic inspections and ensure changes do not invalidate the specification and time standards.

Scrutiny of bonus earnings highlight sudden fluctuations. It is suggested that agreements include a clause stating that: 'Management have the right to reinvestigate the job if the standard time is made to look ridiculous without effort; or the operatives have the right to request an investigation if bonus earnings remain low and these cannot be increased without undue fatigue'.

EFFECTS

Everyone becomes method study conscious and bright ideas pour in. Ensure that method study is applied in its widest field, not on a narrow basis, otherwise bottlenecks may arise due to over-zealousness for Method Study.

The method study approach should be applied with a view to achieving: a smooth flow of work; minimum movement of materials and workers; a high rate of throughput; ease of supervision and administration.

Since the underlying feature in the application of method study is the extent to which the economic justification of investigation exists, its use in the investigation of manufacturing conditions, will be limited by the level of effectiveness engineered into the production processes by the initial planning. The attitude of mind engineered by method study should not be confined to the investigation of existing methods of manufacture but should also be directed to the preparatory production stages to reduce the need for change.

PLANT LAYOUT

Plant layout is the process of obtaining the optimal disposition of the physical facilities of a manufacturing unit.

The conclusive existence in a factory of any one of the strategies of layout design discussed on paper is rare in practice. Whatever the combination of strategy, and apart from the question of conformity to safety standards and general principles of work design and workplace layout, an effective design must satisfy certain important criteria. These can be summarised as follows:

1. *Minimum handling costs.* Can be achieved by the proper choice of handling methods and of the relative balancing and location of facilities for work centres and departments.
2. *Maximum sequential movement.* Permits work to be uni-directional eliminating back-tracking or cross tracking as far as possible.
3. *Minimum capital costs.* Under utilised capacity machines, as well as incurring more capital expenditure, generally occupying more space and having higher operating costs are charged as if they were operating continually under optimised conditions, i.e. by size, feeds, and speeds, and being the correct machine for the job.

The utilisation of facilities is essential and it must be borne in mind that the three vital utilisation indices must be satisfied to ensure minimum capital costs. These are machine activity utilisation—processing; machine capacity utilisation—size; and machine intensity utilisation—feeds and speeds.

Plant layout techniques

These techniques are frequently described as within the method study function, though in practice, firms often set-up plant layout departments or an ad hoc group to plan large scale moves. What is needed is the same analytical approach. Moore[16] shows why layout problems arise and relates them to the range of problems (Figure 20.14).

The development of plant layout techniques have evolved through three stages:

1. *Traditional.* Flow diagrams, charting and models used typically to minimise the distance travelled between work centres.
2. *Graphical and systematic.* These are more objective in application and include travel charting, sequence analysis, and systematic layout planning.
3. *Optimisation models of a mathematical order.*

Traditional techniques have already been adequately emphasised; graphical techniques require some elaboration. Of all those graphical techniques mentioned above and indeed of the three stages of evolution, travel charting is the most effective. Although it does not inherently offer the optimal solution, it does offer relative to one another various layout proposals that can be evaluated quantitatively.

Travel charting was first introduced by Cameron in 1952 and developed by Smith in 1955. It is a tabular record for presenting quantitative data about the movements of people, materials and equipment between any number of places over any given period of time or distance; not forgetting cost valuations that could be unrelated to time. This method enables a large quantity of data to be handled in a concise and rapid way thus:

1. From operation planning sheets, or route cards collate the volume and sequence of operations.
2. Construct a matrix, from process-to-process.
3. Analyse the matrix and determine the most critical moves and the ideal layout.
4. Evaluate moves and make improvements by changing locations relative to one another.
5. Continue alaysis, revision of the matrix and location of work centres or departments until no further improvements are obvious.

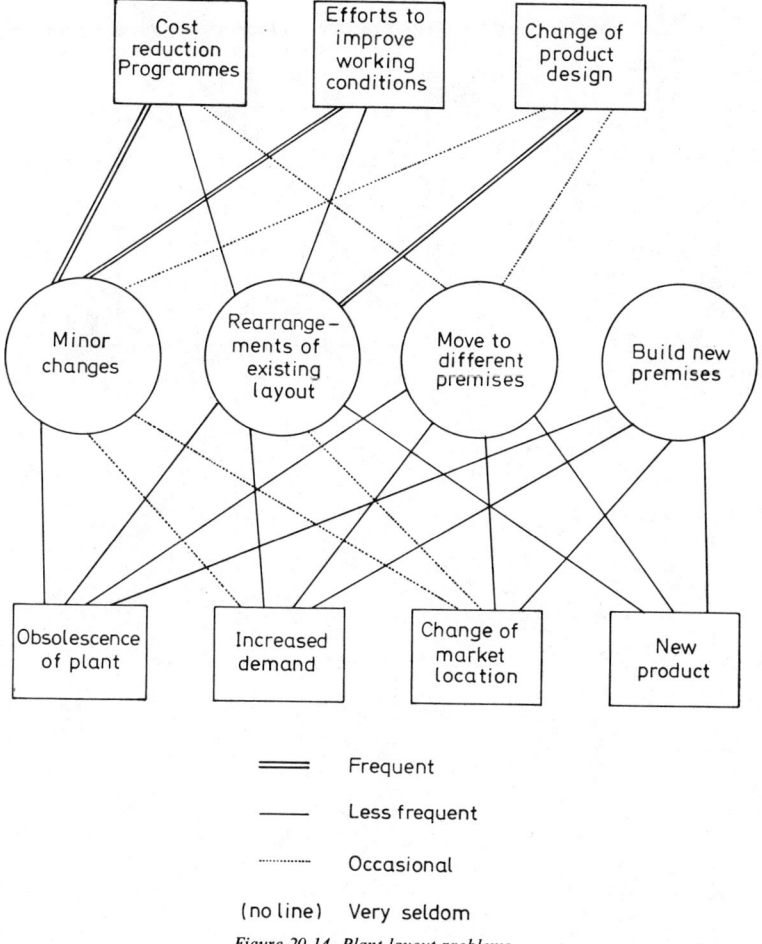

Figure 20.14. Plant layout problems

Gearbox	A	B	C	D	E
Casings/week	10	20	30	40	50
Movement of processes	Store Anneal Turn Grind Inspect Pack	Store Turn Mill Drill Grind Inspect Pack	Store Anneal Mill Drill Mill Turn Grind Inspect Pack	Store Anneal Turn Mill Inspect Pack	Store Anneal Drill Turn Mill Grind Inspect Pack

The units/week are extracted from the above table

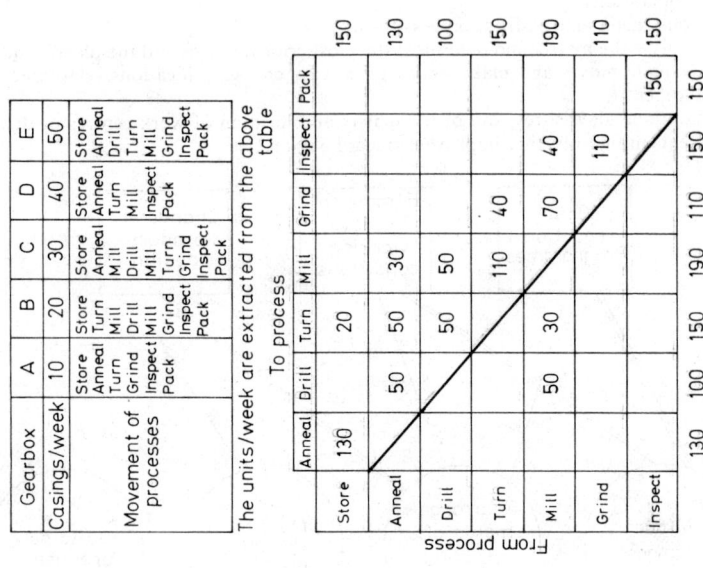

From process \ To process	Anneal	Drill	Turn	Mill	Grind	Inspect	Pack	
Store	130		20					150
Anneal		50	50	30				130
Drill			50	50				100
Turn				110	40			150
Mill		50	30		70	40		190
Grind						110		110
Inspect							150	150
	130	100	150	190	110	150	150	

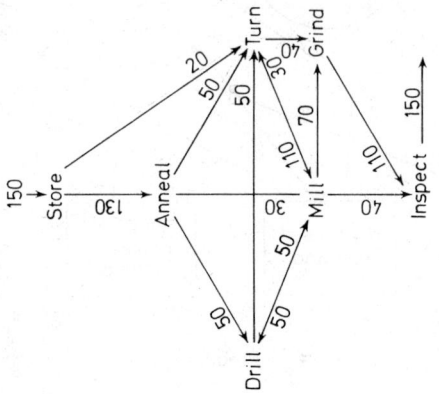

Figure 20.15. Operating planning sheet

EXAMPLE

A machine shop is to be responsible for producing five types of gearbox casing. The operation planning sheets have provided data in the following order of processes, Figure 20.15.

Sequence analysis

Sequence analysis is based upon the analysis of the sequence of operations for parts and products. The steps followed in this type of analysis can be summarised as follows:

1. Collect data concerning sequence of operations, production requirements and unit loads for all parts under consideration. Estimates for area requirements should also be obtained.
2. Develop a sequence summary identifying the work centre each part will go to next.
3. Compile a summary of loads per time period between all work centres.
4. Schematically develop the improved layout. This method represents only a slight improvement over the traditional approach, but is inferior to travel charting.

Systematic layout planning

The following three techniques have been developed by R. E. Muther,[17] an international authority in the work study approach to factory layout. These are P–Q analysis charting; Activity Relationship charting and Flow/Activity Relationship diagrams.

To assist him in identifying certain activities he uses A.S.M.E. symbols in a modified form:

○ Operation or production; subassembly or assembly.

○ Operation or production; processing or fabrication.

⇨ Transport and related activities.

▽ Storage.

☐ Inspection, test, check.

⌂ Services (maintenance, utilities, personnel services).

⇧ Office areas or activities not directly part of the main area.

⬡ Handling.

THE MUTHER MNEMONICS: P–Q ANALYSIS

This rests on the basic sequence of elements for layout planning. Thus:

P   Product (material)
Q   Quantity (volume)
R   Routing (process sequence)
S   Supporting services
T   Time.

The cross-relationship between one defined area and another may be charted in an Activity Relationship Chart (Figure 20.16) with reference to the desired proximity and the reason for this degree of closeness.

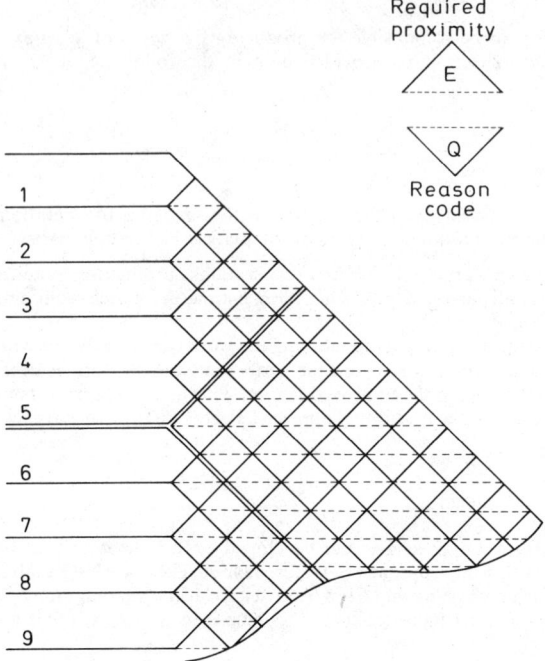

Figure 20.16. *Activity relationship chart*

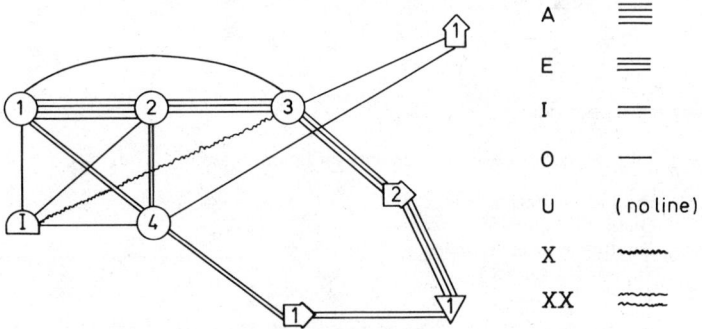

Figure 20.17. *Flow/activity relationship diagram*

The required proximity or relationship is evaluated as follows:

Value	Closeness	Colour
A	Absolutely necessary	☐
E	Especially important	☐
I	Important	☐
O	Ordinarily close	☐
U	Unimportant	☐
X	Undesirable	☐
XX	Extremely undesirable	☐

The reason code is also important, but is not specified, as it must be varied to suit each set of conditions for which the layout is being made. The main use is in the evaluation technique where the people who are in charge of work in the area can weight the particular reasons.

Typical reasons are as follows but this list is in no way exhaustive. Ease of supervision; ease of operation; closeness of materials; closeness of machine; share of common services; hazards of fumes, dust, noise, etc; cost.

Figure 20.17 illustrates a flow/activity relationship diagram in which the number of lines linking activity centres is coded according to closeness requirements.

Schematic analyses

Schematically analysing the relative relationships between work centres is an improvement over other schematic and graphical techniques, and it contains a highly systematic approach and a number of quantitative factors.

The procedure is as follows:

1. Relates activities to each other by a method 'closeness—desired' rating which results in a 'relationship chart'.
2. Establish required areas and configuration for each of these activities.
3. Graphically relate activities and arrange space required to form a basic pattern.
4. Evaluate alternative layouts against objectives and constraints as desired by management.

A rating method is used to evaluate these alternatives and the best alternative selected.

Optimisation models

Numerous mathematical models have been suggested for the optimal solution of the facilities layout problem, but unfortunately all are doomed to failure because of the strong influence of the human element which begins with the customer and continues through—information available—management skill—industrial relations control and feedback—transportation availability. One such model utilises the criteria of minimising the product volume and distance of material handling between all combinations of work centres.[18] Three assumptions however are necessary.

1. Complete interchangeability of individual areas.
2. Independence of the distance between any pair of locations from the direction of movement.
3. Direct proportionality of cost to equivalent distance.

The technique is, broadly, one of arranging volume data into a distance matrix in such a way that it enables the systematic elimination of unfeasable alternatives until an optimal assignment is reached.

ERGONOMICS: ENVIRONMENTAL FACTORS

Metabolism

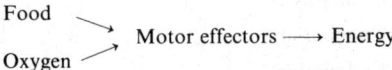

The changes to which the foodstuffs and the oxygen are subjected are summed up under the term metabolism, the principal end products of which are converted into water, carbon dioxide and heat. The foodstuffs are mainly carbohydrates, fat and protein, the

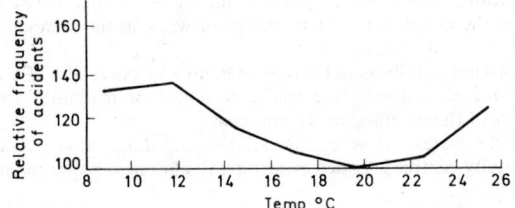

Figure 20.18(a). Accidents in relation to temperature change

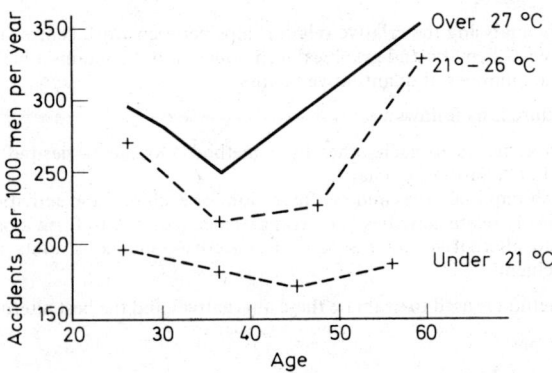

Figure 20.18(b). Accidents in relation to age and temperature

latter are used to maintain body tissues; the former principally as a source of energy—glycogen for muscular activity and blood sugar for the brain (diets can seriously affect the blood sugar content if not controlled).

The energy expended in the complete absence of voluntary muscular activity is known as basal metabolism, which represents the minimum energy expenditure to keep the organism alive.

Temperature and humidity affect workers in industry in a number of ways. When the body temperature rises above normal (36·9°C) the blood vessels in the skin dilate and more blood carrying heat from the deeper tissues passes through the skin. In cold conditions of work the skin blood vessels contract so that the heat in the deeper tissues is conserved.

If the air temperature is so high that the difference in temperature between the skin and air is small, or if the body is producing too much heat by physical or mental exertion, insufficient heat is lost by simple conduction from the skin to the air and body temperature begins to rise. This stimulates the sweat glands and the correct body temperature is then restored by secretion of sweat which on evaporating extracts heat from the skin's surface. In very hot conditions as much as 1 litre of sweat is secreted each hour and all this fluid has to be replaced by drinking water thus some salt is lost.

Industrial studies in relation to temperature are scarce. Vernon and Bedford (1927) shows that there was a slow but steady increase in the time taken to load coal tubs as the effective temperature increased from 19°C to 28°C. Time taken 'resting' also increased more markedly at temperatures above 24°C. The working efficiency at high temperature was 41% less than that at the lower.

On the industrial effects of cold, Bedford (1940) shows in a bicycle chain assembly in which the temperature was reduced from 17·5° to 10°C that the time to complete the task increased by 12%.

Accidents

Osborne and Vernon (1922) show an increase in accidents both with decrease and increase of temperature from an optimum of 19–20·5°C. The increase with cold being somewhat the greater (Figure 20.18a).

A similar effect is shown by Vernon and Bedford (1931) in relation to age of coalminers; below 21°C there is very little difference between the age groups, above it is more pronounced over the age of 30 years (particularly on the over 50's), see Figure 20.18b.

If accidents reflect on any effect of temperature on efficiency, then it might be taken that older men are adversely affected by higher temperatures. The age at which heat will begin to have an adverse effect is not known.

Energy expenditure in heavy work

Methods of measuring energy expenditure directly involve estimating the volume and consumption of expired air while indirect measurements are based upon variations in heart rate. A 'Douglas Bag' consists of a large impermeable envelope carried on the back, into which the whole of expired air is breathed through a mouthpiece, its disadvantage is that capacity is limited to 100 litres so that it can only be used for comparatively short periods before it is full. The 'Kofranji' apparatus and the I.M.P. (integrating motor pneumotachograph) designed by Wolff 1958 are improvements on this method.[19] However the commoner form, of two subjective methods, is to measure heart rate—pulse.

For example, with a sudden sprint to catch a bus, although breathing may start rapidly whilst sprinting, there will likely be a high pulse rate, rapid respiration and probably sweating to get rid of heat, for some time after sitting comfortably in the bus. From this a typical curve of activity can be obtained by pulse rate observation, see Figure 20.19.

In light work, if prolonged activity is studied it will be realised that concentration deteriorates at the point at which performance begins to deteriorate, that is, where an increasing number of pauses in work, errors, missed pick-ups, indicate the end of the effective work period. Murrell defines this as the 'Actile Period'.[20]

Duration of work

The working hours in the early part of this century were frequently quite long and early investigations concentrated on discovering whether these long periods should be broken by a single rest pause. The practice of giving a mid-period break is now reasonably established. However, there is no indication, so far, that more frequent pauses result in an even greater increase in output.

Recent work by Murrell, indicates an improvement in output where activity cycles of 50 min duration followed by 10 min rest were maintained. Experimental research suggests that a decline sets in after 70 min of work in light industry. Whenever irregularities do occur the time has arrived for a break or stimulant.

In determining performance characteristics during a work period, Dudley[21] found that operatives maintain a consistent working pace, but interruptions, due to personal

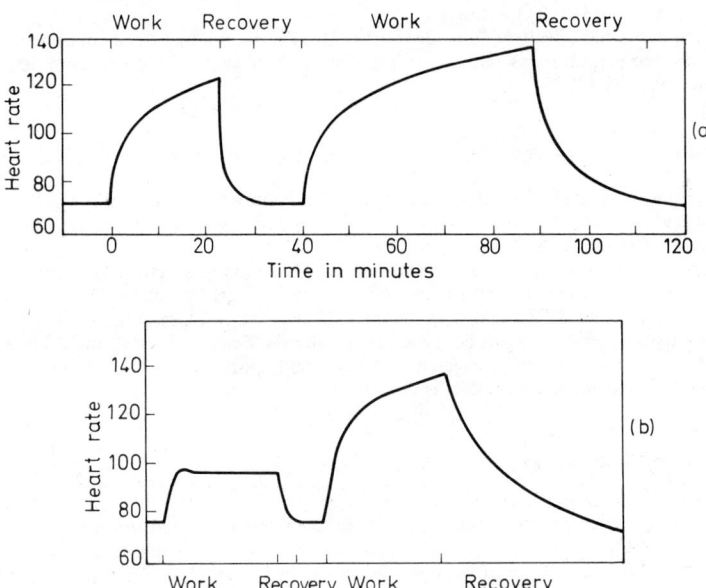

Figure 20.19. (a) The effect on pulse recovery time of continuing heavy work beyond the optimum duration, (b) pulse recovery time with moderate and with heavy work of equal duration

and operational delays and ancillary work, result in changing rates of output. Very little time was lost through operational delays but where such delays did occur they tended to be grouped, as were personal delays, towards the beginning and end of work periods and were mainly responsible for the generally accepted 'late starts' and 'early finishes' at the work station. Dudley concludes that 'it is, therefore, in the analysis of lost time, rather than of operation time, that the need for fatigue, relaxation or other allowances is to be sought' and suggests that the range of relaxation allowances given in practice 'is greatly in excess of time actually lost by reason of personal delays'.

The research studies of Norman and Bahiri[11] into the utilisation of men and machines support Professor Dudley's view that there is a relatively high level of ancillary work being performed in manufacturing industries, i.e. work which does not contribute to the added value activities of manufacture. For example, 'handling' and 'attending' activities in labour utilisation account for over 30% of the working time available, and in addition idle-waiting management account for 5·5% while, idle-labour responsible adds a further 10·5%.

Working patterns

Large numbers of workers on repetitive work are not free to work at their own pace. Especially where performance has been seen to be of a highly consistent pattern, subject to speed restrictions (conveyor, machine, or other worker) and have a maximum period of time in which to complete a specified task.

In the interests of higher rate of output, the time in which an operation has to be performed tends to approximate to the time required to perform the operations, that is, the time allowed for the task tends to be reduced to the minimum. Experience leads to expect quality deterioration and excessive operative frustration and fatigue.

The distinction between paced and unpaced conditions is not too clear; there are many degrees of pacing ranging between situations where:

1. The operative is rigidly paced, and
2. The operative having sufficient time within the work interval to perform his task without undue haste or frustration.

In case (1) the operative may be rigidly paced by a machine in such a way that the time allowed to perform the operation is equal to the time required for its completion and every component must be dealt in the correct manner and to the required specification. In other cases, a few misses may be allowed, faults and lower quality standards tolerated, delays and waiting time accepted. Alternatively, the work itself may be allowed to accumulate at the work station to permit some degree of flexibility on the part of the operative. The use of mechanical handling devices, while facilitating production and output, can too often be readily abused, so that, for example, some conveyor systems, carousels and continuous monorail types in particular become convenient stores for material; in other words mere devices for inefficiency.

When engaged in repetitive work, paced or unpaced, an operative will not perform at a constant rate by collecting and plotting a sufficient number of cycle times in a frequency distribution, a curve of the form shown in Figure 20.20 is likely to result, and while the distribution of cycle times obtained from trained and experienced operatives on repetitive work at their own pace is positively skewed, there is a marked tendency for paced performances to yield a more nearly normal distribution of cycle times. Dudley[21] reveals that 'when trained and experienced workers are engaged on repetitive manual or manually controlled operations (which have been standardised as to the method of working and work content) during the periods of time which are regarded as normal in current practice, their performances display certain characterisations which are common to the wide range of tasks studied'.

These characteristics are as follows:

1. Operation and constituent element times are, within close limits, substantially constant throughout the working day.
2. The range of element times throughout the working day is no greater than that which could be expected to arise from random variations.
3. Variations which occur in the pattern of motion element times are attributable to voluntary changes in the operator's working pace.
4. Personal delays, under the worker's control where these occur, tend to be grouped towards the beginning and end of each work period.
5. Ancillary and operational delays, where these occur tend to be grouped towards the beginning and end of each work period.
6. A slightly greater percentage of time (but not necessarily more time) is lost by reason of personal delays during the afternoon than during the morning work period.

7. Foreign elements result from interpretation of the repetitive work pattern by personal or operational delays.

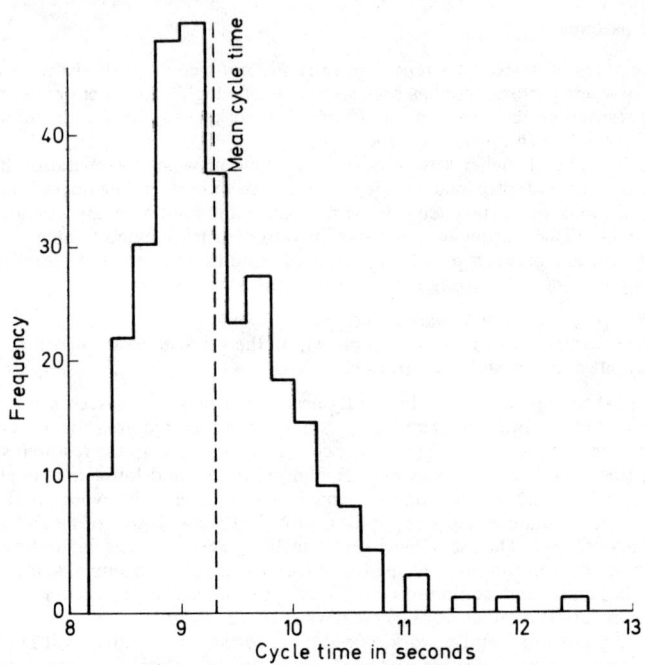

Figure 20.20. Histogram of cycle times in a repetitive task

GROUP TECHNOLOGY

To lay out a plant, department or section within a department, workbench or machine, the production engineer has such aids as process charts and diagrams, two and three dimensional models and travel charts and other quantifiable techniques as described elsewhere. To use these aids successfully, quantitative and qualitative information must be collected and analysed, e.g. to use travel charts, quantities, space dimensions, and costs involved must be known.

Operation layout sheets, as shown in Table 20.5 in the following sub-section, are a source of information issued by the planning engineer. These present quantities in a batch, sequence or route of operations, tools and equipment to be used and in some cases time allowed for each operation. By quantifying the processes involved in the sequence of operations, a layout that appears to be the most effective may emerge. Effective layouts for large batch and flow production work do emerge where it is a simple matter of analysing mass quantities. Batch production, however, particularly small batches and one-off's, and where a high variety exists require a design orientated approach to layout.

Higher costs in batch production, as against flow production, stem from the incidence of set-ups and changes from one batch to another, which prevent an uninterrupted operation of the production line and the costly machine tools on it. Reducing this incidence of set-up changes from one batch to another by simply increasing batch sizes does not afford a solution, since the number of components that have to be held in stock are increased and the high cost is transferred from batch production to storage.

The alternative and most effective solution to date is a matter of design rationalisation or standardisation, and simplification or multiplication of tooling layouts which depend upon whether the machine is conventional or special purpose, and *group technology*, which is family grouping of components. Numerical controlled machine tools and other methods of automating batch production reduces the idle-time in change-overs, but group technology actually reduces the number of production stoppages arising from changeovers.

Group technology originated in 1938 as a purely academic study by Professor Sokolovsky in the USSR. A certain amount was done by the Germans in World War II and by Scania Vabis, a Swedish company in 1948. By 1959, the technique of designing a composite component (Figure 20.21) which incorporated all the shape and feature characteristics of a number of different components was introduced by Mitrafanov, USSR.

The first conference to be held in Britain was in 1962 at which it was intimated that: 'As the use of special purpose machines and automation was extended the production of small quantities of left-overs presented problems'.

Grouping of parts was stated to be the most effective solution, since tooling, method planning, and production of each part was then not considered in isolation but in the context of a group of similar parts.

On the basis of a classification system that had been in operation in one company for about six years, a pilot group technology study was arranged and components that required turning, milling and drilling, totalling ninety-nine different components, were grouped into families (Figures 20.22 and 23). Four machines were installed into a 'Cell': a combination turret lathe, a capstan lathe, a horizontal mill and a radial arm drilling machine, and arranged to process the ninety-nine parts in sequence, attended by three operators. A vital feature of the arrangement was the design of a 'composite component' which embodied all the machining operations required to produce all the different shapes and features; since if the machines were set-up to produce this composite component they would be able to machine any one of the individual parts.

Savings

The savings achieved are listed below:

> Process time for low volume work reduced 35%.
> Setting times decreased 70%.
> Three documents per order instead of twelve.
> Utilisation of lathes, 90%; milling and drilling machines, 12%.
> Manpower utilisation, average 85%.
> Segregation of low volume orders reduced disturbance of normal manufacturing programmes.
> Design of anomalies in the components were removed by rationalisation and standardisation.

On the strength of such savings the company has since extended the application of group technology throughout its manufacturing processes.

The relatively new philosophy of group technology has highlighted the need to rationalise variety, reduce tooling set-up and changeover times. With the aid of Group or Cell type layouts such non-productive time can be reduced. One authority,[22] stated that 'with conventional batch production methods (functional layout) total workpiece throughput times average 100 days due to a large extent to the queueing problems experienced'. The same authority also claimed that 'with the use of the "Cell" system

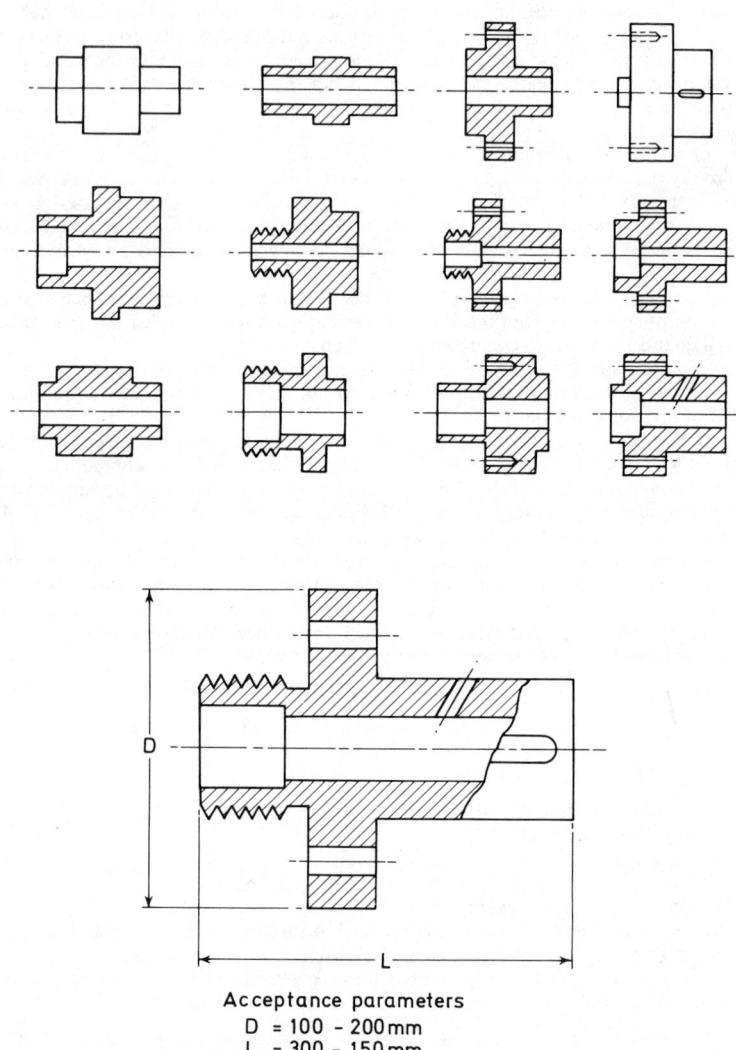

Acceptance parameters
D = 100 – 200 mm
L = 300 – 150 mm

Figure 20.21. Composite component for a production group of three machines: S.C. lathe, mill and drill

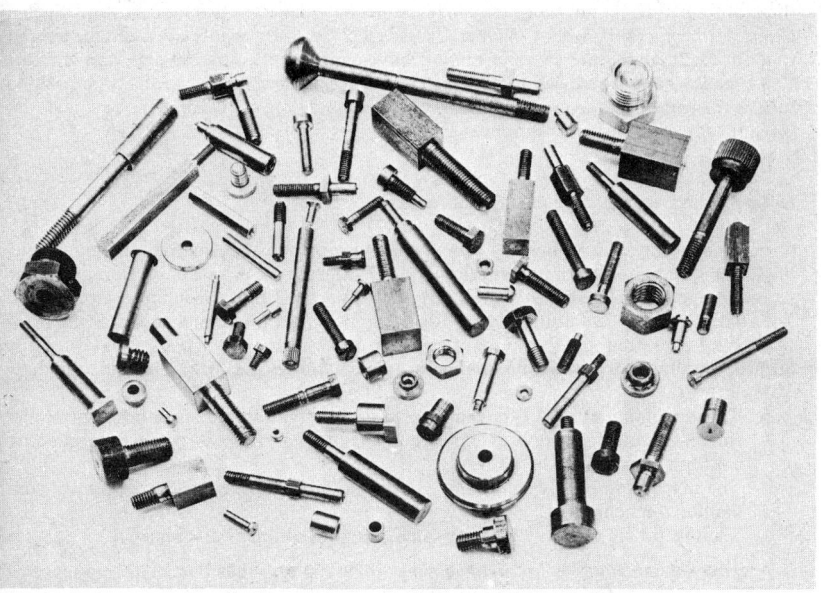

Figure 20.22. Collection of miscellaneous parts

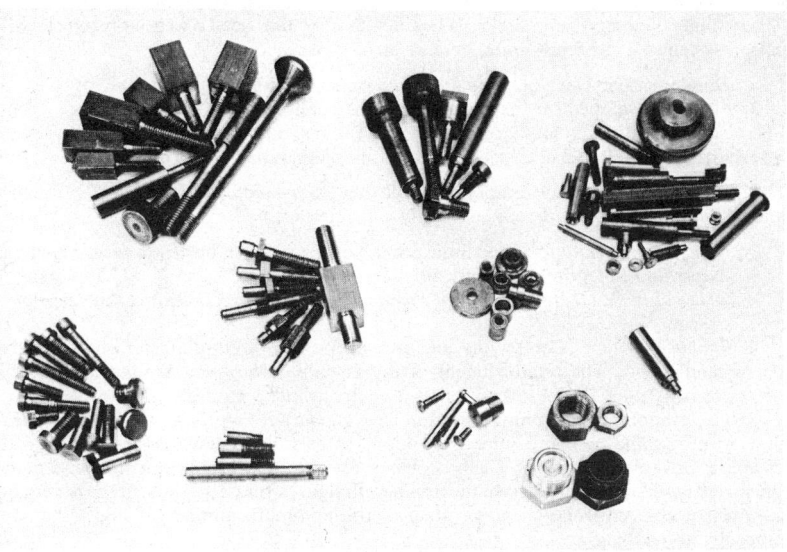

Figure 20.23. Collection of parts into families

these average throughput times may in instances be reduced by as much as 90%'. Such a claim has been confirmed by Walker Crossweller Co. Ltd., whose average throughput time for the manufacture of water mixing valve components, was 100 days and with the Cell type layout was reduced to 10 days. Moreover, the company expects eventually to reduce this time to an average of one week for completing parts from the raw material stage through to finished components.[23]

Defining and constructing a component family

Group technology can be defined as a production method that involves the machining of parts in 'families'. A component family, for example type A, is a collection of similar or related geometrical shapes and/or size, all requiring similar machining operations. Alternatively, a type B family may be dissimilar in shape though related by having some machining operation in common and possibly other similarities, such as materials and accuracy limits. As a production method group technology may be implemented in three different ways:

1. The machining of an A type family on a group of different conventional machines (or on a multi-functional machine equivalent to a group of single function conventional machines).
2. The machining of A type and/or B type families on one or several similar conventional machines.
3. The machining of a B type family on a group of different machines.

A group of machines is an arrangement which enables that particular sequence of operations required for machining of all parts in a given family to be performed within the confines of that group.

Classification systems

The defining of component families is best achieved by the use of a well-designed classification system, e.g. Brisch, Orpitz, Pera or those listed below:

Guildemeister. Developed by a West German machine tool company.
Pittler. Developed for the manufacturer of a range of automatic lathes.
Zafo. Another German designed system for general use.
Viroso. Developed in Czechoslovakia for the analysis of workpiece statistics.

All identical and similar components will then be brought together by classification and three basic types of component family will emerge:

1. *Identical shape and function.* Spur gears, bevel gears and bushes whose individual shapes and functions are identical.
2. *Identical in shape but different in function.* Bearing flanges, sealing flanges, rings, spacers.
3. *Similar in shape.* Centre line and non-centre line multi-diameter shafts, bolts, spindles, etc. The establishment of the machine group and tooling for these is normally more difficult than for the two preceding types of family.

Group technology and computerisation have encouraged a proliferation of classification systems some new and others adaptions of existing systems. Many are used to establish workpiece statistics, such as shape, features, machining operations, surface finish, etc. and are unrestricted in the number of digits, whilst a second category contain as much usable information as possible in the minimum number of digits that are normally of fixed length.

The most common systems are Opitz (Figure 20.24), Brisch mono-code (Figure 20.25), Brisch poly-code, and PERA (Figure 20.28). The Opitz system has a more generalised

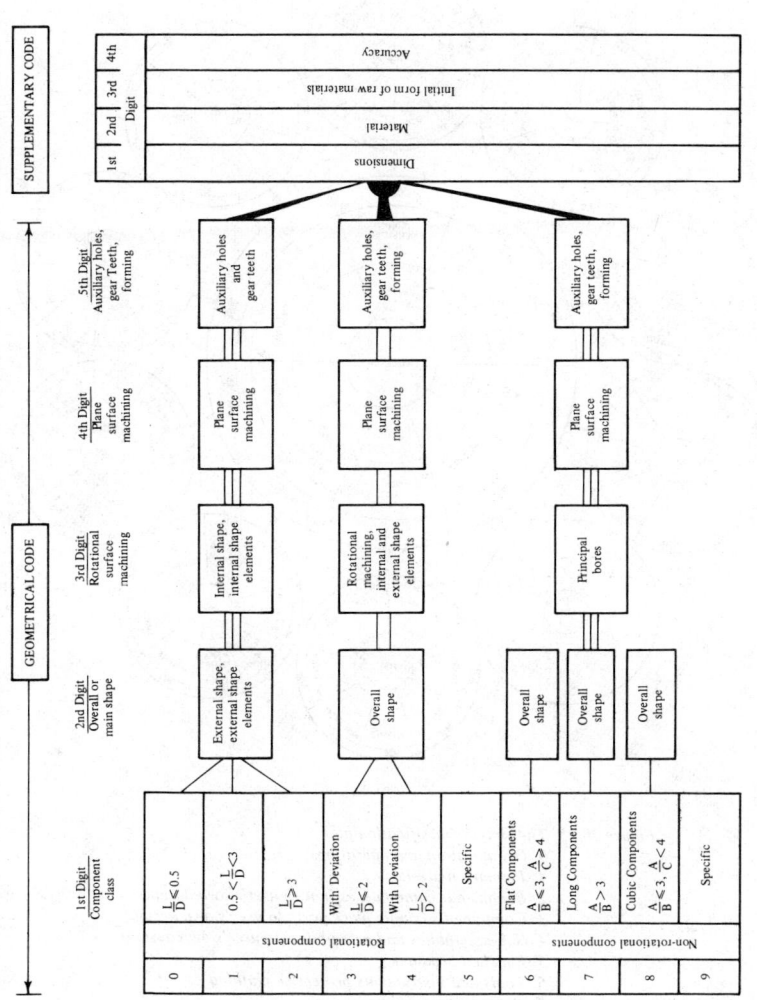

Figure 20.24. The Optiz classification system

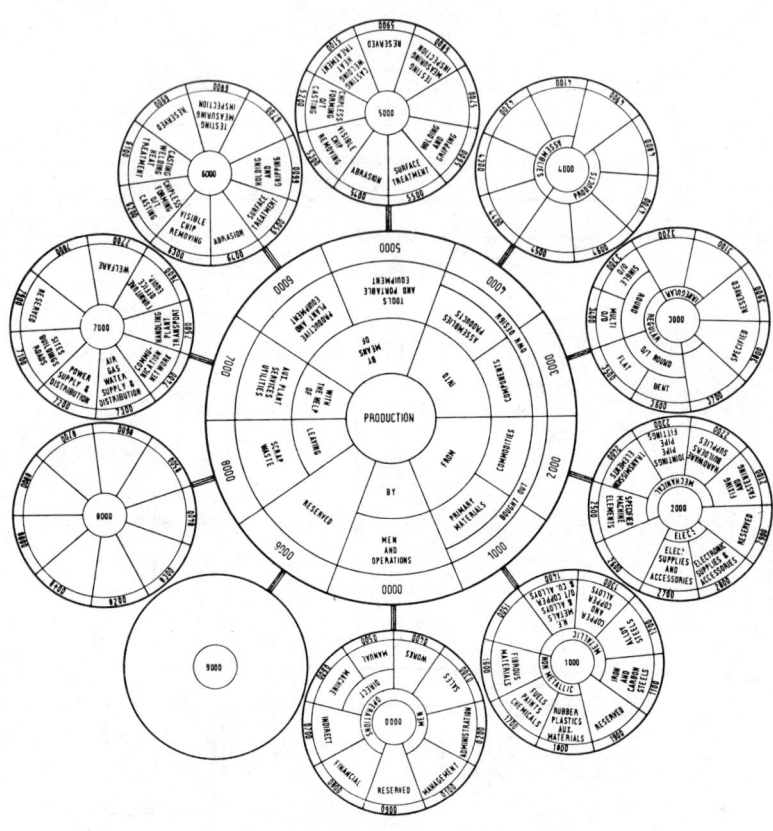

Figure 20.25. The Brisch classification plan
  0. Organisation and operations
  1. Primary materials
  2. Bought-out commodities (not to user's own design)
  3. Components (single piece parts to user's own design)
  4. Sub-assemblies and assemblies (to user's own design)
  5. Finished products
  6. Tools and instruments protective clothing
  7. Plant and machinery
  8. Services
  9. Reserved

General Statement

Company Ref. No.	Serial No.	Total Change Points	Change Point for Maximum Diameter	Change Point for Maximum Length	Workpiece Type	Material	Initial Form	Quantity
60	30	08	03	04	1	0	0	0002

Detailed Statement

Shape and Dimensions			Surface Elements						
Change Points	X Co-ordinates	Y Co-ordinates	Corner Condition	Form	Accuracy	Surface Finish	State of Hardness	Function	Additional Machining Features
01	00000	00175	1	0	3	3	2	6	7
02	00253	00175	3	0	9	7	2	1	1
03	00253	00270	3	0	3	6	2	1	6
04	00453	00270	3	0	9	8	2	1	1
05	00453	00188	1	0	7	8	2	1	1
06	00440	00188	1	0	9	8	2	2	1
07	00440	00125	1	0	3	3	2	6	1
08	00000	00125	1	0	9	8	2	2	1

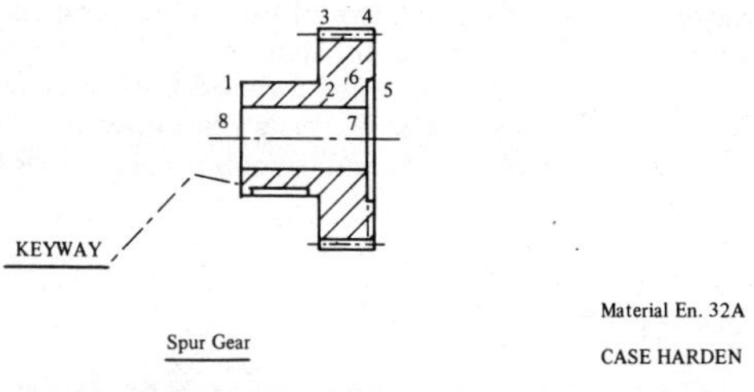
3
4
1
6
2
5
8
7
KEYWAY
Spur Gear
Material En. 32A
CASE HARDEN

Figure 20.26. The Pera system

application. It consists of a 5-digit primary code and a 4-digit supplementary code. The primary code is essentially a geometrical code which groups components by the logical arrangement of shape characteristics and significant features. The supplementary code provides information on component dimensions, type of material and material thickness.

Fixed digital significance exists in certain areas of the code with individual digits describing the same features of all classes of components and each position within a digit has a corresponding meaning. This makes the code relatively easy to memorise and high rates of coding can be achieved. From the information and data contained within the classification number it is possible to construct an approximate picture of the shape and size of the component. It is still necessary, however, to use the component drawing for precise dimensions and therefore the company's unique numbering of components must be incorporated within the total code, e.g.

AB 5919	90005	2110
Drawing number	Primary Code	Supplementary Code

Formation of component family

Experience shows that a pilot study should be initially carried out by examining small areas of simple components first, and gradually progressing to the more complex parts as experience and technique improves. However, the component family analysis should preferably be based on the complete range of components from the products manufactured by the company. This, of course, presupposes that all components have been classified. If not, a preliminary investigation using a representative sample of 10/15% components from the complete component range should be carried out.

The drawings and associated production data of the selected components are collected together, classified and sorted into code number order. The investigation should not be limited to the formation of potential component families, but also to assess the necessary diversity in company operation. This will include the preparation of component family paper work and the scheduling and control of the components into the machine group.

While this investigation is in progress the remainder of the drawings, new designs and obsolescent drawings still liable to be called forward for spares replacement, should be coded. There is no short cut, nevertheless, it is a once only exercise and the effort fully justified, particularly for variety control and the computerisation of production planning and control.

Establishment of component family

The coding provides the first stage in sorting and makes it possible to gather the components into families. If tabulated lists are visually examined the naturally occurring families can be easily determined. These families are normally of the type 'identical in shape and function' and 'identical in shape but different in function', and appear as blocks of near identical blocks of code numbers of the listings. These will be the most obvious component families to begin to develop and establish machine groups.

Finally, recheck the component codes and tabulations and revise where necessary.

Collection of production data

Number and sequence of machining operations, setting times and numbers-off each component within a defined period of time need to be collected. By analysis of the machining operations and sequence it is possible to derive the types of machines required to form the machine group. From the machining and setting times and the numbers-off each component, the potential load on the machine group may be established. Line balancing however cannot be adequately established until the component tooling requirements have been assessed. This is the first stage in 'family' formation. The drawing of each component within the family is examined and the type and number of tools necessary to produce the component are determined.

With the tooling analysis complete the family is established and group layout balanced, i.e. data should now include:

> Geometric shape.
> Maximum and minimum sizes.
> Material type.
> Form and method of holding.
> Tools: type and holding.
> Machine tools type and capacity.

From this information the profile and parameters of the component family are constructed against which the acceptance or non-acceptance of new components into the 'family' can be based.

A point to bear in mind is that once a component family has been formed and integrated into a group layout, it does not necessarily have to remain static. Some components will become obsolete while new components will appear. The more flexibility that can be built into the system the more one can expect to get out of it.

PRODUCTION FLOW ANALYSIS

Production Flow Analysis[24] is a technique for finding the families of components by a systematic analysis of information contained in operation planning sheets. It is based solely on the manufacturing methods and does not consider the design features or shape of components as in Group Technology. It has the more immediate effect of producing savings that would be possible from the application of Group Technology. There are three successive levels of analysis—Factory Flow, Group and Line.

Factory flow analysis

Factory flow analysis is an attempt to find major groups of department size and major families of components which can be completely processed in these departments.

SYSTEMATIC PROCEDURE

1. Divide into departments.
2. Allocate plant and equipment to the departments and obtain utilisation frequency.
3. Determine operation sequence or route process card for each part.
4. Analyse the parts by route.
5. Draw the basic flow chart, e.g. Figure 20.27.
6. Determine which parts are exceptions.
7. Eliminate the exceptions by possible rerouting or purchasing.
8. Check machine loads and forward scheduling.
9. Specify the standard interdepartmental document flow system and construct a basic flow chart, a list of plant for each department and a list of parts processed in each department.

In effect factory flow analysis decides common family groups by process which can generally be changed or revised when necessary at low cost. Once the needs have been

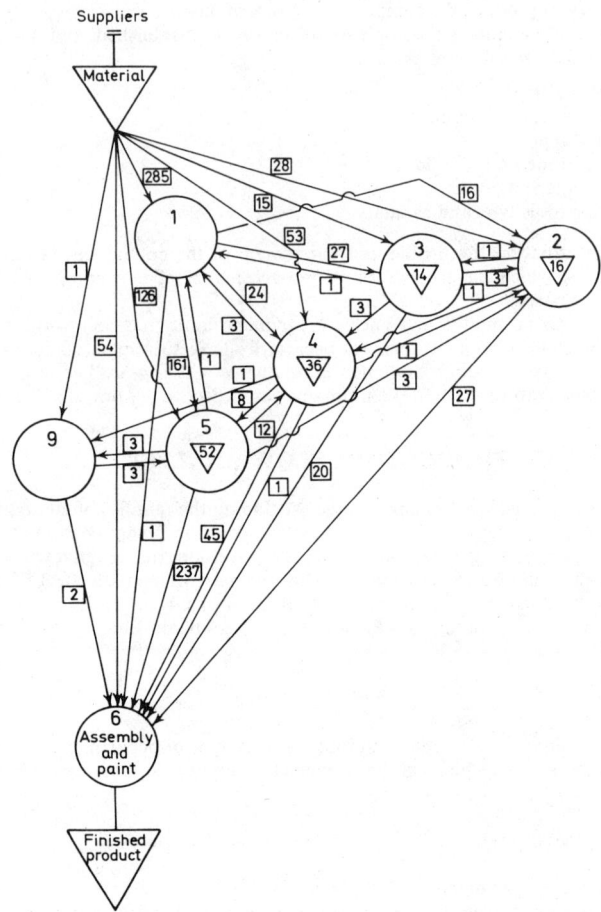

Figure 20.27. Basic flow chart (courtesy 'Production Engineering')
1. Blanks department
2. Sheet metal work
3. Forge
4. Welding department
5. Machine shop
6. Assembly
7. Outside firms

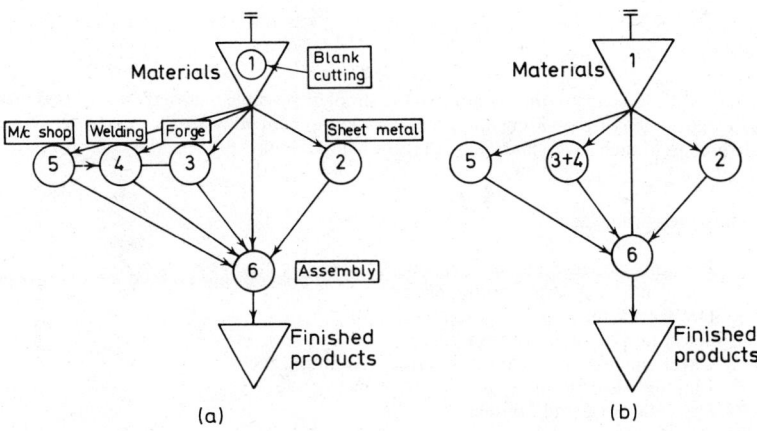

Figure 20.28. Simplified basic flow chart (a) First simplification found by factory flow analysis, (b) Further simplification found possible later

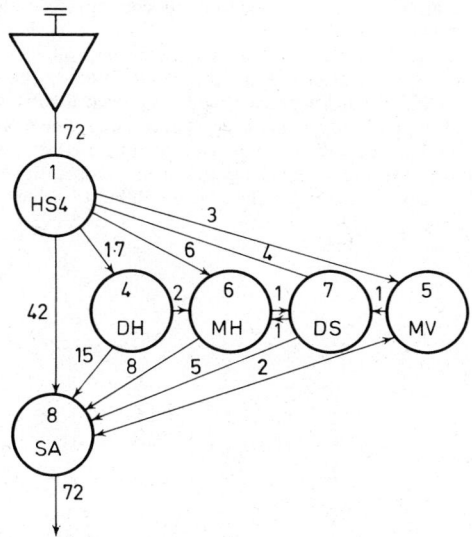

Figure 20.29. Simplified group flow network—Group 2

found, it is only necessary to alter a small proportion of route cards and to move a relatively small number of machines to obtain the advantages of a simple inter-departmental material flow system.

Group analysis

Group analysis is an attempt to achieve the simplest possible material flow system inside each department by dividing the plant and equipment allocated to each department into groups in such a way that each family of components is processed by one group only.

SYSTEMATIC PROCEDURE

1. Renumber operations on operation sequence or process route card.
2. Sort route cards into 'family' packs.
3. Draw pack/machine chart figure.
4. Select families and groups figure.
5. Check machine loads and forward scheduling.
6. Eliminate exceptions as per factory flow analysis.
7. Specify groups and families.
8. Draw final flow system.

Figure 20.28 shows both an intermediate and final stage of analysis for the forge and machining departments.

Line analysis

Line analysis is an attempt to achieve a layout sequence approximately to unidirectional flow (line flow).

With traditional line layout, only one component is produced on the line and the machines are laid out in the sequence in which they are used. The same type of layout can be used if 'line balanced' for batch production, even if some of the machines are used in a different sequence to the majority. The aims of 'line analysis' is therefore to find the simplest and most productive line which can be used.

By constructing a travel chart matrix using machine/operation number frequencies the group analysis flow network can then be simplified to achieve the most effective, flow-line layout (Figure 20.29).

PRODUCTION PLANNING AND CONTROL

R. G. NORMAN

PLANNING CONSTRAINT AND OBJECTIVES

There are two influences on an economical production policy:

1. *A constraint*—to provide goods to a delivery date or at frequencies suitable to customer's requirements, and
2. *An objective*—to utilise the resources so that production costs are kept to a minimum.

Because of these influences production planning is essential.

Delivery promises can be broken if assessments of capacity are wrong; and congestion, bottlenecks, idle time and other excess production costs can arise, if all supplies (including information) are not forthcoming when they are required. Guesswork is not good enough and memory is an unreliable servant. In all but the smallest organisations a person cannot carry all the details of delivery dates, supplies and capacities in his head without making mistakes or neglecting something. Waiting for work, instructions, or tools is one of the biggest contributory causes of high costs—or at least of costs being higher than they should be, i.e. excess costs. Well-thought-out schemes of production planning substitute facts for guesswork and reduce waiting time considerably.

In general, therefore, orders must be delivered to time, and for economical production there must be a minimum of interruption to flow due to lack of work or information, and in order to maintain steady employment a balanced load of work must be maintained between departments.

Typical of many companies today foremen occupy powerful positions. A foreman is often the one man responsible for loading the work and allocating it to specific individuals. Because of the knowledge of operative skills and reliability the foreman decides who should do the job, which results in certain selective jobs being allocated to one or two specific individuals and it is just unfortunate if they are already loaded; the new order is held in delay until the operative can accommodate the work. Failures to meet deliveries is partly due to this form of 'personalised' allocation.

The foreman, with some discretion, decides the size of batch, the production method and the stages of operation. He raises the job card and completes the order of priority, operative's name, time allowed, etc. He also has the responsibility of organising materials, tools and equipment necessary to do the job, as well as filling in the necessary stores requisitions. Because of this stocks of materials, bought-out parts, tools and work-in-progress are held at many stages of production which are generally fragmented and scattered about the works.

A relatively high volume of work-in-progress, often inherent in batch production methods, is generally the result of failure to match loads to operatives, tools and material availability and becomes a burden when it occupies valuable space that could be used for more productive work.

Jigs and tools are often duplicated many times over and retained by individual operative's as their personal property, they consider to have the right to hold them on the

20–53

basis that they devised and manufactured them in order to simplify the production method and reduce process time. Although production methods change and process time reduces, often the original target times remain, tending to add more complications to planning and control.

Where production has to be coordinated between departments other problems emerge; each department periodically experiences either over-, or under-work loading which frequently calls for overtime in one department, while others have insufficient for a normal day's effort and occasionally no work which results in stand-still situations.

Other major causes of a fluctuating load and late deliveries is the dominating position sometimes held by sales representatives. Often they have direct access to foremen and frequently request changes of programme by asking for small (sample) batch orders, and furthermore, expect skilled operatives to be withdrawn from production work to assist in product design and the development of a production method. No planning or control system could operate under such interfering conditions.

One of the essentials in securing a well-defined management is the separation of 'planning' from the 'doing' activities. Work should be planned in advance for each stage of the production process and be factory wide, so that those who supervise the processes are free to fulfil their more important and proper functions of directing and maintaining the 'doing' activities and ensuring that operatives' conform to standard practices and methods.

Systematic planning

Planning decides, prepares and issues schedules for the sequence of manufacturing. It is responsible for the preparation of parts lists, material withdrawals and delivery to stores of finished components and sub-assemblies; long and short-term programming, and resource allocation.

A point of view often expressed is that systematic planning increases clerical staff and paperwork. It is, nevertheless, usual to have some haphazard and fragmented means of recording on scrap pads, paper fragments or carry an excess of duplicated documents. Experience has shown that many companies house a number of offices and personnel that are concerned for a large part of the time transferring information, progressing, and filing orders and related information. The volume of duplicated and filed documents is a matter of serious concern in most firms for it is evidence of a fragmented information system and indicative of a clerical burden under which management labours.

Frequently, the 'flow' of documentation from enquiry to manufacture and despatch is personalised to such an extent that it is difficult to obtain up-to-the-hour progress of production without involving several people. The dangers of such a situation are only too obvious. Relationships with customers are put at risk because of such a fragmented store of information held by individuals. Too often production departments flounder from one week to the next attempting to meet the demands of pressing customers, nearly all of whom receive their goods later than promised merely because too much has been promised without sufficient analysis. The company's internal economic operation depends upon quick access to statistical analyses and controlling facts. Whilst a strong customer relationship is created and supported by individual contact, it must not be at the expense of an economic use of resources.

SUB-OBJECTIVES

1. Relate orders and delivery promises to resources and capacities available or conversely to provide the resources and capacity to meet agreed or acceptable demands.

2. Ensure that material, tools and part components are available when and where required.
3. Preserve a balance and steady flow of work through and between related departments.
4. Provide adequate manufacturing instructions to enable management and supervision to concentrate on production technique and control, and to relieve them of detailed administrative work.
5. Anticipate production delays and difficulties and provide production management with information to counter-act before they arise or become serious bottlenecks.

PREPRODUCTION PLANNING

There are two types of information essential to a company's activities:

1. Permanent customer/order knowledge preferably recorded in a form suitable for statistical analysis and forecasting.
2. Transient demand/supply information in the form of current orders, stock withdrawals, production and despatch, sequencing and loading schedules.

These areas of knowledge and information are the link between company headquarters, sales offices, customers and work's production resources. Both will provide management and customer with pertinent and timely facts about the progress of orders, utilisation of production and servicing resources and the material and stocks positions. A pre-production planning procedure is outlined in Figure 20.30.

A centralised information bureau will provide all that a fragmented system would provide far more economically and speedily. Such a centre would receive, or raise, all sales/works orders and distribute the documents to the appropriate departments concerned with the order. In so doing the centre will provide:

1. An essential overall control of the goods between departments.
2. A library of historical information relating to customers and orders.
3. Periodic analyses of supply and demand essential to senior management in their everyday decision-making and control.
4. Liaison and progressing customer/sales enquiries.

Documentation procedures—an example

WORKS ORDERS

Orders could arrive from these sources:

1. via sales office;
2. directly from the customer to works;
3. as part of an enquiry to the work's services department.

Initially five copies of the works order form (Table 20.6) are prepared by either the sales office (reference S123), or works office (reference W123).

The copies are designated:

Acknowledgement—white.
Accounts office/production—pink.
Planning office/despatch—yellow
Master record—blue.
Sales office record—buff card.

Table 20.4

Customer	Sales	Accounts	Production planning	Production control	Stores	Manufacture	Purchase
	Information bureau capacity schedule-occupied and unoccupied	Invoicing and advice note	Master capacity schedule	Work in progress	Stock position	Capacity schedule	Supp. position
Enquiry →	Seek quotation		Parts required and tools ↓ Operation plan ↓ Estimate—cost ↓ Load to schedule		Parts available		B.O.P. parts supply date
Order (+ Drwg. and spec. if required)	Quote ←	Pricing ←	Delivery date				
	Prepare works order		Detail parts and tools required		Parts and tools available		
	Modification		Decide make or buy ↓ Operations plan (routeing, job cards, inspection tickets, matl. requisition etc.)				B.O.P. supply
Active schedule ←			Schedule for production	Product control schedule	Stores delivery schedule	Manufacture schedules	B.O.P. schedule

Figure 20.30. Pre-production planning procedure

The five works order copies are distributed as follows:

Approvals

Accounts office.   One copy sent to accounts for credit approval, pricing, etc. Returned to the information centre duly signed and dated.

Planning office.   One copy sent to production planning for technical feasibility and delivery statements. Returned to the information centre duly signed and dated.

On receipt of both the accounts office and planning office approvals the information centre up-dates all copies with respect to pricing and deliveries. Production, despatch and acknowledgement copies are then released to the appropriate offices simultaneously.

Action

Production.   The pink copy returned to information centre (after credit approval) from accounts is sent to production planning to initiate manufacture. Planning office will then raise a package of production and/or stores documents to initiate and progess the work. After complete order has been moved into despatch, the planning office will return their work's order copy to the information centre as an indication of production work completed.

Despatch.   The yellow copy returned to the information centre (after delivery acceptance) from planning is sent to despatch to signal goods are on the way and packaging materials and appropriate resources need to be allocated. This copy waits receipt of the goods which may arrive from production and/or stores.

On the receipt of goods, despatch will complete the despatch instruction raised by planning with serial number included and record where necessary package size, weight, works order number and list of documents that will be required for shipment of goods.

The despatch instruction is sent to accounts office, who after rechecking credit, raise the necessary documents, advice notes and invoices (the invoices are held in the office to await instruction for their release), although some copies may be transmitted to despatch for inclusion with goods.

The accounts office forwards the documents and advice notes to despatch who enter the means of transportation and date of shipment after goods are despatched. As goods leave the works, the works order despatch copy is sent to accounts office to signal the release of invoices, a copy invoice being sent to the information centre. The work's order despatch copy is returned to despatch if further items for shipment remain outstanding.

The whole procedure between despatch and the account office is repeated for each part-shipment, until the total order has been despatched. On completion the accounts office forward the work's order despatch copy to the information centre who up-date the work's order master and sales office record with despatch dates, transportation and invoice numbers.

INFORMATION

Acknowledgement.   This copy, after up-dating with pricing and delivery intentions is sent to the customer to confirm order acceptance and terms.

Master record.   Retained at all times in an up-dated condition and filed on order completion as a permanent record.

Sales office record.   This is a copy of the completed master record which is sent to the sales office for their retention.

Production scheduling

Scheduling is the more detailed aspect of planning, it is the stage at which all the activitie of manufacture and despatch are co-ordinated and projected onto a time scale. B

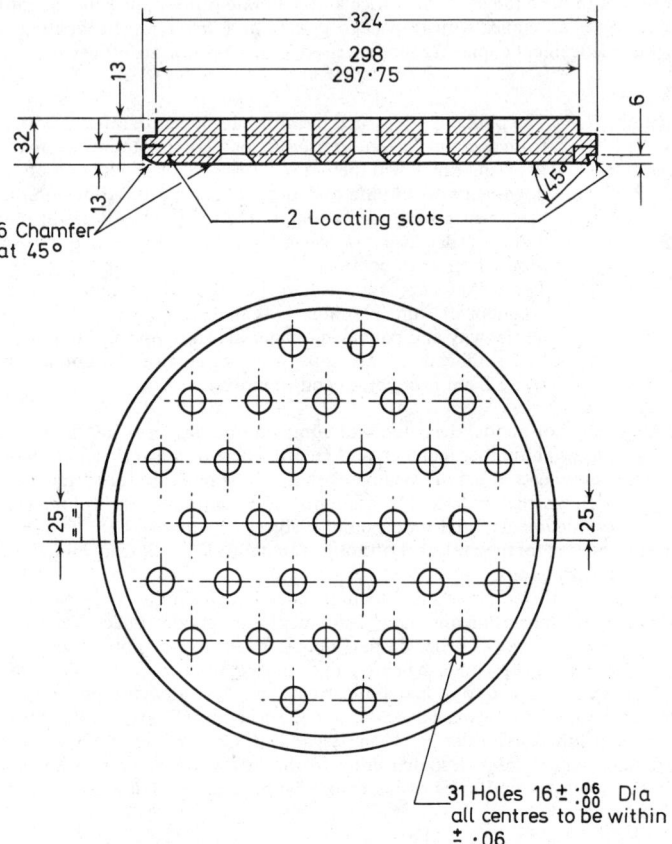

Figure 20.31. Component drawing and specification. Material: graphite. Dimensions in millimetres

referring to component drawings and their related specifications (Figure 20.31) operatio planning schedules and material parts list can be prepared, Tables 20.5 and 20.6.

Operation planning schedules describe the type of operation, sequence, machines an tools, and the materials that are to be used. They are in a form suitable for collatin operations and routes of components to that of final assembly and despatch. 'Th operation sheet therefore constitutes a general statement, about how in the view of th production planning department, the production sequence should be carried out.'[2]

Table 20.5 OPERATION PLANNING SCHEDULE

Component drawing No. SK. 17384 A
Storage code No.

Description top pouring plate (31 holes)
Material: graphite. Size 355 dia. × 102 mm long for 2 components

Date 29/8/56
Revised and retimed 22/8/70

Layout by RGN

	Description of operation	Machine	Special tools and production gauges	Special gauges (insp.)	Notes	Time target	Mins accept
				1 12/2/56 NW		2 29/8/56 SP	3
1	Bandsaw 102 mm billet from bar stock	Bandsaw				5·25	6·3
2	Chuck – face end	No. 10 or		SC (W) 33			
	Turn 324 mm dia. over whole length	No. 10/13		SC (W) 34		14·7	16·2
	Turn 298 mm dia. shoulder to length	Ward		SC (W) 41			
	Turn recess for saw guide				Do *not* c/sink		
	Drill 1 hole 16 mm dia. stage inspection					12·6	15·1
3	Reverse in chuck	No. 10 or					
	Grip on 324 mm dia. – face end	No. 10/13					
	Turn 324 mm dia. o/dia.	Ward					
	Turn 298 mm shoulder to length stage inspection						
4	Bandsaw to saw guide	Bandsaw				2·0	2·62
5	Chuck (3 jaw)	No. 10 or	Fixture 4357–2 F	SC (W) 36 AN		2·1	2·62
	Face to thickness and chamfer	No. 10/13	Plug gauge 4 357–4 G	SC (W) 36-42			
	Stage inspection	Ward	C/sink slip drill 4 357–3 T	SC (W) 32			
6	Drill 30 holes—16 mm dia. and c/sink	Multi-spindle	Drill jib 4 357–1 J			3·15	3·62
7	Mill 25 mm wide slots	Vert. mill	Fixture 4 357–5 F			4·2	5·25
	Final inspection						
					Total	44·1	51·71

Although it indicates the type of machines that could be used for the operation, it does not normally specify precisely to which individual machine in the plant the work should be allocated.

Such schedules may also comprehensively detail how the various parts are to be manufactured, operation by operation, and there will be a separate lay-out for each part. Normally, it will contain the following information:

The part number and the drawing number (if different).
The part number of the assembly on which it is required.
Page reference to master parts list.
Description of the part.
The quantity required for assembly.
Material specification and quantity required.
The operation number for each operation.
The department in which each operation is carried out.
Brief instructions as to how each operation is performed.
The tools required for each operation.
The type of labour required (skilled, semi-skilled, female, etc.).
The type of machine on which the work is to be carried out.
Time allowance.

Each operation in the manufacturing process is thus itemised and described in the operation lay-out. The lay-out performs a dual function. Firstly, it gives manufacturing

Table 20.6

Parts list			Ref. issue No.		
Assembly No.	*Description*	*Batch qty*	*No. per set*	*Used on*	*Date issued*
Part No.	*Part description*	*Finish*	*No. pr. st.*	*Batch total*	*Remarks*

and technical instructions to the factory, and secondly, it provides the information which is necessary for the progressing of production.

It can be ascertained from the lay-out: how long an operation is expected to take; the grade of labour required; and the type of machines which are to be used.

When the information is aggregated to cover the overall master programme, it is possible to tell whether sufficient machine and labour capacity is available in the factory; whether it is necessary to sub-contract on certain items; and in what respects excess capacity is available (if any).

The materials parts list

This generally tabulates all the details which on assembly make up the finished product. As far as possible details will be grouped in sub-assemblies and sub-assemblies into main assemblies. There will generally be a number of parts which are needed in more than one assembly, and these are known as common parts. These common parts will be shown on the parts list under each assembly in which they are required and one of the functions of the clerical production control staff is to isolate and aggregate these so that the total requirements are known.

The master parts list will generally provide the following information in columnar form:

Part number.
Quantity required for assembly.
Description of part.
Part number of assembly on which the item is required.
Material specification (for detail parts).
Dimensions of material required.
An indication as to whether the part is to be factory produced or sub-contracted.

From the parts list the following information can be extracted: how many of each part is to be manufactured and how much of all types of material will be required.

A 'delivery' programme is prepared for each bought-out and internally manufactured component part listed on the materials part list to provide some overall guidance to when an action should begin and end so that the component part is readily available as and when the assembly or despatch processes require it. Such a programme takes the form of a 'Gantt-Chart' from which all individual departmental shop programmes are derived.

The Gantt Chart (designed by Henry L. Gantt, one of the pioneers of scientific management) because of its ease of construction and presentation of facts in relation to time 'is one of the most notable contributions to the art of production management in this generation'.[26] In its original form the chart compares what is to be done with what was planned and indicates within specified time-checks when performance falls short of the plan. An example is shown in Figure 20.32. In constructing a Gantt Chart there are numerous ways of deciding how jobs may be loaded to gain the most benefit of resources available. For instance:

Jobs may be loaded on to machines in the order of priority indicated by the customer delivery dates, normally a constraint.
Jobs may be loaded on the principle of using each machine to the maximum and reducing idle time to a minimum, an objective of resource utilisation.
Jobs may be loaded in the priority indicated by the hours of work contained in the job. For example, job Y might consist of four operations totalling 30 hours; job Z might consist of five operations with a total work time of 42 hours. Adopting this principle job Z could be loaded before job Y, or vice versa.
Jobs may be loaded in reverse order using the completion or delivery date as the starting point. That is to say load the final operation to be completed on the delivery date (or very near to it) and load the remaining operations in preceding order.

Whatever the method used the advantages of such scheduling will control idle time on machines; indicate the extent to which completion dates will be achieved and the extent to which jobs may be completed too soon; this should not be encouraged otherwise it presents a problem of charges on warehousing and handling.

It was mentioned in the section dealing with the master parts list that the quantities of all types of materials required can be extracted from it. The way in which this can be done is to have an over-riding sheet on the Kardex stores record for each type of material on which the requirements of each component can be entered. The total requirements are then entered on the Kardex record proper and internal requisitions are sent to the purchasing department to cover these needs.

It is obviously not necessary or wise to buy all the requirements for a long term programme at once because it would tie up capital unnecessarily. The planning office, therefore, provides not only details of material requirements but also the dates on which they require delivery. The functions of material control and purchase department then become quite distinct in relation to the planning department. The material control section is responsible for stating its requirements to the purchase department and the purchase department is responsible for placing orders and progressing outside suppliers. This division of function should be strictly adhered to, for there is bound to be confusion where two separate departments are urging suppliers independently of each other.

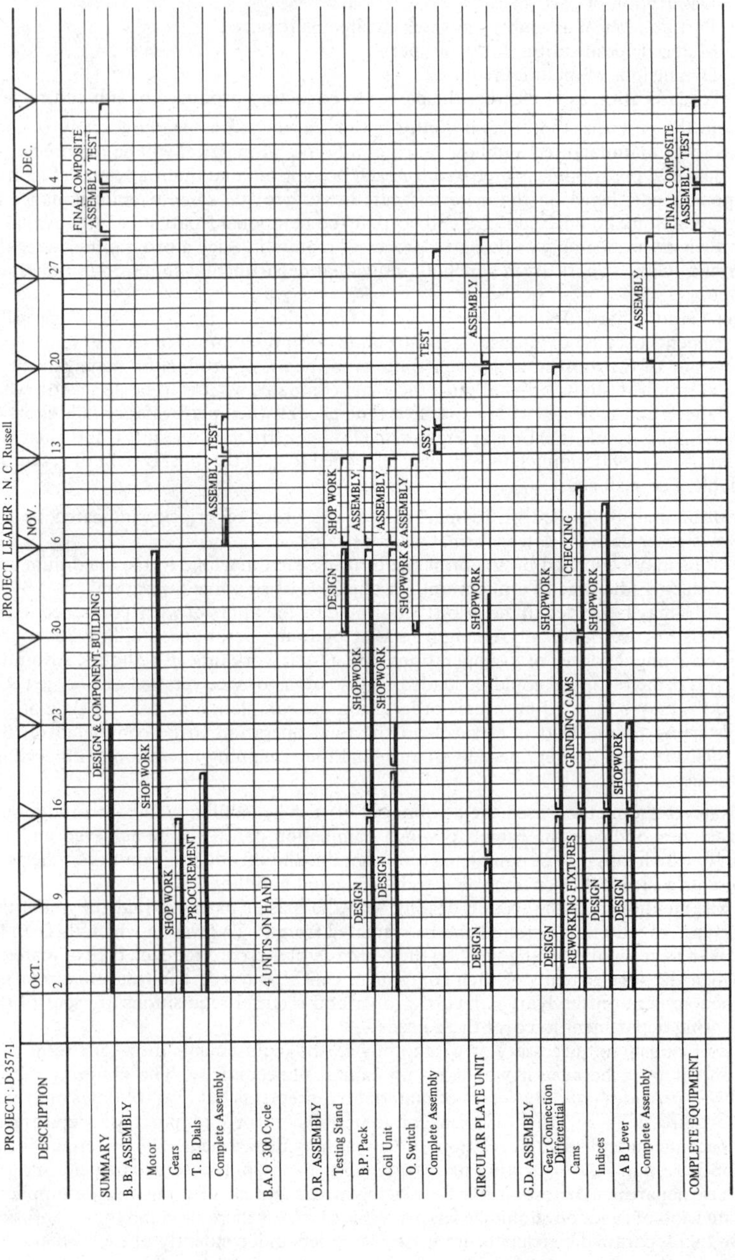

20–62

Figure 20.32. The Gantt Chart (courtesy Pitman & Co.)

Properly designed records in the material control section will enable them to prompt the purchasing office when stocks are dangerously low, but *before* they are exhausted. Also, properly designed records in the purchasing office will enable them to prompt a supplier when deliveries are overdue or in danger of being overdue. The material control department decides the intervals at which they require delivery of the various types of material by reference to the production schedule. A similar routine can be applied to tools required.

Shop instructions

Translating the schedules and lists into shop instructions is mainly a clerical function. In practice, it is at this stage that a certain amount of decentralisation is advisable, and

Figure 20.33. Ormig model AV10 machine

the hour-by-hour machine or operation loading and the actual issue of jobs to operators is done either in or adjacent to the foreman's office.

Typical of many production documentation systems is that it relies entirely upon manual operation. However, as many other documents are necessary for communication, control and progress, integrating all departments and sections concerned with the total manufacture and despatch, it is advisable and more economic to design important and necessary documentation around metal plate or spirit duplicator machines. Such a machine system is the Ormig system (Figure 20.33) by which a master is prepared and is so designed that the constant data is segregated from the variable data which is likewise segregated. In this way the part of the reproduction medium containing the constant information is retained whilst the other part containing the variable data is destroyed after the document set for the particular order has been produced. For repeat orders, only the variable information (date of order, quantity ordered, order number and the like) need be prepared.

A typical factory manufacturing order-set might comprise the following documents:

Operation layout master.
Operation layout cost control copy.
Operation layout progress copy.
Batch route card (which goes with the finished part to the stores and acts as the medium from which the central records are written up).
Material requisition (to authorise the issue of the appropriate material).
Identification card (to accompany and identify the actual work in the factory).
Job card for each operation (for bonus and payroll purposes).

The body of the manufacturing order is designed so that it accommodates the manufacturing instructions (as set out on the operations lay-out). Subject to the materials

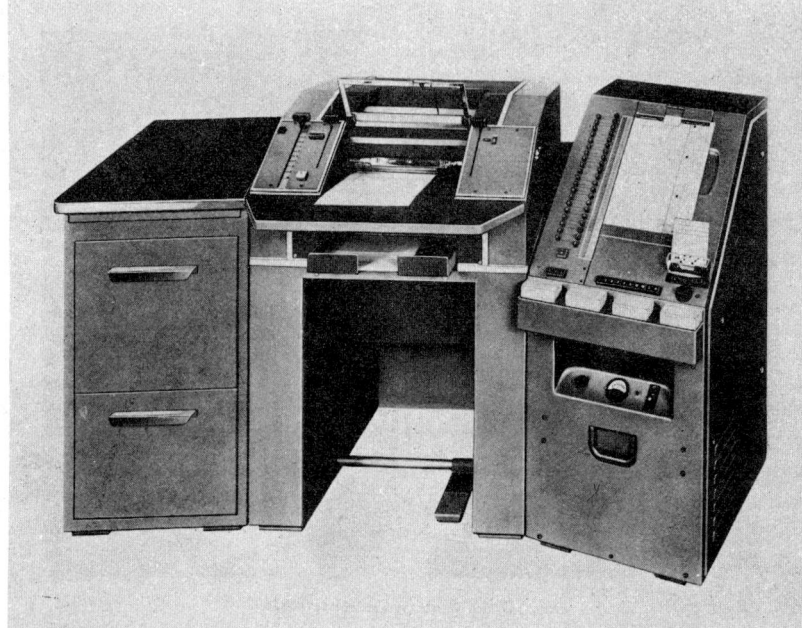

Figure 20.34. Ormig Selectronic N Series machine

being available manufacturing orders are issued some weeks before the programme and this is often done whenever the load of any particular shop becomes dangerously low.

Documentation in every organisation demands a great deal of writing, rewriting, typing and retyping of the whole or parts of the same information on different forms or documents. The risk of errors in transcription can be burdensome and expensive. These form filling problems can be simplified by using a machine which produces from one single typing operation all the necessary forms and documents in a fraction of the time, irrespective of how many different documents with different texts are needed. A typical example is the Ormig Selectronic N Series machine shown in Figure 20.34.

A typical application is invoicing where the order is typed once only and nothing more has to be typed again. The machine will produce complete documentation including: acknowledgement of orders, packing specifications, warehouse requisitions, receipt notes, packing labels, advice notes, delivery instructions, invoices, a list of items which have not been despatched. Documentation of part shipments against an order is obtained instantly and accurately without retyping.

The Ormig parts list schedule, which is typed once only can be used repeatedly for duplicating assembly programmes without retyping.

Production control

Production control is the specific application of recording results and correcting for deviations from planned programmes, schedules and standards. No matter how accurate the standards set or perfect the plans, all factors are not under the control of management, there are in practice wrong decisions and failures to achieve plans and reach standards of performance.

Any deviation from the plan must be investigated and reported to production management, so that action on the deviation can be immediate and not historical. To enable corrective action to be taken to limit and reduce the effect of the mistakes and failures and to prevent recurrence, production control concerns itself with progress and expediting.

PROGRESSING

The carrying out of the plan, which has been stated to be the progress responsibility, entails many duties. Chief of these is the discovery and correction of failures to fulfil the plan which will frequently be due to shortages and delays.

Strict adherence to a careful plan will reduce shortages to a minimum. It will not however eliminate them because however good routine machine maintenance may be, there will inevitably be breakdowns. Also, however close the shop supervision, there will always be scrap and wastage.

Progress clerks are generally responsible for seeing that material and tools are ready at the right time and at the right place so that, as soon as a job is allocated to a worker, all the pre-requisites are at hand for him to start it at once. This becomes little more than a routine matter if work is not released to the shops until material and tools are known to be available.

Visual aids, such as Sched-o-graph and Graphdex are popular in production control systems. Kardex is also used in many organisations where it can provide a more efficient way of doing the job with less clerical effort and in addition gives the control which is helpful and ensures that action is taken at the right time without the necessity to search through records or to keep a system of reminders.

Tool control

A comprehensive tool control scheme is described which covers the ordering and receiving of tools and their subsequent issue; the file is maintained in tool number order.

Figure 20.35 shows the split card which is used. The left-hand side gives a detail of tools ordered on suppliers and of deliveries received against those purchase orders.

The right-hand split card is a record of tools sent to be repaired. When a tool is sent for repair, the details (date, works order number and quantity) are entered on this card, but the quantity is not deducted from the stock balance figure on the insert card. When the tool has been repaired satisfactorily, this fact is recorded by entering the date in the 'date' column. The man or department who carried out the repair is entered in the appropriately headed column and the quantity in the 'to stock' column. If the tool is not satisfactorily repaired, the date and man or department who had the repair in hand are entered in the respective columns and the quantity concerned is entered in the 'to scrap' column. The stock record on the insert card is posted by entering the date and works order number, in the 'date' and 'reference' columns respectively, the quantity in the 'scrapped' column and this quantity is then deducted from the stock balance column.

Issues are dealt with in the following way. For each employee a record is created in the form of a turned-up master card. Each time an employee wishes to draw a tool from the tool stores, a tool requisition which is completed in duplicate, signed and handed to the storekeeper.

The storekeeper issues the tool, initials the requisition and files the original behind the stub of the turned-up master and hands the duplicate back to the man. The requisition is printed on coloured paper and this colour shows through the cut-out on the turned-up master and shows that the employee concerned has a tool on loan.

When the employee returns the tool, the storekeeper removes the original of the requisition from the employee's docket and signs the duplicate requisition as an acknowledgement that the tool has been returned.

If the tool is broken or lost, details are recorded on the employee's card and the original requisition is passed to the clerk maintaining the tool records where the loss or breakage is recorded on the stock card by a posting in the 'lost' column. This figure is, of course, deducted from the stock balance figure.

With this scheme there is a record card for each tool and this card gives details of the original purchases. Where, however, the tool has not been purchased but is on loan from another organisation the card gives full details of the loan. Each time the tool is issued to an operator details of the job for which it is required and of the operator borrowing it are entered on the card.

The signal scheme is very comprehensive. On the left a signal shows that a tool is not available for the following reasons:

O.S.   = With an outside supplier.
Design   = Design being modified.
T.R.M.   = In tool room for adjustment.

On the right a signal shows that tool is out or reserved for the following reasons:

Out   = Out on loan.
Called for   = Will be shortly required for an outside supplier.
Con.   = Sent to contractor.

INFORMATION FEEDBACK

Ideally, the control section should always be able to provide detailed information on the location and state of work of all the materials, equipment and labour in progress.
Method of feedback is as follows:

1. *Mechanical.* If a process is tied to a machine or conveyor, counting or recording devices may be used.
2. *Operator's work record.* The operator may be required to maintain a log, usually verified by the supervisor.
3. *Job ticket with detachable tickets.* Again the operator fills in the quantity and operations carried out. As each operation is completed the appropriate portion of the ticket is removed and either placed in a designated box or collected by the progress chaser.
4. *Walk and count.* This relies on the progress chaser walking round his own sphere of activity and counting the work he sees. Often resorted to and probably the most valuable check of progress.

5. *Bonus record*. If possible a bonus card should be used as this will tend to be filled in promptly and accurately.

One of the important features of the Gantt Chart (Figure 20.32) is that it can be used not only for recording a plan of work but also to show progress is made in completing the work. Progress is recorded by drawing a thick line between the legs of 'starting and finishing'.

A COMPLEMENT TO EDP

Production control creates the necessary conditions for production, calculation, cost accounting and planning of all types. When viewing the whole as *one* single data system within a large organisation, the central function of operation control becomes all too obvious. Proper storage media are required in order to maintain the efficiency and stability of such a system at all times.

A typical system is the Ormig which aims to offer a series of additions to the punched card technique, not only to cover the further expansion of dual purpose punched cards,

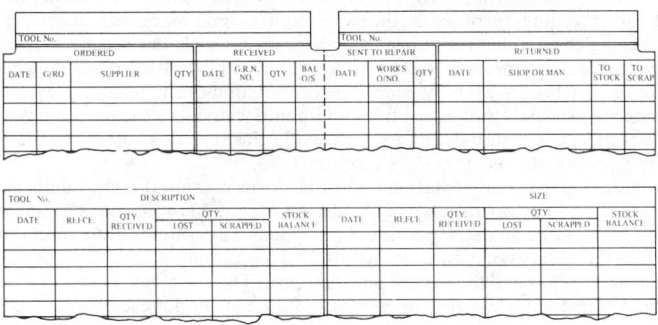

Figure 20.35 Tool control split card

but the production of continuous records capable of being extracted from tub files or external memory devices via tabulating machines and high speed printers.

Normally, works order data assembled from works orders, batch route cards, job cards, material requisitions and identity labels are forwarded in the form of single item tickets to the punched card section after completion of the job, and after costing (by posting the actual data) by the works office. The single item tickets are used as data preparation copies by card punch operators. In other cases the data is posted direct to dual purpose punched cards, thus dispensing with separate job tickets. The punched card then serves as a storage medium for production.

Other common methods are tub or disk files where cards are punched with the reference data by means of a reproducer or card punch. This produces a set of cards to be fed to the data processing system for mechanical calculation. When using these punched cards as storage media they have subsequently to be printed, or, at the very least made intelligible to the human agency with the aid of a card interpreter. This medium can be used in the reverse order, the punched cards being first line selected and subsequently punched with the reference data for the purpose of mechanical calculation.

All such methods call for at least two separate processes, and it is therefore, understandable that a large number of production controllers requires a system which combines line posting and card punching in one machine operation. Furthermore the desired machine has to form an integral part of production control and when producing the storage medium no reliance can be placed on the data processing system for information.

The Ormig system uses a machine that combines the following processes in one single operation. Complete copies such as works order, batch route cards and progress cards are reproduced. Punched cards in the form of material requisition and job tickets are line selected and punched *at the same time*. All relevant single item tickets are supplied by the machine in one batch with no subsequent separating and manual sorting.

The machine is located in the production control department which thus remains independent and flexible. At the same time, the punched cards thus produced are capable of further processing for shop loading or for a material requisition by the computer or the data processing department. The production control department is freed from time consuming routine calculation; see the flow diagram shown in Figure 20.36.

The further processing of the cards is not subject to the organisation having its own data processing installation, as an outside agency can be used. Even small companies can link up with an outside computer service or use a small computing installation at maximum cost efficiency. A schematic diagram for outside data processing is shown in Figure 20.37.

The whole system consists of two units. It will be seen from Figure 20.38 that the recording unit consists of an electric typewriter which is connected via an encoder with a punch for the master punched card. The numerical content (essential for data processing) of the planning layout is recorded on the left of the hectographic master. When depressing a numerical key the figure is taken over by a memory device in the punch and as the typewriter carriage moves forward the next line is punched into the master punched card. The figure is punched in a binary code. The next section of the planning layout master allows space for several typewritten lines and numerical values such as dimensions and reference numbers of materials can be included in this section as often as required, but they do not operate the memory device and punch.

When designing the documentation it must be borne in mind that the number of digits in the heading and any operation line should not exceed 20 as the master punched card has only a capacity of 80 digits per line. Four punching positions are required to show digits 0–9 in the binary code. In other words 20×4 positions equals 80 digits and this is the full capacity per line of the master punched card. The use of horizontal line punching makes for a tremendous increase in capacity of the master punched card. The heading of this master punched card carries all the detail of the planning layout master in the line selection process and is thus fully legible, which avoids the possibility of confusion. The hectographic master and the master punched card are always retained together when filed.

The second unit comprises a Line selection machine; Control cabinet; Memory device; Decoder; Card punch; Card conveyor; Master punched card reading device; Keyboard with digit indicator (to take the variable order data); and Built-in cabinet for printed forms.

The machine operates as follows. The variable data is set on the variable data printing units and on the keyboard, and the planning layout master is inserted on the drum of the machine. The master punched card is fitted into the reading device. The machine is switched to 'DUPLICATION' to produce full copies of the works order, batch route card and progress card. The machine is then reset to line selection and punching, and the punched cards are fed in the following sequence: material requisition; job tickets. All storage media drop out into a special tray in the front of the machine.

After having been line selected, each card is taken by a conveyor into the card punch where the following information is supplied by the memory device to the punch. The variable data fed to the keyboard. The heading read by the master punched card reading device (these values are retained by the memory device until the last card has been dealt with). The punched line, in the proper sequence, from the master punched card.

When the memory device has fed all values to the card punch, punching takes place *en bloc*. The punched card is then moved from the punching position and ejected forward into the tray while the next card is drawn into the punch. When all documents have been reproduced the planning layout master and the master punched card are filed together.

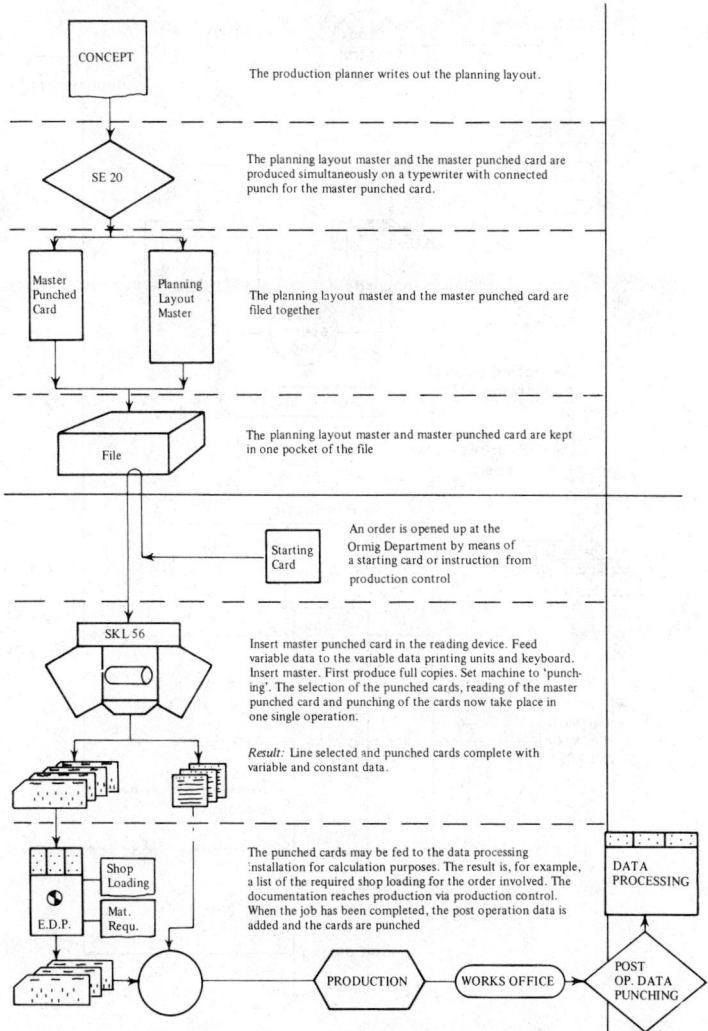

Figure 20.36. Information flow diagram

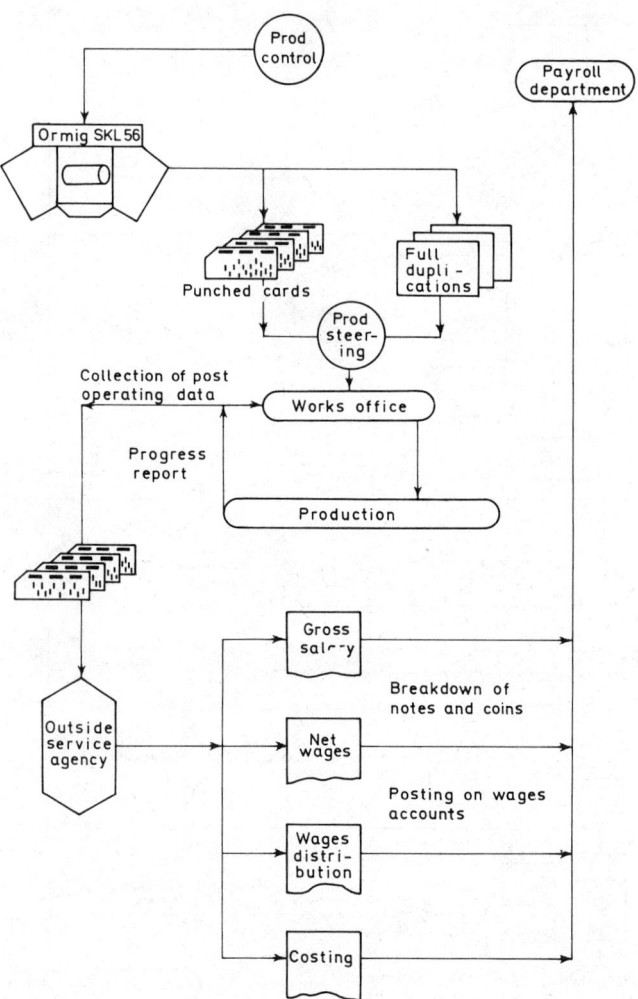

Figure 20.37. Outside data processing with Ormig SKL/56 without internal manual punching of post operation data

In the design of documentation it is in all instances advisable to draw up a card schedule. The layout and division of all forms must be identical and the first essential is the determining of the required positions for the numerical data. The separate recording method, for example, for drawing reference numbers for figure columns, must also be accurately determined. Line selection is then adapted to meet the prevailing practical

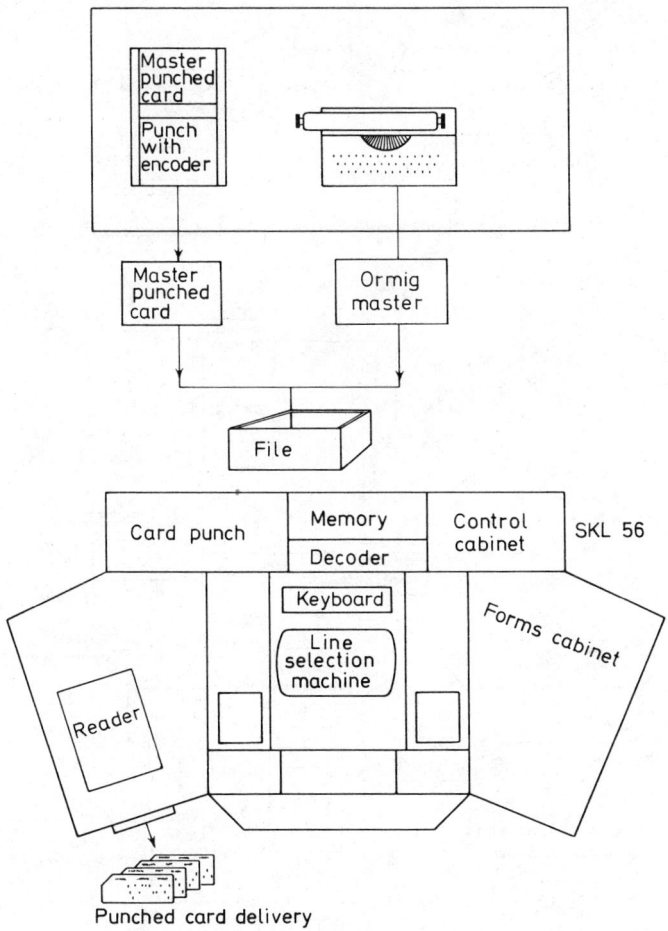

Figure 20.38. Ormig AV System 56/20 using SE20 and SKL56 units

requirements, the machines being subsequently adjusted in accordance with these requirements. Naturally the design of the documentation follows from this preliminary work. When typing a planning layout master punched card is, of course, produced at the same time.

The possibilities provided in co-operation with the data processing installation should be fully implemented as shown in Figure 20.39 and made use of in production control, but a clear distinction should be made between the production and storage media on one hand and the processing of data with expensive and complicated machinery on the other. The full transfer of production planning information to external memory devices involves considerable expenditure and unless results thus obtained represent a great improvement on the inexpensive line selection processes, its use should be carefully and critically examined.

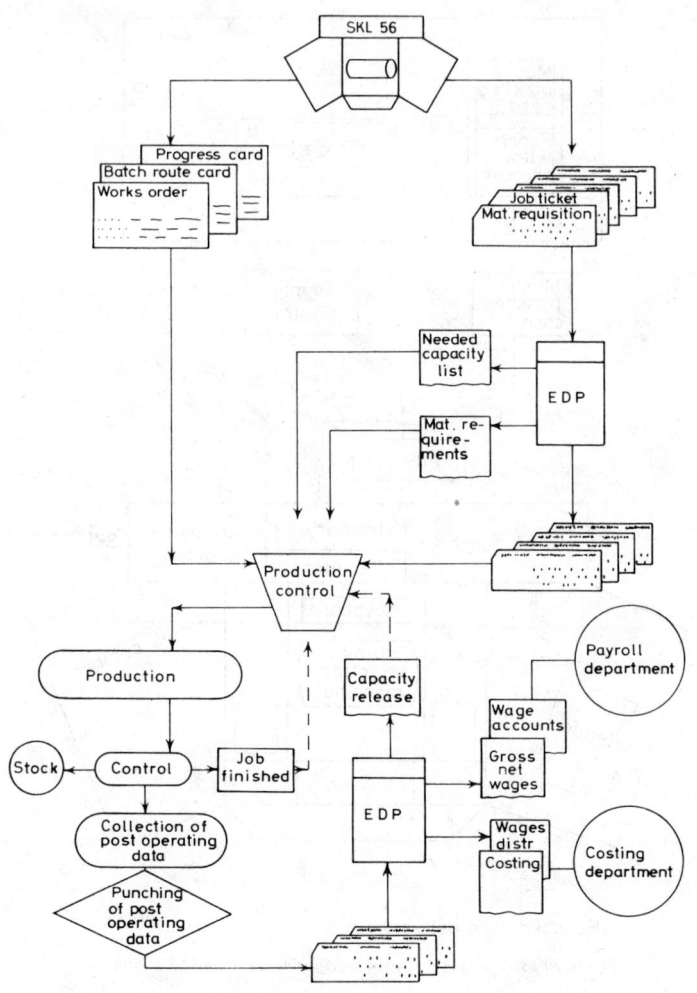

Figure 20.39. Ormig AV System 56/20 a peripheral unit of EDP

Another version is the Ormig SKL 56NL which, instead of the memory punch card, has a paper tape reader for the direct input of paper tape created by systems typewriters. By means of various programme codes, the numeric data of the paper tape are read into the machine and punched into dual purpose cards (80 columns). Paper tapes can be used which fully contain the production plan.

The models described above take numeric programmes only but research is at present being carried out to bring alpha into the scope of the system.

NETWORK ANALYSIS

The Gantt Chart is of limited value when planning and controlling a large project such as construction, maintenance or the manufacture of a complex piece of equipment. It is impossible to deduce from an orthodox Gantt Chart whether 'activity A' must be completed before 'activity B' can start, or that delays between 'activity Y' and 'activity Z' are permissible, but not essential. In such project work the main tasks to be completed are:

1. Determination and specification of each activity concerned in the project.
2. Describing the order in which the activities can be carried out.
3. Preparing a time schedule for each activity.
4. Determining a completion date for the project which will allow high utilisation of resources available and satisfy the customer's wishes.
5. Finding the activities which are important or critical if the completion date is to be met.
6. Controlling the project through its various stages so that its completion date is kept and that resources are economically utilised.

Network analysis is a general term which is used to embrace a whole series of similar planning methods dealing with project control usually denoted by the initial letters of the words in their title, e.g. Critical Path Analysis (CPA); Project Evaluation and Review Technique (PERT). There are numerous systems some of which are listed below:

PERT	Programme Evaluation and Review Technique.
CPS	Critical Path Scheduling.
CPM	Critical Path Method.
CPA	Critical Path Analysis.
LESS	Least Cost Estimating and Scheduling System.
PEP	Programme Evaluation Procedure.
SCANS	Scheduling and Control by Automated Network Systems.
RAMPS	Resource Allocation and Multi-project Scheduling.
WASP	Workshop Analysis Scheduling Programme.

Initially there are significant differences in the use of these techniques but there has been considerable advances in each technique and the differences now are marginal. Network analysis can be carried out manually but is most useful when used in conjunction with a computer. The use of a computer allows resources allocation and minimisation of project cost techniques to be used on projects which would be tedious and complex when carried out manually.

A network depicts the relationship between activities necessary for the completion of a project and by analysing the network the time boundaries of each activity in the project can be determined. The analysis will also reveal the critical or important tasks that have to be completed if a delivery date is to be achieved.[27, 28]

There are three main phases of network analysis:

1. *Planning* or preparation of activity lists, finding the restrictions imposed on each activity and finally the drawing of the network or arrow diagram.
2. *Analysing and scheduling* determining the shortest time to complete the project and the critical activities leading up to that completion date. Preparation of time boundaries for each activity within the project.
3. *Controlling* the project by checking the actual performance on each activity against scheduled and adjusting the schedules to meet any new requirements.

The initial step is to break the project down into its component operations to form a complete list of essential activities known as the 'Activity List'. Activities are often defined to suit planning purposes, e.g. if certain types of labour or specialised machines are in short supply these activities are separated. But the list of activities normally evolves from the general order of completion. In order to evolve a time sequence of activities and their precedence, each activity is subjected to the following questions:

What activities must precede it?
What activities must follow it?
What activities can be done at the same time?

EXAMPLE

Description	Symbol	Activity relationship		Restriction list
		Prerequisite	Postrequisite	
Design	A		Order parts	A < BC
Order parts	B		Scheduling	B < D
Scheduling	C	Design		< precedes

Arrow diagram

A network diagram[29, 32] is made up of two basic elements. An event which must be of a definite recognisable nature and must be a point in time and an activity which is the 'work' or job which leads up to an event, e.g. walk from A to B taking two hours.

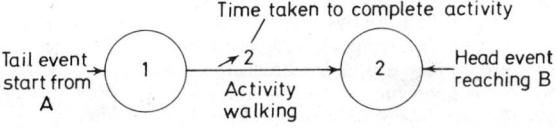

A dummy activity, represented by a dashed line arrow is used in network to show the dependency of one activity on another. They have the same properties of regular activities but require zero time.

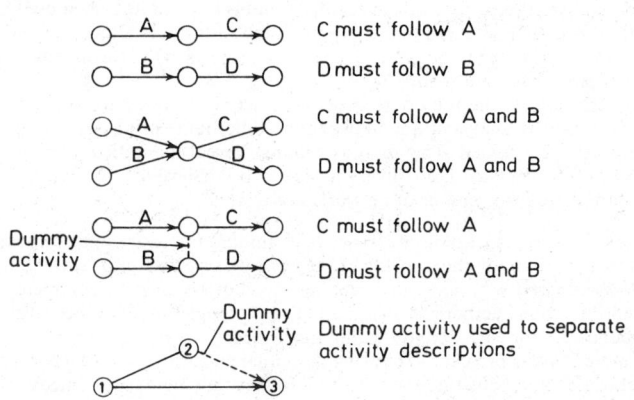

Real networks, however detailed and complex, are merely combinations of the four basic patterns created by simple sequence; divergence; convergence and dummy arrows. It is usual to identify the beginning and end of a network by special activities called 'lead time' and 'end time' activities.

EXAMPLE

There are six activities: A, B, C, D, E, and F.

> E follows B.
> D follows A.
> F follows D and completes the project.
> A is an initial job.
> E follows C.
> C is an initial job.
> F follows E.
> B can start at the same time as A.

The procedure to reconstruct the network is as follows and the solution is shown in Figure 20.40.

> Rearrange data
> 1. A, B, C are initial activities.
> 2. D follows A.
> 3. E follows B, C.
> 4. F follows D, E and completes project.
> Draw the lead time and three initial activities and label them.
> Draw item 2 as this is a simple sequence.
> E cannot follow both B and C unless a dummy is inserted.

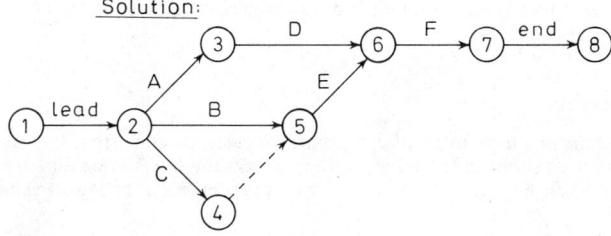

Figure 20.40

Insert dummy. Draw E leaving B or C, depending on the direction in which the dummy arrow points. In this example either direction is permissible.
F follows D and E. Both D and E can terminate in the same node as they do not start at the same node.

Rules for construction of networks

1. Each activity must be uniquely identified by its two end nodes.
2. Each node must have at least one entering and departing arrow except that the lead node and the final node will have only one departing and one entering arrow.
3. Each arrow must point towards its end node.

4. Each node must be numbered uniquely so that each 'head' node is greater than its 'tail' node(s).
5. Each activity must appear once only—this is a consequence of the first two rules.

Numbering networks

A simple four point procedure for numbering nodes, known as Fulkerson's rule, is quoted by Battersby.[28]

To locate the critical path the 'earliest' and 'latest' times must be calculated noting events which have no slack. The first step is to evaluate the duration of each activity and mark this on the diagram. Then calculate earliest event times by the following rule: 'To the earliest times of each immediately preceding event add the duration of the job which connect it and select the highest of the values obtained.'

To calculate latest event times the following rule should be adopted. 'From the latest time of each immediately succeeding event subtract the duration of the job which connects it and select the lowest of the values obtained.'

Battersby suggests showing the earliest event times in ☐ boxes and latest event times in ▽ boxes, above or below the relevant events. Other methods are also available, for example ◒.

The critical path becomes immediately apparent on subtracting the earliest from the latest event times.

Terminology

EARLIEST EVENT TIME

The earliest time at which an event can occur. To calculate, start from the first event giving it zero and proceed to each event in order adding the duration time to each preceding event. If several activities lead into an event, the earliest time is fixed by the *longest chain*. Earliest completion time of the project is obtained from the earliest time of the last event.

LATEST EVENT TIME

The latest time at which an event can occur. To calculate start from the earliest completion time and subtract the duration time obtain the latest event time for the preceding event. When several activities lead into an event select the lowest value.

CRITICAL PATH

The set of activities forming a continuous path which determines the final completion time of the project and whose activities have zero float.

ACTIVITY TIMES

Activities cannot start until their tail events are complete and must finish at the head event, the head and tail event time fix the boundaries for activity times.

EARLIEST START TIME

Obtain from the earliest time of the activity tail event.

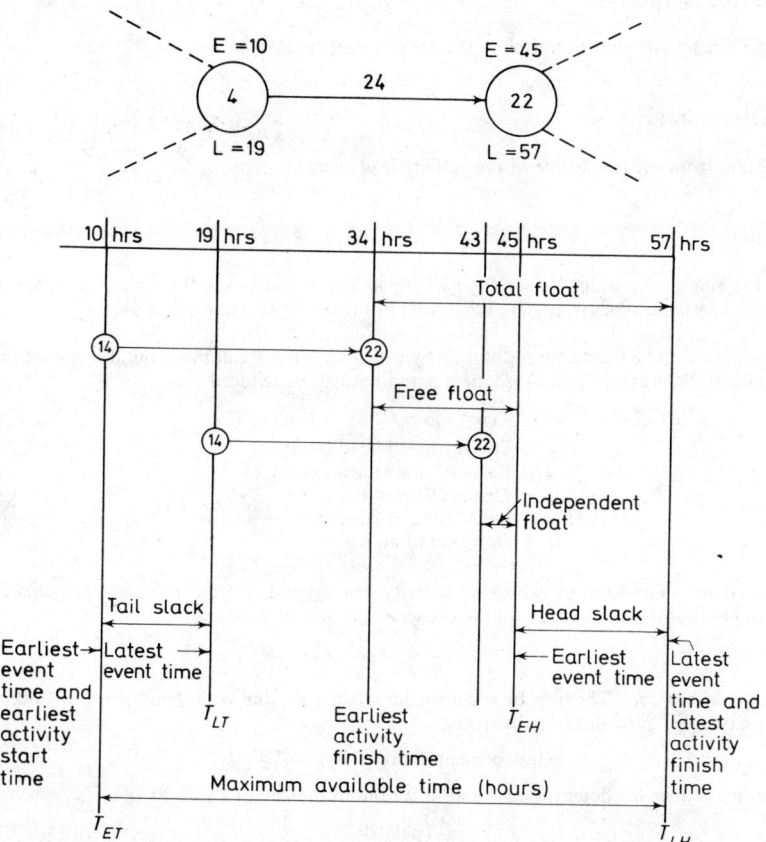

Using this comparison as an example the calculations for the floats in activity 14–22 would be as follows:

Total float = 57 − 10 − 24 = 23
Free float = 45 − 10 − 24 = 11
Independent float = 45 − 19 − 24 = 2

Figure 20.41. Graphical comparison of float

LATEST START TIME

Obtain from subtracting the activity duration time from the latest finish time.

EARLIEST FINISH TIME

Add the activity duration time to the earliest start time.

LATEST FINISH TIME

Obtain from the latest time of the activity lead event.

FLOAT

Is the amount by which an activity can 'move' without affecting the earliest completion date. The main types in use are illustrated in Figure 20.41 and defined as:

Total float. The total amount of time by which an activity can move but if it is used the floats in the previous and subsequent activities may be reduced.

$$\text{Total float} = T_{LH} - T_{ET} - D$$

$$\begin{aligned} T_{LH} &= \text{Latest time at head event} \\ T_{EH} &= \text{Earliest time at head event} \\ T_{ET} &= \text{Earliest time at tail event} \\ T_{LT} &= \text{Latest time at tail event} \\ D &= \text{Duration of activity.} \end{aligned}$$

Free float. The time by which an activity can expand without affecting subsequent activities but if used the float in the earlier stages will be reduced.

$$\text{Free float} = T_{EH} - T_{ET} - D$$

Independent float. The time by which an activity can expand without affecting any other activity either previous or subsequent.

$$\text{Independent float} = T_{EH} - T_{LT} - D$$

The procedure for determining 'float' is set out in the key to Table 20.7.

Table 20.7

Job			Earliest		Latest		Total float
Beginning event	Ending event	Dura-tion	Start	Finish	Start	Finish	
A	B	C	D as calculated	E = C + D	F = G − C	G as calculated	H = F − D = G − E

EXAMPLE

The table on p. **20**–79 lists the activities which together constitute a small engineering project. The table indicates the precedence and restrictions on each activity and Figure 20.42 shows the network diagram obtained for the project.

Activities for small engineering project

Activity	Restriction	Activity duration (Days)
A	A < B, C	4
B	B < D	5
C	C < E	6
D	D < F	5
E	E < F	10
F	No restriction	7

Schedule of activity times and floats

Activity	Duration	Start		Finish		Float		
		E	L	E	L	Total	Free	Ind.
*1–2 (A)	4	0	0	4	4	0	0	0
2–3 (B)	5	4	10	9	15	6	0	0
*2–4 (C)	6	4	4	10	10	0	0	0
3–5 (D)	5	9	15	14	20	6	6	0
*4–5 (E)	10	10	10	20	20	0	0	0
*5–6 (F)	7	20	20	27	27	0	0	0

*Critical activities

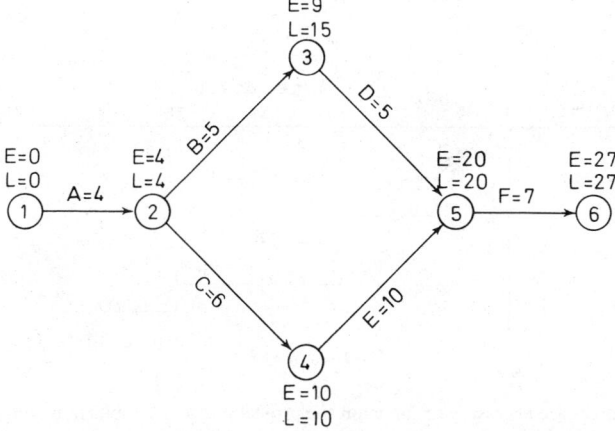

Figure 20.42. Network diagram

The calculations show that the critical path for this project will be through activities A, C, E and F. Earliest completion date for the project from zero start date will be 27 days.

In the above network if activity B was delayed and took 6 days longer than originally forecasted there would be no change in the earliest completion date for the project although activity D would now be critical.

If activity C was delayed by 6 days the project would take 6 days longer to complete, i.e. 33 days.

Scheduling

When the activity durations have been entered on the network and the boundary times for each event and activity calculated, scheduling may be arranged in tabular form as shown in the example or by visual means such as a Sched-U-Graph planning board which is manufactured by Remington Rand.

Resource allocation

Resources employed on a project may be limited, i.e. men or equipment may be restricted either by shortage of necessary skills or lack of capital. To economically carry out a project the use of available resources must be controlled to give the best utilisation. If

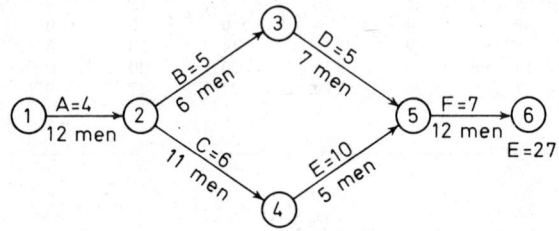

ACTIVITY	D	Time (days)
* A	4	[1 12 men 2]
B	5	[2 6 men 3]
* C	6	[2 11 men 4]
D	5	[3 7 men 5]
* E	10	[4 5 men 5]
* F	7	[5 12 men 6]

Figure 20.43

extra resources are necessary to carry out the project it usually results in paying overtime rates for men or hiring extra equipment at high short term loan costs.

Resources should be allocated by scheduling in such a way as to eliminate any sudden increases or decreases in requirements.[31]

The work involved in each activity may be optimised by the use of work study techniques to evaluate the best method, although the greatest saving in completion time and cost, is obtained by concentrating on the critical path activities.

To allocate the resources effectively the following procedure should be observed:

1. Obtain activity work load, assuming normal resource allocation of men and equipment.

HISTOGRAM

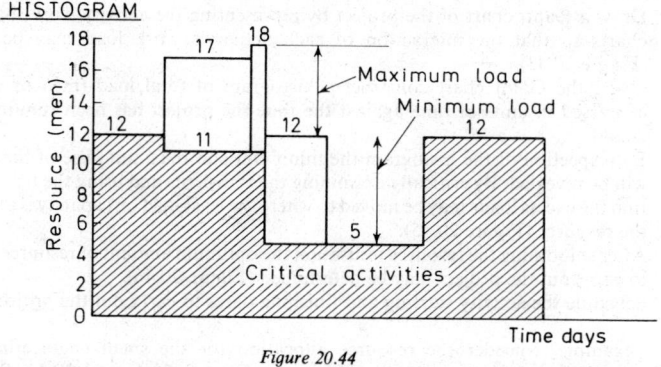

Figure 20.44

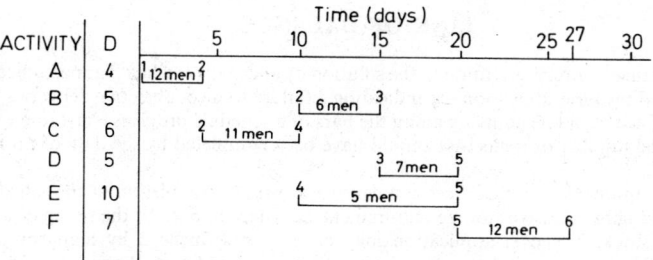

Figure 20.45. Revised Gantt Chart

HISTOGRAM

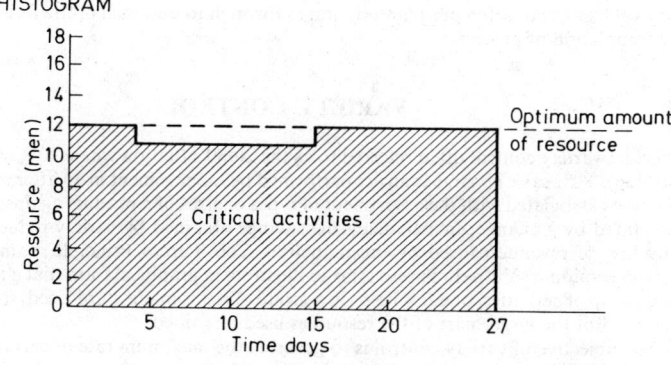

Figure 20.46

2. Draw a Gantt chart of the project by representing the activity work loads as bar charts so that the interaction of each activity's work load may be obtained (Figure 20.43).
3. Using the Gantt chart construct a histogram of total load (men or machines) expressed in units of time against the time the project has been running (Figure 20.44).
4. By inspection of the histogram the minimum and maximum use of the resources will be revealed. By critically examining the transients and using the float information the excess loads may be moved to where the least load on capacity is and smooth the resource (Figure 20.45).
5. After smoothing the resource utilisation the optimum amount of resource necessary to carry out the project may be determined (Figure 20.46).
6. Schedule the activity starting and finishing dates to carry out the project.

For example, consider the resource allocation for the small engineering project previously analysed (Figure 20.42) and prepare Figures 20.43 and 20.44. By moving activities B and D as shown in the revised Gantt chart (Figure 20.45) a smoother resource allocation is obtained (Figure 20.46).

LINE OF BALANCE

Line of balance[32] directs attention to the solution by alternatives when planned schedules of work and resource allocation are indicating a failure to meet objective deliveries. The term is defined as 'a line coursing along the bars of a schedule progress chart to indicate the expected number of items that should have been completed by the date of progress inspection'.

An over-application of resources can cause a too-rapid completion of the scheduled work resulting in excessive storage requirement and more important the tie-up of useful capital in stock. The over-application may have been influenced by many events— shortage of materials and tools, etc. necessary to start or continue other work; or it may have been considered that its completion before the scheduled date may allow resources to be diverted to logging events at a later date. Whatever the cause, without careful examination and forethought production problems gather momentum and the imbalance between resources and work availability increases.

The similarity between line of balance and network analysis suggests a marriage of the two techniques—resource allocation. The planning merits of network analysis integrated with the control features of a line of balance provides a routine for production control from that of the pre-planning stages through to despatch of the finished product or completion of project.

VARIETY CONTROL

Product variety control often referred to as the 20/80 situation where 20% of production produces 80% sales revenue is an act performed by management in an attempt to balance the costs associated with the assortment of types or sizes of a product against the revenue generated by a change; or balancing the savings obtained by variety reduction against any loss of revenue caused by the reduction. The 'Pareto Principle' named after the Italian economist Vilfredo Pareto aims to illustrate in graphical form that a few activities in a group of activities, or a few items in a group of items made, purchased, sold or stored, account for the larger part of the resources used or gained.

The objective of variety control is to generate the maximum rate of earnings from the plant by the determination of the optimum policy between variety reduction (which is the process of eliminating excessive diversity) and diversification. Whilst most industries appear to favour variety reduction without diversification it is important that the two aspects should be examined simultaneously. As variety reduction (or control) may release or create idle capacity by eliminating unpopular products and reduce product machine set-up and change-over times, most of the released capacity will be in the form of 'fixed' cost plant and equipment that cannot be easily disposed of, obviously this would be essential to indicate a reduction in plant overheads. Therefore the creation of

Table 20.8   SALES VOLUME VS. NUMBER OF PRODUCTS

Sales	A	B	C	D	E	Total
			Product group			
No. of products	120	372	163	107	320	1 082
Total sales	642 644	607 264	230 244	615 796	332 737	£2 428 675
over £10 000 sales						
No. of products	21	13	6	11	6	57
% of products	17·5	3·5	3·7	10·3	1·9	5·2
Sales	489 773	227 646	101 705	486 215	101 159	£1 406 498
% of sales	76·2	37·5	44·2	79·0	30·4	58·0
£3 000–£9 999 sales						
No. of products	17	36	11	16	29	109
% of products	14·1	9·8	6·8	15·0	9·1	10·1
Sales	99 509	213 004	62 478	88 067	106 133	£569 193
% of sales	15·4	35·1	27·1	14·2	31·9	23·4
over £3 000 sales						
% of products	31·6	13·2	10·5	25·3	11·0	15·3
% of sales	91·6	72·6	71·6	93·2	62·3	81·4
£100–£2 999 sales						
No. of products	34	191	89	44	145	503
% of products	28·4	51·3	54·5	41·1	45·3	46·5
Sales	52 203	163 373	65 240	40 856	120 984	£442 661
0–£99 sales						
No. of products	48	132	57	36	140	413
% of products	40·0	35·5	35·0	33·6	43·7	38·2
Sales	1 159	3 239	821	658	4 446	£10 323
% of sales	0·81	0·53	0·36	0·11	1·34	0·4
Products with no sales	10	54	31	14	16	127
% of products with no sales	8·3	14·5	19·0	13·1	5·0	11·7
£0–£2 999 sales						
% No. of products	68·4	86·8	89·5	74·7	89·0	84·7
% of total sales	18·4	27·4	28·7	6·8	37·7	18·6

this idle capacity should in turn be utilised by diversification in the wake of a variety reduction exercise.

Product variety is one of the major factors influencing tied-up capital, and the frequency of set-up times.

In a research study carried out in the West Midlands manufacturing industries in 1968,[11] the author together with a colleague, Simcha Bahiri, found that the iron and steel, and electrical industries between them had a product variety of 32 000, whilst the remaining industries of non-ferrous fasteners and motor vehicle components had 15 000 varieties.

In the fastener industry a major problem of variety is the multiplicity of threads, and each attempt to standardise would only add new items to existing lines, rather than eliminate specific thread forms. The introduction of the new metric thread forms (an original attempt to standardise) has shown this to be true.

In one electrical firm employing over 2000 people, five main product groups (10% of the total product groups in the company) representing over 33% of the annual sales revenue, were examined. Table 20.8 shows that 38·2% of the products (items with less than £100 sales per annum) generated only 0·4% of the revenue. For ease of analysis the data was developed by grouping the sales value of each product group into four groups, namely:

> Over £10 000 per annum.
> Over £3000 but less than £10 000 per annum.
> Over £100 but less than £3000 per annum.
> Less than £100 per annum.

Figure 20.47 shows a Pareto graph of the four groups. In the most popular group (£10 000) 5·2% of the number of products contributed 58% of the total sales revenue; and 15·3% of the number of products (£3000) contributed 81·4% of the sales revenue.

Such analyses however only provide one small part of the 'optimised' criteria necessary for effective variety and inventory control. What is being said is that the analysis

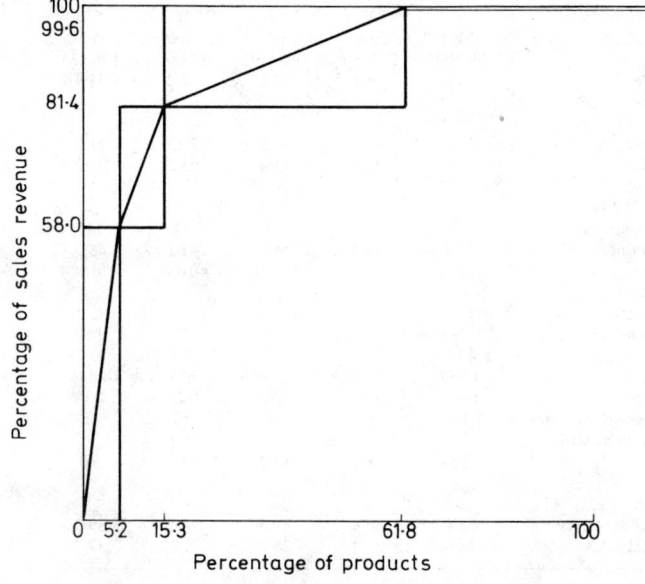

Figure 20.47. Number of products versus sales volume

of sales turnover to quantities produced may be influenced strongly by selling price of product and therefore will not provide a reliable controlling element. Further Pareto analyses need to be made relating production cost and throughput times (turnover times in respect of inventory) to the quantities or items produced. From this trio-analysis an optimised and reliable condition of control can be effected.

Standardisation of products also present similar problems as variety control. Other desirable company objectives highly related to variety control are the reduction in variety and standardisation, the main advantage being that capital tied-up in stocks and work-in progress can be controlled effectively and a fair coverage of a large number of items of plant and equipment can be given by the holding of a small number of spare 'standard' parts, the locations of which are in closer proximity.

A direct example of the benefits of standardisation in the maintenance field, for instance, is the reduction in the variety of lubricants of identical specifications. Purchasing and documentation handling are simplified, savings by bulk purchasing are more readily available.

INVENTORY MANAGEMENT

It is inefficient to devote an equal amount of time and attention to inconsequential items and more vital supplies. Even in a small company the number of items held in stock is likely to be large, say 1000–2000 different varieties, whilst in large companies this number can be very much greater say 5000–20 000 items (not taking into consideration in some cases dimension variance and colour). It might therefore be assumed that the cost of controlling accurately all these items will be high. Some considerable relief can be obtained by using a similar analysis to that suggested in variety control, i.e. Pareto graph. Its application to inventory policy recognises that a small number of production supplies account for the bulk of the total value used.

In a practical situation list the total values of all the stock of all items held in descending order of magnitude of total value; rank the items in order, the biggest value having rank 1, prepare a cumulative table of total values, as follows:

| Part No. | Total value £ | | Rank |
	Individual	Cumulative	
A164	14 260		1
A784	14 000	28 260	2
A326	13 650	41 910	3
A839	13 210	55 120	4
A004	13 050	68 170	5
A453	2 652	506 260	207
A174	104	1 104 000	2 000
and so on.			

From the above table it will be readily possible to calculate what proportion of the total value is attributable to each item of stock, or since this will produce too many figures to deal with, it is usually simpler to calculate the percentage value of stock for each percentile of the number of items of stock as ranked. This will give:

PERCENTAGE OF RANGE COMPARED WITH PERCENTAGE OF VALUE

| Items of range | | Value in % of total | |
Individual	Cumulative	Individual	Cumulative
First 5%		30%	
Next 5%	10%	20%	50%
Next 10%	20%	30%	80%
Next 10%	30%	8%	88%
Next 20%	50%	9%	97%
Next 50%	100%	3%	100%

The implication of this analysis is clear (Figure 20.48); tight control of the first 20% of the range will control 80% of the value of the stock investment. Thus, if there are 5000 different part numbers in stock, control of only 1000 specific items will control the bulk (80% of stock value) and therefore the control problem is grossly reduced.

Inventory control[34, 35, 36]

When supply and demand for goods are not equal over short periods of time, inventories of goods are held to balance these transients. Thus raw materials, goods being processed, finished articles and spare parts for machines may be stocked. When these stocks occur

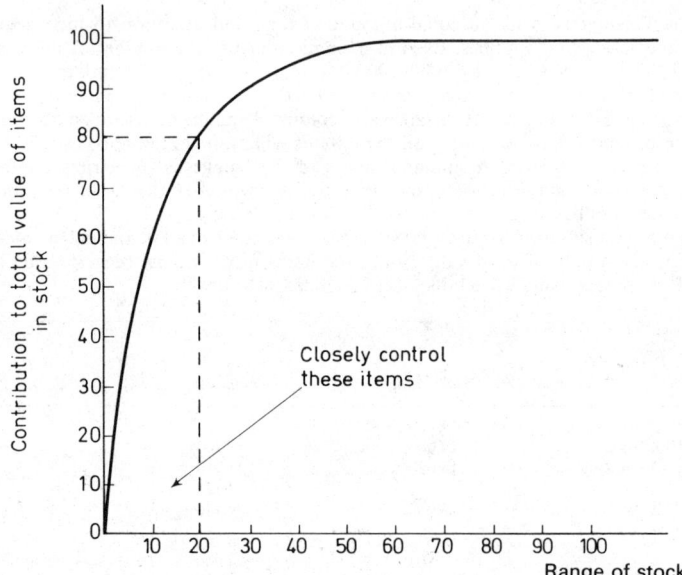

Figure 20.48. Range related to per cent of value

they contribute to the production costs in tied-up capital, storage costs, insurance charges and some wastage costs. These costs may amount to between 10% and 30% of the value of stock held. It is obviously economic to keep stocks as low as possible consistent with satisfying most demands for the goods involved.

In many companies, inventory policy emerges as the balance of view of a number of conflicting interests. The opposing nature of preferences is shown below.

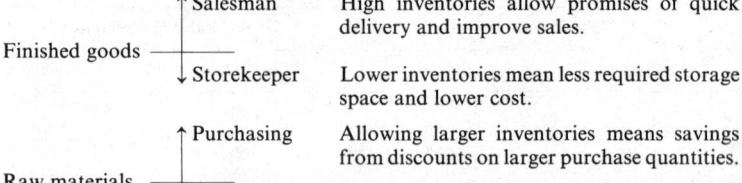

	↑ Salesman	High inventories allow promises of quick delivery and improve sales.
Finished goods		
	↓ Storekeeper	Lower inventories mean less required storage space and lower cost.
	↑ Purchasing	Allowing larger inventories means savings from discounts on larger purchase quantities.
Raw materials		
	↓ Accountant	A large inventory means more capital tied-up in investment in materials.

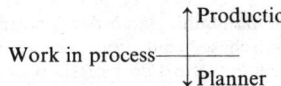

Work in process

Large inventories allow long production runs and reduction in set-up costs.

To ensure delivery dates are achieved and manufacturing capacity fully utilised, batch quantities must keep moving through the process and inventories kept as low as possible.

Purchasing policy is intimately related to inventory policy. Prices of materials tend to fluctuate about a trend line and a purchasing policy to buy materials in constant use

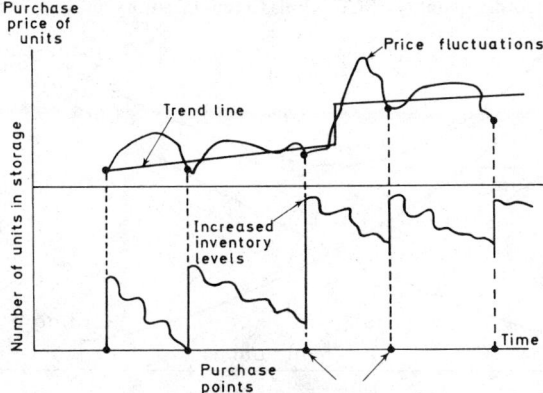

Figure 20.49. Price fluctuations of materials

when the price is below the trend line is often used (Figure 20.49). Such policy will increase inventory levels.

Inventory costs

In order to determine optimum inventory policy the following costs associated with inventory must be obtained.

Price (P).   The value of item or unit or material is its purchase price, or its production cost if produced internally. Quantity discounts can alter the price of external purchases and internal produced items will vary as value is added as they progress through the production cycle.

Holding cost (C_1).   This is the cost of holding items or materials in stock and is dependent upon the materials or items stored, their rate of obsolescence, methods of handling in stores, and the interest charges on capital tied-up by the inventory. Typical costs of holding stock are expressed as a percentage of cost of item or unit of material as follows:

Expected return on investment	10%
Interest on borrowed money	8%
Rent of stores	1%
Cost of stock recording	2%
Insurance of stock	1%
Cost of storekeeping	1%
Losses due to pilfering and wastage	3%
	26%

Order cost (C_2).   Order costs include the fixed cost of maintaining an order department and the variable costs of preparing and executing purchase requisitions. Set-up costs incurred in preparing for a production run and include clerical cost of shop orders, scheduling and expediting. Order costs remain relatively constant irrespective of the quantity involved in the order.

Inventory models

The problems raised by inventory control are; when to order; that is, to determine the reorder level or the interval of time between orders and how much to order; that is, to determine the reorder quantity or a fixed maximum inventory level.

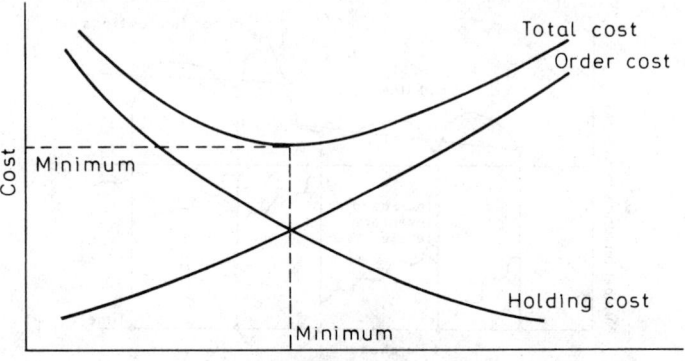

Figure 20.50. Cost of item against quantity

If the costs of inventory are considered it is found that they fall into two categories; holding costs which vary with the size of inventory and order costs which vary with the number of orders. The cost of the material or item is assumed to be constant.

If the total cost of the item or material is plotted against the order quantity a minimum cost point will be found indicating an order quantity that gives minimum cost (Figure 20.50).

ASSUMING CERTAINTY

The following assumptions are used in this model. Demand is known for the item, or can be accurately forecast. Delivery of the order is instantaneous or that lead time from when an order is placed to when it is received is known and is constant.

ECONOMIC ORDER QUANTITY (Q)

Refer to Figures 20.51 and 20.52.

Average inventory level $= \dfrac{Q}{2}$

Demand per year is D

Number of orders per year is $\dfrac{D}{Q}$

Holding costs per unit per year is C_1

Order costs are C_2

Total expected cost $= C_1 \dfrac{Q}{2} + C_2 \dfrac{D}{Q} + DP.$

The total cost is minimised by differentiating with respect to Q.

Thus
$$\frac{d(\text{total cost})}{dQ} = \frac{C_1}{2} - \frac{C_2 D_2}{Q}$$

For minimum cost

$$\frac{d(\text{total cost})}{dQ} = 0 = \frac{C_1}{2} - \frac{C_2 D}{Q^2}$$

$$Q = \sqrt{\frac{2C_2 D}{C_1}}$$

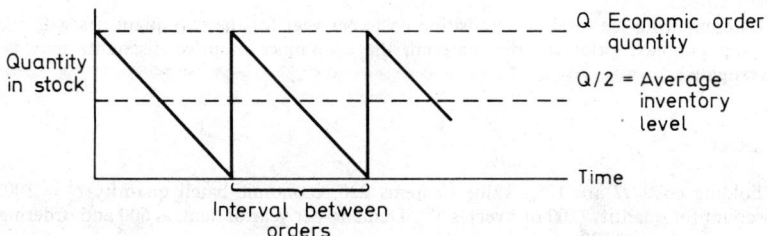

Figure 20.51. Economic order quantity governed by demand

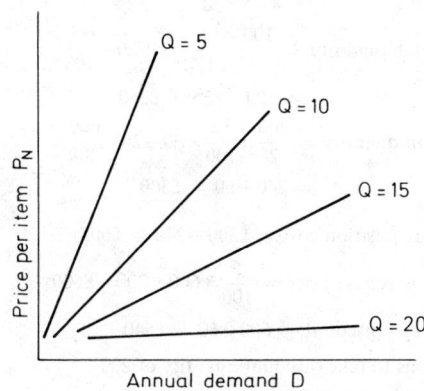

Figure 20.52. Economic order quantity governed by stock

If H is the cost of holding a unit of stock expressed as a fraction of the price of unit of stock then

$$H = \frac{C_1}{P}$$

and

$$Q = \sqrt{\frac{2C_2 D}{H.P.}}$$

For a given stores, C and H may be regarded as constants so that

$$Q = K \sqrt{\frac{D}{P}} \text{ where } K \text{ is a constant.}$$

The interval of time between placing orders of size Q is D/Q years.

Quantity discounts

Price reductions in material, etc., may be obtained by buying large quantities of goods and it may prove that by ordering the fixed quantity Q and minimising costs with respect to holding and ordering costs, the opportunity of further saving from discounts is being missed.

By comparing the total or acquisition costs per year for the two quantities with the varying purchase price for the material, the economics of price discounts may be investigated.

EXAMPLE

If holding costs H are 12%, value of items £20, economic batch quantity $Q = 100$. Discount for quantity (200 or over) is 5%. Demand per year of items is 600 and ordering costs per order are £20.

Acquisition costs per year = holding costs + ordering costs.

$$= \frac{QHP}{2} + \frac{C_2 D}{Q}$$

$$\text{For economic batch quantity} = \frac{100 \times 12 \times 20}{2 \times 100} + 20 \times \frac{600}{100}$$

$$= 120 + 120 = £240$$

$$\text{For price discount quantity} = \frac{200}{2} \times \frac{12}{100} \times 20 + 20 \times \frac{600}{200}$$

$$= 240 + 60 = £300$$

$$\text{Increase in acquisition cost} = £300 - 240 = £60$$

$$\text{Decrease in purchase price} = \frac{5}{100} \times 600 \times 20 = £600$$

$$\text{Saving through discount} = 600 - 60 = £540$$

Hence it is advantageous to take discount quantity of 200.

Economic production quantity

When the products are manufactured by a production process there is a time delay before the stock is built up to a required level (Figure 20.53). During this period there will also be usage of the products and the maximum stock level does not reach the order quantity Q.

ASSUMPTIONS

1. Demand rate is constant.
2. Production rate is constant and equal to or greater than the demand rate.

3. Order cost includes the cost of setting up production lines and procuring material for the production run.
4. Unit costs are constant.

If the manufacturing rate M is just equal to the usage rate D, the items will be used as fast as they are produced. If M is greater than D, inventory is accumulated at the daily rate of

$$\frac{(M-D)}{\text{days}} \text{ per year}$$

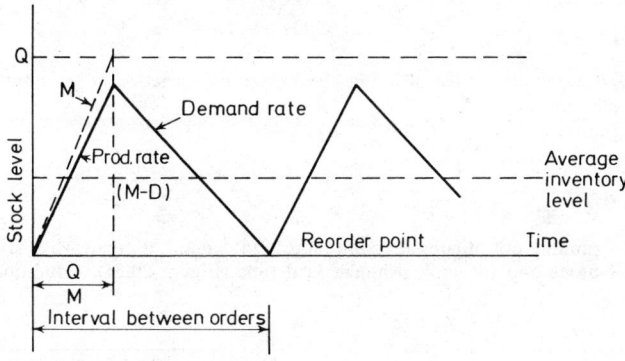

Figure 20.53. *Economic production quantity*

The inventory reaches a maximum after Q/M^1 days where M^1 is the daily production rate or Q/M years.

$$\text{Max. inventory level} = (M-D)\frac{Q}{M}$$

$$= \left(\frac{M}{M} - \frac{D}{M}\right)Q$$

$$= \left(1 - \frac{D}{M}\right)Q$$

$$\text{Average inventory} = \left(1 - \frac{D}{M}\right)\frac{Q}{2}$$

Total costs = ordering costs + holding costs + cost of part.

$$C = \frac{C_2 D}{Q} + C_1 \frac{Q}{2}\left(1 - \frac{D}{M}\right) + DP$$

where C_1 = Holding costs per year per unit

C_2 = Ordering costs per batch

P = Cost of part

Thus

$$\frac{d(\text{total cost})}{dQ} = \frac{-C_2 D}{Q^2} + \frac{C_1}{2} - \frac{C_1}{2}\frac{D}{M} + 0$$

For min. total cost

$$\frac{d(\text{total cost})}{dQ} = 0 = \frac{-C_2 D}{Q_2} + \frac{C_1}{2}\left(1 - \frac{D}{M}\right)$$

$$\frac{C_2 D}{Q^2} = \frac{C_1}{2}\left(1 - \frac{D}{M}\right)$$

$$Q_2 = \frac{2C_2 D}{C_1(1 - D/M)}$$

$$Q = \sqrt{\frac{2C_2 D}{C_1(1 - D/M)}}$$

If $C_1 = HP$, where H = cost of holding expressed as a fraction of the unit cost, and P = cost per unit.

Assuming risk[32]

The risk of running out of supplies continuously in demand is created by variations in the rate of usage and the replenishment lead time (Figure 20.54). Conditions which

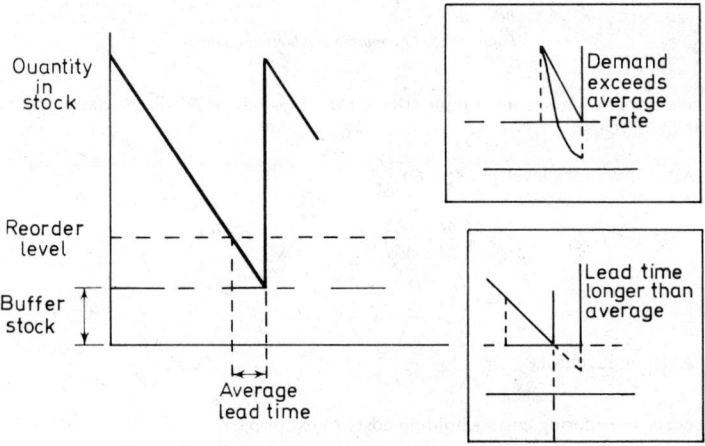

Figure 20.54. Stock and reorder level

contribute to stockout after an order has been placed are: accelerated demand, extended lead time, or a spurt in demand coupled with a delivery delay.

The way of avoiding stock-outs is to hold a buffer stock beyond the amount normally consumed by average usage during average lead time. The problem is determining the size of buffer stock which minimises costs of stock outs, and the holding costs for the buffer stock.

These problems are normally solved by simulation. From the results of the simulation the optimum buffer stock is found and added to the amount of stock consumed during the average lead time gives the reorder level.

Stock ordering procedures[37]

The object of any stock ordering procedure is to control the stock levels kept to that which will minimise inventory costs but maintain an adequate service to the customer and utilise the production equipment economically. Because it is extremely costly and difficult to continuously monitor stock levels it is usual to monitor them at a definite point which may be determined by either the level that the stock has reached or at a predetermined time, e.g. monthly or weekly, etc. These two systems may be classified under the general headings of two bin system (order quantity) and cyclic review.

TWO BIN SYSTEM (Figure 20.55)

When stocks fall to a preset level (reorder level) a fixed quantity (economic order quantity) is ordered. Material is assumed to be held in two bins and when the first is

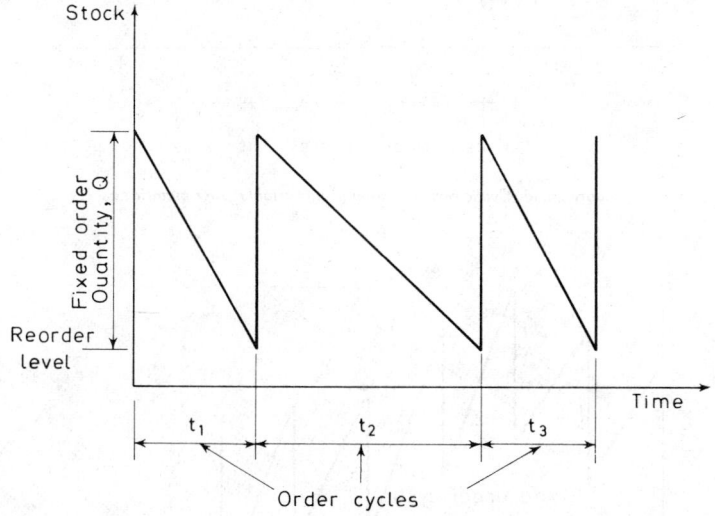

Figure 20.55. Two-bin stock ordering procedure

empty then a fixed quantity order is placed for the replenishment of total stock, and the second bin is used until the new stock is received.

For low value items of comparatively constant consumption the use of the two bin system can often take the place of all stock records leading to substantial administrative economies.

For other than low value stock it is necessary to maintain clerical records in order that the usage, etc. may be investigated. The use of clerical records can introduce a delay factor when the reorder level is reached.

The reorder level is determined by the consumption during the average lead time (time elapsed between stock ordering and stock receiving) under conditions where demand usage and lead time are reasonably constant. When either, or both, of these factors are variable then a buffer stock must be added.

CYCLIC REVIEW

This ordering procedure reviews the stock levels at a *fixed time* and places an order, Figure 20.56. When usage is reasonably constant the order quantity placed may be fixed which is advantageous to the supplier who can easily load his workshops and if internal

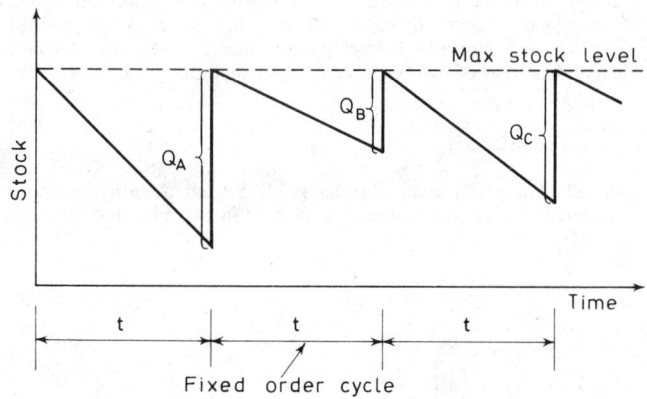

Figure 20.56. Cyclic review system using variable order quantities

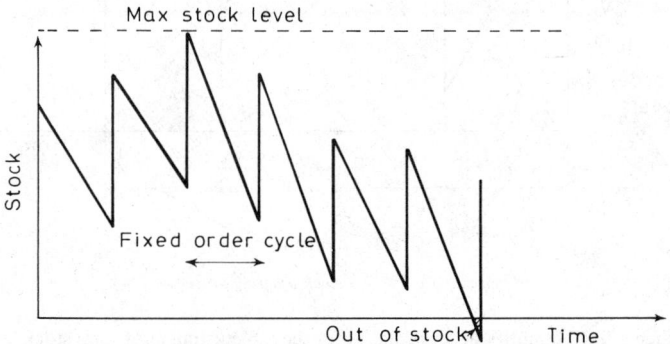

Figure 20.57. Cyclic review system using fixed order quantity

supplying is done, the reorder period can be adjusted to make the most effective use of plant.

If the demand and lead time for the product remains constant then both systems of ordering give the same results. When the demand or lead time are uncertain then the fixed order cycle method may result in high stocks or stock-outs as shown in Figure 20.57.

If variable order quantities are used high stocks may be controlled but stock-outs can be experienced, and the advantages of using economic order quantities are lost.

COVERAGE ANALYSIS

Decisions of stock control are normally based on 'Economic Ordering Quantity' (EOQ) formulae which attempts to balance and minimise the costs incurred. In evaluating such data, an estimate is usually required of the likely costs of placing an order; carrying stock; running out of stock and profit per item.

Whether or not one accepts these formulae, the method of calculation is certainly long and tedious, quantities for every single item need to be derived and it remains equally likely that the stock may increase or decrease when the general aim is to decrease the stock.

One of the main advantages of coverage analysis over EOQ methods lies in its speedy application. Coverage analysis[38] was developed to provide an optimum stock controlling

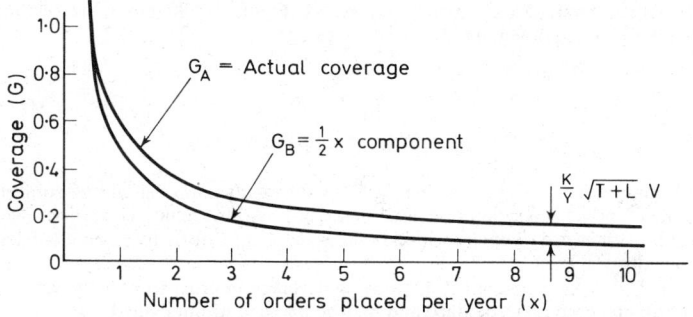

Figure 20.58. Relationship $\frac{1}{2}x$ component and actual coverage

policy, the objective being to 'minimise stock capital over the range of stock items (without affecting overall probability of stock run-out) subject to keeping constant the existing total orders placed for all items'.

The method of coverage analysis is applicable to stock problems where a large number of items are involved (in fact the larger the number, the more effective the technique) and where the demand for the items is relatively stable.

The coverage G, of a stock item is defined as the ratio of the average stock carried to the annual usage of the stock item.

Thus,
$$\text{coverage } G = \frac{\text{average stock, } S}{\text{annual usage, } Y}$$

Accepting $S = Q/2 + K\sqrt{T + LV}$, where Q = quantity; T = periodicity; L = lead time; V = variance; $K = 2.33$ for a 1 in 100 chance of running out.

then
$$G = \frac{Q/2 + K\sqrt{T + LV}}{Y}$$

Number of orders per year,
$$X = \frac{Y}{Q}$$

$$G = \frac{1}{2X} + \frac{K}{Y}\sqrt{T + LV}$$

Thus coverage on a stock item can be regarded as comprising two components:

$\dfrac{1}{2X}$ component, arising from turnover stock

$\dfrac{K}{Y} \sqrt{T} + L\ V$ component, arising from buffer stock.

From the value of G it is possible to clarify the stock items. For example,

$G = \infty$ represents dead stock.
G over 2·0 slow moving stock.
G less than 1·0 normal turnover.
$G = 0$, no stock.

In order that the estimation of the reduction in stock can be carried out the actual coverage must be plotted against the number of orders placed per year. The points produced in the plotting are scattered and it is necessary to plot the $\frac{1}{2}X$ component of coverage also. The $\frac{1}{2}X$ component shows the general trend and the most suitable for the actual coverage can then be estimated. Figure 20.58 shows the relationship between the $\frac{1}{2}X$ component and the actual coverage.

PLANT MAINTENANCE [39]

As with many other activities within firms, plant maintenance should be subjected to management policy consideration and decision.[40] Maintenance is a sheer necessity and the decision is not whether it should take place, but the form in which it will be most economically practised.

BS 3811[41] defines maintenance as 'work undertaken in order to keep or restore every facility—to an acceptable standard and may be planned or unplanned'.

Maintenance in many firms is inhibited by the production methods and production management seems content to accept that maintenance is an on-cost and should be left to be carried out during an annual holiday when it would be most convenient to have a shutdown. Only in an extreme emergency or breakdown will production management cooperate with maintenance staff in allowing examination or repairs to be made.

These 'unplanned emergency maintenance' repairs, are carried out under the urgency of keeping production downtime to a minimum. Patching and unreliable maintenance is encouraged, and rarely is the effort made to analyse and correct the cause of failure.

Production management, particularly those responsible for highly capitalised equipment, too often content themselves with rectification of unforeseen breakdowns as they occur, instead of directing what is necessary to anticipate, or to prevent them occurring. Inadequate maintenance accelerates depreciation and leads to excessive damage and an exceedingly high cost in both repairs and production downtime. Such losses, however, can be prevented.

A utilisation study carried out by work study personnel, or by first line supervision, in the form of an activity sample study, can provide management with the percentage of production downtime spent on breakdowns, even to the extent of time spent on plant being repaired and waiting to be repaired. This information will help management towards the decision in which form maintenance could be most economically practised (Figure 20.59).

Planned maintenance divides into two closely associated parts:

1. *Preventive.* Which is directed towards preservation.
2. *Corrective.* Which is directed towards all predetermined minor and major overhauls and their associated activities.

Preventive maintenance is further divided into two main activities:

1. *Routine attention.* Designed to preserve the assets of the undertaking at a standard of cleanliness and effectiveness, consistent with financial and operating policies. It embraces such activities as: painting, lubricating, rust prevention, regular adjustments, and where justified 'running' replacement of parts.
2. *Routine examination.* Finding incipient faults before they develop into failure.

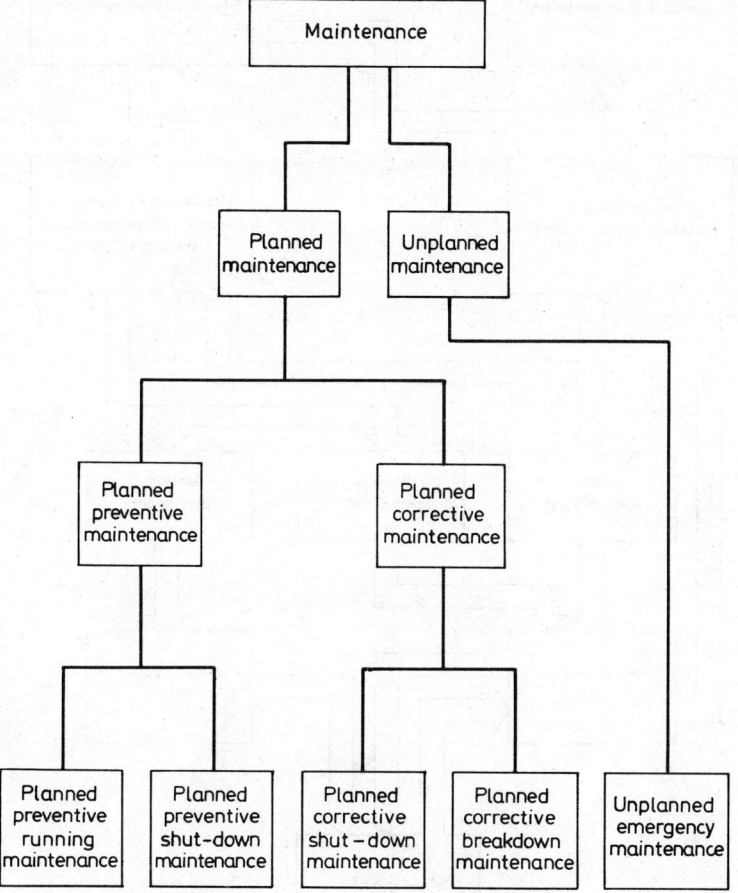

Figure 20.59. Classifications of maintenance

Corrective maintenance is designed to restore a facility to its original standard at periodicities jointly agreed by maintenance and production management.

The essentials of a planned maintenance system are to prepare and execute programmes in respect of:

1. Inspection and examination periodicities.
2. Minor repairs shown necessary by examination.

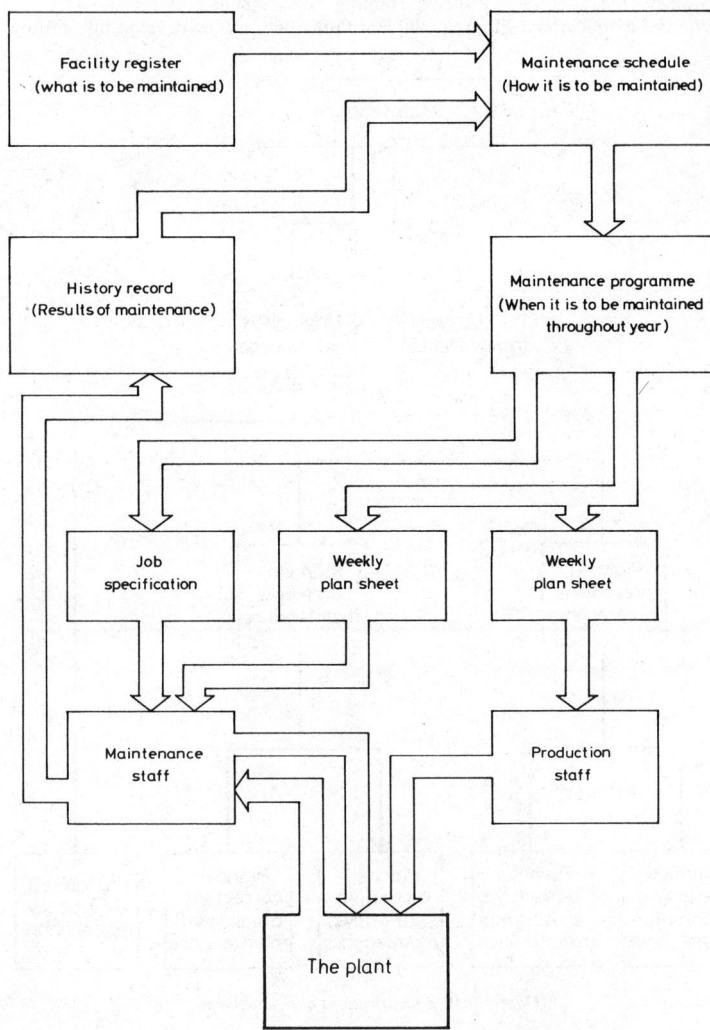

Figure 20.60. How the maintenance control system works

3. Regular corrective maintenance.
4. Inventory of maintenance stores.
5. Disposal or salvage.
6. Manufacture or purchase of new spares.

Also, to maintain a log of all reports made by the inspector and the fitter; and to revise periodicities if economically justified on the basis of statistical records.

Documentation for statistical records, however, is all too often elaborated to provide statistical data for purposes indirectly connected with day to day maintenance work. Such temptation must be resisted otherwise the administration of such a volume of paperwork will develop into 'a slow paper moderated reactor', which becomes critical when fitters complain of too much red tape, and management become frustrated. All that is essential is to provide answers to four questions:

What is to be maintained?
How and when is it to be maintained?
How long will it take?
Is the maintenance adequate and reliable?

Figure 20.60 shows a documentation flow diagram of how a maintenance control system should work.

WHAT IS TO BE MAINTAINED?

In the majority of firms a plant inventory register exists, which is used annually for stock-taking purposes. This register contains information that positively identifies and locates all capital plant and equipment and their most critical parts, e.g. reciprocating parts and seals, etc. An extension of this information would enable a 'facility register' to be compiled, consisting of constructional and technical details and in particular to the critical parts, with reference to drawings, specifications and manufacturers' brochures. This register would then establish 'what has to be maintained'.

HOW AND WHEN IS IT TO BE MAINTAINED?

A maintenance schedule is compiled in consultation between production and maintenance staff to establish how the maintenance is to be effective, i.e. preventive running maintenance, preventive shut-down maintenance, corrective shut-down, or corrective breakdown; and at what periodicities, i.e. daily, monthly, annually, biannually (or equivalent running hours).

A series of job specifications, detailing the various periodical operations required by the itemised plant and critical part are compiled by maintenance staff assisted by work study engineers, who question:

Purpose—what is achieved?	eliminate
Place—where is it done?	
Sequence—when and in what order?	change or combine
Person—who does it?	
Means—how is it done?	simplify

and in addition carry out studies of broad general factors affecting maintenance gangs, 'make ready' and 'put away' activities. For example:

Inclusion on the job card, or instruction of precise details of materials and tools required for the job, intended to reduce visits to stores and cut down waiting time.
Getting to and from the job, changeover and travelling time.
Creation of tea-break facilities nearer to the working area.

Endeavouring to broaden skill and demarcation lines.
Ensuring variety, responsibility and achievement factors are included in the job to provide employee job satisfaction.
Co-ordination of work involving other skills.
The issue of new work and instructions, not necessarily at the beginning of the day.
Arrangement for inclement weather, hold-ups and emergencies.

HOW LONG WILL IT TAKE?

A view held by most maintenance people is that, as their work is so occasional and varied, why go to all the bother of obtaining a time standard for 'an average qualified maintenance fitter' to carry out each element of work described in the job specification. So we find the simple practice of extracting average job times from past records and of estimating times from experience.

Such time standards are unreliable, since they arise from incomplete records and fail to define clearly: job content and the related time; productive and non-productive time; and the average of a wide range of times booked on identical or similar tasks.

What is worse, is that such practices fail to utilise one of the most fruitful sources of improvement, method study. They simply ratify and perpetuate methods which are not the best possible nor the most effective.

Work study engineers assist maintenance management by providing time standards much more reliable than those mentioned above. Firstly, by standardising the method, i.e. broadly described in the job specification and then measuring the work content and contingent delays.

The maintenance programme

After the estimated times have been arrived at for each item of plant it is necessary to draw up a maintenance programme to establish when each item shall receive the specified attention throughout the planning period, which is usually twelve months.

A convenient layout is to list the plant items with their reference numbers, and alongside these a table summarises the mechanical, electrical and lubrication maintenance each requires at various intervals. The maintenance required by each machine is then entered in the appropriate week columns (with due allowance for holiday periods) to ensure that the weekly maintenance load shall be as nearly as possible constant, as well as falling within the specified time limits. Maintenance cycles are commonly based on multiples of three months, and it is therefore convenient to work to a planning year of 48 working weeks. When the scheme is first launched it will be impossible to time and programme more than 15–20% of the total maintenance work done, but as experience is gained the proportion that can be scheduled should increase to 75–80% with a corresponding reduction in emergency maintenance.

The maintenance programme should be prepared in consultation with the production staff, so that all concerned shall be aware of future requirements in running and shutdown maintenance.

IS MAINTENANCE ADEQUATE AND RELIABLE?

As with all management techniques, plant maintenance requires a feed-back of information and this is provided by the tradesman's report on completion of the specified service.
This report includes:

Anticipated failures and defects.
Details of adjustments made.

Failures and action taken.
Cause of breakdown, if known and comments of prevention.
Time taken in carrying out specified service.
Plant down-time.
Name of tradesman and date of service.

The information is examined and dealt with where necessary and then transferred in a summary form on to a history record pertinent to each plant item and the critical parts.

Periodical examination of this history record, otherwise known as a memory storage, will indicate if any revision amendment is necessary to the maintenance programme, method, time and periodicity of examination and repair. It will also indicate if the failures are recurring in the same component parts, an investigation can then be employed to consider new materials, design, etc., value engineering would be most useful in these circumstances. If economically justified and in the interests of minimising maintenance costs, eventually it may only be necessary to maintain on a planned basis the critical parts of equipment rather than religiously maintain all associated parts at the same time. Many planned maintenance systems have been rejected because this revision has never been carried out.

Inventory of maintenance stores

Contents of stores can usually be grouped into four classes:

1. 'Insurance' spares.
2. Wearing parts pertinent to special plant and equipment.
3. Wearing parts of a general nature, i.e. belts, pulleys, packing glands, etc.
4. General hardware, nuts, bolts, sheet metal, pipework, etc.

A stock review procedure should be held at say twelve monthly intervals to ensure that stocks do not exceed the levels agreed. A standard stock list should be published and non-standard items carefully scrutinised for possible standardisation.

Disposal or salvage

While not strictly concerned with the improvements of maintenance standards, properly arranged salvage operations can yield considerable revenue to offset material costs. Recoverable components should be returned to workshop or store.

Manufacture or purchase of new spares

Given adequate workshop facilities and skills, the manufacture of spare parts can often be justified. Careful consideration, however, must be given whether to contract or use direct labour.

Benefits of planned maintenance

These can be summarised as follows:

1. Less production down-time with related benefits to labour and machine utilisation and delivery satisfaction.
2. Improved production control by moving away from random breakdown occurrences.

3. Better industrial relations through the avoidance of involuntary production layoffs or loss of incentive earnings.
4. Fewer large scale repairs and randomness, less need to provide staff cover in 'rush hour' conditions.
5. Lower repair costs due to preventive attention given before anticipated breakdown.
6. Less overtime worked on planned than on unplanned emergency repairs.
7. Postponement or elimination of cash outlay for premature major replacement, because of better conservation of assets.
8. Less capital investment in standby equipment.
9. Better spare parts control leading to improved inventory control.
10. Fewer product rejects as a result of better conditioned plant and equipment.

The use of computers in plant maintenance

In recent years the major oil companies have developed a method of lubrication maintenance for large industrial complexes, which has been called computer planned lubrication maintenance for industry.

The system is based on the previous detailed and routed cards method and follows an analysis of every lubrication point throughout the factory. The disadvantage of the old scheme was that any variation in the positioning of a machine or other alteration could mean many hours of work redrawing the scheme. The application of modern computers, printing out 1500 lines a minute has overcome this. This enables a lubrication plan with full instructions for a lubrication maintenance system to be redrawn in minutes.

HOW THE METHOD WORKS

As with the previous planned lubrication maintenance scheme, an initial survey is first made of all plant, taking into account manufacturers' recommendations, guarantees, environmental operating conditions, and anything that affects performance and running.

A detailed inventory of plant is then made and this is then analysed and broken down into simple daily tasks which take account of the best routes round the factory for the lubricators, operative to follow, manpower requirements, total work loads, frequency and application, etc. Under the old scheme, the route and reminder cards were then produced in two categories short and long term.

In the past it was not easy to keep the system updated. Old plant was scrapped or moved, the labour force was changed, work loads were altered, new grades and lubricants were introduced. This entailed a lot of retiming, reorientation and reprinting. It is at this stage that the computer can produce the answer.

The successful computerisation of an existing manual scheme depends upon two factors:

1. The ability to define the precise location within a factory of each lubrication point.
2. The activity of each lubrication point which should be concise and unambiguously defined.

COMPUTER PROGRAMMING

Each factory is allocated a unique PLM number and this is attached to all lubrication records on the computer file. The PLM number has four digits of which the first three are in the range 1 to 999 and the fourth is a digit check. The factory is divided in areas 1–99 and each one is named. The machines are numbered within each area consecutively from 1 to 999. There is then a smaller dissection and each machine can be divided into a maximum of 99 sections. Each section itself consists of a number of lubrication items.

Each lubrication item is thus uniquely located, not only within the factory but on the whole computer file.

Timings for lubrication operation are available for the computer and can be modified by an efficiency factor to allow for non-normal conditions in any particular area of the factory. Any individual lubrication item requiring an abnormal time has a 'special time' entered against the item on the input computer document. The programme in the planned lubrication maintenance suite, which deals with routing, is designed to allocate a day's work to each maintenance man based on the working time available, and the choice of several combinations of daily, weekly and monthly lubrication schedules.

ADVANTAGES OF THE SYSTEM

With a print out of 1500 lines a minute, the computer can do in two minutes what would have taken 48 hours under the old method. Inventories analysis, routed and updatings can be produced in a fraction of the time previously taken for manual documentation, and therefore a system can be completely updated, at the worst, in a month but usually about ten days. It is easy to record any alteration of grades or machine in areas or throughout. The computer also programmes the tasks to separate jobs requiring special treatment, e.g. items requiring added activity such as removal of guards, tradesman's jobs such as electricians. If machinery is changed to a point where it affects work loads a new works loading sheet can quickly be provided.

An efficiency factor can be applied to downgrade or upgrade the total work load to allow either for age of operator or to allocate the work to smaller or larger numbers of staff. New programmes will be written continuously to improve the service.

Evaluating maintenance productivity

There are so many different approaches to the problem of measuring maintenance productivity that it is hazardous to generalise about them, nevertheless, the maintenance manager in his pursuit for productivity improvement must have available some means of evaluating the effectiveness of his function.

Maintenance is generally considered as an indirect cost of no great consequence in the overall accounting procedure, being acknowledged as an overhead situation and therefore requires no precise control other than budgetary controls. Currently, however, maintenance controls are taking the form of refined and explicit indices to point the directions in which management might be more effective in improving productivity.

Such an index has been developed by Corder[42] and is known as the 'maintenance efficiency index':

$$E = \frac{K}{xC + yL + zW};$$

where

E = Maintenance efficiency index.

C = The total 'pure' maintenance cost comprising total planned maintenance (preventive and corrective) and emergency maintenance expressed as a percentage of the replacement value of the plant and equipment.

L = The total downtime due to maintenance expressed as a percentage of the scheduled production hours.

W = The units of waste caused as a result of maintenance responsibility, expressed as a percentage of the total output.

x = Total cost of maintenance in £'s in base year.

y = Total cost of lost time in £'s due to maintenance causes in base year.

z = Total cost of waste due to maintenance causes in base year.

K = Constant so chosen that the value of the expression is 100 for the base year.

When the value of E calculated, using the above formula for any year, is over 100, this indicates an improvement in the efficiency of the maintenance function. Values less than 100 represent an unfavourable trend.

Corder's index has been found to give good results in some types of industry, but there are others where it is only partly applicable.

OVERTIME INDEX

Work done over the normal working hours costs more and it should, under most circumstances, be kept to a minimum. It is a useful control for maintenance management to keep an open eye on this simple index, which can be expressed as:

$$\frac{\text{Overtime hours}}{\text{Total maintenance man hours}};$$

This index is very useful when the strength of the maintenance department is not subject to variation.

LABOUR PERFORMANCE INDEX

This is applicable where maintenance work is measured through work study techniques and takes the following form:

$$\frac{\text{Standard minutes produced by a section}}{\text{Attendance hours of the section}};$$

METHODS OF REPAIR INDEX

Methods of carrying maintenance work gradually improve. This index shows progress in improving methods by reducing work content and is expressed as:

$$\frac{\text{Standard minutes of work saved}}{\text{Total standard minutes of work produced}};$$

EXCESS STOCKS INDEX

Reductions in spare parts stocks can be effected through proper planning, arithmetical techniques of fixing maximum, minimum and ordering levels of stock, and designing-out maintenance techniques, etc. The effect of all these techniques on the capital tied up in maintenance spares can be studied by the following index:

$$\frac{\text{Cost of maintenance material used}}{\text{Total cost of maintenance stores in hand}};$$

LABOUR UTILISATION INDEX

The effects of planning maintenance work on the utilisation of labour force can be studied using this index, which is expressed as:

$$\frac{\text{Total hours actually worked}}{\text{Total hours doing 'productive' work}};$$

MACHINE MAINTENANCE EFFICIENCY INDEX

The effect of maintenance methods and procedures on the machine availability can be studied by observing the downtime of the machine due to maintenance, expressed as a ratio of the total machine scheduled time.

The machine maintenance efficiency index takes the form:

$$\frac{\text{Total hours of machine downtime due to maintenance}}{\text{Total hours of scheduled production}};$$

The purpose of control indices

It has already been stated that the different control indices help management to guide its decisions along the correct lines in its pursuit for higher productivity. Indices, however, would not serve their full purpose unless they initiate corrective action to be applied, where and when, and if necessary.

The particular indices to be used in any factory depend upon local circumstances and the particular aims of the company concerned. Maintenance management should, therefore, be in a position to design its own indices, which would broadly follow the outlines indicated.

Replacement theory[43]

'Replacement' applies to the replacement of a machine, group of machines or an entire department or layout, by a new replica, by an improved model, or by a totally different machine. The 'replacement analysis' assumes the new or proposed equipment is the best alternative to the old equipment.

Conventional replacement analysis methods exist in the form of 'expert opinion', i.e. estimated machine tool life or depreciation policy, or the 'return on investment' method which provides that a new machine must show savings in direct labour for the first year equal to say 25–30% of the cost of the initial investment plus depreciation.

For example, suppose the replacement of a milling machine costing new £8000 is being considered. Based on an estimate from a machine tool manufacturer a return of 25% on the investment, i.e. £1000 per year for eight years could be obtained. In effect, because there is eight years waiting before the machine begins to pay for itself (ignoring depreciation and inflation) the old machine is shielded from replacement long after it should be replaced. Thus as a result of periodic analyses, by this method, of the old machine against the best new one available it may take many years before the desired replacement signal is obtained. During this time unnecessary costs are incurred as a result of deterioration and obsolescence and apart from the influence of inflation.

Furthermore, by this incomplete method of analysis, it is also possible to replace before it is necessary.

EXPLANATION OF TERMS USED IN REPLACEMENT ANALYSIS[44]

The following should be read in conjunction with the replacement analysis work sheet (Table 20.9) and the chart for determining depreciation value, Figure 20.61.

Table 20.9   REPLACEMENT ANALYSIS WORK SHEET

OLD EQUIPMENT	NEW EQUIPMENT
Manufacturer *Cincinnati Milling Machines Ltd.*	Manufacturer *Cincinnati Milling Machines Ltd.*
Type and size *2 MH. Plain Milling Machine.*	Type and size *205–12 MI with 3 HP. I.O.S.*
Machine No.	Model No.
Year built *1944*	Estimate No.
Department	Date of proposal

Operating cost analysis (next year only)

COST ITEMS	OLD		NEW	
(1) Direct labour				
(2) Indirect labour	£2 572		£1 878	
(3) Fringe benefits				
(4) Maintenance	£30		£10	
(5) Tooling				
(6) Scrap	£100		£25	
(7) Property taxes and insurance				
(8) Down time	£55		£15	
(9) Floor space				
(10) Other costs				
(11) Total operating costs (next year)	£2 757	(A)	£1 928	(B)
(12) Net operating cost favouring new equipment	(A minus B)		£829	

Capital cost analysis (next year only)

OLD EQUIPMENT		NEW EQUIPMENT		
(13) Disposal value (now) £250		(18) Estimated primary		
(14) Disposal value (next year) £225		Service life	*115* yrs.	
(15) Loss of disposal value		(19) Required investment (Y_1) £3 651 Less 30%		
(next year)	£25	Required investment (Y_2) £2556 ←*investment*		
(16) Interest on disposal value		(20) Estimated disposal		*allowance*
10% (of line 13).	£25	value (X)	£365	
		(21) Disposal ratio (X ÷ Y)	*10%*	
		(22) Depreciation		
		chart value (C)	*8·7%*	($Y_2 \times$ C) £222
		(23) Interest on required		
		investment (D)	*10%*	($Y_2 \times$ D) £256
(17) Total old equipment		(24) Total new equipment		
capital cost (next year)	£50	capital cost (next year)		£478
(25) Net capital cost favouring old equipment		(Line 24 minus line 17)	£428	
(26) Savings from replacement next year		(Line 12 minus line 25) Plus £401		

1. *Direct labour.* Such costs are determined primarily by measuring production time in terms of operator wages, including bonuses and enhanced payments for a period of one year (next year). It is necessary to compare the sum of operating cost and capital cost on the old and new equipment for the same base period. Thus we use next year's operations beginning on the machine's installation date.

Special attention should be given to savings resulting from: combining or eliminating operations; improved quality; inventory costs (i.e. less work in progress due to increased

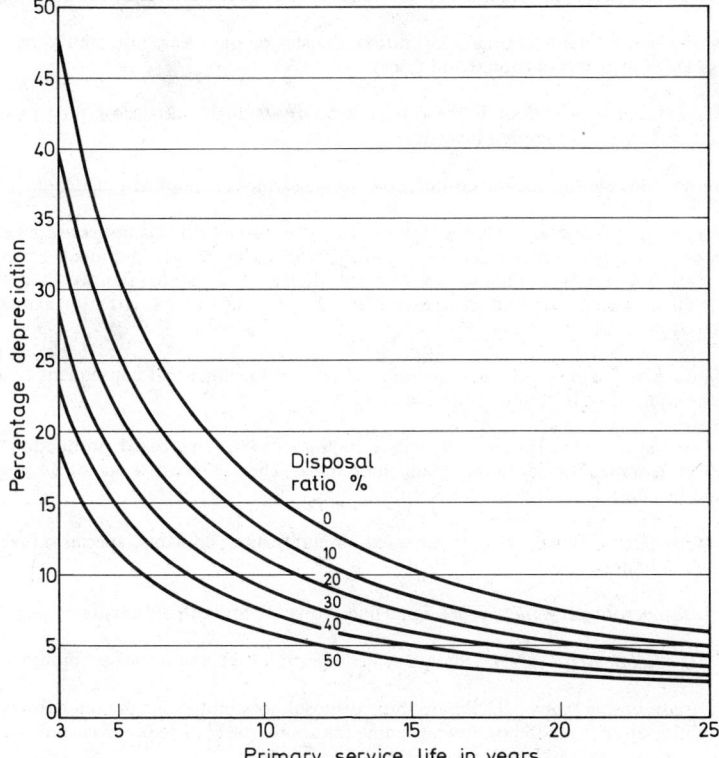

Figure 20.61. Chart for determining depreciation value. Instructions for use:
1. *Locate primary service life (in years) on horizontal axis*
2. *From this point ascend and locate point representing disposal ratio.* (*Estimate point when percentages fall between lines*)
3. *The percent on vertical line directly opposite this point is depreciation chart value*

speed of operation); morale improvement; reduction in operator skill; less setting or breakdown tool time; and reduced maintenance or occurrences of breakdowns.

Estimates on all work should be made on a cost/piece basis including set-up time.

Example of computing cost per piece (simplified):

Old machine: 125 pieces × 8 hours = 1000 pieces per day.
£5.00 labour cost only ÷ 1000 pieces = $0.0\frac{1}{2}$p (0.05) piece.
New Machine: 200 pieces × 8 hours = 1600 pieces per day.
£5.00 ÷ 1600 pieces = 0.035p/piece.

The cost of 1600 pieces is then:
Old machine $1600 \times 0.05 = 80p$.
New machine $1600 \times 0.035 = 56p$.

2. *Indirect labour.* Workshop administration, supervisory, inspection, labourers, etc. (all except sales and administration).

3. *Fringe benefits.* All benefits, pensions and holiday pay, etc. Frequently 15% is used.

4. *Maintenance.* Only ordinary operational costs for one year, not rebuilding costs (these are considered as capital additions).

5. *Tooling and accessories.* If these have a relatively high initial cost they must be considered a part of required investment.

6. *Scrap.* Should include the cost of work spoilage due to equipment inadequacies.

7. *Property taxes and insurance.* Property taxes, fire and casualty insurance are generally computed on a percentage basis, i.e. 3% of the book value of the equipment. However, this may vary with location and type of plant, therefore, use actual figures where these are available. Since tax and insurance rates change frequently this factor should be reviewed periodically.

8. *Down time.* This represents the value of production time and consequent costs to put the equipment back into operation.

9. *Floor space.* This becomes especially important when increased production may result in demands for additional plant floor space. The real estate value of floor space should be considered under these circumstances—otherwise ignore it.

10. *Other costs.* Power, gas, compressed air, lubricants, coolants, special oils, belts, storage of materials, etc.

11. *Total operating cost (next year).* Total lines 1 thro' 10 for both old and new equipment.

12. *Net operating cost A—B.* Should A be larger than B show negative amount.

13. *Disposal value (now).* There are three disposal possibilities for the old equipment: outright scrap; conversion or down-grading for another use; or resale at current market price.

14. *Disposal value.* Value as of the following year.

15. *Loss of disposal value.* Subtract line 14–13.

16. *Interest on disposal value.* If the old equipment were disposed of, or sold, the money could be used as operating capital. As such it would earn a return. A figure of 10% is suggested as a good average for most industries.

If extensive repair or rebuilding of the old equipment is necessary in order to keep it in effective service, such capital costs must be included in the analysis. In such cases the rebuilding cost (capital addition) will extend the effective service life of the old equipment. The cost must then be prorated over the estimated service life extension. To this amount, add the interest for one year which would be earned by the amount of the entire rebuilding cost, if the money were used as operating capital.

17. *Total old equipment capital cost.* (As next year would show.) Add lines 15 and 16.

18. *Estimated primary service life.* This is the estimated length of time in years which the new equipment can profitably be used, without serious challenge by a new piece of equipment. Do not consider the total useful life of the equipment.

19. *Required investment.* The total amount of money necessary to purchase and install the new equipment includes shipping, special insurance and any other cost which may be involved in placing the machine in operation.

20. *Estimated disposal value.* A 'qualified guess' based on past experience and common sense.

21. *Disposal ratio.* This ratio is necessary to obtain the proper depreciation chart value. To express the ratio as the needed percentage, divide line 20 by line 19.

22. *Depreciation chart value.* The chart portrays the geometrical decline in value of the new equipment over the service life. Expressed as a percent of original cost, the curves indicate the next year's equivalent of capital cost chargeable to deterioration and obsolescence. The heavy curved lines represent the disposal values in terms of percentages.

23. *Interest on required investment.* Purchasing the new equipment will tie up a sum of money for a considerable period of time. If invested or used as operating capital, this money would yield interest or income to the company. Therefore, select an interest rate which is the same or close to the current rate the company is earning on its operating capital. A reasonable range would be 1% above bank rate.

24. *Total new equipment capital cost.* As next year would show. Add lines 22 and 23.

25. *Net capital cost favouring old equipment.* Subtract line 17 from 24. The total on line 24 will almost always be larger because of the depreciation chart value (Figure 20.61) and the interest on the investment for one year.

26. *Savings from replacement (penalty for non-replacement) next year.* Subtract line 25—line 12. A plus value indicates replacement should be made, and the greater the amount, the greater the urgency.

VALUE ENGINEERING AND VALUE ANALYSIS

Twin techniques of VE and VA have been developed as methods of critically analysing the value of each part of an assembly under review. The usual cost reduction exercise is to make the existing product by less expensive methods or by using alternative processes, but VE and VA are techniques that investigate systematically costs other than those associated with direct labour and in particular the material costs. Value analysis for instance enquires into the function the product will fulfil and compares the possibility of performing the same function more cheaply. Whilst VA is applied to a fully designed product, the same criteria can be applied to the design of a new product or product concept.

Design Engineers deal adequately with value in their designs, but in the main are inclined to concentrate upon performance and appearance rather than give adequate attention to the relationship between function or purpose and cost.

Competent designers must have in addition to creativity and inventive flair an understanding of production and accounting methods and when all such skills and knowledge are brought together in the design of a product then it can be acknowledged that the designer is practising value engineering.

The technique was initiated by Miles of the General Electric Company, USA[45] from the lead by Henry J. Kaiser who applied prefabricated methods in the construction of Liberty freighters during World War II. It has taken nearly twenty years for this technique to be recognised by the more progressive managements in their endeavours to reduce production costs with consequent effect on the profitability of their concerns.

Organisation for value engineering

There are three possible locations for a value engineer in a company's organisation as follows.

The value engineer could be responsible to the purchasing function which would enable him to be in direct contact with suppliers and sub-contract firms. He would be able to investigate fully the network of suppliers and find the supplier who delivered on time and at a competitive price. For instance, if material purchased at an economical price arrives late for production, this could prove to be most expensive to the company. Nonetheless, if the value engineer was positioned in the purchasing side of the company his field of investigation may tend to be narrow and limited to the purchasing function only.

Of those firms already applying the techniques most approve of the value engineer working in the design office where he has the opportunity to keep up to date with new designs and investigate components that appear to be high priced for their required function. He is available for discussion on production methods and also able to keep up to date with new and improved facilities as they are introduced into the works. Existing components would also come into this field which could be brought to his notice through a suggestion scheme. The men who produce and assemble components often have some practical ideas. If the value engineer was positioned on the design side of the company he could apply acceptable ideas and be able to discuss components with the designers. This would help the designers to find more economical ways of producing components by careful design.

He could be responsible to the financial accountant. Here, where budgets are controlled, he would have an overall view of purchasing, design, and production costs. If it appeared that a component's cost was out of proportion to its function from the figures returned, the value engineer could go to the design office and investigate the component from its beginning through to it being sold as part of the machine or assembly. To be in a position to do this, the value engineer would require a broad knowledge and the ability to learn as he carried out his investigation. For a position in the organisation structure that would give the value engineer a complete freedom to do thorough investigations; the financial side of the company is probably the best 'base'.

Cruickshank[46] recommends a committee consisting of a design engineer with a knowledge of production engineering and work study; technical buyer; methods engineer; cost estimator and production engineer. The committee should be responsible to the works director or his equivalent. A disadvantage of this arrangement is to find a convenient time for all the members to get together and to leave their departmental problems behind thus giving their complete attention to creative thinking and useful speculation.

INVESTIGATION PROCEDURE

Value engineering is simply the application of the questioning technique of method study to the design function in order to provide a desirable product that must satisfy a particular need, fulfil its function efficiently and attain an acceptable standard of quality, i.e. fitness for purpose. It must also be produced at a cost well within the financial means of its potential market.

The basis of the approach is to examine the product, take it apart into its component pieces and question the purpose of each piece to consider the possibility of achieving the function at less cost but without detriment to quality. For control throughout the investigation the following systematic procedure should be applied.

Selection	Component parts that have excessive costs, narrow profit margin, competition in price and quality.
Evaluation	Define the value of the component part in terms of its desirability, i.e. usefulness, importance or intrinsic worth.
	Compare the value of each component part with the rest using cost ratios. A component's cost must be in direct relationship to its function.
Creative thinking and speculation	Develop a range of alternatives to the design, material specification and manufacturing methods. Each alternative must be evaluated by cost against function.
Development	Select the most promising ideas showing best value and highest probability of accomplishment.
Plan and programme	Select specialists and vendors for consultation and establish a programme of investigation to test out each of the ideas showing promise.
Report and execute	Provide encouragement to everyone involved and work with them to overcome problems in applying material and process change.

EXAMPLES OF APPLICATION

Study these features which contribute to the cost of any material, component or process in order to determine whether they are fulfilling their functions economically or whether those functions can be performed more cheaply by other means.

APPEARANCE AND QUALITY

The reason for manufacturing an article is to sell to the customer a product that will fulfil a purpose. Unless it fulfils this purpose it will fail to sell irrespective of good appearance, finish, workmanship, or low price.

Design for function, including performance and quality, is the first consideration. However satisfactory in function, this alone is insufficient to promote large sales. Designers frequently neglect appearance in basic design and then try to achieve eye-appeal with meaningless pieces of chromium plating and do not thereby achieve minimum cost. Good appearance is inherent in good design; it is not just a matter of adding embellishments.

The quality standard is essentially a decision of the board of directors, after correlating a number of considerations: sales regarding market potential; production as to plant suitability and availability; labour of appropriate skill, ability and experience; and the recommendations of the value engineer, to mention only a few.

A competent production engineer can produce an article at any desired quality level between the superlative and shoddy, the designer's proposals can be elegant or crude and between these extremes the required level must be defined, otherwise deviations above or below the necessary standard adds unnecessary cost to the company. The chief inspector should not set standards of quality, his responsibility is to check that the defined quality standards are in fact being maintained, notwithstanding the fact that the real task of maintaining quality is the responsibility of every supervisor and operator. By being involved at the design stage the value engineer investigates material utilisation, dimensions and tolerances and the most effective and economic process of manufacture.

PLAN AND PROGRAMME

Explore the identified alternatives. This requires programmes to cover each of the alternatives thought to be reliable and useful choices. Selection of such specialists is the first requirement. Then provide all pertinent information to these specialists with the proper motivation to propose the contact they believe to meet its application with a breakdown of costs. Assign the final step to purchasing for action.

REPORT AND EXECUTE

Correlate all the furnished details from the specialists and suppliers and report on those having special merit.

To perform a value analysis exercise[47] the precise details of cost and total product value must be determined. Each component cost must be examined to see if it is possible to obtain the same quality at a reduced cost. The breakdown of costs is usually done jointly by accountants and engineers. Any change that is proposed must be examined to see how it affects the cost of the component. It is possible, of course, that an opposite effect might occur whereby itemised costs may increase. It must be emphasised that the aim is not an item for item reduction in cost but an overall reduction.

QUALITY CONTROL

R. G. NORMAN AND W. EDWARDS-SMITH

Inspection is the fundamental aspect which controls quality. The inspection function may consist of a 100% check on the items produced or an acceptance may be based upon the inspection of samples statistically determined. It is essential when producing components in quantity to strike a balance between the highest possible quality and the cost of production. No particular advantage is to be gained from making components to a higher quality than is demanded by the functional requirements of the item concerned. The appropriate quality levels are of course determined by the designer, based upon his experience in this particular field or with the assistance of standards such as BS 4500, which lists suitable tolerances for mating components over a wide range of sizes and applications.

Having laid down a specification for the items being manufactured it is necessary to decide on a procedure which will ensure that the derived quality is achieved. In cases where small batch quantities are involved, 100% inspection may be necessary, but where large quantities are involved it is more usual to adopt sampling techniques. To ensure that the samples inspected are representative of the total production the rules of sampling, based on the probability theory, should be applied.

Practical sampling may be divided into two groups, namely, sampling by variables and sampling by attributes. When the measured dimensions are recorded then inspection is said to be by 'variables', but when examination involves a simple decision to accept or reject, as for example when using a 'go—not go' gauge, then we have inspection by 'attributes'.

Acceptance sampling

In quality control, test data is used to trace back in the manufacturing source, assignable causes of variation, and thereby take immediate steps to remedy these errors; in other words to avoid processing defective material, test data is essential as a guide to subsequent action on batches of items or material received from outside suppliers. Such action may involve rejection or grading if the batch is not to the required specification and quality.

If all the items in the batch are given 100% inspection, then the appropriate action is purely a technical one. If, however, action is taken on the basis of a sample, and one which may not exactly represent the quality of the batch, then there is a risk that the action taken may not be correct. It is the purpose of statistics supplemented by test data to calculate such risks.

This type of statistical sampling describes the distribution and the occurrence of isolated events in a progression, the distribution of such events conform to the 'Poisson' distribution which is based on e (i.e. Naperian Logarithms). This arises in the study of the 'law of natural growth'.

If e is raised to any power Z then $e^Z = (2.718\,3)^Z$. These results can be expressed in the form of an infinite series:

$$e^Z = 1 + Z + \frac{Z^2}{2!} + \frac{Z^3}{3!} + \frac{Z^4}{4!} \cdots \frac{Z^N}{n!}$$

The more general form of the Poisson distribution is:

$$e^{-Z} = \frac{1}{e^Z} \quad \text{and} \quad 1 = e^{-Z} \left(1 + \frac{Z}{1!} + \frac{Z^2}{2!} + \frac{Z^3}{3!} \cdots \right)$$

Thus the solution of each successive stage of the distribution enables the determination of the probabilities of finding the occurrence of the event Z to be calculated. Only the 'average' number of occurrences of event Z is required to calculate the probabilities of observing all the various possible number of occurrences, i.e. the probability of finding (0), (1), (2), (3) . . . defectives in the batches, etc.

Sampling schemes and operating characteristics

A typical inspection specification when using sampling is as follows. A supplier is told to deliver goods in batches of a given size, e.g. $N = 100$. From each batch a certain number of items are drawn at random and inspected, i.e. n = sample number. If the sample contains defectives 'd' to an allowable level, the whole batch is acceptable, if defectives are found beyond the allowable level, then the batch is rejected. Rejection may mean the whole batch is returned to the supplier for 100% inspection.

'Goods inwards' inspection is used to protect the customer from accepting defective goods and satisfaction is not necessarily achieved until the goods are inspected 100%. More usually the customer takes his stand on economic considerations and is aware of the need for protection against abuse and imperfections, but is unwilling to spend money unnecessarily on 100% inspection methods when less investment would satisfy the same guarantees. Thus a balance of risk and cost is demanded by minimising n to ensure that a loss due to an imperfect inspection is not experienced.

Factors to be considered in a favourable sampling scheme are:

1. Risk of sending or accepting defectives; this should be equitably shared by producer and consumer; and
2. the fact that a sampling scheme is a line of defence against a threat whose seriousness varies from time to time.

When it is specified that $N = 100$, $n = 8$, and $d = 2$, the scheme is fixed irrespective of threat, and therefore it must be considered how any specified scheme would operate under different conditions. The following questions may be asked:

What will be the effect when the supplier sends in batches of 'normal' standards?
What will happen when quality falls off?
Would there be a change in the supplier's proportion of defectives?
How will the rate of rejection change as the value of the percentage defectives in producers' goods increases?

To calculate the operating characteristics first of all assume a series of hypothetical values for p the proportion of defectives in the supplied goods, then calculate the expected number of defectives in the stated sample n, this will be nP, and will be the expectation on the probability graph. Finally, refer to the Poisson probability graph

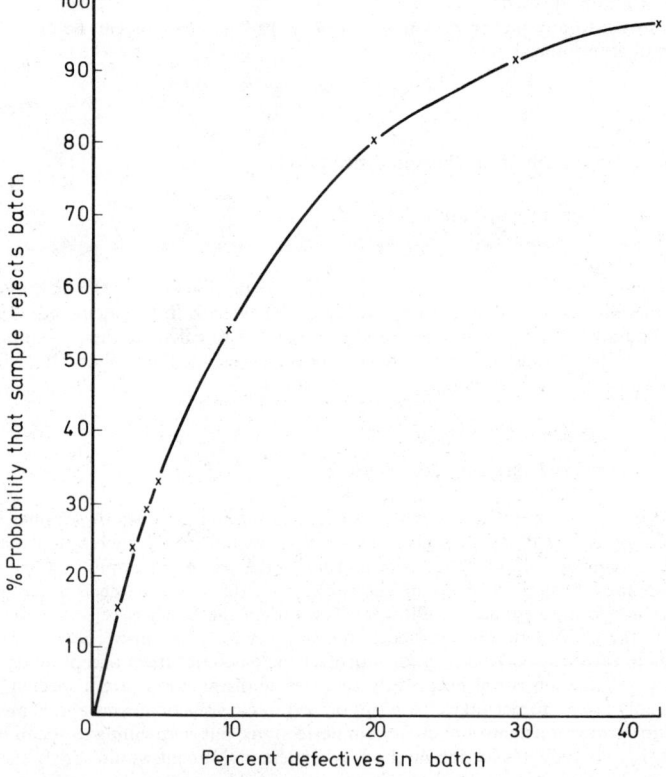

Figure 20.62. Operating characteristics curve (1)

for the probability of one or more (shown by line $c = 1$) defectives occurring for each of the assumed hypothetical values.

Taking the case where the batch size, $N = 100$; sample size, $n = 8$, and allowable number of defectives, $d = 0$, the results would be as shown in Table 20.10 and the curve of operating characteristics shown in Figure 20.62.

Table 20.10

Suppliers proportion defective = p	0·02	0·03	0·04	0·05	0·1	0·2	0·3	0·4
$n = 8$ Expectation of defectives in sample $n\,P$	0·16	0·24	0·32	0·40	0·8	1·6	2·4	3·2
Probability of finding one or more defectives = prob of rejections	0·15	0·24	0·29	0·33	0·54	0·8	0·91	0·96
% of batches rejected by scheme under assumed conditions.	15	24	29	33	54	80	91	96

This scheme, whilst being stringent is not discriminating, even when goods are only 1% defective; 8% of the batches can be rejected. If the customers were prepared to accept a batch which contained no more than 1% defectives then from the customer's

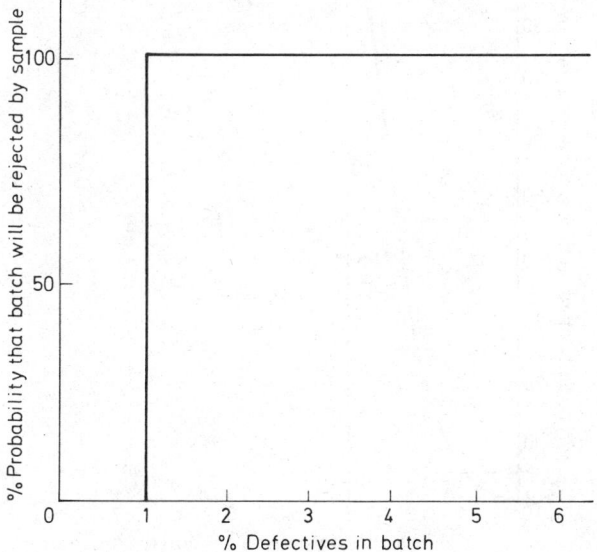

Figure 20.63. Operating characteristics curve (2)

point of view the ideal curve of operating characteristics would be shown in Figure 20.63. This indicates that all batches containing not more than 1% defectives would be certain of being accepted whilst all others would be certain to be rejected. A scheme of this nature is not very realistic, if we are prepared to accept batches containing 1% defective it is ridiculous to reject batches containing 1·1% defectives. From a practical point of view the ideal scheme should have characteristics similar to those shown in Figure 20.64.

The items marked on the percent defective scale are the process average (PA) which is the quality the manufacturer can normally maintain, and lot tolerance percent defective (LTPD) which is a quality the customer would want.

PRODUCER'S RISK

This is the risk taken when a batch of goods of acceptable quality will be rejected by a sampling scheme as a result of a pessimistic-looking sample being drawn.

CONSUMER'S RISK

This is the chance that the goods will be accepted by the scheme as a result of an optimistic sample being drawn from a batch which should otherwise be rejected. Increased stringency does not necessarily bring increased discrimination in a sampling scheme and every scheme favours a particular quality level, i.e. the process average (PA) where

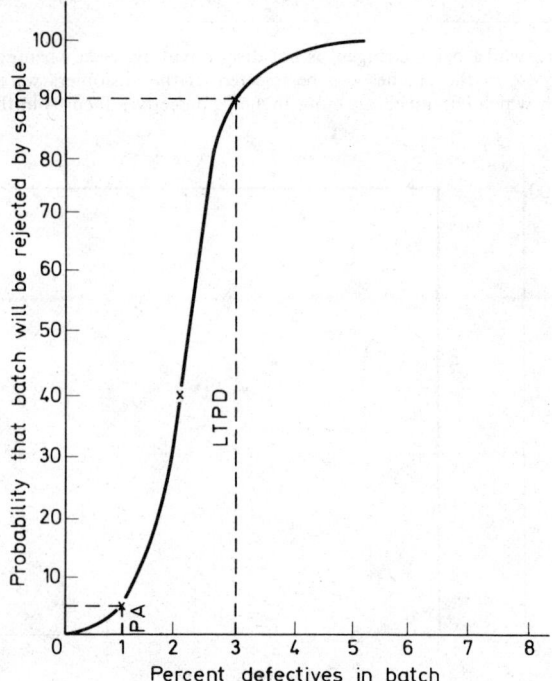

Figure 20.64. Operating characteristics curve (3)

increased stringency has the effect of penalising the producer without contributing to an equitable distribution of the risk.

A sampling scheme is required to give protection which may take the form of protection on every individual 'lot' in order to be sure that the quality of the 'lots' are constant, in this case a 'lot quality protection'.

When the 'lots' are mixed they lose their individuality, and inspection will only give 'average quality protection'.

SCHEMES

The amount of inspection performed in the 'long run' is made up of two distinct parts.

1. Inspection of samples of a batch.
2. 100% inspection of those batches which fail to be accepted by the sampling scheme.

Thus, as suppliers, quality falls off the volume of 100% inspection increases. This increased inspection costs and provides a compelling argument for remedial action.

Calculation of amount of inspection involved

For a given process average 'PA'% defective there will be a precise value for the probability of the allowable number of defectives d being exceeded by chance. This probability is the 'chance' that a 'lot' which is in fact of acceptable quality being rejected by the sampling scheme, i.e. the producer's risk.

The average amount of inspection per 'lot' for a given value of PA% is derived as follows:

Let $I = n = (N-n)R$.

where I = average amount of inspection and R = producer's risks.

EXAMPLE

Assume that the batch size $N = 1000$ units, and the process average (PA) is 1% defectives.

The scheme must ensure a consumer's risk of 10% of accepting batches which contain 5% defectives, in other words, sampling shall reject 90% of such batches. The scheme shall be the most economic possible for this degree of protection under normal conditions.

PROCEDURE

1. Tabulate the data as shown in Table 20.11 based on an assumed number of allowable defectives 'd' per sample.

Table 20.11

Allowed defect No. d	Expectation, e	Sample size, n
0	2·3	46
1	3·9	78
2	5·4	108
3	6·8	136
4	8·0	160
5	9·2	184

2. Since probability of rejection is 0·9 (90%) trace to this value on a left hand vertical scale (using probability paper) as illustrated in Figure 20.65.
3. From this point trace to the right for case 1, i.e. $d = 0$ until curve $c = 1$ is intersected, read off on horizontal base scale the expectation for the conditions.

4. As the 'lot' tolerance *defective* % must not exceed 5% of the sample size for an expected number of defectives:

$$\text{Sample size } n = \frac{e}{5} \times \frac{100}{1}$$

$$\text{Where } e = nP \text{ and } P = \frac{5}{100} \text{ hence } n = \frac{e}{p}$$

$$\text{Sample size } n = \frac{e}{5} \times 100$$

$$\text{Allowed defect No. } 0 = \frac{2\cdot3}{5} \times 100 = 46$$

$$1 = \frac{3\cdot9}{5} \times 100 = 78$$

$$2 = \frac{5\cdot4}{5} \times 100 = 108$$

$$3 = \frac{6\cdot8}{5} \times 100 = 136$$

$$4 = \frac{8\cdot0}{5} \times 100 = 160$$

$$5 = \frac{9\cdot2}{5} \times 100 = 184$$

To find the most economical scheme when work of normal quality is being submitted it is necessary to calculate the average amount of inspection using the expression $I = n + (N+n)R$ and tabulate as shown below in Table 20.12.

Table 20.12

1	2	3	4	5
Sample size	Allowable defectives	Expectation at $PA = nP$	Producers risk R	Average total inspected $n + (N-n)R$
46	0	0·46	0·37	399
78	1	0·78	0·19	253
108	2	1·08	0·10	197
136	3	1·36	0·06	188
160	4	1·6	0·03	175
184	5	1·86	0·01	192

Column 3 is found by multiplying sample size n by the PA expressed as a fraction defective, in this case 0·01. Knowing this value it can be deducted from the curve on the probability paper the producer's risk R for the scheme, entered in Column 4.

It would appear that the most economic procedure would be to submit lots of 1000; inspect sample of 160 with $d = 4$; and to involve overall inspection of 175 per 1000 on average.

OPERATING CHARACTERISTICS

The operating characteristics of the sampling scheme must be examined in order to assess the rate of batch rejection when changes in quality occur.

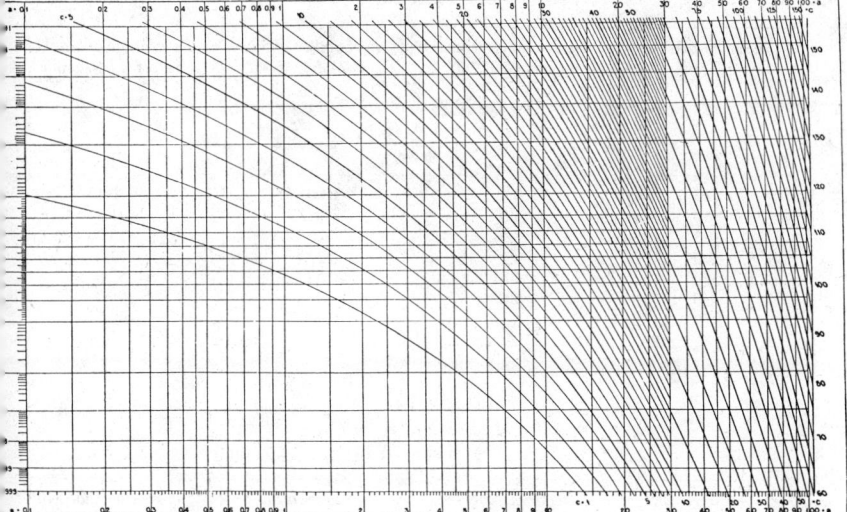

Figure 20.65. *Probability curves showing Poisson's exponential summation*

$$P = 1 - \left[1 + \frac{a}{1!} + \frac{a^2}{2!} + \ldots + \frac{a^{c-1}}{(c-1)!} \right] e^{-a}$$

for the probability P that an event occurs at least c times in a large group of trials for which the average number of occurrences is a. A scale proportional to the normal probability integral is used for P, a logarithmic scale for a

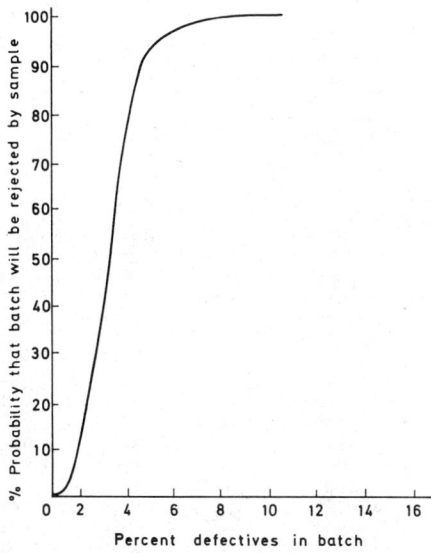

Figure 20.66. *Operating characteristics curve (4)*

Table 20.13

Assumed percent in batch	Defectives expected in sample of 160	Probability of 5 or more defectives	Percentage of batches rejected by sample
1	2	3	4
0·5	0·8	0·001 5	0·15
1·0	1·6	0·013	1·3
1·5	2·4	0·10	10·0
2·0	3·2	0·23	23·0
3·0	4·8	0·52	52·0
6·0	9·6	0·96	96·0
10·0	16	0·999 7	99·97

Table 20.14

Assumed % defectives in batch	Defectives expected in sample of 160 = 1	Probability of five or more defectives	% of batches that will be rejected by samples at the stated no. of defects in batch
0·5	0·8	0·001 5	0·15
1·0	1·6	0·013	1·3
1·5	2·4	0·10	10·0
2·0	3·2	0·23	23·0
2·5	4·0	0·38	38·0
3·0	4·8	0·52	52·0
6·0	9·6	0·96	96·0
10·0	16	0·999 7	99·97

Table 20.15

% Defects in batch	Fraction of batches rejected by sample	Avg. number per batch inspected as reminders	Number inspected as sample	Total avg. number inspected per batch	Fraction of batches not inspected	% Defects in batches after inspection
1	2	3	4	5	6	7
0	0	0	160	160	0·84	0
0·5	0·005	1·2	160	161·2	0·838 8	0·419 4
1·0	0·013	10·92	160	170·92	0·829	0·829
1·5	0·10	84	160	244	0·756	1·034
2·0	0·23	193·2	160	353·2	0·6468	1·293 6
2·5	0·38	319·2	160	479·2	0·520 8	1·302
3·0	0·52	436·8	160	596·8	0·403 2	1·209 6
6·0	0·96	806·4	160	966·4	0·033 6	0·201 6
10·0	0·999 7	839·7	160	999·7	0·000 3	0·003

A series of values for percent defective in a batch are chosen from which the number of defectives expected in, say, a sample of 160, are calculated. Using probability paper, the probability of five or more defectives can be obtained (i.e. probability of rejection) and converting this figure to a percentage will give the percentage of batches that will be rejected by the sample (Table 20.13).

By plotting the values of column 4 against column 1 in the table the curve of operating characteristics will be as shown in Figure 20.66. It can be observed that the rate of rejection increases significantly when batches contain an excess of 1·0% defectives. This offers considerable protection for the customer, but is rather severe on the producer allowing little deviation from his process average.

Average outgoing quality (AOQ)

If batches are 100% inspected and rejected, and the defectives removed, then the proportion of defectives remaining (taken over all the batches which leave the inspection department) will be less than the overall proportion of defectives arriving at inspection.

'Average outgoing quality' is the overall proportion of defectives in a large number of batches as they leave inspection, all the batches having the same proportion 'p' of defectives on arrival at inspection. It is assumed that all defectives found have been removed, and replaced by putting in 'good' items from stock.

Let k = number of batches (large).
n = number of items/batch (large).
p = proportion of defectives in each batch.
P = probability of acceptance of each batch, or, proportion of batches expected to be accepted.

Then:

Number of batches expected to be accepted = Pk
Number of pieces in accepted batches = NPk
Number of defectives accepted = $NPkp$
Total number of outgoing pieces = Nk
Number of outgoing defectives = $NPkP$

$$\text{AOQ} = \frac{NPkp}{Nk} = Pp$$

If batches arriving at inspection contain only a small proportion of defectives (p) few will yield a bad sample and will require 100% inspection. Consequently the AOQ will be marginally less than p. When the proportion is somewhat larger, more batches will be rejected and 100% inspected, and the proportion will be rather less than p. Ultimately as the proportion increases and the number of batches which will have 100 defectives becomes large, p will begin to reduce again. Consequently a graph of AOQ against p will rise initially, reach a maximum and then decrease as p increases.

The maximum value of AOQ is called 'average outgoing quality limit' and is the worst average quality (over a large number of batches) which may reasonably reach the stores, no matter how poor the quality of the batches when they arrive at inspection.

Using data from a previous example: batch size N = 1000 units, sample size n = 160 and allowable defectives d = 4, the results are shown in Table 20.14 and from which the curve of operating characteristic can be constructed.

For any given value of percentage defective in the batch the AOQ, i.e. the fraction of batches rejected by the sample, can be obtained from the operating characteristic (Table 20.15).

From the expression $(N-n)R$ the average number inspected as remainders can be determined (column 3). The values in column 5 are obtained from $Nr(N-n)R$ and that of column 6 from to $[N-(n+(N-n)R)]/N$ and finally the values of column 7 are obtained by multiplying the values of columns 1 and 6. This is the average outgoing quality.

When columns 1 and 7 of Table 20.40 are plotted the average outgoing quality curve is obtained as shown in Figure 20.67.

It is seen from Figure 20.67 that there is a maximum outgoing quality at about 1·31%. This is known as the 'average outgoing quality limit' and occurs when batches contain

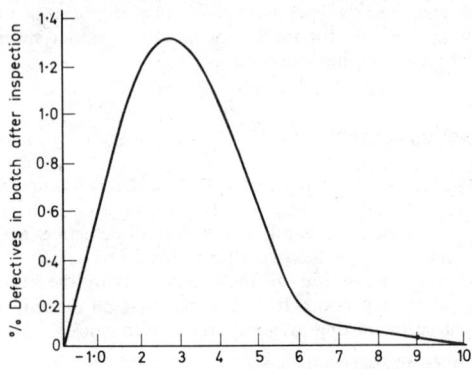

Figure 20.67. Average outgoing quality curve

an average of 2·4% defectives, implying that less control is exerted by inspection under these stated conditions than when any other percentage defects exist.

Double sampling

So far only single sampling schemes have been considered in which the decision to accept or reject a batch was based on the evidence of one sample. Unfortunately this makes no allowance for the possibility of selecting an 'optimistic' or 'pessimistic' sample by pure chance. Furthermore, when first introduced to sampling, prospective users are doubtful about its validity and often like to take a second sample if the first seems to give an 'unsatisfactory' or unexpected result. It is frequently the practice to ignore the less satisfactory result, and it is of course, impossible to determine the cover afforded by such a scheme. If, however, proper rules are adhered to, a 'double sampling' scheme may be used which can lead to less sampling than that required in single sampling plan.

Previous statements indicate that a pessimistic sample may have been drawn and the question arises, should a second sample be taken? If so then the results of the first sample should be incorporated with the results of the second sample, so that a fair assessment of the quantity inspected can be obtained. Any second sample taken after an unfavourable first must be determined in accordance with probability theory, so that the necessary protection built into the scheme does not suffer as a result of taking a second sample.

With this 'second chance' arrangement the schemes can be designed to be more economical in terms of average inspection, and in fact the more chances there are the more economical the scheme, i.e. smaller sample sizes. The procedure for double sampling is shown diagrammatically in Figure 20.68 and is as follows:

1. Inspect first a sample of n_1 items.
2. If the number of defectives found in the first sample does not exceed d_1 (the acceptance number for the first sample), then accept the batch.
3. If the number of defectives found in the first sample exceeds d_2 (the acceptance number for the combined first and second samples), then inspect the remainder of the batch.

4. If the number of defectives found in the first sample exceeds d_1, but does not exceed d_2, then inspect a second sample of n_2 pieces.
5. If the total number of defectives found in the first and second samples combined does not exceed d_2, accept the batch.
6. If the total number of defectives found in the first and second samples combined exceed d_2, then inspect the remainder of the batch.
7. Correct or replace all defective pieces found.

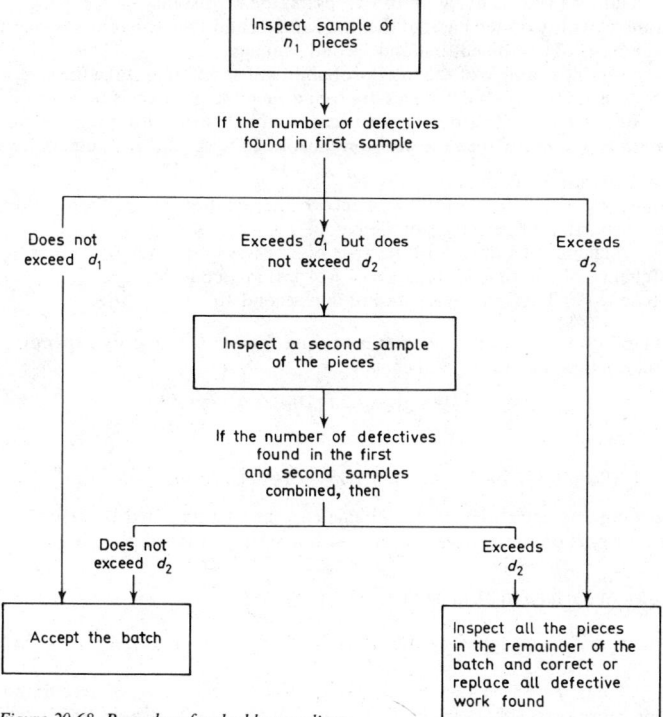

Figure 20.68. Procedure for double samplings

Consumer protection

For both single and double sampling procedures statistical tables have been developed for each of the following two kinds of consumer protection.

LOT QUALITY PROTECTION

In lot quality protection there is prescribed some chosen value of limiting percent defective in the 'lot tolerance defective' and some chosen value for the probability of accepting a submitted batch that has a percent defective equal to the 'lot tolerance percent defective'. This probability is known as the 'consumer's risk'.

AVERAGE QUALITY PROTECTION

In average quality protection there is prescribed some chosen value of average percent defective in the product after inspection (average outgoing quality limit) that shall not be exceeded whatever the level of percent defective in the product submitted to the inspector. Lot sizes of $N = 1$ to $N = 100\,000$ for all practical values of lot tolerance percent defective and AOQ limit up to 10% are catered for.

In the 'lot tolerance' tables for each sampling scheme, the AOQL is derived, and a table of this type is prepared for a wide variety of 'lot sizes' to determine the AOQL for each 'lot size' selected.[48] In addition, the tables show the 'lot tolerance percent defective' based on a consumer's risk of 10% as well as 'double sampling' and 'single sampling' computation.

The 'consumer's risk' is made up of two parts, the probability of accepting a 'lot' of 'lot tolerance' quality on the basis of the first sample, and the probability of accepting a 'lot' on the basis of combined first and second samples.

To give a simple example of the total probability of acceptance, take the case where a sample of 10 is drawn, if no defectives are found then the decision is to accept the 'lot', but if 1, 2 or 3 defectives are found take a second sample of 10; if total defectives in the two samples is 4 or more, then the decision must be to reject the 'lot'; otherwise accept.

Total chance of acceptance $= P_1 + P_2$,
Where $P_1 = q^{10} = (1-p)^{10} =$ chance of acceptance on first sample and P is the proportion of defectives in the batch.
$P_2 =$ (chance of 1 in first 10) (chance of 2 or less in second 10)
$+$ (chance of 2 in first 10) (chance of 1 or less in second 10)
$+$ (chance of 3 in first 10) (chance of 0 in second 10)

The probability of 2, 1 or 0 defectives in second sample is found by expanding to the third term the binomial $(q+p)^{10}$, i.e.

$$(q^{10} + 10q^9 p + 45q^8 p^2 + \ \ldots \ldots)$$

P_2 now becomes:

$$10q^9 p(q^{10} + 10q^9 p + 45q^8 p^2) + 45q^8 p^2 (q^{10} + 10q^9 p) + 120q^7 p^3 q^{10}$$

The operating characteristics of the scheme can then be obtained by assuming values for p, the proportion of defectives in the batch and calculating $P = P_1 + P_2$ when

$$p = 0,\ 0{\cdot}02,\ 0{\cdot}04,\ 0{\cdot}06,\ 0{\cdot}08,\ 0{\cdot}1,\ 0{\cdot}15,\ 0{\cdot}2,\ 0{\cdot}25,\ 0{\cdot}3,\ 0{\cdot}4$$

$$P = P_1 + P_2\ 1,\ 0{\cdot}999,\ 0{\cdot}993,\ 0{\cdot}971,\ 0{\cdot}934,\ 0{\cdot}868,\ 0{\cdot}659,\ 0{\cdot}417,\ 0{\cdot}249,\ 0{\cdot}104,\ 0{\cdot}02$$

Plotting P against p it is seen that there is a risk of about 10% of accepting batches containing more than 30% defectives and a risk of about 10% of rejecting batches containing fewer than 9% defectives.

Sequential sampling

For a batch of 500 components, the average amount of single sampling inspection would be 147 components per batch. In the case of a double sampling scheme, under the same conditions, the total of the two samples would amount to 135 components, a saving of twelve components per batch. Improved economy in average sample size can be achieved by the use of a more sophisticated sequential sampling scheme.

Consider first the case where a batch of components are inspected and classified GOOD or BAD, i.e. inspected in what is essentially a binomial distribution problem $(P+Q)^n = 1$.

To solve this problem by the sequential technique is to first of all lay down two values for the fraction defective, one of which is regarded as GOOD and the other as BAD. If, for example 1% defectives are regarded as GOOD and 5% defectives as BAD in a batch, then the risk of rejecting a batch for which the percentage of defectives was less than 1% should be small, and the chance of accepting a batch where the defectives exceed 5% should also be small.

The chance of a good batch being rejected is the 'producer's risk' and the chance of a bad batch being accepted is the 'consumer's risk'. These four factors:

1. Bad quality
2. Good quality
3. Producer's risk
4. Consumer's risk

are sufficient to determine a sequential sampling plan and the following notation is used:

P_1 = acceptable quality expressed as a fraction defective
α = probability of accepting a batch of acceptable quality
p_2 = unacceptable quality expressed as a fraction defective
β = probability of accepting a batch of unacceptable quality.

For the sequential test one item is selected at random from the batch and inspected, if it is 'bad' it is plotted on the control chart on which the number of defectives are on the ordinate and number of items inspected on the abscissa. The position of the two control lines is derived by:

The upper limit $= Sn + h_2$
The lower limit $= Sn - h_1$.

Where S, h_1 and h_2 are constants and N = number of components inspected.

If P_1, α, P_2 and β are given then the constants h_1, h_2 and S can be calculated from the following expressions:

$$h_1 = \frac{\log \frac{(1-\alpha)}{(\beta)}}{\log \frac{P_2(1-P_1)}{P_1(1-P_2)}} \qquad h_2 = \frac{\log \frac{(1-\beta)}{(\alpha)}}{\log \frac{P_2(1-P_1)}{P_1(1-P_2)}} \qquad S = \frac{\log \frac{(1-P_1)}{(1-P_2)}}{\log \frac{P_2(1-P_1)}{P_1(1-P_2)}}$$

Let

$$g_1 = \log \frac{P_2}{P_1} \qquad g_2 = \frac{(1-P_1)}{(1-P_2)}$$

$$a = \log \frac{(1-\beta)}{(\alpha)} \qquad b = \log \frac{(1-\alpha)}{(\beta)}$$

Then

$$h_1 = \frac{b}{g_1 + g_2} \qquad h_2 = \frac{a}{g_1 + g_2} \qquad S = \frac{g_2}{g_1 + g_2}$$

It will be observed that when $\alpha = \beta$, $h_1 = h_2$.
Having calculated the constants the inspection chart (Figure 20.69) can be constructed.

EXAMPLE

Design a sequential sampling scheme to cover the following case. Two per cent defective is considered acceptable quality and the risk of rejecting a batch as good as this is to be

$p = 0.1$. A batch of 5% defective is considered so bad that the probability of it being accepted is only $p = 0.1$. Plot the operating characteristic, outgoing quality curve and the average sample number curve of the scheme.

To calculate h_1, h_2 and S the characteristic constants of the sequential scheme, given that:

$$P_1 = 0.02; \quad \alpha = 0.10; \quad P_2 = 0.05; \quad \text{and} \quad \beta = 0.10.$$

Since $\alpha = \beta$, $h_1 = h_2$.

Therefore

$$g_1 = \log \frac{P_2}{P_1} = \log \frac{0.05}{0.02} = \log 2.5 = 0.3979;$$

$$g_2 = \log \frac{(1 - P_1)}{(1 - P_2)} = \log \frac{0.98}{0.95} = \log 1.03 = 0.012\,8;$$

$$a = \log \frac{(1 - \beta)}{(\alpha)} = \log \frac{0.90}{0.10} = \log 9 = 0.954;$$

Hence:

$$h_1 = \frac{b}{g_1 + g_2} = \frac{0.954}{0.3979 + 0.0128} = \frac{0.954}{0.4107} = 2.323;$$

and $h_2 = 2.323$

$$S = \frac{g_2}{g_1 + g_2} = \frac{0.0128}{0.4107} = 0.03117;$$

When the sample size $n = 0$ the rejection line passes through point h_2 on the axis for d, the number of defectives and the acceptance line passes through the point h_1 on the

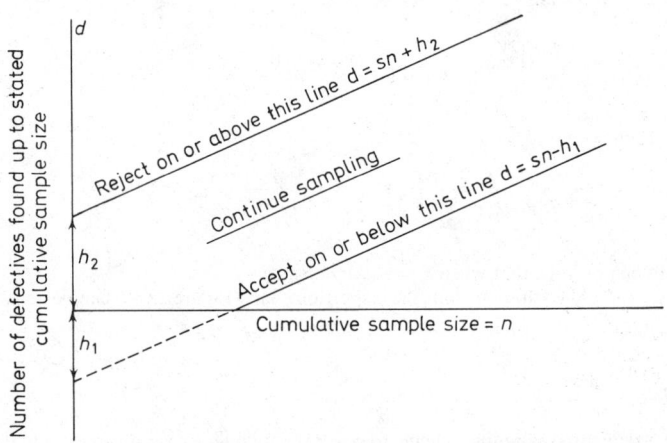

Figure 20.69. Sequential sampling inspection chart (1)

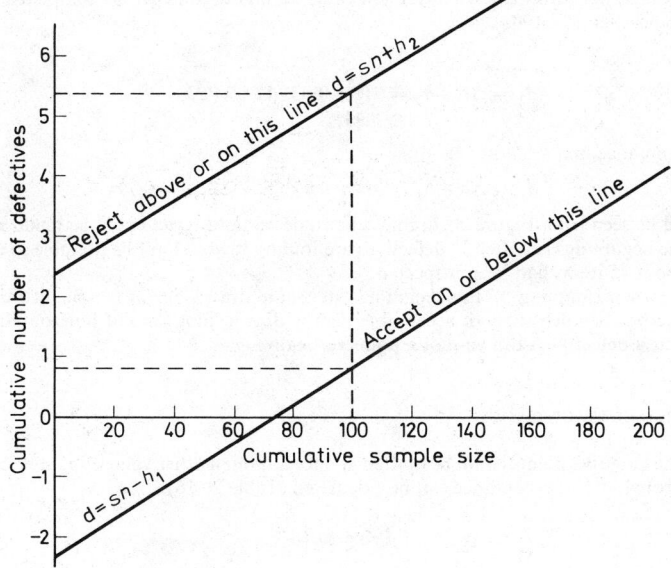

Figure 20.70. Sequential sampling inspection chart (2)

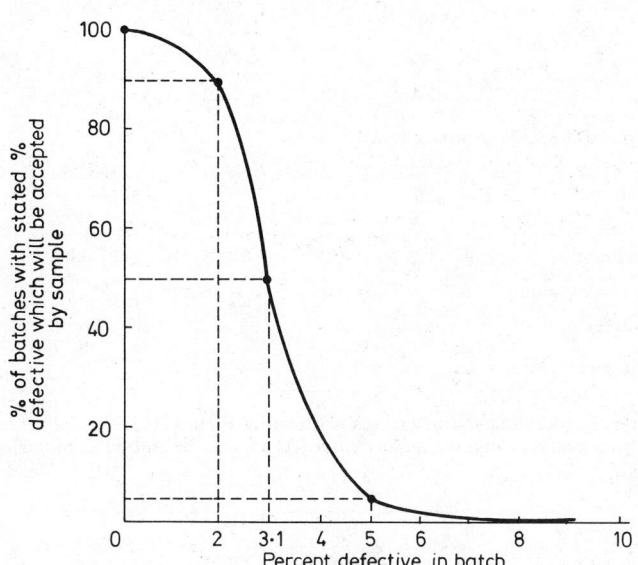

Figure 20.71. Operating characteristics

same axis. By substituting any convenient value for n both lines can be completed fixed; for instance: let $n = 100$.

Rejection line,

$$d_2 = SN + h_2 = 100 \times 0.031\ 17 + 2.323$$
$$= 5.44;$$

and acceptance line

$$d_1 = SN - h_1 = 100 \times 0.031\ 17 - 2.323 = 0.794.$$

It will be seen from Figure 20.70 that while a decision to reject could be made almost from the beginning (i.e. after 32 defectives are found), it would not be possible to accept until about 75 items had been inspected.

Thus, when sampling by this technique, items are drawn one at a time, at random, from the batch. Each time an acceptable item is drawn, plot a point horizontally and each time a defective is drawn plot a point vertically.

OPERATING CHARACTERISTICS

From the data five points could be plotted at once although other values might be chosen if the probability of acceptance can be calculated (Table 20.16).

Table 20.16

Lot fraction defective	Probability of acceptance
0	1
P_1	$1 - \alpha$
S	$\dfrac{h2}{h1 + h2}$
P_2	β
1	0

For example in the scheme just designed

$\alpha = 0.1; \beta = 0.1; h1 = 2.323; h2 = 2.323; P_1 = 0.2; P_2 = 0.05$ and $S = 0.3117$

Lot fraction defective	0	$P_1 = 0.02$	$S = 0.031\ 17$	$P_2 = 0.05$	1
Probability of acceptance	1	$(1 - \alpha) = 0.9$	$\dfrac{h_2}{h_1 + h_2} = 0.5$	$\beta = 0.1$	0
I.E. % of batches accepted is	100	90	50	10	0
When % defectives in lot is	0	2	3.1	5	100

The curve of operating characteristics is shown in Figure 20.71.

Similarly an average outgoing quality curve (AOQ) may be plotted from the following five points:

Lot fraction defective	0	P_1	S	P_2	1
AOQ Fraction defective	0	$(1 - \alpha)P_1$	$\dfrac{Sh_2}{h_1 + h_2}$	βP_2	0

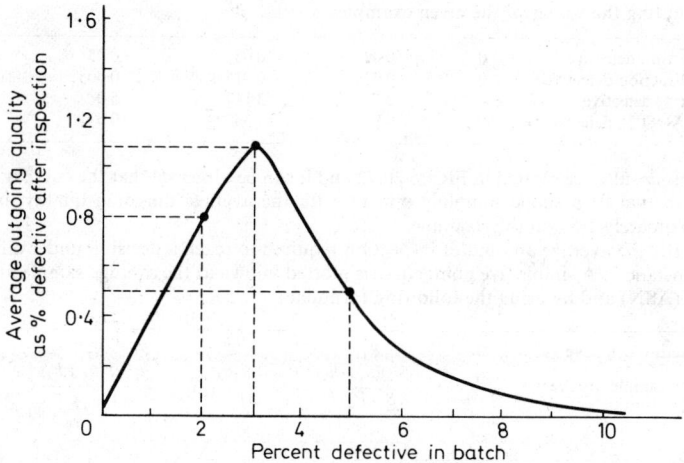

Figure 20.72. Outgoing quality curve

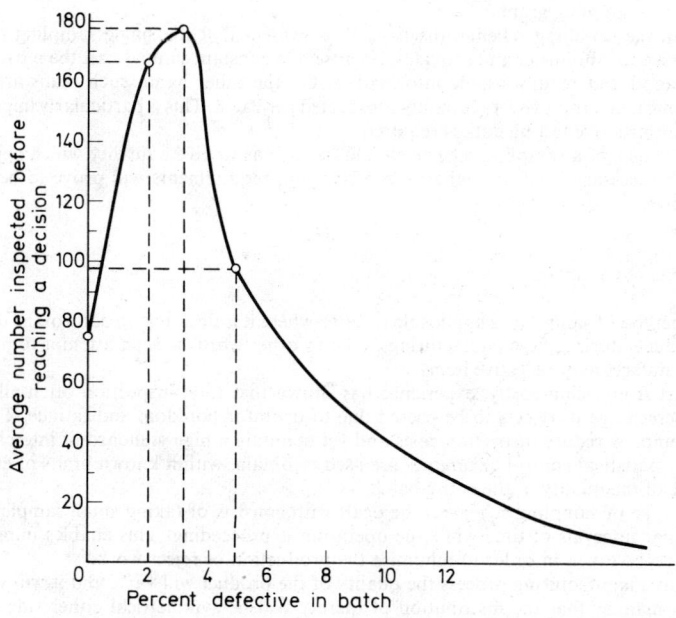

Figure 20.73

Substituting the values of the given example

Lot fraction defective	0	0·09	0·031 17	0·05	1
AOQ Fraction defective	0	0·018	0·015 4	0·005	0
i.e. Lot % defective	0	2	3·117	5·0	100
Gives AOQ % defective	0	0·8	1·54	0·5	0

These results are plotted in Figure 20.72 and it can be observed that the curve is of the form to that of a single sampling scheme with the average outgoing quality limit at approximately 1·6% in this example.

Finally the average amount of inspection required to reach a decision under different circumstances. A simple five point curve is plotted known as the average sample number curve (ASN) and by using the following formulae:

Lot fraction defective	0	P_1	S	P_2	1
Average sample number	$\dfrac{h_1}{s}$	$\dfrac{h_1 - \alpha(h_1 + h_2)}{S - P_1}$	$\dfrac{h_1 h_2}{S(1 - S)}$	$\dfrac{h_2 - \beta(h_1 + h_2)}{P_2 - s}$	$\dfrac{h_2}{1 - s}$

Lot fraction defective	0	0·02	0·031 17	0·05	1
Average sample number	75	167	178	98	2·397
% Defective in lot	0	2	3·117	5	100

The average sample curve is shown in Figure 20.73 from which it is seen that an average of 180 items per batch would be inspected before the batch could be accepted.

In order to reduce calculations, sampling tables[49] show the acceptance and rejection number of defectives for any sample size thus making it unnecessary to construct the sequential sampling graph.

From the sampling schemes discussed it is evident that the single sampling plan is the easiest to administer and provides, because of a constant sample size, the most easily understood and readily usable information. On the other hand, such plans are least economical in terms of average number inspected per batch. This is particularly important where destructive testing data is required.

The choice of a sampling scheme should be such as to give a quality which is just as high as necessary—but no higher; more stringent requirements will prove to be more expensive.

Sampling of variables

This method of sampling is applicable to cases where it is desirable to control the quality of products during the manufacturing cycle. In other words it is an attempt to prevent the manufacture of defective items.

Apart from being costly, experience has shown that total inspection often allows a high percentage of rejects to be passed due to operator boredom and fatigue. Thus in an attempt to reduce inspection costs and yet maintain a high standard of interchangeability, statistical control techniques are used to obtain, within known limits of error, a picture of the quality of the entire batch.

The type of sampling scheme to be dealt with consists of taking small samples 2–10 at regular intervals of time while the operation is proceeding. This enables immediate action to be taken in order to minimise the production of reject work.

In any manufacturing process the quality of the product will vary, and it will vary in such a manner that the distribution of quality will be symmetrical either side of the quality mean.

If many components from a batch, and the frequency with which each dimension occurs are plotted to form a frequency distribution, it will be found that the frequencies group themselves round a central value in a reasonably symmetrical curve approximating to the 'normal' distribution shown in Figure 20.75. When an infinitely large number of

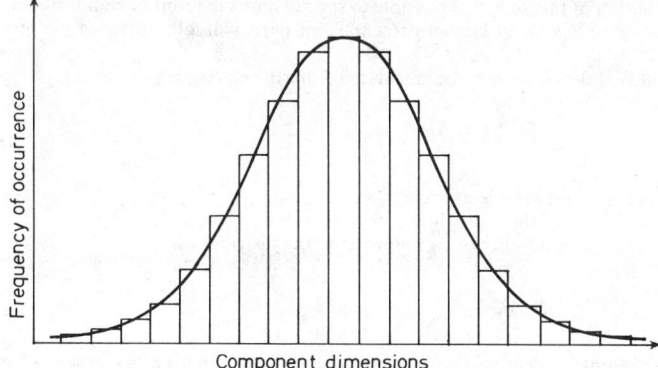

Figure 20.74. Distribution of component sizes in a batch

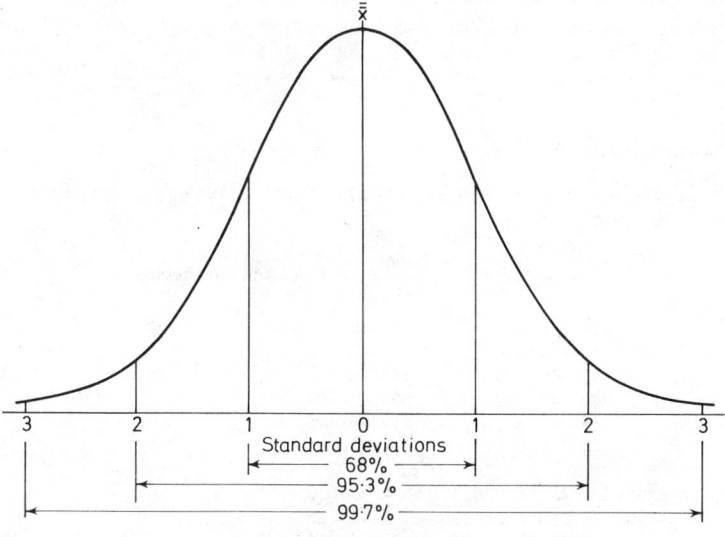

Figure 20.75. Dispersion of population in a normal distribution

components are measured then the distribution loses the stepped characteristics of Figure 20.74 and becomes a smooth 'normal' distribution curve of Figure 20.75.

Dispersion of population in a normal distribution

The central value, $\bar{x}$, in this case is also equal to the mean value. The most convenient measure of dispersion for a normal distribution is the standard deviation; the area under the curve represents total production and the manner in which it is distributed. Thus it will be seen from Figure 20.76 that 68% of total production lies within $\pm 1\sigma$ one standard deviation from the mean and similarly 95·5% of production lies within $\pm 2\sigma$ and 99·7% of production within $\pm 3\sigma$. If the value of standard deviation is found for any distribution of this sort, it is possible to specify limits in terms of standard deviation, outside which only a small known percentage of parts will fall, provided the process is in control.

The standard deviation may be calculated from the equation

$$\sigma = \sqrt{\left\{\frac{\Sigma f(X - \bar{\bar{X}})^2}{N}\right\}}$$

where $\bar{\bar{X}}$ is the grand sample average
X is the size of the individual component
N is the total number of components in the distribution.

EXAMPLE

A typical example of determination of the standard deviation is a case where a hardness test (Rockwell) was taken off a number of gear blanks.

Hardness No.	41	42	43	44	45	46	47	48	49	50	51	Total
Frequency	1	2	5	15	25	30	25	15	5	2	1	126

SUM OF FREQUENCIES

$41 \times 1 + 42 \times 2 + 43 \times 5$ etc. $= 41 + 84 + 215 + 660 + 1125 + 1380 + 1175 + 720 + 245 + 100 + 51 = 5796$

$$\text{Average } \bar{\bar{x}} = \frac{5796}{126} = 46 \text{ Rockwell hardness}$$

$$
\begin{aligned}
(x - \bar{\bar{x}}) = \ & 41 - 46 = 5 & 47 - 46 = 1 \\
& 42 - 46 = 4 & 48 - 46 = 2 \\
& 43 - 46 = 3 & 49 - 46 = 3 \\
& 44 - 46 = 2 & 50 - 46 = 4 \\
& 45 - 46 = 1 & 51 - 46 = 5 \\
& 46 - 46 = 0 &
\end{aligned}
$$

$(x - \bar{\bar{x}})^2 = $ 25, 16, 9, 4, 1, 0, 1, 4, 9, 16, 25

$$
\begin{aligned}
p(x - \bar{\bar{x}})^2 = \ & 25 \times 1 = 25 & 1 \times 25 = 25 \\
& 16 \times 2 = 32 & 4 \times 15 = 60 \\
& 9 \times 5 = 45 & 9 \times 5 = 45 \\
& 4 \times 15 = 60 & 16 \times 2 = 32 \\
& 1 \times 25 = 25 & 25 \times 1 = 25 \\
& 0 \times 0 = 0 & \text{Total} \quad 374
\end{aligned}
$$

Quality Control Chart

In order that close control may be kept on production, data is plotted on a 'quality control chart' which indicates changes in the state of the process. As samples are taken at regular intervals of time from the current production it is possible to keep a close check on any process variability. If, for example, a sample of 4 were inspected it would be impracticable to plot the dimensions of the individual components on the control chart, yet suitable to plot the sample average. The effect of this is to make the distribution of mean sample sizes round the grand sample average more compact, i.e. the dispersion and the sample standard deviation decreases as the sample size increases.

The standard deviation of the sample average becomes:

$$\sigma n = \frac{\sigma}{\sqrt{n}}$$

where n is the sample size. The above equation is often referred to as the 'standard error of the mean'. Thus it follows that the larger the sample size the closer to the mean will

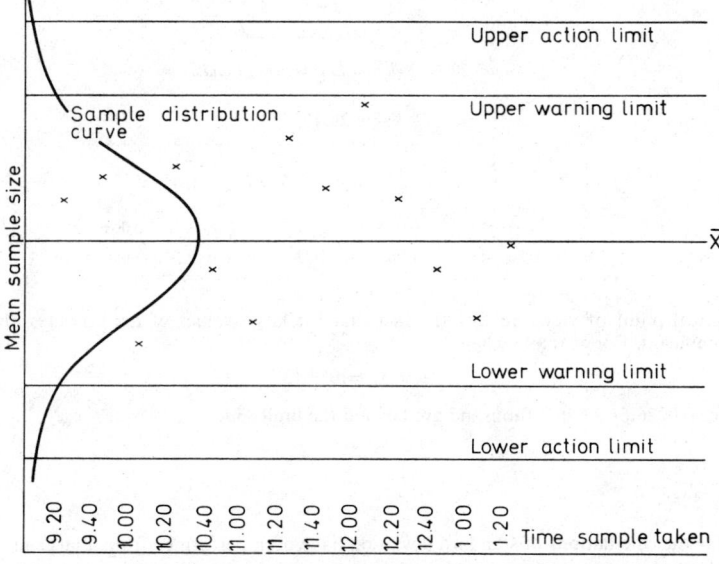

Figure 20.76. Typical mean chart showing action and warning limits

be the $+3\sigma$ control limits on the control chart for 'means' an example of which is shown in Figure 20.76.

Distribution Relationship

In Figure 20.77 the central ordinate is $\bar{X}$ the mean value of X with ordinates at $\bar{\bar{X}}+t\sigma$ and $X-t\sigma$ (i.e. at $\pm t$ times σ) the shaded area to the right and left of $\bar{x}\pm t\sigma$ is proportional to the frequency of observations having values of greater or less than $\bar{x}\pm t\sigma$ as the curve is symmetrical. Table 20.17 shows some of the relationships between t and α for the normal distribution.

Theoretically the normal distribution extend to infinity in both directions of X so that even when t has no value α never reaches zero. This would be ridiculous from a

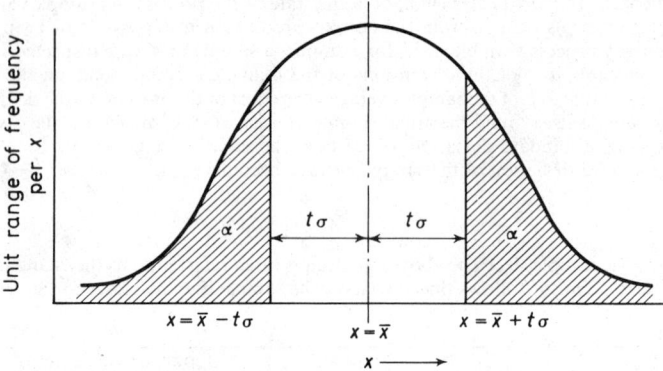

Figure 20.77. Relationships between t and α

Table 20.17

t	1	2		3	
α	0·159	0·023		0·001 3	
α	0·05	0·025	0·01	0·005	0·001
t	1·645	1·96	2·23	2·58	3·09

practical point of view, but for the fact that for large values of t α becomes almost insignificant. For example when

$$t = 3\alpha = 0{\cdot}0013$$

hence 0·0026 or 2·6 in a thousand are beyond the limits 3σ.

The range

So far the calculations of standard deviation have been performed from a large number of observations, in practice however there is often only a small number or a collection of groups of small numbers available in which case the standard deviation may be calculated for each sample and averaged to give σ. It is more convenient (particularly if the sample is less than 20) to use a measure of variation the 'mean range' ϖ. Whilst this 'mean range' is convenient to use, it must be handled carefully for it depends on the number of observations per sample. If the observations are distributed normally, then σ can be calculated according to the standard deviation equation on the previous page.

Where n is very large and the mean range in a very large number of samples 'drawn at random' is related as shown in Table 20.18.

Table 20.18

Number in sample	2	3	4	5	10
Mean range/std. dev. ϖ/σ	1·128	1·693	2·059	2·326	3·078

The assumption of normality is not important in practice, as the relationships do not differ greatly from those given in Table 20.18 provided the samples are 'drawn at random' the departures from normality can be quite large.

The fraction defective

Frequently the proportion of values which are above, below or between certain limits are more important than the whole distribution. For example, in the manufacture of 'gudgeon pins' it is essential to know the proportion of pins which fall outside specified tolerance limits known as the 'fraction defective' and represented by p' when determined for the whole batch and by P for a sample within a batch.

This fraction is often expressed as a percentage and if it is defined by the limits of some measurable quantity distributed normally, then it is related to the mean $\bar{X}$ the standard deviation σ and the limits $\bar{X} \pm t\sigma$. The result has practical use in the setting of design tolerances.

EXAMPLE

A shaft has to be made a certain diameter with tolerances of $+0.000$ and -0.0254 mm. Of those made 0.45 (45%) were defective, i.e. outside the specified limits. The question is, whether this fraction defective is due to the general variability of the lathe or due to other factors which could be corrected by better supervision and control by the operator. (It is generally accepted that lathes do produce diameters with an uncontrollable variation, a characteristic of the state of repair and maintenance of the machine, and often this variation is distributed approximately 'normally'.)

Measurements on 12 samples of 5 shafts produced a mean range of 0.0348 mm leading to an estimate of the standard deviation of 0.0348/2.326 (from $w/\sigma \equiv 2.326$

Table 20.19

Sample size	Value of RPI		
	Low	Medium	High
2	Less than 6	6 to 7	Greater than 7
3	Less than 4	4 to 5	Greater than 5
4	Less than 3	3 to 4	Greater than 4
5 and 6	Less than 2·5	2·5 to 3·5	Greater than 3·5

for a sample of 5) = 0.0150 mm. The most favourable machine setting therefore would be such as to produce a mean diameter of -0.0127 mm so that the tolerance limits would be ± 0.0127 mm relative to the mean and t would be $0.0127/0.0150 = 0.85$.

It can be deduced from Table 20.17 that for a value of 0.85 for t, the corresponding value for α is greater than 0.159; from more complete tables α is shown to be 0.20. Thus, on this basis, the fraction defective would be $2 \times 0.20 = 0.40$ (40%) which is reasonably near the observed fraction of 0.45 (45%). Since the 5 shafts in each sample were made consecutively their range measures an uncontrollable variation indicating that the machine should be overhauled.

On reconditioning the machine, the standard deviation was reduced to 0.0061 mm and produced a value of $t = 0.0127/0.0061 \simeq 2.00$ with approximately 5% defectives (Table 20.17). Whilst this shows a considerable improvement it is not good enough, since changes due to tool wear add to the defectives. There is, however, the possibility of increasing the tolerance range; it may well be that the designer has specified tolerances so that there shall be no misfits. But if the variation was normal, the fraction of shafts and holes at the extremes of the limits would be small and the fraction of misfits in the assembly of shafts and holes, taken at random, would be even smaller. For example, if there were 0·05 defective shafts and holes then the fraction of defective assemblies would be $0·05 \times 0·05 = 0·0025$, and if this small fraction is acceptable, the tolerance limits could be relaxed so that most of the 0·05 previously rejected would be accepted.

In BS 2564[52] the condition of a machine tool is indicated by the ratio:

$$\frac{\text{(total) drawing tolerance}}{\text{Mean range}}$$

and this is referred to as the relative precision index (RPI). Table 20.19 provides suitable values for this ratio classifying as low, medium and high relative precision.

The RPI established for any machine will be of considerable value to the planning engineer because he is able to allocate work to a machine with which it is fully capable of coping.

Range chart

To exercise complete control over a process, a range chart must also be used. This chart records the maximum difference between the largest and smallest components in a sample and incorporates action and warning limits for sample range sizes. As the sample size increases the average range also increases, so the limits on the range chart, when taking larger samples, will embrace a larger variation of size than when taking smaller samples.

Control limits may be set to contain any required percentage of production, but it is usual for the action limits on the 'means chart' to contain 99·8% of the samples ($+3·09$)

Table 20.20 CONTROL CHART LIMITS FOR SAMPLE AVERAGE ($\bar{x}$)

Sample size n	For inner limits a' 0·025	For outer limits A' 0·001
2	1·23	1·94
3	0·67	1·05
4	0·48	0·75
5	0·38	0·59
6	0·32	0·50

Table 20.21 CONTROL CHART LIMITS FOR RANGE (ϖ)

Sample size n	For lower limits D' 0·025	D' 0·025	For upper limits D' 0·975	D' 0·999	For standard deviation dn
2	0·00	0·04	2·81	4·12	1·13
3	0·04	0·18	2·17	2·98	1·69
4	0·1	0·29	1·93	2·57	2·06
5	0·16	0·37	1·81	2·34	2·33
6	0·21	0·42	1·72	2·21	2·53

and for the warning limits to contain 95% ($+1·96$) of the samples. In other words one sample in every 1000 can be expected to fall outside the action limits. Such limits are called the '1 in 1000 control limits' and '1 in 40 warning limits'. Although these limits are calculated from a knowledge of the standard deviation, constants have been evaluated for different sample sizes which enable the limits to be calculated directly from the grand sample average and the sample range without reference to the standard deviation.

When finding control limits a number of samples are checked at regular intervals of time and the sample average x and sample range ω are tabulated. After taking about ten samples the grand sample average $\bar{\bar{x}}$ and the average range, ϖ are found. The data tabulated in Tables 20.20 and 20.21 can then be used to construct the 'means chart' and the range chart respectively.

$$\left. \begin{array}{l} \text{Warning limits} = \bar{\bar{X}} \pm A \; 0 \cdot 025 \; \varpi \text{ and} \\ \text{Action limits} \quad = \bar{\bar{X}} \pm A' \; 0 \cdot 001 \; \varpi \end{array} \right\} \text{Table 20.20}$$

$$\left. \begin{array}{l} \text{Warning limit} = D' \; 0 \cdot 975 \; \varpi \text{ and} \\ \text{Action limit} \quad = D' \; 0 \cdot 999 \; \varpi \end{array} \right\} \text{Table 20.21}$$

Additionally, there are lower action and warning limits for the range chart which are of little use in practice.

These limits are now drawn on a chart similar to that of Figure 20.69 for both the 'averages' and 'ranges'. From the pattern formed it is possible to determine whether the process is stable, i.e. whether it continues with the original degree of accuracy and uniformity. The distribution of points on the 'means chart' will indicate the amount of tool wear, the ability of the setter to set near the mid-tolerance and the effectiveness of the machine to produce a given size without drifting toward either limit. Points on the range chart will show the ability of the machine or operator to repeat consistently. Signs of instability are when a point occurs near the action limit or when two or more points in rapid succession appear near the warning limits.

Sample size

It is usual to use an empirical approach when determining the sample size on the grounds that it seems reasonable to the practical man. It conforms to past practices and should just about keep a reasonable inspection staff busy. Where the articles are measured, the number per sample is usually in the neighbourhood of 5 or 10.

Modified control limits

The statistical limits bear no relation to the design limits, and although the charts described may show the process to be stable, stability is of no use if the components are being produced outside limits. In order that there shall be no appreciable scrap it is necessary that the whole distribution should fall within design limits.

As 99·7% of components fall within 3σ on either side of the mean, allowing the possibility of 0·3% defectives being passed, the production is satisfactory provided the tolerance between top and bottom design limits exceed 6σ. Since samples are taken of n components, the action limits on the mean chart will be $6 \cdot 18\sigma / \sqrt{n}$ apart, and the design tolerance must then equal or exceed the difference between top and bottom control limits. Constants have been calculated to allow critical examination of control charts without first finding the standard deviation. The fundamental requirements are:

1. The mean sample range must not exceed $L \times$ design tolerance.
2. The means chart control limit must lie between values of $A''_{0 \; 001} \varpi$ above the bottom design and below the upper design limit.

Sometimes the process is controlled well within the drawing tolerance and if the statistical limits were worked to, the components would be made to an unnecessarily high standard of perfection.

To make these limits more realistic a drift is allowed on the mean value so that the tails of the distribution centred round the displaced mean fall within the design limits by an amount equal to $\varpi A''_{0\ 001}$ as shown in Figure 20.78.

The modified limits will then be as follows:

    Upper action limit   = Upper design limit $-\varpi A''0\cdot001$
    Uppe. warning limit = Upper design limit $-\varpi A''0\cdot025$

    Lower warning limit = Lower design limit $+\varpi A''0\cdot025$
    Lower action limit   = Lower design limit $+\varpi A''0\cdot001$

The constants are shown in Table 20.22 and are recommended in BS 2564.[50]

Table 20.22

Sample size n	Modified limits		Critical examination of range chart
	Warning A 0·025	Action A 0·001	L
2	1·51	0·80	0·18
3	1·16	0·77	0·27
4	1·02	0·75	0·33
5	0·95	0·73	0·37
6	0·90	0·71	0·41

Control charts for qualitative data

Many characteristics of a product cannot be measured quantitatively, the most extreme case in which no 'measure' of quality is available is when manufactured articles are

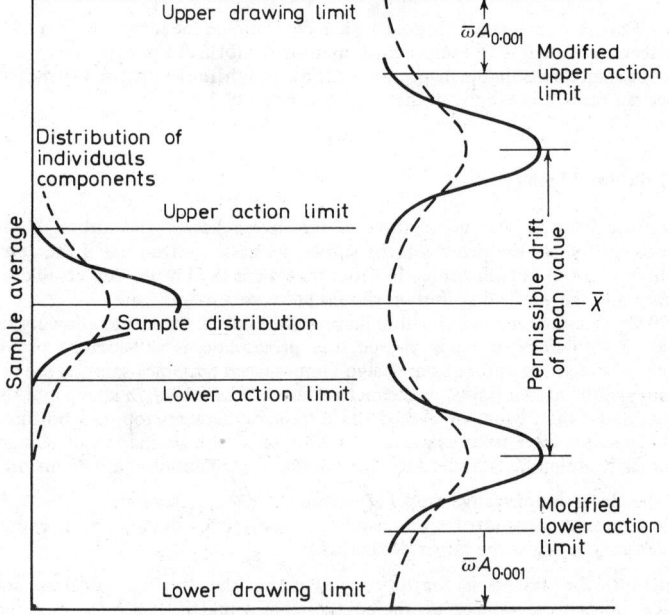

Figure 20.78. Modified control limits

classed as either 'good' or 'bad'. The only information given being the proportion of articles in each class.

An associated problem relates to those articles which, although not completely defective in themselves, may have a number of defects which mar the finished product. Examples of this type are the number of defective rivets in an aircraft wing; the structure may be able to tolerate a number of defects before becoming totally defective. The only practical objective being to ensure that the defects do not increase above a specified level.

Thus we use the word 'defective' to signify that an article fails to reach some standard; while a 'defect' is a fault in an article which is not in itself serious enough to make the article defective.

The number of defectives can be examined by regular sampling, plotting results on a chart, and watching them in relation to probability levels. It should be understood that, the methods used, although theoretically sound, are much less sensitive than those using averages and ranges of measurement, and where technical and economic factors permit, measurement is far superior to qualitative methods of counting the number of defective articles. However, the technique of control by defectives is much simpler to understand and is therefore a good starting point when installing control chart procedures.

CONTROL CHART FOR PROPORTION DEFECTIVE P CHARTS

When items are checked for defective features by using GO and NO GO gauges, control is achieved by using proportion defective charts.

If the mean proportion defective ($\bar{P}$) remains constant, then the proportion of defectives (P) in each sample will vary from sample to sample and will form a Binomial distribution having the following properties:

$$\text{Mean proportion defective} = \bar{P}$$

$$\bar{Q} = 1 - \bar{P}$$

$$\text{Standard deviation of proportion defective} = \sqrt{\left(\frac{\bar{P}\bar{Q}}{n}\right)}$$

For a Normal distribution, only one item in a thousand exceeds the mean by more than 3·090. If n is reasonably large the Normal is a good approximation to the Binomial and hence there is only one chance in a thousand of getting a sample whose proportion defective, P, exceeds

$$\bar{P} + 3 \cdot 09 \sqrt{\left(\frac{\bar{P}\bar{Q}}{n}\right)} \simeq \bar{P} + 3 \sqrt{\left(\frac{\bar{P}\bar{Q}}{n}\right)}$$

Consequently this is taken as the 'action' limit for proportion defective.

Similarly there is only 1 chance in 40 of getting a sample whose proportion defective P, exceeds:

$$\bar{P} + 1 \cdot 96 \sqrt{\left(\frac{\bar{P}\bar{Q}}{n}\right)} \simeq \bar{P} + 2 \sqrt{\left(\frac{\bar{P}\bar{Q}}{n}\right)}$$

If the population proportion defective has not changed there is only 1 chance in 1000 of getting a sample containing a proportion defective greater than the action limit. Thus when such a sample is obtained, it is taken as an indication of an increase in the proportion defective. The process is stopped, and the cause of the change established.

The lower limits on P charts are only of interest in so far as any significant decrease in proportion defective is desirable, and steps should be taken to consolidate the improvement.

P CHARTS. CONSTANT SAMPLE SIZE

If the sample size is constant, or nearly so, then the value of n in the expression:

$$\bar{P} + 3\sqrt{\left(\frac{\bar{P}\bar{Q}}{n}\right)}$$

may be taken to be the average size ($\bar{n}$) of the samples used to set up the charts. Thus:

$$\text{Action limit for} \quad P = \bar{P} + 3\sqrt{\left(\frac{\bar{P}\bar{Q}}{\bar{n}}\right)}$$

$$\text{Warning limit for} \quad P = \bar{P} + 2\sqrt{\left(\frac{\bar{P}\bar{Q}}{\bar{n}}\right)}$$

P CHARTS. VARYING SAMPLE SIZE

If the sample size n varies considerably, then the value of the standard deviation of P i.e. $\sqrt{(\bar{P}\bar{Q}/\bar{n})}$ will vary too. Consequently the control limit values will depend on the sample size and the Control chart will no longer consist of a set of parallel lines.

To deal with this variation of control limit first plot a graph of:

$$\text{Action limit} = \bar{P} + 3\sqrt{\left(\frac{\bar{P}\bar{Q}}{\bar{n}}\right)}$$

$$\text{and Warning limit} = \bar{P} + 2\sqrt{\left(\frac{\bar{P}\bar{Q}}{n}\right)}$$

against sample size "n". Note that n, not $\bar{n}$, is used.

This graph is then used to read off the limits appropriate to any given sample size n as required. The limit lines on the control chart will then move up and down as the sample sizes decrease and increase.

CONTROL CHART FOR NUMBER DEFECTIVE—D CHARTS

When constructing the P chart, the sample size must be reasonably large and it would be hoped that the proportion of defectives in the batch would be small, the Poisson distribution therefore providing a good approximation to the theoretically more correct binomial probabilities of obtaining samples containing any given proportion of defectives. Thus if the sample size is constant a D chart may be used in place of a P chart, by plotting the number of defectives in the sample against the sample number.

The only way in which a reliable estimate can be obtained of whether a change in quality has taken place, is to keep a continuous record of samples from the process and compare new samples with limits based on previous performance. The distribution of the number of defectives will, provided the probability of finding 'bad' articles remains constant, follow a *binomial distribution* with average $\bar{d} = nP$, where n is the sample size and P the average proportion of defectives produced by the process. From this distribution, we can calculate the probability that d, the number of defectives in a sample, will exceed any number of defectives and thus establish warning and action limits for the control chart just as we did for the measurable variables.

Statistical tables are available which give the control limits for values of d up to 15. For larger values of $\bar{d}$ the control limits may be calculated from the approximations:

$$\text{Warning limits} = \bar{d} \pm 1 \cdot 96 \sqrt{(nP(1-P))}$$

$$\text{Action limits} = \bar{d} \pm 3 \cdot 09 \sqrt{(nP(1-P))}$$

Sample size

It has been shown that if for any reason the sample size is changed, the control limits must also be changed as these are dependent upon sample size. If the sample size cannot be kept constant, it is practical to base the limits on the average sample size and indicate the actual sample size on the chart so that points near the limit can be considered in relation to their sample size. If the variation in sample size is greater than $2:1$ it is desirable to use the proportion defective chart.

The smaller the sample size the less sensitive is the chart to changes in quality, thus the sample size should be as large as is practically and economically possible.

PRODUCTIVITY MEASUREMENT

R. G. NORMAN

It is generally acknowledged that 'output per head' in British industry is low in comparison with foreign performances. Maddison cited by Jones and Barnes[53] says that 'British industrial productivity is now below that of all West European countries north of the Alps' and same authors claim that 'throughout industry we are misusing labour, employing too many men to the job and allowing much of their time to be deliberately and systematically wasted'. Many people will agree with such statements for it is obvious that this country's economic situation is such that improvement in productivity is an urgent task. But before comparisons of productivity between nations, regions, industries, firms and departments within such firms, can be reliably drawn, productivity measures need to be examined for their adequacy and meaningfulness.

Productivity and production are often regarded as synonymous; an improvement in production is mistakenly interpreted as an improvement in productivity. Whilst an improvement in production can be obtained by simply increasing the input of resources that are used to convert raw material into finished useful products, e.g. increasing the labour force, working additional hours or by providing more capital and equipment, productivity could very well have been sacrificed. Production statistics show how much is produced while productivity ratios indicate how well the resources have been utilised in achieving the production.

Understandably, unless a manufacturing concern employs productivity measures which are meaningful for that concern and for others in its particular section of industry, the company is in no position to assess its present or potential levels of productive efficiency. Unfortunately, there is little agreement among industrial executives as to what are the relevant criteria in designing productivity indices, and the theorists' are also equally divided.

In this era of technological change and innovation, industry is becoming more and more capital intensive and it would appear that the proportion of labour costs to those of total and realistic manufacturing costs is diminishing. A report produced by the late Ministry of Labour[54] suggests that: 'It may be uneconomic to seek the minimum expenditure in the use of manpower if this detracts from the optimum use of expensive capital equipment. The combination of capital and labour can be varied to give the same output'. If more capital is invested and labour reduced, or maintained at the same level 'output per employee' might rise, but it does not necessarily indicate that the system is more productively applied, especially if it is found that the total unit costs have risen. Laszlo Rostas[55] was aware of this in 1955 when he wrote 'in industries where this (labour) proportion is small and the proportion of other factors in total costs is high, the measurement of labour productivity only may not lead us very far without the measurement of productivity of other factors'. 'Output per head' is dependent upon how well capital, as well as the total input resources of a system iṣ utilised; and while little headway has been made in measuring capital productivity, or even labour and capital productivity in a combined form, comparisons between industries and firms within given industries will continue to be inconsistent and misleading.

As industry becomes more capital intensive it may be argued that it is machinery rather than (or in conjunction with) labour that generated more output per head.[56]

ACCOUNTANTS' 'MEASURES OF PRODUCTIVITY'

There is an abundance of crude financial and efficiency (productivity) measures in existence, unfortunately, many are unable to satisfy the two essential desiderata, namely, that they are sufficiently plausible to appeal to the business entrepreneur and are refined enough for the production engineer-cum-manager.

Many of the productivity measures used in industry today are those developed by accountants and quite naturally are financially orientated. There are firms that attempt to evaluate the worth and effectiveness of business activities by adopting 'financial ratio analyses'; ratios that provide management with valuable pointers to the condition in terms of liquidity, funds and profitability of the company.

Theoretically, there are no limits to the number of ratios that can be derived, Roy Foulke[57] suggests that 500 or more ratios can be made available, and some kind of relationship can be found; the important factor is whether or not the relationship is a useful measure. Such measures are broadly concerned with 'sales return on capital employed' or 'profit to asset' ratios, which have often been referred to as an indication of business efficiency and on occasions by several executives and academics as an indication of productivity change. Some ratio examples are listed below:

$$\text{Current (liquidity) ratio} = \frac{\text{Current asset value}}{\text{Current liabilities}}$$

where current assets include cash, marketable securities, accounts receivable and useful inventories. Current liabilities include accounts outstanding, bank overdrafts, accrued expenses and tax liabilities.

$$\text{Inventory turnover} = \frac{\text{Sales value}}{\text{Average inventory}}$$

a ratio that indicates the number of times the average inventory is turned over in a given period.

$$\text{Return on assets} = \frac{\text{Net profit} + \text{interest}}{\text{Total assets}}$$

a ratio that appraises profitability, but is highly suspect because asset value may well be under valued, and often includes such intangibles as goodwill and patents.

$$\text{Return on sales} = \frac{\text{Net profit}}{\text{Sales}}$$

a ratio that appraises the efficiency of operations, where pricing and the volume of sales may affect its reliability.

Ingham and Harrington[58] of the Centre of Interfirm Comparisons Ltd., developed a pyramid of such ratios under the primary ratio of 'return on capital employed' which was originally devised to help general management to understand the way in which certain business factors determine their firms' return on assets. The framework shown in Figure 20.79 has since effectively been used to present a channel of communication between the technical levels of responsibility and general management. They maintain, however, that their ratios were not conceived as productivity indicators.

In the revitalisation of GEC's operations Arnold Weinstock set out seven key ratios to be used to control the operational aspects of a decentralised complex: profit/capital employed; profit/sales; sales/capital employed; sales/fixed assets; sales/stocks; sales/employee; profits/employee.

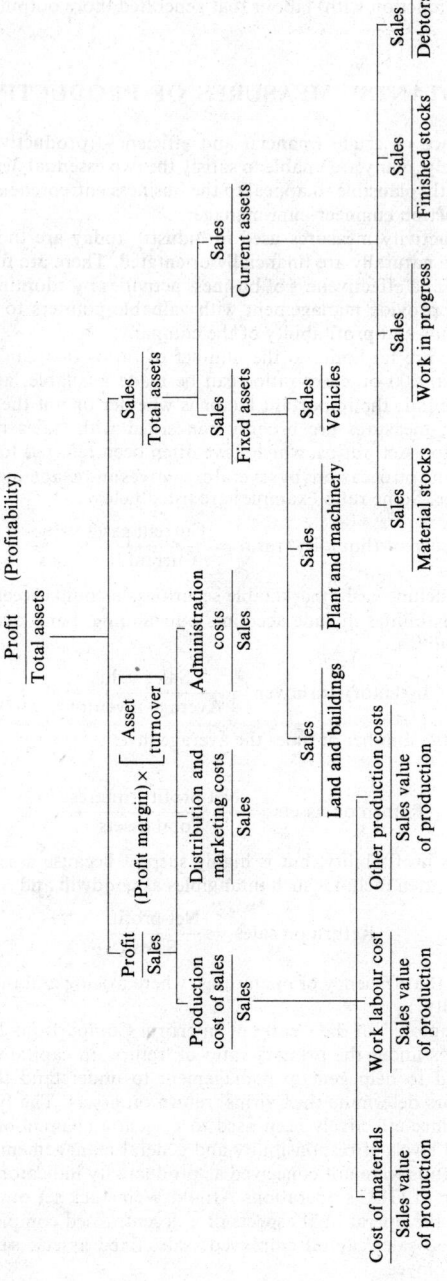

Figure 20.79. Pyramid of ratios

Many publications exist which refer to such ratios and pyramid type frameworks, for example, those used by E. I. Du Pont De Nemours and Company in 1919; also, about four years later, Bliss[59] and his students examined many financial and operating ratios in management used at that time. Dunn and Bradstreet, in whose company Roy Foulke was intimately involved, developed many ratios in order to derive the credit worthiness of companies. It is, therefore, obvious that this form of presenting management information is in no way a new concept, but, however useful comparisons using such ratios might be, they must be treated with great caution; unless it is clearly understood what is being compared the results can be highly misleading.

Professor Witschey[60] is aware of the caution that must be exercised when he points out that: 'Accounting neither strives for nor attains absolute truth. Although it is characterised by a rather elaborate theoretical framework; its results are usually dependent upon judgment. While there are limited objectives in accounting, perhaps its principal purpose is to describe change. Thus in the case of a business entity the central goal of accounting is the determination of income (or expenditure), i.e. change resulting from the efficiency of operations'.

COST ACCOUNTING

The accounting function can be divided into two areas, financial accounting and management accounting.

Financial accounting is concerned with looking after the shareholder's investment. Careful records are kept, usually in a double-entry book-keeping form of assets and liabilities, debtors and creditors, income and expenditure as illustrated by Pickles[61]. At least once a year a trading account with a profit and loss account for the period and a balance sheet of the affairs of the company on a particular day are prepared in a form laid down by the Companies Act 1948 and 1967.

The need for cost accounting

The need for cost accounting systems has arisen out of the growth in the size and complexity of business. With the separation of ownership and management of the public companies there was a need for management to provide information about profits and losses and assets and liabilities. The government intervened by passing various Companies Acts, which required the directors to place before the shareholders a summary of the financial transactions for the period and to keep true and accurate records of the company's affairs. With the increase in competition and the need to produce and sell at low profit margins the need for management accounting is now essential.

The amount of attention paid by engineers to the development of improved bases for decision-making and control, in their respective areas, frequently leaves much to be desired, despite the fact that their work involves and depends upon data developed from costing and budgetary systems. Even though they may occasionally complain about the inadequacy of the data, in fact, the very nature of the information they receive is a function of the adequacy of the costing system from which their information is derived.

Furthermore, developing a costing system is too important a function to be left entirely to the traditional cost accountant, but should involve partnership with the engineering management functions. Indeed, engineers must make themselves acquainted with costing systems and in particular their significance in application.

The main objectives of cost accounting are:

1. To ascertain accurately the costs of products and processes under differing conditions.
2. To provide bases for effective control of costs in every form of economic and engineering activity, and

3. To provide reliable information for sound management decision-making.

Historical costing systems collect data on expenditures and apply them to products at the end of some fiscal period but standard costing attempts to predetermine product costs and to use these as standards for cost control purposes.

However useful this latter technique might be (and even such a technique is often inadequately employed since it is based on comparing actual results, to some budgeted performance) the technique does not reveal the potential utilisation of labour, machines, or resources capacity.

Terminology

The Institute of Cost and Management Accountants,[62] which has given its kind permission to reproduce some of the terminology used in Cost Accountancy defines:

COSTING

This is the techniques and processes of ascertaining costs.

COST ACCOUNTING

The process of accounting for cost from the point at which expenditure is incurred or committed to the establishment of its ultimate relationship with cost centres and cost units. In its widest usage it embraces the preparation of statistical data, the application of cost control methods and the profitability of activities caried out or planned.

COST CENTRE

This term may refer to a department, a group of operatives, an individual machine for which costs may be allocated and used for the purpose of cost control. Examples are a machine shop, a sandblasting plant, the personnel department.

ELEMENTS OF COST

Costs can be divided into direct costs and overheads which are often called indirect costs. The cost of producing an article is made up of direct labour; direct materials; and overheads. The first two of these are direct costs.

The cost of running a firm will include the cost of producing all the products; selling and distribution costs; and administrative and development costs. In order to run the firm successfully, measured by profit, all costs have to be recovered.

Direct costs are those which can be identified with, and allocated to, cost centres and cost units. Direct labour is all the labour used to physically assemble or adapt the materials, components or parts which become part of the product within the process, department, machine operation or other cost centre. Direct materials are all materials components and parts, which become part of the product within a process, department, machine operation or other cost centre.

Overheads are the cost of indirect labour and indirect material and services, which support direct labour and help in the processing of materials. They also represent cost of upkeep of the factory, plant and machinery, general supervision, production control, safety and welfare, and as such cannot be allocated completely to one individual cost centre.

ORDER DOCUMENTATION

The sales office will usually be organised in sections either on a territorial basis, or type of customer, to handle the incoming calls from customers direct or through the sales representative.

The first task is to check the present credit standing of the customer the history of which is traditionally recorded in a sales ledger normally kept by the accounts department. Periodically a black list is submitted to the sales clerk, who exercises control by not processing the order. Where computers are used a white list is produced, listing customers with whom credit can be given.

Simultaneously production planning for the manufacture of such an order must approve the customers' specifications and delivery requirements and ensure that resources are available either by withdrawal from stock, by manufacture or by subcontract.

The next stage is to edit the order and create the necessary documents of the internal works order, if these have not already been prepared by the sales representative.

A straightforward order for goods already finished and in stock can be processed without the preparation of the production document set, by marking the master of the internal order with 'despatch from stock'. On the other hand, an order for goods to be manufactured to specifications laid down by the customer may require considerable discussion between the sales department, production control, production engineering and the purchasing department, before the production document set can be produced.

Production orders

These may be of two types:
1. To meet the customer's own specification (specials).
2. To maintain the stock levels of components and finished parts.

In order to fulfil the specials, the production order is initiated by the production function. The order is itemised and a separate manifold works order issued for each item required. Each order is carefully numbered either using an alphabetical and/or numerical sequence, e.g.

DHG 2756/3. Item 3 on Works Order No. 2756 for double helical gears.
274. 2756.3. Where 274 is the product code for double helical gears.

Full technical details, the quantity to be produced, the customer's name and address, delivery instructions and the date required are generally included.

The material specification of this item will detail and quantify the components indicating whether they are made within the works (MH) or are bought out as finished (BOF) items.

Indirect works order

When it is necessary for a cost centre to carry out indirect work for another cost centre, e.g. maintenance or toolroom which is considered a major activity, it is usual for the production function to issue an indirect works order. This order may be identified by a prefixed letter to indicate that it is not a production order and the numerical sequence may include as suffixes the appropriate cost code and cost centre code, e.g.

Where

W.   Indirect works order.
1624.   Works order number.
26.   Cost code.
16.   Cost centre code.

These orders have as many copies as required by the operations specified the number of which varies with the nature of the task, but copies should always be sent to the initiator and the costing department.

MATERIAL CONTROL

The material control function is responsible for maintaining stocks of materials in accordance with company policy as to the maximum, minimum and reordering levels of stock. When reordering is necessary a purchase requisition is sent to the purchasing department, specifying the material required, with full technical details and the quantity required.

There may already exist an established relationship between the purchasing department and a particular supplier, avoiding the need to call on various suppliers to provide quotations and then to select a supplier on the basis of price, quality, delivery and terms of payment.

A purchase order will then be placed. specifying the material, quantity, delivery and terms agreed. Copies of the purchase order are sent to material control, progress section and the goods inwards section of the stores.

It should be firmly laid down in the policies of the company, that no material purchased by, or on behalf of the company, should be purchased without an official purchase order. Goods accepted because of necessity without an official purchase order should be confirmed by such an order, clearly marked 'confirmation of goods already supplied'.

Receipt and inspection of goods

There are many methods of routing documents related to the receipt of goods. One established method, is that on receipt of the copy purchase order, the goods inwards section will prepare two copies of the 'goods received note', for each item ordered on every purchase order. When the goods are received they are inspected and checked for quantity and quality and for conformation to the technical specifications before they are admitted to stock. Any rejections will be shown on an 'inspection rejection note', which gives details of the supplier, the purchase order, number, date, reasons for rejection and the present location. A copy of the goods received note' with any inspection rejection note is sent to the purchasing department, which initiates the action necessary to obtain replacement or recompensation for sub-standard material.

The accepted stock is moved to its appropriate stock location and the accompanying goods received note is used to record the receipt on the appropriate bin card and is maintained for each different type, shape or size of material held in stock. The bin card is usually divided into two; the top half containing the bin number, location, code number and maximum, minimum and reordering levels. The lower half contains columns for recording receipts and issues and balance quantities.

Purchase invoices are received from suppliers of goods and services, which give details of the quantity of each item supplied, their specification, unit price and gross purchase value. Discounts may be given by the supplier on any or all items and this may be deducted from the gross purchase value of each item separately where discount rates differ, or, in total, if only one rate applies to determine the net purchase value. The individual net purchase values of each item are added for the total net purchase value of the invoice.

The purchasing department will marry up their copy of the goods received note with the purchase invoice from the supplier. It is usual to rubber stamp on a clear space on the invoice or affix a gummed check list, which enables the person inspecting to initial the check he has made as to quantity, price, discounts and calculations.

The invoice with the goods received note will then be passed to the accounts department for payment or to credit the supplier with the invoice total in the 'bought (purchases) ledger'. The invoice is filed, but the goods received note is passed to the stores ledger function, which is part of the costing function for recording the receipt.

Stores ledger maintains a record of every type of material, size, shape in terms of quantity, unit price and value for each receipt and issue. These records constitute a 'perpetual inventory' of all materials in stock as all material documents recording issues or returns to stores are priced and valued from the information they contain. It is possible to check the physical quantity shown on a stores ledger record with that in the appropriate bin in the stores, after making due allowance for delays in the flow of material documents through the system.

The material documents authorising movements in and out of the stores include the following.

MATERIAL REQUISITION

This is a document which authorises and records the issue of materials for use. It may be preprinted as part of the paperwork accompanying a production order. This method provides a better control than the method of individual foremen writing up their own requisitions.

The requisition will state the cost centre for which the material will be used, the quantity, the material description and the purpose for which it is required, quoting the works order number if the material is required for production. All requisitions must be signed by an authorised person, usually the foreman in charge of the section.

MATERIAL RETURNS NOTE

This records the return of unused materials. It is identical in design to a material requisition, distinguished usually by title and colour. It is used to record the return of surplus material of usable quality to the stores.

EXCESS MATERIALS NOTE

A document which records the issue of materials in excess to the normal requirements. In a system of preprinted material requisitions, excess materials for a particular job will be authorised and recorded by the issue of an 'excess materials note'.

MATERIAL TRANSFER NOTE

A document which records the transfer of materials from one store to another, from one cost centre to another from one cost unit to another. Material transfer notes are similar in design to 'material requisitions' but allow for the inclusion of the cost centre or cost unit to which they were originally issued, as well as the cost centre or cost unit to which they are being transferred.

Valuation of material releases

Although receipts of materials are entered in the store ledger records at the actual prices (net, after discounts) shown on the purchase invoices, but these prices are not always used for pricing the issues, except where nil-stock items are purchased specially for a job.

Opinions vary, as to whether it is better to use prices which reflect, what has been paid for the material, or what will be paid for the replacement material necessary to maintain the stock levels. The established methods of pricing issues are as follows.

FIFO (FIRST IN FIRST OUT)

Using actual prices paid, the valuation of the stock released is to be based on the assumption that the oldest receipt of stock is to be the first to be issued. In an inflationary market, where prices are rising, the issues of stock maybe priced at out of date prices, which could be misleading.

LIFO (LAST IN FIRST OUT)

Using actual prices paid, the valuation of the stock released is to be based on the assumption that the latest receipt of stock is to be the first to be issued. In a falling market, where prices are falling, these prices could also be misleading.

REPLACEMENT PRICE

This is the price at which there could be purchased an asset identical to that which is being replaced or revalued. The method is based on the price that will have to be paid to replace it. The trend of the market is therefore taken into account.

WEIGHTED AVERAGE PRICE

This is a price which is calculated by dividing the total cost of materials in the stock from which the material could be drawn, by the total quantity of materials in that stock. This method is popular because of the ease with which it can be applied, using calculating machines or a computer and because averaging smoothes out price fluctuations. It tends to lag behind the current or replacement price, but it does follow the market trend.

STANDARD PRICE

A predetermined price fixed on the basis of the factors which will affect the price within a given period. It therefore attempts to determine the weighted average price which will be paid during the budget period. In systems where standard prices are used for the valuation of stock releases, it is common practice to simplify the stock ledgers by recording only the physical quantities and not to record their valuations. Changes in prices are recorded only for statistical purposes and to modify the standard price in the next budget period.

Material issues analysis sheet

Material documents are priced and extended by quantities to determine their value. Their cost coding is checked and an analysis made on a 'material issue analysis sheet'. This is a document which is a classified record of material issues, returns and transfers.[63] This is prepared weekly, or monthly and is the basic document in the preparation of cost accounts.

LABOUR COST ACCOUNTING

It is necessary, both for wage payment and for cost information, for records to be maintained relating to the activities of all employees. Where detail analysis is required, as in the case of hourly paid workers, or those on a bonus or incentive scheme records of attendance by means of individual clock cards, or similar methods of recording, coupled with records of work done during the hours attended must be existing.

Attendance time

There are many ways of recording attendance time, but the best and most flexible methods are based on the individual. The method chosen should be designed to eliminate as far as possible fraudulent practices, such as one employee making a time record for both himself and another. The actual recording is best achieved by clocking a personal clock card at a clocking station situated as near as possible to the place where the work is normally carried out. This avoids congestion and possibly accidents, which might arise if all the clocking stations were situated together at the works entrance. It also means that payment based on attendance time is made for the time spent at the place of work, as contracted between employee and employer.

Records of work done

The detail required will vary with the nature of the work undertaken. A direct worker will be required to complete a pre-printed job ticket which is part of the production document set. Indirect work may be recorded on an indirect labour cost card prepared in the shop office by the shop clerk on instructions from supervision. There must be strict control over all job tickets to avoid fraudulent booking by operatives on incentive schemes. The indirect labour cost card is usually a different colour from the job ticket to facilitate their segregation.

An indirect worker, who normally works on a number of different tasks such as maintenance, may use a daily or weekly time sheet, which records the nature of the work done, the cost centre for which it was done and the time taken, quoting any relevant indirect works order number. Daily time sheets are to be preferred to weekly ones, which tend to suffer in accuracy because of late compilation.

Indirect workers who are engaged on the same task in the same cost centre need not submit a daily time sheet. It is only necessary for an indirect labour cost card to be completed by the shop clerk at the end of each week. All time records should account for the same number of hours as is shown on the attendance records.

Payroll preparation[64]

The basic records of attendance time will be extended by the wages office at the end of the working week. This extension will show the daily hours worked, divided into ordinary time and overtime. The overtime earnings will be calculated in accordance with national wage agreements. The wages earned by attendance time will also be calculated at the appropriate rates for the grade of worker. If an operative is engaged on an incentive scheme his bonus earnings will be calculated from the records of work done. There may also be an entitlement to other earnings agreed at local, shop or at an individual level. The sum of all his earnings is the gross wage.

Payrolls are compiled on a cost centre basis to facilitate their completion and list out individual and total gross wages, statutory and voluntary deductions and net pay. Statutory deductions include income tax and employee's contributions to national insurance and graduated pensions. Voluntary contributions, which must be agreed in advance by the operative, may include such items as holiday fund, union dues, charitable donations and laundry charges for overalls.

The basic information such as name, clock number, grade, basic rate and voluntary deductions is generally either already printed or stored on punch card or computer tape. The variable information: hours worked, basic and overtime wages, bonus earned, and gross wages obtained from the clock cards and records of work done, is then either added on to the prepared wage sheet, or integrated with the punched cards or computer tape and fed through the computer.

If the wages are calculated manually, the next operation is to calculate from tax-tables, the tax to be deducted. If a computer is used, the tax-tables are programmed into the computer. Each employee has a tax record card, which gives details of his name, clock number and income tax code number.

A weekly entry is made of the gross wages earned and the cumulative gross wages for the tax year is calculated. Against this cumulative gross wage figure is set the tax-free income according to the 'income-tax Table A' and the individual operative's tax code. The difference between these two is the taxable income, and from the 'income-tax Table B' the cumulative tax due to date is shown. The cumulative tax due from the previous week is deducted to determine the tax due to the current week. In certain cases the tax due may be a minus figure, thereby indicating a refund of tax.

The individual's net wages are then calculated and the payroll is totalled. A note and coin analysis is made to facilitate the making up of the wages into individual pay packets. A copy of the payroll showing the details of the gross wages, the separate deductions and the net pay is included in the pay packet. A summary of all the payrolls is made which form the control over gross wages, deductions, net-wages and hours worked.

Payment of wages

This should be carried out at or near the place of work. This avoids the mass congestion of operatives at the wages office and enables supervision to assist in identifying employees so that mistakes and fraud are minimised. It is common practice for operatives to sign some form of receipt, such as a completed pay card in exchange for their wage packet.

Wages analysis sheet

After the payment of the wages, each payroll is analysed. A wages (labour cost) analysis sheet is prepared for each weekly payroll and the summary weekly payroll. These documents are 'classified records of time and/or wages compiled from the labour time records'. The gross wages figures shown on the payrolls are used for balancing this analysis.

OVERHEADS ALLOCATION

Items which constitute overheads can be classified as follows.

Area. In proportion to the area occupied, e.g. rent and rates, allowances should be made for the differences in amenities provided. There is a considerable difference between a central heated office and an open yard, even on the same site.

Cubic space.   In proportion to the space occupied, e.g. heating and ventilation.

Capital value.   Insurance premiums, depreciation of buildings, plant and equipment.

Recording information.   Use of meter and log books for power, telephone, vehicle mileage.

Technical information.   Technical data for use and maintenance requirements, e.g. different machines require different servicing schedules. Certain materials need careful handling and storage. Some processes produce waste, other processes need constant supervision.

Number of personnel.   Canteen and welfare facilities.

A problem arises when not all the employees use the available service, such as the canteen. Can costs be apportioned under such conditions to:
All employees who are permitted to use it?
All employees who eat regularly at the canteen?
All employees who at some time eat at the canteen?

The established authorities favour all employees who are permitted to use it.

Gross wages payable to personnel.   Insurance premium for employer's liability.

Reapportionment of service costs

In apportioning overheads of the factory, certain amounts will be allocated to the service departments, e.g. the stores, works administration and maintenance. The total cost of each of the service departments are apportioned to the production departments, depending upon the use that each production department makes upon the service department.
In practice many considerations have to be taken into account and several principles may have to be considered to apportion the department's cost.

Selling and distribution costs

Market research.   This includes analysing the market and distribution of the products, by fact finding, building up statistical knowledge and making statements, based upon facts and figures.
The costs can be broken down as follows:
Advertising costs.
Selling costs.
Secondary packing, distribution and delivery costs.
Maintaining good customer relationship, necessary for repeat orders.
Sales documentation and the collection of accounts.
Losses due to bad debts.
The cost of money tied up with debtors and finished products stored.

Policy cost

This is the cost which is incurred in the current trading period and is additional to the requirements of manufacturing selling and distribution and the administration of the current output. The policy cost is written off against the profit in the current period, although the expected benefits will accrue in the future.

Administrative and development costs

ADMINISTRATIVE COSTS

These are incurred in the registration of the company, the annual return and the provision of information required by the Companies Act 1948 and 1967, to shareholders, company registry office and the stock exchange. The items are as follows:

The cost of maintaining the Registrar's Office.
Cost of maintaining files and records of the company affairs.
Cost of maintaining the directors.
Cost of providing communication between the office and the works.
Fees payable to the bank, auditor, solicitor and patent office.

DEVELOPMENT COST

The development cost is the cost of the process which begins with the implementation of the decision to produce a new or improved product, or to introduce a new or improved method and ends with the commencement of formal production of that product, or by that method. This cost is also absorbed in the unit cost of the product. The costs are accounted for by the following:

The development and research into new products.
Training.
Statistical and accounting investigations.
Maintaining the method studies department.

RESEARCH COST

The cost of searching for new or improved products, new applications of materials, new or improved methods. If the research is successful the cost is charged against the unit cost of the new product on a conservative basis. If unsuccessful it is written off against profits in the current period.

PREPRODUCTION COST

The preproduction cost is that part of the development cost incurred in making a trial production run preliminary to formal production. The term is often used to cover all activities prior to production, including research, development and trial production runs. This cost is also absorbed in the unit cost of the product.

CONVERSION COST

The sum of direct wages, direct expenses and overheads of converting raw material from one stage of production to the next, or converting the raw material to the finished unit though all the stages of production.

IDLE FACILITIES COST

'The cost of abnormal idleness of fixed assets or available services'. It may be incurred when part of an undertaking is temporarily closed during a period of recession. The fixed overheads applicable to this period are written off against profit in the current period and are not absorbed into the production cost. The reapportionment of other costs are described by Nolan.[63]

OVERHEAD COST ACCOUNTING

The overhead costs may be derived from several sources;
Material issues analysis sheet (indirect materials).
Wages analysis sheet (indirect labour).
Journal vouchers.
Petty cash vouchers.
Bought ledger analysis.

JOURNAL VOUCHERS

These are created in the financial accounts for charges which are of a special nature and which do not arise through the other accounting documents, such as depreciation. The rate at which an asset is written down, depends largely upon the rate allowed by the Inland Revenue. There are two methods which could be adopted. One is the straight-line method which writing off a fixed sum each year, and the other is the reducing balance method; this entails writing off a fixed percentage of the remaining balance each year.

The depreciation charged, is analysed per cost centre by maintaining a plant register. This is usually in the form of a loose-leaf binder. A new sheet is created at the time of acquisition of each item of fixed plant and machinery. It records the full technical specifications, physical location as well as cost details.

PETTY CASH VOUCHERS

These are analysed according to their cost code and cost centre code.

BOUGHT LEDGER ANALYSIS

This is carried out to indicate which invoices relate to capital items or stock items or expenses. The expenses are apportioned to their appropriate cost centre indicating their cost code. The bases of apportionment are predetermined on well established bases.

The indirect material, indirect wages and indirect expenses are posted to the cost ledger. This is a subsidiary ledger whose accounts record those transactions which are included in costs. It is maintained on a cost centre basis and permits the recording of all activities of each cost centre. These include direct wages earned and productive hours worked as well as all expenses in their cost classifications. It is arranged, so that the weekly postings may be made where applicable and periodic and cumulative sub totals computed.

Overhead absorption

Overhead absorption is the allotment of overheads to cost units; the principal methods of overhead absorption are:

PERCENTAGE ON DIRECT WAGES

$$\text{Overhead rate} = \frac{\text{Budget overheads}}{\text{Budget direct wages}} \times 100 = \dots\%$$

Overhead absorption = Actual direct wages × overhead rate.

This is computed on a cost centre basis and is easy to apply. It may give rise to inequities because of the variety of labour grades incurring different amounts of overhead absorption, because of their different direct wage rates. Again, inequities may arise if differences in widely varying overhead costs, such as, between a large expensive plant and benchwork are not reflected in the overheads absorbed.

DIRECT LABOUR HOUR RATE

$$\text{Overhead rate} = \frac{\text{Budget overheads}}{\text{Budget direct labour hours}} = \text{£... per DLH.}$$

Overhead absorption = actual direct wages × overhead rate.

This method introduces the constant factor of time into the absorption rate calculation. It retains the objection of different facilities, but removes the objection of different wage rates. It is considered an equitable method because many overheads accrue on a time basis.

MACHINE HOUR RATE

A calculation in which the overheads are divided into two groups; those that may be identified individually with machines or items of plant, and those of a general nature.

$$\frac{\text{Overhead rate for}}{\text{the cost centre}} = \frac{\text{Budgeted overhead for cost centre}}{\text{Budgeted productive machine hours for cost centre}}$$

$$\frac{\text{Overhead rate for}}{\text{machine 123}} = \frac{\text{Budget overhead (machine 123 only)}}{\text{Budget productive hours (machine 123 only)}}$$

$$\frac{\text{Overhead absorption}}{\text{of machine 123}} = \text{Actual hours of machine 123} \times (\text{machine overhead rate} + \text{cost centre overhead rate})$$

Machines of the same type in the same cost centre should be grouped together, as it would be inequitable to have different rates for the same type of machine in the same cost centre. This could arise, because of different depreciation charges due to machines being of different ages and therefore different capital costs.

In certain exceptional circumstances, such as, an operator working several machines it may be expedient to include what is normally considered to be direct wages of the operator in the overheads.

PLANT HOUR RATE

This is an extension of the machine hour rate to cover large items of plant on which a group of operatives may be working.

PERCENTAGE ON DIRECT MATERIALS

PERCENTAGE ON PRIME COSTS

Both these methods are simple to apply but in practice are found to be inequitable and inaccurate.

The job tickets, which indicate the facilities used in production are designed to permit the inclusion of the appropriate overhead absorption rate. A summary of these for each source of absorption and for each cost centre is required for the cost ledger for each week.

Administration overheads are applied to works cost on varying bases such as percentage on works cost or on a unit basis, but are absorbed on the units produced.

Selling and distribution overheads are absorbed in relation to units sold only on similar bases to administration overheads.

METHODS OF COSTING[65]

Job costing

Job costing is the method employed to determine the cost of a single non-repetitive order. This requires the collation of all the material requisitions and job tickets which have been issued during the production of each component, sub and main-assembly. This calls for a measure of technical knowledge by the cost clerk to ensure that all documents are included.

It is important that the cost of each constituent part of the job should be shown separately to facilitate investigation.

The cost of a component is determined from

1. The priced material requisition, which forms the basis of the materials analysis.
2. The priced job tickets for each operation carried out, which forms the basis of the wages analysis.
3. The application of the appropriate overhead absorption rate in the cost of each operation and each component. In practice there may be various methods of the overhead absorption rate applied, as production may use a variety of production facilities.

Summaries are made of component costs, with the cost of any bought-out finished (BOF) parts and the labour and overheads applicable for both sub and main-assemblies.

The complete works cost requires the addition of the administration overheads and the selling and distribution overheads to determine total cost.

The total cost is compared with the sales value to determine the profit/loss on the order.

Batch costing

Batch costing is applied where batches of components and products are manufactured for stock. The fact that the production is for stock facilitates the standardisation of production processes and unit costs. Where a number of units are scrapped and this is considered normal, the cost incurred will be absorbed in the cost of the good units, which pass the final inspection. The decision as to what is 'normal' is a technical one. Abnormal losses are not absorbed, but represent an offset against any profit being made on the batch.

The costing procedure is the same as for job costing, but cost statements are presented in detail; at the component stage, rather than in total for each product.

Process costing

Process costing differs from both job and batch costing in that the method of production is of a continuous nature in a predetermined sequence and the costs per unit are averaged over a period, which may be a week, or a month. It is used in the manufacture of chemicals, cement and flour milling.

The processing of a unit may involve its transfer in an incomplete state from one process to another. The cost of the raw material entering the original process, together with any added at subsequent processes must be added to the labour and overheads applicable. With continuous operation it is necessary for the technical staff to estimate the state of completion of the units in process at the end of each accounting period. For cost purposes partly finished units are equated to finished units in that process in order to determine the output for the period accurately. To this figure is added the units transferred to the next process. From the total thus obtained, the equivalent units which were in process at the beginning of the period is deducted to determine the output for the period. The total output is divided into the total costs to determine unit costs.

Direct costing[66]

The practice of charging all direct costs to operations, processes, or products leaving all indirect costs to be written off against profits in the period in which they arise.

Absorption costing

This is the practice of charging all costs both fixed and variable to operations, processes, or products.

Historical costing

Historical costing is the ascertainment of costs after they have been incurred.

This method is used in practice, although it has serious limitations, particularly in repetition production. The principal limitation is the comparison of successive batches of production with regard to cost. The variable factors of material used, operator efficiency, changes in production methods may not be clearly defined and may be compensatory in their effects on the unit costs, to be hidden.

Standard costing[67]

The preparation and use of standard costs, their comparison with actual costs and the analysis of variances to their causes and points of incidence.

Under this method, future costs are predetermined and computed in advance of production. A standard cost is a predetermined cost which is calculated from management's standards of efficient operation and the relevant necessary expenditure. It may be used as a basis for price-fixing and for cost control through variance analysis.

Standard costs illustrate what the costs are likely to be, in the manufacturing situation. They provide yard-sticks against which the actual performance and costs are measured. They must be revised in the light of changing circumstances. They provide valuable information with which management may take remedial action to correct adverse tendencies.

PROCEDURE OF ESTABLISHING A STANDARD COST

Select the period of time when standard costs are to be applied and establish whether there are proposals to change material specifications or operation schedules which will occur within the period of time. Also establish the volume of production to determine the overhead rate.

Examine the material specifications to establish standard material quantities required. Establish the standard material cost rate, based upon stock in hand at stock values; orders for materials, where the price is fixed; and the trend of material prices.

Calculate the standard material cost, standard material quantity and standard material cost rate. From the operation schedule find out the number of employees, grade, and number of man/hours required to do the standard operation.

Determine the standard wage rate for each grade of employee. Note that the higher earnings per hour of a man on bonus is reflected in the standard hours allowed for the

Table 20.23

Code SC 2/10		Introduced	
Approved by		Withdrawn	

STANDARD COST CARD

Product GN 2.		Batch of 10	
		£	£
Direct materials			
5732	1·5 kg at £1·00 per kg	1·50	
5768	1·0 kg at £0·50 per kg	0·50	
	Total direct material cost		2·00
Direct labour			
Operation 1.	20 min at £0·30 per hour	0·10	
2.	40 min at £0·60 per hour	0·40	
3.	60 min at £0·75 per hour	0·75	
	Total direct labour cost		1·25
Overheads			
Overhead rate	120 min at £0·50 per hour		1·00
	Total standard cost		4·25

job. Calculate the standard labour cost for each operation; standard time and standard wage rate.

From the production schedule, find out the total volume of production and from the operation sheets, the total number of man/hours and machine/hours to produce the planned production. Establish the forecast of total overhead expenses.

Calculate the overhead rate, based upon the planned production and the man/hours and machine/hours to produce the planned production, following the various methods of apportioning overheads.

Prepare a standard cost card (Table 20.23). A simplified relationship diagram is shown in Figure 20.80.

Marginal costing

This is used to consider the total cost of producing, selling and distributing the last, or marginal unit, exclusive of overheads. It analyses the contribution that the manufacture and sale of the last or marginal unit makes towards the profit of the firm.

It does not normally consider fixed overhead charges, as they do not vary within fixed limits of production output. It therefore confines attention to expenditure, which can be avoided, if production is reduced, or will occur if production is to increase.

The advantages of marginal costing are that it simplifies the cost analysis, permitting more time to be spent on analysing variable costs and it enables management to accept orders in difficult trading periods, if the accepted price covers variable cost and makes a contribution to overheads, as an alternative to take no orders.

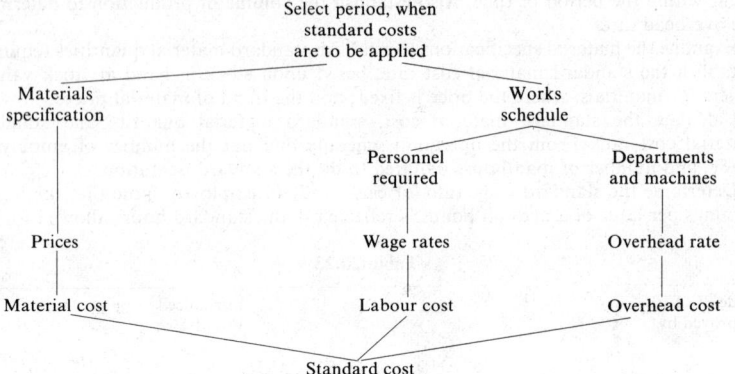

Figure 20.80. Establishment of a standard cost

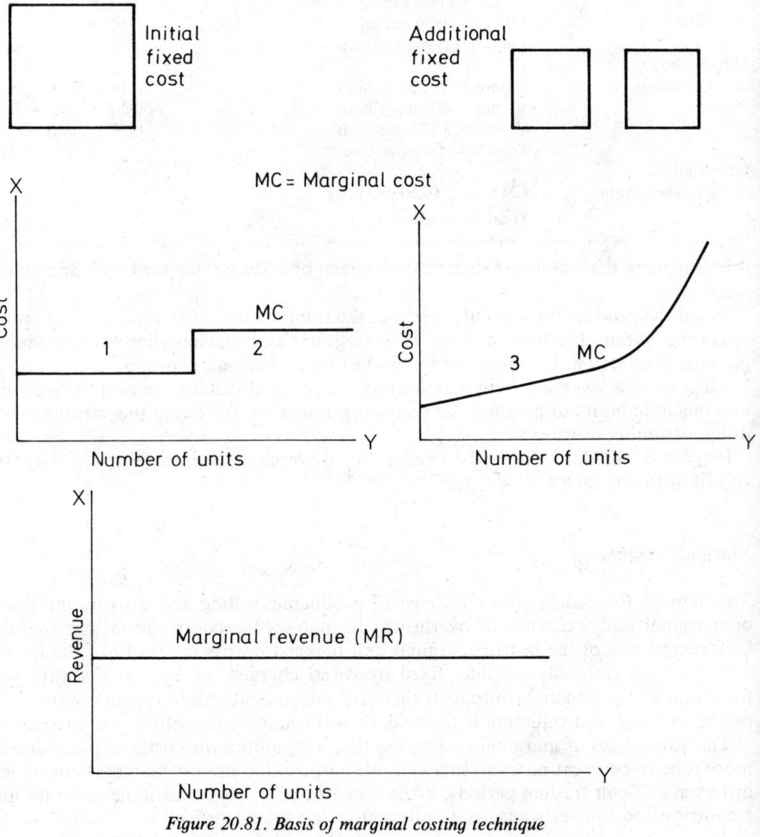

Figure 20.81. Basis of marginal costing technique

The disadvantages are that the economic selling price in the long term cannot be ascertained without regard to fixed costs and the final costs may be difficult to allocate, but they cannot be ignored even in a short period. In many firms the proportion of fixed capital is increasing and the upkeep, represents the greater part of the cost of manufacture. Very few overheads are fixed and attention must be made to the stepping-up of overheads as output increases.

Marginal costing technique can be illustrated by representing squares for fixed costs and graphs, to form areas for variable or marginal cost (Figure 20.81).

As the output is increased, it may be necessary to add further squares, representing additional fixed costs. In the graphs explaining marginal cost a straight line, horizontal to the x ordinate, indicates that marginal cost varies in direct proportion with the quantity produced. A step-up indicates a rise which stays fixed in the marginal cost

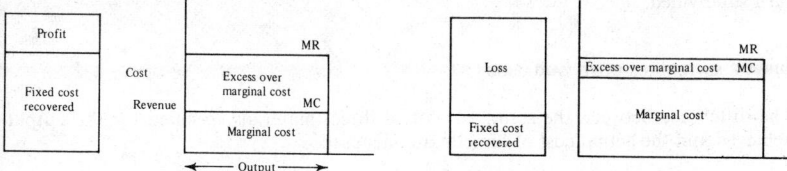

Figure 20.82. Marginal costing technique— long term

Figure 20.83. Marginal costing technique— short term

rate and an upward sloping curve, indicates a falling off in marginal efficiency, or steady rising marginal costs. Marginal revenue can also be shown graphically. A fixed or constant price is represented by a horizontal straight line.

In the long term the area showing the 'excess revenue' over the amount needed to pay the marginal costs, must equal the area of the square or squares, representing the fixed costs (Figure 20.82).

In the short period, the firm in its decision to take or not to take an order could lower it's price to the level or just above the level of the marginal cost. (Figure 20.83).

If a company operates on marginal costs for certain periods, it is essential to obtain a 'good' to an 'excessive price' in other periods because fixed costs have to be recovered if the firm is to exist in the long term.

VARIANCE ANALYSIS

The variances ascertained in any business should be designed to provide the particular information important in the control of the affairs of that business.[67] The principle variances and their relationship are illustrated in Figure 20.84 and are defined below.

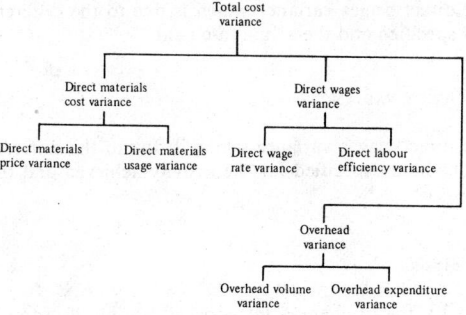

Figure 20.84. Variance analysis

METHODS VARIANCE

The difference between the standard cost of a product or operation produced or performed by the normal method and the standard cost of the product or operation produced by the alternative method actually employed.

TOTAL COST VARIANCE

The difference between the total standard cost value of the output achieved in the period and the total actual cost incurred.

The total cost variance will usually be subdivided into other variances, which are in turn subdivided.

DIRECT MATERIALS COST VARIANCE

The difference between the standard cost of direct materials specified for the output achieved and the actual cost of direct materials used.

DIRECT MATERIALS PRICE VARIANCE

That portion of the direct materials cost variance which is due to the difference between the standard price specified and the actual price paid.

DIRECT MATERIALS USAGE VARIANCE

That portion of the direct materials cost variance which is due to the difference between the standard quantity specified and the actual quantity used.

DIRECT WAGES VARIANCE

The difference between the standard direct wages specified for the activity achieved and the actual wages paid.

DIRECT WAGE RATE VARIANCE

That portion of the direct wages variance which is due to the difference between the standard rate of pay specified and the actual rate paid.

DIRECT LABOUR EFFICIENCY VARIANCE

That portion of the direct wages variance which is due to the difference between the standard direct labour hours specified for the activity achieved and the actual labour hours expended.

Material variance analysis

Price variance could be due to changes in terms of sale by the supplier; quality and therefore price, by the supplier, or transport charges or dock dues.

Usage variance, mix variance and yield variance could all arise from changes in quality of material; storage of material; method of manufacture; volume of production; plant and equipment; operation of plant and equipment; personnel; conditions of personnel; quality of finished product; or inspection control.

MATERIAL VARIANCE EXAMPLE

Standard specification	4 kg per unit.
Standard price	100 p per kg
Actual output	800 units.
Actual material used	3300 kg
Actual price paid	90 p per kg

Variance analysis

Direct materials price variance

$(100 \text{ p} - 90 \text{ p}) \cdot 3300 \text{ kg}$ = £330 (F)

Direct materials usage variance

$((4 \times 800) - 3300) \cdot 100 \text{ p}$ = £100 (A)

Direct materials cost variance

$(4 \text{ kg} \times 800 \text{ units} \times 100 \text{ p})$
$-(3300 \text{ kg} \times 90 \text{ p})$ = £230 (F)

The direct materials price variance is favourable, because the actual price paid is less than the standard price specified.

The direct materials usage variance is adverse, because the actual quantity used is more than the standard quantity specified for the actual output achieved.

The direct materials cost variance is favourable, because the actual cost incurred is less than the standard cost for the output achieved.

Labour variance analysis

Wage rate variance could arise from changes in basic rate of pay; special allowances, which increase the rate for the job, but not necessarily, the earnings and/or SET or similar tax.

Efficiency variance and idle time variance could arise from changes in any of the following: quality of labour; grade of labour; training; supervision; conditions of work; volume of production; flow of work; method of manufacture; supply of power, materials and machine time; quality of materials; plant and equipment; operation of plant and equipment; quality of finished product; inspection control; documentation.

Mix variance could be due to changes in the supply of personnel, supervision, training, nature of the work and/or the supply of work for each grade of employee.

LABOUR VARIANCE EXAMPLE

Standard hours per unit	5 standard hours.
Standard efficiency	100%.
Standard direct wage rate	75 p per hour.
Actual output	800 units.
Actual hours worked	3800 hours.
Actual wage rate paid	82 p per hour.

Variance analysis

Direct wage rate variance
$(75 \text{ p} - 82 \text{ p}) \times 3800 \text{ hours}$ = £266 (A)

Direct labour efficiency variance
$((5 \text{ hours} \times 800 \text{ units} \times 75 \text{ p})$
$- 3800 \text{ hours})$ = £150 (F)

Direct wages variance
$(5 \text{ hours} \times 75 \text{ p} \times 800 \text{ units})$
$-(3800 \times 82 \text{ p})$ = £116 (A)

There has been a reduction in the profit due to the wage rate paid being in excess of the standard wage rate. This reduction has been offset by the actual efficiency being greater than the standard efficiency.

$$\text{Labour efficiency ratio} = \frac{\text{Standard hours produced}}{\text{Actual hours worked}} \times 100$$

$$= \frac{5 \text{ hours} \times 800 \text{ units}}{3800} \times 100$$

$$= 105\%.$$

Overhead variance

This is the difference between the standard cost of overhead absorbed and the actual overhead cost.

Volume variance

Volume variance is that portion of the overhead variance which is the difference between the standard cost of fixed overhead absorbed in the actual output of the period and the standard fixed overhead allowance for the period.

Overhead expenditure variance

Overhead expenditure variance is that portion of the overhead variance which represents the difference between the standard allowance for the output achieved and the actual expenditure incurred.

OVERHEAD COST VARIANCE EXAMPLE

Standard fixed overheads for period	£2000
Standard variable overheads for period	£1600
Standard output for period	1000 units
Actual fixed overheads	£1950
Actual variable overheads	£1310
Actual output	800 units

It is necessary to distinguish between fixed and variable overheads; for instance the fixed overheads remain constant, irrespective of the output, whilst the variable overheads tend to vary with output.

Overhead expenditure variance
= Standard allowance for output − actual overheads

$$= £200 + \left(\frac{800}{1000} \times £1600\right) - (£1950 + £1310) \qquad = £20 \text{ (F)}$$

Overhead volume variance

$$\left(800 \times \frac{£2000}{£1000}\right) - £2000 \qquad\qquad = £400 \text{ (A)}$$

Overhead variance

$$800 \times \frac{£2000 + £1600}{£100} - (£1950 + £1310) \qquad = £380 \text{ (A)}$$

Total cost variance of example

	Standard cost per unit	Standard cost of output	Actual cost	Variances
	£	£	£	£
Direct material	4	3200	2970	230 (F)
Direct wages	3·75	3000	3116	116 (A)
Overheads	3·60	2880	3260	380 (A)
	£11·35	£9080	£9346	£266 (A)

Total cost variance £266 (A).

Total sales margin variance

This is the difference between the standard margin appropriate to the quantity of sales budgeted for a period and the margin between standard cost and the actual selling price of the sales effected.

Sales margin variance due to selling prices

This is that portion of the total margin variance which is due to the difference between the standard price of the quantity of sales effected and the actual price of those sales.

Sales margin variance due to sales quantities (mixture)

That portion of total margin variance which is due to the difference between the budgeted and actual quantities of which the sales mixture is composed, valuing sales at the standard net selling prices and cost of sales at standard.

SALES VARIANCE EXAMPLE

Standard net selling price per unit	£15
Standard cost per unit	£11·35
Standard margin per unit	£ 3·65
Budget sales for the period	1000 units
Actual sales for the period	800 units
Actual selling price per unit	£15·25

Variance analysis

Sales quantity variance

$$(1000 - 800) \times (£15 - £11·35) \qquad = £730 \text{ (A)}$$

Sales price variance

$$(£15 - £15·25) \times 800 \qquad = £200 \text{ (F)}$$

Total sales margin variance

$$(1000 \times £3·65) - 800(£15·25 - £11·35) \qquad = £530 \text{ (A)}$$

There has been a reduction of profit due to the lower sales volume, and this has been offset by a higher sales price.

Presenting variances to management

The method will depend upon:
1. The detail and the period covered and upon the level of management.
2. The need for minimum delay in presenting the information.
3. The presentation, in order that management can assess the information and take corrective action.
4. The clarification of those variances, which are controllable and those which are not.
5. The need to determine the executive, who is responsible for the control of the expense.
6. The use of standard forms to simplify understanding.
7. The breaking down into details of totals.
8. Use of colour, heavy lines, different print, to attract attention to a particular point or variance.
9. The avoidance of all irrelevant information.
10. Explanations, where appropriate.

BUDGETARY CONTROL

A budget is a plan, which is usually historically based, in terms of quantity or money, which sets out the policy for a defined period in the future. Once set, it becomes a target and a measuring device, to compare the actual events, which occur.

Budgetary control is a management technique used to balance the various activities of the company, within the availability of resources and to achieve set objectives for a defined period in the future. Once set, it becomes the control by which the various activities must be regulated.

The main objectives are to plan and control:

Sales, purchases and manufacturing activities, so as to maximise the profit or minimise the loss of the company.

Overheads and to minimise expenditure.

Development and research activities.

Capital structures (shares and debentures) and financial reserves of the organisation (investments in other companies).

Working capital (control of stock, debtors, cash: and creditors, bank overdrafts and tax liabilities).

Liquid resources (cash, overdraft facilities).

Management decisions, by centralising control to the master budget and decentralising responsibility so that each executive becomes responsible for items within his departmental budget.

To prepare a budget, a committee is appointed by the directors which is chaired by the managing director, and the senior accountant, often referred to as the budget officer, is responsible for the budget preparation. Initially, the committee discusses the key factors and determines the initial policy to which each executive, with the help of the budget officer, prepare statements indicating the limiting factors which affect the budget for the department or function.

After reporting back to the committee, the key or limiting factors are reviewed and policy established. The functional budgets are then prepared, with continuous reporting back to the budget committee for decisions. From the functional budgets, a final co-ordinated budget is transcribed into financial accounting terminology and the whole master budget formally approved by the directors.

The limit to the size of the budget, will be due to one or more of the following 'Key Factors':

Sales.	A limited market.
	Shortage of good salesmen.
	Advertising difficulties.
Distribution.	Storage problems.
	Lack of distribution points.
	Heavy or bulky products compared with value, restricts area of distribution.
Production plant.	Restricted capacity.
	Insufficient maintenance.
	Insufficient capital for new plant.
Labour.	General shortage of all grades.
	Shortage of a particular grade.
	Difficult to supervise.
	Restrictive trade union rules and practices.
Materials.	General shortage of materials.
	Shortage of a certain range of materials.
	Transport difficulties in supplying them.
Capital.	Lack of investment in the company.
	Lack of cash to pay current expenses.
	Lack of overdraft or credit facilities.
Management.	Restricted by policies set out by the directors to the budget committee.
	Shortage of executives.

Sections of the master budget

SECTION 1. BUDGET SUMMARIES

These transcribe the functional budgets and sub-divisions into financial accounting terminology, by forecasting for the budget period,

The profit and loss account.
The profit and loss appropriation account and
A balance sheet for the end of the period.

SECTION 2. FUNCTIONAL BUDGETS

Functional budgets are prepared, the order of which depends upon the limiting or 'key factors',

Sales budget.
Selling and distribution cost budget.
Production budget.
Production cost budget.
Plant utilisation budget.
Development and research budget.
Capital expenditure budget.
Purchasing budget.
Personnel budget.
Cash budget.

Each functional budget is sub-divided into defined areas so that there is a budget available for each 'cost centre'. Each functional budget is inter-related with each other as shown in Figure 20.85.

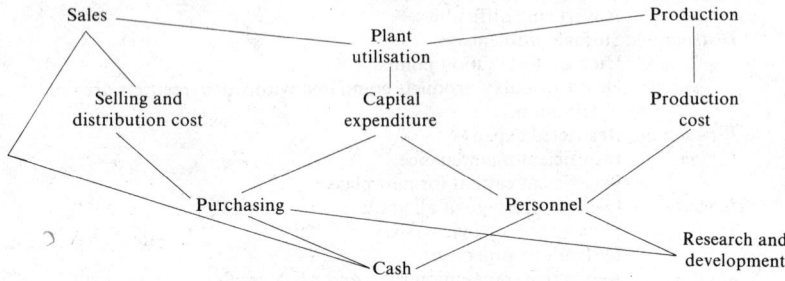

Figure 20.85. Inter-relation of functional budgets

The sales budget

The total sales in money/volume terms, is sub-divided, for example, by products and services; regional or branch areas, and periods of time, depending upon the organisation of production and selling units; selling methods and seasonal fluctuations.

SUB-DIVISION OF THE SALES BUDGET

The sales manager is responsible for the sales budget, but by sub-dividing into areas, the area managers will be responsible for their area sales budgets. These in turn are finally broken down so that each salesman is responsible for his own personal sales budget.

FACTORS TO BE CONSIDERED IN DRAWING UP A SALES BUDGET

These are usually based upon:
  The general economic climate.
    Indications from economic statistics, showing the economic trends.
    Indications from the firm's own statistical records produced from various reports.
    The political situation, since politics influence the general economic climate, at home and abroad.
  The competitive position of the firm.
    The size of the market, the division of the market between the competing firms.
    The prospects of the market, expanding or reducing, or remaining constant in size.
    The factor, or factors, which control the market, i.e. price, quality, credit given, early delivery date, fashion—clothes.
    The advantage that one firm has over another.
  Actions proposed by all the firms competing for the market
    Additions to the sales force, new depots, or new accounts.
    New advertising campaigns.
    Alterations to terms of sale, delivery dates design and packaging.
    The introduction of new products.

The amount of finished stock held by wholesalers retailers and consumers. If a large amount of stock has been sold in the previous period and remains unsold or unused, it affects the current sales, because customers are already stocked up and furthermore customers are reluctant to place future large orders, for they have seen their capital tied up in a slow selling line.

The production capacity of the firm; the firm can only sell what it can produce or purchase from others to resell.

Market research activities and the elasticity of demand for the product. This is an economic theory which explains what will happen to the volume of sales if there is a price change, other things remaining constant.

Production budget

The total production is defined in terms of physical units, standard hours and monetary value, which are analysed into; products and services; manufacturing or process departments, and periods of time.

THE FACTORS TO BE CONSIDERED IN DRAWING UP A PRODUCTION BUDGET ARE BASED UPON:

The sales budget.

Production capacity of the firm, by size of the factory premises; plant—operating time available; personnel available; availability of materials.

The increase or decrease in stock and work in progress.

The capital cash and credit facilities available.

The policies set out by the directors and senior management
    Sub-contracting to and from other firms.
    Purchase of finished or semi finished products.
    Scheduling the work into long or short runs.
    Manufacturing a wide or limited range of products.
    Introduction of shift work.
    Increase pressure for greater production.

The attitudes of management, trade unions, shop stewards and employees to current policies and proposed policies.

Proposed changes to resources within the budget period.

Plant utilisation budget

This lists the total machine load in machine hours for each process, department, group of machines, and an individual machine.

The aims being to determine the utilisation of plant and equipment in order to direct management to take corrective action. For example, by introducing overtime and shift working; transferring work to other machines, or departments, which are less suitable, but capable of doing the work; sub-contracting work, depending upon machine load; purchasing new plant and machinery to cope with planned production within the budget period; selling excess machinery. Also to amend the sales and production budgets by reducing the work load so that existing plant and machinery can cope, or expand by taking-in work to utilise the excess capacity.

Selling and distribution cost budget-lists

The total selling and distribution costs of selling and distributing the sales planned in the sales budget and the long term sales policies set by management can be analysed into the following:

Direct selling costs.
Indirect selling costs—sales office, market research, advertising.
Direct distribution costs—transport and packing.
Indirect distribution costs—warehouse and storage points.

Production cost budget

Using the figures set out in the production budget, the production cost budget analyses the production cost of:

The firm—responsibility of the production manager.
Each factory—responsibility of the factory manager.
Each department—responsibility of the departmental manager.
Each process—responsibility of the process manager.
Each sub-department—responsibility of the department superintendent.
Each machine operation, or activity—responsibility of the supervisor.

into,

Direct labour cost.
Direct materials and services.
Overheads; controllable and uncontrollable.

Depending upon the level of authority and the amount of responsibility delegated, the controllable overheads will be the responsibility of the executive concerned, and the uncontrollable overheads the responsibility of his superior.

The complete production cost budget is a voluminous document and it is essential for each executive to have not only the information at his level, but that of his subordinate. Consequently, the departmental manager should have the production cost budget for his own department, his own department's sub-divisions and the cost centres of each sub-division.

The capital expenditure budget

This lists the planned expenditure on the purchase and the sale of fixed assets for a period which is generally longer than the budget period.

The capital expenditure budget is prepared from:

1. The planned overload on the fixed assets that the firm has currently available while producing the planned output.
2. Current expenditure upon capital equipment which may be affected by an expansionist policy. Certain machinery may be sold and replaced by larger capacity machines that will fit with future purchases in other sections.
3. A list of priorities, and
4. Long term policies aimed at reducing cost. On the basis of a cost investigation to improve efficiency, the firm may or may not justify the purchase of new, or nearly new equipment.

Development and research budget

This is a planned outlay in terms of money, on research and development.

The aims are to limit the activities of the development and research activities, by limiting the amount and supply of money, and direct the development and research activities, to where an allotment of money is available.

Purchasing budget

The purchasing budget lists the total purchases in money/quantity terms analysed in terms of:
  Direct materials and services.
  Indirect materials and services.
  Bought in components of the finished article.
  Finished goods for resale.
This is to enable the firm to achieve the planned sales, production, capital expenditure, development and research and changes in stocks of raw materials, bought in parts, work in progress and finished articles. Taking into account orders already placed by the company and the materials to be manufactured by the firm.
  The aims are to permit the purchasing department to organise itself; establish long term contracts at favourable prices; take advantage of special offers or bulk purchases; record material price changes and keep records; and to define cash requirements for the period.

The personnel budget

The personnel budget lists the total direct and indirect labour required in terms of the number and grades of personnel, the number of working hours and the wage bill per cost centre,
  This is to enable the firm to achieve the planned sales, production, capital expenditure and development and research.
  The aims are to assist in the efficient management of labour by defining requirements for the period and stabilising labour turnover; establishing training needs; providing welfare and medical facilities and defining the wage bill for the period.

The budget period

To obtain the full benefit from budgetary control, the main budget (one year is the natural period) should be supplemented by short period budgets.
  The advantages of a long term budget are that it enables favourable purchase contracts to be negotiated, makes long production runs possible, and introduces specialised plant and labour to generate improved productivity per man/hour.
  The advantages of a short term budget fulfils the need of different personnel, for example, the supervisor needs a weekly budget, the manager needs a monthly budget, and the cashier needs a weekly and monthly budget.
  Long term budgets, such as the capital expenditure budget are generally 3 to 5 years.

Level of attainment

The budgets may be set at various levels of attainment as follows.
  1. *Optimum level.* A level at which the function, or the organisation would be operating with the maximum efficiency possible. This level has a disheartening effect on the morale of the executive and those under him. The attitude is, that the target is impossible and it can never be surpassed. It may result in a falling off in efficiency.
  2. *Efficient level.* A level at which the function, or the organisation would be operating at a highly efficient, but workable level. This is considered to be the level at which the budgets should be set. It encourages efficiency, considered fair and it can be surpassed, giving an incentive to do so.

3. *An easy level.* A level at which the function, or the organisation could without any real effort be reached. This level encourages a go-easy attitude, that no matter what happens, the target will be reached. Therefore a falling off in efficiency and in management control.

Flexible budgets

To avoid the possibility of wrongful decisions based on unreliable budget plans, a flexible budget might be more appropriate. In such a budget certain changing events must be considered; for instance changes in the volume of work; the working period; conditions of work and seasonal variations. If these changes are likely to occur then a 'flexible budget' would be preferred.

Cash budget

This lists the total cash balance or deficit which can be forecasted from the planned receipts and expenditure for the period and shorter periods. The cash budget is prepared from a:

Personnel budget.

Purchasing budget, with adjustments for time-delay in making payments and taking advantage of cash discounts.

Sales budget, with adjustments for credit given to customers, and giving discounts for quick payment.

Cash required for other purposes, tax and dividends.

Cash received from other sources such as, industrial training grants, dividends from investments.

The aims are to ensure sufficient cash is available, indicate the necessity of obtaining additional finance and to explain the basis upon which the decision was made, and indicate surplus funds available for other purposes. The budgets for each function (sales production etc.) are related to the financial statements. Balance sheet × profit and loss account by preparing an estimated balance sheet for the end of the budget period and an estimated profit and loss account for the period of the budget.

Discounted cash flow (DCF)

A method used to evaluate capital or investment projects. All the techniques,[69] net present value (NPV), yield method, sinking fund, terminal values and payback period, introduce:

1. A time period that £1 now is worth more than £1 later, even if depreciation is accounted for, because £1 now can be invested and gain earning interest.
2. The consideration of flows of cash rather than profit.
3. An examination of inter-relationships between one activity and another in a complex business.
4. The consideration of a marginal project.

NPV TECHNIQUE

$$\text{NPV} = X_0 + \frac{X_1}{1+i} + \frac{X_2}{(1+i)^2} + \frac{X_3}{(1+i)^3} + \dots + \frac{X_n}{(1+i)^n}$$

where $X_0 \; X_1 \; X_2 \; X_3 \; ... \; X_n$ are the cash flows in periods of 0 1 2 3 4 (where 0 equals the current period) and i is the discounted rate of interest.

YIELD METHOD

$$C = \frac{X_1}{1+r} + \frac{X_2}{(1+r)^2} + \frac{X_3}{(1+r)^3} + ... + \frac{X_n}{(1+r)^n}$$

C equals capital expenditure; $X_1 \; X_2 \; X_3 \; ... \; X_n$ are the cash flows and r the unknown rate of interest.

The DCF method is used in practice to solve problems of deciding to accept or reject an investment project; of rationing capital between several projects; of making the necessary adjustments and allowances, where the inflows and outflows of cash may alternate within the investment period.

PRODUCTIVITY COSTING

Productivity costing[70] was developed to offer some improvement in methods of cost accounting and cost control, particularly as a means for sound managerial decision making. The intention was to avoid as many as possible of the pitfalls of both conventional overhead absorption and marginal costing. Productivity costing may be considered as an operating overhead expense absorption costing method, in which products absorb materials conversion costs at a rate based on the facilities capacity to produce, rather than on an ability to sell products. Alternatively, it may be considered as a marginal costing system in which due proportions of the fixed costs are absorbed by products according to their usage and occupancy of productive facilities, at a costing rate based on the maximum feasible usage of the facilities.

The most effective way of maximising total earnings is understandably to maximise the rate of generating total earnings per unit of operating cost. Theoretically, this can be achieved by producing the optimum mix of products having a high return on total earnings. However, to determine this maximum it is first necessary to determine the utilisation of facilities and labour, this may be done by activity sampling, by other work measurement techniques, or by recording devices, and must embrace possible shift working policies as well as unavoidable operating interruptions.

An output is generated (or value added) only by the productivity of utilised inputs, and it may be conceptualised that the ratio of utilised inputs to total inputs available is a measure of the efficiency of productivity of the manufacturing system. The operating costs required to add value to throughput materials in order to convert them into saleable or usable products, is the product cost C_d. Since the cost of operating the entire manufacturing system C_s includes the idle and under-utilised costs C_i (which are not allocated to the products in productivity costing), the conversion utilisation would be C_d/C_s.

As a system, productivity costing is compatible with standard, absorption and marginal costing methods, and special systems integrating these concepts may be designed specifically to meet the organisational and functional requirements of any particular enterprise.

The major objectives of industrial-commercial systems is assumed to be the maximisation of total earnings, T (derived from sales-materials purchased) or of the rate of generating total earnings per unit of total conversion costs, C, thereby maximising T/C. Consequently while the primary productivity index is derived from T/C the rate of profit generation is the ratio of profit total conversion, P, to conversion costs, C, or P/C. Since profit is the difference between total earnings and production costs $T - C = P$, the secondary productivity index will be:

$$\frac{P}{C} = \frac{T-C}{C} = \frac{T}{C} - 1.$$

In maximising the primary productivity index T/C, the secondary index P/C is also maximised since it is always unity less than the total earnings productivity.

The potential total earnings output, T_{pot}, generated by a manufacturing system's input C_s, is equal to the total earnings T, that could have been generated if all the inputs had been fully utilised. This potential situation, T_{pot}, can be derived from normal costing records by totalling the product earnings for all products produced $\Sigma T/\Sigma C_d$, and multiplying this by the manufacturing system's input costs, C_s; i.e.

$$\frac{\Sigma T}{\Sigma C_d} \times C_s = T_{pot}.$$

An abundance of productivity indices can be developed quite readily from data available in books of account maintained normally by most enterprises, provided product costs do not include any element of the costs of idle and under-utilised product processing facilities. The total earnings productivity concept has the convenient characteristic that it reveals the profitability of a product or a manufacturing system, something conventional costing practices fail to do. What is just as important is that it provides management with a measure of the potential level of output that could be achieved if all the input factors were fully utilised. This in turn provides an indication of how much productivity improvement is possible rather than have companies continuing to compare productivity levels with past performances.

Productivity costing represents the influence of engineers on cost accounting practices which complements the existing structure of management and financial ratios.

Engineers' measures of productivity

Productivity and efficiency are often regarded synonymous. Engineers speak of efficiency (E) as the ratio O_u/I where O_u = useful output and I = input, both of course in the same units. Clearly this ratio is always less than 1.

$$E = \frac{O_u}{I} < 1$$

For useful output (O_u) to reach its potential useful output (O_p) depends upon how well the input factor(s) is utilised:

$$E = \frac{O_u}{I} = \frac{I - \text{Losses}}{I} = \frac{O_u}{O_p} = \text{Productivity} < 1$$

From this expression we have three productivity ratios:

1. Generation of useful output $\dfrac{\text{Output}}{\text{Input}}$

2. Utilisation of systems input $\dfrac{\text{Effective input}}{\text{Actual input}}$

3. Actual (useful) output to potential (useful) output $\dfrac{\text{Actual output}}{\text{Potential output}}$

Utilisation of resources compared to a standard, such as producing more from a given combination of input factors; or a given useful output from less inputs is most generally regarded as the improvement of productivity, and a low utilisation of input factors as low productivity. Consequently, we define and measure relative productivity levels in comparison with a level achieved in the past; or in comparison with another establishment in the same industry; or in comparison with the national average achieved by another nation.

Standards of efficiency

Since the utilised parts of labour and machines (not forgetting materials) add value to the manufacture of products, then the efficiency with which value is added should be measured. For the engineer this measurement depends upon the concept of a standard. Work measurement supported by method study provides useful standards, which nevertheless, have a built-in element of judgement aggravated by policies of wage payment. The means of establishing these standards varies considerably between industries and between firms within given industries—(and often between departments within given firms). Thus, 'A standard obtained by measuring inappropriate methods of working is no standard. Equally a standard obtained by timing one man on one job is not a standard, only a source of argument. Thus it is essential that an acceptable method is devised before work measurement studies are taken and that several operatives, representative of those doing the same standardised job, are timed in a proper manner before a standard is presented'.[71]

Measured data on which payment-by-results are generally based are determined in many companies by ratefixers, or supervisors, who base their standards of efficiency on extracts from past records, experience, incomplete or out-dated synthetics, or by stopwatch timing one operative doing a task. Methods examination is (when carried out) mainly confined to 'floor-to-floor' activities, rather than concerned over a broader field examining 'door-to-door' activities. Before the time is accepted as the appropriate 'rate for the job', bargaining at shop floor level is carried out, and under this mutual arrangement both the ratefixer and the operative know that the final value for the job will bear little or no relationship to the time it actually takes.

Many trade union officers consider that a ratefixed system has an in-built opportunity for bargaining which tends to favour shop floor workers simply because it enables them to participate in setting their own earnings levels. Nonetheless, a measurement system that depends almost entirely upon mutual agreement provides little opportunity for workers and management to compare the effectiveness of their work with that of a 'representative' standard. The measure of success in any working situation is performance against a standard and in applying work study techniques it is important to realise that there are very few short-cuts and none for the beginner.

In addition to measuring labour performance against some predetermined standard, comparisons should also be made with an ideal, or potential standard. An interesting method that combines potential with standard indices is used in an electrical company in the Midlands. The method makes use of a technical index which is equal to the ideal or minimum work-time (i.e. no interference or breakdown, least setting-time, etc.) divided by the current standard time, multiplied by a labour utilisation index:

Productivity index = technical index − labour index

or

$$\frac{\text{Ideal time}}{\text{Actual time}} = \frac{\text{Ideal time}}{\text{Standard time}} \times \frac{\text{Standard time}}{\text{Actual time}}$$

Other types of utilisation, or productivity measures the engineer uses, are based on:

Volume or physical unit of output e.g. in homogeneous industries.
Horsepower and electricity consumption.
Machine and space utilisation.
Materials utilisation and waste controls.

ACTIVITY SAMPLING

Statistical sampling techniques have been used for the measurement of industrial-commercial activities since 1934, when 'snap-reading' was introduced into textile processing by Tippett;[72] Morrow[73] renamed this form of workmeasurement 'ratio-delay' in order to determine delay allowances in setting time standards, and referred to later as 'work sampling'.[10]

British Standard 3138[6] defines 'activity sampling' as 'A technique in which a large number of instantaneous observations are made over a period of time of a group of machines, processes, or workers. Each observation records what is happening at that instant, and the percentage of observations recorded for a particular activity, or delay, is a measure of the percentage of time during which that activity, or delay, occurs'.

Activity sampling is recognised as a work measurement technique suitably designed to measure the working and non-working time of office, maintenance, stores and other indirect service personnel, as well as those directly employed in productive work. It also permits the measurement of operating time and down-time of machines and equipment.

The purpose of using this technique is to provide management with physical measures of the utilisation of labour and machines, and to provide a basis for analyses, which lead to the reduction of idle capacities caused by inadequate decision and control, inadequate services, and operative 'rest' pauses. As industry becomes more capital intensive the importance of controlling and reducing machine down-time becomes very great.

The advantages of this technique are gained where several observations can be made in one visit to the work area. Even where an observer spends the whole time in the work area, the fact that he is observing several operatives and/or machines almost simultaneously, if need be, is an advantage that cannot be claimed by any other work measurement technique.

It is inevitable however, in all economic sampling studies that complete accuracy cannot be obtained. Accuracy, within defined limits of confidence (as in all work measurement techniques) depends upon the sample size.

Confidence limits

If a number of samples are taken at random from an unchanging mass, the sample results will be distributed in an approximately repeatable pattern known as the 'normal curve of error' and from this together with the theory of probability it has been possible to establish a relationship between the number of observations made, the sample mean and the probable, or sampling error.

When a sample is taken it must be recognised that it is only an approximation, the reliability of which varies according to the number of random observations. One of the great advantages of statistical sampling is that the degree of reliability of the results can be calculated as the study continues. The statistical formula for the calculation on the standard deviation of sampling variations (standard error) is:

$$S = \sqrt{\left\{\frac{p(100-p)}{n}\right\}}$$

where p = the occurrence of the specified activity as a percentage of the total number of observations.

n = the total number of observations.

S = the standard error of p expressed as a percentage of the total.

The properties of the normal curve have been thoroughly worked out and it can be shown that:

682 times in 1000 the true value of p is between $p \pm S$
954 times in 1000 the true value of p is between $p \pm 2S$
997 times in 1000 the true value of p is between $p \pm 3S$

These positions define the limits into which p will fall with three different levels of probability. Obviously, for any given number of observations the higher the required value of probability the wider the margin of error. The most commonly used confidence limits are those based on $\pm 2S$.

Suppose management wished to know the time spent by a machine which was either productive, non-productive but being set-up, being repaired, or idle. These activities could be listed and arrangements made to observe at the end of every tenth minute, over periods totalling fifteen hours, exactly what was taking place. (Care would need to be taken that the time interval for observation did not coincide with any repetitive cycle of operations). Similarly, snap readings must not be taken in such a way that bias is in evidence.

A summary of these readings could be as follows:

Activity	No. of observations	% of total
Productive	39	43
Being set-up	20	22
Being repaired	26	29
Idle	5	6
Total	90	100

The approximate time spent by the machine being set-up is therefore 198 minutes i.e. 22% of 15 hours. How accurate is this?

$$2S = \pm 2 \sqrt{\left\{ \frac{p(100-p)}{n} \right\}}$$

$$= \pm 2 \sqrt{\left\{ \frac{22(100-22)}{90} \right\}} = \pm 8 \cdot 75\%$$

So that the actual time spent by the machine 'being set-up' may be assumed to be between 13·25% and 30·75%, which is probably too indefinite; the higher limit being more than twice the lower.

If an error of not more than 3% was required so that the machine set-up lay between 19% and 25%, then the number of observations would have to be:

$$n = \frac{4p(100-p)}{S}$$

$$n = \frac{(4 \times 22)(100-22)}{3 \times 3} = 762.$$

To keep the error down to 2% the number of observations would be 1716. Even a sampling error of not more than 2% might be inadequate for an activity which occupied only a small percentage of the total time.

The item 'idle' for instance is 6% or on a 2% error basis between 4% and 8% expressed as time between 36 and 72 minutes. The size of sample may not be sufficiently accurate enough to help decide on the controls necessary to eliminate the idle time altogether. It is therefore necessary to calculate for the small percentage factor occurrence.

Summary of procedure

1. Make a preliminary study of the process and decide on the information required.
2. Identify and clearly define the various activities to be studied.
3. Prepare suitable observation sheets, designed in such a way that where appropriate, the various activities, whether of 'men' or 'machines' can be ticked off as they are observed.

4. Decide on the number of observations to be taken over a specified period and calculate the minimum and maximum time interval required (Norman[9]), or if necessary undertake a random selection from random tables.
5. Carry out a trial study and adjust where necessary.
6. Complete the study.
7. Calculate the results and determine further action.

Under conventional application, activity sampling implies two factors; working (p) and not working $(1-p)$. Calculations are normally performed to determine the number of observations (n) required to ensure that the findings of the chosen factors are within close confidence limits of accuracy. For most industrial measurements a confidence level of 95% is considered satisfactory, which means that 'one is confident for 95% of the time that the random observations (n) will represent the facts, and for 5% of the time they will not'.[10]

It is important to realise, however, that observations need not always be made at random. (Tippett[74] says that 'very often the pattern of behaviour of the system is so irregular that randomisation of the observations is unnecessary . . . Randomisation is achieved only at a cost: the cost of handling the hundreds (or thousands) of random numbers, and the cost of the observer making himself available at random instants'.

For example, the operations within a manually controlled process are not necessarily regular, constant and/or repetitive; observations under these conditions might very well be made at regular intervals, providing such regularity does not pick-up cyclic patterns. What is already random cannot be made more random. Only in machine controlled processes, where regular cyclic patterns are more likely to occur does it become important to use random number tables.

Other practical suggestions have been made for determining sample sizes: 'rule 1000' described by Rowe[75] is known to give satisfactory results in industry, as long as there is no desire to measure accurately small percentage activity occurrences.

'Disc-O-Tec' sampling

Disc-O-Tec activity sampling was considered in 1968 to be the first of its kind attempted on a large scale in Britain, and the use of foremen to carry out the observations, in addition to their normal duties, was considered to be quite novel. Foremen were being requested to do something in an ordered fashion, which they themselves do in their every day duties; that is, observe that man and/or machine are working.

For this exercise the foremen were issued with sample study sheets for each working shift hour, or working day whichever method was decided upon in discussion and a maximum of twenty discs, each to represent an operative, or machine, or work station.

During *each hour* the foreman was to withdraw a handful of discs from a pocket (about 4 or 5) at random intervals of his own choice, select a disc number and identify it with 'man' or 'machine', and take a mental 'snap-shot' of the activity, marking the study sheet accordingly. The foreman continued to do this until all discs were transferred from one pocket to the other in one hour. This was repeated for all subsequent hours in his working shift for a period of four normal working weeks.

In order to simplify the analysis of data, the study sheets were designed as data processing cards and arranged to incorporate six major activities embracing 'productive' and 'non-productive' work:

FACILITY (OR MACHINE) SHEET HEADINGS

Forming, removing material and/or joining or separating material.	Machine productive

Setting or inspecting—job change new product.
Tool adjustment—no job change.
Maintenance—planned and unplanned waiting for maintenance, and maintenance in progress. } Machine non-productive
Machine idle—operative not responsible.
Set-up and non set-up.
Machine idle—operative responsible.

MAN (OR LABOUR) SHEET HEADINGS

Operating—man in control of progress.
Attending.
Handling. } Man productive
Servicing—cleaning down, changing tools, doing own inspection, etc.

Waiting at work station—management responsible for work; for service; idle—no work. } Man non-productive
Waiting at, or away from work station—worker responsible.

Bahiri and Martin[56] recognise the significance of truly productive work as distinct from ancillary (or non-productive) work, which may or may not be necessarily entailed in achieving the productive work which alone generates output. They define it in a manufacturing context as: 'work which (whether done by a man or a machine), actually changes the shape, physical characteristics or appearance of production materials, or which joins one material or product to another, or separates one material or product from another during the process of converting throughout materials into saleable (or usable) products'.

It was considered that as the discs were transferred from one pocket to another, ordered stacking might occur, i.e. the last disc to be looked at in one hour's sampling could be the first selected in the hour following. Shuffling was suggested but in order to dispel such fears, tests were carried out using two groups of subjects; twenty in each group. One group sampled with shuffling and the other without. In each case the order in which the discs were sampled was noted. A statistical analysis proved that ordered stacking was highly insignificant at 99%.

In order to show comparisons of utilisation, hour by hour, day by day and shift by shift; the method of recording was divided two ways; by shift hour and time, and by day. Shift hour and time were related to mornings, days, afternoons and nights as shown below and in order to overcome the differences in starting times of each firm, the shift hour (1 to 8) was adopted as the basis for comparison. Comparisons on a 'day' basis presented no problems.

Once the shift hour and time had been recorded on the study sheet, the same sheet was used for the *same hour every day* for the period of sampling. Separate sheets were used for each remaining hour. This meant that each observer chosen to sample in this way was issued with 8 sheets 'man' type and/or eight sheets 'facility' type.

To obtain 'day' trends, the observers were issued with five sheets 'man' type and/or five sheets 'facility' type. Monday was nominated day 1, Tuesday day 2; and so on. Once the day has been recorded on the study sheet, the same sheet was used *every hour during the nominated day,* i.e. if the sheet was recorded day 1 then it was used every hour each Monday for the period of sampling. Separate sheets were issued for day 2, 3 etc. Saturday, Sunday and overtime working were omitted from the study. Tea breaks (morning and afternoon), where these were taken, and the time immediately before and after each break, including lunch and end of shift, were also ignored, that is, work pattern were accepted as they were currently applied.

WORK MEASUREMENT

Work measurement is that aspect of work study which is concerned with the measure ment of the work content of activities. It facilitates accurate planning and programmin, of work, utilisation of labour and machines and cost control; it also serves as the basi

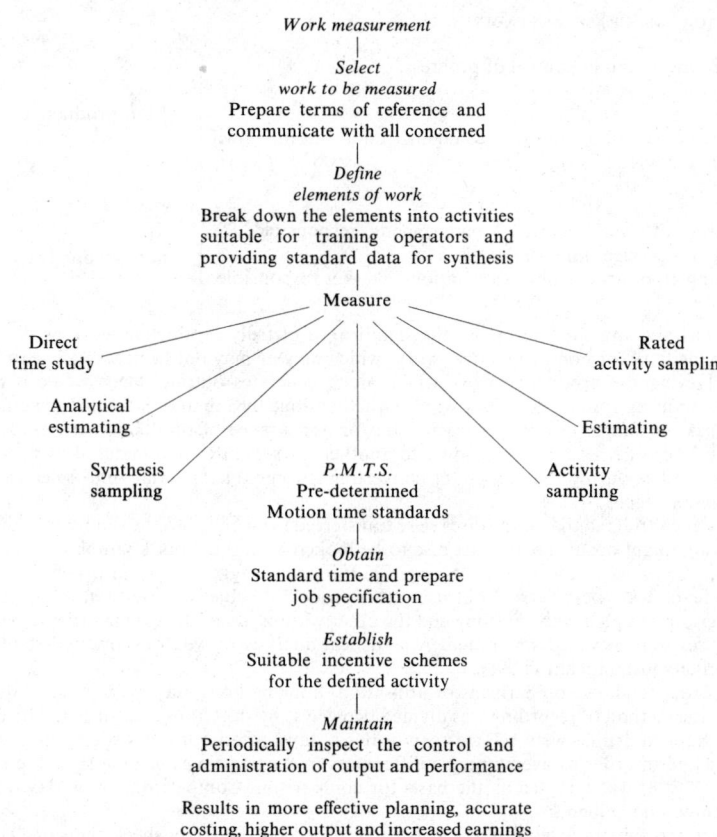

Work measurement

Select
work to be measured
Prepare terms of reference and
communicate with all concerned

Define
elements of work
Break down the elements into activities
suitable for training operators and
providing standard data for synthesis

Measure

Direct
time study

Analytical
estimating

Synthesis
sampling

P.M.T.S.
Pre-determined
Motion time standards

Rated
activity sampling

Estimating

Activity
sampling

Obtain
Standard time and prepare
job specification

Establish
Suitable incentive schemes
for the defined activity

Maintain
Periodically inspect the control and
administration of output and performance

Results in more effective planning, accurate
costing, higher output and increased earnings

Figure 20.86. The systematic approach to work measurement

for incentive schemes which should ensure quality of reward for equivalent skill an effort. A systematic approach to work measurement is shown in Figure 20.86.

Work measurement includes time observation of actual activity; 'rating' of effor and assessing the relaxation allowances required by the operator.

Measurement highlights factors causing delays some of which may originate from 'overwork' and increase operator fatigue and others from 'underwork' which tend to encourage an operator to work ineffectively.

Work measurement is defined as the application of techniques designed to establish the time for a qualified worker to carry out a specified job at a defined level of performance.[6]

Systematic approach

As in method study there are six processes which provide a systematic approach. These are listed in the following paragraphs.

METHODOLOGY

Select.   With due regard to economic justification of study. Activity sampling study, or production study to obtain performance and efficiency levels of manpower and capital equipment.

Measure. Depending upon the nature of work, frequency and job breakdown.

Work at design stage	*Work in progress*
Estimating	Analytical estimating
Analytical estimating	Time study
Synthesis	Synthesis
	Activity sampling
	Production study

Obtain.   Basic time $\dfrac{\text{observed time} \times \text{observed rating}}{\text{standard rating}}$

Selected time

Establish. Allowances and standard time (SMs). Note that policy allowance may be added for wage adjustment thus extending standard time to allowed time (AMs).

Install.   Prepare job specifications and administrative controls, instruct supervisors and clerical personnel involved. Issue standing instructions for scheme and implement financial incentive scheme where applicable. Allow commissioning period.

Maintain. Periodically check controls, amend and modify if agreed and improve.

Select

The selection of activities suitable for work measurement arise from areas of unnecessarily high cost, and the possible reasons for selection may be:

To compare an original method and performance with that of a new and improved method.

To provide a standard time for a new method that has been devised.

To determine the performance index of labour.

To reduce the cost of an operation which appears to be excessive.

To provide time standards for the basis of an incentive scheme.

To provide standard data which can be catalogued and used as synthetics.

Terms of reference must be either provided by management, or prepared by work study investigators after a preliminary survey has been carried out.

Time study

Time study is one of several complementary techniques embraced by work study, and before it is carried out, in an attempt to establish a standard the observer must see that two basic conditions have been satisfied:

1. The operation, or task, has been standardised and causes of fatigue eliminated wherever possible, and

2. The task is being performed in the approved manner by a trained and experienced operative, the length of whose experience depends upon the skill and complexity of the task.

The orthodox procedure for establishing standard time can be illustrated simply as shown in Figure 20.87.

Recording information by time study

An accurate time study of an operation cannot be made unless the study practitioner has thoroughly familiarised himself with the operation and is able to place on record a

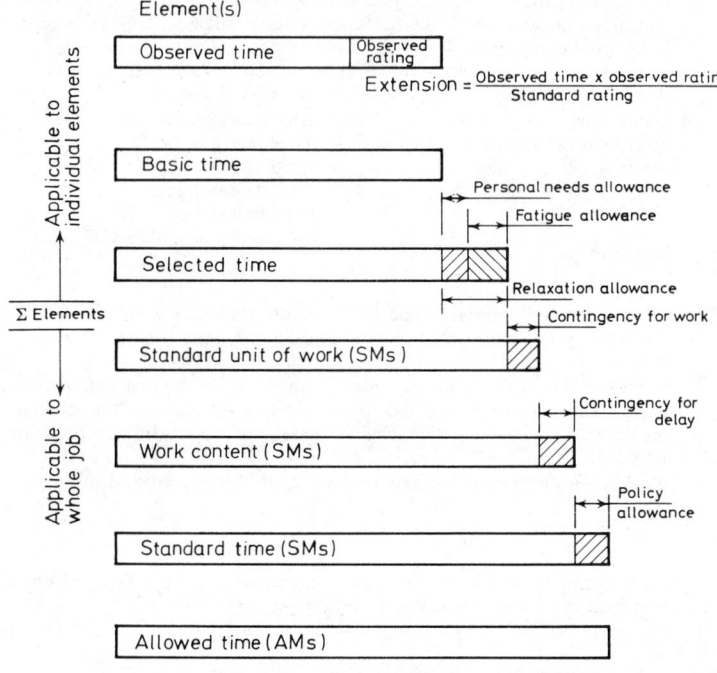

Figure 20.87. Orthodox procedure for establishing standard time

complete and detailed description of both the present and improved methods used in carrying out the operation. The importance of this cannot be over-emphasised.

Comparisons are continually being made of the old and improved methods and the relative time measurements. Having a detailed specification on record assists in determining whether the operation is being correctly performed when bonus payments are fluctuating and are receiving scrutiny from both management and workpeople.

ELEMENTS

Taylor[76] laid down what is still recognised as one of the first steps in the preparation for a time study: 'each job should be carefully sub-divided into elementary operations and each of these should receive the most thorough time study'.

An element is a distinct part of a job usually consisting of a number of basic movements (therbligs grouped together) which need to be identified in order to recognise the method used and conveniently arranged for: observation; measurement; extension (rating adjustment), and analysis.

Currie[77] favours elements of between 10 and 50 centiminutes duration, realising that there is a limit below which accuracy can be maintained in reading a stop watch. There are, nevertheless, elements made conveniently longer than 50 c.min. depending upon the type of job and activities being performed, and where this is necessary, recordings of the consistency of the operatives pace should be made approximately every 50 c.min.

The instant at which one element in a work cycle ends and another begins is called a 'breakpoint', recognised by sight, sound or touch, and should be decisive.

The number of work cycles through which any particular job should be observed varies directly as the amount of variation in the 'basic times' of the element for instance if occasional and variable elements are evident then these must be observed over several occasions. One simple exercise for checking the validity of basic times is shown in ILO.[5] Barnes[15] says that: 'it is necessary to consider the length of work cycle, number of elements in the work cycle, variations in the length of cycle, consistency of the operative and the relation of machine time to handling time'. To this we must add the consistency of the observer in choosing appropriate elements and breakpoints, and his rating conception.

The purpose of breaking a job down into elements is to:

1. Compute synthetic time values by using comparisons of elemental times obtained from different tasks.
2. Train and retrain operatives.
3. Account for any change in consistency that operatives put into a work cycle.
4. Identify compensating elements in the assessment of fatigue allowance per element.
5. Recognise more easily foreign elements that might occur in a work cycle.

Measure the work content

Time observations may be carried out by using one of the following methods. A fly-back stop-watch or a combined fly-back and continuous running stop watch; graduated in centi-minutes i.e., 1/100th part of a minute, and is one of the most accurate and simplest devices for work measurement.

An ordinary pocket watch, wrist watch or clock when there is no need for such accurate results.

Counting. This method is difficult to perform and practice is necessary to ensure uniformity. The exact starting and finishing clock time is essential with counting studies.

Recording instruments such as tape recorders and cine cameras.

'AUTOGRAPH' TIME RECORDER

Amongst the alternative methods that exist is a timing device incorporating a print out. The Gregory-Russell Time Recorder is one example of such a device. Six buttons (coded a to f) allow the observer to distinguish between different elements. The instant at which the observer presses a button is printed in time units to the nearest one tenth of a second. It is necessary to subtract between each numerical recording to obtain the elemental times.

Rating

'Rating is a matter of judgement on the part of the time study engineer'; this is stated by Barnes.[15] Marriott[78] states that: 'It is probably the most difficult part of the procedure,

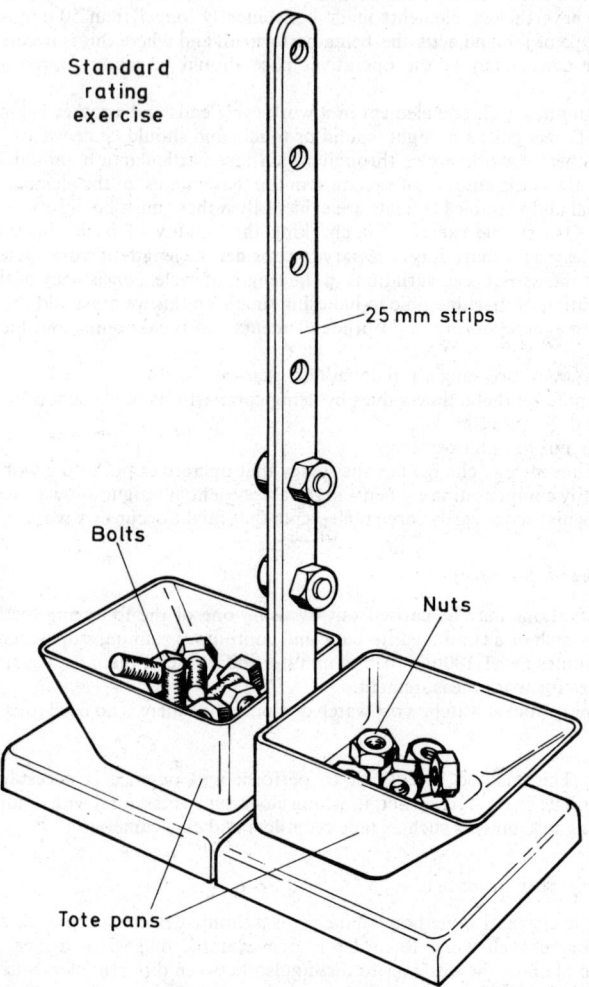

Figure 20.88. Exercise suitable for practitioner training in a metalworking environment and engineering

(time study) mainly because it is entirely subjective and a matter of the observer relying entirely on his judgement'. Sury[79, 80] is also concerned with the procedure of time study particularly the accuracy of the rating process.

Rating is that process during which a time study officer compares speed, effort, and effectiveness of the operative, with his trained concept of standard rate (Figure 20.89).

BS 3138 recommends the 0/100 rating scale where 100 corresponds to 80 and 133, on the 60/80 (Bedeaux) and 100/133 (Metric) scales respectively, and explains that standard rate should not be the supposed rating of a 'timeworker' (i.e. daywork, identified by the lower points of 60 and 100 on the other scales), but the rating to be expected from a worker of average ability and attainment who is motivated to apply himself to his work.

Graham[81] states that 'only two working speeds (rating) have a special meaning—0 which is the stopped condition and 100 which is the standard motivated condition. All other speeds (ratings) are only relative to those two reference points'.

If work study practitioners maintained their rating concepts, then consistency in judgement might be controlled even to the point of eliminating the variations. Matthew[82] supports this view with one qualification, 'within limits of accuracy and consistency'. He found that where groups of time study engineers were trained on established standards of performance they could achieve surprisingly accurate results. Currie[12] also says 'that given the right methods, selection and training of work study men, errors (variations) are minimised and results are well within acceptable limits of accuracy'.

Numerous training aids for rating are available; some establishments favour the concept of a man walking unladen at 4 m.p.h. (6.44 km/h) as standard rate or of dealing 52 cards into four heaps, placed at the corners of a 1 ft. (0·305 m) square, in 38 c.mins. An exercise most suitable for practitioner training in a metalworking and engineering environment is illustrated by Norman[9] (Figure 20.88).

Rating films exist on many forms, the most common having experienced operatives performing the same operation at different speeds. Others are; film loops, multi-image[7] and stepped paced films[83], most of which can be obtained from the Central Film Library, or from ICI, NCB, Pilkington, Hoover, Courtaulds, University of Manchester and The Society for the Advancement of Management (SAM) USA.

Factors affecting rating

Research indicates that there is some interdependence and interaction between rating and relaxation allowances, for example, Dudley[84] states 'that although rating is merely an assessment of actual working pace relative to (standard rate) the establishment of standard pace involves consideration of effort or skill demanded by the nature of the operation . . . . Hence in the process of rating . . . allowances are made for the fact that some jobs are more fatiguing than others and CR allowance (fatigue allowance) also graded accordingly to effort, are additional to these'.

Furthermore, an operative's working pace can be affected by other factors such as, attention required, concentration, confined working areas, abnormal working positions, special clothing, etc. and these are also considered in the Observer's assessment of actual working pace relative to standard rate.

The rating assessment however, does not provide allowance to overcome periods of exertion and fatigue, and therefore, additional allowances are considered necessary, graded accordingly.

Allowances

When making a time study, ineffective time will be seen to occur as judged by the work study man. The manner in which this is decided upon can introduce further subjectiveness into the derivation of standard time.

The ineffective time might be recorded on the time study sheet as:
Rest, smoke, personal needs, etc.
Interruption by supervisor, progress chaser, or others giving job instructions.
Foreign element: not specified in method.
Fumble: deliberate, or lack of skill.
Unoccupied time other than waiting or idle.
Delay: cyclic interference in machine work cycle.
Delay: machine breakdown.
Delay: waiting materials, handling, etc.
Changing tools, cleaning down, clearing waste.
Absent from job.
The above list is by no means exhaustive.

By quantifying the time spent under the general headings of rest, delay—management responsible, delay—operative responsible, unoccupied time, interference time, assessments can then be made as judged necessary by the work study man.

BSI[6] recommends that allowances be given for: personal needs, fatigue, contingency for work, and contingency for delay.

Personal needs allowance is generally accepted as constant within a working group. Barnes[15] says that 'where an operator works eight hours per day without organised rest periods, 2 to 5% (10 to 24 min) per day is all that the average worker will use for

Rating scales in use

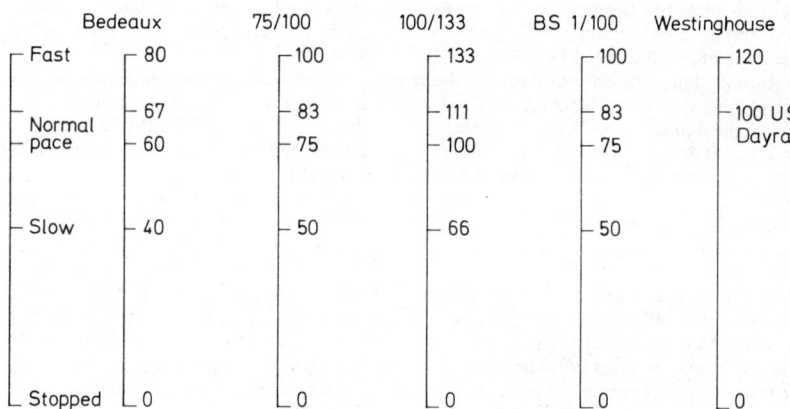

Figure 20.89. Rating scale comparisons

personal needs'. Dudley[21] has also observed that 'personal delays under the workers' control tend to be grouped towards the beginning and end of each working period'. This is supported by current research.

Production studies[5] or activity sampling studies[10] of various types of work can give an indication of the allowances required for personal needs, work contingencies and delay contingencies.

Fatigue allowance, however, is one where judgement is most subjective, although causes of fatigue will have been eliminated wherever possible in standardising the layout of the work and methods of working by systematic method or motion study.

The problem of determining the required amount of time needed to overcome periods of exertion and fatigue that remains, is most complex. Murrell[19] says that 'rest point cannot be determined objectively because it takes some time for lactic acid to accumulate and give a subjective sense of fatigue . . . . Since many industrial jobs vary in work effort . . . for work rates exceeding 350 W the periods of work and rest can be estimated'.

Research in the fields of ergonomics and applied physiological psychology is showing some promise to provide objective means of measuring recovery rates. Some organisations such as AIC, ICI and Courtaulds, have by experience, arrived at tables of points and percentage from which allowances can be built up.

Relaxation allowances

The relaxation allowances given below represent a wide range of industrial conditions and should be regarded as averages. Careful consideration must always be given to each instance before relaxation allowances are fixed.

Policy allowances, e.g., tea breaks, getting to the job after arriving on site, obtaining tools, etc., are excluded.

The following points about these policy allowances should be noted.

They may vary from company to company and from site to site.

They should be measured wherever possible.

They may be credited daily or weekly.

GUIDE TO RELAXATION ALLOWANCES

A.		Basic allowance for personal needs, etc.	Male, $2\frac{1}{2}\%$; Female, 4%.
B.	Posture.	Standing normally	2%.
		Intermittent crouching, bending, etc.	Up to 10% max.
		Sustained awkward muscular positions	As required.
C.	Attention.	When continuous attention is essential	Add up to 7% max.
D.	Conditions.	When work must be done under trying conditions of heat, cold, dust, fumes, noise, vibration, bad light, danger, etc. Normally allowed	Up to 40% max.
E.	Effort.	Up to 18 N force by arm muscles	Nil.

Effort in newtons
22 to 40 add 1%
45 to 62 add 3%
67 to 75 add 7%
80 to 90 add 12%
Add $\frac{1}{2}\%$ to the above figures for every 4·5 N
above 90 N, up to a maximum of 30%.

Note: the approximate equivalent weight handled $= \dfrac{N}{10} kg$

F.	Monotony.	When the work consists of continuous short elements or short cycle operations, Add up to 5%.

Once the allowance has been derived and tabulated, it is added to the selected basic time and the standard time for carrying out the work is obtained, e.g.

$$\text{Standard time} = \text{selected basic time} + \left(\text{basic time} \times \frac{RA}{100} \right)$$

The results are then tabulated on an Element Weighting Sheet.

ELEMENT WEIGHTING SHEET.

Measurement of an operation of work may be carried out on two or more occasions and by two or more work study practitioners. If this does occur an element weighting sheet is prepared and the standard time obtained will be accurate for that particular activity.

The accuracy of work measurement will be reflected in the achievement of target wage obtained by operatives working to given standards of performance.

Job specification

To establish the activity covered by the 'standard time' it is essential that a job specification should be recorded, describing accurately a working situation after method improvements and time measurements have been made, stating all facts relevant to the operation and the control of the operation. In this respect a job specification may become the basis for contract of payment between employer and employee.

Policy allowance

The I.Prod.E. and I.C.W.A.[85] found that an additional allowance was made to the standard time to provide an attractive wage earnings level. 'This wage policy allowance was variable within each firm . . . and introduced to the rate as padding . . . . In the majority of firms, time study rates are not used as standards of measurement, but as means of bolstering up a basic hourly wage to an attractive level . . . '.

The addition of a policy allowance to standard time changes the unit of value to allowed time expressed in allowed minutes (AMs).

It has been shown by Lupton[86] that the time study man may enter into subjective argument just as the ratefixer does in bargaining with workers. This implies misunderstanding of the procedures of work study. *Standard time* is the expert information and the *allowed time* is the management employee negotiated time.

Synthesis

Synthesis is a work measurement technique for building up the time for a job at a defined level of performance by totalling element times obtained previously from time studies on other jobs containing the elements concerned, or from synthetic data.[6]

Synthetics take the form of:

Like elements taken from time studies or film analysis.

Feeds and speeds of machinery.

Predetermined motion time systems.

Data are usually obtained from numerous time studies of recurring elements such as: walking over measured distances, handling machine controls, or assembling components, etc.

To provide synthetics, elements should be selected to ensure 'like' elements are identifiable with each other, e.g. loading and unloading from specified distances. A job might consist of eight types of element, their relationship being:

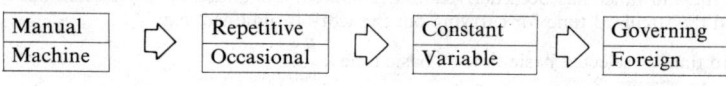

Machine elements can be obtained by calculation, e.g. $N = \dfrac{12S}{\pi D}$

where N = rev/min

$\quad\quad S$ = cutting speed of material.

$\quad\quad D$ = diameter of the rotating part.

The actual speed is dependent upon the available gear progression and feed of the machine extracted from manufacturers tables:

Machine element time taken:
(Lathe and drills)
$$= \frac{\text{length of traverse}}{\text{feed mm per rev.} \times \text{rev/min} \times \text{machine eff.}} \text{ min.}$$

(Milling machines) $= \dfrac{\text{length of traverse}}{\text{No. of cutter teeth} \times \text{feed per tooth} \times \text{rev/min} \times \text{machine eff.}}$

min

Manual elements need to be timed by watch, by cine-camera, or by assigning to each basic motion in the element a pre-determined motion time standard.

Predetermined motion time system (PMTS)

This is a work measurement technique whereby times established for basic human motions (classified according to the nature of the motion and the conditions under which it is made) are used to build up the time for the job at a defined level of performance'.[6]

Most human work is characterised by movement of the body and limbs and this is the basic key to all PMT systems. Manual work can be regarded as consisting of different combinations of a relatively small number of basic motions. By analysis it is possible to construct patterns of basic motions and to construct time values. Gilbreth at the time of his death was endeavouring to evolve time values to each of his 'Therbligs'. However, Segur (1924) is credited with the first system known as 'motion time analysis' and due to his strict secrecy and the market potential, other systems were allowed to be developed. The most prominent being 'work factor' (WF) in 1938, and ten years later 'methods time measurement' (MTM). The various systems are described in published literature (see Refs. 87–97).

The systems are basically similar as they cover similar basic motions, the differences lie in the treatment of time units and levels of performances. Westinghouse Brake and Signal Corporation decided that standard data be developed in 1940, which should conform to:

Methods engineering, i.e. the developments of good methods in advance of their installation, not methods correction.
Universal for both repetitive and non-repetitive work.
Consistent results and answers when analysed by various people.
Employee faith and trust.

Method Time Measurement

Three engineers developed MTM, Maynard, Stegemerton and Schwab. Lowry with Stegemerton and Schwab devised a levelling system (Westinghouse LMS) which took into account factors influencing performance,—skill—effort—and consistency, and weighted them according to their relative effect. Three consultant engineers had to agree the rating of each job film taken at 16 frames per sec. It is considered by members of the MTM association, of which the UK received full status in 1963, that MTM systems, because of the influence of the levelling factor, is 11% tight on British Standard Rate.

In 1949, Cornel University tested and analysed the MTM system and substantially confirmed the MTM findings to the American Society of Mechanical Engineers (ASME). The data collected so far appears to point towards the practicability of defining work elements in terms of codes . . . although errors amounted to 0·000 06 and 0·000 3 min representing in some cases 5%, 10% and even in isolated cases of 20%, they were no considered serious enough as any standard is made up of a composite number of ele ments'. Amendments were made according to many of the suggestions put forth b Cornell University.

MTM is measured in 'Time Measured Units' (TMUs)
$$= 0 \cdot 000 \ 01 \ \text{hr.,} \ 0 \cdot 000 \ 6 \ \text{min.,} \ \text{or} \ 0 \cdot 036 \ \text{seconds}$$

MTM being a non-copyright system is favourably looked upon by numerous com panies who have developed at some time or other simplified versions of MTM data a listed below:

MSD	Master Standard Data Developed by Sirge A Birn & Co.
MCD	Master Clerical Data. Sirge A Birn & Co.
PSD	Primary Standard Data Developed by Urwick, Orr and Partners Ltd
USD	Universal Standard Data Developed by H.B. Maynard and Co. Ltd.
UMS	Universal Maintenance Standard, H.B. Maynard.
UOC	Universal Office Controls, H.B. Maynard.
MMD	Milliminute Data Developed by PA Management Consultants Ltd.
Clerical MMD	PA Management Consultants Ltd.
MODAPTS	Modular Arrangement of Predetermined Time Standards Develope by The Australian Association for Predetermined Time Standards an Research.
GPD	General Purpose Data Developed by the MTM. Association of the United States and Canada.
MTM-2	Developed mainly by the Swedish MTM association in conjunctio with the United Kingdom and the US/Canada Associations. Mad available in 1965 this is now the official simplified system of MTM approved by the International MTM Directorate.

MTM has been defined by the originators as 'a procedure which analyses any manua operation or method into the *basic motions* required to perform it and assigns to each motion a *predetermined time standard* which is determined by the *nature* of the motio and the *conditions* under which it is made'.

MTM-2

One of the major criticisms made of detailed Predetermined Motion Time Systems sinc their development, has been the cost of application in non-repetitive areas, or on wor of lengthy cycle time. In order to overcome this problem and spread the use of PMT over a wider field, several systems of simplified MTM data have been developed. Thi created a problem in that more and more systems, all aiming at the same thing, ye doing so in a different manner, came into being.

A need was recognised for a common system, carrying the full approval of the Inter national MTM Directorate, with provision for instruction and qualification of practi tioners. On the 2nd October 1964, the Managing Board of the International MTM Directorate instructed the International Standing Committee for Applied Research to develop a system that would be:

1. Based on MTM.
2. Capable of combination with other MTM data.
3. Consistent between analysts and areas of application.
4. Easily understood.
5. Descriptive of the method.
6. Universally named.
7. Fast to handle.
8. Specified as to speed of application and accuracy of results.

Development work proceeded rapidly, MTM analyses from various industries wer processed by means of computer, yielding information on motion sequence and fre quencies.

The completed system was given full approval by the Managing Board of the IMD in Munich on the 11th June 1965. MTM-2 has only nine motion categories, some of which are basic MTM motions and some are motion sequencies.

MTM-3

The amount of work involved in analysis was still considerable and as far as variable work areas were concerned MTM2 even when assisted by the collection technique of Tape Data Analysis was considered to be uneconomic, in certain of these areas of work. The project was undertaken in Sweden by K. E. Magnusson and J. Silwer-frost. The objective of the MTM 3 project was to develop a standard time system faster to use than MTM 1 and MTM 2 with similar general applicability throughout all work areas.

The demands made on the system were similar to MTM 2, namely:
1. Elements to be general (independent of company).
2. High speed of analysis unequivocal.
3. Easy to understand.
4. Descriptive of the method.
5. Combinable with other MTM Data.
6. Based on MTM.
7. Specified regarding speed of application and accuracy of results. Particular emphasis being placed on speed of application in relation to accuracy of results.

In order to justify its existence MTM 3 was intended to be at least three times as fast in analysis as MTM 2. It has also been claimed to have achieved an accuracy relationship to MTM 1 of:

+ or − 5% at cycles of approx. 10 min.
+ or − 10% at cycles of approx. $2\frac{1}{2}$ min.

with a confidence limit of 95%.

The purpose of the system is to complement MTM 1 and MTM 2, not to replace them, and therefore create a family of MTM based systems which can be used for either detailed analysis or short cycle operations as with MTM 1 or range to MTM 3 where the achievement of speed of analysis is more important than loss of precision in the time value which would offer an advantage in non-repetitive work areas. MTM 3 has been tested using similar methods to those employed in testing MTM 2.

The types of jobs covered over a range of industrial concerns and consultant companies include, foundry work, sheet metal work, maintenance work and office work. The results indicate that the system has no significant bias in any application area provided highly repetitive elements are excluded. It is necessary however for the analyst to have been previously trained in MTM 2 before training in MTM 3 can be undertaken.

The data totals only 10 elemental values:

MTM 3				
Code	HA	HB	TA	TB
−15	18	34	7	21
−80	34	48	16	29
	SF	18	B	61

The recent addition of MTM 3 to the family of MTM techniques will eventually simplify the analysis still further as with this system even less analysis expression is necessary. More *data blocks* based on MTM 3 will undoubtedly be added to the range available for general use, thereby continually improving the technique range and flexibility in the maintenance field.

SIMPLIFIED PMTS

The variations in time required to perform the same effective motions are small when observations are made of different 'qualified' workers. The composition of data at ICI was obtained by accurate observations based upon a large sample of different operatives in ages and sex and in different localities. In order to bring the time values for all motions to a common level, a rating factor was used and observations were made by a minimum of three practitioners. The times were obtained by means of a synchronous cine-camera with a high speed lens to minimise lighting effects.

There are twenty-three basic motions used in the detailed PMTS of ICI where standardised time values range between 1 and 40 milliminutes. This detailed technique was derived from MTM. Among the second generation of simplified techniques ICI derived SPMTS and based it upon:
1. Simplified MTM.
2. A re-examination of the detailed data.
3. Practical judgment.
4. Validation by comparative trials and the statistical analysis of trials results, achieved by:
    1. Reducing the number of variables of the detailed technique.
    2. Simplifying rules and presentation.
    3. Rounding off values.

LIMITATIONS OF ALL SYSTEMS:

The data can only be used to evaluate uninhibited motions, and will not serve where motions are restrained by the nature of the process or to account for any limiting mental effort involved in a process.

Since the technique is concerned with human motions, it cannot be used to evaluate process times.

LEARNING CURVES

The learning curve[98] approach is most directly useful to production management in that it enables time standards to be set more accurately and provides a basis for planning, analytical estimating and incentive schemes. The setting of time standards at the introduction of a process and sustaining these standards indefinitely is the common practice of most industries even beyond the point where 'standard time' restricts output, because the learning process that both the organisation and individual undergoes, is totally ignored. For example, as skill and dexterity increases, the time standards become correspondingly slacker and result in some rise in bonus until a point is reached when caution overtakes the challenging incentive and the operatives tend to keep their bonus earnings and output reasonably constant, rather than attract the attention of management to the so called 'loose' rate.

Sometimes the learning process is construed to represent only the progress made by the production worker whose learning is limited to improvements in dexterity, whereas learning is accomplished through many other factors. For example:
    Rationalisation, standardisation and simplification in product and process design.
    Increase in batch-size.
    Improved handling to and from the work station.

Reduction in waste production.

Implementation of other cost reduction techniques and continuous application of method study.

According to Jorden[99] the learning curve theorem states that 'every time the production of a product doubles, the new cumulative average cost (hours or some similar

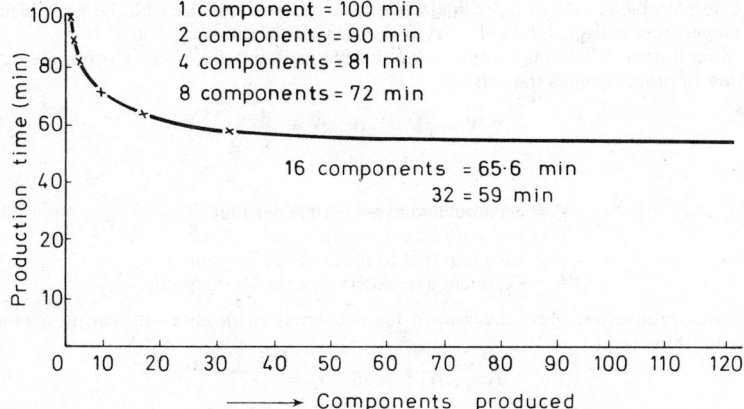

1 component = 100 min
2 components = 90 min
4 components = 81 min
8 components = 72 min

16 components = 65·6 min
32 = 59 min

Figure 20.90. Relationship between production time and components produced

unit of measurement) declines by a fixed percentage of the previous cumulative average. This fixed percentage identifies the learning achieved.' Figure 20.90 attempts to illustrate this theorem using a 90% learning rate.

Experimental method

For the purpose of measuring variations due to different skill requirements appropriate subjects must be selected and the job divided into manipulative elements e.g. touch W,

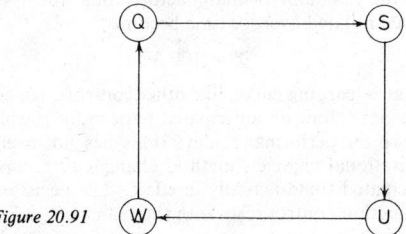

Figure 20.91

switch on Q, turn knob S, touch V, together with reach elements between each successive manipulative element (see Figure 20.91):

Reach from W to Q.

Touch Q and press on toggle switch (spring loaded) reach from Q to S.

Turn knob clockwise through 90° until the curve line is in a specified tolerance band.

Reach from S to U.

Touch a brass peg.

Reach from U to W.

Touch a brass peg.

To accurately record these elements a recording instrument originated and developed by N. T. Welford[100] known as a Sequential Electronic Timer and Recorder (SETAR), recording time intervals in 0·001 sec whenever circuits are made or broken. The recording facility can be extended to eight separate channels and the time intervals recorded directly onto punched tape. The tapes can then be decoded on a computer which lists the time intervals in numerical form.

Considerable scatter in individual cycle time observations need not be a problem if averages from collected data of every five or ten observations are taken.

Many forms of learning/output relationships exist but the most appropriate curve found in practice takes the form of.

$$Y = aN^{-b} \quad \text{or} \quad Y = \frac{a}{N^b}$$

where:

Y = accumulated average time per unit
N = number of units produced.
a = time required to produce the first unit.
b = exponent associated with the learning rate.

When production times are known for two levels of production bearing a binary relationship,

$$Y_1 = aN_1^{-b} \quad \text{and} \quad Y_2 = aN_2^{-b}$$

the exponent 'b' representing the slope or constant relationship between cumulative average cost and units produced, can be determined.

For example:

$$\frac{Y_1}{y_2} = \frac{a/N_1^b}{a/N_2^b} = \frac{N_2^b}{N_1^b} = \left(\frac{2N_1}{N_1}\right)^b$$

where $N_2 = 2N_1$

$$\therefore b = \frac{\log 2N_1}{\log N_1}$$

By knowing that the first unit of manufacture took 100% (or say 100 hours) the expression for the accumulated average time becomes:

$$Y = 100 \, N^b.$$

Rigg[32] explains that 'a learning curve, like other controls, can engender a false sense of security when operations follow an anticipated pattern for a while, but then veer away because the true cause for performance deviations has not been identified.' Similarly, other factors, organisational changes, method changes or component feature changes can obscure an anticipated trend actually in effect. 'The decisions to adopt a probable relationship for production control rests with the manager . . . '.

REFERENCES

1. WILD, RAY, *The Techniques of Production Management*, Holt, Rinehart and Winston, London (1971)
2. PETROV, V. A., *Flowline Group Production Planning*, Business Publications, London (1968)
3. IVANOV, E. K., *Group Production Organisation and Technology*, Business Publications, London (1968)
4. LOCKYER, K. G., *Factory Management*, Pitman, London (1962)
5. INTERNATIONAL LABOUR OFFICE, *Introduction to Work Study*, ILO, Geneva (1960)
6. BS 3138
   Glossary of terms in work study (1969)

7. MUNDELL, M. E., *Motion and Time Study*, Prentice Hall, London (1960)
8. URWICK, L. and BRECK, E. S. L., *The Making of Scientific Management, Vol. 1, Thirteen Pioneers*, Management Publication Trust, London (1945)
9. NORMAN, R. G., *Techniques of Work Study*, Draughtman's and Allied Technicians Assoc., Richmond, Surrey (1963)
10. BARNES, R. M., *Work Sampling*, John Wiley & Sons Inc., New York (1964)
11. NORMAN, R. G. and BAHIRI, S., *Productivity Measurement and Incentives*, Butterworths, London (1972)
12. CURRIE, R. M., *Work Study*, Pitman, London (1963)
13. MAYNARD, H. B., *Industrial Engineering Handbook*, McGraw-Hill, New York (1956)
14. BUFFA, ELWOOD S., *Modern Production Management*, John Wiley & Sons Inc., New York (1965)
15. BARNES, R. M., *Motion and Time Study*, John Wiley & Sons Inc., New York (1955)
16. MOORE, J. M., *Plant Layout and Design*, Macmillan, New York (1962)
17. MUTHER, R. E., *Practical Plant Layout*, McGraw-Hill, New York (1956)
18. EL-RAYAH, T. E. and HOLLIER, R. H., 'A review of plant design techniques, *The International Journal of Production Research*, **8** (1970)
19. MURRELL, K. F. H., *Ergonomics*, Chapman and Hall (1965)
20. MURRELL, K. F. H., 'Data on human performance for engineering designers', *Engineering* (Aug 16 and 23, Sept 6 and 13, Oct 4, 1957)
21. DUDLEY, N. A., *Work Measurement—Some Research Studies*, Macmillan and Co. (1968)
22. MACHINERY, P115/122 (Dec 10 1969)
23. MACHINERY, April 1 1970
24. BURBIDGE, JOHN L., *Production Planning*, Heinemann & Co. (1971)
25. EILON, SAMUEL, *Elements of Production Planning and Control*, Macmillan & Co., New York (1962)
26. CLARK, WALLACE, *The Gantt Chart*, Pitman, London (1963)
27. LOCKYER, K. G., *Introduction to Critical Path Analysis*, Pitman, London (1967)
28. BATTERSBY, A., *Network Analysis*, Macmillan, London (1970)
29. MOORE, J. J. and PHILIPS, C. R., *Project Management with C.P.M. and P.E.R.T.*, Reinhold, New York (1964)
30. MILLIGAN, R. A. and BROOKS, D. F., 'Precedence diagrams for critical path analysis', *The Engineer* (April 5–19 1968)
31. WOODGATE, H. S., *Planning by Network*, Business Publications, London (1967)
32. RIGG, JAMES L., *Production Systems Planning Analysis and Control*, John Wiley & Sons Inc., New York (1970)
33. MAGEE, J. F. and BOODMAN, D. M., *Production Planning and Inventory Control*, McGraw-Hill, London (1967)
34. BRIGGS P. G. et al., *Problems of Stocks and Storage*, Monograph No. 4, Oliver and Boyd, London (1967)
35. SASIENI, M. W., VASPAN, A. and FRIEDMAN, L., *Operation Research—Methods and Problems*, John Wiley & Sons Inc., New York (1959)
36. THOMAS, A. B., *Stock Control in Manufacturing Industries*, Gower Press, London (1968)
37. MORRISON, A., *Storage and Control of Stock*, Pitman, London (1969)
38. MURDOCK, J., *Coverage Analysis*, Conference Proceedings, Cranfield (March 1965)
39. MORROW, L. C., *Maintenance Engineering Handbook* (1957)
40. MILLER, J. E. and BLOOD, J. W., *Modern Maintenance Management*, American Management Association, New York (1963)
41. BS 3811
    Glossary of general terms used in maintenance organisations (1964)
42. CORDER, G. G., *Maintenance Techniques and Outlook*, British Productivity Council, London (1957)
43. TERBORGH, G., *Realistic Depreciation Policy* (1954)
44. MAPI, *MAPI Replacement Manual*, Machinery and Allied Products Institute, London (1950)
45. MILES, L. D., *Techniques of Value Analysis and Engineering*, McGraw-Hill, New York (1961)
46. CRUICKSHANK, R. D., 'Value analyses', *Journal of Work Study and Management* (Feb 1965)
47. BPC, *16 Case Studies of Value Analysis*, British Productivity Council (1961)
48. DODGE, H. F. and RONIG, H. G., *Sampling Inspection Tables, Single and Double Sampling*, John Wiley & Sons Inc. (1959)
49. DEF-131-A
    Sampling procedures and tables for inspection by attributes, H.M.S.O. (1967)
50. BS 2564
    Control chart technique when manufacturing to a specification, with special reference to articles machined to dimensional tolerances (1955)

51. DODGE, H. F. and ROMIG, H. G., *Sampling Inspection Tables,* John Wiley & Sons Inc.
52. RATCLIFFE, J. F., *Elements of Mathematical Statistics,* OUP
53. JONES, G. and BARNES, M., *Britain on Borrowed Time,* Penguin, London (1957)
54. MINISTRY OF LABOUR, *Efficient Use of Manpower,* HMSO (1967)
55. ROSTAS, LAZZLO, *Alternative Productivity Concepts,* Productivity Measurement Concepts Vol. 1, European Productivity Association (EPA), OECD, Paris (1955)
56. BAHIRI, S. and MARTIN, H. W., *Effective Management through Productivity Costing,* Industrial Commercial Techniques (In-Com-Tec), London (1968)
57. FOULKE, ROY, *Practical Financial Statement Analysis,* McGraw-Hill (1968)
58. INGHAM, H. and HARRINGTON, L. TAYLOR, *Productivity Measurement and Incentives, Appendix A, Interfirm Comparisons for Management,* Butterworth (1972)
59. BLISS, J. H., *Financial and Operating Ratios in Management,* Ronald Press, New York (1924)
60. WITSCHEY, R. E., *Accounting Theory and the Accounting Profession,* Prentice-Hall Inc., New Jersey (1966)
61. PICKLES, W., *Accounting,* Pitman, London (1964)
62. INSTITUTE OF COST AND WORKS ACCOUNTS, *Terminology of Cost Accountancy* (1952)
63. NOLAN, R. WARWICK, *An Introduction to Cost Accountancy,* Gee & Co. (1958)
64. GRAHAM, L. R., *Manual of Works Accounting,* Gee & Co. (1967)
65. SIZER, J., *An Insight into Management Accounting,* Penguin, London (1969)
66. WHELDON, H. J., *Cost Accounting and Costing Methods,* Macdonald and Evans, London (1970)
67. BATTY, J., *Standard Costing,* Macdonald and Evans, London (1970)
68. MERRETT, A. J. and SYKES, A., *Finance and Analysis of Capital Projects,* Longmans (1965)
69. I.PROD.E. and ICWA, *An Engineers Guide to Costing,* Institution of Production Engineers (1969)
70. MARTIN, H. W., *Productivity Costing and Control,* Productivity Measurement Review, No. 37, OECD, Paris (May 1964)
71. CORLETT, E. N., NORMAN, R. G. and SNAITH, J. W., 'Measurement and incentives in building maintenance', *Surveyor and Municipal Engineer,* October (1967)
72. TIPPETT, L. H. C., *Statistical Methods in Textile Research—A Snap Reading Method of making Time Studies of Machines and Operatives in Factory Surveys,* Shirley Institute Memoirs, Vol 13 (1934)
73. MORROW, R. L., *Ratios Delay Study,* Time and Motion Economy, Ronald Press Co., New York (1946)
74. TIPPETT, L. H. C., 'Letter to the Editor', *Journal of Time and Motion Study* (March 1959)
75. ROWE, A. J., 'Relative versus absolute errors in delay measurement', *British Management Review,* **13,** No. 1 (1955)
76. TAYLOR, F. W., *Shop Management,* Harper and Bros., New York (1912)
77. CURRIE, R. M., *Measurement of Work,* British Institute of Management (1965)
78. MARRIOTT, R., *Incentive Payment systems,* Staples Press, London (1957)
79. SURY, R. J., 'A comparative study of performance rating systems', *International Journal of Production Research,* **1,** 1 (March 1962)
80. SURY, R. J., 'A survey of time study rating research', *The Production Engineer, Inst.Prod.E.* (Jan 1962)
81. GRAHAM, C. F., *Work Measurement and Cost Control,* Pergamon Press, London (1965)
82. MATTHEW, T. U., 'The accuracy and use of time study', *O.R. Quarterly,* **6,** 1 (1955)
83. NADLER, G., *Work Systems Design: The Ideals Concept,* Irwin Holmwood, Illinois (1967)
84. DUDLEY, N. A., 'Output pattern in repetitive tasks', *The Production Engineer, Inst.Prod.E.* (April 1958)
85. I.PROD.E. and ICWA, 'Measurement of productivity', *Inst.Prod.E.* (1949)
86. LUPTON, T., *A Sociologist looks at Work Study,* Work Study and Industrial Engineering No. 1 (1957)
87. HOLMES, W. G., *Applied Time and Motion Study,* Ronald Press (1938)
88. QUICK, DUNCAN and MALCOLM, *Work Factor Time Standards,* McGraw-Hill (1962)
89. SCHAEFER, N. G., *SAM—ASME Conference,* New York (April 1947)
90. MAYNARD, H. B., STEGMERTON, G. J. and SCHWAB, J. L., *Methods Time Measurement,* McGraw-Hill (1948)
91. KAGER and BOYLE, *Engineered Work Measurement,* Industrial Press (1957)
92. PRESGRAVE, R., BAILEY, G. B. and LOWDEN, J. A., *Basic Motion Time Study,* J. D. Woods & Gordon Ltd., Canada (1950)
93. GOPPINGER, H. C., 'Dimensional motion time studies', *Iron Age* (Jan 8, 1953)
94. LAZARUS, I. F., *A System of Predetermined Human Work Times,* Purdue University Ph.D. Thesis (1952)
95. CURRIE, R. M., *Simplified PMTS,* BIM, London (1965)

96. ICI, 'Basic work data', *British Industry Week* (March 22 1968)
97. EDITOR, *Journal Maintenance Engineering* (April 1966)
98. 'Computation of learning curves', *J. Methods—Time Measurement* (July/Aug 1966)
99. JORDEN, R., *How to Use Learning Curves*, Materials Management Institute, Boston, Mass, USA (1966)
100. MELFORD, N. T., 'An electronic digital recording machine—The SETAR', *J. Science Inst.*, 29, 1

PATENTS, DESIGNS AND TRADE MARKS

H. M. MOSS

In this section the chief provisions of the Acts and Rules relating to Patents, Designs and Trademarks are summarised. Free literature is obtainable from the Patent Office on how to obtain a patent, registered design or trade mark, and how to file a specification. In view of the complexity of patent laws, in order to obtain sound protection for a patent, those not versed in the patent procedure are advised to employ the services of a suitably qualified Patent Agent.

It is a function of the Patent Office to administer the Patent Acts 1949 to 1961 and the Patent Rules 1968.

PATENTS

Letters Patents for inventions are monopolies granted by the Crown for a limited term of years (sixteen) to patentees to make, use, exercise, and vend within the United Kingdom of Great Britain and Northern Ireland and the Isle of Man new and useful inventions.

A patentable invention must be for 'a manner of new manufacture'. It must contain subject-matter or invention, have greater utility than the already known, and not have been known or patented in the United Kingdom within fifty years from the date of applying for a patent.

Applicant for patent. This is any person, who is the true and first inventor, who may apply alone or jointly with one or more persons, company or other corporate body. Any person, company or corporate body may apply for a patent as communicatee for an invention emanating from an inventor residing abroad or under the terms of the International and Colonial Arrangements. A request may be made to the Comptroller on Patents Form 19 (stamped £10) to deal with a dispute between joint applicants.

The personal representative of a deceased inventor may also apply for a patent for the invention of the said inventor.

Employer and employee. It is an implied term in the contract of employment that an employee is a trustee for his employer of any invention made in the course of his duty as employee unless there is an agreement having legal force to the contrary effect. If an employee is specifically engaged to work out the details of an invention or carry it into practice, then any suggestion of invention emanating from the employee during such work is the property of the employer. An application may be made on Patents Form 55 (stamped £8) to request the Comptroller to deal with a dispute between an employer and an employee as to rights in an invention.

An application for patent. This must be accompanied by either a provisional specification or a complete specification, and must be restricted to one invention. An application for patent (Patents Form 1) may be applied for by the inventor, either alone or jointly with one or more persons or on behalf of a company or other corporate body. An application may also be filed by a company alone as assignee of the inventor, who must, however, sign a declaration of assent.

The Comptroller may, if the invention is relevant for defence purposes, prohibit or restrict application and withhold the grant of a patent. It is an offence with a penalty on conviction of a fine or imprisonment, for any person resident in the U.K. to file or cause to be filed any application abroad without the Comptroller's written permission unless an application has been filed in the U.K. and at least six weeks have elapsed since the filing of the application and no directions as to secrecy have been given by the Comptroller. This does not apply when an invention has first been filed abroad by a person resident abroad.

An application for patent under international and colonial arrangements, whereby the priority date of the first patent application filed for the same invention in a country which has a corresponding reciprocal arrangement with this country may be claimed, must be filed in this country within twelve months from the date of the first patent application in such a country, and must be made on Patents Form No. 1 Con.

Publication of invention before applying for patent. An invention may be published with the consent of the inventor by the reading of a paper before a learned society or by being shown at an exhibition certified by the Board of Trade, without affecting the validity of any patents subsequently granted for the invention, provided a patent application for the invention is filed within six months from the said publication.

Divided patent application. If an application for patent covers more than one invention, the Comptroller may call for a division in which case the excised invention may form the basis of a fresh application bearing the same date as the original application.

Post-dating patent applications. If new matter is introduced into a patent application, either in the complete specification or by amendment during its prosecution, the Comptroller may either call for its excision or post-date the application to the date when such new matter was introduced.

The applicant may, before the acceptance of the complete specification request the application to be post-dated to a date not later than six months from the original application date on filing Patents Form 5 stamped £5. Where an application is accompanied by a provisional specification for which the complete specification has not been filed the request must be made within twelve months of the original application date or within fifteen months thereof on the filing of Form 5 stamped £5.

Amendment of application form. An application form may be amended on filing Patents Form 38 stamped £5.

Provisional specification. This must describe the invention and need not necessarily make a reference to a drawing; it should be prepared in duplicate on Patents Form 2 (unstamped) and be filed or posted with the appropriate application form to the Comptroller, Patent Office, 25 Southampton Buildings, Chancery Lane, London, WC2. *Note.* A provisional specification cannot be filed with a patent application made under international or colonial arrangements; thus every convention application must be accompanied by a complete specification.

The specification

The complete specification must describe not only the invention but give detailed information as to the best means for carrying the invention into practice and if capable of being illustrated must be accompanied with drawings to which specific reference must be made in the specification.

The specification must end with a distinct statement of claim setting forth clearly the scope of the invention claimed for which a patent is sought. Claims should not be made for the efficiency, or advantages of the invention. The complete specification must be prepared in duplicate on Patents Form 3 and one copy must bear an impressed stamp for £22 and be filed or posted to the Patents Office.

If a complete specification be not lodged at the Patent Office in the first instance with application Form 1 (it is however obligatory with Form 1 Con as stated above) and a full patent is desired then a complete specification must be filed within twelve months from the date of the patent application or within fifteen months. In the latter case an extension of time for three months (the longest extension obtainable) must be applied for on Patent Form 5 bearing an impressed stamp for £5 as mentioned above.

Amendment of a complete specification may be effected before its acceptance by filing Patents Form 37 (stamped £5).

Cognate inventions.   When the same applicant has two or more pending provisional specifications for inventions which are modifications one of the other and has obtained thereby concurrent provisional protection for the same then such pending provisional applications may be included in a single complete specification provided the modifications shall constitute a single invention.

Chemical substances for food or medicine.   In the case of such inventions the complete specification shall not include claims for a mere admixture of known ingredients.

Drawings for Patent Applications must be prepared on good quality smooth surface white drawing paper measuring 330 mm by 200 to 210 mm or of size A4 with a clear margin of 13 mm around each sheet. With the Comptroller's permission larger sheets of 400 to 420 mm by 330 mm may be used. The drawings shall be in duplicate with lines firmly and evenly drawn in durable very dark marking and no more sheets shall be employed than are necessary. Reference letters and numerals must be bold and exceed 3 mm in height. The drawings must be delivered at the Patent Office without folds breaks or creases.

Official search for novelty.   After the complete specification has been filed and not before, there is made the official search for novelty of the invention as claimed prior to the acceptance of the complete specification.

After the result of such a search is communicated to the applicant the complete specification must be amended within two and a half years of the filing date to overcome the prior citations, if any, and to overcome any other objections raised by the examiner. An extension of time for amending the specification to overcome prior citations may be obtained by filing Patents Form 10 fee £3 for each additional month up to three months.

If, after amendment of the complete specification and after a hearing before the Comptroller he is satisfied that the invention is still partly anticipated he may insert reference to prior specifications or if wholly anticipated may refuse to accept the specification.

The Comptroller has also power to refuse an application for a frivolous invention or one contrary to natural laws or one the use of which would be contrary to law or morality.

A complete specification, in normal cases, must be accepted within twelve months of the date of filing the complete specification but this period may be extended up to fifteen months. Application for extension of time must be made on Patents Form 11 stamped £3 for one month's extension; £6 for 2 months: £9 for 3 months' extension.

Result of official search.   After a complete specification has been accepted any person may make application on Patents Form 8 (stamped 15p) to be informed of the result of the Official Search made in connection with such specification.

Amendment of complete specification (including drawings). Any amendment to the specification subject to the discretionary power of the Court or the Comptroller after acceptance by way of disclaimer, correction, or explanation may be made by the applicant on filing Patents Form 35 (stamped £10) before sealing or after sealing of patent.

Opposition to amendment may be made on Patents Form 36 (stamped £5) and notification to attend hearing before the Comptroller in such proceedings must be made on Form 13 (stamped £3).

Printed Specifications are obtainable priced 50p per copy from the sales branch of the Patent Office.

Opposition to the grant of a patent

This may be made on various specified grounds, such as obtaining, ambiguity, prior publication, prior use, must be made on Patents Form 12 (a £5 stamp) and filed at the Patent Office within three months from the date of publication of the complete specification. No extension of time for filing opposition is allowed beyond such three months. Attendance at hearing before the Comptroller must be notified on Patents Form 13 (stamped £3).

The deviser of the invention

The actual deviser of the whole or a substantial part of an invention may be mentioned in a patent without conferring any rights under the patent provided application is made on Form 14 or 15 (each stamped £2); not later than two months extendable by one month, from the date of publication of the complete specification. Provision is also made for contesting the applicant's claim to be so mentioned and to annul such mention after having been made. Such annulment application must be made on Patents Form 17 (stamped £5).

Grant of patent

After the complete specification has been accepted a patent is sealed on payment of the sealing fee on Form 20 (stamped £8). In normal cases a patent must be sealed within four months of the date of publication of the complete specification an extension of time for sealing a patent may be obtained on filing Patents Form 21 bearing an impressed stamp for £3 for one month's extension, £6 for two months and £9 for three months. In certain exceptional cases the period may be further extended on filing Patents Form 22 bearing an impressed stamp for £3 for the first month and £3 for each subsequent month's extension applied for. An application which has lapsed owing to unintentional failure to pay the sealing fee may be restored by filing Patents Form 32 (stamped £11) within six months of the date when the fee should have been paid. If the Comptroller allows restoration an additional fee of £15 is paid on Patents Form 34 as well as the unpaid fee of £8 on Form 20. Opposition to such sealing may be filed using Patents Form 33 (stamped £5).

Grant of patent to assignee.   Before the sealing of a patent on filing Patents Form 18 (stamped £5) the assignee of an invention may be granted a patent in his name. Where an applicant in a joint patent application has died, the patent may be granted to the survivor. In the case of disputes between joint applicants as to proceedings with the patent application, the Comptroller may allow one or more of the applicants to proceed and grant the patent to him or them. In the case of such dispute Patents Form 19 (stamped £10) should be filed.

Provisional protection is afforded during the period between the date of the patent application and the sealing of the patent. There is no such thing as a provisional patent, only a Provisional Specification which records 'a priority date' for whatever novel matter is disclosed therein.

Joint patentees. Unless specified to the contrary, these are joint owners and if one of the parties dies, his beneficial interest becomes part of his personal estate. Each owner can use the invention for his own profit without accounting to the others, but shall not be entitled to grant a licence or assign his share without the consent of the other owners. Disputes between co-owners may be decided by the Comptroller if application is made on Patents Form 53 or 54 each (stamped £8).

Duplicate patent. If a patent is lost, destroyed or cannot be produced the Comptroller may issue a duplicate patent on receipt of Patents Form 68 (stamped £5).

Register of patents. This is kept in the Patent Office wherein are entered names, addresses and nationalities of Grantees of Patents, notifications of assignments, licences, amendments, extension, and revocations, and all other matters relating to the validity or proprietorship of patents. The fee for inspecting the register, original document, samples or specimens is 15p. A request for alteration of a name or an address for service is made on Patents Form 57 (stamped 50p).

Duration of patent. The term of a patent is sixteen years from the date of filing the complete specification. The duration of a patent may be extended exceptionally up to a further period of ten years beyond its normal life by petition to a High Court.

An application for extension of the term of a patent on the ground of war loss however may be made to the Comptroller of Patents or to the High Court. An application to the Comptroller can be made by Patentee or the Licencee for an extension on Patents Form 27 (stamped £10), which must be filed not more than twelve months nor less than six months before the expiration of the normal term of the patent. Opposition may be lodged within two months of the advertisement of an application in the 'Patents Journal' on Patents Form 28 (stamped £5).

'Licences of right'

A patent (after sealing) may be endorsed 'Licences of Right' by filing Patents Form 42 (stamped £2). Patents so endorsed are subject to the grant of a licence to any person making due application therefor, and thereafter only half the usual fees are payable to keep the patent in force. Objection to such endorsement may be made on Patents Form 45 (stamped £5). An application for settlement of the terms of a licence must be made on Patents Form 43 (stamped £8).

The cancellation of an endorsement may be made by a patentee on filing Patents Form 44 (stamped £5). Attendance at a hearing before the Comptroller in above opposition proceedings must be notified on Form 13 (stamped £3).

Fees are payable to keep a patent in force annually after the fourth year from its date by filing Patents Form 24 having an impressed stamp for £13 in respect of the fifth year, £14 in respect of the sixth year, £16 for the seventh year, £18 for the eighth year, £20 for the ninth year, £24 for the tenth year, £26 for the eleventh year, £28 for the twelfth year, £30 for the thirteenth year, £34 for the fourteenth year, £37 for the fifteenth year and £40 is payable before the expiration of the sixteenth year and in respect of the remainder of the term.

Extension of time for payment of fees up to six months may be obtained on filing

Patents Form 25 (stamped with £3) for one month's extension, £6 for two months' extension and £9 for an extension of three months and so on until £18 is payable for an extension of six months.

In exceptional cases when a patent has been allowed to lapse unintentionally the Comptroller has power to reinstate it on filing Form 29 bearing a stamp for £6 on application within three years of lapsing. Opposition to the reinstatement of a lapsed patent may be made on Patent Form 30 stamped £5. If the Comptroller allows restoration an additional fee of £15 has to be paid on Patents Form 31.

Patents of addition

These are granted for an inclusion in or modification of an invention forming the subject of a patent application, a patent, or a patent of addition. No fees are payable on patents of addition so long as the original patent is in force but only become payable in the event of the original patent being revoked, in which case the patent of addition becomes an independent patent. Application for a patent of addition must be made on Patents Form 1 (stamped £1). Application for the grant of a Patent of Addition in lieu of an independent patent must be made on Patents Form 1 add. (stamped £8).

Fraudulent application for a patent.   Where a patent has been obtained in fraud of the true inventor and has been revoked or the grant of a patent has been refused, a patent may be granted for the invention to the true inventor on his making application therefor and be dated as that of the fraudulent patent or patent application.

Publication or use of the invention subsequent to the fraudulent application will not invalidate the patent subsequently granted to the true inventor.

Correction of clerical errors

A request for the correction of a clerical error in the documents of a Patent Application or in the Patent Register must be made on Patents Form 64 (stamped £3) before sealing or after sealing of the patent.

Notice of opposition to the correction of a clerical error must be made on Patents Form 65 (stamped £4). Notification to attend hearing before the Comptroller in such proceedings must be made on Patents Form 13 (stamped £3).

Exercise of patent rights

Any person having an interest after the expiration of three years from the date of sealing a patent can apply on Patents Form 47 (stamped £8), alleging that the invention has not been put to its fullest practicable use whereby the exercise of Patent Rights has prejudiced the development of industry and can apply thereby for the grant of a compulsory licence under the patent. Notice of opposition to such application is made on Patents Form 51 (stamped £5) the grounds upon which application may be made are as follows:

1. If the invention (being one capable of being worked in the U.K.) is not worked commercially within three years from sealing the patent and no satisfactory reason can be given for such non-working;
2. If the working within the U.K. is prevented or hindered by importation from abroad of the patented article;
3. If the demand for the patented article is not met to an adequate extent on reasonable terms;

4. If by the refusal of the patentee to grant a licence upon reasonable terms trade is prejudicially affected;
5. If any person or trade is unfairly prejudiced by conditions imposed by the patentee on the purchase, hire or use of the patented article; and
6. If it is shown that the existence of the patent being a patent for a process involving the use of materials not protected by the patent or for a substance produced by such a process has been used by the patentee to unfairly prejudice in the U.K. the manufacture use or sale of such materials.

Endorsement etc. on the application of the Crown

At any time after the expiration of three years from the date of sealing of the patent any Government Department may apply to the Comptroller, without having to establish any interest, for an endorsement 'Licences of Right' or for a Grant of a Licence to any person whom the Department may specify in the application to be made on Patents Form 48 (stamped £8).

Assignments and licences

Assignments, licences, and other documents affecting the proprietorship of patents must be entered on the Register of Patents. Application for entry of proprietor or part proprietor must be made on Patents Form 58 or 60, for a mortgage or licence on Forms 59 or 61, and a notification of a document on Form 62. Patents Form 58 must be stamped £2 in respect of one patent if application is made within six months from the date of acquisition of proprietorship; £5 if made within twelve months; and £8 if made after twelve months, and 50p for each additional patent included in this deed. Forms 59, 61, and 62 must each be stamped £2 if filed within six months from the date of the document, acquisition of interest or sealing of the patent whichever is the later and bear a stamp for £5 if made after six months but within twelve months and £8 after twelve months plus 50p for each additional patent included in the same document.

Licences under patents for food or medicine or as part of a surgical or curative device must be applied for on Patents Form 52 (stamped £8).

Revocation of a patent may be obtained on petition to the Court or within twelve months from the date of sealing a patent on making application to the Comptroller on Patents Form 39 (stamped £5). Notification to attend a hearing before the Comptroller in such a proceeding must be made on Patents Form 13 (stamped £3). An offer to surrender the patent after such proceedings have commenced must be made on Patents Form 40. Opposition to the surrender of the patent may be made on Patents Form 41 (stamped £4) and must be filed within one month of the date of the first advertisement of the surrender offer. Notification to attend hearing before the Comptroller in such opposition proceedings must be made on Patents Form 13 (stamped £3).

Appeals from the decisions of the Comptroller are made to an 'Appeal Tribunal' consisting of a judge of the high court.

Certificate of Comptroller

A request for a certificate from the Comptroller as to any entry or matter which he is authorised to make by the Patents Acts or Rules must be made on Patents Form 66 (stamped 75p).

A request for information respecting a patent or patent application must be made on Patents Form 67 (stamped £1.50).

Infringement of patent

No action for infringement of a patent can be commenced until a patent is actually sealed and only damages can be claimed for an infringement committed after the publication of the complete specification. A patentee may obtain an injunction to restrain infringement but cannot recover any damages in respect of infringement if the infringer proves that he is not aware and had no reasonable grounds of supposing that the patent existed. Marking an article with the words 'Patent' or 'Patented' is not notice of the existence of a patent unless the number of the patent is given.

The plaintiff in an infringement action is entitled to relief by way of injunction and damages limited to a maximum of £1000 or at his option to an account of profits in lieu of damages.

One or more invalid claims in a specification are not sufficient to invalidate a patent in an infringement action provided that the invalid claim was framed in good faith, and with reasonable skill and knowledge.

Groundless threats of legal proceedings for infringement of a patent may be restrained and damages recovered. The mere drawing attention to the existence of a patent is not a threat.

A dispute as to infringement of a patent may be determined by the Comptroller if both parties make joint application on Patents Form 56 (stamped £8). The Comptroller may also determine whether any claim is alleged to be infringed.

The Crown. A patent has the like effect as against Her Majesty the Queen as it has against the subject, but Government department or their authorised agent can make use of, or exercise, the invention for the services of the Crown.

Offences

It is not lawful for a patentee to prohibit or restrict a purchaser or licencee from using any article, patented or not, supplied by any person other than the patentee or to make it necessary to acquire from the patentee any article not protected by the patent.

It is a misdemeanour to make any false entry in the Register of Patents. Any person falsely representing an article to be a patented article is liable on conviction to a fine not exceeding £5 for each offence.

The grant of a patent does not authorise the patentee to use the Royal Arms. Unauthorised use of the Royal Arms carries a penalty up to £20 on conviction.

The use of the words 'Patent Office' or a similar designation on business premises, letter paper or other documents renders the user liable on conviction to a fine not exceeding £20.

Any person practising as a Patent Agent unless registered commits an offence punishable on conviction to a fine not exceeding £20 for the first offence and on subsequent convictions to a fine not exceeding £50.

Patent Agents

All business with the Patent Office including the signing of many of the necessary documents, may be transacted through a registered Patent Agent, who must be a British Subject and duly authorised to the satisfaction of the Comptroller of Patents.

A list of Patent Agents may be obtained (price 12½p) from the Chartered Institute of Patent Agents, Staple Inn Buildings, London, WC1.

Development and exploitation of patents

Information on the development and exploitation can be obtained from:
(i) National Research Development Corporation, Kingsgate House, 66–74 Victoria Street, London SW1.
(ii) The Institute of Patentees and Inventors, Abbey House, Victoria Street, London SW1

DESIGNS

The Designs Registry Branch of the Patents Office administers the Registered Design Acts 1949–1961 and the Design Rules 1949–1969.

A Registrable Design under the Designs Act 1949 concerns only the shape, configuration, pattern, or ornament of an article and is judged solely by the eye; it does not cover any principle of construction or any features of shape which are dictated solely by the function the article has to perform. It must not have been published or registered previously in the United Kingdom. (*Note.* Sections 10 and 44 of the Copyright Act 1956 are relevant particularly to designs.) Industrial designs which correspond with 'artistic works' protected under the 1956 Act may be protected under Designs Copyright Act 1968 which amends Section 10. An industrial design is a design which is applied to articles manufactured in quantitiy, i.e. more than fifty articles or in the case of piece goods, a length or piece. Articles are either 'textile' or 'non-textile' with different fee rates. Textile articles include checks or stripes and lace.

Applications for registration. Any person or persons (not necessarily including the author or designer), a firm, company, or body corporate, who is or are the proprietor or proprietors of a new or original design, may apply for registration.

Application for registration of an article not being a textile must be made on Designs Form 2 signed by the applicant or applicants or by a duly appointed agent (Designs Form 1).
Designs Form 2 is stamped with a £8 stamp and must be accompanied by three identical representations or specimens of the design.

International arrangements. For an application for a design registration under International arrangements whereby the priority date of the first design registration filed for the same design in a country which has a corresponding reciprocal arrangement with this country may be claimed; the application must be filed within six months of the first application in such country and must be made on Designs Form 3 also stamped with a £8 stamp if for a non-textile article and accompanied by three identical representations or specimens of the design. A copy of the design filed in the Convention county must be filed not later than three months after the British application. The application form and representations in each case must be filed with or posted to the Comptroller, The Patent Office (Designs Registry), 25 Southampton Buildings, London WC2.

'Set' of designs. Application for registration of a number of articles of the same general character ordinarily on sale or intended to be used together must be made on Designs Form 4 (stamped £16) and accompanied with four representations or specimens. If

application is made under International arrangements, Designs Form 5 must be used, also stamped £16 and accompanied with four representations or specimens.

Lace designs. Application must be made on Designs Form 2 (stamped £2) and accompanied by three representations or specimens of the design. If for a 'set', application is made on Designs Form 4 (stamped £5) and accompanied with four representations or specimens.

Applications for printed or woven designs on textile piece goods must be made on Designs Form (Manchester) 1 (stamped £8) for a textile article other than a check or stripe design. For checks or stripes, application must be made on Designs Form (Manchester) 2 (stamped £2). Each form must be accompanied by an unstamped duplicate together with four identical representations or specimens and be filed with or posted to either the Comptroller, The Patent Office (Designs Registry), or to the Designs Registry, Manchester Branch, Baskerville House, Browncross Street, New Bailey Street, Salford, M3 5FU.

Statement of novelty. A statement of the feature of the design for which novelty is claimed must be filed with every type of design application, except for textile articles, wallpaper, or lace.

Publications of an unregistered design. A design may be published with the consent of the proprietor by being shown at an exhibition certified by the Department of Industry and Trade, without invalidating a subsequent registration provided an application for registration is made within six months from the date of opening of the exhibition.

Registration of same design in respect of other articles. Where a design has already been registered, the proprietor of the design may subsequently register the same design as applied to another article or articles, but the terms of copyright in such subsequent registration will not extend beyond the term of the original registration and any extension thereof.

Extension of time for completion of application. If an application for registration is not completed within twelve months from the date of application by default of the applicant, it becomes abandoned. However, an extension of time up to three months may be obtained by filing Designs Form 8 stamped £3 for one month's extension, £6 for two months, and £9 for three months.

Refusal of registration by Registrar. If, after consideration, there appears to the Registrar to be any objection a statement is sent to the applicant in writing and unless within one month the applicant applies for a hearing, the application is deemed withdrawn. The Registrar will communicate in writing his decision at the hearing to the applicant. If after such a hearing the application for registration is still refused and the applicant desires to appeal to the Appeal Tribunal he must apply on Designs Form 7 (stamped £5) within one month from such hearing and the date when the Registrar complies with the request shall be deemed to be the date of the Registrar's decision for the purpose of an appeal.

Duration of copyright. This is five years from the date of application for registration, which term may be extended for a further period of five years on filing Designs Form 9 (stamped £20) within the term of the first period; and an extension for a further period of five years, making fifteen years in all, may be obtained on filing Designs Form 10 (stamped £30), within the second period. If an extension of time up to six months is required to pay either fee, Designs Form 11 stamped £3 for one month extension, £6 for two months, £8 for three months and so on up to £18 for six months must be filed. The fee for a full term of fifteen years may be paid in advance.

Marking on goods.   The Acts do not require that articles made to a registered design should be marked to indicate this but in proceedings for infringement damages will not be awarded against a defendant proving that at the date of the infringement he was unaware and had no reasonable grounds for supposing that the design was registered. However before delivery on sale of any goods to which a registered design has been applied, the proprietor of such a design should mark each article with the word 'Registered', 'Regd.', or 'RD.', together with the number appearing on the certificate of registration.

If articles are not so marked where requisite, the proprietor of the design is not entitled to recover any penalty or damages for infringement of his copyright unless he proves that all necessary steps were taken to ensure the marking or unless he shows that infringement took place after the infringer had received notice of the existence of the copyright.

Duplicate of certificate of registration.   The Registrar may, in a case where he is satisfied that the certificate has been lost, or destroyed, or in any other case in which he thinks it expedient, furnish one or more copies of the certificate. A duplicate may be issued on filing Designs Form 6 (stamped 75p).

Register of designs.   This is kept at the Patent Office wherein is entered the names and addresses of proprietors, notifications of assignments and transmissions of registered designs and other matters relating thereto. The fee for inspecting the Register or a registered design is 15p.

Manchester register.   Registration of designs for textile articles and matters in connection therewith are entered on the Manchester Register which is kept at Baskerville House, Browncross Street, New Bailey Street, Salford, M3 5FU.

Correction of errors in an application for registration, representation or in the Register may be made by filing Designs Form 18 (stamped £2).

Assignments, mortgages, or licences and notifications.   These must be entered on the Register. A request by a registered proprietor, mortgagee, or licensee must be made on Designs Form 13; a request to enter names of subsequent proprietor, mortgagee or licensee must be made on Designs Form 12 and a request for the notification of a document must be made on Designs Form 14. Each Form must be stamped £2 for one design, plus 50p for each additional design if made within six months from the date of acquisition of the ownership, mortgage or licence, or stamped £5 for one design, plus 50p for each additional design if made after six months and within twelve months of the acquisition; or stamped £8 for one design plus 50p for each additional design if made after twelve months of the acquisition.

Application by a mortgagee or licensee for removal of his name from the Register must be made on Designs Form 15 (stamped 50p) for each design.

Application to enter alteration of name or change in nationality of the registered proprietor must be made on Designs Form 16, stamped 50p for each design.

Application to enter alteration of address or an address for service in the Register must be made on Designs Form 17, stamped 50p for each design.

Application by registered proprietor to cancel entry in Register must be made on Designs Form 19 (unstamped).

Registered designs

Infringement of a registered design.   Any person infringing the copyright in a registered design either by applying the design to any article or exposing the infringing article for sale is liable to an action for the infringement and for an injunction against repetition. The registered proprietor may be awarded damages and an injunction.

No proceedings can be taken in respect of an infringement committed before the date on which the certificate of registration of the design was issued.

Information respecting a registered design. Any person may apply for information respecting a registered design, e.g. if still in force, date of registration, name and address of registered proprietor. If the number of the design registration is known, application is made on Designs Form 20 (stamped £1) or, if unknown, on Designs Form 21 (stamped £4).

Search amongst registered designs. The Registrar may make a search to ascertain if any proposed design is covered in any existing design on the filing of Designs Form 22 (stamped £4).

Certificate of Registrar. A request for a certificate for use in legal proceedings or for other special purposes may be applied for on Designs Forms 23 or 24 (stamped 75p).

Cancellation of a registered design or grant of a compulsory licence. An application for cancellation of a registered design on the grounds that it was not new or original or should have been rejected must be made on Designs Form 26 (stamped £5). After the design has been registered, any interested person may apply to the Registrar for the grant of a compulsory licence in respect of the design on the ground that it is not applied in the United Kingdom by any industrial process or means to the Article of which it is registered to any reasonable circumstance by filing Designs Form 25 (stamped £8). Notification that a hearing before the Registrar in the matter will be attended must be made on Designs Form 27 (stamped £3).

An application for entry of order of the Court in the Register must be made on Designs Form 28 (unstamped).

Crown

Any Government Department and any person authorised in writing by a Government Department may use any registered design in the service of the Crown subject to certain statutory provisions which also define the rights of third parties in respect of such use. Designs relevant to defence purposes may be ordered to be kept secret.

Offences

It is a misdemeanour to make any false entry in the Register of Designs.

Any person falsely representing an article to be the subject of a registered design or who applies the word 'Registered' or any other word implying that the article is the subject of a registered design or applies the word 'Registered' (or similar words or contractions) to an article in which the copyright has expired shall be liable on conviction under the Summary Jurisdiction Acts to a fine not exceeding £5.

Any person failing to comply with the provisions for secrecy of certain designs is liable on conviction to imprisonment and/or a fine.

TRADE MARKS

The Patent Office is responsible for administering the provisions of the Trade Marks Act 1938 and Trade Marks Rules 1938 (as amended) which together govern the registration of Trade Marks in the United Kingdom.

A trade mark is a device, brand, heading, label, ticket, name, signature, word, letter, numeral or any combination thereof used by a person in the pursuit of trade in order to distinguish his goods from similar goods of other traders. Registration of trade marks is not compulsory in this country. Many trade marks in common use are not registered, either because their proprietors do not wish to register them or because they do not qualify for registration. The statutory provision for registration of trade marks is limited to marks used or proposed to be used for goods.

A Registered Trade Mark is a trade mark which is on the Register of trade marks.

The registration of a trade mark confers a statutory monopoly in the use of that mark as related to the goods for which it is registered and the registered owner has the right to sue in the Courts for infringement of his mark.

The Register of trade marks. This is kept at the Patent Office and is a record of all registered trade marks and all matters pertaining thereto. The Register is divided into two parts, Part A and Part B. The fee for inspecting the Register or documents relating to registration is 15p for each quarter of an hour whether the inspection be at the Patent Office, The Manchester Branch thereof, or at the office of the Cutlers Company.

The Manchester Branch of the Register is kept at the Manchester Branch Office, Baskerville House, Browncross Street, New Bailey Street, Salford, M3 5FU, and contains the record and particulars of all marks for 'textile' goods.

The Sheffield Register is kept by The Cutlers' Company in Sheffield and contains particulars of and all matters relating to all marks for 'metal' goods.

Unregistrable features in trade marks. The following may not appear on trade marks seeking to be registered: Pictorial representations of Her Majesty or any members of the Royal Family or any colourable imitations thereof, the Royal or Imperial Crowns, arms, or crests, or Royal, Imperial, or National flags, Admiralty anchor, Royal Air Force Badge, Fleet Air Arm Badge or any device which may be confused with any of the above and the words: 'Anzac', Royal, Imperial, Empire, Dominion, Crown Patent, Patented, Registered, Registered Design, Copyright, Entered at Stationers Hall, to counterfeit this is a forgery (or words to like effect), Red Cross, Geneva Cross, and representations of the Geneva and other crosses in red or of the Swiss Federal Cross in white on a red ground or silver on a red ground or such representations in a similar colour or colours.

No word which is the common or accepted name of a Chemical substance can be registered as a trademark. Where insignia, orders of chivalry, armorial bearings, emblems, decorations, flags of any state, city, society, institution or person appear on a trade mark the Registrar may require justification for their use or proof of consent to their registration and use from persons entitled to give it.

Classification of goods. A trade mark can only be registered in respect of particular goods or classes of goods set out in thirty-four different classes (at one time before 27 July 1938 fifty classes). A list of goods comprised in each class is obtainable from the Patents Office (Sales Branch) 25 Southampton Buildings, London WC2.

Part A of the Register. A mark (other than a certification trade mark) to be registrable in Part A must consist of or contain one of the following essentials: An invented word or words; a word or words having no direct reference to the character or quality of the goods and not being according to its ordinary significance a geographical name or surname; the name of an individual company or firm represented in a particular manner; the signature of the applicant for registration or of some predecessor in his business, or any other distinctive mark upon evidence of its distinctiveness.

Part B of the Register. In this part of the Register the trade mark must be capable of distinguishing goods with which the proprietor of the trade mark is or may be connected in trade from goods in which no such connection exists.

Applicants and applications for registration. Any person, or persons, firm, partnership or body corporate may apply for registration of a trade mark either directly or through an authorised agent on T.M. Form 1 and on filing T.M. Form 2 for an application in either Part A or Part B of the Register, stamped £6, accompanied by four other representations for non-textile or 'metal' goods, referred to below, on T.M. Form 4 and lodged with or posted to the Registrar, The Patent Office, Trades Mark Registry, 25 Southampton Buildings, London WC2.

An application for the registration of a trade mark for textile goods whether with or without other goods in the class of registration in either Part A or Part B must be made on Textile Form 2 in duplicate, one bearing a £6 stamp and accompanied by six additional representations on T.M. Form 4 and filed with or posted to either the Registrar at the Patent Office or the Registrar, Manchester Branch, Trades Mark Registry, Baskerville House, Browncross Street, New Bailey Street, Salford, M3 5FU. In the case of an application relating to 'metal' goods by a person carrying on business in or around six miles of Hallamshire, this may be made to either the Registrar at the Patent Office or the Cutlers' Company of Sheffield, in which case six additional representations each on T.M. Form 4 must accompany the application form.

A mark may be limited in whole or in part to one or more specified colours.

International convention for the protection of industrial property. Application for a priority date under the terms of the International Convention must be made within six months from the date of the first application for registration in a Convention country.

If the Registrar objects to an application he informs the applicant of his objection in writing and unless within one month the applicant applies for a hearing or makes a considered reply in writing, the application is regarded as withdrawn. If the application is to be amended to the satisfaction of the Registrar he shall inform the applicant in writing of the conditions for acceptance; if the applicant still objects he can within one month of the date of the notification apply for a hearing or make a further considered reply. If the applicant does not object he notifies the Registrar and alters his application accordingly. The decision of the Registrar, with or without a hearing if the applicant has communicated his objections or considered reply in writing and has waived the hearing, is written to the applicant who, if he objects to the decision may within one month by filing T.M. Form 5 (stamped £5), request the Registrar to state the grounds of, and the materials used for, his decision.

Appeals. Appeals from Registrar to the Department of Trade and Industry must be made on T.M. Form 30 (stamped £5), accompanied by two copies of the decision of the Registrar.

Appeals to the Court are made by motion informing the Registrar in the usual way and must be given within one month of the date of the decision appealed against.

Preliminary advice on adoption of a new trade mark may be given by the Registrar on filing T.M. Form 29 (stamped £1), and accompanied by duplicate representations of the proposed mark.

Searches. Request for search before making application for registration must be made on T.M. Form 28 (stamped £3), or £4 if Registrar's preliminary advice as to the suitability of the trade mark for registration is also required. This form must be accompanied by two representations of the mark.

Searches among classified trade marks may be made on payment of 15p for every quarter of an hour.

Opposition to application for registration. This may be given by any person within one month from date of advertisement of the application in the Trades Marks Journal on filing T.M. Form 7 in duplicate, one form being stamped £5. Within one month from receipt of notice of opposition, the applicant for registration must file T.M. Form 8 (stamped £3) together with an unstamped duplicate. When a hearing before the Registrar has been appointed notification of attendance by applicant and opponent must be made on T.M. Form 9 (stamped £5).

Opposition to a certification trade mark must be made on T.M. Form 37 (stamped £5) and the counter statement in reply must be made on T.M. Form 38 (stamped £3), each accompanied by unstamped duplicates. Notice of attendance at the hearing must be given on T.M. Form 39 (stamped £5).

Series trade marks. The owner of several trade marks differing only in immaterial particulars may cover such marks in a single registration. Such an application is made on T.M. Form 2 or Textile Form 2 (stamped £6), accompanied by representations of each of the trade marks.

Defensive trade marks. The registered proprietor of a trade mark for an invented word may apply for a defensive trade mark in respect of any goods which he does not actually use but if used by another trader might indicate a connection with the goods of the registered proprietor. A higher registration fee is charged in such cases; application is made on T.M. Form 32 (stamped £9), and accompanied by four other representations on T.M. Form 4; for 'textile' goods six representations are required.

Certification of trade marks. These are registrable in Part A for goods adapted in relation to any other goods to distinguish goods certified by any person in respect of origin, material, quality or other characteristics by filing T.M. Form 6 (stamped £8 for one class and £8 for each additional class), but not exceeding £160 for any number of classes, accompanied by two unstamped duplicates and six representations on T.M. Form 4.

A request for the Registrar's consent to alterations governing the use of the mark must be made on T.M. Form 35, stamped £6 for one registration and 50p for each other registration in the same request.

Application to the Registrar to expunge or vary an entry in the Register must be made on T.M. Form 36 (stamped £6).

For textile trade marks T.M. Form 6 must be accompanied by three unstamped duplicates.

Advertisement of trade mark. After a trade mark has been accepted or in some cases before acceptance, the application is advertised in the Trade Marks Journal for which purposes a printing of the mark must be filed, except in the case of word 'marks' in plain block type of uniform size. No block should be forwarded until it is asked for by the Registrar. If a mark requiring a block is to be registered in more than one class, a separate block must be supplied for each class. The largest space available for the insertion of any single block is 140 mm broad by 190 mm deep. When a block exceeds 50 mm in breadth or depth a charge of 30p is levied for every 25 mm or part thereof beyond 50 mm in either direction.

Registration of trade mark. After a trade mark has been advertised in the Journal for one month, and not opposed or opposition proceedings disposed of, the mark is entered on the Register on filing T.M. Form 10, stamped £8, plus 50p for each mark of a series; or in the case of Certification marks plus £8 for each class not exceeding £160; and for a defensive mark £9. In the case of textile marks the Form must be accompanied by an unstamped duplicate.

An address for service or alteration or cancellation thereof must be given on T.M. Form 33 and stamped 50p for the first entry and 50p for every additional entry with a total limit of £70 for alterations but if filed with the registration fee (T.M. Form 10) there is no charge.

Association between registered marks. Where a mark is registered as associated with any other mark on the Register a fee of 50p is payable on T.M. Form 10 for each mark associated with a newly registered mark. An application to dissolve the association must be made on T.M. Form 19 (stamped £4). Associated marks are assignable and transmissable as a whole and not separately.

Duration of registration. A trade mark is registered for a period of seven years but may be renewed for further periods of fourteen years each.

Application for renewal must be made on T.M. Form 11, (stamped £20 plus 50p for each mark of a series registration and in the case of a certification trade mark £20 for each Class, not exceeding a total of £400). The renewal form must be submitted not more than three months prior to the expiration of the last registration period. If the renewal fee is not paid in due time it may be paid within one month from date of notice received from the Registrar on filing the stamped T.M. Form 11 and T.M. Form 12 (stamped £3). Where a trade mark has been removed from the Register for non-payment of fee it may be restored on filing T.M. Form 11 together with T.M. Form 13 (stamped £8).

Cotton marks. Application for the continuance of a cotton mark in the collection of refused marks (entered before 27 July 1938 and not thereafter) may be made on Cotton Form 6, stamped £8 for each mark in each Class for each period of fourteen years after the date of application and sent to the Registrar, Trade Mark Registry, Baskerville House, Browncross Street, New Bailey Street, Salford, M3 5FU.

Assignment and transmission of registered trade marks. A trade mark in respect of all or part of the goods covered by registration may be assigned together with the goodwill of a business or not.

Application for registration by owner and transferee jointly must be made on T.M. Form 15, but if by transferee alone on T.M. Form 16, each form stamped £2 if the application is made within six months of date of acquisition; £5 after expiration of six months but within twelve months of date of acquisition; and £8 after twelve months of date of acquisition in each case for each additional mark after the first one. The registration of a Corporation as a subsequent proprietor may be extended by filing T.M. Form 14, stamped £3 for two months' extension; £6 not exceeding four months, and £9 not exceeding six months.

Alterations and amendments. Alteration of address on Register must be made on T.M. Form 18 stamped 50p for each entry but no fee is payable if address is changed by public authority.

Change of name or description of proprietor or user must be requested on T.M. Form 21 stamped 50p for each mark involved.

Correction of clerical errors or amendment of application form must be made on T.M. Form 20 (stamped £1.25).

Cancellation or amendment of entry on Register must be made on T.M. Form 22 dealing with cancellation for which there is no fee; T.M. Form 23 which deals with the striking out of goods and is stamped 75p; and T.M. Form 24 for disclaimer or memorandum (stamped £1.25).

Amendment of a registered trade mark. Application must be made on T.M. Form 25, stamped £6 for one mark and £3 for each other mark, accompanpied by four copies of the mark as amended or six copies in the case of a 'textile' or 'metal' mark. Notice of opposition is given on T.M. Form 47 (stamped £5).

A request for entry on the Register of a certificate of the Court relating to the validity of a registered trade mark may be made on T.M. Form 49 stamped £2 for the first mark and 50p for each other mark certified.

Rectification or removal of a registered trade mark. Application must be made on T.M. Form 26 (stamped £6) and opposition to such proceeding must be notified on

T.M. Form 27 (stamped £5). An order of the Court for alteration or rectification of the Register may be made on T.M. Form 48 (stamped £3).

Certificate of the Registrar

A certificate of the Registrar for the registration of a trade mark or a series of marks must be applied for on T.M. Form 31 (stamped 75p).

A certificate of the Keeper of an entry in the Manchester Record relating to one 'textile' mark or to a series of 'textile' marks must be applied for on T.M. Form Textile 5 (stamped 75p).

An application for a certificate of the Registrar for a proposed assignment in respect of any particular goods covered in a registered trade mark must be made on T.M. Form 40 (stamped £6 for the first mark and 50p for every other mark of the same proprietor included in the same assignment). Application for assignment of exclusive rights to different persons in different parts of the United Kingdom must be made on T.M. Form 41 or on T.M. Form 42 if the assignment is in respect of a trade mark registered before 27 July 1938. Each form must be stamped for £6 for the first mark proposed to be assigned and 50p for every other mark of the same proprietor in the assignment. On application for directions by the Registrar for advertisement of assignment of trade marks in use without goodwill, T.M. Form 43 must be filed stamped £3 for one mark and 50p for every other mark assigned with the same devolution of title. If additional time is required to file T.M. Form 43 application for extension is made on T.M. Form 44, (stamped £3 for up to one month, £6 for up to two months, and £9 for up to three months).

Reclassification

Reclassification of goods in respect of registrations effected before 27 July 1938 may be made by filing T.M. Form 45 (stamped £1). Notice of opposition to the reclassification may be given on T.M. Form 46 stamped £5 for one trade mark and 50p for every other mark of the same proprietor having the same trade mark specification.

Registered user

Permitted use of a registered mark may be made by a registered proprietor on T.M. Form 50, (stamped £6 for one mark and 50p for other marks) and variation of such permitted use may be granted on filing T.M. Form 51 (stamped £6 for first mark and 50p for every other mark). Cancellation of entry may be made on T.M. Form 52 or 53 (stamped £3 for one mark and 50p for each additional mark). Notice of intention to intervene in one proceeding is given on T.M. Form 54 (stamped £5).

Non-user of a registered trade mark may be removed from the Register by application to the Court or to the Registrar who may refer the matter to the Court.

INDEX

1

INDEX TO ADVERTISERS